GORDIAN®

Light Commercial Costs with RSMeans data

John Gomes, Senior Editor

2019
38th annual edition

**Numbers in italics are the divisional responsibilities for each editor. Please contact the designated editor directly with any questions.*

RSMeans data from Gordian
Construction Publishers & Consultants
1099 Hingham Street, Suite 201
Rockland, MA 02370
United States of America
1.800.448.8182
RSMeans.com

Printed in the United States of America
ISSN 0896-7601
ISBN 978-1-946872-62-3

0189 $278.99 per copy (in United States)
Price is subject to change without prior notice.

Related Data and Services

Our engineers recommend the following products and services to complement *Light Commercial Costs with RSMeans data:*

Annual Cost Data Books

2019 Open Shop Building Construction Costs with RSMeans data
2019 Residential Costs with RSMeans data

Reference Books

Residential & Light Commercial Construction Standards
Square Foot & Assemblies Estimating Methods
Designing & Building with the IBC
Estimating Building Costs
RSMeans Estimating Handbook
Green Building: Project Planning & Estimating
How to Estimate with RSMeans data
Plan Reading & Material Takeoff
Project Scheduling and Management for Construction

Seminars and In-House Training

Unit Price Estimating
Training for our online estimating solution
Practical Project Management for Construction Professionals
Scheduling with MSProject for Construction Professionals
Mechanical & Electrical Estimating

RSMeans data Online

For access to the latest cost data, an intuitive search, and an easy-to-use estimate builder, take advantage of the time savings available from our online application. To learn more visit: RSMeans.com/2019online.

Enterprise Solutions

Building owners, facility managers, building product manufacturers, and attorneys across the public and private sectors engage with RSMeans data Enterprise to solve unique challenges where trusted construction cost data is critical. To learn more visit: RSMeans.com/Enterprise.

Custom Built Data Sets

Building and Space Models: Quickly plan construction costs across multiple locations based on geography, project size, building system component, product options, and other variables for precise budgeting and cost control.

Predictive Analytics: Accurately plan future builds with custom graphical interactive dashboards, negotiate future costs of tenant build-outs, and identify and compare national account pricing.

Consulting

Building Product Manufacturing Analytics: Validate your claims and assist with new product launches.

Third-Party Legal Resources: Used in cases of construction cost or estimate disputes, construction product failure vs. installation failure, eminent domain, class action construction product liability, and more.

API

For resellers or internal application integration, RSMeans data is offered via API. Deliver Unit, Assembly, and Square Foot Model data within your interface. To learn more about how you can provide your customers with the latest in localized construction cost data visit: RSMeans.com/API.

Table of Contents

Foreword

The Value of RSMeans data from Gordian

Since 1942, RSMeans data has been the industry-standard materials, labor, and equipment cost information database for contractors, facility owners and managers, architects, engineers, and anyone else that requires the latest localized construction cost information. More than 75 years later, the objective remains the same: to provide facility and construction professionals with the most current and comprehensive construction cost database possible.

With the constant influx of new construction methods and materials, in addition to ever-changing labor and material costs, last year's cost data is not reliable for today's designs, estimates, or budgets. Gordian's cost engineers apply real-world construction experience to identify and quantify new building products and methodologies, adjust productivity rates, and adjust costs to local market conditions across the nation. This adds up to more than 22,000 hours in cost research annually. This unparalleled construction cost expertise is why so many facility and construction professionals rely on RSMeans data year over year.

About Gordian

Gordian originated in the spirit of innovation and a strong commitment to helping clients reach and exceed their construction goals. In 1982, Gordian's chairman and founder, Harry H. Mellon, created Job Order Contracting while serving as chief engineer at the Supreme Headquarters Allied Powers Europe. Job Order Contracting is a unique indefinite delivery/indefinite quantity (IDIQ) process, which enables facility owners to complete a substantial number of repair, maintenance, and construction projects with a single, competitively awarded contract. Realizing facility and infrastructure owners across various industries could greatly benefit from the time and cost saving advantages of this innovative construction procurement solution, he established Gordian in 1990.

Continuing the commitment to providing the most relevant and accurate facility and construction data, software, and expertise in the industry, Gordian enhanced the fortitude of its data with the acquisition of RSMeans in 2014. And in an effort to expand its facility management capabilities, Gordian acquired Sightlines, the leading provider of facilities benchmarking data and analysis, in 2015.

Our Offerings

Gordian is the leader in facility and construction cost data, software, and expertise for all phases of the building life cycle. From planning to design, procurement, construction, and operations, Gordian's solutions help clients maximize efficiency, optimize cost savings, and increase building quality with its highly specialized data engineers, software, and unique proprietary data sets.

Our Commitment

At Gordian, we do more than talk about the quality of our data and the usefulness of its application. We stand behind all of our RSMeans data—from historical cost indexes to construction materials and techniques—to craft current costs and predict future trends. If you have any questions about our products or services, please call us toll-free at 800.448.8182 or visit our website at gordian.com.

How the Cost Data Is Built: An Overview

Unit Prices*
All cost data have been divided into 50 divisions according to the MasterFormat® system of classification and numbering.

Assemblies*
The cost data in this section have been organized in an "Assemblies" format. These assemblies are the functional elements of a building and are arranged according to the 7 elements of the UNIFORMAT II classification system. For a complete explanation of a typical "Assembly", see "RSMeans data: Assemblies—How They Work."

*Residential Models**
Model buildings for four classes of construction—economy, average, custom, and luxury—are developed and shown with complete costs per square foot.

*Commercial/Industrial/ Institutional Models**
This section contains complete costs for 77 typical model buildings expressed as costs per square foot.

*Green Commercial/Industrial/ Institutional Models**
This section contains complete costs for 25 green model buildings expressed as costs per square foot.

*References**
This section includes information on Equipment Rental Costs, Crew Listings, Historical Cost Indexes, City Cost Indexes, Location Factors, Reference Tables, and Change Orders, as well as a listing of abbreviations.

- **Equipment Rental Costs:** Included are the average costs to rent and operate hundreds of pieces of construction equipment.
- **Crew Listings**: This section lists all the crews referenced in the cost data. A crew is composed of more than one trade classification and/or the addition of power equipment to any trade classification. Power equipment is included in the cost of the crew. Costs are shown both with bare labor rates and with the installing contractor's overhead and profit added. For each, the total crew cost per eight-hour day and the composite cost per labor-hour are listed.

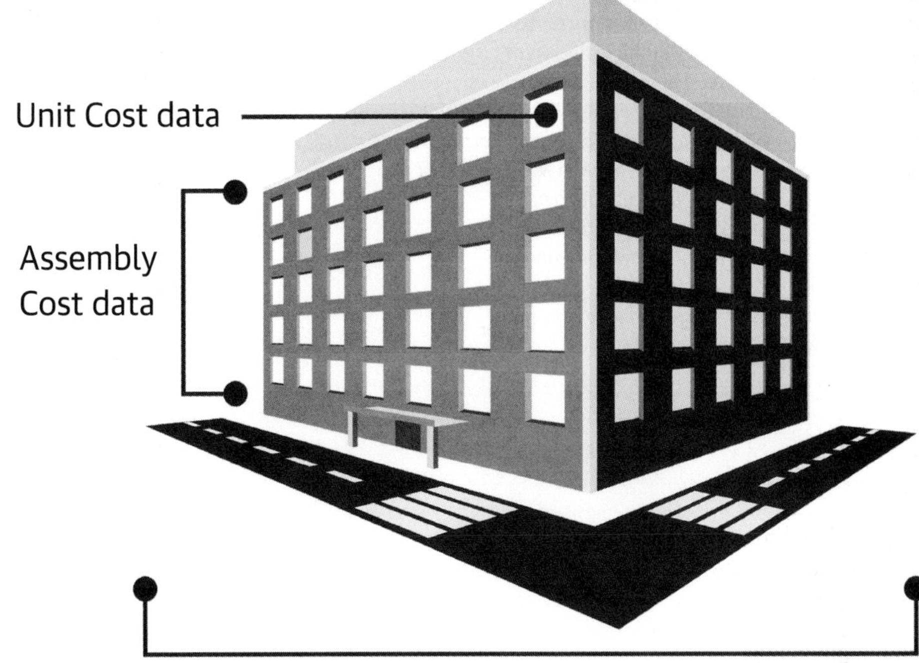

- **Historical Cost Indexes**: These indexes provide you with data to adjust construction costs over time.
- **City Cost Indexes**: All costs in this data set are U.S. national averages. Costs vary by region. You can adjust for this by CSI Division to over 730 cities in 900+ 3-digit zip codes throughout the U.S. and Canada by using this data.
- **Location Factors**: You can adjust total project costs to over 730 cities in 900+ 3-digit zip codes throughout the U.S. and Canada by using the weighted number, which applies across all divisions.
- **Reference Tables**: At the beginning of selected major classifications in the Unit Prices are reference numbers indicators. These numbers refer you to related information in the Reference Section. In this section, you'll find reference tables, explanations, and estimating information that support how we develop the unit price data, technical data, and estimating procedures.
- **Change Orders**: This section includes information on the factors that influence the pricing of change orders.
- **Abbreviations**: A listing of abbreviations used throughout this information, along with the terms they represent, is included.

Index (printed versions only)
A comprehensive listing of all terms and subjects will help you quickly find what you need when you are not sure where it occurs in MasterFormat®.

Conclusion
This information is designed to be as comprehensive and easy to use as possible.

The Construction Specifications Institute (CSI) and Construction Specifications Canada (CSC) have produced the 2016 edition of MasterFormat®, a system of titles and numbers used extensively to organize construction information.

All unit prices in the RSMeans cost data are now arranged in the 50-division MasterFormat® 2016 system.

* Not all information is available in all data sets

Note: The material prices in RSMeans cost data are "contractor's prices." They are the prices that contractors can expect to pay at the lumberyards, suppliers'/distributors' warehouses, etc. Small orders of specialty items would be higher than the costs shown, while very large orders, such as truckload lots, would be less. The variation would depend on the size, timing, and negotiating power of the contractor. The labor costs are primarily for new construction or major renovation rather than repairs or minor alterations. With reasonable exercise of judgment, the figures can be used for any building work.

Estimating with RSMeans data: Unit Prices

Following these steps will allow you to complete an accurate estimate using RSMeans data Unit Prices.

1. Scope Out the Project

- Think through the project and identify the CSI divisions needed in your estimate.
- Identify the individual work tasks that will need to be covered in your estimate.
- The Unit Price data have been divided into 50 divisions according to CSI MasterFormat® 2016.
- In printed versions, the Unit Price Section Table of Contents on page 1 may also be helpful when scoping out your project.
- Experienced estimators find it helpful to begin with Division 2 and continue through completion. Division 1 can be estimated after the full project scope is known.

2. Quantify

- Determine the number of units required for each work task that you identified.
- Experienced estimators include an allowance for waste in their quantities. (Waste is not included in our Unit Price line items unless otherwise stated.)

3. Price the Quantities

- Use the search tools available to locate individual Unit Price line items for your estimate.
- Reference Numbers indicated within a Unit Price section refer to additional information that you may find useful.
- The crew indicates who is performing the work for that task. Crew codes are expanded in the Crew Listings in the Reference Section to include all trades and equipment that comprise the crew.
- The Daily Output is the amount of work the crew is expected to complete in one day.
- The Labor-Hours value is the amount of time it will take for the crew to install one unit of work.
- The abbreviated Unit designation indicates the unit of measure upon which the crew, productivity, and prices are based.
- Bare Costs are shown for materials, labor, and equipment needed to complete the Unit Price line item. Bare costs do not include waste, project overhead, payroll insurance, payroll taxes, main office overhead, or profit.
- The Total Incl O&P cost is the billing rate or invoice amount of the installing contractor or subcontractor who performs the work for the Unit Price line item.

4. Multiply

- Multiply the total number of units needed for your project by the Total Incl O&P cost for each Unit Price line item.
- Be careful that your take off unit of measure matches the unit of measure in the Unit column.
- The price you calculate is an estimate for a completed item of work.
- Keep scoping individual tasks, determining the number of units required for those tasks, matching each task with individual Unit Price line items, and multiplying quantities by Total Incl O&P costs.
- An estimate completed in this manner is priced as if a subcontractor, or set of subcontractors, is performing the work. The estimate does not yet include Project Overhead or Estimate Summary components such as general contractor markups on subcontracted work, general contractor office overhead and profit, contingency, and location factors.

5. Project Overhead

- Include project overhead items from Division 1-General Requirements.
- These items are needed to make the job run. They are typically, but not always, provided by the general contractor. Items include, but are not limited to, field personnel, insurance, performance bond, permits, testing, temporary utilities, field office and storage facilities, temporary scaffolding and platforms, equipment mobilization and demobilization, temporary roads and sidewalks, winter protection, temporary barricades and fencing, temporary security, temporary signs, field engineering and layout, final cleaning, and commissioning.
- Each item should be quantified and matched to individual Unit Price line items in Division 1, then priced and added to your estimate.
- An alternate method of estimating project overhead costs is to apply a percentage of the total project cost—usually 5% to 15% with an average of 10% (see General Conditions).
- Include other project related expenses in your estimate such as:
  - Rented equipment not itemized in the Crew Listings
  - Rubbish handling throughout the project (see section 02 41 19.19)

6. Estimate Summary

- Include sales tax as required by laws of your state or county.
- Include the general contractor's markup on self-performed work, usually 5% to 15% with an average of 10%.
- Include the general contractor's markup on subcontracted work, usually 5% to 15% with an average of 10%.
- Include the general contractor's main office overhead and profit:
  - RSMeans data provides general guidelines on the general contractor's main office overhead (see section 01 31 13.60 and Reference Number R013113-50).
  - Markups will depend on the size of the general contractor's operations, projected annual revenue, the level of risk, and the level of competition in the local area and for this project in particular.
- Include a contingency, usually 3% to 5%, if appropriate.
- Adjust your estimate to the project's location by using the City Cost Indexes or the Location Factors in the Reference Section:
  - Look at the rules in "How to Use the City Cost Indexes" to see how to apply the Indexes for your location.
  - When the proper Index or Factor has been identified for the project's location, convert it to a multiplier by dividing it by 100, then multiply that multiplier by your estimated total cost. The original estimated total cost will now be adjusted up or down from the national average to a total that is appropriate for your location.

Editors' Note:
We urge you to spend time reading and understanding the supporting material. An accurate estimate requires experience, knowledge, and careful calculation. The more you know about how we at RSMeans developed the data, the more accurate your estimate will be. In addition, it is important to take into consideration the reference material such as Equipment Listings, Crew Listings, City Cost Indexes, Location Factors, and Reference Tables.

How to Use the Cost Data: The Details

What's Behind the Numbers? The Development of Cost Data

RSMeans data engineers continually monitor developments in the construction industry in order to ensure reliable, thorough, and up-to-date cost information. While overall construction costs may vary relative to general economic conditions, price fluctuations within the industry are dependent upon many factors. Individual price variations may, in fact, be opposite to overall economic trends. Therefore, costs are constantly tracked and complete updates are performed yearly. Also, new items are frequently added in response to changes in materials and methods.

Costs in U.S. Dollars

All costs represent U.S. national averages and are given in U.S. dollars. The City Cost Index (CCI) with RSMeans data can be used to adjust costs to a particular location. The CCI for Canada can be used to adjust U.S. national averages to local costs in Canadian dollars. No exchange rate conversion is necessary because it has already been factored in.

G The processes or products identified by the green symbol in our publications have been determined to be environmentally responsible and/or resource-efficient solely by RSMeans data engineering staff. The inclusion of the green symbol does not represent compliance with any specific industry association or standard.

Material Costs

RSMeans data engineers contact manufacturers, dealers, distributors, and contractors all across the U.S. and Canada to determine national average material costs. If you have access to current material costs for your specific location, you may wish to make adjustments to reflect differences from the national average. Included within material costs are fasteners for a normal installation. RSMeans data engineers use manufacturers' recommendations, written specifications, and/or standard construction practices for the sizing and spacing of fasteners. Adjustments to material costs may be required for your specific application or location. The manufacturer's warranty is assumed. Extended warranties are not included in the material costs. **Material costs do not include sales tax.**

Labor Costs

Labor costs are based upon a mathematical average of trade-specific wages in 30 major U.S. cities. The type of wage (union, open shop, or residential) is identified on the inside back cover of printed publications or selected by the estimator when using the electronic products. Markups for the wages can also be found on the inside back cover of printed publications and/or under the labor references found in the electronic products.

- If wage rates in your area vary from those used, or if rate increases are expected within a given year, labor costs should be adjusted accordingly.

Labor costs reflect productivity based on actual working conditions. In addition to actual installation, these figures include time spent during a normal weekday on tasks, such as material receiving and handling, mobilization at the site, site movement, breaks, and cleanup.

Productivity data is developed over an extended period so as not to be influenced by abnormal variations and reflects a typical average.

Equipment Costs

Equipment costs include not only rental but also operating costs for equipment under normal use. The operating costs include parts and labor for routine servicing, such as the repair and replacement of pumps, filters, and worn lines. Normal operating expendables, such as fuel, lubricants, tires, and electricity (where applicable), are also included. Extraordinary operating expendables with highly variable wear patterns, such as diamond bits and blades, are excluded. These costs are included under materials. Equipment rental rates are obtained from industry sources throughout North America—contractors, suppliers, dealers, manufacturers, and distributors.

Rental rates can also be treated as reimbursement costs for contractor-owned equipment. Owned equipment costs include depreciation, loan payments, interest, taxes, insurance, storage, and major repairs.

Equipment costs do not include operators' wages.

Equipment Cost/Day—The cost of equipment required for each crew is included in the Crew Listings in the Reference Section (small tools that are considered essential everyday tools are not listed out separately). The Crew Listings itemize specialized tools and heavy equipment along with labor trades. The daily cost of itemized equipment included in a crew is based on dividing the weekly bare rental rate by 5 (number of working days per week), then adding the hourly operating cost times 8 (the number of hours per day). This Equipment Cost/Day is shown in the last column of the Equipment Rental Costs in the Reference Section.

Mobilization, Demobilization—The cost to move construction equipment from an equipment yard or rental company to the job site and back again is not included in equipment costs. Mobilization (to the site) and demobilization (from the site) costs can be found in the Unit Price Section. If a piece of equipment is already at the job site, it is not appropriate to utilize mobilization or demobilization costs again in an estimate.

Overhead and Profit

Total Cost including O&P for the installing contractor is shown in the last column of the Unit Price and/or Assemblies. This figure is the sum of the bare material cost plus 10% for profit, the bare labor cost plus total overhead and profit, and the bare equipment cost plus 10% for profit. Details for the calculation of overhead and profit on labor are shown on the inside back cover of the printed product and in the Reference Section of the electronic product.

General Conditions

Cost data in this data set are presented in two ways: Bare Costs and Total Cost including O&P (Overhead and Profit). General Conditions, or General Requirements, of the contract should also be added to the Total Cost including O&P when applicable. Costs for General Conditions are listed in Division 1 of the Unit Price Section and in the Reference Section.

General Conditions for the installing contractor may range from 0% to 10% of the Total Cost including O&P. For the general or prime contractor, costs for General Conditions may range from 5% to 15% of the Total Cost including O&P, with a figure of 10% as the most typical allowance. If applicable, the Assemblies and Models sections use costs that include the installing contractor's overhead and profit (O&P).

Factors Affecting Costs

Costs can vary depending upon a number of variables. Here's a listing of some factors that affect costs and points to consider.

Quality—The prices for materials and the workmanship upon which productivity is based represent sound construction work. They are also in line with industry standard and manufacturer specifications and are frequently used by federal, state, and local governments.

Overtime—We have made no allowance for overtime. If you anticipate premium time or work beyond normal working hours, be sure to make an appropriate adjustment to your labor costs.

Productivity—The productivity, daily output, and labor-hour figures for each line item are based on an eight-hour work day in daylight hours in moderate temperatures and up to a 14' working height unless otherwise indicated. For work that extends beyond normal work hours or is performed under adverse conditions, productivity may decrease.

Size of Project—The size, scope of work, and type of construction project will have a significant impact on cost. Economies of scale can reduce costs for large projects. Unit costs can often run higher for small projects.

Location—Material prices are for metropolitan areas. However, in dense urban areas, traffic and site storage limitations may increase costs. Beyond a 20-mile radius of metropolitan areas, extra trucking or transportation charges may also increase the material costs slightly. On the other hand, lower wage rates may be in effect. Be sure to consider both of these factors when preparing an estimate, particularly if the job site is located in a central city or remote rural location. In addition, highly specialized subcontract items may require travel and per-diem expenses for mechanics.

Other Factors—

- season of year
- contractor management
- weather conditions
- local union restrictions
- building code requirements
- availability of:
  - adequate energy
  - skilled labor
  - building materials
- owner's special requirements/restrictions
- safety requirements
- environmental considerations
- access

Unpredictable Factors—General business conditions influence "in-place" costs of all items. Substitute materials and construction methods may have to be employed. These may affect the installed cost and/or life cycle costs. Such factors may be difficult to evaluate and cannot necessarily be predicted on the basis of the job's location in a particular section of the country. Thus, where these factors apply, you may find significant but unavoidable cost variations for which you will have to apply a measure of judgment to your estimate.

Rounding of Costs

In printed publications only, all unit prices in excess of $5.00 have been rounded to make them easier to use and still maintain adequate precision of the results.

How Subcontracted Items Affect Costs

A considerable portion of all large construction jobs is usually subcontracted. In fact, the percentage done by subcontractors is constantly increasing and may run over 90%. Since the workers employed by these companies do nothing else but install their particular products, they soon become experts in that line. As a result, installation by these firms is accomplished so efficiently that the total in-place cost, even with the general contractor's overhead and profit, is no more, and often less, than if the principal contractor had handled the installation. Companies that deal with construction specialties are anxious to have their products perform well and, consequently, the installation will be the best possible.

Contingencies

The allowance for contingencies generally provides for unforeseen construction difficulties. On alterations or repair jobs, 20% is not too much. If drawings are final and only field contingencies are being considered, 2% or 3% is probably sufficient and often nothing needs to be added. Contractually, changes in plans will be covered by extras. The contractor should consider inflationary price trends and possible material shortages during the course of the job. These escalation factors are dependent upon both economic conditions and the anticipated time between the estimate and actual construction. If drawings are not complete or approved, or a budget cost is wanted, it is wise to add 5% to 10%. Contingencies, then, are a matter of judgment.

Important Estimating Considerations

The productivity, or daily output, of each craftsman or crew assumes a well-managed job where tradesmen with the proper tools and equipment, along with the appropriate construction materials, are present. Included are daily set-up and cleanup time, break time, and plan layout time. Unless otherwise indicated, time for material movement on site (for items that can be transported by hand) of up to 200' into the building and to the first or second floor is also included. If material has to be transported by other means, over greater distances, or to higher floors, an additional allowance should be considered by the estimator.

While horizontal movement is typically a sole function of distances, vertical transport introduces other variables that can significantly impact productivity. In an occupied building, the use of elevators (assuming access, size, and required protective measures are acceptable) must be understood at the time of the estimate. For new construction, hoist wait and cycle times can easily be 15 minutes and may result in scheduled access extending beyond the normal work day. Finally, all vertical transport will impose strict weight limits likely to preclude the use of any motorized material handling.

The productivity, or daily output, also assumes installation that meets manufacturer/designer/standard specifications. A time allowance for quality control checks, minor adjustments, and any task required to ensure proper function or operation is also included. For items that require connections to services, time is included for positioning, leveling, securing the unit, and making all the necessary connections (and start up where applicable) to ensure a complete installation. Estimating of the services themselves (electrical, plumbing, water, steam, hydraulics, dust collection, etc.) is separate.

In some cases, the estimator must consider the use of a crane and an appropriate crew for the installation of large or heavy items. For those situations where a crane is not included in the assigned crew and as part of the line item cost, then equipment rental costs, mobilization and demobilization costs, and operator and support personnel costs must be considered.

Labor-Hours

The labor-hours expressed in this publication are derived by dividing the total daily labor-hours for the crew by the daily output. Based on average installation time and the assumptions listed above, the labor-hours include: direct labor, indirect labor, and nonproductive time. A typical day for a craftsman might include but is not limited to:

- Direct Work
  - Measuring and layout
  - Preparing materials
  - Actual installation
  - Quality assurance/quality control
- Indirect Work
  - Reading plans or specifications
  - Preparing space
  - Receiving materials
  - Material movement
  - Giving or receiving instruction
  - Miscellaneous
- Non-Work
  - Chatting
  - Personal issues
  - Breaks
  - Interruptions (i.e., sickness, weather, material or equipment shortages, etc.)

If any of the items for a typical day do not apply to the particular work or project situation, the estimator should make any necessary adjustments.

Final Checklist

Estimating can be a straightforward process provided you remember the basics. Here's a checklist of some of the steps you should remember to complete before finalizing your estimate.

Did you remember to:

- factor in the City Cost Index for your locale?
- take into consideration which items have been marked up and by how much?
- mark up the entire estimate sufficiently for your purposes?
- read the background information on techniques and technical matters that could impact your project time span and cost?
- include all components of your project in the final estimate?
- double check your figures for accuracy?
- call RSMeans data engineers if you have any questions about your estimate or the data you've used? Remember, Gordian stands behind all of our products, including our extensive RSMeans data solutions. If you have any questions about your estimate, about the costs you've used from our data, or even about the technical aspects of the job that may affect your estimate, feel free to call the Gordian RSMeans editors at 1.800.448.8182.

Commercial/Industrial/ Institutional Section

Table of Contents

Introduction to the Commercial/ Industrial/Institutional Section

General

The Commercial/Industrial/Institutional section of this manual contains base building costs per square foot of floor area for 50 model buildings. Each model has a table of square foot costs for combinations of exterior wall and framing systems. This table is supplemented by a list of common additives and their unit costs. A breakdown of the component costs used to develop the base cost for the model is included. The total cost derived from the base cost and additives must be modified using the appropriate location factor from the Reference Section.

This section may be used directly to estimate the construction cost of most types of buildings when only the floor area, exterior wall construction, and framing systems are known. To adjust the base cost for components that are different from the model, use the tables from the Assemblies Section.

Building Identification & Model Selection

The building models in this section represent structures by use. Occupancy, however, does not necessarily identify the building, i.e., a restaurant could be a converted warehouse. In all instances, the building should be described and identified by its own physical characteristics. The model selection should also be guided by comparing specifications with the model. In the case of converted use, data from one model may be used to supplement data from another.

Green Models

Consistent with expanding green trends in the design and construction industries, 16 green building models are currently included. Although similar to our standard models in building type and structural system, the green models align with Energy Star requirements and address many of the items necessary to obtain LEED certification. Although our models do not include site-specific information, our design assumption is that they are located in climate zone 5. DOE's eQuest software was used to perform an energy analysis of the model buildings. By reducing energy use, we were able to reduce the size of the service entrances, switchgear, power feeds, and generators. Each of the following building systems was researched and analyzed: building envelope, HVAC, plumbing fixtures, lighting, and electrical service. These systems were targeted because of their impact on energy usage and green building.

Wall and roof insulation was increased to reduce heat loss and a recycled vapor barrier was added to the foundation. White roofs were specified for models with flat roofs. Interior finishes include materials with recycled content and low VOC paint. Stainless steel toilet partitions were selected because the recycled material is easy to maintain. Plumbing fixtures were selected to conserve water. Water closets are low-flow and equipped with auto-sensor valves. Waterless urinals are specified throughout. Faucets use auto-sensor flush valves and are powered by a hydroelectric power unit inside the faucet. Energy-efficient, low-flow water coolers were specified to reduce energy consumption and water usage throughout each building model.

Lighting efficiency was achieved by specifying LED fixtures in place of standard fluorescents. Daylight on/off lighting control systems were incorporated into the models. These controls are equipped with sensors that automatically turn off the lights when sufficient daylight is available. Energy monitoring systems were also included in all green models.

Adjustments

The base cost tables represent the base cost per square foot of floor area for buildings without a basement and without unusual special features. Basement costs and other common additives are listed below the base cost table. Cost adjustments can also be made to the model by using the tables from the Assemblies Section. This table is for example only.

Dimensions

All base cost tables are developed so that measurements can be readily made during the inspection process. Areas are calculated from exterior dimensions and story heights are measured from the top surface of one floor to the top surface of the floor above. Roof areas are measured by horizontal area covered and costs related to inclines are converted with appropriate factors. The precision of measurement is a matter of the user's choice and discretion. For ease in calculation, consideration should be given to measuring in tenths of a foot, i.e., 9' 6" = 9.5 ft. and 9' 4" = 9.3 ft.

Floor Area

The term "Floor Area" as used in this section includes the sum of the floor plate at grade level and above. This dimension is measured from the outside face of the foundation wall. Basement costs are calculated separately. The user must exercise his/her own judgment, where the lowest level floor is slightly below grade, whether to consider it at grade level or make the basement adjustment.

How to Use the Commercial/ Industrial/Institutional Section

The following is a detailed explanation of a sample entry in the Commercial/Industrial/Institutional Square Foot Cost Section. Each bold number below corresponds to the described item in the following list with the appropriate component or cost of the sample entry in parentheses.

Prices listed are costs that include overhead and profit of the installing contractor and additional markups for general conditions and architects' fees.

COMMERCIAL/INDUSTRIAL/ INSTITUTIONAL	M.010 1	Apartment, 1-3 Story 2

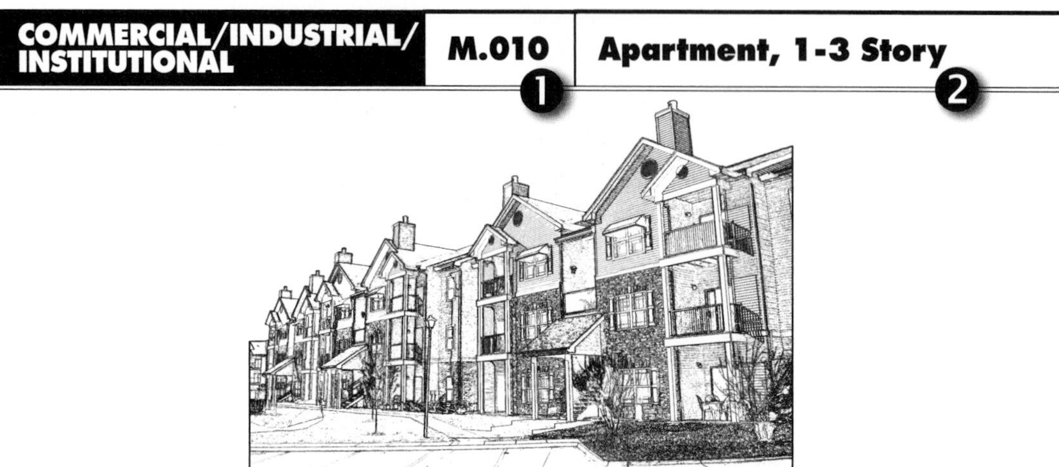

Costs per square foot of floor area

Exterior Wall 3	S.F. Area	8000	12000	15000	19000	22500	25000	29000	32000	36000
	L.F. Perimeter	213	280	330	350	400 4	433	442	480	520
Fiber Cement	Wood Frame	173.15	165.75	162.85	157.60	156.15	155.25	152.45	151.90	150.95
Stone Veneer	Wood Frame	191.20	181.65	177.80	170.15	**168.20** 5	167.00	162.80	162.05	160.80
Brick Veneer	Rigid Steel	182.65	173.85	170.40	163.50	161.85	160.75	157.00	156.35	155.20
E.I.F.S. and Metal Studs	Rigid Steel	177.95	170.05	166.95	161.10	159.50	158.55	155.35	154.80	153.80
Brick Veneer	Reinforced Concrete	201.20	190.75	186.65	178.05	176.05	174.75	170.00	169.20	167.80
Stucco and Concrete Block	Reinforced Concrete	190.45	181.30	177.75	170.65	168.85	167.70	163.85	163.10	161.95
Perimeter Adj., Add or Deduct 6	Per 100 L.F.	14.00	9.40	7.40	5.85	4.95 7	4.45	3.90	3.45	3.10
Story Hgt. Adj., Add or Deduct	Per 1 Ft.	2.20	1.95	1.80	1.55	1.45	1.45	1.30	1.25	1.15

8 ***For Basement**, add $ 34.00 per square foot of basement area*

The above costs were calculated using the basic specifications shown on the facing page. These costs should be adjusted where necessary for design alternatives and owner's requirements.

Common additives 9

Description	Unit	$ Cost
Appliances		
Cooking range, 30" free standing		
1 oven	Ea.	600 - 2775
2 oven	Ea.	1200 - 3800
30" built-in		
1 oven	Ea.	1050 - 2425
2 oven	Ea.	1600 - 3050
Counter top cook tops, 4 burner	Ea.	455 - 2250
Microwave oven	Ea.	255 - 800
Combination range, refrig. & sink, 30" wide	Ea.	1925 - 2025
72" wide	Ea.	2625
Combination range, refrigerator, sink, microwave oven & icemaker	Ea.	6325
Compactor, residential, 4-1 compaction	Ea.	880 - 1450
Dishwasher, built-in, 2 cycles	Ea.	520 - 870
4 cycles	Ea.	650 - 2475
Garbage disposer, sink type	Ea.	193 - 305
Hood for range, 2 speed, vented, 30" wide	Ea.	237 - 1275
42" wide	Ea.	300 - 2400
Refrigerator, no frost 10-12 C.F.	Ea.	575 - 735
18-20 C.F.	Ea.	890 - 2050

Description	Unit	$ Cost
Closed Circuit Surveillance, One station		
Camera and monitor	Ea.	1250
For additional camera stations, add	Ea.	565
Elevators, Hydraulic passenger, 2 stops		
2000# capacity	Ea.	69,400
2500# capacity	Ea.	71,900
3500# capacity	Ea.	76,900
Additional stop, add	Ea.	7000
Emergency Lighting, 25 watt, battery operated		
Lead battery	Ea.	305
Nickel cadmium	Ea.	525
Laundry Equipment		
Dryer, gas, 16 lb. capacity	Ea.	985
30 lb. capacity	Ea.	4000
Washer, 4 cycle	Ea.	1275
Commercial	Ea.	1650
Smoke Detectors		
Ceiling type	Ea.	231
Duct type	Ea.	510

1 Model Number (M.010)

"M" distinguishes this section and stands for model. The number designation is a sequential number. Our newest models are green and designated with the letter "G", followed by the building type number.

2 Type of Building (Apartment, 1-3 Story)

There are 50 different types of commercial/industrial/ institutional buildings highlighted in this section.

3 Exterior Wall Construction and Building Framing Options (Face Brick with Concrete Block Back-up and Open Web Steel Bar Joists)

Three or more commonly used exterior walls and, in most cases, two typical building framing systems are presented for each type of building. The model selected should be based on the actual characteristics of the building being estimated.

4 Total Square Foot of Floor Area and Perimeter Used to Compute Base Costs (22,500 Square Feet and 400 Linear Feet)

Square foot of floor area is the total gross area of all floors at grade and above and does not include a basement. The perimeter in linear feet used for the base cost is generally for a rectangular, economical building shape.

5 Cost per Square Foot of Floor Area ($168.20)

The highlighted cost is for a building of the selected exterior wall and framing system and floor area. Costs for buildings with floor areas other than those calculated may be interpolated between the costs shown.

6 Building Perimeter and Story Height Adjustments

Square foot costs for a building with a perimeter or floor-to-floor story height significantly different from the model used to calculate the base cost may be adjusted (add or deduct) to reflect the actual building geometry.

7 Cost per Square Foot of Floor Area for the Perimeter and/or Height Adjustment ($4.95 for Perimeter Difference and $1.45 for Story Height Difference)

Add (or deduct) $4.95 to the base square foot cost for each 100 feet of perimeter difference between the model and the actual building. Add (or deduct) $1.45 to the base square foot cost for each 1 foot of story height difference between the model and the actual building.

8 Optional Cost per Square Foot of Basement Floor Area ($34.00)

The cost of an unfinished basement for the building being estimated is $34.00 times the gross floor area of the basement.

9 Common Additives

Common components and/or systems used in this type of building are listed. These costs should be added to the total building cost. Additional selections may be found in the Assemblies Section.

How to Use the Commercial/ Industrial/Inst. Section (Continued)

The following is a detailed explanation of the specification and costs for a model building in the Commercial/Industrial/Institutional Square Foot Cost Section. Each bold number below corresponds to the described item in the following list with the appropriate component of the sample entry in parentheses.

Prices listed in the specification are costs that include overhead and profit of the installing contractor but not the general contractor's markup and architect's fees.

Model costs calculated for a 3 story building with 10′ story height and 22,500 square feet of floor area **1**

2 **Apartment, 1-3 Story**

				Unit	Unit Cost	Cost Per S.F.	% Of Sub-Total
A. SUBSTRUCTURE							
1010	Standard Foundations	Poured concrete; strip and spread footings; 4' foundation wall		S.F. Ground	9.15	3.05	
1020	Special Foundations	N/A		—	—	—	
1030	Slab on Grade	4" reinforced concrete		S.F. Slab	5.28	1.76	3.9%
2010	Basement Excavation	Site preparation for slab and trench for foundation wall and footing		S.F. Ground	.28	.09	
2020	Basement Walls	N/A		—	—	—	
B. SHELL							
	B10 Superstructure						
1010	Floor Construction	Wood Beam and joist on wood columns, fireproofed		S.F. Floor	20	13.34	12.5%
1020	Roof Construction	Wood roof, truss, 4/12 slope		S.F. Roof	7.23	2.41	
	B20 Exterior Enclosure						
2010	Exterior Walls	Ashlar stone veneer on wood stud back-up, insulated	*80% of wall*	S.F. Wall	37	15.59	
2020	Exterior Windows	Aluminum horizontal sliding	*20% of wall*	Each	498	3.54	15.6%
2030	Exterior Doors	Steel door , hollow metal with frame		Each	3105	.55	
	B30 Roofing						
3010	Roof Coverings	Asphalt roofing, strip shingles, flashing		S.F. Roof	1.95	.65	0.5%
3020	Roof Openings	N/A		—	—	—	
C. INTERIORS **3**							
1010	Partitions	Gypsum board on wood studs **4**	*10 S.F. Floor/L.F. Partition*	S.F. Partition	6.82	6.06	
1020	Interior Doors	Interior single leaf doors, frames and hardware	*130 S.F. Floor/Door* **6**	Each **7**	1349 **8**	10.23	
1030	Fittings	Residential wall and base wood cabinets, laminate counters	**5**	S.F. Floor	4.82	4.82	
2010	Stair Construction	Wood Stairs with wood rails		Flight	2825	.75	26.4% **9**
3010	Wall Finishes	95% paint, 5% ceramic wall tile		S.F. Surface	1.29	2.29	
3020	Floor Finishes	80% Carpet tile, 10% vinyl composition tile, 10% ceramic tile		S.F. Floor	5.08	5.08	
3030	Ceiling Finishes	Painted gypsum board ceiling on resilient channels		S.F. Ceiling	3.91	3.91	
D. SERVICES							
	D10 Conveying						
1010	Elevators & Lifts	Hydraulic passenger elevator		Each	110,250	4.90	3.9%
1020	Escalators & Moving Walks	N/A		—	—	—	
	D20 Plumbing						
2010	Plumbing Fixtures	Kitchen, bath, laundry and service fixtures, supply and drainage	*1 Fixture/285 S.F. Floor*	Each	1847	6.48	
2020	Domestic Water Distribution	Electric water heater		S.F. Floor	8.39	8.39	12.2%
2040	Rain Water Drainage	Roof drains		S.F. Roof	1.38	.46	
	D30 HVAC						
3010	Energy Supply	Oil fired hot water, baseboard radiation		S.F. Floor	7.56	7.56	
3020	Heat Generating Systems	N/A		—	—	—	
3030	Cooling Generating Systems	Chilled water, air cooled condenser system		S.F. Floor	8.13	8.13	12.5%
3050	Terminal & Package Units	N/A		—	—	—	
3090	Other HVAC Sys. & Equipment	N/A		—	—	—	
	D40 Fire Protection						
4010	Sprinklers	Wet pipe spinkler system, light hazard		S.F. Floor	3.27	3.27	2.6%
4020	Standpipes	N/A		—	—	—	
	D50 Electrical						
5010	Electrical Service/Distribution	800 ampere service, panel board and feeders		S.F. Floor	2.50	2.50	
5020	Lighting & Branch Wiring	Incandescent fixtures, receptacles, switches, A.C. and misc. power		S.F. Floor	6.84	6.84	8.8%
5030	Communications & Security	Addressable alarm systems, emergency lighting, internet and phone wiring		S.F. Floor	1.69	1.69	
5090	Other Electrical Systems	N/A		—	—	—	
E. EQUIPMENT & FURNISHINGS							
1010	Commercial Equipment	N/A		—	—	—	
1020	Institutional Equipment	N/A		—	—	—	1.1 %
1030	Vehicular Equipment	N/A		—	—	—	
1090	Other Equipment	Residential freestanding gas ranges, dishwashers		S.F. Floor	1.42	1.42	
F. SPECIAL CONSTRUCTION							
1020	Integrated Construction	N/A		—	—	—	0.0 %
1040	Special Facilities	N/A		—	—	—	
G. BUILDING SITEWORK	**N/A**						
				10 **Sub-Total**		125.76	**100%**
	CONTRACTOR FEES (General Requirements: 10%, Overhead: 5%, Profit: 10%) **11**				*25%*	31.44	
	ARCHITECT FEES				*7%*	11	
				12 **Total Building Cost**		**168.20**	

1. **Building Description (Model costs are calculated for a 3-story apartment building with a 10' story height and 22,500 square feet of floor area)**
The model highlighted is described in terms of building type, number of stories, typical story height, and square footage.

2. **Type of Building (Apartment, 1-3 Story)**

3. **Division C Interiors (C1020 Interior Doors)**
System costs are presented in divisions according to the 7-division UNIFORMAT II classifications. Each of the component systems is listed.

4. **Specification Highlights (Single leaf solid and hollow core, fire doors)**
All systems in each subdivision are described with the material and proportions used.

5. **Quantity Criteria (130 S.F. Floor/Door)**
The criteria used in determining quantities for the calculations are shown.

6. **Unit (Each)**
The unit of measure shown in this column is the unit of measure of the particular system shown that corresponds to the unit cost.

7. **Unit Cost ($1349)**
The cost per unit of measure of each system subdivision.

8. **Cost per Square Foot ($10.23)**
The cost per square foot for each system is the unit cost of the system times the total number of units, divided by the total square feet of building area.

9. **% of Sub-Total (26.4%)**
The percent of sub-total is the total cost per square foot of all systems in the division divided by the sub-total cost per square foot of the building.

10. **Sub-Total ($125.76)**
The sub-total is the total of all the system costs per square foot.

11. **Project Fees**
(Contractor Fees) (25%); (Architect's Fee) (7%)
Contractor fees to cover the general requirements, overhead, and profit of the general contractor are added as a percentage of the sub-total. An architect's fee, also as a percentage of the sub-total, is also added. These values vary with the building type.

12. **Total Building Cost ($168.20)**
The total building cost per square foot of building area is the sum of the square foot costs of all the systems, the general contractor's general requirements, overhead and profit, and the architect's fee. The total building cost is the amount that appears shaded in the Cost per Square Foot of Floor Area table shown previously.

Costs per square foot of floor area

Exterior Wall	S.F. Area	8000	12000	15000	19000	22500	25000	29000	32000	36000
	L.F. Perimeter	213	280	330	350	400	433	442	480	520
Fiber Cement	Wood Frame	173.15	165.75	162.85	157.60	156.15	155.25	152.45	151.90	150.95
Stone Veneer	Wood Frame	191.20	181.65	177.80	170.15	**168.20**	167.00	162.80	162.05	160.80
Brick Veneer	Rigid Steel	182.65	173.85	170.40	163.50	161.85	160.75	157.00	156.35	155.20
E.I.F.S. and Metal Studs	Rigid Steel	177.95	170.05	166.95	161.10	159.50	158.55	155.35	154.80	153.80
Brick Veneer	Reinforced Concrete	201.20	190.75	186.65	178.05	176.05	174.75	170.00	169.20	167.80
Stucco and Concrete Block	Reinforced Concrete	190.45	181.30	177.75	170.65	168.85	167.70	163.85	163.10	161.95
Perimeter Adj., Add or Deduct	Per 100 L.F.	14.00	9.40	7.40	5.85	4.95	4.45	3.90	3.45	3.10
Story Hgt. Adj., Add or Deduct	Per 1 Ft.	2.20	1.95	1.80	1.55	1.45	1.45	1.30	1.25	1.15
***For Basement**, add $34.00 per square foot of basement area*										

The above costs were calculated using the basic specifications shown on the facing page. These costs should be adjusted where necessary for design alternatives and owner's requirements.

Common additives

Description	Unit	$ Cost
Appliances		
Cooking range, 30" free standing		
1 oven	Ea.	600 - 2775
2 oven	Ea.	1200 - 3800
30" built-in		
1 oven	Ea.	1050 - 2425
2 oven	Ea.	1600 - 3050
Counter top cook tops, 4 burner	Ea.	455 - 2250
Microwave oven	Ea.	255 - 800
Combination range, refrig. & sink, 30" wide	Ea.	1925 - 2025
72" wide	Ea.	2625
Combination range, refrigerator, sink, microwave oven & icemaker	Ea.	6325
Compactor, residential, 4-1 compaction	Ea.	880 - 1450
Dishwasher, built-in, 2 cycles	Ea.	520 - 870
4 cycles	Ea.	650 - 2475
Garbage disposer, sink type	Ea.	193 - 305
Hood for range, 2 speed, vented, 30" wide	Ea.	237 - 1275
42" wide	Ea.	300 - 2400
Refrigerator, no frost 10-12 C.F.	Ea.	575 - 735
18-20 C.F.	Ea.	890 - 2050

Description	Unit	$ Cost
Closed Circuit Surveillance, One station		
Camera and monitor	Ea.	1250
For additional camera stations, add	Ea.	565
Elevators, Hydraulic passenger, 2 stops		
2000# capacity	Ea.	69,400
2500# capacity	Ea.	71,900
3500# capacity	Ea.	76,900
Additional stop, add	Ea.	7000
Emergency Lighting, 25 watt, battery operated		
Lead battery	Ea.	305
Nickel cadmium	Ea.	525
Laundry Equipment		
Dryer, gas, 16 lb. capacity	Ea.	985
30 lb. capacity	Ea.	4000
Washer, 4 cycle	Ea.	1275
Commercial	Ea.	1650
Smoke Detectors		
Ceiling type	Ea.	231
Duct type	Ea.	510

Important: See the Reference Section for Location Factors.

Model costs calculated for a 3 story building with 10′ story height and 22,500 square feet of floor area

Apartment, 1-3 Story

				Unit	Unit Cost	Cost Per S.F.	% Of Sub-Total
A. SUBSTRUCTURE							
1010	Standard Foundations	Poured concrete; strip and spread footings; 4′ foundation wall		S.F. Ground	9.15	3.05	3.9%
1020	Special Foundations	N/A		—	—	—	
1030	Slab on Grade	4″ reinforced concrete		S.F. Slab	5.28	1.76	
2010	Basement Excavation	Site preparation for slab and trench for foundation wall and footing		S.F. Ground	.28	.09	
2020	Basement Walls	N/A		—	—	—	
B. SHELL							
	B10 Superstructure						
1010	Floor Construction	Wood Beam and joist on wood columns, fireproofed		S.F. Floor	20	13.34	12.5%
1020	Roof Construction	Wood roof, truss, 4/12 slope		S.F. Roof	7.23	2.41	
	B20 Exterior Enclosure						
2010	Exterior Walls	Ashlar stone veneer on wood stud back-up, insulated	*80% of wall*	S.F. Wall	37	15.59	15.6%
2020	Exterior Windows	Aluminum horizontal sliding	*20% of wall*	Each	498	3.54	
2030	Exterior Doors	Steel door , hollow metal with frame		Each	3105	.55	
	B30 Roofing						
3010	Roof Coverings	Asphalt roofing, strip shingles, flashing		S.F. Roof	1.95	.65	0.5%
3020	Roof Openings	N/A		—	—	—	
C. INTERIORS							
1010	Partitions	Gypsum board on wood studs	*10 S.F. Floor/L.F. Partition*	S.F. Partition	6.82	6.06	26.4%
1020	Interior Doors	Interior single leaf doors, frames and hardware	*130 S.F. Floor/Door*	Each	1349	10.23	
1030	Fittings	Residential wall and base wood cabinets, laminate counters		S.F. Floor	4.82	4.82	
2010	Stair Construction	Wood Stairs with wood rails		Flight	2825	.75	
3010	Wall Finishes	95% paint, 5% ceramic wall tile		S.F. Surface	1.29	2.29	
3020	Floor Finishes	80% Carpet tile, 10% vinyl composition tile, 10% ceramic tile		S.F. Floor	5.08	5.08	
3030	Ceiling Finishes	Painted gypsum board ceiling on resilient channels		S.F. Ceiling	3.91	3.91	
D. SERVICES							
	D10 Conveying						
1010	Elevators & Lifts	Hydraulic passenger elevator		Each	110,250	4.90	3.9%
1020	Escalators & Moving Walks	N/A		—	—	—	
	D20 Plumbing						
2010	Plumbing Fixtures	Kitchen, bath, laundry and service fixtures, supply and drainage	*1 Fixture/285 S.F. Floor*	Each	1847	6.48	12.2%
2020	Domestic Water Distribution	Electric water heater		S.F. Floor	8.39	8.39	
2040	Rain Water Drainage	Roof drains		S.F. Roof	1.38	.46	
	D30 HVAC						
3010	Energy Supply	Oil fired hot water, baseboard radiation		S.F. Floor	7.56	7.56	12.5%
3020	Heat Generating Systems	N/A		—	—	—	
3030	Cooling Generating Systems	Chilled water, air cooled condenser system		S.F. Floor	8.13	8.13	
3050	Terminal & Package Units	N/A		—	—	—	
3090	Other HVAC Sys. & Equipment	N/A		—	—	—	
	D40 Fire Protection						
4010	Sprinklers	Wet pipe spinkler system, light hazard		S.F. Floor	3.27	3.27	2.6%
4020	Standpipes	N/A		—	—	—	
	D50 Electrical						
5010	Electrical Service/Distribution	800 ampere service, panel board and feeders		S.F. Floor	2.50	2.50	8.8%
5020	Lighting & Branch Wiring	Incandescent fixtures, receptacles, switches, A.C. and misc. power		S.F. Floor	6.84	6.84	
5030	Communications & Security	Addressable alarm systems, emergency lighting, internet and phone wiring		S.F. Floor	1.69	1.69	
5090	Other Electrical Systems	N/A		—	—	—	
E. EQUIPMENT & FURNISHINGS							
1010	Commercial Equipment	N/A		—	—	—	1.1 %
1020	Institutional Equipment	N/A		—	—	—	
1030	Vehicular Equipment	N/A		—	—	—	
1090	Other Equipment	Residential freestanding gas ranges, dishwashers		S.F. Floor	1.42	1.42	
F. SPECIAL CONSTRUCTION							
1020	Integrated Construction	N/A		—	—	—	0.0 %
1040	Special Facilities	N/A		—	—	—	
G. BUILDING SITEWORK		**N/A**					
					Sub-Total	125.76	**100%**
	CONTRACTOR FEES (General Requirements: 10%, Overhead: 5%, Profit: 10%)				*25%*	31.44	
	ARCHITECT FEES				*7%*	11	
					Total Building Cost	**168.20**	

Costs per square foot of floor area

Exterior Wall	S.F. Area	2000	3000	4000	5000	7500	10000	15000	25000	50000
	L.F. Perimeter	188	224	260	300	356	400	504	700	1200
Brick Veneer	Wood Frame	235.85	209.75	196.80	189.40	175.95	**168.55**	161.80	156.10	151.95
	Rigid Steel	262.15	236.25	223.35	216.05	202.75	195.40	188.70	182.95	178.80
Brick Veneer and Concrete Block	Wood Truss	239.00	211.65	198.05	190.40	176.10	168.25	161.05	154.95	150.50
	Bearing Walls	274.95	246.40	232.20	224.20	209.20	200.80	193.20	186.80	182.05
Wood Clapboard	Wood Frame	225.60	201.65	189.70	182.90	170.85	164.25	158.15	153.00	149.30
Vinyl Clapboard	Wood Frame	226.75	202.55	190.45	183.65	171.40	164.75	158.55	153.40	149.55
Perimeter Adj., Add or Deduct	Per 100 L.F.	31.00	20.65	15.45	12.45	8.30	6.30	4.15	2.45	1.20
Story Hgt. Adj., Add or Deduct	Per 1 Ft.	3.60	2.85	2.50	2.35	1.85	1.60	1.30	1.05	0.85

For Basement, add $30.70 per square foot of basement area

The above costs were calculated using the basic specifications shown on the facing page. These costs should be adjusted where necessary for design alternatives and owner's requirements.

Common additives

Description	Unit	$ Cost
Appliances		
Cooking range, 30" free standing		
1 oven	Ea.	600 - 2775
2 oven	Ea.	1200 - 3800
30" built-in		
1 oven	Ea.	1050 - 2425
2 oven	Ea.	1600 - 3050
Counter top cook tops, 4 burner	Ea.	455 - 2250
Microwave oven	Ea.	255 - 800
Combination range, refrig. & sink, 30" wide	Ea.	1925 - 2025
60" wide	Ea.	1925
72" wide	Ea.	2625
Combination range, refrigerator, sink, microwave oven & icemaker	Ea.	6325
Compactor, residential, 4-1 compaction	Ea.	880 - 1450
Dishwasher, built-in, 2 cycles	Ea.	520 - 870
4 cycles	Ea.	650 - 2475
Garbage disposer, sink type	Ea.	193 - 305
Hood for range, 2 speed, vented, 30" wide	Ea.	237 - 1275
42" wide	Ea.	300 - 2400

Description	Unit	$ Cost
Appliances, cont.		
Refrigerator, no frost 10-12 C.F.	Ea.	575 - 735
14-16 C.F.	Ea.	710 - 1200
18-20 C.F.	Ea.	890 - 2050
Laundry Equipment		
Dryer, gas, 16 lb. capacity	Ea.	985
30 lb. capacity	Ea.	4000
Washer, 4 cycle	Ea.	1275
Commercial	Ea.	1650
Sound System		
Amplifier, 250 watts	Ea.	1950
Speaker, ceiling or wall	Ea.	233
Trumpet	Ea.	445

Important: See the Reference Section for Location Factors.

Model costs calculated for a 1 story building with 10′ story height and 10,000 square feet of floor area

Assisted - Senior Living

				Unit	Unit Cost	Cost Per S.F.	% Of Sub-Total
A. SUBSTRUCTURE							
1010	Standard Foundations	Poured concrete strip footings; 4′ foundation wall		S.F. Ground	6.58	6.58	10.2%
1020	Special Foundations	N/A		—	—	—	
1030	Slab on Grade	4″ reinforced concrete with vapor barrier and granular base		S.F. Slab	5.28	5.28	
2010	Basement Excavation	Site preparation for slab and trench for foundation wall and footing		S.F. Ground	.49	.49	
2020	Basement Walls	N/A		—	—	—	
B. SHELL							
	B10 Superstructure						
1010	Floor Construction	Steel column		S.F. Floor	.58	.58	6.4%
1020	Roof Construction	Plywood on wood trusses (pitched); fiberglass batt insulation		S.F. Roof	7.18	7.18	
	B20 Exterior Enclosure						
2010	Exterior Walls	Brick veneer on wood studs, insulated	*80% of wall*	S.F. Wall	23	7.33	10.3%
2020	Exterior Windows	Double hung wood	*20% of wall*	Each	615	3.08	
2030	Exterior Doors	Aluminum and glass; 18 ga. steel flush		Each	3508	2.11	
	B30 Roofing						
3010	Roof Coverings	Asphalt shingles with flashing (pitched); al. gutters & downspouts		S.F. Roof	2.73	2.73	2.2%
3020	Roof Openings	N/A		—	—	—	
C. INTERIORS							
1010	Partitions	Gypsum board on wood studs	*6 S.F. Floor/L.F. Partition*	S.F. Partition	5	6.66	26.3%
1020	Interior Doors	Single leaf wood	*80 S.F. Floor/Door*	Each	691	8.13	
1030	Fittings	N/A		—	—	—	
2010	Stair Construction	N/A		—	—	—	
3010	Wall Finishes	Paint on drywall		S.F. Surface	.79	2.11	
3020	Floor Finishes	80% carpet, 20% ceramic tile		S.F. Floor	7.28	7.28	
3030	Ceiling Finishes	Gypsum board on wood furring, painted		S.F. Ceiling	7.82	7.82	
D. SERVICES							
	D10 Conveying						
1010	Elevators & Lifts	N/A		—	—	—	0.0 %
1020	Escalators & Moving Walks	N/A		—	—	—	
	D20 Plumbing						
2010	Plumbing Fixtures	Kitchen, toilet and service fixtures, supply and drainage	*1 Fixture/200 S.F. Floor*	Each	2968	14.84	14.3%
2020	Domestic Water Distribution	Gas water heater		S.F. Floor	2.55	2.55	
2040	Rain Water Drainage	N/A		—	—	—	
	D30 HVAC						
3010	Energy Supply	N/A		—	—	—	3.4 %
3020	Heat Generating Systems	N/A		—	—	—	
3030	Cooling Generating Systems	N/A		—	—	—	
3050	Terminal & Package Units	Gas fired hot water boiler; terminal package heat/A/C units		S.F. Floor	4.17	4.17	
3090	Other HVAC Sys. & Equipment	N/A		—	—	—	
	D40 Fire Protection						
4010	Sprinklers	Wet pipe sprinkler system		S.F. Floor	3.41	3.41	3.5%
4020	Standpipes	Standpipes		S.F. Floor	.89	.89	
	D50 Electrical						
5010	Electrical Service/Distribution	600 ampere service, panel board, feeders and branch wiring		S.F. Floor	8.04	8.04	21.9%
5020	Lighting & Branch Wiring	Fuorescent & incandescent fixtures, receptacles, switches, A.C. and misc. power		S.F. Floor	10.40	10.40	
5030	Communications & Security	Addressable alarm, fire detection and command center, communication system		S.F. Floor	7.99	7.99	
5090	Other Electrical Systems	Emergency generator, 7.5 kW		S.F. Floor	.22	.22	
E. EQUIPMENT & FURNISHINGS							
1010	Commercial Equipment	Residential washers and dryers		S.F. Floor	.46	.46	1.3%
1020	Institutional Equipment	N/A		—	—	—	
1030	Vehicular Equipment	N/A		—	—	—	
1090	Other Equipment	Residential ranges & refrigerators		S.F. Floor	1.18	1.18	
F. SPECIAL CONSTRUCTION							
1020	Integrated Construction	N/A		—	—	—	0.0 %
1040	Special Facilities	N/A		—	—	—	
G. BUILDING SITEWORK	**N/A**						
					Sub-Total	121.51	**100%**
	CONTRACTOR FEES (General Requirements: 10%, Overhead: 5%, Profit: 10%)				*25%*	30.33	
	ARCHITECT FEES				*11%*	16.71	
					Total Building Cost	**168.55**	

Costs per square foot of floor area

Exterior Wall	S.F. Area	2000	2700	3400	4100	4800	5500	6200	6900	7600
	L.F. Perimeter	180	208	236	256	280	303	317	337	357
Face Brick and Concrete Block	Rigid Steel	253.45	238.60	229.80	222.35	217.70	214.15	210.00	207.50	205.55
	Reinforced Concrete	268.10	253.25	244.45	**236.95**	232.30	228.70	224.60	222.15	220.15
Precast Concrete	Rigid Steel	275.35	257.25	246.60	237.30	231.65	227.30	222.20	219.20	216.70
	Reinforced Concrete	290.40	272.30	261.70	252.45	246.75	242.35	237.30	234.30	231.80
Limestone and Concrete Block	Rigid Steel	283.25	264.15	252.85	243.00	237.00	232.35	226.90	223.70	221.05
	Reinforced Concrete	297.90	278.70	267.45	257.60	251.60	246.95	241.50	238.30	235.65
Perimeter Adj., Add or Deduct	Per 100 L.F.	46.15	34.20	27.20	22.40	19.15	16.70	14.85	13.35	12.10
Story Hgt. Adj., Add or Deduct	Per 1 Ft.	4.35	3.75	3.40	3.00	2.80	2.60	2.45	2.35	2.20
***For Basement**, add $ 32.30 per square foot of basement area*										

The above costs were calculated using the basic specifications shown on the facing page. These costs should be adjusted where necessary for design alternatives and owner's requirements.

Common additives

Description	Unit	$ Cost
Bulletproof Teller Window, 44" x 60"	Ea.	5825
60" x 48"	Ea.	7850
Closed Circuit Surveillance, One station		
Camera and monitor	Ea.	1250
For additional camera stations, add	Ea.	565
Counters, Complete	Station	6075
Door & Frame, 3' x 6'-8", bullet resistant steel		
with vision panel	Ea.	7175 - 9350
Drive-up Window, Drawer & micr., not incl. glass	Ea.	9450 - 12,900
Emergency Lighting, 25 watt, battery operated		
Lead battery	Ea.	305
Nickel cadmium	Ea.	525
Night Depository	Ea.	9875 - 14,800
Package Receiver, painted	Ea.	1900
stainless steel	Ea.	2975
Partitions, Bullet resistant to 8' high	L.F.	325 - 525
Pneumatic Tube Systems, 2 station	Ea.	31,800
With TV viewer	Ea.	60,000

Description	Unit	$ Cost
Service Windows, Pass thru, steel		
24" x 36"	Ea.	4000
48" x 48"	Ea.	3950
72" x 40"	Ea.	6325
Smoke Detectors		
Ceiling type	Ea.	231
Duct type	Ea.	510
Twenty-four Hour Teller		
Automatic deposit cash & memo	Ea.	52,000
Vault Front, Door & frame		
1 hour test, 32"x 78"	Opng.	7700
2 hour test, 32" door	Opng.	9700
40" door	Opng.	10,900
4 hour test, 32" door	Opng.	12,100
40" door	Opng.	13,600
Time lock, two movement, add	Ea.	2375

Important: See the Reference Section for Location Factors.

Model costs calculated for a 1 story building with 14′ story height and 4,100 square feet of floor area

Bank

				Unit	Unit Cost	Cost Per S.F.	% Of Sub-Total
A. SUBSTRUCTURE							
1010	Standard Foundations	Poured concrete; strip and spread footings; 4′ foundation wall		S.F. Ground	10.02	10.02	8.9%
1020	Special Foundations	N/A		–	–	–	
1030	Slab on Grade	4″ reinforced concrete with vapor barrier and granular base		S.F. Slab	5.28	5.28	
2010	Basement Excavation	Site preparation for slab and trench for foundation wall and footing		S.F. Ground	.28	.28	
2020	Basement Walls	N/A		–	–	–	
B. SHELL							
B10 Superstructure							
1010	Floor Construction	Cast-in-place columns		L.F. Column	112	5.46	11.9%
1020	Roof Construction	Cast-in-place concrete flat plate		S.F. Roof	15.40	15.40	
B20 Exterior Enclosure							
2010	Exterior Walls	Face brick with concrete block back-up	*80% of wall*	S.F. Wall	31	21.61	17.3%
2020	Exterior Windows	Horizontal aluminum sliding	*20% of wall*	Each	558	6.50	
2030	Exterior Doors	Double aluminum and glass and hollow metal		Each	4480	2.19	
B30 Roofing							
3010	Roof Coverings	Built-up tar and gravel with flashing; perlite/EPS composite insulation		S.F. Roof	7.54	7.54	4.3%
3020	Roof Openings	N/A		–	–	–	
C. INTERIORS							
1010	Partitions	Gypsum board on metal studs	*20 S.F. of Floor/L.F. Partition*	S.F. Partition	10.22	5.11	14.2%
1020	Interior Doors	Single leaf hollow metal	*200 S.F. Floor/Door*	Each	1201	6.01	
1030	Fittings	N/A		–	–	–	
2010	Stair Construction	N/A		–	–	–	
3010	Wall Finishes	50% vinyl wall covering, 50% paint		S.F. Surface	1.40	1.40	
3020	Floor Finishes	50% carpet tile, 40% vinyl composition tile, 10% quarry tile		S.F. Floor	5.48	5.48	
3030	Ceiling Finishes	Mineral fiber tile on concealed zee bars		S.F. Ceiling	6.86	6.86	
D. SERVICES							
D10 Conveying							
1010	Elevators & Lifts	N/A		–	–	–	0.0 %
1020	Escalators & Moving Walks	N/A		–	–	–	
D20 Plumbing							
2010	Plumbing Fixtures	Toilet and service fixtures, supply and drainage	*1 Fixture/580 S.F. Floor*	Each	5730	9.88	7.4%
2020	Domestic Water Distribution	Gas fired water heater		S.F. Floor	1.62	1.62	
2040	Rain Water Drainage	Roof drains		S.F. Roof	1.43	1.43	
D30 HVAC							
3010	Energy Supply	N/A		–	–	–	7.5 %
3020	Heat Generating Systems	Included in D3050		–	–	–	
3030	Cooling Generating Systems	N/A		–	–	–	
3050	Terminal & Package Units	Single zone rooftop unit, gas heating, electric cooling		S.F. Floor	13.10	13.10	
3090	Other HVAC Sys. & Equipment	N/A		–	–	–	
D40 Fire Protection							
4010	Sprinklers	Wet pipe sprinkler system		S.F. Floor	4.66	4.66	4.4%
4020	Standpipes	Standpipe		S.F. Floor	3.06	3.06	
D50 Electrical							
5010	Electrical Service/Distribution	200 ampere service, panel board and feeders		S.F. Floor	2.58	2.58	11.5%
5020	Lighting & Branch Wiring	Fluorescent fixtures, receptacles, switches, A.C. and misc. power		S.F. Floor	7.19	7.19	
5030	Communications & Security	Alarm systems, internet and phone wiring, emergency lighting, and security television		S.F. Floor	10.12	10.12	
5090	Other Electrical Systems	Emergency generator, 15 kW, Uninterruptible power supply		S.F. Floor	.23	.23	
E. EQUIPMENT & FURNISHINGS							
1010	Commercial Equipment	Automatic teller, drive up window, night depository		S.F. Floor	8.17	8.17	6.3%
1020	Institutional Equipment	Closed circuit TV monitoring system		S.F. Floor	2.95	2.95	
1030	Vehicular Equipment	N/A		–	–	–	
1090	Other Equipment	N/A		–	–	–	
F. SPECIAL CONSTRUCTION							
1020	Integrated Construction	N/A		–	–	–	6.5 %
1040	Special Facilities	Security vault door		S.F. Floor	11.37	11.37	
G. BUILDING SITEWORK	**N/A**						
					Sub-Total	175.50	**100%**
	CONTRACTOR FEES (General Requirements: 10%, Overhead: 5%, Profit: 10%)				*25%*	43.90	
	ARCHITECT FEES				*8%*	17.55	
					Total Building Cost	**236.95**	

Bowling Alley

Costs per square foot of floor area

Exterior Wall	S.F. Area	12000	14000	16000	18000	20000	22000	24000	26000	28000
	L.F. Perimeter	460	491	520	540	566	593	620	645	670
Concrete Block	Steel Roof Deck	172.90	169.85	167.50	165.45	**163.90**	162.60	161.65	160.70	159.95
Decorative Concrete Block	Steel Roof Deck	177.70	174.25	171.55	169.20	167.40	166.00	164.85	163.80	162.95
Face Brick and Concrete Block	Steel Roof Deck	184.55	180.55	177.30	174.50	172.45	170.80	169.50	168.25	167.20
Jumbo Brick and Concrete Block	Steel Roof Deck	181.55	177.75	174.80	172.15	170.25	168.70	167.45	166.30	165.30
Stucco and Concrete Block	Steel Roof Deck	176.05	172.80	170.15	167.90	166.20	164.85	163.75	162.80	161.85
Precast Concrete	Steel Roof Deck	185.80	181.70	178.45	175.50	173.45	171.70	170.30	169.05	168.00
Perimeter Adj., Add or Deduct	Per 100 L.F.	5.10	4.45	3.80	3.35	3.05	2.85	2.55	2.40	2.20
Story Hgt. Adj., Add or Deduct	Per 1 Ft.	1.10	1.10	0.90	0.85	0.80	0.85	0.80	0.80	0.70
***For Basement**, add $32.30 per square foot of basement area*										

The above costs were calculated using the basic specifications shown on the facing page. These costs should be adjusted where necessary for design alternatives and owner's requirements.

Common additives

Description	Unit	$ Cost
Bowling Alleys, Incl. alley, pinsetter,		
scorer, counter & misc. supplies, average	Lane	67,000
For automatic scorer, add (maximum)	Lane	11,500
Emergency Lighting, 25 watt, battery operated		
Lead battery	Ea.	305
Nickel cadmium	Ea.	525
Lockers, Steel, Single tier, 60" or 72"	Opng.	223 - 390
2 tier, 60" or 72" total	Opng.	135 - 167
5 tier, box lockers	Opng.	69.50 - 81.50
Locker bench, lam., maple top only	L.F.	35.50
Pedestals, steel pipe	Ea.	73.50
Seating		
Auditorium chair, all veneer	Ea.	355
Veneer back, padded seat	Ea.	375
Upholstered, spring seat	Ea.	325
Sound System		
Amplifier, 250 watts	Ea.	1950
Speaker, ceiling or wall	Ea.	233
Trumpet	Ea.	445

Important: See the Reference Section for Location Factors.

Model costs calculated for a 1 story building with 14′ story height and 20,000 square feet of floor area

Bowling Alley

				Unit	Unit Cost	Cost Per S.F.	% Of Sub-Total
A. SUBSTRUCTURE							
1010	Standard Foundations	Poured concrete; strip and spread footings; 4′ foundation wall		S.F. Ground	3.43	3.43	
1020	Special Foundations	N/A		—	—	—	
1030	Slab on Grade	4″ reinforced concrete with vapor barrier and granular base		S.F. Slab	5.28	5.28	7.4%
2010	Basement Excavation	Site preparation for slab and trench for foundation wall and footing		S.F. Ground	.49	.49	
2020	Basement Walls	N/A		—	—	—	
B. SHELL							
	B10 Superstructure						
1010	Floor Construction	N/A		—	—	—	7.6 %
1020	Roof Construction	Metal deck on open web steel joists with columns and beams		S.F. Roof	9.45	9.45	
	B20 Exterior Enclosure						
2010	Exterior Walls	Concrete block	*90% of wall*	S.F. Wall	12.76	4.55	
2020	Exterior Windows	Horizontal pivoted	*10% of wall*	Each	2015	3.33	7.1%
2030	Exterior Doors	Double aluminum and glass and hollow metal		Each	3493	.87	
	B30 Roofing						
3010	Roof Coverings	Built-up tar and gravel with flashing; perlite/EPS composite insulation		S.F. Roof	5.77	5.77	4.8%
3020	Roof Openings	Roof hatches		S.F. Roof	.19	.19	
C. INTERIORS							
1010	Partitions	Concrete block	*50 S.F. Floor/L.F. Partition*	S.F. Partition	10.20	2.04	
1020	Interior Doors	Hollow metal single leaf	*1000 S.F. Floor/Door*	Each	1201	1.20	
1030	Fittings	N/A		—	—	—	
2010	Stair Construction	N/A		—	—	—	6.2%
3010	Wall Finishes	Paint and block filler		S.F. Surface	5.68	2.27	
3020	Floor Finishes	Vinyl tile	*25% of floor*	S.F. Floor	2.60	.65	
3030	Ceiling Finishes	Suspended fiberglass board	*25% of area*	S.F. Ceiling	6.10	1.52	
D. SERVICES							
	D10 Conveying						
1010	Elevators & Lifts	N/A		—	—	—	0.0 %
1020	Escalators & Moving Walks	N/A		—	—	—	
	D20 Plumbing						
2010	Plumbing Fixtures	Toilet and service fixtures, supply and drainage	*1 Fixture/2200 S.F. Floor*	Each	3520	1.60	
2020	Domestic Water Distribution	Gas fired water heater		S.F. Floor	.29	.29	2.0%
2040	Rain Water Drainage	Roof drains		S.F. Roof	.63	.63	
	D30 HVAC						
3010	Energy Supply	N/A		—	—	—	
3020	Heat Generating Systems	Included in D3050		—	—	—	
3030	Cooling Generating Systems	N/A		—	—	—	12.0 %
3050	Terminal & Package Units	Single zone rooftop unit, gas heating, electric cooling		S.F. Floor	14.85	14.85	
3090	Other HVAC Sys. & Equipment	N/A		—	—	—	
	D40 Fire Protection						
4010	Sprinklers	Sprinklers, light hazard		S.F. Floor	3.41	3.41	3.3%
4020	Standpipes	Standpipe, wet, Class III		S.F. Floor	.62	.62	
	D50 Electrical						
5010	Electrical Service/Distribution	800 ampere service, panel board and feeders		S.F. Floor	2.15	2.15	
5020	Lighting & Branch Wiring	High efficiency fluorescent fixtures, receptacles, switches, A.C. and misc. power		S.F. Floor	10.27	10.27	11.2%
5030	Communications & Security	Addressable alarm systems and emergency lighting		S.F. Floor	1.30	1.30	
5090	Other Electrical Systems	Emergency generator, 7.5 kW		S.F. Floor	.08	.08	
E. EQUIPMENT & FURNISHINGS							
1010	Commercial Equipment	N/A		—	—	—	
1020	Institutional Equipment	N/A		—	—	—	38.4 %
1030	Vehicular Equipment	N/A		—	—	—	
1090	Other Equipment	Bowling Alley, ballrack, automatic scorers		S.F. Floor	47	47.44	
F. SPECIAL CONSTRUCTION							
1020	Integrated Construction	N/A		—	—	—	0.0 %
1040	Special Facilities	N/A		—	—	—	
G. BUILDING SITEWORK	**N/A**						
					Sub-Total	123.68	**100%**
	CONTRACTOR FEES (General Requirements: 10%, Overhead: 5%, Profit: 10%)				*25%*	30.94	
	ARCHITECT FEES				*6%*	9.28	
					Total Building Cost	**163.90**	

Costs per square foot of floor area

Exterior Wall	S.F. Area	6000	8000	10000	12000	14000	16000	18000	20000	22000
	L.F. Perimeter	320	386	453	475	505	530	570	600	630
Face Brick and Concrete Block	Bearing Walls	165.80	159.85	156.30	150.90	147.50	144.65	143.20	141.60	140.25
	Rigid Steel	168.75	162.80	159.35	154.05	150.65	147.90	146.45	144.85	143.55
Decorative Concrete Block	Bearing Walls	158.40	153.15	150.00	145.45	142.50	140.10	138.85	137.45	136.30
	Rigid Steel	161.35	156.15	153.10	148.55	145.70	143.35	142.10	140.75	139.60
Precast Concrete	Bearing Walls	171.75	165.15	161.35	155.30	151.50	148.35	146.75	144.90	143.40
	Rigid Steel	174.65	168.15	164.35	158.45	154.70	151.55	150.00	148.20	146.70
Perimeter Adj., Add or Deduct	Per 100 L.F.	13.60	10.15	8.15	6.85	5.85	5.15	4.55	4.05	3.75
Story Hgt. Adj., Add or Deduct	Per 1 Ft.	2.25	2.00	1.90	1.65	1.55	1.45	1.40	1.30	1.20
Basement—Not Applicable										

The above costs were calculated using the basic specifications shown on the facing page. These costs should be adjusted where necessary for design alternatives and owner's requirements.

Common additives

Description	Unit	$ Cost
Directory Boards, Plastic, glass covered		
30″ x 20″	Ea.	570
36″ x 48″	Ea.	1500
Aluminum, 24″ x 18″	Ea.	570
36″ x 24″	Ea.	725
48″ x 32″	Ea.	950
48″ x 60″	Ea.	2000
Emergency Lighting, 25 watt, battery operated		
Lead battery	Ea.	305
Nickel cadmium	Ea.	525
Benches, Hardwood	L.F.	137 - 228
Ticket Printer	Ea.	7975
Turnstiles, One way		
4 arm, 46″ dia., manual	Ea.	2475
Electric	Ea.	3375
High security, 3 arm		
65″ dia., manual	Ea.	7275
Electric	Ea.	11,100
Gate with horizontal bars		
65″ dia., 7′ high transit type	Ea.	8800

Important: See the Reference Section for Location Factors.

Model costs calculated for a 1 story building with 14′ story height and 12,000 square feet of floor area

Bus Terminal

				Unit	Unit Cost	Cost Per S.F.	% Of Sub-Total
A. SUBSTRUCTURE							
1010	Standard Foundations	Poured concrete; strip and spread footings; 4′ foundation wall		S.F. Ground	5.58	5.58	11.8%
1020	Special Foundations	N/A		—	—	—	
1030	Slab on Grade	6″ reinforced concrete with vapor barrier and granular base		S.F. Slab	7.56	7.56	
2010	Basement Excavation	Site preparation for slab and trench for foundation wall and footing		S.F. Ground	.28	.28	
2020	Basement Walls	N/A		—	—	—	
B. SHELL							
	B10 Superstructure						
1010	Floor Construction	N/A		—	—	—	5.8 %
1020	Roof Construction	Metal deck on open web steel joists with columns and beams		S.F. Roof	6.58	6.58	
	B20 Exterior Enclosure						
2010	Exterior Walls	Face brick with concrete block back-up	*70% of wall*	S.F. Wall	31	11.99	17.2%
2020	Exterior Windows	Store front	*30% of wall*	Each	32	5.31	
2030	Exterior Doors	Double aluminum and glass		Each	6950	2.32	
	B30 Roofing						
3010	Roof Coverings	Built-up tar and gravel with flashing; perlite/EPS composite insulation		S.F. Roof	6.57	6.57	5.8%
3020	Roof Openings	N/A		—	—	—	
C. INTERIORS							
1010	Partitions	Lightweight concrete block	*15 S.F. Floor/L.F. Partition*	S.F. Partition	10.22	6.81	22.7%
1020	Interior Doors	Hollow metal	*150 S.F. Floor/Door*	Each	1201	.80	
1030	Fittings	N/A		—	—	—	
2010	Stair Construction	N/A		—	—	—	
3010	Wall Finishes	Glazed coating		S.F. Surface	1.93	2.57	
3020	Floor Finishes	Quarry tile and vinyl composition tile		S.F. Floor	9.53	9.53	
3030	Ceiling Finishes	Mineral fiber tile on concealed zee bars		S.F. Ceiling	6.10	6.10	
D. SERVICES							
	D10 Conveying						
1010	Elevators & Lifts	N/A		—	—	—	0.0 %
1020	Escalators & Moving Walks	N/A		—	—	—	
	D20 Plumbing						
2010	Plumbing Fixtures	Toilet and service fixtures, supply and drainage	*1 Fixture/850 S.F. Floor*	Each	7446	8.76	10.6%
2020	Domestic Water Distribution	Electric water heater		S.F. Floor	2.27	2.27	
2040	Rain Water Drainage	Roof drains		S.F. Roof	1.10	1.10	
	D30 HVAC						
3010	Energy Supply	N/A		—	—	—	10.0 %
3020	Heat Generating Systems	Included in D3050		—	—	—	
3030	Cooling Generating Systems	N/A		—	—	—	
3050	Terminal & Package Units	Single zone rooftop unit, gas heating, electric cooling		S.F. Floor	11.40	11.40	
3090	Other HVAC Sys. & Equipment	N/A		—	—	—	
	D40 Fire Protection						
4010	Sprinklers	Wet pipe sprinkler system		S.F. Floor	4.45	4.45	4.8%
4020	Standpipes	Standpipe		S.F. Floor	1.04	1.04	
	D50 Electrical						
5010	Electrical Service/Distribution	400 ampere service, panel board and feeders		S.F. Floor	1.77	1.77	11.3%
5020	Lighting & Branch Wiring	Fluorescent fixtures, receptacles, switches, A.C. and misc. power		S.F. Floor	8.64	8.64	
5030	Communications & Security	Alarm systems and emergency lighting		S.F. Floor	2.03	2.03	
5090	Other Electrical Systems	Emergency generator, 7.5 kW		S.F. Floor	.45	.45	
E. EQUIPMENT & FURNISHINGS							
1010	Commercial Equipment	N/A		—	—	—	0.0 %
1020	Institutional Equipment	N/A		—	—	—	
1030	Vehicular Equipment	N/A		—	—	—	
1090	Other Equipment	N/A		—	—	—	
F. SPECIAL CONSTRUCTION							
1020	Integrated Construction	N/A		—	—	—	0.0 %
1040	Special Facilities	N/A		—	—	—	
G. BUILDING SITEWORK		**N/A**					
					Sub-Total	113.91	**100%**
	CONTRACTOR FEES (General Requirements: 10%, Overhead: 5%, Profit: 10%)				*25%*	28.45	
	ARCHITECT FEES				*6%*	8.54	
					Total Building Cost	**150.90**	

Costs per square foot of floor area

Exterior Wall	S.F. Area	600	800	1000	1200	1600	2000	2400	3000	4000
	L.F. Perimeter	100	114	128	139	164	189	210	235	280
Face Brick and Concrete Block	Rigid Steel	341.30	327.55	319.35	312.50	304.85	300.35	296.40	291.20	286.45
	Bearing Walls	334.25	320.55	312.30	305.45	297.80	293.35	289.30	284.15	279.45
Concrete Block	Rigid Steel	301.10	293.25	288.45	284.55	280.10	277.55	275.25	272.30	269.60
	Bearing Walls	294.10	286.15	281.45	277.45	273.10	270.55	268.15	265.25	262.55
Metal Panel	Rigid Steel	304.50	296.15	291.15	286.90	282.25	279.50	277.05	273.95	271.00
Metal Sandwich Panel	Rigid Steel	306.35	297.55	292.30	287.85	282.95	280.05	277.50	274.15	271.15
Perimeter Adj., Add or Deduct	Per 100 L.F.	94.55	70.95	56.75	47.20	35.50	28.35	23.60	18.85	14.20
Story Hgt. Adj., Add or Deduct	Per 1 Ft.	5.80	5.00	4.50	4.00	3.60	3.25	3.00	2.75	2.45
Basement—Not Applicable										

The above costs were calculated using the basic specifications shown on the facing page. These costs should be adjusted where necessary for design alternatives and owner's requirements.

Common additives

Description	Unit	$ Cost
Air Compressors, Electric		
1-1/2 H.P., standard controls	Ea.	1150
Dual controls	Ea.	1750
5 H.P., 115/230 volt, standard control	Ea.	3225
Dual controls	Ea.	4425
Emergency Lighting, 25 watt, battery operated		
Lead battery	Ea.	305
Nickel cadmium	Ea.	525
Fence, Chain link, 6′ high		
9 ga. wire, galvanized	L.F.	28
6 ga. wire	L.F.	33.50
Gate	Ea.	375
Product Dispenser with vapor recovery for 6 nozzles	Ea.	28,900
Laundry Equipment		
Dryer, gas, 16 lb. capacity	Ea.	985
30 lb. capacity	Ea.	4000
Washer, 4 cycle	Ea.	1275
Commercial	Ea.	1650

Description	Unit	$ Cost
Lockers, Steel, single tier, 60″ or 72″	Opng.	223 - 390
2 tier, 60″ or 72″ total	Opng.	135 - 167
5 tier, box lockers	Opng.	69.50 - 81.50
Locker bench, lam. maple top only	L.F.	35.50
Pedestals, steel pipe	Ea.	73.50
Paving, Bituminous		
Wearing course plus base course	S.Y.	7.20
Safe, Office type, 1 hour rating		
30″ x 18″ x 18″	Ea.	2400
60″ x 36″ x 18″, double door	Ea.	10,200
Sidewalks, Concrete 4″ thick	S.F.	4.76
Yard Lighting, 20′ aluminum pole with 400 watt high pressure sodium fixture	Ea.	3500

Important: See the Reference Section for Location Factors.

Model costs calculated for a 1 story building with 12′ story height and 800 square feet of floor area

Car Wash

			Unit	Unit Cost	Cost Per S.F.	% Of Sub-Total
A. SUBSTRUCTURE						
1010	Standard Foundations	Poured concrete; strip and spread footings; 4′ foundation wall	S.F. Ground	12.99	12.99	8.5%
1020	Special Foundations	N/A	—	—	—	
1030	Slab on Grade	5″ reinforced concrete with vapor barrier and granular base	S.F. Slab	6.90	6.90	
2010	Basement Excavation	Site preparation for slab and trench for foundation wall and footing	S.F. Ground	.95	.95	
2020	Basement Walls	N/A	—	—	—	
B. SHELL						
	B10 Superstructure					
1010	Floor Construction	N/A	—	—	—	3.6 %
1020	Roof Construction	Metal deck, open web steel joists, beams, columns	S.F. Roof	8.82	8.82	
	B20 Exterior Enclosure					
2010	Exterior Walls	Face brick with concrete block back-up *70% of wall*	S.F. Wall	31	36.99	27.4%
2020	Exterior Windows	Horizontal pivoted steel *5% of wall*	Each	802	7.62	
2030	Exterior Doors	Steel overhead hollow metal	Each	4498	22.49	
	B30 Roofing					
3010	Roof Coverings	Built-up tar and gravel with flashing; perlite/EPS composite insulation	S.F. Roof	8.62	8.62	3.5%
3020	Roof Openings	N/A	—	—	—	
C. INTERIORS						
1010	Partitions	Concrete block *20 S.F. Floor/S.F. Partition*	S.F. Partition	9.44	4.72	3.2%
1020	Interior Doors	Hollow metal *600 S.F. Floor/Door*	Each	1201	2.01	
1030	Fittings	N/A	—	—	—	
2010	Stair Construction	N/A	—	—	—	
3010	Wall Finishes	Paint	S.F. Surface	1.22	1.22	
3020	Floor Finishes	N/A	—	—	—	
3030	Ceiling Finishes	N/A	—	—	—	
D. SERVICES						
	D10 Conveying					
1010	Elevators & Lifts	N/A	—	—	—	0.0 %
1020	Escalators & Moving Walks	N/A	—	—	—	
	D20 Plumbing					
2010	Plumbing Fixtures	Toilet and service fixtures, supply and drainage *1 Fixture/160 S.F. Floor*	Each	2656	16.60	24.5%
2020	Domestic Water Distribution	Gas fired water heater	S.F. Floor	39	39	
2040	Rain Water Drainage	Roof drains	S.F. Roof	4.31	4.31	
	D30 HVAC					
3010	Energy Supply	N/A	—	—	—	3.6 %
3020	Heat Generating Systems	N/A	—	—	—	
3030	Cooling Generating Systems	N/A	—	—	—	
3050	Terminal & Package Units	Single zone rooftop unit, gas heating, electric cooling	S.F. Floor	8.88	8.88	
3090	Other HVAC Sys. & Equipment	N/A	—	—	—	
	D40 Fire Protection					
4010	Sprinklers	N/A	—	—	—	0.0 %
4020	Standpipes	N/A	—	—	—	
	D50 Electrical					
5010	Electrical Service/Distribution	400 ampere service, panel board and feeders	S.F. Floor	24	23.69	25.6%
5020	Lighting & Branch Wiring	Fluorescent fixtures, receptacles, switches and misc. power	S.F. Floor	37	37.20	
5030	Communications & Security	N/A	—	—	—	
5090	Other Electrical Systems	Emergency generator, 7.5 kW	S.F. Floor	1.92	1.92	
E. EQUIPMENT & FURNISHINGS						
1010	Commercial Equipment	N/A	—	—	—	0.0 %
1020	Institutional Equipment	N/A	—	—	—	
1030	Vehicular Equipment	N/A	—	—	—	
1090	Other Equipment	N/A	—	—	—	
F. SPECIAL CONSTRUCTION						
1020	Integrated Construction	N/A	—	—	—	0.0 %
1040	Special Facilities	N/A	—	—	—	
G. BUILDING SITEWORK	**N/A**					
				Sub-Total	244.93	**100%**
	CONTRACTOR FEES (General Requirements: 10%, Overhead: 5%, Profit: 10%)			*25%*	61.19	
	ARCHITECT FEES			*7%*	21.43	
				Total Building Cost	**327.55**	

Costs per square foot of floor area

Exterior Wall	S.F. Area	2000	7000	12000	17000	22000	27000	32000	37000	42000
	L.F. Perimeter	180	340	470	540	640	740	762	793	860
Decorative Concrete Block	Wood Arch	255.00	187.85	173.35	162.70	158.70	156.20	151.25	147.95	146.60
	Steel Truss	256.95	189.80	175.30	164.65	160.70	158.15	153.25	149.90	148.60
Stone and Concrete Block	Wood Arch	311.30	218.20	197.85	182.60	176.95	173.35	166.15	161.35	159.40
	Steel Truss	313.30	220.20	199.80	184.55	178.90	175.30	168.10	163.35	161.40
Face Brick and Concrete Block	Wood Arch	279.85	201.25	184.10	171.45	166.80	163.75	157.85	153.85	152.25
	Steel Truss	281.85	203.20	186.10	173.45	168.75	165.75	159.80	155.85	154.25
Perimeter Adj., Add or Deduct	Per 100 L.F.	66.00	18.80	10.95	7.75	6.00	4.85	4.10	3.55	3.15
Story Hgt. Adj., Add or Deduct	Per 1 Ft.	3.95	2.05	1.65	1.35	1.30	1.20	1.00	0.90	0.90
***For Basement**, add $33.10 per square foot of basement area*										

The above costs were calculated using the basic specifications shown on the facing page. These costs should be adjusted where necessary for design alternatives and owner's requirements.

Common additives

Description	Unit	$ Cost
Altar, Wood, custom design, plain	Ea.	3250
Deluxe	Ea.	16,400
Granite or marble, average	Ea.	16,800
Deluxe	Ea.	45,400
Ark, Prefabricated, plain	Ea.	11,700
Deluxe	Ea.	153,500
Baptistry, Fiberglass, incl. plumbing	Ea.	7325 - 11,900
Bells & Carillons, 48 bells	Ea.	1,196,000
24 bells	Ea.	563,000
Confessional, Prefabricated wood		
Single, plain	Ea.	4400
Deluxe	Ea.	10,800
Double, plain	Ea.	8250
Deluxe	Ea.	23,400
Emergency Lighting, 25 watt, battery operated		
Lead battery	Ea.	305
Nickel cadmium	Ea.	525
Lecterns, Wood, plain	Ea.	920
Deluxe	Ea.	6850

Description	Unit	$ Cost
Pews/Benches, Hardwood	L.F.	137 - 228
Pulpits, Prefabricated, hardwood	Ea.	1825 - 15,300
Railing, Hardwood	L.F.	253
Steeples, translucent fiberglass		
30" square, 15' high	Ea.	11,900
25' high	Ea.	13,400
Painted fiberglass, 24" square, 14' high	Ea.	9950
28' high	Ea.	8500
Aluminum		
20' high, 3'- 6" base	Ea.	11,800
35' high, 8'- 0" base	Ea.	41,600
60' high, 14'- 0" base	Ea.	90,000

Important: See the Reference Section for Location Factors.

Model costs calculated for a 1 story building with 24′ story height and 17,000 square feet of floor area

Church

			Unit	Unit Cost	Cost Per S.F.	% Of Sub-Total
A. SUBSTRUCTURE						
1010	Standard Foundations	Poured concrete; strip and spread footings; 4′ foundation wall	S.F. Ground	7.68	7.68	10.5%
1020	Special Foundations	N/A	–	–	–	
1030	Slab on Grade	4″ reinforced concrete with vapor barrier and granular base	S.F. Slab	5.28	5.28	
2010	Basement Excavation	Site preparation for slab and trench for foundation wall and footing	S.F. Ground	.16	.16	
2020	Basement Walls	N/A	–	–	–	
B. SHELL						
	B10 Superstructure					
1010	Floor Construction	N/A	–	–	–	16.5 %
1020	Roof Construction	Wood deck on laminated wood arches	S.F. Roof	19.29	20.58	
	B20 Exterior Enclosure					
2010	Exterior Walls	Face brick with concrete block back-up *80% of wall (adjusted for end walls)*	S.F. Wall	37	22.37	22.6%
2020	Exterior Windows	Aluminum, top hinged, in-swinging and curtain wall panels *20% of wall*	S.F. Window	32	4.91	
2030	Exterior Doors	Double hollow metal swinging, single hollow metal	Each	2380	.84	
	B30 Roofing					
3010	Roof Coverings	Asphalt shingles with flashing; polystyrene insulation	S.F. Roof	6.05	6.05	4.9%
3020	Roof Openings	N/A	–	–	–	
C. INTERIORS						
1010	Partitions	Plaster on metal studs *40 S.F. Floor/L.F. Partitions*	S.F. Partition	14.23	8.54	14.4%
1020	Interior Doors	Hollow metal *400 S.F. Floor/Door*	Each	1201	3	
1030	Fittings	N/A	–	–	–	
2010	Stair Construction	N/A	–	–	–	
3010	Wall Finishes	Paint	S.F. Surface	1.10	1.32	
3020	Floor Finishes	Carpet	S.F. Floor	5.13	5.13	
3030	Ceiling Finishes	N/A	–	–	–	
D. SERVICES						
	D10 Conveying					
1010	Elevators & Lifts	N/A	–	–	–	0.0 %
1020	Escalators & Moving Walks	N/A	–	–	–	
	D20 Plumbing					
2010	Plumbing Fixtures	Kitchen, toilet and service fixtures, supply and drainage *1 Fixture/2430 S.F. Floor*	Each	8043	3.31	3.0%
2020	Domestic Water Distribution	Gas fired hot water heater	S.F. Floor	.39	.39	
2040	Rain Water Drainage	N/A	–	–	–	
	D30 HVAC					
3010	Energy Supply	Oil fired hot water, wall fin radiation	S.F. Floor	9.69	9.69	15.2%
3020	Heat Generating Systems	N/A	–	–	–	
3030	Cooling Generating Systems	N/A	–	–	–	
3050	Terminal & Package Units	Split systems with air cooled condensing units	S.F. Floor	9.27	9.27	
3090	Other HVAC Sys. & Equipment	N/A	–	–	–	
	D40 Fire Protection					
4010	Sprinklers	Wet pipe sprinkler system	S.F. Floor	3.41	3.41	3.7%
4020	Standpipes	Standpipe	S.F. Floor	1.26	1.26	
	D50 Electrical					
5010	Electrical Service/Distribution	400 ampere service, panel board and feeders	S.F. Floor	.75	.75	9.2%
5020	Lighting & Branch Wiring	Fluorescent fixtures, receptacles, switches, A.C. and misc. power	S.F. Floor	7.75	7.75	
5030	Communications & Security	Alarm systems, sound system and emergency lighting	S.F. Floor	2.87	2.87	
5090	Other Electrical Systems	Emergency generator, 7.5 kW	S.F. Floor	.13	.13	
E. EQUIPMENT & FURNISHINGS						
1010	Commercial Equipment	N/A	–	–	–	0.0 %
1020	Institutional Equipment	N/A	–	–	–	
1030	Vehicular Equipment	N/A	–	–	–	
1090	Other Equipment	N/A	–	–	–	
F. SPECIAL CONSTRUCTION						
1020	Integrated Construction	N/A	–	–	–	0.0 %
1040	Special Facilities	N/A	–	–	–	
G. BUILDING SITEWORK	**N/A**					
				Sub-Total	124.69	**100%**
	CONTRACTOR FEES (General Requirements: 10%, Overhead: 5%, Profit: 10%)			*25%*	31.17	
	ARCHITECT FEES			*10%*	15.59	
				Total Building Cost	**171.45**	

Costs per square foot of floor area

Exterior Wall	S.F. Area	2000	4000	6000	8000	12000	15000	18000	20000	22000
	L.F. Perimeter	180	280	340	386	460	535	560	600	640
Stone Ashlar and Concrete Block	Wood Truss	252.85	226.50	211.90	203.00	192.90	189.90	185.50	184.35	183.40
	Steel Joists	257.45	230.50	215.40	206.35	195.85	192.85	188.25	187.10	186.15
Stucco and Concrete Block	Wood Truss	226.30	205.90	195.20	188.80	181.55	179.40	176.35	175.50	174.85
	Steel Joists	230.95	209.85	198.75	192.10	184.55	182.35	179.10	178.25	177.55
Brick Veneer	Wood Frame	235.20	212.80	200.80	193.60	185.35	182.95	179.40	178.50	177.70
Wood Shingles	Wood Frame	224.25	204.25	193.90	187.65	180.65	178.55	175.65	174.80	174.15
Perimeter Adj., Add or Deduct	Per 100 L.F.	43.95	21.95	14.60	10.95	7.25	5.85	4.90	4.40	4.00
Story Hgt. Adj., Add or Deduct	Per 1 Ft.	5.10	4.00	3.20	2.75	2.10	2.00	1.75	1.70	1.70
***For Basement**, add $ 29.20 per square foot of basement area*										

The above costs were calculated using the basic specifications shown on the facing page. These costs should be adjusted where necessary for design alternatives and owner's requirements.

Common additives

Description	Unit	$ Cost
Bar, Front bar	L.F.	445
Back bar	L.F.	355
Booth, Upholstered, custom, straight	L.F.	259 - 480
"L" or "U" shaped	L.F.	268 - 455
Fireplaces, Brick not incl. chimney or foundation, 30" x 24" opening	Ea.	2900
Chimney, standard brick		
Single flue 16" x 20"	V.L.F.	100
20" x 20"	V.L.F.	120
2 flue, 20" x 32"	V.L.F.	186
Lockers, Steel, single tier, 60" or 72"	Opng.	223 - 390
2 tier, 60" or 72" total	Opng.	135 - 167
5 tier, box lockers	Opng.	69.50 - 81.50
Locker bench, lam. maple top only	L.F.	35.50
Pedestals, steel pipe	Ea.	73.50
Refrigerators, Prefabricated, walk-in		
7'-6" High, 6' x 6'	S.F.	137
10' x 10'	S.F.	107
12' x 14'	S.F.	94
12' x 20'	S.F.	134

Description	Unit	$ Cost
Sauna, Prefabricated, complete, 6' x 4'	Ea.	6425
6' x 9'	Ea.	9550
8' x 8'	Ea.	10,500
10' x 12'	Ea.	15,400
Smoke Detectors		
Ceiling type	Ea.	231
Duct type	Ea.	510
Sound System		
Amplifier, 250 watts	Ea.	1950
Speaker, ceiling or wall	Ea.	233
Trumpet	Ea.	445
Steam Bath, Complete, to 140 C.F.	Ea.	3025
To 300 C.F.	Ea.	3425
To 800 C.F.	Ea.	6975
To 2500 C.F.	Ea.	9750
Swimming Pool Complete, gunite	S.F.	101 - 126
Tennis Court, Complete with fence		
Bituminous	Ea.	50,500 - 94,000
Clay	Ea.	54,500 - 97,500

Important: See the Reference Section for Location Factors.

Model costs calculated for a 1 story building with 12′ story height and 6,000 square feet of floor area

Club, Country

			Unit	Unit Cost	Cost Per S.F.	% Of Sub-Total
A. SUBSTRUCTURE						
1010	Standard Foundations	Poured concrete; strip and spread footings; 4′ foundation wall	S.F. Ground	8.18	8.18	
1020	Special Foundations	N/A	—	—	—	
1030	Slab on Grade	4″ reinforced concrete with vapor barrier and granular base	S.F. Slab	5.28	5.28	8.8%
2010	Basement Excavation	Site preparation for slab and trench for foundation wall and footing	S.F. Ground	.28	.28	
2020	Basement Walls	N/A	—	—	—	
B. SHELL						
	B10 Superstructure					
1010	Floor Construction	6 x 6 wood columns	S.F. Floor	.22	.22	4.7%
1020	Roof Construction	Wood truss with plywood sheathing	S.F. Roof	7.22	7.22	
	B20 Exterior Enclosure					
2010	Exterior Walls	Stone ashlar veneer on concrete block *65% of wall*	S.F. Wall	42	18.59	
2020	Exterior Windows	Aluminum horizontal sliding *35% of wall*	Each	498	7.90	19.4%
2030	Exterior Doors	Double aluminum and glass, hollow metal	Each	3932	3.93	
	B30 Roofing					
3010	Roof Coverings	Asphalt shingles	S.F. Roof	16.46	2.90	1.8%
3020	Roof Openings	N/A	—	—	—	
C. INTERIORS						
1010	Partitions	Gypsum board on metal studs, load bearing *14 S.F. Floor/L.F. Partition*	S.F. Partition	7.76	5.54	
1020	Interior Doors	Single leaf wood *140 S.F. Floor/Door*	Each	691	4.93	
1030	Fittings	N/A	—	—	—	
2010	Stair Construction	N/A	—	—	—	16.8%
3010	Wall Finishes	40% vinyl wall covering, 40% paint, 20% ceramic tile	S.F. Surface	2.54	3.63	
3020	Floor Finishes	50% carpet, 30% hardwood tile, 20% ceramic tile	S.F. Floor	8.05	8.05	
3030	Ceiling Finishes	Gypsum plaster on wood furring	S.F. Ceiling	4.22	4.22	
D. SERVICES						
	D10 Conveying					
1010	Elevators & Lifts	N/A	—	—	—	0.0 %
1020	Escalators & Moving Walks	N/A	—	—	—	
	D20 Plumbing					
2010	Plumbing Fixtures	Kitchen, toilet and service fixtures, supply and drainage *1 Fixture/125 S.F. Floor*	Each	2174	17.39	
2020	Domestic Water Distribution	Gas fired hot water	S.F. Floor	9.01	9.01	16.8%
2040	Rain Water Drainage	N/A	—	—	—	
	D30 HVAC					
3010	Energy Supply	N/A	—	—	—	
3020	Heat Generating Systems	Included in D3050	—	—	—	
3030	Cooling Generating Systems	N/A	—	—	—	18.3 %
3050	Terminal & Package Units	Multizone rooftop unit, gas heating, electric cooling	S.F. Floor	29	28.76	
3090	Other HVAC Sys. & Equipment	N/A	—	—	—	
	D40 Fire Protection					
4010	Sprinklers	Wet pipe sprinkler system	S.F. Floor	5.57	5.57	4.7%
4020	Standpipes	Standpipe	S.F. Floor	1.80	1.80	
	D50 Electrical					
5010	Electrical Service/Distribution	400 ampere service, panel board and feeders	S.F. Floor	3.32	3.32	
5020	Lighting & Branch Wiring	High efficiency fluorescent fixtures, receptacles, switches, A.C. and misc. power	S.F. Floor	8.04	8.04	8.6%
5030	Communications & Security	Addressable alarm systems and emergency lighting	S.F. Floor	1.95	1.95	
5090	Other Electrical Systems	Emergency generator, 11.5 kW	S.F. Floor	.24	.24	
E. EQUIPMENT & FURNISHINGS						
1010	Commercial Equipment	N/A	—	—	—	
1020	Institutional Equipment	N/A	—	—	—	0.0 %
1030	Vehicular Equipment	N/A	—	—	—	
1090	Other Equipment	N/A	—	—	—	
F. SPECIAL CONSTRUCTION						
1020	Integrated Construction	N/A	—	—	—	0.0 %
1040	Special Facilities	N/A	—	—	—	
G. BUILDING SITEWORK	**N/A**					
				Sub-Total	156.95	**100%**
	CONTRACTOR FEES (General Requirements: 10%, Overhead: 5%, Profit: 10%)			*25%*	39.25	
	ARCHITECT FEES			*8%*	15.70	
				Total Building Cost	**211.90**	

Costs per square foot of floor area

Exterior Wall	S.F. Area	4000	6000	8000	10000	12000	14000	16000	18000	20000
	L.F. Perimeter	260	340	420	453	460	510	560	610	600
Face Brick and Concrete Block	Bearing Walls	157.95	150.50	146.70	141.05	135.70	134.05	132.85	131.95	129.00
	Rigid Steel	160.80	153.30	149.55	143.90	138.50	136.90	135.70	134.75	131.85
Decorative Concrete Block	Bearing Walls	145.80	139.90	136.90	132.60	128.55	127.25	126.35	125.60	123.40
	Rigid Steel	148.65	142.75	139.75	135.45	131.40	130.05	129.15	128.45	126.25
Tilt-up Concrete Panels	Bearing Walls	140.75	135.50	132.80	129.10	125.55	124.40	123.60	122.95	121.10
	Rigid Steel	143.60	138.35	135.70	131.90	128.35	127.25	126.45	125.80	123.90
Perimeter Adj., Add or Deduct	Per 100 L.F.	18.40	12.25	9.25	7.40	6.10	5.30	4.60	4.10	3.65
Story Hgt. Adj., Add or Deduct	Per 1 Ft.	2.65	2.30	2.15	1.90	1.55	1.50	1.45	1.40	1.25
***For Basement**, add $31.60 per square foot of basement area*										

The above costs were calculated using the basic specifications shown on the facing page. These costs should be adjusted where necessary for design alternatives and owner's requirements.

Common additives

Description	Unit	$ Cost
Bar, Front bar	L.F.	445
Back bar	L.F.	355
Booth, Upholstered, custom straight	L.F.	259 - 480
"L" or "U" shaped	L.F.	268 - 455
Bowling Alleys, incl. alley, pinsetter		
Scorer, counter & misc. supplies, average	Lane	67,000
For automatic scorer, add	Lane	11,500
Emergency Lighting, 25 watt, battery operated		
Lead battery	Ea.	305
Nickel cadmium	Ea.	525
Kitchen Equipment		
Broiler	Ea.	4400
Coffee urn, twin 6 gallon	Ea.	2900
Cooler, 6 ft. long	Ea.	3725
Dishwasher, 10-12 racks per hr.	Ea.	3975
Food warmer	Ea.	785
Freezer, 44 C.F., reach-in	Ea.	5400
Ice cube maker, 50 lb. per day	Ea.	2050
Range with 1 oven	Ea.	3200

Description	Unit	$ Cost
Movie Equipment		
Projector, 35mm	Ea.	13,400 - 18,500
Screen, wall or ceiling hung	S.F.	9.15 - 13.40
Partitions, Folding leaf, wood		
Acoustic type	S.F.	82.50 - 135
Seating		
Auditorium chair, all veneer	Ea.	355
Veneer back, padded seat	Ea.	375
Upholstered, spring seat	Ea.	325
Classroom, movable chair & desk	Set	81 - 171
Lecture hall, pedestal type	Ea.	345 - 645
Sound System		
Amplifier, 250 watts	Ea.	1950
Speaker, ceiling or wall	Ea.	233
Trumpet	Ea.	445
Stage Curtains, Medium weight	S.F.	11.30 - 38.50
Curtain Track, Light duty	L.F.	85.50
Swimming Pools, Complete, gunite	S.F.	101 - 126

Important: See the Reference Section for Location Factors.

Model costs calculated for a 1 story building with 12′ story height and 10,000 square feet of floor area

Community Center

				Unit	Unit Cost	Cost Per S.F.	% Of Sub-Total
A. SUBSTRUCTURE							
1010	Standard Foundations	Poured concrete; strip and spread footings; 4′ foundation wall		S.F. Ground	6.17	6.17	
1020	Special Foundations	N/A		—	—	—	
1030	Slab on Grade	4″ reinforced concrete with vapor barrier and granular base		S.F. Slab	5.28	5.28	11.2%
2010	Basement Excavation	Site preparation for slab and trench for foundation wall and footing		S.F. Ground	.28	.28	
2020	Basement Walls	N/A		—	—	—	
B. SHELL							
	B10 Superstructure						
1010	Floor Construction	N/A		—	—	—	7.3 %
1020	Roof Construction	Metal deck on open web steel joists		S.F. Roof	7.59	7.59	
	B20 Exterior Enclosure						
2010	Exterior Walls	Face brick with concrete block back-up	*80% of wall*	S.F. Wall	31	13.44	
2020	Exterior Windows	Aluminum sliding	*20% of wall*	Each	587	1.99	16.2%
2030	Exterior Doors	Double aluminum and glass and hollow metal		Each	3658	1.47	
	B30 Roofing						
3010	Roof Coverings	Built-up tar and gravel with flashing; perlite/EPS composite insulation		S.F. Roof	6.81	6.81	6.6%
3020	Roof Openings	Roof hatches		S.F. Roof	.11	.11	
C. INTERIORS							
1010	Partitions	Gypsum board on metal studs	*14 S.F. Floor/L.F. Partition*	S.F. Partition	8.09	5.78	
1020	Interior Doors	Single leaf hollow metal	*140 S.F. Floor/Door*	Each	1201	1.20	
1030	Fittings	Toilet partitions, directory board, mailboxes		S.F. Floor	1.75	1.75	
2010	Stair Construction	N/A		—	—	—	20.9%
3010	Wall Finishes	Paint		S.F. Surface	1.70	2.43	
3020	Floor Finishes	50% carpet, 50% vinyl tile		S.F. Floor	4.54	4.54	
3030	Ceiling Finishes	Mineral fiber tile on concealed zee bars		S.F. Ceiling	6.10	6.10	
D. SERVICES							
	D10 Conveying						
1010	Elevators & Lifts	N/A		—	—	—	0.0 %
1020	Escalators & Moving Walks	N/A		—	—	—	
	D20 Plumbing						
2010	Plumbing Fixtures	Kitchen, toilet and service fixtures, supply and drainage	*1 Fixture/910 S.F. Floor*	Each	3103	3.41	
2020	Domestic Water Distribution	Electric water heater		S.F. Floor	11.04	11.04	14.4%
2040	Rain Water Drainage	Roof drains		S.F. Roof	.61	.61	
	D30 HVAC						
3010	Energy Supply	N/A		—	—	—	
3020	Heat Generating Systems	Included in D3050		—	—	—	
3030	Cooling Generating Systems	N/A		—	—	—	10.0 %
3050	Terminal & Package Units	Single zone rooftop unit, gas heating, electric cooling		S.F. Floor	10.46	10.46	
3090	Other HVAC Sys. & Equipment	N/A		—	—	—	
	D40 Fire Protection						
4010	Sprinklers	Wet pipe sprinkler system		S.F. Floor	3.41	3.41	3.3%
4020	Standpipes	N/A		—	—	—	
	D50 Electrical						
5010	Electrical Service/Distribution	200 ampere service, panel board and feeders		S.F. Floor	1.06	1.06	
5020	Lighting & Branch Wiring	High efficiency fluorescent fixtures, receptacles, switches, A.C. and misc. power		S.F. Floor	5.58	5.58	8.0%
5030	Communications & Security	Addressable alarm systems and emergency lighting		S.F. Floor	1.49	1.49	
5090	Other Electrical Systems	Emergency generator, 15 kW		S.F. Floor	.19	.19	
E. EQUIPMENT & FURNISHINGS							
1010	Commercial Equipment	Freezer, chest type		S.F. Floor	.89	.89	
1020	Institutional Equipment	N/A		—	—	—	2.2%
1030	Vehicular Equipment	N/A		—	—	—	
1090	Other Equipment	Kitchen equipment, directory board, mailboxes, built-in coat racks		S.F. Floor	1.41	1.41	
F. SPECIAL CONSTRUCTION							
1020	Integrated Construction	N/A		—	—	—	0.0 %
1040	Special Facilities	N/A		—	—	—	
G. BUILDING SITEWORK	**N/A**						
					Sub-Total	104.49	**100%**
	CONTRACTOR FEES (General Requirements: 10%, Overhead: 5%, Profit: 10%)				*25%*	26.11	
	ARCHITECT FEES				*8%*	10.45	
					Total Building Cost	**141.05**	

Costs per square foot of floor area

Exterior Wall	S.F. Area	10000	12500	15000	17500	20000	22500	25000	30000	40000
	L.F. Perimeter	400	450	500	550	600	625	650	700	800
Fiber Cement	Wood Frame	313.65	310.95	309.10	307.70	306.75	**305.25**	304.00	302.25	299.95
Stone Veneer	Wood Frame	332.75	327.95	324.75	322.40	320.70	318.10	316.00	312.85	308.90
Curtain Wall	Rigid Steel	340.55	335.45	332.10	329.65	327.80	325.00	322.70	319.35	315.15
E.I.F.S.	Rigid Steel	320.15	317.35	315.50	314.15	313.10	311.60	310.30	308.50	306.20
Brick Veneer	Reinforced Concrete	343.05	338.05	334.75	332.30	330.55	327.80	325.55	322.30	318.15
Metal Panel	Reinforced Concrete	341.65	337.05	334.00	331.85	330.25	327.70	325.70	322.70	318.95
Perimeter Adj., Add or Deduct	Per 100 L.F.	6.75	5.35	4.45	3.90	3.35	3.00	2.70	2.20	1.65
Story Hgt. Adj., Add or Deduct	Per 1 Ft.	1.00	0.80	0.80	0.75	0.65	0.65	0.60	0.55	0.40
***For Basement**, add $ 34.70 per square foot of basement area*										

The above costs were calculated using the basic specifications shown on the facing page. These costs should be adjusted where necessary for design alternatives and owner's requirements.

Common additives

Description	Unit	$ Cost
Clock System		
20 room	Ea.	17,900
50 room	Ea.	42,300
Closed Circuit Surveillance, one station		
Camera and monitor	Total	1250
For additional camera stations, add	Ea.	565
For zoom lens - remote control, add	Ea.	2350 - 5875
For automatic iris for low light, add	Ea.	1425
Directory Boards, plastic, glass covered		
30" x 20"	Ea.	570
36" x 48"	Ea.	1500
Aluminum, 24" x 18"	Ea.	570
36" x 24"	Ea.	725
48" x 32"	Ea.	950
48" x 60"	Ea.	2000
Emergency Lighting, 25 watt, battery operated		
Lead battery	Ea.	305
Nickel cadmium	Ea.	525

Description	Unit	$ Cost
Smoke Detectors		
Ceiling type	Ea.	231
Duct type	Ea.	510
Sound System		
Amplifier, 250 watts	Ea.	1950
Speakers, ceiling or wall	Ea.	233
Trumpets	Ea.	445

Important: See the Reference Section for Location Factors.

Model costs calculated for a 1 story building with 16'-6" story height and 22,500 square feet of floor area

Computer Data Center

			Unit	Unit Cost	Cost Per S.F.	% Of Sub-Total
A. SUBSTRUCTURE						
1010	Standard Foundations	Poured concrete; strip and spread footings; 4' foundation wall	S.F. Ground	3.61	3.61	4.0%
1020	Special Foundations	N/A	—	—	—	
1030	Slab on Grade	4" reinforced concrete	S.F. Slab	5.28	5.28	
2010	Basement Excavation	Site preparation for slab and trench for foundation wall and footing	S.F. Ground	.16	.18	
2020	Basement Walls	N/A	—	—	—	
B. SHELL						
	B10 Superstructure					
1010	Floor Construction	Wood columns, fireproofed	S.F. Floor	2.86	2.86	4.6%
1020	Roof Construction	Wood roof truss 4:12 pitch	S.F. Roof	7.55	7.55	
	B20 Exterior Enclosure					
2010	Exterior Walls	Fiber cement siding on wood studs, insulated *95% of wall*	S.F. Wall	17.20	7.49	4.2%
2020	Exterior Windows	Aluminum horizontal sliding *5% of wall*	Each	558	.86	
2030	Exterior Doors	Aluminum and glass single and double doors	Each	4737	1.27	
	B30 Roofing					
3010	Roof Coverings	Asphalt roofing, strip shingles, gutters and downspouts	S.F. Roof	2.10	2.10	1.0%
3020	Roof Openings	Roof hatch	S.F. Roof	.24	.24	
C. INTERIORS						
1010	Partitions	Sound deadening gypsum board on wood studs *15 S.F. Floor/L.F. Partition*	S.F. Partition	8.44	7.79	11.7%
1020	Interior Doors	Solid core wood doors in metal frames *370 S.F. Floor/Door*	Each	923	2.50	
1030	Fittings	Plastic laminate toilet partitions	S.F. Floor	.28	.28	
2010	Stair Construction	N/A	—	—	—	
3010	Wall Finishes	90% Paint, 10% ceramic wall tile	S.F. Surface	1.63	3	
3020	Floor Finishes	65% Carpet, 10% porcelain tile, 10% quarry tile	S.F. Floor	6.29	6.29	
3030	Ceiling Finishes	Acoustic ceiling tiles on suspended channel grid	S.F. Ceiling	6.86	6.86	
D. SERVICES						
	D10 Conveying					
1010	Elevators & Lifts	N/A	—	—	—	0.0 %
1020	Escalators & Moving Walks	N/A	—	—	—	
	D20 Plumbing					
2010	Plumbing Fixtures	Restroom and service fixtures, supply and drainage *1 Fixture/1180 S.F. Floor*	Each	2089	1.77	1.2%
2020	Domestic Water Distribution	Gas fired water heater	S.F. Floor	.88	.88	
2040	Rain Water Drainage	N/A	—	—	—	
	D30 HVAC					
3010	Energy Supply	Hot water reheat system	S.F. Floor	6.27	6.27	41.4%
3020	Heat Generating Systems	Oil fired hot water boiler and pumps	Each	53,375	10.52	
3030	Cooling Generating Systems	Cooling tower and chiller	S.F. Floor	9.52	9.52	
3050	Terminal & Package Units	N/A	—	—	—	
3090	Other HVAC Sys. & Equipment	Plate heat exchanger, ductwork, AHUs and VAV terminals	S.F. Floor	68	68.06	
	D40 Fire Protection					
4010	Sprinklers	85% light hazard wet pipe sprinkler system, 15% preaction system	S.F. Floor	4.14	4.14	2.1%
4020	Standpipes	Standpipes and hose systems	S.F. Floor	.68	.68	
	D50 Electrical					
5010	Electrical Service/Distribution	1200 ampere service, panel boards and feeders	S.F. Floor	3.51	3.51	27.7%
5020	Lighting & Branch Wiring	Fluorescent fixtures, receptacles, switches, A.C. and misc. power	S.F. Floor	15.85	15.85	
5030	Communications & Security	Telephone systems, internet wiring, and addressable alarm systems	S.F. Floor	29	29.30	
5090	Other Electrical Systems	Emergency generator, UPS system with 15 minute pack	S.F. Floor	14.57	14.57	
E. EQUIPMENT & FURNISHINGS						
1010	Commercial Equipment	N/A	—	—	—	0.0 %
1020	Institutional Equipment	N/A	—	—	—	
1030	Vehicular Equipment	N/A	—	—	—	
1090	Other Equipment	N/A	—	—	—	
F. SPECIAL CONSTRUCTION & DEMOLITION						
1020	Integrated Construction	Pedestal access floor	S.F. Floor	4.99	4.99	2.2%
1040	Special Facilities	N/A	—	—	—	
G. BUILDING SITEWORK	**N/A**					
				Sub-Total	228.22	**100%**
	CONTRACTOR FEES (General Requirements: 10%, Overhead: 5%, Profit: 10%)			25%	57.06	
	ARCHITECT FEES			7%	19.97	
				Total Building Cost	**305.25**	

Costs per square foot of floor area

Exterior Wall	S.F. Area	16000	23000	30000	37000	44000	51000	58000	65000	72000
	L.F. Perimeter	597	763	821	954	968	1066	1090	1132	1220
Limestone and Concrete Block	Reinforced Concrete	226.70	216.50	206.45	202.75	196.75	194.60	191.25	189.00	188.00
	Rigid Steel	225.65	215.45	205.40	201.65	195.65	193.50	190.15	187.95	186.90
Face Brick and Concrete Block	Reinforced Concrete	210.70	202.25	194.65	191.65	187.30	185.60	183.15	181.50	180.70
	Rigid Steel	209.60	201.15	193.60	190.60	186.20	184.55	182.10	180.45	179.65
Stone and Concrete Block	Reinforced Concrete	216.55	207.40	198.90	195.70	190.75	188.85	186.10	184.25	183.35
	Rigid Steel	215.45	206.35	197.85	194.65	189.70	187.80	185.05	183.20	182.30
Perimeter Adj., Add or Deduct	Per 100 L.F.	8.05	5.55	4.25	3.45	3.00	2.45	2.20	2.00	1.80
Story Hgt. Adj., Add or Deduct	Per 1 Ft.	2.90	2.50	2.05	1.95	1.70	1.60	1.45	1.40	1.30

***For Basement**, add $ 29.90 per square foot of basement area*

The above costs were calculated using the basic specifications shown on the facing page. These costs should be adjusted where necessary for design alternatives and owner's requirements.

Common additives

Description	Unit	$ Cost
Benches, Hardwood	L.F.	137 - 228
Clock System		
20 room	Ea.	17,900
50 room	Ea.	42,300
Closed Circuit Surveillance, One station		
Camera and monitor	Ea.	1250
For additional camera stations, add	Ea.	565
Directory Boards, Plastic, glass covered		
30" x 20"	Ea.	570
36" x 48"	Ea.	1500
Aluminum, 24" x 18"	Ea.	570
36" x 24"	Ea.	725
48" x 32"	Ea.	950
48" x 60"	Ea.	2000
Emergency Lighting, 25 watt, battery operated		
Lead battery	Ea.	305
Nickel cadmium	Ea.	525

Description	Unit	$ Cost
Flagpoles, Complete		
Aluminum, 20' high	Ea.	1825
40' high	Ea.	4400
70' high	Ea.	11,600
Fiberglass, 23' high	Ea.	1375
39'-5" high	Ea.	3175
59' high	Ea.	7000
Intercom System, 25 station capacity		
Master station	Ea.	3150
Intercom outlets	Ea.	204
Handset	Ea.	575
Safe, Office type, 1 hour rating		
30" x 18" x 18"	Ea.	2400
60" x 36" x 18", double door	Ea.	10,200
Smoke Detectors		
Ceiling type	Ea.	231
Duct type	Ea.	510

Model costs calculated for a 1 story building with 14′ story height and 30,000 square feet of floor area

Courthouse, 1 Story

			Unit	Unit Cost	Cost Per S.F.	% Of Sub-Total
A. SUBSTRUCTURE						
1010	Standard Foundations	Poured concrete; strip and spread footings; 4′ foundation wall	S.F. Ground	4.09	4.09	
1020	Special Foundations	N/A	—	—	—	
1030	Slab on Grade	4″ reinforced concrete with vapor barrier and granular base	S.F. Slab	5.28	5.28	6.2%
2010	Basement Excavation	Site preparation for slab and trench for foundation wall and footing	S.F. Ground	.16	.16	
2020	Basement Walls	N/A	—	—	—	
B. SHELL						
	B10 Superstructure					
1010	Floor Construction	Cast-in-place columns	L.F. Column	66	1.47	14.7%
1020	Roof Construction	Cast-in-place concrete waffle slab	S.F. Roof	21	21.20	
	B20 Exterior Enclosure					
2010	Exterior Walls	Limestone panels with concrete block back-up *75% of wall*	S.F. Wall	62	17.68	
2020	Exterior Windows	Aluminum with insulated glass *25% of wall*	Each	703	2.93	13.8%
2030	Exterior Doors	Double wood	Each	3233	.75	
	B30 Roofing					
3010	Roof Coverings	Built-up tar and gravel with flashing; perlite/EPS composite insulation	S.F. Roof	5.74	5.74	3.8%
3020	Roof Openings	Roof hatches	S.F. Roof	.11	.11	
C. INTERIORS						
1010	Partitions	Plaster on metal studs *10 S.F. Floor/L.F. Partition*	S.F. Partition	12.99	15.59	
1020	Interior Doors	Single leaf wood *100 S.F. Floor/Door*	Each	691	6.91	
1030	Fittings	Toilet partitions	S.F. Floor	.52	.52	
2010	Stair Construction	N/A	—	—	—	33.1%
3010	Wall Finishes	70% paint, 20% wood paneling, 10% vinyl wall covering	S.F. Surface	2.09	5.02	
3020	Floor Finishes	60% hardwood, 20% carpet, 20% terrazzo	S.F. Floor	13.42	13.42	
3030	Ceiling Finishes	Gypsum plaster on metal lath, suspended	S.F. Ceiling	9.69	9.69	
D. SERVICES						
	D10 Conveying					
1010	Elevators & Lifts	N/A	—	—	—	0.0 %
1020	Escalators & Moving Walks	N/A	—	—	—	
	D20 Plumbing					
2010	Plumbing Fixtures	Toilet and service fixtures, supply and drainage *1 Fixture/1110 S.F. Floor*	Each	4618	4.16	
2020	Domestic Water Distribution	Electric hot water heater	S.F. Floor	3.96	3.96	6.2%
2040	Rain Water Drainage	Roof drains	S.F. Roof	1.40	1.40	
	D30 HVAC					
3010	Energy Supply	N/A	—	—	—	
3020	Heat Generating Systems	Included in D3050	—	—	—	
3030	Cooling Generating Systems	N/A	—	—	—	12.5 %
3050	Terminal & Package Units	Multizone unit, gas heating, electric cooling	S.F. Floor	19.25	19.25	
3090	Other HVAC Sys. & Equipment	N/A	—	—	—	
	D40 Fire Protection					
4010	Sprinklers	Wet pipe sprinkler system	S.F. Floor	2.54	2.54	2.0%
4020	Standpipes	Standpipe, wet, Class III	S.F. Floor	.50	.50	
	D50 Electrical					
5010	Electrical Service/Distribution	400 ampere service, panel board and feeders	S.F. Floor	.97	.97	
5020	Lighting & Branch Wiring	High efficiency fluorescent fixtures, receptacles, switches, A.C. and misc. power	S.F. Floor	9.22	9.22	7.7%
5030	Communications & Security	Addressable alarm systems, internet wiring, and emergency lighting	S.F. Floor	1.63	1.63	
5090	Other Electrical Systems	Emergency generator, 11.5 kW	S.F. Floor	.14	.14	
E. EQUIPMENT & FURNISHINGS						
1010	Commercial Equipment	N/A	—	—	—	
1020	Institutional Equipment	N/A	—	—	—	0.0 %
1030	Vehicular Equipment	N/A	—	—	—	
1090	Other Equipment	N/A	—	—	—	
F. SPECIAL CONSTRUCTION						
1020	Integrated Construction	N/A	—	—	—	0.0 %
1040	Special Facilities	N/A	—	—	—	
G. BUILDING SITEWORK		**N/A**				
				Sub-Total	154.33	**100%**
	CONTRACTOR FEES (General Requirements: 10%, Overhead: 5%, Profit: 10%)			*25%*	38.62	
	ARCHITECT FEES			*7%*	13.50	
				Total Building Cost	**206.45**	

Costs per square foot of floor area

Exterior Wall	S.F. Area	2000	5000	7000	10000	12000	15000	18000	21000	25000
	L.F. Perimeter	200	310	360	440	480	520	560	600	660
Tilt-up Concrete Panels	Steel Joists	209.35	179.70	172.25	167.00	164.35	160.90	158.65	157.05	155.65
Decorative Concrete Block	Bearing Walls	215.40	182.75	174.50	168.65	165.65	161.85	159.35	157.50	155.95
Face Brick and Concrete Block	Steel Joists	246.85	202.10	190.65	182.60	178.55	173.25	169.80	167.25	165.05
Stucco and Concrete Block	Wood Truss	207.55	179.60	172.45	167.40	164.75	161.40	159.25	157.60	156.25
Brick Veneer	Wood Frame	217.55	186.30	178.15	**172.45**	169.45	165.65	163.10	161.25	159.75
Wood Clapboard	Wood Frame	204.15	178.00	171.25	166.60	164.10	161.00	159.00	157.50	156.25
Perimeter Adj., Add or Deduct	Per 100 L.F.	26.85	10.75	7.70	5.35	4.45	3.60	3.05	2.55	2.20
Story Hgt. Adj., Add or Deduct	Per 1 Ft.	2.70	1.70	1.40	1.15	1.10	0.95	0.90	0.80	0.75
***For Basement**, add $ 29.60 per square foot of basement area*										

The above costs were calculated using the basic specifications shown on the facing page. These costs should be adjusted where necessary for design alternatives and owner's requirements.

Common additives

Description	Unit	$ Cost
Emergency Lighting, 25 watt, battery operated		
Lead battery	Ea.	305
Nickel cadmium	Ea.	525
Flagpoles, Complete		
Aluminum, 20' high	Ea.	1825
40' high	Ea.	4400
70' high	Ea.	11,600
Fiberglass, 23' high	Ea.	1375
39'-5" high	Ea.	3175
59' high	Ea.	7000
Gym Floor, Incl. sleepers and finish, maple	S.F.	15.60
Intercom System, 25 Station capacity		
Master station	Ea.	3150
Intercom outlets	Ea.	204
Handset	Ea.	575

Description	Unit	$ Cost
Lockers, Steel, single tier, 60" to 72"	Opng.	223 - 390
2 tier, 60" to 72" total	Opng.	135 - 167
5 tier, box lockers	Opng.	69.50 - 81.50
Locker bench, lam. maple top only	L.F.	35.50
Pedestals, steel pipe	Ea.	73.50
Smoke Detectors		
Ceiling type	Ea.	231
Duct type	Ea.	510
Sound System		
Amplifier, 250 watts	Ea.	1950
Speaker, ceiling or wall	Ea.	233
Trumpet	Ea.	445

Important: See the Reference Section for Location Factors.

Model costs calculated for a 1 story building with 12′ story height and 10,000 square feet of floor area

Day Care Center

			Unit	Unit Cost	Cost Per S.F.	% Of Sub-Total
A. SUBSTRUCTURE						
1010	Standard Foundations	Poured concrete; strip and spread footings; 4′ foundation wall	S.F. Ground	5.52	5.52	8.4%
1020	Special Foundations	N/A	—	—	—	
1030	Slab on Grade	4″ concrete with vapor barrier and granular base	S.F. Slab	4.82	4.82	
2010	Basement Excavation	Site preparation for slab and trench for foundation wall and footing	S.F. Ground	.28	.28	
2020	Basement Walls	N/A	—	—	—	
B. SHELL						
	B10 Superstructure					
1010	Floor Construction	Wood beams on columns	S.F. Floor	.22	.22	6.1%
1020	Roof Construction	Wood trusses	S.F. Roof	7.55	7.55	
	B20 Exterior Enclosure					
2010	Exterior Walls	Brick veneer on wood studs *85% of wall*	S.F. Wall	23	10.52	16.9%
2020	Exterior Windows	Window wall *15% of wall*	Each	53	4.20	
2030	Exterior Doors	Aluminum and glass; steel	Each	3493	6.65	
	B30 Roofing					
3010	Roof Coverings	Asphalt shingles, 9″ fiberglass batt insulation, gutters and downspouts	S.F. Roof	4.23	4.23	3.3%
3020	Roof Openings	N/A	—	—	—	
C. INTERIORS						
1010	Partitions	Gypsum board on wood studs *8 S.F. Floor/S.F. Partition*	S.F. Partition	7.47	4.98	15.9%
1020	Interior Doors	Single leaf hollow metal *700 S.F. Floor/Door*	Each	1201	3.21	
1030	Fittings	Toilet partitions	S.F. Floor	.46	.46	
2010	Stair Construction	N/A	—	—	—	
3010	Wall Finishes	Paint	S.F. Surface	1.46	1.95	
3020	Floor Finishes	5% quarry tile, 95% vinyl composition tile	S.F. Floor	3.44	3.44	
3030	Ceiling Finishes	Fiberglass tile on tee grid	S.F. Ceiling	6.10	6.10	
D. SERVICES						
	D10 Conveying					
1010	Elevators & Lifts	N/A	—	—	—	0.0 %
1020	Escalators & Moving Walks	N/A	—	—	—	
	D20 Plumbing					
2010	Plumbing Fixtures	Toilet and service fixtures, supply and drainage *1 Fixture/455 S.F. Floor*	Each	3006	16.89	15.8%
2020	Domestic Water Distribution	Electric water heater	S.F. Floor	3.07	3.07	
2040	Rain Water Drainage	N/A	—	—	—	
	D30 HVAC					
3010	Energy Supply	Oil fired hot water, wall fin radiation	S.F. Floor	10.66	10.66	17.9%
3020	Heat Generating Systems	N/A	—	—	—	
3030	Cooling Generating Systems	N/A	—	—	—	
3050	Terminal & Package Units	Split systems with air cooled condensing units	S.F. Floor	12.03	12.03	
3090	Other HVAC Sys. & Equipment	N/A	—	—	—	
	D40 Fire Protection					
4010	Sprinklers	Sprinkler, light hazard	S.F. Floor	3.41	3.41	3.4%
4020	Standpipes	Standpipe	S.F. Floor	.89	.89	
	D50 Electrical					
5010	Electrical Service/Distribution	200 ampere service, panel board and feeders	S.F. Floor	1.06	1.06	8.1%
5020	Lighting & Branch Wiring	High efficiency fluorescent fixtures, receptacles, switches, A.C. and misc. power	S.F. Floor	7.63	7.63	
5030	Communications & Security	Addressable alarm systems and emergency lighting	S.F. Floor	1.51	1.51	
5090	Other Electrical Systems	Emergency generator, 15 kW	S.F. Floor	.09	.09	
E. EQUIPMENT & FURNISHINGS						
1010	Commercial Equipment	N/A	—	—	—	4.1 %
1020	Institutional Equipment	Cabinets and countertop	S.F. Floor	5.19	5.19	
1030	Vehicular Equipment	N/A	—	—	—	
1090	Other Equipment	N/A	—	—	—	
F. SPECIAL CONSTRUCTION						
1020	Integrated Construction	N/A	—	—	—	0.0 %
1040	Special Facilities	N/A	—	—	—	
G. BUILDING SITEWORK		**N/A**				
				Sub-Total	126.56	**100%**
	CONTRACTOR FEES (General Requirements: 10%, Overhead: 5%, Profit: 10%)			*25%*	31.65	
	ARCHITECT FEES			*9%*	14.24	
				Total Building Cost	**172.45**	

Costs per square foot of floor area

Exterior Wall	S.F. Area	12000	18000	24000	30000	36000	42000	48000	54000	60000
	L.F. Perimeter	460	580	713	730	826	880	965	1006	1045
Concrete Block	Rigid Steel	133.00	125.80	122.60	118.15	116.60	114.85	113.95	112.70	111.70
	Bearing Walls	130.70	123.50	120.30	115.85	114.30	112.55	111.65	110.40	109.40
Precast Concrete	Rigid Steel	142.15	133.50	129.70	123.95	122.05	119.85	118.70	117.15	115.85
Metal Panel	Rigid Steel	135.10	127.55	124.20	119.45	117.80	115.95	115.00	113.70	112.65
Face Brick and Common Brick	Rigid Steel	151.65	141.50	137.05	130.00	127.75	125.05	123.70	121.80	120.20
Tilt-up Concrete Panels	Rigid Steel	134.60	127.20	123.85	119.20	117.55	115.75	114.80	113.50	112.45
Perimeter Adj., Add or Deduct	Per 100 L.F.	5.35	3.60	2.70	2.10	1.75	1.55	1.30	1.15	1.10
Story Hgt. Adj., Add or Deduct	Per 1 Ft.	0.80	0.65	0.60	0.45	0.40	0.40	0.35	0.40	0.35
For Basement, add $ 30.20 per square foot of basement area										

The above costs were calculated using the basic specifications shown on the facing page. These costs should be adjusted where necessary for design alternatives and owner's requirements.

Common additives

Description	Unit	$ Cost
Clock System		
20 room	Ea.	17,900
50 room	Ea.	42,300
Dock Bumpers, Rubber blocks		
4-1/2" thick, 10" high, 14" long	Ea.	83
24" long	Ea.	135
36" long	Ea.	345
12" high, 14" long	Ea.	99
24" long	Ea.	117
36" long	Ea.	170
6" thick, 10" high, 14" long	Ea.	101
24" long	Ea.	168
36" long	Ea.	279
20" high, 11" long	Ea.	184
Dock Boards, Heavy		
60" x 60" Aluminum, 5,000# cap.	Ea.	1550
9000# cap.	Ea.	1550
15,000# cap.	Ea.	1475

Description	Unit	$ Cost
Dock Levelers, Hinged 10 ton cap.		
6' x 8'	Ea.	5950
7' x 8'	Ea.	8750
Partitions, Woven wire, 10 ga., 1-1/2" mesh		
4' wide x 7' high	Ea.	199
8' high	Ea.	219
10' High	Ea.	274
Platform Lifter, Portable, 6'x 6'		
3000# cap.	Ea.	10,800
4000# cap.	Ea.	13,300
Fixed, 6' x 8', 5000# cap.	Ea.	11,100

Important: See the Reference Section for Location Factors.

Model costs calculated for a 1 story building with 20′ story height and 30,000 square feet of floor area

Factory, 1 Story

				Unit	Unit Cost	Cost Per S.F.	% Of Sub-Total
A. SUBSTRUCTURE							
1010	Standard Foundations	Poured concrete; strip and spread footings; 4′ foundation wall		S.F. Ground	3.78	3.78	
1020	Special Foundations	N/A		—	—	—	
1030	Slab on Grade	4″ reinforced concrete with vapor barrier and granular base		S.F. Slab	6.90	6.90	12.3%
2010	Basement Excavation	Site preparation for slab and trench for foundation wall and footing		S.F. Ground	.16	.16	
2020	Basement Walls	N/A		—	—	—	
B. SHELL							
	B10 Superstructure						
1010	Floor Construction	N/A		—	—	—	10.6 %
1020	Roof Construction	Metal deck, open web steel joists, beams and columns		S.F. Roof	9.38	9.38	
	B20 Exterior Enclosure						
2010	Exterior Walls	Concrete block	*75% of wall*	S.F. Wall	8.36	3.05	
2020	Exterior Windows	Industrial horizontal pivoted steel	*25% of wall*	Each	827	3.14	8.9%
2030	Exterior Doors	Double aluminum and glass, hollow metal, steel overhead		Each	3293	1.64	
	B30 Roofing						
3010	Roof Coverings	Built-up tar and gravel with flashing; perlite/EPS composite insulation		S.F. Roof	5.93	5.93	7.2%
3020	Roof Openings	Roof hatches		S.F. Roof	.43	.43	
C. INTERIORS							
1010	Partitions	Concrete block	*60 S.F. Floor/L.F. Partition*	S.F. Partition	8.40	1.68	
1020	Interior Doors	Single leaf hollow metal and fire doors	*600 S.F. Floor/Door*	Each	1201	2.01	
1030	Fittings	Toilet partitions		S.F. Floor	1.02	1.02	
2010	Stair Construction	N/A		—	—	—	8.4%
3010	Wall Finishes	Paint		S.F. Surface	4.45	1.78	
3020	Floor Finishes	Vinyl composition tile	*10%of floor*	S.F. Floor	2.80	.28	
3030	Ceiling Finishes	Fiberglass board on exposed grid system	*10% of area*	S.F. Ceiling	6.86	.69	
D. SERVICES							
	D10 Conveying						
1010	Elevators & Lifts	N/A		—	—	—	0.0 %
1020	Escalators & Moving Walks	N/A		—	—	—	
	D20 Plumbing						
2010	Plumbing Fixtures	Toilet and service fixtures, supply and drainage	*1 Fixture/1000 S.F. Floor*	Each	4980	4.98	
2020	Domestic Water Distribution	Gas fired water heater		S.F. Floor	.67	.67	7.7%
2040	Rain Water Drainage	Roof drains		S.F. Roof	1.17	1.17	
	D30 HVAC						
3010	Energy Supply	Oil fired hot water, unit heaters		S.F. Floor	9.15	9.15	
3020	Heat Generating Systems	N/A		—	—	—	
3030	Cooling Generating Systems	Chilled water, air cooled condenser system		S.F. Floor	10.73	10.73	22.5%
3050	Terminal & Package Units	N/A		—	—	—	
3090	Other HVAC Sys. & Equipment	N/A		—	—	—	
	D40 Fire Protection						
4010	Sprinklers	Sprinklers, ordinary hazard		S.F. Floor	3.88	3.88	5.1%
4020	Standpipes	Standpipe, wet, Class III		S.F. Floor	.64	.64	
	D50 Electrical						
5010	Electrical Service/Distribution	600 ampere service, panel board and feeders		S.F. Floor	1.01	1.01	
5020	Lighting & Branch Wiring	High intensity discharge fixtures, receptacles, switches, A.C. and misc. power		S.F. Floor	12.28	12.28	16.7%
5030	Communications & Security	Addressable alarm systems and emergency lighting		S.F. Floor	1.44	1.44	
5090	Other Electrical Systems	N/A		—	—	—	
E. EQUIPMENT & FURNISHINGS							
1010	Commercial Equipment	N/A		—	—	—	
1020	Institutional Equipment	N/A		—	—	—	0.6 %
1030	Vehicular Equipment	Dock shelters		S.F. Floor	.49	.49	
1090	Other Equipment	N/A		—	—	—	
F. SPECIAL CONSTRUCTION							
1020	Integrated Construction	N/A		—	—	—	0.0 %
1040	Special Facilities	N/A		—	—	—	
G. BUILDING SITEWORK	**N/A**						
					Sub-Total	88.31	**100%**
	CONTRACTOR FEES (General Requirements: 10%, Overhead: 5%, Profit: 10%)				*25%*	22.11	
	ARCHITECT FEES				*7%*	7.73	
					Total Building Cost	**118.15**	

Costs per square foot of floor area

Exterior Wall	S.F. Area	4000	4500	5000	5500	6000	6500	7000	7500	8000
	L.F. Perimeter	260	280	300	310	320	336	353	370	386
Face Brick and Concrete Block	Steel Joists	185.55	181.90	179.00	175.10	**171.90**	169.90	168.35	167.00	165.70
	Bearing Walls	181.75	178.05	175.15	171.25	168.05	166.10	164.50	163.15	161.85
Decorative Concrete Block	Steel Joists	172.30	169.20	166.70	163.65	161.05	159.35	158.10	156.95	155.85
	Bearing Walls	168.45	165.35	162.85	159.75	157.20	155.50	154.20	153.10	152.00
Limestone and Concrete Block	Steel Joists	205.75	201.20	197.60	192.60	188.45	185.95	184.00	182.35	180.65
	Bearing Walls	201.90	197.40	193.75	188.75	184.60	182.15	180.15	178.50	176.80
Perimeter Adj., Add or Deduct	Per 100 L.F.	19.85	17.60	15.80	14.40	13.20	12.25	11.30	10.55	9.85
Story Hgt. Adj., Add or Deduct	Per 1 Ft.	2.65	2.55	2.40	2.35	2.25	2.15	2.05	2.00	1.95

For Basement, add $36.80 per square foot of basement area

The above costs were calculated using the basic specifications shown on the facing page. These costs should be adjusted where necessary for design alternatives and owner's requirements.

Common additives

Description	Unit	$ Cost
Appliances		
Cooking range, 30" free standing		
1 oven	Ea.	600 - 2775
2 oven	Ea.	1200 - 3800
30" built-in		
1 oven	Ea.	1050 - 2425
2 oven	Ea.	1600 - 3050
Counter top cook tops, 4 burner	Ea.	455 - 2250
Microwave oven	Ea.	255 - 800
Combination range, refrig. & sink, 30" wide	Ea.	1925 - 2025
60" wide	Ea.	1925
72" wide	Ea.	2625
Combination range, refrigerator, sink microwave oven & icemaker	Ea.	6325
Compactor, residential, 4-1 compaction	Ea.	880 - 1450
Dishwasher, built-in, 2 cycles	Ea.	520 - 870
4 cycles	Ea.	650 - 2475
Garbage disposer, sink type	Ea.	193 - 305
Hood for range, 2 speed, vented, 30" wide	Ea.	237 - 1275
42" wide	Ea.	300 - 2400

Description	Unit	$ Cost
Appliances, cont.		
Refrigerator, no frost 10-12 C.F.	Ea.	575 - 735
14-16 C.F.	Ea.	710 - 1200
18-20 C.F.	Ea.	890 - 2050
Lockers, Steel, single tier, 60" or 72"	Opng.	223 - 390
2 tier, 60" or 72" total	Opng.	135 - 167
5 tier, box lockers	Opng.	69.50 - 81.50
Locker bench, lam. maple top only	L.F.	35.50
Pedestals, steel pipe	Ea.	73.50
Sound System		
Amplifier, 250 watts	Ea.	1950
Speaker, ceiling or wall	Ea.	233
Trumpet	Ea.	445

Important: See the Reference Section for Location Factors.

Model costs calculated for a 1 story building with 14′ story height and 6,000 square feet of floor area

Fire Station, 1 Story

				Unit	Unit Cost	Cost Per S.F.	% Of Sub-Total
A. SUBSTRUCTURE							
1010	Standard Foundations	Poured concrete; strip and spread footings; 4′ foundation wall		S.F. Ground	6.79	6.79	
1020	Special Foundations	N/A		—	—	—	
1030	Slab on Grade	6″ reinforced concrete with vapor barrier and granular base		S.F. Slab	6.90	6.90	11.1%
2010	Basement Excavation	Site preparation for slab and trench for foundation wall and footing		S.F. Ground	.49	.49	
2020	Basement Walls	N/A		—	—	—	
B. SHELL							
B10 Superstructure							
1010	Floor Construction	N/A		—	—	—	8.2 %
1020	Roof Construction	Metal deck, open web steel joists, beams on columns		S.F. Roof	10.49	10.49	
B20 Exterior Enclosure							
2010	Exterior Walls	Face brick with concrete block back-up	*75% of wall*	S.F. Wall	31	17.30	
2020	Exterior Windows	Aluminum insulated glass	*10% of wall*	Each	827	1.93	19.4%
2030	Exterior Doors	Single aluminum and glass, overhead, hollow metal	*15% of wall*	S.F. Door	49	5.47	
B30 Roofing							
3010	Roof Coverings	Built-up tar and gravel with flashing; perlite/EPS composite insulation		S.F. Roof	7.15	7.15	5.8%
3020	Roof Openings	Skylights, roof hatches		S.F. Roof	.19	.19	
C. INTERIORS							
1010	Partitions	Concrete block	*17 S.F. Floor/L.F. Partition*	S.F. Partition	9.15	5.38	
1020	Interior Doors	Single leaf hollow metal	*500 S.F. Floor/Door*	Each	1201	2.40	
1030	Fittings	Toilet partitions		S.F. Floor	.52	.52	
2010	Stair Construction	N/A		—	—	—	13.6%
3010	Wall Finishes	Paint		S.F. Surface	2.86	3.36	
3020	Floor Finishes	50% vinyl tile, 50% paint		S.F. Floor	2.28	2.28	
3030	Ceiling Finishes	Fiberglass board on exposed grid, suspended	*50% of area*	S.F. Ceiling	6.86	3.44	
D. SERVICES							
D10 Conveying							
1010	Elevators & Lifts	N/A		—	—	—	0.0 %
1020	Escalators & Moving Walks	N/A		—	—	—	
D20 Plumbing							
2010	Plumbing Fixtures	Kitchen, toilet and service fixtures, supply and drainage	*1 Fixture/375 S.F. Floor*	Each	3930	10.48	
2020	Domestic Water Distribution	Gast fired water heater		S.F. Floor	3.37	3.37	12.1%
2040	Rain Water Drainage	Roof drains		S.F. Roof	1.61	1.61	
D30 HVAC							
3010	Energy Supply	N/A		—	—	—	
3020	Heat Generating Systems	Included in D3050		—	—	—	
3030	Cooling Generating Systems	N/A		—	—	—	18.1 %
3050	Terminal & Package Units	Rooftop multizone unit system		S.F. Floor	23	23.10	
3090	Other HVAC Sys. & Equipment	N/A		—	—	—	
D40 Fire Protection							
4010	Sprinklers	Wet pipe sprinkler system		S.F. Floor	4.66	4.66	4.8%
4020	Standpipes	Standpipe, wet, Class III		S.F. Floor	1.49	1.49	
D50 Electrical							
5010	Electrical Service/Distribution	200 ampere service, panel board and feeders		S.F. Floor	1.55	1.55	
5020	Lighting & Branch Wiring	High efficiency fluorescent fixtures, receptacles, switches, A.C. and misc. power		S.F. Floor	5.04	5.04	6.7%
5030	Communications & Security	Addressable alarm systems		S.F. Floor	1.95	1.95	
5090	Other Electrical Systems	N/A		—	—	—	
E. EQUIPMENT & FURNISHINGS							
1010	Commercial Equipment	N/A		—	—	—	
1020	Institutional Equipment	N/A		—	—	—	0.0 %
1030	Vehicular Equipment	N/A		—	—	—	
1090	Other Equipment	N/A		—	—	—	
F. SPECIAL CONSTRUCTION							
1020	Integrated Construction	N/A		—	—	—	0.0 %
1040	Special Facilities	N/A		—	—	—	
G. BUILDING SITEWORK	**N/A**						
					Sub-Total	127.34	**100%**
	CONTRACTOR FEES (General Requirements: 10%, Overhead: 5%, Profit: 10%)				*25%*	31.83	
	ARCHITECT FEES				*8%*	12.73	
					Total Building Cost	**171.90**	

Costs per square foot of floor area

Exterior Wall	S.F. Area	4000	5000	6000	8000	10000	12000	14000	16000	18000
	L.F. Perimeter	180	205	230	260	300	340	353	386	420
Wood Clapboard	Wood Frame	198.25	190.10	184.75	176.45	**172.15**	169.15	165.85	164.15	162.95
Aluminum Clapboard	Wood Frame	195.15	187.35	182.15	174.20	170.05	167.25	164.15	162.50	161.35
Wood Board and Batten	Wood Frame	197.60	189.55	184.25	175.95	171.70	168.75	165.50	163.80	162.60
Face Brick and Concrete Block	Bearing Walls	211.00	201.35	194.95	184.40	179.05	175.55	171.00	168.90	167.40
Stucco and Concrete Block	Bearing Walls	194.70	186.45	181.10	172.65	168.25	165.25	161.90	160.15	158.90
Decorative Concrete Block	Wood Joists	203.30	194.75	189.10	180.05	175.45	172.35	168.70	166.90	165.60
Perimeter Adj., Add or Deduct	Per 100 L.F.	15.90	12.80	10.65	7.95	6.30	5.35	4.55	4.05	3.55
Story Hgt. Adj., Add or Deduct	Per 1 Ft.	2.10	2.00	1.85	1.55	1.40	1.35	1.20	1.20	1.10
***For Basement**, add $ 22.80 per square foot of basement area*										

The above costs were calculated using the basic specifications shown on the facing page. These costs should be adjusted where necessary for design alternatives and owner's requirements.

Common additives

Description	Unit	$ Cost
Appliances		
Cooking range, 30" free standing		
1 oven	Ea.	600 - 2775
2 oven	Ea.	1200 - 3800
30" built-in		
1 oven	Ea.	1050 - 2425
2 oven	Ea.	1600 - 3050
Counter top cook tops, 4 burner	Ea.	455 - 2250
Microwave oven	Ea.	255 - 800
Combination range, refrig. & sink, 30" wide	Ea.	1925 - 2025
60" wide	Ea.	1925
72" wide	Ea.	2625
Combination range, refrigerator, sink, microwave oven & icemaker	Ea.	6325
Compactor, residential, 4-1 compaction	Ea.	880 - 1450
Dishwasher, built-in, 2 cycles	Ea.	520 - 870
4 cycles	Ea.	650 - 2475
Garbage disposer, sink type	Ea.	193 - 305
Hood for range, 2 speed, vented, 30" wide	Ea.	237 - 1275
42" wide	Ea.	300 - 2400

Description	Unit	$ Cost
Appliances, cont.		
Refrigerator, no frost 10-12 C.F.	Ea.	575 - 735
14-16 C.F.	Ea.	710 - 1200
18-20 C.F.	Ea.	890 - 2050
Elevators, Hydraulic passenger, 2 stops		
1500# capacity	Ea.	68,400
2500# capacity	Ea.	71,900
3500# capacity	Ea.	76,900
Laundry Equipment		
Dryer, gas, 16 lb. capacity	Ea.	985
30 lb. capacity	Ea.	4000
Washer, 4 cycle	Ea.	1275
Commercial	Ea.	1650
Sound System		
Amplifier, 250 watts	Ea.	1950
Speaker, ceiling or wall	Ea.	233
Trumpet	Ea.	445

Important: See the Reference Section for Location Factors.

Model costs calculated for a 2 story building with 10′ story height and 10,000 square feet of floor area

Fraternity/Sorority House

				Unit	Unit Cost	Cost Per S.F.	% Of Sub-Total
A. SUBSTRUCTURE							
1010	Standard Foundations	Poured concrete; strip and spread footings; 4′ foundation wall		S.F. Ground	6.90	3.45	
1020	Special Foundations	N/A		—	—	—	
1030	Slab on Grade	4″ reinforced concrete with vapor barrier and granular base		S.F. Slab	5.28	2.64	5.0%
2010	Basement Excavation	Site preparation for slab and trench for foundation wall and footing		S.F. Ground	.49	.25	
2020	Basement Walls	N/A		—	—	—	
B. SHELL							
	B10 Superstructure						
1010	Floor Construction	Plywood on wood joists		S.F. Floor	24	11.87	11.0%
1020	Roof Construction	Plywood on wood rafters (pitched)		S.F. Roof	3.63	2.03	
	B20 Exterior Enclosure						
2010	Exterior Walls	Cedar bevel siding on wood studs, insulated	*80% of wall*	S.F. Wall	13.10	6.29	
2020	Exterior Windows	Double hung wood	*20% of wall*	Each	615	2.95	9.1%
2030	Exterior Doors	Solid core wood		Each	3172	2.23	
	B30 Roofing						
3010	Roof Coverings	Asphalt shingles with flashing (pitched); rigid fiberglass insulation		S.F. Roof	4.92	2.46	1.9%
3020	Roof Openings	N/A		—	—	—	
C. INTERIORS							
1010	Partitions	Gypsum board on wood studs	*25 S.F. Floor/L.F. Partition*	S.F. Partition	7.81	2.50	
1020	Interior Doors	Single leaf wood	*200 S.F. Floor/Door*	Each	691	3.46	
1030	Fittings	N/A		—	—	—	
2010	Stair Construction	Wood		Flight	2825	.85	15.3%
3010	Wall Finishes	Paint		S.F. Surface	.78	.50	
3020	Floor Finishes	10% hardwood, 70% carpet, 20% ceramic tile		S.F. Floor	7.78	7.78	
3030	Ceiling Finishes	Gypsum board on wood furring		S.F. Ceiling	4.22	4.22	
D. SERVICES							
	D10 Conveying						
1010	Elevators & Lifts	One hydraulic passenger elevator		Each	85,600	8.56	6.8%
1020	Escalators & Moving Walks	N/A		—	—	—	
	D20 Plumbing						
2010	Plumbing Fixtures	Kitchen toilet and service fixtures, supply and drainage	*1 Fixture/150 S.F. Floor*	Each	2199	14.66	
2020	Domestic Water Distribution	Gas fired water heater		S.F. Floor	3.60	3.60	14.5%
2040	Rain Water Drainage	N/A		—	—	—	
	D30 HVAC						
3010	Energy Supply	Oil fired hot water, baseboard radiation		S.F. Floor	7.51	7.51	
3020	Heat Generating Systems	N/A		—	—	—	
3030	Cooling Generating Systems	N/A		—	—	—	11.3%
3050	Terminal & Package Units	Split system with air cooled condensing unit		S.F. Floor	6.74	6.74	
3090	Other HVAC Sys. & Equipment	N/A		—	—	—	
	D40 Fire Protection						
4010	Sprinklers	Wet pipe sprinkler system		S.F. Floor	3.64	3.64	3.8%
4020	Standpipes	Standpipe		S.F. Floor	1.10	1.10	
	D50 Electrical						
5010	Electrical Service/Distribution	600 ampere service, panel board and feeders		S.F. Floor	3.74	3.74	
5020	Lighting & Branch Wiring	High efficiency fluorescent fixtures, receptacles, switches, A.C. and misc. power		S.F. Floor	9.41	9.41	21.4%
5030	Communications & Security	Addressable alarm, communication system, internet and phone wiring, and generator set		S.F. Floor	13.66	13.66	
5090	Other Electrical Systems	Emergency generator, 7.5 kW		S.F. Floor	.22	.22	
E. EQUIPMENT & FURNISHINGS							
1010	Commercial Equipment	N/A		—	—	—	
1020	Institutional Equipment	N/A		—	—	—	0.0 %
1030	Vehicular Equipment	N/A		—	—	—	
1090	Other Equipment	N/A		—	—	—	
F. SPECIAL CONSTRUCTION							
1020	Integrated Construction	N/A		—	—	—	0.0 %
1040	Special Facilities	N/A		—	—	—	
G. BUILDING SITEWORK	**N/A**						
					Sub-Total	126.32	**100%**
	CONTRACTOR FEES (General Requirements: 10%, Overhead: 5%, Profit: 10%)				*25%*	31.62	
	ARCHITECT FEES				*9%*	14.21	
					Total Building Cost	**172.15**	

Costs per square foot of floor area

Exterior Wall	S.F. Area	4000	6000	8000	10000	12000	14000	16000	18000	20000
	L.F. Perimeter	260	320	384	424	460	484	510	540	576
Wood Board and Batten	Wood Frame	206.40	178.70	165.20	**155.80**	149.40	144.45	140.80	138.05	135.95
Brick Veneer	Wood Frame	213.40	184.50	170.30	160.40	153.55	148.20	144.20	141.30	139.10
Aluminum Clapboard	Wood Frame	199.85	173.35	160.30	151.55	145.55	140.90	137.60	135.00	133.10
Face Brick and Concrete Block	Wood Truss	219.90	189.70	175.05	164.55	157.25	151.45	147.20	144.10	141.80
Limestone and Concrete Block	Wood Truss	240.30	206.50	190.15	177.85	169.30	162.30	157.25	153.50	150.85
Stucco and Concrete Block	Wood Truss	204.00	176.70	163.30	154.15	147.85	142.95	139.45	136.75	134.70
Perimeter Adj., Add or Deduct	Per 100 L.F.	12.15	8.10	6.00	4.85	4.10	3.50	3.05	2.70	2.45
Story Hgt. Adj., Add or Deduct	Per 1 Ft.	1.80	1.50	1.30	1.15	1.10	0.90	0.90	0.85	0.85
***For Basement**, add $30.50 per square foot of basement area*										

The above costs were calculated using the basic specifications shown on the facing page. These costs should be adjusted where necessary for design alternatives and owner's requirements.

Common additives

Description	Unit	$ Cost
Autopsy Table, Standard	Ea.	11,800
Deluxe	Ea.	19,500
Directory Boards, Plastic, glass covered		
30" x 20"	Ea.	570
36" x 48"	Ea.	1500
Aluminum, 24" x 18"	Ea.	570
36" x 24"	Ea.	725
48" x 32"	Ea.	950
48" x 60"	Ea.	2000
Emergency Lighting, 25 watt, battery operated		
Lead battery	Ea.	305
Nickel cadmium	Ea.	525
Mortuary Refrigerator, End operated		
Two capacity	Ea.	10,000
Six capacity	Ea.	18,600

Description	Unit	$ Cost
Planters, Precast concrete		
48" diam., 24" high	Ea.	775
7" diam., 36" high	Ea.	1850
Fiberglass, 36" diam., 24" high	Ea.	845
60" diam., 24" high	Ea.	1300
Smoke Detectors		
Ceiling type	Ea.	231
Duct type	Ea.	510

Important: See the Reference Section for Location Factors.

Model costs calculated for a 1 story building with 12′ story height and 10,000 square feet of floor area

Funeral Home

				Unit	Unit Cost	Cost Per S.F.	% Of Sub-Total
A. SUBSTRUCTURE							
1010	Standard Foundations	Poured concrete; strip and spread footings; 4′ foundation wall		S.F. Ground	5.38	5.38	9.6%
1020	Special Foundations	N/A		—	—	—	
1030	Slab on Grade	4″ reinforced concrete with vapor barrier and granular base		S.F. Slab	5.28	5.28	
2010	Basement Excavation	Site preparation for slab and trench for foundation wall and footing		S.F. Ground	.28	.28	
2020	Basement Walls	N/A		—	—	—	
B. SHELL							
	B10 Superstructure						
1010	Floor Construction	Wood beams on columns		S.F. Floor	.78	.78	5.2%
1020	Roof Construction	Plywood on wood truss		S.F. Roof	5.12	5.12	
	B20 Exterior Enclosure						
2010	Exterior Walls	1″ x 4″ vertical T. & G. redwood siding on wood studs	*90% of wall*	S.F. Wall	16.14	7.39	10.1%
2020	Exterior Windows	Double hung wood	*10% of wall*	Each	491	2.09	
2030	Exterior Doors	Wood swinging double doors, single leaf hollow metal		Each	3375	2.02	
	B30 Roofing						
3010	Roof Coverings	Single ply membrane, fully adhered; polyisocyanurate sheets		S.F. Roof	5.82	5.82	5.1%
3020	Roof Openings	N/A		—	—	—	
C. INTERIORS							
1010	Partitions	Gypsum board on wood studs with sound deadening board	*15 S.F. Floor/L.F. Partition*	S.F. Partition	9	4.80	24.1%
1020	Interior Doors	Single leaf wood	*150 S.F. Floor/Door*	Each	691	4.60	
1030	Fittings	N/A		—	—	—	
2010	Stair Construction	N/A		—	—	—	
3010	Wall Finishes	50% wallpaper, 25% wood paneling, 25% paint		S.F. Surface	3.19	3.40	
3020	Floor Finishes	70% carpet, 30% ceramic tile		S.F. Floor	8.63	8.63	
3030	Ceiling Finishes	Fiberglass board on exposed grid, suspended		S.F. Ceiling	6.10	6.10	
D. SERVICES							
	D10 Conveying						
1010	Elevators & Lifts	N/A		—	—	—	0.0 %
1020	Escalators & Moving Walks	N/A		—	—	—	
	D20 Plumbing						
2010	Plumbing Fixtures	Toilet and service fixtures, supply and drainage	*1 Fixture/770 S.F. Floor*	Each	2780	3.61	18.5%
2020	Domestic Water Distribution	Electric water heater		S.F. Floor	17.52	17.52	
2040	Rain Water Drainage	Roof drain		S.F. Roof			
	D30 HVAC						
3010	Energy Supply	N/A		—	—	—	15.2 %
3020	Heat Generating Systems	Included in D3030		—	—	—	
3030	Cooling Generating Systems	Multizone rooftop unit, gas heating, electric cooling		S.F. Floor	17.35	17.35	
3050	Terminal & Package Units	N/A		—	—	—	
3090	Other HVAC Sys. & Equipment	N/A		—	—	—	
	D40 Fire Protection						
4010	Sprinklers	Wet pipe sprinkler system		S.F. Floor	3.41	3.41	3.8%
4020	Standpipes	Standpipe, wet, Class III		S.F. Floor	.89	.89	
	D50 Electrical						
5010	Electrical Service/Distribution	400 ampere service, panel board and feeders		S.F. Floor	1.85	1.85	8.7%
5020	Lighting & Branch Wiring	High efficiency fluorescent fixtures, receptacles, switches, A.C. and misc. power		S.F. Floor	6.45	6.45	
5030	Communications & Security	Addressable alarm systems and emergency lighting		S.F. Floor	1.51	1.51	
5090	Other Electrical Systems	Emergency generator, 15 kW		S.F. Floor	.09	.09	
E. EQUIPMENT & FURNISHINGS							
1010	Commercial Equipment	N/A		—	—	—	0.0 %
1020	Institutional Equipment	N/A		—	—	—	
1030	Vehicular Equipment	N/A		—	—	—	
1090	Other Equipment	N/A		—	—	—	
F. SPECIAL CONSTRUCTION							
1020	Integrated Construction	N/A		—	—	—	0.0 %
1040	Special Facilities	N/A		—	—	—	
G. BUILDING SITEWORK		**N/A**					
					Sub-Total	114.37	**100%**
	CONTRACTOR FEES (General Requirements: 10%, Overhead: 5%, Profit: 10%)				25%	28.56	
	ARCHITECT FEES				9%	12.87	
					Total Building Cost	**155.80**	

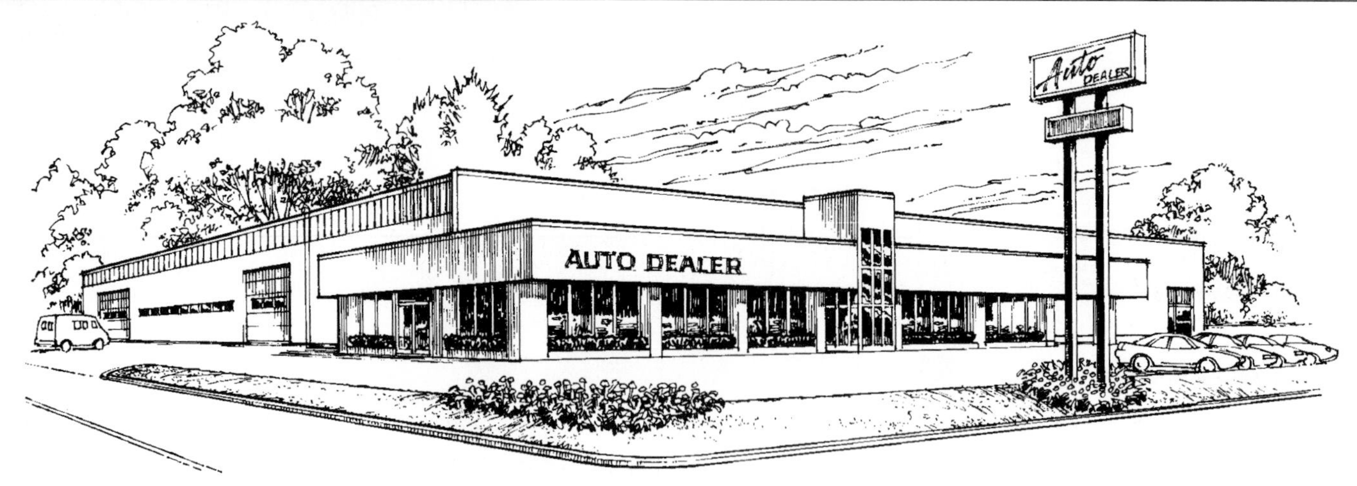

Costs per square foot of floor area

Exterior Wall	S.F. Area	12000	14000	16000	19000	21000	23000	26000	28000	30000
	L.F. Perimeter	440	474	510	556	583	607	648	670	695
E.I.F.S. and Concrete Block	Rigid Steel	127.80	125.10	123.20	120.80	**119.45**	118.15	116.85	115.90	115.20
Tilt-up Concrete Panels	Rigid Steel	125.55	123.05	121.20	119.00	117.70	116.50	115.25	114.40	113.80
Face Brick and Concrete Block	Bearing Walls	127.35	124.15	121.85	119.00	117.40	115.85	114.25	113.10	112.35
	Rigid Steel	134.50	131.35	129.00	126.15	124.50	123.00	121.40	120.30	119.50
Stucco and Concrete Block	Bearing Walls	119.35	116.75	114.95	112.65	111.30	110.15	108.80	107.90	107.30
	Rigid Steel	126.55	123.95	122.10	119.80	118.50	117.25	116.00	115.05	114.45
Perimeter Adj., Add or Deduct	Per 100 L.F.	6.70	5.80	5.05	4.25	3.85	3.55	3.10	2.90	2.75
Story Hgt. Adj., Add or Deduct	Per 1 Ft.	1.45	1.35	1.20	1.20	1.10	1.10	0.95	0.95	1.00
***For Basement**, add $34.50 per square foot of basement area*										

The above costs were calculated using the basic specifications shown on the facing page. These costs should be adjusted where necessary for design alternatives and owner's requirements.

Common additives

Description	Unit	$ Cost
Emergency Lighting, 25 watt, battery operated		
Lead battery	Ea.	305
Nickel cadmium	Ea.	525
Smoke Detectors		
Ceiling type	Ea.	231
Duct type	Ea.	510
Sound System		
Amplifier, 250 watts	Ea.	1950
Speaker, ceiling or wall	Ea.	233
Trumpet	Ea.	445

Model costs calculated for a 1 story building with 14′ story height and 21,000 square feet of floor area

Garage, Auto Sales

			Unit	Unit Cost	Cost Per S.F.	% Of Sub-Total
A. SUBSTRUCTURE						
1010	Standard Foundations	Poured concrete; strip and spread footings; 4′ foundation wall	S.F. Ground	3.60	3.60	
1020	Special Foundations	N/A	—	—	—	
1030	Slab on Grade	4″ reinforced concrete with vapor barrier and granular base	S.F. Slab	7.56	7.56	12.8%
2010	Basement Excavation	Site preparation for slab and trench for foundation wall and footing	S.F. Ground	.28	.28	
2020	Basement Walls	N/A	—	—	—	
B. SHELL						
	B10 Superstructure					
1010	Floor Construction	N/A	—	—	—	13.0 %
1020	Roof Construction	Metal deck, open web steel joists, beams, columns	S.F. Roof	11.62	11.62	
	B20 Exterior Enclosure					
2010	Exterior Walls	E.I.F.S. *70% of wall*	S.F. Wall	16.72	4.55	
2020	Exterior Windows	Window wall *30% of wall*	Each	56	6.55	16.0%
2030	Exterior Doors	Double aluminum and glass, hollow metal, steel overhead	Each	3710	3.18	
	B30 Roofing					
3010	Roof Coverings	Built-up tar and gravel with flashing; perlite/EPS composite insulation	S.F. Roof	6.72	6.72	7.5%
3020	Roof Openings	N/A	—	—	—	
C. INTERIORS						
1010	Partitions	Gypsum board on metal studs *28 S.F. Floor/L.F. Partition*	S.F. Partition	4.81	2.06	
1020	Interior Doors	Hollow metal *280 S.F. Floor/Door*	Each	1201	4.29	
1030	Fittings	N/A	—	—	—	
2010	Stair Construction	N/A	—	—	—	14.1%
3010	Wall Finishes	Paint	S.F. Surface	1.05	.90	
3020	Floor Finishes	50% vinyl tile, 50% paint	S.F. Floor	2.28	2.28	
3030	Ceiling Finishes	Fiberglass board on exposed grid, suspended *50% of area*	S.F. Ceiling	6.10	3.06	
D. SERVICES						
	D10 Conveying					
1010	Elevators & Lifts	N/A	—	—	—	0.0 %
1020	Escalators & Moving Walks	N/A	—	—	—	
	D20 Plumbing					
2010	Plumbing Fixtures	Toilet and service fixtures, supply and drainage *1 Fixture/1500 S.F. Floor*	Each	2910	1.94	
2020	Domestic Water Distribution	Gas fired water heater	S.F. Floor	2.25	2.25	7.3%
2040	Rain Water Drainage	Roof drains	S.F. Roof	2.29	2.29	
	D30 HVAC					
3010	Energy Supply	N/A	—	—	—	
3020	Heat Generating Systems	N/A	—	—	—	
3030	Cooling Generating Systems	N/A	—	—	—	11.1 %
3050	Terminal & Package Units	Single zone rooftop unit, gas heating, electric cooling	S.F. Floor	9.21	9.21	
3090	Other HVAC Sys. & Equipment	Underfloor garage exhaust system	S.F. Floor	.70	.70	
	D40 Fire Protection					
4010	Sprinklers	Wet pipe sprinkler system	S.F. Floor	4.45	4.45	5.5%
4020	Standpipes	Standpipe	S.F. Floor	.46	.46	
	D50 Electrical					
5010	Electrical Service/Distribution	200 ampere service, panel board and feeders	S.F. Floor	.46	.46	
5020	Lighting & Branch Wiring	T-8 fluorescent fixtures, receptacles, switches, A.C. and misc. power	S.F. Floor	6.25	6.25	11.2%
5030	Communications & Security	Addressable alarm systems, partial internet wiring and emergency lighting	S.F. Floor	3.20	3.20	
5090	Other Electrical Systems	Emergency generator, 7.5 kW	S.F. Floor	.07	.07	
E. EQUIPMENT & FURNISHINGS						
1010	Commercial Equipment	N/A	—	—	—	
1020	Institutional Equipment	N/A	—	—	—	1.5 %
1030	Vehicular Equipment	Hoists, compressor, fuel pump	S.F. Floor	1.37	1.37	
1090	Other Equipment	N/A	—	—	—	
F. SPECIAL CONSTRUCTION						
1020	Integrated Construction	N/A	—	—	—	0.0 %
1040	Special Facilities	N/A	—	—	—	
G. BUILDING SITEWORK	**N/A**					
				Sub-Total	89.30	**100%**
	CONTRACTOR FEES (General Requirements: 10%, Overhead: 5%, Profit: 10%)			*25%*	22.34	
	ARCHITECT FEES			*7%*	7.81	
				Total Building Cost	**119.45**	

Costs per square foot of floor area

Exterior Wall	S.F. Area	2000	4000	6000	8000	10000	12000	14000	16000	18000
	L.F. Perimeter	180	260	340	420	450	500	550	575	600
Concrete Block	Wood Joists	151.50	133.25	127.20	124.10	119.85	117.85	116.40	114.50	113.10
	Steel Joists	160.70	140.25	133.40	130.00	**125.30**	123.10	121.40	119.40	117.85
Cast in Place Concrete	Wood Joists	160.55	140.10	133.25	129.80	124.75	122.40	120.65	118.50	116.75
	Steel Joists	171.20	147.80	140.05	136.15	130.55	127.95	126.00	123.60	121.70
Stucco	Wood Frame	149.55	132.20	126.50	123.50	119.40	117.50	116.05	114.20	112.80
Metal Panel	Rigid Steel	173.25	149.05	141.00	136.95	131.25	128.60	126.60	124.10	122.20
Perimeter Adj., Add or Deduct	Per 100 L.F.	24.25	12.10	8.05	6.10	4.85	4.05	3.45	3.05	2.75
Story Hgt. Adj., Add or Deduct	Per 1 Ft.	2.10	1.50	1.25	1.20	1.05	0.95	0.85	0.80	0.75

***For Basement**, add $32.50 per square foot of basement area*

The above costs were calculated using the basic specifications shown on the facing page. These costs should be adjusted where necessary for design alternatives and owner's requirements.

Common additives

Description	Unit	$ Cost
Air Compressors		
Electric 1-1/2 H.P., standard controls	Ea.	1150
Dual controls	Ea.	1750
5 H.P. 115/230 Volt, standard controls	Ea.	3225
Dual controls	Ea.	4425
Product Dispenser		
with vapor recovery for 6 nozzles	Ea.	28,900
Lifts, Single post		
8000# cap., swivel arm	Ea.	9725
Two post, adjustable frames, 12,000# cap.	Ea.	5350
24,000# cap.	Ea.	13,300
30,000# cap.	Ea.	63,000
Four post, roll on ramp, 25,000# cap.	Ea.	24,900
Lockers, Steel, single tier, 60" or 72"	Opng.	223 - 390
2 tier, 60" or 72" total	Opng.	135 - 167
5 tier, box lockers	Opng.	69.50 - 81.50
Locker bench, lam. maple top only	L.F.	35.50
Pedestals, steel pipe	Ea.	73.50
Lube Equipment		
3 reel type, with pumps, no piping	Ea.	13,200
Spray Painting Booth, 26' long, complete	Ea.	13,900

Important: See the Reference Section for Location Factors.

Model costs calculated for a 1 story building with 14′ story height and 10,000 square feet of floor area

Garage, Repair

				Unit	Unit Cost	Cost Per S.F.	% Of Sub-Total
A. SUBSTRUCTURE							
1010	Standard Foundations	Poured concrete; strip and spread footings; 4′ foundation wall		S.F. Ground	5.44	5.44	
1020	Special Foundations	N/A		—	—	—	
1030	Slab on Grade	6″ reinforced concrete with vapor barrier and granular base		S.F. Slab	7.56	7.56	14.3%
2010	Basement Excavation	Site preparation for slab and trench for foundation wall and footing		S.F. Ground	.28	.28	
2020	Basement Walls	N/A		—	—	—	
B. SHELL							
	B10 Superstructure						
1010	Floor Construction	N/A		—	—	—	6.0 %
1020	Roof Construction	Metal deck on open web steel joists		S.F. Roof	5.55	5.55	
	B20 Exterior Enclosure						
2010	Exterior Walls	Concrete block	*80% of wall*	S.F. Wall	12.74	6.42	
2020	Exterior Windows	Hopper type commercial steel	*5% of wall*	Each	498	1.05	10.7%
2030	Exterior Doors	Steel overhead and hollow metal	*15% of wall*	S.F. Door	26	2.43	
	B30 Roofing						
3010	Roof Coverings	Built-up tar and gravel; perlite/EPS composite insulation		S.F. Roof	6.58	6.58	7.1%
3020	Roof Openings	Skylight		S.F. Roof	.02	.02	
C. INTERIORS							
1010	Partitions	Concrete block	*50 S.F. Floor/L.F. Partition*	S.F. Partition	22	4.40	
1020	Interior Doors	Single leaf hollow metal	*3000 S.F. Floor/Door*	Each	1201	.40	
1030	Fittings	Toilet partitions		S.F. Floor	.16	.16	
2010	Stair Construction	N/A		—	—	—	9.9%
3010	Wall Finishes	Paint		S.F. Surface	6.78	2.71	
3020	Floor Finishes	90% metallic floor hardener, 10% vinyl composition tile		S.F. Floor	1.15	1.15	
3030	Ceiling Finishes	Gypsum board on wood joists in office and washrooms	*10% of area*	S.F. Ceiling	4.10	.41	
D. SERVICES							
	D10 Conveying						
1010	Elevators & Lifts	N/A		—	—	—	0.0 %
1020	Escalators & Moving Walks	N/A		—	—	—	
	D20 Plumbing						
2010	Plumbing Fixtures	Toilet and service fixtures, supply and drainage	*1 Fixture/500 S.F. Floor*	Each	1625	3.25	
2020	Domestic Water Distribution	Gas fired water heater		S.F. Floor	.66	.66	6.6%
2040	Rain Water Drainage	Roof drains		S.F. Roof	2.22	2.22	
	D30 HVAC						
3010	Energy Supply	N/A		—	—	—	
3020	Heat Generating Systems	N/A		—	—	—	
3030	Cooling Generating Systems	N/A		—	—	—	11.2 %
3050	Terminal & Package Units	Single zone AC unit		S.F. Floor	9.45	9.45	
3090	Other HVAC Sys. & Equipment	Garage exhaust system		S.F. Floor	.96	.96	
	D40 Fire Protection						
4010	Sprinklers	Sprinklers, ordinary hazard		S.F. Floor	4.45	4.45	5.8%
4020	Standpipes	Standpipe		S.F. Floor	.97	.97	
	D50 Electrical						
5010	Electrical Service/Distribution	200 ampere service, panel board and feeders		S.F. Floor	.43	.43	
5020	Lighting & Branch Wiring	T-8 fluorescent fixtures, receptacles, switches, A.C. and misc. power		S.F. Floor	7.75	7.75	12.6%
5030	Communications & Security	Addressable alarm systems, partial internet wiring and emergency lighting		S.F. Floor	3.44	3.44	
5090	Other Electrical Systems	Emergency generator, 15 kW		S.F. Floor	.09	.09	
E. EQUIPMENT & FURNISHINGS							
1010	Commercial Equipment	N/A		—	—	—	
1020	Institutional Equipment	N/A		—	—	—	15.7 %
1030	Vehicular Equipment	Hoists		S.F. Floor	14.59	14.59	
1090	Other Equipment	N/A		—	—	—	
F. SPECIAL CONSTRUCTION							
1020	Integrated Construction	N/A		—	—	—	0.0 %
1040	Special Facilities	N/A		—	—	—	
G. BUILDING SITEWORK	**N/A**						
					Sub-Total	92.82	**100%**
	CONTRACTOR FEES (General Requirements: 10%, Overhead: 5%, Profit: 10%)				*25%*	23.20	
	ARCHITECT FEES				*8%*	9.28	
					Total Building Cost	**125.30**	

Costs per square foot of floor area

Exterior Wall	S.F. Area	600	800	1000	1200	1400	1600	1800	2000	2200
	L.F. Perimeter	100	120	126	140	153	160	170	180	190
Face Brick and Concrete Block	Wood Truss	278.25	261.00	240.50	231.75	**224.85**	217.05	212.10	208.25	205.10
	Steel Joists	297.55	279.35	257.70	248.40	241.15	232.95	227.75	223.65	220.25
Metal Sandwich Panel	Rigid Steel	266.80	250.10	230.90	222.45	215.90	208.60	203.95	200.35	197.30
Face Brick and Glazed Block	Steel Joists	317.20	297.00	272.55	262.15	254.00	244.70	238.85	234.30	230.45
Aluminum Clapboard	Wood Frame	239.55	225.35	209.80	202.80	197.35	191.45	187.70	184.75	182.20
Wood Clapboard	Wood Frame	240.75	226.45	210.70	203.65	198.20	192.20	188.35	185.35	182.85
Perimeter Adj., Add or Deduct	Per 100 L.F.	119.85	89.90	71.95	59.90	51.40	44.95	40.00	35.95	32.70
Story Hgt. Adj., Add or Deduct	Per 1 Ft.	7.90	7.05	5.95	5.45	5.15	4.70	4.50	4.30	4.00
Basement—Not Applicable										

The above costs were calculated using the basic specifications shown on the facing page. These costs should be adjusted where necessary for design alternatives and owner's requirements.

Common additives

Description	Unit	$ Cost
Air Compressors		
Electric 1-1/2 H.P., standard controls	Ea.	1150
Dual controls	Ea.	1750
5 H.P. 115/230 volt, standard controls	Ea.	3225
Dual controls	Ea.	4425
Product Dispenser		
with vapor recovery for 6 nozzles	Ea.	28,900
Lifts, Single post		
8000# cap. swivel arm	Ea.	9725
Two post, adjustable frames, 12,000# cap.	Ea.	5350
24,000# cap.	Ea.	13,300
30,000# cap.	Ea.	63,000
Four post, roll on ramp, 25,000# cap.	Ea.	24,900
Lockers, Steel, single tier, 60" or 72"	Opng.	223 - 390
2 tier, 60" or 72" total	Opng.	135 - 167
5 tier, box lockers	Ea.	69.50 - 81.50
Locker bench, lam. maple top only	L.F.	35.50
Pedestals, steel pipe	Ea.	73.50
Lube Equipment		
3 reel type, with pumps, no piping	Ea.	13,200

Important: See the Reference Section for Location Factors.

Model costs calculated for a 1 story building with 12′ story height and 1,400 square feet of floor area

Garage, Service Station

				Unit	Unit Cost	Cost Per S.F.	% Of Sub-Total
A. SUBSTRUCTURE							
1010	Standard Foundations	Poured concrete; strip and spread footings; 4′ foundation wall		S.F. Ground	11.63	11.63	12.1%
1020	Special Foundations	N/A		—	—	—	
1030	Slab on Grade	6″ reinforced concrete with vapor barrier and granular base		S.F. Slab	7.56	7.56	
2010	Basement Excavation	Site preparation for slab and trench for foundation wall and footing		S.F. Ground	.95	.95	
2020	Basement Walls	N/A		—	—	—	
B. SHELL							
	B10 Superstructure						
1010	Floor Construction	N/A		—	—	—	4.6 %
1020	Roof Construction	Plywood on wood trusses		S.F. Roof	7.61	7.61	
	B20 Exterior Enclosure						
2010	Exterior Walls	Face brick with concrete block back-up	*60% of wall*	S.F. Wall	31	24.32	30.3%
2020	Exterior Windows	Store front and metal top hinged outswinging	*20% of wall*	Each	46	11.94	
2030	Exterior Doors	Steel overhead, aluminum & glass and hollow metal	*20% of wall*	Each	54	14.16	
	B30 Roofing						
3010	Roof Coverings	Asphalt shingles with flashing; perlite/EPS composite insulation		S.F. Roof	5.08	5.08	3.0%
3020	Roof Openings	N/A		—	—	—	
C. INTERIORS							
1010	Partitions	Concrete block	*25 S.F. Floor/L.F. Partition*	S.F. Partition	7.31	2.34	7.3%
1020	Interior Doors	Single leaf hollow metal	*700 S.F. Floor/Door*	Each	1201	1.72	
1030	Fittings	Toilet partitions		S.F. Floor	2.19	2.19	
2010	Stair Construction	N/A		—	—	—	
3010	Wall Finishes	Paint		S.F. Surface	5.39	3.45	
3020	Floor Finishes	Vinyl composition tile	*35% of floor area*	S.F. Floor	2.86	1	
3030	Ceiling Finishes	Painted gypsum board on furring in sales area & washrooms	*35% of floor area*	S.F. Ceiling	4.22	1.47	
D. SERVICES							
	D10 Conveying						
1010	Elevators & Lifts	N/A		—	—	—	0.0 %
1020	Escalators & Moving Walks	N/A		—	—	—	
	D20 Plumbing						
2010	Plumbing Fixtures	Toilet and service fixtures, supply and drainage	*1 Fixture/235 S.F. Floor*	Each	2143	9.12	8.3%
2020	Domestic Water Distribution	Gas fired water heater		S.F. Floor	4.74	4.74	
2040	Rain Water Drainage	N/A		—	—	—	
	D30 HVAC						
3010	Energy Supply	N/A		—	—	—	14.4 %
3020	Heat Generating Systems	N/A		—	—	—	
3030	Cooling Generating Systems	N/A		—	—	—	
3050	Terminal & Package Units	Single zone AC unit		S.F. Floor	18.26	18.26	
3090	Other HVAC Sys. & Equipment	Underfloor exhaust system		S.F. Floor	5.71	5.71	
	D40 Fire Protection						
4010	Sprinklers	Wet pipe sprinkler system		S.F. Floor	9.44	9.44	8.3%
4020	Standpipes	Standpipe		S.F. Floor	4.36	4.36	
	D50 Electrical						
5010	Electrical Service/Distribution	200 ampere service, panel board and feeders		S.F. Floor	4.57	4.57	11.7%
5020	Lighting & Branch Wiring	High efficiency fluorescent fixtures, receptacles, switches, A.C. and misc. power		S.F. Floor	5.88	5.88	
5030	Communications & Security	Addressable alarm systems and emergency lighting		S.F. Floor	8.40	8.40	
5090	Other Electrical Systems	Emergency generator		S.F. Floor	.66	.66	
E. EQUIPMENT & FURNISHINGS							
1010	Commercial Equipment	N/A		—	—	—	0.0 %
1020	Institutional Equipment	N/A		—	—	—	
1030	Vehicular Equipment	N/A		—	—	—	
1090	Other Equipment	N/A		—	—	—	
F. SPECIAL CONSTRUCTION							
1020	Integrated Construction	N/A		—	—	—	0.0 %
1040	Special Facilities	N/A		—	—	—	
G. BUILDING SITEWORK		**N/A**					
					Sub-Total	166.56	**100%**
	CONTRACTOR FEES (General Requirements: 10%, Overhead: 5%, Profit: 10%)				*25%*	41.63	
	ARCHITECT FEES				*8%*	16.66	
					Total Building Cost	**224.85**	

Gymnasium

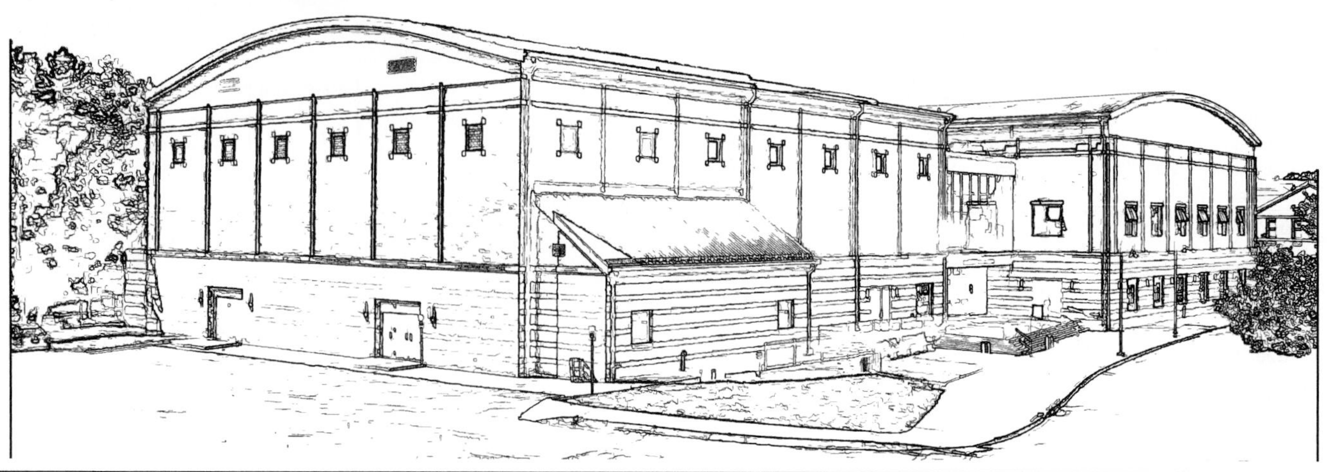

Costs per square foot of floor area

Exterior Wall	S.F. Area	12000	16000	20000	25000	30000	35000	40000	45000	50000
	L.F. Perimeter	440	520	600	700	708	780	841	910	979
Concrete Block	Wood Arch	173.15	168.75	**166.20**	164.05	160.25	158.95	157.75	157.00	156.30
	Rigid Steel	180.55	176.10	173.50	171.35	167.45	166.20	164.95	164.15	163.55
Face Brick and Concrete Block	Wood Arch	194.90	188.00	183.95	180.65	174.20	172.15	170.25	169.00	167.95
	Rigid Steel	201.80	194.95	190.85	187.55	181.10	179.05	177.10	175.90	174.80
Metal Sandwich Panel	Wood Arch	183.10	177.55	174.25	171.60	166.60	164.95	163.45	162.45	161.65
	Rigid Steel	193.40	187.55	184.00	181.15	175.75	174.00	172.35	171.25	170.40
Perimeter Adj., Add or Deduct	Per 100 L.F.	6.60	4.95	3.90	3.15	2.60	2.25	1.95	1.75	1.60
Story Hgt. Adj., Add or Deduct	Per 1 Ft.	0.95	0.85	0.75	0.75	0.55	0.60	0.55	0.50	0.55
Basement—Not Applicable										

The above costs were calculated using the basic specifications shown on the facing page. These costs should be adjusted where necessary for design alternatives and owner's requirements.

Common additives

Description	Unit	$ Cost
Bleachers, Telescoping, manual		
To 15 tier	Seat	158 - 216
16-20 tier	Seat	315 - 405
21-30 tier	Seat	330 - 520
For power operation, add	Seat	64 - 103
Gym Divider Curtain, Mesh top		
Manual roll-up	S.F.	12.80
Gym Mats		
2" naugahyde covered	S.F.	5.75
2" nylon	S.F.	9.20
1-1/2" wall pads	S.F.	8.60
1" wrestling mats	S.F.	5.60
Scoreboard		
Basketball, one side	Ea.	3550 - 13,800
Basketball Backstop		
Wall mtd., 6' extended, fixed	Ea.	2675 - 3175
Swing up, wall mtd.	Ea.	2725 - 4275

Description	Unit	$ Cost
Lockers, Steel, single tier, 60" or 72"	Opng.	223 - 390
2 tier, 60" or 72" total	Opng.	135 - 167
5 tier, box lockers	Opng.	69.50 - 81.50
Locker bench, lam. maple top only	L.F.	35.50
Pedestals, steel pipe	Ea.	73.50
Sound System		
Amplifier, 250 watts	Ea.	1950
Speaker, ceiling or wall	Ea.	233
Trumpet	Ea.	445
Emergency Lighting, 25 watt, battery operated		
Lead battery	Ea.	305
Nickel cadmium	Ea.	525

Important: See the Reference Section for Location Factors.

Model costs calculated for a 1 story building with 25′ story height and 20,000 square feet of floor area

Gymnasium

				Unit	Unit Cost	Cost Per S.F.	% Of Sub-Total
A. SUBSTRUCTURE							
1010	Standard Foundations	Poured concrete; strip and spread footings; 4′ foundation wall		S.F. Ground	3.54	3.54	7.2%
1020	Special Foundations	N/A		—	—	—	
1030	Slab on Grade	4″ reinforced concrete with vapor barrier and granular base		S.F. Slab	5.28	5.28	
2010	Basement Excavation	Site preparation for slab and trench for foundation wall and footing		S.F. Ground	.16	.16	
2020	Basement Walls	N/A		—	—	—	
B. SHELL							
	B10 Superstructure						
1010	Floor Construction	Steel column		S.F. Floor	2.24	2.24	17.3%
1020	Roof Construction	Wood deck on laminated wood arches		S.F. Roof	19.25	19.25	
	B20 Exterior Enclosure						
2010	Exterior Walls	Reinforced concrete block (end walls included)	*90% of wall*	S.F. Wall	11.20	7.56	10.6%
2020	Exterior Windows	Metal horizontal pivoted	*10% of wall*	Each	650	4.87	
2030	Exterior Doors	Aluminum and glass, hollow metal, steel overhead		Each	2263	.68	
	B30 Roofing						
3010	Roof Coverings	EPDM, 60 mils, fully adhered; polyisocyanurate insulation		S.F. Roof	5.25	5.25	4.2%
3020	Roof Openings	N/A		—	—	—	
C. INTERIORS							
1010	Partitions	Concrete block	*50 S.F. Floor/L.F. Partition*	S.F. Partition	9.15	1.83	19.8%
1020	Interior Doors	Single leaf hollow metal	*500 S.F. Floor/Door*	Each	1201	2.40	
1030	Fittings	Toilet partitions		S.F. Floor	.31	.31	
2010	Stair Construction	N/A		—	—	—	
3010	Wall Finishes	50% paint, 50% ceramic tile		S.F. Surface	9.50	3.80	
3020	Floor Finishes	90% hardwood, 10% ceramic tile		S.F. Floor	15.22	15.22	
3030	Ceiling Finishes	Mineral fiber tile on concealed zee bars	*15% of area*	S.F. Ceiling	6.86	1.03	
D. SERVICES							
	D10 Conveying						
1010	Elevators & Lifts	N/A		—	—	—	0.0 %
1020	Escalators & Moving Walks	N/A		—	—	—	
	D20 Plumbing						
2010	Plumbing Fixtures	Toilet and service fixtures, supply and drainage	*1 Fixture/515 S.F. Floor*	Each	4779	9.28	11.8%
2020	Domestic Water Distribution	Electric water heater		S.F. Floor	5.40	5.40	
2040	Rain Water Drainage	N/A		—	—	—	
	D30 HVAC						
3010	Energy Supply	N/A		—	—	—	9.2 %
3020	Heat Generating Systems	Included in D3050		—	—	—	
3030	Cooling Generating Systems	N/A		—	—	—	
3050	Terminal & Package Units	Single zone rooftop unit, gas heating, electric cooling		S.F. Floor	11.40	11.40	
3090	Other HVAC Sys. & Equipment	N/A		—	—	—	
	D40 Fire Protection						
4010	Sprinklers	Wet pipe sprinkler system		S.F. Floor	3.41	3.41	3.6%
4020	Standpipes	Standpipe		S.F. Floor	1.05	1.05	
	D50 Electrical						
5010	Electrical Service/Distribution	400 ampere service, panel board and feeders		S.F. Floor	1.06	1.06	10.3%
5020	Lighting & Branch Wiring	High efficiency fluorescent fixtures, receptacles, switches, A.C. and misc. power		S.F. Floor	8.65	8.65	
5030	Communications & Security	Addressable alarm systems, sound system and emergency lighting		S.F. Floor	2.82	2.82	
5090	Other Electrical Systems	Emergency generator, 7.5 kW		S.F. Floor	.22	.22	
E. EQUIPMENT & FURNISHINGS							
1010	Commercial Equipment	N/A		—	—	—	6.1 %
1020	Institutional Equipment	N/A		—	—	—	
1030	Vehicular Equipment	N/A		—	—	—	
1090	Other Equipment	Bleachers, sauna, weight room		S.F. Floor	7.53	7.53	
F. SPECIAL CONSTRUCTION							
1020	Integrated Construction	N/A		—	—	—	0.0 %
1040	Special Facilities	N/A		—	—	—	
G. BUILDING SITEWORK	**N/A**						
					Sub-Total	124.24	**100%**
	CONTRACTOR FEES (General Requirements: 10%, Overhead: 5%, Profit: 10%)				*25%*	31.09	
	ARCHITECT FEES				*7%*	10.87	
					Total Building Cost	**166.20**	

Costs per square foot of floor area

Exterior Wall	S.F. Area	1000	2000	3000	4000	5000	10000	15000	20000	25000
	L.F. Perimeter	126	179	219	253	283	400	490	568	632
Decorative Concrete Block	Rigid Steel	278.80	236.70	**221.05**	212.55	207.10	195.15	190.35	187.80	185.95
	Bearing Walls	300.00	250.35	231.20	220.80	213.95	198.50	192.20	188.75	186.30
Face Brick and Concrete Block	Rigid Steel	328.75	272.00	249.65	237.30	229.20	210.55	202.85	198.55	195.50
	Bearing Walls	326.25	269.00	246.45	233.95	225.70	206.80	199.00	194.70	191.60
Metal Sandwich Panel	Rigid Steel	301.35	252.70	234.05	223.80	217.20	202.15	196.05	192.75	190.35
Precast Concrete	Bearing Walls	294.00	246.10	227.75	217.80	211.25	196.60	190.70	187.45	185.10
Perimeter Adj., Add or Deduct	Per 100 L.F.	39.20	19.65	13.05	9.85	7.90	3.85	2.60	1.95	1.55
Story Hgt. Adj., Add or Deduct	Per 1 Ft.	1.90	1.35	1.05	0.95	0.85	0.55	0.50	0.40	0.40
***For Basement**, add $31.80 per square foot of basement area*										

The above costs were calculated using the basic specifications shown on the facing page. These costs should be adjusted where necessary for design alternatives and owner's requirements.

Common additives

Description	Unit	$ Cost
Closed Circuit Surveillance, One station		
Camera and monitor	Ea.	1250
For additional camera stations, add	Ea.	565
Emergency Lighting, 25 watt, battery operated		
Lead battery	Ea.	305
Nickel cadmium	Ea.	525
Laundry Equipment		
Dryers, coin operated 30 lb.	Ea.	4000
Double stacked	Ea.	9300
50 lb.	Ea.	5300
Dry cleaner 20 lb.	Ea.	41,600
30 lb.	Ea.	65,000
Washers, coin operated	Ea.	1650
Washer/extractor 20 lb.	Ea.	7900
30 lb.	Ea.	12,200
50 lb.	Ea.	15,300
75 lb.	Ea.	23,800
Smoke Detectors		
Ceiling type	Ea.	231
Duct type	Ea.	510

Important: See the Reference Section for Location Factors.

Model costs calculated for a 1 story building with 12′ story height and 3,000 square feet of floor area

Laundromat

				Unit	Unit Cost	Cost Per S.F.	% Of Sub-Total
A. SUBSTRUCTURE							
1010	Standard Foundations	Poured concrete; strip and spread footings; 4′ foundation wall		S.F. Ground	8.34	8.34	
1020	Special Foundations	N/A		—	—	—	
1030	Slab on Grade	5″ reinforced concrete with vapor barrier and granular base		S.F. Slab	5.80	5.80	8.9%
2010	Basement Excavation	Site preparation for slab and trench for foundation wall and footing		S.F. Ground	.49	.49	
2020	Basement Walls	N/A		—	—	—	
B. SHELL							
	B10 Superstructure						
1010	Floor Construction	Steel column fireproofing		S.F. Floor	.85	.85	6.6%
1020	Roof Construction	Metal deck, open web steel joists, beams, columns		S.F. Roof	10.04	10.04	
	B20 Exterior Enclosure						
2010	Exterior Walls	Decorative concrete block	*90% of wall*	S.F. Wall			
2020	Exterior Windows	Store front	*10% of wall*	Each	72	6.34	5.9%
2030	Exterior Doors	Double aluminum and glass		Each	5113	3.42	
	B30 Roofing						
3010	Roof Coverings	Built-up tar and gravel with flashing; perlite/EPS composite insulation		S.F. Roof	8.67	8.67	5.2%
3020	Roof Openings	N/A		—	—	—	
C. INTERIORS							
1010	Partitions	Gypsum board on metal studs	*60 S.F. Floor/L.F. Partition*	S.F. Partition	26	4.26	
1020	Interior Doors	Single leaf wood	*750 S.F. Floor/Door*	Each	691	.92	
1030	Fittings	N/A		—	—	—	
2010	Stair Construction	N/A		—	—	—	7.1%
3010	Wall Finishes	Paint		S.F. Surface	.78	.26	
3020	Floor Finishes	Vinyl composition tile		S.F. Floor	2.27	2.27	
3030	Ceiling Finishes	Fiberglass board on exposed grid system		S.F. Ceiling	4.10	4.10	
D. SERVICES							
	D10 Conveying						
1010	Elevators & Lifts	N/A		—	—	—	0.0 %
1020	Escalators & Moving Walks	N/A		—	—	—	
	D20 Plumbing						
2010	Plumbing Fixtures	Toilet and service fixtures, supply and drainage	*1 Fixture/600 S.F. Floor*	Each	7086	11.81	
2020	Domestic Water Distribution	Gas fired hot water heater		S.F. Floor	54	54.02	41.0%
2040	Rain Water Drainage	Roof drains		S.F. Roof	1.87	1.87	
	D30 HVAC						
3010	Energy Supply	N/A		—	—	—	
3020	Heat Generating Systems	Included in D3050		—	—	—	
3030	Cooling Generating Systems	N/A		—	—	—	5.3 %
3050	Terminal & Package Units	Rooftop single zone unit systems		S.F. Floor	8.70	8.70	
3090	Other HVAC Sys. & Equipment	N/A		—	—	—	
	D40 Fire Protection						
4010	Sprinklers	Sprinkler, ordinary hazard		S.F. Floor	5.15	5.15	3.1%
4020	Standpipes	N/A		—	—	—	
	D50 Electrical						
5010	Electrical Service/Distribution	200 ampere service, panel board and feeders		S.F. Floor	3.91	3.91	
5020	Lighting & Branch Wiring	High efficiency fluorescent fixtures, receptacles, switches, A.C. and misc. power		S.F. Floor	20	20.42	16.9%
5030	Communications & Security	Addressable alarm systems and emergency lighting		S.F. Floor	3.09	3.09	
5090	Other Electrical Systems	Emergency generator, 7.5 kW		S.F. Floor	.52	.52	
E. EQUIPMENT & FURNISHINGS							
1010	Commercial Equipment	N/A		—	—	—	
1020	Institutional Equipment	N/A		—	—	—	0.0 %
1030	Vehicular Equipment	N/A		—	—	—	
1090	Other Equipment	N/A		—	—	—	
F. SPECIAL CONSTRUCTION							
1020	Integrated Construction	N/A		—	—	—	0.0 %
1040	Special Facilities	N/A		—	—	—	
G. BUILDING SITEWORK	**N/A**						
					Sub-Total	165.25	**100%**
	CONTRACTOR FEES (General Requirements: 10%, Overhead: 5%, Profit: 10%)				*25%*	41.34	
	ARCHITECT FEES				*7%*	14.46	
					Total Building Cost	**221.05**	

Costs per square foot of floor area

Exterior Wall	S.F. Area	7000	10000	13000	16000	19000	22000	25000	28000	31000
	L.F. Perimeter	240	300	336	386	411	435	472	510	524
Face Brick and Concrete Block	Reinforced Concrete	190.90	181.80	173.85	170.35	165.85	**162.45**	160.75	159.50	157.15
	Rigid Steel	189.90	180.75	172.85	169.35	164.75	161.40	159.75	158.40	156.10
Limestone and Concrete Block	Reinforced Concrete	222.05	207.10	194.70	189.00	182.00	176.85	174.15	172.10	168.60
	Rigid Steel	214.80	202.50	191.60	186.85	180.50	175.80	173.45	171.70	168.40
Precast Concrete	Reinforced Concrete	231.45	215.25	201.75	195.60	187.95	182.20	179.35	177.05	173.25
	Rigid Steel	230.40	214.25	200.70	194.55	186.90	181.20	178.25	176.05	172.20
Perimeter Adj., Add or Deduct	Per 100 L.F.	22.85	15.95	12.25	9.95	8.35	7.25	6.40	5.70	5.15
Story Hgt. Adj., Add or Deduct	Per 1 Ft.	3.45	3.00	2.65	2.40	2.15	2.00	1.95	1.75	1.75
***For Basement**, add $45.40 per square foot of basement area*										

The above costs were calculated using the basic specifications shown on the facing page. These costs should be adjusted where necessary for design alternatives and owner's requirements.

Common additives

Description	Unit	$ Cost
Carrels Hardwood	Ea.	715 - 1900
Closed Circuit Surveillance, One station		
Camera and monitor	Ea.	1250
For additional camera stations, add	Ea.	565
Elevators, Hydraulic passenger, 2 stops		
1500# capacity	Ea.	68,400
2500# capacity	Ea.	71,900
3500# capacity	Ea.	76,900
Emergency Lighting, 25 watt, battery operated		
Lead battery	Ea.	305
Nickel cadmium	Ea.	525
Flagpoles, Complete		
Aluminum, 20′ high	Ea.	1825
40′ high	Ea.	4400
70′ high	Ea.	11,600
Fiberglass, 23′ high	Ea.	1375
39′-5″ high	Ea.	3175
59′ high	Ea.	7000

Description	Unit	$ Cost
Library Furnishings		
Bookshelf, 90″ high, 10″ shelf double face	L.F.	253
single face	L.F.	172
Charging desk, built-in with counter		
Plastic laminated top	L.F.	595
Reading table, laminated		
top 60″ x 36″	Ea.	510

Important: See the Reference Section for Location Factors.

Model costs calculated for a 2 story building with 14′ story height and 22,000 square feet of floor area

Library

				Unit	Unit Cost	Cost Per S.F.	% Of Sub-Total
A. SUBSTRUCTURE							
1010	Standard Foundations	Poured concrete; strip and spread footings; 4′ foundation wall		S.F. Ground	6.96	3.48	5.2%
1020	Special Foundations	N/A		—	—	—	
1030	Slab on Grade	4″ reinforced concrete with vapor barrier and granular base		S.F. Slab	5.28	2.64	
2010	Basement Excavation	Site preparation for slab and trench for foundation wall and footing		S.F. Ground	.28	.14	
2020	Basement Walls	N/A		—	—	—	
B. SHELL							
	B10 Superstructure						
1010	Floor Construction	Concrete waffle slab		S.F. Floor	26	13.15	19.6%
1020	Roof Construction	Concrete waffle slab		S.F. Roof	21	10.41	
	B20 Exterior Enclosure						
2010	Exterior Walls	Face brick with concrete block back-up	*90% of wall*	S.F. Wall	31	15.30	15.7%
2020	Exterior Windows	Window wall	*10% of wall*	Each	54	2.98	
2030	Exterior Doors	Double aluminum and glass, single leaf hollow metal		Each	6950	.63	
	B30 Roofing						
3010	Roof Coverings	Single ply membrane, EPDM, fully adhered; perlite/EPS composite insulation		S.F. Roof	5.42	2.71	2.3%
3020	Roof Openings	Roof hatches		S.F. Roof	.10	.05	
C. INTERIORS							
1010	Partitions	Gypsum board on metal studs	*30 S.F. Floor/L.F. Partition*	S.F. Partition	11.53	4.61	16.8%
1020	Interior Doors	Single leaf wood	*300 S.F. Floor/Door*	Each	1201	4	
1030	Fittings	N/A		—	—	—	
2010	Stair Construction	Concrete filled metal pan		Flight	9650	.88	
3010	Wall Finishes	Paint		S.F. Surface	.80	.64	
3020	Floor Finishes	50% carpet, 50% vinyl tile		S.F. Floor	3.97	3.97	
3030	Ceiling Finishes	Mineral fiber on concealed zee bars		S.F. Ceiling	6.10	6.10	
D. SERVICES							
	D10 Conveying						
1010	Elevators & Lifts	One hydraulic passenger elevator		Each	88,880	4.04	3.4%
1020	Escalators & Moving Walks	N/A		—	—	—	
	D20 Plumbing						
2010	Plumbing Fixtures	Toilet and service fixtures, supply and drainage	*1 Fixture/1835 S.F. Floor*	Each	7891	4.30	5.4%
2020	Domestic Water Distribution	Gas fired water heater		S.F. Floor	1.42	1.42	
2040	Rain Water Drainage	Roof drains		S.F. Roof	1.46	.73	
	D30 HVAC						
3010	Energy Supply	N/A		—	—	—	16.9 %
3020	Heat Generating Systems	Included in D3050		—	—	—	
3030	Cooling Generating Systems	N/A		—	—	—	
3050	Terminal & Package Units	Multizone unit, gas heating, electric cooling		S.F. Floor	20	20.35	
3090	Other HVAC Sys. & Equipment	N/A		—	—	—	
	D40 Fire Protection						
4010	Sprinklers	Wet pipe sprinkler system		S.F. Floor	2.91	2.91	3.3%
4020	Standpipes	Standpipe		S.F. Floor	1.01	1.01	
	D50 Electrical						
5010	Electrical Service/Distribution	400 ampere service, panel board and feeders		S.F. Floor	1.12	1.12	11.5%
5020	Lighting & Branch Wiring	High efficiency fluorescent fixtures, receptacles, switches, A.C. and misc. power		S.F. Floor	10.64	10.64	
5030	Communications & Security	Addressable alarm systems, internet wiring, and emergency lighting		S.F. Floor	2.02	2.02	
5090	Other Electrical Systems	Emergency generator, 7.5 kW, Uninterruptible power supply		S.F. Floor	.10	.10	
E. EQUIPMENT & FURNISHINGS							
1010	Commercial Equipment	N/A		—	—	—	0.0 %
1020	Institutional Equipment	N/A		—	—	—	
1030	Vehicular Equipment	N/A		—	—	—	
1090	Other Equipment	N/A		—	—	—	
F. SPECIAL CONSTRUCTION							
1020	Integrated Construction	N/A		—	—	—	0.0 %
1040	Special Facilities	N/A		—	—	—	
G. BUILDING SITEWORK	**N/A**						
					Sub-Total	120.33	**100%**
	CONTRACTOR FEES (General Requirements: 10%, Overhead: 5%, Profit: 10%)				*25%*	30.09	
	ARCHITECT FEES				*8%*	12.03	
					Total Building Cost	**162.45**	

Costs per square foot of floor area

Exterior Wall	S.F. Area	4000	5500	7000	8500	10000	11500	13000	14500	16000
	L.F. Perimeter	280	320	380	440	453	503	510	522	560
Vinyl Clapboard	Wood Frame	181.15	172.75	169.70	167.80	163.50	162.30	159.35	157.20	156.45
Stone Veneer	Wood Frame	185.80	175.00	**171.15**	168.70	163.05	161.55	157.65	154.85	153.90
Fiber Cement	Rigid Steel	185.25	176.45	173.25	171.25	166.80	165.50	162.40	160.15	159.35
E.I.F.S.	Rigid Steel	186.80	177.75	174.50	172.50	167.80	166.50	163.35	161.00	160.15
Precast Concrete	Reinforced Concrete	224.65	210.30	205.30	202.15	194.55	192.55	187.35	183.60	182.35
Brick Veneer	Reinforced Concrete	237.00	221.10	215.55	212.05	203.65	201.40	195.60	191.40	190.05
Perimeter Adj., Add or Deduct	Per 100 L.F.	15.60	11.25	8.90	7.35	6.20	5.40	4.85	4.30	4.00
Story Hgt. Adj., Add or Deduct	Per 1 Ft.	2.05	1.75	1.60	1.55	1.35	1.25	1.15	1.05	1.05
***For Basement**, add $29.50 per square foot of basement area*										

The above costs were calculated using the basic specifications shown on the facing page. These costs should be adjusted where necessary for design alternatives and owner's requirements.

Common additives

Description	Unit	$ Cost
Cabinets, Hospital, base		
Laminated plastic	L.F.	540
Stainless steel	L.F.	905
Counter top, laminated plastic	L.F.	84.50
Stainless steel	L.F.	217
For drop-in sink, add	Ea.	1600
Nurses station, door type		
Laminated plastic	L.F.	605
Enameled steel	L.F.	585
Stainless steel	L.F.	1075
Wall cabinets, laminated plastic	L.F.	400
Enameled steel	L.F.	470
Stainless steel	L.F.	870

Description	Unit	$ Cost
Directory Boards, Plastic, glass covered		
30" x 20"	Ea.	570
36" x 48"	Ea.	1500
Aluminum, 24" x 18"	Ea.	570
36" x 24"	Ea.	725
48" x 32"	Ea.	950
48" x 60"	Ea.	2000
Heat Therapy Unit		
Humidified, 26" x 78" x 28"	Ea.	4750
Smoke Detectors		
Ceiling type	Ea.	231
Duct type	Ea.	510
Tables, Examining, vinyl top with base cabinets	Ea.	1525 - 6500
Utensil Washer, Sanitizer	Ea.	9950
X-Ray, Mobile	Ea.	19,400 - 106,500

Important: See the Reference Section for Location Factors.

Model costs calculated for a 1 story building with 10′ story height and 7,000 square feet of floor area

Medical Office, 1 Story

				Unit	Unit Cost	Cost Per S.F.	% Of Sub-Total
A. SUBSTRUCTURE							
1010	Standard Foundations	Poured concrete; strip and spread footings; 4′ foundation wall		S.F. Ground	7.24	7.24	
1020	Special Foundations	N/A		—	—	—	
1030	Slab on Grade	4″ reinforced slab on grade		S.F. Slab	5.28	5.28	10.2%
2010	Basement Excavation	Site preparation for slab and trench for foundation wall and footing		S.F. Ground	.28	.28	
2020	Basement Walls	N/A		—	—	—	
B. SHELL							
B10 Superstructure							
1010	Floor Construction	Wood columns, fireproofed		S.F. Floor	3.13	3.13	8.5%
1020	Roof Construction	Wood roof trusses, 4:12 slope		S.F. Roof	7.55	7.55	
B20 Exterior Enclosure							
2010	Exterior Walls	Ashlar stone veneer on wood stud back-up	*70% of wall*	S.F. Wall	39	14.76	
2020	Exterior Windows	Double hung, insulated wood windows	*30% of wall*	Each	615	5.89	18.1%
2030	Exterior Doors	Single leaf solid core wood doors		Each	2380	2.04	
B30 Roofing							
3010	Roof Coverings	Asphalt shingles with flashing, gutters and downspouts		S.F. Roof	2.54	2.54	2.0%
3020	Roof Openings	N/A		—	—	—	
C. INTERIORS							
1010	Partitions	Fire rated gypsum board on wood studs	*6 S.F. Floor/L.F. Partition*	S.F. Partition	7.19	9.58	
1020	Interior Doors	Single leaf solid core wood doors	*240 S.F. Floor/Door*	Each	1201	5.01	
1030	Fittings	N/A		—	—	—	
2010	Stair Construction	N/A		—	—	—	21.6%
3010	Wall Finishes	Paint on drywall		S.F. Surface	.90	2.41	
3020	Floor Finishes	50% carpet , 50% vinyl composition tile		S.F. Floor	3.97	3.97	
3030	Ceiling Finishes	Acoustic ceiling tiles on suspended support system		S.F. Ceiling	6.15	6.15	
D. SERVICES							
D10 Conveying							
1010	Elevators & Lifts	N/A		—	—	—	0.0 %
1020	Escalators & Moving Walks	N/A		—	—	—	
D20 Plumbing							
2010	Plumbing Fixtures	Restroom, exam room and service fixtures, supply and drainage	*1 Fixture/350 S.F. Floor*	Each	1960	5.60	
2020	Domestic Water Distribution	Gas fired water heater		S.F. Floor	2.27	2.27	6.3%
2040	Rain Water Drainage	Roof drains		—	—	—	
D30 HVAC							
3010	Energy Supply	N/A		—	—	—	
3020	Heat Generating Systems	Included in D3050		—	—	—	
3030	Cooling Generating Systems	N/A		—	—	—	11.9 %
3050	Terminal & Package Units	Multizone unit, gas heating , electric cooling		S.F. Floor	14.95	14.95	
3090	Other HVAC Sys. & Equipment	N/A		—	—	—	
D40 Fire Protection							
4010	Sprinklers	Wet pipe spinkler system		S.F. Floor	4.66	4.66	4.7%
4020	Standpipes	Standpipes and hose systems, with pumps		S.F. Floor	1.28	1.28	
D50 Electrical							
5010	Electrical Service/Distribution	200 Ampere service, panel boards and feeders		S.F. Floor	1.54	1.54	
5020	Lighting & Branch Wiring	Fluorescent fixtures, receptacles, switches, A.C. and misc. power		S.F. Floor	7.23	7.23	13.0%
5030	Communications & Security	Alarm systems, emergency lighting, internet and phone wiring		S.F. Floor	7.54	7.54	
5090	Other Electrical Systems	N/A		—	—	—	
E. EQUIPMENT & FURNISHINGS							
1010	Commercial Equipment	N/A		—	—	—	
1020	Institutional Equipment	Exam room equipment, cabinets and countertops		S.F. Floor	4.71	4.71	3.7 %
1030	Vehicular Equipment	N/A		—	—	—	
1090	Other Equipment	N/A		—	—	—	
F. SPECIAL CONSTRUCTION							
1020	Integrated Construction	N/A		—	—	—	0.0 %
1040	Special Facilities	N/A		—	—	—	
G. BUILDING SITEWORK	**N/A**						
					Sub-Total	125.61	**100%**
	CONTRACTOR FEES (General Requirements: 10%, Overhead: 5%, Profit: 10%)				*25%*	31.41	
	ARCHITECT FEES				*9%*	14.13	
					Total Building Cost	**171.15**	

Costs per square foot of floor area

Exterior Wall	S.F. Area	4000	5500	7000	8500	10000	11500	13000	14500	16000
	L.F. Perimeter	180	210	240	270	286	311	336	361	386
Vinyl Clapboard	Wood Frame	214.65	205.45	200.25	196.85	193.20	191.15	189.70	188.45	187.55
Stone Veneer	Wood Frame	232.65	220.60	213.65	209.35	204.35	201.65	199.60	198.00	196.70
Fiber Cement	Rigid Steel	216.55	206.80	**201.25**	197.75	193.75	191.70	190.10	188.80	187.75
E.I.F.S.	Rigid Steel	220.50	210.55	204.85	201.25	197.20	195.05	193.40	192.10	191.10
Precast Concrete	Reinforced Concrete	261.55	245.80	236.85	231.10	224.40	220.95	218.20	216.15	214.40
Brick Veneer	Reinforced Concrete	228.70	216.30	209.20	204.70	199.55	196.80	194.70	193.05	191.75
Perimeter Adj., Add or Deduct	Per 100 L.F.	22.80	16.60	13.05	10.70	9.15	8.00	7.05	6.35	5.70
Story Hgt. Adj., Add or Deduct	Per 1 Ft.	2.55	2.20	1.90	1.85	1.65	1.60	1.50	1.45	1.30
***For Basement**, add $32.80 per square foot of basement area*										

The above costs were calculated using the basic specifications shown on the facing page. These costs should be adjusted where necessary for design alternatives and owner's requirements.

Common additives

Description	Unit	$ Cost
Cabinets, Hospital, base		
Laminated plastic	L.F.	540
Stainless steel	L.F.	905
Counter top, laminated plastic	L.F.	84.50
Stainless steel	L.F.	217
For drop-in sink, add	Ea.	1600
Nurses station, door type		
Laminated plastic	L.F.	605
Enameled steel	L.F.	585
Stainless steel	L.F.	1075
Wall cabinets, laminated plastic	L.F.	400
Enameled steel	L.F.	470
Stainless steel	L.F.	870
Elevators, Hydraulic passenger, 2 stops		
1500# capacity	Ea.	68,400
2500# capacity	Ea.	71,900
3500# capacity	Ea.	76,900

Description	Unit	$ Cost
Directory Boards, Plastic, glass covered		
30" x 20"	Ea.	570
36" x 48"	Ea.	1500
Aluminum, 24" x 18"	Ea.	570
36" x 24"	Ea.	725
48" x 32"	Ea.	950
48" x 60"	Ea.	2000
Emergency Lighting, 25 watt, battery operated		
Lead battery	Ea.	305
Nickel cadmium	Ea.	525
Heat Therapy Unit		
Humidified, 26" x 78" x 28"	Ea.	4750
Smoke Detectors		
Ceiling type	Ea.	231
Duct type	Ea.	510
Tables, Examining, vinyl top		
with base cabinets	Ea.	1525 - 6500
Utensil Washer, Sanitizer	Ea.	9950
X-Ray, Mobile	Ea.	19,400 - 106,500

Important: See the Reference Section for Location Factors.

Model costs calculated for a 2 story building with 10′ story height and 7,000 square feet of floor area

Medical Office, 2 Story

				Unit	Unit Cost	Cost Per S.F.	% Of Sub-Total
A. SUBSTRUCTURE							
1010	Standard Foundations	Poured concrete; strip and spread footings; 4′ foundation wall		S.F. Ground	10.48	5.24	
1020	Special Foundations	N/A		–	–	–	
1030	Slab on Grade	4″ reinforced slab on grade		S.F. Slab	5.28	2.64	5.5%
2010	Basement Excavation	Site preparation for slab and trench for foundation wall and footing		S.F. Ground	.49	.25	
2020	Basement Walls	N/A		–	–	–	
B. SHELL							
	B10 Superstructure						
1010	Floor Construction	Open web bar joist, slab form, fireproofed columns		S.F. Floor	16.56	8.28	9.0%
1020	Roof Construction	Metal deck, open web steel joists, beams, columns		S.F. Roof	10.06	5.03	
	B20 Exterior Enclosure						
2010	Exterior Walls	Fiber cement siding on metal studs	*70% of wall*	S.F. Wall	15	7.20	
2020	Exterior Windows	Aluminum sliding windows	*30% of wall*	Each	498	6.83	12.2%
2030	Exterior Doors	Aluminum & glass entrance doors		Each	6950	3.97	
	B30 Roofing						
3010	Roof Coverings	Single ply membrane, stone ballast, rigid insulation		S.F. Roof	8.76	4.38	3.7%
3020	Roof Openings	Roof and smoke hatches		S.F. Roof	2.12	1.06	
C. INTERIORS							
1010	Partitions	Fire rated gypsum board on metal studs	*6 S.F. Floor/L.F. Partition*	S.F. Partition	5.31	7.08	
1020	Interior Doors	Single leaf solid core wood doors	*240 S.F. Floor/Door*	Each	1201	5.67	
1030	Fittings	N/A		–	–	–	
2010	Stair Construction	Concrete filled metal pan, with rails		Flight	15,375	10.98	24.6%
3010	Wall Finishes	Paint on drywall		S.F. Surface	.93	2.49	
3020	Floor Finishes	50% carpet , 50% vinyl composition tile		S.F. Floor	3.97	3.97	
3030	Ceiling Finishes	Acoustic ceiling tiles on suspended support system		S.F. Ceiling	6.15	6.15	
D. SERVICES							
	D10 Conveying						
1010	Elevators & Lifts	One hydraulic elevator		Each	112,070	16.01	10.8%
1020	Escalators & Moving Walks	N/A		–	–	–	
	D20 Plumbing						
2010	Plumbing Fixtures	Restroom, exam room and service fixtures, supply and drainage	*1 Fixture/350 S.F. Floor*	Each	2065	5.90	
2020	Domestic Water Distribution	Gas fired water heater		S.F. Floor	2.27	2.27	6.4%
2040	Rain Water Drainage	Roof drains		S.F. Roof	2.54	1.27	
	D30 HVAC						
3010	Energy Supply	N/A		–	–	–	
3020	Heat Generating Systems	Included in D3050		–	–	–	
3030	Cooling Generating Systems	N/A		–	–	–	10.1 %
3050	Terminal & Package Units	Multizone unit, gas heating , electric cooling		S.F. Floor	14.95	14.95	
3090	Other HVAC Sys. & Equipment	N/A		–	–	–	
	D40 Fire Protection						
4010	Sprinklers	Wet pipe spinkler system		S.F. Floor	3.64	3.64	3.8%
4020	Standpipes	Standpipes and hose systems, with pumps		S.F. Floor	1.94	1.94	
	D50 Electrical						
5010	Electrical Service/Distribution	400 Ampere service, panel boards and feeders		S.F. Floor	2.27	2.27	
5020	Lighting & Branch Wiring	Fluorescent fixtures, receptacles, switches, A.C. and misc. power		S.F. Floor	7.23	7.23	11.5%
5030	Communications & Security	Alarm systems, emergency lighting, internet and phone wiring		S.F. Floor	7.54	7.54	
5090	Other Electrical Systems	N/A		–	–	–	
E. EQUIPMENT & FURNISHINGS							
1010	Commercial Equipment	N/A		–	–	–	
1020	Institutional Equipment	Exam room equipment, cabinets and countertops		S.F. Floor	3.47	3.47	2.3 %
1030	Vehicular Equipment	N/A		–	–	–	
1090	Other Equipment	N/A		–	–	–	
F. SPECIAL CONSTRUCTION							
1020	Integrated Construction	N/A		–	–	–	0.0 %
1040	Special Facilities	N/A		–	–	–	
G. BUILDING SITEWORK	**N/A**						
					Sub-Total	147.71	**100%**
	CONTRACTOR FEES (General Requirements: 10%, Overhead: 5%, Profit: 10%)				25%	36.92	
	ARCHITECT FEES				9%	16.62	
					Total Building Cost	**201.25**	

Motel, 1 Story

Costs per square foot of floor area

Exterior Wall	S.F. Area	2000	3000	4000	6000	8000	10000	12000	14000	16000
	L.F. Perimeter	240	260	280	380	480	560	580	660	740
Brick Veneer	Wood Frame	191.75	168.65	157.10	150.70	**147.45**	144.50	139.95	138.90	138.10
Aluminum Clapboard	Wood Frame	175.40	156.80	147.50	142.05	139.25	136.90	133.40	132.45	131.85
Wood Clapboard	Wood Frame	174.90	156.45	147.25	141.75	139.00	136.65	133.20	132.25	131.65
Wood Shingles	Wood Frame	178.35	159.00	149.25	143.65	140.75	138.30	134.60	133.65	132.95
Concrete Block	Wood Truss	174.35	156.05	146.90	141.50	138.75	136.45	132.95	132.10	131.40
Face Brick and Concrete Block	Wood Truss	200.35	174.90	162.10	155.20	151.75	148.55	143.45	142.30	141.45
Perimeter Adj., Add or Deduct	Per 100 L.F.	25.75	17.15	12.85	8.55	6.45	5.10	4.25	3.65	3.25
Story Hgt. Adj., Add or Deduct	Per 1 Ft.	4.70	3.35	2.75	2.45	2.35	2.20	1.90	1.90	1.80
***For Basement**, add $ 22.50 per square foot of basement area*										

The above costs were calculated using the basic specifications shown on the facing page. These costs should be adjusted where necessary for design alternatives and owner's requirements.

Common additives

Description	Unit	$ Cost
Closed Circuit Surveillance, One station		
Camera and monitor	Ea.	1250
For additional camera stations, add	Ea.	565
Emergency Lighting, 25 watt, battery operated		
Lead battery	Ea.	305
Nickel cadmium	Ea.	525
Laundry Equipment		
Dryer, gas, 16 lb. capacity	Ea.	985
30 lb. capacity	Ea.	4000
Washer, 4 cycle	Ea.	1275
Commercial	Ea.	1650
Sauna, Prefabricated, complete		
6′ x 4′	Ea.	6425
6′ x 6′	Ea.	8600
6′ x 9′	Ea.	9550
8′ x 8′	Ea.	10,500
8′ x 10′	Ea.	11,500
10′ x 12′	Ea.	15,400
Smoke Detectors		
Ceiling type	Ea.	231
Duct type	Ea.	510

Description	Unit	$ Cost
Swimming Pools, Complete, gunite	S.F.	101 - 126
TV Antenna, Master system, 12 outlet	Outlet	229
30 outlet	Outlet	325
100 outlet	Outlet	360

Important: See the Reference Section for Location Factors.

Model costs calculated for a 1 story building with 9′ story height and 8,000 square feet of floor area

Motel, 1 Story

				Unit	Unit Cost	Cost Per S.F.	% Of Sub-Total
A. SUBSTRUCTURE							
1010	Standard Foundations	Poured concrete; strip and spread footings; 4′ foundation wall		S.F. Ground	6.68	6.68	
1020	Special Foundations	N/A		—	—	—	
1030	Slab on Grade	4″ reinforced concrete with vapor barrier and granular base		S.F. Slab	5.28	5.28	11.1%
2010	Basement Excavation	Site preparation for slab and trench for foundation wall and footing		S.F. Ground	.28	.28	
2020	Basement Walls	N/A		—	—	—	
B. SHELL							
	B10 Superstructure						
1010	Floor Construction	N/A		—	—	—	6.5 %
1020	Roof Construction	Plywood on wood trusses		S.F. Roof	7.22	7.22	
	B20 Exterior Enclosure						
2010	Exterior Walls	Face brick on wood studs with sheathing, insulation and paper	*80% of wall*	S.F. Wall	23	10.13	
2020	Exterior Windows	Wood double hung	*20% of wall*	Each	491	3.79	18.6%
2030	Exterior Doors	Wood solid core		Each	2380	6.55	
	B30 Roofing						
3010	Roof Coverings	Asphalt shingles with flashing (pitched); rigid fiberglass insulation		S.F. Roof	4.26	4.26	3.9%
3020	Roof Openings	N/A		—	—	—	
C. INTERIORS							
1010	Partitions	Gypsum bd. and sound deadening bd. on wood studs	*9 S.F. Floor/L.F. Partition*	S.F. Partition	9.66	8.59	
1020	Interior Doors	Single leaf hollow core wood	*300 S.F. Floor/Door*	Each	618	2.06	
1030	Fittings	N/A		—	—	—	
2010	Stair Construction	N/A		—	—	—	22.1%
3010	Wall Finishes	90% paint, 10% ceramic tile		S.F. Surface	1.43	2.54	
3020	Floor Finishes	85% carpet, 15% ceramic tile		S.F. Floor	6.92	6.92	
3030	Ceiling Finishes	Painted gypsum board on furring		S.F. Ceiling	4.22	4.22	
D. SERVICES							
	D10 Conveying						
1010	Elevators & Lifts	N/A		—	—	—	0.0 %
1020	Escalators & Moving Walks	N/A		—	—	—	
	D20 Plumbing						
2010	Plumbing Fixtures	Toilet and service fixtures, supply and drainage	*1 Fixture/90 S.F. Floor*	Each	1741	19.34	
2020	Domestic Water Distribution	Gas fired water heater		S.F. Floor	2.10	2.10	19.4%
2040	Rain Water Drainage	N/A		—	—	—	
	D30 HVAC						
3010	Energy Supply	N/A		—	—	—	
3020	Heat Generating Systems	Included in D3050		—	—	—	
3030	Cooling Generating Systems	N/A		—	—	—	2.8 %
3050	Terminal & Package Units	Through the wall electric heating and cooling units		S.F. Floor	3.07	3.07	
3090	Other HVAC Sys. & Equipment	N/A		—	—	—	
	D40 Fire Protection						
4010	Sprinklers	Wet pipe sprinkler system		S.F. Floor	4.66	4.66	5.2%
4020	Standpipes	Standpipe, wet, Class III		S.F. Floor	1.12	1.12	
	D50 Electrical						
5010	Electrical Service/Distribution	200 ampere service, panel board and feeders		S.F. Floor	1.25	1.25	
5020	Lighting & Branch Wiring	Fluorescent fixtures, receptacles, switches and misc. power		S.F. Floor	6.50	6.50	9.4%
5030	Communications & Security	Addressable alarm systems		S.F. Floor	2.65	2.65	
5090	Other Electrical Systems	N/A		—	—	—	
E. EQUIPMENT & FURNISHINGS							
1010	Commercial Equipment	Laundry equipment		S.F. Floor	1.03	1.03	
1020	Institutional Equipment	N/A		—	—	—	0.9%
1030	Vehicular Equipment	N/A		—	—	—	
1090	Other Equipment	N/A		—	—	—	
F. SPECIAL CONSTRUCTION							
1020	Integrated Construction	N/A		—	—	—	0.0 %
1040	Special Facilities	N/A		—	—	—	
G. BUILDING SITEWORK	**N/A**						
					Sub-Total	110.24	**100%**
	CONTRACTOR FEES (General Requirements: 10%, Overhead: 5%, Profit: 10%)				*25%*	27.56	
	ARCHITECT FEES				*7%*	9.65	
					Total Building Cost	**147.45**	

Costs per square foot of floor area

Exterior Wall	S.F. Area	25000	37000	49000	61000	73000	81000	88000	96000	104000
	L.F. Perimeter	433	593	606	720	835	911	978	1054	1074
Decorative Concrete Block	Wood Joists	162.45	159.70	155.15	154.15	153.50	153.10	152.85	152.65	151.85
	Precast Concrete	183.20	179.20	**173.80**	172.40	171.50	170.90	170.60	170.25	169.35
Stucco and Concrete Block	Wood Joists	167.10	163.00	157.20	155.75	154.75	154.25	153.90	153.65	152.60
	Precast Concrete	186.85	182.15	175.95	174.20	173.15	172.50	172.15	171.75	170.70
Wood Clapboard	Wood Frame	161.15	158.50	154.20	153.25	152.60	152.20	152.05	151.75	151.05
Brick Veneer	Wood Frame	166.90	162.85	157.10	155.60	154.70	154.15	153.85	153.50	152.55
Perimeter Adj., Add or Deduct	Per 100 L.F.	4.30	2.90	2.15	1.75	1.45	1.30	1.25	1.15	1.00
Story Hgt. Adj., Add or Deduct	Per 1 Ft.	1.75	1.65	1.20	1.15	1.10	1.15	1.10	1.10	1.00
***For Basement**, add $ 29.50 per square foot of basement area*										

The above costs were calculated using the basic specifications shown on the facing page. These costs should be adjusted where necessary for design alternatives and owner's requirements.

Common additives

Description	Unit	$ Cost
Closed Circuit Surveillance, One station		
Camera and monitor	Ea.	1250
For additional camera station, add	Ea.	565
Elevators, Hydraulic passenger, 2 stops		
1500# capacity	Ea.	68,400
2500# capacity	Ea.	71,900
3500# capacity	Ea.	76,900
Additional stop, add	Ea.	7000
Emergency Lighting, 25 watt, battery operated		
Lead battery	Ea.	305
Nickel cadmium	Ea.	525
Laundry Equipment		
Dryer, gas, 16 lb. capacity	Ea.	985
30 lb. capacity	Ea.	4000
Washer, 4 cycle	Ea.	1275
Commercial	Ea.	1650

Description	Unit	$ Cost
Sauna, Prefabricated, complete		
6' x 4'	Ea.	6425
6' x 6'	Ea.	8600
6' x 9'	Ea.	9550
8' x 8'	Ea.	10,500
8' x 10'	Ea.	11,500
10' x 12'	Ea.	15,400
Smoke Detectors		
Ceiling type	Ea.	231
Duct type	Ea.	510
Swimming Pools, Complete, gunite	S.F.	101 - 126
TV Antenna, Master system, 12 outlet	Outlet	229
30 outlet	Outlet	325
100 outlet	Outlet	360

Important: See the Reference Section for Location Factors.

Model costs calculated for a 3 story building with 9′ story height and 49,000 square feet of floor area

Motel, 2-3 Story

				Unit	Unit Cost	Cost Per S.F.	% Of Sub-Total
A. SUBSTRUCTURE							
1010	Standard Foundations	Poured concrete; strip and spread footings; 4′ foundation wall		S.F. Ground	5.43	1.81	2.8%
1020	Special Foundations	N/A		—	—	—	
1030	Slab on Grade	4″ reinforced concrete with vapor barrier and granular base		S.F. Slab	5.28	1.76	
2010	Basement Excavation	Site preparation for slab and trench for foundation wall and footing		S.F. Ground	.28	.09	
2020	Basement Walls	N/A		—	—	—	
B. SHELL							
	B10 Superstructure						
1010	Floor Construction	Precast concrete plank		S.F. Floor	14.15	9.43	10.1%
1020	Roof Construction	Precast concrete plank		S.F. Roof	11.58	3.86	
	B20 Exterior Enclosure						
2010	Exterior Walls	Decorative concrete block	*85% of wall*	S.F. Wall	18.25	5.18	12.7%
2020	Exterior Windows	Aluminum sliding	*15% of wall*	Each	558	1.87	
2030	Exterior Doors	Aluminum and glass doors and entrance with transom		Each	2923	9.65	
	B30 Roofing						
3010	Roof Coverings	Built-up tar and gravel with flashing; perlite/EPS composite insulation		S.F. Roof	6.30	2.10	1.6%
3020	Roof Openings	Roof hatches		S.F. Roof	.15	.05	
C. INTERIORS							
1010	Partitions	Concrete block	*7 S.F. Floor/L.F. Partition*	S.F. Partition	19.71	22.52	37.7%
1020	Interior Doors	Wood hollow core	*70 S.F. Floor/Door*	Each	618	8.82	
1030	Fittings	N/A		—	—	—	
2010	Stair Construction	Concrete filled metal pan		Flight	15,375	3.76	
3010	Wall Finishes	90% paint, 10% ceramic tile		S.F. Surface	1.43	3.26	
3020	Floor Finishes	85% carpet, 5% vinyl composition tile, 10% ceramic tile		S.F. Floor	6.92	6.92	
3030	Ceiling Finishes	Textured finish		S.F. Ceiling	4.22	4.22	
D. SERVICES							
	D10 Conveying						
1010	Elevators & Lifts	Two hydraulic passenger elevators		Each	110,005	4.49	3.4%
1020	Escalators & Moving Walks	N/A		—	—	—	
	D20 Plumbing						
2010	Plumbing Fixtures	Toilet and service fixtures, supply and drainage	*1 Fixture/180 S.F. Floor*	Each	4084	22.69	18.8%
2020	Domestic Water Distribution	Gas fired water heater		S.F. Floor	1.36	1.36	
2040	Rain Water Drainage	Roof drains		S.F. Roof	1.71	.57	
	D30 HVAC						
3010	Energy Supply	N/A		—	—	—	2.1 %
3020	Heat Generating Systems	Included in D3050		—	—	—	
3030	Cooling Generating Systems	N/A		—	—	—	
3050	Terminal & Package Units	Through the wall electric heating and cooling units		S.F. Floor	2.77	2.77	
3090	Other HVAC Sys. & Equipment	N/A		—	—	—	
	D40 Fire Protection						
4010	Sprinklers	Sprinklers, wet, light hazard		S.F. Floor	2.74	2.74	2.3%
4020	Standpipes	Standpipe, wet, Class III		S.F. Floor	.26	.26	
	D50 Electrical						
5010	Electrical Service/Distribution	800 ampere service, panel board and feeders		S.F. Floor	1.47	1.47	8.1%
5020	Lighting & Branch Wiring	Fluorescent fixtures, receptacles, switches and misc. power		S.F. Floor	7.25	7.25	
5030	Communications & Security	Addressable alarm systems and emergency lighting		S.F. Floor	1.80	1.80	
5090	Other Electrical Systems	Emergency generator, 7.5 kW		S.F. Floor	.13	.13	
E. EQUIPMENT & FURNISHINGS							
1010	Commercial Equipment	Commercial laundry equipment		S.F. Floor	.35	.35	0.3%
1020	Institutional Equipment	N/A		—	—	—	
1030	Vehicular Equipment	N/A		—	—	—	
1090	Other Equipment	N/A		—	—	—	
F. SPECIAL CONSTRUCTION							
1020	Integrated Construction	N/A		—	—	—	0.0 %
1040	Special Facilities	N/A		—	—	—	
G. BUILDING SITEWORK	**N/A**						
					Sub-Total	131.18	**100%**
	CONTRACTOR FEES (General Requirements: 10%, Overhead: 5%, Profit: 10%)				*25%*	32.78	
	ARCHITECT FEES				*6%*	9.84	
					Total Building Cost	**173.80**	

Costs per square foot of floor area

Exterior Wall	S.F. Area	9000	10000	12000	13000	14000	15000	16000	18000	20000
	L.F. Perimeter	385	410	440	460	480	500	510	547	583
Decorative Concrete Block	Steel Joists	186.95	182.00	**173.05**	169.95	167.35	165.05	162.40	158.95	156.20
Concrete Block	Steel Joists	180.05	175.45	167.15	164.25	161.85	159.70	157.30	154.15	151.50
Face Brick and Concrete Block	Steel Joists	196.70	191.40	181.40	177.95	175.20	172.60	169.70	165.90	162.85
Precast Concrete	Steel Joists	223.10	216.65	204.00	199.75	196.25	193.10	189.25	184.55	180.70
Tilt-up Concrete Panels	Steel Joists	184.00	179.20	170.55	167.45	164.95	162.70	160.20	156.85	154.15
Metal Sandwich Panel	Steel Joists	182.15	177.40	168.90	165.95	163.50	161.25	158.80	155.55	152.95
Perimeter Adj., Add or Deduct	Per 100 L.F.	11.15	10.05	8.35	7.65	7.20	6.70	6.30	5.65	5.05
Story Hgt. Adj., Add or Deduct	Per 1 Ft.	1.75	1.65	1.50	1.40	1.40	1.30	1.30	1.25	1.20
Basement—Not Applicable										

The above costs were calculated using the basic specifications shown on the facing page. These costs should be adjusted where necessary for design alternatives and owner's requirements.

Common additives

Description	Unit	$ Cost
Emergency Lighting, 25 watt, battery operated		
Lead battery	Ea.	305
Nickel cadmium	Ea.	525
Seating		
Auditorium chair, all veneer	Ea.	355
Veneer back, padded seat	Ea.	375
Upholstered, spring seat	Ea.	325
Classroom, movable chair & desk	Set	81 - 171
Lecture hall, pedestal type	Ea.	345 - 645
Smoke Detectors		
Ceiling type	Ea.	231
Duct type	Ea.	510
Sound System		
Amplifier, 250 watts	Ea.	1950
Speaker, ceiling or wall	Ea.	233
Trumpet	Ea.	445

Important: See the Reference Section for Location Factors.

Model costs calculated for a 1 story building with 20′ story height and 12,000 square feet of floor area

Movie Theater

				Unit	Unit Cost	Cost Per S.F.	% Of Sub-Total
A. SUBSTRUCTURE							
1010	Standard Foundations	Poured concrete; strip and spread footings; 4′ foundation wall		S.F. Ground	4.34	4.34	
1020	Special Foundations	N/A		—	—	—	
1030	Slab on Grade	4″ reinforced concrete with vapor barrier and granular base		S.F. Slab	5.28	5.28	7.7%
2010	Basement Excavation	Site preparation for slab and trench for foundation wall and footing		S.F. Ground	.28	.28	
2020	Basement Walls	N/A		—	—	—	
B. SHELL							
	B10 Superstructure						
1010	Floor Construction	Open web steel joists, slab form, concrete	*mezzanine 2250 S.F.*	S.F. Floor	11.50	2.15	9.7%
1020	Roof Construction	Metal deck on open web steel joists		S.F. Roof	10.41	10.41	
	B20 Exterior Enclosure						
2010	Exterior Walls	Decorative concrete block	*80% of wall*	S.F. Wall	16.20	11.88	
2020	Exterior Windows	Window wall	*20% of wall*	Each	56	8.23	16.9%
2030	Exterior Doors	Sliding mallfront aluminum and glass and hollow metal		Each	3521	1.76	
	B30 Roofing						
3010	Roof Coverings	Built-up tar and gravel with flashing; perlite/EPS composite insulation		S.F. Roof	6.27	6.27	4.8%
3020	Roof Openings	N/A		—	—	—	
C. INTERIORS							
1010	Partitions	Concrete block	*40 S.F. Floor/L.F. Partition*	S.F. Partition	9.15	3.66	
1020	Interior Doors	Single leaf hollow metal	*705 S.F. Floor/Door*	Each	1201	1.71	
1030	Fittings	Toilet partitions		S.F. Floor	.65	.65	
2010	Stair Construction	Concrete filled metal pan		Flight	18,425	3.07	19.7%
3010	Wall Finishes	Paint		S.F. Surface	3.90	3.12	
3020	Floor Finishes	Carpet 50%, ceramic tile 5%	*50% of area*	S.F. Floor	12.88	6.44	
3030	Ceiling Finishes	Mineral fiber tile on concealed zee runners suspended		S.F. Ceiling	6.86	6.86	
D. SERVICES							
	D10 Conveying						
1010	Elevators & Lifts	N/A		—	—	—	0.0 %
1020	Escalators & Moving Walks	N/A		—	—	—	
	D20 Plumbing						
2010	Plumbing Fixtures	Toilet and service fixtures, supply and drainage	*1 Fixture/500 S.F. Floor*	Each	5295	10.59	
2020	Domestic Water Distribution	Gas fired water heater		S.F. Floor	.55	.55	9.7%
2040	Rain Water Drainage	Roof drains		S.F. Roof	1.46	1.46	
	D30 HVAC						
3010	Energy Supply	N/A		—	—	—	
3020	Heat Generating Systems	Included in D3050		—	—	—	
3030	Cooling Generating Systems	N/A		—	—	—	8.1 %
3050	Terminal & Package Units	Single zone rooftop unit, gas heating, electric cooling		S.F. Floor	10.46	10.46	
3090	Other HVAC Sys. & Equipment	N/A		—	—	—	
	D40 Fire Protection						
4010	Sprinklers	Wet pipe sprinkler system		S.F. Floor	4.06	4.06	3.8%
4020	Standpipes	Standpipe		S.F. Floor	.90	.90	
	D50 Electrical						
5010	Electrical Service/Distribution	400 ampere service, panel board and feeders		S.F. Floor	1.67	1.67	
5020	Lighting & Branch Wiring	High efficiency fluorescent fixtures, receptacles, switches, A.C. and misc. power		S.F. Floor	6.38	6.38	8.1%
5030	Communications & Security	Addressable alarm systems, sound system and emergency lighting		S.F. Floor	2.28	2.28	
5090	Other Electrical Systems	Emergency generator, 7.5 kW		S.F. Floor	.18	.18	
E. EQUIPMENT & FURNISHINGS							
1010	Commercial Equipment	N/A		—	—	—	
1020	Institutional Equipment	Projection equipment, screen		S.F. Floor	9.88	9.88	11.4 %
1030	Vehicular Equipment	N/A		—	—	—	
2010	Fixed Furnishings	Movie theater seating		S.F. Floor	4.87	4.87	
F. SPECIAL CONSTRUCTION							
1020	Integrated Construction	N/A		—	—	—	0.0 %
1040	Special Facilities	N/A		—	—	—	
G. BUILDING SITEWORK	**N/A**						
					Sub-Total	129.39	**100%**
	CONTRACTOR FEES (General Requirements: 10%, Overhead: 5%, Profit: 10%)				*25%*	32.34	
	ARCHITECT FEES				*7%*	11.32	
					Total Building Cost	**173.05**	

Costs per square foot of floor area

Exterior Wall	S.F. Area	10000	15000	20000	25000	30000	35000	40000	45000	50000
	L.F. Perimeter	286	370	453	457	513	568	624	680	735
Precast Concrete	Bearing Walls	230.95	220.95	215.85	206.85	204.10	202.10	200.55	199.45	198.55
	Rigid Steel	231.90	221.95	216.95	207.95	205.25	203.25	201.70	200.60	199.70
Face Brick and Concrete Block	Bearing Walls	211.20	203.95	200.25	194.25	192.35	190.95	189.85	189.05	188.40
	Rigid Steel	213.05	205.85	202.20	196.25	194.30	192.95	191.85	191.10	190.45
Stucco and Concrete Block	Bearing Walls	206.55	199.90	196.60	191.30	189.55	188.30	187.30	186.60	186.05
	Steel Joists	208.40	201.80	198.50	193.25	191.55	190.30	189.30	188.65	188.00
Perimeter Adj., Add or Deduct	Per 100 L.F.	18.85	12.55	9.50	7.55	6.30	5.40	4.80	4.20	3.75
Story Hgt. Adj., Add or Deduct	Per 1 Ft.	4.75	4.15	3.85	3.05	2.90	2.70	2.60	2.55	2.40
***For Basement**, add $30.70 per square foot of basement area*										

The above costs were calculated using the basic specifications shown on the facing page. These costs should be adjusted where necessary for design alternatives and owner's requirements.

Common additives

Description	Unit	$ Cost
Beds, Manual	Ea.	945 - 3100
Elevators, Hydraulic passenger, 2 stops		
1500# capacity	Ea.	68,400
2500# capacity	Ea.	71,900
3500# capacity	Ea.	76,900
Emergency Lighting, 25 watt, battery operated		
Lead battery	Ea.	305
Nickel cadmium	Ea.	525
Intercom System, 25 station capacity		
Master station	Ea.	3150
Intercom outlets	Ea.	204
Handset	Ea.	575
Kitchen Equipment		
Broiler	Ea.	4400
Coffee urn, twin 6 gallon	Ea.	2900
Cooler, 6 ft. long	Ea.	3725
Dishwasher, 10-12 racks per hr.	Ea.	3975
Food warmer	Ea.	785
Freezer, 44 C.F., reach-in	Ea.	5400

Description	Unit	$ Cost
Kitchen Equipment, cont.		
Ice cube maker, 50 lb. per day	Ea.	2050
Range with 1 oven	Ea.	3200
Laundry Equipment		
Dryer, gas, 16 lb. capacity	Ea.	985
30 lb. capacity	Ea.	4000
Washer, 4 cycle	Ea.	1275
Commercial	Ea.	1650
Nurses Call System		
Single bedside call station	Ea.	263
Pillow speaker	Ea.	320
Refrigerator, Prefabricated, walk-in		
7'-6" high, 6' x 6'	S.F.	137
10' x 10'	S.F.	107
12' x 14'	S.F.	94
12' x 20'	S.F.	134
TV Antenna, Master system, 12 outlet	Outlet	229
30 outlet	Outlet	325
100 outlet	Outlet	360
Whirlpool Bath, Mobile, 18" x 24" x 60"	Ea.	6700
X-Ray, Mobile	Ea.	19,400 - 106,500

Important: See the Reference Section for Location Factors.

Model costs calculated for a 2 story building with 10′ story height and 25,000 square feet of floor area

Nursing Home

				Unit	Unit Cost	Cost Per S.F.	% Of Sub-Total
A. SUBSTRUCTURE							
1010	Standard Foundations	Poured concrete; strip and spread footings; 4′ foundation wall		S.F. Ground	4.24	2.12	
1020	Special Foundations	N/A		—	—	—	
1030	Slab on Grade	4″ reinforced concrete with vapor barrier and granular base		S.F. Slab	5.28	2.64	3.3%
2010	Basement Excavation	Site preparation for slab and trench for foundation wall and footing		S.F. Ground	.28	.14	
2020	Basement Walls	N/A		—	—	—	
B. SHELL							
	B10 Superstructure						
1010	Floor Construction	Pre-cast double tees with concrete topping		S.F. Floor	16.38	8.19	10.4%
1020	Roof Construction	Pre-cast double tees		S.F. Roof	14.72	7.36	
	B20 Exterior Enclosure						
2010	Exterior Walls	Precast concrete panels		S.F. Wall	60	18.62	
2020	Exterior Windows	Wood double hung	*15% of wall*	Each	558	2.04	14.3%
2030	Exterior Doors	Double aluminum & glass doors, single leaf hollow metal		Each	3294	.66	
	B30 Roofing						
3010	Roof Coverings	Built-up tar and gravel with flashing; perlite/EPS composite insulation		S.F. Roof	6.26	3.13	2.1%
3020	Roof Openings	Roof hatches		S.F. Roof	.10	.05	
C. INTERIORS							
1010	Partitions	Gypsum board on metal studs	*8 S.F. Floor/L.F. Partition*	S.F. Partition	8.62	8.62	
1020	Interior Doors	Single leaf wood	*80 S.F. Floor/Door*	Each	691	8.64	
1030	Fittings	N/A		—	—	—	
2010	Stair Construction	Concrete stair		Flight	15,375	2.46	20.3%
3010	Wall Finishes	50% vinyl wall coverings, 45% paint, 5% ceramic tile		S.F. Surface	1.71	3.42	
3020	Floor Finishes	95% vinyl tile, 5% ceramic tile		S.F. Floor	3.24	3.24	
3030	Ceiling Finishes	Painted gypsum board		S.F. Ceiling	3.91	3.91	
D. SERVICES							
	D10 Conveying						
1010	Elevators & Lifts	One hydraulic hospital elevator		Each	112,000	4.48	3.0%
1020	Escalators & Moving Walks	N/A		—	—	—	
	D20 Plumbing						
2010	Plumbing Fixtures	Kitchen, toilet and service fixtures, supply and drainage	*1 Fixture/230 S.F. Floor*	Each	8275	35.98	
2020	Domestic Water Distribution	Oil fired water heater		S.F. Floor	1.44	1.44	25.6%
2040	Rain Water Drainage	Roof drains		S.F. Roof	1.42	.71	
	D30 HVAC						
3010	Energy Supply	Oil fired hot water, wall fin radiation		S.F. Floor	7.94	7.94	
3020	Heat Generating Systems	N/A		—	—	—	
3030	Cooling Generating Systems	N/A		—	—	—	10.0%
3050	Terminal & Package Units	Split systems with air cooled condensing units		S.F. Floor	6.97	6.97	
3090	Other HVAC Sys. & Equipment	N/A		—	—	—	
	D40 Fire Protection						
4010	Sprinklers	Sprinkler, light hazard		S.F. Floor	2.91	2.91	2.3%
4020	Standpipes	Standpipe		S.F. Floor	.47	.47	
	D50 Electrical						
5010	Electrical Service/Distribution	800 ampere service, panel board and feeders		S.F. Floor	1.55	1.55	
5020	Lighting & Branch Wiring	High efficiency fluorescent fixtures, receptacles, switches, A.C. and misc. power		S.F. Floor	9.44	9.44	8.7%
5030	Communications & Security	Addressable alarm systems and emergency lighting		S.F. Floor	1.39	1.39	
5090	Other Electrical Systems	Emergency generator, 15 kW		S.F. Floor	.56	.56	
E. EQUIPMENT & FURNISHINGS							
1010	Commercial Equipment	N/A		—	—	—	
1020	Institutional Equipment	N/A		—	—	—	0.0 %
1030	Vehicular Equipment	N/A		—	—	—	
1090	Other Equipment	N/A		—	—	—	
F. SPECIAL CONSTRUCTION							
1020	Integrated Construction	N/A		—	—	—	0.0 %
1040	Special Facilities	N/A		—	—	—	
G. BUILDING SITEWORK	**N/A**						
					Sub-Total	149.08	**100%**
	CONTRACTOR FEES (General Requirements: 10%, Overhead: 5%, Profit: 10%)				*25%*	37.27	
	ARCHITECT FEES				*11%*	20.50	
					Total Building Cost	**206.85**	

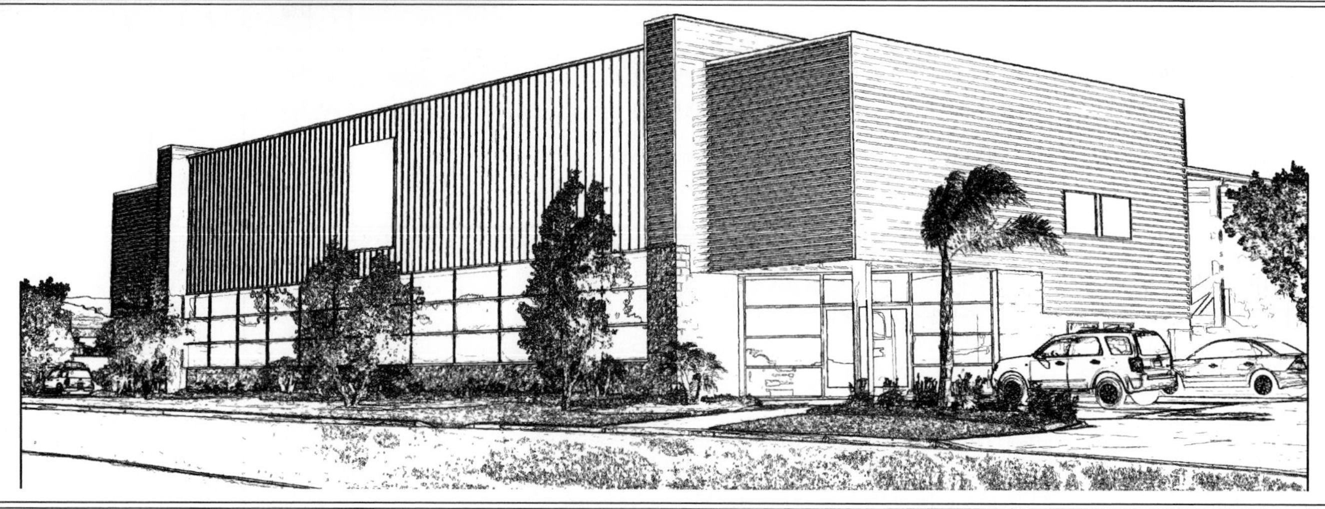

Costs per square foot of floor area

Exterior Wall	S.F. Area	2000	3000	5000	7000	9000	12000	15000	20000	25000
	L.F. Perimeter	220	260	320	360	420	480	520	640	700
Vinyl Clapboard	Wood Frame	201.60	181.00	162.55	153.05	149.00	144.20	140.55	138.35	135.70
Stone Veneer	Wood Frame	232.85	205.70	180.90	167.90	162.55	155.80	150.65	147.75	144.00
Fiber Cement	Rigid Steel	217.60	195.20	174.95	164.55	160.10	154.80	150.75	148.40	145.40
E.I.F.S.	Rigid Steel	220.00	197.10	176.35	165.65	161.15	155.65	151.50	149.05	146.00
Precast Concrete	Reinforced Concrete	287.95	252.00	218.60	**200.85**	193.65	184.45	177.25	173.35	168.05
Brick Veneer	Reinforced Concrete	293.60	256.90	222.90	204.90	197.55	188.20	180.95	176.95	171.55
Perimeter Adj., Add or Deduct	Per 100 L.F.	26.05	17.50	10.40	7.45	5.80	4.25	3.45	2.60	2.10
Story Hgt. Adj., Add or Deduct	Per 1 Ft.	2.75	2.20	1.55	1.30	1.15	0.95	0.85	0.75	0.70
***For Basement**, add $31.10 per square foot of basement area*										

The above costs were calculated using the basic specifications shown on the facing page. These costs should be adjusted where necessary for design alternatives and owner's requirements.

Common additives

Description	Unit	$ Cost
Closed circuit surveillance, one station		
Camera and monitor	Ea.	1250
For additional camera stations, add	Ea.	565
Directory boards, plastic, glass covered		
30" x 20"	Ea.	570
36" x 48"	Ea.	1500
Aluminum, 24" x 18"	Ea.	570
36" x 24"	Ea.	725
48" x 32"	Ea.	950
48" x 60"	Ea.	2000
Electronic, wall mounted	S.F.	4600
Free standing	S.F.	4050
Pedestal access floor system w/plastic laminate cover		
Computer room, less than 6000 SF	S.F.	22
Greater than 6000 SF	S.F.	21.50
Office, greater than 6000 S.F.	S.F.	17.55

Description	Unit	$ Cost
Security access systems		
Metal detectors, wand type	Ea.	138
Walk-through portal type, single-zone	Ea.	3950
Multi-zone	Ea.	5250
X-ray equipment		
Desk top, for mail, small packages	Ea.	3925
Conveyer type, including monitor, minimum	Ea.	17,700
Maximum	Ea.	30,700
Explosive detection equipment		
Hand held, battery operated	Ea.	28,100
Walk-through portal type	Ea.	48,400
Uninterruptible power supply, 15 kVA/12.75 kW	kW	1.23

Important: See the Reference Section for Location Factors.

Model costs calculated for a 1 story building with 12′ story height and 7,000 square feet of floor area

Office, 1 Story

				Unit	Unit Cost	Cost Per S.F.	% Of Sub-Total
A. SUBSTRUCTURE							
1010	Standard Foundations	Poured concrete; strip and spread footings; 4′ foundation wall		S.F. Ground	7.34	7.34	8.6%
1020	Special Foundations	N/A		—	—	—	
1030	Slab on Grade	4″ reinforced concrete		S.F. Slab	5.28	5.28	
2010	Basement Excavation	Site preparation for slab and trench for foundation wall and footing		S.F. Ground	.28	.28	
2020	Basement Walls	N/A		—	—	—	
B. SHELL							
	B10 Superstructure						
1010	Floor Construction	Cast in place concrete columns		S.F. Floor	2.39	2.39	11.4%
1020	Roof Construction	Concrete beam and slab		S.F. Roof	14.80	14.80	
	B20 Exterior Enclosure						
2010	Exterior Walls	Precast concrete panels, insulated	*80% of wall*	S.F. Wall	60	29.66	24.8%
2020	Exterior Windows	Aluminum awning windows	*20% of wall*	Each	703	3.77	
2030	Exterior Doors	Aluminum and glass single and double doors, hollow metal doors		Each	4458	3.83	
	B30 Roofing						
3010	Roof Coverings	Single ply membrane, stone ballast, rigid insulation		S.F. Roof	8.11	8.11	5.9%
3020	Roof Openings	Smoke hatches		S.F. Roof	.72	.72	
C. INTERIORS							
1010	Partitions	Gypsum board on metal studs	*20 S.F. Floor/L.F. Partition*	S.F. Partition	5.74	2.87	12.8%
1020	Interior Doors	70% Single leaf solid wood doors, 30% hollow metal single leaf	*320 S.F. Floor/Door*	Each	1201	3.77	
1030	Fittings	Plastic laminate toilet partitions		S.F. Floor	.37	.37	
2010	Stair Construction	N/A		—	—	—	
3010	Wall Finishes	90% Paint, 10% ceramic wall tile		S.F. Surface	1.18	1.18	
3020	Floor Finishes	65% Carpet, 10% porcelain tile, 10% quarry tile		S.F. Floor	4.99	4.99	
3030	Ceiling Finishes	Acoustic ceiling tiles on suspended channel grid		S.F. Ceiling	6.10	6.10	
D. SERVICES							
	D10 Conveying						
1010	Elevators & Lifts	N/A		—	—	—	0.0 %
1020	Escalators & Moving Walks	N/A		—	—	—	
	D20 Plumbing						
2010	Plumbing Fixtures	Restroom and service fixtures, supply and drainage	*1 Fixture/700 S.F. Floor*	Each	2100	3	3.6%
2020	Domestic Water Distribution	Gas fired water heater		S.F. Floor	1.68	1.68	
2040	Rain Water Drainage	Roof drains		S.F. Roof	.74	.74	
	D30 HVAC						
3010	Energy Supply	N/A		—	—	—	14.0 %
3020	Heat Generating Systems	Included in D3050		—	—	—	
3030	Cooling Generating Systems	N/A		—	—	—	
3050	Terminal & Package Units	Multizone gas heating, electric cooling unit		S.F. Floor	21	21	
3090	Other HVAC Sys. & Equipment	N/A		—	—	—	
	D40 Fire Protection						
4010	Sprinklers	Wet pipe light hazard sprinkler system		S.F. Floor	3.41	3.41	3.3%
4020	Standpipes	Standpipes		S.F. Floor	1.53	1.53	
	D50 Electrical						
5010	Electrical Service/Distribution	400 ampere service, panel boards and feeders		S.F. Floor	4.69	4.69	15.5%
5020	Lighting & Branch Wiring	Fluorescent fixtures, receptacles, switches, A.C. and misc. power		S.F. Floor	11.76	11.76	
5030	Communications & Security	Telephone systems, internet wiring, and addressable alarm systems		S.F. Floor	6.87	6.87	
5090	Other Electrical Systems	Emergency generator, 7.5 kW		S.F. Floor			
E. EQUIPMENT & FURNISHINGS							
1010	Commercial Equipment	N/A		—	—	—	0.0 %
1020	Institutional Equipment	N/A		—	—	—	
1030	Vehicular Equipment	N/A		—	—	—	
1090	Other Equipment	N/A		—	—	—	
F. SPECIAL CONSTRUCTION							
1020	Integrated Construction	N/A		—	—	—	0.0 %
1040	Special Facilities	N/A		—	—	—	
G. BUILDING SITEWORK		**N/A**					
					Sub-Total	150.14	**100%**
	CONTRACTOR FEES (General Requirements: 10%, Overhead: 5%, Profit: 10%)				*25%*	37.57	
	ARCHITECT FEES				*7%*	13.14	
					Total Building Cost	**200.85**	

Costs per square foot of floor area

Exterior Wall	S.F. Area	5000	8000	12000	16000	20000	35000	50000	65000	80000
	L.F. Perimeter	220	260	310	330	360	440	550	600	675
Vinyl Clapboard	Wood Frame	198.90	177.70	165.55	157.20	152.85	144.15	141.40	138.85	137.55
Stone Veneer	Wood Frame	236.00	205.15	187.40	174.65	**168.05**	154.90	150.80	146.75	144.80
Fiber Cement	Rigid Steel	203.15	180.35	167.30	158.20	153.45	144.05	141.10	138.25	136.90
E.I.F.S.	Rigid Steel	208.55	185.00	171.50	162.05	157.15	147.35	144.30	141.40	139.95
Precast Concrete	Reinforced Concrete	282.45	240.20	215.90	198.00	188.90	170.30	164.65	158.85	156.15
Brick Veneer	Reinforced Concrete	273.05	234.20	211.80	195.45	187.10	170.05	164.95	159.60	157.15
Perimeter Adj., Add or Deduct	Per 100 L.F.	24.25	15.15	10.10	7.60	6.05	3.50	2.45	1.90	1.50
Story Hgt. Adj., Add or Deduct	Per 1 Ft.	3.30	2.40	1.90	1.55	1.30	1.00	0.90	0.70	0.65
For Basement, add $32.30 per square foot of basement area										

The above costs were calculated using the basic specifications shown on the facing page. These costs should be adjusted where necessary for design alternatives and owner's requirements.

Common additives

Description	Unit	$ Cost
Closed circuit surveillance, one station		
Camera and monitor	Ea.	1250
For additional camera stations, add	Ea.	565
Directory boards, plastic, glass covered		
30" x 20"	Ea.	570
36" x 48"	Ea.	1500
Aluminum, 24" x 18"	Ea.	570
36" x 24"	Ea.	725
48" x 32"	Ea.	950
48" x 60"	Ea.	2000
Electronic, wall mounted	S.F.	4600
Free standing	S.F.	4050
Escalators, 10' rise, 32" wide, glass balustrade	Ea.	142,600
Metal balustrade	Ea.	151,100
48" wide, glass balustrade	Ea.	150,100
Metal balustrade	Ea.	161,100
Pedestal access floor system w/plastic laminate cover		
Computer room, less than 6000 SF	S.F.	22
Greater than 6000 SF	S.F.	21.50
Office, greater than 6000 S.F.	S.F.	17.55

Description	Unit	$ Cost
Security access systems		
Metal detectors, wand type	Ea.	138
Walk-through portal type, single-zone	Ea.	3950
Multi-zone	Ea.	5250
X-ray equipment		
Desk top, for mail, small packages	Ea.	3925
Conveyer type, including monitor, minimum	Ea.	17,700
Maximum	Ea.	30,700
Explosive detection equipment		
Hand held, battery operated	Ea.	28,100
Walk-through portal type	Ea.	48,400
Uninterruptible power supply, 15 kVA/12.75 kW	kW	1.23

Important: See the Reference Section for Location Factors.

Model costs calculated for a 3 story building with 12′ story height and 20,000 square feet of floor area

Office, 2-4 Story

				Unit	Unit Cost	Cost Per S.F.	% Of Sub-Total
A. SUBSTRUCTURE							
1010	Standard Foundations	Poured concrete; strip and spread footings; 4′ foundation wall		S.F. Ground	9.09	3.03	
1020	Special Foundations	N/A		–	–	–	
1030	Slab on Grade	4″ reinforced slab on grade		S.F. Slab	5.28	1.76	3.9%
2010	Basement Excavation	Site preparation for slab and trench for foundation wall and footing		S.F. Ground	.28	.09	
2020	Basement Walls	N/A		–	–	–	
B. SHELL							
	B10 Superstructure						
1010	Floor Construction	Wood Beam and joist on wood columns, fireproofed		S.F. Floor	21	14.01	13.1%
1020	Roof Construction	Wood roof, truss, 4/12 slope		S.F. Roof	7.53	2.51	
	B20 Exterior Enclosure						
2010	Exterior Walls	Ashlar stone veneer on wood stud back-up, insulated	*80% of wall*	S.F. Wall	36	18.68	
2020	Exterior Windows	Aluminum awning type windows	*20% of wall*	Each	703	3.96	19.1%
2030	Exterior Doors	Aluminum and glass single and double doors, hollow metal doors		Each	4458	1.34	
	B30 Roofing						
3010	Roof Coverings	Asphalt shingles with flashing, gutters and downspouts		S.F. Roof	2.25	.75	0.6%
3020	Roof Openings	N/A		–	–	–	
C. INTERIORS							
1010	Partitions	Gypsum board on wood studs	*20 S.F. Floor/L.F. Partition*	S.F. Partition	7.35	2.94	
1020	Interior Doors	Single leaf solid core wood doors, hollow metal egress doors	*315 S.F. Floor/Door*	Each	1201	3.85	
1030	Fittings	Plastic laminate toilet partitions		S.F. Floor	.21	.21	
2010	Stair Construction	Concrete filled metal pan with rails		Flight	12,375	4.34	18.7%
3010	Wall Finishes	Paint on drywall		S.F. Surface	1.31	1.05	
3020	Floor Finishes	60% Carpet tile, 30% vinyl composition tile, 10% ceramic tile		S.F. Floor	4.99	4.99	
3030	Ceiling Finishes	Acoustic ceiling tiles on suspended support system		S.F. Ceiling	6.10	6.10	
D. SERVICES							
	D10 Conveying						
1010	Elevators & Lifts	Two hydraulic passenger elevators		Each	117,600	11.76	9.4%
1020	Escalators & Moving Walks	N/A		–	–	–	
	D20 Plumbing						
2010	Plumbing Fixtures	Restroom and service fixtures, supply and drainage	*1 Fixture/1050 S.F. Floor*	Each	2159	2.05	
2020	Domestic Water Distribution	Gas fired water heater		S.F. Floor	.59	.59	2.1%
2040	Rain Water Drainage	Roof drains		S.F. Roof			
	D30 HVAC						
3010	Energy Supply	N/A		–	–	–	
3020	Heat Generating Systems	Included in D3050		–	–	–	
3030	Cooling Generating Systems	N/A		–	–	–	13.3 %
3050	Terminal & Package Units	Multizone unit, gas heating , electric cooling		S.F. Floor	16.65	16.65	
3090	Other HVAC Sys. & Equipment	N/A		–	–	–	
	D40 Fire Protection						
4010	Sprinklers	Wet pipe spinkler system		S.F. Floor	3.55	3.55	3.6%
4020	Standpipes	Standpipes and hose systems		S.F. Floor	.99	.99	
	D50 Electrical						
5010	Electrical Service/Distribution	1000 Ampere service, panel boards and feeders		S.F. Floor	3.42	3.42	
5020	Lighting & Branch Wiring	Fluorescent fixtures, receptacles, switches, A.C. and misc. power		S.F. Floor	11.33	11.33	16.3%
5030	Communications & Security	Addressable alarm systems, emergency lighting, internet and phone wiring		S.F. Floor	5.71	5.71	
5090	Other Electrical Systems	Uninterruptible power supply with battery pack, 12.75kW		S.F. Floor			
E. EQUIPMENT & FURNISHINGS							
1010	Commercial Equipment	N/A		–	–	–	
1020	Institutional Equipment	N/A		–	–	–	0.0 %
1030	Vehicular Equipment	N/A		–	–	–	
1090	Other Equipment	N/A		–	–	–	
F. SPECIAL CONSTRUCTION							
1020	Integrated Construction	N/A		–	–	–	0.0 %
1040	Special Facilities	N/A		–	–	–	
G. BUILDING SITEWORK	**N/A**						
					Sub-Total	125.66	**100%**
	CONTRACTOR FEES (General Requirements: 10%, Overhead: 5%, Profit: 10%)				*25%*	31.39	
	ARCHITECT FEES				*7%*	11	
					Total Building Cost	**168.05**	

M.485 Outpatient Surgery Center

Costs per square foot of floor area

Exterior Wall	S.F. Area	5000	7500	10000	12500	15000	17500	20000	22500	25000
	L.F. Perimeter	284	350	410	450	490	530	570	600	632
Brick Veneer	Wood Truss	458.70	446.45	439.70	434.20	430.45	427.80	425.85	423.95	422.40
Brick Veneer	Rigid Steel	459.50	447.70	441.20	435.90	432.40	429.85	428.00	426.15	424.70
Decorative Concrete Block	Rigid Steel	462.80	450.40	443.65	438.00	434.30	431.65	429.65	427.70	426.15
Decorative Concrete Block	Bearing Walls	454.10	443.25	437.30	432.50	429.25	427.00	425.30	423.60	422.30
Tilt-up Concrete Panels	Rigid Steel	454.85	443.85	437.85	433.00	429.75	427.40	425.65	423.95	422.65
E.I.F.S. and Concrete Block	Rigid Steel	454.70	443.70	437.70	432.85	429.60	427.30	425.55	423.85	422.50
Perimeter Adj., Add or Deduct	Per 100 L.F.	18.95	12.60	9.45	7.55	6.30	5.40	4.80	4.15	3.75
Story Hgt. Adj., Add or Deduct	Per 1 Ft.	2.80	2.30	2.10	1.75	1.70	1.55	1.45	1.25	1.25
***For Basement**, add $ 29.50 per square foot of basement area*										

The above costs were calculated using the basic specifications shown on the facing page. These costs should be adjusted where necessary for design alternatives and owner's requirements.

Common additives

Description	Unit	$ Cost
Cabinets, base, door units, metal	L.F.	345
Drawer units	L.F.	690
Tall storage cabinets, open, 7′ high	L.F.	660
With glazed doors	L.F.	1050
Wall cabinets, metal, 12-1/2″ deep, open	L.F.	293
With doors	L.F.	525
Counter top, laminated plastic, no backsplash	L.F.	84.50
Stainless steel counter top	L.F.	217
For drop-in stainless 43″ x 21″ sink, add	Ea.	1600
Nurse Call Station		
Single bedside call station	Ea.	263
Ceiling speaker station	Ea.	171
Emergency call station	Ea.	174
Pillow speaker	Ea.	320
Standard call button	Ea.	214
Master control station for 20 stations	Total	5800

Description	Unit	$ Cost
Directory Boards, Plastic, glass covered		
Plastic, glass covered, 30″ x 20″	Ea.	570
36″ x 48″	Ea.	1500
Building directory, alum., black felt panels, 1 door, 24″ x 18″	Ea.	570
36″ x 24″	Ea.	725
48″ x 32″	Ea.	950
48″ x 60″	Ea.	2000
Tables, Examining, vinyl top		
with base cabinets	Ea.	1525
Utensil washer-sanitizer	Ea.	9950
X-ray, mobile, minimum	Ea.	19,400

Important: See the Reference Section for Location Factors.

Model costs calculated for a 1 story building with 15'-4" story height and 12,500 square feet of floor area

Outpatient Surgery Center

			Unit	Unit Cost	Cost Per S.F.	% Of Sub-Total
A. SUBSTRUCTURE						
1010	Standard Foundations	Poured concrete; strip and spread footings; 4' foundation wall	S.F. Ground	4.60	4.60	3.5%
1020	Special Foundations	N/A	—	—	—	
1030	Slab on Grade	4" reinforced concrete with vapor barrier and granular base	S.F. Slab	6.40	6.40	
2010	Basement Excavation	Site preparation for slab and trench for foundation wall and footing	S.F. Ground	.28	.28	
2020	Basement Walls	N/A	—	—	—	
B. SHELL						
	B10 Superstructure					
1010	Floor Construction	N/A	—	—	—	2.4 %
1020	Roof Construction	Steel Joists and girders on columns	S.F. Roof	7.51	7.51	
	B20 Exterior Enclosure					
2010	Exterior Walls	Brick veneer *85% of wall*	S.F. Wall	18.21	8.54	6.4%
2020	Exterior Windows	Aluminum frame, fixed *15% of wall*	Each	827	4.56	
2030	Exterior Doors	Aluminum and glass entry; steel flush doors	Each	8994	7.19	
	B30 Roofing					
3010	Roof Coverings	Single-ply membrane, loose laid and ballasted; polyisocyanurate insulation	S.F. Roof	6.36	6.36	2.0%
3020	Roof Openings	Roof hatches	S.F. Roof	.11	.11	
C. INTERIORS						
1010	Partitions	Gypsum board on metal studs *7 S.F. Floor/L.F. Partition*	S.F. Partition	7.36	10.52	12.2%
1020	Interior Doors	Single and double leaf fire doors *200 S.F. Floor/Door*	Each	1333	6.67	
1030	Fittings	Stainless steel toilet partitions, steel lockers	S.F. Floor	1.68	1.68	
2010	Stair Construction	N/A	—	—	—	
3010	Wall Finishes	80% Paint, 10% vinyl wall covering, 10% ceramic tile	S.F. Surface	1.92	5.49	
3020	Floor Finishes	10% Carpet, 80% vinyl sheet, 10% ceramic tile	S.F. Floor	10.08	10.08	
3030	Ceiling Finishes	Mineral fiber tile and plastic coated tile on suspension system	S.F. Ceiling	4.37	4.37	
D. SERVICES						
	D10 Conveying					
1010	Elevators & Lifts	N/A	—	—	—	0.0 %
1020	Escalators & Moving Walks	N/A	—	—	—	
	D20 Plumbing					
2010	Plumbing Fixtures	Medical, patient & specialty fixtures, supply and drainage *1 fixture/350 S.F. Floor*	Each	3388	9.68	4.5%
2020	Domestic Water Distribution	Electric water heater	S.F. Floor	2.46	2.46	
2040	Rain Water Drainage	Roof drains	S.F. Roof	2.24	2.24	
	D30 HVAC					
3010	Energy Supply	Hot water reheat for surgery	S.F. Floor	11.05	11.05	39.2%
3020	Heat Generating Systems	Boiler	Each	60,925	6.91	
3030	Cooling Generating Systems	N/A	—	—	—	
3050	Terminal & Package Units	N/A	—	—	—	
3090	Other HVAC Sys. & Equipment	Rooftop AC, CAV, roof vents, heat exchanger, surgical air curtains	S.F. Floor	107	106.50	
	D40 Fire Protection					
4010	Sprinklers	Wet pipe sprinkler system	S.F. Floor	3.41	3.41	1.6%
4020	Standpipes	Standpipe	S.F. Floor	1.69	1.69	
	D50 Electrical					
5010	Electrical Service/Distribution	1200 ampere service, panel board and feeders	S.F. Floor	6.29	6.29	12.1%
5020	Lighting & Branch Wiring	T-8 fluorescent light fixtures, receptacles, switches, A.C. and misc. power	S.F. Floor	16.02	16.02	
5030	Communications & Security	Addressable alarm system, internet wiring, communications system, emergency lighting	S.F. Floor	9.58	9.58	
5090	Other Electrical Systems	Emergency generator, 300 kW with fuel tank	S.F. Floor	6.52	6.52	
E. EQUIPMENT & FURNISHINGS						
1010	Commercial Equipment	N/A	—	—	—	16.1 %
1020	Institutional Equipment	Medical gas system, cabinetry, scrub sink, sterilizer	S.F. Floor	50	50.06	
1030	Vehicular Equipment	N/A	—	—	—	
1090	Other Equipment	Ice maker, washing machine	S.F. Floor	1.01	1.01	
F. SPECIAL CONSTRUCTION & DEMOLITION						
1020	Integrated Construction	N/A	—	—	—	0.0 %
1040	Special Facilities	N/A	—	—	—	
G. BUILDING SITEWORK	**N/A**					
				Sub-Total	317.78	**100%**
	CONTRACTOR FEES (General Requirements: 10%, Overhead: 5%, Profit: 5%)			20%	79.47	
	ARCHITECT FEES			9%	35.75	
				Total Building Cost	**433**	

Costs per square foot of floor area

Exterior Wall	S.F. Area	7000	9000	11000	13000	15000	17000	19000	21000	23000
	L.F. Perimeter	240	280	303	325	354	372	397	422	447
Limestone and Concrete Block	Bearing Walls	279.20	261.30	**246.20**	235.55	228.80	222.15	217.75	214.25	211.30
	Reinforced Concrete	293.40	275.50	260.45	249.85	243.15	236.55	232.15	228.60	225.70
Face Brick and Concrete Block	Bearing Walls	251.75	236.40	224.10	215.55	209.90	204.65	201.05	198.15	195.75
	Reinforced Concrete	265.95	250.60	238.40	229.80	224.25	219.00	215.40	212.55	210.15
Decorative Concrete Block	Bearing Walls	242.20	227.70	216.45	208.60	203.35	198.55	195.20	192.55	190.35
	Reinforced Concrete	256.35	241.90	230.75	222.85	217.65	212.95	209.55	206.95	204.70
Perimeter Adj., Add or Deduct	Per 100 L.F.	34.60	26.85	21.95	18.55	16.15	14.20	12.75	11.50	10.50
Story Hgt. Adj., Add or Deduct	Per 1 Ft.	6.25	5.70	4.95	4.55	4.30	4.00	3.85	3.65	3.55
***For Basement**, add $ 27.50 per square foot of basement area*										

The above costs were calculated using the basic specifications shown on the facing page. These costs should be adjusted where necessary for design alternatives and owner's requirements.

Common additives

Description	Unit	$ Cost
Cells Prefabricated, 5'-6' wide,		
7'-8' high, 7'-8' deep	Ea.	12,200
Elevators, Hydraulic passenger, 2 stops		
1500# capacity	Ea.	68,400
2500# capacity	Ea.	71,900
3500# capacity	Ea.	76,900
Emergency Lighting, 25 watt, battery operated		
Lead battery	Ea.	305
Nickel cadmium	Ea.	525
Flagpoles, Complete		
Aluminum, 20' high	Ea.	1825
40' high	Ea.	4400
70' high	Ea.	11,600
Fiberglass, 23' high	Ea.	1375
39'-5" high	Ea.	3175
59' high	Ea.	7000

Description	Unit	$ Cost
Lockers, Steel, Single tier, 60" to 72"	Opng.	223 - 390
2 tier, 60" or 72" total	Opng.	135 - 167
5 tier, box lockers	Opng.	69.50 - 81.50
Locker bench, lam. maple top only	L.F.	35.50
Pedestals, steel pipe	Ea.	73.50
Safe, Office type, 1 hour rating		
30" x 18" x 18"	Ea.	2400
60" x 36" x 18", double door	Ea.	10,200
Shooting Range, Incl. bullet traps,		
target provisions, and contols,		
not incl. structural shell	Ea.	58,000
Smoke Detectors		
Ceiling type	Ea.	231
Duct type	Ea.	510
Sound System		
Amplifier, 250 watts	Ea.	1950
Speaker, ceiling or wall	Ea.	233
Trumpet	Ea.	445

Important: See the Reference Section for Location Factors.

Model costs calculated for a 2 story building with 12′ story height and 11,000 square feet of floor area

Police Station

			Unit	Unit Cost	Cost Per S.F.	% Of Sub-Total
A. SUBSTRUCTURE						
1010	Standard Foundations	Poured concrete; strip and spread footings; 4′ foundation wall	S.F. Ground	7.14	3.57	3.5%
1020	Special Foundations	N/A	—	—	—	
1030	Slab on Grade	4″ reinforced concrete with vapor barrier and granular base	S.F. Slab	5.28	2.64	
2010	Basement Excavation	Site preparation for slab and trench for foundation wall and footing	S.F. Ground	.28	.14	
2020	Basement Walls	N/A	—	—	—	
B. SHELL						
	B10 Superstructure					
1010	Floor Construction	Open web steel joists, slab form, concrete	S.F. Floor	12.14	6.07	4.7%
1020	Roof Construction	Metal deck on open web steel joists	S.F. Roof	4.86	2.43	
	B20 Exterior Enclosure					
2010	Exterior Walls	Limestone with concrete block back-up *80% of wall*	S.F. Wall	62	32.53	25.0%
2020	Exterior Windows	Metal horizontal sliding *20% of wall*	Each	1141	10.06	
2030	Exterior Doors	Hollow metal	Each	3454	2.51	
	B30 Roofing					
3010	Roof Coverings	Built-up tar and gravel with flashing; perlite/EPS composite insulation	S.F. Roof	7.22	3.61	2.0%
3020	Roof Openings	N/A	—	—	—	
C. INTERIORS						
1010	Partitions	Concrete block *20 S.F. Floor/L.F. Partition*	S.F. Partition	14	7	18.1%
1020	Interior Doors	Single leaf kalamein fire door *200 S.F. Floor/Door*	Each	1201	6.01	
1030	Fittings	Toilet partitions	S.F. Floor	.95	.95	
2010	Stair Construction	Concrete filled metal pan	Flight	18,425	5.03	
3010	Wall Finishes	90% paint, 10% ceramic tile	S.F. Surface	3.55	3.55	
3020	Floor Finishes	70% vinyl composition tile, 20% carpet tile, 10% ceramic tile	S.F. Floor	4.11	4.11	
3030	Ceiling Finishes	Mineral fiber tile on concealed zee bars	S.F. Ceiling	6.10	6.10	
D. SERVICES						
	D10 Conveying					
1010	Elevators & Lifts	One hydraulic passenger elevator	Each	85,580	7.78	4.3%
1020	Escalators & Moving Walks	N/A	—	—	—	
	D20 Plumbing					
2010	Plumbing Fixtures	Toilet and service fixtures, supply and drainage *1 Fixture/580 S.F. Floor*	Each	4524	7.80	7.9%
2020	Domestic Water Distribution	Oil fired water heater	S.F. Floor	4.89	4.89	
2040	Rain Water Drainage	Roof drains	S.F. Roof	3.10	1.55	
	D30 HVAC					
3010	Energy Supply	N/A	—	—	—	11.6 %
3020	Heat Generating Systems	N/A	—	—	—	
3030	Cooling Generating Systems	N/A	—	—	—	
3050	Terminal & Package Units	Multizone HVAC air cooled system	S.F. Floor	21	21	
3090	Other HVAC Sys. & Equipment	N/A	—	—	—	
	D40 Fire Protection					
4010	Sprinklers	Wet pipe sprinkler system	S.F. Floor	3.64	3.64	2.6%
4020	Standpipes	Standpipe	S.F. Floor	1.08	1.08	
	D50 Electrical					
5010	Electrical Service/Distribution	400 ampere service, panel board and feeders	S.F. Floor	1.93	1.93	11.4%
5020	Lighting & Branch Wiring	T-8 fluorescent fixtures, receptacles, switches, A.C. and misc. power	S.F. Floor	11.36	11.36	
5030	Communications & Security	Addressable alarm systems, internet wiring, intercom and emergency lighting	S.F. Floor	6.99	6.99	
5090	Other Electrical Systems	Emergency generator, 15 kW	S.F. Floor	.24	.24	
E. EQUIPMENT & FURNISHINGS						
1010	Commercial Equipment	N/A	—	—	—	8.7 %
1020	Institutional Equipment	Lockers, detention rooms, cells, gasoline dispensers	S.F. Floor	13.12	13.12	
1030	Vehicular Equipment	Gasoline dispenser system	S.F. Floor	2.63	2.63	
1090	Other Equipment	N/A	—	—	—	
F. SPECIAL CONSTRUCTION						
1020	Integrated Construction	N/A	—	—	—	0.0 %
1040	Special Facilities	N/A	—	—	—	
G. BUILDING SITEWORK		**N/A**				
				Sub-Total	180.69	**100%**
	CONTRACTOR FEES (General Requirements: 10%, Overhead: 5%, Profit: 10%)			25%	45.18	
	ARCHITECT FEES			9%	20.33	
				Total Building Cost	**246.20**	

Costs per square foot of floor area

Exterior Wall	S.F. Area	600	750	925	1150	1450	1800	2100	2600	3200
	L.F. Perimeter	100	112	128	150	164	182	200	216	240
Metal Panel and Wood Studs	Wood Truss	78.40	73.40	70.15	67.75	63.10	**59.85**	58.20	55.00	52.80
Aluminum Clapboard	Wood Truss	72.15	67.65	64.75	62.60	58.45	55.50	54.05	51.20	49.25
Wood Clapboard	Wood Truss	77.20	72.35	69.20	66.90	62.30	59.15	57.55	54.40	52.30
Plywood Siding	Wood Truss	72.05	67.55	64.60	62.45	58.30	55.40	53.95	51.10	49.20
Perimeter Adj., Add or Deduct	Per 100 L.F.	41.50	33.20	26.95	21.65	17.20	13.80	11.90	9.55	7.80
Story Hgt. Adj., Add or Deduct	Per 1 Ft.	3.00	2.60	2.45	2.30	2.00	1.80	1.70	1.45	1.40
Basement—Not Applicable										

The above costs were calculated using the basic specifications shown on the facing page. These costs should be adjusted where necessary for design alternatives and owner's requirements.

Common additives

Description	Unit	$ Cost
Slab on Grade		
4" thick, non-industrial, non-reinforced	S.F.	4.82
Reinforced	S.F.	5.28
Light industrial, non-reinforced	S.F.	5.91
Reinforced	S.F.	6.40
Industrial, non-reinforced	S.F.	9.95
Reinforced	S.F.	10.43
6" thick, non-industrial, non-reinforced	S.F.	5.79
Reinforced	S.F.	6.46
Light industrial, non-reinforced	S.F.	6.91
Reinforced	S.F.	7.56
Heavy industrial, non-reinforced	S.F.	12.90
Reinforced	S.F.	13.55
Insulation		
Fiberboard, low density, 1" thick, R2.78	S.F.	1.38
Foil on reinforced scrim, single bubble air space, R8.8	C.S.F.	63
Vinyl faced fiberglass, 1-1/2" thick, R5	S.F.	0.87
Vinyl/scrim/foil (VSF), 1-1/2" thick, R5	S.F.	1.07

Description	Unit	$ Cost
Roof ventilation, ridge vent, aluminum	L.F.	7.60
Mushroom vent, aluminum	Ea.	91.50
Gutter, aluminum, 5" K type, enameled	L.F.	7.80
Steel, half round or box, 5", enameled	L.F.	7.05
Wood, clear treated cedar, fir or hemlock, 4" x 5"	L.F.	29
Downspout, aluminum, 3" x 4", enameled	L.F.	6.45
Steel, round, 4" diameter	L.F.	6.30

Important: See the Reference Section for Location Factors.

Model costs calculated for a 1 story building with 14′ story height and 1,800 square feet of floor area

Post Frame Barn

			Unit	Unit Cost	Cost Per S.F.	% Of Sub-Total
A. SUBSTRUCTURE						
1010	Standard Foundations	Included in B2010 Exterior Walls	–	–	–	0.0%
1020	Special Foundations	N/A	–	–	–	
1030	Slab on Grade	N/A	–	–	–	
2010	Basement Excavation	N/A	–	–	–	
2020	Basement Walls	N/A	–	–	–	
B. SHELL						
	B10 Superstructure					
1010	Floor Construction	N/A	–	–	–	16.3 %
1020	Roof Construction	Wood truss, 4/12 pitch	S.F. Roof	7.22	7.22	
	B20 Exterior Enclosure					
2010	Exterior Walls	2 x 6 Wood framing, 6 x 6 wood posts in concrete; steel siding panels	S.F. Wall	10.89	14.65	54.3%
2020	Exterior Windows	Steel windows, fixed	Each	507	3.99	
2030	Exterior Doors	Steel overhead, hollow metal doors	Each	2440	5.42	
	B30 Roofing					
3010	Roof Coverings	Metal panel roof	S.F. Roof	4.13	4.13	9.3%
3020	Roof Openings	N/A	–	–	–	
C. INTERIORS						
1010	Partitions	N/A	–	–	–	0.0 %
1020	Interior Doors	N/A	–	–	–	
1030	Fittings	N/A	–	–	–	
2010	Stair Construction	N/A	–	–	–	
3010	Wall Finishes	N/A	–	–	–	
3020	Floor Finishes	N/A	–	–	–	
3030	Ceiling Finishes	N/A	–	–	–	
D. SERVICES						
	D10 Conveying					
1010	Elevators & Lifts	N/A	–	–	–	0.0 %
1020	Escalators & Moving Walks	N/A	–	–	–	
	D20 Plumbing					
2010	Plumbing Fixtures	N/A	–	–	–	0.0 %
2020	Domestic Water Distribution	N/A	–	–	–	
2040	Rain Water Drainage	N/A	–	–	–	
	D30 HVAC					
3010	Energy Supply	N/A	–	–	–	0.0 %
3020	Heat Generating Systems	N/A	–	–	–	
3030	Cooling Generating Systems	N/A	–	–	–	
3050	Terminal & Package Units	N/A	–	–	–	
3090	Other HVAC Sys. & Equipment	N/A	–	–	–	
	D40 Fire Protection					
4010	Sprinklers	N/A	–	–	–	0.0 %
4020	Standpipes	N/A	–	–	–	
	D50 Electrical					
5010	Electrical Service/Distribution	60 Ampere service, panel board & feeders	S.F. Floor	1.84	1.84	20.1%
5020	Lighting & Branch Wiring	High bay fixtures, receptacles & switches	S.F. Floor	7.07	7.07	
5030	Communications & Security	N/A	–	–	–	
5090	Other Electrical Systems	N/A	–	–	–	
E. EQUIPMENT & FURNISHINGS						
1010	Commercial Equipment	N/A	–	–	–	0.0 %
1020	Institutional Equipment	N/A	–	–	–	
1030	Vehicular Equipment	N/A	–	–	–	
1090	Other Equipment	N/A	–	–	–	
F. SPECIAL CONSTRUCTION & DEMOLITION						
1020	Integrated Construction	N/A	–	–	–	0.0 %
1040	Special Facilities	N/A	–	–	–	
G. BUILDING SITEWORK		**N/A**				
				Sub-Total	44.32	**100%**
	CONTRACTOR FEES (General Requirements: 10%, Overhead: 5%, Profit: 5%)			20%	11.10	
	ARCHITECT FEES			8%	4.43	
				Total Building Cost	**59.85**	

Costs per square foot of floor area

Exterior Wall	S.F. Area	5000	7000	9000	11000	13000	15000	17000	19000	21000
	L.F. Perimeter	300	380	420	468	486	513	540	580	620
Face Brick and Concrete Block	Rigid Steel	167.90	158.40	149.35	144.20	**138.75**	135.20	132.55	130.95	129.70
	Bearing Walls	168.25	158.70	149.55	144.40	138.75	135.25	132.50	130.95	129.65
Limestone and Concrete Block	Rigid Steel	195.90	183.75	171.10	164.10	156.20	151.15	147.35	145.20	143.50
	Bearing Walls	196.30	184.00	171.30	164.25	156.25	151.20	147.35	145.20	143.45
Decorative Concrete Block	Rigid Steel	158.15	149.55	141.80	137.35	132.65	129.65	127.35	125.95	124.95
	Bearing Walls	158.50	149.85	141.95	137.45	132.70	129.70	127.35	125.95	124.90
Perimeter Adj., Add or Deduct	Per 100 L.F.	16.95	12.15	9.50	7.75	6.50	5.70	4.95	4.50	4.05
Story Hgt. Adj., Add or Deduct	Per 1 Ft.	2.70	2.40	2.15	1.95	1.65	1.50	1.40	1.35	1.35
***For Basement**, add $28.70 per square foot of basement area*										

The above costs were calculated using the basic specifications shown on the facing page. These costs should be adjusted where necessary for design alternatives and owner's requirements.

Common additives

Description	Unit	$ Cost
Closed Circuit Surveillance, One station		
Camera and monitor	Ea.	1250
For additional camera stations, add	Ea.	565
Emergency Lighting, 25 watt, battery operated		
Lead battery	Ea.	305
Nickel cadmium	Ea.	525
Flagpoles, Complete		
Aluminum, 20′ high	Ea.	1825
40′ high	Ea.	4400
70′ high	Ea.	11,600
Fiberglass, 23′ high	Ea.	1375
39′-5″ high	Ea.	3175
59′ high	Ea.	7000
Mail Boxes, Horizontal, key lock, 15″ x 6″ x 5″	Ea.	62
Double 15″ x 12″ x 5″	Ea.	91
Quadruple 15″ x 12″ x 10″	Ea.	146
Vertical, 6″ x 5″ x 15″, aluminum	Ea.	64.50
Bronze	Ea.	68
Steel, enameled	Ea.	64.50
Scales, Dial type, 5 ton cap.		
8′ x 6′ platform	Ea.	13,200
9′ x 7′ platform	Ea.	15,200
Smoke Detectors		
Ceiling type	Ea.	231
Duct type	Ea.	510

Important: See the Reference Section for Location Factors.

Model costs calculated for a 1 story building with 14′ story height and 13,000 square feet of floor area

Post Office

			Unit	Unit Cost	Cost Per S.F.	% Of Sub-Total
A. SUBSTRUCTURE						
1010	Standard Foundations	Poured concrete; strip and spread footings; 4′ foundation wall	S.F. Ground	4.54	4.54	9.9%
1020	Special Foundations	N/A	–	–	–	
1030	Slab on Grade	4″ reinforced concrete with vapor barrier and granular base	S.F. Slab	5.28	5.28	
2010	Basement Excavation	Site preparation for slab and trench for foundation wall and footing	S.F. Ground	.28	.28	
2020	Basement Walls	N/A	–	–	–	
B. SHELL						
	B10 Superstructure					
1010	Floor Construction	Steel column fireproofing	L.F. Column	33	.24	9.1%
1020	Roof Construction	Metal deck, open web steel joists, columns	S.F. Roof	9.03	9.03	
	B20 Exterior Enclosure					
2010	Exterior Walls	Face brick with concrete block back-up *80% of wall*	S.F. Wall	31	12.93	17.5%
2020	Exterior Windows	Double strength window glass *20% of wall*	Each	703	3.20	
2030	Exterior Doors	Double aluminum & glass, single aluminum, hollow metal, steel overhead	Each	3647	1.69	
	B30 Roofing					
3010	Roof Coverings	Built-up tar and gravel with flashing; perlite/EPS composite insulation	S.F. Roof	6.47	6.47	6.4%
3020	Roof Openings	N/A	–	–	–	
C. INTERIORS						
1010	Partitions	Concrete block *15 S.F. Floor/L.F. Partition*	S.F. Partition	9.15	7.32	24.0%
1020	Interior Doors	Single leaf hollow metal *150 S.F. Floor/Door*	Each	1201	8.01	
1030	Fittings	Toilet partitions, cabinets, shelving, lockers	S.F. Floor	1.64	1.64	
2010	Stair Construction	N/A	–	–	–	
3010	Wall Finishes	Paint	S.F. Surface	2.17	3.47	
3020	Floor Finishes	50% vinyl tile, 50% paint	S.F. Floor	2.28	2.28	
3030	Ceiling Finishes	Mineral fiber tile on concealed zee bars *25% of area*	S.F. Ceiling	6.86	1.72	
D. SERVICES						
	D10 Conveying					
1010	Elevators & Lifts	N/A	–	–	–	0.0 %
1020	Escalators & Moving Walks	N/A	–	–	–	
	D20 Plumbing					
2010	Plumbing Fixtures	Toilet and service fixtures, supply and drainage *1 Fixture/1180 S.F. Floor*	Each	2159	1.83	4.0%
2020	Domestic Water Distribution	Gas fired water heater	S.F. Floor	.51	.51	
2040	Rain Water Drainage	Roof drains	S.F. Roof	1.71	1.71	
	D30 HVAC					
3010	Energy Supply	N/A	–	–	–	8.8 %
3020	Heat Generating Systems	Included in D3050	–	–	–	
3030	Cooling Generating Systems	N/A	–	–	–	
3050	Terminal & Package Units	Single zone, gas heating, electric cooling	S.F. Floor	8.94	8.94	
3090	Other HVAC Sys. & Equipment	N/A	–	–	–	
	D40 Fire Protection					
4010	Sprinklers	Wet pipe sprinkler system	S.F. Floor	4.45	4.45	5.3%
4020	Standpipes	Standpipe	S.F. Floor	.97	.97	
	D50 Electrical					
5010	Electrical Service/Distribution	600 ampere service, panel board and feeders	S.F. Floor	2.31	2.31	15.0%
5020	Lighting & Branch Wiring	T-8 fluorescent fixtures, receptacles, switches, A.C. and misc. power	S.F. Floor	8.41	8.41	
5030	Communications & Security	Addressable alarm systems, internet wiring and emergency lighting	S.F. Floor	4.44	4.44	
5090	Other Electrical Systems	Emergency generator, 15 kW	S.F. Floor	.14	.14	
E. EQUIPMENT & FURNISHINGS						
1010	Commercial Equipment	N/A	–	–	–	0.0 %
1020	Institutional Equipment	N/A	–	–	–	
1030	Vehicular Equipment	N/A	–	–	–	
1090	Other Equipment	N/A	–	–	–	
F. SPECIAL CONSTRUCTION						
1020	Integrated Construction	N/A	–	–	–	0.0 %
1040	Special Facilities	N/A	–	–	–	
G. BUILDING SITEWORK		**N/A**				
				Sub-Total	101.81	**100%**
	CONTRACTOR FEES (General Requirements: 10%, Overhead: 5%, Profit: 10%)			*25%*	25.49	
	ARCHITECT FEES			*9%*	11.45	
				Total Building Cost	**138.75**	

M.510 Racquetball Court

Costs per square foot of floor area

Exterior Wall	S.F. Area	5000	10000	15000	21000	25000	30000	40000	50000	60000
	L.F. Perimeter	287	400	500	600	660	700	834	900	1000
Face Brick and Concrete Block	Rigid Steel	237.95	203.55	190.80	182.15	178.30	**173.20**	168.85	164.30	162.05
	Bearing Walls	235.95	201.10	188.20	179.35	175.50	170.35	165.90	161.25	158.95
Concrete Block	Rigid Steel	206.25	181.45	172.35	166.30	163.70	160.35	157.35	154.35	152.80
Brick Veneer	Rigid Steel	223.95	193.85	182.70	175.20	171.85	167.55	163.80	159.90	158.00
Metal Panel	Rigid Steel	195.65	174.10	166.25	161.10	158.85	156.05	153.50	151.00	149.75
Metal Sandwich Panel	Rigid Steel	247.10	209.95	196.10	186.70	182.45	176.95	172.20	167.15	164.70
Perimeter Adj., Add or Deduct	Per 100 L.F.	29.05	14.50	9.70	6.90	5.80	4.85	3.60	2.85	2.45
Story Hgt. Adj., Add or Deduct	Per 1 Ft.	5.95	4.15	3.45	2.90	2.70	2.45	2.15	1.80	1.70
***For Basement**, add $ 27.70 per square foot of basement area*										

The above costs were calculated using the basic specifications shown on the facing page. These costs should be adjusted where necessary for design alternatives and owner's requirements.

Common additives

Description	Unit	$ Cost
Bar, Front Bar	L.F.	445
Back Bar	L.F.	355
Booth, Upholstered, custom straight	L.F.	259 - 480
"L" or "U" shaped	L.F.	268 - 455
Bleachers, Telescoping, manual		
To 15 tier	Seat	158 - 216
21-30 tier	Seat	330 - 520
Courts		
Ceiling	Court	9800
Floor	Court	16,900
Walls	Court	28,200
Emergency Lighting, 25 watt, battery operated		
Lead battery	Ea.	305
Nickel cadmium	Ea.	525
Kitchen Equipment		
Broiler	Ea.	4400
Cooler, 6 ft. long, reach-in	Ea.	3725
Dishwasher, 10-12 racks per hr.	Ea.	3975
Food warmer, counter 1.2 KW	Ea.	785
Freezer, reach-in, 44 C.F.	Ea.	5400
Ice cube maker, 50 lb. per day	Ea.	2050

Description	Unit	$ Cost
Lockers, Steel, single tier, 60" or 72"	Opng.	223 - 390
2 tier, 60" or 72" total	Opng.	135 - 167
5 tier, box lockers	Opng.	69.50 - 81.50
Locker bench, lam. maple top only	L.F.	35.50
Pedestals, steel pipe	Ea.	73.50
Sauna, Prefabricated, complete		
6' x 4'	Ea.	6425
6' x 9'	Ea.	9550
8' x 8'	Ea.	10,500
8' x 10'	Ea.	11,500
10' x 12'	Ea.	15,400
Sound System		
Amplifier, 250 watts	Ea.	1950
Speaker, ceiling or wall	Ea.	233
Trumpet	Ea.	445
Steam Bath, Complete, to 140 C.F.	Ea.	3025
To 300 C.F.	Ea.	3425
To 800 C.F.	Ea.	6975
To 2500 C.F.	Ea.	9750

Important: See the Reference Section for Location Factors.

Model costs calculated for a 2 story building with 12′ story height and 30,000 square feet of floor area

Racquetball Court

			Unit	Unit Cost	Cost Per S.F.	% Of Sub-Total
A. SUBSTRUCTURE						
1010	Standard Foundations	Poured concrete; strip and spread footings; 4′ foundation wall	S.F. Ground	5.86	2.93	5.1%
1020	Special Foundations	N/A	—	—	—	
1030	Slab on Grade	5″ reinforced concrete with vapor barrier and granular base	S.F. Slab	6.90	3.45	
2010	Basement Excavation	Site preparation for slab and trench for foundation wall and footing	S.F. Ground	.28	.14	
2020	Basement Walls	N/A	—	—	—	
B. SHELL						
	B10 Superstructure					
1010	Floor Construction	Open web steel joists, slab form, concrete, columns *50% of area*	S.F. Floor	18.14	9.07	10.0%
1020	Roof Construction	Metal deck on open web steel joists, columns	S.F. Roof	7.54	3.77	
	B20 Exterior Enclosure					
2010	Exterior Walls	Face brick with concrete block back-up *95% of wall*	S.F. Wall	31	16.33	15.3%
2020	Exterior Windows	Storefront *5% of wall*	S.F. Window	100	2.79	
2030	Exterior Doors	Aluminum and glass and hollow metal	Each	4033	.54	
	B30 Roofing					
3010	Roof Coverings	Built-up tar and gravel with flashing; perlite/EPS composite insulation	S.F. Roof	6.64	3.32	2.6%
3020	Roof Openings	Roof hatches	S.F. Roof	.16	.08	
C. INTERIORS						
1010	Partitions	Concrete block, gypsum board on metal studs *25 S.F. Floor/L.F. Partition*	S.F. Partition	13.83	5.53	13.7%
1020	Interior Doors	Single leaf hollow metal *810 S.F. Floor/Door*	Each	1201	1.48	
1030	Fittings	Toilet partitions	S.F. Floor	.41	.41	
2010	Stair Construction	Concrete filled metal pan	Flight	18,425	1.84	
3010	Wall Finishes	Paint	S.F. Surface	1.33	1.06	
3020	Floor Finishes	80% carpet tile, 20% ceramic tile *50% of floor area*	S.F. Floor	6.28	3.14	
3030	Ceiling Finishes	Mineral fiber tile on concealed zee bars *60% of area*	S.F. Ceiling	6.86	4.11	
D. SERVICES						
	D10 Conveying					
1010	Elevators & Lifts	N/A	—	—	—	0.0 %
1020	Escalators & Moving Walks	N/A	—	—	—	
	D20 Plumbing					
2010	Plumbing Fixtures	Kitchen, bathroom and service fixtures, supply and drainage *1 Fixture/1000 S.F. Floor*	Each	3660	3.66	5.9%
2020	Domestic Water Distribution	Gas fired water heater	S.F. Floor	3.60	3.60	
2040	Rain Water Drainage	Roof drains	S.F. Roof	.72	.36	
	D30 HVAC					
3010	Energy Supply	N/A	—	—	—	21.6 %
3020	Heat Generating Systems	Included in D3050	—	—	—	
3030	Cooling Generating Systems	N/A	—	—	—	
3050	Terminal & Package Units	Multizone unit, gas heating, electric cooling	S.F. Floor	28	27.70	
3090	Other HVAC Sys. & Equipment	N/A	—	—	—	
	D40 Fire Protection					
4010	Sprinklers	Sprinklers, light hazard	S.F. Floor	3.17	3.17	2.8%
4020	Standpipes	Standpipe	S.F. Floor	.40	.40	
	D50 Electrical					
5010	Electrical Service/Distribution	400 ampere service, panel board and feeders	S.F. Floor	.84	.84	7.5%
5020	Lighting & Branch Wiring	Fluorescent and high intensity discharge fixtures, receptacles, switches, A.C. and misc. power	S.F. Floor	7.59	7.59	
5030	Communications & Security	Addressable alarm systems and emergency lighting	S.F. Floor	1.08	1.08	
5090	Other Electrical Systems	Emergency generator, 15 kW	S.F. Floor	.06	.06	
E. EQUIPMENT & FURNISHINGS						
1010	Commercial Equipment	N/A	—	—	—	15.5 %
1020	Institutional Equipment	N/A	—	—	—	
1030	Vehicular Equipment	N/A	—	—	—	
1090	Other Equipment	Courts, sauna baths	S.F. Floor	19.86	19.86	
F. SPECIAL CONSTRUCTION						
1020	Integrated Construction	N/A	—	—	—	0.0 %
1040	Special Facilities	N/A	—	—	—	
G. BUILDING SITEWORK	**N/A**					
				Sub-Total	128.31	**100%**
	CONTRACTOR FEES (General Requirements: 10%, Overhead: 5%, Profit: 10%)			*25%*	32.06	
	ARCHITECT FEES			*8%*	12.83	
				Total Building Cost	**173.20**	

Costs per square foot of floor area

Exterior Wall	S.F. Area	2000	2800	3500	4200	5000	5800	6500	7200	8000
	L.F. Perimeter	180	212	240	268	300	314	336	344	368
Wood Clapboard	Wood Frame	216.70	205.30	199.60	195.80	**192.80**	189.00	187.20	184.80	183.50
Brick Veneer	Wood Frame	225.50	213.10	206.90	202.80	199.50	195.30	193.40	190.70	189.30
Stone Veneer	Rigid Steel	259.70	242.95	234.60	229.05	224.60	218.60	215.95	212.05	210.15
Fiber Cement	Rigid Steel	223.85	212.75	207.20	203.60	200.70	197.00	195.30	193.00	191.75
E.I.F.S.	Reinforced Concrete	249.60	236.50	229.90	225.65	222.15	217.65	215.60	212.75	211.20
Stucco	Reinforced Concrete	250.35	237.15	230.50	226.15	222.65	218.10	216.05	213.15	211.55
Perimeter Adj., Add or Deduct	Per 100 L.F.	25.15	17.95	14.40	12.00	10.00	8.70	7.75	7.00	6.30
Story Hgt. Adj., Add or Deduct	Per 1 Ft.	2.60	2.15	1.95	1.80	1.65	1.55	1.45	1.35	1.25
***For Basement**, add $ 33.70 per square foot of basement area*										

The above costs were calculated using the basic specifications shown on the facing page. These costs should be adjusted where necessary for design alternatives and owner's requirements.

Common additives

Description	Unit	$ Cost
Bar, Front Bar	L.F.	445
Back bar	L.F.	355
Booth, Upholstered, custom straight	L.F.	259 - 480
"L" or "U" shaped	L.F.	268 - 455
Cupola, Stock unit, redwood		
30" square, 37" high, aluminum roof	Ea.	810
Copper roof	Ea.	455
Fiberglass, 5'-0" base, 63" high	Ea.	4350 - 4700
6'-0" base, 63" high	Ea.	5125 - 7575
Decorative Wood Beams, Non load bearing		
Rough sawn, 4" x 6"	L.F.	8.10
4" x 8"	L.F.	9.15
4" x 10"	L.F.	11.65
4" x 12"	L.F.	12.80
8" x 8"	L.F.	13.30
Emergency Lighting, 25 watt, battery operated		
Lead battery	Ea.	305
Nickel cadmium	Ea.	525

Description	Unit	$ Cost
Fireplace, Brick, not incl. chimney or foundation		
30" x 29" opening	Ea.	2900
Chimney, standard brick		
Single flue, 16" x 20"	V.L.F.	100
20" x 20"	V.L.F.	120
2 Flue, 20" x 24"	V.L.F.	148
20" x 32"	V.L.F.	186
Kitchen Equipment		
Broiler	Ea.	4400
Coffee urn, twin 6 gallon	Ea.	2900
Cooler, 6 ft. long	Ea.	3725
Dishwasher, 10-12 racks per hr.	Ea.	3975
Food warmer, counter, 1.2 KW	Ea.	785
Freezer, 44 C.F., reach-in	Ea.	5400
Ice cube maker, 50 lb. per day	Ea.	2050
Range with 1 oven	Ea.	3200
Refrigerators, Prefabricated, walk-in		
7'-6" high, 6' x 6'	S.F.	137
10' x 10'	S.F.	107
12' x 14'	S.F.	94
12' x 20'	S.F.	134

Important: See the Reference Section for Location Factors.

Model costs calculated for a 1 story building with 12′ story height and 5,000 square feet of floor area

Restaurant

			Unit	Unit Cost	Cost Per S.F.	% Of Sub-Total
A. SUBSTRUCTURE						
1010	Standard Foundations	Poured concrete; strip and spread footings; 4′ foundation wall	S.F. Ground	8.97	8.97	
1020	Special Foundations	N/A	—	—	—	
1030	Slab on Grade	4″ reinforced concrete	S.F. Slab	5.28	5.28	10.2%
2010	Basement Excavation	Site preparation for slab and trench for foundation wall and footing	S.F. Ground	.49	.49	
2020	Basement Walls	N/A	—	—	—	
B. SHELL						
	B10 Superstructure					
1010	Floor Construction	Wood columns	S.F. Floor	.17	.42	5.3%
1020	Roof Construction	Wood roof truss, 4:12 slope	S.F. Roof	6.45	7.22	
	B20 Exterior Enclosure					
2010	Exterior Walls	Wood siding on wood studs, insulated *70% of wall*	S.F. Wall	23	11.68	
2020	Exterior Windows	Curtain glazing wall system *30% of wall*	Each	42	3.86	15.4%
2030	Exterior Doors	Aluminum & glass double doors, hollow metal doors	Each	6667	6.68	
	B30 Roofing					
3010	Roof Coverings	Asphalt strip shingles, gutters and downspouts	S.F. Roof	14.05	2.24	1.6%
3020	Roof Openings	N/A	—	—	—	
C. INTERIORS						
1010	Partitions	Fire rated gypsum board on wood studs *25 S.F. Floor/L.F. Partition*	S.F. Partition	6.40	2.56	
1020	Interior Doors	Single leaf hollow core wood doors *1000 S.F. Floor/Door*	Each	618	.62	
1030	Fittings	Plastic laminate toilet partitions	S.F. Floor	.87	.87	
2010	Stair Construction	N/A	—	—	—	13.1%
3010	Wall Finishes	80% Paint, 20% ceramic tile	S.F. Surface	2.56	2.05	
3020	Floor Finishes	70% Carpet tile, 30% quarry tile	S.F. Floor	8.52	8.52	
3030	Ceiling Finishes	Gypsum board ceiling on wood furring	S.F. Ceiling	4.21	4.21	
D. SERVICES						
	D10 Conveying					
1010	Elevators & Lifts	N/A	—	—	—	0.0 %
1020	Escalators & Moving Walks	N/A	—	—	—	
	D20 Plumbing					
2010	Plumbing Fixtures	Restroom, kitchen and service fixtures, supply and drainage *1 Fixture/350 S.F. Floor*	Each	3106	8.75	
2020	Domestic Water Distribution	Gas fired water heater	S.F. Floor	5.57	5.57	9.9%
2040	Rain Water Drainage	Roof drains	—	—	—	
	D30 HVAC					
3010	Energy Supply	N/A	—	—	—	
3020	Heat Generating Systems	Included in D3050	—	—	—	
3030	Cooling Generating Systems	N/A	—	—	—	25.3 %
3050	Terminal & Package Units	Multizone unit gas heating, electric cooling, kitchen exhaust system	S.F. Floor	37	36.50	
3090	Other HVAC Sys. & Equipment	N/A	—	—	—	
	D40 Fire Protection					
4010	Sprinklers	Wet pipe sprinkler system	S.F. Floor	8.64	8.64	7.5%
4020	Standpipes	Standpipe	S.F. Floor	2.15	2.15	
	D50 Electrical					
5010	Electrical Service/Distribution	400 Ampere service, panelboards and feeders	S.F. Floor	4.43	4.43	
5020	Lighting & Branch Wiring	Fluorescent fixtures, receptacles, switches, a.c. and misc. power	S.F. Floor	8.93	8.93	11.7%
5030	Communications & Security	Addressable alarm system	S.F. Floor	3.49	3.49	
5090	Other Electrical Systems	N/A	—	—	—	
E. EQUIPMENT & FURNISHINGS						
1010	Commercial Equipment	N/A	—	—	—	
1020	Institutional Equipment	N/A	—	—	—	0.0 %
1030	Vehicular Equipment	N/A	—	—	—	
1090	Other Equipment	N/A	—	—	—	
F. SPECIAL CONSTRUCTION						
1020	Integrated Construction	N/A	—	—	—	0.0 %
1040	Special Facilities	N/A	—	—	—	
G. BUILDING SITEWORK		**N/A**				
				Sub-Total	144.13	**100%**
	CONTRACTOR FEES (General Requirements: 10%, Overhead: 5%, Profit: 10%)			*25%*	36.06	
	ARCHITECT FEES			*7%*	12.61	
				Total Building Cost	**192.80**	

Costs per square foot of floor area

Exterior Wall	S.F. Area	2000	2800	3500	4000	5000	5800	6500	7200	8000
	L.F. Perimeter	180	212	240	260	300	314	336	344	368
Wood Clapboard	Wood Frame	215.75	208.45	204.80	203.00	200.45	197.50	196.25	194.25	193.30
Brick Veneer	Wood Frame	218.10	210.40	206.60	204.65	202.00	198.85	197.50	195.45	194.45
Stone Veneer	Rigid Steel	248.10	236.80	231.10	228.30	224.35	219.70	217.80	214.65	213.25
Fiber Cement	Rigid Steel	217.95	211.45	208.10	206.50	204.20	201.55	200.40	198.65	197.80
E.I.F.S.	Reinforced Concrete	241.05	232.70	228.50	226.40	223.50	220.05	218.60	216.40	215.25
Stucco	Reinforced Concrete	241.70	233.25	229.00	226.90	223.95	220.45	218.95	216.70	215.65
Perimeter Adj., Add or Deduct	Per 100 L.F.	25.30	18.10	14.50	12.75	10.15	8.75	7.85	7.05	6.40
Story Hgt. Adj., Add or Deduct	Per 1 Ft.	2.65	2.30	2.15	2.05	1.85	1.65	1.60	1.50	1.50
Basement—Not Applicable										

The above costs were calculated using the basic specifications shown on the facing page. These costs should be adjusted where necessary for design alternatives and owner's requirements.

Common additives

Description	Unit	$ Cost
Bar, Front Bar	L.F.	445
Back bar	L.F.	355
Booth, Upholstered, custom straight	L.F.	259 - 480
"L" or "U" shaped	L.F.	268 - 455
Drive-up Window	Ea.	9450 - 12,900
Emergency Lighting, 25 watt, battery operated		
Lead battery	Ea.	305
Nickel cadmium	Ea.	525
Kitchen Equipment		
Broiler	Ea.	4400
Coffee urn, twin 6 gallon	Ea.	2900
Cooler, 6 ft. long	Ea.	3725
Dishwasher, 10-12 racks per hr.	Ea.	3975
Food warmer, counter, 1.2 KW	Ea.	785
Freezer, 44 C.F., reach-in	Ea.	5400
Ice cube maker, 50 lb. per day	Ea.	2050
Range with 1 oven	Ea.	3200

Description	Unit	$ Cost
Refrigerators, Prefabricated, walk-in		
7'-6" High, 6' x 6'	S.F.	137
10' x 10'	S.F.	107
12' x 14'	S.F.	94
12' x 20'	S.F.	134
Serving		
Counter top (Stainless steel)	L.F.	217
Base cabinets	L.F.	540 - 905
Sound System		
Amplifier, 250 watts	Ea.	1950
Speaker, ceiling or wall	Ea.	233
Trumpet	Ea.	445
Storage		
Shelving	S.F.	9.65
Washing		
Stainless steel counter	L.F.	217

Important: See the Reference Section for Location Factors.

Model costs calculated for a 1 story building with 10′ story height and 4,000 square feet of floor area

Restaurant, Fast Food

				Unit	Unit Cost	Cost Per S.F.	% Of Sub-Total
A. SUBSTRUCTURE							
1010	Standard Foundations	Poured concrete; strip and spread footings; 4′ foundation wall		S.F. Ground	8.47	8.47	
1020	Special Foundations	N/A		—	—	—	
1030	Slab on Grade	4″ reinforced concrete		S.F. Slab	5.28	5.28	9.5%
2010	Basement Excavation	Site preparation for slab and trench for foundation wall and footing		S.F. Ground	.49	.49	
2020	Basement Walls	N/A		—	—	—	
B. SHELL							
	B10 Superstructure						
1010	Floor Construction	Wood columns		S.F. Floor	.42	.42	3.2%
1020	Roof Construction	Flat wood rafter roof		S.F. Roof	4.35	4.35	
	B20 Exterior Enclosure						
2010	Exterior Walls	Wood siding on wood studs, insulated	*70% of wall*	S.F. Wall	20	9.28	
2020	Exterior Windows	Curtain glazing wall system	*30% of wall*	Each	50	4.49	16.3%
2030	Exterior Doors	Aluminum & glass double doors, hollow metal doors		Each	5355	10.72	
	B30 Roofing						
3010	Roof Coverings	Single ply membrane, stone ballast, rigid insulation		S.F. Roof	7.77	7.77	5.6%
3020	Roof Openings	Roof and smoke hatches		S.F. Roof	.63	.63	
C. INTERIORS							
1010	Partitions	Water resistant gypsum board on wood studs	*25 S.F. Floor/L.F. Partition*	S.F. Partition	6.43	2.57	
1020	Interior Doors	Single leaf hollow core wood doors	*1000 S.F. Floor/Door*	Each	618	.62	
1030	Fittings	Stainless steel toilet partitions		S.F. Floor	1.54	1.54	
2010	Stair Construction	N/A		—	—	—	19.2%
3010	Wall Finishes	90% Paint, 10% ceramic tile		S.F. Surface	1.88	1.50	
3020	Floor Finishes	Quarry tile		S.F. Floor	16.51	16.51	
3030	Ceiling Finishes	Acoustic ceiling tiles on suspended support grid		S.F. Ceiling	6.10	6.10	
D. SERVICES							
	D10 Conveying						
1010	Elevators & Lifts	N/A		—	—	—	0.0 %
1020	Escalators & Moving Walks	N/A		—	—	—	
	D20 Plumbing						
2010	Plumbing Fixtures	Restroom, kitchen and service fixtures, supply and drainage	*1 Fixture/330 S.F. Floor*	Each	2751	8.26	
2020	Domestic Water Distribution	Gas fired water heater		S.F. Floor	3.98	3.98	9.2%
2040	Rain Water Drainage	Roof drains		S.F. Roof	1.65	1.65	
	D30 HVAC						
3010	Energy Supply	N/A		—	—	—	
3020	Heat Generating Systems	Included in D3050		—	—	—	
3030	Cooling Generating Systems	N/A		—	—	—	12.0 %
3050	Terminal & Package Units	Single zone unit gas heating, electric cooling		S.F. Floor	18.11	18.11	
3090	Other HVAC Sys. & Equipment	N/A		—	—	—	
	D40 Fire Protection						
4010	Sprinklers	Wet pipe sprinkler system, ordinary hazard		S.F. Floor	8.64	8.64	7.2%
4020	Standpipes	Standpipe		S.F. Floor	2.24	2.24	
	D50 Electrical						
5010	Electrical Service/Distribution	400 Ampere service, panelboards and feeders		S.F. Floor	4.64	4.64	
5020	Lighting & Branch Wiring	Fluorescent fixtures, receptacles, switches, A.C. and misc. power		S.F. Floor	9.68	9.68	12.4%
5030	Communications & Security	Addressable alarm, emergency lighting		S.F. Floor	4.36	4.36	
5090	Other Electrical Systems	N/A		—	—	—	
E. EQUIPMENT & FURNISHINGS							
1010	Commercial Equipment	N/A		—	—	—	
1020	Institutional Equipment	N/A		—	—	—	5.4 %
1030	Vehicular Equipment	N/A		—	—	—	
1090	Other Equipment	Prefabricated walk-in refrigerators		S.F. Floor	8.06	8.06	
F. SPECIAL CONSTRUCTION							
1020	Integrated Construction	N/A		—	—	—	0.0 %
1040	Special Facilities	N/A		—	—	—	
G. BUILDING SITEWORK	**N/A**						
					Sub-Total	150.36	**100%**
	CONTRACTOR FEES (General Requirements: 10%, Overhead: 5%, Profit: 10%)				*25%*	37.60	
	ARCHITECT FEES				*8%*	15.04	
					Total Building Cost	**203**	

M.550 Rink, Hockey/Indoor Soccer

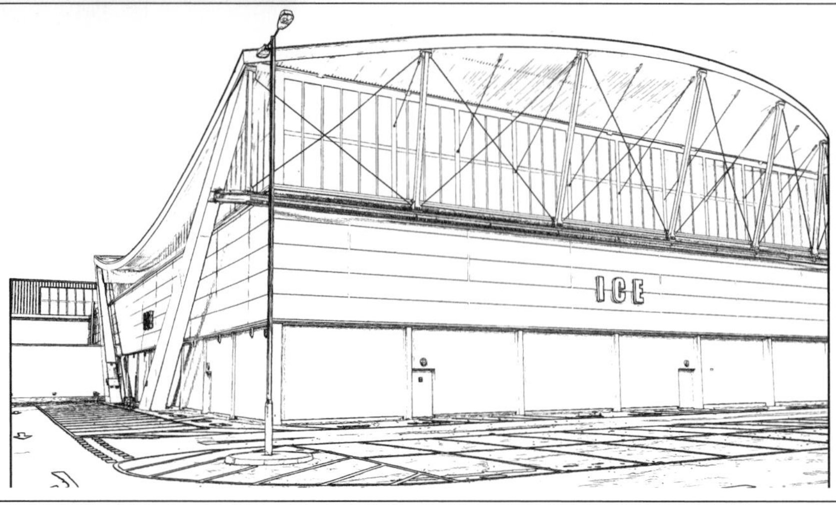

Costs per square foot of floor area

Exterior Wall	S.F. Area	10000	15000	20000	25000	30000	35000	40000	45000	50000
	L.F. Perimeter	450	500	600	700	740	822	890	920	966
Face Brick and Concrete Block	Steel Joists	195.85	178.70	173.40	170.20	165.45	163.65	161.85	159.35	157.75
	Wood Truss	207.50	191.60	187.00	184.15	179.70	178.05	176.35	173.95	172.45
Concrete Block	Rigid Steel	189.60	178.15	174.55	172.30	**169.20**	167.95	166.75	165.15	164.10
	Wood Truss	185.50	175.30	172.30	170.45	167.60	166.50	165.45	163.95	162.95
Metal Panel	Rigid Steel	188.10	177.10	173.50	171.35	168.35	167.20	166.00	164.45	163.45
Metal Sandwich Panel	Rigid Steel	192.15	180.15	176.25	173.95	170.60	169.30	168.05	166.35	165.15
Perimeter Adj., Add or Deduct	Per 100 L.F.	13.10	8.75	6.50	5.20	4.35	3.75	3.25	2.90	2.60
Story Hgt. Adj., Add or Deduct	Per 1 Ft.	2.05	1.50	1.40	1.30	1.15	1.10	1.05	0.90	0.85
Basement—Not Applicable										

The above costs were calculated using the basic specifications shown on the facing page. These costs should be adjusted where necessary for design alternatives and owner's requirements.

Common additives

Description	Unit	$ Cost
Bar, Front Bar	L.F.	445
Back bar	L.F.	355
Booth, Upholstered, custom straight	L.F.	259 - 480
"L" or "U" shaped	L.F.	268 - 455
Bleachers, Telescoping, manual		
To 15 tier	Seat	158 - 216
16-20 tier	Seat	315 - 405
21-30 tier	Seat	330 - 520
For power operation, add	Seat	64 - 103
Emergency Lighting, 25 watt, battery operated		
Lead battery	Ea.	305
Nickel cadmium	Ea.	525
Lockers, Steel, single tier, 60" or 72"	Opng.	223 - 390
2 tier, 60" or 72" total	Opng.	135 - 167
5 tier, box lockers	Opng.	69.50 - 81.50
Locker bench, lam. maple top only	L.F.	35.50
Pedestals, steel pipe	Ea.	73.50

Description	Unit	$ Cost
Rink		
Dasher boards & top guard	Ea.	189,000
Mats, rubber	S.F.	24.50
Score Board	Ea.	16,800 - 28,800

Important: See the Reference Section for Location Factors.

Model costs calculated for a 1 story building with 24′ story height and 30,000 square feet of floor area

Rink, Hockey/Indoor Soccer

			Unit	Unit Cost	Cost Per S.F.	% Of Sub-Total
A. SUBSTRUCTURE						
1010	Standard Foundations	Poured concrete; strip and spread footings; 4′ foundation wall	S.F. Ground	3.28	3.28	7.8%
1020	Special Foundations	N/A	–	–	–	
1030	Slab on Grade	6″ reinforced concrete with vapor barrier and granular base	S.F. Slab	6.46	6.46	
2010	Basement Excavation	Site preparation for slab and trench for foundation wall and footing	S.F. Ground	.16	.16	
2020	Basement Walls	N/A	–	–	–	
B. SHELL						
	B10 Superstructure					
1010	Floor Construction	Wide flange beams and columns	S.F. Floor	39	27.50	21.7%
1020	Roof Construction	(incl. in B1010)				
	B20 Exterior Enclosure					
2010	Exterior Walls	Concrete block *95% of wall*	S.F. Wall	14.85	8.35	8.7%
2020	Exterior Windows	Store front *5% of wall*	Each	49	1.45	
2030	Exterior Doors	Aluminum and glass, hollow metal, overhead	Each	4268	1.15	
	B30 Roofing					
3010	Roof Coverings	Elastomeric neoprene membrane with flashing; perlite/EPS composite insulation	S.F. Roof	4.92	4.92	4.1%
3020	Roof Openings	Roof hatches	S.F. Roof	.29	.29	
C. INTERIORS						
1010	Partitions	Concrete block *140 S.F. Floor/L.F. Partition*	S.F. Partition	9.22	.79	6.5%
1020	Interior Doors	Hollow metal *2500 S.F. Floor/Door*	Each	1201	.48	
1030	Fittings	N/A	–	–	–	
2010	Stair Construction	N/A	–	–	–	
3010	Wall Finishes	Paint	S.F. Surface	11.55	1.98	
3020	Floor Finishes	80% rubber mat, 20% paint *50% of floor area*	S.F. Floor	8.64	4.32	
3030	Ceiling Finishes	Mineral fiber tile on concealed zee bar *10% of area*	S.F. Ceiling	6.86	.69	
D. SERVICES						
	D10 Conveying					
1010	Elevators & Lifts	N/A	–	–	–	0.0 %
1020	Escalators & Moving Walks	N/A	–	–	–	
	D20 Plumbing					
2010	Plumbing Fixtures	Toilet and service fixtures, supply and drainage *1 Fixture/1070 S.F. Floor*	Each	6613	6.18	8.5%
2020	Domestic Water Distribution	Oil fired water heater	S.F. Floor	3.60	3.60	
2040	Rain Water Drainage	Roof drains	S.F. Roof	.98	.98	
	D30 HVAC					
3010	Energy Supply	Oil fired hot water, unit heaters *10% of area*	S.F. Floor	.83	.83	11.2%
3020	Heat Generating Systems	N/A	–	–	–	
3030	Cooling Generating Systems	N/A	–	–	–	
3050	Terminal & Package Units	Single zone, electric cooling *90% of area*	S.F. Floor	13.37	13.37	
3090	Other HVAC Sys. & Equipment	N/A	–	–	–	
	D40 Fire Protection					
4010	Sprinklers	N/A	–	–	–	0.5 %
4020	Standpipes	Standpipe	S.F. Floor	.69	.69	
	D50 Electrical					
5010	Electrical Service/Distribution	600 ampere service, panel board and feeders	S.F. Floor	1.41	1.41	8.7%
5020	Lighting & Branch Wiring	High intensity discharge and fluorescent fixtures, receptacles, switches, A.C. and misc. power	S.F. Floor	7.35	7.35	
5030	Communications & Security	Addressable alarm systems, emergency lighting and public address	S.F. Floor	2.02	2.02	
5090	Other Electrical Systems	Emergency generator	S.F. Floor	.22	.22	
E. EQUIPMENT & FURNISHINGS						
1010	Commercial Equipment	N/A	–	–	–	0.0 %
1020	Institutional Equipment	N/A	–	–	–	
1030	Vehicular Equipment	N/A	–	–	–	
1090	Other Equipment	N/A	–	–	–	
F. SPECIAL CONSTRUCTION						
1020	Integrated Construction	N/A	–	–	–	22.2 %
1040	Special Facilities	Dasher boards and rink (including ice making system)	S.F. Floor	28	28.03	
G. BUILDING SITEWORK	**N/A**					
				Sub-Total	126.50	**100%**
	CONTRACTOR FEES (General Requirements: 10%, Overhead: 5%, Profit: 10%)			*25%*	31.63	
	ARCHITECT FEES			*7%*	11.07	
				Total Building Cost	**169.20**	

Costs per square foot of floor area

Exterior Wall	S.F. Area	25000	30000	35000	40000	45000	50000	55000	60000	65000
	L.F. Perimeter	900	1050	1200	1350	1510	1650	1800	1970	2100
Fiber Cement	Rigid Steel	150.65	149.50	148.60	148.00	147.75	147.15	146.80	146.80	146.30
Metal Panel	Rigid Steel	150.55	149.40	148.50	147.85	**147.60**	147.05	146.70	146.65	146.15
E.I.F.S.	Rigid Steel	153.80	152.60	151.70	151.05	150.80	150.20	149.90	149.85	149.35
Tilt-up Concrete Panels	Reinforced Concrete	199.65	197.60	196.05	194.90	194.45	193.30	192.75	192.85	191.85
Brick Veneer	Reinforced Concrete	174.05	172.55	171.50	170.75	170.35	169.55	169.15	169.20	168.55
Stone Veneer	Reinforced Concrete	223.00	220.25	218.25	216.75	216.20	214.65	213.90	214.15	212.80
Perimeter Adj., Add or Deduct	Per 100 L.F.	3.00	2.50	2.20	1.90	1.60	1.50	1.35	1.25	1.15
Story Hgt. Adj., Add or Deduct	Per 1 Ft.	1.15	1.10	1.10	1.10	1.00	1.00	1.05	1.05	1.05
***For Basement**, add $24.90 per square foot of basement area*										

The above costs were calculated using the basic specifications shown on the facing page. These costs should be adjusted where necessary for design alternatives and owner's requirements.

Common additives

Description	Unit	$ Cost
Bleachers, Telescoping, manual		
To 15 tier	Seat	158 - 216
16-20 tier	Seat	315 - 405
21-30 tier	Seat	330 - 520
For power operation, add	Seat	64 - 103
Carrels Hardwood	Ea.	715 - 1900
Clock System		
20 room	Ea.	17,900
50 room	Ea.	42,300
Emergency Lighting, 25 watt, battery operated		
Lead battery	Ea.	305
Nickel cadmium	Ea.	525
Flagpoles, Complete		
Aluminum, 20' high	Ea.	1825
40' high	Ea.	4400
Fiberglass, 23' high	Ea.	1375
39'-5" high	Ea.	3175
Kitchen Equipment		
Broiler	Ea.	4400
Cooler, 6 ft. long, reach-in	Ea.	3725

Description	Unit	$ Cost
Kitchen Equipment, cont.		
Dishwasher, 10-12 racks per hr.	Ea.	3975
Food warmer, counter, 1.2 KW	Ea.	785
Freezer, 44 C.F., reach-in	Ea.	5400
Ice cube maker, 50 lb. per day	Ea.	2050
Range with 1 oven	Ea.	3200
Lockers, Steel, single tier, 60" to 72"	Opng.	223 - 390
2 tier, 60" to 72" total	Opng.	135 - 167
5 tier, box lockers	Opng.	69.50 - 81.50
Locker bench, lam. maple top only	L.F.	35.50
Pedestals, steel pipe	Ea.	73.50
Seating		
Auditorium chair, all veneer	Ea.	355
Veneer back, padded seat	Ea.	375
Upholstered, spring seat	Ea.	325
Classroom, movable chair & desk	Set	81 - 171
Lecture hall, pedestal type	Ea.	345 - 645
Sound System		
Amplifier, 250 watts	Ea.	1950
Speaker, ceiling or wall	Ea.	233
Trumpet	Ea.	445

Important: See the Reference Section for Location Factors.

Model costs calculated for a 1 story building with 15′ story height and 45,000 square feet of floor area

School, Elementary

			Unit	Unit Cost	Cost Per S.F.	% Of Sub-Total
A. SUBSTRUCTURE						
1010	Standard Foundations	Poured concrete; strip and spread footings; 4′ foundation wall	S.F. Ground	4.76	4.76	9.4%
1020	Special Foundations	N/A	—	—	—	
1030	Slab on Grade	4″ reinforced concrete	S.F. Slab	5.28	5.28	
2010	Basement Excavation	Site preparation for slab and trench for foundation wall and footing	S.F. Ground	.16	.28	
2020	Basement Walls	N/A	—	—	—	
B. SHELL						
	B10 Superstructure					
1010	Floor Construction	Fireproofing for structural columns	S.F. Floor	.94	.94	9.9%
1020	Roof Construction	Metal deck, open web steel joists, beams, columns	S.F. Roof	10.04	10.04	
	B20 Exterior Enclosure					
2010	Exterior Walls	Textured metal panel on metal stud, insulated *70% of wall*	S.F. Wall	16.21	5.71	11.3%
2020	Exterior Windows	Curtain wall system, thermo-break frame, aluminum awning windows *30% of wall*	Each	759	5.91	
2030	Exterior Doors	Aluminum & glass double doors, hollow metal doors	Each	4893	.87	
	B30 Roofing					
3010	Roof Coverings	Single ply membrane, stone ballast, rigid insulation	S.F. Roof	7.75	7.75	7.2%
3020	Roof Openings	Roof and smoke hatches	S.F. Roof	2496	.22	
C. INTERIORS						
1010	Partitions	Gypsum board on metal studs, sound attentuation insulation *20 S.F. Floor/L.F. Partition*	S.F. Partition	5.20	2.60	19.3%
1020	Interior Doors	Hollow metal doors *700 S.F. Floor/Door*	Each	1201	1.72	
1030	Fittings	Chalkboards, lockers, toilet partitions	S.F. Floor	1.24	1.24	
2010	Stair Construction	N/A	—	—	—	
3010	Wall Finishes	90% Paint, 10% ceramic tile	S.F. Surface	1.71	1.71	
3020	Floor Finishes	10% Carpet, 10% terrazzo, 60% VCT, 20% wood	S.F. Floor	7.14	7.14	
3030	Ceiling Finishes	Acoustic ceiling tiles on suspended channel grid	S.F. Ceiling	6.86	6.86	
D. SERVICES						
	D10 Conveying					
1010	Elevators & Lifts	N/A	—	—	—	0.0 %
1020	Escalators & Moving Walks	N/A	—	—	—	
	D20 Plumbing					
2010	Plumbing Fixtures	Restroom, kitchen, and service fixtures, supply and drainage *1 Fixture/325 S.F. Floor*	Each	2181	6.71	7.6%
2020	Domestic Water Distribution	Gas fired water heater	S.F. Floor	.70	.70	
2040	Rain Water Drainage	Roof drains	S.F. Roof	1.02	1.02	
	D30 HVAC					
3010	Energy Supply	Forced hot water heating system, fin tube radiation	S.F. Floor	9.69	9.69	18.9%
3020	Heat Generating Systems	N/A	—	—	—	
3030	Cooling Generating Systems	N/A	—	—	—	
3050	Terminal & Package Units	Split system with air cooled condensing unit	S.F. Floor	11.19	11.19	
3090	Other HVAC Sys. & Equipment	N/A	—	—	—	
	D40 Fire Protection					
4010	Sprinklers	Wet pipe spinkler system	S.F. Floor	2.54	2.54	2.7%
4020	Standpipes	Standpipe	S.F. Floor	.42	.42	
	D50 Electrical					
5010	Electrical Service/Distribution	800 Ampere service, panelboards and feeders	S.F. Floor	1.05	1.05	13.4%
5020	Lighting & Branch Wiring	Fluorescent fixtures, receptacles, switches, A.C. and misc. power	S.F. Floor	10.09	10.09	
5030	Communications & Security	Addressable alarm system, emergency lighting, internet & phone wiring	S.F. Floor	3.58	3.58	
5090	Other Electrical Systems	Emergency generator, 15kW	S.F. Floor	.09	.09	
E. EQUIPMENT & FURNISHINGS						
1010	Commercial Equipment	N/A	—	—	—	0.2 %
1020	Institutional Equipment	Laboratory equipment	S.F. Floor	.25	.25	
1030	Vehicular Equipment	N/A	—	—	—	
1090	Other Equipment	N/A	—	—	—	
F. SPECIAL CONSTRUCTION						
1020	Integrated Construction	N/A	—	—	—	0.0 %
1040	Special Facilities	N/A	—	—	—	
G. BUILDING SITEWORK		**N/A**				
				Sub-Total	110.36	**100%**
	CONTRACTOR FEES (General Requirements: 10%, Overhead: 5%, Profit: 10%)			*25%*	27.58	
	ARCHITECT FEES			*7%*	9.66	
				Total Building Cost	**147.60**	

For customer support on your Light Commercial Costs with RSMeans data, call 800.448.8182.

Costs per square foot of floor area

Exterior Wall	S.F. Area	50000	65000	80000	95000	110000	125000	140000	155000	170000
	L.F. Perimeter	850	1060	1280	1490	1700	1920	2140	2340	2560
Fiber Cement	Rigid Steel	155.30	153.50	152.45	151.70	**151.15**	150.75	150.50	150.15	149.95
Metal Panel	Rigid Steel	155.95	154.10	153.10	152.30	151.70	151.35	151.10	150.70	150.50
E.I.F.S.	Rigid Steel	158.60	156.75	155.70	154.95	154.40	154.00	153.75	153.35	153.15
Tilt-up Concrete Panels	Reinforced Concrete	189.80	187.10	185.80	184.70	183.80	183.35	183.05	182.45	182.20
Brick Veneer	Reinforced Concrete	179.80	177.50	176.35	175.35	174.65	174.15	173.90	173.35	173.20
Stone Veneer	Reinforced Concrete	209.00	205.60	203.95	202.45	201.30	200.70	200.35	199.50	199.25
Perimeter Adj., Add or Deduct	Per 100 L.F.	3.00	2.30	1.95	1.55	1.35	1.25	1.05	0.95	0.90
Story Hgt. Adj., Add or Deduct	Per 1 Ft.	1.20	1.10	1.15	1.10	1.05	1.10	1.10	1.05	1.10

***For Basement**, add $34.00 per square foot of basement area*

The above costs were calculated using the basic specifications shown on the facing page. These costs should be adjusted where necessary for design alternatives and owner's requirements.

Common additives

Description	Unit	$ Cost
Bleachers, Telescoping, manual		
To 15 tier	Seat	158 - 216
16-20 tier	Seat	315 - 405
21-30 tier	Seat	330 - 520
For power operation, add	Seat	64 - 103
Carrels Hardwood	Ea.	715 - 1900
Clock System		
20 room	Ea.	17,900
50 room	Ea.	42,300
Elevators, Hydraulic passenger, 2 stops		
1500# capacity	Ea.	68,400
2500# capacity	Ea.	71,900
Emergency Lighting, 25 watt, battery operated		
Lead battery	Ea.	305
Nickel cadmium	Ea.	525
Flagpoles, Complete		
Aluminum, 20' high	Ea.	1825
40' high	Ea.	4400
Fiberglass, 23' high	Ea.	1375
39'-5" high	Ea.	3175

Description	Unit	$ Cost
Kitchen Equipment		
Broiler	Ea.	4400
Cooler, 6 ft. long, reach-in	Ea.	3725
Dishwasher, 10-12 racks per hr.	Ea.	3975
Food warmer, counter, 1.2 KW	Ea.	785
Freezer, 44 C.F., reach-in	Ea.	5400
Lockers, Steel, single tier, 60" to 72"	Opng.	223 - 390
2 tier, 60" to 72" total	Opng.	135 - 167
5 tier, box lockers	Opng.	69.50 - 81.50
Locker bench, lam. maple top only	L.F.	35.50
Pedestals, steel pipe	Ea.	73.50
Seating		
Auditorium chair, all veneer	Ea.	355
Veneer back, padded seat	Ea.	375
Upholstered, spring seat	Ea.	325
Classroom, movable chair & desk	Set	81 - 171
Lecture hall, pedestal type	Ea.	345 - 645
Sound System		
Amplifier, 250 watts	Ea.	1950
Speaker, ceiling or wall	Ea.	233
Trumpet	Ea.	445

Important: See the Reference Section for Location Factors.

Model costs calculated for a 2 story building with 15′ story height and 110,000 square feet of floor area

School, Jr High, 2-3 Story

			Unit	Unit Cost	Cost Per S.F.	% Of Sub-Total
A. SUBSTRUCTURE						
1010	Standard Foundations	Poured concrete; strip and spread footings; 4′ foundation wall	S.F. Ground	4.76	2.38	
1020	Special Foundations	N/A	—	—	—	
1030	Slab on Grade	4″ reinforced concrete	S.F. Slab	5.28	2.64	4.6%
2010	Basement Excavation	Site preparation for slab and trench for foundation wall and footing	S.F. Ground	.16	.14	
2020	Basement Walls	N/A	—	—	—	
B. SHELL						
	B10 Superstructure					
1010	Floor Construction	Open web steel joists, slab form, concrete, fireproofed steel columns	S.F. Floor	15.04	7.52	11.1%
1020	Roof Construction	Metal deck, open web steel joists, beams, columns	S.F. Roof	10.06	5.03	
	B20 Exterior Enclosure					
2010	Exterior Walls	Fiber cement siding on wood studs, insulated *75% of wall*	S.F. Wall	15.01	5.22	
2020	Exterior Windows	Curtain wall system, thermo-break frame *25% of wall*	Each	55	6.36	11.0%
2030	Exterior Doors	Aluminum & glass double doors, hollow metal doors, overhead doors	Each	3376	.89	
	B30 Roofing					
3010	Roof Coverings	Single ply membrane, stone ballast, rigid insulation	S.F. Roof	7.58	3.79	3.5%
3020	Roof Openings	Roof and smoke hatches	S.F. Roof	.36	.18	
C. INTERIORS						
1010	Partitions	Gypsum board on metal studs, sound attentuation insulation *20 S.F. Floor/L.F. Partition*	S.F. Partition	5.95	3.57	
1020	Interior Doors	Hollow metal doors *750 S.F. Floor/Door*	Each	1201	1.60	
1030	Fittings	Chalkboards, lockers, toilet partitions	S.F. Floor	7.32	7.32	
2010	Stair Construction	Cement filled metal pan, picket rail	Flight	12,375	.90	24.6%
3010	Wall Finishes	90% Paint, 10% ceramic tile	S.F. Surface	1.67	2	
3020	Floor Finishes	10% Carpet, 10% terrazzo, 70% VCT, 10% wood	S.F. Floor	6.30	6.30	
3030	Ceiling Finishes	Acoustic ceiling tiles on suspended channel grid	S.F. Ceiling	6.10	6.10	
D. SERVICES						
	D10 Conveying					
1010	Elevators & Lifts	Hydraulic passenger elevator	Each	85,800	.78	0.7%
1020	Escalators & Moving Walks	N/A	—	—	—	
	D20 Plumbing					
2010	Plumbing Fixtures	Restroom, kitchen, and service fixtures, supply and drainage *1 Fixture/350 S.F. Floor*	Each	2051	5.86	
2020	Domestic Water Distribution	Gas fired water heater	S.F. Floor	.71	.71	6.2%
2040	Rain Water Drainage	Roof drains	S.F. Roof	.84	.42	
	D30 HVAC					
3010	Energy Supply	N/A	—	—	—	
3020	Heat Generating Systems	Included in D3050	—	—	—	
3030	Cooling Generating Systems	N/A	—	—	—	17.0 %
3050	Terminal & Package Units	Single zone unit gas heating, electric cooling	S.F. Floor	19.25	19.25	
3090	Other HVAC Sys. & Equipment	N/A	—	—	—	
	D40 Fire Protection					
4010	Sprinklers	Wet pipe spinkler system	S.F. Floor	2.20	2.20	2.3%
4020	Standpipes	Standpipe	S.F. Floor	.42	.42	
	D50 Electrical					
5010	Electrical Service/Distribution	1600 Ampere service, panelboards and feeders	S.F. Floor	.96	.96	
5020	Lighting & Branch Wiring	Fluorescent fixtures, receptacles, switches, a.c. and misc. power	S.F. Floor	9.38	9.38	13.1%
5030	Communications & Security	Addressable alarm system, emergency lighting, internet & phone wiring	S.F. Floor	4.12	4.12	
5090	Other Electrical Systems	Emergency generator, 100kW	S.F. Floor	.33	.33	
E. EQUIPMENT & FURNISHINGS						
1010	Commercial Equipment	N/A	—	—	—	
1020	Institutional Equipment	Laboratory equipment	S.F. Floor	1.89	1.89	5.9 %
1030	Vehicular Equipment	N/A	—	—	—	
1090	Other Equipment	Gym equipment, bleachers, scoreboards	S.F. Floor	4.74	4.74	
F. SPECIAL CONSTRUCTION						
1020	Integrated Construction	N/A	—	—	—	0.0 %
1040	Special Facilities	N/A	—	—	—	
G. BUILDING SITEWORK	**N/A**					
				Sub-Total	113	**100%**
	CONTRACTOR FEES (General Requirements: 10%, Overhead: 5%, Profit: 10%)			*25%*	28.26	
	ARCHITECT FEES			*7%*	9.89	
				Total Building Cost	**151.15**	

Costs per square foot of floor area

Exterior Wall	S.F. Area	1000	2000	3000	4000	6000	8000	10000	12000	15000
	L.F. Perimeter	126	179	219	253	310	358	400	438	490
Wood Clapboard	Wood Frame	156.15	130.55	120.15	**114.35**	107.80	104.00	101.45	99.60	97.65
Face Brick	Wood Frame	173.00	141.10	127.80	120.30	111.70	106.70	103.35	100.95	98.30
Stucco and Concrete Block	Rigid Steel	175.75	146.55	134.60	127.80	120.10	115.65	112.60	110.45	108.10
	Bearing Walls	170.80	141.60	129.65	122.85	115.20	110.70	107.70	105.50	103.15
Metal Sandwich Panel	Rigid Steel	188.25	155.45	141.80	134.05	125.20	120.05	116.60	114.05	111.40
Precast Concrete	Rigid Steel	250.75	199.85	178.10	165.45	150.85	142.30	136.45	132.15	127.60
Perimeter Adj., Add or Deduct	Per 100 L.F.	42.50	21.25	14.20	10.65	7.05	5.30	4.25	3.50	2.85
Story Hgt. Adj., Add or Deduct	Per 1 Ft.	3.20	2.25	1.90	1.55	1.25	1.10	1.00	0.95	0.80

***For Basement**, add $24.80 per square foot of basement area*

The above costs were calculated using the basic specifications shown on the facing page. These costs should be adjusted where necessary for design alternatives and owner's requirements.

Common additives

Description	Unit	$ Cost
Check Out Counter		
Single belt	Ea.	4025
Double belt	Ea.	5325
Emergency Lighting, 25 watt, battery operated		
Lead battery	Ea.	305
Nickel cadmium	Ea.	525
Refrigerators, Prefabricated, walk-in		
7'-6" high, 6' x 6'	S.F.	137
10' x 10'	S.F.	107
12' x 14'	S.F.	94
12' x 20'	S.F.	134
Refrigerated Food Cases		
Dairy, multi deck, 12' long	Ea.	13,200
Delicatessen case, single deck, 12' long	Ea.	9825
Multi deck, 18 S.F. shelf display	Ea.	9200
Freezer, self-contained chest type, 30 C.F.	Ea.	5125
Glass door upright, 78 C.F.	Ea.	10,600
Refrigerated Food Cases, cont.		
Frozen food, chest type, 12' long	Ea.	8875
Glass door reach-in, 5 door	Ea.	11,600
Island case 12' long, single deck	Ea.	8700
Multi deck	Ea.	10,400
Meat cases, 12' long, single deck	Ea.	8925
Multi deck	Ea.	11,400
Produce, 12' long single deck	Ea.	7625
Multi deck	Ea.	9625
Safe, Office type, 1 hour rating		
30" x 18" x 18"	Ea.	2400
60" x 36" x 18", double door	Ea.	10,200
Smoke Detectors		
Ceiling type	Ea.	231
Duct type	Ea.	510
Sound System		
Amplifier, 250 watts	Ea.	1950
Speaker, ceiling or wall	Ea.	233
Trumpet	Ea.	445

Important: See the Reference Section for Location Factors.

Model costs calculated for a 1 story building with 12' story height and 4,000 square feet of floor area

Store, Convenience

			Unit	Unit Cost	Cost Per S.F.	% Of Sub-Total
A. SUBSTRUCTURE						
1010	Standard Foundations	Poured concrete; strip and spread footings; 4' foundation wall	S.F. Ground	6.05	6.05	13.8%
1020	Special Foundations	N/A	–	–	–	
1030	Slab on Grade	4" reinforced concrete with vapor barrier and granular base	S.F. Slab	5.28	5.28	
2010	Basement Excavation	Site preparation for slab and trench for foundation wall and footing	S.F. Ground	.49	.49	
2020	Basement Walls	N/A	–	–	–	
B. SHELL						
	B10 Superstructure					
1010	Floor Construction	N/A	–	–	–	8.8 %
1020	Roof Construction	Wood truss with plywood sheathing	S.F. Roof	7.22	7.22	
	B20 Exterior Enclosure					
2010	Exterior Walls	Wood siding on wood studs, insulated *80% of wall*	S.F. Wall	9.52	5.78	19.0%
2020	Exterior Windows	Storefront *20% of wall*	Each	48	7.34	
2030	Exterior Doors	Double aluminum and glass, solid core wood	Each	4212	3.16	
	B30 Roofing					
3010	Roof Coverings	Asphalt shingles; rigid fiberglass insulation	S.F. Roof	3.59	3.59	4.2%
3020	Roof Openings	N/A	–	–	–	
C. INTERIORS						
1010	Partitions	Gypsum board on wood studs *60 S.F. Floor/L.F. Partition*	S.F. Partition	12.66	2.11	12.5%
1020	Interior Doors	Single leaf wood, hollow metal *1300 S.F. Floor/Door*	Each	1663	1.28	
1030	Fittings	N/A	–	–	–	
2010	Stair Construction	N/A	–	–	–	
3010	Wall Finishes	Paint	S.F. Surface	1.08	.36	
3020	Floor Finishes	Vinyl composition tile	S.F. Floor	2.84	2.84	
3030	Ceiling Finishes	Mineral fiber tile on wood furring	S.F. Ceiling	4.10	4.10	
D. SERVICES						
	D10 Conveying					
1010	Elevators & Lifts	N/A	–	–	–	0.0 %
1020	Escalators & Moving Walks	N/A	–	–	–	
	D20 Plumbing					
2010	Plumbing Fixtures	Toilet and service fixtures, supply and drainage *1 Fixture/1000 S.F. Floor*	Each	3830	3.83	6.4%
2020	Domestic Water Distribution	Gas fired water heater	S.F. Floor	1.66	1.66	
2040	Rain Water Drainage	N/A	–	–	–	
	D30 HVAC					
3010	Energy Supply	N/A	–	–	–	8.3 %
3020	Heat Generating Systems	Included in D3050	–	–	–	
3030	Cooling Generating Systems	N/A	–	–	–	
3050	Terminal & Package Units	Single zone rooftop unit, gas heating, electric cooling	S.F. Floor	7.12	7.12	
3090	Other HVAC Sys. & Equipment	N/A	–	–	–	
	D40 Fire Protection					
4010	Sprinklers	Sprinkler, ordinary hazard	S.F. Floor	4.66	4.66	8.1%
4020	Standpipes	Standpipe	S.F. floor	2.24	2.24	
	D50 Electrical					
5010	Electrical Service/Distribution	200 ampere service, panel board and feeders	S.F. Floor	2.92	2.92	18.8%
5020	Lighting & Branch Wiring	High efficiency fluorescent fixtures, receptacles, switches, A.C. and misc. power	S.F. Floor	10.03	10.03	
5030	Communications & Security	Addressable alarm systems and emergency lighting	S.F. Floor	2.76	2.76	
5090	Other Electrical Systems	Emergency generator, 7.5 kW	S.F. Floor	.39	.39	
E. EQUIPMENT & FURNISHINGS						
1010	Commercial Equipment	N/A	–	–	–	0.0 %
1020	Institutional Equipment	N/A	–	–	–	
1030	Vehicular Equipment	N/A	–	–	–	
1090	Other Equipment	N/A	–	–	–	
F. SPECIAL CONSTRUCTION						
1020	Integrated Construction	N/A	–	–	–	0.0 %
1040	Special Facilities	N/A	–	–	–	
G. BUILDING SITEWORK		**N/A**				
				Sub-Total	85.49	**100%**
	CONTRACTOR FEES (General Requirements: 10%, Overhead: 5%, Profit: 10%)			*25%*	21.38	
	ARCHITECT FEES			*7%*	7.48	
				Total Building Cost	**114.35**	

Costs per square foot of floor area

Exterior Wall	S.F. Area	50000	65000	80000	95000	110000	125000	140000	155000	170000
	L.F. Perimeter	920	1065	1167	1303	1333	1433	1533	1633	1733
Face Brick and Concrete Block	Reinforced Concrete	129.85	127.45	125.45	124.40	**122.80**	122.10	121.45	121.05	120.65
	Rigid Steel	101.60	99.25	97.25	96.15	94.50	93.85	93.25	92.80	92.45
Decorative Concrete Block	Reinforced Concrete	126.00	124.00	122.45	121.55	120.25	119.75	119.20	118.85	118.55
	Steel Joists	97.75	95.80	94.20	93.35	92.05	91.45	90.95	90.65	90.35
Precast Concrete	Reinforced Concrete	130.35	127.90	125.90	124.75	123.10	122.45	121.75	121.35	120.95
	Steel Joists	102.15	99.65	97.65	96.55	94.90	94.20	93.55	93.10	92.75
Perimeter Adj., Add or Deduct	Per 100 L.F.	1.80	1.40	1.15	1.00	0.85	0.75	0.70	0.55	0.55
Story Hgt. Adj., Add or Deduct	Per 1 Ft.	0.90	0.80	0.75	0.75	0.60	0.60	0.60	0.50	0.55
***For Basement**, add $24.50 per square foot of basement area*										

The above costs were calculated using the basic specifications shown on the facing page. These costs should be adjusted where necessary for design alternatives and owner's requirements.

Common additives

Description	Unit	$ Cost
Closed Circuit Surveillance, One station		
Camera and monitor	Ea.	1250
For additional camera stations, add	Ea.	565
Directory Boards, Plastic, glass covered		
30" x 20"	Ea.	570
36" x 48"	Ea.	1500
Aluminum, 24" x 18"	Ea.	570
36" x 24"	Ea.	725
48" x 32"	Ea.	950
48" x 60"	Ea.	2000
Emergency Lighting, 25 watt, battery operated		
Lead battery	Ea.	305
Nickel cadmium	Ea.	525
Safe, Office type, 1 hour rating		
30" x 18" x 18"	Ea.	2400
60" x 36" x 18", double door	Ea.	10,200
Sound System		
Amplifier, 250 watts	Ea.	1950
Speaker, ceiling or wall	Ea.	233
Trumpet	Ea.	445

Important: See the Reference Section for Location Factors.

Model costs calculated for a 1 story building with 14′ story height and 110,000 square feet of floor area

Store, Department, 1 Story

				Unit	Unit Cost	Cost Per S.F.	% Of Sub-Total
A. SUBSTRUCTURE							
1010	Standard Foundations	Poured concrete; strip and spread footings; 4′ foundation wall		S.F. Ground	1.85	1.85	8.0%
1020	Special Foundations	N/A		—	—	—	
1030	Slab on Grade	4″ reinforced concrete with vapor barrier and granular base		S.F. Slab	5.28	5.28	
2010	Basement Excavation	Site preparation for slab and trench for foundation wall and footing		S.F. Ground	.27	.27	
2020	Basement Walls	N/A		—	—	—	
B. SHELL							
	B10 Superstructure						
1010	Floor Construction	Cast-in-place concrete columns		L.F. Column	66	.84	34.7%
1020	Roof Construction	Precast concrete beam and plank, concrete columns		S.F. Roof	31	31.34	
	B20 Exterior Enclosure						
2010	Exterior Walls	Face brick with concrete block back-up	*90% of wall*	S.F. Wall	31	4.68	6.8%
2020	Exterior Windows	Storefront	*10% of wall*	Each	67	1.14	
2030	Exterior Doors	Sliding electric operated entrance, hollow metal		Each	8657	.47	
	B30 Roofing						
3010	Roof Coverings	Built-up tar and gravel with flashing; perlite/EPS composite insulation		S.F. Roof	5.34	5.34	5.8%
3020	Roof Openings	Roof hatches		S.F. Roof	.07	.07	
C. INTERIORS							
1010	Partitions	Gypsum board on metal studs	*60 S.F. Floor/L.F. Partition*	S.F. Partition	10.08	1.68	18.3%
1020	Interior Doors	Single leaf hollow metal	*600 S.F. Floor/Door*	Each	1201	2.01	
1030	Fittings	N/A		—	—	—	
2010	Stair Construction	N/A		—	—	—	
3010	Wall Finishes	Paint		S.F. Surface	.78	.26	
3020	Floor Finishes	50% ceramic tile, 50% carpet tile		S.F. Floor	8.36	8.36	
3030	Ceiling Finishes	Mineral fiber board on exposed grid system, suspended		S.F. Ceiling	4.65	4.65	
D. SERVICES							
	D10 Conveying						
1010	Elevators & Lifts	N/A		—	—	—	0.0 %
1020	Escalators & Moving Walks	N/A		—	—	—	
	D20 Plumbing						
2010	Plumbing Fixtures	Toilet and service fixtures, supply and drainage	*1 Fixture/4075 S.F. Floor*	Each	7294	1.79	3.4%
2020	Domestic Water Distribution	Gas fired water heater		S.F. Floor	.43	.43	
2040	Rain Water Drainage	Roof drains		S.F. Roof	.90	.90	
	D30 HVAC						
3010	Energy Supply	N/A		—	—	—	8.9 %
3020	Heat Generating Systems	Included in D3050		—	—	—	
3030	Cooling Generating Systems	N/A		—	—	—	
3050	Terminal & Package Units	Single zone unit gas heating, electric cooling		S.F. Floor	8.26	8.26	
3090	Other HVAC Sys. & Equipment	N/A		—	—	—	
	D40 Fire Protection						
4010	Sprinklers	Sprinklers, light hazard		S.F. Floor	2.54	2.54	3.1%
4020	Standpipes	Standpipe		S.F. Floor	.29	.29	
	D50 Electrical						
5010	Electrical Service/Distribution	1200 ampere service, panel board and feeders		S.F. Floor	.71	.71	11.0%
5020	Lighting & Branch Wiring	High efficiency fluorescent fixtures, receptacles, switches, A.C. and misc. power		S.F. Floor	8.27	8.27	
5030	Communications & Security	Addressable alarm systems, internet wiring and emergency lighting		S.F. Floor	1.18	1.18	
5090	Other Electrical Systems	Emergency generator, 7.5 kW		S.F. Floor	.04	.04	
E. EQUIPMENT & FURNISHINGS							
1010	Commercial Equipment	N/A		—	—	—	0.0 %
1020	Institutional Equipment	N/A		—	—	—	
1030	Vehicular Equipment	N/A		—	—	—	
1090	Other Equipment	N/A		—	—	—	
F. SPECIAL CONSTRUCTION							
1020	Integrated Construction	N/A		—	—	—	0.0 %
1040	Special Facilities	N/A		—	—	—	
G. BUILDING SITEWORK	**N/A**						
					Sub-Total	92.65	**100%**
	CONTRACTOR FEES (General Requirements: 10%, Overhead: 5%, Profit: 10%)				*25%*	23.20	
	ARCHITECT FEES				*6%*	6.95	
					Total Building Cost	**122.80**	

Costs per square foot of floor area

Exterior Wall	S.F. Area	4000	6000	8000	10000	12000	15000	18000	20000	22000
	L.F. Perimeter	260	340	360	410	440	490	540	565	594
Vinyl Clapboard	Wood Frame	130.10	119.90	**110.95**	107.15	103.70	100.45	98.30	97.00	96.00
Fiber Cement	Wood Frame	132.20	121.95	112.85	109.00	105.55	102.30	100.05	98.75	97.75
E.I.F.S. and Metal Studs	Steel Joists	141.45	130.80	121.20	117.15	113.55	110.10	107.75	106.35	105.30
Stone Veneer	Rigid Steel	183.25	168.10	152.20	145.95	140.00	134.35	130.55	128.20	126.50
Brick Veneer	Reinforced Concrete	173.30	159.45	145.35	139.70	134.35	129.35	125.90	123.90	122.30
Stucco	Reinforced Concrete	197.40	180.80	162.85	155.95	149.25	142.90	138.70	136.05	134.05
Perimeter Adj., Add or Deduct	Per 100 L.F.	13.05	8.75	6.50	5.20	4.40	3.50	2.90	2.65	2.40
Story Hgt. Adj., Add or Deduct	Per 1 Ft.	1.40	1.35	1.00	0.95	0.90	0.75	0.65	0.65	0.65
***For Basement**, add $34.50 per square foot of basement area*										

The above costs were calculated using the basic specifications shown on the facing page. These costs should be adjusted where necessary for design alternatives and owner's requirements.

Common additives

Description	Unit	$ Cost
Emergency Lighting, 25 watt, battery operated		
Lead battery	Ea.	305
Nickel cadmium	Ea.	525
Safe, Office type, 1 hour rating		
30" x 18" x 18"	Ea.	2400
60" x 36" x 18", double door	Ea.	10,200
Smoke Detectors		
Ceiling type	Ea.	231
Duct type	Ea.	510
Sound System		
Amplifier, 250 watts	Ea.	1950
Speaker, ceiling or wall	Ea.	233
Trumpet	Ea.	445

Important: See the Reference Section for Location Factors.

Model costs calculated for a 1 story building with 14′ story height and 8,000 square feet of floor area

Store, Retail

				Unit	Unit Cost	Cost Per S.F.	% Of Sub-Total
A. SUBSTRUCTURE							
1010	Standard Foundations	Poured concrete; strip and spread footings; 4′ foundation wall		S.F. Ground	5.55	5.55	13.5%
1020	Special Foundations	N/A		—	—	—	
1030	Slab on Grade	4″ reinforced concrete		S.F. Slab	5.28	5.28	
2010	Basement Excavation	Site preparation for slab and trench for foundation wall and footing		S.F. Ground	.28	.28	
2020	Basement Walls	N/A		—	—	—	
B. SHELL							
	B10 Superstructure						
1010	Floor Construction	Wood columns,		S.F. Floor	.42	.42	9.7%
1020	Roof Construction	Wood roof truss 4:12 pitch		S.F. Roof	7.55	7.55	
	B20 Exterior Enclosure						
2010	Exterior Walls	Vinyl siding on wood studs, insulated	*85% of wall*	S.F. Wall	13.33	7.14	13.3%
2020	Exterior Windows	Storefront glazing	*15% of wall*	Each	48	2.11	
2030	Exterior Doors	Aluminum and glass single doors, hollow metal doors		Each	3289	1.65	
	B30 Roofing						
3010	Roof Coverings	Asphalt roofing, strip shingles, gutters and downspouts		S.F. Roof	3.64	3.64	4.6%
3020	Roof Openings	Roof hatches		S.F. Roof	.14	.14	
C. INTERIORS							
1010	Partitions	Sound deadening gypsum board on wood studs	*60 S.F. Floor/L.F. Partition*	S.F. Partition	9.12	1.52	16.3%
1020	Interior Doors	Hollow metal doors	*1150 S.F. Floor/Door*	Each	1201	1.05	
1030	Fittings	Toilet partitions		S.F. Floor	.19	.19	
2010	Stair Construction	N/A		—	—	—	
3010	Wall Finishes	90% Paint, 10% ceramic tile		S.F. Surface	2.73	.91	
3020	Floor Finishes	Vinyl composition tile		S.F. Floor	2.84	2.84	
3030	Ceiling Finishes	Acoustic ceiling tiles on suspended channel grid		S.F. Ceiling	6.86	6.86	
D. SERVICES							
	D10 Conveying						
1010	Elevators & Lifts	N/A		—	—	—	0.0 %
1020	Escalators & Moving Walks	N/A		—	—	—	
	D20 Plumbing						
2010	Plumbing Fixtures	Restroom and service fixtures, supply and drainage	*1 Fixture/1350 S.F. Floor*	Each	2733	2.05	8.6%
2020	Domestic Water Distribution	Gas fired water heater		S.F. Floor	3.48	3.48	
2040	Rain Water Drainage	Roof Drains		S.F. Roof	1.53	1.53	
	D30 HVAC						
3010	Energy Supply	N/A		—	—	—	10.0 %
3020	Heat Generating Systems	Included in D3050		—	—	—	
3030	Cooling Generating Systems	N/A		—	—	—	
3050	Terminal & Package Units	Single zone unit gas heating, electric cooling		S.F. Floor	8.26	8.26	
3090	Other HVAC Sys. & Equipment	N/A		—	—	—	
	D40 Fire Protection						
4010	Sprinklers	Wet pipe spinkler system		S.F. Floor	4.45	4.45	6.8%
4020	Standpipes	Standpipe		S.F. Floor	1.12	1.12	
	D50 Electrical						
5010	Electrical Service/Distribution	400 Ampere service, panelboards and feeders		S.F. Floor	2.65	2.65	17.3%
5020	Lighting & Branch Wiring	Flourescent fixtures, receptacles, A.C. and misc. power		S.F. Floor	9.88	9.88	
5030	Communications & Security	Addressable alarm system, emergency lighting		S.F. Floor	1.65	1.65	
5090	Other Electrical Systems	N/A		—	—	—	
E. EQUIPMENT & FURNISHINGS							
1010	Commercial Equipment	N/A		—	—	—	0.0 %
1020	Institutional Equipment	N/A		—	—	—	
1030	Vehicular Equipment	N/A		—	—	—	
1090	Other Equipment	N/A		—	—	—	
F. SPECIAL CONSTRUCTION							
1020	Integrated Construction	N/A		—	—	—	0.0 %
1040	Special Facilities	N/A		—	—	—	
G. BUILDING SITEWORK		**N/A**					
					Sub-Total	82.20	**100%**
	CONTRACTOR FEES (General Requirements: 10%, Overhead: 5%, Profit: 10%)				25%	20.53	
	ARCHITECT FEES				8%	8.22	
					Total Building Cost	**110.95**	

Supermarket

Costs per square foot of floor area

Exterior Wall	S.F. Area	12000	16000	20000	26000	32000	38000	44000	52000	60000
	L.F. Perimeter	450	510	570	650	716	780	840	920	1020
Stucco	Reinforced Concrete	182.70	167.45	158.30	149.05	142.25	137.55	133.95	130.40	128.50
Curtain Wall	Rigid Steel	145.35	134.30	127.70	121.05	116.25	112.85	**110.35**	107.80	106.45
E.I.F.S.	Rigid Steel	123.95	116.45	111.95	107.50	104.35	102.15	100.50	98.95	98.00
Metal Panel	Rigid Steel	121.30	113.85	109.35	105.00	101.85	99.70	98.05	96.45	95.55
Tilt-up Concrete Panels	Reinforced Concrete	135.30	126.70	121.55	116.50	112.80	110.25	108.30	106.45	105.40
Brick Veneer	Reinforced Concrete	192.40	175.70	165.65	155.55	148.05	142.85	138.85	135.00	132.90
Perimeter Adj., Add or Deduct	Per 100 L.F.	18.00	13.45	10.70	8.25	6.75	5.65	4.85	4.15	3.65
Story Hgt. Adj., Add or Deduct	Per 1 Ft.	1.55	1.25	1.10	1.00	0.90	0.80	0.80	0.70	0.70
***For Basement**, add $24.90 per square foot of basement area*										

The above costs were calculated using the basic specifications shown on the facing page. These costs should be adjusted where necessary for design alternatives and owner's requirements.

Common additives

Description	Unit	$ Cost
Check Out Counter		
Single belt	Ea.	4025
Double belt	Ea.	5325
Scanner, registers, guns & memory 2 lanes	Ea.	19,000
10 lanes	Ea.	180,500
Power take away	Ea.	6550
Emergency Lighting, 25 watt, battery operated		
Lead battery	Ea.	305
Nickel cadmium	Ea.	525
Refrigerators, Prefabricated, walk-in		
7'-6" High, 6' x 6'	S.F.	137
10' x 10'	S.F.	107
12' x 14'	S.F.	94
12' x 20'	S.F.	134
Refrigerated Food Cases		
Dairy, multi deck, 12' long	Ea.	13,200
Delicatessen case, single deck, 12' long	Ea.	9825
Multi deck, 18 S.F. shelf display	Ea.	9200
Freezer, self-contained chest type, 30 C.F.	Ea.	5125
Glass door upright, 78 C.F.	Ea.	10,600

Description	Unit	$ Cost
Refrigerated Food Cases, cont.		
Frozen food, chest type, 12' long	Ea.	8875
Glass door reach in, 5 door	Ea.	11,600
Island case 12' long, single deck	Ea.	8700
Multi deck	Ea.	10,400
Meat case 12' long, single deck	Ea.	8925
Multi deck	Ea.	11,400
Produce, 12' long single deck	Ea.	7625
Multi deck	Ea.	9625
Safe, Office type, 1 hour rating		
30" x 18" x 18"	Ea.	2400
60" x 36" x 18", double door	Ea.	10,200
Smoke Detectors		
Ceiling type	Ea.	231
Duct type	Ea.	510
Sound System		
Amplifier, 250 watts	Ea.	1950
Speaker, ceiling or wall	Ea.	233
Trumpet	Ea.	445

Important: See the Reference Section for Location Factors.

Model costs calculated for a 1 story building with 18′ story height and 44,000 square feet of floor area

Supermarket

			Unit	Unit Cost	Cost Per S.F.	% Of Sub-Total
A. SUBSTRUCTURE						
1010	Standard Foundations	Poured concrete; strip and spread footings; 4′ foundation wall	S.F. Ground	2.49	2.49	
1020	Special Foundations	N/A	—	—	—	
1030	Slab on Grade	4″ reinforced slab on grade	S.F. Slab	5.28	5.28	9.6%
2010	Basement Excavation	Site preparation for slab and trench for foundation wall and footing	S.F. Ground	.16	.16	
2020	Basement Walls	N/A	—	—	—	
B. SHELL						
	B10 Superstructure					
1010	Floor Construction	N/A	—	—	—	12.2 %
1020	Roof Construction	Metal deck on open web steel joints, columns	S.F. Roof	10.04	10.04	
	B20 Exterior Enclosure					
2010	Exterior Walls	N/A	—	—	—	
2020	Exterior Windows	Curtain wall glazing system, thermo-break frame *100% of wall*	Each	48	16.43	21.7 %
2030	Exterior Doors	Glass sliding entrance doors, overhead doors, hollow metal doors	Each	5935	1.48	
	B30 Roofing					
3010	Roof Coverings	Single ply membrane, rigid insulation, gravel stop	S.F. Roof	6.47	6.47	8.0%
3020	Roof Openings	Roof and smoke hatches	S.F. Roof	.14	.14	
C. INTERIORS						
1010	Partitions	Concrete partitions, gypsum board on metal stud *50 S.F. Floor/L.F. Partition*	S.F. Partition	8.54	2.05	
1020	Interior Doors	Hollow metal doors *2000 S.F. Floor/Door*	Each	1201	.60	
1030	Fittings	Stainless steel toilet partitions	S.F. Floor	.11	.11	
2010	Stair Construction	N/A	—	—	—	15.3%
3010	Wall Finishes	Paint	S.F. Surface	1.83	.88	
3020	Floor Finishes	Vinyl composition tile	S.F. Floor	2.84	2.84	
3030	Ceiling Finishes	Acoustic ceiling tiles on suspended channel grid	S.F. Ceiling	6.10	6.10	
D. SERVICES						
	D10 Conveying					
1010	Elevators & Lifts	N/A	—	—	—	0.0 %
1020	Escalators & Moving Walks	N/A	—	—	—	
	D20 Plumbing					
2010	Plumbing Fixtures	Restroom, food prep and service fixtures, supply and drainage *1 Fixture/1750 S.F. Floor*	Each	2112	1.20	
2020	Domestic Water Distribution	Gas fired water heater	S.F. Floor	.27	.27	3.1%
2040	Rain Water Drainage	Roof drains	S.F. Roof	1.12	1.12	
	D30 HVAC					
3010	Energy Supply	N/A	—	—	—	
3020	Heat Generating Systems	Included in D3050	—	—	—	
3030	Cooling Generating Systems	N/A	—	—	—	5.6 %
3050	Terminal & Package Units	Single zone unit gas heating, electric cooling	S.F. Floor	4.66	4.66	
3090	Other HVAC Sys. & Equipment	N/A	—	—	—	
	D40 Fire Protection					
4010	Sprinklers	Wet pipe sprinkler system	S.F. Floor	3.41	3.41	4.8%
4020	Standpipes	Standpipe	S.F. Floor	.58	.58	
	D50 Electrical					
5010	Electrical Service/Distribution	1600 ampere service, panel board and feeders	S.F. Floor	2.44	2.44	
5020	Lighting & Branch Wiring	Fluorescent fixtures, receptacles, switches, A.C. and misc. power	S.F. Floor	11.54	11.54	19.6%
5030	Communications & Security	Addressable alarm system, emergency lighting, internet & phone wiring	S.F. Floor	2.14	2.14	
5090	Other Electrical Systems	Emergency generator, 15kW	S.F. Floor	.05	.05	
E. EQUIPMENT & FURNISHINGS						
1010	Commercial Equipment	N/A	—	—	—	
1020	Institutional Equipment	N/A	—	—	—	0.0 %
1030	Vehicular Equipment	N/A	—	—	—	
1090	Other Equipment	N/A	—	—	—	
F. SPECIAL CONSTRUCTION						
1020	Integrated Construction	N/A	—	—	—	0.0 %
1040	Special Facilities	N/A	—	—	—	
G. BUILDING SITEWORK		**N/A**				
				Sub-Total	82.48	**100%**
	CONTRACTOR FEES (General Requirements: 10%, Overhead: 5%, Profit: 10%)			*25%*	20.65	
	ARCHITECT FEES			*7%*	7.22	
				Total Building Cost	**110.35**	

Telephone Exchange

Costs per square foot of floor area

Exterior Wall	S.F. Area	2000	3000	4000	5000	6000	7000	8000	9000	10000
	L.F. Perimeter	180	220	260	286	320	353	368	397	425
Face Brick and Concrete Block	Rigid Steel	218.65	193.85	181.50	171.40	166.00	161.85	156.70	154.20	152.10
	Bearing Walls	208.65	186.65	175.70	166.45	161.60	157.90	153.00	150.75	148.85
Limestone and Concrete Block	Rigid Steel	250.30	221.50	207.20	194.70	188.30	183.35	176.75	173.70	171.10
	Bearing Walls	250.35	218.75	203.00	189.75	182.75	177.45	170.45	167.15	164.40
Decorative Concrete Block	Rigid Steel	205.90	183.45	172.30	163.30	158.45	154.75	150.20	147.90	146.05
	Bearing Walls	200.90	178.50	167.30	158.30	153.45	149.75	145.20	142.90	141.10
Perimeter Adj., Add or Deduct	Per 100 L.F.	47.65	31.75	23.85	19.10	15.90	13.65	11.90	10.60	9.55
Story Hgt. Adj., Add or Deduct	Per 1 Ft.	5.55	4.50	3.95	3.55	3.25	3.15	2.80	2.70	2.60
***For Basement**, add $36.00 per square foot of basement area*										

The above costs were calculated using the basic specifications shown on the facing page. These costs should be adjusted where necessary for design alternatives and owner's requirements.

Common additives

Description	Unit	$ Cost
Emergency Lighting, 25 watt, battery operated		
Lead battery	Ea.	305
Nickel cadmium	Ea.	525
Smoke Detectors		
Ceiling type	Ea.	231
Duct type	Ea.	510
Emergency Generators, complete system, gas		
15 kw	Ea.	18,600
85 kw	Ea.	44,100
170 kw	Ea.	105,500
Diesel, 50 kw	Ea.	26,100
150 kw	Ea.	53,500
350 kw	Ea.	78,000

Model costs calculated for a 1 story building with 12′ story height and 5,000 square feet of floor area

Telephone Exchange

				Unit	Unit Cost	Cost Per S.F.	% Of Sub-Total
A. SUBSTRUCTURE							
1010	Standard Foundations	Poured concrete; strip and spread footings; 4′ foundation wall		S.F. Ground	6.95	6.95	
1020	Special Foundations	N/A		—	—	—	
1030	Slab on Grade	4″ reinforced concrete with vapor barrier and granular base		S.F. Slab	5.28	5.28	10.4%
2010	Basement Excavation	Site preparation for slab and trench for foundation wall and footing		S.F. Ground	.49	.61	
2020	Basement Walls	N/A		—	—	—	
B. SHELL							
	B10 Superstructure						
1010	Floor Construction	Steel column fireproofing		S.F. Floor	.73	.73	8.7%
1020	Roof Construction	Metal deck, open web steel joists, beams, columns		S.F. Roof	10.04	10.04	
	B20 Exterior Enclosure						
2010	Exterior Walls	Face brick with concrete block back-up	*80% of wall*	S.F. Wall	31	16.97	
2020	Exterior Windows	Outward projecting steel	*20% of wall*	Each	1730	11.88	25.0%
2030	Exterior Doors	Single aluminum glass with transom		Each	5130	2.06	
	B30 Roofing						
3010	Roof Coverings	Built-up tar and gravel with flashing; perlite/EPS composite insulation		S.F. Roof	7.03	7.03	5.7%
3020	Roof Openings	N/A		—	—	—	
C. INTERIORS							
1010	Partitions	Double layer gypsum board on metal studs	*15 S.F. Floor/L.F. Partition*	S.F. Partition	5.94	3.96	
1020	Interior Doors	Single leaf hollow metal	*150 S.F. Floor/Door*	Each	1201	8.01	
1030	Fittings	Toilet partitions		S.F. Floor	1.22	1.22	
2010	Stair Construction	N/A		—	—	—	22.7%
3010	Wall Finishes	Paint		S.F. Surface	2.03	2.70	
3020	Floor Finishes	90% carpet, 10% terrazzo		S.F. Floor	6.10	6.10	
3030	Ceiling Finishes	Fiberglass board on exposed grid system, suspended		S.F. Ceiling	6.10	6.10	
D. SERVICES							
	D10 Conveying						
1010	Elevators & Lifts	N/A		—	—	—	0.0 %
1020	Escalators & Moving Walks	N/A		—	—	—	
	D20 Plumbing						
2010	Plumbing Fixtures	Kitchen, toilet and service fixtures, supply and drainage	*1 Fixture/715 S.F. Floor*	Each	3825	5.35	
2020	Domestic Water Distribution	Gas fired water heater		S.F. Floor	2.44	2.44	7.7%
2040	Rain Water Drainage	Roof drains		S.F. Roof	1.72	1.72	
	D30 HVAC						
3010	Energy Supply	N/A		—	—	—	
3020	Heat Generating Systems	Included in D3050		—	—	—	
3030	Cooling Generating Systems	N/A		—	—	—	6.4 %
3050	Terminal & Package Units	Single zone unit, gas heating, electric cooling		S.F. Floor	7.94	7.94	
3090	Other HVAC Sys. & Equipment	N/A		—	—	—	
	D40 Fire Protection						
4010	Sprinklers	Wet pipe sprinkler system		S.F. Floor	5.15	5.15	5.9%
4020	Standpipes	Standpipe		S.F. Floor	2.15	2.15	
	D50 Electrical						
5010	Electrical Service/Distribution	200 ampere service, panel board and feeders		S.F. Floor	1.85	1.85	
5020	Lighting & Branch Wiring	High efficiency fluorescent fixtures, receptacles, switches, A.C. and misc. power		S.F. Floor	4.22	4.22	7.4%
5030	Communications & Security	Addressable alarm systems and emergency lighting		S.F. Floor	2.89	2.89	
5090	Other Electrical Systems	Emergency generator, 15 kW		S.F. Floor	.19	.19	
E. EQUIPMENT & FURNISHINGS							
1010	Commercial Equipment	N/A		—	—	—	
1020	Institutional Equipment	N/A		—	—	—	0.0 %
1030	Vehicular Equipment	N/A		—	—	—	
1090	Other Equipment	N/A		—	—	—	
F. SPECIAL CONSTRUCTION							
1020	Integrated Construction	N/A		—	—	—	0.0 %
1040	Special Facilities	N/A		—	—	—	
G. BUILDING SITEWORK	**N/A**						
					Sub-Total	123.54	**100%**
	CONTRACTOR FEES (General Requirements: 10%, Overhead: 5%, Profit: 10%)				*25%*	30.87	
	ARCHITECT FEES				*11%*	16.99	
					Total Building Cost	**171.40**	

Costs per square foot of floor area

Exterior Wall	S.F. Area	5000	6500	8000	9500	11000	14000	17500	21000	24000
	L.F. Perimeter	300	360	386	396	435	510	550	620	680
Face Brick and Concrete Block	Steel Joists	162.75	156.55	149.40	143.25	**140.80**	137.20	132.50	130.55	129.30
	Wood Joists	155.15	149.20	142.50	136.70	134.35	130.95	126.55	124.70	123.55
Stone and Concrete Block	Steel Joists	170.45	163.65	155.60	148.55	145.80	141.85	136.55	134.30	132.95
	Wood Joists	162.85	156.30	148.65	142.00	139.40	135.65	130.60	128.45	127.10
Brick Veneer	Wood Frame	150.05	144.55	138.40	133.10	131.00	127.85	123.90	122.15	121.10
E.I.F.S.	Wood Frame	147.95	142.60	136.70	131.70	129.60	126.60	122.80	121.15	120.10
Perimeter Adj., Add or Deduct	Per 100 L.F.	15.45	11.90	9.65	8.10	6.95	5.50	4.40	3.65	3.25
Story Hgt. Adj., Add or Deduct	Per 1 Ft.	2.80	2.55	2.25	1.90	1.80	1.70	1.45	1.35	1.30
***For Basement**, add $29.60 per square foot of basement area*										

The above costs were calculated using the basic specifications shown on the facing page. These costs should be adjusted where necessary for design alternatives and owner's requirements.

Common additives

Description	Unit	$ Cost
Directory Boards, Plastic, glass covered		
30" x 20"	Ea.	570
36" x 48"	Ea.	1500
Aluminum, 24" x 18"	Ea.	570
36" x 24"	Ea.	725
48" x 32"	Ea.	950
48" x 60"	Ea.	2000
Emergency Lighting, 25 watt, battery operated		
Lead battery	Ea.	305
Nickel cadmium	Ea.	525
Flagpoles, Complete		
Aluminum, 20' high	Ea.	1825
40' high	Ea.	4400
70' high	Ea.	11,600
Fiberglass, 23' high	Ea.	1375
39'-5" high	Ea.	3175
59' high	Ea.	7000
Safe, Office type, 1 hour rating		
30" x 18" x 18"	Ea.	2400
60" x 36" x 18", double door	Ea.	10,200

Description	Unit	$ Cost
Smoke Detectors		
Ceiling type	Ea.	231
Duct type	Ea.	510
Vault Front, Door & frame		
1 Hour test, 32" x 78"	Opng.	7700
2 Hour test, 32" door	Opng.	9700
40" door	Opng.	10,900
4 Hour test, 32" door	Opng.	12,100
40" door	Opng.	13,600
Time lock movement; two movement	Ea.	2375

Important: See the Reference Section for Location Factors.

Model costs calculated for a 1 story building with 12′ story height and 11,000 square feet of floor area

Town Hall, 1 Story

				Unit	Unit Cost	Cost Per S.F.	% Of Sub-Total
A. SUBSTRUCTURE							
1010	Standard Foundations	Poured concrete; strip and spread footings; 4′ foundation wall		S.F. Ground	4.77	4.77	10.0%
1020	Special Foundations	N/A		—	—	—	
1030	Slab on Grade	4″ reinforced concrete with vapor barrier and granular base		S.F. Slab	5.28	5.28	
2010	Basement Excavation	Site preparation for slab and trench for foundation wall and footing		S.F. Ground	.28	.31	
2020	Basement Walls	N/A		—	—	—	
B. SHELL							
	B10 Superstructure						
1010	Floor Construction	Steel column fireproofing		S.F. Floor	.25	.25	7.6%
1020	Roof Construction	Metal deck, open web steel joists, beams, interior columns		S.F. Roof	7.64	7.64	
	B20 Exterior Enclosure						
2010	Exterior Walls	Face brick with concrete block back-up	*70% of wall*	S.F. Wall	31	10.26	15.5%
2020	Exterior Windows	Metal outward projecting	*30% of wall*	Each	703	4.35	
2030	Exterior Doors	Metal and glass with transom		Each	3864	1.40	
	B30 Roofing						
3010	Roof Coverings	Built-up tar and gravel with flashing; perlite/EPS composite insulation		S.F. Roof	6.56	6.56	6.4%
3020	Roof Openings	N/A		—	—	—	
C. INTERIORS							
1010	Partitions	Gypsum board on metal studs	*20 S.F. Floor/L.F. Partition*	S.F. Partition	8.94	4.47	24.0%
1020	Interior Doors	Wood solid core	*200 S.F. Floor/Door*	Each	691	3.46	
1030	Fittings	Toilet partitions		S.F. Floor	.47	.47	
2010	Stair Construction	N/A		—	—	—	
3010	Wall Finishes	90% paint, 10% ceramic tile		S.F. Surface	1.43	1.43	
3020	Floor Finishes	70% carpet tile, 15% terrazzo, 15% vinyl composition tile		S.F. Floor	8.12	8.12	
3030	Ceiling Finishes	Mineral fiber tile on concealed zee bars		S.F. Ceiling	6.86	6.86	
D. SERVICES							
	D10 Conveying						
1010	Elevators & Lifts	N/A		—	—	—	0.0 %
1020	Escalators & Moving Walks	N/A		—	—	—	
	D20 Plumbing						
2010	Plumbing Fixtures	Kitchen, toilet and service fixtures, supply and drainage	*1 Fixture/500 S.F. Floor*	Each	2555	5.11	9.4%
2020	Domestic Water Distribution	Gas fired water heater		S.F. Floor	1.89	1.89	
2040	Rain Water Drainage	Roof drains		S.F. Roof	2.66	2.66	
	D30 HVAC						
3010	Energy Supply	N/A		—	—	—	8.7 %
3020	Heat Generating Systems	Included in D3050		—	—	—	
3030	Cooling Generating Systems	N/A		—	—	—	
3050	Terminal & Package Units	Multizone unit, gas heating, electric cooling		S.F. Floor	8.94	8.94	
3090	Other HVAC Sys. & Equipment	N/A		—	—	—	
	D40 Fire Protection						
4010	Sprinklers	Wet pipe sprinkler system		S.F. Floor	3.41	3.41	4.1%
4020	Standpipes	Standpipe, wet, Class III		S.F. Floor	.82	.82	
	D50 Electrical						
5010	Electrical Service/Distribution	400 ampere service, panel board and feeders		S.F. Floor	1.93	1.93	14.4%
5020	Lighting & Branch Wiring	High efficiency fluorescent fixtures, receptacles, switches, A.C. and misc. power		S.F. Floor	9.64	9.64	
5030	Communications & Security	Addressable alarm systems, internet wiring, and emergency lighting		S.F. Floor	3.10	3.10	
5090	Other Electrical Systems	Emergency generator, 15 kW		S.F. Floor	.17	.17	
E. EQUIPMENT & FURNISHINGS							
1010	Commercial Equipment	N/A		—	—	—	0.0 %
1020	Institutional Equipment	N/A		—	—	—	
1030	Vehicular Equipment	N/A		—	—	—	
1090	Other Equipment	N/A		—	—	—	
F. SPECIAL CONSTRUCTION							
1020	Integrated Construction	N/A		—	—	—	0.0 %
1040	Special Facilities	N/A		—	—	—	
G. BUILDING SITEWORK		**N/A**					
					Sub-Total	103.30	**100%**
	CONTRACTOR FEES (General Requirements: 10%, Overhead: 5%, Profit: 10%)				25%	25.88	
	ARCHITECT FEES				9%	11.62	
					Total Building Cost	**140.80**	

Costs per square foot of floor area

Exterior Wall	S.F. Area	8000	10000	12000	15000	18000	24000	28000	35000	40000
	L.F. Perimeter	206	233	260	300	320	360	393	451	493
Face Brick and Concrete Block	Rigid Steel	215.50	203.80	196.10	188.20	180.85	171.70	168.25	164.05	161.95
	Reinforced Concrete	219.05	207.35	199.60	191.70	184.35	175.25	171.75	167.55	165.50
Stone and Concrete Block	Rigid Steel	225.35	212.75	204.35	195.85	**187.70**	177.45	173.60	168.95	166.70
	Reinforced Concrete	228.85	216.25	207.90	199.40	191.20	180.95	177.10	172.45	170.20
Limestone and Concrete Block	Rigid Steel	240.40	226.15	216.70	207.10	197.35	185.30	180.85	175.45	172.75
	Reinforced Concrete	246.10	231.85	222.40	212.75	203.05	191.00	186.50	181.10	178.45
Perimeter Adj., Add or Deduct	Per 100 L.F.	23.65	18.90	15.75	12.60	10.55	7.90	6.80	5.35	4.70
Story Hgt. Adj., Add or Deduct	Per 1 Ft.	3.55	3.25	3.00	2.80	2.45	2.10	1.95	1.80	1.70
***For Basement**, add $28.80 per square foot of basement area*										

The above costs were calculated using the basic specifications shown on the facing page. These costs should be adjusted where necessary for design alternatives and owner's requirements.

Common additives

Description	Unit	$ Cost
Directory Boards, Plastic, glass covered		
30" x 20"	Ea.	570
36" x 48"	Ea.	1500
Aluminum, 24" x 18"	Ea.	570
36" x 24"	Ea.	725
48" x 32"	Ea.	950
48" x 60"	Ea.	2000
Elevators, Hydraulic passenger, 2 stops		
1500# capacity	Ea.	68,400
2500# capacity	Ea.	71,900
3500# capacity	Ea.	76,900
Additional stop, add	Ea.	7000
Emergency Lighting, 25 watt, battery operated		
Lead battery	Ea.	305
Nickel cadmium	Ea.	525

Description	Unit	$ Cost
Flagpoles, Complete		
Aluminum, 20' high	Ea.	1825
40' high	Ea.	4400
70' high	Ea.	11,600
Fiberglass, 23' high	Ea.	1375
39'-5" high	Ea.	3175
59' high	Ea.	7000
Safe, Office type, 1 hour rating		
30" x 18" x 18"	Ea.	2400
60" x 36" x 18", double door	Ea.	10,200
Smoke Detectors		
Ceiling type	Ea.	231
Duct type	Ea.	510
Vault Front, Door & frame		
1 Hour test, 32" x 78"	Opng.	7700
2 Hour test, 32" door	Opng.	9700
40" door	Opng.	10,900
4 Hour test, 32" door	Opng.	12,100
40" door	Opng.	13,600
Time lock movement; two movement	Ea.	2375

Important: See the Reference Section for Location Factors.

Model costs calculated for a 3 story building with 12′ story height and 18,000 square feet of floor area

Town Hall, 2-3 Story

				Unit	Unit Cost	Cost Per S.F.	% Of Sub-Total
A. SUBSTRUCTURE							
1010	Standard Foundations	Poured concrete; strip and spread footings; 4′ foundation wall		S.F. Ground	8.58	2.86	3.5%
1020	Special Foundations	N/A		—	—	—	
1030	Slab on Grade	4″ reinforced concrete with vapor barrier and granular base		S.F. Slab	5.28	1.76	
2010	Basement Excavation	Site preparation for slab and trench for foundation wall and footing		S.F. Ground	.49	.25	
2020	Basement Walls	N/A		—	—	—	
B. SHELL							
	B10 Superstructure						
1010	Floor Construction	Open web steel joists, slab form, concrete, wide flange steel columns		S.F. Floor	27	17.87	15.0%
1020	Roof Construction	Metal deck, open web steel joists, beams, interior columns		S.F. Roof	8.37	2.79	
	B20 Exterior Enclosure						
2010	Exterior Walls	Stone with concrete block back-up	*70% of wall*	S.F. Wall	42	18.84	18.7%
2020	Exterior Windows	Metal outward projecting	*10% of wall*	Each	703	5.86	
2030	Exterior Doors	Metal and glass with transoms		Each	3658	1.02	
	B30 Roofing						
3010	Roof Coverings	Built-up tar and gravel with flashing; perlite/EPS composite insulation		S.F. Roof	7.17	2.39	1.7%
3020	Roof Openings	N/A		—	—	—	
C. INTERIORS							
1010	Partitions	Gypsum board on metal studs	*20 S.F. Floor/L.F. Partition*	S.F. Partition	9.98	4.99	24.2%
1020	Interior Doors	Wood solid core	*200 S.F. Floor/Door*	Each	691	3.46	
1030	Fittings	Toilet partitions		S.F. Floor	.29	.29	
2010	Stair Construction	Concrete filled metal pan		Flight	18,425	8.18	
3010	Wall Finishes	90% paint, 10% ceramic tile		S.F. Surface	1.43	1.43	
3020	Floor Finishes	70% carpet tile, 15% terrazzo, 15% vinyl composition tile		S.F. Floor	8.12	8.12	
3030	Ceiling Finishes	Mineral fiber tile on concealed zee bars		S.F. Ceiling	6.86	6.86	
D. SERVICES							
	D10 Conveying						
1010	Elevators & Lifts	Two hydraulic elevators		Each	117,540	13.06	9.5%
1020	Escalators & Moving Walks	N/A		—	—	—	
	D20 Plumbing						
2010	Plumbing Fixtures	Toilet and service fixtures, supply and drainage	*1 Fixture/1385 S.F. Floor*	Each	6759	4.88	4.7%
2020	Domestic Water Distribution	Gas fired water heater		S.F. Floor	.65	.65	
2040	Rain Water Drainage	Roof drains		S.F. Roof	2.94	.98	
	D30 HVAC						
3010	Energy Supply	N/A		—	—	—	6.5 %
3020	Heat Generating Systems	Included in D3050		—	—	—	
3030	Cooling Generating Systems	N/A		—	—	—	
3050	Terminal & Package Units	Multizone unit, gas heating, electric cooling		S.F. Floor	8.94	8.94	
3090	Other HVAC Sys. & Equipment	N/A		—	—	—	
	D40 Fire Protection						
4010	Sprinklers	Sprinklers, light hazard		S.F. Floor	3.08	3.08	4.7%
4020	Standpipes	Standpipe, wet, Class III		S.F. Floor	3.40	3.40	
	D50 Electrical						
5010	Electrical Service/Distribution	400 ampere service, panel board and feeders		S.F. Floor	1.62	1.62	11.4%
5020	Lighting & Branch Wiring	High efficiency fluorescent fixtures, receptacles, switches, A.C. and misc. power		S.F. Floor	10.73	10.73	
5030	Communications & Security	Addressable alarm systems, internet wiring, and emergency lighting		S.F. Floor	3.22	3.22	
5090	Other Electrical Systems	Emergency generator, 15 kW		S.F. Floor	.20	.20	
E. EQUIPMENT & FURNISHINGS							
1010	Commercial Equipment	N/A		—	—	—	0.0 %
1020	Institutional Equipment	N/A		—	—	—	
1030	Vehicular Equipment	N/A		—	—	—	
1090	Other Equipment	N/A		—	—	—	
F. SPECIAL CONSTRUCTION							
1020	Integrated Construction	N/A		—	—	—	0.0 %
1040	Special Facilities	N/A		—	—	—	
G. BUILDING SITEWORK		**N/A**					
					Sub-Total	137.73	**100%**
	CONTRACTOR FEES (General Requirements: 10%, Overhead: 5%, Profit: 10%)				25%	34.48	
	ARCHITECT FEES				9%	15.49	
					Total Building Cost	**187.70**	

Costs per square foot of floor area

Exterior Wall	S.F. Area	3500	5000	7500	9000	11000	12500	15000	20000	25000
	L.F. Perimeter	240	290	350	390	430	460	510	580	650
E.I.F.S.	Wood Truss	167.80	158.95	150.75	148.30	145.40	143.80	141.95	138.95	137.20
	Steel Joists	178.15	168.95	160.35	157.75	154.75	153.15	151.15	148.00	146.10
Wood Clapboard	Wood Truss	171.05	162.30	**154.20**	151.70	148.90	147.35	145.55	142.60	140.85
Brick Veneer	Wood Truss	180.75	170.55	160.80	157.85	154.45	152.60	150.35	146.70	144.55
Face Brick and Concrete Block	Wood Truss	184.25	173.00	162.15	158.90	155.05	152.95	150.50	146.30	143.80
Tilt-up Concrete Panels	Steel Joists	170.85	162.55	154.90	152.60	149.90	148.55	146.75	144.05	142.40
Perimeter Adj., Add or Deduct	Per 100 L.F.	12.35	8.65	5.75	4.75	3.90	3.50	2.90	2.20	1.75
Story Hgt. Adj., Add or Deduct	Per 1 Ft.	1.55	1.30	1.00	0.90	0.85	0.85	0.75	0.65	0.55
***For Basement**, add $31.10 per square foot of basement area*										

The above costs were calculated using the basic specifications shown on the facing page. These costs should be adjusted where necessary for design alternatives and owner's requirements.

Common additives

Description	Unit	$ Cost
Closed circuit surveillance, one station		
Camera and monitor	Ea.	1250
For additional camera stations, add	Ea.	565
Directory boards, plastic, glass covered		
30″ x 20″	Ea.	570
36″ x 48″	Ea.	1500
Aluminum, 24″ x 18″	Ea.	570
36″ x 24″	Ea.	725
48″ x 32″	Ea.	950
48″ x 60″	Ea.	2000
Electronic, wall mounted	S.F.	4600
Free standing	S.F.	4050
Kennel fencing		
Kennel fencing, 1-1/2″ mesh, 6′ long, 3′-6″ wide, 6′-2″ high	Ea.	750
12′ long	Ea.	975
Top covers, 1-1/2″ mesh, 6′ long	Ea.	204
12′ long	Ea.	274

Description	Unit	$ Cost
Kennel doors		
2 way, swinging type, 13″ x 19″ opening	Opng.	192
17″ x 29″ opening	Opng.	238
9″ x 9″ opening, electronic with accessories	Opng.	262

Important: See the Reference Section for Location Factors.

Model costs calculated for a 1 story building with 12′ story height and 7,500 square feet of floor area

Veterinary Hospital

			Unit	Unit Cost	Cost Per S.F.	% Of Sub-Total
A. SUBSTRUCTURE						
1010	Standard Foundations	Poured concrete; strip and spread footings; 4′ foundation wall	S.F. Ground	5.63	5.63	
1020	Special Foundations	N/A	—	—	—	
1030	Slab on Grade	4″ and 5″ reinforced concrete with vapor barrier and granular base	S.F. Slab	5.28	5.28	10.5%
2010	Basement Excavation	Site preparation for slab and trench for foundation wall and footing	S.F. Ground	.49	.92	
2020	Basement Walls	N/A	—	—	—	
B. SHELL						
	B10 Superstructure					
1010	Floor Construction	6 x 6 wood columns	S.F. Floor	.22	.22	6.9%
1020	Roof Construction	Wood roof truss with plywood sheathing	S.F. Roof	7.55	7.55	
	B20 Exterior Enclosure					
2010	Exterior Walls	Cedar bevel siding on wood studs, insulated *85% of wall*	S.F. Wall	13.26	6.31	
2020	Exterior Windows	Wood double hung and awning *15% of wall*	Each	1129	3.86	11.5%
2030	Exterior Doors	Aluminum and glass	Each	3522	2.81	
	B30 Roofing					
3010	Roof Coverings	Ashphalt shingles with flashing (pitched); al. gutters and downspouts	S.F. Roof	5.22	5.22	4.6%
3020	Roof Openings	N/A	—	—	—	
C. INTERIORS						
1010	Partitions	Gypsum board on wood studs *8 S.F. Floor/L.F. Partition*	S.F. Partition	4.81	6.01	
1020	Interior Doors	Solid core wood *180 S.F. Floor/Door*	Each	714	1.99	
1030	Fittings	Lockers	S.F. Floor	.45	.45	
2010	Stair Construction	N/A	—	—	—	18.6%
3010	Wall Finishes	Paint on drywall	S.F. Surface	.94	2.35	
3020	Floor Finishes	90% vinyl composition tile, 5% ceramic tile, 5% sealed concrete	S.F. Floor	3.45	3.45	
3030	Ceiling Finishes	90% acoustic tile, 10% gypsum board	S.F. Ceiling	6.77	6.77	
D. SERVICES						
	D10 Conveying					
1010	Elevators & Lifts	N/A	—	—	—	0.0 %
1020	Escalators & Moving Walks	N/A	—	—	—	
	D20 Plumbing					
2010	Plumbing Fixtures	Toilet and service fixtures, supply and drainage *1 Fixture/1075 S.F. Floor*	Each	1736	3.10	
2020	Domestic Water Distribution	Electric water heater	S.F. Floor	1.60	1.60	4.2%
2040	Rain Water Drainage	N/A	—	—	—	
	D30 HVAC					
3010	Energy Supply	N/A	—	—	—	
3020	Heat Generating Systems	N/A	—	—	—	
3030	Cooling Generating Systems	N/A	—	—	—	5.0 %
3050	Terminal & Package Units	Split system with air cooled condensing units	S.F. Floor	5.65	5.65	
3090	Other HVAC Sys. & Equipment	N/A	—	—	—	
	D40 Fire Protection					
4010	Sprinklers	Wet pipe sprinkler system	S.F. Floor	3.41	3.41	3.0%
4020	Standpipes	N/A	—	—	—	
	D50 Electrical					
5010	Electrical Service/Distribution	800 ampere service, panel board and feeders	S.F. Floor	6.32	6.32	
5020	Lighting & Branch Wiring	High efficiency fluorescent fixtures, receptacles, switches, A.C. and misc. power	S.F. Floor	11.59	11.59	19.9%
5030	Communications & Security	Addressable alarm systems, internet wiring, and emergency lighting	S.F. Floor	4.60	4.60	
5090	Other Electrical Systems	Emergency generator, 100 kW	S.F. Floor	.06	.06	
E. EQUIPMENT & FURNISHINGS						
1010	Commercial Equipment	N/A	—	—	—	
1020	Institutional Equipment	Eye wash station	S.F. Floor	15.97	15.97	14.1 %
1030	Vehicular Equipment	N/A	—	—	—	
1090	Other Equipment	Countertop, plastic laminate	S.F. Floor			
F. SPECIAL CONSTRUCTION & DEMOLITION						
1020	Integrated Construction	N/A	—	—	—	0.0 %
1040	Special Facilities	N/A	—	—	—	
G. BUILDING SITEWORK	**N/A**					
				Sub-Total	113.15	**100%**
	CONTRACTOR FEES (General Requirements: 10%, Overhead: 5%, Profit: 10%)			*25%*	28.32	
	ARCHITECT FEES			*9%*	12.73	
				Total Building Cost	**154.20**	

Costs per square foot of floor area

Exterior Wall	S.F. Area	10000	15000	20000	25000	30000	35000	40000	50000	60000
	L.F. Perimeter	410	500	600	640	700	766	833	966	1000
Metal Panel	Rigid Steel	118.15	109.05	105.00	100.45	98.00	96.35	95.25	93.55	91.05
Pre-Engineered Metal Building	Rigid Steel	117.05	106.80	102.15	96.90	94.10	92.20	90.95	89.00	86.10
E.I.F.S.	Rigid Steel	118.75	109.55	105.45	100.80	98.35	96.70	95.55	93.85	91.30
Brick Veneer	Reinforced Concrete	163.45	147.60	140.60	132.15	**127.70**	124.85	122.75	119.70	114.95
Precast Concrete	Reinforced Concrete	192.05	170.50	161.00	149.25	143.15	139.20	136.30	132.20	125.40
Tilt-up Concrete Panels	Reinforced Concrete	143.00	130.95	125.65	119.35	116.10	113.90	112.40	110.10	106.60
Perimeter Adj., Add or Deduct	Per 100 L.F.	8.50	5.70	4.15	3.40	2.80	2.45	2.05	1.65	1.40
Story Hgt. Adj., Add or Deduct	Per 1 Ft.	1.00	0.85	0.70	0.65	0.60	0.55	0.45	0.50	0.45
***For Basement**, add $27.70 per square foot of basement area*										

The above costs were calculated using the basic specifications shown on the facing page. These costs should be adjusted where necessary for design alternatives and owner's requirements.

Common additives

Description	Unit	$ Cost
Dock Leveler, 10 ton cap.		
6′ x 8′	Ea.	5950
7′ x 8′	Ea.	8750
Emergency Lighting, 25 watt, battery operated		
Lead battery	Ea.	305
Nickel cadmium	Ea.	525
Fence, Chain link, 6′ high		
9 ga. wire	L.F.	28
6 ga. wire	L.F.	33.50
Gate	Ea.	375
Flagpoles, Complete		
Aluminum, 20′ high	Ea.	1825
40′ high	Ea.	4400
70′ high	Ea.	11,600
Fiberglass, 23′ high	Ea.	1375
39′-5″ high	Ea.	3175
59′ high	Ea.	7000
Paving, Bituminous		
Wearing course plus base course	S.Y.	7.20
Sidewalks, Concrete 4″ thick	S.F.	4.76

Description	Unit	$ Cost
Sound System		
Amplifier, 250 watts	Ea.	1950
Speaker, ceiling or wall	Ea.	233
Trumpet	Ea.	445
Yard Lighting, 20′ aluminum pole with 400 watt high pressure sodium fixture	Ea.	3500

Important: See the Reference Section for Location Factors.

Model costs calculated for a 1 story building with 24′ story height and 30,000 square feet of floor area

Warehouse

			Unit	Unit Cost	Cost Per S.F.	% Of Sub-Total
A. SUBSTRUCTURE						
1010	Standard Foundations	Poured concrete; strip and spread footings; 4′ foundation wall	S.F. Ground	3.20	3.20	
1020	Special Foundations	N/A	—	—	—	
1030	Slab on Grade	5″ reinforced concrete	S.F. Slab	5.80	5.80	9.6%
2010	Basement Excavation	Site preparation for slab and trench for foundation wall and footing	S.F. Ground	.16	.16	
2020	Basement Walls	N/A	—	—	—	
B. SHELL						
	B10 Superstructure					
1010	Floor Construction	Cast in place concrete columns, beams, and slab, fireproofed	S.F. Floor	65	6.50	26.2%
1020	Roof Construction	Precast double T with 2″ topping	S.F. Roof	18.49	18.49	
	B20 Exterior Enclosure					
2010	Exterior Walls	Face brick with concrete block back-up *98% of wall*	S.F. Wall	36	19.97	
2020	Exterior Windows	Aluminum horizontal sliding *2% of wall*	Each	498	.38	22.8%
2030	Exterior Doors	Aluminum and glass entry doors, hollow steel doors, overhead doors	Each	3785	1.39	
	B30 Roofing					
3010	Roof Coverings	Single ply membrane, stone ballast, rigid insulation	S.F. Roof	6.62	6.62	7.4%
3020	Roof Openings	Roof and smoke hatches	S.F. Roof	.40	.40	
C. INTERIORS						
1010	Partitions	Concrete block partitions, gypsum board on metal stud *100 S.F. Floor/L.F. Partition*	S.F. Partition	16	1.28	
1020	Interior Doors	Single leaf wood hollow core doors, metal doors *2500 S.F. Floor/Door*	Each	1201	.48	
1030	Fittings	N/A	—	—	—	
2010	Stair Construction	Steel grate type with rails	Flight	14,475	.96	6.1%
3010	Wall Finishes	Paint	S.F. Surface	4.31	.69	
3020	Floor Finishes	90% hardener, 10% vinyl composition tile	S.F. Floor	1.77	1.77	
3030	Ceiling Finishes	Acoustic ceiling tiles on suspended channel grid	S.F. Ceiling	6.86	.69	
D. SERVICES						
	D10 Conveying					
1010	Elevators & Lifts	N/A	—	—	—	0.0 %
1020	Escalators & Moving Walks	N/A	—	—	—	
	D20 Plumbing					
2010	Plumbing Fixtures	Restroom and service fixtures, supply and drainage *1 Fixture/2500 S.F. Floor*	Each	1400	.56	
2020	Domestic Water Distribution	Gas fired water heater	S.F. Floor	.24	.24	1.5%
2040	Rain Water Drainage	Roof drains	S.F. Roof	.60	.60	
	D30 HVAC					
3010	Energy Supply	N/A	—	—	—	
3020	Heat Generating Systems	Ventilation with heat system	Each	144,700	5.21	
3030	Cooling Generating Systems	N/A	—	—	—	6.3 %
3050	Terminal & Package Units	Single zone unit gas heating, electric cooling	S.F. Floor	.82	.82	
3090	Other HVAC Sys. & Equipment	N/A	—	—	—	
	D40 Fire Protection					
4010	Sprinklers	Wet pipe sprinkler system, ordinary hazard	S.F. Floor	3.94	3.94	4.6%
4020	Standpipes	Standpipe	S.F. Floor	.50	.50	
	D50 Electrical					
5010	Electrical Service/Distribution	200 ampere service, panel board and feeders	S.F. Floor	.54	.54	
5020	Lighting & Branch Wiring	Fluorescent fixtures, receptacles, switches, A.C. and misc. power	S.F. Floor	4.13	4.13	7.6%
5030	Communications & Security	Addressable alarm system	S.F. Floor	2.63	2.63	
5090	Other Electrical Systems	N/A	—	—	—	
E. EQUIPMENT & FURNISHINGS						
1010	Commercial Equipment	N/A	—	—	—	
1020	Institutional Equipment	N/A	—	—	—	1.9 %
1030	Vehicular Equipment	Dock boards and levelers	S.F. Floor	1.80	1.80	
1090	Other Equipment	N/A	—	—	—	
F. SPECIAL CONSTRUCTION						
1020	Integrated Construction	Shipping and receiving air curtain	S.F. Floor	5.74	5.74	6.0%
1040	Special Facilities	N/A	—	—	—	
G. BUILDING SITEWORK		**N/A**				
				Sub-Total	95.49	**100%**
	CONTRACTOR FEES (General Requirements: 10%, Overhead: 5%, Profit: 10%)			*25%*	23.85	
	ARCHITECT FEES			*7%*	8.36	
				Total Building Cost	**127.70**	

Costs per square foot of floor area

Exterior Wall	S.F. Area	2000	3000	5000	8000	12000	20000	30000	50000	100000
	L.F. Perimeter	180	220	300	420	580	900	1300	2100	4100
Concrete Block	Rigid Steel	102.65	90.20	80.40	74.85	71.75	69.35	68.05	67.05	66.40
	Reinforced Concrete	111.85	100.10	90.90	85.60	82.70	80.40	79.25	78.25	77.55
Metal Sandwich Panel	Rigid Steel	102.85	90.40	80.55	74.95	71.90	69.45	68.20	67.20	66.45
Tilt-up Concrete Panels	Reinforced Concrete	104.45	92.30	82.75	77.25	74.30	71.90	70.65	69.70	68.95
Precast Concrete	Rigid Steel	128.20	111.10	97.45	89.85	85.50	82.15	80.45	79.00	78.05
	Reinforced Concrete	164.40	146.15	131.65	123.40	118.90	115.25	113.40	111.90	110.85
Perimeter Adj., Add or Deduct	Per 100 L.F.	19.80	13.25	7.90	5.00	3.35	2.00	1.35	0.80	0.40
Story Hgt. Adj., Add or Deduct	Per 1 Ft.	1.30	1.10	0.90	0.80	0.75	0.65	0.70	0.60	0.55
Basement—Not Applicable										

The above costs were calculated using the basic specifications shown on the facing page. These costs should be adjusted where necessary for design alternatives and owner's requirements.

Common additives

Description	Unit	$ Cost
Dock Leveler, 10 ton cap.		
6′ x 8′	Ea.	5950
7′ x 8′	Ea.	8750
Emergency Lighting, 25 watt, battery operated		
Lead battery	Ea.	305
Nickel cadmium	Ea.	525
Fence, Chain link, 6′ high		
9 ga. wire	L.F.	28
6 ga. wire	L.F.	33.50
Gate	Ea.	375
Flagpoles, Complete		
Aluminum, 20′ high	Ea.	1825
40′ high	Ea.	4400
70′ high	Ea.	11,600
Fiberglass, 23′ high	Ea.	1375
39′-5″ high	Ea.	3175
59′ high	Ea.	7000
Paving, Bituminous		
Wearing course plus base course	S.Y.	7.20
Sidewalks, Concrete 4″ thick	S.F.	4.76

Description	Unit	$ Cost
Sound System		
Amplifier, 250 watts	Ea.	1950
Speaker, ceiling or wall	Ea.	233
Trumpet	Ea.	445
Yard Lighting,		
20′ aluminum pole with 400 watt high pressure sodium fixture	Ea.	3500

Important: See the Reference Section for Location Factors.

Model costs calculated for a 1 story building with 12′ story height and 20,000 square feet of floor area

Warehouse Self Storage

				Unit	Unit Cost	Cost Per S.F.	% Of Sub-Total
A. SUBSTRUCTURE							
1010	Standard Foundations	Poured concrete; strip and spread footings; 4′ foundation wall		S.F. Ground	5.78	5.78	24.6%
1020	Special Foundations	N/A		—	—	—	
1030	Slab on Grade	4″ reinforced concrete with vapor barrier and granular base		S.F. Slab	6.40	6.40	
2010	Basement Excavation	Site preparation for slab and trench for foundation wall and footing		S.F. Ground	.28	.56	
2020	Basement Walls	N/A		—	—	—	
B. SHELL							
B10 Superstructure							
1010	Floor Construction	Steel column fireproofing		S.F. Floor	.63	.63	13.8%
1020	Roof Construction	Metal deck, open web steel joists, beams, columns		S.F. Roof	6.54	6.54	
B20 Exterior Enclosure							
2010	Exterior Walls	Concrete block	*53% of wall*	S.F. Wall	21	5.91	24.5%
2020	Exterior Windows	Aluminum projecting	*1% of wall*	Each	886	.09	
2030	Exterior Doors	Steel overhead, hollow metal	*46% of wall*	Each	1650	6.69	
B30 Roofing							
3010	Roof Coverings	Metal panel roof; perlite/EPS composite insulation		S.F. Roof	6.80	6.80	13.1%
3020	Roof Openings	N/A		—	—	—	
C. INTERIORS							
1010	Partitions	Metal panels on metal studs	*11 S.F. Floor/L.F. Partition*	S.F. Partition	2.66	2.90	5.7%
1020	Interior Doors	Single leaf hollow metal	*20,000 S.F. Floor/Door*	Each	600	.06	
1030	Fittings	N/A		—	—	—	
2010	Stair Construction	N/A		—	—	—	
3010	Wall Finishes	N/A		—	—	—	
3020	Floor Finishes	N/A		—	—	—	
3030	Ceiling Finishes	N/A		—	—	—	
D. SERVICES							
D10 Conveying							
1010	Elevators & Lifts	N/A		—	—	—	0.0 %
1020	Escalators & Moving Walks	N/A		—	—	—	
D20 Plumbing							
2010	Plumbing Fixtures	Toilet and service fixtures, supply and drainage	*1 Fixture/10,000 S.F. Floor*	Each	2600	.26	2.4%
2020	Domestic Water Distribution	Gas fired water heater		S.F. Floor	.33	.33	
2040	Rain Water Drainage	Roof drain		S.F. Roof	.67	.67	
D30 HVAC							
3010	Energy Supply	N/A		—	—	—	2.3 %
3020	Heat Generating Systems	N/A		—	—	—	
3030	Cooling Generating Systems	N/A		—	—	—	
3050	Terminal & Package Units	Single zone rooftop unit		S.F. Floor	1.21	1.21	
3090	Other HVAC Sys. & Equipment	N/A		—	—	—	
D40 Fire Protection							
4010	Sprinklers	N/A		—	—	—	0.0 %
4020	Standpipes	N/A		—	—	—	
D50 Electrical							
5010	Electrical Service/Distribution	60 ampere service, panel board and feeders		S.F. Floor	.36	.36	13.5%
5020	Lighting & Branch Wiring	High bay fixtures, receptacles, switches and misc. power		S.F. Floor	5.35	5.35	
5030	Communications & Security	Addressable alarm systems		S.F. Floor	1.30	1.30	
5090	Other Electrical Systems	N/A		—	—	—	
E. EQUIPMENT & FURNISHINGS							
1010	Commercial Equipment	N/A		—	—	—	0.0 %
1020	Institutional Equipment	N/A		—	—	—	
1030	Vehicular Equipment	N/A		—	—	—	
1090	Other Equipment	N/A		—	—	—	
F. SPECIAL CONSTRUCTION							
1020	Integrated Construction	N/A		—	—	—	0.0 %
1040	Special Facilities	N/A		—	—	—	
G. BUILDING SITEWORK		**N/A**					
					Sub-Total	51.84	**100%**
	CONTRACTOR FEES (General Requirements: 10%, Overhead: 5%, Profit: 10%)				*25%*	12.97	
	ARCHITECT FEES				*7%*	4.54	
					Total Building Cost	**69.35**	

Green Commercial/Industrial/ Institutional Section

Table of Contents

Costs per square foot of floor area

Exterior Wall	S.F. Area	2000	2700	3400	4100	4800	5500	6200	6900	7600
	L.F. Perimeter	180	208	236	256	280	303	317	337	357
Face Brick and Concrete Block	Rigid Steel	283.55	265.10	254.35	245.20	239.50	235.15	230.25	227.25	224.80
	Reinforced Concrete	298.60	280.15	269.35	260.20	254.55	250.20	245.25	242.25	239.80
Precast Concrete	Rigid Steel	285.15	264.20	251.95	241.50	235.05	230.00	224.35	220.85	218.05
	Reinforced Concrete	316.00	295.00	282.75	272.30	265.80	260.85	255.10	251.70	248.90
Limestone and Concrete Block	Rigid Steel	310.60	288.50	275.55	264.50	257.60	252.35	246.35	242.65	239.70
	Reinforced Concrete	324.05	301.90	288.95	277.85	271.05	265.80	259.70	256.05	253.10
Perimeter Adj., Add or Deduct	Per 100 L.F.	52.40	38.85	30.80	25.55	21.85	19.05	16.90	15.20	13.75
Story Hgt. Adj., Add or Deduct	Per 1 Ft.	5.05	4.35	3.90	3.50	3.30	3.10	2.90	2.70	2.65
***For Basement**, add $ 34.87 per square foot of basement area*										

The above costs were calculated using the basic specifications shown on the facing page. These costs should be adjusted where necessary for design alternatives and owner's requirements.

Common additives

Description	Unit	$ Cost
Bulletproof Teller Window, 44" x 60"	Ea.	5825
60" x 48"	Ea.	7850
Closed Circuit Surveillance, one station, camera & monitor	Ea.	1250
For additional camera stations, add	Ea.	565
Counters, complete, dr & frame, 3' x 6'-8", bullet resist stl with vision panel	Ea.	7175 - 9350
Drive-up Window, drawer & micr., not incl. glass	Ea.	9450 - 12,900
Night Depository	Ea.	9875 - 14,800
Package Receiver, painted	Ea.	1900
Stainless steel	Ea.	2975
Partitions, bullet resistant to 8' high	L.F.	325 - 525
Pneumatic Tube Systems, 2 station	Ea.	31,800
With TV viewer	Ea.	60,000
Service Windows, pass thru, steel, 24" x 36"	Ea.	4000
48" x 48"	Ea.	3950
Twenty-four Hour Teller, automatic deposit cash & memo	Ea.	52,000
Vault Front, door & frame, 2 hour test, 32" door	Opng.	9700
4 hour test, 40" door	Opng.	13,600
Time lock, two movement, add	Ea.	2375

Description	Unit	$ Cost
Commissioning Fees, sustainable commercial construction	S.F.	0.24 - 3.04
Energy Modelling Fees, banks to 10,000 SF	Ea.	11,000
Green Bldg Cert Fees for comm construction project reg	Project	900
Photovoltaic Pwr Sys, grid connected, 20 kW (~2400 SF), roof	Ea.	250,600
Green Roofs, 6" soil depth, w/treated wd edging & sedum mats	S.F.	12.09
10" Soil depth, with treated wood edging & sedum mats	S.F.	13.58
Greywater Recovery Systems, prepackaged comm, 1530 gal	Ea.	35,190
Rainwater Harvest Sys, prepckged comm, 10,000 gal, sys contrller	Ea.	37,050
20,000 gal. w/system controller	Ea.	60,450
30,000 gal. w/system controller	Ea.	96,175
Solar Domestic HW, closed loop, add-on sys, ext heat exchanger	Ea.	10,050
Draindown, hot water system, 120 gal tank	Ea.	13,125

Important: See the Reference Section for Location Factors.

Model costs calculated for a 1 story building with 14′ story height and 4,100 square feet of floor area

G Bank

				Unit	Unit Cost	Cost Per S.F.	% Of Sub-Total
A. SUBSTRUCTURE							
1010	Standard Foundations	Poured concrete; strip and spread footings		S.F. Ground	4.18	4.18	8.6%
1020	Special Foundations	N/A		—	—	—	
1030	Slab on Grade	4″ reinforced concrete with recycled vapor barrier and granular base		S.F. Slab	5.35	5.35	
2010	Basement Excavation	Site preparation for slab and trench for foundation wall and footing		S.F. Ground	.28	.28	
2020	Basement Walls	4′ Foundation wall		L.F. Wall	85	6.36	
B. SHELL							
	B10 Superstructure						
1010	Floor Construction	Cast-in-place columns		L.F. Column	5.46	5.46	11.1%
1020	Roof Construction	Cast-in-place concrete flat plate		S.F. Roof	15.40	15.40	
	B20 Exterior Enclosure						
2010	Exterior Walls	Face brick with concrete block backup	*80% of wall*	S.F. Wall	36	25.45	18.2%
2020	Exterior Windows	Horizontal aluminum sliding	*20% of wall*	Each	558	6.50	
2030	Exterior Doors	Double aluminum and glass and hollow metal, low VOC paint		Each	4457	2.17	
	B30 Roofing						
3010	Roof Coverings	Single-ply TPO membrane, 60 mils, heat welded seams w/4″ thk. R20 insul.		S.F. Roof	8.38	8.38	4.5%
3020	Roof Openings	N/A		—	—	—	
C. INTERIORS							
1010	Partitions	Gypsum board on metal studs w/sound attenuation	*20 SF of Flr./LF Part.*	S.F. Partition	11.68	5.84	13.9%
1020	Interior Doors	Single leaf hollow metal, low VOC paint	*200 S.F. Floor/Door*	Each	1201	6.01	
1030	Fittings	N/A		—	—	—	
2010	Stair Construction	N/A		—	—	—	
3010	Wall Finishes	50% vinyl wall covering, 50% paint, low VOC		S.F. Surface	1.48	1.48	
3020	Floor Finishes	50% carpet tile, 40% vinyl composition tile, recycled, 10% quarry tile		S.F. Floor	5.83	5.83	
3030	Ceiling Finishes	Mineral fiber tile on concealed zee bars		S.F. Ceiling	6.86	6.86	
D. SERVICES							
	D10 Conveying						
1010	Elevators & Lifts	N/A		—	—	—	0.0 %
1020	Escalators & Moving Walks	N/A		—	—	—	
	D20 Plumbing						
2010	Plumbing Fixtures	Toilet low flow, auto sensor and service fixt., supply and drainage	*1 Fixt./580 SF Flr.*	Each	5997	10.34	6.8%
2020	Domestic Water Distribution	Tankless, on demand water heaters, natural gas/propane		S.F. Floor	.97	.97	
2040	Rain Water Drainage	Roof drains		S.F. Roof	1.43	1.43	
	D30 HVAC						
3010	Energy Supply	N/A		—	—	—	6.2 %
3020	Heat Generating Systems	Included in D3050		—	—	—	
3040	Distribution Systems	Enthalpy heat recovery packages		Each	15,500	3.78	
3050	Terminal & Package Units	Single zone rooftop air conditioner		S.F. Floor	7.80	7.80	
3090	Other HVAC Sys. & Equipment	N/A		—	—	—	
	D40 Fire Protection						
4010	Sprinklers	Wet pipe sprinkler system		S.F. Floor	4.66	4.66	4.1%
4020	Standpipes	Standpipe		S.F. Floor	3.06	3.06	
	D50 Electrical						
5010	Electrical Service/Distribution	200 ampere service, panel board and feeders		S.F. Floor	1.41	1.41	13.8%
5020	Lighting & Branch Wiring	LED fixtures, daylt. dimming & ltg. on/off control, receptacles, switches, A.C.		S.F. Floor	10.53	10.53	
5030	Communications & Security	Alarm systems, internet/phone wiring, and security television		S.F. Floor	10.11	10.11	
5090	Other Electrical Systems	Emergency generator, 15 kW, UPS, energy monitoring systems		S.F. Floor	3.87	3.87	
E. EQUIPMENT & FURNISHINGS							
1010	Commercial Equipment	Automatic teller, drive up window, night depository		S.F. Floor	8.17	8.17	6.8%
1020	Institutional Equipment	Closed circuit TV monitoring system		S.F. Floor	2.95	2.95	
1090	Other Equipment	Waste handling recycling tilt truck		S.F. Floor	1.51	1.51	
2020	Moveable Furnishings	No smoking signage		S.F. Floor	.03	.03	
F. SPECIAL CONSTRUCTION							
1020	Integrated Construction	N/A		—	—	—	6.1 %
1040	Special Facilities	Security vault door		S.F. Floor	11.37	11.37	
G. BUILDING SITEWORK	**N/A**						
					Sub-Total	187.54	**100%**
	CONTRACTOR FEES (General Requirements: 10%, Overhead: 5%, Profit: 10%)				*25%*	46.87	
	ARCHITECT FEES				*11%*	25.79	
					Total Building Cost	**260.20**	

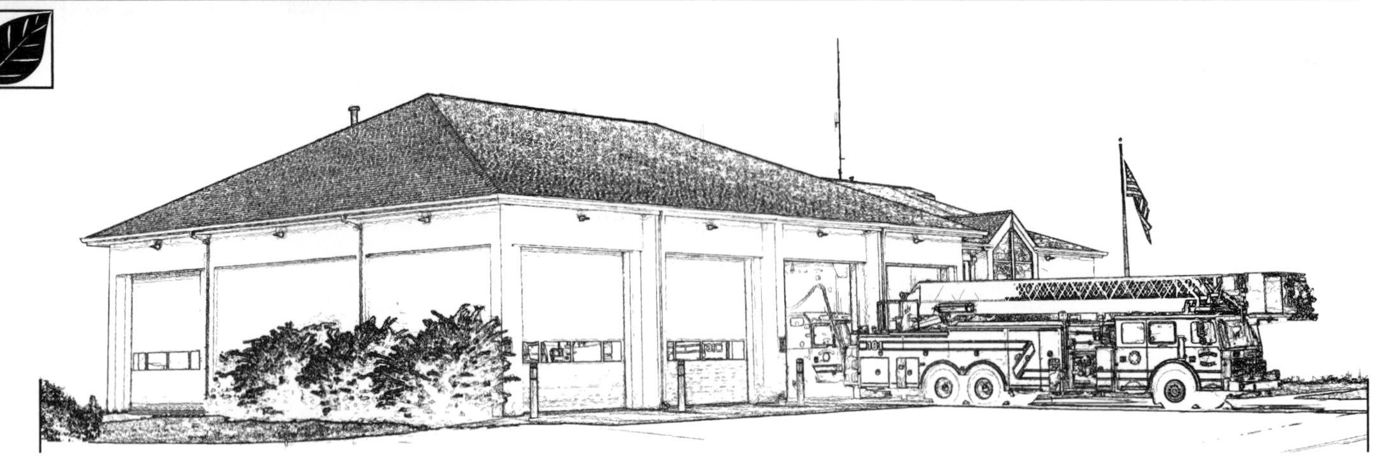

Costs per square foot of floor area

Exterior Wall	S.F. Area	4000	4500	5000	5500	6000	6500	7000	7500	8000
	L.F. Perimeter	260	280	300	310	320	336	353	370	386
Face Brick and Concrete Block	Steel Joists	214.05	209.85	206.60	202.25	**198.65**	196.45	194.65	193.10	191.60
	Bearing Walls	206.95	202.75	199.50	195.15	191.60	189.30	187.55	186.00	184.55
Decorative Concrete Block	Steel Joists	199.20	195.65	192.90	189.35	186.45	184.60	183.15	181.85	180.55
	Bearing Walls	193.45	189.95	187.20	183.65	180.80	178.90	177.45	176.10	174.85
Limestone and Concrete Block	Steel Joists	230.80	225.90	222.10	216.75	212.45	209.70	207.70	205.80	204.05
	Bearing Walls	225.10	220.20	216.35	211.05	206.70	204.00	201.95	200.10	198.35
Perimeter Adj., Add or Deduct	Per 100 L.F.	22.70	20.20	18.20	16.55	15.20	13.95	13.00	12.15	11.35
Story Hgt. Adj., Add or Deduct	Per 1 Ft.	3.00	2.95	2.80	2.60	2.50	2.40	2.35	2.25	2.25
***For Basement**, add $ 39.73 per square foot of basement area*										

The above costs were calculated using the basic specifications shown on the facing page. These costs should be adjusted where necessary for design alternatives and owner's requirements.

Common additives

Description	Unit	$ Cost
Appliances, cooking range, 30" free standing		
1 oven	Ea.	600 - 2775
2 oven	Ea.	1200 - 3800
Microwave oven	Ea.	255 - 800
Compactor, residential, 4-1 compaction	Ea.	880 - 1450
Dishwasher, built-in, 2 cycles	Ea.	520 - 870
4 cycles	Ea.	650 - 2475
Garbage disposer, sink type	Ea.	193 - 305
Hood for range, 2 speed, vented, 30" wide	Ea.	237 - 1275
Refrigerator, no frost 10-12 C.F.	Ea.	575 - 735
14-16 C.F.	Ea.	710 - 1200
18-20 C.F.	Ea.	890 - 2050
Lockers, Steel, single tier, 60" or 72"	Opng.	223 - 390
Locker bench, lam. maple top only	L.F.	35.50
Pedestals, steel pipe	Ea.	73.50
Sound System, amplifier, 250 watts	Ea.	1950
Speaker, ceiling or wall	Ea.	233
Trumpet	Ea.	445

Description	Unit	$ Cost
Commissioning Fees, sustainable commercial construction	S.F.	0.24 - 3.04
Energy Modelling Fees, Fire Stations to 10,000 SF	Ea.	11,000
Green Bldg Cert Fees for comm construction project reg	Project	900
Photovoltaic Pwr Sys, grid connected, 20 kW (~2400 SF), roof	Ea.	250,600
Green Roofs, 6" soil depth, w/treated wd edging & sedum mats	S.F.	12.09
10" Soil depth, with treated wood edging & sedum mats	S.F.	13.58
Greywater Recovery Systems, prepackaged comm, 3060 gal.	Ea.	47,695
4590 gal.	Ea.	59,835
Rainwater Harvest Sys, prepckged comm, 10,000 gal, sys contrller	Ea.	37,050
20,000 gal. w/system controller	Ea.	60,450
30,000 gal. w/system controller	Ea.	96,175
Solar Domestic HW, closed loop, add-on sys, ext heat exchanger	Ea.	10,050
Drainback, hot water system, 120 gal tank	Ea.	12,875

Important: See the Reference Section for Location Factors.

Model costs calculated for a 1 story building with 14′ story height and 6,000 square feet of floor area

G Fire Station, 1 Story

			Unit	Unit Cost	Cost Per S.F.	% Of Sub-Total
A. SUBSTRUCTURE						
1010	Standard Foundations	Poured concrete; strip and spread footings	S.F. Ground	3.32	3.32	
1020	Special Foundations	N/A	–	–	–	
1030	Slab on Grade	6″ reinforced concrete with recycled vapor barrier and granular base	S.F. Slab	6.94	6.94	10.4%
2010	Basement Excavation	Site preparation for slab and trench for foundation wall and footing	S.F. Ground	.49	.49	
2020	Basement Walls	4′ foundation wall	L.F. Wall	85	4.54	
B. SHELL						
	B10 Superstructure					
1010	Floor Construction	N/A	–	–	–	6.9 %
1020	Roof Construction	Metal deck, open web steel joists, beams on columns	S.F. Roof	10.08	10.08	
	B20 Exterior Enclosure					
2010	Exterior Walls	Face brick with concrete block backup *75% of wall*	S.F. Wall	36	20.38	
2020	Exterior Windows	Aluminum insulated glass *10% of wall*	Each	827	1.93	18.9%
2030	Exterior Doors	Single aluminum and glass, overhead, hollow metal, low VOC paint *15% of wall*	S.F. Door	3633	5.44	
	B30 Roofing					
3010	Roof Coverings	Single-ply TPO membrane, 60 mil, with flashing, R-20 insulation and roof edges	S.F. Roof	7.99	7.99	5.6%
3020	Roof Openings	Skylights, roof hatches	S.F. Roof	.19	.19	
C. INTERIORS						
1010	Partitions	Concrete block w/foamed-in insulation *17 S.F. Floor/LF Part.*	S.F. Partition	10.59	6.23	
1020	Interior Doors	Single leaf hollow metal, low VOC paint *500 S.F. Floor/Door*	Each	1201	2.40	
1030	Fittings	Toilet partitions	S.F. Floor	.52	.52	
2010	Stair Construction	N/A	–	–	–	12.3%
3010	Wall Finishes	Paint, low VOC	S.F. Surface	2.25	2.65	
3020	Floor Finishes	50% vinyl tile with recycled content, 50% paint, low VOC	S.F. Floor	2.79	2.79	
3030	Ceiling Finishes	Mineral board acoustic ceiling tiles, concealed grid, suspended *50% of area*	S.F. Ceiling	3.43	3.44	
D. SERVICES						
	D10 Conveying					
1010	Elevators & Lifts	N/A	–	–	–	0.0 %
1020	Escalators & Moving Walks	N/A	–	–	–	
	D20 Plumbing					
2010	Plumbing Fixtures	Kitchen, toilet, low flow, auto sensor & service fixt., supply & drain. *1 Fixt./375 SF Flr.*	Each	4335	11.56	
2020	Domestic Water Distribution	Water heater, tankless, on-demand, natural gas/propane	S.F. Floor	2.06	2.06	10.3%
2040	Rain Water Drainage	Roof drains	S.F. Roof	1.61	1.61	
	D30 HVAC					
3010	Energy Supply	N/A	–	–	–	
3020	Heat Generating Systems	Included in D3050	–	–	–	
3040	Distribution Systems	Enthalpy heat recovery packages	Each	15,500	2.59	16.2 %
3050	Terminal & Package Units	Multizone rooftop air conditioner, SEER 14	S.F. Floor	21	21.29	
3090	Other HVAC Sys. & Equipment	N/A	–	–	–	
	D40 Fire Protection					
4010	Sprinklers	Wet pipe sprinkler system	S.F. Floor	4.66	4.66	4.2%
4020	Standpipes	Standpipe, wet, Class III	S.F. Floor	1.49	1.49	
	D50 Electrical					
5010	Electrical Service/Distribution	200 ampere service, panel board and feeders	S.F. Floor	1.33	1.33	
5020	Lighting & Branch Wiring	LED fixt., daylt. dimming & ltg. on/off control, receptacles, switches, A.C. & misc. pwr.	S.F. Floor	14.05	14.05	14.6%
5030	Communications & Security	Addressable alarm systems	S.F. Floor	1.95	1.95	
5090	Other Electrical Systems	Energy monitoring systems	S.F. Floor	4.16	4.16	
E. EQUIPMENT & FURNISHINGS						
1010	Commercial Equipment	N/A	–	–	–	
1020	Institutional Equipment	N/A	–	–	–	0.7 %
1090	Other Equipment	Waste handling recycling tilt truck	S.F. Floor	1.03	1.03	
2020	Moveable Furnishings	No smoking signage	S.F. Floor	.04	.04	
F. SPECIAL CONSTRUCTION						
1020	Integrated Construction	N/A	–	–	–	0.0 %
1040	Special Facilities	N/A	–	–	–	
G. BUILDING SITEWORK	**N/A**					
				Sub-Total	147.15	**100%**
	CONTRACTOR FEES (General Requirements: 10%, Overhead: 5%, Profit: 10%)			*25%*	36.78	
	ARCHITECT FEES			*8%*	14.72	
				Total Building Cost	**198.65**	

Costs per square foot of floor area

Exterior Wall	S.F. Area	4000	6000	8000	10000	12000	14000	16000	18000	20000
	L.F. Perimeter	260	320	384	424	460	484	510	540	576
Wood Board and Batten	Wood Frame	162.95	151.65	146.25	141.70	138.40	135.60	133.60	132.10	131.15
Brick Veneer	Wood Frame	170.95	158.45	152.50	147.40	143.75	140.55	138.30	136.65	135.50
Aluminum Clapboard	Wood Frame	157.95	147.80	142.90	138.90	136.10	133.65	131.95	130.60	129.80
Face Brick and Concrete Block	Wood Truss	184.90	170.60	163.75	157.75	153.45	149.65	147.00	145.05	143.75
Limestone and Concrete Block	Wood Truss	203.60	185.95	177.60	170.00	164.50	159.60	156.20	153.70	152.05
Stucco and Concrete Block	Wood Truss	192.95	176.90	169.30	162.50	157.60	153.25	150.15	147.90	146.50
Perimeter Adj., Add or Deduct	Per 100 L.F.	14.00	9.35	7.00	5.55	4.65	4.05	3.50	3.10	2.80
Story Hgt. Adj., Add or Deduct	Per 1 Ft.	1.90	1.50	1.35	1.20	1.10	1.00	0.90	0.85	0.85
***For Basement**, add $ 32.89 per square foot of basement area*										

The above costs were calculated using the basic specifications shown on the facing page. These costs should be adjusted where necessary for design alternatives and owner's requirements.

Common additives

Description	Unit	$ Cost
Autopsy Table, standard	Ea.	11,800
Deluxe	Ea.	19,500
Directory Boards, plastic, glass covered, 30" x 20"	Ea.	570
36" x 48"	Ea.	1500
Aluminum, 24" x 18"	Ea.	570
36" x 24"	Ea.	725
48" x 32"	Ea.	950
Mortuary Refrigerator, end operated		
Two capacity	Ea.	10,000
Six capacity	Ea.	18,600
Planters, precast concrete, 48" diam., 24" high	Ea.	775
7" diam., 36" high	Ea.	1850
Fiberglass, 36" diam., 24" high	Ea.	845
60" diam., 24" high	Ea.	1300

Description	Unit	$ Cost
Commissioning Fees, sustainable commercial construction	S.F.	0.24 - 3.04
Energy Modelling Fees, commercial buildings to 10,000 SF	Ea.	11,000
Greater than 10,000 SF add	S.F.	0.04
Green Bldg Cert Fees for comm construction project reg	Project	900
Photovoltaic Pwr Sys, grid connected, 20 kW (~2400 SF), roof	Ea.	250,600
Green Roofs, 6" soil depth, w/treated wd edging & sedum mats	S.F.	12.09
10" Soil depth, with treated wood edging & sedum mats	S.F.	13.58
Greywater Recovery Systems, prepackaged comm, 3060 gal.	Ea.	47,695
4590 gal.	Ea.	59,835
Rainwater Harvest Sys, prepckged comm, 10,000 gal, sys contrller	Ea.	37,050
20,000 gal. w/system controller	Ea.	60,450
30,000 gal. w/system controller	Ea.	96,175
Solar Domestic HW, closed loop, add-on sys, ext heat exchanger	Ea.	10,050
Drainback, hot water system, 120 gal tank	Ea.	12,875
Draindown, hot water system, 120 gal tank	Ea.	13,125

Model costs calculated for a 1 story building with 12′ story height and 10,000 square feet of floor area

G Funeral Home

				Unit	Unit Cost	Cost Per S.F.	% Of Sub-Total
A. SUBSTRUCTURE							
1010	Standard Foundations	Poured concrete; strip and spread footings		S.F. Ground	1.81	1.81	10.5%
1020	Special Foundations	N/A		—	—	—	
1030	Slab on Grade	4″ reinforced concrete with recycled plastic vapor barrier and granular base		S.F. Slab	5.35	5.35	
2010	Basement Excavation	Site preparation for slab and trench for foundation wall and footing		S.F. Ground	.28	.28	
2020	Basement Walls	4′ foundation wall		L.F. Wall	76	3.52	
B. SHELL							
B10 Superstructure							
1010	Floor Construction	N/A		—	—	—	3.5 %
1020	Roof Construction	Plywood on wood truss		S.F. Roof	3.62	3.62	
B20 Exterior Enclosure							
2010	Exterior Walls	1″ x 4″ vertical T. & G. redwood siding on 2x6 wood studs with insul.	*90% of wall*	S.F. Wall	16.79	7.69	11.3%
2020	Exterior Windows	Double hung wood	*10% of wall*	Each	491	2.09	
2030	Exterior Doors	Wood swinging double doors, single leaf hollow metal, low VOC paint		Each	3375	2.02	
B30 Roofing							
3010	Roof Coverings	Single ply membrane, TPO, 45 mil, fully adhered; polyisocyanurate sheets		S.F. Roof	5.97	5.97	5.7%
3020	Roof Openings	N/A		—	—	—	
C. INTERIORS							
1010	Partitions	Gypsum board on wood studs with sound deadening board	*15 SF Flr./LF Part.*	S.F. Partition	9.36	4.99	25.4%
1020	Interior Doors	Single leaf wood, low VOC paint	*150 S.F. Floor/Door*	Each	691	4.60	
1030	Fittings	N/A		—	—	—	
2010	Stair Construction	N/A		—	—	—	
3010	Wall Finishes	50% wallpaper, 25% wood paneling, 25% paint (low VOC)		S.F. Surface	3.23	3.44	
3020	Floor Finishes	70% carpet tile, 30% ceramic tile		S.F. Floor	7.24	7.24	
3030	Ceiling Finishes	Fiberglass board on exposed grid, suspended		S.F. Ceiling	6.10	6.10	
D. SERVICES							
D10 Conveying							
1010	Elevators & Lifts	N/A		—	—	—	0.0 %
1020	Escalators & Moving Walks	N/A		—	—	—	
D20 Plumbing							
2010	Plumbing Fixtures	Toilets, low flow, auto sensor, urinals & service fixt., supply & drain.	*1 Fixt./770 SF Flr.*	Each	3303	4.29	5.5%
2020	Domestic Water Distribution	Electric water heater, point-of-use		S.F. Floor	.45	.45	
2040	Rain Water Drainage	Roof drain		S.F. Roof	.93	.93	
D30 HVAC							
3010	Energy Supply	N/A		—	—	—	19.3 %
3020	Heat Generating Systems	Included in D3050		—	—	—	
3040	Distribution Systems	Enthalpy heat recovery packages		Each	39,100	3.91	
3050	Terminal & Package Units	Multizone rooftop air conditioner, SEER 14		S.F. Floor	16.14	16.14	
3090	Other HVAC Sys. & Equipment	N/A		—	—	—	
D40 Fire Protection							
4010	Sprinklers	Wet pipe sprinkler system		S.F. Floor	3.41	3.41	4.1%
4020	Standpipes	Standpipe, wet, Class III		S.F. Floor	.89	.89	
D50 Electrical							
5010	Electrical Service/Distribution	400 ampere service, panel board and feeders		S.F. Floor	1.49	1.49	14.0%
5020	Lighting & Branch Wiring	LED light fixtures, daylt. dimming control & ltg. on/off sys., recept., switches, & A.C. power		S.F. Floor	9.66	9.66	
5030	Communications & Security	Addressable alarm systems		S.F. Floor	1.51	1.51	
5090	Other Electrical Systems	Emergency generator, 15 kW and energy monitoring systems		S.F. Floor	1.91	1.91	
E. EQUIPMENT & FURNISHINGS							
1010	Commercial Equipment	N/A		—	—	—	0.6 %
1020	Institutional Equipment	N/A		—	—	—	
1090	Other Equipment	Waste handling recycling tilt truck		S.F. Floor	.62	.62	
2020	Moveable Furnishings	No smoking signage		S.F. Floor	.04	.04	
F. SPECIAL CONSTRUCTION							
1020	Integrated Construction	N/A		—	—	—	0.0 %
1040	Special Facilities	N/A		—	—	—	
G. BUILDING SITEWORK		**N/A**					
					Sub-Total	103.97	**100%**
	CONTRACTOR FEES (General Requirements: 10%, Overhead: 5%, Profit: 10%)				*25%*	26.03	
	ARCHITECT FEES				*9%*	11.70	
					Total Building Cost	**141.70**	

Costs per square foot of floor area

Exterior Wall	S.F. Area	7000	10000	13000	16000	19000	22000	25000	28000	31000
	L.F. Perimeter	240	300	336	386	411	435	472	510	524
Face Brick and Concrete Block	Reinforced Concrete	205.05	194.75	185.90	181.95	176.95	173.25	171.35	169.95	167.40
	Rigid Steel	204.10	193.80	184.95	181.00	175.95	172.30	170.40	168.95	166.40
Limestone and Concrete Block	Reinforced Concrete	229.95	215.90	203.85	198.40	191.50	186.40	183.85	181.80	178.35
	Rigid Steel	226.90	213.75	202.20	197.05	190.35	185.45	182.95	181.05	177.65
Precast Concrete	Reinforced Concrete	237.20	222.20	209.25	203.50	196.10	190.55	187.80	185.65	181.90
	Rigid Steel	236.20	221.25	208.30	202.50	195.15	189.60	186.85	184.65	180.90
Perimeter Adj., Add or Deduct	Per 100 L.F.	24.80	17.35	13.35	10.85	9.10	7.95	6.95	6.20	5.55
Story Hgt. Adj., Add or Deduct	Per 1 Ft.	3.75	3.25	2.85	2.65	2.35	2.15	2.10	1.95	1.85
***For Basement**, add $48.97 per square foot of basement area*										

The above costs were calculated using the basic specifications shown on the facing page. These costs should be adjusted where necessary for design alternatives and owner's requirements.

Common additives

Description	Unit	$ Cost
Carrels Hardwood	Ea.	715 - 1900
Closed Circuit Surveillance, one station, camera and monitor	Ea.	1250
For additional camera stations, add	Ea.	565
Elevators, hydraulic passenger, 2 stops, 1500# capacity	Ea.	68,400
2500# capacity	Ea.	71,900
3500# capacity	Ea.	76,900
Library Furnishings, bookshelf, 90" high, 10" shelf double face	L.F.	253
Single face	L.F.	172
Charging desk, built-in with counter		
Plastic laminated top	L.F.	595
Reading table, laminated		
Top 60" x 36"	Ea.	510

Description	Unit	$ Cost
Commissioning Fees, sustainable institutional construction	S.F.	0.58 - 2.47
Energy Modelling Fees, commercial buildings to 10,000 SF	Ea.	11,000
Greater than 10,000 SF add	S.F.	0.04
Green Bldg Cert Fees for comm construction project reg	Project	900
Photovoltaic Pwr Sys, grid connected, 20 kW (~2400 SF), roof	Ea.	250,600
Green Roofs, 6" soil depth, w/treated wd edging & sedum mats	S.F.	12.09
10" Soil depth, with treated wood edging & sedum mats	S.F.	13.58
Greywater Recovery Systems, prepackaged comm, 3060 gal.	Ea.	47,695
4590 gal.	Ea.	59,835
Rainwater Harvest Sys, prepckged comm, 10,000 gal, sys contrller	Ea.	37,050
20,000 gal. w/system controller	Ea.	60,450
30,000 gal. w/system controller	Ea.	96,175
Solar Domestic HW, closed loop, add-on sys, ext heat exchanger	Ea.	10,050
Drainback, hot water system, 120 gal tank	Ea.	12,875
Draindown, hot water system, 120 gal tank	Ea.	13,125

Important: See the Reference Section for Location Factors.

Model costs calculated for a 2 story building with 14′ story height and 22,000 square feet of floor area

				Unit	Unit Cost	Cost Per S.F.	% Of Sub-Total
A. SUBSTRUCTURE							
1010	Standard Foundations	Poured concrete; strip and spread footings		S.F. Ground	3.88	1.94	5.0%
1020	Special Foundations	N/A		—	—	—	
1030	Slab on Grade	4″ reinforced concrete with recycled vapor barrier and granular base		S.F. Slab	5.35	2.68	
2010	Basement Excavation	Site preparation for slab and trench for foundation wall and footing		S.F. Ground	.28	.14	
2020	Basement Walls	4′ foundation wall		L.F. Wall	85	1.68	
B. SHELL							
B10 Superstructure							
1010	Floor Construction	Concrete waffle slab		S.F. Floor	28	13.87	18.9%
1020	Roof Construction	Concrete waffle slab		S.F. Roof	21	10.41	
B20 Exterior Enclosure							
2010	Exterior Walls	Face brick with concrete block backup	*90% of wall*	S.F. Wall	34	17.11	16.1%
2020	Exterior Windows	Window wall	*10% of wall*	Each	54	2.98	
2030	Exterior Doors	Double aluminum and glass, single leaf hollow metal, low VOC paint		Each	6950	.63	
B30 Roofing							
3010	Roof Coverings	Single-ply TPO membrane, 60 mil, heat welded w/R-20 insulation		S.F. Roof	7.08	3.54	2.8%
3020	Roof Openings	Roof hatches		S.F. Roof	.10	.05	
C. INTERIORS							
1010	Partitions	Gypsum board on metal studs w/sound attenuation insul.	*30 SF Flr./LF Part.*	S.F. Partition	12.98	5.19	16.5%
1020	Interior Doors	Single leaf wood, low VOC paint	*300 S.F. Floor/Door*	Each	691	2.31	
1030	Fittings	N/A		—	—	—	
2010	Stair Construction	Concrete filled metal pan		Flight	9650	.88	
3010	Wall Finishes	Paint, low VOC		S.F. Surface	1.85	1.48	
3020	Floor Finishes	50% carpet tile, 50% vinyl composition tile, recycled content		S.F. Floor	4.40	4.40	
3030	Ceiling Finishes	Mineral fiber on concealed zee bars		S.F. Ceiling	6.86	6.86	
D. SERVICES							
D10 Conveying							
1010	Elevators & Lifts	One hydraulic passenger elevator		Each	88,880	4.04	3.1%
1020	Escalators & Moving Walks	N/A		—	—	—	
D20 Plumbing							
2010	Plumbing Fixtures	Toilet, low flow, auto sensor & service fixt., supply & drain.	*1 Fixt./1835 SF.Flr.*	Each	8331	4.54	4.5%
2020	Domestic Water Distribution	Tankless, on demand water heater, gas/propane		S.F. Floor	.56	.56	
2040	Rain Water Drainage	Roof drains		S.F. Roof	1.46	.73	
D30 HVAC							
3010	Energy Supply	N/A		—	—	—	16.2 %
3020	Heat Generating Systems	Included in D3050		—	—	—	
3040	Distribution Systems	Enthalpy heat recovery packages		Each	39,100	1.77	
3050	Terminal & Package Units	Multizone rooftop air conditioner, SEER 14		S.F. Floor	19.05	19.05	
3090	Other HVAC Sys. & Equipment	N/A		—	—	—	
D40 Fire Protection							
4010	Sprinklers	Wet pipe sprinkler system		S.F. Floor	2.91	2.91	3.1%
4020	Standpipes	Standpipe		S.F. Floor	1.01	1.01	
D50 Electrical							
5010	Electrical Service/Distribution	400 ampere service, panel board and feeders		S.F. Floor	.52	.52	13.5%
5020	Lighting & Branch Wiring	LED fixtures, daylt. dim., ltg. on/off, recept., switches, and A.C. power		S.F. Floor	13.58	13.58	
5030	Communications & Security	Addressable alarm systems, internet wiring, and emergency lighting		S.F. Floor	2.02	2.02	
5090	Other Electrical Systems	Emergency generator, 7.5 kW, UPS, energy monitoring systems		S.F. Floor	1.17	1.17	
E. EQUIPMENT & FURNISHINGS							
1010	Commercial Equipment	N/A		—	—	—	0.2 %
1020	Institutional Equipment	N/A		—	—	—	
1090	Other Equipment	Waste handling recycling tilt truck		S.F. Floor	.28	.28	
2020	Moveable Furnishings	No smoking signage		—	—	—	
F. SPECIAL CONSTRUCTION							
1020	Integrated Construction	N/A		—	—	—	0.0 %
1040	Special Facilities	N/A		—	—	—	
G. BUILDING SITEWORK		**N/A**					
					Sub-Total	128.33	**100%**
	CONTRACTOR FEES (General Requirements: 10%, Overhead: 5%, Profit: 10%)				*25%*	32.09	
	ARCHITECT FEES				*8%*	12.83	
					Total Building Cost	**173.25**	

Costs per square foot of floor area

Exterior Wall	S.F. Area	4000	5500	7000	8500	10000	11500	13000	14500	16000
	L.F. Perimeter	280	320	380	440	453	503	510	522	560
Face Brick and Concrete Block	Steel Joists	249.45	238.85	234.85	232.20	227.10	225.45	221.95	219.40	218.50
	Wood Truss	246.80	236.10	**232.10**	229.50	224.40	222.80	219.25	216.70	215.80
Stucco and Concrete Block	Steel Joists	254.25	242.45	238.05	235.20	229.50	227.75	223.80	220.95	219.95
	Wood Truss	253.50	241.65	237.30	234.45	228.70	226.95	223.00	220.15	219.15
Brick Veneer	Wood Truss	238.45	229.05	225.45	223.05	218.60	217.20	214.10	211.95	211.10
Wood Clapboard	Wood Frame	230.20	222.35	219.25	217.25	213.70	212.40	210.00	208.20	207.50
Perimeter Adj., Add or Deduct	Per 100 L.F.	17.85	12.90	10.20	8.35	7.05	6.20	5.50	4.95	4.45
Story Hgt. Adj., Add or Deduct	Per 1 Ft.	3.70	3.00	2.80	2.65	2.35	2.30	2.05	1.90	1.85
***For Basement**, add $31.90 per square foot of basement area*										

The above costs were calculated using the basic specifications shown on the facing page. These costs should be adjusted where necessary for design alternatives and owner's requirements.

Common additives

Description	Unit	$ Cost
Cabinets, hospital, base, laminated plastic	L.F.	540
Counter top, laminated plastic	L.F.	84.50
Nurses station, door type , laminated plastic	L.F.	605
Wall cabinets, laminated plastic	L.F.	400
Directory Boards, plastic, glass covered, 30" x 20"	Ea.	570
36" x 48"	Ea.	1500
Aluminum, 36" x 24"	Ea.	725
48" x 32"	Ea.	950
Heat Therapy Unit, humidified, 26" x 78" x 28"	Ea.	4750
Tables, examining, vinyl top, with base cabinets	Ea.	1525 - 6500
Utensil Washer, Sanitizer	Ea.	9950
X-Ray, Mobile	Ea.	19,400 - 106,500

Description	Unit	$ Cost
Commissioning Fees, sustainable commercial construction	S.F.	0.24 - 3.04
Energy Modelling Fees, commercial buildings to 10,000 SF	Ea.	11,000
Greater than 10,000 SF add	S.F.	0.04
Green Bldg Cert Fees for comm construction project reg	Project	900
Photovoltaic Pwr Sys, grid connected, 20 kW (~2400 SF), roof	Ea.	250,600
Green Roofs, 6" soil depth, w/treated wd edging & sedum mats	S.F.	12.09
10" Soil depth, with treated wood edging & sedum mats	S.F.	13.58
Greywater Recovery Systems, prepackaged comm, 3060 gal.	Ea.	47,695
4590 gal.	Ea.	59,835
Rainwater Harvest Sys, prepckged comm, 10,000 gal, sys contrller	Ea.	37,050
20,000 gal. w/system controller	Ea.	60,450
30,000 gal. w/system controller	Ea.	96,175
Solar Domestic HW, closed loop, add-on sys, ext heat exchanger	Ea.	10,050
Drainback, hot water system, 120 gal tank	Ea.	12,875
Draindown, hot water system, 120 gal tank	Ea.	13,125

Important: See the Reference Section for Location Factors.

Model costs calculated for a 1 story building with 10′ story height and 7,000 square feet of floor area

G Medical Office, 1 Story

				Unit	Unit Cost	Cost Per S.F.	% Of Sub-Total
A. SUBSTRUCTURE							
1010	Standard Foundations	Poured concrete; strip and spread footings		S.F. Ground	2.12	2.12	
1020	Special Foundations	N/A		—	—	—	
1030	Slab on Grade	4″ reinforced concrete with recycled vapor barrier and granular base		S.F. Slab	5.35	5.35	7.3%
2010	Basement Excavation	Site preparation for slab and trench for foundation wall and footing		S.F. Ground	.28	.28	
2020	Basement Walls	4′ foundation wall		L.F. Wall	85	4.61	
B. SHELL							
	B10 Superstructure						
1010	Floor Construction	N/A		—	—	—	4.2 %
1020	Roof Construction	Plywood on wood trusses		S.F. Roof	7.22	7.22	
	B20 Exterior Enclosure						
2010	Exterior Walls	Face brick with concrete block backup	*70% of wall*	S.F. Wall	38	14.45	
2020	Exterior Windows	Wood double hung	*30% of wall*	Each	615	5.89	14.3%
2030	Exterior Doors	Aluminum and glass doors and entrance with transoms		Each	2380	4.08	
	B30 Roofing						
3010	Roof Coverings	Asphalt shingles with flashing (Pitched); rigid fiber glass insulation, gutters		S.F. Roof	2.99	2.99	1.8%
3020	Roof Openings	N/A		—	—	—	
C. INTERIORS							
1010	Partitions	Gypsum bd. & sound deadening bd. on wood studs w/insul.	*6 SF Flr./LF Part.*	S.F. Partition	9.09	12.12	
1020	Interior Doors	Single leaf wood, low VOC paint	*60 S.F. Floor/Door*	Each	691	11.51	
1030	Fittings	N/A		—	—	—	
2010	Stair Construction	N/A		—	—	—	22.1%
3010	Wall Finishes	50% paint, low VOC, 50% vinyl wall covering		S.F. Surface	1.39	3.70	
3020	Floor Finishes	50% carpet tile, 50% vinyl composition tile, recycled content		S.F. Floor	4.10	4.10	
3030	Ceiling Finishes	Mineral fiber tile on concealed zee bars		S.F. Ceiling	6.15	6.15	
D. SERVICES							
	D10 Conveying						
1010	Elevators & Lifts	N/A		—	—	—	0.0 %
1020	Escalators & Moving Walks	N/A		—	—	—	
	D20 Plumbing						
2010	Plumbing Fixtures	Toilet, low flow, auto sensor, exam room & service fixt., supply & drain.	*1 Fixt./195 SF Flr.*	Each	4826	24.75	
2020	Domestic Water Distribution	Gas fired tankless water heater		S.F. Floor	2.77	2.77	16.9%
2040	Rain Water Drainage	Roof drains		S.F. Roof	1.25	1.25	
	D30 HVAC						
3010	Energy Supply	N/A		—	—	—	
3020	Heat Generating Systems	Included in D3050		—	—	—	
3040	Distribution Systems	Enthalpy heat recovery packages		Each	14,075	2.01	9.1 %
3050	Terminal & Package Units	Multizone rooftop air conditioner, SEER 14		S.F. Floor	13.45	13.45	
3090	Other HVAC Sys. & Equipment	N/A		—	—	—	
	D40 Fire Protection						
4010	Sprinklers	Wet pipe sprinkler system		S.F. Floor	4.66	4.66	3.5%
4020	Standpipes	Standpipe		S.F. Floor	1.28	1.28	
	D50 Electrical						
5010	Electrical Service/Distribution	200 ampere service, panel board and feeders		S.F. Floor	1.25	1.25	
5020	Lighting & Branch Wiring	LED fixtures, daylt. dim., ltg. on/off, recept., switches, and A.C. power		S.F. Floor	12.19	12.19	14.1%
5030	Communications & Security	Alarm systems, internet & phone wiring, intercom system, & emerg. ltg.		S.F. Floor	7.54	7.54	
5090	Other Electrical Systems	Emergency generator, 7.5 kW, and energy monitoring systems		S.F. Floor	3.06	3.06	
E. EQUIPMENT & FURNISHINGS							
1010	Commercial Equipment	N/A		—	—	—	
1020	Institutional Equipment	Exam room casework and countertops		S.F. Floor	10.60	10.60	6.8 %
1090	Other Equipment	Waste handling recycling tilt truck		S.F. Floor	.88	.88	
2020	Moveable Furnishings	No smoking signage		S.F. Floor	.10	.10	
F. SPECIAL CONSTRUCTION							
1020	Integrated Construction	N/A		—	—	—	0.0 %
1040	Special Facilities	N/A		—	—	—	
G. BUILDING SITEWORK		**N/A**					
					Sub-Total	170.36	**100%**
	CONTRACTOR FEES (General Requirements: 10%, Overhead: 5%, Profit: 10%)				*25%*	42.57	
	ARCHITECT FEES				*9%*	19.17	
					Total Building Cost	**232.10**	

Costs per square foot of floor area

Exterior Wall	S.F. Area	4000	5500	7000	8500	10000	11500	13000	14500	16000
	L.F. Perimeter	180	210	240	270	286	311	336	361	386
Face Brick and Concrete Block	Steel Joists	281.10	269.75	263.30	259.15	254.35	251.85	250.00	248.45	247.20
	Wood Joists	282.65	271.25	264.85	260.65	255.90	253.40	251.50	249.95	248.75
Stucco and Concrete Block	Steel Joists	265.55	256.55	251.40	248.15	244.50	242.50	241.05	239.85	238.85
	Wood Joists	267.05	258.05	252.95	249.65	246.00	244.00	242.55	241.35	240.40
Brick Veneer	Wood Frame	271.55	261.90	256.40	252.85	248.90	246.75	245.15	243.85	242.80
Wood Clapboard	Wood Frame	267.25	258.25	253.10	249.80	246.10	244.15	242.70	241.45	240.50
Perimeter Adj., Add or Deduct	Per 100 L.F.	32.15	23.40	18.40	15.10	12.85	11.20	9.90	8.90	8.10
Story Hgt. Adj., Add or Deduct	Per 1 Ft.	4.80	4.05	3.60	3.30	3.05	2.85	2.70	2.65	2.60
***For Basement**, add $ 35.45 per square foot of basement area*										

The above costs were calculated using the basic specifications shown on the facing page. These costs should be adjusted where necessary for design alternatives and owner's requirements.

Common additives

Description	Unit	$ Cost
Cabinets, hospital, base, laminated plastic	L.F.	540
Counter top, laminated plastic	L.F.	84.50
Nurses station, door type , laminated plastic	L.F.	605
Wall cabinets, laminated plastic	L.F.	400
Elevators, hydraulic passenger, 2 stops, 2500# capacity	Ea.	71,900
3500# capacity	Ea.	76,900
Directory Boards, plastic, glass covered, 30" x 20"	Ea.	570
36" x 48"	Ea.	1500
Aluminum, 36" x 24"	Ea.	725
48" x 32"	Ea.	950
Heat Therapy Unit, humidified, 26" x 78" x 28"	Ea.	4750
Tables, examining, vinyl top, with base cabinets	Ea.	1525 - 6500
Utensil Washer, Sanitizer	Ea.	9950
X-Ray, Mobile	Ea.	19,400 - 106,500

Description	Unit	$ Cost
Commissioning Fees, sustainable commercial construction	S.F.	0.24 - 3.04
Energy Modelling Fees, commercial buildings to 10,000 SF	Ea.	11,000
Greater than 10,000 SF add	S.F.	0.04
Green Bldg Cert Fees for comm construction project reg	Project	900
Photovoltaic Pwr Sys, grid connected, 20 kW (~2400 SF), roof	Ea.	250,600
Green Roofs, 6" soil depth, w/treated wd edging & sedum mats	S.F.	12.09
10" Soil depth, with treated wood edging & sedum mats	S.F.	13.58
Greywater Recovery Systems, prepackaged comm, 3060 gal.	Ea.	47,695
4590 gal.	Ea.	59,835
Rainwater Harvest Sys, prepckged comm, 10,000 gal, sys contrller	Ea.	37,050
20,000 gal. w/system controller	Ea.	60,450
30,000 gal. w/system controller	Ea.	96,175
Solar Domestic HW, closed loop, add-on sys, ext heat exchanger	Ea.	10,050
Drainback, hot water system, 120 gal tank	Ea.	12,875
Draindown, hot water system, 120 gal tank	Ea.	13,125

Important: See the Reference Section for Location Factors.

Model costs calculated for a 2 story building with 10' story height and 7,000 square feet of floor area

G Medical Office, 2 Story

				Unit	Unit Cost	Cost Per S.F.	% Of Sub-Total
A. SUBSTRUCTURE							
1010	Standard Foundations	Poured concrete; strip and spread footings		S.F. Ground	3.36	1.68	4.1%
1020	Special Foundations	N/A		—	—	—	
1030	Slab on Grade	4" reinforced concrete with vapor barrier and granular base		S.F. Slab	5.35	2.68	
2010	Basement Excavation	Site preparation for slab and trench for foundation wall and footing		S.F. Ground	.49	.25	
2020	Basement Walls	4' foundation wall		L.F. Wall	85	2.91	
B. SHELL							
	B10 Superstructure						
1010	Floor Construction	Open web steel joists, slab form, concrete, columns		S.F. Floor	13.32	6.66	5.3%
1020	Roof Construction	Metal deck, open web steel joists, beams, columns		S.F. Roof	6.08	3.04	
	B20 Exterior Enclosure						
2010	Exterior Walls	Stucco on concrete block		S.F. Wall	18.25	8.76	9.5%
2020	Exterior Windows	Outward projecting metal	*30% of wall*	Each	498	6.83	
2030	Exterior Doors	Aluminum and glass doors with transoms		Each	6950	1.99	
	B30 Roofing						
3010	Roof Coverings	Built-up tar and gravel with flashing; perlite/EPS composite insulation		S.F. Roof	8.32	4.16	2.3%
3020	Roof Openings	Roof hatches		S.F. Roof	.32	.16	
C. INTERIORS							
1010	Partitions	Gypsum bd. & acous. insul. on metal studs	*6 SF Floor/LF Part.*	S.F. Partition	7.43	9.90	21.5%
1020	Interior Doors	Single leaf wood	*60 S.F. Floor/Door*	Each	691	11.51	
1030	Fittings	N/A		—	—	—	
2010	Stair Construction	Concrete filled metal pan		Flight	15,375	4.40	
3010	Wall Finishes	45% paint, 50% vinyl wall coating, 5% ceramic tile		S.F. Surface	1.39	3.70	
3020	Floor Finishes	50% carpet, 50% vinyl composition tile		S.F. Floor	4.10	4.10	
3030	Ceiling Finishes	Mineral fiber tile on concealed zee bars		S.F. Ceiling	6.15	6.15	
D. SERVICES							
	D10 Conveying						
1010	Elevators & Lifts	One hydraulic hospital elevator		Each	112,070	16.01	8.7%
1020	Escalators & Moving Walks	N/A		—	—	—	
	D20 Plumbing						
2010	Plumbing Fixtures	Toilet, exam room & service fixt., supply & drain.	*1 Fixt./160 SF Flr.*	Each	4496	28.10	17.5%
2020	Domestic Water Distribution	Gas fired water heater		S.F. Floor	2.77	2.77	
2040	Rain Water Drainage	Roof drains		S.F. Roof	2.72	1.36	
	D30 HVAC						
3010	Energy Supply	N/A		—	—	—	8.4 %
3020	Heat Generating Systems	Included in D3050		—	—	—	
3040	Distribution Systems	Enthalpy heat recovery packages		Each	14,075	2.01	
3050	Terminal & Package Units	Multizone rooftop air conditioner, SEER 14		S.F. Floor	13.45	13.45	
3090	Other HVAC Sys. & Equipment	N/A		—	—	—	
	D40 Fire Protection						
4010	Sprinklers	Wet pipe sprinkler system		S.F. Floor	3.64	3.64	3.0%
4020	Standpipes	Standpipe		S.F. Floor	1.94	1.94	
	D50 Electrical						
5010	Electrical Service/Distribution	400 ampere service, panel board and feeders		S.F. Floor	1.54	1.54	13.7%
5020	Lighting & Branch Wiring	LED fixtures, receptacles, switches, A.C. and misc. power		S.F. Floor	13.03	13.03	
5030	Communications & Security	Alarm system, internet & phone wiring, intercom system, & emergency ltg.		S.F. Floor	7.54	7.54	
5090	Other Electrical Systems	Emergency generator, 7.5 kW		S.F. Floor	3.15	3.15	
E. EQUIPMENT & FURNISHINGS							
1010	Commercial Equipment	N/A		—	—	—	6.0 %
1020	Institutional Equipment	Exam room casework and contertops		S.F. Floor	10.60	10.60	
1090	Other Equipment	Waste handling recycling tilt truck		S.F. Floor	.50	.50	
2020	Moveable Furnishings	No smoking signage		S.F. Floor	.02	.02	
F. SPECIAL CONSTRUCTION							
1020	Integrated Construction	N/A		—	—	—	0.0 %
1040	Special Facilities	N/A		—	—	—	
G. BUILDING SITEWORK	**N/A**						
					Sub-Total	184.54	**100%**
	CONTRACTOR FEES (General Requirements: 10%, Overhead: 5%, Profit: 10%)				25%	46.10	
	ARCHITECT FEES				9%	20.76	
					Total Building Cost	**251.40**	

Costs per square foot of floor area

Exterior Wall	S.F. Area	2000	3000	4000	6000	8000	10000	12000	14000	16000
	L.F. Perimeter	240	260	280	380	480	560	580	660	740
Brick Veneer	Wood Frame	219.00	193.10	180.15	173.10	**169.50**	166.20	161.05	159.85	159.05
Aluminum Clapboard	Wood Frame	203.30	181.70	170.95	164.80	161.65	158.90	154.75	153.70	153.00
Wood Clapboard	Wood Frame	202.85	181.30	170.65	164.50	161.40	158.65	154.55	153.50	152.85
Wood Shingles	Wood Frame	206.35	183.85	172.70	166.35	163.15	160.30	155.95	154.85	154.15
Precast Concrete	Wood Truss	205.80	183.50	172.40	166.15	162.90	160.05	155.70	154.70	153.95
Face Brick and Concrete Block	Wood Truss	233.85	203.75	188.80	180.90	176.90	173.15	167.05	165.70	164.75
Perimeter Adj., Add or Deduct	Per 100 L.F.	29.75	19.75	14.80	9.85	7.45	5.95	4.95	4.30	3.70
Story Hgt. Adj., Add or Deduct	Per 1 Ft.	4.70	3.35	2.70	2.45	2.35	2.20	1.85	1.85	1.75
***For Basement**, add $ 24.33 per square foot of basement area*										

The above costs were calculated using the basic specifications shown on the facing page. These costs should be adjusted where necessary for design alternatives and owner's requirements.

Common additives

Description	Unit	$ Cost
Closed Circuit Surveillance, one station, camera & monitor	Ea.	1250
For additional camera stations, add	Ea.	565
Laundry Equipment, dryer, gas, 30 lb. capacity	Ea.	4000
Washer, commercial	Ea.	1650
Sauna, prefabricated, complete, 6′ x 9′	Ea.	9550
8′ x 8′	Ea.	10,500
Swimming Pools, Complete, gunite	S.F.	101 - 126
TV Antenna, Master system, 12 outlet	Outlet	229
30 outlet	Outlet	325
100 outlet	Outlet	360

Description	Unit	$ Cost
Commissioning Fees, sustainable commercial construction	S.F.	0.24 - 3.04
Energy Modelling Fees, commercial buildings to 10,000 SF	Ea.	11,000
Greater than 10,000 SF add	S.F.	0.04
Green Bldg Cert Fees for comm construction project reg	Project	900
Photovoltaic Pwr Sys, grid connected, 20 kW (~2400 SF), roof	Ea.	250,600
Green Roofs, 6″ soil depth, w/treated wd edging & sedum mats	S.F.	12.09
10″ Soil depth, with treated wood edging & sedum mats	S.F.	13.58
Greywater Recovery Systems, prepackaged comm, 3060 gal.	Ea.	47,695
4590 gal.	Ea.	59,835
Rainwater Harvest Sys, prepckged comm, 10,000 gal, sys contrller	Ea.	37,050
20,000 gal. w/system controller	Ea.	60,450
30,000 gal. w/system controller	Ea.	96,175
Solar Domestic HW, closed loop, add-on sys, ext heat exchanger	Ea.	10,050
Drainback, hot water system, 120 gal tank	Ea.	12,875
Draindown, hot water system, 120 gal tank	Ea.	13,125

Important: See the Reference Section for Location Factors.

Model costs calculated for a 1 story building with 9' story height and 8,000 square feet of floor area

G Motel, 1 Story

			Unit	Unit Cost	Cost Per S.F.	% Of Sub-Total
A. SUBSTRUCTURE						
1010	Standard Foundations	Poured concrete; strip and spread footings	S.F. Ground	3.19	3.19	12.5%
1020	Special Foundations	N/A	–	–	–	
1030	Slab on Grade	4" reinforced concrete with recycled vapor barrier and granular base	S.F. Slab	5.35	5.35	
2010	Basement Excavation	Site preparation for slab and trench for foundation wall and footing	S.F. Ground	.28	.28	
2020	Basement Walls	4' foundation wall	L.F. Wall	85	6.99	
B. SHELL						
	B10 Superstructure					
1010	Floor Construction	N/A	–	–	–	5.7 %
1020	Roof Construction	Plywood on wood trusses	S.F. Roof	7.22	7.22	
	B20 Exterior Enclosure					
2010	Exterior Walls	Face brick on wood studs with sheathing, insulation and paper *80% of wall*	S.F. Wall	24	10.17	16.2%
2020	Exterior Windows	Wood double hung *20% of wall*	Each	491	3.79	
2030	Exterior Doors	Wood solid core	Each	2380	6.55	
	B30 Roofing					
3010	Roof Coverings	Asphalt strip shingles, 210-235 lbs/Sq., R-10 insul. & weather barrier-recycled	S.F. Roof	8.06	8.06	6.4%
3020	Roof Openings	N/A	–	–	–	
C. INTERIORS						
1010	Partitions	Gypsum bd. and sound deadening bd. on wood studs *9 SF Flr./LF Part.*	S.F. Partition	9.74	8.66	18.4%
1020	Interior Doors	Single leaf hollow core wood, low VOC paint *300 S.F. Floor/Door*	Each	623	2.08	
1030	Fittings	N/A	–	–	–	
2010	Stair Construction	N/A	–	–	–	
3010	Wall Finishes	90% paint, low VOC, 10% ceramic tile	S.F. Surface	1.58	2.80	
3020	Floor Finishes	85% carpet tile, 15% ceramic tile	S.F. Floor	5.45	5.45	
3030	Ceiling Finishes	Painted gypsum board on furring, low VOC paint	S.F. Ceiling	4.31	4.31	
D. SERVICES						
	D10 Conveying					
1010	Elevators & Lifts	N/A	–	–	–	0.0 %
1020	Escalators & Moving Walks	N/A	–	–	–	
	D20 Plumbing					
2010	Plumbing Fixtures	Toilet, low flow, auto sensor, & service fixt., supply & drainage *1 Fixt./90 SF Flr.*	Each	2093	23.25	20.1%
2020	Domestic Water Distribution	Gas fired tankless water heater	S.F. Floor	.38	.38	
2040	Rain Water Drainage	Roof Drains	S.F. Roof	1.88	1.88	
	D30 HVAC					
3010	Energy Supply	N/A	–	–	–	2.4 %
3020	Heat Generating Systems	Included in D3050	–	–	–	
3030	Cooling Generating Systems	N/A	–	–	–	
3050	Terminal & Package Units	Through the wall electric heating and cooling units	S.F. Floor	3.07	3.07	
3090	Other HVAC Sys. & Equipment	N/A	–	–	–	
	D40 Fire Protection					
4010	Sprinklers	Wet pipe sprinkler system	S.F. Floor	4.66	4.66	4.6%
4020	Standpipes	Standpipe, wet, Class III	S.F. Floor	1.12	1.12	
	D50 Electrical					
5010	Electrical Service/Distribution	200 ampere service, panel board and feeders	S.F. Floor	1.23	1.23	12.2%
5020	Lighting & Branch Wiring	LED fixt., daylit. dim., ltg. on/off control, recept., switches and misc. pwr.	S.F. Floor	9.59	9.59	
5030	Communications & Security	Addressable alarm systems	S.F. Floor	2.65	2.65	
5090	Other Electrical Systems	Emergency generator, 7.5 kW, and energy monitoring systems	S.F. Floor	2.02	2.02	
E. EQUIPMENT & FURNISHINGS						
1010	Commercial Equipment	Laundry equipment	S.F. Floor	1.03	1.03	1.5%
1020	Institutional Equipment	N/A	–	–	–	
1090	Other Equipment	Waste handling recycling tilt truck	S.F. Floor	.77	.77	
2020	Moveable Furnishings	No smoking signage	S.F. Floor	.16	.16	
F. SPECIAL CONSTRUCTION						
1020	Integrated Construction	N/A	–	–	–	0.0 %
1040	Special Facilities	N/A	–	–	–	
G. BUILDING SITEWORK	**N/A**					
				Sub-Total	126.71	**100%**
	CONTRACTOR FEES (General Requirements:10%, Overhead: 5%, Profit:10%)			*25%*	31.70	
	ARCHITECT FEES			*7%*	11.09	
				Total Building Cost	**169.50**	

Costs per square foot of floor area

Exterior Wall	S.F. Area	25000	37000	49000	61000	73000	81000	88000	96000	104000
	L.F. Perimeter	433	593	606	720	835	911	978	1054	1074
Decorative Concrete Block	Wood Joists	183.80	179.50	173.25	171.70	170.70	170.05	169.70	169.35	168.30
	Precast Concrete	199.60	195.25	**189.05**	187.50	186.50	185.85	185.50	185.15	184.15
Stucco and Concrete Block	Wood Joists	183.90	179.50	173.10	171.50	170.50	169.90	169.50	169.10	168.05
	Precast Concrete	200.55	196.15	189.75	188.15	187.15	186.50	186.10	185.75	184.70
Wood Clapboard	Wood Frame	180.05	175.90	170.35	168.90	167.95	167.35	167.05	166.65	165.75
Brick Veneer	Wood Frame	183.80	179.45	173.05	171.45	170.40	169.80	169.45	169.10	168.00
Perimeter Adj., Add or Deduct	Per 100 L.F.	5.40	3.65	2.75	2.20	1.85	1.70	1.60	1.45	1.30
Story Hgt. Adj., Add or Deduct	Per 1 Ft.	1.95	1.75	1.35	1.30	1.30	1.30	1.30	1.20	1.15

***For Basement**, add $31.90 per square foot of basement area*

The above costs were calculated using the basic specifications shown on the facing page. These costs should be adjusted where necessary for design alternatives and owner's requirements.

Common additives

Description	Unit	$ Cost
Closed Circuit Surveillance, one station, camera & monitor	Ea.	1250
For additional camera station, add	Ea.	565
Elevators, hydraulic passenger, 2 stops, 2500# capacity	Ea.	71,900
3500# capacity	Ea.	76,900
Additional stop, add	Ea.	7000
Laundry Equipment, dryer, gas, 30 lb. capacity	Ea.	4000
Washer, commercial	Ea.	1650
Sauna, prefabricated, complete, 6′ x 9′	Ea.	9550
8′ x 8′	Ea.	10,500
Swimming Pools, Complete, gunite	S.F.	101 - 126
TV Antenna, Master system, 12 outlet	Outlet	229
30 outlet	Outlet	325
100 outlet	Outlet	360

Description	Unit	$ Cost
Commissioning Fees, sustainable commercial construction	S.F.	0.24 - 3.04
Energy Modelling Fees, commercial buildings to 10,000 SF	Ea.	11,000
Greater than 10,000 SF add	S.F.	0.04
Green Bldg Cert Fees for comm construction project reg	Project	900
Photovoltaic Pwr Sys, grid connected, 20 kW (~2400 SF), roof	Ea.	250,600
Green Roofs, 6″ soil depth, w/treated wd edging & sedum mats	S.F.	12.09
10″ Soil depth, with treated wood edging & sedum mats	S.F.	13.58
Greywater Recovery Systems, prepackaged comm, 3060 gal.	Ea.	47,695
4590 gal.	Ea.	59,835
Rainwater Harvest Sys, prepckged comm, 10,000 gal, sys contrller	Ea.	37,050
20,000 gal. w/system controller	Ea.	60,450
30,000 gal. w/system controller	Ea.	96,175
Solar Domestic HW, closed loop, add-on sys, ext heat exchanger	Ea.	10,050
Drainback, hot water system, 120 gal tank	Ea.	12,875
Draindown, hot water system, 120 gal tank	Ea.	13,125

Model costs calculated for a 3 story building with 9′ story height and 49,000 square feet of floor area

G Motel, 2-3 Story

			Unit	Unit Cost	Cost Per S.F.	% Of Sub-Total
A. SUBSTRUCTURE						
1010	Standard Foundations	Poured concrete; strip and spread footings	S.F. Ground	1.50	.50	3.1%
1020	Special Foundations	N/A	—	—	—	
1030	Slab on Grade	4″ reinforced concrete with recycled vapor barrier and granular base	S.F. Slab	5.35	1.78	
2010	Basement Excavation	Site preparation for slab and trench for foundation wall and footing	S.F. Ground	.16	.05	
2020	Basement Walls	4′ foundation wall	L.F. Wall	85	2.10	
B. SHELL						
	B10 Superstructure					
1010	Floor Construction	Precast concrete plank	S.F. Floor	14.15	9.43	9.3%
1020	Roof Construction	Precast concrete plank	S.F. Roof	11.58	3.86	
	B20 Exterior Enclosure					
2010	Exterior Walls	Decorative concrete block *85% of wall*	S.F. Wall	22	6.26	12.4%
2020	Exterior Windows	Aluminum sliding *15% of wall*	Each	558	1.87	
2030	Exterior Doors	Aluminum and glass doors and entrance with transom	Each	2880	9.52	
	B30 Roofing					
3010	Roof Coverings	Single-ply TPO membrane, 60 mils w/R-20 insul.	S.F. Roof	7.59	2.53	1.8%
3020	Roof Openings	Roof hatches	S.F. Roof	.15	.05	
C. INTERIORS						
1010	Partitions	Concrete block w/foamed-in insul. *7 SF Flr./LF Part.*	S.F. Partition	21	24.23	35.1%
1020	Interior Doors	Wood hollow core, low VOC paint *70 S.F. Floor/Door*	Each	623	8.90	
1030	Fittings	N/A	—	—	—	
2010	Stair Construction	Concrete filled metal pan	Flight	15,375	3.76	
3010	Wall Finishes	90% paint, low VOC, 10% ceramic tile	S.F. Surface	1.57	3.59	
3020	Floor Finishes	85% carpet tile, 5% vinyl composition tile, recycled content, 10% ceramic tile	S.F. Floor	5.45	5.45	
3030	Ceiling Finishes	Textured finish	S.F. Ceiling	4.11	4.11	
D. SERVICES						
	D10 Conveying					
1010	Elevators & Lifts	Two hydraulic passenger elevators	Each	110,005	4.49	3.1%
1020	Escalators & Moving Walks	N/A	—	—	—	
	D20 Plumbing					
2010	Plumbing Fixtures	Toilet, low flow, auto sensor & service fixtures, supply & drain. *1 Fixt./180 SF Flr.*	Each	5188	28.82	20.8%
2020	Domestic Water Distribution	Gas fired, tankless water heater	S.F. Floor	.28	.28	
2040	Rain Water Drainage	Roof drains	S.F. Roof	1.71	.57	
	D30 HVAC					
3010	Energy Supply	N/A	—	—	—	1.9 %
3020	Heat Generating Systems	Included in D3050	—	—	—	
3030	Cooling Generating Systems	N/A	—	—	—	
3050	Terminal & Package Units	Through the wall electric heating and cooling units	S.F. Floor	2.77	2.77	
3090	Other HVAC Sys. & Equipment	N/A	—	—	—	
	D40 Fire Protection					
4010	Sprinklers	Sprinklers, wet, light hazard	S.F. Floor	2.74	2.74	2.1%
4020	Standpipes	Standpipe, wet, Class III	S.F. Floor	.26	.26	
	D50 Electrical					
5010	Electrical Service/Distribution	800 ampere service, panel board and feeders	S.F. Floor	1.15	1.15	10.0%
5020	Lighting & Branch Wiring	LED fixtures, daylt. dim., ltg. on/off, recept., switches and misc. power	S.F. Floor	10.34	10.34	
5030	Communications & Security	Addressable alarm systems and emergency lighting	S.F. Floor	1.80	1.80	
5090	Other Electrical Systems	Emergency generator, 7.5 kW, energy monitoring systems	S.F. Floor	1.02	1.02	
E. EQUIPMENT & FURNISHINGS						
1010	Commercial Equipment	Commercial laundry equipment	S.F. Floor	.35	.35	0.3%
1020	Institutional Equipment	N/A	—	—	—	
1090	Other Equipment	Waste handling recycling tilt truck	S.F. Floor	.08	.08	
2020	Moveable Furnishings	No smoking signage	S.F. Floor	.02	.02	
F. SPECIAL CONSTRUCTION						
1020	Integrated Construction	N/A	—	—	—	0.0 %
1040	Special Facilities	N/A	—	—	—	
G. BUILDING SITEWORK		**N/A**				
				Sub-Total	142.68	**100%**
	CONTRACTOR FEES (General Requirements: 10%, Overhead: 5%, Profit: 10%)			*25%*	35.67	
	ARCHITECT FEES			*6%*	10.70	
				Total Building Cost	**189.05**	

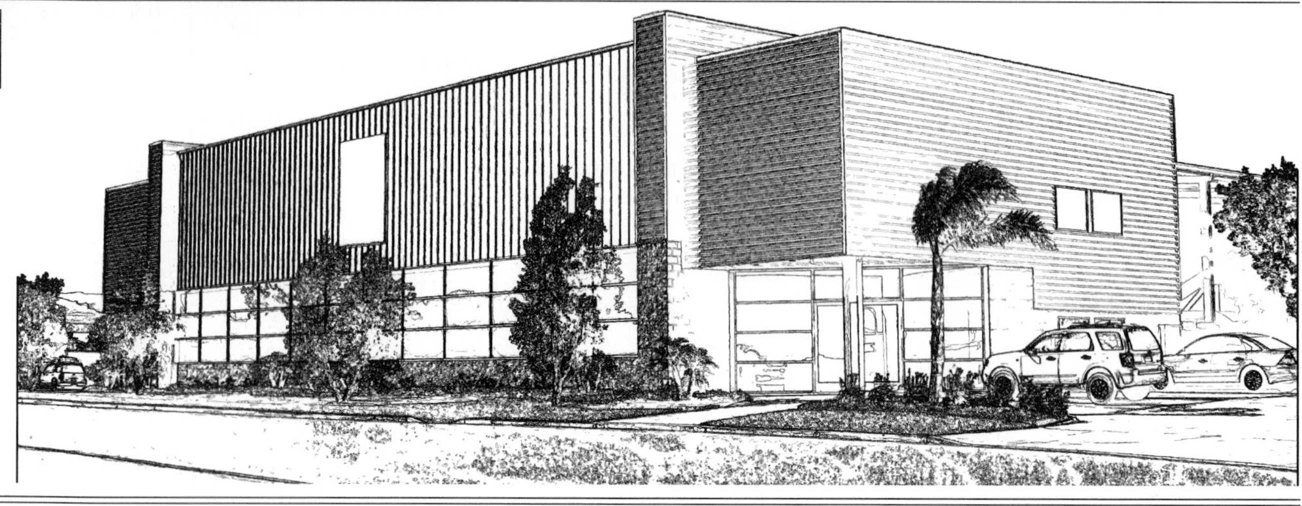

Costs per square foot of floor area

Exterior Wall	S.F. Area	2000	3000	5000	7000	9000	12000	15000	20000	25000
	L.F. Perimeter	220	260	320	360	420	480	520	640	700
Wood Clapboard	Wood Truss	217.95	197.25	178.50	168.85	164.65	159.70	155.95	153.65	150.95
Brick Veneer	Wood Truss	223.80	201.75	181.65	171.25	166.80	161.45	157.35	154.95	152.00
Face Brick and Concrete Block	Wood Truss	240.35	214.90	191.55	179.30	174.20	167.85	163.00	160.15	156.65
	Steel Roof Deck	256.40	229.40	204.55	191.50	186.05	179.35	174.15	171.20	167.40
E.I.F.S. and Metal Studs	Steel Roof Deck	236.00	214.30	194.55	**184.40**	180.00	174.75	170.85	168.45	165.60
Tilt-up Concrete Panels	Steel Roof Deck	242.65	219.75	198.95	188.10	183.55	177.95	173.75	171.25	168.20
Perimeter Adj., Add or Deduct	Per 100 L.F.	28.40	19.00	11.35	8.10	6.40	4.75	3.75	2.85	2.30
Story Hgt. Adj., Add or Deduct	Per 1 Ft.	3.65	2.85	2.10	1.65	1.55	1.35	1.15	1.05	0.90
***For Basement**, add $33.62 per square foot of basement area*										

The above costs were calculated using the basic specifications shown on the facing page. These costs should be adjusted where necessary for design alternatives and owner's requirements.

Common additives

Description	Unit	$ Cost
Closed Circuit Surveillance, one station, camera & monitor	Ea.	1250
For additional camera stations, add	Ea.	565
Directory boards, plastic, glass covered, 30" x 20"	Ea.	570
36" x 48"	Ea.	1500
Aluminum, 36" x 24"	Ea.	725
48" x 32"	Ea.	950
Electronic, wall mounted	S.F.	4600
Pedestal access flr sys w/PLAM cover, comp rm,		
Less than 6000 SF	S.F.	22
Greater than 6000 SF	S.F.	21.50
Office, greater than 6000 S.F.	S.F.	17.55
Uninterruptible power supply, 15 kVA/12.75 kW	kW	1.23

Description	Unit	$ Cost
Commissioning Fees, sustainable commercial construction	S.F.	0.24 - 3.04
Energy Modelling Fees, commercial buildings to 10,000 SF	Ea.	11,000
Greater than 10,000 SF add	S.F.	0.04
Green Bldg Cert Fees for comm construction project reg	Project	900
Photovoltaic Pwr Sys, grid connected, 20 kW (~2400 SF), roof	Ea.	250,600
Green Roofs, 6" soil depth, w/treated wd edging & sedum mats	S.F.	12.09
10" Soil depth, with treated wood edging & sedum mats	S.F.	13.58
Greywater Recovery Systems, prepackaged comm, 3060 gal.	Ea.	47,695
4590 gal.	Ea.	59,835
Rainwater Harvest Sys, prepckged comm, 10,000 gal, sys contrller	Ea.	37,050
20,000 gal. w/system controller	Ea.	60,450
30,000 gal. w/system controller	Ea.	96,175
Solar Domestic HW, closed loop, add-on sys, ext heat exchanger	Ea.	10,050
Drainback, hot water system, 120 gal tank	Ea.	12,875
Draindown, hot water system, 120 gal tank	Ea.	13,125

Important: See the Reference Section for Location Factors.

Model costs calculated for a 1 story building with 12′ story height and 7,000 square feet of floor area

G Office, 1 Story

				Unit	Unit Cost	Cost Per S.F.	% Of Sub-Total
A. SUBSTRUCTURE							
1010	Standard Foundations	Poured concrete; strip and spread footings		S.F. Ground	2.76	2.76	
1020	Special Foundations	N/A		—	—	—	
1030	Slab on Grade	4″ reinforced concrete with recycled vapor barrier and granular base		S.F. Slab	5.35	5.35	9.4%
2010	Basement Excavation	Site preparation for slab and trench for foundation wall and footing		S.F. Ground	.28	.28	
2020	Basement Walls	4′ foundation wall		L.F. Wall	161	4.58	
B. SHELL							
B10 Superstructure							
1010	Floor Construction	N/A		—	—	—	7.3 %
1020	Roof Construction	Steel joists, girders & deck on columns		S.F. Roof	10.07	10.07	
B20 Exterior Enclosure							
2010	Exterior Walls	E.I.F.S. on metal studs		S.F. Wall	16.69	8.24	
2020	Exterior Windows	Aluminum outward projecting	*20% of wall*	Each	703	3.77	11.5%
2030	Exterior Doors	Aluminum and glass, hollow metal		Each	4443	3.81	
B30 Roofing							
3010	Roof Coverings	Single-ply TPO membrane, 45 mils, heat welded seams w/R-20 insul.		S.F. Roof	12.36	12.36	9.1%
3020	Roof Openings	Roof hatch		S.F. Roof	.20	.20	
C. INTERIORS							
1010	Partitions	Gyp. bd. on mtl. studs w/sound attenuation	*20 SF Flr/LF Part.*	S.F. Partition	9.86	4.93	
1020	Interior Doors	Single leaf hollow metal, low VOC paint	*200 S.F. Floor/Door*	Each	1201	6.01	
1030	Fittings	Toilet partitions		S.F. Floor	.88	.88	
2010	Stair Construction	N/A		—	—	—	18.2%
3010	Wall Finishes	60% vinyl wall covering, 40% paint, low VOC		S.F. Surface	1.57	1.57	
3020	Floor Finishes	60% carpet tile, 30% vinyl composition tile, recycled content, 10% ceramic tile		S.F. Floor	4.89	4.89	
3030	Ceiling Finishes	Mineral fiber tile on concealed zee bars		S.F. Ceiling	6.86	6.86	
D. SERVICES							
D10 Conveying							
1010	Elevators & Lifts	N/A		—	—	—	0.0 %
1020	Escalators & Moving Walks	N/A		—	—	—	
D20 Plumbing							
2010	Plumbing Fixtures	Toilet, low flow, auto sensor, & service fixt., supply & drain.	*1 Fixt./1320 SF Flr.*	Each	9557	7.24	
2020	Domestic Water Distribution	Tankless, on-demand water heater, gas/propane		S.F. Floor	.88	.88	6.4%
2040	Rain Water Drainage	Roof drains		S.F. Roof	.77	.77	
D30 HVAC							
3010	Energy Supply	N/A		—	—	—	
3020	Heat Generating Systems	Included in D3050		—	—	—	
3040	Distribution Systems	Enthalpy heat recovery packages		Each	10,325	1.48	15.1 %
3050	Terminal & Package Units	Multizone rooftop air conditioner, SEER 14		S.F. Floor	19.35	19.35	
3090	Other HVAC Sys. & Equipment	N/A		—	—	—	
D40 Fire Protection							
4010	Sprinklers	Sprinkler system, light hazard		S.F. Floor	3.41	3.41	3.6%
4020	Standpipes	Standpipes and hose systems		S.F. Floor	1.53	1.53	
D50 Electrical							
5010	Electrical Service/Distribution	400 ampere service, panel board and feeders		S.F. Floor	1.92	1.92	
5020	Lighting & Branch Wiring	LED fixtures, daylt. dim., ltg. on/off, recept., switches, and A.C. power		S.F. Floor	14.55	14.55	18.6%
5030	Communications & Security	Addressable alarm systems, internet and phone wiring, and emergency lighting		S.F. Floor	6.87	6.87	
5090	Other Electrical Systems	Emergency generator, 7.5 kW, energy monitoring systems		S.F. Floor	2.35	2.35	
E. EQUIPMENT & FURNISHINGS							
1010	Commercial Equipment	N/A		—	—	—	
1020	Institutional Equipment	N/A		—	—	—	0.7 %
1090	Other Equipment	Waste handling recycling tilt truck		S.F. Floor	.88	.88	
2020	Moveable Furnishings	No smoking signage		S.F. Floor	.05	.05	
F. SPECIAL CONSTRUCTION							
1020	Integrated Construction	N/A		—	—	—	0.0 %
1040	Special Facilities	N/A		—	—	—	
G. BUILDING SITEWORK	**N/A**						
					Sub-Total	137.84	**100%**
	CONTRACTOR FEES (General Requirements: 10%, Overhead: 5%, Profit: 10%)				*25%*	34.50	
	ARCHITECT FEES				*7%*	12.06	
					Total Building Cost	**184.40**	

Costs per square foot of floor area

Exterior Wall	S.F. Area	5000	8000	12000	16000	20000	35000	50000	65000	80000
	L.F. Perimeter	220	260	310	330	360	440	550	600	675
Face Brick and Concrete Block	Wood Joists	260.05	227.55	209.00	195.70	188.80	175.10	170.85	166.55	164.60
	Steel Joists	266.90	234.35	215.80	202.50	195.65	181.85	177.65	173.40	171.45
Curtain Wall	Rigid Steel	301.35	261.30	238.35	221.60	212.95	195.65	190.35	184.95	182.40
	Reinforced Concrete	302.80	262.75	239.80	223.05	214.45	197.10	191.75	186.40	183.90
Wood Clapboard	Wood Frame	218.10	193.55	179.60	169.85	164.80	154.80	151.65	148.65	147.20
Brick Veneer	Wood Frame	226.65	199.90	184.60	173.90	168.30	157.20	153.80	150.45	148.80
Perimeter Adj., Add or Deduct	Per 100 L.F.	42.00	26.25	17.50	13.10	10.60	5.95	4.15	3.25	2.60
Story Hgt. Adj., Add or Deduct	Per 1 Ft.	6.90	5.10	4.00	3.25	2.85	1.95	1.65	1.50	1.30

***For Basement**, add $34.87 per square foot of basement area*

The above costs were calculated using the basic specifications shown on the facing page. These costs should be adjusted where necessary for design alternatives and owner's requirements.

Common additives

Description	Unit	$ Cost
Closed Circuit Surveillance, one station, camera and monitor	Ea.	1250
For additional camera stations, add	Ea.	565
Directory Boards, plastic, glass covered, 30" x 20"	Ea.	570
36" x 48"	Ea.	1500
Aluminum, 36" x 24"	Ea.	725
48" x 32"	Ea.	950
Electronic, wall mounted	S.F.	4600
Escalators, 10' rise, 32" wide, glass balustrade	Ea.	142,600
48" wide, glass balustrade	Ea.	150,100
Pedestal access flr sys w/PLAM cover, comp rm,		
Less than 6000 SF	S.F.	22
Greater than 6000 SF	S.F.	21.50
Office, greater than 6000 S.F.	S.F.	17.55
Uninterruptible power supply, 15 kVA/12.75 kW	kW	1.23

Description	Unit	$ Cost
Commissioning Fees, sustainable commercial construction	S.F.	0.24 - 3.04
Energy Modelling Fees, commercial buildings to 10,000 SF	Ea.	11,000
Greater than 10,000 SF add	S.F.	0.04
Green Bldg Cert Fees for comm construction project reg	Project	900
Photovoltaic Pwr Sys, grid connected, 20 kW (~2400 SF), roof	Ea.	250,600
Green Roofs, 6" soil depth, w/treated wd edging & sedum mats	S.F.	12.09
10" Soil depth, with treated wood edging & sedum mats	S.F.	13.58
Greywater Recovery Systems, prepackaged comm, 3060 gal.	Ea.	47,695
4590 gal.	Ea.	59,835
Rainwater Harvest Sys, prepckged comm, 10,000 gal, sys contrller	Ea.	37,050
20,000 gal. w/system controller	Ea.	60,450
30,000 gal. w/system controller	Ea.	96,175
Solar Domestic HW, closed loop, add-on sys, ext heat exchanger	Ea.	10,050
Drainback, hot water system, 120 gal tank	Ea.	12,875
Draindown, hot water system, 120 gal tank	Ea.	13,125

Important: See the Reference Section for Location Factors.

Model costs calculated for a 3 story building with 12′ story height and 20,000 square feet of floor area

G Office, 2-4 Story

			Unit	Unit Cost	Cost Per S.F.	% Of Sub-Total
A. SUBSTRUCTURE						
1010	Standard Foundations	Poured concrete; strip and spread footings	S.F. Ground	7.53	2.51	4.2%
1020	Special Foundations	N/A	—	—	—	
1030	Slab on Grade	4″ reinforced concrete with recycled vapor barrier and granular base	S.F. Slab	5.35	1.78	
2010	Basement Excavation	Site preparation for slab and trench for foundation wall and footing	S.F. Ground	.28	.09	
2020	Basement Walls	4′ foundation wall	L.F. Wall	161	1.79	
B. SHELL						
	B10 Superstructure					
1010	Floor Construction	Open web steel joists, slab form, concrete, columns	S.F. Floor	18.53	12.35	10.1%
1020	Roof Construction	Metal deck, open web steel joists, columns	S.F. Roof	7.17	2.39	
	B20 Exterior Enclosure					
2010	Exterior Walls	Face brick with concrete block backup *80% of wall*	S.F. Wall	36	18.87	16.5%
2020	Exterior Windows	Aluminum outward projecting *20% of wall*	Each	703	3.96	
2030	Exterior Doors	Aluminum and glass, hollow metal	Each	4443	1.34	
	B30 Roofing					
3010	Roof Coverings	Single-ply, TPO membrane, 60 mils, heat welded seams w/R-20 insul.	S.F. Roof	8.37	2.79	1.9%
3020	Roof Openings	N/A	—	—	—	
C. INTERIORS						
1010	Partitions	Gypsum board on metal studs *20 SF Flr/LF Part.*	S.F. Partition	9.95	3.98	20.7%
1020	Interior Doors	Single leaf hollow metal, low VOC paint *200 S.F. Floor/Door*	Each	1201	6.01	
1030	Fittings	Toilet partitions	S.F. Floor	1.84	1.84	
2010	Stair Construction	Concrete filled metal pan	Flight	15,375	5.38	
3010	Wall Finishes	60% vinyl wall covering, 40% paint, low VOC	S.F. Surface	1.56	1.25	
3020	Floor Finishes	60% carpet tile, 30% vinyl composition tile, recycled content, 10% ceramic tile	S.F. Floor	4.89	4.89	
3030	Ceiling Finishes	Mineral fiber tile on concealed zee bars	S.F. Ceiling	6.86	6.86	
D. SERVICES						
	D10 Conveying					
1010	Elevators & Lifts	Two hydraulic passenger elevators	Each	117,600	11.76	8.0%
1020	Escalators & Moving Walks	N/A	—	—	—	
	D20 Plumbing					
2010	Plumbing Fixtures	Toilet, low flow, auto sensor, & service fixt., supply & drain. *1 Fixt./1320 SF Flr.*	Each	5650	4.28	3.6%
2020	Domestic Water Distribution	Gas fired, tankless water heater	S.F. Floor	.20	.20	
2040	Rain Water Drainage	Roof drains	S.F. Roof	2.55	.85	
	D30 HVAC					
3010	Energy Supply	N/A	—	—	—	12.5 %
3020	Heat Generating Systems	Included in D3050	—	—	—	
3040	Distribution Systems	Enthalpy heat recovery packages	Each	18,375	2.76	
3050	Terminal & Package Units	Multizone rooftop air conditioner. SEER 14	S.F. Floor	15.59	15.59	
3090	Other HVAC Sys. & Equipment	N/A	—	—	—	
	D40 Fire Protection					
4010	Sprinklers	Wet pipe sprinkler system	S.F. Floor	3.55	3.55	3.1%
4020	Standpipes	Standpipes and hose systems	S.F. Floor	.99	.99	
	D50 Electrical					
5010	Electrical Service/Distribution	1000 ampere service, panel board and feeders	S.F. Floor	2.69	2.69	19.1%
5020	Lighting & Branch Wiring	LED fixtures, daylt. dim., ltg. on/off, recept., switches, and A.C. power	S.F. Floor	15.19	15.19	
5030	Communications & Security	Addressable alarm systems, internet and phone wiring, & emergency ltg.	S.F. Floor	8.72	8.72	
5090	Other Electrical Systems	Emergency generator, 7.5 kW, UPS, energy monitoring systems	S.F. Floor	1.27	1.27	
E. EQUIPMENT & FURNISHINGS						
1010	Commercial Equipment	N/A	—	—	—	0.2 %
1020	Institutional Equipment	N/A	—	—	—	
1090	Other Equipment	Waste handling recycling tilt truck	S.F. Floor	.31	.31	
2020	Moveable Furnishings	No smoking signage	S.F. Floor	.02	.02	
F. SPECIAL CONSTRUCTION						
1020	Integrated Construction	N/A	—	—	—	0.0 %
1040	Special Facilities	N/A	—	—	—	
G. BUILDING SITEWORK	**N/A**					
			Sub-Total		146.26	**100%**
	CONTRACTOR FEES (General Requirements: 10%, Overhead: 5%, Profit: 10%)			*25%*	36.59	
	ARCHITECT FEES			*7%*	12.80	
			Total Building Cost		**195.65**	

Costs per square foot of floor area

Exterior Wall	S.F. Area	7000	9000	11000	13000	15000	17000	19000	21000	23000
	L.F. Perimeter	240	280	303	325	354	372	397	422	447
Limestone and Concrete Block	Bearing Walls	260.50	245.80	233.30	224.45	218.95	213.45	209.80	206.90	204.45
	Reinforced Concrete	280.35	265.55	253.05	244.25	238.75	233.20	229.60	226.65	224.25
Face Brick and Concrete Block	Bearing Walls	241.60	228.60	218.05	210.65	205.95	201.30	198.30	195.80	193.70
	Reinforced Concrete	261.40	248.40	237.90	230.50	225.70	221.15	218.05	215.55	213.50
Decorative Concrete Block	Bearing Walls	229.95	218.00	208.75	202.15	197.90	193.90	191.15	188.95	187.15
	Reinforced Concrete	249.75	237.80	228.50	221.95	217.70	213.70	210.95	208.75	206.90
Perimeter Adj., Add or Deduct	Per 100 L.F.	29.00	22.55	18.40	15.70	13.50	11.90	10.70	9.60	8.75
Story Hgt. Adj., Add or Deduct	Per 1 Ft.	5.05	4.50	4.00	3.70	3.45	3.10	3.05	2.90	2.80

***For Basement**, add $ 29.65 per square foot of basement area*

The above costs were calculated using the basic specifications shown on the facing page. These costs should be adjusted where necessary for design alternatives and owner's requirements.

Common additives

Description	Unit	$ Cost
Cells Prefabricated, 5'-6' x 7'-8", 7'-8' high	Ea.	12,200
Elevators, hydraulic passenger, 2 stops, 2500# capacity	Ea.	71,900
3500# capacity	Ea.	76,900
Lockers, Steel, Single tier, 60" to 72"	Opng.	223 - 390
Locker bench, lam. maple top only	L.F.	35.50
Pedestals, steel pipe	Ea.	73.50
Safe, Office type, 1 hr rating, 60"x36"x18", double door	Ea.	10,200
Shooting Range, incl. bullet traps,		
Target provisions & controls, excl. struct shell	Ea.	58,000
Sound System, amplifier, 250 watts	Ea.	1950
Speaker, ceiling or wall	Ea.	233
Trumpet	Ea.	445

Description	Unit	$ Cost
Commissioning Fees, sustainable commercial construction	S.F.	0.24 - 3.04
Energy Modelling Fees, commercial buildings to 10,000 SF	Ea.	11,000
Greater than 10,000 SF add	S.F.	0.04
Green Bldg Cert Fees for comm construction project reg	Project	900
Photovoltaic Pwr Sys, grid connected, 20 kW (~2400 SF), roof	Ea.	250,600
Green Roofs, 6" soil depth, w/treated wd edging & sedum mats	S.F.	12.09
10" Soil depth, with treated wood edging & sedum mats	S.F.	13.58
Greywater Recovery Systems, prepackaged comm, 3060 gal.	Ea.	47,695
4590 gal.	Ea.	59,835
Rainwater Harvest Sys, prepckged comm, 10,000 gal, sys contrller	Ea.	37,050
20,000 gal. w/system controller	Ea.	60,450
30,000 gal. w/system controller	Ea.	96,175
Solar Domestic HW, closed loop, add-on sys, ext heat exchanger	Ea.	10,050
Drainback, hot water system, 120 gal tank	Ea.	12,875
Draindown, hot water system, 120 gal tank	Ea.	13,125

Important: See the Reference Section for Location Factors.

Model costs calculated for a 2 story building with 12′ story height and 11,000 square feet of floor area

G Police Station

			Unit	Unit Cost	Cost Per S.F.	% Of Sub-Total
A. SUBSTRUCTURE						
1010	Standard Foundations	Poured concrete; strip and spread footings	S.F. Ground	2.84	1.42	4.3%
1020	Special Foundations	N/A	—	—	—	
1030	Slab on Grade	4″ reinforced concrete with recycled vapor barrier and granular base	S.F. Slab	5.35	2.68	
2010	Basement Excavation	Site preparation for slab and trench for foundation wall and footing	S.F. Ground	.28	.14	
2020	Basement Walls	4′ foundation wall	L.F. Wall	85	3.12	
B. SHELL						
	B10 Superstructure					
1010	Floor Construction	Open web steel joists, slab form, concrete	S.F. Floor	12.14	6.07	5.0%
1020	Roof Construction	Metal deck on open web steel joists	S.F. Roof	4.86	2.43	
	B20 Exterior Enclosure					
2010	Exterior Walls	Limestone with concrete block backup *80% of wall*	S.F. Wall	58	30.41	21.8%
2020	Exterior Windows	Metal horizontal sliding *20% of wall*	Each	498	4.39	
2030	Exterior Doors	Hollow metal	Each	3431	2.50	
	B30 Roofing					
3010	Roof Coverings	Single-ply TPO membrane, 60 mils, heat welded w/R-20 insul.	S.F. Roof	8.06	4.03	2.4%
3020	Roof Openings	N/A	—	—	—	
C. INTERIORS						
1010	Partitions	Concrete block w/foamed-in insul. *20 SF Flr./LF Part.*	S.F. Partition	10.60	5.30	17.4%
1020	Interior Doors	Single leaf kalamein fire door, low VOC paint *200 S.F. Floor/Door*	Each	1201	6.01	
1030	Fittings	Toilet partitions	S.F. Floor	.95	.95	
2010	Stair Construction	Concrete filled metal pan	Flight	18,425	3.35	
3010	Wall Finishes	90% paint, low VOC, 10% ceramic tile	S.F. Surface	2.64	2.64	
3020	Floor Finishes	70% vinyl composition tile, recycled content, 20% carpet tile, 10% ceramic tile	S.F. Floor	4.72	4.72	
3030	Ceiling Finishes	Mineral fiber tile on concealed zee bars	S.F. Ceiling	6.86	6.86	
D. SERVICES						
	D10 Conveying					
1010	Elevators & Lifts	One hydraulic passenger elevator	Each	85,580	7.78	4.5%
1020	Escalators & Moving Walks	N/A	—	—	—	
	D20 Plumbing					
2010	Plumbing Fixtures	Toilet, low flow, auto sensor, & service fixt., supply & drainage *1 Fixt./580 SF Flr.*	Each	5533	9.54	7.2%
2020	Domestic Water Distribution	Electric water heater, point-of-use, energy saver	S.F. Floor	1.17	1.17	
2040	Rain Water Drainage	Roof drains	S.F. Roof	3.10	1.55	
	D30 HVAC					
3010	Energy Supply	N/A	—	—	—	12.1 %
3020	Heat Generating Systems	N/A	—	—	—	
3040	Distribution Systems	Enthalpy heat recovery packages	Each	15,500	1.41	
3050	Terminal & Package Units	Multizone rooftop air conditioner, SEER 14	S.F. Floor	19.35	19.35	
3090	Other HVAC Sys. & Equipment	N/A	—	—	—	
	D40 Fire Protection					
4010	Sprinklers	Wet pipe sprinkler system	S.F. Floor	3.64	3.64	2.8%
4020	Standpipes	Standpipe	S.F. Floor	1.08	1.08	
	D50 Electrical					
5010	Electrical Service/Distribution	400 ampere service, panel board and feeders	S.F. Floor	.96	.96	12.8%
5020	Lighting & Branch Wiring	LED fixtures, daylt. dim., ltg. on/off cntrl., recept., switches, & A.C. pwr.	S.F. Floor	12.35	12.35	
5030	Communications & Security	Addressable alarm systems, internet wiring, intercom &emergency ltg.	S.F. Floor	6.99	6.99	
5090	Other Electrical Systems	Emergency generator, 15 kW, energy monitoring systems	S.F. Floor	1.65	1.65	
E. EQUIPMENT & FURNISHINGS						
1020	Institutional Equipment	Lockers, detention rooms, cells, gasoline dispensers	S.F. Floor	13.12	13.12	9.5%
1030	Vehicular Equipment	Gasoline dispenser system	S.F. Floor	2.63	2.63	
1090	Other Equipment	Waste handling recycling tilt truck	S.F. Floor	.56	.56	
2020	Moveable Furnishings	No smoking signage	S.F. Floor	.04	.04	
F. SPECIAL CONSTRUCTION						
1020	Integrated Construction	N/A	—	—	—	0.0 %
1040	Special Facilities	N/A	—	—	—	
G. BUILDING SITEWORK		**N/A**				

	Sub-Total	171.21	**100%**
CONTRACTOR FEES (General Requirements: 10%, Overhead: 5%, Profit: 10%)	*25%*	42.83	
ARCHITECT FEES	*9%*	19.26	
	Total Building Cost	**233.30**	

Costs per square foot of floor area

Exterior Wall	S.F. Area	2000	2800	3500	4200	5000	5800	6500	7200	8000
	L.F. Perimeter	180	212	240	268	300	314	336	344	368
Wood Clapboard	Wood Frame	265.15	251.35	244.55	239.95	**236.30**	231.75	229.60	226.70	225.05
Brick Veneer	Wood Frame	275.90	260.40	252.75	247.55	243.45	238.20	235.75	232.40	230.55
Face Brick and Concrete Block	Wood Joists	288.90	271.35	262.65	256.75	252.10	246.00	243.25	239.25	237.25
	Steel Joists	282.70	265.15	256.50	250.60	245.95	239.80	237.05	233.10	231.05
Stucco and Concrete Block	Wood Joists	278.70	263.15	255.45	250.20	246.15	240.90	238.45	235.05	233.25
	Steel Joists	269.85	254.35	246.70	241.45	237.35	232.10	229.70	226.30	224.45
Perimeter Adj., Add or Deduct	Per 100 L.F.	30.50	21.80	17.40	14.50	12.20	10.55	9.40	8.45	7.65
Story Hgt. Adj., Add or Deduct	Per 1 Ft.	3.25	2.75	2.50	2.30	2.15	1.90	1.85	1.70	1.70

For Basement, add $36.39 per square foot of basement area

The above costs were calculated using the basic specifications shown on the facing page. These costs should be adjusted where necessary for design alternatives and owner's requirements.

Common additives

Description	Unit	$ Cost
Bar, Front Bar	L.F.	445
Back bar	L.F.	355
Booth, Upholstered, custom straight	L.F.	259 - 480
"L" or "U" shaped	L.F.	268 - 455
Fireplace, brick, excl. chimney or foundation, 30"x29" opening	Ea.	2900
Chimney, standard brick, single flue, 16" x 20"	V.L.F.	100
2 Flue, 20" x 24"	V.L.F.	148
Kitchen Equipment		
Broiler	Ea.	4400
Coffee urn, twin 6 gallon	Ea.	2900
Cooler, 6 ft. long	Ea.	3725
Dishwasher, 10-12 racks per hr.	Ea.	3975
Food warmer, counter, 1.2 KW	Ea.	785
Freezer, 44 C.F., reach-in	Ea.	5400
Ice cube maker, 50 lb. per day	Ea.	2050
Range with 1 oven	Ea.	3200
Refrigerators, prefabricated, walk-in, 7'-6" high, 6' x 6'	S.F.	137
10' x 10'	S.F.	107
12' x 14'	S.F.	94
12' x 20'	S.F.	134

Description	Unit	$ Cost
Commissioning Fees, sustainable commercial construction	S.F.	0.24 - 3.04
Energy Modelling Fees, commercial buildings to 10,000 SF	Ea.	11,000
Greater than 10,000 SF add	S.F.	0.04
Green Bldg Cert Fees for comm construction project reg	Project	900
Photovoltaic Pwr Sys, grid connected, 20 kW (~2400 SF), roof	Ea.	250,600
Green Roofs, 6" soil depth, w/treated wd edging & sedum mats	S.F.	12.09
10" Soil depth, with treated wood edging & sedum mats	S.F.	13.58
Greywater Recovery Systems, prepackaged comm, 3060 gal.	Ea.	47,695
4590 gal.	Ea.	59,835
Rainwater Harvest Sys, prepckged comm, 10,000 gal, sys contrller	Ea.	37,050
20,000 gal. w/system controller	Ea.	60,450
30,000 gal. w/system controller	Ea.	96,175
Solar Domestic HW, closed loop, add-on sys, ext heat exchanger	Ea.	10,050
Drainback, hot water system, 120 gal tank	Ea.	12,875
Draindown, hot water system, 120 gal tank	Ea.	13,125

Important: See the Reference Section for Location Factors.

Model costs calculated for a 1 story building with 12′ story height and 5,000 square feet of floor area

G Restaurant

				Unit	Unit Cost	Cost Per S.F.	% Of Sub-Total
A. SUBSTRUCTURE							
1010	Standard Foundations	Poured concrete; strip and spread footings		S.F. Ground	2.60	2.60	7.7%
1020	Special Foundations	N/A		—	—	—	
1030	Slab on Grade	4″ reinforced concrete with recycled vapor barrier and granular base		S.F. Slab	5.35	5.35	
2010	Basement Excavation	Site preparation for slab and trench for foundation wall and footing		S.F. Ground	.49	.49	
2020	Basement Walls	4′ foundation wall		L.F. Wall	85	5.10	
B. SHELL							
	B10 Superstructure						
1010	Floor Construction	Wood columns		S.F. Floor	.83	.83	5.0%
1020	Roof Construction	Plywood on wood truss (pitched)		S.F. Roof	8.07	8.07	
	B20 Exterior Enclosure						
2010	Exterior Walls	Cedar siding on wood studs with insulation	*70% of wall*	S.F. Wall	15.73	7.93	14.1%
2020	Exterior Windows	Storefront windows	*30% of wall*	Each	51	10.33	
2030	Exterior Doors	Aluminum and glass doors and entrance with transom		Each	6658	6.67	
	B30 Roofing						
3010	Roof Coverings	Cedar shingles with flashing (pitched); rigid fiberglass insulation		S.F. Roof	6.97	6.97	4.0%
3020	Roof Openings	Skylights		S.F. Roof	.10	.10	
C. INTERIORS							
1010	Partitions	Gypsum board on wood studs	*25 SF Floor/LF Part.*	S.F. Partition	11.53	4.61	14.9%
1020	Interior Doors	Hollow core wood, low VOC paint	*250 S.F. Floor/Door*	Each	623	2.49	
1030	Fittings	Toilet partitions		S.F. Floor	1.22	1.22	
2010	Stair Construction	N/A		—	—	—	
3010	Wall Finishes	75% paint, low VOC, 25% ceramic tile		S.F. Surface	2.50	2	
3020	Floor Finishes	65% carpet, 35% quarry tile		S.F. Floor	9.08	9.08	
3030	Ceiling Finishes	Mineral fiber tile on concealed zee bars		S.F. Ceiling	6.86	6.86	
D. SERVICES							
	D10 Conveying						
1010	Elevators & Lifts	N/A		—	—	—	0.0 %
1020	Escalators & Moving Walks	N/A		—	—	—	
	D20 Plumbing						
2010	Plumbing Fixtures	Kitchen, toilet, low flow, auto sensor, & service fixt., supply & drain.	*1 Fixt./335 SF Flr.*	Each	3770	10.62	8.7%
2020	Domestic Water Distribution	Tankless, on-demand water heater gas/propane		S.F. Floor	2.58	2.58	
2040	Rain Water Drainage	Roof drains		S.F. Roof	2.14	2.14	
	D30 HVAC						
3010	Energy Supply	N/A		—	—	—	26.7 %
3020	Heat Generating Systems	Included in D3050		—	—	—	
3040	Distribution Systems	Enthalpy heat recovery packages and kitchen exhaust/make-up air system		Each	44,375	8.88	
3050	Terminal & Package Units	Multizone rooftop air conditioner, SEER 14		S.F. Floor	38	38.35	
3090	Other HVAC Sys. & Equipment	N/A		—	—	—	
	D40 Fire Protection						
4010	Sprinklers	Sprinklers, light hazard and ordinary hazard kitchen		S.F. Floor	8.64	8.64	6.1%
4020	Standpipes	Standpipe		S.F. Floor	2.15	2.15	
	D50 Electrical						
5010	Electrical Service/Distribution	400 ampere service, panel board and feeders		S.F. Floor	2.20	2.20	12.1%
5020	Lighting & Branch Wiring	LED fixtures, daylt. dim., ltg. on/off, receptacles, switches, and A.C. power		S.F. Floor	12.27	12.27	
5030	Communications & Security	Addressable alarm systems and emergency lighting		S.F. Floor	3.49	3.49	
5090	Other Electrical Systems	Emergency generator, 15 kW and energy monitoring systems		S.F. Floor	3.35	3.35	
E. EQUIPMENT & FURNISHINGS							
1010	Commercial Equipment	N/A		—	—	—	0.7 %
1020	Institutional Equipment	N/A		—	—	—	
1090	Other Equipment	Waste handling recycling tilt truck		S.F. Floor	1.24	1.24	
2020	Moveable Furnishings	No smoking signage		S.F. Floor	.06	.06	
F. SPECIAL CONSTRUCTION							
1020	Integrated Construction	N/A		—	—	—	0.0 %
1040	Special Facilities	N/A		—	—	—	
G. BUILDING SITEWORK	**N/A**						
					Sub-Total	176.67	**100%**
	CONTRACTOR FEES (General Requirements: 10%, Overhead: 5%, Profit: 10%)				*25%*	44.17	
	ARCHITECT FEES				*7%*	15.46	
					Total Building Cost	**236.30**	

Costs per square foot of floor area

Exterior Wall	S.F. Area	25000	30000	35000	40000	45000	50000	55000	60000	65000
	L.F. Perimeter	900	1050	1200	1350	1510	1650	1800	1970	2100
Face Brick and Concrete Block	Rigid Steel	192.15	190.00	188.50	187.30	186.65	185.65	185.10	184.95	184.15
	Bearing Walls	184.25	182.35	181.05	180.05	179.50	178.60	178.10	178.10	177.30
Stucco and Concrete Block	Rigid Steel	185.75	183.75	182.35	181.30	180.65	179.75	179.25	179.15	178.35
	Bearing Walls	179.35	177.30	175.90	174.85	174.25	173.35	172.85	172.65	172.00
Decorative Concrete Block	Rigid Steel	185.60	183.65	182.20	181.15	180.50	179.65	179.10	178.95	178.25
	Bearing Walls	179.20	177.20	175.80	174.70	174.10	173.20	172.70	172.50	171.80
Perimeter Adj., Add or Deduct	Per 100 L.F.	4.75	3.95	3.35	2.95	2.60	2.35	2.10	1.95	1.85
Story Hgt. Adj., Add or Deduct	Per 1 Ft.	1.75	1.75	1.65	1.65	1.65	1.65	1.60	1.65	1.60
***For Basement**, add $ 26.89 per square foot of basement area*										

The above costs were calculated using the basic specifications shown on the facing page. These costs should be adjusted where necessary for design alternatives and owner's requirements.

Common additives

Description	Unit	$ Cost
Bleachers, Telescoping, manual, 16-20 tier	Seat	315 - 405
21-30 tier	Seat	330 - 520
For power operation, add	Seat	64 - 103
Carrels Hardwood	Ea.	715 - 1900
Clock System, 20 room	Ea.	17,900
50 room	Ea.	42,300
Kitchen Equipment		
Broiler	Ea.	4400
Cooler, 6 ft. long, reach-in	Ea.	3725
Dishwasher, 10-12 racks per hr.	Ea.	3975
Food warmer, counter, 1.2 KW	Ea.	785
Freezer, 44 C.F., reach-in	Ea.	5400
Ice cube maker, 50 lb. per day	Ea.	2050
Range with 1 oven	Ea.	3200
Lockers, Steel, single tier, 60" to 72"	Opng.	223 - 390
2 tier, 60" to 72" total	Opng.	135 - 167
Locker bench, lam. maple top only	L.F.	35.50
Pedestals, steel pipe	Ea.	73.50
Seating, auditorium chair, veneer back, padded seat	Ea.	375
Classroom, movable chair & desk	Set	81 - 171
Lecture hall, pedestal type	Ea.	345 - 645

Description	Unit	$ Cost
Sound System, amplifier, 250 watts	Ea.	1950
Speaker, ceiling or wall	Ea.	233
Commissioning Fees, sustainable institutional construction	S.F.	0.58 - 2.47
Energy Modelling Fees, academic buildings to 10,000 SF	Ea.	9000
Greater than 10,000 SF add	S.F.	0.20
Green Bldg Cert Fees for school construction project reg	Project	900
Photovoltaic Pwr Sys, grid connected, 20 kW (~2400 SF), roof	Ea.	250,600
Green Roofs, 6" soil depth, w/treated wd edging & sedum mats	S.F.	12.09
10" Soil depth, with treated wood edging & sedum mats	S.F.	13.58
Greywater Recovery Systems, prepackaged comm, 3060 gal.	Ea.	47,695
4590 gal.	Ea.	59,835
Rainwater Harvest Sys, prepckged comm, 10,000 gal, sys contrller	Ea.	37,050
20,000 gal. w/system controller	Ea.	60,450
30,000 gal. w/system controller	Ea.	96,175
Solar Domestic HW, closed loop, add-on sys, ext heat exchanger	Ea.	10,050
Drainback, hot water system, 120 gal tank	Ea.	12,875
Draindown, hot water system, 120 gal tank	Ea.	13,125

Important: See the Reference Section for Location Factors.

Model costs calculated for a 1 story building with 15′ story height and 45,000 square feet of floor area

G School, Elementary

				Unit	Unit Cost	Cost Per S.F.	% Of Sub-Total
A. SUBSTRUCTURE							
1010	Standard Foundations	Poured concrete; strip and spread footings		S.F. Ground	4.96	4.96	11.4%
1020	Special Foundations	N/A		—	—	—	
1030	Slab on Grade	4″ reinforced concrete with recycled vapor barrier and granular base		S.F. Slab	5.35	5.35	
2010	Basement Excavation	Site preparation for slab and trench for foundation wall and footing		S.F. Ground	.16	.16	
2020	Basement Walls	4′ foundation wall		L.F. Wall	85	4.85	
B. SHELL							
	B10 Superstructure						
1010	Floor Construction	N/A		—	—	—	3.3 %
1020	Roof Construction	Metal deck on open web steel joists		S.F. Roof	4.41	4.41	
	B20 Exterior Enclosure						
2010	Exterior Walls	Face brick with concrete block backup	*70% of wall*	S.F. Wall	36	12.82	13.6%
2020	Exterior Windows	Steel outward projecting	*25% of wall*	Each	703	4.62	
2030	Exterior Doors	Metal and glass	*5% of wall*	Each	4870	.87	
	B30 Roofing						
3010	Roof Coverings	Single-ply TPO membrane, 60 mils, w/flashing; polyiso. insulation		S.F. Roof	9.76	9.76	7.3%
3020	Roof Openings	N/A		—	—	—	
C. INTERIORS							
1010	Partitions	Concrete block w/foamed-in insul.	*20 SF Flr./LF Part.*	S.F. Partition	10.88	5.44	20.0%
1020	Interior Doors	Single leaf kalamein fire doors, low VOC paint	*700 S.F. Floor/Door*	Each	1201	1.72	
1030	Fittings	Toilet partitions		S.F. Floor	2	2	
2010	Stair Construction	N/A		—	—	—	
3010	Wall Finishes	75% paint, low VOC, 15% glazed coating, 10% ceramic tile		S.F. Surface	4.55	4.55	
3020	Floor Finishes	65% vinyl composition tile, recycled content, 25% carpet tile, 10% terrazzo		S.F. Floor	6.29	6.29	
3030	Ceiling Finishes	Mineral fiber tile on concealed zee bars		S.F. Ceiling	6.86	6.86	
D. SERVICES							
	D10 Conveying						
1010	Elevators & Lifts	N/A		—	—	—	0.0 %
1020	Escalators & Moving Walks	N/A		—	—	—	
	D20 Plumbing						
2010	Plumbing Fixtures	Kitchen, toilet, low flow, auto sensor, & service fixt., supply & drain.	*1 Fixt./625 SF Flr.*	Each	9550	15.28	12.7%
2020	Domestic Water Distribution	Gas fired, tankless water heater		S.F. Floor	.15	.15	
2040	Rain Water Drainage	Roof drains		S.F. Roof	1.58	1.58	
	D30 HVAC						
3010	Energy Supply	N/A		—	—	—	14.4 %
3020	Heat Generating Systems	N/A		—	—	—	
3040	Distribution Systems	Enthalpy heat recovery packages		Each	32,100	.71	
3050	Terminal & Package Units	Multizone rooftop air conditioner		S.F. Floor	18.55	18.55	
3090	Other HVAC Sys. & Equipment	N/A		—	—	—	
	D40 Fire Protection						
4010	Sprinklers	Sprinklers, light hazard		S.F. Floor	2.54	2.54	2.2%
4020	Standpipes	Standpipe		S.F. Floor	.42	.42	
	D50 Electrical						
5010	Electrical Service/Distribution	800 ampere service, panel board and feeders		S.F. Floor	.81	.81	14.9%
5020	Lighting & Branch Wiring	LED fixtures, daylt. dim., ltg. on/off, recept., switches, and A.C. power		S.F. Floor	13.24	13.24	
5030	Communications & Security	Addressable alarm systems, internet wiring, comm. systems & emerg. ltg.		S.F. Floor	4.92	4.92	
5090	Other Electrical Systems	Emergency generator, 15 kW, energy monitoring systems		S.F. Floor	1	1	
E. EQUIPMENT & FURNISHINGS							
1010	Commercial Equipment	N/A		—	—	—	0.3 %
1020	Institutional Equipment	Chalkboards		S.F. Floor	.25	.25	
1090	Other Equipment	Waste handling recycling tilt truck		S.F. Floor	.08	.08	
2020	Moveable Furnishings	No smoking signage		S.F. Floor	.01	.01	
F. SPECIAL CONSTRUCTION							
1020	Integrated Construction	N/A		—	—	—	0.0 %
1040	Special Facilities	N/A		—	—	—	
G. BUILDING SITEWORK	**N/A**						
					Sub-Total	134.20	**100%**
	CONTRACTOR FEES (General Requirements: 10%, Overhead: 5%, Profit: 10%)				*25%*	33.56	
	ARCHITECT FEES				*7%*	11.74	
					Total Building Cost	**179.50**	

Costs per square foot of floor area

Exterior Wall	S.F. Area	50000	65000	80000	95000	110000	125000	140000	155000	170000
	L.F. Perimeter	850	1060	1280	1490	1700	1920	2140	2340	2560
Face Brick and Concrete Block	Rigid Steel	186.45	184.15	182.95	181.95	181.15	180.75	180.50	179.95	179.75
	Bearing Walls	178.95	176.60	175.35	174.40	173.65	173.20	172.95	172.35	172.20
Stone and Concrete Block	Rigid Steel	192.45	189.90	188.60	187.50	186.65	186.20	185.85	185.25	185.05
	Bearing Walls	184.90	182.30	181.00	179.90	179.05	178.60	178.30	177.65	177.50
Decorative Concrete Block	Rigid Steel	179.80	177.75	176.70	175.80	175.15	174.75	174.50	174.00	173.90
	Bearing Walls	172.50	170.45	169.35	168.45	167.80	167.40	167.20	166.65	166.50
Perimeter Adj., Add or Deduct	Per 100 L.F.	4.05	3.10	2.50	2.15	1.85	1.60	1.40	1.30	1.20
Story Hgt. Adj., Add or Deduct	Per 1 Ft.	2.05	1.95	1.90	1.85	1.85	1.85	1.80	1.75	1.80

***For Basement**, add $36.70 per square foot of basement area*

The above costs were calculated using the basic specifications shown on the facing page. These costs should be adjusted where necessary for design alternatives and owner's requirements.

Common additives

Description	Unit	$ Cost
Bleachers, Telescoping, manual, 16-20 tier	Seat	315 - 405
21-30 tier	Seat	330 - 520
For power operation, add	Seat	64 - 103
Carrels Hardwood	Ea.	715 - 1900
Clock System, 20 room	Ea.	17,900
50 room	Ea.	42,300
Elevators, hydraulic passenger, 2 stops, 2500# capacity	Ea.	71,900
3500# capacity	Ea.	76,900
Kitchen Equipment		
Broiler	Ea.	4400
Cooler, 6 ft. long, reach-in	Ea.	3725
Dishwasher, 10-12 racks per hr.	Ea.	3975
Food warmer, counter, 1.2 KW	Ea.	785
Freezer, 44 C.F., reach-in	Ea.	5400
Lockers, Steel, single tier, 60" to 72"	Opng.	223 - 390
2 tier, 60" to 72" total	Opng.	135 - 167
Locker bench, lam. maple top only	L.F.	35.50
Pedestals, steel pipe	Ea.	73.50

Description	Unit	$ Cost
Seating, auditorium chair, veneer back, padded seat	Ea.	375
Upholstered, spring seat	Ea.	325
Classroom, movable chair & desk	Set	81 - 171
Lecture hall, pedestal type	Ea.	345 - 645
Sound System, amplifier, 250 watts	Ea.	1950
Speaker, ceiling or wall	Ea.	233
Commissioning Fees, sustainable institutional construction	S.F.	0.58 - 2.47
Energy Modelling Fees, academic buildings to 10,000 SF	Ea.	9000
Greater than 10,000 SF add	S.F.	0.20
Green Bldg Cert Fees for school construction project reg	Project	900
Photovoltaic Pwr Sys, grid connected, 20 kW (~2400 SF), roof	Ea.	250,600
Green Roofs, 6" soil depth, w/treated wd edging & sedum mats	S.F.	12.09
10" Soil depth, with treated wood edging & sedum mats	S.F.	13.58
Greywater Recovery Systems, prepackaged comm, 3060 gal.	Ea.	47,695
4590 gal.	Ea.	59,835
Rainwater Harvest Sys, prepckged comm, 10,000 gal, sys contrller	Ea.	37,050
20,000 gal. w/system controller	Ea.	60,450
30,000 gal. w/system controller	Ea.	96,175
Solar Domestic HW, closed loop, add-on sys, ext heat exchanger	Ea.	10,050
Drainback, hot water system, 120 gal tank	Ea.	12,875
Draindown, hot water system, 120 gal tank	Ea.	13,125

Important: See the Reference Section for Location Factors.

Model costs calculated for a 2 story building with 15′ story height and 110,000 square feet of floor area

G School, Jr High, 2-3 Story

				Unit	Unit Cost	Cost Per S.F.	% Of Sub-Total
A. SUBSTRUCTURE							
1010	Standard Foundations	Poured concrete; strip and spread footings		S.F. Ground	2.46	1.23	3.9%
1020	Special Foundations	N/A		—	—	—	
1030	Slab on Grade	4″ reinforced concrete with recycled vapor barrier and granular base		S.F. Slab	5.35	2.68	
2010	Basement Excavation	Site preparation for slab and trench for foundation wall and footing		S.F. Ground	.16	.08	
2020	Basement Walls	4′ foundation wall		L.F. Wall	85	1.31	
B. SHELL							
	B10 Superstructure						
1010	Floor Construction	Open web steel joists, slab form, concrete, columns		S.F. Floor	28	13.86	14.5%
1020	Roof Construction	Metal deck, open web steel joists, columns		S.F. Roof	11.64	5.82	
	B20 Exterior Enclosure						
2010	Exterior Walls	Face brick with concrete block backup	*75% of wall*	S.F. Wall	36	12.66	15.0%
2020	Exterior Windows	Window wall	*25% of wall*	Each	61	6.73	
2030	Exterior Doors	Double aluminum & glass		Each	3338	.88	
	B30 Roofing						
3010	Roof Coverings	Single-ply TPO membrane & standing seam metal; polyiso. insulation		S.F. Roof	13.78	6.89	5.1%
3020	Roof Openings	Roof hatches		S.F. Roof	.06	.03	
C. INTERIORS							
1010	Partitions	Concrete block w/foamed-in insul.	*20 SF Flr./LF Part.*	S.F. Partition	11.67	7	22.9%
1020	Interior Doors	Single leaf kalamein fire doors, low VOC paint	*750 S.F. Floor/Door*	Each	1201	1.60	
1030	Fittings	Toilet partitions, chalkboards		S.F. Floor	1.73	1.73	
2010	Stair Construction	Concrete filled metal pan		Flight	15,375	.84	
3010	Wall Finishes	50% paint, low VOC, 40% glazed coatings, 10% ceramic tile		S.F. Surface	3.54	4.25	
3020	Floor Finishes	50% vinyl comp. tile, recycled content, 30% carpet tile, 20% terrrazzo		S.F. Floor	8.71	8.71	
3030	Ceiling Finishes	Mineral fiberboard on concealed zee bars		S.F. Ceiling	6.86	6.86	
D. SERVICES							
	D10 Conveying						
1010	Elevators & Lifts	One hydraulic passenger elevator		Each	85,800	.78	0.6%
1020	Escalators & Moving Walks	N/A		—	—	—	
	D20 Plumbing						
2010	Plumbing Fixtures	Kitchen, toilet, low flow, auto sensor, & service fixt., supply & drain.	*1 Fixt./1170 SF Flr.*	Each	7137	6.10	5.4%
2020	Domestic Water Distribution	Gas fired, tankless water heater		S.F. Floor	.35	.35	
2040	Rain Water Drainage	Roof drains		S.F. Roof	1.84	.92	
	D30 HVAC						
3010	Energy Supply	N/A		—	—	—	14.2 %
3020	Heat Generating Systems	Included in D3050		—	—	—	
3040	Distribution Systems	Enthalpy heat recovery packages		Each	39,100	.71	
3050	Terminal & Package Units	Multizone rooftop air conditioner, SEER 14		S.F. Floor	18.55	18.55	
3090	Other HVAC Sys. & Equipment	N/A		—	—	—	
	D40 Fire Protection						
4010	Sprinklers	Sprinklers, light hazard	*10% of area*	S.F. Floor	2.20	2.20	1.9%
4020	Standpipes	Standpipe, wet, Class III		S.F. Floor	.42	.42	
	D50 Electrical						
5010	Electrical Service/Distribution	1600 ampere service, panel board and feeders		S.F. Floor	.71	.71	13.5%
5020	Lighting & Branch Wiring	LED fixtures, daylt. dim., ltg. on/off, recept., switches, and A.C. power		S.F. Floor	12.09	12.09	
5030	Communications & Security	Addressable alarm systems, internet wiring, comm. systems and emerg. ltg.		S.F. Floor	4.67	4.67	
5090	Other Electrical Systems	Emergency generator, 100 kW, and energy monitoring systems		S.F. Floor	.76	.76	
E. EQUIPMENT & FURNISHINGS							
1010	Commercial Equipment	N/A		—	—	—	3.0 %
1020	Institutional Equipment	Laboratory casework and counters		S.F. Floor	2.70	2.70	
1090	Other Equipment	Waste handlg. recyc. tilt truck, built-in athletic equip., bleachers & backstps.		S.F. Floor	1.32	1.32	
2020	Moveable Furnishings	No smoking signage		S.F. Floor	.02	.02	
F. SPECIAL CONSTRUCTION							
1020	Integrated Construction	N/A		—	—	—	0.0 %
1040	Special Facilities	N/A		—	—	—	
G. BUILDING SITEWORK		**N/A**					
					Sub-Total	135.46	**100%**
	CONTRACTOR FEES (General Requirements: 10%, Overhead: 5%, Profit: 10%)				*25%*	33.84	
	ARCHITECT FEES				*7%*	11.85	
					Total Building Cost	**181.15**	

Costs per square foot of floor area

Exterior Wall	S.F. Area	12000	16000	20000	26000	32000	38000	44000	52000	60000
	L.F. Perimeter	450	510	570	650	716	780	840	920	1020
Face Brick and Concrete Block	Rigid Steel	174.15	163.60	157.35	150.95	146.45	143.25	140.85	138.60	137.25
	Bearing Walls	169.10	158.55	152.30	145.90	141.40	138.20	135.85	133.50	132.20
Stucco and Concrete Block	Rigid Steel	164.45	155.35	149.95	144.45	140.65	137.95	135.95	134.00	132.85
	Bearing Walls	162.10	152.95	147.55	142.05	138.30	135.60	133.55	131.65	130.45
Precast Concrete	Rigid Steel	190.15	177.20	169.50	161.60	156.00	152.00	149.05	146.10	144.50
Metal Sandwich Panel	Rigid Steel	174.10	163.50	157.25	150.90	146.45	143.25	140.85	138.55	137.25
Perimeter Adj., Add or Deduct	Per 100 L.F.	10.45	7.80	6.15	4.80	3.85	3.30	2.85	2.35	2.05
Story Hgt. Adj., Add or Deduct	Per 1 Ft.	2.20	1.85	1.65	1.40	1.30	1.15	1.10	1.00	1.00

***For Basement**, add $26.94 per square foot of basement area*

The above costs were calculated using the basic specifications shown on the facing page. These costs should be adjusted where necessary for design alternatives and owner's requirements.

Common additives

Description	Unit	$ Cost
Check Out Counter, single belt	Ea.	4025
Scanner, registers, guns & memory 10 lanes	Ea.	180,500
Power take away	Ea.	6550
Refrigerators, prefabricated, walk-in, 7'-6" High, 10' x 10'	S.F.	107
12' x 14'	S.F.	94
12' x 20'	S.F.	134
Refrigerated Food Cases, dairy, multi deck, 12' long	Ea.	13,200
Delicatessen case, multi deck, 18 S.F. shelf display	Ea.	9200
Freezer, glass door upright, 78 C.F.	Ea.	10,600
Frozen food, chest type, 12' long	Ea.	8875
Glass door reach in, 5 door	Ea.	11,600
Island case 12' long, multi deck	Ea.	10,400
Meat case, 12' long, multi deck	Ea.	11,400
Produce, 12' long, multi deck	Ea.	9625
Safe, Office type, 1 hr rating, 60"x36"x18", double door	Ea.	10,200

Description	Unit	$ Cost
Sound System, amplifier, 250 watts	Ea.	1950
Speaker, ceiling or wall	Ea.	233
Commissioning Fees, sustainable commercial construction	S.F.	0.24 - 3.04
Energy Modelling Fees, commercial buildings to 10,000 SF	Ea.	11,000
Greater than 10,000 SF add	S.F.	0.04
Green Bldg Cert Fees for comm construction project reg	Project	900
Photovoltaic Pwr Sys, grid connected, 20 kW (~2400 SF), roof	Ea.	250,600
Green Roofs, 6" soil depth, w/treated wd edging & sedum mats	S.F.	12.09
10" Soil depth, with treated wood edging & sedum mats	S.F.	13.58
Greywater Recovery Systems, prepackaged comm, 3060 gal.	Ea.	47,695
4590 gal.	Ea.	59,835
Rainwater Harvest Sys, prepckged comm, 10,000 gal, sys contrller	Ea.	37,050
20,000 gal. w/system controller	Ea.	60,450
30,000 gal. w/system controller	Ea.	96,175
Solar Domestic HW, closed loop, add-on sys, ext heat exchanger	Ea.	10,050
Drainback, hot water system, 120 gal tank	Ea.	12,875
Draindown, hot water system, 120 gal tank	Ea.	13,125

Important: See the Reference Section for Location Factors.

Model costs calculated for a 1 story building with 18′ story height and 44,000 square feet of floor area

				Unit	Unit Cost	Cost Per S.F.	
A. SUBSTRUCTURE							
1010	Standard Foundations	Poured concrete; strip and spread footings		S.F. Ground	1.12	1.12	8.1%
1020	Special Foundations	N/A		—	—	—	
1030	Slab on Grade	4″ reinforced concrete with recycled vapor barrier and granular base		S.F. Slab	5.35	5.35	
2010	Basement Excavation	Site preparation for slab and trench for foundation wall and footing		S.F. Ground	.16	.16	
2020	Basement Walls	4′ foundation wall		L.F. Wall	85	1.63	
B. SHELL							
	B10 Superstructure						
1010	Floor Construction	N/A		—	—	—	7.8 %
1020	Roof Construction	Metal deck, open web steel joists, beams, interior columns		S.F. Roof	7.96	7.96	
	B20 Exterior Enclosure						
2010	Exterior Walls	Face brick with concrete block backup	*85% of wall*	S.F. Wall	36	10.63	15.0%
2020	Exterior Windows	Storefront windows	*15% of wall*	Each	61	3.15	
2030	Exterior Doors	Sliding entrance doors with electrical operator, hollow metal		Each	5918	1.48	
	B30 Roofing						
3010	Roof Coverings	Single-ply TPO membrane, 60 mils, heat welded seams w/R-20 insul.		S.F. Roof	6.45	6.45	6.5%
3020	Roof Openings	Roof hatches		S.F. Roof	.14	.14	
C. INTERIORS							
1010	Partitions	50% CMU w/foamed-in insul., 50% gyp. bd. on mtl. studs	*50 SF Flr./LF Part.*	S.F. Partition	8.88	2.13	15.5%
1020	Interior Doors	Single leaf hollow metal, low VOC paint	*2000 S.F. Floor/Door*	Each	1201	.60	
1030	Fittings	N/A		—	—	—	
2010	Stair Construction	N/A		—	—	—	
3010	Wall Finishes	Paint, low VOC		S.F. Surface	5.10	2.45	
3020	Floor Finishes	Vinyl composition tile, recycled content		S.F. Floor	3.71	3.71	
3030	Ceiling Finishes	Mineral fiber tile on concealed zee bars		S.F. Ceiling	6.86	6.86	
D. SERVICES							
	D10 Conveying						
1010	Elevators & Lifts	N/A		—	—	—	0.0 %
1020	Escalators & Moving Walks	N/A		—	—	—	
	D20 Plumbing						
2010	Plumbing Fixtures	Toilet, low flow, auto sensor, & service fixt., supply & drain.	*1 Fixt./1820 SF Flr.*	Each	2512	1.38	2.7%
2020	Domestic Water Distribution	Point of use water heater, electric, energy saver		S.F. Floor	.20	.20	
2040	Rain Water Drainage	Roof drains		S.F. Roof	1.12	1.12	
	D30 HVAC						
3010	Energy Supply	N/A		—	—	—	4.8 %
3020	Heat Generating Systems	Included in D3050		—	—	—	
3040	Distribution Systems	Enthalpy heat recovery packages		Each	32,100	.80	
3050	Terminal & Package Units	Single zone rooftop air conditioner		S.F. Floor	4.04	4.04	
3090	Other HVAC Sys. & Equipment	N/A		—	—	—	
	D40 Fire Protection						
4010	Sprinklers	Sprinklers, light hazard		S.F. Floor	3.41	3.41	3.9%
4020	Standpipes	Standpipe		S.F. Floor	.58	.58	
	D50 Electrical						
5010	Electrical Service/Distribution	1600 ampere service, panel board and feeders		S.F. Floor	1.78	1.78	35.5%
5020	Lighting & Branch Wiring	LED fixtures, daylt. dim., ltg. on/off, receptacles, switches, and A.C.power		S.F. Floor	30	29.73	
5030	Communications & Security	Addressable alarm systems, partial internet wiring and emergency lighting		S.F. Floor	3.51	3.51	
5090	Other Electrical Systems	Emergency generator, 15 kW and energy monitoring systems		S.F. Floor	1.01	1.01	
E. EQUIPMENT & FURNISHINGS							
1010	Commercial Equipment	N/A		—	—	—	0.2 %
1020	Institutional Equipment	N/A		—	—	—	
1090	Other Equipment	Waste handling recycling tilt truck		S.F. Floor	.14	.14	
2020	Moveable Furnishings	No smoking signage		S.F. Floor	.02	.02	
F. SPECIAL CONSTRUCTION							
1020	Integrated Construction	N/A		—	—	—	0.0 %
1040	Special Facilities	N/A		—	—	—	
G. BUILDING SITEWORK	**N/A**						
					Sub-Total	101.54	**100%**
	CONTRACTOR FEES (General Requirements: 10%, Overhead: 5%, Profit: 10%)				*25%*	25.42	
	ARCHITECT FEES				*7%*	8.89	
					Total Building Cost	**135.85**	

Costs per square foot of floor area

Exterior Wall	S.F. Area	10000	15000	20000	25000	30000	35000	40000	50000	60000
	L.F. Perimeter	410	500	600	640	700	766	833	966	1000
Tilt-up Concrete Panels	Rigid Steel	150.00	139.50	134.80	129.85	127.10	125.35	124.00	122.15	119.50
Face Brick and Concrete Block	Bearing Walls	170.60	155.50	148.75	141.10	137.05	134.45	132.40	129.60	125.35
Concrete Block	Rigid Steel	149.85	139.45	134.70	129.75	127.00	125.25	123.90	122.05	119.40
	Bearing Walls	145.80	135.35	130.60	125.70	**122.95**	121.15	119.85	117.95	115.30
Metal Panel	Rigid Steel	150.65	140.05	135.25	130.25	127.45	125.70	124.25	122.40	119.70
Metal Sandwich Panel	Rigid Steel	157.55	145.60	140.25	134.55	131.35	129.35	127.80	125.65	122.55
Perimeter Adj., Add or Deduct	Per 100 L.F.	8.60	5.80	4.30	3.45	2.90	2.45	2.20	1.70	1.40
Story Hgt. Adj., Add or Deduct	Per 1 Ft.	1.05	0.85	0.75	0.65	0.60	0.60	0.55	0.45	0.40
***For Basement**, add $29.91 per square foot of basement area*										

The above costs were calculated using the basic specifications shown on the facing page. These costs should be adjusted where necessary for design alternatives and owner's requirements.

Common additives

Description	Unit	$ Cost
Dock Leveler, 10 ton cap., 6′ x 8′	Ea.	5950
7′ x 8′	Ea.	8750
Fence, Chain link, 6′ high, 9 ga. wire	L.F.	28
6 ga. wire	L.F.	33.50
Gate	Ea.	375
Paving, Bituminous, wearing course plus base course	S.Y.	7.20
Sidewalks, Concrete 4″ thick	S.F.	4.76
Sound System, amplifier, 250 watts	Ea.	1950
Speaker, ceiling or wall	Ea.	233
Yard Lighting, 20′ aluminum pole, 400W, HP sodium fixture	Ea.	3500

Description	Unit	$ Cost
Commissioning Fees, sustainable industrial construction	S.F.	0.20 - 1.06
Energy Modelling Fees, commercial buildings to 10,000 SF	Ea.	11,000
Greater than 10,000 SF add	S.F.	0.04
Green Bldg Cert Fees for comm construction project reg	Project	900
Photovoltaic Pwr Sys, grid connected, 20 kW (~2400 SF), roof	Ea.	250,600
Green Roofs, 6″ soil depth, w/treated wd edging & sedum mats	S.F.	12.09
10″ Soil depth, with treated wood edging & sedum mats	S.F.	13.58
Greywater Recovery Systems, prepackaged comm, 3060 gal.	Ea.	47,695
4590 gal.	Ea.	59,835
Rainwater Harvest Sys, prepckged comm, 10,000 gal, sys contrller	Ea.	37,050
20,000 gal. w/system controller	Ea.	60,450
30,000 gal. w/system controller	Ea.	96,175
Solar Domestic HW, closed loop, add-on sys, ext heat exchanger	Ea.	10,050
Drainback, hot water system, 120 gal tank	Ea.	12,875
Draindown, hot water system, 120 gal tank	Ea.	13,125

Important: See the Reference Section for Location Factors.

Model costs calculated for a 1 story building with 24′ story height and 30,000 square feet of floor area

G Warehouse

				Unit	Unit Cost	Cost Per S.F.	% Of Sub-Total
A. SUBSTRUCTURE							
1010	Standard Foundations	Poured concrete; strip and spread footings		S.F. Ground	1.43	1.43	18.3%
1020	Special Foundations	N/A		—	—	—	
1030	Slab on Grade	5″ reinforced concrete with recycled vapor barrier and granular base		S.F. Slab	12.69	12.69	
2010	Basement Excavation	Site preparation for slab and trench for foundation wall and footing		S.F. Ground	.16	.16	
2020	Basement Walls	4′ foundation wall		L.F. Wall	85	2.58	
B. SHELL							
	B10 Superstructure						
1010	Floor Construction	Mezzanine: open web steel joists, slab form, concrete beams, columns	*10% of area*	S.F. Floor	2.12	2.12	10.1%
1020	Roof Construction	Metal deck, open web steel joists, beams, columns		S.F. Roof	7.19	7.19	
	B20 Exterior Enclosure						
2010	Exterior Walls	Concrete block	*95% of wall*	S.F. Wall	16.56	8.81	11.1%
2020	Exterior Windows	N/A		—	—	—	
2030	Exterior Doors	Steel overhead, hollow metal	*5% of wall*	Each	3768	1.38	
	B30 Roofing						
3010	Roof Coverings	Single-ply TPO membrane, 60 mils, heat welded w/R-20 insul.		S.F. Roof	6.60	6.60	7.6%
3020	Roof Openings	Roof hatches and skylight		S.F. Roof	.40	.40	
C. INTERIORS							
1010	Partitions	Concrete block w/foamed-in insul. (office and washrooms)	*100 SF Flr./LF Part.*	S.F. Partition	10.63	.85	7.3%
1020	Interior Doors	Single leaf hollow metal, low VOC paint	*5000 S.F. Floor/Door*	Each	1201	.24	
1030	Fittings	N/A		—	—	—	
2010	Stair Construction	Steel gate with rails		Flight	14,475	.96	
3010	Wall Finishes	Paint, low VOC		S.F. Surface	13.13	2.10	
3020	Floor Finishes	90% hardener, 10% vinyl composition tile, recycled content		S.F. Floor	1.86	1.86	
3030	Ceiling Finishes	Suspended mineral tile on zee channels in office area	*10% of area*	S.F. Ceiling	.69	.69	
D. SERVICES							
	D10 Conveying						
1010	Elevators & Lifts	N/A		—	—	—	0.0 %
1020	Escalators & Moving Walks	N/A		—	—	—	
	D20 Plumbing						
2010	Plumbing Fixtures	Toilet, low flow, auto sensor & service fixt., supply & drainage	*1 Fixt./2500 SF Flr.*	Each	4450	1.78	3.9%
2020	Domestic Water Distribution	Tankless, on-demand water heater, gas/propane		S.F. Floor	.42	.42	
2040	Rain Water Drainage	Roof drains		S.F. Roof	1.35	1.35	
	D30 HVAC						
3010	Energy Supply	N/A		—	—	—	6.4 %
3020	Heat Generating Systems	Ventilation with heat		Each	144,700	5.21	
3030	Cooling Generating Systems	N/A		—	—	—	
3050	Terminal & Package Units	Single zone rooftop air conditioner, SEER 14		S.F. Floor	.68	.68	
3090	Other HVAC Sys. & Equipment	N/A		—	—	—	
	D40 Fire Protection						
4010	Sprinklers	Sprinklers, ordinary hazard		S.F. Floor	3.94	3.94	4.8%
4020	Standpipes	Standpipe		S.F. Floor	.50	.50	
	D50 Electrical						
5010	Electrical Service/Distribution	200 ampere service, panel board and feeders		S.F. Floor	.28	.28	22.0%
5020	Lighting & Branch Wiring	LED fixtures, daylt. dim., ltg. on/off, receptacles, switches, and A.C. power		S.F. Floor	16.11	16.11	
5030	Communications & Security	Addressable alarm systems		S.F. Floor	2.63	2.63	
5090	Other Electrical Systems	Energy monitoring systems		S.F. Floor	1.18	1.18	
E. EQUIPMENT & FURNISHINGS							
1010	Commercial Equipment	N/A		—	—	—	2.2 %
1030	Vehicular Equipment	Dock boards, dock levelers		S.F. Floor	1.80	1.80	
1090	Other Equipment	Waste handling recycling tilt truck		S.F. Floor	.21	.21	
2020	Moveable Furnishings	No smoking signage		S.F. Floor	.02	.02	
F. SPECIAL CONSTRUCTION							
1020	Integrated Construction	Shipping & receiving air curtain		S.F. Floor	5.74	5.74	6.2%
1040	Special Facilities	N/A		—	—	—	
G. BUILDING SITEWORK		**N/A**					
					Sub-Total	91.91	**100%**
	CONTRACTOR FEES (General Requirements: 10%, Overhead: 5%, Profit: 10%)				*25%*	23	
	ARCHITECT FEES				*7%*	8.04	
					Total Building Cost	**122.95**	

Assemblies Section

Table of Contents

Table of Contents

Table of Contents

Table of Contents

RSMeans data: Assemblies—How They Work

Assemblies estimating provides a fast and reasonably accurate way to develop construction costs. An assembly is the grouping of individual work items—with appropriate quantities—to provide a cost for a major construction component in a convenient unit of measure.

An assemblies estimate is often used during early stages of design development to compare the cost impact of various design alternatives on total building cost.

Assemblies estimates are also used as an efficient tool to verify construction estimates.

Assemblies estimates do not require a completed design or detailed drawings. Instead, they are based on the general size of the structure and other known parameters of the project. The degree of accuracy of an assemblies estimate is generally within +/- 15%.

Most assemblies consist of three major elements: a graphic, the system components, and the cost data itself. The **Graphic** is a visual representation showing the typical appearance of the assembly

---

1 Unique 12-character Identifier

Our assemblies are identified by a **unique 12-character identifier**. The assemblies are numbered using UNIFORMAT II, ASTM Standard E1557. The first 5 characters represent this system to Level 3. The last 7 characters represent further breakdown in order to arrange items in understandable groups of similar tasks. Line numbers are consistent across all of our publications, so a line number in any assemblies data set will always refer to the same item.

2 Narrative Descriptions

Our assemblies descriptions appear in two formats: narrative and table. **Narrative descriptions** are shown in a hierarchical structure to make them readable. In order to read a complete description, read up through the indents to the top of the section. Include everything that is above and to the left that is not contradicted by information below.

Narrative Format

C20 Stairs

C2010 Stair Construction

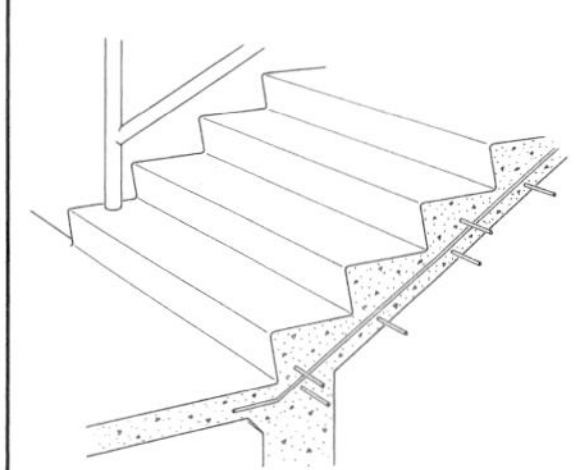

General Design: See reference section for code requirements. Maximum height between landings is 12′; usual stair angle is 20° to 50° with 30° to 35° best. Usual relation of riser to treads is:
Riser + tread = 17.5.
2x (Riser) + tread = 25.
Riser x tread = 70 or 75.
Maximum riser height is 7″ for commercial, 8-1/4″ for residential.
Usual riser height is 6-1/2″ to 7-1/4″.

Minimum tread width is 11″ for commercial and 9″ for residential.

For additional information please see reference section.

Cost Per Flight: Table below lists the cost per flight for 4′-0″ wide stairs. Side walls are not included. Railings are included.

C2010 110 (1)	**Stairs**	COST PER FLIGHT		
		MAT.	INST.	TOTAL
0470	Stairs, C.I.P. concrete, w/o landing, 12 risers, w/o nosing	1,500	2,075	3,575
0480	With nosing	2,450	2,250	4,700
0550	W/landing, 12 risers, w/o nosing	1,675	2,550	4,225
0560	With nosing	2,625	2,725	5,350
0570	(2) 16 risers, w/o nosing	2,050	3,225	5,275
0580	With nosing	3,325	3,475	6,800
0590	20 risers, w/o nosing	2,400	3,875	6,275
0600	With nosing	4,025	4,175	8,200
0610	24 risers, w/o nosing	2,775	4,550	7,325
0620	With nosing	4,725	4,925	9,650
0630	Steel, grate type w/nosing & rails, 12 risers, w/o landing	6,125	970	7,095
0640	With landing	8,400	1,325	9,725
0660	16 risers, with landing	10,500	1,675	12,175
0680	20 risers, with landing	12,500	1,975	14,475
0700	24 risers, with landing	14,500	2,300	16,800
0710	Metal pan stairs for concrete in-fill, picket rail, 12 risers, w/o landing	8,050	970	9,020
0720	With landing	10,900	1,475	12,375
0740	16 risers, with landing	13,600	1,775	15,375
0760	20 risers, with landing	16,300	2,125	18,425
0780	24 risers, with landing	19,000	2,425	21,425
0790	Cast iron tread & pipe rail, 12 risers, w/o landing	8,100	970	9,070
0800	With landing	11,000	1,475	12,475
1120	Wood, prefab box type, oak treads, wood rails 3′-6″ wide, 14 risers	2,400	425	2,825
1150	Prefab basement type, oak treads, wood rails 3′-0″ wide, 14 risers	1,150	105	1,255

For supplemental customizable square foot estimating forms, visit: **RSMeans.com/2019books**

in question. It is frequently accompanied by additional explanatory technical information describing the class of items. The **System Components** is a listing of the individual tasks that make up the assembly, including the quantity and unit of measure for each item, along with the cost of material and installation. The **Assemblies data** below lists prices for other similar systems with dimensional and/or size variations.

All of our assemblies costs represent the cost for the installing contractor. An allowance for profit has been added to all material, labor, and equipment rental costs. A markup for labor burdens, including workers' compensation, fixed overhead, and business overhead, is included with installation costs.

The information in RSMeans cost data represents a "national average" cost. This data should be modified to the project location using the **City Cost Indexes** or **Location Factors** tables found in the Reference Section.

Table Format

A20 Basement Construction

A2020 Basement Walls

The Foundation Bearing Wall System includes: forms up to 16′ high (four uses); 3,000 p.s.i. concrete placed and vibrated; and form removal with breaking form ties and patching walls. The wall systems list walls from 6″ to 16″ thick and are designed with minimum reinforcement.

Excavation and backfill are not included.

Please see the reference section for further design and cost information.

A2020 110 — Walls, Cast in Place ❸

	WALL HEIGHT (FT.)	PLACING METHOD	CONCRETE (C.Y. per L.F.)	REINFORCING (LBS. per L.F.)	WALL THICKNESS (IN.)	COST PER L.F. MAT.	COST PER L.F. INST.	COST PER L.F. TOTAL
5000	8′	direct chute	.148	6.6	6	38	87	125
5020			.199	9.6	8	46.50	92.50	139
5040			.250	12	10	55	91	146
5061	❹		.296	14.39	12	62.50	93	155.50
6020		direct chute	.248	12	8	58	112	170
6040			.307	14.99	10	68	114	182
6061			.370	17.99	12	78	116	194
7220	12′	pumped	.298	14.39	8	70	139	209
7240			.369	17.99	10	81.50	142	223.50
7260			.444	21.59	12	94	146	240
9220	16′	pumped	.397	19.19	8	93	185	278
9240			.492	23.99	10	109	189	298
9260			.593	28.79	12	125	195	320

❸ Unit of Measure

All RSMeans data: Assemblies include a typical **Unit of Measure** used for estimating that item. For instance, for continuous footings or foundation walls the unit is linear feet (L.F.). For spread footings the unit is each (Ea.). The estimator needs to take special care that the unit in the data matches the unit in the takeoff. Abbreviations and unit conversions can be found in the Reference Section.

❹ Table Descriptions

Table descriptions work similar to Narrative Descriptions, except that if there is a blank in the column at a particular line number, read up to the description above in the same column.

RSMeans data: Assemblies—How They Work (Continued)

Sample Estimate

This sample demonstrates the elements of an estimate, including a tally of the RSMeans data lines. Published assemblies costs include all markups for labor burden and profit for the installing contractor. This estimate adds a summary of the markups applied by a general contractor on the installing contractor's work. These figures represent the total cost to the owner. The location factor with RSMeans data is applied at the bottom of the estimate to adjust the cost of the work to a specific location.

Project Name:	**Interior Fit-out, ABC Office**			
Location:	**Anywhere, USA**	**Date: 1/1/2019**		**OPN**
Assembly Number	**Description**	**Qty.**	**Unit**	**Subtotal**
C1010 124 1200 (1)	Wood partition, 2 x 4 @ 16" OC w/5/8" FR gypsum board	560	S.F.	$2,497.60
C1020 114 1800	Metal door & frame, flush hollow core, 3'-0" x 7'-0"	2	Ea.	$2,410.00
C3010 230 0080	Painting, brushwork, primer & 2 coats	1,120	S.F.	$1,232.00
C3020 410 0140	Carpet, tufted, nylon, roll goods, 12' wide, 26 oz	240	S.F.	$849.60
C3030 210 6000	Acoustic ceilings, 24" x 48" tile, tee grid suspension	200	S.F.	$1,220.00
D5020 125 0560	Receptacles incl plate, box, conduit, wire, 20 A duplex	8	Ea.	$2,060.00
D5020 125 0720	Light switch incl plate, box, conduit, wire, 20 A single pole	2	Ea.	$501.00
D5020 210 0560	Fluorescent fixtures, recess mounted, 20 per 1000 SF	200	S.F.	$2,000.00
	Assembly Subtotal			**$12,770.20**
	Sales Tax @ (2)	**5 %**		**$ 319.26**
	General Requirements @ (3)	**7 %**		**$ 893.91**
	Subtotal A			**$13,983.37**
	GC Overhead @ (4)	**5 %**		**$ 699.17**
	Subtotal B			**$14,682.54**
	GC Profit @ (5)	**5 %**		**$ 734.13**
	Subtotal C			**$15,416.66**
	Adjusted by Location Factor (6)	**113.9**		**$ 17,559.58**
	Architects Fee @ (7)	**8 %**		**$ 1,404.77**
	Contingency @ (8)	**15 %**		**$ 2,633.94**
	Project Total Cost			**$ 21,598.28**

This estimate is based on an interactive spreadsheet. You are free to download it and adjust it to your methodology.
A copy of this spreadsheet is available at **RSMeans.com/2019books.**

1 Work Performed

The body of the estimate shows the RSMeans data selected, including line numbers, a brief description of each item, its takeoff quantity and unit, and the total installed cost, including the installing contractor's overhead and profit.

2 Sales Tax

If the work is subject to state or local sales taxes, the amount must be added to the estimate. In a conceptual estimate it can be assumed that one half of the total represents material costs. Therefore, apply the sales tax rate to 50% of the assembly subtotal.

3 General Requirements

This item covers project-wide needs provided by the general contractor. These items vary by project but may include temporary facilities and utilities, security, testing, project cleanup, etc. In assemblies estimates a percentage is used—typically between 5% and 15% of project cost.

4 General Contractor Overhead

This entry represents the general contractor's markup on all work to cover project administration costs.

5 General Contractor Profit

This entry represents the GC's profit on all work performed. The value included here can vary widely by project and is influenced by the GC's perception of the project's financial risk and market conditions.

6 Location Factor

RSMeans published data are based on national average costs. If necessary, adjust the total cost of the project using a location factor from the "Location Factor" table or the "City Cost Indexes" table found in the Reference Section. Use location factors if the work is general, covering the work of multiple trades. If the work is by a single trade (e.g., masonry) use the more specific data found in the City Cost Indexes.

To adjust costs by location factors, multiply the base cost by the factor and divide by 100.

7 Architect's Fee

If appropriate, add the design cost to the project estimate. These fees vary based on project complexity and size. Typical design and engineering fees can be found in the Reference Section.

8 Contingency

A factor for contingency may be added to any estimate to represent the cost of unknowns that may occur between the time that the estimate is performed and the time the project is constructed. The amount of the allowance will depend on the stage of design at which the estimate is done, as well as the contractor's assessment of the risk involved.

A10 Foundations

A1010 Standard Foundations

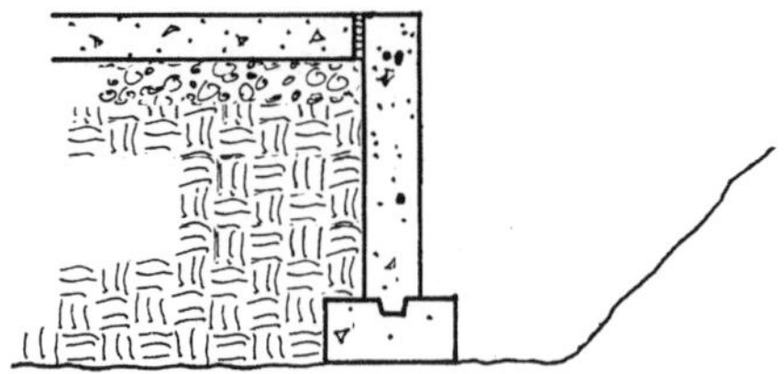

The Foundation Bearing Wall System includes: forms up to 6′ high (four uses); 3,000 p.s.i. concrete placed and vibrated; and form removal with breaking form ties and patching walls. The wall systems list walls from 6″ to 16″ thick and are designed with minimum reinforcement.

Excavation and backfill are not included.

Please see the reference section for further design and cost information.

A1010 105 Wall Foundations

	WALL HEIGHT (FT.)	PLACING METHOD	CONCRETE (C.Y. per L.F.)	REINFORCING (LBS. per L.F.)	WALL THICKNESS (IN.)	COST PER L.F.		
						MAT.	INST.	TOTAL
1500	4′	direct chute	.074	3.3	6	19	43.50	62.50
1520			.099	4.8	8	23	44.50	67.50
1540			.123	6.0	10	27	45.50	72.50
1560			.148	7.2	12	31.50	46.50	78
1580			.173	8.1	14	35	47.50	82.50
1600			.197	9.44	16	39	48.50	87.50
3000	6′	direct chute	.111	4.95	6	28.50	65	93.50
3020			.149	7.20	8	35	67	102
3040			.184	9.00	10	40.50	68	108.50
3060			.222	10.8	12	47	70	117

A1010 Standard Foundations

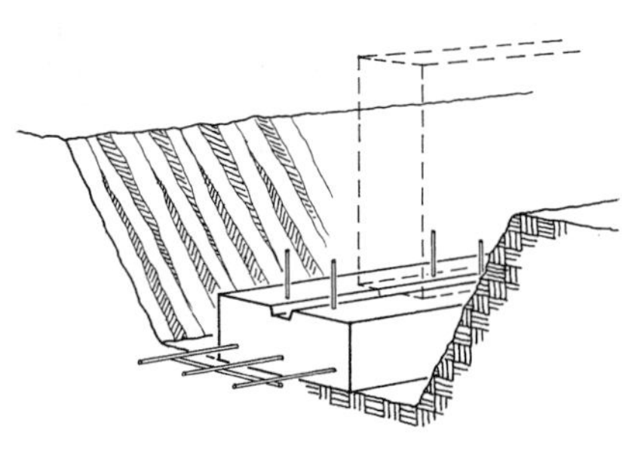

The Strip Footing System includes: excavation; hand trim; all forms needed for footing placement; forms for 2″ x 6″ keyway (four uses); dowels; and 3,000 p.s.i. concrete.

The footing size required varies for different soils. Soil bearing capacities are listed for 3 KSF and 6 KSF. Depths of the system range from 8″ and deeper. Widths range from 16″ and wider. Smaller strip footings may not require reinforcement.

Please see the reference section for further design and cost information.

A1010 110	Strip Footings	COST PER L.F.		
		MAT.	INST.	TOTAL
2100	Strip footing, load 2.6 KLF, soil capacity 3 KSF, 16″ wide x 8″ deep, plain	8.25	9.30	17.55
2300	Load 3.9 KLF, soil capacity 3 KSF, 24″ wide x 8″ deep, plain	10.55	10.50	21.05
2500	Load 5.1 KLF, soil capacity 3 KSF, 24″ wide x 12″ deep, reinf.	18.90	20	38.90
2700	Load 11.1 KLF, soil capacity 6 KSF, 24″ wide x 12″ deep, reinf.	18.90	20	38.90
2900	Load 6.8 KLF, soil capacity 3 KSF, 32″ wide x 12″ deep, reinf.	23	21.50	44.50
3100	Load 14.8 KLF, soil capacity 6 KSF, 32″ wide x 12″ deep, reinf.	23	22	45
3300	Load 9.3 KLF, soil capacity 3 KSF, 40″ wide x 12″ deep, reinf.	27	23.50	50.50
3500	Load 18.4 KLF, soil capacity 6 KSF, 40″ wide x 12″ deep, reinf.	27	23.50	50.50
4500	Load 10 KLF, soil capacity 3 KSF, 48″ wide x 16″ deep, reinf.	38.50	29.50	68
4700	Load 22 KLF, soil capacity 6 KSF, 48″ wide, 16″ deep, reinf.	39	30.50	69.50
5700	Load 15 KLF, soil capacity 3 KSF, 72″ wide x 20″ deep, reinf.	66.50	43	109.50
5900	Load 33 KLF, soil capacity 6 KSF, 72″ wide x 20″ deep, reinf.	70	46	116

A1010 Standard Foundations

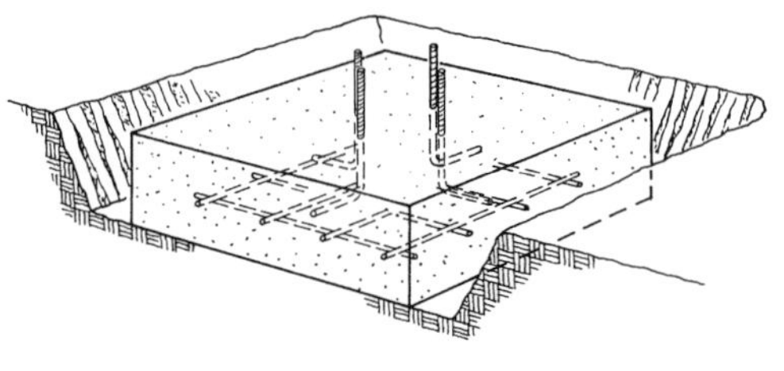

The Spread Footing System includes: excavation; backfill; forms (four uses); all reinforcement; 3,000 p.s.i. concrete (chute placed); and float finish.

Footing systems are priced per individual unit. The Expanded System Listing at the bottom shows various footing sizes. It is assumed that excavation is done by a truck mounted hydraulic excavator with an operator and oiler.

Backfill is with a dozer, and compaction by air tamp. The excavation and backfill equipment is assumed to operate at 30 C.Y. per hour.

Please see the reference section for further design and cost information.

A1010 210	Spread Footings	COST EACH		
		MAT.	INST.	TOTAL
7090	Spread footings, 3000 psi concrete, chute delivered			
7100	Load 25K, soil capacity 3 KSF, 3'-0" sq. x 12" deep	68	103	171
7150	Load 50K, soil capacity 3 KSF, 4'-6" sq. x 12" deep	145	179	324
7200	Load 50K, soil capacity 6 KSF, 3'-0" sq. x 12" deep	68	103	171
7250	Load 75K, soil capacity 3 KSF, 5'-6" sq. x 13" deep	231	254	485
7300	Load 75K, soil capacity 6 KSF, 4'-0" sq. x 12" deep	118	153	271
7350	Load 100K, soil capacity 3 KSF, 6'-0" sq. x 14" deep	293	305	598
7410	Load 100K, soil capacity 6 KSF, 4'-6" sq. x 15" deep	179	211	390
7450	Load 125K, soil capacity 3 KSF, 7'-0" sq. x 17" deep	465	440	905
7500	Load 125K, soil capacity 6 KSF, 5'-0" sq. x 16" deep	231	254	485
7550	Load 150K, soil capacity 3 KSF 7'-6" sq. x 18" deep	565	510	1,075
7610	Load 150K, soil capacity 6 KSF, 5'-6" sq. x 18" deep	310	320	630
7650	Load 200K, soil capacity 3 KSF, 8'-6" sq. x 20" deep	805	675	1,480
7700	Load 200K, soil capacity 6 KSF, 6'-0" sq. x 20" deep	405	400	805
7750	Load 300K, soil capacity 3 KSF, 10'-6" sq. x 25" deep	1,475	1,100	2,575
7810	Load 300K, soil capacity 6 KSF, 7'-6" sq. x 25" deep	770	665	1,435
7850	Load 400K, soil capacity 3 KSF, 12'-6" sq. x 28" deep	2,350	1,650	4,000
7900	Load 400K, soil capacity 6 KSF, 8'-6" sq. x 27" deep	1,075	865	1,940
8010	Load 500K, soil capacity 6 KSF, 9'-6" sq. x 30" deep	1,475	1,125	2,600
8100	Load 600K, soil capacity 6 KSF, 10'-6" sq. x 33" deep	2,000	1,450	3,450
8200	Load 700K, soil capacity 6 KSF, 11'-6" sq. x 36" deep	2,550	1,775	4,325
8300	Load 800K, soil capacity 6 KSF, 12'-0" sq. x 37" deep	2,875	1,950	4,825
8400	Load 900K, soil capacity 6 KSF, 13'-0" sq. x 39" deep	3,550	2,325	5,875
8500	Load 1000K, soil capacity 6 KSF, 13'-6" sq. x 41" deep	4,025	2,550	6,575

A1010 320	Foundation Dampproofing	COST PER L.F.		
		MAT.	INST.	TOTAL
1000	Foundation dampproofing, bituminous, 1 coat, 4' high	1	4.08	5.08
1400	8' high	2	8.15	10.15
1800	12' high	3	12.65	15.65
2000	2 coats, 4' high	2	5.05	7.05
2400	8' high	4	10.10	14.10
2800	12' high	6	15.50	21.50
3000	Asphalt with fibers, 1/16" thick, 4' high	1.76	5.05	6.81
3400	8' high	3.52	10.10	13.62
3800	12' high	5.30	15.50	20.80
4000	1/8" thick, 4' high	3.08	6.05	9.13
4400	8' high	6.15	12.10	18.25
4800	12' high	9.25	18.50	27.75

A1010 Standard Foundations

A1010 320	Foundation Dampproofing	COST PER L.F.		
		MAT.	INST.	TOTAL
5000	Asphalt coated board and mastic, 1/4" thick, 4' high	5.25	5.50	10.75
5400	8' high	10.50	10.95	21.45
5800	12' high	15.70	16.85	32.55
6000	1/2" thick, 4' high	7.85	7.60	15.45
6400	8' high	15.70	15.25	30.95
6800	12' high	23.50	23.50	47
7000	Cementitious coating, on walls, 1/8" thick coating, 4' high	2.96	7.75	10.71
7400	8' high	5.90	15.50	21.40
7800	12' high	8.90	23.50	32.40
8000	Cementitious/metallic slurry, 4 coat, 1/2"thick, 2' high	1.47	8	9.47
8400	4' high	2.94	16	18.94
8800	6' high	4.41	24	28.41

A1030 Slab on Grade

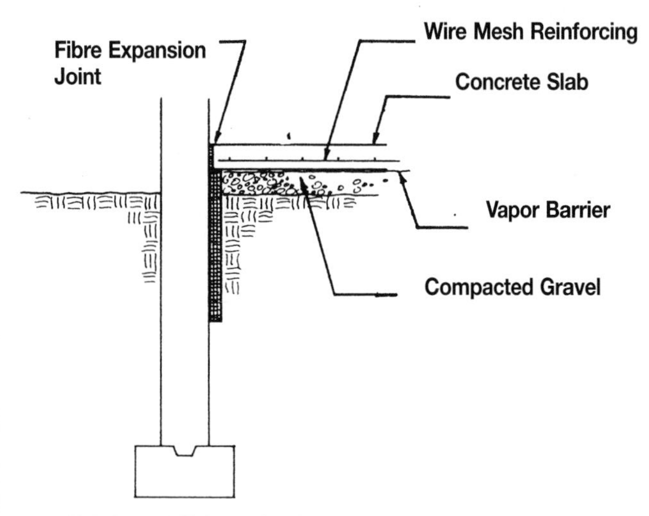

Reinforced Slab on Grade

A Slab on Grade system includes fine grading; 6″ of compacted gravel; vapor barrier; 3500 p.s.i. concrete; bituminous fiber expansion joint; all necessary edge forms 4 uses; steel trowel finish; and sprayed on membrane curing compound. Wire mesh reinforcing used in all reinforced slabs.

Non-industrial slabs are for foot traffic only with negligible abrasion. Light industrial slabs are for pneumatic wheels and light abrasion. Industrial slabs are for solid rubber wheels and moderate abrasion. Heavy industrial slabs are for steel wheels and severe abrasion.

A1030 120	Plain & Reinforced	COST PER S.F.		
		MAT.	INST.	TOTAL
2220	Slab on grade, 4″ thick, non industrial, non reinforced	2.37	2.45	4.82
2240	Reinforced	2.48	2.80	5.28
2260	Light industrial, non reinforced	2.89	3.02	5.91
2280	Reinforced	3.07	3.33	6.40
2300	Industrial, non reinforced	3.70	6.25	9.95
2320	Reinforced	3.88	6.55	10.43
3340	5″ thick, non industrial, non reinforced	2.79	2.52	5.31
3360	Reinforced	2.97	2.83	5.80
3380	Light industrial, non reinforced	3.32	3.09	6.41
3400	Reinforced	3.50	3.40	6.90
3420	Heavy industrial, non reinforced	4.70	7.50	12.20
3440	Reinforced	4.80	7.85	12.65
4460	6″ thick, non industrial, non reinforced	3.33	2.46	5.79
4480	Reinforced	3.59	2.87	6.46
4500	Light industrial, non reinforced	3.88	3.03	6.91
4520	Reinforced	4.22	3.34	7.56
4540	Heavy industrial, non reinforced	5.30	7.60	12.90
4560	Reinforced	5.55	8	13.55
5580	7″ thick, non industrial, non reinforced	3.75	2.55	6.30
5600	Reinforced	4.09	2.99	7.08
5620	Light industrial, non reinforced	4.31	3.12	7.43
5640	Reinforced	4.65	3.56	8.21
5660	Heavy industrial, non reinforced	5.70	7.50	13.20
5680	Reinforced	5.95	7.90	13.85
6700	8″ thick, non industrial, non reinforced	4.17	2.59	6.76
6720	Reinforced	4.46	2.96	7.42
6740	Light industrial, non reinforced	4.74	3.16	7.90
6760	Reinforced	5.05	3.53	8.58
6780	Heavy industrial, non reinforced	6.15	7.60	13.75
6800	Reinforced	6.60	8	14.60

A2010 Basement Excavation

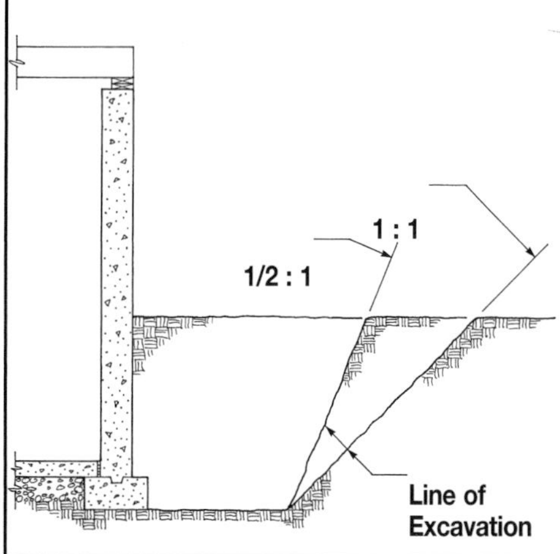

Pricing Assumptions: Two-thirds of excavation is by 2-1/2 C.Y. wheel mounted front end loader and one-third by 1-1/2 C.Y. hydraulic excavator.

Two-mile round trip haul by 12 C.Y. tandem trucks is included for excavation wasted and storage of suitable fill from excavated soil. For excavation in clay, all is wasted and the cost of suitable backfill with two-mile haul is included.

Sand and gravel assumes 15% swell and compaction; common earth assumes 25% swell and 15% compaction; clay assumes 40% swell and 15% compaction (non-clay).

In general, the following items are accounted for in the costs in the table below.

1. Excavation for building or other structure to depth and extent indicated.
2. Backfill compacted in place.
3. Haul of excavated waste.
4. Replacement of unsuitable material with bank run gravel.

Note: Additional excavation and fill beyond this line of general excavation for the building (as required for isolated spread footings, strip footings, etc.) are included in the cost of the appropriate component systems.

A2010 110	Building Excavation & Backfill	COST PER S.F.		
		MAT.	INST.	TOTAL
2220	Excav & fill, 1000 S.F., 4' deep sand, gravel, or common earth, on site storage		.95	.95
2240	Off site storage		1.41	1.41
2260	Clay excavation, bank run gravel borrow for backfill	2.05	2.17	4.22
2280	8' deep, sand, gravel, or common earth, on site storage		5.65	5.65
2300	Off site storage		12.40	12.40
2320	Clay excavation, bank run gravel borrow for backfill	8.30	10.20	18.50
2340	16' deep, sand, gravel, or common earth, on site storage		15.20	15.20
2350	Off site storage		30.50	30.50
2360	Clay excavation, bank run gravel borrow for backfill	23	26	49
3380	4000 S.F., 4' deep, sand, gravel, or common earth, on site storage		.49	.49
3400	Off site storage		.97	.97
3420	Clay excavation, bank run gravel borrow for backfill	1.05	1.10	2.15
3440	8' deep, sand, gravel, or common earth, on site storage		3.99	3.99
3460	Off site storage		6.95	6.95
3480	Clay excavation, bank run gravel borrow for backfill	3.79	6.10	9.89
3500	16' deep, sand, gravel, or common earth, on site storage		9.40	9.40
3520	Off site storage		18.75	18.75
3540	Clay, excavation, bank run gravel borrow for backfill	10.15	14.50	24.65
4560	10,000 S.F., 4' deep, sand, gravel, or common earth, on site storage		.28	.28
4580	Off site storage		.57	.57
4600	Clay excavation, bank run gravel borrow for backfill	.64	.68	1.32
4620	8' deep, sand, gravel, or common earth, on site storage		3.46	3.46
4640	Off site storage		5.25	5.25
4660	Clay excavation, bank run gravel borrow for backfill	2.30	4.75	7.05
4680	16' deep, sand, gravel, or common earth, on site storage		7.70	7.70
4700	Off site storage		13.15	13.15
4720	Clay excavation, bank run gravel borrow for backfill	6.10	10.85	16.95
5740	30,000 S.F., 4' deep, sand, gravel, or common earth, on site storage		.16	.16
5760	Off site storage		.33	.33
5780	Clay excavation, bank run gravel borrow for backfill	.37	.39	.76
5860	16' deep, sand, gravel, or common earth, on site storage		6.65	6.65
5880	Off site storage		9.60	9.60
5900	Clay excavation, bank run gravel borrow for backfill	3.38	8.40	11.78
6910	100,000 S.F., 4' deep, sand, gravel, or common earth, on site storage		.10	.10
6920	Off site storage		.20	.20
6930	Clay excavation, bank run gravel borrow for backfill	.18	.19	.37
6940	8' deep, sand, gravel, or common earth, on site storage		2.90	2.90
6950	Off site storage		3.43	3.43
6960	Clay excavation, bank run gravel borrow for backfill	.70	3.29	3.99
6970	16' deep, sand, gravel, or common earth, on site storage		6	6
6980	Off site storage		7.60	7.60
6990	Clay excavation, bank run gravel borrow for backfill	1.82	7	8.82

A2020 Basement Walls

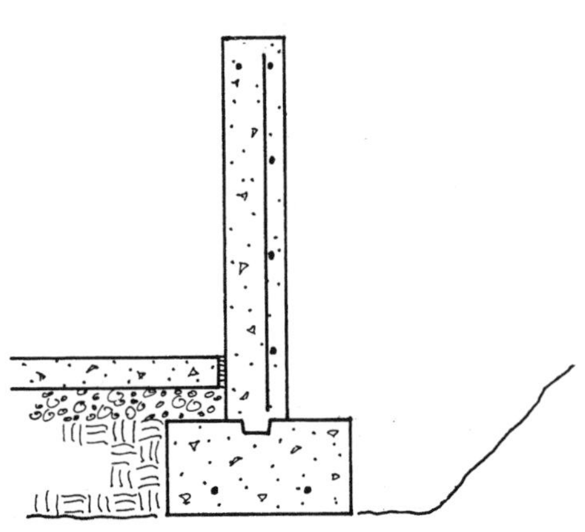

The Foundation Bearing Wall System includes: forms up to 16′ high (four uses); 3,000 p.s.i. concrete placed and vibrated; and form removal with breaking form ties and patching walls. The wall systems list walls from 6″ to 16″ thick and are designed with minimum reinforcement.

Excavation and backfill are not included.

Please see the reference section for further design and cost information.

A2020 110 Walls, Cast in Place

	WALL HEIGHT (FT.)	PLACING METHOD	CONCRETE (C.Y. per L.F.)	REINFORCING (LBS. per L.F.)	WALL THICKNESS (IN.)	COST PER L.F.		
						MAT.	INST.	TOTAL
5000	8′	direct chute	.148	6.6	6	38	87	125
5020			.199	9.6	8	46.50	92.50	139
5040			.250	12	10	55	91	146
5061			.296	14.39	12	62.50	93	155.50
6020	10′	direct chute	.248	12	8	58	112	170
6040			.307	14.99	10	68	114	182
6061			.370	17.99	12	78	116	194
7220	12′	pumped	.298	14.39	8	70	139	209
7240			.369	17.99	10	81.50	142	223.50
7260			.444	21.59	12	94	146	240
9220	16′	pumped	.397	19.19	8	93	185	278
9240			.492	23.99	10	109	189	298
9260			.593	28.79	12	125	195	320

B1010 Floor Construction

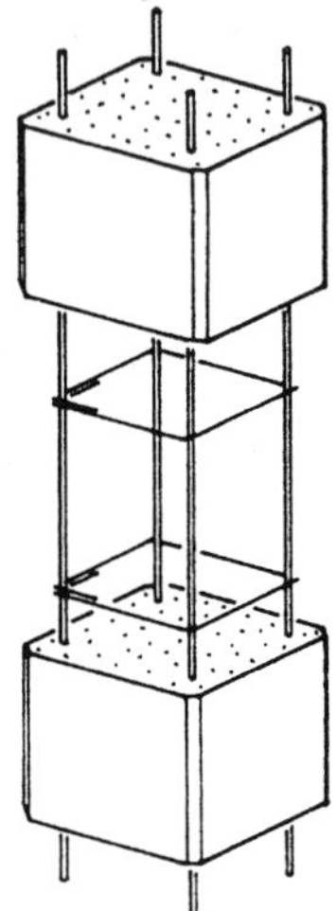

General: It is desirable for purposes of consistency and simplicity to maintain constant column sizes throughout building height. To do this, concrete strength may be varied (higher strength concrete at lower stories and lower strength concrete at upper stories), as well as varying the amount of reinforcing.

The table provides probably minimum column sizes with related costs and weight per lineal foot of story height.

B1010 203 C.I.P. Column, Square Tied

	LOAD (KIPS)	STORY HEIGHT (FT.)	COLUMN SIZE (IN.)	COLUMN WEIGHT (P.L.F.)	CONCRETE STRENGTH (PSI)	COST PER V.L.F.		
						MAT.	INST.	TOTAL
0640	100	10	10	96	4000	11.75	41	52.75
0680		12	10	97	4000	11.95	41.50	53.45
0700		14	12	142	4000	15.70	50	65.70
0840	200	10	12	140	4000	17.15	52	69.15
0860		12	12	142	4000	17.45	52	69.45
0900		14	14	196	4000	20.50	58.50	79
0920	300	10	14	192	4000	21.50	60	81.50
0960		12	14	194	4000	22	60.50	82.50
0980		14	16	253	4000	25.50	66	91.50
1020	400	10	16	248	4000	27	68	95
1060		12	16	251	4000	27.50	68.50	96
1080		14	16	253	4000	28	69.50	97.50
1200	500	10	18	315	4000	32	76	108
1250		12	20	394	4000	38.50	86.50	125
1300		14	20	397	4000	39.50	87.50	127
1350	600	10	20	388	4000	40.50	88.50	129
1400		12	20	394	4000	41	89.50	130.50
1600		14	20	397	4000	42	90.50	132.50
3400	900	10	24	560	4000	55	110	165
3800		12	24	567	4000	56	112	168
4000		14	24	571	4000	57	113	170
7300	300	10	14	192	6000	22	59.50	81.50
7500		12	14	194	6000	22.50	60	82.50
7600		14	14	196	6000	23	60.50	83.50
8000	500	10	16	248	6000	28.50	68	96.50
8050		12	16	251	6000	29	68.50	97.50
8100		14	16	253	6000	29.50	69.50	99
8200	600	10	18	315	6000	33	76.50	109.50
8300		12	18	319	6000	33.50	77.50	111
8400		14	18	321	6000	34	78	112
8800	800	10	20	388	6000	40.50	86.50	127
8900		12	20	394	6000	41	87.50	128.50
9000		14	20	397	6000	42	88.50	130.50

B1010 Floor Construction

B1010 203 — C.I.P. Column, Square Tied

	LOAD (KIPS)	STORY HEIGHT (FT.)	COLUMN SIZE (IN.)	COLUMN WEIGHT (P.L.F.)	CONCRETE STRENGTH (PSI)	COST PER V.L.F.		
						MAT.	INST.	TOTAL
9100	900	10	20	388	6000	43.50	90.50	134
9300		12	20	394	6000	44.50	91.50	136
9600		14	20	397	6000	45	92.50	137.50

B1010 204 — C.I.P. Column, Square Tied-Minimum Reinforcing

	LOAD (KIPS)	STORY HEIGHT (FT.)	COLUMN SIZE (IN.)	COLUMN WEIGHT (P.L.F.)	CONCRETE STRENGTH (PSI)	COST PER V.L.F.		
						MAT.	INST.	TOTAL
9913	150	10-14	12	135	4000	14.05	44	58.05
9918	300	10-14	16	240	4000	23	57.50	80.50
9924	500	10-14	20	375	4000	34.50	81.50	116
9930	700	10-14	24	540	4000	48	102	150
9936	1000	10-14	28	740	4000	63	123	186
9942	1400	10-14	32	965	4000	77.50	139	216.50
9948	1800	10-14	36	1220	4000	97	162	259

B1010 Floor Construction

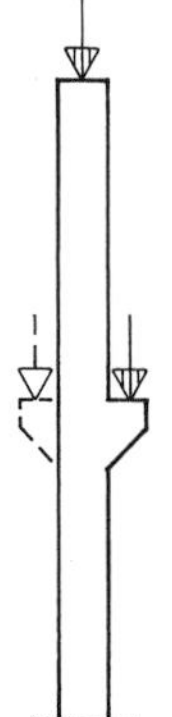

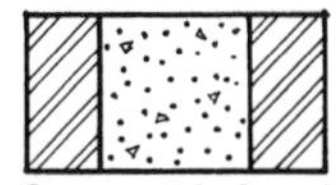

Concentric Load

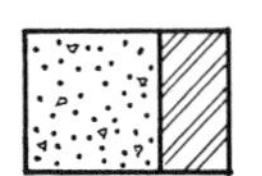

Eccentric Load

General: Data presented here is for plant produced members transported 50 miles to 100 miles to the site and erected.

Design and pricing assumptions:
Normal wt. concrete, f'c = 5 KSI

Main reinforcement, fy = 60 KSI
Ties, fy = 40 KSI

Minimum design eccentricity, 0.1t.

Concrete encased structural steel haunches are assumed where practical; otherwise galvanized rebar haunches are assumed.

Base plates are integral with columns.

Foundation anchor bolts, nuts and washers are included in price.

B1010 206 Tied, Concentric Loaded Precast Concrete Columns

	LOAD (KIPS)	STORY HEIGHT (FT.)	COLUMN SIZE (IN.)	COLUMN WEIGHT (P.L.F.)	LOAD LEVELS	COST PER V.L.F.		
						MAT.	INST.	TOTAL
0560	100	10	12x12	164	2	236	10.45	246.45
0570		12	12x12	162	2	230	8.70	238.70
0580		14	12x12	161	2	230	8.55	238.55
0590	150	10	12x12	166	3	230	8.45	238.45
0600		12	12x12	169	3	230	7.55	237.55
0610		14	12x12	162	3	230	7.40	237.40
0620	200	10	12x12	168	4	230	8.90	238.90
0630		12	12x12	170	4	230	8.10	238.10
0640		14	12x12	220	4	229	8	237

B1010 207 Tied, Eccentric Loaded Precast Concrete Columns

	LOAD (KIPS)	STORY HEIGHT (FT.)	COLUMN SIZE (IN.)	COLUMN WEIGHT (P.L.F.)	LOAD LEVELS	COST PER V.L.F.		
						MAT.	INST.	TOTAL
1130	100	10	12x12	161	2	226	10.45	236.45
1140		12	12x12	159	2	226	8.70	234.70
1150		14	12x12	159	2	226	8.55	234.55
1390	600	10	18x18	385	4	350	8.90	358.90
1400		12	18x18	380	4	350	8.10	358.10
1410		14	18x18	375	4	350	8	358
1480	800	10	20x20	490	4	320	8.90	328.90
1490		12	20x20	480	4	320	8.10	328.10
1500		14	20x20	475	4	320	8	328

B1010 Floor Construction

(A) Wide Flange

(B) Pipe

(C) Pipe, Concrete Filled

(D) Square Tube

(E) Square Tube Concrete Filled

(F) Rectangular Tube

(G) Rectangular Tube, Concrete Filled

General: The following pages provide data for seven types of steel columns: wide flange, round pipe, round pipe concrete filled, square tube, square tube concrete filled, rectangular tube and rectangular tube concrete filled.

Design Assumptions: Loads are concentric; wide flange and round pipe bearing capacity is for 36 KSI steel. Square and rectangular tubing bearing capacity is for 46 KSI steel.

The effective length factor K=1.1 is used for determining column values in the tables. K=1.1 is within a frequently used range for pinned connections with cross bracing.

How To Use Tables:

a. Steel columns usually extend through two or more stories to minimize splices. Determine floors with splices.

b. Enter Table No. below with load to column at the splice. Use the unsupported height.

c. Determine the column type desired by price or design.

Cost:

a. Multiply number of columns at the desired level by the total height of the column by the cost/VLF.

b. Repeat the above for all tiers.

Please see the reference section for further design and cost information.

B1010 208 Steel Columns

	LOAD (KIPS)	UNSUPPORTED HEIGHT (FT.)	WEIGHT (P.L.F.)	SIZE (IN.)	TYPE	COST PER V.L.F.		
						MAT.	INST.	TOTAL
1000	25	10	13	4	A	23	8.90	31.90
1020			7.58	3	B	13.30	8.90	22.20
1040			15	3-1/2	C	16.25	8.90	25.15
1120			20	4x3	G	18.80	8.90	27.70
1200		16	16	5	A	26	6.65	32.65
1220			10.79	4	B	17.50	6.65	24.15
1240			36	5-1/2	C	24.50	6.65	31.15
1320			64	8x6	G	38	6.65	44.65
1600	50	10	16	5	A	28	8.90	36.90
1620			14.62	5	B	25.50	8.90	34.40
1640			24	4-1/2	C	19.45	8.90	28.35
1720			28	6x3	G	25	8.90	33.90
1800		16	24	8	A	39	6.65	45.65
1840			36	5-1/2	C	24.50	6.65	31.15
1920			64	8x6	G	38	6.65	44.65
2000		20	28	8	A	43	6.65	49.65
2040			49	6-5/8	C	30.50	6.65	37.15
2120			64	8x6	G	36	6.65	42.65
2200	75	10	20	6	A	35	8.90	43.90
2240			36	4-1/2	C	48.50	8.90	57.40
2320			35	6x4	G	28	8.90	36.90
2400		16	31	8	A	50.50	6.65	57.15
2440			49	6-5/8	C	32	6.65	38.65
2520			64	8x6	G	38	6.65	44.65
2600		20	31	8	A	47.50	6.65	54.15
2640			81	8-5/8	C	46	6.65	52.65
2720			64	8x6	G	36	6.65	42.65
2800	100	10	24	8	A	42	8.90	50.90
2840			35	4-1/2	C	48.50	8.90	57.40
2920			46	8x4	G	34.50	8.90	43.40
3000		16	31	8	A	50.50	6.65	57.15
3040			56	6-5/8	C	47.50	6.65	54.15
3120			64	8x6	G	38	6.65	44.65

B1010 Floor Construction

B1010 208 Steel Columns

	LOAD (KIPS)	UNSUPPORTED HEIGHT (FT.)	WEIGHT (P.L.F.)	SIZE (IN.)	TYPE	COST PER V.L.F.		
						MAT.	INST.	TOTAL
3200	100	20	40	8	A	61.50	6.65	68.15
3240			81	8-5/8	C	46	6.65	52.65
3320			70	8x6	G	51.50	6.65	58.15
3400	125	10	31	8	A	54.50	8.90	63.40
3440			81	8	C	52	8.90	60.90
3520			64	8x6	G	41	8.90	49.90
3600		16	40	8	A	65	6.65	71.65
3640			81	8	C	48.50	6.65	55.15
3720			64	8x6	G	38	6.65	44.65
3800		20	48	8	A	74	6.65	80.65
3840			81	8	C	46	6.65	52.65
3920			60	8x6	G	51.50	6.65	58.15
4000	150	10	35	8	A	61.50	8.90	70.40
4040			81	8-5/8	C	52	8.90	60.90
4120			64	8x6	G	41	8.90	49.90
4200		16	45	10	A	73	6.65	79.65
4240			81	8-5/8	C	48.50	6.65	55.15
4320			70	8x6	G	54.50	6.65	61.15
4400		20	49	10	A	75.50	6.65	82.15
4440			123	10-3/4	C	65.50	6.65	72.15
4520			86	10x6	G	51	6.65	57.65
4600	200	10	45	10	A	79	8.90	87.90
4640			81	8-5/8	C	52	8.90	60.90
4720			70	8x6	G	58.50	8.90	67.40
4800		16	49	10	A	79.50	6.65	86.15
4840			123	10-3/4	C	69	6.65	75.65
4920			85	10x6	G	63	6.65	69.65
5200	300	10	61	14	A	107	8.90	115.90
5240			169	12-3/4	C	91.50	8.90	100.40
5320			86	10x6	G	87.50	8.90	96.40
5400		16	72	12	A	117	6.65	123.65
5440			169	12-3/4	C	85	6.65	91.65
5600		20	79	12	A	121	6.65	127.65
5640			169	12-3/4	C	80.50	6.65	87.15
5800	400	10	79	12	A	139	8.90	147.90
5840			178	12-3/4	C	119	8.90	127.90
6000		16	87	12	A	141	6.65	147.65
6040			178	12-3/4	C	111	6.65	117.65
6400	500	10	99	14	A	174	8.90	182.90
6600		16	109	14	A	177	6.65	183.65
6800		20	120	12	A	184	6.65	190.65
7000	600	10	120	12	A	211	8.90	219.90
7200		16	132	14	A	214	6.65	220.65
7400		20	132	14	A	203	6.65	209.65
7600	700	10	136	12	A	239	8.90	247.90
7800		16	145	14	A	235	6.65	241.65
8000		20	145	14	A	223	6.65	229.65
8200	800	10	145	14	A	254	8.90	262.90
8300		16	159	14	A	258	6.65	264.65
8400		20	176	14	A	271	6.65	277.65
8800	900	10	159	14	A	279	8.90	287.90
8900		16	176	14	A	286	6.65	292.65
9000		20	193	14	A	297	6.65	303.65

B1010 Floor Construction

B1010 208 Steel Columns

	LOAD (KIPS)	UNSUPPORTED HEIGHT (FT.)	WEIGHT (P.L.F.)	SIZE (IN.)	TYPE	COST PER V.L.F.		
						MAT.	INST.	TOTAL
9100	1000	10	176	14	A	310	8.90	318.90
9200		16	193	14	A	315	6.65	321.65
9300		20	211	14	A	325	6.65	331.65

B1010 Floor Construction

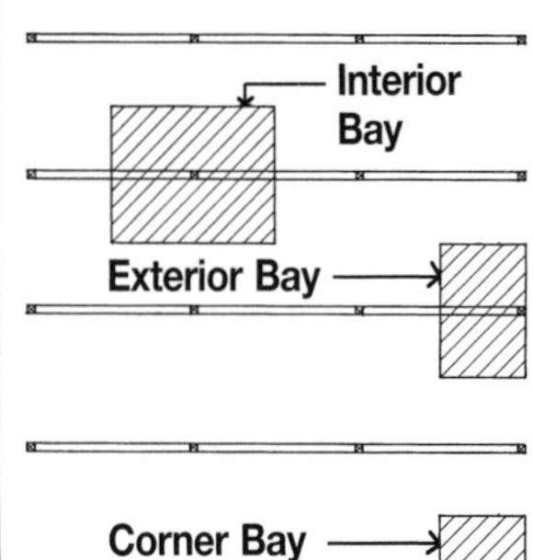

Description: Table below lists costs of columns per S.F. of bay for wood columns of various sizes and unsupported heights and the maximum allowable total load per S.F. by bay size.

Design Assumptions: Columns are concentrically loaded and are not subject to bending.

Fiber stress (f) is 1200 psi maximum.

Modulus of elasticity is 1,760,000. Use table to factor load capacity figures for modulus of elasticity other than 1,760,000.

The cost of columns per S.F. of exterior bay is proportional to the area supported. For exterior bays, multiply the costs below by two. For corner bays, multiply the cost by four.

Modulus of Elasticity	Factor
1,210,000 psi	0.69
1,320,000 psi	0.75
1,430,000 psi	0.81
1,540,000 psi	0.87
1,650,000 psi	0.94
1,760,000 psi	1.00

B1010 210 Wood Columns

	NOMINAL COLUMN SIZE (IN.)	BAY SIZE (FT.)	UNSUPPORTED HEIGHT (FT.)	MATERIAL (BF per M.S.F.)	TOTAL LOAD (P.S.F.)	COST PER S.F.		
						MAT.	INST.	TOTAL
1000	4 x 4	10 x 8	8	133	100	.27	.23	.50
1050			10	167	60	.34	.28	.62
1200		10 x 10	8	106	80	.21	.18	.39
1250			10	133	50	.27	.23	.50
1400		10 x 15	8	71	50	.14	.12	.26
1450			10	88	30	.18	.15	.33
1600		15 x 15	8	47	30	.10	.08	.18
1650			10	59	15	.12	.10	.22
2000	6 x 6	10 x 15	8	160	230	.24	.25	.49
2050			10	200	210	.30	.32	.62
2200		15 x 15	8	107	150	.16	.17	.33
2250			10	133	140	.20	.21	.41
2400		15 x 20	8	80	110	.12	.13	.25
2450			10	100	100	.15	.16	.31
2500			12	120	70	.18	.19	.37
2600		20 x 20	8	60	80	.09	.09	.18
2650			10	75	70	.11	.12	.23
2800		20 x 25	8	48	60	.07	.08	.15
2850			10	60	50	.09	.09	.18
3400	8 x 8	20 x 20	8	107	160	.17	.16	.33
3450			10	133	160	.22	.20	.42
3600		20 x 25	8	85	130	.17	.14	.31
3650			10	107	130	.22	.18	.40
3800		25 x 25	8	68	100	.13	.09	.22
3850			10	85	100	.16	.12	.28
4200	10 x 10	20 x 25	8	133	210	.26	.18	.44
4250			10	167	210	.32	.23	.55
4400		25 x 25	8	107	160	.21	.15	.36
4450			10	133	160	.26	.18	.44
4700	12 x 12	20 x 25	8	192	310	.32	.24	.56
4750			10	240	310	.40	.31	.71
4900		25 x 25	8	154	240	.26	.20	.46
4950			10	192	240	.32	.24	.56

B1010 Floor Construction

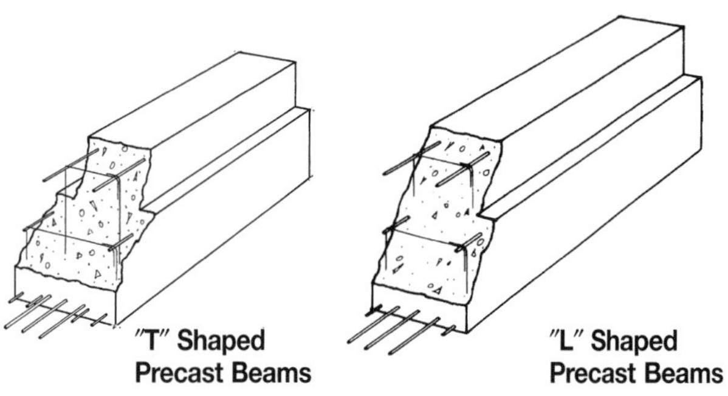

B1010 214 "T" Shaped Precast Beams

	SPAN (FT.)	SUPERIMPOSED LOAD (K.L.F.)	SIZE W X D (IN.)	BEAM WEIGHT (P.L.F.)	TOTAL LOAD (K.L.F.)	COST PER L.F.		
						MAT.	INST.	TOTAL
2300	15	2.8	12x16	260	3.06	253	14.70	267.70
2500		8.37	12x28	515	8.89	320	15.90	335.90
8900	45	3.34	12x60	1165	4.51	500	9.45	509.45
9900		6.5	24x60	1915	8.42	525	13.20	538.20

B1010 215 "L" Shaped Precast Beams

	SPAN (FT.)	SUPERIMPOSED LOAD (K.L.F.)	SIZE W X D (IN.)	BEAM WEIGHT (P.L.F.)	TOTAL LOAD (K.L.F.)	COST PER L.F.		
						MAT.	INST.	TOTAL
2250	15	2.58	12x16	230	2.81	246	14.70	260.70
2400		5.92	12x24	370	6.29	260	14.70	274.70
4000	25	2.64	12x28	435	3.08	281	9.50	290.50
4450		6.44	18x36	790	7.23	340	11.35	351.35
5300	30	2.80	12x36	565	3.37	315	8.60	323.60
6400		8.66	24x44	1245	9.90	400	12.40	412.40

B1010 Floor Construction

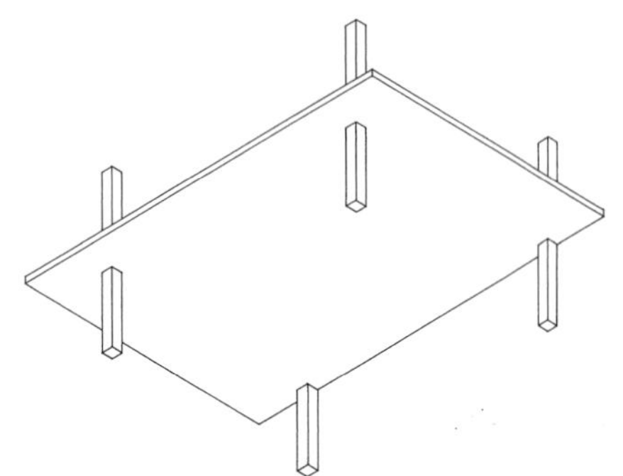

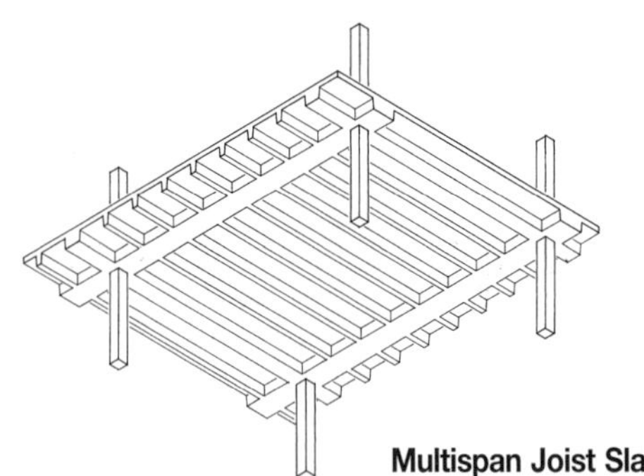
Multispan Joist Slab

General: Flat Plates: Solid uniform depth concrete two way slab without drops or interior beams. Primary design limit is shear at columns.

General: Combination of thin concrete slab and monolithic ribs at uniform spacing to reduce dead weight and increase rigidity.

B1010 223 Cast in Place Flat Plate

	BAY SIZE (FT.)	SUPERIMPOSED LOAD (P.S.F.)	MINIMUM COL. SIZE (IN.)	SLAB THICKNESS (IN.)	TOTAL LOAD (P.S.F.)	COST PER S.F.		
						MAT.	INST.	TOTAL
3000	15 x 20	40	14	7	127	5.70	7.40	13.10
3400		75	16	7-1/2	169	6.10	7.55	13.65
3600		125	22	8-1/2	231	6.75	7.80	14.55
3800		175	24	8-1/2	281	6.80	7.80	14.60
4200	20 x 20	40	16	7	127	5.75	7.40	13.15
4400		75	20	7-1/2	175	6.15	7.55	13.70
4600		125	24	8-1/2	231	6.75	7.75	14.50
5000		175	24	8-1/2	281	6.80	7.80	14.60
5600	20 x 25	40	18	8-1/2	146	6.70	7.80	14.50
6000		75	20	9	188	6.95	7.85	14.80
6400		125	26	9-1/2	244	7.50	8.10	15.60
6600		175	30	10	300	7.80	8.20	16
7000	25 x 25	40	20	9	152	6.95	7.85	14.80
7400		75	24	9-1/2	194	7.35	8.05	15.40
7600		125	30	10	250	7.85	8.20	16.05

B1010 226 Cast in Place Multispan Joist Slab

	BAY SIZE (FT.)	SUPERIMPOSED LOAD (P.S.F.)	MINIMUM COL. SIZE (IN.)	RIB DEPTH (IN.)	TOTAL LOAD (P.S.F.)	COST PER S.F.		
						MAT.	INST.	TOTAL
2000	15 x 15	40	12	8	115	8.70	9.05	17.75
2100		75	12	8	150	8.75	9.10	17.85
2200		125	12	8	200	8.90	9.20	18.10
2300		200	14	8	275	9.05	9.55	18.60
2600	15 x 20	40	12	8	115	8.85	9.05	17.90
2800		75	12	8	150	9	9.55	18.55
3000		125	14	8	200	9.25	9.75	19
3300		200	16	8	275	9.60	9.85	19.45
3600	20 x 20	40	12	10	120	9.05	8.90	17.95
3900		75	14	10	155	9.35	9.40	18.75
4000		125	16	10	205	9.40	9.55	18.95
4100		200	18	10	280	9.75	10	19.75
6200	30 x 30	40	14	14	131	9.90	9.35	19.25
6400		75	18	14	166	10.10	9.65	19.75
6600		125	20	14	216	10.65	10.15	20.80
6700		200	24	16	297	11.25	10.50	21.75

B1010 Floor Construction

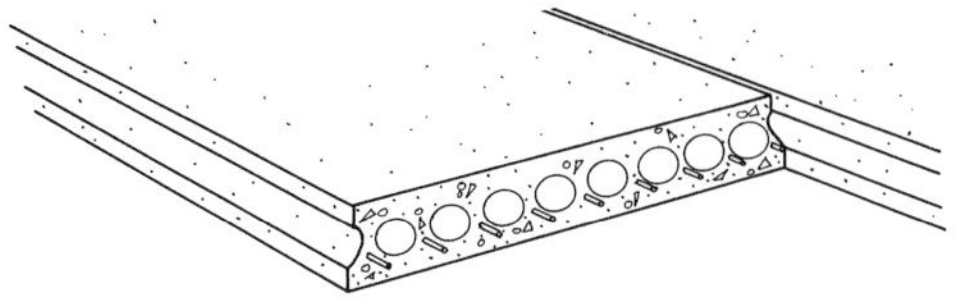

Precast Plank with No Topping

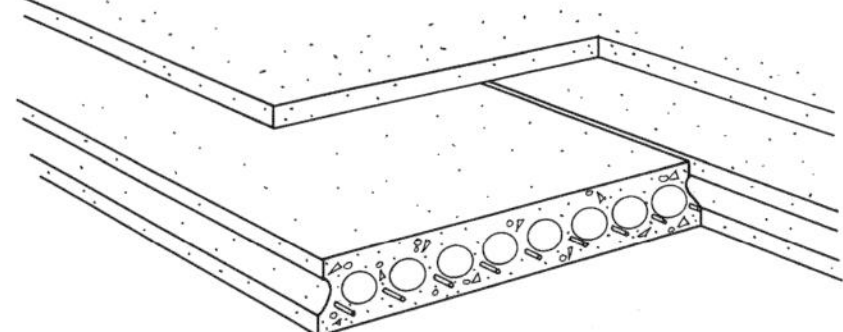

Precast Plank with 2″ Concrete Topping

B1010 229 Precast Plank with No Topping

	SPAN (FT.)	SUPERIMPOSED LOAD (P.S.F.)	TOTAL DEPTH (IN.)	DEAD LOAD (P.S.F.)	TOTAL LOAD (P.S.F.)	COST PER S.F.		
						MAT.	INST.	TOTAL
0720	10	40	4	50	90	9.30	2.48	11.78
0750		75	6	50	125	9.45	2.12	11.57
0770		100	6	50	150	9.45	2.12	11.57
0800	15	40	6	50	90	9.45	2.12	11.57
0820		75	6	50	125	9.45	2.12	11.57
0850		100	6	50	150	9.45	2.12	11.57
0950	25	40	6	50	90	9.45	2.12	11.57
0970		75	8	55	130	12.30	1.85	14.15
1000		100	8	55	155	12.30	1.85	14.15
1200	30	40	8	55	95	12.30	1.85	14.15
1300		75	8	55	130	12.30	1.85	14.15
1400		100	10	70	170	10.95	1.65	12.60
1500	40	40	10	70	110	10.95	1.65	12.60
1600		75	12	70	145	12.65	1.49	14.14
1700	45	40	12	70	110	12.65	1.49	14.14

B1010 230 Precast Plank with 2″ Concrete Topping

	SPAN (FT.)	SUPERIMPOSED LOAD (P.S.F.)	TOTAL DEPTH (IN.)	DEAD LOAD (P.S.F.)	TOTAL LOAD (P.S.F.)	COST PER S.F.		
						MAT.	INST.	TOTAL
2000	10	40	6	75	115	10.50	4.57	15.07
2100		75	8	75	150	10.65	4.21	14.86
2200		100	8	75	175	10.65	4.21	14.86
2500	15	40	8	75	115	10.65	4.21	14.86
2600		75	8	75	150	10.65	4.21	14.86
2700		100	8	75	175	10.65	4.21	14.86
3100	25	40	8	75	115	10.65	4.21	14.86
3200		75	8	75	150	10.65	4.21	14.86
3300		100	10	80	180	13.50	3.94	17.44
3400	30	40	10	80	120	13.50	3.94	17.44
3500		75	10	80	155	13.50	3.94	17.44
3600		100	10	80	180	13.50	3.94	17.44
4000	40	40	12	95	135	12.15	3.74	15.89
4500		75	14	95	170	13.85	3.58	17.43
5000	45	40	14	95	135	13.85	3.58	17.43

B1010 Floor Construction

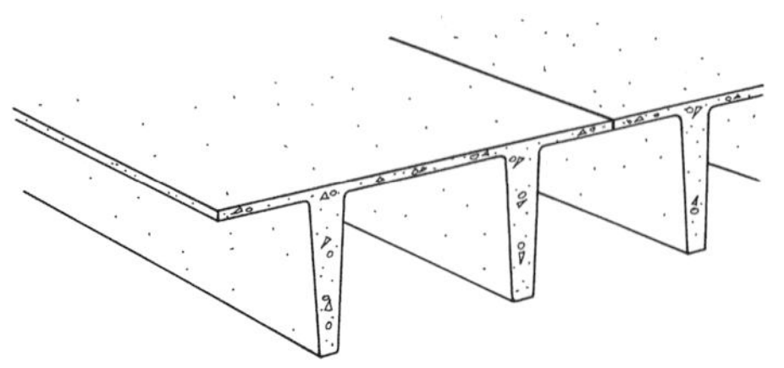

Most widely used for moderate span floors and roofs. At shorter spans, they tend to be competitive with hollow core slabs. They are also used as wall panels.

B1010 234 Precast Double "T" Beams with No Topping

	SPAN (FT.)	SUPERIMPOSED LOAD (P.S.F.)	DBL. "T" SIZE D (IN.) W (FT.)	CONCRETE "T" TYPE	TOTAL LOAD (P.S.F.)	COST PER S.F. MAT.	INST.	TOTAL
4300	50	30	20x8	Lt. Wt.	66	13.60	1.15	14.75
4400		40	20x8	Lt. Wt.	76	13.65	1.05	14.70
4500		50	20x8	Lt. Wt.	86	13.75	1.28	15.03
4600		75	20x8	Lt. Wt.	111	13.85	1.48	15.33
5600	70	30	32x10	Lt. Wt.	78	13.60	.79	14.39
5750		40	32x10	Lt. Wt.	88	13.70	.98	14.68
5900		50	32x10	Lt. Wt.	98	13.75	1.13	14.88
6000		75	32x10	Lt. Wt.	123	13.85	1.38	15.23
6100		100	32x10	Lt. Wt.	148	14.10	1.90	16
6200	80	30	32x10	Lt. Wt.	78	13.75	1.13	14.88
6300		40	32x10	Lt. Wt.	88	13.85	1.38	15.23
6400		50	32x10	Lt. Wt.	98	14	1.64	15.64

B1010 235 Precast Double "T" Beams With 2″ Topping

	SPAN (FT.)	SUPERIMPOSED LOAD (P.S.F.)	DBL. "T" SIZE D (IN.) W (FT.)	CONCRETE "T" TYPE	TOTAL LOAD (P.S.F.)	COST PER S.F. MAT.	INST.	TOTAL
7100	40	30	18x8	Reg. Wt.	120	11.65	2.60	14.25
7200		40	20x8	Reg. Wt.	130	13.75	2.49	16.24
7300		50	20x8	Reg. Wt.	140	13.85	2.71	16.56
7400		75	20x8	Reg. Wt.	165	13.90	2.82	16.72
7500		100	20x8	Reg. Wt.	190	14.05	3.14	17.19
7550	50	30	24x8	Reg. Wt.	120	13.80	2.60	16.40
7600		40	24x8	Reg. Wt.	130	13.85	2.69	16.54
7750		50	24x8	Reg. Wt.	140	13.85	2.71	16.56
7800		75	24x8	Reg. Wt.	165	14	3.03	17.03
7900		100	32x10	Reg. Wt.	189	13.85	2.52	16.37

B1010 Floor Construction

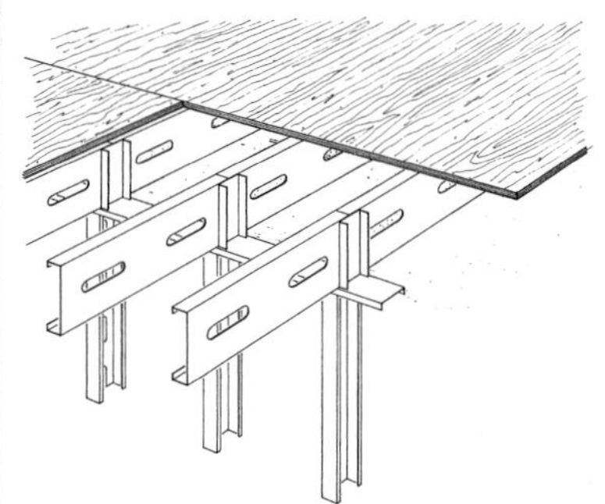

Description: Table below lists costs for light gauge CEE or PUNCHED DOUBLE joists to suit the span and loading with the minimum thickness subfloor required by the joist spacing.

Design Assumptions:

Maximum live load deflection is 1/360 of the clear span.

Maximum total load deflection is 1/240 of the clear span.

Bending strength is 20,000 psi.

8% allowance has been added to framing quantities for overlaps, double joists at openings under partitions, etc.; 5% added to glued & nailed subfloor for waste.

Maximum span is in feet and is the unsupported clear span.

B1010 244 Light Gauge Steel Floor Systems

	SPAN (FT.)	SUPERIMPOSED LOAD (P.S.F.)	FRAMING DEPTH (IN.)	FRAMING SPAC. (IN.)	TOTAL LOAD (P.S.F.)	COST PER S.F.		
						MAT.	INST.	TOTAL
1500	15	40	8	16	54	2.77	1.57	4.34
1550			8	24	54	2.92	1.62	4.54
1600		65	10	16	80	3.59	1.82	5.41
1650			10	24	80	3.57	1.80	5.37
1700		75	10	16	90	3.59	1.82	5.41
1750			10	24	90	3.57	1.80	5.37
1800		100	10	16	116	4.65	2.22	6.87
1850			10	24	116	3.57	1.80	5.37
1900		125	10	16	141	4.65	2.22	6.87
1950			10	24	141	3.57	1.80	5.37
2500	20	40	8	16	55	3.17	1.73	4.90
2550			8	24	55	2.92	1.62	4.54
2600		65	8	16	80	3.66	1.93	5.59
2650			10	24	80	3.58	1.82	5.40
2700		75	10	16	90	3.39	1.74	5.13
2750			12	24	90	3.98	1.65	5.63
2800		100	10	16	115	3.39	1.74	5.13
2850			10	24	116	4.57	2.03	6.60
2900		125	12	16	142	5.25	1.98	7.23
2950			12	24	141	4.96	2.13	7.09
3500	25	40	10	16	55	3.59	1.82	5.41
3550			10	24	55	3.57	1.80	5.37
3600		65	10	16	81	4.65	2.22	6.87
3650			12	24	81	4.40	1.76	6.16
3700		75	12	16	92	5.25	1.98	7.23
3750			12	24	91	4.96	2.13	7.09
3800		100	12	16	117	5.90	2.15	8.05
3850		125	12	16	143	6.75	2.71	9.46
4500	30	40	12	16	57	5.25	1.98	7.23
4550			12	24	56	4.38	1.76	6.14
4600		65	12	16	82	5.85	2.14	7.99
4650		75	12	16	92	5.85	2.14	7.99

B1010 Floor Construction

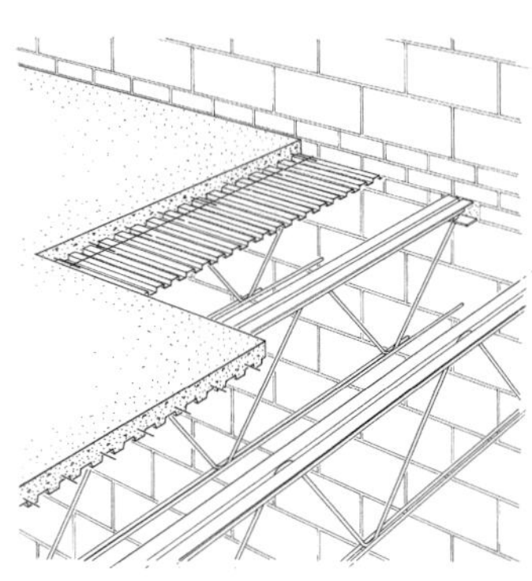

Description: Table below lists cost per S.F. for a floor system on bearing walls using open web steel joists, galvanized steel slab form and 2-1/2" concrete slab reinforced with welded wire fabric.

Design and Pricing Assumptions:
Concrete f'c = 3 KSI placed by pump.
WWF 6 x 6 – W1.4 x W1.4 (10 x 10)
Joists are spaced as shown.
Slab form is 28 gauge galvanized.
Joists costs include appropriate bridging. Deflection is limited to 1/360 of the span. Screeds and steel trowel finish.

Design Loads	Min.	Max.
Joists	3.0 PSF	7.6 PSF
Slab Form	1.0	1.0
2-1/2" Concrete	27.0	27.0
Ceiling	3.0	3.0
Misc.	9.0	9.4
	43.0 PSF	48.0 PSF

B1010 246 — Deck & Joists on Bearing Walls

	SPAN (FT.)	SUPERIMPOSED LOAD (P.S.F.)	JOIST SPACING FT. - IN.	DEPTH (IN.)	TOTAL LOAD (P.S.F.)	COST PER S.F.		
						MAT.	INST.	TOTAL
1050	20	40	2-0	14-1/2	83	6.40	3.31	9.71
1070		65	2-0	16-1/2	109	6.85	3.50	10.35
1100		75	2-0	16-1/2	119	6.85	3.50	10.35
1120		100	2-0	18-1/2	145	6.90	3.51	10.41
1150		125	1-9	18-1/2	170	7.50	3.71	11.21
1170	25	40	2-0	18-1/2	84	6.90	3.51	10.41
1200		65	2-0	20-1/2	109	7.10	3.58	10.68
1220		75	2-0	20-1/2	119	7.55	3.74	11.29
1250		100	2-0	22-1/2	145	7.70	3.77	11.47
1270		125	1-9	22-1/2	170	8.70	4.12	12.82
1300	30	40	2-0	22-1/2	84	7.65	3.37	11.02
1320		65	2-0	24-1/2	110	8.25	3.51	11.76
1350		75	2-0	26-1/2	121	8.55	3.58	12.13
1370		100	2-0	26-1/2	146	9.10	3.70	12.80
1400		125	2-0	24-1/2	172	10.60	4.05	14.65
1420	35	40	2-0	26-1/2	85	10.05	3.92	13.97
1450		65	2-0	28-1/2	111	10.35	4	14.35
1470		75	2-0	28-1/2	121	10.35	4	14.35
1500		100	1-11	28-1/2	147	11.25	4.19	15.44
1520		125	1-8	28-1/2	172	12.30	4.45	16.75
1550	5/8" gyp. fireproof.							
1560	On metal furring, add					.87	2.85	3.72

B1010 Floor Construction

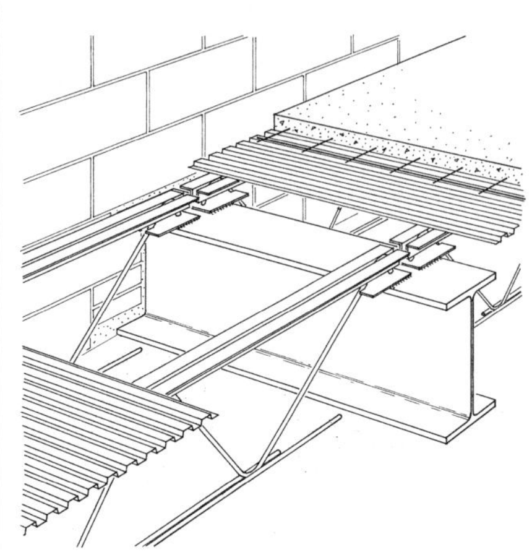

Table below lists costs for a floor system on exterior bearing walls and interior columns and beams using open web steel joists, galvanized steel slab form, 2-1/2″ concrete slab reinforced with welded wire fabric.

Design and Pricing Assumptions:
- Structural Steel is A36.
- Concrete f'c = 3 KSI placed by pump.
- WWF 6 x 6 – W1.4 x W1.4 (10 x 10)
- Columns are 12′ high.
- Building is 4 bays long by 4 bays wide.
- Joists are 2′ O.C. ± and span the long direction of the bay.

Joists at columns have bottom chords extended and are connected to columns.

Slab form is 28 gauge galvanized. Column costs in table are for columns to support 1 floor plus roof loading in a 2-story building; however, column costs are from ground floor to 2nd floor only. Joist costs include appropriate bridging. Deflection is limited to 1/360 of the span. Screeds and steel trowel finish.

Design Loads	Min.	Max.
S.S & Joists	4.4 PSF	11.5 PSF
Slab Form	1.0	1.0
2-1/2″ Concrete	27.0	27.0
Ceiling	3.0	3.0
Misc.	7.6	5.5
	43.0 PSF	48.0 PSF

B1010 248 Steel Joists on Beam & Wall

	BAY SIZE (FT.)	SUPERIMPOSED LOAD (P.S.F.)	DEPTH (IN.)	TOTAL LOAD (P.S.F.)	COLUMN ADD	COST PER S.F.		
						MAT.	INST.	TOTAL
1720	25x25	40	23	84		8.45	3.78	12.23
1730					columns	.41	.10	.51
1750	25x25	65	29	110		8.90	3.89	12.79
1760					columns	.41	.10	.51
1770	25x25	75	26	120		9.85	4.18	14.03
1780					columns	.48	.11	.59
1800	25x25	100	29	145		11.05	4.56	15.61
1810					columns	.48	.11	.59
1820	25x25	125	29	170		11.60	4.70	16.30
1830					columns	.53	.13	.66
2020	30x30	40	29	84		9.65	3.74	13.39
2030					columns	.37	.09	.46
2050	30x30	65	29	110		10.90	4.03	14.93
2060					columns	.37	.09	.46
2070	30x30	75	32	120		11.20	4.10	15.30
2080					columns	.43	.10	.53
2100	30x30	100	35	145		12.35	4.37	16.72
2110					columns	.50	.12	.62
2120	30x30	125	35	172		14.25	4.82	19.07
2130					columns	.60	.14	.74
2170	30x35	40	29	85		10.85	4.02	14.87
2180					columns	.36	.08	.44
2200	30x35	65	29	111		12.90	4.49	17.39
2210					columns	.41	.10	.51
2220	30x35	75	32	121		12.90	4.49	17.39
2230					columns	.42	.10	.52
2250	30x35	100	35	148		13.10	4.54	17.64
2260					columns	.51	.13	.64
2270	30x35	125	38	173		14.65	4.91	19.56
2280					columns	.52	.13	.65
2320	35x35	40	32	85		11.15	4.09	15.24
2330					columns	.37	.09	.46
2350	35x35	65	35	111		13.55	4.65	18.20
2360					columns	.44	.11	.55
2370	35x35	75	35	121		13.85	4.72	18.57
2380					columns	.44	.11	.55

B1010 Floor Construction

B1010 248 Steel Joists on Beam & Wall

	BAY SIZE (FT.)	SUPERIMPOSED LOAD (P.S.F.)	DEPTH (IN.)	TOTAL LOAD (P.S.F.)	COLUMN ADD	COST PER S.F.		
						MAT.	INST.	TOTAL
2400	35x35	100	38	148		14.40	4.86	19.26
2410					columns	.54	.13	.67
2460	5/8 gyp. fireproof.							
2475	On metal furring, add					.87	2.85	3.72

B1010 Floor Construction

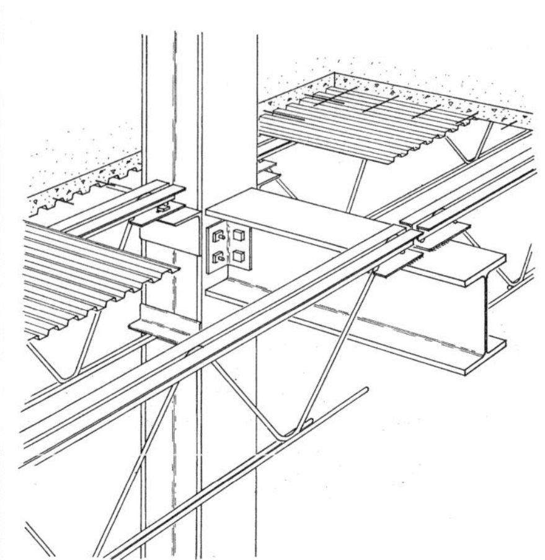

Table below lists costs for a floor system on steel columns and beams using open web steel joists, galvanized steel slab form, and 2-1/2″ concrete slab reinforced with welded wire fabric.

Design and Pricing Assumptions:

Structural Steel is A36.
Concrete f'c = 3 KSI placed by pump.
WWF 6 x 6 – W1.4 x W1.4 (10 x 10)
Columns are 12′ high.
Building is 4 bays long by 4 bays wide.
Joists are 2′ O.C. ± and span the long direction of the bay.

Joists at columns have bottom chords extended and are connected to columns.

Slab form is 28 gauge galvanized.
Column costs in table are for columns to support 1 floor plus roof loading in a 2-story building; however, column costs are from ground floor to 2nd floor only. Joist costs include appropriate bridging.
Deflection is limited to 1/360 of the span.
Screeds and steel trowel finish.

Design Loads	Min.	Max.
S.S. & Joists	6.3 PSF	15.3 PSF
Slab Form	1.0	1.0
2-1/2″ Concrete	27.0	27.0
Ceiling	3.0	3.0
Misc.	5.7	1.7
	43.0 PSF	48.0 PSF

B1010 250 Steel Joists, Beams & Slab on Columns

	BAY SIZE (FT.)	SUPERIMPOSED LOAD (P.S.F.)	DEPTH (IN.)	TOTAL LOAD (P.S.F.)	COLUMN ADD	COST PER S.F.		
						MAT.	INST.	TOTAL
2350	15x20	40	17	83		8.80	3.80	12.60
2400					column	1.37	.33	1.70
2450	15x20	65	19	108		9.70	4.03	13.73
2500					column	1.37	.33	1.70
2550	15x20	75	19	119		10.05	4.15	14.20
2600					column	1.50	.36	1.86
2650	15x20	100	19	144		10.70	4.33	15.03
2700					column	1.50	.36	1.86
2750	15x20	125	19	170		11.20	4.45	15.65
2800					column	2	.48	2.48
2850	20x20	40	19	83		9.55	3.97	13.52
2900					column	1.12	.27	1.39
2950	20x20	65	23	109		10.45	4.24	14.69
3000					column	1.50	.36	1.86
3100	20x20	75	26	119		11.05	4.38	15.43
3200					column	1.50	.36	1.86
3400	20x20	100	23	144		11.45	4.48	15.93
3450					column	1.50	.36	1.86
3500	20x20	125	23	170		12.70	4.83	17.53
3600					column	1.80	.43	2.23
3700	20x25	40	44	83		10.15	4.17	14.32
3800					column	1.20	.29	1.49
3900	20x25	65	26	110		11.10	4.42	15.52
4000					column	1.20	.29	1.49
4100	20x25	75	26	120		11.60	4.57	16.17
4200					column	1.44	.35	1.79
4300	20x25	100	26	145		12.25	4.74	16.99
4400					column	1.44	.35	1.79
4500	20x25	125	29	170		13.60	5.15	18.75
4600					column	1.68	.41	2.09
4700	25x25	40	23	84		11	4.36	15.36
4800					column	1.15	.28	1.43
4900	25x25	65	29	110		11.65	4.53	16.18
5000					column	1.15	.28	1.43
5100	25x25	75	26	120		12.80	4.87	17.67
5200					column	1.34	.32	1.66

B1010 Floor Construction

B1010 250 Steel Joists, Beams & Slab on Columns

	BAY SIZE (FT.)	SUPERIMPOSED LOAD (P.S.F.)	DEPTH (IN.)	TOTAL LOAD (P.S.F.)	COLUMN ADD	COST PER S.F.		
						MAT.	INST.	TOTAL
5300	25x25	100	29	145		14.20	5.30	19.50
5400					column	1.34	.32	1.66
5500	25x25	125	32	170		15	5.50	20.50
5600					column	1.48	.36	1.84
5700	25x30	40	29	84		12.55	4.39	16.94
5800					column	1.12	.27	1.39
5900	25x30	65	29	110		13.05	4.51	17.56
6000					column	1.12	.27	1.39
6050	25x30	75	29	120		13.40	4.62	18.02
6100					column	1.24	.30	1.54
6150	25x30	100	29	145		14.50	4.87	19.37
6200					column	1.24	.30	1.54
6250	25x30	125	32	170		16.25	5.25	21.50
6300					column	1.42	.35	1.77
6350	30x30	40	29	84		12.55	4.40	16.95
6400					column	1.03	.25	1.28
6500	30x30	65	29	110		14.20	4.80	19
6600					column	1.03	.25	1.28
6700	30x30	75	32	120		14.50	4.87	19.37
6800					column	1.19	.28	1.47
6900	30x30	100	35	145		16.05	5.25	21.30
7000					column	1.38	.34	1.72
7100	30x30	125	35	172		18.25	5.75	24
7200					column	1.54	.38	1.92
7300	30x35	40	29	85		14.10	4.77	18.87
7400					column	.88	.21	1.09
7500	30x35	65	29	111		16.35	5.30	21.65
7600					column	1.14	.28	1.42
7700	30x35	75	32	121		16.35	5.30	21.65
7800					column	1.16	.28	1.44
7900	30x35	100	35	148		16.85	5.45	22.30
8000					column	1.42	.35	1.77
8100	30x35	125	38	173		18.65	5.85	24.50
8200					column	1.45	.35	1.80
8300	35x35	40	32	85		14.45	4.86	19.31
8400					column	1.02	.25	1.27
8500	35x35	65	35	111		17.20	5.50	22.70
8600					column	1.22	.29	1.51
9300	35x35	75	38	121		17.60	5.60	23.20
9400					column	1.22	.29	1.51
9500	35x35	100	38	148		18.95	5.95	24.90
9600					column	1.51	.36	1.87
9750	35x35	125	41	173		19.55	6.05	25.60
9800					column	1.53	.37	1.90
9810	5/8 gyp. fireproof.							
9815	On metal furring, add					.87	2.85	3.72

B1010 Floor Construction

Description: Table B1010 258 lists S.F. costs for steel deck and concrete slabs for various spans.

Description: Table B1010 261 lists the S.F. costs for wood joists and a minimum thickness plywood subfloor.

Description: Table B1010 264 lists the S.F. costs, total load, and member sizes, for various bay sizes and loading conditions.

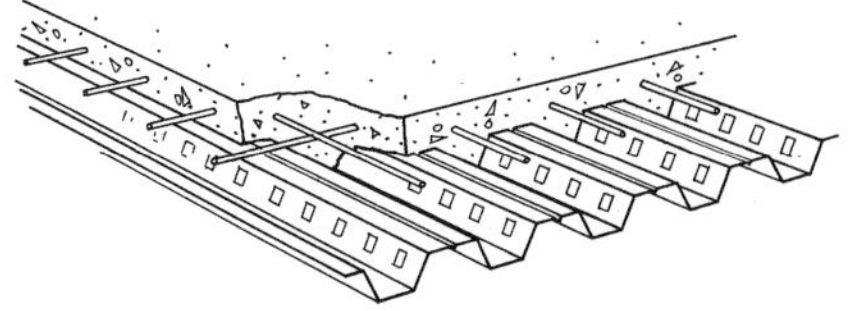

Metal Deck/Concrete Fill

Wood Joist

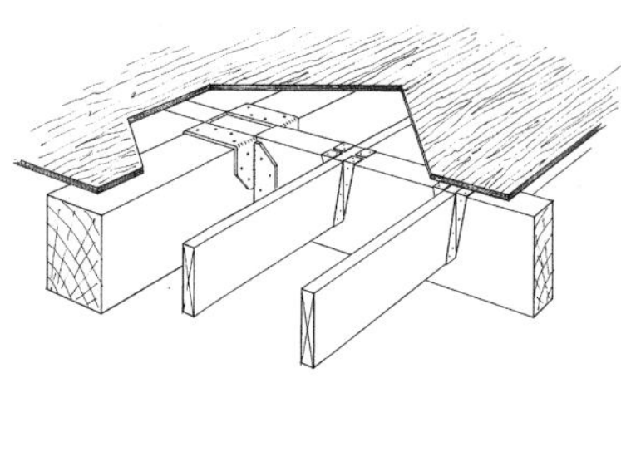

Wood Beam and Joist

B1010 258 Metal Deck/Concrete Fill

	SUPERIMPOSED LOAD (P.S.F.)	DECK SPAN (FT.)	DECK GAGE DEPTH	SLAB THICKNESS (IN.)	TOTAL LOAD (P.S.F.)	COST PER S.F.		
						MAT.	INST.	TOTAL
0900	125	6	22 1-1/2	4	164	3.38	2.17	5.55
0920		7	20 1-1/2	4	164	3.70	2.29	5.99
0950		8	20 1-1/2	4	165	3.70	2.29	5.99
0970		9	18 1-1/2	4	165	4.31	2.29	6.60
1000		10	18 2	4	165	4.94	2.41	7.35
1020		11	18 3	5	169	5.15	2.60	7.75

B1010 261 Wood Joist

		COST PER S.F.		
		MAT.	INST.	TOTAL
2900	2"x8", 12" O.C.	1.91	1.73	3.64
2950	16" O.C.	1.62	1.48	3.10
3000	24" O.C.	1.67	1.38	3.05
3300	2"x10", 12" O.C.	2.50	1.98	4.48
3350	16" O.C.	2.06	1.66	3.72
3400	24" O.C.	1.95	1.49	3.44
3700	2"x12", 12" O.C.	2.94	2	4.94
3750	16" O.C.	2.39	1.68	4.07
3800	24" O.C.	2.18	1.50	3.68
4100	2"x14", 12" O.C.	3.56	2.18	5.74
4150	16" O.C.	2.86	1.82	4.68
4200	24" O.C.	2.50	1.60	4.10
7101	Note: Subfloor cost is included in these prices.			

B1010 264 Wood Beam & Joist

	BAY SIZE (FT.)	SUPERIMPOSED LOAD (P.S.F.)	GIRDER BEAM (IN.)	JOISTS (IN.)	TOTAL LOAD (P.S.F.)	COST PER S.F.		
						MAT.	INST.	TOTAL
2000	15x15	40	8 x 12 4 x 12	2 x 6 @ 16	53	5.70	3.56	9.26
2050		75	8 x 16 4 x 16	2 x 8 @ 16	90	8	3.79	11.79
2100		125	12 x 16 6 x 16	2 x 8 @ 12	144	12.40	4.77	17.17
2150		200	14 x 22 12 x 16	2 x 10 @ 12	227	22	6.55	28.55
3000	20x20	40	10 x 14 10 x 12	2 x 8 @ 16	63	8.50	3.60	12.10
3050		75	12 x 16 8 x 16	2 x 10 @ 16	102	13.25	4.28	17.53
3100		125	14 x 22 12 x 16	2 x 10 @ 12	163	17.95	5.35	23.30

B1010 Floor Construction

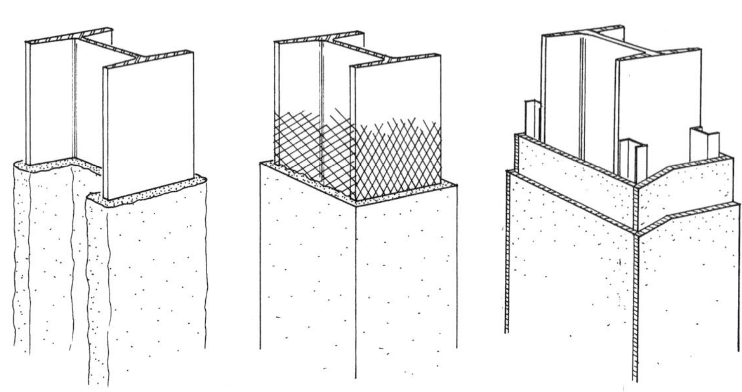

Listed below are costs per V.L.F. for fireproofing by material, column size, thickness and fire rating. Weights listed are for the fireproofing material only.

B1010 720 Steel Column Fireproofing

	ENCASEMENT SYSTEM	COLUMN SIZE (IN.)	THICKNESS (IN.)	FIRE RATING (HRS.)	WEIGHT (P.L.F.)	COST PER V.L.F.		
						MAT.	INST.	TOTAL
3000	Concrete	8	1	1	110	7.20	30.50	37.70
3300		14	1	1	258	12.15	44	56.15
3400			2	3	325	14.55	50	64.55
3450	Gypsum board	8	1/2	2	8	3.78	19.85	23.63
3550	1 layer	14	1/2	2	18	4.09	21	25.09
3600	Gypsum board	8	1	3	14	5.25	25.50	30.75
3650	1/2" fire rated	10	1	3	17	5.60	27	32.60
3700	2 layers	14	1	3	22	5.80	27.50	33.30
3750	Gypsum board	8	1-1/2	3	23	7	32	39
3800	1/2" fire rated	10	1-1/2	3	27	7.85	35.50	43.35
3850	3 layers	14	1-1/2	3	35	8.70	38.50	47.20
3900	Sprayed fiber	8	1-1/2	2	6.3	4.34	6.45	10.79
3950	Direct application		2	3	8.3	6	8.90	14.90
4050		10	1-1/2	2	7.9	5.25	7.80	13.05
4200	Sprayed fiber	14	1-1/2	2	10.8	6.50	9.65	16.15

B1020 Roof Construction

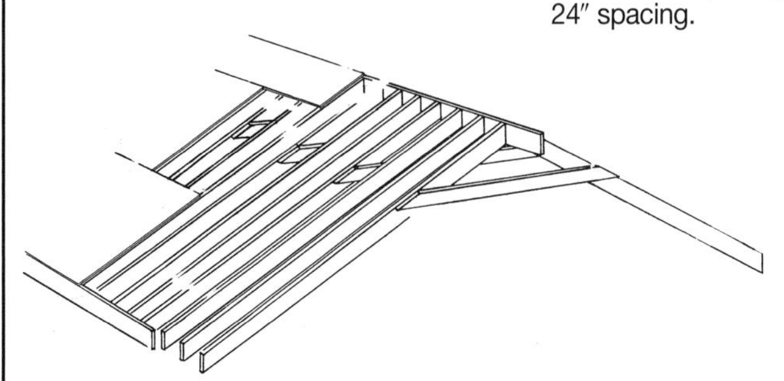

The table below lists prices per S.F. for roof rafters and sheathing by nominal size and spacing. Sheathing is 5/16″ CDX for 12″ and 16″ spacing and 3/8″ CDX for 24″ spacing.

Factors for Converting Inclined to Horizontal

Roof Slope	Approx. Angle	Factor	Roof Slope	Approx. Angle	Factor
Flat	0°	1.000	12 in 12	45.0°	1.414
1 in 12	4.8°	1.003	13 in 12	47.3°	1.474
2 in 12	9.5°	1.014	14 in 12	49.4°	1.537
3 in 12	14.0°	1.031	15 in 12	51.3°	1.601
4 in 12	18.4°	1.054	16 in 12	53.1°	1.667
5 in 12	22.6°	1.083	17 in 12	54.8°	1.734
6 in 12	26.6°	1.118	18 in 12	56.3°	1.803
7 in 12	30.3°	1.158	19 in 12	57.7°	1.873
8 in 12	33.7°	1.202	20 in 12	59.0°	1.943
9 in 12	36.9°	1.250	21 in 12	60.3°	2.015
10 in 12	39.8°	1.302	22 in 12	61.4°	2.088
11 in 12	42.5°	1.357	23 in 12	62.4°	2.162

B1020 102	Wood/Flat or Pitched	COST PER S.F.		
		MAT.	INST.	TOTAL
2500	Flat rafter, 2″x4″, 12″ O.C.	1.19	1.56	2.75
2550	16″ O.C.	1.07	1.34	2.41
2600	24″ O.C.	.97	1.15	2.12
2900	2″x6″, 12″ O.C.	1.55	1.57	3.12
2950	16″ O.C.	1.34	1.35	2.69
3000	24″ O.C.	1.15	1.15	2.30
3300	2″x8″, 12″ O.C.	1.85	1.69	3.54
3350	16″ O.C.	1.56	1.44	3
3400	24″ O.C.	1.30	1.22	2.52
3700	2″x10″, 12″ O.C.	2.44	1.94	4.38
3750	16″ O.C.	2	1.62	3.62
3800	24″ O.C.	1.58	1.33	2.91
4100	2″x12″, 12″ O.C.	2.88	1.96	4.84
4150	16″ O.C.	2.55	1.80	4.35
4200	24″ O.C.	1.81	1.34	3.15
4500	2″x14″, 12″ O.C.	3.50	2.86	6.36
4550	16″ O.C.	2.80	2.32	5.12
4600	24″ O.C.	2.13	1.80	3.93
4900	3″x6″, 12″ O.C.	3.20	1.67	4.87
4950	16″ O.C.	2.57	1.42	3.99
5000	24″ O.C.	1.98	1.20	3.18
5300	3″x8″, 12″ O.C.	4.26	1.86	6.12
5350	16″ O.C.	3.36	1.56	4.92
5400	24″ O.C.	2.50	1.29	3.79
5700	3″x10″, 12″ O.C.	5.15	2.11	7.26
5750	16″ O.C.	4.04	1.76	5.80
5800	24″ O.C.	2.95	1.42	4.37
6100	3″x12″, 12″ O.C.	5.95	2.55	8.50
6150	16″ O.C.	4.65	2.08	6.73
6200	24″ O.C.	3.35	1.64	4.99
7001	Wood truss, 4 in 12 slope, 24″ O.C., 24′ to 29′ span	3.93	2.23	6.16
7100	30′ to 43′ span	4.76	2.46	7.22
7200	44′ to 60′ span	4.58	2.23	6.81

B1020 Roof Construction

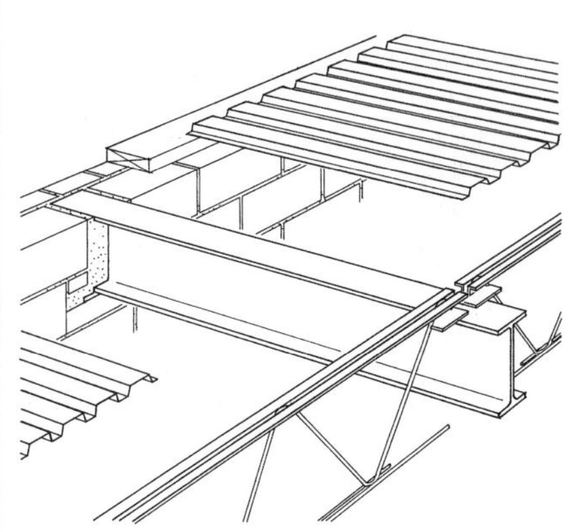

Table below lists the cost per S.F. for a roof system with steel columns, beams, and deck using open web steel joists and 1-1/2" galvanized metal deck. Perimeter of system is supported on bearing walls.

Design and Pricing Assumptions:
Columns are 18' high.
Joists are 5'-0" O.C. and span the long direction of the bay.

Joists at columns have bottom chords extended and are connected to columns. Column costs are not included but are listed separately per S.F. of floor.

Roof deck is 1-1/2", 22 gauge galvanized steel. Joist cost includes appropriate bridging. Deflection is limited to 1/240 of the span. Fireproofing is not included.

Costs/S.F. are based on a building 4 bays long and 4 bays wide.

Design Loads	**Min.**	**Max.**
Joists & Beams	3 PSF	5 PSF
Deck	2	2
Insulation	3	3
Roofing	6	6
Misc.	6	6
Total Dead Load	20 PSF	22 PSF

B1020 108	**Steel Joists, Beams & Deck on Columns & Walls**							
	BAY SIZE (FT.)	SUPERIMPOSED LOAD (P.S.F.)	DEPTH (IN.)	TOTAL LOAD (P.S.F.)	COLUMN ADD	COST PER S.F.		
						MAT.	INST.	TOTAL
3000	25x25	20	18	40		4.42	1.17	5.59
3100					columns	.39	.07	.46
3200		30	22	50		4.73	1.23	5.96
3300					columns	.52	.10	.62
3400		40	20	60		5.20	1.37	6.57
3500					columns	.52	.10	.62
3600	25x30	20	22	40		4.66	1.11	5.77
3700					columns	.43	.08	.51
3800		30	20	50		5.25	1.24	6.49
3900					columns	.43	.08	.51
4000		40	25	60		5.45	1.27	6.72
4100					columns	.52	.10	.62
4200	30x30	20	25	42		5.10	1.19	6.29
4300					columns	.36	.07	.43
4400		30	22	52		5.55	1.29	6.84
4500					columns	.43	.08	.51
4600		40	28	62		5.80	1.33	7.13
4700					columns	.43	.08	.51
4800	30x35	20	22	42		5.30	1.24	6.54
4900					columns	.37	.07	.44
5000		30	28	52		5.60	1.29	6.89
5100					columns	.37	.07	.44
5200		40	25	62		6.05	1.40	7.45
5300					columns	.43	.08	.51
5400	35x35	20	28	42		5.30	1.24	6.54
5500					columns	.32	.06	.38
5600		30	25	52		6.45	1.48	7.93
5700					columns	.37	.07	.44
5800		40	28	62		6.55	1.49	8.04
5900					columns	.41	.08	.49

B1020 Roof Construction

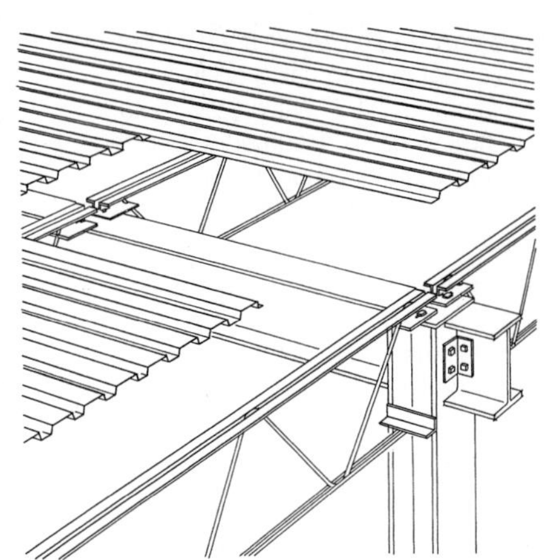

Description: Table below lists the cost per S.F. for a roof system with steel columns, beams, and deck, using open web steel joists and 1-1/2" galvanized metal deck.

Design and Pricing Assumptions:
Columns are 18' high.
Building is 4 bays long by 4 bays wide.
Joists are 5'-0" O.C. and span the long direction of the bay.
Joists at columns have bottom chords extended and are connected to columns.
Column costs are not included but are listed separately per S.F. of floor.

Roof deck is 1-1/2", 22 gauge galvanized steel. Joist cost includes appropriate bridging. Deflection is limited to 1/240 of the span. Fireproofing is not included.

Design Loads	Min.	Max.
Joists & Beams	3 PSF	5 PSF
Deck	2	2
Insulation	3	3
Roofing	6	6
Misc.	6	6
Total Dead Load	20 PSF	22 PSF

B1020 112 Steel Joists, Beams, & Deck on Columns

	BAY SIZE (FT.)	SUPERIMPOSED LOAD (P.S.F.)	DEPTH (IN.)	TOTAL LOAD (P.S.F.)	COLUMN ADD	COST PER S.F.		
						MAT.	INST.	TOTAL
1100	15x20	20	16	40		4.37	1.16	5.53
1200					columns	2.24	.43	2.67
1500		40	18	60		4.86	1.29	6.15
1600					columns	2.24	.43	2.67
2300	20x25	20	18	40		4.96	1.30	6.26
2400					columns	1.35	.26	1.61
2700		40	20	60		5.55	1.45	7
2800					columns	1.80	.35	2.15
2900	25x25	20	18	40		5.80	1.45	7.25
3000					columns	1.08	.21	1.29
3100		30	22	50		6	1.49	7.49
3200					columns	1.44	.28	1.72
3300		40	20	60		6.65	1.67	8.32
3400					columns	1.44	.28	1.72
3500	25x30	20	22	40		5.65	1.31	6.96
3600					columns	1.20	.24	1.44
3900		40	25	60		6.80	1.55	8.35
4000					columns	1.44	.28	1.72
4100	30x30	20	25	42		6.40	1.45	7.85
4200					columns	1	.20	1.20
4300		30	22	52		7.05	1.59	8.64
4400					columns	1.20	.24	1.44
4500		40	28	62		7.40	1.65	9.05
4600					columns	1.20	.24	1.44
5300	35x35	20	28	42		6.95	1.57	8.52
5400					columns	.88	.17	1.05
5500		30	25	52		7.80	1.73	9.53
5600					columns	1.03	.20	1.23
5700		40	28	62		8.40	1.87	10.27
5800					columns	1.13	.22	1.35

B1020 Roof Construction

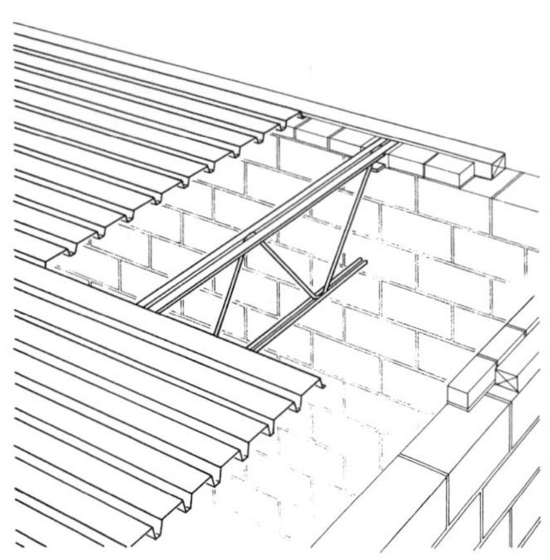

Description: Table below lists cost per S.F. for a roof system using open web steel joists and 1-1/2″ galvanized metal deck. The system is assumed supported on bearing walls or other suitable support. Costs for the supports are not included.

Design and Pricing Assumptions:
Joists are 5′-0″ O.C.
Roof deck is 1-1/2″, 22 gauge galvanized.

B1020 116 Steel Joists & Deck on Bearing Walls

	BAY SIZE (FT.)	SUPERIMPOSED LOAD (P.S.F.)	DEPTH (IN.)	TOTAL LOAD (P.S.F.)		COST PER S.F.		
						MAT.	INST.	TOTAL
1100	20	20	13-1/2	40		2.85	.85	3.70
1200		30	15-1/2	50		2.91	.87	3.78
1300		40	15-1/2	60		3.11	.94	4.05
1400	25	20	17-1/2	40		3.12	.94	4.06
1500		30	17-1/2	50		3.35	1.01	4.36
1600		40	19-1/2	60		3.39	1.02	4.41
1700	30	20	19-1/2	40		3.38	1.02	4.40
1800		30	21-1/2	50		3.46	1.05	4.51
1900		40	23-1/2	60		3.71	1.13	4.84
2000	35	20	23-1/2	40		3.67	.93	4.60
2100		30	25-1/2	50		3.77	.95	4.72
2200		40	25-1/2	60		4	1	5
2300	40	20	25-1/2	41		4.05	1.02	5.07
2400		30	25-1/2	51		4.30	1.07	5.37
2500		40	25-1/2	61		4.44	1.11	5.55
2600	45	20	27-1/2	41		4.85	1.20	6.05
2700		30	31-1/2	51		5.10	1.27	6.37
2800		40	31-1/2	61		5.40	1.34	6.74
2900	50	20	29-1/2	42		5.40	1.34	6.74
3000		30	31-1/2	52		5.90	1.45	7.35
3100		40	31-1/2	62		6.25	1.54	7.79
3200	60	20	37-1/2	42		6.45	1.75	8.20
3300		30	37-1/2	52		7.10	1.94	9.04
3400		40	37-1/2	62		7.10	1.94	9.04
3500	70	20	41-1/2	42		7.10	1.94	9.04
3600		30	41-1/2	52		7.60	2.07	9.67
3700		40	41-1/2	64		9.20	2.50	11.70
3800	80	20	45-1/2	44		8.95	2.43	11.38
3900		30	45-1/2	54		8.95	2.43	11.38
4000		40	45-1/2	64		9.85	2.68	12.53
4400	100	20	57-1/2	44		8.80	2.31	11.11
4500		30	57-1/2	54		10.45	2.74	13.19
4600		40	57-1/2	65		11.55	3.02	14.57
4700	125	20	69-1/2	44		10.45	2.74	13.19
4800		30	69-1/2	56		12.15	3.17	15.32
4900		40	69-1/2	67		13.80	3.61	17.41

B1020 Roof Construction

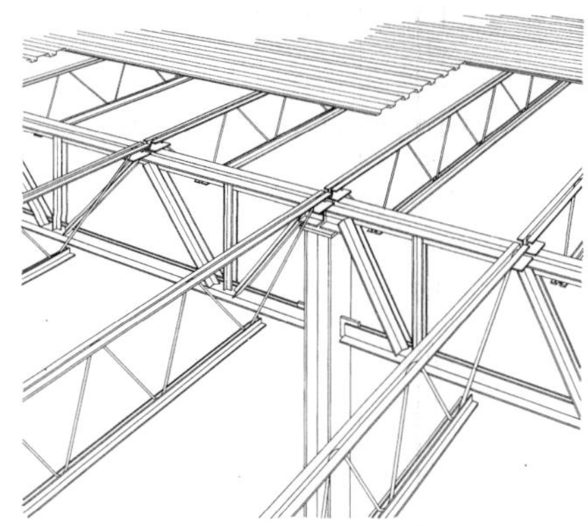

Description: Table below lists costs for a roof system supported on exterior bearing walls and interior columns. Costs include bracing, joist girders, open web steel joists and 1-1/2" galvanized metal deck.

Design and Pricing Assumptions:
Columns are 18′ high.
Joists are 5′-0″ O.C.
Joist girders and joists have bottom chords connected to columns. Roof deck is 1-1/2", 22 gauge galvanized steel. Costs include bridging and bracing. Deflection is limited to 1/240 of the span.

Fireproofing is not included.
Costs/S.F. are based on a building 4 bays long and 4 bays wide.
Costs for bearing walls are not included.

Column costs are not included but are listed separately per S.F. of floor.

B1020 120 Steel Joists, Joist Girders & Deck on Columns & Walls

	BAY SIZE (FT.) GIRD X JOISTS	SUPERIMPOSED LOAD (P.S.F.)	DEPTH (IN.)	TOTAL LOAD (P.S.F.)	COLUMN ADD	COST PER S.F.		
						MAT.	INST.	TOTAL
2350	30x35	20	32-1/2	40		4.09	1.16	5.25
2400					columns	.37	.07	.44
2550		40	36-1/2	60		4.57	1.28	5.85
2600					columns	.43	.08	.51
3000	35x35	20	36-1/2	40		4.45	1.44	5.89
3050					columns	.32	.06	.38
3200		40	36-1/2	60		4.97	1.58	6.55
3250					columns	.41	.08	.49
3300	35x40	20	36-1/2	40		4.51	1.30	5.81
3350					columns	.32	.06	.38
3500		40	36-1/2	60		5.05	1.46	6.51
3550					columns	.36	.07	.43
3900	40x40	20	40-1/2	41		5.20	1.47	6.67
3950					columns	.31	.06	.37
4100		40	40-1/2	61		5.80	1.60	7.40
4150					columns	.31	.06	.37
5100	45x50	20	52-1/2	41		5.85	1.66	7.51
5150					columns	.22	.04	.26
5300		40	52-1/2	61		7.10	1.97	9.07
5350					columns	.32	.06	.38
5400	50x45	20	56-1/2	41		5.65	1.61	7.26
5450					columns	.22	.04	.26
5600		40	56-1/2	61		7.25	2.02	9.27
5650					columns	.30	.06	.36
5700	50x50	20	56-1/2	42		6.35	1.80	8.15
5750					columns	.23	.04	.27
5900		40	59	64		7.40	2.21	9.61
5950					columns	.32	.06	.38
6300	60x50	20	62-1/2	43		6.40	1.85	8.25
6350					columns	.36	.07	.43
6500		40	71	65		7.55	2.28	9.83
6550					columns	.48	.10	.58

B1020 Roof Construction

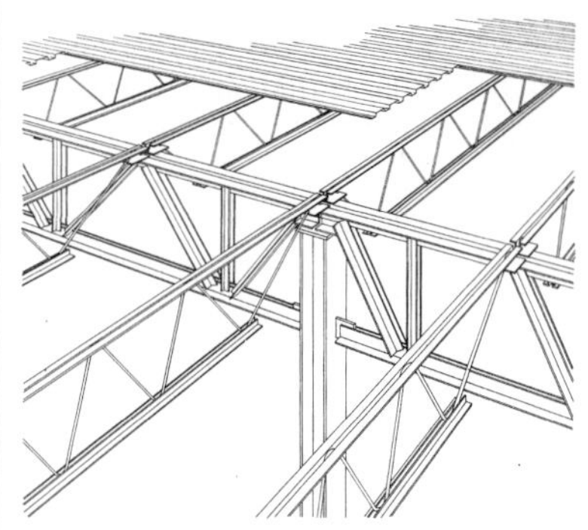

Description: Table below lists the cost per S.F. for a roof system supported on columns. Costs include joist girders, open web steel joists and 1-1/2" galvanized metal deck.

Design and Pricing Assumptions:
Columns are 18′ high.
Joists are 5′-0" O.C.
Joist girders and joists have bottom chords connected to columns. Roof deck is 1-1/2", 22 gauge galvanized steel. Costs include bridging and bracing. Deflection is limited to 1/240 of the span. Fireproofing is not included.

Costs/S.F. are based on a building 4 bays long and 4 bays wide.

Costs for columns are not included, but are listed separately per S.F. of area.

B1020 124 Steel Joists & Joist Girders on Columns

	BAY SIZE (FT.) GIRD X JOISTS	SUPERIMPOSED LOAD (P.S.F.)	DEPTH (IN.)	TOTAL LOAD (P.S.F.)	COLUMN ADD	COST PER S.F.		
						MAT.	INST.	TOTAL
2000	30x30	20	17-1/2	40		4.10	1.39	5.49
2050					columns	1.20	.23	1.43
2100		30	17-1/2	50		4.43	1.50	5.93
2150					columns	1.20	.23	1.43
2200		40	21-1/2	60		4.55	1.53	6.08
2250					columns	1.20	.23	1.43
2300	30x35	20	32-1/2	40		4.38	1.30	5.68
2350					columns	1.03	.19	1.22
2500		40	36-1/2	60		4.91	1.43	6.34
2550					columns	1.20	.23	1.43
3200	35x40	20	36-1/2	40		5.65	1.76	7.41
3250					columns	.90	.17	1.07
3400		40	36-1/2	60		6.35	1.93	8.28
3450					columns	.99	.19	1.18
3800	40x40	20	40-1/2	41		5.70	1.77	7.47
3850					columns	.87	.17	1.04
4000		40	40-1/2	61		6.40	1.94	8.34
4050					columns	.87	.17	1.04
4100	40x45	20	40-1/2	41		6.15	1.99	8.14
4150					columns	.77	.15	.92
4200		30	40-1/2	51		6.65	2.13	8.78
4250					columns	.77	.15	.92
4300		40	40-1/2	61		7.40	2.31	9.71
4350					columns	.87	.17	1.04
5000	45x50	20	52-1/2	41		6.45	2.02	8.47
5050					columns	.62	.12	.74
5200		40	52-1/2	61		7.90	2.37	10.27
5250					columns	.80	.15	.95
5300	50x45	20	56-1/2	41		6.30	1.99	8.29
5350					columns	.62	.12	.74
5500		40	56-1/2	61		8.10	2.44	10.54
5550					columns	.80	.15	.95
5600	50x50	20	56-1/2	42		7	2.17	9.17
5650					columns	.63	.12	.75
5800		40	59	64		8.10	2.60	10.70
5850					columns	.88	.17	1.05

B1020 310 Canopies

		COST PER S.F.		
		MAT.	INST.	TOTAL
0100	Canopies, wall hung, prefinished aluminum, 8′ x 10′	34.50	18.25	52.75

B2010 Exterior Walls

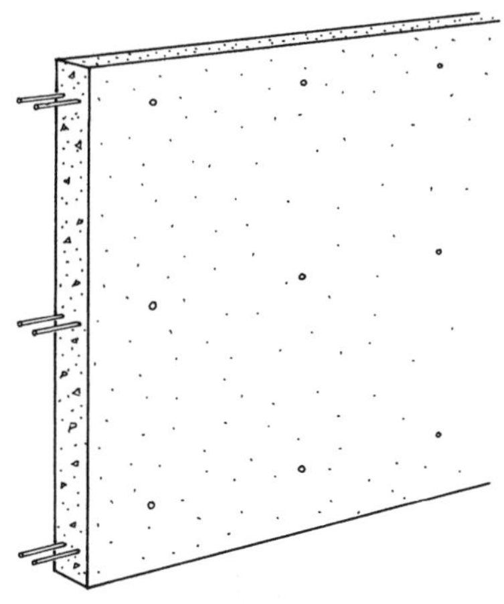

The table below describes a concrete wall system for exterior closure. There are several types of wall finishes priced from plain finish to a finish with 3/4" rustication strip.

Design Assumptions:
Conc. f'c = 3000 to 5000 psi
Reinf. fy = 60,000 psi

B2010 101	Cast In Place Concrete	COST PER S.F.		
		MAT.	INST.	TOTAL
2100	Conc wall reinforced, 8' high, 6" thick, plain finish, 3000 PSI	5.30	13.75	19.05
2700	Aged wood liner, 3000 PSI	6.50	15.75	22.25
3000	Sand blast light 1 side, 3000 PSI	5.90	15.65	21.55
3700	3/4" bevel rustication strip, 3000 PSI	5.40	14.65	20.05
4000	8" thick, plain finish, 3000 PSI	6.30	14.15	20.45
4100	4000 PSI	6.40	14.15	20.55
4300	Rub concrete 1 side, 3000 PSI	6.30	16.90	23.20
4550	8" thick, aged wood liner, 3000 PSI	7.55	16.15	23.70
4750	Sand blast light 1 side, 3000 PSI	6.90	16.05	22.95
5300	3/4" bevel rustication strip, 3000 PSI	6.45	15.05	21.50
5600	10" thick, plain finish, 3000 PSI	7.25	14.50	21.75
5900	Rub concrete 1 side, 3000 PSI	7.25	17.25	24.50
6200	Aged wood liner, 3000 PSI	8.50	16.50	25
6500	Sand blast light 1 side, 3000 PSI	7.85	16.40	24.25
7100	3/4" bevel rustication strip, 3000 PSI	7.40	15.40	22.80
7400	12" thick, plain finish, 3000 PSI	8.40	14.90	23.30
7700	Rub concrete 1 side, 3000 PSI	8.40	17.65	26.05
8000	Aged wood liner, 3000 PSI	9.65	16.90	26.55
8300	Sand blast light 1 side, 3000 PSI	9	16.80	25.80
8900	3/4" bevel rustication strip, 3000 PSI	8.55	15.80	24.35

B2010 Exterior Walls

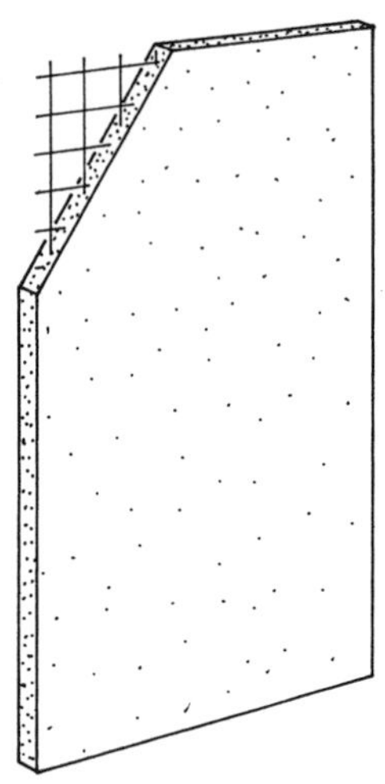

Precast concrete wall panels are either solid or insulated with plain, colored or textured finishes. Transportation is an important cost factor. Prices below are based on delivery within fifty miles of a plant. Engineering data is available from fabricators to assist with construction details. Usual minimum job size for economical use of panels is about 5000 S.F. Small jobs can double the prices below. For large, highly repetitive jobs, deduct up to 15% from the prices below.

B2010 102 Flat Precast Concrete

	THICKNESS (IN.)	PANEL SIZE (FT.)	FINISHES	RIGID INSULATION (IN)	TYPE	COST PER S.F.		
						MAT.	INST.	TOTAL
3000	4	5x18	smooth gray	none	low rise	16.80	5.65	22.45
3050		6x18				14.05	4.75	18.80
3100		8x20				28	2.30	30.30
3150		12x20				26.50	2.17	28.67
3200	6	5x18	smooth gray	2	low rise	17.80	6.20	24
3250		6x18				15.05	5.25	20.30
3300		8x20				29.50	2.90	32.40
3350		12x20				26.50	2.66	29.16
3400	8	5x18	smooth gray	2	low rise	38	3.62	41.62
3450		6x18				36	3.45	39.45
3500		8x20				33	3.19	36.19
3550		12x20				30	2.94	32.94
3600	4	4x8	white face	none	low rise	64	3.28	67.28
3650		8x8				48	3.79	51.79
3700		10x10				42	2.16	44.16
3750		20x10				38	1.95	39.95
3800	5	4x8	white face	none	low rise	65	3.36	68.36
3850		8x8				49	2.53	51.53
3900		10x10				43.50	2.24	45.74
3950		20x20				40	2.05	42.05
4000	6	4x8	white face	none	low rise	67.50	3.47	70.97
4050		8x8				51.50	2.64	54.14
4100		10x10				45	2.32	47.32
4150		20x10				41.50	2.13	43.63
4200	6	4x8	white face	2	low rise	68.50	4.05	72.55
4250		8x8				52.50	3.22	55.72
4300		10x10				46.50	2.90	49.40
4350		20x10				41.50	2.13	43.63
4400	7	4x8	white face	none	low rise	69	3.55	72.55
4450		8x8				53	2.73	55.73
4500		10x10				47.50	2.45	49.95
4550		20x10				43.50	2.24	45.74

B2010 Exterior Walls

B2010 102 Flat Precast Concrete

	THICKNESS (IN.)	PANEL SIZE (FT.)	FINISHES	RIGID INSULATION (IN)	TYPE	COST PER S.F.		
						MAT.	INST.	TOTAL
4600	7	4x8	white face	2	low rise	70.50	4.13	74.63
4650		8x8				54.50	3.31	57.81
4700		10x10				49	3.03	52.03
4750		20x10				44.50	2.82	47.32
4800	8	4x8	white face	none	low rise	70.50	3.63	74.13
4850		8x8				54.50	2.80	57.30
4900		10x10				49	2.52	51.52
4950		20x10				45	2.32	47.32
5000	8	4x8	white face	2	low rise	72	4.21	76.21
5050		8x8				55.50	3.38	58.88
5100		10x10				56.50	3.59	60.09
5150		20x10				56.50	3.59	60.09

B2010 103 Fluted Window or Mullion Precast Concrete

	THICKNESS (IN.)	PANEL SIZE (FT.)	FINISHES	RIGID INSULATION (IN)	TYPE	COST PER S.F.		
						MAT.	INST.	TOTAL
5200	4	4x8	smooth gray	none	high rise	38.50	13	51.50
5250		8x8				27.50	9.25	36.75
5300		10x10				53	4.38	57.38
5350		20x10				46.50	3.84	50.34
5400	5	4x8	smooth gray	none	high rise	39	13.20	52.20
5450		8x8				28.50	9.60	38.10
5500		10x10				55.50	4.57	60.07
5550		20x10				49	4.04	53.04
5600	6	4x8	smooth gray	none	high rise	40	13.60	53.60
5650		8x8				29.50	9.90	39.40
5700		10x10				57	4.70	61.70
5750		20x10				50.50	4.17	54.67
5800	6	4x8	smooth gray	2	high rise	41.50	14.15	55.65
5850		8x8				30.50	10.50	41
5900		10x10				58	5.30	63.30
5950		20x10				52	4.75	56.75
6000	7	4x8	smooth gray	none	high rise	41	13.85	54.85
6050		8x8				30	10.20	40.20
6100		10x10				59.50	4.90	64.40
6150		20x10				52	4.30	56.30
6200	7	4x8	smooth gray	2	high rise	42	14.45	56.45
6250		8x8				31.50	10.75	42.25
6300		10x10				60.50	5.50	66
6350		20x10				53.50	4.88	58.38
6400	8	4x8	smooth gray	none	high rise	41.50	14.05	55.55
6450		8x8				31	10.45	41.45
6500		10x10				61.50	5.05	66.55
6550		20x10				54.50	4.52	59.02
6600	8	4x8	smooth gray	2	high rise	42.50	14.60	57.10
6650		8x8				32	11.05	43.05
6700		10x10				62.50	5.65	68.15
6750		20x10				56	5.10	61.10

B2010 Exterior Walls

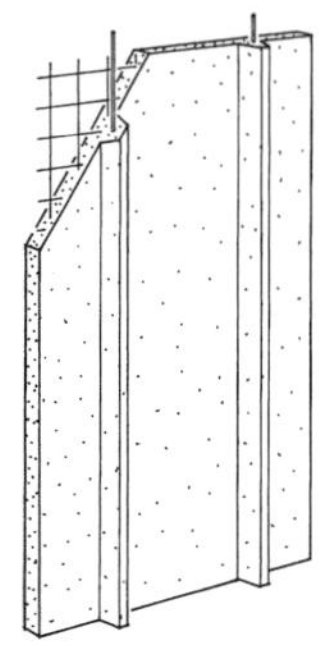

Ribbed Precast Panel

B2010 104	Ribbed Precast Concrete							
	THICKNESS (IN.)	PANEL SIZE (FT.)	FINISHES	RIGID INSULATION(IN.)	TYPE	COST PER S.F.		
						MAT.	INST.	TOTAL
6800	4	4x8	aggregate	none	high rise	46	12.95	58.95
6850		8x8				33.50	9.50	43
6900		10x10				57	3.44	60.44
6950		20x10				50.50	3.04	53.54
7000	5	4x8	aggregate	none	high rise	46.50	13.20	59.70
7050		8x8				34.50	9.75	44.25
7100		10x10				59.50	3.58	63.08
7150		20x10				53	3.20	56.20
7200	6	4x8	aggregate	none	high rise	47.50	13.45	60.95
7250		8x8				35.50	10	45.50
7300		10x10				61.50	3.69	65.19
7350		20x10				55	3.31	58.31
7400	6	4x8	aggregate	2	high rise	49	14.05	63.05
7450		8x8				36.50	10.60	47.10
7500		10x10				62.50	4.27	66.77
7550		20x10				56	3.89	59.89
7600	7	4x8	aggregate	none	high rise	49	13.75	62.75
7650		8x8				36.50	10.25	46.75
7700		10x10				63.50	3.81	67.31
7750		20x10				56.50	3.40	59.90
7800	7	4x8	aggregate	2	high rise	50	14.35	64.35
7850		8x8				37.50	10.85	48.35
7900		10x10				64.50	4.39	68.89
7950		20x10				57.50	3.98	61.48
8000	8	4x8	aggregate	none	high rise	49.50	14	63.50
8050		8x8				37.50	10.55	48.05
8100		10x10				65.50	3.94	69.44
8150		20x10				59	3.56	62.56
8200	8	4x8	aggregate	2	high rise	51	14.60	65.60
8250		8x8				38.50	11.15	49.65
8300		10x10				67	5.85	72.85
8350		20x10				60.50	4.14	64.64

B2010 Exterior Walls

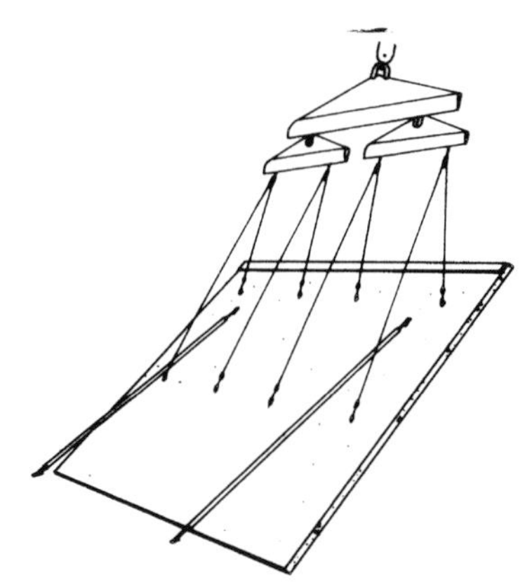

The advantage of tilt up construction is in the low cost of forms and placing of concrete and reinforcing. Tilt up has been used for several types of buildings, including warehouses, stores, offices, and schools. The panels are cast in forms on the ground, or floor slab. Most jobs use 5-1/2″ thick solid reinforced concrete panels.

Design Assumptions:
Conc. f'c = 3000 psi
Reinf. fy = 60,000

B2010 106	Tilt-Up Concrete Panel	COST PER S.F.		
		MAT.	INST.	TOTAL
3200	Tilt-up conc panels, broom finish, 5-1/2″ thick, 3000 PSI	4.75	5.15	9.90
3250	5000 PSI	4.93	5.05	9.98
3300	6″ thick, 3000 PSI	5.20	5.30	10.50
3350	5000 PSI	5.40	5.20	10.60
3400	7-1/2″ thick, 3000 PSI	6.65	5.50	12.15
3450	5000 PSI	6.90	5.40	12.30
3500	8″ thick, 3000 PSI	7.10	5.65	12.75
3550	5000 PSI	7.40	5.55	12.95
3700	Steel trowel finish, 5-1/2″ thick, 3000 PSI	4.75	5.25	10
3750	5000 PSI	4.93	5.15	10.08
3800	6″ thick, 3000 PSI	5.20	5.35	10.55
3850	5000 PSI	5.40	5.25	10.65
3900	7-1/2″ thick, 3000 PSI	6.65	5.55	12.20
3950	5000 PSI	6.90	5.45	12.35
4000	8″ thick, 3000 PSI	7.10	5.70	12.80
4050	5000 PSI	7.40	5.60	13
4200	Exp. aggregate finish, 5-1/2″ thick, 3000 PSI	5.10	5.30	10.40
4250	5000 PSI	5.25	5.20	10.45
4300	6″ thick, 3000 PSI	5.55	5.45	11
4350	5000 PSI	5.75	5.35	11.10
4400	7-1/2″ thick, 3000 PSI	7	5.65	12.65
4450	5000 PSI	7.25	5.55	12.80
4500	8″ thick, 3000 PSI	7.45	5.75	13.20
4550	5000 PSI	7.75	5.65	13.40
4600	Exposed aggregate & vert. rustication 5-1/2″ thick, 3000 PSI	7.50	6.65	14.15
4650	5000 PSI	7.70	6.55	14.25
4700	6″ thick, 3000 PSI	7.95	6.75	14.70
4750	5000 PSI	8.15	6.65	14.80
4800	7-1/2″ thick, 3000 PSI	9.40	6.95	16.35
4850	5000 PSI	9.70	6.85	16.55
4900	8″ thick, 3000 PSI	9.90	7.10	17
4950	5000 PSI	10.20	7	17.20
5000	Vertical rib & light sandblast, 5-1/2″ thick, 3000 PSI	7.25	8.70	15.95
5100	6″ thick, 3000 PSI	7.70	8.85	16.55
5200	7-1/2″ thick, 3000 PSI	9.15	9.05	18.20
5300	8″ thick, 3000 PSI	9.65	9.15	18.80
6000	Broom finish w/2″ polystyrene insulation, 6″ thick, 3000 PSI	4.52	6.45	10.97
6100	Broom finish 2″ fiberplank insulation, 6″ thick, 3000 PSI	5.10	6.40	11.50
6200	Exposed aggregate w/2″ polystyrene insulation, 6″ thick, 3000 PSI	4.74	6.45	11.19
6300	Exposed aggregate 2″ fiberplank insulation, 6″ thick, 3000 PSI	5.30	6.40	11.70

B2010 Exterior Walls

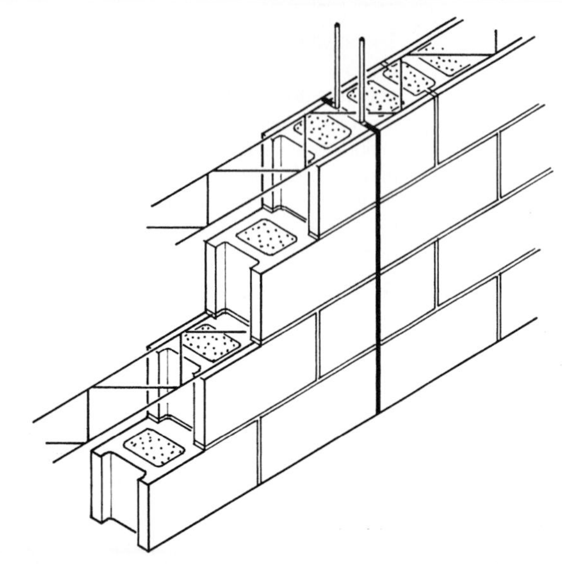

Exterior concrete block walls are defined in the following terms; structural reinforcement, weight, percent solid, size, strength and insulation. Within each of these categories, two to four variations are shown. No costs are included for brick shelf or relieving angles.

B2010 109 Concrete Block Wall - Regular Weight

	TYPE	SIZE (IN.)	STRENGTH (P.S.I.)	CORE FILL		COST PER S.F.		
						MAT.	INST.	TOTAL
1200	Hollow	4x8x16	2,000	none		2.29	5.50	7.79
1250			4,500	none		2.83	5.50	8.33
1400		8x8x16	2,000	perlite		4.64	6.70	11.34
1410				styrofoam		4.64	6.30	10.94
1440				none		3.13	6.30	9.43
1450			4,500	perlite		5.70	6.70	12.40
1460				styrofoam		5.70	6.30	12
1490				none		4.20	6.30	10.50
2000	75% solid	4x8x16	2,000	none		2.77	5.60	8.37
2100		6x8x16	2,000	perlite		4.15	6.15	10.30
2500	Solid	4x8x16	2,000	none		2.58	5.70	8.28
2700		8x8x16	2,000	none		4.39	6.55	10.94

B2010 110 Concrete Block Wall - Lightweight

	TYPE	SIZE (IN.)	WEIGHT (P.C.F.)	CORE FILL		COST PER S.F.		
						MAT.	INST.	TOTAL
3100	Hollow	8x4x16	105	perlite		4.54	6.30	10.84
3110				styrofoam		4.54	5.90	10.44
3200		4x8x16	105	none		2.43	5.40	7.83
3250			85	none		3.08	5.30	8.38
3300		6x8x16	105	perlite		4.37	6.10	10.47
3310				styrofoam		4.69	5.80	10.49
3340				none		3.35	5.80	9.15
3400		8x8x16	105	perlite		5.60	6.55	12.15
3410				styrofoam		5.60	6.15	11.75
3440				none		4.07	6.15	10.22
3450			85	perlite		5.70	6.40	12.10
4000	75% solid	4x8x16	105	none		3.51	5.45	8.96
4050			85	none		3.80	5.35	9.15
4100		6x8x16	105	perlite		5.40	6	11.40
4500	Solid	4x8x16	105	none		2.27	5.65	7.92
4700		8x8x16	105	none		4.39	6.45	10.84

B2010 Exterior Walls

B2010 111 Reinforced Concrete Block Wall - Regular Weight

	TYPE	SIZE (IN.)	STRENGTH (P.S.I.)	VERT. REINF & GROUT SPACING		COST PER S.F.		
						MAT.	INST.	TOTAL
5200	Hollow	4x8x16	2,000	#4 @ 48"		2.45	6.10	8.55
5300		6x8x16	2,000	#4 @ 48"		3.29	6.60	9.89
5330				#5 @ 32"		3.53	6.85	10.38
5340				#5 @ 16"		4.04	7.80	11.84
5350			4,500	#4 @ 28"		3.77	6.60	10.37
5390				#5 @ 16"		4.52	7.80	12.32
5400		8x8x16	2,000	#4 @ 48"		3.57	7.05	10.62
5430				#5 @ 32"		3.75	7.45	11.20
5440				#5 @ 16"		4.35	8.60	12.95
5450		8x8x16	4,500	#4 @ 48"		4.55	7.15	11.70
5490				#5 @ 16"		5.40	8.60	14
5500		12x8x16	2,000	#4 @ 48"		5.45	9.05	14.50
5540				#5 @ 16"		6.65	10.60	17.25
6100	75% solid	6x8x16	2,000	#4 @ 48"		3.79	6.50	10.29
6140				#5 @ 16"		4.26	7.40	11.66
6150			4,500	#4 @ 48"		4.42	6.50	10.92
6190				#5 @ 16"		4.89	7.40	12.29
6200		8x8x16	2,000	#4 @ 48"		3.93	7	10.93
6230				#5 @ 32"		4.12	7.25	11.37
6240				#5 @ 16"		4.44	8.10	12.54
6250			4,500	#4 @ 48"		5.35	7	12.35
6280				#5 @ 32"		5.50	7.25	12.75
6290				#5 @ 16"		5.85	8.10	13.95
6500	Solid-double	2-4x8x16	2,000	#4 @ 48" E.W.		6.10	13.05	19.15
6530	Wythe			#5 @ 16" E.W.		6.85	13.85	20.70
6550			4,500	#4 @ 48" E.W.		8.70	12.95	21.65
6580				#5 @ 16" E.W.		9.50	13.75	23.25

B2010 112 Reinforced Concrete Block Wall - Lightweight

	TYPE	SIZE (IN.)	WEIGHT (P.C.F.)	VERT REINF. & GROUT SPACING		COST PER S.F.		
						MAT.	INST.	TOTAL
7100	Hollow	8x4x16	105	#4 @ 48"		3.38	6.75	10.13
7140				#5 @ 16"		4.25	8.20	12.45
7150			85	#4 @ 48"		5	6.65	11.65
7190				#5 @ 16"		5.90	8.10	14
7400		8x8x16	105	#4 @ 48"		4.42	7	11.42
7440				#5 @ 16"		5.30	8.45	13.75
7450		8x8x16	85	#4 @ 48"		4.55	6.85	11.40
7490				#5 @ 16"		5.40	8.30	13.70
7800		8x8x24	105	#4 @ 48"		3.38	7.55	10.93
7840				#5 @ 16"		4.25	9	13.25
7850			85	#4 @ 48"		7.70	6.45	14.15
7890				#5 @ 16"		8.55	7.90	16.45
8100	75% solid	6x8x16	105	#4 @ 48"		5.05	6.35	11.40
8130				#5 @ 32"		5.25	6.55	11.80
8150			85	#4 @ 48"		5.20	6.20	11.40
8180				#5 @ 32"		5.35	6.40	11.75
8200		8x8x16	105	#4 @ 48"		6.15	6.80	12.95
8230				#5 @ 32"		6.35	7.05	13.40
8250			85	#4 @ 48"		5.30	6.65	11.95
8280				#5 @ 32"		5.50	6.90	12.40

B2010 Exterior Walls

B2010 112 Reinforced Concrete Block Wall - Lightweight

	TYPE	SIZE (IN.)	WEIGHT (P.C.F.)	VERT REINF. & GROUT SPACING		COST PER S.F.		
						MAT.	INST.	TOTAL
8500	Solid-double	2-4x8x16	105	#4 @ 48"		5.45	12.95	18.40
8530	Wythe		105	#5 @ 16"		6.20	13.75	19.95
8600		2-6x8x16	105	#4 @ 48"		8.55	13.80	22.35
8630				#5 @ 16"		9.30	14.60	23.90
8650			85	#4 @ 48"		14	13.30	27.30

B2010 Exterior Walls

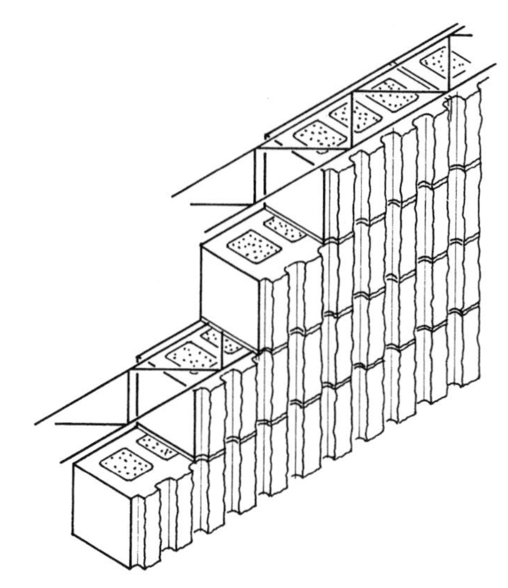

Exterior split ribbed block walls are defined in the following terms; structural reinforcement, weight, percent solid, size, number of ribs and insulation. Within each of these categories two to four variations are shown. No costs are included for brick shelf or relieving angles. Costs include control joints every 20′ and horizontal reinforcing.

B2010 113 Split Ribbed Block Wall - Regular Weight

	TYPE	SIZE (IN.)	RIBS	CORE FILL		COST PER S.F.		
						MAT.	INST.	TOTAL
1220	Hollow	4x8x16	4	none		4.62	6.85	11.47
1250			8	none		5	6.85	11.85
1280			16	none		5.40	6.95	12.35
1430		8x8x16	8	perlite		8.55	8.10	16.65
1440				styrofoam		8.35	7.70	16.05
1450				none		7.05	7.70	14.75
1530		12x8x16	8	perlite		10.60	10.80	21.40
1540				styrofoam		9.85	10.05	19.90
1550				none		8.10	10.05	18.15
2120	75% solid	4x8x16	4	none		5.80	6.95	12.75
2150			8	none		6.35	6.95	13.30
2180			16	none		6.85	7.05	13.90
2520	Solid	4x8x16	4	none		6.60	7.05	13.65
2550			8	none		7.80	7.05	14.85
2580			16	none		7.80	7.15	14.95

B2010 115 Reinforced Split Ribbed Block Wall - Regular Weight

	TYPE	SIZE (IN.)	RIBS	VERT. REINF. & GROUT SPACING		COST PER S.F.		
						MAT.	INST.	TOTAL
5200	Hollow	4x8x16	4	#4 @ 48″		4.78	7.45	12.23
5230			8	#4 @ 48″		5.20	7.45	12.65
5260			16	#4 @ 48″		5.55	7.55	13.10
5430		8x8x16	8	#4 @ 48″		7.40	8.55	15.95
5440				#5 @ 32″		7.65	8.85	16.50
5450				#5 @ 16″		8.25	10	18.25
5530		12x8x16	8	#4 @ 48″		8.60	10.90	19.50
5540				#5 @ 32″		9	11.25	20.25
5550				#5 @ 16″		9.80	12.45	22.25
6230	75% solid	6x8x16	8	#4 @ 48″		7.35	7.85	15.20
6240				#5 @ 32″		7.50	8.05	15.55
6250				#5 @ 16″		7.80	8.75	16.55
6330		8x8x16	8	#4 @ 48″		9	8.45	17.45
6340				#5 @ 32″		9.20	8.70	17.90
6350				#5 @ 16″		9.50	9.55	19.05

B2010 Exterior Walls

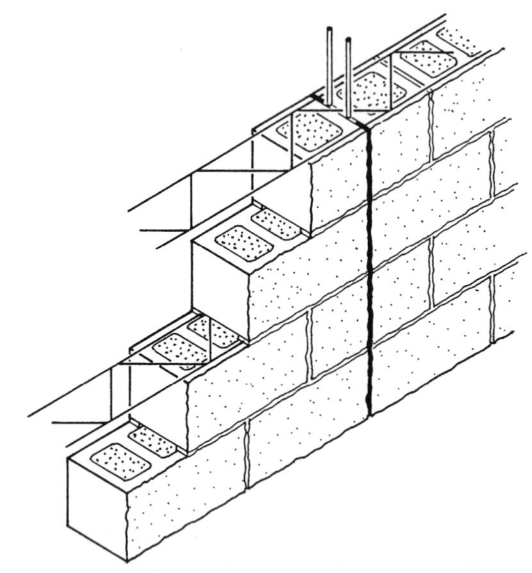

Exterior split face block walls are defined in the following terms; structural reinforcement, weight, percent solid, size, scores and insulation. Within each of these categories two to four variations are shown. No costs are included for brick shelf or relieving angles. Costs include control joints every 20′ and horizontal reinforcing.

B2010 117 Split Face Block Wall - Regular Weight

	TYPE	SIZE (IN.)	SCORES	CORE FILL		COST PER S.F.		
						MAT.	INST.	TOTAL
1200	Hollow	8x4x16	0	perlite		8.05	8.90	16.95
1210				styrofoam		8.05	8.50	16.55
1240				none		6.55	8.50	15.05
1250			1	perlite		8.55	8.90	17.45
1260				styrofoam		8.55	8.50	17.05
1290				none		7.05	8.50	15.55
1300		12x4x16	0	perlite		10.50	10.15	20.65
1310				styrofoam		9.75	9.40	19.15
1340				none		8	9.40	17.40
1350			1	perlite		11.05	10.15	21.20
1360				styrofoam		10.30	9.40	19.70
1390				none		8.55	9.40	17.95
1400		4x8x16	0	none		4.38	6.75	11.13
1450			1	none		4.75	6.85	11.60
1500		6x8x16	0	perlite		5.95	7.75	13.70
1510				styrofoam		6.25	7.45	13.70
1540				none		4.92	7.45	12.37
1550			1	perlite		6.35	7.90	14.25
1560				styrofoam		6.65	7.60	14.25
1590				none		5.35	7.60	12.95
1600		8x8x16	0	perlite		7.35	8.35	15.70
1610				styrofoam		7.35	7.95	15.30
1640				none		5.85	7.95	13.80
1650		8x8x16	1	perlite		7.85	8.50	16.35
1660				styrofoam		7.85	8.10	15.95
1690				none		6.35	8.10	14.45
1700		12x8x16	0	perlite		9.50	11	20.50
1705			0	perlite		11.05	13.30	24.35
1710				styrofoam		8.75	10.25	19
1740				none		7	10.25	17.25

B2010 Exterior Walls

B2010 117 Split Face Block Wall - Regular Weight

	TYPE	SIZE (IN.)	SCORES	CORE FILL		COST PER S.F.		
						MAT.	INST.	TOTAL
1750	Hollow	12x8x16	1	perlite		10.05	11.20	21.25
1760				styrofoam		9.30	10.45	19.75
1790				none		7.55	10.45	18
1800	75% solid	8x4x16	0	perlite		8.80	8.90	17.70
1840				none		8.05	8.70	16.75
1850			1	perlite		9.40	8.90	18.30
1890				none		8.65	8.70	17.35
2000		4x8x16	0	none		5.50	6.85	12.35
2050			1	none		5.95	6.95	12.90
2400	Solid	8x4x16	0	none		9	8.85	17.85
2450			1	none		9.75	8.85	18.60
2900		12x8x16	0	none		9.85	10.65	20.50
2950			1	none		10.70	10.85	21.55

B2010 119 Reinforced Split Face Block Wall - Regular Weight

	TYPE	SIZE (IN.)	SCORES	VERT. REINF. & GROUT SPACING		COST PER S.F.		
						MAT.	INST.	TOTAL
5200	Hollow	8x4x16	0	#4 @ 48″		6.90	9.35	16.25
5210				#5 @ 32″		7.15	9.65	16.80
5240				#5 @ 16″		7.75	10.80	18.55
5250			1	#4 @ 48″		7.40	9.35	16.75
5260				#5 @ 32″		7.65	9.65	17.30
5290				#5 @ 16″		8.25	10.80	19.05
5700		12x8x16	0	#4 @ 48″		7.50	11.10	18.60
5710				#5 @ 32″		7.90	11.45	19.35
5740				#5 @ 16″		8.70	12.65	21.35
5750			1	#4 @ 48″		8.05	11.30	19.35
5760				#5 @ 32″		8.45	11.65	20.10
5790				#5 @ 16″		9.25	12.85	22.10
6000	75% solid	8x4x16	0	#4 @ 48″		8.20	9.30	17.50
6010				#5 @ 32″		8.40	9.55	17.95
6040				#5 @ 16″		8.70	10.40	19.10
6050			1	#4 @ 48″		8.80	9.30	18.10
6060				#5 @ 32″		9	9.55	18.55
6090				#5 @ 16″		9.30	10.40	19.70
6100		12x4x16	0	#4 @ 48″		10	10.30	20.30
6110				#5 @ 32″		10.25	10.55	20.80
6140				#5 @ 16″		10.70	11.30	22
6150			1	#4 @ 48″		10.75	10.30	21.05
6160				#5 @ 32″		11	10.55	21.55
6190				#5 @ 16″		11.50	11.50	23
6700	Solid-double	2-4x8x16	0	#4 @ 48″ E.W.		13.40	15.55	28.95
6710	Wythe			#5 @ 32″ E.W.		13.65	15.70	29.35
6740				#5 @ 16″ E.W.		14.15	16.35	30.50
6750			1	#4 @ 48″ E.W.		14.50	15.75	30.25
6760				#5 @ 32″ E.W.		14.75	15.90	30.65
6790				#5 @ 16″ E.W.		15.25	16.55	31.80
6800		2-6x8x16	0	#4 @ 48″ E.W.		14.90	17.10	32
6810				#5 @ 32″ E.W.		15.15	17.25	32.40
6840				#5 @ 16″ E.W.		15.65	17.90	33.55
6850			1	#4 @ 48″ E.W.		16.10	17.40	33.50
6860				#5 @ 32″ E.W.		16.35	17.55	33.90
6890				#5 @ 16″ E.W.		16.85	18.20	35.05

B2010 Exterior Walls

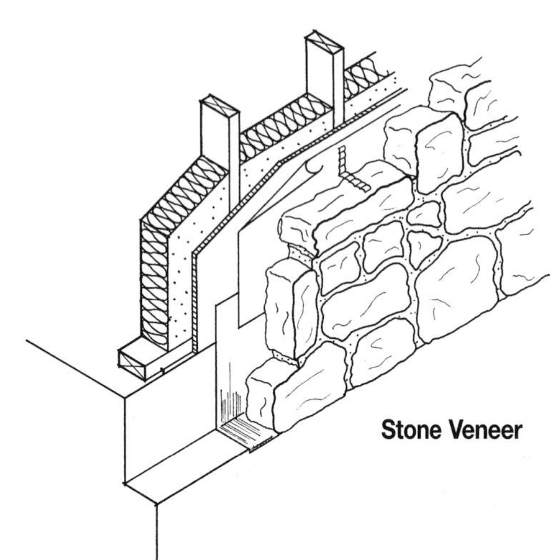

The table below lists costs per S.F. for stone veneer walls on various backup using different stone.

B2010 128	Stone Veneer	COST PER S.F.		
		MAT.	INST.	TOTAL
2000	Ashlar veneer, 4", 2" x 4" stud backup, 16" O.C., 8' high, low priced stone	15.60	19.55	35.15
2050	2" x 6" stud backup, 16" O.C.	17.55	23	40.55
2100	Metal stud backup, 8' high, 16" O.C.	16.40	19.85	36.25
2150	24" O.C.	16.05	19.40	35.45
2200	Conc. block backup, 4" thick	17.20	23.50	40.70
2300	6" thick	17.90	23.50	41.40
2350	8" thick	18.05	24	42.05
2400	10" thick	18.60	25	43.60
2500	12" thick	19.80	27	46.80
3100	High priced stone, wood stud backup, 10' high, 16" O.C.	22	22.50	44.50
3200	Metal stud backup, 10' high, 16" O.C.	23	22.50	45.50
3250	24" O.C.	22.50	22	44.50
3300	Conc. block backup, 10' high, 4" thick	23.50	26	49.50
3350	6" thick	24.50	26	50.50
3400	8" thick	25.50	28	53.50
3450	10" thick	25	28	53
3500	12" thick	26.50	29.50	56
4000	Indiana limestone 2" thk., sawn finish, wood stud backup, 10' high, 16" O.C	34	11.85	45.85
4100	Metal stud backup, 10' high, 16" O.C.	35.50	15	50.50
4150	24" O.C.	35.50	15	50.50
4200	Conc. block backup, 4" thick	35.50	15.55	51.05
4250	6" thick	36	15.80	51.80
4300	8" thick	36.50	16.30	52.80
4350	10" thick	37	17.55	54.55
4400	12" thick	38	19.30	57.30
4450	2" thick, smooth finish, wood stud backup, 8' high, 16" O.C.	34	11.85	45.85
4550	Metal stud backup, 8' high, 16" O.C.	34.50	11.95	46.45
4600	24" O.C.	34.50	11.55	46.05
4650	Conc. block backup, 4" thick	35.50	15.40	50.90
4700	6" thick	36	15.60	51.60
4750	8" thick	36.50	16.10	52.60
4800	10" thick	37	17.55	54.55
4850	12" thick	38	19.30	57.30
5350	4" thick, smooth finish, wood stud backup, 8' high, 16" O.C.	35	11.85	46.85
5450	Metal stud backup, 8' high, 16" O.C.	35.50	12.15	47.65
5500	24" O.C.	35.50	11.70	47.20
5550	Conc. block backup, 4" thick	36.50	15.55	52.05
5600	6" thick	37	15.80	52.80

B2010 Exterior Walls

B2010 128	Stone Veneer	COST PER S.F.		
		MAT.	INST.	TOTAL
5650	8" thick	41.50	20	61.50
5700	10" thick	38	17.55	55.55
5750	12" thick	39	19.30	58.30
6000	Granite, gray or pink, 2" thick, wood stud backup, 8' high, 16" O.C.	34.50	21.50	56
6100	Metal studs, 8' high, 16" O.C.	35	22	57
6150	24" O.C.	35	21.50	56.50
6200	Conc. block backup, 4" thick	36	25	61
6250	6" thick	36.50	25.50	62
6300	8" thick	37	26	63
6350	10" thick	37.50	27	64.50
6400	12" thick	38.50	29	67.50
6900	4" thick, wood stud backup, 8' high, 16" O.C.	45	24.50	69.50
7000	Metal studs, 8' high, 16" O.C.	45.50	25	70.50
7050	24" O.C.	45.50	24.50	70
7100	Conc. block backup, 4" thick	46.50	28.50	75
7150	6" thick	47	28.50	75.50
7200	8" thick	47.50	29	76.50
7250	10" thick	48	30.50	78.50
7300	12" thick	49	32	81

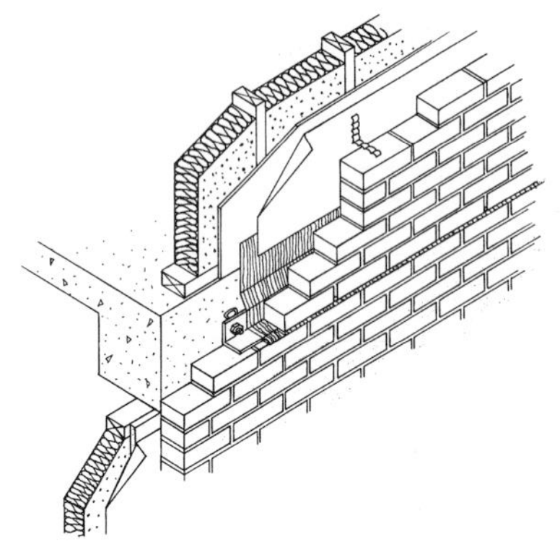

Exterior brick veneer/stud backup walls are defined in the following terms: type of brick and studs, stud spacing and bond. All systems include a back-up wall, a control joint every 20′, a brick shelf every 12′ of height, ties to the backup and the necessary dampproofing, flashing and insulation.

B2010 129 Brick Veneer/Wood Stud Backup

	FACE BRICK	STUD BACKUP	STUD SPACING (IN.)	BOND	FACE	COST PER S.F.		
						MAT.	INST.	TOTAL
1100	Standard	2x4-wood	16	running		7.65	15.25	22.90
1120				common		8.70	17.20	25.90
1140				Flemish		9.45	20	29.45
1160				English		10.35	21	31.35
1400		2x6-wood	16	running		8.25	15.20	23.45
1420				common		9.05	17.30	26.35
1440				Flemish		9.80	20	29.80
1460				English		10.70	21.50	32.20
1700	Glazed	2x4-wood	16	running		24.50	15.75	40.25
1720				common		29	17.95	46.95
1740				Flemish		32	21	53
1760				English		35.50	22.50	58
2300	Engineer	2x4-wood	16	running		7.35	13.65	21
2320				common		8.35	15.25	23.60
2340				Flemish		9	17.95	26.95
2360				English		9.85	18.70	28.55
2900	Roman	2x4-wood	16	running		10.75	14	24.75
2920				common		12.45	15.75	28.20
2940				Flemish		13.55	18.30	31.85
2960				English		15.05	19.60	34.65
4100	Norwegian	2x4-wood	16	running		7.95	10.95	18.90
4120				common		9	12.10	21.10
4140				Flemish		9.75	14	23.75
4160				English		10.70	14.60	25.30

B2010 130 Brick Veneer/Metal Stud Backup

	FACE BRICK	STUD BACKUP	STUD SPACING (IN.)	BOND	FACE	COST PER S.F.		
						MAT.	INST.	TOTAL
5050	Standard	16 ga x 6″LB	16	running		9.15	16	25.15
5100		25ga.x6″NLB	24	running		7.20	14.90	22.10
5120				common		8.20	16.85	25.05
5140				Flemish		8.60	19.10	27.70
5160				English		9.85	21	30.85
5200		20ga.x3-5/8″NLB	16	running		7.30	15.50	22.80
5220				common		8.30	17.45	25.75
5240				Flemish		9.05	20.50	29.55
5260				English		9.95	21.50	31.45

B2010 Exterior Walls

B2010 130 Brick Veneer/Metal Stud Backup

	FACE BRICK	STUD BACKUP	STUD SPACING (IN.)	BOND	FACE	COST PER S.F.		
						MAT.	INST.	TOTAL
5400	Standard	16ga.x3-5/8"LB	16	running		8.05	15.75	23.80
5420				common		9.10	17.70	26.80
5440				Flemish		9.85	20.50	30.35
5460				English		10.75	21.50	32.25
5500			24	running		7.70	15.30	23
5520				common		8.75	17.25	26
5540				Flemish		9.50	20	29.50
5560				English		10.40	21.50	31.90
5700	Glazed	25ga.x6"NLB	24	running		24	15.40	39.40
5720				common		28.50	17.60	46.10
5740				Flemish		31.50	21	52.50
5760				English		35	22	57
5800		20ga.x3-5/8"NLB	24	running		24	15.55	39.55
5820				common		28.50	17.75	46.25
5840				Flemish		31.50	21	52.50
5860				English		35	22.50	57.50
6000		16ga.x3-5/8"LB	16	running		25	16.25	41.25
6020				common		29.50	18.45	47.95
6040				Flemish		32.50	21.50	54
6060				English		36	23	59
6300	Engineer	25ga.x6"NLB	24	running		6.90	13.30	20.20
6320				common		7.85	14.90	22.75
6340				Flemish		8.50	17.60	26.10
6360				English		9.35	18.35	27.70
6400		20ga.x3-5/8"NLB	16	running		7	13.90	20.90
6420				common		7.95	15.50	23.45
6440				Flemish		8.60	18.20	26.80
6460				English		9.45	18.95	28.40
6900	Roman	25ga.x6"NLB	24	running		10.25	13.65	23.90
6920				common		11.95	15.35	27.30
6940				Flemish		13.05	17.95	31
6960				English		14.55	19.25	33.80
7000		20ga.x3-5/8"NLB	16	running		10.35	14.25	24.60
7020				common		12.05	16	28.05
7040				Flemish		13.15	18.55	31.70
7060				English		14.65	19.85	34.50
7500	Norman	25ga.x6"NLB	24	running		8.20	11.65	19.85
7520				common		9.45	13	22.45
7540				Flemish		19	15.15	34.15
7560				English		11.35	15.95	27.30
7600		20ga.x3-5/8"NLB	24	running		8.15	11.80	19.95
7620				common		9.40	13.15	22.55
7640				Flemish		18.95	15.30	34.25
7660				English		11.30	16.10	27.40
8100	Norwegian	25ga.x6"NLB	24	running		7.45	10.60	18.05
8120				common		8.50	11.75	20.25
8140				Flemish		9.25	13.65	22.90
8160				English		10.20	14.25	24.45
8400		16ga.x3-5/8"LB	16	running		8.35	11.45	19.80
8420				common		9.40	12.60	22
8440				Flemish		10.15	14.50	24.65
8460				English		11.10	15.10	26.20

B2010 Exterior Walls

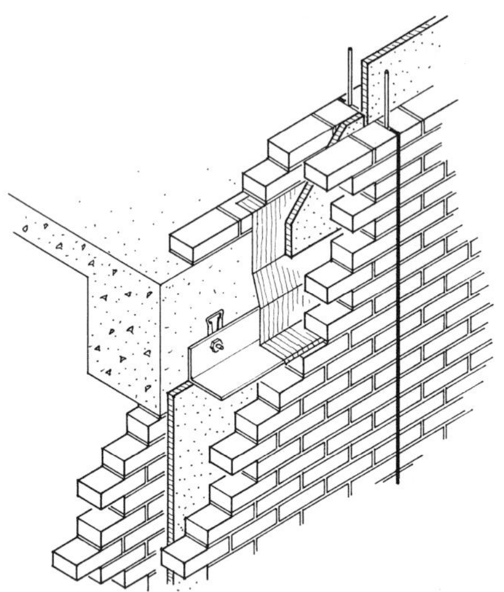

Exterior brick face cavity walls are defined in the following terms: cavity treatment, type of face brick, backup masonry, total thickness and insulation. Seven types of face brick are shown with various types of backup. All systems include a brick shelf, ties to the backups and necessary dampproofing, flashing, and control joints every 20′.

B2010 134 Brick Face Cavity Wall

	FACE BRICK	BACKUP MASONRY	TOTAL THICKNESS (IN.)	CAVITY INSULATION		COST PER S.F.		
						MAT.	INST.	TOTAL
1000	Standard	4″ common brick	10	polystyrene		11.20	23.50	34.70
1020				none		10.90	23	33.90
1040		6″ SCR brick	12	polystyrene		13.30	21	34.30
1060				none		13	20.50	33.50
1080		4″ conc. block	10	polystyrene		8.35	18.95	27.30
1100				none		8.05	18.30	26.35
1120		6″ conc. block	12	polystyrene		9.10	19.35	28.45
1140				none		8.80	18.70	27.50
1160		4″ L.W. block	10	polystyrene		8.50	18.85	27.35
1180				none		8.20	18.20	26.40
1200		6″ L.W. block	12	polystyrene		9.45	19.25	28.70
1220				none		9.15	18.60	27.75
1240		4″ glazed block	10	polystyrene		19.40	20.50	39.90
1260				none		19.10	19.65	38.75
1280		6″ glazed block	12	polystyrene		19.45	20.50	39.95
1300				none		19.15	19.65	38.80
1320		4″ clay tile	10	polystyrene		12.60	18.25	30.85
1340				none		12.30	17.60	29.90
1360		4″ glazed tile	10	polystyrene		21	24	45
1380				none		20.50	23.50	44
1500	Glazed	4″ common brick	10	polystyrene		28	24	52
1520				none		28	23.50	51.50
1580		4″ conc. block	10	polystyrene		25.50	19.45	44.95
1600				none		25	18.80	43.80
1660		4″ L.W. block	10	polystyrene		25.50	19.35	44.85
1680				none		25	18.70	43.70
1740		4″ glazed block	10	polystyrene		36.50	21	57.50
1760				none		36	20	56
1820		4″ clay tile	10	polystyrene		29.50	18.75	48.25
1840				none		29.50	18.10	47.60
1860		4″ glazed tile	10	polystyrene		33.50	19.25	52.75
1880				none		33.50	18.60	52.10
2000	Engineer	4″ common brick	10	polystyrene		10.90	22	32.90
2020				none		10.60	21.50	32.10

B2010 Exterior Walls

B2010 134 Brick Face Cavity Wall

	FACE BRICK	BACKUP MASONRY	TOTAL THICKNESS (IN.)	CAVITY INSULATION		COST PER S.F.		
						MAT.	INST.	TOTAL
2080	Engineer	4" conc. block	10	polystyrene		8.05	17.35	25.40
2100				none		7.75	16.70	24.45
2160		4" L.W. block	10	polystyrene		8.20	17.25	25.45
2180				none		7.90	16.60	24.50
2240		4" glazed block	10	polystyrene		19.10	18.70	37.80
2260				none		18.80	18.05	36.85
2320		4" clay tile	10	polystyrene		12.30	16.65	28.95
2340				none		12	16	28
2360		4" glazed tile	10	polystyrene		20.50	22.50	43
2380				none		20	22	42
2500	Roman	4" common brick	10	polystyrene		14.25	22.50	36.75
2520				none		13.95	21.50	35.45
2580		4" conc. block	10	polystyrene		11.45	17.70	29.15
2600				none		11.15	17.05	28.20
2660		4" L.W. block	10	polystyrene		11.60	17.60	29.20
2680				none		11.30	16.95	28.25
2740		4" glazed block	10	polystyrene		22.50	19.05	41.55
2760				none		22	18.40	40.40
2820		4" clay tile	10	polystyrene		15.70	17	32.70
2840				none		15.40	16.35	31.75
2860		4" glazed tile	10	polystyrene		24	23	47
2880				none		23.50	22	45.50
3000	Norman	4" common brick	10	polystyrene		12.20	20.50	32.70
3020				none		11.90	19.70	31.60
3080		4" conc. block	10	polystyrene		9.40	15.70	25.10
3100				none		9.10	15.05	24.15
3160		4" L.W. block	10	polystyrene		9.55	15.60	25.15
3180				none		9.25	14.95	24.20
3240		4" glazed block	10	polystyrene		20.50	17.05	37.55
3260				none		20	16.40	36.40
3320		4" clay tile	10	polystyrene		13.65	15	28.65
3340				none		13.35	14.35	27.70
3360		4" glazed tile	10	polystyrene		22	21	43
3380				none		21.50	20	41.50
3500	Norwegian	4" common brick	10	polystyrene		11.45	19.30	30.75
3520				none		11.15	18.65	29.80
3580		4" conc. block	10	polystyrene		8.65	14.65	23.30
3600				none		8.35	14	22.35
3660		4" L.W. block	10	polystyrene		8.80	14.55	23.35
3680				none		8.50	13.90	22.40
3740		4" glazed block	10	polystyrene		19.70	16	35.70
3760				none		19.40	15.35	34.75
3820		4" clay tile	10	polystyrene		12.90	13.95	26.85
3840				none		12.60	13.30	25.90
3860		4" glazed tile	10	polystyrene		21	19.75	40.75
3880				none		21	19.10	40.10
4000	Utility	4" common brick	10	polystyrene		12.05	18.30	30.35
4020				none		11.75	17.65	29.40
4080		4" conc. block	10	polystyrene		9.25	13.65	22.90
4100				none		8.95	13	21.95
4160		4" L.W. block	10	polystyrene		9.40	13.55	22.95
4180				none		9.10	12.90	22

B2010 Exterior Walls

B2010 134 Brick Face Cavity Wall

	FACE BRICK	BACKUP MASONRY	TOTAL THICKNESS (IN.)	CAVITY INSULATION		COST PER S.F.		
						MAT.	INST.	TOTAL
4240	Utility	4" glazed block	10	polystyrene		20.50	15	35.50
4260				none		20	14.35	34.35
4320		4" clay tile	10	polystyrene		13.50	12.95	26.45
4340				none		13.20	12.30	25.50
4360		4" glazed tile	10	polystyrene		21.50	18.75	40.25
4380				none		21.50	18.10	39.60

B2010 135 Brick Face Cavity Wall - Insulated Backup

	FACE BRICK	BACKUP MASONRY	TOTAL THICKNESS (IN.)	BACKUP CORE FILL		COST PER S.F.		
						MAT.	INST.	TOTAL
5100	Standard	6" conc. block	10	perlite		9.80	19	28.80
5120				styrofoam		10.15	18.70	28.85
5180		6" L.W. block	10	perlite		10.15	18.90	29.05
5200				styrofoam		10.50	18.60	29.10
5260		6" glazed block	10	perlite		20.50	20.50	41
5280				styrofoam		21	19.95	40.95
5340		6" clay tile	10	none		16.40	18.10	34.50
5360		8" clay tile	12	none		19	18.70	37.70
5600	Glazed	6" conc. block	10	perlite		27	19.50	46.50
5620				styrofoam		27	19.20	46.20
5680		6" L.W. block	10	perlite		27	19.40	46.40
5700				styrofoam		27.50	19.10	46.60
5760		6" glazed block	10	perlite		37.50	21	58.50
5780				styrofoam		38	20.50	58.50
5840		6" clay tile	10	none		33.50	18.60	52.10
5860		8" clay tile	8	none		36	19.20	55.20
6100	Engineer	6" conc. block	10	perlite		9.50	17.40	26.90
6120				styrofoam		9.85	17.10	26.95
6180		6" L.W. block	10	perlite		9.85	17.30	27.15
6200				styrofoam		10.20	17	27.20
6260		6" glazed block	10	perlite		20	18.65	38.65
6280				styrofoam		20.50	18.35	38.85
6340		6" clay tile	10	none		16.10	16.50	32.60
6360		8" clay tile	12	none		18.70	17.10	35.80
6600	Roman	6" conc. block	10	perlite		12.90	17.75	30.65
6620				styrofoam		13.20	17.45	30.65
6680		6" L.W. block	10	perlite		13.25	17.65	30.90
6700				styrofoam		13.55	17.35	30.90
6760		6" glazed block	10	perlite		23.50	19	42.50
6780				styrofoam		24	18.70	42.70
6840		6" clay tile	10	none		19.50	16.85	36.35
6860		8" clay tile	12	none		22	17.45	39.45
7100	Norman	6" conc. block	10	perlite		10.85	15.75	26.60
7120				styrofoam		11.15	15.45	26.60
7180		6" L.W. block	10	perlite		11.20	15.65	26.85
7200				styrofoam		11.50	15.35	26.85
7260		6" glazed block	10	perlite		21.50	17	38.50
7280				styrofoam		22	16.70	38.70
7340		6" clay tile	10	none		17.45	14.85	32.30
7360		8" clay tile	12	none		20	15.45	35.45

B2010 Exterior Walls

B2010 135 Brick Face Cavity Wall - Insulated Backup

	FACE BRICK	BACKUP MASONRY	TOTAL THICKNESS (IN.)	BACKUP CORE FILL		COST PER S.F.		
						MAT.	INST.	TOTAL
7600	Norwegian	6″ conc. block	10	perlite		10.10	14.70	24.80
7620				styrofoam		10.40	14.40	24.80
7680		6″ L.W. block	10	perlite		10.45	14.60	25.05
7700				styrofoam		10.75	14.30	25.05
7760		6″ glazed block	10	perlite		20.50	15.95	36.45
7780				styrofoam		21	15.65	36.65
7840		6″ clay tile	10	none		16.70	13.80	30.50
7860		8″ clay tile	12	none		19.25	14.40	33.65
8100	Utility	6″ conc. block	10	perlite		10.70	13.70	24.40
8120				styrofoam		11	13.40	24.40
8180		6″ L.W. block	10	perlite		11.05	13.60	24.65
8200				styrofoam		11.35	13.30	24.65
8220		8″ L.W. block	12	perlite		12.25	14.05	26.30
8240				styrofoam		12.25	13.65	25.90
8260		6″ glazed block	10	perlite		21.50	14.95	36.45
8280				styrofoam		21.50	14.65	36.15
8300		8″ glazed block	12	perlite		23.50	15.50	39
8340		6″ clay tile	10	none		17.30	12.80	30.10
8360		8″ clay tile	12	none		19.85	13.40	33.25

B2010 Exterior Walls

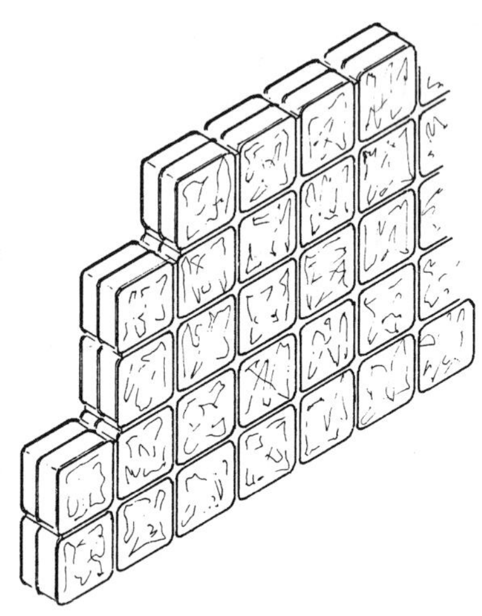

The table below lists costs per S.F. for glass block walls. Included in the costs are the following special accessories required for glass block walls.

Glass block accessories required for proper installation.

Wall ties: Galvanized double steel mesh full length of joint.

Fiberglass expansion joint at sides and top.

Silicone caulking: One gallon does 95 L.F.

Oakum: One lb. does 30 L.F.

Asphalt emulsion: One gallon does 600 L.F.

If block are not set in wall chase, use 2′-0″ long wall anchors at 2′-0″ O.C.

B2010 140	Glass Block	COST PER S.F.		
		MAT.	INST.	TOTAL
2300	Glass block 4″ thick, 6″x6″ plain, under 1,000 S.F.	25.50	22	47.50
2400	1,000 to 5,000 S.F.	25	19.25	44.25
2500	Over 5,000 S.F.	24.50	18.10	42.60
2600	Solar reflective, under 1,000 S.F.	36	30	66
2700	1,000 to 5,000 S.F.	35	26	61
2800	Over 5,000 S.F.	34.50	24.50	59
3500	8″x8″ plain, under 1,000 S.F.	14.85	16.60	31.45
3600	1,000 to 5,000 S.F.	14.55	14.35	28.90
3700	Over 5,000 S.F.	14.10	12.95	27.05
3800	Solar reflective, under 1,000 S.F.	20.50	22.50	43
3900	1,000 to 5,000 S.F.	20.50	19.15	39.65
4000	Over 5,000 S.F.	19.65	17.20	36.85
5000	12″x12″ plain, under 1,000 S.F.	24	15.40	39.40
5100	1,000 to 5,000 S.F.	23.50	12.95	36.45
5200	Over 5,000 S.F.	22.50	11.85	34.35
5300	Solar reflective, under 1,000 S.F.	34	20.50	54.50
5400	1,000 to 5,000 S.F.	33	17.20	50.20
5600	Over 5,000 S.F.	31.50	15.65	47.15
5800	3″ thinline, 6″x6″ plain, under 1,000 S.F.	26.50	22	48.50
5900	Over 5,000 S.F.	25.50	18.10	43.60
6000	Solar reflective, under 1,000 S.F.	37.50	30	67.50
6100	Over 5,000 S.F.	36	24.50	60.50
6200	8″x8″ plain, under 1,000 S.F.	14.75	16.60	31.35
6300	Over 5,000 S.F.	14.55	12.95	27.50
6400	Solar reflective, under 1,000 S.F.	20.50	22.50	43
6500	Over 5,000 S.F.	20.50	17.20	37.70

B2010 Exterior Walls

The table below lists costs for metal siding of various descriptions, not including the steel frame, or the structural steel, of a building. Costs are per S.F. including all accessories and insulation.

For steel frame support see System B2010 154.

B2010 146	Metal Siding Panel	COST PER S.F.		
		MAT.	INST.	TOTAL
1400	Metal siding aluminum panel, corrugated, .024" thick, natural	3.43	3.61	7.04
1450	Painted	3.60	3.61	7.21
1500	.032" thick, natural	3.68	3.61	7.29
1550	Painted	4.26	3.61	7.87
1600	Ribbed 4" pitch, .032" thick, natural	3.71	3.61	7.32
1650	Painted	4.31	3.61	7.92
1700	.040" thick, natural	4.28	3.61	7.89
1750	Painted	4.84	3.61	8.45
1800	.050" thick, natural	4.80	3.61	8.41
1850	Painted	5.45	3.61	9.06
1900	Ribbed 8" pitch, .032" thick, natural	3.54	3.46	7
1950	Painted	4.13	3.48	7.61
2000	.040" thick, natural	4.11	3.48	7.59
2050	Painted	4.65	3.52	8.17
2100	.050" thick, natural	4.63	3.50	8.13
2150	Painted	5.30	3.53	8.83
3000	Steel, corrugated or ribbed, 29 Ga., .0135" thick, galvanized	2.25	3.27	5.52
3050	Colored	3.21	3.30	6.51
3100	26 Ga., .0179" thick, galvanized	2.38	3.28	5.66
3150	Colored	3.28	3.31	6.59
3200	24 Ga., .0239" thick, galvanized	2.82	3.30	6.12
3250	Colored	3.54	3.33	6.87
3300	22 Ga., .0299" thick, galvanized	3.45	3.32	6.77
3350	Colored	4.24	3.35	7.59
3400	20 Ga., .0359" thick, galvanized	3.45	3.32	6.77
3450	Colored	4.49	3.54	8.03
4100	Sandwich panels, factory fab., 1" polystyrene, steel core, 26 Ga., galv.	5.95	5.15	11.10
4200	Colored, 1 side	7.70	5.15	12.85
4300	2 sides	9.35	5.15	14.50
4400	2" polystyrene core, 26 Ga., galvanized	10.10	5.15	15.25
4500	Colored, 1 side	8.80	5.15	13.95
4600	2 sides	10.45	5.15	15.60
4700	22 Ga., baked enamel exterior	13.50	5.40	18.90
4800	Polyvinyl chloride exterior	14.30	5.40	19.70
5100	Textured aluminum, 4' x 8' x 5/16" plywood backing, single face	4.52	3.06	7.58
5200	Double face	5.85	3.06	8.91

B2010 Exterior Walls

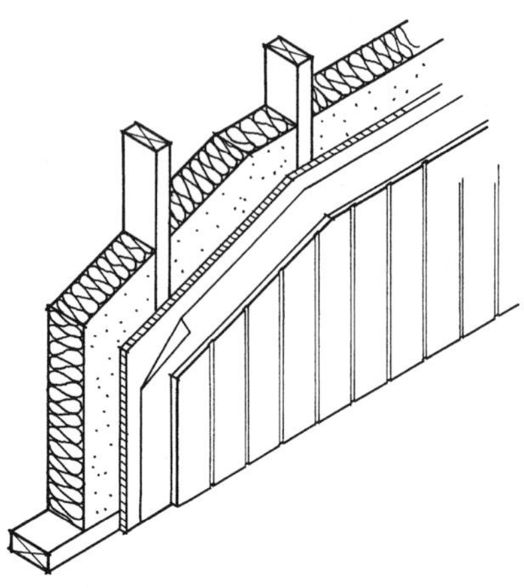

The table below lists costs per S.F. for exterior walls with wood siding. A variety of systems are presented using both wood and metal studs at 16" and 24" O.C.

B2010 148	Panel, Shingle & Lap Siding	COST PER S.F.		
		MAT.	INST.	TOTAL
1400	Wood siding w/2" x 4" studs, 16" O.C., insul. wall, 5/8" text 1-11 fir ply.	3.60	5.25	8.85
1450	5/8" text 1-11 cedar plywood	4.98	5.05	10.03
1500	1" x 4" vert T.&G. redwood	9.65	6.05	15.70
1600	1" x 8" vert T.&G. redwood	10.90	5.25	16.15
1650	1" x 5" rabbetted cedar bev. siding	6.95	5.45	12.40
1700	1" x 6" cedar drop siding	7.05	5.50	12.55
1750	1" x 12" rough sawn cedar	3.55	5.30	8.85
1800	1" x 12" sawn cedar, 1" x 4" battens	7.35	4.94	12.29
1850	1" x 10" redwood shiplap siding	7.20	5.10	12.30
1900	18" no. 1 red cedar shingles, 5-1/2" exposed	4.97	6.90	11.87
1950	6" exposed	4.77	6.70	11.47
2000	6-1/2" exposed	4.56	6.45	11.01
2100	7" exposed	4.36	6.25	10.61
2150	7-1/2" exposed	4.15	6	10.15
3000	8" wide aluminum siding	4.58	4.70	9.28
3150	8" plain vinyl siding	3.34	4.78	8.12
3250	8" insulated vinyl siding	3.69	5.30	8.99
3300				
3400	2" x 6" studs, 16" O.C., insul. wall, w/ 5/8" text 1-11 fir plywood	4.11	5.40	9.51
3500	5/8" text 1-11 cedar plywood	5.50	5.40	10.90
3600	1" x 4" vert T.&G. redwood	10.15	6.20	16.35
3700	1" x 8" vert T.&G. redwood	11.40	5.40	16.80
3800	1" x 5" rabbetted cedar bev siding	7.50	5.60	13.10
3900	1" x 6" cedar drop siding	7.60	5.65	13.25
4000	1" x 12" rough sawn cedar	4.06	5.45	9.51
4200	1" x 12" sawn cedar, 1" x 4" battens	7.85	5.10	12.95
4500	1" x 10" redwood shiplap siding	7.70	5.25	12.95
4550	18" no. 1 red cedar shingles, 5-1/2" exposed	5.50	7.05	12.55
4600	6" exposed	5.30	6.85	12.15
4650	6-1/2" exposed	5.05	6.60	11.65
4700	7" exposed	4.87	6.40	11.27
4750	7-1/2" exposed	4.66	6.15	10.81
4800	8" wide aluminum siding	5.10	4.85	9.95
4850	8" plain vinyl siding	4.99	8.35	13.34
4900	8" insulated vinyl siding	5.35	8.85	14.20
4950	8" fiber cement siding	4.99	9	13.99
5000	2" x 6" studs, 24" O.C., insul. wall, 5/8" text 1-11, fir plywood	3.91	5.05	8.96
5050	5/8" text 1-11 cedar plywood	5.30	5.05	10.35
5100	1" x 4" vert T.&G. redwood	11.10	9.30	20.40
5150	1" x 8" vert T.&G. redwood	11.20	5.10	16.30

B2010 Exterior Walls

B2010 148	Panel, Shingle & Lap Siding	COST PER S.F.		
		MAT.	INST.	TOTAL
5200	1″ x 5″ rabbetted cedar bev siding	7.30	5.30	12.60
5250	1″ x 6″ cedar drop siding	7.40	5.30	12.70
5300	1″ x 12″ rough sawn cedar	3.86	5.10	8.96
5400	1″ x 12″ sawn cedar, 1″ x 4″ battens	7.65	4.77	12.42
5450	1″ x 10″ redwood shiplap siding	7.50	4.92	12.42
5500	18″ no. 1 red cedar shingles, 5-1/2″ exposed	5.30	6.75	12.05
5550	6″ exposed	5.10	6.50	11.60
5650	7″ exposed	4.67	6.05	10.72
5700	7-1/2″ exposed	4.46	5.85	10.31
5750	8″ wide aluminum siding	4.89	4.53	9.42
5800	8″ plain vinyl siding	3.65	4.61	8.26
5850	8″ insulated vinyl siding	4	5.10	9.10
5875	8″ fiber cement siding	4.79	8.70	13.49
5900	3-5/8″ metal studs, 16 Ga. 16″ O.C. insul. wall, 5/8″ text 1-11 fir plywood	4.36	5.50	9.86
5950	5/8″ text 1-11 cedar plywood	5.75	5.50	11.25
6000	1″ x 4″ vert T.&G. redwood	10.40	6.35	16.75
6050	1″ x 8″ vert T.&G. redwood	11.65	5.55	17.20
6100	1″ x 5″ rabbetted cedar bev siding	7.75	5.75	13.50
6150	1″ x 6″ cedar drop siding	7.85	5.75	13.60
6200	1″ x 12″ rough sawn cedar	4.31	5.55	9.86
6250	1″ x 12″ sawn cedar, 1″ x 4″ battens	8.10	5.20	13.30
6300	1″ x 10″ redwood shiplap siding	7.95	5.35	13.30
6350	18″ no. 1 red cedar shingles, 5-1/2″ exposed	5.75	7.20	12.95
6500	6″ exposed	5.55	6.95	12.50
6550	6-1/2″ exposed	5.30	6.75	12.05
6600	7″ exposed	5.10	6.50	11.60
6650	7-1/2″ exposed	5.10	6.35	11.45
6700	8″ wide aluminum siding	5.35	4.98	10.33
6750	8″ plain vinyl siding	4.10	4.88	8.98
6800	8″ insulated vinyl siding	4.45	5.40	9.85
7000	3-5/8″ metal studs, 16 Ga. 24″ O.C. insul wall, 5/8″ text 1-11 fir plywood	4.01	4.90	8.91
7050	5/8″ text 1-11 cedar plywood	5.40	4.90	10.30
7100	1″ x 4″ vert T.&G. redwood	10.05	5.90	15.95
7150	1″ x 8″ vert T.&G. redwood	11.30	5.10	16.40
7200	1″ x 5″ rabbetted cedar bev siding	7.40	5.30	12.70
7250	1″ x 6″ cedar drop siding	7.45	5.35	12.80
7300	1″ x 12″ rough sawn cedar	3.96	5.15	9.11
7350	1″ x 12″ sawn cedar 1″ x 4″ battens	7.75	4.78	12.53
7400	1″ x 10″ redwood shiplap siding	7.60	4.93	12.53
7450	18″ no. 1 red cedar shingles, 5-1/2″ exposed	5.60	7	12.60
7500	6″ exposed	5.40	6.75	12.15
7550	6-1/2″ exposed	5.20	6.50	11.70
7600	7″ exposed	4.97	6.30	11.27
7650	7-1/2″ exposed	4.56	5.85	10.41
7700	8″ wide aluminum siding	4.99	4.54	9.53
7750	8″ plain vinyl siding	3.75	4.62	8.37
7800	8″ insul. vinyl siding	4.10	5.15	9.25

B2010 Exterior Walls

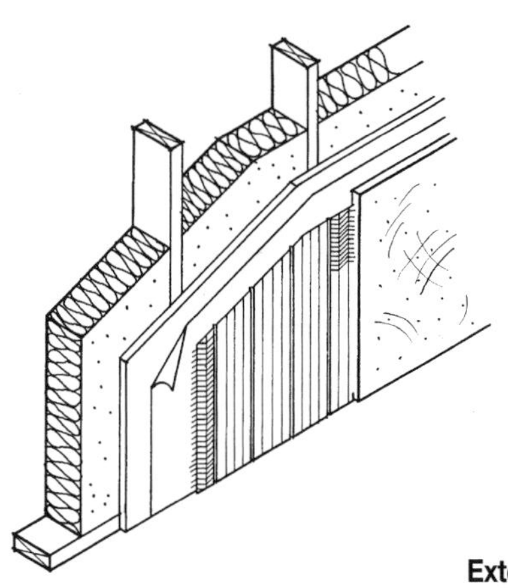

Exterior Stucco Wall

The table below lists costs for some typical stucco walls including all the components as demonstrated in the component block below. Prices are presented for backup walls using wood studs, metal studs and CMU.

B2010 151	Stucco Wall	COST PER S.F.		
		MAT.	INST.	TOTAL
2100	Cement stucco, 7/8" th., plywood sheathing, stud wall, 2" x 4", 16" O.C.	3.23	8.70	11.93
2200	24" O.C.	3.11	8.45	11.56
2300	2" x 6", 16" O.C.	3.74	8.85	12.59
2400	24" O.C.	3.54	8.55	12.09
2500	No sheathing, metal lath on stud wall, 2" x 4", 16" O.C.	2.31	7.60	9.91
2600	24" O.C.	2.19	7.35	9.54
2700	2" x 6", 16" O.C.	2.82	7.75	10.57
2800	24" O.C.	2.62	7.45	10.07
2900	1/2" gypsum sheathing, 3-5/8" metal studs, 16" O.C.	3.35	7.95	11.30
2950	24" O.C.	3	7.50	10.50
3000	Cement stucco, 5/8" th., 2 coats on std. CMU block, 8"x 16", 8" thick	3.68	10.40	14.08
3100	10" thick	5.30	10.65	15.95
3200	12" thick	5.45	12.25	17.70
3300	Std. light Wt. block 8" x 16", 8" thick	4.57	10.20	14.77
3400	10" thick	5.30	10.45	15.75
3500	12" thick	5.55	12	17.55
3600	3 coat stucco, self furring metal lath 3.4 Lb/SY, on 8" x 16", 8" thick	4.64	12.40	17.04
3700	10" thick	5.35	11.30	16.65
3800	12" thick	8.65	15.15	23.80
3900	Lt. Wt. block, 8" thick	4.63	10.85	15.48
4000	10" thick	5.40	11.10	16.50
4100	12" thick	5.60	12.65	18.25

B2010 152	E.I.F.S.	COST PER S.F.		
		MAT.	INST.	TOTAL
5100	E.I.F.S., plywood sheathing, stud wall, 2" x 4", 16" O.C., 1" EPS	3.89	8.80	12.69
5110	2" EPS	4.18	8.80	12.98
5120	3" EPS	4.48	8.80	13.28
5130	4" EPS	4.78	8.80	13.58
5140	2" x 6", 16" O.C., 1" EPS	4.40	8.95	13.35
5150	2" EPS	4.88	9.05	13.93
5160	3" EPS	4.99	8.95	13.94
5170	4" EPS	5.30	8.95	14.25
5180	Cement board sheathing, 3-5/8" metal studs, 16" O.C., 1" EPS	4.66	10.85	15.51
5190	2" EPS	4.95	10.85	15.80
5200	3" EPS	5.25	10.85	16.10
5210	4" EPS	6.40	10.85	17.25
5220	6" metal studs, 16" O.C., 1" EPS	5.20	10.90	16.10
5230	2" EPS	5.70	11	16.70

B2010 Exterior Walls

B2010 152	E.I.F.S.	COST PER S.F.		
		MAT.	INST.	TOTAL
5240	3" EPS	5.80	10.90	16.70
5250	4" EPS	7	10.90	17.90
5260	CMU block, 8" x 8" x 16", 1" EPS	5.05	12.35	17.40
5270	2" EPS	6.90	16.15	23.05
5280	3" EPS	5.65	12.35	18
5290	4" EPS	6.85	13.70	20.55
5300	8" x 10" x 16", 1" EPS	6.65	12.60	19.25
5310	2" EPS	6.95	12.60	19.55
5320	3" EPS	7.25	12.60	19.85
5330	4" EPS	7.55	12.60	20.15
5340	8" x 12" x 16", 1" EPS	6.80	14.20	21
5350	2" EPS	7.10	14.20	21.30
5360	3" EPS	7.40	14.20	21.60
5370	4" EPS	7.70	14.20	21.90

B2010 Exterior Walls

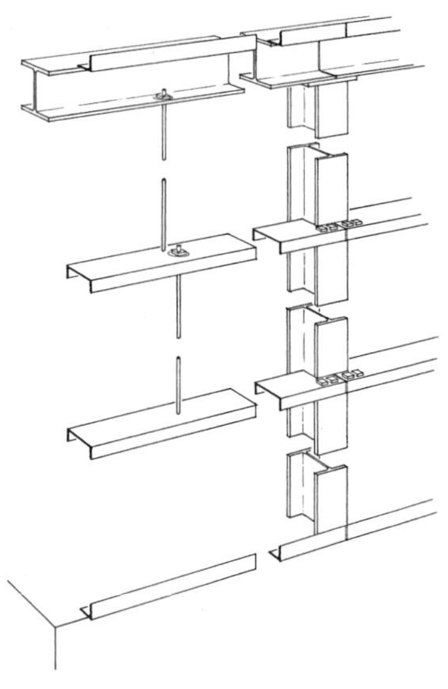

Description: The table below lists costs, $/S.F., for channel girts with sag rods and connector angles top and bottom for various column spacings, building heights and wind loads. Additive costs are shown for wind columns.

How to Use this Table: Add the cost of girts, sag rods, and angles to the framing costs of steel buildings clad in metal or composition siding. If the column spacing is in excess of the column spacing shown, use intermediate wind columns. Additive costs are shown under "Wind Columns".

B2010 154	Metal Siding Support							
	BLDG. HEIGHT (FT.)	WIND LOAD (P.S.F.)	COL. SPACING (FT.)		INTERMEDIATE COLUMNS	COST PER S.F.		
						MAT.	INST.	TOTAL
3000	18	20	20			2.40	3.19	5.59
3100					wind cols.	1.24	.24	1.48
3200		20	25			2.63	3.24	5.87
3300					wind cols.	.99	.19	1.18
3400		20	30			2.90	3.29	6.19
3500					wind cols.	.83	.16	.99
3600		20	35			3.19	3.35	6.54
3700					wind cols.	.71	.14	.85
3800		30	20			2.66	3.25	5.91
3900					wind cols.	1.24	.24	1.48
4000		30	25			2.90	3.29	6.19
4100					wind cols.	.99	.19	1.18
4200		30	30			3.21	3.35	6.56
4300					wind cols.	1.12	.21	1.33
4600	30	20	20			2.33	2.26	4.59
4700					wind cols.	1.75	.34	2.09
4800		20	25			2.62	2.31	4.93
4900					wind cols.	1.40	.27	1.67
5000		20	30			2.95	2.38	5.33
5100					wind cols.	1.38	.27	1.65
5200		20	35			3.30	2.44	5.74
5300					wind cols.	1.37	.27	1.64
5400		30	20			2.65	2.31	4.96
5500					wind cols.	2.07	.40	2.47
5600		30	25			2.94	2.37	5.31
5700					wind cols.	1.98	.38	2.36
5800		30	30			3.32	2.45	5.77
5900					wind cols.	1.86	.36	2.22

B2010 Exterior Walls

B2010 160	Pole Barn Exterior Wall	COST PER S.F.		
		MAT.	INST.	TOTAL
2000	Pole barn exterior wall, pressure treated pole in concrete, 8′ O.C.			
3000	Steel siding, 8′ eave	4.66	8.55	13.21
3050	10′ eave	4.46	7.90	12.36
3100	12′ eave	4.22	7.20	11.42
3150	14′ eave	4.10	6.80	10.90
3200	16′ eave	4	6.45	10.45
4000	Aluminum siding, 8′ eave	4.48	8.55	13.03
4050	10′ eave	4.22	7.75	11.97
4100	12′ eave	4.04	7.20	11.24
4150	14′ eave	3.92	6.80	10.72
4200	16′ eave	3.82	6.45	10.27
5000	Wood siding, 8′ eave	5.20	7.60	12.80
5050	10′ eave	4.94	6.75	11.69
5100	12′ eave	4.76	6.20	10.96
5150	14′ eave	4.64	5.80	10.44
5200	16′ eave	4.54	5.50	10.04
6000	Plywood siding, 8′ eave	3.94	7.55	11.49
6050	10′ eave	3.68	6.75	10.43
6100	12′ eave	3.50	6.20	9.70
6150	14′ eave	3.38	5.80	9.18
6200	16′ eave	3.28	5.45	8.73

B2020 Exterior Windows

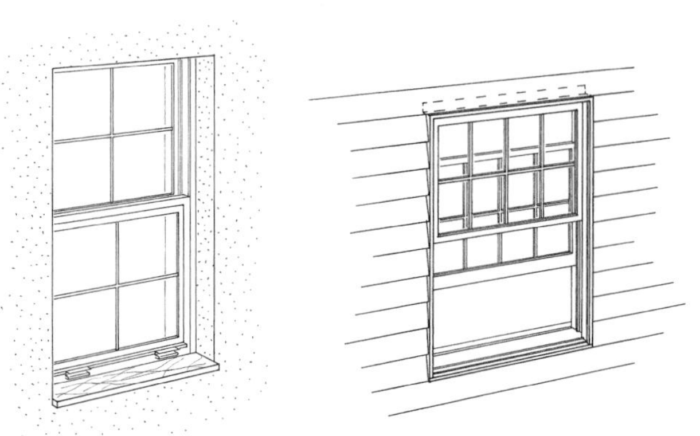

The table below lists window systems by material, type and size. Prices between sizes listed can be interpolated with reasonable accuracy. Prices include frame, hardware, and casing as illustrated in the component block below.

B2020 102 Wood Windows

	MATERIAL	TYPE	GLAZING	SIZE	DETAIL	COST PER UNIT MAT.	COST PER UNIT INST.	COST PER UNIT TOTAL
3000	Wood	double hung	std. glass	2′-8″ x 4′-6″		253	223	476
3050				3′-0″ x 5′-6″		330	255	585
3100			insul. glass	2′-8″ x 4′-6″		268	223	491
3150				3′-0″ x 5′-6″		360	255	615
3200		sliding	std. glass	3′-4″ x 2′-7″		335	185	520
3250				4′-4″ x 3′-3″		345	203	548
3300				5′-4″ x 6′-0″		440	244	684
3350			insul. glass	3′-4″ x 2′-7″		405	217	622
3400				4′-4″ x 3′-3″		420	236	656
3450				5′-4″ x 6′-0″		535	278	813
3500		awning	std. glass	2′-10″ x 1′-9″		234	111	345
3600				4′-4″ x 2′-8″		375	107	482
3700			insul. glass	2′-10″ x 1′-9″		285	128	413
3800				4′-4″ x 2′-8″		460	118	578
3900		casement	std. glass	1′-10″ x 3′-2″	1 lite	345	146	491
3950				4′-2″ x 4′-2″	2 lite	610	194	804
4000				5′-11″ x 5′-2″	3 lite	910	250	1,160
4050				7′-11″ x 6′-3″	4 lite	1,300	300	1,600
4100				9′-11″ x 6′-3″	5 lite	1,700	350	2,050
4150			insul. glass	1′-10″ x 3′-2″	1 lite	345	146	491
4200				4′-2″ x 4′-2″	2 lite	625	194	819
4250				5′-11″ x 5′-2″	3 lite	960	250	1,210
4300				7′-11″ x 6′-3″	4 lite	1,375	300	1,675
4350				9′-11″ x 6′-3″	5 lite	1,700	350	2,050
4400		picture	std. glass	4′-6″ x 4′-6″		480	251	731
4450				5′-8″ x 4′-6″		540	279	819
4500		picture	insul. glass	4′-6″ x 4′-6″		585	292	877
4550				5′-8″ x 4′-6″		655	325	980
4600		fixed bay	std. glass	8′ x 5′		1,675	460	2,135
4650				9′-9″ x 5′-4″		1,200	645	1,845
4700			insul. glass	8′ x 5′		2,250	460	2,710
4750				9′-9″ x 5′-4″		1,300	645	1,945
4800		casement bay	std. glass	8′ x 5′		1,725	530	2,255
4850			insul. glass	8′ x 5′		1,875	530	2,405
4900		vert. bay	std. glass	8′ x 5′		1,875	530	2,405
4950			insul. glass	8′ x 5′		1,975	530	2,505

B2020 Exterior Windows

B2020 104 Steel Windows

	MATERIAL	TYPE	GLAZING	SIZE	DETAIL	COST PER UNIT		
						MAT.	INST.	TOTAL
5000	Steel	double hung	1/4″ tempered	2′-8″ x 4′-6″		900	170	1,070
5050				3′-4″ x 5′-6″		1,375	259	1,634
5100			insul. glass	2′-8″ x 4′-6″		940	196	1,136
5150				3′-4″ x 5′-6″		1,425	299	1,724
5200		horz. pivoted	std. glass	2′ x 2′		277	56.50	333.50
5250				3′ x 3′		625	127	752
5300				4′ x 4′		1,100	226	1,326
5350				6′ x 4′		1,675	340	2,015
5400			insul. glass	2′ x 2′		292	65.50	357.50
5450				3′ x 3′		655	147	802
5500				4′ x 4′		1,175	262	1,437
5550				6′ x 4′		1,750	390	2,140
5600		picture window	std. glass	3′ x 3′		380	127	507
5650				6′ x 4′		1,025	340	1,365
5700			insul. glass	3′ x 3′		415	147	562
5750				6′ x 4′		1,100	390	1,490
5800		industrial security	std. glass	2′-9″ x 4′-1″		835	159	994
5850				4′-1″ x 5′-5″		1,650	315	1,965
5900			insul. glass	2′-9″ x 4′-1″		875	183	1,058
5950				4′-1″ x 5′-5″		1,725	360	2,085
6000		comm. projected	std. glass	3′-9″ x 5′-5″		1,350	288	1,638
6050				6′-9″ x 4′-1″		1,825	390	2,215
6100			insul. glass	3′-9″ x 5′-5″		1,400	330	1,730
6150				6′-9″ x 4′-1″		1,925	450	2,375
6200		casement	std. glass	4′-2″ x 4′-2″	2 lite	1,200	246	1,446
6250			insul. glass	4′-2″ x 4′-2″		1,250	284	1,534
6300			std. glass	5′-11″ x 5′-2″	3 lite	2,525	435	2,960
6350			insul. glass	5′-11″ x 5′-2″		2,625	500	3,125

B2020 106 Aluminum Windows

	MATERIAL	TYPE	GLAZING	SIZE	DETAIL	COST PER UNIT		
						MAT.	INST.	TOTAL
6400	Aluminum	double hung	insul. glass	3′-0″ x 4′-0″		535	115	650
6450				4′-5″ x 5′-3″		440	144	584
6500			insul. glass	3′-1″ x 3′-2″		480	138	618
6550				4′-5″ x 5′-3″		530	173	703
6600		sliding	std. glass	3′ x 2′		242	115	357
6650				5′ x 3′		370	128	498
6700				8′ x 4′		395	192	587
6750				9′ x 5′		595	288	883
6800			insul. glass	3′ x 2′		260	115	375
6850				5′ x 3′		430	128	558
6900				8′ x 4′		635	192	827
6950				9′ x 5′		935	288	1,223
7000		single hung	std. glass	2′ x 3′		236	115	351
7050				2′-8″ x 6′-8″		410	144	554
7100				3′-4″ x 5′-0″		340	128	468
7150			insul. glass	2′ x 3′		286	115	401
7200				2′-8″ x 6′-8″		520	144	664
7250				3′-4″ x 5′		370	128	498
7300		double hung	std. glass	2′ x 3′		325	85	410
7350				2′-8″ x 6′-8″		965	252	1,217

B2020 Exterior Windows

B2020 106 Aluminum Windows

	MATERIAL	TYPE	GLAZING	SIZE	DETAIL	COST PER UNIT		
						MAT.	INST.	TOTAL
7400	Aluminum	double hung		3'-4" x 5'		905	236	1,141
7450			insul. glass	2' x 3'		350	98	448
7500				2'-8" x 6'-8"		1,025	291	1,316
7550				3'-4" x 5'-0"		965	272	1,237
7600		casement	std. glass	3'-1" x 3'-2"		276	138	414
7650				4'-5" x 5'-3"		670	335	1,005
7700			insul. glass	3'-1" x 3'-2"		310	159	469
7750				4'-5" x 5'-3"		755	385	1,140
7800		hinged swing	std. glass	3' x 4'		600	170	770
7850				4' x 5'		995	283	1,278
7900			insul. glass	3' x 4'		640	196	836
7950				4' x 5'		1,075	325	1,400
8200		picture unit	std. glass	2'-0" x 3'-0"		179	85	264
8250				2'-8" x 6'-8"		530	252	782
8300				3'-4" x 5'-0"		495	236	731
8350			insul. glass	2'-0" x 3'-0"		201	98	299
8400				2'-8" x 6'-8"		595	291	886
8450				3'-4" x 5'-0"		555	272	827
8500		awning type	std. glass	3'-0" x 3'-0"	2 lite	480	82	562
8550				3'-0" x 4'-0"	3 lite	550	115	665
8600				3'-0" x 5'-4"	4 lite	660	115	775
8650				4'-0" x 5'-4"	4 lite	730	128	858
8700			insul. glass	3'-0" x 3'-0"	2 lite	510	82	592
8750				3'-0" x 4'-0"	3 lite	635	115	750
8800				3'-0" x 5'-4"	4 lite	785	115	900
8850				4'-0" x 5'-4"	4 lite	865	128	993

B2020 108 Vinyl Clad Windows

	MATERIAL	TYPE	GLAZING	SIZE	DETAIL	COST PER UNIT		
						MAT.	INST.	TOTAL
1000	Vinyl clad	casement	insul. glass	1'-4" x 4'-0"		455	118	573
1010				2'-0" x 3'-0"		380	118	498
1020				2'-0" x 4'-0"		430	124	554
1030				2'-0" x 5'-0"		485	131	616
1040				2'-0" x 6'-0"		515	131	646
1050				2'-4" x 4'-0"		485	131	616
1060				2'-6" x 5'-0"		615	128	743
1070				3'-0" x 5'-0"		840	131	971
1080				4'-0" x 3'-0"		910	131	1,041
1090				4'-0" x 4'-0"		790	131	921
1100				4'-8" x 4'-0"		865	131	996
1110				4'-8" x 5'-0"		980	153	1,133
1120				4'-8" x 6'-0"		1,100	153	1,253
1130				6'-0" x 4'-0"		1,000	153	1,153
1140				6'-0" x 5'-0"		1,100	153	1,253
3000		double-hung	insul. glass	2'-0" x 4'-0"		485	114	599
3050				2'-0" x 5'-0"		530	114	644
3100				2'-4" x 4'-0"		505	118	623
3150				2'-4" x 4'-8"		485	118	603
3200				2'-4" x 6'-0"		530	118	648
3250				2'-6" x 4'-0"		480	118	598
3300				2'-8" x 4'-0"		700	153	853

B2020 Exterior Windows

B2020 108 Vinyl Clad Windows

	MATERIAL	TYPE	GLAZING	SIZE	DETAIL	COST PER UNIT		
						MAT.	INST.	TOTAL
3350	Aluminum	double hung	insul. glass	2'-8" x 5'-0"		560	131	691
3375				2'-8" x 6'-0"		555	131	686
3400				3'-0 x 3'-6"		445	118	563
3450				3'-0 x 4'-0"		510	124	634
3500				3'-0 x 4'-8"		560	124	684
3550				3'-0 x 5'-0"		570	131	701
3600				3'-0 x 6'-0"		510	131	641
3700				4'-0 x 5'-0"		700	140	840
3800				4'-0 x 6'-0"		865	140	1,005

B2020 210 Tubular Aluminum Framing

		COST/S.F. OPNG.		
		MAT.	INST.	TOTAL
1100	Alum flush tube frame, for 1/4" glass, 1-3/4"x4", 5'x6'opng, no inter horiz	15.55	10.45	26
1150	One intermediate horizontal	21	12.50	33.50
1200	Two intermediate horizontals	26.50	14.50	41
1250	5' x 20' opening, three intermediate horizontals	14.30	8.95	23.25
1400	1-3/4" x 4-1/2", 5' x 6' opening, no intermediate horizontals	17.60	10.45	28.05
1450	One intermediate horizontal	23	12.50	35.50
1500	Two intermediate horizontals	29	14.50	43.50
1550	5' x 20' opening, three intermediate horizontals	15.75	8.95	24.70
1700	For insulating glass, 2"x4-1/2", 5'x6' opening, no intermediate horizontals	17.20	11.10	28.30
1750	One intermediate horizontal	22	13.25	35.25
1800	Two intermediate horizontals	27	15.35	42.35
1850	5' x 20' opening, three intermediate horizontals	14.80	9.50	24.30
2000	Thermal break frame, 2-1/4"x4-1/2", 5'x6'opng, no intermediate horizontals	17.50	11.25	28.75
2050	One intermediate horizontal	23.50	13.80	37.30
2100	Two intermediate horizontals	29.50	16.35	45.85
2150	5' x 20' opening, three intermediate horizontals	15.95	9.95	25.90

B2020 Exterior Windows

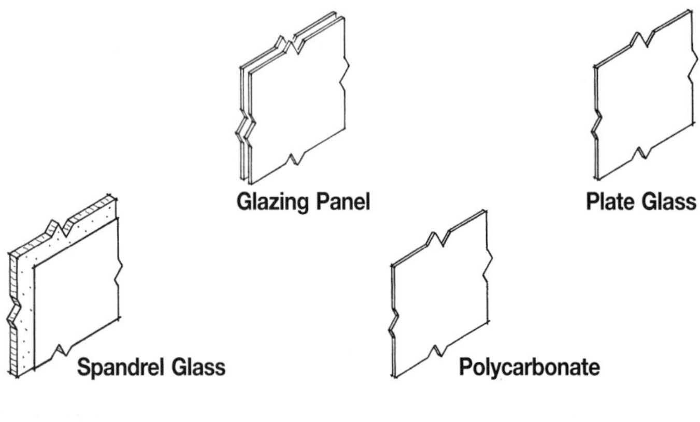

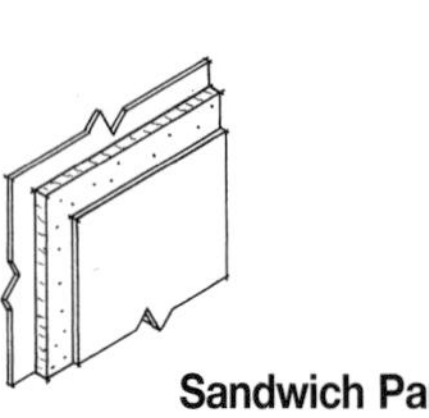

Sandwich Panel

The table below lists costs of curtain wall and spandrel panels per S.F. Costs do not include structural framing used to hang the panels.

B2020 220	Curtain Wall Panels	COST PER S.F.		
		MAT.	INST.	TOTAL
1000	Glazing panel, insulating, 1/2" thick, 2 lites 1/8" float, clear	11.45	10.60	22.05
1100	Tinted	15.95	10.60	26.55
1200	5/8" thick units, 2 lites 3/16" float, clear	15.50	11.20	26.70
1300	Tinted	15.60	11.20	26.80
1400	1" thick units, 2 lites, 1/4" float, clear	19.45	13.45	32.90
1500	Tinted	26.50	13.45	39.95
1600	Heat reflective film inside	31	11.85	42.85
1700	Light and heat reflective glass, tinted	35.50	11.85	47.35
2000	Plate glass, 1/4" thick, clear	11.15	8.40	19.55
2050	Tempered	7.85	8.40	16.25
2100	Tinted	10.55	8.40	18.95
2200	3/8" thick, clear	12.55	13.45	26
2250	Tempered	19.80	13.45	33.25
2300	Tinted	18.45	13.45	31.90
2400	1/2" thick, clear	21	18.30	39.30
2450	Tempered	29.50	18.30	47.80
2500	Tinted	32.50	18.30	50.80
2600	3/4" thick, clear	41.50	29	70.50
2650	Tempered	48.50	29	77.50
3000	Spandrel glass, panels, 1/4" plate glass insul w/fiberglass, 1" thick	19.35	8.40	27.75
3100	2" thick	24	8.40	32.40
3200	Galvanized steel backing, add	6.75		6.75
3300	3/8" plate glass, 1" thick	33.50	8.40	41.90
3400	2" thick	38	8.40	46.40
4000	Polycarbonate, masked, clear or colored, 1/8" thick	17.35	5.90	23.25
4100	3/16" thick	21.50	6.10	27.60
4200	1/4" thick	21	6.50	27.50
4300	3/8" thick	32.50	6.70	39.20
5000	Facing panel, textured alum., 4' x 8' x 5/16" plywood backing, sgl face	4.52	3.06	7.58
5100	Double face	5.85	3.06	8.91
5200	4' x 10' x 5/16" plywood backing, single face	4.79	3.06	7.85
5300	Double face	6.40	3.06	9.46
5400	4' x 12' x 5/16" plywood backing, single face	4.84	3.06	7.90
5500	Sandwich panel, 22 Ga. galv., both sides 2" insulation, enamel exterior	13.50	5.40	18.90
5600	Polyvinylidene fluoride exterior finish	14.30	5.40	19.70
5700	26 Ga., galv. both sides, 1" insulation, colored 1 side	7.70	5.15	12.85
5800	Colored 2 sides	9.35	5.15	14.50

B2030 Exterior Doors

B2030 110 Glazed Doors, Steel or Aluminum

	MATERIAL	TYPE	DOORS	SPECIFICATION	OPENING	COST PER OPNG.		
						MAT.	INST.	TOTAL
5600	St. Stl. & glass	revolving	stock unit	manual oper.	6'-0" x 7'-0"	55,000	7,625	62,625
5650				auto Cntrls.	6'-10" x 7'-0"	72,000	8,225	80,225
5700	Bronze	revolving	stock unit	manual oper.	6'-10" x 7'-0"	55,000	15,300	70,300
5750				auto Cntrls.	6'-10" x 7'-0"	72,000	15,900	87,900
5800	St. Stl. & glass	balanced	standard	economy	3'-0" x 7'-0"	11,200	1,275	12,475
5850				premium	3'-0" x 7'-0"	18,600	1,700	20,300
6300	Alum. & glass	w/o transom	narrow stile	w/panic Hrdwre.	3'-0" x 7'-0"	2,750	880	3,630
6350				dbl. door, Hrdwre.	6'-0" x 7'-0"	4,650	1,475	6,125
6400			wide stile	hdwre.	3'-0" x 7'-0"	3,000	865	3,865
6450				dbl. door, Hdwre.	6'-0" x 7'-0"	6,150	1,725	7,875
6500			full vision	hdwre.	3'-0" x 7'-0"	3,750	1,400	5,150
6550				dbl. door, Hdwre.	6'-0" x 7'-0"	5,475	1,950	7,425
6600			non-standard	hdwre.	3'-0" x 7'-0"	3,100	865	3,965
6650				dbl. door, Hdwre.	6'-0" x 7'-0"	6,200	1,725	7,925
6700			bronze fin.	hdwre.	3'-0" x 7'-0"	2,575	865	3,440
6750				dbl. door, Hrdwre.	6'-0" x 7'-0"	5,150	1,725	6,875
6800			black fin.	hdwre.	3'-0" x 7'-0"	2,950	865	3,815
6850				dbl. door, Hdwre.	6'-0" x 7'-0"	5,900	1,725	7,625
6900		w/transom	narrow stile	hdwre.	3'-0" x 10'-0"	3,250	1,000	4,250
6950				dbl. door, Hdwre.	6'-0" x 10'-0"	5,175	1,775	6,950
7000			wide stile	hdwre.	3'-0" x 10'-0"	3,800	1,200	5,000
7050				dbl. door, Hdwre.	6'-0" x 10'-0"	5,850	2,100	7,950
7100			full vision	hdwre.	3'-0" x 10'-0"	4,075	1,325	5,400
7150				dbl. door, Hdwre.	6'-0" x 10'-0"	6,175	2,300	8,475
7200			non-standard	hdwre.	3'-0" x 10'-0"	3,175	940	4,115
7250				dbl. door, Hdwre.	6'-0" x 10'-0"	6,350	1,875	8,225
7300			bronze fin.	hdwre.	3'-0" x 10'-0"	2,650	940	3,590
7350				dbl. door, Hdwre.	6'-0" x 10'-0"	5,275	1,875	7,150
7400			black fin.	hdwre.	3'-0" x 10'-0"	3,025	940	3,965
7450				dbl. door, Hdwre.	6'-0" x 10'-0"	6,050	1,875	7,925
7500		revolving	stock design	minimum	6'-10" x 7'-0"	37,400	3,075	40,475
7550				average	6'-0" x 7'-0"	43,000	3,825	46,825
7600				maximum	6'-10" x 7'-0"	48,400	5,125	53,525
7650				min., automatic	6'-10" x 7'-0"	54,500	3,675	58,175
7700				avg., automatic	6'-10" x 7'-0"	60,000	4,425	64,425
7750				max., automatic	6'-10" x 7'-0"	65,500	5,725	71,225
7800		balanced	standard	economy	3'-0" x 7'-0"	7,750	1,275	9,025
7850				premium	3'-0" x 7'-0"	9,475	1,650	11,125
7900		mall front	sliding panels	alum. fin.	16'-0" x 9'-0"	4,450	775	5,225
7950					24'-0" x 9'-0"	6,400	1,450	7,850
8000				bronze fin.	16'-0" x 9'-0"	5,175	905	6,080
8050					24'-0" x 9'-0"	7,450	1,700	9,150
8100			fixed panels	alum. fin.	48'-0" x 9'-0"	11,500	1,125	12,625
8150				bronze fin.	48'-0" x 9'-0"	13,400	1,300	14,700
8200		sliding entrance	5' x 7' door	electric oper.	12'-0" x 7'-6"	9,700	1,450	11,150
8250		sliding patio	temp. glass	economy	6'-0" x 7'-0"	1,925	256	2,181
8300			temp. glass	economy	12'-0" x 7'-0"	4,625	340	4,965
8350				premium	6'-0" x 7'-0"	2,900	385	3,285
8400					12'-0" x 7'-0"	6,950	510	7,460

B2030 Exterior Doors

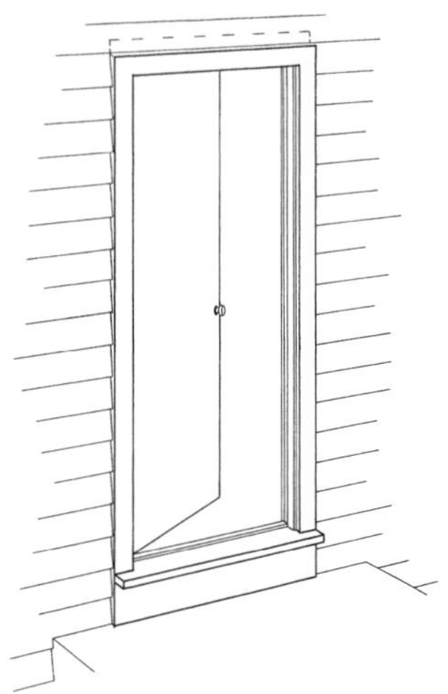

Costs are listed for exterior door systems by material, type and size. Prices between sizes listed can be interpolated with reasonable accuracy. Prices are per opening for a complete door system including frame as illustrated in the component block.

B2030 210 Wood Doors

	MATERIAL	TYPE	DOORS	SPECIFICATION	OPENING	COST PER OPNG. MAT.	INST.	TOTAL
2350	Birch	solid core	single door	hinged	2'-6" x 6'-8"	2,100	280	2,380
2400					2'-6" x 7'-0"	2,100	280	2,380
2450					2'-8" x 7'-0"	2,175	280	2,455
2500					3'-0" x 7'-0"	2,100	280	2,380
2550			double door		2'-6" x 6'-8"	4,050	510	4,560
2600					2'-6" x 7'-0"	4,050	510	4,560
2650					2'-8" x 7'-0"	4,225	510	4,735
2700					3'-0" x 7'-0"	3,925	445	4,370
2750	Wood	combination	storm & screen		3'-0" x 6'-8"	355	68.50	423.50
2800					3'-0" x 7'-0"	390	75.50	465.50
2850		overhead	panels, H.D.	manual oper.	8'-0" x 8'-0"	1,375	510	1,885
2900					10'-0" x 10'-0"	1,925	570	2,495
2950					12'-0" x 12'-0"	2,650	685	3,335
3000					14'-0" x 14'-0"	4,125	790	4,915
3050					20'-0" x 16'-0"	6,325	1,575	7,900
3100				electric oper.	8'-0" x 8'-0"	2,600	765	3,365
3150					10'-0" x 10'-0"	3,150	825	3,975
3200					12'-0" x 12'-0"	3,875	940	4,815
3250					14'-0" x 14'-0"	5,350	1,050	6,400
3300					20'-0" x 16'-0"	7,675	2,075	9,750

B2030 220 Steel Doors

	MATERIAL	TYPE	DOORS	SPECIFICATION	OPENING	COST PER OPNG. MAT.	INST.	TOTAL
3350	Steel 18 Ga.	hollow metal	1 door w/frame	no label	2'-6" x 7'-0"	2,500	289	2,789
3400					2'-8" x 7'-0"	2,500	289	2,789
3450					3'-0" x 7'-0"	2,525	310	2,835
3500					3'-6" x 7'-0"	2,800	305	3,105
3550		hollow metal	1 door w/frame	no label	4'-0" x 8'-0"	2,950	305	3,255
3600			2 doors w/frame	no label	5'-0" x 7'-0"	4,900	535	5,435
3650					5'-4" x 7'-0"	4,900	535	5,435
3700					6'-0" x 7'-0"	4,900	535	5,435
3750					7'-0" x 7'-0"	5,500	565	6,065
3800					8'-0" x 8'-0"	5,800	565	6,365

B2030 Exterior Doors

B2030 220 Steel Doors

	MATERIAL	TYPE	DOORS	SPECIFICATION	OPENING	COST PER OPNG.		
						MAT.	INST.	TOTAL
3850	Steel 18 Ga.	Hollow metal	1 door w/frame	"A" label	2'-6" x 7'-0"	2,925	350	3,275
3900					2'-8" x 7'-0"	2,925	350	3,275
3950					3'-0" x 7'-0"	2,925	350	3,275
4000					3'-6" x 7'-0"	3,100	360	3,460
4050					4'-0" x 8'-0"	3,275	375	3,650
4100			2 doors w/frame	"A" label	5'-0" x 7'-0"	5,725	645	6,370
4150					5'-4" x 7'-0"	5,725	655	6,380
4200					6'-0" x 7'-0"	5,725	655	6,380
4250					7'-0" x 7'-0"	6,075	665	6,740
4300					8'-0" x 8'-0"	6,075	670	6,745
4350	Steel 24 Ga.	overhead	sectional	manual oper.	8'-0" x 8'-0"	1,125	510	1,635
4400					10'-0" x 10'-0"	1,475	570	2,045
4450					12'-0" x 12'-0"	1,725	685	2,410
4500					20'-0" x 14'-0"	4,050	1,475	5,525
4550				electric oper.	8'-0" x 8'-0"	2,350	765	3,115
4600					10'-0" x 10'-0"	2,700	825	3,525
4650					12'-0" x 12'-0"	2,950	940	3,890
4700					20'-0" x 14'-0"	5,400	1,975	7,375
4750	Steel	overhead	rolling	manual oper.	8'-0" x 8'-0"	790	720	1,510
4800					10'-0" x 10'-0"	1,200	820	2,020
4850					12'-0" x 12'-0"	1,300	960	2,260
4900					14'-0" x 14'-0"	3,275	1,450	4,725
4950					20'-0" x 12'-0"	2,275	1,275	3,550
5000					20'-0" x 16'-0"	4,050	1,925	5,975
5050				electric oper.	8'-0" x 8'-0"	2,100	950	3,050
5100					10'-0" x 10'-0"	2,500	1,050	3,550
5150					12'-0" x 12'-0"	2,600	1,200	3,800
5200					14'-0" x 14'-0"	4,575	1,675	6,250
5250					20'-0" x 12'-0"	3,750	1,500	5,250
5300					20'-0" x 16'-0"	5,525	2,150	7,675
5350				fire rated	10'-0" x 10'-0"	2,375	1,050	3,425
5400			rolling grille	manual oper.	10'-0" x 10'-0"	2,925	1,150	4,075
5450					15'-0" x 8'-0"	3,275	1,450	4,725
5500		vertical lift	1 door w/frame	motor operator	16"-0" x 16"-0"	24,600	4,500	29,100
5550					32'-0" x 24'-0"	56,000	3,000	59,000

B2030 230 Aluminum Doors

	MATERIAL	TYPE	DOORS	SPECIFICATION	OPENING	COST PER OPNG.		
						MAT.	INST.	TOTAL
6000	Aluminum	combination	storm & screen	hinged	3'-0" x 6'-8"	340	73	413
6050					3'-0" x 7'-0"	375	80.50	455.50
6100		overhead	rolling grille	manual oper.	12'-0" x 12'-0"	4,750	2,025	6,775
6150				motor oper.	12'-0" x 12'-0"	6,175	2,250	8,425
6200	Alum. & Fbrgls.	overhead	heavy duty	manual oper.	12'-0" x 12'-0"	3,725	685	4,410
6250				electric oper.	12'-0" x 12'-0"	4,950	940	5,890

B3010 Roof Coverings

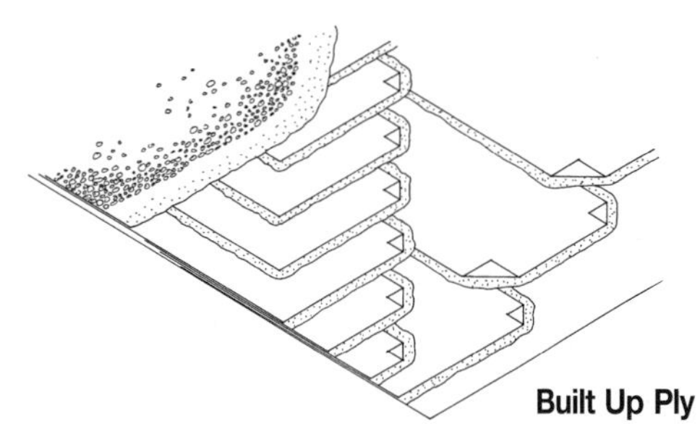

Built Up Ply

Multiple ply roofing is the most popular covering for minimum pitch roofs.

B3010 105	Built-Up	COST PER S.F.		
		MAT.	INST.	TOTAL
1200	Asphalt flood coat w/gravel; not incl. insul, flash., nailers			
1300				
1400	Asphalt base sheets & 3 plies #15 asphalt felt, mopped	1.18	1.70	2.88
1500	On nailable deck	1.21	1.78	2.99
1600	4 plies #15 asphalt felt, mopped	1.60	1.87	3.47
1700	On nailable deck	1.42	1.96	3.38
1800	Coated glass base sheet, 2 plies glass (type IV), mopped	1.27	1.70	2.97
1900	For 3 plies	1.53	1.87	3.40
2000	On nailable deck	1.44	1.96	3.40
2300	4 plies glass fiber felt (type IV), mopped	1.88	1.87	3.75
2400	On nailable deck	1.70	1.96	3.66
2500	Organic base sheet & 3 plies #15 organic felt, mopped	1.20	1.87	3.07
2600	On nailable deck	1.22	1.96	3.18
2700	4 plies #15 organic felt, mopped	1.52	1.70	3.22
2750				
2800	Asphalt flood coat, smooth surface, not incl. insul, flash., nailers			
2900	Asphalt base sheet & 3 plies #15 asphalt felt, mopped	1.21	1.56	2.77
3000	On nailable deck	1.12	1.63	2.75
3100	Coated glass fiber base sheet & 2 plies glass fiber felt, mopped	1.18	1.50	2.68
3200	On nailable deck	1.12	1.56	2.68
3300	For 3 plies, mopped	1.44	1.63	3.07
3400	On nailable deck	1.35	1.70	3.05
3700	4 plies glass fiber felt (type IV), mopped	1.70	1.63	3.33
3800	On nailable deck	1.61	1.70	3.31
3900	Organic base sheet & 3 plies #15 organic felt, mopped	1.22	1.56	2.78
4000	On nailable decks	1.13	1.63	2.76
4100	4 plies #15 organic felt, mopped	1.43	1.70	3.13
4200	Coal tar pitch with gravel surfacing			
4300	4 plies #15 tarred felt, mopped	2.28	1.78	4.06
4400	3 plies glass fiber felt (type IV), mopped	1.89	1.96	3.85
4500	Coated glass fiber base sheets 2 plies glass fiber felt, mopped	1.94	1.96	3.90
4600	On nailable decks	1.70	2.08	3.78
4800	3 plies glass fiber felt (type IV), mopped	2.61	1.78	4.39
4900	On nailable decks	2.37	1.87	4.24

B3010 Roof Coverings

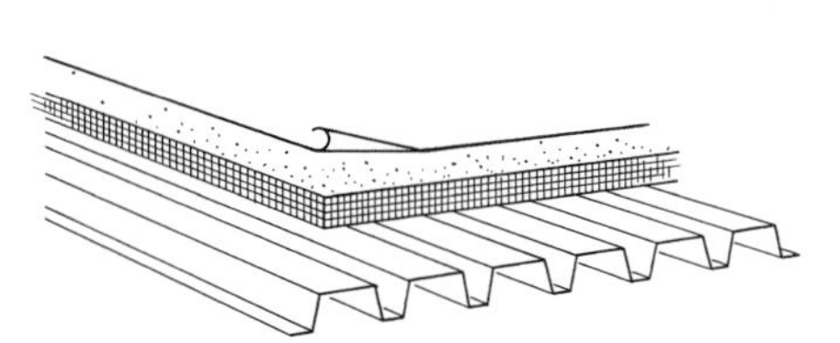

Fully Adhered

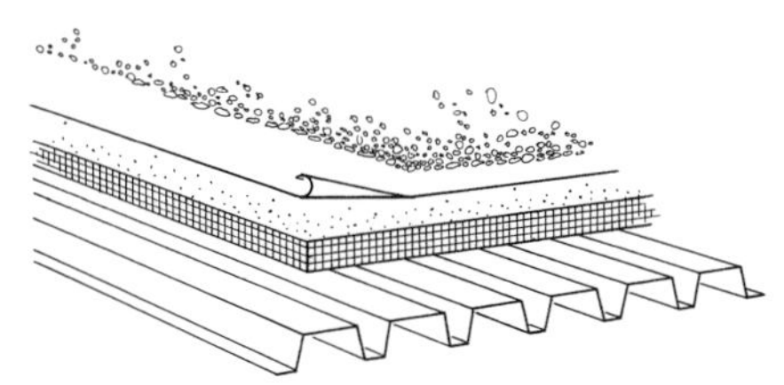

Ballasted

The systems listed below reflect only the cost for the single ply membrane.

B3010 120	Single Ply Membrane	COST PER S.F.		
		MAT.	INST.	TOTAL
1000	CSPE (Chlorosulfonated polyethylene), 45 mils, plate attached	2.65	.69	3.34
2000	EPDM (Ethylene propylene diene monomer), 45 mils, fully adhered	1.24	.93	2.17
4000	Modified bit., SBS modified, granule surface cap sheet, mopped, 150 mils	1.49	1.86	3.35
4500	APP modified, granule surface cap sheet, torched, 180 mils	1.06	1.20	2.26
6000	Reinforced PVC, 48 mils, loose laid and ballasted with stone	1.31	.47	1.78
6200	Fully adhered with adhesive	1.66	.93	2.59

B3010 130	Preformed Metal Roofing	COST PER S.F.		
		MAT.	INST.	TOTAL
0200	Corrugated roofing, aluminum, mill finish, .0175" thick, .272 P.S.F.	1.12	1.63	2.75
0250	.0215" thick, .334 P.S.F.	1.39	1.63	3.02

B3010 135	Formed Metal	COST PER S.F.		
		MAT.	INST.	TOTAL
1000	Batten seam, formed copper roofing, 3" min slope, 16 oz., 1.2 P.S.F.	14.10	5.30	19.40
1100	18 oz., 1.35 P.S.F.	15.85	5.85	21.70
2000	Zinc copper alloy, 3" min slope, .020" thick, .88 P.S.F.	16.35	4.88	21.23
3000	Flat seam, copper, 1/4" min. slope, 16 oz., 1.2 P.S.F.	10.35	4.88	15.23
3100	18 oz., 1.35 P.S.F.	11.60	5.10	16.70
5000	Standing seam, copper, 2-1/2" min. slope, 16 oz., 1.25 P.S.F.	11.10	4.48	15.58
5100	18 oz., 1.40 P.S.F.	12.60	4.88	17.48
6000	Zinc copper alloy, 2-1/2" min. slope, .020" thick, .87 P.S.F.	15.85	4.88	20.73
6100	.032" thick, 1.39 P.S.F.	20.50	5.30	25.80

B3010 Roof Coverings

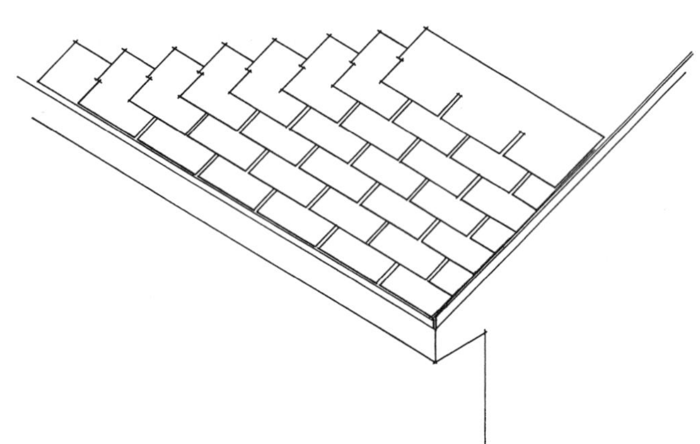

Shingles and tiles are practical in applications where the roof slope is more than 3-1/2" per foot of rise. Table below lists the various materials and the weight per S.F.

B3010 140	Shingle & Tile	COST PER S.F.		
		MAT.	INST.	TOTAL
1095	Asphalt roofing			
1100	Strip shingles, 4" slope, inorganic class A 210-235 lb./sq.	.93	1.02	1.95
1150	Organic, class C, 235-240 lb./sq.	1.12	1.11	2.23
1200	Premium laminated multi-layered, class A, 260-300 lb./sq.	1.66	1.52	3.18
1545	Metal roofing			
1550	Alum., shingles, colors, 3" min slope, .019" thick, 0.4 PSF	2.87	1.10	3.97
1850	Steel, colors, 3" min slope, 26 gauge, 1.0 PSF	4.77	2.28	7.05
2795	Slate roofing			
2800	Slate roofing, 4" min. slope, shingles, 3/16" thick, 8.0 PSF	6.25	2.85	9.10
3495	Wood roofing			
3500	4" min slope, cedar shingles, 16" x 5", 5" exposure 1.6 PSF	3.37	2.21	5.58
4000	Shakes, 18", 8-1/2" exposure, 2.8 PSF	3.17	2.72	5.89
5095	Tile roofing			
5100	Aluminum, mission, 3" min slope, .019" thick, 0.65 PSF	9.20	2.13	11.33
6000	Clay tile, flat shingle, interlocking, 15", 166 pcs/sq, fireflashed blend	5.30	2.50	7.80
6002				

B3010 Roof Coverings

B3010 320	Roof Deck Rigid Insulation	COST PER S.F.		
		MAT.	INST.	TOTAL
0100	Fiberboard low density			
0150	1" thick R2.78	.77	.61	1.38
0300	1 1/2" thick R4.17	1.11	.61	1.72
0350	2" thick R5.56	1.40	.61	2.01
0370	Fiberboard high density, 1/2" thick R1.3	.42	.52	.94
0380	1" thick R2.5	.75	.61	1.36
0390	1 1/2" thick R3.8	1.04	.61	1.65
0410	Fiberglass, 3/4" thick R2.78	.80	.52	1.32
0450	15/16" thick R3.70	1.02	.52	1.54
0550	1-5/16" thick R5.26	1.69	.52	2.21
0650	2 7/16" thick R10	2.02	.61	2.63
1510	Polyisocyanurate 2#/CF density, 1" thick	.60	.42	1.02
1550	1 1/2" thick	.78	.45	1.23
1600	2" thick	.99	.49	1.48
1650	2 1/2" thick	1.24	.50	1.74
1700	3" thick	1.35	.52	1.87
1750	3 1/2" thick	2.17	.52	2.69
1800	Tapered for drainage	.69	.42	1.11
1810	Expanded polystyrene, 1#/CF density, 3/4" thick R2.89	.34	.40	.74
1820	2" thick R7.69	.71	.45	1.16
1830	15 PSI compressive strength, 1" thick, R5	.74	.40	1.14
1835	2" thick R10	.92	.45	1.37
1840	3" thick R15	1.72	.52	2.24
2550	40 PSI compressive strength, 1" thick R5	1.08	.40	1.48
2600	2" thick R10	1.94	.45	2.39
2650	3" thick R15	2.75	.52	3.27
2700	4" thick R20	3.57	.52	4.09
2750	Tapered for drainage	1.08	.42	1.50
2810	60 PSI compressive strength, 1" thick R5	1.28	.41	1.69
2850	2" thick R10	2.31	.46	2.77
2900	Tapered for drainage	1.23	.42	1.65

B3010 410 Base Flashing

	TYPE	DESCRIPTION	SPECIFICATION	REGLET	COUNTER FLASHING	COST PER L.F.		
						MAT.	INST.	TOTAL
1000	Aluminum	mill finish	.019" thick	alum. .025	alum. .032	13.95	14.55	28.50
1100			.050" thick	.025	.032	15.40	14.55	29.95
1200		fabric, 2 sides	.004" thick	.025	.032	14.10	12.70	26.80
1300			.016" thick	.025	.032	14.35	12.70	27.05
1400		mastic, 2 sides	.004" thick	.025	.032	14.10	12.70	26.80
1500			.016" thick	.025	.032	14.60	12.70	27.30
1700	Copper	sheets, plain	16 oz.	copper 16 oz.	copper 16 oz.	31	15.45	46.45
1800			20 oz.	16 oz.	16 oz.	34	15.65	49.65
1900			24 oz.	16 oz.	16 oz.	38.50	15.85	54.35
2000			32 oz.	16 oz.	16 oz.	44	16.05	60.05
2200	Sheets, Metal	galvanized	20 Ga.	galv. 24 Ga.	galv. 24 Ga.	12.85	33	45.85
2300			24 Ga.	24 Ga.	24 Ga.	12.20	26.50	38.70
2400			30 Ga.	24 Ga.	24 Ga.	11.60	19.85	31.45
2600	Rubber	butyl	1/32" thick	galv. 24 Ga.	galv. 24 Ga.	13.25	12.95	26.20
2700			1/16" thick	24 Ga.	24 Ga.	14.45	12.95	27.40
2800		neoprene	1/16" thick	24 Ga.	24 Ga.	13.65	12.95	26.60
2900			1/8" thick	24 Ga.	24 Ga.	17.55	12.95	30.50

B3010 Roof Coverings

B3010 420 Roof Edges

	EDGE TYPE	DESCRIPTION	SPECIFICATION	FACE HEIGHT		COST PER L.F.		
						MAT.	INST.	TOTAL
1000	Aluminum	mill finish	.050" thick	4"		15.25	9.10	24.35
1100				6"		15.85	9.40	25.25
1300		duranodic	.050" thick	4"		16.35	9.10	25.45
1400				6"		16.80	9.40	26.20
1600		painted	.050" thick	4"		16.35	9.10	25.45
2000	Copper	plain	16 oz.	4"		35.50	9.10	44.60
2100				6"		45	9.40	54.40
2300			20 oz.	4"		44.50	10.40	54.90
2400				6"		50.50	10.05	60.55
2700	Sheet Metal	galvanized	20 Ga.	4"		18.45	10.80	29.25
2800				6"		18.60	10.80	29.40
3000			24 Ga.	4"		15.30	9.10	24.40
3100				6"		15.40	9.10	24.50

B3010 430 Flashing

	MATERIAL	BACKING	SIDES	SPECIFICATION	QUANTITY	COST PER S.F.		
						MAT.	INST.	TOTAL
0040	Aluminum	none		.019"		1.56	3.32	4.88
0050				.032"		1.50	3.32	4.82
0300		fabric	2	.004"		1.74	1.46	3.20
0400		mastic		.004"		1.74	1.46	3.20
0700	Copper	none		16 oz.	<500 lbs.	8.90	4.19	13.09
0800				24 oz.	<500 lbs.	16.25	4.59	20.84
3500	PVC black	none		.010"		.30	1.69	1.99
3700				.030"		.36	1.69	2.05
4200	Neoprene			1/16"		2.93	1.69	4.62
4500	Stainless steel	none		.015"	<500 lbs.	7.35	4.19	11.54
4600	Copper clad				>2000 lbs.	7.50	3.11	10.61
5000	Plain			32 ga.		3.75	3.11	6.86
5009								

B3010 Roof Coverings

B3010 610 — Gutters

	SECTION	MATERIAL	THICKNESS	SIZE	FINISH	COST PER L.F.		
						MAT.	INST.	TOTAL
0050	Box	aluminum	.027"	5"	enameled	3.31	4.75	8.06
0100					mill	3.22	4.75	7.97
0200			.032"	5"	enameled	3.93	4.60	8.53
0500		copper	16 Oz.	5"	lead coated	20.50	4.60	25.10
0600					mill	9.25	4.60	13.85
1000		steel galv.	28 Ga.	5"	enameled	2.45	4.60	7.05
1200			26 Ga.	5"	mill	2.74	4.60	7.34
1800		vinyl		4"	colors	1.50	4.46	5.96
1900				5"	colors	1.80	4.46	6.26
2300		hemlock or fir		4"x5"	treated	24	5.10	29.10
3000	Half round	copper	16 Oz.	4"	lead coated	17.30	4.60	21.90
3100					mill	9.25	4.60	13.85
3600		steel galv.	28 Ga.	5"	enameled	2.45	4.60	7.05
4102		stainless steel		5"	mill	11.05	4.60	15.65
5000		vinyl		4"	white	1.57	4.46	6.03
5002								

B3010 620 — Downspouts

	MATERIALS	SECTION	SIZE	FINISH	THICKNESS	COST PER V.L.F.		
						MAT.	INST.	TOTAL
0100	Aluminum	rectangular	2"x3"	embossed mill	.020"	1.02	3.02	4.04
0150				enameled	.020"	1.51	3.02	4.53
0250			3"x4"	enameled	.024"	2.35	4.10	6.45
0300		round corrugated	3"	enameled	.020"	2.31	3.02	5.33
0350			4"	enameled	.025"	3.56	4.10	7.66
0500	Copper	rectangular corr.	2"x3"	mill	16 Oz.	9.05	3.02	12.07
0600		smooth		mill	16 Oz.	12.50	3.02	15.52
0700		rectangular corr.	3"x4"	mill	16 Oz.	10.30	3.96	14.26
1300	Steel	rectangular corr.	2"x3"	galvanized	28 Ga.	2.45	3.02	5.47
1350				epoxy coated	24 Ga.	2.71	3.02	5.73
1400		smooth		galvanized	28 Ga.	4.44	3.02	7.46
1450		rectangular corr.	3"x4"	galvanized	28 Ga.	2.46	3.96	6.42
1500				epoxy coated	24 Ga.	3.28	3.96	7.24
1550		smooth		galvanized	28 Ga.	4.81	3.96	8.77
1600		round corrugated	2"	galvanized	28 Ga.	2.32	3.02	5.34
1650			3"	galvanized	28 Ga.	2.32	3.02	5.34
1700			4"	galvanized	28 Ga.	2.33	3.96	6.29
1750			5"	galvanized	28 Ga.	4.02	4.42	8.44
2552	S.S. tubing sch.5	rectangular	3"x4"	mill		133	3.96	136.96
2702								

B3010 630 — Gravel Stop

	MATERIALS	SECTION	SIZE	FINISH	THICKNESS	COST PER L.F.		
						MAT.	INST.	TOTAL
5100	Aluminum	extruded	4"	mill	.050"	7.25	3.96	11.21
5200			4"	duranodic	.050"	8.35	3.96	12.31
5300			8"	mill	.050"	8.55	4.60	13.15
5400			8"	duranodic	.050"	10.35	4.60	14.95
6000			12"-2 pc.	duranodic	.050"	12.20	5.75	17.95
6100	Stainless	formed	6"	mill	24 Ga.	17.15	4.25	21.40

B3020 Roof Openings

Roof Hatch

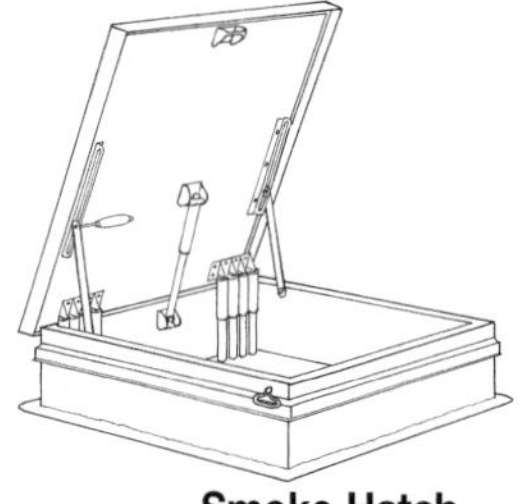
Smoke Hatch

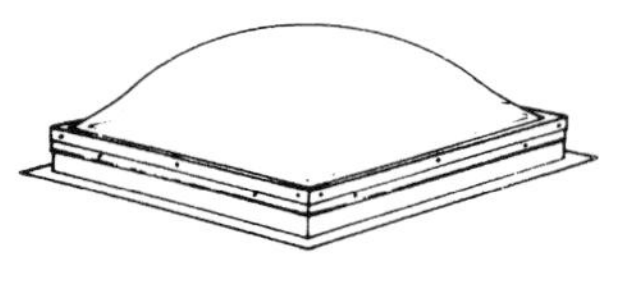
Skylight

B3020 110	Skylights	COST PER S.F.		
		MAT.	INST.	TOTAL
5100	Skylights, plastic domes, insul curbs, nom. size to 10 S.F., single glaze	33	12.20	45.20
5200	Double glazing	37.50	15	52.50
5300	10 S.F. to 20 S.F., single glazing	34.50	4.94	39.44
5400	Double glazing	32	6.20	38.20
5500	20 S.F. to 30 S.F., single glazing	26	4.20	30.20
5600	Double glazing	30	4.94	34.94
5700	30 S.F. to 65 S.F., single glazing	24.50	3.20	27.70
5800	Double glazing	32	4.20	36.20
6000	Sandwich panels fiberglass, 1-9/16" thick, 2 S.F. to 10 S.F.	20.50	9.75	30.25
6100	10 S.F. to 18 S.F.	18.60	7.35	25.95
6200	2-3/4" thick, 25 S.F. to 40 S.F.	29.50	6.60	36.10
6300	40 S.F. to 70 S.F.	24.50	5.90	30.40

B3020 210	Hatches	COST PER OPNG.		
		MAT.	INST.	TOTAL
0200	Roof hatches with curb and 1" fiberglass insulation, 2'-6"x3'-0", aluminum	965	195	1,160
0300	Galvanized steel 165 lbs.	940	195	1,135
0400	Primed steel 164 lbs.	690	195	885
0500	2'-6"x4'-6" aluminum curb and cover, 150 lbs.	1,175	217	1,392
0600	Galvanized steel 220 lbs.	1,050	217	1,267
0650	Primed steel 218 lbs.	1,100	217	1,317
0800	2'x6"x8'-0" aluminum curb and cover, 260 lbs.	2,325	296	2,621
0900	Galvanized steel, 360 lbs.	2,100	296	2,396
0950	Primed steel 358 lbs.	1,325	296	1,621
1200	For plexiglass panels, add to the above	525		525
2100	Smoke hatches, unlabeled not incl. hand winch operator, 2'-6"x3', galv	1,125	236	1,361
2200	Plain steel, 160 lbs.	835	236	1,071
2400	2'-6"x8'-0",galvanized steel, 360 lbs.	2,300	325	2,625
2500	Plain steel, 350 lbs.	1,450	325	1,775
3000	4'-0"x8'-0", double leaf low profile, aluminum cover, 359 lb.	3,500	244	3,744
3100	Galvanized steel 475 lbs.	3,025	244	3,269
3200	High profile, aluminum cover, galvanized curb, 361 lbs.	3,375	244	3,619

C1010 Partitions

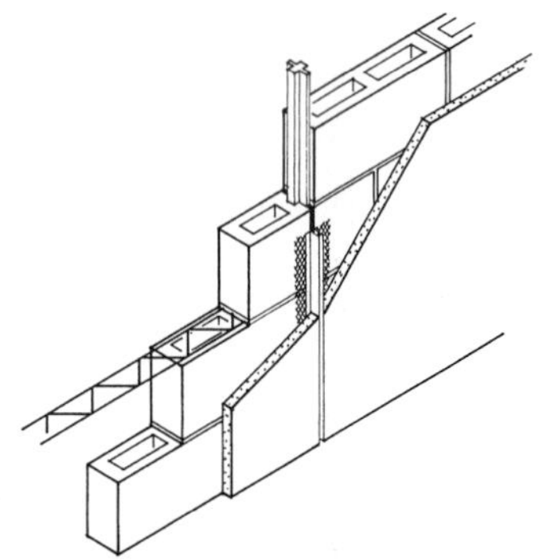

The Concrete Block Partition Systems are defined by weight and type of block, thickness, type of finish and number of sides finished. System components include joint reinforcing on alternate courses and vertical control joints.

C1010 102 Concrete Block Partitions - Regular Weight

	TYPE	THICKNESS (IN.)	TYPE FINISH	SIDES FINISHED		COST PER S.F.		
						MAT.	INST.	TOTAL
1000	Hollow	4	none	0		2.29	5.50	7.79
1010			gyp. plaster 2 coat	1		3.09	8.15	11.24
1020				2		3.89	10.85	14.74
1200			portland - 3 coat	1		2.85	8.60	11.45
1400			5/8" drywall	1		2.95	7.40	10.35
1500		6	none	0		3	5.90	8.90
1510			gyp. plaster 2 coat	1		3.80	8.55	12.35
1520				2		4.60	11.25	15.85
1700			portland - 3 coat	1		3.56	9	12.56
1900			5/8" drywall	1		3.66	7.80	11.46
1910				2		4.32	9.70	14.02
2000		8	none	0		3.13	6.30	9.43
2010			gyp. plaster 2 coat	1		3.93	8.95	12.88
2020			gyp. plaster 2 coat	2		4.73	11.65	16.38
2200			portland - 3 coat	1		3.69	9.40	13.09
2400			5/8" drywall	1		3.79	8.25	12.04
2410				2		4.45	10.15	14.60
2500		10	none	0		3.73	6.60	10.33
2510			gyp. plaster 2 coat	1		4.53	9.25	13.78
2520				2		5.35	11.95	17.30
2700			portland - 3 coat	1		4.29	9.70	13.99
2900			5/8" drywall	1		4.39	8.50	12.89
2910				2		5.05	10.45	15.50
3000	Solid	2	none	0		2.01	5.45	7.46
3010			gyp. plaster	1		2.81	8.10	10.91
3020				2		3.61	10.80	14.41
3200			portland - 3 coat	1		2.57	8.55	11.12
3400			5/8" drywall	1		2.67	7.35	10.02
3410				2		3.33	9.25	12.58
3500		4	none	0		2.58	5.70	8.28
3510			gyp. plaster	1		3.50	8.35	11.85
3520				2		4.18	11.05	15.23
3700			portland - 3 coat	1		3.14	8.80	11.94
3900			5/8" drywall	1		3.24	7.60	10.84
3910				2		3.90	9.50	13.40

C1010 Partitions

C1010 102 Concrete Block Partitions - Regular Weight

	TYPE	THICKNESS (IN.)	TYPE FINISH	SIDES FINISHED		COST PER S.F.		
						MAT.	INST.	TOTAL
4010	Solid	6	gyp. plaster	1		4.16	8.80	12.96
4020				2		4.96	11.50	16.46
4200			portland - 3 coat	1		3.92	9.25	13.17
4400			5/8" drywall	1		4.02	8.05	12.07
4410				2		4.68	9.95	14.63

C1010 104 Concrete Block Partitions - Lightweight

	TYPE	THICKNESS (IN.)	TYPE FINISH	SIDES FINISHED		COST PER S.F.		
						MAT.	INST.	TOTAL
5000	Hollow	4	none	0		2.43	5.40	7.83
5010			gyp. plaster	1		3.23	8.05	11.28
5020				2		4.03	10.75	14.78
5200			portland - 3 coat	1		2.99	8.50	11.49
5400			5/8" drywall	1		3.09	7.30	10.39
5410				2		3.75	9.20	12.95
5500		6	none	0		3.35	5.80	9.15
5520			gyp. plaster	2		4.95	11.15	16.10
5700			portland - 3 coat	1		3.91	8.90	12.81
5900			5/8" drywall	1		4.01	7.70	11.71
5910				2		4.67	9.60	14.27
6000	Hollow	8	none	0		4.07	6.15	10.22
6010			gyp. plaster	1		4.87	8.80	13.67
6020				2		5.65	11.50	17.15
6200			portland - 3 coat	1		4.63	9.25	13.88
6400			5/8" drywall	1		4.73	8.10	12.83
6410				2		5.40	10	15.40
6500		10	none	0		4.83	6.45	11.28
6700			portland - 3 coat	1		5.40	9.55	14.95
6900			5/8" drywall	1		5.50	8.35	13.85
6910				2		6.15	10.30	16.45
7000	Solid	4	none	0		2.27	5.65	7.92
7010			gyp. plaster	1		3.07	8.30	11.37
7020				2		3.87	11	14.87
7200			portland - 3 coat	1		2.83	8.75	11.58
7400			5/8" drywall	1		2.93	7.55	10.48
7410				2		3.59	9.45	13.04
7500		6	none	0		3.82	6.05	9.87
7510			gyp. plaster	1		4.78	8.75	13.53
7520				2		5.40	11.40	16.80
7700			portland - 3 coat	1		4.38	9.15	13.53
7900			5/8" drywall	1		4.48	7.95	12.43
7910				2		5.15	9.85	15
8000		8	none	0		4.39	6.45	10.84
8010			gyp. plaster	1		5.20	9.10	14.30
8020				2		6	11.80	17.80
8200			portland - 3 coat	1		4.95	9.55	14.50
8400			5/8" drywall	1		5.05	8.40	13.45
8410				2		5.70	10.30	16

C1010 Partitions

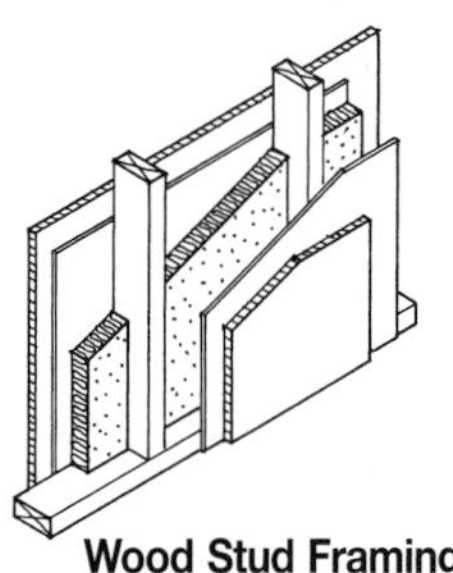

Wood Stud Framing

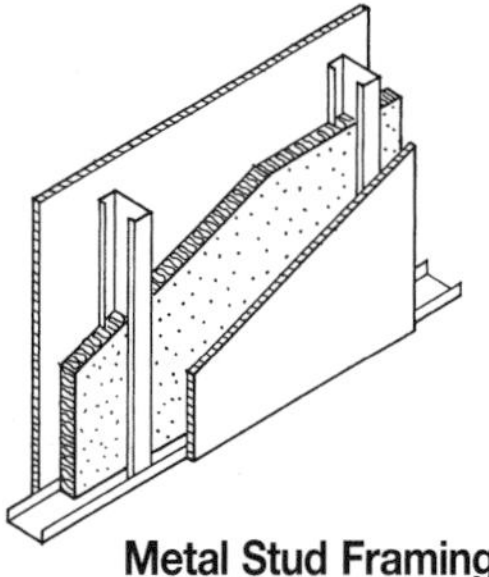

Metal Stud Framing

The Drywall Partitions/Stud Framing Systems are defined by type of drywall and number of layers, type and spacing of stud framing, and treatment on the opposite face. Components include taping and finishing.

Cost differences between regular and fire resistant drywall are negligible, and terminology is interchangeable. In some cases fiberglass insulation is included for additional sound deadening.

C1010 124 Drywall Partitions/Wood Stud Framing

	FACE LAYER	BASE LAYER	FRAMING	OPPOSITE FACE	INSULATION	COST PER S.F.		
						MAT.	INST.	TOTAL
1200	5/8" FR drywall	none	2 x 4, @ 16" O.C.	same	0	1.39	3.07	4.46
1250				5/8" reg. drywall	0	1.38	3.07	4.45
1300				nothing	0	.94	2.05	2.99
1400		1/4" SD gypsum	2 x 4 @ 16" O.C.	same	1-1/2" fiberglass	2.75	4.72	7.47
1425					Sound attenuation	2.99	4.77	7.76
1450				5/8" FR drywall	1-1/2" fiberglass	2.29	4.15	6.44
1475					Sound attenuation	2.53	4.20	6.73
1500				nothing	1-1/2" fiberglass	1.84	3.13	4.97
1600		resil. channels	2 x 4 @ 16", O.C.	same	1-1/2" fiberglass	2.18	6	8.18
1650				5/8" FR drywall	1-1/2" fiberglass	2.01	4.79	6.80
1700				nothing	1-1/2" fiberglass	1.56	3.77	5.33
1800		5/8" FR drywall	2 x 4 @ 24" O.C.	same	0	2.08	3.88	5.96
1850				5/8" FR drywall	0	1.68	3.37	5.05
1900				nothing	0	1.23	2.35	3.58
2200		5/8" FR drywall	2 rows-2 x 4	same	2" fiberglass	3.23	5.60	8.83
2250			16"O.C.	5/8" FR drywall	2" fiberglass	2.83	5.10	7.93
2300				nothing	2" fiberglass	2.38	4.09	6.47
2400	5/8" WR drywall	none	2 x 4, @ 16" O.C.	same	0	1.69	3.07	4.76
2450				5/8" FR drywall	0	1.54	3.07	4.61
2500				nothing	0	1.09	2.05	3.14
2600		5/8" FR drywall	2 x 4, @ 24" O.C.	same	0	2.38	3.88	6.26
2650				5/8" FR drywall	0	1.83	3.37	5.20
2700				nothing	0	1.38	2.35	3.73
2800	5/8" VF drywall	none	2 x 4, @ 16" O.C.	same	0	2.31	3.31	5.62
2850				5/8" FR drywall	0	1.85	3.19	5.04
2900				nothing	0	1.40	2.17	3.57
3000		5/8" FR drywall	2 x 4 , 24" O.C.	same	0	3	4.12	7.12
3050				5/8" FR drywall	0	2.14	3.49	5.63
3100				nothing	0	1.69	2.47	4.16
3200	1/2" reg drywall	3/8" reg drywall	2 x 4, @ 16" O.C.	same	0	2.13	4.09	6.22
3250				5/8" FR drywall	0	1.76	3.58	5.34
3300				nothing	0	1.31	2.56	3.87

C1010 Partitions

C1010 126 Drywall Partitions/Metal Stud Framing

	FACE LAYER	BASE LAYER	FRAMING	OPPOSITE FACE	INSULATION	COST PER S.F.		
						MAT.	INST.	TOTAL
5200	5/8" FR drywall	none	1-5/8" @ 24" O.C.	same	0	1.10	2.71	3.81
5250				5/8" reg. drywall	0	1.09	2.71	3.80
5300				nothing	0	.65	1.69	2.34
5310			2-1/2" @ 24" O.C.	same	0	1.17	2.72	3.89
5320				5/8" reg. drywall	0	1.16	2.72	3.88
5330				nothing	0	.72	1.70	2.42
5400			3-5/8" @ 24" O.C.	same	0	1.20	2.73	3.93
5450				5/8" reg. drywall	0	1.19	2.73	3.92
5500				nothing	0	.75	1.71	2.46
5530		1/4" SD gypsum	1-5/8" @ 16" O.C.	same	0	2.10	4.22	6.32
5535				5/8" FR drywall	0	1.18	3.08	4.26
5540				nothing	0	1.19	2.63	3.82
5545			2-1/2" @ 16" O.C.	same	0	2.19	4.23	6.42
5550				5/8" FR drywall	0	1.27	3.09	4.36
5555				nothing	0	1.28	2.64	3.92
5560			3-5/8" @ 16" O.C.	same	0	2.23	4.25	6.48
5565				5/8" FR drywall	0	1.31	3.11	4.42
5570				nothing	0	1.32	2.66	3.98
5600			1-5/8" @ 24" O.C.	same	0	2.02	3.85	5.87
5650				5/8" FR drywall	0	1.56	3.28	4.84
5700				nothing	0	1.11	2.26	3.37
5800			2-1/2" @ 24" O.C.	same	0	2.09	3.86	5.95
5850				5/8" FR drywall	0	1.63	3.29	4.92
5900				nothing	0	1.18	2.27	3.45
5910			3-5/8" @ 24" O.C.	same	0	2.12	3.87	5.99
5920				5/8" FR drywall	0	1.66	3.30	4.96
5930				nothing	0	1.21	2.28	3.49
6000		5/8" FR drywall	2-1/2" @ 16" O.C.	same	0	2.15	4.37	6.52
6050				5/8" FR drywall	0	1.75	3.86	5.61
6100				nothing	0	1.30	2.84	4.14
6110			3-5/8" @ 16" O.C.	same	0	2.11	4.13	6.24
6120				5/8" FR drywall	0	1.71	3.62	5.33
6130				nothing	0	1.26	2.60	3.86
6170			2-1/2" @ 24" O.C.	same	0	1.97	3.74	5.71
6180				5/8" FR drywall	0	1.57	3.23	4.80
6190				nothing	0	1.12	2.21	3.33
6200			3-5/8" @ 24" O.C.	same	0	2	3.75	5.75
6250				5/8"FR drywall	3-1/2" fiberglass	2.12	3.62	5.74
6300				nothing	0	1.15	2.22	3.37
6310	5/8" WR drywall	none	1-5/8" @ 16" O.C.	same	0	1.48	3.08	4.56
6320				5/8" WR drywall	0	2.03	3.59	5.62
6330				nothing	0	1.43	2.57	4
6340			2-1/2" @ 16" O.C.	same	0	2.67	4.11	6.78
6350				5/8" WR drywall	0	2.12	3.60	5.72
6360				nothing	0	1.52	2.58	4.10
6370			3-5/8" @ 16" O.C.	same	0	2.71	4.13	6.84
6380				5/8" WR drywall	0	2.16	3.62	5.78
6390				nothing	0	1.56	2.60	4.16

C1010 Partitions

C1010 126 Drywall Partitions/Metal Stud Framing

	FACE LAYER	BASE LAYER	FRAMING	OPPOSITE FACE	INSULATION	COST PER S.F.		
						MAT.	INST.	TOTAL
6400	5/8" WR drywall	none	1-5/8" @ 24" O.C.	same	0	1.40	2.71	4.11
6450				5/8" WR drywall	0	1.25	2.71	3.96
6500				nothing	0	.80	1.69	2.49
6510			2-1/2" @ 24" O.C.	same	0	2.57	3.74	6.31
6520				5/8" WR drywall	0	1.32	2.72	4.04
6530				nothing	0	.87	1.70	2.57
6600			3-5/8" @ 24" O.C.	same	0	1.50	2.73	4.23
6650				5/8" WR drywall	0	1.35	2.73	4.08
6700				nothing	0	.90	1.71	2.61
6800		5/8" FR drywall	2-1/2" @ 16" O.C.	same	0	2.45	4.37	6.82
6850				5/8" FR drywall	0	1.90	3.86	5.76
6900				nothing	0	1.45	2.84	4.29
6910			3-5/8" @ 16" O.C.	same	0	2.41	4.13	6.54
6920				5/8" FR drywall	0	1.86	3.62	5.48
6930				nothing	0	1.41	2.60	4.01
6940			2-1/2" @ 24" O.C.	same	0	1.97	3.74	5.71
6950				5/8" FR drywall	0	1.72	3.23	4.95
6960				nothing	0	1.27	2.21	3.48
7000			3-5/8" @ 24" O.C.	same	0	2.30	3.75	6.05
7050				5/8"FR drywall	3-1/2" fiberglass	2.27	3.62	5.89
7100				nothing	0	1.30	2.22	3.52
7110	5/8" VF drywall	none	1-5/8" @ 16" O.C.	same	0	2.10	3.32	5.42
7120				5/8" FR drywall	0	1.73	3.59	5.32
7130				nothing	0	1.28	2.57	3.85
7140			3-5/8" @ 16" O.C.	same	0	2.41	4.13	6.54
7150				5/8" FR drywall	0	1.86	3.62	5.48
7160				nothing	0	1.41	2.60	4.01
7200		none	1-5/8" @ 24" O.C.	same	0	2.02	2.95	4.97
7250				5/8" FR drywall	0	1.56	2.83	4.39
7300				nothing	0	1.11	1.81	2.92
7400			3-5/8" @ 24" O.C.	same	0	2.12	2.97	5.09
7450				5/8" FR drywall	0	1.66	2.85	4.51
7500				nothing	0	1.21	1.83	3.04
7600		5/8" FR drywall	2-1/2" @ 16" O.C.	same	0	3.07	4.61	7.68
7650				5/8" FR drywall	0	2.21	3.98	6.19
7700				nothing	0	1.76	2.96	4.72
7710			3-5/8" @ 16" O.C.	same	0	3.03	4.37	7.40
7720				5/8" FR drywall	0	2.17	3.74	5.91
7730				nothing	0	1.72	2.72	4.44
7740			2-1/2" @ 24" O.C.	same	0	2.89	3.98	6.87
7750				5/8" FR drywall	0	2.03	3.35	5.38
7760				nothing	0	1.58	2.33	3.91
7800			3-5/8" @ 24" O.C.	same	0	2.92	3.99	6.91
7850				5/8"FR drywall	3-1/2" fiberglass	2.58	3.74	6.32
7900				nothing	0	1.61	2.34	3.95

C1010 Partitions

C1010 128	Drywall Components	COST PER S.F.		
		MAT.	INST.	TOTAL
0060	Metal studs, 24" O.C. including track, load bearing, 20 gage, 2-1/2"	.70	.96	1.66
0080	3-5/8"	.84	.98	1.82
0100	4"	.75	1	1.75
0120	6"	1.12	1.02	2.14
0140	Metal studs, 24" O.C. including track, load bearing, 18 gage, 2-1/2"	.70	.96	1.66
0160	3-5/8"	.84	.98	1.82
0180	4"	.75	1	1.75
0200	6"	1.12	1.02	2.14
0220	16 gage, 2-1/2"	.81	1.09	1.90
0240	3-5/8"	.94	1.12	2.06
0260	4"	.98	1.14	2.12
0280	6"	1.24	1.17	2.41
0300	Non load bearing, 25 gage, 1-5/8"	.20	.67	.87
0340	3-5/8"	.30	.69	.99
0360	4"	.34	.69	1.03
0380	6"	.42	.71	1.13
0400	20 gage, 2-1/2"	.33	.85	1.18
0420	3-5/8"	.38	.87	1.25
0440	4"	.46	.87	1.33
0460	6"	.55	.88	1.43
0540	Wood studs including blocking, shoe and double top plate, 2"x4", 12" O.C.	.61	1.28	1.89
0560	16" O.C.	.49	1.03	1.52
0580	24" O.C.	.38	.82	1.20
0600	2"x6", 12" O.C.	1.03	1.47	2.50
0620	16" O.C.	.84	1.14	1.98
0640	24" O.C.	.64	.89	1.53
0642	Furring one side only, steel channels, 3/4", 12" O.C.	.43	2.09	2.52
0644	16" O.C.	.38	1.85	2.23
0646	24" O.C.	.26	1.40	1.66
0647	1-1/2" , 12" O.C.	.58	2.34	2.92
0648	16" O.C.	.52	2.05	2.57
0649	24" O.C.	.35	1.61	1.96
0650	Wood strips, 1" x 3", on wood, 12" O.C.	.49	.93	1.42
0651	16" O.C.	.37	.70	1.07
0652	On masonry, 12" O.C.	.52	1.04	1.56
0653	16" O.C.	.39	.78	1.17
0654	On concrete, 12" O.C.	.52	1.97	2.49
0655	16" O.C.	.39	1.48	1.87
0665	Gypsum board, one face only, exterior sheathing, 1/2"	.51	.91	1.42
0680	Interior, fire resistant, 1/2"	.41	.51	.92
0700	5/8"	.40	.51	.91
0720	Sound deadening board 1/4"	.46	.57	1.03
0740	Standard drywall 3/8"	.40	.51	.91
0760	1/2"	.37	.51	.88
0780	5/8"	.39	.51	.90
0800	Tongue & groove coreboard 1"	.88	2.13	3.01
0820	Water resistant, 1/2"	.46	.51	.97
0840	5/8"	.55	.51	1.06
0860	Add for the following:, foil backing	.20		.20
0880	Fiberglass insulation, 3-1/2"	.52	.38	.90
0900	6"	.67	.38	1.05
0920	Rigid insulation 1"	.62	.51	1.13
0940	Resilient furring @ 16" O.C.	.24	1.61	1.85
0960	Taping and finishing	.05	.51	.56
0980	Texture spray	.04	.61	.65
1000	Thin coat plaster	.13	.66	.79
1050	Sound wall framing, 2x6 plates, 2x4 staggered studs, 12" O.C.	.77	1.27	2.04

C1010 Partitions

C1010 144	Plaster Partition Components	COST PER S.F.		
		MAT.	INST.	TOTAL
0060	Metal studs, 16" O.C., including track, non load bearing, 25 gage, 1-5/8"	.37	1.05	1.42
0080	2-1/2"	.37	1.05	1.42
0100	3-1/4"	.41	1.07	1.48
0120	3-5/8"	.41	1.07	1.48
0140	4"	.46	1.08	1.54
0160	6"	.57	1.09	1.66
0180	Load bearing, 20 gage, 2-1/2"	.95	1.33	2.28
0200	3-5/8"	1.14	1.35	2.49
0220	4"	1.18	1.39	2.57
0240	6"	1.51	1.41	2.92
0260	16 gage 2-1/2"	1.12	1.51	2.63
0280	3-5/8"	1.29	1.56	2.85
0300	4"	1.35	1.58	2.93
0320	6"	1.70	1.60	3.30
0340	Wood studs, including blocking, shoe and double plate, 2"x4" , 12" O.C.	.61	1.28	1.89
0360	16" O.C.	.49	1.03	1.52
0380	24" O.C.	.38	.82	1.20
0400	2"x6" , 12" O.C.	1.03	1.47	2.50
0420	16" O.C.	.84	1.14	1.98
0440	24" O.C.	.64	.89	1.53
0460	Furring one face only, steel channels, 3/4" , 12" O.C.	.43	2.09	2.52
0480	16" O.C.	.38	1.85	2.23
0500	24" O.C.	.26	1.40	1.66
0520	1-1/2" , 12" O.C.	.58	2.34	2.92
0540	16" O.C.	.52	2.05	2.57
0560	24"O.C.	.35	1.61	1.96
0580	Wood strips 1"x3", on wood., 12" O.C.	.49	.93	1.42
0600	16"O.C.	.37	.70	1.07
0620	On masonry, 12" O.C.	.52	1.04	1.56
0640	16" O.C.	.39	.78	1.17
0660	On concrete, 12" O.C.	.52	1.97	2.49
0680	16" O.C.	.39	1.48	1.87
0700	Gypsum lath. plain or perforated, nailed to studs, 3/8" thick	.39	.64	1.03
0720	1/2" thick	.33	.68	1.01
0740	Clipped to studs, 3/8" thick	.39	.73	1.12
0760	1/2" thick	.33	.78	1.11
0780	Metal lath, diamond painted, nailed to wood studs, 2.5 lb.	.50	.64	1.14
0800	3.4 lb.	.55	.68	1.23
0820	Screwed to steel studs, 2.5 lb.	.50	.68	1.18
0840	3.4 lb.	.53	.73	1.26
0860	Rib painted, wired to steel, 2.75 lb	.44	.73	1.17
0880	3.4 lb	.58	.78	1.36
0900	4.0 lb	.69	.84	1.53
0910				
0920	Gypsum plaster, 2 coats	.46	2.55	3.01
0940	3 coats	.65	3.08	3.73
0960	Perlite or vermiculite plaster, 2 coats	.74	2.90	3.64
0980	3 coats	.79	3.61	4.40
1000	Stucco, 3 coats, 1" thick, on wood framing	1.03	5.10	6.13
1020	On masonry	.43	3.97	4.40
1100	Metal base galvanized and painted 2-1/2" high	.71	2.05	2.76

C1010 Partitions

Folding Accordian

Folding Leaf

Movable and Borrow Lites

Corrugated Panel

C1010 205	Partitions	COST PER S.F.		
		MAT.	INST.	TOTAL
0360	Folding accordion, vinyl covered, acoustical, 3 lb. S.F., 17 ft max. hgt	31	10.25	41.25
0380	5 lb. per S.F. 27 ft max height	42.50	10.80	53.30
0400	5.5 lb. per S.F., 17 ft. max height	50	11.40	61.40
0420	Commercial, 1.75 lb per S.F., 8 ft. max height	28.50	4.55	33.05
0440	2.0 Lb per S.F., 17 ft. max height	29.50	6.85	36.35
0460	Industrial, 4.0 lb. per S.F. 27 ft. max height	44	13.65	57.65
0480	Vinyl clad wood or steel, electric operation 6 psf	62	6.40	68.40
0500	Wood, non acoustic, birch or mahogany	33.50	3.42	36.92
0560	Folding leaf, alum framed acoustical 12′ high., 5.5 lb/S.F. standard trim	47	17.10	64.10
0580	Premium trim	57	34	91
0600	6.5 lb. per S.F., standard trim	49.50	17.10	66.60
0620	Premium trim	61	34	95
0640	Steel acoustical, 7.5 per S.F., vinyl faced, standard trim	70	17.10	87.10
0660	Premium trim	85.50	34	119.50
0680	Wood acoustic type, vinyl faced to 18′ high 6 psf, economy trim	65.50	17.10	82.60
0700	Standard trim	78.50	23	101.50
0720	Premium trim	101	34	135
0740	Plastic lam. or hardwood faced, standard trim	67.50	17.10	84.60
0760	Premium trim	72	34	106
0780	Wood, low acoustical type to 12 ft. high 4.5 psf	49	20.50	69.50
0840	Demountable, trackless wall, cork finish, semi acous, 1-5/8″ th, unsealed	42.50	3.15	45.65
0860	Sealed	47.50	5.40	52.90
0880	Acoustic, 2″ thick, unsealed	40.50	3.36	43.86
0900	Sealed	62	4.55	66.55
0920	In-plant modular office system, w/prehung steel door			
0940	3″ thick honeycomb core panels			
0960	12′ x 12′, 2 wall	16.35	.63	16.98
0970	4 wall	16.65	.84	17.49
0980	16′ x 16′, 2 wall	17.95	.45	18.40
0990	4 wall	11.30	.45	11.75
1000	Gypsum, demountable, 3″ to 3-3/4″ thick x 9′ high, vinyl clad	7.40	2.39	9.79
1020	Fabric clad	18.45	2.61	21.06
1040	1.75 system, vinyl clad hardboard, paper honeycomb core panel			
1060	1-3/4″ to 2-1/2″ thick x 9′ high	12.45	2.39	14.84
1080	Unitized gypsum panel system, 2″ to 2-1/2″ thick x 9′ high			
1100	Vinyl clad gypsum	16	2.39	18.39
1120	Fabric clad gypsum	26.50	2.61	29.11
1140	Movable steel walls, modular system			
1160	Unitized panels, 48″ wide x 9′ high			
1180	Baked enamel, pre-finished	18.10	1.90	20
1200	Fabric clad	26	2.03	28.03
1203				
1300	Metal panel partition, load bearing studs, 16 gage, corrugate/ribbed panel,	3.99	4.05	8.04

C1020 Interior Doors

Sliding Panel-Mall Front

Rolling Overhead Steel Door

C1020 102	Special Doors	COST PER OPNG.		
		MAT.	INST.	TOTAL
2500	Single leaf, wood, 3'-0"x7'-0"x1 3/8", birch, solid core	500	191	691
2510	Hollow core	435	183	618
2530	Hollow core, lauan	420	183	603
2540	Louvered pine	590	183	773
2550	Paneled pine	645	183	828
2600	Hollow metal, comm. quality, flush, 3'-0"x7'-0"x1-3/8"	1,000	201	1,201
2650	3'-0"x10'-0" openings with panel	1,450	260	1,710
2700	Metal fire, comm. quality, 3'-0"x7'-0"x1-3/8"	1,225	209	1,434
3200	Double leaf, wood, hollow core, 2 - 3'-0"x7'-0"x1-3/8"	675	350	1,025
3300	Hollow metal, comm. quality, B label, 2'-3'-0"x7'-0"x1-3/8"	2,300	440	2,740
3400	6'-0"x10'-0" opening, with panel	2,750	545	3,295
3500	Double swing door system, 12'-0"x7'-0", mill finish	10,100	2,700	12,800
3700	Black finish	10,200	2,775	12,975
3800	Sliding entrance door and system mill finish	10,900	2,550	13,450
3900	Bronze finish	12,000	2,775	14,775
4000	Black finish	12,500	2,875	15,375
4100	Sliding panel mall front, 16'x9' opening, mill finish	4,450	775	5,225
4200	Bronze finish	5,775	1,000	6,775
4300	Black finish	7,125	1,250	8,375
4400	24'x9' opening mill finish	6,400	1,450	7,850
4500	Bronze finish	8,325	1,875	10,200
4600	Black finish	10,200	2,325	12,525
4700	48'x9' opening mill finish	11,500	1,125	12,625
4800	Bronze finish	15,000	1,475	16,475
4900	Black finish	18,400	1,800	20,200
5000	Rolling overhead steel door, manual, 8' x 8' high	790	720	1,510
5100	10' x 10' high	1,200	820	2,020
5200	20' x 10' high	3,300	1,150	4,450
5300	12' x 12' high	1,300	960	2,260
5400	Motor operated, 8' x 8' high	2,100	950	3,050
5500	10' x 10' high	2,500	1,050	3,550
5600	20' x 10' high	4,600	1,375	5,975
5700	12' x 12' high	2,600	1,200	3,800
5800	Roll up grille, aluminum, manual, 10' x 10' high, mill finish	3,300	1,400	4,700
5900	Bronze anodized	5,100	1,400	6,500
6000	Motor operated, 10' x 10' high, mill finish	4,725	1,625	6,350
6100	Bronze anodized	6,525	1,625	8,150
6200	Steel, manual, 10' x 10' high	2,925	1,150	4,075
6300	15' x 8' high	3,275	1,450	4,725
6400	Motor operated, 10' x 10' high	4,350	1,375	5,725
6500	15' x 8' high	4,700	1,675	6,375
8970	Counter door, rolling, 6' high, 14' wide, aluminum	3,250	730	3,980

C1030 Fittings

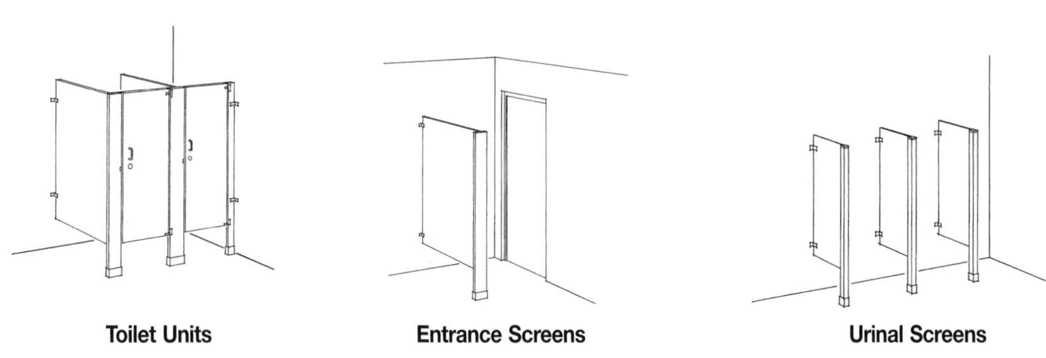

C1030 110	Toilet Partitions	COST PER UNIT		
		MAT.	INST.	TOTAL
0380	Toilet partitions, cubicles, ceiling hung, marble	1,825	485	2,310
0400	Painted metal	615	256	871
0420	Plastic laminate	610	256	866
0430	Phenolic	990	256	1,246
0440	Polymer plastic	1,175	256	1,431
0460	Stainless steel	1,275	256	1,531
0480	Handicap addition	510		510
0520	Floor and ceiling anchored, marble	1,900	390	2,290
0540	Painted metal	675	205	880
0560	Plastic laminate	855	205	1,060
0570	Phenolic	945	205	1,150
0580	Polymer plastic	1,325	205	1,530
0600	Stainless steel	1,375	205	1,580
0620	Handicap addition	395		395
0660	Floor mounted marble	1,150	325	1,475
0680	Painted metal	685	146	831
0700	Plastic laminate	600	146	746
0710	Phenolic	820	146	966
0720	Polymer plastic	950	146	1,096
0740	Stainless steel	1,400	146	1,546
0760	Handicap addition	365		365
0780	Juvenile deduction	47.50		47.50
0820	Floor mounted with headrail marble	1,325	325	1,650
0840	Painted metal	435	171	606
0860	Plastic laminate	920	171	1,091
0870	Phenolic	825	171	996
0880	Polymer plastic	945	171	1,116
0900	Stainless steel	1,175	171	1,346
0920	Handicap addition	360		360
0960	Wall hung, painted metal	735	146	881
1020	Stainless steel	1,900	146	2,046
1040	Handicap addition	360		360
1080	Entrance screens, floor mounted, 54″ high, marble	755	108	863
1100	Painted metal	265	68.50	333.50
1140	Stainless steel	1,050	68.50	1,118.50
1150	Polymer plastic	550	171	721
1300	Urinal screens, floor mounted, 24″ wide, plastic laminate	227	128	355
1320	Marble	750	150	900
1330	Polymer plastic	475	171	646
1340	Painted metal	252	128	380
1380	Stainless steel	640	128	768
1428	Wall mounted wedge type, painted metal	156	102	258

C1030 Fittings

C1030 110	Toilet Partitions	COST PER UNIT		
		MAT.	INST.	TOTAL
1460	Stainless steel	665	102	767
1500	Partitions, shower stall, single wall, painted steel, 2'-8" x 2'-8"	990	230	1,220
1510	Fiberglass, 2'-8" x 2'-8"	840	255	1,095
1520	Double wall, enameled steel, 2'-8" x 2'-8"	1,225	230	1,455
1530	Stainless steel, 2'-8" x 2'-8"	2,600	230	2,830
1560	Tub enclosure, sliding panels, tempered glass, aluminum frame	435	287	722
1570	Chrome/brass frame, clear glass	1,275	385	1,660

C1030 310	Storage Specialties, EACH	COST EACH		
		MAT.	INST.	TOTAL
0200	Lockers, steel, 1-tier, 5' to 6' high, per opng, 1 wide, knock down constr.	182	41	223
0210	3 wide	340	48	388
0220	2- tier	113	22	135
0235	Two person	370	28.50	398.50
0240	Duplex	375	28.50	403.50
0245	Welded, athletic type, ventilated, 1- tier	365	76.50	441.50
0600	Shelving, metal industrial, braced, 3' wide, 1' deep	26.50	9.85	36.35
0610	2' deep	42	10.45	52.45

C1030 510	Identifying/Visual Aid Specialties, EACH	COST EACH		
		MAT.	INST.	TOTAL
0100	Control boards, magnetic, porcelain finish, framed, 24" x 18"	219	128	347
0110	96" x 48"	1,175	205	1,380
0120	Directory boards, outdoor, black plastic, 36" x 24"	765	510	1,275
0130	36" x 36"	880	685	1,565
0140	Indoor, economy, open faced, 18" x 24"	193	146	339
0510	Street, reflective alum., dbl. face, 4 way, w/bracket	215	34	249
0520	Letters, cast aluminum, 1/2" deep, 4" high	33	28.50	61.50
0530	1" deep, 10" high	65.50	28.50	94
0540	Plaques, cast aluminum, 20" x 30"	2,075	256	2,331
0550	Cast bronze, 36" x 48"	5,400	510	5,910

C1030 520	Identifying/Visual Aid Specialties, S.F.	COST PER S.F.		
		MAT.	INST.	TOTAL
0100	Bulletin board, cork sheets, no frame, 1/4" thick	1.76	3.53	5.29
0120	Aluminum frame, 1/4" thick, 3' x 5'	10.45	4.29	14.74
0200	Chalkboards, wall hung, alum, frame & chalktrough	13.95	2.26	16.21
0210	Wood frame & chalktrough	12.70	2.45	15.15
0220	Sliding board, one board with back panel	66	2.19	68.19
0230	Two boards with back panel	98	2.19	100.19
0240	Liquid chalk type, alum. frame & chalktrough	15.50	2.26	17.76
0250	Wood frame & chalktrough	35	2.26	37.26

C1030 710	Bath and Toilet Accessories, EACH	COST EACH		
		MAT.	INST.	TOTAL
0100	Specialties, bathroom accessories, st. steel, curtain rod, 5' long, 1" diam	30	39.50	69.50
0120	Dispenser, towel, surface mounted	47.50	32	79.50
0140	Grab bar, 1-1/4" diam., 12" long	33	21.50	54.50
0160	Mirror, framed with shelf, 18" x 24"	204	25.50	229.50
0170	72" x 24"	296	85.50	381.50
0180	Toilet tissue dispenser, surface mounted, single roll	19.45	17.10	36.55
0200	Towel bar, 18" long	46.50	22.50	69
0300	Medicine cabinets, sliding mirror doors, 20" x 16" x 4-3/4", unlighted	139	73	212
0310	24" x 19" x 8-1/2", lighted	238	102	340

C1030 Fittings

C1030 730	Bath and Toilet Accessories, L.F.	COST PER L.F.		
		MAT.	INST.	TOTAL
0100	Partitions, hospital curtain, ceiling hung, polyester oxford cloth	27	5	32
0110	Designer oxford cloth	16.50	6.30	22.80

C1030 830	Fabricated Cabinets, EACH	COST EACH		
		MAT.	INST.	TOTAL
0110	Household, base, hardwood, one top drawer & one door below x 12″ wide	350	41.50	391.50
0115	24″ wide	480	46	526
0120	Four drawer x 24″ wide	445	46	491
0130	Wall, hardwood, 30″ high with one door x 12″ wide	298	46.50	344.50
0140	Two doors x 48″ wide	650	55.50	705.50

C1030 830	Fabricated Counters, L.F.	COST PER L.F.		
		MAT.	INST.	TOTAL
0150	Counter top-laminated plastic, stock, economy	19.25	17.10	36.35
0160	Custom-square edge, 7/8″ thick	17.45	38.50	55.95
0170	School, counter, wood, 32″ high	290	51	341
0180	Metal, 84″ high	685	68.50	753.50

C1030 910	Other Fittings, EACH	COST EACH		
		MAT.	INST.	TOTAL
0500	Mail boxes, horizontal, rear loaded, aluminum, 5″ x 6″ x 15″ deep	47	15.05	62.05
0510	Front loaded, aluminum, 10″ x 12″ x 15″ deep	105	25.50	130.50
0520	Vertical, front loaded, aluminum, 15″ x 5″ x 6″ deep	49.50	15.05	64.55
0530	Bronze, duranodic finish	53	15.05	68.05
0540	Letter slot, post office	132	64	196
0550	Mail counter, window, post office, with grille	620	256	876
0700	Turnstiles, one way, 4′ arm, 46″ diam., manual	2,275	205	2,480
0710	Electric	2,525	855	3,380
0720	3 arm, 5′-5″ diam. & 7′ high, manual	6,250	1,025	7,275
0730	Electric	9,375	1,700	11,075

C2010 Stair Construction

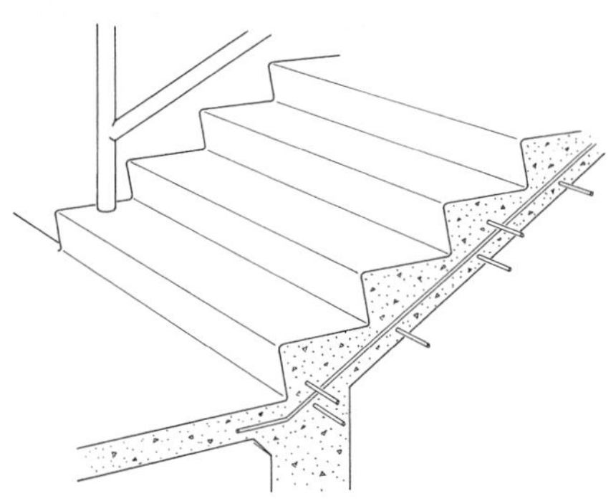

General Design: See reference section for code requirements. Maximum height between landings is 12′; usual stair angle is 20° to 50° with 30° to 35° best. Usual relation of riser to treads is:

Riser + tread = 17.5.
2x (Riser) + tread = 25.
Riser x tread = 70 or 75.

Maximum riser height is 7″ for commercial, 8-1/4″ for residential.
Usual riser height is 6-1/2″ to 7-1/4″.

Minimum tread width is 11″ for commercial and 9″ for residential.

For additional information please see reference section.

Cost Per Flight: Table below lists the cost per flight for 4′-0″ wide stairs.
Side walls are not included.
Railings are included.

C2010 110	Stairs	COST PER FLIGHT		
		MAT.	INST.	TOTAL
0470	Stairs, C.I.P. concrete, w/o landing, 12 risers, w/o nosing	1,500	2,075	3,575
0480	With nosing	2,450	2,250	4,700
0550	W/landing, 12 risers, w/o nosing	1,675	2,550	4,225
0560	With nosing	2,625	2,725	5,350
0570	16 risers, w/o nosing	2,050	3,225	5,275
0580	With nosing	3,325	3,475	6,800
0590	20 risers, w/o nosing	2,400	3,875	6,275
0600	With nosing	4,025	4,175	8,200
0610	24 risers, w/o nosing	2,775	4,550	7,325
0620	With nosing	4,725	4,925	9,650
0630	Steel, grate type w/nosing & rails, 12 risers, w/o landing	6,125	970	7,095
0640	With landing	8,400	1,325	9,725
0660	16 risers, with landing	10,500	1,675	12,175
0680	20 risers, with landing	12,500	1,975	14,475
0700	24 risers, with landing	14,500	2,300	16,800
0710	Metal pan stairs for concrete in-fill, picket rail, 12 risers, w/o landing	8,050	970	9,020
0720	With landing	10,900	1,475	12,375
0740	16 risers, with landing	13,600	1,775	15,375
0760	20 risers, with landing	16,300	2,125	18,425
0780	24 risers, with landing	19,000	2,425	21,425
0790	Cast iron tread & pipe rail, 12 risers, w/o landing	8,100	970	9,070
0800	With landing	11,000	1,475	12,475
1120	Wood, prefab box type, oak treads, wood rails 3′-6″ wide, 14 risers	2,400	425	2,825
1150	Prefab basement type, oak treads, wood rails 3′-0″ wide, 14 risers	1,150	105	1,255

C3010 Wall Finishes

C3010 230	Paint & Covering	COST PER S.F.		
		MAT.	INST.	TOTAL
0060	Painting, interior on plaster and drywall, brushwork, primer & 1 coat	.14	.66	.80
0080	Primer & 2 coats	.22	.88	1.10
0100	Primer & 3 coats	.30	1.08	1.38
0120	Walls & ceilings, roller work, primer & 1 coat	.14	.44	.58
0140	Primer & 2 coats	.22	.57	.79
0160	Woodwork incl. puttying, brushwork, primer & 1 coat	.14	.96	1.10
0180	Primer & 2 coats	.22	1.27	1.49
0200	Primer & 3 coats	.30	1.72	2.02
0260	Cabinets and casework, enamel, primer & 1 coat	.15	1.08	1.23
0280	Primer & 2 coats	.23	1.32	1.55
0300	Masonry or concrete, latex, brushwork, primer & 1 coat	.32	.90	1.22
0320	Primer & 2 coats	.42	1.28	1.70
0340	Addition for block filler	.20	1.10	1.30
0380	Fireproof paints, intumescent, 1/8" thick 3/4 hour	2.19	.88	3.07
0400	3/16" thick 1 hour	4.88	1.32	6.20
0420	7/16" thick 2 hour	6.30	3.07	9.37
0440	1-1/16" thick 3 hour	10.25	6.15	16.40
0500	Gratings, primer & 1 coat	.38	1.35	1.73
0600	Pipes over 12" diameter	.85	4.30	5.15
0700	Structural steel, brushwork, light framing 300-500 S.F./Ton	.10	1.49	1.59
0720	Heavy framing 50-100 S.F./Ton	.10	.74	.84
0740	Spraywork, light framing 300-500 S.F./Ton	.11	.33	.44
0760	Heavy framing 50-100 S.F./Ton	.11	.37	.48
0800	Varnish, interior wood trim, no sanding sealer & 1 coat	.08	1.08	1.16
0820	Hardwood floor, no sanding 2 coats	.17	.23	.40
0840	Wall coatings, acrylic glazed coatings, minimum	.41	.82	1.23
0860	Maximum	.86	1.41	2.27
0880	Epoxy coatings, solvent based	.52	.82	1.34
0900	Water based	.36	2.53	2.89
0940	Exposed epoxy aggregate, troweled on, 1/16" to 1/4" aggregate, topping mix	.81	1.83	2.64
0960	Integral mix	1.74	3.31	5.05
0980	1/2" to 5/8" aggregate, topping mix	1.56	3.31	4.87
1000	Integral mix	2.72	5.40	8.12
1020	1" aggregate, topping mix	2.77	4.78	7.55
1040	Integral mix	4.21	7.85	12.06
1060	Sprayed on, topping mix	.64	1.46	2.10
1080	Water based	1.36	2.97	4.33
1100	High build epoxy 50 mil, solvent based	.84	1.10	1.94
1120	Water based	1.50	4.53	6.03
1140	Laminated epoxy with fiberglass solvent based	.95	1.46	2.41
1160	Water based	1.73	2.97	4.70
1180	Sprayed perlite or vermiculite 1/16" thick, solvent based	.31	.15	.46
1200	Water based	.96	.67	1.63
1260	Wall coatings, vinyl plastic, solvent based	.44	.59	1.03
1280	Water based	1.08	1.79	2.87
1300	Urethane on smooth surface, 2 coats, solvent based	.39	.38	.77
1320	Water based	.67	.65	1.32
1340	3 coats, solvent based	.48	.51	.99
1360	Water based	1.05	.92	1.97
1380	Ceramic-like glazed coating, cementitious, solvent based	.54	.98	1.52
1400	Water based	1.07	1.25	2.32
1420	Resin base, solvent based	.37	.67	1.04
1440	Water based	.65	1.30	1.95
1460	Wall coverings, aluminum foil	1.13	1.57	2.70
1480	Copper sheets, .025" thick, phenolic backing	7.85	1.80	9.65
1500	Vinyl backing	6.05	1.80	7.85

C3010 Wall Finishes

C3010 230	Paint & Covering	COST PER S.F.		
		MAT.	INST.	TOTAL
1520	Cork tiles, 12"x12", light or dark, 3/16" thick	4.68	1.80	6.48
1540	5/16" thick	3.66	1.84	5.50
1560	Basketweave, 1/4" thick	3.73	1.80	5.53
1580	Natural, non-directional, 1/2" thick	7.30	1.80	9.10
1600	12"x36", granular, 3/16" thick	1.46	1.12	2.58
1620	1" thick	1.86	1.17	3.03
1640	12"x12", polyurethane coated, 3/16" thick	4.52	1.80	6.32
1660	5/16" thick	6.50	1.84	8.34
1661	Paneling, prefinished plywood, birch	1.33	2.44	3.77
1662	Mahogany, African	2.73	2.56	5.29
1663	Philippine (lauan)	.72	2.05	2.77
1664	Oak or cherry	2.18	2.56	4.74
1665	Rosewood	3.48	3.20	6.68
1666	Teak	3.55	2.56	6.11
1667	Chestnut	5.85	2.73	8.58
1668	Pecan	2.76	2.56	5.32
1669	Walnut	2.74	2.56	5.30
1670	Wood board, knotty pine, finished	2.65	4.24	6.89
1671	Rough sawn cedar	4.15	4.24	8.39
1672	Redwood	5.95	4.24	10.19
1673	Aromatic cedar	3.12	4.55	7.67
1680	Cork wallpaper, paper backed, natural	1.76	.90	2.66
1700	Color	3.18	.90	4.08
1720	Gypsum based, fabric backed, minimum	.86	.54	1.40
1740	Average	1.42	.60	2.02
1760	Small quantities	.86	.68	1.54
1780	Vinyl wall covering, fabric back, light weight	1.50	.68	2.18
1800	Medium weight	1.09	.90	1.99
1820	Heavy weight	1.54	.99	2.53
1840	Wall paper, double roll, solid pattern, avg. workmanship	.65	.68	1.33
1860	Basic pattern, avg. workmanship	1.34	.81	2.15
1880	Basic pattern, quality workmanship	2.35	.99	3.34
1900	Grass cloths with lining paper, minimum	1.53	1.08	2.61
1920	Maximum	3.60	1.24	4.84
1940	Ceramic tile, thin set, 4-1/4" x 4-1/4"	2.85	4.31	7.16
1960	12" x 12"	5.15	5.10	10.25

C3010 235	Paint Trim	COST PER L.F.		
		MAT.	INST.	TOTAL
2040	Painting, wood trim, to 6" wide, enamel, primer & 1 coat	.15	.54	.69
2060	Primer & 2 coats	.23	.68	.91
2080	Misc. metal brushwork, ladders	.75	5.40	6.15
2100	Pipes, to 4" dia.	.10	1.13	1.23
2120	6" to 8" dia.	.21	2.27	2.48
2140	10" to 12" dia.	.66	3.39	4.05
2160	Railings, 2" pipe	.27	2.69	2.96
2180	Handrail, single	.19	1.08	1.27
2185	Caulking & Sealants, Polyurethane, In place, 1 or 2 component, 1/2" X 1/4"	.40	1.70	2.10

C3020 Floor Finishes

C3020 410	Tile & Covering	COST PER S.F.		
		MAT.	INST.	TOTAL
0060	Carpet tile, nylon, fusion bonded, 18" x 18" or 24" x 24", 24 oz.	3.83	.64	4.47
0080	35 oz.	4.44	.64	5.08
0100	42 oz.	5.75	.64	6.39
0140	Carpet, tufted, nylon, roll goods, 12' wide, 26 oz.	2.85	.69	3.54
0160	36 oz.	4.44	.69	5.13
0180	Woven, wool, 36 oz.	12.45	.73	13.18
0200	42 oz.	13.65	.73	14.38
0220	Padding, add to above, 2.7 density	.73	.34	1.07
0240	13.0 density	.99	.34	1.33
0260	Composition flooring, acrylic, 1/4" thick	1.99	4.94	6.93
0280	3/8" thick	2.64	5.75	8.39
0300	Epoxy, 3/8" thick	3.54	3.81	7.35
0320	1/2" thick	5.05	5.25	10.30
0340	Epoxy terrazzo, granite chips	7.15	6	13.15
0360	Recycled porcelain	11	8.05	19.05
0380	Mastic, hot laid, 1-1/2" thick, minimum	5.25	3.72	8.97
0400	Maximum	6.70	4.94	11.64
0420	Neoprene 1/4" thick, minimum	5.15	4.72	9.87
0440	Maximum	7.10	6	13.10
0460	Polyacrylate with ground granite 1/4", granite chips	4.24	3.50	7.74
0480	Recycled porcelain	7.80	5.35	13.15
0500	Polyester with colored quartz chips 1/16", minimum	3.85	2.42	6.27
0520	Maximum	6.10	3.81	9.91
0540	Polyurethane with vinyl chips, clear	8.50	2.42	10.92
0560	Pigmented	12.35	2.99	15.34
0600	Concrete topping, granolithic concrete, 1/2" thick	.42	4.13	4.55
0620	1" thick	.84	4.23	5.07
0640	2" thick	1.68	4.87	6.55
0660	Heavy duty 3/4" thick, minimum	.57	6.40	6.97
0680	Maximum	1	7.60	8.60
0700	For colors, add to above, minimum	.50	1.47	1.97
0720	Maximum	.83	1.62	2.45
0740	Exposed aggregate finish, minimum	.22	.77	.99
0760	Maximum	.37	1.03	1.40
0780	Abrasives, .25 P.S.F. add to above, minimum	.65	.57	1.22
0800	Maximum	.94	.57	1.51
0820	Dust on coloring, add, minimum	.50	.37	.87
0840	Maximum	.83	.77	1.60
0860	Floor coloring using 0.6 psf powdered color, 1/2" integral, minimum	5.15	4.13	9.28
0880	Maximum	5.50	4.13	9.63
0900	Dustproofing, add, minimum	.20	.25	.45
0920	Maximum	.72	.37	1.09
0930	Paint	.42	1.28	1.70
0940	Hardeners, metallic add, minimum	.47	.57	1.04
0960	Maximum	1.42	.84	2.26
0980	Non-metallic, minimum	.18	.57	.75
1000	Maximum	.54	.84	1.38
1020	Integral topping and finish, 1:1:2 mix, 3/16" thick	.14	2.43	2.57
1040	1/2" thick	.38	2.56	2.94
1060	3/4" thick	.57	2.87	3.44
1080	1" thick	.76	3.25	4.01
1100	Terrazzo, minimum	4.08	16	20.08
1120	Maximum	7.55	20	27.55
1340	Cork tile, minimum	5.80	1.47	7.27
1360	Maximum	8.35	1.47	9.82
1380	Polyethylene, in rolls, minimum	4.61	1.69	6.30
1400	Maximum	7.95	1.69	9.64
1420	Polyurethane, thermoset, minimum	6.30	4.64	10.94

C3020 Floor Finishes

C3020 410	Tile & Covering	COST PER S.F.		
		MAT.	INST.	TOTAL
1440	Maximum	7.15	9.25	16.40
1460	Rubber, sheet goods, minimum	9.95	3.86	13.81
1480	Maximum	13.15	5.15	18.30
1500	Tile, minimum	6.50	1.16	7.66
1520	Maximum	11.80	1.69	13.49
1540	Synthetic turf, minimum	5.10	2.21	7.31
1560	1/2" thick	5.60	2.44	8.04
1580	Vinyl, composition tile, minimum	1.34	.93	2.27
1600	Maximum	1.91	.93	2.84
1620	Vinyl tile, 3/32", minimum	4.09	.93	5.02
1640	Maximum	3.54	.93	4.47
1660	Sheet goods, plain pattern/colors	4.92	1.85	6.77
1680	Intricate pattern/colors	8.05	2.32	10.37
1720	Tile, ceramic natural clay	6.50	4.48	10.98
1730	Marble, synthetic 12"x12"x5/8"	12.60	13.65	26.25
1740	Ceramic tile, floors, porcelain type, 1 color, color group 2, 1" x 1"	7.15	4.48	11.63
1760	Ceramic flr porcelain 1 color, color group 2 2" x 2" epoxy grout	8.40	5.30	13.70
1800	Quarry tile, mud set, minimum	9.10	5.85	14.95
1820	Maximum	10.60	7.45	18.05
1840	Thin set, deduct		1.17	1.17
1850	Tile, natural stone, marble, in mortar bed, 12" x 12" x 3/8" thick	16.75	21.50	38.25
1860	Terrazzo precast, minimum	5.70	6.20	11.90
1880	Maximum	13.45	6.20	19.65
1900	Non-slip, minimum	27.50	15	42.50
1920	Maximum	28	20	48
2020	Wood block, end grain factory type, natural finish, 2" thick	5.70	4.10	9.80
2040	Fir, vertical grain, 1"x4", no finish, minimum	3.67	2.01	5.68
2060	Maximum	3.91	2.01	5.92
2080	Prefinished white oak, prime grade, 2-1/4" wide	6.45	3.01	9.46
2100	3-1/4" wide	6.30	2.77	9.07
2120	Maple strip, sanded and finished, minimum	5.80	4.37	10.17
2140	Maximum	6.65	4.37	11.02
2160	Oak strip, sanded and finished, minimum	4.09	4.37	8.46
2180	Maximum	4.98	4.37	9.35
2200	Parquetry, sanded and finished, plain pattern	6.25	4.56	10.81
2220	Intricate pattern	11.55	6.45	18
2260	Add for sleepers on concrete, treated, 24" O.C., 1"x2"	2.03	2.63	4.66
2280	1"x3"	2.13	2.05	4.18
2300	2"x4"	1.15	1.03	2.18
2340	Underlayment, plywood, 3/8" thick	1.16	.68	1.84
2350	1/2" thick	1.36	.71	2.07
2360	5/8" thick	1.50	.73	2.23
2370	3/4" thick	1.62	.79	2.41
2380	Particle board, 3/8" thick	.44	.68	1.12
2390	1/2" thick	.46	.71	1.17
2400	5/8" thick	.59	.73	1.32
2410	3/4" thick	.74	.79	1.53
2420	Hardboard, 4' x 4', .215" thick	.76	.68	1.44
9200	Vinyl, composition tile, 12" x 12" x 1/8" thick, recycled content	2.78	.93	3.71

C3020 600	Bases, Curbs & Trim	COST PER UNIT		
		MAT.	INST.	TOTAL
0050	1/8" vinyl base, 2-1/2" H, straight or cove, std. colors	.77	1.47	2.24
0055	4" H	1.30	1.47	2.77
0060	6" H	1.69	1.47	3.16
0065	Corners, 2-1/2" H	2.45	1.47	3.92
0070	4" H	3.33	1.47	4.80
0075	6" H	3.15	1.47	4.62

C3020 Floor Finishes

C3020 600	Bases, Curbs & Trim	COST PER UNIT		
		MAT.	INST.	TOTAL
0080	1/8" rubber base, 2-1/2" H, straight or cove, std. colors	1.22	1.47	2.69
0085	4" H	1.43	1.47	2.90
0090	6" H	2.11	1.47	3.58
0095	Corners, 2-1/2" H	2.73	1.47	4.20
0100	4" H	2.82	1.47	4.29
0105	6" H	3.50	1.47	4.97

C3030 Ceiling Finishes

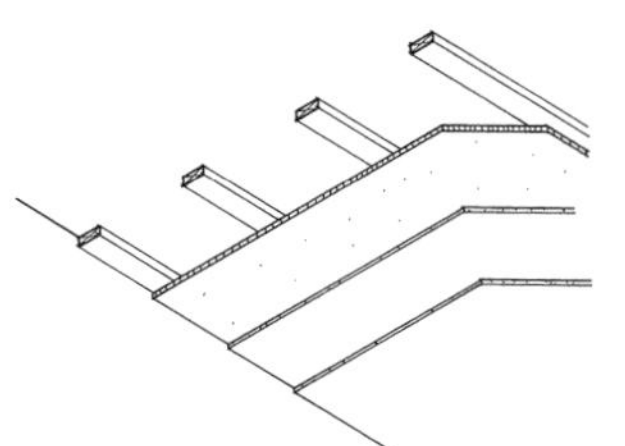

2 Coats of Plaster on Gypsum Lath on Wood Furring

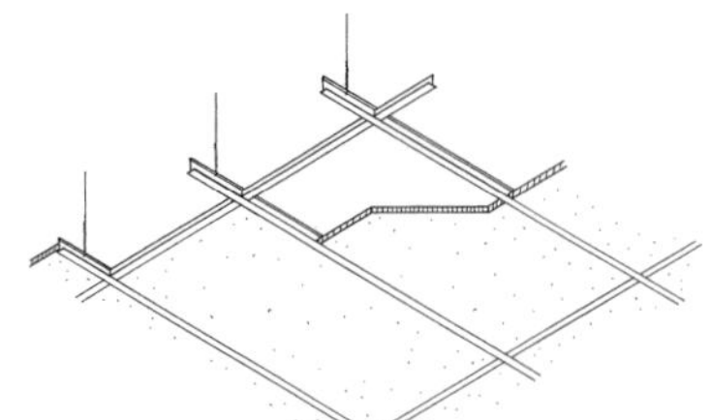

Fiberglass Board on Exposed Suspended Grid System

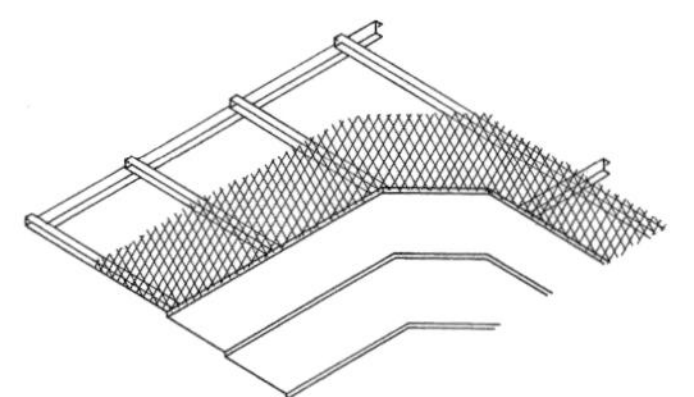

Plaster and Metal Lath on Metal Furring

C3030 105 Plaster Ceilings

	TYPE	LATH	FURRING	SUPPORT		COST PER S.F.		
						MAT.	INST.	TOTAL
2400	2 coat gypsum	3/8" gypsum	1"x3" wood, 16" O.C.	wood		1.36	5.35	6.71
2500	Painted			masonry		1.38	5.45	6.83
2600				concrete		1.38	6.10	7.48
2700	3 coat gypsum	3.4# metal	1"x3" wood, 16" O.C.	wood		1.71	5.70	7.41
2800	Painted			masonry		1.73	5.80	7.53
2900				concrete		1.73	6.40	8.13
3000	2 coat perlite	3/8" gypsum	1"x3" wood, 16" O.C.	wood		1.64	5.55	7.19
3100	Painted			masonry		1.66	5.65	7.31
3200				concrete		1.66	6.25	7.91
3300	3 coat perlite	3.4# metal	1"x3" wood, 16" O.C.	wood		1.80	5.70	7.50
3400	Painted			masonry		1.82	5.80	7.62
3500				concrete		1.82	6.40	8.22
3600	2 coat gypsum	3/8" gypsum	3/4" CRC, 12" O.C.	1-1/2" CRC, 48"O.C.		1.42	6.60	8.02
3700	Painted		3/4" CRC, 16" O.C.	1-1/2" CRC, 48"O.C.		1.37	5.95	7.32
3800			3/4" CRC, 24" O.C.	1-1/2" CRC, 48"O.C.		1.25	5.40	6.65
3900	2 coat perlite	3/8" gypsum	3/4" CRC, 12" O.C.	1-1/2" CRC, 48"O.C		1.70	7.05	8.75
4000	Painted		3/4" CRC, 16" O.C.	1-1/2" CRC, 48"O.C.		1.65	6.40	8.05
4100			3/4" CRC, 24" O.C.	1-1/2" CRC, 48"O.C.		1.53	5.85	7.38
4200	3 coat gypsum	3.4# metal	3/4" CRC, 12" O.C.	1-1/2" CRC, 36" O.C.		1.94	8.45	10.39
4300	Painted		3/4" CRC, 16" O.C.	1-1/2" CRC, 36" O.C.		1.89	7.80	9.69
4400			3/4" CRC, 24" O.C.	1-1/2" CRC, 36" O.C.		1.77	7.30	9.07
4500	3 coat perlite	3.4# metal	3/4" CRC, 12" O.C.	1-1/2" CRC,36" O.C.		2.08	9.30	11.38
4600	Painted		3/4" CRC, 16" O.C.	1-1/2" CRC, 36" O.C.		2.03	8.65	10.68
4700			3/4" CRC, 24" O.C.	1-1/2" CRC, 36" O.C.		1.91	8.10	10.01

C3030 110 Drywall Ceilings

	TYPE	FINISH	FURRING	SUPPORT		COST PER S.F.		
						MAT.	INST.	TOTAL
4800	1/2" F.R. drywall	painted and textured	1"x3" wood, 16" O.C.	wood		1.01	3.21	4.22
4900				masonry		1.03	3.31	4.34
5000				concrete		1.03	3.94	4.97
5100	5/8" F.R. drywall	painted and textured	1"x3" wood, 16" O.C.	wood		1	3.21	4.21
5200				masonry		1.02	3.31	4.33
5300				concrete		1.02	3.94	4.96

C3030 Ceiling Finishes

C3030 110 Drywall Ceilings

	TYPE	FINISH	FURRING	SUPPORT		COST PER S.F. MAT.	INST.	TOTAL
5400	1/2" F.R. drywall	painted and textured	7/8" resil. channels	24" O.C.		.79	3.12	3.91
5500			1"x2" wood	stud clips		.99	3.02	4.01
5602	1/2" F.R. drywall	painted	1-5/8" metal studs	24" O.C.		.84	2.78	3.62
5700	5/8" F.R. drywall	painted and textured	1-5/8"metal studs	24" O.C.		.83	2.78	3.61
5702								

C3030 140 Plaster Ceiling Components

		COST PER S.F. MAT.	INST.	TOTAL
0060	Plaster, gypsum incl. finish			
0080	3 coats	.65	3.37	4.02
0100	Perlite, incl. finish, 2 coats	.74	3.37	4.11
0120	3 coats	.79	4.20	4.99
0140	Thin coat on drywall	.13	.66	.79
0200	Lath, gypsum, 3/8" thick	.39	.89	1.28
0220	1/2" thick	.33	.93	1.26
0240	5/8" thick	.35	1.09	1.44
0260	Metal, diamond, 2.5 lb.	.50	.73	1.23
0280	3.4 lb.	.55	.78	1.33
0300	Flat rib, 2.75 lb.	.44	.73	1.17
0320	3.4 lb.	.58	.78	1.36
0440	Furring, steel channels, 3/4" galvanized , 12" O.C.	.43	2.34	2.77
0460	16" O.C.	.38	1.69	2.07
0480	24" O.C.	.26	1.17	1.43
0500	1-1/2" galvanized , 12" O.C.	.58	2.59	3.17
0520	16" O.C.	.52	1.89	2.41
0540	24" O.C.	.35	1.26	1.61
0560	Wood strips, 1"x3", on wood, 12" O.C.	.49	1.46	1.95
0580	16" O.C.	.37	1.10	1.47
0600	24" O.C.	.25	.73	.98
0620	On masonry, 12" O.C.	.52	1.60	2.12
0640	16" O.C.	.39	1.20	1.59
0660	24" O.C.	.26	.80	1.06
0680	On concrete, 12" O.C.	.52	2.44	2.96
0700	16" O.C.	.39	1.83	2.22
0720	24" O.C.	.26	1.22	1.48
0722				
0740				
0940	Paint on plaster or drywall, roller work, primer + 1 coat	.14	.44	.58
0960	Primer + 2 coats	.22	.57	.79

C3030 210 Acoustical Ceilings

	TYPE	TILE	GRID	SUPPORT		COST PER S.F. MAT.	INST.	TOTAL
5800	5/8" fiberglass board	24" x 48"	tee	suspended		2.56	1.54	4.10
5900		24" x 24"	tee	suspended		2.82	1.69	4.51
6000	3/4" fiberglass board	24" x 48"	tee	suspended		4.53	1.57	6.10
6100		24" x 24"	tee	suspended		4.79	1.72	6.51
6500	5/8" mineral fiber	12" x 12"	1"x3" wood, 12" O.C.	wood		2.98	3.17	6.15
6600				masonry		3.01	3.31	6.32
6700				concrete		3.01	4.15	7.16
6800	3/4" mineral fiber	12" x 12"	1"x3" wood, 12" O.C.	wood		3.89	3.17	7.06
6900				masonry		3.89	3.17	7.06
7000				concrete		3.89	3.17	7.06
7100	3/4"mineral fiber on	12" x 12"	25 ga. channels	runners		4.30	3.98	8.28

C3030 Ceiling Finishes

C3030 210 Acoustical Ceilings

	TYPE	TILE	GRID	SUPPORT		COST PER S.F. MAT.	INST.	TOTAL
7102	5/8" F.R. drywall							
7200 7201 7202	5/8" plastic coated Mineral fiber	12" x 12"		adhesive backed		2.94	1.71	4.65
7300 7301 7302	3/4" plastic coated Mineral fiber	12" x 12"		adhesive backed		3.85	1.71	5.56
7400 7401 7402	3/4" mineral fiber	12" x 12"	conceal 2" bar & channels	suspended		2.99	3.87	6.86

C3030 240 Acoustical Ceiling Components

		COST PER S.F. MAT.	INST.	TOTAL
2480	Ceiling boards, eggcrate, acrylic, 1/2" x 1/2" x 1/2" cubes	2.09	1.02	3.11
2500	Polystyrene, 3/8" x 3/8" x 1/2" cubes	1.82	1	2.82
2520	1/2" x 1/2" x 1/2" cubes	1.88	1.02	2.90
2540	Fiberglass boards, plain, 5/8" thick	1.44	.82	2.26
2560	3/4" thick	3.41	.85	4.26
2580	Grass cloth faced, 3/4" thick	3.36	1.02	4.38
2600	1" thick	4.06	1.06	5.12
2620	Luminous panels, prismatic, acrylic	3.33	1.28	4.61
2640	Polystyrene	1.88	1.28	3.16
2660	Flat or ribbed, acrylic	4.91	1.28	6.19
2680	Polystyrene	2.62	1.28	3.90
2700	Drop pan, white, acrylic	6.25	1.28	7.53
2720	Polystyrene	4.99	1.28	6.27
2740	Mineral fiber boards, 5/8" thick, standard	1.10	.76	1.86
2760	Plastic faced	2.89	1.28	4.17
2780	2 hour rating	1.44	.76	2.20
2800	Perforated aluminum sheets, .024 thick, corrugated painted	3.06	1.05	4.11
2820	Plain	5.35	1.02	6.37
3080	Mineral fiber, plastic coated, 12" x 12" or 12" x 24", 5/8" thick	2.49	1.71	4.20
3100	3/4" thick	3.40	1.71	5.11
3120	Fire rated, 3/4" thick, plain faced	1.57	1.71	3.28
3140	Mylar faced	2.37	1.71	4.08
3160	Add for ceiling primer	.13		.13
3180	Add for ceiling cement	.45		.45
3240	Suspension system, furring, 1" x 3" wood 12" O.C.	.49	1.46	1.95
3260	T bar suspension system, 2' x 4' grid	.88	.64	1.52
3280	2' x 2' grid	1.14	.79	1.93
3300	Concealed Z bar suspension system 12" module	1.05	.99	2.04
3320	Add to above for 1-1/2" carrier channels 4' O.C.	.13	1.09	1.22
3340	Add to above for carrier channels for recessed lighting	.23	1.11	1.34

D1010 Elevators and Lifts

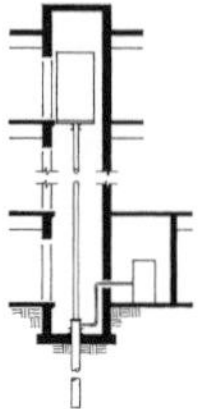

The hydraulic elevator obtains its motion from the movement of liquid under pressure in the piston connected to the car bottom. These pistons can provide travel to a maximum rise of 70′ and are sized for the intended load. As the rise reaches the upper limits the cost tends to exceed that of a geared electric unit.

D1010 110	Hydraulic	COST EACH		
		MAT.	INST.	TOTAL
1300	Pass. elev., 1500 lb., 2 Floors, 100 FPM	52,000	16,400	68,400
1400	5 Floors, 100 FPM	89,000	43,500	132,500
1600	2000 lb., 2 Floors, 100 FPM	53,000	16,400	69,400
1700	5 floors, 100 FPM	90,000	43,500	133,500
1900	2500 lb., 2 Floors, 100 FPM	55,500	16,400	71,900
2000	5 floors, 100 FPM	92,500	43,500	136,000
2200	3000 lb., 2 Floors, 100 FPM	57,000	16,400	73,400
2300	5 floors, 100 FPM	94,000	43,500	137,500
2500	3500 lb., 2 Floors, 100 FPM	60,500	16,400	76,900
2600	5 floors, 100 FPM	97,500	43,500	141,000
2800	4000 lb., 2 Floors, 100 FPM	62,000	16,400	78,400
2900	5 floors, 100 FPM	99,000	43,500	142,500
3100	4500 lb., 2 Floors, 100 FPM	65,000	16,400	81,400
3200	5 floors, 100 FPM	102,000	43,500	145,500
4000	Hospital elevators, 3500 lb., 2 Floors, 100 FPM	88,000	16,400	104,400
4100	5 floors, 100 FPM	133,000	43,500	176,500
4300	4000 lb., 2 Floors, 100 FPM	88,000	16,400	104,400
4400	5 floors, 100 FPM	133,000	43,500	176,500
4600	4500 lb., 2 Floors, 100 FPM	96,000	16,400	112,400
4800	5 floors, 100 FPM	141,500	43,500	185,000
4900	5000 lb., 2 Floors, 100 FPM	100,000	16,400	116,400
5000	5 floors, 100 FPM	145,000	43,500	188,500
6700	Freight elevators (Class "B"), 3000 lb., 2 Floors, 50 FPM	118,500	20,900	139,400
6800	5 floors, 100 FPM	187,000	54,500	241,500
7000	4000 lb., 2 Floors, 50 FPM	124,000	20,900	144,900
7100	5 floors, 100 FPM	192,000	54,500	246,500
7500	10,000 lb., 2 Floors, 50 FPM	148,500	20,900	169,400
7600	5 floors, 100 FPM	216,500	54,500	271,000
8100	20,000 lb., 2 Floors, 50 FPM	176,500	20,900	197,400
8200	5 Floors, 100 FPM	245,000	54,500	299,500

Minimum Plumbing Fixture Requirements

Classification	Occupancy	Description	Water		Lavatories		Bathtubs/Showers	Drinking Fountains	Other
			Male	Female	Male	Female			
Assembly	A-1	Theaters and other buildings for the performing arts and motion pictures	1:125	1:65	1:200			1:500	1 Service Sink
	A-2	Nightclubs, bars, taverns dance halls	1:40		1:75			1:500	1 Service Sink
		Restaurants, banquet halls, food courts	1:75		1:200			1:500	1 Service Sink
	A-3	Auditorium w/o permanent seating, art galleries, exhibition halls, museums, lecture halls, libraries, arcades & gymnasiums	1:125	1:65	1:200			1:500	1 Service Sink
		Passenger terminals and transportation facilities	1:500		1:750			1:1000	1 Service Sink
		Places of worship and other religious services	1:150	1:75	1:200			1:1000	1 Service Sink
	A-4	Indoor sporting events and activities, coliseums, arenas, skating rinks, pools, and tennis courts	1:75 for the first 1500, then 1:120 for the remainder	1:40 for the first 1520, then 1:60 for the remainder	1:200	1:150		1:1000	1 Service Sink
	A-5	Outdoor sporting events and activities, stadiums, amusement parks, bleachers, grandstands	1:75 for the first 1500, then 1:120 for the remainder	1:40 for the first 1520, then 1:60 for the remainder	1:200	1:150		1:1000	1 Service Sink
Business	B	Buildings for the transaction of business, professional services, other services involving merchandise, office buildings, banks, light industrial	1:25 for the first 50, then 1:50 for the remainder		1:40 for the first 80, then 1:80 for the remainder			1:100	1 Service Sink
Educational	E	Educational Facilities	1:50		1:50			1:100	1 Service Sink
Factory and industrial	F-1 and F-2	Structures in which occupants are engaged in work fabricating, assembly or processing of products or materials	1:100		1:100		See *International Plumbing Code*	1:400	1 Service Sink
Institutional	I-1	Residential Care	1:10		1:10		1:8	1:100	1 Service Sink
	I-2	Hospitals, ambulatory nursing home care recipient	1 per room		1 per room		1:15	1:100	1 Service Sink
		Employees, other than residential care	1:25		1:35			1:100	
		Visitors, other than residential care	1:75		1:100			1:500	
	I-3	Prisons	1 per cell		1 per cell		1:15	1:100	1 Service Sink
		Reformatories, detention and correction centers	1:15		1:15		1:15	1:100	1 Service Sink
		Employees	1:25		1:35			1:100	
	I-4	Adult and child day care	1:15		1:15		1	1:100	1 Service Sink
Mercantile	M	Retail stores, service stations, shops, salesrooms, markets and shopping centers	1:500		1:750			1:1000	1 Service Sink
Residential	R-1	Hotels, Motels, boarding houses (transient)	1 per sleeping unit		1 per sleeping unit		1 per sleeping unit		1 Service Sink
	R-2	Dormitories, fraternities, sororities and boarding houses (not transient)	1:10		1:10		1:8	1:100	1 Service Sink
		Apartment House	1 per dwelling unit		1 per dwelling unit		1 per dwelling unit		1 Kitchen sink per dwelling; 1 clothes washer connection per 20 dwellings
	R-3	1 and 2 Family Dwellings	1 per dwelling unit		1:10		1 per dwelling unit		1 Kitchen sink per dwelling; 1 clothes washer connection per dwelling
	R-3	Congregate living facilities w/<16 people	1:10		1:10		1:8	1:100	1 Service Sink
	R-4	Congregate living facilities w/<16 people	1:10		1:10		1:8	1:100	1 Service Sink
Storage	S-1 and S-2	Structures for the storage of good, warehouses, storehouses and freight depots, low and moderate hazard	1:100		1:100		See *International Plumbing Code*	1;1000	1 Service Sink

Table 2902.1

D2010 Plumbing Fixtures

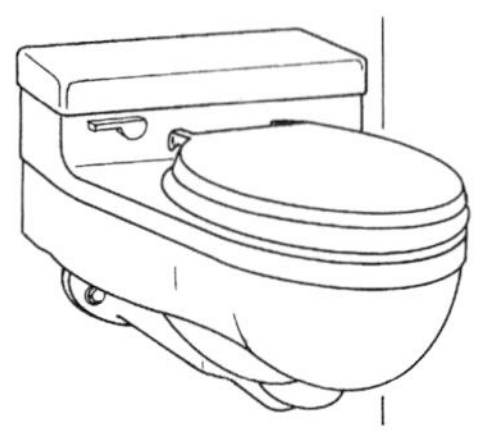
One Piece Wall Hung Water Closet

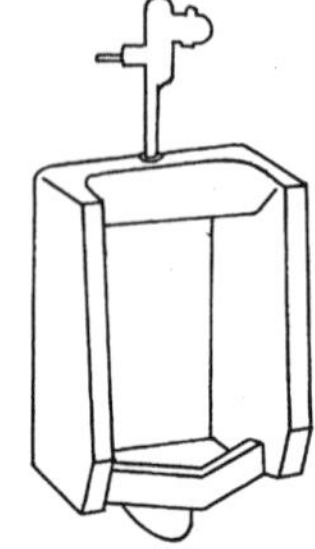
Wall Hung Urinal

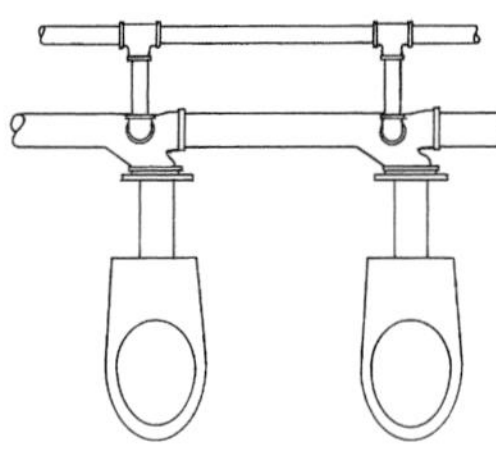
Side By Side Water Closet Group

Floor Mount Water Closet

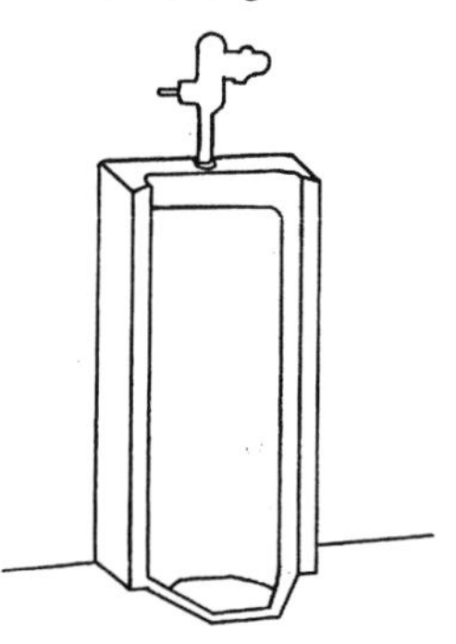
Stall Type Urinal

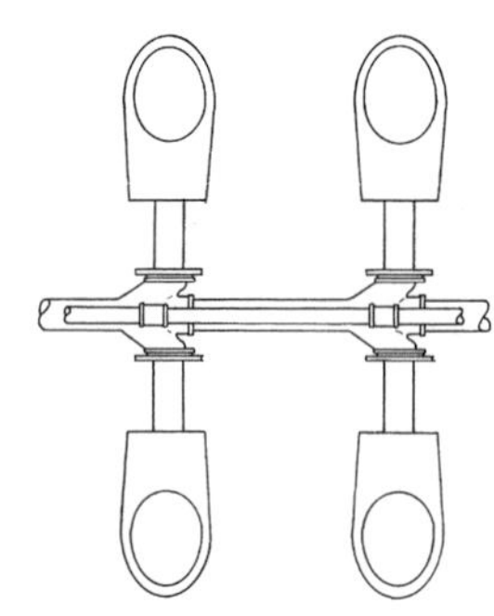
Back to Back Water Closet Group

D2010 110	Water Closet Systems	COST EACH		
		MAT.	INST.	TOTAL
1800	Water closet, vitreous china			
1840	Tank type, wall hung			
1880	Close coupled two piece	1,825	625	2,450
1920	Floor mount, one piece	1,375	665	2,040
1960	One piece low profile	1,325	665	1,990
2000	Two piece close coupled	705	665	1,370
2040	Bowl only with flush valve			
2080	Wall hung	2,650	710	3,360
2120	Floor mount	885	675	1,560
2122				
2160	Floor mount, ADA compliant with 18″ high bowl	895	695	1,590

D2010 120	Water Closets, Group	COST EACH		
		MAT.	INST.	TOTAL
1760	Water closets, battery mount, wall hung, side by side, first closet	2,775	730	3,505
1800	Each additional water closet, add	2,675	690	3,365
3000	Back to back, first pair of closets	4,675	970	5,645
3100	Each additional pair of closets, back to back	4,600	955	5,555
9000	Back to back, first pair of closets, auto sensor flush valve, 1.28 gpf	5,100	1,050	6,150
9100	Ea additional pair of cls, back to back, auto sensor flush valve, 1.28 gpf	4,900	985	5,885

D2010 210	Urinal Systems	COST EACH		
		MAT.	INST.	TOTAL
2000	Urinal, vitreous china, wall hung	600	700	1,300
2040	Stall type	1,350	840	2,190

D2010 Plumbing Fixtures

Systems are complete with trim and rough-in (supply, waste and vent) to connect to supply branches and waste mains.

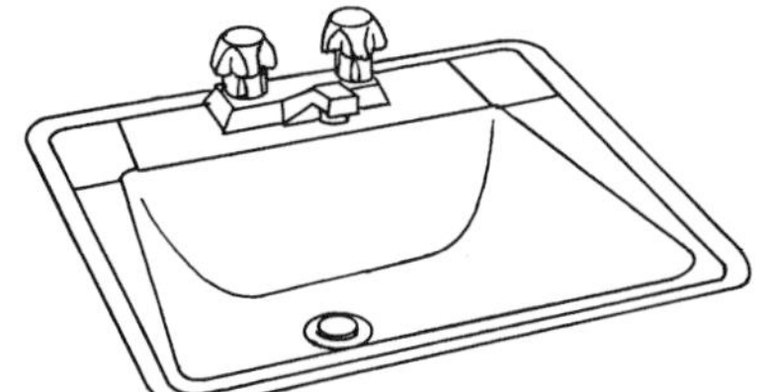

Vanity Top

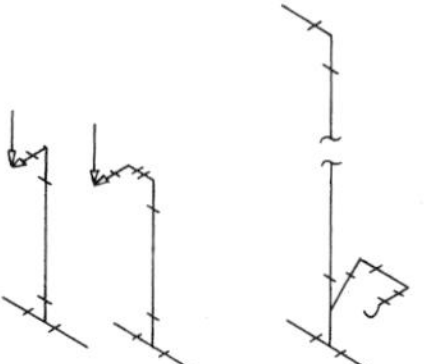

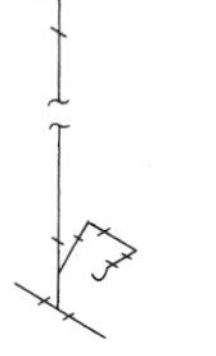

Supply Waste/Vent

Wall Hung

D2010 310	Lavatory Systems	COST EACH		
		MAT.	INST.	TOTAL
1560	Lavatory w/trim, vanity top, PE on CI, 20" x 18", Vanity top by others.	600	625	1,225
1600	19" x 16" oval	400	625	1,025
1640	18" round	710	625	1,335
1680	Cultured marble, 19" x 17"	410	625	1,035
1720	25" x 19"	440	625	1,065
1760	Stainless, self-rimming, 25" x 22"	615	625	1,240
1800	17" x 22"	605	625	1,230
1840	Steel enameled, 20" x 17"	410	645	1,055
1880	19" round	450	645	1,095
1920	Vitreous china, 20" x 16"	505	655	1,160
1960	19" x 16"	510	655	1,165
2000	22" x 13"	510	655	1,165
2040	Wall hung, PE on CI, 18" x 15"	915	690	1,605
2080	19" x 17"	965	690	1,655
2120	20" x 18"	800	690	1,490
2160	Vitreous china, 18" x 15"	705	710	1,415
2200	19" x 17"	665	710	1,375
2240	24" x 20"	795	710	1,505
2300	20" x 27", handicap	1,575	765	2,340

D2010 Plumbing Fixtures

Systems are complete with trim and rough-in (supply, waste and vent) to connect to supply branches and waste mains.

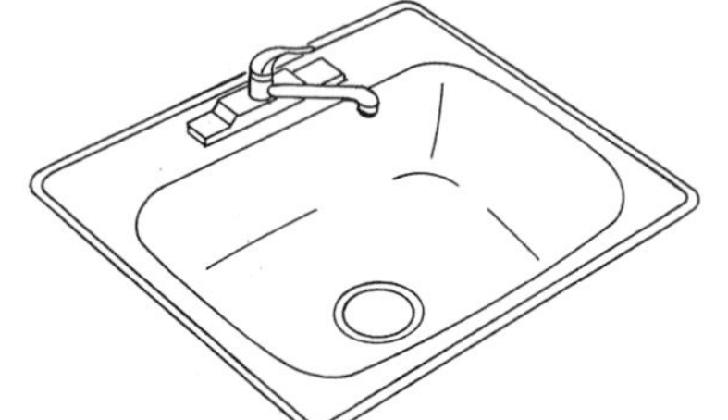

Countertop
Single Bowl

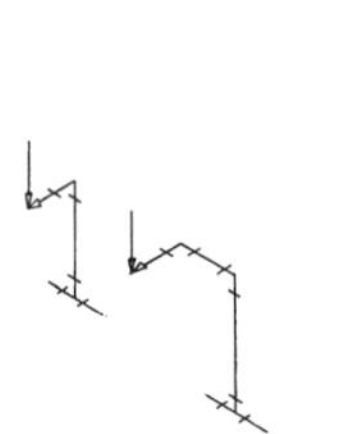

Supply

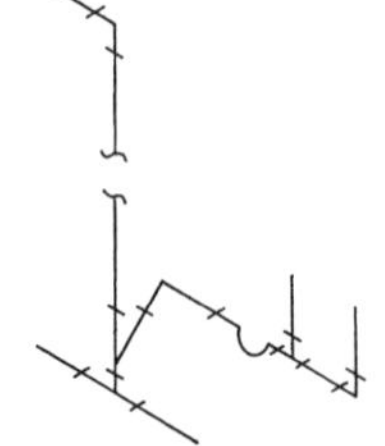

Waste/Vent

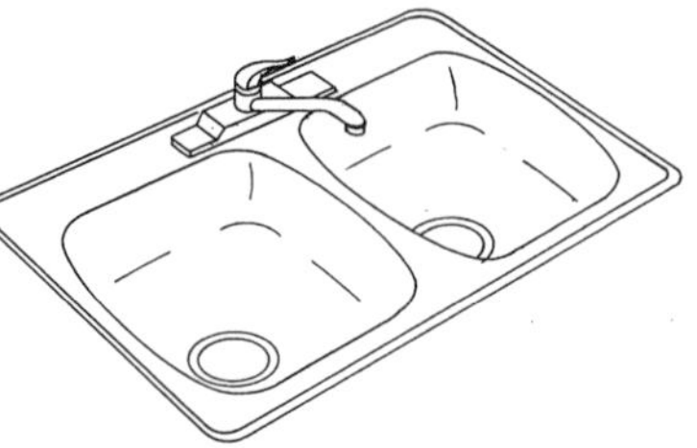

Countertop
Double Bowl

D2010 410	Kitchen Sink Systems	COST EACH		
		MAT.	INST.	TOTAL
1720	Kitchen sink w/trim, countertop, PE on CI, 24"x21", single bowl	660	685	1,345
1760	30" x 21" single bowl	1,200	685	1,885
1800	32" x 21" double bowl	765	740	1,505
1880	Stainless steel, 19" x 18" single bowl	995	685	1,680
1920	25" x 22" single bowl	1,075	685	1,760
1960	33" x 22" double bowl	1,425	740	2,165
2000	43" x 22" double bowl	1,600	750	2,350
2040	44" x 22" triple bowl	1,625	780	2,405
2080	44" x 24" corner double bowl	1,200	750	1,950
2120	Steel, enameled, 24" x 21" single bowl	900	685	1,585
2160	32" x 21" double bowl	945	740	1,685
2240	Raised deck, PE on CI, 32" x 21", dual level, double bowl	850	940	1,790
2280	42" x 21" dual level, triple bowl	1,425	1,025	2,450
2281				
2282				

D2010 Plumbing Fixtures

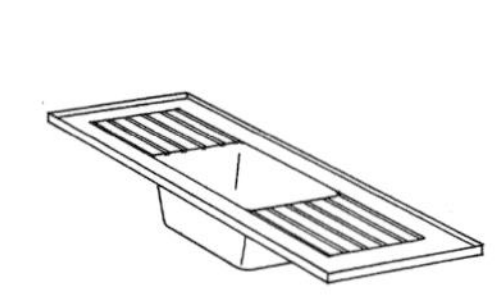

Laboratory Sink

Service Sink

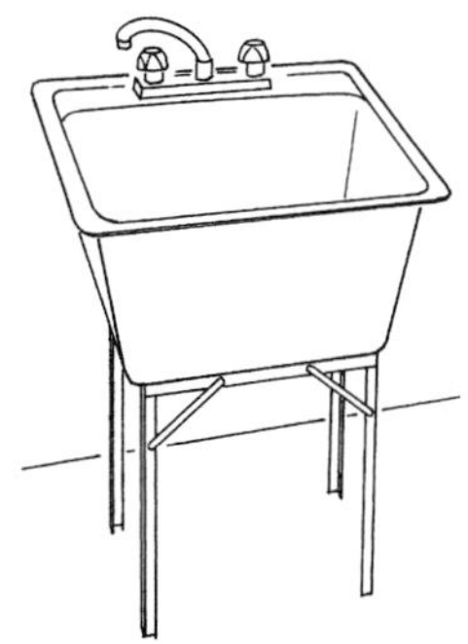

Single Compartment Sink

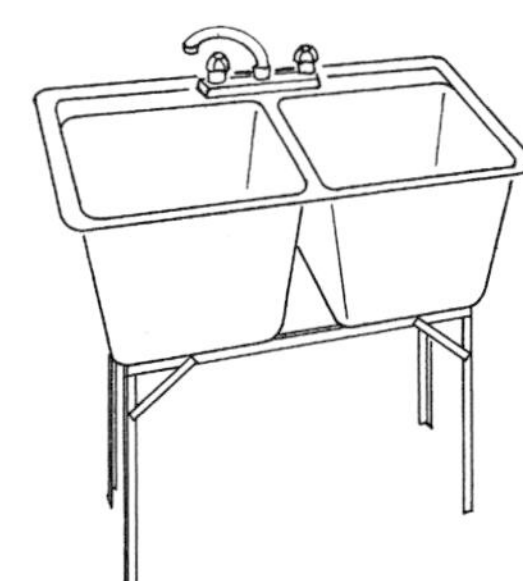

Double Compartment Sink

Systems are complete with trim and rough-in (supply, waste and vent) to connect to supply branches and waste mains.

D2010 420	Laundry Sink Systems	COST EACH		
		MAT.	INST.	TOTAL
1740	Laundry sink w/trim, PE on CI, black iron frame			
1760	24" x 20", single compartment	975	675	1,650
1840	48" x 21" double compartment	1,225	730	1,955
1920	Molded stone, on wall, 22" x 21" single compartment	500	675	1,175
1960	45"x 21" double compartment	715	730	1,445
2040	Plastic, on wall or legs, 18" x 23" single compartment	480	660	1,140
2080	20" x 24" single compartment	500	660	1,160
2120	36" x 23" double compartment	580	710	1,290
2160	40" x 24" double compartment	660	710	1,370

D2010 430	Laboratory Sink Systems	COST EACH		
		MAT.	INST.	TOTAL
1580	Laboratory sink w/trim,			
1590	Stainless steel, single bowl,			
1600	Double drainboard, 54" x 24" O.D.	1,550	880	2,430
1640	Single drainboard, 47" x 24"O.D.	1,150	880	2,030
1670	Stainless steel, double bowl,			
1680	70" x 24" O.D.	1,675	880	2,555
1750	Polyethylene, single bowl,			
1760	Flanged, 14-1/2" x 14-1/2" O.D.	540	790	1,330
1800	18-1/2" x 18-1/2" O.D.	660	790	1,450
1840	23-1/2" x 20-1/2" O.D.	685	790	1,475
1920	Polypropylene, cup sink, oval, 7" x 4" O.D.	430	700	1,130
1960	10" x 4-1/2" O.D.	460	700	1,160
1961				

D2010 440	Service Sink Systems	COST EACH		
		MAT.	INST.	TOTAL
4260	Service sink w/trim, PE on CI, corner floor, 28" x 28", w/rim guard	2,225	885	3,110
4300	Wall hung w/rim guard, 22" x 18"	2,575	1,025	3,600
4340	24" x 20"	2,650	1,025	3,675
4380	Vitreous china, wall hung 22" x 20"	2,575	1,025	3,600

D2010 Plumbing Fixtures

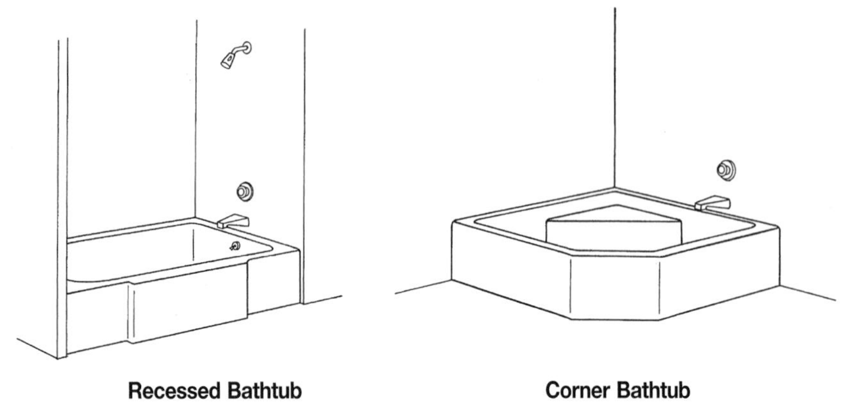

Recessed Bathtub

Corner Bathtub

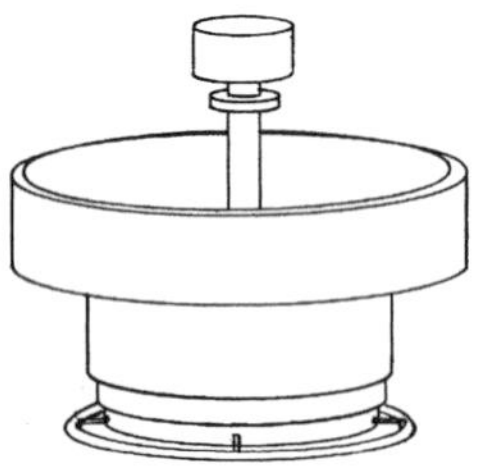

Circular Wash Fountain

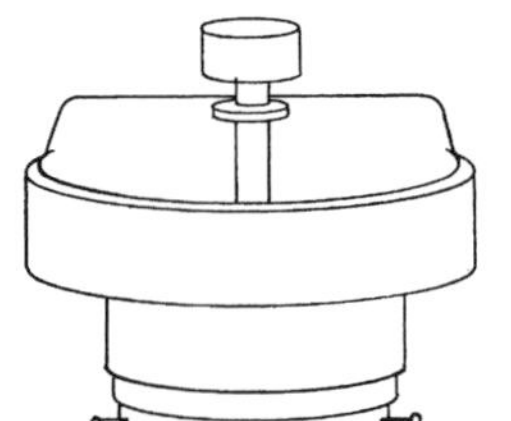

Semi-Circular Wash Fountain

Systems are complete with trim and rough-in (supply, waste and vent) to connect to supply branches and waste mains.

D2010 510	Bathtub Systems	COST EACH		
		MAT.	INST.	TOTAL
2000	Bathtub, recessed, P.E. on CI., 48" x 42"	3,600	780	4,380
2040	72" x 36"	3,675	870	4,545
2080	Mat bottom, 5' long	1,825	755	2,580
2120	5'-6" long	2,450	780	3,230
2160	Corner, 48" x 42"	3,600	755	4,355
2200	Formed steel, enameled, 4'-6" long	1,050	695	1,745

D2010 610	Group Wash Fountain Systems	COST EACH		
		MAT.	INST.	TOTAL
1740	Group wash fountain, precast terrazzo			
1760	Circular, 36" diameter	9,100	1,100	10,200
1800	54" diameter	11,000	1,200	12,200
1840	Semi-circular, 36" diameter	7,525	1,100	8,625
1880	54" diameter	10,600	1,200	11,800
1960	Stainless steel, circular, 36" diameter	7,775	1,025	8,800
2000	54" diameter	9,600	1,125	10,725
2040	Semi-circular, 36" diameter	6,400	1,025	7,425
2080	54" diameter	8,300	1,125	9,425
2160	Thermoplastic, circular, 36" diameter	5,550	835	6,385
2200	54" diameter	6,350	965	7,315
2240	Semi-circular, 36" diameter	5,250	835	6,085
2280	54" diameter	6,400	965	7,365

D2010 Plumbing Fixtures

Square Shower Stall

Corner Angle Shower Stall

Systems are complete with trim and rough-in (supply, waste and vent) to connect to supply branches and waste mains.

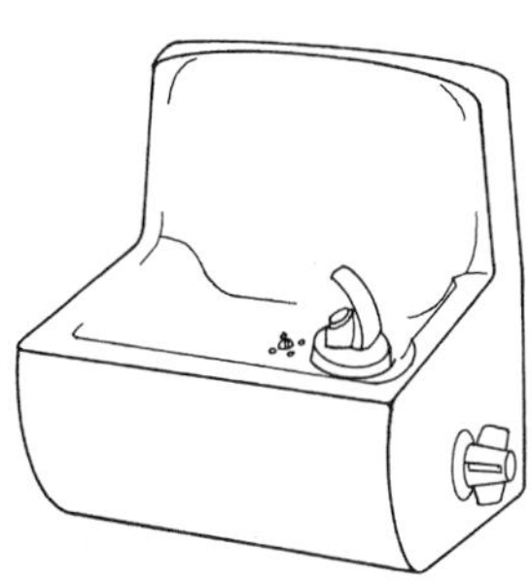

Wall Mounted, Low Back

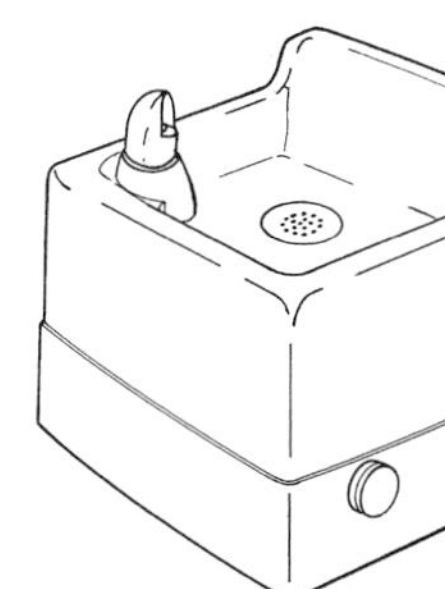

Wall Mounted, No Back

D2010 710	Shower Systems	COST EACH		
		MAT.	INST.	TOTAL
1560	Shower, stall, baked enamel, molded stone receptor, 30" square	1,600	720	2,320
1600	32" square	1,625	730	2,355
1640	Terrazzo receptor, 32" square	1,800	730	2,530
1680	36" square	1,950	740	2,690
1720	36" corner angle	2,075	305	2,380
1800	Fiberglass one piece, three walls, 32" square	715	710	1,425
1840	36" square	775	710	1,485
1880	Polypropylene, molded stone receptor, 30" square	1,050	1,050	2,100
1920	32" square	1,050	1,050	2,100
1960	Built-in head, arm, bypass, stops and handles	123	272	395

D2010 810	Drinking Fountain Systems	COST EACH		
		MAT.	INST.	TOTAL
1740	Drinking fountain, one bubbler, wall mounted			
1760	Non recessed			
1800	Bronze, no back	1,375	415	1,790
1840	Cast iron, enameled, low back	1,525	415	1,940
1880	Fiberglass, 12" back	2,575	415	2,990
1920	Stainless steel, no back	1,350	415	1,765
1960	Semi-recessed, poly marble	1,275	415	1,690
2040	Stainless steel	1,675	415	2,090
2080	Vitreous china	1,225	415	1,640
2120	Full recessed, poly marble	2,100	415	2,515
2200	Stainless steel	1,925	415	2,340
2240	Floor mounted, pedestal type, aluminum	2,950	560	3,510
2320	Bronze	2,625	560	3,185
2360	Stainless steel	2,425	560	2,985

D2010 Plumbing Fixtures

D2010 820	Water Cooler Systems	COST EACH		
		MAT.	INST.	TOTAL
1840	Water cooler, electric, wall hung, 8.2 G.P.H.	1,250	530	1,780
1880	Dual height, 14.3 G.P.H.	1,875	545	2,420
1920	Wheelchair type, 7.5 G.P.H.	1,325	530	1,855
1960	Semi recessed, 8.1 G.P.H.	1,150	530	1,680
2000	Full recessed, 8 G.P.H.	2,775	570	3,345
2040	Floor mounted, 14.3 G.P.H.	1,325	460	1,785
2080	Dual height, 14.3 G.P.H.	1,600	560	2,160
2120	Refrigerated compartment type, 1.5 G.P.H.	1,925	460	2,385

D2010 Plumbing Fixtures

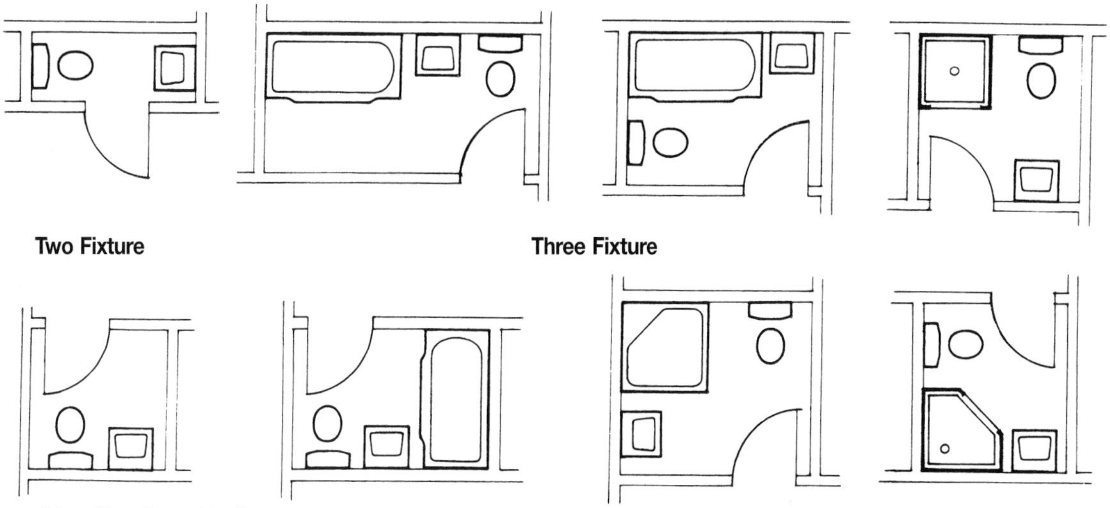

Two Fixture

Three Fixture

*Common wall is with adjacent bathroom

D2010 920	Two Fixture Bathroom, Two Wall Plumbing	COST EACH		
		MAT.	INST.	TOTAL
1180	Bathroom, lavatory & water closet, 2 wall plumbing, stand alone	1,875	1,725	3,600
1200	Share common plumbing wall*	1,700	1,500	3,200

D2010 922	Two Fixture Bathroom, One Wall Plumbing	COST EACH		
		MAT.	INST.	TOTAL
2220	Bathroom, lavatory & water closet, one wall plumbing, stand alone	1,775	1,550	3,325
2240	Share common plumbing wall*	1,425	1,325	2,750

D2010 924	Three Fixture Bathroom, One Wall Plumbing	COST EACH		
		MAT.	INST.	TOTAL
1150	Bathroom, three fixture, one wall plumbing			
1160	Lavatory, water closet & bathtub			
1170	Stand alone	3,175	2,025	5,200
1180	Share common plumbing wall *	2,725	1,450	4,175

D2010 926	Three Fixture Bathroom, Two Wall Plumbing	COST EACH		
		MAT.	INST.	TOTAL
2130	Bathroom, three fixture, two wall plumbing			
2140	Lavatory, water closet & bathtub			
2160	Stand alone	3,200	2,050	5,250
2180	Long plumbing wall common *	2,850	1,625	4,475
3610	Lavatory, bathtub & water closet			
3620	Stand alone	3,475	2,325	5,800
3640	Long plumbing wall common *	3,250	2,100	5,350
4660	Water closet, corner bathtub & lavatory			
4680	Stand alone	5,000	2,075	7,075
4700	Long plumbing wall common *	4,525	1,575	6,100
6100	Water closet, stall shower & lavatory			
6120	Stand alone	3,550	2,325	5,875
6140	Long plumbing wall common *	3,325	2,150	5,475
7060	Lavatory, corner stall shower & water closet			
7080	Stand alone	3,775	2,050	5,825
7100	Short plumbing wall common *	3,225	1,375	4,600

D2020 Domestic Water Distribution

Installation includes piping and fittings within 10′ of heater. Gas and oil fired heaters require vent piping (not included with these units).

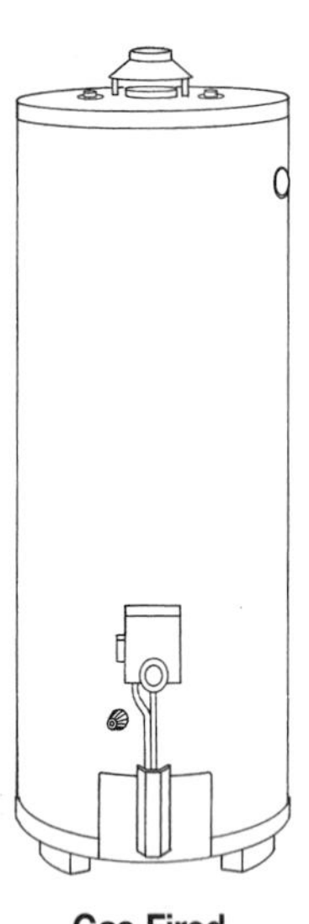

Gas Fired

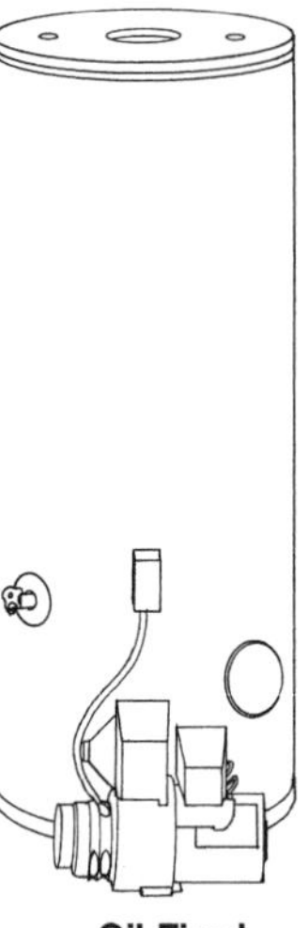

Oil Fired

D2020 220	Gas Fired Water Heaters - Residential Systems	COST EACH		
		MAT.	INST.	TOTAL
2200	Gas fired water heater, residential, 100°F rise			
2260	30 gallon tank, 32 GPH	2,675	1,225	3,900
2300	40 gallon tank, 32 GPH	2,875	1,375	4,250
2340	50 gallon tank, 63 GPH	3,025	1,375	4,400
2380	75 gallon tank, 63 GPH	4,375	1,550	5,925
2420	100 gallon tank, 63 GPH	4,575	1,625	6,200

D2020 230	Oil Fired Water Heaters - Residential Systems	COST EACH		
		MAT.	INST.	TOTAL
2200	Oil fired water heater, residential, 100°F rise			
2220	30 gallon tank, 103 GPH	2,100	1,125	3,225
2260	50 gallon tank, 145 GPH	2,525	1,275	3,800
2300	70 gallon tank, 164 GPH	3,800	1,425	5,225
2340	85 gallon tank, 181 GPH	12,400	1,475	13,875
2344				

D2020 Domestic Water Distribution

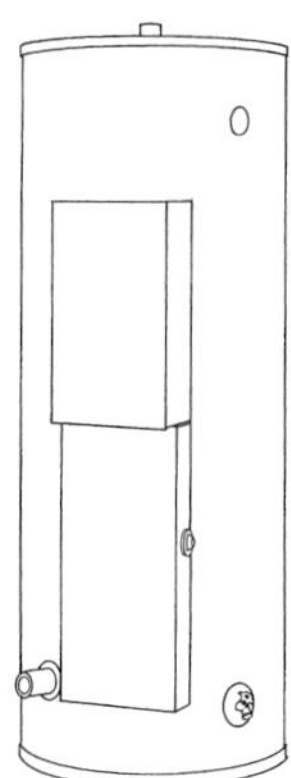

Systems below include piping and fittings within 10′ of heater. Electric water heaters do not require venting.

D2020 240	Electric Water Heaters - Commercial Systems	COST EACH		
		MAT.	INST.	TOTAL
1800	Electric water heater, commercial, 100°F rise			
1820	50 gallon tank, 9 KW 37 GPH	8,900	1,025	9,925
1860	80 gal, 12 KW 49 GPH	10,700	1,275	11,975
1900	36 KW 147 GPH	14,700	1,375	16,075
1940	120 gal, 36 KW 147 GPH	16,600	1,475	18,075
1980	150 gal, 120 KW 490 GPH	52,000	1,575	53,575
2020	200 gal, 120 KW 490 GPH	51,000	1,625	52,625
2060	250 gal, 150 KW 615 GPH	56,500	1,900	58,400
2100	300 gal, 180 KW 738 GPH	64,500	2,000	66,500
2140	350 gal, 30 KW 123 GPH	48,700	2,150	50,850
2180	180 KW 738 GPH	68,000	2,150	70,150
2220	500 gal, 30 KW 123 GPH	59,500	2,525	62,025
2260	240 KW 984 GPH	100,500	2,525	103,025
2300	700 gal, 30 KW 123 GPH	77,000	2,775	79,775
2340	300 KW 1230 GPH	114,500	2,775	117,275
2380	1000 gal, 60 KW 245 GPH	93,000	3,850	96,850
2420	480 KW 1970 GPH	151,000	3,850	154,850
2460	1500 gal, 60 KW 245 GPH	127,500	4,725	132,225
2500	480 KW 1970 GPH	187,000	4,725	191,725

D2020 Domestic Water Distribution

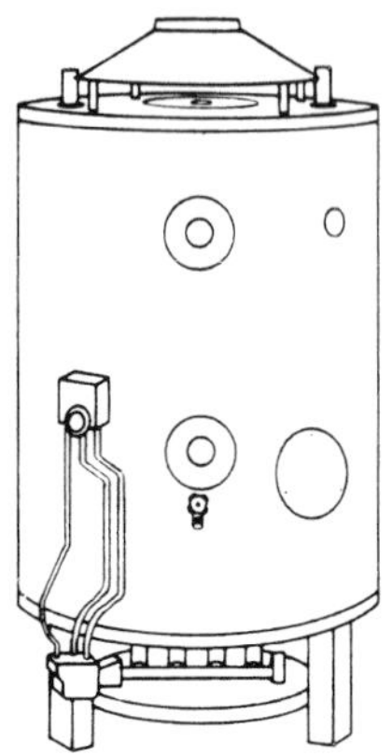

Units may be installed in multiples for increased capacity.

Included below is the heater with self-energizing gas controls, safety pilots, insulated jacket, hi-limit aquastat and pressure relief valve.

Installation includes piping and fittings within 10′ of heater. Gas heaters require vent piping (not included in these prices).

D2020 250	Gas Fired Water Heaters - Commercial Systems	COST EACH		
		MAT.	INST.	TOTAL
1760	Gas fired water heater, commercial, 100°F rise			
1780	75.5 MBH input, 63 GPH	5,575	1,575	7,150
1820	95 MBH input, 86 GPH	9,850	1,575	11,425
1860	100 MBH input, 91 GPH	10,100	1,650	11,750
1900	115 MBH input, 110 GPH	10,200	1,700	11,900
1980	155 MBH input, 150 GPH	12,700	1,900	14,600
2020	175 MBH input, 168 GPH	13,000	2,025	15,025
2060	200 MBH input, 192 GPH	13,600	2,300	15,900
2100	240 MBH input, 230 GPH	14,100	2,500	16,600
2140	300 MBH input, 278 GPH	15,500	2,850	18,350
2180	390 MBH input, 374 GPH	18,300	2,900	21,200
2220	500 MBH input, 480 GPH	24,700	3,125	27,825
2260	600 MBH input, 576 GPH	28,400	3,375	31,775

D2020 Domestic Water Distribution

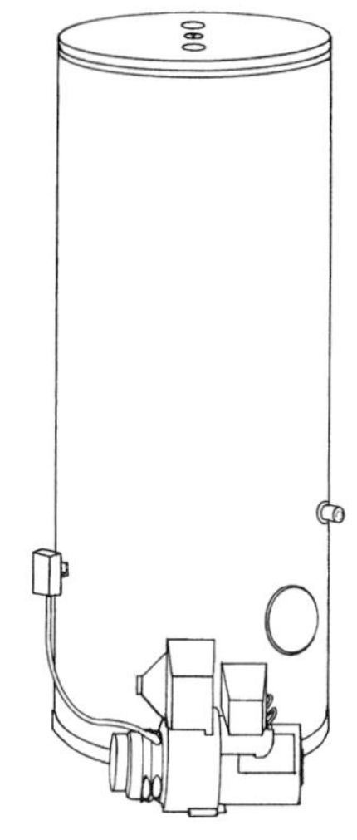

Units may be installed in multiples for increased capacity.

Included below is the heater, wired-in flame retention burners, cadmium cell primary controls, hi-limit controls, ASME pressure relief valves, draft controls, and insulated jacket.

Oil fired water heater systems include piping and fittings within 10′ of heater. Oil fired heaters require vent piping (not included in these systems).

D2020 260	Oil Fired Water Heaters - Commercial Systems	COST EACH		
		MAT.	INST.	TOTAL
1800	Oil fired water heater, commercial, 100°F rise			
1820	140 gal., 140 MBH input, 134 GPH	30,300	1,350	31,650
1900	140 gal., 255 MBH input, 247 GPH	32,800	1,700	34,500
1940	140 gal., 270 MBH input, 259 GPH	40,200	1,925	42,125
1980	140 gal., 400 MBH input, 384 GPH	41,500	2,200	43,700
2060	140 gal., 720 MBH input, 691 GPH	44,100	2,300	46,400
2100	221 gal., 300 MBH input, 288 GPH	58,000	2,525	60,525
2140	221 gal., 600 MBH input, 576 GPH	64,500	2,550	67,050
2180	221 gal., 800 MBH input, 768 GPH	65,500	2,675	68,175
2220	201 gal., 1000 MBH input, 960 GPH	67,000	2,600	69,600
2260	201 gal., 1250 MBH input, 1200 GPH	69,000	2,650	71,650
2300	201 gal., 1500 MBH input, 1441 GPH	74,500	2,725	77,225
2340	411 gal., 600 MBH input, 576 GPH	75,000	2,800	77,800
2380	411 gal., 800 MBH input, 768 GPH	78,000	2,850	80,850
2420	411 gal., 1000 MBH input, 960 GPH	81,000	3,300	84,300
2460	411 gal., 1250 MBH input, 1200 GPH	82,500	3,375	85,875
2500	397 gal., 1500 MBH input, 1441 GPH	88,000	3,500	91,500
2540	397 gal., 1750 MBH input, 1681 GPH	90,000	3,600	93,600

D2040 Rain Water Drainage

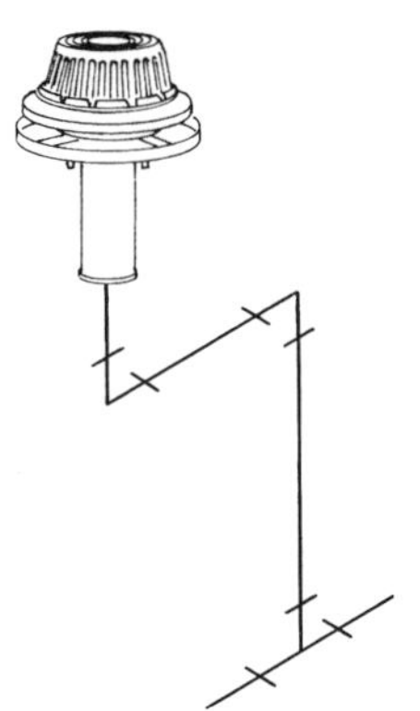

Design Assumptions: Vertical conductor size is based on a maximum rate of rainfall of 4″ per hour. To convert roof area to other rates multiply "Max. S.F. Roof Area" shown by four and divide the result by desired local rate. The answer is the local roof area that may be handled by the indicated pipe diameter.

Basic cost is for roof drain, 10′ of vertical leader and 10′ of horizontal, plus connection to the main.

Pipe Dia.	Max. S.F. Roof Area	Gallons per Min.
2″	544	23
3″	1610	67
4″	3460	144
5″	6280	261
6″	10,200	424
8″	22,000	913

D2040 210	Roof Drain Systems	COST EACH		
		MAT.	INST.	TOTAL
1880	Roof drain, DWV PVC, 2″ diam., piping, 10′ high	380	590	970
1920	For each additional foot add	9.75	17.95	27.70
1960	3″ diam., 10′ high	405	690	1,095
2000	For each additional foot add	9.55	20	29.55
2040	4″ diam., 10′ high	480	770	1,250
2080	For each additional foot add	11.50	22	33.50
2120	5″ diam., 10′ high	1,800	880	2,680
2160	For each additional foot add	41.50	24.50	66
2200	6″ diam., 10′ high	1,575	975	2,550
2240	For each additional foot add	19.70	27	46.70
2280	8″ diam., 10′ high	3,350	1,600	4,950
2320	For each additional foot add	39	32	71
3940	C.I., soil, single hub, service wt., 2″ diam. piping, 10′ high	825	645	1,470
3980	For each additional foot add	18.75	16.85	35.60
4120	3″ diam., 10′ high	955	700	1,655
4160	For each additional foot add	22	17.65	39.65
4200	4″ diam., 10′ high	1,450	765	2,215
4240	For each additional foot add	37.50	19.25	56.75
4280	5″ diam., 10′ high	1,950	785	2,735
4320	For each additional foot add	48	20	68
4360	6″ diam., 10′ high	2,175	845	3,020
4400	For each additional foot add	46	21	67
4440	8″ diam., 10′ high	4,175	1,775	5,950
4480	For each additional foot add	70	36	106
6040	Steel galv. sch 40 threaded, 2″ diam. piping, 10′ high	840	625	1,465
6080	For each additional foot add	13.80	16.55	30.35
6120	3″ diam., 10′ high	1,425	895	2,320
6160	For each additional foot add	23	24.50	47.50
6200	4″ diam., 10′ high	2,075	1,150	3,225
6240	For each additional foot add	26.50	29.50	56
6280	5″ diam., 10′ high	2,150	945	3,095
6320	For each additional foot add	35	28.50	63.50
6360	6″ diam, 10′ high	2,600	1,150	3,750
6400	For each additional foot add	38.50	36.50	75
6440	8″ diam., 10′ high	4,825	1,700	6,525
6480	For each additional foot add	57.50	41.50	99

D3010 Energy Supply

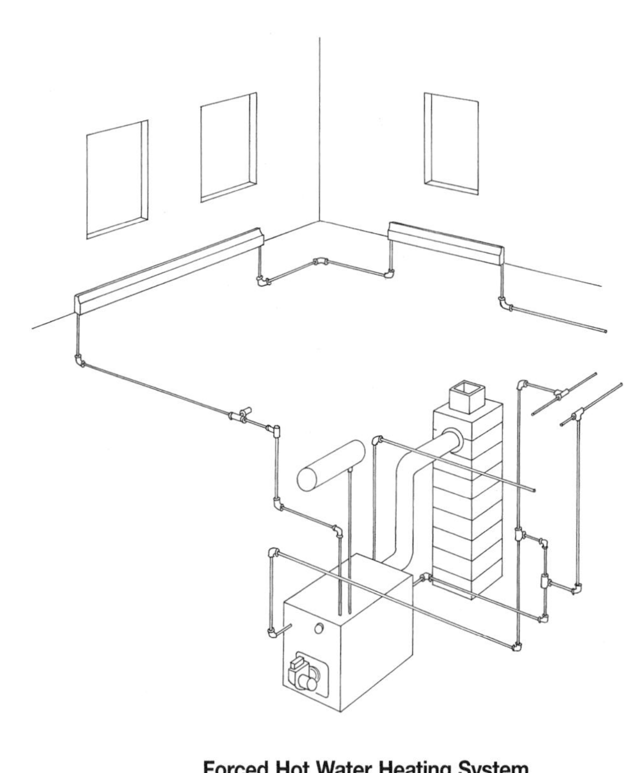

Forced Hot Water Heating System

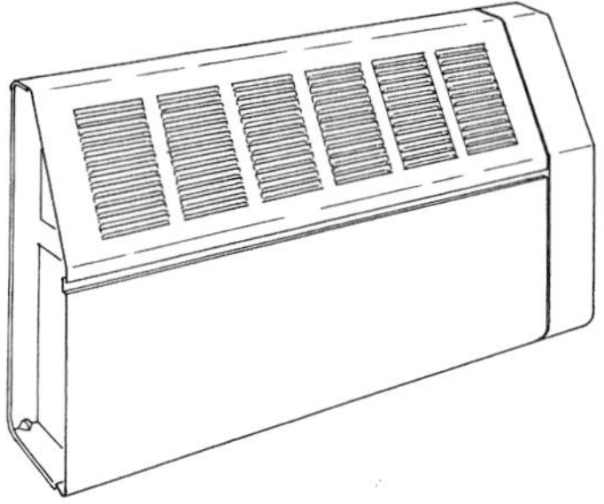

Fin Tube Radiation

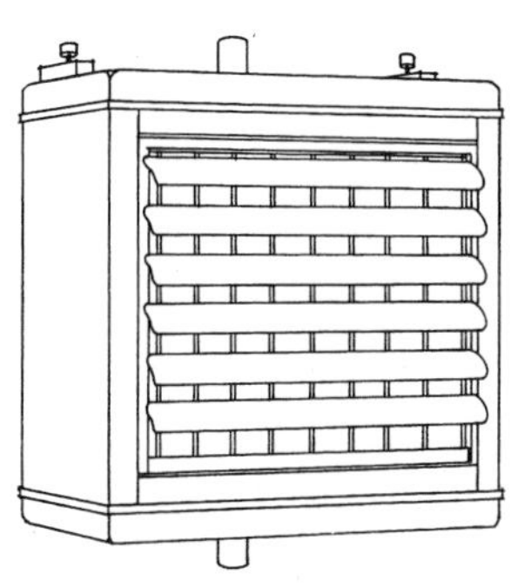

Terminal Unit Heater

D3010 510	Apartment Building Heating - Fin Tube Radiation	COST PER S.F.		
		MAT.	INST.	TOTAL
1740	Heating systems, fin tube radiation, forced hot water			
1800	10,000 S.F. area, 100,000 C.F. volume	3.39	3.44	6.83
1840	20,000 S.F. area, 200,000 C.F. volume	3.70	3.86	7.56
1880	30,000 S.F. area, 300,000 C.F. volume	3.60	3.75	7.35

D3010 520	Commercial Building Heating - Fin Tube Radiation	COST PER S.F.		
		MAT.	INST.	TOTAL
1940	Heating systems, fin tube radiation, forced hot water			
1960	1,000 S.F. bldg, one floor	19.40	12.80	32.20
2000	10,000 S.F., 100,000 C.F., total two floors	4.81	4.88	9.69
2040	100,000 S.F., 1,000,000 C.F., total three floors	2.19	2.17	4.36
2080	1,000,000 S.F., 10,000,000 C.F., total five floors	1.15	1.15	2.30

D3010 530	Commercial Bldg. Heating - Terminal Unit Heaters	COST PER S.F.		
		MAT.	INST.	TOTAL
1860	Heating systems, terminal unit heaters, forced hot water			
1880	1,000 S.F. bldg., one floor	19	11.70	30.70
1920	10,000 S.F. bldg., 100,000 C.F. total two floors	4.30	4.02	8.32
1960	100,000 S.F. bldg., 1,000,000 C.F. total three floors	2.23	1.98	4.21
2000	1,000,000 S.F. bldg., 10,000,000 C.F. total five floors	1.47	1.29	2.76

D3020 Heat Generating Systems

Small Electric Boiler Systems Considerations:

1. Terminal units are fin tube baseboard radiation rated at 720 BTU/hr with 200° water temperature or 820 BTU/hr steam.
2. Primary use being for residential or smaller supplementary areas, the floor levels are based on 7-1/2″ ceiling heights.
3. All distribution piping is copper for boilers through 205 MBH. All piping for larger systems is steel pipe.

Large Electric Boiler Systems Considerations:

1. Terminal units are all unit heaters of the same size. Quantities are varied to accommodate total requirements.
2. All air is circulated through the heaters a minimum of three times per hour.
3. As the capacities are adequate for commercial use, gross output rating by floor levels are based on 10′ ceiling height.
4. All distribution piping is black steel pipe.

D3020 102	Small Heating Systems, Hydronic, Electric Boilers	COST PER S.F.		
		MAT.	INST.	TOTAL
1100	Small heating systems, hydronic, electric boilers			
1120	Steam, 1 floor, 1480 S.F., 61 M.B.H.	7.97	7.43	15.40
1160	3,000 S.F., 123 M.B.H.	6.15	6.50	12.65
1200	5,000 S.F., 205 M.B.H.	5.05	6	11.05
1240	2 floors, 12,400 S.F., 512 M.B.H.	4.10	5.95	10.05
1280	3 floors, 24,800 S.F., 1023 M.B.H.	4.69	5.90	10.59
1320	34,750 S.F., 1,433 M.B.H.	4.31	5.70	10.01
1360	Hot water, 1 floor, 1,000 S.F., 41 M.B.H.	13.40	4.16	17.56
1400	2,500 S.F., 103 M.B.H.	9.20	7.40	16.60
1440	2 floors, 4,850 S.F., 205 M.B.H.	8.55	8.85	17.40
1480	3 floors, 9,700 S.F., 410 M.B.H.	9.05	9.15	18.20

D3020 104	Large Heating Systems, Hydronic, Electric Boilers	COST PER S.F.		
		MAT.	INST.	TOTAL
1230	Large heating systems, hydronic, electric boilers			
1240	9,280 S.F., 135 K.W., 461 M.B.H., 1 floor	4.55	2.76	7.31
1280	14,900 S.F., 240 K.W., 820 M.B.H., 2 floors	5.95	4.46	10.41
1320	18,600 S.F., 296 K.W., 1,010 M.B.H., 3 floors	5.90	4.83	10.73
1360	26,100 S.F., 420 K.W., 1,432 M.B.H., 4 floors	5.80	4.76	10.56
1400	39,100 S.F., 666 K.W., 2,273 M.B.H., 4 floors	4.95	3.98	8.93
1440	57,700 S.F., 900 K.W., 3,071 M.B.H., 5 floors	4.71	3.89	8.60
1480	111,700 S.F., 1,800 K.W., 6,148 M.B.H., 6 floors	4.15	3.35	7.50
1520	149,000 S.F., 2,400 K.W., 8,191 M.B.H., 8 floors	4.13	3.35	7.48
1560	223,300 S.F., 3,600 K.W., 12,283 M.B.H., 14 floors	4.39	3.80	8.19

D3020 Heat Generating Systems

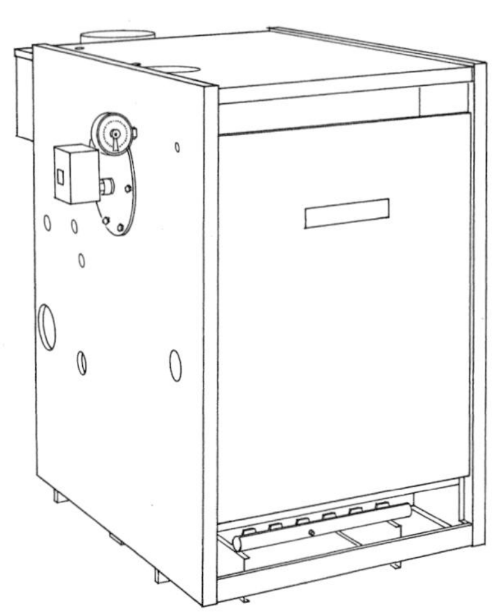

Boiler Selection: The maximum allowable working pressures are limited by ASME "Code for Heating Boilers" to 15 PSI for steam and 160 PSI for hot water heating boilers, with a maximum temperature limitation of 250°F. Hot water boilers are generally rated for a working pressure of 30 PSI. High pressure boilers are governed by the ASME "Code for Power Boilers" which is used almost universally for boilers operating over 15 PSIG. High pressure boilers used for a combination of heating/process loads are usually designed for 150 PSIG.

Boiler ratings are usually indicated as either Gross or Net Output. The Gross Load is equal to the Net Load plus a piping and pickup allowance. When this allowance cannot be determined, divide the gross output rating by 1.25 for a value equal to or greater than the next heat loss requirement of the building.

Table below lists installed cost per boiler and includes insulating jacket, standard controls, burner and safety controls. Costs do not include piping or boiler base pad. Outputs are Gross.

D3020 106	Boilers, Hot Water & Steam	COST EACH		
		MAT.	INST.	TOTAL
0600	Boiler, electric, steel, hot water, 12 K.W., 41 M.B.H.	5,725	1,300	7,025
0620	30 K.W., 103 M.B.H.	6,275	1,400	7,675
0640	60 K.W., 205 M.B.H.	6,775	1,525	8,300
0660	120 K.W., 410 M.B.H.	7,325	1,875	9,200
0680	210 K.W., 716 M.B.H.	8,825	2,800	11,625
0700	510 K.W., 1,739 M.B.H.	24,500	5,200	29,700
0720	720 K.W., 2,452 M.B.H.	30,300	5,875	36,175
0740	1,200 K.W., 4,095 M.B.H.	38,800	6,725	45,525
0760	2,100 K.W., 7,167 M.B.H.	71,000	8,475	79,475
0780	3,600 K.W., 12,283 M.B.H.	104,000	14,300	118,300
0820	Steam, 6 K.W., 20.5 M.B.H.	4,725	1,400	6,125
0840	24 K.W., 81.8 M.B.H.	5,625	1,525	7,150
0860	60 K.W., 205 M.B.H.	7,775	1,675	9,450
0880	150 K.W., 512 M.B.H.	11,200	2,600	13,800
0900	510 K.W., 1,740 M.B.H.	36,000	6,350	42,350
0920	1,080 K.W., 3,685 M.B.H.	44,800	9,150	53,950
0940	2,340 K.W., 7,984 M.B.H.	95,500	14,300	109,800
0980	Gas, cast iron, hot water, 80 M.B.H.	2,125	1,500	3,625
1000	100 M.B.H.	2,500	1,600	4,100
1020	163 M.B.H.	3,175	2,175	5,350
1040	280 M.B.H.	4,375	2,425	6,800
1060	544 M.B.H.	9,450	4,275	13,725
1080	1,088 M.B.H.	14,300	5,450	19,750
1100	2,000 M.B.H.	22,700	8,500	31,200
1120	2,856 M.B.H.	32,600	10,900	43,500
1140	4,720 M.B.H.	78,500	15,000	93,500
1160	6,970 M.B.H.	137,000	24,400	161,400
1180	For steam systems under 2,856 M.B.H., add 8%			
1520	Oil, cast iron, hot water, 109 M.B.H.	2,450	1,825	4,275
1540	173 M.B.H.	3,125	2,175	5,300
1560	236 M.B.H.	3,975	2,550	6,525
1580	1,084 M.B.H.	11,700	5,775	17,475
1600	1,600 M.B.H.	18,500	8,300	26,800
1620	2,480 M.B.H.	22,000	10,600	32,600
1640	3,550 M.B.H.	29,500	12,700	42,200
1660	Steam systems same price as hot water			

D3020 Heat Generating Systems

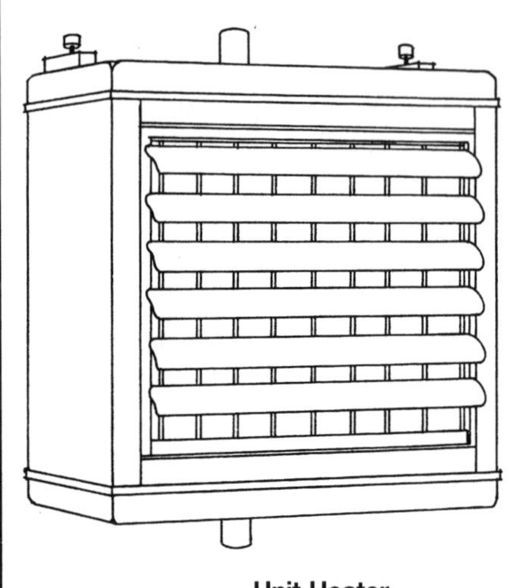

Unit Heater

Fossil Fuel Boiler Systems Considerations:

1. Terminal units are horizontal unit heaters. Quantities are varied to accommodate total heat loss per building.
2. Unit heater selection was determined by their capacity to circulate the building volume a minimum of three times per hour in addition to the BTU output.
3. Systems shown are forced hot water. Steam boilers cost slightly more than hot water boilers. However, this is compensated for by the smaller size or fewer terminal units required with steam.
4. Floor levels are based on 10′ story heights.
5. MBH requirements are gross boiler output.

D3020 108	Heating Systems, Unit Heaters	COST PER S.F.		
		MAT.	INST.	TOTAL
1260	Heating systems, hydronic, fossil fuel, terminal unit heaters,			
1280	Cast iron boiler, gas, 80 M.B.H., 1,070 S.F. bldg.	12.24	8.71	20.95
1320	163 M.B.H., 2,140 S.F. bldg.	8.10	5.90	14
1360	544 M.B.H., 7,250 S.F. bldg.	5.85	4.17	10.02
1400	1,088 M.B.H., 14,500 S.F. bldg.	4.95	3.96	8.91
1440	3,264 M.B.H., 43,500 S.F. bldg.	3.98	3	6.98
1480	5,032 M.B.H., 67,100 S.F. bldg.	4.24	3.08	7.32
1520	Oil, 109 M.B.H., 1,420 S.F. bldg.	13.55	7.75	21.30
1560	235 M.B.H., 3,150 S.F. bldg.	7.85	5.50	13.35
1600	940 M.B.H., 12,500 S.F. bldg.	5.80	3.61	9.41
1640	1,600 M.B.H., 21,300 S.F. bldg.	5.10	3.49	8.59
1680	2,480 M.B.H., 33,100 S.F. bldg.	4.71	3.13	7.84
1720	3,350 M.B.H., 44,500 S.F. bldg.	4.56	3.21	7.77
1760	Coal, 148 M.B.H., 1,975 S.F. bldg.	96.50	5.05	101.55
1800	300 M.B.H., 4,000 S.F. bldg.	53	3.98	56.98
1840	2,360 M.B.H., 31,500 S.F. bldg.	12	3.20	15.20

D3020 Heat Generating Systems

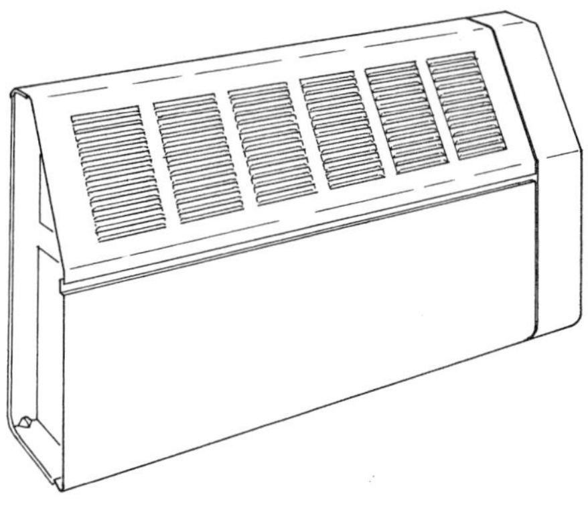

Fin Tube Radiator

Fossil Fuel Boiler System Considerations:

1. Terminal units are commercial steel fin tube radiation. Quantities are varied to accommodate total heat loss per building.
2. Systems shown are forced hot water. Steam boilers cost slightly more than hot water boilers. However, this is compensated for by the smaller size or fewer terminal units required with steam.
3. Floor levels are based on 10′ story heights.
4. MBH requirements are gross boiler output.

D3020 110	Heating System, Fin Tube Radiation	COST PER S.F.		
		MAT.	INST.	TOTAL
3230	Heating systems, hydronic, fossil fuel, fin tube radiation			
3240	Cast iron boiler, gas, 80 MBH, 1,070 S.F. bldg.	14.95	13.18	28.13
3280	169 M.B.H., 2,140 S.F. bldg.	9.35	8.35	17.70
3320	544 M.B.H., 7,250 S.F. bldg.	7.85	7.15	15
3360	1,088 M.B.H., 14,500 S.F. bldg.	7.05	7.05	14.10
3400	3,264 M.B.H., 43,500 S.F. bldg.	6.25	6.20	12.45
3440	5,032 M.B.H., 67,100 S.F. bldg.	6.55	6.25	12.80
3480	Oil, 109 M.B.H., 1,420 S.F. bldg.	18.40	14.40	32.80
3520	235 M.B.H., 3,150 S.F. bldg.	9.85	8.40	18.25
3560	940 M.B.H., 12,500 S.F. bldg.	7.95	6.70	14.65
3600	1,600 M.B.H., 21,300 S.F. bldg.	7.40	6.70	14.10
3640	2,480 M.B.H., 33,100 S.F. bldg.	7	6.35	13.35
3680	3,350 M.B.H., 44,500 S.F. bldg.	6.85	6.40	13.25
3720	Coal, 148 M.B.H., 1,975 S.F. bldg.	98.50	7.95	106.45
3760	300 M.B.H., 4,000 S.F. bldg.	55	6.85	61.85
3800	2,360 M.B.H., 31,500 S.F. bldg.	14.20	6.35	20.55
4080	Steel boiler, oil, 97 M.B.H., 1,300 S.F. bldg.	13	11.85	24.85
4120	315 M.B.H., 4,550 S.F. bldg.	6.95	6.05	13
4160	525 M.B.H., 7,000 S.F. bldg.	9.70	6.90	16.60
4200	1,050 M.B.H., 14,000 S.F. bldg.	8.90	6.90	15.80
4240	2,310 M.B.H., 30,800 S.F. bldg.	7.55	6.40	13.95
4280	3,150 M.B.H., 42,000 S.F. bldg.	7.40	6.45	13.85

D3030 Cooling Generating Systems

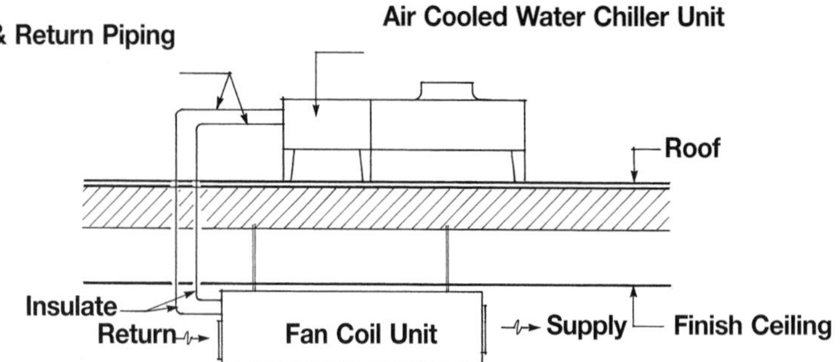

Design Assumptions: The chilled water, air cooled systems priced, utilize reciprocating hermetic compressors and propeller-type condenser fans. Piping with pumps and expansion tanks is included based on a two pipe system. No ducting is included and the fan-coil units are cooling only. Water treatment and balancing are not included. Chilled water piping is insulated. Area distribution is through the use of multiple fan coil units. Fewer but larger fan coil units with duct distribution would be approximately the same S.F. cost.

D3030 110	Chilled Water, Air Cooled Condenser Systems	COST PER S.F.		
		MAT.	INST.	TOTAL
1180	Packaged chiller, air cooled, with fan coil unit			
1200	Apartment corridors, 3,000 S.F., 5.50 ton	7.77	7.37	15.14
1240	6,000 S.F., 11.00 ton	6.20	5.85	12.05
1280	10,000 S.F., 18.33 ton	5.25	4.41	9.66
1320	20,000 S.F., 36.66 ton	4.03	3.19	7.22
1360	40,000 S.F., 73.33 ton	5	2.99	7.99
1440	Banks and libraries, 3,000 S.F., 12.50 ton	11.15	8.50	19.65
1480	6,000 S.F., 25.00 ton	10.05	6.90	16.95
1520	10,000 S.F., 41.66 ton	7.75	4.79	12.54
1560	20,000 S.F., 83.33 ton	8.35	4.05	12.40
1600	40,000 S.F., 167 ton*			
1680	Bars and taverns, 3,000 S.F., 33.25 ton	18.70	9.90	28.60
1720	6,000 S.F., 66.50 ton	16.40	7.50	23.90
1760	10,000 S.F., 110.83 ton	12.75	2.19	14.94
1800	20,000 S.F., 220 ton*			
1840	40,000 S.F., 440 ton*			
1920	Bowling alleys, 3,000 S.F., 17.00 ton	13.90	9.50	23.40
1960	6,000 S.F., 34.00 ton	10.50	6.65	17.15
2000	10,000 S.F., 56.66 ton	8.85	4.74	13.59
2040	20,000 S.F., 113.33 ton	9.30	4.09	13.39
2080	40,000 S.F., 227 ton*			
2160	Department stores, 3,000 S.F., 8.75 ton	10.70	8.15	18.85
2200	6,000 S.F., 17.50 ton	7.80	6.25	14.05
2240	10,000 S.F., 29.17 ton	6.30	4.50	10.80
2280	20,000 S.F., 58.33 ton	5.15	3.28	8.43
2320	40,000 S.F., 116.66 ton	6.10	3.08	9.18
2400	Drug stores, 3,000 S.F., 20.00 ton	15.65	9.75	25.40
2440	6000 S.F., 40.00 ton	12.25	7.15	19.40
2480	10,000 S.F., 66.66 ton	12.75	5.80	18.55
2520	20,000 S.F., 133.33 ton	11	4.29	15.29
2560	40,000 S.F., 267 ton*			
2640	Factories, 2,000 S.F., 10.00 ton	9.90	8.20	18.10
2680	6,000 S.F., 20.00 ton	8.80	6.60	15.40
2720	10,000 S.F., 33.33 ton	6.75	4.59	11.34
2760	20,000 S.F., 66.66 ton	7.35	3.85	11.20
2800	40,000 S.F., 133.33 ton	6.60	3.15	9.75
2880	Food supermarkets, 3,000 S.F., 8.50 ton	10.50	8.15	18.65
2920	6,000 S.F., 17.00 ton	7.70	6.20	13.90
2960	10,000 S.F., 28.33 ton	5.95	4.37	10.32
3000	20,000 S.F., 56.66 ton	4.93	3.22	8.15

D3030 Cooling Generating Systems

D3030 110	Chilled Water, Air Cooled Condenser Systems	COST PER S.F. MAT.	COST PER S.F. INST.	COST PER S.F. TOTAL
3040	40,000 S.F., 113.33 ton	5.95	3.09	9.04
3120	Medical centers, 3,000 S.F., 7.00 ton	9.55	7.95	17.50
3160	6,000 S.F., 14.00 ton	6.95	6.05	13
3200	10,000 S.F., 23.33 ton	6.10	4.58	10.68
3240	20,000 S.F., 46.66 ton	4.78	3.35	8.13
3280	40,000 S.F., 93.33 ton	5.50	3.04	8.54
3360	Offices, 3,000 S.F., 9.50 ton	9.55	8.10	17.65
3400	6,000 S.F., 19.00 ton	8.60	6.50	15.10
3440	10,000 S.F., 31.66 ton	6.55	4.55	11.10
3480	20,000 S.F., 63.33 ton	7.40	3.93	11.33
3520	40,000 S.F., 126.66 ton	6.55	3.19	9.74
3600	Restaurants, 3,000 S.F., 15.00 ton	12.40	8.80	21.20
3640	6,000 S.F., 30.00 ton	10.10	6.70	16.80
3680	10,000 S.F., 50.00 ton	8.65	4.85	13.50
3720	20,000 S.F., 100.00 ton	9.65	4.32	13.97
3760	40,000 S.F., 200 ton*			
3840	Schools and colleges, 3,000 S.F., 11.50 ton	10.65	8.40	19.05
3880	6,000 S.F., 23.00 ton	9.55	6.75	16.30
3920	10,000 S.F., 38.33 ton	7.35	4.71	12.06
3960	20,000 S.F., 76.66 ton	8.25	4.11	12.36

D3030 Cooling Generating Systems

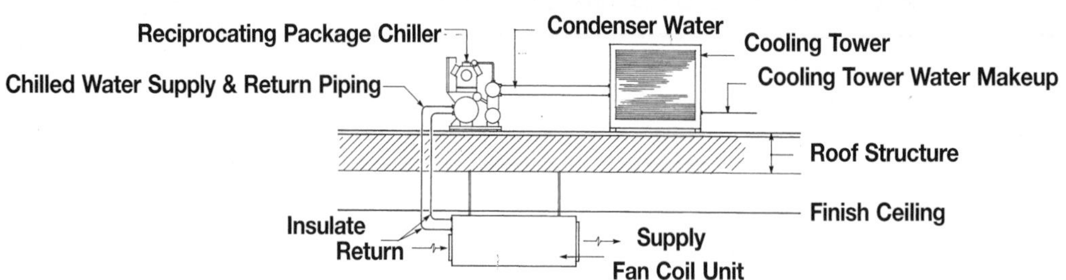

General: Water cooled chillers are available in the same sizes as air cooled units. They are also available in larger capacities.

Design Assumptions: The chilled water systems with water cooled condenser include reciprocating hermetic compressors, water cooling tower, pumps, piping and expansion tanks and are based on a two pipe system. Chilled water piping is insulated. No ducts are included and fan-coil units are cooling only. Area distribution is through use of multiple fan coil units. Fewer but larger fan coil units with duct distribution would be approximately the same S.F. cost. Water treatment and balancing are not included.

D3030 115	Chilled Water, Cooling Tower Systems	COST PER S.F.		
		MAT.	INST.	TOTAL
1300	Packaged chiller, water cooled, with fan coil unit			
1320	Apartment corridors, 4,000 S.F., 7.33 ton	7.43	7.23	14.66
1360	6,000 S.F., 11.00 ton	6.05	6.15	12.20
1400	10,000 S.F., 18.33 ton	6	4.59	10.59
1440	20,000 S.F., 26.66 ton	4.56	3.32	7.88
1480	40,000 S.F., 73.33 ton	5.55	3.11	8.66
1520	60,000 S.F., 110.00 ton	5.40	3.29	8.69
1600	Banks and libraries, 4,000 S.F., 16.66 ton	12.90	7.70	20.60
1640	6,000 S.F., 25.00 ton	10.55	6.80	17.35
1680	10,000 S.F., 41.66 ton	8.45	5	13.45
1720	20,000 S.F., 83.33 ton	9.45	4.24	13.69
1760	40,000 S.F., 166.66 ton	8.80	5.85	14.65
1800	60,000 S.F., 250.00 ton	9.05	6.20	15.25
1880	Bars and taverns, 4,000 S.F., 44.33 ton	19.70	9.50	29.20
1920	6,000 S.F., 66.50 ton	21	9.45	30.45
1960	10,000 S.F., 110.83 ton	19	7.25	26.25
2000	20,000 S.F., 221.66 ton	20	7.90	27.90
2040	40,000 S.F., 440 ton*			
2080	60,000 S.F., 660 ton*			
2160	Bowling alleys, 4,000 S.F., 22.66 ton	14.75	8.65	23.40
2200	6,000 S.F., 34.00 ton	12.40	7.30	19.70
2240	10,000 S.F., 56.66 ton	10.30	5.40	15.70
2280	20,000 S.F., 113.33 ton	10.70	4.67	15.37
2320	40,000 S.F., 226.66 ton	11.50	5.80	17.30
2360	60,000 S.F., 340 ton			
2440	Department stores, 4,000 S.F., 11.66 ton	8.10	7.65	15.75
2480	6,000 S.F., 17.50 ton	9.55	6.30	15.85
2520	10,000 S.F., 29.17 ton	6.85	4.61	11.46
2560	20,000 S.F., 58.33 ton	5.40	3.48	8.88
2600	40,000 S.F., 116.66 ton	6.50	3.33	9.83
2640	60,000 S.F., 175.00 ton	7.85	5.70	13.55
2720	Drug stores, 4,000 S.F., 26.66 ton	15.50	8.65	24.15
2760	6,000 S.F., 40.00 ton	13.35	7.40	20.75
2800	10,000 S.F., 66.66 ton	13.65	6.40	20.05
2840	20,000 S.F., 133.33 ton	11.35	4.98	16.33
2880	40,000 S.F., 266.67 ton	12.05	6.55	18.60
2920	60,000 S.F., 400 ton*			
3000	Factories, 4,000 S.F., 13.33 ton	11.20	7.40	18.60
3040	6,000 S.F., 20.00 ton	9.45	6.50	15.95
3080	10,000 S.F., 33.33 ton	7.40	4.74	12.14

D3030 Cooling Generating Systems

D3030 115	Chilled Water, Cooling Tower Systems	COST PER S.F.		
		MAT.	INST.	TOTAL
3120	20,000 S.F., 66.66 ton	7.65	4.10	11.75
3160	40,000 S.F., 133.33 ton	6.75	3.50	10.25
3200	60,000 S.F., 200.00 ton	8.20	6	14.20
3280	Food supermarkets, 4,000 S.F., 11.33 ton	8	7.65	15.65
3320	6,000 S.F., 17.00 ton	8.45	6.35	14.80
3360	10,000 S.F., 28.33 ton	6.70	4.59	11.29
3400	20,000 S.F., 56.66 ton	5.50	3.50	9
3440	40,000 S.F., 113.33 ton	6.45	3.35	9.80
3480	60,000 S.F., 170.00 ton	7.80	5.70	13.50
3560	Medical centers, 4.000 S.F., 9.33 ton	6.95	7	13.95
3600	6,000 S.F., 14.00 ton	8.35	6.10	14.45
3640	10,000 S.F., 23.33 ton	6.40	4.53	10.93
3680	20,000 S.F., 46.66 ton	5	3.42	8.42
3720	40,000 S.F., 93.33 ton	6.05	3.26	9.31
3760	60,000 S.F., 140.00 ton	6.80	5.80	12.60
3840	Offices, 4,000 S.F., 12.66 ton	10.85	7.35	18.20
3880	6,000 S.F., 19.00 ton	9.45	6.65	16.10
3920	10,000 S.F., 31.66 ton	7.40	4.80	12.20
3960	20,000 S.F., 63.33 ton	7.60	4.11	11.71
4000	40,000 S.F., 126.66 ton	7.75	5.55	13.30
4040	60,000 S.F., 190.00 ton	7.95	5.90	13.85
4120	Restaurants, 4,000 S.F., 20.00 ton	13.20	8.05	21.25
4160	6,000 S.F., 30.00 ton	11.20	6.85	18.05
4200	10,000 S.F., 50.00 ton	9.50	5.20	14.70
4240	20,000 S.F., 100.00 ton	10.55	4.61	15.16
4280	40,000 S.F., 200.00 ton	9.20	5.60	14.80
4320	60,000 S.F., 300.00 ton	10.05	6.40	16.45
4400	Schools and colleges, 4,000 S.F., 15.33 ton	12.25	7.60	19.85
4440	6,000 S.F., 23.00 ton	10.05	6.70	16.75
4480	10,000 S.F., 38.33 ton	8	4.93	12.93
4520	20,000 S.F., 76.66 ton	9.15	4.24	13.39
4560	40,000 S.F., 153.33 ton	8.35	5.70	14.05
4600	60,000 S.F., 230.00 ton	8.35	6	14.35
4603				

D3050 Terminal & Package Units

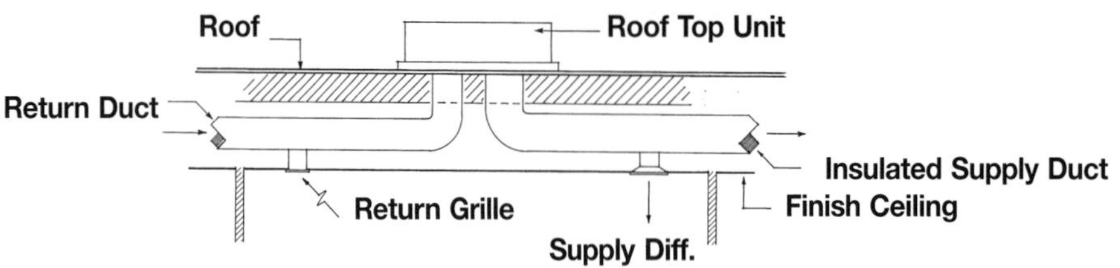

System Description: Rooftop single zone units are electric cooling and gas heat. Duct systems are low velocity, galvanized steel supply and return. Price variations between sizes are due to several factors. Jumps in the cost of the rooftop unit occur when the manufacturer shifts from the largest capacity unit on a small frame to the smallest capacity on the next larger frame, or changes from one compressor to two. As the unit capacity increases for larger areas the duct distribution grows in proportion. For most applications there is a tradeoff point where it is less expensive and more efficient to utilize smaller units with short simple distribution systems. Larger units also require larger initial supply and return ducts which can create a space problem. Supplemental heat may be desired in colder locations. The table below is based on one unit supplying the area listed. The 10,000 S.F. unit for bars and taverns is not listed because a nominal 110 ton unit would be required and this is above the normal single zone rooftop capacity.

*Size would suggest multiple units.

D3050 150	Rooftop Single Zone Unit Systems	COST PER S.F.		
		MAT.	INST.	TOTAL
1260	Rooftop, single zone, air conditioner			
1280	Apartment corridors, 500 S.F., .92 ton	3.60	3.40	7
1320	1,000 S.F., 1.83 ton	3.57	3.39	6.96
1360	1500 S.F., 2.75 ton	2.33	2.79	5.12
1400	3,000 S.F., 5.50 ton	2.32	2.69	5.01
1440	5,000 S.F., 9.17 ton	2.39	2.40	4.79
1480	10,000 S.F., 18.33 ton	3.52	2.27	5.79
1560	Banks or libraries, 500 S.F., 2.08 ton	8.15	7.70	15.85
1600	1,000 S.F., 4.17 ton	5.30	6.35	11.65
1640	1,500 S.F., 6.25 ton	5.25	6.10	11.35
1680	3,000 S.F., 12.50 ton	5.45	5.45	10.90
1720	5,000 S.F., 20.80 ton	7.95	5.15	13.10
1760	10,000 S.F., 41.67 ton	6.25	5.15	11.40
1840	Bars and taverns, 500 S.F. 5.54 ton	12.35	10.30	22.65
1880	1,000 S.F., 11.08 ton	12.80	8.55	21.35
1920	1,500 S.F., 16.62 ton	12.25	8	20.25
1960	3,000 S.F., 33.25 ton	16.15	7.70	23.85
2000	5,000 S.F., 55.42 ton	13.80	7.70	21.50
2040	10,000 S.F., 110.83 ton*			
2080	Bowling alleys, 500 S.F., 2.83 ton	7.20	8.65	15.85
2120	1,000 S.F., 5.67 ton	7.15	8.30	15.45
2160	1,500 S.F., 8.50 ton	7.35	7.40	14.75
2200	3,000 S.F., 17.00 ton	7.10	7.15	14.25
2240	5,000 S.F., 28.33 ton	9.05	6.95	16
2280	10,000 S.F., 56.67 ton	7.90	6.95	14.85
2360	Department stores, 500 S.F., 1.46 ton	5.70	5.40	11.10
2400	1,000 S.F., 2.92 ton	3.71	4.45	8.16
2480	3,000 S.F., 8.75 ton	3.80	3.83	7.63
2520	5,000 S.F., 14.58 ton	3.65	3.66	7.31
2560	10,000 S.F., 29.17 ton	4.67	3.59	8.26
2640	Drug stores, 500 S.F., 3.33 ton	8.50	10.15	18.65
2680	1,000 S.F., 6.67 ton	8.40	9.75	18.15
2720	1,500 S.F., 10.00 ton	8.70	8.75	17.45
2760	3,000 S.F., 20.00 ton	12.80	8.25	21.05
2800	5,000 S.F., 33.33 ton	10.70	8.20	18.90
2840	10,000 S.F., 66.67 ton	9.30	8.20	17.50

D3050 Terminal & Package Units

D3050 150	Rooftop Single Zone Unit Systems	COST PER S.F.		
		MAT.	INST.	TOTAL
2920	Factories, 500 S.F., 1.67 ton	6.50	6.15	12.65
3000	1,500 S.F., 5.00 ton	4.20	4.88	9.08
3040	3,000 S.F., 10.00 ton	4.34	4.36	8.70
3080	5,000 S.F., 16.67 ton	4.17	4.19	8.36
3120	10,000 S.F., 33.33 ton	5.35	4.10	9.45
3200	Food supermarkets, 500 S.F., 1.42 ton	5.55	5.25	10.80
3240	1,000 S.F., 2.83 ton	3.58	4.30	7.88
3280	1,500 S.F., 4.25 ton	3.58	4.16	7.74
3320	3,000 S.F., 8.50 ton	3.69	3.72	7.41
3360	5,000 S.F., 14.17 ton	3.55	3.57	7.12
3400	10,000 S.F., 28.33 ton	4.54	3.49	8.03
3480	Medical centers, 500 S.F., 1.17 ton	4.56	4.33	8.89
3520	1,000 S.F., 2.33 ton	4.56	4.31	8.87
3560	1,500 S.F., 3.50 ton	2.97	3.56	6.53
3640	5,000 S.F., 11.67 ton	3.04	3.06	6.10
3680	10,000 S.F., 23.33 ton	4.47	2.89	7.36
3760	Offices, 500 S.F., 1.58 ton	6.20	5.85	12.05
3800	1,000 S.F., 3.17 ton	4.04	4.84	8.88
3840	1,500 S.F., 4.75 ton	3.99	4.64	8.63
3880	3,000 S.F., 9.50 ton	4.12	4.14	8.26
3920	5,000 S.F., 15.83 ton	3.96	3.98	7.94
3960	10,000 S.F., 31.67 ton	5.05	3.89	8.94
4000	Restaurants, 500 S.F., 2.50 ton	9.75	9.25	19
4040	1,000 S.F., 5.00 ton	6.30	7.35	13.65
4080	1,500 S.F., 7.50 ton	6.50	6.55	13.05
4120	3,000 S.F., 15.00 ton	6.25	6.30	12.55
4160	5,000 S.F., 25.00 ton	9.60	6.20	15.80
4200	10,000 S.F., 50.00 ton	6.95	6.15	13.10
4240	Schools and colleges, 500 S.F., 1.92 ton	7.50	7.10	14.60
4280	1,000 S.F., 3.83 ton	4.88	5.85	10.73
4360	3,000 S.F., 11.50 ton	4.99	5	9.99
4400	5,000 S.F., 19.17 ton	7.35	4.74	12.09
4441	10,000 S.F., 38.33 ton	5.75	4.71	10.46

D3050 Terminal & Package Units

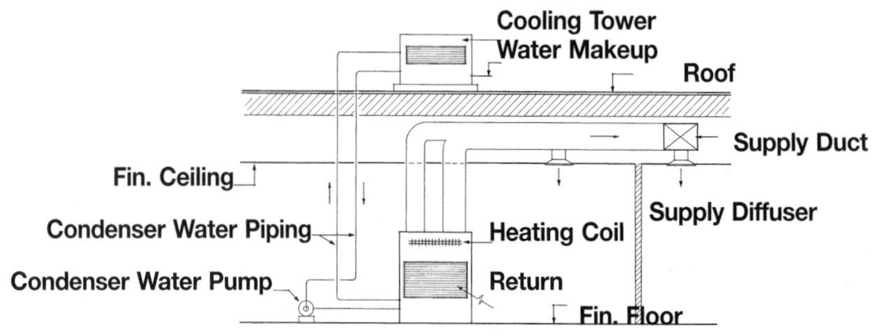

System Description: Self-contained, single package water cooled units include cooling tower, pump, piping allowance. Systems for 1000 S.F. and up include duct and diffusers to provide for even distribution of air. Smaller units distribute air through a supply air plenum, which is integral with the unit.

Returns are not ducted and supplies are not insulated.

Hot water or steam heating coils are included but piping to boiler and the boiler itself is not included.

Where local codes or conditions permit single pass cooling for the smaller units, deduct 10%.

D3050 160	Self-contained, Water Cooled Unit Systems	COST PER S.F.		
		MAT.	INST.	TOTAL
1280	Self-contained, water cooled unit	3.52	2.68	6.20
1300	Apartment corridors, 500 S.F., .92 ton	3.50	2.70	6.20
1320	1,000 S.F., 1.83 ton	3.51	2.67	6.18
1360	3,000 S.F., 5.50 ton	2.77	2.25	5.02
1400	5,000 S.F., 9.17 ton	2.67	2.10	4.77
1440	10,000 S.F., 18.33 ton	3.85	1.88	5.73
1520	Banks or libraries, 500 S.F., 2.08 ton	7.70	2.64	10.34
1560	1,000 S.F., 4.17 ton	6.30	5.10	11.40
1600	3,000 S.F., 12.50 ton	6.05	4.77	10.82
1640	5,000 S.F., 20.80 ton	8.70	4.27	12.97
1680	10,000 S.F., 41.66 ton	7.05	4.28	11.33
1760	Bars and taverns, 500 S.F., 5.54 ton	15.85	4.46	20.31
1800	1,000 S.F., 11.08 ton	15.65	8.10	23.75
1840	3,000 S.F., 33.25 ton	20.50	6.45	26.95
1880	5,000 S.F., 55.42 ton	17.20	6.80	24
1920	10,000 S.F., 110.00 ton	16.70	6.70	23.40
2000	Bowling alleys, 500 S.F., 2.83 ton	10.50	3.61	14.11
2040	1,000 S.F., 5.66 ton	8.55	6.95	15.50
2080	3,000 S.F., 17.00 ton	11.90	5.85	17.75
2120	5,000 S.F., 28.33 ton	10.75	5.65	16.40
2160	10,000 S.F., 56.66 ton	9	5.80	14.80
2200	Department stores, 500 S.F., 1.46 ton	5.40	1.86	7.26
2240	1,000 S.F., 2.92 ton	4.40	3.57	7.97
2280	3,000 S.F., 8.75 ton	4.24	3.34	7.58
2320	5,000 S.F., 14.58 ton	6.10	3	9.10
2360	10,000 S.F., 29.17 ton	5.55	2.89	8.44
2440	Drug stores, 500 S.F., 3.33 ton	12.30	4.25	16.55
2480	1,000 S.F., 6.66 ton	10.05	8.20	18.25
2520	3,000 S.F., 20.00 ton	13.95	6.85	20.80
2560	5,000 S.F., 33.33 ton	14.65	6.80	21.45
2600	10,000 S.F., 66.66 ton	10.60	6.85	17.45
2680	Factories, 500 S.F., 1.66 ton	6.15	2.12	8.27
2720	1,000 S.F. 3.37 ton	5.05	4.13	9.18
2760	3,000 S.F., 10.00 ton	4.84	3.82	8.66
2800	5,000 S.F., 16.66 ton	7	3.43	10.43
2840	10,000 S.F., 33.33 ton	6.30	3.31	9.61
2920	Food supermarkets, 500 S.F., 1.42 ton	5.25	1.80	7.05
2960	1,000 S.F., 2.83 ton	5.45	4.12	9.57

D3050 Terminal & Package Units

D3050 160	Self-contained, Water Cooled Unit Systems	COST PER S.F.		
		MAT.	INST.	TOTAL
3000	3,000 S.F., 8.50 ton	4.28	3.47	7.75
3040	5,000 S.F., 14.17 ton	4.12	3.25	7.37
3080	10,000 S.F., 28.33 ton	5.40	2.81	8.21
3160	Medical centers, 500 S.F., 1.17 ton	4.30	1.49	5.79
3200	1,000 S.F., 2.33 ton	4.47	3.40	7.87
3240	3,000 S.F., 7.00 ton	3.51	2.86	6.37
3280	5,000 S.F., 11.66 ton	3.38	2.67	6.05
3320	10,000 S.F., 23.33 ton	4.88	2.40	7.28
3400	Offices, 500 S.F., 1.58 ton	5.85	2.02	7.87
3440	1,000 S.F., 3.17 ton	6.05	4.62	10.67
3480	3,000 S.F., 9.50 ton	4.60	3.63	8.23
3520	5,000 S.F., 15.83 ton	6.65	3.26	9.91
3560	10,000 S.F., 31.67 ton	6	3.15	9.15
3640	Restaurants, 500 S.F., 2.50 ton	9.25	3.19	12.44
3680	1,000 S.F., 5.00 ton	7.55	6.15	13.70
3720	3,000 S.F., 15.00 ton	7.25	5.75	13
3760	5,000 S.F., 25.00 ton	10.50	5.15	15.65
3800	10,000 S.F., 50.00 ton	6.50	4.74	11.24
3880	Schools and colleges, 500 S.F., 1.92 ton	7.10	2.44	9.54
3920	1,000 S.F., 3.83 ton	5.80	4.70	10.50
3960	3,000 S.F., 11.50 ton	5.55	4.39	9.94
4000	5,000 S.F., 19.17 ton	8.05	3.94	11.99
4040	10,000 S.F., 38.33 ton	6.45	3.94	10.39

D3050 Terminal & Package Units

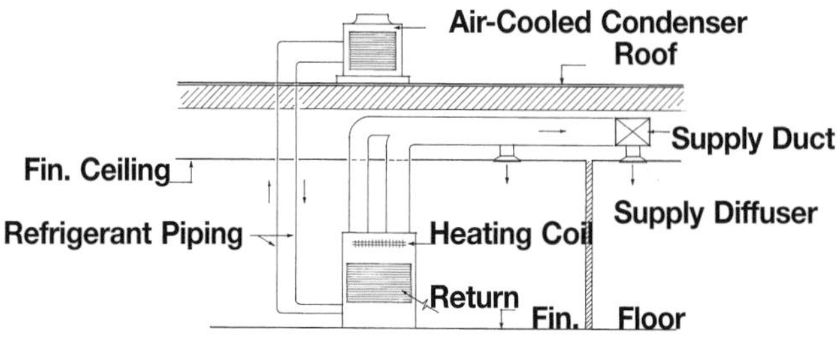

System Description: Self-contained air cooled units with remote air cooled condenser and interconnecting tubing. Systems for 1000 S.F. and up include duct and diffusers. Smaller units distribute air directly.

Returns are not ducted and supplies are not insulated.

Potential savings may be realized by using a single zone rooftop system or through-the-wall unit, especially in the smaller capacities, if the application permits.

Hot water or steam heating coils are included but piping to boiler and the boiler itself is not included.

Condenserless models are available for 15% less where remote refrigerant source is available.

D3050 165	Self-contained, Air Cooled Unit Systems	COST PER S.F.		
		MAT.	INST.	TOTAL
1300	Self-contained, air cooled unit			
1320	Apartment corridors, 500 S.F., .92 ton	5.40	3.55	8.95
1360	1,000 S.F., 1.83 ton	5.35	3.52	8.87
1400	3,000 S.F., 5.50 ton	4.92	3.22	8.14
1440	5,000 S.F., 9.17 ton	4.01	3.08	7.09
1480	10,000 S.F., 18.33 ton	3.40	2.84	6.24
1560	Banks or libraries, 500 S.F., 2.08 ton	11.75	4.56	16.31
1600	1,000 S.F., 4.17 ton	11.10	7.30	18.40
1640	3,000 S.F., 12.50 ton	9.10	7	16.10
1680	5,000 S.F., 20.80 ton	7.85	6.45	14.30
1720	10,000 S.F., 41.66 ton	8.65	6.30	14.95
1800	Bars and taverns, 500 S.F., 5.54 ton	27.50	9.95	37.45
1840	1,000 S.F., 11.08 ton	24	14.10	38.10
1880	3,000 S.F., 33.25 ton	23	12.25	35.25
1920	5,000 S.F., 55.42 ton	23	12.30	35.30
1960	10,000 S.F., 110.00 ton	23.50	12.25	35.75
2040	Bowling alleys, 500 S.F., 2.83 ton	16.15	6.25	22.40
2080	1,000 S.F., 5.66 ton	15.20	10	25.20
2120	3,000 S.F., 17.00 ton	10.55	8.80	19.35
2160	5,000 S.F., 28.33 ton	11.90	8.60	20.50
2200	10,000 S.F., 56.66 ton	12.05	8.60	20.65
2240	Department stores, 500 S.F., 1.46 ton	8.30	3.20	11.50
2280	1,000 S.F., 2.92 ton	7.80	5.15	12.95
2320	3,000 S.F., 8.75 ton	6.35	4.91	11.26
2360	5,000 S.F., 14.58 ton	6.35	4.91	11.26
2400	10,000 S.F., 29.17 ton	6.15	4.42	10.57
2480	Drug stores, 500 S.F., 3.33 ton	19	7.30	26.30
2520	1,000 S.F., 6.66 ton	17.85	11.75	29.60
2560	3,000 S.F., 20.00 ton	12.55	10.35	22.90
2600	5,000 S.F., 33.33 ton	14	10.10	24.10
2640	10,000 S.F., 66.66 ton	14.30	10.10	24.40
2720	Factories, 500 S.F., 1.66 ton	9.65	3.70	13.35
2760	1,000 S.F., 3.33 ton	9	5.90	14.90
2800	3,000 S.F., 10.00 ton	7.30	5.60	12.90
2840	5,000 S.F., 16.66 ton	6.15	5.15	11.30
2880	10,000 S.F., 33.33 ton	7	5.05	12.05

D3050 Terminal & Package Units

D3050 165	Self-contained, Air Cooled Unit Systems	COST PER S.F.		
		MAT.	INST.	TOTAL
2960	Food supermarkets, 500 S.F., 1.42 ton	8.10	3.11	11.21
3000	1,000 S.F., 2.83 ton	8.25	5.40	13.65
3040	3,000 S.F., 8.50 ton	7.60	4.98	12.58
3080	5,000 S.F., 14.17 ton	6.20	4.76	10.96
3120	10,000 S.F., 28.33 ton	5.95	4.29	10.24
3200	Medical centers, 500 S.F., 1.17 ton	6.70	2.57	9.27
3240	1,000 S.F., 2.33 ton	6.85	4.48	11.33
3280	3,000 S.F., 7.00 ton	6.25	4.10	10.35
3320	5,000 S.F., 16.66 ton	5.10	3.92	9.02
3360	10,000 S.F., 23.33 ton	4.34	3.61	7.95
3440	Offices, 500 S.F., 1.58 ton	9.05	3.49	12.54
3480	1,000 S.F., 3.16 ton	9.30	6.10	15.40
3520	3,000 S.F., 9.50 ton	6.95	5.35	12.30
3560	5,000 S.F., 15.83 ton	5.95	4.92	10.87
3600	10,000 S.F., 31.66 ton	6.65	4.81	11.46
3680	Restaurants, 500 S.F., 2.50 ton	14.20	5.50	19.70
3720	1,000 S.F., 5.00 ton	13.40	8.80	22.20
3760	3,000 S.F., 15.00 ton	10.95	8.40	19.35
3800	5,000 S.F., 25.00 ton	10.45	7.60	18.05
3840	10,000 S.F., 50.00 ton	11.15	7.55	18.70
3920	Schools and colleges, 500 S.F., 1.92 ton	10.90	4.21	15.11
3960	1,000 S.F., 3.83 ton	10.30	6.75	17.05
4000	3,000 S.F., 11.50 ton	8.35	6.45	14.80
4040	5,000 S.F., 19.17 ton	7.10	5.95	13.05
4080	10,000 S.F., 38.33 ton	7.95	5.80	13.75

D3050 Terminal & Package Units

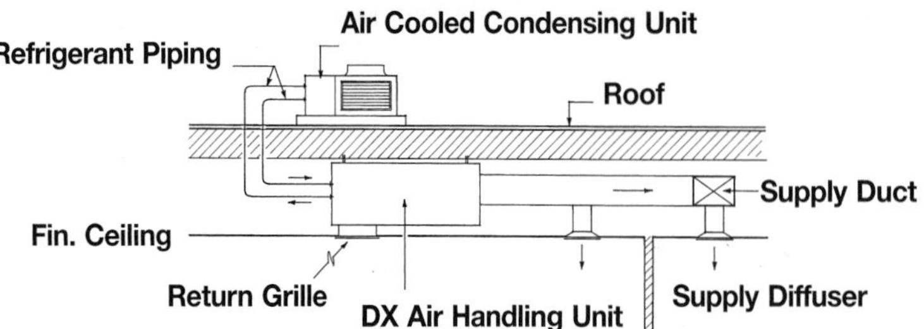

General: Split systems offer several important advantages which should be evaluated when a selection is to be made. They provide a greater degree of flexibility in component selection which permits an accurate match-up of the proper equipment size and type with the particular needs of the building. This allows for maximum use of modern energy saving concepts in heating and cooling. Outdoor installation of the air cooled condensing unit allows space savings in the building and also isolates the equipment operating sounds from building occupants.

Design Assumptions: The systems below are comprised of a direct expansion air handling unit and air cooled condensing unit with interconnecting copper tubing. Ducts and diffusers are also included for distribution of air. Systems are priced for cooling only. Heat can be added as desired either by putting hot water/steam coils into the air unit or into the duct supplying the particular area of need. Gas fired duct furnaces are also available. Refrigerant liquid line is insulated.

D3050 170	Split Systems With Air Cooled Condensing Units	COST PER S.F.		
		MAT.	INST.	TOTAL
1260	Split system, air cooled condensing unit			
1280	Apartment corridors, 1,000 S.F., 1.83 ton	2.33	2.23	4.56
1320	2,000 S.F., 3.66 ton	1.79	2.23	4.02
1360	5,000 S.F., 9.17 ton	1.66	2.78	4.44
1400	10,000 S.F., 18.33 ton	1.82	3.01	4.83
1440	20,000 S.F., 36.66 ton	2.17	3.06	5.23
1520	Banks and libraries, 1,000 S.F., 4.17 ton	4.08	5.10	9.18
1560	2,000 S.F., 8.33 ton	3.78	6.35	10.13
1600	5,000 S.F., 20.80 ton	4.13	6.85	10.98
1640	10,000 S.F., 41.66 ton	4.93	6.95	11.88
1680	20,000 S.F., 83.32 ton	5.95	7.05	13
1760	Bars and taverns, 1,000 S.F., 11.08 ton	8.55	9.95	18.50
1800	2,000 S.F., 22.16 ton	11.15	11.65	22.80
1840	5,000 S.F., 55.42 ton	10.85	10.95	21.80
1880	10,000 S.F., 110.84 ton	13.65	11.15	24.80
1920	20,000 S.F., 220 ton*			
2000	Bowling alleys, 1,000 S.F., 5.66 ton	5.65	9.50	15.15
2040	2,000 S.F., 11.33 ton	5.15	8.60	13.75
2080	5,000 S.F., 28.33 ton	5.60	9.30	14.90
2120	10,000 S.F., 56.66 ton	6.70	9.45	16.15
2160	20,000 S.F., 113.32 ton	9.20	9.95	19.15
2320	Department stores, 1,000 S.F., 2.92 ton	2.84	3.49	6.33
2360	2,000 S.F., 5.83 ton	2.90	4.89	7.79
2400	5,000 S.F., 14.58 ton	2.65	4.43	7.08
2440	10,000 S.F., 29.17 ton	2.89	4.78	7.67
2480	20,000 S.F., 58.33 ton	3.45	4.87	8.32
2560	Drug stores, 1,000 S.F., 6.66 ton	6.65	11.20	17.85
2600	2,000 S.F., 13.32 ton	6.05	10.10	16.15
2640	5,000 S.F., 33.33 ton	7.90	11.10	19
2680	10,000 S.F., 66.66 ton	8.15	11.30	19.45
2720	20,000 S.F., 133.32 ton*			
2800	Factories, 1,000 S.F., 3.33 ton	3.25	3.99	7.24
2840	2,000 S.F., 6.66 ton	3.32	5.60	8.92
2880	5,000 S.F., 16.66 ton	3.31	5.45	8.76
2920	10,000 S.F., 33.33 ton	3.95	5.55	9.50
2960	20,000 S.F., 66.66 ton	4.08	5.65	9.73
3040	Food supermarkets, 1,000 S.F., 2.83 ton	2.76	3.40	6.16
3080	2,000 S.F., 5.66 ton	2.81	4.75	7.56

D3050 Terminal & Package Units

D3050 170	Split Systems With Air Cooled Condensing Units	COST PER S.F.		
		MAT.	INST.	TOTAL
3120	5,000 S.F., 14.66 ton	2.56	4.30	6.86
3160	10,000 S.F., 28.33 ton	2.80	4.64	7.44
3200	20,000 S.F., 56.66 ton	3.35	4.73	8.08
3280	Medical centers, 1,000 S.F., 2.33 ton	2.43	2.75	5.18
3320	2,000 S.F., 4.66 ton	2.32	3.91	6.23
3360	5,000 S.F., 11.66 ton	2.11	3.54	5.65
3400	10,000 S.F., 23.33 ton	2.31	3.82	6.13
3440	20,000 S.F., 46.66 ton	2.75	3.89	6.64
3520	Offices, 1,000 S.F., 3.17 ton	3.08	3.79	6.87
3560	2,000 S.F., 6.33 ton	3.15	5.30	8.45
3600	5,000 S.F., 15.83 ton	2.88	4.81	7.69
3640	10,000 S.F., 31.66 ton	3.14	5.20	8.34
3680	20,000 S.F., 63.32 ton	3.87	5.40	9.27
3760	Restaurants, 1,000 S.F., 5.00 ton	4.98	8.40	13.38
3800	2,000 S.F., 10.00 ton	4.54	7.60	12.14
3840	5,000 S.F., 25.00 ton	4.96	8.20	13.16
3880	10,000 S.F., 50.00 ton	5.95	8.35	14.30
3920	20,000 S.F., 100.00 ton	8.15	8.80	16.95
4000	Schools and colleges, 1,000 S.F., 3.83 ton	3.75	4.67	8.42
4040	2,000 S.F., 7.66 ton	3.47	5.80	9.27
4080	5,000 S.F., 19.17 ton	3.81	6.30	10.11
4120	10,000 S.F., 38.33 ton	4.54	6.40	10.94
4160	20,000 S.F., 76.66 ton	4.69	6.50	11.19

D3090 Other HVAC Systems/Equip

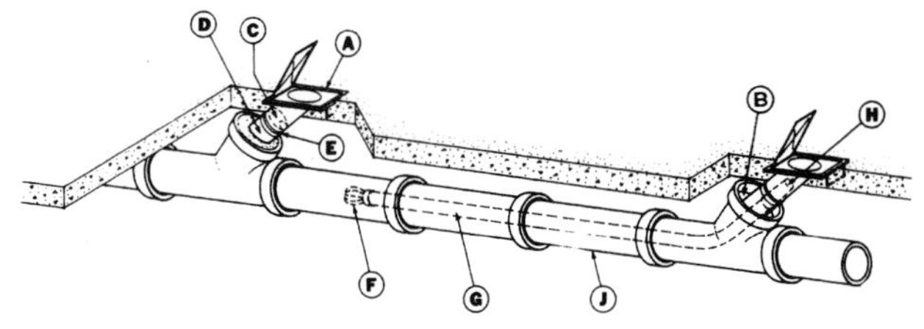

Cast Iron Garage Exhaust System

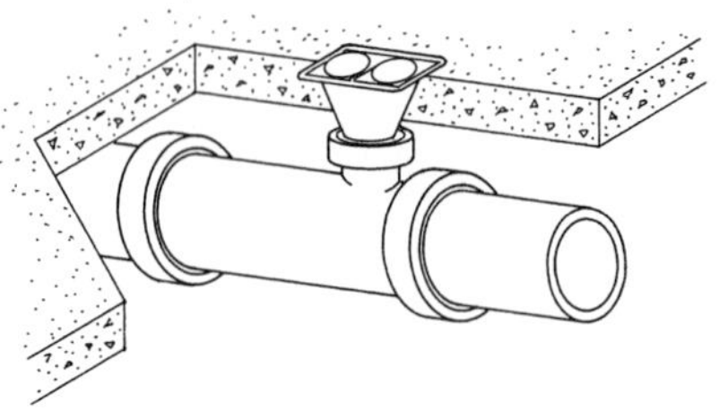

Dual Exhaust System

D3090 320	Garage Exhaust Systems	COST PER BAY		
		MAT.	INST.	TOTAL
1040	Garage, single 3" exhaust outlet, cars & light trucks, one bay	4,775	1,600	6,375
1060	Additional bays up to seven bays	1,200	415	1,615
1500	4" outlet, trucks, one bay	4,800	1,600	6,400
1520	Additional bays up to six bays	1,225	415	1,640
1600	5" outlet, diesel trucks, one bay	5,075	1,600	6,675
1650	Additional single bays up to six bays	1,650	485	2,135
1700	Two adjoining bays	5,075	1,600	6,675
2000	Dual exhaust, 3" outlets, pair of adjoining bays	6,025	2,000	8,025
2100	Additional pairs of adjoining bays	1,825	485	2,310

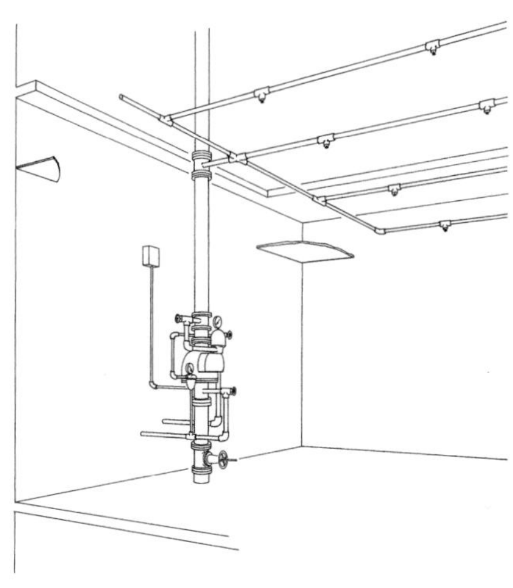

Dry Pipe System: A system employing automatic sprinklers attached to a piping system containing air under pressure, the release of which from the opening of sprinklers permits the water pressure to open a valve known as a "dry pipe valve". The water then flows into the piping system and out the opened sprinklers.

All areas are assumed to be open.

D4010 310	Dry Pipe Sprinkler Systems	COST PER S.F.		
		MAT.	INST.	TOTAL
0520	Dry pipe sprinkler systems, steel, black, sch. 40 pipe			
0530	Light hazard, one floor, 500 S.F.	11.30	5.25	16.55
0560	1000 S.F.	6.10	3.06	9.16
0580	2000 S.F.	5.90	3.10	9
0600	5000 S.F.	3.08	2.09	5.17
0620	10,000 S.F.	2.17	1.74	3.91
0640	50,000 S.F.	1.54	1.52	3.06
0660	Each additional floor, 500 S.F.	2.22	2.55	4.77
0680	1000 S.F.	1.85	2.09	3.94
0700	2000 S.F.	1.88	1.96	3.84
0720	5000 S.F.	1.57	1.66	3.23
0740	10,000 S.F.	1.44	1.52	2.96
0760	50,000 S.F.	1.28	1.34	2.62
1000	Ordinary hazard, one floor, 500 S.F.	11.50	5.30	16.80
1020	1000 S.F.	6.35	3.09	9.44
1040	2000 S.F.	6.10	3.21	9.31
1060	5000 S.F.	3.60	2.23	5.83
1080	10,000 S.F.	2.90	2.30	5.20
1100	50,000 S.F.	2.55	2.15	4.70
1140	Each additional floor, 500 S.F.	2.44	2.61	5.05
1160	1000 S.F.	2.26	2.34	4.60
1180	2000 S.F.	2.28	2.13	4.41
1200	5000 S.F.	2.16	1.83	3.99
1220	10,000 S.F.	1.95	1.81	3.76
1240	50,000 S.F.	1.92	1.56	3.48
1500	Extra hazard, one floor, 500 S.F.	15.50	6.60	22.10
1520	1000 S.F.	9.45	4.75	14.20
1540	2000 S.F.	6.70	4.15	10.85
1560	5000 S.F.	4.07	3.11	7.18
1580	10,000 S.F.	4.37	2.97	7.34
1600	50,000 S.F.	4.72	2.83	7.55
1660	Each additional floor, 500 S.F.	3.31	3.23	6.54
1680	1000 S.F.	3.24	3.06	6.30
1700	2000 S.F.	3.08	3.09	6.17
1720	5000 S.F.	2.65	2.68	5.33
1740	10,000 S.F.	3.32	2.44	5.76
1760	50,000 S.F.	3.36	2.34	5.70
2020	Grooved steel, black, sch. 40 pipe, light hazard, one floor, 2000 S.F.	5.85	2.65	8.50

D4010 Sprinklers

D4010 310	Dry Pipe Sprinkler Systems	COST PER S.F.		
		MAT.	INST.	TOTAL
2060	10,000 S.F.	2.23	1.53	3.76
2100	Each additional floor, 2000 S.F.	2.04	1.57	3.61
2150	10,000 S.F.	1.50	1.31	2.81
2200	Ordinary hazard, one floor, 2000 S.F.	6.15	2.81	8.96
2250	10,000 S.F.	2.75	1.94	4.69
2300	Each additional floor, 2000 S.F.	2.32	1.73	4.05
2350	10,000 S.F.	2.02	1.73	3.75
2400	Extra hazard, one floor, 2000 S.F.	6.95	3.55	10.50
2450	10,000 S.F.	3.93	2.50	6.43
2500	Each additional floor, 2000 S.F.	3.32	2.55	5.87
2550	10,000 S.F.	3.02	2.20	5.22
3050	Grooved steel, black, sch. 10 pipe, light hazard, one floor, 2000 S.F.	5.75	2.62	8.37
3100	10,000 S.F.	2.16	1.50	3.66
3150	Each additional floor, 2000 S.F.	1.95	1.54	3.49
3200	10,000 S.F.	1.43	1.28	2.71
3250	Ordinary hazard, one floor, 2000 S.F.	6.05	2.79	8.84
3300	10,000 S.F.	2.63	1.90	4.53
3350	Each additional floor, 2000 S.F.	2.23	1.71	3.94
3400	10,000 S.F.	1.90	1.69	3.59
3450	Extra hazard, one floor, 2000 S.F.	6.85	3.52	10.37
3500	10,000 S.F.	3.69	2.46	6.15
3550	Each additional floor, 2000 S.F.	3.24	2.52	5.76
3600	10,000 S.F.	2.90	2.17	5.07
4050	Copper tubing, type M, light hazard, one floor, 2000 S.F.	6.25	2.62	8.87
4100	10,000 S.F.	2.80	1.52	4.32
4150	Each additional floor, 2000 S.F.	2.46	1.57	4.03
4200	10,000 S.F.	2.07	1.31	3.38
4250	Ordinary hazard, one floor, 2000 S.F.	6.65	2.93	9.58
4300	10,000 S.F.	3.42	1.78	5.20
4350	Each additional floor, 2000 S.F.	3.22	1.83	5.05
4400	10,000 S.F.	2.63	1.54	4.17
4450	Extra hazard, one floor, 2000 S.F.	7.65	3.55	11.20
4500	10,000 S.F.	5.90	2.66	8.56
4550	Each additional floor, 2000 S.F.	4.02	2.55	6.57
4600	10,000 S.F.	4.32	2.34	6.66
5050	Copper tubing, type M, T-drill system, light hazard, one floor			
5060	2000 S.F.	6.25	2.46	8.71
5100	10,000 S.F.	2.65	1.27	3.92
5150	Each additional floor, 2000 S.F.	2.45	1.41	3.86
5200	10,000 S.F.	1.92	1.06	2.98
5250	Ordinary hazard, one floor, 2000 S.F.	6.45	2.52	8.97
5300	10,000 S.F.	3.31	1.60	4.91
5350	Each additional floor, 2000 S.F.	2.63	1.44	4.07
5400	10,000 S.F.	2.45	1.32	3.77
5450	Extra hazard, one floor, 2000 S.F.	7.10	2.94	10.04
5500	10,000 S.F.	4.96	1.95	6.91
5550	Each additional floor, 2000 S.F.	3.46	1.94	5.40
5600	10,000 S.F.	3.40	1.63	5.03

Reference - System Classification

System Classification

Rules for installation of sprinkler systems vary depending on the classification of occupancy falling into one of three categories as follows:

Light Hazard Occupancy

The protection area allotted per sprinkler should not exceed 225 S.F., with the maximum distance between lines and sprinklers on lines being 15′. The sprinklers do not need to be staggered. Branch lines should not exceed eight sprinklers on either side of a cross main. Each large area requiring more than 100 sprinklers and without a sub-dividing partition should be supplied by feed mains or risers sized for ordinary hazard occupancy.
Maximum system area = 52,000 S.F.

Included in this group are:

Churches
Clubs
Educational
Hospitals
Institutional
Libraries
(except large stack rooms)
Museums
Nursing Homes
Offices
Residential
Restaurants
Theaters and Auditoriums
(except stages and prosceniums)
Unused Attics

Ordinary Hazard Occupancy

The protection area allotted per sprinkler shall not exceed 130 S.F. of noncombustible ceiling and 130 S.F. of combustible ceiling. The maximum allowable distance between sprinkler lines and sprinklers on line is 15′. Sprinklers shall be staggered if the distance between heads exceeds 12′. Branch lines should not exceed eight sprinklers on either side of a cross main.
Maximum system area = 52,000 S.F.

Included in this group are:

Group 1
Automotive Parking and Showrooms
Bakeries
Beverage manufacturing
Canneries
Dairy Products Manufacturing/Processing
Electronic Plans
Glass and Glass Products Manufacturing
Laundries
Restaurant Service Areas

Group 2
Cereal Mills
Chemical Plants—Ordinary
Confectionery Products
Distilleries
Dry Cleaners
Feed Mills
Horse Stables
Leather Goods Manufacturing
Libraries—Large Stack Room Areas
Machine Shops
Metal Working
Mercantile
Paper and Pulp Mills
Paper Process Plants
Piers and Wharves
Post Offices
Printing and Publishing
Repair Garages
Stages
Textile Manufacturing
Tire Manufacturing
Tobacco Products Manufacturing
Wood Machining
Wood Product Assembly

Extra Hazard Occupancy

The protection area allotted per sprinkler shall not exceed 100 S.F. of noncombustible ceiling and 100 S.F. of combustible ceiling. The maximum allowable distance between lines and between sprinklers on lines is 12′. Sprinklers on alternate lines shall be staggered if the distance between sprinklers on lines exceeds 8′. Branch lines should not exceed six sprinklers on either side of a cross main.
Maximum system area:
Design by pipe schedule = 25,000 S.F.
Design by hydraulic calculation = 40,000 S.F.

Included in this group are:

Group 1
Aircraft hangars
Combustible Hydraulic Fluid Use Area
Die Casting
Metal Extruding
Plywood/Particle Board Manufacturing
Printing (inks with flash points < 100 degrees F
Rubber Reclaiming, Compounding, Drying, Milling, Vulcanizing
Saw Mills
Textile Picking, Opening, Blending, Garnetting, Carding, Combing of Cotton, Synthetics, Wood Shoddy, or Burlap
Upholstering with Plastic Foams

Group 2
Asphalt Saturating
Flammable Liquids Spraying
Flow Coating
Manufactured/Modular Home Building Assemblies (where finished enclosure is present and has combustible interiors)
Open Oil Quenching
Plastics Processing
Solvent Cleaning
Varnish and Paint Dipping

Reference - Sprinkler Systems (Automatic)

Sprinkler systems may be classified by type as follows:

1. **Wet Pipe System.** A system employing automatic sprinklers attached to a piping system containing water and connected to a water supply so that water discharges immediately from sprinklers opened by a fire.
2. **Dry Pipe System.** A system employing automatic sprinklers attached to a piping system containing air under pressure, the release of which as from the opening of sprinklers permits the water pressure to open a valve known as a "dry pipe valve". The water then flows into the piping system and out the opened sprinklers.
3. **Pre-Action System.** A system employing automatic sprinklers attached to a piping system containing air that may or may not be under pressure, with a supplemental heat responsive system of generally more sensitive characteristics than the automatic sprinklers themselves, installed in the same areas as the sprinklers; actuation of the heat responsive system, as from a fire, opens a valve which permits water to flow into the sprinkler piping system and to be discharged from any sprinklers which may be open.
4. **Deluge System.** A system employing open sprinklers attached to a piping system connected to a water supply through a valve which is opened by the operation of a heat responsive system installed in the same areas as the sprinklers. When this valve opens, water flows into the piping system and discharges from all sprinklers attached thereto.
5. **Combined Dry Pipe and Pre-Action Sprinkler System.** A system employing automatic sprinklers attached to a piping system containing air under pressure with a supplemental heat responsive system of generally more sensitive characteristics than the automatic sprinklers themselves, installed in the same areas as the sprinklers; operation of the heat responsive system, as from a fire, actuates tripping devices which open dry pipe valves simultaneously and without loss of air pressure in the system. Operation of the heat responsive system also opens approved air exhaust valves at the end of the feed main which facilitates the filling of the system with water which usually precedes the opening of sprinklers. The heat responsive system also serves as an automatic fire alarm system.
6. **Limited Water Supply System.** A system employing automatic sprinklers and conforming to these standards but supplied by a pressure tank of limited capacity.
7. **Chemical Systems.** Systems using halon, carbon dioxide, dry chemical or high expansion foam as selected for special requirements. Agent may extinguish flames by chemically inhibiting flame propagation, suffocate flames by excluding oxygen, interrupting chemical action of oxygen uniting with fuel or sealing and cooling the combustion center.
8. **Firecycle System.** Firecycle is a fixed fire protection sprinkler system utilizing water as its extinguishing agent. It is a time delayed, recycling, preaction type which automatically shuts the water off when heat is reduced below the detector operating temperature and turns the water back on when that temperature is exceeded. The system senses a fire condition through a closed circuit electrical detector system which controls water flow to the fire automatically. Batteries supply up to 90 hour emergency power supply for system operation. The piping system is dry (until water is required) and is monitored with pressurized air. Should any leak in the system piping occur, an alarm will sound, but water will not enter the system until heat is sensed by a firecycle detector.

Area coverage sprinkler systems may be laid out and fed from the supply in any one of several patterns as shown below. It is desirable, if possible, to utilize a central feed and achieve a shorter flow path from the riser to the furthest sprinkler. This permits use of the smallest sizes of pipe possible with resulting savings.

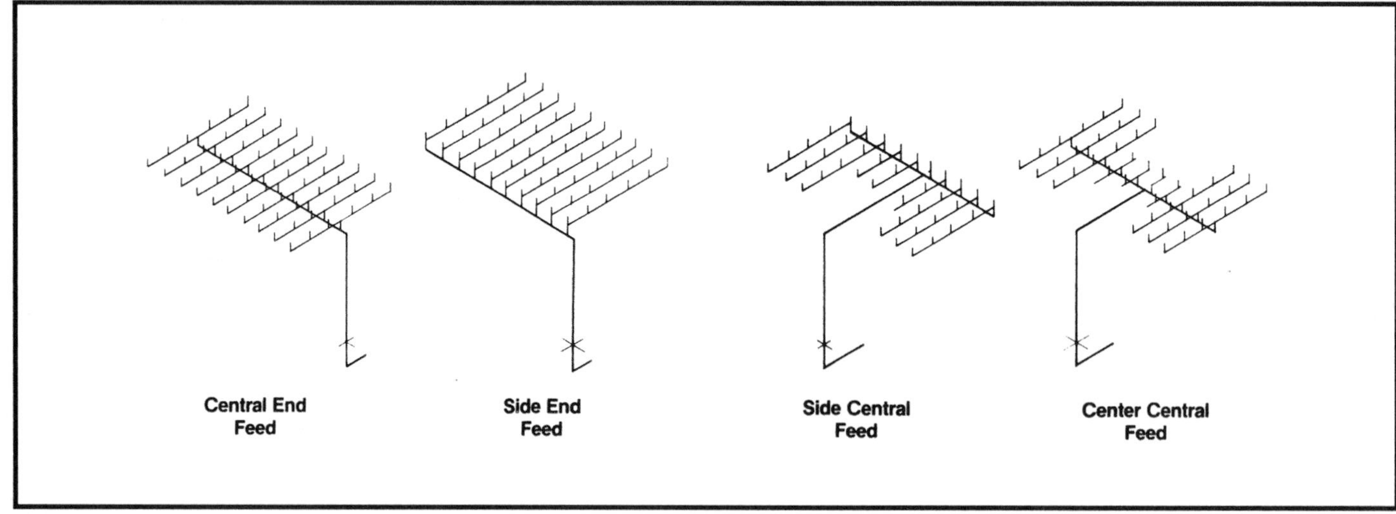

D4010 Sprinklers

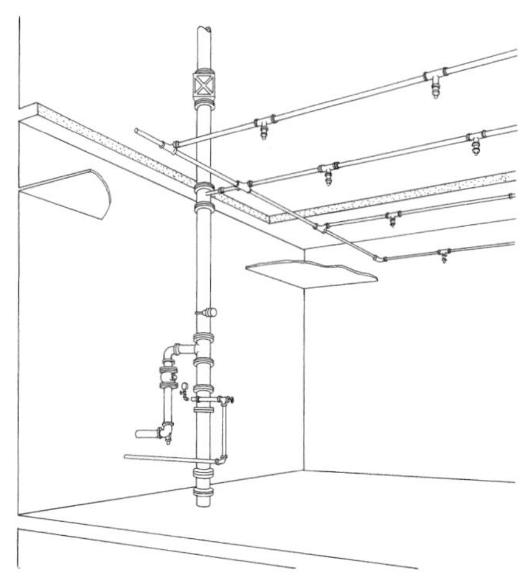

Wet Pipe System. A system employing automatic sprinklers attached to a piping system containing water and connected to a water supply so that water discharges immediately from sprinklers opened by heat from a fire.

All areas are assumed to be open.

D4010 410	Wet Pipe Sprinkler Systems	COST PER S.F.		
		MAT.	INST.	TOTAL
0520	Wet pipe sprinkler systems, steel, black, sch. 40 pipe			
0530	Light hazard, one floor, 500 S.F.	3.03	2.70	5.73
0560	1000 S.F.	6.30	2.77	9.07
0580	2000 S.F.	5.75	2.79	8.54
0600	5000 S.F.	2.71	1.95	4.66
0620	10,000 S.F.	1.74	1.67	3.41
0640	50,000 S.F.	1.06	1.48	2.54
0660	Each additional floor, 500 S.F.	1.25	2.29	3.54
0680	1000 S.F.	1.29	2.12	3.41
0700	2000 S.F.	1.29	1.93	3.22
0720	5000 S.F.	.95	1.64	2.59
0740	10,000 S.F.	.90	1.50	2.40
0760	50,000 S.F.	.70	1.16	1.86
1000	Ordinary hazard, one floor, 500 S.F.	3.36	2.89	6.25
1020	1000 S.F.	6.35	2.72	9.07
1040	2000 S.F.	5.90	2.90	8.80
1060	5000 S.F.	3.07	2.08	5.15
1080	10,000 S.F.	2.23	2.22	4.45
1100	50,000 S.F.	1.81	2.07	3.88
1140	Each additional floor, 500 S.F.	1.66	2.58	4.24
1160	1000 S.F.	1.31	2.09	3.40
1180	2000 S.F.	1.50	2.10	3.60
1200	5000 S.F.	1.53	1.99	3.52
1220	10,000 S.F.	1.39	2.06	3.45
1240	50,000 S.F.	1.30	1.82	3.12
1500	Extra hazard, one floor, 500 S.F.	12.15	4.46	16.61
1520	1000 S.F.	8.10	3.85	11.95
1540	2000 S.F.	6.20	3.94	10.14
1560	5000 S.F.	3.72	3.41	7.13
1580	10,000 S.F.	3.41	3.26	6.67
1600	50,000 S.F.	3.90	3.09	6.99
1660	Each additional floor, 500 S.F.	2.14	3.19	5.33
1680	1000 S.F.	2.07	3.02	5.09
1700	2000 S.F.	1.91	3.05	4.96
1720	5000 S.F.	1.59	2.68	4.27
1740	10,000 S.F.	2.16	2.45	4.61
1760	50,000 S.F.	2.15	2.32	4.47
2020	Grooved steel, black, sch. 40 pipe, light hazard, one floor, 2000 S.F.	5.85	2.34	8.19

D4010 Sprinklers

D4010 410	Wet Pipe Sprinkler Systems	COST PER S.F.		
		MAT.	INST.	TOTAL
2060	10,000 S.F.	2.29	1.51	3.80
2100	Each additional floor, 2000 S.F.	1.45	1.54	2.99
2150	10,000 S.F.	.96	1.29	2.25
2200	Ordinary hazard, one floor, 2000 S.F.	5.95	2.50	8.45
2250	10,000 S.F.	2.08	1.86	3.94
2300	Each additional floor, 2000 S.F.	1.54	1.70	3.24
2350	10,000 S.F.	1.24	1.70	2.94
2400	Extra hazard, one floor, 2000 S.F.	6.40	3.23	9.63
2450	10,000 S.F.	2.86	2.40	5.26
2500	Each additional floor, 2000 S.F.	2.15	2.51	4.66
2550	10,000 S.F.	1.84	2.15	3.99
3050	Grooved steel, black, sch. 10 pipe, light hazard, one floor, 2000 S.F.	5.75	2.31	8.06
3100	10,000 S.F.	1.73	1.43	3.16
3150	Each additional floor, 2000 S.F.	1.36	1.51	2.87
3200	10,000 S.F.	.89	1.26	2.15
3250	Ordinary hazard, one floor, 2000 S.F.	5.85	2.48	8.33
3300	10,000 S.F.	1.96	1.82	3.78
3350	Each additional floor, 2000 S.F.	1.45	1.68	3.13
3400	10,000 S.F.	1.12	1.66	2.78
3450	Extra hazard, one floor, 2000 S.F.	6.30	3.20	9.50
3500	10,000 S.F.	2.62	2.36	4.98
3550	Each additional floor, 2000 S.F.	2.07	2.48	4.55
3600	10,000 S.F.	1.72	2.12	3.84
4050	Copper tubing, type M, light hazard, one floor, 2000 S.F.	6.25	2.31	8.56
4100	10,000 S.F.	2.37	1.45	3.82
4150	Each additional floor, 2000 S.F.	1.87	1.54	3.41
4200	10,000 S.F.	1.53	1.29	2.82
4250	Ordinary hazard, one floor, 2000 S.F.	6.50	2.62	9.12
4300	10,000 S.F.	2.75	1.70	4.45
4350	Each additional floor, 2000 S.F.	2.12	1.74	3.86
4400	10,000 S.F.	1.85	1.51	3.36
4450	Extra hazard, one floor, 2000 S.F.	7.10	3.23	10.33
4500	10,000 S.F.	4.77	2.56	7.33
4550	Each additional floor, 2000 S.F.	2.85	2.51	5.36
4600	10,000 S.F.	3.14	2.29	5.43
5050	Copper tubing, type M, T-drill system, light hazard, one floor			
5060	2000 S.F.	6.25	2.15	8.40
5100	10,000 S.F.	2.22	1.20	3.42
5150	Each additional floor, 2000 S.F.	1.86	1.38	3.24
5200	10,000 S.F.	1.38	1.04	2.42
5250	Ordinary hazard, one floor, 2000 S.F.	6.25	2.21	8.46
5300	10,000 S.F.	2.64	1.52	4.16
5350	Each additional floor, 2000 S.F.	1.85	1.41	3.26
5400	10,000 S.F.	1.80	1.36	3.16
5450	Extra hazard, one floor, 2000 S.F.	6.50	2.62	9.12
5500	10,000 S.F.	3.85	1.85	5.70
5550	Each additional floor, 2000 S.F.	2.38	1.94	4.32
5600	10,000 S.F.	2.22	1.58	3.80

D4020 Standpipes

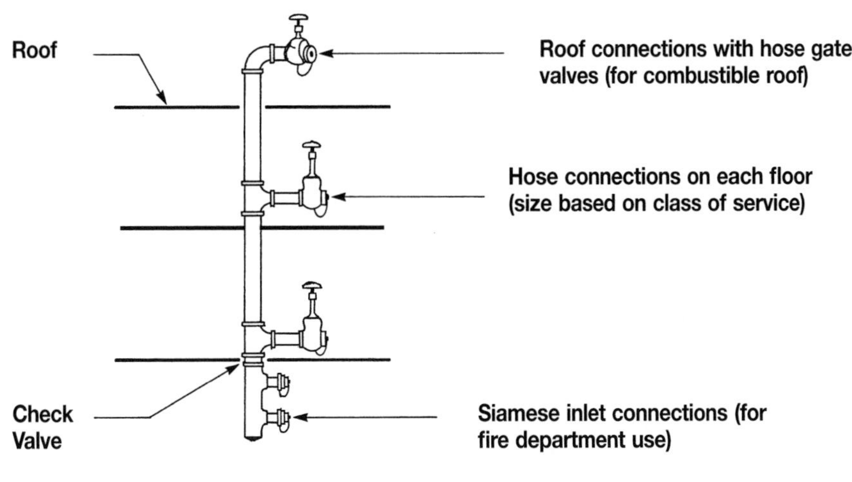

D4020 310	Wet Standpipe Risers, Class I	COST PER FLOOR		
		MAT.	INST.	TOTAL
0550	Wet standpipe risers, Class I, steel, black, sch. 40, 10′ height			
0560	4″ diameter pipe, one floor	5,900	2,925	8,825
0580	Additional floors	1,450	910	2,360
0600	6″ diameter pipe, one floor	9,675	5,000	14,675
0620	Additional floors	2,550	1,425	3,975
0640	8″ diameter pipe, one floor	14,500	6,050	20,550
0660	Additional floors	3,700	1,725	5,425

D4020 310	Wet Standpipe Risers, Class II	COST PER FLOOR		
		MAT.	INST.	TOTAL
1030	Wet standpipe risers, Class II, steel, black, sch. 40, 10′ height			
1040	2″ diameter pipe, one floor	2,500	1,050	3,550
1060	Additional floors	940	410	1,350
1080	2-1/2″ diameter pipe, one floor	3,525	1,550	5,075
1100	Additional floors	1,025	475	1,500

D4020 310	Wet Standpipe Risers, Class III	COST PER FLOOR		
		MAT.	INST.	TOTAL
1530	Wet standpipe risers, Class III, steel, black, sch. 40, 10′ height			
1540	4″ diameter pipe, one floor	6,025	2,925	8,950
1560	Additional floors	1,250	760	2,010
1580	6″ diameter pipe, one floor	9,825	5,000	14,825
1600	Additional floors	2,625	1,425	4,050
1620	8″ diameter pipe, one floor	14,700	6,050	20,750
1640	Additional floors	3,775	1,725	5,500

D4020 Standpipes

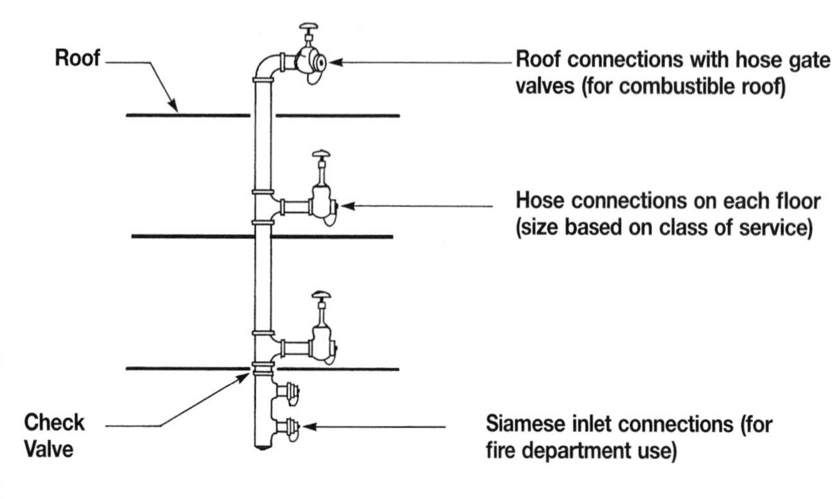

D4020 330	Dry Standpipe Risers, Class I	COST PER FLOOR		
		MAT.	INST.	TOTAL
0530	Dry standpipe risers, Class I, steel, black, sch. 40, 10′ height			
0540	4″ diameter pipe, one floor	3,725	2,375	6,100
0560	Additional floors	1,325	860	2,185
0580	6″ diameter pipe, one floor	7,400	4,000	11,400
0600	Additional floors	2,425	1,375	3,800
0620	8″ diameter pipe, one floor	11,300	4,850	16,150
0640	Additional floors	3,575	1,675	5,250

D4020 330	Dry Standpipe Risers, Class II	COST PER FLOOR		
		MAT.	INST.	TOTAL
1030	Dry standpipe risers, Class II, steel, black, sch. 40, 10′ height			
1040	2″ diameter pipe, one floor	2,250	1,100	3,350
1060	Additional floors	810	360	1,170
1080	2-1/2″ diameter pipe, one floor	2,925	1,275	4,200
1100	Additional floors	880	425	1,305

D4020 330	Dry Standpipe Risers, Class III	COST PER FLOOR		
		MAT.	INST.	TOTAL
1530	Dry standpipe risers, Class III, steel, black, sch. 40, 10′ height			
1540	4″ diameter pipe, one floor	3,775	2,325	6,100
1560	Additional floors	1,125	775	1,900
1580	6″ diameter pipe, one floor	7,475	4,000	11,475
1600	Additional floors	2,500	1,375	3,875
1620	8″ diameter pipe, one floor	11,400	4,850	16,250
1640	Additional floors	3,650	1,675	5,325

D4020 Standpipes

D4020 410	Fire Hose Equipment	COST EACH		
		MAT.	INST.	TOTAL
0100	Adapters, reducing, 1 piece, FxM, hexagon, cast brass, 2-1/2" x 1-1/2"	74.50		74.50
0200	Pin lug, 1-1/2" x 1"	53		53
0250	3" x 2-1/2"	163		163
0300	For polished chrome, add 75% mat.			
0400	Cabinets, D.S. glass in door, recessed, steel box, not equipped			
0500	Single extinguisher, steel door & frame	154	131	285
0550	Stainless steel door & frame	243	131	374
0600	Valve, 2-1/2" angle, steel door & frame	206	87.50	293.50
0650	Aluminum door & frame	249	87.50	336.50
0700	Stainless steel door & frame	335	87.50	422.50
0750	Hose rack assy, 2-1/2" x 1-1/2" valve & 100' hose, steel door & frame	385	175	560
0800	Aluminum door & frame	565	175	740
0850	Stainless steel door & frame	755	175	930
0900	Hose rack assy & extinguisher,2-1/2"x1-1/2" valve & hose,steel door & frame	320	210	530
0950	Aluminum	725	210	935
1000	Stainless steel	700	210	910
1550	Compressor, air, dry pipe system, automatic, 200 gal., 3/4 H.P.	1,300	450	1,750
1600	520 gal., 1 H.P.	1,625	450	2,075
1650	Alarm, electric pressure switch (circuit closer)	122	22.50	144.50
2500	Couplings, hose, rocker lug, cast brass, 1-1/2"	73.50		73.50
2550	2-1/2"	57.50		57.50
3000	Escutcheon plate, for angle valves, polished brass, 1-1/2"	16.90		16.90
3050	2-1/2"	25.50		25.50
3500	Fire pump, electric, w/controller, fittings, relief valve			
3550	4" pump, 30 HP, 500 GPM	16,700	3,075	19,775
3600	5" pump, 40 H.P., 1000 G.P.M.	19,000	3,500	22,500
3650	5" pump, 100 H.P., 1000 G.P.M.	26,500	3,875	30,375
3700	For jockey pump system, add	3,300	525	3,825
5000	Hose, per linear foot, synthetic jacket, lined,			
5100	300 lb. test, 1-1/2" diameter	3.54	.40	3.94
5150	2-1/2" diameter	6.60	.48	7.08
5200	500 lb. test, 1-1/2" diameter	2.74	.40	3.14
5250	2-1/2" diameter	6.35	.48	6.83
5500	Nozzle, plain stream, polished brass, 1-1/2" x 10"	67		67
5550	2-1/2" x 15" x 13/16" or 1-1/2"	118		118
5600	Heavy duty combination adjustable fog and straight stream w/handle 1-1/2"	465		465
5650	2-1/2" direct connection	525		525
6000	Rack, for 1-1/2" diameter hose 100 ft. long, steel	104	52.50	156.50
6050	Brass	159	52.50	211.50
6500	Reel, steel, for 50 ft. long 1-1/2" diameter hose	158	75	233
6550	For 75 ft. long 2-1/2" diameter hose	350	75	425
7050	Siamese, w/plugs & chains, polished brass, sidewalk, 4" x 2-1/2" x 2-1/2"	800	420	1,220
7100	6" x 2-1/2" x 2-1/2"	855	525	1,380
7200	Wall type, flush, 4" x 2-1/2" x 2-1/2"	905	210	1,115
7250	6" x 2-1/2" x 2-1/2"	1,000	228	1,228
7300	Projecting, 4" x 2-1/2" x 2-1/2"	600	210	810
7350	6" x 2-1/2" x 2-1/2"	1,000	228	1,228
7400	For chrome plate, add 15% mat.			
8000	Valves, angle, wheel handle, 300 Lb., rough brass, 1-1/2"	130	48.50	178.50
8050	2-1/2"	236	83	319
8100	Combination pressure restricting, 1-1/2"	108	48.50	156.50
8150	2-1/2"	234	83	317
8200	Pressure restricting, adjustable, satin brass, 1-1/2"	365	48.50	413.50
8250	2-1/2"	435	83	518
8300	Hydrolator, vent and drain, rough brass, 1-1/2"	133	48.50	181.50
8350	2-1/2"	133	48.50	181.50
8400	Cabinet assy, incls. adapter, rack, hose, and nozzle	990	315	1,305

D4090 Other Fire Protection Systems

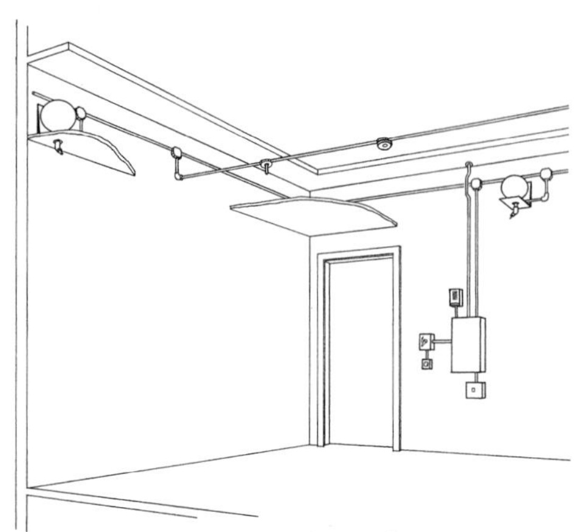

General: Automatic fire protection (suppression) systems other than water sprinklers may be desired for special environments, high risk areas, isolated locations or unusual hazards. Some typical applications would include:

Paint dip tanks
Securities vaults
Electronic data processing
Tape and data storage
Transformer rooms
Spray booths
Petroleum storage
High rack storage

Piping and wiring costs are dependent on the individual application and must be added to the component costs shown below.

All areas are assumed to be open.

D4090 910	Fire Suppression Unit Components	COST EACH		
		MAT.	INST.	TOTAL
0020	Detectors with brackets			
0040	Fixed temperature heat detector	50	85	135
0060	Rate of temperature rise detector	57.50	74.50	132
0080	Ion detector (smoke) detector	135	96	231
0200	Extinguisher agent			
0240	200 lb FM200, container	7,375	265	7,640
0280	75 lb carbon dioxide cylinder	1,475	177	1,652
0320	Dispersion nozzle			
0340	FM200 1-1/2" dispersion nozzle	210	42	252
0380	Carbon dioxide 3" x 5" dispersion nozzle	169	32.50	201.50
0420	Control station			
0440	Single zone control station with batteries	1,075	595	1,670
0470	Multizone (4) control station with batteries	3,225	1,200	4,425
0500	Electric mechanical release	1,075	184	1,259
0550	Manual pull station	91.50	98	189.50
0640	Battery standby power 10" x 10" x 17"	430	149	579
0740	Bell signalling device	149	74.50	223.50

D4090 920	FM200 Systems	COST PER C.F.		
		MAT.	INST.	TOTAL
0820	Average FM200 system, minimum			1.97
0840	Maximum			3.92

D5010 Electrical Service/Distribution

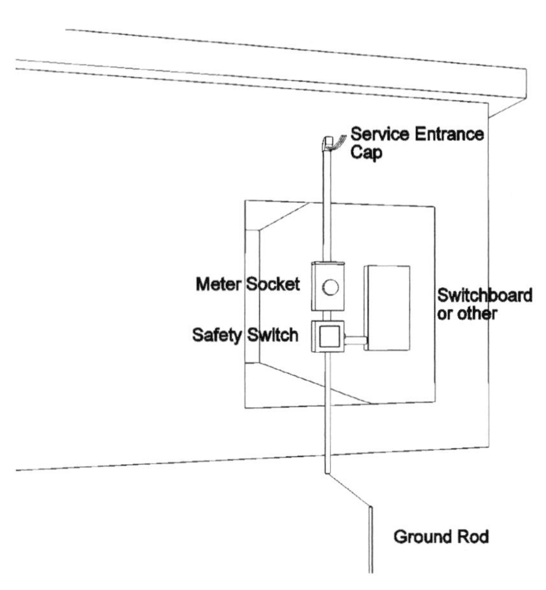

D5010 120	Overhead Electric Service, 3 Phase - 4 Wire	COST EACH		
		MAT.	INST.	TOTAL
0200	Service installation, includes breakers, metering, 20' conduit & wire			
0220	3 phase, 4 wire, 120/208 volts, 60 A	515	970	1,485
0240	100 A	690	1,100	1,790
0245	100 A w/circuit breaker	1,475	1,350	2,825
0280	200 A	1,175	1,500	2,675
0285	200 A w/circuit breaker	3,175	1,900	5,075
0320	400 A	2,375	3,025	5,400
0325	400 A w/circuit breaker	5,475	3,775	9,250
0360	600 A	4,000	4,475	8,475
0365	600 A, w/switchboard	8,350	5,675	14,025
0400	800 A	6,150	5,325	11,475
0405	800 A, w/switchboard	10,500	6,675	17,175
0440	1000 A	7,575	6,575	14,150
0445	1000 A, w/switchboard	12,800	8,075	20,875
0480	1200 A	10,000	7,500	17,500
0485	1200 A, w/groundfault switchboard	32,600	9,300	41,900
0520	1600 A	12,600	9,825	22,425
0525	1600 A, w/groundfault switchboard	37,100	11,700	48,800
0560	2000 A	16,500	11,900	28,400
0565	2000 A, w/groundfault switchboard	44,100	14,700	58,800
0610	1 phase, 3 wire, 120/240 volts, 100 A (no safety switch)	160	535	695
0615	100 A w/load center	330	1,025	1,355
0620	200 A	370	740	1,110
0625	200 A w/load center	705	1,575	2,280

D5010 Electrical Service/Distribution

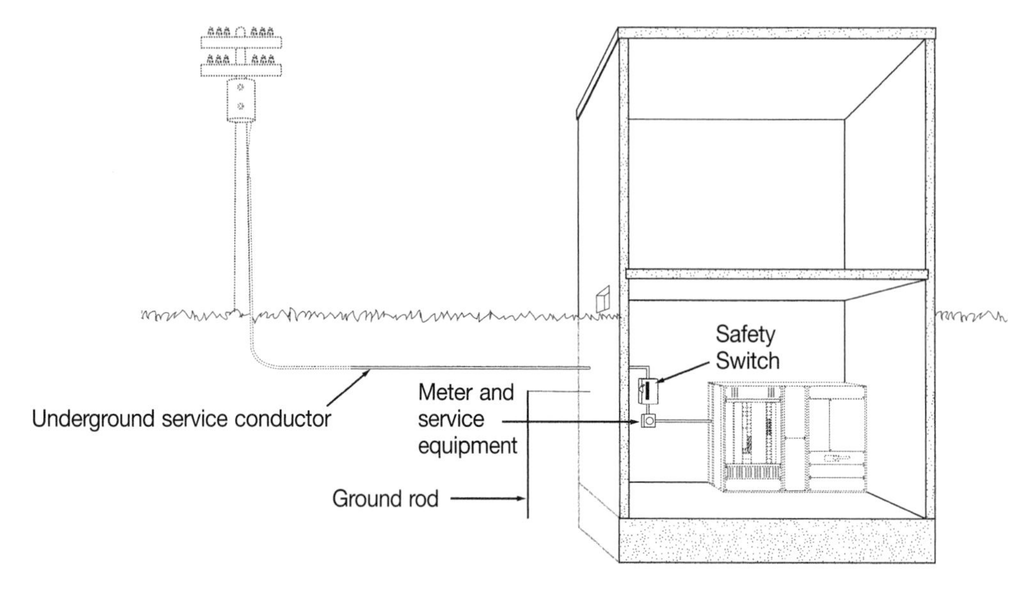

D5010 130	Underground Electric Service	COST EACH		
		MAT.	INST.	TOTAL
0950	Underground electric service including excavation, backfill and compaction			
1000	3 phase, 4 wire, 277/480 volts, 2000 A	32,000	18,900	50,900
1050	2000 A w/groundfault switchboard	58,500	21,100	79,600
1100	1600 A	25,200	15,300	40,500
1150	1600 A w/groundfault switchboard	49,700	17,200	66,900
1200	1200 A	19,500	12,100	31,600
1250	1200 A w/groundfault switchboard	42,100	13,900	56,000
1400	800 A	12,800	10,000	22,800
1450	800 A w/switchboard	17,200	11,400	28,600
1500	600 A	8,400	9,225	17,625
1550	600 A w/switchboard	12,800	10,400	23,200
1600	1 phase, 3 wire, 120/240 volts, 200 A	2,975	3,125	6,100
1650	200 A w/load center	3,325	3,950	7,275
1700	100 A	2,400	2,550	4,950
1750	100 A w/load center	2,725	3,375	6,100

D5010 Electrical Service/Distribution

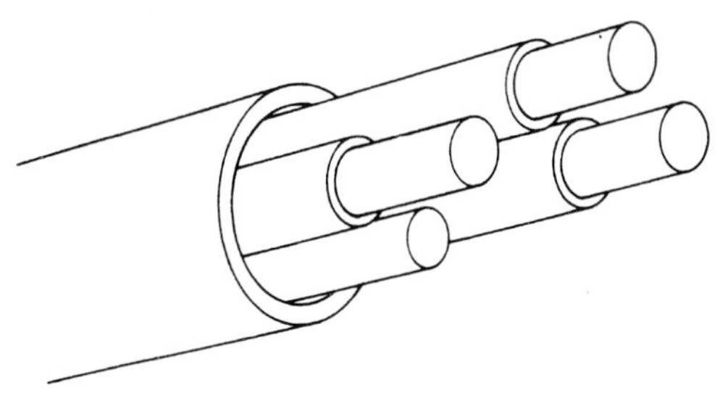

D5010 230	Feeder Installation	COST PER L.F.		
		MAT.	INST.	TOTAL
0200	Feeder installation 600 V, including RGS conduit and XHHW wire, 60 A	7.20	10.95	18.15
0240	100 A	8.85	14.20	23.05
0280	200 A	22.50	21.50	44
0320	400 A	45	43	88
0360	600 A	73	70.50	143.50
0400	800 A	94.50	83.50	178
0440	1000 A	117	109	226
0480	1200 A	146	141	287
0520	1600 A	189	167	356
0560	2000 A	234	218	452
1200	Branch installation 600 V, including EMT conduit and THW wire, 15 A	1.22	5.35	6.57
1240	20 A	1.49	5.65	7.14
1280	30 A	1.83	5.90	7.73
1320	50 A	3.57	8.20	11.77
1360	65 A	3.87	8.70	12.57
1400	85 A	6.35	10.05	16.40
1440	100 A	7.40	10.25	17.65
1480	130 A	8.70	12	20.70
1520	150 A	11.75	13.75	25.50
1560	200 A	12.75	15.70	28.45

D5010 Electrical Service/Distribution

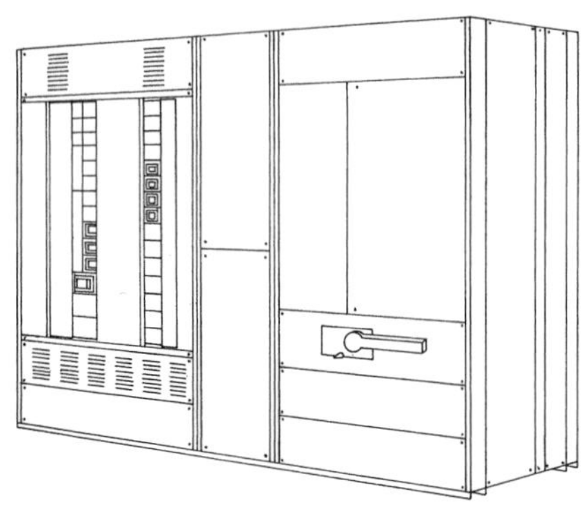

D5010 240	Switchgear	COST PER EACH		
		MAT.	INST.	TOTAL
0190	Switchgear installation, including switchboard, panels, & circuit breaker			
0200	120/208 V, 3 phase, 400 A	8,825	2,525	11,350
0240	600 A	11,600	2,850	14,450
0280	800 A	14,900	3,250	18,150
0300	1000 A	18,400	3,650	22,050
0320	1200 A	19,300	4,075	23,375
0360	1600 A	31,400	4,725	36,125
0400	2000 A	37,900	5,150	43,050
0500	277/480 V, 3 phase, 400 A	11,900	4,675	16,575
0520	600 A	16,500	5,425	21,925
0540	800 A	19,100	5,950	25,050
0560	1000 A	24,200	6,650	30,850
0580	1200 A	26,100	7,375	33,475
0600	1600 A	38,800	8,125	46,925
0620	2000 A	48,600	9,000	57,600

D5010 Electrical Service/Distribution

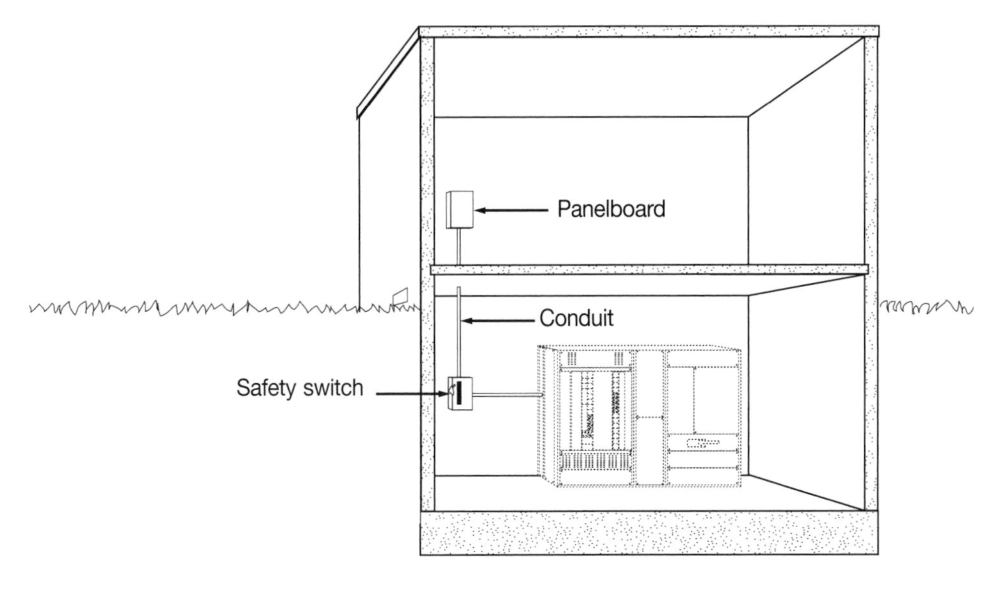

D5010 250	Panelboard	COST EACH		
		MAT.	INST.	TOTAL
0900	Panelboards, NQOD, 4 wire, 120/208 volts w/conductor & conduit			
1000	100 A, 0 stories, 0′ horizontal	2,075	1,275	3,350
1020	1 stories, 25′ horizontal	2,450	1,825	4,275
1040	5 stories, 50′ horizontal	3,200	2,900	6,100
1060	10 stories, 75′ horizontal	4,075	4,125	8,200
1080	225A, 0 stories, 0′ horizontal	3,750	1,650	5,400
2000	1 stories, 25′ horizontal	4,800	2,625	7,425
2020	5 stories, 50′ horizontal	6,875	4,550	11,425
2040	10 stories, 75′ horizontal	9,300	6,800	16,100
2060	400A, 0 stories, 0′ horizontal	5,500	2,475	7,975
2080	1 stories, 25′ horizontal	6,925	4,075	11,000
3000	5 stories, 50′ horizontal	9,775	7,200	16,975
3020	10 stories, 75′ horizontal	16,400	14,100	30,500
3040	600 A, 0 stories, 0′ horizontal	8,125	2,975	11,100
3060	1 stories, 25′ horizontal	10,500	5,575	16,075
3080	5 stories, 50′ horizontal	15,100	10,700	25,800
4000	10 stories, 75′ horizontal	26,700	21,100	47,800
4010	Panelboards, NEHB, 4 wire, 277/480 volts w/conductor, conduit, & safety switch			
4020	100 A, 0 stories, 0′ horizontal, includes safety switch	3,175	1,750	4,925
4040	1 stories, 25′ horizontal	3,550	2,275	5,825
4060	5 stories, 50′ horizontal	4,300	3,350	7,650
4080	10 stories, 75′ horizontal	5,175	4,600	9,775
5000	225 A, 0 stories, 0′ horizontal	4,700	2,100	6,800
5020	1 stories, 25′ horizontal	5,750	3,100	8,850
5040	5 stories, 50′ horizontal	7,825	5,025	12,850
5060	10 stories, 75′ horizontal	10,200	7,275	17,475
5080	400 A, 0 stories, 0′ horizontal	7,425	3,250	10,675
6000	1 stories, 25′ horizontal	8,850	4,850	13,700
6020	5 stories, 50′ horizontal	11,700	8,000	19,700
6040	10 stories, 75′ horizontal	18,300	14,900	33,200
6060	600 A, 0 stories, 0′ horizontal	10,600	4,125	14,725
6080	1 stories, 25′ horizontal	12,900	6,725	19,625
7000	5 stories, 50′ horizontal	17,500	11,900	29,400
7020	10 stories, 75′ horizontal	29,200	22,200	51,400

D5020 Lighting and Branch Wiring

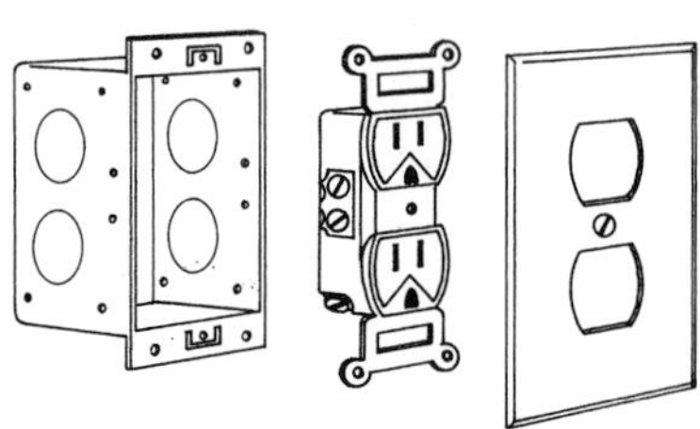

Duplex Wall Receptacle

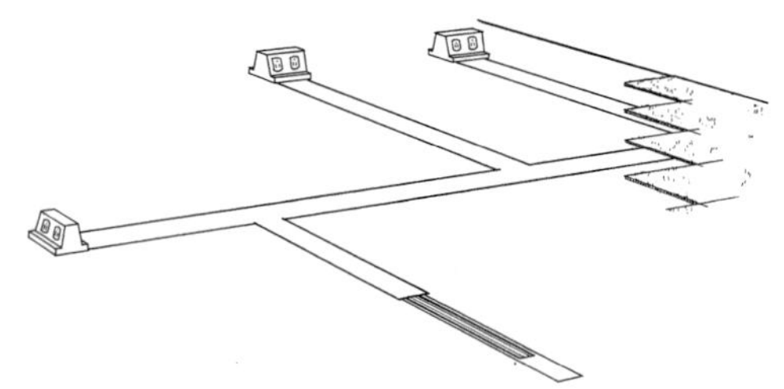

Undercarpet Receptacle System

D5020 110	Receptacle (by Wattage)	COST PER S.F.		
		MAT.	INST.	TOTAL
0190	Receptacles include plate, box, conduit, wire & transformer when required			
0200	2.5 per 1000 S.F., .3 watts per S.F.	.36	1.33	1.69
0240	With transformer	.46	1.40	1.86
0280	4 per 1000 S.F., .5 watts per S.F.	.42	1.55	1.97
0320	With transformer	.57	1.66	2.23
0360	5 per 1000 S.F., .6 watts per S.F.	.14	.62	.76
0400	With transformer	.69	1.96	2.65
0440	8 per 1000 S.F., .9 watts per S.F.	.53	2.03	2.56
0480	With transformer	.84	2.24	3.08
0520	10 per 1000 S.F., 1.2 watts per S.F.	.57	2.19	2.76
0560	With transformer	.96	2.45	3.41
0600	16.5 per 1000 S.F., 2.0 watts per S.F.	.67	2.75	3.42
0640	With transformer	1.31	3.19	4.50
0680	20 per 1000 S.F., 2.4 watts per S.F.	.70	3	3.70
0720	With transformer	1.47	3.53	5

D5020 115	Receptacles, Floor	COST PER S.F.		
		MAT.	INST.	TOTAL
0200	Receptacle systems, underfloor duct, 5′ on center, low density	7.45	2.93	10.38
0240	High density	8.05	3.77	11.82
0280	7′ on center, low density	5.90	2.51	8.41
0320	High density	6.50	3.35	9.85
0400	Poke thru fittings, low density	1.33	1.42	2.75
0440	High density	2.65	2.83	5.48
0520	Telepoles, using Romex, low density	1.57	.88	2.45
0560	High density	3.13	1.77	4.90
0600	Using EMT, low density	1.65	1.17	2.82
0640	High density	3.31	2.34	5.65
0720	Conduit system with floor boxes, low density	1.45	1.01	2.46
0760	High density	2.92	2.01	4.93
0840	Undercarpet power system, 3 conductor with 5 conductor feeder, low density	1.48	.34	1.82
0880	High density	2.98	.72	3.70

D5020 Lighting and Branch Wiring

Description: Table D5020 130 includes the cost for switch, plate, box, conduit in slab or EMT exposed and copper wire. Add 20% for exposed conduit.

No power required for switches.

Federal energy guidelines recommend the maximum lighting area controlled per switch shall not exceed 1000 S.F. and that areas over 500 S.F. shall be so controlled that total illumination can be reduced by at least 50%.

D5020 130	Wall Switch by Sq. Ft.	COST PER S.F.		
		MAT.	INST.	TOTAL
0200	Wall switches, 1.0 per 1000 S.F.	.05	.21	.26
0240	1.2 per 1000 S.F.	.06	.25	.31
0280	2.0 per 1000 S.F.	.08	.34	.42
0320	2.5 per 1000 S.F.	.09	.43	.52
0360	5.0 per 1000 S.F.	.22	.92	1.14
0400	10.0 per 1000 S.F.	.46	1.86	2.32

D5020 135	Miscellaneous Power	COST PER S.F.		
		MAT.	INST.	TOTAL
0200	Miscellaneous power, to .5 watts	.03	.11	.14
0240	.8 watts	.04	.14	.18
0280	1 watt	.06	.20	.26
0320	1.2 watts	.07	.24	.31
0360	1.5 watts	.08	.28	.36
0400	1.8 watts	.10	.32	.42
0440	2 watts	.11	.39	.50
0480	2.5 watts	.13	.47	.60
0520	3 watts	.17	.56	.73

D5020 140	Central A. C. Power (by Wattage)	COST PER S.F.		
		MAT.	INST.	TOTAL
0200	Central air conditioning power, 1 watt	.07	.23	.30
0220	2 watts	.08	.26	.34
0240	3 watts	.15	.39	.54
0280	4 watts	.16	.40	.56
0320	6 watts	.24	.53	.77
0360	8 watts	.34	.58	.92
0400	10 watts	.46	.68	1.14

D5020 Lighting and Branch Wiring

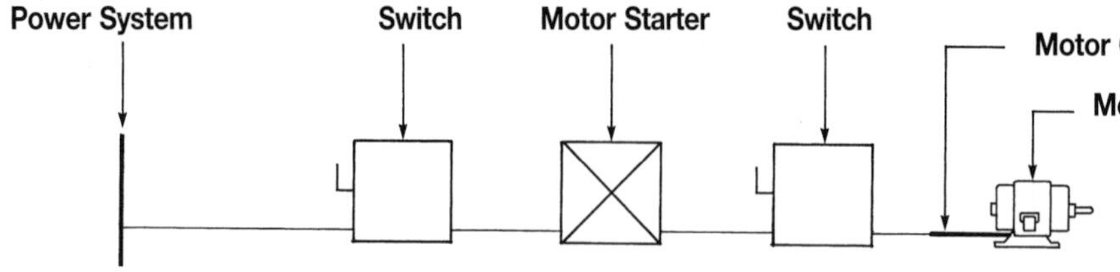

Motor Installation

D5020 145	Motor Installation	COST EACH		
		MAT.	INST.	TOTAL
0200	Motor installation, single phase, 115V, 1/3 HP motor size	470	905	1,375
0240	1 HP motor size	490	905	1,395
0280	2 HP motor size	540	960	1,500
0320	3 HP motor size	620	980	1,600
0360	230V, 1 HP motor size	470	920	1,390
0400	2 HP motor size	510	920	1,430
0440	3 HP motor size	575	985	1,560
0520	Three phase, 200V, 1-1/2 HP motor size	550	1,000	1,550
0560	3 HP motor size	650	1,100	1,750
0600	5 HP motor size	655	1,225	1,880
0640	7-1/2 HP motor size	685	1,250	1,935
0680	10 HP motor size	1,125	1,550	2,675
0720	15 HP motor size	1,550	1,725	3,275
0760	20 HP motor size	1,950	2,000	3,950
0800	25 HP motor size	2,025	2,025	4,050
0840	30 HP motor size	3,125	2,375	5,500
0880	40 HP motor size	3,875	2,800	6,675
0920	50 HP motor size	6,825	3,275	10,100
0960	60 HP motor size	7,100	3,450	10,550
1000	75 HP motor size	8,775	3,950	12,725
1040	100 HP motor size	25,100	4,650	29,750
1080	125 HP motor size	25,600	5,125	30,725
1120	150 HP motor size	29,200	6,000	35,200
1160	200 HP motor size	29,800	7,100	36,900
1240	230V, 1-1/2 HP motor size	525	995	1,520
1280	3 HP motor size	625	1,075	1,700
1320	5 HP motor size	630	1,200	1,830
1360	7-1/2 HP motor size	630	1,200	1,830
1400	10 HP motor size	985	1,475	2,460
1440	15 HP motor size	1,200	1,625	2,825
1480	20 HP motor size	1,825	1,950	3,775
1520	25 HP motor size	1,950	2,000	3,950
1560	30 HP motor size	1,975	2,025	4,000
1600	40 HP motor size	3,725	2,725	6,450
1640	50 HP motor size	3,925	2,875	6,800
1680	60 HP motor size	6,850	3,275	10,125
1720	75 HP motor size	7,975	3,725	11,700
1760	100 HP motor size	8,875	4,125	13,000
1800	125 HP motor size	26,000	4,800	30,800
1840	150 HP motor size	26,600	5,450	32,050
1880	200 HP motor size	27,500	6,075	33,575
1960	460V, 2 HP motor size	655	1,000	1,655
2000	5 HP motor size	755	1,100	1,855
2040	10 HP motor size	725	1,200	1,925
2080	15 HP motor size	995	1,375	2,370
2120	20 HP motor size	1,050	1,475	2,525

D5020 Lighting and Branch Wiring

D5020 145	Motor Installation	COST EACH		
		MAT.	INST.	TOTAL
2160	25 HP motor size	1,200	1,550	2,750
2200	30 HP motor size	1,550	1,675	3,225
2240	40 HP motor size	1,900	1,800	3,700
2280	50 HP motor size	2,125	2,000	4,125
2320	60 HP motor size	3,250	2,350	5,600
2360	75 HP motor size	3,750	2,600	6,350
2400	100 HP motor size	4,150	2,900	7,050
2440	125 HP motor size	7,075	3,300	10,375
2480	150 HP motor size	8,725	3,675	12,400
2520	200 HP motor size	9,675	4,150	13,825
2600	575V, 2 HP motor size	655	1,000	1,655
2640	5 HP motor size	755	1,100	1,855
2680	10 HP motor size	725	1,200	1,925
2720	20 HP motor size	995	1,375	2,370
2760	25 HP motor size	1,050	1,475	2,525
2800	30 HP motor size	1,550	1,675	3,225
2840	50 HP motor size	1,650	1,750	3,400
2880	60 HP motor size	3,225	2,325	5,550
2920	75 HP motor size	3,250	2,350	5,600
2960	100 HP motor size	3,750	2,600	6,350
3000	125 HP motor size	6,825	3,225	10,050
3040	150 HP motor size	7,075	3,300	10,375
3080	200 HP motor size	8,775	3,725	12,500

D5020 155	Motor Feeder	COST PER L.F.		
		MAT.	INST.	TOTAL
0200	Motor feeder systems, single phase, feed up to 115V 1HP or 230V 2 HP	2.08	7.05	9.13
0240	115V 2HP, 230V 3HP	2.18	7.15	9.33
0280	115V 3HP	2.48	7.45	9.93
0360	Three phase, feed to 200V 3HP, 230V 5HP, 460V 10HP, 575V 10HP	2.18	7.55	9.73
0440	200V 5HP, 230V 7.5HP, 460V 15HP, 575V 20HP	2.34	7.75	10.09
0520	200V 10HP, 230V 10HP, 460V 30HP, 575V 30HP	2.79	8.20	10.99
0600	200V 15HP, 230V 15HP, 460V 40HP, 575V 50HP	3.86	9.35	13.21
0680	200V 20HP, 230V 25HP, 460V 50HP, 575V 60HP	5.10	11.85	16.95
0760	200V 25HP, 230V 30HP, 460V 60HP, 575V 75HP	6.05	12.05	18.10
0840	200V 30HP	6.70	12.45	19.15
0920	230V 40HP, 460V 75HP, 575V 100HP	9.05	13.60	22.65
1000	200V 40HP	10.50	14.55	25.05
1080	230V 50HP, 460V 100HP, 575V 125HP	12.45	16.05	28.50
1160	200V 50HP, 230V 60HP, 460V 125HP, 575V 150HP	14.45	17.05	31.50
1240	200V 60HP, 460V 150HP	18.45	20	38.45
1320	230V 75HP, 575V 200HP	21	21	42
1400	200V 75HP	23.50	21.50	45
1480	230V 100HP, 460V 200HP	29.50	25	54.50
1560	200V 100HP	41	31	72
1640	230V 125HP	41	31	72
1720	200V 125HP, 230V 150HP	46.50	39	85.50
1800	200V 150HP	58	42	100
1880	200V 200HP	74.50	44	118.50
1960	230V 200HP	68.50	46	114.50

D5020 Lighting and Branch Wiring

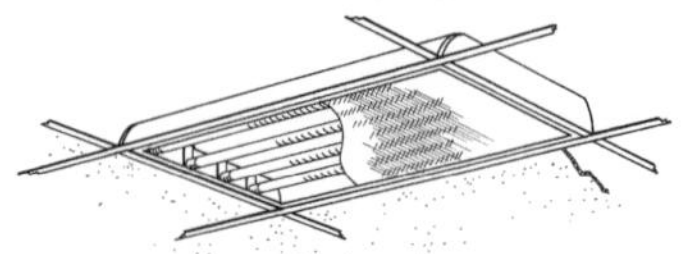

Fluorescent Fixture

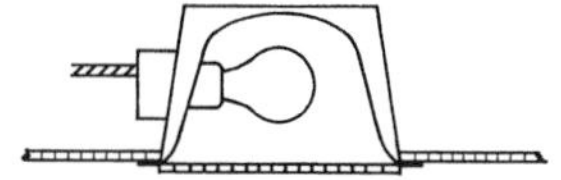

Incandescent Fixture

D5020 210	Fluorescent Fixtures (by Wattage)	COST PER S.F.		
		MAT.	INST.	TOTAL
0190	Fluorescent fixtures recess mounted in ceiling			
0195	T12, standard 40 watt lamps			
0200	1 watt per S.F., 20 FC, 5 fixtures @40 watts per 1000 S.F.	.71	1.75	2.46
0240	2 watt per S.F., 40 FC, 10 fixtures @40 watt per 1000 S.F.	1.42	3.43	4.85
0280	3 watt per S.F., 60 FC, 15 fixtures @40 watt per 1000 S.F	2.13	5.20	7.33
0320	4 watt per S.F., 80 FC, 20 fixtures @40 watt per 1000 S.F.	2.84	6.90	9.74
0400	5 watt per S.F., 100 FC, 25 fixtures @40 watt per 1000 S.F.	3.55	8.65	12.20
0402				
0450	T8, energy saver 32 watt lamps			
0500	0.8 watt per S.F., 20 FC, 5 fixtures @32 watt per 1000 S.F.	.78	1.75	2.53
0520	1.6 watt per S.F., 40 FC, 10 fixtures @32 watt per 1000 S.F.	1.55	3.43	4.98
0540	2.4 watt per S.F., 60 FC, 15 fixtures @ 32 watt per 1000 S.F	2.32	5.20	7.52
0560	3.2 watt per S.F., 80 FC, 20 fixtures @32 watt per 1000 S.F.	3.10	6.90	10
0580	4 watt per S.F., 100 FC, 25 fixtures @32 watt per 1000 S.F.	3.88	8.65	12.53

D5020 216	Incandescent Fixture (by Wattage)	COST PER S.F.		
		MAT.	INST.	TOTAL
0190	Incandescent fixture recess mounted, type A			
0200	1 watt per S.F., 8 FC, 6 fixtures per 1000 S.F.	.97	1.40	2.37
0240	2 watt per S.F., 16 FC, 12 fixtures per 1000 S.F.	1.92	2.82	4.74
0280	3 watt per S.F., 24 FC, 18 fixtures, per 1000 S.F.	2.87	4.16	7.03
0320	4 watt per S.F., 32 FC, 24 fixtures per 1000 S.F.	3.84	5.60	9.44
0400	5 watt per S.F., 40 FC, 30 fixtures per 1000 S.F.	4.80	7	11.80

D5020 226	H.I.D. Fixture, High Bay, 16′ (by Wattage)	COST PER S.F.		
		MAT.	INST.	TOTAL
0190	High intensity discharge fixture, 16′ above work plane			
0240	1 watt/S.F., type E, 42 FC, 1 fixture/1000 S.F.	1.02	1.47	2.49
0280	Type G, 52 FC, 1 fixture/1000 S.F.	1.02	1.47	2.49
0320	Type C, 54 FC, 2 fixture/1000 S.F.	1.20	1.64	2.84
0440	2 watt/S.F., type E, 84 FC, 2 fixture/1000 S.F.	2.05	2.99	5.04
0480	Type G, 105 FC, 2 fixture/1000 S.F.	2.05	2.99	5.04
0520	Type C, 108 FC, 4 fixture/1000 S.F.	2.40	3.29	5.69
0640	3 watt/S.F., type E, 126 FC, 3 fixture/1000 S.F.	3.06	4.46	7.52
0680	Type G, 157 FC, 3 fixture/1000 S.F.	3.06	4.46	7.52
0720	Type C, 162 FC, 6 fixture/1000 S.F.	3.60	4.92	8.52
0840	4 watt/S.F., type E, 168 FC, 4 fixture/1000 S.F.	4.10	6	10.10
0880	Type G, 210 FC, 4 fixture/1000 S.F.	4.10	6	10.10
0920	Type C, 243 FC, 9 fixture/1000 S.F.	5.25	6.85	12.10
1040	5 watt/S.F., type E, 210 FC, 5 fixture/1000 S.F.	5.10	7.45	12.55
1080	Type G, 262 FC, 5 fixture/1000 S.F.	5.10	7.45	12.55
1120	Type C, 297 FC, 11 fixture/1000 S.F.	6.45	8.50	14.95

D5020 Lighting and Branch Wiring

D5020 234	H.I.D. Fixture, High Bay, 30′ (by Wattage)	COST PER S.F.		
		MAT.	INST.	TOTAL
0190	High intensity discharge fixture, 30′ above work plane			
0240	1 watt/S.F., type E, 37 FC, 1 fixture/1000 S.F.	1.13	1.87	3
0280	Type G, 45 FC., 1 fixture/1000 S.F.	1.13	1.87	3
0320	Type F, 50 FC, 1 fixture/1000 S.F.	.96	1.46	2.42
0440	2 watt/S.F., type E, 74 FC, 2 fixtures/1000 S.F.	2.26	3.71	5.97
0480	Type G, 92 FC, 2 fixtures/1000 S.F.	2.26	3.71	5.97
0520	Type F, 100 FC, 2 fixtures/1000 S.F.	1.93	2.97	4.90
0640	3 watt/S.F., type E, 110 FC, 3 fixtures/1000 S.F.	3.41	5.65	9.06
0680	Type G, 138 FC, 3 fixtures/1000 S.F.	3.41	5.65	9.06
0720	Type F, 150 FC, 3 fixtures/1000 S.F.	2.90	4.41	7.31
0840	4 watt/S.F., type E, 148 FC, 4 fixtures/1000 S.F.	4.54	7.50	12.04
0880	Type G, 185 FC, 4 fixtures/1000 S.F.	4.54	7.50	12.04
0920	Type F, 200 FC, 4 fixtures/1000 S.F.	3.87	5.95	9.82
1040	5 watt/S.F., type E, 185 FC, 5 fixtures/1000 S.F.	5.65	9.35	15
1080	Type G, 230 FC, 5 fixtures/1000 S.F.	5.65	9.35	15
1120	Type F, 250 FC, 5 fixtures/1000 S.F.	4.83	7.40	12.23

D5020 Lighting and Branch Wiring

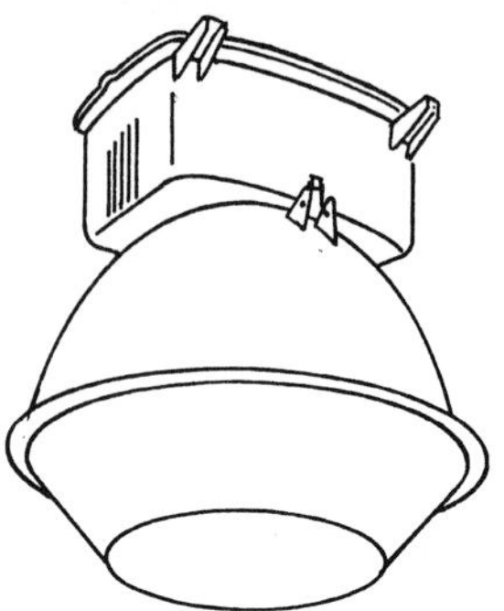

LOW BAY FIXTURES

J. Metal halide 250 watt

K. High pressure sodium 150 watt

D5020 238	H.I.D. Fixture, Low Bay, 8'-10' (by Wattage)	COST PER S.F.		
		MAT.	INST.	TOTAL
0190	High intensity discharge fixture, 8'-10' above work plane			
0240	1 watt/S.F., type J, 30 FC, 4 fixtures/1000 S.F.	2.33	3.04	5.37
0280	Type K, 29 FC, 5 fixtures/1000 S.F.	2.39	2.73	5.12
0400	2 watt/S.F., type J, 52 FC, 7 fixtures/1000 S.F.	4.16	5.55	9.71
0440	Type K, 63 FC, 11 fixtures/1000 S.F.	5.15	5.75	10.90
0560	3 watt/S.F., type J, 81 FC, 11 fixtures/1000 S.F.	6.40	8.40	14.80
0600	Type K, 92 FC, 16 fixtures/1000 S.F.	7.55	8.45	16
0720	4 watt/S.F., type J, 103 FC, 14 fixtures/1000 S.F.	8.30	11.05	19.35
0760	Type K, 127 FC, 22 fixtures/1000 S.F.	10.35	11.45	21.80
0880	5 watt/S.F., type J, 133 FC, 18 fixtures/1000 S.F.	10.60	13.95	24.55
0920	Type K, 155 FC, 27 fixtures/1000 S.F.	12.75	14.20	26.95

D5020 242	H.I.D. Fixture, Low Bay, 16' (by Wattage)	COST PER S.F.		
		MAT.	INST.	TOTAL
0190	High intensity discharge fixture, mounted 16' above work plane			
0240	1 watt/S.F., type J, 28 FC, 4 fixt./1000 S.F.	2.45	3.44	5.89
0280	Type K, 27 FC, 5 fixt./1000 S.F.	2.73	3.85	6.58
0400	2 watt/S.F., type J, 48 FC, 7 fixt/1000 S.F.	4.49	6.65	11.14
0440	Type K, 58 FC, 11 fixt/1000 S.F.	5.80	7.90	13.70
0560	3 watt/S.F., type J, 75 FC, 11 fixt/1000 S.F.	6.95	10.10	17.05
0600	Type K, 85 FC, 16 fixt/1000 S.F.	8.55	11.75	20.30
0720	4 watt/S.F., type J, 95 FC, 14 fixt/1000 S.F.	8.95	13.30	22.25
0760	Type K, 117 FC, 22 fixt/1000 S.F.	11.65	15.75	27.40
0880	5 watt/S.F., type J, 122 FC, 18 fixt/1000 S.F.	11.45	16.75	28.20
0920	Type K, 143 FC, 27 fixt/1000 S.F.	14.40	19.60	34

D5030 Communications and Security

D5030 910	Communication & Alarm Systems	COST EACH		
		MAT.	INST.	TOTAL
0200	Communication & alarm systems, includes outlets, boxes, conduit & wire			
0210	Sound system, 6 outlets	5,800	7,525	13,325
0220	12 outlets	8,250	12,000	20,250
0240	30 outlets	14,500	22,800	37,300
0280	100 outlets	43,300	76,000	119,300
0320	Fire detection systems, non-addressable, 12 detectors	3,350	6,250	9,600
0360	25 detectors	5,950	10,600	16,550
0400	50 detectors	11,700	20,900	32,600
0440	100 detectors	22,000	38,100	60,100
0450	Addressable type, 12 detectors	5,025	6,300	11,325
0452	25 detectors	9,150	10,700	19,850
0456	100 detectors	34,100	38,400	72,500
0480	Intercom systems, 6 stations	4,200	5,150	9,350
0560	25 stations	13,300	19,700	33,000
0640	100 stations	51,500	72,500	124,000
0680	Master clock systems, 6 rooms	4,500	8,400	12,900
0720	12 rooms	7,200	14,300	21,500
0760	20 rooms	10,100	20,300	30,400
0800	30 rooms	16,700	37,400	54,100
0840	50 rooms	27,200	63,000	90,200
0920	Master TV antenna systems, 6 outlets	2,300	5,325	7,625
0960	12 outlets	4,250	9,900	14,150
1000	30 outlets	14,300	22,900	37,200
1040	100 outlets	52,500	74,500	127,000

D5090 Other Electrical Systems

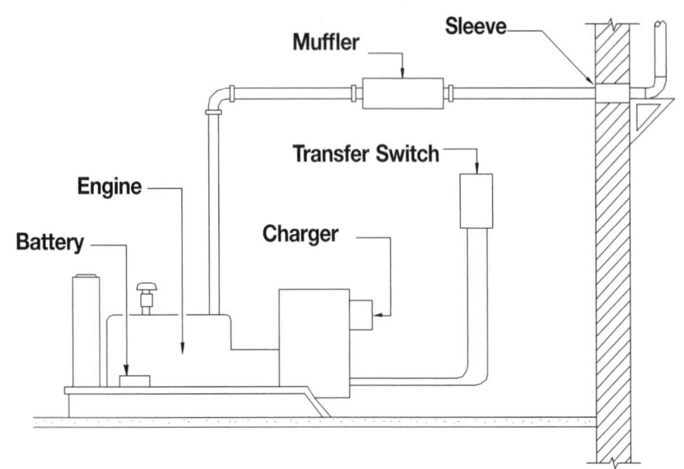

Description: System below tabulates the installed cost for generators by kW. Included in costs are battery, charger, muffler, and transfer switch.

No conduit, wire, or terminations included.

D5090 210	Generators (by kW)	COST PER kW		
		MAT.	INST.	TOTAL
0190	Generator sets, include battery, charger, muffler & transfer switch			
0200	Gas/gasoline operated, 3 phase, 4 wire, 277/480 volt, 7.5 kW	1,275	260	1,535
0240	11.5 kW	1,175	198	1,373
0280	20 kW	800	127	927
0320	35 kW	545	84	629
0360	80 kW	390	102	492
0400	100 kW	340	99	439
0440	125 kW	560	93	653
0480	185 kW	500	70.50	570.50
0560	Diesel engine with fuel tank, 30 kW	410	97.50	507.50
0600	50 kW	445	77	522
0760	150 kW	315	41.50	356.50
0840	200 kW	250	33.50	283.50
0880	250 kW	214	28	242
0960	350 kW	200	23	223
1040	500 kW	217	17.95	234.95

E1010 Commercial Equipment

E1010 110	Security/Vault, EACH	COST EACH		
		MAT.	INST.	TOTAL
0100	Bank equipment, drive up window, drawer & mike, no glazing, economy	8,300	1,150	9,450
0110	Deluxe	10,600	2,300	12,900
0120	Night depository, economy	8,725	1,150	9,875
0130	Deluxe	12,500	2,300	14,800
0140	Pneumatic tube systems, 2 station, standard	29,100	2,650	31,750
0150	Teller, automated, 24 hour, single unit	49,100	2,650	51,750
0160	Teller window, bullet proof glazing, 44" x 60"	4,975	840	5,815
0170	Pass through, painted steel, 72" x 40"	4,875	1,450	6,325
0300	Safe, office type, 1 hr. rating, 34" x 20" x 20"	2,400		2,400
0310	4 hr. rating, 62" x 33" x 20"	10,700		10,700
0320	Data storage, 4 hr. rating, 63" x 44" x 16"	15,900		15,900
0330	Jewelers, 63" x 44" x 16"	15,700		15,700
0340	Money, "B" label, 9" x 14" x 14"	625		625
0350	Tool and torch resistive, 24" x 24" x 20"	9,225	201	9,426
0500	Security gates-scissors type, painted steel, single, 6' high, 5-1/2' wide	257	288	545
0510	Double gate, 7-1/2' high, 14' wide	685	575	1,260

E1010 510	Mercantile Equipment, EACH	COST EACH		
		MAT.	INST.	TOTAL
0015	Barber equipment, chair, hydraulic, economy	660	21.50	681.50
0020	Deluxe	4,225	32	4,257
0100	Checkout counter, single belt	3,950	80	4,030
0110	Double belt, power take-away	5,225	89	5,314
0200	Display cases, freestanding, glass and aluminum, 3'-6" x 3' x 1'-0" deep	1,525	128	1,653
0220	Wall mounted, glass and aluminum, 3' x 4' x 1'-4" deep	2,700	205	2,905
0320	Frozen food, chest type, 12 ft. long	8,550	330	8,880

E1010 610	Laundry/Dry Cleaning, EACH	COST EACH		
		MAT.	INST.	TOTAL
0100	Laundry equipment, dryers, gas fired, residential, 16 lb. capacity	790	196	986
0110	Commercial, 30 lb. capacity, single	3,800	196	3,996
0120	Dry cleaners, electric, 20 lb. capacity	37,900	3,700	41,600
0130	30 lb. capacity	60,000	4,925	64,925
0140	Ironers, commercial, 120" with canopy, 8 roll	208,000	10,500	218,500
0150	Institutional, 110", single roll	37,600	2,975	40,575
0160	Washers, residential, 4 cycle	1,075	196	1,271
0170	Commercial, coin operated, deluxe	4,075	196	4,271

E1020 Institutional Equipment

E1020 110	Ecclesiastical Equipment, EACH	COST EACH		
		MAT.	INST.	TOTAL
0090	Church equipment, altar, wood, custom, plain	2,875	365	3,240
0100	Granite, custom, deluxe	40,500	4,850	45,350
0110	Baptistry, fiberglass, economy	6,125	1,200	7,325
0120	Bells & carillons, keyboard operation	20,500	7,100	27,600
0130	Confessional, wood, single, economy	3,550	855	4,405
0140	Double, deluxe	20,800	2,550	23,350
0150	Steeples, translucent fiberglass, 30″ square, 15′ high	10,500	1,425	11,925

E1020 130	Ecclesiastical Equipment, L.F.	COST PER L.F.		
		MAT.	INST.	TOTAL
0100	Arch. equip., church equip. pews, bench type, hardwood, economy	111	25.50	136.50
0110	Deluxe	194	34	228

E1020 210	Library Equipment, EACH	COST EACH		
		MAT.	INST.	TOTAL
0110	Library equipment, carrels, metal, economy	300	102	402
0120	Hardwood, deluxe	1,775	128	1,903

E1020 230	Library Equipment, L.F.	COST PER L.F.		
		MAT.	INST.	TOTAL
0100	Library equipment, book shelf, metal, single face, 90″ high x 10″ shelf	129	42.50	171.50
0110	Double face, 90″ high x 10″ shelf	435	92.50	527.50
0120	Charging desk, built-in, with counter, plastic laminate	520	73	593

E1020 310	Theater and Stage Equipment, EACH	COST EACH		
		MAT.	INST.	TOTAL
0200	Movie equipment, changeover, economy	560		560
0210	Film transport, incl. platters and autowind, economy	6,100		6,100
0220	Lamphouses, incl. rectifiers, xenon, 1000W	8,075	298	8,373
0230	4000W	12,900	395	13,295
0240	Projector mechanisms, 35 mm, economy	13,400		13,400
0250	Deluxe	18,500		18,500
0260	Sound systems, incl. amplifier, single, economy	4,050	660	4,710
0270	Dual, Dolby/super sound	20,700	1,500	22,200
0280	Projection screens, wall hung, manual operation, 50 S.F., economy	355	103	458
0290	Electric operation, 100 S.F., deluxe	3,075	510	3,585

E1020 320	Theater and Stage Equipment, S.F.	COST PER S.F.		
		MAT.	INST.	TOTAL
0090	Movie equipment, projection screens, rigid in wall, acrylic, 1/4″ thick	51.50	5.15	56.65
0100	1/2″ thick	59	7.75	66.75
0110	Stage equipment, curtains, velour, medium weight	9.60	1.71	11.31
0120	Silica based yarn, fireproof	18.05	20.50	38.55
0130	Stages, portable with steps, folding legs, 8″ high	49		49
0140	Telescoping platforms, aluminum, deluxe	59.50	26.50	86

E1020 330	Theater and Stage Equipment, L.F.	COST PER L.F.		
		MAT.	INST.	TOTAL
0100	Stage equipment, curtain track, heavy duty	74.50	57	131.50
0110	Lights, border, quartz, colored	205	30	235

E1020 Institutional Equipment

E1020 610	Detention Equipment, EACH	COST PER EACH		
		MAT.	INST.	TOTAL
0110	Detention equipment, cell front rolling door, 7/8" bars, 5' x 7' high	5,975	1,225	7,200
0120	Cells, prefab., including front, 5' x 7' x 7' deep	10,600	1,625	12,225
0130	Doors and frames, 3' x 7', single plate	5,525	605	6,130
0140	Double plate	6,725	605	7,330
0150	Toilet apparatus, incl wash basin	3,825	795	4,620
0160	Visitor cubicle, vision panel, no intercom	3,725	1,225	4,950

E1020 710	Laboratory Equipment, EACH	COST PER EACH		
		MAT.	INST.	TOTAL
0110	Laboratory equipment, glassware washer, distilled water, economy	7,575	410	7,985
0120	Deluxe	16,400	740	17,140
0140	Radio isotope	19,500		19,500

E1020 720	Laboratory Equipment, S.F.	COST PER S.F.		
		MAT.	INST.	TOTAL
0100	Arch. equip., lab equip., counter tops, acid proof, economy	54	12.50	66.50
0110	Stainless steel	219	12.50	231.50

E1020 730	Laboratory Equipment, L.F.	COST PER L.F.		
		MAT.	INST.	TOTAL
0110	Laboratory equipment, cabinets, wall, open	242	51	293
0120	Base, drawer units	635	57	692
0130	Fume hoods, not incl. HVAC, economy	625	190	815
0140	Deluxe incl. fixtures	1,200	425	1,625

E1020 810	Medical Equipment, EACH	COST EACH		
		MAT.	INST.	TOTAL
0100	Dental equipment, central suction system, economy	1,175	490	1,665
0110	Compressor-air, deluxe	9,550	1,050	10,600
0120	Chair, hydraulic, economy	2,650	1,050	3,700
0130	Deluxe	4,700	2,100	6,800
0140	Drill console with accessories, economy	3,125	330	3,455
0150	Deluxe	5,625	330	5,955
0160	X-ray unit, portable	2,975	131	3,106
0170	Panoramic unit	18,800	875	19,675
0300	Medical equipment, autopsy table, standard	11,200	590	11,790
0310	Deluxe	18,500	980	19,480
0320	Incubators, economy	3,675		3,675
0330	Deluxe	15,500		15,500
0700	Station, scrub-surgical, single, economy	6,075	196	6,271
0710	Dietary, medium, with ice	23,800		23,800
0720	Sterilizers, general purpose, single door, 20" x 20" x 28"	14,000		14,000
0730	Floor loading, double door, 28" x 67" x 52"	253,000		253,000
0740	Surgery tables, standard	16,400	820	17,220
0750	Deluxe	28,200	1,150	29,350
0770	Tables, standard, with base cabinets, economy	1,175	340	1,515
0780	Deluxe	6,000	510	6,510
0790	X-ray, mobile, economy	19,400		19,400
0800	Stationary, deluxe	307,500		307,500

E1030 Vehicular Equipment

E1030 110	Vehicular Service Equipment, EACH	COST EACH		
		MAT.	INST.	TOTAL
0110	Automotive equipment, compressors, electric, 1-1/2 H.P., std. controls	540	610	1,150
0120	5 H.P., dual controls	3,500	915	4,415
0130	Hoists, single post, 4 ton capacity, swivel arms	7,450	2,275	9,725
0140	Dual post, 12 ton capacity, adjustable frame	12,800	480	13,280
0150	Lube equipment, 3 reel type, with pumps	11,400	1,825	13,225
0160	Product dispenser, 6 nozzles, w/vapor recovery, not incl. piping, installed	28,900		28,900
0800	Scales, dial type, built in floor, 5 ton capacity, 8′ x 6′ platform	10,100	3,075	13,175
0810	10 ton capacity, 9′ x 7′ platform	9,825	4,400	14,225
0820	Truck (including weigh bridge), 20 ton capacity, 24′ x 10′	14,000	5,125	19,125

E1030 210	Parking Control Equipment, EACH	COST EACH		
		MAT.	INST.	TOTAL
0110	Parking equipment, automatic gates, 8 ft. arm, one way	3,525	1,075	4,600
0120	Traffic detectors, single treadle	2,200	495	2,695
0130	Booth for attendant, economy	7,725		7,725
0140	Deluxe	30,500		30,500
0150	Ticket printer/dispenser, rate computing	9,450	850	10,300
0160	Key station on pedestal	715	290	1,005

E1030 310	Loading Dock Equipment, EACH	COST EACH		
		MAT.	INST.	TOTAL
0110	Dock bumpers, rubber blocks, 4-1/2″ thick, 10″ high, 14″ long	63.50	19.70	83.20
0120	6″ thick, 20″ high, 11″ long	144	39.50	183.50
0130	Dock boards, H.D., 5′ x 5′, aluminum, 5000 lb. capacity	1,550		1,550
0140	16,000 lb. capacity	1,475		1,475
0150	Dock levelers, hydraulic, 7′ x 8′, 10 ton capacity	6,250	1,200	7,450
0160	Dock lifters, platform, 6′ x 6′, portable, 3000 lb. capacity	10,800		10,800
0170	Dock shelters, truck, scissor arms, economy	2,350	510	2,860
0180	Deluxe	2,650	1,025	3,675

E1090 Other Equipment

E1090 210	Solid Waste Handling Equipment, EACH	COST EACH		
		MAT.	INST.	TOTAL
0110	Waste handling, compactors, single bag, 250 lbs./hr., hand fed	18,100	380	18,480
0120	Heavy duty industrial, 5 C.Y. capacity	39,000	1,825	40,825
0130	Incinerator, electric, 100 lbs./hr., economy	75,000	2,875	77,875
0140	Gas, 2000 lbs./hr., deluxe	450,000	21,200	471,200
0150	Shredder, no baling, 35 tons/hr.	355,500		355,500
0160	Incl. baling, 50 tons/day	710,500		710,500

E1090 350	Food Service Equipment, EACH	COST EACH		
		MAT.	INST.	TOTAL
0110	Kitchen equipment, bake oven, single deck	6,425	133	6,558
0120	Broiler, without oven	4,275	133	4,408
0130	Commercial dish washer, semiautomatic, 50 racks/hr.	7,050	815	7,865
0140	Automatic, 275 racks/hr.	31,800	3,550	35,350
0150	Cooler, beverage, reach-in, 6 ft. long	3,550	177	3,727
0160	Food warmer, counter, 1.65 kw	785		785
0170	Fryers, with submerger, single	1,525	151	1,676
0180	Double	2,775	212	2,987
0185	Ice maker, 1000 lb. per day, with bin	5,800	1,050	6,850
0190	Kettles, steam jacketed, 20 gallons	9,350	206	9,556
0200	Range, restaurant type, burners, 2 ovens and 24" griddle	5,950	177	6,127
0210	Range hood, incl. carbon dioxide system, elect. stove	2,275	355	2,630
0220	Gas stove	2,600	355	2,955

E1090 360	Food Service Equipment, S.F.	COST PER S.F.		
		MAT.	INST.	TOTAL
0110	Refrigerators, prefab, walk-in, 7'-6" high, 6' x 6'	118	18.70	136.70
0120	12' x 20'	125	9.35	134.35

E1090 410	Residential Equipment, EACH	COST EACH		
		MAT.	INST.	TOTAL
0110	Arch. equip., appliances, range, cook top, 4 burner, economy	355	99	454
0120	Built in, single oven 30" wide, economy	945	99	1,044
0130	Standing, single oven-21" wide, economy	520	80	600
0135	Free standing, 30" wide, 1 oven, average	1,050	101	1,151
0140	Double oven-30" wide, deluxe	3,725	80	3,805
0150	Compactor, residential, economy	780	102	882
0160	Deluxe	1,275	171	1,446
0170	Dish washer, built-in, 2 cycles, economy	335	184	519
0180	4 or more cycles, deluxe	2,100	370	2,470
0190	Garbage disposer, sink type, economy	119	74	193
0200	Deluxe	229	74	303
0210	Refrigerator, no frost, 10 to 12 C.F., economy	495	80	575
0220	21 to 29 C.F., deluxe	2,700	267	2,967
0300	Washing machine, automatic	1,500	590	2,090

E1090 610	School Equipment, EACH	COST EACH		
		MAT.	INST.	TOTAL
0110	School equipment, basketball backstops, wall mounted, wood, fixed	1,775	900	2,675
0120	Suspended type, electrically operated	7,650	1,750	9,400
0130	Bleachers-telescoping, manual operation, 15 tier, economy (per seat)	130	27.50	157.50
0140	Power operation, 30 tier, deluxe (per seat)	575	50	625
0150	Weight lifting gym, universal, economy	330	800	1,130
0160	Deluxe	16,500	1,600	18,100
0170	Scoreboards, basketball, 1 side, economy	2,775	780	3,555
0180	4 sides, deluxe	18,000	10,800	28,800

E1090 Other Equipment

E1090 610	School Equipment, EACH	COST EACH		
		MAT.	INST.	TOTAL
0800	Vocational shop equipment, benches, metal	500	205	705
0810	Wood	810	205	1,015
0820	Dust collector, not incl. ductwork, 6′ diam.	6,075	520	6,595
0830	Planer, 13″ x 6″	1,325	256	1,581

E1090 620	School Equipment, S.F.	COST PER S.F.		
		MAT.	INST.	TOTAL
0110	School equipment, gym mats, naugahyde cover, 2″ thick	5.75		5.75
0120	Wrestling, 1″ thick, heavy duty	5.60		5.60

E1090 810	Athletic, Recreational, and Therapeutic Equipment, EACH	COST EACH		
		MAT.	INST.	TOTAL
0110	Sauna, prefabricated, incl. heater and controls, 7′ high, 6′ x 4′	5,775	655	6,430
0120	10′ x 12′	13,900	1,450	15,350
0130	Heaters, wall mounted, to 200 C.F.	1,050		1,050
0140	Floor standing, to 1000 C.F., 12500 W	3,950	198	4,148
0610	Shooting range incl. bullet traps, controls, separators, ceilings, economy	50,500	6,725	57,225
0620	Deluxe	69,500	11,300	80,800
0650	Sport court, squash, regulation, in existing building, economy			45,000
0660	Deluxe			50,000
0670	Racketball, regulation, in existing building, economy	48,700	9,075	57,775
0680	Deluxe	52,500	18,100	70,600
0700	Swimming pool equipment, diving stand, stainless steel, 1 meter	11,500	380	11,880
0710	3 meter	19,100	2,550	21,650
0720	Diving boards, 16 ft. long, aluminum	4,825	380	5,205
0730	Fiberglass	3,875	380	4,255
0740	Filter system, sand, incl. pump, 6000 gal./hr.	2,425	655	3,080
0750	Lights, underwater, 12 volt with transformer, 300W	400	595	995
0760	Slides, fiberglass with aluminum handrails & ladder, 6′ high, straight	4,150	640	4,790
0780	12′ high, straight with platform	17,400	855	18,255

E1090 820	Athletic, Recreational, and Therapeutic Equipment, S.F.	COST PER S.F.		
		MAT.	INST.	TOTAL
0110	Swimming pools, residential, vinyl liner, metal sides	25	7.05	32.05
0120	Concrete sides	30	12.45	42.45
0130	Gunite shell, plaster finish, 350 S.F.	55.50	26	81.50
0140	800 S.F.	44.50	14.95	59.45
0150	Motel, gunite shell, plaster finish	68.50	33	101.50
0160	Municipal, gunite shell, tile finish, formed gutters	263	56.50	319.50

E2010 Fixed Furnishings

E2010 310	Window Treatment, EACH	COST EACH		
		MAT.	INST.	TOTAL
0110	Furnishings, blinds, exterior, aluminum, louvered, 1'-4" wide x 3'-0" long	220	51	271
0120	1'-4" wide x 6'-8" long	395	57	452
0130	Hemlock, solid raised, 1-'4" wide x 3'-0" long	96	51	147
0140	1'-4" wide x 6'-9" long	164	57	221
0150	Polystyrene, louvered, 1'-3" wide x 3'-3" long	43	51	94
0160	1'-3" wide x 6'-8" long	76.50	57	133.50
0200	Interior, wood folding panels, louvered, 7" x 20" (per pair)	105	30	135
0210	18" x 40" (per pair)	152	30	182

E2010 320	Window Treatment, S.F.	COST PER S.F.		
		MAT.	INST.	TOTAL
0110	Furnishings, blinds-interior, venetian-aluminum, stock, 2" slats	5.40	.87	6.27
0120	Custom, 1" slats, deluxe	6.50	.87	7.37
0130	Vertical, PVC or cloth, T&B track, economy	9.80	1.11	10.91
0140	Deluxe	20	1.28	21.28
0150	Draperies, unlined, economy	34		34
0160	Lightproof, deluxe	67		67
0510	Shades, mylar, wood roller, single layer, non-reflective	3.37	.75	4.12
0520	Metal roller, triple layer, heat reflective	9.70	.75	10.45
0530	Vinyl, light weight, 4 ga.	.95	.75	1.70
0540	Heavyweight, 6 ga.	2.94	.75	3.69
0550	Vinyl coated cotton, lightproof decorator shades	6.90	.75	7.65
0560	Woven aluminum, 3/8" thick, light and fireproof	9.10	1.46	10.56

E2010 420	Fixed Floor Grilles and Mats, S.F.	COST PER S.F.		
		MAT.	INST.	TOTAL
0110	Floor mats, recessed, inlaid black rubber, 3/8" thick, solid	29.50	2.59	32.09
0120	Colors, 1/2" thick, perforated	42	2.59	44.59
0130	Link-including nosings, steel-galvanized, 3/8" thick	32	2.59	34.59
0140	Vinyl, in colors	29.50	2.59	32.09

E2010 510	Fixed Multiple Seating, EACH	COST EACH		
		MAT.	INST.	TOTAL
0110	Seating, painted steel, upholstered, economy	165	29.50	194.50
0120	Deluxe	555	36.50	591.50
0400	Seating, lecture hall, pedestal type, economy	298	46.50	344.50
0410	Deluxe	575	70.50	645.50
0500	Auditorium chair, veneer construction	310	46.50	356.50
0510	Fully upholstered, spring seat	280	46.50	326.50

E2020 Moveable Furnishings

E2020 210	Furnishings/EACH	COST EACH		
		MAT.	INST.	TOTAL
0200	Hospital furniture, beds, manual, economy	945		945
0210	Deluxe	3,100		3,100
0220	All electric, economy	2,075		2,075
0230	Deluxe	4,100		4,100
0240	Patient wall systems, no utilities, economy, per room	1,575		1,575
0250	Deluxe, per room	2,200		2,200
0300	Hotel furnishings, standard room set, economy, per room	2,800		2,800
0310	Deluxe, per room	9,350		9,350
0500	Office furniture, standard employee set, economy, per person	655		655
0510	Deluxe, per person	2,725		2,725
0550	Posts, portable, pedestrian traffic control, economy	174		174
0560	Deluxe	261		261
0700	Restaurant furniture, booth, molded plastic, stub wall and 2 seats, economy	395	256	651
0710	Deluxe	1,725	340	2,065
0720	Upholstered seats, foursome, single-economy	895	102	997
0730	Foursome, double-deluxe	2,225	171	2,396

E2020 220	Furniture and Accessories, L.F.	COST PER L.F.		
		MAT.	INST.	TOTAL
0210	Dormitory furniture, desk top (built-in),laminated plastc, 24"deep, economy	53.50	20.50	74
0220	30" deep, deluxe	298	25.50	323.50
0230	Dressing unit, built-in, economy	228	85.50	313.50
0240	Deluxe	685	128	813
0310	Furnishings, cabinets, hospital, base, laminated plastic	440	102	542
0320	Stainless steel	805	102	907
0330	Countertop, laminated plastic, no backsplash	59	25.50	84.50
0340	Stainless steel	191	25.50	216.50
0350	Nurses station, door type, laminated plastic	505	102	607
0360	Stainless steel	970	102	1,072
0710	Restaurant furniture, bars, built-in, back bar	251	102	353
0720	Front bar	345	102	447
0910	Wardrobes & coatracks, standing, steel, single pedestal, 30" x 18" x 63"	164		164
0920	Double face rack, 39" x 26" x 70"	143		143
0930	Wall mounted rack, steel frame & shelves, 12" x 15" x 26"	71	7.30	78.30
0940	12" x 15" x 50"	43	3.81	46.81

F1010 Special Structures

F1010 120	Air-Supported Structures, S.F.	COST PER S.F.		
		MAT.	INST.	TOTAL
0110	Air supported struc., polyester vinyl fabric, 24oz., warehouse, 5000 S.F.	30.50	.32	30.82
0120	50,000 S.F.	14.35	.26	14.61
0130	Tennis, 7,200 S.F.	26	.27	26.27
0140	24,000 S.F.	19.10	.27	19.37
0150	Woven polyethylene, 6 oz., shelter, 3,000 S.F.	18.25	.54	18.79
0160	24,000 S.F.	13.65	.27	13.92
0170	Teflon coated fiberglass, stadium cover, economy	66.50	.14	66.64
0180	Deluxe	79	.19	79.19
0190	Air supported storage tank covers, reinf. vinyl fabric, 12 oz., 400 S.F.	27	.45	27.45
0200	18,000 S.F.	9.85	.41	10.26

F1010 210	Pre-Engineered Structures, EACH	COST EACH		
		MAT.	INST.	TOTAL
0600	Radio towers, guyed, 40 lb. section, 50′ high, 70 MPH basic wind speed	3,050	1,150	4,200
0610	90 lb. section, 400′ high, wind load 70 MPH basic wind speed	41,400	13,500	54,900
0620	Self supporting, 60′ high, 70 MPH basic wind speed	4,825	2,375	7,200
0630	190′ high, wind load 90 MPH basic wind speed	30,900	9,475	40,375
0700	Shelters, aluminum frame, acrylic glazing, 8′ high, 3′ x 9′	3,400	1,000	4,400
0710	9′ x 12′	8,425	1,575	10,000

F1010 320	Other Special Structures, S.F.	COST PER S.F.		
		MAT.	INST.	TOTAL
0110	Swimming pool enclosure, transluscent, freestanding, economy	55	5.10	60.10
0120	Deluxe	101	14.65	115.65
0510	Tension structures, steel frame, polyester vinyl fabric, 12,000 S.F.	18.90	2.43	21.33
0520	20,800 S.F.	18.50	2.19	20.69

F1010 330	Special Structures, EACH	COST EACH		
		MAT.	INST.	TOTAL
0110	Kiosks, round, 5′ diam., 7′ high, aluminum wall, illuminated	25,700		25,700
0120	Rectangular, 5′ x 9′, 1″ insulated dbl. wall fiberglass, 7′-6″ high	26,200		26,200
0220	Silos, steel prefab, 30,000 gal., painted, economy	24,500	4,675	29,175
0230	Epoxy-lined, deluxe	50,500	9,325	59,825

F1010 340	Special Structures, S.F.	COST PER S.F.		
		MAT.	INST.	TOTAL
0110	Comfort stations, prefab, mobile on steel frame, economy	209		209
0120	Permanent on concrete slab, deluxe	208	44	252
0210	Domes, bulk storage, wood framing, wood decking, 50′ diam.	75.50	1.44	76.94
0220	116′ diam.	36.50	1.66	38.16
0230	Steel framing, metal decking, 150′ diam.	37	9.70	46.70
0240	400′ diam.	30	7.40	37.40
0250	Geodesic, wood framing, wood panels, 30′ diam.	38.50	1.61	40.11
0260	60′ diam.	24.50	1.01	25.51
0270	Aluminum framing, acrylic panels, 40′ diam.			69.50
0280	Aluminum panels, 400′ diam.			22.50
0310	Garden house, prefab, wood, shell only, 48 S.F.	58	5.10	63.10
0320	200 S.F.	33.50	21.50	55
0410	Greenhouse, shell-stock, residential, lean-to, 8′-6″ long x 3′-10″ wide	48.50	30	78.50
0420	Freestanding, 8′-6″ long x 13′-6″ wide	50	9.50	59.50
0430	Commercial-truss frame, under 2000 S.F., deluxe			14.40
0440	Over 5,000 S.F., economy			13.40
0450	Institutional-rigid frame, under 500 S.F., deluxe			32.50
0460	Over 2,000 S.F., economy			12.20
0510	Hangar, prefab, galv. roof and walls, bottom rolling doors, economy	14.90	3.84	18.74
0520	Electric bifolding doors, deluxe	13.75	5	18.75

F1020 Integrated Construction

F1020 110	Integrated Construction, EACH	COST EACH		
		MAT.	INST.	TOTAL
0110	Integrated ceilings, radiant electric, 2' x 4' panel, manila finish	305	24	329
0120	ABS plastic finish	123	46	169

F1020 120	Integrated Construction, S.F.	COST PER S.F.		
		MAT.	INST.	TOTAL
0110	Integrated ceilings, Luminaire, suspended, 5' x 5' modules, 50% lighted	5.65	7.35	13
0120	100% lighted	7.35	13.20	20.55
0130	Dimensionaire, 2' x 4' module tile system, no air bar	2.71	2.05	4.76
0140	With air bar, deluxe	3.98	2.05	6.03
0220	Pedestal access floor pkg., inc. stl. pnls, peds. & stringers, w/vinyl cov.	30.50	2.73	33.23
0230	With high pressure laminate covering	28.50	2.73	31.23
0240	With carpet covering	30	2.73	32.73
0250	Aluminum panels, no stringers, no covering	37	2.05	39.05

F1020 250	Special Purpose Room, EACH	COST EACH		
		MAT.	INST.	TOTAL
0110	Portable booth, acoustical, 27 db 1000 hz., 15 S.F. floor	4,425		4,425
0120	55 S.F. flr.	9,100		9,100

F1020 260	Special Purpose Room, S.F.	COST PER S.F.		
		MAT.	INST.	TOTAL
0110	Anechoic chambers, 7' high, 100 cps cutoff, 25 S.F.			2,975
0120	200 cps cutoff, 100 S.F.			1,550
0130	Audiometric rooms, under 500 S.F.	66	21	87
0140	Over 500 S.F.	63	17.10	80.10
0300	Darkrooms, shell, not including door, 240 S.F., 8' high	30.50	8.55	39.05
0310	64 S.F., 12' high	78	16	94
0510	Music practice room, modular, perforated steel, under 500 S.F.	41	14.65	55.65
0520	Over 500 S.F.	35	12.80	47.80

F1020 330	Special Construction, L.F.	COST PER L.F.		
		MAT.	INST.	TOTAL
0110	Spec. const., air curtains, shipping & receiving, 8'high x 5'wide, economy	915	115	1,030
0120	20' high x 8' wide, heated, deluxe	3,875	287	4,162
0130	Customer entrance, 10' high x 5' wide, economy	1,200	115	1,315
0140	12' high x 4' wide, heated, deluxe	2,600	144	2,744

F1030 Special Construction Systems

F1030 120	Sound, Vibration, and Seismic Construction, S.F.	COST PER S.F.		
		MAT.	INST.	TOTAL
0020	Special construction, acoustical, enclosure, 4" thick, 8 psf panels	34.50	21.50	56
0030	Reverb chamber, 4" thick, parallel walls	49	25.50	74.50
0110	Sound absorbing panels, 2'-6" x 8', painted metal	13	7.15	20.15
0120	Vinyl faced	10.15	6.40	16.55
0130	Flexible transparent curtain, clear	7.85	8	15.85
0140	With absorbing foam, 75% coverage	11	8	19
0150	Strip entrance, 2/3 overlap	8.45	12.75	21.20
0160	Full overlap	10.70	15	25.70
0200	Audio masking system, plenum mounted, over 10,000 S.F.	.77	.27	1.04
0210	Ceiling mounted, under 5,000 S.F.	1.32	.50	1.82

F1030 210	Radiation Protection, EACH	COST EACH		
		MAT.	INST.	TOTAL
0110	Shielding, lead x-ray protection, radiography room, 1/16" lead, economy	12,400	3,925	16,325
0120	Deluxe	14,900	6,550	21,450
0210	Deep therapy x-ray, 1/4" lead, economy	34,700	12,300	47,000
0220	Deluxe	42,800	16,400	59,200

F1030 220	Radiation Protection, S.F.	COST PER S.F.		
		MAT.	INST.	TOTAL
0110	Shielding, lead, gypsum board, 5/8" thick, 1/16" lead	17.50	6.30	23.80
0120	1/8" lead	27.50	7.20	34.70
0130	Lath, 1/16" thick	12.20	7.30	19.50
0140	1/8" thick	25	8.20	33.20
0150	Radio frequency, galvanized steel, prefab type, economy	5.90	2.73	8.63
0160	Radio frequency, door, copper/wood laminate, 4" x 7'	10.05	7.30	17.35

F1030 910	Other Special Construction Systems, EACH	COST EACH		
		MAT.	INST.	TOTAL
0110	Disappearing stairways, folding, pine, 8'-6" ceiling	194	128	322
0220	Automatic electric, wood, 8' to 9' ceiling	10,100	1,025	11,125
0300	Fireplace prefabricated, freestanding or wall hung, painted	1,775	395	2,170
0310	Stainless steel	3,550	570	4,120
0320	Woodburning stoves, cast iron, economy, less than 1500 S.F. htg. area	1,525	790	2,315
0330	Greater than 2000 S.F. htg. area	3,075	1,275	4,350

F1040 Special Facilities

F1040 210	Ice Rinks, EACH	COAT EACH		
		MAT.	INST.	TOTAL
0100	Ice skating rink, 85′ x 200′, 55° system, 5 mos., 100 ton			652,000
0110	90° system, 12 mos., 135 ton			737,500
0120	Dash boards, acrylic screens, polyethylene coated plywood	159,000	30,000	189,000
0130	Fiberglass and aluminum construction	176,000	30,000	206,000

F1040 510	Liquid & Gas Storage Tanks, EACH	COST EACH		
		MAT.	INST.	TOTAL
0100	Tanks, steel, ground level, 100,000 gal.			229,500
0110	10,000,000 gal.			5,352,000
0120	Elevated water, 50,000 gal.		158,500	409,000
0130	1,000,000 gal.		739,500	1,806,000
0150	Cypress wood, ground level, 3,000 gal.	10,400	9,550	19,950
0160	Redwood, ground level, 45,000 gal.	82,500	25,900	108,400

F1040 910	Special Construction, EACH	COST EACH		
		MAT.	INST.	TOTAL
0110	Special construction, bowling alley incl. pinsetter, scorer etc., economy	53,000	10,200	63,200
0120	Deluxe	64,500	11,400	75,900
0130	For automatic scorer, economy, add	6,900		6,900
0140	Deluxe, add	11,500		11,500
0300	Control tower, modular, 12′ x 10′, incl. instrumentation, economy			891,500
0310	Deluxe			1,396,000
0400	Garage costs, residential, prefab, wood, single car economy	6,675	1,025	7,700
0410	Two car deluxe	16,400	2,050	18,450
0500	Hangars, prefab, galv. steel, bottom rolling doors, economy (per plane)	19,000	4,275	23,275
0510	Electrical bi-folding doors, deluxe (per plane)	14,700	5,850	20,550

G1030 Site Earthwork

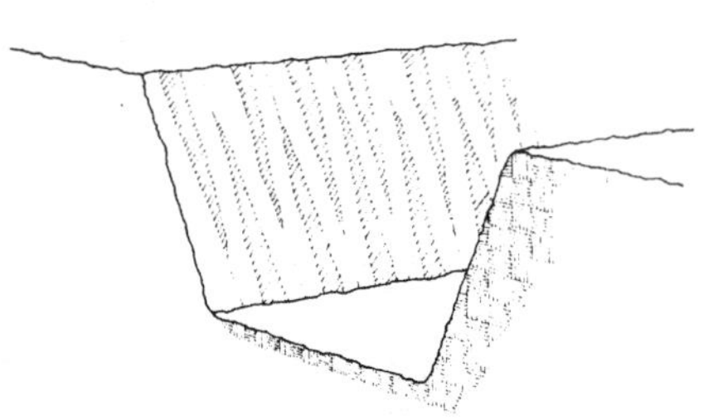

Trenching Systems are shown on a cost per linear foot basis. The systems include: excavation; backfill and removal of spoil; and compaction for various depths and trench bottom widths. The backfill has been reduced to accommodate a pipe of suitable diameter and bedding.

The slope for trench sides varies from none to 1:1.

The Expanded System Listing shows Trenching Systems that range from 2′ to 12′ in width. Depths range from 2′ to 25′.

G1030 805	Trenching Common Earth	COST PER L.F.		
		MAT.	INST.	TOTAL
1310	Trenching, common earth, no slope, 2′ wide, 2′ deep, 3/8 C.Y. bucket	.63	1.63	2.26
1330	4′ deep, 3/8 C.Y. bucket	1.16	3.22	4.38
1360	10′ deep, 1 C.Y. bucket	3.51	6.50	10.01
1400	4′ wide, 2′ deep, 3/8 C.Y. bucket	1.40	3.26	4.66
1420	4′ deep, 1/2 C.Y. bucket	2.40	5.35	7.75
1450	10′ deep, 1 C.Y. bucket	8.10	12.40	20.50
1480	18′ deep, 2-1/2 C.Y. bucket	12.40	19.15	31.55
1520	6′ wide, 6′ deep, 5/8 C.Y. bucket w/trench box	8.45	11.85	20.30
1540	10′ deep, 1 C.Y. bucket	10.85	15.30	26.15
1570	20′ deep, 3-1/2 C.Y. bucket	21	23	44
1640	8′ wide, 12′ deep, 1-1/2 C.Y. bucket w/trench box	15.95	20.50	36.45
1680	24′ deep, 3-1/2 C.Y. bucket	33.50	35.50	69
1730	10′ wide, 20′ deep, 3-1/2 C.Y. bucket w/trench box	27.50	33.50	61
1740	24′ deep, 3-1/2 C.Y. bucket	40	41	81
3500	1 to 1 slope, 2′ wide, 2′ deep, 3/8 C.Y. bucket	1.16	3.22	4.38
3540	4′ deep, 3/8 C.Y. bucket	3.34	9.70	13.04
3600	10′ deep, 1 C.Y. bucket	20.50	39	59.50
3800	4′ wide, 2′ deep, 3/8 C.Y. bucket	1.94	4.87	6.81
3840	4′ deep, 1/2 C.Y. bucket	4.44	10.85	15.29
3900	10′ deep, 1 C.Y. bucket	28	44	72
4030	6′ wide, 6′ deep, 5/8 C.Y. bucket w/trench box	16.55	23	39.55
4050	10′ deep, 1 C.Y. bucket	25	36	61
4080	20′ deep, 3-1/2 C.Y. bucket	84.50	99	183.50
4500	8′ wide, 12′ deep, 1-1/2 C.Y. bucket w/trench box	38	53.50	91.50
4650	24′ deep, 3-1/2 C.Y. bucket	132	148	280
4800	10′ wide, 20′ deep, 3-1/2 C.Y. bucket w/trench box	79.50	105	184.50
4850	24′ deep, 3-1/2 C.Y. bucket	141	156	297

G1030 Site Earthwork

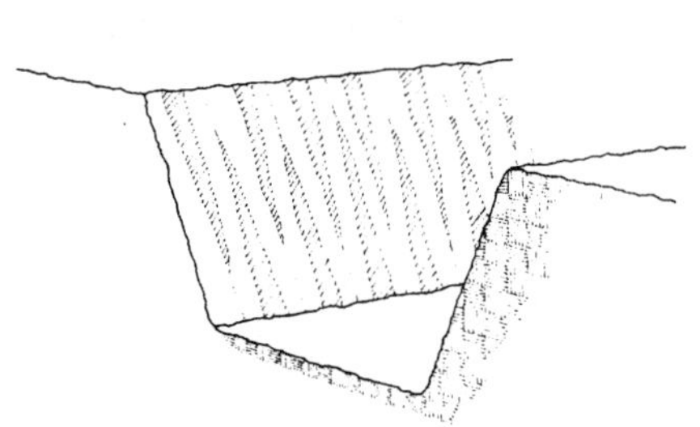

Trenching Systems are shown on a cost per linear foot basis. The systems include: excavation; backfill and removal of spoil; and compaction for various depths and trench bottom widths. The backfill has been reduced to accommodate a pipe of suitable diameter and bedding.

The slope for trench sides varies from none to 1:1.

The Expanded System Listing shows Trenching Systems that range from 2′ to 12′ in width. Depths range from 2′ to 25′.

G1030 807	Trenching Sand & Gravel	COST PER L.F.		
		MAT.	INST.	TOTAL
1310	Trenching, sand & gravel, no slope, 2′ wide, 2′ deep, 3/8 C.Y. bucket	.57	1.51	2.08
1320	3′ deep, 3/8 C.Y. bucket	.91	2.51	3.42
1330	4′ deep, 3/8 C.Y. bucket	1.06	2.99	4.05
1340	6′ deep, 3/8 C.Y. bucket	1.65	3.53	5.20
1350	8′ deep, 1/2 C.Y. bucket	2.19	4.65	6.85
1360	10′ deep, 1 C.Y. bucket	2.35	4.83	7.20
1400	4′ wide, 2′ deep, 3/8 C.Y. bucket	1.27	3	4.27
1410	3′ deep, 3/8 C.Y. bucket	1.78	4.49	6.25
1420	4′ deep, 1/2 C.Y. bucket	2.20	4.99	7.20
1430	6′ deep, 1/2 C.Y. bucket	4.05	6.60	10.65
1440	8′ deep, 1/2 C.Y. bucket	6.10	9.35	15.45
1450	10′ deep, 1 C.Y. bucket	5.75	9.10	14.85
1460	12′ deep, 1 C.Y. bucket	7.25	11.40	18.65
1470	15′ deep, 1-1/2 C.Y. bucket	8.60	12.95	21.50
1480	18′ deep, 2-1/2 C.Y. bucket	10.25	13.75	24
1520	6′ wide, 6′ deep, 5/8 C.Y. bucket w/trench box	7.70	10.95	18.65
1530	8′ deep, 3/4 C.Y. bucket	10.30	14.25	24.50
1540	10′ deep, 1 C.Y. bucket	9.45	13.70	23
1550	12′ deep, 1-1/2 C.Y. bucket	10.60	15	25.50
1560	16′ deep, 2 C.Y. bucket	14.80	18.55	33.50
1570	20′ deep, 3-1/2 C.Y. bucket	18.25	21.50	40
1580	24′ deep, 3-1/2 C.Y. bucket	23	26	49
1640	8′ wide, 12′ deep, 1-1/2 C.Y. bucket w/trench box	14.80	19.40	34
1650	15′ deep, 1-1/2 C.Y. bucket	18	24.50	42.50
1660	18′ deep, 2-1/2 C.Y. bucket	22	26.50	48.50
1680	24′ deep, 3-1/2 C.Y. bucket	31	34	65
1730	10′ wide, 20′ deep, 3-1/2 C.Y. bucket w/trench box	31	34	65
1740	24′ deep, 3-1/2 C.Y. bucket	38.50	42	80.50
1780	12′ wide, 20′ deep, 3-1/2 C.Y. bucket w/trench box	37.50	40.50	78
1790	25′ deep, 3-1/2 C.Y. bucket	48.50	52	101
1800	1/2:1 slope, 2′ wide, 2′ deep, 3/8 C.Y. bucket	.82	2.25	3.07
1810	3′ deep, 3/8 C.Y. bucket	1.38	3.94	5.30
1820	4′ deep, 3/8 C.Y. bucket	2.05	6	8.05
1840	6′ deep, 3/8 C.Y. bucket	4.03	8.80	12.85
1860	8′ deep, 1/2 C.Y. bucket	6.40	14	20.50
1880	10′ deep, 1 C.Y. bucket	8.05	17.10	25
2300	4′ wide, 2′ deep, 3/8 C.Y. bucket	1.53	3.74	5.25
2310	3′ deep, 3/8 C.Y. bucket	2.33	6.15	8.50
2320	4′ deep, 1/2 C.Y. bucket	3.14	7.55	10.70
2340	6′ deep, 1/2 C.Y. bucket	6.90	11.70	18.60
2360	8′ deep, 1/2 C.Y. bucket	11.95	18.75	30.50
2380	10′ deep, 1 C.Y. bucket	12.80	20.50	33.50
2400	12′ deep, 1 C.Y. bucket	21.50	34.50	56
2430	15′ deep, 1-1/2 C.Y. bucket	24.50	37.50	62

G1030 Site Earthwork

G1030 807	Trenching Sand & Gravel	COST PER L.F.		
		MAT.	INST.	TOTAL
2460	18' deep, 2-1/2 C.Y. bucket	40.50	56	96.50
2840	6' wide, 6' deep, 5/8 C.Y. bucket w/trench box	11.35	16.40	28
2860	8' deep, 3/4 C.Y. bucket	16.75	24	41
2880	10' deep, 1 C.Y. bucket	17.05	25.50	42.50
2900	12' deep, 1-1/2 C.Y. bucket	21	30.50	51.50
2940	16' deep, 2 C.Y. bucket	34	44	78
2980	20' deep, 3-1/2 C.Y. bucket	48.50	57.50	106
3020	24' deep, 3-1/2 C.Y. bucket	68.50	79.50	148
3100	8' wide, 12' deep, 1-1/4 C.Y. bucket w/trench box	25	35	60
3120	15' deep, 1-1/2 C.Y. bucket	34	48.50	82.50
3140	18' deep, 2-1/2 C.Y. bucket	46.50	58.50	105
3180	24' deep, 3-1/2 C.Y. bucket	76.50	87	164
3270	10' wide, 20' deep, 3-1/2 C.Y. bucket w/trench box	61	70.50	132
3280	24' deep, 3-1/2 C.Y. bucket	84	95.50	180
3370	12' wide, 20' deep, 3-1/2 C.Y. bucket w/trench box	68	78.50	147
3380	25' deep, 3-1/2 C.Y. bucket	98	110	208
3500	1:1 slope, 2' wide, 2' deep, 3/8 C.Y. bucket	1.76	3.83	5.60
3520	3' deep, 3/8 C.Y. bucket	1.93	5.60	7.55
3540	4' deep, 3/8 C.Y. bucket	3.04	9	12.05
3560	6' deep, 3/8 C.Y. bucket	4.03	8.80	12.85
3580	8' deep, 1/2 C.Y. bucket	10.60	23.50	34
3600	10' deep, 1 C.Y. bucket	13.75	29.50	43.50
3800	4' wide, 2' deep, 3/8 C.Y. bucket	1.78	4.49	6.25
3820	3' deep, 3/8 C.Y. bucket	2.89	7.85	10.75
3840	4' deep, 1/2 C.Y. bucket	4.07	10.15	14.20
3860	6' deep, 1/2 C.Y. bucket	9.75	16.75	26.50
3880	8' deep, 1/2 C.Y. bucket	17.80	28	46
3900	10' deep, 1 C.Y. bucket	19.80	32.50	52.50
3920	12' deep, 1 C.Y. bucket	28.50	46	74.50
3940	15' deep, 1-1/2 C.Y. bucket	40.50	62.50	103
3960	18' deep, 2-1/2 C.Y. bucket	56	77	133
4030	6' wide, 6' deep, 5/8 C.Y. bucket w/trench box	15	22	37
4040	8' deep, 3/4 C.Y. bucket	23	33.50	56.50
4050	10' deep, 1 C.Y. bucket	24.50	37.50	62
4060	12' deep, 1-1/2 C.Y. bucket	31	46	77
4070	16' deep, 2 C.Y. bucket	53.50	69.50	123
4080	20' deep, 3-1/2 C.Y. bucket	78.50	94	173
4090	24' deep, 3-1/2 C.Y. bucket	114	133	247
4500	8' wide, 12' deep, 1-1/2 C.Y. bucket w/trench box	35.50	50.50	86
4550	15' deep, 1-1/2 C.Y. bucket	50	73	123
4600	18' deep, 2-1/2 C.Y. bucket	70.50	90.50	161
4650	24' deep, 3-1/2 C.Y. bucket	122	141	263
4800	10' wide, 20' deep, 3-1/2 C.Y. bucket w/trench box	91	107	198
4850	24' deep, 3-1/2 C.Y. bucket	130	149	279
4950	12' wide, 20' deep, 3-1/2 C.Y. bucket w/trench box	97.50	113	211
4980	25' deep, 3-1/2 C.Y. bucket	147	168	315

G1030 815	Pipe Bedding	COST PER L.F.		
		MAT.	INST.	TOTAL
1440	Pipe bedding, side slope 0 to 1, 1' wide, pipe size 6" diameter	1.70	.80	2.50
1460	2' wide, pipe size 8" diameter	3.68	1.73	5.41
1500	Pipe size 12" diameter	3.84	1.81	5.65
1600	4' wide, pipe size 20" diameter	9.20	4.33	13.53
1660	Pipe size 30" diameter	9.65	4.54	14.19
1680	6' wide, pipe size 32" diameter	16.50	7.80	24.30
1740	8' wide, pipe size 60" diameter	32.50	15.40	47.90
1780	12' wide, pipe size 84" diameter	60	28.50	88.50

G2010 Roadways

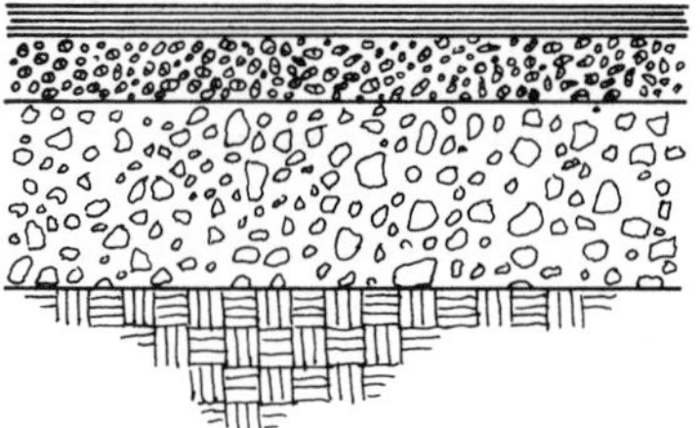

The Bituminous Roadway Systems are listed for pavement thicknesses between 3-l/2″ and 7″ and crushed stone bases from 3″ to 22″ in depth. Systems costs are expressed per linear foot for varying widths of two and multi-lane roads. Earth moving is not included. Granite curbs and line painting are added as required system components.

G2010 232	Bituminous Roadways Crushed Stone	COST PER L.F.		
		MAT.	INST.	TOTAL
1050	Bitum. roadway, two lanes, 3-1/2″ th. pvmt., 3″ th. crushed stone,24′ wide	78.50	46.50	125
1100	28′ wide	85.50	47.50	133
1150	32′ wide	93	54	147
1300	9″ thick crushed stone, 24′ wide	94	49	143
1350	28′ wide	104	53	157
1400	32′ wide	114	57	171
1420	12″ thick crushed stone, 24′ wide	102	51	153
1550	4″ thick pavement., 4″ thick crushed stone, 24′ wide	86	45.50	131.50
1600	28′ wide	94.50	49	143.50
1650	32′ wide	103	52.50	155.50
1800	10″ thick crushed stone, 24′ wide	102	50	152
1850	28′ wide	113	53.50	166.50
1900	32′ wide	124	59	183
3050	5″ thick pavement, 6″ thick crushed stone, 24′ wide	102	49.50	151.50
3100	28′ wide	114	54	168
3150	32′ wide	125	58	183
3300	14″ thick crushed stone, 24′ wide	123	55	178
3350	28′ wide	138	60	198
3400	32′ wide	153	65.50	218.50

G2020 Parking Lots

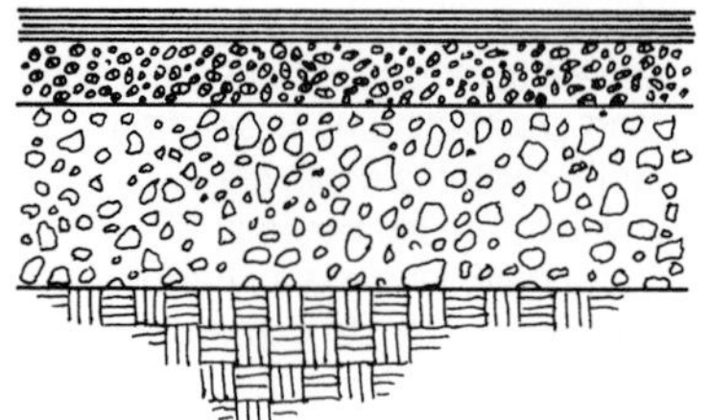

The Parking Lot System includes: compacted bank-run gravel; fine grading with a grader and roller; and bituminous concrete wearing course. All Parking Lot systems are on a cost per car basis. There are three basic types of systems: 90° angle, 60° angle, and 45° angle. The gravel base is compacted to 98%. Final stall design and lay-out of the parking lot with precast bumpers, sealcoating and white paint is also included.

The Expanded System Listing shows the three basic parking lot types with various depths of both gravel base and wearing course. The gravel base depths range from 6″ to 10″. The bituminous paving wearing course varies from a depth of 3″ to 6″.

G2020 210	Parking Lots Gravel Base	COST PER CAR		
		MAT.	INST.	TOTAL
1500	Parking lot, 90° angle parking, 3″ bituminous paving, 6″ gravel base	780	390	1,170
1540	10″ gravel base	880	450	1,330
1560	4″ bituminous paving, 6″ gravel base	985	410	1,395
1600	10″ gravel base	1,075	470	1,545
1620	6″ bituminous paving, 6″ gravel base	1,325	440	1,765
1660	10″ gravel base	1,425	505	1,930
1800	60° angle parking, 3″ bituminous paving, 6″ gravel base	780	390	1,170
1840	10″ gravel base	880	450	1,330
1860	4″ bituminous paving, 6″ gravel base	985	410	1,395
1900	10″ gravel base	1,075	470	1,545
1920	6″ bituminous paving, 6″ gravel base	1,325	440	1,765
1960	10″ gravel base	1,425	505	1,930
2200	45° angle parking, 3″ bituminous paving, 6″ gravel base	795	400	1,195
2240	10″ gravel base	900	465	1,365
2260	4″ bituminous paving, 6″ gravel base	1,000	425	1,425
2300	10″ gravel base	1,100	485	1,585
2320	6″ bituminous paving, 6″ gravel base	1,350	455	1,805
2360	10″ gravel base	1,450	515	1,965

G2040 Site Development

G2040 810	Flagpoles	COST EACH		
		MAT.	INST.	TOTAL
0110	Flagpoles, on grade, aluminum, tapered, 20' high	1,200	615	1,815
0120	70' high	10,100	1,550	11,650
0140	59' high	5,625	1,375	7,000
0150	Concrete, internal halyard, 20' high	1,550	495	2,045
0160	100' high	20,100	1,225	21,325

G2040 950	Other Site Development, EACH	COST EACH		
		MAT.	INST.	TOTAL
0110	Grandstands, permanent, closed deck, steel, economy (per seat)			29
0120	Deluxe (per seat)			32.50
0130	Composite design, economy (per seat)			40.50
0140	Deluxe (per seat)			97.50

G3030 Storm Sewer

G3030 210	Manholes & Catch Basins	COST PER EACH		
		MAT.	INST.	TOTAL
1920	Manhole/catch basin, brick, 4' I.D. riser, 4' deep	1,425	1,575	3,000
1980	10' deep	3,025	3,725	6,750
3200	Block, 4' I.D. riser, 4' deep	1,225	1,275	2,500
3260	10' deep	2,575	3,075	5,650
4620	Concrete, cast-in-place, 4' I.D. riser, 4' deep	1,375	2,200	3,575
4680	10' deep	3,225	5,250	8,475
5820	Concrete, precast, 4' I.D. riser, 4' deep	1,725	1,125	2,850
5880	10' deep	3,275	2,625	5,900
6200	6' I.D. riser, 4' deep	3,725	1,725	5,450
6260	10' deep	7,350	4,075	11,425

G4020 Site Lighting

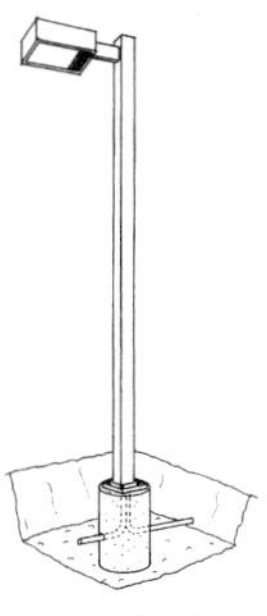

Light Pole

G4020 210	Light Pole (Installed)	COST EACH		
		MAT.	INST.	TOTAL
0200	Light pole, aluminum, 20' high, 1 arm bracket	1,450	1,050	2,500
0240	2 arm brackets	1,600	1,050	2,650
0280	3 arm brackets	1,775	1,100	2,875
0320	4 arm brackets	1,925	1,100	3,025
0360	30' high, 1 arm bracket	2,550	1,325	3,875
0400	2 arm brackets	2,700	1,325	4,025
0440	3 arm brackets	2,875	1,375	4,250
0480	4 arm brackets	3,025	1,375	4,400
0680	40' high, 1 arm bracket	3,325	1,775	5,100
0720	2 arm brackets	3,475	1,775	5,250
0760	3 arm brackets	3,650	1,825	5,475
0800	4 arm brackets	3,800	1,825	5,625
0840	Steel, 20' high, 1 arm bracket	1,475	1,125	2,600
0880	2 arm brackets	1,600	1,125	2,725
0920	3 arm brackets	1,500	1,150	2,650
0960	4 arm brackets	1,600	1,150	2,750
1000	30' high, 1 arm bracket	2,050	1,425	3,475
1040	2 arm brackets	2,175	1,425	3,600
1080	3 arm brackets	2,075	1,475	3,550
1120	4 arm brackets	2,175	1,475	3,650
1320	40' high, 1 arm bracket	2,625	1,900	4,525
1360	2 arm brackets	2,750	1,900	4,650
1400	3 arm brackets	2,650	1,950	4,600
1440	4 arm brackets	2,750	1,950	4,700

Unit Price Section

Table of Contents

Table of Contents (cont.)

RSMeans data: Unit Prices—How They Work

All RSMeans data: Unit Prices are organized in the same way.

03 30 Cast-In-Place Concrete

03 30 53 – Miscellaneous Cast-In-Place Concrete

03 30 53.40 Concrete In Place		Crew	Daily Output	Labor-Hours	Unit	2019 Bare Costs Material	Labor	Equipment	Total	Total Incl O&P
0010	CONCRETE IN PLACE R033053-10									
0020	Including forms (4 uses), Grade 60 rebar, concrete (Portland cement R033105-10									
0050	Type I), placement and finishing unless otherwise indicated R033105-20									
0500	Chimney foundations (5000 psi), over 5 C.Y. R033105-50	C-14C	32.22	3.476	C.Y.	178	127	.83	305.83	405
0510	(3500 psi), under 5 C.Y. R033105-70	"	23.71	4.724	"	206	173	1.13	380.13	515
3540	Equipment pad (3000 psi), 3' x 3' x 6" thick	C-14H	45	1.067	Ea.	46.50	40.50	.59	87.59	118
3550	4' x 4' x 6" thick		30	1.600		72.50	60.50	.88	133.88	180
3560	5' x 5' x 8" thick		18	2.667		132	101	1.47	234.47	315
3570	6' x 6' x 8" thick		14	3.429		181	129	1.89	311.89	415
3580	8' x 8' x 10" thick		8	6		385	226	3.30	614.30	800
3590	10' x 10' x 12" thick		5	9.600		665	360	5.30	1,030.30	1,325
3800	Footings (3000 psi), spread under 1 C.Y.	C-14C	28	4	C.Y.	193	147	.96	340.96	455
3825	1 C.Y. to 5 C.Y.		43	2.605		227	95.50	.63	323.13	410
3850	Over 5 C.Y. R033105-80		75	1.493		212	54.50	.36	266.86	325

It is important to understand the structure of RSMeans data: Unit Prices so that you can find information easily and use it correctly.

1 Line Numbers

Line Numbers consist of 12 characters, which identify a unique location in the database for each task. The first 6 or 8 digits conform to the Construction Specifications Institute MasterFormat® 2016. The remainder of the digits are a further breakdown in order to arrange items in understandable groups of similar tasks. Line numbers are consistent across all of our publications, so a line number in any of our products will always refer to the same item of work.

2 Descriptions

Descriptions are shown in a hierarchical structure to make them readable. In order to read a complete description, read up through the indents to the top of the section. Include everything that is above and to the left that is not contradicted by information below. For instance, the complete description for line 03 30 53.40 3550 is "Concrete in place, including forms (4 uses), Grade 60 rebar, concrete (Portland cement Type 1), placement and finishing unless otherwise indicated; Equipment pad (3000 psi), 4' × 4' × 6" thick."

3 RSMeans data

When using **RSMeans data**, it is important to read through an entire section to ensure that you use the data that most closely matches your work. Note that sometimes there is additional information shown in the section that may improve your price. There are frequently lines that further describe, add to, or adjust data for specific situations.

4 Reference Information

Gordian's RSMeans engineers have created **reference** information to assist you in your estimate. **If** there is information that applies to a section, it will be indicated at the start of the section. The Reference Section is located in the back of the data set.

5 Crews

Crews include labor and/or equipment necessary to accomplish each task. In this case, Crew C-14H is used. Gordian's RSMeans staff selects a crew to represent the workers and equipment that are typically used for that task. In this case, Crew C-14H consists of one carpenter foreman (outside), two carpenters, one rodman, one laborer, one cement finisher, and one gas engine vibrator. Details of all crews can be found in the Reference Section.

Crews - Open Shop

Crew No.	Bare Costs		Incl. Subs O&P		Cost Per Labor-Hour	
Crew C-14H	Hr.	Daily	Hr.	Daily	Bare Costs	Incl. O&P
1 Carpenter Foreman (outside)	$40.75	$326.00	$67.35	$538.80	$37.73	$62.13
2 Carpenters	38.75	620.00	64.05	1024.80		
1 Rodman (reinf.)	40.60	324.80	67.15	537.20		
1 Laborer	30.35	242.80	50.15	401.20		
1 Cement Finisher	37.15	297.20	60.05	480.40		
1 Gas Engine Vibrator		26.55		29.20	0.55	0.61
48 L.H., Daily Totals		$1837.35		$3011.61	$38.28	$62.74

6 Daily Output

The **Daily Output** is the amount of work that the crew can do in a normal 8-hour workday, including mobilization, layout, movement of materials, and cleanup. In this case, crew C-14H can install thirty 4' × 4' × 6" thick concrete pads in a day. Daily output is variable and based on many factors, including the size of the job, location, and environmental conditions. RSMeans data represents work done in daylight (or adequate lighting) and temperate conditions.

7 Labor-Hours

The figure in the **Labor-Hours** column is the amount of labor required to perform one unit of work–in this case the amount of labor required to construct one 4' × 4' equipment pad. This figure is calculated by dividing the number of hours of labor in the crew by the daily output (48 labor-hours divided by 30 pads = 1.6 hours of labor per pad). Multiply 1.6 times 60 to see the value in minutes: 60 × 1.6 = 96 minutes. Note: the labor-hour figure is not dependent on the crew size. A change in crew size will result in a corresponding change in daily output, but the labor-hours per unit of work will not change.

8 Unit of Measure

All RSMeans data: Unit Prices include the typical **Unit of Measure** used for estimating that item. For concrete-in-place the typical unit is cubic yards (C.Y.) or each (Ea.). For installing broadloom carpet it is square yard and for gypsum board it is square foot. The estimator needs to take special care that the unit in the data matches the unit in the take-off. Unit conversions may be found in the Reference Section.

9 Bare Costs

Bare Costs are the costs of materials, labor, and equipment that the installing contractor pays. They represent the cost, in U.S. dollars, for one unit of work. They do not include any markups for profit or labor burden.

10 Bare Total

The **Total column** represents the total bare cost for the installing contractor in U.S. dollars. In this case, the sum of $72.50 for material + $60.50 for labor + $.88 for equipment is $133.88.

11 Total Incl O&P

The **Total Incl O&P column** is the total cost, including overhead and profit, that the installing contractor will charge the customer. This represents the cost of materials plus 10% profit, the cost of labor plus labor burden and 10% profit, and the cost of equipment plus 10% profit. It does not include the general contractor's overhead and profit. Note: See the inside back cover of the printed product or the Reference Section of the electronic product for details on how the labor burden is calculated.

National Average

The RSMeans data in our print publications represent a "national average" cost. This data should be modified to the project location using the ***City Cost Indexes*** *or* ***Location Factors*** *tables found in the Reference Section. Use the Location Factors to adjust estimate totals if the project covers multiple trades. Use the City Cost Indexes (CCI) for single trade projects or projects where a more detailed analysis is required. All figures in the two tables are derived from the same research. The last row of data in the CCI—the weighted average—is the same as the numbers reported for each location in the location factor table.*

RSMeans data: Unit Prices—How They Work (Continued)

Project Name: Pre-Engineered Steel Building		Architect: As Shown						
Location:	**Anywhere, USA**						**01/01/19**	**OPN**
1 **Line Number**	**Description**	**Qty**	**Unit**	**Material**	**Labor**	**Equipment**	**SubContract**	**Estimate Total**
03 30 53.40 3940	Strip footing, 12" x 24", reinforced	15	C.Y.	$2,460.00	$1,282.50	$8.40	$0.00	
03 30 53.40 3950	Strip footing, 12" x 36", reinforced	34	C.Y.	$5,372.00	$2,329.00	$15.30	$0.00	
03 11 13.65 3000	Concrete slab edge forms	500	L.F.	$155.00	$910.00	$0.00	$0.00	
03 22 11.10 0200	Welded wire fabric reinforcing	150	C.S.F.	$3,000.00	$3,150.00	$0.00	$0.00	
03 31 13.35 0300	Ready mix concrete, 4000 psi for slab on grade	278	C.Y.	$35,584.00	$0.00	$0.00	$0.00	
03 31 13.70 4300	Place, strike off & consolidate concrete slab	278	C.Y.	$0.00	$3,864.20	$133.44	$0.00	
03 35 13.30 0250	Machine float & trowel concrete slab	15,000	S.F.	$0.00	$7,350.00	$300.00	$0.00	
03 15 16.20 0140	Cut control joints in concrete slab	950	L.F.	$47.50	$313.50	$57.00	$0.00	
03 39 23.13 0300	Sprayed concrete curing membrane	150	C.S.F.	$1,890.00	$765.00	$0.00	$0.00	
Division 03	**Subtotal**			**$48,508.50**	**$19,964.20**	**$514.14**	**$0.00**	**$68,986.84**
08 36 13.10 2650	Manual 10' x 10' steel sectional overhead door	8	Ea.	$10,600.00	$2,760.00	$0.00	$0.00	
08 36 13.10 2860	Insulation and steel back panel for OH door	800	S.F.	$4,120.00	$0.00	$0.00	$0.00	
Division 08	**Subtotal**			**$14,720.00**	**$2,760.00**	**$0.00**	**$0.00**	**$17,480.00**
13 34 19.50 1100	Pre-Engineered Steel Building, 100' x 150' x 24'	15,000	SF Flr.	$0.00	$0.00	$0.00	$322,500.00	
13 34 19.50 6050	Framing for PESB door opening, 3' x 7'	4	Opng.	$0.00	$0.00	$0.00	$1,960.00	
13 34 19.50 6100	Framing for PESB door opening, 10' x 10'	8	Opng.	$0.00	$0.00	$0.00	$8,400.00	
13 34 19.50 6200	Framing for PESB window opening, 4' x 3'	6	Opng.	$0.00	$0.00	$0.00	$2,940.00	
13 34 19.50 5750	PESB door, 3' x 7', single leaf	4	Opng.	$2,700.00	$532.00	$0.00	$0.00	
13 34 19.50 7750	PESB sliding window, 4' x 3' with screen	6	Opng.	$2,910.00	$306.00	$44.10	$0.00	
13 34 19.50 6550	PESB gutter, eave type, 26 ga., painted	300	L.F.	$2,160.00	$621.00	$0.00	$0.00	
13 34 19.50 8650	PESB roof vent, 12" wide x 10' long	15	Ea.	$562.50	$2,490.00	$0.00	$0.00	
13 34 19.50 6900	PESB insulation, vinyl faced, 4" thick	27,400	S.F.	$12,330.00	$7,398.00	$0.00	$0.00	
Division 13	**Subtotal**			**$20,662.50**	**$11,347.00**	**$44.10**	**$335,800.00**	**$367,853.60**
		Subtotal		**$83,891.00**	**$34,071.20**	**$558.24**	**$335,800.00**	**$454,320.44**
Division 01 2	**General Requirements @ 7%**			**5,872.37**	**2,384.98**	**39.08**	**23,506.00**	
		Estimate Subtotal		**$89,763.37**	**$36,456.18**	**$597.32**	**$359,306.00**	**$454,320.44**
	3	**Sales Tax @ 5%**		**4,488.17**		**29.87**	**8,982.65**	
		Subtotal A		**94,251.54**	**36,456.18**	**627.18**	**368,288.65**	
	4	**GC O & P**		**9,425.15**	**24,097.54**	**62.72**	**36,828.87**	
		Subtotal B		**103,676.69**	**60,553.72**	**689.90**	**405,117.52**	**$570,037.83**
	5	**Contingency @ 5%**						**28,501.89**
		Subtotal C						**$598,539.72**
	6	**Bond @ $12/1000 +10% O&P**						**7,900.72**
		Subtotal D						**$606,440.45**
	7	**Location Adjustment Factor**			**113.90**			**84,295.22**
		Grand Total						**$690,735.67**

This estimate is based on an interactive spreadsheet. You are free to download it and adjust it to your methodology. A copy of this spreadsheet is available at **RSMeans.com/2019books.**

Sample Estimate

This sample demonstrates the elements of an estimate, including a tally of the RSMeans data lines and a summary of the markups on a contractor's work to arrive at a total cost to the owner. The Location Factor with RSMeans data is added at the bottom of the estimate to adjust the cost of the work to a specific location.

---

1 Work Performed

The body of the estimate shows the RSMeans data selected, including the line number, a brief description of each item, its take-off unit and quantity, and the bare costs of materials, labor, and equipment. This estimate also includes a column titled "SubContract." This data is taken from the column "Total Incl O&P" and represents the total that a subcontractor would charge a general contractor for the work, including the sub's markup for overhead and profit.

2 Division 1, General Requirements

This is the first division numerically but the last division estimated. Division 1 includes project-wide needs provided by the general contractor. These requirements vary by project but may include temporary facilities and utilities, security, testing, project cleanup, etc. For small projects a percentage can be used—typically between 5% and 15% of project cost. For large projects the costs may be itemized and priced individually.

3 Sales Tax

If the work is subject to state or local sales taxes, the amount must be added to the estimate. Sales tax may be added to material costs, equipment costs, and subcontracted work. In this case, sales tax was added in all three categories. It was assumed that approximately half the subcontracted work would be material cost, so the tax was applied to 50% of the subcontract total.

4 GC O&P

This entry represents the general contractor's markup on material, labor, equipment, and subcontractor costs. Our standard markup on materials, equipment, and subcontracted work is 10%. In this estimate, the markup on the labor performed by the GC's workers uses "Skilled Workers Average" shown in Column F on the table "Installing Contractor's Overhead & Profit," which can be found on the inside back cover of the printed product or in the Reference Section of the electronic product.

5 Contingency

A factor for contingency may be added to any estimate to represent the cost of unknowns that may occur between the time that the estimate is performed and the time the project is constructed. The amount of the allowance will depend on the stage of design at which the estimate is done and the contractor's assessment of the risk involved. Refer to section 01 21 16.50 for contingency allowances.

6 Bonds

Bond costs should be added to the estimate. The figures here represent a typical performance bond, ensuring the owner that if the general contractor does not complete the obligations in the construction contract the bonding company will pay the cost for completion of the work.

7 Location Adjustment

Published prices are based on national average costs. If necessary, adjust the total cost of the project using a location factor from the "Location Factor" table or the "City Cost Index" table. Use location factors if the work is general, covering multiple trades. If the work is by a single trade (e.g., masonry) use the more specific data found in the "City Cost Indexes."

Estimating Tips

01 20 00 Price and Payment Procedures

- Allowances that should be added to estimates to cover contingencies and job conditions that are not included in the national average material and labor costs are shown in Section 01 21.
- When estimating historic preservation projects (depending on the condition of the existing structure and the owner's requirements), a 15–20% contingency or allowance is recommended, regardless of the stage of the drawings.

01 30 00 Administrative Requirements

- Before determining a final cost estimate, it is good practice to review all the items listed in Subdivisions 01 31 and 01 32 to make final adjustments for items that may need customizing to specific job conditions.
- Requirements for initial and periodic submittals can represent a significant cost to the General Requirements of a job. Thoroughly check the submittal specifications when estimating a project to determine any costs that should be included.

01 40 00 Quality Requirements

- All projects will require some degree of quality control. This cost is not included in the unit cost of construction listed in each division. Depending upon the terms of the contract, the various costs of inspection and testing can be the responsibility of either the owner or the contractor. Be sure to include the required costs in your estimate.

01 50 00 Temporary Facilities and Controls

- Barricades, access roads, safety nets, scaffolding, security, and many more requirements for the execution of a safe project are elements of direct cost. These costs can easily be overlooked when preparing an estimate. When looking through the major classifications of this subdivision, determine which items apply to each division in your estimate.
- Construction equipment rental costs can be found in the Reference Section in Section 01 54 33. Operators' wages are not included in equipment rental costs.
- Equipment mobilization and demobilization costs are not included in equipment rental costs and must be considered separately.
- The cost of small tools provided by the installing contractor for his workers is covered in the "Overhead" column on the "Installing Contractor's Overhead and Profit" table that lists labor trades, base rates, and markups. Therefore, it is included in the "Total Incl. O&P" cost of any unit price line item.

01 70 00 Execution and Closeout Requirements

- When preparing an estimate, thoroughly read the specifications to determine the requirements for Contract Closeout. Final cleaning, record documentation, operation and maintenance data, warranties and bonds, and spare parts and maintenance materials can all be elements of cost for the completion of a contract. Do not overlook these in your estimate.

Reference Numbers

Reference numbers are shown at the beginning of some major classifications. These numbers refer to related items in the Reference Section. The reference information may be an estimating procedure, an alternate pricing method, or technical information.

Note: Not all subdivisions listed here necessarily appear. ■

01 11 Summary of Work

01 11 31 – Professional Consultants

	01 11 31.10 Architectural Fees		Crew	Daily Output	Labor-Hours	Unit	2019 Bare Costs Material	Labor	Equipment	Total	Total Incl O&P
0010	**ARCHITECTURAL FEES**	R011110-10									
0020	For new construction										
0060	Minimum					Project				4.90%	4.90%
0090	Maximum									16%	16%
0100	For alteration work, to $500,000, add to new construction fee									50%	50%
0150	Over $500,000, add to new construction fee									25%	25%
2000	For "Greening" of building	G				↓				3%	3%

	01 11 31.20 Construction Management Fees	Crew	Daily Output	Labor-Hours	Unit	Material	Labor	Equipment	Total	Total Incl O&P
0010	**CONSTRUCTION MANAGEMENT FEES**									
0060	For work to $100,000				Project				10%	10%
0070	To $250,000								9%	9%
0090	To $1,000,000				↓				6%	6%

	01 11 31.30 Engineering Fees		Crew	Daily Output	Labor-Hours	Unit	Material	Labor	Equipment	Total	Total Incl O&P
0010	**ENGINEERING FEES**	R011110-30									
1200	Structural, minimum					Project				1%	1%
1300	Maximum					"				2.50%	2.50%

	01 11 31.75 Renderings	Crew	Daily Output	Labor-Hours	Unit	Material	Labor	Equipment	Total	Total Incl O&P
0010	**RENDERINGS** Color, matted, 20" x 30", eye level,									
0020	1 building, minimum				Ea.	2,225			2,225	2,450
0050	Average					3,175			3,175	3,500
0100	Maximum				↓	5,125			5,125	5,625

01 21 Allowances

01 21 16 – Contingency Allowances

	01 21 16.50 Contingencies	Crew	Daily Output	Labor-Hours	Unit	Material	Labor	Equipment	Total	Total Incl O&P
0010	**CONTINGENCIES**, Add to estimate									
0020	Conceptual stage				Project				20%	20%
0150	Final working drawing stage				"				3%	3%

01 21 53 – Factors Allowance

	01 21 53.60 Security Factors		Crew	Daily Output	Labor-Hours	Unit	Material	Labor	Equipment	Total	Total Incl O&P
0010	**SECURITY FACTORS**	R012153-60									
0100	Additional costs due to security requirements										
0110	Daily search of personnel, supplies, equipment and vehicles										
0120	Physical search, inventory and doc of assets, at entry					Costs		30%			
0130	At entry and exit							50%			
0140	Physical search, at entry							6.25%			
0150	At entry and exit							12.50%			
0160	Electronic scan search, at entry							2%			
0170	At entry and exit							4%			
0180	Visual inspection only, at entry							.25%			
0190	At entry and exit							.50%			
0200	ID card or display sticker only, at entry							.12%			
0210	At entry and exit					↓		.25%			
0220	Day 1 as described below, then visual only for up to 5 day job duration										
0230	Physical search, inventory and doc of assets, at entry					Costs		5%			
0240	At entry and exit							10%			
0250	Physical search, at entry							1.25%			
0260	At entry and exit							2.50%			
0270	Electronic scan search, at entry							.42%			
0280	At entry and exit					↓		.83%			

01 21 Allowances

01 21 53 – Factors Allowance

01 21 53.60 Security Factors		Crew	Daily Output	Labor-Hours	Unit	Material	2019 Bare Costs Labor	Equipment	Total	Total Incl O&P
0290	Day 1 as described below, then visual only for 6-10 day job duration									
0300	Physical search, inventory and doc of assets, at entry				Costs		2.50%			
0310	At entry and exit						5%			
0320	Physical search, at entry						.63%			
0330	At entry and exit						1.25%			
0340	Electronic scan search, at entry						.21%			
0350	At entry and exit				↓		.42%			
0360	Day 1 as described below, then visual only for 11-20 day job duration									
0370	Physical search, inventory and doc of assets, at entry				Costs		1.25%			
0380	At entry and exit						2.50%			
0390	Physical search, at entry						.31%			
0400	At entry and exit						.63%			
0410	Electronic scan search, at entry						.10%			
0420	At entry and exit				↓		.21%			
0430	Beyond 20 days, costs are negligible									
0440	Escort required to be with tradesperson during work effort				Costs		6.25%			

01 21 63 – Taxes

01 21 63.10 Taxes			Crew	Daily Output	Labor-Hours	Unit	Material	Labor	Equipment	Total	Total Incl O&P
0010	**TAXES**	R012909-80									
0020	Sales tax, State, average					%	5.08%				
0050	Maximum	R012909-85					7.50%				
0200	Social Security, on first $118,500 of wages							7.65%			
0300	Unemployment, combined Federal and State, minimum	R012909-86						.60%			
0350	Average							9.60%			
0400	Maximum					↓		12%			

01 31 Project Management and Coordination

01 31 13 – Project Coordination

01 31 13.30 Insurance			Crew	Daily Output	Labor-Hours	Unit	Material	Labor	Equipment	Total	Total Incl O&P
0010	**INSURANCE**	R013113-40									
0020	Builders risk, standard, minimum					Job				.24%	.24%
0050	Maximum									.64%	.64%
0200	All-risk type, minimum									.25%	.25%
0250	Maximum	R013113-60				↓				.62%	.62%
0400	Contractor's equipment floater, minimum					Value				.50%	.50%
0450	Maximum					"				1.50%	1.50%
0600	Public liability, average					Job				2.02%	2.02%
0800	Workers' compensation & employer's liability, average										
0850	by trade, carpentry, general					Payroll		11.97%			
0900	Clerical							.38%			
0950	Concrete							10.84%			
1000	Electrical							4.91%			
1050	Excavation							7.81%			
1100	Glazing							11.29%			
1150	Insulation							10.07%			
1200	Lathing							7.58%			
1250	Masonry							13.40%			
1300	Painting & decorating							10.44%			
1350	Pile driving							12.06%			
1400	Plastering					↓		10.24%			

01 31 Project Management and Coordination

01 31 13 – Project Coordination

01 31 13.30 Insurance		Crew	Daily Output	Labor-Hours	Unit	Material	2019 Bare Costs Labor	Equipment	Total	Total Incl O&P
1450	Plumbing				Payroll		5.77%			
1500	Roofing						27.34%			
1550	Sheet metal work (HVAC)						7.56%			
1600	Steel erection, structural						17.21%			
1650	Tile work, interior ceramic						8.10%			
1700	Waterproofing, brush or hand caulking						6.05%			
1800	Wrecking						15.48%			
2000	Range of 35 trades in 50 states, excl. wrecking & clerical, min.						1.37%			
2100	Average						10.60%			
2200	Maximum				↓		120.29%			

01 31 13.40 Main Office Expense		Crew	Daily Output	Labor-Hours	Unit	Material	Labor	Equipment	Total	Total Incl O&P
0010	**MAIN OFFICE EXPENSE** Average for General Contractors R013113-50									
0020	As a percentage of their annual volume									
0125	Annual volume under $1,000,000				% Vol.				17.50%	
0145	Up to $2,500,000								8%	
0150	Up to $4,000,000								6.80%	
0200	Up to $7,000,000				↓				5.60%	

01 31 13.80 Overhead and Profit		Crew	Daily Output	Labor-Hours	Unit	Material	Labor	Equipment	Total	Total Incl O&P
0010	**OVERHEAD & PROFIT** Allowance to add to items in this									
0020	book that do not include Subs O&P, average				%				25%	
0100	Allowance to add to items in this book that R013113-55									
0110	do include Subs O&P, minimum				%				5%	5%
0150	Average								10%	10%
0200	Maximum								15%	15%
0290	Typical, by size of project, under $50,000								40%	
0310	$50,000 to $100,000								35%	
0320	$100,000 to $500,000								25%	
0330	$500,000 to $1,000,000				↓				20%	

01 41 Regulatory Requirements

01 41 26 – Permit Requirements

01 41 26.50 Permits		Crew	Daily Output	Labor-Hours	Unit	Material	Labor	Equipment	Total	Total Incl O&P
0010	**PERMITS**									
0020	Rule of thumb, most cities, minimum				Job				.50%	.50%
0100	Maximum				"				2%	2%

01 45 Quality Control

01 45 23 – Testing and Inspecting Services

01 45 23.50 Testing		Crew	Daily Output	Labor-Hours	Unit	Material	Labor	Equipment	Total	Total Incl O&P
0010	**TESTING** and Inspecting Services									
0015	For concrete building costing $1,000,000, minimum				Project				4,725	5,200
0020	Maximum								38,000	41,800
0050	Steel building, minimum								4,725	5,200
0070	Maximum				↓				14,800	16,300
0600	Concrete testing, aggregates, abrasion, ASTM C 131				Ea.				136	150
0650	Absorption, ASTM C 127								42	46
1800	Compressive test, cylinder, delivered to lab, ASTM C 39								22	25
1900	Picked up by lab, minimum								65	72
1950	Average				↓				108	119

01 45 Quality Control

01 45 23 – Testing and Inspecting Services

01 45 23.50 Testing		Crew	Daily Output	Labor-Hours	Unit	Material	2019 Bare Costs Labor	Equipment	Total	Total Incl O&P
2000	Maximum				Ea.				150	166
2200	Compressive strength, cores (not incl. drilling), ASTM C 42								45	50
2250	Core drilling, 4" diameter (plus technician), up to 6" thick	B-89A	14	1.143		.78	40	8	48.78	75.50
2260	Technician for core drilling				Hr.				50	55
2300	Patching core holes				Ea.				22	24
9000	Thermographic testing, for bldg envelope heat loss, average 2,000 S.F. G				"				500	500

01 52 Construction Facilities

01 52 13 – Field Offices and Sheds

01 52 13.20 Office and Storage Space		Crew	Daily Output	Labor-Hours	Unit	Material	Labor	Equipment	Total	Total Incl O&P
0010	**OFFICE AND STORAGE SPACE**									
0020	Office trailer, furnished, no hookups, 20' x 8', buy	2 Skwk	1	16	Ea.	9,050	630		9,680	11,000
0250	Rent per month					201			201	222
0300	32' x 8', buy	2 Skwk	.70	22.857		14,500	900		15,400	17,400
0350	Rent per month					254			254	280
0700	For air conditioning, rent per month, add					51			51	56
1200	Storage boxes, 20' x 8', buy	2 Skwk	1.80	8.889		3,250	350		3,600	4,150
1250	Rent per month					85			85	93.50

01 52 13.40 Field Office Expense		Crew	Daily Output	Labor-Hours	Unit	Material	Labor	Equipment	Total	Total Incl O&P
0010	**FIELD OFFICE EXPENSE**									
0100	Office equipment rental average				Month	207			207	227
0120	Office supplies, average					85			85	93.50
0140	Telephone bill; avg. bill/month incl. long dist.					87			87	95.50
0160	Lights & HVAC					162			162	179

01 54 Construction Aids

01 54 16 – Temporary Hoists

01 54 16.50 Weekly Forklift Crew		Crew	Daily Output	Labor-Hours	Unit	Material	Labor	Equipment	Total	Total Incl O&P
0010	**WEEKLY FORKLIFT CREW**									
0100	All-terrain forklift, 45' lift, 35' reach, 9000 lb. capacity	A-3P	.20	40	Week		1,550	2,325	3,875	5,100

01 54 19 – Temporary Cranes

01 54 19.50 Daily Crane Crews		Crew	Daily Output	Labor-Hours	Unit	Material	Labor	Equipment	Total	Total Incl O&P
0010	**DAILY CRANE CREWS** for small jobs, portal to portal R015433-15									
0100	12-ton truck-mounted hydraulic crane	A-3H	1	8	Day		335	710	1,045	1,325
0200	25-ton	A-3I	1	8			335	785	1,120	1,425
0300	40-ton	A-3J	1	8			335	1,275	1,610	1,950
0400	55-ton	A-3K	1	16			625	1,450	2,075	2,600
0500	80-ton	A-3L	1	16			625	2,200	2,825	3,450
0900	If crane is needed on a Saturday, Sunday or Holiday									
0910	At time-and-a-half, add				Day		50%			
0920	At double time, add				"		100%			

01 54 19.60 Monthly Tower Crane Crew		Crew	Daily Output	Labor-Hours	Unit	Material	Labor	Equipment	Total	Total Incl O&P
0010	**MONTHLY TOWER CRANE CREW**, excludes concrete footing									
0100	Static tower crane, 130' high, 106' jib, 6200 lb. capacity	A-3N	.05	176	Month		7,375	26,000	33,375	40,700

01 54 23 – Temporary Scaffolding and Platforms

01 54 23.60 Pump Staging

	01 54 23.60 Pump Staging	Crew	Daily Output	Labor-Hours	Unit	2019 Bare Costs Material	Labor	Equipment	Total	Total Incl O&P
0010	**PUMP STAGING**, Aluminum R015423-20									
0200	24′ long pole section, buy				Ea.	390			390	425
0300	18′ long pole section, buy					315			315	350
0400	12′ long pole section, buy					203			203	223
0500	6′ long pole section, buy					119			119	130
0600	6′ long splice joint section, buy					79.50			79.50	87.50
0700	Pump jack, buy					156			156	171
0900	Foldable brace, buy					62			62	68.50
1000	Workbench/back safety rail support, buy					83			83	91
1100	Scaffolding planks/workbench, 14″ wide x 24′ long, buy					725			725	800
1200	Plank end safety rail, buy					360			360	395
1250	Safety net, 22′ long, buy					375			375	415
1300	System in place, 50′ working height, per use based on 50 uses	2 Carp	84.80	.189	C.S.F.	6.55	7.30		13.85	19.30
1400	100 uses		84.80	.189		3.27	7.30		10.57	15.70
1500	150 uses		84.80	.189		2.19	7.30		9.49	14.50

01 54 23.70 Scaffolding

	01 54 23.70 Scaffolding	Crew	Daily Output	Labor-Hours	Unit	Material	Labor	Equipment	Total	Total Incl O&P
0010	**SCAFFOLDING** R015423-10									
0015	Steel tube, regular, no plank, labor only to erect & dismantle									
0091	Building exterior, wall face, 1 to 5 stories, 6′-4″ x 5′ frames	3 Clab	8	3	C.S.F.		91		91	150
0201	6 to 12 stories	4 Clab	8	4			121		121	201
0310	13 to 20 stories	5 Carp	8	5			194		194	320
0461	Building interior, wall face area, up to 16′ high	3 Clab	12	2			60.50		60.50	100
0561	16′ to 40′ high		10	2.400			73		73	120
0801	Building interior floor area, up to 30′ high		150	.160	C.C.F.		4.86		4.86	8
0901	Over 30′ high	4 Clab	160	.200	″		6.05		6.05	10.05
0906	Complete system for face of walls, no plank, material only rent/mo				C.S.F.	33			33	36.50
0908	Interior spaces, no plank, material only rent/mo				C.C.F.	3.80			3.80	4.18
0910	Steel tubular, heavy duty shoring, buy									
0920	Frames 5′ high 2′ wide				Ea.	92			92	101
0925	5′ high 4′ wide					97.50			97.50	107
0930	6′ high 2′ wide					99			99	109
0935	6′ high 4′ wide					116			116	128
0940	Accessories									
0945	Cross braces				Ea.	17.10			17.10	18.80
0950	U-head, 8″ x 8″					18.90			18.90	21
0955	J-head, 4″ x 8″					13.75			13.75	15.15
0960	Base plate, 8″ x 8″					15.15			15.15	16.65
0965	Leveling jack					36			36	39.50
1000	Steel tubular, regular, buy									
1100	Frames 3′ high 5′ wide				Ea.	89.50			89.50	98.50
1150	5′ high 5′ wide					108			108	119
1200	6′-4″ high 5′ wide					106			106	117
1350	7′-6″ high 6′ wide					157			157	173
1500	Accessories, cross braces					18.05			18.05	19.85
1550	Guardrail post					19.90			19.90	22
1600	Guardrail 7′ section					8.15			8.15	8.95
1650	Screw jacks & plates					25.50			25.50	28.50
1700	Sidearm brackets					22.50			22.50	25
1750	8″ casters					37.50			37.50	41
1800	Plank 2″ x 10″ x 16′-0″					67			67	74
1900	Stairway section					284			284	310
1910	Stairway starter bar					32.50			32.50	35.50

01 54 23 – Temporary Scaffolding and Platforms

01 54 23.70 Scaffolding		Crew	Daily Output	Labor-Hours	Unit	2019 Bare Costs Material	Labor	Equipment	Total	Total Incl O&P
1920	Stairway inside handrail				Ea.	53			53	58.50
1930	Stairway outside handrail					84.50			84.50	93
1940	Walk-thru frame guardrail				↓	42			42	46
2000	Steel tubular, regular, rent/mo.									
2100	Frames 3′ high 5′ wide				Ea.	4.45			4.45	4.90
2150	5′ high 5′ wide					4.45			4.45	4.90
2200	6′-4″ high 5′ wide					5.40			5.40	5.95
2250	7′-6″ high 6′ wide					9.85			9.85	10.85
2500	Accessories, cross braces					.89			.89	.98
2550	Guardrail post					.89			.89	.98
2600	Guardrail 7′ section					.89			.89	.98
2650	Screw jacks & plates					1.78			1.78	1.96
2700	Sidearm brackets					1.78			1.78	1.96
2750	8″ casters					7.10			7.10	7.80
2800	Outrigger for rolling tower					2.67			2.67	2.94
2850	Plank 2″ x 10″ x 16′-0″					9.85			9.85	10.85
2900	Stairway section					31.50			31.50	34.50
2940	Walk-thru frame guardrail				↓	2.22			2.22	2.44
3000	Steel tubular, heavy duty shoring, rent/mo.									
3250	5′ high 2′ & 4′ wide				Ea.	8.35			8.35	9.20
3300	6′ high 2′ & 4′ wide					8.35			8.35	9.20
3500	Accessories, cross braces					.89			.89	.98
3600	U-head, 8″ x 8″					2.46			2.46	2.71
3650	J-head, 4″ x 8″					2.46			2.46	2.71
3700	Base plate, 8″ x 8″					.89			.89	.98
3750	Leveling jack					2.46			2.46	2.71
5700	Planks, 2″ x 10″ x 16′-0″, labor only to erect & remove to 50′ H	3 Carp	72	.333			12.90		12.90	21.50
5800	Over 50′ high	4 Carp	80	.400	↓		15.50		15.50	25.50

01 54 23.75 Scaffolding Specialties		Crew	Daily Output	Labor-Hours	Unit	Material	Labor	Equipment	Total	Total Incl O&P
0010	**SCAFFOLDING SPECIALTIES**									
5000	Motorized work platform, mast climber									
5050	Base unit, 50′ W, less than 100′ tall, rent/mo				Ea.	3,375			3,375	3,725
5100	Less than 200′ tall, rent/mo					3,925			3,925	4,300
5150	Less than 300′ tall, rent/mo					4,525			4,525	5,000
5200	Less than 400′ tall, rent/mo				↓	5,275			5,275	5,800
5250	Set up and demob, per unit, less than 100′ tall	B-68F	16.60	1.446	C.S.F.		58	16.65	74.65	115
5300	Less than 200′ tall		25	.960			38.50	11.05	49.55	76
5350	Less than 300′ tall		30	.800			32	9.20	41.20	63.50
5400	Less than 400′ tall	↓	33	.727	↓		29	8.35	37.35	57.50
5500	Mobilization (price includes freight in and out) per unit	B-34N	1	16	Ea.		605	670	1,275	1,725

01 54 23.80 Staging Aids		Crew	Daily Output	Labor-Hours	Unit	Material	Labor	Equipment	Total	Total Incl O&P
0010	**STAGING AIDS** and fall protection equipment									
0110	Cost each per day, based on 250 days use				Day	.23			.23	.26
0200	Guard post, buy				Ea.	55			55	60.50
0210	Cost each per day, based on 250 days use				Day	.22			.22	.24
0300	End guard chains, buy per pair				Pair	46.50			46.50	51
0310	Cost per set per day, based on 250 days use				Day	.23			.23	.26
1010	Cost each per day, based on 250 days use				″	.05			.05	.05
1100	Wood bracket, buy				Ea.	25.50			25.50	28
1110	Cost each per day, based on 250 days use				Day	.10			.10	.11
2010	Cost per pair per day, based on 250 days use					.55			.55	.61
3010	Cost each per day, based on 250 days use				↓	.23			.23	.26

01 54 Construction Aids

01 54 23 – Temporary Scaffolding and Platforms

01 54 23.80 Staging Aids		Crew	Daily Output	Labor-Hours	Unit	2019 Bare Costs Material	Labor	Equipment	Total	Total Incl O&P
3100	Aluminum scaffolding plank, 20" wide x 24' long, buy				Ea.	735			735	810
3110	Cost each per day, based on 250 days use				Day	2.94			2.94	3.24
4010	Cost each per day, based on 250 days use				"	.67			.67	.74
4100	Rope for safety line, 5/8" x 100' nylon, buy				Ea.	56.50			56.50	62
4110	Cost each per day, based on 250 days use				Day	.23			.23	.25
4200	Permanent U-Bolt roof anchor, buy				Ea.	30.50			30.50	33.50
4300	Temporary (one use) roof ridge anchor, buy				"	6.40			6.40	7.05
5000	Installation (setup and removal) of staging aids									
5010	Sidewall staging bracket	2 Carp	64	.250	Ea.		9.70		9.70	16
5020	Guard post with 2 wood rails	"	64	.250			9.70		9.70	16
5030	End guard chains, set	1 Carp	64	.125			4.84		4.84	8
5100	Roof shingling bracket		96	.083			3.23		3.23	5.35
5200	Ladder jack		64	.125			4.84		4.84	8
5300	Wood plank, 2" x 10" x 16'	2 Carp	80	.200			7.75		7.75	12.80
5310	Aluminum scaffold plank, 20" x 24'	"	40	.400			15.50		15.50	25.50
5410	Safety rope	1 Carp	40	.200			7.75		7.75	12.80
5420	Permanent U-Bolt roof anchor (install only)	2 Carp	40	.400			15.50		15.50	25.50
5430	Temporary roof ridge anchor (install only)	1 Carp	64	.125			4.84		4.84	8

01 54 36 – Equipment Mobilization

01 54 36.50 Mobilization		Crew	Daily Output	Labor-Hours	Unit	Material	Labor	Equipment	Total	Total Incl O&P
0010	**MOBILIZATION** (Use line item again for demobilization) R015436-50									
0015	Up to 25 mi. haul dist. (50 mi. RT for mob/demob crew)									
1200	Small equipment, placed in rear of, or towed by pickup truck	A-3A	4	2	Ea.		77.50	30.50	108	161
1300	Equipment hauled on 3-ton capacity towed trailer	A-3Q	2.67	3			116	55.50	171.50	252
1400	20-ton capacity	B-34U	2	8			294	218	512	725
1500	40-ton capacity	B-34N	2	8			300	335	635	865
1600	50-ton capacity	B-34V	1	24			925	920	1,845	2,525
1700	Crane, truck-mounted, up to 75 ton (driver only)	1 Eqhv	4	2			84		84	138
2500	For each additional 5 miles haul distance, add						10%	10%		
3000	For large pieces of equipment, allow for assembly/knockdown									
3100	For mob/demob of micro-tunneling equip, see Section 33 05 23.19									

01 56 Temporary Barriers and Enclosures

01 56 13 – Temporary Air Barriers

01 56 13.60 Tarpaulins		Crew	Daily Output	Labor-Hours	Unit	Material	Labor	Equipment	Total	Total Incl O&P
0010	**TARPAULINS**									
0020	Cotton duck, 10-13.13 oz./S.Y., 6' x 8'				S.F.	.86			.86	.95
0050	30' x 30'					.60			.60	.66
0200	Reinforced polyethylene 3 mils thick, white					.04			.04	.04
0300	4 mils thick, white, clear or black					.13			.13	.14
0730	Polyester reinforced w/integral fastening system, 11 mils thick					.19			.19	.21

01 56 16 – Temporary Dust Barriers

01 56 16.10 Dust Barriers, Temporary		Crew	Daily Output	Labor-Hours	Unit	Material	Labor	Equipment	Total	Total Incl O&P
0010	**DUST BARRIERS, TEMPORARY**									
0020	Spring loaded telescoping pole & head, to 12', erect and dismantle	1 Clab	240	.033	Ea.		1.01		1.01	1.67
0025	Cost per day (based upon 250 days)				Day	.22			.22	.24
0030	To 21', erect and dismantle	1 Clab	240	.033	Ea.		1.01		1.01	1.67
0035	Cost per day (based upon 250 days)				Day	.68			.68	.74
0040	Accessories, caution tape reel, erect and dismantle	1 Clab	480	.017	Ea.		.51		.51	.84
0045	Cost per day (based upon 250 days)				Day	.32			.32	.35

01 56 Temporary Barriers and Enclosures

01 56 16 – Temporary Dust Barriers

01 56 16.10 Dust Barriers, Temporary		Crew	Daily Output	Labor-Hours	Unit	2019 Bare Costs Material	Labor	Equipment	Total	Total Incl O&P
0060	Foam rail and connector, erect and dismantle	1 Clab	240	.033	Ea.		1.01		1.01	1.67
0065	Cost per day (based upon 250 days)				Day	.10			.10	.11
0070	Caution tape	1 Clab	384	.021	C.L.F.	2.69	.63		3.32	4
0080	Zipper, standard duty		60	.133	Ea.	7.40	4.05		11.45	14.85
0090	Heavy duty		48	.167	"	9.95	5.05		15	19.25
0100	Polyethylene sheet, 4 mil		37	.216	Sq.	2.66	6.55		9.21	13.80
0110	6 mil		37	.216	"	3.87	6.55		10.42	15.10
1000	Dust partition, 6 mil polyethylene, 1" x 3" frame	2 Carp	2000	.008	S.F.	.33	.31		.64	.87
1080	2" x 4" frame	"	2000	.008	"	.35	.31		.66	.90
1085	Negative air machine, 1800 CFM				Ea.	910			910	1,000
1090	Adhesive strip application, 2" width	1 Clab	192	.042	C.L.F.	6.90	1.26		8.16	9.65

01 66 Product Storage and Handling Requirements

01 66 19 – Material Handling

01 66 19.10 Material Handling		Crew	Daily Output	Labor-Hours	Unit	Material	Labor	Equipment	Total	Total Incl O&P
0010	**MATERIAL HANDLING**									
0020	Above 2nd story, via stairs, per C.Y. of material per floor	2 Clab	145	.110	C.Y.		3.35		3.35	5.55
0030	Via elevator, per C.Y. of material		240	.067			2.02		2.02	3.34
0050	Distances greater than 200', per C.Y. of material per each addl 200'		300	.053			1.62		1.62	2.67

01 71 Examination and Preparation

01 71 23 – Field Engineering

01 71 23.13 Construction Layout		Crew	Daily Output	Labor-Hours	Unit	Material	Labor	Equipment	Total	Total Incl O&P
0010	**CONSTRUCTION LAYOUT**									
1100	Crew for layout of building, trenching or pipe laying, 2 person crew	A-6	1	16	Day		620	45.50	665.50	1,075
1200	3 person crew	A-7	1	24	"		995	45.50	1,040.50	1,700

01 74 Cleaning and Waste Management

01 74 13 – Progress Cleaning

01 74 13.20 Cleaning Up		Crew	Daily Output	Labor-Hours	Unit	Material	Labor	Equipment	Total	Total Incl O&P
0010	**CLEANING UP**									
0020	After job completion, allow, minimum				Job				.30%	.30%
0040	Maximum				"				1%	1%

01 76 Protecting Installed Construction

01 76 13 – Temporary Protection of Installed Construction

01 76 13.20 Temporary Protection		Crew	Daily Output	Labor-Hours	Unit	Material	Labor	Equipment	Total	Total Incl O&P
0010	**TEMPORARY PROTECTION**									
0020	Flooring, 1/8" tempered hardboard, taped seams	2 Carp	1500	.011	S.F.	.46	.41		.87	1.19
0030	Peel away carpet protection	1 Clab	3200	.003	"	.13	.08		.21	.27

Division Notes

		CREW	DAILY OUTPUT	LABOR-HOURS	UNIT	BARE COSTS MAT.	BARE COSTS LABOR	BARE COSTS EQUIP.	BARE COSTS TOTAL	TOTAL INCL O&P

Division 2 Existing Conditions

Estimating Tips

02 30 00 Subsurface Investigation

In preparing estimates on structures involving earthwork or foundations, all information concerning soil characteristics should be obtained. Look particularly for hazardous waste, evidence of prior dumping of debris, and previous stream beds.

02 40 00 Demolition and Structure Moving

The costs shown for selective demolition do not include rubbish handling or disposal. These items should be estimated separately using RSMeans data or other sources.

- Historic preservation often requires that the contractor remove materials from the existing structure, rehab them, and replace them. The estimator must be aware of any related measures and precautions that must be taken when doing selective demolition and cutting and patching. Requirements may include special handling and storage, as well as security.
- In addition to Subdivision 02 41 00, you can find selective demolition items in each division. Example: Roofing demolition is in Division 7.
- Absent of any other specific reference, an approximate demolish-in-place cost can be obtained by halving the new-install labor cost. To remove for reuse, allow the entire new-install labor figure.

02 40 00 Building Deconstruction

This section provides costs for the careful dismantling and recycling of most low-rise building materials.

02 50 00 Containment of Hazardous Waste

This section addresses on-site hazardous waste disposal costs.

02 80 00 Hazardous Material Disposal/Remediation

This subdivision includes information on hazardous waste handling, asbestos remediation, lead remediation, and mold remediation. See reference numbers R028213-20 and R028319-60 for further guidance in using these unit price lines.

02 90 00 Monitoring Chemical Sampling, Testing Analysis

This section provides costs for on-site sampling and testing hazardous waste.

Reference Numbers

Reference numbers are shown at the beginning of some major classifications. These numbers refer to related items in the Reference Section. The reference information may be an estimating procedure, an alternate pricing method, or technical information.

Note: Not all subdivisions listed here necessarily appear. ■

02 21 Surveys

02 21 13 – Site Surveys

02 21 13.09 Topographical Surveys		Crew	Daily Output	Labor-Hours	Unit	Material	2019 Bare Costs Labor	Equipment	Total	Total Incl O&P
0010	TOPOGRAPHICAL SURVEYS									
0020	Topographical surveying, conventional, minimum	A-7	3.30	7.273	Acre	23	300	13.75	336.75	540
0100	Maximum	A-8	.60	53.333	"	61	2,175	75.50	2,311.50	3,725

02 21 13.13 Boundary and Survey Markers		Crew	Daily Output	Labor-Hours	Unit	Material	Labor	Equipment	Total	Total Incl O&P
0010	BOUNDARY AND SURVEY MARKERS									
0300	Lot location and lines, large quantities, minimum	A-7	2	12	Acre	35.50	500	22.50	558	890
0320	Average	"	1.25	19.200		61.50	795	36.50	893	1,425
0400	Small quantities, maximum	A-8	1	32	↓	76	1,300	45.50	1,421.50	2,275
0600	Monuments, 3′ long	A-7	10	2.400	Ea.	32	99.50	4.54	136.04	205
0800	Property lines, perimeter, cleared land	"	1000	.024	L.F.	.08	1	.05	1.13	1.79
0900	Wooded land	A-8	875	.037	"	.10	1.49	.05	1.64	2.63

02 32 Geotechnical Investigations

02 32 13 – Subsurface Drilling and Sampling

02 32 13.10 Boring and Exploratory Drilling		Crew	Daily Output	Labor-Hours	Unit	Material	Labor	Equipment	Total	Total Incl O&P
0010	BORING AND EXPLORATORY DRILLING									
0020	Borings, initial field stake out & determination of elevations	A-6	1	16	Day		620	45.50	665.50	1,075
0100	Drawings showing boring details				Total		335		335	425
0200	Report and recommendations from P.E.						775		775	970
0300	Mobilization and demobilization	B-55	4	4	↓		129	251	380	490
0350	For over 100 miles, per added mile		450	.036	Mile		1.15	2.23	3.38	4.34
0600	Auger holes in earth, no samples, 2-1/2″ diameter		78.60	.204	L.F.		6.55	12.75	19.30	25
0800	Cased borings in earth, with samples, 2-1/2″ diameter		55.50	.288	"	17.05	9.30	18.05	44.40	54
1400	Borings, earth, drill rig and crew with truck mounted auger	↓	1	16	Day		515	1,000	1,515	1,950
1500	For inner city borings, add, minimum								10%	10%
1510	Maximum								20%	20%

02 41 Demolition

02 41 13 – Selective Site Demolition

02 41 13.17 Demolish, Remove Pavement and Curb		Crew	Daily Output	Labor-Hours	Unit	Material	Labor	Equipment	Total	Total Incl O&P
0010	DEMOLISH, REMOVE PAVEMENT AND CURB R024119-10									
5010	Pavement removal, bituminous roads, up to 3″ thick	B-38	690	.035	S.Y.		1.15	1.62	2.77	3.69
5050	4″-6″ thick		420	.057			1.89	2.67	4.56	6.05
5100	Bituminous driveways		640	.038			1.24	1.75	2.99	3.97
5200	Concrete to 6″ thick, hydraulic hammer, mesh reinforced		255	.094			3.12	4.39	7.51	10
5300	Rod reinforced	↓	200	.120	↓		3.98	5.60	9.58	12.70
5600	With hand held air equipment, bituminous, to 6″ thick	B-39	1900	.025	S.F.		.78	.12	.90	1.42
5700	Concrete to 6″ thick, no reinforcing		1600	.030			.92	.15	1.07	1.68
5800	Mesh reinforced		1400	.034			1.05	.17	1.22	1.92
5900	Rod reinforced	↓	765	.063	↓		1.93	.31	2.24	3.52
6000	Curbs, concrete, plain	B-6	360	.067	L.F.		2.21	.89	3.10	4.62
6100	Reinforced		275	.087			2.89	1.16	4.05	6.05
6200	Granite		360	.067			2.21	.89	3.10	4.62
6300	Bituminous	↓	528	.045	↓		1.51	.61	2.12	3.15

02 41 13.23 Utility Line Removal		Crew	Daily Output	Labor-Hours	Unit	Material	Labor	Equipment	Total	Total Incl O&P
0010	UTILITY LINE REMOVAL									
0015	No hauling, abandon catch basin or manhole	B-6	7	3.429	Ea.		114	45.50	159.50	238
0020	Remove existing catch basin or manhole, masonry		4	6			199	80	279	420
0030	Catch basin or manhole frames and covers, stored	↓	13	1.846	↓		61	24.50	85.50	128

02 41 Demolition

02 41 13 – Selective Site Demolition

02 41 13.23 Utility Line Removal		Crew	Daily Output	Labor-Hours	Unit	Material	2019 Bare Costs Labor	Equipment	Total	Total Incl O&P
0040	Remove and reset	B-6	7	3.429	Ea.		114	45.50	159.50	238
2900	Pipe removal, sewer/water, no excavation, 12″ diameter	↓	175	.137	L.F.		4.55	1.83	6.38	9.50
2930	15″-18″ diameter	B-12Z	150	.160			5.50	9.60	15.10	19.55
2960	21″-24″ diameter	″	120	.200	↓		6.85	12	18.85	24.50

02 41 13.30 Minor Site Demolition		Crew	Daily Output	Labor-Hours	Unit	Material	Labor	Equipment	Total	Total Incl O&P
0010	**MINOR SITE DEMOLITION** R024119-10									
1000	Masonry walls, block, solid	B-5	1800	.022	C.F.		.73	.80	1.53	2.08
1200	Brick, solid		900	.044			1.46	1.60	3.06	4.17
1400	Stone, with mortar		900	.044			1.46	1.60	3.06	4.17
1500	Dry set	↓	1500	.027	↓		.88	.96	1.84	2.49
4000	Sidewalk removal, bituminous, 2″ thick	B-6	350	.069	S.Y.		2.27	.91	3.18	4.76
4010	2-1/2″ thick		325	.074			2.45	.99	3.44	5.10
4050	Brick, set in mortar		185	.130			4.30	1.73	6.03	9
4100	Concrete, plain, 4″		160	.150			4.97	2	6.97	10.40
4110	Plain, 5″		140	.171			5.70	2.29	7.99	11.85
4120	Plain, 6″		120	.200			6.65	2.67	9.32	13.90
4200	Mesh reinforced, concrete, 4″		150	.160			5.30	2.13	7.43	11.10
4210	5″ thick		131	.183			6.05	2.44	8.49	12.70
4220	6″ thick	↓	112	.214	↓		7.10	2.86	9.96	14.85
4300	Slab on grade removal, plain	B-5	45	.889	C.Y.		29	32	61	83
4310	Mesh reinforced		33	1.212			40	43.50	83.50	114
4320	Rod reinforced	↓	25	1.600			52.50	57.50	110	150
4400	For congested sites or small quantities, add up to								200%	200%
4450	For disposal on site, add	B-11A	232	.069			2.45	5.55	8	10.15
4500	To 5 miles, add	B-34D	76	.105	↓		3.66	8.35	12.01	15.25

02 41 13.60 Selective Demolition Fencing		Crew	Daily Output	Labor-Hours	Unit	Material	Labor	Equipment	Total	Total Incl O&P
0010	**SELECTIVE DEMOLITION FENCING** R024119-10									
1600	Fencing, barbed wire, 3 strand	2 Clab	430	.037	L.F.		1.13		1.13	1.87
1650	5 strand	″	280	.057			1.73		1.73	2.87
1700	Chain link, posts & fabric, 8′-10′ high, remove only	B-6	445	.054	↓		1.79	.72	2.51	3.74

02 41 16 – Structure Demolition

02 41 16.13 Building Demolition		Crew	Daily Output	Labor-Hours	Unit	Material	Labor	Equipment	Total	Total Incl O&P
0010	**BUILDING DEMOLITION** Large urban projects, incl. 20 mi. haul R024119-10									
0500	Small bldgs, or single bldgs, no salvage included, steel	B-3	14800	.003	C.F.		.11	.16	.27	.35
0600	Concrete	″	11300	.004	″		.14	.21	.35	.47
0605	Concrete, plain	B-5	33	1.212	C.Y.		40	43.50	83.50	114
0610	Reinforced		25	1.600			52.50	57.50	110	150
0615	Concrete walls		34	1.176			38.50	42.50	81	110
0620	Elevated slabs	↓	26	1.538	↓		50.50	55.50	106	145
0650	Masonry	B-3	14800	.003	C.F.		.11	.16	.27	.35
0700	Wood	″	14800	.003			.11	.16	.27	.35
0750	For buildings with no interior walls, deduct				↓				30%	30%
1000	Demolition single family house, one story, wood 1600 S.F.	B-3	1	48	Ea.		1,625	2,325	3,950	5,250
1020	3200 S.F.		.50	96			3,250	4,675	7,925	10,500
1200	Demolition two family house, two story, wood 2400 S.F.		.67	71.964			2,450	3,500	5,950	7,875
1220	4200 S.F.		.38	128			4,350	6,225	10,575	14,000
1300	Demolition three family house, three story, wood 3200 S.F.		.50	96			3,250	4,675	7,925	10,500
1320	5400 S.F.	↓	.30	160			5,425	7,775	13,200	17,500
5000	For buildings with no interior walls, deduct				↓				30%	30%

02 41 Demolition

02 41 16 – Structure Demolition

	02 41 16.17 Building Demolition Footings and Foundations	Crew	Daily Output	Labor-Hours	Unit	Material	2019 Bare Costs Labor	Equipment	Total	Total Incl O&P
0010	**BUILDING DEMOLITION FOOTINGS AND FOUNDATIONS** R024119-10									
0200	Floors, concrete slab on grade,									
0240	4" thick, plain concrete	B-13L	5000	.003	S.F.		.13	.40	.53	.66
0280	Reinforced, wire mesh		4000	.004			.17	.50	.67	.83
0300	Rods		4500	.004			.15	.44	.59	.74
0400	6" thick, plain concrete		4000	.004			.17	.50	.67	.83
0420	Reinforced, wire mesh		3200	.005			.21	.62	.83	1.03
0440	Rods		3600	.004			.19	.55	.74	.92
1000	Footings, concrete, 1' thick, 2' wide		300	.053	L.F.		2.24	6.65	8.89	10.95
1080	1'-6" thick, 2' wide		250	.064			2.68	8	10.68	13.20
1120	3' wide		200	.080			3.36	10	13.36	16.50
1200	Average reinforcing, add								10%	10%
2000	Walls, block, 4" thick	B-13L	8000	.002	S.F.		.08	.25	.33	.41
2040	6" thick		6000	.003			.11	.33	.44	.55
2080	8" thick		4000	.004			.17	.50	.67	.83
2100	12" thick		3000	.005			.22	.67	.89	1.10
2400	Concrete, plain concrete, 6" thick		4000	.004			.17	.50	.67	.83
2420	8" thick		3500	.005			.19	.57	.76	.94
2440	10" thick		3000	.005			.22	.67	.89	1.10
2500	12" thick		2500	.006			.27	.80	1.07	1.32
2600	For average reinforcing, add								10%	10%
4000	For congested sites or small quantities, add up to								200%	200%
4200	Add for disposal, on site	B-11A	232	.069	C.Y.		2.45	5.55	8	10.15
4250	To five miles	B-30	220	.109	"		4.01	9.30	13.31	16.80

02 41 19 – Selective Demolition

	02 41 19.13 Selective Building Demolition	Crew	Daily Output	Labor-Hours	Unit	Material	Labor	Equipment	Total	Total Incl O&P
0010	**SELECTIVE BUILDING DEMOLITION**									
0020	Costs related to selective demolition of specific building components									
0025	are included under Common Work Results (XX 05)									
0030	in the component's appropriate division.									

	02 41 19.16 Selective Demolition, Cutout	Crew	Daily Output	Labor-Hours	Unit	Material	Labor	Equipment	Total	Total Incl O&P
0010	**SELECTIVE DEMOLITION, CUTOUT** R024119-10									
0020	Concrete, elev. slab, light reinforcement, under 6 C.F.	B-9	65	.615	C.F.		18.90	3.61	22.51	35.50
0050	Light reinforcing, over 6 C.F.		75	.533	"		16.40	3.13	19.53	30.50
0200	Slab on grade to 6" thick, not reinforced, under 8 S.F.		85	.471	S.F.		14.45	2.76	17.21	27
0250	8-16 S.F.		175	.229	"		7.05	1.34	8.39	13.10
0255	For over 16 S.F. see Line 02 41 16.17 0400									
0600	Walls, not reinforced, under 6 C.F.	B-9	60	.667	C.F.		20.50	3.91	24.41	38.50
0650	6-12 C.F.	"	80	.500	"		15.40	2.94	18.34	28.50
0655	For over 12 C.F. see Line 02 41 16.17 2500									
1000	Concrete, elevated slab, bar reinforced, under 6 C.F.	B-9	45	.889	C.F.		27.50	5.20	32.70	51
1050	Bar reinforced, over 6 C.F.		50	.800	"		24.50	4.70	29.20	45.50
1200	Slab on grade to 6" thick, bar reinforced, under 8 S.F.		75	.533	S.F.		16.40	3.13	19.53	30.50
1250	8-16 S.F.		150	.267	"		8.20	1.57	9.77	15.25
1255	For over 16 S.F. see Line 02 41 16.17 0440									
1400	Walls, bar reinforced, under 6 C.F.	B-9	50	.800	C.F.		24.50	4.70	29.20	45.50
1450	6-12 C.F.	"	70	.571	"		17.55	3.35	20.90	32.50
1455	For over 12 C.F. see Lines 02 41 16.17 2500 and 2600									
2000	Brick, to 4 S.F. opening, not including toothing									
2040	4" thick	B-9	30	1.333	Ea.		41	7.85	48.85	76.50
2060	8" thick		18	2.222			68.50	13.05	81.55	127

02 41 Demolition

02 41 19 – Selective Demolition

02 41 19.16 Selective Demolition, Cutout		Crew	Daily Output	Labor-Hours	Unit	Material	2019 Bare Costs Labor	Equipment	Total	Total Incl O&P
2080	12″ thick	B-9	10	4	Ea.		123	23.50	146.50	229
2400	Concrete block, to 4 S.F. opening, 2″ thick		35	1.143			35	6.70	41.70	65.50
2420	4″ thick		30	1.333			41	7.85	48.85	76.50
2440	8″ thick		27	1.481			45.50	8.70	54.20	85
2460	12″ thick		24	1.667			51.50	9.80	61.30	95.50
2600	Gypsum block, to 4 S.F. opening, 2″ thick		80	.500			15.40	2.94	18.34	28.50
2620	4″ thick		70	.571			17.55	3.35	20.90	32.50
2640	8″ thick		55	.727			22.50	4.27	26.77	41.50
2800	Terra cotta, to 4 S.F. opening, 4″ thick		70	.571			17.55	3.35	20.90	32.50
2840	8″ thick		65	.615			18.90	3.61	22.51	35.50
2880	12″ thick		50	.800			24.50	4.70	29.20	45.50
3000	Toothing masonry cutouts, brick, soft old mortar	1 Brhe	40	.200	V.L.F.		6.15		6.15	10.25
3100	Hard mortar		30	.267			8.20		8.20	13.65
3200	Block, soft old mortar		70	.114			3.51		3.51	5.85
3400	Hard mortar		50	.160			4.91		4.91	8.20
6000	Walls, interior, not including re-framing,									
6010	openings to 5 S.F.									
6100	Drywall to 5/8″ thick	1 Clab	24	.333	Ea.		10.10		10.10	16.70
6200	Paneling to 3/4″ thick		20	.400			12.15		12.15	20
6300	Plaster, on gypsum lath		20	.400			12.15		12.15	20
6340	On wire lath		14	.571			17.35		17.35	28.50
7000	Wood frame, not including re-framing, openings to 5 S.F.									
7200	Floors, sheathing and flooring to 2″ thick	1 Clab	5	1.600	Ea.		48.50		48.50	80
7310	Roofs, sheathing to 1″ thick, not including roofing		6	1.333			40.50		40.50	67
7410	Walls, sheathing to 1″ thick, not including siding		7	1.143			34.50		34.50	57.50

02 41 19.19 Selective Demolition		Crew	Daily Output	Labor-Hours	Unit	Material	Labor	Equipment	Total	Total Incl O&P
0010	**SELECTIVE DEMOLITION**, Rubbish Handling R024119-10									
0020	The following are to be added to the demolition prices									
0050	The following are components for a complete chute system									
0100	Top chute circular steel, 4′ long, 18″ diameter R024119-30	B-1C	15	1.600	Ea.	273	49.50	30	352.50	415
0102	23″ diameter		15	1.600		296	49.50	30	375.50	440
0104	27″ diameter		15	1.600		320	49.50	30	399.50	465
0106	30″ diameter		15	1.600		340	49.50	30	419.50	490
0108	33″ diameter		15	1.600		365	49.50	30	444.50	515
0110	36″ diameter		15	1.600		385	49.50	30	464.50	540
0112	Regular chute, 18″ diameter		15	1.600		207	49.50	30	286.50	345
0114	23″ diameter		15	1.600		227	49.50	30	306.50	365
0116	27″ diameter		15	1.600		252	49.50	30	331.50	390
0118	30″ diameter		15	1.600		263	49.50	30	342.50	405
0120	33″ diameter		15	1.600		296	49.50	30	375.50	440
0122	36″ diameter		15	1.600		320	49.50	30	399.50	465
0124	Control door chute, 18″ diameter		15	1.600		385	49.50	30	464.50	540
0126	23″ diameter		15	1.600		415	49.50	30	494.50	570
0128	27″ diameter		15	1.600		430	49.50	30	509.50	590
0130	30″ diameter		15	1.600		455	49.50	30	534.50	615
0132	33″ diameter		15	1.600		480	49.50	30	559.50	640
0134	36″ diameter		15	1.600		500	49.50	30	579.50	665
0136	Chute liners, 14 ga., 18″-30″ diameter		15	1.600		214	49.50	30	293.50	350
0138	33″-36″ diameter		15	1.600		268	49.50	30	347.50	410
0140	17% thinner chute, 30″ diameter		15	1.600		226	49.50	30	305.50	365
0142	33% thinner chute, 30″ diameter		15	1.600		168	49.50	30	247.50	300
0144	Top chute cover	1 Clab	24	.333		146	10.10		156.10	178

02 41 19.19 Selective Demolition		Crew	Daily Output	Labor-Hours	Unit	Material	2019 Bare Costs Labor	Equipment	Total	Total Incl O&P
0146	Door chute cover	1 Clab	24	.333	Ea.	146	10.10		156.10	178
0148	Top chute trough	2 Clab	12	1.333		490	40.50		530.50	605
0150	Bolt down frame & counter weights, 250 lb.	B-1	4	6		4,200	186		4,386	4,925
0152	500 lb.		4	6		6,225	186		6,411	7,150
0154	750 lb.		4	6		9,825	186		10,011	11,100
0156	1000 lb.		2.67	8.989		10,900	279		11,179	12,500
0158	1500 lb.		2.67	8.989		13,700	279		13,979	15,600
0160	Chute warning light system, 5 stories	B-1C	4	6		8,450	186	113	8,749	9,725
0162	10 stories	"	2	12		14,600	370	226	15,196	17,000
0164	Dust control device for dumpsters	1 Clab	8	1		152	30.50		182.50	218
0166	Install or replace breakaway cord		8	1		26.50	30.50		57	79
0168	Install or replace warning sign		16	.500		10.35	15.20		25.55	36.50
0600	Dumpster, weekly rental, 1 dump/week, 6 C.Y. capacity (2 tons)				Week	415			415	455
0700	10 C.Y. capacity (3 tons)					480			480	530
0725	20 C.Y. capacity (5 tons) R024119-20					565			565	625
0800	30 C.Y. capacity (7 tons)					730			730	800
0840	40 C.Y. capacity (10 tons)					775			775	850
2000	Load, haul, dump and return, 0'-50' haul, hand carried	2 Clab	24	.667	C.Y.		20		20	33.50
2005	Wheeled		37	.432			13.10		13.10	21.50
2040	0'-100' haul, hand carried		16.50	.970			29.50		29.50	48.50
2045	Wheeled		25	.640			19.40		19.40	32
2050	Forklift	A-3R	25	.320			12.40	5.75	18.15	27
2080	Haul and return, add per each extra 100' haul, hand carried	2 Clab	35.50	.451			13.70		13.70	22.50
2085	Wheeled		54	.296			9		9	14.85
2120	For travel in elevators, up to 10 floors, add		140	.114			3.47		3.47	5.75
2130	0'-50' haul, incl. up to 5 riser stairs, hand carried		23	.696			21		21	35
2135	Wheeled		35	.457			13.85		13.85	23
2140	6-10 riser stairs, hand carried		22	.727			22		22	36.50
2145	Wheeled		34	.471			14.30		14.30	23.50
2150	11-20 riser stairs, hand carried		20	.800			24.50		24.50	40
2155	Wheeled		31	.516			15.65		15.65	26
2160	21-40 riser stairs, hand carried		16	1			30.50		30.50	50
2165	Wheeled		24	.667			20		20	33.50
2170	0'-100' haul, incl. 5 riser stairs, hand carried		15	1.067			32.50		32.50	53.50
2175	Wheeled		23	.696			21		21	35
2180	6-10 riser stairs, hand carried		14	1.143			34.50		34.50	57.50
2185	Wheeled		21	.762			23		23	38
2190	11-20 riser stairs, hand carried		12	1.333			40.50		40.50	67
2195	Wheeled		18	.889			27		27	44.50
2200	21-40 riser stairs, hand carried		8	2			60.50		60.50	100
2205	Wheeled		12	1.333			40.50		40.50	67
2210	Haul and return, add per each extra 100' haul, hand carried		35.50	.451			13.70		13.70	22.50
2215	Wheeled		54	.296			9		9	14.85
2220	For each additional flight of stairs, up to 5 risers, add		550	.029	Flight		.88		.88	1.46
2225	6-10 risers, add		275	.058			1.77		1.77	2.92
2230	11-20 risers, add		138	.116			3.52		3.52	5.80
2235	21-40 risers, add		69	.232			7.05		7.05	11.65
3000	Loading & trucking, including 2 mile haul, chute loaded	B-16	45	.711	C.Y.		22.50	12.60	35.10	51.50
3040	Hand loading truck, 50' haul	"	48	.667			21.50	11.80	33.30	48
3080	Machine loading truck	B-17	120	.267			8.95	5.50	14.45	21
5000	Haul, per mile, up to 8 C.Y. truck	B-34B	1165	.007			.24	.49	.73	.93
5100	Over 8 C.Y. truck	"	1550	.005			.18	.37	.55	.70

02 41 Demolition

02 41 19 – Selective Demolition

02 41 19.20 Selective Demolition, Dump Charges

	02 41 19.20 Selective Demolition, Dump Charges	Crew	Daily Output	Labor-Hours	Unit	Material	2019 Bare Costs Labor	Equipment	Total	Total Incl O&P
0010	SELECTIVE DEMOLITION, DUMP CHARGES R024119-10									
0020	Dump charges, typical urban city, tipping fees only									
0100	Building construction materials				Ton	74			74	81
0200	Trees, brush, lumber					63			63	69.50
0300	Rubbish only					63			63	69.50
0500	Reclamation station, usual charge				↓	74			74	81

02 41 19.21 Selective Demolition, Gutting

	02 41 19.21 Selective Demolition, Gutting	Crew	Daily Output	Labor-Hours	Unit	Material	Labor	Equipment	Total	Total Incl O&P
0010	SELECTIVE DEMOLITION, GUTTING R024119-10									
0020	Building interior, including disposal, dumpster fees not included									
0500	Residential building									
0560	Minimum	B-16	400	.080	SF Flr.		2.56	1.42	3.98	5.80
0580	Maximum	"	360	.089	"		2.84	1.58	4.42	6.40
0900	Commercial building									
1000	Minimum	B-16	350	.091	SF Flr.		2.92	1.62	4.54	6.60
1020	Maximum	"	250	.128	"		4.09	2.27	6.36	9.25

02 82 Asbestos Remediation

02 82 13 – Asbestos Abatement

02 82 13.39 Asbestos Remediation Plans and Methods

	02 82 13.39 Asbestos Remediation Plans and Methods	Crew	Daily Output	Labor-Hours	Unit	Material	Labor	Equipment	Total	Total Incl O&P
0010	ASBESTOS REMEDIATION PLANS AND METHODS									
0100	Building Survey-Commercial Building				Ea.				2,200	2,400
0200	Asbestos Abatement Remediation Plan				"				1,350	1,475

02 82 13.41 Asbestos Abatement Equipment

	02 82 13.41 Asbestos Abatement Equipment	Crew	Daily Output	Labor-Hours	Unit	Material	Labor	Equipment	Total	Total Incl O&P
0010	ASBESTOS ABATEMENT EQUIPMENT R028213-20									
0011	Equipment and supplies, buy									
0200	Air filtration device, 2000 CFM				Ea.	800			800	880
0250	Large volume air sampling pump, minimum					325			325	355
0260	Maximum					345			345	380
0300	Airless sprayer unit, 2 gun					2,500			2,500	2,750
0350	Light stand, 500 watt				↓	38.50			38.50	42.50
0400	Personal respirators									
0410	Negative pressure, 1/2 face, dual operation, minimum				Ea.	27			27	29.50
0420	Maximum					30.50			30.50	33.50
0450	P.A.P.R., full face, minimum					680			680	745
0460	Maximum					1,425			1,425	1,575
0470	Supplied air, full face, including air line, minimum					375			375	410
0480	Maximum					530			530	580
0500	Personnel sampling pump					243			243	268
1500	Power panel, 20 unit, including GFI					445			445	490
1600	Shower unit, including pump and filters					1,075			1,075	1,200
1700	Supplied air system (type C)					3,500			3,500	3,825
1750	Vacuum cleaner, HEPA, 16 gal., stainless steel, wet/dry					1,125			1,125	1,250
1760	55 gallon					1,425			1,425	1,575
1800	Vacuum loader, 9-18 ton/hr.					98,500			98,500	108,000
1900	Water atomizer unit, including 55 gal. drum					290			290	320
2000	Worker protection, whole body, foot, head cover & gloves, plastic					8.25			8.25	9.10
2500	Respirator, single use					29.50			29.50	32.50
2550	Cartridge for respirator					4.30			4.30	4.73
2570	Glove bag, 7 mil, 50" x 64"					8			8	8.80
2580	10 mil, 44" x 60"				↓	8.90			8.90	9.80

02 82 Asbestos Remediation

02 82 13 – Asbestos Abatement

02 82 13.41 Asbestos Abatement Equipment

	02 82 13.41 Asbestos Abatement Equipment	Crew	Daily Output	Labor-Hours	Unit	2019 Bare Costs Material	Labor	Equipment	Total	Total Incl O&P
6000	Disposable polyethylene bags, 6 mil, 3 C.F.				Ea.	.90			.90	.99
6300	Disposable fiber drums, 3 C.F.					18.50			18.50	20.50
6400	Pressure sensitive caution labels, 3" x 5"					2.98			2.98	3.28
6450	11" x 17"					7.70			7.70	8.45
6500	Negative air machine, 1800 CFM					910			910	1,000

02 82 13.42 Preparation of Asbestos Containment Area

	02 82 13.42 Preparation of Asbestos Containment Area	Crew	Daily Output	Labor-Hours	Unit	Material	Labor	Equipment	Total	Total Incl O&P
0010	**PREPARATION OF ASBESTOS CONTAINMENT AREA**									
0100	Pre-cleaning, HEPA vacuum and wet wipe, flat surfaces	A-9	12000	.005	S.F.	.01	.22		.23	.39
0200	Protect carpeted area, 2 layers 6 mil poly on 3/4" plywood	"	1000	.064		2.06	2.61		4.67	6.65
0300	Separation barrier, 2" x 4" @ 16", 1/2" plywood ea. side, 8' high	2 Carp	400	.040		2.87	1.55		4.42	5.70
0310	12' high		320	.050		2.82	1.94		4.76	6.30
0320	16' high		200	.080		2.79	3.10		5.89	8.15
0400	Personnel decontam. chamber, 2" x 4" @ 16", 3/4" ply ea. side		280	.057		3.43	2.21		5.64	7.45
0450	Waste decontam. chamber, 2" x 4" studs @ 16", 3/4" ply ea. side		360	.044		3.43	1.72		5.15	6.60
0500	Cover surfaces with polyethylene sheeting									
0501	Including glue and tape									
0550	Floors, each layer, 6 mil	A-9	8000	.008	S.F.	.04	.33		.37	.59
0551	4 mil		9000	.007		.03	.29		.32	.52
0560	Walls, each layer, 6 mil		6000	.011		.04	.43		.47	.77
0561	4 mil		7000	.009		.03	.37		.40	.66
0570	For heights above 14', add						20%			
0575	For heights above 20', add						30%			
0580	For fire retardant poly, add					100%				
0590	For large open areas, deduct					10%	20%			
0600	Seal floor penetrations with foam firestop to 36 sq. in.	2 Carp	200	.080	Ea.	13.20	3.10		16.30	19.60
0610	36 sq. in. to 72 sq. in.		125	.128		26.50	4.96		31.46	37
0615	72 sq. in. to 144 sq. in.		80	.200		53	7.75		60.75	71
0620	Wall penetrations, to 36 sq. in.		180	.089		13.20	3.44		16.64	20
0630	36 sq. in. to 72 sq. in.		100	.160		26.50	6.20		32.70	39.50
0640	72 sq. in. to 144 sq. in.		60	.267		53	10.35		63.35	75
0800	Caulk seams with latex	1 Carp	230	.035	L.F.	.17	1.35		1.52	2.42
0900	Set up neg. air machine, 1-2k CFM/25 M.C.F. volume	1 Asbe	4.30	1.860	Ea.		75.50		75.50	128
0950	Set up and remove portable shower unit	2 Asbe	4	4	"		163		163	274

02 82 13.43 Bulk Asbestos Removal

	02 82 13.43 Bulk Asbestos Removal	Crew	Daily Output	Labor-Hours	Unit	Material	Labor	Equipment	Total	Total Incl O&P
0010	**BULK ASBESTOS REMOVAL**									
0020	Includes disposable tools and 2 suits and 1 respirator filter/day/worker									
0100	Beams, W 10 x 19	A-9	235	.272	L.F.	.71	11.10		11.81	19.50
0110	W 12 x 22		210	.305		.79	12.40		13.19	22
0120	W 14 x 26		180	.356		.93	14.50		15.43	25.50
0130	W 16 x 31		160	.400		1.04	16.30		17.34	28.50
0140	W 18 x 40		140	.457		1.19	18.65		19.84	33
0150	W 24 x 55		110	.582		1.52	23.50		25.02	41.50
0160	W 30 x 108		85	.753		1.96	30.50		32.46	53.50
0170	W 36 x 150		72	.889		2.32	36		38.32	63.50
0200	Boiler insulation		480	.133	S.F.	.42	5.45		5.87	9.60
0210	With metal lath, add				%				50%	50%
0300	Boiler breeching or flue insulation	A-9	520	.123	S.F.	.32	5		5.32	8.80
0310	For active boiler, add				%				100%	100%
0400	Duct or AHU insulation	A-10B	440	.073	S.F.	.19	2.97		3.16	5.20
0500	Duct vibration isolation joints, up to 24 sq. in. duct	A-9	56	1.143	Ea.	2.98	46.50		49.48	82
0520	25 sq. in. to 48 sq. in. duct		48	1.333		3.47	54.50		57.97	95.50
0530	49 sq. in. to 76 sq. in. duct		40	1.600		4.17	65		69.17	115

02 82 Asbestos Remediation

02 82 13 – Asbestos Abatement

02 82 13.43 Bulk Asbestos Removal

	02 82 13.43 Bulk Asbestos Removal	Crew	Daily Output	Labor-Hours	Unit	Material	Labor	Equipment	Total	Total Incl O&P
						2019 Bare Costs				
0600	Pipe insulation, air cell type, up to 4" diameter pipe	A-9	900	.071	L.F.	.19	2.90		3.09	5.10
0610	4" to 8" diameter pipe		800	.080		.21	3.26		3.47	5.75
0620	10" to 12" diameter pipe		700	.091		.24	3.73		3.97	6.55
0630	14" to 16" diameter pipe		550	.116		.30	4.74		5.04	8.35
0650	Over 16" diameter pipe		650	.098	S.F.	.26	4.01		4.27	7.05
0700	With glove bag up to 3" diameter pipe		200	.320	L.F.	9.50	13.05		22.55	32.50
1000	Pipe fitting insulation up to 4" diameter pipe		320	.200	Ea.	.52	8.15		8.67	14.30
1100	6" to 8" diameter pipe		304	.211		.55	8.60		9.15	15.05
1110	10" to 12" diameter pipe		192	.333		.87	13.60		14.47	24
1120	14" to 16" diameter pipe		128	.500		1.30	20.50		21.80	36
1130	Over 16" diameter pipe		176	.364	S.F.	.95	14.80		15.75	26
1200	With glove bag, up to 8" diameter pipe		75	.853	L.F.	6.50	35		41.50	65.50
2000	Scrape foam fireproofing from flat surface		2400	.027	S.F.	.07	1.09		1.16	1.91
2100	Irregular surfaces		1200	.053		.14	2.17		2.31	3.81
3000	Remove cementitious material from flat surface		1800	.036		.09	1.45		1.54	2.54
3100	Irregular surface		1000	.064		.12	2.61		2.73	4.52
4000	Scrape acoustical coating/fireproofing, from ceiling		3200	.020		.05	.82		.87	1.43
5000	Remove VAT and mastic from floor by hand		2400	.027		.07	1.09		1.16	1.91
5100	By machine	A-11	4800	.013		.03	.54	.01	.58	.97
5150	For 2 layers, add				%				50%	50%
6000	Remove contaminated soil from crawl space by hand	A-9	400	.160	C.F.	.42	6.50		6.92	11.45
6100	With large production vacuum loader	A-12	700	.091	"	.24	3.73	1.03	5	7.70
7000	Radiator backing, not including radiator removal	A-9	1200	.053	S.F.	.14	2.17		2.31	3.81
8000	Cement-asbestos transite board and cement wall board	2 Asbe	1000	.016		.15	.65		.80	1.27
8100	Transite shingle siding	A-10B	750	.043		.21	1.74		1.95	3.16
8200	Shingle roofing	"	2000	.016		.08	.65		.73	1.19
8250	Built-up, no gravel, non-friable	B-2	1400	.029		.08	.88		.96	1.54
8260	Bituminous flashing	1 Rofc	300	.027		.08	.89		.97	1.70
8300	Asbestos millboard, flat board and VAT contaminated plywood	2 Asbe	1000	.016		.08	.65		.73	1.19
9000	For type B (supplied air) respirator equipment, add				%				10%	10%

02 82 13.44 Demolition In Asbestos Contaminated Area

	Description	Crew	Daily Output	Labor-Hours	Unit	Material	Labor	Equipment	Total	Total Incl O&P
0010	**DEMOLITION IN ASBESTOS CONTAMINATED AREA**									
0200	Ceiling, including suspension system, plaster and lath	A-9	2100	.030	S.F.	.08	1.24		1.32	2.18
0210	Finished plaster, leaving wire lath		585	.109		.28	4.46		4.74	7.80
0220	Suspended acoustical tile		3500	.018		.05	.75		.80	1.31
0230	Concealed tile grid system		3000	.021		.06	.87		.93	1.52
0240	Metal pan grid system		1500	.043		.11	1.74		1.85	3.05
0250	Gypsum board		2500	.026		.07	1.04		1.11	1.83
0260	Lighting fixtures up to 2' x 4'		72	.889	Ea.	2.32	36		38.32	63.50
0400	Partitions, non load bearing									
0410	Plaster, lath, and studs	A-9	690	.093	S.F.	.88	3.78		4.66	7.30
0450	Gypsum board and studs	"	1390	.046	"	.12	1.88		2	3.29
9000	For type B (supplied air) respirator equipment, add				%				10%	10%

02 82 13.45 OSHA Testing

	Description	Crew	Daily Output	Labor-Hours	Unit	Material	Labor	Equipment	Total	Total Incl O&P
0010	**OSHA TESTING**									
0100	Certified technician, minimum				Day				200	220
0110	Maximum								300	330
0121	Industrial hygienist, minimum	1 Asbe	1.75	4.571			186		186	315
0130	Maximum								400	440
0200	Asbestos sampling and PCM analysis, NIOSH 7400, minimum	1 Asbe	8	1	Ea.	16.20	40.50		56.70	86.50
0210	Maximum		4	2		28	81.50		109.50	168
1000	Cleaned area samples		8	1		203	40.50		243.50	292

02 82 Asbestos Remediation

02 82 13 – Asbestos Abatement

02 82 13.45 OSHA Testing		Crew	Daily Output	Labor-Hours	Unit	Material	2019 Bare Costs Labor	Equipment	Total	Total Incl O&P
1100	PCM air sample analysis, NIOSH 7400, minimum	1 Asbe	8	1	Ea.	15.80	40.50		56.30	86
1110	Maximum		4	2		2.34	81.50		83.84	140
1200	TEM air sample analysis, NIOSH 7402, minimum								80	106
1210	Maximum								360	450

02 82 13.46 Decontamination of Asbestos Containment Area		Crew	Daily Output	Labor-Hours	Unit	Material	Labor	Equipment	Total	Total Incl O&P
0010	**DECONTAMINATION OF ASBESTOS CONTAINMENT AREA**									
0100	Spray exposed substrate with surfactant (bridging)									
0200	Flat surfaces	A-9	6000	.011	S.F.	.41	.43		.84	1.18
0250	Irregular surfaces		4000	.016	"	.46	.65		1.11	1.61
0300	Pipes, beams, and columns		2000	.032	L.F.	.60	1.30		1.90	2.86
1000	Spray encapsulate polyethylene sheeting		8000	.008	S.F.	.41	.33		.74	1
1100	Roll down polyethylene sheeting		8000	.008	"		.33		.33	.55
1500	Bag polyethylene sheeting		400	.160	Ea.	1.01	6.50		7.51	12.10
2000	Fine clean exposed substrate, with nylon brush		2400	.027	S.F.		1.09		1.09	1.83
2500	Wet wipe substrate		4800	.013			.54		.54	.92
2600	Vacuum surfaces, fine brush		6400	.010			.41		.41	.69
3000	Structural demolition									
3100	Wood stud walls	A-9	2800	.023	S.F.		.93		.93	1.57
3500	Window manifolds, not incl. window replacement		4200	.015			.62		.62	1.05
3600	Plywood carpet protection		2000	.032			1.30		1.30	2.20
4000	Remove custom decontamination facility	A-10A	8	3	Ea.	15.15	123		138.15	223
4100	Remove portable decontamination facility	3 Asbe	12	2	"	14.30	81.50		95.80	153
5000	HEPA vacuum, shampoo carpeting	A-9	4800	.013	S.F.	.11	.54		.65	1.04
9000	Final cleaning of protected surfaces	A-10A	8000	.003	"		.12		.12	.21

02 82 13.47 Asbestos Waste Pkg., Handling, and Disp.		Crew	Daily Output	Labor-Hours	Unit	Material	Labor	Equipment	Total	Total Incl O&P
0010	**ASBESTOS WASTE PACKAGING, HANDLING, AND DISPOSAL**									
0100	Collect and bag bulk material, 3 C.F. bags, by hand	A-9	400	.160	Ea.	.90	6.50		7.40	12
0200	Large production vacuum loader	A-12	880	.073		1.01	2.96	.82	4.79	7
1000	Double bag and decontaminate	A-9	960	.067		.90	2.72		3.62	5.55
2000	Containerize bagged material in drums, per 3 C.F. drum	"	800	.080		18.50	3.26		21.76	26
3000	Cart bags 50′ to dumpster	2 Asbe	400	.040			1.63		1.63	2.74
5000	Disposal charges, not including haul, minimum				C.Y.				61	67
5020	Maximum				"				355	395
9000	For type B (supplied air) respirator equipment, add				%				10%	10%

02 82 13.48 Asbestos Encapsulation With Sealants		Crew	Daily Output	Labor-Hours	Unit	Material	Labor	Equipment	Total	Total Incl O&P
0010	**ASBESTOS ENCAPSULATION WITH SEALANTS**									
0100	Ceilings and walls, minimum	A-9	21000	.003	S.F.	.42	.12		.54	.67
0110	Maximum		10600	.006		.52	.25		.77	.98
0200	Columns and beams, minimum		13300	.005		.42	.20		.62	.79
0210	Maximum		5325	.012		.63	.49		1.12	1.52
0300	Pipes to 12″ diameter including minor repairs, minimum		800	.080	L.F.	.51	3.26		3.77	6.05
0310	Maximum		400	.160	"	1.21	6.50		7.71	12.35

02 83 19 – Lead-Based Paint Remediation

02 83 19.21 Lead Paint Remediation Plans and Methods

	02 83 19.21 Lead Paint Remediation Plans and Methods	Crew	Daily Output	Labor-Hours	Unit	2019 Bare Costs Material	Labor	Equipment	Total	Total Incl O&P
0010	LEAD PAINT REMEDIATION PLANS AND METHODS									
0100	Building Survey-Commercial Building				Ea.				2,050	2,250
0200	Lead Abatement Remediation Plan								1,225	1,350
0300	Lead Paint Testing, AAS Analysis								51	56
0400	Lead Paint Testing, X-Ray Fluorescence								51	56

02 83 19.23 Encapsulation of Lead-Based Paint

	02 83 19.23 Encapsulation of Lead-Based Paint	Crew	Daily Output	Labor-Hours	Unit	Material	Labor	Equipment	Total	Total Incl O&P
0010	ENCAPSULATION OF LEAD-BASED PAINT									
0020	Interior, brushwork, trim, under 6"	1 Pord	240	.033	L.F.	2.37	1.09		3.46	4.40
0030	6"-12" wide		180	.044		3.09	1.46		4.55	5.80
0040	Balustrades		300	.027		1.95	.88		2.83	3.58
0050	Pipe to 4" diameter		500	.016		1.17	.53		1.70	2.15
0060	To 8" diameter		375	.021		1.54	.70		2.24	2.84
0070	To 12" diameter		250	.032		2.28	1.05		3.33	4.23
0080	To 16" diameter		170	.047		3.39	1.55		4.94	6.25
0090	Cabinets, ornate design		200	.040	S.F.	2.92	1.31		4.23	5.35
0100	Simple design		250	.032	"	2.34	1.05		3.39	4.29
0110	Doors, 3' x 7', both sides, incl. frame & trim									
0120	Flush	1 Pord	6	1.333	Ea.	29	44		73	104
0130	French, 10-15 lite		3	2.667		5.95	87.50		93.45	150
0140	Panel		4	2		36	65.50		101.50	148
0150	Louvered		2.75	2.909		33	95.50		128.50	193
0160	Windows, per interior side, per 15 S.F.									
0170	1-6 lite	1 Pord	14	.571	Ea.	20.50	18.75		39.25	53
0180	7-10 lite		7.50	1.067		22.50	35		57.50	82.50
0190	12 lite		5.75	1.391		30.50	45.50		76	109
0200	Radiators		8	1		72	33		105	134
0210	Grilles, vents		275	.029	S.F.	2.05	.96		3.01	3.83
0220	Walls, roller, drywall or plaster		1000	.008		.59	.26		.85	1.08
0230	With spunbonded reinforcing fabric		720	.011		.69	.37		1.06	1.36
0240	Wood		800	.010		.72	.33		1.05	1.33
0250	Ceilings, roller, drywall or plaster		900	.009		.69	.29		.98	1.24
0260	Wood		700	.011		.83	.38		1.21	1.52
0270	Exterior, brushwork, gutters and downspouts		300	.027	L.F.	1.95	.88		2.83	3.58
0280	Columns		400	.020	S.F.	1.44	.66		2.10	2.66
0290	Spray, siding		600	.013	"	.97	.44		1.41	1.79
0300	Miscellaneous									
0310	Electrical conduit, brushwork, to 2" diameter	1 Pord	500	.016	L.F.	1.17	.53		1.70	2.15
0320	Brick, block or concrete, spray		500	.016	S.F.	1.17	.53		1.70	2.15
0330	Steel, flat surfaces and tanks to 12"		500	.016		1.17	.53		1.70	2.15
0340	Beams, brushwork		400	.020		1.44	.66		2.10	2.66
0350	Trusses		400	.020		1.44	.66		2.10	2.66

02 83 19.26 Removal of Lead-Based Paint

	02 83 19.26 Removal of Lead-Based Paint	Crew	Daily Output	Labor-Hours	Unit	Material	Labor	Equipment	Total	Total Incl O&P
0010	REMOVAL OF LEAD-BASED PAINT R028319-60									
0011	By chemicals, per application									
0050	Baseboard, to 6" wide	1 Pord	64	.125	L.F.	.79	4.11		4.90	7.60
0070	To 12" wide		32	.250	"	1.49	8.20		9.69	15.10
0200	Balustrades, one side		28	.286	S.F.	1.46	9.40		10.86	16.95
1400	Cabinets, simple design		32	.250		1.33	8.20		9.53	14.90
1420	Ornate design		25	.320		1.56	10.50		12.06	18.90
1600	Cornice, simple design		60	.133		1.59	4.38		5.97	8.90
1620	Ornate design		20	.400		5.50	13.15		18.65	27.50
2800	Doors, one side, flush		84	.095		2.08	3.13		5.21	7.40

02 83 Lead Remediation

02 83 19 – Lead-Based Paint Remediation

02 83 19.26 Removal of Lead-Based Paint		Crew	Daily Output	Labor-Hours	Unit	2019 Bare Costs Material	Labor	Equipment	Total	Total Incl O&P
2820	Two panel	1 Pord	80	.100	S.F.	1.33	3.29		4.62	6.85
2840	Four panel		45	.178		1.40	5.85		7.25	11.10
2880	For trim, one side, add		64	.125	L.F.	.76	4.11		4.87	7.60
3000	Fence, picket, one side		30	.267	S.F.	1.33	8.75		10.08	15.80
3200	Grilles, one side, simple design		30	.267		1.34	8.75		10.09	15.80
3220	Ornate design		25	.320		1.44	10.50		11.94	18.80
3240	Handrails		90	.089	L.F.	1.34	2.92		4.26	6.25
4400	Pipes, to 4" diameter		90	.089		2.08	2.92		5	7.05
4420	To 8" diameter		50	.160		4.16	5.25		9.41	13.15
4440	To 12" diameter		36	.222		6.25	7.30		13.55	18.80
4460	To 16" diameter		20	.400		8.30	13.15		21.45	30.50
4500	For hangers, add		40	.200	Ea.	2.52	6.55		9.07	13.50
4800	Siding		90	.089	S.F.	1.32	2.92		4.24	6.25
5000	Trusses, open		55	.145	SF Face	1.99	4.78		6.77	10.05
6200	Windows, one side only, double-hung, 1/1 light, 24" x 48" high		4	2	Ea.	25	65.50		90.50	136
6220	30" x 60" high		3	2.667		33	87.50		120.50	180
6240	36" x 72" high		2.50	3.200		39.50	105		144.50	216
6280	40" x 80" high		2	4		49	131		180	269
6400	Colonial window, 6/6 light, 24" x 48" high		2	4		49	131		180	269
6420	30" x 60" high		1.50	5.333		64.50	175		239.50	360
6440	36" x 72" high		1	8		97	263		360	535
6480	40" x 80" high		1	8		97	263		360	535
6600	8/8 light, 24" x 48" high		2	4		49	131		180	269
6620	40" x 80" high		1	8		97	263		360	535
6800	12/12 light, 24" x 48" high		1	8		97	263		360	535
6820	40" x 80" high		.75	10.667		130	350		480	720
6840	Window frame & trim items, included in pricing above									
7000	Hand scraping and HEPA vacuum, less than 4 S.F.	1 Pord	8	1	Ea.	.90	33		33.90	55
8000	Collect and bag bulk material, 3 C.F. bags, by hand	"	30	.267	"	.90	8.75		9.65	15.35

02 87 Biohazard Remediation

02 87 13 – Mold Remediation

02 87 13.16 Mold Remediation Preparation and Containment		Crew	Daily Output	Labor-Hours	Unit	Material	Labor	Equipment	Total	Total Incl O&P
0010	**MOLD REMEDIATION PREPARATION AND CONTAINMENT**									
6010	Preparation of mold containment area									
6100	Pre-cleaning, HEPA vacuum and wet wipe, flat surfaces	A-9	12000	.005	S.F.	.01	.22		.23	.39
6300	Separation barrier, 2" x 4" @ 16", 1/2" plywood ea. side, 8' high	2 Carp	400	.040		3.47	1.55		5.02	6.40
6310	12' high		320	.050		3.47	1.94		5.41	7
6320	16' high		200	.080		2.38	3.10		5.48	7.70
6400	Personnel decontam. chamber, 2" x 4" @ 16", 3/4" ply ea. side		280	.057		4.50	2.21		6.71	8.60
6450	Waste decontam. chamber, 2" x 4" studs @ 16", 3/4" ply each side		360	.044		4.51	1.72		6.23	7.80
6500	Cover surfaces with polyethylene sheeting									
6501	Including glue and tape									
6550	Floors, each layer, 6 mil	A-9	8000	.008	S.F.	.04	.33		.37	.59
6551	4 mil		9000	.007		.03	.29		.32	.52
6560	Walls, each layer, 6 mil		6000	.011		.04	.43		.47	.77
6561	4 mil		7000	.009		.03	.37		.40	.66
6570	For heights above 14', add						20%			
6575	For heights above 20', add						30%			
6580	For fire retardant poly, add					100%				
6590	For large open areas, deduct					10%	20%			

02 87 13 – Mold Remediation

02 87 13.16 Mold Remediation Preparation and Containment		Crew	Daily Output	Labor-Hours	Unit	Material	Labor	Equipment	Total	Total Incl O&P
6600	Seal floor penetrations with foam firestop to 36 sq. in.	2 Carp	200	.080	Ea.	13.20	3.10		16.30	19.60
6610	36 sq. in. to 72 sq. in.		125	.128		26.50	4.96		31.46	37
6615	72 sq. in. to 144 sq. in.		80	.200		53	7.75		60.75	71
6620	Wall penetrations, to 36 sq. in.		180	.089		13.20	3.44		16.64	20
6630	36 sq. in. to 72 sq. in.		100	.160		26.50	6.20		32.70	39.50
6640	72 sq. in. to 144 sq. in.		60	.267		53	10.35		63.35	75
6800	Caulk seams with latex caulk	1 Carp	230	.035	L.F.	.17	1.35		1.52	2.42
6900	Set up neg. air machine, 1-2k CFM/25 M.C.F. volume	1 Asbe	4.30	1.860	Ea.		75.50		75.50	128

(2019 Bare Costs: Material, Labor, Equipment, Total)

02 87 13.33 Removal and Disposal of Materials With Mold		Crew	Daily Output	Labor-Hours	Unit	Material	Labor	Equipment	Total	Total Incl O&P
0010	**REMOVAL AND DISPOSAL OF MATERIALS WITH MOLD**									
0015	Demolition in mold contaminated area									
0200	Ceiling, including suspension system, plaster and lath	A-9	2100	.030	S.F.	.08	1.24		1.32	2.18
0210	Finished plaster, leaving wire lath		585	.109		.28	4.46		4.74	7.80
0220	Suspended acoustical tile		3500	.018		.05	.75		.80	1.31
0230	Concealed tile grid system		3000	.021		.06	.87		.93	1.52
0240	Metal pan grid system		1500	.043		.11	1.74		1.85	3.05
0250	Gypsum board		2500	.026		.07	1.04		1.11	1.83
0255	Plywood		2500	.026		.07	1.04		1.11	1.83
0260	Lighting fixtures up to 2′ x 4′		72	.889	Ea.	2.32	36		38.32	63.50
0400	Partitions, non load bearing									
0410	Plaster, lath, and studs	A-9	690	.093	S.F.	.88	3.78		4.66	7.30
0450	Gypsum board and studs		1390	.046		.12	1.88		2	3.29
0465	Carpet & pad		1390	.046		.12	1.88		2	3.29
0600	Pipe insulation, air cell type, up to 4″ diameter pipe		900	.071	L.F.	.19	2.90		3.09	5.10
0610	4″ to 8″ diameter pipe		800	.080		.21	3.26		3.47	5.75
0620	10″ to 12″ diameter pipe		700	.091		.24	3.73		3.97	6.55
0630	14″ to 16″ diameter pipe		550	.116		.30	4.74		5.04	8.35
0650	Over 16″ diameter pipe		650	.098	S.F.	.26	4.01		4.27	7.05
9000	For type B (supplied air) respirator equipment, add				%				10%	10%

Division Notes

	CREW	DAILY OUTPUT	LABOR-HOURS	UNIT	BARE COSTS MAT.	BARE COSTS LABOR	BARE COSTS EQUIP.	BARE COSTS TOTAL	TOTAL INCL O&P

Division 3

Concrete

Estimating Tips

General

- Carefully check all the plans and specifications. Concrete often appears on drawings other than structural drawings, including mechanical and electrical drawings for equipment pads. The cost of cutting and patching is often difficult to estimate. See Subdivision 03 81 for Concrete Cutting, Subdivision 02 41 19.16 for Cutout Demolition, Subdivision 03 05 05.10 for Concrete Demolition, and Subdivision 02 41 19.19 for Rubbish Handling (handling, loading, and hauling of debris).
- Always obtain concrete prices from suppliers near the job site. A volume discount can often be negotiated, depending upon competition in the area. Remember to add for waste, particularly for slabs and footings on grade.

03 10 00 Concrete Forming and Accessories

- A primary cost for concrete construction is forming. Most jobs today are constructed with prefabricated forms. The selection of the forms best suited for the job and the total square feet of forms required for efficient concrete forming and placing are key elements in estimating concrete construction. Enough forms must be available for erection to make efficient use of the concrete placing equipment and crew.
- Concrete accessories for forming and placing depend upon the systems used. Study the plans and specifications to ensure that all special accessory requirements have been included in the cost estimate, such as anchor bolts, inserts, and hangers.
- Included within costs for forms-in-place are all necessary bracing and shoring.

03 20 00 Concrete Reinforcing

- Ascertain that the reinforcing steel supplier has included all accessories, cutting, bending, and an allowance for lapping, splicing, and waste. A good rule of thumb is 10% for lapping, splicing, and waste. Also, 10% waste should be allowed for welded wire fabric.
- The unit price items in the subdivisions for Reinforcing In Place, Glass Fiber Reinforcing, and Welded Wire Fabric include the labor to install accessories such as beam and slab bolsters, high chairs, and bar ties and tie wire. The material cost for these accessories is not included; they may be obtained from the Accessories Subdivisions.

03 30 00 Cast-In-Place Concrete

- When estimating structural concrete, pay particular attention to requirements for concrete additives, curing methods, and surface treatments. Special consideration for climate, hot or cold, must be included in your estimate. Be sure to include requirements for concrete placing equipment and concrete finishing.
- For accurate concrete estimating, the estimator must consider each of the following major components individually: forms, reinforcing steel, ready-mix concrete, placement of the concrete, and finishing of the top surface. For faster estimating, Subdivision 03 30 53.40 for Concrete-In-Place can be used; here, various items of concrete work are presented that include the costs of all five major components (unless specifically stated otherwise).

03 40 00 Precast Concrete
03 50 00 Cast Decks and Underlayment

- The cost of hauling precast concrete structural members is often an important factor. For this reason, it is important to get a quote from the nearest supplier. It may become economically feasible to set up precasting beds on the site if the hauling costs are prohibitive.

Reference Numbers

Reference numbers are shown at the beginning of some major classifications. These numbers refer to related items in the Reference Section. The reference information may be an estimating procedure, an alternate pricing method, or technical information.

Note: Not all subdivisions listed here necessarily appear. ■

03 01 Maintenance of Concrete

03 01 30 – Maintenance of Cast-In-Place Concrete

03 01 30.64 Floor Patching		Crew	Daily Output	Labor-Hours	Unit	2019 Bare Costs Material	Labor	Equipment	Total	Total Incl O&P
0010	**FLOOR PATCHING**									
0012	Floor patching, 1/4" thick, small areas, regular	1 Cefi	170	.047	S.F.	3.38	1.75		5.13	6.55
0100	Epoxy	"	100	.080	"	8.60	2.97		11.57	14.25

03 05 Common Work Results for Concrete

03 05 05 – Selective Demolition for Concrete

03 05 05.10 Selective Demolition, Concrete

		Crew	Daily Output	Labor-Hours	Unit	Material	Labor	Equipment	Total	Total Incl O&P
0010	**SELECTIVE DEMOLITION, CONCRETE** R024119-10									
0012	Excludes saw cutting, torch cutting, loading or hauling									
0050	Break into small pieces, reinf. less than 1% of cross-sectional area	B-9	24	1.667	C.Y.		51.50	9.80	61.30	95.50
0060	Reinforcing 1% to 2% of cross-sectional area		16	2.500			77	14.70	91.70	143
0070	Reinforcing more than 2% of cross-sectional area		8	5			154	29.50	183.50	287
0150	Remove whole pieces, up to 2 tons per piece	E-18	36	1.111	Ea.		46.50	24	70.50	106
0160	2-5 tons per piece		30	1.333			55.50	29	84.50	128
0170	5-10 tons per piece		24	1.667			69.50	36	105.50	159
0180	10-15 tons per piece		18	2.222			92.50	48.50	141	212
0250	Precast unit embedded in masonry, up to 1 C.F.	D-1	16	1			33.50		33.50	56
0260	1-2 C.F.		12	1.333			45		45	75
0270	2-5 C.F.		10	1.600			54		54	90
0280	5-10 C.F.		8	2			67.50		67.50	112

03 05 13 – Basic Concrete Materials

03 05 13.85 Winter Protection

		Crew	Daily Output	Labor-Hours	Unit	Material	Labor	Equipment	Total	Total Incl O&P
0010	**WINTER PROTECTION**									
0012	For heated ready mix, add				C.Y.	5.35			5.35	5.90
0100	Temporary heat to protect concrete, 24 hours	2 Clab	50	.320	M.S.F.	204	9.70		213.70	240
0200	Temporary shelter for slab on grade, wood frame/polyethylene sheeting									
0201	Build or remove, light framing for short spans	2 Carp	10	1.600	M.S.F.	315	62		377	445
0210	Large framing for long spans	"	3	5.333	"	420	207		627	800
0710	Electrically heated pads, 15 watts/S.F., 20 uses				S.F.	.56			.56	.61

03 11 Concrete Forming

03 11 13 – Structural Cast-In-Place Concrete Forming

03 11 13.45 Forms In Place, Footings

		Crew	Daily Output	Labor-Hours	Unit	Material	Labor	Equipment	Total	Total Incl O&P
0010	**FORMS IN PLACE, FOOTINGS** R031113-40									
0020	Continuous wall, plywood, 1 use	C-1	375	.085	SFCA	7	2.92		9.92	12.55
0150	4 use	"	485	.066	"	2.28	2.26		4.54	6.25
1500	Keyway, 4 use, tapered wood, 2" x 4" R031113-60	1 Carp	530	.015	L.F.	.23	.58		.81	1.22
1550	2" x 6"	"	500	.016	"	.36	.62		.98	1.41
5000	Spread footings, job-built lumber, 1 use	C-1	305	.105	SFCA	2.31	3.59		5.90	8.50
5150	4 use	"	414	.077	"	.75	2.64		3.39	5.20

03 11 13.65 Forms In Place, Slab On Grade

		Crew	Daily Output	Labor-Hours	Unit	Material	Labor	Equipment	Total	Total Incl O&P
0010	**FORMS IN PLACE, SLAB ON GRADE** R031113-40									
1000	Bulkhead forms w/keyway, wood, 6" high, 1 use R031113-60	C-1	510	.063	L.F.	1.13	2.15		3.28	4.80
1400	Bulkhead form for slab, 4-1/2" high, exp metal, incl. keyway & stakes G		1200	.027		.94	.91		1.85	2.54
1410	5-1/2" high G		1100	.029		1.18	1		2.18	2.95
1420	7-1/2" high G		960	.033		1.41	1.14		2.55	3.44
1430	9-1/2" high G		840	.038		1.53	1.30		2.83	3.84
2000	Curb forms, wood, 6" to 12" high, on grade, 1 use		215	.149	SFCA	1.94	5.10		7.04	10.60

03 11 Concrete Forming

03 11 13 – Structural Cast-In-Place Concrete Forming

03 11 13.65 Forms In Place, Slab On Grade		Crew	Daily Output	Labor-Hours	Unit	Material	2019 Bare Costs Labor	Equipment	Total	Total Incl O&P
2150	4 use	C-1	275	.116	SFCA	.63	3.98		4.61	7.30
3000	Edge forms, wood, 4 use, on grade, to 6″ high		600	.053	L.F.	.31	1.82		2.13	3.36
3050	7″ to 12″ high		435	.074	SFCA	.73	2.52		3.25	4.97
4000	For slab blockouts, to 12″ high, 1 use		200	.160	L.F.	.84	5.45		6.29	9.95
4100	Plastic (extruded), to 6″ high, multiple use, on grade		800	.040	″	6.65	1.37		8.02	9.55
8760	Void form, corrugated fiberboard, 4″ x 12″, 4′ long G		3000	.011	S.F.	3.43	.37		3.80	4.38
8770	6″ x 12″, 4′ long		3000	.011		4.59	.37		4.96	5.65
8780	1/4″ thick hardboard protective cover for void form	2 Carp	1500	.011		.69	.41		1.10	1.44

03 11 13.85 Forms In Place, Walls		Crew	Daily Output	Labor-Hours	Unit	Material	Labor	Equipment	Total	Total Incl O&P
0010	**FORMS IN PLACE, WALLS** R031113-10									
0100	Box out for wall openings, to 16″ thick, to 10 S.F.	C-2	24	2	Ea.	27	69		96	144
0150	Over 10 S.F. (use perimeter) R031113-40	″	280	.171	L.F.	2.32	5.90		8.22	12.35
0250	Brick shelf, 4″ w, add to wall forms, use wall area above shelf									
0260	1 use R031113-60	C-2	240	.200	SFCA	2.51	6.90		9.41	14.20
0350	4 use		300	.160	″	1	5.50		6.50	10.25
0500	Bulkhead, wood with keyway, 1 use, 2 piece		265	.181	L.F.	2.23	6.25		8.48	12.80
0600	Bulkhead forms with keyway, 1 piece expanded metal, 8″ wall G	C-1	1000	.032		1.41	1.09		2.50	3.36
0610	10″ wall G		800	.040		1.53	1.37		2.90	3.95
0620	12″ wall G		525	.061		1.84	2.09		3.93	5.50
2000	Wall, job-built plywood, to 8′ high, 1 use	C-2	370	.130	SFCA	2.78	4.47		7.25	10.45
2050	2 use		435	.110		1.77	3.80		5.57	8.25
2100	3 use		495	.097		1.29	3.34		4.63	6.95
2150	4 use		505	.095		1.05	3.27		4.32	6.60
2400	Over 8′ to 16′ high, 1 use		280	.171		3.08	5.90		8.98	13.20
2450	2 use		345	.139		1.36	4.79		6.15	9.45
2500	3 use		375	.128		.97	4.41		5.38	8.35
2550	4 use		395	.122		.79	4.18		4.97	7.80
7800	Modular prefabricated plywood, based on 20 uses of purchased									
7820	forms, and 4 uses of bracing lumber									
7860	To 8′ high	C-2	800	.060	SFCA	1.13	2.07		3.20	4.68
8060	Over 8′ to 16′ high	″	600	.080	″	1.19	2.75		3.94	5.90

03 11 19 – Insulating Concrete Forming

03 11 19.10 Insulating Forms, Left In Place		Crew	Daily Output	Labor-Hours	Unit	Material	Labor	Equipment	Total	Total Incl O&P
0010	**INSULATING FORMS, LEFT IN PLACE**									
0020	Forms includes layout, excludes rebar, embedments, bucks for openings,									
0030	scaffolding, wall bracing, concrete, and concrete placing.									
0040	S.F. is for exterior face but includes forms for both faces of wall									
0100	Straight blocks or panels, molded, walls up to 4′ high									
0110	4″ core wall	4 Carp	1984	.016	S.F.	3.61	.63		4.24	5
0120	6″ core wall		1808	.018		3.63	.69		4.32	5.10
0130	8″ core wall		1536	.021		3.76	.81		4.57	5.45
0140	10″ core wall		1152	.028		4.29	1.08		5.37	6.50
0150	12″ core wall		992	.032		4.87	1.25		6.12	7.40
0200	90 degree corner blocks or panels, molded, walls up to 4′ high									
0210	4″ core wall	4 Carp	1880	.017	S.F.	3.72	.66		4.38	5.20
0220	6″ core wall		1708	.019		3.74	.73		4.47	5.30
0230	8″ core wall		1324	.024		3.87	.94		4.81	5.80
0240	10″ core wall		987	.032		4.47	1.26		5.73	7
0250	12″ core wall		884	.036		4.79	1.40		6.19	7.55
0300	45 degree corner blocks or panels, molded, walls up to 4′ high									
0310	4″ core wall	4 Carp	1880	.017	S.F.	3.92	.66		4.58	5.40
0320	6″ core wall		1712	.019		4.05	.72		4.77	5.65

03 11 Concrete Forming

03 11 19 – Insulating Concrete Forming

03 11 19.10 Insulating Forms, Left In Place		Crew	Daily Output	Labor-Hours	Unit	Material	2019 Bare Costs Labor	Equipment	Total	Total Incl O&P
0330	8″ core wall	4 Carp	1324	.024	S.F.	4.16	.94		5.10	6.15
0400	T blocks or panels, molded, walls up to 4′ high									
0420	6″ core wall	4 Carp	1540	.021	S.F.	4.77	.81		5.58	6.60
0430	8″ core wall		1325	.024		4.82	.94		5.76	6.85
0440	Non-standard corners or Ts requiring trimming & strapping		192	.167		4.87	6.45		11.32	16.05
0500	Radius blocks or panels, molded, walls up to 4′ high, 6″ core wall									
0520	5′ to 10′ diameter, molded blocks or panels	4 Carp	2400	.013	S.F.	5.90	.52		6.42	7.30
0530	10′ to 15′ diameter, requiring trimming and strapping, add		500	.064		4.43	2.48		6.91	8.95
0540	15′-1″ to 30′ diameter, requiring trimming and strapping, add		1200	.027		4.43	1.03		5.46	6.60
0550	30′-1″ to 60′ diameter, requiring trimming and strapping, add		1600	.020		4.43	.78		5.21	6.15
0560	60′-1″ to 100′ diameter, requiring trimming and strapping, add		2800	.011		4.43	.44		4.87	5.60
0600	Additional labor for blocks/panels in higher walls (excludes scaffolding)									
0610	4′-1″ to 9′-4″ high, add						10%			
0620	9′-5″ to 12′-0″ high, add						20%			
0630	12′-1″ to 20′-0″ high, add						35%			
0640	Over 20′-0″ high, add						55%			
0700	Taper block or panels, molded, single course									
0720	6″ core wall	4 Carp	1600	.020	S.F.	3.90	.78		4.68	5.55
0730	8″ core wall	″	1392	.023	″	3.95	.89		4.84	5.80
0800	ICF brick ledge (corbel) block or panels, molded, single course									
0820	6″ core wall	4 Carp	1200	.027	S.F.	4.30	1.03		5.33	6.45
0830	8″ core wall	″	1152	.028	″	4.39	1.08		5.47	6.60
0900	ICF curb (shelf) block or panels, molded, single course									
0930	8″ core wall	4 Carp	688	.047	S.F.	3.71	1.80		5.51	7.05
0940	10″ core wall	″	544	.059	″	4.24	2.28		6.52	8.45
0950	Wood form to hold back concrete to form shelf, 8″ high	2 Carp	400	.040	L.F.	.85	1.55		2.40	3.50
1000	ICF half height block or panels, molded, single course									
1010	4″ core wall	4 Carp	1248	.026	S.F.	4.70	.99		5.69	6.80
1020	6″ core wall		1152	.028		4.72	1.08		5.80	7
1030	8″ core wall		942	.034		4.77	1.32		6.09	7.45
1040	10″ core wall		752	.043		5.25	1.65		6.90	8.55
1050	12″ core wall		648	.049		5.20	1.91		7.11	8.85
1100	ICF half height block/panels, made by field sawing full height block/panels									
1110	4″ core wall	4 Carp	800	.040	S.F.	1.81	1.55		3.36	4.55
1120	6″ core wall		752	.043		1.82	1.65		3.47	4.73
1130	8″ core wall		600	.053		1.88	2.07		3.95	5.50
1140	10″ core wall		496	.065		2.15	2.50		4.65	6.50
1150	12″ core wall		400	.080		2.44	3.10		5.54	7.80
1200	Additional insulation inserted into forms between ties									
1210	1 layer (2″ thick)	4 Carp	14000	.002	S.F.	1.03	.09		1.12	1.28
1220	2 layers (4″ thick)		7000	.005		2.06	.18		2.24	2.56
1230	3 layers (6″ thick)		4622	.007		3.09	.27		3.36	3.84
1300	EPS window/door bucks, molded, permanent									
1310	4″ core wall (9″ wide)	2 Carp	200	.080	L.F.	3.46	3.10		6.56	8.90
1320	6″ core wall (11″ wide)		200	.080		3.55	3.10		6.65	9
1330	8″ core wall (13″ wide)		176	.091		3.72	3.52		7.24	9.90
1340	10″ core wall (15″ wide)		152	.105		4.84	4.08		8.92	12.05
1350	12″ core wall (17″ wide)		152	.105		5.40	4.08		9.48	12.70
1360	2″ x 6″ temporary buck bracing (includes installing and removing)		400	.040		.71	1.55		2.26	3.34
1400	Wood window/door bucks (instead of EPS bucks), permanent									
1410	4″ core wall (9″ wide)	2 Carp	400	.040	L.F.	1.43	1.55		2.98	4.14
1420	6″ core wall (11″ wide)		400	.040		1.81	1.55		3.36	4.55
1430	8″ core wall (13″ wide)		350	.046		7.55	1.77		9.32	11.25

03 11 Concrete Forming

03 11 19 – Insulating Concrete Forming

03 11 19.10 Insulating Forms, Left In Place		Crew	Daily Output	Labor-Hours	Unit	Material	Labor	Equipment	Total	Total Incl O&P
						2019 Bare Costs				
1440	10″ core wall (15″ wide)	2 Carp	300	.053	L.F.	10.40	2.07		12.47	14.85
1450	12″ core wall (17″ wide)		300	.053		10.40	2.07		12.47	14.85
1460	2″ x 6″ temporary buck bracing (includes installing and removing)		800	.020		.71	.78		1.49	2.06
1500	ICF alignment brace (incl. stiff-back, diagonal kick-back, work platform									
1501	bracket & guard rail post), fastened to one face of wall forms @ 6′ O.C.									
1510	1st tier up to 10′ tall									
1520	Rental of ICF alignment brace set, per set				Week	10			10	11
1530	Labor (includes installing & removing)	2 Carp	30	.533	Ea.		20.50		20.50	34
1560	2nd tier from 10′ to 20′ tall (excludes mason's scaffolding up to 10′ high)									
1570	Rental of ICF alignment brace set, per set				Week	10			10	11
1580	Labor (includes installing & removing)	4 Carp	30	1.067	Ea.		41.50		41.50	68.50
1600	2″ x 10″ wood plank for work platform, 16′ long									
1610	Plank material cost pro-rated over 20 uses				Ea.	1.15			1.15	1.26
1620	Labor (includes installing & removing)	2 Carp	48	.333	″		12.90		12.90	21.50
1700	2″ x 4″ lumber for top & middle rails for work platform									
1710	Railing material cost pro-rated over 20 uses				Ea.	.02			.02	.02
1720	Labor (includes installing & removing)	2 Carp	2400	.007	L.F.		.26		.26	.43
1800	ICF accessories									
1810	Wire clip to secure forms in place	2 Carp	2100	.008	Ea.	.39	.30		.69	.92
1820	Masonry anchor embedment (excludes ties by mason)		1600	.010		4.23	.39		4.62	5.30
1830	Ledger anchor embedment (excludes timber hanger & screws)		128	.125		7.55	4.84		12.39	16.35
1900	See section 01 54 23.70 for mason's scaffolding components									
1910	See section 03 15 19.05 for anchor bolt sleeves									
1920	See section 03 15 19.10 for anchor bolts									
1930	See section 03 15 19.20 for dovetail anchor components									
1940	See section 03 15 19.30 for embedded inserts									
1950	See section 03 21 05.10 for rebar accessories									
1960	See section 03 21 11.60 for reinforcing bars in place									
1970	See section 03 31 13.35 for ready-mix concrete material									
1980	See section 03 31 13.70 for placement and consolidation of concrete									
1990	See section 06 05 23.60 for timber connectors									

03 11 23 – Permanent Stair Forming

03 11 23.75 Forms In Place, Stairs		Crew	Daily Output	Labor-Hours	Unit	Material	Labor	Equipment	Total	Total Incl O&P
0010	**FORMS IN PLACE, STAIRS** R031113-40									
0015	(Slant length x width), 1 use	C-2	165	.291	S.F.	6.10	10		16.10	23.50
0150	4 use R031113-60		190	.253		2.15	8.70		10.85	16.80
2000	Stairs, cast on sloping ground (length x width), 1 use		220	.218		2.43	7.50		9.93	15.15
2025	2 use		232	.207		1.34	7.10		8.44	13.25
2050	3 use		244	.197		.97	6.75		7.72	12.30
2100	4 use		256	.188		.79	6.45		7.24	11.55

03 15 Concrete Accessories

03 15 05 – Concrete Forming Accessories

03 15 05.85 Stair Tread Inserts		Crew	Daily Output	Labor-Hours	Unit	Material	Labor	Equipment	Total	Total Incl O&P
						2019 Bare Costs				
0010	**STAIR TREAD INSERTS**									
0105	Cast nosing insert, abrasive surface, pre-drilled, includes screws									
0110	Aluminum, 3" wide x 3' long	1 Cefi	32	.250	Ea.	56	9.30		65.30	76.50
0120	4' long		31	.258		73	9.60		82.60	96
0130	5' long		30	.267		88.50	9.90		98.40	114
0135	Extruded nosing insert, black abrasive strips, continuous anchor									
0140	Aluminum, 3" wide x 3' long	1 Cefi	64	.125	Ea.	35.50	4.64		40.14	47
0150	4' long		60	.133		49	4.95		53.95	62
0160	5' long		56	.143		64.50	5.30		69.80	79.50
0165	Extruded nosing insert, black abrasive strips, pre-drilled, incl. screws									
0170	Aluminum, 3" wide x 3' long	1 Cefi	32	.250	Ea.	44.50	9.30		53.80	64
0180	4' long		31	.258		59	9.60		68.60	80.50
0190	5' long		30	.267		79.50	9.90		89.40	104

03 15 13 – Waterstops

03 15 13.50 Waterstops

Line	Description	Crew	Daily Output	Labor-Hours	Unit	Material	Labor	Equipment	Total	Total Incl O&P
0010	**WATERSTOPS**, PVC and Rubber									
0020	PVC, ribbed 3/16" thick, 4" wide	1 Carp	155	.052	L.F.	1.43	2		3.43	4.88
0050	6" wide		145	.055		2.47	2.14		4.61	6.25
0500	With center bulb, 6" wide, 3/16" thick		135	.059		2.56	2.30		4.86	6.60
0550	3/8" thick		130	.062		4.59	2.38		6.97	9
0600	9" wide x 3/8" thick		125	.064		7.80	2.48		10.28	12.70

03 15 16 – Concrete Construction Joints

03 15 16.20 Control Joints, Saw Cut

Line	Description	Crew	Daily Output	Labor-Hours	Unit	Material	Labor	Equipment	Total	Total Incl O&P
0010	**CONTROL JOINTS, SAW CUT**									
0100	Sawcut control joints in green concrete									
0120	1" depth	C-27	2000	.008	L.F.	.04	.30	.06	.40	.58
0140	1-1/2" depth		1800	.009		.05	.33	.06	.44	.66
0160	2" depth		1600	.010		.07	.37	.07	.51	.76
0180	Sawcut joint reservoir in cured concrete									
0182	3/8" wide x 3/4" deep, with single saw blade	C-27	1000	.016	L.F.	.05	.59	.11	.75	1.14
0184	1/2" wide x 1" deep, with double saw blades		900	.018		.10	.66	.12	.88	1.31
0186	3/4" wide x 1-1/2" deep, with double saw blades		800	.020		.21	.74	.14	1.09	1.58
0190	Water blast joint to wash away laitance, 2 passes	C-29	2500	.003			.10	.03	.13	.19
0200	Air blast joint to blow out debris and air dry, 2 passes	C-28	2000	.004			.15	.01	.16	.25
0300	For backer rod, see Section 07 91 23.10									
0342	For joint sealant, see Section 07 92 13.20									
0900	For replacement of joint sealant, see Section 07 01 90.81									

03 15 19 – Cast-In Concrete Anchors

03 15 19.05 Anchor Bolt Accessories

Line	Description	Crew	Daily Output	Labor-Hours	Unit	Material	Labor	Equipment	Total	Total Incl O&P
0010	**ANCHOR BOLT ACCESSORIES**									
0015	For anchor bolts set in fresh concrete, see Section 03 15 19.10									
8150	Anchor bolt sleeve, plastic, 1" diameter bolts	1 Carp	60	.133	Ea.	13.60	5.15		18.75	23.50
8500	1-1/2" diameter		28	.286		20	11.05		31.05	40.50
8600	2" diameter		24	.333		19.95	12.90		32.85	43.50
8650	3" diameter		20	.400		36.50	15.50		52	65.50

03 15 19.10 Anchor Bolts

Line	Description	Crew	Daily Output	Labor-Hours	Unit	Material	Labor	Equipment	Total	Total Incl O&P
0010	**ANCHOR BOLTS**									
0015	Made from recycled materials									
0025	Single bolts installed in fresh concrete, no templates									
0030	Hooked w/nut and washer, 1/2" diameter, 8" long G	1 Carp	132	.061	Ea.	1.48	2.35		3.83	5.50
0040	12" long G		131	.061		1.64	2.37		4.01	5.70

03 15 Concrete Accessories

03 15 19 – Cast-In Concrete Anchors

03 15 19.10 Anchor Bolts			Crew	Daily Output	Labor-Hours	Unit	2019 Bare Costs Material	Labor	Equipment	Total	Total Incl O&P
0050	5/8" diameter, 8" long	G	1 Carp	129	.062	Ea.	4.10	2.40		6.50	8.45
0060	12" long	G		127	.063		5.05	2.44		7.49	9.60
0070	3/4" diameter, 8" long	G		127	.063		5.05	2.44		7.49	9.60
0080	12" long	G		125	.064		6.30	2.48		8.78	11.05
0090	2-bolt pattern, including job-built 2-hole template, per set										
0100	J-type, incl. hex nut & washer, 1/2" diameter x 6" long	G	1 Carp	21	.381	Set	5.30	14.75		20.05	30.50
0110	12" long	G		21	.381		5.95	14.75		20.70	31
0120	18" long	G		21	.381		6.95	14.75		21.70	32
0130	3/4" diameter x 8" long	G		20	.400		12.75	15.50		28.25	39.50
0140	12" long	G		20	.400		15.30	15.50		30.80	42.50
0150	18" long	G		20	.400		19.05	15.50		34.55	46.50
0160	1" diameter x 12" long	G		19	.421		24.50	16.30		40.80	54
0170	18" long	G		19	.421		29	16.30		45.30	59
0180	24" long	G		19	.421		35	16.30		51.30	65.50
0190	36" long	G		18	.444		47	17.20		64.20	80
0200	1-1/2" diameter x 18" long	G		17	.471		43	18.25		61.25	77.50
0210	24" long	G		16	.500		51	19.40		70.40	88
0300	L-type, incl. hex nut & washer, 3/4" diameter x 12" long	G		20	.400		15.90	15.50		31.40	43
0310	18" long	G		20	.400		19.60	15.50		35.10	47
0320	24" long	G		20	.400		23.50	15.50		39	51
0330	30" long	G		20	.400		29	15.50		44.50	57.50
0340	36" long	G		20	.400		32.50	15.50		48	61.50
0350	1" diameter x 12" long	G		19	.421		24	16.30		40.30	53.50
0360	18" long	G		19	.421		29	16.30		45.30	59
0370	24" long	G		19	.421		35.50	16.30		51.80	66
0380	30" long	G		19	.421		41.50	16.30		57.80	72.50
0390	36" long	G		18	.444		47	17.20		64.20	80
0400	42" long	G		18	.444		56.50	17.20		73.70	90.50
0410	48" long	G		18	.444		63	17.20		80.20	98
0420	1-1/4" diameter x 18" long	G		18	.444		37	17.20		54.20	69
0430	24" long	G		18	.444		43.50	17.20		60.70	76
0440	30" long	G		17	.471		49.50	18.25		67.75	84.50
0450	36" long	G		17	.471		56	18.25		74.25	91.50
1000	4-bolt pattern, including job-built 4-hole template, per set										
1100	J-type, incl. hex nut & washer, 1/2" diameter x 6" long	G	1 Carp	19	.421	Set	7.95	16.30		24.25	35.50
1110	12" long	G		19	.421		9.25	16.30		25.55	37
1120	18" long	G		18	.444		11.20	17.20		28.40	41
1130	3/4" diameter x 8" long	G		17	.471		23	18.25		41.25	55
1140	12" long	G		17	.471		28	18.25		46.25	60.50
1150	18" long	G		17	.471		35.50	18.25		53.75	69
1160	1" diameter x 12" long	G		16	.500		46.50	19.40		65.90	83.50
1170	18" long	G		15	.533		56	20.50		76.50	95.50
1180	24" long	G		15	.533		67.50	20.50		88	108
1190	36" long	G		15	.533		91.50	20.50		112	134
1200	1-1/2" diameter x 18" long	G		13	.615		83.50	24		107.50	132
1210	24" long	G		12	.667		99	26		125	152
1300	L-type, incl. hex nut & washer, 3/4" diameter x 12" long	G		17	.471		29	18.25		47.25	62
1310	18" long	G		17	.471		36.50	18.25		54.75	70
1320	24" long	G		17	.471		44	18.25		62.25	78.50
1330	30" long	G		16	.500		55	19.40		74.40	92.50
1340	36" long	G		16	.500		62.50	19.40		81.90	101
1350	1" diameter x 12" long	G		16	.500		45.50	19.40		64.90	82
1360	18" long	G		15	.533		56	20.50		76.50	95.50

03 15 Concrete Accessories

03 15 19 – Cast-In Concrete Anchors

03 15 19.10 Anchor Bolts		Crew	Daily Output	Labor-Hours	Unit	2019 Bare Costs Material	Labor	Equipment	Total	Total Incl O&P
1370	24″ long G	1 Carp	15	.533	Set	68	20.50		88.50	109
1380	30″ long G		15	.533		80	20.50		100.50	122
1390	36″ long G		15	.533		91	20.50		111.50	134
1400	42″ long G		14	.571		110	22		132	158
1410	48″ long G		14	.571		123	22		145	173
1420	1-1/4″ diameter x 18″ long G		14	.571		71	22		93	115
1430	24″ long G		14	.571		84	22		106	129
1440	30″ long G		13	.615		96.50	24		120.50	146
1450	36″ long G		13	.615		109	24		133	160

03 21 Reinforcement Bars

03 21 11 – Plain Steel Reinforcement Bars

03 21 11.60 Reinforcing In Place

		Crew	Daily Output	Labor-Hours	Unit	Material	Labor	Equipment	Total	Total Incl O&P
0010	**REINFORCING IN PLACE**, 50-60 ton lots, A615 Grade 60 R032110-10									
0020	Includes labor, but not material cost, to install accessories									
0030	Made from recycled materials									
0500	Footings, #4 to #7 G	4 Rodm	2.10	15.238	Ton	1,025	620		1,645	2,150
0550	#8 to #18 R032110-20 G		3.60	8.889		1,025	360		1,385	1,725
0700	Walls, #3 to #7 G		3	10.667		1,025	435		1,460	1,850
0750	#8 to #18 G		4	8		1,025	325		1,350	1,650
0900	For other than 50-60 ton lots									
1000	Under 10 ton job, #3 to #7, add					25%	10%			
1010	#8 to #18, add					20%	10%			
1050	10-50 ton job, #3 to #7, add					10%				
1060	#8 to #18, add					5%				
1100	60-100 ton job, #3 to #7, deduct					5%				
1110	#8 to #18, deduct					10%				
1150	Over 100 ton job, #3 to #7, deduct					10%				
1160	#8 to #18, deduct					15%				
2400	Dowels, 2 feet long, deformed, #3 G	2 Rodm	520	.031	Ea.	.43	1.25		1.68	2.54
2410	#4 R032110-40 G		480	.033		.76	1.35		2.11	3.07
2420	#5 G		435	.037		1.18	1.49		2.67	3.77
2430	#6 R032110-50 G		360	.044		1.70	1.80		3.50	4.85
2600	Dowel sleeves for CIP concrete, 2-part system R032110-70									
2610	Sleeve base, plastic, for 5/8″ smooth dowel sleeve, fasten to edge form	1 Rodm	200	.040	Ea.	.55	1.62		2.17	3.30
2615	Sleeve, plastic, 12″ long, for 5/8″ smooth dowel, snap onto base		400	.020		1.46	.81		2.27	2.95
2620	Sleeve base, for 3/4″ smooth dowel sleeve R032110-80		175	.046		.54	1.86		2.40	3.66
2625	Sleeve, 12″ long, for 3/4″ smooth dowel		350	.023		1.27	.93		2.20	2.94
2630	Sleeve base, for 1″ smooth dowel sleeve		150	.053		.70	2.17		2.87	4.35
2635	Sleeve, 12″ long, for 1″ smooth dowel		300	.027		1.44	1.08		2.52	3.37
2700	Dowel caps, visual warning only, plastic, #3 to #8	2 Rodm	800	.020		.32	.81		1.13	1.69
2720	#8 to #18		750	.021		.84	.87		1.71	2.35
2750	Impalement protective, plastic, #4 to #9		800	.020		1.16	.81		1.97	2.62

03 21 21 – Composite Reinforcement Bars

03 21 21.11 Glass Fiber-Reinforced Polymer Reinf. Bars

		Crew	Daily Output	Labor-Hours	Unit	Material	Labor	Equipment	Total	Total Incl O&P
0010	**GLASS FIBER-REINFORCED POLYMER REINFORCEMENT BARS**									
0020	Includes labor, but not material cost, to install accessories									
0050	#2 bar, .043 lb./L.F.	4 Rodm	9500	.003	L.F.	.41	.14		.55	.68
0100	#3 bar, .092 lb./L.F.		9300	.003		.65	.14		.79	.95
0150	#4 bar, .160 lb./L.F.		9100	.004		.94	.14		1.08	1.27

03 21 Reinforcement Bars

03 21 21 – Composite Reinforcement Bars

03 21 21.11 Glass Fiber-Reinforced Polymer Reinf. Bars		Crew	Daily Output	Labor-Hours	Unit	2019 Bare Costs Material	Labor	Equipment	Total	Total Incl O&P
0200	#5 bar, .258 lb./L.F.	4 Rodm	8700	.004	L.F.	1.12	.15		1.27	1.48
0250	#6 bar, .372 lb./L.F.		8300	.004		1.88	.16		2.04	2.33
0300	#7 bar, .497 lb./L.F.		7900	.004		2.54	.16		2.70	3.06
0350	#8 bar, .620 lb./L.F.		7400	.004		2.44	.18		2.62	2.97
0400	#9 bar, .800 lb./L.F.		6800	.005		4.23	.19		4.42	4.97
0450	#10 bar, 1.08 lb./L.F.		5800	.006		3.83	.22		4.05	4.58
0500	For bends, add per bend				Ea.	1.61			1.61	1.77

03 22 Fabric and Grid Reinforcing

03 22 11 – Plain Welded Wire Fabric Reinforcing

03 22 11.10 Plain Welded Wire Fabric		Crew	Daily Output	Labor-Hours	Unit	Material	Labor	Equipment	Total	Total Incl O&P
0010	**PLAIN WELDED WIRE FABRIC** ASTM A185 R032205-30									
0020	Includes labor, but not material cost, to install accessories									
0030	Made from recycled materials									
0050	Sheets									
0100	6 x 6 - W1.4 x W1.4 (10 x 10) 21 lb./C.S.F. G	2 Rodm	35	.457	C.S.F.	15.90	18.55		34.45	48
0300	6 x 6 - W2.9 x W2.9 (6 x 6) 42 lb./C.S.F. G		29	.552		26	22.50		48.50	65.50
0500	4 x 4 - W1.4 x W1.4 (10 x 10) 31 lb./C.S.F. G		31	.516		24	21		45	60.50
0750	Rolls									
0900	2 x 2 - #12 galv. for gunite reinforcing G	2 Rodm	6.50	2.462	C.S.F.	70.50	100		170.50	243

03 22 13 – Galvanized Welded Wire Fabric Reinforcing

03 22 13.10 Galvanized Welded Wire Fabric		Crew	Daily Output	Labor-Hours	Unit	Material	Labor	Equipment	Total	Total Incl O&P
0010	**GALVANIZED WELDED WIRE FABRIC**									
0100	Add to plain welded wire pricing for galvanized welded wire				Lb.	.24			.24	.27

03 22 16 – Epoxy-Coated Welded Wire Fabric Reinforcing

03 22 16.10 Epoxy-Coated Welded Wire Fabric		Crew	Daily Output	Labor-Hours	Unit	Material	Labor	Equipment	Total	Total Incl O&P
0010	**EPOXY-COATED WELDED WIRE FABRIC**									
0100	Add to plain welded wire pricing for epoxy-coated welded wire				Lb.	.42			.42	.46

03 23 Stressed Tendon Reinforcing

03 23 05 – Prestressing Tendons

03 23 05.50 Prestressing Steel		Crew	Daily Output	Labor-Hours	Unit	Material	Labor	Equipment	Total	Total Incl O&P
0010	**PRESTRESSING STEEL** R034136-90									
3000	Slabs on grade, 0.5-inch diam. non-bonded strands, HDPE sheathed,									
3050	attached dead-end anchors, loose stressing-end anchors									
3100	25′ x 30′ slab, strands @ 36″ OC, placing	2 Rodm	2940	.005	S.F.	.66	.22		.88	1.09
3105	Stressing	C-4A	3750	.004			.17	.01	.18	.30
3110	42″ OC, placing	2 Rodm	3200	.005		.58	.20		.78	.98
3115	Stressing	C-4A	4040	.004			.16	.01	.17	.28
3120	48″ OC, placing	2 Rodm	3510	.005		.51	.19		.70	.87
3125	Stressing	C-4A	4390	.004			.15	.01	.16	.25
3150	25′ x 40′ slab, strands @ 36″ OC, placing	2 Rodm	3370	.005		.64	.19		.83	1.02
3155	Stressing	C-4A	4360	.004			.15	.01	.16	.26
3160	42″ OC, placing	2 Rodm	3760	.004		.55	.17		.72	.90
3165	Stressing	C-4A	4820	.003			.13	.01	.14	.23
3170	48″ OC, placing	2 Rodm	4090	.004		.49	.16		.65	.80
3175	Stressing	C-4A	5190	.003			.13	.01	.14	.22
3200	30′ x 30′ slab, strands @ 36″ OC, placing	2 Rodm	3260	.005		.63	.20		.83	1.03

03 23 Stressed Tendon Reinforcing

03 23 05 – Prestressing Tendons

03 23 05.50 Prestressing Steel		Crew	Daily Output	Labor-Hours	Unit	Material	Labor	Equipment	Total	Total Incl O&P
						2019 Bare Costs				
3205	Stressing	C-4A	4190	.004	S.F.		.16	.01	.17	.27
3210	42" OC, placing	2 Rodm	3530	.005		.57	.18		.75	.93
3215	Stressing	C-4A	4500	.004			.14	.01	.15	.25
3220	48" OC, placing	2 Rodm	3840	.004		.51	.17		.68	.84
3225	Stressing	C-4A	4850	.003			.13	.01	.14	.23
3230	30' x 40' slab, strands @ 36" OC, placing	2 Rodm	3780	.004		.61	.17		.78	.95
3235	Stressing	C-4A	4920	.003			.13	.01	.14	.23
3240	42" OC, placing	2 Rodm	4190	.004		.53	.16		.69	.85
3245	Stressing	C-4A	5410	.003			.12	.01	.13	.21
3250	48" OC, placing	2 Rodm	4520	.004		.48	.14		.62	.77
3255	Stressing	C-4A	5790	.003			.11	.01	.12	.20
3260	30' x 50' slab, strands @ 36" OC, placing	2 Rodm	4300	.004		.57	.15		.72	.88
3265	Stressing	C-4A	5650	.003			.11	.01	.12	.20
3270	42" OC, placing	2 Rodm	4720	.003		.51	.14		.65	.79
3275	Stressing	C-4A	6150	.003			.11	.01	.12	.18
3280	48" OC, placing	2 Rodm	5240	.003		.45	.12		.57	.69
3285	Stressing	C-4A	6760	.002	↓		.10	.01	.11	.17

03 24 Fibrous Reinforcing

03 24 05 – Reinforcing Fibers

03 24 05.30 Synthetic Fibers		Crew	Daily Output	Labor-Hours	Unit	Material	Labor	Equipment	Total	Total Incl O&P
0010	**SYNTHETIC FIBERS**									
0100	Synthetic fibers, add to concrete				Lb.	4.71			4.71	5.20
0110	1-1/2 lb./C.Y.				C.Y.	7.30			7.30	8

03 24 05.70 Steel Fibers		Crew	Daily Output	Labor-Hours	Unit	Material	Labor	Equipment	Total	Total Incl O&P
0010	**STEEL FIBERS**									
0140	ASTM A850, Type V, continuously deformed, 1-1/2" long x 0.045" diam.									
0150	Add to price of ready mix concrete G				Lb.	1.25			1.25	1.38
0205	Alternate pricing, dosing at 5 lb./C.Y., add to price of RMC G				C.Y.	6.25			6.25	6.90
0210	10 lb./C.Y. G					12.50			12.50	13.75
0215	15 lb./C.Y. G					18.75			18.75	20.50
0220	20 lb./C.Y. G					25			25	27.50
0225	25 lb./C.Y. G					31.50			31.50	34.50
0230	30 lb./C.Y. G					37.50			37.50	41.50
0235	35 lb./C.Y. G					44			44	48
0240	40 lb./C.Y. G					50			50	55
0250	50 lb./C.Y. G					62.50			62.50	69
0275	75 lb./C.Y. G					94			94	103
0300	100 lb./C.Y. G				↓	125			125	138

03 30 Cast-In-Place Concrete

03 30 53 – Miscellaneous Cast-In-Place Concrete

03 30 53.40 Concrete In Place		Crew	Daily Output	Labor-Hours	Unit	Material	Labor (2019 Bare Costs)	Equipment	Total	Total Incl O&P
0010	**CONCRETE IN PLACE** R033053-10									
0020	Including forms (4 uses), Grade 60 rebar, concrete (Portland cement R033105-10									
0050	Type I), placement and finishing unless otherwise indicated R033105-20									
0500	Chimney foundations (5000 psi), over 5 C.Y. R033105-50	C-14C	32.22	3.476	C.Y.	178	127	.83	305.83	405
0510	(3500 psi), under 5 C.Y. R033105-70	"	23.71	4.724	"	206	173	1.13	380.13	515
3540	Equipment pad (3000 psi), 3′ x 3′ x 6″ thick	C-14H	45	1.067	Ea.	46.50	40.50	.59	87.59	118
3550	4′ x 4′ x 6″ thick		30	1.600		72.50	60.50	.88	133.88	180
3560	5′ x 5′ x 8″ thick		18	2.667		132	101	1.47	234.47	315
3570	6′ x 6′ x 8″ thick		14	3.429		181	129	1.89	311.89	415
3580	8′ x 8′ x 10″ thick		8	6		385	226	3.30	614.30	800
3590	10′ x 10′ x 12″ thick		5	9.600		665	360	5.30	1,030.30	1,325
3800	Footings (3000 psi), spread under 1 C.Y.	C-14C	28	4	C.Y.	193	147	.96	340.96	455
3825	1 C.Y. to 5 C.Y.		43	2.605		227	95.50	.63	323.13	410
3850	Over 5 C.Y. R033105-80		75	1.493		212	54.50	.36	266.86	325
3900	Footings, strip (3000 psi), 18″ x 9″, unreinforced	C-14L	40	2.400		149	86.50	.67	236.17	305
3920	18″ x 9″, reinforced R033105-85	C-14C	35	3.200		174	117	.77	291.77	385
3925	20″ x 10″, unreinforced	C-14L	45	2.133		146	77	.60	223.60	288
3930	20″ x 10″, reinforced	C-14C	40	2.800		166	103	.67	269.67	350
3935	24″ x 12″, unreinforced	C-14L	55	1.745		143	63	.49	206.49	263
3940	24″ x 12″, reinforced	C-14C	48	2.333		164	85.50	.56	250.06	320
3945	36″ x 12″, unreinforced	C-14L	70	1.371		139	49.50	.38	188.88	235
3950	36″ x 12″, reinforced	C-14C	60	1.867		158	68.50	.45	226.95	286
4000	Foundation mat (3000 psi), under 10 C.Y.		38.67	2.896		231	106	.70	337.70	430
4050	Over 20 C.Y.		56.40	1.986		205	73	.48	278.48	345
4520	Handicap access ramp (4000 psi), railing both sides, 3′ wide	C-14H	14.58	3.292	L.F.	370	124	1.81	495.81	610
4525	5′ wide		12.22	3.928		380	148	2.16	530.16	665
4530	With 6″ curb and rails both sides, 3′ wide		8.55	5.614		380	212	3.09	595.09	770
4535	5′ wide		7.31	6.566		385	248	3.61	636.61	840
4751	Slab on grade (3500 psi), incl. troweled finish, not incl. forms									
4760	or reinforcing, over 10,000 S.F., 4″ thick	C-14F	3425	.021	S.F.	1.62	.74	.01	2.37	2.99
4820	6″ thick	"	3350	.021	"	2.37	.75	.01	3.13	3.85
5000	Slab on grade (3000 psi), incl. broom finish, not incl. forms									
5001	or reinforcing, 4″ thick	C-14G	2873	.019	S.F.	1.60	.67	.01	2.28	2.88
5010	6″ thick		2590	.022		2.51	.75	.01	3.27	3.99
5020	8″ thick		2320	.024		3.28	.83	.01	4.12	4.97
6800	Stairs (3500 psi), not including safety treads, free standing, 3′-6″ wide	C-14H	83	.578	LF Nose	6.25	22	.32	28.57	43.50
6850	Cast on ground		125	.384	"	5.30	14.50	.21	20.01	30
7000	Stair landings, free standing		200	.240	S.F.	5	9.05	.13	14.18	20.50
7050	Cast on ground		475	.101	"	4.03	3.81	.06	7.90	10.80

03 31 Structural Concrete

03 31 13 – Heavyweight Structural Concrete

03 31 13.25 Concrete, Hand Mix		Crew	Daily Output	Labor-Hours	Unit	Material	Labor (2019 Bare Costs)	Equipment	Total	Total Incl O&P
0010	**CONCRETE, HAND MIX** for small quantities or remote areas									
0050	Includes bulk local aggregate, bulk sand, bagged Portland									
0060	cement (Type I) and water, using gas powered cement mixer									
0125	2500 psi	C-30	135	.059	C.F.	4.03	1.80	1.20	7.03	8.70
0130	3000 psi		135	.059		4.35	1.80	1.20	7.35	9.10
0135	3500 psi		135	.059		4.54	1.80	1.20	7.54	9.30
0140	4000 psi		135	.059		4.77	1.80	1.20	7.77	9.55
0145	4500 psi		135	.059		5.05	1.80	1.20	8.05	9.85

03 31 Structural Concrete

03 31 13 – Heavyweight Structural Concrete

03 31 13.25 Concrete, Hand Mix		Crew	Daily Output	Labor-Hours	Unit	Material	2019 Bare Costs Labor	Equipment	Total	Total Incl O&P
0150	5000 psi	C-30	135	.059	C.F.	5.40	1.80	1.20	8.40	10.20
0300	Using pre-bagged dry mix and wheelbarrow (80-lb. bag = 0.6 C.F.)									
0340	4000 psi	1 Clab	48	.167	C.F.	7.95	5.05		13	17.10

03 31 13.30 Concrete, Volumetric Site-Mixed

		Crew	Daily Output	Labor-Hours	Unit	Material	Labor	Equipment	Total	Total Incl O&P
0010	**CONCRETE, VOLUMETRIC SITE-MIXED**									
0015	Mixed on-site in volumetric truck									
0020	Includes local aggregate, sand, Portland cement (Type I) and water									
0025	Excludes all additives and treatments									
0100	3000 psi, 1 C.Y. mixed and discharged				C.Y.	215			215	236
0110	2 C.Y.					159			159	174
0120	3 C.Y.					140			140	154
0130	4 C.Y.					124			124	136
0140	5 C.Y.					114			114	125
0200	For truck holding/waiting time past first 2 on-site hours, add				Hr.	97			97	107
0210	For trip charge beyond first 20 miles, each way, add				Mile	3.65			3.65	4.02
0220	For each additional increase of 500 psi, add				Ea.	4.59			4.59	5.05

03 31 13.35 Heavyweight Concrete, Ready Mix

		Crew	Daily Output	Labor-Hours	Unit	Material	Labor	Equipment	Total	Total Incl O&P
0010	**HEAVYWEIGHT CONCRETE, READY MIX**, delivered R033105-10									
0012	Includes local aggregate, sand, Portland cement (Type I) and water									
0015	Excludes all additives and treatments R033105-20									
0020	2000 psi				C.Y.	110			110	121
0100	2500 psi R033105-30					113			113	125
0150	3000 psi					124			124	136
0200	3500 psi R033105-40					125			125	137
0300	4000 psi					128			128	141
0350	4500 psi R033105-50					132			132	145
0400	5000 psi					139			139	153
0411	6000 psi					143			143	157
0412	8000 psi					150			150	166
0413	10,000 psi					158			158	174
0414	12,000 psi					166			166	182
1000	For high early strength (Portland cement Type III), add					10%				
1300	For winter concrete (hot water), add					5.35			5.35	5.90
1410	For mid-range water reducer, add					4.01			4.01	4.41
1420	For high-range water reducer/superplasticizer, add					6.25			6.25	6.90
1430	For retarder, add					3.25			3.25	3.58
1440	For non-Chloride accelerator, add					6.40			6.40	7.05
1450	For Chloride accelerator, per 1%, add					3.91			3.91	4.30
1460	For fiber reinforcing, synthetic (1 lb./C.Y.), add					8.05			8.05	8.85
1500	For Saturday delivery, add					8.25			8.25	9.10
1510	For truck holding/waiting time past 1st hour per load, add				Hr.	105			105	115
1520	For short load (less than 4 C.Y.), add per load				Ea.	92			92	101
2000	For all lightweight aggregate, add				C.Y.	45%				

03 31 13.70 Placing Concrete

		Crew	Daily Output	Labor-Hours	Unit	Material	Labor	Equipment	Total	Total Incl O&P
0010	**PLACING CONCRETE**									
0020	Includes labor and equipment to place, level (strike off) and consolidate									
1400	Elevated slabs, less than 6" thick, pumped	C-20	140	.457	C.Y.		14.95	6.40	21.35	31.50
1450	With crane and bucket R033105-70	C-7	95	.758			25	11.15	36.15	54
1500	6" to 10" thick, pumped	C-20	160	.400			13.10	5.60	18.70	27.50
1550	With crane and bucket	C-7	110	.655			21.50	9.65	31.15	46
1600	Slabs over 10" thick, pumped	C-20	180	.356			11.65	4.99	16.64	24.50
1650	With crane and bucket	C-7	130	.554			18.35	8.15	26.50	39

03 31 Structural Concrete

03 31 13 – Heavyweight Structural Concrete

03 31 13.70 Placing Concrete		Crew	Daily Output	Labor-Hours	Unit	Material	2019 Bare Costs Labor	Equipment	Total	Total Incl O&P
1900	Footings, continuous, shallow, direct chute	C-6	120	.400	C.Y.		12.75	.44	13.19	21.50
1950	Pumped	C-20	150	.427			13.95	6	19.95	29.50
2000	With crane and bucket	C-7	90	.800			26.50	11.80	38.30	56.50
2400	Footings, spread, under 1 C.Y., direct chute	C-6	55	.873			28	.97	28.97	46.50
2600	Over 5 C.Y., direct chute		120	.400			12.75	.44	13.19	21.50
2900	Foundation mats, over 20 C.Y., direct chute		350	.137			4.36	.15	4.51	7.35
4300	Slab on grade, up to 6" thick, direct chute		110	.436			13.90	.48	14.38	23.50
4350	Pumped	C-20	130	.492			16.10	6.90	23	34
4400	With crane and bucket	C-7	110	.655			21.50	9.65	31.15	46
4900	Walls, 8" thick, direct chute	C-6	90	.533			16.95	.59	17.54	28.50
4950	Pumped	C-20	100	.640			21	8.95	29.95	44.50
5000	With crane and bucket	C-7	80	.900			30	13.25	43.25	63.50
5050	12" thick, direct chute	C-6	100	.480			15.25	.53	15.78	25.50
5100	Pumped	C-20	110	.582			19.05	8.15	27.20	40.50
5200	With crane and bucket	C-7	90	.800			26.50	11.80	38.30	56.50
5600	Wheeled concrete dumping, add to placing costs above									
5610	Walking cart, 50' haul, add	C-18	32	.281	C.Y.		8.60	1.82	10.42	16.20
5620	150' haul, add		24	.375			11.45	2.42	13.87	21.50
5700	250' haul, add		18	.500			15.30	3.23	18.53	29
5800	Riding cart, 50' haul, add	C-19	80	.113			3.44	1.22	4.66	7.05
5810	150' haul, add		60	.150			4.59	1.63	6.22	9.40
5900	250' haul, add		45	.200			6.10	2.17	8.27	12.50
6000	Concrete in-fill for pan-type metal stairs and landings. Manual placement									
6010	includes up to 50' horizontal haul from point of concrete discharge.									
6100	Stair pan treads, 2" deep									
6110	Flights in 1st floor level up/down from discharge point	C-8A	3200	.015	S.F.		.49		.49	.81
6120	2nd floor level		2500	.019			.63		.63	1.04
6130	3rd floor level		2000	.024			.79		.79	1.30
6140	4th floor level		1800	.027			.88		.88	1.44
6200	Intermediate stair landings, pan-type 4" deep									
6210	Flights in 1st floor level up/down from discharge point	C-8A	2000	.024	S.F.		.79		.79	1.30
6220	2nd floor level		1500	.032			1.05		1.05	1.73
6230	3rd floor level		1200	.040			1.32		1.32	2.16
6240	4th floor level		1000	.048			1.58		1.58	2.59

03 35 Concrete Finishing

03 35 13 – High-Tolerance Concrete Floor Finishing

03 35 13.30 Finishing Floors, High Tolerance		Crew	Daily Output	Labor-Hours	Unit	Material	Labor	Equipment	Total	Total Incl O&P
0010	**FINISHING FLOORS, HIGH TOLERANCE**									
0012	Finishing of fresh concrete flatwork requires that concrete									
0013	first be placed, struck off & consolidated									
0015	Basic finishing for various unspecified flatwork									
0100	Bull float only	C-10	4000	.006	S.F.		.21		.21	.34
0125	Bull float & manual float		2000	.012			.42		.42	.68
0150	Bull float, manual float & broom finish, w/edging & joints		1850	.013			.45		.45	.74
0200	Bull float, manual float & manual steel trowel		1265	.019			.66		.66	1.08
0210	For specified Random Access Floors in ACI Classes 1, 2, 3 and 4 to achieve									
0215	Composite Overall Floor Flatness and Levelness values up to FF35/FL25									
0250	Bull float, machine float & machine trowel (walk-behind)	C-10C	1715	.014	S.F.		.49	.02	.51	.82
0300	Power screed, bull float, machine float & trowel (walk-behind)	C-10D	2400	.010			.35	.05	.40	.63
0350	Power screed, bull float, machine float & trowel (ride-on)	C-10E	4000	.006			.21	.06	.27	.41

03 35 Concrete Finishing

03 35 13 – High-Tolerance Concrete Floor Finishing

03 35 13.30 Finishing Floors, High Tolerance		Crew	Daily Output	Labor-Hours	Unit	Material	2019 Bare Costs Labor	Equipment	Total	Total Incl O&P
0352	For specified Random Access Floors in ACI Classes 5, 6, 7 and 8 to achieve									
0354	Composite Overall Floor Flatness and Levelness values up to FF50/FL50									
0356	Add for two-dimensional restraightening after power float	C-10	6000	.004	S.F.		.14		.14	.23
0358	For specified Random or Defined Access Floors in ACI Class 9 to achieve									
0360	Composite Overall Floor Flatness and Levelness values up to FF100/FL100									
0362	Add for two-dimensional restraightening after bull float & power float	C-10	3000	.008	S.F.		.28		.28	.45
0364	For specified Superflat Defined Access Floors in ACI Class 9 to achieve									
0366	Minimum Floor Flatness and Levelness values of FF100/FL100									
0368	Add for 2-dim'l restraightening after bull float, power float, power trowel	C-10	2000	.012	S.F.		.42		.42	.68

03 35 23 – Exposed Aggregate Concrete Finishing

03 35 23.30 Finishing Floors, Exposed Aggregate		Crew	Daily Output	Labor-Hours	Unit	Material	Labor	Equipment	Total	Total Incl O&P
0010	**FINISHING FLOORS, EXPOSED AGGREGATE**									
1600	Exposed local aggregate finish, seeded on fresh concrete, 3 lb./S.F.	1 Cefi	625	.013	S.F.	.20	.48		.68	.99
1650	4 lb./S.F.	″	465	.017	″	.34	.64		.98	1.40

03 35 29 – Tooled Concrete Finishing

03 35 29.30 Finishing Floors, Tooled		Crew	Daily Output	Labor-Hours	Unit	Material	Labor	Equipment	Total	Total Incl O&P
0010	**FINISHING FLOORS, TOOLED**									
4400	Stair finish, fresh concrete, float finish	1 Cefi	275	.029	S.F.		1.08		1.08	1.75
4500	Steel trowel finish	″	200	.040	″		1.49		1.49	2.40

03 35 29.60 Finishing Walls		Crew	Daily Output	Labor-Hours	Unit	Material	Labor	Equipment	Total	Total Incl O&P
0010	**FINISHING WALLS**									
0020	Break ties and patch voids	1 Cefi	540	.015	S.F.	.04	.55		.59	.94
0050	Burlap rub with grout	″	450	.018		.04	.66		.70	1.12
0300	Bush hammer, green concrete	B-39	1000	.048			1.47	.23	1.70	2.69
0350	Cured concrete	″	650	.074			2.27	.36	2.63	4.14
0500	Acid etch	1 Cefi	575	.014		.13	.52		.65	.98
0850	Grind form fins flush	1 Clab	700	.011	L.F.		.35		.35	.57

03 35 33 – Stamped Concrete Finishing

03 35 33.50 Slab Texture Stamping		Crew	Daily Output	Labor-Hours	Unit	Material	Labor	Equipment	Total	Total Incl O&P
0010	**SLAB TEXTURE STAMPING**									
0050	Stamping requires that concrete first be placed, struck off, consolidated,									
0060	bull floated and free of bleed water. Decorative stamping tasks include:									
0100	Step 1 - first application of dry shake colored hardener	1 Cefi	6400	.001	S.F.	.42	.05		.47	.54
0110	Step 2 - bull float		6400	.001			.05		.05	.08
0130	Step 3 - second application of dry shake colored hardener		6400	.001		.21	.05		.26	.31
0140	Step 4 - bull float, manual float & steel trowel	3 Cefi	1280	.019			.70		.70	1.13
0150	Step 5 - application of dry shake colored release agent	1 Cefi	6400	.001		.10	.05		.15	.19
0160	Step 6 - place, tamp & remove mats	3 Cefi	2400	.010		.82	.37		1.19	1.51
0170	Step 7 - touch up edges, mat joints & simulated grout lines	1 Cefi	1280	.006			.23		.23	.38
0300	Alternate stamping estimating method includes all tasks above	4 Cefi	800	.040		1.55	1.49		3.04	4.10
0400	Step 8 - pressure wash @ 3000 psi after 24 hours	1 Cefi	1600	.005			.19		.19	.30
0500	Step 9 - roll 2 coats cure/seal compound when dry	″	800	.010		.63	.37		1	1.29

03 35 43 – Polished Concrete Finishing

03 35 43.10 Polished Concrete Floors		Crew	Daily Output	Labor-Hours	Unit	Material	Labor	Equipment	Total	Total Incl O&P
0010	**POLISHED CONCRETE FLOORS** R033543-10									
0015	Processing of cured concrete to include grinding, honing,									
0020	and polishing of interior floors with 22″ segmented diamond									
0025	planetary floor grinder (2 passes in different directions per grit)									
0100	Removal of pre-existing coatings, dry, with carbide discs using									
0105	dry vacuum pick-up system, final hand sweeping									

03 35 Concrete Finishing

03 35 43 – Polished Concrete Finishing

03 35 43.10 Polished Concrete Floors		Crew	Daily Output	Labor-Hours	Unit	Material	2019 Bare Costs Labor	Equipment	Total	Total Incl O&P
0110	Glue, adhesive or tar	J-4	1.60	15	M.S.F.	21	525	159	705	1,050
0120	Paint, epoxy, 1 coat		3.60	6.667		21	233	70.50	324.50	480
0130	2 coats		1.80	13.333		21	465	141	627	935
0200	Grinding and edging, wet, including wet vac pick-up and auto									
0205	scrubbing between grit changes									
0210	40-grit diamond/metal matrix	J-4A	1.60	20	M.S.F.	42.50	675	325	1,042.50	1,500
0220	80-grit diamond/metal matrix		2	16		42.50	540	260	842.50	1,200
0230	120-grit diamond/metal matrix		2.40	13.333		42.50	450	216	708.50	1,025
0240	200-grit diamond/metal matrix		2.80	11.429		42.50	385	185	612.50	880
0300	Spray on dye or stain (1 coat)	1 Cefi	16	.500		243	18.60		261.60	297
0400	Spray on densifier/hardener (2 coats)	"	8	1		325	37		362	415
0410	Auto scrubbing after 2nd coat, when dry	J-4B	16	.500			15.20	16.55	31.75	43.50
0500	Honing and edging, wet, including wet vac pick-up and auto									
0505	scrubbing between grit changes									
0510	100-grit diamond/resin matrix	J-4A	2.80	11.429	M.S.F.	42.50	385	185	612.50	880
0520	200-grit diamond/resin matrix	"	2.80	11.429	"	42.50	385	185	612.50	880
0530	Dry, including dry vacuum pick-up system, final hand sweeping									
0540	400-grit diamond/resin matrix	J-4A	2.80	11.429	M.S.F.	42.50	385	185	612.50	880
0600	Polishing and edging, dry, including dry vac pick-up and hand									
0605	sweeping between grit changes									
0610	800-grit diamond/resin matrix	J-4A	2.80	11.429	M.S.F.	42.50	385	185	612.50	880
0620	1500-grit diamond/resin matrix		2.80	11.429		42.50	385	185	612.50	880
0630	3000-grit diamond/resin matrix		2.80	11.429		42.50	385	185	612.50	880
0700	Auto scrubbing after final polishing step	J-4B	16	.500			15.20	16.55	31.75	43.50

03 39 Concrete Curing

03 39 13 – Water Concrete Curing

03 39 13.50 Water Curing		Crew	Daily Output	Labor-Hours	Unit	Material	Labor	Equipment	Total	Total Incl O&P
0010	**WATER CURING**									
0015	With burlap, 4 uses assumed, 7.5 oz.	2 Clab	55	.291	C.S.F.	15.30	8.85		24.15	31.50
0100	10 oz.	"	55	.291	"	27.50	8.85		36.35	45
0400	Curing blankets, 1" to 2" thick, buy				S.F.	.25			.25	.28

03 39 23 – Membrane Concrete Curing

03 39 23.13 Chemical Compound Membrane Concrete Curing		Crew	Daily Output	Labor-Hours	Unit	Material	Labor	Equipment	Total	Total Incl O&P
0010	**CHEMICAL COMPOUND MEMBRANE CONCRETE CURING**									
0300	Sprayed membrane curing compound	2 Clab	95	.168	C.S.F.	12.60	5.10		17.70	22.50

03 39 23.23 Sheet Membrane Concrete Curing		Crew	Daily Output	Labor-Hours	Unit	Material	Labor	Equipment	Total	Total Incl O&P
0010	**SHEET MEMBRANE CONCRETE CURING**									
0200	Curing blanket, burlap/poly, 2-ply	2 Clab	70	.229	C.S.F.	21.50	6.95		28.45	35

03 41 Precast Structural Concrete

03 41 13 – Precast Concrete Hollow Core Planks

03 41 13.50 Precast Slab Planks		Crew	Daily Output	Labor-Hours	Unit	Material	Labor	Equipment	Total	Total Incl O&P
						2019 Bare Costs				
0010	**PRECAST SLAB PLANKS** R034105-30									
0020	Prestressed roof/floor members, grouted, solid, 4" thick	C-11	2400	.023	S.F.	8.45	.94	.84	10.23	11.80
0050	6" thick		2800	.020		8.60	.80	.72	10.12	11.55
0100	Hollow, 8" thick		3200	.018		11.15	.70	.63	12.48	14.15
0150	10" thick		3600	.016		9.95	.62	.56	11.13	12.60
0200	12" thick		4000	.014		11.50	.56	.51	12.57	14.15

03 41 23 – Precast Concrete Stairs

03 41 23.50 Precast Stairs		Crew	Daily Output	Labor-Hours	Unit	Material	Labor	Equipment	Total	Total Incl O&P
0010	**PRECAST STAIRS**									
0020	Precast concrete treads on steel stringers, 3' wide	C-12	75	.640	Riser	153	24.50	6.30	183.80	215
0300	Front entrance, 5' wide with 48" platform, 2 risers		16	3	Flight	675	115	29.50	819.50	960
0350	5 risers		12	4		1,075	153	39.50	1,267.50	1,475
0500	6' wide, 2 risers		15	3.200		745	122	31.50	898.50	1,050
0550	5 risers		11	4.364		1,175	167	43	1,385	1,625
0700	7' wide, 2 risers		14	3.429		935	131	33.50	1,099.50	1,275
1200	Basement entrance stairwell, 6 steps, incl. steel bulkhead door	B-51	22	2.182		1,875	68.50	8.90	1,952.40	2,175
1250	14 steps	"	11	4.364		3,125	137	17.75	3,279.75	3,675

03 48 Precast Concrete Specialties

03 48 43 – Precast Concrete Trim

03 48 43.40 Precast Lintels		Crew	Daily Output	Labor-Hours	Unit	Material	Labor	Equipment	Total	Total Incl O&P
0010	**PRECAST LINTELS**, smooth gray, prestressed, stock units only									
0800	4" wide x 8" high x 4' long	D-10	28	1.143	Ea.	35.50	42.50	13.30	91.30	124
0850	8' long		24	1.333		81.50	49.50	15.50	146.50	189
1000	6" wide x 8" high x 4' long		26	1.231		54	45.50	14.30	113.80	151
1050	10' long		22	1.455		132	54	16.90	202.90	253

03 48 43.90 Precast Window Sills		Crew	Daily Output	Labor-Hours	Unit	Material	Labor	Equipment	Total	Total Incl O&P
0010	**PRECAST WINDOW SILLS**									
0600	Precast concrete, 4" tapers to 3", 9" wide	D-1	70	.229	L.F.	20.50	7.70		28.20	35.50
0650	11" wide	"	60	.267	"	33	9		42	51.50

03 54 Cast Underlayment

03 54 13 – Gypsum Cement Underlayment

03 54 13.50 Poured Gypsum Underlayment		Crew	Daily Output	Labor-Hours	Unit	Material	Labor	Equipment	Total	Total Incl O&P
0010	**POURED GYPSUM UNDERLAYMENT**									
0400	Underlayment, gypsum based, self-leveling 2500 psi, pumped, 1/2" thick	C-8	24000	.002	S.F.	.44	.08	.04	.56	.66
0500	3/4" thick		20000	.003		.66	.10	.04	.80	.94
0600	1" thick		16000	.004		.88	.12	.05	1.05	1.23
1400	Hand placed, 1/2" thick	C-18	450	.020		.44	.61	.13	1.18	1.64
1500	3/4" thick	"	300	.030		.66	.92	.19	1.77	2.46

03 63 Epoxy Grouting

03 63 05 – Grouting of Dowels and Fasteners

03 63 05.10 Epoxy Only		Crew	Daily Output	Labor-Hours	Unit	Material	2019 Bare Costs Labor	Equipment	Total	Total Incl O&P
0010	**EPOXY ONLY**									
1500	Chemical anchoring, epoxy cartridge, excludes layout, drilling, fastener									
1530	For fastener 3/4" diam. x 6" embedment	2 Skwk	72	.222	Ea.	5.20	8.75		13.95	20.50
1535	1" diam. x 8" embedment		66	.242		7.75	9.55		17.30	24.50
1540	1-1/4" diam. x 10" embedment		60	.267		15.55	10.50		26.05	34.50
1545	1-3/4" diam. x 12" embedment		54	.296		26	11.70		37.70	48
1550	14" embedment		48	.333		31	13.15		44.15	56
1555	2" diam. x 12" embedment		42	.381		41.50	15.05		56.55	70.50
1560	18" embedment		32	.500		52	19.75		71.75	90

03 82 Concrete Boring

03 82 13 – Concrete Core Drilling

03 82 13.10 Core Drilling

		Crew	Daily Output	Labor-Hours	Unit	Material	Labor	Equipment	Total	Total Incl O&P
0010	**CORE DRILLING**									
0015	Includes bit cost, layout and set-up time									
0020	Reinforced concrete slab, up to 6" thick									
0100	1" diameter core	B-89A	17	.941	Ea.	.18	33	6.60	39.78	62
0150	For each additional inch of slab thickness in same hole, add		1440	.011		.03	.39	.08	.50	.76
0200	2" diameter core		16.50	.970		.29	34	6.80	41.09	64
0250	For each additional inch of slab thickness in same hole, add		1080	.015		.05	.52	.10	.67	1.02
0300	3" diameter core		16	1		.39	35	7	42.39	66
0350	For each additional inch of slab thickness in same hole, add		720	.022		.07	.78	.16	1.01	1.53
0500	4" diameter core		15	1.067		.51	37	7.50	45.01	70.50
0550	For each additional inch of slab thickness in same hole, add		480	.033		.08	1.16	.23	1.47	2.28
0700	6" diameter core		14	1.143		.78	40	8	48.78	75.50
0750	For each additional inch of slab thickness in same hole, add		360	.044		.13	1.55	.31	1.99	3.05
0900	8" diameter core		13	1.231		1.09	43	8.65	52.74	81.50
0950	For each additional inch of slab thickness in same hole, add		288	.056		.18	1.94	.39	2.51	3.84
1760	For horizontal holes, add to above						20%	20%		
1770	Prestressed hollow core plank, 8" thick									
1780	1" diameter core	B-89A	17.50	.914	Ea.	.24	32	6.40	38.64	60.50
1790	For each additional inch of plank thickness in same hole, add		3840	.004		.03	.15	.03	.21	.30
1794	2" diameter core		17.25	.928		.39	32.50	6.50	39.39	61
1796	For each additional inch of plank thickness in same hole, add		2880	.006		.05	.19	.04	.28	.41
1800	3" diameter core		17	.941		.52	33	6.60	40.12	62.50
1810	For each additional inch of plank thickness in same hole, add		1920	.008		.07	.29	.06	.42	.61
1820	4" diameter core		16.50	.970		.68	34	6.80	41.48	64.50
1830	For each additional inch of plank thickness in same hole, add		1280	.013		.08	.44	.09	.61	.91
1840	6" diameter core		15.50	1.032		1.04	36	7.25	44.29	68.50
1850	For each additional inch of plank thickness in same hole, add		960	.017		.13	.58	.12	.83	1.23
1860	8" diameter core		15	1.067		1.45	37	7.50	45.95	71.50
1870	For each additional inch of plank thickness in same hole, add		768	.021		.18	.73	.15	1.06	1.57

03 82 16 – Concrete Drilling

03 82 16.10 Concrete Impact Drilling

		Crew	Daily Output	Labor-Hours	Unit	Material	Labor	Equipment	Total	Total Incl O&P
0010	**CONCRETE IMPACT DRILLING**									
0020	Includes bit cost, layout and set-up time, no anchors									
0050	Up to 4" deep in concrete/brick floors/walls									
0100	Holes, 1/4" diameter	1 Carp	75	.107	Ea.	.07	4.13		4.20	6.95
0150	For each additional inch of depth in same hole, add		430	.019		.02	.72		.74	1.21
0200	3/8" diameter		63	.127		.05	4.92		4.97	8.20

03 82 Concrete Boring

03 82 16 – Concrete Drilling

03 82 16.10 Concrete Impact Drilling		Crew	Daily Output	Labor-Hours	Unit	Material	Labor	Equipment	Total	Total Incl O&P
						2019 Bare Costs				
0250	For each additional inch of depth in same hole, add	1 Carp	340	.024	Ea.	.01	.91		.92	1.53
0300	1/2" diameter		50	.160		.06	6.20		6.26	10.30
0350	For each additional inch of depth in same hole, add		250	.032		.01	1.24		1.25	2.07
0400	5/8" diameter		48	.167		.10	6.45		6.55	10.80
0450	For each additional inch of depth in same hole, add		240	.033		.03	1.29		1.32	2.16
0500	3/4" diameter		45	.178		.13	6.90		7.03	11.55
0550	For each additional inch of depth in same hole, add		220	.036		.03	1.41		1.44	2.37
0600	7/8" diameter		43	.186		.18	7.20		7.38	12.10
0650	For each additional inch of depth in same hole, add		210	.038		.05	1.48		1.53	2.49
0700	1" diameter		40	.200		.17	7.75		7.92	13
0750	For each additional inch of depth in same hole, add		190	.042		.04	1.63		1.67	2.75
0800	1-1/4" diameter		38	.211		.28	8.15		8.43	13.80
0850	For each additional inch of depth in same hole, add		180	.044		.07	1.72		1.79	2.93
0900	1-1/2" diameter		35	.229		.51	8.85		9.36	15.20
0950	For each additional inch of depth in same hole, add	▼	165	.048	▼	.13	1.88		2.01	3.25
1000	For ceiling installations, add						40%			

Estimating Tips

04 05 00 Common Work Results for Masonry

- The terms mortar and grout are often used interchangeably—and incorrectly. Mortar is used to bed masonry units, seal the entry of air and moisture, provide architectural appearance, and allow for size variations in the units. Grout is used primarily in reinforced masonry construction and to bond the masonry to the reinforcing steel. Common mortar types are M (2500 psi), S (1800 psi), N (750 psi), and O (350 psi), and they conform to ASTM C270. Grout is either fine or coarse and conforms to ASTM C476, and in-place strengths generally exceed 2500 psi. Mortar and grout are different components of masonry construction and are placed by entirely different methods. An estimator should be aware of their unique uses and costs.
- Mortar is included in all assembled masonry line items. The mortar cost, part of the assembled masonry material cost, includes all ingredients, all labor, and all equipment required. Please see reference number R040513-10.
- Waste, specifically the loss/droppings of mortar and the breakage of brick and block, is included in all unit cost lines that include mortar and masonry units in this division. A factor of 25% is added for mortar and 3% for brick and concrete masonry units.
- Scaffolding or staging is not included in any of the Division 4 costs. Refer to Subdivision 01 54 23 for scaffolding and staging costs.

04 20 00 Unit Masonry

- The most common types of unit masonry are brick and concrete masonry. The major classifications of brick are building brick (ASTM C62), facing brick (ASTM C216), glazed brick, fire brick, and pavers. Many varieties of texture and appearance can exist within these classifications, and the estimator would be wise to check local custom and availability within the project area. For repair and remodeling jobs, matching the existing brick may be the most important criteria.
- Brick and concrete block are priced by the piece and then converted into a price per square foot of wall. Openings less than two square feet are generally ignored by the estimator because any savings in units used are offset by the cutting and trimming required.
- It is often difficult and expensive to find and purchase small lots of historic brick. Costs can vary widely. Many design issues affect costs, selection of mortar mix, and repairs or replacement of masonry materials. Cleaning techniques must be reflected in the estimate.
- All masonry walls, whether interior or exterior, require bracing. The cost of bracing walls during construction should be included by the estimator, and this bracing must remain in place until permanent bracing is complete. Permanent bracing of masonry walls is accomplished by masonry itself, in the form of pilasters or abutting wall corners, or by anchoring the walls to the structural frame. Accessories in the form of anchors, anchor slots, and ties are used, but their supply and installation can be by different trades. For instance, anchor slots on spandrel beams and columns are supplied and welded in place by the steel fabricator, but the ties from the slots into the masonry are installed by the bricklayer. Regardless of the installation method, the estimator must be certain that these accessories are accounted for in pricing.

Reference Numbers

Reference numbers are shown at the beginning of some major classifications. These numbers refer to related items in the Reference Section. The reference information may be an estimating procedure, an alternate pricing method, or technical information.

Note: Not all subdivisions listed here necessarily appear. ■

04 01 Maintenance of Masonry

04 01 20 – Maintenance of Unit Masonry

04 01 20.20 Pointing Masonry		Crew	Daily Output	Labor-Hours	Unit	2019 Bare Costs Material	Labor	Equipment	Total	Total Incl O&P
0010	POINTING MASONRY									
0300	Cut and repoint brick, hard mortar, running bond	1 Bric	80	.100	S.F.	.56	3.68		4.24	6.75
0320	Common bond		77	.104		.56	3.82		4.38	6.95
0360	Flemish bond		70	.114		.59	4.20		4.79	7.65
0400	English bond		65	.123		.59	4.52		5.11	8.20
0600	Soft old mortar, running bond		100	.080		.56	2.94		3.50	5.50
0620	Common bond		96	.083		.56	3.06		3.62	5.70
0640	Flemish bond		90	.089		.59	3.27		3.86	6.10
0680	English bond		82	.098		.59	3.59		4.18	6.65
0700	Stonework, hard mortar		140	.057	L.F.	.74	2.10		2.84	4.32
0720	Soft old mortar		160	.050	″	.74	1.84		2.58	3.88
1000	Repoint, mask and grout method, running bond		95	.084	S.F.	.74	3.09		3.83	5.95
1020	Common bond		90	.089		.74	3.27		4.01	6.25
1040	Flemish bond		86	.093		.78	3.42		4.20	6.55
1060	English bond		77	.104		.78	3.82		4.60	7.20
2000	Scrub coat, sand grout on walls, thin mix, brushed		120	.067		3.49	2.45		5.94	7.90
2020	Troweled		98	.082		4.85	3		7.85	10.35

04 01 20.30 Pointing CMU

		Crew	Daily Output	Labor-Hours	Unit	Material	Labor	Equipment	Total	Total Incl O&P
0010	POINTING CMU									
0300	Cut and repoint block, hard mortar, running bond	1 Bric	190	.042	S.F.	.23	1.55		1.78	2.83
0310	Stacked bond		200	.040		.23	1.47		1.70	2.70
0600	Soft old mortar, running bond		230	.035		.23	1.28		1.51	2.38
0610	Stacked bond		245	.033		.23	1.20		1.43	2.25

04 01 30 – Unit Masonry Cleaning

04 01 30.60 Brick Washing

		Crew	Daily Output	Labor-Hours	Unit	Material	Labor	Equipment	Total	Total Incl O&P
0010	BRICK WASHING R040130-10									
0012	Acid cleanser, smooth brick surface	1 Bric	560	.014	S.F.	.05	.53		.58	.94
0050	Rough brick		400	.020		.07	.74		.81	1.31
0060	Stone, acid wash		600	.013		.09	.49		.58	.92
1000	Muriatic acid, price per gallon in 5 gallon lots				Gal.	10.95			10.95	12.05

04 05 Common Work Results for Masonry

04 05 05 – Selective Demolition for Masonry

04 05 05.10 Selective Demolition

		Crew	Daily Output	Labor-Hours	Unit	Material	Labor	Equipment	Total	Total Incl O&P
0010	SELECTIVE DEMOLITION R024119-10									
0200	Bond beams, 8″ block with #4 bar	2 Clab	32	.500	L.F.		15.20		15.20	25
0300	Concrete block walls, unreinforced, 2″ thick		1200	.013	S.F.		.40		.40	.67
0310	4″ thick		1150	.014			.42		.42	.70
0320	6″ thick		1100	.015			.44		.44	.73
0330	8″ thick		1050	.015			.46		.46	.76
0340	10″ thick		1000	.016			.49		.49	.80
0360	12″ thick		950	.017			.51		.51	.84
0380	Reinforced alternate courses, 2″ thick		1130	.014			.43		.43	.71
0390	4″ thick		1080	.015			.45		.45	.74
0400	6″ thick		1035	.015			.47		.47	.78
0410	8″ thick		990	.016			.49		.49	.81
0420	10″ thick		940	.017			.52		.52	.85
0430	12″ thick		890	.018			.55		.55	.90
0440	Reinforced alternate courses & vertically 48″ OC, 4″ thick		900	.018			.54		.54	.89
0450	6″ thick		850	.019			.57		.57	.94

04 05 05.10 Selective Demolition		Crew	Daily Output	Labor-Hours	Unit	Material	2019 Bare Costs Labor	Equipment	Total	Total Incl O&P
0460	8″ thick	2 Clab	800	.020	S.F.		.61		.61	1
0480	10″ thick		750	.021			.65		.65	1.07
0490	12″ thick		700	.023			.69		.69	1.15
1000	Chimney, 16″ x 16″, soft old mortar	1 Clab	55	.145	C.F.		4.41		4.41	7.30
1020	Hard mortar		40	.200			6.05		6.05	10.05
1030	16″ x 20″, soft old mortar		55	.145			4.41		4.41	7.30
1040	Hard mortar		40	.200			6.05		6.05	10.05
1050	16″ x 24″, soft old mortar		55	.145			4.41		4.41	7.30
1060	Hard mortar		40	.200			6.05		6.05	10.05
1080	20″ x 20″, soft old mortar		55	.145			4.41		4.41	7.30
1100	Hard mortar		40	.200			6.05		6.05	10.05
1110	20″ x 24″, soft old mortar		55	.145			4.41		4.41	7.30
1120	Hard mortar		40	.200			6.05		6.05	10.05
1140	20″ x 32″, soft old mortar		55	.145			4.41		4.41	7.30
1160	Hard mortar		40	.200			6.05		6.05	10.05
1200	48″ x 48″, soft old mortar		55	.145			4.41		4.41	7.30
1220	Hard mortar		40	.200			6.05		6.05	10.05
1250	Metal, high temp steel jacket, 24″ diameter	E-2	130	.369	V.L.F.		15.45	13.05	28.50	41
1260	60″ diameter	″	60	.800			33.50	28.50	62	88.50
1280	Flue lining, up to 12″ x 12″	1 Clab	200	.040			1.21		1.21	2.01
1282	Up to 24″ x 24″		150	.053			1.62		1.62	2.67
2000	Columns, 8″ x 8″, soft old mortar		48	.167			5.05		5.05	8.35
2020	Hard mortar		40	.200			6.05		6.05	10.05
2060	16″ x 16″, soft old mortar		16	.500			15.20		15.20	25
2100	Hard mortar		14	.571			17.35		17.35	28.50
2140	24″ x 24″, soft old mortar		8	1			30.50		30.50	50
2160	Hard mortar		6	1.333			40.50		40.50	67
2200	36″ x 36″, soft old mortar		4	2			60.50		60.50	100
2220	Hard mortar		3	2.667			81		81	134
2230	Alternate pricing method, soft old mortar		30	.267	C.F.		8.10		8.10	13.35
2240	Hard mortar		23	.348	″		10.55		10.55	17.45
3000	Copings, precast or masonry, to 8″ wide									
3020	Soft old mortar	1 Clab	180	.044	L.F.		1.35		1.35	2.23
3040	Hard mortar	″	160	.050	″		1.52		1.52	2.51
3100	To 12″ wide									
3120	Soft old mortar	1 Clab	160	.050	L.F.		1.52		1.52	2.51
3140	Hard mortar	″	140	.057	″		1.73		1.73	2.87
4000	Fireplace, brick, 30″ x 24″ opening									
4020	Soft old mortar	1 Clab	2	4	Ea.		121		121	201
4040	Hard mortar		1.25	6.400			194		194	320
4100	Stone, soft old mortar		1.50	5.333			162		162	267
4120	Hard mortar		1	8			243		243	400
4150	Up to 48″ fireplace, 15′ chimney and foundation		.28	28.571			865		865	1,425
4400	Premanufactured, up to 48″	2 Clab	14	1.143			34.50		34.50	57.50
5000	Veneers, brick, soft old mortar	1 Clab	140	.057	S.F.		1.73		1.73	2.87
5020	Hard mortar		125	.064			1.94		1.94	3.21
5050	Glass block, up to 4″ thick		500	.016			.49		.49	.80
5100	Granite and marble, 2″ thick		180	.044			1.35		1.35	2.23
5120	4″ thick		170	.047			1.43		1.43	2.36
5140	Stone, 4″ thick		180	.044			1.35		1.35	2.23
5160	8″ thick		175	.046			1.39		1.39	2.29
5400	Alternate pricing method, stone, 4″ thick		60	.133	C.F.		4.05		4.05	6.70
5420	8″ thick		85	.094			2.86		2.86	4.72

04 05 05 – Selective Demolition for Masonry

04 05 05.10 Selective Demolition		Crew	Daily Output	Labor-Hours	Unit	Material	2019 Bare Costs Labor	Equipment	Total	Total Incl O&P
5450	Solid masonry	1 Clab	130	.062	C.F.		1.87		1.87	3.09
5460	Stone or precast sills, treads, copings		130	.062			1.87		1.87	3.09
5470	Solid stone or precast		110	.073			2.21		2.21	3.65
5500	Remove and reset steel lintel	1 Bric	40	.200	L.F.		7.35		7.35	12.25
5600	Vent box removal	1 Clab	50	.160	S.F.		4.86		4.86	8
5700	Remove block pilaster for fence, 6′ high		2.33	3.433	Ea.		104		104	172
5800	Remove 12″ x 12″ step flashing from mortar joints		240	.033	C.F.		1.01		1.01	1.67

04 05 13 – Masonry Mortaring

04 05 13.10 Cement		Crew	Daily Output	Labor-Hours	Unit	Material	Labor	Equipment	Total	Total Incl O&P
0010	**CEMENT** R040513-10									
0100	Masonry, 70 lb. bag, T.L. lots				Bag	14.30			14.30	15.75
0150	L.T.L. lots					15.20			15.20	16.70
0200	White, 70 lb. bag, T.L. lots					18.05			18.05	19.85
0250	L.T.L. lots					20			20	22

04 05 16 – Masonry Grouting

04 05 16.30 Grouting		Crew	Daily Output	Labor-Hours	Unit	Material	Labor	Equipment	Total	Total Incl O&P
0010	**GROUTING**									
0011	Bond beams & lintels, 8″ deep, 6″ thick, 0.15 C.F./L.F.	D-4	1480	.027	L.F.	.78	.86	.08	1.72	2.38
0020	8″ thick, 0.2 C.F./L.F.		1400	.029		1.27	.91	.09	2.27	3
0050	10″ thick, 0.25 C.F./L.F.		1200	.033		1.30	1.06	.10	2.46	3.32
0060	12″ thick, 0.3 C.F./L.F.		1040	.038		1.57	1.22	.12	2.91	3.89
0200	Concrete block cores, solid, 4″ thk., by hand, 0.067 C.F./S.F. of wall	D-8	1100	.036	S.F.	.35	1.25		1.60	2.46
0210	6″ thick, pumped, 0.175 C.F./S.F.	D-4	720	.056		.91	1.77	.17	2.85	4.13
0250	8″ thick, pumped, 0.258 C.F./S.F.		680	.059		1.35	1.87	.18	3.40	4.80
0300	10″ thick, pumped, 0.340 C.F./S.F.		660	.061		1.77	1.93	.19	3.89	5.35
0350	12″ thick, pumped, 0.422 C.F./S.F.		640	.063		2.20	1.99	.20	4.39	5.95

04 05 19 – Masonry Anchorage and Reinforcing

04 05 19.05 Anchor Bolts		Crew	Daily Output	Labor-Hours	Unit	Material	Labor	Equipment	Total	Total Incl O&P
0010	**ANCHOR BOLTS**									
0015	Installed in fresh grout in CMU bond beams or filled cores, no templates									
0020	Hooked, with nut and washer, 1/2″ diameter, 8″ long	1 Bric	132	.061	Ea.	1.48	2.23		3.71	5.35
0030	12″ long		131	.061		1.64	2.24		3.88	5.55
0040	5/8″ diameter, 8″ long		129	.062		4.10	2.28		6.38	8.30
0050	12″ long		127	.063		5.05	2.31		7.36	9.40
0060	3/4″ diameter, 8″ long		127	.063		5.05	2.31		7.36	9.40
0070	12″ long		125	.064		6.30	2.35		8.65	10.85

04 05 19.16 Masonry Anchors		Crew	Daily Output	Labor-Hours	Unit	Material	Labor	Equipment	Total	Total Incl O&P
0010	**MASONRY ANCHORS**									
0020	For brick veneer, galv., corrugated, 7/8″ x 7″, 22 ga.	1 Bric	10.50	.762	C	15.75	28		43.75	64
0100	24 ga.		10.50	.762		10.25	28		38.25	58
0150	16 ga.		10.50	.762		31	28		59	80.50
0200	Buck anchors, galv., corrugated, 16 ga., 2″ bend, 8″ x 2″		10.50	.762		64.50	28		92.50	118
0250	8″ x 3″		10.50	.762		69.50	28		97.50	123
0660	Cavity wall, Z-type, galvanized, 6″ long, 1/8″ diam.		10.50	.762		24	28		52	73
0670	3/16″ diameter		10.50	.762		33.50	28		61.50	83
0680	1/4″ diameter		10.50	.762		40	28		68	90.50
0850	8″ long, 3/16″ diameter		10.50	.762		26.50	28		54.50	75.50
0855	1/4″ diameter		10.50	.762		48.50	28		76.50	99.50
1000	Rectangular type, galvanized, 1/4″ diameter, 2″ x 6″		10.50	.762		76	28		104	130
1050	4″ x 6″		10.50	.762		91.50	28		119.50	147
1100	3/16″ diameter, 2″ x 6″		10.50	.762		49	28		77	101

04 05 Common Work Results for Masonry

04 05 19 – Masonry Anchorage and Reinforcing

04 05 19.16 Masonry Anchors		Crew	Daily Output	Labor-Hours	Unit	2019 Bare Costs Material	Labor	Equipment	Total	Total Incl O&P
1150	4" x 6"	1 Bric	10.50	.762	C	56	28		84	109
1500	Rigid partition anchors, plain, 8" long, 1" x 1/8"		10.50	.762		242	28		270	315
1550	1" x 1/4"		10.50	.762		289	28		317	365
1580	1-1/2" x 1/8"		10.50	.762		258	28		286	330
1600	1-1/2" x 1/4"		10.50	.762		330	28		358	410
1650	2" x 1/8"		10.50	.762		305	28		333	380
1700	2" x 1/4"		10.50	.762		415	28		443	500

04 05 19.26 Masonry Reinforcing Bars		Crew	Daily Output	Labor-Hours	Unit	Material	Labor	Equipment	Total	Total Incl O&P
0010	**MASONRY REINFORCING BARS** R040519-50									
0015	Steel bars A615, placed horiz., #3 & #4 bars	1 Bric	450	.018	Lb.	.51	.65		1.16	1.66
0050	Placed vertical, #3 & #4 bars		350	.023		.51	.84		1.35	1.97
0060	#5 & #6 bars		650	.012		.51	.45		.96	1.32
0200	Joint reinforcing, regular truss, to 6" wide, mill std galvanized		30	.267	C.L.F.	26	9.80		35.80	45
0250	12" wide		20	.400		28.50	14.70		43.20	56
0400	Cavity truss with drip section, to 6" wide		30	.267		24	9.80		33.80	42.50
0450	12" wide		20	.400		27.50	14.70		42.20	54.50

04 05 23 – Masonry Accessories

04 05 23.13 Masonry Control and Expansion Joints		Crew	Daily Output	Labor-Hours	Unit	Material	Labor	Equipment	Total	Total Incl O&P
0010	**MASONRY CONTROL AND EXPANSION JOINTS**									
0020	Rubber, for double wythe 8" minimum wall (Brick/CMU)	1 Bric	400	.020	L.F.	2.32	.74		3.06	3.78
0025	"T" shaped		320	.025		1.19	.92		2.11	2.84
0030	Cross-shaped for CMU units		280	.029		1.69	1.05		2.74	3.61
0050	PVC, for double wythe 8" minimum wall (Brick/CMU)		400	.020		1.69	.74		2.43	3.09
0120	"T" shaped		320	.025		.91	.92		1.83	2.53
0160	Cross-shaped for CMU units		280	.029		1.18	1.05		2.23	3.05

04 21 Clay Unit Masonry

04 21 13 – Brick Masonry

04 21 13.13 Brick Veneer Masonry		Crew	Daily Output	Labor-Hours	Unit	Material	Labor	Equipment	Total	Total Incl O&P
0010	**BRICK VENEER MASONRY**, T.L. lots, excl. scaff., grout & reinforcing									
0015	Material costs incl. 3% brick and 25% mortar waste									
2000	Standard, sel. common, 4" x 2-2/3" x 8" (6.75/S.F.)	D-8	230	.174	S.F.	4.32	5.95		10.27	14.70
2020	Red, 4" x 2-2/3" x 8", running bond		220	.182		4.15	6.25		10.40	14.95
2050	Full header every 6th course (7.88/S.F.)		185	.216		4.84	7.40		12.24	17.65
2100	English, full header every 2nd course (10.13/S.F.)		140	.286		6.20	9.80		16	23
2150	Flemish, alternate header every course (9.00/S.F.)		150	.267		5.50	9.15		14.65	21.50
2200	Flemish, alt. header every 6th course (7.13/S.F.)		205	.195		4.38	6.70		11.08	15.95
2250	Full headers throughout (13.50/S.F.)		105	.381		8.25	13.10		21.35	31
2300	Rowlock course (13.50/S.F.)		100	.400		8.25	13.75		22	32
2350	Rowlock stretcher (4.50/S.F.)		310	.129		2.79	4.43		7.22	10.45
2400	Soldier course (6.75/S.F.)		200	.200		4.15	6.85		11	16
2450	Sailor course (4.50/S.F.)		290	.138		2.79	4.74		7.53	10.95
2600	Buff or gray face, running bond (6.75/S.F.)		220	.182		4.39	6.25		10.64	15.25
2700	Glazed face brick, running bond		210	.190		12.85	6.55		19.40	25
2750	Full header every 6th course (7.88/S.F.)		170	.235		15	8.10		23.10	30
3000	Jumbo, 6" x 4" x 12" running bond (3.00/S.F.)		435	.092		5.35	3.16		8.51	11.15
3050	Norman, 4" x 2-2/3" x 12" running bond (4.5/S.F.)		320	.125		6.60	4.29		10.89	14.40
3100	Norwegian, 4" x 3-1/5" x 12" (3.75/S.F.)		375	.107		5.75	3.66		9.41	12.45
3150	Economy, 4" x 4" x 8" (4.50/S.F.)		310	.129		4.55	4.43		8.98	12.40
3200	Engineer, 4" x 3-1/5" x 8" (5.63/S.F.)		260	.154		3.91	5.30		9.21	13.10

04 21 Clay Unit Masonry

04 21 13 – Brick Masonry

04 21 13.13 Brick Veneer Masonry

	04 21 13.13 Brick Veneer Masonry	Crew	Daily Output	Labor-Hours	Unit	Material	Labor	Equipment	Total	Total Incl O&P
						2019 Bare Costs				
3250	Roman, 4" x 2" x 12" (6.00/S.F.)	D-8	250	.160	S.F.	7.20	5.50		12.70	17.10
3300	S.C.R., 6" x 2-2/3" x 12" (4.50/S.F.)		310	.129		6.30	4.43		10.73	14.35
3350	Utility, 4" x 4" x 12" (3.00/S.F.)		360	.111		5.15	3.81		8.96	12
3360	For less than truck load lots, add					.10%				
3400	For cavity wall construction, add						15%			
3450	For stacked bond, add						10%			
3500	For interior veneer construction, add						15%			
3550	For curved walls, add						30%			

04 21 13.14 Thin Brick Veneer

	04 21 13.14 Thin Brick Veneer	Crew	Daily Output	Labor-Hours	Unit	Material	Labor	Equipment	Total	Total Incl O&P
0010	**THIN BRICK VENEER**									
0015	Material costs incl. 3% brick and 25% mortar waste									
0020	On & incl. metal panel support sys, modular, 2-2/3" x 5/8" x 8", red	D-7	92	.174	S.F.	9.50	5.50		15	19.35
0100	Closure, 4" x 5/8" x 8"		110	.145		9.25	4.62		13.87	17.60
0110	Norman, 2-2/3" x 5/8" x 12"		110	.145		10.05	4.62		14.67	18.55
0120	Utility, 4" x 5/8" x 12"		125	.128		8.95	4.06		13.01	16.40
0130	Emperor, 4" x 3/4" x 16"		175	.091		10.15	2.90		13.05	15.85
0140	Super emperor, 8" x 3/4" x 16"		195	.082		11.50	2.60		14.10	16.85
0150	For L shaped corners with 4" return, add				L.F.	9.25			9.25	10.20
0200	On masonry/plaster back-up, modular, 2-2/3" x 5/8" x 8", red	D-7	137	.117	S.F.	4.55	3.71		8.26	11
0210	Closure, 4" x 5/8" x 8"		165	.097		4.30	3.08		7.38	9.70
0220	Norman, 2-2/3" x 5/8" x 12"		165	.097		5.15	3.08		8.23	10.60
0230	Utility, 4" x 5/8" x 12"		185	.086		4.03	2.74		6.77	8.85
0240	Emperor, 4" x 3/4" x 16"		260	.062		5.20	1.95		7.15	8.85
0250	Super emperor, 8" x 3/4" x 16"		285	.056		6.55	1.78		8.33	10.05
0260	For L shaped corners with 4" return, add				L.F.	9.25			9.25	10.20
0270	For embedment into pre-cast concrete panels, add				S.F.	14.40			14.40	15.85

04 21 13.15 Chimney

	04 21 13.15 Chimney	Crew	Daily Output	Labor-Hours	Unit	Material	Labor	Equipment	Total	Total Incl O&P
0010	**CHIMNEY**, excludes foundation, scaffolding, grout and reinforcing									
0100	Brick, 16" x 16", 8" flue	D-1	18.20	.879	V.L.F.	25.50	29.50		55	77.50
0150	16" x 20" with one 8" x 12" flue		16	1		40	33.50		73.50	100
0200	16" x 24" with two 8" x 8" flues		14	1.143		58.50	38.50		97	129
0250	20" x 20" with one 12" x 12" flue		13.70	1.168		49	39.50		88.50	120
0300	20" x 24" with two 8" x 12" flues		12	1.333		66.50	45		111.50	148
0350	20" x 32" with two 12" x 12" flues		10	1.600		86.50	54		140.50	186

04 21 13.18 Columns

	04 21 13.18 Columns	Crew	Daily Output	Labor-Hours	Unit	Material	Labor	Equipment	Total	Total Incl O&P
0010	**COLUMNS**, solid, excludes scaffolding, grout and reinforcing									
0050	Brick, 8" x 8", 9 brick/V.L.F.	D-1	56	.286	V.L.F.	5.40	9.65		15.05	22
0100	12" x 8", 13.5 brick/V.L.F.		37	.432		8.10	14.60		22.70	33.50
0200	12" x 12", 20 brick/V.L.F.		25	.640		12	21.50		33.50	49
0300	16" x 12", 27 brick/V.L.F.		19	.842		16.25	28.50		44.75	65.50
0400	16" x 16", 36 brick/V.L.F.		14	1.143		21.50	38.50		60	88.50
0500	20" x 16", 45 brick/V.L.F.		11	1.455		27	49		76	112
0600	20" x 20", 56 brick/V.L.F.		9	1.778		33.50	60		93.50	137
0700	24" x 20", 68 brick/V.L.F.		7	2.286		41	77		118	174
0800	24" x 24", 81 brick/V.L.F.		6	2.667		48.50	90		138.50	204
1000	36" x 36", 182 brick/V.L.F.		3	5.333		109	180		289	420

04 21 13.30 Oversized Brick

	04 21 13.30 Oversized Brick	Crew	Daily Output	Labor-Hours	Unit	Material	Labor	Equipment	Total	Total Incl O&P
0010	**OVERSIZED BRICK**, excludes scaffolding, grout and reinforcing									
0100	Veneer, 4" x 2.25" x 16"	D-8	387	.103	S.F.	5.25	3.55		8.80	11.70
0102	8" x 2.25" x 16", multicell		265	.151		17.25	5.20		22.45	27.50
0105	4" x 2.75" x 16"		412	.097		5.65	3.33		8.98	11.75
0107	8" x 2.75" x 16", multicell		295	.136		17.20	4.65		21.85	26.50

04 21 Clay Unit Masonry

04 21 13 – Brick Masonry

04 21 13.30 Oversized Brick		Crew	Daily Output	Labor-Hours	Unit	Material	2019 Bare Costs Labor	Equipment	Total	Total Incl O&P
0110	4" x 4" x 16"	D-8	460	.087	S.F.	3.74	2.99		6.73	9.10
0120	4" x 8" x 16"		533	.075		4.49	2.58		7.07	9.25
0122	4" x 8" x 16" multicell		327	.122		15.45	4.20		19.65	24
0125	Loadbearing, 6" x 4" x 16", grouted and reinforced		387	.103		11.75	3.55		15.30	18.85
0130	8" x 4" x 16", grouted and reinforced		327	.122		12.55	4.20		16.75	21
0132	10" x 4" x 16", grouted and reinforced		327	.122		25.50	4.20		29.70	35
0135	6" x 8" x 16", grouted and reinforced		440	.091		14.80	3.12		17.92	21.50
0140	8" x 8" x 16", grouted and reinforced		400	.100		15.90	3.43		19.33	23
0145	Curtainwall/reinforced veneer, 6" x 4" x 16"		387	.103		16.95	3.55		20.50	24.50
0150	8" x 4" x 16"		327	.122		20	4.20		24.20	29.50
0152	10" x 4" x 16"		327	.122		28	4.20		32.20	38
0155	6" x 8" x 16"		440	.091		20.50	3.12		23.62	27.50
0160	8" x 8" x 16"	↓	400	.100		28.50	3.43		31.93	37
0200	For 1 to 3 slots in face, add					15%				
0210	For 4 to 7 slots in face, add					25%				
0220	For bond beams, add					20%				
0230	For bullnose shapes, add					20%				
0240	For open end knockout, add					10%				
0250	For white or gray color group, add					10%				
0260	For 135 degree corner, add				↓	250%				

04 21 13.35 Common Building Brick

		Crew	Daily Output	Labor-Hours	Unit	Material	Labor	Equipment	Total	Total Incl O&P
0010	**COMMON BUILDING BRICK**, C62, T.L. lots, material only R042110-20									
0020	Standard				M	565			565	625
0050	Select				"	530			530	585

04 21 13.45 Face Brick

		Crew	Daily Output	Labor-Hours	Unit	Material	Labor	Equipment	Total	Total Incl O&P
0010	**FACE BRICK** Material Only, C216, T.L. lots R042110-20									
0300	Standard modular, 4" x 2-2/3" x 8"				M	505			505	560
2170	For less than truck load lots, add					15			15	16.50
2180	For buff or gray brick, add				↓	16			16	17.60

04 22 Concrete Unit Masonry

04 22 10 – Concrete Masonry Units

04 22 10.11 Autoclave Aerated Concrete Block

		Crew	Daily Output	Labor-Hours	Unit	Material	Labor	Equipment	Total	Total Incl O&P
0010	**AUTOCLAVE AERATED CONCRETE BLOCK**, excl. scaffolding, grout & reinforcing									
0050	Solid, 4" x 8" x 24", incl. mortar G	D-8	600	.067	S.F.	1.61	2.29		3.90	5.60
0060	6" x 8" x 24" G		600	.067		2.38	2.29		4.67	6.45
0070	8" x 8" x 24" G		575	.070		3.19	2.39		5.58	7.50
0080	10" x 8" x 24" G		575	.070		3.88	2.39		6.27	8.25
0090	12" x 8" x 24" G	↓	550	.073	↓	4.79	2.50		7.29	9.40

04 22 10.14 Concrete Block, Back-Up

		Crew	Daily Output	Labor-Hours	Unit	Material	Labor	Equipment	Total	Total Incl O&P
0010	**CONCRETE BLOCK, BACK-UP**, C90, 2000 psi									
0020	Normal weight, 8" x 16" units, tooled joint 1 side									
0050	Not-reinforced, 2000 psi, 2" thick	D-8	475	.084	S.F.	1.62	2.89		4.51	6.60
0200	4" thick		460	.087		1.92	2.99		4.91	7.10
0300	6" thick		440	.091		2.57	3.12		5.69	8.05
0350	8" thick		400	.100		2.69	3.43		6.12	8.65
0400	10" thick	↓	330	.121		3.21	4.16		7.37	10.50
0450	12" thick	D-9	310	.155		4.30	5.20		9.50	13.45
1000	Reinforced, alternate courses, 4" thick	D-8	450	.089		2.10	3.05		5.15	7.40
1100	6" thick	↓	430	.093	↓	2.76	3.19		5.95	8.35

04 22 Concrete Unit Masonry

04 22 10 – Concrete Masonry Units

04 22 10.14 Concrete Block, Back-Up

	04 22 10.14 Concrete Block, Back-Up	Crew	Daily Output	Labor-Hours	Unit	2019 Bare Costs Material	Labor	Equipment	Total	Total Incl O&P
1150	8″ thick	D-8	395	.101	S.F.	2.89	3.48		6.37	9
1200	10″ thick		320	.125		3.39	4.29		7.68	10.85
1250	12″ thick	D-9	300	.160		4.49	5.40		9.89	13.95

04 22 10.16 Concrete Block, Bond Beam

	04 22 10.16 Concrete Block, Bond Beam	Crew	Daily Output	Labor-Hours	Unit	Material	Labor	Equipment	Total	Total Incl O&P
0010	**CONCRETE BLOCK, BOND BEAM**, C90, 2000 psi									
0020	Not including grout or reinforcing									
0125	Regular block, 6″ thick	D-8	584	.068	L.F.	2.78	2.35		5.13	7
0130	8″ high, 8″ thick	″	565	.071		2.90	2.43		5.33	7.25
0150	12″ thick	D-9	510	.094		4.07	3.17		7.24	9.80
0525	Lightweight, 6″ thick	D-8	592	.068		3.03	2.32		5.35	7.20

04 22 10.19 Concrete Block, Insulation Inserts

	04 22 10.19 Concrete Block, Insulation Inserts	Crew	Daily Output	Labor-Hours	Unit	Material	Labor	Equipment	Total	Total Incl O&P
0010	**CONCRETE BLOCK, INSULATION INSERTS**									
0100	Styrofoam, plant installed, add to block prices									
0200	8″ x 16″ units, 6″ thick				S.F.	1.22			1.22	1.34
0250	8″ thick					1.37			1.37	1.51
0300	10″ thick					1.42			1.42	1.56
0350	12″ thick					1.58			1.58	1.74
0500	8″ x 8″ units, 8″ thick					1.21			1.21	1.33
0550	12″ thick					1.44			1.44	1.58

04 22 10.23 Concrete Block, Decorative

	04 22 10.23 Concrete Block, Decorative	Crew	Daily Output	Labor-Hours	Unit	Material	Labor	Equipment	Total	Total Incl O&P
0010	**CONCRETE BLOCK, DECORATIVE**, C90, 2000 psi									
0020	Embossed, simulated brick face									
0100	8″ x 16″ units, 4″ thick	D-8	400	.100	S.F.	2.92	3.43		6.35	8.90
0200	8″ thick		340	.118		3.17	4.04		7.21	10.25
0250	12″ thick		300	.133		5.30	4.58		9.88	13.45
0400	Embossed both sides									
0500	8″ thick	D-8	300	.133	S.F.	4.60	4.58		9.18	12.70
0550	12″ thick	″	275	.145	″	5.85	4.99		10.84	14.75
1000	Fluted high strength									
1100	8″ x 16″ x 4″ thick, flutes 1 side,	D-8	345	.116	S.F.	4.11	3.98		8.09	11.20
1150	Flutes 2 sides		335	.119		4.87	4.10		8.97	12.20
1200	8″ thick		300	.133		5.55	4.58		10.13	13.75
1250	For special colors, add					.67			.67	.73
1400	Deep grooved, smooth face									
1450	8″ x 16″ x 4″ thick	D-8	345	.116	S.F.	2.70	3.98		6.68	9.60
1500	8″ thick	″	300	.133	″	4.20	4.58		8.78	12.25
2000	Formblock, incl. inserts & reinforcing									
2100	8″ x 16″ x 8″ thick	D-8	345	.116	S.F.	3.68	3.98		7.66	10.70
2150	12″ thick	″	310	.129	″	4.89	4.43		9.32	12.80
2500	Ground face									
2600	8″ x 16″ x 4″ thick	D-8	345	.116	S.F.	4.44	3.98		8.42	11.55
2650	6″ thick		325	.123		5.05	4.23		9.28	12.60
2700	8″ thick		300	.133		5.60	4.58		10.18	13.85
2750	12″ thick	D-9	265	.181		6.50	6.10		12.60	17.30
2900	For special colors, add, minimum					15%				
2950	For special colors, add, maximum					45%				
4000	Slump block									
4100	4″ face height x 16″ x 4″ thick	D-1	165	.097	S.F.	4.81	3.27		8.08	10.75
4150	6″ thick		160	.100		6.25	3.37		9.62	12.45
4200	8″ thick		155	.103		7.05	3.48		10.53	13.55
4250	10″ thick		140	.114		12.85	3.86		16.71	20.50
4300	12″ thick		130	.123		13.05	4.15		17.20	21.50

04 22 10 – Concrete Masonry Units

04 22 10.23 Concrete Block, Decorative		Crew	Daily Output	Labor-Hours	Unit	Material	2019 Bare Costs Labor	Equipment	Total	Total Incl O&P
4400	6" face height x 16" x 6" thick	D-1	155	.103	S.F.	6.15	3.48		9.63	12.60
4450	8" thick		150	.107		9.05	3.60		12.65	15.95
4500	10" thick		130	.123		13.85	4.15		18	22
4550	12" thick		120	.133		14.35	4.50		18.85	23.50
5000	Split rib profile units, 1" deep ribs, 8 ribs									
5100	8" x 16" x 4" thick	D-8	345	.116	S.F.	4.16	3.98		8.14	11.25
5150	6" thick		325	.123		4.73	4.23		8.96	12.25
5200	8" thick		300	.133		5.85	4.58		10.43	14.05
5250	12" thick	D-9	275	.175		6.75	5.90		12.65	17.25
5400	For special deeper colors, 4" thick, add					1.39			1.39	1.53
5450	12" thick, add					1.39			1.39	1.53
5600	For white, 4" thick, add					1.39			1.39	1.53
5650	6" thick, add					1.39			1.39	1.53
5700	8" thick, add					1.45			1.45	1.59
5750	12" thick, add					1.45			1.45	1.59
6000	Split face									
6100	8" x 16" x 4" thick	D-8	350	.114	S.F.	3.90	3.92		7.82	10.85
6150	6" thick		325	.123		4.40	4.23		8.63	11.90
6200	8" thick		300	.133		5.25	4.58		9.83	13.40
6250	12" thick	D-9	270	.178		6.30	6		12.30	16.95
6300	For scored, add					.38			.38	.42
6400	For special deeper colors, 4" thick, add					.67			.67	.73
6450	6" thick, add					.81			.81	.89
6500	8" thick, add					.81			.81	.89
6550	12" thick, add					.87			.87	.96
6650	For white, 4" thick, add					1.39			1.39	1.53
6700	6" thick, add					1.37			1.37	1.50
6750	8" thick, add					1.38			1.38	1.52
6800	12" thick, add					1.39			1.39	1.53
7000	Scored ground face, 2 to 5 scores									
7100	8" x 16" x 4" thick	D-8	340	.118	S.F.	9.10	4.04		13.14	16.75
7150	6" thick		310	.129		10.20	4.43		14.63	18.65
7200	8" thick		290	.138		11.50	4.74		16.24	20.50
7250	12" thick	D-9	265	.181		15.90	6.10		22	27.50
8000	Hexagonal face profile units, 8" x 16" units									
8100	4" thick, hollow	D-8	340	.118	S.F.	3.90	4.04		7.94	11.05
8200	Solid		340	.118		4.97	4.04		9.01	12.20
8300	6" thick, hollow		310	.129		4.16	4.43		8.59	12
8350	8" thick, hollow		290	.138		4.70	4.74		9.44	13.05
8500	For stacked bond, add						26%			
8550	For high rise construction, add per story	D-8	67.80	.590	M.S.F.		20.50		20.50	34
8600	For scored block, add					10%				
8650	For honed or ground face, per face, add				Ea.	1.21			1.21	1.33
8700	For honed or ground end, per end, add				"	1.19			1.19	1.31
8750	For bullnose block, add					10%				
8800	For special color, add					13%				

04 22 10.24 Concrete Block, Exterior		Crew	Daily Output	Labor-Hours	Unit	Material	Labor	Equipment	Total	Total Incl O&P
0010	**CONCRETE BLOCK, EXTERIOR**, C90, 2000 psi									
0020	Reinforced alt courses, tooled joints 2 sides									
0100	Normal weight, 8" x 16" x 6" thick	D-8	395	.101	S.F.	2.56	3.48		6.04	8.60
0200	8" thick		360	.111		3.96	3.81		7.77	10.70
0250	10" thick		290	.138		4.42	4.74		9.16	12.75

04 22 10 – Concrete Masonry Units

04 22 10.24 Concrete Block, Exterior		Crew	Daily Output	Labor-Hours	Unit	2019 Bare Costs Material	Labor	Equipment	Total	Total Incl O&P
0300	12″ thick	D-9	250	.192	S.F.	5.30	6.50		11.80	16.65

04 22 10.26 Concrete Block Foundation Wall

		Crew	Daily Output	Labor-Hours	Unit	Material	Labor	Equipment	Total	Total Incl O&P
0010	**CONCRETE BLOCK FOUNDATION WALL**, C90/C145									
0050	Normal-weight, cut joints, horiz joint reinf, no vert reinf.									
0200	Hollow, 8″ x 16″ x 6″ thick	D-8	455	.088	S.F.	3.03	3.02		6.05	8.40
0250	8″ thick		425	.094		3.16	3.23		6.39	8.85
0300	10″ thick		350	.114		3.66	3.92		7.58	10.55
0350	12″ thick	D-9	300	.160		4.76	5.40		10.16	14.25
0500	Solid, 8″ x 16″ block, 6″ thick	D-8	440	.091		3.35	3.12		6.47	8.90
0550	8″ thick	″	415	.096		4.30	3.31		7.61	10.25
0600	12″ thick	D-9	350	.137		6	4.63		10.63	14.30

04 22 10.32 Concrete Block, Lintels

		Crew	Daily Output	Labor-Hours	Unit	Material	Labor	Equipment	Total	Total Incl O&P
0010	**CONCRETE BLOCK, LINTELS**, C90, normal weight									
0100	Including grout and horizontal reinforcing									
0200	8″ x 8″ x 8″, 1 #4 bar	D-4	300	.133	L.F.	4.10	4.25	.42	8.77	12
0250	2 #4 bars		295	.136		4.34	4.32	.43	9.09	12.45
0400	8″ x 16″ x 8″, 1 #4 bar		275	.145		4.13	4.63	.46	9.22	12.75
0450	2 #4 bars		270	.148		4.36	4.72	.47	9.55	13.15
1000	12″ x 8″ x 8″, 1 #4 bar		275	.145		5.65	4.63	.46	10.74	14.40
1100	2 #4 bars		270	.148		5.85	4.72	.47	11.04	14.80
1150	2 #5 bars		270	.148		6.10	4.72	.47	11.29	15.10
1200	2 #6 bars		265	.151		6.45	4.81	.47	11.73	15.55
1500	12″ x 16″ x 8″, 1 #4 bar		250	.160		7.05	5.10	.50	12.65	16.80
1600	2 #3 bars		245	.163		7.05	5.20	.51	12.76	17
1650	2 #4 bars		245	.163		7.30	5.20	.51	13.01	17.20
1700	2 #5 bars		240	.167		7.55	5.30	.52	13.37	17.75

04 22 10.33 Lintel Block

		Crew	Daily Output	Labor-Hours	Unit	Material	Labor	Equipment	Total	Total Incl O&P
0010	**LINTEL BLOCK**									
3481	Lintel block 6″ x 8″ x 8″	D-1	300	.053	Ea.	1.38	1.80		3.18	4.52
3501	6″ x 16″ x 8″		275	.058		2.13	1.96		4.09	5.60
3521	8″ x 8″ x 8″		275	.058		1.29	1.96		3.25	4.69
3561	8″ x 16″ x 8″		250	.064		2	2.16		4.16	5.80

04 22 10.34 Concrete Block, Partitions

		Crew	Daily Output	Labor-Hours	Unit	Material	Labor	Equipment	Total	Total Incl O&P
0010	**CONCRETE BLOCK, PARTITIONS**, excludes scaffolding									
1000	Lightweight block, tooled joints, 2 sides, hollow									
1100	Not reinforced, 8″ x 16″ x 4″ thick	D-8	440	.091	S.F.	1.95	3.12		5.07	7.35
1150	6″ thick		410	.098		2.79	3.35		6.14	8.65
1200	8″ thick		385	.104		3.44	3.57		7.01	9.75
1250	10″ thick		370	.108		4.11	3.71		7.82	10.70
1300	12″ thick	D-9	350	.137		4.29	4.63		8.92	12.40
4000	Regular block, tooled joints, 2 sides, hollow									
4100	Not reinforced, 8″ x 16″ x 4″ thick	D-8	430	.093	S.F.	1.82	3.19		5.01	7.30
4150	6″ thick		400	.100		2.48	3.43		5.91	8.40
4200	8″ thick		375	.107		2.59	3.66		6.25	8.95
4250	10″ thick		360	.111		3.11	3.81		6.92	9.75
4300	12″ thick	D-9	340	.141		4.21	4.76		8.97	12.60

04 23 Glass Unit Masonry

04 23 13 – Vertical Glass Unit Masonry

04 23 13.10 Glass Block		Crew	Daily Output	Labor-Hours	Unit	2019 Bare Costs Material	Labor	Equipment	Total	Total Incl O&P
0010	**GLASS BLOCK**									
0100	Plain, 4" thick, under 1,000 S.F., 6" x 6"	D-8	115	.348	S.F.	23.50	11.95		35.45	45.50
0150	8" x 8"		160	.250		13.30	8.60		21.90	29
0160	end block		160	.250		62.50	8.60		71.10	83.50
0170	90 degree corner		160	.250		66.50	8.60		75.10	88
0180	45 degree corner		160	.250		60	8.60		68.60	80.50
0200	12" x 12"		175	.229		21.50	7.85		29.35	37
0210	4" x 8"		160	.250		36.50	8.60		45.10	54.50
0220	6" x 8"		160	.250		23	8.60		31.60	39.50
0700	For solar reflective blocks, add					100%				
1000	Thinline, plain, 3-1/8" thick, under 1,000 S.F., 6" x 6"	D-8	115	.348	S.F.	24	11.95		35.95	46.50
1050	8" x 8"		160	.250		13.25	8.60		21.85	29
1400	For cleaning block after installation (both sides), add		1000	.040		.16	1.37		1.53	2.47

04 24 Adobe Unit Masonry

04 24 16 – Manufactured Adobe Unit Masonry

04 24 16.06 Adobe Brick		Crew	Daily Output	Labor-Hours	Unit	Material	Labor	Equipment	Total	Total Incl O&P
0010	**ADOBE BRICK**, Semi-stabilized, with cement mortar									
0060	Brick, 10" x 4" x 14", 2.6/S.F. G	D-8	560	.071	S.F.	4.75	2.45		7.20	9.30
0080	12" x 4" x 16", 2.3/S.F. G		580	.069		6.95	2.37		9.32	11.60
0100	10" x 4" x 16", 2.3/S.F. G		590	.068		6.50	2.33		8.83	11.05
0120	8" x 4" x 16", 2.3/S.F. G		560	.071		4.96	2.45		7.41	9.55
0140	4" x 4" x 16", 2.3/S.F. G		540	.074		4.85	2.54		7.39	9.60
0160	6" x 4" x 16", 2.3/S.F. G		540	.074		4.38	2.54		6.92	9.05
0180	4" x 4" x 12", 3.0/S.F. G		520	.077		5.60	2.64		8.24	10.60
0200	8" x 4" x 12", 3.0/S.F. G		520	.077		4.36	2.64		7	9.20

04 27 Multiple-Wythe Unit Masonry

04 27 10 – Multiple-Wythe Masonry

04 27 10.10 Cornices		Crew	Daily Output	Labor-Hours	Unit	Material	Labor	Equipment	Total	Total Incl O&P
0010	**CORNICES**									
0110	Face bricks, 12 brick/S.F.	D-1	30	.533	SF Face	5.75	18		23.75	36.50
0150	15 brick/S.F.	"	23	.696	"	6.90	23.50		30.40	46.50
04 27 10.30 Brick Walls										
0010	**BRICK WALLS**, including mortar, excludes scaffolding									
0800	Face brick, 4" thick wall, 6.75 brick/S.F.	D-8	215	.186	S.F.	4.07	6.40		10.47	15.15
0850	Common brick, 4" thick wall, 6.75 brick/S.F.		240	.167		4.48	5.70		10.18	14.50
0900	8" thick, 13.50 brick/S.F.		135	.296		9.20	10.15		19.35	27
1000	12" thick, 20.25 brick/S.F.		95	.421		13.85	14.45		28.30	39.50
1050	16" thick, 27.00 brick/S.F.		75	.533		18.70	18.30		37	51
1200	Reinforced, face brick, 4" thick wall, 6.75 brick/S.F.		210	.190		4.24	6.55		10.79	15.55
1220	Common brick, 4" thick wall, 6.75 brick/S.F.		235	.170		4.65	5.85		10.50	14.85
1250	8" thick, 13.50 brick/S.F.		130	.308		9.55	10.55		20.10	28
1260	8" thick, 2.25 brick/S.F.		130	.308		1.71	10.55		12.26	19.50
1300	12" thick, 20.25 brick/S.F.		90	.444		14.35	15.25		29.60	41.50
1350	16" thick, 27.00 brick/S.F.		70	.571		19.40	19.60		39	54

04 27 Multiple-Wythe Unit Masonry

04 27 10 – Multiple-Wythe Masonry

04 27 10.40 Steps		Crew	Daily Output	Labor-Hours	Unit	Material	Labor	Equipment	Total	Total Incl O&P
						2019 Bare Costs				
0010	**STEPS**									
0012	Entry steps, select common brick	D-1	.30	53.333	M	530	1,800		2,330	3,575

04 41 Dry-Placed Stone

04 41 10 – Dry Placed Stone

04 41 10.10 Rough Stone Wall

			Crew	Daily Output	Labor-Hours	Unit	Material	Labor	Equipment	Total	Total Incl O&P
0011	**ROUGH STONE WALL**, Dry										
0012	Dry laid (no mortar), under 18" thick	G	D-1	60	.267	C.F.	13.90	9		22.90	30.50
0100	Random fieldstone, under 18" thick	G	D-12	60	.533		13.90	18		31.90	45.50
0150	Over 18" thick	G	"	63	.508	↓	16.65	17.15		33.80	47
0500	Field stone veneer	G	D-8	120	.333	S.F.	12.65	11.45		24.10	33
0600	Rubble stone walls, in mortar bed, up to 18" thick	G	D-11	75	.320	C.F.	16.75	11.10		27.85	37

04 43 Stone Masonry

04 43 10 – Masonry with Natural and Processed Stone

04 43 10.45 Granite

		Crew	Daily Output	Labor-Hours	Unit	Material	Labor	Equipment	Total	Total Incl O&P
0010	**GRANITE**, cut to size									
0050	Veneer, polished face, 3/4" to 1-1/2" thick									
0150	Low price, gray, light gray, etc.	D-10	130	.246	S.F.	29	9.10	2.86	40.96	50
0180	Medium price, pink, brown, etc.		130	.246		30	9.10	2.86	41.96	51.50
0220	High price, red, black, etc.	↓	130	.246	↓	42.50	9.10	2.86	54.46	65.50
0300	1-1/2" to 2-1/2" thick, veneer									
0350	Low price, gray, light gray, etc.	D-10	130	.246	S.F.	29	9.10	2.86	40.96	50.50
0500	Medium price, pink, brown, etc.		130	.246		34	9.10	2.86	45.96	55.50
0550	High price, red, black, etc.	↓	130	.246	↓	52.50	9.10	2.86	64.46	76.50
0700	2-1/2" to 4" thick, veneer									
0750	Low price, gray, light gray, etc.	D-10	110	.291	S.F.	39	10.80	3.38	53.18	64
0850	Medium price, pink, brown, etc.		110	.291		44.50	10.80	3.38	58.68	70.50
0950	High price, red, black, etc.	↓	110	.291		63.50	10.80	3.38	77.68	91.50
1000	For bush hammered finish, deduct					5%				
1050	Coarse rubbed finish, deduct					10%				
1100	Honed finish, deduct					5%				
1150	Thermal finish, deduct				↓	18%				
2500	Steps, copings, etc., finished on more than one surface									
2550	Low price, gray, light gray, etc.	D-10	50	.640	C.F.	92.50	23.50	7.45	123.45	149
2575	Medium price, pink, brown, etc.		50	.640		120	23.50	7.45	150.95	180
2600	High price, red, black, etc.	↓	50	.640	↓	148	23.50	7.45	178.95	210
2800	Pavers, 4" x 4" x 4" blocks, split face and joints									
2850	Low price, gray, light gray, etc.	D-11	80	.300	S.F.	13.15	10.40		23.55	32
2875	Medium price, pink, brown, etc.		80	.300		21	10.40		31.40	41
2900	High price, red, black, etc.	↓	80	.300		29	10.40		39.40	49.50
4000	Soffits, 2" thick, low price, gray, light gray	D-13	35	1.371		38	49.50	10.65	98.15	136
4050	Medium price, pink, brown, etc.		35	1.371		65	49.50	10.65	125.15	165
4100	High price, red, black, etc.		35	1.371		91.50	49.50	10.65	151.65	195
4200	Low price, gray, light gray, etc.		35	1.371		63.50	49.50	10.65	123.65	164
4250	Medium price, pink, brown, etc.		35	1.371		91.50	49.50	10.65	151.65	194
4300	High price, red, black, etc.	↓	35	1.371	↓	119	49.50	10.65	179.15	225
5000	Reclaimed or antique									
5010	Treads, up to 12" wide	D-10	100	.320	L.F.	42.50	11.85	3.72	58.07	70

04 43 10 – Masonry with Natural and Processed Stone

04 43 10.45 Granite		Crew	Daily Output	Labor-Hours	Unit	2019 Bare Costs Material	Labor	Equipment	Total	Total Incl O&P
5020	Up to 18" wide	D-10	100	.320	L.F.	40	11.85	3.72	55.57	67.50
5030	Capstone, size varies		50	.640		29	23.50	7.45	59.95	79.50
5040	Posts		30	1.067	V.L.F.	30	39.50	12.40	81.90	112

04 43 10.55 Limestone

Code	Description	Crew	Daily Output	Labor-Hours	Unit	Material	Labor	Equipment	Total	Total Incl O&P
0010	**LIMESTONE**, cut to size									
0020	Veneer facing panels									
0500	Texture finish, light stick, 4-1/2" thick, 5' x 12'	D-4	300	.133	S.F.	20.50	4.25	.42	25.17	30
0750	5" thick, 5' x 14' panels	D-10	275	.116		21.50	4.31	1.35	27.16	32.50
1000	Sugarcube finish, 2" thick, 3' x 5' panels		275	.116		28.50	4.31	1.35	34.16	40
1050	3" thick, 4' x 9' panels		275	.116		25.50	4.31	1.35	31.16	36.50
1200	4" thick, 5' x 11' panels		275	.116		29.50	4.31	1.35	35.16	41
1400	Sugarcube, textured finish, 4-1/2" thick, 5' x 12'		275	.116		32	4.31	1.35	37.66	43.50
1450	5" thick, 5' x 14' panels		275	.116		33	4.31	1.35	38.66	45
2000	Coping, sugarcube finish, top & 2 sides		30	1.067	C.F.	66.50	39.50	12.40	118.40	153
2100	Sills, lintels, jambs, trim, stops, sugarcube finish, simple		20	1.600		66.50	59.50	18.60	144.60	193
2150	Detailed		20	1.600		66.50	59.50	18.60	144.60	193
2300	Steps, extra hard, 14" wide, 6" rise		50	.640	L.F.	28.50	23.50	7.45	59.45	78.50
3000	Quoins, plain finish, 6" x 12" x 12"	D-12	25	1.280	Ea.	39.50	43		82.50	116
3050	6" x 16" x 24"	"	25	1.280	"	53	43		96	130

04 43 10.60 Marble

Code	Description	Crew	Daily Output	Labor-Hours	Unit	Material	Labor	Equipment	Total	Total Incl O&P
0011	**MARBLE**, ashlar, split face, +/- 4" thick, random									
0040	Lengths 1' to 4' & heights 2" to 7-1/2", average	D-8	175	.229	S.F.	17.80	7.85		25.65	32.50
1000	Facing, polished finish, cut to size, 3/4" to 7/8" thick									
1050	Carrara or equal	D-10	130	.246	S.F.	22	9.10	2.86	33.96	42.50
1100	Arabescato or equal	"	130	.246	"	38.50	9.10	2.86	50.46	61
2200	Window sills, 6" x 3/4" thick	D-1	85	.188	L.F.	11.95	6.35		18.30	24
2500	Flooring, polished tiles, 12" x 12" x 3/8" thick									
2510	Thin set, Giallo Solare or equal	D-11	90	.267	S.F.	16.95	9.25		26.20	34
2600	Sky Blue or equal		90	.267		15.20	9.25		24.45	32
2700	Mortar bed, Giallo Solare or equal		65	.369		16.95	12.80		29.75	40
2740	Sky Blue or equal		65	.369		15.20	12.80		28	38.50
2780	Travertine, 3/8" thick, Sierra or equal	D-10	130	.246		9.25	9.10	2.86	21.21	28.50
2790	Silver or equal	"	130	.246		25.50	9.10	2.86	37.46	47
3500	Thresholds, 3' long, 7/8" thick, 4" to 5" wide, plain	D-12	24	1.333	Ea.	34	45		79	113
3550	Beveled		24	1.333	"	76	45		121	159
3700	Window stools, polished, 7/8" thick, 5" wide		85	.376	L.F.	21	12.70		33.70	44.50

04 43 10.75 Sandstone or Brownstone

Code	Description	Crew	Daily Output	Labor-Hours	Unit	Material	Labor	Equipment	Total	Total Incl O&P
0011	**SANDSTONE OR BROWNSTONE**									
0100	Sawed face veneer, 2-1/2" thick, to 2' x 4' panels	D-10	130	.246	S.F.	21	9.10	2.86	32.96	42
0150	4" thick, to 3'-6" x 8' panels		100	.320		21	11.85	3.72	36.57	47
0300	Split face, random sizes		100	.320		13.65	11.85	3.72	29.22	39
0350	Cut stone trim (limestone)									
0360	Ribbon stone, 4" thick, 5' pieces	D-8	120	.333	Ea.	161	11.45		172.45	196
0370	Cove stone, 4" thick, 5' pieces		105	.381		162	13.10		175.10	200
0380	Cornice stone, 10" to 12" wide		90	.444		200	15.25		215.25	246
0390	Band stone, 4" thick, 5' pieces		145	.276		106	9.45		115.45	133
0410	Window and door trim, 3" to 4" wide		160	.250		90.50	8.60		99.10	114
0420	Key stone, 18" long		60	.667		93	23		116	140

04 43 10.80 Slate

Code	Description	Crew	Daily Output	Labor-Hours	Unit	Material	Labor	Equipment	Total	Total Incl O&P
0010	**SLATE**									
3100	Stair landings, 1" thick, black, clear	D-1	65	.246	S.F.	21	8.30		29.30	37
3200	Ribbon	"	65	.246	"	23	8.30		31.30	39.50

04 43 Stone Masonry

04 43 10 – Masonry with Natural and Processed Stone

04 43 10.80 Slate		Crew	Daily Output	Labor-Hours	Unit	Material	2019 Bare Costs Labor	Equipment	Total	Total Incl O&P
3500	Stair treads, sand finish, 1" thick x 12" wide									
3600	3 L.F. to 6 L.F.	D-10	120	.267	L.F.	25	9.90	3.10	38	47.50
3700	Ribbon, sand finish, 1" thick x 12" wide									
3750	To 6 L.F.	D-10	120	.267	L.F.	21	9.90	3.10	34	43

04 43 10.85 Window Sill		Crew	Daily Output	Labor-Hours	Unit	Material	Labor	Equipment	Total	Total Incl O&P
0010	**WINDOW SILL**									
0020	Bluestone, thermal top, 10" wide, 1-1/2" thick	D-1	85	.188	S.F.	8.90	6.35		15.25	20.50
0050	2" thick		75	.213	"	9.25	7.20		16.45	22
0100	Cut stone, 5" x 8" plain		48	.333	L.F.	12.60	11.25		23.85	32.50
0200	Face brick on edge, brick, 8" wide		80	.200		5.55	6.75		12.30	17.35
0400	Marble, 9" wide, 1" thick		85	.188		9	6.35		15.35	20.50
0900	Slate, colored, unfading, honed, 12" wide, 1" thick		85	.188		9.30	6.35		15.65	21
0950	2" thick	↓	70	.229	↓	8.95	7.70		16.65	22.50

04 51 Flue Liner Masonry

04 51 10 – Clay Flue Lining

04 51 10.10 Flue Lining		Crew	Daily Output	Labor-Hours	Unit	Material	Labor	Equipment	Total	Total Incl O&P
0010	**FLUE LINING**, including mortar									
0020	Clay, 8" x 8"	D-1	125	.128	V.L.F.	6.10	4.32		10.42	13.90
0100	8" x 12"		103	.155		8.75	5.25		14	18.35
0200	12" x 12"		93	.172		11.95	5.80		17.75	23
0300	12" x 18"		84	.190		23.50	6.40		29.90	36.50
0400	18" x 18"		75	.213		30.50	7.20		37.70	46
0500	20" x 20"		66	.242		45.50	8.20		53.70	63.50
0600	24" x 24"		56	.286		58.50	9.65		68.15	80.50
1000	Round, 18" diameter		66	.242		40	8.20		48.20	57.50
1100	24" diameter	↓	47	.340	↓	79	11.50		90.50	106

04 57 Masonry Fireplaces

04 57 10 – Brick or Stone Fireplaces

04 57 10.10 Fireplace		Crew	Daily Output	Labor-Hours	Unit	Material	Labor	Equipment	Total	Total Incl O&P
0010	**FIREPLACE**									
0100	Brick fireplace, not incl. foundations or chimneys									
0110	30" x 29" opening, incl. chamber, plain brickwork	D-1	.40	40	Ea.	585	1,350		1,935	2,900
0200	Fireplace box only (110 brick)	"	2	8	"	162	270		432	630
0300	For elaborate brickwork and details, add					35%	35%			
0400	For hearth, brick & stone, add	D-1	2	8	Ea.	211	270		481	680
0410	For steel, damper, cleanouts, add		4	4		15.45	135		150.45	242
0600	Plain brickwork, incl. metal circulator		.50	32	↓	905	1,075		1,980	2,800
0800	Face brick only, standard size, 8" x 2-2/3" x 4"	↓	.30	53.333	M	595	1,800		2,395	3,650
0900	Stone fireplace, fieldstone, add				SF Face	9.05			9.05	10
1000	Cut stone, add				"	8.80			8.80	9.65

04 72 Cast Stone Masonry

04 72 10 – Cast Stone Masonry Features

04 72 10.10 Coping		Crew	Daily Output	Labor-Hours	Unit	Material	Labor	Equipment	Total	Total Incl O&P
						2019 Bare Costs				
0010	**COPING**, stock units									
0050	Precast concrete, 10" wide, 4" tapers to 3-1/2", 8" wall	D-1	75	.213	L.F.	24	7.20		31.20	38.50
0100	12" wide, 3-1/2" tapers to 3", 10" wall		70	.229		26	7.70		33.70	41.50
0110	14" wide, 4" tapers to 3-1/2", 12" wall		65	.246		30	8.30		38.30	47
0150	16" wide, 4" tapers to 3-1/2", 14" wall		60	.267		32	9		41	50.50
0300	Limestone for 12" wall, 4" thick		90	.178		16.80	6		22.80	28.50
0350	6" thick		80	.200		24	6.75		30.75	38
0500	Marble, to 4" thick, no wash, 9" wide		90	.178		12.70	6		18.70	24
0550	12" wide		80	.200		19.35	6.75		26.10	33
0700	Terra cotta, 9" wide		90	.178		7.75	6		13.75	18.50
0750	12" wide		80	.200		8.70	6.75		15.45	21
0800	Aluminum, for 12" wall		80	.200		9.20	6.75		15.95	21.50

04 72 20 – Cultured Stone Veneer

04 72 20.10 Cultured Stone Veneer Components

		Crew	Daily Output	Labor-Hours	Unit	Material	Labor	Equipment	Total	Total Incl O&P
0010	**CULTURED STONE VENEER COMPONENTS**									
0110	On wood frame and sheathing substrate, random sized cobbles, corner stones	D-8	70	.571	V.L.F.	10.55	19.60		30.15	44
0120	Field stones		140	.286	S.F.	7.60	9.80		17.40	25
0130	Random sized flats, corner stones		70	.571	V.L.F.	10.70	19.60		30.30	44.50
0140	Field stones		140	.286	S.F.	8.85	9.80		18.65	26
0150	Horizontal lined ledgestones, corner stones		75	.533	V.L.F.	10.55	18.30		28.85	42
0160	Field stones		150	.267	S.F.	7.60	9.15		16.75	23.50
0170	Random shaped flats, corner stones		65	.615	V.L.F.	10.55	21		31.55	46.50
0180	Field stones		150	.267	S.F.	7.55	9.15		16.70	23.50
0190	Random shaped/textured face, corner stones		65	.615	V.L.F.	10.55	21		31.55	46.50
0200	Field stones		130	.308	S.F.	7.50	10.55		18.05	26
0210	Random shaped river rock, corner stones		65	.615	V.L.F.	10.55	21		31.55	46.50
0220	Field stones		130	.308	S.F.	7.50	10.55		18.05	26
0240	On concrete or CMU substrate, random sized cobbles, corner stones		70	.571	V.L.F.	9.70	19.60		29.30	43
0250	Field stones		140	.286	S.F.	7.20	9.80		17	24.50
0260	Random sized flats, corner stones		70	.571	V.L.F.	9.85	19.60		29.45	43.50
0270	Field stones		140	.286	S.F.	8.45	9.80		18.25	25.50
0280	Horizontal lined ledgestones, corner stones		75	.533	V.L.F.	9.70	18.30		28	41
0290	Field stones		150	.267	S.F.	7.20	9.15		16.35	23
0300	Random shaped flats, corner stones		70	.571	V.L.F.	9.70	19.60		29.30	43
0310	Field stones		140	.286	S.F.	7.15	9.80		16.95	24
0320	Random shaped/textured face, corner stones		65	.615	V.L.F.	9.70	21		30.70	45.50
0330	Field stones		130	.308	S.F.	7.10	10.55		17.65	25.50
0340	Random shaped river rock, corner stones		65	.615	V.L.F.	9.70	21		30.70	45.50
0350	Field stones		130	.308	S.F.	7.10	10.55		17.65	25.50
0360	Cultured stone veneer, #15 felt weather resistant barrier	1 Clab	3700	.002	Sq.	5.40	.07		5.47	6.05
0370	Expanded metal lath, diamond, 2.5 lb./S.Y., galvanized	1 Lath	85	.094	S.Y.	3.83	3.59		7.42	10
0390	Water table or window sill, 18" long	1 Bric	80	.100	Ea.	10	3.68		13.68	17.15

Division Notes

		CREW	DAILY OUTPUT	LABOR-HOURS	UNIT	BARE COSTS MAT.	BARE COSTS LABOR	BARE COSTS EQUIP.	BARE COSTS TOTAL	TOTAL INCL O&P

Estimating Tips

05 05 00 Common Work Results for Metals

- Nuts, bolts, washers, connection angles, and plates can add a significant amount to both the tonnage of a structural steel job and the estimated cost. As a rule of thumb, add 10% to the total weight to account for these accessories.
- Type 2 steel construction, commonly referred to as "simple construction," consists generally of field-bolted connections with lateral bracing supplied by other elements of the building, such as masonry walls or x-bracing. The estimator should be aware, however, that shop connections may be accomplished by welding or bolting. The method may be particular to the fabrication shop and may have an impact on the estimated cost.

05 10 00 Structural Steel

- Steel items can be obtained from two sources: a fabrication shop or a metals service center. Fabrication shops can fabricate items under more controlled conditions than crews in the field can. They are also more efficient and can produce items more economically. Metal service centers serve as a source of long mill shapes to both fabrication shops and contractors.
- Most line items in this structural steel subdivision, and most items in 05 50 00 Metal Fabrications, are indicated as being shop fabricated. The bare material cost for these shop fabricated items is the "Invoice Cost" from the shop and includes the mill base price of steel plus mill extras, transportation to the shop, shop drawings and detailing where warranted, shop fabrication and handling, sandblasting and a shop coat of primer paint, all necessary structural bolts, and delivery to the job site. The bare labor cost and bare equipment cost for these shop fabricated items are for field installation or erection.
- Line items in Subdivision 05 12 23.40 Lightweight Framing, and other items scattered in Division 5, are indicated as being field fabricated. The bare material cost for these field fabricated items is the "Invoice Cost" from the metals service center and includes the mill base price of steel plus mill extras, transportation to the metals service center, material handling, and delivery of long lengths of mill shapes to the job site. Material costs for structural bolts and welding rods should be added to the estimate. The bare labor cost and bare equipment cost for these items are for both field fabrication and field installation or erection, and include time for cutting, welding, and drilling in the fabricated metal items. Drilling into concrete and fasteners to fasten field fabricated items to other work is not included and should be added to the estimate.

05 20 00 Steel Joist Framing

- In any given project the total weight of open web steel joists is determined by the loads to be supported and the design. However, economies can be realized in minimizing the amount of labor used to place the joists. This is done by maximizing the joist spacing and therefore minimizing the number of joists required to be installed on the job. Certain spacings and locations may be required by the design, but in other cases maximizing the spacing and keeping it as uniform as possible will keep the costs down.

05 30 00 Steel Decking

- The takeoff and estimating of a metal deck involve more than the area of the floor or roof and the type of deck specified or shown on the drawings. Many different sizes and types of openings may exist. Small openings for individual pipes or conduits may be drilled after the floor/roof is installed, but larger openings may require special deck lengths as well as reinforcing or structural support. The estimator should determine who will be supplying this reinforcing. Additionally, some deck terminations are part of the deck package, such as screed angles and pour stops, and others will be part of the steel contract, such as angles attached to structural members and cast-in-place angles and plates. The estimator must ensure that all pieces are accounted for in the complete estimate.

05 50 00 Metal Fabrications

- The most economical steel stairs are those that use common materials, standard details, and most importantly, a uniform and relatively simple method of field assembly. Commonly available A36/A992 channels and plates are very good choices for the main stringers of the stairs, as are angles and tees for the carrier members. Risers and treads are usually made by specialty shops, and it is most economical to use a typical detail in as many places as possible. The stairs should be pre-assembled and shipped directly to the site. The field connections should be simple and straightforward enough to be accomplished efficiently, and with minimum equipment and labor.

Reference Numbers

Reference numbers are shown at the beginning of some major classifications. These numbers refer to related items in the Reference Section. The reference information may be an estimating procedure, an alternate pricing method, or technical information.

Note: Not all subdivisions listed here necessarily appear. ■

05 01 Maintenance of Metals

05 01 10 – Maintenance of Structural Metal Framing

05 01 10.51 Cleaning of Structural Metal Framing		Crew	Daily Output	Labor-Hours	Unit	Material	2019 Bare Costs Labor	Equipment	Total	Total Incl O&P
0010	**CLEANING OF STRUCTURAL METAL FRAMING**									
6125	Steel surface treatments, PDCA guidelines									
6235	Com'l blast (SSPC-SP6), loose scale, fine pwder rust, 2.0#/S.F. sand	E-11	1200	.027	S.F.	.36	.91	.17	1.44	2.13
6240	Tight mill scale, little/no rust, 3.0#/S.F. sand		1000	.032		.54	1.09	.20	1.83	2.67
6245	Exist coat blistered/pitted, 4.0#/S.F. sand	↓	875	.037	↓	.72	1.24	.23	2.19	3.18

05 05 Common Work Results for Metals

05 05 05 – Selective Demolition for Metals

05 05 05.10 Selective Demolition, Metals		Crew	Daily Output	Labor-Hours	Unit	Material	Labor	Equipment	Total	Total Incl O&P
0010	**SELECTIVE DEMOLITION, METALS** R024119-10									
0015	Excludes shores, bracing, cutting, loading, hauling, dumping									
0020	Remove nuts only up to 3/4" diameter	1 Sswk	480	.017	Ea.		.69		.69	1.20
0030	7/8" to 1-1/4" diameter		240	.033			1.38		1.38	2.40
0040	1-3/8" to 2" diameter		160	.050			2.07		2.07	3.60
0060	Unbolt and remove structural bolts up to 3/4" diameter		240	.033			1.38		1.38	2.40
0070	7/8" to 2" diameter		160	.050			2.07		2.07	3.60
0140	Light weight framing members, remove whole or cut up, up to 20 lb.	↓	240	.033			1.38		1.38	2.40
0150	21-40 lb.	2 Sswk	210	.076			3.16		3.16	5.50
0160	41-80 lb.	3 Sswk	180	.133			5.55		5.55	9.60
0170	81-120 lb.	4 Sswk	150	.213			8.85		8.85	15.35
0230	Structural members, remove whole or cut up, up to 500 lb.	E-19	48	.500			20.50	18.10	38.60	55
0240	1/4-2 tons	E-18	36	1.111			46.50	24	70.50	106
0250	2-5 tons	E-24	30	1.067			44	19.40	63.40	97
0260	5-10 tons	E-20	24	2.667			110	55	165	249
0270	10-15 tons	E-2	18	2.667			112	94.50	206.50	296
0340	Fabricated item, remove whole or cut up, up to 20 lb.	1 Sswk	96	.083			3.45		3.45	6
0350	21-40 lb.	2 Sswk	84	.190			7.90		7.90	13.70
0360	41-80 lb.	3 Sswk	72	.333			13.80		13.80	24
0370	81-120 lb.	4 Sswk	60	.533			22		22	38.50
0380	121-500 lb.	E-19	48	.500	↓		20.50	18.10	38.60	55
0500	Steel roof decking, uncovered, bare	B-2	5000	.008	S.F.		.25		.25	.41

05 05 13 – Shop-Applied Coatings for Metal

05 05 13.50 Paints and Protective Coatings		Crew	Daily Output	Labor-Hours	Unit	Material	Labor	Equipment	Total	Total Incl O&P
0010	**PAINTS AND PROTECTIVE COATINGS**									
5900	Galvanizing structural steel in shop, under 1 ton				Ton	560			560	615
5950	1 ton to 20 tons					505			505	555
6000	Over 20 tons				↓	470			470	515

05 05 19 – Post-Installed Concrete Anchors

05 05 19.10 Chemical Anchors		Crew	Daily Output	Labor-Hours	Unit	Material	Labor	Equipment	Total	Total Incl O&P
0010	**CHEMICAL ANCHORS**									
0020	Includes layout & drilling									
1430	Chemical anchor, w/rod & epoxy cartridge, 3/4" diameter x 9-1/2" long	B-89A	27	.593	Ea.	10.90	20.50	4.15	35.55	51
1435	1" diameter x 11-3/4" long		24	.667		18.80	23.50	4.67	46.97	64
1440	1-1/4" diameter x 14" long		21	.762		37.50	26.50	5.35	69.35	91.50
1445	1-3/4" diameter x 15" long		20	.800		72	28	5.60	105.60	132
1450	18" long		17	.941		86	33	6.60	125.60	156
1455	2" diameter x 18" long		16	1		114	35	7	156	191
1460	24" long	↓	15	1.067	↓	148	37	7.50	192.50	233

05 05 19.20 Expansion Anchors		Crew	Daily Output	Labor-Hours	Unit	Material	Labor	Equipment	Total	Total Incl O&P
						2019 Bare Costs				
0010	**EXPANSION ANCHORS**									
0100	Anchors for concrete, brick or stone, no layout and drilling									
0200	Expansion shields, zinc, 1/4" diameter, 1-5/16" long, single G	1 Carp	90	.089	Ea.	.45	3.44		3.89	6.20
0300	1-3/8" long, double G		85	.094		.59	3.65		4.24	6.70
0400	3/8" diameter, 1-1/2" long, single G		85	.094		.70	3.65		4.35	6.80
0500	2" long, double G		80	.100		1.18	3.88		5.06	7.70
0600	1/2" diameter, 2-1/16" long, single G		80	.100		1.37	3.88		5.25	7.90
0700	2-1/2" long, double G		75	.107		2.17	4.13		6.30	9.25
0800	5/8" diameter, 2-5/8" long, single G		75	.107		2.19	4.13		6.32	9.25
0900	2-3/4" long, double G		70	.114		2.97	4.43		7.40	10.55
1000	3/4" diameter, 2-3/4" long, single G		70	.114		3.78	4.43		8.21	11.45
1100	3-15/16" long, double G	↓	65	.123	↓	5.55	4.77		10.32	14
2100	Hollow wall anchors for gypsum wall board, plaster or tile									
2500	3/16" diameter, short G	1 Carp	150	.053	Ea.	.52	2.07		2.59	3.99
3000	Toggle bolts, bright steel, 1/8" diameter, 2" long G		85	.094		.26	3.65		3.91	6.35
3100	4" long G		80	.100		.29	3.88		4.17	6.70
3200	3/16" diameter, 3" long G		80	.100		.34	3.88		4.22	6.75
3300	6" long G		75	.107		.45	4.13		4.58	7.35
3400	1/4" diameter, 3" long G		75	.107		.46	4.13		4.59	7.35
3500	6" long G		70	.114		.59	4.43		5.02	7.95
3600	3/8" diameter, 3" long G		70	.114		.86	4.43		5.29	8.25
3700	6" long G		60	.133		1.71	5.15		6.86	10.45
3800	1/2" diameter, 4" long G		60	.133		1.68	5.15		6.83	10.40
3900	6" long G	↓	50	.160	↓	2.25	6.20		8.45	12.75
4000	Nailing anchors									
4100	Nylon nailing anchor, 1/4" diameter, 1" long	1 Carp	3.20	2.500	C	24	97		121	187
4200	1-1/2" long		2.80	2.857		25.50	111		136.50	212
4300	2" long		2.40	3.333		28	129		157	245
4400	Metal nailing anchor, 1/4" diameter, 1" long G		3.20	2.500		22.50	97		119.50	185
4500	1-1/2" long G		2.80	2.857		29.50	111		140.50	216
4600	2" long G	↓	2.40	3.333	↓	36.50	129		165.50	254
5000	Screw anchors for concrete, masonry,									
5100	stone & tile, no layout or drilling included									
5700	Lag screw shields, 1/4" diameter, short G	1 Carp	90	.089	Ea.	.48	3.44		3.92	6.25
5800	Long G		85	.094		.55	3.65		4.20	6.65
5900	3/8" diameter, short G		85	.094		.73	3.65		4.38	6.85
6000	Long G		80	.100		.93	3.88		4.81	7.40
6100	1/2" diameter, short G		80	.100		1.04	3.88		4.92	7.55
6200	Long G		75	.107		1.38	4.13		5.51	8.35
6300	5/8" diameter, short G		70	.114		1.56	4.43		5.99	9
6400	Long G		65	.123		2.08	4.77		6.85	10.20
6600	Lead, #6 & #8, 3/4" long G		260	.031		.19	1.19		1.38	2.18
6700	#10 - #14, 1-1/2" long G		200	.040		.40	1.55		1.95	3
6800	#16 & #18, 1-1/2" long G		160	.050		.42	1.94		2.36	3.66
6900	Plastic, #6 & #8, 3/4" long		260	.031		.05	1.19		1.24	2.03
7000	#8 & #10, 7/8" long		240	.033		.06	1.29		1.35	2.20
7100	#10 & #12, 1" long		220	.036		.08	1.41		1.49	2.42
7200	#14 & #16, 1-1/2" long		160	.050		.07	1.94		2.01	3.28
8950	Self-drilling concrete screw, hex washer head, 3/16" diam. x 1-3/4" long G		300	.027		.19	1.03		1.22	1.92
8960	2-1/4" long G		250	.032		.21	1.24		1.45	2.28
8970	Phillips flat head, 3/16" diam. x 1-3/4" long G		300	.027		.19	1.03		1.22	1.92
8980	2-1/4" long G	↓	250	.032	↓	.21	1.24		1.45	2.28

05 05 Common Work Results for Metals

05 05 21 – Fastening Methods for Metal

05 05 21.15 Drilling Steel		Crew	Daily Output	Labor-Hours	Unit	2019 Bare Costs Material	Labor	Equipment	Total	Total Incl O&P
0010	**DRILLING STEEL**									
1910	Drilling & layout for steel, up to 1/4" deep, no anchor									
1920	Holes, 1/4" diameter	1 Sswk	112	.071	Ea.	.09	2.96		3.05	5.25
1925	For each additional 1/4" depth, add		336	.024		.09	.99		1.08	1.81
1930	3/8" diameter		104	.077		.09	3.19		3.28	5.65
1935	For each additional 1/4" depth, add		312	.026		.09	1.06		1.15	1.94
1940	1/2" diameter		96	.083		.10	3.45		3.55	6.10
1945	For each additional 1/4" depth, add		288	.028		.10	1.15		1.25	2.11
1950	5/8" diameter		88	.091		.17	3.77		3.94	6.75
1955	For each additional 1/4" depth, add		264	.030		.17	1.26		1.43	2.36
1960	3/4" diameter		80	.100		.19	4.15		4.34	7.40
1965	For each additional 1/4" depth, add		240	.033		.19	1.38		1.57	2.61
1970	7/8" diameter		72	.111		.27	4.61		4.88	8.30
1975	For each additional 1/4" depth, add		216	.037		.27	1.54		1.81	2.96
1980	1" diameter		64	.125		.25	5.20		5.45	9.25
1985	For each additional 1/4" depth, add		192	.042		.25	1.73		1.98	3.27
1990	For drilling up, add						40%			

05 05 23 – Metal Fastenings

05 05 23.10 Bolts and Hex Nuts			Crew	Daily Output	Labor-Hours	Unit	Material	Labor	Equipment	Total	Total Incl O&P
0010	**BOLTS & HEX NUTS**, Steel, A307										
0100	1/4" diameter, 1/2" long	G	1 Sswk	140	.057	Ea.	.06	2.37		2.43	4.18
0200	1" long	G		140	.057		.07	2.37		2.44	4.19
0300	2" long	G		130	.062		.10	2.55		2.65	4.53
0400	3" long	G		130	.062		.15	2.55		2.70	4.59
0500	4" long	G		120	.067		.17	2.76		2.93	4.97
0600	3/8" diameter, 1" long	G		130	.062		.14	2.55		2.69	4.58
0700	2" long	G		130	.062		.18	2.55		2.73	4.62
0800	3" long	G		120	.067		.24	2.76		3	5.05
0900	4" long	G		120	.067		.30	2.76		3.06	5.10
1000	5" long	G		115	.070		.38	2.88		3.26	5.40
1100	1/2" diameter, 1-1/2" long	G		120	.067		.48	2.76		3.24	5.30
1200	2" long	G		120	.067		.56	2.76		3.32	5.40
1300	4" long	G		115	.070		.92	2.88		3.80	6
1400	6" long	G		110	.073		1.30	3.01		4.31	6.70
1500	8" long	G		105	.076		1.73	3.16		4.89	7.40
1600	5/8" diameter, 1-1/2" long	G		120	.067		.85	2.76		3.61	5.75
1700	2" long	G		120	.067		.94	2.76		3.70	5.85
1800	4" long	G		115	.070		1.35	2.88		4.23	6.50
1900	6" long	G		110	.073		1.73	3.01		4.74	7.15
2000	8" long	G		105	.076		2.56	3.16		5.72	8.30
2100	10" long	G		100	.080		3.22	3.32		6.54	9.30
2200	3/4" diameter, 2" long	G		120	.067		1.13	2.76		3.89	6.05
2300	4" long	G		110	.073		1.61	3.01		4.62	7
2400	6" long	G		105	.076		2.07	3.16		5.23	7.80
2500	8" long	G		95	.084		3.12	3.49		6.61	9.50
2600	10" long	G		85	.094		4.09	3.90		7.99	11.25
2700	12" long	G		80	.100		4.79	4.15		8.94	12.45
2800	1" diameter, 3" long	G		105	.076		2.91	3.16		6.07	8.70
2900	6" long	G		90	.089		4.24	3.68		7.92	11.05
3000	12" long	G		75	.107		7.60	4.42		12.02	16
3100	For galvanized, add						75%				
3200	For stainless, add						350%				

05 05 23 – Metal Fastenings

05 05 23.30 Lag Screws		Crew	Daily Output	Labor-Hours	Unit	Material	Labor	Equipment	Total	Total Incl O&P
						2019 Bare Costs				
0010	LAG SCREWS									
0020	Steel, 1/4" diameter, 2" long G	1 Carp	200	.040	Ea.	.10	1.55		1.65	2.67
0100	3/8" diameter, 3" long G		150	.053		.28	2.07		2.35	3.73
0200	1/2" diameter, 3" long G		130	.062		.69	2.38		3.07	4.70
0300	5/8" diameter, 3" long G	↓	120	.067	↓	1.31	2.58		3.89	5.70

05 05 23.50 Powder Actuated Tools and Fasteners		Crew	Daily Output	Labor-Hours	Unit	Material	Labor	Equipment	Total	Total Incl O&P
0010	POWDER ACTUATED TOOLS & FASTENERS									
0020	Stud driver, .22 caliber, single shot				Ea.	162			162	178
0100	.27 caliber, semi automatic, strip				"	460			460	505
0300	Powder load, single shot, .22 cal, power level 2, brown				C	6.50			6.50	7.15
0400	Strip, .27 cal, power level 4, red					9.30			9.30	10.20
0600	Drive pin, .300 x 3/4" long G	1 Carp	4.80	1.667		4.43	64.50		68.93	112
0700	.300 x 3" long with washer G	"	4	2	↓	13.50	77.50		91	143

05 05 23.55 Rivets		Crew	Daily Output	Labor-Hours	Unit	Material	Labor	Equipment	Total	Total Incl O&P
0010	RIVETS									
0100	Aluminum rivet & mandrel, 1/2" grip length x 1/8" diameter G	1 Carp	4.80	1.667	C	8.05	64.50		72.55	116
0200	3/16" diameter G		4	2		11	77.50		88.50	140
0300	Aluminum rivet, steel mandrel, 1/8" diameter G		4.80	1.667		10.35	64.50		74.85	118
0400	3/16" diameter G		4	2		16.80	77.50		94.30	147
0500	Copper rivet, steel mandrel, 1/8" diameter G		4.80	1.667		10.20	64.50		74.70	118
0800	Stainless rivet & mandrel, 1/8" diameter G		4.80	1.667		23.50	64.50		88	133
0900	3/16" diameter G		4	2		39	77.50		116.50	171
1000	Stainless rivet, steel mandrel, 1/8" diameter G		4.80	1.667		15.50	64.50		80	124
1100	3/16" diameter G		4	2		26	77.50		103.50	157
1200	Steel rivet and mandrel, 1/8" diameter G		4.80	1.667		8.15	64.50		72.65	116
1300	3/16" diameter G	↓	4	2	↓	11.50	77.50		89	141
1400	Hand riveting tool, standard				Ea.	75.50			75.50	83
1500	Deluxe					425			425	470
1600	Power riveting tool, standard					540			540	595
1700	Deluxe				↓	1,550			1,550	1,700

05 05 23.70 Structural Blind Bolts		Crew	Daily Output	Labor-Hours	Unit	Material	Labor	Equipment	Total	Total Incl O&P
0010	STRUCTURAL BLIND BOLTS									
0100	1/4" diameter x 1/4" grip G	1 Sswk	240	.033	Ea.	1.76	1.38		3.14	4.34
0150	1/2" grip G		216	.037		1.87	1.54		3.41	4.72
0200	3/8" diameter x 1/2" grip G		232	.034		3.46	1.43		4.89	6.30
0250	3/4" grip G		208	.038		3.33	1.59		4.92	6.45
0300	1/2" diameter x 1/2" grip G		224	.036		6.15	1.48		7.63	9.35
0350	3/4" grip G		200	.040		8.35	1.66		10.01	12.10
0400	5/8" diameter x 3/4" grip G		216	.037		11.10	1.54		12.64	14.90
0450	1" grip G	↓	192	.042	↓	15.50	1.73		17.23	20

05 12 23.10 Ceiling Supports			Crew	Daily Output	Labor-Hours	Unit	Material	Labor	Equipment	Total	Total Incl O&P
							2019 Bare Costs				
0010	**CEILING SUPPORTS**										
1000	Entrance door/folding partition supports, shop fabricated	G	E-4	60	.533	L.F.	26.50	22.50	1.59	50.59	70
1100	Linear accelerator door supports	G		14	2.286		120	96	6.85	222.85	305
1200	Lintels or shelf angles, hung, exterior hot dipped galv.	G		267	.120		18.05	5.05	.36	23.46	29
1250	Two coats primer paint instead of galv.	G		267	.120		15.65	5.05	.36	21.06	26.50
1400	Monitor support, ceiling hung, expansion bolted	G		4	8	Ea.	420	335	24	779	1,075
1450	Hung from pre-set inserts	G		6	5.333		450	224	15.95	689.95	905
1600	Motor supports for overhead doors	G		4	8		213	335	24	572	840
1700	Partition support for heavy folding partitions, without pocket	G		24	1.333	L.F.	60	56	3.99	119.99	167
1750	Supports at pocket only	G		12	2.667		120	112	7.95	239.95	335
2000	Rolling grilles & fire door supports	G		34	.941		51.50	39.50	2.81	93.81	128
2100	Spider-leg light supports, expansion bolted to ceiling slab	G		8	4	Ea.	172	168	11.95	351.95	495
2150	Hung from pre-set inserts	G		12	2.667	"	185	112	7.95	304.95	405
2400	Toilet partition support	G		36	.889	L.F.	60	37.50	2.66	100.16	133
2500	X-ray travel gantry support	G		12	2.667	"	206	112	7.95	325.95	430

05 12 23.15 Columns, Lightweight			Crew	Daily Output	Labor-Hours	Unit	Material	Labor	Equipment	Total	Total Incl O&P
0010	**COLUMNS, LIGHTWEIGHT**										
8000	Lally columns, to 8′, 3-1/2″ diameter		2 Carp	24	.667	Ea.	45	26		71	92
8080	4″ diameter		"	20	.800	"	64	31		95	121

05 12 23.17 Columns, Structural			Crew	Daily Output	Labor-Hours	Unit	Material	Labor	Equipment	Total	Total Incl O&P
0010	**COLUMNS, STRUCTURAL**	R051223-10									
0015	Made from recycled materials										
0020	Shop fab'd for 100-ton, 1-2 story project, bolted connections										
0800	Steel, concrete filled, extra strong pipe, 3-1/2″ diameter		E-2	660	.073	L.F.	43.50	3.05	2.57	49.12	56
0830	4″ diameter			780	.062		48.50	2.58	2.18	53.26	60.50
0890	5″ diameter			1020	.047		58	1.97	1.67	61.64	68.50
0930	6″ diameter			1200	.040		76.50	1.67	1.42	79.59	89
0940	8″ diameter			1100	.044		76.50	1.83	1.54	79.87	89.50
1100	For galvanizing, add					Lb.	.25			.25	.28
1300	For web ties, angles, etc., add per added lb.		1 Sswk	945	.008		1.32	.35		1.67	2.06
1500	Steel pipe, extra strong, no concrete, 3″ to 5″ diameter	G	E-2	16000	.003		1.32	.13	.11	1.56	1.79
1600	6″ to 12″ diameter	G		14000	.003		1.32	.14	.12	1.58	1.83
5100	Structural tubing, rect., 5″ to 6″ wide, light section	G		8000	.006		1.32	.25	.21	1.78	2.11
5200	Heavy section	G		12000	.004		1.32	.17	.14	1.63	1.90
8090	For projects 75 to 99 tons, add					%	10%				
8092	50 to 74 tons, add						20%				
8094	25 to 49 tons, add						30%	10%			
8096	10 to 24 tons, add						50%	25%			
8098	2 to 9 tons, add						75%	50%			
8099	Less than 2 tons, add						100%	100%			

05 12 23.18 Corner Guards			Crew	Daily Output	Labor-Hours	Unit	Material	Labor	Equipment	Total	Total Incl O&P
0010	**CORNER GUARDS**										

05 12 23.45 Lintels			Crew	Daily Output	Labor-Hours	Unit	Material	Labor	Equipment	Total	Total Incl O&P
0010	**LINTELS**										
0015	Made from recycled materials										
0020	Plain steel angles, shop fabricated, under 500 lb.	G	1 Bric	550	.015	Lb.	1.02	.53		1.55	2.01
0100	500 to 1,000 lb.	G		640	.013	"	.99	.46		1.45	1.86
2000	Steel angles, 3-1/2″ x 3″, 1/4″ thick, 2′-6″ long	G		47	.170	Ea.	14.25	6.25		20.50	26
2100	4′-6″ long	G		26	.308		25.50	11.30		36.80	47.50
2600	4″ x 3-1/2″, 1/4″ thick, 5′-0″ long	G		21	.381		33	14		47	59.50
2700	9′-0″ long	G		12	.667		59	24.50		83.50	106

05 12 23.65 Plates

05 12 23.65 Plates			Crew	Daily Output	Labor-Hours	Unit	2019 Bare Costs Material	Labor	Equipment	Total	Total Incl O&P
0010	**PLATES**										
0015	Made from recycled materials										
0020	For connections & stiffener plates, shop fabricated										
0050	1/8" thick (5.1 lb./S.F.)	G				S.F.	6.75			6.75	7.40
0100	1/4" thick (10.2 lb./S.F.)	G					13.50			13.50	14.85
0300	3/8" thick (15.3 lb./S.F.)	G					20			20	22
0400	1/2" thick (20.4 lb./S.F.)	G					27			27	29.50
0450	3/4" thick (30.6 lb./S.F.)	G					40.50			40.50	44.50
0500	1" thick (40.8 lb./S.F.)	G					54			54	59.50
2000	Steel plate, warehouse prices, no shop fabrication										
2100	1/4" thick (10.2 lb./S.F.)	G				S.F.	7.55			7.55	8.30

05 12 23.77 Structural Steel Projects

05 12 23.77 Structural Steel Projects			Crew	Daily Output	Labor-Hours	Unit	Material	Labor	Equipment	Total	Total Incl O&P
0010	**STRUCTURAL STEEL PROJECTS** R050516-30										
0015	Made from recycled materials										
0020	Shop fab'd for 100-ton, 1-2 story project, bolted connections										
0201	Apartments, nursing homes, etc., 1 to 2 stories R050523-10	G	E-2	6.45	7.442	Ton	2,650	310	263	3,223	3,725
0701	Offices, hospitals, etc., steel bearing, 1 to 2 stories	G		6.45	7.442		2,650	310	263	3,223	3,725
3101	Roof trusses, simple connections	G		8.13	5.904		3,700	247	209	4,156	4,725
3200	Moment/composite connections	G	E-5	8.30	8.675		4,500	360	216	5,076	5,825
5390	For projects 75 to 99 tons, add R051223-20						10%				
5392	50 to 74 tons, add						20%				
5394	25 to 49 tons, add R051223-25						30%	10%			
5396	10 to 24 tons, add						50%	25%			
5398	2 to 9 tons, add R051223-30						75%	50%			
5399	Less than 2 tons, add						100%	100%			

05 12 23.78 Structural Steel Secondary Members

05 12 23.78 Structural Steel Secondary Members		Crew	Daily Output	Labor-Hours	Unit	Material	Labor	Equipment	Total	Total Incl O&P
0010	**STRUCTURAL STEEL SECONDARY MEMBERS**									
0015	Made from recycled materials									
0020	Shop fabricated for 20-ton girt/purlin framing package, materials only									
0100	Girts/purlins, C/Z-shapes, includes clips and bolts									
0110	6" x 2-1/2" x 2-1/2", 16 ga., 3.0 lb./L.F.				L.F.	3.57			3.57	3.93
0115	14 ga., 3.5 lb./L.F.					4.16			4.16	4.58
0120	8" x 2-3/4" x 2-3/4", 16 ga., 3.4 lb./L.F.					4.04			4.04	4.45
0125	14 ga., 4.1 lb./L.F.					4.88			4.88	5.35
0130	12 ga., 5.6 lb./L.F.					6.65			6.65	7.35
0135	10" x 3-1/2" x 3-1/2", 14 ga., 4.7 lb./L.F.					5.60			5.60	6.15
0140	12 ga., 6.7 lb./L.F.					7.95			7.95	8.75
0145	12" x 3-1/2" x 3-1/2", 14 ga., 5.3 lb./L.F.					6.30			6.30	6.95
0150	12 ga., 7.4 lb./L.F.					8.80			8.80	9.70
0200	Eave struts, C-shape, includes clips and bolts									
0210	6" x 4" x 3", 16 ga., 3.1 lb./L.F.				L.F.	3.69			3.69	4.06
0215	14 ga., 3.9 lb./L.F.					4.64			4.64	5.10
0220	8" x 4" x 3", 16 ga., 3.5 lb./L.F.					4.16			4.16	4.58
0225	14 ga., 4.4 lb./L.F.					5.25			5.25	5.75
0230	12 ga., 6.2 lb./L.F.					7.35			7.35	8.10
0235	10" x 5" x 3", 14 ga., 5.2 lb./L.F.					6.20			6.20	6.80
0240	12 ga., 7.3 lb./L.F.					8.70			8.70	9.55
0245	12" x 5" x 4", 14 ga., 6.0 lb./L.F.					7.15			7.15	7.85
0250	12 ga., 8.4 lb./L.F.					10			10	11
0300	Rake/base angle, excludes concrete drilling and expansion anchors									
0310	2" x 2", 14 ga., 1.0 lb./L.F.	2 Sswk	640	.025	L.F.	1.19	1.04		2.23	3.11
0315	3" x 2", 14 ga., 1.3 lb./L.F.		535	.030		1.55	1.24		2.79	3.85

05 12 Structural Steel Framing

05 12 23 – Structural Steel for Buildings

05 12 23.78 Structural Steel Secondary Members		Crew	Daily Output	Labor-Hours	Unit	2019 Bare Costs Material	Labor	Equipment	Total	Total Incl O&P
0320	3″ x 3″, 14 ga., 1.6 lb./L.F.	2 Sswk	500	.032	L.F.	1.90	1.33		3.23	4.39
0325	4″ x 3″, 14 ga., 1.8 lb./L.F.		480	.033		2.14	1.38		3.52	4.76
0600	Installation of secondary members, erection only									
0610	Girts, purlins, eave struts, 16 ga., 6″ deep	E-18	100	.400	Ea.		16.70	8.70	25.40	38
0615	8″ deep		80	.500			21	10.85	31.85	48
0620	14 ga., 6″ deep		80	.500			21	10.85	31.85	48
0625	8″ deep		65	.615			25.50	13.35	38.85	58.50
0630	10″ deep		55	.727			30.50	15.80	46.30	69.50
0635	12″ deep		50	.800			33.50	17.40	50.90	76.50
0640	12 ga., 8″ deep		50	.800			33.50	17.40	50.90	76.50
0645	10″ deep		45	.889			37	19.30	56.30	84.50
0650	12″ deep		40	1			41.50	21.50	63	95.50
0900	For less than 20-ton job lots									
0905	For 15 to 19 tons, add				%	10%				
0910	For 10 to 14 tons, add					25%				
0915	For 5 to 9 tons, add					50%	50%	50%		
0920	For 1 to 4 tons, add					75%	75%	75%		
0925	For less than 1 ton, add					100%	100%	100%		

05 12 23.79 Structural Steel		Crew	Daily Output	Labor-Hours	Unit	Material	Labor	Equipment	Total	Total Incl O&P
0010	**STRUCTURAL STEEL** R050516-30									
0020	Shop fab'd for 100-ton, 1-2 story project, bolted conn's.									
0050	Beams, W 6 x 9 R050521-20 G	E-2	720	.067	L.F.	14.25	2.79	2.36	19.40	23
0100	W 8 x 10 G		720	.067		15.85	2.79	2.36	21	25
0200	Columns, W 6 x 15 R051223-10 G		540	.089		26	3.72	3.14	32.86	38.50
0250	W 8 x 31 G		540	.089		53.50	3.72	3.14	60.36	68.50
7990	For projects 75 to 99 tons, add R051223-20				All	10%				
7992	50 to 75 tons, add					20%				
7994	25 to 49 tons, add R051223-25					30%	10%			
7996	10 to 24 tons, add					50%	25%			
7998	2 to 9 tons, add R051223-30					75%	50%			
7999	Less than 2 tons, add					100%	100%			

05 15 Wire Rope Assemblies

05 15 16 – Steel Wire Rope Assemblies

05 15 16.70 Temporary Cable Safety Railing		Crew	Daily Output	Labor-Hours	Unit	Material	Labor	Equipment	Total	Total Incl O&P
0010	**TEMPORARY CABLE SAFETY RAILING**, Each 100′ strand incl.									
0020	2 eyebolts, 1 turnbuckle, 100′ cable, 2 thimbles, 6 clips									
0025	Made from recycled materials									
0100	One strand using 1/4″ cable & accessories G	2 Sswk	4	4	C.L.F.	158	166		324	460
0200	1/2″ cable & accessories G	″	2	8	″	340	330		670	950

05 21 Steel Joist Framing

05 21 19 – Open Web Steel Joist Framing

05 21 19.10 Open Web Joists			Crew	Daily Output	Labor-Hours	Unit	Material	Labor	Equipment	Total	Total Incl O&P
							2019 Bare Costs				
0010	**OPEN WEB JOISTS**										
0015	Made from recycled materials										
0050	K series, 40-ton lots, horiz. bridging, spans to 30', shop primer	G	E-7	12	6	Ton	1,675	250	157	2,082	2,425
0130	8K1, 5.1 lb./L.F.	G		1200	.060	L.F.	4.25	2.50	1.57	8.32	10.70
0140	10K1, 5.0 lb./L.F.	G		1200	.060		4.16	2.50	1.57	8.23	10.65
0160	12K3, 5.7 lb./L.F.	G		1500	.048		4.75	2	1.26	8.01	10.05
0180	14K3, 6.0 lb./L.F.	G		1500	.048		5	2	1.26	8.26	10.35
0200	16K3, 6.3 lb./L.F.	G		1800	.040		5.25	1.67	1.05	7.97	9.80
0220	16K6, 8.1 lb./L.F.	G		1800	.040		6.75	1.67	1.05	9.47	11.45
0240	18K5, 7.7 lb./L.F.	G		2000	.036		6.40	1.50	.94	8.84	10.70
0260	18K9, 10.2 lb./L.F.	G		2000	.036		8.50	1.50	.94	10.94	13
0440	K series, 30' to 50' spans	G		17	4.235	Ton	1,625	177	111	1,913	2,225
0500	20K5, 8.2 lb./L.F.	G		2000	.036	L.F.	6.70	1.50	.94	9.14	11
0520	20K9, 10.8 lb./L.F.	G		2000	.036		8.80	1.50	.94	11.24	13.35
0540	22K5, 8.8 lb./L.F.	G		2000	.036		7.20	1.50	.94	9.64	11.55
0560	22K9, 11.3 lb./L.F.	G		2000	.036		9.25	1.50	.94	11.69	13.80
0580	24K6, 9.7 lb./L.F.	G		2200	.033		7.90	1.37	.86	10.13	12
0600	24K10, 13.1 lb./L.F.	G		2200	.033		10.70	1.37	.86	12.93	15.05
0620	26K6, 10.6 lb./L.F.	G		2200	.033		8.65	1.37	.86	10.88	12.80
0640	26K10, 13.8 lb./L.F.	G		2200	.033		11.25	1.37	.86	13.48	15.70
0660	28K8, 12.7 lb./L.F.	G		2400	.030		10.35	1.25	.79	12.39	14.45
0680	28K12, 17.1 lb./L.F.	G		2400	.030		13.95	1.25	.79	15.99	18.40
0700	30K8, 13.2 lb./L.F.	G		2400	.030		10.80	1.25	.79	12.84	14.90
0720	30K12, 17.6 lb./L.F.	G		2400	.030		14.35	1.25	.79	16.39	18.85
0800	For less than 40-ton job lots										
0802	For 30 to 39 tons, add					%	10%				
0804	20 to 29 tons, add						20%				
0806	10 to 19 tons, add						30%				
0807	5 to 9 tons, add						50%	25%			
0808	1 to 4 tons, add						75%	50%			
0809	Less than 1 ton, add						100%	100%			
6200	For shop prime paint other than mfrs. standard, add						20%				
6400	Individual steel bearing plate, 6" x 6" x 1/4" with J-hook	G	1 Bric	160	.050	Ea.	7.95	1.84		9.79	11.75

05 31 Steel Decking

05 31 13 – Steel Floor Decking

05 31 13.50 Floor Decking			Crew	Daily Output	Labor-Hours	Unit	Material	Labor	Equipment	Total	Total Incl O&P
0010	**FLOOR DECKING**										
0015	Made from recycled materials										
5100	Non-cellular composite decking, galvanized, 1-1/2" deep, 16 ga.	G	E-4	3500	.009	S.F.	4.17	.38	.03	4.58	5.30
5120	18 ga.	G		3650	.009		2.99	.37	.03	3.39	3.95
5140	20 ga.	G		3800	.008		3.05	.35	.03	3.43	3.99
5200	2" deep, 22 ga.	G		3860	.008		2.05	.35	.02	2.42	2.89
5300	20 ga.	G		3600	.009		2.93	.37	.03	3.33	3.90
5400	18 ga.	G		3380	.009		3.11	.40	.03	3.54	4.14
5500	16 ga.	G		3200	.010		4.12	.42	.03	4.57	5.30
5700	3" deep, 22 ga.	G		3200	.010		2.23	.42	.03	2.68	3.22
5800	20 ga.	G		3000	.011		3.22	.45	.03	3.70	4.37
5900	18 ga.	G		2850	.011		3.11	.47	.03	3.61	4.28
6000	16 ga.	G		2700	.012		4.68	.50	.04	5.22	6.05

05 31 Steel Decking

05 31 23 – Steel Roof Decking

05 31 23.50 Roof Decking			Crew	Daily Output	Labor-Hours	Unit	2019 Bare Costs Material	Labor	Equipment	Total	Total Incl O&P
0010	**ROOF DECKING**										
0015	Made from recycled materials										
2100	Open type, 1-1/2″ deep, Type B, wide rib, galv., 22 ga., under 50 sq.	G	E-4	4500	.007	S.F.	2.20	.30	.02	2.52	2.96
2600	20 ga., under 50 squares	G		3865	.008		2.58	.35	.02	2.95	3.46
2900	18 ga., under 50 squares	G		3800	.008		3.31	.35	.03	3.69	4.28
3050	16 ga., under 50 squares	G		3700	.009		4.47	.36	.03	4.86	5.55
3200	3″ deep, Type N, 22 ga., under 50 squares	G		3600	.009		3.23	.37	.03	3.63	4.23
3300	20 ga., under 50 squares	G		3400	.009		3.48	.39	.03	3.90	4.53
3400	18 ga., under 50 squares	G		3200	.010		4.51	.42	.03	4.96	5.70
3500	16 ga., under 50 squares	G		3000	.011		5.95	.45	.03	6.43	7.35
3700	4-1/2″ deep, Type J, 20 ga., over 50 squares	G		2700	.012		3.92	.50	.04	4.46	5.20
4100	6″ deep, Type H, 18 ga., over 50 squares	G		2000	.016		6.60	.67	.05	7.32	8.50
4500	7-1/2″ deep, Type H, 18 ga., over 50 squares	G		1690	.019		7.85	.79	.06	8.70	10.10

05 31 33 – Steel Form Decking

05 31 33.50 Form Decking			Crew	Daily Output	Labor-Hours	Unit	Material	Labor	Equipment	Total	Total Incl O&P
0010	**FORM DECKING**										
0015	Made from recycled materials										
6100	Slab form, steel, 28 ga., 9/16″ deep, Type UFS, uncoated	G	E-4	4000	.008	S.F.	1.66	.34	.02	2.02	2.44
6200	Galvanized	G		4000	.008		1.47	.34	.02	1.83	2.23
6220	24 ga., 1″ deep, Type UF1X, uncoated	G		3900	.008		1.59	.34	.02	1.95	2.38
6240	Galvanized	G		3900	.008		1.87	.34	.02	2.23	2.69
6300	24 ga., 1-5/16″ deep, Type UFX, uncoated	G		3800	.008		1.69	.35	.03	2.07	2.50
6400	Galvanized	G		3800	.008		1.99	.35	.03	2.37	2.83
6500	22 ga., 1-5/16″ deep, uncoated	G		3700	.009		2.15	.36	.03	2.54	3.02
6600	Galvanized	G		3700	.009		2.19	.36	.03	2.58	3.07
6700	22 ga., 2″ deep, uncoated	G		3600	.009		2.78	.37	.03	3.18	3.74
6800	Galvanized	G		3600	.009		2.73	.37	.03	3.13	3.68

05 41 Structural Metal Stud Framing

05 41 13 – Load-Bearing Metal Stud Framing

05 41 13.05 Bracing			Crew	Daily Output	Labor-Hours	Unit	Material	Labor	Equipment	Total	Total Incl O&P
0010	**BRACING**, shear wall X-bracing, per 10′ x 10′ bay, one face										
0015	Made of recycled materials										
0120	Metal strap, 20 ga. x 4″ wide	G	2 Carp	18	.889	Ea.	18.90	34.50		53.40	78
0130	6″ wide	G		18	.889		30.50	34.50		65	90.50
0160	18 ga. x 4″ wide	G		16	1		31	39		70	98
0170	6″ wide	G		16	1		46	39		85	115
0410	Continuous strap bracing, per horizontal row on both faces										
0420	Metal strap, 20 ga. x 2″ wide, studs 12″ OC	G	1 Carp	7	1.143	C.L.F.	55	44.50		99.50	134
0430	16″ OC	G		8	1		55	39		94	125
0440	24″ OC	G		10	.800		55	31		86	112
0450	18 ga. x 2″ wide, studs 12″ OC	G		6	1.333		75	51.50		126.50	168
0460	16″ OC	G		7	1.143		75	44.50		119.50	156
0470	24″ OC	G		8	1		75	39		114	147

05 41 13.10 Bridging			Crew	Daily Output	Labor-Hours	Unit	Material	Labor	Equipment	Total	Total Incl O&P
0010	**BRIDGING**, solid between studs w/1-1/4″ leg track, per stud bay										
0015	Made from recycled materials										
0200	Studs 12″ OC, 18 ga. x 2-1/2″ wide	G	1 Carp	125	.064	Ea.	.90	2.48		3.38	5.10
0210	3-5/8″ wide	G		120	.067		1.16	2.58		3.74	5.55
0220	4″ wide	G		120	.067		1.22	2.58		3.80	5.60

05 41 13 – Load-Bearing Metal Stud Framing

05 41 13.10 Bridging		Crew	Daily Output	Labor-Hours	Unit	Material	2019 Bare Costs Labor	Equipment	Total	Total Incl O&P
0230	6" wide	G	1 Carp	115	.070	Ea.	1.60	2.70		4.30	6.20
0240	8" wide	G		110	.073		1.85	2.82		4.67	6.70
0300	16 ga. x 2-1/2" wide	G		115	.070		1.15	2.70		3.85	5.70
0310	3-5/8" wide	G		110	.073		1.39	2.82		4.21	6.20
0320	4" wide	G		110	.073		1.48	2.82		4.30	6.30
0330	6" wide	G		105	.076		1.89	2.95		4.84	6.95
0340	8" wide	G		100	.080		2.61	3.10		5.71	7.95
1200	Studs 16" OC, 18 ga. x 2-1/2" wide	G		125	.064		1.16	2.48		3.64	5.35
1210	3-5/8" wide	G		120	.067		1.49	2.58		4.07	5.90
1220	4" wide	G		120	.067		1.57	2.58		4.15	6
1230	6" wide	G		115	.070		2.05	2.70		4.75	6.70
1240	8" wide	G		110	.073		2.38	2.82		5.20	7.25
1300	16 ga. x 2-1/2" wide	G		115	.070		1.47	2.70		4.17	6.10
1310	3-5/8" wide	G		110	.073		1.78	2.82		4.60	6.60
1320	4" wide	G		110	.073		1.90	2.82		4.72	6.75
1330	6" wide	G		105	.076		2.43	2.95		5.38	7.55
1340	8" wide	G		100	.080		3.35	3.10		6.45	8.80
2200	Studs 24" OC, 18 ga. x 2-1/2" wide	G		125	.064		1.67	2.48		4.15	5.95
2210	3-5/8" wide	G		120	.067		2.15	2.58		4.73	6.65
2220	4" wide	G		120	.067		2.27	2.58		4.85	6.75
2230	6" wide	G		115	.070		2.96	2.70		5.66	7.70
2240	8" wide	G		110	.073		3.44	2.82		6.26	8.45
2300	16 ga. x 2-1/2" wide	G		115	.070		2.12	2.70		4.82	6.80
2310	3-5/8" wide	G		110	.073		2.58	2.82		5.40	7.50
2320	4" wide	G		110	.073		2.75	2.82		5.57	7.70
2330	6" wide	G		105	.076		3.51	2.95		6.46	8.75
2340	8" wide	G	↓	100	.080	↓	4.85	3.10		7.95	10.45
3000	Continuous bridging, per row										
3100	16 ga. x 1-1/2" channel thru studs 12" OC	G	1 Carp	6	1.333	C.L.F.	49	51.50		100.50	140
3110	16" OC	G		7	1.143		49	44.50		93.50	127
3120	24" OC	G		8.80	.909		49	35		84	112
4100	2" x 2" angle x 18 ga., studs 12" OC	G		7	1.143		76	44.50		120.50	157
4110	16" OC	G		9	.889		76	34.50		110.50	141
4120	24" OC	G		12	.667		76	26		102	126
4200	16 ga., studs 12" OC	G		5	1.600		95.50	62		157.50	207
4210	16" OC	G		7	1.143		95.50	44.50		140	178
4220	24" OC	G	↓	10	.800	↓	95.50	31		126.50	156

05 41 13.25 Framing, Boxed Headers/Beams			Crew	Daily Output	Labor-Hours	Unit	Material	Labor	Equipment	Total	Total Incl O&P
0010	**FRAMING, BOXED HEADERS/BEAMS**										
0015	Made from recycled materials										
0200	Double, 18 ga. x 6" deep	G	2 Carp	220	.073	L.F.	5.65	2.82		8.47	10.85
0210	8" deep	G		210	.076		5.85	2.95		8.80	11.30
0220	10" deep	G		200	.080		7	3.10		10.10	12.80
0230	12" deep	G		190	.084		7.60	3.26		10.86	13.80
0300	16 ga. x 8" deep	G		180	.089		6.70	3.44		10.14	13.10
0310	10" deep	G		170	.094		8.50	3.65		12.15	15.40
0320	12" deep	G		160	.100		9.25	3.88		13.13	16.60
0400	14 ga. x 10" deep	G		140	.114		9.20	4.43		13.63	17.40
0410	12" deep	G		130	.123		10.05	4.77		14.82	18.95
1210	Triple, 18 ga. x 8" deep	G		170	.094		8.45	3.65		12.10	15.35
1220	10" deep	G		165	.097		10	3.76		13.76	17.20
1230	12" deep	G	↓	160	.100	↓	10.95	3.88		14.83	18.45

05 41 Structural Metal Stud Framing

05 41 13 – Load-Bearing Metal Stud Framing

05 41 13.25 Framing, Boxed Headers/Beams		Crew	Daily Output	Labor-Hours	Unit	Material	Labor	Equipment	Total	Total Incl O&P
						2019 Bare Costs				
1300 16 ga. x 8" deep	G	2 Carp	145	.110	L.F.	9.75	4.28		14.03	17.80
1310 10" deep	G		140	.114		12.30	4.43		16.73	21
1320 12" deep	G		135	.119		13.45	4.59		18.04	22.50
1400 14 ga. x 10" deep	G		115	.139		12.50	5.40		17.90	22.50
1410 12" deep	G		110	.145		13.85	5.65		19.50	24.50

05 41 13.30 Framing, Stud Walls		Crew	Daily Output	Labor-Hours	Unit	Material	Labor	Equipment	Total	Total Incl O&P
0010 **FRAMING, STUD WALLS** w/top & bottom track, no openings,										
0020 Headers, beams, bridging or bracing										
0025 Made from recycled materials										
4100 8' high walls, 18 ga. x 2-1/2" wide, studs 12" OC	G	2 Carp	54	.296	L.F.	9.10	11.50		20.60	29
4110 16" OC	G		77	.208		7.25	8.05		15.30	21.50
4120 24" OC	G		107	.150		5.40	5.80		11.20	15.55
4130 3-5/8" wide, studs 12" OC	G		53	.302		10.85	11.70		22.55	31.50
4140 16" OC	G		76	.211		8.70	8.15		16.85	23
4150 24" OC	G		105	.152		6.55	5.90		12.45	16.95
4160 4" wide, studs 12" OC	G		52	.308		11.30	11.90		23.20	32
4170 16" OC	G		74	.216		9.05	8.40		17.45	24
4180 24" OC	G		103	.155		6.80	6		12.80	17.45
4190 6" wide, studs 12" OC	G		51	.314		14.40	12.15		26.55	36
4200 16" OC	G		73	.219		11.55	8.50		20.05	27
4210 24" OC	G		101	.158		8.70	6.15		14.85	19.75
4220 8" wide, studs 12" OC	G		50	.320		16.40	12.40		28.80	38.50
4230 16" OC	G		72	.222		13.20	8.60		21.80	29
4240 24" OC	G		100	.160		9.95	6.20		16.15	21
4300 16 ga. x 2-1/2" wide, studs 12" OC	G		47	.340		10.75	13.20		23.95	34
4310 16" OC	G		68	.235		8.50	9.10		17.60	24.50
4320 24" OC	G		94	.170		6.25	6.60		12.85	17.75
4330 3-5/8" wide, studs 12" OC	G		46	.348		12.35	13.50		25.85	36
4340 16" OC	G		66	.242		9.85	9.40		19.25	26.50
4350 24" OC	G		92	.174		7.30	6.75		14.05	19.15
4360 4" wide, studs 12" OC	G		45	.356		12.90	13.80		26.70	37
4370 16" OC	G		65	.246		10.25	9.55		19.80	27
4380 24" OC	G		90	.178		7.60	6.90		14.50	19.75
4390 6" wide, studs 12" OC	G		44	.364		16.25	14.10		30.35	41.50
4400 16" OC	G		64	.250		12.95	9.70		22.65	30.50
4410 24" OC	G		88	.182		9.65	7.05		16.70	22.50
4420 8" wide, studs 12" OC	G		43	.372		19.70	14.40		34.10	45.50
4430 16" OC	G		63	.254		15.65	9.85		25.50	33.50
4440 24" OC	G		86	.186		11.60	7.20		18.80	24.50
5100 10' high walls, 18 ga. x 2-1/2" wide, studs 12" OC	G		54	.296		10.90	11.50		22.40	31
5110 16" OC	G		77	.208		8.60	8.05		16.65	23
5120 24" OC	G		107	.150		6.30	5.80		12.10	16.55
5130 3-5/8" wide, studs 12" OC	G		53	.302		13	11.70		24.70	33.50
5140 16" OC	G		76	.211		10.30	8.15		18.45	25
5150 24" OC	G		105	.152		7.60	5.90		13.50	18.10
5160 4" wide, studs 12" OC	G		52	.308		13.55	11.90		25.45	34.50
5170 16" OC	G		74	.216		10.75	8.40		19.15	25.50
5180 24" OC	G		103	.155		7.95	6		13.95	18.65
5190 6" wide, studs 12" OC	G		51	.314		17.25	12.15		29.40	39
5200 16" OC	G		73	.219		13.70	8.50		22.20	29
5210 24" OC	G		101	.158		10.15	6.15		16.30	21.50
5220 8" wide, studs 12" OC	G		50	.320		19.65	12.40		32.05	42

05 41 13.30 Framing, Stud Walls			Crew	Daily Output	Labor-Hours	Unit	Material	2019 Bare Costs Labor	Equipment	Total	Total Incl O&P
5230	16" OC	G	2 Carp	72	.222	L.F.	15.60	8.60		24.20	31.50
5240	24" OC	G		100	.160		11.60	6.20		17.80	23
5300	16 ga. x 2-1/2" wide, studs 12" OC	G		47	.340		13	13.20		26.20	36.50
5310	16" OC	G		68	.235		10.20	9.10		19.30	26.50
5320	24" OC	G		94	.170		7.35	6.60		13.95	19
5330	3-5/8" wide, studs 12" OC	G		46	.348		14.90	13.50		28.40	39
5340	16" OC	G		66	.242		11.75	9.40		21.15	28.50
5350	24" OC	G		92	.174		8.55	6.75		15.30	20.50
5360	4" wide, studs 12" OC	G		45	.356		15.55	13.80		29.35	40
5370	16" OC	G		65	.246		12.25	9.55		21.80	29
5380	24" OC	G		90	.178		8.95	6.90		15.85	21
5390	6" wide, studs 12" OC	G		44	.364		19.55	14.10		33.65	45
5400	16" OC	G		64	.250		15.40	9.70		25.10	33
5410	24" OC	G		88	.182		11.30	7.05		18.35	24
5420	8" wide, studs 12" OC	G		43	.372		23.50	14.40		37.90	50
5430	16" OC	G		63	.254		18.70	9.85		28.55	37
5440	24" OC	G		86	.186		13.65	7.20		20.85	27
6190	12' high walls, 18 ga. x 6" wide, studs 12" OC	G		41	.390		20	15.10		35.10	47
6200	16" OC	G		58	.276		15.80	10.70		26.50	35
6210	24" OC	G		81	.198		11.55	7.65		19.20	25.50
6220	8" wide, studs 12" OC	G		40	.400		23	15.50		38.50	50.50
6230	16" OC	G		57	.281		18	10.90		28.90	38
6240	24" OC	G		80	.200		13.20	7.75		20.95	27.50
6390	16 ga. x 6" wide, studs 12" OC	G		35	.457		23	17.70		40.70	54.50
6400	16" OC	G		51	.314		17.90	12.15		30.05	39.50
6410	24" OC	G		70	.229		12.95	8.85		21.80	29
6420	8" wide, studs 12" OC	G		34	.471		28	18.25		46.25	60.50
6430	16" OC	G		50	.320		21.50	12.40		33.90	44.50
6440	24" OC	G		69	.232		15.65	9		24.65	32
6530	14 ga. x 3-5/8" wide, studs 12" OC	G		34	.471		21.50	18.25		39.75	53.50
6540	16" OC	G		48	.333		16.80	12.90		29.70	40
6550	24" OC	G		65	.246		12.05	9.55		21.60	29
6560	4" wide, studs 12" OC	G		33	.485		22.50	18.80		41.30	56
6570	16" OC	G		47	.340		17.65	13.20		30.85	41.50
6580	24" OC	G		64	.250		12.70	9.70		22.40	30
6730	12 ga. x 3-5/8" wide, studs 12" OC	G		31	.516		30	20		50	65.50
6740	16" OC	G		43	.372		23	14.40		37.40	49.50
6750	24" OC	G		59	.271		16.20	10.50		26.70	35
6760	4" wide, studs 12" OC	G		30	.533		31.50	20.50		52	69
6770	16" OC	G		42	.381		24.50	14.75		39.25	51.50
6780	24" OC	G		58	.276		17.30	10.70		28	36.50
7390	16' high walls, 16 ga. x 6" wide, studs 12" OC	G		33	.485		29.50	18.80		48.30	63.50
7400	16" OC	G		48	.333		23	12.90		35.90	46.50
7410	24" OC	G		67	.239		16.25	9.25		25.50	33
7420	8" wide, studs 12" OC	G		32	.500		36	19.40		55.40	71.50
7430	16" OC	G		47	.340		28	13.20		41.20	52.50
7440	24" OC	G		66	.242		19.70	9.40		29.10	37
7560	14 ga. x 4" wide, studs 12" OC	G		31	.516		29	20		49	65
7570	16" OC	G		45	.356		22.50	13.80		36.30	48
7580	24" OC	G		61	.262		16	10.15		26.15	34.50
7590	6" wide, studs 12" OC	G		30	.533		37	20.50		57.50	74.50
7600	16" OC	G		44	.364		28.50	14.10		42.60	55
7610	24" OC	G		60	.267		20	10.35		30.35	39.50

05 41 Structural Metal Stud Framing

05 41 13 – Load-Bearing Metal Stud Framing

05 41 13.30 Framing, Stud Walls			Crew	Daily Output	Labor-Hours	Unit	Material	Labor	Equipment	Total	Total Incl O&P
							2019 Bare Costs				
7760	12 ga. x 4" wide, studs 12" OC	G	2 Carp	29	.552	L.F.	41.50	21.50		63	81
7770	16" OC	G		40	.400		31.50	15.50		47	60.50
7780	24" OC	G		55	.291		22	11.25		33.25	43
7790	6" wide, studs 12" OC	G		28	.571		52	22		74	94
7800	16" OC	G		39	.410		40	15.90		55.90	70.50
7810	24" OC	G		54	.296		28	11.50		39.50	49.50
8590	20' high walls, 14 ga. x 6" wide, studs 12" OC	G		29	.552		45	21.50		66.50	85
8600	16" OC	G		42	.381		35	14.75		49.75	63
8610	24" OC	G		57	.281		24.50	10.90		35.40	45
8620	8" wide, studs 12" OC	G		28	.571		52	22		74	93.50
8630	16" OC	G		41	.390		40	15.10		55.10	69.50
8640	24" OC	G		56	.286		28.50	11.05		39.55	50
8790	12 ga. x 6" wide, studs 12" OC	G		27	.593		64.50	23		87.50	109
8800	16" OC	G		37	.432		49	16.75		65.75	81.50
8810	24" OC	G		51	.314		34	12.15		46.15	57.50
8820	8" wide, studs 12" OC	G		26	.615		78.50	24		102.50	126
8830	16" OC	G		36	.444		60	17.20		77.20	94.50
8840	24" OC	G	↓	50	.320	↓	41.50	12.40		53.90	66.50

05 42 Cold-Formed Metal Joist Framing

05 42 13 – Cold-Formed Metal Floor Joist Framing

05 42 13.05 Bracing			Crew	Daily Output	Labor-Hours	Unit	Material	Labor	Equipment	Total	Total Incl O&P
0010	**BRACING**, continuous, per row, top & bottom										
0015	Made from recycled materials										
0120	Flat strap, 20 ga. x 2" wide, joists at 12" OC	G	1 Carp	4.67	1.713	C.L.F.	57.50	66.50		124	174
0130	16" OC	G		5.33	1.501		55.50	58		113.50	157
0140	24" OC	G		6.66	1.201		53.50	46.50		100	136
0150	18 ga. x 2" wide, joists at 12" OC	G		4	2		74	77.50		151.50	210
0160	16" OC	G		4.67	1.713		73	66.50		139.50	190
0170	24" OC	G	↓	5.33	1.501	↓	71.50	58		129.50	175

05 42 13.10 Bridging			Crew	Daily Output	Labor-Hours	Unit	Material	Labor	Equipment	Total	Total Incl O&P
0010	**BRIDGING**, solid between joists w/1-1/4" leg track, per joist bay										
0015	Made from recycled materials										
0230	Joists 12" OC, 18 ga. track x 6" wide	G	1 Carp	80	.100	Ea.	1.60	3.88		5.48	8.15
0240	8" wide	G		75	.107		1.85	4.13		5.98	8.90
0250	10" wide	G		70	.114		2.32	4.43		6.75	9.85
0260	12" wide	G		65	.123		2.63	4.77		7.40	10.80
0330	16 ga. track x 6" wide	G		70	.114		1.89	4.43		6.32	9.40
0340	8" wide	G		65	.123		2.61	4.77		7.38	10.75
0350	10" wide	G		60	.133		2.93	5.15		8.08	11.80
0360	12" wide	G		55	.145		3.37	5.65		9.02	13
0440	14 ga. track x 8" wide	G		60	.133		2.95	5.15		8.10	11.80
0450	10" wide	G		55	.145		4.05	5.65		9.70	13.75
0460	12" wide	G		50	.160		4.68	6.20		10.88	15.40
0550	12 ga. track x 10" wide	G		45	.178		5.95	6.90		12.85	17.95
0560	12" wide	G		40	.200		5.65	7.75		13.40	19
1230	16" OC, 18 ga. track x 6" wide	G		80	.100		2.05	3.88		5.93	8.65
1240	8" wide	G		75	.107		2.38	4.13		6.51	9.45
1250	10" wide	G		70	.114		2.97	4.43		7.40	10.55
1260	12" wide	G		65	.123		3.37	4.77		8.14	11.60
1330	16 ga. track x 6" wide	G	↓	70	.114	↓	2.43	4.43		6.86	9.95

05 42 13 – Cold-Formed Metal Floor Joist Framing

05 42 13.10 Bridging			Crew	Daily Output	Labor-Hours	Unit	2019 Bare Costs Material	Labor	Equipment	Total	Total Incl O&P
1340	8″ wide	G	1 Carp	65	.123	Ea.	3.35	4.77		8.12	11.60
1350	10″ wide	G		60	.133		3.76	5.15		8.91	12.70
1360	12″ wide	G		55	.145		4.32	5.65		9.97	14.05
1440	14 ga. track x 8″ wide	G		60	.133		3.78	5.15		8.93	12.70
1450	10″ wide	G		55	.145		5.20	5.65		10.85	15
1460	12″ wide	G		50	.160		6	6.20		12.20	16.85
1550	12 ga. track x 10″ wide	G		45	.178		7.60	6.90		14.50	19.75
1560	12″ wide	G		40	.200		7.25	7.75		15	21
2230	24″ OC, 18 ga. track x 6″ wide	G		80	.100		2.96	3.88		6.84	9.65
2240	8″ wide	G		75	.107		3.44	4.13		7.57	10.65
2250	10″ wide	G		70	.114		4.30	4.43		8.73	12.05
2260	12″ wide	G		65	.123		4.87	4.77		9.64	13.25
2330	16 ga. track x 6″ wide	G		70	.114		3.51	4.43		7.94	11.15
2340	8″ wide	G		65	.123		4.85	4.77		9.62	13.25
2350	10″ wide	G		60	.133		5.45	5.15		10.60	14.55
2360	12″ wide	G		55	.145		6.25	5.65		11.90	16.20
2440	14 ga. track x 8″ wide	G		60	.133		5.45	5.15		10.60	14.55
2450	10″ wide	G		55	.145		7.50	5.65		13.15	17.55
2460	12″ wide	G		50	.160		8.70	6.20		14.90	19.80
2550	12 ga. track x 10″ wide	G		45	.178		11	6.90		17.90	23.50
2560	12″ wide	G		40	.200		10.50	7.75		18.25	24.50

05 42 13.25 Framing, Band Joist			Crew	Daily Output	Labor-Hours	Unit	Material	Labor	Equipment	Total	Total Incl O&P
0010	**FRAMING, BAND JOIST** (track) fastened to bearing wall										
0015	Made from recycled materials										
0220	18 ga. track x 6″ deep	G	2 Carp	1000	.016	L.F.	1.30	.62		1.92	2.45
0230	8″ deep	G		920	.017		1.51	.67		2.18	2.77
0240	10″ deep	G		860	.019		1.89	.72		2.61	3.27
0320	16 ga. track x 6″ deep	G		900	.018		1.54	.69		2.23	2.84
0330	8″ deep	G		840	.019		2.13	.74		2.87	3.56
0340	10″ deep	G		780	.021		2.39	.79		3.18	3.94
0350	12″ deep	G		740	.022		2.75	.84		3.59	4.41
0430	14 ga. track x 8″ deep	G		750	.021		2.40	.83		3.23	4.01
0440	10″ deep	G		720	.022		3.31	.86		4.17	5.05
0450	12″ deep	G		700	.023		3.82	.89		4.71	5.65
0540	12 ga. track x 10″ deep	G		670	.024		4.84	.93		5.77	6.85
0550	12″ deep	G		650	.025		4.61	.95		5.56	6.65

05 42 13.30 Framing, Boxed Headers/Beams			Crew	Daily Output	Labor-Hours	Unit	Material	Labor	Equipment	Total	Total Incl O&P
0010	**FRAMING, BOXED HEADERS/BEAMS**										
0015	Made from recycled materials										
0200	Double, 18 ga. x 6″ deep	G	2 Carp	220	.073	L.F.	5.65	2.82		8.47	10.85
0210	8″ deep	G		210	.076		5.85	2.95		8.80	11.30
0220	10″ deep	G		200	.080		7	3.10		10.10	12.80
0230	12″ deep	G		190	.084		7.60	3.26		10.86	13.80
0300	16 ga. x 8″ deep	G		180	.089		6.70	3.44		10.14	13.10
0310	10″ deep	G		170	.094		8.50	3.65		12.15	15.40
0320	12″ deep	G		160	.100		9.25	3.88		13.13	16.60
0400	14 ga. x 10″ deep	G		140	.114		9.20	4.43		13.63	17.40
0410	12″ deep	G		130	.123		10.05	4.77		14.82	18.95
0500	12 ga. x 10″ deep	G		110	.145		12.10	5.65		17.75	22.50
0510	12″ deep	G		100	.160		13.40	6.20		19.60	25
1210	Triple, 18 ga. x 8″ deep	G		170	.094		8.45	3.65		12.10	15.35
1220	10″ deep	G		165	.097		10	3.76		13.76	17.20

05 42 Cold-Formed Metal Joist Framing

05 42 13 – Cold-Formed Metal Floor Joist Framing

05 42 13.30 Framing, Boxed Headers/Beams			Crew	Daily Output	Labor-Hours	Unit	Material	Labor	Equipment	Total	Total Incl O&P
							2019 Bare Costs				
1230	12″ deep	G	2 Carp	160	.100	L.F.	10.95	3.88		14.83	18.45
1300	16 ga. x 8″ deep	G		145	.110		9.75	4.28		14.03	17.80
1310	10″ deep	G		140	.114		12.30	4.43		16.73	21
1320	12″ deep	G		135	.119		13.45	4.59		18.04	22.50
1400	14 ga. x 10″ deep	G		115	.139		13.30	5.40		18.70	23.50
1410	12″ deep	G		110	.145		14.65	5.65		20.30	25.50
1500	12 ga. x 10″ deep	G		90	.178		17.70	6.90		24.60	31
1510	12″ deep	G		85	.188		19.60	7.30		26.90	33.50

05 42 13.40 Framing, Joists

Line	Description		Crew	Daily Output	Labor-Hours	Unit	Material	Labor	Equipment	Total	Total Incl O&P
0010	**FRAMING, JOISTS**, no band joists (track), web stiffeners, headers,										
0020	Beams, bridging or bracing										
0025	Made from recycled materials										
0030	Joists (2″ flange) and fasteners, materials only										
0220	18 ga. x 6″ deep	G				L.F.	1.76			1.76	1.94
0230	8″ deep	G					1.87			1.87	2.06
0240	10″ deep	G					2.22			2.22	2.44
0320	16 ga. x 6″ deep	G					2.16			2.16	2.38
0330	8″ deep	G					2.33			2.33	2.56
0340	10″ deep	G					3.01			3.01	3.31
0350	12″ deep	G					3.41			3.41	3.75
0430	14 ga. x 8″ deep	G					2.93			2.93	3.22
0440	10″ deep	G					3.37			3.37	3.71
0450	12″ deep	G					3.83			3.83	4.22
0540	12 ga. x 10″ deep	G					4.90			4.90	5.40
0550	12″ deep	G					5.60			5.60	6.15
1010	Installation of joists to band joists, beams & headers, labor only										
1220	18 ga. x 6″ deep		2 Carp	110	.145	Ea.		5.65		5.65	9.30
1230	8″ deep			90	.178			6.90		6.90	11.40
1240	10″ deep			80	.200			7.75		7.75	12.80
1320	16 ga. x 6″ deep			95	.168			6.55		6.55	10.80
1330	8″ deep			70	.229			8.85		8.85	14.65
1340	10″ deep			60	.267			10.35		10.35	17.10
1350	12″ deep			55	.291			11.25		11.25	18.65
1430	14 ga. x 8″ deep			65	.246			9.55		9.55	15.75
1440	10″ deep			45	.356			13.80		13.80	23
1450	12″ deep			35	.457			17.70		17.70	29.50
1540	12 ga. x 10″ deep			40	.400			15.50		15.50	25.50
1550	12″ deep			30	.533			20.50		20.50	34

05 42 13.45 Framing, Web Stiffeners

Line	Description		Crew	Daily Output	Labor-Hours	Unit	Material	Labor	Equipment	Total	Total Incl O&P
0010	**FRAMING, WEB STIFFENERS** at joist bearing, fabricated from										
0020	Stud piece (1-5/8″ flange) to stiffen joist (2″ flange)										
0025	Made from recycled materials										
2120	For 6″ deep joist, with 18 ga. x 2-1/2″ stud	G	1 Carp	120	.067	Ea.	.92	2.58		3.50	5.30
2130	3-5/8″ stud	G		110	.073		1.08	2.82		3.90	5.85
2140	4″ stud	G		105	.076		1.12	2.95		4.07	6.10
2150	6″ stud	G		100	.080		1.42	3.10		4.52	6.65
2160	8″ stud	G		95	.084		1.61	3.26		4.87	7.15
2220	8″ deep joist, with 2-1/2″ stud	G		120	.067		1.23	2.58		3.81	5.65
2230	3-5/8″ stud	G		110	.073		1.45	2.82		4.27	6.25
2240	4″ stud	G		105	.076		1.50	2.95		4.45	6.55
2250	6″ stud	G		100	.080		1.90	3.10		5	7.20
2260	8″ stud	G		95	.084		2.16	3.26		5.42	7.75

05 42 Cold-Formed Metal Joist Framing

05 42 13 – Cold-Formed Metal Floor Joist Framing

05 42 13.45 Framing, Web Stiffeners			Crew	Daily Output	Labor-Hours	Unit	2019 Bare Costs Material	Labor	Equipment	Total	Total Incl O&P
2320	10" deep joist, with 2-1/2" stud	G	1 Carp	110	.073	Ea.	1.53	2.82		4.35	6.35
2330	3-5/8" stud	G		100	.080		1.79	3.10		4.89	7.05
2340	4" stud	G		95	.084		1.86	3.26		5.12	7.45
2350	6" stud	G		90	.089		2.36	3.44		5.80	8.30
2360	8" stud	G		85	.094		2.67	3.65		6.32	9
2420	12" deep joist, with 2-1/2" stud	G		110	.073		1.84	2.82		4.66	6.70
2430	3-5/8" stud	G		100	.080		2.16	3.10		5.26	7.50
2440	4" stud	G		95	.084		2.24	3.26		5.50	7.85
2450	6" stud	G		90	.089		2.84	3.44		6.28	8.80
2460	8" stud	G		85	.094		3.22	3.65		6.87	9.60
3130	For 6" deep joist, with 16 ga. x 3-5/8" stud	G		100	.080		1.27	3.10		4.37	6.50
3140	4" stud	G		95	.084		1.32	3.26		4.58	6.85
3150	6" stud	G		90	.089		1.65	3.44		5.09	7.50
3160	8" stud	G		85	.094		2.02	3.65		5.67	8.25
3230	8" deep joist, with 3-5/8" stud	G		100	.080		1.70	3.10		4.80	6.95
3240	4" stud	G		95	.084		1.77	3.26		5.03	7.35
3250	6" stud	G		90	.089		2.21	3.44		5.65	8.15
3260	8" stud	G		85	.094		2.71	3.65		6.36	9.05
3330	10" deep joist, with 3-5/8" stud	G		85	.094		2.11	3.65		5.76	8.35
3340	4" stud	G		80	.100		2.19	3.88		6.07	8.80
3350	6" stud	G		75	.107		2.74	4.13		6.87	9.85
3360	8" stud	G		70	.114		3.35	4.43		7.78	11
3430	12" deep joist, with 3-5/8" stud	G		85	.094		2.54	3.65		6.19	8.85
3440	4" stud	G		80	.100		2.64	3.88		6.52	9.30
3450	6" stud	G		75	.107		3.30	4.13		7.43	10.50
3460	8" stud	G		70	.114		4.04	4.43		8.47	11.75
4230	For 8" deep joist, with 14 ga. x 3-5/8" stud	G		90	.089		2.10	3.44		5.54	8
4240	4" stud	G		85	.094		2.21	3.65		5.86	8.50
4250	6" stud	G		80	.100		2.79	3.88		6.67	9.45
4260	8" stud	G		75	.107		3.15	4.13		7.28	10.30
4330	10" deep joist, with 3-5/8" stud	G		75	.107		2.61	4.13		6.74	9.70
4340	4" stud	G		70	.114		2.74	4.43		7.17	10.30
4350	6" stud	G		65	.123		3.45	4.77		8.22	11.70
4360	8" stud	G		60	.133		3.90	5.15		9.05	12.85
4430	12" deep joist, with 3-5/8" stud	G		75	.107		3.14	4.13		7.27	10.30
4440	4" stud	G		70	.114		3.30	4.43		7.73	10.95
4450	6" stud	G		65	.123		4.16	4.77		8.93	12.50
4460	8" stud	G		60	.133		4.70	5.15		9.85	13.70
5330	For 10" deep joist, with 12 ga. x 3-5/8" stud	G		65	.123		3.75	4.77		8.52	12.05
5340	4" stud	G		60	.133		4	5.15		9.15	12.95
5350	6" stud	G		55	.145		5.05	5.65		10.70	14.85
5360	8" stud	G		50	.160		6.10	6.20		12.30	16.95
5430	12" deep joist, with 3-5/8" stud	G		65	.123		4.52	4.77		9.29	12.85
5440	4" stud	G		60	.133		4.82	5.15		9.97	13.85
5450	6" stud	G		55	.145		6.10	5.65		11.75	16
5460	8" stud	G	↓	50	.160	↓	7.35	6.20		13.55	18.30

05 42 23 – Cold-Formed Metal Roof Joist Framing

05 42 23.05 Framing, Bracing			Crew	Daily Output	Labor-Hours	Unit	Material	Labor	Equipment	Total	Total Incl O&P
0010	**FRAMING, BRACING**										
0015	Made from recycled materials										
0020	Continuous bracing, per row										
0100	16 ga. x 1-1/2" channel thru rafters/trusses @ 16" OC	G	1 Carp	4.50	1.778	C.L.F.	49	69		118	168

05 42 Cold-Formed Metal Joist Framing

05 42 23 – Cold-Formed Metal Roof Joist Framing

05 42 23.05 Framing, Bracing		Crew	Daily Output	Labor-Hours	Unit	Material	2019 Bare Costs Labor	Equipment	Total	Total Incl O&P
0120	24" OC	G	1 Carp	6	1.333	C.L.F.	49	51.50		100.50	140
0300	2" x 2" angle x 18 ga., rafters/trusses @ 16" OC	G		6	1.333		76	51.50		127.50	169
0320	24" OC	G		8	1		76	39		115	148
0400	16 ga., rafters/trusses @ 16" OC	G		4.50	1.778		95.50	69		164.50	219
0420	24" OC	G		6.50	1.231		95.50	47.50		143	184

05 42 23.10 Framing, Bridging			Crew	Daily Output	Labor-Hours	Unit	Material	Labor	Equipment	Total	Total Incl O&P
0010	**FRAMING, BRIDGING**										
0015	Made from recycled materials										
0020	Solid, between rafters w/1-1/4" leg track, per rafter bay										
1200	Rafters 16" OC, 18 ga. x 4" deep	G	1 Carp	60	.133	Ea.	1.57	5.15		6.72	10.25
1210	6" deep	G		57	.140		2.05	5.45		7.50	11.25
1220	8" deep	G		55	.145		2.38	5.65		8.03	11.90
1230	10" deep	G		52	.154		2.97	5.95		8.92	13.10
1240	12" deep	G		50	.160		3.37	6.20		9.57	13.95
2200	24" OC, 18 ga. x 4" deep	G		60	.133		2.27	5.15		7.42	11.05
2210	6" deep	G		57	.140		2.96	5.45		8.41	12.25
2220	8" deep	G		55	.145		3.44	5.65		9.09	13.10
2230	10" deep	G		52	.154		4.30	5.95		10.25	14.60
2240	12" deep	G		50	.160		4.87	6.20		11.07	15.60

05 42 23.50 Framing, Parapets			Crew	Daily Output	Labor-Hours	Unit	Material	Labor	Equipment	Total	Total Incl O&P
0010	**FRAMING, PARAPETS**										
0015	Made from recycled materials										
0100	3' high installed on 1st story, 18 ga. x 4" wide studs, 12" OC	G	2 Carp	100	.160	L.F.	5.70	6.20		11.90	16.50
0110	16" OC	G		150	.107		4.85	4.13		8.98	12.20
0120	24" OC	G		200	.080		4.01	3.10		7.11	9.50
0200	6" wide studs, 12" OC	G		100	.160		7.30	6.20		13.50	18.30
0210	16" OC	G		150	.107		6.25	4.13		10.38	13.70
0220	24" OC	G		200	.080		5.15	3.10		8.25	10.80
1100	Installed on 2nd story, 18 ga. x 4" wide studs, 12" OC	G		95	.168		5.70	6.55		12.25	17.05
1110	16" OC	G		145	.110		4.85	4.28		9.13	12.40
1120	24" OC	G		190	.084		4.01	3.26		7.27	9.80
1200	6" wide studs, 12" OC	G		95	.168		7.30	6.55		13.85	18.85
1210	16" OC	G		145	.110		6.25	4.28		10.53	13.90
1220	24" OC	G		190	.084		5.15	3.26		8.41	11.10
2100	Installed on gable, 18 ga. x 4" wide studs, 12" OC	G		85	.188		5.70	7.30		13	18.30
2110	16" OC	G		130	.123		4.85	4.77		9.62	13.25
2120	24" OC	G		170	.094		4.01	3.65		7.66	10.45
2200	6" wide studs, 12" OC	G		85	.188		7.30	7.30		14.60	20
2210	16" OC	G		130	.123		6.25	4.77		11.02	14.75
2220	24" OC	G		170	.094		5.15	3.65		8.80	11.75

05 42 23.60 Framing, Roof Rafters			Crew	Daily Output	Labor-Hours	Unit	Material	Labor	Equipment	Total	Total Incl O&P
0010	**FRAMING, ROOF RAFTERS**										
0015	Made from recycled materials										
0100	Boxed ridge beam, double, 18 ga. x 6" deep	G	2 Carp	160	.100	L.F.	5.65	3.88		9.53	12.60
0110	8" deep	G		150	.107		5.85	4.13		9.98	13.25
0120	10" deep	G		140	.114		7	4.43		11.43	15
0130	12" deep	G		130	.123		7.60	4.77		12.37	16.30
0200	16 ga. x 6" deep	G		150	.107		6.40	4.13		10.53	13.90
0210	8" deep	G		140	.114		6.70	4.43		11.13	14.70
0220	10" deep	G		130	.123		8.50	4.77		13.27	17.25
0230	12" deep	G		120	.133		9.25	5.15		14.40	18.75
1100	Rafters, 2" flange, material only, 18 ga. x 6" deep	G					1.76			1.76	1.94

05 42 Cold-Formed Metal Joist Framing

05 42 23 – Cold-Formed Metal Roof Joist Framing

05 42 23.60 Framing, Roof Rafters		Crew	Daily Output	Labor-Hours	Unit	Material (2019 Bare Costs)	Labor (2019 Bare Costs)	Equipment (2019 Bare Costs)	Total (2019 Bare Costs)	Total Incl O&P
1110	8" deep	G				L.F.	1.87			1.87	2.06
1120	10" deep	G					2.22			2.22	2.44
1130	12" deep	G					2.55			2.55	2.81
1200	16 ga. x 6" deep	G					2.16			2.16	2.38
1210	8" deep	G					2.33			2.33	2.56
1220	10" deep	G					3.01			3.01	3.31
1230	12" deep	G					3.41			3.41	3.75
2100	Installation only, ordinary rafter to 4:12 pitch, 18 ga. x 6" deep		2 Carp	35	.457	Ea.		17.70		17.70	29.50
2110	8" deep			30	.533			20.50		20.50	34
2120	10" deep			25	.640			25		25	41
2130	12" deep			20	.800			31		31	51
2200	16 ga. x 6" deep			30	.533			20.50		20.50	34
2210	8" deep			25	.640			25		25	41
2220	10" deep			20	.800			31		31	51
2230	12" deep			15	1.067			41.50		41.50	68.50
8100	Add to labor, ordinary rafters on steep roofs							25%			
8110	Dormers & complex roofs							50%			
8200	Hip & valley rafters to 4:12 pitch							25%			
8210	Steep roofs							50%			
8220	Dormers & complex roofs							75%			
8300	Hip & valley jack rafters to 4:12 pitch							50%			
8310	Steep roofs							75%			
8320	Dormers & complex roofs							100%			

05 42 23.70 Framing, Soffits and Canopies		Crew	Daily Output	Labor-Hours	Unit	Material	Labor	Equipment	Total	Total Incl O&P
0010	**FRAMING, SOFFITS & CANOPIES**									
0015	Made from recycled materials									
0130	Continuous ledger track @ wall, studs @ 16" OC, 18 ga. x 4" wide	G	2 Carp	535	.030	L.F.	1.05	1.16		2.21	3.07
0140	6" wide	G		500	.032		1.36	1.24		2.60	3.55
0150	8" wide	G		465	.034		1.58	1.33		2.91	3.94
0160	10" wide	G		430	.037		1.98	1.44		3.42	4.56
0230	Studs @ 24" OC, 18 ga. x 4" wide	G		800	.020		1	.78		1.78	2.38
0240	6" wide	G		750	.021		1.30	.83		2.13	2.80
0250	8" wide	G		700	.023		1.51	.89		2.40	3.12
0260	10" wide	G		650	.025		1.89	.95		2.84	3.66
1000	Horizontal soffit and canopy members, material only										
1030	1-5/8" flange studs, 18 ga. x 4" deep	G				L.F.	1.34			1.34	1.48
1040	6" deep	G					1.70			1.70	1.87
1050	8" deep	G					1.93			1.93	2.13
1140	2" flange joists, 18 ga. x 6" deep	G					2.02			2.02	2.22
1150	8" deep	G					2.14			2.14	2.35
1160	10" deep	G					2.53			2.53	2.79
4030	Installation only, 18 ga., 1-5/8" flange x 4" deep		2 Carp	130	.123	Ea.		4.77		4.77	7.90
4040	6" deep			110	.145			5.65		5.65	9.30
4050	8" deep			90	.178			6.90		6.90	11.40
4140	2" flange, 18 ga. x 6" deep			110	.145			5.65		5.65	9.30
4150	8" deep			90	.178			6.90		6.90	11.40
4160	10" deep			80	.200			7.75		7.75	12.80
6010	Clips to attach fascia to rafter tails, 2" x 2" x 18 ga. angle	G	1 Carp	120	.067		.90	2.58		3.48	5.25
6020	16 ga. angle	G	"	100	.080		1.13	3.10		4.23	6.35

05 44 13.60 Framing, Roof Trusses		Crew	Daily Output	Labor-Hours	Unit	Material	2019 Bare Costs Labor	Equipment	Total	Total Incl O&P
0010	**FRAMING, ROOF TRUSSES**									
0015	Made from recycled materials									
0020	Fabrication of trusses on ground, Fink (W) or King Post, to 4:12 pitch									
0120	18 ga. x 4" chords, 16' span G	2 Carp	12	1.333	Ea.	62.50	51.50		114	155
0130	20' span G		11	1.455		78.50	56.50		135	179
0140	24' span G		11	1.455		94	56.50		150.50	196
0150	28' span G		10	1.600		110	62		172	223
0160	32' span G		10	1.600		125	62		187	240
0250	6" chords, 28' span G		9	1.778		139	69		208	267
0260	32' span G		9	1.778		159	69		228	289
0270	36' span G		8	2		179	77.50		256.50	325
0280	40' span G		8	2		199	77.50		276.50	345
1120	5:12 to 8:12 pitch, 18 ga. x 4" chords, 16' span G		10	1.600		71.50	62		133.50	181
1130	20' span G		9	1.778		89.50	69		158.50	213
1140	24' span G		9	1.778		108	69		177	232
1150	28' span G		8	2		125	77.50		202.50	266
1160	32' span G		8	2		143	77.50		220.50	286
1250	6" chords, 28' span G		7	2.286		159	88.50		247.50	320
1260	32' span G		7	2.286		182	88.50		270.50	345
1270	36' span G		6	2.667		204	103		307	395
1280	40' span G		6	2.667		227	103		330	420
2120	9:12 to 12:12 pitch, 18 ga. x 4" chords, 16' span G		8	2		89.50	77.50		167	227
2130	20' span G		7	2.286		112	88.50		200.50	269
2140	24' span G		7	2.286		134	88.50		222.50	294
2150	28' span G		6	2.667		157	103		260	345
2160	32' span G		6	2.667		179	103		282	370
2250	6" chords, 28' span G		5	3.200		199	124		323	425
2260	32' span G		5	3.200		227	124		351	455
2270	36' span G		4	4		256	155		411	535
2280	40' span G		4	4		284	155		439	565
5120	Erection only of roof trusses, to 4:12 pitch, 16' span	F-6	48	.833			30	9.85	39.85	60.50
5130	20' span		46	.870			31.50	10.25	41.75	63
5140	24' span		44	.909			33	10.70	43.70	66
5150	28' span		42	.952			34.50	11.25	45.75	69
5160	32' span		40	1			36	11.80	47.80	72.50
5170	36' span		38	1.053			38	12.40	50.40	76
5180	40' span		36	1.111			40	13.10	53.10	80.50
5220	5:12 to 8:12 pitch, 16' span		42	.952			34.50	11.25	45.75	69
5230	20' span		40	1			36	11.80	47.80	72.50
5240	24' span		38	1.053			38	12.40	50.40	76
5250	28' span		36	1.111			40	13.10	53.10	80.50
5260	32' span		34	1.176			42.50	13.85	56.35	85.50
5270	36' span		32	1.250			45	14.75	59.75	90.50
5280	40' span		30	1.333			48	15.70	63.70	97
5320	9:12 to 12:12 pitch, 16' span		36	1.111			40	13.10	53.10	80.50
5330	20' span		34	1.176			42.50	13.85	56.35	85.50
5340	24' span		32	1.250			45	14.75	59.75	90.50
5350	28' span		30	1.333			48	15.70	63.70	97
5360	32' span		28	1.429			51.50	16.85	68.35	104
5370	36' span		26	1.538			55.50	18.15	73.65	111
5380	40' span		24	1.667			60	19.65	79.65	121

05 51 Metal Stairs

05 51 13 – Metal Pan Stairs

05 51 13.50 Pan Stairs			Crew	Daily Output	Labor-Hours	Unit	2019 Bare Costs Material	Labor	Equipment	Total	Total Incl O&P
0010	**PAN STAIRS**, shop fabricated, steel stringers										
0015	Made from recycled materials										
1700	Pre-erected, steel pan tread, 3′-6″ wide, 2 line pipe rail	G	E-2	87	.552	Riser	535	23	19.50	577.50	645
1800	With flat bar picket rail	G	"	87	.552	"	595	23	19.50	637.50	715

05 51 19 – Metal Grating Stairs

05 51 19.50 Grating Stairs

Line	Description		Crew	Daily Output	Labor-Hours	Unit	Material	Labor	Equipment	Total	Total Incl O&P
0010	**GRATING STAIRS**, shop fabricated, steel stringers, safety nosing on treads										
0015	Made from recycled materials										
0020	Grating tread and pipe railing, 3′-6″ wide	G	E-4	35	.914	Riser	400	38.50	2.73	441.23	510
0100	4′-0″ wide	G	"	30	1.067	"	405	45	3.19	453.19	525

05 51 23 – Metal Fire Escapes

05 51 23.50 Fire Escape Stairs

Line	Description		Crew	Daily Output	Labor-Hours	Unit	Material	Labor	Equipment	Total	Total Incl O&P
0010	**FIRE ESCAPE STAIRS**, portable										
0100	Portable ladder					Ea.	123			123	135

05 51 33 – Metal Ladders

05 51 33.13 Vertical Metal Ladders

Line	Description		Crew	Daily Output	Labor-Hours	Unit	Material	Labor	Equipment	Total	Total Incl O&P
0010	**VERTICAL METAL LADDERS**, shop fabricated										
0015	Made from recycled materials										
0020	Steel, 20″ wide, bolted to concrete, with cage	G	E-4	50	.640	V.L.F.	85	27	1.91	113.91	142
0100	Without cage	G		85	.376		41.50	15.80	1.13	58.43	74.50
0300	Aluminum, bolted to concrete, with cage	G		50	.640		134	27	1.91	162.91	196
0400	Without cage	G		85	.376		50.50	15.80	1.13	67.43	84

05 51 33.23 Alternating Tread Ladders

Line	Description	Crew	Daily Output	Labor-Hours	Unit	Material	Labor	Equipment	Total	Total Incl O&P
0010	**ALTERNATING TREAD LADDERS**, shop fabricated									
0015	Made from recycled materials									
0800	Alternating tread ladders, 68-degree angle of incline									
0810	8′ vertical rise, steel, 149 lb., standard paint color	B-68G	3	5.333	Ea.	2,175	221	92	2,488	2,875
0820	Non-standard paint color		3	5.333		2,625	221	92	2,938	3,350
0830	Galvanized		3	5.333		2,625	221	92	2,938	3,350
0840	Stainless		3	5.333		3,800	221	92	4,113	4,650
0850	Aluminum, 87 lb.		3	5.333		2,775	221	92	3,088	3,525
1010	10′ vertical rise, steel, 181 lb., standard paint color		2.75	5.818		2,725	241	100	3,066	3,525
1020	Non-standard paint color		2.75	5.818		3,125	241	100	3,466	3,975
1030	Galvanized		2.75	5.818		3,175	241	100	3,516	4,025
1040	Stainless		2.75	5.818		4,575	241	100	4,916	5,550
1050	Aluminum, 103 lb.		2.75	5.818		3,350	241	100	3,691	4,200
1210	12′ vertical rise, steel, 245 lb., standard paint color		2.50	6.400		3,200	265	110	3,575	4,075
1220	Non-standard paint color		2.50	6.400		3,650	265	110	4,025	4,575
1230	Galvanized		2.50	6.400		3,725	265	110	4,100	4,675
1240	Stainless		2.50	6.400		5,350	265	110	5,725	6,475
1250	Aluminum, 103 lb.		2.50	6.400		3,800	265	110	4,175	4,750
1410	14′ vertical rise, steel, 281 lb., standard paint color		2.25	7.111		3,650	295	123	4,068	4,675
1420	Non-standard paint color		2.25	7.111		4,150	295	123	4,568	5,225
1430	Galvanized		2.25	7.111		4,275	295	123	4,693	5,350
1440	Stainless		2.25	7.111		6,125	295	123	6,543	7,400
1450	Aluminum, 136 lb.		2.25	7.111		4,550	295	123	4,968	5,650
1610	16′ vertical rise, steel, 317 lb., standard paint color		2	8		4,125	330	138	4,593	5,275
1620	Non-standard paint color		2	8		4,650	330	138	5,118	5,850
1630	Galvanized		2	8		4,825	330	138	5,293	6,025
1640	Stainless		2	8		6,900	330	138	7,368	8,325
1650	Aluminum, 153 lb.		2	8		5,050	330	138	5,518	6,300

05 52 13.50 Railings, Pipe			Crew	Daily Output	Labor-Hours	Unit	2019 Bare Costs Material	Labor	Equipment	Total	Total Incl O&P
0010	**RAILINGS, PIPE**, shop fab'd, 3'-6" high, posts @ 5' OC										
0015	Made from recycled materials										
0020	Aluminum, 2 rail, satin finish, 1-1/4" diameter	G	E-4	160	.200	L.F.	52	8.40	.60	61	72
0030	Clear anodized	G		160	.200		63.50	8.40	.60	72.50	85
0040	Dark anodized	G		160	.200		70.50	8.40	.60	79.50	92.50
0080	1-1/2" diameter, satin finish	G		160	.200		61	8.40	.60	70	82
0090	Clear anodized	G		160	.200		68.50	8.40	.60	77.50	90.50
0100	Dark anodized	G		160	.200		75.50	8.40	.60	84.50	98
0140	Aluminum, 3 rail, 1-1/4" diam., satin finish	G		137	.234		67.50	9.80	.70	78	92.50
0150	Clear anodized	G		137	.234		84	9.80	.70	94.50	110
0160	Dark anodized	G		137	.234		92.50	9.80	.70	103	120
0200	1-1/2" diameter, satin finish	G		137	.234		80	9.80	.70	90.50	106
0210	Clear anodized	G		137	.234		91	9.80	.70	101.50	118
0220	Dark anodized	G		137	.234		100	9.80	.70	110.50	128
0500	Steel, 2 rail, on stairs, primed, 1-1/4" diameter	G		160	.200		29.50	8.40	.60	38.50	47.50
0520	1-1/2" diameter	G		160	.200		32	8.40	.60	41	50
0540	Galvanized, 1-1/4" diameter	G		160	.200		39	8.40	.60	48	58
0560	1-1/2" diameter	G		160	.200		45.50	8.40	.60	54.50	65
0580	Steel, 3 rail, primed, 1-1/4" diameter	G		137	.234		43	9.80	.70	53.50	65
0600	1-1/2" diameter	G		137	.234		46.50	9.80	.70	57	69
0620	Galvanized, 1-1/4" diameter	G		137	.234		60	9.80	.70	70.50	84
0640	1-1/2" diameter	G		137	.234		72	9.80	.70	82.50	97
0700	Stainless steel, 2 rail, 1-1/4" diam., #4 finish	G		137	.234		131	9.80	.70	141.50	162
0720	High polish	G		137	.234		213	9.80	.70	223.50	252
0740	Mirror polish	G		137	.234		265	9.80	.70	275.50	310
0760	Stainless steel, 3 rail, 1-1/2" diam., #4 finish	G		120	.267		197	11.20	.80	209	237
0770	High polish	G		120	.267		330	11.20	.80	342	380
0780	Mirror finish	G		120	.267		400	11.20	.80	412	460
0900	Wall rail, alum. pipe, 1-1/4" diam., satin finish	G		213	.150		25	6.30	.45	31.75	39
0905	Clear anodized	G		213	.150		31.50	6.30	.45	38.25	46
0910	Dark anodized	G		213	.150		37	6.30	.45	43.75	52
0915	1-1/2" diameter, satin finish	G		213	.150		28	6.30	.45	34.75	42
0920	Clear anodized	G		213	.150		34.50	6.30	.45	41.25	49.50
0925	Dark anodized	G		213	.150		44	6.30	.45	50.75	59.50
0930	Steel pipe, 1-1/4" diameter, primed	G		213	.150		17.25	6.30	.45	24	30.50
0935	Galvanized	G		213	.150		25	6.30	.45	31.75	39
0940	1-1/2" diameter	G		176	.182		18.15	7.65	.54	26.34	34
0945	Galvanized	G		213	.150		25	6.30	.45	31.75	39
0955	Stainless steel pipe, 1-1/2" diam., #4 finish	G		107	.299		105	12.55	.89	118.44	139
0960	High polish	G		107	.299		214	12.55	.89	227.44	258
0965	Mirror polish	G	↓	107	.299	↓	253	12.55	.89	266.44	300
2000	2-line pipe rail (1-1/2" T&B) with 1/2" pickets @ 4-1/2" OC,										
2005	attached handrail on brackets										
2010	42" high aluminum, satin finish, straight & level	G	E-4	120	.267	L.F.	277	11.20	.80	289	325
2050	42" high steel, primed, straight & level	G	"	120	.267		148	11.20	.80	160	183
4000	For curved and level rails, add						10%	10%			
4100	For sloped rails for stairs, add					↓	30%	30%			

05 55 Metal Stair Treads and Nosings

05 55 13 – Metal Stair Treads

05 55 13.50 Stair Treads		Crew	Daily Output	Labor-Hours	Unit	Material	2019 Bare Costs Labor	Equipment	Total	Total Incl O&P
0010	**STAIR TREADS**, stringers and bolts not included									
3000	Diamond plate treads, steel, 1/8" thick									
3005	Open riser, black enamel									
3010	9" deep x 36" long G	2 Sswk	48	.333	Ea.	102	13.80		115.80	137
3020	42" long G		48	.333		108	13.80		121.80	143
3030	48" long G		48	.333		113	13.80		126.80	148
3040	11" deep x 36" long G		44	.364		108	15.05		123.05	145
3050	42" long G		44	.364		114	15.05		129.05	152
3060	48" long G		44	.364		122	15.05		137.05	160
3110	Galvanized, 9" deep x 36" long G		48	.333		162	13.80		175.80	202
3120	42" long G		48	.333		172	13.80		185.80	213
3130	48" long G		48	.333		183	13.80		196.80	225
3140	11" deep x 36" long G		44	.364		168	15.05		183.05	211
3150	42" long G		44	.364		183	15.05		198.05	227
3160	48" long G	↓	44	.364	↓	190	15.05		205.05	235
3200	Closed riser, black enamel									
3210	12" deep x 36" long G	2 Sswk	40	.400	Ea.	123	16.60		139.60	165
3220	42" long G		40	.400		134	16.60		150.60	176
3230	48" long G		40	.400		141	16.60		157.60	184
3240	Galvanized, 12" deep x 36" long G		40	.400		195	16.60		211.60	243
3250	42" long G		40	.400		212	16.60		228.60	263
3260	48" long G	↓	40	.400	↓	223	16.60		239.60	274
4000	Bar grating treads									
4005	Steel, 1-1/4" x 3/16" bars, anti-skid nosing, black enamel									
4010	8-5/8" deep x 30" long G	2 Sswk	48	.333	Ea.	54	13.80		67.80	83.50
4020	36" long G		48	.333		63	13.80		76.80	93.50
4030	48" long G		48	.333		97	13.80		110.80	131
4040	10-15/16" deep x 36" long G		44	.364		70	15.05		85.05	103
4050	48" long G		44	.364		100	15.05		115.05	136
4060	Galvanized, 8-5/8" deep x 30" long G		48	.333		62	13.80		75.80	92
4070	36" long G		48	.333		73.50	13.80		87.30	105
4080	48" long G		48	.333		108	13.80		121.80	143
4090	10-15/16" deep x 36" long G		44	.364		86	15.05		101.05	121
4100	48" long G	↓	44	.364	↓	111	15.05		126.05	148
4200	Aluminum, 1-1/4" x 3/16" bars, serrated, with nosing									
4210	7-5/8" deep x 18" long G	2 Sswk	52	.308	Ea.	49.50	12.75		62.25	76.50
4220	24" long G		52	.308		59	12.75		71.75	86.50
4230	30" long G		52	.308		68	12.75		80.75	97
4240	36" long G		52	.308		176	12.75		188.75	216
4250	8-13/16" deep x 18" long G		48	.333		67	13.80		80.80	97.50
4260	24" long G		48	.333		99	13.80		112.80	133
4270	30" long G		48	.333		114	13.80		127.80	150
4280	36" long G		48	.333		194	13.80		207.80	237
4290	10" deep x 18" long G		44	.364		130	15.05		145.05	169
4300	30" long G		44	.364		172	15.05		187.05	215
4310	36" long G	↓	44	.364	↓	209	15.05		224.05	256
5000	Channel grating treads									
5005	Steel, 14 ga., 2-1/2" thick, galvanized									
5010	9" deep x 36" long G	2 Sswk	48	.333	Ea.	103	13.80		116.80	137
5020	48" long G	"	48	.333	"	138	13.80		151.80	176

05 55 Metal Stair Treads and Nosings

05 55 19 – Metal Stair Tread Covers

05 55 19.50 Stair Tread Covers for Renovation		Crew	Daily Output	Labor-Hours	Unit	2019 Bare Costs Material	Labor	Equipment	Total	Total Incl O&P
0010	**STAIR TREAD COVERS FOR RENOVATION**									
0205	Extruded tread cover with nosing, pre-drilled, includes screws									
0210	Aluminum with black abrasive strips, 9″ wide x 3′ long	1 Carp	24	.333	Ea.	105	12.90		117.90	137
0220	4′ long		22	.364		139	14.10		153.10	176
0230	5′ long		20	.400		182	15.50		197.50	226
0240	11″ wide x 3′ long		24	.333		142	12.90		154.90	178
0250	4′ long		22	.364		191	14.10		205.10	234
0260	5′ long		20	.400		233	15.50		248.50	282
0305	Black abrasive strips with yellow front strips									
0310	Aluminum, 9″ wide x 3′ long	1 Carp	24	.333	Ea.	123	12.90		135.90	157
0320	4′ long		22	.364		164	14.10		178.10	204
0330	5′ long		20	.400		208	15.50		223.50	255
0340	11″ wide x 3′ long		24	.333		154	12.90		166.90	191
0350	4′ long		22	.364		198	14.10		212.10	242
0360	5′ long		20	.400		255	15.50		270.50	305
0405	Black abrasive strips with photoluminescent front strips									
0410	Aluminum, 9″ wide x 3′ long	1 Carp	24	.333	Ea.	154	12.90		166.90	191
0420	4′ long		22	.364		189	14.10		203.10	232
0430	5′ long		20	.400		236	15.50		251.50	285
0440	11″ wide x 3′ long		24	.333		168	12.90		180.90	207
0450	4′ long		22	.364		224	14.10		238.10	270
0460	5′ long		20	.400		280	15.50		295.50	335

05 58 Formed Metal Fabrications

05 58 13 – Column Covers

05 58 13.05 Column Covers

			Crew	Daily Output	Labor-Hours	Unit	Material	Labor	Equipment	Total	Total Incl O&P
0010	**COLUMN COVERS**										
0015	Made from recycled materials										
0020	Excludes structural steel, light ga. metal framing, misc. metals, sealants										
0100	Round covers, 2 halves with 2 vertical joints for backer rod and sealant										
0110	Up to 12′ high, no horizontal joints										
0120	12″ diameter, 0.125″ aluminum, anodized/painted finish	G	2 Sswk	32	.500	V.L.F.	31	20.50		51.50	70.50
0130	Type 304 stainless steel, 16 gauge, #4 brushed finish	G		32	.500		44.50	20.50		65	85
0140	Type 316 stainless steel, 16 gauge, #4 brushed finish	G		32	.500		55	20.50		75.50	96.50
0150	18″ diameter, aluminum	G		32	.500		47	20.50		67.50	87.50
0160	Type 304 stainless steel	G		32	.500		66.50	20.50		87	109
0170	Type 316 stainless steel	G		32	.500		83	20.50		103.50	127
0180	24″ diameter, aluminum	G		32	.500		62.50	20.50		83	105
0190	Type 304 stainless steel	G		32	.500		89	20.50		109.50	134
0200	Type 316 stainless steel	G		32	.500		110	20.50		130.50	157
0210	30″ diameter, aluminum	G		30	.533		78	22		100	125
0220	Type 304 stainless steel	G		30	.533		111	22		133	161
0230	Type 316 stainless steel	G		30	.533		138	22		160	191
0240	36″ diameter, aluminum	G		30	.533		93.50	22		115.50	142
0250	Type 304 stainless steel	G		30	.533		133	22		155	185
0260	Type 316 stainless steel	G		30	.533		166	22		188	221
0400	Up to 24′ high, 2 stacked sections with 1 horizontal joint										
0410	18″ diameter, aluminum	G	2 Sswk	28	.571	V.L.F.	49	23.50		72.50	95
0450	Type 304 stainless steel	G		28	.571		70	23.50		93.50	118
0460	Type 316 stainless steel	G		28	.571		87	23.50		110.50	137

05 58 Formed Metal Fabrications

05 58 13 – Column Covers

05 58 13.05 Column Covers			Crew	Daily Output	Labor-Hours	Unit	2019 Bare Costs Material	Labor	Equipment	Total	Total Incl O&P
0470	24" diameter, aluminum	G	2 Sswk	28	.571	V.L.F.	65.50	23.50		89	113
0480	Type 304 stainless steel	G		28	.571		93	23.50		116.50	144
0490	Type 316 stainless steel	G		28	.571		116	23.50		139.50	168
0500	30" diameter, aluminum	G		24	.667		82	27.50		109.50	138
0510	Type 304 stainless steel	G		24	.667		117	27.50		144.50	176
0520	Type 316 stainless steel	G		24	.667		145	27.50		172.50	207
0530	36" diameter, aluminum	G		24	.667		98.50	27.50		126	156
0540	Type 304 stainless steel	G		24	.667		140	27.50		167.50	202
0550	Type 316 stainless steel	G		24	.667		174	27.50		201.50	239

05 58 25 – Formed Lamp Posts

05 58 25.40 Lamp Posts			Crew	Daily Output	Labor-Hours	Unit	Material	Labor	Equipment	Total	Total Incl O&P
0010	**LAMP POSTS**										
0020	Aluminum, 7' high, stock units, post only	G	1 Carp	16	.500	Ea.	82.50	19.40		101.90	123
0100	Mild steel, plain	G	"	16	.500	"	75	19.40		94.40	115

05 71 Decorative Metal Stairs

05 71 13 – Fabricated Metal Spiral Stairs

05 71 13.50 Spiral Stairs			Crew	Daily Output	Labor-Hours	Unit	Material	Labor	Equipment	Total	Total Incl O&P
0010	**SPIRAL STAIRS**										
1805	Shop fabricated, custom ordered										
1810	Aluminum, 5'-0" diameter, plain units	G	E-4	45	.711	Riser	730	30	2.13	762.13	860
1820	Fancy units	G		45	.711		1,300	30	2.13	1,332.13	1,500
1900	Cast iron, 4'-0" diameter, plain units	G		45	.711		735	30	2.13	767.13	860
1920	Fancy units	G		25	1.280		1,275	53.50	3.83	1,332.33	1,500
3100	Spiral stair kits, 12 stacking risers to fit exact floor height										

05 73 Decorative Metal Railings

05 73 16 – Wire Rope Decorative Metal Railings

05 73 16.10 Cable Railings			Crew	Daily Output	Labor-Hours	Unit	Material	Labor	Equipment	Total	Total Incl O&P
0010	**CABLE RAILINGS**, with 316 stainless steel 1 x 19 cable, 3/16" diameter										
0015	Made from recycled materials										
0100	1-3/4" diameter stainless steel posts x 42" high, cables 4" OC	G	2 Sswk	25	.640	L.F.	46.50	26.50		73	97

05 73 23 – Ornamental Railings

05 73 23.50 Railings, Ornamental			Crew	Daily Output	Labor-Hours	Unit	Material	Labor	Equipment	Total	Total Incl O&P
0010	**RAILINGS, ORNAMENTAL**, 3'-6" high, posts @ 6' OC										
0020	Bronze or stainless, hand forged, plain	G	2 Sswk	24	.667	L.F.	265	27.50		292.50	340
0100	Fancy	G		18	.889		525	37		562	640
0200	Aluminum, panelized, plain	G		24	.667		13.25	27.50		40.75	62.50
0300	Fancy	G		18	.889		25.50	37		62.50	92
0400	Wrought iron, hand forged, plain	G		24	.667		95.50	27.50		123	153
0500	Fancy	G		18	.889		230	37		267	315
0550	Steel, panelized, plain	G		24	.667		22.50	27.50		50	72.50
0560	Fancy	G		18	.889		33	37		70	100
0600	Composite metal/wood/glass, plain			18	.889		145	37		182	223
0700	Fancy			12	1.333		289	55.50		344.50	415

Division Notes

		CREW	DAILY OUTPUT	LABOR-HOURS	UNIT	BARE COSTS MAT.	LABOR	EQUIP.	TOTAL	TOTAL INCL O&P

Estimating Tips

06 05 00 Common Work Results for Wood, Plastics, and Composites

- Common to any wood-framed structure are the accessory connector items such as screws, nails, adhesives, hangers, connector plates, straps, angles, and hold-downs. For typical wood-framed buildings, such as residential projects, the aggregate total for these items can be significant, especially in areas where seismic loading is a concern. For floor and wall framing, the material cost is based on 10 to 25 lbs. of accessory connectors per MBF. Hold-downs, hangers, and other connectors should be taken off by the piece.

  Included with material costs are fasteners for a normal installation. Gordian's RSMeans engineers use manufacturers' recommendations, written specifications, and/or standard construction practice for the sizing and spacing of fasteners. Prices for various fasteners are shown for informational purposes only. Adjustments should be made if unusual fastening conditions exist.

06 10 00 Carpentry

- Lumber is a traded commodity and therefore sensitive to supply and demand in the marketplace. Even with "budgetary" estimating of wood-framed projects, it is advisable to call local suppliers for the latest market pricing.
- The common quantity unit for wood-framed projects is "thousand board feet" (MBF). A board foot is a volume of wood—1" x 1' x 1' or 144 cubic inches. Board-foot quantities are generally calculated using nominal material dimensions—dressed sizes are ignored. Board foot per lineal foot of any stick of lumber can be calculated by dividing the nominal cross-sectional area by 12. As an example, 2,000 lineal feet of 2 x 12 equates to 4 MBF by dividing the nominal area, 2 x 12, by 12, which equals 2, and multiplying that by 2,000 to give 4,000 board feet. This simple rule applies to all nominal dimensioned lumber.
- Waste is an issue of concern at the quantity takeoff for any area of construction. Framing lumber is sold in even foot lengths, i.e., 8', 10', 12', 14', 16', and depending on spans, wall heights, and the grade of lumber, waste is inevitable. A rule of thumb for lumber waste is 5–10% depending on material quality and the complexity of the framing.
- Wood in various forms and shapes is used in many projects, even where the main structural framing is steel, concrete, or masonry. Plywood as a back-up partition material and 2x boards used as blocking and cant strips around roof edges are two common examples. The estimator should ensure that the costs of all wood materials are included in the final estimate.

06 20 00 Finish Carpentry

- It is necessary to consider the grade of workmanship when estimating labor costs for erecting millwork and an interior finish. In practice, there are three grades: premium, custom, and economy. The RSMeans daily output for base and case moldings is in the range of 200 to 250 L.F. per carpenter per day. This is appropriate for most average custom-grade projects. For premium projects, an adjustment to productivity of 25–50% should be made, depending on the complexity of the job.

Reference Numbers

Reference numbers are shown at the beginning of some major classifications. These numbers refer to related items in the Reference Section. The reference information may be an estimating procedure, an alternate pricing method, or technical information.

Note: Not all subdivisions listed here necessarily appear. ■

06 05 05 – Selective Demolition for Wood, Plastics, and Composites

06 05 05.10 Selective Demolition Wood Framing		Crew	Daily Output	Labor-Hours	Unit	Material	2019 Bare Costs Labor	Equipment	Total	Total Incl O&P
0010	SELECTIVE DEMOLITION WOOD FRAMING R024119-10									
0100	Timber connector, nailed, small	1 Clab	96	.083	Ea.		2.53		2.53	4.18
0110	Medium		60	.133			4.05		4.05	6.70
0120	Large		48	.167			5.05		5.05	8.35
0130	Bolted, small		48	.167			5.05		5.05	8.35
0140	Medium		32	.250			7.60		7.60	12.55
0150	Large		24	.333			10.10		10.10	16.70
2958	Beams, 2" x 6"	2 Clab	1100	.015	L.F.		.44		.44	.73
2960	2" x 8"		825	.019			.59		.59	.97
2965	2" x 10"		665	.024			.73		.73	1.21
2970	2" x 12"		550	.029			.88		.88	1.46
2972	2" x 14"		470	.034			1.03		1.03	1.71
2975	4" x 8"	B-1	413	.058			1.80		1.80	2.98
2980	4" x 10"		330	.073			2.26		2.26	3.73
2985	4" x 12"		275	.087			2.71		2.71	4.47
3000	6" x 8"		275	.087			2.71		2.71	4.47
3040	6" x 10"		220	.109			3.38		3.38	5.60
3080	6" x 12"		185	.130			4.02		4.02	6.65
3120	8" x 12"		140	.171			5.30		5.30	8.80
3160	10" x 12"		110	.218			6.75		6.75	11.20
3162	Alternate pricing method		1.10	21.818	M.B.F.		675		675	1,125
3170	Blocking, in 16" OC wall framing, 2" x 4"	1 Clab	600	.013	L.F.		.40		.40	.67
3172	2" x 6"		400	.020			.61		.61	1
3174	In 24" OC wall framing, 2" x 4"		600	.013			.40		.40	.67
3176	2" x 6"		400	.020			.61		.61	1
3178	Alt method, wood blocking removal from wood framing		.40	20	M.B.F.		605		605	1,000
3179	Wood blocking removal from steel framing		.36	22.222	"		675		675	1,125
3180	Bracing, let in, 1" x 3", studs 16" OC		1050	.008	L.F.		.23		.23	.38
3181	Studs 24" OC		1080	.007			.22		.22	.37
3182	1" x 4", studs 16" OC		1050	.008			.23		.23	.38
3183	Studs 24" OC		1080	.007			.22		.22	.37
3184	1" x 6", studs 16" OC		1050	.008			.23		.23	.38
3185	Studs 24" OC		1080	.007			.22		.22	.37
3186	2" x 3", studs 16" OC		800	.010			.30		.30	.50
3187	Studs 24" OC		830	.010			.29		.29	.48
3188	2" x 4", studs 16" OC		800	.010			.30		.30	.50
3189	Studs 24" OC		830	.010			.29		.29	.48
3190	2" x 6", studs 16" OC		800	.010			.30		.30	.50
3191	Studs 24" OC		830	.010			.29		.29	.48
3192	2" x 8", studs 16" OC		800	.010			.30		.30	.50
3193	Studs 24" OC		830	.010			.29		.29	.48
3194	"T" shaped metal bracing, studs at 16" OC		1060	.008			.23		.23	.38
3195	Studs at 24" OC		1200	.007			.20		.20	.33
3196	Metal straps, studs at 16" OC		1200	.007			.20		.20	.33
3197	Studs at 24" OC		1240	.006			.20		.20	.32
3200	Columns, round, 8' to 14' tall		40	.200	Ea.		6.05		6.05	10.05
3202	Dimensional lumber sizes	2 Clab	1.10	14.545	M.B.F.		440		440	730
3250	Blocking, between joists	1 Clab	320	.025	Ea.		.76		.76	1.25
3252	Bridging, metal strap, between joists		320	.025	Pr.		.76		.76	1.25
3254	Wood, between joists		320	.025	"		.76		.76	1.25
3260	Door buck, studs, header & access., 8' high 2" x 4" wall, 3' wide		32	.250	Ea.		7.60		7.60	12.55
3261	4' wide		32	.250			7.60		7.60	12.55
3262	5' wide		32	.250			7.60		7.60	12.55

06 05 05 – Selective Demolition for Wood, Plastics, and Composites

06 05 05.10 Selective Demolition Wood Framing		Crew	Daily Output	Labor-Hours	Unit	Material	2019 Bare Costs Labor	Equipment	Total	Total Incl O&P
3263	6′ wide	1 Clab	32	.250	Ea.		7.60		7.60	12.55
3264	8′ wide		30	.267			8.10		8.10	13.35
3265	10′ wide		30	.267			8.10		8.10	13.35
3266	12′ wide		30	.267			8.10		8.10	13.35
3267	2″ x 6″ wall, 3′ wide		32	.250			7.60		7.60	12.55
3268	4′ wide		32	.250			7.60		7.60	12.55
3269	5′ wide		32	.250			7.60		7.60	12.55
3270	6′ wide		32	.250			7.60		7.60	12.55
3271	8′ wide		30	.267			8.10		8.10	13.35
3272	10′ wide		30	.267			8.10		8.10	13.35
3273	12′ wide		30	.267			8.10		8.10	13.35
3274	Window buck, studs, header & access., 8′ high 2″ x 4″ wall, 2′ wide		24	.333			10.10		10.10	16.70
3275	3′ wide		24	.333			10.10		10.10	16.70
3276	4′ wide		24	.333			10.10		10.10	16.70
3277	5′ wide		24	.333			10.10		10.10	16.70
3278	6′ wide		24	.333			10.10		10.10	16.70
3279	7′ wide		24	.333			10.10		10.10	16.70
3280	8′ wide		22	.364			11.05		11.05	18.25
3281	10′ wide		22	.364			11.05		11.05	18.25
3282	12′ wide		22	.364			11.05		11.05	18.25
3283	2″ x 6″ wall, 2′ wide		24	.333			10.10		10.10	16.70
3284	3′ wide		24	.333			10.10		10.10	16.70
3285	4′ wide		24	.333			10.10		10.10	16.70
3286	5′ wide		24	.333			10.10		10.10	16.70
3287	6′ wide		24	.333			10.10		10.10	16.70
3288	7′ wide		24	.333			10.10		10.10	16.70
3289	8′ wide		22	.364			11.05		11.05	18.25
3290	10′ wide		22	.364			11.05		11.05	18.25
3291	12′ wide		22	.364			11.05		11.05	18.25
3360	Deck or porch decking		825	.010	L.F.		.29		.29	.49
3400	Fascia boards, 1″ x 6″		500	.016			.49		.49	.80
3440	1″ x 8″		450	.018			.54		.54	.89
3480	1″ x 10″		400	.020			.61		.61	1
3490	2″ x 6″		450	.018			.54		.54	.89
3500	2″ x 8″		400	.020			.61		.61	1
3510	2″ x 10″		350	.023			.69		.69	1.15
3610	Furring, on wood walls or ceiling		4000	.002	S.F.		.06		.06	.10
3620	On masonry or concrete walls or ceiling		1200	.007	"		.20		.20	.33
3800	Headers over openings, 2 @ 2″ x 6″		110	.073	L.F.		2.21		2.21	3.65
3840	2 @ 2″ x 8″		100	.080			2.43		2.43	4.01
3880	2 @ 2″ x 10″		90	.089			2.70		2.70	4.46
3885	Alternate pricing method		.26	30.651	M.B.F.		930		930	1,525
3920	Joists, 1″ x 4″		1250	.006	L.F.		.19		.19	.32
3930	1″ x 6″		1135	.007			.21		.21	.35
3940	1″ x 8″		1000	.008			.24		.24	.40
3950	1″ x 10″		895	.009			.27		.27	.45
3960	1″ x 12″		765	.010			.32		.32	.52
4200	2″ x 4″	2 Clab	1000	.016			.49		.49	.80
4230	2″ x 6″		970	.016			.50		.50	.83
4240	2″ x 8″		940	.017			.52		.52	.85
4250	2″ x 10″		910	.018			.53		.53	.88
4280	2″ x 12″		880	.018			.55		.55	.91
4281	2″ x 14″		850	.019			.57		.57	.94

	06 05 05.10 Selective Demolition Wood Framing	Crew	Daily Output	Labor-Hours	Unit	2019 Bare Costs Material	Labor	Equipment	Total	Total Incl O&P
4282	Composite joists, 9-1/2"	2 Clab	960	.017	L.F.		.51		.51	.84
4283	11-7/8"		930	.017			.52		.52	.86
4284	14"		897	.018			.54		.54	.89
4285	16"		865	.019			.56		.56	.93
4290	Wood joists, alternate pricing method		1.50	10.667	M.B.F.		325		325	535
4500	Open web joist, 12" deep		500	.032	L.F.		.97		.97	1.60
4505	14" deep		475	.034			1.02		1.02	1.69
4510	16" deep		450	.036			1.08		1.08	1.78
4520	18" deep		425	.038			1.14		1.14	1.89
4530	24" deep		400	.040			1.21		1.21	2.01
4550	Ledger strips, 1" x 2"	1 Clab	1200	.007			.20		.20	.33
4560	1" x 3"		1200	.007			.20		.20	.33
4570	1" x 4"		1200	.007			.20		.20	.33
4580	2" x 2"		1100	.007			.22		.22	.36
4590	2" x 4"		1000	.008			.24		.24	.40
4600	2" x 6"		1000	.008			.24		.24	.40
4601	2" x 8 or 2" x 10"		800	.010			.30		.30	.50
4602	4" x 6"		600	.013			.40		.40	.67
4604	4" x 8"		450	.018			.54		.54	.89
5400	Posts, 4" x 4"	2 Clab	800	.020			.61		.61	1
5405	4" x 6"		550	.029			.88		.88	1.46
5410	4" x 8"		440	.036			1.10		1.10	1.82
5425	4" x 10"		390	.041			1.25		1.25	2.06
5430	4" x 12"		350	.046			1.39		1.39	2.29
5440	6" x 6"		400	.040			1.21		1.21	2.01
5445	6" x 8"		350	.046			1.39		1.39	2.29
5450	6" x 10"		320	.050			1.52		1.52	2.51
5455	6" x 12"		290	.055			1.67		1.67	2.77
5480	8" x 8"		300	.053			1.62		1.62	2.67
5500	10" x 10"		240	.067			2.02		2.02	3.34
5660	T&G floor planks		2	8	M.B.F.		243		243	400
5682	Rafters, ordinary, 16" OC, 2" x 4"		880	.018	S.F.		.55		.55	.91
5683	2" x 6"		840	.019			.58		.58	.96
5684	2" x 8"		820	.020			.59		.59	.98
5685	2" x 10"		820	.020			.59		.59	.98
5686	2" x 12"		810	.020			.60		.60	.99
5687	24" OC, 2" x 4"		1170	.014			.42		.42	.69
5688	2" x 6"		1117	.014			.43		.43	.72
5689	2" x 8"		1091	.015			.45		.45	.74
5690	2" x 10"		1091	.015			.45		.45	.74
5691	2" x 12"		1077	.015			.45		.45	.75
5795	Rafters, ordinary, 2" x 4" (alternate method)		862	.019	L.F.		.56		.56	.93
5800	2" x 6" (alternate method)		850	.019			.57		.57	.94
5840	2" x 8" (alternate method)		837	.019			.58		.58	.96
5855	2" x 10" (alternate method)		825	.019			.59		.59	.97
5865	2" x 12" (alternate method)		812	.020			.60		.60	.99
5870	Sill plate, 2" x 4"	1 Clab	1170	.007			.21		.21	.34
5871	2" x 6"		780	.010			.31		.31	.51
5872	2" x 8"		586	.014			.41		.41	.68
5873	Alternate pricing method		.78	10.256	M.B.F.		310		310	515
5885	Ridge board, 1" x 4"	2 Clab	900	.018	L.F.		.54		.54	.89
5886	1" x 6"		875	.018			.56		.56	.92
5887	1" x 8"		850	.019			.57		.57	.94

06 05 05 – Selective Demolition for Wood, Plastics, and Composites

06 05 05.10 Selective Demolition Wood Framing		Crew	Daily Output	Labor-Hours	Unit	Material	2019 Bare Costs Labor	Equipment	Total	Total Incl O&P
5888	1" x 10"	2 Clab	825	.019	L.F.		.59		.59	.97
5889	1" x 12"		800	.020			.61		.61	1
5890	2" x 4"		900	.018			.54		.54	.89
5892	2" x 6"		875	.018			.56		.56	.92
5894	2" x 8"		850	.019			.57		.57	.94
5896	2" x 10"		825	.019			.59		.59	.97
5898	2" x 12"		800	.020			.61		.61	1
6050	Rafter tie, 1" x 4"		1250	.013			.39		.39	.64
6052	1" x 6"		1135	.014			.43		.43	.71
6054	2" x 4"		1000	.016			.49		.49	.80
6056	2" x 6"		970	.016			.50		.50	.83
6070	Sleepers, on concrete, 1" x 2"	1 Clab	4700	.002			.05		.05	.09
6075	1" x 3"		4000	.002			.06		.06	.10
6080	2" x 4"		3000	.003			.08		.08	.13
6085	2" x 6"		2600	.003			.09		.09	.15
6086	Sheathing from roof, 5/16"	2 Clab	1600	.010	S.F.		.30		.30	.50
6088	3/8"		1525	.010			.32		.32	.53
6090	1/2"		1400	.011			.35		.35	.57
6092	5/8"		1300	.012			.37		.37	.62
6094	3/4"		1200	.013			.40		.40	.67
6096	Board sheathing from roof		1400	.011			.35		.35	.57
6100	Sheathing, from walls, 1/4"		1200	.013			.40		.40	.67
6110	5/16"		1175	.014			.41		.41	.68
6120	3/8"		1150	.014			.42		.42	.70
6130	1/2"		1125	.014			.43		.43	.71
6140	5/8"		1100	.015			.44		.44	.73
6150	3/4"		1075	.015			.45		.45	.75
6152	Board sheathing from walls		1500	.011			.32		.32	.54
6158	Subfloor/roof deck, with boards		2200	.007			.22		.22	.36
6159	Subfloor/roof deck, with tongue & groove boards		2000	.008			.24		.24	.40
6160	Plywood, 1/2" thick		768	.021			.63		.63	1.04
6162	5/8" thick		760	.021			.64		.64	1.06
6164	3/4" thick		750	.021			.65		.65	1.07
6165	1-1/8" thick		720	.022			.67		.67	1.11
6166	Underlayment, particle board, 3/8" thick	1 Clab	780	.010			.31		.31	.51
6168	1/2" thick		768	.010			.32		.32	.52
6170	5/8" thick		760	.011			.32		.32	.53
6172	3/4" thick		750	.011			.32		.32	.54
6200	Stairs and stringers, straight run	2 Clab	40	.400	Riser		12.15		12.15	20
6240	With platforms, winders or curves	"	26	.615	"		18.70		18.70	31
6300	Components, tread	1 Clab	110	.073	Ea.		2.21		2.21	3.65
6320	Riser		80	.100	"		3.04		3.04	5
6390	Stringer, 2" x 10"		260	.031	L.F.		.93		.93	1.54
6400	2" x 12"		260	.031			.93		.93	1.54
6410	3" x 10"		250	.032			.97		.97	1.60
6420	3" x 12"		250	.032			.97		.97	1.60
6590	Wood studs, 2" x 3"	2 Clab	3076	.005			.16		.16	.26
6600	2" x 4"		2000	.008			.24		.24	.40
6640	2" x 6"		1600	.010			.30		.30	.50
6720	Wall framing, including studs, plates and blocking, 2" x 4"	1 Clab	600	.013	S.F.		.40		.40	.67
6740	2" x 6"		480	.017	"		.51		.51	.84
6750	Headers, 2" x 4"		1125	.007	L.F.		.22		.22	.36
6755	2" x 6"		1125	.007			.22		.22	.36

06 05 05 – Selective Demolition for Wood, Plastics, and Composites

06 05 05.10 Selective Demolition Wood Framing

	06 05 05.10 Selective Demolition Wood Framing	Crew	Daily Output	Labor-Hours	Unit	Material	2019 Bare Costs Labor	Equipment	Total	Total Incl O&P
6760	2" x 8"	1 Clab	1050	.008	L.F.		.23		.23	.38
6765	2" x 10"		1050	.008			.23		.23	.38
6770	2" x 12"		1000	.008			.24		.24	.40
6780	4" x 10"		525	.015			.46		.46	.76
6785	4" x 12"		500	.016			.49		.49	.80
6790	6" x 8"		560	.014			.43		.43	.72
6795	6" x 10"		525	.015			.46		.46	.76
6797	6" x 12"		500	.016			.49		.49	.80
7000	Trusses									
7050	12' span	2 Clab	74	.216	Ea.		6.55		6.55	10.85
7150	24' span	F-3	66	.606			21.50	7.15	28.65	43.50
7200	26' span		64	.625			22	7.35	29.35	45
7250	28' span		62	.645			23	7.60	30.60	46.50
7300	30' span		58	.690			24.50	8.15	32.65	49.50
7350	32' span		56	.714			25.50	8.40	33.90	51.50
7400	34' span		54	.741			26.50	8.75	35.25	53
7450	36' span		52	.769			27.50	9.05	36.55	55.50
8000	Soffit, T&G wood	1 Clab	520	.015	S.F.		.47		.47	.77
8010	Hardboard, vinyl or aluminum	"	640	.013			.38		.38	.63
8030	Plywood	2 Carp	315	.051			1.97		1.97	3.25
9500	See Section 02 41 19.19 for rubbish handling									

06 05 05.20 Selective Demolition Millwork and Trim

	06 05 05.20 Selective Demolition Millwork and Trim	Crew	Daily Output	Labor-Hours	Unit	Material	2019 Bare Costs Labor	Equipment	Total	Total Incl O&P
0010	**SELECTIVE DEMOLITION MILLWORK AND TRIM** R024119-10									
1000	Cabinets, wood, base cabinets, per L.F.	2 Clab	80	.200	L.F.		6.05		6.05	10.05
1020	Wall cabinets, per L.F.	"	80	.200	"		6.05		6.05	10.05
1060	Remove and reset, base cabinets	2 Carp	18	.889	Ea.		34.50		34.50	57
1070	Wall cabinets		20	.800			31		31	51
1072	Oven cabinet, 7' high		11	1.455			56.50		56.50	93
1074	Cabinet door, up to 2' high	1 Clab	66	.121			3.68		3.68	6.10
1076	2' - 4' high	"	46	.174			5.30		5.30	8.70
1100	Steel, painted, base cabinets	2 Clab	60	.267	L.F.		8.10		8.10	13.35
1120	Wall cabinets		60	.267	"		8.10		8.10	13.35
1200	Casework, large area		320	.050	S.F.		1.52		1.52	2.51
1220	Selective		200	.080	"		2.43		2.43	4.01
1500	Counter top, straight runs		200	.080	L.F.		2.43		2.43	4.01
1510	L, U or C shapes		120	.133	"		4.05		4.05	6.70
2000	Paneling, 4' x 8' sheets		2000	.008	S.F.		.24		.24	.40
2100	Boards, 1" x 4"		700	.023			.69		.69	1.15
2120	1" x 6"		750	.021			.65		.65	1.07
2140	1" x 8"		800	.020			.61		.61	1
3000	Trim, baseboard, to 6" wide		1200	.013	L.F.		.40		.40	.67
3040	Greater than 6" and up to 12" wide		1000	.016			.49		.49	.80
3080	Remove and reset, minimum	2 Carp	400	.040			1.55		1.55	2.56
3090	Maximum	"	300	.053			2.07		2.07	3.42
3100	Ceiling trim	2 Clab	1000	.016			.49		.49	.80
3120	Chair rail		1200	.013			.40		.40	.67
3140	Railings with balusters		240	.067			2.02		2.02	3.34
3160	Wainscoting		700	.023	S.F.		.69		.69	1.15
4000	Curtain rod	1 Clab	80	.100	L.F.		3.04		3.04	5

06 05 23 – Wood, Plastic, and Composite Fastenings

06 05 23.10 Nails		Crew	Daily Output	Labor-Hours	Unit	Material	Labor	Equipment	Total	Total Incl O&P
						2019 Bare Costs				
0010	**NAILS**, material only, based upon 50# box purchase									
0020	Copper nails, plain				Lb.	11.45			11.45	12.60
0400	Stainless steel, plain					8.70			8.70	9.55
0500	Box, 3d to 20d, bright					1.46			1.46	1.61
0520	Galvanized					2.39			2.39	2.63
0600	Common, 3d to 60d, plain					1.16			1.16	1.28
0700	Galvanized					2.40			2.40	2.64
0800	Aluminum					11			11	12.10
1000	Annular or spiral thread, 4d to 60d, plain					3.28			3.28	3.61
1200	Galvanized					3.07			3.07	3.38
1400	Drywall nails, plain					1.82			1.82	2
1600	Galvanized					1.88			1.88	2.07
1800	Finish nails, 4d to 10d, plain					1.26			1.26	1.39
2000	Galvanized					1.88			1.88	2.07
2100	Aluminum					8.10			8.10	8.90
2300	Flooring nails, hardened steel, 2d to 10d, plain					3.63			3.63	3.99
2400	Galvanized					4.03			4.03	4.43
2500	Gypsum lath nails, 1-1/8", 13 ga. flathead, blued					3.27			3.27	3.60
2600	Masonry nails, hardened steel, 3/4" to 3" long, plain					2.39			2.39	2.63
2700	Galvanized					3.90			3.90	4.29
2900	Roofing nails, threaded, galvanized					2.78			2.78	3.06
3100	Aluminum					7.55			7.55	8.30
3300	Compressed lead head, threaded, galvanized					3.04			3.04	3.34
3600	Siding nails, plain shank, galvanized					2.54			2.54	2.79
3800	Aluminum					5.85			5.85	6.40
5000	Add to prices above for cement coating					.14			.14	.15
5200	Zinc or tin plating					.26			.26	.29
5500	Vinyl coated sinkers, 8d to 16d				↓	2.63			2.63	2.89

06 05 23.50 Wood Screws		Crew	Daily Output	Labor-Hours	Unit	Material	Labor	Equipment	Total	Total Incl O&P
0010	**WOOD SCREWS**									
0020	#8, 1" long, steel				C	4.71			4.71	5.20
0100	Brass					12.20			12.20	13.40
0200	#8, 2" long, steel					8.15			8.15	9
0300	Brass					25			25	27
0400	#10, 1" long, steel					3.38			3.38	3.72
0500	Brass					15.65			15.65	17.25
0600	#10, 2" long, steel					5.40			5.40	5.95
0700	Brass					27			27	29.50
0800	#10, 3" long, steel					9.05			9.05	9.95
1000	#12, 2" long, steel					7.60			7.60	8.35
1100	Brass					35			35	38.50
1500	#12, 3" long, steel					11.25			11.25	12.40
2000	#12, 4" long, steel				↓	25			25	27.50

06 05 23.60 Timber Connectors		Crew	Daily Output	Labor-Hours	Unit	Material	Labor	Equipment	Total	Total Incl O&P
0010	**TIMBER CONNECTORS**									
0020	Add up cost of each part for total cost of connection									
0100	Connector plates, steel, with bolts, straight	2 Carp	75	.213	Ea.	28	8.25		36.25	44
0110	Tee, 7 ga.		50	.320		40	12.40		52.40	64.50
0120	T- Strap, 14 ga., 12" x 8" x 2"		50	.320		40	12.40		52.40	64.50
0150	Anchor plates, 7 ga., 9" x 7"	↓	75	.213		28	8.25		36.25	44
0200	Bolts, machine, sq. hd. with nut & washer, 1/2" diameter, 4" long	1 Carp	140	.057		.92	2.21		3.13	4.67
0300	7-1/2" long	↓	130	.062	↓	1.71	2.38		4.09	5.80

06 05 23 – Wood, Plastic, and Composite Fastenings

06 05 23.60 Timber Connectors		Crew	Daily Output	Labor-Hours	Unit	2019 Bare Costs Material	Labor	Equipment	Total	Total Incl O&P
0500	3/4" diameter, 7-1/2" long	1 Carp	130	.062	Ea.	3.12	2.38		5.50	7.35
0610	Machine bolts, w/nut, washer, 3/4" diameter, 15" L, HD's & beam hangers		95	.084		5.80	3.26		9.06	11.80
0720	Machine bolts, sq. hd. w/nut & wash		150	.053	Lb.	3.79	2.07		5.86	7.60
0800	Drilling bolt holes in timber, 1/2" diameter		450	.018	Inch		.69		.69	1.14
0900	1" diameter		350	.023	"		.89		.89	1.46
1100	Framing anchor, angle, 3" x 3" x 1-1/2", 12 ga.		175	.046	Ea.	2.44	1.77		4.21	5.60
1150	Framing anchors, 18 ga., 4-1/2" x 2-3/4"		175	.046		2.44	1.77		4.21	5.60
1160	Framing anchors, 18 ga., 4-1/2" x 3"		175	.046		2.44	1.77		4.21	5.60
1170	Clip anchors plates, 18 ga., 12" x 1-1/8"		175	.046		2.44	1.77		4.21	5.60
1250	Holdowns, 3 ga. base, 10 ga. body		8	1		40.50	39		79.50	109
1260	Holdowns, 7 ga. 11-1/16" x 3-1/4"		8	1		40.50	39		79.50	109
1270	Holdowns, 7 ga. 14-3/8" x 3-1/8"		8	1		40.50	39		79.50	109
1275	Holdowns, 12 ga. 8" x 2-1/2"		8	1		40.50	39		79.50	109
1300	Joist and beam hangers, 18 ga. galv., for 2" x 4" joist		175	.046		.77	1.77		2.54	3.78
1400	2" x 6" to 2" x 10" joist		165	.048		1.46	1.88		3.34	4.72
1600	16 ga. galv., 3" x 6" to 3" x 10" joist		160	.050		2.91	1.94		4.85	6.40
1700	3" x 10" to 3" x 14" joist		160	.050		4.31	1.94		6.25	7.95
1800	4" x 6" to 4" x 10" joist		155	.052		3.36	2		5.36	7
1900	4" x 10" to 4" x 14" joist		155	.052		4.81	2		6.81	8.60
2000	Two-2" x 6" to two-2" x 10" joists		150	.053		3.89	2.07		5.96	7.70
2100	Two-2" x 10" to two-2" x 14" joists		150	.053		4.35	2.07		6.42	8.20
2300	3/16" thick, 6" x 8" joist		145	.055		70	2.14		72.14	80.50
2400	6" x 10" joist		140	.057		73	2.21		75.21	83.50
2500	6" x 12" joist		135	.059		74	2.30		76.30	85.50
2700	1/4" thick, 6" x 14" joist		130	.062		77	2.38		79.38	89
2900	Plywood clips, extruded aluminum H clip, for 3/4" panels					.23			.23	.25
3000	Galvanized 18 ga. back-up clip					.18			.18	.20
3200	Post framing, 16 ga. galv. for 4" x 4" base, 2 piece	1 Carp	130	.062		15.35	2.38		17.73	21
3300	Cap		130	.062		24.50	2.38		26.88	31
3500	Rafter anchors, 18 ga. galv., 1-1/2" wide, 5-1/4" long		145	.055		.45	2.14		2.59	4.03
3600	10-3/4" long		145	.055		1.38	2.14		3.52	5.05
3800	Shear plates, 2-5/8" diameter		120	.067		2.87	2.58		5.45	7.45
3900	4" diameter		115	.070		2.83	2.70		5.53	7.55
4000	Sill anchors, embedded in concrete or block, 25-1/2" long		115	.070		13.65	2.70		16.35	19.45
4100	Spike grids, 3" x 6"		120	.067		1.27	2.58		3.85	5.65
4400	Split rings, 2-1/2" diameter		120	.067		2.83	2.58		5.41	7.40
4500	4" diameter		110	.073		3.27	2.82		6.09	8.25
4550	Tie plate, 20 ga., 7" x 3-1/8"		110	.073		3.27	2.82		6.09	8.25
4560	5" x 4-1/8"		110	.073		3.27	2.82		6.09	8.25
4575	Twist straps, 18 ga., 12" x 1-1/4"		110	.073		3.27	2.82		6.09	8.25
4580	16" x 1-1/4"		110	.073		3.27	2.82		6.09	8.25
4600	Strap ties, 20 ga., 2-1/16" wide, 12-13/16" long		180	.044		.89	1.72		2.61	3.83
4700	16 ga., 1-3/8" wide, 12" long		180	.044		.89	1.72		2.61	3.83
4800	1-1/4" wide, 21-5/8" long		160	.050		2.73	1.94		4.67	6.20
5000	Toothed rings, 2-5/8" or 4" diameter		90	.089		2.39	3.44		5.83	8.35
5200	Truss plates, nailed, 20 ga., up to 32' span		17	.471	Truss	15.15	18.25		33.40	46.50
5400	Washers, 2" x 2" x 1/8"				Ea.	.48			.48	.53
5500	3" x 3" x 3/16"				"	1.29			1.29	1.42

06 05 23.80 Metal Bracing		Crew	Daily Output	Labor-Hours	Unit	Material	Labor	Equipment	Total	Total Incl O&P
0010	**METAL BRACING** R051223-20									
0302	Let-in, "T" shaped, 22 ga. galv. steel, studs at 16" OC	1 Carp	580	.014	L.F.	.82	.53		1.35	1.78
0402	Studs at 24" OC		600	.013		.82	.52		1.34	1.75

06 05 Common Work Results for Wood, Plastics, and Composites

06 05 23 – Wood, Plastic, and Composite Fastenings

06 05 23.80 Metal Bracing		Crew	Daily Output	Labor-Hours	Unit	Material	2019 Bare Costs Labor	Equipment	Total	Total Incl O&P
0502	Steel straps, 16 ga. galv. steel, studs at 16" OC	1 Carp	600	.013	L.F.	1.06	.52		1.58	2.02
0602	Studs at 24" OC	↓	620	.013	↓	1.06	.50		1.56	2

06 11 Wood Framing

06 11 10 – Framing with Dimensional, Engineered or Composite Lumber

06 11 10.01 Forest Stewardship Council Certification		Crew	Daily Output	Labor-Hours	Unit	Material	Labor	Equipment	Total	Total Incl O&P
0010	**FOREST STEWARDSHIP COUNCIL CERTIFICATION**									
0020	For Forest Stewardship Council (FSC) cert dimension lumber, add [G]					65%				

06 11 10.02 Blocking		Crew	Daily Output	Labor-Hours	Unit	Material	Labor	Equipment	Total	Total Incl O&P
0010	**BLOCKING**									
1790	Bolted to concrete									
1798	Ledger board, 2" x 4"	1 Carp	180	.044	L.F.	4.82	1.72		6.54	8.15
1800	2" x 6"		160	.050		5.10	1.94		7.04	8.80
1810	4" x 6"		140	.057		8.90	2.21		11.11	13.45
1820	4" x 8"	↓	120	.067	↓	10.05	2.58		12.63	15.30
1950	Miscellaneous, to wood construction									
2000	2" x 4"	1 Carp	250	.032	L.F.	.43	1.24		1.67	2.52
2005	Pneumatic nailed		305	.026		.43	1.02		1.45	2.15
2050	2" x 6"		222	.036		.72	1.40		2.12	3.10
2055	Pneumatic nailed		271	.030		.73	1.14		1.87	2.69
2100	2" x 8"		200	.040		.97	1.55		2.52	3.63
2105	Pneumatic nailed		244	.033		.98	1.27		2.25	3.18
2150	2" x 10"		178	.045		1.45	1.74		3.19	4.48
2155	Pneumatic nailed		217	.037		1.47	1.43		2.90	3.97
2200	2" x 12"		151	.053		1.83	2.05		3.88	5.40
2205	Pneumatic nailed	↓	185	.043	↓	1.85	1.68		3.53	4.80
2300	To steel construction									
2320	2" x 4"	1 Carp	208	.038	L.F.	.43	1.49		1.92	2.93
2340	2" x 6"		180	.044		.72	1.72		2.44	3.65
2360	2" x 8"		158	.051		.97	1.96		2.93	4.31
2380	2" x 10"		136	.059		1.45	2.28		3.73	5.35
2400	2" x 12"	↓	109	.073	↓	1.83	2.84		4.67	6.70

06 11 10.04 Wood Bracing		Crew	Daily Output	Labor-Hours	Unit	Material	Labor	Equipment	Total	Total Incl O&P
0010	**WOOD BRACING**									
0012	Let-in, with 1" x 6" boards, studs @ 16" OC	1 Carp	150	.053	L.F.	.79	2.07		2.86	4.29
0202	Studs @ 24" OC	"	230	.035	"	.79	1.35		2.14	3.10

06 11 10.06 Bridging		Crew	Daily Output	Labor-Hours	Unit	Material	Labor	Equipment	Total	Total Incl O&P
0010	**BRIDGING**									
0012	Wood, for joists 16" OC, 1" x 3"	1 Carp	130	.062	Pr.	.71	2.38		3.09	4.72
0017	Pneumatic nailed		170	.047		.79	1.82		2.61	3.88
0102	2" x 3" bridging		130	.062		.72	2.38		3.10	4.73
0107	Pneumatic nailed		170	.047		.75	1.82		2.57	3.84
0302	Steel, galvanized, 18 ga., for 2" x 10" joists at 12" OC		130	.062		1.86	2.38		4.24	6
0352	16" OC		135	.059		1.75	2.30		4.05	5.75
0402	24" OC		140	.057		2.57	2.21		4.78	6.50
0602	For 2" x 14" joists at 16" OC		130	.062		1.88	2.38		4.26	6
0902	Compression type, 16" OC, 2" x 8" joists		200	.040		1.29	1.55		2.84	3.98
1002	2" x 12" joists	↓	200	.040	↓	1.28	1.55		2.83	3.97

06 11 10 – Framing with Dimensional, Engineered or Composite Lumber

06 11 10.10 Beam and Girder Framing		Crew	Daily Output	Labor-Hours	Unit	2019 Bare Costs Material	Labor	Equipment	Total	Total Incl O&P
0010	**BEAM AND GIRDER FRAMING**									
1000	Single, 2" x 6"	2 Carp	700	.023	L.F.	.72	.89		1.61	2.25
1005	Pneumatic nailed		812	.020		.73	.76		1.49	2.06
1020	2" x 8"		650	.025		.97	.95		1.92	2.65
1025	Pneumatic nailed		754	.021		.98	.82		1.80	2.44
1040	2" x 10"		600	.027		1.45	1.03		2.48	3.31
1045	Pneumatic nailed		696	.023		1.47	.89		2.36	3.08
1060	2" x 12"		550	.029		1.83	1.13		2.96	3.87
1065	Pneumatic nailed		638	.025		1.85	.97		2.82	3.64
1080	2" x 14"		500	.032		2.32	1.24		3.56	4.60
1085	Pneumatic nailed		580	.028		2.34	1.07		3.41	4.34
1100	3" x 8"		550	.029		2.95	1.13		4.08	5.10
1120	3" x 10"		500	.032		3.70	1.24		4.94	6.10
1140	3" x 12"		450	.036		4.37	1.38		5.75	7.10
1160	3" x 14"		400	.040		5.05	1.55		6.60	8.10
1170	4" x 6"	F-3	1100	.036		3.15	1.29	.43	4.87	6.05
1180	4" x 8"		1000	.040		4.29	1.42	.47	6.18	7.60
1200	4" x 10"		950	.042		5.05	1.49	.50	7.04	8.60
1220	4" x 12"		900	.044		5.55	1.58	.52	7.65	9.30
1240	4" x 14"		850	.047		6.55	1.67	.55	8.77	10.60
1250	6" x 8"		525	.076		8.05	2.70	.90	11.65	14.30
1260	6" x 10"		500	.080		7	2.84	.94	10.78	13.45
1290	8" x 12"		300	.133		16.85	4.73	1.57	23.15	28
1300	Treated, single, 2" x 4"	2 Carp	700	.023		.61	.89		1.50	2.13
1320	2" x 6"		700	.023		.77	.89		1.66	2.30
1340	2" x 8"		650	.025		1.14	.95		2.09	2.83
1360	2" x 10"		600	.027		1.42	1.03		2.45	3.27
1380	2" x 12"		550	.029		2.10	1.13		3.23	4.17
1400	2" x 14"		500	.032		2.60	1.24		3.84	4.90
1420	3" x 8"		550	.029		3.48	1.13		4.61	5.70
1440	3" x 10"		500	.032		4.61	1.24		5.85	7.10
1460	3" x 12"		450	.036		.12	1.38		1.50	2.41
1480	3" x 14"		400	.040		6.30	1.55		7.85	9.50
1500	4" x 8"	F-3	1000	.040		4.01	1.42	.47	5.90	7.30
1520	4" x 10"		950	.042		5.80	1.49	.50	7.79	9.45
1540	4" x 12"		900	.044		6.85	1.58	.52	8.95	10.75
1560	4" x 14"		850	.047		10.60	1.67	.55	12.82	15.05
2000	Double, 2" x 6"	2 Carp	625	.026		1.44	.99		2.43	3.23
2005	Pneumatic nailed		725	.022		1.46	.86		2.32	3.02
2020	2" x 8"		575	.028		1.95	1.08		3.03	3.92
2025	Pneumatic nailed		667	.024		1.97	.93		2.90	3.70
2040	2" x 10"		550	.029		2.90	1.13		4.03	5.05
2045	Pneumatic nailed		638	.025		2.93	.97		3.90	4.84
2060	2" x 12"		525	.030		3.66	1.18		4.84	6
2065	Pneumatic nailed		610	.026		3.69	1.02		4.71	5.75
2080	2" x 14"		475	.034		4.64	1.31		5.95	7.25
2085	Pneumatic nailed		551	.029		4.68	1.13		5.81	7
3000	Triple, 2" x 6"		550	.029		2.16	1.13		3.29	4.24
3005	Pneumatic nailed		638	.025		2.19	.97		3.16	4.02
3020	2" x 8"		525	.030		2.92	1.18		4.10	5.15
3025	Pneumatic nailed		609	.026		2.95	1.02		3.97	4.93
3040	2" x 10"		500	.032		4.35	1.24		5.59	6.85

06 11 Wood Framing

06 11 10 – Framing with Dimensional, Engineered or Composite Lumber

06 11 10.10 Beam and Girder Framing		Crew	Daily Output	Labor-Hours	Unit	2019 Bare Costs Material	Labor	Equipment	Total	Total Incl O&P
3045	Pneumatic nailed	2 Carp	580	.028	L.F.	4.40	1.07		5.47	6.60
3060	2″ x 12″		475	.034		5.50	1.31		6.81	8.20
3065	Pneumatic nailed		551	.029		5.55	1.13		6.68	7.95
3080	2″ x 14″		450	.036		7.55	1.38		8.93	10.60
3085	Pneumatic nailed		522	.031		7	1.19		8.19	9.65

06 11 10.12 Ceiling Framing

Line	Description	Crew	Daily Output	Labor-Hours	Unit	Material	Labor	Equipment	Total	Total Incl O&P
0010	**CEILING FRAMING**									
6000	Suspended, 2″ x 3″	2 Carp	1000	.016	L.F.	.42	.62		1.04	1.48
6050	2″ x 4″		900	.018		.43	.69		1.12	1.61
6100	2″ x 6″		800	.020		.72	.78		1.50	2.07
6150	2″ x 8″		650	.025		.97	.95		1.92	2.65

06 11 10.14 Posts and Columns

Line	Description	Crew	Daily Output	Labor-Hours	Unit	Material	Labor	Equipment	Total	Total Incl O&P
0010	**POSTS AND COLUMNS**									
0100	4″ x 4″	2 Carp	390	.041	L.F.	1.86	1.59		3.45	4.68
0150	4″ x 6″		275	.058		3.15	2.25		5.40	7.20
0200	4″ x 8″		220	.073		4.29	2.82		7.11	9.40
0250	6″ x 6″		215	.074		5.45	2.88		8.33	10.75
0300	6″ x 8″		175	.091		8.05	3.54		11.59	14.70
0350	6″ x 10″		150	.107		7	4.13		11.13	14.55

06 11 10.18 Joist Framing

Line	Description	Crew	Daily Output	Labor-Hours	Unit	Material	Labor	Equipment	Total	Total Incl O&P
0010	**JOIST FRAMING**									
2000	Joists, 2″ x 4″	2 Carp	1250	.013	L.F.	.43	.50		.93	1.29
2005	Pneumatic nailed		1438	.011		.43	.43		.86	1.18
2100	2″ x 6″		1250	.013		.72	.50		1.22	1.61
2105	Pneumatic nailed		1438	.011		.73	.43		1.16	1.51
2150	2″ x 8″		1100	.015		.97	.56		1.53	2
2155	Pneumatic nailed		1265	.013		.98	.49		1.47	1.89
2200	2″ x 10″		900	.018		1.45	.69		2.14	2.74
2205	Pneumatic nailed		1035	.015		1.47	.60		2.07	2.60
2250	2″ x 12″		875	.018		1.83	.71		2.54	3.18
2255	Pneumatic nailed		1006	.016		1.85	.62		2.47	3.05
2300	2″ x 14″		770	.021		2.32	.81		3.13	3.88
2305	Pneumatic nailed		886	.018		2.34	.70		3.04	3.73
2350	3″ x 6″		925	.017		2.09	.67		2.76	3.41
2400	3″ x 10″		780	.021		3.70	.79		4.49	5.40
2450	3″ x 12″		600	.027		4.37	1.03		5.40	6.50
2500	4″ x 6″		800	.020		3.15	.78		3.93	4.74
2550	4″ x 10″		600	.027		5.05	1.03		6.08	7.25
2600	4″ x 12″		450	.036		5.55	1.38		6.93	8.40
2605	Sister joist, 2″ x 6″		800	.020		.72	.78		1.50	2.07
2606	Pneumatic nailed		960	.017		.73	.65		1.38	1.87
3000	Composite wood joist 9-1/2″ deep		.90	17.778	M.L.F.	1,775	690		2,465	3,100
3010	11-1/2″ deep		.88	18.182		2,025	705		2,730	3,400
3020	14″ deep		.82	19.512		2,600	755		3,355	4,100
3030	16″ deep		.78	20.513		4,125	795		4,920	5,875
4000	Open web joist 12″ deep		.88	18.182		3,625	705		4,330	5,175
4002	Per linear foot		880	.018	L.F.	3.64	.70		4.34	5.15
4004	Treated, per linear foot		880	.018	″	4.58	.70		5.28	6.20
4010	14″ deep		.82	19.512	M.L.F.	3,875	755		4,630	5,525
4012	Per linear foot		820	.020	L.F.	3.88	.76		4.64	5.50
4014	Treated, per linear foot		820	.020	″	4.98	.76		5.74	6.75
4020	16″ deep		.78	20.513	M.L.F.	3,850	795		4,645	5,575

06 11 Wood Framing

06 11 10 – Framing with Dimensional, Engineered or Composite Lumber

06 11 10.18 Joist Framing		Crew	Daily Output	Labor-Hours	Unit	Material	2019 Bare Costs Labor	Equipment	Total	Total Incl O&P
4022	Per linear foot	2 Carp	780	.021	L.F.	3.86	.79		4.65	5.55
4024	Treated, per linear foot		780	.021	"	5.10	.79		5.89	6.95
4030	18" deep		.74	21.622	M.L.F.	4,200	840		5,040	5,975
4032	Per linear foot		740	.022	L.F.	4.19	.84		5.03	6
4034	Treated, per linear foot		740	.022	"	5.60	.84		6.44	7.55
6000	Composite rim joist, 1-1/4" x 9-1/2"		.90	17.778	M.L.F.	2,050	690		2,740	3,425
6010	1-1/4" x 11-1/2"		.88	18.182		2,300	705		3,005	3,700
6020	1-1/4" x 14-1/2"		.82	19.512		3,000	755		3,755	4,550
6030	1-1/4" x 16-1/2"		.78	20.513		2,875	795		3,670	4,475
06 11 10.24 Miscellaneous Framing										
0010	**MISCELLANEOUS FRAMING**									
2000	Firestops, 2" x 4"	2 Carp	780	.021	L.F.	.43	.79		1.22	1.78
2005	Pneumatic nailed R061110-30		952	.017		.43	.65		1.08	1.55
2100	2" x 6"		600	.027		.72	1.03		1.75	2.50
2105	Pneumatic nailed		732	.022		.73	.85		1.58	2.20
5000	Nailers, treated, wood construction, 2" x 4"		800	.020		.60	.78		1.38	1.94
5005	Pneumatic nailed		960	.017		.61	.65		1.26	1.74
5100	2" x 6"		750	.021		.76	.83		1.59	2.20
5105	Pneumatic nailed		900	.018		.77	.69		1.46	1.98
5120	2" x 8"		700	.023		1.12	.89		2.01	2.70
5125	Pneumatic nailed		840	.019		1.14	.74		1.88	2.47
5200	Steel construction, 2" x 4"		750	.021		.60	.83		1.43	2.03
5220	2" x 6"		700	.023		.76	.89		1.65	2.30
5240	2" x 8"		650	.025		1.13	.95		2.08	2.82
7000	Rough bucks, treated, for doors or windows, 2" x 6"		400	.040		.76	1.55		2.31	3.39
7005	Pneumatic nailed		480	.033		.77	1.29		2.06	2.97
7100	2" x 8"		380	.042		1.12	1.63		2.75	3.94
7105	Pneumatic nailed		456	.035		1.14	1.36		2.50	3.50
8000	Stair stringers, 2" x 10"		130	.123		1.45	4.77		6.22	9.50
8100	2" x 12"		130	.123		1.83	4.77		6.60	9.90
8150	3" x 10"		125	.128		3.70	4.96		8.66	12.25
8200	3" x 12"		125	.128		4.37	4.96		9.33	13
06 11 10.26 Partitions										
0010	**PARTITIONS**									
0020	Single bottom and double top plate, no waste, std. & better lumber									
0180	2" x 4" studs, 8' high, studs 12" OC	2 Carp	80	.200	L.F.	4.69	7.75		12.44	17.95
0185	12" OC, pneumatic nailed		96	.167		4.75	6.45		11.20	15.90
0200	16" OC		100	.160		3.84	6.20		10.04	14.45
0205	16" OC, pneumatic nailed		120	.133		3.88	5.15		9.03	12.80
0300	24" OC		125	.128		2.98	4.96		7.94	11.50
0305	24" OC, pneumatic nailed		150	.107		3.02	4.13		7.15	10.15
0380	10' high, studs 12" OC		80	.200		5.55	7.75		13.30	18.90
0385	12" OC, pneumatic nailed		96	.167		5.60	6.45		12.05	16.85
0400	16" OC		100	.160		4.48	6.20		10.68	15.15
0405	16" OC, pneumatic nailed		120	.133		4.53	5.15		9.68	13.55
0500	24" OC		125	.128		3.41	4.96		8.37	11.95
0505	24" OC, pneumatic nailed		150	.107		3.45	4.13		7.58	10.65
0580	12' high, studs 12" OC		65	.246		6.40	9.55		15.95	23
0585	12" OC, pneumatic nailed		78	.205		6.45	7.95		14.40	20.50
0600	16" OC		80	.200		5.10	7.75		12.85	18.45
0605	16" OC, pneumatic nailed		96	.167		5.20	6.45		11.65	16.40
0700	24" OC		100	.160		3.84	6.20		10.04	14.45

06 11 10 – Framing with Dimensional, Engineered or Composite Lumber

06 11 10.26 Partitions		Crew	Daily Output	Labor-Hours	Unit	Material	2019 Bare Costs Labor	Equipment	Total	Total Incl O&P
0705	24" OC, pneumatic nailed	2 Carp	120	.133	L.F.	3.88	5.15		9.03	12.80
0780	2" x 6" studs, 8' high, studs 12" OC		70	.229		7.95	8.85		16.80	23.50
0785	12" OC, pneumatic nailed		84	.190		8.05	7.40		15.45	21
0800	16" OC		90	.178		6.50	6.90		13.40	18.55
0805	16" OC, pneumatic nailed		108	.148		6.55	5.75		12.30	16.70
0900	24" OC		115	.139		5.05	5.40		10.45	14.45
0905	24" OC, pneumatic nailed		138	.116		5.10	4.49		9.59	13.05
0980	10' high, studs 12" OC		70	.229		9.40	8.85		18.25	25
0985	12" OC, pneumatic nailed		84	.190		9.50	7.40		16.90	22.50
1000	16" OC		90	.178		7.60	6.90		14.50	19.75
1005	16" OC, pneumatic nailed		108	.148		7.65	5.75		13.40	17.95
1100	24" OC		115	.139		5.75	5.40		11.15	15.25
1105	24" OC, pneumatic nailed		138	.116		5.85	4.49		10.34	13.85
1180	12' high, studs 12" OC		55	.291		10.80	11.25		22.05	30.50
1185	12" OC, pneumatic nailed		66	.242		10.95	9.40		20.35	27.50
1200	16" OC		70	.229		8.65	8.85		17.50	24
1205	16" OC, pneumatic nailed		84	.190		8.75	7.40		16.15	22
1300	24" OC		90	.178		6.50	6.90		13.40	18.55
1305	24" OC, pneumatic nailed		108	.148		6.55	5.75		12.30	16.70
1400	For horizontal blocking, 2" x 4", add		600	.027		.43	1.03		1.46	2.18
1500	2" x 6", add		600	.027		.72	1.03		1.75	2.50
1600	For openings, add		250	.064			2.48		2.48	4.10
1700	Headers for above openings, material only, add				M.B.F.	730			730	805

06 11 10.28 Porch or Deck Framing		Crew	Daily Output	Labor-Hours	Unit	Material	Labor	Equipment	Total	Total Incl O&P
0010	PORCH OR DECK FRAMING									
0100	Treated lumber, posts or columns, 4" x 4"	2 Carp	390	.041	L.F.	1.25	1.59		2.84	4
0110	4" x 6"		275	.058		1.95	2.25		4.20	5.85
0120	4" x 8"		220	.073		4.06	2.82		6.88	9.10
0130	Girder, single, 4" x 4"		675	.024		1.25	.92		2.17	2.89
0140	4" x 6"		600	.027		1.95	1.03		2.98	3.85
0150	4" x 8"		525	.030		4.06	1.18		5.24	6.40
0160	Double, 2" x 4"		625	.026		1.24	.99		2.23	3
0170	2" x 6"		600	.027		1.57	1.03		2.60	3.44
0180	2" x 8"		575	.028		2.32	1.08		3.40	4.33
0190	2" x 10"		550	.029		2.89	1.13		4.02	5.05
0200	2" x 12"		525	.030		4.28	1.18		5.46	6.65
0210	Triple, 2" x 4"		575	.028		1.86	1.08		2.94	3.82
0220	2" x 6"		550	.029		2.35	1.13		3.48	4.45
0230	2" x 8"		525	.030		3.48	1.18		4.66	5.80
0240	2" x 10"		500	.032		4.34	1.24		5.58	6.80
0250	2" x 12"		475	.034		6.40	1.31		7.71	9.20
0260	Ledger, bolted 4' OC, 2" x 4"		400	.040		.77	1.55		2.32	3.40
0270	2" x 6"		395	.041		.92	1.57		2.49	3.60
0280	2" x 8"		390	.041		1.29	1.59		2.88	4.04
0290	2" x 10"		385	.042		1.56	1.61		3.17	4.37
0300	2" x 12"		380	.042		2.24	1.63		3.87	5.15
0310	Joists, 2" x 4"		1250	.013		.62	.50		1.12	1.50
0320	2" x 6"		1250	.013		.78	.50		1.28	1.68
0330	2" x 8"		1100	.015		1.16	.56		1.72	2.21
0340	2" x 10"		900	.018		1.45	.69		2.14	2.73
0350	2" x 12"		875	.018		1.73	.71		2.44	3.08
0360	Railings and trim, 1" x 4"	1 Carp	300	.027		.53	1.03		1.56	2.29

06 11 10 – Framing with Dimensional, Engineered or Composite Lumber

06 11 10.28 Porch or Deck Framing		Crew	Daily Output	Labor-Hours	Unit	2019 Bare Costs Material	Labor	Equipment	Total	Total Incl O&P
0370	2″ x 2″	1 Carp	300	.027	L.F.	.42	1.03		1.45	2.17
0380	2″ x 4″		300	.027		.61	1.03		1.64	2.38
0390	2″ x 6″		300	.027		.77	1.03		1.80	2.55
0400	Decking, 1″ x 4″		275	.029	S.F.	2.92	1.13		4.05	5.05
0410	2″ x 4″		300	.027		2.05	1.03		3.08	3.96
0420	2″ x 6″		320	.025		1.65	.97		2.62	3.42
0430	5/4″ x 6″		320	.025		2.17	.97		3.14	3.98
0440	Balusters, square, 2″ x 2″	2 Carp	660	.024	L.F.	.42	.94		1.36	2.01
0450	Turned, 2″ x 2″		420	.038		.56	1.48		2.04	3.05
0460	Stair stringer, 2″ x 10″		130	.123		1.45	4.77		6.22	9.50
0470	2″ x 12″		130	.123		1.73	4.77		6.50	9.80
0480	Stair treads, 1″ x 4″		140	.114		2.92	4.43		7.35	10.50
0490	2″ x 4″		140	.114		.62	4.43		5.05	8
0500	2″ x 6″		160	.100		.93	3.88		4.81	7.40
0510	5/4″ x 6″		160	.100		1.01	3.88		4.89	7.50
0520	Turned handrail post, 4″ x 4″		64	.250	Ea.	37	9.70		46.70	56.50
0530	Lattice panel, 4′ x 8′, 1/2″		1600	.010	S.F.	.73	.39		1.12	1.44
0535	3/4″		1600	.010	″	1.08	.39		1.47	1.83
0540	Cedar, posts or columns, 4″ x 4″		390	.041	L.F.	3.76	1.59		5.35	6.75
0550	4″ x 6″		275	.058		5.85	2.25		8.10	10.20
0560	4″ x 8″		220	.073		11.15	2.82		13.97	16.95
0800	Decking, 1″ x 4″		550	.029		2.84	1.13		3.97	4.98
0810	2″ x 4″		600	.027		5.80	1.03		6.83	8.05
0820	2″ x 6″		640	.025		10.55	.97		11.52	13.20
0830	5/4″ x 6″		640	.025		6.65	.97		7.62	8.95
0840	Railings and trim, 1″ x 4″		600	.027		2.84	1.03		3.87	4.83
0860	2″ x 4″		600	.027		5.80	1.03		6.83	8.05
0870	2″ x 6″		600	.027		10.55	1.03		11.58	13.30
0920	Stair treads, 1″ x 4″		140	.114		2.84	4.43		7.27	10.40
0930	2″ x 4″		140	.114		5.80	4.43		10.23	13.65
0940	2″ x 6″		160	.100		10.55	3.88		14.43	18
0950	5/4″ x 6″		160	.100		6.65	3.88		10.53	13.75
0980	Redwood, posts or columns, 4″ x 4″		390	.041		6.55	1.59		8.14	9.85
0990	4″ x 6″		275	.058		12.80	2.25		15.05	17.85
1000	4″ x 8″		220	.073		23.50	2.82		26.32	30.50
1240	Decking, 1″ x 4″	1 Carp	275	.029	S.F.	4.04	1.13		5.17	6.30
1260	2″ x 6″		340	.024		7.60	.91		8.51	9.85
1270	5/4″ x 6″		320	.025		4.85	.97		5.82	6.95
1280	Railings and trim, 1″ x 4″	2 Carp	600	.027	L.F.	1.19	1.03		2.22	3.02
1310	2″ x 6″		600	.027		7.60	1.03		8.63	10.10
1420	Alternative decking, wood/plastic composite, 5/4″ x 6″ G		640	.025		3.43	.97		4.40	5.40
1440	1″ x 4″ square edge fir		550	.029		2.93	1.13		4.06	5.10
1450	1″ x 4″ tongue and groove fir		450	.036		1.54	1.38		2.92	3.97
1460	1″ x 4″ mahogany		550	.029		2.09	1.13		3.22	4.16
1462	5/4″ x 6″ PVC		550	.029		3.35	1.13		4.48	5.55
1465	Framing, porch or deck, alt deck fastening, screws, add	1 Carp	240	.033	S.F.		1.29		1.29	2.13
1470	Accessories, joist hangers, 2″ x 4″		160	.050	Ea.	.77	1.94		2.71	4.05
1480	2″ x 6″ through 2″ x 12″		150	.053		1.46	2.07		3.53	5.05
1530	Post footing, incl excav, backfill, tube form & concrete, 4′ deep, 8″ diam.	F-7	12	2.667		17.95	92		109.95	172
1540	10″ diameter		11	2.909		24.50	101		125.50	193
1550	12″ diameter		10	3.200		31.50	111		142.50	218

06 11 10 – Framing with Dimensional, Engineered or Composite Lumber

06 11 10.30 Roof Framing		Crew	Daily Output	Labor-Hours	Unit	Material	Labor	Equipment	Total	Total Incl O&P
						2019 Bare Costs				
0010	**ROOF FRAMING**									
1900	Rough fascia, 2" x 6"	2 Carp	250	.064	L.F.	.72	2.48		3.20	4.89
2000	2" x 8"		225	.071		.97	2.76		3.73	5.60
2100	2" x 10"		180	.089		1.45	3.44		4.89	7.30
5000	Rafters, to 4 in 12 pitch, 2" x 6", ordinary		1000	.016		.72	.62		1.34	1.81
5020	On steep roofs		800	.020		.72	.78		1.50	2.07
5040	On dormers or complex roofs		590	.027		.72	1.05		1.77	2.53
5060	2" x 8", ordinary		950	.017		.97	.65		1.62	2.15
5080	On steep roofs		750	.021		.97	.83		1.80	2.44
5100	On dormers or complex roofs		540	.030		.97	1.15		2.12	2.97
5120	2" x 10", ordinary		630	.025		1.45	.98		2.43	3.23
5140	On steep roofs		495	.032		1.45	1.25		2.70	3.67
5160	On dormers or complex roofs		425	.038		1.45	1.46		2.91	4.01
5180	2" x 12", ordinary		575	.028		1.83	1.08		2.91	3.79
5200	On steep roofs		455	.035		1.83	1.36		3.19	4.26
5220	On dormers or complex roofs		395	.041		1.83	1.57		3.40	4.60
5300	Hip and valley rafters, 2" x 6", ordinary		760	.021		.72	.82		1.54	2.14
5320	On steep roofs		585	.027		.72	1.06		1.78	2.54
5340	On dormers or complex roofs		510	.031		.72	1.22		1.94	2.80
5360	2" x 8", ordinary		720	.022		.97	.86		1.83	2.49
5380	On steep roofs		545	.029		.97	1.14		2.11	2.95
5400	On dormers or complex roofs		470	.034		.97	1.32		2.29	3.25
5420	2" x 10", ordinary		570	.028		1.45	1.09		2.54	3.40
5440	On steep roofs		440	.036		1.45	1.41		2.86	3.93
5460	On dormers or complex roofs		380	.042		1.45	1.63		3.08	4.30
5480	2" x 12", ordinary		525	.030		1.83	1.18		3.01	3.96
5500	On steep roofs		410	.039		1.83	1.51		3.34	4.51
5520	On dormers or complex roofs		355	.045		1.83	1.75		3.58	4.90
5540	Hip and valley jacks, 2" x 6", ordinary		600	.027		.72	1.03		1.75	2.50
5560	On steep roofs		475	.034		.72	1.31		2.03	2.95
5580	On dormers or complex roofs		410	.039		.72	1.51		2.23	3.29
5600	2" x 8", ordinary		490	.033		.97	1.27		2.24	3.16
5620	On steep roofs		385	.042		.97	1.61		2.58	3.73
5640	On dormers or complex roofs		335	.048		.97	1.85		2.82	4.13
5660	2" x 10", ordinary		450	.036		1.45	1.38		2.83	3.88
5680	On steep roofs		350	.046		1.45	1.77		3.22	4.53
5700	On dormers or complex roofs		305	.052		1.45	2.03		3.48	4.96
5720	2" x 12", ordinary		375	.043		1.83	1.65		3.48	4.74
5740	On steep roofs		295	.054		1.83	2.10		3.93	5.50
5760	On dormers or complex roofs		255	.063		1.83	2.43		4.26	6.05
5780	Rafter tie, 1" x 4", #3		800	.020		.53	.78		1.31	1.86
5790	2" x 4", #3		800	.020		.43	.78		1.21	1.75
5800	Ridge board, #2 or better, 1" x 6"		600	.027		.79	1.03		1.82	2.58
5820	1" x 8"		550	.029		1.35	1.13		2.48	3.34
5840	1" x 10"		500	.032		1.71	1.24		2.95	3.93
5860	2" x 6"		500	.032		.72	1.24		1.96	2.84
5880	2" x 8"		450	.036		.97	1.38		2.35	3.35
5900	2" x 10"		400	.040		1.45	1.55		3	4.16
5920	Roof cants, split, 4" x 4"		650	.025		1.86	.95		2.81	3.63
5940	6" x 6"		600	.027		5.45	1.03		6.48	7.70
5960	Roof curbs, untreated, 2" x 6"		520	.031		.72	1.19		1.91	2.76
5980	2" x 12"		400	.040		1.83	1.55		3.38	4.57

06 11 Wood Framing

06 11 10 – Framing with Dimensional, Engineered or Composite Lumber

06 11 10.30	Roof Framing	Crew	Daily Output	Labor-Hours	Unit	2019 Bare Costs Material	Labor	Equipment	Total	Total Incl O&P
6000	Sister rafters, 2″ x 6″	2 Carp	800	.020	L.F.	.72	.78		1.50	2.07
6020	2″ x 8″		640	.025		.97	.97		1.94	2.67
6040	2″ x 10″		535	.030		1.45	1.16		2.61	3.52
6060	2″ x 12″		455	.035		1.83	1.36		3.19	4.26

06 11 10.32 Sill and Ledger Framing

		Crew	Daily Output	Labor-Hours	Unit	Material	Labor	Equipment	Total	Total Incl O&P
0010	**SILL AND LEDGER FRAMING**									
0020	Extruded polystyrene sill sealer, 5-1/2″ wide	1 Carp	1600	.005	L.F.	.14	.19		.33	.47
2000	Ledgers, nailed, 2″ x 4″	2 Carp	755	.021		.43	.82		1.25	1.83
2050	2″ x 6″		600	.027		.72	1.03		1.75	2.50
2100	Bolted, not including bolts, 3″ x 6″		325	.049		2.07	1.91		3.98	5.45
2150	3″ x 12″		233	.069		4.34	2.66		7	9.20
2600	Mud sills, redwood, construction grade, 2″ x 4″		895	.018		2.30	.69		2.99	3.68
2620	2″ x 6″		780	.021		3.45	.79		4.24	5.10
4000	Sills, 2″ x 4″		600	.027		.42	1.03		1.45	2.17
4050	2″ x 6″		550	.029		.71	1.13		1.84	2.64
4080	2″ x 8″		500	.032		.96	1.24		2.20	3.11
4100	2″ x 10″		450	.036		1.43	1.38		2.81	3.86
4120	2″ x 12″		400	.040		1.81	1.55		3.36	4.55
4200	Treated, 2″ x 4″		550	.029		.60	1.13		1.73	2.51
4220	2″ x 6″		500	.032		.75	1.24		1.99	2.87
4240	2″ x 8″		450	.036		1.11	1.38		2.49	3.50
4280	2″ x 12″		350	.046		2.07	1.77		3.84	5.20
4400	4″ x 4″		450	.036		1.20	1.38		2.58	3.60
4420	4″ x 6″		350	.046		1.87	1.77		3.64	4.99
4460	4″ x 8″		300	.053		3.96	2.07		6.03	7.80
4480	4″ x 10″		260	.062		5.75	2.38		8.13	10.25

06 11 10.34 Sleepers

		Crew	Daily Output	Labor-Hours	Unit	Material	Labor	Equipment	Total	Total Incl O&P
0010	**SLEEPERS**									
0100	On concrete, treated, 1″ x 2″	2 Carp	2350	.007	L.F.	.31	.26		.57	.78
0150	1″ x 3″		2000	.008		.48	.31		.79	1.04
0200	2″ x 4″		1500	.011		.65	.41		1.06	1.39
0250	2″ x 6″		1300	.012		.85	.48		1.33	1.72

06 11 10.36 Soffit and Canopy Framing

		Crew	Daily Output	Labor-Hours	Unit	Material	Labor	Equipment	Total	Total Incl O&P
0010	**SOFFIT AND CANOPY FRAMING**									
1000	Canopy or soffit framing, 1″ x 4″	2 Carp	900	.018	L.F.	.53	.69		1.22	1.72
1020	1″ x 6″		850	.019		.79	.73		1.52	2.08
1040	1″ x 8″		750	.021		1.35	.83		2.18	2.85
1100	2″ x 4″		620	.026		.43	1		1.43	2.12
1120	2″ x 6″		560	.029		.72	1.11		1.83	2.62
1140	2″ x 8″		500	.032		.97	1.24		2.21	3.12
1200	3″ x 4″		500	.032		1.22	1.24		2.46	3.40
1220	3″ x 6″		400	.040		2.09	1.55		3.64	4.86
1240	3″ x 10″		300	.053		3.70	2.07		5.77	7.50

06 11 10.38 Treated Lumber Framing Material

		Crew	Daily Output	Labor-Hours	Unit	Material	Labor	Equipment	Total	Total Incl O&P
0010	**TREATED LUMBER FRAMING MATERIAL**									
0100	2″ x 4″				M.B.F.	895			895	980
0110	2″ x 6″					750			750	825
0120	2″ x 8″					835			835	920
0130	2″ x 10″					830			830	915
0140	2″ x 12″					1,025			1,025	1,125
0200	4″ x 4″					900			900	990
0210	4″ x 6″					935			935	1,025

06 11 10 – Framing with Dimensional, Engineered or Composite Lumber

06 11 10.38 Treated Lumber Framing Material		Crew	Daily Output	Labor-Hours	Unit	2019 Bare Costs Material	Labor	Equipment	Total	Total Incl O&P
0220	4" x 8"				M.B.F.	1,475			1,475	1,625

06 11 10.40 Wall Framing

		Crew	Daily Output	Labor-Hours	Unit	Material	Labor	Equipment	Total	Total Incl O&P
0010	**WALL FRAMING**									
2000	Headers over openings, 2" x 6"	2 Carp	360	.044	L.F.	.72	1.72		2.44	3.64
2005	2" x 6", pneumatic nailed R061110-30		432	.037		.73	1.44		2.17	3.17
2050	2" x 8"		340	.047		.97	1.82		2.79	4.08
2055	2" x 8", pneumatic nailed		408	.039		.98	1.52		2.50	3.59
2100	2" x 10"		320	.050		1.45	1.94		3.39	4.80
2105	2" x 10", pneumatic nailed		384	.042		1.47	1.61		3.08	4.28
2150	2" x 12"		300	.053		1.83	2.07		3.90	5.45
2155	2" x 12", pneumatic nailed		360	.044		1.85	1.72		3.57	4.88
2180	4" x 8"		260	.062		4.29	2.38		6.67	8.65
2185	4" x 8", pneumatic nailed		312	.051		4.31	1.99		6.30	8
2190	4" x 10"		240	.067		5.05	2.58		7.63	9.80
2195	4" x 10", pneumatic nailed		288	.056		5.10	2.15		7.25	9.15
2200	4" x 12"		190	.084		5.55	3.26		8.81	11.50
2205	4" x 12", pneumatic nailed		228	.070		5.60	2.72		8.32	10.65
2240	6" x 10"		165	.097		7	3.76		10.76	13.90
2245	6" x 10", pneumatic nailed		198	.081		7.05	3.13		10.18	12.95
2250	6" x 12"		140	.114		8.85	4.43		13.28	17.05
2255	6" x 12", pneumatic nailed		168	.095		8.90	3.69		12.59	15.90
5000	Plates, untreated, 2" x 3"		850	.019		.42	.73		1.15	1.67
5005	2" x 3", pneumatic nailed		1020	.016		.42	.61		1.03	1.47
5020	2" x 4"		800	.020		.43	.78		1.21	1.75
5025	2" x 4", pneumatic nailed		960	.017		.43	.65		1.08	1.54
5040	2" x 6"		750	.021		.72	.83		1.55	2.16
5045	2" x 6", pneumatic nailed		900	.018		.73	.69		1.42	1.94
5060	Treated, 2" x 3"		850	.019		.52	.73		1.25	1.78
5065	2" x 3", treated, pneumatic nailed		1020	.016		.53	.61		1.14	1.58
5080	2" x 4"		800	.020		.60	.78		1.38	1.94
5085	2" x 4", treated, pneumatic nailed		960	.017		.61	.65		1.26	1.74
5100	2" x 6"		750	.021		.76	.83		1.59	2.20
5102	2 x 8		725	.022		1.12	.86		1.98	2.65
5103	2" x 6", treated, pneumatic nailed		900	.018		.77	.69		1.46	1.98
5104	2 x 10		700	.023		1.40	.89		2.29	3
5105	2" x 12"		670	.024		2.08	.93		3.01	3.82
5107	2" x 8", pneumatic nailed		870	.018		1.14	.71		1.85	2.43
5108	2" x 10", pneumatic nailed		840	.019		1.42	.74		2.16	2.78
5109	2" x 12", pneumatic nailed		804	.020		2.10	.77		2.87	3.58
5120	Studs, 8' high wall, 2" x 3"		1200	.013		.42	.52		.94	1.31
5125	2" x 3", pneumatic nailed		1440	.011		.42	.43		.85	1.18
5140	2" x 4"		1100	.015		.43	.56		.99	1.40
5146	2" x 4", pneumatic nailed		1320	.012		.43	.47		.90	1.25
5160	2" x 6"		1000	.016		.72	.62		1.34	1.81
5166	2" x 6", pneumatic nailed		1200	.013		.73	.52		1.25	1.65
5180	3" x 4"		800	.020		1.22	.78		2	2.63
5185	3" x 4", pneumatic nailed		960	.017		1.23	.65		1.88	2.43
5200	Installed on second story, 2" x 3"		1170	.014		.42	.53		.95	1.34
5205	2" x 3", pneumatic nailed		1404	.011		.42	.44		.86	1.20
5220	2" x 4"		1015	.016		.43	.61		1.04	1.48
5225	2" x 4", pneumatic nailed		1218	.013		.43	.51		.94	1.31
5240	2" x 6"		890	.018		.72	.70		1.42	1.94

06 11 10 – Framing with Dimensional, Engineered or Composite Lumber

06 11 10.40 Wall Framing		Crew	Daily Output	Labor-Hours	Unit	Material	Labor	Equipment	Total	Total Incl O&P
						2019 Bare Costs				
5245	2″ x 6″, pneumatic nailed	2 Carp	1080	.015	L.F.	.73	.57		1.30	1.75
5260	3″ x 4″		800	.020		1.22	.78		2	2.63
5265	3″ x 4″, pneumatic nailed		960	.017		1.23	.65		1.88	2.43
5280	Installed on dormer or gable, 2″ x 3″		1045	.015		.42	.59		1.01	1.44
5285	2″ x 3″, pneumatic nailed		1254	.013		.42	.49		.91	1.29
5300	2″ x 4″		905	.018		.43	.69		1.12	1.60
5305	2″ x 4″, pneumatic nailed		1086	.015		.43	.57		1	1.41
5320	2″ x 6″		800	.020		.72	.78		1.50	2.07
5325	2″ x 6″, pneumatic nailed		960	.017		.73	.65		1.38	1.87
5340	3″ x 4″		700	.023		1.22	.89		2.11	2.81
5345	3″ x 4″, pneumatic nailed		840	.019		1.23	.74		1.97	2.58
5360	6′ high wall, 2″ x 3″		970	.016		.42	.64		1.06	1.52
5365	2″ x 3″, pneumatic nailed		1164	.014		.42	.53		.95	1.35
5380	2″ x 4″		850	.019		.43	.73		1.16	1.68
5385	2″ x 4″, pneumatic nailed		1020	.016		.43	.61		1.04	1.47
5400	2″ x 6″		740	.022		.72	.84		1.56	2.17
5405	2″ x 6″, pneumatic nailed		888	.018		.73	.70		1.43	1.95
5420	3″ x 4″		600	.027		1.22	1.03		2.25	3.06
5425	3″ x 4″, pneumatic nailed		720	.022		1.23	.86		2.09	2.78
5440	Installed on second story, 2″ x 3″		950	.017		.42	.65		1.07	1.54
5445	2″ x 3″, pneumatic nailed		1140	.014		.42	.54		.96	1.37
5460	2″ x 4″		810	.020		.43	.77		1.20	1.74
5465	2″ x 4″, pneumatic nailed		972	.016		.43	.64		1.07	1.52
5480	2″ x 6″		700	.023		.72	.89		1.61	2.25
5485	2″ x 6″, pneumatic nailed		840	.019		.73	.74		1.47	2.02
5500	3″ x 4″		550	.029		1.22	1.13		2.35	3.21
5505	3″ x 4″, pneumatic nailed		660	.024		1.23	.94		2.17	2.91
5520	Installed on dormer or gable, 2″ x 3″		850	.019		.42	.73		1.15	1.67
5525	2″ x 3″, pneumatic nailed		1020	.016		.42	.61		1.03	1.47
5540	2″ x 4″		720	.022		.43	.86		1.29	1.89
5545	2″ x 4″, pneumatic nailed		864	.019		.43	.72		1.15	1.66
5560	2″ x 6″		620	.026		.72	1		1.72	2.44
5565	2″ x 6″, pneumatic nailed		744	.022		.73	.83		1.56	2.18
5580	3″ x 4″		480	.033		1.22	1.29		2.51	3.48
5585	3″ x 4″, pneumatic nailed		576	.028		1.23	1.08		2.31	3.14
5600	3′ high wall, 2″ x 3″		740	.022		.42	.84		1.26	1.84
5605	2″ x 3″, pneumatic nailed		888	.018		.42	.70		1.12	1.62
5620	2″ x 4″		640	.025		.43	.97		1.40	2.07
5625	2″ x 4″, pneumatic nailed		768	.021		.43	.81		1.24	1.80
5640	2″ x 6″		550	.029		.72	1.13		1.85	2.65
5645	2″ x 6″, pneumatic nailed		660	.024		.73	.94		1.67	2.35
5660	3″ x 4″		440	.036		1.22	1.41		2.63	3.68
5665	3″ x 4″, pneumatic nailed		528	.030		1.23	1.17		2.40	3.30
5680	Installed on second story, 2″ x 3″		700	.023		.42	.89		1.31	1.92
5685	2″ x 3″, pneumatic nailed		840	.019		.42	.74		1.16	1.69
5700	2″ x 4″		610	.026		.43	1.02		1.45	2.15
5705	2″ x 4″, pneumatic nailed		732	.022		.43	.85		1.28	1.87
5720	2″ x 6″		520	.031		.72	1.19		1.91	2.76
5725	2″ x 6″, pneumatic nailed		624	.026		.73	.99		1.72	2.44
5740	3″ x 4″		430	.037		1.22	1.44		2.66	3.73
5745	3″ x 4″, pneumatic nailed		516	.031		1.23	1.20		2.43	3.35
5760	Installed on dormer or gable, 2″ x 3″		625	.026		.42	.99		1.41	2.10
5765	2″ x 3″, pneumatic nailed	↓	750	.021	↓	.42	.83		1.25	1.84

06 11 Wood Framing

06 11 10 – Framing with Dimensional, Engineered or Composite Lumber

06 11 10.40 Wall Framing		Crew	Daily Output	Labor-Hours	Unit	Material (2019 Bare Costs)	Labor (2019 Bare Costs)	Equipment (2019 Bare Costs)	Total (2019 Bare Costs)	Total Incl O&P
5780	2″ x 4″	2 Carp	545	.029	L.F.	.43	1.14		1.57	2.35
5785	2″ x 4″, pneumatic nailed		654	.024		.43	.95		1.38	2.04
5800	2″ x 6″		465	.034		.72	1.33		2.05	2.99
5805	2″ x 6″, pneumatic nailed		558	.029		.73	1.11		1.84	2.64
5820	3″ x 4″		380	.042		1.22	1.63		2.85	4.05
5825	3″ x 4″, pneumatic nailed	↓	456	.035	↓	1.23	1.36		2.59	3.61
8250	For second story & above, add						5%			
8300	For dormer & gable, add						15%			

06 11 10.42 Furring

		Crew	Daily Output	Labor-Hours	Unit	Material	Labor	Equipment	Total	Total Incl O&P
0010	**FURRING**									
0012	Wood strips, 1″ x 2″, on walls, on wood	1 Carp	550	.015	L.F.	.27	.56		.83	1.23
0015	On wood, pneumatic nailed		710	.011		.27	.44		.71	1.02
0300	On masonry		495	.016		.29	.63		.92	1.36
0400	On concrete		260	.031		.29	1.19		1.48	2.29
0600	1″ x 3″, on walls, on wood		550	.015		.44	.56		1	1.42
0605	On wood, pneumatic nailed		710	.011		.44	.44		.88	1.21
0700	On masonry		495	.016		.48	.63		1.11	1.56
0800	On concrete		260	.031		.48	1.19		1.67	2.49
0850	On ceilings, on wood		350	.023		.44	.89		1.33	1.95
0855	On wood, pneumatic nailed		450	.018		.44	.69		1.13	1.63
0900	On masonry		320	.025		.48	.97		1.45	2.12
0950	On concrete	↓	210	.038	↓	.48	1.48		1.96	2.96

06 11 10.44 Grounds

		Crew	Daily Output	Labor-Hours	Unit	Material	Labor	Equipment	Total	Total Incl O&P
0010	**GROUNDS**									
0020	For casework, 1″ x 2″ wood strips, on wood	1 Carp	330	.024	L.F.	.27	.94		1.21	1.85
0100	On masonry		285	.028		.29	1.09		1.38	2.12
0200	On concrete		250	.032		.29	1.24		1.53	2.37
0400	For plaster, 3/4″ deep, on wood		450	.018		.27	.69		.96	1.44
0500	On masonry		225	.036		.29	1.38		1.67	2.60
0600	On concrete		175	.046		.29	1.77		2.06	3.25
0700	On metal lath	↓	200	.040	↓	.29	1.55		1.84	2.88

06 12 Structural Panels

06 12 10 – Structural Insulated Panels

06 12 10.10 OSB Faced Panels

			Crew	Daily Output	Labor-Hours	Unit	Material	Labor	Equipment	Total	Total Incl O&P
0010	**OSB FACED PANELS**										
0100	Structural insul. panels, 7/16″ OSB both faces, EPS insul., 3-5/8″ T	G	F-3	2075	.019	S.F.	3.74	.68	.23	4.65	5.50
0110	5-5/8″ thick	G		1725	.023		4.20	.82	.27	5.29	6.30
0120	7-3/8″ thick	G		1425	.028		5.30	1	.33	6.63	7.85
0130	9-3/8″ thick	G		1125	.036		5.65	1.26	.42	7.33	8.75
0140	7/16″ OSB one face, EPS insul., 3-5/8″ thick	G		2175	.018		3.89	.65	.22	4.76	5.60
0150	5-5/8″ thick	G		1825	.022		4.51	.78	.26	5.55	6.55
0160	7-3/8″ thick	G		1525	.026		4.77	.93	.31	6.01	7.15
0170	9-3/8″ thick	G		1225	.033		5.55	1.16	.38	7.09	8.45
0190	7/16″ OSB - 1/2″ GWB faces, EPS insul., 3-5/8″ T	G		2075	.019		3.56	.68	.23	4.47	5.30
0200	5-5/8″ thick	G		1725	.023		4.87	.82	.27	5.96	7
0210	7-3/8″ thick	G		1425	.028		4.85	1	.33	6.18	7.35
0220	9-3/8″ thick	G		1125	.036		5.45	1.26	.42	7.13	8.50
0240	7/16″ OSB - 1/2″ MRGWB faces, EPS insul., 3-5/8″ T	G		2075	.019		4.38	.68	.23	5.29	6.20
0250	5-5/8″ thick	G	↓	1725	.023	↓	4.39	.82	.27	5.48	6.50

06 12 Structural Panels

06 12 10 – Structural Insulated Panels

06 12 10.10 OSB Faced Panels			Crew	Daily Output	Labor-Hours	Unit	Material	2019 Bare Costs Labor	Equipment	Total	Total Incl O&P
0260	7-3/8″ thick	G	F-3	1425	.028	S.F.	4.86	1	.33	6.19	7.35
0270	9-3/8″ thick	G		1125	.036		5.55	1.26	.42	7.23	8.65
0300	For 1/2″ GWB added to OSB skin, add	G					1.44			1.44	1.58
0310	For 1/2″ MRGWB added to OSB skin, add	G					1.63			1.63	1.79
0320	For one T1-11 skin, add to OSB-OSB	G					2.06			2.06	2.27
0330	For one 19/32″ CDX skin, add to OSB-OSB	G					1.54			1.54	1.69
0500	Structural insulated panel, 7/16″ OSB both sides, straw core										
0510	4-3/8″ T, walls (w/sill, splines, plates)	G	F-6	2400	.017	S.F.	7.50	.60	.20	8.30	9.45
0520	Floors (w/splines)	G		2400	.017		7.50	.60	.20	8.30	9.45
0530	Roof (w/splines)	G		2400	.017		7.50	.60	.20	8.30	9.45
0550	7-7/8″ T, walls (w/sill, splines, plates)	G		2400	.017		11.70	.60	.20	12.50	14.05
0560	Floors (w/splines)	G		2400	.017		11.70	.60	.20	12.50	14.05
0570	Roof (w/splines)	G		2400	.017		11.70	.60	.20	12.50	14.05

06 12 19 – Composite Shearwall Panels

06 12 19.10 Steel and Wood Composite Shearwall Panels		Crew	Daily Output	Labor-Hours	Unit	Material	Labor	Equipment	Total	Total Incl O&P
0010	**STEEL & WOOD COMPOSITE SHEARWALL PANELS**									
0020	Anchor bolts, 36″ long (must be placed in wet concrete)	1 Carp	150	.053	Ea.	40.50	2.07		42.57	48
0030	On concrete, 2″ x 4″ & 2″ x 6″ walls, 7′-10′ high, 360 lb. shear, 12″ wide	2 Carp	8	2		475	77.50		552.50	655
0040	715 lb. shear, 15″ wide		8	2		570	77.50		647.50	755
0050	1860 lb. shear, 18″ wide		8	2		635	77.50		712.50	825
0060	2780 lb. shear, 21″ wide		8	2		740	77.50		817.50	945
0070	3790 lb. shear, 24″ wide		8	2		795	77.50		872.50	1,000
0080	2″ x 6″ walls, 11′-13′ high, 1180 lb. shear, 18″ wide		6	2.667		725	103		828	970
0090	1555 lb. shear, 21″ wide		6	2.667		740	103		843	980
0100	2280 lb. shear, 24″ wide		6	2.667		885	103		988	1,150
0110	For installing above on wood floor frame, add									
0120	Coupler nuts, threaded rods, bolts, shear transfer plate kit	1 Carp	16	.500	Ea.	60.50	19.40		79.90	98.50
0130	Framing anchors, angle (2 required)	″	96	.083	″	2.44	3.23		5.67	8.05
0140	For blocking see Section 06 11 10.02									
0150	For installing above, first floor to second floor, wood floor frame, add									
0160	Add stack option to first floor wall panel				Ea.	66.50			66.50	73
0170	Threaded rods, bolts, shear transfer plate kit	1 Carp	16	.500		76.50	19.40		95.90	116
0180	Framing anchors, angle (2 required)	″	96	.083		2.44	3.23		5.67	8.05
0190	For blocking see section 06 11 10.02									
0200	For installing stacked panels, balloon framing									
0210	Add stack option to first floor wall panel				Ea.	66.50			66.50	73
0220	Threaded rods, bolts kit	1 Carp	16	.500	″	43	19.40		62.40	79

06 13 Heavy Timber Construction

06 13 23 – Heavy Timber Framing

06 13 23.10 Heavy Framing		Crew	Daily Output	Labor-Hours	Unit	Material	Labor	Equipment	Total	Total Incl O&P
0010	**HEAVY FRAMING**									
0020	Beams, single 6″ x 10″	2 Carp	1.10	14.545	M.B.F.	1,550	565		2,115	2,650
0100	Single 8″ x 16″		1.20	13.333		1,900	515		2,415	2,950
0200	Built from 2″ lumber, multiple 2″ x 14″		.90	17.778		985	690		1,675	2,225
0210	Built from 3″ lumber, multiple 3″ x 6″		.70	22.857		1,375	885		2,260	3,000
0220	Multiple 3″ x 8″		.80	20		1,475	775		2,250	2,900
0230	Multiple 3″ x 10″		.90	17.778		1,475	690		2,165	2,775
0240	Multiple 3″ x 12″		1	16		1,450	620		2,070	2,625
0250	Built from 4″ lumber, multiple 4″ x 6″		.80	20		1,575	775		2,350	3,000

06 13 Heavy Timber Construction

06 13 23 – Heavy Timber Framing

06 13 23.10 Heavy Framing		Crew	Daily Output	Labor-Hours	Unit	Material	Labor	Equipment	Total	Total Incl O&P
						2019 Bare Costs				
0260	Multiple 4″ x 8″	2 Carp	.90	17.778	M.B.F.	1,600	690		2,290	2,900
0270	Multiple 4″ x 10″		1	16		1,500	620		2,120	2,675
0280	Multiple 4″ x 12″		1.10	14.545		1,375	565		1,940	2,450
0290	Columns, structural grade, 1500f, 4″ x 4″		.60	26.667		1,850	1,025		2,875	3,725
0300	6″ x 6″		.65	24.615		1,375	955		2,330	3,075
0400	8″ x 8″		.70	22.857		1,500	885		2,385	3,100
0500	10″ x 10″		.75	21.333		1,750	825		2,575	3,300
0600	12″ x 12″		.80	20		1,525	775		2,300	2,950
0800	Floor planks, 2″ thick, T&G, 2″ x 6″		1.05	15.238		1,575	590		2,165	2,700
0900	2″ x 10″		1.10	14.545		1,575	565		2,140	2,675
1100	3″ thick, 3″ x 6″		1.05	15.238		1,625	590		2,215	2,750
1200	3″ x 10″		1.10	14.545		1,650	565		2,215	2,725
1400	Girders, structural grade, 12″ x 12″		.80	20		1,525	775		2,300	2,950
1500	10″ x 16″		1	16		2,525	620		3,145	3,800
2300	Roof purlins, 4″ thick, structural grade	↓	1.05	15.238	↓	1,600	590		2,190	2,725

06 15 Wood Decking

06 15 16 – Wood Roof Decking

06 15 16.10 Solid Wood Roof Decking		Crew	Daily Output	Labor-Hours	Unit	Material	Labor	Equipment	Total	Total Incl O&P
0010	**SOLID WOOD ROOF DECKING**									
0350	Cedar planks, 2″ thick	2 Carp	350	.046	S.F.	6.85	1.77		8.62	10.45
0400	3″ thick		320	.050		10.25	1.94		12.19	14.45
0500	4″ thick		250	.064		13.65	2.48		16.13	19.10
0550	6″ thick		200	.080		20.50	3.10		23.60	27.50
0650	Douglas fir, 2″ thick		350	.046		3.07	1.77		4.84	6.30
0700	3″ thick		320	.050		4.60	1.94		6.54	8.25
0800	4″ thick		250	.064		6.15	2.48		8.63	10.85
0850	6″ thick		200	.080		9.20	3.10		12.30	15.20
0950	Hemlock, 2″ thick		350	.046		3.12	1.77		4.89	6.35
1000	3″ thick		320	.050		4.68	1.94		6.62	8.35
1100	4″ thick		250	.064		6.25	2.48		8.73	10.95
1150	6″ thick		200	.080		9.35	3.10		12.45	15.40
1250	Western white spruce, 2″ thick		350	.046		1.99	1.77		3.76	5.10
1300	3″ thick		320	.050		2.99	1.94		4.93	6.50
1400	4″ thick		250	.064		3.99	2.48		6.47	8.50
1450	6″ thick	↓	200	.080	↓	6	3.10		9.10	11.70

06 15 23 – Laminated Wood Decking

06 15 23.10 Laminated Roof Deck		Crew	Daily Output	Labor-Hours	Unit	Material	Labor	Equipment	Total	Total Incl O&P
0010	**LAMINATED ROOF DECK**									
0020	Pine or hemlock, 3″ thick	2 Carp	425	.038	S.F.	5.65	1.46		7.11	8.60
0100	4″ thick		325	.049		7.35	1.91		9.26	11.20
0300	Cedar, 3″ thick		425	.038		7.70	1.46		9.16	10.90
0400	4″ thick		325	.049		10.30	1.91		12.21	14.50
0600	Fir, 3″ thick		425	.038		6.60	1.46		8.06	9.65
0700	4″ thick	↓	325	.049	↓	8.40	1.91		10.31	12.40

06 16 Sheathing

06 16 13 – Insulating Sheathing

06 16 13.10 Insulating Sheathing			Crew	Daily Output	Labor-Hours	Unit	Material	2019 Bare Costs Labor	Equipment	Total	Total Incl O&P
0010	INSULATING SHEATHING										
0020	Expanded polystyrene, 1#/C.F. density, 3/4" thick, R2.89	G	2 Carp	1400	.011	S.F.	.36	.44		.80	1.13
0030	1" thick, R3.85	G		1300	.012		.43	.48		.91	1.26
0040	2" thick, R7.69	G		1200	.013		.70	.52		1.22	1.62
0050	Extruded polystyrene, 15 psi compressive strength, 1" thick, R5	G		1300	.012		.72	.48		1.20	1.58
0060	2" thick, R10	G		1200	.013		.89	.52		1.41	1.82
0070	Polyisocyanurate, 2#/C.F. density, 3/4" thick	G		1400	.011		.66	.44		1.10	1.45
0080	1" thick	G		1300	.012		.60	.48		1.08	1.45
0090	1-1/2" thick	G		1250	.013		.76	.50		1.26	1.65
0100	2" thick	G	↓	1200	.013	↓	.95	.52		1.47	1.89

06 16 23 – Subflooring

06 16 23.10 Subfloor			Crew	Daily Output	Labor-Hours	Unit	Material	Labor	Equipment	Total	Total Incl O&P
0010	SUBFLOOR	R061636-20									
0011	Plywood, CDX, 1/2" thick		2 Carp	1500	.011	SF Flr.	.63	.41		1.04	1.38
0015	Pneumatic nailed			1860	.009		.63	.33		.96	1.25
0100	5/8" thick			1350	.012		.78	.46		1.24	1.62
0105	Pneumatic nailed			1674	.010		.78	.37		1.15	1.47
0200	3/4" thick			1250	.013		.94	.50		1.44	1.85
0205	Pneumatic nailed			1550	.010		.94	.40		1.34	1.69
0300	1-1/8" thick, 2-4-1 including underlayment			1050	.015		2.12	.59		2.71	3.31
0440	With boards, 1" x 6", S4S, laid regular			900	.018		1.71	.69		2.40	3.02
0450	1" x 8", laid regular			1000	.016		2.15	.62		2.77	3.38
0460	Laid diagonal			850	.019		2.15	.73		2.88	3.57
0500	1" x 10", laid regular			1100	.015		2.14	.56		2.70	3.29
0600	Laid diagonal			900	.018	↓	2.14	.69		2.83	3.50
1500	OSB, 5/8" thick			1330	.012	S.F.	.54	.47		1.01	1.36
1600	3/4" thick		↓	1230	.013	"	.67	.50		1.17	1.57
8990	Subfloor adhesive, 3/8" bead		1 Carp	2300	.003	L.F.	.11	.13		.24	.34

06 16 26 – Underlayment

06 16 26.10 Wood Product Underlayment			Crew	Daily Output	Labor-Hours	Unit	Material	Labor	Equipment	Total	Total Incl O&P
0010	WOOD PRODUCT UNDERLAYMENT										
0015	Plywood, underlayment grade, 1/4" thick		2 Carp	1500	.011	S.F.	.95	.41		1.36	1.72
0018	Pneumatic nailed			1860	.009		.95	.33		1.28	1.59
0030	3/8" thick			1500	.011		1.05	.41		1.46	1.84
0070	Pneumatic nailed			1860	.009		1.05	.33		1.38	1.71
0100	1/2" thick			1450	.011		1.24	.43		1.67	2.07
0105	Pneumatic nailed			1798	.009		1.24	.34		1.58	1.93
0200	5/8" thick			1400	.011		1.36	.44		1.80	2.23
0205	Pneumatic nailed			1736	.009		1.36	.36		1.72	2.09
0300	3/4" thick			1300	.012		1.47	.48		1.95	2.41
0305	Pneumatic nailed			1612	.010		1.47	.38		1.85	2.26
0500	Particle board, 3/8" thick	G		1500	.011		.40	.41		.81	1.12
0505	Pneumatic nailed	G		1860	.009		.40	.33		.73	.99
0600	1/2" thick	G		1450	.011		.42	.43		.85	1.17
0605	Pneumatic nailed	G		1798	.009		.42	.34		.76	1.03
0800	5/8" thick	G		1400	.011		.54	.44		.98	1.32
0805	Pneumatic nailed	G		1736	.009		.54	.36		.90	1.18
0900	3/4" thick	G		1300	.012		.67	.48		1.15	1.53
0905	Pneumatic nailed	G		1612	.010		.67	.38		1.05	1.38
1100	Hardboard, underlayment grade, 4' x 4', .215" thick	G	↓	1500	.011	↓	.69	.41		1.10	1.44

06 16 Sheathing

06 16 33 – Wood Board Sheathing

06 16 33.10 Board Sheathing		Crew	Daily Output	Labor-Hours	Unit	Material	2019 Bare Costs Labor	Equipment	Total	Total Incl O&P
0009	BOARD SHEATHING									
0010	Roof, 1" x 6" boards, laid horizontal	2 Carp	725	.022	S.F.	1.71	.86		2.57	3.29
0020	On steep roof		520	.031		1.71	1.19		2.90	3.85
0040	On dormers, hips, & valleys		480	.033		1.71	1.29		3	4.01
0050	Laid diagonal		650	.025		1.71	.95		2.66	3.46
0070	1" x 8" boards, laid horizontal		875	.018		2.15	.71		2.86	3.53
0080	On steep roof		635	.025		2.15	.98		3.13	3.97
0090	On dormers, hips, & valleys		580	.028		2.20	1.07		3.27	4.19
0100	Laid diagonal		725	.022		2.15	.86		3.01	3.77
0110	Skip sheathing, 1" x 4", 7" OC	1 Carp	1200	.007		.66	.26		.92	1.15
0120	1" x 6", 9" OC		1450	.006		.79	.21		1	1.21
0180	T&G sheathing/decking, 1" x 6"		1000	.008		1.78	.31		2.09	2.47
0190	2" x 6"		1000	.008		3.87	.31		4.18	4.77
0200	Walls, 1" x 6" boards, laid regular	2 Carp	650	.025		1.71	.95		2.66	3.46
0210	Laid diagonal		585	.027		1.71	1.06		2.77	3.63
0220	1" x 8" boards, laid regular		765	.021		2.15	.81		2.96	3.70
0230	Laid diagonal		650	.025		2.15	.95		3.10	3.94

06 16 36 – Wood Panel Product Sheathing

06 16 36.10 Sheathing		Crew	Daily Output	Labor-Hours	Unit	Material	Labor	Equipment	Total	Total Incl O&P
0010	SHEATHING R061636-20									
0012	Plywood on roofs, CDX									
0030	5/16" thick	2 Carp	1600	.010	S.F.	.59	.39		.98	1.29
0035	Pneumatic nailed		1952	.008		.59	.32		.91	1.18
0050	3/8" thick		1525	.010		.61	.41		1.02	1.35
0055	Pneumatic nailed		1860	.009		.61	.33		.94	1.23
0100	1/2" thick		1400	.011		.63	.44		1.07	1.43
0105	Pneumatic nailed		1708	.009		.63	.36		.99	1.30
0200	5/8" thick		1300	.012		.78	.48		1.26	1.65
0205	Pneumatic nailed		1586	.010		.78	.39		1.17	1.51
0300	3/4" thick		1200	.013		.94	.52		1.46	1.88
0305	Pneumatic nailed		1464	.011		.94	.42		1.36	1.73
0500	Plywood on walls, with exterior CDX, 3/8" thick		1200	.013		.61	.52		1.13	1.53
0505	Pneumatic nailed		1488	.011		.61	.42		1.03	1.37
0600	1/2" thick		1125	.014		.63	.55		1.18	1.61
0605	Pneumatic nailed		1395	.011		.63	.44		1.07	1.43
0700	5/8" thick		1050	.015		.78	.59		1.37	1.84
0705	Pneumatic nailed		1302	.012		.78	.48		1.26	1.65
0800	3/4" thick		975	.016		.94	.64		1.58	2.08
0805	Pneumatic nailed		1209	.013		.94	.51		1.45	1.88
0900	For exterior C-C grade plywood, add					15%				
0920	For application to metal studs, joists, rafters, add						20%			
1000	For shear wall construction, add						20%			
1200	For structural 1 exterior plywood, add				S.F.	10%				
3000	Wood fiber, regular, no vapor barrier, 1/2" thick	2 Carp	1200	.013		.64	.52		1.16	1.55
3100	5/8" thick		1200	.013		.71	.52		1.23	1.63
3300	No vapor barrier, in colors, 1/2" thick		1200	.013		.81	.52		1.33	1.74
3400	5/8" thick		1200	.013		.85	.52		1.37	1.79
3600	With vapor barrier one side, white, 1/2" thick		1200	.013		.62	.52		1.14	1.53
3700	Vapor barrier 2 sides, 1/2" thick		1200	.013		.85	.52		1.37	1.79
3800	Asphalt impregnated, 25/32" thick		1200	.013		.32	.52		.84	1.20
3850	Intermediate, 1/2" thick		1200	.013		.38	.52		.90	1.27
4500	Oriented strand board, on roof, 7/16" thick G		1460	.011		.47	.42		.89	1.22

06 16 Sheathing

06 16 36 – Wood Panel Product Sheathing

06 16 36.10 Sheathing			Crew	Daily Output	Labor-Hours	Unit	2019 Bare Costs Material	Labor	Equipment	Total	Total Incl O&P
4505	Pneumatic nailed	G	2 Carp	1780	.009	S.F.	.47	.35		.82	1.10
4550	1/2" thick	G		1400	.011		.47	.44		.91	1.25
4555	Pneumatic nailed	G		1736	.009		.47	.36		.83	1.11
4600	5/8" thick	G		1300	.012		.64	.48		1.12	1.49
4605	Pneumatic nailed	G		1586	.010		.64	.39		1.03	1.35
4610	On walls, 7/16" thick	G		1200	.013		.47	.52		.99	1.37
4615	Pneumatic nailed	G		1488	.011		.47	.42		.89	1.21
4620	1/2" thick	G		1195	.013		.47	.52		.99	1.38
4625	Pneumatic nailed	G		1325	.012		.47	.47		.94	1.29
4630	5/8" thick	G		1050	.015		.64	.59		1.23	1.68
4635	Pneumatic nailed	G		1302	.012		.64	.48		1.12	1.49
4700	Oriented strand board, factory laminated W.R. barrier, on roof, 1/2" thick	G		1400	.011		.77	.44		1.21	1.58
4705	Pneumatic nailed	G		1736	.009		.77	.36		1.13	1.44
4720	5/8" thick	G		1300	.012		.93	.48		1.41	1.81
4725	Pneumatic nailed	G		1586	.010		.93	.39		1.32	1.67
4730	5/8" thick, T&G	G		1150	.014		1.24	.54		1.78	2.25
4735	Pneumatic nailed, T&G	G		1400	.011		1.24	.44		1.68	2.09
4740	On walls, 7/16" thick	G		1200	.013		.71	.52		1.23	1.63
4745	Pneumatic nailed	G		1488	.011		.71	.42		1.13	1.47
4750	1/2" thick	G		1195	.013		.77	.52		1.29	1.71
4755	Pneumatic nailed	G		1325	.012		.77	.47		1.24	1.62
4800	Joint sealant tape, 3-1/2"			7600	.002	L.F.	.30	.08		.38	.47
4810	Joint sealant tape, 6"			7600	.002	"	.41	.08		.49	.59

06 16 43 – Gypsum Sheathing

06 16 43.10 Gypsum Sheathing		Crew	Daily Output	Labor-Hours	Unit	Material	Labor	Equipment	Total	Total Incl O&P
0010	**GYPSUM SHEATHING**									
0020	Gypsum, weatherproof, 1/2" thick	2 Carp	1125	.014	S.F.	.46	.55		1.01	1.42
0040	With embedded glass mats	"	1100	.015	"	.71	.56		1.27	1.71

06 17 Shop-Fabricated Structural Wood

06 17 33 – Wood I-Joists

06 17 33.10 Wood and Composite I-Joists		Crew	Daily Output	Labor-Hours	Unit	Material	Labor	Equipment	Total	Total Incl O&P
0010	**WOOD AND COMPOSITE I-JOISTS**									
0100	Plywood webs, incl. bridging & blocking, panels 24" OC									
1200	15' to 24' span, 50 psf live load	F-5	2400	.013	SF Flr.	1.99	.45		2.44	2.94
1300	55 psf live load		2250	.014		2.29	.48		2.77	3.32
1400	24' to 30' span, 45 psf live load		2600	.012		2.93	.42		3.35	3.91
1500	55 psf live load		2400	.013		4.68	.45		5.13	5.90

06 17 53 – Shop-Fabricated Wood Trusses

06 17 53.10 Roof Trusses		Crew	Daily Output	Labor-Hours	Unit	Material	Labor	Equipment	Total	Total Incl O&P
0010	**ROOF TRUSSES**									
5000	Common wood, 2" x 4" metal plate connected, 24" OC, 4/12 slope									
5010	1' overhang, 12' span	F-5	55	.582	Ea.	36.50	19.70		56.20	73
5050	20' span	F-6	62	.645		74	23.50	7.60	105.10	128
5100	24' span		60	.667		78	24	7.85	109.85	134
5150	26' span		57	.702		78	25.50	8.25	111.75	136
5200	28' span		53	.755		98.50	27	8.90	134.40	163
5240	30' span		51	.784		107	28.50	9.25	144.75	175
5250	32' span		50	.800		114	29	9.45	152.45	183
5280	34' span		48	.833		107	30	9.85	146.85	178

06 17 Shop-Fabricated Structural Wood

06 17 53 – Shop-Fabricated Wood Trusses

06 17 53.10 Roof Trusses		Crew	Daily Output	Labor-Hours	Unit	Material	2019 Bare Costs Labor	Equipment	Total	Total Incl O&P
5350	8/12 pitch, 1′ overhang, 20′ span	F-6	57	.702	Ea.	89	25.50	8.25	122.75	148
5400	24′ span		55	.727		104	26	8.55	138.55	167
5450	26′ span		52	.769		110	27.50	9.05	146.55	177
5500	28′ span		49	.816		123	29.50	9.60	162.10	195
5550	32′ span		45	.889		144	32	10.50	186.50	224
5600	36′ span		41	.976		170	35	11.50	216.50	258
5650	38′ span		40	1		193	36	11.80	240.80	285
5700	40′ span	↓	40	1	↓	201	36	11.80	248.80	293

06 18 Glued-Laminated Construction

06 18 13 – Glued-Laminated Beams

06 18 13.10 Laminated Beams		Crew	Daily Output	Labor-Hours	Unit	Material	Labor	Equipment	Total	Total Incl O&P
0010	**LAMINATED BEAMS**									
0050	3-1/2″ x 18″	F-3	480	.083	L.F.	28	2.96	.98	31.94	36.50
0100	5-1/4″ x 11-7/8″		450	.089		27.50	3.15	1.05	31.70	36.50
0150	5-1/4″ x 16″		360	.111		38.50	3.94	1.31	43.75	50.50
0200	5-1/4″ x 18″		290	.138		42	4.90	1.63	48.53	56
0250	5-1/4″ x 24″		220	.182		61.50	6.45	2.14	70.09	80.50
0300	7″ x 11-7/8″		320	.125		42	4.44	1.47	47.91	55
0350	7″ x 16″		260	.154		58	5.45	1.81	65.26	75
0400	7″ x 18″	↓	210	.190		67.50	6.75	2.25	76.50	88
0500	For premium appearance, add to L.F. prices					5%				
0550	For industrial type, deduct					15%				
0600	For stain and varnish, add					5%				
0650	For 3/4″ laminations, add				↓	25%				

06 18 13.20 Laminated Framing		Crew	Daily Output	Labor-Hours	Unit	Material	Labor	Equipment	Total	Total Incl O&P
0010	**LAMINATED FRAMING**									
0020	30 lb., short term live load, 15 lb. dead load									
0200	Straight roof beams, 20′ clear span, beams 8′ OC	F-3	2560	.016	SF Flr.	2.20	.55	.18	2.93	3.54
0300	Beams 16′ OC		3200	.013		1.61	.44	.15	2.20	2.67
0500	40′ clear span, beams 8′ OC		3200	.013		4.17	.44	.15	4.76	5.50
0600	Beams 16′ OC	↓	3840	.010		3.45	.37	.12	3.94	4.55
0800	60′ clear span, beams 8′ OC	F-4	2880	.014		7.15	.49	.34	7.98	9.05
0900	Beams 16′ OC	″	3840	.010		5.35	.37	.26	5.98	6.80
1100	Tudor arches, 30′ to 40′ clear span, frames 8′ OC	F-3	1680	.024		9.30	.85	.28	10.43	11.95
1200	Frames 16′ OC	″	2240	.018		7.25	.63	.21	8.09	9.30
1400	50′ to 60′ clear span, frames 8′ OC	F-4	2200	.018		10	.65	.45	11.10	12.55
1500	Frames 16′ OC		2640	.015		8.55	.54	.37	9.46	10.70
1700	Radial arches, 60′ clear span, frames 8′ OC		1920	.021		9.40	.74	.51	10.65	12.10
1800	Frames 16′ OC		2880	.014		7.50	.49	.34	8.33	9.45
2000	100′ clear span, frames 8′ OC		1600	.025		9.65	.89	.61	11.15	12.80
2100	Frames 16′ OC		2400	.017		8.55	.59	.41	9.55	10.85
2300	120′ clear span, frames 8′ OC		1440	.028		12.85	.99	.68	14.52	16.50
2400	Frames 16′ OC	↓	1920	.021		11.75	.74	.51	13	14.70
2600	Bowstring trusses, 20′ OC, 40′ clear span	F-3	2400	.017		5.80	.59	.20	6.59	7.60
2700	60′ clear span	F-4	3600	.011		5.25	.39	.27	5.91	6.70
2800	100′ clear span		4000	.010		7.40	.35	.25	8	9
2900	120′ clear span	↓	3600	.011		7.85	.39	.27	8.51	9.60
3100	For premium appearance, add to S.F. prices					5%				
3300	For industrial type, deduct					15%				
3500	For stain and varnish, add				↓	5%				

06 18 Glued-Laminated Construction

06 18 13 – Glued-Laminated Beams

06 18 13.20 Laminated Framing		Crew	Daily Output	Labor-Hours	Unit	Material	2019 Bare Costs Labor	Equipment	Total	Total Incl O&P
3900	For 3/4" laminations, add to straight				SF Flr.	25%				
4100	Add to curved					15%				
4300	Alternate pricing method: (use nominal footage of									
4310	components). Straight beams, camber less than 6"	F-3	3.50	11.429	M.B.F.	3,150	405	135	3,690	4,275
4400	Columns, including hardware		2	20		3,375	710	236	4,321	5,125
4600	Curved members, radius over 32'		2.50	16		3,450	570	189	4,209	4,950
4700	Radius 10' to 32'		3	13.333		3,425	475	157	4,057	4,725
4900	For complicated shapes, add maximum					100%				
5100	For pressure treating, add to straight					35%				
5200	Add to curved					45%				
6000	Laminated veneer members, southern pine or western species									
6050	1-3/4" wide x 5-1/2" deep	2 Carp	480	.033	L.F.	3.49	1.29		4.78	5.95
6100	9-1/2" deep		480	.033		4.36	1.29		5.65	6.95
6150	14" deep		450	.036		7.55	1.38		8.93	10.60
6200	18" deep		450	.036		10.40	1.38		11.78	13.75
6300	Parallel strand members, southern pine or western species									
6350	1-3/4" wide x 9-1/4" deep	2 Carp	480	.033	L.F.	4.65	1.29		5.94	7.25
6400	11-1/4" deep		450	.036		5.10	1.38		6.48	7.90
6450	14" deep		400	.040		7.55	1.55		9.10	10.85
6500	3-1/2" wide x 9-1/4" deep		480	.033		16.05	1.29		17.34	19.80
6550	11-1/4" deep		450	.036		19.90	1.38		21.28	24.50
6600	14" deep		400	.040		23	1.55		24.55	27.50
6650	7" wide x 9-1/4" deep		450	.036		33.50	1.38		34.88	39
6700	11-1/4" deep		420	.038		42	1.48		43.48	49
6750	14" deep		400	.040		50	1.55		51.55	57.50

06 22 Millwork

06 22 13 – Standard Pattern Wood Trim

06 22 13.10 Millwork

		Crew	Daily Output	Labor-Hours	Unit	Material	Labor	Equipment	Total	Total Incl O&P
0010	**MILLWORK** R061110-30									
0020	Rule of thumb, milled material equals rough lumber cost x 3									
1020	1" x 12", custom birch				L.F.	4.70			4.70	5.15
1040	Cedar					5.55			5.55	6.10
1060	Oak					5.25			5.25	5.75
1080	Redwood					4.79			4.79	5.25
1100	Southern yellow pine					4.11			4.11	4.52
1120	Sugar pine					6.30			6.30	6.95
1140	Teak					34			34	37.50
1160	Walnut					7.60			7.60	8.40
1180	White pine					5.40			5.40	5.95

06 22 13.15 Moldings, Base

		Crew	Daily Output	Labor-Hours	Unit	Material	Labor	Equipment	Total	Total Incl O&P
0010	**MOLDINGS, BASE**									
5100	Classic profile, 5/8" x 5-1/2", finger jointed and primed	1 Carp	250	.032	L.F.	1.63	1.24		2.87	3.84
5105	Poplar		240	.033		1.90	1.29		3.19	4.22
5110	Red oak		220	.036		2.51	1.41		3.92	5.10
5115	Maple		220	.036		3.79	1.41		5.20	6.50
5120	Cherry		220	.036		4.49	1.41		5.90	7.25
5125	3/4" x 7-1/2", finger jointed and primed		250	.032		2.02	1.24		3.26	4.27
5130	Poplar		240	.033		2.61	1.29		3.90	5
5135	Red oak		220	.036		3.63	1.41		5.04	6.30
5140	Maple		220	.036		5.05	1.41		6.46	7.90

06 22 13.15 Moldings, Base		Crew	Daily Output	Labor-Hours	Unit	2019 Bare Costs Material	Labor	Equipment	Total	Total Incl O&P
5145	Cherry	1 Carp	220	.036	L.F.	6	1.41		7.41	8.95
5150	Modern profile, 5/8" x 3-1/2", finger jointed and primed		250	.032		.97	1.24		2.21	3.12
5155	Poplar		240	.033		1.04	1.29		2.33	3.27
5160	Red oak		220	.036		1.71	1.41		3.12	4.21
5165	Maple		220	.036		2.65	1.41		4.06	5.25
5170	Cherry		220	.036		2.92	1.41		4.33	5.55
5175	Ogee profile, 7/16" x 3", finger jointed and primed		250	.032		.68	1.24		1.92	2.80
5180	Poplar		240	.033		.77	1.29		2.06	2.98
5185	Red oak		220	.036		.97	1.41		2.38	3.40
5200	9/16" x 3-1/2", finger jointed and primed		250	.032		.67	1.24		1.91	2.79
5205	Pine		240	.033		1.21	1.29		2.50	3.46
5210	Red oak		220	.036		2.89	1.41		4.30	5.50
5215	9/16" x 4-1/2", red oak		220	.036		4.33	1.41		5.74	7.10
5220	5/8" x 3-1/2", finger jointed and primed		250	.032		1.01	1.24		2.25	3.16
5225	Poplar		240	.033		1.04	1.29		2.33	3.27
5230	Red oak		220	.036		1.71	1.41		3.12	4.21
5235	Maple		220	.036		2.65	1.41		4.06	5.25
5240	Cherry		220	.036		2.87	1.41		4.28	5.50
5245	5/8" x 4", finger jointed and primed		250	.032		1.26	1.24		2.50	3.43
5250	Poplar		240	.033		1.34	1.29		2.63	3.60
5255	Red oak		220	.036		1.92	1.41		3.33	4.44
5260	Maple		220	.036		2.97	1.41		4.38	5.60
5265	Cherry		220	.036		3.11	1.41		4.52	5.75
5270	Rectangular profile, oak, 3/8" x 1-1/4"		260	.031		1.27	1.19		2.46	3.36
5275	1/2" x 2-1/2"		255	.031		2.25	1.22		3.47	4.48
5280	1/2" x 3-1/2"		250	.032		2.83	1.24		4.07	5.15
5285	1" x 6"		240	.033		4.09	1.29		5.38	6.65
5290	1" x 8"		240	.033		5.05	1.29		6.34	7.75
5295	Pine, 3/8" x 1-3/4"		260	.031		.50	1.19		1.69	2.52
5300	7/16" x 2-1/2"		255	.031		.79	1.22		2.01	2.88
5305	1" x 6"		240	.033		.73	1.29		2.02	2.93
5310	1" x 8"		240	.033		.94	1.29		2.23	3.16
5315	Shoe, 1/2" x 3/4", primed		260	.031		.53	1.19		1.72	2.55
5320	Pine		240	.033		.35	1.29		1.64	2.51
5325	Poplar		240	.033		.42	1.29		1.71	2.59
5330	Red oak		220	.036		.56	1.41		1.97	2.94
5335	Maple		220	.036		.84	1.41		2.25	3.25
5340	Cherry		220	.036		.85	1.41		2.26	3.26
5345	11/16" x 1-1/2", pine		240	.033		.74	1.29		2.03	2.94
5350	Caps, 11/16" x 1-3/8", pine		240	.033		.61	1.29		1.90	2.80
5355	3/4" x 1-3/4", finger jointed and primed		260	.031		.88	1.19		2.07	2.94
5360	Poplar		240	.033		1.05	1.29		2.34	3.28
5365	Red oak		220	.036		1.34	1.41		2.75	3.80
5370	Maple		220	.036		1.65	1.41		3.06	4.14
5375	Cherry		220	.036		3.34	1.41		4.75	6
5380	Combination base & shoe, 9/16" x 3-1/2" & 1/2" x 3/4", pine		125	.064		1.56	2.48		4.04	5.80
5385	Three piece oak, 6" high		80	.100		6	3.88		9.88	13
5390	Including 3/4" x 1" base shoe		70	.114		6.50	4.43		10.93	14.45
5395	Flooring cant strip, 3/4" x 3/4", pre-finished pine		260	.031		.45	1.19		1.64	2.46
5400	For pre-finished, stain and clear coat, add					.57			.57	.63
5405	Clear coat only, add					.44			.44	.48

06 22 13 – Standard Pattern Wood Trim

06 22 13.30 Moldings, Casings		Crew	Daily Output	Labor-Hours	Unit	Material	Labor	Equipment	Total	Total Incl O&P
						2019 Bare Costs				
0010	**MOLDINGS, CASINGS**									
0085	Apron, 9/16" x 2-1/2", pine	1 Carp	250	.032	L.F.	1.44	1.24		2.68	3.63
0090	5/8" x 2-1/2", pine		250	.032		1.68	1.24		2.92	3.90
0110	5/8" x 3-1/2", pine		220	.036		1.99	1.41		3.40	4.52
0300	Band, 11/16" x 1-1/8", pine		270	.030		.76	1.15		1.91	2.73
0310	11/16" x 1-1/2", finger jointed and primed		270	.030		.76	1.15		1.91	2.73
0320	Pine		270	.030		1.02	1.15		2.17	3.02
0330	11/16" x 1-3/4", finger jointed and primed		270	.030		.95	1.15		2.10	2.94
0350	Pine		270	.030		1.03	1.15		2.18	3.03
0355	Beaded, 3/4" x 3-1/2", finger jointed and primed		220	.036		.97	1.41		2.38	3.40
0360	Poplar		220	.036		1.16	1.41		2.57	3.60
0365	Red oak		220	.036		1.71	1.41		3.12	4.21
0370	Maple		220	.036		2.65	1.41		4.06	5.25
0375	Cherry		220	.036		3.21	1.41		4.62	5.85
0380	3/4" x 4", finger jointed and primed		220	.036		1.25	1.41		2.66	3.70
0385	Poplar		220	.036		1.65	1.41		3.06	4.14
0390	Red oak		220	.036		2.35	1.41		3.76	4.91
0395	Maple		220	.036		2.68	1.41		4.09	5.30
0400	Cherry		220	.036		3.85	1.41		5.26	6.55
0405	3/4" x 5-1/2", finger jointed and primed		200	.040		1.49	1.55		3.04	4.20
0410	Poplar		200	.040		2.01	1.55		3.56	4.77
0415	Red oak		200	.040		2.95	1.55		4.50	5.80
0420	Maple		200	.040		3.89	1.55		5.44	6.85
0425	Cherry		200	.040		4.47	1.55		6.02	7.50
0430	Classic profile, 3/4" x 2-3/4", finger jointed and primed		250	.032		.87	1.24		2.11	3.01
0435	Poplar		250	.032		1.05	1.24		2.29	3.20
0440	Red oak		250	.032		1.54	1.24		2.78	3.74
0445	Maple		250	.032		2.18	1.24		3.42	4.45
0450	Cherry		250	.032		2.51	1.24		3.75	4.81
0455	Fluted, 3/4" x 3-1/2", poplar		220	.036		1.18	1.41		2.59	3.63
0460	Red oak		220	.036		1.72	1.41		3.13	4.22
0465	Maple		220	.036		3.15	1.41		4.56	5.80
0470	Cherry		220	.036		3.21	1.41		4.62	5.85
0475	3/4" x 4", poplar		220	.036		1.45	1.41		2.86	3.92
0480	Red oak		220	.036		2.05	1.41		3.46	4.58
0485	Maple		220	.036		2.68	1.41		4.09	5.30
0490	Cherry		220	.036		3.85	1.41		5.26	6.55
0495	3/4" x 5-1/2", poplar		200	.040		2.01	1.55		3.56	4.77
0500	Red oak		200	.040		3.07	1.55		4.62	5.95
0505	Maple		200	.040		3.89	1.55		5.44	6.85
0510	Cherry		200	.040		4.47	1.55		6.02	7.50
0515	3/4" x 7-1/2", poplar		190	.042		1.28	1.63		2.91	4.11
0520	Red oak		190	.042		3.96	1.63		5.59	7.05
0525	Maple		190	.042		6.75	1.63		8.38	10.15
0530	Cherry		190	.042		8.10	1.63		9.73	11.60
0535	3/4" x 9-1/2", poplar		180	.044		4.16	1.72		5.88	7.40
0540	Red oak		180	.044		6.60	1.72		8.32	10.10
0545	Maple		180	.044		10.80	1.72		12.52	14.70
0550	Cherry		180	.044		11.75	1.72		13.47	15.80
0555	Modern profile, 9/16" x 2-1/4", poplar		250	.032		.84	1.24		2.08	2.97
0560	Red oak		250	.032		.94	1.24		2.18	3.08
0565	11/16" x 2-1/2", finger jointed & primed		250	.032		.86	1.24		2.10	2.99

06 22 13 – Standard Pattern Wood Trim

06 22 13.30 Moldings, Casings		Crew	Daily Output	Labor-Hours	Unit	Material	Labor	Equipment	Total	Total Incl O&P
						2019 Bare Costs				
0570	Pine	1 Carp	250	.032	L.F.	1.38	1.24		2.62	3.57
0575	3/4" x 2-1/2", poplar		250	.032		.93	1.24		2.17	3.07
0580	Red oak		250	.032		1.24	1.24		2.48	3.41
0585	Maple		250	.032		1.91	1.24		3.15	4.15
0590	Cherry		250	.032		2.73	1.24		3.97	5.05
0595	Mullion, 5/16" x 2", pine		270	.030		.91	1.15		2.06	2.90
0600	9/16" x 2-1/2", finger jointed and primed		250	.032		.98	1.24		2.22	3.13
0605	Pine		250	.032		1.35	1.24		2.59	3.53
0610	Red oak		250	.032		3.37	1.24		4.61	5.75
0615	1-1/16" x 3-3/4", red oak		220	.036		7	1.41		8.41	10.05
0620	Ogee, 7/16" x 2-1/2", poplar		250	.032		.74	1.24		1.98	2.86
0625	Red oak		250	.032		.90	1.24		2.14	3.04
0630	9/16" x 2-1/4", finger jointed and primed		250	.032		.60	1.24		1.84	2.71
0635	Poplar		250	.032		.62	1.24		1.86	2.73
0640	Red oak		250	.032		.83	1.24		2.07	2.96
0645	11/16" x 2-1/2", finger jointed and primed		250	.032		.72	1.24		1.96	2.84
0700	Pine		250	.032		1.42	1.24		2.66	3.61
0701	Red oak		250	.032		3.28	1.24		4.52	5.65
0730	11/16" x 3-1/2", finger jointed and primed		220	.036		1.37	1.41		2.78	3.84
0750	Pine		220	.036		1.85	1.41		3.26	4.36
0755	3/4" x 2-1/2", finger jointed and primed		250	.032		.75	1.24		1.99	2.87
0760	Poplar		250	.032		.96	1.24		2.20	3.10
0765	Red oak		250	.032		1.29	1.24		2.53	3.47
0770	Maple		250	.032		1.88	1.24		3.12	4.12
0775	Cherry		250	.032		2.36	1.24		3.60	4.64
0780	3/4" x 3-1/2", finger jointed and primed		220	.036		.98	1.41		2.39	3.41
0785	Poplar		220	.036		1.18	1.41		2.59	3.63
0790	Red oak		220	.036		1.72	1.41		3.13	4.22
0795	Maple		220	.036		2.65	1.41		4.06	5.25
0800	Cherry		220	.036		3.26	1.41		4.67	5.90
4700	Square profile, 1" x 1", teak		215	.037		2.31	1.44		3.75	4.92
4800	Rectangular profile, 1" x 3", teak		200	.040		6.75	1.55		8.30	10

06 22 13.35 Moldings, Ceilings		Crew	Daily Output	Labor-Hours	Unit	Material	Labor	Equipment	Total	Total Incl O&P
0010	**MOLDINGS, CEILINGS**									
0600	Bed, 9/16" x 1-3/4", pine	1 Carp	270	.030	L.F.	1.21	1.15		2.36	3.23
0650	9/16" x 2", pine		270	.030		1.22	1.15		2.37	3.24
0710	9/16" x 1-3/4", oak		270	.030		2.38	1.15		3.53	4.52
1200	Cornice, 9/16" x 1-3/4", pine		270	.030		.99	1.15		2.14	2.99
1300	9/16" x 2-1/4", pine		265	.030		1.32	1.17		2.49	3.38
1350	Cove, 1/2" x 2-1/4", poplar		265	.030		1.17	1.17		2.34	3.21
1360	Red oak		265	.030		1.56	1.17		2.73	3.64
1370	Hard maple		265	.030		1.67	1.17		2.84	3.76
1380	Cherry		265	.030		2.28	1.17		3.45	4.43
2400	9/16" x 1-3/4", pine		270	.030		1	1.15		2.15	3
2401	Oak		270	.030		.99	1.15		2.14	2.99
2500	11/16" x 2-3/4", pine		265	.030		1.90	1.17		3.07	4.02
2510	Crown, 5/8" x 5/8", poplar		300	.027		.47	1.03		1.50	2.23
2520	Red oak		300	.027		.55	1.03		1.58	2.31
2530	Hard maple		300	.027		.76	1.03		1.79	2.54
2540	Cherry		300	.027		.80	1.03		1.83	2.59
2600	9/16" x 3-5/8", pine		250	.032		2.13	1.24		3.37	4.39
2700	11/16" x 4-1/4", pine		250	.032		3.02	1.24		4.26	5.35

06 22 13 – Standard Pattern Wood Trim

06 22 13.35 Moldings, Ceilings		Crew	Daily Output	Labor-Hours	Unit	Material	Labor	Equipment	Total	Total Incl O&P
						2019 Bare Costs				
2705	Oak	1 Carp	250	.032	L.F.	6.40	1.24		7.64	9.10
2710	3/4" x 1-3/4", poplar		270	.030		.75	1.15		1.90	2.72
2720	Red oak		270	.030		1.11	1.15		2.26	3.12
2730	Hard maple		270	.030		1.46	1.15		2.61	3.50
2740	Cherry		270	.030		1.79	1.15		2.94	3.87
2750	3/4" x 2", poplar		270	.030		.99	1.15		2.14	2.99
2760	Red oak		270	.030		1.32	1.15		2.47	3.35
2770	Hard maple		270	.030		1.96	1.15		3.11	4.05
2780	Cherry		270	.030		1.99	1.15		3.14	4.09
2790	3/4" x 2-3/4", poplar		265	.030		1.07	1.17		2.24	3.10
2800	Red oak		265	.030		1.65	1.17		2.82	3.74
2810	Hard maple		265	.030		2.24	1.17		3.41	4.39
2820	Cherry		265	.030		2.54	1.17		3.71	4.72
2830	3/4" x 3-1/2", poplar		250	.032		1.38	1.24		2.62	3.56
2840	Red oak		250	.032		2.05	1.24		3.29	4.30
2850	Hard maple		250	.032		3	1.24		4.24	5.35
2860	Cherry		250	.032		3.12	1.24		4.36	5.50
2870	FJP poplar		250	.032		1.05	1.24		2.29	3.20
2880	3/4" x 5", poplar		245	.033		2.01	1.27		3.28	4.30
2890	Red oak		245	.033		2.99	1.27		4.26	5.40
2900	Hard maple		245	.033		3.95	1.27		5.22	6.45
2910	Cherry		245	.033		4.77	1.27		6.04	7.35
2920	FJP poplar		245	.033		1.47	1.27		2.74	3.70
2930	3/4" x 6-1/4", poplar		240	.033		2.42	1.29		3.71	4.79
2940	Red oak		240	.033		3.64	1.29		4.93	6.15
2950	Hard maple		240	.033		4.96	1.29		6.25	7.60
2960	Cherry		240	.033		5.90	1.29		7.19	8.65
2970	7/8" x 8-3/4", poplar		220	.036		4.50	1.41		5.91	7.25
2980	Red oak		220	.036		6.15	1.41		7.56	9.10
2990	Hard maple		220	.036		8.50	1.41		9.91	11.70
3000	Cherry		220	.036		10.05	1.41		11.46	13.40
3010	1" x 7-1/4", poplar		220	.036		4.57	1.41		5.98	7.35
3020	Red oak		220	.036		6.15	1.41		7.56	9.10
3030	Hard maple		220	.036		9.10	1.41		10.51	12.35
3040	Cherry		220	.036		10	1.41		11.41	13.35
3050	1-1/16" x 4-1/4", poplar		250	.032		2.56	1.24		3.80	4.86
3060	Red oak		250	.032		2.82	1.24		4.06	5.15
3070	Hard maple		250	.032		4.30	1.24		5.54	6.80
3080	Cherry		250	.032		5.35	1.24		6.59	7.95
3090	Dentil crown, 3/4" x 5", poplar		250	.032		2.01	1.24		3.25	4.26
3100	Red oak		250	.032		2.99	1.24		4.23	5.35
3110	Hard maple		250	.032		3.93	1.24		5.17	6.35
3120	Cherry		250	.032		4.54	1.24		5.78	7.05
3130	Dentil piece for above, 1/2" x 1/2", poplar		300	.027		3.31	1.03		4.34	5.35
3140	Red oak		300	.027		3.29	1.03		4.32	5.35
3150	Hard maple		300	.027		4.26	1.03		5.29	6.40
3160	Cherry		300	.027		4.10	1.03		5.13	6.20

06 22 13.40 Moldings, Exterior		Crew	Daily Output	Labor-Hours	Unit	Material	Labor	Equipment	Total	Total Incl O&P
0010	**MOLDINGS, EXTERIOR**									
0100	Band board, cedar, rough sawn, 1" x 2"	1 Carp	300	.027	L.F.	.69	1.03		1.72	2.47
0110	1" x 3"		300	.027		1.03	1.03		2.06	2.84
0120	1" x 4"		250	.032		1.37	1.24		2.61	3.55

06 22 13.40 Moldings, Exterior		Crew	Daily Output	Labor-Hours	Unit	2019 Bare Costs Material	Labor	Equipment	Total	Total Incl O&P
0130	1" x 6"	1 Carp	250	.032	L.F.	2.05	1.24		3.29	4.31
0140	1" x 8"		225	.036		2.74	1.38		4.12	5.30
0150	1" x 10"		225	.036		3.40	1.38		4.78	6
0160	1" x 12"		200	.040		4.09	1.55		5.64	7.05
0240	STK, 1" x 2"		300	.027		.46	1.03		1.49	2.22
0250	1" x 3"		300	.027		.50	1.03		1.53	2.26
0260	1" x 4"		250	.032		.82	1.24		2.06	2.95
0270	1" x 6"		250	.032		1.34	1.24		2.58	3.52
0280	1" x 8"		225	.036		2.29	1.38		3.67	4.80
0290	1" x 10"		225	.036		2.81	1.38		4.19	5.35
0300	1" x 12"		200	.040		4.42	1.55		5.97	7.40
0310	Pine, #2, 1" x 2"		300	.027		.29	1.03		1.32	2.03
0320	1" x 3"		300	.027		.46	1.03		1.49	2.21
0330	1" x 4"		250	.032		.56	1.24		1.80	2.67
0340	1" x 6"		250	.032		.82	1.24		2.06	2.96
0350	1" x 8"		225	.036		1.38	1.38		2.76	3.80
0360	1" x 10"		225	.036		1.76	1.38		3.14	4.22
0370	1" x 12"		200	.040		2.20	1.55		3.75	4.98
0380	D & better, 1" x 2"		300	.027		.50	1.03		1.53	2.26
0390	1" x 3"		300	.027		.65	1.03		1.68	2.42
0400	1" x 4"		250	.032		.83	1.24		2.07	2.96
0410	1" x 6"		250	.032		1.10	1.24		2.34	3.26
0420	1" x 8"		225	.036		1.63	1.38		3.01	4.07
0430	1" x 10"		225	.036		2.20	1.38		3.58	4.70
0440	1" x 12"		200	.040		2.71	1.55		4.26	5.55
0450	Redwood, clear all heart, 1" x 2"		300	.027		.65	1.03		1.68	2.42
0460	1" x 3"		300	.027		.96	1.03		1.99	2.77
0470	1" x 4"		250	.032		1.22	1.24		2.46	3.39
0480	1" x 6"		252	.032		1.81	1.23		3.04	4.03
0490	1" x 8"		225	.036		2.41	1.38		3.79	4.93
0500	1" x 10"		225	.036		4.02	1.38		5.40	6.70
0510	1" x 12"		200	.040		4.85	1.55		6.40	7.90
0530	Corner board, cedar, rough sawn, 1" x 2"		225	.036		.69	1.38		2.07	3.04
0540	1" x 3"		225	.036		1.03	1.38		2.41	3.41
0550	1" x 4"		200	.040		1.37	1.55		2.92	4.06
0560	1" x 6"		200	.040		2.05	1.55		3.60	4.82
0570	1" x 8"		200	.040		2.74	1.55		4.29	5.55
0580	1" x 10"		175	.046		3.40	1.77		5.17	6.65
0590	1" x 12"		175	.046		4.09	1.77		5.86	7.45
0670	STK, 1" x 2"		225	.036		.46	1.38		1.84	2.79
0680	1" x 3"		225	.036		.50	1.38		1.88	2.83
0690	1" x 4"		200	.040		.80	1.55		2.35	3.44
0700	1" x 6"		200	.040		1.34	1.55		2.89	4.03
0710	1" x 8"		200	.040		2.29	1.55		3.84	5.10
0720	1" x 10"		175	.046		2.81	1.77		4.58	6
0730	1" x 12"		175	.046		4.42	1.77		6.19	7.80
0740	Pine, #2, 1" x 2"		225	.036		.29	1.38		1.67	2.60
0750	1" x 3"		225	.036		.46	1.38		1.84	2.78
0760	1" x 4"		200	.040		.56	1.55		2.11	3.18
0770	1" x 6"		200	.040		.82	1.55		2.37	3.47
0780	1" x 8"		200	.040		1.38	1.55		2.93	4.08
0790	1" x 10"		175	.046		1.76	1.77		3.53	4.87
0800	1" x 12"		175	.046		2.20	1.77		3.97	5.35

06 22 13.40 Moldings, Exterior		Crew	Daily Output	Labor-Hours	Unit	Material	Labor	Equipment	Total	Total Incl O&P
						2019 Bare Costs				
0810	D & better, 1" x 2"	1 Carp	225	.036	L.F.	.50	1.38		1.88	2.83
0820	1" x 3"		225	.036		.65	1.38		2.03	2.99
0830	1" x 4"		200	.040		.83	1.55		2.38	3.47
0840	1" x 6"		200	.040		1.10	1.55		2.65	3.77
0850	1" x 8"		200	.040		1.63	1.55		3.18	4.35
0860	1" x 10"		175	.046		2.20	1.77		3.97	5.35
0870	1" x 12"		175	.046		2.71	1.77		4.48	5.90
0880	Redwood, clear all heart, 1" x 2"		225	.036		.65	1.38		2.03	2.99
0890	1" x 3"		225	.036		.96	1.38		2.34	3.34
0900	1" x 4"		200	.040		1.22	1.55		2.77	3.90
0910	1" x 6"		200	.040		1.81	1.55		3.36	4.56
0920	1" x 8"		200	.040		2.41	1.55		3.96	5.20
0930	1" x 10"		175	.046		4.02	1.77		5.79	7.35
0940	1" x 12"		175	.046		4.85	1.77		6.62	8.30
0950	Cornice board, cedar, rough sawn, 1" x 2"		330	.024		.69	.94		1.63	2.31
0960	1" x 3"		290	.028		1.03	1.07		2.10	2.90
0970	1" x 4"		250	.032		1.37	1.24		2.61	3.55
0980	1" x 6"		250	.032		2.05	1.24		3.29	4.31
0990	1" x 8"		200	.040		2.74	1.55		4.29	5.55
1000	1" x 10"		180	.044		3.40	1.72		5.12	6.60
1010	1" x 12"		180	.044		4.09	1.72		5.81	7.35
1020	STK, 1" x 2"		330	.024		.46	.94		1.40	2.06
1030	1" x 3"		290	.028		.50	1.07		1.57	2.32
1040	1" x 4"		250	.032		.82	1.24		2.06	2.95
1050	1" x 6"		250	.032		1.34	1.24		2.58	3.52
1060	1" x 8"		200	.040		2.29	1.55		3.84	5.10
1070	1" x 10"		180	.044		2.81	1.72		4.53	5.95
1080	1" x 12"		180	.044		4.42	1.72		6.14	7.70
1500	Pine, #2, 1" x 2"		330	.024		.29	.94		1.23	1.87
1510	1" x 3"		290	.028		.31	1.07		1.38	2.11
1600	1" x 4"		250	.032		.56	1.24		1.80	2.67
1700	1" x 6"		250	.032		.82	1.24		2.06	2.96
1800	1" x 8"		200	.040		1.38	1.55		2.93	4.08
1900	1" x 10"		180	.044		1.76	1.72		3.48	4.79
2000	1" x 12"		180	.044		2.20	1.72		3.92	5.25
2020	D & better, 1" x 2"		330	.024		.50	.94		1.44	2.10
2030	1" x 3"		290	.028		.65	1.07		1.72	2.48
2040	1" x 4"		250	.032		.83	1.24		2.07	2.96
2050	1" x 6"		250	.032		1.10	1.24		2.34	3.26
2060	1" x 8"		200	.040		1.63	1.55		3.18	4.35
2070	1" x 10"		180	.044		2.20	1.72		3.92	5.25
2080	1" x 12"		180	.044		2.71	1.72		4.43	5.85
2090	Redwood, clear all heart, 1" x 2"		330	.024		.65	.94		1.59	2.26
2100	1" x 3"		290	.028		.96	1.07		2.03	2.83
2110	1" x 4"		250	.032		1.22	1.24		2.46	3.39
2120	1" x 6"		250	.032		1.81	1.24		3.05	4.05
2130	1" x 8"		200	.040		2.41	1.55		3.96	5.20
2140	1" x 10"		180	.044		4.02	1.72		5.74	7.25
2150	1" x 12"		180	.044		4.85	1.72		6.57	8.20
2160	3 piece, 1" x 2", 1" x 4", 1" x 6", rough sawn cedar		80	.100		4.13	3.88		8.01	10.95
2180	STK cedar		80	.100		2.62	3.88		6.50	9.30
2200	#2 pine		80	.100		1.67	3.88		5.55	8.25
2210	D & better pine		80	.100		2.43	3.88		6.31	9.05

06 22 13.40 Moldings, Exterior		Crew	Daily Output	Labor-Hours	Unit	Material	Labor	Equipment	Total	Total Incl O&P
						2019 Bare Costs				
2220	Clear all heart redwood	1 Carp	80	.100	L.F.	3.68	3.88		7.56	10.45
2230	1" x 8", 1" x 10", 1" x 12", rough sawn cedar		65	.123		10.20	4.77		14.97	19.10
2240	STK cedar		65	.123		9.45	4.77		14.22	18.30
2300	#2 pine		65	.123		5.30	4.77		10.07	13.75
2320	D & better pine		65	.123		6.50	4.77		11.27	15.05
2330	Clear all heart redwood		65	.123		11.25	4.77		16.02	20.50
2340	Door/window casing, cedar, rough sawn, 1" x 2"		275	.029		.69	1.13		1.82	2.62
2350	1" x 3"		275	.029		1.03	1.13		2.16	2.99
2360	1" x 4"		250	.032		1.37	1.24		2.61	3.55
2370	1" x 6"		250	.032		2.05	1.24		3.29	4.31
2380	1" x 8"		230	.035		2.74	1.35		4.09	5.25
2390	1" x 10"		230	.035		3.40	1.35		4.75	5.95
2395	1" x 12"		210	.038		4.09	1.48		5.57	6.95
2410	STK, 1" x 2"		275	.029		.46	1.13		1.59	2.37
2420	1" x 3"		275	.029		.50	1.13		1.63	2.41
2430	1" x 4"		250	.032		.82	1.24		2.06	2.95
2440	1" x 6"		250	.032		1.34	1.24		2.58	3.52
2450	1" x 8"		230	.035		2.29	1.35		3.64	4.75
2460	1" x 10"		230	.035		2.81	1.35		4.16	5.30
2470	1" x 12"		210	.038		4.42	1.48		5.90	7.30
2550	Pine, #2, 1" x 2"		275	.029		.29	1.13		1.42	2.18
2560	1" x 3"		275	.029		.46	1.13		1.59	2.36
2570	1" x 4"		250	.032		.56	1.24		1.80	2.67
2580	1" x 6"		250	.032		.82	1.24		2.06	2.96
2590	1" x 8"		230	.035		1.38	1.35		2.73	3.75
2600	1" x 10"		230	.035		1.76	1.35		3.11	4.17
2610	1" x 12"		210	.038		2.20	1.48		3.68	4.86
2620	Pine, D & better, 1" x 2"		275	.029		.50	1.13		1.63	2.41
2630	1" x 3"		275	.029		.65	1.13		1.78	2.57
2640	1" x 4"		250	.032		.83	1.24		2.07	2.96
2650	1" x 6"		250	.032		1.10	1.24		2.34	3.26
2660	1" x 8"		230	.035		1.63	1.35		2.98	4.02
2670	1" x 10"		230	.035		2.20	1.35		3.55	4.65
2680	1" x 12"		210	.038		2.71	1.48		4.19	5.40
2690	Redwood, clear all heart, 1" x 2"		275	.029		.65	1.13		1.78	2.57
2695	1" x 3"		275	.029		.96	1.13		2.09	2.92
2710	1" x 4"		250	.032		1.22	1.24		2.46	3.39
2715	1" x 6"		250	.032		1.81	1.24		3.05	4.05
2730	1" x 8"		230	.035		2.41	1.35		3.76	4.88
2740	1" x 10"		230	.035		4.02	1.35		5.37	6.65
2750	1" x 12"		210	.038		4.85	1.48		6.33	7.80
3500	Bellyband, pine, 11/16" x 4-1/4"		250	.032		2.84	1.24		4.08	5.15
3610	Brickmold, pine, 1-1/4" x 2"		200	.040		2.11	1.55		3.66	4.88
3620	FJP, 1-1/4" x 2"		200	.040		.99	1.55		2.54	3.65
5100	Fascia, cedar, rough sawn, 1" x 2"		275	.029		.69	1.13		1.82	2.62
5110	1" x 3"		275	.029		1.03	1.13		2.16	2.99
5120	1" x 4"		250	.032		1.37	1.24		2.61	3.55
5200	1" x 6"		250	.032		2.05	1.24		3.29	4.31
5300	1" x 8"		230	.035		2.74	1.35		4.09	5.25
5310	1" x 10"		230	.035		3.40	1.35		4.75	5.95
5320	1" x 12"		210	.038		4.09	1.48		5.57	6.95
5400	2" x 4"		220	.036		1.11	1.41		2.52	3.55
5500	2" x 6"		220	.036		1.66	1.41		3.07	4.16

06 22 13 – Standard Pattern Wood Trim

06 22 13.40 Moldings, Exterior		Crew	Daily Output	Labor-Hours	Unit	Material	Labor	Equipment	Total	Total Incl O&P
5600	2" x 8"	1 Carp	200	.040	L.F.	2.21	1.55		3.76	4.99
5700	2" x 10"		180	.044		2.75	1.72		4.47	5.90
5800	2" x 12"		170	.047		6.55	1.82		8.37	10.20
6120	STK, 1" x 2"		275	.029		.46	1.13		1.59	2.37
6130	1" x 3"		275	.029		.50	1.13		1.63	2.41
6140	1" x 4"		250	.032		.82	1.24		2.06	2.95
6150	1" x 6"		250	.032		1.34	1.24		2.58	3.52
6160	1" x 8"		230	.035		2.29	1.35		3.64	4.75
6170	1" x 10"		230	.035		2.81	1.35		4.16	5.30
6180	1" x 12"		210	.038		4.42	1.48		5.90	7.30
6185	2" x 2"		260	.031		.75	1.19		1.94	2.79
6190	Pine, #2, 1" x 2"		275	.029		.29	1.13		1.42	2.18
6200	1" x 3"		275	.029		.46	1.13		1.59	2.36
6210	1" x 4"		250	.032		.56	1.24		1.80	2.67
6220	1" x 6"		250	.032		.82	1.24		2.06	2.96
6230	1" x 8"		230	.035		1.38	1.35		2.73	3.75
6240	1" x 10"		230	.035		1.76	1.35		3.11	4.17
6250	1" x 12"		210	.038		2.20	1.48		3.68	4.86
6260	D & better, 1" x 2"		275	.029		.50	1.13		1.63	2.41
6270	1" x 3"		275	.029		.65	1.13		1.78	2.57
6280	1" x 4"		250	.032		.83	1.24		2.07	2.96
6290	1" x 6"		250	.032		1.10	1.24		2.34	3.26
6300	1" x 8"		230	.035		1.63	1.35		2.98	4.02
6310	1" x 10"		230	.035		2.20	1.35		3.55	4.65
6312	1" x 12"		210	.038		2.71	1.48		4.19	5.40
6330	Southern yellow, 1-1/4" x 5"		240	.033		3.07	1.29		4.36	5.50
6340	1-1/4" x 6"		240	.033		2.60	1.29		3.89	4.99
6350	1-1/4" x 8"		215	.037		3.68	1.44		5.12	6.45
6360	1-1/4" x 12"		190	.042		5.35	1.63		6.98	8.60
6370	Redwood, clear all heart, 1" x 2"		275	.029		.65	1.13		1.78	2.57
6380	1" x 3"		275	.029		1.22	1.13		2.35	3.20
6390	1" x 4"		250	.032		1.22	1.24		2.46	3.39
6400	1" x 6"		250	.032		1.81	1.24		3.05	4.05
6410	1" x 8"		230	.035		2.41	1.35		3.76	4.88
6420	1" x 10"		230	.035		4.02	1.35		5.37	6.65
6430	1" x 12"		210	.038		4.85	1.48		6.33	7.80
6440	1-1/4" x 5"		240	.033		1.90	1.29		3.19	4.22
6450	1-1/4" x 6"		240	.033		2.27	1.29		3.56	4.62
6460	1-1/4" x 8"		215	.037		3.60	1.44		5.04	6.35
6470	1-1/4" x 12"		190	.042		7.25	1.63		8.88	10.70
6580	Frieze, cedar, rough sawn, 1" x 2"		275	.029		.69	1.13		1.82	2.62
6590	1" x 3"		275	.029		1.03	1.13		2.16	2.99
6600	1" x 4"		250	.032		1.37	1.24		2.61	3.55
6610	1" x 6"		250	.032		2.05	1.24		3.29	4.31
6620	1" x 8"		250	.032		2.74	1.24		3.98	5.05
6630	1" x 10"		225	.036		3.40	1.38		4.78	6
6640	1" x 12"		200	.040		4.05	1.55		5.60	7
6650	STK, 1" x 2"		275	.029		.46	1.13		1.59	2.37
6660	1" x 3"		275	.029		.50	1.13		1.63	2.41
6670	1" x 4"		250	.032		.82	1.24		2.06	2.95
6680	1" x 6"		250	.032		1.34	1.24		2.58	3.52
6690	1" x 8"		250	.032		2.29	1.24		3.53	4.57
6700	1" x 10"		225	.036		2.81	1.38		4.19	5.35

Material, Labor, Equipment and Total columns: 2019 Bare Costs

06 22 13 – Standard Pattern Wood Trim

06 22 13.40 Moldings, Exterior		Crew	Daily Output	Labor-Hours	Unit	Material	Labor	Equipment	Total	Total Incl O&P
						2019 Bare Costs				
6710	1″ x 12″	1 Carp	200	.040	L.F.	4.42	1.55		5.97	7.40
6790	Pine, #2, 1″ x 2″		275	.029		.29	1.13		1.42	2.18
6800	1″ x 3″		275	.029		.46	1.13		1.59	2.36
6810	1″ x 4″		250	.032		.56	1.24		1.80	2.67
6820	1″ x 6″		250	.032		.82	1.24		2.06	2.96
6830	1″ x 8″		250	.032		1.38	1.24		2.62	3.57
6840	1″ x 10″		225	.036		1.76	1.38		3.14	4.22
6850	1″ x 12″		200	.040		2.20	1.55		3.75	4.98
6860	D & better, 1″ x 2″		275	.029		.50	1.13		1.63	2.41
6870	1″ x 3″		275	.029		.65	1.13		1.78	2.57
6880	1″ x 4″		250	.032		.83	1.24		2.07	2.96
6890	1″ x 6″		250	.032		1.10	1.24		2.34	3.26
6900	1″ x 8″		250	.032		1.63	1.24		2.87	3.84
6910	1″ x 10″		225	.036		2.20	1.38		3.58	4.70
6920	1″ x 12″		200	.040		2.71	1.55		4.26	5.55
6930	Redwood, clear all heart, 1″ x 2″		275	.029		.65	1.13		1.78	2.57
6940	1″ x 3″		275	.029		.96	1.13		2.09	2.92
6950	1″ x 4″		250	.032		1.22	1.24		2.46	3.39
6960	1″ x 6″		250	.032		1.81	1.24		3.05	4.05
6970	1″ x 8″		250	.032		2.41	1.24		3.65	4.70
6980	1″ x 10″		225	.036		4.02	1.38		5.40	6.70
6990	1″ x 12″		200	.040		4.85	1.55		6.40	7.90
7000	Grounds, 1″ x 1″, cedar, rough sawn		300	.027		.35	1.03		1.38	2.10
7010	STK		300	.027		.28	1.03		1.31	2.02
7020	Pine, #2		300	.027		.18	1.03		1.21	1.91
7030	D & better		300	.027		.31	1.03		1.34	2.05
7050	Redwood		300	.027		.40	1.03		1.43	2.15
7060	Rake/verge board, cedar, rough sawn, 1″ x 2″		225	.036		.69	1.38		2.07	3.04
7070	1″ x 3″		225	.036		1.03	1.38		2.41	3.41
7080	1″ x 4″		200	.040		1.37	1.55		2.92	4.06
7090	1″ x 6″		200	.040		2.05	1.55		3.60	4.82
7100	1″ x 8″		190	.042		2.74	1.63		4.37	5.70
7110	1″ x 10″		190	.042		3.40	1.63		5.03	6.45
7120	1″ x 12″		180	.044		4.09	1.72		5.81	7.35
7130	STK, 1″ x 2″		225	.036		.46	1.38		1.84	2.79
7140	1″ x 3″		225	.036		.50	1.38		1.88	2.83
7150	1″ x 4″		200	.040		.82	1.55		2.37	3.46
7160	1″ x 6″		200	.040		1.34	1.55		2.89	4.03
7170	1″ x 8″		190	.042		2.29	1.63		3.92	5.20
7180	1″ x 10″		190	.042		2.81	1.63		4.44	5.80
7190	1″ x 12″		180	.044		4.42	1.72		6.14	7.70
7200	Pine, #2, 1″ x 2″		225	.036		.29	1.38		1.67	2.60
7210	1″ x 3″		225	.036		.46	1.38		1.84	2.78
7220	1″ x 4″		200	.040		.56	1.55		2.11	3.18
7230	1″ x 6″		200	.040		.82	1.55		2.37	3.47
7240	1″ x 8″		190	.042		1.38	1.63		3.01	4.22
7250	1″ x 10″		190	.042		1.76	1.63		3.39	4.64
7260	1″ x 12″		180	.044		2.20	1.72		3.92	5.25
7340	D & better, 1″ x 2″		225	.036		.50	1.38		1.88	2.83
7350	1″ x 3″		225	.036		.65	1.38		2.03	2.99
7360	1″ x 4″		200	.040		.83	1.55		2.38	3.47
7370	1″ x 6″		200	.040		1.10	1.55		2.65	3.77
7380	1″ x 8″		190	.042		1.63	1.63		3.26	4.49

06 22 13.40 Moldings, Exterior		Crew	Daily Output	Labor-Hours	Unit	Material	Labor	Equipment	Total	Total Incl O&P
						2019 Bare Costs				
7390	1" x 10"	1 Carp	190	.042	L.F.	2.20	1.63		3.83	5.10
7400	1" x 12"		180	.044		2.71	1.72		4.43	5.85
7410	Redwood, clear all heart, 1" x 2"		225	.036		.65	1.38		2.03	2.99
7420	1" x 3"		225	.036		.96	1.38		2.34	3.34
7430	1" x 4"		200	.040		1.22	1.55		2.77	3.90
7440	1" x 6"		200	.040		1.81	1.55		3.36	4.56
7450	1" x 8"		190	.042		2.41	1.63		4.04	5.35
7460	1" x 10"		190	.042		4.02	1.63		5.65	7.10
7470	1" x 12"		180	.044		4.85	1.72		6.57	8.20
7480	2" x 4"		200	.040		2.33	1.55		3.88	5.15
7490	2" x 6"		182	.044		3.48	1.70		5.18	6.65
7500	2" x 8"		165	.048		4.64	1.88		6.52	8.20
7630	Soffit, cedar, rough sawn, 1" x 2"	2 Carp	440	.036		.69	1.41		2.10	3.09
7640	1" x 3"		440	.036		1.03	1.41		2.44	3.46
7650	1" x 4"		420	.038		1.37	1.48		2.85	3.94
7660	1" x 6"		420	.038		2.05	1.48		3.53	4.70
7670	1" x 8"		420	.038		2.74	1.48		4.22	5.45
7680	1" x 10"		400	.040		3.40	1.55		4.95	6.30
7690	1" x 12"		400	.040		4.09	1.55		5.64	7.05
7700	STK, 1" x 2"		440	.036		.46	1.41		1.87	2.84
7710	1" x 3"		440	.036		.50	1.41		1.91	2.88
7720	1" x 4"		420	.038		.82	1.48		2.30	3.34
7730	1" x 6"		420	.038		1.34	1.48		2.82	3.91
7740	1" x 8"		420	.038		2.29	1.48		3.77	4.96
7750	1" x 10"		400	.040		2.81	1.55		4.36	5.65
7760	1" x 12"		400	.040		4.42	1.55		5.97	7.40
7770	Pine, #2, 1" x 2"		440	.036		.29	1.41		1.70	2.65
7780	1" x 3"		440	.036		.46	1.41		1.87	2.83
7790	1" x 4"		420	.038		.56	1.48		2.04	3.06
7800	1" x 6"		420	.038		.82	1.48		2.30	3.35
7810	1" x 8"		420	.038		1.38	1.48		2.86	3.96
7820	1" x 10"		400	.040		1.76	1.55		3.31	4.50
7830	1" x 12"		400	.040		2.20	1.55		3.75	4.98
7840	D & better, 1" x 2"		440	.036		.50	1.41		1.91	2.88
7850	1" x 3"		440	.036		.65	1.41		2.06	3.04
7860	1" x 4"		420	.038		.83	1.48		2.31	3.35
7870	1" x 6"		420	.038		1.10	1.48		2.58	3.65
7880	1" x 8"		420	.038		1.63	1.48		3.11	4.23
7890	1" x 10"		400	.040		2.20	1.55		3.75	4.98
7900	1" x 12"		400	.040		2.71	1.55		4.26	5.55
7910	Redwood, clear all heart, 1" x 2"		440	.036		.65	1.41		2.06	3.04
7920	1" x 3"		440	.036		.96	1.41		2.37	3.39
7930	1" x 4"		420	.038		1.22	1.48		2.70	3.78
7940	1" x 6"		420	.038		1.81	1.48		3.29	4.44
7950	1" x 8"		420	.038		2.41	1.48		3.89	5.10
7960	1" x 10"		400	.040		4.02	1.55		5.57	7
7970	1" x 12"		400	.040		4.85	1.55		6.40	7.90
8050	Trim, crown molding, pine, 11/16" x 4-1/4"	1 Carp	250	.032		3.96	1.24		5.20	6.40
8060	Back band, 11/16" x 1-1/16"		250	.032		.89	1.24		2.13	3.03
8070	Insect screen frame stock, 1-1/16" x 1-3/4"		395	.020		2.24	.78		3.02	3.76
8080	Dentils, 2-1/2" x 2-1/2" x 4", 6" OC		30	.267		1.29	10.35		11.64	18.50
8100	Fluted, 5-1/2"		165	.048		5.10	1.88		6.98	8.70
8110	Stucco bead, 1-3/8" x 1-5/8"		250	.032		2.35	1.24		3.59	4.63

06 22 13.45 Moldings, Trim		Crew	Daily Output	Labor-Hours	Unit	Material	Labor	Equipment	Total	Total Incl O&P
						2019 Bare Costs				
0010	**MOLDINGS, TRIM**									
0200	Astragal, stock pine, 11/16" x 1-3/4"	1 Carp	255	.031	L.F.	1.25	1.22		2.47	3.39
0250	1-5/16" x 2-3/16"		240	.033		2.63	1.29		3.92	5
0800	Chair rail, stock pine, 5/8" x 2-1/2"		270	.030		1.54	1.15		2.69	3.59
0900	5/8" x 3-1/2"		240	.033		2.31	1.29		3.60	4.67
1000	Closet pole, stock pine, 1-1/8" diameter		200	.040		1.16	1.55		2.71	3.84
1100	Fir, 1-5/8" diameter		200	.040		2.05	1.55		3.60	4.82
1150	Corner, inside, 5/16" x 1"		225	.036		.34	1.38		1.72	2.65
1160	Outside, 1-1/16" x 1-1/16"		240	.033		1.35	1.29		2.64	3.62
1161	1-5/16" x 1-5/16"		240	.033		1.61	1.29		2.90	3.90
3300	Half round, stock pine, 1/4" x 1/2"		270	.030		.25	1.15		1.40	2.18
3350	1/2" x 1"	↓	255	.031	↓	.64	1.22		1.86	2.71
3400	Handrail, fir, single piece, stock, hardware not included									
3450	1-1/2" x 1-3/4"	1 Carp	80	.100	L.F.	2.13	3.88		6.01	8.75
3470	Pine, 1-1/2" x 1-3/4"		80	.100		1.99	3.88		5.87	8.60
3500	1-1/2" x 2-1/2"		76	.105		2.71	4.08		6.79	9.75
3600	Lattice, stock pine, 1/4" x 1-1/8"		270	.030		.41	1.15		1.56	2.35
3700	1/4" x 1-3/4"		250	.032		.62	1.24		1.86	2.73
3800	Miscellaneous, custom, pine, 1" x 1"		270	.030		.45	1.15		1.60	2.40
3850	1" x 2"		265	.030		.90	1.17		2.07	2.92
3900	1" x 3"		240	.033		1.35	1.29		2.64	3.62
4100	Birch or oak, nominal 1" x 1"		240	.033		.39	1.29		1.68	2.56
4200	Nominal 1" x 3"		215	.037		1.18	1.44		2.62	3.67
4400	Walnut, nominal 1" x 1"		215	.037		.64	1.44		2.08	3.08
4500	Nominal 1" x 3"		200	.040		1.91	1.55		3.46	4.66
4700	Teak, nominal 1" x 1"		215	.037		2.85	1.44		4.29	5.50
4800	Nominal 1" x 3"		200	.040		8.55	1.55		10.10	11.95
4900	Quarter round, stock pine, 1/4" x 1/4"		275	.029		.25	1.13		1.38	2.14
4950	3/4" x 3/4"		255	.031	↓	.53	1.22		1.75	2.59
5600	Wainscot moldings, 1-1/8" x 9/16", 2' high, minimum		76	.105	S.F.	12.55	4.08		16.63	20.50
5700	Maximum	↓	65	.123	"	16.30	4.77		21.07	26

06 22 13.50 Moldings, Window and Door		Crew	Daily Output	Labor-Hours	Unit	Material	Labor	Equipment	Total	Total Incl O&P
0010	**MOLDINGS, WINDOW AND DOOR**									
2800	Door moldings, stock, decorative, 1-1/8" wide, plain	1 Carp	17	.471	Set	48	18.25		66.25	82.50
2900	Detailed		17	.471	"	111	18.25		129.25	152
2960	Clear pine door jamb, no stops, 11/16" x 4-9/16"		240	.033	L.F.	4.93	1.29		6.22	7.55
3150	Door trim set, 1 head and 2 sides, pine, 2-1/2" wide		12	.667	Opng.	24	26		50	69
3170	3-1/2" wide		11	.727	"	31.50	28		59.50	81
3250	Glass beads, stock pine, 3/8" x 1/2"		275	.029	L.F.	.36	1.13		1.49	2.26
3270	3/8" x 7/8"		270	.030		.40	1.15		1.55	2.34
4850	Parting bead, stock pine, 3/8" x 3/4"		275	.029		.47	1.13		1.60	2.38
4870	1/2" x 3/4"		255	.031		.42	1.22		1.64	2.47
5000	Stool caps, stock pine, 11/16" x 3-1/2"		200	.040		2.40	1.55		3.95	5.20
5100	1-1/16" x 3-1/4"		150	.053	↓	3.76	2.07		5.83	7.55
5300	Threshold, oak, 3' long, inside, 5/8" x 3-5/8"		32	.250	Ea.	12.95	9.70		22.65	30.50
5400	Outside, 1-1/2" x 7-5/8"	↓	16	.500	"	47.50	19.40		66.90	84
5900	Window trim sets, including casings, header, stops,									
5910	stool and apron, 2-1/2" wide, FJP	1 Carp	13	.615	Opng.	33	24		57	75.50
5950	Pine		10	.800		38.50	31		69.50	93.50
6000	Oak	↓	6	1.333	↓	78	51.50		129.50	172

06 22 Millwork

06 22 13 – Standard Pattern Wood Trim

06 22 13.60 Moldings, Soffits		Crew	Daily Output	Labor-Hours	Unit	2019 Bare Costs Material	Labor	Equipment	Total	Total Incl O&P
0010	**MOLDINGS, SOFFITS**									
0200	Soffits, pine, 1″ x 4″	2 Carp	420	.038	L.F.	.52	1.48		2	3.02
0210	1″ x 6″		420	.038		.79	1.48		2.27	3.30
0220	1″ x 8″		420	.038		1.34	1.48		2.82	3.92
0230	1″ x 10″		400	.040		1.70	1.55		3.25	4.43
0240	1″ x 12″		400	.040		2.14	1.55		3.69	4.92
0250	STK cedar, 1″ x 4″		420	.038		.78	1.48		2.26	3.30
0260	1″ x 6″		420	.038		1.30	1.48		2.78	3.87
0270	1″ x 8″		420	.038		2.25	1.48		3.73	4.92
0280	1″ x 10″		400	.040		2.75	1.55		4.30	5.60
0290	1″ x 12″		400	.040		4.36	1.55		5.91	7.35
1000	Exterior AC plywood, 1/4″ thick		400	.040	S.F.	1	1.55		2.55	3.66
1050	3/8″ thick		400	.040		1.05	1.55		2.60	3.72
1100	1/2″ thick		400	.040		1.24	1.55		2.79	3.92
1150	Polyvinyl chloride, white, solid	1 Carp	230	.035		2.19	1.35		3.54	4.64
1160	Perforated	″	230	.035		2.19	1.35		3.54	4.64
1170	Accessories, "J" channel 5/8″	2 Carp	700	.023	L.F.	.51	.89		1.40	2.02

06 25 Prefinished Paneling

06 25 13 – Prefinished Hardboard Paneling

06 25 13.10 Paneling, Hardboard

			Crew	Daily Output	Labor-Hours	Unit	Material	Labor	Equipment	Total	Total Incl O&P
0010	**PANELING, HARDBOARD**										
0050	Not incl. furring or trim, hardboard, tempered, 1/8″ thick	G	2 Carp	500	.032	S.F.	.46	1.24		1.70	2.56
0100	1/4″ thick	G		500	.032		.64	1.24		1.88	2.75
0300	Tempered pegboard, 1/8″ thick	G		500	.032		.44	1.24		1.68	2.53
0400	1/4″ thick	G		500	.032		.70	1.24		1.94	2.82
0600	Untempered hardboard, natural finish, 1/8″ thick	G		500	.032		.44	1.24		1.68	2.53
0700	1/4″ thick	G		500	.032		.52	1.24		1.76	2.62
0900	Untempered pegboard, 1/8″ thick	G		500	.032		.49	1.24		1.73	2.59
1000	1/4″ thick	G		500	.032		.48	1.24		1.72	2.58
1200	Plastic faced hardboard, 1/8″ thick	G		500	.032		.62	1.24		1.86	2.73
1300	1/4″ thick	G		500	.032		.87	1.24		2.11	3.01
1500	Plastic faced pegboard, 1/8″ thick	G		500	.032		.64	1.24		1.88	2.75
1600	1/4″ thick	G		500	.032		.80	1.24		2.04	2.93
1800	Wood grained, plain or grooved, 1/8″ thick	G		500	.032		.65	1.24		1.89	2.77
1900	1/4″ thick	G		425	.038		1.42	1.46		2.88	3.97
2100	Moldings, wood grained MDF			500	.032	L.F.	.41	1.24		1.65	2.50
2200	Pine			425	.038	″	1.41	1.46		2.87	3.96

06 25 16 – Prefinished Plywood Paneling

06 25 16.10 Paneling, Plywood

		Crew	Daily Output	Labor-Hours	Unit	Material	Labor	Equipment	Total	Total Incl O&P
0010	**PANELING, PLYWOOD**									
2400	Plywood, prefinished, 1/4″ thick, 4′ x 8′ sheets									
2410	with vertical grooves. Birch faced, economy	2 Carp	500	.032	S.F.	1.49	1.24		2.73	3.69
2420	Average		420	.038		1.21	1.48		2.69	3.77
2430	Custom		350	.046		1.11	1.77		2.88	4.15
2600	Mahogany, African		400	.040		2.48	1.55		4.03	5.30
2700	Philippine (Lauan)		500	.032		.65	1.24		1.89	2.77
2900	Oak		500	.032		1.35	1.24		2.59	3.54
3000	Cherry		400	.040		1.98	1.55		3.53	4.74
3200	Rosewood		320	.050		3.16	1.94		5.10	6.70

06 25 Prefinished Paneling

06 25 16 – Prefinished Plywood Paneling

06 25 16.10 Paneling, Plywood		Crew	Daily Output	Labor-Hours	Unit	2019 Bare Costs Material	Labor	Equipment	Total	Total Incl O&P
3400	Teak	2 Carp	400	.040	S.F.	3.23	1.55		4.78	6.10
3600	Chestnut		375	.043		5.35	1.65		7	8.60
3800	Pecan		400	.040		2.51	1.55		4.06	5.30
3900	Walnut, average		500	.032		2.57	1.24		3.81	4.88
3950	Custom		400	.040		2.49	1.55		4.04	5.30
4000	Plywood, prefinished, 3/4" thick, stock grades, economy		320	.050		1.42	1.94		3.36	4.76
4100	Average		224	.071		4.98	2.77		7.75	10.10
4300	Architectural grade, custom		224	.071		5.35	2.77		8.12	10.50
4400	Luxury		160	.100		5.45	3.88		9.33	12.40
4600	Plywood, "A" face, birch, VC, 1/2" thick, natural		450	.036		1.39	1.38		2.77	3.81
4700	Select		450	.036		1.92	1.38		3.30	4.39
4900	Veneer core, 3/4" thick, natural		320	.050		2.27	1.94		4.21	5.70
5000	Select		320	.050		2.63	1.94		4.57	6.10
5200	Lumber core, 3/4" thick, natural		320	.050		3.14	1.94		5.08	6.65
5500	Plywood, knotty pine, 1/4" thick, A2 grade		450	.036		1.59	1.38		2.97	4.03
5600	A3 grade		450	.036		2.10	1.38		3.48	4.59
5800	3/4" thick, veneer core, A2 grade		320	.050		2.33	1.94		4.27	5.75
5900	A3 grade		320	.050		2.40	1.94		4.34	5.85
6100	Aromatic cedar, 1/4" thick, plywood		400	.040		2.26	1.55		3.81	5.05
6200	1/4" thick, particle board		400	.040		1.16	1.55		2.71	3.84

06 25 26 – Panel System

06 25 26.10 Panel Systems		Crew	Daily Output	Labor-Hours	Unit	Material	Labor	Equipment	Total	Total Incl O&P
0010	PANEL SYSTEMS									
0100	Raised panel, eng. wood core w/wood veneer, std., paint grade	2 Carp	300	.053	S.F.	12.10	2.07		14.17	16.70
0110	Oak veneer		300	.053		24.50	2.07		26.57	30.50
0120	Maple veneer		300	.053		30	2.07		32.07	36.50
0130	Cherry veneer		300	.053		36.50	2.07		38.57	44
0300	Class I fire rated, paint grade		300	.053		13.80	2.07		15.87	18.55
0310	Oak veneer		300	.053		30.50	2.07		32.57	37
0320	Maple veneer		300	.053		41.50	2.07		43.57	49
0330	Cherry veneer		300	.053		49.50	2.07		51.57	58
0510	Beadboard, 5/8" MDF, standard, primed		300	.053		9.50	2.07		11.57	13.85
0520	Oak veneer, unfinished		300	.053		14.90	2.07		16.97	19.80
0530	Maple veneer, unfinished		300	.053		16	2.07		18.07	21
0610	Rustic paneling, 5/8" MDF, standard, maple veneer, unfinished		300	.053		18.60	2.07		20.67	24

06 26 Board Paneling

06 26 13 – Profile Board Paneling

06 26 13.10 Paneling, Boards		Crew	Daily Output	Labor-Hours	Unit	Material	Labor	Equipment	Total	Total Incl O&P
0010	PANELING, BOARDS									
6400	Wood board paneling, 3/4" thick, knotty pine	2 Carp	300	.053	S.F.	2.04	2.07		4.11	5.65
6500	Rough sawn cedar		300	.053		3.40	2.07		5.47	7.15
6700	Redwood, clear, 1" x 4" boards		300	.053		5.05	2.07		7.12	8.95
6900	Aromatic cedar, closet lining, boards		275	.058		2.46	2.25		4.71	6.45

06 43 Wood Stairs and Railings

06 43 13 – Wood Stairs

06 43 13.20 Prefabricated Wood Stairs		Crew	Daily Output	Labor-Hours	Unit	2019 Bare Costs Material	Labor	Equipment	Total	Total Incl O&P
0010	**PREFABRICATED WOOD STAIRS**									
0100	Box stairs, prefabricated, 3′-0″ wide									
0110	Oak treads, up to 14 risers	2 Carp	39	.410	Riser	98.50	15.90		114.40	136
0600	With pine treads for carpet, up to 14 risers	″	39	.410	″	63.50	15.90		79.40	96.50
1100	For 4′ wide stairs, add				Flight	25%				
1550	Stairs, prefabricated stair handrail with balusters	1 Carp	30	.267	L.F.	82	10.35		92.35	108
1700	Basement stairs, prefabricated, pine treads									
1710	Pine risers, 3′ wide, up to 14 risers	2 Carp	52	.308	Riser	63.50	11.90		75.40	89.50
4000	Residential, wood, oak treads, prefabricated		1.50	10.667	Flight	1,275	415		1,690	2,075
4200	Built in place		.44	36.364	″	2,325	1,400		3,725	4,875
4400	Spiral, oak, 4′-6″ diameter, unfinished, prefabricated,									
4500	incl. railing, 9′ high	2 Carp	1.50	10.667	Flight	2,675	415		3,090	3,600
06 43 13.40 Wood Stair Parts										
0010	**WOOD STAIR PARTS**									
0020	Pin top balusters, 1-1/4″, oak, 34″	1 Carp	96	.083	Ea.	5.20	3.23		8.43	11.05
0030	38″		96	.083		5.80	3.23		9.03	11.70
0040	42″		96	.083		6.30	3.23		9.53	12.30
0050	Poplar, 34″		96	.083		3.35	3.23		6.58	9.05
0060	38″		96	.083		5.90	3.23		9.13	11.85
0070	42″		96	.083		6.55	3.23		9.78	12.55
0080	Maple, 34″		96	.083		4.96	3.23		8.19	10.80
0090	38″		96	.083		5.60	3.23		8.83	11.50
0100	42″		96	.083		7.15	3.23		10.38	13.20
0130	Primed, 34″		96	.083		3.68	3.23		6.91	9.40
0140	38″		96	.083		4.22	3.23		7.45	10
0150	42″		96	.083		4.86	3.23		8.09	10.70
0180	Box top balusters, 1-1/4″, oak, 34″		60	.133		8.70	5.15		13.85	18.10
0190	38″		60	.133		11.80	5.15		16.95	21.50
0200	42″		60	.133		12.90	5.15		18.05	23
0210	Poplar, 34″		60	.133		7.70	5.15		12.85	17
0220	38″		60	.133		8.30	5.15		13.45	17.70
0230	42″		60	.133		9.15	5.15		14.30	18.60
0240	Maple, 34″		60	.133		9.75	5.15		14.90	19.30
0250	38″		60	.133		10.70	5.15		15.85	20.50
0260	42″		60	.133		11.80	5.15		16.95	21.50
0290	Primed, 34″		60	.133		8.25	5.15		13.40	17.60
0300	38″		60	.133		12.05	5.15		17.20	22
0310	42″		60	.133		11.35	5.15		16.50	21
0340	Square balusters, cut from lineal stock, pine, 1-1/16″ x 1-1/16″		180	.044	L.F.	1.30	1.72		3.02	4.28
0350	1-5/16″ x 1-5/16″		180	.044		1.99	1.72		3.71	5.05
0360	1-5/8″ x 1-5/8″		180	.044		2.55	1.72		4.27	5.65
0370	Turned newel, oak, 3-1/2″ square, 48″ high		8	1	Ea.	104	39		143	178
0380	62″ high		8	1		100	39		139	174
0390	Poplar, 3-1/2″ square, 48″ high		8	1		46.50	39		85.50	115
0400	62″ high		8	1		57	39		96	127
0410	Maple, 3-1/2″ square, 48″ high		8	1		64.50	39		103.50	135
0420	62″ high		8	1		79.50	39		118.50	152
0430	Square newel, oak, 3-1/2″ square, 48″ high		8	1		54.50	39		93.50	124
0440	58″ high		8	1		65.50	39		104.50	136
0450	Poplar, 3-1/2″ square, 48″ high		8	1		42	39		81	110
0460	58″ high		8	1		51	39		90	121
0470	Maple, 3″ square, 48″ high		8	1		53.50	39		92.50	123

06 43 13 – Wood Stairs

06 43 13.40 Wood Stair Parts		Crew	Daily Output	Labor-Hours	Unit	Material	Labor	Equipment	Total	Total Incl O&P
						2019 Bare Costs				
0480	58″ high	1 Carp	8	1	Ea.	67.50	39		106.50	138
0490	Railings, oak, economy		96	.083	L.F.	9.30	3.23		12.53	15.60
0500	Average		96	.083		15.05	3.23		18.28	22
0510	Custom		96	.083		18	3.23		21.23	25
0520	Maple, economy		96	.083		11.75	3.23		14.98	18.30
0530	Average		96	.083		13.40	3.23		16.63	20
0540	Custom		96	.083		21	3.23		24.23	28.50
0550	Oak, for bending rail, economy		48	.167		26.50	6.45		32.95	39.50
0560	Average		48	.167		29.50	6.45		35.95	43
0570	Custom		48	.167		33	6.45		39.45	47
0580	Maple, for bending rail, economy		48	.167		28.50	6.45		34.95	42
0590	Average		48	.167		31.50	6.45		37.95	45
0600	Custom		48	.167		34.50	6.45		40.95	48.50
0610	Risers, oak, 3/4″ x 8″, 36″ long		80	.100	Ea.	12.90	3.88		16.78	20.50
0620	42″ long		70	.114		15.05	4.43		19.48	24
0630	48″ long		63	.127		17.20	4.92		22.12	27
0640	54″ long		56	.143		19.35	5.55		24.90	30.50
0650	60″ long		50	.160		21.50	6.20		27.70	34
0660	72″ long		42	.190		26	7.40		33.40	40.50
0670	Poplar, 3/4″ x 8″, 36″ long		80	.100		12.65	3.88		16.53	20.50
0680	42″ long		71	.113		14.75	4.37		19.12	23.50
0690	48″ long		63	.127		16.85	4.92		21.77	26.50
0700	54″ long		56	.143		18.95	5.55		24.50	30
0710	60″ long		50	.160		21	6.20		27.20	33.50
0720	72″ long		42	.190		25.50	7.40		32.90	40
0730	Pine, 1″ x 8″, 36″ long		80	.100		4.03	3.88		7.91	10.85
0740	42″ long		70	.114		4.70	4.43		9.13	12.45
0750	48″ long		63	.127		5.35	4.92		10.27	14.05
0760	54″ long		56	.143		6.05	5.55		11.60	15.80
0770	60″ long		50	.160		6.70	6.20		12.90	17.65
0780	72″ long		42	.190		8.05	7.40		15.45	21
0790	Treads, oak, no returns, 1-1/32″ x 11-1/2″ x 36″ long		32	.250		29	9.70		38.70	48
0800	42″ long		32	.250		34	9.70		43.70	53.50
0810	48″ long		32	.250		39	9.70		48.70	58.50
0820	54″ long		32	.250		43.50	9.70		53.20	64
0830	60″ long		32	.250		48.50	9.70		58.20	69.50
0840	72″ long		32	.250		58	9.70		67.70	80
0850	Mitred return one end, 1-1/32″ x 11-1/2″ x 36″ long		24	.333		38.50	12.90		51.40	64
0860	42″ long		24	.333		45	12.90		57.90	71
0870	48″ long		24	.333		51.50	12.90		64.40	78
0880	54″ long		24	.333		57.50	12.90		70.40	85
0890	60″ long		24	.333		64	12.90		76.90	92
0900	72″ long		24	.333		77	12.90		89.90	106
0910	Mitred return two ends, 1-1/32″ x 11-1/2″ x 36″ long		12	.667		48.50	26		74.50	96
0920	42″ long		12	.667		56.50	26		82.50	105
0930	48″ long		12	.667		64.50	26		90.50	114
0940	54″ long		12	.667		73	26		99	123
0950	60″ long		12	.667		81	26		107	132
0960	72″ long		12	.667		97	26		123	150
0970	Starting step, oak, 48″, bullnose		8	1		171	39		210	252
0980	Double end bullnose		8	1		259	39		298	350
1030	Skirt board, pine, 1″ x 10″		55	.145	L.F.	1.70	5.65		7.35	11.15
1040	1″ x 12″		52	.154	″	2.14	5.95		8.09	12.20

06 43 Wood Stairs and Railings

06 43 13 – Wood Stairs

06 43 13.40 Wood Stair Parts		Crew	Daily Output	Labor-Hours	Unit	2019 Bare Costs Material	Labor	Equipment	Total	Total Incl O&P
1050	Oak landing tread, 1-1/16" thick	1 Carp	54	.148	S.F.	7.90	5.75		13.65	18.20
1060	Oak cove molding		96	.083	L.F.	.99	3.23		4.22	6.45
1070	Oak stringer molding		96	.083	"	3.95	3.23		7.18	9.70
1090	Rail bolt, 5/16" x 3-1/2"		48	.167	Ea.	3.50	6.45		9.95	14.55
1100	5/16" x 4-1/2"		48	.167		3.03	6.45		9.48	14.05
1120	Newel post anchor		16	.500		12.90	19.40		32.30	46
1130	Tapered plug, 1/2"		240	.033		.99	1.29		2.28	3.22
1140	1"		240	.033		.98	1.29		2.27	3.21

06 43 16 – Wood Railings

06 43 16.10 Wood Handrails and Railings		Crew	Daily Output	Labor-Hours	Unit	Material	Labor	Equipment	Total	Total Incl O&P
0010	**WOOD HANDRAILS AND RAILINGS**									
0020	Custom design, architectural grade, hardwood, plain	1 Carp	38	.211	L.F.	12.65	8.15		20.80	27.50
0100	Shaped		30	.267		52.50	10.35		62.85	74.50
0300	Stock interior railing with spindles 4" OC, 4' long		40	.200		48	7.75		55.75	65.50
0400	8' long		48	.167		48	6.45		54.45	63

06 44 Ornamental Woodwork

06 44 19 – Wood Grilles

06 44 19.10 Grilles		Crew	Daily Output	Labor-Hours	Unit	Material	Labor	Equipment	Total	Total Incl O&P
0010	**GRILLES** and panels, hardwood, sanded									
0020	2' x 4' to 4' x 8', custom designs, unfinished, economy	1 Carp	38	.211	S.F.	57.50	8.15		65.65	76.50
0050	Average		30	.267		71.50	10.35		81.85	95.50
0100	Custom		19	.421		75	16.30		91.30	110

06 44 33 – Wood Mantels

06 44 33.10 Fireplace Mantels		Crew	Daily Output	Labor-Hours	Unit	Material	Labor	Equipment	Total	Total Incl O&P
0010	**FIREPLACE MANTELS**									
0015	6" molding, 6' x 3'-6" opening, plain, paint grade	1 Carp	5	1.600	Opng.	465	62		527	610
0100	Ornate, oak		5	1.600		635	62		697	800
0300	Prefabricated pine, colonial type, stock, deluxe		2	4		1,900	155		2,055	2,325
0400	Economy		3	2.667		820	103		923	1,075

06 44 33.20 Fireplace Mantel Beam		Crew	Daily Output	Labor-Hours	Unit	Material	Labor	Equipment	Total	Total Incl O&P
0010	**FIREPLACE MANTEL BEAM**									
0020	Rough texture wood, 4" x 8"	1 Carp	36	.222	L.F.	8.70	8.60		17.30	24
0100	4" x 10"		35	.229	"	11	8.85		19.85	27
0300	Laminated hardwood, 2-1/4" x 10-1/2" wide, 6' long		5	1.600	Ea.	108	62		170	221
0400	8' long		5	1.600	"	151	62		213	268
0600	Brackets for above, rough sawn		12	.667	Pr.	10.45	26		36.45	54
0700	Laminated		12	.667	"	21.50	26		47.50	66

06 44 39 – Wood Posts and Columns

06 44 39.10 Decorative Beams		Crew	Daily Output	Labor-Hours	Unit	Material	Labor	Equipment	Total	Total Incl O&P
0010	**DECORATIVE BEAMS**									
0020	Rough sawn cedar, non-load bearing, 4" x 4"	2 Carp	180	.089	L.F.	1.75	3.44		5.19	7.65
0100	4" x 6"		170	.094		1.88	3.65		5.53	8.10
0200	4" x 8"		160	.100		2.49	3.88		6.37	9.15
0300	4" x 10"		150	.107		4.35	4.13		8.48	11.65
0400	4" x 12"		140	.114		5	4.43		9.43	12.80
0500	8" x 8"		130	.123		4.89	4.77		9.66	13.30
0600	Plastic beam, "hewn finish", 6" x 2"		240	.067		3.57	2.58		6.15	8.20
0601	6" x 4"		220	.073		3.94	2.82		6.76	9

06 44 Ornamental Woodwork

06 44 39 – Wood Posts and Columns

06 44 39.20 Columns		Crew	Daily Output	Labor-Hours	Unit	2019 Bare Costs Material	Labor	Equipment	Total	Total Incl O&P
0010	**COLUMNS**									
0050	Aluminum, round colonial, 6" diameter	2 Carp	80	.200	V.L.F.	18.80	7.75		26.55	33.50
0100	8" diameter		62.25	.257		18.65	9.95		28.60	37
0200	10" diameter		55	.291		23	11.25		34.25	43.50
0250	Fir, stock units, hollow round, 6" diameter		80	.200		26.50	7.75		34.25	42
0300	8" diameter		80	.200		34	7.75		41.75	50.50
0350	10" diameter		70	.229		42	8.85		50.85	61
0360	12" diameter		65	.246		61.50	9.55		71.05	83.50
0400	Solid turned, to 8' high, 3-1/2" diameter		80	.200		11.60	7.75		19.35	25.50
0500	4-1/2" diameter		75	.213		12.60	8.25		20.85	27.50
0600	5-1/2" diameter		70	.229		16.95	8.85		25.80	33.50
0800	Square columns, built-up, 5" x 5"		65	.246		35	9.55		44.55	54.50
0900	Solid, 3-1/2" x 3-1/2"		130	.123		10.30	4.77		15.07	19.25
1600	Hemlock, tapered, T&G, 12" diam., 10' high		100	.160		51	6.20		57.20	66.50
1700	16' high		65	.246		85.50	9.55		95.05	110
1900	14" diameter, 10' high		100	.160		103	6.20		109.20	124
2000	18' high		65	.246		106	9.55		115.55	132
2200	18" diameter, 12' high		65	.246		179	9.55		188.55	213
2300	20' high		50	.320		134	12.40		146.40	168
2500	20" diameter, 14' high		40	.400		194	15.50		209.50	240
2600	20' high		35	.457		185	17.70		202.70	233
2800	For flat pilasters, deduct					33%				
3000	For splitting into halves, add				Ea.	115			115	126
4000	Rough sawn cedar posts, 4" x 4"	2 Carp	250	.064	V.L.F.	4.37	2.48		6.85	8.90
4100	4" x 6"		235	.068		8.90	2.64		11.54	14.15
4200	6" x 6"		220	.073		15.20	2.82		18.02	21.50
4300	8" x 8"		200	.080		19.15	3.10		22.25	26

06 48 Wood Frames

06 48 13 – Exterior Wood Door Frames

06 48 13.10 Exterior Wood Door Frames and Accessories		Crew	Daily Output	Labor-Hours	Unit	2019 Bare Costs Material	Labor	Equipment	Total	Total Incl O&P
0010	**EXTERIOR WOOD DOOR FRAMES AND ACCESSORIES**									
0400	Exterior frame, incl. ext. trim, pine, 5/4 x 4-9/16" deep	2 Carp	375	.043	L.F.	6.75	1.65		8.40	10.15
0420	5-3/16" deep		375	.043		8.25	1.65		9.90	11.85
0440	6-9/16" deep		375	.043		10.05	1.65		11.70	13.80
0600	Oak, 5/4 x 4-9/16" deep		350	.046		12.65	1.77		14.42	16.85
0620	5-3/16" deep		350	.046		13.90	1.77		15.67	18.25
0640	6-9/16" deep		350	.046		18.70	1.77		20.47	23.50
1000	Sills, 8/4 x 8" deep, oak, no horns		100	.160		7.15	6.20		13.35	18.10
1020	2" horns		100	.160		21.50	6.20		27.70	34
1040	3" horns		100	.160		21.50	6.20		27.70	34
1100	8/4 x 10" deep, oak, no horns		90	.178		6.55	6.90		13.45	18.60
1120	2" horns		90	.178		27	6.90		33.90	41
1140	3" horns		90	.178		27	6.90		33.90	41
2000	Wood frame & trim, ext., colonial, 3' opng., fluted pilasters, flat head		22	.727	Ea.	505	28		533	600
2010	Dentil head		21	.762		600	29.50		629.50	710
2020	Ram's head		20	.800		675	31		706	795
2100	5'-4" opening, in-swing, fluted pilasters, flat head		17	.941		450	36.50		486.50	560
2120	Ram's head		15	1.067		1,325	41.50		1,366.50	1,550
2140	Out-swing, fluted pilasters, flat head		17	.941		480	36.50		516.50	585

06 48 Wood Frames

06 48 13 – Exterior Wood Door Frames

06 48 13.10 Exterior Wood Door Frames and Accessories		Crew	Daily Output	Labor-Hours	Unit	Material	Labor	Equipment	Total	Total Incl O&P
						2019 Bare Costs				
2160	Ram's head	2 Carp	15	1.067	Ea.	1,525	41.50		1,566.50	1,750
2400	6'-0" opening, in-swing, fluted pilasters, flat head		16	1		470	39		509	580
2420	Ram's head		10	1.600		1,500	62		1,562	1,750
2460	Out-swing, fluted pilasters, flat head		16	1		480	39		519	595
2480	Ram's head		10	1.600		1,525	62		1,587	1,775
2600	For two sidelights, flat head, add		30	.533	Opng.	300	20.50		320.50	365
2620	Ram's head, add		20	.800	"	910	31		941	1,050
2700	Custom birch frame, 3'-0" opening		16	1	Ea.	235	39		274	320
2750	6'-0" opng.		16	1		390	39		429	495
2900	Exterior, modern, plain trim, 3' opng., in-swing, FJP		26	.615		48	24		72	92.50
2920	Fir		24	.667		56.50	26		82.50	105
2940	Oak		22	.727		66.50	28		94.50	120

06 48 16 – Interior Wood Door Frames

06 48 16.10 Interior Wood Door Jamb and Frames		Crew	Daily Output	Labor-Hours	Unit	Material	Labor	Equipment	Total	Total Incl O&P
0010	**INTERIOR WOOD DOOR JAMB AND FRAMES**									
3000	Interior frame, pine, 11/16" x 3-5/8" deep	2 Carp	375	.043	L.F.	4.16	1.65		5.81	7.30
3020	4-9/16" deep		375	.043		4.75	1.65		6.40	8
3200	Oak, 11/16" x 3-5/8" deep		350	.046		2.78	1.77		4.55	6
3220	4-9/16" deep		350	.046		11.15	1.77		12.92	15.20
3240	5-3/16" deep		350	.046		18.20	1.77		19.97	23
3400	Walnut, 11/16" x 3-5/8" deep		350	.046		9.55	1.77		11.32	13.45
3420	4-9/16" deep		350	.046		10.85	1.77		12.62	14.85
3440	5-3/16" deep		350	.046		9.85	1.77		11.62	13.75
3600	Pocket door frame		16	1	Ea.	101	39		140	175
3800	Threshold, oak, 5/8" x 3-5/8" deep		200	.080	L.F.	3.57	3.10		6.67	9.05
3820	4-5/8" deep		190	.084		4.16	3.26		7.42	10
3840	5-5/8" deep		180	.089		6.65	3.44		10.09	13

06 49 Wood Screens and Exterior Wood Shutters

06 49 19 – Exterior Wood Shutters

06 49 19.10 Shutters, Exterior		Crew	Daily Output	Labor-Hours	Unit	Material	Labor	Equipment	Total	Total Incl O&P
0010	**SHUTTERS, EXTERIOR**									
0012	Aluminum, louvered, 1'-4" wide, 3'-0" long	1 Carp	10	.800	Pr.	200	31		231	271
0200	4'-0" long		10	.800		241	31		272	315
0300	5'-4" long		10	.800		291	31		322	370
0400	6'-8" long		9	.889		360	34.50		394.50	450
1000	Pine, louvered, primed, each 1'-2" wide, 3'-3" long		10	.800		267	31		298	345
1100	4'-7" long		10	.800		294	31		325	375
1250	Each 1'-4" wide, 3'-0" long		10	.800		288	31		319	365
1350	5'-3" long		10	.800		375	31		406	465
1500	Each 1'-6" wide, 3'-3" long		10	.800		279	31		310	355
1600	4'-7" long		10	.800		370	31		401	460
1620	Cedar, louvered, 1'-2" wide, 5'-7" long		10	.800		345	31		376	430
1630	Each 1'-4" wide, 2'-2" long		10	.800		185	31		216	254
1640	3'-0" long		10	.800		240	31		271	315
1650	3'-3" long		10	.800		251	31		282	325
1660	3'-11" long		10	.800		290	31		321	370
1670	4'-3" long		10	.800		320	31		351	405
1680	5'-3" long		10	.800		395	31		426	485
1690	5'-11" long		10	.800		415	31		446	510

06 49 Wood Screens and Exterior Wood Shutters

06 49 19 – Exterior Wood Shutters

06 49 19.10 Shutters, Exterior		Crew	Daily Output	Labor-Hours	Unit	2019 Bare Costs Material	Labor	Equipment	Total	Total Incl O&P
1700	Door blinds, 6'-9" long, each 1'-3" wide	1 Carp	9	.889	Pr.	465	34.50		499.50	570
1710	1'-6" wide		9	.889		420	34.50		454.50	515
2500	Polystyrene, solid raised panel, each 1'-4" wide, 3'-3" long		10	.800		87.50	31		118.50	147
2600	3'-11" long		10	.800		93	31		124	153
2700	4'-7" long		10	.800		105	31		136	167
2800	5'-3" long		10	.800		120	31		151	183
2900	6'-8" long		9	.889		149	34.50		183.50	221
4500	Polystyrene, louvered, each 1'-2" wide, 3'-3" long		10	.800		39	31		70	94
4600	4'-7" long		10	.800		45.50	31		76.50	101
4750	5'-3" long		10	.800		53.50	31		84.50	110
4850	6'-8" long		9	.889		69.50	34.50		104	134
6000	Vinyl, louvered, each 1'-2" x 4'-7" long		10	.800		61.50	31		92.50	119
6200	Each 1'-4" x 6'-8" long	↓	9	.889	↓	82.50	34.50		117	148
8000	PVC exterior rolling shutters									
8100	including crank control	1 Carp	8	1	Ea.	760	39		799	900
8500	Insulative - 6' x 6'-8" stock unit	"	8	1	"	1,075	39		1,114	1,275

06 51 Structural Plastic Shapes and Plates

06 51 13 – Plastic Lumber

06 51 13.12 Structural Plastic Lumber		Crew	Daily Output	Labor-Hours	Unit	Material	Labor	Equipment	Total	Total Incl O&P
0010	**STRUCTURAL PLASTIC LUMBER**									
1320	Plastic lumber, posts or columns, 4" x 4"	2 Carp	390	.041	L.F.	11.05	1.59		12.64	14.80
1325	4" x 6"		275	.058		13.55	2.25		15.80	18.65
1330	4" x 8"		220	.073		19.80	2.82		22.62	26.50
1340	Girder, single, 4" x 4"		675	.024		11.05	.92		11.97	13.65
1345	4" x 6"		600	.027		13.55	1.03		14.58	16.60
1350	4" x 8"		525	.030		19.80	1.18		20.98	24
1352	Double, 2" x 4"		625	.026		8.45	.99		9.44	10.95
1354	2" x 6"		600	.027		9.95	1.03		10.98	12.65
1356	2" x 8"		575	.028		17.05	1.08		18.13	20.50
1358	2" x 10"		550	.029		24.50	1.13		25.63	29
1360	2" x 12"		525	.030		25.50	1.18		26.68	30
1362	Triple, 2" x 4"		575	.028		12.70	1.08		13.78	15.75
1364	2" x 6"		550	.029		14.95	1.13		16.08	18.30
1366	2" x 8"		525	.030		25.50	1.18		26.68	30
1368	2" x 10"		500	.032		36.50	1.24		37.74	42.50
1370	2" x 12"		475	.034		38.50	1.31		39.81	44.50
1372	Ledger, bolted 4' OC, 2" x 4"		400	.040		4.38	1.55		5.93	7.40
1374	2" x 6"		550	.029		5.05	1.13		6.18	7.40
1376	2" x 8"		390	.041		8.65	1.59		10.24	12.15
1378	2" x 10"		385	.042		12.30	1.61		13.91	16.20
1380	2" x 12"		380	.042		12.90	1.63		14.53	16.90
1382	Joists, 2" x 4"		1250	.013		4.23	.50		4.73	5.50
1384	2" x 6"		1250	.013		4.99	.50		5.49	6.30
1386	2" x 8"		1100	.015		8.55	.56		9.11	10.35
1388	2" x 10"		500	.032		12.30	1.24		13.54	15.60
1390	2" x 12"	↓	875	.018		12.80	.71		13.51	15.25
1392	Railings and trim, 5/4" x 4"	1 Carp	300	.027		4.95	1.03		5.98	7.15
1394	2" x 2"		300	.027		2.10	1.03		3.13	4.02
1396	2" x 4"		300	.027		4.21	1.03		5.24	6.35
1398	2" x 6"	↓	300	.027	↓	4.95	1.03		5.98	7.15

06 63 Plastic Railings

06 63 10 – Plastic (PVC) Railings

06 63 10.10 Plastic Railings		Crew	Daily Output	Labor-Hours	Unit	2019 Bare Costs Material	Labor	Equipment	Total	Total Incl O&P
0010	**PLASTIC RAILINGS**									
0100	Horizontal PVC handrail with balusters, 3-1/2" wide, 36" high	1 Carp	96	.083	L.F.	29.50	3.23		32.73	38
0150	42" high		96	.083		29.50	3.23		32.73	38
0200	Angled PVC handrail with balusters, 3-1/2" wide, 36" high		72	.111		20	4.31		24.31	29
0250	42" high		72	.111		29.50	4.31		33.81	39.50
0300	Post sleeve for 4 x 4 post		96	.083		15.70	3.23		18.93	22.50
0400	Post cap for 4 x 4 post, flat profile		48	.167	Ea.	13.55	6.45		20	25.50
0450	Newel post style profile		48	.167		26.50	6.45		32.95	39.50
0500	Raised corbeled profile		48	.167		37	6.45		43.45	51
0550	Post base trim for 4 x 4 post		96	.083		21.50	3.23		24.73	29

06 65 Plastic Trim

06 65 10 – PVC Trim

06 65 10.10 PVC Trim, Exterior		Crew	Daily Output	Labor-Hours	Unit	Material	Labor	Equipment	Total	Total Incl O&P
0010	**PVC TRIM, EXTERIOR**									
0100	Cornerboards, 5/4" x 6" x 6"	1 Carp	240	.033	L.F.	6.95	1.29		8.24	9.75
0110	Door/window casing, 1" x 4"		200	.040		1.63	1.55		3.18	4.35
0120	1" x 6"		200	.040		2.11	1.55		3.66	4.88
0130	1" x 8"		195	.041		2.98	1.59		4.57	5.90
0140	1" x 10"		195	.041		3.84	1.59		5.43	6.85
0150	1" x 12"		190	.042		4.20	1.63		5.83	7.30
0160	5/4" x 4"		195	.041		1.81	1.59		3.40	4.62
0170	5/4" x 6"		195	.041		2.78	1.59		4.37	5.70
0180	5/4" x 8"		190	.042		3.66	1.63		5.29	6.75
0190	5/4" x 10"		190	.042		4.64	1.63		6.27	7.80
0200	5/4" x 12"		185	.043		5.35	1.68		7.03	8.65
0210	Fascia, 1" x 4"		250	.032		1.63	1.24		2.87	3.84
0220	1" x 6"		250	.032		2.11	1.24		3.35	4.37
0230	1" x 8"		225	.036		2.98	1.38		4.36	5.55
0240	1" x 10"		225	.036		3.84	1.38		5.22	6.50
0250	1" x 12"		200	.040		4.20	1.55		5.75	7.20
0260	5/4" x 4"		240	.033		1.81	1.29		3.10	4.12
0270	5/4" x 6"		240	.033		2.78	1.29		4.07	5.20
0280	5/4" x 8"		215	.037		3.66	1.44		5.10	6.40
0290	5/4" x 10"		215	.037		4.64	1.44		6.08	7.50
0300	5/4" x 12"		190	.042		5.35	1.63		6.98	8.60
0310	Frieze, 1" x 4"		250	.032		1.63	1.24		2.87	3.84
0320	1" x 6"		250	.032		2.11	1.24		3.35	4.37
0330	1" x 8"		225	.036		2.98	1.38		4.36	5.55
0340	1" x 10"		225	.036		3.84	1.38		5.22	6.50
0350	1" x 12"		200	.040		4.20	1.55		5.75	7.20
0360	5/4" x 4"		240	.033		1.81	1.29		3.10	4.12
0370	5/4" x 6"		240	.033		2.78	1.29		4.07	5.20
0380	5/4" x 8"		215	.037		3.66	1.44		5.10	6.40
0390	5/4" x 10"		215	.037		4.64	1.44		6.08	7.50
0400	5/4" x 12"		190	.042		5.35	1.63		6.98	8.60
0410	Rake, 1" x 4"		200	.040		1.63	1.55		3.18	4.35
0420	1" x 6"		200	.040		2.11	1.55		3.66	4.88
0430	1" x 8"		190	.042		2.98	1.63		4.61	6
0440	1" x 10"		190	.042		3.84	1.63		5.47	6.90
0450	1" x 12"		180	.044		4.20	1.72		5.92	7.45

06 65 Plastic Trim

06 65 10 – PVC Trim

06 65 10.10 PVC Trim, Exterior		Crew	Daily Output	Labor-Hours	Unit	Material	Labor	Equipment	Total	Total Incl O&P
						2019 Bare Costs				
0460	5/4" x 4"	1 Carp	195	.041	L.F.	1.81	1.59		3.40	4.62
0470	5/4" x 6"		195	.041		2.78	1.59		4.37	5.70
0480	5/4" x 8"		185	.043		3.66	1.68		5.34	6.80
0490	5/4" x 10"		185	.043		4.64	1.68		6.32	7.85
0500	5/4" x 12"		175	.046		5.35	1.77		7.12	8.85
0510	Rake trim, 1" x 4"		225	.036		1.63	1.38		3.01	4.07
0520	1" x 6"		225	.036		2.11	1.38		3.49	4.60
0560	5/4" x 4"		220	.036		1.81	1.41		3.22	4.32
0570	5/4" x 6"	↓	220	.036		2.78	1.41		4.19	5.40
0610	Soffit, 1" x 4"	2 Carp	420	.038		1.63	1.48		3.11	4.23
0620	1" x 6"		420	.038		2.11	1.48		3.59	4.76
0630	1" x 8"		420	.038		2.98	1.48		4.46	5.70
0640	1" x 10"		400	.040		3.84	1.55		5.39	6.80
0650	1" x 12"		400	.040		4.20	1.55		5.75	7.20
0660	5/4" x 4"		410	.039		1.81	1.51		3.32	4.49
0670	5/4" x 6"		410	.039		2.78	1.51		4.29	5.55
0680	5/4" x 8"		410	.039		3.66	1.51		5.17	6.55
0690	5/4" x 10"		390	.041		4.64	1.59		6.23	7.75
0700	5/4" x 12"	↓	390	.041	↓	5.35	1.59		6.94	8.55

06 80 Composite Fabrications

06 80 10 – Composite Decking

06 80 10.10 Woodgrained Composite Decking		Crew	Daily Output	Labor-Hours	Unit	Material	Labor	Equipment	Total	Total Incl O&P
0010	**WOODGRAINED COMPOSITE DECKING**									
0100	Woodgrained composite decking, 1" x 6"	2 Carp	640	.025	L.F.	4.13	.97		5.10	6.15
0110	Grooved edge		660	.024		4.30	.94		5.24	6.25
0120	2" x 6"		640	.025		3.95	.97		4.92	5.95
0130	Encased, 1" x 6"		640	.025		4.35	.97		5.32	6.40
0140	Grooved edge		660	.024		4.52	.94		5.46	6.50
0150	2" x 6"	↓	640	.025	↓	5.35	.97		6.32	7.50

06 81 Composite Railings

06 81 10 – Encased Railings

06 81 10.10 Encased Composite Railings		Crew	Daily Output	Labor-Hours	Unit	Material	Labor	Equipment	Total	Total Incl O&P
0010	**ENCASED COMPOSITE RAILINGS**									
0100	Encased composite railing, 6' long, 36" high, incl. balusters	1 Carp	16	.500	Ea.	147	19.40		166.40	193
0110	42" high, incl. balusters		16	.500		219	19.40		238.40	273
0120	8' long, 36" high, incl. balusters		12	.667		165	26		191	224
0130	42" high, incl. balusters		12	.667		163	26		189	222
0140	Accessories, post sleeve, 4" x 4", 39" long		32	.250		28.50	9.70		38.20	47
0150	96" long		24	.333		75.50	12.90		88.40	105
0160	6" x 6", 39" long		32	.250		54	9.70		63.70	75.50
0170	96" long		24	.333		157	12.90		169.90	194
0180	Accessories, post skirt, 4" x 4"		96	.083		5.30	3.23		8.53	11.15
0190	6" x 6"		96	.083		6.90	3.23		10.13	12.95
0200	Post cap, 4" x 4", flat		48	.167		9.05	6.45		15.50	20.50
0210	Pyramid		48	.167		7.75	6.45		14.20	19.25
0220	Post cap, 6" x 6", flat		48	.167		12.85	6.45		19.30	25
0230	Pyramid	↓	48	.167	↓	10.65	6.45		17.10	22.50

Division Notes

		CREW	DAILY OUTPUT	LABOR-HOURS	UNIT	BARE COSTS MAT.	BARE COSTS LABOR	BARE COSTS EQUIP.	BARE COSTS TOTAL	TOTAL INCL O&P

Estimating Tips

07 10 00 Dampproofing and Waterproofing

- Be sure of the job specifications before pricing this subdivision. The difference in cost between waterproofing and dampproofing can be great. Waterproofing will hold back standing water. Dampproofing prevents the transmission of water vapor. Also included in this section are vapor retarding membranes.

07 20 00 Thermal Protection

- Insulation and fireproofing products are measured by area, thickness, volume, or R-value. Specifications may give only what the specific R-value should be in a certain situation. The estimator may need to choose the type of insulation to meet that R-value.

07 30 00 Steep Slope Roofing
07 40 00 Roofing and Siding Panels

- Many roofing and siding products are bought and sold by the square. One square is equal to an area that measures 100 square feet.

  This simple change in unit of measure could create a large error if the estimator is not observant. Accessories necessary for a complete installation must be figured into any calculations for both material and labor.

07 50 00 Membrane Roofing
07 60 00 Flashing and Sheet Metal
07 70 00 Roofing and Wall Specialties and Accessories

- The items in these subdivisions compose a roofing system. No one component completes the installation, and all must be estimated. Built-up or single-ply membrane roofing systems are made up of many products and installation trades. Wood blocking at roof perimeters or penetrations, parapet coverings, reglets, roof drains, gutters, downspouts, sheet metal flashing, skylights, smoke vents, and roof hatches all need to be considered along with the roofing material. Several different installation trades will need to work together on the roofing system. Inherent difficulties in the scheduling and coordination of various trades must be accounted for when estimating labor costs.

07 90 00 Joint Protection

- To complete the weather-tight shell, the sealants and caulkings must be estimated. Where different materials meet—at expansion joints, at flashing penetrations, and at hundreds of other locations throughout a construction project—caulking and sealants provide another line of defense against water penetration. Often, an entire system is based on the proper location and placement of caulking or sealants. The detailed drawings that are included as part of a set of architectural plans show typical locations for these materials. When caulking or sealants are shown at typical locations, this means the estimator must include them for all the locations where this detail is applicable. Be careful to keep different types of sealants separate, and remember to consider backer rods and primers if necessary.

Reference Numbers

Reference numbers are shown at the beginning of some major classifications. These numbers refer to related items in the Reference Section. The reference information may be an estimating procedure, an alternate pricing method, or technical information.

Note: Not all subdivisions listed here necessarily appear. ■

07 01 Operation and Maint. of Thermal and Moisture Protection

07 01 50 – Maintenance of Membrane Roofing

07 01 50.10 Roof Coatings		Crew	Daily Output	Labor-Hours	Unit	Material	2019 Bare Costs Labor	Equipment	Total	Total Incl O&P
0010	**ROOF COATINGS**									
0012	Asphalt, brush grade, material only				Gal.	9.05			9.05	9.95
0200	Asphalt base, fibered aluminum coating G					9.15			9.15	10.05
0300	Asphalt primer, 5 gal.				↓	7.60			7.60	8.35
0600	Coal tar pitch, 200 lb. barrels				Ton	1,275			1,275	1,425
0700	Tar roof cement, 5 gal. lots				Gal.	14.55			14.55	16
0800	Glass fibered roof & patching cement, 5 gal.				"	9.15			9.15	10.05
0900	Reinforcing glass membrane, 450 S.F./roll				Ea.	61			61	67
1000	Neoprene roof coating, 5 gal., 2 gal./sq.				Gal.	32			32	35
1100	Roof patch & flashing cement, 5 gal.					9.10			9.10	10
1200	Roof resaturant, glass fibered, 3 gal./sq.					9.65			9.65	10.60
1600	Reflective roof coating, white, elastomeric, approx. 50 S.F./gal. G				↓	17.40			17.40	19.15

07 01 90 – Maintenance of Joint Protection

07 01 90.81 Joint Sealant Replacement		Crew	Daily Output	Labor-Hours	Unit	Material	Labor	Equipment	Total	Total Incl O&P
0010	**JOINT SEALANT REPLACEMENT**									
0050	Control joints in concrete floors/slabs									
0100	Option 1 for joints with hard dry sealant									
0110	Step 1: Sawcut to remove 95% of old sealant									
0112	1/4" wide x 1/2" deep, with single saw blade	C-27	4800	.003	L.F.	.02	.12	.02	.16	.25
0114	3/8" wide x 3/4" deep, with single saw blade		4000	.004		.03	.15	.03	.21	.30
0116	1/2" wide x 1" deep, with double saw blades		3600	.004		.06	.16	.03	.25	.36
0118	3/4" wide x 1-1/2" deep, with double saw blades	↓	3200	.005		.12	.19	.03	.34	.47
0120	Step 2: Water blast joint faces and edges	C-29	2500	.003			.10	.03	.13	.19
0130	Step 3: Air blast joint faces and edges	C-28	2000	.004			.15	.01	.16	.25
0140	Step 4: Sand blast joint faces and edges	E-11	2000	.016			.54	.10	.64	1.04
0150	Step 5: Air blast joint faces and edges	C-28	2000	.004	↓		.15	.01	.16	.25
0200	Option 2 for joints with soft pliable sealant									
0210	Step 1: Plow joint with rectangular blade	B-62	2600	.009	L.F.		.31	.07	.38	.57
0220	Step 2: Sawcut to re-face joint faces									
0222	1/4" wide x 1/2" deep, with single saw blade	C-27	2400	.007	L.F.	.02	.25	.05	.32	.47
0224	3/8" wide x 3/4" deep, with single saw blade		2000	.008		.04	.30	.06	.40	.58
0226	1/2" wide x 1" deep, with double saw blades		1800	.009		.08	.33	.06	.47	.69
0228	3/4" wide x 1-1/2" deep, with double saw blades	↓	1600	.010		.16	.37	.07	.60	.86
0230	Step 3: Water blast joint faces and edges	C-29	2500	.003			.10	.03	.13	.19
0240	Step 4: Air blast joint faces and edges	C-28	2000	.004			.15	.01	.16	.25
0250	Step 5: Sand blast joint faces and edges	E-11	2000	.016			.54	.10	.64	1.04
0260	Step 6: Air blast joint faces and edges	C-28	2000	.004	↓		.15	.01	.16	.25
0290	For saw cutting new control joints, see Section 03 15 16.20									
8910	For backer rod, see Section 07 91 23.10									

07 05 Common Work Results for Thermal and Moisture Protection

07 05 05 – Selective Demolition for Thermal and Moisture Protection

07 05 05.10 Selective Demo., Thermal and Moist. Protection		Crew	Daily Output	Labor-Hours	Unit	Material	Labor	Equipment	Total	Total Incl O&P
0010	**SELECTIVE DEMO., THERMAL AND MOISTURE PROTECTION**									
0020	Caulking/sealant, to 1" x 1" joint R024119-10	1 Clab	600	.013	L.F.		.40		.40	.67
0120	Downspouts, including hangers		350	.023	"		.69		.69	1.15
0220	Flashing, sheet metal		290	.028	S.F.		.84		.84	1.38
0420	Gutters, aluminum or wood, edge hung		240	.033	L.F.		1.01		1.01	1.67
0520	Built-in		100	.080	"		2.43		2.43	4.01
0620	Insulation, air/vapor barrier	↓	3500	.002	S.F.		.07		.07	.11

07 05 05 – Selective Demolition for Thermal and Moisture Protection

07 05 05.10 Selective Demo., Thermal and Moist. Protection		Crew	Daily Output	Labor-Hours	Unit	Material	2019 Bare Costs Labor	Equipment	Total	Total Incl O&P
0670	Batts or blankets	1 Clab	1400	.006	C.F.		.17		.17	.29
0720	Foamed or sprayed in place	2 Clab	1000	.016	B.F.		.49		.49	.80
0770	Loose fitting	1 Clab	3000	.003	C.F.		.08		.08	.13
0870	Rigid board		3450	.002	B.F.		.07		.07	.12
1120	Roll roofing, cold adhesive		12	.667	Sq.		20		20	33.50
1170	Roof accessories, adjustable metal chimney flashing		9	.889	Ea.		27		27	44.50
1325	Plumbing vent flashing		32	.250	"		7.60		7.60	12.55
1375	Ridge vent strip, aluminum		310	.026	L.F.		.78		.78	1.29
1620	Skylight to 10 S.F.		8	1	Ea.		30.50		30.50	50
2120	Roof edge, aluminum soffit and fascia		570	.014	L.F.		.43		.43	.70
2170	Concrete coping, up to 12" wide	2 Clab	160	.100			3.04		3.04	5
2220	Drip edge	1 Clab	1000	.008			.24		.24	.40
2270	Gravel stop		950	.008			.26		.26	.42
2370	Sheet metal coping, up to 12" wide		240	.033			1.01		1.01	1.67
2470	Roof insulation board, over 2" thick	B-2	7800	.005	B.F.		.16		.16	.26
2520	Up to 2" thick	"	3900	.010	S.F.		.32		.32	.52
2620	Roof ventilation, louvered gable vent	1 Clab	16	.500	Ea.		15.20		15.20	25
2670	Remove, roof hatch	G-3	15	2.133			78.50		78.50	130
2675	Rafter vents	1 Clab	960	.008			.25		.25	.42
2720	Soffit vent and/or fascia vent		575	.014	L.F.		.42		.42	.70
2775	Soffit vent strip, aluminum, 3" to 4" wide		160	.050			1.52		1.52	2.51
2820	Roofing accessories, shingle moulding, to 1" x 4"		1600	.005			.15		.15	.25
2870	Cant strip	B-2	2000	.020			.62		.62	1.02
2920	Concrete block walkway	1 Clab	230	.035			1.06		1.06	1.74
3070	Roofing, felt paper, #15		70	.114	Sq.		3.47		3.47	5.75
3125	#30 felt		30	.267	"		8.10		8.10	13.35
3170	Asphalt shingles, 1 layer	B-2	3500	.011	S.F.		.35		.35	.58
3180	2 layers		1750	.023	"		.70		.70	1.16
3370	Modified bitumen		26	1.538	Sq.		47.50		47.50	78
3420	Built-up, no gravel, 3 ply		25	1.600			49		49	81.50
3470	4 ply		21	1.905			58.50		58.50	97
3620	5 ply		1600	.025	S.F.		.77		.77	1.27
3720	5 ply, with gravel		890	.045			1.38		1.38	2.28
3725	Loose gravel removal		5000	.008			.25		.25	.41
3730	Embedded gravel removal		2000	.020			.62		.62	1.02
3870	Fiberglass sheet		1200	.033			1.02		1.02	1.69
4120	Slate shingles		1900	.021			.65		.65	1.07
4170	Ridge shingles, clay or slate		2000	.020	L.F.		.62		.62	1.02
4320	Single ply membrane, attached at seams		52	.769	Sq.		23.50		23.50	39
4370	Ballasted		75	.533			16.40		16.40	27
4420	Fully adhered		39	1.026			31.50		31.50	52
4550	Roof hatch, 2'-6" x 3'-0"	1 Clab	10	.800	Ea.		24.50		24.50	40
4670	Wood shingles	B-2	2200	.018	S.F.		.56		.56	.92
4820	Sheet metal roofing	"	2150	.019			.57		.57	.95
4970	Siding, horizontal wood clapboards	1 Clab	380	.021			.64		.64	1.06
5025	Exterior insulation finish system	"	120	.067			2.02		2.02	3.34
5070	Tempered hardboard, remove and reset	1 Carp	380	.021			.82		.82	1.35
5120	Tempered hardboard sheet siding	"	375	.021			.83		.83	1.37
5170	Metal, corner strips	1 Clab	850	.009	L.F.		.29		.29	.47
5225	Horizontal strips		444	.018	S.F.		.55		.55	.90
5320	Vertical strips		400	.020			.61		.61	1
5520	Wood shingles		350	.023			.69		.69	1.15
5620	Stucco siding		360	.022			.67		.67	1.11

07 05 Common Work Results for Thermal and Moisture Protection

07 05 05 – Selective Demolition for Thermal and Moisture Protection

07 05 05.10 Selective Demo., Thermal and Moist. Protection		Crew	Daily Output	Labor-Hours	Unit	2019 Bare Costs Material	Labor	Equipment	Total	Total Incl O&P
5670	Textured plywood	1 Clab	725	.011	S.F.		.33		.33	.55
5720	Vinyl siding		510	.016			.48		.48	.79
5770	Corner strips		900	.009	L.F.		.27		.27	.45
5870	Wood, boards, vertical		400	.020	S.F.		.61		.61	1
5880	Steel siding, corrugated/ribbed		402.50	.020	"		.60		.60	1
5920	Waterproofing, protection/drain board	2 Clab	3900	.004	B.F.		.12		.12	.21
5970	Over 1/2" thick		1750	.009	S.F.		.28		.28	.46
6020	To 1/2" thick		2000	.008	"		.24		.24	.40

07 11 Dampproofing

07 11 13 – Bituminous Dampproofing

07 11 13.10 Bituminous Asphalt Coating

		Crew	Daily Output	Labor-Hours	Unit	Material	Labor	Equipment	Total	Total Incl O&P
0010	**BITUMINOUS ASPHALT COATING**									
0030	Brushed on, below grade, 1 coat	1 Rofc	665	.012	S.F.	.23	.40		.63	.97
0100	2 coat		500	.016		.45	.53		.98	1.46
0300	Sprayed on, below grade, 1 coat		830	.010		.23	.32		.55	.83
0400	2 coat		500	.016		.44	.53		.97	1.45
0500	Asphalt coating, with fibers				Gal.	9.15			9.15	10.05
0600	Troweled on, asphalt with fibers, 1/16" thick	1 Rofc	500	.016	S.F.	.40	.53		.93	1.40
0700	1/8" thick		400	.020		.70	.67		1.37	1.98
1000	1/2" thick		350	.023		2.29	.76		3.05	3.89

07 11 16 – Cementitious Dampproofing

07 11 16.20 Cementitious Parging

		Crew	Daily Output	Labor-Hours	Unit	Material	Labor	Equipment	Total	Total Incl O&P
0010	**CEMENTITIOUS PARGING**									
0020	Portland cement, 2 coats, 1/2" thick	D-1	250	.064	S.F.	.39	2.16		2.55	4.03
0100	Waterproofed Portland cement, 1/2" thick, 2 coats	"	250	.064	"	6.55	2.16		8.71	10.80

07 19 Water Repellents

07 19 19 – Silicone Water Repellents

07 19 19.10 Silicone Based Water Repellents

		Crew	Daily Output	Labor-Hours	Unit	Material	Labor	Equipment	Total	Total Incl O&P
0010	**SILICONE BASED WATER REPELLENTS**									
0020	Water base liquid, roller applied	2 Rofc	7000	.002	S.F.	.42	.08		.50	.61
0200	Silicone or stearate, sprayed on CMU, 1 coat	1 Rofc	4000	.002		.40	.07		.47	.56
0300	2 coats	"	3000	.003		.80	.09		.89	1.04

07 21 Thermal Insulation

07 21 13 – Board Insulation

07 21 13.10 Rigid Insulation

			Crew	Daily Output	Labor-Hours	Unit	Material	Labor	Equipment	Total	Total Incl O&P
0010	**RIGID INSULATION**, for walls										
0020	Fiberboard, 3/4" thick, R2.08	G	1 Carp	1100	.007	S.F.	.42	.28		.70	.93
0025	1" thick, R2.78	G		800	.010		.56	.39		.95	1.26
0030	2" thick, R5.26	G		730	.011		1.09	.42		1.51	1.90
0040	Fiberglass, 1.5#/C.F., unfaced, 1" thick, R4.1	G		1000	.008		.34	.31		.65	.88
0060	1-1/2" thick, R6.2	G		1000	.008		.40	.31		.71	.95
0080	2" thick, R8.3	G		1000	.008		.50	.31		.81	1.06
0120	3" thick, R12.4	G		800	.010		.60	.39		.99	1.30
0370	3#/C.F., unfaced, 1" thick, R4.3	G		1000	.008		.56	.31		.87	1.13

07 21 Thermal Insulation

07 21 13 – Board Insulation

07 21 13.10 Rigid Insulation			Crew	Daily Output	Labor-Hours	Unit	Material	Labor	Equipment	Total	Total Incl O&P
							2019 Bare Costs				
0390	1-1/2″ thick, R6.5	G	1 Carp	1000	.008	S.F.	.80	.31		1.11	1.39
0400	2″ thick, R8.7	G		890	.009		1.07	.35		1.42	1.76
0420	2-1/2″ thick, R10.9	G		800	.010		1.11	.39		1.50	1.86
0440	3″ thick, R13	G		800	.010		1.63	.39		2.02	2.43
0520	Foil faced, 1″ thick, R4.3	G		1000	.008		.84	.31		1.15	1.43
0540	1-1/2″ thick, R6.5	G		1000	.008		1.25	.31		1.56	1.89
0560	2″ thick, R8.7	G		890	.009		1.58	.35		1.93	2.32
0580	2-1/2″ thick, R10.9	G		800	.010		1.86	.39		2.25	2.69
0600	3″ thick, R13	G		800	.010		2.09	.39		2.48	2.94
1600	Isocyanurate, 4′ x 8′ sheet, foil faced, both sides										
1610	1/2″ thick	G	1 Carp	800	.010	S.F.	.31	.39		.70	.98
1620	5/8″ thick	G		800	.010		.55	.39		.94	1.25
1630	3/4″ thick	G		800	.010		.46	.39		.85	1.15
1640	1″ thick	G		800	.010		.61	.39		1	1.31
1650	1-1/2″ thick	G		730	.011		.67	.42		1.09	1.44
1660	2″ thick	G		730	.011		.90	.42		1.32	1.69
1670	3″ thick	G		730	.011		2.84	.42		3.26	3.82
1680	4″ thick	G		730	.011		2.56	.42		2.98	3.52
1700	Perlite, 1″ thick, R2.77	G		800	.010		.48	.39		.87	1.17
1750	2″ thick, R5.55	G		730	.011		.78	.42		1.20	1.56
1900	Extruded polystyrene, 25 psi compressive strength, 1″ thick, R5	G		800	.010		.56	.39		.95	1.26
1940	2″ thick, R10	G		730	.011		1.12	.42		1.54	1.93
1960	3″ thick, R15	G		730	.011		1.61	.42		2.03	2.47
2100	Expanded polystyrene, 1″ thick, R3.85	G		800	.010		.27	.39		.66	.94
2120	2″ thick, R7.69	G		730	.011		.54	.42		.96	1.29
2140	3″ thick, R11.49	G		730	.011		.81	.42		1.23	1.59
2360	Fiberboard, low density, 1/2″ thick, R1.39	G		750	.011		.35	.41		.76	1.07
2400	Wood fiber, 1″ thick, R3.85	G		1000	.008		.56	.31		.87	1.13
2410	2″ thick, R7.7	G		1000	.008		1.09	.31		1.40	1.71
2680	Mineral fiberboard, rigid, 1″ thick, R4.2	G		800	.010		.56	.39		.95	1.26
2700	2″ thick, R8.4	G		730	.011		1.09	.42		1.51	1.90

07 21 13.13 Foam Board Insulation

			Crew	Daily Output	Labor-Hours	Unit	Material	Labor	Equipment	Total	Total Incl O&P
0010	**FOAM BOARD INSULATION**										
0600	Polystyrene, expanded, 1″ thick, R4	G	1 Carp	680	.012	S.F.	.27	.46		.73	1.05
0700	2″ thick, R8	G	″	675	.012	″	.54	.46		1	1.35

07 21 16 – Blanket Insulation

07 21 16.10 Blanket Insulation for Floors/Ceilings

			Crew	Daily Output	Labor-Hours	Unit	Material	Labor	Equipment	Total	Total Incl O&P
0010	**BLANKET INSULATION FOR FLOORS/CEILINGS**										
0020	Including spring type wire fasteners										
2000	Fiberglass, blankets or batts, paper or foil backing										
2100	3-1/2″ thick, R13	G	1 Carp	700	.011	S.F.	.41	.44		.85	1.18
2150	6-1/4″ thick, R19	G		600	.013		.50	.52		1.02	1.40
2210	9-1/2″ thick, R30	G		500	.016		.73	.62		1.35	1.82
2220	12″ thick, R38	G		475	.017		1.06	.65		1.71	2.25
3000	Unfaced, 3-1/2″ thick, R13	G		600	.013		.33	.52		.85	1.21
3010	6-1/4″ thick, R19	G		500	.016		.39	.62		1.01	1.45
3020	9-1/2″ thick, R30	G		450	.018		.61	.69		1.30	1.81
3030	12″ thick, R38	G		425	.019		.74	.73		1.47	2.02

07 21 16.20 Blanket Insulation for Walls

			Crew	Daily Output	Labor-Hours	Unit	Material	Labor	Equipment	Total	Total Incl O&P
0010	**BLANKET INSULATION FOR WALLS**										
0020	Kraft faced fiberglass, 3-1/2″ thick, R11, 15″ wide	G	1 Carp	1350	.006	S.F.	.30	.23		.53	.71
0030	23″ wide	G		1600	.005		.30	.19		.49	.65

07 21 Thermal Insulation

07 21 16 – Blanket Insulation

07 21 16.20 Blanket Insulation for Walls			Crew	Daily Output	Labor-Hours	Unit	Material	2019 Bare Costs Labor	Equipment	Total	Total Incl O&P
0060	R13, 11″ wide	G	1 Carp	1150	.007	S.F.	.34	.27		.61	.82
0110	R15, 11″ wide	G		1150	.007		.53	.27		.80	1.03
0140	6″ thick, R19, 11″ wide	G		1150	.007		.43	.27		.70	.92
0182	R21, 11″ wide	G		1150	.007		.66	.27		.93	1.18
0184	15″ wide	G		1350	.006		.66	.23		.89	1.11
0186	23″ wide	G		1600	.005		.66	.19		.85	1.05
0188	9″ thick, R30, 11″ wide	G		985	.008		.73	.31		1.04	1.32
0200	15″ wide	G		1150	.007		.73	.27		1	1.25
0230	12″ thick, R38, 11″ wide	G		985	.008		1.06	.31		1.37	1.69
0240	15″ wide	G		1150	.007		1.06	.27		1.33	1.62
0410	Foil faced fiberglass, 3-1/2″ thick, R13, 11″ wide	G		1150	.007		.47	.27		.74	.97
0420	15″ wide	G		1350	.006		.47	.23		.70	.90
0442	R15, 11″ wide	G		1150	.007		.48	.27		.75	.98
0444	15″ wide	G		1350	.006		.48	.23		.71	.91
0448	6″ thick, R19, 11″ wide	G		1150	.007		.61	.27		.88	1.12
0460	15″ wide	G		1350	.006		.61	.23		.84	1.05
0482	R21, 11″ wide	G		1150	.007		.64	.27		.91	1.15
0484	15″ wide	G		1350	.006		.64	.23		.87	1.08
0488	9″ thick, R30, 11″ wide	G		985	.008		.93	.31		1.24	1.54
0500	15″ wide	G		1150	.007		.93	.27		1.20	1.47
0560	12″ thick, R38, 11″ wide	G		985	.008		1.16	.31		1.47	1.80
0570	15″ wide	G		1150	.007		1.16	.27		1.43	1.73
0620	Unfaced fiberglass, 3-1/2″ thick, R13, 11″ wide	G		1150	.007		.33	.27		.60	.81
0820	15″ wide	G		1350	.006		.33	.23		.56	.74
0832	R15, 11″ wide	G		1150	.007		.46	.27		.73	.96
0838	6″ thick, R19, 11″ wide	G		1150	.007		.39	.27		.66	.88
0860	15″ wide	G		1150	.007		.39	.27		.66	.88
0882	R21, 11″ wide	G		1150	.007		.68	.27		.95	1.20
0886	15″ wide	G		1350	.006		.68	.23		.91	1.13
0890	9″ thick, R30, 11″ wide	G		985	.008		.61	.31		.92	1.19
0900	15″ wide	G		1150	.007		.61	.27		.88	1.12
0930	12″ thick, R38, 11″ wide	G		985	.008		.74	.31		1.05	1.33
0940	15″ wide	G		1000	.008		.74	.31		1.05	1.32
1300	Wall or ceiling insulation, mineral wool batts										
1320	3-1/2″ thick, R15	G	1 Carp	1600	.005	S.F.	.79	.19		.98	1.19
1340	5-1/2″ thick, R23	G		1600	.005		1.24	.19		1.43	1.69
1380	7-1/4″ thick, R30	G		1350	.006		1.64	.23		1.87	2.18
1700	Non-rigid insul., recycled blue cotton fiber, unfaced batts, R13, 16″ wide	G		1600	.005		1.10	.19		1.29	1.53
1710	R19, 16″ wide	G		1600	.005		1.35	.19		1.54	1.81
1850	Friction fit wire insulation supports, 16″ OC			960	.008	Ea.	.07	.32		.39	.61

07 21 19 – Foamed In Place Insulation

07 21 19.10 Masonry Foamed In Place Insulation			Crew	Daily Output	Labor-Hours	Unit	Material	Labor	Equipment	Total	Total Incl O&P
0010	**MASONRY FOAMED IN PLACE INSULATION**										
0100	Amino-plast foam, injected into block core, 6″ block	G	G-2A	6000	.004	Ea.	.17	.12	.10	.39	.51
0110	8″ block	G		5000	.005		.20	.14	.12	.46	.61
0120	10″ block	G		4000	.006		.25	.18	.16	.59	.76
0130	12″ block	G		3000	.008		.34	.24	.21	.79	1.03
0140	Injected into cavity wall	G		13000	.002	B.F.	.06	.05	.05	.16	.22
0150	Preparation, drill holes into mortar joint every 4 V.L.F., 5/8″ diameter		1 Clab	960	.008	Ea.		.25		.25	.42
0160	7/8″ diameter			680	.012			.36		.36	.59
0170	Patch drilled holes, 5/8″ diameter			1800	.004		.04	.13		.17	.26
0180	7/8″ diameter			1200	.007		.05	.20		.25	.38

07 21 Thermal Insulation

07 21 23 – Loose-Fill Insulation

07 21 23.10 Poured Loose-Fill Insulation

	07 21 23.10 Poured Loose-Fill Insulation		Crew	Daily Output	Labor-Hours	Unit	Material	2019 Bare Costs Labor	Equipment	Total	Total Incl O&P
0010	**POURED LOOSE-FILL INSULATION**										
0020	Cellulose fiber, R3.8 per inch	G	1 Carp	200	.040	C.F.	.70	1.55		2.25	3.33
0021	4" thick	G		1000	.008	S.F.	.17	.31		.48	.69
0022	6" thick	G		800	.010	"	.28	.39		.67	.95
0080	Fiberglass wool, R4 per inch	G		200	.040	C.F.	.70	1.55		2.25	3.33
0081	4" thick	G		600	.013	S.F.	.24	.52		.76	1.11
0082	6" thick	G		400	.020	"	.34	.78		1.12	1.65
0100	Mineral wool, R3 per inch	G		200	.040	C.F.	.53	1.55		2.08	3.14
0101	4" thick	G		600	.013	S.F.	.17	.52		.69	1.04
0102	6" thick	G		400	.020	"	.27	.78		1.05	1.57
0300	Polystyrene, R4 per inch	G		200	.040	C.F.	1.51	1.55		3.06	4.22
0301	4" thick	G		600	.013	S.F.	.50	.52		1.02	1.40
0302	6" thick	G		400	.020	"	.76	.78		1.54	2.11
0400	Perlite, R2.78 per inch	G		200	.040	C.F.	5.30	1.55		6.85	8.40
0401	4" thick	G		1000	.008	S.F.	1.77	.31		2.08	2.46
0402	6" thick	G		800	.010	"	2.66	.39		3.05	3.57

07 21 23.20 Masonry Loose-Fill Insulation

	07 21 23.20 Masonry Loose-Fill Insulation		Crew	Daily Output	Labor-Hours	Unit	Material	Labor	Equipment	Total	Total Incl O&P
0010	**MASONRY LOOSE-FILL INSULATION**, vermiculite or perlite										
0100	In cores of concrete block, 4" thick wall, .115 C.F./S.F.	G	D-1	4800	.003	S.F.	.61	.11		.72	.86
0700	Foamed in place, urethane in 2-5/8" cavity	G	G-2A	1035	.023		1.35	.69	.60	2.64	3.34
0800	For each 1" added thickness, add	G	"	2372	.010		.51	.30	.26	1.07	1.39

07 21 26 – Blown Insulation

07 21 26.10 Blown Insulation

	07 21 26.10 Blown Insulation		Crew	Daily Output	Labor-Hours	Unit	Material	Labor	Equipment	Total	Total Incl O&P
0010	**BLOWN INSULATION** Ceilings, with open access										
0020	Cellulose, 3-1/2" thick, R13	G	G-4	5000	.005	S.F.	.24	.15	.06	.45	.58
0030	5-3/16" thick, R19	G		3800	.006		.35	.20	.08	.63	.80
0050	6-1/2" thick, R22	G		3000	.008		.45	.25	.10	.80	1.02
0100	8-11/16" thick, R30	G		2600	.009		.61	.29	.12	1.02	1.27
0120	10-7/8" thick, R38	G		1800	.013		.78	.41	.17	1.36	1.73
1000	Fiberglass, 5.5" thick, R11	G		3800	.006		.24	.20	.08	.52	.67
1050	6" thick, R12	G		3000	.008		.34	.25	.10	.69	.89
1100	8.8" thick, R19	G		2200	.011		.42	.34	.14	.90	1.18
1200	10" thick, R22	G		1800	.013		.49	.41	.17	1.07	1.41
1300	11.5" thick, R26	G		1500	.016		.58	.50	.21	1.29	1.69
1350	13" thick, R30	G		1400	.017		.68	.53	.22	1.43	1.86
1450	16" thick, R38	G		1145	.021		.86	.65	.27	1.78	2.32
1500	20" thick, R49	G		920	.026		1.14	.81	.34	2.29	2.97

07 21 27 – Reflective Insulation

07 21 27.10 Reflective Insulation Options

	07 21 27.10 Reflective Insulation Options		Crew	Daily Output	Labor-Hours	Unit	Material	Labor	Equipment	Total	Total Incl O&P
0010	**REFLECTIVE INSULATION OPTIONS**										
0020	Aluminum foil on reinforced scrim	G	1 Carp	19	.421	C.S.F.	15	16.30		31.30	43.50
0100	Reinforced with woven polyolefin	G		19	.421		22.50	16.30		38.80	51.50
0500	With single bubble air space, R8.8	G		15	.533		26.50	20.50		47	63
0600	With double bubble air space, R9.8	G		15	.533		31.50	20.50		52	68.50

07 21 29 – Sprayed Insulation

07 21 29.10 Sprayed-On Insulation

	07 21 29.10 Sprayed-On Insulation		Crew	Daily Output	Labor-Hours	Unit	Material	Labor	Equipment	Total	Total Incl O&P
0010	**SPRAYED-ON INSULATION**										
0300	Closed cell, spray polyurethane foam, 2 lb./C.F. density										
0310	1" thick	G	G-2A	6000	.004	S.F.	.51	.12	.10	.73	.89
0320	2" thick	G		3000	.008		1.03	.24	.21	1.48	1.78

07 21 Thermal Insulation

07 21 29 – Sprayed Insulation

07 21 29.10 Sprayed-On Insulation			Crew	Daily Output	Labor-Hours	Unit	2019 Bare Costs Material	Labor	Equipment	Total	Total Incl O&P
0330	3" thick	G	G-2A	2000	.012	S.F.	1.54	.36	.31	2.21	2.66
0335	3-1/2" thick	G		1715	.014		1.80	.41	.36	2.57	3.11
0340	4" thick	G		1500	.016		2.06	.47	.42	2.95	3.55
0350	5" thick	G		1200	.020		2.57	.59	.52	3.68	4.44
0355	5-1/2" thick	G		1090	.022		2.83	.65	.57	4.05	4.88
0360	6" thick	G		1000	.024		3.08	.71	.62	4.41	5.35

07 22 Roof and Deck Insulation

07 22 16 – Roof Board Insulation

07 22 16.10 Roof Deck Insulation			Crew	Daily Output	Labor-Hours	Unit	Material	Labor	Equipment	Total	Total Incl O&P
0010	**ROOF DECK INSULATION**, fastening excluded										
0016	Asphaltic cover board, fiberglass lined, 1/8" thick		1 Rofc	1400	.006	S.F.	.48	.19		.67	.87
0018	1/4" thick			1400	.006		.97	.19		1.16	1.41
0020	Fiberboard low density, 1/2" thick, R1.39	G		1300	.006		.35	.21		.56	.76
0030	1" thick, R2.78	G		1040	.008		.59	.26		.85	1.11
0080	1-1/2" thick, R4.17	G		1040	.008		.90	.26		1.16	1.45
0100	2" thick, R5.56	G		1040	.008		1.16	.26		1.42	1.74
0110	Fiberboard high density, 1/2" thick, R1.3	G		1300	.006		.27	.21		.48	.67
0120	1" thick, R2.5	G		1040	.008		.57	.26		.83	1.09
0130	1-1/2" thick, R3.8	G		1040	.008		.84	.26		1.10	1.38
0200	Fiberglass, 3/4" thick, R2.78	G		1300	.006		.62	.21		.83	1.05
0400	15/16" thick, R3.70	G		1300	.006		.82	.21		1.03	1.27
0460	1-1/16" thick, R4.17	G		1300	.006		1.08	.21		1.29	1.56
0600	1-5/16" thick, R5.26	G		1300	.006		1.43	.21		1.64	1.94
0650	2-1/16" thick, R8.33	G		1040	.008		1.60	.26		1.86	2.22
0700	2-7/16" thick, R10	G		1040	.008		1.73	.26		1.99	2.36
0800	Gypsum cover board, fiberglass mat facer, 1/4" thick			1400	.006		.48	.19		.67	.87
0810	1/2" thick			1300	.006		.59	.21		.80	1.02
0820	5/8" thick			1200	.007		.62	.22		.84	1.08
0830	Primed fiberglass mat facer, 1/4" thick			1400	.006		.50	.19		.69	.89
0840	1/2" thick			1300	.006		.58	.21		.79	1.01
0850	5/8" thick			1200	.007		.61	.22		.83	1.07
1650	Perlite, 1/2" thick, R1.32	G		1365	.006		.27	.20		.47	.65
1655	3/4" thick, R2.08	G		1040	.008		.38	.26		.64	.88
1660	1" thick, R2.78	G		1040	.008		.55	.26		.81	1.07
1670	1-1/2" thick, R4.17	G		1040	.008		.81	.26		1.07	1.35
1680	2" thick, R5.56	G		910	.009		1.10	.29		1.39	1.74
1685	2-1/2" thick, R6.67	G		910	.009		1.45	.29		1.74	2.13
1690	Tapered for drainage	G		1040	.008	B.F.	1.05	.26		1.31	1.62
1700	Polyisocyanurate, 2#/C.F. density, 3/4" thick	G		1950	.004	S.F.	.50	.14		.64	.80
1705	1" thick	G		1820	.004		.44	.15		.59	.75
1715	1-1/2" thick	G		1625	.005		.60	.16		.76	.96
1725	2" thick	G		1430	.006		.79	.19		.98	1.21
1735	2-1/2" thick	G		1365	.006		1.02	.20		1.22	1.47
1745	3" thick	G		1300	.006		1.12	.21		1.33	1.60
1755	3-1/2" thick	G		1300	.006		1.86	.21		2.07	2.42
1765	Tapered for drainage	G		1820	.004	B.F.	.52	.15		.67	.84
1900	Extruded polystyrene										
1910	15 psi compressive strength, 1" thick, R5	G	1 Rofc	1950	.004	S.F.	.56	.14		.70	.87
1920	2" thick, R10	G		1625	.005		.73	.16		.89	1.10
1930	3" thick, R15	G		1300	.006		1.46	.21		1.67	1.97

07 22 16.10 Roof Deck Insulation			Crew	Daily Output	Labor-Hours	Unit	2019 Bare Costs Material	Labor	Equipment	Total	Total Incl O&P
1932	4" thick, R20	G	1 Rofc	1300	.006	S.F.	1.96	.21		2.17	2.53
1934	Tapered for drainage	G		1950	.004	B.F.	.51	.14		.65	.81
1940	25 psi compressive strength, 1" thick, R5	G		1950	.004	S.F.	1.12	.14		1.26	1.48
1942	2" thick, R10	G		1625	.005		2.13	.16		2.29	2.64
1944	3" thick, R15	G		1300	.006		3.25	.21		3.46	3.94
1946	4" thick, R20	G		1300	.006		4.48	.21		4.69	5.30
1948	Tapered for drainage	G		1950	.004	B.F.	.56	.14		.70	.87
1950	40 psi compressive strength, 1" thick, R5	G		1950	.004	S.F.	.87	.14		1.01	1.21
1952	2" thick, R10	G		1625	.005		1.65	.16		1.81	2.12
1954	3" thick, R15	G		1300	.006		2.39	.21		2.60	3
1956	4" thick, R20	G		1300	.006		3.13	.21		3.34	3.82
1958	Tapered for drainage	G		1820	.004	B.F.	.87	.15		1.02	1.23
1960	60 psi compressive strength, 1" thick, R5	G		1885	.004	S.F.	1.05	.14		1.19	1.42
1962	2" thick, R10	G		1560	.005		2	.17		2.17	2.50
1964	3" thick, R15	G		1270	.006		3.26	.21		3.47	3.96
1966	4" thick, R20	G		1235	.006		4.04	.22		4.26	4.84
1968	Tapered for drainage	G		1820	.004	B.F.	1.01	.15		1.16	1.38
2010	Expanded polystyrene, 1#/C.F. density, 3/4" thick, R2.89	G		1950	.004	S.F.	.20	.14		.34	.47
2020	1" thick, R3.85	G		1950	.004		.27	.14		.41	.55
2100	2" thick, R7.69	G		1625	.005		.54	.16		.70	.89
2110	3" thick, R11.49	G		1625	.005		.81	.16		.97	1.19
2120	4" thick, R15.38	G		1625	.005		1.08	.16		1.24	1.49
2130	5" thick, R19.23	G		1495	.005		1.35	.18		1.53	1.81
2140	6" thick, R23.26	G		1495	.005		1.62	.18		1.80	2.10
2150	Tapered for drainage	G		1950	.004	B.F.	.53	.14		.67	.83
2400	Composites with 2" EPS										
2410	1" fiberboard	G	1 Rofc	1325	.006	S.F.	1.47	.20		1.67	1.98
2420	7/16" oriented strand board	G		1040	.008		1.24	.26		1.50	1.82
2430	1/2" plywood	G		1040	.008		1.47	.26		1.73	2.08
2440	1" perlite	G		1040	.008		1.16	.26		1.42	1.74
2450	Composites with 1-1/2" polyisocyanurate										
2460	1" fiberboard	G	1 Rofc	1040	.008	S.F.	1.07	.26		1.33	1.64
2470	1" perlite	G		1105	.007		.94	.24		1.18	1.47
2480	7/16" oriented strand board	G		1040	.008		.85	.26		1.11	1.40
3000	Fastening alternatives, coated screws, 2" long			3744	.002	Ea.	.06	.07		.13	.20
3010	4" long			3120	.003		.11	.09		.20	.27
3020	6" long			2675	.003		.19	.10		.29	.39
3030	8" long			2340	.003		.28	.11		.39	.52
3040	10" long			1872	.004		.47	.14		.61	.78
3050	Pre-drill and drive wedge spike, 2-1/2"			1248	.006		.40	.21		.61	.83
3060	3-1/2"			1101	.007		.58	.24		.82	1.08
3070	4-1/2"			936	.009		.65	.29		.94	1.24
3075	3" galvanized deck plates			7488	.001		.09	.04		.13	.16
3080	Spot mop asphalt		G-1	295	.190	Sq.	5.80	5.95	1.74	13.49	19
3090	Full mop asphalt		"	192	.292	"	11.65	9.10	2.67	23.42	32

07 24 Exterior Insulation and Finish Systems

07 24 13 – Polymer-Based Exterior Insulation and Finish System

07 24 13.10 Exterior Insulation and Finish Systems			Crew	Daily Output	Labor-Hours	Unit	Material	Labor	Equipment	Total	Total Incl O&P
							2019 Bare Costs				
0010	EXTERIOR INSULATION AND FINISH SYSTEMS										
0095	Field applied, 1" EPS insulation	G	J-1	390	.103	S.F.	1.80	3.52	.35	5.67	8.10
0100	With 1/2" cement board sheathing	G		268	.149		2.59	5.10	.50	8.19	11.75
0105	2" EPS insulation	G		390	.103		2.07	3.52	.35	5.94	8.40
0110	With 1/2" cement board sheathing	G		268	.149		2.86	5.10	.50	8.46	12.05
0115	3" EPS insulation	G		390	.103		2.34	3.52	.35	6.21	8.70
0120	With 1/2" cement board sheathing	G		268	.149		3.13	5.10	.50	8.73	12.35
0125	4" EPS insulation	G		390	.103		2.61	3.52	.35	6.48	9
0130	With 1/2" cement board sheathing	G		268	.149		4.19	5.10	.50	9.79	13.50
0140	Premium finish add			1265	.032		.41	1.08	.11	1.60	2.34
0150	Heavy duty reinforcement add			914	.044		.60	1.50	.15	2.25	3.27
0160	2.5#/S.Y. metal lath substrate add		1 Lath	75	.107	S.Y.	3.83	4.07		7.90	10.75
0170	3.4#/S.Y. metal lath substrate add		"	75	.107	"	4.39	4.07		8.46	11.40
0180	Color or texture change		J-1	1265	.032	S.F.	.81	1.08	.11	2	2.78
0190	With substrate leveling base coat		1 Plas	530	.015		.82	.55		1.37	1.80
0210	With substrate sealing base coat		1 Pord	1224	.007		.15	.21		.36	.52
0370	V groove shape in panel face					L.F.	.69			.69	.76
0380	U groove shape in panel face					"	.85			.85	.94
0440	For higher than one story, add							25%			

07 25 Weather Barriers

07 25 10 – Weather Barriers or Wraps

07 25 10.10 Weather Barriers		Crew	Daily Output	Labor-Hours	Unit	Material	Labor	Equipment	Total	Total Incl O&P
0010	WEATHER BARRIERS									
0400	Asphalt felt paper, #15	1 Carp	37	.216	Sq.	5.40	8.40		13.80	19.80
0401	Per square foot	"	3700	.002	S.F.	.05	.08		.13	.20
0450	Housewrap, exterior, spun bonded polypropylene									
0470	Small roll	1 Carp	3800	.002	S.F.	.16	.08		.24	.31
0480	Large roll	"	4000	.002	"	.14	.08		.22	.28
2100	Asphalt felt roof deck vapor barrier, class 1 metal decks	1 Rofc	37	.216	Sq.	22.50	7.20		29.70	37.50
2200	For all other decks	"	37	.216		17	7.20		24.20	32
2800	Asphalt felt, 50% recycled content, 15 lb., 4 sq./roll	1 Carp	36	.222		5.90	8.60		14.50	21
2810	30 lb., 2 sq./roll	"	36	.222		10	8.60		18.60	25.50
3000	Building wrap, spun bonded polyethylene	2 Carp	8000	.002	S.F.	.17	.08		.25	.32

07 26 Vapor Retarders

07 26 10 – Above-Grade Vapor Retarders

07 26 10.10 Vapor Retarders			Crew	Daily Output	Labor-Hours	Unit	Material	Labor	Equipment	Total	Total Incl O&P
0010	VAPOR RETARDERS										
0020	Aluminum and kraft laminated, foil 1 side	G	1 Carp	37	.216	Sq.	14.30	8.40		22.70	29.50
0100	Foil 2 sides	G		37	.216		15.10	8.40		23.50	30.50
0600	Polyethylene vapor barrier, standard, 2 mil	G		37	.216		1.56	8.40		9.96	15.55
0700	4 mil	G		37	.216		2.66	8.40		11.06	16.80
0900	6 mil	G		37	.216		3.87	8.40		12.27	18.10
1200	10 mil	G		37	.216		9.15	8.40		17.55	24
1800	Reinf. waterproof, 2 mil polyethylene backing, 1 side			37	.216		10.90	8.40		19.30	26
1900	2 sides			37	.216		14.15	8.40		22.55	29.50

07 27 Air Barriers

07 27 13 – Modified Bituminous Sheet Air Barriers

07 27 13.10 Modified Bituminous Sheet Air Barrier		Crew	Daily Output	Labor-Hours	Unit	Material	2019 Bare Costs Labor	Equipment	Total	Total Incl O&P
0010	MODIFIED BITUMINOUS SHEET AIR BARRIER									
0100	SBS modified sheet laminated to polyethylene sheet, 40 mils, 4" wide	1 Carp	1200	.007	L.F.	.33	.26		.59	.79
0120	6" wide		1100	.007		.45	.28		.73	.96
0140	9" wide		1000	.008		.62	.31		.93	1.20
0160	12" wide		900	.009		.80	.34		1.14	1.45
0180	18" wide	2 Carp	1700	.009	S.F.	.75	.36		1.11	1.42
0200	36" wide	"	1800	.009		.73	.34		1.07	1.37
0220	Adhesive for above	1 Carp	1400	.006		.32	.22		.54	.72

07 27 26 – Fluid-Applied Membrane Air Barriers

07 27 26.10 Fluid Applied Membrane Air Barrier

		Crew	Daily Output	Labor-Hours	Unit	Material	Labor	Equipment	Total	Total Incl O&P
0010	FLUID APPLIED MEMBRANE AIR BARRIER									
0100	Spray applied vapor barrier, 25 S.F./gallon	1 Pord	1375	.006	S.F.	.02	.19		.21	.33

07 31 Shingles and Shakes

07 31 13 – Asphalt Shingles

07 31 13.10 Asphalt Roof Shingles

		Crew	Daily Output	Labor-Hours	Unit	Material	Labor	Equipment	Total	Total Incl O&P
0010	ASPHALT ROOF SHINGLES									
0100	Standard strip shingles									
0150	Inorganic, class A, 25 year	1 Rofc	5.50	1.455	Sq.	79	48.50		127.50	174
0155	Pneumatic nailed		7	1.143		79	38		117	156
0200	30 year		5	1.600		96	53.50		149.50	203
0205	Pneumatic nailed		6.25	1.280		96	42.50		138.50	183
0250	Standard laminated multi-layered shingles									
0300	Class A, 240-260 lb./square	1 Rofc	4.50	1.778	Sq.	110	59.50		169.50	228
0305	Pneumatic nailed		5.63	1.422		110	47.50		157.50	207
0350	Class A, 250-270 lb./square		4	2		110	66.50		176.50	242
0355	Pneumatic nailed		5	1.600		110	53.50		163.50	218
0400	Premium, laminated multi-layered shingles									
0450	Class A, 260-300 lb./square	1 Rofc	3.50	2.286	Sq.	145	76		221	298
0455	Pneumatic nailed		4.37	1.831		145	61		206	270
0500	Class A, 300-385 lb./square		3	2.667		274	89		363	460
0505	Pneumatic nailed		3.75	2.133		274	71		345	430
0800	#15 felt underlayment		64	.125		5.40	4.17		9.57	13.50
0825	#30 felt underlayment		58	.138		10.55	4.60		15.15	19.90
0850	Self adhering polyethylene and rubberized asphalt underlayment		22	.364		79	12.15		91.15	109
0900	Ridge shingles		330	.024	L.F.	2.28	.81		3.09	3.97
0905	Pneumatic nailed		412.50	.019	"	2.28	.65		2.93	3.68
1000	For steep roofs (7 to 12 pitch or greater), add						50%			

07 31 16 – Metal Shingles

07 31 16.10 Aluminum Shingles

		Crew	Daily Output	Labor-Hours	Unit	Material	Labor	Equipment	Total	Total Incl O&P
0010	ALUMINUM SHINGLES									
0020	Mill finish, .019" thick	1 Carp	5	1.600	Sq.	228	62		290	355
0100	.020" thick	"	5	1.600		260	62		322	390
0300	For colors, add					22			22	24
0600	Ridge cap, .024" thick	1 Carp	170	.047	L.F.	4.02	1.82		5.84	7.45
0700	End wall flashing, .024" thick		170	.047		2.30	1.82		4.12	5.55
0900	Valley section, .024" thick		170	.047		4.03	1.82		5.85	7.45
1000	Starter strip, .024" thick		400	.020		1.84	.78		2.62	3.30
1200	Side wall flashing, .024" thick		170	.047		2.24	1.82		4.06	5.45
1500	Gable flashing, .024" thick		400	.020		1.79	.78		2.57	3.25

07 31 Shingles and Shakes

07 31 26 – Slate Shingles

07 31 26.10 Slate Roof Shingles

	07 31 26.10 Slate Roof Shingles	Crew	Daily Output	Labor-Hours	Unit	Material	Labor	Equipment	Total	Total Incl O&P
0010	SLATE ROOF SHINGLES R073126-20									
0100	Buckingham Virginia black, 3/16" - 1/4" thick G	1 Rots	1.75	4.571	Sq.	560	153		713	890
0200	1/4" thick G		1.75	4.571		560	153		713	890
0900	Pennsylvania black, Bangor, #1 clear G		1.75	4.571		495	153		648	820
1200	Vermont, unfading, green, mottled green G		1.75	4.571		510	153		663	835
1300	Semi-weathering green & gray G		1.75	4.571		405	153		558	720
1400	Purple G		1.75	4.571		435	153		588	755
1500	Black or gray G		1.75	4.571		490	153		643	810
2700	Ridge shingles, slate		200	.040	L.F.	10.15	1.34		11.49	13.60

07 31 29 – Wood Shingles and Shakes

07 31 29.13 Wood Shingles

	07 31 29.13 Wood Shingles	Crew	Daily Output	Labor-Hours	Unit	Material	Labor	Equipment	Total	Total Incl O&P
0010	WOOD SHINGLES R061110-30									
0012	16" No. 1 red cedar shingles, 5" exposure, on roof	1 Carp	2.50	3.200	Sq.	296	124		420	530
0015	Pneumatic nailed		3.25	2.462		296	95.50		391.50	485
0200	7-1/2" exposure, on walls		2.05	3.902		197	151		348	465
0205	Pneumatic nailed		2.67	2.996		197	116		313	410
0300	18" No. 1 red cedar perfections, 5-1/2" exposure, on roof		2.75	2.909		254	113		367	465
0305	Pneumatic nailed		3.57	2.241		254	87		341	425
0500	7-1/2" exposure, on walls		2.25	3.556		187	138		325	435
0505	Pneumatic nailed		2.92	2.740		187	106		293	380
0600	Resquared and rebutted, 5-1/2" exposure, on roof		3	2.667		286	103		389	485
0605	Pneumatic nailed		3.90	2.051		286	79.50		365.50	445
0900	7-1/2" exposure, on walls		2.45	3.265		210	127		337	440
0905	Pneumatic nailed		3.18	2.516		210	97.50		307.50	390
1000	Add to above for fire retardant shingles					55.50			55.50	61
1060	Preformed ridge shingles	1 Carp	400	.020	L.F.	3.95	.78		4.73	5.65
2000	White cedar shingles, 16" long, extras, 5" exposure, on roof		2.40	3.333	Sq.	199	129		328	435
2005	Pneumatic nailed		3.12	2.564		199	99.50		298.50	385
2050	5" exposure on walls		2	4		199	155		354	475
2055	Pneumatic nailed		2.60	3.077		199	119		318	415
2100	7-1/2" exposure, on walls		2	4		142	155		297	410
2105	Pneumatic nailed		2.60	3.077		142	119		261	355
2150	"B" grade, 5" exposure on walls		2	4		172	155		327	445
2155	Pneumatic nailed		2.60	3.077		172	119		291	385
2300	For #15 organic felt underlayment on roof, 1 layer, add		64	.125		5.40	4.84		10.24	13.95
2400	2 layers, add		32	.250		10.80	9.70		20.50	28
2600	For steep roofs (7/12 pitch or greater), add to above						50%			
2700	Panelized systems, No.1 cedar shingles on 5/16" CDX plywood									
2800	On walls, 8' strips, 7" or 14" exposure	2 Carp	700	.023	S.F.	6.45	.89		7.34	8.55
3000	Ridge shakes or shingle, wood	1 Carp	280	.029	L.F.	4.79	1.11		5.90	7.10

07 31 29.16 Wood Shakes

	07 31 29.16 Wood Shakes	Crew	Daily Output	Labor-Hours	Unit	Material	Labor	Equipment	Total	Total Incl O&P
0010	WOOD SHAKES									
1100	Hand-split red cedar shakes, 1/2" thick x 24" long, 10" exp. on roof	1 Carp	2.50	3.200	Sq.	300	124		424	535
1105	Pneumatic nailed		3.25	2.462		300	95.50		395.50	490
1110	3/4" thick x 24" long, 10" exp. on roof		2.25	3.556		300	138		438	560
1115	Pneumatic nailed		2.92	2.740		300	106		406	505
1200	1/2" thick, 18" long, 8-1/2" exp. on roof		2	4		279	155		434	560
1205	Pneumatic nailed		2.60	3.077		279	119		398	500
1210	3/4" thick x 18" long, 8-1/2" exp. on roof		1.80	4.444		279	172		451	590
1215	Pneumatic nailed		2.34	3.419		279	132		411	525
1255	10" exposure on walls		2	4		269	155		424	550

07 31 Shingles and Shakes

07 31 29 – Wood Shingles and Shakes

07 31 29.16 Wood Shakes		Crew	Daily Output	Labor-Hours	Unit	2019 Bare Costs Material	Labor	Equipment	Total	Total Incl O&P
1260	10" exposure on walls, pneumatic nailed	1 Carp	2.60	3.077	Sq.	269	119		388	495
1700	Add to above for fire retardant shakes, 24" long					55.50			55.50	61
1800	18" long				↓	55.50			55.50	61
1810	Ridge shakes	1 Carp	350	.023	L.F.	5.75	.89		6.64	7.80

07 32 Roof Tiles

07 32 13 – Clay Roof Tiles

07 32 13.10 Clay Tiles		Crew	Daily Output	Labor-Hours	Unit	Material	Labor	Equipment	Total	Total Incl O&P
0010	**CLAY TILES**, including accessories									
0300	Flat shingle, interlocking, 15", 166 pcs./sq., fireflashed blend	3 Rots	6	4	Sq.	475	134		609	760
0500	Terra cotta red		6	4		495	134		629	780
0600	Roman pan and top, 18", 102 pcs./sq., fireflashed blend	↓	5.50	4.364		500	146		646	815
0640	Terra cotta red	1 Rots	2.40	3.333		560	112		672	815
1100	Barrel mission tile, 18", 166 pcs./sq., fireflashed blend	3 Rots	5.50	4.364		415	146		561	720
1140	Terra cotta red		5.50	4.364		420	146		566	725
1700	Scalloped edge flat shingle, 14", 145 pcs./sq., fireflashed blend		6	4		1,150	134		1,284	1,525
1800	Terra cotta red	↓	6	4		1,050	134		1,184	1,400
3010	#15 felt underlayment	1 Rofc	64	.125		5.40	4.17		9.57	13.50
3020	#30 felt underlayment		58	.138		10.55	4.60		15.15	19.90
3040	Polyethylene and rubberized asph. underlayment	↓	22	.364	↓	79	12.15		91.15	109

07 32 16 – Concrete Roof Tiles

07 32 16.10 Concrete Tiles		Crew	Daily Output	Labor-Hours	Unit	Material	Labor	Equipment	Total	Total Incl O&P
0010	**CONCRETE TILES**									
0020	Corrugated, 13" x 16-1/2", 90 per sq., 950 lb./sq.									
0050	Earthtone colors, nailed to wood deck	1 Rots	1.35	5.926	Sq.	104	199		303	475
0150	Blues		1.35	5.926		105	199		304	475
0200	Greens		1.35	5.926		105	199		304	475
0250	Premium colors	↓	1.35	5.926	↓	105	199		304	475
0500	Shakes, 13" x 16-1/2", 90 per sq., 950 lb./sq.									
0600	All colors, nailed to wood deck	1 Rots	1.50	5.333	Sq.	131	179		310	470
1500	Accessory pieces, ridge & hip, 10" x 16-1/2", 8 lb. each	"	120	.067	Ea.	3.78	2.24		6.02	8.20
1700	Rake, 6-1/2" x 16-3/4", 9 lb. each					3.78			3.78	4.16
1800	Mansard hip, 10" x 16-1/2", 9.2 lb. each					3.78			3.78	4.16
1900	Hip starter, 10" x 16-1/2", 10.5 lb. each					10.60			10.60	11.65
2000	3 or 4 way apex, 10" each side, 11.5 lb. each				↓	12.10			12.10	13.30

07 32 19 – Metal Roof Tiles

07 32 19.10 Metal Roof Tiles		Crew	Daily Output	Labor-Hours	Unit	Material	Labor	Equipment	Total	Total Incl O&P
0010	**METAL ROOF TILES**									
0020	Accessories included, .032" thick aluminum, mission tile	1 Carp	2.50	3.200	Sq.	830	124		954	1,125
0200	Spanish tiles	"	3	2.667	"	550	103		653	775

07 41 Roof Panels

07 41 13 – Metal Roof Panels

07 41 13.10 Aluminum Roof Panels			Crew	Daily Output	Labor-Hours	Unit	Material	2019 Bare Costs Labor	Equipment	Total	Total Incl O&P
0010	ALUMINUM ROOF PANELS										
0020	Corrugated or ribbed, .0155" thick, natural		G-3	1200	.027	S.F.	1.02	.98		2	2.75
0300	Painted		"	1200	.027	"	1.48	.98		2.46	3.26

07 41 13.20 Steel Roofing Panels			Crew	Daily Output	Labor-Hours	Unit	Material	Labor	Equipment	Total	Total Incl O&P
0010	STEEL ROOFING PANELS										
0012	Corrugated or ribbed, on steel framing, 30 ga. galv		G-3	1100	.029	S.F.	1.71	1.07		2.78	3.65
0100	28 ga.			1050	.030		1.49	1.12		2.61	3.50
0300	26 ga.			1000	.032		2.09	1.18		3.27	4.25
0400	24 ga.			950	.034		3.03	1.24		4.27	5.40
0510	Painted, including fasteners, 18 ga.	G		850	.038		4.20	1.39		5.59	6.90
0520	20 ga.	G		875	.037		3.25	1.35		4.60	5.80
0530	22 ga.	G		900	.036		4.50	1.31		5.81	7.10
0600	Colored, 28 ga.			1050	.030		1.87	1.12		2.99	3.92
0700	26 ga.			1000	.032		2.11	1.18		3.29	4.27
0710	Flat profile, 1-3/4" standing seams, 10" wide, standard finish, 26 ga.			1000	.032		3.98	1.18		5.16	6.35
0715	24 ga.			950	.034		4.64	1.24		5.88	7.15
0720	22 ga.			900	.036		5.75	1.31		7.06	8.45
0725	Zinc aluminum alloy finish, 26 ga.			1000	.032		3.19	1.18		4.37	5.45
0730	24 ga.			950	.034		3.75	1.24		4.99	6.20
0735	22 ga.			900	.036		4.28	1.31		5.59	6.90
0740	12" wide, standard finish, 26 ga.			1000	.032		4.01	1.18		5.19	6.35
0745	24 ga.			950	.034		5.20	1.24		6.44	7.80
0750	Zinc aluminum alloy finish, 26 ga.			1000	.032		4.50	1.18		5.68	6.90
0755	24 ga.			950	.034		3.75	1.24		4.99	6.20
0840	Flat profile, 1" x 3/8" batten, 12" wide, standard finish, 26 ga.			1000	.032		3.50	1.18		4.68	5.80
0845	24 ga.			950	.034		4.12	1.24		5.36	6.60
0850	22 ga.			900	.036		4.95	1.31		6.26	7.60
0855	Zinc aluminum alloy finish, 26 ga.			1000	.032		3.38	1.18		4.56	5.65
0860	24 ga.			950	.034		3.82	1.24		5.06	6.25
0865	22 ga.			900	.036		4.37	1.31		5.68	7
0870	16-1/2" wide, standard finish, 24 ga.			950	.034		4.07	1.24		5.31	6.55
0875	22 ga.			900	.036		4.58	1.31		5.89	7.20
0880	Zinc aluminum alloy finish, 24 ga.			950	.034		3.55	1.24		4.79	5.95
0885	22 ga.			900	.036		4	1.31		5.31	6.55
0890	Flat profile, 2" x 2" batten, 12" wide, standard finish, 26 ga.			1000	.032		4.06	1.18		5.24	6.40
0895	24 ga.			950	.034		4.88	1.24		6.12	7.40
0900	22 ga.			900	.036		5.95	1.31		7.26	8.70
0905	Zinc aluminum alloy finish, 26 ga.			1000	.032		3.82	1.18		5	6.15
0910	24 ga.			950	.034		4.32	1.24		5.56	6.80
0915	22 ga.			900	.036		5	1.31		6.31	7.65
0920	16-1/2" wide, standard finish, 24 ga.			950	.034		4.47	1.24		5.71	6.95
0925	22 ga.			900	.036		5.20	1.31		6.51	7.85
0930	Zinc aluminum alloy finish, 24 ga.			950	.034		4.02	1.24		5.26	6.45
0935	22 ga.			900	.036		4.59	1.31		5.90	7.20
0950	Box rib roof panels, painted, including fasteners, 18 ga.	G		850	.038		3.78	1.39		5.17	6.45
0960	20 ga.	G		875	.037		3.31	1.35		4.66	5.85
0970	22 ga.	G		900	.036		4.31	1.31		5.62	6.90
1000	4" rib panel, painted, including fasteners, 22 ga.	G		900	.036		3.30	1.31		4.61	5.80
1010	20 ga.	G		875	.037		3.22	1.35		4.57	5.75
1020	18 ga.	G		850	.038		6.50	1.39		7.89	9.45
1050	On substrate, 2" standing seam panel, painted, 22 ga.	G		900	.036		6	1.31		7.31	8.75
1060	24 ga.	G	↓	950	.034	↓	5.55	1.24		6.79	8.15

07 41 Roof Panels

07 41 13 – Metal Roof Panels

07 41 13.20 Steel Roofing Panels		Crew	Daily Output	Labor-Hours	Unit	Material	2019 Bare Costs Labor	Equipment	Total	Total Incl O&P
1070	26 ga. G	G-3	1000	.032	S.F.	2.77	1.18		3.95	5
1200	Ridge, galvanized, 10" wide G	↓	800	.040	L.F.	3.24	1.47		4.71	6
1203	14" wide G	2 Shee	316	.051		3.89	2.19		6.08	7.90
1205	18" wide G	"	308	.052		4.54	2.25		6.79	8.70
1210	20" wide G	G-3	750	.043	↓	4.31	1.57		5.88	7.35

07 41 33 – Plastic Roof Panels

07 41 33.10 Fiberglass Panels		Crew	Daily Output	Labor-Hours	Unit	Material	Labor	Equipment	Total	Total Incl O&P
0010	**FIBERGLASS PANELS**									
0012	Corrugated panels, roofing, 8 oz./S.F.	G-3	1000	.032	S.F.	2.60	1.18		3.78	4.81
0100	12 oz./S.F.		1000	.032		4.62	1.18		5.80	7.05
0300	Corrugated siding, 6 oz./S.F.		880	.036		2.02	1.34		3.36	4.44
0400	8 oz./S.F.		880	.036		2.60	1.34		3.94	5.10
0500	Fire retardant		880	.036		4.05	1.34		5.39	6.70
0600	12 oz. siding, textured		880	.036		3.95	1.34		5.29	6.55
0700	Fire retardant		880	.036		4.57	1.34		5.91	7.25
0900	Flat panels, 6 oz./S.F., clear or colors		880	.036		2.58	1.34		3.92	5.05
1100	Fire retardant, class A		880	.036		3.64	1.34		4.98	6.20
1300	8 oz./S.F., clear or colors	↓	880	.036	↓	2.48	1.34		3.82	4.95

07 42 Wall Panels

07 42 13 – Metal Wall Panels

07 42 13.20 Aluminum Siding		Crew	Daily Output	Labor-Hours	Unit	Material	Labor	Equipment	Total	Total Incl O&P
0011	**ALUMINUM SIDING**									
6040	0.024" thick smooth white single 8" wide	2 Carp	515	.031	S.F.	2.99	1.20		4.19	5.30
6060	Double 4" pattern		515	.031		2.98	1.20		4.18	5.25
6080	Double 5" pattern		550	.029		2.77	1.13		3.90	4.91
6120	Embossed white, 8" wide		515	.031		2.80	1.20		4	5.05
6140	Double 4" pattern		515	.031		2.88	1.20		4.08	5.15
6160	Double 5" pattern		550	.029		2.90	1.13		4.03	5.05
6170	Vertical, embossed white, 12" wide		590	.027		2.91	1.05		3.96	4.94
6320	0.019" thick, insulated, smooth white, 8" wide		515	.031		2.54	1.20		3.74	4.78
6340	Double 4" pattern		515	.031		2.52	1.20		3.72	4.76
6360	Double 5" pattern		550	.029		2.53	1.13		3.66	4.64
6400	Embossed white, 8" wide		515	.031		2.93	1.20		4.13	5.20
6420	Double 4" pattern		515	.031		2.95	1.20		4.15	5.25
6440	Double 5" pattern		550	.029		2.95	1.13		4.08	5.10
6500	Shake finish 10" wide white		550	.029		3.19	1.13		4.32	5.35
6600	Vertical pattern, 12" wide, white	↓	590	.027		2.42	1.05		3.47	4.40
6640	For colors add				↓	.15			.15	.17
6700	Accessories, white									
6720	Starter strip 2-1/8"	2 Carp	610	.026	L.F.	.48	1.02		1.50	2.21
6740	Sill trim		450	.036		.67	1.38		2.05	3.02
6760	Inside corner		610	.026		1.73	1.02		2.75	3.58
6780	Outside corner post		610	.026		3.71	1.02		4.73	5.75
6800	Door & window trim	↓	440	.036		.65	1.41		2.06	3.05
6820	For colors add					.15			.15	.17
6900	Soffit & fascia 1' overhang solid	2 Carp	110	.145		4.52	5.65		10.17	14.25
6920	Vented		110	.145		4.55	5.65		10.20	14.30
6940	2' overhang solid		100	.160		6.60	6.20		12.80	17.50
6960	Vented	↓	100	.160	↓	6.60	6.20		12.80	17.50

07 42 Wall Panels

07 42 13 – Metal Wall Panels

07 42 13.30 Steel Siding		Crew	Daily Output	Labor-Hours	Unit	Material	2019 Bare Costs Labor	Equipment	Total	Total Incl O&P
0010	**STEEL SIDING**									
0020	Beveled, vinyl coated, 8" wide	1 Carp	265	.030	S.F.	1.79	1.17		2.96	3.90
0050	10" wide	"	275	.029		1.96	1.13		3.09	4.02
0080	Galv, corrugated or ribbed, on steel frame, 30 ga.	G-3	800	.040		1.24	1.47		2.71	3.80
0100	28 ga.		795	.040		1.35	1.48		2.83	3.94
0300	26 ga.		790	.041		1.75	1.49		3.24	4.40
0400	24 ga.		785	.041		2.33	1.50		3.83	5.05
0600	22 ga.		770	.042		2.41	1.53		3.94	5.20
0700	Colored, corrugated/ribbed, on steel frame, 10 yr. finish, 28 ga.		800	.040		2.02	1.47		3.49	4.66
0900	26 ga.		795	.040		2.08	1.48		3.56	4.74
1000	24 ga.		790	.041		2.32	1.49		3.81	5
1020	20 ga.		785	.041		2.95	1.50		4.45	5.75

07 46 Siding

07 46 23 – Wood Siding

07 46 23.10 Wood Board Siding

		Crew	Daily Output	Labor-Hours	Unit	Material	Labor	Equipment	Total	Total Incl O&P
0010	**WOOD BOARD SIDING**									
2000	Board & batten, cedar, "B" grade, 1" x 10"	1 Carp	375	.021	S.F.	6.65	.83		7.48	8.65
2200	Redwood, clear, vertical grain, 1" x 10"		375	.021		5.40	.83		6.23	7.25
2400	White pine, #2 & better, 1" x 10"		375	.021		3.67	.83		4.50	5.40
2410	White pine, #2 & better, 1" x 12"		420	.019		3.67	.74		4.41	5.25
3200	Wood, cedar bevel, A grade, 1/2" x 6"		295	.027		4.43	1.05		5.48	6.60
3300	1/2" x 8"		330	.024		7.75	.94		8.69	10.10
3500	3/4" x 10", clear grade		375	.021		7.20	.83		8.03	9.25
3600	"B" grade		375	.021		4.10	.83		4.93	5.90
3800	Cedar, rough sawn, 1" x 4", A grade, natural		220	.036		7.50	1.41		8.91	10.60
3900	Stained		220	.036		7.65	1.41		9.06	10.75
4100	1" x 12", board & batten, #3 & Btr., natural		420	.019		4.79	.74		5.53	6.45
4200	Stained		420	.019		5.15	.74		5.89	6.85
4400	1" x 8" channel siding, #3 & Btr., natural		330	.024		4.92	.94		5.86	6.95
4500	Stained		330	.024		5.05	.94		5.99	7.10
4700	Redwood, clear, beveled, vertical grain, 1/2" x 4"		220	.036		4.87	1.41		6.28	7.70
4750	1/2" x 6"		295	.027		4.99	1.05		6.04	7.25
4800	1/2" x 8"		330	.024		5.45	.94		6.39	7.55
5000	3/4" x 10"		375	.021		5.10	.83		5.93	6.95
5200	Channel siding, 1" x 10", B grade		375	.021		4.65	.83		5.48	6.45
5250	Redwood, T&G boards, B grade, 1" x 4"		220	.036		6.85	1.41		8.26	9.90
5270	1" x 8"		330	.024		8	.94		8.94	10.35
5400	White pine, rough sawn, 1" x 8", natural		330	.024		2.51	.94		3.45	4.31
5500	Stained		330	.024		2.40	.94		3.34	4.19
5600	T&G, 1" x 8"		330	.024		2.54	.94		3.48	4.34

07 46 29 – Plywood Siding

07 46 29.10 Plywood Siding Options

		Crew	Daily Output	Labor-Hours	Unit	Material	Labor	Equipment	Total	Total Incl O&P
0010	**PLYWOOD SIDING OPTIONS**									
0900	Plywood, medium density overlaid, 3/8" thick	2 Carp	750	.021	S.F.	1.38	.83		2.21	2.89
1000	1/2" thick		700	.023		1.57	.89		2.46	3.19
1100	3/4" thick		650	.025		1.93	.95		2.88	3.70
1600	Texture 1-11, cedar, 5/8" thick, natural		675	.024		2.62	.92		3.54	4.40
1700	Factory stained		675	.024		2.92	.92		3.84	4.73
1900	Texture 1-11, fir, 5/8" thick, natural		675	.024		1.36	.92		2.28	3.02

07 46 29 – Plywood Siding

07 46 29.10 Plywood Siding Options		Crew	Daily Output	Labor-Hours	Unit	Material	Labor (2019 Bare Costs)	Equipment	Total	Total Incl O&P
2000	Factory stained	2 Carp	675	.024	S.F.	1.93	.92		2.85	3.64
2050	Texture 1-11, S.Y.P., 5/8" thick, natural		675	.024		1.45	.92		2.37	3.12
2100	Factory stained		675	.024		1.53	.92		2.45	3.20
2200	Rough sawn cedar, 3/8" thick, natural		675	.024		1.28	.92		2.20	2.93
2300	Factory stained		675	.024		1.59	.92		2.51	3.27
2500	Rough sawn fir, 3/8" thick, natural		675	.024		.93	.92		1.85	2.54
2600	Factory stained		675	.024		1.10	.92		2.02	2.73
2800	Redwood, textured siding, 5/8" thick		675	.024		2.03	.92		2.95	3.75

07 46 33 – Plastic Siding

07 46 33.10 Vinyl Siding		Crew	Daily Output	Labor-Hours	Unit	Material	Labor (2019 Bare Costs)	Equipment	Total	Total Incl O&P
0010	**VINYL SIDING**									
3995	Clapboard profile, woodgrain texture, .048 thick, double 4	2 Carp	495	.032	S.F.	1.08	1.25		2.33	3.26
4000	Double 5		550	.029		1.08	1.13		2.21	3.05
4005	Single 8		495	.032		1.39	1.25		2.64	3.60
4010	Single 10		550	.029		1.67	1.13		2.80	3.70
4015	.044 thick, double 4		495	.032		1.07	1.25		2.32	3.25
4020	Double 5		550	.029		1.09	1.13		2.22	3.06
4025	.042 thick, double 4		495	.032		1.09	1.25		2.34	3.27
4030	Double 5		550	.029		1.09	1.13		2.22	3.06
4035	Cross sawn texture, .040 thick, double 4		495	.032		.74	1.25		1.99	2.89
4040	Double 5		550	.029		.67	1.13		1.80	2.60
4045	Smooth texture, .042 thick, double 4		495	.032		.81	1.25		2.06	2.97
4050	Double 5		550	.029		.80	1.13		1.93	2.75
4055	Single 8		495	.032		.81	1.25		2.06	2.97
4060	Cedar texture, .044 thick, double 4		495	.032		1.16	1.25		2.41	3.35
4065	Double 6		600	.027		1.17	1.03		2.20	3
4070	Dutch lap profile, woodgrain texture, .048 thick, double 5		550	.029		1.10	1.13		2.23	3.08
4075	.044 thick, double 4.5		525	.030		1.08	1.18		2.26	3.14
4080	.042 thick, double 4.5		525	.030		.93	1.18		2.11	2.98
4085	.040 thick, double 4.5		525	.030		.74	1.18		1.92	2.77
4100	Shake profile, 10" wide		400	.040		3.81	1.55		5.36	6.75
4105	Vertical pattern, .046 thick, double 5		550	.029		1.58	1.13		2.71	3.60
4110	.044 thick, triple 3		550	.029		1.79	1.13		2.92	3.83
4115	.040 thick, triple 4		550	.029		1.73	1.13		2.86	3.77
4120	.040 thick, triple 2.66		550	.029		1.80	1.13		2.93	3.85
4125	Insulation, fan folded extruded polystyrene, 1/4"		2000	.008		.29	.31		.60	.83
4130	3/8"		2000	.008		.32	.31		.63	.86
4135	Accessories, J channel, 5/8" pocket		700	.023	L.F.	.52	.89		1.41	2.03
4140	3/4" pocket		695	.023		.56	.89		1.45	2.08
4145	1-1/4" pocket		680	.024		1.05	.91		1.96	2.66
4150	Flexible, 3/4" pocket		600	.027		2.44	1.03		3.47	4.39
4155	Under sill finish trim		500	.032		.57	1.24		1.81	2.67
4160	Vinyl starter strip		700	.023		.70	.89		1.59	2.23
4165	Aluminum starter strip		700	.023		.30	.89		1.19	1.79
4170	Window casing, 2-1/2" wide, 3/4" pocket		510	.031		1.79	1.22		3.01	3.98
4175	Outside corner, woodgrain finish, 4" face, 3/4" pocket		700	.023		2.16	.89		3.05	3.84
4180	5/8" pocket		700	.023		2.17	.89		3.06	3.85
4185	Smooth finish, 4" face, 3/4" pocket		700	.023		2.15	.89		3.04	3.83
4190	7/8" pocket		690	.023		2.02	.90		2.92	3.72
4195	1-1/4" pocket		700	.023		1.42	.89		2.31	3.03
4200	Soffit and fascia, 1' overhang, solid		120	.133		4.84	5.15		9.99	13.90
4205	Vented		120	.133		4.84	5.15		9.99	13.90

07 46 Siding

07 46 33 – Plastic Siding

07 46 33.10 Vinyl Siding		Crew	Daily Output	Labor-Hours	Unit	2019 Bare Costs Material	Labor	Equipment	Total	Total Incl O&P
4207	18″ overhang, solid	2 Carp	110	.145	L.F.	5.65	5.65		11.30	15.50
4208	Vented		110	.145		5.65	5.65		11.30	15.50
4210	2′ overhang, solid		100	.160		6.40	6.20		12.60	17.30
4215	Vented		100	.160		6.40	6.20		12.60	17.30
4217	3′ overhang, solid		100	.160		8	6.20		14.20	19.05
4218	Vented		100	.160		8	6.20		14.20	19.05
4220	Colors for siding and soffits, add				S.F.	.15			.15	.17
4225	Colors for accessories and trim, add				L.F.	.31			.31	.34

07 46 33.20 Polypropylene Siding		Crew	Daily Output	Labor-Hours	Unit	Material	Labor	Equipment	Total	Total Incl O&P
0010	**POLYPROPYLENE SIDING**									
4090	Shingle profile, random grooves, double 7	2 Carp	400	.040	S.F.	3.47	1.55		5.02	6.40
4092	Cornerpost for above	1 Carp	365	.022	L.F.	13	.85		13.85	15.70
4095	Triple 5	2 Carp	400	.040	S.F.	3.48	1.55		5.03	6.40
4097	Cornerpost for above	1 Carp	365	.022	L.F.	12.35	.85		13.20	14.95
5000	Staggered butt, double 7″	2 Carp	400	.040	S.F.	3.68	1.55		5.23	6.60
5002	Cornerpost for above	1 Carp	365	.022	L.F.	13.25	.85		14.10	15.95
5010	Half round, double 6-1/4″	2 Carp	360	.044	S.F.	4.05	1.72		5.77	7.30
5020	Shake profile, staggered butt, double 9″	″	510	.031	″	3.81	1.22		5.03	6.20
5022	Cornerpost for above	1 Carp	365	.022	L.F.	9.90	.85		10.75	12.30
5030	Straight butt, double 7″	2 Carp	400	.040	S.F.	4.05	1.55		5.60	7
5032	Cornerpost for above	1 Carp	365	.022	L.F.	13.90	.85		14.75	16.70
6000	Accessories, J channel, 5/8″ pocket	2 Carp	700	.023		.52	.89		1.41	2.03
6010	3/4″ pocket		695	.023		.56	.89		1.45	2.08
6020	1-1/4″ pocket		680	.024		1.05	.91		1.96	2.66
6030	Aluminum starter strip		700	.023		.30	.89		1.19	1.79

07 46 46 – Fiber Cement Siding

07 46 46.10 Fiber Cement Siding		Crew	Daily Output	Labor-Hours	Unit	Material	Labor	Equipment	Total	Total Incl O&P
0010	**FIBER CEMENT SIDING**									
0020	Lap siding, 5/16″ thick, 6″ wide, 4-3/4″ exposure, smooth texture	2 Carp	415	.039	S.F.	1.31	1.49		2.80	3.91
0025	Woodgrain texture		415	.039		1.31	1.49		2.80	3.91
0030	7-1/2″ wide, 6-1/4″ exposure, smooth texture		425	.038		1.82	1.46		3.28	4.42
0035	Woodgrain texture		425	.038		1.82	1.46		3.28	4.42
0040	8″ wide, 6-3/4″ exposure, smooth texture		425	.038		1.33	1.46		2.79	3.88
0045	Rough sawn texture		425	.038		1.33	1.46		2.79	3.88
0050	9-1/2″ wide, 8-1/4″ exposure, smooth texture		440	.036		1.31	1.41		2.72	3.77
0055	Woodgrain texture		440	.036		1.31	1.41		2.72	3.77
0060	12″ wide, 10-3/8″ exposure, smooth texture		455	.035		2.12	1.36		3.48	4.58
0065	Woodgrain texture		455	.035		2.12	1.36		3.48	4.58
0070	Panel siding, 5/16″ thick, smooth texture		750	.021		1.22	.83		2.05	2.71
0075	Stucco texture		750	.021		1.22	.83		2.05	2.71
0080	Grooved woodgrain texture		750	.021		1.22	.83		2.05	2.71
0085	V - grooved woodgrain texture		750	.021		1.22	.83		2.05	2.71
0088	Shingle siding, 48″ x 15-1/4″ panels, 7″ exposure		700	.023		4.30	.89		5.19	6.20
0090	Wood starter strip		400	.040	L.F.	.44	1.55		1.99	3.05

07 51 13.10 Built-Up Roofing Components

	07 51 13.10 Built-Up Roofing Components	Crew	Daily Output	Labor-Hours	Unit	2019 Bare Costs Material	Labor	Equipment	Total	Total Incl O&P
0010	BUILT-UP ROOFING COMPONENTS									
0012	Asphalt saturated felt, #30, 2 sq./roll	1 Rofc	58	.138	Sq.	10.55	4.60		15.15	19.90
0200	#15, 4 sq./roll, plain or perforated, not mopped		58	.138		5.40	4.60		10	14.25
0250	Perforated		58	.138		5.40	4.60		10	14.25
0300	Roll roofing, smooth, #65		15	.533		10.55	17.80		28.35	43.50
0500	#90		12	.667		38	22		60	82
0520	Mineralized		12	.667		35.50	22		57.50	79
0540	D.C. (double coverage), 19" selvage edge		10	.800		48	26.50		74.50	101
0580	Adhesive (lap cement)				Gal.	8.65			8.65	9.50

07 51 13.13 Cold-Applied Built-Up Asphalt Roofing

	07 51 13.13 Cold-Applied Built-Up Asphalt Roofing	Crew	Daily Output	Labor-Hours	Unit	Material	Labor	Equipment	Total	Total Incl O&P
0010	COLD-APPLIED BUILT-UP ASPHALT ROOFING									
0020	3 ply system, installation only (components listed below)	G-5	50	.800	Sq.		24.50	3.62	28.12	48
0100	Spunbond poly. fabric, 1.35 oz./S.Y., 36"W, 10.8 sq./roll				Ea.	135			135	149
0500	Base & finish coat, 3 gal./sq., 5 gal./can				Gal.	8.45			8.45	9.30
0600	Coating, ceramic granules, 1/2 sq./bag				Ea.	23.50			23.50	26
0700	Aluminum, 2 gal./sq.				Gal.	12.10			12.10	13.30
0800	Emulsion, fibered or non-fibered, 4 gal./sq.				″	6.40			6.40	7.05

07 51 13.20 Built-Up Roofing Systems

	07 51 13.20 Built-Up Roofing Systems	Crew	Daily Output	Labor-Hours	Unit	Material	Labor	Equipment	Total	Total Incl O&P
0010	BUILT-UP ROOFING SYSTEMS									
0120	Asphalt flood coat with gravel/slag surfacing, not including									
0140	Insulation, flashing or wood nailers									
0200	Asphalt base sheet, 3 plies #15 asphalt felt, mopped	G-1	22	2.545	Sq.	107	79.50	23.50	210	288
0350	On nailable decks		21	2.667		110	83.50	24.50	218	299
0500	4 plies #15 asphalt felt, mopped		20	2.800		145	87.50	25.50	258	345
0550	On nailable decks		19	2.947		129	92	27	248	340
0700	Coated glass base sheet, 2 plies glass (type IV), mopped		22	2.545		116	79.50	23.50	219	297
0850	3 plies glass, mopped		20	2.800		139	87.50	25.50	252	340
0950	On nailable decks		19	2.947		131	92	27	250	340
1100	4 plies glass fiber felt (type IV), mopped		20	2.800		171	87.50	25.50	284	375
1150	On nailable decks		19	2.947		154	92	27	273	365
1200	Coated & saturated base sheet, 3 plies #15 asph. felt, mopped		20	2.800		119	87.50	25.50	232	320
1250	On nailable decks		19	2.947		111	92	27	230	320
1300	4 plies #15 asphalt felt, mopped		22	2.545		138	79.50	23.50	241	320
2000	Asphalt flood coat, smooth surface									
2200	Asphalt base sheet & 3 plies #15 asphalt felt, mopped	G-1	24	2.333	Sq.	110	73	21.50	204.50	277
2400	On nailable decks		23	2.435		102	76	22.50	200.50	275
2600	4 plies #15 asphalt felt, mopped		24	2.333		129	73	21.50	223.50	298
2700	On nailable decks		23	2.435		121	76	22.50	219.50	296
2900	Coated glass fiber base sheet, mopped, and 2 plies of									
2910	glass fiber felt (type IV)	G-1	25	2.240	Sq.	108	70	20.50	198.50	268
3100	On nailable decks		24	2.333		102	73	21.50	196.50	268
3200	3 plies, mopped		23	2.435		131	76	22.50	229.50	305
3300	On nailable decks		22	2.545		123	79.50	23.50	226	305
3800	4 plies glass fiber felt (type IV), mopped		23	2.435		154	76	22.50	252.50	335
3900	On nailable decks		22	2.545		146	79.50	23.50	249	330
4000	Coated & saturated base sheet, 3 plies #15 asph. felt, mopped		24	2.333		111	73	21.50	205.50	278
4200	On nailable decks		23	2.435		103	76	22.50	201.50	276
4300	4 plies #15 organic felt, mopped		22	2.545		130	79.50	23.50	233	315
4500	Coal tar pitch with gravel/slag surfacing									
4600	4 plies #15 tarred felt, mopped	G-1	21	2.667	Sq.	207	83.50	24.50	315	405
4800	3 plies glass fiber felt (type IV), mopped	″	19	2.947	″	172	92	27	291	385
5000	Coated glass fiber base sheet, and 2 plies of									

07 51 Built-Up Bituminous Roofing

07 51 13 – Built-Up Asphalt Roofing

07 51 13.20 Built-Up Roofing Systems		Crew	Daily Output	Labor-Hours	Unit	2019 Bare Costs Material	Labor	Equipment	Total	Total Incl O&P
5010	glass fiber felt (type IV), mopped	G-1	19	2.947	Sq.	177	92	27	296	390
5300	On nailable decks		18	3.111		155	97.50	28.50	281	380
5600	4 plies glass fiber felt (type IV), mopped		21	2.667		237	83.50	24.50	345	440
5800	On nailable decks		20	2.800		215	87.50	25.50	328	425

07 51 13.30 Cants

07 51 13.30 Cants		Crew	Daily Output	Labor-Hours	Unit	Material	Labor	Equipment	Total	Total Incl O&P
0010	**CANTS**									
0012	Lumber, treated, 4" x 4" cut diagonally	1 Rofc	325	.025	L.F.	1.96	.82		2.78	3.64
0300	Mineral or fiber, trapezoidal, 1" x 4" x 48"		325	.025		.31	.82		1.13	1.82
0400	1-1/2" x 5-5/8" x 48"		325	.025		.49	.82		1.31	2.02

07 51 13.40 Felts

07 51 13.40 Felts		Crew	Daily Output	Labor-Hours	Unit	Material	Labor	Equipment	Total	Total Incl O&P
0010	**FELTS**									
0012	Glass fibered roofing felt, #15, not mopped	1 Rofc	58	.138	Sq.	9.95	4.60		14.55	19.25
0300	Base sheet, #80, channel vented		58	.138		44.50	4.60		49.10	57.50
0400	#70, coated		58	.138		18.45	4.60		23.05	29
0500	Cap, #87, mineral surfaced		58	.138		84	4.60		88.60	101
0600	Flashing membrane, #65		16	.500		10.55	16.70		27.25	41.50
0800	Coal tar fibered, #15, no mopping		58	.138		18	4.60		22.60	28
0900	Asphalt felt, #15, 4 sq./roll, no mopping		58	.138		5.40	4.60		10	14.25
1100	#30, 2 sq./roll		58	.138		10.55	4.60		15.15	19.90
1200	Double coated, #33		58	.138		11.40	4.60		16	21
1400	#40, base sheet		58	.138		11.25	4.60		15.85	20.50
1450	Coated and saturated		58	.138		12.50	4.60		17.10	22
1500	Tarred felt, organic, #15, 4 sq. rolls		58	.138		13.50	4.60		18.10	23
1550	#30, 2 sq. roll		58	.138		25.50	4.60		30.10	37
1700	Add for mopping above felts, per ply, asphalt, 24 lb./sq.	G-1	192	.292		11.65	9.10	2.67	23.42	32
1800	Coal tar mopping, 30 lb./sq.		186	.301		19.30	9.40	2.76	31.46	41
1900	Flood coat, with asphalt, 60 lb./sq.		60	.933		29	29	8.55	66.55	94
2000	With coal tar, 75 lb./sq.		56	1		48	31.50	9.15	88.65	120

07 52 Modified Bituminous Membrane Roofing

07 52 13 – Atactic-Polypropylene-Modified Bituminous Membrane Roofing

07 52 13.10 APP Modified Bituminous Membrane

07 52 13.10 APP Modified Bituminous Membrane		Crew	Daily Output	Labor-Hours	Unit	Material	Labor	Equipment	Total	Total Incl O&P
0010	**APP MODIFIED BITUMINOUS MEMBRANE** R075213-30									
0020	Base sheet, #15 glass fiber felt, nailed to deck	1 Rofc	58	.138	Sq.	11.50	4.60		16.10	21
0030	Spot mopped to deck	G-1	295	.190		15.80	5.95	1.74	23.49	30
0040	Fully mopped to deck	"	192	.292		21.50	9.10	2.67	33.27	43.50
0050	#15 organic felt, nailed to deck	1 Rofc	58	.138		6.95	4.60		11.55	15.95
0060	Spot mopped to deck	G-1	295	.190		11.20	5.95	1.74	18.89	25
0070	Fully mopped to deck	"	192	.292		17	9.10	2.67	28.77	38
2100	APP mod., smooth surf. cap sheet, poly. reinf., torched, 160 mils	G-5	2100	.019	S.F.	.80	.58	.09	1.47	2.02
2150	170 mils		2100	.019		.78	.58	.09	1.45	2
2200	Granule surface cap sheet, poly. reinf., torched, 180 mils		2000	.020		.96	.61	.09	1.66	2.26
2250	Smooth surface flashing, torched, 160 mils		1260	.032		.80	.97	.14	1.91	2.79
2300	170 mils		1260	.032		.78	.97	.14	1.89	2.77
2350	Granule surface flashing, torched, 180 mils		1260	.032		.96	.97	.14	2.07	2.97
2400	Fibrated aluminum coating	1 Rofc	3800	.002		.09	.07		.16	.23
2450	Seam heat welding	"	205	.039	L.F.	.09	1.30		1.39	2.45

07 52 Modified Bituminous Membrane Roofing

07 52 16 – Styrene-Butadiene-Styrene Modified Bituminous Membrane Roofing

07 52 16.10 SBS Modified Bituminous Membrane		Crew	Daily Output	Labor-Hours	Unit	Material	Labor	Equipment	Total	Total Incl O&P
						2019 Bare Costs				
0010	**SBS MODIFIED BITUMINOUS MEMBRANE**									
0080	Mod. bit. rfng., SBS mod, gran surf. cap sheet, poly. reinf.									
0650	120 to 149 mils thick	G-1	2000	.028	S.F.	1.35	.88	.26	2.49	3.35
0750	150 to 160 mils	"	2000	.028		1.78	.88	.26	2.92	3.82
1150	For reflective granules, add					.71	.79	.28	1.78	2.54
1600	Smooth surface cap sheet, mopped, 145 mils	G-1	2100	.027		.83	.83	.24	1.90	2.69
1620	Lightweight base sheet, fiberglass reinforced, 35 to 47 mil		2100	.027		.29	.83	.24	1.36	2.10
1625	Heavyweight base/ply sheet, reinforced, 87 to 120 mil thick		2100	.027		.92	.83	.24	1.99	2.79
1650	Granulated walkpad, 180 to 220 mils	1 Rofc	400	.020		1.85	.67		2.52	3.24
1700	Smooth surface flashing, 145 mils	G-1	1260	.044		.83	1.39	.41	2.63	3.87
1800	150 mils		1260	.044		.51	1.39	.41	2.31	3.52
1900	Granular surface flashing, 150 mils		1260	.044		.73	1.39	.41	2.53	3.76
2000	160 mils		1260	.044		.76	1.39	.41	2.56	3.80
2010	Elastomeric asphalt primer	1 Rofc	2600	.003		.17	.10		.27	.38
2015	Roofing asphalt, 30 lb./square	G-1	19000	.003		.15	.09	.03	.27	.36
2020	Cold process adhesive, 20 to 30 mils thick	1 Rofc	750	.011		.25	.36		.61	.92
2025	Self adhering vapor retarder, 30 to 45 mils thick	G-5	2150	.019		1.05	.57	.08	1.70	2.27
2050	Seam heat welding	1 Rofc	205	.039	L.F.	.09	1.30		1.39	2.45

07 53 Elastomeric Membrane Roofing

07 53 16 – Chlorosulfonate-Polyethylene Roofing

07 53 16.10 Chlorosulfonated Polyethylene Roofing

		Crew	Daily Output	Labor-Hours	Unit	Material	Labor	Equipment	Total	Total Incl O&P
0010	**CHLOROSULFONATED POLYETHYLENE ROOFING**									
0800	Chlorosulfonated polyethylene (CSPE)									
0900	45 mils, heat welded seams, plate attachment	G-5	35	1.143	Sq.	241	35	5.20	281.20	335
1100	Heat welded seams, plate attachment and ballasted		26	1.538		257	47	6.95	310.95	375
1200	60 mils, heat welded seams, plate attachment		35	1.143		315	35	5.20	355.20	415
1300	Heat welded seams, plate attachment and ballasted		26	1.538		330	47	6.95	383.95	455

07 53 23 – Ethylene-Propylene-Diene-Monomer Roofing

07 53 23.20 Ethylene-Propylene-Diene-Monomer Roofing

		Crew	Daily Output	Labor-Hours	Unit	Material	Labor	Equipment	Total	Total Incl O&P
0010	**ETHYLENE-PROPYLENE-DIENE-MONOMER ROOFING (EPDM)**									
3500	Ethylene-propylene-diene-monomer (EPDM), 45 mils, 0.28 psf									
3600	Loose-laid & ballasted with stone (10 psf)	G-5	51	.784	Sq.	91	24	3.55	118.55	147
3700	Mechanically attached		35	1.143		79	35	5.20	119.20	156
3800	Fully adhered with adhesive		26	1.538		113	47	6.95	166.95	216
4500	60 mils, 0.40 psf									
4600	Loose-laid & ballasted with stone (10 psf)	G-5	51	.784	Sq.	108	24	3.55	135.55	166
4700	Mechanically attached		35	1.143		94.50	35	5.20	134.70	173
4800	Fully adhered with adhesive		26	1.538		128	47	6.95	181.95	233
4810	45 mil, 0.28 psf, membrane only					48.50			48.50	53.50
4820	60 mil, 0.40 psf, membrane only					63			63	69
4850	Seam tape for membrane, 3" x 100' roll				Ea.	49.50			49.50	54.50
4900	Batten strips, 10' sections					4.04			4.04	4.44
4910	Cover tape for batten strips, 6" x 100' roll					193			193	212
4930	Plate anchors				M	81.50			81.50	90
4970	Adhesive for fully adhered systems, 60 S.F./gal.				Gal.	20.50			20.50	23

07 53 Elastomeric Membrane Roofing

07 53 29 – Polyisobutylene Roofing

07 53 29.10 Polyisobutylene Roofing		Crew	Daily Output	Labor-Hours	Unit	2019 Bare Costs Material	Labor	Equipment	Total	Total Incl O&P
0010	**POLYISOBUTYLENE ROOFING**									
7500	Polyisobutylene (PIB), 100 mils, 0.57 psf									
7600	Loose-laid & ballasted with stone/gravel (10 psf)	G-5	51	.784	Sq.	214	24	3.55	241.55	282
7700	Partially adhered with adhesive		35	1.143		255	35	5.20	295.20	350
7800	Hot asphalt attachment		35	1.143		243	35	5.20	283.20	335
7900	Fully adhered with contact cement		26	1.538		264	47	6.95	317.95	385

07 54 Thermoplastic Membrane Roofing

07 54 19 – Polyvinyl-Chloride Roofing

07 54 19.10 Polyvinyl-Chloride Roofing (PVC)

Line	Description	Crew	Daily Output	Labor-Hours	Unit	Material	Labor	Equipment	Total	Total Incl O&P
0010	**POLYVINYL-CHLORIDE ROOFING (PVC)**									
8200	Heat welded seams									
8700	Reinforced, 48 mils, 0.33 psf									
8750	Loose-laid & ballasted with stone/gravel (12 psf)	G-5	51	.784	Sq.	119	24	3.55	146.55	178
8800	Mechanically attached		35	1.143		105	35	5.20	145.20	185
8850	Fully adhered with adhesive		26	1.538		151	47	6.95	204.95	258
8860	Reinforced, 60 mils, 0.40 psf									
8870	Loose-laid & ballasted with stone/gravel (12 psf)	G-5	51	.784	Sq.	121	24	3.55	148.55	180
8880	Mechanically attached		35	1.143		107	35	5.20	147.20	187
8890	Fully adhered with adhesive		26	1.538		153	47	6.95	206.95	260

07 54 23 – Thermoplastic-Polyolefin Roofing

07 54 23.10 Thermoplastic Polyolefin Roofing (T.P.O.)

Line	Description	Crew	Daily Output	Labor-Hours	Unit	Material	Labor	Equipment	Total	Total Incl O&P
0010	**THERMOPLASTIC POLYOLEFIN ROOFING (T.P.O.)**									
0100	45 mil, loose laid & ballasted with stone (1/2 ton/sq.)	G-5	51	.784	Sq.	94	24	3.55	121.55	150
0120	Fully adhered		25	1.600		78	48.50	7.25	133.75	182
0140	Mechanically attached		34	1.176		77.50	36	5.35	118.85	156
0160	Self adhered		35	1.143		77.50	35	5.20	117.70	154
0180	60 mil membrane, heat welded seams, ballasted		50	.800		105	24.50	3.62	133.12	164
0200	Fully adhered		25	1.600		89	48.50	7.25	144.75	194
0220	Mechanically attached		34	1.176		93	36	5.35	134.35	172
0240	Self adhered		35	1.143		105	35	5.20	145.20	184

07 54 30 – Ketone Ethylene Ester Roofing

07 54 30.10 Ketone Ethylene Ester Roofing

Line	Description	Crew	Daily Output	Labor-Hours	Unit	Material	Labor	Equipment	Total	Total Incl O&P
0010	**KETONE ETHYLENE ESTER ROOFING**									
0100	Ketone ethylene ester roofing, 50 mil, fully adhered	G-5	26	1.538	Sq.	217	47	6.95	270.95	330
0120	Mechanically attached		35	1.143		140	35	5.20	180.20	223
0140	Ballasted with stone		51	.784		154	24	3.55	181.55	217
0160	50 mil, fleece backed, adhered w/hot asphalt	G-1	26	2.154		160	67.50	19.75	247.25	320
0180	Accessories, pipe boot	1 Rofc	32	.250	Ea.	27	8.35		35.35	44.50
0200	Pre-formed corners		32	.250	"	9.70	8.35		18.05	26
0220	Ketone clad metal, including up to 4 bends		330	.024	S.F.	4.03	.81		4.84	5.90
0240	Walkway pad	2 Rofc	800	.020	"	4.23	.67		4.90	5.85
0260	Stripping material	1 Rofc	310	.026	L.F.	.96	.86		1.82	2.62

07 57 Coated Foamed Roofing

07 57 13 – Sprayed Polyurethane Foam Roofing

07 57 13.10 Sprayed Polyurethane Foam Roofing (S.P.F.)		Crew	Daily Output	Labor-Hours	Unit	Material	2019 Bare Costs Labor	Equipment	Total	Total Incl O&P
0010	SPRAYED POLYURETHANE FOAM ROOFING (S.P.F.)									
0100	Primer for metal substrate (when required)	G-2A	3000	.008	S.F.	.50	.24	.21	.95	1.20
0200	Primer for non-metal substrate (when required)		3000	.008		.19	.24	.21	.64	.86
0300	Closed cell spray, polyurethane foam, 3 lb./C.F. density, 1", R6.7		15000	.002		.64	.05	.04	.73	.84
0400	2", R13.4		13125	.002		1.29	.05	.05	1.39	1.56
0500	3", R18.6		11485	.002		1.93	.06	.05	2.04	2.29
0550	4", R24.8		10080	.002		2.58	.07	.06	2.71	3.02
0700	Spray-on silicone coating	↓	2500	.010		1.27	.28	.25	1.80	2.16
0800	Warranty 5-20 year manufacturer's								.15	.15
0900	Warranty 20 year, no dollar limit				↓				.20	.20

07 58 Roll Roofing

07 58 10 – Asphalt Roll Roofing

07 58 10.10 Roll Roofing		Crew	Daily Output	Labor-Hours	Unit	Material	Labor	Equipment	Total	Total Incl O&P
0010	ROLL ROOFING									
0100	Asphalt, mineral surface									
0200	1 ply #15 organic felt, 1 ply mineral surfaced									
0300	Selvage roofing, lap 19", nailed & mopped	G-1	27	2.074	Sq.	71	65	19	155	217
0400	3 plies glass fiber felt (type IV), 1 ply mineral surfaced									
0500	Selvage roofing, lapped 19", mopped	G-1	25	2.240	Sq.	124	70	20.50	214.50	287
0600	Coated glass fiber base sheet, 2 plies of glass fiber									
0700	Felt (type IV), 1 ply mineral surfaced selvage									
0800	Roofing, lapped 19", mopped	G-1	25	2.240	Sq.	133	70	20.50	223.50	296
0900	On nailable decks	"	24	2.333	"	121	73	21.50	215.50	289
1000	3 plies glass fiber felt (type III), 1 ply mineral surfaced									
1100	Selvage roofing, lapped 19", mopped	G-1	25	2.240	Sq.	124	70	20.50	214.50	287

07 61 Sheet Metal Roofing

07 61 13 – Standing Seam Sheet Metal Roofing

07 61 13.10 Standing Seam Sheet Metal Roofing, Field Fab.		Crew	Daily Output	Labor-Hours	Unit	Material	Labor	Equipment	Total	Total Incl O&P
0010	STANDING SEAM SHEET METAL ROOFING, FIELD FABRICATED									
0400	Copper standing seam roofing, over 10 squares, 16 oz., 125 lb./sq.	1 Shee	1.30	6.154	Sq.	1,000	266		1,266	1,550
0600	18 oz., 140 lb./sq.	"	1.20	6.667		1,125	289		1,414	1,725
1200	For abnormal conditions or small areas, add					25%	100%			
1300	For lead-coated copper, add				↓	25%				

07 61 16 – Batten Seam Sheet Metal Roofing

07 61 16.10 Batten Seam Sheet Metal Roofing, Field Fab.		Crew	Daily Output	Labor-Hours	Unit	Material	Labor	Equipment	Total	Total Incl O&P
0010	BATTEN SEAM SHEET METAL ROOFING, FIELD FABRICATED									
0012	Copper batten seam roofing, over 10 sq., 16 oz., 130 lb./sq.	1 Shee	1.10	7.273	Sq.	1,275	315		1,590	1,925
0020	Lead batten seam roofing, 5 lb./S.F.		1.20	6.667		1,725	289		2,014	2,375
0100	Zinc/copper alloy batten seam roofing, .020" thick		1.20	6.667		1,475	289		1,764	2,100
0200	Copper roofing, batten seam, over 10 sq., 18 oz., 145 lb./sq.		1	8		1,425	345		1,770	2,150
0800	Zinc, copper alloy roofing, batten seam, .027" thick		1.15	6.957		1,925	300		2,225	2,625
0900	.032" thick		1.10	7.273		1,950	315		2,265	2,675
1000	.040" thick	↓	1.05	7.619	↓	2,500	330		2,830	3,300

07 61 Sheet Metal Roofing

07 61 19 – Flat Seam Sheet Metal Roofing

07 61 19.10 Flat Seam Sheet Metal Roofing, Field Fabricated		Crew	Daily Output	Labor-Hours	Unit	Material	2019 Bare Costs Labor	Equipment	Total	Total Incl O&P
0010	**FLAT SEAM SHEET METAL ROOFING, FIELD FABRICATED**									
0900	Copper flat seam roofing, over 10 squares, 16 oz., 115 lb./sq.	1 Shee	1.20	6.667	Sq.	940	289		1,229	1,500
0950	18 oz., 130 lb./sq.		1.15	6.957		1,050	300		1,350	1,650
1000	20 oz., 145 lb./sq.		1.10	7.273		1,250	315		1,565	1,900
1008	Zinc flat seam roofing, .020" thick		1.20	6.667		1,275	289		1,564	1,875
1010	.027" thick		1.15	6.957		1,650	300		1,950	2,300
1020	.032" thick		1.12	7.143		1,675	310		1,985	2,350
1030	.040" thick		1.05	7.619		2,125	330		2,455	2,875
1100	Lead flat seam roofing, 5 lb./S.F.	↓	1.30	6.154	↓	1,475	266		1,741	2,075

07 65 Flexible Flashing

07 65 10 – Sheet Metal Flashing

07 65 10.10 Sheet Metal Flashing and Counter Flashing		Crew	Daily Output	Labor-Hours	Unit	Material	2019 Bare Costs Labor	Equipment	Total	Total Incl O&P
0010	**SHEET METAL FLASHING AND COUNTER FLASHING**									
0011	Including up to 4 bends									
0020	Aluminum, mill finish, .013" thick	1 Rofc	145	.055	S.F.	.86	1.84		2.70	4.27
0030	.016" thick		145	.055		1.01	1.84		2.85	4.43
0060	.019" thick		145	.055		1.42	1.84		3.26	4.88
0100	.032" thick		145	.055		1.36	1.84		3.20	4.82
0200	.040" thick		145	.055		2.30	1.84		4.14	5.85
0300	.050" thick		145	.055	↓	2.75	1.84		4.59	6.35
0325	Mill finish 5" x 7" step flashing, .016" thick		1920	.004	Ea.	.15	.14		.29	.42
0350	Mill finish 12" x 12" step flashing, .016" thick	↓	1600	.005	"	.56	.17		.73	.92
0400	Painted finish, add				S.F.	.33			.33	.36
1000	Mastic-coated 2 sides, .005" thick	1 Rofc	330	.024		1.83	.81		2.64	3.47
1100	.016" thick		330	.024		2.02	.81		2.83	3.68
1600	Copper, 16 oz. sheets, under 1000 lb.		115	.070		8.10	2.32		10.42	13.10
1700	Over 4000 lb.		155	.052		8.10	1.72		9.82	12
1900	20 oz. sheets, under 1000 lb.		110	.073		10.75	2.43		13.18	16.25
2000	Over 4000 lb.		145	.055		10.20	1.84		12.04	14.55
2200	24 oz. sheets, under 1000 lb.		105	.076		14.75	2.54		17.29	21
2500	32 oz. sheets, under 1000 lb.		100	.080	↓	19.60	2.67		22.27	26.50
2700	W shape for valleys, 16 oz., 24" wide		100	.080	L.F.	16.25	2.67		18.92	22.50
5800	Lead, 2.5 lb./S.F., up to 12" wide		135	.059	S.F.	6.20	1.98		8.18	10.35
5900	Over 12" wide		135	.059		3.99	1.98		5.97	7.95
8900	Stainless steel sheets, 32 ga.		155	.052		3.41	1.72		5.13	6.85
9000	28 ga.		155	.052		4.75	1.72		6.47	8.35
9100	26 ga.		155	.052		4.70	1.72		6.42	8.25
9200	24 ga.	↓	155	.052	↓	5.25	1.72		6.97	8.85
9290	For mechanically keyed flashing, add					40%				
9320	Steel sheets, galvanized, 20 ga.	1 Rofc	130	.062	S.F.	1.20	2.05		3.25	5.05
9322	22 ga.		135	.059		1.26	1.98		3.24	4.96
9324	24 ga.		140	.057		.96	1.91		2.87	4.50
9326	26 ga.		148	.054		.84	1.80		2.64	4.18
9328	28 ga.		155	.052		.72	1.72		2.44	3.91
9340	30 ga.		160	.050		.61	1.67		2.28	3.68
9400	Terne coated stainless steel, .015" thick, 28 ga.		155	.052		8.20	1.72		9.92	12.10
9500	.018" thick, 26 ga.		155	.052		9.15	1.72		10.87	13.15
9600	Zinc and copper alloy (brass), .020" thick		155	.052		10.10	1.72		11.82	14.25
9700	.027" thick	↓	155	.052	↓	12.50	1.72		14.22	16.85

07 65 Flexible Flashing

07 65 10 – Sheet Metal Flashing

07 65 10.10 Sheet Metal Flashing and Counter Flashing		Crew	Daily Output	Labor-Hours	Unit	Material	Labor (2019 Bare Costs)	Equipment (2019 Bare Costs)	Total (2019 Bare Costs)	Total Incl O&P
9800	.032" thick	1 Rofc	155	.052	S.F.	15.80	1.72		17.52	20.50
9900	.040" thick		155	.052		20.50	1.72		22.22	25.50

07 65 12 – Fabric and Mastic Flashings

07 65 12.10 Fabric and Mastic Flashing and Counter Flashing

		Crew	Daily Output	Labor-Hours	Unit	Material	Labor	Equipment	Total	Total Incl O&P
0010	**FABRIC AND MASTIC FLASHING AND COUNTER FLASHING**									
1300	Asphalt flashing cement, 5 gallon				Gal.	9.85			9.85	10.85
4900	Fabric, asphalt-saturated cotton, specification grade	1 Rofc	35	.229	S.Y.	3.26	7.60		10.86	17.35
5000	Utility grade		35	.229		1.48	7.60		9.08	15.40
5300	Close-mesh fabric, saturated, 17 oz./S.Y.		35	.229		2.19	7.60		9.79	16.15
5500	Fiberglass, resin-coated		35	.229		1.12	7.60		8.72	15
8500	Shower pan, bituminous membrane, 7 oz.		155	.052	S.F.	1.98	1.72		3.70	5.30

07 65 13 – Laminated Sheet Flashing

07 65 13.10 Laminated Sheet Flashing

		Crew	Daily Output	Labor-Hours	Unit	Material	Labor	Equipment	Total	Total Incl O&P
0010	**LAMINATED SHEET FLASHING**, Including up to 4 bends									
0500	Aluminum, fabric-backed 2 sides, mill finish, .004" thick	1 Rofc	330	.024	S.F.	1.58	.81		2.39	3.20
0700	.005" thick		330	.024		1.82	.81		2.63	3.46
0750	Mastic-backed, self adhesive		460	.017		3.38	.58		3.96	4.77
0800	Mastic-coated 2 sides, .004" thick		330	.024		1.58	.81		2.39	3.20
2800	Copper, paperbacked 1 side, 2 oz.		330	.024		2.17	.81		2.98	3.85
2900	3 oz.		330	.024		3.18	.81		3.99	4.96
3100	Paperbacked 2 sides, 2 oz.		330	.024		2.38	.81		3.19	4.08
3150	3 oz.		330	.024		2.24	.81		3.05	3.92
3200	5 oz.		330	.024		3.42	.81		4.23	5.20
3400	Mastic-backed 2 sides, copper, 2 oz.		330	.024		2.03	.81		2.84	3.69
3500	3 oz.		330	.024		2.53	.81		3.34	4.24
3700	5 oz.		330	.024		3.84	.81		4.65	5.70
3800	Fabric-backed 2 sides, copper, 2 oz.		330	.024		2.05	.81		2.86	3.72
4000	3 oz.		330	.024		2.83	.81		3.64	4.57
4100	5 oz.		330	.024		3.96	.81		4.77	5.80
4300	Copper-clad stainless steel, .015" thick, under 500 lb.		115	.070		6.70	2.32		9.02	11.55
4400	Over 2000 lb.		155	.052		6.80	1.72		8.52	10.60
4600	.018" thick, under 500 lb.		100	.080		8.05	2.67		10.72	13.65
4700	Over 2000 lb.		145	.055		7.85	1.84		9.69	11.90
8550	Shower pan, 3 ply copper and fabric, 3 oz.		155	.052		4.05	1.72		5.77	7.55
8600	7 oz.		155	.052		4.87	1.72		6.59	8.45
9300	Stainless steel, paperbacked 2 sides, .005" thick		330	.024		3.96	.81		4.77	5.80

07 65 19 – Plastic Sheet Flashing

07 65 19.10 Plastic Sheet Flashing and Counter Flashing

		Crew	Daily Output	Labor-Hours	Unit	Material	Labor	Equipment	Total	Total Incl O&P
0010	**PLASTIC SHEET FLASHING AND COUNTER FLASHING**									
7300	Polyvinyl chloride, black, 10 mil	1 Rofc	285	.028	S.F.	.27	.94		1.21	1.99
7400	20 mil		285	.028		.26	.94		1.20	1.98
7600	30 mil		285	.028		.33	.94		1.27	2.05
7700	60 mil		285	.028		.86	.94		1.80	2.64
7900	Black or white for exposed roofs, 60 mil		285	.028		1.25	.94		2.19	3.07
8060	PVC tape, 5" x 45 mils, for joint covers, 100 L.F./roll				Ea.	180			180	198
8850	Polyvinyl chloride, 30 mil	1 Rofc	160	.050	S.F.	1.53	1.67		3.20	4.69

07 65 Flexible Flashing

07 65 23 – Rubber Sheet Flashing

07 65 23.10 Rubber Sheet Flashing and Counter Flashing		Crew	Daily Output	Labor-Hours	Unit	Material	2019 Bare Costs Labor	Equipment	Total	Total Incl O&P
0010	**RUBBER SHEET FLASHING AND COUNTER FLASHING**									
4750	EPDM Cured	1 Rofc	285	.028	S.F.	1.25	.94		2.19	3.07
4800	Uncured		285	.028	"	1.76	.94		2.70	3.63
4810	EPDM 90 mils, 1" diameter pipe flashing		32	.250	Ea.	20.50	8.35		28.85	37.50
4820	2" diameter		30	.267		20	8.90		28.90	38
4830	3" diameter		28	.286		21.50	9.55		31.05	41
4840	4" diameter		24	.333		29	11.10		40.10	52
4850	6" diameter		22	.364		29	12.15		41.15	54
7150	Neoprene, 60 mil		285	.028	S.F.	2.27	.94		3.21	4.19
7160	Self-curing		285	.028		2.54	.94		3.48	4.48
7170	Uncured		285	.028		2.18	.94		3.12	4.09
8100	Rubber, butyl, 1/32" thick		285	.028		2.27	.94		3.21	4.19
8200	1/16" thick		285	.028		3.37	.94		4.31	5.40
8300	Neoprene, cured, 1/16" thick		285	.028		2.66	.94		3.60	4.62
8400	1/8" thick		285	.028		6.15	.94		7.09	8.50

07 65 26 – Self-Adhering Sheet Flashing

07 65 26.10 Self-Adhering Sheet or Roll Flashing		Crew	Daily Output	Labor-Hours	Unit	Material	Labor	Equipment	Total	Total Incl O&P
0010	**SELF-ADHERING SHEET OR ROLL FLASHING**									
0020	Self-adhered flashing, 25 mil cross laminated HDPE, 4" wide	1 Rofc	960	.008	L.F.	.20	.28		.48	.72
0040	6" wide		896	.009		.31	.30		.61	.88
0060	9" wide		832	.010		.46	.32		.78	1.08
0080	12" wide		768	.010		.61	.35		.96	1.30

07 71 Roof Specialties

07 71 16 – Manufactured Counterflashing Systems

07 71 16.10 Roof Drain Boot		Crew	Daily Output	Labor-Hours	Unit	Material	Labor	Equipment	Total	Total Incl O&P
0010	**ROOF DRAIN BOOT**									
0100	Cast iron, 4" diameter	1 Shee	125	.064	L.F.	45.50	2.77		48.27	54.50
0300	4" x 3"		125	.064		62	2.77		64.77	73
0400	5" x 4"		125	.064		124	2.77		126.77	142

07 71 16.20 Pitch Pockets, Variable Sizes		Crew	Daily Output	Labor-Hours	Unit	Material	Labor	Equipment	Total	Total Incl O&P
0010	**PITCH POCKETS, VARIABLE SIZES**									
0100	Adjustable, 4" to 7", welded corners, 4" deep	1 Rofc	48	.167	Ea.	19.50	5.55		25.05	31.50
0200	Side extenders, 6"	"	240	.033	"	2.87	1.11		3.98	5.15

07 71 19 – Manufactured Gravel Stops and Fasciae

07 71 19.10 Gravel Stop		Crew	Daily Output	Labor-Hours	Unit	Material	Labor	Equipment	Total	Total Incl O&P
0010	**GRAVEL STOP**									
0020	Aluminum, .050" thick, 4" face height, mill finish	1 Shee	145	.055	L.F.	6.55	2.39		8.94	11.20
0080	Duranodic finish		145	.055		7.60	2.39		9.99	12.30
0100	Painted		145	.055		7.60	2.39		9.99	12.30
1200	Copper, 16 oz., 3" face height		145	.055		25	2.39		27.39	31.50
1300	6" face height		135	.059		34	2.57		36.57	41.50
1350	Galv steel, 24 ga., 4" leg, plain, with continuous cleat, 4" face		145	.055		6.60	2.39		8.99	11.25
1360	6" face height		145	.055		6.70	2.39		9.09	11.35
1500	Polyvinyl chloride, 6" face height		135	.059		5.80	2.57		8.37	10.65
1800	Stainless steel, 24 ga., 6" face height		135	.059		15.60	2.57		18.17	21.50

07 71 19 – Manufactured Gravel Stops and Fasciae

07 71 19.30 Fascia		Crew	Daily Output	Labor-Hours	Unit	Material	Labor	Equipment	Total	Total Incl O&P
						2019 Bare Costs				
0010	FASCIA									
0100	Aluminum, reverse board and batten, .032" thick, colored, no furring incl.	1 Shee	145	.055	S.F.	7.15	2.39		9.54	11.80
0200	Residential type, aluminum	1 Carp	200	.040	L.F.	2.13	1.55		3.68	4.90
0220	Vinyl	"	200	.040	"	2.25	1.55		3.80	5.05
0300	Steel, galv and enameled, stock, no furring, long panels	1 Shee	145	.055	S.F.	5.55	2.39		7.94	10.05
0600	Short panels	"	115	.070	"	5.25	3.01		8.26	10.80

07 71 23 – Manufactured Gutters and Downspouts

07 71 23.10 Downspouts		Crew	Daily Output	Labor-Hours	Unit	Material	Labor	Equipment	Total	Total Incl O&P
0010	DOWNSPOUTS									
0020	Aluminum, embossed, .020" thick, 2" x 3"	1 Shee	190	.042	L.F.	.93	1.82		2.75	4.04
0100	Enameled		190	.042		1.37	1.82		3.19	4.53
0300	.024" thick, 2" x 3"		180	.044		2.14	1.92		4.06	5.55
0400	3" x 4"		140	.057		2.14	2.47		4.61	6.45
0600	Round, corrugated aluminum, 3" diameter, .020" thick		190	.042		2.10	1.82		3.92	5.35
0700	4" diameter, .025" thick		140	.057		3.24	2.47		5.71	7.65
0900	Wire strainer, round, 2" diameter		155	.052	Ea.	1.90	2.23		4.13	5.80
1000	4" diameter		155	.052		2.42	2.23		4.65	6.35
1200	Rectangular, perforated, 2" x 3"		145	.055		2.34	2.39		4.73	6.55
1300	3" x 4"		145	.055		3.35	2.39		5.74	7.65
1500	Copper, round, 16 oz., stock, 2" diameter		190	.042	L.F.	8.50	1.82		10.32	12.35
1600	3" diameter		190	.042		9.05	1.82		10.87	12.95
1800	4" diameter		145	.055		9.75	2.39		12.14	14.70
1900	5" diameter		130	.062		14.85	2.66		17.51	21
2100	Rectangular, corrugated copper, stock, 2" x 3"		190	.042		8.20	1.82		10.02	12.05
2200	3" x 4"		145	.055		9.35	2.39		11.74	14.25
2400	Rectangular, plain copper, stock, 2" x 3"		190	.042		11.35	1.82		13.17	15.50
2500	3" x 4"		145	.055		14.15	2.39		16.54	19.55
2700	Wire strainers, rectangular, 2" x 3"		145	.055	Ea.	17.40	2.39		19.79	23
2800	3" x 4"		145	.055		18.20	2.39		20.59	24
3000	Round, 2" diameter		145	.055		7.20	2.39		9.59	11.85
3100	3" diameter		145	.055		9.20	2.39		11.59	14.05
3300	4" diameter		145	.055		12.10	2.39		14.49	17.25
3400	5" diameter		115	.070		23	3.01		26.01	30.50
3600	Lead-coated copper, round, stock, 2" diameter		190	.042	L.F.	23	1.82		24.82	28.50
3700	3" diameter		190	.042		23	1.82		24.82	28.50
3900	4" diameter		145	.055		24	2.39		26.39	30.50
4000	5" diameter, corrugated		130	.062		24	2.66		26.66	31
4200	6" diameter, corrugated		105	.076		31.50	3.30		34.80	40.50
4300	Rectangular, corrugated, stock, 2" x 3"		190	.042		15.15	1.82		16.97	19.65
4500	Plain, stock, 2" x 3"		190	.042		24.50	1.82		26.32	30
4600	3" x 4"		145	.055		34.50	2.39		36.89	42
4800	Steel, galvanized, round, corrugated, 2" or 3" diameter, 28 ga.		190	.042		2.11	1.82		3.93	5.35
4900	4" diameter, 28 ga.		145	.055		2.12	2.39		4.51	6.30
5700	Rectangular, corrugated, 28 ga., 2" x 3"		190	.042		2.23	1.82		4.05	5.45
5800	3" x 4"		145	.055		2.24	2.39		4.63	6.40
6000	Rectangular, plain, 28 ga., galvanized, 2" x 3"		190	.042		4.04	1.82		5.86	7.45
6100	3" x 4"		145	.055		4.37	2.39		6.76	8.75
6300	Epoxy painted, 24 ga., corrugated, 2" x 3"		190	.042		2.46	1.82		4.28	5.75
6400	3" x 4"		145	.055		2.98	2.39		5.37	7.25
6600	Wire strainers, rectangular, 2" x 3"		145	.055	Ea.	17.50	2.39		19.89	23
6700	3" x 4"		145	.055		19.40	2.39		21.79	25.50
6900	Round strainers, 2" or 3" diameter		145	.055		4.38	2.39		6.77	8.80

07 71 23 – Manufactured Gutters and Downspouts

07 71 23.10 Downspouts		Crew	Daily Output	Labor-Hours	Unit	Material	Labor	Equipment	Total	Total Incl O&P
						2019 Bare Costs				
7000	4" diameter	1 Shee	145	.055	Ea.	6.05	2.39		8.44	10.65
8200	Vinyl, rectangular, 2" x 3"		210	.038	L.F.	1.94	1.65		3.59	4.87
8300	Round, 2-1/2"	↓	220	.036	"	1.40	1.57		2.97	4.15

07 71 23.20 Downspout Elbows

Line	Description	Crew	Daily Output	Labor-Hours	Unit	Material	Labor	Equipment	Total	Total Incl O&P
0010	**DOWNSPOUT ELBOWS**									
0020	Aluminum, embossed, 2" x 3", .020" thick	1 Shee	100	.080	Ea.	.93	3.46		4.39	6.75
0100	Enameled		100	.080		2.17	3.46		5.63	8.15
0200	Embossed, 3" x 4", .025" thick		100	.080		2.98	3.46		6.44	9.05
0300	Enameled		100	.080		4	3.46		7.46	10.15
0400	Embossed, corrugated, 3" diameter, .020" thick		100	.080		3.13	3.46		6.59	9.20
0500	4" diameter, .025" thick		100	.080		6.70	3.46		10.16	13.10
0600	Copper, 16 oz., 2" diameter		100	.080		10.15	3.46		13.61	16.90
0700	3" diameter		100	.080		9.95	3.46		13.41	16.70
0800	4" diameter		100	.080		13.90	3.46		17.36	21
1000	Rectangular, 2" x 3" corrugated		100	.080		9.50	3.46		12.96	16.20
1100	3" x 4" corrugated		100	.080		15.40	3.46		18.86	22.50
1300	Vinyl, 2-1/2" diameter, 45 or 75 degree bend		100	.080		3.99	3.46		7.45	10.15
1400	Tee Y junction	↓	75	.107	↓	13.10	4.62		17.72	22

07 71 23.30 Gutters

Line	Description	Crew	Daily Output	Labor-Hours	Unit	Material	Labor	Equipment	Total	Total Incl O&P
0010	**GUTTERS**									
0012	Aluminum, stock units, 5" K type, .027" thick, plain	1 Shee	125	.064	L.F.	2.84	2.77		5.61	7.70
0100	Enameled		125	.064		2.92	2.77		5.69	7.80
0300	5" K type, .032" thick, plain		125	.064		3.59	2.77		6.36	8.55
0400	Enameled		125	.064		3.58	2.77		6.35	8.55
0700	Copper, half round, 16 oz., stock units, 4" wide		125	.064		8.45	2.77		11.22	13.85
0900	5" wide		125	.064		7.20	2.77		9.97	12.55
1000	6" wide		118	.068		11.40	2.94		14.34	17.40
1200	K type, 16 oz., stock, 5" wide		125	.064		8.40	2.77		11.17	13.85
1300	6" wide		125	.064		8.80	2.77		11.57	14.25
1500	Lead coated copper, 16 oz., half round, stock, 4" wide		125	.064		15.75	2.77		18.52	22
1600	6" wide		118	.068		18.65	2.94		21.59	25.50
1800	K type, stock, 5" wide		125	.064		18.65	2.77		21.42	25
1900	6" wide		125	.064		18.85	2.77		21.62	25
2100	Copper clad stainless steel, K type, 5" wide		125	.064		7.85	2.77		10.62	13.25
2200	6" wide		125	.064		10.05	2.77		12.82	15.65
2400	Steel, galv, half round or box, 28 ga., 5" wide, plain		125	.064		2.16	2.77		4.93	7
2500	Enameled		125	.064		2.23	2.77		5	7.05
2700	26 ga., stock, 5" wide		125	.064		2.49	2.77		5.26	7.35
2800	6" wide	↓	125	.064		2.67	2.77		5.44	7.55
3000	Vinyl, O.G., 4" wide	1 Carp	115	.070		1.36	2.70		4.06	5.95
3100	5" wide		115	.070		1.64	2.70		4.34	6.25
3200	4" half round, stock units	↓	115	.070	↓	1.43	2.70		4.13	6.05
3250	Joint connectors				Ea.	3.06			3.06	3.37
3300	Wood, clear treated cedar, fir or hemlock, 3" x 4"	1 Carp	100	.080	L.F.	11.20	3.10		14.30	17.45
3400	4" x 5"	"	100	.080	"	22	3.10		25.10	29
5000	Accessories, end cap, K type, aluminum 5"	1 Shee	625	.013	Ea.	.74	.55		1.29	1.73
5010	6"		625	.013		1.52	.55		2.07	2.59
5020	Copper, 5"		625	.013		3.45	.55		4	4.72
5030	6"		625	.013		3.67	.55		4.22	4.96
5040	Lead coated copper, 5"		625	.013		13.05	.55		13.60	15.25
5050	6"		625	.013		13.95	.55		14.50	16.25
5060	Copper clad stainless steel, 5"	↓	625	.013	↓	3.55	.55		4.10	4.83

07 71 23 – Manufactured Gutters and Downspouts

07 71 23.30 Gutters		Crew	Daily Output	Labor-Hours	Unit	Material	2019 Bare Costs Labor	Equipment	Total	Total Incl O&P
5070	6"	1 Shee	625	.013	Ea.	3.55	.55		4.10	4.83
5080	Galvanized steel, 5"		625	.013		1.50	.55		2.05	2.57
5090	6"		625	.013		2.59	.55		3.14	3.77
5100	Vinyl, 4"	1 Carp	625	.013		6.40	.50		6.90	7.85
5110	5"	"	625	.013		6.75	.50		7.25	8.25
5120	Half round, copper, 4"	1 Shee	625	.013		4.62	.55		5.17	6
5130	5"		625	.013		4.95	.55		5.50	6.35
5140	6"		625	.013		8.20	.55		8.75	9.95
5150	Lead coated copper, 5"		625	.013		14.70	.55		15.25	17.10
5160	6"		625	.013		22	.55		22.55	25
5170	Copper clad stainless steel, 5"		625	.013		4.85	.55		5.40	6.25
5180	6"		625	.013		4.57	.55		5.12	5.95
5190	Galvanized steel, 5"		625	.013		2.66	.55		3.21	3.85
5200	6"		625	.013		3.31	.55		3.86	4.56
5210	Outlet, aluminum, 2" x 3"		420	.019		.68	.82		1.50	2.12
5220	3" x 4"		420	.019		1.07	.82		1.89	2.55
5230	2-3/8" round		420	.019		.62	.82		1.44	2.05
5240	Copper, 2" x 3"		420	.019		7.75	.82		8.57	9.85
5250	3" x 4"		420	.019		8.70	.82		9.52	10.95
5260	2-3/8" round		420	.019		4.76	.82		5.58	6.60
5270	Lead coated copper, 2" x 3"		420	.019		26	.82		26.82	30
5280	3" x 4"		420	.019		29.50	.82		30.32	34
5290	2-3/8" round		420	.019		26	.82		26.82	30
5300	Copper clad stainless steel, 2" x 3"		420	.019		7.25	.82		8.07	9.35
5310	3" x 4"		420	.019		8.25	.82		9.07	10.45
5320	2-3/8" round		420	.019		4.76	.82		5.58	6.60
5330	Galvanized steel, 2" x 3"		420	.019		3.77	.82		4.59	5.50
5340	3" x 4"		420	.019		5.80	.82		6.62	7.70
5350	2-3/8" round		420	.019		4.71	.82		5.53	6.55
5360	K type mitres, aluminum		65	.123		4.89	5.35		10.24	14.25
5370	Copper		65	.123		16	5.35		21.35	26.50
5380	Lead coated copper		65	.123		55.50	5.35		60.85	70
5390	Copper clad stainless steel		65	.123		27.50	5.35		32.85	39.50
5400	Galvanized steel		65	.123		26.50	5.35		31.85	38.50
5420	Half round mitres, copper		65	.123		68	5.35		73.35	84
5430	Lead coated copper		65	.123		90.50	5.35		95.85	108
5440	Copper clad stainless steel		65	.123		57	5.35		62.35	71.50
5450	Galvanized steel		65	.123		32	5.35		37.35	44
5460	Vinyl mitres and outlets		65	.123		10.80	5.35		16.15	21
5470	Sealant		940	.009	L.F.	.01	.37		.38	.62
5480	Soldering		96	.083	"	.27	3.61		3.88	6.30

07 71 23.35 Gutter Guard		Crew	Daily Output	Labor-Hours	Unit	Material	Labor	Equipment	Total	Total Incl O&P
0010	**GUTTER GUARD**									
0020	6" wide strip, aluminum mesh	1 Carp	500	.016	L.F.	2.54	.62		3.16	3.81
0100	Vinyl mesh	"	500	.016	"	2.88	.62		3.50	4.19

07 71 26 – Reglets

07 71 26.10 Reglets and Accessories		Crew	Daily Output	Labor-Hours	Unit	Material	Labor	Equipment	Total	Total Incl O&P
0010	**REGLETS AND ACCESSORIES**									
0020	Reglet, aluminum, .025" thick, in parapet	1 Carp	225	.036	L.F.	1.80	1.38		3.18	4.26
0300	16 oz. copper		225	.036		6.75	1.38		8.13	9.75
0400	Galvanized steel, 24 ga.		225	.036		1.24	1.38		2.62	3.64
0600	Stainless steel, .020" thick		225	.036		3.92	1.38		5.30	6.60

07 71 Roof Specialties

07 71 26 – Reglets

07 71 26.10 Reglets and Accessories		Crew	Daily Output	Labor-Hours	Unit	Material	Labor	Equipment	Total	Total Incl O&P
						2019 Bare Costs				
0900	Counter flashing for above, 12" wide, .032" aluminum	1 Shee	150	.053	L.F.	2.18	2.31		4.49	6.25
1200	16 oz. copper		150	.053		6.25	2.31		8.56	10.75
1300	Galvanized steel, 26 ga.		150	.053		1.26	2.31		3.57	5.20
1500	Stainless steel, .020" thick		150	.053		6.35	2.31		8.66	10.85

07 71 43 – Drip Edge

07 71 43.10 Drip Edge, Rake Edge, Ice Belts		Crew	Daily Output	Labor-Hours	Unit	Material	Labor	Equipment	Total	Total Incl O&P
0010	**DRIP EDGE, RAKE EDGE, ICE BELTS**									
0020	Aluminum, .016" thick, 5" wide, mill finish	1 Carp	400	.020	L.F.	.59	.78		1.37	1.93
0100	White finish		400	.020		.66	.78		1.44	2.01
0200	8" wide, mill finish		400	.020		1.53	.78		2.31	2.96
0300	Ice belt, 28" wide, mill finish		100	.080		7.90	3.10		11	13.80
0310	Vented, mill finish		400	.020		2.26	.78		3.04	3.77
0320	Painted finish		400	.020		2.53	.78		3.31	4.06
0400	Galvanized, 5" wide		400	.020		.61	.78		1.39	1.95
0500	8" wide, mill finish		400	.020		.84	.78		1.62	2.20
0510	Rake edge, aluminum, 1-1/2" x 1-1/2"		400	.020		.34	.78		1.12	1.65
0520	3-1/2" x 1-1/2"		400	.020		.44	.78		1.22	1.76

07 72 Roof Accessories

07 72 26 – Ridge Vents

07 72 26.10 Ridge Vents and Accessories		Crew	Daily Output	Labor-Hours	Unit	Material	Labor	Equipment	Total	Total Incl O&P
0010	**RIDGE VENTS AND ACCESSORIES**									
2300	Ridge vent strip, mill finish	1 Shee	155	.052	L.F.	3.94	2.23		6.17	8.05

07 72 33 – Roof Hatches

07 72 33.10 Roof Hatch Options		Crew	Daily Output	Labor-Hours	Unit	Material	Labor	Equipment	Total	Total Incl O&P
0010	**ROOF HATCH OPTIONS**									
0500	2'-6" x 3', aluminum curb and cover	G-3	10	3.200	Ea.	880	118		998	1,150
0520	Galvanized steel curb and aluminum cover		10	3.200		850	118		968	1,125
0540	Galvanized steel curb and cover		10	3.200		630	118		748	885
0600	2'-6" x 4'-6", aluminum curb and cover		9	3.556		1,075	131		1,206	1,400
0800	Galvanized steel curb and aluminum cover		9	3.556		960	131		1,091	1,275
0900	Galvanized steel curb and cover		9	3.556		1,000	131		1,131	1,325
1100	4' x 4' aluminum curb and cover		8	4		1,725	147		1,872	2,150
1120	Galvanized steel curb and aluminum cover		8	4		1,775	147		1,922	2,200
1140	Galvanized steel curb and cover		8	4		1,125	147		1,272	1,475
1200	2'-6" x 8'-0", aluminum curb and cover		6.60	4.848		2,125	179		2,304	2,625
1400	Galvanized steel curb and aluminum cover		6.60	4.848		1,900	179		2,079	2,400
1500	Galvanized steel curb and cover		6.60	4.848		1,200	179		1,379	1,625
1800	For plexiglass panels, 2'-6" x 3'-0", add to above					475			475	525

07 72 53 – Snow Guards

07 72 53.10 Snow Guard Options		Crew	Daily Output	Labor-Hours	Unit	Material	Labor	Equipment	Total	Total Incl O&P
0010	**SNOW GUARD OPTIONS**									
0100	Slate & asphalt shingle roofs, fastened with nails	1 Rofc	160	.050	Ea.	12.40	1.67		14.07	16.65
0200	Standing seam metal roofs, fastened with set screws		48	.167		17.70	5.55		23.25	29.50
0300	Surface mount for metal roofs, fastened with solder		48	.167		7.50	5.55		13.05	18.30
0400	Double rail pipe type, including pipe		130	.062	L.F.	34	2.05		36.05	41

07 72 Roof Accessories

07 72 80 – Vents

07 72 80.30 Vent Options		Crew	Daily Output	Labor-Hours	Unit	2019 Bare Costs Material	Labor	Equipment	Total	Total Incl O&P
0010	**VENT OPTIONS**									
0020	Plastic, for insulated decks, 1 per M.S.F.	1 Rofc	40	.200	Ea.	20.50	6.65		27.15	34.50
0100	Heavy duty		20	.400		49.50	13.35		62.85	78.50
0300	Aluminum		30	.267		21.50	8.90		30.40	40
0800	Polystyrene baffles, 12" wide for 16" OC rafter spacing	1 Carp	90	.089		.49	3.44		3.93	6.25
0900	For 24" OC rafter spacing	"	110	.073		.79	2.82		3.61	5.55

07 76 Roof Pavers

07 76 16 – Roof Decking Pavers

07 76 16.10 Roof Pavers and Supports		Crew	Daily Output	Labor-Hours	Unit	Material	Labor	Equipment	Total	Total Incl O&P
0010	**ROOF PAVERS AND SUPPORTS**									
1000	Roof decking pavers, concrete blocks, 2" thick, natural	1 Clab	115	.070	S.F.	3.56	2.11		5.67	7.40
1100	Colors		115	.070	"	3.82	2.11		5.93	7.70
1200	Support pedestal, bottom cap		960	.008	Ea.	2.64	.25		2.89	3.32
1300	Top cap		960	.008		4.84	.25		5.09	5.70
1400	Leveling shims, 1/16"		1920	.004		1.21	.13		1.34	1.54
1500	1/8"		1920	.004		1.21	.13		1.34	1.54
1600	Buffer pad		960	.008		2.52	.25		2.77	3.19
1700	PVC legs (4" SDR 35)		2880	.003	Inch	.14	.08		.22	.29
2000	Alternate pricing method, system in place		101	.079	S.F.	7.15	2.40		9.55	11.80

07 81 Applied Fireproofing

07 81 16 – Cementitious Fireproofing

07 81 16.10 Sprayed Cementitious Fireproofing		Crew	Daily Output	Labor-Hours	Unit	Material	Labor	Equipment	Total	Total Incl O&P
0010	**SPRAYED CEMENTITIOUS FIREPROOFING**									
0050	Not including canvas protection, normal density									
0100	Per 1" thick, on flat plate steel	G-2	3000	.008	S.F.	.58	.26	.04	.88	1.12
0200	Flat decking		2400	.010		.58	.33	.05	.96	1.24
0400	Beams		1500	.016		.58	.52	.08	1.18	1.59
0500	Corrugated or fluted decks		1250	.019		.87	.63	.10	1.60	2.10
0700	Columns, 1-1/8" thick		1100	.022		.65	.71	.11	1.47	2.02
0800	2-3/16" thick		700	.034		1.39	1.12	.18	2.69	3.57
0900	For canvas protection, add		5000	.005		.11	.16	.03	.30	.41
1000	Not including canvas protection, high density									
1100	Per 1" thick, on flat plate steel	G-2	3000	.008	S.F.	2.25	.26	.04	2.55	2.96
1110	On flat decking		2400	.010		2.25	.33	.05	2.63	3.08
1120	On beams		1500	.016		2.25	.52	.08	2.85	3.43
1130	Corrugated or fluted decks		1250	.019		2.25	.63	.10	2.98	3.62
1140	Columns, 1-1/8" thick		1100	.022		2.53	.71	.11	3.35	4.09
1150	2-3/16" thick		1100	.022		5.05	.71	.11	5.87	6.85
1170	For canvas protection, add		5000	.005		.11	.16	.03	.30	.41
1200	Not including canvas protection, retrofitting									
1210	Per 1" thick, on flat plate steel	G-2	1500	.016	S.F.	.51	.52	.08	1.11	1.51
1220	On flat decking		1200	.020		.51	.65	.10	1.26	1.75
1230	On beams		750	.032		.51	1.04	.17	1.72	2.45
1240	Corrugated or fluted decks		625	.038		.76	1.25	.20	2.21	3.12
1250	Columns, 1-1/8" thick		550	.044		.57	1.42	.23	2.22	3.22
1260	2-3/16" thick		500	.048		1.15	1.57	.25	2.97	4.11
1400	Accessories, preliminary spattered texture coat		4500	.005		.05	.17	.03	.25	.37

07 81 Applied Fireproofing

07 81 16 – Cementitious Fireproofing

07 81 16.10 Sprayed Cementitious Fireproofing		Crew	Daily Output	Labor-Hours	Unit	Material	2019 Bare Costs Labor	Equipment	Total	Total Incl O&P
1410	Bonding agent	1 Plas	1000	.008	S.F.	.10	.29		.39	.58

07 84 Firestopping

07 84 13 – Penetration Firestopping

07 84 13.10 Firestopping

Line	Description	Crew	Daily Output	Labor-Hours	Unit	Material	2019 Bare Costs Labor	Equipment	Total	Total Incl O&P
0010	**FIRESTOPPING** R078413-30									
0100	Metallic piping, non insulated									
0110	Through walls, 2″ diameter	1 Carp	16	.500	Ea.	3.95	19.40		23.35	36.50
0120	4″ diameter		14	.571		6.90	22		28.90	44
0130	6″ diameter		12	.667		10.15	26		36.15	53.50
0140	12″ diameter		10	.800		20	31		51	73.50
0150	Through floors, 2″ diameter		32	.250		2.01	9.70		11.71	18.20
0160	4″ diameter		28	.286		3.53	11.05		14.58	22
0170	6″ diameter		24	.333		5.15	12.90		18.05	27
0180	12″ diameter		20	.400		10.35	15.50		25.85	37
0190	Metallic piping, insulated									
0200	Through walls, 2″ diameter	1 Carp	16	.500	Ea.	6.90	19.40		26.30	39.50
0210	4″ diameter		14	.571		10.15	22		32.15	47.50
0220	6″ diameter		12	.667		13.60	26		39.60	57.50
0230	12″ diameter		10	.800		23	31		54	76.50
0240	Through floors, 2″ diameter		32	.250		3.53	9.70		13.23	19.90
0250	4″ diameter		28	.286		5.15	11.05		16.20	24
0260	6″ diameter		24	.333		6.95	12.90		19.85	29
0270	12″ diameter		20	.400		10.60	15.50		26.10	37
0280	Non metallic piping, non insulated									
0290	Through walls, 2″ diameter	1 Carp	12	.667	Ea.	49	26		75	96.50
0300	4″ diameter		10	.800		97	31		128	158
0310	6″ diameter		8	1		183	39		222	266
0330	Through floors, 2″ diameter		16	.500		32.50	19.40		51.90	68
0340	4″ diameter		6	1.333		61	51.50		112.50	153
0350	6″ diameter		6	1.333		118	51.50		169.50	216
0370	Ductwork, insulated & non insulated, round									
0380	Through walls, 6″ diameter	1 Carp	12	.667	Ea.	13.60	26		39.60	57.50
0390	12″ diameter		10	.800		23	31		54	76.50
0400	18″ diameter		8	1		28	39		67	95
0410	Through floors, 6″ diameter		16	.500		6.95	19.40		26.35	39.50
0420	12″ diameter		14	.571		11.85	22		33.85	49.50
0430	18″ diameter		12	.667		14.40	26		40.40	58.50
0440	Ductwork, insulated & non insulated, rectangular									
0450	With stiffener/closure angle, through walls, 6″ x 12″	1 Carp	8	1	Ea.	18.15	39		57.15	84
0460	12″ x 24″		6	1.333		35.50	51.50		87	125
0470	24″ x 48″		4	2		76	77.50		153.50	212
0480	With stiffener/closure angle, through floors, 6″ x 12″		10	.800		9.05	31		40.05	61
0490	12″ x 24″		8	1		18.15	39		57.15	84
0500	24″ x 48″		6	1.333		38	51.50		89.50	128
0510	Multi trade openings									
0520	Through walls, 6″ x 12″	1 Carp	2	4	Ea.	49.50	155		204.50	310
0530	12″ x 24″	″	1	8		172	310		482	700
0540	24″ x 48″	2 Carp	1	16		590	620		1,210	1,675
0550	48″ x 96″	″	.75	21.333		2,200	825		3,025	3,800
0560	Through floors, 6″ x 12″	1 Carp	2	4		46.50	155		201.50	305

07 84 Firestopping

07 84 13 – Penetration Firestopping

07 84 13.10 Firestopping		Crew	Daily Output	Labor-Hours	Unit	Material	2019 Bare Costs Labor	Equipment	Total	Total Incl O&P
0570	12" x 24"	1 Carp	1	8	Ea.	63	310		373	580
0580	24" x 48"	2 Carp	.75	21.333		151	825		976	1,550
0590	48" x 96"	"	.50	32		360	1,250		1,610	2,450
0600	Structural penetrations, through walls									
0610	Steel beams, W8 x 10	1 Carp	8	1	Ea.	75	39		114	147
0620	W12 x 14		6	1.333		109	51.50		160.50	206
0630	W21 x 44		5	1.600		147	62		209	264
0640	W36 x 135		3	2.667		241	103		344	435
0650	Bar joists, 18" deep		6	1.333		44.50	51.50		96	135
0660	24" deep		6	1.333		59	51.50		110.50	151
0670	36" deep		5	1.600		89	62		151	200
0680	48" deep		4	2		119	77.50		196.50	258
0690	Construction joints, floor slab at exterior wall									
0700	Precast, brick, block or drywall exterior									
0710	2" wide joint	1 Carp	125	.064	L.F.	9.85	2.48		12.33	14.95
0720	4" wide joint	"	75	.107	"	16.35	4.13		20.48	25
0730	Metal panel, glass or curtain wall exterior									
0740	2" wide joint	1 Carp	40	.200	L.F.	10.65	7.75		18.40	24.50
0750	4" wide joint	"	25	.320	"	16.65	12.40		29.05	39
0760	Floor slab to drywall partition									
0770	Flat joint	1 Carp	100	.080	L.F.	14.55	3.10		17.65	21
0780	Fluted joint		50	.160		16.15	6.20		22.35	28
0790	Etched fluted joint		75	.107		20.50	4.13		24.63	29.50
0800	Floor slab to concrete/masonry partition									
0810	Flat joint	1 Carp	75	.107	L.F.	14.50	4.13		18.63	23
0820	Fluted joint	"	50	.160	"	16.55	6.20		22.75	28.50
0830	Concrete/CMU wall joints									
0840	1" wide	1 Carp	100	.080	L.F.	20.50	3.10		23.60	28
0850	2" wide		75	.107		25.50	4.13		29.63	35
0860	4" wide		50	.160		35.50	6.20		41.70	49.50
0870	Concrete/CMU floor joints									
0880	1" wide	1 Carp	200	.040	L.F.	20.50	1.55		22.05	25
0890	2" wide		150	.053		25.50	2.07		27.57	31.50
0900	4" wide		100	.080		36.50	3.10		39.60	45

07 91 Preformed Joint Seals

07 91 13 – Compression Seals

07 91 13.10 Compression Seals		Crew	Daily Output	Labor-Hours	Unit	Material	Labor	Equipment	Total	Total Incl O&P
0010	COMPRESSION SEALS									
4900	O-ring type cord, 1/4"	1 Bric	472	.017	L.F.	.41	.62		1.03	1.49
4910	1/2"		440	.018		1.06	.67		1.73	2.28
4920	3/4"		424	.019		2.05	.69		2.74	3.42
4930	1"		408	.020		4.15	.72		4.87	5.75
4940	1-1/4"		384	.021		7.90	.77		8.67	10
4950	1-1/2"		368	.022		9.70	.80		10.50	12
4960	1-3/4"		352	.023		13.60	.84		14.44	16.35
4970	2"		344	.023		22.50	.85		23.35	26

07 91 Preformed Joint Seals

07 91 16 – Joint Gaskets

07 91 16.10 Joint Gaskets		Crew	Daily Output	Labor-Hours	Unit	Material	2019 Bare Costs Labor	Equipment	Total	Total Incl O&P
0010	**JOINT GASKETS**									
4400	Joint gaskets, neoprene, closed cell w/adh, 1/8″ x 3/8″	1 Bric	240	.033	L.F.	.33	1.22		1.55	2.40
4500	1/4″ x 3/4″		215	.037		.64	1.37		2.01	2.98
4700	1/2″ x 1″		200	.040		1.51	1.47		2.98	4.11
4800	3/4″ x 1-1/2″		165	.048		1.65	1.78		3.43	4.79

07 91 23 – Backer Rods

07 91 23.10 Backer Rods

Code	Description	Crew	Daily Output	Labor-Hours	Unit	Material	Labor	Equipment	Total	Total Incl O&P
0010	**BACKER RODS**									
0030	Backer rod, polyethylene, 1/4″ diameter	1 Bric	4.60	1.739	C.L.F.	2.42	64		66.42	110
0050	1/2″ diameter		4.60	1.739		4.09	64		68.09	112
0070	3/4″ diameter		4.60	1.739		6.95	64		70.95	115
0090	1″ diameter		4.60	1.739		12.45	64		76.45	121

07 91 26 – Joint Fillers

07 91 26.10 Joint Fillers

Code	Description	Crew	Daily Output	Labor-Hours	Unit	Material	Labor	Equipment	Total	Total Incl O&P
0010	**JOINT FILLERS**									
4360	Butyl rubber filler, 1/4″ x 1/4″	1 Bric	290	.028	L.F.	.23	1.01		1.24	1.94
4365	1/2″ x 1/2″		250	.032		.91	1.18		2.09	2.96
4370	1/2″ x 3/4″		210	.038		1.37	1.40		2.77	3.83
4375	3/4″ x 3/4″		230	.035		2.05	1.28		3.33	4.38
4380	1″ x 1″		180	.044		2.73	1.63		4.36	5.70
4390	For coloring, add					12%				
4980	Polyethylene joint backing, 1/4″ x 2″	1 Bric	2.08	3.846	C.L.F.	13.85	141		154.85	251
4990	1/4″ x 6″		1.28	6.250	″	29.50	230		259.50	420
5600	Silicone, room temp vulcanizing foam seal, 1/4″ x 1/2″		1312	.006	L.F.	.46	.22		.68	.88
5610	1/2″ x 1/2″		656	.012		.92	.45		1.37	1.77
5620	1/2″ x 3/4″		442	.018		1.38	.67		2.05	2.63
5630	3/4″ x 3/4″		328	.024		2.08	.90		2.98	3.78
5640	1/8″ x 1″		1312	.006		.46	.22		.68	.88
5650	1/8″ x 3″		442	.018		1.38	.67		2.05	2.63
5670	1/4″ x 3″		295	.027		2.77	1		3.77	4.71
5680	1/4″ x 6″		148	.054		5.55	1.99		7.54	9.40
5690	1/2″ x 6″		82	.098		11.10	3.59		14.69	18.20
5700	1/2″ x 9″		52.50	.152		16.60	5.60		22.20	27.50
5710	1/2″ x 12″		33	.242		22	8.90		30.90	39.50

07 92 Joint Sealants

07 92 13 – Elastomeric Joint Sealants

07 92 13.20 Caulking and Sealant Options

Code	Description	Crew	Daily Output	Labor-Hours	Unit	Material	Labor	Equipment	Total	Total Incl O&P
0010	**CAULKING AND SEALANT OPTIONS**									
0050	Latex acrylic based, bulk				Gal.	30			30	33
0055	Bulk in place 1/4″ x 1/4″ bead	1 Bric	300	.027	L.F.	.09	.98		1.07	1.73
0060	1/4″ x 3/8″		294	.027		.16	1		1.16	1.84
0065	1/4″ x 1/2″		288	.028		.21	1.02		1.23	1.93
0075	3/8″ x 3/8″		284	.028		.24	1.04		1.28	1.99
0080	3/8″ x 1/2″		280	.029		.32	1.05		1.37	2.10
0085	3/8″ x 5/8″		276	.029		.39	1.07		1.46	2.21
0095	3/8″ x 3/4″		272	.029		.47	1.08		1.55	2.32
0100	1/2″ x 1/2″		275	.029		.42	1.07		1.49	2.24
0105	1/2″ x 5/8″		269	.030		.53	1.09		1.62	2.40

07 92 Joint Sealants

07 92 13 – Elastomeric Joint Sealants

07 92 13.20 Caulking and Sealant Options		Crew	Daily Output	Labor-Hours	Unit	Material	Labor	Equipment	Total	Total Incl O&P
						2019 Bare Costs				
0110	1/2" x 3/4"	1 Bric	263	.030	L.F.	.63	1.12		1.75	2.55
0115	1/2" x 7/8"		256	.031		.74	1.15		1.89	2.72
0120	1/2" x 1"		250	.032		.84	1.18		2.02	2.89
0125	3/4" x 3/4"		244	.033		.95	1.21		2.16	3.05
0130	3/4" x 1"		225	.036		1.26	1.31		2.57	3.57
0135	1" x 1"		200	.040		1.68	1.47		3.15	4.30
0190	Cartridges				Gal.	33			33	36
0200	11 fl. oz. cartridge				Ea.	2.82			2.82	3.10
0500	1/4" x 1/2"	1 Bric	288	.028	L.F.	.23	1.02		1.25	1.95
0600	1/2" x 1/2"		275	.029		.46	1.07		1.53	2.29
0800	3/4" x 3/4"		244	.033		1.04	1.21		2.25	3.15
0900	3/4" x 1"		225	.036		1.38	1.31		2.69	3.70
1000	1" x 1"		200	.040		1.73	1.47		3.20	4.35
1400	Butyl based, bulk				Gal.	37.50			37.50	41
1500	Cartridges				"	37.50			37.50	41
1700	1/4" x 1/2", 154 L.F./gal.	1 Bric	288	.028	L.F.	.24	1.02		1.26	1.97
1800	1/2" x 1/2", 77 L.F./gal.	"	275	.029	"	.49	1.07		1.56	2.31
2300	Polysulfide compounds, 1 component, bulk				Gal.	84.50			84.50	93
2400	Cartridges				"	131			131	144
2600	1 or 2 component, in place, 1/4" x 1/4", 308 L.F./gal.	1 Bric	300	.027	L.F.	.27	.98		1.25	1.93
2700	1/2" x 1/4", 154 L.F./gal.		288	.028		.55	1.02		1.57	2.30
2900	3/4" x 3/8", 68 L.F./gal.		272	.029		1.24	1.08		2.32	3.16
3000	1" x 1/2", 38 L.F./gal.		250	.032		2.22	1.18		3.40	4.40
3200	Polyurethane, 1 or 2 component				Gal.	55.50			55.50	61
3300	Cartridges				"	68			68	75
3500	Bulk, in place, 1/4" x 1/4"	1 Bric	300	.027	L.F.	.18	.98		1.16	1.83
3655	1/2" x 1/4"		288	.028		.36	1.02		1.38	2.10
3800	3/4" x 3/8"		272	.029		.82	1.08		1.90	2.70
3900	1" x 1/2"		250	.032		1.44	1.18		2.62	3.55
4100	Silicone rubber, bulk				Gal.	58.50			58.50	64
4200	Cartridges				"	52.50			52.50	58

07 92 19 – Acoustical Joint Sealants

07 92 19.10 Acoustical Sealant		Crew	Daily Output	Labor-Hours	Unit	Material	Labor	Equipment	Total	Total Incl O&P
0010	**ACOUSTICAL SEALANT**									
0020	Acoustical sealant, elastomeric, cartridges				Ea.	8.60			8.60	9.45
0025	In place, 1/4" x 1/4"	1 Bric	300	.027	L.F.	.35	.98		1.33	2.02
0030	1/4" x 1/2"		288	.028		.70	1.02		1.72	2.47
0035	1/2" x 1/2"		275	.029		1.40	1.07		2.47	3.32
0040	1/2" x 3/4"		263	.030		2.11	1.12		3.23	4.18
0045	3/4" x 3/4"		244	.033		3.16	1.21		4.37	5.50
0050	1" x 1"		200	.040		5.60	1.47		7.07	8.65

Division Notes

		CREW	DAILY OUTPUT	LABOR-HOURS	UNIT	BARE COSTS MAT.	BARE COSTS LABOR	BARE COSTS EQUIP.	BARE COSTS TOTAL	TOTAL INCL O&P

Estimating Tips

08 10 00 Doors and Frames

All exterior doors should be addressed for their energy conservation (insulation and seals).

- Most metal doors and frames look alike, but there may be significant differences among them. When estimating these items, be sure to choose the line item that most closely compares to the specification or door schedule requirements regarding:
  - □ type of metal
  - □ metal gauge
  - □ door core material
  - □ fire rating
  - □ finish
- Wood and plastic doors vary considerably in price. The primary determinant is the veneer material. Lauan, birch, and oak are the most common veneers. Other variables include the following:
  - □ hollow or solid core
  - □ fire rating
  - □ flush or raised panel
  - □ finish
- Door pricing includes bore for cylindrical locksets and mortise for hinges.

08 30 00 Specialty Doors and Frames

- There are many varieties of special doors, and they are usually priced per each. Add frames, hardware, or operators required for a complete installation.

08 40 00 Entrances, Storefronts, and Curtain Walls

- Glazed curtain walls consist of the metal tube framing and the glazing material. The cost data in this subdivision is presented for the metal tube framing alone or the composite wall. If your estimate requires a detailed takeoff of the framing, be sure to add the glazing cost and any tints.

08 50 00 Windows

- Steel windows are unglazed and aluminum can be glazed or unglazed. Some metal windows are priced without glass. Refer to 08 80 00 Glazing for glass pricing. The grade C indicates commercial grade windows, usually ASTM C-35.
- All wood windows and vinyl are priced preglazed. The glazing is insulating glass. Add the cost of screens and grills if required and not already included.

08 70 00 Hardware

- Hardware costs add considerably to the cost of a door. The most efficient method to determine the hardware requirements for a project is to review the door and hardware schedule together. One type of door may have different hardware, depending on the door usage.
- Door hinges are priced by the pair, with most doors requiring 1-1/2 pairs per door. The hinge prices do not include installation labor because it is included in door installation.

  Hinges are classified according to the frequency of use, base material, and finish.

08 80 00 Glazing

- Different openings require different types of glass. The most common types are:
  - □ float
  - □ tempered
  - □ insulating
  - □ impact-resistant
  - □ ballistic-resistant
- Most exterior windows are glazed with insulating glass. Entrance doors and window walls, where the glass is less than 18" from the floor, are generally glazed with tempered glass. Interior windows and some residential windows are glazed with float glass.
- Coastal communities require the use of impact-resistant glass, dependent on wind speed.
- The insulation or 'u' value is a strong consideration, along with solar heat gain, to determine total energy efficiency.

Reference Numbers

Reference numbers are shown at the beginning of some major classifications. These numbers refer to related items in the Reference Section. The reference information may be an estimating procedure, an alternate pricing method, or technical information.

Note: Not all subdivisions listed here necessarily appear. ■

08 01 53 – Operation and Maintenance of Plastic Windows

08 01 53.81 Solid Vinyl Replacement Windows			Crew	Daily Output	Labor-Hours	Unit	2019 Bare Costs Material	Labor	Equipment	Total	Total Incl O&P
0010	**SOLID VINYL REPLACEMENT WINDOWS** R085313-20										
0020	Double-hung, insulated glass, up to 83 united inches	G	2 Carp	8	2	Ea.	181	77.50		258.50	325
0040	84 to 93	G		8	2		203	77.50		280.50	350
0060	94 to 101	G		6	2.667		203	103		306	395
0080	102 to 111	G		6	2.667		213	103		316	405
0100	112 to 120	G		6	2.667		232	103		335	425
0120	For each united inch over 120, add	G		800	.020	Inch	2.85	.78		3.63	4.42
0140	Casement windows, one operating sash, 42 to 60 united inches	G		8	2	Ea.	244	77.50		321.50	395
0160	61 to 70	G		8	2		270	77.50		347.50	425
0180	71 to 80	G		8	2		305	77.50		382.50	465
0200	81 to 96	G		8	2		330	77.50		407.50	495
0220	Two operating sash, 58 to 78 united inches	G		8	2		600	77.50		677.50	795
0240	79 to 88	G		8	2		680	77.50		757.50	880
0260	89 to 98	G		8	2		690	77.50		767.50	890
0280	99 to 108	G		6	2.667		735	103		838	975
0300	109 to 121	G		6	2.667		805	103		908	1,050
0320	Two operating, one fixed sash, 73 to 108 united inches	G		8	2		560	77.50		637.50	750
0340	109 to 118	G		8	2		615	77.50		692.50	805
0360	119 to 128	G		6	2.667		720	103		823	960
0380	129 to 138	G		6	2.667		775	103		878	1,025
0400	139 to 156	G		6	2.667		940	103		1,043	1,200
0420	Four operating sash, 98 to 118 united inches	G		8	2		1,175	77.50		1,252.50	1,400
0440	119 to 128	G		8	2		1,225	77.50		1,302.50	1,475
0460	129 to 138	G		6	2.667		1,275	103		1,378	1,575
0480	139 to 148	G		6	2.667		1,325	103		1,428	1,625
0500	149 to 168	G		6	2.667		1,425	103		1,528	1,750
0520	169 to 178	G		6	2.667		1,475	103		1,578	1,800
0560	Fixed picture window, up to 63 united inches	G		8	2		163	77.50		240.50	310
0580	64 to 83	G		8	2		196	77.50		273.50	345
0600	84 to 101	G		8	2		247	77.50		324.50	400
0620	For each united inch over 101, add	G		900	.018	Inch	3	.69		3.69	4.44
0800	Cellulose fiber insulation, poured into sash balance cavity	G	1 Carp	36	.222	C.F.	.70	8.60		9.30	15
0820	Silicone caulking at perimeter	G	"	800	.010	L.F.	.17	.39		.56	.83
2000	Impact resistant replacement windows										
2005	Laminated glass, 120 MPH rating, measure in united inches										
2010	Installation labor does not cover any rework of the window opening										
2020	Double-hung, insulated glass, up to 101 united inches		2 Carp	8	2	Ea.	530	77.50		607.50	710
2025	For each united inch over 101, add			80	.200	Inch	3.31	7.75		11.06	16.45
2100	Casement windows, impact resistant, up to 60 united inches			8	2	Ea.	490	77.50		567.50	670
2120	61 to 70			8	2		515	77.50		592.50	700
2130	71 to 80			8	2		545	77.50		622.50	730
2140	81 to 100			8	2		560	77.50		637.50	745
2150	For each united inch over 100, add			80	.200	Inch	5.05	7.75		12.80	18.35
2200	Awning windows, impact resistant, up to 60 united inches			8	2	Ea.	500	77.50		577.50	680
2220	61 to 70			8	2		525	77.50		602.50	705
2230	71 to 80			8	2		535	77.50		612.50	720
2240	For each united inch over 80, add			80	.200	Inch	4.75	7.75		12.50	18.05
2300	Picture windows, impact resistant, up to 63 united inches			8	2	Ea.	360	77.50		437.50	530
2320	63 to 83			8	2		390	77.50		467.50	560
2330	84 to 101			8	2		425	77.50		502.50	600
2340	For each united inch over 101, add			80	.200	Inch	3.19	7.75		10.94	16.30

08 05 05 – Selective Demolition for Openings

08 05 05.10 Selective Demolition Doors	Crew	Daily Output	Labor-Hours	Unit	Material	2019 Bare Costs Labor	Equipment	Total	Total Incl O&P
0010 **SELECTIVE DEMOLITION DOORS** R024119-10									
0200 Doors, exterior, 1-3/4" thick, single, 3' x 7' high	1 Clab	16	.500	Ea.		15.20		15.20	25
0202 Doors, exterior, 3' - 6' wide x 7' high		12	.667			20		20	33.50
0210 3' x 8' high		10	.800			24.50		24.50	40
0215 Double, 3' x 8' high		6	1.333			40.50		40.50	67
0220 Double, 6' x 7' high		12	.667			20		20	33.50
0500 Interior, 1-3/8" thick, single, 3' x 7' high		20	.400			12.15		12.15	20
0520 Double, 6' x 7' high		16	.500			15.20		15.20	25
0700 Bi-folding, 3' x 6'-8" high		20	.400			12.15		12.15	20
0720 6' x 6'-8" high		18	.444			13.50		13.50	22.50
0900 Bi-passing, 3' x 6'-8" high		16	.500			15.20		15.20	25
0940 6' x 6'-8" high		14	.571			17.35		17.35	28.50
0960 Interior metal door 1-3/4" thick, 3'-0" x 6'-8"		18	.444			13.50		13.50	22.50
0980 Interior metal door 1-3/4" thick, 3'-0" x 7'-0"		18	.444			13.50		13.50	22.50
1000 Interior wood door 1-3/4" thick, 3'-0" x 6'-8"		20	.400			12.15		12.15	20
1020 Interior wood door 1-3/4" thick, 3'-0" x 7'-0"	↓	20	.400			12.15		12.15	20
1100 Door demo, floor door	2 Sswk	5	3.200			133		133	230
1500 Remove and reset, hollow core	1 Carp	8	1			39		39	64
1520 Solid		6	1.333			51.50		51.50	85.50
2000 Frames, including trim, metal	↓	8	1			39		39	64
2200 Wood	2 Carp	32	.500	↓		19.40		19.40	32
2201 Alternate pricing method	1 Carp	200	.040	L.F.		1.55		1.55	2.56
3000 Special doors, counter doors	2 Carp	6	2.667	Ea.		103		103	171
3001 Special doors, counter doors	2 Clab	6	2.667			81		81	134
3200 Floor door (trap type), or access type	2 Carp	8	2			77.50		77.50	128
3300 Glass, sliding, including frames		12	1.333			51.50		51.50	85.50
3400 Overhead, commercial, 12' x 12' high		4	4			155		155	256
3500 Residential, 9' x 7' high		8	2			77.50		77.50	128
3540 16' x 7' high		7	2.286			88.50		88.50	146
3600 Remove and reset, small		4	4			155		155	256
3620 Large	↓	2.50	6.400			248		248	410
3660 Remove and reset elec. garage door opener	1 Carp	8	1			39		39	64
3902 Cafe/bar swing door	2 Clab	8	2			60.50		60.50	100
4000 Residential lockset, exterior	1 Carp	28	.286			11.05		11.05	18.30
4010 Residential lockset, exterior w/deadbolt		26	.308			11.90		11.90	19.70
4020 Residential lockset, interior		30	.267			10.35		10.35	17.10
4200 Deadbolt lock		32	.250			9.70		9.70	16
4224 Pocket door, no frame	↓	8	1			39		39	64
5500 Remove door closer	1 Clab	10	.800			24.50		24.50	40
5520 Remove push/pull plate		20	.400			12.15		12.15	20
5540 Remove door bolt		30	.267			8.10		8.10	13.35
5560 Remove kick/mop plate		20	.400			12.15		12.15	20
5590 Remove mail slot		45	.178			5.40		5.40	8.90
5595 Remove peep hole	↓	45	.178			5.40		5.40	8.90
5600 Remove door sidelight	1 Carp	6	1.333			51.50		51.50	85.50
5604 Remove 9' x 7' garage door	2 Carp	9.50	1.684			65.50		65.50	108
5624 16' x 7' garage door		8	2			77.50		77.50	128
5626 9' x 7' swing up garage door		7	2.286			88.50		88.50	146
5628 16' x 7' swing up garage door	↓	7	2.286			88.50		88.50	146
5630 Garage door demolition, remove garage door track	1 Carp	6	1.333			51.50		51.50	85.50
5644 Remove overhead door opener	1 Clab	5	1.600			48.50		48.50	80
6334 Remove shower door unit		16	.500			15.20		15.20	25
6384 Remove French door unit	↓	8	1	↓		30.50		30.50	50

08 05 Common Work Results for Openings

08 05 05 – Selective Demolition for Openings

08 05 05.10 Selective Demolition Doors		Crew	Daily Output	Labor-Hours	Unit	Material	2019 Bare Costs Labor	Equipment	Total	Total Incl O&P
7100	Remove double swing pneumatic doors, openers and sensors	2 Skwk	.50	32	Opng.		1,250		1,250	2,100
7570	Remove shock absorbing door	2 Sswk	1.90	8.421	"		350		350	605

08 05 05.20 Selective Demolition of Windows

Code	Description	Crew	Daily Output	Labor-Hours	Unit	Material	Labor	Equipment	Total	Total Incl O&P
0010	**SELECTIVE DEMOLITION OF WINDOWS** R024119-10									
0200	Aluminum, including trim, to 12 S.F.	1 Clab	16	.500	Ea.		15.20		15.20	25
0240	To 25 S.F.		11	.727			22		22	36.50
0280	To 50 S.F.		5	1.600			48.50		48.50	80
0320	Storm windows/screens, to 12 S.F.		27	.296			9		9	14.85
0360	To 25 S.F.		21	.381			11.55		11.55	19.10
0400	To 50 S.F.		16	.500			15.20		15.20	25
0600	Glass, up to 10 S.F./window		200	.040	S.F.		1.21		1.21	2.01
0620	Over 10 S.F./window		150	.053	"		1.62		1.62	2.67
2000	Wood, including trim, to 12 S.F.		22	.364	Ea.		11.05		11.05	18.25
2020	To 25 S.F.		18	.444			13.50		13.50	22.50
2060	To 50 S.F.		13	.615			18.70		18.70	31
2065	To 180 S.F.		8	1			30.50		30.50	50
4300	Remove bay/bow window	2 Carp	6	2.667			103		103	171
4410	Remove skylight, plstc domes, flush/curb mtd	"	395	.041	S.F.		1.57		1.57	2.59
4420	Remove skylight, plstc/glass up to 2' x 3'	1 Carp	15	.533	Ea.		20.50		20.50	34
4440	Remove skylight, plstc/glass up to 4' x 6'	2 Carp	10	1.600			62		62	102
4480	Remove roof window up to 3' x 4'	1 Carp	8	1			39		39	64
4500	Remove roof window up to 4' x 6'	2 Carp	6	2.667			103		103	171
5020	Remove and reset window, up to a 2' x 2' window	1 Carp	6	1.333			51.50		51.50	85.50
5040	Up to a 3' x 3' window		4	2			77.50		77.50	128
5080	Up to a 4' x 5' window		2	4			155		155	256
6000	Screening only	1 Clab	4000	.002	S.F.		.06		.06	.10

08 11 Metal Doors and Frames

08 11 16 – Aluminum Doors and Frames

08 11 16.10 Entrance Doors

Code	Description	Crew	Daily Output	Labor-Hours	Unit	Material	Labor	Equipment	Total	Total Incl O&P
0010	**ENTRANCE DOORS** and frame, aluminum, narrow stile									
0011	Including standard hardware, clear finish, no glass									
0012	Top and bottom offset pivots, 1/4" beveled glass stops, threshold									
0013	Dead bolt lock with inside thumb screw, standard push pull									
0020	3'-0" x 7'-0" opening	2 Sswk	2	8	Ea.	935	330		1,265	1,600
0025	Anodizing aluminum entr. door & frame, add					113			113	125
0030	3'-6" x 7'-0" opening	2 Sswk	2	8		875	330		1,205	1,550
0100	3'-0" x 10'-0" opening, 3' high transom		1.80	8.889		1,325	370		1,695	2,125
0200	3'-6" x 10'-0" opening, 3' high transom		1.80	8.889		1,375	370		1,745	2,175
0280	5'-0" x 7'-0" opening		2	8		1,600	330		1,930	2,350
0300	6'-0" x 7'-0" opening		1.30	12.308		1,125	510		1,635	2,125
0301	6'-0" x 7'-0" opening		1.30	12.308	Pr.	1,125	510		1,635	2,125
0400	6'-0" x 10'-0" opening, 3' high transom		1.10	14.545	"	1,500	605		2,105	2,675
0520	3'-0" x 7'-0" opening, wide stile		2	8	Ea.	900	330		1,230	1,575
0540	3'-6" x 7'-0" opening		2	8		990	330		1,320	1,675
0560	5'-0" x 7'-0" opening		2	8		1,325	330		1,655	2,050
0580	6'-0" x 7'-0" opening		1.30	12.308	Pr.	1,525	510		2,035	2,575
0600	7'-0" x 7'-0" opening		1	16	"	2,325	665		2,990	3,700
1200	For non-standard size, add				Leaf	80%				
1250	For installation of non-standard size, add						20%			
1300	Light bronze finish, add				Leaf	36%				

08 11 Metal Doors and Frames

08 11 16 – Aluminum Doors and Frames

08 11 16.10 Entrance Doors		Crew	Daily Output	Labor-Hours	Unit	Material	Labor (2019 Bare Costs)	Equipment (2019 Bare Costs)	Total (2019 Bare Costs)	Total Incl O&P
1400	Dark bronze finish, add				Leaf	25%				
1500	For black finish, add					40%				
1600	Concealed panic device, add					1,000			1,000	1,100
1700	Electric striker release, add				Opng.	325			325	355
1800	Floor check, add				Leaf	705			705	780
1900	Concealed closer, add				″	535			535	590

08 11 63 – Metal Screen and Storm Doors and Frames

08 11 63.23 Aluminum Screen and Storm Doors and Frames		Crew	Daily Output	Labor-Hours	Unit	Material	Labor	Equipment	Total	Total Incl O&P
0010	**ALUMINUM SCREEN AND STORM DOORS AND FRAMES**									
0020	Combination storm and screen									
0420	Clear anodic coating, 2′-8″ wide	2 Carp	14	1.143	Ea.	220	44.50		264.50	315
0440	3′-0″ wide	″	14	1.143	″	191	44.50		235.50	283
0500	For 7′-0″ door height, add					8%				
1020	Mill finish, 2′-8″ wide	2 Carp	14	1.143	Ea.	250	44.50		294.50	350
1040	3′-0″ wide	″	14	1.143		269	44.50		313.50	370
1100	For 7′-0″ door, add					8%				
1520	White painted, 2′-8″ wide	2 Carp	14	1.143		256	44.50		300.50	355
1540	3′-0″ wide		14	1.143		310	44.50		354.50	415
1541	Storm door, painted, alum., insul., 6′-8″ x 2′-6″ wide		14	1.143		184	44.50		228.50	275
1545	2′-8″ wide		14	1.143		291	44.50		335.50	395
1600	For 7′-0″ door, add					8%				
1800	Aluminum screen door, 6′-8″ x 2′-8″ wide	2 Carp	14	1.143		209	44.50		253.50	305
1810	3′-0″ wide	″	14	1.143		283	44.50		327.50	385

08 12 Metal Frames

08 12 13 – Hollow Metal Frames

08 12 13.13 Standard Hollow Metal Frames		Crew	Daily Output	Labor-Hours	Unit	Material	Labor	Equipment	Total	Total Incl O&P
0010	**STANDARD HOLLOW METAL FRAMES**									
0020	16 ga., up to 5-3/4″ jamb depth									
0025	3′-0″ x 6′-8″ single G	2 Carp	16	1	Ea.	153	39		192	232
0028	3′-6″ wide, single G		16	1		263	39		302	355
0030	4′-0″ wide, single G		16	1		288	39		327	380
0040	6′-0″ wide, double G		14	1.143		236	44.50		280.50	335
0045	8′-0″ wide, double G		14	1.143		229	44.50		273.50	325
0100	3′-0″ x 7′-0″ single G		16	1		205	39		244	289
0110	3′-6″ wide, single G		16	1		208	39		247	293
0112	4′-0″ wide, single G		16	1		256	39		295	345
0140	6′-0″ wide, double G		14	1.143		237	44.50		281.50	335
0145	8′-0″ wide, double G		14	1.143		247	44.50		291.50	345
1000	16 ga., up to 4-7/8″ deep, 3′-0″ x 7′-0″ single G		16	1		191	39		230	274
1140	6′-0″ wide, double G		14	1.143		216	44.50		260.50	310
1200	16 ga., 8-3/4″ deep, 3′-0″ x 7′-0″ single G		16	1		206	39		245	291
1240	6′-0″ wide, double G		14	1.143		275	44.50		319.50	375
2800	14 ga., up to 3-7/8″ deep, 3′-0″ x 7′-0″ single G		16	1		265	39		304	355
2840	6′-0″ wide, double G		14	1.143		310	44.50		354.50	420
3000	14 ga., up to 5-3/4″ deep, 3′-0″ x 6′-8″ single G		16	1		160	39		199	240
3002	3′-6″ wide, single G		16	1		257	39		296	345
3005	4′-0″ wide, single G		16	1		167	39		206	248
3600	up to 5-3/4″ jamb depth, 4′-0″ x 7′-0″ single G		15	1.067		255	41.50		296.50	350
3620	6′-0″ wide, double G		12	1.333		190	51.50		241.50	295

08 12 13.13 Standard Hollow Metal Frames			Crew	Daily Output	Labor-Hours	Unit	Material	Labor	Equipment	Total	Total Incl O&P
							2019 Bare Costs				
3640	8′-0″ wide, double	G	2 Carp	12	1.333	Ea.	300	51.50		351.50	415
3700	8′-0″ high, 4′-0″ wide, single	G		15	1.067		290	41.50		331.50	390
3740	8′-0″ wide, double	G		12	1.333		340	51.50		391.50	455
4000	6-3/4″ deep, 4′-0″ x 7′-0″ single	G		15	1.067		282	41.50		323.50	380
4020	6′-0″ wide, double	G		12	1.333		320	51.50		371.50	435
4040	8′-0″ wide, double	G		12	1.333		241	51.50		292.50	350
4100	8′-0″ high, 4′-0″ wide, single	G		15	1.067		189	41.50		230.50	277
4140	8′-0″ wide, double	G		12	1.333		440	51.50		491.50	570
4400	8-3/4″ deep, 4′-0″ x 7′-0″, single	G		15	1.067		410	41.50		451.50	520
4440	8′-0″ wide, double	G		12	1.333		440	51.50		491.50	565
4500	4′-0″ x 8′-0″, single	G		15	1.067		465	41.50		506.50	585
4540	8′-0″ wide, double	G		12	1.333		475	51.50		526.50	610
4900	For welded frames, add						51.50			51.50	56.50
5380	Steel frames, KD, 14 ga., "B" label, to 5-3/4″ throat, to 3′-0″ x 7′-0″		2 Carp	15	1.067		233	41.50		274.50	325
5400	14 ga., "B" label, up to 5-3/4″ deep, 4′-0″ x 7′-0″ single	G		15	1.067		180	41.50		221.50	267
5440	8′-0″ wide, double	G		12	1.333		235	51.50		286.50	345
5800	6-3/4″ deep, 7′-0″ high, 4′-0″ wide, single	G		15	1.067		180	41.50		221.50	268
5840	8′-0″ wide, double	G		12	1.333		330	51.50		381.50	450
6200	8-3/4″ deep, 4′-0″ x 7′-0″ single	G		15	1.067		266	41.50		307.50	360
6240	8′-0″ wide, double	G		12	1.333		395	51.50		446.50	520
6300	For "A" label use same price as "B" label										
6400	For baked enamel finish, add						30%	15%			
6500	For galvanizing, add						20%				
6600	For hospital stop, add					Ea.	244			244	268
6620	For hospital stop, stainless steel, add					"	113			113	124
7900	Transom lite frames, fixed, add		2 Carp	155	.103	S.F.	45	4		49	56
8000	Movable, add		"	130	.123	"	57.50	4.77		62.27	71.50

08 12 13.15 Borrowed Lites		Crew	Daily Output	Labor-Hours	Unit	Material	Labor	Equipment	Total	Total Incl O&P
0010	**BORROWED LITES**									
0100	Hollow metal section 20 ga., 3-1/2″ with glass stop	2 Carp	100	.160	L.F.	18.75	6.20		24.95	31
0110	3-3/4″ with glass stop		100	.160		8.60	6.20		14.80	19.70
0120	4″ with glass stop		100	.160		8.75	6.20		14.95	19.90
0130	4-5/8″ with glass stop		100	.160		9.05	6.20		15.25	20
0140	4-7/8″ with glass stop		100	.160		9.20	6.20		15.40	20.50
0150	5″ with glass stop		100	.160		10.05	6.20		16.25	21.50
0160	5-3/8″ with glass stop		100	.160		11.80	6.20		18	23.50
0300	Hollow metal section 18 ga., 3-1/2″ with glass stop		80	.200		9.30	7.75		17.05	23
0310	3-3/4″ with glass stop		80	.200		9.50	7.75		17.25	23.50
0320	4″ with glass stop		80	.200		13.20	7.75		20.95	27.50
0330	4-5/8″ with glass stop		80	.200		10.10	7.75		17.85	24
0340	4-7/8″ with glass stop		80	.200		10.35	7.75		18.10	24
0350	5″ with glass stop		80	.200		10.45	7.75		18.20	24.50
0360	5-3/8″ with glass stop		80	.200		10.45	7.75		18.20	24.50
0370	6-5/8″ with glass stop		80	.200		11.65	7.75		19.40	25.50
0380	6-7/8″ with glass stop		80	.200		11.80	7.75		19.55	26
0390	7-1/4″ with glass stop		80	.200		12.30	7.75		20.05	26.50
0500	Mullion section 18 ga., 3-1/2″ with double stop		50	.320		21.50	12.40		33.90	44.50
0510	3-3/4″ with double stop		50	.320		22	12.40		34.40	45
0530	4-5/8″ with double stop		50	.320		23.50	12.40		35.90	46.50
0540	4-7/8″ with double stop		50	.320		24	12.40		36.40	47
0550	5″ with double stop		50	.320		24	12.40		36.40	47
0560	5-3/8″ with double stop		50	.320		25	12.40		37.40	48

08 12 Metal Frames

08 12 13 – Hollow Metal Frames

08 12 13.15 Borrowed Lites		Crew	Daily Output	Labor-Hours	Unit	2019 Bare Costs Material	Labor	Equipment	Total	Total Incl O&P
0570	6-5/8" with double stop	2 Carp	50	.320	L.F.	27	12.40		39.40	50.50
0580	6-7/8" with double stop		50	.320		27.50	12.40		39.90	51
0590	7-1/4" with double stop		50	.320		28	12.40		40.40	51.50
1000	Assembled frame 2'-0" x 2'-0" 16 ga., 3-3/4" with glass stop		10	1.600	Ea.	204	62		266	325
1010	3'-0" x 2'-0"		10	1.600		204	62		266	325
1020	4'-0" x 2'-0"		10	1.600		205	62		267	330
1030	5'-0" x 2'-0"		10	1.600		211	62		273	335
1040	6'-0" x 2'-0"		8	2		305	77.50		382.50	465
1050	7'-0" x 2'-0"		8	2		305	77.50		382.50	465
1060	8'-0" x 2'-0"		8	2		305	77.50		382.50	465
1100	2'-0" x 7'-0"		8	2		305	77.50		382.50	465
1110	3'-0" x 7'-0"		8	2		305	77.50		382.50	465
1120	4'-0" x 7'-0"		8	2		305	77.50		382.50	465
1130	5'-0" x 7'-0"		8	2		305	77.50		382.50	465
1140	6'-0" x 7'-0"		6	2.667		330	103		433	535
1150	7'-0" x 7'-0"		6	2.667		330	103		433	535
1160	8'-0" x 7'-0"		6	2.667		330	103		433	535
1200	Assembled frame 2'-0" x 2'-0" 16 ga., 4-7/8" with glass stop		10	1.600		205	62		267	330
1210	3'-0" x 2'-0"		10	1.600		205	62		267	330
1220	4'-0" x 2'-0"		8	2		205	77.50		282.50	355
1230	5'-0" x 2'-0"		8	2		305	77.50		382.50	465
1240	6'-0" x 2'-0"		8	2		305	77.50		382.50	465
1250	7'-0" x 2'-0"		8	2		305	77.50		382.50	465
1260	8'-0" x 2'-0"		8	2		305	77.50		382.50	465
1300	2'-0" x 7'-0"		8	2		305	77.50		382.50	465
1310	3'-0" x 7'-0"		8	2		305	77.50		382.50	465
1320	4'-0" x 7'-0"		8	2		305	77.50		382.50	465
1330	5'-0" x 7'-0"		6	2.667		305	103		408	505
1340	6'-0" x 7'-0"		6	2.667		330	103		433	535
1350	7'-0" x 7'-0"		6	2.667		340	103		443	545
1360	8'-0" x 7'-0"		6	2.667		340	103		443	545
1400	Assembled frame 2'-0" x 3'-0" 16 ga., 6-1/8" with glass stop		6	2.667		206	103		309	395
1410	3'-0" x 3'-0"		8	2		205	77.50		282.50	355
1420	4'-0" x 3'-0"		8	2		205	77.50		282.50	355
1430	5'-0" x 3'-0"		6	2.667		305	103		408	505
1440	6'-0" x 3'-0"		6	2.667		305	103		408	505
1450	7'-0" x 3'-0"		6	2.667		305	103		408	505
1460	8'-0" x 3'-0"		10	1.600		305	62		367	435

08 12 13.18 Transoms and Sidelights		Crew	Daily Output	Labor-Hours	Unit	Material	Labor	Equipment	Total	Total Incl O&P
0010	**TRANSOMS & SIDELIGHTS**									
0160	Transom sash, 6-3/4" thick, 3'-0" x 1'-4"	2 Carp	11	1.455	Ea.	188	56.50		244.50	300
0180	3'-4" x 1'-4"		11	1.455		191	56.50		247.50	305
0200	6'-0" x 1'-4"		6	2.667		219	103		322	410
0220	4-3/4" thick, 3'-0" x 1'-4"		11	1.455		179	56.50		235.50	290
0240	3'-4" x 1'-4"		11	1.455		179	56.50		235.50	290
0260	6'-0" x 1'-4"		6	2.667		205	103		308	395

08 12 13.25 Channel Metal Frames		Crew	Daily Output	Labor-Hours	Unit	Material	Labor	Equipment	Total	Total Incl O&P
0010	**CHANNEL METAL FRAMES**									
0020	Steel channels with anchors and bar stops									
0100	6" channel @ 8.2#/L.F., 3' x 7' door, weighs 150# G	E-4	13	2.462	Ea.	238	103	7.35	348.35	450
0200	8" channel @ 11.5#/L.F., 6' x 8' door, weighs 275# G		9	3.556		435	149	10.65	594.65	750
0300	8' x 12' door, weighs 400# G		6.50	4.923		635	207	14.70	856.70	1,075

08 12 Metal Frames

08 12 13 – Hollow Metal Frames

08 12 13.25 Channel Metal Frames			Crew	Daily Output	Labor-Hours	Unit	2019 Bare Costs Material	Labor	Equipment	Total	Total Incl O&P
0400	10" channel @ 15.3#/L.F., 10' x 10' door, weighs 500#	G	E-4	6	5.333	Ea.	795	224	15.95	1,034.95	1,275
0500	12' x 12' door, weighs 600#	G		5.50	5.818		950	244	17.40	1,211.40	1,500
0600	12" channel @ 20.7#/L.F., 12' x 12' door, weighs 825#	G		4.50	7.111		1,300	298	21.50	1,619.50	2,000
0700	12' x 16' door, weighs 1000#	G		4	8		1,575	335	24	1,934	2,350
0800	For frames without bar stops, light sections, deduct						15%				
0900	Heavy sections, deduct						10%				

08 13 Metal Doors

08 13 13 – Hollow Metal Doors

08 13 13.13 Standard Hollow Metal Doors			Crew	Daily Output	Labor-Hours	Unit	Material	Labor	Equipment	Total	Total Incl O&P
0010	**STANDARD HOLLOW METAL DOORS** R081313-20										
0015	Flush, full panel, hollow core										
0017	When noted doors are prepared but do not include glass or louvers										
0020	1-3/8" thick, 20 ga., 2'-0" x 6'-8"	G	2 Carp	20	.800	Ea.	345	31		376	430
0040	2'-8" x 6'-8"	G		18	.889		355	34.50		389.50	445
0060	3'-0" x 6'-8"	G		17	.941		360	36.50		396.50	455
0100	3'-0" x 7'-0"	G		17	.941		370	36.50		406.50	465
0120	For vision lite, add						108			108	119
0140	For narrow lite, add						114			114	126
0320	Half glass, 20 ga., 2'-0" x 6'-8"	G	2 Carp	20	.800		610	31		641	720
0340	2'-8" x 6'-8"	G		18	.889		620	34.50		654.50	735
0360	3'-0" x 6'-8"	G		17	.941		610	36.50		646.50	730
0400	3'-0" x 7'-0"	G		17	.941		400	36.50		436.50	505
0410	1-3/8" thick, 18 ga., 2'-0" x 6'-8"	G		20	.800		410	31		441	500
0420	3'-0" x 6'-8"	G		17	.941		415	36.50		451.50	515
0425	3'-0" x 7'-0"	G		17	.941		435	36.50		471.50	540
0450	For vision lite, add						108			108	119
0452	For narrow lite, add						114			114	126
0460	Half glass, 18 ga., 2'-0" x 6'-8"	G	2 Carp	20	.800		570	31		601	675
0465	2'-8" x 6'-8"	G		18	.889		590	34.50		624.50	700
0470	3'-0" x 6'-8"	G		17	.941		575	36.50		611.50	695
0475	3'-0" x 7'-0"	G		17	.941		595	36.50		631.50	715
0500	Hollow core, 1-3/4" thick, full panel, 20 ga., 2'-8" x 6'-8"	G		18	.889		445	34.50		479.50	545
0520	3'-0" x 6'-8"	G		17	.941		440	36.50		476.50	545
0640	3'-0" x 7'-0"	G		17	.941		470	36.50		506.50	575
0680	4'-0" x 7'-0"	G		15	1.067		640	41.50		681.50	775
0700	4'-0" x 8'-0"	G		13	1.231		745	47.50		792.50	895
1000	18 ga., 2'-8" x 6'-8"	G		17	.941		490	36.50		526.50	600
1020	3'-0" x 6'-8"	G		16	1		490	39		529	605
1120	3'-0" x 7'-0"	G		17	.941		495	36.50		531.50	605
1180	4'-0" x 7'-0"	G		14	1.143		765	44.50		809.50	920
1200	4'-0" x 8'-0"	G		17	.941		870	36.50		906.50	1,025
1212	For vision lite, add						108			108	119
1214	For narrow lite, add						114			114	126
1230	Half glass, 20 ga., 2'-8" x 6'-8"	G	2 Carp	20	.800		585	31		616	695
1240	3'-0" x 6'-8"	G		18	.889		585	34.50		619.50	695
1260	3'-0" x 7'-0"	G		18	.889		600	34.50		634.50	715
1280	Embossed panel, 1-3/4" thick, poly core, 20 ga., 3'-0" x 7'-0"	G		18	.889		420	34.50		454.50	520
1290	Half glass, 1-3/4" thick, poly core, 20 ga., 3'-0" x 7'-0"	G		18	.889		620	34.50		654.50	735
1320	18 ga., 2'-8" x 6'-8"	G		18	.889		645	34.50		679.50	765
1340	3'-0" x 6'-8"	G		17	.941		640	36.50		676.50	760

08 13 13 – Hollow Metal Doors

08 13 13.13 Standard Hollow Metal Doors		Crew	Daily Output	Labor-Hours	Unit	Material	2019 Bare Costs Labor	Equipment	Total	Total Incl O&P
1360	3'-0" x 7'-0" G	2 Carp	17	.941	Ea.	650	36.50		686.50	775
1380	4'-0" x 7'-0" G		15	1.067		800	41.50		841.50	950
1400	4'-0" x 8'-0" G		14	1.143		900	44.50		944.50	1,075
1500	Flush full panel, 16 ga., steel hollow core									
1520	2'-0" x 6'-8" G	2 Carp	20	.800	Ea.	575	31		606	680
1530	2'-8" x 6'-8" G		20	.800		575	31		606	680
1540	3'-0" x 6'-8" G		20	.800		555	31		586	660
1560	2'-8" x 7'-0" G		18	.889		585	34.50		619.50	700
1570	3'-0" x 7'-0" G		18	.889		565	34.50		599.50	680
1580	3'-6" x 7'-0" G		18	.889		670	34.50		704.50	795
1590	4'-0" x 7'-0" G		18	.889		735	34.50		769.50	865
1600	2'-8" x 8'-0" G		18	.889		720	34.50		754.50	845
1620	3'-0" x 8'-0" G		18	.889		735	34.50		769.50	860
1630	3'-6" x 8'-0" G		18	.889		815	34.50		849.50	950
1640	4'-0" x 8'-0" G		18	.889		860	34.50		894.50	1,000
1650	1-13/16", 14 ga., 2'-8" x 7'-0" G		10	1.600		1,150	62		1,212	1,350
1670	3'-0" x 7'-0" G		10	1.600		1,100	62		1,162	1,325
1690	3'-6" x 7'-0" G		10	1.600		1,225	62		1,287	1,450
1700	4'-0" x 7'-0" G		10	1.600		1,300	62		1,362	1,525
1720	Insulated, 1-3/4" thick, full panel, 18 ga., 3'-0" x 6'-8" G		15	1.067		485	41.50		526.50	605
1740	2'-8" x 7'-0" G		16	1		505	39		544	625
1760	3'-0" x 7'-0" G		15	1.067		500	41.50		541.50	620
1800	4'-0" x 8'-0" G		13	1.231		775	47.50		822.50	935
1820	Half glass, 18 ga., 3'-0" x 6'-8" G		16	1		635	39		674	765
1840	2'-8" x 7'-0" G		17	.941		665	36.50		701.50	795
1860	3'-0" x 7'-0" G		16	1		695	39		734	830
1900	4'-0" x 8'-0" G		14	1.143		560	44.50		604.50	690
2000	For vision lite, add					108			108	119
2010	For narrow lite, add					114			114	126
8100	For bottom louver, add					226			226	249
8110	For baked enamel finish, add					30%	15%			
8120	For galvanizing, add					20%				
8190	Commercial door, flush steel, fire door see section 08 13 13 15									
8270	For dutch door with shelf, add to standard door				Ea.	300			300	330

08 13 13.15 Metal Fire Doors		Crew	Daily Output	Labor-Hours	Unit	Material	Labor	Equipment	Total	Total Incl O&P
0010	**METAL FIRE DOORS** R081313-20									
0015	Steel, flush, "B" label, 90 minutes									
0020	Full panel, 20 ga., 2'-0" x 6'-8"	2 Carp	20	.800	Ea.	425	31		456	520
0040	2'-8" x 6'-8"		18	.889		440	34.50		474.50	540
0060	3'-0" x 6'-8"		17	.941		440	36.50		476.50	545
0080	3'-0" x 7'-0"		17	.941		460	36.50		496.50	565
0140	18 ga., 3'-0" x 6'-8"		16	1		505	39		544	620
0160	2'-8" x 7'-0"		17	.941		535	36.50		571.50	650
0180	3'-0" x 7'-0"		16	1		520	39		559	635
0200	4'-0" x 7'-0"		15	1.067		655	41.50		696.50	790
0220	For "A" label, 3 hour, 18 ga., use same price as "B" label									
0240	For vision lite, add				Ea.	165			165	181
0300	Full panel, 16 ga., 2'-0" x 6'-8"	2 Carp	20	.800		315	31		346	400
0310	2'-8" x 6'-8"		18	.889		555	34.50		589.50	665
0320	3'-0" x 6'-8"		17	.941		445	36.50		481.50	550
0350	2'-8" x 7'-0"		17	.941		575	36.50		611.50	690
0360	3'-0" x 7'-0"		16	1		555	39		594	680

08 13 13 – Hollow Metal Doors

08 13 13.15 Metal Fire Doors		Crew	Daily Output	Labor-Hours	Unit	2019 Bare Costs Material	Labor	Equipment	Total	Total Incl O&P
0370	4'-0" x 7'-0"	2 Carp	15	1.067	Ea.	720	41.50		761.50	865
0520	Flush, "B" label, 90 minutes, egress core, 20 ga., 2'-0" x 6'-8"		18	.889		670	34.50		704.50	795
0540	2'-8" x 6'-8"		17	.941		675	36.50		711.50	800
0560	3'-0" x 6'-8"		16	1		675	39		714	810
0580	3'-0" x 7'-0"		16	1		705	39		744	840
0640	Flush, "A" label, 3 hour, egress core, 18 ga., 3'-0" x 6'-8"		15	1.067		735	41.50		776.50	875
0660	2'-8" x 7'-0"		16	1		770	39		809	910
0680	3'-0" x 7'-0"		15	1.067		755	41.50		796.50	900
0700	4'-0" x 7'-0"	↓	14	1.143	↓	915	44.50		959.50	1,075

08 13 13.20 Residential Steel Doors		Crew	Daily Output	Labor-Hours	Unit	Material	Labor	Equipment	Total	Total Incl O&P
0010	**RESIDENTIAL STEEL DOORS**									
2510	Bi-passing closet, incl. hardware, no frame or trim incl.									
2511	Mirrored, metal frame, 4'-0" x 6'-8"	2 Carp	10	1.600	Opng.	265	62		327	395
2512	5'-0" wide		10	1.600		267	62		329	395
2513	6'-0" wide		10	1.600		295	62		357	425
2514	7'-0" wide		9	1.778		287	69		356	430
2515	8'-0" wide		9	1.778		480	69		549	640
2611	Mirrored, metal, 4'-0" x 8'-0"		10	1.600		315	62		377	450
2612	5'-0" wide		10	1.600		325	62		387	455
2613	6'-0" wide		10	1.600		330	62		392	465
2614	7'-0" wide		9	1.778		355	69		424	505
2615	8'-0" wide	↓	9	1.778	↓	445	69		514	605

08 13 13.25 Doors Hollow Metal			Crew	Daily Output	Labor-Hours	Unit	Material	Labor	Equipment	Total	Total Incl O&P
0010	**DOORS HOLLOW METAL**										
0500	Exterior, commercial, flush, 20 ga., 1-3/4" x 7'-0" x 2'-6" wide	G	2 Carp	15	1.067	Ea.	295	41.50		336.50	395
0530	2'-8" wide	G		15	1.067		293	41.50		334.50	395
0560	3'-0" wide	G		14	1.143		294	44.50		338.50	400
1000	18 ga., 1-3/4" x 7'-0" x 2'-6" wide	G		15	1.067		320	41.50		361.50	425
1030	2'-8" wide	G		15	1.067		325	41.50		366.50	430
1060	3'-0" wide	G		14	1.143		315	44.50		359.50	425
1500	16 ga., 1-3/4" x 7'-0" x 2'-6" wide	G		15	1.067		375	41.50		416.50	485
1530	2'-8" wide	G		15	1.067		375	41.50		416.50	485
1560	3'-0" wide	G		14	1.143		375	44.50		419.50	490
1590	3'-6" wide	G		14	1.143		715	44.50		759.50	860
2900	Fire door, "A" label, 18 gauge, 1-3/4" x 2'-6" x 7'-0"	G		15	1.067		805	41.50		846.50	955
2930	2'-8" wide	G		15	1.067		700	41.50		741.50	840
2960	3'-0" wide	G		14	1.143		725	44.50		769.50	875
2990	3'-6" wide	G		14	1.143		475	44.50		519.50	600
3100	"B" label, 2'-6" wide	G		15	1.067		585	41.50		626.50	715
3130	2'-8" wide	G		15	1.067		585	41.50		626.50	715
3160	3'-0" wide	G	↓	14	1.143	↓	590	44.50		634.50	720

08 13 16 – Aluminum Doors

08 13 16.10 Commercial Aluminum Doors		Crew	Daily Output	Labor-Hours	Unit	Material	Labor	Equipment	Total	Total Incl O&P
0010	**COMMERCIAL ALUMINUM DOORS**, flush, no glazing									
5000	Flush panel doors, pair of 2'-6" x 7'-0"	2 Sswk	2	8	Pr.	1,325	330		1,655	2,025
5050	3'-0" x 7'-0", single		2.50	6.400	Ea.	780	265		1,045	1,325
5100	Pair of 3'-0" x 7'-0"		2	8	Pr.	1,375	330		1,705	2,075
5150	3'-6" x 7'-0", single	↓	2.50	6.400	Ea.	905	265		1,170	1,450

08 14 Wood Doors

08 14 13 – Carved Wood Doors

08 14 13.10 Types of Wood Doors, Carved		Crew	Daily Output	Labor-Hours	Unit	Material	2019 Bare Costs Labor	Equipment	Total	Total Incl O&P
0010	**TYPES OF WOOD DOORS, CARVED**									
3000	Solid wood, 1-3/4" thick stile and rail									
3020	Mahogany, 3'-0" x 7'-0", six panel	2 Carp	14	1.143	Ea.	1,150	44.50		1,194.50	1,350
3030	With two lites		10	1.600		3,275	62		3,337	3,700
3040	3'-6" x 8'-0", six panel		10	1.600		1,725	62		1,787	2,000
3050	With two lites		8	2		3,000	77.50		3,077.50	3,425
3100	Pine, 3'-0" x 7'-0", six panel		14	1.143		550	44.50		594.50	680
3110	With two lites		10	1.600		870	62		932	1,050
3120	3'-6" x 8'-0", six panel		10	1.600		1,075	62		1,137	1,300
3130	With two lites		8	2		2,050	77.50		2,127.50	2,375
3200	Red oak, 3'-0" x 7'-0", six panel		14	1.143		1,400	44.50		1,444.50	1,600
3210	With two lites		10	1.600		2,600	62		2,662	2,950
3220	3'-6" x 8'-0", six panel		10	1.600		2,925	62		2,987	3,300
3230	With two lites		8	2		3,475	77.50		3,552.50	3,950
4000	Hand carved door, mahogany									
4020	3'-0" x 7'-0", simple design	2 Carp	14	1.143	Ea.	1,825	44.50		1,869.50	2,075
4030	Intricate design		11	1.455		3,800	56.50		3,856.50	4,275
4040	3'-6" x 8'-0", simple design		10	1.600		2,900	62		2,962	3,275
4050	Intricate design		8	2		3,800	77.50		3,877.50	4,300
4400	For custom finish, add					595			595	655
4600	Side light, mahogany, 7'-0" x 1'-6" wide, 4 lites	2 Carp	18	.889		1,150	34.50		1,184.50	1,300
4610	6 lites		14	1.143		2,625	44.50		2,669.50	2,950
4620	8'-0" x 1'-6" wide, 4 lites		14	1.143		2,200	44.50		2,244.50	2,500
4630	6 lites		10	1.600		2,150	62		2,212	2,475
4640	Side light, oak, 7'-0" x 1'-6" wide, 4 lites		18	.889		1,250	34.50		1,284.50	1,425
4650	6 lites		14	1.143		2,175	44.50		2,219.50	2,475
4660	8'-0" x 1'-6" wide, 4 lites		14	1.143		1,300	44.50		1,344.50	1,525
4670	6 lites		10	1.600		2,175	62		2,237	2,500

08 14 16 – Flush Wood Doors

08 14 16.09 Smooth Wood Doors		Crew	Daily Output	Labor-Hours	Unit	Material	Labor	Equipment	Total	Total Incl O&P
0010	**SMOOTH WOOD DOORS**									
0085	3'-0" x 6'-8"	1 Carp	16	.500	Ea.	57	19.40		76.40	94.50
0108	3'-0" x 7'-0"	2 Carp	16	1	"	214	39		253	299
0112	Pair of 3'-0" x 7'-0"		9	1.778	Pr.	140	69		209	268
0206	2'-6" x 7'-0"		16	1	Ea.	130	39		169	207
0210	3'-0" x 7'-0"		16	1	"	165	39		204	246
0214	Pair of 3'-0" x 7'-0"		9	1.778	Pr.	272	69		341	415
2205	1-3/4", 3'-0" x 7'-0"		13	1.231	Ea.	202	47.50		249.50	300
4045	3'-0" x 8'-0"	1 Carp	8	1		495	39		534	610
5000	Wood doors, for vision lite, add					108			108	119
5010	Wood doors, for narrow lite, add					114			114	126
5015	Wood doors, for bottom (or top) louver, add					226			226	249

08 14 16.10 Wood Doors Decorator		Crew	Daily Output	Labor-Hours	Unit	Material	Labor	Equipment	Total	Total Incl O&P
0010	**WOOD DOORS DECORATOR**									
1800	Exterior, flush, solid wood core, birch 1-3/4" x 2'-6" x 7'-0"	2 Carp	15	1.067	Ea.	325	41.50		366.50	430
1820	2'-8" wide		15	1.067		370	41.50		411.50	475
1840	3'-0" wide		14	1.143		370	44.50		414.50	485
1900	Oak faced, 1-3/4" x 2'-6" x 7'-0"		15	1.067		215	41.50		256.50	305
1920	2'-8" wide		15	1.067		465	41.50		506.50	585
1940	3'-0" wide		14	1.143		480	44.50		524.50	605
2100	Walnut faced, 1-3/4" x 2'-6" x 7'-0"		15	1.067		415	41.50		456.50	525
2120	2'-8" wide		15	1.067		430	41.50		471.50	540

08 14 Wood Doors

08 14 16 – Flush Wood Doors

08 14 16.10 Wood Doors Decorator		Crew	Daily Output	Labor-Hours	Unit	Material	Labor	Equipment	Total	Total Incl O&P
						2019 Bare Costs				
2140	3'-0" wide	2 Carp	14	1.143	Ea.	455	44.50		499.50	575

08 14 33 – Stile and Rail Wood Doors

08 14 33.10 Wood Doors Paneled		Crew	Daily Output	Labor-Hours	Unit	Material	Labor	Equipment	Total	Total Incl O&P
0010	**WOOD DOORS PANELED**									
0020	Interior, six panel, hollow core, 1-3/8" thick									
0040	Molded hardboard, 2'-0" x 6'-8"	2 Carp	17	.941	Ea.	64	36.50		100.50	131
0060	2'-6" x 6'-8"		17	.941		66.50	36.50		103	134
0070	2'-8" x 6'-8"		17	.941		69.50	36.50		106	137
0080	3'-0" x 6'-8"		17	.941		76.50	36.50		113	145
0140	Embossed print, molded hardboard, 2'-0" x 6'-8"		17	.941		66.50	36.50		103	134
0160	2'-6" x 6'-8"		17	.941		66.50	36.50		103	134
0180	3'-0" x 6'-8"		17	.941		76.50	36.50		113	145
0540	Six panel, solid, 1-3/8" thick, pine, 2'-0" x 6'-8"		15	1.067		161	41.50		202.50	246
0560	2'-6" x 6'-8"		14	1.143		159	44.50		203.50	248
0580	3'-0" x 6'-8"		13	1.231		154	47.50		201.50	249
1020	Two panel, bored rail, solid, 1-3/8" thick, pine, 1'-6" x 6'-8"		16	1		253	39		292	340
1040	2'-0" x 6'-8"		15	1.067		330	41.50		371.50	430
1060	2'-6" x 6'-8"		14	1.143		375	44.50		419.50	490
1340	Two panel, solid, 1-3/8" thick, fir, 2'-0" x 6'-8"		15	1.067		166	41.50		207.50	252
1360	2'-6" x 6'-8"		14	1.143		225	44.50		269.50	320
1380	3'-0" x 6'-8"		13	1.231		405	47.50		452.50	525
1740	Five panel, solid, 1-3/8" thick, fir, 2'-0" x 6'-8"		15	1.067		300	41.50		341.50	400
1760	2'-6" x 6'-8"		14	1.143		395	44.50		439.50	510
1780	3'-0" x 6'-8"		13	1.231		395	47.50		442.50	515

08 14 33.20 Wood Doors Residential		Crew	Daily Output	Labor-Hours	Unit	Material	Labor	Equipment	Total	Total Incl O&P
0010	**WOOD DOORS RESIDENTIAL**									
1420	6'-8" x 3'-0" wide, fir	2 Carp	16	1	Ea.	530	39		569	645
1800	Lauan, solid core, 1-3/4" x 7'-0" x 2'-4" wide		16	1		144	39		183	222
1810	2'-6" wide		15	1.067		150	41.50		191.50	235
1820	2'-8" wide		9	1.778		168	69		237	298
1830	3'-0" wide		16	1		184	39		223	266
1840	3'-4" wide		16	1		270	39		309	360
1850	Pair of 3'-0" wide		15	1.067	Pr.	365	41.50		406.50	475
1930	For dutch door with shelf, add					140%				
2920	Hardboard, primed 7'-0" x 4'-0" wide	2 Carp	12	1.333	Ea.	205	51.50		256.50	310
2930	6'-0" wide		10	1.600		199	62		261	320
3100	7'-0" x 3'-0" wide		12	1.333		310	51.50		361.50	425
3120	6'-0" wide		10	1.600		450	62		512	595
3140	For oak or ash, add					125%				
9000	Passage doors, flush, no frame, birch, solid core, 1-3/8" x 2'-4" x 7'-0"	2 Carp	16	1	Ea.	131	39		170	209
9020	2'-8" wide		16	1		139	39		178	217
9040	3'-0" wide		16	1		154	39		193	234
9060	3'-4" wide		15	1.067		292	41.50		333.50	390
9080	Pair of 3'-0" wide		9	1.778	Pr.	305	69		374	450
9100	Lauan, solid core, 1-3/8" x 7'-0" x 2'-4" wide		16	1	Ea.	172	39		211	253
9120	2'-8" wide		16	1		154	39		193	233
9140	3'-0" wide		16	1		203	39		242	287
9160	3'-4" wide		15	1.067		216	41.50		257.50	305
9180	Pair of 3'-0" wide		9	1.778	Pr.	405	69		474	560
9200	Hardboard, solid core, 1-3/8" x 7'-0" x 2'-4" wide		16	1	Ea.	175	39		214	257
9220	2'-8" wide		16	1		182	39		221	265
9240	3'-0" wide		16	1		188	39		227	271

08 14 Wood Doors

08 14 33 – Stile and Rail Wood Doors

08 14 33.20 Wood Doors Residential		Crew	Daily Output	Labor-Hours	Unit	Material	Labor	Equipment	Total	Total Incl O&P
						2019 Bare Costs				
9260	3'-4" wide	2 Carp	15	1.067	Ea.	360	41.50		401.50	465

08 14 35 – Torrified Doors

08 14 35.10 Torrified Exterior Doors

		Crew	Daily Output	Labor-Hours	Unit	Material	Labor	Equipment	Total	Total Incl O&P
0010	**TORRIFIED EXTERIOR DOORS**									
0020	Wood doors made from torrified wood, exterior									
0030	All doors require a finish be applied, all glass is insulated									
0040	All doors require pilot holes for all fasteners									
0100	6 panel, paint grade poplar, 1-3/4" x 3'-0" x 6'-8"	2 Carp	12	1.333	Ea.	1,375	51.50		1,426.50	1,575
0120	Half glass 3'-0" x 6'-8"	"	12	1.333		1,475	51.50		1,526.50	1,700
0200	Side lite, full glass, 1-3/4" x 1'-2" x 6'-8"					970			970	1,075
0220	Side lite, half glass, 1-3/4" x 1'-2" x 6'-8"					950			950	1,050
0300	Raised face, 2 panel, paint grade poplar, 1-3/4" x 3'-0" x 7'-0"	2 Carp	12	1.333		1,375	51.50		1,426.50	1,575
0320	Side lite, raised face, half glass, 1-3/4" x 1'-2" x 7'-0"					1,100			1,100	1,225
0500	6 panel, Fir, 1-3/4" x 3'-0" x 6'-8"	2 Carp	12	1.333		1,950	51.50		2,001.50	2,225
0520	Half glass 3'-0" x 6'-8"	"	12	1.333		1,925	51.50		1,976.50	2,200
0600	Side lite, full glass, 1-3/4" x 1'-2" x 6'-8"					980			980	1,075
0620	Side lite, half glass, 1-3/4" x 1'-2" x 6'-8"					1,000			1,000	1,100
0700	6 panel, Mahogany, 1-3/4" x 3'-0" x 6'-8"	2 Carp	12	1.333		2,175	51.50		2,226.50	2,475
0800	Side lite, full glass, 1-3/4" x 1'-2" x 6'-8"					1,100			1,100	1,200
0820	Side lite, half glass, 1-3/4" x 1'-2" x 6'-8"				↓	1,100			1,100	1,200

08 14 40 – Interior Cafe Doors

08 14 40.10 Cafe Style Doors

		Crew	Daily Output	Labor-Hours	Unit	Material	Labor	Equipment	Total	Total Incl O&P
0010	**CAFE STYLE DOORS**									
6520	Interior cafe doors, 2'-6" opening, stock, panel pine	2 Carp	16	1	Ea.	415	39		454	520
6540	3'-0" opening	"	16	1	"	460	39		499	570
6550	Louvered pine									
6560	2'-6" opening	2 Carp	16	1	Ea.	335	39		374	430
8000	3'-0" opening		16	1		360	39		399	460
8010	2'-6" opening, hardwood		16	1		365	39		404	465
8020	3'-0" opening	↓	16	1	↓	420	39		459	525

08 16 Composite Doors

08 16 13 – Fiberglass Doors

08 16 13.10 Entrance Doors, Fibrous Glass

			Crew	Daily Output	Labor-Hours	Unit	Material	Labor	Equipment	Total	Total Incl O&P
0010	**ENTRANCE DOORS, FIBROUS GLASS**										
0020	Exterior, fiberglass, door, 2'-8" wide x 6'-8" high	G	2 Carp	15	1.067	Ea.	278	41.50		319.50	375
0040	3'-0" wide x 6'-8" high	G		15	1.067		280	41.50		321.50	380
0060	3'-0" wide x 7'-0" high	G		15	1.067		500	41.50		541.50	620
0080	3'-0" wide x 6'-8" high, with two lites	G		15	1.067		355	41.50		396.50	460
0100	3'-0" wide x 8'-0" high, with two lites	G		15	1.067		545	41.50		586.50	670
0110	Half glass, 3'-0" wide x 6'-8" high	G		15	1.067		465	41.50		506.50	580
0120	3'-0" wide x 6'-8" high, low E	G		15	1.067		500	41.50		541.50	620
0130	3'-0" wide x 8'-0" high	G		15	1.067		600	41.50		641.50	730
0140	3'-0" wide x 8'-0" high, low E	G	↓	15	1.067		680	41.50		721.50	815
0150	Side lights, 1'-0" wide x 6'-8" high	G					291			291	320
0160	1'-0" wide x 6'-8" high, low E	G					298			298	325
0180	1'-0" wide x 6'-8" high, full glass	G					340			340	375
0190	1'-0" wide x 6'-8" high, low E	G				↓	380			380	415

08 16 Composite Doors

08 16 14 – French Doors

08 16 14.10 Exterior Doors With Glass Lites		Crew	Daily Output	Labor-Hours	Unit	Material	2019 Bare Costs Labor	Equipment	Total	Total Incl O&P
0010	EXTERIOR DOORS WITH GLASS LITES									
0020	French, Fir, 1-3/4", 3'-0" wide x 6'-8" high	2 Carp	12	1.333	Ea.	635	51.50		686.50	785
0025	Double		12	1.333		1,275	51.50		1,326.50	1,475
0030	Maple, 1-3/4", 3'-0" wide x 6'-8" high		12	1.333		715	51.50		766.50	870
0035	Double		12	1.333		1,425	51.50		1,476.50	1,650
0040	Cherry, 1-3/4", 3'-0" wide x 6'-8" high		12	1.333		835	51.50		886.50	1,000
0045	Double		12	1.333		1,675	51.50		1,726.50	1,900
0100	Mahogany, 1-3/4", 3'-0" wide x 8'-0" high		10	1.600		855	62		917	1,050
0105	Double		10	1.600		1,725	62		1,787	1,975
0110	Fir, 1-3/4", 3'-0" wide x 8'-0" high		10	1.600		1,275	62		1,337	1,500
0115	Double		10	1.600		2,575	62		2,637	2,925
0120	Oak, 1-3/4", 3'-0" wide x 8'-0" high		10	1.600		1,925	62		1,987	2,225
0125	Double		10	1.600		3,850	62		3,912	4,350

08 17 Integrated Door Opening Assemblies

08 17 13 – Integrated Metal Door Opening Assemblies

08 17 13.10 Hollow Metal Doors and Frames		Crew	Daily Output	Labor-Hours	Unit	Material	Labor	Equipment	Total	Total Incl O&P
0010	HOLLOW METAL DOORS AND FRAMES									
0100	Prehung, flush, 18 ga., 1-3/4" x 6'-8" x 2'-8" wide	1 Carp	16	.500	Ea.	650	19.40		669.40	745
0120	3'-0" wide		15	.533		655	20.50		675.50	755
0130	3'-6" wide		13	.615		715	24		739	830
0140	4'-0" wide		10	.800		760	31		791	885
0300	Double, 1-3/4" x 3'-0" x 6'-8"		5	1.600		1,350	62		1,412	1,575
0350	3'-0" x 8'-0"		4	2		1,350	77.50		1,427.50	1,625

08 17 13.20 Stainless Steel Doors and Frames		Crew	Daily Output	Labor-Hours	Unit	Material	Labor	Equipment	Total	Total Incl O&P
0010	STAINLESS STEEL DOORS AND FRAMES									
0020	Stainless steel (304) prehung 24 ga. 2'-6" x 6'-8" door w/16 g frame G	2 Carp	6	2.667	Ea.	1,825	103		1,928	2,175
0025	2'-8" x 6'-8" G		6	2.667		1,900	103		2,003	2,250
0030	3'-0" x 6'-8" G		6	2.667		1,975	103		2,078	2,350
0040	3'-0" x 7'-0" G		6	2.667		2,050	103		2,153	2,425
0050	4'-0" x 7'-0" G		5	3.200		2,225	124		2,349	2,650
0100	Stainless steel (316) prehung 24 ga. 2'-6" x 6'-8" door w/16 g frame G		6	2.667		2,500	103		2,603	2,925
0110	2'-8" x 6'-8" G		6	2.667		2,325	103		2,428	2,725
0120	3'-0" x 6'-8" G		6	2.667		2,425	103		2,528	2,825
0150	3'-0" x 7'-0" G		6	2.667		2,550	103		2,653	2,975
0160	4'-0" x 7'-0" G		6	2.667		3,175	103		3,278	3,675
0300	Stainless steel (304) prehung 18 ga. 2'-6" x 6'-8" door w/16 g frame G		6	2.667		2,875	103		2,978	3,325
0310	2'-8" x 6'-8" G		6	2.667		2,900	103		3,003	3,375
0320	3'-0" x 6'-8" G		6	2.667		2,950	103		3,053	3,425
0350	3'-0" x 7'-0" G		6	2.667		3,000	103		3,103	3,475
0360	4'-0" x 7'-0" G		5	3.200		3,150	124		3,274	3,650
0500	Stainless steel, prehung door, foam core, 14 ga., 3'-0" x 7'-0" G		5	3.200		3,150	124		3,274	3,650
0600	Stainless steel, prehung double door, foam core, 14 ga., 3'-0" x 7'-0" G		4	4		6,175	155		6,330	7,050

08 17 23 – Integrated Wood Door Opening Assemblies

08 17 23.10 Pre-Hung Doors		Crew	Daily Output	Labor-Hours	Unit	Material	Labor	Equipment	Total	Total Incl O&P
0010	PRE-HUNG DOORS									
0300	Exterior, wood, comb. storm & screen, 6'-9" x 2'-6" wide	2 Carp	15	1.067	Ea.	257	41.50		298.50	350
0320	2'-8" wide		15	1.067		345	41.50		386.50	450
0340	3'-0" wide		15	1.067		325	41.50		366.50	425
0360	For 7'-0" high door, add					41			41	45

08 17 23 – Integrated Wood Door Opening Assemblies

08 17 23.10 Pre-Hung Doors		Crew	Daily Output	Labor-Hours	Unit	2019 Bare Costs Material	Labor	Equipment	Total	Total Incl O&P
1600	Entrance door, flush, birch, solid core									
1620	4-5/8" solid jamb, 1-3/4" x 6'-8" x 2'-8" wide	2 Carp	16	1	Ea.	300	39		339	395
1640	3'-0" wide		16	1		400	39		439	505
1642	5-5/8" jamb		16	1		345	39		384	445
1680	For 7'-0" high door, add					25			25	27.50
2000	Entrance door, colonial, 6 panel pine									
2020	4-5/8" solid jamb, 1-3/4" x 6'-8" x 2'-8" wide	2 Carp	16	1	Ea.	655	39		694	790
2040	3'-0" wide	"	16	1		690	39		729	820
2060	For 7'-0" high door, add					56.50			56.50	62.50
2200	For 5-5/8" solid jamb, add					44.50			44.50	49
2230	French style, exterior, 1 lite, 1-3/4" x 3'-0" x 6'-8"	1 Carp	14	.571		675	22		697	780
2235	9 lites	"	14	.571		710	22		732	815
2245	15 lites	2 Carp	14	1.143		770	44.50		814.50	920
2250	Double, 15 lites, 2'-0" x 6'-8", 4'-0" opening		7	2.286	Pr.	1,300	88.50		1,388.50	1,575
2260	2'-6" x 6'-8", 5'-0" opening		7	2.286		1,425	88.50		1,513.50	1,700
2280	3'-0" x 6'-8", 6'-0" opening		7	2.286		1,600	88.50		1,688.50	1,925
2430	3'-0" x 7'-0", 15 lites		14	1.143	Ea.	1,000	44.50		1,044.50	1,175
2432	Two 3'-0" x 7'-0"		7	2.286	Pr.	2,100	88.50		2,188.50	2,450
2435	3'-0" x 8'-0"		14	1.143	Ea.	1,075	44.50		1,119.50	1,250
2437	Two, 3'-0" x 8'-0"		7	2.286	Pr.	2,225	88.50		2,313.50	2,575
4000	Interior, passage door, 4-5/8" solid jamb									
4400	Lauan, flush, solid core, 1-3/8" x 6'-8" x 2'-6" wide	2 Carp	17	.941	Ea.	195	36.50		231.50	275
4420	2'-8" wide		17	.941		195	36.50		231.50	275
4440	3'-0" wide		16	1		212	39		251	297
4600	Hollow core, 1-3/8" x 6'-8" x 2'-6" wide		17	.941		134	36.50		170.50	209
4620	2'-8" wide		17	.941		137	36.50		173.50	211
4640	3'-0" wide		16	1		140	39		179	218
4700	For 7'-0" high door, add					41.50			41.50	46
5000	Birch, flush, solid core, 1-3/8" x 6'-8" x 2'-6" wide	2 Carp	17	.941		300	36.50		336.50	390
5020	2'-8" wide		17	.941		209	36.50		245.50	291
5040	3'-0" wide		16	1		325	39		364	425
5200	Hollow core, 1-3/8" x 6'-8" x 2'-6" wide		17	.941		247	36.50		283.50	335
5220	2'-8" wide		17	.941		280	36.50		316.50	370
5240	3'-0" wide		16	1		266	39		305	355
5280	For 7'-0" high door, add					35.50			35.50	39
5500	Hardboard paneled, 1-3/8" x 6'-8" x 2'-6" wide	2 Carp	17	.941		152	36.50		188.50	228
5520	2'-8" wide		17	.941		164	36.50		200.50	241
5540	3'-0" wide		16	1		162	39		201	242
6000	Pine paneled, 1-3/8" x 6'-8" x 2'-6" wide		17	.941		275	36.50		311.50	360
6020	2'-8" wide		17	.941		293	36.50		329.50	380
7600	Oak, 6 panel, 1-3/4" x 6'-8" x 3'-0"	1 Carp	17	.471		915	18.25		933.25	1,025
8200	Birch, flush, solid core, 1-3/4" x 6'-8" x 2'-4" wide		17	.471		234	18.25		252.25	288
8220	2'-6" wide		17	.471		227	18.25		245.25	280
8240	2'-8" wide		17	.471		235	18.25		253.25	288
8260	3'-0" wide		16	.500		239	19.40		258.40	295
8280	3'-6" wide		15	.533		375	20.50		395.50	445
8500	Pocket door frame with lauan, flush, hollow core, 1-3/8" x 3'-0" x 6'-8"		17	.471		295	18.25		313.25	355

08 31 Access Doors and Panels

08 31 13 – Access Doors and Frames

08 31 13.10 Types of Framed Access Doors

	08 31 13.10 Types of Framed Access Doors	Crew	Daily Output	Labor-Hours	Unit	2019 Bare Costs Material	Labor	Equipment	Total	Total Incl O&P
0010	TYPES OF FRAMED ACCESS DOORS									
1000	Fire rated door with lock									
1100	Metal, 12″ x 12″	1 Carp	10	.800	Ea.	161	31		192	228
1150	18″ x 18″		9	.889		216	34.50		250.50	295
1200	24″ x 24″		9	.889		345	34.50		379.50	435
1250	24″ x 36″		8	1		350	39		389	450
1300	24″ x 48″		8	1		400	39		439	505
1350	36″ x 36″		7.50	1.067		515	41.50		556.50	635
1400	48″ x 48″		7.50	1.067		670	41.50		711.50	805
1600	Stainless steel, 12″ x 12″		10	.800		340	31		371	425
1650	18″ x 18″		9	.889		375	34.50		409.50	465
1700	24″ x 24″		9	.889		550	34.50		584.50	660
1750	24″ x 36″		8	1		655	39		694	785
2000	Flush door for finishing									
2100	Metal 8″ x 8″	1 Carp	10	.800	Ea.	35.50	31		66.50	90
2150	12″ x 12″	″	10	.800	″	37	31		68	92
3000	Recessed door for acoustic tile									
3100	Metal, 12″ x 12″	1 Carp	4.50	1.778	Ea.	51.50	69		120.50	171
3150	12″ x 24″		4.50	1.778		108	69		177	233
3200	24″ x 24″		4	2		98	77.50		175.50	236
3250	24″ x 36″		4	2		162	77.50		239.50	305
4000	Recessed door for drywall									
4100	Metal 12″ x 12″	1 Carp	6	1.333	Ea.	74.50	51.50		126	168
4150	12″ x 24″		5.50	1.455		119	56.50		175.50	224
4200	24″ x 36″		5	1.600		167	62		229	286
6000	Standard door									
6100	Metal, 8″ x 8″	1 Carp	10	.800	Ea.	28.50	31		59.50	82.50
6150	12″ x 12″		10	.800		31	31		62	85
6200	18″ x 18″		9	.889		35	34.50		69.50	95.50
6250	24″ x 24″		9	.889		50.50	34.50		85	113
6300	24″ x 36″		8	1		87	39		126	160
6350	36″ x 36″		8	1		99.50	39		138.50	173
6500	Stainless steel, 8″ x 8″		10	.800		79.50	31		110.50	139
6550	12″ x 12″		10	.800		98.50	31		129.50	159
6600	18″ x 18″		9	.889		189	34.50		223.50	265
6650	24″ x 24″		9	.889		286	34.50		320.50	370

08 31 13.20 Bulkhead/Cellar Doors

	08 31 13.20 Bulkhead/Cellar Doors	Crew	Daily Output	Labor-Hours	Unit	Material	Labor	Equipment	Total	Total Incl O&P
0010	BULKHEAD/CELLAR DOORS									
0020	Steel, not incl. sides, 44″ x 62″	1 Carp	5.50	1.455	Ea.	655	56.50		711.50	815
0100	52″ x 73″		5.10	1.569		825	61		886	1,000
0500	With sides and foundation plates, 57″ x 45″ x 24″		4.70	1.702		895	66		961	1,100
0600	42″ x 49″ x 51″		4.30	1.860		590	72		662	770

08 31 13.30 Commercial Floor Doors

	08 31 13.30 Commercial Floor Doors	Crew	Daily Output	Labor-Hours	Unit	Material	Labor	Equipment	Total	Total Incl O&P
0010	COMMERCIAL FLOOR DOORS									
0020	Aluminum tile, steel frame, one leaf, 2′ x 2′ opng.	2 Sswk	3.50	4.571	Opng.	625	189		814	1,025
0021	Aluminum tile, steel frame, one leaf, 2′ x 2′ opng.	L-4	3.50	4.571		625	156		781	950
0050	3′-6″ x 3′-6″ opening	2 Sswk	3.50	4.571		1,325	189		1,514	1,800
0051	3′-6″ x 3′-6″ opening	L-4	3.50	4.571		1,325	156		1,481	1,725
0500	Double leaf, 4′ x 4′ opening	2 Sswk	3	5.333		1,625	221		1,846	2,150
0501	Double leaf, 4′ x 4′ opening	L-4	3	5.333		1,625	183		1,808	2,075
0550	5′ x 5′ opening	2 Sswk	3	5.333		2,775	221		2,996	3,450
0551	5′ x 5′ opening	L-4	3	5.333		2,775	183		2,958	3,375

08 31 Access Doors and Panels

08 31 13 – Access Doors and Frames

08 31 13.35 Industrial Floor Doors		Crew	Daily Output	Labor-Hours	Unit	2019 Bare Costs Material	Labor	Equipment	Total	Total Incl O&P
0010	**INDUSTRIAL FLOOR DOORS**									
0021	Steel 300 psf L.L., single leaf, 2' x 2', 175#	L-4	6	2.667	Opng.	755	91.50		846.50	985
0051	3' x 3' opening, 300#		5.50	2.909		1,100	99.50		1,199.50	1,375
0301	Double leaf, 4' x 4' opening, 455#		5	3.200		2,475	110		2,585	2,900
0351	5' x 5' opening, 645#		4.50	3.556		2,975	122		3,097	3,475
1001	Aluminum, 300 psf L.L., single leaf, 2' x 2', 60#		6	2.667		780	91.50		871.50	1,000
1051	3' x 3' opening, 100#		5.50	2.909		935	99.50		1,034.50	1,200
1501	Double leaf, 4' x 4' opening, 160#		5	3.200		2,425	110		2,535	2,850
1551	5' x 5' opening, 235#		4.50	3.556		2,850	122		2,972	3,350
2001	Aluminum, 150 psf L.L., single leaf, 2' x 2', 60#		6	2.667		725	91.50		816.50	950
2051	3' x 3' opening, 95#		5.50	2.909		1,050	99.50		1,149.50	1,325
2501	Double leaf, 4' x 4' opening, 150#		5	3.200		1,475	110		1,585	1,800
2551	5' x 5' opening, 230#	↓	4.50	3.556	↓	1,975	122		2,097	2,375

08 32 Sliding Glass Doors

08 32 13 – Sliding Aluminum-Framed Glass Doors

08 32 13.10 Sliding Aluminum Doors		Crew	Daily Output	Labor-Hours	Unit	Material	Labor	Equipment	Total	Total Incl O&P
0010	**SLIDING ALUMINUM DOORS**									
0350	Aluminum, 5/8" tempered insulated glass, 6' wide									
0400	Premium	2 Carp	4	4	Ea.	1,600	155		1,755	2,025
0450	Economy		4	4		855	155		1,010	1,200
0500	8' wide, premium		3	5.333		1,775	207		1,982	2,300
0550	Economy		3	5.333		1,525	207		1,732	2,025
0600	12' wide, premium		2.50	6.400		3,100	248		3,348	3,825
0650	Economy		2.50	6.400		1,550	248		1,798	2,100
4000	Aluminum, baked on enamel, temp glass, 6'-8" x 10'-0" wide		4	4		1,125	155		1,280	1,500
4020	Insulating glass, 6'-8" x 6'-0" wide		4	4		975	155		1,130	1,325
4040	8'-0" wide		3	5.333		1,150	207		1,357	1,625
4060	10'-0" wide		2	8		1,450	310		1,760	2,100
4080	Anodized, temp glass, 6'-8" x 6'-0" wide		4	4		470	155		625	775
4100	8'-0" wide		3	5.333		595	207		802	995
4120	10'-0" wide	↓	2	8	↓	695	310		1,005	1,275
5000	Aluminum sliding glass door system									
5010	Sliding door 4' wide opening single side	2 Carp	2	8	Ea.	6,200	310		6,510	7,325
5015	8' wide opening single side		2	8		9,325	310		9,635	10,700
5020	Telescoping glass door system, 4' wide opening biparting		2	8		5,275	310		5,585	6,300
5025	8' wide opening biparting		2	8		6,200	310		6,510	7,325
5030	Folding glass door, 4' wide opening biparting		2	8		8,275	310		8,585	9,600
5035	8' wide opening biparting		2	8		10,300	310		10,610	11,900
5040	ICU-CCU sliding telescoping glass door, 4' x 7', single side opening		2	8		3,150	310		3,460	3,975
5045	8' x 7', single side opening	↓	2	8		4,800	310		5,110	5,775
7000	Electric swing door operator and control, single door w/sensors	1 Carp	4	2		2,875	77.50		2,952.50	3,275
7005	Double door w/sensors		2	4		5,775	155		5,930	6,600
7010	Electric folding door operator and control, single door w/sensors		4	2		6,750	77.50		6,827.50	7,550
7015	Bi-folding door		4	2		7,975	77.50		8,052.50	8,900
7020	Electric swing door operator and control, single door	↓	4	2	↓	2,875	77.50		2,952.50	3,300

08 32 Sliding Glass Doors

08 32 19 – Sliding Wood-Framed Glass Doors

08 32 19.10 Sliding Wood Doors		Crew	Daily Output	Labor-Hours	Unit	Material	2019 Bare Costs Labor	Equipment	Total	Total Incl O&P
0010	SLIDING WOOD DOORS									
0020	Wood, tempered insul. glass, 6′ wide, premium	2 Carp	4	4	Ea.	1,500	155		1,655	1,900
0100	Economy		4	4		1,250	155		1,405	1,625
0150	8′ wide, wood, premium		3	5.333		1,925	207		2,132	2,450
0200	Economy		3	5.333		1,550	207		1,757	2,075
0235	10′ wide, wood, premium		2.50	6.400		2,750	248		2,998	3,425
0240	Economy		2.50	6.400		2,350	248		2,598	2,975
0250	12′ wide, wood, premium		2.50	6.400		3,225	248		3,473	3,925
0300	Economy		2.50	6.400		2,575	248		2,823	3,225

08 32 19.15 Sliding Glass Vinyl-Clad Wood Doors

		Crew	Daily Output	Labor-Hours	Unit	Material	Labor	Equipment	Total	Total Incl O&P
0010	SLIDING GLASS VINYL-CLAD WOOD DOORS									
0020	Glass, sliding vinyl-clad, insul. glass, 6′-0″ x 6′-8″ G	2 Carp	4	4	Opng.	1,575	155		1,730	1,975
0025	6′-0″ x 6′-10″ high G		4	4		1,750	155		1,905	2,175
0030	6′-0″ x 8′-0″ high G		4	4		2,125	155		2,280	2,600
0050	5′-0″ x 6′-8″ high G		4	4		1,575	155		1,730	2,000
0100	8′-0″ x 6′-10″ high G		4	4		2,100	155		2,255	2,550
0150	8′-0″ x 8′-0″ high G		4	4		2,400	155		2,555	2,900
0500	4 leaf, 9′-0″ x 6′-10″ high G		3	5.333		3,500	207		3,707	4,200
0550	9′-0″ x 8′-0″ high G		3	5.333		3,925	207		4,132	4,675
0600	12′-0″ x 6′-10″ high G		3	5.333		4,200	207		4,407	4,975
0650	12′-0″ x 8′-0″ high G		3	5.333		4,200	207		4,407	4,975

08 33 Coiling Doors and Grilles

08 33 13 – Coiling Counter Doors

08 33 13.10 Counter Doors, Coiling Type

		Crew	Daily Output	Labor-Hours	Unit	Material	Labor	Equipment	Total	Total Incl O&P
0010	COUNTER DOORS, COILING TYPE									
0020	Manual, incl. frame and hardware, galv. stl., 4′ roll-up, 6′ long	2 Carp	2	8	Opng.	1,325	310		1,635	1,950
0300	Galvanized steel, UL label		1.80	8.889		1,325	345		1,670	2,025
0600	Stainless steel, 4′ high roll-up, 6′ long		2	8		2,200	310		2,510	2,925
0700	10′ long		1.80	8.889		2,650	345		2,995	3,475
2000	Aluminum, 4′ high, 4′ long		2.20	7.273		1,675	282		1,957	2,325
2020	6′ long		2	8		2,000	310		2,310	2,700
2040	8′ long		1.90	8.421		2,300	325		2,625	3,100
2060	10′ long		1.80	8.889		2,300	345		2,645	3,100
2080	14′ long		1.40	11.429		2,950	445		3,395	3,975
2100	6′ high, 4′ long		2	8		2,050	310		2,360	2,750
2120	6′ long		1.60	10		1,800	390		2,190	2,625
2140	10′ long		1.40	11.429		2,375	445		2,820	3,325

08 33 16 – Coiling Counter Grilles

08 33 16.10 Coiling Grilles

		Crew	Daily Output	Labor-Hours	Unit	Material	Labor	Equipment	Total	Total Incl O&P
0010	COILING GRILLES									
2020	Aluminum, manual operated, mill finish	2 Sswk	82	.195	S.F.	30	8.10		38.10	47
2040	Bronze anodized		82	.195	″	46	8.10		54.10	65
2060	Steel, manual operated, 10′ x 10′ high		1	16	Opng.	2,675	665		3,340	4,075
2080	15′ x 8′ high		.80	20	″	2,975	830		3,805	4,725
3000	For safety edge bottom bar, electric, add				L.F.	50			50	55.50
8000	For motor operation, add	2 Sswk	5	3.200	Opng.	1,300	133		1,433	1,650

08 33 23.10 Coiling Service Doors		Crew	Daily Output	Labor-Hours	Unit	Material	2019 Bare Costs Labor	Equipment	Total	Total Incl O&P
0010	**COILING SERVICE DOORS** Steel, manual, 20 ga., incl. hardware									
0051	8′ x 8′ high	L-4	1.60	10	Ea.	715	340		1,055	1,350
0101	10′ x 10′ high	"	1.40	11.429		1,100	390		1,490	1,850
0130	12′ x 12′ high, standard	2 Sswk	1.20	13.333		2,125	555		2,680	3,300
0160	10′ x 20′ high, standard	"	.50	32		2,600	1,325		3,925	5,150
0201	20′ x 10′ high	L-4	1	16		3,000	550		3,550	4,225
0301	12′ x 12′ high	"	1.20	13.333		1,175	455		1,630	2,050
0303	Doors, rolling service, steel, manual, 20ga., 12′ x 12′, incl. hardware	2 Sswk	.92	17.391		1,175	720		1,895	2,550
0401	20′ x 12′ high	L-4	.90	17.778		2,075	610		2,685	3,300
0420	14′ x 12′ high	2 Sswk	.80	20		2,850	830		3,680	4,575
0430	14′ x 13′ high	"	.80	20		3,025	830		3,855	4,775
0501	14′ x 14′ high	L-4	.80	20		2,975	685		3,660	4,425
0540	14′ x 20′ high	2 Sswk	.56	28.571	Opng.	3,700	1,175		4,875	6,125
0550	18′ x 18′ high	"	.60	26.667	"	2,850	1,100		3,950	5,050
0601	20′ x 16′ high	L-4	.60	26.667	Ea.	3,675	915		4,590	5,575
0700	10′ x 20′ high	2 Sswk	.50	32	"	2,600	1,325		3,925	5,150
0750	16′ x 24′ high	"	.48	33.333	Opng.	5,150	1,375		6,525	8,075
1001	12′ x 12′, crank operated, crank on door side	L-4	.80	20	Ea.	1,750	685		2,435	3,075
1101	Crank thru wall	"	.70	22.857		2,050	780		2,830	3,550
1300	For vision panel, add					350			350	385
1600	3′ x 7′ pass door within rolling steel door, new construction					1,725			1,725	1,900
1701	Existing construction	L-4	2	8		2,025	274		2,299	2,675
2001	Class A fire doors, manual, 20 ga., 8′ x 8′ high		1.40	11.429		1,575	390		1,965	2,375
2101	10′ x 10′ high		1.10	14.545		2,150	500		2,650	3,200
2201	20′ x 10′ high		.80	20		4,375	685		5,060	5,975
2301	12′ x 12′ high		1	16		3,375	550		3,925	4,650
2304	Overhead door, roll up, fire rated 12′ x 14′	2 Sswk	.90	17.778		3,950	735		4,685	5,625
2401	20′ x 12′ high	L-4	.80	20		4,750	685		5,435	6,375
2501	14′ x 14′ high		.60	26.667		3,700	915		4,615	5,600
2601	20′ x 16′ high		.50	32		5,825	1,100		6,925	8,250
2701	10′ x 20′ high		.40	40		4,600	1,375		5,975	7,325
2730	Manual steel rollup fire doors	2 Sswk	160	.100	S.F.	37	4.15		41.15	48
2740	For motor operated, add	"	1	16	Ea.	755	665		1,420	1,975
3000	For 18 ga. doors, add				S.F.	1.10			1.10	1.21
3300	For enamel finish, add				"	2.76			2.76	3.04
3600	For safety edge bottom bar, pneumatic, add				L.F.	22.50			22.50	25
3700	Electric, add					41.50			41.50	45.50
4000	For weatherstripping, extruded rubber, jambs, add					18.75			18.75	20.50
4100	Hood, add					8.90			8.90	9.80
4200	Sill, add					8.90			8.90	9.80
4501	Motor operators, to 14′ x 14′ opening	L-4	5	3.200	Ea.	1,175	110		1,285	1,475
4601	Over 14′ x 14′, jack shaft type	"	5	3.200		1,325	110		1,435	1,650
4700	For fire door, additional fusible link, add					74			74	81.50
5100	Radio control operator to 12′ x 12′ door	2 Sswk	3	5.333		805	221		1,026	1,275
5120	Receiver to 12′ x 12′ door		13	1.231		241	51		292	355
5140	Transmitter to 12′ x 12′ door		49	.327		55	13.55		68.55	84

08 34 Special Function Doors

08 34 36 – Darkroom Doors

08 34 36.10 Various Types of Darkroom Doors		Crew	Daily Output	Labor-Hours	Unit	Material	2019 Bare Costs Labor	Equipment	Total	Total Incl O&P
0010	**VARIOUS TYPES OF DARKROOM DOORS**									
0015	Revolving, standard, 2 way, 36" diameter	2 Carp	3.10	5.161	Opng.	3,050	200		3,250	3,700
0020	41" diameter		3.10	5.161		3,500	200		3,700	4,175
0050	3 way, 51" diameter		1.40	11.429		4,050	445		4,495	5,175
2000	Hinged safety, 2 way, 41" diameter		2.30	6.957		3,975	270		4,245	4,825
2500	3 way, 51" diameter		1.40	11.429		4,750	445		5,195	5,950
3000	Pop out safety, 2 way, 41" diameter		3.10	5.161		4,200	200		4,400	4,950
4000	3 way, 51" diameter		1.40	11.429		5,250	445		5,695	6,475
5000	Wheelchair-type, pop out, 51" diameter	↓	1.40	11.429	↓	5,850	445		6,295	7,175

08 34 53 – Security Doors and Frames

08 34 53.20 Steel Door		Crew	Daily Output	Labor-Hours	Unit	Material	Labor	Equipment	Total	Total Incl O&P
0010	**STEEL DOOR** flush with ballistic core and welded frame both 14 ga.									
0120	UL 752 Level 3, 1-3/4", 3'-0" x 7'-0"	2 Carp	1.50	10.667	Opng.	2,625	415		3,040	3,575
0125	1-3/4", 3'-6" x 7'-0"		1.50	10.667		2,900	415		3,315	3,850
0130	1-3/4", 4'-0" x 7'-0"		1.20	13.333		3,100	515		3,615	4,250
0150	UL 752 Level 8, 1-3/4", 3'-0" x 7'-0"		1.50	10.667		13,800	415		14,215	15,900
0155	1-3/4", 3'-6" x 7'-0"		1.50	10.667		15,000	415		15,415	17,300
0160	1-3/4", 4'-0" x 7'-0"		1.20	13.333		15,000	515		15,515	17,400
1000	Safe room sliding door and hardware, 1-3/4", 3'-0" x 7'-0" UL 752 Level 3		.50	32		23,900	1,250		25,150	28,400
1050	Safe room swinging door and hardware, 1-3/4", 3'-0" x 7'-0" UL 752 Level 3	↓	.50	32	↓	28,700	1,250		29,950	33,700

08 34 53.30 Wood Ballistic Doors		Crew	Daily Output	Labor-Hours	Unit	Material	Labor	Equipment	Total	Total Incl O&P
0010	**WOOD BALLISTIC DOORS** with frames and hardware									
0050	Wood, 1-3/4", 3'-0" x 7'-0" UL 752 Level 3	2 Carp	1.50	10.667	Opng.	2,625	415		3,040	3,575

08 34 56 – Security Gates

08 34 56.10 Gates		Crew	Daily Output	Labor-Hours	Unit	Material	Labor	Equipment	Total	Total Incl O&P
0010	**GATES**									
0015	Driveway gates include mounting hardware									
0500	Wood, security gate, driveway, dual, 10' wide	H-4	.80	25	Opng.	3,800	905		4,705	5,675
0505	12' wide		.80	25		3,675	905		4,580	5,525
0510	15' wide		.80	25		4,350	905		5,255	6,300
0600	Steel, security gate, driveway, single, 10' wide		.80	25		1,750	905		2,655	3,450
0605	12' wide		.80	25		2,050	905		2,955	3,750
0620	Steel, security gate, driveway, dual, 12' wide		.80	25		2,125	905		3,030	3,825
0625	14' wide		.80	25		2,400	905		3,305	4,150
0630	16' wide		.80	25		2,475	905		3,380	4,225
0700	Aluminum, security gate, driveway, dual, 10' wide		.80	25		3,225	905		4,130	5,050
0705	12' wide		.80	25		3,925	905		4,830	5,825
0710	16' wide	↓	.80	25	↓	4,550	905		5,455	6,500
1000	Security gate, driveway, opener 12 VDC				Ea.	835			835	920
1010	Wireless					1,925			1,925	2,125
1020	Security gate, driveway, opener 24 VDC					1,450			1,450	1,600
1030	Wireless					1,625			1,625	1,775
1040	Security gate, driveway, opener 12 VDC, solar panel 10 watt	1 Elec	2	4		465	182		647	810
1050	20 watt	"	2	4	↓	610	182		792	970

08 34 73 – Sound Control Door Assemblies

08 34 73.10 Acoustical Doors		Crew	Daily Output	Labor-Hours	Unit	Material	Labor	Equipment	Total	Total Incl O&P
0010	**ACOUSTICAL DOORS**									
0020	Including framed seals, 3' x 7', wood, 40 STC rating	2 Carp	1.50	10.667	Ea.	3,375	415		3,790	4,400
0100	Steel, 41 STC rating		1.50	10.667		3,750	415		4,165	4,800
0200	45 STC rating		1.50	10.667		4,075	415		4,490	5,150
0300	48 STC rating	↓	1.50	10.667	↓	4,750	415		5,165	5,900

08 34 Special Function Doors

08 34 73 – Sound Control Door Assemblies

08 34 73.10 Acoustical Doors		Crew	Daily Output	Labor-Hours	Unit	2019 Bare Costs Material	Labor	Equipment	Total	Total Incl O&P
0400	52 STC rating	2 Carp	1.50	10.667	Ea.	5,275	415		5,690	6,475

08 36 Panel Doors

08 36 13 – Sectional Doors

08 36 13.10 Overhead Commercial Doors

		Crew	Daily Output	Labor-Hours	Unit	Material	Labor	Equipment	Total	Total Incl O&P
0010	**OVERHEAD COMMERCIAL DOORS**									
1000	Stock, sectional, heavy duty, wood, 1-3/4" thick, 8' x 8' high	2 Carp	2	8	Ea.	1,250	310		1,560	1,875
1100	10' x 10' high		1.80	8.889		1,750	345		2,095	2,500
1200	12' x 12' high		1.50	10.667		2,400	415		2,815	3,325
1300	Chain hoist, 14' x 14' high		1.30	12.308		3,750	475		4,225	4,925
1400	12' x 16' high		1	16		3,725	620		4,345	5,100
1500	20' x 8' high		1.30	12.270		2,950	475		3,425	4,025
1600	20' x 16' high		.65	24.615		5,750	955		6,705	7,900
1800	Center mullion openings, 8' high		4	4		1,325	155		1,480	1,725
1900	20' high	↓	2	8	↓	2,175	310		2,485	2,900
2100	For medium duty custom door, deduct					5%	5%			
2150	For medium duty stock doors, deduct					10%	5%			
2300	Fiberglass and aluminum, heavy duty, sectional, 12' x 12' high	2 Carp	1.50	10.667	Ea.	3,400	415		3,815	4,400
2450	Chain hoist, 20' x 20' high		.50	32		7,725	1,250		8,975	10,600
2600	Steel, 24 ga. sectional, manual, 8' x 8' high		2	8		1,025	310		1,335	1,625
2650	10' x 10' high		1.80	8.889		1,325	345		1,670	2,050
2700	12' x 12' high		1.50	10.667		1,575	415		1,990	2,400
2800	Chain hoist, 20' x 14' high	↓	.70	22.857	↓	3,675	885		4,560	5,525
2850	For 1-1/4" rigid insulation and 26 ga. galv.									
2860	back panel, add				S.F.	5.15			5.15	5.65
2900	For electric trolley operator, 1/3 HP, to 12' x 12', add	1 Carp	2	4	Ea.	1,125	155		1,280	1,475
2950	Over 12' x 12', 1/2 HP, add		1	8		1,225	310		1,535	1,850
2980	Overhead, for row of clear lites, add	↓	1	8	↓	166	310		476	695

08 38 Traffic Doors

08 38 13 – Flexible Strip Doors

08 38 13.10 Flexible Transparent Strip Doors

		Crew	Daily Output	Labor-Hours	Unit	Material	Labor	Equipment	Total	Total Incl O&P
0010	**FLEXIBLE TRANSPARENT STRIP DOORS**									
0100	12" strip width, 2/3 overlap	3 Shee	135	.178	SF Surf	7.65	7.70		15.35	21
0200	Full overlap		115	.209		9.75	9.05		18.80	25.50
0220	8" strip width, 1/2 overlap		140	.171		6.35	7.40		13.75	19.25
0240	Full overlap	↓	120	.200	↓	7.95	8.65		16.60	23
0300	Add for suspension system, header mount				L.F.	9.35			9.35	10.25
0400	Wall mount				"	9.65			9.65	10.60

08 38 16 – Flexible Traffic Doors

08 38 16.10 Rubber Doors

		Crew	Daily Output	Labor-Hours	Unit	Material	Labor	Equipment	Total	Total Incl O&P
0010	**RUBBER DOORS**									
0015	Incl. frame and hardware, 14" x 16" vision panel,									
0020	48" wear panel, pair of 2'-6" x 7'	2 Sswk	1.60	10	Ea.	2,275	415		2,690	3,225
0040	Pair of 3' x 7'		1.50	10.667		2,625	440		3,065	3,675
0060	Pair of 4' x 7'	↓	1.40	11.429	↓	3,100	475		3,575	4,225

08 38 Traffic Doors

08 38 19 – Rigid Traffic Doors

08 38 19.20 Double Acting Swing Doors		Crew	Daily Output	Labor-Hours	Unit	2019 Bare Costs Material	Labor	Equipment	Total	Total Incl O&P
0010	**DOUBLE ACTING SWING DOORS**									
0020	Including frame, closer, hardware and vision panel									
1000	Polymer, 7′-0″ high, 4′-0″ wide	2 Carp	4.20	3.810	Pr.	2,100	148		2,248	2,575
1025	6′-0″ wide		4	4		2,225	155		2,380	2,700
1050	6′-8″ wide		4	4		2,425	155		2,580	2,925
2000	3/4″ thick, stainless steel									
2010	Stainless steel, 7′ high opening, 4′ wide	2 Carp	4	4	Pr.	2,625	155		2,780	3,150
2050	7′ wide	"	3.80	4.211	"	2,800	163		2,963	3,375

08 38 19.30 Shock Absorbing Doors		Crew	Daily Output	Labor-Hours	Unit	Material	Labor	Equipment	Total	Total Incl O&P
0010	**SHOCK ABSORBING DOORS**									
0020	Rigid, no frame, 1-1/2″ thick, 5′ x 7′	2 Sswk	1.90	8.421	Opng.	1,550	350		1,900	2,300
0100	8′ x 8′		1.80	8.889		2,025	370		2,395	2,875
0500	Flexible, no frame, insulated, .16″ thick, economy, 5′ x 7′		2	8		1,775	330		2,105	2,525
0600	Deluxe		1.90	8.421		2,650	350		3,000	3,525
1000	8′ x 8′ opening, economy		2	8		2,800	330		3,130	3,650
1100	Deluxe		1.90	8.421		3,550	350		3,900	4,500

08 41 Entrances and Storefronts

08 41 13 – Aluminum-Framed Entrances and Storefronts

08 41 13.20 Tube Framing		Crew	Daily Output	Labor-Hours	Unit	Material	Labor	Equipment	Total	Total Incl O&P
0010	**TUBE FRAMING**, For window walls and storefronts, aluminum stock									
0050	Plain tube frame, mill finish, 1-3/4″ x 1-3/4″	2 Glaz	103	.155	L.F.	10.75	5.95		16.70	21.50
0150	1-3/4″ x 4″		98	.163		14.25	6.25		20.50	26
0200	1-3/4″ x 4-1/2″		95	.168		17.35	6.45		23.80	29.50
0250	2″ x 6″		89	.180		25.50	6.90		32.40	39.50
0350	4″ x 4″		87	.184		28.50	7.05		35.55	43
0400	4-1/2″ x 4-1/2″		85	.188		30.50	7.20		37.70	45.50
0450	Glass bead		240	.067		3.32	2.55		5.87	7.85
1000	Flush tube frame, mill finish, 1/4″ glass, 1-3/4″ x 4″, open header		80	.200		14.40	7.65		22.05	28.50
1050	Open sill		82	.195		11.70	7.45		19.15	25
1100	Closed back header		83	.193		20.50	7.35		27.85	34.50
1150	Closed back sill		85	.188		19.70	7.20		26.90	33.50
1160	Tube fmg., spandrel cover both sides, alum 1″ wide	1 Sswk	85	.094	S.F.	97.50	3.90		101.40	114
1170	Tube fmg., spandrel cover both sides, alum 2″ wide	"	85	.094	"	38.50	3.90		42.40	49.50
1200	Vertical mullion, one piece	2 Glaz	75	.213	L.F.	20.50	8.15		28.65	36
1250	Two piece		73	.219		22	8.40		30.40	38.50
1300	90° or 180° vertical corner post		75	.213		33.50	8.15		41.65	50.50
1400	1-3/4″ x 4-1/2″, open header		80	.200		17	7.65		24.65	31.50
1450	Open sill		82	.195		14	7.45		21.45	27.50
1500	Closed back header		83	.193		20.50	7.35		27.85	34.50
1550	Closed back sill		85	.188		19.65	7.20		26.85	33.50
1600	Vertical mullion, one piece		75	.213		22	8.15		30.15	37.50
1650	Two piece		73	.219		23.50	8.40		31.90	39.50
1700	90° or 180° vertical corner post		75	.213		24	8.15		32.15	39.50
2000	Flush tube frame, mil fin.,ins. glass w/thml brk, 2″ x 4-1/2″, open header		75	.213		16.40	8.15		24.55	31.50
2050	Open sill		77	.208		13.80	7.95		21.75	28.50
2100	Closed back header		78	.205		14.95	7.85		22.80	29.50
2150	Closed back sill		80	.200		15.80	7.65		23.45	30
2200	Vertical mullion, one piece		70	.229		16.75	8.75		25.50	33
2250	Two piece		68	.235		18.05	9		27.05	34.50

08 41 Entrances and Storefronts

08 41 13 – Aluminum-Framed Entrances and Storefronts

08 41 13.20 Tube Framing		Crew	Daily Output	Labor-Hours	Unit	2019 Bare Costs Material	Labor	Equipment	Total	Total Incl O&P
2300	90° or 180° vertical corner post	2 Glaz	70	.229	L.F.	18.85	8.75		27.60	35.50
5000	Flush tube frame, mill fin., thermal brk., 2-1/4" x 4-1/2", open header		74	.216		16.65	8.25		24.90	32
5050	Open sill		75	.213		14.70	8.15		22.85	29.50
5100	Vertical mullion, one piece		69	.232		18.95	8.85		27.80	35.50
5150	Two piece		67	.239		21.50	9.15		30.65	39
5200	90° or 180° vertical corner post		69	.232		18.55	8.85		27.40	35
6980	Door stop (snap in)		380	.042		3.73	1.61		5.34	6.75
7000	For joints, 90°, clip type, add				Ea.	25.50			25.50	28.50
7050	Screw spline joint, add					24			24	26.50
7100	For joint other than 90°, add					50			50	55
8000	For bronze anodized aluminum, add					15%				
8050	For stainless steel materials, add					350%				
8100	For monumental grade, add					53%				
8150	For steel stiffener, add	2 Glaz	200	.080	L.F.	11.55	3.06		14.61	17.75
8200	For 2 to 5 stories, add per story				Story		8%			

08 42 Entrances

08 42 26 – All-Glass Entrances

08 42 26.10 Swinging Glass Doors		Crew	Daily Output	Labor-Hours	Unit	Material	Labor	Equipment	Total	Total Incl O&P
0010	**SWINGING GLASS DOORS**									
0020	Including hardware, 1/2" thick, tempered, 3' x 7' opening	2 Glaz	2	8	Opng.	2,350	305		2,655	3,100
0100	6' x 7' opening	"	1.40	11.429	"	4,625	435		5,060	5,825

08 42 36 – Balanced Door Entrances

08 42 36.10 Balanced Entrance Doors		Crew	Daily Output	Labor-Hours	Unit	Material	Labor	Equipment	Total	Total Incl O&P
0010	**BALANCED ENTRANCE DOORS**									
0020	Hardware & frame, alum. & glass, 3' x 7', econ.	2 Sswk	.90	17.778	Ea.	7,050	735		7,785	9,025
0150	Premium	"	.70	22.857	"	8,600	945		9,545	11,100

08 43 Storefronts

08 43 13 – Aluminum-Framed Storefronts

08 43 13.10 Aluminum-Framed Entrance Doors and Frames		Crew	Daily Output	Labor-Hours	Unit	Material	Labor	Equipment	Total	Total Incl O&P
0010	**ALUMINUM-FRAMED ENTRANCE DOORS AND FRAMES**									
0015	Standard hardware and glass stops but no glass									
0020	Entrance door, 3' x 7' opening, clear anodized finish	2 Sswk	7	2.286	Opng.	660	94.50		754.50	895
0200	3'-6" x 7'-0", mill finish		7	2.286		775	94.50		869.50	1,025
0220	Bronze finish		7	2.286		930	94.50		1,024.50	1,200
0240	Black finish		7	2.286		970	94.50		1,064.50	1,250
0500	6' x 7' opening, clear finish		6	2.667		1,025	111		1,136	1,325
0520	Bronze finish		6	2.667		1,125	111		1,236	1,425
0540	Black finish		6	2.667		1,225	111		1,336	1,550
0600	Door frame for above doors 3'-0" x 7'-0", mill finish		6	2.667		465	111		576	700
0620	Bronze finish		6	2.667		530	111		641	775
0640	Black finish		6	2.667		595	111		706	840
0700	3'-6" x 7'-0", mill finish		6	2.667		380	111		491	610
0720	Bronze finish		6	2.667		380	111		491	610
0740	Black finish		6	2.667		380	111		491	610
0800	6'-0" x 7'-0", mill finish		6	2.667		380	111		491	610
0820	Bronze finish		6	2.667		385	111		496	615
0840	Black finish		6	2.667		420	111		531	655

08 43 Storefronts

08 43 13 – Aluminum-Framed Storefronts

08 43 13.10 Aluminum-Framed Entrance Doors and Frames		Crew	Daily Output	Labor-Hours	Unit	2019 Bare Costs Material	Labor	Equipment	Total	Total Incl O&P
8000	For 8′ high doors, add				Ea.	200			200	220

08 43 13.20 Storefront Systems

		Crew	Daily Output	Labor-Hours	Unit	Material	Labor	Equipment	Total	Total Incl O&P
0010	**STOREFRONT SYSTEMS**, aluminum frame clear 3/8″ plate glass									
0020	incl. 3′ x 7′ door with hardware (400 sq. ft. max. wall)									
0500	Wall height to 12′ high, commercial grade	2 Glaz	150	.107	S.F.	25.50	4.08		29.58	34.50
0600	Institutional grade		130	.123		31.50	4.71		36.21	42.50
0700	Monumental grade		115	.139		81	5.30		86.30	98
1000	6′ x 7′ door with hardware, commercial grade		135	.119		75.50	4.53		80.03	90.50
1100	Institutional grade		115	.139		58	5.30		63.30	73
1200	Monumental grade		100	.160		112	6.10		118.10	133
1500	For bronze anodized finish, add					15%				
1600	For black anodized finish, add					36%				
1700	For stainless steel framing, add to monumental					78%				

08 43 29 – Sliding Storefronts

08 43 29.10 Sliding Panels

		Crew	Daily Output	Labor-Hours	Unit	Material	Labor	Equipment	Total	Total Incl O&P
0010	**SLIDING PANELS**									
0020	Mall fronts, aluminum & glass, 15′ x 9′ high	2 Glaz	1.30	12.308	Opng.	4,050	470		4,520	5,225
0100	24′ x 9′ high		.70	22.857		5,800	875		6,675	7,850
0200	48′ x 9′ high, with fixed panels		.90	17.778		10,500	680		11,180	12,600
0500	For bronze finish, add					17%				

08 45 Translucent Wall and Roof Assemblies

08 45 10 – Translucent Roof Assemblies

08 45 10.10 Skyroofs

		Crew	Daily Output	Labor-Hours	Unit	Material	Labor	Equipment	Total	Total Incl O&P
0010	**SKYROOFS**									
1200	Skylights, circular, clear, double glazed acrylic									
1230	30″ diameter	2 Carp	3	5.333	Ea.	3,025	207		3,232	3,675
1250	60″ diameter		3	5.333		4,025	207		4,232	4,800
1290	96″ diameter		2	8		5,050	310		5,360	6,050
1330	3′-0″ x 12′-0″	G-3	3	10.667		5,050	395		5,445	6,200
1350	4′-0″ x 12′-0″		3	10.667		5,550	395		5,945	6,750
1390	5′-0″ x 12′-0″		2	16		6,050	590		6,640	7,625
1430	Square, 3′ x 3′		3	10.667		6,050	395		6,445	7,300
1440	4′ x 4′		3	10.667		7,050	395		7,445	8,425
1450	5′ x 5′, glass must be field installed		3	10.667		8,075	395		8,470	9,525
1460	6′ x 6′, glass must be field installed		2	16		10,100	590		10,690	12,100

08 51 Metal Windows

08 51 13 – Aluminum Windows

08 51 13.10 Aluminum Sash

		Crew	Daily Output	Labor-Hours	Unit	Material	Labor	Equipment	Total	Total Incl O&P
0010	**ALUMINUM SASH**									
0020	Stock, grade C, glaze & trim not incl., casement	2 Sswk	200	.080	S.F.	42.50	3.32		45.82	53
0050	Double-hung		200	.080		42.50	3.32		45.82	52.50
0100	Fixed casement		200	.080		18.50	3.32		21.82	26.50
0150	Picture window		200	.080		19.85	3.32		23.17	28
0200	Projected window		200	.080		38	3.32		41.32	48
0250	Single-hung		200	.080		16.95	3.32		20.27	24.50
0300	Sliding		200	.080		22.50	3.32		25.82	31

08 51 13 – Aluminum Windows

08 51 13.10 Aluminum Sash		Crew	Daily Output	Labor-Hours	Unit	2019 Bare Costs Material	Labor	Equipment	Total	Total Incl O&P
1000	Mullions for above, tubular	2 Sswk	240	.067	L.F.	6.55	2.76		9.31	12

08 51 13.20 Aluminum Windows

		Crew	Daily Output	Labor-Hours	Unit	Material	Labor	Equipment	Total	Total Incl O&P
0010	**ALUMINUM WINDOWS**, incl. frame and glazing, commercial grade									
1000	Stock units, casement, 3′-1″ x 3′-2″ opening	2 Sswk	10	1.600	Ea.	385	66.50		451.50	540
1040	Insulating glass	″	10	1.600		525	66.50		591.50	690
1050	Add for storms					123			123	136
1600	Projected, with screen, 3′-1″ x 3′-2″ opening	2 Sswk	10	1.600		365	66.50		431.50	515
1650	Insulating glass	″	10	1.600		390	66.50		456.50	545
1700	Add for storms					120			120	132
2000	4′-5″ x 5′-3″ opening	2 Sswk	8	2		400	83		483	585
2050	Insulating glass	″	8	2		480	83		563	675
2100	Add for storms					129			129	142
2500	Enamel finish windows, 3′-1″ x 3′-2″	2 Sswk	10	1.600		370	66.50		436.50	520
2550	Insulating glass		10	1.600		340	66.50		406.50	490
2600	4′-5″ x 5′-3″		8	2		415	83		498	605
2700	Insulating glass		8	2		445	83		528	635
3000	Single-hung, 2′ x 3′ opening, enameled, standard glazed		10	1.600		214	66.50		280.50	350
3100	Insulating glass		10	1.600		260	66.50		326.50	400
3300	2′-8″ x 6′-8″ opening, standard glazed		8	2		375	83		458	555
3400	Insulating glass		8	2		475	83		558	665
3700	3′-4″ x 5′-0″ opening, standard glazed		9	1.778		310	73.50		383.50	470
3800	Insulating glass		9	1.778		335	73.50		408.50	500
4000	Sliding aluminum, 3′ x 2′ opening, standard glazed		10	1.600		220	66.50		286.50	355
4100	Insulating glass		10	1.600		236	66.50		302.50	375
4300	5′ x 3′ opening, standard glazed		9	1.778		335	73.50		408.50	500
4400	Insulating glass		9	1.778		390	73.50		463.50	560
4600	8′ x 4′ opening, standard glazed		6	2.667		360	111		471	585
4700	Insulating glass		6	2.667		575	111		686	825
5000	9′ x 5′ opening, standard glazed		4	4		540	166		706	885
5100	Insulating glass		4	4		850	166		1,016	1,225
5500	Sliding, with thermal barrier and screen, 6′ x 4′, 2 track		8	2		725	83		808	945
5700	4 track		8	2		910	83		993	1,150
6000	For above units with bronze finish, add					15%				
6200	For installation in concrete openings, add					8%				

08 51 13.30 Impact Resistant Aluminum Windows

		Crew	Daily Output	Labor-Hours	Unit	Material	Labor	Equipment	Total	Total Incl O&P
0010	**IMPACT RESISTANT ALUMINUM WINDOWS**, incl. frame and glazing									
0100	Single-hung, impact resistant, 2′-8″ x 5′-0″	2 Carp	9	1.778	Ea.	1,250	69		1,319	1,500
0120	3′-0″ x 5′-0″		9	1.778		1,375	69		1,444	1,625
0130	4′-0″ x 5′-0″		9	1.778		1,475	69		1,544	1,725
0250	Horizontal slider, impact resistant, 5′-5″ x 5′-2″		9	1.778		1,625	69		1,694	1,925

08 51 23 – Steel Windows

08 51 23.10 Steel Sash

		Crew	Daily Output	Labor-Hours	Unit	Material	Labor	Equipment	Total	Total Incl O&P
0010	**STEEL SASH** Custom units, glazing and trim not included									
0100	Casement, 100% vented	2 Sswk	200	.080	S.F.	68	3.32		71.32	80.50
0200	50% vented		200	.080		55.50	3.32		58.82	67
0300	Fixed		200	.080		29.50	3.32		32.82	38.50
1000	Projected, commercial, 40% vented		200	.080		53	3.32		56.32	64
1100	Intermediate, 50% vented		200	.080		60.50	3.32		63.82	73
1500	Industrial, horizontally pivoted		200	.080		56	3.32		59.32	67.50
1600	Fixed		200	.080		32.50	3.32		35.82	42
2000	Industrial security sash, 50% vented		200	.080		60.50	3.32		63.82	72.50
2100	Fixed		200	.080		49	3.32		52.32	60

08 51 Metal Windows

08 51 23 – Steel Windows

08 51 23.10 Steel Sash

	08 51 23.10 Steel Sash	Crew	Daily Output	Labor-Hours	Unit	Material	Labor	Equipment	Total	Total Incl O&P
2500	Picture window	2 Sswk	200	.080	S.F.	31	3.32		34.32	40.50
3000	Double-hung		200	.080		61	3.32		64.32	73
5000	Mullions for above, open interior face		240	.067	L.F.	10.75	2.76		13.51	16.65
5100	With interior cover		240	.067	"	17.80	2.76		20.56	24.50
6000	Double glazing for above, add	2 Glaz	200	.080	S.F.	13.25	3.06		16.31	19.65
6100	Triple glazing for above, add	"	85	.188	"	12.65	7.20		19.85	26

08 51 23.20 Steel Windows

	08 51 23.20 Steel Windows	Crew	Daily Output	Labor-Hours	Unit	Material	Labor	Equipment	Total	Total Incl O&P
0010	**STEEL WINDOWS** Stock, including frame, trim and insul. glass									
1000	Custom units, double-hung, 2′-8″ x 4′-6″ opening	2 Sswk	12	1.333	Ea.	730	55.50		785.50	895
1100	2′-4″ x 3′-9″ opening		12	1.333		600	55.50		655.50	755
1500	Commercial projected, 3′-9″ x 5′-5″ opening		10	1.600		1,250	66.50		1,316.50	1,525
1600	6′-9″ x 4′-1″ opening		7	2.286		1,675	94.50		1,769.50	2,000
2000	Intermediate projected, 2′-9″ x 4′-1″ opening		12	1.333		710	55.50		765.50	875
2100	4′-1″ x 5′-5″ opening		10	1.600		1,450	66.50		1,516.50	1,725

08 51 23.40 Basement Utility Windows

	08 51 23.40 Basement Utility Windows	Crew	Daily Output	Labor-Hours	Unit	Material	Labor	Equipment	Total	Total Incl O&P
0010	**BASEMENT UTILITY WINDOWS**									
0015	1′-3″ x 2′-8″	1 Carp	16	.500	Ea.	147	19.40		166.40	194
1100	1′-7″ x 2′-8″		16	.500		149	19.40		168.40	196
1200	1′-11″ x 2′-8″		14	.571		154	22		176	206

08 51 66 – Metal Window Screens

08 51 66.10 Screens

	08 51 66.10 Screens	Crew	Daily Output	Labor-Hours	Unit	Material	Labor	Equipment	Total	Total Incl O&P
0010	**SCREENS**									
0020	For metal sash, aluminum or bronze mesh, flat screen	2 Sswk	1200	.013	S.F.	4.54	.55		5.09	5.95
0500	Wicket screen, inside window	"	1000	.016	"	6.90	.66		7.56	8.75
0600	Residential, aluminum mesh and frame, 2′ x 3′	2 Carp	32	.500	Ea.	17.70	19.40		37.10	51.50
0610	Rescreen		50	.320		14.10	12.40		26.50	36
0620	3′ x 5′		32	.500		59.50	19.40		78.90	97.50
0630	Rescreen		45	.356		36.50	13.80		50.30	63.50
0640	4′ x 8′		25	.640		91	25		116	141
0650	Rescreen		40	.400		54	15.50		69.50	85
0660	Patio door		25	.640		207	25		232	269
0680	Rescreening		1600	.010	S.F.	2.67	.39		3.06	3.58
0800	Security screen, aluminum frame with stainless steel cloth	2 Sswk	1200	.013		24.50	.55		25.05	28
0900	Steel grate, painted, on steel frame		1600	.010		13.60	.41		14.01	15.65
1000	Screens for solar louvers		160	.100		25.50	4.15		29.65	35

08 52 Wood Windows

08 52 10 – Plain Wood Windows

08 52 10.10 Wood Windows

	08 52 10.10 Wood Windows	Crew	Daily Output	Labor-Hours	Unit	Material	Labor	Equipment	Total	Total Incl O&P
0010	**WOOD WINDOWS**, including frame, screens and grilles									
0020	Residential, stock units									
0050	Awning type, double insulated glass, 2′-10″ x 1′-9″ opening	2 Carp	12	1.333	Opng.	231	51.50		282.50	340
0100	2′-10″ x 6′-0″ opening	1 Carp	8	1		560	39		599	680
0200	4′-0″ x 3′-6″ single pane		10	.800		375	31		406	465
0300	6′ x 5′ single pane		8	1	Ea.	515	39		554	630
1000	Casement, 2′-0″ x 3′-4″ high	2 Carp	20	.800		258	31		289	335
1020	2′-0″ x 4′-0″		18	.889		273	34.50		307.50	355
1040	2′-0″ x 5′-0″		17	.941		320	36.50		356.50	410
1060	2′-0″ x 6′-0″		16	1		315	39		354	415
1080	4′-0″ x 3′-4″		15	1.067		595	41.50		636.50	725

08 52 10 – Plain Wood Windows

08 52 10.10 Wood Windows		Crew	Daily Output	Labor-Hours	Unit	Material	2019 Bare Costs Labor	Equipment	Total	Total Incl O&P
1100	4′-0″ x 4′-0″	2 Carp	15	1.067	Ea.	670	41.50		711.50	805
1120	4′-0″ x 5′-0″		14	1.143		755	44.50		799.50	905
1140	4′-0″ x 6′-0″		12	1.333		850	51.50		901.50	1,025
1600	Casement units, 8′ x 5′, with screens, double insulated glass		2.50	6.400	Opng.	1,500	248		1,748	2,050
1700	Low E glass		2.50	6.400		1,650	248		1,898	2,200
2300	Casements, including screens, 2′-0″ x 3′-4″, double insulated glass		11	1.455		277	56.50		333.50	400
2400	Low E glass		11	1.455		277	56.50		333.50	400
2600	2 lite, 4′-0″ x 4′-0″, double insulated glass		9	1.778		505	69		574	670
2700	Low E glass		9	1.778		520	69		589	685
2900	3 lite, 5′-2″ x 5′-0″, double insulated glass		7	2.286		765	88.50		853.50	985
3000	Low E glass		7	2.286		810	88.50		898.50	1,025
3200	4 lite, 7′-0″ x 5′-0″, double insulated glass		6	2.667		1,100	103		1,203	1,375
3300	Low E glass		6	2.667		1,175	103		1,278	1,450
3500	5 lite, 8′-6″ x 5′-0″, double insulated glass		5	3.200		1,450	124		1,574	1,800
3600	Low E glass		5	3.200		1,450	124		1,574	1,800
3800	For removable wood grilles, diamond pattern, add				Leaf	39.50			39.50	43.50
3900	Rectangular pattern, add				"	39			39	43
4000	Bow, fixed lites, 8′ x 5′, double insulated glass	2 Carp	3	5.333	Opng.	1,450	207		1,657	1,950
4100	Low E glass	"	3	5.333	"	1,975	207		2,182	2,525
4150	6′-0″ x 5′-0″	1 Carp	8	1	Ea.	1,275	39		1,314	1,500
4300	Fixed lites, 9′-9″ x 5′-0″, double insulated glass	2 Carp	2	8	Opng.	1,000	310		1,310	1,600
4400	Low E glass	"	2	8	"	1,100	310		1,410	1,700
5000	Bow, casement, 8′-1″ x 4′-8″ high	3 Carp	8	3	Ea.	1,600	116		1,716	1,975
5020	9′-6″ x 4′-8″		8	3		1,750	116		1,866	2,125
5040	8′-1″ x 5′-1″		8	3		1,950	116		2,066	2,350
5060	9′-6″ x 5′-1″		6	4		1,950	155		2,105	2,400
5080	8′-1″ x 6′-0″		6	4		1,950	155		2,105	2,400
5100	9′-6″ x 6′-0″		6	4		2,050	155		2,205	2,500
5800	Skylights, hatches, vents, and sky roofs, see Section 08 62 13.00									

08 52 10.20 Awning Window		Crew	Daily Output	Labor-Hours	Unit	Material	Labor	Equipment	Total	Total Incl O&P
0010	**AWNING WINDOW**, Including frame, screens and grilles									
0100	34″ x 22″, insulated glass	1 Carp	10	.800	Ea.	278	31		309	355
0200	Low E glass		10	.800		315	31		346	395
0300	40″ x 28″, insulated glass		9	.889		315	34.50		349.50	400
0400	Low E glass		9	.889		345	34.50		379.50	435
0500	48″ x 36″, insulated glass		8	1		475	39		514	585
0600	Low E glass		8	1		500	39		539	615
4000	Impact windows, minimum, add					60%				
4010	Impact windows, maximum, add					160%				

08 52 10.30 Wood Windows		Crew	Daily Output	Labor-Hours	Unit	Material	Labor	Equipment	Total	Total Incl O&P
0010	**WOOD WINDOWS**, double-hung									
0020	Including frame, double insulated glass, screens and grilles									
0040	Double-hung, 2′-2″ x 3′-4″ high	2 Carp	15	1.067	Ea.	218	41.50		259.50	310
0060	2′-2″ x 4′-4″		14	1.143		236	44.50		280.50	330
0080	2′-6″ x 3′-4″		13	1.231		227	47.50		274.50	330
0100	2′-6″ x 4′-0″		12	1.333		236	51.50		287.50	345
0120	2′-6″ x 4′-8″		12	1.333		254	51.50		305.50	365
0140	2′-10″ x 3′-4″		10	1.600		229	62		291	355
0160	2′-10″ x 4′-0″		10	1.600		254	62		316	380
0180	3′-7″ x 3′-4″		9	1.778		261	69		330	400
0200	3′-7″ x 5′-4″		9	1.778		295	69		364	440
0220	3′-10″ x 5′-4″		8	2		520	77.50		597.50	705

08 52 Wood Windows

08 52 10 – Plain Wood Windows

08 52 10.30 Wood Windows		Crew	Daily Output	Labor-Hours	Unit	2019 Bare Costs Material	Labor	Equipment	Total	Total Incl O&P
3800	Triple glazing for above, add				Ea.	25%				

08 52 10.40 Casement Window		Crew	Daily Output	Labor-Hours	Unit	Material	Labor	Equipment	Total	Total Incl O&P
0010	**CASEMENT WINDOW**, including frame, screen and grilles									
0100	2′-0″ x 3′-0″ H, double insulated glass G	1 Carp	10	.800	Ea.	281	31		312	360
0150	Low E glass G		10	.800		276	31		307	355
0200	2′-0″ x 4′-6″ high, double insulated glass G		9	.889		395	34.50		429.50	490
0250	Low E glass G		9	.889		410	34.50		444.50	505
0260	Casement 4′-2″ x 4′-2″ double insulated glass G		11	.727		925	28		953	1,075
0270	4′-0″ x 4′-0″ Low E glass G		11	.727		555	28		583	655
0290	6′-4″ x 5′-7″ Low E glass G		9	.889		1,175	34.50		1,209.50	1,350
0300	2′-4″ x 6′-0″ high, double insulated glass G		8	1		450	39		489	560
0350	Low E glass G		8	1		445	39		484	550
0600	3′-0″ x 5′-0″ G		8	1		700	39		739	840
0700	4′-0″ x 3′-0″ G		8	1		770	39		809	910
0710	4′-0″ x 4′-0″ G		8	1		660	39		699	790
0720	4′-8″ x 4′-0″ G		8	1		725	39		764	865
0730	4′-8″ x 5′-0″ G		6	1.333		830	51.50		881.50	1,000
0740	4′-8″ x 6′-0″ G		6	1.333		935	51.50		986.50	1,100
0750	6′-0″ x 4′-0″ G		6	1.333		850	51.50		901.50	1,025
0800	6′-0″ x 5′-0″ G		6	1.333		935	51.50		986.50	1,100
0900	5′-6″ x 5′-6″ G	2 Carp	15	1.067		1,500	41.50		1,541.50	1,725
2000	Bay, casement units, 8′ x 5′, w/screens, double insulated glass		2.50	6.400	Opng.	1,650	248		1,898	2,200
2100	Low E glass		2.50	6.400	″	1,725	248		1,973	2,300
8190	For installation, add per leaf				Ea.		15%			
8200	For multiple leaf units, deduct for stationary sash									
8220	2′ high				Ea.	24.50			24.50	26.50
8240	4′-6″ high					27.50			27.50	30
8260	6′ high					36.50			36.50	40
8300	Impact windows, minimum, add					60%				
8310	Impact windows, maximum, add					160%				

08 52 10.50 Double-Hung		Crew	Daily Output	Labor-Hours	Unit	Material	Labor	Equipment	Total	Total Incl O&P
0010	**DOUBLE-HUNG**, Including frame, screens and grilles									
0100	2′-0″ x 3′-0″ high, low E insul. glass G	1 Carp	10	.800	Ea.	193	31		224	263
0200	3′-0″ x 4′-0″ high, double insulated glass G		9	.889		292	34.50		326.50	375
0300	4′-0″ x 4′-6″ high, low E insulated glass G		8	1		340	39		379	435

08 52 10.55 Picture Window		Crew	Daily Output	Labor-Hours	Unit	Material	Labor	Equipment	Total	Total Incl O&P
0010	**PICTURE WINDOW**, Including frame and grilles									
0100	3′-6″ x 4′-0″ high, double insulated glass	2 Carp	12	1.333	Ea.	435	51.50		486.50	565
0150	Low E glass		12	1.333		445	51.50		496.50	575
0200	4′-0″ x 4′-6″ high, double insulated glass		11	1.455		550	56.50		606.50	700
0250	Low E glass		11	1.455		535	56.50		591.50	680
0300	5′-0″ x 4′-0″ high, double insulated glass		11	1.455		585	56.50		641.50	735
0350	Low E glass		11	1.455		610	56.50		666.50	765
0400	6′-0″ x 4′-6″ high, double insulated glass		10	1.600		640	62		702	805
0450	Low E glass		10	1.600		645	62		707	810

08 52 10.65 Wood Sash		Crew	Daily Output	Labor-Hours	Unit	Material	Labor	Equipment	Total	Total Incl O&P
0010	**WOOD SASH**, Including glazing but not trim									
0050	Custom, 5′-0″ x 4′-0″, 1″ double glazed, 3/16″ thick lites	2 Carp	3.20	5	Ea.	234	194		428	580
0100	1/4″ thick lites		5	3.200		266	124		390	500
0200	1″ thick, triple glazed		5	3.200		435	124		559	685
0300	7′-0″ x 4′-6″ high, 1″ double glazed, 3/16″ thick lites		4.30	3.721		435	144		579	715
0400	1/4″ thick lites		4.30	3.721		495	144		639	780

08 52 Wood Windows

08 52 10 – Plain Wood Windows

08 52 10.65 Wood Sash			Crew	Daily Output	Labor-Hours	Unit	Material	2019 Bare Costs Labor	Equipment	Total	Total Incl O&P
0500	1" thick, triple glazed		2 Carp	4.30	3.721	Ea.	565	144		709	860
0600	8'-6" x 5'-0" high, 1" double glazed, 3/16" thick lites			3.50	4.571		590	177		767	940
0700	1/4" thick lites			3.50	4.571		645	177		822	1,000
0800	1" thick, triple glazed			3.50	4.571		685	177		862	1,050
0900	Window frames only, based on perimeter length					L.F.	4.17			4.17	4.59

08 52 10.70 Sliding Windows			Crew	Daily Output	Labor-Hours	Unit	Material	Labor	Equipment	Total	Total Incl O&P
0010	**SLIDING WINDOWS**										
0100	3'-0" x 3'-0" high, double insulated	G	1 Carp	10	.800	Ea.	297	31		328	375
0120	Low E glass	G		10	.800		325	31		356	405
0200	4'-0" x 3'-6" high, double insulated	G		9	.889		395	34.50		429.50	490
0220	Low E glass	G		9	.889		390	34.50		424.50	485
0300	6'-0" x 5'-0" high, double insulated	G		8	1		515	39		554	630
0320	Low E glass	G		8	1		580	39		619	700
6000	Sliding, insulating glass, including screens,										
6100	3'-0" x 3'-0"		2 Carp	6.50	2.462	Ea.	330	95.50		425.50	525
6200	4'-0" x 3'-6"			6.30	2.540		335	98.50		433.50	535
6300	5'-0" x 4'-0"			6	2.667		420	103		523	630

08 52 13 – Metal-Clad Wood Windows

08 52 13.10 Awning Windows, Metal-Clad			Crew	Daily Output	Labor-Hours	Unit	Material	Labor	Equipment	Total	Total Incl O&P
0010	**AWNING WINDOWS, METAL-CLAD**										
2000	Metal-clad, awning deluxe, double insulated glass, 34" x 22"		1 Carp	9	.889	Ea.	253	34.50		287.50	335
2050	36" x 25"			9	.889		278	34.50		312.50	360
2100	40" x 22"			9	.889		295	34.50		329.50	380
2150	40" x 30"			9	.889		345	34.50		379.50	435
2200	48" x 28"			8	1		355	39		394	455
2250	60" x 36"			8	1		375	39		414	480

08 52 13.20 Casement Windows, Metal-Clad			Crew	Daily Output	Labor-Hours	Unit	Material	Labor	Equipment	Total	Total Incl O&P
0010	**CASEMENT WINDOWS, METAL-CLAD**										
0100	Metal-clad, deluxe, dbl. insul. glass, 2'-0" x 3'-0" high	G	1 Carp	10	.800	Ea.	296	31		327	375
0120	2'-0" x 4'-0" high	G		9	.889		320	34.50		354.50	410
0130	2'-0" x 5'-0" high	G		8	1		345	39		384	440
0140	2'-0" x 6'-0" high	G		8	1		385	39		424	490
0300	Metal-clad, casement, bldrs mdl, 6'-0" x 4'-0", dbl. insul. glass, 3 panels		2 Carp	10	1.600		1,225	62		1,287	1,450
0310	9'-0" x 4'-0", 4 panels			8	2		1,550	77.50		1,627.50	1,825
0320	10'-0" x 5'-0", 5 panels			7	2.286		2,125	88.50		2,213.50	2,475
0330	12'-0" x 6'-0", 6 panels			6	2.667		2,700	103		2,803	3,150

08 52 13.30 Double-Hung Windows, Metal-Clad			Crew	Daily Output	Labor-Hours	Unit	Material	Labor	Equipment	Total	Total Incl O&P
0010	**DOUBLE-HUNG WINDOWS, METAL-CLAD**										
0100	Metal-clad, deluxe, dbl. insul. glass, 2'-6" x 3'-0" high	G	1 Carp	10	.800	Ea.	286	31		317	365
0120	3'-0" x 3'-6" high	G		10	.800		325	31		356	410
0140	3'-0" x 4'-0" high	G		9	.889		345	34.50		379.50	430
0160	3'-0" x 4'-6" high	G		9	.889		360	34.50		394.50	450
0180	3'-0" x 5'-0" high	G		8	1		385	39		424	490
0200	3'-6" x 6'-0" high	G		8	1		470	39		509	580

08 52 13.35 Picture and Sliding Windows Metal-Clad			Crew	Daily Output	Labor-Hours	Unit	Material	Labor	Equipment	Total	Total Incl O&P
0010	**PICTURE AND SLIDING WINDOWS METAL-CLAD**										
2000	Metal-clad, dlx picture, dbl. insul. glass, 4'-0" x 4'-0" high		2 Carp	12	1.333	Ea.	380	51.50		431.50	505
2100	4'-0" x 6'-0" high			11	1.455		565	56.50		621.50	715
2200	5'-0" x 6'-0" high			10	1.600		625	62		687	785
2300	6'-0" x 6'-0" high			10	1.600		715	62		777	885
2400	Metal-clad, dlx sliding, dbl. insul. glass, 3'-0" x 3'-0" high	G	1 Carp	10	.800		330	31		361	415

08 52 13 – Metal-Clad Wood Windows

08 52 13.35 Picture and Sliding Windows Metal-Clad		Crew	Daily Output	Labor-Hours	Unit	Material	Labor	Equipment	Total	Total Incl O&P
						2019 Bare Costs				
2420	4'-0" x 3'-6" high G	1 Carp	9	.889	Ea.	400	34.50		434.50	495
2440	5'-0" x 4'-0" high G		9	.889		480	34.50		514.50	585
2460	6'-0" x 5'-0" high G	↓	8	1	↓	755	39		794	895

08 52 13.40 Bow and Bay Windows, Metal-Clad

		Crew	Daily Output	Labor-Hours	Unit	Material	Labor	Equipment	Total	Total Incl O&P
0010	**BOW AND BAY WINDOWS, METAL-CLAD**									
0100	Metal-clad, deluxe, dbl. insul. glass, 8'-0" x 5'-0" high, 4 panels	2 Carp	10	1.600	Ea.	1,700	62		1,762	1,975
0120	10'-0" x 5'-0" high, 5 panels		8	2		1,825	77.50		1,902.50	2,150
0140	10'-0" x 6'-0" high, 5 panels		7	2.286		2,175	88.50		2,263.50	2,525
0160	12'-0" x 6'-0" high, 6 panels		6	2.667		2,950	103		3,053	3,425
0400	Double-hung, bldrs. model, bay, 8' x 4' high, dbl. insul. glass		10	1.600		1,350	62		1,412	1,600
0440	Low E glass		10	1.600		1,450	62		1,512	1,700
0460	9'-0" x 5'-0" high, dbl. insul. glass		6	2.667		1,450	103		1,553	1,775
0480	Low E glass		6	2.667		1,550	103		1,653	1,875
0500	Metal-clad, deluxe, dbl. insul. glass, 7'-0" x 4'-0" high		10	1.600		1,300	62		1,362	1,525
0520	8'-0" x 4'-0" high		8	2		1,350	77.50		1,427.50	1,600
0540	8'-0" x 5'-0" high		7	2.286		1,400	88.50		1,488.50	1,675
0560	9'-0" x 5'-0" high	↓	6	2.667	↓	1,475	103		1,578	1,800

08 52 16 – Plastic-Clad Wood Windows

08 52 16.10 Bow Window

		Crew	Daily Output	Labor-Hours	Unit	Material	Labor	Equipment	Total	Total Incl O&P
0010	**BOW WINDOW** including frames, screens, and grilles									
0020	End panels operable									
1000	Bow type, casement, wood, bldrs. mdl., 8' x 5' dbl. insul. glass, 4 panel	2 Carp	10	1.600	Ea.	1,575	62		1,637	1,850
1050	Low E glass		10	1.600		1,325	62		1,387	1,575
1100	10'-0" x 5'-0", dbl. insul. glass, 6 panels		6	2.667		1,375	103		1,478	1,675
1200	Low E glass, 6 panels		6	2.667		1,475	103		1,578	1,800
1300	Vinyl-clad, bldrs. model, dbl. insul. glass, 6'-0" x 4'-0", 3 panel		10	1.600		1,050	62		1,112	1,250
1340	9'-0" x 4'-0", 4 panel		8	2		1,425	77.50		1,502.50	1,700
1380	10'-0" x 6'-0", 5 panels		7	2.286		2,350	88.50		2,438.50	2,750
1420	12'-0" x 6'-0", 6 panels		6	2.667		3,100	103		3,203	3,575
2000	Bay window, 8' x 5', dbl. insul. glass		10	1.600		1,925	62		1,987	2,225
2050	Low E glass		10	1.600		2,325	62		2,387	2,675
2100	12'-0" x 6'-0", dbl. insul. glass, 6 panels		6	2.667		2,400	103		2,503	2,825
2200	Low E glass		6	2.667		3,275	103		3,378	3,775
2280	6'-0" x 4'-0"		11	1.455		1,275	56.50		1,331.50	1,500
2300	Vinyl-clad, premium, dbl. insul. glass, 8'-0" x 5'-0"		10	1.600		1,800	62		1,862	2,100
2340	10'-0" x 5'-0"		8	2		2,425	77.50		2,502.50	2,800
2380	10'-0" x 6'-0"		7	2.286		2,825	88.50		2,913.50	3,275
2420	12'-0" x 6'-0"		6	2.667		3,375	103		3,478	3,900
3300	Vinyl-clad, premium, dbl. insul. glass, 7'-0" x 4'-6"		10	1.600		1,400	62		1,462	1,650
3340	8'-0" x 4'-6"		8	2		1,425	77.50		1,502.50	1,700
3380	8'-0" x 5'-0"		7	2.286		1,500	88.50		1,588.50	1,800
3420	9'-0" x 5'-0"	↓	6	2.667	↓	1,550	103		1,653	1,875

08 52 16.15 Awning Window Vinyl-Clad

		Crew	Daily Output	Labor-Hours	Unit	Material	Labor	Equipment	Total	Total Incl O&P
0010	**AWNING WINDOW VINYL-CLAD** including frames, screens, and grilles									
0240	Vinyl-clad, 34" x 22"	1 Carp	10	.800	Ea.	269	31		300	345
0280	36" x 28"		9	.889		310	34.50		344.50	395
0300	36" x 36"		9	.889		345	34.50		379.50	435
0340	40" x 22"		10	.800		296	31		327	375
0360	48" x 28"		8	1		375	39		414	475
0380	60" x 36"	↓	8	1	↓	515	39		554	635

08 52 Wood Windows

08 52 16 – Plastic-Clad Wood Windows

08 52 16.30 Palladian Windows

Line	Description		Crew	Daily Output	Labor-Hours	Unit	Material	Labor	Equipment	Total	Total Incl O&P
							2019 Bare Costs				
0010	**PALLADIAN WINDOWS**										
0020	Vinyl-clad, double insulated glass, including frame and grilles										
0040	3′-2″ x 2′-6″ high		2 Carp	11	1.455	Ea.	1,275	56.50		1,331.50	1,500
0060	3′-2″ x 4′-10″			11	1.455		1,775	56.50		1,831.50	2,050
0080	3′-2″ x 6′-4″			10	1.600		1,775	62		1,837	2,050
0100	4′-0″ x 4′-0″			10	1.600		1,575	62		1,637	1,850
0120	4′-0″ x 5′-4″		3 Carp	10	2.400		1,900	93		1,993	2,250
0140	4′-0″ x 6′-0″			9	2.667		1,900	103		2,003	2,275
0160	4′-0″ x 7′-4″			9	2.667		2,150	103		2,253	2,525
0180	5′-5″ x 4′-10″			9	2.667		2,300	103		2,403	2,725
0200	5′-5″ x 6′-10″			9	2.667		2,650	103		2,753	3,100
0220	5′-5″ x 7′-9″			9	2.667		2,825	103		2,928	3,275
0240	6′-0″ x 7′-11″			8	3		3,600	116		3,716	4,175
0260	8′-0″ x 6′-0″			8	3		3,200	116		3,316	3,725

08 52 16.35 Double-Hung Window

Line	Description		Crew	Daily Output	Labor-Hours	Unit	Material	Labor	Equipment	Total	Total Incl O&P
0010	**DOUBLE-HUNG WINDOW** including frames, screens, and grilles										
0300	Vinyl-clad, premium, double insulated glass, 2′-6″ x 3′-0″	G	1 Carp	10	.800	Ea.	345	31		376	425
0305	2′-6″ x 4′-0″	G		10	.800		380	31		411	465
0400	3′-0″ x 3′-6″	G		10	.800		345	31		376	430
0500	3′-0″ x 4′-0″	G		9	.889		405	34.50		439.50	500
0600	3′-0″ x 4′-6″	G		9	.889		430	34.50		464.50	525
0700	3′-0″ x 5′-0″	G		8	1		460	39		499	570
0790	3′-4″ x 5′-0″	G		8	1		450	39		489	560
0800	3′-6″ x 6′-0″	G		8	1		520	39		559	635
0820	4′-0″ x 5′-0″	G		7	1.143		575	44.50		619.50	710
0830	4′-0″ x 6′-0″	G		7	1.143		730	44.50		774.50	875

08 52 16.40 Transom Windows

Line	Description		Crew	Daily Output	Labor-Hours	Unit	Material	Labor	Equipment	Total	Total Incl O&P
0010	**TRANSOM WINDOWS**										
0050	Vinyl-clad, premium, dbl. insul. glass, 32″ x 8″		1 Carp	16	.500	Ea.	193	19.40		212.40	244
0100	36″ x 8″			16	.500		209	19.40		228.40	261
0110	36″ x 12″			16	.500		224	19.40		243.40	278
0200	44″ x 48″			12	.667		595	26		621	700
1000	Vinyl-clad, premium, dbl. insul. glass, 4′-0″ x 4′-0″		2 Carp	12	1.333		530	51.50		581.50	670
1100	4′-0″ x 6′-0″			11	1.455		980	56.50		1,036.50	1,175
1200	5′-0″ x 6′-0″			10	1.600		1,075	62		1,137	1,300
1300	6′-0″ x 6′-0″			10	1.600		1,125	62		1,187	1,325

08 52 50 – Window Accessories

08 52 50.10 Window Grille or Muntin

Line	Description	Crew	Daily Output	Labor-Hours	Unit	Material	Labor	Equipment	Total	Total Incl O&P
0010	**WINDOW GRILLE OR MUNTIN**, snap in type									
0020	Standard pattern interior grilles									
2000	Wood, awning window, glass size, 28″ x 16″ high	1 Carp	30	.267	Ea.	31	10.35		41.35	51
2060	44″ x 24″ high		32	.250		45	9.70		54.70	65.50
2100	Casement, glass size, 20″ x 36″ high		30	.267		34	10.35		44.35	54.50
2180	20″ x 56″ high		32	.250		46.50	9.70		56.20	67
2200	Double-hung, glass size, 16″ x 24″ high		24	.333	Set	57.50	12.90		70.40	84.50
2280	32″ x 32″ high		34	.235	″	141	9.10		150.10	170
2500	Picture, glass size, 48″ x 48″ high		30	.267	Ea.	129	10.35		139.35	159
2580	60″ x 68″ high		28	.286	″	201	11.05		212.05	239
2600	Sliding, glass size, 14″ x 36″ high		24	.333	Set	38	12.90		50.90	63.50
2680	36″ x 36″ high		22	.364	″	45.50	14.10		59.60	73.50

08 52 Wood Windows

08 52 69 – Wood Storm Windows

08 52 69.10 Storm Windows			Crew	Daily Output	Labor-Hours	Unit	2019 Bare Costs Material	Labor	Equipment	Total	Total Incl O&P
0010	**STORM WINDOWS**, aluminum residential										
0300	Basement, mill finish, incl. fiberglass screen										
0320	1'-10" x 1'-0" high	G	2 Carp	30	.533	Ea.	36	20.50		56.50	74
0340	2'-9" x 1'-6" high	G		30	.533		39	20.50		59.50	77
0360	3'-4" x 2'-0" high	G		30	.533		46.50	20.50		67	85
1600	Double-hung, combination, storm & screen										
1700	Custom, clear anodic coating, 2'-0" x 3'-5" high		2 Carp	30	.533	Ea.	99.50	20.50		120	144
1720	2'-6" x 5'-0" high			28	.571		125	22		147	175
1740	4'-0" x 6'-0" high			25	.640		237	25		262	300
1800	White painted, 2'-0" x 3'-5" high			30	.533		116	20.50		136.50	162
1820	2'-6" x 5'-0" high			28	.571		176	22		198	231
1840	4'-0" x 6'-0" high			25	.640		288	25		313	355
2000	Clear anodic coating, 2'-0" x 3'-5" high	G		30	.533		97	20.50		117.50	141
2020	2'-6" x 5'-0" high	G		28	.571		125	22		147	175
2040	4'-0" x 6'-0" high	G		25	.640		136	25		161	190
2400	White painted, 2'-0" x 3'-5" high	G		30	.533		96	20.50		116.50	140
2420	2'-6" x 5'-0" high	G		28	.571		100	22		122	147
2440	4'-0" x 6'-0" high	G		25	.640		116	25		141	169
2600	Mill finish, 2'-0" x 3'-5" high	G		30	.533		91	20.50		111.50	134
2620	2'-6" x 5'-0" high	G		28	.571		96	22		118	143
2640	4'-0" x 6'-8" high	G		25	.640		116	25		141	169
4000	Picture window, storm, 1 lite, white or bronze finish										
4020	4'-6" x 4'-6" high		2 Carp	25	.640	Ea.	142	25		167	198
4040	5'-8" x 4'-6" high			20	.800		157	31		188	224
4400	Mill finish, 4'-6" x 4'-6" high			25	.640		142	25		167	198
4420	5'-8" x 4'-6" high			20	.800		162	31		193	229
4600	3 lite, white or bronze finish										
4620	4'-6" x 4'-6" high		2 Carp	25	.640	Ea.	166	25		191	224
4640	5'-8" x 4'-6" high			20	.800		176	31		207	244
4800	Mill finish, 4'-6" x 4'-6" high			25	.640		162	25		187	219
4820	5'-8" x 4'-6" high			20	.800		166	31		197	234
6000	Sliding window, storm, 2 lite, white or bronze finish										
6020	3'-4" x 2'-7" high		2 Carp	28	.571	Ea.	145	22		167	197
6040	4'-4" x 3'-3" high			25	.640		161	25		186	218
6060	5'-4" x 6'-0" high			20	.800		217	31		248	290
9000	Interior storm window										
9100	Storm window interior glass		1 Glaz	107	.075	S.F.	6.30	2.86		9.16	11.65

08 53 Plastic Windows

08 53 13 – Vinyl Windows

08 53 13.10 Solid Vinyl Windows		Crew	Daily Output	Labor-Hours	Unit	Material	Labor	Equipment	Total	Total Incl O&P
0010	**SOLID VINYL WINDOWS**									
0020	Double-hung, including frame and screen, 2'-0" x 2'-6"	2 Carp	15	1.067	Ea.	206	41.50		247.50	296
0040	2'-0" x 3'-6"		14	1.143		216	44.50		260.50	310
0060	2'-6" x 4'-6"		13	1.231		257	47.50		304.50	360
0080	3'-0" x 4'-0"		10	1.600		229	62		291	355
0100	3'-0" x 4'-6"		9	1.778		294	69		363	440
0120	3'-6" x 4'-6"		8	2		310	77.50		387.50	470
0140	3'-6" x 6'-0"		7	2.286		385	88.50		473.50	570

08 53 Plastic Windows

08 53 13 – Vinyl Windows

08 53 13.20 Vinyl Single-Hung Windows		Crew	Daily Output	Labor-Hours	Unit	2019 Bare Costs Material	Labor	Equipment	Total	Total Incl O&P
0010	**VINYL SINGLE-HUNG WINDOWS**, insulated glass									
0020	Grids, low E, J fin, extension jambs									
0130	25" x 41" G	2 Carp	20	.800	Ea.	204	31		235	276
0140	25" x 49" G		18	.889		214	34.50		248.50	293
0150	25" x 57" G		17	.941		228	36.50		264.50	310
0160	25" x 65" G		16	1		247	39		286	335
0170	29" x 41" G		18	.889		206	34.50		240.50	283
0180	29" x 53" G		18	.889		225	34.50		259.50	305
0190	29" x 57" G		17	.941		238	36.50		274.50	325
0200	29" x 65" G		16	1		248	39		287	335
0210	33" x 41" G		20	.800		220	31		251	293
0220	33" x 53" G		18	.889		239	34.50		273.50	320
0230	33" x 57" G		17	.941		248	36.50		284.50	335
0240	33" x 65" G		16	1		258	39		297	350
0250	37" x 41" G		20	.800		251	31		282	325
0260	37" x 53" G		18	.889		275	34.50		309.50	355
0270	37" x 57" G		17	.941		285	36.50		321.50	375
0280	37" x 65" G		16	1		300	39		339	395

08 53 13.30 Vinyl Double-Hung Windows		Crew	Daily Output	Labor-Hours	Unit	Material	Labor	Equipment	Total	Total Incl O&P
0010	**VINYL DOUBLE-HUNG WINDOWS**, insulated glass									
0100	Grids, low E, J fin, ext. jambs, 21" x 53" G	2 Carp	18	.889	Ea.	216	34.50		250.50	294
0102	21" x 37" G		18	.889		229	34.50		263.50	310
0104	21" x 41" G		18	.889		239	34.50		273.50	320
0106	21" x 49" G		18	.889		249	34.50		283.50	330
0110	21" x 57" G		17	.941		268	36.50		304.50	355
0120	21" x 65" G		16	1		283	39		322	375
0128	25" x 37" G		20	.800		238	31		269	315
0130	25" x 41" G		20	.800		248	31		279	325
0140	25" x 49" G		18	.889		255	34.50		289.50	340
0145	25" x 53" G		18	.889		274	34.50		308.50	355
0150	25" x 57" G		17	.941		278	36.50		314.50	365
0160	25" x 65" G		16	1		293	39		332	385
0162	25" x 69" G		16	1		297	39		336	390
0164	25" x 77" G		16	1		320	39		359	415
0168	29" x 37" G		18	.889		249	34.50		283.50	330
0170	29" x 41" G		18	.889		263	34.50		297.50	345
0172	29" x 49" G		18	.889		268	34.50		302.50	350
0180	29" x 53" G		18	.889		278	34.50		312.50	360
0190	29" x 57" G		17	.941		293	36.50		329.50	380
0200	29" x 65" G		16	1		305	39		344	400
0202	29" x 69" G		16	1		345	39		384	445
0205	29" x 77" G		16	1		335	39		374	430
0208	33" x 37" G		20	.800		268	31		299	345
0210	33" x 41" G		20	.800		269	31		300	345
0215	33" x 49" G		20	.800		283	31		314	360
0220	33" x 53" G		18	.889		288	34.50		322.50	370
0230	33" x 57" G		17	.941		297	36.50		333.50	385
0240	33" x 65" G		16	1		315	39		354	415
0242	33" x 69" G		16	1		325	39		364	420
0246	33" x 77" G		16	1		345	39		384	445
0250	37" x 41" G		20	.800		289	31		320	370
0255	37" x 49" G		20	.800		297	31		328	375

08 53 Plastic Windows

08 53 13 – Vinyl Windows

08 53 13.30 Vinyl Double-Hung Windows

	08 53 13.30 Vinyl Double-Hung Windows		Crew	Daily Output	Labor-Hours	Unit	Material	2019 Bare Costs Labor	Equipment	Total	Total Incl O&P
0260	37″ x 53″	G	2 Carp	18	.889	Ea.	315	34.50		349.50	400
0270	37″ x 57″	G		17	.941		340	36.50		376.50	435
0280	37″ x 65″	G		16	1		355	39		394	455
0282	37″ x 69″	G		16	1		360	39		399	460
0286	37″ x 77″	G		16	1		375	39		414	475
0300	Solid vinyl, average quality, double insulated glass, 2′-0″ x 3′-0″	G	1 Carp	10	.800		293	31		324	370
0310	3′-0″ x 4′-0″	G		9	.889		217	34.50		251.50	295
0330	Premium, double insulated glass, 2′-6″ x 3′-0″	G		10	.800		278	31		309	355
0340	3′-0″ x 3′-6″	G		9	.889		310	34.50		344.50	395
0350	3′-0″ x 4′-0″	G		9	.889		335	34.50		369.50	425
0360	3′-0″ x 4′-6″	G		9	.889		340	34.50		374.50	430
0370	3′-0″ x 5′-0″	G		8	1		365	39		404	465
0380	3′-6″ x 6′-0″	G		8	1		410	39		449	515

08 53 13.40 Vinyl Casement Windows

	08 53 13.40 Vinyl Casement Windows		Crew	Daily Output	Labor-Hours	Unit	Material	Labor	Equipment	Total	Total Incl O&P
0010	**VINYL CASEMENT WINDOWS**, insulated glass										
0015	Grids, low E, J fin, extension jambs, screens										
0100	One lite, 21″ x 41″	G	2 Carp	20	.800	Ea.	320	31		351	400
0110	21″ x 47″	G		20	.800		335	31		366	420
0120	21″ x 53″	G		20	.800		355	31		386	440
0128	24″ x 35″	G		19	.842		305	32.50		337.50	390
0130	24″ x 41″	G		19	.842		325	32.50		357.50	415
0140	24″ x 47″	G		19	.842		340	32.50		372.50	430
0150	24″ x 53″	G		19	.842		370	32.50		402.50	460
0158	28″ x 35″	G		19	.842		315	32.50		347.50	405
0160	28″ x 41″	G		19	.842		330	32.50		362.50	415
0170	28″ x 47″	G		19	.842		365	32.50		397.50	455
0180	28″ x 53″	G		19	.842		390	32.50		422.50	485
0184	28″ x 59″	G		19	.842		400	32.50		432.50	495
0188	Two lites, 33″ x 35″	G		18	.889		530	34.50		564.50	640
0190	33″ x 41″	G		18	.889		530	34.50		564.50	635
0200	33″ x 47″	G		18	.889		555	34.50		589.50	665
0210	33″ x 53″	G		18	.889		570	34.50		604.50	685
0212	33″ x 59″	G		18	.889		625	34.50		659.50	745
0215	33″ x 72″	G		18	.889		645	34.50		679.50	760
0220	41″ x 41″	G		18	.889		565	34.50		599.50	675
0230	41″ x 47″	G		18	.889		600	34.50		634.50	715
0240	41″ x 53″	G		17	.941		645	36.50		681.50	770
0242	41″ x 59″	G		17	.941		680	36.50		716.50	810
0246	41″ x 72″	G		17	.941		710	36.50		746.50	840
0250	47″ x 41″	G		17	.941		575	36.50		611.50	695
0260	47″ x 47″	G		17	.941		605	36.50		641.50	725
0270	47″ x 53″	G		17	.941		645	36.50		681.50	765
0272	47″ x 59″	G		17	.941		715	36.50		751.50	850
0280	56″ x 41″	G		15	1.067		620	41.50		661.50	750
0290	56″ x 47″	G		15	1.067		645	41.50		686.50	775
0300	56″ x 53″	G		15	1.067		715	41.50		756.50	860
0302	56″ x 59″	G		15	1.067		740	41.50		781.50	885
0310	56″ x 72″	G		15	1.067		795	41.50		836.50	945
0340	Solid vinyl, premium, double insulated glass, 2′-0″ x 3′-0″ high	G	1 Carp	10	.800		272	31		303	350
0360	2′-0″ x 4′-0″ high	G		9	.889		300	34.50		334.50	385
0380	2′-0″ x 5′-0″ high	G		8	1		350	39		389	450

08 53 Plastic Windows

08 53 13 – Vinyl Windows

08 53 13.50 Vinyl Picture Windows			Crew	Daily Output	Labor-Hours	Unit	Material	2019 Bare Costs Labor	Equipment	Total	Total Incl O&P
0010	**VINYL PICTURE WINDOWS**, insulated glass										
0120	Grids, low E, J fin, ext. jambs, 47″ x 35″		2 Carp	12	1.333	Ea.	300	51.50		351.50	415
0130	47″ x 41″			12	1.333		455	51.50		506.50	590
0140	47″ x 47″			12	1.333		345	51.50		396.50	465
0150	47″ x 53″			11	1.455		370	56.50		426.50	500
0160	71″ x 35″			11	1.455		445	56.50		501.50	585
0170	71″ x 41″			11	1.455		465	56.50		521.50	610
0180	71″ x 47″		↓	11	1.455	↓	500	56.50		556.50	645

08 54 Composite Windows

08 54 13 – Fiberglass Windows

08 54 13.10 Fiberglass Single-Hung Windows			Crew	Daily Output	Labor-Hours	Unit	Material	Labor	Equipment	Total	Total Incl O&P
0010	**FIBERGLASS SINGLE-HUNG WINDOWS**										
0100	Grids, low E, 18″ x 24″	G	2 Carp	18	.889	Ea.	380	34.50		414.50	470
0110	18″ x 40″	G		17	.941		365	36.50		401.50	465
0130	24″ x 40″	G		20	.800		390	31		421	475
0230	36″ x 36″	G		17	.941		410	36.50		446.50	515
0250	36″ x 48″	G		20	.800		445	31		476	540
0260	36″ x 60″	G		18	.889		495	34.50		529.50	600
0280	36″ x 72″	G		16	1		530	39		569	645
0290	48″ x 40″	G	↓	16	1	↓	530	39		569	645
08 54 13.30 Fiberglass Slider Windows											
0010	**FIBERGLASS SLIDER WINDOWS**										
0100	Grids, low E, 36″ x 24″	G	2 Carp	20	.800	Ea.	385	31		416	470
0110	36″ x 36″	G	″	20	.800	″	385	31		416	470
08 54 13.50 Fiberglass Bay Windows											
0010	**FIBERGLASS BAY WINDOWS**										
0150	48″ x 36″	G	2 Carp	11	1.455	Ea.	1,125	56.50		1,181.50	1,350

08 61 Roof Windows

08 61 13 – Metal Roof Windows

08 61 13.10 Roof Windows		Crew	Daily Output	Labor-Hours	Unit	Material	Labor	Equipment	Total	Total Incl O&P
0010	**ROOF WINDOWS**, fixed high perf tmpd glazing, metallic framed									
0020	46″ x 21-1/2″, Flashed for shingled roof	1 Carp	8	1	Ea.	277	39		316	370
0100	46″ x 28″		8	1		310	39		349	405
0125	57″ x 44″		6	1.333		375	51.50		426.50	500
0130	72″ x 28″		7	1.143		375	44.50		419.50	490
0150	Fixed, laminated tempered glazing, 46″ x 21-1/2″		8	1		465	39		504	580
0175	46″ x 28″		8	1		515	39		554	635
0200	57″ x 44″		6	1.333		485	51.50		536.50	620
0500	Vented flashing set for shingled roof, 46″ x 21-1/2″		7	1.143		465	44.50		509.50	590
0525	46″ x 28″		6	1.333		515	51.50		566.50	655
0550	57″ x 44″		5	1.600		650	62		712	815
0560	72″ x 28″		5	1.600		650	62		712	815
0575	Flashing set for low pitched roof, 46″ x 21-1/2″		7	1.143		540	44.50		584.50	670
0600	46″ x 28″		7	1.143		590	44.50		634.50	725
0625	57″ x 44″		5	1.600		735	62		797	905
0650	Flashing set for curb, 46″ x 21-1/2″	↓	7	1.143	↓	605	44.50		649.50	745

08 61 Roof Windows

08 61 13 – Metal Roof Windows

08 61 13.10 Roof Windows		Crew	Daily Output	Labor-Hours	Unit	Material	Labor	Equipment	Total	Total Incl O&P
						2019 Bare Costs				
0675	46″ x 28″	1 Carp	7	1.143	Ea.	660	44.50		704.50	800
0700	57″ x 44″	↓	5	1.600	↓	805	62		867	990

08 61 16 – Wood Roof Windows

08 61 16.16 Roof Windows, Wood Framed		Crew	Daily Output	Labor-Hours	Unit	Material	Labor	Equipment	Total	Total Incl O&P
0010	**ROOF WINDOWS, WOOD FRAMED**									
5600	Roof window incl. frame, flashing, double insulated glass & screens,									
5610	complete unit, 22″ x 38″	2 Carp	3	5.333	Ea.	780	207		987	1,200
5650	2′-5″ x 3′-8″		3.20	5		985	194		1,179	1,400
5700	3′-5″ x 4′-9″	↓	3.40	4.706	↓	1,100	182		1,282	1,500

08 62 Unit Skylights

08 62 13 – Domed Unit Skylights

08 62 13.10 Domed Skylights			Crew	Daily Output	Labor-Hours	Unit	Material	Labor	Equipment	Total	Total Incl O&P
0010	**DOMED SKYLIGHTS**										
0020	Skylight, fixed dome type, 22″ x 22″	G	G-3	12	2.667	Ea.	216	98		314	400
0030	22″ x 46″	G		10	3.200		271	118		389	495
0040	30″ x 30″	G		12	2.667		285	98		383	480
0050	30″ x 46″	G		10	3.200		375	118		493	605
0110	Fixed, double glazed, 22″ x 27″	G		12	2.667		275	98		373	470
0120	22″ x 46″	G		10	3.200		291	118		409	515
0130	44″ x 46″	G		10	3.200		430	118		548	665
0210	Operable, double glazed, 22″ x 27″	G		12	2.667		405	98		503	610
0220	22″ x 46″	G		10	3.200		485	118		603	730
0230	44″ x 46″	G	↓	10	3.200	↓	925	118		1,043	1,225

08 62 13.20 Skylights			Crew	Daily Output	Labor-Hours	Unit	Material	Labor	Equipment	Total	Total Incl O&P
0010	**SKYLIGHTS**, flush or curb mounted										
2120	Ventilating insulated plexiglass dome with										
2130	curb mounting, 36″ x 36″	G	G-3	12	2.667	Ea.	480	98		578	690
2150	52″ x 52″	G		12	2.667		665	98		763	900
2160	28″ x 52″	G		10	3.200		500	118		618	740
2170	36″ x 52″	G	↓	10	3.200		540	118		658	790
2180	For electric opening system, add	G					320			320	350
2210	Operating skylight, with thermopane glass, 24″ x 48″	G	G-3	10	3.200		595	118		713	850
2220	32″ x 48″	G	″	9	3.556	↓	620	131		751	900
2310	Non venting insulated plexiglass dome skylight with										
2320	Flush mount 22″ x 46″	G	G-3	15.23	2.101	Ea.	335	77.50		412.50	500
2330	30″ x 30″	G		16	2		310	73.50		383.50	460
2340	46″ x 46″	G		13.91	2.301		570	84.50		654.50	765
2350	Curb mount 22″ x 46″	G		15.23	2.101		365	77.50		442.50	530
2360	30″ x 30″	G		16	2		405	73.50		478.50	565
2370	46″ x 46″	G		13.91	2.301		625	84.50		709.50	830
2381	Non-insulated flush mount 22″ x 46″			15.23	2.101		228	77.50		305.50	380
2382	30″ x 30″			16	2		205	73.50		278.50	350
2383	46″ x 46″			13.91	2.301		385	84.50		469.50	560
2384	Curb mount 22″ x 46″			15.23	2.101		191	77.50		268.50	340
2385	30″ x 30″		↓	16	2		188	73.50		261.50	330
4000	Skylight, solar tube kit, incl. dome, flashing, diffuser, 1 pipe, 10″ diam.	G	1 Carp	2	4		305	155		460	590
4010	14″ diam.	G		2	4		405	155		560	700
4020	21″ diam.	G		2	4		485	155		640	790
4030	Accessories for, 1′ long x 9″ diam. pipe	G	↓	24	.333	↓	50	12.90		62.90	76.50

08 62 Unit Skylights

08 62 13 – Domed Unit Skylights

08 62 13.20 Skylights		Crew	Daily Output	Labor-Hours	Unit	Material	2019 Bare Costs Labor	Equipment	Total	Total Incl O&P
4040	2′ long x 9″ diam. pipe G	1 Carp	24	.333	Ea.	42.50	12.90		55.40	68.50
4050	4′ long x 9″ diam. pipe G		20	.400		74	15.50		89.50	107
4060	1′ long x 13″ diam. pipe G		24	.333		70.50	12.90		83.40	99
4070	2′ long x 13″ diam. pipe G		24	.333		55	12.90		67.90	82
4080	4′ long x 13″ diam. pipe G		20	.400		102	15.50		117.50	138
4090	6.5″ turret ext. for 21″ diam. pipe G		16	.500		109	19.40		128.40	152
4100	12′ long x 21″ diam. flexible pipe G		12	.667		90	26		116	141
4110	45 degree elbow, 10″ G		16	.500		151	19.40		170.40	198
4120	14″ G		16	.500		84.50	19.40		103.90	125
4130	Interior decorative ring, 9″ G		20	.400		47.50	15.50		63	78
4140	13″ G		20	.400		70	15.50		85.50	102

08 63 Metal-Framed Skylights

08 63 13 – Domed Metal-Framed Skylights

08 63 13.20 Skylight Rigid Metal-Framed		Crew	Daily Output	Labor-Hours	Unit	Material	Labor	Equipment	Total	Total Incl O&P
0010	**SKYLIGHT RIGID METAL-FRAMED** skylight framing is aluminum									
0050	Fixed acrylic double domes, curb mount, 25-1/2″ x 25-1/2″	G-3	10	3.200	Ea.	212	118		330	430
0060	25-1/2″ x 33-1/2″		10	3.200		246	118		364	465
0070	25-1/2″ x 49-1/2″		6	5.333		281	196		477	635
0080	33-1/2″ x 33-1/2″		8	4		300	147		447	575
0090	37-1/2″ x 25-1/2″		8	4		254	147		401	525
0100	37-1/2″ x 37-1/2″		8	4		249	147		396	520
0110	37-1/2″ x 49-1/2″		6	5.333		425	196		621	795
0120	49-1/2″ x 33-1/2″		6	5.333		380	196		576	745
0130	49-1/2″ x 49-1/2″		6	5.333		470	196		666	845
1000	Fixed tempered glass, curb mount, 17-1/2″ x 33-1/2″		10	3.200		173	118		291	385
1020	17-1/2″ x 49-1/2″		6	5.333		193	196		389	540
1030	25-1/2″ x 25-1/2″		10	3.200		173	118		291	385
1040	25-1/2″ x 33-1/2″		10	3.200		204	118		322	420
1050	25-1/2″ x 37-1/2″		8	4		214	147		361	480
1060	25-1/2″ x 49-1/2″		6	5.333		226	196		422	575
1070	25-1/2″ x 73-1/2″		6	5.333		370	196		566	730
1080	33-1/2″ x 33-1/2″		8	4		225	147		372	490
2000	Manual vent tempered glass & screen, curb, 25-1/2″ x 25-1/2″		10	3.200		410	118		528	645
2020	25-1/2″ x 37-1/2″		10	3.200		470	118		588	710
2030	25-1/2″ x 49-1/2″		8	4		505	147		652	800
2040	33-1/2″ x 33-1/2″		8	4		540	147		687	840
2050	33-1/2″ x 49-1/2″		6	5.333		680	196		876	1,075
2060	37-1/2″ x 37-1/2″		6	5.333		620	196		816	1,000
2070	49-1/2″ x 49-1/2″		6	5.333		795	196		991	1,200
3000	Electric vent tempered glass, curb mount, 25-1/2″ x 25-1/2″		10	3.200		1,000	118		1,118	1,325
3020	25-1/2″ x 37-1/2″		10	3.200		1,100	118		1,218	1,400
3030	25-1/2″ x 49-1/2″		8	4		1,175	147		1,322	1,525
3040	33-1/2″ x 33-1/2″		8	4		1,175	147		1,322	1,550
3050	33-1/2″ x 49-1/2″		6	5.333		1,250	196		1,446	1,725
3060	37-1/2″ x 37-1/2″		6	5.333		1,250	196		1,446	1,700
3070	49-1/2″ x 49-1/2″		6	5.333		1,350	196		1,546	1,825

08 71 Door Hardware

08 71 13 – Automatic Door Operators

08 71 13.10 Automatic Openers Commercial		Crew	Daily Output	Labor-Hours	Unit	Material	Labor (2019 Bare Costs)	Equipment	Total	Total Incl O&P
0010	AUTOMATIC OPENERS COMMERCIAL									
0020	Pneumatic door opener incl. motion sens, control box, tubing, compressor									
0050	For single swing door, per opening	2 Skwk	.80	20	Ea.	4,725	790		5,515	6,500
0100	Pair, per opening		.50	32	Opng.	7,700	1,250		8,950	10,600
1000	For single sliding door, per opening		.60	26.667		5,200	1,050		6,250	7,450
1300	Bi-parting pair		.50	32		7,800	1,250		9,050	10,700
1420	Electronic door opener incl. motion sens, 12 V control box, motor									
1450	For single swing door, per opening	2 Skwk	.80	20	Opng.	3,850	790		4,640	5,550
1500	Pair, per opening		.50	32		7,450	1,250		8,700	10,300
1600	For single sliding door, per opening		.60	26.667		4,900	1,050		5,950	7,150
1700	Bi-parting pair		.50	32		5,575	1,250		6,825	8,250
1750	Handicap actuator buttons, 2, incl. 12 V DC wiring, add	1 Carp	1.50	5.333	Pr.	490	207		697	880
2000	Electric panic button for door	2 Skwk	.50	32	Opng.	207	1,250		1,457	2,325

08 71 20 – Hardware

08 71 20.10 Bolts, Flush

Line	Description	Crew	Daily Output	Labor-Hours	Unit	Material	Labor	Equipment	Total	Total Incl O&P
0010	BOLTS, FLUSH									
0020	Standard, concealed	1 Carp	7	1.143	Ea.	23.50	44.50		68	99
0800	Automatic fire exit	"	5	1.600		177	62		239	297
1600	Electrified dead bolt	1 Elec	3	2.667		176	122		298	390
2000	Brass, UL rated, with lever extension	1 Carp	40	.200		36	7.75		43.75	52.50
2500	Surface bolt, 6″ long, with strike		32	.250		28.50	9.70		38.20	47.50
3000	Barrel, brass, 2″ long		40	.200		5.25	7.75		13	18.55
3020	4″ long		40	.200		13.20	7.75		20.95	27.50
3060	6″ long		40	.200		24.50	7.75		32.25	40

08 71 20.15 Hardware

Line	Description	Crew	Daily Output	Labor-Hours	Unit	Material	Labor	Equipment	Total	Total Incl O&P
0010	HARDWARE									
0020	Average hardware cost									
1000	Door hardware, apartment, interior	1 Carp	4	2	Door	545	77.50		622.50	730
1300	Average, door hardware, motel/hotel interior, with access card		4	2		690	77.50		767.50	885
1500	Hospital bedroom, average quality		4	2		705	77.50		782.50	905
2000	High quality		3	2.667		710	103		813	950
2100	Pocket door		6	1.333	Ea.	113	51.50		164.50	210
2250	School, single exterior, incl. lever, incl. panic device		3	2.667	Door	1,575	103		1,678	1,900
2500	Single interior, regular use, lever included		3	2.667		575	103		678	805
2550	Ave., door hdwe., school, classroom, ANSI F84, lever handle		3	2.667		810	103		913	1,075
2600	Ave., door hdwe. set, school, classroom, ANSI F88, incl. lever		3	2.667		875	103		978	1,125
2850	Stairway, single interior		3	2.667		635	103		738	865
3100	Double exterior, with panic device		2	4	Pr.	2,975	155		3,130	3,525
6020	Add for fire alarm door holder, electro-magnetic	1 Elec	4	2	Ea.	91.50	91		182.50	250

08 71 20.20 Door Protectors

Line	Description	Crew	Daily Output	Labor-Hours	Unit	Material	Labor	Equipment	Total	Total Incl O&P
0010	DOOR PROTECTORS									
0020	1-3/4″ x 3/4″ U channel	2 Carp	80	.200	L.F.	24.50	7.75		32.25	40
0021	1-3/4″ x 1-1/4″ U channel		80	.200	"	20.50	7.75		28.25	35.50
1000	Tear drop, spring stl., 8″ high x 19″ long		15	1.067	Ea.	104	41.50		145.50	183
1010	8″ high x 32″ long		15	1.067		128	41.50		169.50	210
1100	Tear drop, stainless stl., 8″ high x 19″ long		15	1.067		345	41.50		386.50	450
1200	8″ high x 32″ long		15	1.067		355	41.50		396.50	460

08 71 20.30 Door Closers		Crew	Daily Output	Labor-Hours	Unit	Material	2019 Bare Costs Labor	Equipment	Total	Total Incl O&P
0010	**DOOR CLOSERS** adjustable backcheck, multiple mounting									
0015	and rack and pinion									
0020	Standard Regular Arm	1 Carp	6	1.333	Ea.	204	51.50		255.50	310
0023	Install new door closer	"	4.61	1.735		204	67.50		271.50	335
0025	Door closer, adjustable backset, 3-way mount					204			204	225
0040	Hold open arm	1 Carp	6	1.333		101	51.50		152.50	197
0100	Fusible link		6.50	1.231		178	47.50		225.50	275
0210	Light duty, regular arm		6	1.333		122	51.50		173.50	220
0220	Parallel arm		6	1.333		144	51.50		195.50	245
0230	Hold open arm		6	1.333		130	51.50		181.50	229
0240	Fusible link arm		6	1.333		159	51.50		210.50	261
0250	Medium duty, regular arm		6	1.333		138	51.50		189.50	238
0500	Surface mount regular arm		6.50	1.231		156	47.50		203.50	250
0550	Fusible link		6.50	1.231		157	47.50		204.50	251
1000	3′-8″ wide					160			160	176
1500	Concealed closers, normal use, head, pivot hung, interior	1 Carp	5.50	1.455		410	56.50		466.50	545
1510	Exterior		5.50	1.455		450	56.50		506.50	590
1520	Overhead concealed, all sizes, regular arm		5.50	1.455		233	56.50		289.50	350
1525	Concealed arm		5	1.600		345	62		407	480
1530	Concealed in door, all sizes, regular arm		5.50	1.455		360	56.50		416.50	490
1535	Concealed arm		5	1.600		278	62		340	405
1550	Floor		2.20	3.636		292	141		433	555
1560	Floor concealed, all sizes, single acting		2.20	3.636		570	141		711	860
1565	Double acting		2.20	3.636		560	141		701	850
1570	Interior, floor, offset pivot, single acting		3.50	2.286		915	88.50		1,003.50	1,150
1590	Exterior		3.50	2.286		1,000	88.50		1,088.50	1,250
1610	Hold open arm		6	1.333		455	51.50		506.50	585
1620	Double acting, standard arm		6	1.333		850	51.50		901.50	1,025
1630	Hold open arm		6	1.333		855	51.50		906.50	1,025
1640	Floor, center hung, single acting, bottom arm		6	1.333		455	51.50		506.50	585
1650	Double acting		6	1.333		525	51.50		576.50	660
1660	Offset hung, single acting, bottom arm		6	1.333		625	51.50		676.50	770
2000	Backcheck and adjustable power, hinge face mount									
5000	For cast aluminum cylinder, deduct				Ea.	38.50			38.50	42.50
5010	For delayed action surface mounted, add	1 Carp	6	1.333		39.50	51.50		91	129
5040	For delayed action, add					52.50			52.50	58
5080	For fusible link arm, add					46.50			46.50	51
5120	For shock absorbing arm, add					53.50			53.50	59
5160	For spring power adjustment, add					43			43	47
6000	Closer-holder, hinge face mount, all sizes, exposed arm	1 Carp	6.50	1.231		226	47.50		273.50	325
6500	Electro magnetic closer/holder									
6510	Single point, no detector	1 Carp	4	2	Ea.	560	77.50		637.50	745
6515	Including detector		4	2		730	77.50		807.50	930
6520	Multi-point, no detector		4	2		990	77.50		1,067.50	1,225
6524	Including detector		4	2		1,500	77.50		1,577.50	1,775
6550	Electric automatic operators									
6555	Operator	1 Carp	4	2	Ea.	2,100	77.50		2,177.50	2,450
6570	Wall plate actuator		4	2		252	77.50		329.50	405
8000	Surface mounted, standard duty, parallel arm, primed, traditional		6	1.333		203	51.50		254.50	310
8030	Light duty		6	1.333		167	51.50		218.50	269
8042	Extra duty parallel arm		6	1.333		126	51.50		177.50	225
8044	Hold open arm		6	1.333		160	51.50		211.50	262

08 71 20.30 Door Closers

08 71 20.30 Door Closers		Crew	Daily Output	Labor-Hours	Unit	Material	Labor	Equipment	Total	Total Incl O&P
						2019 Bare Costs				
8046	Positive stop arm	1 Carp	6	1.333	Ea.	237	51.50		288.50	345
8050	Heavy duty		6	1.333		247	51.50		298.50	360
8052	Heavy duty, regular arm		6	1.333		239	51.50		290.50	350
8054	Top jamb mount		6	1.333		239	51.50		290.50	350
8056	Extra duty parallel arm		6	1.333		241	51.50		292.50	350
8058	Hold open arm		6	1.333		294	51.50		345.50	410
8060	Positive stop arm		6	1.333		263	51.50		314.50	375
8062	Fusible link arm		6	1.333		284	51.50		335.50	400
8080	Universal heavy duty, regular arm		6	1.333		251	51.50		302.50	360
8084	Parallel arm		6	1.333		251	51.50		302.50	360
8088	Extra duty, parallel arm		6	1.333		282	51.50		333.50	395
8090	Hold open arm		6	1.333		292	51.50		343.50	405
8094	Positive stop arm		6	1.333		310	51.50		361.50	425
8100	Standard duty, parallel arm, modern		6	1.333		282	51.50		333.50	395
8150	Heavy duty		6	1.333		320	51.50		371.50	435

08 71 20.36 Panic Devices

08 71 20.36 Panic Devices		Crew	Daily Output	Labor-Hours	Unit	Material	Labor	Equipment	Total	Total Incl O&P
0010	**PANIC DEVICES** R087110-10									
0015	Touch bars various styles									
0040	Single door exit only, rim device, wide stile									
0050	Economy US28	1 Carp	5	1.600	Ea.	286	62		348	415
0060	Standard duty US28		5	1.600		785	62		847	965
0065	US26D		5	1.600		805	62		867	985
0070	US10		5	1.600		840	62		902	1,025
0075	US3		5	1.600		955	62		1,017	1,150
0080	Night latch, economy US28		5	1.600		355	62		417	490
0085	Standard duty, night latch US28		5	1.600		885	62		947	1,075
0090	US26D		5	1.600		1,050	62		1,112	1,250
0095	US10		5	1.600		1,050	62		1,112	1,250
0100	US3		5	1.600		1,200	62		1,262	1,425
0500	Single door exit only, rim device, narrow stile									
0540	Economy US28	1 Carp	5	1.600	Ea.	455	62		517	600
0550	Standard duty US28		5	1.600		840	62		902	1,025
0565	US26D		5	1.600		795	62		857	975
0570	US10		5	1.600		860	62		922	1,050
0575	US3		5	1.600		1,075	62		1,137	1,275
0580	Economy with night latch US28		5	1.600		345	62		407	480
0585	Standard duty US28		5	1.600		960	62		1,022	1,150
0590	US26D		5	1.600		1,150	62		1,212	1,375
0595	US10		5	1.600		965	62		1,027	1,150
0600	US32D		5	1.600		1,275	62		1,337	1,500
1040	Single door exit only, surface vertical rod									
1050	Surface vertical rod, economy US28	1 Carp	4	2	Ea.	520	77.50		597.50	705
1060	Surface vertical rod, standard duty US28		4	2		1,100	77.50		1,177.50	1,325
1065	US26D		4	2		1,250	77.50		1,327.50	1,500
1070	US10		4	2		1,175	77.50		1,252.50	1,400
1075	US3		4	2		1,225	77.50		1,302.50	1,475
1080	US32D		4	2		1,300	77.50		1,377.50	1,550
1500	Single door exit, rim, narrow stile, economy, surface vertical rod									
1540	Surface vertical rod, narrow, economy US28	1 Carp	4	2	Ea.	525	77.50		602.50	705
2040	Single door exit only, concealed vertical rod									
2042	These devices will require separate and extra door preparation									
2060	Concealed rod, standard duty US28	1 Carp	4	2	Ea.	1,075	77.50		1,152.50	1,300

08 71 20.36 Panic Devices

	08 71 20.36 Panic Devices	Crew	Daily Output	Labor-Hours	Unit	2019 Bare Costs Material	Labor	Equipment	Total	Total Incl O&P
2065	US26D	1 Carp	4	2	Ea.	1,200	77.50		1,277.50	1,450
2070	US10		4	2		1,300	77.50		1,377.50	1,575
2560	Concealed rod, narrow, standard duty US28		4	2		1,200	77.50		1,277.50	1,450
2565	US26D		4	2		1,425	77.50		1,502.50	1,700
2567	US26		4	2		1,500	77.50		1,577.50	1,775
2570	US10		4	2		1,250	77.50		1,327.50	1,500
2575	US3		4	2		1,450	77.50		1,527.50	1,725
3040	Single door exit only, mortise device, wide stile									
3042	These devices must be combined with a mortise lock									
3060	Mortise, standard duty US28	1 Carp	4	2	Ea.	1,125	77.50		1,202.50	1,375
3065	US26D		4	2		1,125	77.50		1,202.50	1,350
3067	US26		4	2		1,350	77.50		1,427.50	1,600
3070	US10		4	2		1,225	77.50		1,302.50	1,475
3075	US3		4	2		1,475	77.50		1,552.50	1,725
3080	US32D		4	2		1,175	77.50		1,252.50	1,425
3500	Single door exit only, mortise device, narrow stile									
3510	These devices must be combined with a mortise lock									
3540	Mortise, economy US28	1 Carp	4	2	Ea.	520	77.50		597.50	700

08 71 20.40 Lockset

	08 71 20.40 Lockset	Crew	Daily Output	Labor-Hours	Unit	Material	Labor	Equipment	Total	Total Incl O&P
0010	**LOCKSET**, Standard duty									
0020	Non-keyed, passage, w/sect. trim	1 Carp	12	.667	Ea.	79.50	26		105.50	130
0100	Privacy		12	.667		81.50	26		107.50	132
0400	Keyed, single cylinder function		10	.800		151	31		182	217
0420	Hotel (see also Section 08 71 20.15)		8	1		213	39		252	299
0500	Lever handled, keyed, single cylinder function		10	.800		140	31		171	205
0600	Bedroom, bathroom and inner office doors		10	.800		148	31		179	214
0900	Apartment, office and corridor doors		10	.800		210	31		241	282
1400	Keyed, single cylinder function		10	.800		192	31		223	263

08 71 20.42 Mortise Locksets

	08 71 20.42 Mortise Locksets	Crew	Daily Output	Labor-Hours	Unit	Material	Labor	Equipment	Total	Total Incl O&P
0010	**MORTISE LOCKSETS**, Comm., wrought knobs & full escutcheon trim									
0015	Assumes mortise is cut									
0020	Non-keyed, passage, Grade 3	1 Carp	9	.889	Ea.	167	34.50		201.50	240
0030	Grade 1		8	1		435	39		474	540
0040	Privacy set, Grade 3		9	.889		177	34.50		211.50	252
0050	Grade 1		8	1		475	39		514	590
0100	Keyed, office/entrance/apartment, Grade 2		8	1		206	39		245	291
0110	Grade 1		7	1.143		540	44.50		584.50	670
0120	Single cylinder, typical, Grade 3		8	1		197	39		236	281
0130	Grade 1		7	1.143		540	44.50		584.50	665
0200	Hotel, room, Grade 3		7	1.143		198	44.50		242.50	291
0210	Grade 1 (see also Section 08 71 20.15)		6	1.333		540	51.50		591.50	680
0300	Double cylinder, Grade 3		8	1		236	39		275	325
0310	Grade 1		7	1.143		555	44.50		599.50	685
1000	Wrought knobs and sectional trim, non-keyed, passage, Grade 3		10	.800		136	31		167	200
1010	Grade 1		9	.889		435	34.50		469.50	535
1040	Privacy, Grade 3		10	.800		157	31		188	224
1050	Grade 1		9	.889		490	34.50		524.50	595
1100	Keyed, entrance, office/apartment, Grade 3		9	.889		234	34.50		268.50	315
1103	Install lockset		6.92	1.156		234	45		279	330
1110	Grade 1		8	1		560	39		599	680
1120	Single cylinder, Grade 3		9	.889		236	34.50		270.50	315
1130	Grade 1		8	1		520	39		559	640

08 71 20.42 Mortise Locksets		Crew	Daily Output	Labor-Hours	Unit	Material	2019 Bare Costs Labor	Equipment	Total	Total Incl O&P
2000	Cast knobs and full escutcheon trim									
2010	Non-keyed, passage, Grade 3	1 Carp	9	.889	Ea.	287	34.50		321.50	370
2020	Grade 1		8	1		395	39		434	500
2040	Privacy, Grade 3		9	.889		335	34.50		369.50	420
2050	Grade 1		8	1		455	39		494	565
2120	Keyed, single cylinder, Grade 3		8	1		345	39		384	445
2123	Mortise lock		6.15	1.301		345	50.50		395.50	465
2130	Grade 1		7	1.143		545	44.50		589.50	675
3000	Cast knob and sectional trim, non-keyed, passage, Grade 3		10	.800		220	31		251	293
3010	Grade 1		10	.800		400	31		431	490
3040	Privacy, Grade 3		10	.800		231	31		262	305
3050	Grade 1		10	.800		465	31		496	560
3100	Keyed, office/entrance/apartment, Grade 3		9	.889		262	34.50		296.50	345
3110	Grade 1		9	.889		600	34.50		634.50	715
3120	Single cylinder, Grade 3		9	.889		271	34.50		305.50	355
3130	Grade 1		9	.889		550	34.50		584.50	660
3190	For re-core cylinder, add					90			90	99
5000	Wrought steel case, brass base, knob US26D									
5020	Closet, non-keyed passage	1 Carp	8	1	Ea.	300	39		339	395
5040	Bath/bedroom, keyed		8	1		330	39		369	430
5060	Entrance, keyed		8	1		390	39		429	495
5080	Classroom, outside keyed		8	1		410	39		449	520
5100	Storeroom, keyed		8	1		415	39		454	520
5120	Front door, keyed		8	1		410	39		449	515
5140	Dormitory/exit, keyed		8	1		410	39		449	515

08 71 20.44 Anti-Ligature Locksets Grade 1		Crew	Daily Output	Labor-Hours	Unit	Material	Labor	Equipment	Total	Total Incl O&P
0010	**ANTI-LIGATURE LOCKSETS GRADE 1**									
0100	Anti-ligature cylindrical locksets									
0500	Arch handle passage set, US32D	1 Carp	8	1	Ea.	525	39		564	645
0510	privacy set		8	1		585	39		624	705
0520	classroom set		8	1		600	39		639	725
0530	storeroom set		8	1		600	39		639	725
0550	Lever handle passage set US32D		8	1		430	39		469	535
0560	Hospital/privacy set		8	1		445	39		484	555
0570	Office set		8	1		440	39		479	550
0580	Classroom set		8	1		465	39		504	580
0590	Storeroom set		8	1		440	39		479	550
0600	Exit set		8	1		500	39		539	615
0610	Entry set		8	1		465	39		504	580
0620	Asylum set		8	1		485	39		524	600
0630	Back to back dummy set		8	1		310	39		349	405
0690	Anti-ligature mortise locksets									
0700	Knob handle privacy set, US32D	1 Carp	8	1	Ea.	610	39		649	740
0710	Office set		8	1		635	39		674	765
0720	Classroom set		8	1		635	39		674	765
0730	Hotel set		8	1		650	39		689	780
0740	Apartment set		8	1		650	39		689	780
0750	Institutional privacy set		8	1		695	39		734	830
0760	Lever handle office set		8	1		590	39		629	715
0770	Classroom set sectional trim		8	1		590	39		629	715
0780	Hotel set		8	1		610	39		649	735
0790	Apartment set		8	1		610	39		649	735

08 71 20.44 Anti-Ligature Locksets Grade 1		Crew	Daily Output	Labor-Hours	Unit	Material	Labor	Equipment	Total	Total Incl O&P
						2019 Bare Costs				
0800	Institutional privacy set	1 Carp	8	1	Ea.	655	39		694	790
0810	Office set US32D		8	1		675	39		714	810
0820	Classroom set escutcheon trim		8	1		755	39		794	895
0830	Hotel set escutcheon trim		8	1		770	39		809	915
0840	Apartment set escutcheon trim		8	1		770	39		809	915
0850	Institutional privacy set escutcheon trim		8	1		770	39		809	915
08 71 20.45 Peepholes										
0010	**PEEPHOLES**									
2010	Peephole	1 Carp	32	.250	Ea.	9.15	9.70		18.85	26
2020	Peephole, wide view	"	32	.250	"	11.10	9.70		20.80	28
08 71 20.50 Door Stops										
0010	**DOOR STOPS**									
0020	Holder & bumper, floor or wall	1 Carp	32	.250	Ea.	36.50	9.70		46.20	56
1300	Wall bumper, 4" diameter, with rubber pad, aluminum		32	.250		13.85	9.70		23.55	31.50
1600	Door bumper, floor type, aluminum		32	.250		3.46	9.70		13.16	19.80
1620	Brass		32	.250		11.70	9.70		21.40	29
1630	Bronze		32	.250		19.80	9.70		29.50	38
1900	Plunger type, door mounted		32	.250		30	9.70		39.70	49
2500	Holder, floor type, aluminum		32	.250		20.50	9.70		30.20	38.50
2520	Wall type, aluminum		32	.250		34.50	9.70		44.20	53.50
2530	Overhead type, bronze		32	.250		114	9.70		123.70	141
2540	Plunger type, aluminum		32	.250		30	9.70		39.70	49
2560	Brass		32	.250		45	9.70		54.70	65.50
3000	Electromagnetic, wall mounted, US3		3	2.667		292	103		395	490
3020	Floor mounted, US3		3	2.667		345	103		448	550
4000	Doorstop, ceiling mounted		3	2.667		18.85	103		121.85	192
4030	Doorstop, header		3	2.667		44	103		147	220
08 71 20.55 Push-Pull Plates										
0010	**PUSH-PULL PLATES**									
0090	Push plate, 0.050 thick, 3" x 12", aluminum	1 Carp	12	.667	Ea.	6	26		32	49
0100	4" x 16"		12	.667		14.75	26		40.75	59
0110	6" x 16"		12	.667		8.05	26		34.05	51.50
0120	8" x 16"		12	.667		9.50	26		35.50	53
0200	Push plate, 0.050 thick, 3" x 12", brass		12	.667		13.80	26		39.80	57.50
0210	4" x 16"		12	.667		17.25	26		43.25	61.50
0220	6" x 16"		12	.667		26.50	26		52.50	71.50
0230	8" x 16"		12	.667		34.50	26		60.50	80.50
0250	Push plate, 0.050 thick, 3" x 12", satin brass		12	.667		14	26		40	58
0260	4" x 16"		12	.667		17.45	26		43.45	61.50
0270	6" x 16"		12	.667		26.50	26		52.50	72
0280	8" x 16"		12	.667		35	26		61	81
0490	Push plate, 0.050 thick, 3" x 12", bronze		12	.667		16.65	26		42.65	61
0500	4" x 16"		12	.667		26.50	26		52.50	71.50
0510	6" x 16"		12	.667		31.50	26		57.50	77.50
0520	8" x 16"		12	.667		38	26		64	84.50
0600	Push plate, antimicrobial copper alloy finish, 3.5" x 15"		13	.615		20.50	24		44.50	62
0610	4" x 16"		13	.615		25	24		49	67
0620	6" x 16"		13	.615		26.50	24		50.50	68.50
0630	6" x 20"		13	.615		36	24		60	79
0740	Push plate, 0.050 thick, 3" x 12", stainless steel		12	.667		11.75	26		37.75	55.50
0750	4" x 16"		12	.667		28.50	26		54.50	74
0760	6" x 16"		12	.667		18.45	26		44.45	63

08 71 20 – Hardware

08 71 20.55 Push-Pull Plates		Crew	Daily Output	Labor-Hours	Unit	Material	Labor	Equipment	Total	Total Incl O&P
0780	8" x 16"	1 Carp	12	.667	Ea.	23.50	26		49.50	68.50
0790	Push plate, 0.050 thick, 3" x 12", satin stainless steel		12	.667		6.65	26		32.65	50
0810	4" x 16"		12	.667		8.15	26		34.15	51.50
0820	6" x 16"		12	.667		11.60	26		37.60	55.50
0830	8" x 16"		12	.667		15.70	26		41.70	60
0980	Pull plate, 0.050 thick, 3" x 12", aluminum		12	.667		25.50	26		51.50	70.50
1000	4" x 16"		12	.667		27.50	26		53.50	72.50
1050	Pull plate, 0.050 thick, 3" x 12", brass		12	.667		39.50	26		65.50	86
1060	4" x 16"		12	.667		35	26		61	81
1080	Pull plate, 0.050 thick, 3" x 12", bronze		12	.667		50	26		76	97.50
1100	4" x 16"		12	.667		59	26		85	108
1180	Pull plate, 0.050 thick, 3" x 12", stainless steel		12	.667		46	26		72	93
1200	4" x 16"		12	.667		61.50	26		87.50	111
1250	Pull plate, 0.050 thick, 3" x 12", chrome		12	.667		46	26		72	93
1270	4" x 16"		12	.667		48	26		74	95
1500	Pull handle and push bar, aluminum		11	.727		126	28		154	186
2000	Bronze		10	.800		170	31		201	238

08 71 20.60 Entrance Locks		Crew	Daily Output	Labor-Hours	Unit	Material	Labor	Equipment	Total	Total Incl O&P
0010	**ENTRANCE LOCKS**									
0015	Cylinder, grip handle deadlocking latch	1 Carp	9	.889	Ea.	184	34.50		218.50	259
0020	Deadbolt		8	1		195	39		234	279
0100	Push and pull plate, dead bolt		8	1		239	39		278	325
0900	For handicapped lever, add					156			156	171

08 71 20.65 Thresholds		Crew	Daily Output	Labor-Hours	Unit	Material	Labor	Equipment	Total	Total Incl O&P
0010	**THRESHOLDS**									
0011	Threshold 3' long saddles aluminum	1 Carp	48	.167	L.F.	9.95	6.45		16.40	21.50
0100	Aluminum, 8" wide, 1/2" thick		12	.667	Ea.	51.50	26		77.50	99.50
0500	Bronze		60	.133	L.F.	47	5.15		52.15	60.50
0600	Bronze, panic threshold, 5" wide, 1/2" thick		12	.667	Ea.	165	26		191	225
0700	Rubber, 1/2" thick, 5-1/2" wide		20	.400		42.50	15.50		58	72.50
0800	2-3/4" wide		20	.400		51.50	15.50		67	82
1950	ADA compliant thresholds									
2000	Threshold, wood oak 3-1/2" wide x 24" long	1 Carp	12	.667	Ea.	12.20	26		38.20	56
2010	3-1/2" wide x 36" long		12	.667		16.40	26		42.40	60.50
2020	3-1/2" wide x 48" long		12	.667		22	26		48	66.50
2030	4-1/2" wide x 24" long		12	.667		14.10	26		40.10	58
2040	4-1/2" wide x 36" long		12	.667		20.50	26		46.50	65.50
2050	4-1/2" wide x 48" long		12	.667		27	26		53	72.50
2060	6-1/2" wide x 24" long		12	.667		19.60	26		45.60	64
2070	6-1/2" wide x 36" long		12	.667		29.50	26		55.50	75
2080	6-1/2" wide x 48" long		12	.667		39	26		65	85.50
2090	Threshold, wood cherry 3-1/2" wide x 24" long		12	.667		14.60	26		40.60	58.50
2100	3-1/2" wide x 36" long		12	.667		33.50	26		59.50	79.50
2110	3-1/2" wide x 48" long		12	.667		38.50	26		64.50	85
2120	4-1/2" wide x 24" long		12	.667		18.60	26		44.60	63
2130	4-1/2" wide x 36" long		12	.667		37.50	26		63.50	83.50
2140	4-1/2" wide x 48" long		12	.667		45.50	26		71.50	92.50
2150	6-1/2" wide x 24" long		12	.667		26.50	26		52.50	71.50
2160	6-1/2" wide x 36" long		12	.667		50.50	26		76.50	98
2170	6-1/2" wide x 48" long		12	.667		61	26		87	110
2180	Threshold, wood walnut 3-1/2" wide x 24" long		12	.667		17.55	26		43.55	62
2190	3-1/2" wide x 36" long		12	.667		35	26		61	81

08 71 Door Hardware

08 71 20 – Hardware

08 71 20.65 Thresholds

08 71 20.65 Thresholds		Crew	Daily Output	Labor-Hours	Unit	2019 Bare Costs Material	Labor	Equipment	Total	Total Incl O&P
2200	3-1/2" wide x 48" long	1 Carp	12	.667	Ea.	40.50	26		66.50	87
2210	4-1/2" wide x 24" long		12	.667		28	26		54	73
2220	4-1/2" wide x 36" long		12	.667		47	26		73	94
2230	4-1/2" wide x 48" long		12	.667		57.50	26		83.50	106
2240	6-1/2" wide x 24" long		12	.667		44.50	26		70.50	91.50
2250	6-1/2" wide x 36" long		12	.667		70.50	26		96.50	120
2260	6-1/2" wide x 48" long		12	.667		92	26		118	144
2300	Threshold, aluminum 4" wide x 36" long		12	.667		25.50	26		51.50	70.50
2310	4" wide x 48" long		12	.667		30.50	26		56.50	76
2320	4" wide x 72" long		12	.667		46	26		72	93
2330	5" wide x 36" long		12	.667		33.50	26		59.50	79.50
2340	5" wide x 48" long		12	.667		39.50	26		65.50	86
2350	5" wide x 72" long		12	.667		74	26		100	124
2360	6" wide x 36" long		12	.667		46	26		72	93
2370	6" wide x 48" long		12	.667		49	26		75	96.50
2380	6" wide x 72" long		12	.667		71.50	26		97.50	121
2390	7" wide x 36" long		12	.667		50	26		76	97.50
2400	7" wide x 48" long		12	.667		67	26		93	117
2410	7" wide x 72" long		12	.667		95.50	26		121.50	148
2500	Threshold, ramp, aluminum or rubber 24" x 24"		12	.667		190	26		216	252

08 71 20.75 Door Hardware Accessories

08 71 20.75	Door Hardware Accessories	Crew	Daily Output	Labor-Hours	Unit	Material	Labor	Equipment	Total	Total Incl O&P
0010	**DOOR HARDWARE ACCESSORIES**									
0050	Door closing coordinator, 36" (for paired openings up to 56")	1 Carp	8	1	Ea.	131	39		170	208
0060	48" (for paired openings up to 84")		8	1		139	39		178	217
0070	56" (for paired openings up to 96")		8	1		149	39		188	228
1000	Knockers, brass, standard		16	.500		66	19.40		85.40	105
1100	Deluxe		10	.800		94.50	31		125.50	155
2000	Torsion springs for overhead doors									
2050	1-3/4" diam., 32" long, 0.243" spring	1 Carp	8	1	Ea.	55.50	39		94.50	125
2060	1-3/4" diam., 32" long, 0.250" spring		8	1		50	39		89	119
2070	2" diam., 32" long, 0.243" spring		8	1		43	39		82	112
2080	2" diam., 32" long, 0.253" spring		8	1		51.50	39		90.50	121
4500	Rubber door silencers		540	.015		.44	.57		1.01	1.43

08 71 20.80 Hasps

08 71 20.80	Hasps	Crew	Daily Output	Labor-Hours	Unit	Material	Labor	Equipment	Total	Total Incl O&P
0010	**HASPS**, steel assembly									
0015	3"	1 Carp	26	.308	Ea.	5.70	11.90		17.60	26
0020	4-1/2"		13	.615		6.85	24		30.85	47
0040	6"		12.50	.640		10.10	25		35.10	52

08 71 20.90 Hinges

08 71 20.90	Hinges	Crew	Daily Output	Labor-Hours	Unit	Material	Labor	Equipment	Total	Total Incl O&P
0010	**HINGES**									
0012	Full mortise, avg. freq., steel base, USP, 4-1/2" x 4-1/2"				Pr.	46			46	50.50
0100	5" x 5", USP					66.50			66.50	73
0200	6" x 6", USP					139			139	153
0400	Brass base, 4-1/2" x 4-1/2", US10					66			66	72.50
0500	5" x 5", US10					76			76	83.50
0600	6" x 6", US10					169			169	186
0800	Stainless steel base, 4-1/2" x 4-1/2", US32					87.50			87.50	96.50
0900	For non removable pin, add (security item)				Ea.	5.65			5.65	6.20
0910	For floating pin, driven tips, add					3.28			3.28	3.61
0930	For hospital type tip on pin, add					14.45			14.45	15.85
0940	For steeple type tip on pin, add					20			20	22
0950	Full mortise, high frequency, steel base, 3-1/2" x 3-1/2", US26D				Pr.	31.50			31.50	34.50

08 71 Door Hardware

08 71 20 – Hardware

08 71 20.90 Hinges

08 71 20.90 Hinges		Crew	Daily Output	Labor-Hours	Unit	Material	Labor	Equipment	Total	Total Incl O&P
						2019 Bare Costs				
1000	4-1/2" x 4-1/2", USP				Pr.	66.50			66.50	73.50
1100	5" x 5", USP					54			54	59
1200	6" x 6", USP					140			140	155
1400	Brass base, 3-1/2" x 3-1/2", US4					53.50			53.50	59
1430	4-1/2" x 4-1/2", US10					78			78	85.50
1500	5" x 5", US10					128			128	141
1600	6" x 6", US10					174			174	191
1800	Stainless steel base, 4-1/2" x 4-1/2", US32					107			107	118
1810	5" x 4-1/2", US32					143			143	157
1930	For hospital type tip on pin, add				Ea.	14.60			14.60	16.05
1950	Full mortise, low frequency, steel base, 3-1/2" x 3-1/2", US26D				Pr.	29.50			29.50	32.50
2000	4-1/2" x 4-1/2", USP					24.50			24.50	27
2100	5" x 5", USP					50.50			50.50	55.50
2200	6" x 6", USP					97.50			97.50	107
2300	4-1/2" x 4-1/2", US3					17.95			17.95	19.75
2310	5" x 5", US3					43			43	47.50
2400	Brass bass, 4-1/2" x 4-1/2", US10					57			57	62.50
2500	5" x 5", US10					81.50			81.50	89.50
2800	Stainless steel base, 4-1/2" x 4-1/2", US32					76.50			76.50	84.50
8000	Install hinge	1 Carp	34	.235			9.10		9.10	15.05

08 71 20.91 Special Hinges

08 71 20.91 Special Hinges		Crew	Daily Output	Labor-Hours	Unit	Material	Labor	Equipment	Total	Total Incl O&P
0010	**SPECIAL HINGES**									
8000	Continuous hinges									
8010	Steel, piano, 2" x 72"	1 Carp	20	.400	Ea.	23.50	15.50		39	51.50
8020	Brass, piano, 1-1/16" x 30"		30	.267		9.20	10.35		19.55	27
8030	Acrylic, piano, 1-3/4" x 12"		40	.200		15.25	7.75		23	29.50
8040	Aluminum, door, standard duty, 7'		3	2.667		122	103		225	305
8050	Heavy duty, 7'		3	2.667		139	103		242	325
8060	8'		3	2.667		162	103		265	350
8070	Steel, door, heavy duty, 7'		3	2.667		204	103		307	395
8080	8'		3	2.667		255	103		358	450
8090	Stainless steel, door, heavy duty, 7'		3	2.667		260	103		363	455
8100	8'		3	2.667		293	103		396	490
9200	Continuous geared hinge, aluminum, full mortise, standard duty, 83"		3	2.667		151	103		254	335
9250	Continuous geared hinge, aluminum, full mortise, heavy duty, 83"		3	2.667		222	103		325	415

08 71 20.92 Mortised Hinges

08 71 20.92 Mortised Hinges		Crew	Daily Output	Labor-Hours	Unit	Material	Labor	Equipment	Total	Total Incl O&P
0010	**MORTISED HINGES**									
0200	Average frequency, steel plated, ball bearing, 3-1/2" x 3-1/2"				Pr.	28			28	31
0300	Bronze, ball bearing					33			33	36
0900	High frequency, steel plated, ball bearing					106			106	117
1100	Bronze, ball bearing					106			106	117
1300	Average frequency, steel plated, ball bearing, 4-1/2" x 4-1/2"					36.50			36.50	40
1500	Bronze, ball bearing, to 36" wide					37.50			37.50	41
1700	Low frequency, steel, plated, plain bearing					20			20	22
1900	Bronze, plain bearing					28			28	30.50

08 71 20.95 Kick Plates

08 71 20.95 Kick Plates		Crew	Daily Output	Labor-Hours	Unit	Material	Labor	Equipment	Total	Total Incl O&P
0010	**KICK PLATES**									
0020	Stainless steel, .050", 16 ga., 8" x 28", US32	1 Carp	15	.533	Ea.	41.50	20.50		62	80
0030	8" x 30"		15	.533		45.50	20.50		66	84
0040	8" x 34"		15	.533		51	20.50		71.50	90
0050	10" x 28"		15	.533		84	20.50		104.50	126
0060	10" x 30"		15	.533		90.50	20.50		111	134

08 71 20.95 Kick Plates		Crew	Daily Output	Labor-Hours	Unit	Material	2019 Bare Costs Labor	Equipment	Total	Total Incl O&P
0070	10" x 34"	1 Carp	15	.533	Ea.	101	20.50		121.50	146
0080	Mop/Kick, 4" x 28"		15	.533		37.50	20.50		58	75.50
0090	4" x 30"		15	.533		39.50	20.50		60	77.50
0100	4" x 34"		15	.533		45	20.50		65.50	83.50
0110	6" x 28"		15	.533		43	20.50		63.50	81
0120	6" x 30"		15	.533		52	20.50		72.50	91
0130	6" x 34"		15	.533		58.50	20.50		79	98.50
0500	Bronze, .050", 8" x 28"		15	.533		74	20.50		94.50	115
0510	8" x 30"		15	.533		71	20.50		91.50	112
0520	8" x 34"		15	.533		73.50	20.50		94	115
0530	10" x 28"		15	.533		76.50	20.50		97	119
0540	10" x 30"		15	.533		83	20.50		103.50	126
0550	10" x 34"		15	.533		93	20.50		113.50	136
0560	Mop/Kick, 4" x 28"		15	.533		36	20.50		56.50	73.50
0570	4" x 30"		15	.533		39.50	20.50		60	77.50
0580	4" x 34"		15	.533		40.50	20.50		61	78.50
0590	6" x 28"		15	.533		50	20.50		70.50	89
0600	6" x 30"		15	.533		51.50	20.50		72	90.50
0610	6" x 34"		15	.533		62.50	20.50		83	103
1000	Acrylic, .125", 8" x 26"		15	.533		27.50	20.50		48	64
1010	8" x 36"		15	.533		38.50	20.50		59	76
1020	8" x 42"		15	.533		44.50	20.50		65	83
1030	10" x 26"		15	.533		34.50	20.50		55	72
1040	10" x 36"		15	.533		48.50	20.50		69	87
1050	10" x 42"		15	.533		56	20.50		76.50	95.50
1060	Mop/Kick, 4" x 26"		15	.533		16.55	20.50		37.05	52
1070	4" x 36"		15	.533		23	20.50		43.50	59
1080	4" x 42"		15	.533		26.50	20.50		47	63
1090	6" x 26"		15	.533		24.50	20.50		45	60.50
1100	6" x 36"		15	.533		33.50	20.50		54	71
1110	6" x 42"		15	.533		39.50	20.50		60	77.50
1220	Brass, .050", 8" x 26"		15	.533		71	20.50		91.50	112
1230	8" x 36"		15	.533		70	20.50		90.50	111
1240	8" x 42"		15	.533		86	20.50		106.50	129
1250	10" x 26"		15	.533		71.50	20.50		92	113
1260	10" x 36"		15	.533		91.50	20.50		112	135
1270	10" x 42"		15	.533		106	20.50		126.50	150
1320	Mop/Kick, 4" x 26"		15	.533		26	20.50		46.50	63
1330	4" x 36"		15	.533		36.50	20.50		57	74
1340	4" x 42"		15	.533		42.50	20.50		63	80.50
1350	6" x 26"		15	.533		39	20.50		59.50	77
1360	6" x 36"		15	.533		48.50	20.50		69	87
1370	6" x 42"		15	.533		56	20.50		76.50	96
1800	Aluminum, .050", 8" x 26"		15	.533		39.50	20.50		60	77.50
1810	8" x 36"		15	.533		36.50	20.50		57	74.50
1820	8" x 42"		15	.533		43.50	20.50		64	81.50
1830	10" x 26"		15	.533		33	20.50		53.50	70.50
1840	10" x 36"		15	.533		46	20.50		66.50	84.50
1850	10" x 42"		15	.533		63.50	20.50		84	104
1860	Mop/Kick, 4" x 26"		15	.533		16.30	20.50		36.80	52
1870	4" x 36"		15	.533		21.50	20.50		42	57.50
1880	4" x 42"		15	.533		26	20.50		46.50	63
1890	6" x 26"		15	.533		24	20.50		44.50	60

08 71 Door Hardware

08 71 20 – Hardware

08 71 20.95 Kick Plates		Crew	Daily Output	Labor-Hours	Unit	Material	Labor	Equipment	Total	Total Incl O&P
						2019 Bare Costs				
1900	6″ x 36″	1 Carp	15	.533	Ea.	32.50	20.50		53	69.50
1910	6″ x 42″		15	.533		37.50	20.50		58	75.50

08 71 21 – Astragals

08 71 21.10 Exterior Mouldings, Astragals

Line	Description	Crew	Daily Output	Labor-Hours	Unit	Material	Labor	Equipment	Total	Total Incl O&P
0010	**EXTERIOR MOULDINGS, ASTRAGALS**									
4170	Astragal for double doors, aluminum	1 Carp	4	2	Opng.	39	77.50		116.50	171
4174	Bronze	″	4	2	″	53	77.50		130.50	187

08 71 25 – Weatherstripping

08 71 25.10 Mechanical Seals, Weatherstripping

Line	Description	Crew	Daily Output	Labor-Hours	Unit	Material	Labor	Equipment	Total	Total Incl O&P
0010	**MECHANICAL SEALS, WEATHERSTRIPPING**									
1000	Doors, wood frame, interlocking, for 3′ x 7′ door, zinc	1 Carp	3	2.667	Opng.	49	103		152	225
1100	Bronze		3	2.667		62	103		165	239
1300	6′ x 7′ opening, zinc		2	4		59.50	155		214.50	320
1400	Bronze		2	4		71	155		226	335
1700	Wood frame, spring type, bronze									
1800	3′ x 7′ door	1 Carp	7.60	1.053	Opng.	25.50	41		66.50	95.50
1900	6′ x 7′ door		7	1.143		32	44.50		76.50	109
1920	Felt, 3′ x 7′ door		14	.571		4.50	22		26.50	41.50
1930	6′ x 7′ door		13	.615		5.20	24		29.20	45
1950	Rubber, 3′ x 7′ door		7.60	1.053		11.55	41		52.55	80
1951	Rubber, 3′ x 7′ door		119	.067	L.F.	.68	2.61		3.29	5.05
1960	6′ x 7′ door		7	1.143	Opng.	13.40	44.50		57.90	87.50
2200	Metal frame, spring type, bronze									
2300	3′ x 7′ door	1 Carp	3	2.667	Opng.	51	103		154	227
2400	6′ x 7′ door	″	2.50	3.200	″	55	124		179	266
2500	For stainless steel, spring type, add					133%				
2700	Metal frame, extruded sections, 3′ x 7′ door, aluminum	1 Carp	3	2.667	Opng.	28.50	103		131.50	203
2800	Bronze		3	2.667		84	103		187	264
3100	6′ x 7′ door, aluminum		1.50	5.333		35	207		242	380
3200	Bronze		1.50	5.333		145	207		352	500
3500	Threshold weatherstripping									
3650	Door sweep, flush mounted, aluminum	1 Carp	25	.320	Ea.	21	12.40		33.40	43.50
3700	Vinyl		25	.320		18.75	12.40		31.15	41
5000	Garage door bottom weatherstrip, 12′ aluminum, clear		14	.571		26.50	22		48.50	65.50
5010	Bronze		14	.571		94	22		116	140
5050	Bottom protection, rubber		14	.571		48.50	22		70.50	90
5100	Threshold		14	.571		70	22		92	114

08 75 Window Hardware

08 75 10 – Window Handles and Latches

08 75 10.10 Handles and Latches

Line	Description	Crew	Daily Output	Labor-Hours	Unit	Material	Labor	Equipment	Total	Total Incl O&P
0010	**HANDLES AND LATCHES**									
1000	Handles, surface mounted, aluminum	1 Carp	24	.333	Ea.	5.90	12.90		18.80	28
1020	Brass		24	.333		5.80	12.90		18.70	28
1040	Chrome		24	.333		8.65	12.90		21.55	31
1200	Window handles window crank ADA		24	.333		16.40	12.90		29.30	39.50
1500	Recessed, aluminum		12	.667		3.52	26		29.52	46.50
1520	Brass		12	.667		5.35	26		31.35	48.50
1540	Chrome		12	.667		4.43	26		30.43	47.50
2000	Latches, aluminum		20	.400		4.29	15.50		19.79	30

08 75 Window Hardware

08 75 10 – Window Handles and Latches

08 75 10.10 Handles and Latches		Crew	Daily Output	Labor-Hours	Unit	Material	2019 Bare Costs Labor	Equipment	Total	Total Incl O&P
2020	Brass	1 Carp	20	.400	Ea.	5.60	15.50		21.10	31.50
2040	Chrome	↓	20	.400	↓	3.65	15.50		19.15	29.50

08 75 10.15 Window Opening Control		Crew	Daily Output	Labor-Hours	Unit	Material	Labor	Equipment	Total	Total Incl O&P
0010	**WINDOW OPENING CONTROL**									
0015	Window stops									
0020	For double-hung window	1 Carp	35	.229	Ea.	39.50	8.85		48.35	58
0030	Cam action for sliding window		35	.229		4.42	8.85		13.27	19.50
0040	Thumb screw for sliding window	↓	35	.229	↓	4.77	8.85		13.62	19.90
0100	Window guards, child safety bars for single or double-hung									
0110	14″ to 17″ wide max vert opening 26″	1 Carp	16	.500	Ea.	66.50	19.40		85.90	105
0120	17″ to 23″ wide max vert opening 26″		16	.500		76	19.40		95.40	116
0130	23″ to 36″ wide max vert opening 26″		16	.500		86.50	19.40		105.90	127
0140	35″ to 58″ wide max vert opening 26″		16	.500		100	19.40		119.40	142
0150	58″ to 90″ wide max vert opening 26″		16	.500		135	19.40		154.40	180
0160	73″ to 120″ wide max vert opening 26″	↓	16	.500	↓	236	19.40		255.40	292

08 75 30 – Weatherstripping

08 75 30.10 Mechanical Weather Seals		Crew	Daily Output	Labor-Hours	Unit	Material	Labor	Equipment	Total	Total Incl O&P
0010	**MECHANICAL WEATHER SEALS**, Window, double-hung, 3′ x 5′									
0020	Zinc	1 Carp	7.20	1.111	Opng.	22.50	43		65.50	96
0100	Bronze		7.20	1.111		43	43		86	119
0200	Vinyl V strip		7	1.143		11.10	44.50		55.60	85
0500	As above but heavy duty, zinc		4.60	1.739		21	67.50		88.50	134
0600	Bronze	↓	4.60	1.739	↓	77.50	67.50		145	196

08 79 Hardware Accessories

08 79 20 – Door Accessories

08 79 20.10 Door Hardware Accessories		Crew	Daily Output	Labor-Hours	Unit	Material	Labor	Equipment	Total	Total Incl O&P
0010	**DOOR HARDWARE ACCESSORIES**									
0140	Door bolt, surface, 4″	1 Carp	32	.250	Ea.	16.20	9.70		25.90	34
0160	Door latch	″	12	.667	″	8.75	26		34.75	52
0200	Sliding closet door									
0220	Track and hanger, single	1 Carp	10	.800	Ea.	69.50	31		100.50	128
0240	Double		8	1		84.50	39		123.50	157
0260	Door guide, single		48	.167		32	6.45		38.45	45.50
0280	Double		48	.167		42.50	6.45		48.95	57.50
0600	Deadbolt and lock cover plate, brass or stainless steel		30	.267		31	10.35		41.35	51
0620	Hole cover plate, brass or chrome		35	.229		8.50	8.85		17.35	24
2240	Mortise lockset, passage, lever handle		9	.889		166	34.50		200.50	240
4000	Security chain, standard		18	.444		13.15	17.20		30.35	43
4100	Deluxe	↓	18	.444	↓	46.50	17.20		63.70	79.50

08 81 Glass Glazing

08 81 10 – Float Glass

08 81 10.10 Various Types and Thickness of Float Glass		Crew	Daily Output	Labor-Hours	Unit	2019 Bare Costs Material	Labor	Equipment	Total	Total Incl O&P
0010	**VARIOUS TYPES AND THICKNESS OF FLOAT GLASS** R088110-10									
0020	3/16" plain	2 Glaz	130	.123	S.F.	7.10	4.71		11.81	15.55
0200	Tempered, clear		130	.123		7.35	4.71		12.06	15.85
0300	Tinted		130	.123		7	4.71		11.71	15.45
0600	1/4" thick, clear, plain		120	.133		10.15	5.10		15.25	19.55
0700	Tinted		120	.133		9.60	5.10		14.70	18.95
0800	Tempered, clear		120	.133		7.10	5.10		12.20	16.25
0900	Tinted		120	.133		11.95	5.10		17.05	21.50
1600	3/8" thick, clear, plain		75	.213		11.45	8.15		19.60	26
1700	Tinted		75	.213		16.80	8.15		24.95	32
1800	Tempered, clear		75	.213		18	8.15		26.15	33.50
1900	Tinted		75	.213		20.50	8.15		28.65	36.50
2200	1/2" thick, clear, plain		55	.291		18.90	11.15		30.05	39.50
2300	Tinted		55	.291		29.50	11.15		40.65	51
2400	Tempered, clear		55	.291		27	11.15		38.15	48
2500	Tinted		55	.291		28	11.15		39.15	49
2800	5/8" thick, clear, plain		45	.356		29.50	13.60		43.10	55
2900	Tempered, clear		45	.356		33.50	13.60		47.10	59.50
8900	For low emissivity coating for 3/16" & 1/4" only, add to above					18%				

08 81 17 – Fire Glass

08 81 17.10 Fire Resistant Glass

Code	Description	Crew	Daily Output	Labor-Hours	Unit	Material	Labor	Equipment	Total	Total Incl O&P
0010	**FIRE RESISTANT GLASS**									
0020	Fire glass minimum	2 Glaz	40	.400	S.F.	41.50	15.30		56.80	71
0030	Mid range		40	.400		84.50	15.30		99.80	118
0050	High end		40	.400		370	15.30		385.30	430

08 81 20 – Vision Panels

08 81 20.10 Full Vision

Code	Description	Crew	Daily Output	Labor-Hours	Unit	Material	Labor	Equipment	Total	Total Incl O&P
0010	**FULL VISION**, window system with 3/4" glass mullions									
0020	Up to 10' high	H-2	130	.185	S.F.	69	6.60		75.60	87
0100	10' to 20' high, minimum		110	.218		74	7.75		81.75	94
0150	Average		100	.240		79.50	8.55		88.05	101
0200	Maximum		80	.300		89	10.70		99.70	116

08 81 25 – Glazing Variables

08 81 25.10 Applications of Glazing

Code	Description	Crew	Daily Output	Labor-Hours	Unit	Material	Labor	Equipment	Total	Total Incl O&P
0010	**APPLICATIONS OF GLAZING**									
0600	For glass replacement, add				S.F.		100%			
0700	For gasket settings, add				L.F.	6.30			6.30	6.95
0900	For sloped glazing, add				S.F.		26%			
2000	Fabrication, polished edges, 1/4" thick				Inch	.60			.60	.66
2100	1/2" thick					1.42			1.42	1.56
2500	Mitered edges, 1/4" thick					1.42			1.42	1.56
2600	1/2" thick					2.34			2.34	2.57

08 81 30 – Insulating Glass

08 81 30.10 Reduce Heat Transfer Glass

Code	Description	Crew	Daily Output	Labor-Hours	Unit	Material	Labor	Equipment	Total	Total Incl O&P
0010	**REDUCE HEAT TRANSFER GLASS**									
0015	2 lites 1/8" float, 1/2" thk under 15 S.F.									
0100	Tinted G	2 Glaz	95	.168	S.F.	14.50	6.45		20.95	26.50
0280	Double glazed, 5/8" thk unit, 3/16" float, 15 to 30 S.F., clear		90	.178		14.10	6.80		20.90	26.50
0400	1" thk, dbl. glazed, 1/4" float, 30 to 70 S.F., clear G		75	.213		17.70	8.15		25.85	33
0500	Tinted G		75	.213		24	8.15		32.15	40

08 81 Glass Glazing

08 81 30 – Insulating Glass

08 81 30.10 Reduce Heat Transfer Glass		Crew	Daily Output	Labor-Hours	Unit	Material	Labor	Equipment	Total	Total Incl O&P
						2019 Bare Costs				
2000	Both lites, light & heat reflective G	2 Glaz	85	.188	S.F.	32.50	7.20		39.70	47.50
2500	Heat reflective, film inside, 1" thick unit, clear G		85	.188		28.50	7.20		35.70	43
2600	Tinted G		85	.188		30.50	7.20		37.70	45.50
3000	Film on weatherside, clear, 1/2" thick unit G		95	.168		20	6.45		26.45	32.50
3100	5/8" thick unit G		90	.178		21	6.80		27.80	34
3200	1" thick unit G		85	.188		28	7.20		35.20	43
3350	Clear heat reflective film on inside G	1 Glaz	50	.160		13.55	6.10		19.65	25
3360	Heat reflective film on inside, tinted G		25	.320		14.40	12.25		26.65	36
3370	Heat reflective film on the inside, metalized G		20	.400		17.05	15.30		32.35	44
5000	Spectrally selective film, on ext., blocks solar gain/allows 70% of light G	2 Glaz	95	.168		15.45	6.45		21.90	27.50

08 81 55 – Window Glass

08 81 55.10 Sheet Glass		Crew	Daily Output	Labor-Hours	Unit	Material	Labor	Equipment	Total	Total Incl O&P
0010	**SHEET GLASS** (window), clear float, stops, putty bed									
0015	1/8" thick, clear float	2 Glaz	480	.033	S.F.	3.92	1.27		5.19	6.40
0500	3/16" thick, clear		480	.033		6.45	1.27		7.72	9.20
0600	Tinted		480	.033		8.05	1.27		9.32	10.95
0700	Tempered		480	.033		9.85	1.27		11.12	12.95

08 81 65 – Wire Glass

08 81 65.10 Glass Reinforced With Wire		Crew	Daily Output	Labor-Hours	Unit	Material	Labor	Equipment	Total	Total Incl O&P
0010	**GLASS REINFORCED WITH WIRE**									
0012	1/4" thick rough obscure	2 Glaz	135	.119	S.F.	27	4.53		31.53	37.50
1000	Polished wire, 1/4" thick, diamond, clear		135	.119		30.50	4.53		35.03	41
1500	Pinstripe, obscure		135	.119		56	4.53		60.53	69

08 83 Mirrors

08 83 13 – Mirrored Glass Glazing

08 83 13.10 Mirrors		Crew	Daily Output	Labor-Hours	Unit	Material	Labor	Equipment	Total	Total Incl O&P
0010	**MIRRORS**, No frames, wall type, 1/4" plate glass, polished edge									
0100	Up to 5 S.F.	2 Glaz	125	.128	S.F.	9.85	4.90		14.75	18.85
0200	Over 5 S.F.		160	.100		9.45	3.83		13.28	16.70
0500	Door type, 1/4" plate glass, up to 12 S.F.		160	.100		9.35	3.83		13.18	16.60
1000	Float glass, up to 10 S.F., 1/8" thick		160	.100		6.20	3.83		10.03	13.15
1100	3/16" thick		150	.107		7.70	4.08		11.78	15.15
1500	12" x 12" wall tiles, square edge, clear		195	.082		2.68	3.14		5.82	8.10
1600	Veined		195	.082		6.55	3.14		9.69	12.35
2010	Bathroom, unframed, laminated		160	.100		15.05	3.83		18.88	23

08 84 Plastic Glazing

08 84 10 – Plexiglass Glazing

08 84 10.10 Plexiglass Acrylic		Crew	Daily Output	Labor-Hours	Unit	Material	Labor	Equipment	Total	Total Incl O&P
0010	**PLEXIGLASS ACRYLIC**, clear, masked,									
0020	1/8" thick, cut sheets	2 Glaz	170	.094	S.F.	12.20	3.60		15.80	19.30
0200	Full sheets		195	.082		5.95	3.14		9.09	11.70
0500	1/4" thick, cut sheets		165	.097		14.30	3.71		18.01	22
0600	Full sheets		185	.086		9.70	3.31		13.01	16.10
0900	3/8" thick, cut sheets		155	.103		22	3.95		25.95	30.50
1000	Full sheets		180	.089		17.50	3.40		20.90	25
1300	1/2" thick, cut sheets		135	.119		33	4.53		37.53	43.50

08 84 Plastic Glazing

08 84 10 – Plexiglass Glazing

08 84 10.10 Plexiglass Acrylic		Crew	Daily Output	Labor-Hours	Unit	Material	2019 Bare Costs Labor	Equipment	Total	Total Incl O&P
1400	Full sheets	2 Glaz	150	.107	S.F.	19.80	4.08		23.88	28.50
1700	3/4" thick, cut sheets		115	.139		33	5.30		38.30	45
1800	Full sheets		130	.123		18.90	4.71		23.61	29
2100	1" thick, cut sheets		105	.152		37	5.85		42.85	50
2200	Full sheets		125	.128		23	4.90		27.90	33
3000	Colored, 1/8" thick, cut sheets		170	.094		21	3.60		24.60	29
3200	Full sheets		195	.082		12	3.14		15.14	18.35
3500	1/4" thick, cut sheets		165	.097		23	3.71		26.71	31.50
3600	Full sheets		185	.086		15.60	3.31		18.91	22.50
4000	Mirrors, untinted, cut sheets, 1/8" thick		185	.086		14.95	3.31		18.26	22
4200	1/4" thick	↓	180	.089	↓	17.90	3.40		21.30	25.50

08 84 20 – Polycarbonate

08 84 20.10 Thermoplastic		Crew	Daily Output	Labor-Hours	Unit	Material	Labor	Equipment	Total	Total Incl O&P
0010	**THERMOPLASTIC**, clear, masked, cut sheets									
0020	1/8" thick	2 Glaz	170	.094	S.F.	15.80	3.60		19.40	23.50
0500	3/16" thick		165	.097		19.35	3.71		23.06	27.50
1000	1/4" thick		155	.103		18.95	3.95		22.90	27.50
1500	3/8" thick	↓	150	.107	↓	29.50	4.08		33.58	39

08 85 Glazing Accessories

08 85 10 – Miscellaneous Glazing Accessories

08 85 10.10 Glazing Gaskets		Crew	Daily Output	Labor-Hours	Unit	Material	Labor	Equipment	Total	Total Incl O&P
0010	**GLAZING GASKETS**, Neoprene for glass tongued mullion									
0015	1/4" glass	2 Glaz	200	.080	L.F.	.94	3.06		4	6.10
0020	3/8"		200	.080		1.21	3.06		4.27	6.40
0040	1/2"		200	.080		1.34	3.06		4.40	6.50
0060	3/4"		180	.089		1.90	3.40		5.30	7.70
0080	1"	↓	180	.089		1.57	3.40		4.97	7.35
1000	Glazing compound, wood	1 Glaz	58	.138		1.26	5.30		6.56	10.10
1005	Glazing compound, per window, up to 30 L.F.				↓	1.26			1.26	1.38
1006	Glazing compound, per window, up to 30 L.F.				Ea.	37.50			37.50	41.50
1020	Metal	1 Glaz	58	.138	L.F.	1.26	5.30		6.56	10.10

08 85 20 – Structural Glazing Sealants

08 85 20.10 Structural Glazing Adhesives		Crew	Daily Output	Labor-Hours	Unit	Material	Labor	Equipment	Total	Total Incl O&P
0010	**STRUCTURAL GLAZING ADHESIVES**									
0050	Structural glazing adhesive, 1/4" x 1/4" joint seal	1 Glaz	200	.040	L.F.	14.35	1.53		15.88	18.25

08 87 Glazing Surface Films

08 87 13 – Solar Control Films

08 87 13.10 Solar Films On Glass		Crew	Daily Output	Labor-Hours	Unit	Material	Labor	Equipment	Total	Total Incl O&P
0010	**SOLAR FILMS ON GLASS** (glass not included)									
1000	Bronze, 20% VLT	H-2	180	.133	S.F.	1.63	4.75		6.38	9.60
1020	50% VLT		180	.133		1.83	4.75		6.58	9.80
1050	Neutral, 20% VLT		180	.133		1.83	4.75		6.58	9.80
1100	Silver, 15% VLT		180	.133		1.06	4.75		5.81	8.95
1120	35% VLT		180	.133		3.07	4.75		7.82	11.20
1150	68% VLT		180	.133		.41	4.75		5.16	8.25
3000	One way mirror with night vision 5% in and 15% out VLT	↓	180	.133	↓	3.08	4.75		7.83	11.20

08 87 Glazing Surface Films

08 87 16 – Glass Safety Films

08 87 16.10 Safety Films On Glass		Crew	Daily Output	Labor-Hours	Unit	Material	2019 Bare Costs Labor	Equipment	Total	Total Incl O&P
0010	**SAFETY FILMS ON GLASS** (glass not included)									
0015	Safety film helps hold glass together when broken									
0050	Safety film, clear, 2 mil	H-2	180	.133	S.F.	1.23	4.75		5.98	9.15
0100	Safety film, clear, 4 mil		180	.133		1.84	4.75		6.59	9.80
0110	Safety film, tinted, 4 mil		180	.133		3.52	4.75		8.27	11.65
0150	Safety film, black tint, 8 mil		180	.133		2.27	4.75		7.02	10.30

08 87 26 – Bird Control Film

08 87 26.10 Bird Control Film		Crew	Daily Output	Labor-Hours	Unit	Material	Labor	Equipment	Total	Total Incl O&P
0010	**BIRD CONTROL FILM**									
0050	Patterned, adhered to glass	2 Glaz	180	.089	S.F.	6.85	3.40		10.25	13.15
0200	Decals small, adhered to glass	1 Glaz	50	.160	Ea.	2.52	6.10		8.62	12.80
0250	Decals large, adhered to glass		50	.160	"	8	6.10		14.10	18.85
0300	Decals set of 4, adhered to glass		25	.320	Set	7.20	12.25		19.45	28
0400	Bird control film tape		10	.800	Roll	20.50	30.50		51	73

08 87 33 – Electrically Tinted Window Film

08 87 33.20 Window Film		Crew	Daily Output	Labor-Hours	Unit	Material	Labor	Equipment	Total	Total Incl O&P
0010	**WINDOW FILM** adhered on glass (glass not included)									
0015	Film is pre-wired and can be trimmed									
0100	Window film 4 S.F. adhered on glass	1 Glaz	200	.040	S.F.	224	1.53		225.53	250
0120	6 S.F.		180	.044		335	1.70		336.70	375
0160	9 S.F.		170	.047		505	1.80		506.80	560
0180	10 S.F.		150	.053		570	2.04		572.04	635
0200	12 S.F.		144	.056		670	2.13		672.13	740
0220	14 S.F.		140	.057		780	2.19		782.19	865
0240	16 S.F.		128	.063		895	2.39		897.39	985
0260	18 S.F.		126	.063		1,000	2.43		1,002.43	1,100
0280	20 S.F.		120	.067		1,125	2.55		1,127.55	1,225

08 87 53 – Security Films On Glass

08 87 53.10 Security Film Adhered On Glass		Crew	Daily Output	Labor-Hours	Unit	Material	Labor	Equipment	Total	Total Incl O&P
0010	**SECURITY FILM ADHERED ON GLASS** (glass not included)									
0020	Security film, clear, 7 mil	1 Glaz	200	.040	S.F.	1.11	1.53		2.64	3.74
0030	8 mil		200	.040		2	1.53		3.53	4.72
0040	9 mil		200	.040		1.10	1.53		2.63	3.73
0050	10 mil		200	.040		1.89	1.53		3.42	4.60
0060	12 mil		200	.040		1.87	1.53		3.40	4.58
0075	15 mil		200	.040		2.36	1.53		3.89	5.10
0100	Security film, sealed with structural adhesive, 7 mil		180	.044		5.65	1.70		7.35	9
0110	8 mil		180	.044		6.55	1.70		8.25	10
0140	14 mil		180	.044		7.20	1.70		8.90	10.70

08 88 52 – Prefabricated Glass Block Windows

08 88 52.10 Prefabricated Glass Block Windows

	08 88 52.10 Prefabricated Glass Block Windows	Crew	Daily Output	Labor-Hours	Unit	Material	Labor	Equipment	Total	Total Incl O&P
						2019 Bare Costs				
0010	PREFABRICATED GLASS BLOCK WINDOWS									
0015	Includes frame and silicone seal									
0020	Glass Block Window 16" x 8"	2 Glaz	6	2.667	Ea.	166	102		268	350
0025	16" x 16"		6	2.667		190	102		292	375
0030	16" x 24"		6	2.667		220	102		322	410
0050	16" x 48"		4	4		355	153		508	640
0100	Glass Block Window 24" x 8"		6	2.667		179	102		281	365
0110	24" x 16"		6	2.667		229	102		331	420
0120	24" x 24"		6	2.667		289	102		391	490
0150	24" x 48"		6	2.667		425	102		527	640
0200	Glass Block Window 32" x 8"		6	2.667		179	102		281	365
0210	32" x 16"		6	2.667		263	102		365	455
0220	32" x 24"		4	4		320	153		473	605
0250	32" x 48"		4	4		520	153		673	820
0300	Glass Block Window 40" x 8"		6	2.667		241	102		343	435
0310	40" x 16"		6	2.667		299	102		401	500
0320	40" x 24"		6	2.667		370	102		472	575
0350	40" x 48"		6	2.667		610	102		712	840
0400	Glass Block Window 48" x 8"		6	2.667		270	102		372	465
0410	48" x 16"		6	2.667		355	102		457	560
0420	48" x 24"		6	2.667		425	102		527	640
0450	48" x 48"		6	2.667		695	102		797	930
0500	Glass Block Window 56" x 8"		6	2.667		286	102		388	485
0510	56" x 16"		4	4		405	153		558	695
0520	56" x 24"		4	4		500	153		653	800
0550	56" x 48"	↓	4	4	↓	845	153		998	1,175

08 88 52.20 Prefab. Glass Block Windows Hurricane Resistant

	08 88 52.20 Prefab. Glass Block Windows Hurricane Resistant	Crew	Daily Output	Labor-Hours	Unit	Material	Labor	Equipment	Total	Total Incl O&P
0010	PREFABRICATED GLASS BLOCK WINDOWS HURRICANE RESISTANT									
0015	Includes frame and silicone seal									
3020	Glass Block Window 16" x 16"	2 Glaz	6	2.667	Ea.	425	102		527	640
3030	16" x 24"		6	2.667		510	102		612	730
3040	16" x 32"		4	4		590	153		743	900
3050	16" x 40"		4	4		635	153		788	950
3060	16" x 48"		4	4		670	153		823	990
3070	16" x 56"		4	4		835	153		988	1,175
3080	Glass Block Window 24" x 24"		6	2.667		590	102		692	820
3090	24" x 32"		4	4		670	153		823	990
3100	24" x 40"		4	4		835	153		988	1,175
3110	24" x 48"		4	4		875	153		1,028	1,225
3120	24" x 56"		4	4		910	153		1,063	1,250
3130	Glass Block Window 32" x 32"		4	4		750	153		903	1,075
3140	32" x 40"		4	4		855	153		1,008	1,200
3150	32" x 48"		4	4		995	153		1,148	1,350
3160	32" x 56"		4	4		1,150	153		1,303	1,525
3200	Glass Block Window 40" x 40"		4	4		995	153		1,148	1,350
3210	40" x 48"		3	5.333		1,200	204		1,404	1,650
3220	40" x 56"		3	5.333		1,350	204		1,554	1,800
3260	Glass Block Window 48" x 48"		3	5.333		1,400	204		1,604	1,875
3270	48" x 56"		3	5.333		1,550	204		1,754	2,050
3310	Glass Block Window 56" x 56"	↓	3	5.333	↓	1,725	204		1,929	2,225

08 88 Special Function Glazing

08 88 52 – Prefabricated Glass Block Windows

08 88 52.30 Prefabricated Acrylic Block Casement Windows		Crew	Daily Output	Labor-Hours	Unit	2019 Bare Costs Material	Labor	Equipment	Total	Total Incl O&P
0010	**PREFABRICATED ACRYLIC BLOCK CASEMENT WINDOWS**									
0015	Windows include frame and nail fin.									
3050	6" x 6" x 1-3/4" Block Casement Window 2 x 2 Pattern	2 Glaz	6	2.667	Ea.	455	102		557	670
3055	2 x 3		6	2.667		475	102		577	690
3060	2 x 4		6	2.667		515	102		617	735
3065	2 x 5		6	2.667		575	102		677	800
3070	2 x 6		6	2.667		605	102		707	835
3075	2 x 7		4	4		720	153		873	1,050
3080	2 x 8		4	4		750	153		903	1,075
3085	2' x 9'		4	4		820	153		973	1,150
3090	3 x 3		4	4		500	153		653	800
3095	3 x 4		4	4		550	153		703	855
3100	3 x 5		4	4		615	153		768	925
3110	3 x 6		4	4		655	153		808	970
3115	3 x 7		4	4		735	153		888	1,050
3120	3 x 8		4	4		815	153		968	1,150
3125	3 x 9		3	5.333		880	204		1,084	1,300
3130	4 x 4		3	5.333		640	204		844	1,050
3135	4 x 5		3	5.333		685	204		889	1,075
3140	4 x 6		3	5.333		740	204		944	1,150
3145	4 x 7		3	5.333		765	204		969	1,175
3150	4 x 8		3	5.333		880	204		1,084	1,300
3500	8" x 8" x 1-3/4" Block Casement Window 2 x 2 Pattern		6	2.667		420	102		522	630
3510	2 x 3		6	2.667		465	102		567	680
3515	2 x 4		6	2.667		525	102		627	745
3520	2 x 5		6	2.667		590	102		692	820
3525	2 x 6		4	4		660	153		813	975
3530	2 x 7		4	4		690	153		843	1,000
3535	2 x 8		3	5.333		755	204		959	1,175
3540	2 x 9		3	5.333		825	204		1,029	1,250
3545	2 x 10		3	5.333		790	204		994	1,200
3550	2 x 11		3	5.333		935	204		1,139	1,350
3555	3 x 3		4	4		520	153		673	820
3560	3 x 4		4	4		575	153		728	885
3565	3 x 5		4	4		680	153		833	1,000
3570	3 x 6		4	4		745	153		898	1,075
3575	3 x 7		4	4		860	153		1,013	1,200
3580	3 x 8		3	5.333		965	204		1,169	1,400
3585	3 x 9		3	5.333		875	204		1,079	1,300
3590	4 x 4		3	5.333		710	204		914	1,125
3595	4 x 5		3	5.333		875	204		1,079	1,300
3600	4 x 6		3	5.333		940	204		1,144	1,350

08 88 56 – Ballistics-Resistant Glazing

08 88 56.10 Laminated Glass		Crew	Daily Output	Labor-Hours	Unit	Material	Labor	Equipment	Total	Total Incl O&P
0010	**LAMINATED GLASS**									
0020	Clear float .03" vinyl 1/4" thick	2 Glaz	90	.178	S.F.	13.05	6.80		19.85	25.50
0100	3/8" thick		78	.205		28.50	7.85		36.35	44.50
0200	.06" vinyl, 1/2" thick		65	.246		28	9.40		37.40	46.50
1000	5/8" thick		90	.178		31	6.80		37.80	45
2000	Bullet-resisting, 1-3/16" thick, to 15 S.F.		16	1		121	38.50		159.50	196
2100	Over 15 S.F.		16	1		154	38.50		192.50	232
2200	2" thick, to 15 S.F.		10	1.600		152	61		213	268

08 88 Special Function Glazing

08 88 56 – Ballistics-Resistant Glazing

08 88 56.10 Laminated Glass		Crew	Daily Output	Labor-Hours	Unit	2019 Bare Costs Material	Labor	Equipment	Total	Total Incl O&P
2300	Over 15 S.F.	2 Glaz	10	1.600	S.F.	130	61		191	244
2500	2-1/4" thick, to 15 S.F.		12	1.333		208	51		259	315
2600	Over 15 S.F.		12	1.333		197	51		248	300
2700	Level 2 (.357 magnum), NIJ and UL		12	1.333		61.50	51		112.50	152
2750	Level 3A (.44 magnum) NIJ, UL 3		12	1.333		67	51		118	158
2800	Level 4 (AK-47) NIJ, UL 7 & 8		12	1.333		84.50	51		135.50	177
2850	Level 5 (M-16) UL		12	1.333		94.50	51		145.50	188
2900	Level 3 (7.62 Armor Piercing) NIJ, UL 4 & 5	↓	12	1.333	↓	112	51		163	207

08 91 Louvers

08 91 16 – Operable Wall Louvers

08 91 16.10 Movable Blade Louvers		Crew	Daily Output	Labor-Hours	Unit	Material	Labor	Equipment	Total	Total Incl O&P
0010	**MOVABLE BLADE LOUVERS**									
0100	PVC, commercial grade, 12" x 12"	1 Shee	20	.400	Ea.	82	17.30		99.30	119
0110	16" x 16"		14	.571		95.50	24.50		120	146
0120	18" x 18"		14	.571		98	24.50		122.50	149
0130	24" x 24"		14	.571		140	24.50		164.50	195
0140	30" x 30"		12	.667		169	29		198	234
0150	36" x 36"		12	.667		199	29		228	266
0300	Stainless steel, commercial grade, 12" x 12"		20	.400		225	17.30		242.30	276
0310	16" x 16"		14	.571		272	24.50		296.50	340
0320	18" x 18"		14	.571		288	24.50		312.50	355
0330	20" x 20"		14	.571		330	24.50		354.50	405
0340	24" x 24"		14	.571		395	24.50		419.50	475
0350	30" x 30"		12	.667		440	29		469	535
0360	36" x 36"	↓	12	.667	↓	535	29		564	640

08 91 19 – Fixed Louvers

08 91 19.10 Aluminum Louvers		Crew	Daily Output	Labor-Hours	Unit	Material	Labor	Equipment	Total	Total Incl O&P
0010	**ALUMINUM LOUVERS**									
0020	Aluminum with screen, residential, 8" x 8"	1 Carp	38	.211	Ea.	21	8.15		29.15	36.50
0100	12" x 12"		38	.211		16.50	8.15		24.65	31.50
0200	12" x 18"		35	.229		22	8.85		30.85	38.50
0250	14" x 24"		30	.267		30	10.35		40.35	50
0300	18" x 24"		27	.296		33	11.50		44.50	55.50
0500	24" x 30"		24	.333		63.50	12.90		76.40	91.50
0700	Triangle, adjustable, small		20	.400		60.50	15.50		76	92
0800	Large		15	.533		85	20.50		105.50	128
2100	Midget, aluminum, 3/4" deep, 1" diameter		85	.094		.94	3.65		4.59	7.10
2150	3" diameter		60	.133		3.07	5.15		8.22	11.95
2200	4" diameter		50	.160		5.75	6.20		11.95	16.60
2250	6" diameter	↓	30	.267		3.80	10.35		14.15	21.50
3000	PVC, commercial grade, 12" x 12"	1 Shee	20	.400		96.50	17.30		113.80	135
3010	12" x 18"		20	.400		109	17.30		126.30	149
3020	12" x 24"		20	.400		146	17.30		163.30	190
3030	14" x 24"		20	.400		153	17.30		170.30	197
3100	Aluminum, commercial grade, 12" x 12"		20	.400		166	17.30		183.30	211
3110	24" x 24"		16	.500		228	21.50		249.50	287
3120	24" x 36"		16	.500		325	21.50		346.50	395
3130	36" x 24"		16	.500		340	21.50		361.50	405
3140	36" x 36"	↓	14	.571	↓	420	24.50		444.50	505

08 91 Louvers

08 91 19 – Fixed Louvers

08 91 19.10 Aluminum Louvers		Crew	Daily Output	Labor-Hours	Unit	2019 Bare Costs Material	Labor	Equipment	Total	Total Incl O&P
3150	36″ x 48″	1 Shee	10	.800	Ea.	510	34.50		544.50	620
3160	48″ x 36″		10	.800		485	34.50		519.50	595
3170	48″ x 48″		10	.800		445	34.50		479.50	550
3180	60″ x 48″		8	1		505	43.50		548.50	625
3190	60″ x 60″		8	1		730	43.50		773.50	875

08 91 19.20 Steel Louvers

		Crew	Daily Output	Labor-Hours	Unit	Material	Labor	Equipment	Total	Total Incl O&P
0010	**STEEL LOUVERS**									
3300	Galvanized steel, fixed blades, commercial grade, 18″ x 18″	1 Shee	20	.400	Ea.	159	17.30		176.30	204
3310	24″ x 24″		20	.400		213	17.30		230.30	263
3320	24″ x 36″		16	.500		277	21.50		298.50	340
3330	36″ x 24″		16	.500		283	21.50		304.50	345
3340	36″ x 36″		14	.571		350	24.50		374.50	425
3350	36″ x 48″		10	.800		410	34.50		444.50	510
3360	48″ x 36″		10	.800		400	34.50		434.50	500
3370	48″ x 48″		10	.800		495	34.50		529.50	605
3380	60″ x 48″		10	.800		540	34.50		574.50	650
3390	60″ x 60″		10	.800		645	34.50		679.50	770

08 91 26 – Door Louvers

08 91 26.10 Steel Louvers, 18 Gauge, Fixed Blade

		Crew	Daily Output	Labor-Hours	Unit	Material	Labor	Equipment	Total	Total Incl O&P
0010	**STEEL LOUVERS, 18 GAUGE, FIXED BLADE**									
0050	12″ x 12″, with powder coat	1 Carp	20	.400	Ea.	97	15.50		112.50	133
0055	18″ x 12″		20	.400		101	15.50		116.50	137
0060	18″ x 18″		20	.400		108	15.50		123.50	145
0065	24″ x 12″		20	.400		120	15.50		135.50	158
0070	24″ x 18″		20	.400		127	15.50		142.50	165
0075	24″ x 24″		20	.400		160	15.50		175.50	202
0100	12″ x 12″, galvanized		20	.400		80	15.50		95.50	114
0105	18″ x 12″		20	.400		84.50	15.50		100	119
0115	24″ x 12″		20	.400		107	15.50		122.50	143
0125	24″ x 24″		20	.400		147	15.50		162.50	188
0300	6″ x 12″, painted		20	.400		46.50	15.50		62	76.50
0320	10″ x 16″		20	.400		72.50	15.50		88	106
0340	12″ x 22″		20	.400		105	15.50		120.50	141
0360	14″ x 14″		20	.400		106	15.50		121.50	142
0400	18″ x 18″		20	.400		82.50	15.50		98	117
0420	20″ x 26″		20	.400		143	15.50		158.50	184
0440	22″ x 22″		20	.400		136	15.50		151.50	175
0460	26″ x 26″		20	.400		220	15.50		235.50	268
3000	Fire rated									
3010	12″ x 12″	1 Carp	10	.800	Ea.	178	31		209	247
3050	18″ x 18″		10	.800		261	31		292	340
3100	24″ x 12″		10	.800		259	31		290	335
3120	24″ x 18″		10	.800		293	31		324	375
3130	24″ x 24″		10	.800		293	31		324	370

08 91 26.20 Aluminum Louvers

		Crew	Daily Output	Labor-Hours	Unit	Material	Labor	Equipment	Total	Total Incl O&P
0010	**ALUMINUM LOUVERS**, 18 ga., fixed blade, clear anodized									
0050	6″ x 12″	1 Carp	20	.400	Ea.	64	15.50		79.50	96
0060	10″ x 10″		20	.400		63.50	15.50		79	95.50
0065	10″ x 14″		20	.400		74	15.50		89.50	107
0080	12″ x 22″		20	.400		101	15.50		116.50	137
0090	14″ x 14″		20	.400		90.50	15.50		106	126
0100	18″ x 10″		20	.400		75	15.50		90.50	108

08 91 Louvers

08 91 26 – Door Louvers

08 91 26.20 Aluminum Louvers		Crew	Daily Output	Labor-Hours	Unit	Material	Labor	Equipment	Total	Total Incl O&P
						2019 Bare Costs				
0110	18" x 18"	1 Carp	20	.400	Ea.	128	15.50		143.50	167
0120	22" x 22"		20	.400		164	15.50		179.50	207
0130	26" x 26"	↓	20	.400	↓	211	15.50		226.50	258

08 95 Vents

08 95 13 – Soffit Vents

08 95 13.10 Wall Louvers		Crew	Daily Output	Labor-Hours	Unit	Material	Labor	Equipment	Total	Total Incl O&P
0010	**WALL LOUVERS**									
2400	Under eaves vent, aluminum, mill finish, 16" x 4"	1 Carp	48	.167	Ea.	1.95	6.45		8.40	12.85
2500	16" x 8"	"	48	.167	"	2.92	6.45		9.37	13.90

08 95 16 – Wall Vents

08 95 16.10 Louvers		Crew	Daily Output	Labor-Hours	Unit	Material	Labor	Equipment	Total	Total Incl O&P
0010	**LOUVERS**									
0020	Redwood, 2'-0" diameter, full circle	1 Carp	16	.500	Ea.	294	19.40		313.40	355
0100	Half circle		16	.500		214	19.40		233.40	267
0200	Octagonal		16	.500		206	19.40		225.40	259
0300	Triangular, 5/12 pitch, 5'-0" at base		16	.500		630	19.40		649.40	720
1000	Rectangular, 1'-4" x 1'-3"		16	.500		27.50	19.40		46.90	62
1100	Rectangular, 1'-4" x 1'-8"		16	.500		42	19.40		61.40	78
1200	1'-4" x 2'-2"		15	.533		49.50	20.50		70	88.50
1300	1'-9" x 2'-2"		15	.533		59.50	20.50		80	99
1400	2'-3" x 2'-2"		14	.571		79.50	22		101.50	124
1700	2'-4" x 2'-11"		13	.615		79.50	24		103.50	127
2000	Aluminum, 12" x 16"		25	.320		24	12.40		36.40	47
2010	16" x 20"		25	.320		32	12.40		44.40	55.50
2020	24" x 30"		25	.320		65.50	12.40		77.90	92.50
2100	6' triangle		12	.667		181	26		207	242
3100	Round, 2'-2" diameter		16	.500		195	19.40		214.40	247
7000	Vinyl gable vent, 8" x 8"		38	.211		17.65	8.15		25.80	33
7020	12" x 12"		38	.211		31.50	8.15		39.65	48
7080	12" x 18"		35	.229		40.50	8.85		49.35	59
7200	18" x 24"	↓	30	.267	↓	60	10.35		70.35	83

Estimating Tips

General

- Room Finish Schedule: A complete set of plans should contain a room finish schedule. If one is not available, it would be well worth the time and effort to obtain one.

09 20 00 Plaster and Gypsum Board

- Lath is estimated by the square yard plus a 5% allowance for waste. Furring, channels, and accessories are measured by the linear foot. An extra foot should be allowed for each accessory miter or stop.
- Plaster is also estimated by the square yard. Deductions for openings vary by preference, from zero deduction to 50% of all openings over 2 feet in width. The estimator should allow one extra square foot for each linear foot of horizontal interior or exterior angle located below the ceiling level. Also, double the areas of small radius work.
- Drywall accessories, studs, track, and acoustical caulking are all measured by the linear foot. Drywall taping is figured by the square foot. Gypsum wallboard is estimated by the square foot. No material deductions should be made for door or window openings under 32 S.F.

09 60 00 Flooring

- Tile and terrazzo areas are taken off on a square foot basis. Trim and base materials are measured by the linear foot. Accent tiles are listed per each. Two basic methods of installation are used. Mud set is approximately 30% more expensive than thin set. The cost of grout is included with tile unit price lines unless otherwise noted. In terrazzo work, be sure to include the linear footage of embedded decorative strips, grounds, machine rubbing, and power cleanup.
- Wood flooring is available in strip, parquet, or block configuration. The latter two types are set in adhesives with quantities estimated by the square foot. The laying pattern will influence labor costs and material waste. In addition to the material and labor for laying wood floors, the estimator must make allowances for sanding and finishing these areas, unless the flooring is prefinished.
- Sheet flooring is measured by the square yard. Roll widths vary, so consideration should be given to use the most economical width, as waste must be figured into the total quantity. Consider also the installation methods available—direct glue down or stretched. Direct glue-down installation is assumed with sheet carpet unit price lines unless otherwise noted.

09 70 00 Wall Finishes

- Wall coverings are estimated by the square foot. The area to be covered is measured—length by height of the wall above the baseboards—to calculate the square footage of each wall. This figure is divided by the number of square feet in the single roll which is being used. Deduct, in full, the areas of openings such as doors and windows. Where a pattern match is required allow 25–30% waste.

09 80 00 Acoustic Treatment

- Acoustical systems fall into several categories. The takeoff of these materials should be by the square foot of area with a 5% allowance for waste. Do not forget about scaffolding, if applicable, when estimating these systems.

09 90 00 Painting and Coating

- A major portion of the work in painting involves surface preparation. Be sure to include cleaning, sanding, filling, and masking costs in the estimate.
- Protection of adjacent surfaces is not included in painting costs. When considering the method of paint application, an important factor is the amount of protection and masking required. These must be estimated separately and may be the determining factor in choosing the method of application.

Reference Numbers

Reference numbers are shown at the beginning of some major classifications. These numbers refer to related items in the Reference Section. The reference information may be an estimating procedure, an alternate pricing method, or technical information.

Note: Not all subdivisions listed here necessarily appear. ■

09 05 Common Work Results for Finishes

09 05 05 – Selective Demolition for Finishes

09 05 05.10 Selective Demolition, Ceilings

	09 05 05.10 Selective Demolition, Ceilings	Crew	Daily Output	Labor-Hours	Unit	Material	2019 Bare Costs Labor	Equipment	Total	Total Incl O&P
0010	**SELECTIVE DEMOLITION, CEILINGS** R024119-10									
0200	Ceiling, gypsum wall board, furred and nailed or screwed	2 Clab	800	.020	S.F.		.61		.61	1
1000	Plaster, lime and horse hair, on wood lath, incl. lath		700	.023			.69		.69	1.15
1200	Suspended ceiling, mineral fiber, 2′ x 2′ or 2′ x 4′		1500	.011			.32		.32	.54
1250	On suspension system, incl. system		1200	.013			.40		.40	.67
1500	Tile, wood fiber, 12″ x 12″, glued		900	.018			.54		.54	.89
1540	Stapled		1500	.011			.32		.32	.54
2000	Wood, tongue and groove, 1″ x 4″		1000	.016			.49		.49	.80
2040	1″ x 8″		1100	.015			.44		.44	.73
2400	Plywood or wood fiberboard, 4′ x 8′ sheets		1200	.013			.40		.40	.67
2500	Remove & refinish textured ceiling	1 Plas	222	.036		.03	1.31		1.34	2.18

09 05 05.20 Selective Demolition, Flooring

	09 05 05.20 Selective Demolition, Flooring	Crew	Daily Output	Labor-Hours	Unit	Material	Labor	Equipment	Total	Total Incl O&P
0010	**SELECTIVE DEMOLITION, FLOORING**									
0200	Brick with mortar	2 Clab	475	.034	S.F.		1.02		1.02	1.69
0400	Carpet, bonded, including surface scraping		2000	.008			.24		.24	.40
0480	Tackless		9000	.002			.05		.05	.09
0550	Carpet tile, releasable adhesive		5000	.003			.10		.10	.16
0560	Permanent adhesive		1850	.009			.26		.26	.43
0800	Resilient, sheet goods		1400	.011			.35		.35	.57
0850	Vinyl or rubber cove base	1 Clab	1000	.008	L.F.		.24		.24	.40
0860	Vinyl or rubber cove base, molded corner	"	1000	.008	Ea.		.24		.24	.40
0870	For glued and caulked installation, add to labor						50%			
0900	Vinyl composition tile, 12″ x 12″	2 Clab	1000	.016	S.F.		.49		.49	.80
2000	Tile, ceramic, thin set		675	.024			.72		.72	1.19
2020	Mud set		625	.026			.78		.78	1.28
3000	Wood, block, on end	1 Carp	400	.020			.78		.78	1.28
3200	Parquet		450	.018			.69		.69	1.14
3400	Strip flooring, interior, 2-1/4″ x 25/32″ thick		325	.025			.95		.95	1.58
3500	Exterior, porch flooring, 1″ x 4″		220	.036			1.41		1.41	2.33
3800	Subfloor, tongue and groove, 1″ x 6″		325	.025			.95		.95	1.58
3820	1″ x 8″		430	.019			.72		.72	1.19
3840	1″ x 10″		520	.015			.60		.60	.99
4000	Plywood, nailed		600	.013			.52		.52	.85
4100	Glued and nailed		400	.020			.78		.78	1.28
4200	Hardboard, 1/4″ thick		760	.011			.41		.41	.67
9050	For grinding concrete floors, see Section 03 35 43.10									

09 05 05.30 Selective Demolition, Walls and Partitions

	09 05 05.30 Selective Demolition, Walls and Partitions	Crew	Daily Output	Labor-Hours	Unit	Material	Labor	Equipment	Total	Total Incl O&P
0010	**SELECTIVE DEMOLITION, WALLS AND PARTITIONS** R024119-10									
0020	Walls, concrete, reinforced	B-39	120	.400	C.F.		12.25	1.96	14.21	22.50
0025	Plain	"	160	.300	"		9.20	1.47	10.67	16.80
1000	Gypsum wallboard, nailed or screwed	1 Clab	1000	.008	S.F.		.24		.24	.40
1010	2 layers		400	.020			.61		.61	1
1500	Fiberboard, nailed		900	.009			.27		.27	.45
1568	Plenum barrier, sheet lead		300	.027			.81		.81	1.34
2200	Metal or wood studs, finish 2 sides, fiberboard	B-1	520	.046			1.43		1.43	2.37
2250	Lath and plaster		260	.092			2.86		2.86	4.73
2300	Gypsum wallboard		520	.046			1.43		1.43	2.37
2350	Plywood		450	.053			1.65		1.65	2.73
2800	Paneling, 4′ x 8′ sheets	1 Clab	475	.017			.51		.51	.84
3000	Plaster, lime and horsehair, on wood lath		400	.020			.61		.61	1
3020	On metal lath		335	.024			.72		.72	1.20
3450	Plaster, interior gypsum, acoustic, or cement		60	.133	S.Y.		4.05		4.05	6.70

09 05 Common Work Results for Finishes

09 05 05 – Selective Demolition for Finishes

09 05 05.30 Selective Demolition, Walls and Partitions		Crew	Daily Output	Labor-Hours	Unit	Material	2019 Bare Costs Labor	Equipment	Total	Total Incl O&P
3500	Stucco, on masonry	1 Clab	145	.055	S.Y.		1.67		1.67	2.77
3510	Commercial 3-coat		80	.100			3.04		3.04	5
3520	Interior stucco		25	.320			9.70		9.70	16.05
3760	Tile, ceramic, on walls, thin set		300	.027	S.F.		.81		.81	1.34
3765	Mud set		250	.032	"		.97		.97	1.60

09 22 Supports for Plaster and Gypsum Board

09 22 03 – Fastening Methods for Finishes

09 22 03.20 Drilling Plaster/Drywall

		Crew	Daily Output	Labor-Hours	Unit	Material	Labor	Equipment	Total	Total Incl O&P
0010	**DRILLING PLASTER/DRYWALL**									
1100	Drilling & layout for drywall/plaster walls, up to 1" deep, no anchor									
1200	Holes, 1/4" diameter	1 Carp	150	.053	Ea.	.01	2.07		2.08	3.43
1300	3/8" diameter		140	.057		.01	2.21		2.22	3.67
1400	1/2" diameter		130	.062		.01	2.38		2.39	3.95
1500	3/4" diameter		120	.067		.02	2.58		2.60	4.29
1600	1" diameter		110	.073		.02	2.82		2.84	4.68
1700	1-1/4" diameter		100	.080		.04	3.10		3.14	5.15
1800	1-1/2" diameter		90	.089		.06	3.44		3.50	5.75
1900	For ceiling installations, add						40%			

09 22 13 – Metal Furring

09 22 13.13 Metal Channel Furring

		Crew	Daily Output	Labor-Hours	Unit	Material	Labor	Equipment	Total	Total Incl O&P
0010	**METAL CHANNEL FURRING**									
0030	Beams and columns, 7/8" hat channels, galvanized, 12" OC	1 Lath	155	.052	S.F.	.43	1.97		2.40	3.64
0050	16" OC		170	.047		.35	1.80		2.15	3.27
0070	24" OC		185	.043		.23	1.65		1.88	2.91
0100	Ceilings, on steel, 7/8" hat channels, galvanized, 12" OC		210	.038		.39	1.45		1.84	2.77
0300	16" OC		290	.028		.35	1.05		1.40	2.07
0400	24" OC		420	.019		.23	.73		.96	1.43
0600	1-5/8" hat channels, galvanized, 12" OC		190	.042		.53	1.61		2.14	3.17
0700	16" OC		260	.031		.47	1.17		1.64	2.41
0900	24" OC		390	.021		.32	.78		1.10	1.61
0930	7/8" hat channels with sound isolation clips, 12" OC		120	.067		1.70	2.54		4.24	5.95
0940	16" OC		100	.080		1.28	3.05		4.33	6.30
0950	24" OC		165	.048		.85	1.85		2.70	3.92
0960	1-5/8" hat channels, galvanized, 12" OC		110	.073		1.84	2.77		4.61	6.50
0970	16" OC		100	.080		1.38	3.05		4.43	6.45
0980	24" OC		155	.052		.92	1.97		2.89	4.18
1000	Walls, 7/8" hat channels, galvanized, 12" OC		235	.034		.39	1.30		1.69	2.52
1200	16" OC		265	.030		.35	1.15		1.50	2.23
1300	24" OC		350	.023		.23	.87		1.10	1.66
1500	1-5/8" hat channels, galvanized, 12" OC		210	.038		.53	1.45		1.98	2.92
1600	16" OC		240	.033		.47	1.27		1.74	2.57
1800	24" OC		305	.026		.32	1		1.32	1.96
1920	7/8" hat channels with sound isolation clips, 12" OC		125	.064		1.70	2.44		4.14	5.80
1940	16" OC		100	.080		1.28	3.05		4.33	6.30
1950	24" OC		150	.053		.85	2.03		2.88	4.21
1960	1-5/8" hat channels, galvanized, 12" OC		115	.070		1.84	2.65		4.49	6.30
1970	16" OC		95	.084		1.38	3.21		4.59	6.65
1980	24" OC		140	.057		.92	2.18		3.10	4.52
3000	Z Furring, walls, 1" deep, 25 ga., 24" OC		350	.023		1.36	.87		2.23	2.90

09 22 Supports for Plaster and Gypsum Board

09 22 13 – Metal Furring

09 22 13.13 Metal Channel Furring		Crew	Daily Output	Labor-Hours	Unit	Material	2019 Bare Costs Labor	Equipment	Total	Total Incl O&P
3010	48" OC	1 Lath	700	.011	S.F.	.68	.44		1.12	1.45
3020	1-1/2" deep, 24" OC		345	.023		1.59	.88		2.47	3.16
3030	48" OC		695	.012		.79	.44		1.23	1.58
3040	2" deep, 24" OC		340	.024		1.91	.90		2.81	3.54
3050	48" OC		690	.012		.95	.44		1.39	1.76
3060	1" deep, 20 ga., 24" OC		350	.023		2.25	.87		3.12	3.87
3070	48" OC		700	.011		1.12	.44		1.56	1.93
3080	1-1/2" deep, 24" OC		345	.023		2.59	.88		3.47	4.26
3090	48" OC		695	.012		1.29	.44		1.73	2.13
4000	2" deep, 24" OC		340	.024		3.18	.90		4.08	4.93
4010	48" OC	↓	690	.012	↓	1.59	.44		2.03	2.46

09 22 16 – Non-Structural Metal Framing

09 22 16.13 Non-Structural Metal Stud Framing

		Crew	Daily Output	Labor-Hours	Unit	Material	Labor	Equipment	Total	Total Incl O&P
0010	**NON-STRUCTURAL METAL STUD FRAMING**									
1600	Non-load bearing, galv., 8' high, 25 ga. 1-5/8" wide, 16" OC	1 Carp	619	.013	S.F.	.27	.50		.77	1.12
1610	24" OC		950	.008		.20	.33		.53	.76
1620	2-1/2" wide, 16" OC		613	.013		.35	.51		.86	1.23
1630	24" OC		938	.009		.26	.33		.59	.84
1640	3-5/8" wide, 16" OC		600	.013		.40	.52		.92	1.29
1650	24" OC		925	.009		.30	.34		.64	.88
1660	4" wide, 16" OC		594	.013		.44	.52		.96	1.35
1670	24" OC		925	.009		.33	.34		.67	.91
1680	6" wide, 16" OC		588	.014		.55	.53		1.08	1.47
1690	24" OC		906	.009		.41	.34		.75	1.02
1700	20 ga. studs, 1-5/8" wide, 16" OC		494	.016		.34	.63		.97	1.42
1710	24" OC		763	.010		.26	.41		.67	.95
1720	2-1/2" wide, 16" OC		488	.016		.43	.64		1.07	1.52
1730	24" OC		750	.011		.33	.41		.74	1.04
1740	3-5/8" wide, 16" OC		481	.017		.49	.64		1.13	1.61
1750	24" OC		738	.011		.37	.42		.79	1.10
1760	4" wide, 16" OC		475	.017		.60	.65		1.25	1.74
1770	24" OC		738	.011		.45	.42		.87	1.18
1780	6" wide, 16" OC		469	.017		.72	.66		1.38	1.88
1790	24" OC		725	.011		.54	.43		.97	1.30
2000	Non-load bearing, galv., 10' high, 25 ga. 1-5/8" wide, 16" OC		495	.016		.25	.63		.88	1.32
2100	24" OC		760	.011		.19	.41		.60	.87
2200	2-1/2" wide, 16" OC		490	.016		.34	.63		.97	1.42
2250	24" OC		750	.011		.25	.41		.66	.95
2300	3-5/8" wide, 16" OC		480	.017		.37	.65		1.02	1.48
2350	24" OC		740	.011		.28	.42		.70	.99
2400	4" wide, 16" OC		475	.017		.42	.65		1.07	1.54
2450	24" OC		740	.011		.31	.42		.73	1.03
2500	6" wide, 16" OC		470	.017		.52	.66		1.18	1.66
2550	24" OC		725	.011		.38	.43		.81	1.13
2600	20 ga. studs, 1-5/8" wide, 16" OC		395	.020		.32	.78		1.10	1.66
2650	24" OC		610	.013		.24	.51		.75	1.10
2700	2-1/2" wide, 16" OC		390	.021		.41	.79		1.20	1.76
2750	24" OC		600	.013		.30	.52		.82	1.18
2800	3-5/8" wide, 16" OC		385	.021		.47	.81		1.28	1.85
2850	24" OC		590	.014		.35	.53		.88	1.25
2900	4" wide, 16" OC		380	.021		.57	.82		1.39	1.97
2950	24" OC	↓	590	.014	↓	.42	.53		.95	1.33

09 22 16 – Non-Structural Metal Framing

09 22 16.13 Non-Structural Metal Stud Framing		Crew	Daily Output	Labor-Hours	Unit	Material	2019 Bare Costs Labor	Equipment	Total	Total Incl O&P
3000	6" wide, 16" OC	1 Carp	375	.021	S.F.	.68	.83		1.51	2.12
3050	24" OC		580	.014		.50	.53		1.03	1.43
3060	Non-load bearing, galv., 12' high, 25 ga. 1-5/8" wide, 16" OC		413	.019		.24	.75		.99	1.51
3070	24" OC		633	.013		.18	.49		.67	1
3080	2-1/2" wide, 16" OC		408	.020		.32	.76		1.08	1.61
3090	24" OC		625	.013		.23	.50		.73	1.08
3100	3-5/8" wide, 16" OC		400	.020		.36	.78		1.14	1.67
3110	24" OC		617	.013		.26	.50		.76	1.12
3120	4" wide, 16" OC		396	.020		.40	.78		1.18	1.73
3130	24" OC		617	.013		.29	.50		.79	1.15
3140	6" wide, 16" OC		392	.020		.50	.79		1.29	1.86
3150	24" OC		604	.013		.36	.51		.87	1.25
3160	20 ga. studs, 1-5/8" wide, 16" OC		329	.024		.31	.94		1.25	1.90
3170	24" OC		508	.016		.23	.61		.84	1.26
3180	2-1/2" wide, 16" OC		325	.025		.39	.95		1.34	2.01
3190	24" OC		500	.016		.28	.62		.90	1.33
3200	3-5/8" wide, 16" OC		321	.025		.45	.97		1.42	2.09
3210	24" OC		492	.016		.33	.63		.96	1.40
3220	4" wide, 16" OC		317	.025		.54	.98		1.52	2.22
3230	24" OC		492	.016		.40	.63		1.03	1.47
3240	6" wide, 16" OC		313	.026		.65	.99		1.64	2.36
3250	24" OC		483	.017		.47	.64		1.11	1.58
3260	Non-load bearing, galv., 16' high, 25 ga. 4" wide, 12" OC		195	.041		.51	1.59		2.10	3.19
3270	16" OC		275	.029		.40	1.13		1.53	2.30
3280	24" OC		400	.020		.29	.78		1.07	1.60
3290	6" wide, 12" OC		190	.042		.64	1.63		2.27	3.40
3300	16" OC		280	.029		.50	1.11		1.61	2.38
3310	24" OC		400	.020		.37	.78		1.15	1.68
3320	20 ga. studs, 2-1/2" wide, 12" OC		180	.044		.50	1.72		2.22	3.40
3330	16" OC		254	.032		.39	1.22		1.61	2.45
3340	24" OC		390	.021		.29	.79		1.08	1.62
3350	3-5/8" wide, 12" OC		170	.047		.57	1.82		2.39	3.64
3360	16" OC		251	.032		.45	1.24		1.69	2.54
3370	24" OC		384	.021		.33	.81		1.14	1.69
3380	4" wide, 12" OC		170	.047		.69	1.82		2.51	3.77
3390	16" OC		247	.032		.55	1.26		1.81	2.67
3400	24" OC		384	.021		.40	.81		1.21	1.77
3410	6" wide, 12" OC		175	.046		.83	1.77		2.60	3.85
3420	16" OC		245	.033		.66	1.27		1.93	2.81
3430	24" OC		400	.020		.48	.78		1.26	1.81
3440	Non-load bearing, galv., 20' high, 25 ga. 6" wide, 12" OC		125	.064		.63	2.48		3.11	4.79
3450	16" OC		220	.036		.49	1.41		1.90	2.87
3460	24" OC		360	.022		.35	.86		1.21	1.81
3470	20 ga. studs, 4" wide, 12" OC		120	.067		.68	2.58		3.26	5
3480	16" OC		215	.037		.54	1.44		1.98	2.98
3490	6" wide, 12" OC		115	.070		.82	2.70		3.52	5.35
3500	16" OC		215	.037		.64	1.44		2.08	3.08
3510	24" OC	↓	331	.024	↓	.46	.94		1.40	2.06
5000	For load bearing studs, see Section 05 41 13.30									

09 22 Supports for Plaster and Gypsum Board

09 22 26 – Suspension Systems

09 22 26.13 Ceiling Suspension Systems		Crew	Daily Output	Labor-Hours	Unit	2019 Bare Costs Material	Labor	Equipment	Total	Total Incl O&P
0010	**CEILING SUSPENSION SYSTEMS** for gypsum board or plaster									
8000	Suspended ceilings, including carriers									
8200	1-1/2" carriers, 24" OC with:									
8300	7/8" channels, 16" OC	1 Lath	275	.029	S.F.	.56	1.11		1.67	2.40
8320	24" OC		310	.026		.44	.98		1.42	2.07
8400	1-5/8" channels, 16" OC		205	.039		.68	1.49		2.17	3.15
8420	24" OC		250	.032		.52	1.22		1.74	2.54
8600	2" carriers, 24" OC with:									
8700	7/8" channels, 16" OC	1 Lath	250	.032	S.F.	.66	1.22		1.88	2.69
8720	24" OC		285	.028		.54	1.07		1.61	2.32
8800	1-5/8" channels, 16" OC		190	.042		.78	1.61		2.39	3.45
8820	24" OC		225	.036		.63	1.36		1.99	2.87

09 22 36 – Lath

09 22 36.13 Gypsum Lath

		Crew	Daily Output	Labor-Hours	Unit	Material	Labor	Equipment	Total	Total Incl O&P
0010	**GYPSUM LATH** R092000-50									
0020	Plain or perforated, nailed, 3/8" thick	1 Lath	85	.094	S.Y.	3.15	3.59		6.74	9.25
0100	1/2" thick		80	.100		2.70	3.82		6.52	9.10
0300	Clipped to steel studs, 3/8" thick		75	.107		3.15	4.07		7.22	10
0400	1/2" thick		70	.114		2.70	4.36		7.06	9.95
1500	For ceiling installations, add		216	.037			1.41		1.41	2.27
1600	For columns and beams, add		170	.047			1.80		1.80	2.89

09 22 36.23 Metal Lath

		Crew	Daily Output	Labor-Hours	Unit	Material	Labor	Equipment	Total	Total Incl O&P
0010	**METAL LATH** R092000-50									
3600	2.5 lb. diamond painted, on wood framing, on walls	1 Lath	85	.094	S.Y.	4.11	3.59		7.70	10.30
3700	On ceilings		75	.107		4.11	4.07		8.18	11.05
4200	3.4 lb. diamond painted, wired to steel framing		75	.107		4.49	4.07		8.56	11.50
4300	On ceilings		60	.133		4.49	5.10		9.59	13.15
5100	Rib lath, painted, wired to steel, on walls, 2.5 lb.		75	.107		3.61	4.07		7.68	10.50
5200	3.4 lb.		70	.114		4.75	4.36		9.11	12.25
5700	Suspended ceiling system, incl. 3.4 lb. diamond lath, painted		15	.533		4.35	20.50		24.85	38
5800	Galvanized		15	.533		4.56	20.50		25.06	38

09 22 36.43 Security Mesh

		Crew	Daily Output	Labor-Hours	Unit	Material	Labor	Equipment	Total	Total Incl O&P
0010	**SECURITY MESH**, expanded metal, flat, screwed to framing									
0100	On walls, 3/4", 1.76 lb./S.F.	2 Carp	1500	.011	S.F.	2.08	.41		2.49	2.97
0110	1-1/2", 1.14 lb./S.F.		1600	.010		1.56	.39		1.95	2.36
0200	On ceilings, 3/4", 1.76 lb./S.F.		1350	.012		2.08	.46		2.54	3.05
0210	1-1/2", 1.14 lb./S.F.		1450	.011		1.56	.43		1.99	2.43

09 22 36.83 Accessories, Plaster

		Crew	Daily Output	Labor-Hours	Unit	Material	Labor	Equipment	Total	Total Incl O&P
0010	**ACCESSORIES, PLASTER**									
0020	Casing bead, expanded flange, galvanized	1 Lath	2.70	2.963	C.L.F.	53	113		166	240
0200	Foundation weep screed, galvanized	"	2.70	2.963		53	113		166	240
0900	Channels, cold rolled, 16 ga., 3/4" deep, galvanized					39			39	42.50
1200	1-1/2" deep, 16 ga., galvanized					52.50			52.50	58
1620	Corner bead, expanded bullnose, 3/4" radius, #10, galvanized	1 Lath	2.60	3.077		26	117		143	218
1650	#1, galvanized		2.55	3.137		48.50	120		168.50	247
1670	Expanded wing, 2-3/4" wide, #1, galvanized		2.65	3.019		40.50	115		155.50	230
1700	Inside corner (corner rite), 3" x 3", painted		2.60	3.077		21	117		138	212
1750	Strip-ex, 4" wide, painted		2.55	3.137		25	120		145	221
1800	Expansion joint, 3/4" grounds, limited expansion, galv., 1 piece		2.70	2.963		70.50	113		183.50	260
2100	Extreme expansion, galvanized, 2 piece		2.60	3.077		140	117		257	345

09 23 Gypsum Plastering

09 23 13 – Acoustical Gypsum Plastering

09 23 13.10 Perlite or Vermiculite Plaster		Crew	Daily Output	Labor-Hours	Unit	Material	Labor (2019 Bare Costs)	Equipment	Total	Total Incl O&P
0010	**PERLITE OR VERMICULITE PLASTER** R092000-50									
0020	In 100 lb. bags, under 200 bags				Bag	18.45			18.45	20.50
0300	2 coats, no lath included, on walls	J-1	92	.435	S.Y.	6.05	14.90	1.47	22.42	33
0400	On ceilings		79	.506		6.05	17.35	1.71	25.11	37
0900	3 coats, no lath included, on walls		74	.541		6.50	18.55	1.83	26.88	39.50
1000	On ceilings	↓	63	.635		6.50	22	2.15	30.65	45
1700	For irregular or curved surfaces, add to above						30%			
1800	For columns and beams, add to above						50%			
1900	For soffits, add to ceiling prices				↓		40%			

09 23 20 – Gypsum Plaster

09 23 20.10 Gypsum Plaster On Walls and Ceilings		Crew	Daily Output	Labor-Hours	Unit	Material	Labor	Equipment	Total	Total Incl O&P
0010	**GYPSUM PLASTER ON WALLS AND CEILINGS** R092000-50									
0020	80# bag, less than 1 ton				Bag	15.70			15.70	17.30
0300	2 coats, no lath included, on walls	J-1	105	.381	S.Y.	3.78	13.05	1.29	18.12	27
0400	On ceilings		92	.435		3.78	14.90	1.47	20.15	30.50
0900	3 coats, no lath included, on walls		87	.460		5.40	15.75	1.55	22.70	33.50
1000	On ceilings	↓	78	.513	↓	5.40	17.60	1.73	24.73	36.50
1600	For irregular or curved surfaces, add						30%			
1800	For columns & beams, add						50%			

09 24 Cement Plastering

09 24 23 – Cement Stucco

09 24 23.40 Stucco		Crew	Daily Output	Labor-Hours	Unit	Material	Labor	Equipment	Total	Total Incl O&P
0010	**STUCCO** R092000-50									
0015	3 coats 7/8" thick, float finish, with mesh, on wood frame	J-2	63	.762	S.Y.	8.45	26.50	2.15	37.10	55
0100	On masonry construction, no mesh incl.	J-1	67	.597		3.55	20.50	2.02	26.07	39.50
0150	2 coats, 5/8" thick, float finish, no lath incl.	"	110	.364		3.44	12.45	1.23	17.12	25.50
0300	For trowel finish, add	1 Plas	170	.047			1.71		1.71	2.79
0600	For coloring, add	J-1	685	.058		.42	2	.20	2.62	3.95
0700	For special texture, add		200	.200		1.47	6.85	.68	9	13.55
1000	Stucco, with bonding agent, 3 coats, on walls, no mesh incl.		200	.200		4.64	6.85	.68	12.17	17.05
1200	Ceilings		180	.222		3.79	7.60	.75	12.14	17.45
1300	Beams		80	.500		3.79	17.15	1.69	22.63	34
1500	Columns	↓	100	.400		3.79	13.70	1.35	18.84	28
1600	Mesh, galvanized, nailed to wood, 1.8 lb.	1 Lath	60	.133		7.35	5.10		12.45	16.30
1800	3.6 lb.		55	.145		4.24	5.55		9.79	13.60
1900	Wired to steel, galvanized, 1.8 lb.		53	.151		7.35	5.75		13.10	17.35
2100	3.6 lb.	↓	50	.160	↓	4.24	6.10		10.34	14.45

09 25 Other Plastering

09 25 23 – Lime Based Plastering

09 25 23.10 Venetian Plaster		Crew	Daily Output	Labor-Hours	Unit	2019 Bare Costs Material	Labor	Equipment	Total	Total Incl O&P
0010	**VENETIAN PLASTER**									
0100	Walls, 1 coat primer, roller applied	1 Plas	950	.008	S.F.	.17	.31		.48	.69
0200	Plaster, 3 coats, incl. sanding	2 Plas	700	.023		.41	.83		1.24	1.81
0210	For pigment, light colors add per S.F. plaster					.02			.02	.02
0220	For pigment, dark colors add per S.F. plaster					.04			.04	.04
0300	For sealer/wax coat incl. burnishing, add	1 Plas	300	.027		.38	.97		1.35	1.99

09 26 Veneer Plastering

09 26 13 – Gypsum Veneer Plastering

09 26 13.20 Blueboard

Line	Description	Crew	Daily Output	Labor-Hours	Unit	Material	Labor	Equipment	Total	Total Incl O&P
0010	**BLUEBOARD** For use with thin coat									
0100	plaster application see Section 09 26 13.80									
1000	3/8" thick, on walls or ceilings, standard, no finish included	2 Carp	1900	.008	S.F.	.35	.33		.68	.93
1100	With thin coat plaster finish		875	.018		.46	.71		1.17	1.68
1400	On beams, columns, or soffits, standard, no finish included		675	.024		.40	.92		1.32	1.96
1450	With thin coat plaster finish		475	.034		.52	1.31		1.83	2.73
3000	1/2" thick, on walls or ceilings, standard, no finish included		1900	.008		.35	.33		.68	.93
3100	With thin coat plaster finish		875	.018		.46	.71		1.17	1.68
3300	Fire resistant, no finish included		1900	.008		.35	.33		.68	.93
3400	With thin coat plaster finish		875	.018		.46	.71		1.17	1.68
3450	On beams, columns, or soffits, standard, no finish included		675	.024		.40	.92		1.32	1.96
3500	With thin coat plaster finish		475	.034		.52	1.31		1.83	2.73
3700	Fire resistant, no finish included		675	.024		.40	.92		1.32	1.96
3800	With thin coat plaster finish		475	.034		.52	1.31		1.83	2.73
5000	5/8" thick, on walls or ceilings, fire resistant, no finish included		1900	.008		.36	.33		.69	.94
5100	With thin coat plaster finish		875	.018		.47	.71		1.18	1.69
5500	On beams, columns, or soffits, no finish included		675	.024		.41	.92		1.33	1.98
5600	With thin coat plaster finish		475	.034		.53	1.31		1.84	2.74
6000	For high ceilings, over 8' high, add		3060	.005			.20		.20	.34
6500	For distribution costs 3 stories and above, add per story		6100	.003			.10		.10	.17

09 26 13.80 Thin Coat Plaster

Line	Description	Crew	Daily Output	Labor-Hours	Unit	Material	Labor	Equipment	Total	Total Incl O&P
0010	**THIN COAT PLASTER** R092000-50									
0012	1 coat veneer, not incl. lath	J-1	3600	.011	S.F.	.11	.38	.04	.53	.79
1000	In 50 lb. bags				Bag	15.35			15.35	16.90

09 29 Gypsum Board

09 29 10 – Gypsum Board Panels

09 29 10.30 Gypsum Board

Line	Description	Crew	Daily Output	Labor-Hours	Unit	Material	Labor	Equipment	Total	Total Incl O&P
0010	**GYPSUM BOARD** on walls & ceilings R092910-10									
0100	Nailed or screwed to studs unless otherwise noted									
0110	1/4" thick, on walls or ceilings, standard, no finish included	2 Carp	1330	.012	S.F.	.40	.47		.87	1.21
0115	1/4" thick, on walls or ceilings, flexible, no finish included		1050	.015		.54	.59		1.13	1.57
0117	1/4" thick, on columns or soffits, flexible, no finish included		1050	.015		.54	.59		1.13	1.57
0130	1/4" thick, standard, no finish included, less than 800 S.F.		510	.031		.40	1.22		1.62	2.45
0150	3/8" thick, on walls, standard, no finish included		2000	.008		.36	.31		.67	.91
0200	On ceilings, standard, no finish included		1800	.009		.36	.34		.70	.97
0250	On beams, columns, or soffits, no finish included		675	.024		.36	.92		1.28	1.92
0300	1/2" thick, on walls, standard, no finish included		2000	.008		.34	.31		.65	.88
0350	Taped and finished (level 4 finish)		965	.017		.39	.64		1.03	1.49

09 29 10.30 Gypsum Board		Crew	Daily Output	Labor-Hours	Unit	Material	2019 Bare Costs Labor	Equipment	Total	Total Incl O&P
0390	With compound skim coat (level 5 finish)	2 Carp	775	.021	S.F.	.44	.80		1.24	1.80
0400	Fire resistant, no finish included		2000	.008		.37	.31		.68	.92
0450	Taped and finished (level 4 finish)		965	.017		.42	.64		1.06	1.52
0490	With compound skim coat (level 5 finish)		775	.021		.47	.80		1.27	1.84
0500	Water resistant, no finish included		2000	.008		.42	.31		.73	.97
0550	Taped and finished (level 4 finish)		965	.017		.47	.64		1.11	1.57
0590	With compound skim coat (level 5 finish)		775	.021		.52	.80		1.32	1.89
0600	Prefinished, vinyl, clipped to studs		900	.018		.51	.69		1.20	1.70
0700	Mold resistant, no finish included		2000	.008		.43	.31		.74	.98
0710	Taped and finished (level 4 finish)		965	.017		.48	.64		1.12	1.58
0720	With compound skim coat (level 5 finish)		775	.021		.53	.80		1.33	1.90
1000	On ceilings, standard, no finish included		1800	.009		.34	.34		.68	.94
1050	Taped and finished (level 4 finish)		765	.021		.39	.81		1.20	1.77
1090	With compound skim coat (level 5 finish)		610	.026		.44	1.02		1.46	2.16
1100	Fire resistant, no finish included		1800	.009		.37	.34		.71	.98
1150	Taped and finished (level 4 finish)		765	.021		.42	.81		1.23	1.80
1195	With compound skim coat (level 5 finish)		610	.026		.47	1.02		1.49	2.20
1200	Water resistant, no finish included		1800	.009		.42	.34		.76	1.03
1250	Taped and finished (level 4 finish)		765	.021		.47	.81		1.28	1.85
1290	With compound skim coat (level 5 finish)		610	.026		.52	1.02		1.54	2.25
1310	Mold resistant, no finish included		1800	.009		.43	.34		.77	1.04
1320	Taped and finished (level 4 finish)		765	.021		.48	.81		1.29	1.86
1330	With compound skim coat (level 5 finish)		610	.026		.53	1.02		1.55	2.26
1350	Sag resistant, no finish included		1600	.010		.37	.39		.76	1.05
1360	Taped and finished (level 4 finish)		765	.021		.42	.81		1.23	1.80
1370	With compound skim coat (level 5 finish)		610	.026		.47	1.02		1.49	2.20
1500	On beams, columns, or soffits, standard, no finish included		675	.024		.39	.92		1.31	1.95
1550	Taped and finished (level 4 finish)		540	.030		.39	1.15		1.54	2.33
1590	With compound skim coat (level 5 finish)		475	.034		.44	1.31		1.75	2.64
1600	Fire resistant, no finish included		675	.024		.37	.92		1.29	1.93
1650	Taped and finished (level 4 finish)		540	.030		.42	1.15		1.57	2.36
1690	With compound skim coat (level 5 finish)		475	.034		.47	1.31		1.78	2.68
1700	Water resistant, no finish included		675	.024		.48	.92		1.40	2.05
1750	Taped and finished (level 4 finish)		540	.030		.47	1.15		1.62	2.41
1790	With compound skim coat (level 5 finish)		475	.034		.52	1.31		1.83	2.73
1800	Mold resistant, no finish included		675	.024		.49	.92		1.41	2.06
1810	Taped and finished (level 4 finish)		540	.030		.48	1.15		1.63	2.42
1820	With compound skim coat (level 5 finish)		475	.034		.53	1.31		1.84	2.74
1850	Sag resistant, no finish included		675	.024		.43	.92		1.35	1.99
1860	Taped and finished (level 4 finish)		540	.030		.42	1.15		1.57	2.36
1870	With compound skim coat (level 5 finish)		475	.034		.47	1.31		1.78	2.68
2000	5/8" thick, on walls, standard, no finish included		2000	.008		.35	.31		.66	.90
2050	Taped and finished (level 4 finish)		965	.017		.40	.64		1.04	1.50
2090	With compound skim coat (level 5 finish)		775	.021		.45	.80		1.25	1.81
2100	Fire resistant, no finish included		2000	.008		.36	.31		.67	.91
2150	Taped and finished (level 4 finish)		965	.017		.41	.64		1.05	1.51
2195	With compound skim coat (level 5 finish)		775	.021		.46	.80		1.26	1.83
2200	Water resistant, no finish included		2000	.008		.50	.31		.81	1.06
2250	Taped and finished (level 4 finish)		965	.017		.55	.64		1.19	1.66
2290	With compound skim coat (level 5 finish)		775	.021		.60	.80		1.40	1.98
2300	Prefinished, vinyl, clipped to studs		900	.018		.83	.69		1.52	2.05
2510	Mold resistant, no finish included		2000	.008		.49	.31		.80	1.05
2520	Taped and finished (level 4 finish)	↓	965	.017	↓	.54	.64		1.18	1.65

09 29 10.30 Gypsum Board		Crew	Daily Output	Labor-Hours	Unit	Material	Labor	Equipment	Total	Total Incl O&P
						2019 Bare Costs				
2530	With compound skim coat (level 5 finish)	2 Carp	775	.021	S.F.	.59	.80		1.39	1.97
3000	On ceilings, standard, no finish included		1800	.009		.35	.34		.69	.96
3050	Taped and finished (level 4 finish)		765	.021		.40	.81		1.21	1.78
3090	With compound skim coat (level 5 finish)		615	.026		.45	1.01		1.46	2.16
3100	Fire resistant, no finish included		1800	.009		.36	.34		.70	.97
3150	Taped and finished (level 4 finish)		765	.021		.41	.81		1.22	1.79
3190	With compound skim coat (level 5 finish)		615	.026		.46	1.01		1.47	2.18
3200	Water resistant, no finish included		1800	.009		.50	.34		.84	1.12
3250	Taped and finished (level 4 finish)		765	.021		.55	.81		1.36	1.94
3290	With compound skim coat (level 5 finish)		615	.026		.60	1.01		1.61	2.33
3300	Mold resistant, no finish included		1800	.009		.49	.34		.83	1.11
3310	Taped and finished (level 4 finish)		765	.021		.54	.81		1.35	1.93
3320	With compound skim coat (level 5 finish)		615	.026		.59	1.01		1.60	2.32
3500	On beams, columns, or soffits, no finish included		675	.024		.40	.92		1.32	1.96
3550	Taped and finished (level 4 finish)		475	.034		.46	1.31		1.77	2.66
3590	With compound skim coat (level 5 finish)		380	.042		.52	1.63		2.15	3.27
3600	Fire resistant, no finish included		675	.024		.41	.92		1.33	1.98
3650	Taped and finished (level 4 finish)		475	.034		.47	1.31		1.78	2.68
3690	With compound skim coat (level 5 finish)		380	.042		.46	1.63		2.09	3.21
3700	Water resistant, no finish included		675	.024		.58	.92		1.50	2.15
3750	Taped and finished (level 4 finish)		475	.034		.60	1.31		1.91	2.82
3790	With compound skim coat (level 5 finish)		380	.042		.63	1.63		2.26	3.39
3800	Mold resistant, no finish included		675	.024		.56	.92		1.48	2.14
3810	Taped and finished (level 4 finish)		475	.034		.59	1.31		1.90	2.81
3820	With compound skim coat (level 5 finish)		380	.042		.62	1.63		2.25	3.38
4000	Fireproofing, beams or columns, 2 layers, 1/2" thick, incl finish		330	.048		.83	1.88		2.71	4.03
4010	Mold resistant		330	.048		.95	1.88		2.83	4.16
4050	5/8" thick		300	.053		.81	2.07		2.88	4.32
4060	Mold resistant		300	.053		1.07	2.07		3.14	4.60
4100	3 layers, 1/2" thick		225	.071		1.25	2.76		4.01	5.95
4110	Mold resistant		225	.071		1.43	2.76		4.19	6.10
4150	5/8" thick		210	.076		1.22	2.95		4.17	6.20
4160	Mold resistant		210	.076		1.61	2.95		4.56	6.65
5200	For work over 8' high, add		3060	.005			.20		.20	.34
5270	For textured spray, add	2 Lath	1600	.010		.04	.38		.42	.65
5300	For distribution cost 3 stories and above, add per story	2 Carp	6100	.003			.10		.10	.17
5350	For finishing inner corners, add		950	.017	L.F.	.10	.65		.75	1.19
5355	For finishing outer corners, add		1250	.013		.23	.50		.73	1.08
5500	For acoustical sealant, add per bead	1 Carp	500	.016		.04	.62		.66	1.07
5550	Sealant, 1 quart tube				Ea.	7.10			7.10	7.80
6000	Gypsum sound dampening panels									
6010	1/2" thick on walls, multi-layer, lightweight, no finish included	2 Carp	1500	.011	S.F.	2.22	.41		2.63	3.12
6015	Taped and finished (level 4 finish)		725	.022		2.27	.86		3.13	3.90
6020	With compound skim coat (level 5 finish)		580	.028		2.32	1.07		3.39	4.32
6025	5/8" thick on walls, for wood studs, no finish included		1500	.011		2.24	.41		2.65	3.14
6030	Taped and finished (level 4 finish)		725	.022		2.29	.86		3.15	3.93
6035	With compound skim coat (level 5 finish)		580	.028		2.34	1.07		3.41	4.34
6040	For metal stud, no finish included		1500	.011		2.31	.41		2.72	3.22
6045	Taped and finished (level 4 finish)		725	.022		2.36	.86		3.22	4
6050	With compound skim coat (level 5 finish)		580	.028		2.41	1.07		3.48	4.42
6055	Abuse resist, no finish included		1500	.011		4.06	.41		4.47	5.15
6060	Taped and finished (level 4 finish)		725	.022		4.11	.86		4.97	5.95
6065	With compound skim coat (level 5 finish)		580	.028		4.16	1.07		5.23	6.35

09 29 10 – Gypsum Board Panels

09 29 10.30 Gypsum Board		Crew	Daily Output	Labor-Hours	Unit	2019 Bare Costs Material	Labor	Equipment	Total	Total Incl O&P
6070	Shear rated, no finish included	2 Carp	1500	.011	S.F.	5.35	.41		5.76	6.60
6075	Taped and finished (level 4 finish)		725	.022		5.40	.86		6.26	7.35
6080	With compound skim coat (level 5 finish)		580	.028		5.45	1.07		6.52	7.75
6085	For SCIF applications, no finish included		1500	.011		5.15	.41		5.56	6.35
6090	Taped and finished (level 4 finish)		725	.022		5.20	.86		6.06	7.10
6095	With compound skim coat (level 5 finish)		580	.028		5.25	1.07		6.32	7.50
6100	1-3/8" thick on walls, THX certified, no finish included		1500	.011		9.30	.41		9.71	10.95
6105	Taped and finished (level 4 finish)		725	.022		9.35	.86		10.21	11.70
6110	With compound skim coat (level 5 finish)		580	.028		9.40	1.07		10.47	12.10
6115	5/8" thick on walls, score & snap installation, no finish included		2000	.008		1.96	.31		2.27	2.67
6120	Taped and finished (level 4 finish)		965	.017		2.01	.64		2.65	3.27
6125	With compound skim coat (level 5 finish)		775	.021		2.06	.80		2.86	3.59
7020	5/8" thick on ceilings, for wood joists, no finish included		1200	.013		2.24	.52		2.76	3.31
7025	Taped and finished (level 4 finish)		510	.031		2.29	1.22		3.51	4.53
7030	With compound skim coat (level 5 finish)		410	.039		2.34	1.51		3.85	5.05
7035	For metal joists, no finish included		1200	.013		2.31	.52		2.83	3.39
7040	Taped and finished (level 4 finish)		510	.031		2.36	1.22		3.58	4.60
7045	With compound skim coat (level 5 finish)		410	.039		2.41	1.51		3.92	5.15
7050	Abuse resist, no finish included		1200	.013		4.06	.52		4.58	5.30
7055	Taped and finished (level 4 finish)		510	.031		4.11	1.22		5.33	6.55
7060	With compound skim coat (level 5 finish)		410	.039		4.16	1.51		5.67	7.10
7065	Shear rated, no finish included		1200	.013		5.35	.52		5.87	6.75
7070	Taped and finished (level 4 finish)		510	.031		5.40	1.22		6.62	7.95
7075	With compound skim coat (level 5 finish)		410	.039		5.45	1.51		6.96	8.50
7080	For SCIF applications, no finish included		1200	.013		5.15	.52		5.67	6.50
7085	Taped and finished (level 4 finish)		510	.031		5.20	1.22		6.42	7.70
7090	With compound skim coat (level 5 finish)		410	.039		5.25	1.51		6.76	8.25
8010	5/8" thick on ceilings, score & snap installation, no finish included		1600	.010		1.96	.39		2.35	2.80
8015	Taped and finished (level 4 finish)		680	.024		2.01	.91		2.92	3.72
8020	With compound skim coat (level 5 finish)		545	.029		2.06	1.14		3.20	4.15

09 29 10.50 High Abuse Gypsum Board		Crew	Daily Output	Labor-Hours	Unit	Material	Labor	Equipment	Total	Total Incl O&P
0010	**HIGH ABUSE GYPSUM BOARD**, fiber reinforced, nailed or									
0100	screwed to studs unless otherwise noted									
0110	1/2" thick, on walls, no finish included	2 Carp	1800	.009	S.F.	.81	.34		1.15	1.46
0120	Taped and finished (level 4 finish)		870	.018		.86	.71		1.57	2.12
0130	With compound skim coat (level 5 finish)		700	.023		.91	.89		1.80	2.46
0150	On ceilings, no finish included		1620	.010		.81	.38		1.19	1.52
0160	Taped and finished (level 4 finish)		690	.023		.86	.90		1.76	2.43
0170	With compound skim coat (level 5 finish)		550	.029		.91	1.13		2.04	2.86
0210	5/8" thick, on walls, no finish included		1800	.009		.82	.34		1.16	1.47
0220	Taped and finished (level 4 finish)		870	.018		.87	.71		1.58	2.13
0230	With compound skim coat (level 5 finish)		700	.023		.92	.89		1.81	2.47
0250	On ceilings, no finish included		1620	.010		.82	.38		1.20	1.53
0260	Taped and finished (level 4 finish)		690	.023		.87	.90		1.77	2.44
0270	With compound skim coat (level 5 finish)		550	.029		.92	1.13		2.05	2.87
0310	5/8" thick, on walls, very high impact, no finish included		1800	.009		.91	.34		1.25	1.57
0320	Taped and finished (level 4 finish)		870	.018		.96	.71		1.67	2.23
0330	With compound skim coat (level 5 finish)		700	.023		1.01	.89		1.90	2.57
0350	On ceilings, no finish included		1620	.010		.91	.38		1.29	1.63
0360	Taped and finished (level 4 finish)		690	.023		.96	.90		1.86	2.54
0370	With compound skim coat (level 5 finish)		550	.029		1.01	1.13		2.14	2.97
0400	High abuse, gypsum core, paper face									

09 29 Gypsum Board

09 29 10 – Gypsum Board Panels

09 29 10.50 High Abuse Gypsum Board		Crew	Daily Output	Labor-Hours	Unit	Material	Labor	Equipment	Total	Total Incl O&P
0410	1/2" thick, on walls, no finish included	2 Carp	1800	.009	S.F.	.65	.34		.99	1.29
0420	Taped and finished (level 4 finish)		870	.018		.70	.71		1.41	1.95
0430	With compound skim coat (level 5 finish)		700	.023		.75	.89		1.64	2.28
0450	On ceilings, no finish included		1620	.010		.65	.38		1.03	1.35
0460	Taped and finished (level 4 finish)		690	.023		.70	.90		1.60	2.26
0470	With compound skim coat (level 5 finish)		550	.029		.75	1.13		1.88	2.68
0510	5/8" thick, on walls, no finish included		1800	.009		.69	.34		1.03	1.33
0520	Taped and finished (level 4 finish)		870	.018		.74	.71		1.45	1.99
0530	With compound skim coat (level 5 finish)		700	.023		.79	.89		1.68	2.33
0550	On ceilings, no finish included		1620	.010		.69	.38		1.07	1.39
0560	Taped and finished (level 4 finish)		690	.023		.74	.90		1.64	2.30
0570	With compound skim coat (level 5 finish)		550	.029		.79	1.13		1.92	2.73
1000	For high ceilings, over 8' high, add		2750	.006			.23		.23	.37
1010	For distribution cost 3 stories and above, add per story	▼	5500	.003	▼		.11		.11	.19

09 29 15 – Gypsum Board Accessories

09 29 15.10 Accessories, Gypsum Board		Crew	Daily Output	Labor-Hours	Unit	Material	Labor	Equipment	Total	Total Incl O&P
0010	**ACCESSORIES, GYPSUM BOARD**									
0020	Casing bead, galvanized steel	1 Carp	2.90	2.759	C.L.F.	24	107		131	204
0100	Vinyl		3	2.667		23	103		126	196
0300	Corner bead, galvanized steel, 1" x 1"		4	2		15.65	77.50		93.15	145
0400	1-1/4" x 1-1/4"		3.50	2.286		16.95	88.50		105.45	165
0600	Vinyl		4	2		20	77.50		97.50	150
0900	Furring channel, galv. steel, 7/8" deep, standard		2.60	3.077		35	119		154	236
1000	Resilient		2.55	3.137		26.50	122		148.50	231
1100	J trim, galvanized steel, 1/2" wide		3	2.667		22.50	103		125.50	196
1120	5/8" wide	▼	2.95	2.712	▼	31.50	105		136.50	209
1160	Screws #6 x 1" A				M	11			11	12.10
1170	#6 x 1-5/8" A				"	14.60			14.60	16.05
1500	Z stud, galvanized steel, 1-1/2" wide	1 Carp	2.60	3.077	C.L.F.	42.50	119		161.50	244

09 30 Tiling

09 30 13 – Ceramic Tiling

09 30 13.45 Ceramic Tile Accessories		Crew	Daily Output	Labor-Hours	Unit	Material	Labor	Equipment	Total	Total Incl O&P
0010	**CERAMIC TILE ACCESSORIES**									
0100	Spacers, 1/8"				C	1.99			1.99	2.19
1310	Sealer for natural stone tile, installed	1 Tilf	650	.012	S.F.	.05	.44		.49	.77

09 30 29 – Metal Tiling

09 30 29.10 Metal Tile		Crew	Daily Output	Labor-Hours	Unit	Material	Labor	Equipment	Total	Total Incl O&P
0010	**METAL TILE** 4' x 4' sheet, 24 ga., tile pattern, nailed									
0200	Stainless steel	2 Carp	512	.031	S.F.	28	1.21		29.21	32.50
0400	Aluminized steel	"	512	.031	"	20.50	1.21		21.71	24.50

09 30 95 – Tile & Stone Setting Materials and Specialties

09 30 95.10 Moisture Resistant, Anti-Fracture Membrane		Crew	Daily Output	Labor-Hours	Unit	Material	Labor	Equipment	Total	Total Incl O&P
0010	**MOISTURE RESISTANT, ANTI-FRACTURE MEMBRANE**									
0200	Elastomeric membrane, 1/16" thick	D-7	275	.058	S.F.	1.13	1.85		2.98	4.22

Costs in the Material, Labor, Equipment and Total columns are 2019 Bare Costs.

09 31 13 – Thin-Set Ceramic Tiling

09 31 13.10 Thin-Set Ceramic Tile		Crew	Daily Output	Labor-Hours	Unit	2019 Bare Costs Material	Labor	Equipment	Total	Total Incl O&P
0010	**THIN-SET CERAMIC TILE**									
0020	Backsplash, average grade tiles	1 Tilf	50	.160	S.F.	2.77	5.75		8.52	12.30
0022	Custom grade tiles		50	.160		5.55	5.75		11.30	15.35
0024	Luxury grade tiles		50	.160		11.10	5.75		16.85	21.50
0026	Economy grade tiles		50	.160		2.54	5.75		8.29	12.05
0100	Base, using 1′ x 4″ high piece with 1″ x 1″ tiles	D-7	128	.125	L.F.	5	3.97		8.97	11.90
0300	For 6″ high base, 1″ x 1″ tile face, add					1.23			1.23	1.36
0400	For 2″ x 2″ tile face, add to above					.72			.72	.79
0700	Cove base, 4-1/4″ x 4-1/4″	D-7	128	.125		4.21	3.97		8.18	11.05
1000	6″ x 4-1/4″ high		137	.117		4.26	3.71		7.97	10.70
1300	Sanitary cove base, 6″ x 4-1/4″ high		124	.129		4.68	4.09		8.77	11.75
1600	6″ x 6″ high		117	.137		5.35	4.34		9.69	12.90
1800	Bathroom accessories, average (soap dish, toothbrush holder)		82	.195	Ea.	9.70	6.20		15.90	20.50
1900	Bathtub, 5′, rec. 4-1/4″ x 4-1/4″ tile wainscot, adhesive set 6′ high		2.90	5.517		173	175		348	470
2100	7′ high wainscot		2.50	6.400		202	203		405	550
2200	8′ high wainscot		2.20	7.273		231	231		462	625
2500	Bullnose trim, 4-1/4″ x 4-1/4″		128	.125	L.F.	4.24	3.97		8.21	11.05
2800	2″ x 6″		124	.129	″	4.25	4.09		8.34	11.30
3300	Ceramic tile, porcelain type, 1 color, color group 2, 1″ x 1″		183	.087	S.F.	6.50	2.77		9.27	11.65
3310	2″ x 2″ or 2″ x 1″		190	.084		6.35	2.67		9.02	11.30
3350	For random blend, 2 colors, add					1.01			1.01	1.11
3360	4 colors, add					1.51			1.51	1.66
4300	Specialty tile, 4-1/4″ x 4-1/4″ x 1/2″, decorator finish	D-7	183	.087		12.80	2.77		15.57	18.60
4500	Add for epoxy grout, 1/16″ joint, 1″ x 1″ tile		800	.020		.69	.63		1.32	1.78
4600	2″ x 2″ tile		820	.020		.63	.62		1.25	1.69
4610	Add for epoxy grout, 1/8″ joint, 8″ x 8″ x 3/8″ tile, add		900	.018		1.42	.56		1.98	2.47
4800	Pregrouted sheets, walls, 4-1/4″ x 4-1/4″, 6″ x 4-1/4″									
4810	and 8-1/2″ x 4-1/4″, 4 S.F. sheets, silicone grout	D-7	240	.067	S.F.	5.70	2.12		7.82	9.65
5100	Floors, unglazed, 2 S.F. sheets,									
5110	urethane adhesive	D-7	180	.089	S.F.	2.22	2.82		5.04	7
5400	Walls, interior, 4-1/4″ x 4-1/4″ tile		190	.084		2.59	2.67		5.26	7.15
5500	6″ x 4-1/4″ tile		190	.084		3.17	2.67		5.84	7.80
5700	8-1/2″ x 4-1/4″ tile		190	.084		5.70	2.67		8.37	10.60
5800	6″ x 6″ tile		175	.091		3.63	2.90		6.53	8.70
5810	8″ x 8″ tile		170	.094		5.20	2.99		8.19	10.50
5820	12″ x 12″ tile		160	.100		4.69	3.17		7.86	10.25
5830	16″ x 16″ tile		150	.107		5.15	3.38		8.53	11.10
6000	Decorated wall tile, 4-1/4″ x 4-1/4″, color group 1		270	.059		3.63	1.88		5.51	7
6100	Color group 4		180	.089		53.50	2.82		56.32	63.50
9300	Ceramic tiles, recycled glass, standard colors, 2″ x 2″ thru 6″ x 6″ G		190	.084		22.50	2.67		25.17	29
9310	6″ x 6″ G		175	.091		22.50	2.90		25.40	29
9320	8″ x 8″ G		170	.094		23.50	2.99		26.49	31
9330	12″ x 12″ G		160	.100		23.50	3.17		26.67	31
9340	Earthtones, 2″ x 2″ to 4″ x 8″ G		190	.084		26	2.67		28.67	33.50
9350	6″ x 6″ G		175	.091		26	2.90		28.90	33.50
9360	8″ x 8″ G		170	.094		27	2.99		29.99	35
9370	12″ x 12″ G		160	.100		27	3.17		30.17	35
9380	Deep colors, 2″ x 2″ to 4″ x 8″ G		190	.084		31	2.67		33.67	38.50
9390	6″ x 6″ G		175	.091		31	2.90		33.90	38.50
9400	8″ x 8″ G		170	.094		32.50	2.99		35.49	40.50
9410	12″ x 12″ G		160	.100		32.50	3.17		35.67	40.50

09 31 Thin-Set Tiling

09 31 33 – Thin-Set Stone Tiling

09 31 33.10 Tiling, Thin-Set Stone		Crew	Daily Output	Labor-Hours	Unit	Material	2019 Bare Costs Labor	Equipment	Total	Total Incl O&P
0010	TILING, THIN-SET STONE									
3000	Floors, natural clay, random or uniform, color group 1	D-7	183	.087	S.F.	4.63	2.77		7.40	9.60
3100	Color group 2		183	.087		5.90	2.77		8.67	11
3255	Floors, glazed, 6" x 6", color group 1		300	.053		5.30	1.69		6.99	8.60
3260	8" x 8" tile		300	.053		5.30	1.69		6.99	8.55
3270	12" x 12" tile		290	.055		6.75	1.75		8.50	10.20
3280	16" x 16" tile		280	.057		7.40	1.81		9.21	11.10
3281	18" x 18" tile		270	.059		7.25	1.88		9.13	11.05
3282	20" x 20" tile		260	.062		7.15	1.95		9.10	11
3283	24" x 24" tile		250	.064		9.10	2.03		11.13	13.35
3285	Border, 6" x 12" tile		200	.080		10.95	2.54		13.49	16.15
3290	3" x 12" tile		200	.080		12.60	2.54		15.14	18

09 32 Mortar-Bed Tiling

09 32 13 – Mortar-Bed Ceramic Tiling

09 32 13.10 Ceramic Tile		Crew	Daily Output	Labor-Hours	Unit	Material	Labor	Equipment	Total	Total Incl O&P
0010	CERAMIC TILE									
0050	Base, using 1' x 4" high pc. with 1" x 1" tiles	D-7	82	.195	L.F.	5.30	6.20		11.50	15.80
0600	Cove base, 4-1/4" x 4-1/4" high		91	.176		4.35	5.60		9.95	13.80
0900	6" x 4-1/4" high		100	.160		4.40	5.10		9.50	13.05
1200	Sanitary cove base, 6" x 4-1/4" high		93	.172		4.82	5.45		10.27	14.10
1500	6" x 6" high		84	.190		5.50	6.05		11.55	15.80
2400	Bullnose trim, 4-1/4" x 4-1/4"		82	.195		4.34	6.20		10.54	14.75
2700	2" x 6" bullnose trim		84	.190		4.33	6.05		10.38	14.50
6210	Wall tile, 4-1/4" x 4-1/4", better grade	1 Tilf	50	.160	S.F.	9.55	5.75		15.30	19.75
6240	2" x 2"		50	.160		7.55	5.75		13.30	17.55
6250	6" x 6"		55	.145		10.20	5.20		15.40	19.65
6260	8" x 8"		60	.133		9.45	4.79		14.24	18.15
6600	Crystalline glazed, 4-1/4" x 4-1/4", plain	D-7	100	.160		4.48	5.10		9.58	13.15
6700	4-1/4" x 4-1/4", scored tile		100	.160		6.55	5.10		11.65	15.45
6900	6" x 6" plain		93	.172		5.80	5.45		11.25	15.15
7000	For epoxy grout, 1/16" joints, 4-1/4" tile, add		800	.020		.41	.63		1.04	1.47
7200	For tile set in dry mortar, add		1735	.009			.29		.29	.47
7300	For tile set in Portland cement mortar, add		290	.055		.18	1.75		1.93	3.02

09 32 16 – Mortar-Bed Quarry Tiling

09 32 16.10 Quarry Tile		Crew	Daily Output	Labor-Hours	Unit	Material	Labor	Equipment	Total	Total Incl O&P
0010	QUARRY TILE									
0100	Base, cove or sanitary, to 5" high, 1/2" thick	D-7	110	.145	L.F.	6.40	4.62		11.02	14.50
0300	Bullnose trim, red, 6" x 6" x 1/2" thick		120	.133		5.40	4.23		9.63	12.80
0400	4" x 4" x 1/2" thick		110	.145		4.97	4.62		9.59	12.90
0600	4" x 8" x 1/2" thick, using 8" as edge		130	.123		5.40	3.91		9.31	12.25
0700	Floors, 1,000 S.F. lots, red, 4" x 4" x 1/2" thick		120	.133	S.F.	8.80	4.23		13.03	16.55
0900	6" x 6" x 1/2" thick		140	.114		8.25	3.63		11.88	14.95
1000	4" x 8" x 1/2" thick		130	.123		6.80	3.91		10.71	13.75
1300	For waxed coating, add					.76			.76	.84
1500	For non-standard colors, add					.46			.46	.51
1600	For abrasive surface, add					.53			.53	.58
1800	Brown tile, imported, 6" x 6" x 3/4"	D-7	120	.133		7.30	4.23		11.53	14.85
1900	8" x 8" x 1"		110	.145		9.65	4.62		14.27	18.05
2100	For thin set mortar application, deduct		700	.023			.73		.73	1.17

09 32 Mortar-Bed Tiling

09 32 16 – Mortar-Bed Quarry Tiling

09 32 16.10 Quarry Tile		Crew	Daily Output	Labor-Hours	Unit	2019 Bare Costs Material	Labor	Equipment	Total	Total Incl O&P
2700	Stair tread, 6" x 6" x 3/4", plain	D-7	50	.320	S.F.	7.30	10.15		17.45	24.50
2800	Abrasive		47	.340		8.65	10.80		19.45	27
3000	Wainscot, 6" x 6" x 1/2", thin set, red		105	.152		6.40	4.84		11.24	14.85
3100	Non-standard colors		105	.152		6.40	4.84		11.24	14.85
3300	Window sill, 6" wide, 3/4" thick		90	.178	L.F.	8.75	5.65		14.40	18.70
3400	Corners		80	.200	Ea.	6.05	6.35		12.40	16.90

09 32 23 – Mortar-Bed Glass Mosaic Tiling

09 32 23.10 Glass Mosaics		Crew	Daily Output	Labor-Hours	Unit	Material	Labor	Equipment	Total	Total Incl O&P
0010	**GLASS MOSAICS** 3/4" tile on 12" sheets, standard grout									
1020	1" tile on 12" sheets, opalescent finish	D-7	73	.219	S.F.	18.55	6.95		25.50	31.50
1040	1" x 2" tile on 12" sheet, blend		73	.219		21	6.95		27.95	34
1060	2" tile on 12" sheet, blend		73	.219		17.35	6.95		24.30	30.50
1080	5/8" x random tile, linear, on 12" sheet, blend		73	.219		25	6.95		31.95	38
1600	Dots on 12" sheet		73	.219		25	6.95		31.95	38.50
1700	For glass mosaic tiles set in dry mortar, add		290	.055		.45	1.75		2.20	3.32
1720	For glass mosaic tiles set in Portland cement mortar, add		290	.055		.01	1.75		1.76	2.83
1730	For polyblend sanded tile grout		96.15	.166	Lb.	2.19	5.30		7.49	10.90

09 34 Waterproofing-Membrane Tiling

09 34 13 – Waterproofing-Membrane Ceramic Tiling

09 34 13.10 Ceramic Tile Waterproofing Membrane		Crew	Daily Output	Labor-Hours	Unit	Material	Labor	Equipment	Total	Total Incl O&P
0010	**CERAMIC TILE WATERPROOFING MEMBRANE**									
0020	On floors, including thinset									
0030	Fleece laminated polyethylene grid, 1/8" thick	D-7	250	.064	S.F.	2.29	2.03		4.32	5.80
0040	5/16" thick	"	250	.064	"	2.60	2.03		4.63	6.15
0050	On walls, including thinset									
0060	Fleece laminated polyethylene sheet, 8 mil thick	D-7	480	.033	S.F.	2.29	1.06		3.35	4.23
0070	Accessories, including thinset									
0080	Joint and corner sheet, 4 mils thick, 5" wide	1 Tilf	240	.033	L.F.	1.35	1.20		2.55	3.41
0090	7-1/4" wide		180	.044		1.71	1.60		3.31	4.46
0100	10" wide		120	.067		2.08	2.39		4.47	6.15
0110	Pre-formed corners, inside		32	.250	Ea.	7.85	9		16.85	23
0120	Outside		32	.250		7.65	9		16.65	23
0130	2" flanged floor drain with 6" stainless steel grate		16	.500		370	17.95		387.95	440
0140	EPS, sloped shower floor		480	.017	S.F.	5.55	.60		6.15	7.05
0150	Curb		32	.250	L.F.	14.05	9		23.05	30

09 35 Chemical-Resistant Tiling

09 35 13 – Chemical-Resistant Ceramic Tiling

09 35 13.10 Chemical-Resistant Ceramic Tiling		Crew	Daily Output	Labor-Hours	Unit	Material	Labor	Equipment	Total	Total Incl O&P
0010	**CHEMICAL-RESISTANT CERAMIC TILING**									
0100	4-1/4" x 4-1/4" x 1/4", 1/8" joint	D-7	130	.123	S.F.	12.15	3.91		16.06	19.65
0200	6" x 6" x 1/2" thick		120	.133		9.85	4.23		14.08	17.70
0300	8" x 8" x 1/2" thick		110	.145		11	4.62		15.62	19.55
0400	4-1/4" x 4-1/4" x 1/4", 1/4" joint		130	.123		12.95	3.91		16.86	20.50
0500	6" x 6" x 1/2" thick		120	.133		11	4.23		15.23	18.95
0600	8" x 8" x 1/2" thick		110	.145		11.70	4.62		16.32	20.50
0700	4-1/4" x 4-1/4" x 1/4", 3/8" joint		130	.123		13.70	3.91		17.61	21.50
0800	6" x 6" x 1/2" thick		120	.133		12	4.23		16.23	20

09 35 Chemical-Resistant Tiling

09 35 13 – Chemical-Resistant Ceramic Tiling

09 35 13.10 Chemical-Resistant Ceramic Tiling		Crew	Daily Output	Labor-Hours	Unit	Material	Labor	Equipment	Total	Total Incl O&P
0900	8" x 8" x 1/2" thick	D-7	110	.145	S.F.	12.90	4.62		17.52	21.50

09 35 16 – Chemical-Resistant Quarry Tiling

09 35 16.10 Chemical-Resistant Quarry Tiling		Crew	Daily Output	Labor-Hours	Unit	Material	Labor	Equipment	Total	Total Incl O&P
0010	**CHEMICAL-RESISTANT QUARRY TILING**									
0100	4" x 8" x 1/2" thick, 1/8" joint	D-7	130	.123	S.F.	11.35	3.91		15.26	18.80
0200	6" x 6" x 1/2" thick		120	.133		11.40	4.23		15.63	19.40
0300	8" x 8" x 1/2" thick		110	.145		10.45	4.62		15.07	18.95
0400	4" x 8" x 1/2" thick, 1/4" joint		130	.123		12.65	3.91		16.56	20
0500	6" x 6" x 1/2" thick		120	.133		12.55	4.23		16.78	20.50
0600	8" x 8" x 1/2" thick		110	.145		11.10	4.62		15.72	19.70
0700	4" x 8" x 1/2" thick, 3/8" joint		130	.123		13.80	3.91		17.71	21.50
0800	6" x 6" x 1/2" thick		120	.133		13.55	4.23		17.78	22
0900	8" x 8" x 1/2" thick		110	.145		12.35	4.62		16.97	21

09 51 Acoustical Ceilings

09 51 13 – Acoustical Panel Ceilings

09 51 13.10 Ceiling, Acoustical Panel		Crew	Daily Output	Labor-Hours	Unit	Material	Labor	Equipment	Total	Total Incl O&P
0010	**CEILING, ACOUSTICAL PANEL**									
0100	Fiberglass boards, film faced, 2' x 2' or 2' x 4', 5/8" thick	1 Carp	625	.013	S.F.	1.31	.50		1.81	2.26
0120	3/4" thick		600	.013		3.10	.52		3.62	4.26
0130	3" thick, thermal, R11		450	.018		3.88	.69		4.57	5.40

09 51 14 – Acoustical Fabric-Faced Panel Ceilings

09 51 14.10 Ceiling, Acoustical Fabric-Faced Panel		Crew	Daily Output	Labor-Hours	Unit	Material	Labor	Equipment	Total	Total Incl O&P
0010	**CEILING, ACOUSTICAL FABRIC-FACED PANEL**									
0100	Glass cloth faced fiberglass, 3/4" thick	1 Carp	500	.016	S.F.	3.05	.62		3.67	4.38
0120	1" thick		485	.016		3.69	.64		4.33	5.10
0130	1-1/2" thick, nubby face		475	.017		2.73	.65		3.38	4.08

09 51 23 – Acoustical Tile Ceilings

09 51 23.10 Suspended Acoustic Ceiling Tiles		Crew	Daily Output	Labor-Hours	Unit	Material	Labor	Equipment	Total	Total Incl O&P
0010	**SUSPENDED ACOUSTIC CEILING TILES**, not including									
0100	suspension system									
1110	Mineral fiber tile, lay-in, 2' x 2' or 2' x 4', 5/8" thick, fine texture	1 Carp	625	.013	S.F.	.83	.50		1.33	1.73
1115	Rough textured		625	.013		.75	.50		1.25	1.65
1125	3/4" thick, fine textured		600	.013		2.17	.52		2.69	3.24
1130	Rough textured		600	.013		1.91	.52		2.43	2.95
1135	Fissured		600	.013		2.15	.52		2.67	3.22
1150	Tegular, 5/8" thick, fine textured		470	.017		1.15	.66		1.81	2.36
1155	Rough textured		470	.017		1.37	.66		2.03	2.60
1165	3/4" thick, fine textured		450	.018		2.38	.69		3.07	3.76
1170	Rough textured		450	.018		1.57	.69		2.26	2.87
1175	Fissured		450	.018		2.40	.69		3.09	3.78
1185	For plastic film face, add					.80			.80	.88
1190	For fire rating, add					.53			.53	.58
5020	66-78% recycled content, 3/4" thick G	1 Carp	600	.013		2.08	.52		2.60	3.14
5040	Mylar, 42% recycled content, 3/4" thick G		600	.013		5.10	.52		5.62	6.45
6000	Remove & install new ceiling tiles, min fiber, 2' x 2' or 2' x 4', 5/8"thk.		335	.024		.83	.93		1.76	2.44

09 51 Acoustical Ceilings

09 51 23 – Acoustical Tile Ceilings

09 51 23.30 Suspended Ceilings, Complete		Crew	Daily Output	Labor-Hours	Unit	Material	2019 Bare Costs Labor	Equipment	Total	Total Incl O&P
0010	**SUSPENDED CEILINGS, COMPLETE**, incl. standard									
0100	suspension system but not incl. 1-1/2" carrier channels									
0600	Fiberglass ceiling board, 2' x 4' x 5/8", plain faced	1 Carp	500	.016	S.F.	2.11	.62		2.73	3.34
0700	Offices, 2' x 4' x 3/4"		380	.021		3.90	.82		4.72	5.65
1800	Tile, Z bar suspension, 5/8" mineral fiber tile		150	.053		2.18	2.07		4.25	5.80
1900	3/4" mineral fiber tile		150	.053		2.45	2.07		4.52	6.10

09 51 53 – Direct-Applied Acoustical Ceilings

09 51 53.10 Ceiling Tile

Line	Description	Crew	Daily Output	Labor-Hours	Unit	Material	Labor	Equipment	Total	Total Incl O&P
0010	**CEILING TILE**, stapled or cemented									
0100	12" x 12" or 12" x 24", not including furring									
0600	Mineral fiber, vinyl coated, 5/8" thick	1 Carp	300	.027	S.F.	2.26	1.03		3.29	4.20
0700	3/4" thick		300	.027		3.09	1.03		4.12	5.10
0900	Fire rated, 3/4" thick, plain faced		300	.027		1.43	1.03		2.46	3.28
1000	Plastic coated face		300	.027		2.15	1.03		3.18	4.08
1200	Aluminum faced, 5/8" thick, plain		300	.027		1.88	1.03		2.91	3.78
3300	For flameproofing, add					.07			.07	.08
3400	For sculptured 3 dimensional, add					.33			.33	.36
3900	For ceiling primer, add					.12			.12	.13
4000	For ceiling cement, add					.41			.41	.45

09 53 Acoustical Ceiling Suspension Assemblies

09 53 23 – Metal Acoustical Ceiling Suspension Assemblies

09 53 23.30 Ceiling Suspension Systems

Line	Description	Crew	Daily Output	Labor-Hours	Unit	Material	Labor	Equipment	Total	Total Incl O&P
0010	**CEILING SUSPENSION SYSTEMS** for boards and tile									
0050	Class A suspension system, 15/16" T bar, 2' x 4' grid	1 Carp	800	.010	S.F.	.80	.39		1.19	1.52
0300	2' x 2' grid		650	.012		1.04	.48		1.52	1.93
0310	25% recycled steel, 2' x 4' grid G		800	.010		.84	.39		1.23	1.57
0320	2' x 2' grid G		650	.012		1.05	.48		1.53	1.95
0350	For 9/16" grid, add					.16			.16	.18
0360	For fire rated grid, add					.09			.09	.10
0370	For colored grid, add					.22			.22	.24
0400	Concealed Z bar suspension system, 12" module	1 Carp	520	.015		.95	.60		1.55	2.04
0600	1-1/2" carrier channels, 4' OC, add		470	.017		.11	.66		.77	1.22
0650	1-1/2" x 3-1/2" channels		470	.017		.30	.66		.96	1.42
0700	Carrier channels for ceilings with									
0900	recessed lighting fixtures, add	1 Carp	460	.017	S.F.	.21	.67		.88	1.34
3000	Seismic ceiling bracing, IBC Site Class D, Occupancy Category II									
3050	For ceilings less than 2500 S.F.									
3060	Seismic clips at attached walls	1 Carp	180	.044	Ea.	1.15	1.72		2.87	4.11
3100	For ceilings greater than 2500 S.F., add									
3120	Seismic clips, joints at cross tees	1 Carp	120	.067	Ea.	1.88	2.58		4.46	6.35
3140	At cross tees and mains, mains field cut	"	60	.133	"	1.88	5.15		7.03	10.60
3200	Compression posts, telescopic, attached to structure above									
3210	To 30" high	1 Carp	26	.308	Ea.	39.50	11.90		51.40	63
3220	30" to 48" high		25.50	.314		42	12.15		54.15	66.50
3230	48" to 84" high		25	.320		51	12.40		63.40	76.50
3240	84" to 102" high		24.50	.327		57.50	12.65		70.15	84.50
3250	102" to 120" high		24	.333		83	12.90		95.90	113
3260	120" to 144" high		24	.333		93	12.90		105.90	124
3300	Stabilizer bars									

09 53 Acoustical Ceiling Suspension Assemblies

09 53 23 – Metal Acoustical Ceiling Suspension Assemblies

09 53 23.30 Ceiling Suspension Systems		Crew	Daily Output	Labor-Hours	Unit	2019 Bare Costs Material	Labor	Equipment	Total	Total Incl O&P
3310	12″ long	1 Carp	240	.033	Ea.	1.11	1.29		2.40	3.35
3320	24″ long		235	.034		1.04	1.32		2.36	3.32
3330	36″ long		230	.035		1.01	1.35		2.36	3.35
3340	48″ long	↓	220	.036	↓	.81	1.41		2.22	3.23
3400	Wire support for light fixtures, per L.F. height to structure above									
3410	Less than 10 lb.	1 Carp	400	.020	L.F.	.30	.78		1.08	1.61
3420	10 lb. to 56 lb.		240	.033	″	.61	1.29		1.90	2.80
5000	Wire hangers, #12 wire	↓	300	.027	Ea.	.07	1.03		1.10	1.79

09 54 Specialty Ceilings

09 54 16 – Luminous Ceilings

09 54 16.10 Ceiling, Luminous		Crew	Daily Output	Labor-Hours	Unit	Material	Labor	Equipment	Total	Total Incl O&P
0010	**CEILING, LUMINOUS**									
0020	Translucent lay-in panels, 2′ x 2′	1 Carp	500	.016	S.F.	23	.62		23.62	26.50
0030	2′ x 6′	″	500	.016	″	17.95	.62		18.57	21

09 54 23 – Linear Metal Ceilings

09 54 23.10 Metal Ceilings		Crew	Daily Output	Labor-Hours	Unit	Material	Labor	Equipment	Total	Total Incl O&P
0010	**METAL CEILINGS**									
0015	Solid alum. planks, 3-1/4″ x 12′, open reveal	1 Carp	500	.016	S.F.	2.37	.62		2.99	3.63
0020	Closed reveal		500	.016		3.03	.62		3.65	4.35
0030	7-1/4″ x 12′, open reveal		500	.016		4.04	.62		4.66	5.45
0040	Closed reveal		500	.016		5.15	.62		5.77	6.70
0050	Metal, open cell, 2′ x 2′, 6″ cell		500	.016		8.65	.62		9.27	10.55
0060	8″ cell		500	.016		9.60	.62		10.22	11.55
0070	2′ x 4′, 6″ cell		500	.016		5.75	.62		6.37	7.35
0080	8″ cell	↓	500	.016	↓	5.75	.62		6.37	7.35

09 54 26 – Suspended Wood Ceilings

09 54 26.10 Wood Ceilings		Crew	Daily Output	Labor-Hours	Unit	Material	Labor	Equipment	Total	Total Incl O&P
0010	**WOOD CEILINGS**									
1000	4″-6″ wood slats on heavy duty 15/16″ T-bar grid	2 Carp	250	.064	S.F.	26	2.48		28.48	32.50

09 54 33 – Decorative Panel Ceilings

09 54 33.20 Metal Panel Ceilings		Crew	Daily Output	Labor-Hours	Unit	Material	Labor	Equipment	Total	Total Incl O&P
0010	**METAL PANEL CEILINGS**									
0020	Lay-in or screwed to furring, not including grid									
0100	Tin ceilings, 2′ x 2′ or 2′ x 4′, bare steel finish	2 Carp	300	.053	S.F.	2.17	2.07		4.24	5.80
0120	Painted white finish		300	.053	″	3.54	2.07		5.61	7.30
0140	Copper, chrome or brass finish		300	.053	L.F.	6.30	2.07		8.37	10.35
0200	Cornice molding, 2-1/2″ to 3-1/2″ wide, 4′ long, bare steel finish		200	.080	S.F.	2.13	3.10		5.23	7.45
0220	Painted white finish		200	.080		2.82	3.10		5.92	8.20
0240	Copper, chrome or brass finish		200	.080		3.94	3.10		7.04	9.45
0320	5″ to 6-1/2″ wide, 4′ long, bare steel finish		150	.107		3.39	4.13		7.52	10.55
0340	Painted white finish		150	.107		3.90	4.13		8.03	11.15
0360	Copper, chrome or brass finish		150	.107		5.85	4.13		9.98	13.30
0420	Flat molding, 3-1/2″ to 5″ wide, 4′ long, bare steel finish		250	.064		3.81	2.48		6.29	8.30
0440	Painted white finish		250	.064		3.98	2.48		6.46	8.50
0460	Copper, chrome or brass finish	↓	250	.064	↓	7.15	2.48		9.63	12

09 61 Flooring Treatment

09 61 19 – Concrete Floor Staining

09 61 19.40 Floors, Interior		Crew	Daily Output	Labor-Hours	Unit	Material	Labor	Equipment	Total	Total Incl O&P
						2019 Bare Costs				
0010	FLOORS, INTERIOR									
0300	Acid stain and sealer									
0310	Stain, one coat	1 Pord	650	.012	S.F.	.15	.40		.55	.82
0320	Two coats		570	.014		.29	.46		.75	1.08
0330	Acrylic sealer, one coat		2600	.003		.23	.10		.33	.42
0340	Two coats	↓	1400	.006	↓	.45	.19		.64	.81

09 62 Specialty Flooring

09 62 19 – Laminate Flooring

09 62 19.10 Floating Floor		Crew	Daily Output	Labor-Hours	Unit	Material	Labor	Equipment	Total	Total Incl O&P
0010	FLOATING FLOOR									
8300	Floating floor, laminate, wood pattern strip, complete	1 Clab	133	.060	S.F.	4.67	1.83		6.50	8.15
8310	Components, T&G wood composite strips					4.18			4.18	4.60
8320	Film					.18			.18	.20
8330	Foam					.27			.27	.30
8340	Adhesive					.47			.47	.52
8350	Installation kit				↓	.20			.20	.22
8360	Trim, 2" wide x 3' long				L.F.	4.39			4.39	4.83
8370	Reducer moulding				"	5.80			5.80	6.35

09 62 23 – Bamboo Flooring

09 62 23.10 Flooring, Bamboo			Crew	Daily Output	Labor-Hours	Unit	Material	Labor	Equipment	Total	Total Incl O&P
0010	FLOORING, BAMBOO										
8600	Flooring, wood, bamboo strips, unfinished, 5/8" x 4" x 3'	G	1 Carp	255	.031	S.F.	5.85	1.22		7.07	8.40
8610	5/8" x 4" x 4'	G		275	.029		6.05	1.13		7.18	8.50
8620	5/8" x 4" x 6'	G		295	.027		6.65	1.05		7.70	9.05
8630	Finished, 5/8" x 4" x 3'	G		255	.031		6.40	1.22		7.62	9.05
8640	5/8" x 4" x 4'	G		275	.029		6.75	1.13		7.88	9.25
8650	5/8" x 4" x 6'	G		295	.027	↓	5.05	1.05		6.10	7.30
8660	Stair treads, unfinished, 1-1/16" x 11-1/2" x 4'	G		18	.444	Ea.	56	17.20		73.20	90
8670	Finished, 1-1/16" x 11-1/2" x 4'	G		18	.444		86	17.20		103.20	123
8680	Stair risers, unfinished, 5/8" x 7-1/2" x 4'	G		18	.444		20.50	17.20		37.70	51.50
8690	Finished, 5/8" x 7-1/2" x 4'	G		18	.444		39	17.20		56.20	71.50
8700	Stair nosing, unfinished, 6' long	G		16	.500		46	19.40		65.40	82.50
8710	Finished, 6' long	G	↓	16	.500	↓	43.50	19.40		62.90	80

09 62 29 – Cork Flooring

09 62 29.10 Cork Tile Flooring			Crew	Daily Output	Labor-Hours	Unit	Material	Labor	Equipment	Total	Total Incl O&P
0010	CORK TILE FLOORING										
2200	Cork tile, standard finish, 1/8" thick	G	1 Tilf	315	.025	S.F.	5.30	.91		6.21	7.25
2250	3/16" thick	G		315	.025		5.95	.91		6.86	8
2300	5/16" thick	G		315	.025		6.50	.91		7.41	8.60
2350	1/2" thick	G		315	.025		7.05	.91		7.96	9.20
2500	Urethane finish, 1/8" thick	G		315	.025		5.20	.91		6.11	7.15
2550	3/16" thick	G		315	.025		7.20	.91		8.11	9.35
2600	5/16" thick	G		315	.025		7.50	.91		8.41	9.70
2650	1/2" thick	G	↓	315	.025	↓	7.60	.91		8.51	9.80

09 63 Masonry Flooring

09 63 13 – Brick Flooring

09 63 13.10 Miscellaneous Brick Flooring			Crew	Daily Output	Labor-Hours	Unit	Material	Labor	Equipment	Total	Total Incl O&P
							2019 Bare Costs				
0010	**MISCELLANEOUS BRICK FLOORING**										
0020	Acid-proof shales, red, 8″ x 3-3/4″ x 1-1/4″ thick		D-7	.43	37.209	M	735	1,175		1,910	2,700
0050	2-1/4″ thick		D-1	.40	40		1,025	1,350		2,375	3,400
0200	Acid-proof clay brick, 8″ x 3-3/4″ x 2-1/4″ thick	G		.40	40		1,025	1,350		2,375	3,375
0250	9″ x 4-1/2″ x 3″	G		95	.168	S.F.	4.47	5.70		10.17	14.35
0260	Cast ceramic, pressed, 4″ x 8″ x 1/2″, unglazed		D-7	100	.160		7	5.10		12.10	15.90
0270	Glazed			100	.160		9.30	5.10		14.40	18.45
0280	Hand molded flooring, 4″ x 8″ x 3/4″, unglazed			95	.168		9.20	5.35		14.55	18.75
0290	Glazed			95	.168		11.60	5.35		16.95	21.50
0300	8″ hexagonal, 3/4″ thick, unglazed			85	.188		10.10	5.95		16.05	21
0310	Glazed			85	.188		18.25	5.95		24.20	29.50
0450	Acid-proof joints, 1/4″ wide		D-1	65	.246		1.61	8.30		9.91	15.60
0500	Pavers, 8″ x 4″, 1″ to 1-1/4″ thick, red		D-7	95	.168		4.06	5.35		9.41	13.05
0510	Ironspot		″	95	.168		5.75	5.35		11.10	14.90
0540	1-3/8″ to 1-3/4″ thick, red		D-1	95	.168		3.92	5.70		9.62	13.75
0560	Ironspot			95	.168		5.70	5.70		11.40	15.70
0580	2-1/4″ thick, red			90	.178		3.99	6		9.99	14.40
0590	Ironspot			90	.178		6.20	6		12.20	16.80
0700	Paver, adobe brick, 6″ x 12″, 1/2″ joint	G		42	.381		1.46	12.85		14.31	23
0710	Mexican red, 12″ x 12″	G	1 Tilf	48	.167		1.86	6		7.86	11.70
0720	Saltillo, 12″ x 12″	G	″	48	.167		1.50	6		7.50	11.30
0800	For sidewalks and patios with pavers, see Section 32 14 16.10										
0870	For epoxy joints, add		D-1	600	.027	S.F.	3.08	.90		3.98	4.89
0880	For Furan underlayment, add		″	600	.027		2.55	.90		3.45	4.31
0890	For waxed surface, steam cleaned, add		A-1H	1000	.008		.21	.24	.07	.52	.71

09 63 40 – Stone Flooring

09 63 40.10 Marble		Crew	Daily Output	Labor-Hours	Unit	Material	Labor	Equipment	Total	Total Incl O&P
0010	**MARBLE**									
0020	Thin gauge tile, 12″ x 6″, 3/8″, white Carara	D-7	60	.267	S.F.	17.25	8.45		25.70	32.50
0100	Travertine		60	.267		9.05	8.45		17.50	23.50
0200	12″ x 12″ x 3/8″, thin set, floors		60	.267		11.45	8.45		19.90	26.50
0300	On walls		52	.308		10	9.75		19.75	27
1000	Marble threshold, 4″ wide x 36″ long x 5/8″ thick, white		60	.267	Ea.	11.45	8.45		19.90	26.50

09 63 40.20 Slate Tile		Crew	Daily Output	Labor-Hours	Unit	Material	Labor	Equipment	Total	Total Incl O&P
0010	**SLATE TILE**									
0020	Vermont, 6″ x 6″ x 1/4″ thick, thin set	D-7	180	.089	S.F.	7.80	2.82		10.62	13.15

09 64 Wood Flooring

09 64 16 – Wood Block Flooring

09 64 16.10 End Grain Block Flooring		Crew	Daily Output	Labor-Hours	Unit	Material	Labor	Equipment	Total	Total Incl O&P
0010	**END GRAIN BLOCK FLOORING**									
0020	End grain flooring, coated, 2″ thick	1 Carp	295	.027	S.F.	3.69	1.05		4.74	5.80
0400	Natural finish, 1″ thick, fir		125	.064		3.81	2.48		6.29	8.30
0600	1-1/2″ thick, pine		125	.064		3.74	2.48		6.22	8.20
0700	2″ thick, pine		125	.064		5.15	2.48		7.63	9.80

09 64 23 – Wood Parquet Flooring

09 64 23.10 Wood Parquet		Crew	Daily Output	Labor-Hours	Unit	Material	Labor	Equipment	Total	Total Incl O&P
0010	**WOOD PARQUET** flooring									
5200	Parquetry, 5/16″ thk, no finish, oak, plain pattern	1 Carp	160	.050	S.F.	5.45	1.94		7.39	9.20
5300	Intricate pattern		100	.080		10.25	3.10		13.35	16.40

09 64 Wood Flooring

09 64 23 – Wood Parquet Flooring

09 64 23.10 Wood Parquet		Crew	Daily Output	Labor-Hours	Unit	Material	Labor	Equipment	Total	Total Incl O&P
						2019 Bare Costs				
5500	Teak, plain pattern	1 Carp	160	.050	S.F.	6.50	1.94		8.44	10.35
5600	Intricate pattern		100	.080		11	3.10		14.10	17.20
5650	13/16" thick, select grade oak, plain pattern		160	.050		12.45	1.94		14.39	16.90
5700	Intricate pattern		100	.080		17.75	3.10		20.85	24.50
5800	Custom parquetry, including finish, plain pattern		100	.080		17.50	3.10		20.60	24.50
5900	Intricate pattern		50	.160		25.50	6.20		31.70	38.50
6700	Parquetry, prefinished white oak, 5/16" thick, plain pattern		160	.050		8.55	1.94		10.49	12.60
6800	Intricate pattern		100	.080		8.95	3.10		12.05	14.95
7000	Walnut or teak, parquetry, plain pattern		160	.050		8.65	1.94		10.59	12.75
7100	Intricate pattern	▼	100	.080	▼	16.80	3.10		19.90	23.50
7200	Acrylic wood parquet blocks, 12" x 12" x 5/16",									
7210	Irradiated, set in epoxy	1 Carp	160	.050	S.F.	10.65	1.94		12.59	14.90

09 64 29 – Wood Strip and Plank Flooring

09 64 29.10 Wood		Crew	Daily Output	Labor-Hours	Unit	Material	Labor	Equipment	Total	Total Incl O&P
0010	**WOOD**									
0020	Fir, vertical grain, 1" x 4", not incl. finish, grade B & better	1 Carp	255	.031	S.F.	3.55	1.22		4.77	5.90
0100	Grade C & better		255	.031		3.34	1.22		4.56	5.70
0300	Flat grain, 1" x 4", not incl. finish, grade B & better		255	.031		4.04	1.22		5.26	6.45
0400	Grade C & better		255	.031		3.89	1.22		5.11	6.30
4000	Maple, strip, 25/32" x 2-1/4", not incl. finish, select		170	.047		5.05	1.82		6.87	8.55
4100	#2 & better		170	.047		5.05	1.82		6.87	8.55
4300	33/32" x 3-1/4", not incl. finish, #1 grade		170	.047		5.80	1.82		7.62	9.40
4400	#2 & better	▼	170	.047	▼	5.20	1.82		7.02	8.70
4600	Oak, white or red, 25/32" x 2-1/4", not incl. finish									
4700	#1 common	1 Carp	170	.047	S.F.	3.50	1.82		5.32	6.85
4900	Select quartered, 2-1/4" wide		170	.047		4.38	1.82		6.20	7.85
5000	Clear		170	.047		4.31	1.82		6.13	7.75
6100	Prefinished, white oak, prime grade, 2-1/4" wide		170	.047		5.90	1.82		7.72	9.45
6200	3-1/4" wide		185	.043		5.70	1.68		7.38	9.05
6400	Ranch plank		145	.055		6.70	2.14		8.84	10.95
6500	Hardwood blocks, 9" x 9", 25/32" thick		160	.050		7.65	1.94		9.59	11.60
7400	Yellow pine, 3/4" x 3-1/8", T&G, C & better, not incl. finish	▼	200	.040		1.73	1.55		3.28	4.46
7500	Refinish wood floor, sand, 2 coats poly, wax, soft wood	1 Clab	400	.020		.22	.61		.83	1.24
7600	Hardwood		130	.062		.22	1.87		2.09	3.33
7800	Sanding and finishing, 2 coats polyurethane	▼	295	.027	▼	.22	.82		1.04	1.60
7900	Subfloor and underlayment, see Section 06 16									
8015	Transition molding, 2-1/4" wide, 5' long	1 Carp	19.20	.417	Ea.	21.50	16.15		37.65	50

09 65 Resilient Flooring

09 65 10 – Resilient Tile Underlayment

09 65 10.10 Latex Underlayment		Crew	Daily Output	Labor-Hours	Unit	Material	Labor	Equipment	Total	Total Incl O&P
0010	**LATEX UNDERLAYMENT**									
3600	Latex underlayment, 1/8" thk., cementitious for resilient flooring	1 Tilf	160	.050	S.F.	1.25	1.80		3.05	4.27
4000	Liquid, fortified				Gal.	32			32	35.50

09 65 13 – Resilient Base and Accessories

09 65 13.13 Resilient Base		Crew	Daily Output	Labor-Hours	Unit	Material	Labor	Equipment	Total	Total Incl O&P
0010	**RESILIENT BASE**									
0690	1/8" vinyl base, 2-1/2" H, straight or cove, standard colors	1 Tilf	315	.025	L.F.	.70	.91		1.61	2.24
0700	4" high		315	.025		1.18	.91		2.09	2.77
0710	6" high	▼	315	.025	▼	1.54	.91		2.45	3.16

09 65 13 – Resilient Base and Accessories

09 65 13.13 Resilient Base		Crew	Daily Output	Labor-Hours	Unit	Material	Labor	Equipment	Total	Total Incl O&P
						2019 Bare Costs				
0720	Corners, 2-1/2" high	1 Tilf	315	.025	Ea.	2.23	.91		3.14	3.92
0730	4" high		315	.025		3.03	.91		3.94	4.80
0740	6" high		315	.025		2.86	.91		3.77	4.62
0800	1/8" rubber base, 2-1/2" H, straight or cove, standard colors		315	.025	L.F.	1.11	.91		2.02	2.69
1100	4" high		315	.025		1.30	.91		2.21	2.90
1110	6" high		315	.025		1.91	.91		2.82	3.58
1150	Corners, 2-1/2" high		315	.025	Ea.	2.48	.91		3.39	4.20
1153	4" high		315	.025		2.56	.91		3.47	4.29
1155	6" high		315	.025		3.18	.91		4.09	4.97
1450	For premium color/finish add					50%				
1500	Millwork profile	1 Tilf	315	.025	L.F.	6.30	.91		7.21	8.40

09 65 13.37 Vinyl Transition Strips		Crew	Daily Output	Labor-Hours	Unit	Material	Labor	Equipment	Total	Total Incl O&P
0010	**VINYL TRANSITION STRIPS**									
0100	Various mats. to various mats., adhesive applied, 1/4" to 1/8"	1 Tilf	315	.025	L.F.	1.48	.91		2.39	3.10
0105	0.08" to 1/8"		315	.025		1.34	.91		2.25	2.94
0110	0.08" to 1/4"		315	.025		1.46	.91		2.37	3.08
0115	1/4" to 3/8"		315	.025		1.36	.91		2.27	2.97
0120	1/4" to 1/2"		315	.025		1.36	.91		2.27	2.97
0125	1/4" to 0.08"		315	.025		1.46	.91		2.37	3.08
0200	Vinyl wheeled trans. strips, carpet to var. mats., 1/4" to 1/8" x 2-1/2"		315	.025		5.05	.91		5.96	7
0205	1/4" to 1/8" x 4"		315	.025		6.15	.91		7.06	8.20
0210	Various mats. to various mats. 1/4" to 0.08" x 2-1/2"		315	.025		5.05	.91		5.96	7.05
0215	Carpet to various materials, 1/4" to flush x 2-1/2"		315	.025		4.22	.91		5.13	6.10
0220	1/4" to flush x 4"		315	.025		6.50	.91		7.41	8.60
0225	Various materials to resilient, 3/8" to 1/8" x 2-1/2"		315	.025		4.22	.91		5.13	6.10
0230	Carpet to various materials, 3/8" to 1/4" x 2-1/2"		315	.025		5.80	.91		6.71	7.85
0235	1/4" to 1/4" x 2-1/2"		315	.025		6.50	.91		7.41	8.60
0240	Various materials to resilient, 1/8" to 1/8" x 2-1/2"		315	.025		4.91	.91		5.82	6.85
0245	Various materials to var. mats., 1/8" to flush x 2-1/2"		315	.025		3.29	.91		4.20	5.10
0250	3/8" to flush x 4"		315	.025		6.60	.91		7.51	8.75
0255	1/2" to flush x 4"		315	.025		8.90	.91		9.81	11.25
0260	Various materials to resilient, 1/8" to 0.08" x 2-1/2"		315	.025		3.81	.91		4.72	5.65
0265	0.08" to 0.08" x 2-1/2"		315	.025		3.62	.91		4.53	5.45
0270	3/8" to 0.08" x 2-1/2"		315	.025		3.62	.91		4.53	5.45

09 65 16 – Resilient Sheet Flooring

09 65 16.10 Rubber and Vinyl Sheet Flooring		Crew	Daily Output	Labor-Hours	Unit	Material	Labor	Equipment	Total	Total Incl O&P
0010	**RUBBER AND VINYL SHEET FLOORING**									
5500	Linoleum, sheet goods G	1 Tilf	360	.022	S.F.	3.60	.80		4.40	5.25
5900	Rubber, sheet goods, 36" wide, 1/8" thick		120	.067		9.05	2.39		11.44	13.80
5950	3/16" thick		100	.080		10.15	2.87		13.02	15.80
6000	1/4" thick		90	.089		11.95	3.19		15.14	18.30
8000	Vinyl sheet goods, backed, .065" thick, plain pattern/colors		250	.032		4.47	1.15		5.62	6.75
8050	Intricate pattern/colors		200	.040		3.87	1.44		5.31	6.60
8100	.080" thick, plain pattern/colors		230	.035		4.20	1.25		5.45	6.65
8150	Intricate pattern/colors		200	.040		6.60	1.44		8.04	9.60
8200	.125" thick, plain pattern/colors		230	.035		3.93	1.25		5.18	6.35
8250	Intricate pattern/colors		200	.040		7.30	1.44		8.74	10.35
8700	Adhesive cement, 1 gallon per 200 to 300 S.F.				Gal.	31.50			31.50	34.50
8800	Asphalt primer, 1 gallon per 300 S.F.					14.85			14.85	16.35
8900	Emulsion, 1 gallon per 140 S.F.					19.10			19.10	21

09 65 Resilient Flooring

09 65 19 – Resilient Tile Flooring

09 65 19.19 Vinyl Composition Tile Flooring		Crew	Daily Output	Labor-Hours	Unit	Material	Labor	Equipment	Total	Total Incl O&P
						2019 Bare Costs				
0010	**VINYL COMPOSITION TILE FLOORING**									
7000	Vinyl composition tile, 12″ x 12″, 1/16″ thick	1 Tilf	500	.016	S.F.	1.22	.57		1.79	2.27
7050	Embossed		500	.016		2.67	.57		3.24	3.87
7100	Marbleized		500	.016		2.67	.57		3.24	3.87
7150	Solid		500	.016		3.45	.57		4.02	4.73
7200	3/32″ thick, embossed		500	.016		1.54	.57		2.11	2.62
7250	Marbleized		500	.016		3.07	.57		3.64	4.31
7300	Solid		500	.016		2.86	.57		3.43	4.08
7350	1/8″ thick, marbleized		500	.016		2.45	.57		3.02	3.63
7400	Solid		500	.016		1.74	.57		2.31	2.84
7450	Conductive		500	.016		5.90	.57		6.47	7.45

09 65 19.23 Vinyl Tile Flooring

Code	Description	Crew	Daily Output	Labor-Hours	Unit	Material	Labor	Equipment	Total	Total Incl O&P
0010	**VINYL TILE FLOORING**									
7500	Vinyl tile, 12″ x 12″, 3/32″ thick, standard colors/patterns	1 Tilf	500	.016	S.F.	3.72	.57		4.29	5
7550	1/8″ thick, standard colors/patterns		500	.016		5.20	.57		5.77	6.65
7600	1/8″ thick, premium colors/patterns		500	.016		7	.57		7.57	8.65
7650	Solid colors		500	.016		3.22	.57		3.79	4.47
7700	Marbleized or Travertine pattern		500	.016		6.20	.57		6.77	7.80
7750	Florentine pattern		500	.016		6.65	.57		7.22	8.30
7800	Premium colors/patterns		500	.016		6.55	.57		7.12	8.15

09 65 19.33 Rubber Tile Flooring

Code	Description	Crew	Daily Output	Labor-Hours	Unit	Material	Labor	Equipment	Total	Total Incl O&P
0010	**RUBBER TILE FLOORING**									
6050	Rubber tile, marbleized colors, 12″ x 12″, 1/8″ thick	1 Tilf	400	.020	S.F.	5.90	.72		6.62	7.65
6100	3/16″ thick		400	.020		8.30	.72		9.02	10.25
6300	Special tile, plain colors, 1/8″ thick		400	.020		8.35	.72		9.07	10.35
6350	3/16″ thick		400	.020		10.30	.72		11.02	12.45

09 65 33 – Conductive Resilient Flooring

09 65 33.10 Conductive Rubber and Vinyl Flooring

Code	Description	Crew	Daily Output	Labor-Hours	Unit	Material	Labor	Equipment	Total	Total Incl O&P
0010	**CONDUCTIVE RUBBER AND VINYL FLOORING**									
1700	Conductive flooring, rubber tile, 1/8″ thick	1 Tilf	315	.025	S.F.	7.15	.91		8.06	9.30
1800	Homogeneous vinyl tile, 1/8″ thick	″	315	.025	″	6.95	.91		7.86	9.10

09 65 66 – Resilient Athletic Flooring

09 65 66.10 Resilient Athletic Flooring

Code	Description	Crew	Daily Output	Labor-Hours	Unit	Material	Labor	Equipment	Total	Total Incl O&P
0010	**RESILIENT ATHLETIC FLOORING**									
1000	Recycled rubber rolled goods, for weight rooms, 3/8″ thk.	1 Tilf	315	.025	S.F.	2.80	.91		3.71	4.55
1050	Interlocking 2′ x 2′ squares, rubber, 1/4″ thk.		310	.026		1.90	.93		2.83	3.59
1055	5/16″ thk.		310	.026		2.31	.93		3.24	4.04
1060	3/8″ thk.		300	.027		3.32	.96		4.28	5.20
1065	1/2″ thk.		310	.026		3.67	.93		4.60	5.55
2000	Vinyl sheet flooring, 1/4″ thk.		315	.025		4.89	.91		5.80	6.85

09 66 13 – Portland Cement Terrazzo Flooring

09 66 13.10 Portland Cement Terrazzo		Crew	Daily Output	Labor-Hours	Unit	2019 Bare Costs Material	Labor	Equipment	Total	Total Incl O&P
0010	**PORTLAND CEMENT TERRAZZO**, cast-in-place									
0020	Cove base, 6" high, 16 ga. zinc	1 Mstz	20	.400	L.F.	3.35	14.15		17.50	26.50
0101	Curb, 6" high and 6" wide	J-3	12	1.333		6.55	44	26.50	77.05	107
0300	Divider strip for floors, 14 ga., 1-1/4" deep, zinc	1 Mstz	375	.021		1.46	.76		2.22	2.83
0400	Brass		375	.021		2.60	.76		3.36	4.08
0600	Heavy top strip 1/4" thick, 1-1/4" deep, zinc		300	.027		2.23	.94		3.17	3.97
1200	For thin set floors, 16 ga., 1/2" x 1/2", zinc		350	.023		1.36	.81		2.17	2.81
1500	Floor, bonded to concrete, 1-3/4" thick, gray cement	J-3	75	.213	S.F.	3.71	7.05	4.23	14.99	20
1600	White cement, mud set		75	.213		4.40	7.05	4.23	15.68	21
1800	Not bonded, 3" total thickness, gray cement		70	.229		4.58	7.55	4.53	16.66	22
1900	White cement, mud set		70	.229		5.55	7.55	4.53	17.63	23.50
4300	Stone chips, onyx gemstone, per 50 lb. bag				Bag	19			19	21

09 66 16 – Terrazzo Floor Tile

09 66 16.10 Tile or Terrazzo Base		Crew	Daily Output	Labor-Hours	Unit	Material	Labor	Equipment	Total	Total Incl O&P
0010	**TILE OR TERRAZZO BASE**									
0020	Scratch coat only	1 Mstz	150	.053	S.F.	.48	1.89		2.37	3.58
0500	Scratch and brown coat only	"	75	.107	"	.97	3.78		4.75	7.15

09 66 16.13 Portland Cement Terrazzo Floor Tile		Crew	Daily Output	Labor-Hours	Unit	Material	Labor	Equipment	Total	Total Incl O&P
0010	**PORTLAND CEMENT TERRAZZO FLOOR TILE**									
1200	Floor tiles, non-slip, 1" thick, 12" x 12"	D-1	60	.267	S.F.	25	9		34	42.50
1300	1-1/4" thick, 12" x 12"		60	.267		26	9		35	43.50
1500	16" x 16"		50	.320		28	10.80		38.80	49
1600	1-1/2" thick, 16" x 16"		45	.356		25.50	12		37.50	48

09 66 16.16 Plastic Matrix Terrazzo Floor Tile		Crew	Daily Output	Labor-Hours	Unit	Material	Labor	Equipment	Total	Total Incl O&P
0010	**PLASTIC MATRIX TERRAZZO FLOOR TILE**									
0100	12" x 12", 3/16" thick, floor tiles w/marble chips	1 Tilf	500	.016	S.F.	7.50	.57		8.07	9.20
0200	12" x 12", 3/16" thick, floor tiles w/glass chips		500	.016		8.05	.57		8.62	9.80
0300	12" x 12", 3/16" thick, floor tiles w/recycled content		500	.016		6.35	.57		6.92	7.95

09 66 16.30 Terrazzo, Precast		Crew	Daily Output	Labor-Hours	Unit	Material	Labor	Equipment	Total	Total Incl O&P
0010	**TERRAZZO, PRECAST**									
0020	Base, 6" high, straight	1 Mstz	70	.114	L.F.	12.60	4.05		16.65	20.50
0100	Cove		60	.133		17.25	4.72		21.97	26.50
0300	8" high, straight		60	.133		16.55	4.72		21.27	26
0400	Cove		50	.160		24.50	5.65		30.15	36
0600	For white cement, add					.59			.59	.65
0700	For 16 ga. zinc toe strip, add					2.20			2.20	2.42
0900	Curbs, 4" x 4" high	1 Mstz	40	.200		42.50	7.10		49.60	58.50
1000	8" x 8" high		30	.267		46.50	9.45		55.95	66.50
4800	Wainscot, 12" x 12" x 1" tiles		12	.667	S.F.	9.70	23.50		33.20	48.50
4900	16" x 16" x 1-1/2" tiles		8	1	"	18.65	35.50		54.15	77.50

09 66 23 – Resinous Matrix Terrazzo Flooring

09 66 23.13 Polyacrylate Mod. Cementitious Terrazzo Flr.		Crew	Daily Output	Labor-Hours	Unit	Material	Labor	Equipment	Total	Total Incl O&P
0010	**POLYACRYLATE MODIFIED CEMENTITIOUS TERRAZZO FLOORING**									
3150	Polyacrylate, 1/4" thick, granite chips	C-6	735	.065	S.F.	3.85	2.08	.07	6	7.75
3170	Recycled porcelain		480	.100		5.05	3.18	.11	8.34	10.95
3200	3/8" thick, granite chips		620	.077		5.10	2.46	.09	7.65	9.75
3220	Recycled porcelain		480	.100		7.10	3.18	.11	10.39	13.15

09 66 23.16 Epoxy-Resin Terrazzo Flooring		Crew	Daily Output	Labor-Hours	Unit	Material	Labor	Equipment	Total	Total Incl O&P
0010	**EPOXY-RESIN TERRAZZO FLOORING**									
1800	Epoxy terrazzo, 1/4" thick, chemical resistant, granite chips	J-3	200	.080	S.F.	6.50	2.64	1.59	10.73	13.15
1900	Recycled porcelain		150	.107		10	3.52	2.11	15.63	19.05

09 66 Terrazzo Flooring

09 66 23 – Resinous Matrix Terrazzo Flooring

09 66 23.16 Epoxy-Resin Terrazzo Flooring		Crew	Daily Output	Labor-Hours	Unit	2019 Bare Costs Material	Labor	Equipment	Total	Total Incl O&P
2650	Epoxy terrazzo, 3/8" thick, marble chips	J-3	200	.080	S.F.	5.50	2.64	1.59	9.73	12.05
2675	Glass or mother of pearl	↓	200	.080	↓	7.05	2.64	1.59	11.28	13.75

09 66 33 – Conductive Terrazzo Flooring

09 66 33.19 Conductive Plastic-Matrix Terrazzo Flooring		Crew	Daily Output	Labor-Hours	Unit	Material	Labor	Equipment	Total	Total Incl O&P
0010	**CONDUCTIVE PLASTIC-MATRIX TERRAZZO FLOORING**									
3300	Conductive, 1/4" thick, granite chips	C-6	450	.107	S.F.	8.35	3.39	.12	11.86	14.95
3330	Recycled porcelain		305	.157		11	5	.17	16.17	20.50
3350	3/8" thick, granite chips		365	.132		11.60	4.18	.15	15.93	19.80
3370	Recycled porcelain		255	.188		14.05	6	.21	20.26	25.50
3450	Granite, conductive, 1/4" thick, 20% chip		695	.069		10.20	2.20	.08	12.48	14.90
3470	50% chip		420	.114		13.15	3.64	.13	16.92	20.50
3500	3/8" thick, 20% chip		695	.069		14.90	2.20	.08	17.18	20
3520	50% chip	↓	380	.126	↓	18	4.02	.14	22.16	26.50

09 67 Fluid-Applied Flooring

09 67 13 – Elastomeric Liquid Flooring

09 67 13.13 Elastomeric Liquid Flooring		Crew	Daily Output	Labor-Hours	Unit	Material	Labor	Equipment	Total	Total Incl O&P
0010	**ELASTOMERIC LIQUID FLOORING**									
0020	Cementitious acrylic, 1/4" thick	C-6	520	.092	S.F.	1.81	2.94	.10	4.85	6.95
0200	Methyl methachrylate, 1/4" thick	C-8A	3000	.016	"	7.20	.53		7.73	8.75

09 67 23 – Resinous Flooring

09 67 23.23 Resinous Flooring		Crew	Daily Output	Labor-Hours	Unit	Material	Labor	Equipment	Total	Total Incl O&P
0010	**RESINOUS FLOORING**									
1200	Heavy duty epoxy topping, 1/4" thick,									
1300	500 to 1,000 S.F.	C-6	420	.114	S.F.	6.25	3.64	.13	10.02	13.05
1500	1,000 to 2,000 S.F.		450	.107		5.20	3.39	.12	8.71	11.50
1600	Over 10,000 S.F.	↓	480	.100	↓	4.82	3.18	.11	8.11	10.65

09 67 26 – Quartz Flooring

09 67 26.26 Quartz Flooring		Crew	Daily Output	Labor-Hours	Unit	Material	Labor	Equipment	Total	Total Incl O&P
0010	**QUARTZ FLOORING**									
0600	Epoxy, with colored quartz chips, broadcast, 3/8" thick	C-6	675	.071	S.F.	3.22	2.26	.08	5.56	7.35
0700	1/2" thick		490	.098		4.59	3.12	.11	7.82	10.30
0900	Troweled, minimum		560	.086		3.65	2.73	.10	6.48	8.60
1000	Maximum	↓	480	.100	↓	6.20	3.18	.11	9.49	12.15

09 68 Carpeting

09 68 05 – Carpet Accessories

09 68 05.11 Flooring Transition Strip		Crew	Daily Output	Labor-Hours	Unit	Material	Labor	Equipment	Total	Total Incl O&P
0010	**FLOORING TRANSITION STRIP**									
0107	Clamp down brass divider, 12' strip, vinyl to carpet	1 Tilf	31.25	.256	Ea.	14.65	9.20		23.85	31
0117	Vinyl to hard surface	"	31.25	.256	"	14.65	9.20		23.85	31

09 68 10 – Carpet Pad

09 68 10.10 Commercial Grade Carpet Pad		Crew	Daily Output	Labor-Hours	Unit	Material	Labor	Equipment	Total	Total Incl O&P
0010	**COMMERCIAL GRADE CARPET PAD**									
9000	Sponge rubber pad, 20 oz./sq. yd.	1 Tilf	150	.053	S.Y.	4.75	1.91		6.66	8.35
9100	40 to 62 oz./sq. yd.		150	.053		8.80	1.91		10.71	12.75
9200	Felt pad, 20 oz./sq. yd.	↓	150	.053	↓	6.20	1.91		8.11	9.90

09 68 Carpeting

09 68 10 – Carpet Pad

09 68 10.10 Commercial Grade Carpet Pad		Crew	Daily Output	Labor-Hours	Unit	Material	2019 Bare Costs Labor	Equipment	Total	Total Incl O&P
9300	32 to 56 oz./sq. yd.	1 Tilf	150	.053	S.Y.	11.60	1.91		13.51	15.85
9400	Bonded urethane pad, 2.7 density		150	.053		5.95	1.91		7.86	9.65
9500	13.0 density		150	.053		8.05	1.91		9.96	12
9600	Prime urethane pad, 2.7 density		150	.053		3.59	1.91		5.50	7.05
9700	13.0 density		150	.053		6.70	1.91		8.61	10.50

09 68 13 – Tile Carpeting

09 68 13.10 Carpet Tile		Crew	Daily Output	Labor-Hours	Unit	Material	Labor	Equipment	Total	Total Incl O&P
0010	**CARPET TILE**									
0100	Tufted nylon, 18" x 18", hard back, 20 oz.	1 Tilf	80	.100	S.Y.	28	3.59		31.59	36.50
0110	26 oz.		80	.100		25.50	3.59		29.09	34
0200	Cushion back, 20 oz.		80	.100		25	3.59		28.59	33.50
0210	26 oz.		80	.100		30.50	3.59		34.09	39.50
6000	Electrostatic dissapative carpet tile, 24" x 24", 24 oz.		80	.100		38	3.59		41.59	47.50
6100	Electrostatic dissapative carpet tile for access floors, 24" x 24", 24 oz.		80	.100		47.50	3.59		51.09	58.50

09 68 16 – Sheet Carpeting

09 68 16.10 Sheet Carpet		Crew	Daily Output	Labor-Hours	Unit	Material	Labor	Equipment	Total	Total Incl O&P
0010	**SHEET CARPET**									
0700	Nylon, level loop, 26 oz., light to medium traffic	1 Tilf	75	.107	S.Y.	23.50	3.83		27.33	31.50
0720	28 oz., light to medium traffic		75	.107		32.50	3.83		36.33	41.50
0900	32 oz., medium traffic		75	.107		39.50	3.83		43.33	49.50
1100	40 oz., medium to heavy traffic		75	.107		49	3.83		52.83	60
2920	Nylon plush, 30 oz., medium traffic		75	.107		28	3.83		31.83	37
3000	36 oz., medium traffic		75	.107		36.50	3.83		40.33	46
3100	42 oz., medium to heavy traffic		70	.114		46	4.10		50.10	57
3200	46 oz., medium to heavy traffic		70	.114		53	4.10		57.10	64.50
3300	54 oz., heavy traffic		70	.114		60	4.10		64.10	72.50
4110	Wool, level loop, 40 oz., medium traffic		70	.114		116	4.10		120.10	134
4500	50 oz., medium to heavy traffic		70	.114		111	4.10		115.10	129
4700	Patterned, 32 oz., medium to heavy traffic		70	.114		102	4.10		106.10	119
4900	48 oz., heavy traffic		70	.114		111	4.10		115.10	130
5000	For less than full roll (approx. 1500 S.F.), add					25%				
5100	For small rooms, less than 12' wide, add						25%			
5200	For large open areas (no cuts), deduct						25%			
5600	For bound carpet baseboard, add	1 Tilf	300	.027	L.F.	1.83	.96		2.79	3.56
5610	For stairs, not incl. price of carpet, add	"	30	.267	Riser		9.55		9.55	15.45
5620	For borders and patterns, add to labor						18%			
8950	For tackless, stretched installation, add padding from 09 68 10.10 to above									
9850	For brand-named specific fiber, add				S.Y.	25%				

09 68 20 – Athletic Carpet

09 68 20.10 Indoor Athletic Carpet		Crew	Daily Output	Labor-Hours	Unit	Material	Labor	Equipment	Total	Total Incl O&P
0010	**INDOOR ATHLETIC CARPET**									
3700	Polyethylene, in rolls, no base incl., landscape surfaces	1 Tilf	275	.029	S.F.	4.19	1.04		5.23	6.30
3800	Nylon action surface, 1/8" thick		275	.029		4	1.04		5.04	6.10
3900	1/4" thick		275	.029		5.75	1.04		6.79	8.05
4000	3/8" thick		275	.029		7.25	1.04		8.29	9.65

09 72 Wall Coverings

09 72 13 – Cork Wall Coverings

09 72 13.10 Covering, Cork Wall		Crew	Daily Output	Labor-Hours	Unit	Material (2019 Bare Costs)	Labor (2019 Bare Costs)	Equipment (2019 Bare Costs)	Total (2019 Bare Costs)	Total Incl O&P
0010	COVERING, CORK WALL									
0600	Cork tiles, light or dark, 12" x 12" x 3/16"	1 Pape	240	.033	S.F.	4.25	1.10		5.35	6.50
0700	5/16" thick		235	.034		3.33	1.13		4.46	5.50
0900	1/4" basket weave		240	.033		3.39	1.10		4.49	5.55
1000	1/2" natural, non-directional pattern		240	.033		6.65	1.10		7.75	9.10
1100	3/4" natural, non-directional pattern		240	.033		11.70	1.10		12.80	14.65
1200	Granular surface, 12" x 36", 1/2" thick		385	.021		1.33	.69		2.02	2.58
1300	1" thick		370	.022		1.69	.71		2.40	3.03
1500	Polyurethane coated, 12" x 12" x 3/16" thick		240	.033		4.11	1.10		5.21	6.30
1600	5/16" thick		235	.034		5.90	1.13		7.03	8.35
1800	Cork wallpaper, paperbacked, natural		480	.017		1.60	.55		2.15	2.66
1900	Colors		480	.017		2.89	.55		3.44	4.08

09 72 16 – Vinyl-Coated Fabric Wall Coverings

09 72 16.13 Flexible Vinyl Wall Coverings

		Crew	Daily Output	Labor-Hours	Unit	Material	Labor	Equipment	Total	Total Incl O&P
0010	FLEXIBLE VINYL WALL COVERINGS									
3000	Vinyl wall covering, fabric-backed, lightweight, type 1 (12-15 oz./S.Y.)	1 Pape	640	.013	S.F.	1.36	.41		1.77	2.18
3300	Medium weight, type 2 (20-24 oz./S.Y.)		480	.017		.99	.55		1.54	1.99
3400	Heavy weight, type 3 (28 oz./S.Y.)		435	.018		1.40	.61		2.01	2.53
3600	Adhesive, 5 gal. lots (18 S.Y./gal.)				Gal.	12.90			12.90	14.20

09 72 16.16 Rigid-Sheet Vinyl Wall Coverings

		Crew	Daily Output	Labor-Hours	Unit	Material	Labor	Equipment	Total	Total Incl O&P
0010	RIGID-SHEET VINYL WALL COVERINGS									
0100	Acrylic, modified, semi-rigid PVC, .028" thick	2 Carp	330	.048	S.F.	1.30	1.88		3.18	4.54
0110	.040" thick	"	320	.050	"	1.96	1.94		3.90	5.35

09 72 19 – Textile Wall Coverings

09 72 19.10 Textile Wall Covering

		Crew	Daily Output	Labor-Hours	Unit	Material	Labor	Equipment	Total	Total Incl O&P
0010	**TEXTILE WALL COVERING**, including sizing; add 10-30% waste @ takeoff									
0020	Silk	1 Pape	640	.013	S.F.	4.75	.41		5.16	5.95
0030	Cotton		640	.013		7.05	.41		7.46	8.45
0040	Linen		640	.013		1.91	.41		2.32	2.78
0050	Blend		640	.013		3.27	.41		3.68	4.28
0060	Linen wall covering, paper backed									
0070	Flame treatment				S.F.	1.09			1.09	1.20
0080	Stain resistance treatment					1.92			1.92	2.11
0100	Grass cloths with lining paper G	1 Pape	400	.020		1.39	.66		2.05	2.61
0110	Premium texture/color G	"	350	.023		3.27	.76		4.03	4.84

09 72 20 – Natural Fiber Wall Covering

09 72 20.10 Natural Fiber Wall Covering

		Crew	Daily Output	Labor-Hours	Unit	Material	Labor	Equipment	Total	Total Incl O&P
0010	**NATURAL FIBER WALL COVERING**, including sizing; add 10-30% waste @ takeoff									
0015	Bamboo	1 Pape	640	.013	S.F.	2.19	.41		2.60	3.09
0030	Burlap		640	.013		1.84	.41		2.25	2.70
0045	Jute		640	.013		1.32	.41		1.73	2.13
0060	Sisal		640	.013		1.73	.41		2.14	2.58

09 72 23 – Wallpapering

09 72 23.10 Wallpaper

		Crew	Daily Output	Labor-Hours	Unit	Material	Labor	Equipment	Total	Total Incl O&P
0010	**WALLPAPER** including sizing; add 10-30% waste @ takeoff									
0050	Aluminum foil	1 Pape	275	.029	S.F.	1.03	.96		1.99	2.70
0100	Copper sheets, .025" thick, vinyl backing		240	.033		5.50	1.10		6.60	7.85
0300	Phenolic backing		240	.033		7.10	1.10		8.20	9.65
2400	Gypsum-based, fabric-backed, fire resistant									
2500	for masonry walls, 21 oz./S.Y.	1 Pape	800	.010	S.F.	.78	.33		1.11	1.40

09 72 Wall Coverings

09 72 23 – Wallpapering

09 72 23.10 Wallpaper		Crew	Daily Output	Labor-Hours	Unit	Material	2019 Bare Costs Labor	Equipment	Total	Total Incl O&P
2600	Average	1 Pape	720	.011	S.F.	1.29	.37		1.66	2.02
2700	Small quantities		640	.013		.78	.41		1.19	1.54
3700	Wallpaper, average workmanship, solid pattern, low cost paper		640	.013		.59	.41		1	1.33
3900	Basic patterns (matching required), avg. cost paper		535	.015		1.22	.49		1.71	2.15
4000	Paper at $85 per double roll, quality workmanship		435	.018		2.14	.61		2.75	3.34

09 74 Flexible Wood Sheets

09 74 16 – Flexible Wood Veneers

09 74 16.10 Veneer, Flexible Wood		Crew	Daily Output	Labor-Hours	Unit	Material	Labor	Equipment	Total	Total Incl O&P
0010	**VENEER, FLEXIBLE WOOD**									
0100	Flexible wood veneer, 1/32" thick, plain woods	1 Pape	100	.080	S.F.	2.46	2.64		5.10	7.05
0110	Exotic woods	"	95	.084	"	3.71	2.78		6.49	8.65

09 77 Special Wall Surfacing

09 77 30 – Fiberglass Reinforced Panels

09 77 30.10 Fiberglass Reinforced Plastic Panels		Crew	Daily Output	Labor-Hours	Unit	Material	Labor	Equipment	Total	Total Incl O&P
0010	**FIBERGLASS REINFORCED PLASTIC PANELS**, .090" thick									
0020	On walls, adhesive mounted, embossed surface	2 Carp	640	.025	S.F.	1.21	.97		2.18	2.93
0030	Smooth surface		640	.025		1.49	.97		2.46	3.24
0040	Fire rated, embossed surface		640	.025		2.16	.97		3.13	3.98
0050	Nylon rivet mounted, on drywall, embossed surface		480	.033		1.17	1.29		2.46	3.42
0060	Smooth surface		480	.033		1.45	1.29		2.74	3.73
0070	Fire rated, embossed surface		480	.033		2.15	1.29		3.44	4.50
0080	On masonry, embossed surface		320	.050		1.17	1.94		3.11	4.49
0090	Smooth surface		320	.050		1.45	1.94		3.39	4.80
0100	Fire rated, embossed surface		320	.050		2.15	1.94		4.09	5.55
0110	Nylon rivet and adhesive mounted, on drywall, embossed surface		240	.067		1.33	2.58		3.91	5.75
0120	Smooth surface		240	.067		1.35	2.58		3.93	5.75
0130	Fire rated, embossed surface		240	.067		2.31	2.58		4.89	6.80
0140	On masonry, embossed surface		190	.084		1.33	3.26		4.59	6.85
0150	Smooth surface		190	.084		1.35	3.26		4.61	6.90
0160	Fire rated, embossed surface		190	.084		2.31	3.26		5.57	7.95
0170	For moldings, add	1 Carp	250	.032	L.F.	.28	1.24		1.52	2.36
0180	On ceilings, for lay in grid system, embossed surface		400	.020	S.F.	1.21	.78		1.99	2.61
0190	Smooth surface		400	.020		1.49	.78		2.27	2.92
0200	Fire rated, embossed surface		400	.020		2.16	.78		2.94	3.66

09 77 43 – Panel Systems

09 77 43.20 Slatwall Panels and Accessories		Crew	Daily Output	Labor-Hours	Unit	Material	Labor	Equipment	Total	Total Incl O&P
0010	**SLATWALL PANELS AND ACCESSORIES**									
0100	Slatwall panel, 4' x 8' x 3/4" T, MDF, paint grade	1 Carp	500	.016	S.F.	1.56	.62		2.18	2.74
0110	Melamine finish		500	.016		2.09	.62		2.71	3.32
0120	High pressure plastic laminate finish		500	.016		3.33	.62		3.95	4.68
0125	Wood veneer		500	.016		3.72	.62		4.34	5.10
0130	Aluminum channel inserts, add					2.63			2.63	2.89
0200	Accessories, corner forms, 8' L				L.F.	5.25			5.25	5.75
0210	T-connector, 8' L					7.40			7.40	8.15
0220	J-mold, 8' L					1.27			1.27	1.40
0230	Edge cap, 8' L					1.70			1.70	1.87
0240	Finish end cap, 8' L					3.21			3.21	3.53

09 77 Special Wall Surfacing

09 77 43 – Panel Systems

09 77 43.20 Slatwall Panels and Accessories		Crew	Daily Output	Labor-Hours	Unit	Material	Labor	Equipment	Total	Total Incl O&P
						2019 Bare Costs				
0300	Display hook, metal, 4" L				Ea.	.49			.49	.54
0310	6" L					.55			.55	.61
0320	8" L					.59			.59	.65
0330	10" L					.57			.57	.63
0340	12" L					.54			.54	.59
0350	Acrylic, 4" L					1.27			1.27	1.40
0360	6" L					.98			.98	1.08
0400	Waterfall hanger, metal, 12"-16"					3.71			3.71	4.08
0410	Acrylic					11.45			11.45	12.60
0500	Shelf bracket, metal, 8"					1.88			1.88	2.07
0510	10"					2.03			2.03	2.23
0520	12"					2.25			2.25	2.48
0530	14"					2.47			2.47	2.72
0540	16"					2.81			2.81	3.09
0550	Acrylic, 8"					3.75			3.75	4.13
0560	10"					4.03			4.03	4.43
0570	12"					4.58			4.58	5.05
0580	14"					5.90			5.90	6.50
0600	Shelf, acrylic, 12" x 16" x 1/4"					22.50			22.50	24.50
0610	12" x 24" x 1/4"				↓	25			25	27.50

09 81 Acoustic Insulation

09 81 13 – Acoustic Board Insulation

09 81 13.10 Acoustic Board Insulation		Crew	Daily Output	Labor-Hours	Unit	Material	Labor	Equipment	Total	Total Incl O&P
0010	**ACOUSTIC BOARD INSULATION**									
0020	Cellulose fiber board, 1/2" thk.	1 Carp	800	.010	S.F.	.89	.39		1.28	1.62

09 81 16 – Acoustic Blanket Insulation

09 81 16.10 Sound Attenuation Blanket		Crew	Daily Output	Labor-Hours	Unit	Material	Labor	Equipment	Total	Total Incl O&P
0010	**SOUND ATTENUATION BLANKET**									
0020	Blanket, 1" thick	1 Carp	925	.009	S.F.	.26	.34		.60	.84
0500	1-1/2" thick		920	.009		.33	.34		.67	.92
1000	2" thick		915	.009		.42	.34		.76	1.02
1500	3" thick	↓	910	.009	↓	.62	.34		.96	1.24
2000	Wall hung, STC 18-21, 1" thick, 4' x 20'	2 Carp	22	.727	Ea.	545	28		573	645
2010	10' x 20'	"	19	.842		1,375	32.50		1,407.50	1,550
2020	Wall hung, STC 27-28, 3" thick, 4' x 20'	3 Carp	12	2		575	77.50		652.50	765
2030	10' x 20'	"	9	2.667	↓	1,450	103		1,553	1,750

09 84 Acoustic Room Components

09 84 13 – Fixed Sound-Absorptive Panels

09 84 13.10 Fixed Panels		Crew	Daily Output	Labor-Hours	Unit	Material	Labor	Equipment	Total	Total Incl O&P
0010	**FIXED PANELS** Perforated steel facing, painted with									
0100	Fiberglass or mineral filler, no backs, 2-1/4" thick, modular									
0200	space units, ceiling or wall hung, white or colored	1 Carp	100	.080	S.F.	9.40	3.10		12.50	15.45
0300	Fiberboard sound deadening panels, 1/2" thick	"	600	.013	"	.34	.52		.86	1.22
0500	Fiberglass panels, 4' x 8' x 1" thick, with									
0600	glass cloth face for walls, cemented	1 Carp	155	.052	S.F.	9.20	2		11.20	13.40
0700	1-1/2" thick, dacron covered, inner aluminum frame,									
0710	wall mounted	1 Carp	300	.027	S.F.	9.05	1.03		10.08	11.65

09 91 Painting

09 91 03 – Paint Restoration

09 91 03.20 Sanding		Crew	Daily Output	Labor-Hours	Unit	Material	2019 Bare Costs Labor	Equipment	Total	Total Incl O&P
0010	**SANDING** and puttying interior trim, compared to									
0100	Painting 1 coat, on quality work				L.F.		100%			
0300	Medium work						50%			
0400	Industrial grade						25%			
0500	Surface protection, placement and removal									
0510	Surface protection, placement and removal, basic drop cloths	1 Pord	6400	.001	S.F.		.04		.04	.07
0520	Masking with paper		800	.010		.07	.33		.40	.62
0530	Volume cover up (using plastic sheathing or building paper)		16000	.001			.02		.02	.03

09 91 03.30 Exterior Surface Preparation

Line	Description	Crew	Daily Output	Labor-Hours	Unit	Material	Labor	Equipment	Total	Total Incl O&P
0010	**EXTERIOR SURFACE PREPARATION**									
0015	Doors, per side, not incl. frames or trim									
0020	Scrape & sand									
0030	Wood, flush	1 Pord	616	.013	S.F.		.43		.43	.70
0040	Wood, detail		496	.016			.53		.53	.87
0050	Wood, louvered		280	.029			.94		.94	1.54
0060	Wood, overhead		616	.013			.43		.43	.70
0070	Wire brush									
0080	Metal, flush	1 Pord	640	.013	S.F.		.41		.41	.67
0090	Metal, detail		520	.015			.51		.51	.83
0100	Metal, louvered		360	.022			.73		.73	1.20
0110	Metal or fibr., overhead		640	.013			.41		.41	.67
0120	Metal, roll up		560	.014			.47		.47	.77
0130	Metal, bulkhead		640	.013			.41		.41	.67
0140	Power wash, based on 2500 lb. operating pressure									
0150	Metal, flush	A-1H	2240	.004	S.F.		.11	.03	.14	.22
0160	Metal, detail		2120	.004			.11	.04	.15	.23
0170	Metal, louvered		2000	.004			.12	.04	.16	.24
0180	Metal or fibr., overhead		2400	.003			.10	.03	.13	.20
0190	Metal, roll up		2400	.003			.10	.03	.13	.20
0200	Metal, bulkhead		2200	.004			.11	.03	.14	.22
0400	Windows, per side, not incl. trim									
0410	Scrape & sand									
0420	Wood, 1-2 lite	1 Pord	320	.025	S.F.		.82		.82	1.35
0430	Wood, 3-6 lite		280	.029			.94		.94	1.54
0440	Wood, 7-10 lite		240	.033			1.09		1.09	1.79
0450	Wood, 12 lite		200	.040			1.31		1.31	2.15
0460	Wood, Bay/Bow		320	.025			.82		.82	1.35
0470	Wire brush									
0480	Metal, 1-2 lite	1 Pord	480	.017	S.F.		.55		.55	.90
0490	Metal, 3-6 lite		400	.020			.66		.66	1.08
0500	Metal, Bay/Bow		480	.017			.55		.55	.90
0510	Power wash, based on 2500 lb. operating pressure									
0520	1-2 lite	A-1H	4400	.002	S.F.		.06	.02	.08	.11
0530	3-6 lite		4320	.002			.06	.02	.08	.11
0540	7-10 lite		4240	.002			.06	.02	.08	.11
0550	12 lite		4160	.002			.06	.02	.08	.12
0560	Bay/Bow		4400	.002			.06	.02	.08	.11
0600	Siding, scrape and sand, light=10-30%, med.=30-70%									
0610	Heavy=70-100% of surface to sand									
0650	Texture 1-11, light	1 Pord	480	.017	S.F.		.55		.55	.90
0660	Med.		440	.018			.60		.60	.98
0670	Heavy		360	.022			.73		.73	1.20

09 91 03.30 Exterior Surface Preparation		Crew	Daily Output	Labor-Hours	Unit	Material	2019 Bare Costs Labor	Equipment	Total	Total Incl O&P
0680	Wood shingles, shakes, light	1 Pord	440	.018	S.F.		.60		.60	.98
0690	Med.		360	.022			.73		.73	1.20
0700	Heavy		280	.029			.94		.94	1.54
0710	Clapboard, light		520	.015			.51		.51	.83
0720	Med.		480	.017			.55		.55	.90
0730	Heavy	↓	400	.020	↓		.66		.66	1.08
0740	Wire brush									
0750	Aluminum, light	1 Pord	600	.013	S.F.		.44		.44	.72
0760	Med.		520	.015			.51		.51	.83
0770	Heavy	↓	440	.018	↓		.60		.60	.98
0780	Pressure wash, based on 2500 lb. operating pressure									
0790	Stucco	A-1H	3080	.003	S.F.		.08	.02	.10	.16
0800	Aluminum or vinyl		3200	.003			.08	.02	.10	.16
0810	Siding, masonry, brick & block	↓	2400	.003	↓		.10	.03	.13	.20
1300	Miscellaneous, wire brush									
1310	Metal, pedestrian gate	1 Pord	100	.080	S.F.		2.63		2.63	4.30
1320	Aluminum chain link, both sides		250	.032			1.05		1.05	1.72
1400	Existing galvanized surface, clean and prime, prep for painting	↓	380	.021	↓	.13	.69		.82	1.27
8000	For chemical washing, see Section 04 01 30									

09 91 03.40 Interior Surface Preparation		Crew	Daily Output	Labor-Hours	Unit	Material	Labor	Equipment	Total	Total Incl O&P
0010	**INTERIOR SURFACE PREPARATION**									
0020	Doors, per side, not incl. frames or trim									
0030	Scrape & sand									
0040	Wood, flush	1 Pord	616	.013	S.F.		.43		.43	.70
0050	Wood, detail		496	.016			.53		.53	.87
0060	Wood, louvered	↓	280	.029	↓		.94		.94	1.54
0070	Wire brush									
0080	Metal, flush	1 Pord	640	.013	S.F.		.41		.41	.67
0090	Metal, detail		520	.015			.51		.51	.83
0100	Metal, louvered	↓	360	.022	↓		.73		.73	1.20
0110	Hand wash									
0120	Wood, flush	1 Pord	2160	.004	S.F.		.12		.12	.20
0130	Wood, detail		2000	.004			.13		.13	.22
0140	Wood, louvered		1360	.006			.19		.19	.32
0150	Metal, flush		2160	.004			.12		.12	.20
0160	Metal, detail		2000	.004			.13		.13	.22
0170	Metal, louvered	↓	1360	.006	↓		.19		.19	.32
0400	Windows, per side, not incl. trim									
0410	Scrape & sand									
0420	Wood, 1-2 lite	1 Pord	360	.022	S.F.		.73		.73	1.20
0430	Wood, 3-6 lite		320	.025			.82		.82	1.35
0440	Wood, 7-10 lite		280	.029			.94		.94	1.54
0450	Wood, 12 lite		240	.033			1.09		1.09	1.79
0460	Wood, Bay/Bow	↓	360	.022	↓		.73		.73	1.20
0470	Wire brush									
0480	Metal, 1-2 lite	1 Pord	520	.015	S.F.		.51		.51	.83
0490	Metal, 3-6 lite		440	.018			.60		.60	.98
0500	Metal, Bay/Bow	↓	520	.015	↓		.51		.51	.83
0600	Walls, sanding, light=10-30%, medium=30-70%,									
0610	heavy=70-100% of surface to sand									
0650	Walls, sand									
0660	Gypsum board or plaster, light	1 Pord	3077	.003	S.F.		.09		.09	.14

09 91 03 – Paint Restoration

09 91 03.40 Interior Surface Preparation		Crew	Daily Output	Labor-Hours	Unit	Material	Labor	Equipment	Total	Total Incl O&P
						2019 Bare Costs				
0670	Gypsum board or plaster, medium	1 Pord	2160	.004	S.F.		.12		.12	.20
0680	Gypsum board or plaster, heavy		923	.009			.28		.28	.47
0690	Wood, T&G, light		2400	.003			.11		.11	.18
0700	Wood, T&G, medium		1600	.005			.16		.16	.27
0710	Wood, T&G, heavy		800	.010			.33		.33	.54
0720	Walls, wash									
0730	Gypsum board or plaster	1 Pord	3200	.003	S.F.		.08		.08	.13
0740	Wood, T&G		3200	.003			.08		.08	.13
0750	Masonry, brick & block, smooth		2800	.003			.09		.09	.15
0760	Masonry, brick & block, coarse		2000	.004			.13		.13	.22
8000	For chemical washing, see Section 04 01 30									
8020	For sand blasting, see Sections 03 35 29.60 and 05 01 10.51									

09 91 03.41 Scrape After Fire Damage		Crew	Daily Output	Labor-Hours	Unit	Material	Labor	Equipment	Total	Total Incl O&P
0010	**SCRAPE AFTER FIRE DAMAGE**									
0050	Boards, 1" x 4"	1 Pord	336	.024	L.F.		.78		.78	1.28
0060	1" x 6"		260	.031			1.01		1.01	1.66
0070	1" x 8"		207	.039			1.27		1.27	2.08
0080	1" x 10"		174	.046			1.51		1.51	2.47
0500	Framing, 2" x 4"		265	.030			.99		.99	1.62
0510	2" x 6"		221	.036			1.19		1.19	1.95
0520	2" x 8"		190	.042			1.38		1.38	2.27
0530	2" x 10"		165	.048			1.59		1.59	2.61
0540	2" x 12"		144	.056			1.83		1.83	2.99
1000	Heavy framing, 3" x 4"		226	.035			1.16		1.16	1.90
1010	4" x 4"		210	.038			1.25		1.25	2.05
1020	4" x 6"		191	.042			1.38		1.38	2.25
1030	4" x 8"		165	.048			1.59		1.59	2.61
1040	4" x 10"		144	.056			1.83		1.83	2.99
1060	4" x 12"		131	.061			2.01		2.01	3.29
2900	For sealing, light damage		825	.010	S.F.	.15	.32		.47	.69
2920	Heavy damage		460	.017	"	.33	.57		.90	1.30

09 91 13 – Exterior Painting

09 91 13.30 Fences		Crew	Daily Output	Labor-Hours	Unit	Material	Labor	Equipment	Total	Total Incl O&P
0010	**FENCES** R099100-20									
0100	Chain link or wire metal, one side, water base									
0110	Roll & brush, first coat	1 Pord	960	.008	S.F.	.07	.27		.34	.53
0120	Second coat		1280	.006		.07	.21		.28	.41
0130	Spray, first coat		2275	.004		.07	.12		.19	.27
0140	Second coat		2600	.003		.07	.10		.17	.25
0150	Picket, water base									
0160	Roll & brush, first coat	1 Pord	865	.009	S.F.	.08	.30		.38	.58
0170	Second coat		1050	.008		.08	.25		.33	.49
0180	Spray, first coat		2275	.004		.08	.12		.20	.27
0190	Second coat		2600	.003		.08	.10		.18	.25
0200	Stockade, water base									
0210	Roll & brush, first coat	1 Pord	1040	.008	S.F.	.08	.25		.33	.49
0220	Second coat		1200	.007		.08	.22		.30	.44
0230	Spray, first coat		2275	.004		.08	.12		.20	.27
0240	Second coat		2600	.003		.08	.10		.18	.25

09 91 Painting

09 91 13 – Exterior Painting

09 91 13.42 Miscellaneous, Exterior

	09 91 13.42 Miscellaneous, Exterior	Crew	Daily Output	Labor-Hours	Unit	Material	2019 Bare Costs Labor	Equipment	Total	Total Incl O&P
0010	**MISCELLANEOUS, EXTERIOR**									
0015	For painting metals, see Section 09 97 13.23									
0100	Railing, ext., decorative wood, incl. cap & baluster									
0110	Newels & spindles @ 12" OC									
0120	Brushwork, stain, sand, seal & varnish									
0130	First coat	1 Pord	90	.089	L.F.	.90	2.92		3.82	5.75
0140	Second coat	"	120	.067	"	.90	2.19		3.09	4.58
0150	Rough sawn wood, 42" high, 2" x 2" verticals, 6" OC									
0160	Brushwork, stain, each coat	1 Pord	90	.089	L.F.	.31	2.92		3.23	5.10
0170	Wrought iron, 1" rail, 1/2" sq. verticals									
0180	Brushwork, zinc chromate, 60" high, bars 6" OC									
0190	Primer	1 Pord	130	.062	L.F.	.86	2.02		2.88	4.25
0200	Finish coat		130	.062		1.14	2.02		3.16	4.57
0210	Additional coat		190	.042		1.33	1.38		2.71	3.73
0220	Shutters or blinds, single panel, 2' x 4', paint all sides									
0230	Brushwork, primer	1 Pord	20	.400	Ea.	.72	13.15		13.87	22.50
0240	Finish coat, exterior latex		20	.400		.58	13.15		13.73	22
0250	Primer & 1 coat, exterior latex		13	.615		1.13	20		21.13	34.50
0260	Spray, primer		35	.229		1.04	7.50		8.54	13.45
0270	Finish coat, exterior latex		35	.229		1.23	7.50		8.73	13.65
0280	Primer & 1 coat, exterior latex		20	.400		1.12	13.15		14.27	22.50
0290	For louvered shutters, add				S.F.	10%				
0300	Stair stringers, exterior, metal									
0310	Roll & brush, zinc chromate, to 14", each coat	1 Pord	320	.025	L.F.	.38	.82		1.20	1.77
0320	Rough sawn wood, 4" x 12"									
0330	Roll & brush, exterior latex, each coat	1 Pord	215	.037	L.F.	.09	1.22		1.31	2.09
0340	Trellis/lattice, 2" x 2" @ 3" OC with 2" x 8" supports									
0350	Spray, latex, per side, each coat	1 Pord	475	.017	S.F.	.09	.55		.64	1
0450	Decking, ext., sealer, alkyd, brushwork, sealer coat		1140	.007		.10	.23		.33	.49
0460	1st coat		1140	.007		.12	.23		.35	.51
0470	2nd coat		1300	.006		.09	.20		.29	.43
0500	Paint, alkyd, brushwork, primer coat		1140	.007		.11	.23		.34	.50
0510	1st coat		1140	.007		.14	.23		.37	.54
0520	2nd coat		1300	.006		.10	.20		.30	.44
0600	Sand paint, alkyd, brushwork, 1 coat		150	.053		.14	1.75		1.89	3.02

09 91 13.60 Siding Exterior

	09 91 13.60 Siding Exterior	Crew	Daily Output	Labor-Hours	Unit	Material	Labor	Equipment	Total	Total Incl O&P
0010	**SIDING EXTERIOR**, Alkyd (oil base)									
0450	Steel siding, oil base, paint 1 coat, brushwork	2 Pord	2015	.008	S.F.	.11	.26		.37	.55
0500	Spray		4550	.004		.17	.12		.29	.38
0800	Paint 2 coats, brushwork		1300	.012		.23	.40		.63	.91
1000	Spray		2750	.006		.15	.19		.34	.48
1200	Stucco, rough, oil base, paint 2 coats, brushwork		1300	.012		.23	.40		.63	.91
1400	Roller		1625	.010		.24	.32		.56	.79
1600	Spray		2925	.005		.25	.18		.43	.57
1800	Texture 1-11 or clapboard, oil base, primer coat, brushwork		1300	.012		.14	.40		.54	.82
2000	Spray		4550	.004		.14	.12		.26	.35
2100	Paint 1 coat, brushwork		1300	.012		.17	.40		.57	.84
2200	Spray		4550	.004		.17	.12		.29	.37
2400	Paint 2 coats, brushwork		810	.020		.33	.65		.98	1.42
2600	Spray		2600	.006		.37	.20		.57	.73
3000	Stain 1 coat, brushwork		1520	.011		.10	.35		.45	.68
3200	Spray		5320	.003		.12	.10		.22	.29

09 91 13 – Exterior Painting

09 91 13.60 Siding Exterior		Crew	Daily Output	Labor-Hours	Unit	Material	2019 Bare Costs Labor	Equipment	Total	Total Incl O&P
3400	Stain 2 coats, brushwork	2 Pord	950	.017	S.F.	.21	.55		.76	1.14
4000	Spray		3050	.005		.23	.17		.40	.53
4200	Wood shingles, oil base primer coat, brushwork		1300	.012		.13	.40		.53	.81
4400	Spray		3900	.004		.12	.13		.25	.36
4600	Paint 1 coat, brushwork		1300	.012		.14	.40		.54	.81
4800	Spray		3900	.004		.17	.13		.30	.41
5000	Paint 2 coats, brushwork		810	.020		.28	.65		.93	1.36
5200	Spray		2275	.007		.26	.23		.49	.67
5800	Stain 1 coat, brushwork		1500	.011		.10	.35		.45	.68
6000	Spray		3900	.004		.10	.13		.23	.33
6500	Stain 2 coats, brushwork		950	.017		.21	.55		.76	1.14
7000	Spray		2660	.006		.29	.20		.49	.64
8000	For latex paint, deduct					10%				
8100	For work over 12′ H, from pipe scaffolding, add						15%			
8200	For work over 12′ H, from extension ladder, add						25%			
8300	For work over 12′ H, from swing staging, add						35%			

09 91 13.62 Siding, Misc.		Crew	Daily Output	Labor-Hours	Unit	Material	Labor	Equipment	Total	Total Incl O&P
0010	**SIDING, MISC.**, latex paint									
0100	Aluminum siding									
0110	Brushwork, primer	2 Pord	2275	.007	S.F.	.07	.23		.30	.45
0120	Finish coat, exterior latex		2275	.007		.06	.23		.29	.44
0130	Primer & 1 coat exterior latex		1300	.012		.13	.40		.53	.81
0140	Primer & 2 coats exterior latex		975	.016		.19	.54		.73	1.09
0150	Mineral fiber shingles									
0160	Brushwork, primer	2 Pord	1495	.011	S.F.	.14	.35		.49	.74
0170	Finish coat, industrial enamel		1495	.011		.19	.35		.54	.79
0180	Primer & 1 coat enamel		810	.020		.33	.65		.98	1.43
0190	Primer & 2 coats enamel		540	.030		.52	.97		1.49	2.17
0200	Roll, primer		1625	.010		.16	.32		.48	.71
0210	Finish coat, industrial enamel		1625	.010		.21	.32		.53	.76
0220	Primer & 1 coat enamel		975	.016		.37	.54		.91	1.28
0230	Primer & 2 coats enamel		650	.025		.57	.81		1.38	1.95
0240	Spray, primer		3900	.004		.12	.13		.25	.36
0250	Finish coat, industrial enamel		3900	.004		.17	.13		.30	.41
0260	Primer & 1 coat enamel		2275	.007		.30	.23		.53	.70
0270	Primer & 2 coats enamel		1625	.010		.47	.32		.79	1.04
0280	Waterproof sealer, first coat		4485	.004		.13	.12		.25	.33
0290	Second coat		5235	.003		.12	.10		.22	.29
0300	Rough wood incl. shingles, shakes or rough sawn siding									
0310	Brushwork, primer	2 Pord	1280	.013	S.F.	.14	.41		.55	.83
0320	Finish coat, exterior latex		1280	.013		.10	.41		.51	.78
0330	Primer & 1 coat exterior latex		960	.017		.25	.55		.80	1.17
0340	Primer & 2 coats exterior latex		700	.023		.35	.75		1.10	1.61
0350	Roll, primer		2925	.005		.19	.18		.37	.50
0360	Finish coat, exterior latex		2925	.005		.12	.18		.30	.43
0370	Primer & 1 coat exterior latex		1790	.009		.32	.29		.61	.83
0380	Primer & 2 coats exterior latex		1300	.012		.44	.40		.84	1.14
0390	Spray, primer		3900	.004		.16	.13		.29	.40
0400	Finish coat, exterior latex		3900	.004		.10	.13		.23	.32
0410	Primer & 1 coat exterior latex		2600	.006		.26	.20		.46	.61
0420	Primer & 2 coats exterior latex		2080	.008		.35	.25		.60	.80
0430	Waterproof sealer, first coat		4485	.004		.23	.12		.35	.44

09 91 13.62 Siding, Misc.		Crew	Daily Output	Labor-Hours	Unit	2019 Bare Costs Material	Labor	Equipment	Total	Total Incl O&P
0440	Second coat	2 Pord	4485	.004	S.F.	.13	.12		.25	.33
0450	Smooth wood incl. butt, T&G, beveled, drop or B&B siding									
0460	Brushwork, primer	2 Pord	2325	.007	S.F.	.10	.23		.33	.48
0470	Finish coat, exterior latex		1280	.013		.10	.41		.51	.78
0480	Primer & 1 coat exterior latex		800	.020		.21	.66		.87	1.31
0490	Primer & 2 coats exterior latex		630	.025		.31	.83		1.14	1.71
0500	Roll, primer		2275	.007		.12	.23		.35	.51
0510	Finish coat, exterior latex		2275	.007		.11	.23		.34	.50
0520	Primer & 1 coat exterior latex		1300	.012		.23	.40		.63	.91
0530	Primer & 2 coats exterior latex		975	.016		.34	.54		.88	1.25
0540	Spray, primer		4550	.004		.09	.12		.21	.29
0550	Finish coat, exterior latex		4550	.004		.10	.12		.22	.29
0560	Primer & 1 coat exterior latex		2600	.006		.18	.20		.38	.53
0570	Primer & 2 coats exterior latex		1950	.008		.28	.27		.55	.75
0580	Waterproof sealer, first coat		5230	.003		.13	.10		.23	.30
0590	Second coat		5980	.003		.13	.09		.22	.28
0600	For oil base paint, add					10%				

09 91 13.70 Doors and Windows, Exterior		Crew	Daily Output	Labor-Hours	Unit	Material	Labor	Equipment	Total	Total Incl O&P
0010	**DOORS AND WINDOWS, EXTERIOR** R099100-10									
0100	Door frames & trim, only									
0110	Brushwork, primer	1 Pord	512	.016	L.F.	.07	.51		.58	.91
0120	Finish coat, exterior latex		512	.016		.07	.51		.58	.92
0130	Primer & 1 coat, exterior latex		300	.027		.14	.88		1.02	1.58
0135	2 coats, exterior latex, both sides		15	.533	Ea.	6.45	17.50		23.95	35.50
0140	Primer & 2 coats, exterior latex		265	.030	L.F.	.21	.99		1.20	1.85
0150	Doors, flush, both sides, incl. frame & trim									
0160	Roll & brush, primer	1 Pord	10	.800	Ea.	4.94	26.50		31.44	48.50
0170	Finish coat, exterior latex		10	.800		5.50	26.50		32	49
0180	Primer & 1 coat, exterior latex		7	1.143		10.45	37.50		47.95	73
0190	Primer & 2 coats, exterior latex		5	1.600		15.90	52.50		68.40	104
0200	Brushwork, stain, sealer & 2 coats polyurethane		4	2		31	65.50		96.50	143
0210	Doors, French, both sides, 10-15 lite, incl. frame & trim									
0220	Brushwork, primer	1 Pord	6	1.333	Ea.	2.47	44		46.47	74
0230	Finish coat, exterior latex		6	1.333		2.74	44		46.74	74.50
0240	Primer & 1 coat, exterior latex		3	2.667		5.20	87.50		92.70	149
0250	Primer & 2 coats, exterior latex		2	4		7.80	131		138.80	224
0260	Brushwork, stain, sealer & 2 coats polyurethane		2.50	3.200		11.40	105		116.40	185
0270	Doors, louvered, both sides, incl. frame & trim									
0280	Brushwork, primer	1 Pord	7	1.143	Ea.	4.94	37.50		42.44	67
0290	Finish coat, exterior latex		7	1.143		5.50	37.50		43	67.50
0300	Primer & 1 coat, exterior latex		4	2		10.45	65.50		75.95	119
0310	Primer & 2 coats, exterior latex		3	2.667		15.60	87.50		103.10	160
0320	Brushwork, stain, sealer & 2 coats polyurethane		4.50	1.778		31	58.50		89.50	130
0330	Doors, panel, both sides, incl. frame & trim									
0340	Roll & brush, primer	1 Pord	6	1.333	Ea.	4.94	44		48.94	77
0350	Finish coat, exterior latex		6	1.333		5.50	44		49.50	77.50
0360	Primer & 1 coat, exterior latex		3	2.667		10.45	87.50		97.95	154
0370	Primer & 2 coats, exterior latex		2.50	3.200		15.60	105		120.60	189
0380	Brushwork, stain, sealer & 2 coats polyurethane		3	2.667		31	87.50		118.50	178
0400	Windows, per ext. side, based on 15 S.F.									
0410	1 to 6 lite									
0420	Brushwork, primer	1 Pord	13	.615	Ea.	.98	20		20.98	34

09 91 13.70 Doors and Windows, Exterior		Crew	Daily Output	Labor-Hours	Unit	Material	2019 Bare Costs Labor	Equipment	Total	Total Incl O&P
0430	Finish coat, exterior latex	1 Pord	13	.615	Ea.	1.08	20		21.08	34
0440	Primer & 1 coat, exterior latex		8	1		2.06	33		35.06	56.50
0450	Primer & 2 coats, exterior latex		6	1.333		3.08	44		47.08	75
0460	Stain, sealer & 1 coat varnish		7	1.143		4.49	37.50		41.99	66.50
0470	7 to 10 lite									
0480	Brushwork, primer	1 Pord	11	.727	Ea.	.98	24		24.98	40
0490	Finish coat, exterior latex		11	.727		1.08	24		25.08	40
0500	Primer & 1 coat, exterior latex		7	1.143		2.06	37.50		39.56	64
0510	Primer & 2 coats, exterior latex		5	1.600		3.08	52.50		55.58	89.50
0520	Stain, sealer & 1 coat varnish		6	1.333		4.49	44		48.49	76.50
0530	12 lite									
0540	Brushwork, primer	1 Pord	10	.800	Ea.	.98	26.50		27.48	44
0550	Finish coat, exterior latex		10	.800		1.08	26.50		27.58	44
0560	Primer & 1 coat, exterior latex		6	1.333		2.06	44		46.06	74
0570	Primer & 2 coats, exterior latex		5	1.600		3.08	52.50		55.58	89.50
0580	Stain, sealer & 1 coat varnish		6	1.333		4.38	44		48.38	76.50
0590	For oil base paint, add					10%				

09 91 13.80 Trim, Exterior		Crew	Daily Output	Labor-Hours	Unit	Material	Labor	Equipment	Total	Total Incl O&P
0010	**TRIM, EXTERIOR**									
0100	Door frames & trim (see Doors, interior or exterior)									
0110	Fascia, latex paint, one coat coverage									
0120	1" x 4", brushwork	1 Pord	640	.013	L.F.	.02	.41		.43	.69
0130	Roll		1280	.006		.02	.21		.23	.37
0140	Spray		2080	.004		.02	.13		.15	.23
0150	1" x 6" to 1" x 10", brushwork		640	.013		.08	.41		.49	.75
0160	Roll		1230	.007		.08	.21		.29	.44
0170	Spray		2100	.004		.06	.13		.19	.28
0180	1" x 12", brushwork		640	.013		.08	.41		.49	.75
0190	Roll		1050	.008		.08	.25		.33	.50
0200	Spray		2200	.004		.06	.12		.18	.27
0210	Gutters & downspouts, metal, zinc chromate paint									
0220	Brushwork, gutters, 5", first coat	1 Pord	640	.013	L.F.	.40	.41		.81	1.11
0230	Second coat		960	.008		.38	.27		.65	.87
0240	Third coat		1280	.006		.31	.21		.52	.68
0250	Downspouts, 4", first coat		640	.013		.40	.41		.81	1.11
0260	Second coat		960	.008		.38	.27		.65	.87
0270	Third coat		1280	.006		.31	.21		.52	.68
0280	Gutters & downspouts, wood									
0290	Brushwork, gutters, 5", primer	1 Pord	640	.013	L.F.	.07	.41		.48	.74
0300	Finish coat, exterior latex		640	.013		.06	.41		.47	.74
0310	Primer & 1 coat exterior latex		400	.020		.14	.66		.80	1.23
0320	Primer & 2 coats exterior latex		325	.025		.21	.81		1.02	1.55
0330	Downspouts, 4", primer		640	.013		.07	.41		.48	.74
0340	Finish coat, exterior latex		640	.013		.06	.41		.47	.74
0350	Primer & 1 coat exterior latex		400	.020		.14	.66		.80	1.23
0360	Primer & 2 coats exterior latex		325	.025		.10	.81		.91	1.44
0370	Molding, exterior, up to 14" wide									
0380	Brushwork, primer	1 Pord	640	.013	L.F.	.08	.41		.49	.76
0390	Finish coat, exterior latex		640	.013		.08	.41		.49	.76
0400	Primer & 1 coat exterior latex		400	.020		.16	.66		.82	1.26
0410	Primer & 2 coats exterior latex		315	.025		.16	.83		.99	1.55
0420	Stain & fill		1050	.008		.13	.25		.38	.55

09 91 Painting

09 91 13 – Exterior Painting

09 91 13.80 Trim, Exterior		Crew	Daily Output	Labor-Hours	Unit	Material	Labor	Equipment	Total	Total Incl O&P
						2019 Bare Costs				
0430	Shellac	1 Pord	1850	.004	L.F.	.15	.14		.29	.40
0440	Varnish		1275	.006		.11	.21		.32	.46

09 91 13.90 Walls, Masonry (CMU), Exterior

Code	Description	Crew	Daily Output	Labor-Hours	Unit	Material	Labor	Equipment	Total	Total Incl O&P
0010	**WALLS, MASONRY (CMU), EXTERIOR**									
0360	Concrete masonry units (CMU), smooth surface									
0370	Brushwork, latex, first coat	1 Pord	640	.013	S.F.	.06	.41		.47	.74
0380	Second coat		960	.008		.05	.27		.32	.50
0390	Waterproof sealer, first coat		736	.011		.28	.36		.64	.89
0400	Second coat		1104	.007		.28	.24		.52	.70
0410	Roll, latex, paint, first coat		1465	.005		.07	.18		.25	.37
0420	Second coat		1790	.004		.06	.15		.21	.30
0430	Waterproof sealer, first coat		1680	.005		.28	.16		.44	.57
0440	Second coat		2060	.004		.28	.13		.41	.52
0450	Spray, latex, paint, first coat		1950	.004		.06	.13		.19	.28
0460	Second coat		2600	.003		.05	.10		.15	.22
0470	Waterproof sealer, first coat		2245	.004		.28	.12		.40	.50
0480	Second coat		2990	.003		.28	.09		.37	.45
0490	Concrete masonry unit (CMU), porous									
0500	Brushwork, latex, first coat	1 Pord	640	.013	S.F.	.12	.41		.53	.80
0510	Second coat		960	.008		.06	.27		.33	.52
0520	Waterproof sealer, first coat		736	.011		.28	.36		.64	.89
0530	Second coat		1104	.007		.28	.24		.52	.70
0540	Roll latex, first coat		1465	.005		.09	.18		.27	.39
0550	Second coat		1790	.004		.06	.15		.21	.30
0560	Waterproof sealer, first coat		1680	.005		.28	.16		.44	.57
0570	Second coat		2060	.004		.28	.13		.41	.52
0580	Spray latex, first coat		1950	.004		.07	.13		.20	.29
0590	Second coat		2600	.003		.05	.10		.15	.22
0600	Waterproof sealer, first coat		2245	.004		.28	.12		.40	.50
0610	Second coat		2990	.003		.28	.09		.37	.45

09 91 23 – Interior Painting

09 91 23.20 Cabinets and Casework

Code	Description	Crew	Daily Output	Labor-Hours	Unit	Material	Labor	Equipment	Total	Total Incl O&P
0010	**CABINETS AND CASEWORK**									
1000	Primer coat, oil base, brushwork	1 Pord	650	.012	S.F.	.07	.40		.47	.73
2000	Paint, oil base, brushwork, 1 coat		650	.012		.12	.40		.52	.79
2500	2 coats		400	.020		.23	.66		.89	1.33
3000	Stain, brushwork, wipe off		650	.012		.10	.40		.50	.77
4000	Shellac, 1 coat, brushwork		650	.012		.13	.40		.53	.80
4500	Varnish, 3 coats, brushwork, sand after 1st coat		325	.025		.28	.81		1.09	1.62
5000	For latex paint, deduct					10%				

09 91 23.33 Doors and Windows, Interior Alkyd (Oil Base)

Code	Description	Crew	Daily Output	Labor-Hours	Unit	Material	Labor	Equipment	Total	Total Incl O&P
0010	**DOORS AND WINDOWS, INTERIOR ALKYD (OIL BASE)**									
0500	Flush door & frame, 3′ x 7′, oil, primer, brushwork	1 Pord	10	.800	Ea.	4.04	26.50		30.54	47.50
1000	Paint, 1 coat		10	.800		4.24	26.50		30.74	47.50
1200	2 coats		6	1.333		4.66	44		48.66	76.50
1400	Stain, brushwork, wipe off		18	.444		2.19	14.60		16.79	26.50
1600	Shellac, 1 coat, brushwork		25	.320		2.67	10.50		13.17	20
1800	Varnish, 3 coats, brushwork, sand after 1st coat		9	.889		5.80	29		34.80	54.50
2000	Panel door & frame, 3′ x 7′, oil, primer, brushwork		6	1.333		2.47	44		46.47	74
2200	Paint, 1 coat		6	1.333		4.24	44		48.24	76
2400	2 coats		3	2.667		10.95	87.50		98.45	155
2600	Stain, brushwork, panel door, 3′ x 7′, not incl. frame		16	.500		2.19	16.45		18.64	29.50

09 91 Painting

09 91 23 – Interior Painting

09 91 23.33 Doors and Windows, Interior Alkyd (Oil Base)		Crew	Daily Output	Labor-Hours	Unit	2019 Bare Costs Material	Labor	Equipment	Total	Total Incl O&P
2800	Shellac, 1 coat, brushwork	1 Pord	22	.364	Ea.	2.67	11.95		14.62	22.50
3000	Varnish, 3 coats, brushwork, sand after 1st coat		7.50	1.067		5.80	35		40.80	64
3020	French door, incl. 3′ x 7′, 6 lites, frame & trim									
3022	Paint, 1 coat, over existing paint	1 Pord	5	1.600	Ea.	8.50	52.50		61	95.50
3024	2 coats, over existing paint		5	1.600		16.45	52.50		68.95	104
3026	Primer & 1 coat		3.50	2.286		13.45	75		88.45	138
3028	Primer & 2 coats		3	2.667		22	87.50		109.50	167
3032	Varnish or polyurethane, 1 coat		5	1.600		8.40	52.50		60.90	95.50
3034	2 coats, sanding between		3	2.667		16.80	87.50		104.30	162
4400	Windows, including frame and trim, per side									
4600	Colonial type, 6/6 lites, 2′ x 3′, oil, primer, brushwork	1 Pord	14	.571	Ea.	.39	18.75		19.14	31
5800	Paint, 1 coat		14	.571		.67	18.75		19.42	31
6000	2 coats		9	.889		1.30	29		30.30	49.50
6200	3′ x 5′ opening, 6/6 lites, primer coat, brushwork		12	.667		.98	22		22.98	37
6400	Paint, 1 coat		12	.667		1.67	22		23.67	38
6600	2 coats		7	1.143		3.25	37.50		40.75	65
6800	4′ x 8′ opening, 6/6 lites, primer coat, brushwork		8	1		2.08	33		35.08	56.50
7000	Paint, 1 coat		8	1		3.57	33		36.57	58
7200	2 coats		5	1.600		6.95	52.50		59.45	93.50
8000	Single lite type, 2′ x 3′, oil base, primer coat, brushwork		33	.242		.39	7.95		8.34	13.50
8200	Paint, 1 coat		33	.242		.67	7.95		8.62	13.80
8400	2 coats		20	.400		1.30	13.15		14.45	23
8600	3′ x 5′ opening, primer coat, brushwork		20	.400		.98	13.15		14.13	22.50
8800	Paint, 1 coat		20	.400		1.67	13.15		14.82	23.50
8900	2 coats		13	.615		3.25	20		23.25	36.50
9200	4′ x 8′ opening, primer coat, brushwork		14	.571		2.08	18.75		20.83	33
9400	Paint, 1 coat		14	.571		3.57	18.75		22.32	34.50
9600	2 coats		8	1		6.95	33		39.95	61.50

09 91 23.35 Doors and Windows, Interior Latex		Crew	Daily Output	Labor-Hours	Unit	Material	Labor	Equipment	Total	Total Incl O&P
0010	**DOORS & WINDOWS, INTERIOR LATEX**									
0100	Doors, flush, both sides, incl. frame & trim									
0110	Roll & brush, primer	1 Pord	10	.800	Ea.	4.22	26.50		30.72	47.50
0120	Finish coat, latex		10	.800		5.60	26.50		32.10	49
0130	Primer & 1 coat latex		7	1.143		9.85	37.50		47.35	72.50
0140	Primer & 2 coats latex		5	1.600		15.10	52.50		67.60	103
0160	Spray, both sides, primer		20	.400		4.44	13.15		17.59	26.50
0170	Finish coat, latex		20	.400		5.90	13.15		19.05	28
0180	Primer & 1 coat latex		11	.727		10.40	24		34.40	50.50
0190	Primer & 2 coats latex		8	1		16	33		49	71.50
0200	Doors, French, both sides, 10-15 lite, incl. frame & trim									
0210	Roll & brush, primer	1 Pord	6	1.333	Ea.	2.11	44		46.11	74
0220	Finish coat, latex		6	1.333		2.81	44		46.81	74.50
0230	Primer & 1 coat latex		3	2.667		4.92	87.50		92.42	148
0240	Primer & 2 coats latex		2	4		7.55	131		138.55	223
0260	Doors, louvered, both sides, incl. frame & trim									
0270	Roll & brush, primer	1 Pord	7	1.143	Ea.	4.22	37.50		41.72	66
0280	Finish coat, latex		7	1.143		5.60	37.50		43.10	67.50
0290	Primer & 1 coat, latex		4	2		9.60	65.50		75.10	119
0300	Primer & 2 coats, latex		3	2.667		15.45	87.50		102.95	160
0320	Spray, both sides, primer		20	.400		4.44	13.15		17.59	26.50
0330	Finish coat, latex		20	.400		5.90	13.15		19.05	28
0340	Primer & 1 coat, latex		11	.727		10.40	24		34.40	50.50

09 91 23.35 Doors and Windows, Interior Latex		Crew	Daily Output	Labor-Hours	Unit	Material	Labor	Equipment	Total	Total Incl O&P
0350	Primer & 2 coats, latex	1 Pord	8	1	Ea.	16.35	33		49.35	72
0360	Doors, panel, both sides, incl. frame & trim									
0370	Roll & brush, primer	1 Pord	6	1.333	Ea.	4.44	44		48.44	76.50
0380	Finish coat, latex		6	1.333		5.60	44		49.60	77.50
0390	Primer & 1 coat, latex		3	2.667		9.85	87.50		97.35	154
0400	Primer & 2 coats, latex		2.50	3.200		15.45	105		120.45	189
0420	Spray, both sides, primer		10	.800		4.44	26.50		30.94	48
0430	Finish coat, latex		10	.800		5.90	26.50		32.40	49.50
0440	Primer & 1 coat, latex		5	1.600		10.40	52.50		62.90	97.50
0450	Primer & 2 coats, latex		4	2		16.35	65.50		81.85	126
0460	Windows, per interior side, based on 15 S.F.									
0470	1 to 6 lite									
0480	Brushwork, primer	1 Pord	13	.615	Ea.	.83	20		20.83	34
0490	Finish coat, enamel		13	.615		1.11	20		21.11	34
0500	Primer & 1 coat enamel		8	1		1.94	33		34.94	56
0510	Primer & 2 coats enamel		6	1.333		3.05	44		47.05	75
0530	7 to 10 lite									
0540	Brushwork, primer	1 Pord	11	.727	Ea.	.83	24		24.83	40
0550	Finish coat, enamel		11	.727		1.11	24		25.11	40
0560	Primer & 1 coat enamel		7	1.143		1.94	37.50		39.44	63.50
0570	Primer & 2 coats enamel		5	1.600		3.05	52.50		55.55	89.50
0590	12 lite									
0600	Brushwork, primer	1 Pord	10	.800	Ea.	.83	26.50		27.33	44
0610	Finish coat, enamel		10	.800		1.11	26.50		27.61	44
0620	Primer & 1 coat enamel		6	1.333		1.94	44		45.94	73.50
0630	Primer & 2 coats enamel		5	1.600		3.05	52.50		55.55	89.50
0650	For oil base paint, add					10%				

09 91 23.39 Doors and Windows, Interior Latex, Zero Voc			Crew	Daily Output	Labor-Hours	Unit	Material	Labor	Equipment	Total	Total Incl O&P
0010	**DOORS & WINDOWS, INTERIOR LATEX, ZERO VOC**										
0100	Doors flush, both sides, incl. frame & trim										
0110	Roll & brush, primer	G	1 Pord	10	.800	Ea.	6.40	26.50		32.90	50
0120	Finish coat, latex	G		10	.800		6.95	26.50		33.45	50.50
0130	Primer & 1 coat latex	G		7	1.143		13.35	37.50		50.85	76
0140	Primer & 2 coats latex	G		5	1.600		19.90	52.50		72.40	108
0160	Spray, both sides, primer	G		20	.400		6.75	13.15		19.90	29
0170	Finish coat, latex	G		20	.400		7.30	13.15		20.45	29.50
0180	Primer & 1 coat latex	G		11	.727		14.15	24		38.15	54.50
0190	Primer & 2 coats latex	G		8	1		21	33		54	77
0200	Doors, French, both sides, 10-15 lite, incl. frame & trim										
0210	Roll & brush, primer	G	1 Pord	6	1.333	Ea.	3.21	44		47.21	75
0220	Finish coat, latex	G		6	1.333		3.47	44		47.47	75.50
0230	Primer & 1 coat latex	G		3	2.667		6.70	87.50		94.20	150
0240	Primer & 2 coats latex	G		2	4		9.95	131		140.95	226
0360	Doors, panel, both sides, incl. frame & trim										
0370	Roll & brush, primer	G	1 Pord	6	1.333	Ea.	6.75	44		50.75	79
0380	Finish coat, latex	G		6	1.333		6.95	44		50.95	79
0390	Primer & 1 coat, latex	G		3	2.667		13.35	87.50		100.85	158
0400	Primer & 2 coats, latex	G		2.50	3.200		20.50	105		125.50	195
0420	Spray, both sides, primer	G		10	.800		6.75	26.50		33.25	50.50
0430	Finish coat, latex	G		10	.800		7.30	26.50		33.80	51
0440	Primer & 1 coat, latex	G		5	1.600		14.15	52.50		66.65	102
0450	Primer & 2 coats, latex	G		4	2		21.50	65.50		87	132

09 91 23.39 Doors and Windows, Interior Latex, Zero Voc		Crew	Daily Output	Labor-Hours	Unit	2019 Bare Costs Material	Labor	Equipment	Total	Total Incl O&P
0460	Windows, per interior side, based on 15 S.F.									
0470	1 to 6 lite									
0480	Brushwork, primer G	1 Pord	13	.615	Ea.	1.27	20		21.27	34.50
0490	Finish coat, enamel G		13	.615		1.37	20		21.37	34.50
0500	Primer & 1 coat enamel G		8	1		2.64	33		35.64	57
0510	Primer & 2 coats enamel G		6	1.333		4.01	44		48.01	76

09 91 23.40 Floors, Interior		Crew	Daily Output	Labor-Hours	Unit	Material	Labor	Equipment	Total	Total Incl O&P
0010	**FLOORS, INTERIOR**									
0100	Concrete paint, latex									
0110	Brushwork									
0120	1st coat	1 Pord	975	.008	S.F.	.15	.27		.42	.61
0130	2nd coat		1150	.007		.10	.23		.33	.48
0140	3rd coat		1300	.006		.08	.20		.28	.42
0150	Roll									
0160	1st coat	1 Pord	2600	.003	S.F.	.20	.10		.30	.39
0170	2nd coat		3250	.002		.12	.08		.20	.26
0180	3rd coat		3900	.002		.09	.07		.16	.21
0190	Spray									
0200	1st coat	1 Pord	2600	.003	S.F.	.17	.10		.27	.36
0210	2nd coat		3250	.002		.09	.08		.17	.23
0220	3rd coat		3900	.002		.08	.07		.15	.19

09 91 23.52 Miscellaneous, Interior		Crew	Daily Output	Labor-Hours	Unit	Material	Labor	Equipment	Total	Total Incl O&P
0010	**MISCELLANEOUS, INTERIOR**									
2400	Floors, conc./wood, oil base, primer/sealer coat, brushwork	2 Pord	1950	.008	S.F.	.09	.27		.36	.54
2450	Roller		5200	.003		.09	.10		.19	.27
2600	Spray		6000	.003		.09	.09		.18	.24
2650	Paint 1 coat, brushwork		1950	.008		.11	.27		.38	.56
2800	Roller		5200	.003		.11	.10		.21	.29
2850	Spray		6000	.003		.12	.09		.21	.27
3000	Stain, wood floor, brushwork, 1 coat		4550	.004		.10	.12		.22	.30
3200	Roller		5200	.003		.11	.10		.21	.29
3250	Spray		6000	.003		.11	.09		.20	.26
3400	Varnish, wood floor, brushwork		4550	.004		.09	.12		.21	.29
3450	Roller		5200	.003		.10	.10		.20	.28
3600	Spray		6000	.003		.10	.09		.19	.25
3800	Grilles, per side, oil base, primer coat, brushwork	1 Pord	520	.015		.13	.51		.64	.97
3850	Spray		1140	.007		.14	.23		.37	.53
3880	Paint 1 coat, brushwork		520	.015		.22	.51		.73	1.08
3900	Spray		1140	.007		.25	.23		.48	.65
3920	Paint 2 coats, brushwork		325	.025		.43	.81		1.24	1.80
3940	Spray		650	.012		.50	.40		.90	1.21
4500	Louvers, 1 side, primer, brushwork		524	.015		.09	.50		.59	.92
4520	Paint 1 coat, brushwork		520	.015		.10	.51		.61	.94
4530	Spray		1140	.007		.11	.23		.34	.50
4540	Paint 2 coats, brushwork		325	.025		.20	.81		1.01	1.54
4550	Spray		650	.012		.22	.40		.62	.90
4560	Paint 3 coats, brushwork		270	.030		.29	.97		1.26	1.91
4570	Spray		500	.016		.33	.53		.86	1.22
5000	Pipe, 1"-4" diameter, primer or sealer coat, oil base, brushwork	2 Pord	1250	.013	L.F.	.09	.42		.51	.79
5100	Spray		2165	.007		.09	.24		.33	.50
5200	Paint 1 coat, brushwork		1250	.013		.12	.42		.54	.82
5300	Spray		2165	.007		.10	.24		.34	.51

09 91 23.52 Miscellaneous, Interior		Crew	Daily Output	Labor-Hours	Unit	2019 Bare Costs Material	Labor	Equipment	Total	Total Incl O&P
5350	Paint 2 coats, brushwork	2 Pord	775	.021	L.F.	.21	.68		.89	1.34
5400	Spray		1240	.013		.23	.42		.65	.94
5450	5"-8" diameter, primer or sealer coat, brushwork		620	.026		.19	.85		1.04	1.60
5500	Spray		1085	.015		.31	.48		.79	1.13
5550	Paint 1 coat, brushwork		620	.026		.32	.85		1.17	1.74
5600	Spray		1085	.015		.35	.48		.83	1.18
5650	Paint 2 coats, brushwork		385	.042		.42	1.37		1.79	2.70
5700	Spray		620	.026		.46	.85		1.31	1.90
5750	9"-12" diameter, primer or sealer coat, brushwork		415	.039		.28	1.27		1.55	2.38
5800	Spray		725	.022		.38	.73		1.11	1.61
5850	Paint 1 coat, brushwork		415	.039		.32	1.27		1.59	2.42
6000	Spray		725	.022		.36	.73		1.09	1.58
6200	Paint 2 coats, brushwork		260	.062		.62	2.02		2.64	4
6250	Spray		415	.039		.69	1.27		1.96	2.83
6300	13"-16" diameter, primer or sealer coat, brushwork		310	.052		.38	1.70		2.08	3.20
6350	Spray		540	.030		.42	.97		1.39	2.05
6400	Paint 1 coat, brushwork		310	.052		.43	1.70		2.13	3.25
6450	Spray		540	.030		.48	.97		1.45	2.11
6500	Paint 2 coats, brushwork		195	.082		.83	2.70		3.53	5.35
6550	Spray		310	.052		.93	1.70		2.63	3.80
6600	Radiators, per side, primer, brushwork	1 Pord	520	.015	S.F.	.09	.51		.60	.93
6620	Paint, 1 coat		520	.015		.08	.51		.59	.92
6640	2 coats		340	.024		.20	.77		.97	1.49
6660	3 coats		283	.028		.29	.93		1.22	1.84
7000	Trim, wood, incl. puttying, under 6" wide									
7200	Primer coat, oil base, brushwork	1 Pord	650	.012	L.F.	.03	.40		.43	.70
7250	Paint, 1 coat, brushwork		650	.012		.06	.40		.46	.72
7400	2 coats		400	.020		.11	.66		.77	1.20
7450	3 coats		325	.025		.16	.81		.97	1.50
7500	Over 6" wide, primer coat, brushwork		650	.012		.07	.40		.47	.73
7550	Paint, 1 coat, brushwork		650	.012		.11	.40		.51	.78
7600	2 coats		400	.020		.22	.66		.88	1.32
7650	3 coats		325	.025		.32	.81		1.13	1.67
8000	Cornice, simple design, primer coat, oil base, brushwork		650	.012	S.F.	.07	.40		.47	.73
8250	Paint, 1 coat		650	.012		.11	.40		.51	.78
8300	2 coats		400	.020		.22	.66		.88	1.32
8350	Ornate design, primer coat		350	.023		.07	.75		.82	1.30
8400	Paint, 1 coat		350	.023		.11	.75		.86	1.35
8450	2 coats		400	.020		.22	.66		.88	1.32
8600	Balustrades, primer coat, oil base, brushwork		520	.015		.07	.51		.58	.90
8650	Paint, 1 coat		520	.015		.11	.51		.62	.95
8700	2 coats		325	.025		.22	.81		1.03	1.56
8900	Trusses and wood frames, primer coat, oil base, brushwork		800	.010		.07	.33		.40	.61
8950	Spray		1200	.007		.07	.22		.29	.43
9000	Paint 1 coat, brushwork		750	.011		.11	.35		.46	.69
9200	Spray		1200	.007		.12	.22		.34	.50
9220	Paint 2 coats, brushwork		500	.016		.22	.53		.75	1.10
9240	Spray		600	.013		.24	.44		.68	.98
9260	Stain, brushwork, wipe off		600	.013		.10	.44		.54	.83
9280	Varnish, 3 coats, brushwork		275	.029		.28	.96		1.24	1.87
9350	For latex paint, deduct					10%				

09 91 Painting

09 91 23 – Interior Painting

	09 91 23.62 Electrostatic Painting	Crew	Daily Output	Labor-Hours	Unit	Material	2019 Bare Costs Labor	Equipment	Total	Total Incl O&P
0010	**ELECTROSTATIC PAINTING**									
0100	In shop									
0200	Flat surfaces (lockers, casework, elevator doors, etc.)									
0300	One coat	1 Pord	200	.040	S.F.	.63	1.31		1.94	2.84
0400	Two coats	"	120	.067	"	.92	2.19		3.11	4.60
0500	Irregular surfaces (furniture, door frames, etc.)									
0600	One coat	1 Pord	150	.053	S.F.	.63	1.75		2.38	3.56
0700	Two coats	"	100	.080	"	.92	2.63		3.55	5.30
0800	On site									
0900	Flat surfaces (lockers, casework, elevator doors, etc.)									
1000	One coat	1 Pord	150	.053	S.F.	.63	1.75		2.38	3.56
1100	Two coats	"	100	.080	"	.92	2.63		3.55	5.30
1200	Irregular surfaces (furniture, door frames, etc.)									
1300	One coat	1 Pord	115	.070	S.F.	.63	2.29		2.92	4.43
1400	Two coats	"	70	.114	"	.92	3.75		4.67	7.15

09 91 23.72 Walls and Ceilings, Interior

	09 91 23.72 Walls and Ceilings, Interior	Crew	Daily Output	Labor-Hours	Unit	Material	Labor	Equipment	Total	Total Incl O&P
0010	**WALLS AND CEILINGS, INTERIOR**									
0100	Concrete, drywall or plaster, latex, primer or sealer coat									
0200	Smooth finish, brushwork	1 Pord	1150	.007	S.F.	.06	.23		.29	.44
0240	Roller		1350	.006		.06	.19		.25	.39
0280	Spray		2750	.003		.06	.10		.16	.22
0300	Sand finish, brushwork		975	.008		.06	.27		.33	.51
0340	Roller		1150	.007		.06	.23		.29	.44
0380	Spray		2275	.004		.06	.12		.18	.25
0400	Paint 1 coat, smooth finish, brushwork		1200	.007		.07	.22		.29	.44
0440	Roller		1300	.006		.07	.20		.27	.41
0480	Spray		2275	.004		.06	.12		.18	.26
0500	Sand finish, brushwork		1050	.008		.07	.25		.32	.48
0540	Roller		1600	.005		.07	.16		.23	.35
0580	Spray		2100	.004		.02	.13		.15	.24
0800	Paint 2 coats, smooth finish, brushwork		680	.012		.15	.39		.54	.79
0840	Roller		800	.010		.15	.33		.48	.70
0880	Spray		1625	.005		.13	.16		.29	.41
0900	Sand finish, brushwork		605	.013		.15	.43		.58	.87
0940	Roller		1020	.008		.15	.26		.41	.58
0980	Spray		1700	.005		.13	.15		.28	.40
1200	Paint 3 coats, smooth finish, brushwork		510	.016		.22	.52		.74	1.08
1240	Roller		650	.012		.22	.40		.62	.90
1280	Spray		850	.009		.20	.31		.51	.73
1300	Sand finish, brushwork		454	.018		.33	.58		.91	1.31
1340	Roller		680	.012		.35	.39		.74	1.02
1380	Spray		1133	.007		.30	.23		.53	.71
1600	Glaze coating, 2 coats, spray, clear		1200	.007		.56	.22		.78	.98
1640	Multicolor		1200	.007		.87	.22		1.09	1.31
1660	Painting walls, complete, including surface prep, primer &									
1670	2 coats finish, on drywall or plaster, with roller	1 Pord	325	.025	S.F.	.22	.81		1.03	1.56
1700	For oil base paint, add					10%				
1800	For ceiling installations, add						25%			
2000	Masonry or concrete block, primer/sealer, latex paint									
2100	Primer, smooth finish, brushwork	1 Pord	1000	.008	S.F.	.15	.26		.41	.60
2110	Roller		1150	.007		.11	.23		.34	.49
2180	Spray		2400	.003		.10	.11		.21	.29

09 91 23.72 Walls and Ceilings, Interior		Crew	Daily Output	Labor-Hours	Unit	Material	Labor	Equipment	Total	Total Incl O&P
						2019 Bare Costs				
2200	Sand finish, brushwork	1 Pord	850	.009	S.F.	.11	.31		.42	.63
2210	Roller		975	.008		.11	.27		.38	.56
2280	Spray		2050	.004		.10	.13		.23	.32
2400	Finish coat, smooth finish, brush		1100	.007		.09	.24		.33	.48
2410	Roller		1300	.006		.09	.20		.29	.42
2480	Spray		2400	.003		.07	.11		.18	.26
2500	Sand finish, brushwork		950	.008		.09	.28		.37	.54
2510	Roller		1090	.007		.09	.24		.33	.48
2580	Spray		2040	.004		.07	.13		.20	.29
2800	Primer plus one finish coat, smooth brush		525	.015		.31	.50		.81	1.16
2810	Roller		615	.013		.20	.43		.63	.92
2880	Spray		1200	.007		.17	.22		.39	.55
2900	Sand finish, brushwork		450	.018		.20	.58		.78	1.18
2910	Roller		515	.016		.20	.51		.71	1.06
2980	Spray		1025	.008		.17	.26		.43	.61
3200	Primer plus 2 finish coats, smooth, brush		355	.023		.28	.74		1.02	1.52
3210	Roller		415	.019		.28	.63		.91	1.35
3280	Spray		800	.010		.25	.33		.58	.81
3300	Sand finish, brushwork		305	.026		.28	.86		1.14	1.72
3310	Roller		350	.023		.28	.75		1.03	1.54
3380	Spray		675	.012		.25	.39		.64	.91
3600	Glaze coating, 3 coats, spray, clear		900	.009		.80	.29		1.09	1.36
3620	Multicolor		900	.009		1.06	.29		1.35	1.65
4000	Block filler, 1 coat, brushwork		425	.019		.13	.62		.75	1.15
4100	Silicone, water repellent, 2 coats, spray	↓	2000	.004		.48	.13		.61	.75
4120	For oil base paint, add					10%				
8200	For work 8'-15' H, add						10%			
8300	For work over 15' H, add						20%			
8400	For light textured surfaces, add						10%			
8410	Heavy textured, add				↓		25%			

09 91 23.74 Walls and Ceilings, Interior, Zero VOC Latex			Crew	Daily Output	Labor-Hours	Unit	Material	Labor	Equipment	Total	Total Incl O&P
0010	**WALLS AND CEILINGS, INTERIOR, ZERO VOC LATEX**										
0100	Concrete, dry wall or plaster, latex, primer or sealer coat										
0200	Smooth finish, brushwork	G	1 Pord	1150	.007	S.F.	.08	.23		.31	.46
0240	Roller	G		1350	.006		.08	.19		.27	.41
0280	Spray	G		2750	.003		.06	.10		.16	.23
0300	Sand finish, brushwork	G		975	.008		.08	.27		.35	.53
0340	Roller	G		1150	.007		.09	.23		.32	.47
0380	Spray	G		2275	.004		.07	.12		.19	.27
0400	Paint 1 coat, smooth finish, brushwork	G		1200	.007		.09	.22		.31	.46
0440	Roller	G		1300	.006		.09	.20		.29	.43
0480	Spray	G		2275	.004		.08	.12		.20	.28
0500	Sand finish, brushwork	G		1050	.008		.09	.25		.34	.51
0540	Roller	G		1600	.005		.09	.16		.25	.37
0580	Spray	G		2100	.004		.08	.13		.21	.30
0800	Paint 2 coats, smooth finish, brushwork	G		680	.012		.18	.39		.57	.83
0840	Roller	G		800	.010		.19	.33		.52	.75
0880	Spray	G		1625	.005		.16	.16		.32	.44
0900	Sand finish, brushwork	G		605	.013		.18	.43		.61	.91
0940	Roller	G		1020	.008		.19	.26		.45	.63
0980	Spray	G		1700	.005		.16	.15		.31	.43
1200	Paint 3 coats, smooth finish, brushwork	G	↓	510	.016	↓	.27	.52		.79	1.14

09 91 Painting

09 91 23 – Interior Painting

09 91 23.74 Walls and Ceilings, Interior, Zero VOC Latex		Crew	Daily Output	Labor-Hours	Unit	Material	Labor	Equipment	Total	Total Incl O&P
						2019 Bare Costs				
1240	Roller G	1 Pord	650	.012	S.F.	.28	.40		.68	.97
1280	Spray G	↓	850	.009		.24	.31		.55	.78
1800	For ceiling installations, add G						25%			
8200	For work 8' - 15' H, add						10%			
8300	For work over 15' H, add				↓		20%			

09 91 23.75 Dry Fall Painting		Crew	Daily Output	Labor-Hours	Unit	Material	Labor	Equipment	Total	Total Incl O&P
0010	DRY FALL PAINTING R099100-20									
0100	Sprayed on walls, gypsum board or plaster									
0220	One coat	1 Pord	2600	.003	S.F.	.08	.10		.18	.26
0250	Two coats		1560	.005		.16	.17		.33	.46
0280	Concrete or textured plaster, one coat		1560	.005		.08	.17		.25	.37
0310	Two coats		1300	.006		.16	.20		.36	.51
0340	Concrete block, one coat		1560	.005		.08	.17		.25	.37
0370	Two coats		1300	.006		.16	.20		.36	.51
0400	Wood, one coat		877	.009		.08	.30		.38	.58
0430	Two coats	↓	650	.012	↓	.16	.40		.56	.84
0440	On ceilings, gypsum board or plaster									
0470	One coat	1 Pord	1560	.005	S.F.	.08	.17		.25	.37
0500	Two coats		1300	.006		.16	.20		.36	.51
0530	Concrete or textured plaster, one coat		1560	.005		.08	.17		.25	.37
0560	Two coats		1300	.006		.16	.20		.36	.51
0570	Structural steel, bar joists or metal deck, one coat		1560	.005		.08	.17		.25	.37
0580	Two coats	↓	1040	.008	↓	.16	.25		.41	.59

09 93 Staining and Transparent Finishing

09 93 23 – Interior Staining and Finishing

09 93 23.10 Varnish		Crew	Daily Output	Labor-Hours	Unit	Material	Labor	Equipment	Total	Total Incl O&P
0010	VARNISH									
0012	1 coat + sealer, on wood trim, brush, no sanding included	1 Pord	400	.020	S.F.	.07	.66		.73	1.16
0020	1 coat + sealer, on wood trim, brush, no sanding included, no VOC		400	.020		.22	.66		.88	1.32
0100	Hardwood floors, 2 coats, no sanding included, roller	↓	1890	.004	↓	.15	.14		.29	.40

09 96 High-Performance Coatings

09 96 23 – Graffiti-Resistant Coatings

09 96 23.10 Graffiti-Resistant Treatments		Crew	Daily Output	Labor-Hours	Unit	Material	Labor	Equipment	Total	Total Incl O&P
0010	GRAFFITI-RESISTANT TREATMENTS, sprayed on walls									
0100	Non-sacrificial, permanent non-stick coating, clear, on metals	1 Pord	2000	.004	S.F.	2.03	.13		2.16	2.45
0200	Concrete		2000	.004		2.31	.13		2.44	2.76
0300	Concrete block		2000	.004		2.99	.13		3.12	3.51
0400	Brick		2000	.004		3.39	.13		3.52	3.94
0500	Stone		2000	.004		3.39	.13		3.52	3.94
0600	Unpainted wood		2000	.004		3.91	.13		4.04	4.52
2000	Semi-permanent cross linking polymer primer, on metals		2000	.004		.61	.13		.74	.90
2100	Concrete		2000	.004		.74	.13		.87	1.03
2200	Concrete block		2000	.004		.92	.13		1.05	1.23
2300	Brick		2000	.004		.74	.13		.87	1.03
2400	Stone		2000	.004		.74	.13		.87	1.03
2500	Unpainted wood		2000	.004		1.02	.13		1.15	1.35
3000	Top coat, on metals	↓	2000	.004	↓	.56	.13		.69	.83

09 96 High-Performance Coatings

09 96 23 – Graffiti-Resistant Coatings

09 96 23.10 Graffiti-Resistant Treatments		Crew	Daily Output	Labor-Hours	Unit	Material	2019 Bare Costs Labor	Equipment	Total	Total Incl O&P
3100	Concrete	1 Pord	2000	.004	S.F.	.64	.13		.77	.92
3200	Concrete block		2000	.004		.89	.13		1.02	1.20
3300	Brick		2000	.004		.74	.13		.87	1.04
3400	Stone		2000	.004		.74	.13		.87	1.04
3500	Unpainted wood		2000	.004		.89	.13		1.02	1.20
5000	Sacrificial, water based, on metal		2000	.004		.33	.13		.46	.58
5100	Concrete		2000	.004		.33	.13		.46	.58
5200	Concrete block		2000	.004		.33	.13		.46	.58
5300	Brick		2000	.004		.33	.13		.46	.58
5400	Stone		2000	.004		.33	.13		.46	.58
5500	Unpainted wood		2000	.004		.33	.13		.46	.58
8000	Cleaner for use after treatment									
8100	Towels or wipes, per package of 30				Ea.	.64			.64	.70
8200	Aerosol spray, 24 oz. can				"	18.35			18.35	20
8500	Graffiti removal with chemicals									
8510	Brush on, spray rinse off, on brick, masonry or stone	A-1H	200	.040	S.F.	.33	1.21	.37	1.91	2.78
8520	On smooth concrete	"	400	.020		.27	.61	.19	1.07	1.50
8530	Wipe on, wipe off, on plastic or painted metal	1 Clab	1500	.005		.64	.16		.80	.97
8540	Brush on, wipe off, on painted wood	"	500	.016		.35	.49		.84	1.19

09 96 46 – Intumescent Painting

09 96 46.10 Coatings, Intumescent		Crew	Daily Output	Labor-Hours	Unit	Material	Labor	Equipment	Total	Total Incl O&P
0010	**COATINGS, INTUMESCENT**, spray applied									
0100	On exterior structural steel, 0.25" d.f.t.	1 Pord	475	.017	S.F.	.49	.55		1.04	1.45
0150	0.51" d.f.t.		350	.023		.49	.75		1.24	1.77
0200	0.98" d.f.t.		280	.029		.49	.94		1.43	2.08
0300	On interior structural steel, 0.108" d.f.t.		300	.027		.47	.88		1.35	1.94
0350	0.310" d.f.t.		150	.053		.47	1.75		2.22	3.38
0400	0.670" d.f.t.		100	.080		.47	2.63		3.10	4.81

09 96 53 – Elastomeric Coatings

09 96 53.10 Coatings, Elastomeric		Crew	Daily Output	Labor-Hours	Unit	Material	Labor	Equipment	Total	Total Incl O&P
0010	**COATINGS, ELASTOMERIC**									
0020	High build, water proof, one coat system									
0100	Concrete, brush	1 Pord	650	.012	S.F.	.29	.40		.69	.98

09 96 56 – Epoxy Coatings

09 96 56.20 Wall Coatings		Crew	Daily Output	Labor-Hours	Unit	Material	Labor	Equipment	Total	Total Incl O&P
0010	**WALL COATINGS**									
0100	Acrylic glazed coatings, matte	1 Pord	525	.015	S.F.	.37	.50		.87	1.23
0200	Gloss		305	.026		.78	.86		1.64	2.27
0300	Epoxy coatings, solvent based		525	.015		.47	.50		.97	1.34
0400	Water based		170	.047		.33	1.55		1.88	2.89
0600	Exposed aggregate, troweled on, 1/16" to 1/4", solvent based		235	.034		.74	1.12		1.86	2.64
0700	Water based (epoxy or polyacrylate)		130	.062		1.58	2.02		3.60	5.05
0900	1/2" to 5/8" aggregate, solvent based		130	.062		1.42	2.02		3.44	4.87
1000	Water based		80	.100		2.47	3.29		5.76	8.10
1200	1" aggregate size, solvent based		90	.089		2.52	2.92		5.44	7.55
1300	Water based		55	.145		3.83	4.78		8.61	12.05
1500	Exposed aggregate, sprayed on, 1/8" aggregate, solvent based		295	.027		.58	.89		1.47	2.10
1600	Water based		145	.055		1.24	1.81		3.05	4.33
1800	High build epoxy, 50 mil, solvent based		390	.021		.76	.67		1.43	1.94
1900	Water based		95	.084		1.36	2.77		4.13	6.05
2100	Laminated epoxy with fiberglass, solvent based		295	.027		.86	.89		1.75	2.41

09 96 High-Performance Coatings

09 96 56 – Epoxy Coatings

09 96 56.20 Wall Coatings		Crew	Daily Output	Labor-Hours	Unit	Material	2019 Bare Costs Labor	Equipment	Total	Total Incl O&P
2200	Water based	1 Pord	145	.055	S.F.	1.57	1.81		3.38	4.70
2400	Sprayed perlite or vermiculite, 1/16" thick, solvent based		2935	.003		.28	.09		.37	.46
2500	Water based		640	.013		.87	.41		1.28	1.63
2700	Vinyl plastic wall coating, solvent based		735	.011		.40	.36		.76	1.03
2800	Water based		240	.033		.98	1.09		2.07	2.87
3000	Urethane on smooth surface, 2 coats, solvent based		1135	.007		.35	.23		.58	.77
3100	Water based		665	.012		.61	.40		1.01	1.32
3300	3 coat, solvent based		840	.010		.44	.31		.75	.99
3400	Water based		470	.017		.95	.56		1.51	1.97
3600	Ceramic-like glazed coating, cementitious, solvent based		440	.018		.49	.60		1.09	1.52
3700	Water based		345	.023		.97	.76		1.73	2.32
3900	Resin base, solvent based		640	.013		.34	.41		.75	1.04
4000	Water based	↓	330	.024	↓	.59	.80		1.39	1.95

09 97 Special Coatings

09 97 13 – Steel Coatings

09 97 13.23 Exterior Steel Coatings		Crew	Daily Output	Labor-Hours	Unit	Material	Labor	Equipment	Total	Total Incl O&P
0010	**EXTERIOR STEEL COATINGS**									
6101	Cold galvanizing, brush in field	1 Pord	1100	.007	S.F.	.21	.24		.45	.62
6510	Paints & protective coatings, sprayed in field									
6520	Alkyds, primer	2 Psst	3600	.004	S.F.	.09	.15		.24	.36
6540	Gloss topcoats		3200	.005		.09	.17		.26	.40
6560	Silicone alkyd		3200	.005		.17	.17		.34	.49
6610	Epoxy, primer		3000	.005		.25	.18		.43	.60
6630	Intermediate or topcoat		2800	.006		.28	.19		.47	.65
6650	Enamel coat		2800	.006		.35	.19		.54	.73
6700	Epoxy ester, primer		2800	.006		.49	.19		.68	.88
6720	Topcoats		2800	.006		.23	.19		.42	.60
6810	Latex primer		3600	.004		.07	.15		.22	.33
6830	Topcoats		3200	.005		.07	.17		.24	.37
6910	Universal primers, one part, phenolic, modified alkyd		2000	.008		.39	.27		.66	.90
6940	Two part, epoxy spray		2000	.008		.40	.27		.67	.92
7000	Zinc rich primers, self cure, spray, inorganic		1800	.009		.86	.30		1.16	1.47
7010	Epoxy, spray, organic	↓	1800	.009	↓	.28	.30		.58	.84
7020	Above one story, spray painting simple structures, add						25%			
7030	Intricate structures, add						50%			

09 97 35 – Dry Erase Coatings

09 97 35.10 Dry Erase Coatings		Crew	Daily Output	Labor-Hours	Unit	Material	Labor	Equipment	Total	Total Incl O&P
0010	**DRY ERASE COATINGS**									
0020	Dry erase coatings, clear, roller applied	1 Pord	1325	.006	S.F.	2.09	.20		2.29	2.63

Estimating Tips

General

- The items in this division are usually priced per square foot or each.
- Many items in Division 10 require some type of support system or special anchors that are not usually furnished with the item. The required anchors must be added to the estimate in the appropriate division.
- Some items in Division 10, such as lockers, may require assembly before installation. Verify the amount of assembly required. Assembly can often exceed installation time.

10 20 00 Interior Specialties

- Support angles and blocking are not included in the installation of toilet compartments, shower/dressing compartments, or cubicles. Appropriate line items from Division 5 or 6 may need to be added to support the installations.
- Toilet partitions are priced by the stall. A stall consists of a side wall, pilaster, and door with hardware. Toilet tissue holders and grab bars are extra.
- The required acoustical rating of a folding partition can have a significant impact on costs. Verify the sound transmission coefficient rating of the panel priced against the specification requirements.
- Grab bar installation does not include supplemental blocking or backing to support the required load. When grab bars are installed at an existing facility, provisions must be made to attach the grab bars to a solid structure.

Reference Numbers

Reference numbers are shown at the beginning of some major classifications. These numbers refer to related items in the Reference Section. The reference information may be an estimating procedure, an alternate pricing method, or technical information.

Note: Not all subdivisions listed here necessarily appear. ■

10 11 13 – Chalkboards

10 11 13.13 Fixed Chalkboards		Crew	Daily Output	Labor-Hours	Unit	Material	2019 Bare Costs Labor	Equipment	Total	Total Incl O&P
0010	**FIXED CHALKBOARDS** Porcelain enamel steel									
3900	Wall hung									
4000	Aluminum frame and chalktrough									
4300	3′ x 5′	2 Carp	15	1.067	Ea.	310	41.50		351.50	415
4600	4′ x 12′	"	13	1.231	"	565	47.50		612.50	700
4700	Wood frame and chalktrough									
4800	3′ x 4′	2 Carp	16	1	Ea.	211	39		250	296
5300	4′ x 8′	"	13	1.231	"	370	47.50		417.50	490
5400	Liquid chalk, white porcelain enamel, wall hung									
5420	Deluxe units, aluminum trim and chalktrough									
5450	4′ x 4′	2 Carp	16	1	Ea.	305	39		344	400
5550	4′ x 12′	"	12	1.333	"	630	51.50		681.50	780
5700	Wood trim and chalktrough									
5900	4′ x 4′	2 Carp	16	1	Ea.	755	39		794	895
6200	4′ x 8′	"	14	1.143	"	1,025	44.50		1,069.50	1,200

10 11 13.43 Portable Chalkboards		Crew	Daily Output	Labor-Hours	Unit	Material	Labor	Equipment	Total	Total Incl O&P
0010	**PORTABLE CHALKBOARDS**									
0100	Freestanding, reversible									
0120	Economy, wood frame, 4′ x 6′									
0140	Chalkboard both sides				Ea.	710			710	780
0160	Chalkboard one side, cork other side				"	630			630	695
0200	Standard, lightweight satin finished aluminum, 4′ x 6′									
0220	Chalkboard both sides				Ea.	690			690	760
0240	Chalkboard one side, cork other side				"	610			610	670
0300	Deluxe, heavy duty extruded aluminum, 4′ x 6′									
0320	Chalkboard both sides				Ea.	1,150			1,150	1,275
0340	Chalkboard one side, cork other side				"	1,200			1,200	1,300

10 11 16 – Markerboards

10 11 16.53 Electronic Markerboards		Crew	Daily Output	Labor-Hours	Unit	Material	Labor	Equipment	Total	Total Incl O&P
0010	**ELECTRONIC MARKERBOARDS**									
0100	Wall hung or free standing, 3′ x 4′ to 4′ x 6′	2 Carp	8	2	S.F.	86	77.50		163.50	223
0150	5′ x 6′ to 4′ x 8′		8	2	"	62.50	77.50		140	197
0500	Interactive projection module for existing whiteboards	↓	8	2	Ea.	1,300	77.50		1,377.50	1,550

10 11 23 – Tackboards

10 11 23.10 Fixed Tackboards		Crew	Daily Output	Labor-Hours	Unit	Material	Labor	Equipment	Total	Total Incl O&P
0010	**FIXED TACKBOARDS**									
0020	Cork sheets, unbacked, no frame, 1/4″ thick	2 Carp	290	.055	S.F.	1.60	2.14		3.74	5.30
2120	Prefabricated, 1/4″ cork, 3′ x 5′ with aluminum frame		16	1	Ea.	142	39		181	220
2140	Wood frame	↓	16	1	"	165	39		204	246
2300	Glass enclosed cabinets, alum., cork panel, hinged doors									
2600	4′ x 7′, 3 door	2 Carp	10	1.600	Ea.	1,875	62		1,937	2,175

10 13 Directories

10 13 10 – Building Directories

10 13 10.10 Directory Boards		Crew	Daily Output	Labor-Hours	Unit	2019 Bare Costs Material	Labor	Equipment	Total	Total Incl O&P
0010	**DIRECTORY BOARDS**									
0050	Plastic, glass covered, 30″ x 20″	2 Carp	3	5.333	Ea.	210	207		417	570
0100	36″ x 48″		2	8		905	310		1,215	1,500
0900	Outdoor, weatherproof, black plastic, 36″ x 24″		2	8		695	310		1,005	1,275
1000	36″ x 36″		1.50	10.667		800	415		1,215	1,575

10 14 Signage

10 14 19 – Dimensional Letter Signage

10 14 19.10 Plaques		Crew	Daily Output	Labor-Hours	Unit	Material	Labor	Equipment	Total	Total Incl O&P
0012	**PLAQUES**									
3900	Plaques, custom, 20″ x 30″, for up to 450 letters, cast aluminum	2 Carp	4	4	Ea.	1,900	155		2,055	2,325
4000	Cast bronze		4	4		1,950	155		2,105	2,375
4200	30″ x 36″, up to 900 letters, cast aluminum		3	5.333		2,775	207		2,982	3,400
4300	Cast bronze		3	5.333		4,150	207		4,357	4,900
5100	Exit signs, 24 ga. alum., 14″ x 12″ surface mounted	1 Carp	30	.267		51.50	10.35		61.85	73.50
5200	10″ x 7″	″	20	.400		30	15.50		45.50	58.50
6400	Replacement sign faces, 6″ or 8″	1 Clab	50	.160		65.50	4.86		70.36	80.50
8000	Internally illuminated, custom									
8100	On pedestal, 84″ x 30″ x 12″	L-7	.50	52	Ea.	11,500	1,725		13,225	15,500

10 14 23 – Panel Signage

10 14 23.13 Engraved Panel Signage		Crew	Daily Output	Labor-Hours	Unit	Material	Labor	Equipment	Total	Total Incl O&P
0010	**ENGRAVED PANEL SIGNAGE**, interior									
1010	Flexible door sign, adhesive back, w/Braille, 5/8″ letters, 4″ x 4″	1 Clab	32	.250	Ea.	33.50	7.60		41.10	49.50
1050	6″ x 6″		32	.250		48.50	7.60		56.10	65.50
1100	8″ x 2″		32	.250		34.50	7.60		42.10	50.50
1150	8″ x 4″		32	.250		43	7.60		50.60	60
1200	8″ x 8″		32	.250		69	7.60		76.60	88.50
1250	12″ x 2″		32	.250		35	7.60		42.60	51
1300	12″ x 6″		32	.250		41	7.60		48.60	57.50
1350	12″ x 12″		32	.250		153	7.60		160.60	182
1500	Graphic symbols, 2″ x 2″		32	.250		11.95	7.60		19.55	25.50
1550	6″ x 6″		32	.250		31	7.60		38.60	46.50
1600	8″ x 8″		32	.250		38.50	7.60		46.10	55
2010	Corridor, stock acrylic, 2-sided, with mounting bracket, 2″ x 8″	1 Carp	24	.333		29.50	12.90		42.40	54
2020	2″ x 10″		24	.333		34.50	12.90		47.40	59.50
2050	3″ x 8″		24	.333		27	12.90		39.90	51
2060	3″ x 10″		24	.333		38.50	12.90		51.40	63.50
2070	3″ x 12″		24	.333		39	12.90		51.90	64
2100	4″ x 8″		24	.333		23.50	12.90		36.40	47.50
2110	4″ x 10″		24	.333		40.50	12.90		53.40	66
2120	4″ x 12″		24	.333		50	12.90		62.90	76.50
7000	Wayfinding signage, custom									
7010	Plastic, flexible, 1/8″ thick, incl. mounting									
7020	6″ x 6″	1 Carp	32	.250	Ea.	16.85	9.70		26.55	34.50
7030	6″ x 12″		32	.250		36	9.70		45.70	55.50
7040	6″ x 24″		28	.286		63	11.05		74.05	88
7050	8″ x 8″		32	.250		28	9.70		37.70	46.50
7060	8″ x 16″		28	.286		60	11.05		71.05	84.50
7070	8″ x 24″		26	.308		91	11.90		102.90	120
7080	12″ x 12″		28	.286		70.50	11.05		81.55	96

10 14 Signage

10 14 23 – Panel Signage

10 14 23.13 Engraved Panel Signage		Crew	Daily Output	Labor-Hours	Unit	2019 Bare Costs Material	Labor	Equipment	Total	Total Incl O&P
7090	12″ x 24″	1 Carp	24	.333	Ea.	133	12.90		145.90	169
7100	12″ x 36″		20	.400		202	15.50		217.50	248
7210	Weather resistant, engraved and color filled									
7220	8″ x 8″	1 Carp	32	.250	Ea.	38.50	9.70		48.20	58.50
7230	8″ x 16″		28	.286		61	11.05		72.05	85.50
7240	8″ x 24″		26	.308		91.50	11.90		103.40	121
7250	12″ x 12″		28	.286		68.50	11.05		79.55	94
7260	12″ x 24″		24	.333		120	12.90		132.90	154
7270	12″ x 36″		20	.400		201	15.50		216.50	247
7280	16″ x 32″	2 Carp	60	.267		222	10.35		232.35	261
7290	16″ x 48″	″	60	.267		350	10.35		360.35	400
7320	For engraved letters, 1/2″ high, add					.70			.70	.77
7330	1″ high, add					.82			.82	.90
7340	2″ high, add					1.65			1.65	1.82
7350	3″ high, add					2.24			2.24	2.46
7360	For engraved graphic symbols, 3″ high, add					14.20			14.20	15.65
7370	7″ high, add					17.75			17.75	19.50
7380	10″ high, add					23.50			23.50	26
9990	Replace interior labels	1 Carp	12	.667		26	26		52	71.50
9991	Replace interior plaques	″	12	.667		84.50	26		110.50	136

10 14 26 – Post and Panel/Pylon Signage

10 14 26.10 Post and Panel Signage		Crew	Daily Output	Labor-Hours	Unit	Material	Labor	Equipment	Total	Total Incl O&P
0010	**POST AND PANEL SIGNAGE**, Alum., incl. 2 posts,									
0011	2-sided panel (blank), concrete footings per post									
0025	24″ W x 18″ H	B-1	25	.960	Ea.	900	30		930	1,050
0050	36″ W x 24″ H		25	.960		955	30		985	1,100
0100	57″ W x 36″ H		25	.960		2,025	30		2,055	2,275
0150	81″ W x 46″ H		25	.960		2,225	30		2,255	2,500

10 14 43 – Photoluminescent Signage

10 14 43.10 Photoluminescent Signage		Crew	Daily Output	Labor-Hours	Unit	Material	Labor	Equipment	Total	Total Incl O&P
0010	**PHOTOLUMINESCENT SIGNAGE**									
0011	Photoluminescent exit sign, plastic, 7″x10″	1 Carp	32	.250	Ea.	28	9.70		37.70	46.50
0022	Polyester		32	.250		15.10	9.70		24.80	32.50
0033	Aluminum		32	.250		53	9.70		62.70	74
0044	Plastic, 10″x14″		32	.250		59	9.70		68.70	81
0055	Polyester		32	.250		35	9.70		44.70	54.50
0066	Aluminum		32	.250		70	9.70		79.70	93
0111	Directional exit sign, plastic, 10″x14″		32	.250		58	9.70		67.70	79.50
0122	Polyester		32	.250		45.50	9.70		55.20	66
0133	Aluminum		32	.250		67.50	9.70		77.20	90.50
0153	Kick plate exit sign, 10″x34″		28	.286		261	11.05		272.05	305
0173	Photoluminescent tape, 1″x60′		300	.027	L.F.	.54	1.03		1.57	2.30
0183	2″x60′		290	.028	″	1.17	1.07		2.24	3.06
0211	Photoluminescent directional arrows		42	.190	Ea.	1.29	7.40		8.69	13.60

10 17 Telephone Specialties

10 17 16 – Telephone Enclosures

10 17 16.10 Commercial Telephone Enclosures		Crew	Daily Output	Labor-Hours	Unit	2019 Bare Costs Material	Labor	Equipment	Total	Total Incl O&P
0010	**COMMERCIAL TELEPHONE ENCLOSURES**									
0300	Shelf type, wall hung, recessed	2 Carp	5	3.200	Ea.	825	124		949	1,100
0400	Surface mount	″	5	3.200	″	1,675	124		1,799	2,025

10 21 Compartments and Cubicles

10 21 13 – Toilet Compartments

10 21 13.13 Metal Toilet Compartments		Crew	Daily Output	Labor-Hours	Unit	Material	Labor	Equipment	Total	Total Incl O&P
0010	**METAL TOILET COMPARTMENTS**									
0110	Cubicles, ceiling hung									
0200	Powder coated steel	2 Carp	4	4	Ea.	560	155		715	870
0500	Stainless steel	″	4	4		1,150	155		1,305	1,525
0600	For handicap units, add					465			465	510
0900	Floor and ceiling anchored									
1000	Powder coated steel	2 Carp	5	3.200	Ea.	615	124		739	880
1300	Stainless steel	″	5	3.200		1,250	124		1,374	1,575
1400	For handicap units, add					355			355	395
1610	Floor anchored									
1700	Powder coated steel	2 Carp	7	2.286	Ea.	620	88.50		708.50	830
2000	Stainless steel	″	7	2.286		1,275	88.50		1,363.50	1,550
2100	For handicap units, add					335			335	365
2200	For juvenile units, deduct					43.50			43.50	47.50
2450	Floor anchored, headrail braced									
2500	Powder coated steel	2 Carp	6	2.667	Ea.	395	103		498	605
2804	Stainless steel	″	6	2.667		1,075	103		1,178	1,350
2900	For handicap units, add					325			325	360
3000	Wall hung partitions, powder coated steel	2 Carp	7	2.286		670	88.50		758.50	880
3300	Stainless steel	″	7	2.286		1,725	88.50		1,813.50	2,050
3400	For handicap units, add					325			325	360
4000	Screens, entrance, floor mounted, 58″ high, 48″ wide									
4200	Powder coated steel	2 Carp	15	1.067	Ea.	241	41.50		282.50	335
4500	Stainless steel	″	15	1.067	″	955	41.50		996.50	1,125
4650	Urinal screen, 18″ wide									
4704	Powder coated steel	2 Carp	6.15	2.602	Ea.	226	101		327	415
5004	Stainless steel	″	6.15	2.602	″	535	101		636	750
5100	Floor mounted, headrail braced									
5300	Powder coated steel	2 Carp	8	2	Ea.	229	77.50		306.50	380
5600	Stainless steel	″	8	2	″	580	77.50		657.50	770
5750	Pilaster, flush									
5800	Powder coated steel	2 Carp	10	1.600	Ea.	280	62		342	410
6100	Stainless steel		10	1.600		615	62		677	780
6300	Post braced, powder coated steel		10	1.600		168	62		230	287
6800	Powder coated steel		10	1.600		173	62		235	293
7800	Wedge type, powder coated steel		10	1.600		142	62		204	258
8100	Stainless steel		10	1.600		605	62		667	765

10 21 13.16 Plastic-Laminate-Clad Toilet Compartments		Crew	Daily Output	Labor-Hours	Unit	Material	Labor	Equipment	Total	Total Incl O&P
0010	**PLASTIC-LAMINATE-CLAD TOILET COMPARTMENTS**									
0110	Cubicles, ceiling hung									
0300	Plastic laminate on particle board	2 Carp	4	4	Ea.	555	155		710	865
0600	For handicap units, add				″	465			465	510
0900	Floor and ceiling anchored									
1100	Plastic laminate on particle board	2 Carp	5	3.200	Ea.	775	124		899	1,050

10 21 13 – Toilet Compartments

10 21 13.16 Plastic-Laminate-Clad Toilet Compartments		Crew	Daily Output	Labor-Hours	Unit	Material	Labor	Equipment	Total	Total Incl O&P
						2019 Bare Costs				
1400	For handicap units, add				Ea.	355			355	395
1610	Floor mounted									
1800	Plastic laminate on particle board	2 Carp	7	2.286	Ea.	545	88.50		633.50	745
2450	Floor mounted, headrail braced									
2600	Plastic laminate on particle board	2 Carp	6	2.667	Ea.	835	103		938	1,100
3400	For handicap units, add					325			325	360
4300	Entrance screen, floor mtd., plas. lam., 58" high, 48" wide	2 Carp	15	1.067		515	41.50		556.50	635
4800	Urinal screen, 18" wide, ceiling braced, plastic laminate		8	2		213	77.50		290.50	360
5400	Floor mounted, headrail braced		8	2		207	77.50		284.50	355
5900	Pilaster, flush, plastic laminate		10	1.600		425	62		487	570
6400	Post braced, plastic laminate		10	1.600		254	62		316	380
6700	Wall hung, bracket supported									
6900	Plastic laminate on particle board	2 Carp	10	1.600	Ea.	96	62		158	207
7450	Flange supported									
7500	Plastic laminate on particle board	2 Carp	10	1.600	Ea.	246	62		308	370

10 21 13.19 Plastic Toilet Compartments		Crew	Daily Output	Labor-Hours	Unit	Material	Labor	Equipment	Total	Total Incl O&P
0010	**PLASTIC TOILET COMPARTMENTS**									
0110	Cubicles, ceiling hung									
0250	Phenolic	2 Carp	4	4	Ea.	900	155		1,055	1,250
0260	Polymer plastic	"	4	4		1,050	155		1,205	1,425
0600	For handicap units, add					465			465	510
0900	Floor and ceiling anchored									
1050	Phenolic	2 Carp	5	3.200	Ea.	860	124		984	1,150
1060	Polymer plastic	"	5	3.200		1,200	124		1,324	1,525
1400	For handicap units, add					355			355	395
1610	Floor mounted									
1750	Phenolic	2 Carp	7	2.286	Ea.	745	88.50		833.50	965
1760	Polymer plastic	"	7	2.286		865	88.50		953.50	1,100
2100	For handicap units, add					335			335	365
2200	For juvenile units, deduct					43.50			43.50	47.50
2450	Floor mounted, headrail braced									
2550	Phenolic	2 Carp	6	2.667	Ea.	750	103		853	995
3600	Polymer plastic		6	2.667		855	103		958	1,125
3810	Entrance screen, polymer plastic, flr. mtd., 48"x58"		6	2.667		500	103		603	720
3820	Entrance screen, polymer plastic, flr. to clg pilaster, 48"x58"		6	2.667		595	103		698	825
6110	Urinal screen, polymer plastic, pilaster flush, 18" w		6	2.667		435	103		538	645
7110	Wall hung		6	2.667		740	103		843	985
7710	Flange mounted		6	2.667		925	103		1,028	1,200

10 21 13.40 Stone Toilet Compartments		Crew	Daily Output	Labor-Hours	Unit	Material	Labor	Equipment	Total	Total Incl O&P
0010	**STONE TOILET COMPARTMENTS**									
0100	Cubicles, ceiling hung, marble	2 Marb	2	8	Ea.	1,650	291		1,941	2,300
0600	For handicap units, add					465			465	510
0800	Floor & ceiling anchored, marble	2 Marb	2.50	6.400		1,725	233		1,958	2,300
1400	For handicap units, add					355			355	395
1600	Floor mounted, marble	2 Marb	3	5.333		1,050	194		1,244	1,475
2400	Floor mounted, headrail braced, marble	"	3	5.333		1,225	194		1,419	1,650
2900	For handicap units, add					325			325	360
4100	Entrance screen, floor mounted marble, 58" high, 48" wide	2 Marb	9	1.778		690	64.50		754.50	865
4600	Urinal screen, 18" wide, ceiling braced, marble	D-1	6	2.667		800	90		890	1,025
5100	Floor mounted, headrail braced									
5200	Marble	D-1	6	2.667	Ea.	680	90		770	900
5700	Pilaster, flush, marble		9	1.778		890	60		950	1,075

10 21 13 – Toilet Compartments

10 21 13.40 Stone Toilet Compartments		Crew	Daily Output	Labor-Hours	Unit	Material	Labor	Equipment	Total	Total Incl O&P
						2019 Bare Costs				
6200	Post braced, marble	D-1	9	1.778	Ea.	870	60		930	1,050

10 21 14 – Toilet Compartment Components

10 21 14.13 Metal Toilet Compartment Components

		Crew	Daily Output	Labor-Hours	Unit	Material	Labor	Equipment	Total	Total Incl O&P
0010	**METAL TOILET COMPARTMENT COMPONENTS**									
0100	Pilasters									
0110	Overhead braced, powder coated steel, 7" wide x 82" high	2 Carp	22.20	.721	Ea.	69.50	28		97.50	123
0120	Stainless steel		22.20	.721		130	28		158	189
0130	Floor braced, powder coated steel, 7" wide x 70" high		23.30	.687		108	26.50		134.50	163
0140	Stainless steel		23.30	.687		223	26.50		249.50	290
0150	Ceiling hung, powder coated steel, 7" wide x 83" high		13.30	1.203		144	46.50		190.50	236
0160	Stainless steel		13.30	1.203		285	46.50		331.50	390
0170	Wall hung, powder coated steel, 3" wide x 58" high		18.90	.847		134	33		167	202
0180	Stainless steel	↓	18.90	.847	↓	190	33		223	263
0200	Panels									
0210	Powder coated steel, 31" wide x 58" high	2 Carp	18.90	.847	Ea.	133	33		166	200
0220	Stainless steel		18.90	.847		335	33		368	425
0230	Powder coated steel, 53" wide x 58" high		18.90	.847		166	33		199	236
0240	Stainless steel		18.90	.847		400	33		433	495
0250	Powder coated steel, 63" wide x 58" high		18.90	.847		178	33		211	249
0260	Stainless steel	↓	18.90	.847	↓	485	33		518	590
0300	Doors									
0310	Powder coated steel, 24" wide x 58" high	2 Carp	14.10	1.135	Ea.	126	44		170	212
0320	Stainless steel		14.10	1.135		325	44		369	430
0330	Powder coated steel, 26" wide x 58" high		14.10	1.135		147	44		191	235
0340	Stainless steel		14.10	1.135		330	44		374	440
0350	Powder coated steel, 28" wide x 58" high		14.10	1.135		172	44		216	263
0360	Stainless steel		14.10	1.135		345	44		389	455
0370	Powder coated steel, 36" wide x 58" high		14.10	1.135		183	44		227	275
0380	Stainless steel	↓	14.10	1.135	↓	390	44		434	505
0400	Headrails									
0410	For powder coated steel, 62" long	2 Carp	65	.246	Ea.	21	9.55		30.55	39
0420	Stainless steel		65	.246		21	9.55		30.55	39
0430	For powder coated steel, 84" long		50	.320		30.50	12.40		42.90	54
0440	Stainless steel		50	.320		30.50	12.40		42.90	54
0450	For powder coated steel, 120" long		30	.533		39.50	20.50		60	77.50
0460	Stainless steel	↓	30	.533	↓	41.50	20.50		62	79.50

10 21 14.16 Plastic-Lam Clad Toilet Compart. Components

		Crew	Daily Output	Labor-Hours	Unit	Material	Labor	Equipment	Total	Total Incl O&P
0010	**PLASTIC-LAMINATE CLAD TOILET COMPARTMENT COMPONENTS**									
0100	Pilasters									
0110	Overhead braced, 7" wide x 82" high	2 Carp	22.20	.721	Ea.	102	28		130	158
0130	Floor anchored, 7" wide x 70" high		23.30	.687		111	26.50		137.50	166
0150	Ceiling hung, 7" wide x 83" high		13.30	1.203		124	46.50		170.50	213
0180	Wall hung, 3" wide x 58" high	↓	18.90	.847	↓	96.50	33		129.50	160
0200	Panels									
0210	31" wide x 58" high	2 Carp	18.90	.847	Ea.	165	33		198	236
0230	51" wide x 58" high		18.90	.847		207	33		240	282
0250	63" wide x 58" high	↓	18.90	.847	↓	242	33		275	320
0300	Doors									
0310	24" wide x 58" high	2 Carp	14.10	1.135	Ea.	152	44		196	240
0330	26" wide x 58" high		14.10	1.135		157	44		201	245
0350	28" wide x 58" high		14.10	1.135		162	44		206	251
0370	36" wide x 58" high	↓	14.10	1.135	↓	179	44		223	270

10 21 14 – Toilet Compartment Components

10 21 14.16 Plastic-Lam Clad Toilet Compart. Components		Crew	Daily Output	Labor-Hours	Unit	2019 Bare Costs Material	Labor	Equipment	Total	Total Incl O&P
0400	Headrails									
0410	62" long	2 Carp	65	.246	Ea.	26.50	9.55		36.05	45
0430	84" long		60	.267		30.50	10.35		40.85	51
0450	120" long		30	.533		39.50	20.50		60	77.50

10 21 14.19 Plastic Toilet Compartment Components		Crew	Daily Output	Labor-Hours	Unit	Material	Labor	Equipment	Total	Total Incl O&P
0010	**PLASTIC TOILET COMPARTMENT COMPONENTS**									
0100	Pilasters									
0110	Overhead braced, polymer plastic, 7" wide x 82" high	2 Carp	22.20	.721	Ea.	114	28		142	171
0120	Phenolic		22.20	.721		140	28		168	200
0130	Floor braced, polymer plastic, 7" wide x 70" high		23.30	.687		180	26.50		206.50	242
0140	Phenolic		23.30	.687		134	26.50		160.50	191
0150	Ceiling hung, polymer plastic, 7" wide x 83" high		13.30	1.203		174	46.50		220.50	269
0160	Phenolic		13.30	1.203		163	46.50		209.50	256
0180	Wall hung, phenolic, 3" wide x 58" high		18.90	.847		105	33		138	169
0200	Panels									
0203	Polymer plastic, 18" wide x 55" high	2 Carp	18.90	.847	Ea.	310	33		343	395
0206	Phenolic, 18" wide x 58" high		18.90	.847		258	33		291	340
0210	Polymer plastic, 31" wide x 55" high		18.90	.847		300	33		333	390
0220	Phenolic, 31" wide x 58" high		18.90	.847		277	33		310	360
0223	Polymer plastic, 48" wide x 55" high		18.90	.847		535	33		568	645
0226	Phenolic, 48" wide x 58" high		18.90	.847		465	33		498	565
0230	Polymer plastic, 51" wide x 55" high		18.90	.847		450	33		483	550
0240	Phenolic, 51" wide x 58" high		18.90	.847		400	33		433	495
0250	Polymer plastic, 63" wide x 55" high		18.90	.847		615	33		648	730
0260	Phenolic, 63" wide x 58" high		18.90	.847		425	33		458	525
0300	Doors									
0310	Polymer plastic, 24" wide x 55" high	2 Carp	14.10	1.135	Ea.	217	44		261	310
0320	Phenolic, 24" wide x 58" high		14.10	1.135		310	44		354	420
0330	Polymer plastic, 26" wide x 55" high		14.10	1.135		240	44		284	335
0340	Phenolic, 26" wide x 58" high		14.10	1.135		330	44		374	440
0350	Polymer plastic, 28" wide x 55" high		14.10	1.135		261	44		305	360
0360	Phenolic, 28" wide x 58" high		14.10	1.135		350	44		394	460
0370	Polymer plastic, 36" wide x 55" high		14.10	1.135		310	44		354	415
0380	Phenolic, 36" wide x 58" high		14.10	1.135		495	44		539	620
0400	Headrails									
0410	For polymer plastic, 62" long	2 Carp	65	.246	Ea.	23	9.55		32.55	41
0420	Phenolic		65	.246		22.50	9.55		32.05	41
0430	For polymer plastic, 84" long		50	.320		31.50	12.40		43.90	55.50
0440	Phenolic		50	.320		34	12.40		46.40	58
0450	For polymer plastic, 120" long		30	.533		41	20.50		61.50	79
0460	Phenolic		30	.533		40.50	20.50		61	78.50

10 22 Partitions

10 22 19 – Demountable Partitions

10 22 19.43 Demountable Composite Partitions		Crew	Daily Output	Labor-Hours	Unit	2019 Bare Costs Material	Labor	Equipment	Total	Total Incl O&P
0010	**DEMOUNTABLE COMPOSITE PARTITIONS**, add for doors									
0100	Do not deduct door openings from total L.F.									
0900	Demountable gypsum system on 2″ to 2-1/2″									
1000	Steel studs, 9′ high, 3″ to 3-3/4″ thick									
1200	Vinyl-clad gypsum	2 Carp	48	.333	L.F.	60.50	12.90		73.40	88
1300	Fabric clad gypsum		44	.364		151	14.10		165.10	190
1500	Steel clad gypsum		40	.400		169	15.50		184.50	211
1600	1.75 system, aluminum framing, vinyl-clad hardboard,									
1800	Paper honeycomb core panel, 1-3/4″ to 2-1/2″ thick									
1900	9′ high	2 Carp	48	.333	L.F.	102	12.90		114.90	134
2100	7′ high		60	.267		91.50	10.35		101.85	117
2200	5′ high		80	.200		77	7.75		84.75	98
2250	Unitized gypsum system									
2300	Unitized panel, 9′ high, 2″ to 2-1/2″ thick									
2350	Vinyl-clad gypsum	2 Carp	48	.333	L.F.	131	12.90		143.90	166
2400	Fabric clad gypsum	″	44	.364	″	216	14.10		230.10	261
2500	Unitized mineral fiber system									
2510	Unitized panel, 9′ high, 2-1/4″ thick, aluminum frame									
2550	Vinyl-clad mineral fiber	2 Carp	48	.333	L.F.	130	12.90		142.90	165
2600	Fabric clad mineral fiber	″	44	.364	″	194	14.10		208.10	238
2800	Movable steel walls, modular system									
2900	Unitized panels, 9′ high, 48″ wide									
3100	Baked enamel, pre-finished	2 Carp	60	.267	L.F.	148	10.35		158.35	180
3200	Fabric clad steel	″	56	.286	″	213	11.05		224.05	253
5500	For acoustical partitions, add, unsealed				S.F.	2.38			2.38	2.62
5550	Sealed				″	11.10			11.10	12.20
5700	For doors, see Section 08 16									
5800	For door hardware, see Section 08 71									
6100	In-plant modular office system, w/prehung hollow core door									
6200	3″ thick polystyrene core panels									
6250	12′ x 12′, 2 wall	2 Clab	3.80	4.211	Ea.	4,950	128		5,078	5,650
6300	4 wall		1.90	8.421		7,550	256		7,806	8,750
6350	16′ x 16′, 2 wall		3.60	4.444		8,150	135		8,285	9,200
6400	4 wall		1.80	8.889		10,200	270		10,470	11,700

10 22 23 – Portable Partitions, Screens, and Panels

10 22 23.13 Wall Screens		Crew	Daily Output	Labor-Hours	Unit	Material	Labor	Equipment	Total	Total Incl O&P
0010	**WALL SCREENS**, divider panels, free standing, fiber core									
0020	Fabric face straight									
1500	6′-0″ high	2 Carp	125	.128	L.F.	112	4.96		116.96	131
1600	6′-0″ long, 5′-0″ high		162	.099		112	3.83		115.83	129
3200	Economical panels, fabric face, 4′-0″ long, 5′-0″ high		132	.121		60.50	4.70		65.20	74.50
3250	6′-0″ high		112	.143		67	5.55		72.55	82.50
3300	5′-0″ long, 5′-0″ high		150	.107		61.50	4.13		65.63	75
3350	6′-0″ high		125	.128		54.50	4.96		59.46	68
3450	Acoustical panels, 60 to 90 NRC, 3′-0″ long, 5′-0″ high		90	.178		78.50	6.90		85.40	98
3550	6′-0″ high		75	.213		80	8.25		88.25	102
3600	5′-0″ long, 5′-0″ high		150	.107		65.50	4.13		69.63	79.50
3650	6′-0″ high		125	.128		69	4.96		73.96	84
3700	6′-0″ long, 5′-0″ high		162	.099		53.50	3.83		57.33	65.50
3750	6′-0″ high		138	.116		83	4.49		87.49	99
3800	Economy acoustical panels, 40 NRC, 4′-0″ long, 5′-0″ high		132	.121		60.50	4.70		65.20	74.50
3850	6′-0″ high		112	.143		67	5.55		72.55	82.50

10 22 Partitions

10 22 23 – Portable Partitions, Screens, and Panels

10 22 23.13 Wall Screens		Crew	Daily Output	Labor-Hours	Unit	Material	Labor	Equipment	Total	Total Incl O&P
						2019 Bare Costs				
3900	5'-0" long, 6'-0" high	2 Carp	125	.128	L.F.	54.50	4.96		59.46	68
3950	6'-0" long, 5'-0" high	↓	162	.099	↓	50	3.83		53.83	62

10 22 33 – Accordion Folding Partitions

10 22 33.10 Partitions, Accordion Folding		Crew	Daily Output	Labor-Hours	Unit	Material	Labor	Equipment	Total	Total Incl O&P
0010	**PARTITIONS, ACCORDION FOLDING**									
0100	Vinyl covered, over 150 S.F., frame not included									
0300	Residential, 1.25 lb./S.F., 8' maximum height	2 Carp	300	.053	S.F.	22.50	2.07		24.57	28.50
0400	Commercial, 1.75 lb./S.F., 8' maximum height		225	.071		25.50	2.76		28.26	33
0900	Acoustical, 3 lb./S.F., 17' maximum height		100	.160		28	6.20		34.20	41.50
1200	5 lb./S.F., 20' maximum height		95	.168		39	6.55		45.55	53.50
1500	Vinyl-clad wood or steel, electric operation, 5.0 psf		160	.100		56	3.88		59.88	68.50
1900	Wood, non-acoustic, birch or mahogany, to 10' high	↓	300	.053	↓	30.50	2.07		32.57	37

10 22 39 – Folding Panel Partitions

10 22 39.10 Partitions, Folding Panel		Crew	Daily Output	Labor-Hours	Unit	Material	Labor	Equipment	Total	Total Incl O&P
0010	**PARTITIONS, FOLDING PANEL**, acoustic, wood									
0100	Vinyl faced, to 18' high, 6 psf, economy trim	2 Carp	60	.267	S.F.	60	10.35		70.35	82.50
0150	Standard trim		45	.356		71.50	13.80		85.30	102
0200	Premium trim		30	.533		92	20.50		112.50	135
0400	Plastic laminate or hardwood finish, standard trim		60	.267		61.50	10.35		71.85	84.50
0500	Premium trim		30	.533		65.50	20.50		86	106
0600	Wood, low acoustical type, 4.5 psf, to 14' high	↓	50	.320	↓	44.50	12.40		56.90	69.50

10 26 Wall and Door Protection

10 26 16 – Bumper Guards

10 26 16.10 Guard, Bumper		Crew	Daily Output	Labor-Hours	Unit	Material	Labor	Equipment	Total	Total Incl O&P
0010	**GUARD, BUMPER**									
1200	Bed bumper, vinyl acrylic, alum. retainer, 21" long	1 Carp	10	.800	Ea.	42	31		73	97.50
1300	53" long with aligner		9	.889	"	106	34.50		140.50	174
1400	Bumper, vinyl cover, alum. retain., cush. mnt., 1-1/2" x 2-3/4"		80	.100	L.F.	14.30	3.88		18.18	22
1500	2" x 4-1/4"		80	.100		21	3.88		24.88	29.50
1600	Surface mounted, 1-3/4" x 3-5/8"	↓	80	.100	↓	12.05	3.88		15.93	19.65

10 26 16.16 Protective Corridor Handrails		Crew	Daily Output	Labor-Hours	Unit	Material	Labor	Equipment	Total	Total Incl O&P
0010	**PROTECTIVE CORRIDOR HANDRAILS**									
3000	Handrail/bumper, vinyl cover, alum. retainer									
3010	Bracket mounted, flat rail, 5-1/2"	1 Carp	80	.100	L.F.	18.40	3.88		22.28	27
3100	6-1/2"		80	.100		23	3.88		26.88	31.50
3200	Bronze bracket, 1-3/4" diam. rail	↓	80	.100	↓	16.60	3.88		20.48	24.50

10 26 23 – Protective Wall Covering

10 26 23.10 Wall Covering, Protective		Crew	Daily Output	Labor-Hours	Unit	Material	Labor	Equipment	Total	Total Incl O&P
0010	**WALL COVERING, PROTECTIVE**									
0400	Rub rail, vinyl, adhesive mounted	1 Carp	185	.043	L.F.	9.75	1.68		11.43	13.50
0500	Neoprene, aluminum backing, 1-1/2" x 2"		110	.073		9.15	2.82		11.97	14.75
1000	Trolley rail, PVC, clipped to wall, 5" high		185	.043		8.95	1.68		10.63	12.60
1050	8" high	↓	180	.044		15.30	1.72		17.02	19.70
1700	Bumper rail, stainless steel, flat bar on brackets, 4" x 1/4"	2 Skwk	120	.133	↓	38	5.25		43.25	51
1775	Wall end guard, stainless steel, 16 ga., 36" tall, screwed to studs	1 Skwk	30	.267	Ea.	26	10.50		36.50	46
2000	Crash rail, vinyl cover, alum. retainer, 1" x 4"	1 Carp	110	.073	L.F.	11.10	2.82		13.92	16.90
2100	1" x 8"		90	.089		18.85	3.44		22.29	26
2150	Vinyl inserts, aluminum plate, 1" x 2-1/2"	↓	110	.073	↓	15	2.82		17.82	21

10 26 Wall and Door Protection

10 26 23 – Protective Wall Covering

10 26 23.10 Wall Covering, Protective		Crew	Daily Output	Labor-Hours	Unit	2019 Bare Costs Material	Labor	Equipment	Total	Total Incl O&P
2200	1″ x 5″	1 Carp	90	.089	L.F.	24	3.44		27.44	32

10 26 23.13 Impact Resistant Wall Protection

		Crew	Daily Output	Labor-Hours	Unit	Material	Labor	Equipment	Total	Total Incl O&P
0010	**IMPACT RESISTANT WALL PROTECTION**									
0100	Vinyl wall protection, complete instl. incl. panels, trim, adhesive									
0110	.040″ thk., std. colors	2 Carp	320	.050	S.F.	15.55	1.94		17.49	20.50
0120	.040″ thk., element patterns		300	.053		17.25	2.07		19.32	22.50
0130	.060″ thk., std. colors		300	.053		17.90	2.07		19.97	23
0140	.060″ thk., element patterns		280	.057		18.75	2.21		20.96	24
1750	Wallguard stainless steel baseboard, 12″ tall, adhesive applied	2 Skwk	260	.062	L.F.	32.50	2.43		34.93	40
1775	Wall protection, stainless steel, 16 ga., 48″ x 36″ tall, screwed to studs	″	500	.032	S.F.	7.85	1.26		9.11	10.75

10 26 33 – Door and Frame Protection

10 26 33.10 Protection, Door and Frame

		Crew	Daily Output	Labor-Hours	Unit	Material	Labor	Equipment	Total	Total Incl O&P
0010	**PROTECTION, DOOR AND FRAME**									
0100	Door frame guard, vinyl, 3″ x 3″ x 4′	1 Carp	30	.267	Ea.	85.50	10.35		95.85	112
0110	Door frame guard, vinyl, 3″ x 3″ x 8′		26	.308		133	11.90		144.90	166
0120	Door frame guard, stainless steel, 3″ x 3″ x 4′		30	.267		425	10.35		435.35	480
0501	Steel door track/wheel guard, 4′-0″ high	2 Carp	12	1.333		110	51.50		161.50	207

10 28 Toilet, Bath, and Laundry Accessories

10 28 13 – Toilet Accessories

10 28 13.13 Commercial Toilet Accessories

		Crew	Daily Output	Labor-Hours	Unit	Material	Labor	Equipment	Total	Total Incl O&P
0010	**COMMERCIAL TOILET ACCESSORIES**									
0200	Curtain rod, stainless steel, 5′ long, 1″ diameter	1 Carp	13	.615	Ea.	27	24		51	69.50
0300	1-1/4″ diameter		13	.615		29.50	24		53.50	71.50
0400	Diaper changing station, horizontal, wall mounted, plastic		10	.800		244	31		275	320
0500	Dispenser units, combined soap & towel dispensers,									
0510	Mirror and shelf, flush mounted	1 Carp	10	.800	Ea.	365	31		396	455
0600	Towel dispenser and waste receptacle,									
0610	18 gallon capacity	1 Carp	10	.800	Ea.	340	31		371	425
0800	Grab bar, straight, 1-1/4″ diameter, stainless steel, 18″ long		24	.333		30	12.90		42.90	54.50
1100	36″ long		20	.400		34	15.50		49.50	63
1105	42″ long		20	.400		38.50	15.50		54	67.50
1120	Corner, 36″ long		20	.400		86.50	15.50		102	121
2010	Tub/shower/toilet, 2-wall, 36″ x 24″		12	.667		94	26		120	147
3000	Mirror, with stainless steel 3/4″ square frame, 18″ x 24″		20	.400		47	15.50		62.50	77
3100	36″ x 24″		15	.533		117	20.50		137.50	163
3300	72″ x 24″		6	1.333		291	51.50		342.50	405
3500	With 5″ stainless steel shelf, 18″ x 24″		20	.400		185	15.50		200.50	230
3600	36″ x 24″		15	.533		242	20.50		262.50	300
3800	72″ x 24″		6	1.333		269	51.50		320.50	380
4100	Mop holder strip, stainless steel, 5 holders, 48″ long		20	.400		80	15.50		95.50	114
4200	Napkin/tampon dispenser, recessed		15	.533		555	20.50		575.50	645
4220	Semi-recessed		6.50	1.231		365	47.50		412.50	485
4250	Napkin receptacle, recessed		6.50	1.231		171	47.50		218.50	268
4300	Robe hook, single, regular		96	.083		19.95	3.23		23.18	27.50
4400	Heavy duty, concealed mounting		56	.143		22.50	5.55		28.05	33.50
4600	Soap dispenser, chrome, surface mounted, liquid		20	.400		50.50	15.50		66	81
5000	Recessed stainless steel, liquid		10	.800		158	31		189	225
5600	Shelf, stainless steel, 5″ wide, 18 ga., 24″ long		24	.333		76	12.90		88.90	106
6100	Toilet tissue dispenser, surface mounted, SS, single roll		30	.267		17.65	10.35		28	36.50

10 28 Toilet, Bath, and Laundry Accessories

10 28 13 – Toilet Accessories

10 28 13.13 Commercial Toilet Accessories		Crew	Daily Output	Labor-Hours	Unit	Material	2019 Bare Costs Labor	Equipment	Total	Total Incl O&P
6200	Double roll	1 Carp	24	.333	Ea.	21.50	12.90		34.40	45.50
6240	Plastic, twin/jumbo dbl. roll		24	.333		30.50	12.90		43.40	55
6400	Towel bar, stainless steel, 18" long		23	.348		42	13.50		55.50	69
6500	30" long		21	.381		51	14.75		65.75	80.50
6700	Towel dispenser, stainless steel, surface mounted		16	.500		43	19.40		62.40	79.50
6800	Flush mounted, recessed		10	.800		109	31		140	171
6900	Plastic, touchless, battery operated		16	.500		107	19.40		126.40	150
7400	Tumbler holder, for tumbler only		30	.267		27	10.35		37.35	46.50
7410	Tumbler holder, recessed		20	.400		7.85	15.50		23.35	34
7500	Soap, tumbler & toothbrush		30	.267		19.80	10.35		30.15	39
7510	Tumbler & toothbrush holder		20	.400		13.30	15.50		28.80	40
8000	Waste receptacles, stainless steel, with top, 13 gallon		10	.800		305	31		336	385

10 28 16 – Bath Accessories

10 28 16.20 Medicine Cabinets

		Crew	Daily Output	Labor-Hours	Unit	Material	Labor	Equipment	Total	Total Incl O&P
0010	**MEDICINE CABINETS**									
0020	With mirror, sst frame, 16" x 22", unlighted	1 Carp	14	.571	Ea.	96.50	22		118.50	143
0100	Wood frame		14	.571		136	22		158	187
0300	Sliding mirror doors, 20" x 16" x 4-3/4", unlighted		7	1.143		126	44.50		170.50	212
0400	24" x 19" x 8-1/2", lighted		5	1.600		216	62		278	340
0600	Triple door, 30" x 32", unlighted, plywood body		7	1.143		355	44.50		399.50	465
0700	Steel body		7	1.143		355	44.50		399.50	465
0900	Oak door, wood body, beveled mirror, single door		7	1.143		176	44.50		220.50	266
1000	Double door		6	1.333		370	51.50		421.50	490
1200	Hotel cabinets, stainless, with lower shelf, unlighted		10	.800		217	31		248	290
1300	Lighted		5	1.600		335	62		397	470

10 28 19 – Tub and Shower Enclosures

10 28 19.10 Partitions, Shower

		Crew	Daily Output	Labor-Hours	Unit	Material	Labor	Equipment	Total	Total Incl O&P
0010	**PARTITIONS, SHOWER** floor mounted, no plumbing									
0400	Cabinet, one piece, fiberglass, 32" x 32"	2 Carp	5	3.200	Ea.	620	124		744	890
0420	36" x 36"		5	3.200		600	124		724	865
0440	36" x 48"		5	3.200		1,375	124		1,499	1,725
0460	Acrylic, 32" x 32"		5	3.200		355	124		479	595
0480	36" x 36"		5	3.200		1,075	124		1,199	1,375
0500	36" x 48"		5	3.200		1,300	124		1,424	1,625
0520	Shower door for above, clear plastic, 24" wide	1 Carp	8	1		186	39		225	268
0540	28" wide		8	1		246	39		285	335
0560	Tempered glass, 24" wide		8	1		241	39		280	330
0580	28" wide		8	1		281	39		320	375
2400	Glass stalls, with doors, no receptors, chrome on brass	2 Shee	3	5.333		1,650	231		1,881	2,175
2700	Anodized aluminum	"	4	4		1,325	173		1,498	1,725
3200	Receptors, precast terrazzo, 32" x 32"	2 Marb	14	1.143		360	41.50		401.50	465
3300	48" x 34"		9.50	1.684		470	61.50		531.50	615
3500	Plastic, simulated terrazzo receptor, 32" x 32"		14	1.143		170	41.50		211.50	257
3600	32" x 48"		12	1.333		300	48.50		348.50	410
3800	Precast concrete, colors, 32" x 32"		14	1.143		250	41.50		291.50	345
3900	48" x 48"		8	2		277	73		350	425
4100	Shower doors, economy plastic, 24" wide	1 Shee	9	.889		135	38.50		173.50	213
4200	Tempered glass door, economy		8	1		295	43.50		338.50	395
4400	Folding, tempered glass, aluminum frame		6	1.333		435	57.50		492.50	570
4700	Deluxe, tempered glass, chrome on brass frame, 42" to 44"		8	1		435	43.50		478.50	550
4800	39" to 48" wide		1	8		665	345		1,010	1,300
4850	On anodized aluminum frame, obscure glass		2	4		580	173		753	920

10 28 Toilet, Bath, and Laundry Accessories

10 28 19 – Tub and Shower Enclosures

10 28 19.10 Partitions, Shower		Crew	Daily Output	Labor-Hours	Unit	Material	Labor	Equipment	Total	Total Incl O&P
						2019 Bare Costs				
4900	Clear glass	1 Shee	1	8	Ea.	655	345		1,000	1,300
5100	Shower enclosure, tempered glass, anodized alum. frame									
5120	2 panel & door, corner unit, 32″ x 32″	1 Shee	2	4	Ea.	970	173		1,143	1,350
5140	Neo-angle corner unit, 16″ x 24″ x 16″	″	2	4		1,150	173		1,323	1,525
5200	Shower surround, 3 wall, polypropylene, 32″ x 32″	1 Carp	4	2		615	77.50		692.50	805
5220	PVC, 32″ x 32″		4	2		400	77.50		477.50	570
5240	Fiberglass		4	2		395	77.50		472.50	565
5250	2 wall, polypropylene, 32″ x 32″		4	2		310	77.50		387.50	475
5270	PVC		4	2		375	77.50		452.50	540
5290	Fiberglass		4	2		375	77.50		452.50	545
5300	Tub doors, tempered glass & frame, obscure glass	1 Shee	8	1		222	43.50		265.50	315
5400	Clear glass		6	1.333		515	57.50		572.50	660
5600	Chrome plated, brass frame, obscure glass		8	1		291	43.50		334.50	390
5700	Clear glass		6	1.333		715	57.50		772.50	885
5900	Tub/shower enclosure, temp. glass, alum. frame, obscure glass		2	4		395	173		568	720
6200	Clear glass		1.50	5.333		825	231		1,056	1,300
6500	On chrome-plated brass frame, obscure glass		2	4		545	173		718	885
6600	Clear glass		1.50	5.333		1,175	231		1,406	1,650
6800	Tub surround, 3 wall, polypropylene	1 Carp	4	2		253	77.50		330.50	405
6900	PVC		4	2		370	77.50		447.50	535
7000	Fiberglass, obscure glass		4	2		390	77.50		467.50	555
7100	Clear glass		3	2.667		655	103		758	895

10 31 Manufactured Fireplaces

10 31 13 – Manufactured Fireplace Chimneys

10 31 13.10 Fireplace Chimneys

Line	Description	Crew	Daily Output	Labor-Hours	Unit	Material	Labor	Equipment	Total	Total Incl O&P
0010	**FIREPLACE CHIMNEYS**									
0500	Chimney dbl. wall, all stainless, over 8′-6″, 7″ diam., add to fireplace	1 Carp	33	.242	V.L.F.	88	9.40		97.40	113
0600	10″ diameter, add to fireplace		32	.250		111	9.70		120.70	138
0700	12″ diameter, add to fireplace		31	.258		173	10		183	207
0800	14″ diameter, add to fireplace		30	.267		223	10.35		233.35	262
1000	Simulated brick chimney top, 4′ high, 16″ x 16″		10	.800	Ea.	455	31		486	550
1100	24″ x 24″		7	1.143	″	560	44.50		604.50	690

10 31 23 – Prefabricated Fireplaces

10 31 23.10 Fireplace, Prefabricated

Line	Description	Crew	Daily Output	Labor-Hours	Unit	Material	Labor	Equipment	Total	Total Incl O&P
0010	**FIREPLACE, PREFABRICATED**, free standing or wall hung									
0100	With hood & screen, painted	1 Carp	1.30	6.154	Ea.	1,625	238		1,863	2,175
0150	Average		1	8		1,800	310		2,110	2,500
0200	Stainless steel		.90	8.889		3,225	345		3,570	4,125
1500	Simulated logs, gas fired, 40,000 BTU, 2′ long, manual safety pilot		7	1.143	Set	540	44.50		584.50	670
1600	Adjustable flame remote pilot		6	1.333		1,275	51.50		1,326.50	1,475
1700	Electric, 1,500 BTU, 1′-6″ long, incandescent flame		7	1.143		262	44.50		306.50	360
1800	1,500 BTU, LED flame		6	1.333		345	51.50		396.50	465
2000	Fireplace, built-in, 36″ hearth, radiant		1.30	6.154	Ea.	720	238		958	1,200
2100	Recirculating, small fan		1	8		910	310		1,220	1,500
2150	Large fan		.90	8.889		2,025	345		2,370	2,800
2200	42″ hearth, radiant		1.20	6.667		1,125	258		1,383	1,650
2300	Recirculating, small fan		.90	8.889		1,250	345		1,595	1,950
2350	Large fan		.80	10		1,450	390		1,840	2,225
2400	48″ hearth, radiant		1.10	7.273		2,375	282		2,657	3,075

10 31 Manufactured Fireplaces

10 31 23 – Prefabricated Fireplaces

10 31 23.10 Fireplace, Prefabricated		Crew	Daily Output	Labor-Hours	Unit	Material	2019 Bare Costs Labor	Equipment	Total	Total Incl O&P
2500	Recirculating, small fan	1 Carp	.80	10	Ea.	2,900	390		3,290	3,850
2550	Large fan		.70	11.429		2,600	445		3,045	3,575
3000	See through, including doors		.80	10		2,350	390		2,740	3,250
3200	Corner (2 wall)	↓	1	8	↓	3,275	310		3,585	4,100

10 43 Emergency Aid Specialties

10 43 13 – Defibrillator Cabinets

10 43 13.05 Defibrillator Cabinets		Crew	Daily Output	Labor-Hours	Unit	Material	Labor	Equipment	Total	Total Incl O&P
0010	**DEFIBRILLATOR CABINETS**, not equipped, stainless steel									
0050	Defibrillator cabinet, stainless steel with strobe & alarm 12" x 27"	1 Carp	10	.800	Ea.	450	31		481	545
0100	Automatic External Defibrillator	"	30	.267	"	1,375	10.35		1,385.35	1,550

10 44 Fire Protection Specialties

10 44 13 – Fire Protection Cabinets

10 44 13.53 Fire Equipment Cabinets		Crew	Daily Output	Labor-Hours	Unit	Material	Labor	Equipment	Total	Total Incl O&P
0010	**FIRE EQUIPMENT CABINETS**, not equipped, 20 ga. steel box									
0040	Recessed, D.S. glass in door, box size given									
1000	Portable extinguisher, single, 8" x 12" x 27", alum. door & frame	Q-12	8	2	Ea.	135	79.50		214.50	279
1100	Steel door and frame		8	2		140	79.50		219.50	285
2700	Fire blanket & extinguisher cab, inc blanket, rec stl., 14" x 40" x 8"		7	2.286		217	91		308	390
2800	Fire blanket cab, inc blanket, surf mtd, stl, 15"x10"x5", w/pwdr coat fin	↓	8	2	↓	101	79.50		180.50	242
3000	Hose rack assy., 1-1/2" valve & 100' hose, 24" x 40" x 5-1/2"									
3100	Aluminum door and frame	Q-12	6	2.667	Ea.	510	106		616	735
3200	Steel door and frame	"	6	2.667	"	268	106		374	470

10 44 16 – Fire Extinguishers

10 44 16.13 Portable Fire Extinguishers		Crew	Daily Output	Labor-Hours	Unit	Material	Labor	Equipment	Total	Total Incl O&P
0010	**PORTABLE FIRE EXTINGUISHERS**									
0140	CO_2, with hose and "H" horn, 10 lb.				Ea.	295			295	325
1000	Dry chemical, pressurized									
1040	Standard type, portable, painted, 2-1/2 lb.				Ea.	40.50			40.50	44.50
1080	10 lb.					86.50			86.50	95.50
1100	20 lb.					138			138	152
1120	30 lb.					435			435	480
2000	ABC all purpose type, portable, 2-1/2 lb.					23.50			23.50	26
2080	9-1/2 lb.					51.50			51.50	56.50
3500	Halotron 1, 2-1/2 lb.					132			132	145
3600	5 lb.					226			226	249
3700	60 lb.				↓	440			440	480

10 44 16.16 Wheeled Fire Extinguisher Units		Crew	Daily Output	Labor-Hours	Unit	Material	Labor	Equipment	Total	Total Incl O&P
0010	**WHEELED FIRE EXTINGUISHER UNITS**									
0350	CO_2, portable, with swivel horn									
0360	Wheeled type, cart mounted, 50 lb.				Ea.	1,225			1,225	1,350
0400	100 lb.				"	4,400			4,400	4,850
2200	ABC all purpose type									
2300	Wheeled, 45 lb.				Ea.	780			780	860
2360	150 lb.				"	2,000			2,000	2,200

10 51 13 – Metal Lockers

10 51 13.10 Lockers		Crew	Daily Output	Labor-Hours	Unit	Material	2019 Bare Costs Labor	Equipment	Total	Total Incl O&P
0011	**LOCKERS** steel, baked enamel, knock down construction									
0012	Body- 24ga, door- 16 or 18ga									
0110	1-tier locker, 12" x 15" x 72"	1 Shee	20	.400	Ea.	267	17.30		284.30	325
0120	18" x 15" x 72"		20	.400		236	17.30		253.30	288
0130	12" x 18" x 72"		20	.400		207	17.30		224.30	257
0140	18" x 18" x 72"		20	.400		281	17.30		298.30	340
0410	2- tier, 12" x 15" x 36"		30	.267		253	11.55		264.55	297
0420	18" x 15" x 36"		30	.267		238	11.55		249.55	280
0430	12" x 18" x 36"		30	.267		310	11.55		321.55	365
0440	18" x 18" x 36"		30	.267		262	11.55		273.55	305
0450	3- tier 12" x 15" x 24"		30	.267		288	11.55		299.55	335
0455	12" x 18" x 24"		30	.267		297	11.55		308.55	345
0480	4- tier 12" x 15" x 18"		30	.267		370	11.55		381.55	425
0485	12" x 18" x 18"		30	.267		385	11.55		396.55	445
0500	Two person, 18" x 15" x 72"		20	.400		335	17.30		352.30	400
0510	18" x 18" x 72"		20	.400		305	17.30		322.30	365
0520	Duplex, 15" x 15" x 72"		20	.400		340	17.30		357.30	405
0530	15" x 21" x 72"		20	.400		380	17.30		397.30	450
1100	1-tier, 12" x 18" x 72"		7.50	1.067		335	46		381	440
1110	18" x 15" x 72"		7.50	1.067		550	46		596	680
1115	18" x 18" x 72"		7.50	1.067		560	46		606	690
1130	2-tier, 12" x 15" x 36"		7.50	1.067		575	46		621	705
1135	12" x 18" x 36"		7.50	1.067		610	46		656	750
1140	18" x 15" x 36"		7.50	1.067		665	46		711	805
1145	18" x 18" x 36"		7.50	1.067		675	46		721	820
1160	3-tier, 12" x 18" x 24"		7.50	1.067		780	46		826	935
1180	4-tier, 12" x 18" x 18"		7.50	1.067		595	46		641	730
1190	5H box, 12" x 18" x 14.4"		7.50	1.067		610	46		656	745
1195	18" x 18" x 14.4"		7.50	1.067		785	46		831	935
1196	6H box, 12" x 15" x 12"		7.50	1.067		525	46		571	650
1197	12" x 18" x 12"		7.50	1.067		595	46		641	730
1198	18" x 18" x 12"		7.50	1.067		740	46		786	890
3250	Rack w/24 wire mesh baskets		1.50	5.333	Set	405	231		636	830
3260	30 baskets		1.25	6.400		390	277		667	890
3270	36 baskets		.95	8.421		455	365		820	1,100
3280	42 baskets		.80	10		455	435		890	1,225
3600	For hanger rods, add				Ea.	2.53			2.53	2.78
3650	For number plate kit, 100 plates #1 - #100, add	1 Shee	4	2		84	86.50		170.50	237
3700	For locker base, closed front panel		90	.089		5.50	3.85		9.35	12.45
3710	End panel, bolted		36	.222		8.60	9.60		18.20	25.50
3800	For sloping top, 12" wide		24	.333		31	14.45		45.45	58
3810	15" wide		24	.333		31.50	14.45		45.95	58.50
3820	18" wide		24	.333		35	14.45		49.45	62.50
3850	Sloping top end panel, 12" deep		72	.111		16.55	4.81		21.36	26
3860	15" deep		72	.111		16.40	4.81		21.21	26
3870	18" deep		72	.111		17.50	4.81		22.31	27.50
3900	For finish end panels, steel, 60" high, 15" deep		12	.667		30	29		59	81
3910	72" high, 12" deep		12	.667		27	29		56	78
3920	18" deep		12	.667		42.50	29		71.50	94.50
5000	For "ready to assemble" lockers,									
5010	Add to labor						75%			
5020	Deduct from material					20%				

10 55 Postal Specialties

10 55 23 – Mail Boxes

10 55 23.10 Commercial Mail Boxes		Crew	Daily Output	Labor-Hours	Unit	Material	2019 Bare Costs Labor	Equipment	Total	Total Incl O&P
0010	**COMMERCIAL MAIL BOXES**									
0020	Horiz., key lock, 5"H x 6"W x 15"D, alum., rear load	1 Carp	34	.235	Ea.	42.50	9.10		51.60	62
0100	Front loading		34	.235		42.50	9.10		51.60	62
0200	Double, 5"H x 12"W x 15"D, rear loading		26	.308		65	11.90		76.90	91
0300	Front loading		26	.308		79	11.90		90.90	107
0500	Quadruple, 10"H x 12"W x 15"D, rear loading		20	.400		109	15.50		124.50	146
0600	Front loading		20	.400		95.50	15.50		111	131
1600	Vault type, horizontal, for apartments, 4" x 5"		34	.235		26.50	9.10		35.60	44.50
1700	Alphabetical directories, 120 names	↓	10	.800	↓	104	31		135	165

10 56 Storage Assemblies

10 56 13 – Metal Storage Shelving

10 56 13.10 Shelving

10 56 13.10 Shelving		Crew	Daily Output	Labor-Hours	Unit	Material	Labor	Equipment	Total	Total Incl O&P
0010	**SHELVING**									
0020	Metal, industrial, cross-braced, 3' W, 12" D	1 Sswk	175	.046	SF Shlf	8.05	1.89		9.94	12.20
0100	24" D		330	.024		6.35	1		7.35	8.75
2200	Wide span, 1600 lb. capacity per shelf, 6' W, 24" D		380	.021		7.40	.87		8.27	9.65
2400	36" D	↓	440	.018	↓	6.30	.75		7.05	8.25
3000	Residential, vinyl covered wire, wardrobe, 12" D	1 Carp	195	.041	L.F.	15.30	1.59		16.89	19.50
3100	16" D		195	.041		11.65	1.59		13.24	15.50
3200	Standard, 6" D		195	.041		3.80	1.59		5.39	6.80
3300	9" D		195	.041		6.70	1.59		8.29	10.05
3400	12" D		195	.041		7.95	1.59		9.54	11.35
3500	16" D		195	.041		18.10	1.59		19.69	22.50
3600	20" D		195	.041	↓	21	1.59		22.59	26
3700	Support bracket	↓	80	.100	Ea.	7.30	3.88		11.18	14.45

10 57 Wardrobe and Closet Specialties

10 57 23 – Closet and Utility Shelving

10 57 23.19 Wood Closet and Utility Shelving

10 57 23.19 Wood Closet and Utility Shelving		Crew	Daily Output	Labor-Hours	Unit	Material	Labor	Equipment	Total	Total Incl O&P
0010	**WOOD CLOSET AND UTILITY SHELVING**									
0020	Pine, clear grade, no edge band, 1" x 8"	1 Carp	115	.070	L.F.	3.61	2.70		6.31	8.45
0100	1" x 10"		110	.073		4.49	2.82		7.31	9.60
0200	1" x 12"		105	.076		5.40	2.95		8.35	10.85
0600	Plywood, 3/4" thick with lumber edge, 12" wide		75	.107		1.92	4.13		6.05	8.95
0700	24" wide		70	.114	↓	3.39	4.43		7.82	11.05
0900	Bookcase, clear grade pine, shelves 12" OC, 8" deep, per S.F. shelf		70	.114	S.F.	11.70	4.43		16.13	20
1000	12" deep shelves		65	.123	"	17.60	4.77		22.37	27.50
1200	Adjustable closet rod and shelf, 12" wide, 3' long		20	.400	Ea.	13.25	15.50		28.75	40
1300	8' long		15	.533	"	25.50	20.50		46	62
1500	Prefinished shelves with supports, stock, 8" wide		75	.107	L.F.	5.80	4.13		9.93	13.20
1600	10" wide	↓	70	.114	"	5.50	4.43		9.93	13.35

10 73 Protective Covers

10 73 13 – Awnings

10 73 13.10 Awnings, Fabric		Crew	Daily Output	Labor-Hours	Unit	Material	2019 Bare Costs Labor	Equipment	Total	Total Incl O&P
0010	**AWNINGS, FABRIC**									
0020	Including acrylic canvas and frame, standard design									
0100	Door and window, slope, 3′ high, 4′ wide	1 Carp	4.50	1.778	Ea.	755	69		824	945
0110	6′ wide		3.50	2.286		970	88.50		1,058.50	1,225
0120	8′ wide		3	2.667		1,200	103		1,303	1,475
0200	Quarter round convex, 4′ wide		3	2.667		1,175	103		1,278	1,475
0210	6′ wide		2.25	3.556		1,525	138		1,663	1,900
0220	8′ wide		1.80	4.444		1,875	172		2,047	2,325
0300	Dome, 4′ wide		7.50	1.067		455	41.50		496.50	570
0310	6′ wide		3.50	2.286		1,025	88.50		1,113.50	1,275
0320	8′ wide		2	4		1,800	155		1,955	2,250
0350	Elongated dome, 4′ wide		1.33	6.015		1,700	233		1,933	2,250
0360	6′ wide		1.11	7.207		2,025	279		2,304	2,675
0370	8′ wide		1	8		2,375	310		2,685	3,125
1000	Entry or walkway, peak, 12′ long, 4′ wide	2 Carp	.90	17.778		5,425	690		6,115	7,125
1010	6′ wide		.60	26.667		8,375	1,025		9,400	10,900
1020	8′ wide		.40	40		11,600	1,550		13,150	15,300
1100	Radius with dome end, 4′ wide		1.10	14.545		4,125	565		4,690	5,450
1110	6′ wide		.70	22.857		6,625	885		7,510	8,775
1120	8′ wide		.50	32		9,425	1,250		10,675	12,500
2000	Retractable lateral arm awning, manual									
2010	To 12′ wide, 8′-6″ projection	2 Carp	1.70	9.412	Ea.	1,225	365		1,590	1,950
2020	To 14′ wide, 8′-6″ projection		1.10	14.545		1,425	565		1,990	2,500
2030	To 19′ wide, 8′-6″ projection		.85	18.824		1,950	730		2,680	3,350
2040	To 24′ wide, 8′-6″ projection		.67	23.881		2,450	925		3,375	4,225
2050	Motor for above, add	1 Carp	2.67	3		1,050	116		1,166	1,350
3000	Patio/deck canopy with frame									
3010	12′ wide, 12′ projection	2 Carp	2	8	Ea.	1,725	310		2,035	2,400
3020	16′ wide, 14′ projection	″	1.20	13.333		2,700	515		3,215	3,825
9000	For fire retardant canvas, add					7%				
9010	For lettering or graphics, add					35%				
9020	For painted or coated acrylic canvas, deduct					8%				
9030	For translucent or opaque vinyl canvas, add					10%				
9040	For 6 or more units, deduct					20%	15%			

10 73 16 – Canopies

10 73 16.20 Metal Canopies		Crew	Daily Output	Labor-Hours	Unit	Material	Labor	Equipment	Total	Total Incl O&P
0010	**METAL CANOPIES**									
0020	Wall hung, .032″, aluminum, prefinished, 8′ x 10′	K-2	1.30	18.462	Ea.	2,500	730	189	3,419	4,225
0300	8′ x 20′	″	1.10	21.818	″	4,125	865	224	5,214	6,250
2300	Aluminum entrance canopies, flat soffit, .032″									
2500	3′-6″ x 4′-0″, clear anodized	2 Carp	4	4	Ea.	1,050	155		1,205	1,425

10 74 Manufactured Exterior Specialties

10 74 23 – Cupolas

10 74 23.10 Wood Cupolas		Crew	Daily Output	Labor-Hours	Unit	Material	2019 Bare Costs Labor	Equipment	Total	Total Incl O&P
0010	**WOOD CUPOLAS**									
0020	Stock units, pine, painted, 18" sq., 28" high, alum. roof	1 Carp	4.10	1.951	Ea.	266	75.50		341.50	420
0100	Copper roof		3.80	2.105		292	81.50		373.50	455
0300	23" square, 33" high, aluminum roof		3.70	2.162		460	84		544	645
0400	Copper roof		3.30	2.424		615	94		709	830
0600	30" square, 37" high, aluminum roof		3.70	2.162		610	84		694	810
0700	Copper roof		3.30	2.424		740	94		834	970
0900	Hexagonal, 31" wide, 46" high, copper roof		4	2		980	77.50		1,057.50	1,200
1000	36" wide, 50" high, copper roof		3.50	2.286		1,750	88.50		1,838.50	2,075
1200	For deluxe stock units, add to above					25%				
1400	For custom built units, add to above					50%	50%			

10 74 46 – Window Wells

10 74 46.10 Area Window Wells		Crew	Daily Output	Labor-Hours	Unit	Material	Labor	Equipment	Total	Total Incl O&P
0010	**AREA WINDOW WELLS**, Galvanized steel									
0020	20 ga., 3'-2" wide, 1' deep	1 Sswk	29	.276	Ea.	17.45	11.45		28.90	39
0100	2' deep		23	.348		30	14.40		44.40	58.50
0300	16 ga., 3'-2" wide, 1' deep		29	.276		23.50	11.45		34.95	45.50
0400	3' deep		23	.348		48	14.40		62.40	78
0600	Welded grating for above, 15 lb., painted		45	.178		95.50	7.35		102.85	118
0700	Galvanized		45	.178		125	7.35		132.35	150
0900	Translucent plastic cap for above		60	.133		21	5.55		26.55	32.50

10 75 Flagpoles

10 75 16 – Ground-Set Flagpoles

10 75 16.10 Flagpoles		Crew	Daily Output	Labor-Hours	Unit	Material	Labor	Equipment	Total	Total Incl O&P
0010	**FLAGPOLES**, ground set									
0050	Not including base or foundation									
0100	Aluminum, tapered, ground set 20' high	K-1	2	8	Ea.	1,100	291	123	1,514	1,825
0200	25' high		1.70	9.412		1,150	345	145	1,640	2,000
0300	30' high		1.50	10.667		1,400	390	164	1,954	2,375
0500	40' high		1.20	13.333		3,075	485	205	3,765	4,400

Estimating Tips

General

- The items in this division are usually priced per square foot or each. Many of these items are purchased by the owner for installation by the contractor. Check the specifications for responsibilities and include time for receiving, storage, installation, and mechanical and electrical hookups in the appropriate divisions.
- Many items in Division 11 require some type of support system that is not usually furnished with the item. Examples of these systems include blocking for the attachment of casework and support angles for ceiling-hung projection screens. The required blocking or supports must be added to the estimate in the appropriate division.
- Some items in Division 11 may require assembly or electrical hookups. Verify the amount of assembly required or the need for a hard electrical connection and add the appropriate costs.

Reference Numbers

Reference numbers are shown at the beginning of some major classifications. These numbers refer to related items in the Reference Section. The reference information may be an estimating procedure, an alternate pricing method, or technical information.

11 13 Loading Dock Equipment

11 13 13 – Loading Dock Bumpers

11 13 13.10 Dock Bumpers		Crew	Daily Output	Labor-Hours	Unit	Material	Labor	Equipment	Total	Total Incl O&P
						2019 Bare Costs				
0010	**DOCK BUMPERS** Bolts not included									
0020	2" x 6" to 4" x 8", average	1 Carp	.30	26.667	M.B.F.	1,425	1,025		2,450	3,275

11 30 Residential Equipment

11 30 13 – Residential Appliances

11 30 13.15 Cooking Equipment		Crew	Daily Output	Labor-Hours	Unit	Material	Labor	Equipment	Total	Total Incl O&P
0010	**COOKING EQUIPMENT**									
0020	Cooking range, 30" free standing, 1 oven, minimum	2 Clab	10	1.600	Ea.	470	48.50		518.50	600
0050	Maximum		4	4		2,325	121		2,446	2,775
0150	2 oven, minimum		10	1.600		1,025	48.50		1,073.50	1,200
0200	Maximum		10	1.600		3,400	48.50		3,448.50	3,800
0350	Built-in, 30" wide, 1 oven, minimum	1 Elec	6	1.333		855	61		916	1,050
0400	Maximum	2 Carp	2	8		1,750	310		2,060	2,425
0500	2 oven, conventional, minimum		4	4		1,250	155		1,405	1,600
0550	1 conventional, 1 microwave, maximum		2	8		2,325	310		2,635	3,050
0700	Free standing, 1 oven, 21" wide range, minimum	2 Clab	10	1.600		475	48.50		523.50	600
0750	21" wide, maximum	"	4	4		670	121		791	940
0900	Countertop cooktops, 4 burner, standard, minimum	1 Elec	6	1.333		325	61		386	455
0950	Maximum		3	2.667		1,850	122		1,972	2,250
1050	As above, but with grill and griddle attachment, minimum		6	1.333		1,425	61		1,486	1,675
1100	Maximum		3	2.667		3,925	122		4,047	4,525
1250	Microwave oven, minimum		4	2		96	91		187	255
1300	Maximum		2	4		455	182		637	800
5380	Oven, built-in, standard		4	2		935	91		1,026	1,175
5390	Deluxe		2	4		3,150	182		3,332	3,750

11 30 13.16 Refrigeration Equipment		Crew	Daily Output	Labor-Hours	Unit	Material	Labor	Equipment	Total	Total Incl O&P
0010	**REFRIGERATION EQUIPMENT**									
2000	Deep freeze, 15 to 23 C.F., minimum	2 Clab	10	1.600	Ea.	640	48.50		688.50	785
2050	Maximum		5	3.200		800	97		897	1,050
2200	30 C.F., minimum		8	2		795	60.50		855.50	975
2250	Maximum		3	5.333		890	162		1,052	1,250
5200	Icemaker, automatic, 20 lbs./day	1 Plum	7	1.143		1,400	51.50		1,451.50	1,600
5350	51 lbs./day	"	2	4		1,400	179		1,579	1,850
5450	Refrigerator, no frost, 6 C.F.	2 Clab	15	1.067		380	32.50		412.50	470
5500	Refrigerator, no frost, 10 C.F. to 12 C.F., minimum		10	1.600		450	48.50		498.50	575
5600	Maximum		6	2.667		545	81		626	735
5750	14 C.F. to 16 C.F., minimum		9	1.778		565	54		619	710
5800	Maximum		5	3.200		945	97		1,042	1,200
5950	18 C.F. to 20 C.F., minimum		8	2		715	60.50		775.50	890
6000	Maximum		4	4		1,675	121		1,796	2,050
6150	21 C.F. to 29 C.F., minimum		7	2.286		1,100	69.50		1,169.50	1,325
6200	Maximum		3	5.333		2,450	162		2,612	2,975
6790	Energy-star qualified, 18 C.F., minimum (G)	2 Carp	4	4		515	155		670	820
6795	Maximum (G)		2	8		1,700	310		2,010	2,375
6797	21.7 C.F., minimum (G)		4	4		1,025	155		1,180	1,375
6799	Maximum (G)		4	4		1,750	155		1,905	2,200

11 30 13.17 Kitchen Cleaning Equipment		Crew	Daily Output	Labor-Hours	Unit	Material	Labor	Equipment	Total	Total Incl O&P
0010	**KITCHEN CLEANING EQUIPMENT**									
2750	Dishwasher, built-in, 2 cycles, minimum	L-1	4	2.500	Ea.	305	113		418	520
2800	Maximum		2	5		455	225		680	870

11 30 Residential Equipment

11 30 13 – Residential Appliances

11 30 13.17 Kitchen Cleaning Equipment			Crew	Daily Output	Labor-Hours	Unit	Material	Labor	Equipment	Total	Total Incl O&P
							2019 Bare Costs				
2950	4 or more cycles, minimum		L-1	4	2.500	Ea.	420	113		533	650
2960	Average			4	2.500		560	113		673	800
3000	Maximum			2	5		1,900	225		2,125	2,475
3100	Energy-star qualified, minimum	G		4	2.500		405	113		518	630
3110	Maximum	G		2	5		1,775	225		2,000	2,325

11 30 13.18 Waste Disposal Equipment

Code	Description		Crew	Daily Output	Labor-Hours	Unit	Material	Labor	Equipment	Total	Total Incl O&P
0010	**WASTE DISPOSAL EQUIPMENT**										
1750	Compactor, residential size, 4 to 1 compaction, minimum		1 Carp	5	1.600	Ea.	710	62		772	880
1800	Maximum		″	3	2.667		1,150	103		1,253	1,450
3300	Garbage disposal, sink type, minimum		L-1	10	1		109	45		154	193
3350	Maximum		″	10	1		208	45		253	305

11 30 13.19 Kitchen Ventilation Equipment

Code	Description		Crew	Daily Output	Labor-Hours	Unit	Material	Labor	Equipment	Total	Total Incl O&P
0010	**KITCHEN VENTILATION EQUIPMENT**										
4150	Hood for range, 2 speed, vented, 30″ wide, minimum		L-3	5	2	Ea.	96	80		176	237
4200	Maximum			3	3.333		965	134		1,099	1,275
4300	42″ wide, minimum			5	2		155	80		235	300
4330	Custom			5	2		1,625	80		1,705	1,925
4350	Maximum			3	3.333		1,975	134		2,109	2,400
4500	For ventless hood, 2 speed, add						17.80			17.80	19.60
4650	For vented 1 speed, deduct from maximum						64			64	70.50

11 30 13.24 Washers

Code	Description		Crew	Daily Output	Labor-Hours	Unit	Material	Labor	Equipment	Total	Total Incl O&P
0010	**WASHERS**										
6650	Washing machine, automatic, minimum		1 Plum	3	2.667	Ea.	590	120		710	845
6700	Maximum			1	8		1,350	360		1,710	2,100
6750	Energy star qualified, front loading, minimum	G		3	2.667		850	120		970	1,125
6760	Maximum	G		1	8		1,825	360		2,185	2,625
6764	Top loading, minimum	G		3	2.667		650	120		770	910
6766	Maximum	G		3	2.667		1,000	120		1,120	1,300

11 30 13.25 Dryers

Code	Description		Crew	Daily Output	Labor-Hours	Unit	Material	Labor	Equipment	Total	Total Incl O&P
0010	**DRYERS**										
6770	Electric, front loading, energy-star qualified, minimum	G	L-2	3	5.333	Ea.	460	181		641	810
6780	Maximum	G	″	2	8		1,500	271		1,771	2,100
7450	Vent kits for dryers		1 Carp	10	.800		46.50	31		77.50	102

11 30 15 – Miscellaneous Residential Appliances

11 30 15.13 Sump Pumps

Code	Description		Crew	Daily Output	Labor-Hours	Unit	Material	Labor	Equipment	Total	Total Incl O&P
0010	**SUMP PUMPS**										
6400	Cellar drainer, pedestal, 1/3 HP, molded PVC base		1 Plum	3	2.667	Ea.	141	120		261	350
6450	Solid brass		″	2	4	″	239	179		418	555

11 30 15.23 Water Heaters

Code	Description		Crew	Daily Output	Labor-Hours	Unit	Material	Labor	Equipment	Total	Total Incl O&P
0010	**WATER HEATERS**										
6900	Electric, glass lined, 30 gallon, minimum		L-1	5	2	Ea.	895	90		985	1,125
6950	Maximum			3	3.333		1,250	150		1,400	1,625
7100	80 gallon, minimum			2	5		1,650	225		1,875	2,200
7150	Maximum			1	10		2,300	450		2,750	3,300
7180	Gas, glass lined, 30 gallon, minimum		2 Plum	5	3.200		1,600	144		1,744	1,975
7220	Maximum			3	5.333		2,225	239		2,464	2,850
7260	50 gallon, minimum			2.50	6.400		1,725	287		2,012	2,375
7300	Maximum			1.50	10.667		2,400	480		2,880	3,400
7310	Water heater, see also Section 22 33 30.13										

11 30 Residential Equipment

11 30 15 – Miscellaneous Residential Appliances

11 30 15.43 Air Quality		Crew	Daily Output	Labor-Hours	Unit	Material	2019 Bare Costs Labor	Equipment	Total	Total Incl O&P
0010	**AIR QUALITY**									
2450	Dehumidifier, portable, automatic, 15 pint	1 Elec	4	2	Ea.	208	91		299	380
2550	40 pint		3.75	2.133		240	97.50		337.50	425
3550	Heater, electric, built-in, 1250 watt, ceiling type, minimum		4	2		119	91		210	280
3600	Maximum		3	2.667		185	122		307	400
3700	Wall type, minimum		4	2		199	91		290	370
3750	Maximum		3	2.667		194	122		316	410
3900	1500 watt wall type, with blower		4	2		189	91		280	355
3950	3000 watt		3	2.667		490	122		612	740
4850	Humidifier, portable, 8 gallons/day					164			164	180
5000	15 gallons/day					198			198	218

11 30 33 – Retractable Stairs

11 30 33.10 Disappearing Stairway		Crew	Daily Output	Labor-Hours	Unit	Material	2019 Bare Costs Labor	Equipment	Total	Total Incl O&P
0010	**DISAPPEARING STAIRWAY** No trim included									
0020	One piece, yellow pine, 8′-0″ ceiling	2 Carp	4	4	Ea.	241	155		396	520
0030	9′-0″ ceiling		4	4		276	155		431	560
0040	10′-0″ ceiling		3	5.333		238	207		445	600
0050	11′-0″ ceiling		3	5.333		315	207		522	685
0060	12′-0″ ceiling		3	5.333		355	207		562	730
0100	Custom grade, pine, 8′-6″ ceiling, minimum	1 Carp	4	2		268	77.50		345.50	420
0150	Average		3.50	2.286		236	88.50		324.50	405
0200	Maximum		3	2.667		295	103		398	495
0500	Heavy duty, pivoted, from 7′-7″ to 12′-10″ floor to floor		3	2.667		1,375	103		1,478	1,675
0600	16′-0″ ceiling		2	4		1,550	155		1,705	1,950
0800	Economy folding, pine, 8′-6″ ceiling		4	2		176	77.50		253.50	320
0900	9′-6″ ceiling		4	2		201	77.50		278.50	350
1100	Automatic electric, aluminum, floor to floor height, 8′ to 9′	2 Carp	1	16		9,200	620		9,820	11,100

11 32 Unit Kitchens

11 32 13 – Metal Unit Kitchens

11 32 13.10 Commercial Unit Kitchens		Crew	Daily Output	Labor-Hours	Unit	Material	2019 Bare Costs Labor	Equipment	Total	Total Incl O&P
0010	**COMMERCIAL UNIT KITCHENS**									
1500	Combination range, refrigerator and sink, 30″ wide, minimum	L-1	2	5	Ea.	1,400	225		1,625	1,925
1550	Maximum		1	10		1,150	450		1,600	2,025
1570	60″ wide, average		1.40	7.143		1,275	320		1,595	1,925
1590	72″ wide, average		1.20	8.333		1,800	375		2,175	2,625
1640	Combination range, refrigerator, sink, microwave									
1660	Oven and ice maker	L-1	.80	12.500	Ea.	4,900	565		5,465	6,325

11 41 13 – Refrigerated Food Storage Cases

11 41 13.20 Refrigerated Food Storage Equipment		Crew	Daily Output	Labor-Hours	Unit	2019 Bare Costs Material	Labor	Equipment	Total	Total Incl O&P
0010	**REFRIGERATED FOOD STORAGE EQUIPMENT**									
4600	Freezer, pre-fab, 8′ x 8′ w/refrigeration	2 Carp	.45	35.556	Ea.	10,600	1,375		11,975	13,900
4620	8′ x 12′		.35	45.714		11,400	1,775		13,175	15,400
4640	8′ x 16′		.25	64		14,000	2,475		16,475	19,500
4660	8′ x 20′		.17	94.118		19,900	3,650		23,550	27,800
4680	Reach-in, 1 compartment	Q-1	4	4		3,225	162		3,387	3,825
4685	Energy star rated G	R-18	7.80	3.333		3,250	133		3,383	3,800
4700	2 compartment	Q-1	3	5.333		4,600	215		4,815	5,400
4705	Energy star rated G	R-18	6.20	4.194		3,500	168		3,668	4,125
4710	3 compartment	Q-1	3	5.333		6,125	215		6,340	7,075
4715	Energy star rated G	R-18	5.60	4.643		5,100	186		5,286	5,900
8320	Refrigerator, reach-in, 1 compartment		7.80	3.333		3,450	133		3,583	4,000
8325	Energy star rated G		7.80	3.333		3,250	133		3,383	3,800
8330	2 compartment		6.20	4.194		4,000	168		4,168	4,700
8335	Energy star rated G		6.20	4.194		3,500	168		3,668	4,125
8340	3 compartment		5.60	4.643		5,725	186		5,911	6,600
8345	Energy star rated G		5.60	4.643		5,100	186		5,286	5,900
8350	Pre-fab, with refrigeration, 8′ x 8′	2 Carp	.45	35.556		6,975	1,375		8,350	9,950
8360	8′ x 12′		.35	45.714		8,500	1,775		10,275	12,300
8370	8′ x 16′		.25	64		11,500	2,475		13,975	16,700
8380	8′ x 20′		.17	94.118		14,400	3,650		18,050	21,800
8390	Pass-thru/roll-in, 1 compartment	R-18	7.80	3.333		3,700	133		3,833	4,275
8400	2 compartment		6.24	4.167		5,700	167		5,867	6,550
8410	3 compartment		5.60	4.643		8,725	186		8,911	9,900
8420	Walk-in, alum, door & floor only, no refrig, 6′ x 6′ x 7′-6″	2 Carp	1.40	11.429		6,625	445		7,070	8,000
8430	10′ x 6′ x 7′-6″		.55	29.091		9,700	1,125		10,825	12,600
8440	12′ x 14′ x 7′-6″		.25	64		13,400	2,475		15,875	18,900
8450	12′ x 20′ x 7′-6″		.17	94.118		13,000	3,650		16,650	20,300
8460	Refrigerated cabinets, mobile					5,550			5,550	6,100
8470	Refrigerator/freezer, reach-in, 1 compartment	R-18	5.60	4.643		6,550	186		6,736	7,500
8480	2 compartment	″	4.80	5.417		7,450	217		7,667	8,550

11 41 33 – Foodservice Shelving

11 41 33.20 Metal Food Storage Shelving		Crew	Daily Output	Labor-Hours	Unit	Material	Labor	Equipment	Total	Total Incl O&P
0010	**METAL FOOD STORAGE SHELVING**									
8600	Stainless steel shelving, louvered 4-tier, 20″ x 3′	1 Clab	6	1.333	Ea.	1,475	40.50		1,515.50	1,700
8605	20″ x 4′		6	1.333		1,650	40.50		1,690.50	1,900
8610	20″ x 6′		6	1.333		1,775	40.50		1,815.50	2,025
8615	24″ x 3′		6	1.333		2,050	40.50		2,090.50	2,325
8620	24″ x 4′		6	1.333		2,500	40.50		2,540.50	2,825
8625	24″ x 6′		6	1.333		3,450	40.50		3,490.50	3,875
8630	Flat 4-tier, 20″ x 3′		6	1.333		1,200	40.50		1,240.50	1,400
8635	20″ x 4′		6	1.333		1,400	40.50		1,440.50	1,625
8640	20″ x 5′		6	1.333		1,600	40.50		1,640.50	1,825
8645	24″ x 3′		6	1.333		910	40.50		950.50	1,075
8650	24″ x 4′		6	1.333		2,325	40.50		2,365.50	2,625
8655	24″ x 6′		6	1.333		2,825	40.50		2,865.50	3,175
8700	Galvanized shelving, louvered 4-tier, 20″ x 3′		6	1.333		805	40.50		845.50	950
8705	20″ x 4′		6	1.333		915	40.50		955.50	1,075
8710	20″ x 6′		6	1.333		945	40.50		985.50	1,125
8715	24″ x 3′		6	1.333		750	40.50		790.50	890
8720	24″ x 4′		6	1.333		945	40.50		985.50	1,125
8725	24″ x 6′		6	1.333		1,350	40.50		1,390.50	1,575

11 41 Foodservice Storage Equipment

11 41 33 – Foodservice Shelving

11 41 33.20 Metal Food Storage Shelving		Crew	Daily Output	Labor-Hours	Unit	Material	2019 Bare Costs Labor	Equipment	Total	Total Incl O&P
8730	Flat 4-tier, 20" x 3'	1 Clab	6	1.333	Ea.	745	40.50		785.50	885
8735	20" x 4'		6	1.333		650	40.50		690.50	775
8740	20" x 6'		6	1.333		975	40.50		1,015.50	1,150
8745	24" x 3'		6	1.333		720	40.50		760.50	860
8750	24" x 4'		6	1.333		565	40.50		605.50	685
8755	24" x 6'		6	1.333		870	40.50		910.50	1,025
8760	Stainless steel dunnage rack, 24" x 3'		8	1		340	30.50		370.50	425
8765	24" x 4'		8	1		420	30.50		450.50	515
8770	Galvanized dunnage rack, 24" x 3'		8	1		141	30.50		171.50	205
8775	24" x 4'		8	1		191	30.50		221.50	260

11 42 Food Preparation Equipment

11 42 10 – Commercial Food Preparation Equipment

11 42 10.10 Choppers, Mixers and Misc. Equipment

Code	Description	Crew	Daily Output	Labor-Hours	Unit	Material	Labor	Equipment	Total	Total Incl O&P
0010	**CHOPPERS, MIXERS AND MISC. EQUIPMENT**									
1700	Choppers, 5 pounds	R-18	7	3.714	Ea.	3,425	149		3,574	4,000
1720	16 pounds		5	5.200		2,525	208		2,733	3,125
1740	35 to 40 pounds		4	6.500		3,625	260		3,885	4,400
1840	Coffee brewer, 5 burners	1 Plum	3	2.667		955	120		1,075	1,250
1860	Single, 3 gallon	"	3	2.667		1,600	120		1,720	1,975
3850	40 quarts	L-7	5.40	4.815		9,150	160		9,310	10,400
4040	80 quarts		3.90	6.667		19,400	222		19,622	21,700
4100	Floor type, 20 quarts		15	1.733		3,725	57.50		3,782.50	4,200
4120	60 quarts		14	1.857		10,900	62		10,962	12,100
4140	80 quarts		12	2.167		18,000	72		18,072	19,900
4160	140 quarts		8.60	3.023		26,800	101		26,901	29,700
6700	Peelers, small	R-18	8	3.250		1,975	130		2,105	2,375
6720	Large	"	6	4.333		3,125	174		3,299	3,725
6800	Pulper/extractor, close coupled, 5 HP	1 Plum	1.90	4.211		2,475	189		2,664	3,025
8580	Slicer with table	R-18	9	2.889		4,150	116		4,266	4,775

11 43 Food Delivery Carts and Conveyors

11 43 13 – Food Delivery Carts

11 43 13.10 Mobile Carts, Racks and Trays

Code	Description	Crew	Daily Output	Labor-Hours	Unit	Material	Labor	Equipment	Total	Total Incl O&P
0010	**MOBILE CARTS, RACKS AND TRAYS**									
1650	Cabinet, heated, 1 compartment, reach-in	R-18	5.60	4.643	Ea.	3,700	186		3,886	4,375
1655	Pass-thru roll-in		5.60	4.643		3,400	186		3,586	4,025
1660	2 compartment, reach-in		4.80	5.417		5,950	217		6,167	6,900
1670	Mobile					3,650			3,650	4,025
2000	Hospital food cart, hot and cold service, 20 tray capacity					15,600			15,600	17,200
6850	Mobile rack w/pan slide					1,375			1,375	1,525
9180	Tray and silver dispenser, mobile	1 Clab	16	.500		695	15.20		710.20	790

11 44 Food Cooking Equipment

11 44 13 – Commercial Ranges

11 44 13.10 Cooking Equipment		Crew	Daily Output	Labor-Hours	Unit	Material	2019 Bare Costs Labor	Equipment	Total	Total Incl O&P
0010	COOKING EQUIPMENT									
5170	Energy star rated, 50 lb. capacity G	R-18	4	6.500	Ea.	5,325	260		5,585	6,275
5175	85 lb. capacity G	"	4	6.500		8,475	260		8,735	9,750
6200	Iced tea brewer	1 Plum	3.44	2.326		675	104		779	915
9160	Pop-up, 2 slot					650			650	715

11 46 Food Dispensing Equipment

11 46 16 – Service Line Equipment

11 46 16.10 Commercial Food Dispensing Equipment

		Crew	Daily Output	Labor-Hours	Unit	Material	Labor	Equipment	Total	Total Incl O&P
0010	COMMERCIAL FOOD DISPENSING EQUIPMENT									
1050	Butter pat dispenser	1 Clab	13	.615	Ea.	850	18.70		868.70	970
1100	Bread dispenser, counter top		13	.615		760	18.70		778.70	865
1900	Cup and glass dispenser, drop in		4	2		565	60.50		625.50	725
1920	Disposable cup, drop in		16	.500		665	15.20		680.20	755
2650	Dish dispenser, drop in, 12"		11	.727		2,600	22		2,622	2,875
2660	Mobile		10	.800		2,825	24.50		2,849.50	3,150
3600	Well, hot food, built-in, rectangular, 12" x 20"	R-30	10	2.600		735	94.50		829.50	960
3610	Circular, 7 qt.		10	2.600		405	94.50		499.50	600
3620	Refrigerated, 2 compartments		10	2.600		3,200	94.50		3,294.50	3,650
3630	3 compartments		9	2.889		4,075	105		4,180	4,650
3640	4 compartments		8	3.250		4,675	118		4,793	5,350
5700	Hot chocolate dispenser	1 Plum	4	2		1,150	89.50		1,239.50	1,400
6300	Juice dispenser, concentrate	R-18	4.50	5.778		1,925	231		2,156	2,475
6690	Milk dispenser, bulk, 2 flavor	R-30	8	3.250		2,000	118		2,118	2,400
6695	3 flavor	"	8	3.250		2,525	118		2,643	2,975
8800	Serving counter, straight	1 Carp	40	.200	L.F.	1,475	7.75		1,482.75	1,650
8820	Curved section	"	30	.267	"	2,175	10.35		2,185.35	2,400
8825	Solid surface, see Section 12 36 61.16									
9100	Soft serve ice cream machine, medium	R-18	11	2.364	Ea.	7,725	94.50		7,819.50	8,650
9110	Large	"	9	2.889	"	21,900	116		22,016	24,300

11 46 83 – Ice Machines

11 46 83.10 Commercial Ice Equipment

		Crew	Daily Output	Labor-Hours	Unit	Material	Labor	Equipment	Total	Total Incl O&P
0010	COMMERCIAL ICE EQUIPMENT									
5800	Ice cube maker, 50 lbs./day	Q-1	6	2.667	Ea.	1,700	108		1,808	2,050
6050	500 lbs./day		4	4		2,675	162		2,837	3,200
6090	1000 lbs./day, with bin		1	16		5,275	645		5,920	6,850
6100	Ice flakers, 300 lbs./day		1.60	10		3,350	405		3,755	4,350
6120	600 lbs./day		.95	16.842		4,050	680		4,730	5,575
6130	1000 lbs./day		.75	21.333		4,900	860		5,760	6,825
6140	2000 lbs./day		.65	24.615		21,900	995		22,895	25,700
6160	Ice storage bin, 500 pound capacity	Q-5	1	16		1,075	665		1,740	2,300
6180	1000 pound	"	.56	28.571		2,950	1,175		4,125	5,200

11 48 Foodservice Cleaning and Disposal Equipment

11 48 13 – Commercial Dishwashers

11 48 13.10 Dishwashers		Crew	Daily Output	Labor-Hours	Unit	Material	Labor	Equipment	Total	Total Incl O&P
						2019 Bare Costs				
0010	**DISHWASHERS**									
2950	Dishwasher hood, canopy type	L-3A	10	1.200	L.F.	1,175	50		1,225	1,350
2960	Pant leg type	"	2.50	4.800	Ea.	7,800	200		8,000	8,900
6750	Pot sink, 3 compartment	1 Plum	7.25	1.103	L.F.	960	49.50		1,009.50	1,125
6760	Pot washer, low temp wash/rinse		1.60	5	Ea.	5,400	224		5,624	6,325
6770	High pressure wash, high temperature rinse		1.20	6.667		37,700	299		37,999	41,900
9170	Trash compactor, small, up to 125 lb. compacted weight	L-4	4	4		21,400	137		21,537	23,800
9175	Large, up to 175 lb. compacted weight	"	3	5.333		28,900	183		29,083	32,100

11 52 Audio-Visual Equipment

11 52 13 – Projection Screens

11 52 13.10 Projection Screens, Wall or Ceiling Hung

		Crew	Daily Output	Labor-Hours	Unit	Material	Labor	Equipment	Total	Total Incl O&P
0010	**PROJECTION SCREENS, WALL OR CEILING HUNG**, matte white									
0100	Manually operated, economy	2 Carp	500	.032	S.F.	6.45	1.24		7.69	9.15
0300	Intermediate		450	.036		7.90	1.38		9.28	11
0400	Deluxe		400	.040		9.85	1.55		11.40	13.40

11 81 Facility Maintenance Equipment

11 81 19 – Vacuum Cleaning Systems

11 81 19.10 Vacuum Cleaning

		Crew	Daily Output	Labor-Hours	Unit	Material	Labor	Equipment	Total	Total Incl O&P
0010	**VACUUM CLEANING**									
0020	Central, 3 inlet, residential	1 Skwk	.90	8.889	Total	1,200	350		1,550	1,900
0400	5 inlet system, residential		.50	16		1,725	630		2,355	2,950
0600	7 inlet system, commercial		.40	20		2,200	790		2,990	3,725
0800	9 inlet system, residential		.30	26.667		4,175	1,050		5,225	6,350
4010	Rule of thumb: First 1200 S.F., installed								1,425	1,575
4020	For each additional S.F., add				S.F.				.26	.26

Estimating Tips

General

- The items in this division are usually priced per square foot or each. Most of these items are purchased by the owner and installed by the contractor. Do not assume the items in Division 12 will be purchased and installed by the contractor. Check the specifications for responsibilities and include receiving, storage, installation, and mechanical and electrical hookups in the appropriate divisions.
- Some items in this division require some type of support system that is not usually furnished with the item. Examples of these systems include blocking for the attachment of casework and heavy drapery rods. The required blocking must be added to the estimate in the appropriate division.

Reference Numbers

Reference numbers are shown at the beginning of some major classifications. These numbers refer to related items in the Reference Section. The reference information may be an estimating procedure, an alternate pricing method, or technical information.

12 21 Window Blinds

12 21 13 – Horizontal Louver Blinds

12 21 13.13 Metal Horizontal Louver Blinds		Crew	Daily Output	Labor-Hours	Unit	Material	Labor	Equipment	Total	Total Incl O&P
						2019 Bare Costs				
0010	**METAL HORIZONTAL LOUVER BLINDS**									
0020	Horizontal, 1" aluminum slats, solid color, stock	1 Carp	590	.014	S.F.	5.90	.53		6.43	7.35

12 21 13.33 Vinyl Horizontal Louver Blinds		Crew	Daily Output	Labor-Hours	Unit	Material	Labor	Equipment	Total	Total Incl O&P
0010	**VINYL HORIZONTAL LOUVER BLINDS**									
0100	2" composite, 48" wide, 48" high	1 Carp	30	.267	Ea.	95	10.35		105.35	121
0120	72" high		29	.276		128	10.70		138.70	159
0140	96" high		28	.286		174	11.05		185.05	210
0200	60" wide, 60" high		27	.296		133	11.50		144.50	165
0220	72" high		25	.320		152	12.40		164.40	189
0240	96" high		24	.333		228	12.90		240.90	273
0300	72" wide, 72" high		25	.320		199	12.40		211.40	240
0320	96" high		23	.348		274	13.50		287.50	325
0400	96" wide, 96" high		20	.400		445	15.50		460.50	515
1000	2" faux wood, 48" wide, 48" high		30	.267		83.50	10.35		93.85	109
1020	72" high		29	.276		109	10.70		119.70	137
1040	96" high		28	.286		129	11.05		140.05	160
1300	72" wide, 72" high		25	.320		128	12.40		140.40	162
1320	96" high		23	.348		192	13.50		205.50	234
1400	96" wide, 96" high		20	.400		252	15.50		267.50	305

12 22 Curtains and Drapes

12 22 16 – Drapery Track and Accessories

12 22 16.10 Drapery Hardware		Crew	Daily Output	Labor-Hours	Unit	Material	Labor	Equipment	Total	Total Incl O&P
0010	**DRAPERY HARDWARE**									
0030	Standard traverse, per foot, minimum	1 Carp	59	.136	L.F.	6.40	5.25		11.65	15.75
0100	Maximum		51	.157	"	16.10	6.10		22.20	28
0200	Decorative traverse, 28" to 48", minimum		22	.364	Ea.	26	14.10		40.10	52
0220	Maximum		21	.381		57.50	14.75		72.25	87.50
0300	48" to 84", minimum		20	.400		28	15.50		43.50	56.50
0320	Maximum		19	.421		73.50	16.30		89.80	108
0400	66" to 120", minimum		18	.444		38.50	17.20		55.70	71
0420	Maximum		17	.471		112	18.25		130.25	153
0500	84" to 156", minimum		16	.500		49.50	19.40		68.90	86.50
0520	Maximum		15	.533		153	20.50		173.50	202
0600	130" to 240", minimum		14	.571		72	22		94	116
0620	Maximum		13	.615		212	24		236	273
0700	Slide rings, each, minimum					1.34			1.34	1.47
0720	Maximum					2.18			2.18	2.40
4000	Traverse rods, adjustable, 28" to 48"	1 Carp	22	.364		32.50	14.10		46.60	59
4020	48" to 84"		20	.400		41	15.50		56.50	70.50
4040	66" to 120"		18	.444		49.50	17.20		66.70	83
4060	84" to 156"		16	.500		56.50	19.40		75.90	94
4080	100" to 180"		14	.571		66	22		88	109
4100	228" to 312"		13	.615		91.50	24		115.50	141
4500	Curtain rod, 28" to 48", single		22	.364		10.35	14.10		24.45	35
4510	Double		22	.364		17.65	14.10		31.75	43
4520	48" to 86", single		20	.400		17.75	15.50		33.25	45
4530	Double		20	.400		29.50	15.50		45	58
4540	66" to 120", single		18	.444		29.50	17.20		46.70	61
4550	Double		18	.444		46.50	17.20		63.70	79.50
4600	Valance, pinch pleated fabric, 12" deep, up to 54" long, minimum					42.50			42.50	46.50

12 22 Curtains and Drapes

12 22 16 – Drapery Track and Accessories

12 22 16.10 Drapery Hardware		Crew	Daily Output	Labor-Hours	Unit	2019 Bare Costs Material	Labor	Equipment	Total	Total Incl O&P
4610	Maximum				Ea.	105			105	116
4620	Up to 77" long, minimum					81.50			81.50	89.50
4630	Maximum					170			170	187
5000	Stationary rods, first 2'					8.65			8.65	9.50

12 22 16.20 Blast Curtains		Crew	Daily Output	Labor-Hours	Unit	Material	Labor	Equipment	Total	Total Incl O&P
0010	**BLAST CURTAINS** per L.F. horizontal opening width, off-white or gray fabric									
0100	Blast curtains, drapery system, complete, including hardware, minimum	1 Carp	10.25	.780	L.F.	253	30		283	330
0120	Average		10.25	.780		264	30		294	340
0140	Maximum		10.25	.780		284	30		314	360

12 23 Interior Shutters

12 23 10 – Wood Interior Shutters

12 23 10.10 Wood Interior Shutters		Crew	Daily Output	Labor-Hours	Unit	Material	Labor	Equipment	Total	Total Incl O&P
0010	**WOOD INTERIOR SHUTTERS**, louvered									
0200	Two panel, 27" wide, 36" high	1 Carp	5	1.600	Set	165	62		227	284
0300	33" wide, 36" high		5	1.600		213	62		275	335
0500	47" wide, 36" high		5	1.600		286	62		348	415
1000	Four panel, 27" wide, 36" high		5	1.600		157	62		219	274
1100	33" wide, 36" high		5	1.600		201	62		263	325
1300	47" wide, 36" high		5	1.600		268	62		330	395
1400	Plantation shutters, 16" x 48"		5	1.600	Ea.	184	62		246	305
1450	16" x 96"		4	2		300	77.50		377.50	460
1460	36" x 96"		3	2.667		665	103		768	900

12 23 10.13 Wood Panels		Crew	Daily Output	Labor-Hours	Unit	Material	Labor	Equipment	Total	Total Incl O&P
0010	**WOOD PANELS**									
3000	Wood folding panels with movable louvers, 7" x 20" each	1 Carp	17	.471	Pr.	95	18.25		113.25	135
3300	8" x 28" each		17	.471		95	18.25		113.25	135
3450	9" x 36" each		17	.471		109	18.25		127.25	150
3600	10" x 40" each		17	.471		119	18.25		137.25	161
4000	Fixed louver type, stock units, 8" x 20" each		17	.471		97	18.25		115.25	137
4150	10" x 28" each		17	.471		82	18.25		100.25	120
4300	12" x 36" each		17	.471		97	18.25		115.25	137
4450	18" x 40" each		17	.471		139	18.25		157.25	182
5000	Insert panel type, stock, 7" x 20" each		17	.471		28	18.25		46.25	60.50
5150	8" x 28" each		17	.471		51	18.25		69.25	86
5300	9" x 36" each		17	.471		64.50	18.25		82.75	101
5450	10" x 40" each		17	.471		69.50	18.25		87.75	107
5600	Raised panel type, stock, 10" x 24" each		17	.471		118	18.25		136.25	160
5650	12" x 26" each		17	.471		118	18.25		136.25	160
5700	14" x 30" each		17	.471		130	18.25		148.25	173
5750	16" x 36" each		17	.471		144	18.25		162.25	188
6000	For custom built pine, add					22%				
6500	For custom built hardwood blinds, add					42%				

12 24 Window Shades

12 24 13 – Roller Window Shades

12 24 13.10 Shades		Crew	Daily Output	Labor-Hours	Unit	Material	2019 Bare Costs Labor	Equipment	Total	Total Incl O&P
0010	**SHADES**									
0020	Basswood, roll-up, stain finish, 3/8″ slats	1 Carp	300	.027	S.F.	20.50	1.03		21.53	24.50
5011	Insulative shades G		125	.064		16.45	2.48		18.93	22
6011	Solar screening, fiberglass G	↓	85	.094	↓	7.85	3.65		11.50	14.70
8011	Interior insulative shutter									
8111	Stock unit, 15″ x 60″ G	1 Carp	17	.471	Pr.	17.70	18.25		35.95	49.50

12 32 Manufactured Wood Casework

12 32 23 – Hardwood Casework

12 32 23.10 Manufactured Wood Casework, Stock Units

12 32 23.10		Crew	Daily Output	Labor-Hours	Unit	Material	Labor	Equipment	Total	Total Incl O&P
0010	**MANUFACTURED WOOD CASEWORK, STOCK UNITS**									
0700	Kitchen base cabinets, hardwood, not incl. counter tops,									
0710	24″ deep, 35″ high, prefinished									
0800	One top drawer, one door below, 12″ wide	2 Carp	24.80	.645	Ea.	315	25		340	390
0820	15″ wide		24	.667		330	26		356	405
0840	18″ wide		23.30	.687		350	26.50		376.50	430
0860	21″ wide		22.70	.705		360	27.50		387.50	440
0880	24″ wide		22.30	.717		435	28		463	525
1000	Four drawers, 12″ wide		24.80	.645		325	25		350	395
1020	15″ wide		24	.667		325	26		351	405
1040	18″ wide		23.30	.687		360	26.50		386.50	440
1060	24″ wide		22.30	.717		405	28		433	490
1200	Two top drawers, two doors below, 27″ wide		22	.727		445	28		473	535
1220	30″ wide		21.40	.748		505	29		534	605
1240	33″ wide		20.90	.766		525	29.50		554.50	630
1260	36″ wide		20.30	.788		550	30.50		580.50	650
1280	42″ wide		19.80	.808		580	31.50		611.50	685
1300	48″ wide		18.90	.847		620	33		653	735
1500	Range or sink base, two doors below, 30″ wide		21.40	.748		420	29		449	515
1520	33″ wide		20.90	.766		450	29.50		479.50	545
1540	36″ wide		20.30	.788		475	30.50		505.50	570
1560	42″ wide		19.80	.808		500	31.50		531.50	600
1580	48″ wide	↓	18.90	.847		525	33		558	630
1800	For sink front units, deduct					182			182	201
2000	Corner base cabinets, 36″ wide, standard	2 Carp	18	.889		745	34.50		779.50	875
2100	Lazy Susan with revolving door	″	16.50	.970	↓	1,000	37.50		1,037.50	1,150
4000	Kitchen wall cabinets, hardwood, 12″ deep with two doors									
4050	12″ high, 30″ wide	2 Carp	24.80	.645	Ea.	284	25		309	350
4400	15″ high, 30″ wide		24	.667		281	26		307	355
4420	33″ wide		23.30	.687		355	26.50		381.50	435
4440	36″ wide		22.70	.705		345	27.50		372.50	425
4450	42″ wide		22.70	.705		390	27.50		417.50	475
4700	24″ high, 30″ wide		23.30	.687		375	26.50		401.50	460
4720	36″ wide		22.70	.705		410	27.50		437.50	495
4740	42″ wide		22.30	.717		292	28		320	365
5000	30″ high, one door, 12″ wide		22	.727		271	28		299	345
5020	15″ wide		21.40	.748		284	29		313	365
5040	18″ wide		20.90	.766		310	29.50		339.50	395
5060	24″ wide		20.30	.788		365	30.50		395.50	450
5300	Two doors, 27″ wide		19.80	.808		405	31.50		436.50	495
5320	30″ wide	↓	19.30	.829	↓	425	32		457	520

12 32 23 – Hardwood Casework

12 32 23.10 Manufactured Wood Casework, Stock Units		Crew	Daily Output	Labor-Hours	Unit	Material	2019 Bare Costs Labor	Equipment	Total	Total Incl O&P
5340	36" wide	2 Carp	18.80	.851	Ea.	480	33		513	580
5360	42" wide		18.50	.865		525	33.50		558.50	630
5380	48" wide		18.40	.870		590	33.50		623.50	705
6000	Corner wall, 30" high, 24" wide		18	.889		420	34.50		454.50	515
6050	30" wide		17.20	.930		430	36		466	535
6100	36" wide		16.50	.970		495	37.50		532.50	605
6500	Revolving Lazy Susan		15.20	1.053		132	41		173	213
7000	Broom cabinet, 84" high, 24" deep, 18" wide		10	1.600		775	62		837	955
7500	Oven cabinets, 84" high, 24" deep, 27" wide		8	2		1,175	77.50		1,252.50	1,425
7750	Valance board trim		396	.040	L.F.	18.80	1.57		20.37	23
7780	Toe kick trim	1 Carp	256	.031	"	3.55	1.21		4.76	5.90
7790	Base cabinet corner filler		16	.500	Ea.	48.50	19.40		67.90	85.50
7800	Cabinet filler, 3" x 24"		20	.400		19.75	15.50		35.25	47
7810	3" x 30"		20	.400		24.50	15.50		40	52.50
7820	3" x 42"		18	.444		34.50	17.20		51.70	66.50
7830	3" x 80"		16	.500		66	19.40		85.40	105
7850	Cabinet panel		50	.160	S.F.	10.80	6.20		17	22
9000	For deluxe models of all cabinets, add					40%				
9500	For custom built in place, add					25%	10%			
9558	Rule of thumb, kitchen cabinets not including									
9560	appliances & counter top, minimum	2 Carp	30	.533	L.F.	211	20.50		231.50	266
9600	Maximum	"	25	.640	"	445	25		470	530

12 32 23.30 Manufactured Wood Casework Vanities		Crew	Daily Output	Labor-Hours	Unit	Material	Labor	Equipment	Total	Total Incl O&P
0010	**MANUFACTURED WOOD CASEWORK VANITIES**									
8000	Vanity bases, 2 doors, 30" high, 21" deep, 24" wide	2 Carp	20	.800	Ea.	370	31		401	460
8050	30" wide		16	1		445	39		484	550
8100	36" wide		13.33	1.200		400	46.50		446.50	515
8150	48" wide		11.43	1.400		595	54		649	745
9000	For deluxe models of all vanities, add to above					40%				
9500	For custom built in place, add to above					25%	10%			

12 32 23.35 Manufactured Wood Casework Hardware		Crew	Daily Output	Labor-Hours	Unit	Material	Labor	Equipment	Total	Total Incl O&P
0010	**MANUFACTURED WOOD CASEWORK HARDWARE**									
1000	Catches, minimum	1 Carp	235	.034	Ea.	1.49	1.32		2.81	3.82
1020	Average		119.40	.067		4.28	2.60		6.88	9
1040	Maximum		80	.100		9	3.88		12.88	16.30
2000	Door/drawer pulls, handles									
2200	Handles and pulls, projecting, metal, minimum	1 Carp	48	.167	Ea.	5.30	6.45		11.75	16.55
2220	Average		42	.190		8.15	7.40		15.55	21
2240	Maximum		36	.222		10.95	8.60		19.55	26.50
2300	Wood, minimum		48	.167		5.45	6.45		11.90	16.70
2320	Average		42	.190		7.30	7.40		14.70	20.50
2340	Maximum		36	.222		10.10	8.60		18.70	25.50
2600	Flush, metal, minimum		48	.167		5.45	6.45		11.90	16.70
2620	Average		42	.190		7.40	7.40		14.80	20.50
2640	Maximum		36	.222		10.20	8.60		18.80	25.50
3000	Drawer tracks/glides, minimum		48	.167	Pr.	9.20	6.45		15.65	21
3020	Average		32	.250		16.60	9.70		26.30	34.50
3040	Maximum		24	.333		27	12.90		39.90	51
4000	Cabinet hinges, minimum		160	.050		3.16	1.94		5.10	6.70
4020	Average		95.24	.084		5.30	3.26		8.56	11.20
4040	Maximum		68	.118		13.85	4.56		18.41	23

12 36 Countertops

12 36 16 – Metal Countertops

12 36 16.10 Stainless Steel Countertops		Crew	Daily Output	Labor-Hours	Unit	Material	2019 Bare Costs Labor	Equipment	Total	Total Incl O&P
0010	**STAINLESS STEEL COUNTERTOPS**									
3200	Stainless steel, custom	1 Carp	24	.333	S.F.	183	12.90		195.90	223

12 36 19 – Wood Countertops

12 36 19.10 Maple Countertops

		Crew	Daily Output	Labor-Hours	Unit	Material	Labor	Equipment	Total	Total Incl O&P
0010	**MAPLE COUNTERTOPS**									
2900	Solid, laminated, 1-1/2" thick, no splash	1 Carp	28	.286	L.F.	91.50	11.05		102.55	119
3000	With square splash		28	.286	"	109	11.05		120.05	138
3400	Recessed cutting block with trim, 16" x 20" x 1"		8	1	Ea.	111	39		150	186
3410	Replace cutting block only		8	1	"	92	39		131	165

12 36 23 – Plastic Countertops

12 36 23.13 Plastic-Laminate-Clad Countertops

		Crew	Daily Output	Labor-Hours	Unit	Material	Labor	Equipment	Total	Total Incl O&P
0010	**PLASTIC-LAMINATE-CLAD COUNTERTOPS**									
0020	Stock, 24" wide w/backsplash, minimum	1 Carp	30	.267	L.F.	17.50	10.35		27.85	36.50
0100	Maximum		25	.320		41	12.40		53.40	65.50
0300	Custom plastic, 7/8" thick, aluminum molding, no splash		30	.267		36	10.35		46.35	56.50
0400	Cove splash		30	.267		46	10.35		56.35	67.50
0600	1-1/4" thick, no splash		28	.286		39	11.05		50.05	61
0700	Square splash		28	.286		44	11.05		55.05	67
0900	Square edge, plastic face, 7/8" thick, no splash		30	.267		34.50	10.35		44.85	55
1000	With splash		30	.267		42	10.35		52.35	63
1200	For stainless channel edge, 7/8" thick, add					3.84			3.84	4.22
1300	1-1/4" thick, add					4.51			4.51	4.96
1500	For solid color suede finish, add					5.90			5.90	6.50
1700	For end splash, add				Ea.	22			22	24
1900	For cut outs, standard, add, minimum	1 Carp	32	.250		17.70	9.70		27.40	35.50
2000	Maximum		8	1		7.30	39		46.30	72
2010	Cut out in backsplash for elec. wall outlet		38	.211			8.15		8.15	13.50
2020	Cut out for sink		20	.400			15.50		15.50	25.50
2030	Cut out for stove top		18	.444			17.20		17.20	28.50
2100	Postformed, including backsplash and front edge		30	.267	L.F.	12.60	10.35		22.95	31
2110	Mitred, add		12	.667	Ea.		26		26	42.50
2200	Built-in place, 25" wide, plastic laminate		25	.320	L.F.	61.50	12.40		73.90	88.50

12 36 33 – Tile Countertops

12 36 33.10 Ceramic Tile Countertops

		Crew	Daily Output	Labor-Hours	Unit	Material	Labor	Equipment	Total	Total Incl O&P
0010	**CERAMIC TILE COUNTERTOPS**									
2300	Ceramic tile mosaic	1 Carp	25	.320	L.F.	41.50	12.40		53.90	66.50

12 36 40 – Stone Countertops

12 36 40.10 Natural Stone Countertops

		Crew	Daily Output	Labor-Hours	Unit	Material	Labor	Equipment	Total	Total Incl O&P
0010	**NATURAL STONE COUNTERTOPS**									
2500	Marble, stock, with splash, 1/2" thick, minimum	1 Bric	17	.471	L.F.	51.50	17.30		68.80	86
2700	3/4" thick, maximum	"	13	.615	"	130	22.50		152.50	181

12 36 61 – Simulated Stone Countertops

12 36 61.16 Solid Surface Countertops

		Crew	Daily Output	Labor-Hours	Unit	Material	Labor	Equipment	Total	Total Incl O&P
0010	**SOLID SURFACE COUNTERTOPS**, Acrylic polymer									
0020	Pricing for orders of 100 L.F. or greater									
0100	25" wide, solid colors	2 Carp	28	.571	L.F.	53.50	22		75.50	95.50
0200	Patterned colors		28	.571		68	22		90	112
0300	Premium patterned colors		28	.571		90.50	22		112.50	136
0400	With silicone attached 4" backsplash, solid colors		27	.593		62.50	23		85.50	107
0500	Patterned colors		27	.593		79.50	23		102.50	126

12 36 Countertops

12 36 61 – Simulated Stone Countertops

12 36 61.16 Solid Surface Countertops		Crew	Daily Output	Labor-Hours	Unit	Material	2019 Bare Costs Labor	Equipment	Total	Total Incl O&P
0600	Premium patterned colors	2 Carp	27	.593	L.F.	99	23		122	147
0700	With hard seam attached 4" backsplash, solid colors		23	.696		62.50	27		89.50	114
0800	Patterned colors		23	.696		79.50	27		106.50	132
0900	Premium patterned colors		23	.696		99	27		126	154
1000	Pricing for order of 51-99 L.F.									
1100	25" wide, solid colors	2 Carp	24	.667	L.F.	61.50	26		87.50	110
1200	Patterned colors		24	.667		78.50	26		104.50	129
1300	Premium patterned colors		24	.667		104	26		130	158
1400	With silicone attached 4" backsplash, solid colors		23	.696		72	27		99	124
1500	Patterned colors		23	.696		91.50	27		118.50	146
1600	Premium patterned colors		23	.696		114	27		141	170
1700	With hard seam attached 4" backsplash, solid colors		20	.800		72	31		103	131
1800	Patterned colors		20	.800		91.50	31		122.50	152
1900	Premium patterned colors		20	.800		114	31		145	176
2000	Pricing for order of 1-50 L.F.									
2100	25" wide, solid colors	2 Carp	20	.800	L.F.	72	31		103	131
2200	Patterned colors		20	.800		92	31		123	152
2300	Premium patterned colors		20	.800		122	31		153	186
2400	With silicone attached 4" backsplash, solid colors		19	.842		84.50	32.50		117	147
2500	Patterned colors		19	.842		107	32.50		139.50	172
2600	Premium patterned colors		19	.842		134	32.50		166.50	201
2700	With hard seam attached 4" backsplash, solid colors		15	1.067		84.50	41.50		126	162
2800	Patterned colors		15	1.067		107	41.50		148.50	187
2900	Premium patterned colors		15	1.067		134	41.50		175.50	216
3000	Sinks, pricing for order of 100 or greater units									
3100	Single bowl, hard seamed, solid colors, 13" x 17"	1 Carp	3	2.667	Ea.	385	103		488	595
3200	10" x 15"		7	1.143		178	44.50		222.50	269
3300	Cutouts for sinks		8	1			39		39	64
3400	Sinks, pricing for order of 51-99 units									
3500	Single bowl, hard seamed, solid colors, 13" x 17"	1 Carp	2.55	3.137	Ea.	445	122		567	690
3600	10" x 15"		6	1.333		205	51.50		256.50	310
3700	Cutouts for sinks		7	1.143			44.50		44.50	73
3800	Sinks, pricing for order of 1-50 units									
3900	Single bowl, hard seamed, solid colors, 13" x 17"	1 Carp	2	4	Ea.	520	155		675	830
4000	10" x 15"		4.55	1.758		241	68		309	380
4100	Cutouts for sinks		5.25	1.524			59		59	97.50
4200	Cooktop cutouts, pricing for 100 or greater units		4	2		26	77.50		103.50	157
4300	51-99 units		3.40	2.353		30	91		121	184
4400	1-50 units		3	2.667		35.50	103		138.50	210

12 36 61.17 Solid Surface Vanity Tops		Crew	Daily Output	Labor-Hours	Unit	Material	Labor	Equipment	Total	Total Incl O&P
0010	**SOLID SURFACE VANITY TOPS**									
0015	Solid surface, center bowl, 17" x 19"	1 Carp	12	.667	Ea.	173	26		199	233
0020	19" x 25"		12	.667		191	26		217	253
0030	19" x 31"		12	.667		224	26		250	289
0040	19" x 37"		12	.667		261	26		287	330
0050	22" x 25"		10	.800		335	31		366	420
0060	22" x 31"		10	.800		390	31		421	480
0070	22" x 37"		10	.800		455	31		486	550
0080	22" x 43"		10	.800		525	31		556	625
0090	22" x 49"		10	.800		570	31		601	675
0110	22" x 55"		8	1		420	39		459	525
0120	22" x 61"		8	1		475	39		514	590

12 36 Countertops

12 36 61 – Simulated Stone Countertops

12 36 61.17 Solid Surface Vanity Tops			Crew	Daily Output	Labor-Hours	Unit	Material	Labor	Equipment	Total	Total Incl O&P
							2019 Bare Costs				
0220	Double bowl, 22" x 61"		1 Carp	8	1	Ea.	535	39		574	655
0230	Double bowl, 22" x 73"		↓	8	1	↓	1,025	39		1,064	1,200
0240	For aggregate colors, add						35%				
0250	For faucets and fittings, see Section 22 41 39.10										
12 36 61.19 Quartz Agglomerate Countertops											
0010	**QUARTZ AGGLOMERATE COUNTERTOPS**										
0100	25" wide, 4" backsplash, color group A, minimum		2 Carp	15	1.067	L.F.	68.50	41.50		110	144
0110	Maximum			15	1.067		95.50	41.50		137	174
0120	Color group B, minimum			15	1.067		73.50	41.50		115	150
0130	Maximum			15	1.067		99.50	41.50		141	178
0140	Color group C, minimum			15	1.067		82	41.50		123.50	159
0150	Maximum			15	1.067		118	41.50		159.50	198
0160	Color group D, minimum			15	1.067		89	41.50		130.50	167
0170	Maximum		↓	15	1.067	↓	122	41.50		163.50	203

12 48 Rugs and Mats

12 48 13 – Entrance Floor Mats and Frames

12 48 13.13 Entrance Floor Mats			Crew	Daily Output	Labor-Hours	Unit	Material	Labor	Equipment	Total	Total Incl O&P
0010	**ENTRANCE FLOOR MATS**										
2000	Recycled rubber tire tile, 12" x 12" x 3/8" thick	G	1 Clab	125	.064	S.F.	11.30	1.94		13.24	15.65
2510	Natural cocoa fiber, 1/2" thick	G		125	.064		7.95	1.94		9.89	11.95
2520	3/4" thick	G		125	.064		6.85	1.94		8.79	10.70
2530	1" thick	G	↓	125	.064	↓	13.60	1.94		15.54	18.15

12 93 Interior Public Space Furnishings

12 93 13 – Bicycle Racks

12 93 13.10 Bicycle Racks			Crew	Daily Output	Labor-Hours	Unit	Material	Labor	Equipment	Total	Total Incl O&P
0010	**BICYCLE RACKS**										
0020	Single side, grid, 1-5/8" OD stl. pipe, w/1/2" bars, galv, 5 bike cap		2 Clab	10	1.600	Ea.	320	48.50		368.50	430
0025	Powder coat finish			10	1.600		315	48.50		363.50	430
0030	Single side, grid, 1-5/8" OD stl. pipe, w/1/2" bars, galv, 9 bike cap			8	2		400	60.50		460.50	540
0035	Powder coat finish			8	2		425	60.50		485.50	570
0040	Single side, grid, 1-5/8" OD stl. pipe, w/1/2" bars, galv, 18 bike cap			4	4		520	121		641	770
0045	Powder coat finish			4	4		535	121		656	785
0050	S curve, 1-7/8" OD stl. pipe, 11 ga., galv, 5 bike cap			10	1.600		159	48.50		207.50	255
0055	Powder coat finish			10	1.600		159	48.50		207.50	255
0060	S curve, 1-7/8" OD stl. pipe, 11 ga., galv, 7 bike cap			9	1.778		201	54		255	310
0065	Powder coat finish			9	1.778		218	54		272	330
0070	S curve, 1-7/8" OD stl. pipe, 11 ga., galv, 9 bike cap			8	2		315	60.50		375.50	445
0075	Powder coat finish			8	2		305	60.50		365.50	435
0080	S curve, 1-7/8" OD stl. pipe, 11 ga., galv, 11 bike cap			6	2.667		420	81		501	600
0085	Powder coat finish		↓	6	2.667	↓	395	81		476	565

12 93 23 – Trash and Litter Receptacles

12 93 23.10 Trash Receptacles			Crew	Daily Output	Labor-Hours	Unit	Material	Labor	Equipment	Total	Total Incl O&P
0010	**TRASH RECEPTACLES**										
0500	Recycled plastic, var colors, round, 32 gal., 28" x 38" high	G	2 Clab	5	3.200	Ea.	685	97		782	915
0510	32 gal., 31" x 32" high	G	"	5	3.200	"	765	97		862	1,000

Estimating Tips

General

- The items and systems in this division are usually estimated, purchased, supplied, and installed as a unit by one or more subcontractors. The estimator must ensure that all parties are operating from the same set of specifications and assumptions, and that all necessary items are estimated and will be provided. Many times the complex items and systems are covered, but the more common ones, such as excavation or a crane, are overlooked for the very reason that everyone assumes nobody could miss them. The estimator should be the central focus and be able to ensure that all systems are complete.
- It is important to consider factors such as site conditions, weather, shape and size of building, as well as labor availability as they may impact the overall cost of erecting special structures and systems included in this division.
- Another area where problems can develop in this division is at the interface between systems. The estimator must ensure, for instance, that anchor bolts, nuts, and washers are estimated and included for the air-supported structures and pre-engineered buildings to be bolted to their foundations. Utility supply is a common area where essential items or pieces of equipment can be missed or overlooked because each subcontractor may feel it is another's responsibility. The estimator should also be aware of certain items which may be supplied as part of a package but installed by others, and ensure that the installing contractor's estimate includes the cost of installation. Conversely, the estimator must also ensure that items are not costed by two different subcontractors, resulting in an inflated overall estimate.

13 30 00 Special Structures

- The foundations and floor slab, as well as rough mechanical and electrical, should be estimated, as this work is required for the assembly and erection of the structure. Generally, as noted in the data set, the pre-engineered building comes as a shell. Pricing is based on the size and structural design parameters stated in the reference section. Additional features, such as windows and doors with their related structural framing, must also be included by the estimator. Here again, the estimator must have a clear understanding of the scope of each portion of the work and all the necessary interfaces.

Reference Numbers

Reference numbers are shown at the beginning of some major classifications. These numbers refer to related items in the Reference Section. The reference information may be an estimating procedure, an alternate pricing method, or technical information.

Note: Not all subdivisions listed here necessarily appear. ■

13 11 13 – Below-Grade Swimming Pools

13 11 13.50 Swimming Pools		Crew	Daily Output	Labor-Hours	Unit	2019 Bare Costs Material	Labor	Equipment	Total	Total Incl O&P
0010	**SWIMMING POOLS** Residential in-ground, vinyl lined									
0020	Concrete sides, w/equip, sand bottom	B-52	300	.187	SF Surf	27	6.20	2	35.20	42.50
0100	Metal or polystyrene sides R131113-20	B-14	410	.117		22.50	3.76	.78	27.04	32
0200	Add for vermiculite bottom					1.73			1.73	1.90
0500	Gunite bottom and sides, white plaster finish									
0600	12′ x 30′ pool	B-52	145	.386	SF Surf	50.50	12.85	4.14	67.49	81.50
0720	16′ x 32′ pool		155	.361		45.50	12.05	3.88	61.43	74
0750	20′ x 40′ pool		250	.224		40.50	7.45	2.40	50.35	59.50
0810	Concrete bottom and sides, tile finish									
0820	12′ x 30′ pool	B-52	80	.700	SF Surf	51	23.50	7.50	82	103
0830	16′ x 32′ pool		95	.589		42	19.65	6.35	68	86
0840	20′ x 40′ pool		130	.431		33.50	14.35	4.62	52.47	65.50
1100	Motel, gunite with plaster finish, incl. medium									
1150	capacity filtration & chlorination	B-52	115	.487	SF Surf	62	16.20	5.25	83.45	101
1200	Municipal, gunite with plaster finish, incl. high									
1250	capacity filtration & chlorination	B-52	100	.560	SF Surf	80.50	18.65	6	105.15	126
1600	For water heating system, see Section 23 52 28.10									
1700	Filtration and deck equipment only, as % of total				Total				20%	20%
1800	Deck equipment, rule of thumb, 20′ x 40′ pool				SF Pool				1.18	1.30
3000	Painting pools, preparation + 3 coats, 20′ x 40′ pool, epoxy	2 Pord	.33	48.485	Total	1,750	1,600		3,350	4,525
3100	Rubber base paint, 18 gallons	″	.33	48.485		1,350	1,600		2,950	4,100
3500	42′ x 82′ pool, 75 gallons, epoxy paint	3 Pord	.14	171		7,425	5,625		13,050	17,400
3600	Rubber base paint	″	.14	171		5,600	5,625		11,225	15,400

13 11 23 – On-Grade Swimming Pools

13 11 23.50 Swimming Pools		Crew	Daily Output	Labor-Hours	Unit	Material	Labor	Equipment	Total	Total Incl O&P
0010	**SWIMMING POOLS** Residential above ground, steel construction									
0100	Round, 15′ diam.	B-80A	3	8	Ea.	810	243	82	1,135	1,375
0120	18′ diam.		2.50	9.600		910	291	98.50	1,299.50	1,600
0140	21′ diam.		2	12		1,025	365	123	1,513	1,850
0160	24′ diam.		1.80	13.333		1,125	405	137	1,667	2,075
0180	27′ diam.		1.50	16		1,325	485	164	1,974	2,425
0200	30′ diam.		1	24		1,450	730	246	2,426	3,050
0220	Oval, 12′ x 24′		2.30	10.435		1,500	315	107	1,922	2,300
0240	15′ x 30′		1.80	13.333		2,075	405	137	2,617	3,100
0260	18′ x 33′		1	24		2,350	730	246	3,326	4,075

13 11 46 – Swimming Pool Accessories

13 11 46.50 Swimming Pool Equipment		Crew	Daily Output	Labor-Hours	Unit	Material	Labor	Equipment	Total	Total Incl O&P
0010	**SWIMMING POOL EQUIPMENT**									
0020	Diving stand, stainless steel, 3 meter	2 Carp	.40	40	Ea.	17,300	1,550		18,850	21,700
0300	1 meter		2.70	5.926		10,500	230		10,730	11,900
0600	Diving boards, 16′ long, aluminum		2.70	5.926		4,375	230		4,605	5,200
0700	Fiberglass		2.70	5.926		3,525	230		3,755	4,250
0800	14′ long, aluminum		2.70	5.926		4,125	230		4,355	4,925
0850	Fiberglass		2.70	5.926		3,500	230		3,730	4,225
1100	Bulkhead, movable, PVC, 8′-2″ wide	2 Clab	8	2		2,775	60.50		2,835.50	3,150
1120	7′-9″ wide		8	2		2,675	60.50		2,735.50	3,050
1140	7′-3″ wide		8	2		2,175	60.50		2,235.50	2,500
1160	6′-9″ wide		8	2		2,175	60.50		2,235.50	2,500
1200	Ladders, heavy duty, stainless steel, 2 tread	2 Carp	7	2.286		910	88.50		998.50	1,150
1500	4 tread	″	6	2.667		790	103		893	1,050
2100	Lights, underwater, 12 volt, with transformer, 300 watt	1 Elec	1	8		365	365		730	995
2200	110 volt, 500 watt, standard	″	1	8		320	365		685	945

13 11 Swimming Pools

13 11 46 – Swimming Pool Accessories

13 11 46.50 Swimming Pool Equipment		Crew	Daily Output	Labor-Hours	Unit	2019 Bare Costs Material	Labor	Equipment	Total	Total Incl O&P
3000	Pool covers, reinforced vinyl	3 Clab	1800	.013	S.F.	1.24	.40		1.64	2.03
3100	Vinyl, for winter, 400 S.F. max pool surface		3200	.008		.24	.23		.47	.64
3200	With water tubes, 400 S.F. max pool surface		3000	.008		.38	.24		.62	.82
3250	Sealed air bubble polyethylene solar blanket, 16 mils					.36			.36	.40
3300	Slides, tubular, fiberglass, aluminum handrails & ladder, 5′-0″, straight	2 Carp	1.60	10	Ea.	3,775	390		4,165	4,800
3320	8′-0″, curved	″	3	5.333	″	7,450	207		7,657	8,550

13 12 Fountains

13 12 13 – Exterior Fountains

13 12 13.10 Outdoor Fountains		Crew	Daily Output	Labor-Hours	Unit	Material	Labor	Equipment	Total	Total Incl O&P
0010	**OUTDOOR FOUNTAINS**									
0100	Outdoor fountain, 48″ high with bowl and figures	2 Clab	2	8	Ea.	460	243		703	905
0200	Commercial, concrete or cast stone, 40-60″ H, simple		2	8		845	243		1,088	1,325
0220	Average		2	8		1,625	243		1,868	2,200
0240	Ornate		2	8		3,700	243		3,943	4,475
0260	Metal, 72″ high		2	8		1,575	243		1,818	2,125
0280	90″ high		2	8		2,125	243		2,368	2,750
0300	120″ high		2	8		5,150	243		5,393	6,050
0320	Resin or fiberglass, 40-60″ H, wall type		2	8		595	243		838	1,050
0340	Waterfall type		2	8		1,000	243		1,243	1,500

13 12 23 – Interior Fountains

13 12 23.10 Indoor Fountains		Crew	Daily Output	Labor-Hours	Unit	Material	Labor	Equipment	Total	Total Incl O&P
0010	**INDOOR FOUNTAINS**									
0100	Commercial, floor type, resin or fiberglass, lighted, cascade type	2 Clab	2	8	Ea.	365	243		608	805
0120	Tiered type		2	8		430	243		673	875
0140	Waterfall type		2	8		276	243		519	705

13 17 Tubs and Pools

13 17 13 – Hot Tubs

13 17 13.10 Redwood Hot Tub System		Crew	Daily Output	Labor-Hours	Unit	Material	Labor	Equipment	Total	Total Incl O&P
0010	**REDWOOD HOT TUB SYSTEM**									
7050	4′ diameter x 4′ deep	Q-1	1	16	Ea.	3,200	645		3,845	4,575
7150	6′ diameter x 4′ deep		.80	20		4,900	810		5,710	6,700
7200	8′ diameter x 4′ deep		.80	20		7,175	810		7,985	9,225

13 17 33 – Whirlpool Tubs

13 17 33.10 Whirlpool Bath		Crew	Daily Output	Labor-Hours	Unit	Material	Labor	Equipment	Total	Total Incl O&P
0010	**WHIRLPOOL BATH**									
6000	Whirlpool, bath with vented overflow, molded fiberglass									
6100	66″ x 36″ x 24″	Q-1	1	16	Ea.	935	645		1,580	2,075
6400	72″ x 36″ x 21″		1	16		1,225	645		1,870	2,400
6500	60″ x 34″ x 21″		1	16		1,250	645		1,895	2,425

13 24 Special Activity Rooms

13 24 16 – Saunas

13 24 16.50 Saunas and Heaters		Crew	Daily Output	Labor-Hours	Unit	Material	Labor	Equipment	Total	Total Incl O&P
0010	**SAUNAS AND HEATERS**									
0020	Prefabricated, incl. heater & controls, 7′ high, 6′ x 4′, C/C	L-7	2.20	11.818	Ea.	5,250	395		5,645	6,425
0050	6′ x 4′, C/P		2	13		4,350	435		4,785	5,500
0400	6′ x 5′, C/C		2	13		5,425	435		5,860	6,675
0450	6′ x 5′, C/P		2	13		5,150	435		5,585	6,375
0600	6′ x 6′, C/C		1.80	14.444		7,100	480		7,580	8,600
0650	6′ x 6′, C/P		1.80	14.444		5,450	480		5,930	6,800
0800	6′ x 9′, C/C		1.60	16.250		7,850	540		8,390	9,550
0850	6′ x 9′, C/P		1.60	16.250		6,450	540		6,990	8,000
1000	8′ x 12′, C/C		1.10	23.636		10,700	785		11,485	13,100
1050	8′ x 12′, C/P		1.10	23.636		8,850	785		9,635	11,000
1200	8′ x 8′, C/C		1.40	18.571		8,600	620		9,220	10,500
1250	8′ x 8′, C/P		1.40	18.571		7,675	620		8,295	9,450
1400	8′ x 10′, C/C		1.20	21.667		9,400	720		10,120	11,500
1450	8′ x 10′, C/P		1.20	21.667		8,325	720		9,045	10,400
1600	10′ x 12′, C/C		1	26		12,600	865		13,465	15,400
1650	10′ x 12′, C/P		1	26		13,600	865		14,465	16,400
1700	Door only, cedar, 2′ x 6′, w/ 1′ x 4′ tempered insulated glass window	2 Carp	3.40	4.706		625	182		807	990
1800	Prehung, incl. jambs, pulls & hardware	"	12	1.333		600	51.50		651.50	745
2500	Heaters only (incl. above), wall mounted, to 200 C.F.					960			960	1,050
2750	To 300 C.F.					1,050			1,050	1,150
3000	Floor standing, to 720 C.F., 10,000 watts, w/controls	1 Elec	3	2.667		2,475	122		2,597	2,925
3250	To 1,000 C.F., 16,000 watts	"	3	2.667		3,600	122		3,722	4,150

13 24 26 – Steam Baths

13 24 26.50 Steam Baths and Components		Crew	Daily Output	Labor-Hours	Unit	Material	Labor	Equipment	Total	Total Incl O&P
0010	**STEAM BATHS AND COMPONENTS**									
0020	Heater, timer & head, single, to 140 C.F.	1 Plum	1.20	6.667	Ea.	2,300	299		2,599	3,025
0500	To 300 C.F.	"	1.10	7.273		2,625	325		2,950	3,425
2700	Conversion unit for residential tub, including door					3,900			3,900	4,275

13 34 Fabricated Engineered Structures

13 34 13 – Glazed Structures

13 34 13.13 Greenhouses		Crew	Daily Output	Labor-Hours	Unit	Material	Labor	Equipment	Total	Total Incl O&P
0010	**GREENHOUSES**, Shell only, stock units, not incl. 2′ stub walls,									
0020	foundation, floors, heat or compartments									
0300	Residential type, free standing, 8′-6″ long x 7′-6″ wide	2 Carp	59	.271	SF Flr.	24	10.50		34.50	43.50
0400	10′-6″ wide		85	.188		44	7.30		51.30	60.50
0600	13′-6″ wide		108	.148		45.50	5.75		51.25	59.50
0700	17′-0″ wide		160	.100		44.50	3.88		48.38	55.50
0900	Lean-to type, 3′-10″ wide		34	.471		44	18.25		62.25	78.50
1000	6′-10″ wide		58	.276		28.50	10.70		39.20	49
1100	Wall mounted to existing window, 3′ x 3′	1 Carp	4	2	Ea.	2,125	77.50		2,202.50	2,450
1120	4′ x 5′	"	3	2.667	"	2,525	103		2,628	2,950
1200	Deluxe quality, free standing, 7′-6″ wide	2 Carp	55	.291	SF Flr.	83	11.25		94.25	110
1220	10′-6″ wide		81	.198		63	7.65		70.65	82
1240	13′-6″ wide		104	.154		55	5.95		60.95	71
1260	17′-0″ wide		150	.107		45.50	4.13		49.63	57
1400	Lean-to type, 3′-10″ wide		31	.516		104	20		124	148
1420	6′-10″ wide		55	.291		72.50	11.25		83.75	98.50
1440	8′-0″ wide		97	.165		64.50	6.40		70.90	81.50

13 34 Fabricated Engineered Structures

13 34 13 – Glazed Structures

13 34 13.13 Greenhouses		Crew	Daily Output	Labor-Hours	Unit	Material	Labor (2019 Bare Costs)	Equipment (2019 Bare Costs)	Total (2019 Bare Costs)	Total Incl O&P
1500	Commercial, custom, truss frame, incl. equip., plumbing, elec.,									
1550	benches and controls, under 2,000 S.F.				SF Flr.	13.10			13.10	14.40
1700	Over 5,000 S.F.				"	12.20			12.20	13.40

13 34 13.19 Swimming Pool Enclosures		Crew	Daily Output	Labor-Hours	Unit	Material	Labor	Equipment	Total	Total Incl O&P
0010	**SWIMMING POOL ENCLOSURES** Translucent, free standing									
0020	not including foundations, heat or light									
0200	Economy	2 Carp	200	.080	SF Hor.	50	3.10		53.10	60
0600	Deluxe	"	70	.229	"	92	8.85		100.85	116

13 34 63 – Natural Fiber Construction

13 34 63.50 Straw Bale Construction		Crew	Daily Output	Labor-Hours	Unit	Material	Labor	Equipment	Total	Total Incl O&P
0010	**STRAW BALE CONSTRUCTION**									
2020	Straw bales in walls w/modified post and beam frame G	2 Carp	320	.050	S.F.	6.50	1.94		8.44	10.35

Division Notes

		CREW	DAILY OUTPUT	LABOR-HOURS	UNIT	BARE COSTS MAT.	LABOR	EQUIP.	TOTAL	TOTAL INCL O&P

Division 14 Conveying Equipment

Estimating Tips

General

- Many products in Division 14 will require some type of support or blocking for installation not included with the item itself. Examples are supports for conveyors or tube systems, attachment points for lifts, and footings for hoists or cranes. Add these supports in the appropriate division.

14 10 00 Dumbwaiters
14 20 00 Elevators

- Dumbwaiters and elevators are estimated and purchased in a method similar to buying a car. The manufacturer has a base unit with standard features. Added to this base unit price will be whatever options the owner or specifications require. Increased load capacity, additional vertical travel, additional stops, higher speed, and cab finish options are items to be considered. When developing an estimate for dumbwaiters and elevators, remember that some items needed by the installers may have to be included as part of the general contract.

  Examples are:
  - ☐ shaftway
  - ☐ rail support brackets
  - ☐ machine room
  - ☐ electrical supply
  - ☐ sill angles
  - ☐ electrical connections
  - ☐ pits
  - ☐ roof penthouses
  - ☐ pit ladders

  Check the job specifications and drawings before pricing.
- Installation of elevators and handicapped lifts in historic structures can require significant additional costs. The associated structural requirements may involve cutting into and repairing finishes, moldings, flooring, etc. The estimator must account for these special conditions.

14 30 00 Escalators and Moving Walks

- Escalators and moving walks are specialty items installed by specialty contractors. There are numerous options associated with these items. For specific options, contact a manufacturer or contractor. In a method similar to estimating dumbwaiters and elevators, you should verify the extent of general contract work and add items as necessary.

14 40 00 Lifts
14 90 00 Other Conveying Equipment

- Products such as correspondence lifts, chutes, and pneumatic tube systems, as well as other items specified in this subdivision, may require trained installers. The general contractor might not have any choice as to who will perform the installation or when it will be performed. Long lead times are often required for these products, making early decisions in scheduling necessary.

Reference Numbers

Reference numbers are shown at the beginning of some major classifications. These numbers refer to related items in the Reference Section. The reference information may be an estimating procedure, an alternate pricing method, or technical information.

Note: Not all subdivisions listed here necessarily appear. ■

14 11 Manual Dumbwaiters

14 11 10 – Hand Operated Dumbwaiters

14 11 10.20 Manual Dumbwaiters		Crew	Daily Output	Labor-Hours	Unit	2019 Bare Costs Material	Labor	Equipment	Total	Total Incl O&P
0010	**MANUAL DUMBWAITERS**									
0020	2 stop, hand powered, up to 75 lb. capacity	2 Elev	.75	21.333	Ea.	3,300	1,300		4,600	5,750
0100	76 lb. capacity and up		.50	32	"	6,950	1,950		8,900	10,900
0300	For each additional stop, add	↓	.75	21.333	Stop	1,150	1,300		2,450	3,375

14 12 Electric Dumbwaiters

14 12 10 – Dumbwaiters

14 12 10.10 Electric Dumbwaiters		Crew	Daily Output	Labor-Hours	Unit	Material	Labor	Equipment	Total	Total Incl O&P
0010	**ELECTRIC DUMBWAITERS**									
0020	2 stop, up to 75 lb. capacity	2 Elev	.13	123	Ea.	8,575	7,550		16,125	21,700
0100	76 lb. capacity and up		.11	145	"	23,100	8,925		32,025	39,900
0600	For each additional stop, add	↓	.54	29.630	Stop	3,450	1,825		5,275	6,750

14 21 Electric Traction Elevators

14 21 33 – Electric Traction Residential Elevators

14 21 33.20 Residential Elevators		Crew	Daily Output	Labor-Hours	Unit	Material	Labor	Equipment	Total	Total Incl O&P
0010	**RESIDENTIAL ELEVATORS**									
7000	Residential, cab type, 1 floor, 2 stop, economy model	2 Elev	.20	80	Ea.	10,800	4,900		15,700	19,800
7100	Custom model		.10	160		18,100	9,800		27,900	36,000
7200	2 floor, 3 stop, economy model		.12	133		16,000	8,175		24,175	30,900
7300	Custom model	↓	.06	267	↓	26,000	16,300		42,300	55,000

14 42 Wheelchair Lifts

14 42 13 – Inclined Wheelchair Lifts

14 42 13.10 Inclined Wheelchair Lifts and Stairclimbers		Crew	Daily Output	Labor-Hours	Unit	Material	Labor	Equipment	Total	Total Incl O&P
0010	**INCLINED WHEELCHAIR LIFTS AND STAIRCLIMBERS**									
7700	Stair climber (chair lift), single seat, minimum	2 Elev	1	16	Ea.	5,000	980		5,980	7,100
7800	Maximum	"	.20	80	"	6,825	4,900		11,725	15,500

Estimating Tips

Pipe for fire protection and all uses is located in Subdivisions 21 11 13 and 22 11 13.

The labor adjustment factors listed in Subdivision 22 01 02.20 also apply to Division 21.

Many, but not all, areas in the U.S. require backflow protection in the fire system. Insurance underwriters may have specific requirements for the type of materials to be installed or design requirements based on the hazard to be protected. Local jurisdictions may have requirements not covered by code. It is advisable to be aware of any special conditions.

For your reference, the following is a list of the most applicable Fire Codes and Standards, which may be purchased from the NFPA, 1 Batterymarch Park, Quincy, MA 02169-7471.

- NFPA 1: Uniform Fire Code
- NFPA 10: Portable Fire Extinguishers
- NFPA 11: Low-, Medium-, and High-Expansion Foam
- NFPA 12: Carbon Dioxide Extinguishing Systems (Also companion 12A)
- NFPA 13: Installation of Sprinkler Systems (Also companion 13D, 13E, and 13R)
- NFPA 14: Installation of Standpipe and Hose Systems
- NFPA 15: Water Spray Fixed Systems for Fire Protection
- NFPA 16: Installation of Foam-Water Sprinkler and Foam-Water Spray Systems
- NFPA 17: Dry Chemical Extinguishing Systems (Also companion 17A)
- NFPA 18: Wetting Agents
- NFPA 20: Installation of Stationary Pumps for Fire Protection
- NFPA 22: Water Tanks for Private Fire Protection
- NFPA 24: Installation of Private Fire Service Mains and their Appurtenances
- NFPA 25: Inspection, Testing and Maintenance of Water-Based Fire Protection

Reference Numbers

Reference numbers are shown at the beginning of some major classifications. These numbers refer to related items in the Reference Section. The reference information may be an estimating procedure, an alternate pricing method, or technical information.

Note: "Powered in part by CINX™, based on liscensed proprietary information of Harrison Publishing House, Inc."

Note: Trade Service, in part, has been used as a reference source for some of the material prices used in Division 21.

21 05 Common Work Results for Fire Suppression

21 05 23 – General-Duty Valves for Water-Based Fire-Suppression Piping

	21 05 23.50 General-Duty Valves	Crew	Daily Output	Labor-Hours	Unit	2019 Bare Costs Material	Labor	Equipment	Total	Total Incl O&P
0010	**GENERAL-DUTY VALVES**, for water-based fire suppression									
6200	Valves and components									
6210	Wet alarm, includes									
6220	retard chamber, trim, gauges, alarm line strainer									
6260	3″ size	Q-12	3	5.333	Ea.	1,850	212		2,062	2,375
6280	4″ size	″	2	8	″	2,200	320		2,520	2,950
6400	Dry alarm, includes									
6405	retard chamber, trim, gauges, alarm line strainer									
6410	1-1/2″ size	Q-12	3	5.333	Ea.	4,825	212		5,037	5,675
6420	2″ size		3	5.333		4,825	212		5,037	5,675
6430	3″ size		3	5.333		4,900	212		5,112	5,750
6440	4″ size		2	8		5,250	320		5,570	6,300
6450	6″ size	Q-13	3	10.667		6,050	425		6,475	7,350
6460	8″ size	″	3	10.667		8,875	425		9,300	10,500

21 05 53 – Identification For Fire-Suppression Piping and Equipment

	21 05 53.50 Identification	Crew	Daily Output	Labor-Hours	Unit	Material	Labor	Equipment	Total	Total Incl O&P
0010	**IDENTIFICATION**, for fire suppression piping and equipment									
3010	Plates and escutcheons for identification of fire dept. service/connections									
3100	Wall mount, round, aluminum									
3110	4″	1 Plum	96	.083	Ea.	24.50	3.74		28.24	33
3120	6″	″	96	.083	″	63.50	3.74		67.24	76
3200	Wall mount, round, cast brass									
3210	2-1/2″	1 Plum	70	.114	Ea.	58.50	5.15		63.65	73
3220	3″		70	.114		68.50	5.15		73.65	83.50
3230	4″		70	.114		117	5.15		122.15	137
3240	6″		70	.114		137	5.15		142.15	158
3300	Wall mount, square, cast brass									
3310	2-1/2″	1 Plum	70	.114	Ea.	156	5.15		161.15	180
3320	3″	″	70	.114	″	166	5.15		171.15	190
3400	Wall mount, cast brass, multiple outlets									
3410	rect. 2 way	Q-1	5	3.200	Ea.	216	129		345	450
3420	rect. 3 way		4	4		505	162		667	825
3430	rect. 4 way		4	4		635	162		797	965
3440	square 4 way		4	4		630	162		792	955
3450	rect. 6 way		3	5.333		870	215		1,085	1,300
3500	Base mount, free standing fdc, cast brass									
3510	4″	1 Plum	60	.133	Ea.	117	6		123	139
3520	6″	″	60	.133	″	156	6		162	182

21 11 Facility Fire-Suppression Water-Service Piping

21 11 16 – Facility Fire Hydrants

	21 11 16.50 Fire Hydrants for Buildings	Crew	Daily Output	Labor-Hours	Unit	Material	Labor	Equipment	Total	Total Incl O&P
0010	**FIRE HYDRANTS FOR BUILDINGS**									
3750	Hydrants, wall, w/caps, single, flush, polished brass									
3800	2-1/2″ x 2-1/2″	Q-12	5	3.200	Ea.	260	127		387	495
3840	2-1/2″ x 3″		5	3.200		520	127		647	780
3860	3″ x 3″		4.80	3.333		415	133		548	680
3900	For polished chrome, add					20%				

21 11 Facility Fire-Suppression Water-Service Piping

21 11 19 – Fire-Department Connections

21 11 19.50 Connections for the Fire-Department		Crew	Daily Output	Labor-Hours	Unit	2019 Bare Costs Material	Labor	Equipment	Total	Total Incl O&P
0010	**CONNECTIONS FOR THE FIRE-DEPARTMENT**									
7140	Standpipe connections, wall, w/plugs & chains									
7160	Single, flush, brass, 2-1/2" x 2-1/2"	Q-12	5	3.200	Ea.	193	127		320	420
7180	2-1/2" x 3"	"	5	3.200	"	202	127		329	430
7240	For polished chrome, add					15%				
7280	Double, flush, polished brass									
7300	2-1/2" x 2-1/2" x 4"	Q-12	5	3.200	Ea.	820	127		947	1,125
8800	Free standing siamese unit, polished brass, two way									
8820	2-1/2" x 2-1/2" x 4"	Q-12	2.50	6.400	Ea.	725	255		980	1,225

21 12 Fire-Suppression Standpipes

21 12 13 – Fire-Suppression Hoses and Nozzles

21 12 13.50 Fire Hoses and Nozzles

		Crew	Daily Output	Labor-Hours	Unit	Material	Labor	Equipment	Total	Total Incl O&P
0010	**FIRE HOSES AND NOZZLES**									
0200	Adapters, rough brass, straight hose threads									
1400	Couplings, sngl. & dbl. jacket, pin lug or rocker lug, cast brass									
1410	1-1/2"				Ea.	66.50			66.50	73.50
1420	2-1/2"				"	52.50			52.50	57.50
1500	For polished brass, add					20%				
1520	For polished chrome, add					40%				
2200	Hose, less couplings									
2260	Synthetic jacket, lined, 300 lb. test, 1-1/2" diameter	Q-12	2600	.006	L.F.	3.22	.25		3.47	3.94
2270	2" diameter		2200	.007		2.51	.29		2.80	3.24
2280	2-1/2" diameter		2200	.007		6	.29		6.29	7.10
2290	3" diameter		2200	.007		3.17	.29		3.46	3.97
2360	High strength, 500 lb. test, 1-1/2" diameter		2600	.006		2.49	.25		2.74	3.14
2380	2-1/2" diameter		2200	.007		5.75	.29		6.04	6.85
5600	Nozzles, brass									
5620	Adjustable fog, 3/4" booster line				Ea.	139			139	153
5630	1" booster line					141			141	155
5640	1-1/2" leader line					109			109	120
5660	2-1/2" direct connection					201			201	221
5680	2-1/2" playpipe nozzle					235			235	258
5780	For chrome plated, add					8%				
5850	Electrical fire, adjustable fog, no shock									
5900	1-1/2"				Ea.	505			505	555
5920	2-1/2"					830			830	910
5980	For polished chrome, add					6%				
6200	Heavy duty, comb. adj. fog and str. stream, with handle									
6210	1" booster line				Ea.	196			196	216
6240	1-1/2"					425			425	465
6260	2-1/2", for playpipe					670			670	735
6280	2-1/2" direct connection					475			475	525
6300	2-1/2" playpipe combination					565			565	625

21 12 16 – Fire-Suppression Hose Reels

21 12 16.50 Fire-Suppression Hose Reels

		Crew	Daily Output	Labor-Hours	Unit	Material	Labor	Equipment	Total	Total Incl O&P
0010	**FIRE-SUPPRESSION HOSE REELS**									
2990	Hose reel, swinging, for 1-1/2" polyester neoprene lined hose									
3000	50' long	Q-12	14	1.143	Ea.	144	45.50		189.50	233
3020	100' long		14	1.143		264	45.50		309.50	365

21 12 Fire-Suppression Standpipes

21 12 16 – Fire-Suppression Hose Reels

21 12 16.50 Fire-Suppression Hose Reels		Crew	Daily Output	Labor-Hours	Unit	Material	2019 Bare Costs Labor	Equipment	Total	Total Incl O&P
3060	For 2-1/2″ cotton rubber hose, 75′ long	Q-12	14	1.143	Ea.	320	45.50		365.50	425
3100	150′ long	↓	14	1.143	↓	315	45.50		360.50	425

21 12 19 – Fire-Suppression Hose Racks

21 12 19.50 Fire Hose Racks

		Crew	Daily Output	Labor-Hours	Unit	Material	Labor	Equipment	Total	Total Incl O&P
0010	**FIRE HOSE RACKS**									
2600	Hose rack, swinging, for 1-1/2″ diameter hose,									
2620	Enameled steel, 50′ and 75′ lengths of hose	Q-12	20	.800	Ea.	71	32		103	131
2640	100′ and 125′ lengths of hose		20	.800		94	32		126	157
2680	Chrome plated, 50′ and 75′ lengths of hose		20	.800		75.50	32		107.50	136
2700	100′ and 125′ lengths of hose	↓	20	.800	↓	144	32		176	212

21 12 23 – Fire-Suppression Hose Valves

21 12 23.70 Fire Hose Valves

		Crew	Daily Output	Labor-Hours	Unit	Material	Labor	Equipment	Total	Total Incl O&P
0010	**FIRE HOSE VALVES**									
0020	Angle, combination pressure adjust/restricting, rough brass									
0030	1-1/2″	1 Spri	12	.667	Ea.	98	29.50		127.50	157
0040	2-1/2″	″	7	1.143	″	213	50.50		263.50	315
0042	Nonpressure adjustable/restricting, rough brass									
0044	1-1/2″	1 Spri	12	.667	Ea.	49	29.50		78.50	103
0046	2-1/2″	″	7	1.143	″	128	50.50		178.50	224
0050	For polished brass, add					30%				
0060	For polished chrome, add					40%				

21 13 Fire-Suppression Sprinkler Systems

21 13 13 – Wet-Pipe Sprinkler Systems

21 13 13.50 Wet-Pipe Sprinkler System Components

		Crew	Daily Output	Labor-Hours	Unit	Material	Labor	Equipment	Total	Total Incl O&P
0010	**WET-PIPE SPRINKLER SYSTEM COMPONENTS**									
1100	Alarm, electric pressure switch (circuit closer)	1 Spri	26	.308	Ea.	111	13.60		124.60	145
1140	For explosion proof, max 20 psi, contacts close or open		26	.308		690	13.60		703.60	785
1220	Water motor, complete with gong	↓	4	2	↓	455	88.50		543.50	645
1900	Flexible sprinkler head connectors									
1910	Braided stainless steel hose with mounting bracket									
1920	1/2″ and 3/4″ outlet size									
1940	40″ length	1 Spri	30	.267	Ea.	78	11.80		89.80	105
1960	60″ length	″	22	.364	″	89.50	16.10		105.60	125
1982	May replace hard-pipe armovers									
1984	for wet and pre-action systems.									
2600	Sprinkler heads, not including supply piping									
3700	Standard spray, pendent or upright, brass, 135°F to 286°F									
3730	1/2″ NPT, 7/16″ orifice	1 Spri	16	.500	Ea.	16.35	22		38.35	54.50
5600	Concealed, complete with cover plate									
5620	1/2″ NPT, 1/2″ orifice, 135°F to 212°F	1 Spri	9	.889	Ea.	25.50	39.50		65	92.50

21 13 16 – Dry-Pipe Sprinkler Systems

21 13 16.50 Dry-Pipe Sprinkler System Components

		Crew	Daily Output	Labor-Hours	Unit	Material	Labor	Equipment	Total	Total Incl O&P
0010	**DRY-PIPE SPRINKLER SYSTEM COMPONENTS**									
0600	Accelerator	1 Spri	8	1	Ea.	825	44.50		869.50	985
0800	Air compressor for dry pipe system, automatic, complete									
0820	30 gal. system capacity, 3/4 HP	1 Spri	1.30	6.154	Ea.	1,175	272		1,447	1,750
2600	Sprinkler heads, not including supply piping									
2640	Dry, pendent, 1/2″ orifice, 3/4″ or 1″ NPT									

21 13 Fire-Suppression Sprinkler Systems

21 13 16 – Dry-Pipe Sprinkler Systems

21 13 16.50 Dry-Pipe Sprinkler System Components		Crew	Daily Output	Labor-Hours	Unit	Material	2019 Bare Costs Labor	Equipment	Total	Total Incl O&P
2700	15-1/4″ to 18″ length	1 Spri	14	.571	Ea.	158	25.50		183.50	215

21 21 Carbon-Dioxide Fire-Extinguishing Systems

21 21 16 – Carbon-Dioxide Fire-Extinguishing Equipment

21 21 16.50 CO2 Fire Extinguishing System		Crew	Daily Output	Labor-Hours	Unit	Material	2019 Bare Costs Labor	Equipment	Total	Total Incl O&P
0010	**CO_2 FIRE EXTINGUISHING SYSTEM**									
0100	Control panel, single zone with batteries (2 zones det., 1 suppr.)	1 Elec	1	8	Ea.	970	365		1,335	1,675
0150	Multizone (4) with batteries (8 zones det., 4 suppr.)	″	.50	16		2,925	730		3,655	4,425
1000	Dispersion nozzle, CO_2, 3″ x 5″	1 Plum	18	.444		154	19.95		173.95	202
2000	Extinguisher, CO_2 system, high pressure, 75 lb. cylinder	Q-1	6	2.667		1,350	108		1,458	1,650
2100	100 lb. cylinder	″	5	3.200		1,750	129		1,879	2,125
3000	Electro/mechanical release	L-1	4	2.500		985	113		1,098	1,250
3400	Manual pull station	1 Plum	6	1.333		83	60		143	190
4000	Pneumatic damper release	″	8	1		163	45		208	253

21 22 Clean-Agent Fire-Extinguishing Systems

21 22 16 – Clean-Agent Fire-Extinguishing Equipment

21 22 16.50 Clean-Agent Extinguishing Systems		Crew	Daily Output	Labor-Hours	Unit	Material	2019 Bare Costs Labor	Equipment	Total	Total Incl O&P
0010	**CLEAN-AGENT EXTINGUISHING SYSTEMS**									
0020	FM200 fire extinguishing system									
1100	Dispersion nozzle FM200, 1-1/2″	1 Plum	14	.571	Ea.	191	25.50		216.50	252
2400	Extinguisher, FM200 system, filled, with mounting bracket									
2460	26 lb. container	Q-1	8	2	Ea.	2,025	81		2,106	2,350
2480	44 lb. container		7	2.286		2,675	92.50		2,767.50	3,100
2500	63 lb. container		6	2.667		3,200	108		3,308	3,700
2520	101 lb. container		5	3.200		5,275	129		5,404	6,000
2540	196 lb. container		4	4		6,700	162		6,862	7,650
6000	FM200 system, simple nozzle layout, with broad dispersion				C.F.	1.79			1.79	1.97
6010	Extinguisher, FM200 system, filled, with mounting bracket									
6020	Complex nozzle layout and/or including underfloor dispersion				C.F.	3.56			3.56	3.92
6100	20,000 C.F. 2 exits, 8′ clng					2.13			2.13	2.34
6200	100,000 C.F. 4 exits, 8′ clng					1.92			1.92	2.11
6300	250,000 C.F. 6 exits, 8′ clng					1.62			1.62	1.78
7010	HFC-227ea fire extinguishing system									
7100	Cylinders with clean-agent									
7110	Does not include pallete jack/fork lift rental fees									
7120	70 lb. cyl, w/35 lb. agent, no solenoid	Q-12	14	1.143	Ea.	2,425	45.50		2,470.50	2,750
7130	70 lb. cyl w/70 lb. agent, no solenoid		10	1.600		3,175	63.50		3,238.50	3,600
7140	70 lb. cyl w/35 lb. agent, w/solenoid		14	1.143		3,550	45.50		3,595.50	3,975
7150	70 lb. cyl w/70 lb. agent, w/solenoid		10	1.600		4,400	63.50		4,463.50	4,925
7220	250 lb. cyl, w/125 lb. agent, no solenoid		8	2		5,600	79.50		5,679.50	6,275
7230	250 lb. cyl, w/250 lb. agent, no solenoid		5	3.200		5,600	127		5,727	6,350
7240	250 lb. cyl, w/125 lb. agent, w/solenoid		8	2		7,525	79.50		7,604.50	8,400
7250	250 lb. cyl, w/250 lb. agent, w/solenoid		5	3.200		10,800	127		10,927	12,100
7320	560 lb. cyl, w/300 lb. agent, no solenoid		4	4		10,300	159		10,459	11,600
7330	560 lb. cyl, w/560 lb. agent, no solenoid		2.50	6.400		15,400	255		15,655	17,400
7340	560 lb. cyl, w/300 lb. agent, w/solenoid		4	4		13,900	159		14,059	15,600
7350	560 lb. cyl, w/560 lb. agent, w/solenoid		2.50	6.400		21,200	255		21,455	23,700
7420	1,200 lb. cyl, w/600 lb. agent, no solenoid	Q-13	4	8		19,400	320		19,720	21,900

21 22 Clean-Agent Fire-Extinguishing Systems

21 22 16 – Clean-Agent Fire-Extinguishing Equipment

21 22 16.50 Clean-Agent Extinguishing Systems		Crew	Daily Output	Labor-Hours	Unit	2019 Bare Costs Material	Labor	Equipment	Total	Total Incl O&P
7430	1,200 lb. cyl, w/1,200 lb. agent, no solenoid	Q-13	3	10.667	Ea.	31,900	425		32,325	35,800
7440	1,200 lb. cyl, w/600 lb. agent, w/solenoid		4	8		49,400	320		49,720	55,000
7450	1,200 lb. cyl, w/1,200 lb. agent, w/solenoid	↓	3	10.667	↓	76,000	425		76,425	84,000
7500	Accessories									
7510	Dispersion nozzle	1 Spri	16	.500	Ea.	98	22		120	145
7520	Agent release panel	1 Elec	4	2		62	91		153	217
7530	Maintenance switch	"	6	1.333		39.50	61		100.50	143
7540	Solenoid valve, 12v dc	1 Spri	8	1		234	44.50		278.50	330
7550	12v ac		8	1		510	44.50		554.50	640
7560	12v dc, explosion proof	↓	8	1	↓	395	44.50		439.50	510

Estimating Tips

22 10 00 Plumbing Piping and Pumps

This subdivision is primarily basic pipe and related materials. The pipe may be used by any of the mechanical disciplines, i.e., plumbing, fire protection, heating, and air conditioning.

Note: CPVC plastic piping approved for fire protection is located in 21 11 13.

- The labor adjustment factors listed in Subdivision 22 01 02.20 apply throughout Divisions 21, 22, and 23. CAUTION: the correct percentage may vary for the same items. For example, the percentage add for the basic pipe installation should be based on the maximum height that the installer must install for that particular section. If the pipe is to be located 14' above the floor but it is suspended on threaded rod from beams, the bottom flange of which is 18' high (4' rods), then the height is actually 18' and the add is 20%. The pipe cover, however, does not have to go above the 14' and so the add should be 10%.
- Most pipe is priced first as straight pipe with a joint (coupling, weld, etc.) every 10' and a hanger usually every 10'. There are exceptions with hanger spacing such as for cast iron pipe (5') and plastic pipe (3 per 10'). Following each type of pipe there are several lines listing sizes and the amount to be subtracted to delete couplings and hangers. This is for pipe that is to be buried or supported together on trapeze hangers. The reason that the couplings are deleted is that these runs are usually long, and frequently longer lengths of pipe are used. By deleting the couplings, the estimator is expected to look up and add back the correct reduced number of couplings.
- When preparing an estimate, it may be necessary to approximate the fittings. Fittings usually run between 25% and 50% of the cost of the pipe. The lower percentage is for simpler runs, and the higher number is for complex areas, such as mechanical rooms.
- For historic restoration projects, the systems must be as invisible as possible, and pathways must be sought for pipes, conduit, and ductwork. While installations in accessible spaces (such as basements and attics) are relatively straightforward to estimate, labor costs may be more difficult to determine when delivery systems must be concealed.

22 40 00 Plumbing Fixtures

- Plumbing fixture costs usually require two lines: the fixture itself and its "rough-in, supply, and waste."
- In the Assemblies Section (Plumbing D2010) for the desired fixture, the System Components Group at the center of the page shows the fixture on the first line. The rest of the list (fittings, pipe, tubing, etc.) will total up to what we refer to in the Unit Price section as "Rough-in, supply, waste, and vent." Note that for most fixtures we allow a nominal 5' of tubing to reach from the fixture to a main or riser.
- Remember that gas- and oil-fired units need venting.

Reference Numbers

Reference numbers are shown at the beginning of some major classifications. These numbers refer to related items in the Reference Section. The reference information may be an estimating procedure, an alternate pricing method, or technical information.

Note: Not all subdivisions listed here necessarily appear. ■

Note: "Powered in part by CINX™, based on liscensed proprietary information of Harrison Publishing House, Inc."

Note: Trade Service, in part, has been used as a reference source for some of the material prices used in Division 22.

22 01 02.10 Boilers, General		Crew	Daily Output	Labor-Hours	Unit	Material	2019 Bare Costs Labor	Equipment	Total	Total Incl O&P
0010	**BOILERS, GENERAL**, Prices do not include flue piping, elec. wiring,									
0020	gas or oil piping, boiler base, pad, or tankless unless noted									

22 01 02.20 Labor Adjustment Factors		Crew	Daily Output	Labor-Hours	Unit	Material	2019 Bare Costs Labor	Equipment	Total	Total Incl O&P
0010	**LABOR ADJUSTMENT FACTORS** (For Div. 21, 22 and 23) R220102-20									
0100	Labor factors: The below are reasonable suggestions, but									
0110	each project must be evaluated for its own peculiarities, and									
0120	the adjustments be increased or decreased depending on the									
0130	severity of the special conditions.									
1000	Add to labor for elevated installation (Above floor level)									
1080	10′ to 14.5′ high						10%			
1100	15′ to 19.5′ high						20%			
1120	20′ to 24.5′ high						25%			
1140	25′ to 29.5′ high						35%			
1160	30′ to 34.5′ high						40%			
1180	35′ to 39.5′ high						50%			
1200	40′ and higher						55%			
2000	Add to labor for crawl space									
2100	3′ high						40%			
2140	4′ high						30%			
3000	Add to labor for multi-story building									
3010	For new construction (No elevator available)									
3100	Add for floors 3 thru 10						5%			
3110	Add for floors 11 thru 15						10%			
3120	Add for floors 16 thru 20						15%			
3130	Add for floors 21 thru 30						20%			
3140	Add for floors 31 and up						30%			
3170	For existing structure (Elevator available)									
3180	Add for work on floor 3 and above						2%			
4000	Add to labor for working in existing occupied buildings									
4100	Hospital						35%			
4140	Office building						25%			
4180	School						20%			
4220	Factory or warehouse						15%			
4260	Multi dwelling						15%			
5000	Add to labor, miscellaneous									
5100	Cramped shaft						35%			
5140	Congested area						15%			
5180	Excessive heat or cold						30%			
9000	Labor factors: The above are reasonable suggestions, but									
9010	each project should be evaluated for its own peculiarities.									
9100	Other factors to be considered are:									
9140	Movement of material and equipment through finished areas									
9180	Equipment room									
9220	Attic space									
9260	No service road									
9300	Poor unloading/storage area									
9340	Congested site area/heavy traffic									

22 05 05 – Selective Demolition for Plumbing

22 05 05.10 Plumbing Demolition		Crew	Daily Output	Labor-Hours	Unit	Material	2019 Bare Costs Labor	Equipment	Total	Total Incl O&P
0010	**PLUMBING DEMOLITION**									
1020	Fixtures, including 10′ piping									
1101	Bathtubs, cast iron	1 Clab	4	2	Ea.		60.50		60.50	100
1121	Fiberglass		6	1.333			40.50		40.50	67
1141	Steel		5	1.600			48.50		48.50	80
1200	Lavatory, wall hung	1 Plum	10	.800			36		36	59
1221	Counter top	1 Clab	16	.500			15.20		15.20	25
1301	Sink, single compartment		16	.500			15.20		15.20	25
1321	Double		10	.800			24.50		24.50	40
1400	Water closet, floor mounted	1 Plum	8	1			45		45	73.50
1421	Wall mounted	1 Clab	7	1.143			34.50		34.50	57.50
2001	Piping, metal, to 1-1/2″ diameter		200	.040	L.F.		1.21		1.21	2.01
2051	2″ thru 3-1/2″ diameter		150	.053			1.62		1.62	2.67
2101	4″ thru 6″ diameter		100	.080			2.43		2.43	4.01
2160	Plastic pipe with fittings, up thru 1-1/2″ diameter	1 Plum	250	.032			1.44		1.44	2.36
2162	2″ thru 3″ diameter	″	200	.040			1.79		1.79	2.94
2164	4″ thru 6″ diameter	Q-1	200	.080			3.23		3.23	5.30
2166	8″ thru 14″ diameter		150	.107			4.31		4.31	7.05
2168	16″ diameter		100	.160			6.45		6.45	10.60
3000	Submersible sump pump	1 Plum	24	.333	Ea.		14.95		14.95	24.50
6000	Remove and reset fixtures, easy access		6	1.333			60		60	98
6100	Difficult access		4	2			89.50		89.50	147

22 05 23 – General-Duty Valves for Plumbing Piping

22 05 23.20 Valves, Bronze		Crew	Daily Output	Labor-Hours	Unit	Material	Labor	Equipment	Total	Total Incl O&P
0010	**VALVES, BRONZE**									
1300	Ball									
1398	Threaded, 150 psi									
1450	1/2″	1 Plum	22	.364	Ea.	15.15	16.30		31.45	43.50
1460	3/4″		20	.400		28.50	17.95		46.45	61
1470	1″		19	.421		31	18.90		49.90	65
1480	1-1/4″		15	.533		52.50	24		76.50	97
1490	1-1/2″		13	.615		70.50	27.50		98	123
1500	2″		11	.727		82.50	32.50		115	144
1522	Solder the same price as threaded									
1600	Butterfly, 175 psi, full port, solder or threaded ends									
1610	Stainless steel disc and stem									
1620	1/4″	1 Plum	24	.333	Ea.	20	14.95		34.95	47
1630	3/8″		24	.333		16.20	14.95		31.15	42.50
1640	1/2″		22	.364		17.30	16.30		33.60	46
1650	3/4″		20	.400		28	17.95		45.95	60
1660	1″		19	.421		34	18.90		52.90	68.50
1670	1-1/4″		15	.533		55	24		79	100
1680	1-1/2″		13	.615		71	27.50		98.50	124
1690	2″		11	.727		90	32.50		122.50	153
1750	Check, swing, class 150, regrinding disc, threaded									
1850	1/2″	1 Plum	24	.333	Ea.	69.50	14.95		84.45	101
1860	3/4″		20	.400		91	17.95		108.95	130
1870	1″		19	.421		143	18.90		161.90	188
1880	1-1/4″		15	.533		206	24		230	266
1900	2″		11	.727		315	32.50		347.50	405
2850	Gate, N.R.S., soldered, 125 psi									
2920	1/2″	1 Plum	24	.333	Ea.	59.50	14.95		74.45	90

22 05 23 – General-Duty Valves for Plumbing Piping

22 05 23.20 Valves, Bronze		Crew	Daily Output	Labor-Hours	Unit	2019 Bare Costs Material	Labor	Equipment	Total	Total Incl O&P
2940	3/4"	1 Plum	20	.400	Ea.	68	17.95		85.95	105
2950	1"		19	.421		76.50	18.90		95.40	116
2960	1-1/4"		15	.533		142	24		166	196
2970	1-1/2"		13	.615		163	27.50		190.50	225
2980	2"		11	.727		186	32.50		218.50	258
4250	Threaded, class 150									
4340	3/4"	1 Plum	20	.400	Ea.	79	17.95		96.95	117
4350	1"		19	.421		106	18.90		124.90	148
4360	1-1/4"		15	.533		144	24		168	198
4370	1-1/2"		13	.615		182	27.50		209.50	246
5600	Relief, pressure & temperature, self-closing, ASME, threaded									
5640	3/4"	1 Plum	28	.286	Ea.	210	12.80		222.80	252
5650	1"		24	.333		335	14.95		349.95	395
5660	1-1/4"		20	.400		680	17.95		697.95	780
6400	Pressure, water, ASME, threaded									
6440	3/4"	1 Plum	28	.286	Ea.	144	12.80		156.80	179
6450	1"		24	.333		305	14.95		319.95	360
6460	1-1/4"		20	.400		470	17.95		487.95	550
6470	1-1/2"		18	.444		660	19.95		679.95	765
6480	2"		16	.500		955	22.50		977.50	1,075
6490	2-1/2"		15	.533		4,250	24		4,274	4,725
6900	Reducing, water pressure									
6920	300 psi to 25-75 psi, threaded or sweat									
6940	1/2"	1 Plum	24	.333	Ea.	460	14.95		474.95	530
6950	3/4"		20	.400		460	17.95		477.95	540
6960	1"		19	.421		715	18.90		733.90	815
8350	Tempering, water, sweat connections									
8400	1/2"	1 Plum	24	.333	Ea.	105	14.95		119.95	141
8440	3/4"	"	20	.400	"	144	17.95		161.95	189
8800	Water heater water & gas safety shut off									
8810	Protection against a leaking water heater									
8814	Shut off valve	1 Plum	16	.500	Ea.	193	22.50		215.50	249
8818	Water heater dam		32	.250		32	11.20		43.20	53.50
8822	Gas control wiring harness		32	.250		22	11.20		33.20	43
8830	Whole house flood safety shut off									
8834	Connections									
8838	3/4" NPT	1 Plum	12	.667	Ea.	950	30		980	1,100
8842	1" NPT		11	.727		975	32.50		1,007.50	1,125
8846	1-1/4" NPT		10	.800		1,000	36		1,036	1,150

22 05 76 – Facility Drainage Piping Cleanouts

22 05 76.20 Cleanout Tees		Crew	Daily Output	Labor-Hours	Unit	Material	Labor	Equipment	Total	Total Incl O&P
0010	**CLEANOUT TEES**									
0100	Cast iron, B&S, with countersunk plug									
0220	3" pipe size	1 Plum	3.60	2.222	Ea.	154	99.50		253.50	335
0240	4" pipe size	"	3.30	2.424	"	240	109		349	440
0500	For round smooth access cover, same price									
4000	Plastic, tees and adapters. Add plugs									
4010	ABS, DWV									
4020	Cleanout tee, 1-1/2" pipe size	1 Plum	15	.533	Ea.	11.55	24		35.55	52.50

22 07 Plumbing Insulation

22 07 19 – Plumbing Piping Insulation

22 07 19.10 Piping Insulation		Crew	Daily Output	Labor-Hours	Unit	Material	Labor	Equipment	Total	Total Incl O&P
						2019 Bare Costs				
0010	**PIPING INSULATION**									
0600	Pipe covering (price copper tube one size less than IPS)									
6600	Fiberglass, with all service jacket									
6840	1" wall, 1/2" iron pipe size G	Q-14	240	.067	L.F.	.88	2.44		3.32	5.10
6860	3/4" iron pipe size G		230	.070		.96	2.55		3.51	5.35
6870	1" iron pipe size G		220	.073		1.03	2.66		3.69	5.60
6900	2" iron pipe size G		200	.080		1.61	2.93		4.54	6.70
6910	2-1/2" iron pipe size G		190	.084		1.65	3.08		4.73	7
6920	3" iron pipe size G		180	.089		1.78	3.26		5.04	7.45
6940	4" iron pipe size G		150	.107		2.36	3.91		6.27	9.20
6960	6" iron pipe size G		120	.133		2.82	4.88		7.70	11.30
7800	For fiberglass with standard canvas jacket, deduct					5%				
7802	For fittings, add 3 L.F. for each fitting									
7804	plus 4 L.F. for each flange of the fitting									
7879	Rubber tubing, flexible closed cell foam									
8300	3/4" wall, 1/4" iron pipe size G	1 Asbe	90	.089	L.F.	.92	3.62		4.54	7.10
8330	1/2" iron pipe size G		89	.090		1.11	3.66		4.77	7.35
8340	3/4" iron pipe size G		89	.090		1.81	3.66		5.47	8.15
8350	1" iron pipe size G		88	.091		2.08	3.70		5.78	8.55
8360	1-1/4" iron pipe size G		87	.092		2.42	3.74		6.16	8.95
8380	2" iron pipe size G		86	.093		3.57	3.79		7.36	10.35
8390	2-1/2" iron pipe size G		86	.093		4.27	3.79		8.06	11.10
8400	3" iron pipe size G		85	.094		4.82	3.83		8.65	11.75
8420	4" iron pipe size G		80	.100		6	4.07		10.07	13.45
8440	6" iron pipe size G		80	.100		8.70	4.07		12.77	16.45
8444	1" wall, 1/2" iron pipe size G		86	.093		2.71	3.79		6.50	9.40
8445	3/4" iron pipe size G		84	.095		3.29	3.88		7.17	10.15
8446	1" iron pipe size G		84	.095		3.43	3.88		7.31	10.30
8447	1-1/4" iron pipe size G		82	.098		3.76	3.97		7.73	10.85
8448	1-1/2" iron pipe size G		82	.098		5.25	3.97		9.22	12.50
8449	2" iron pipe size G		80	.100		6.50	4.07		10.57	14
8450	2-1/2" iron pipe size G		80	.100		7.50	4.07		11.57	15.10
8456	Rubber insulation tape, 1/8" x 2" x 30' G				Ea.	21.50			21.50	23.50

22 11 Facility Water Distribution

22 11 13 – Facility Water Distribution Piping

22 11 13.23 Pipe/Tube, Copper		Crew	Daily Output	Labor-Hours	Unit	Material	Labor	Equipment	Total	Total Incl O&P
0010	**PIPE/TUBE, COPPER**, Solder joints									
1000	Type K tubing, couplings & clevis hanger assemblies 10' OC									
1180	3/4" diameter	1 Plum	74	.108	L.F.	8.55	4.85		13.40	17.35
1200	1" diameter		66	.121		12.70	5.45		18.15	23
1240	1-1/2" diameter		50	.160		18.15	7.20		25.35	32
1260	2" diameter		40	.200		27	8.95		35.95	44
1280	2-1/2" diameter	Q-1	60	.267		41	10.75		51.75	62.50
1300	3" diameter	"	54	.296		55	11.95		66.95	80
2000	Type L tubing, couplings & clevis hanger assemblies 10' OC									
2140	1/2" diameter	1 Plum	81	.099	L.F.	3.38	4.43		7.81	10.95
2180	3/4" diameter		76	.105		4.39	4.72		9.11	12.55
2200	1" diameter		68	.118		7	5.30		12.30	16.40
2220	1-1/4" diameter		58	.138		11.65	6.20		17.85	23
2240	1-1/2" diameter		52	.154		10.45	6.90		17.35	23

22 11 13.23 Pipe/Tube, Copper

	22 11 13.23 Pipe/Tube, Copper	Crew	Daily Output	Labor-Hours	Unit	2019 Bare Costs Material	Labor	Equipment	Total	Total Incl O&P
2260	2″ diameter	1 Plum	42	.190	L.F.	18.05	8.55		26.60	34
2280	2-1/2″ diameter	Q-1	62	.258		25.50	10.40		35.90	45
2300	3″ diameter		56	.286		43.50	11.55		55.05	66.50
2320	3-1/2″ diameter		43	.372		54	15.05		69.05	83.50
2340	4″ diameter		39	.410		62	16.55		78.55	95
3000	Type M tubing, couplings & clevis hanger assemblies 10′ OC									
3140	1/2″ diameter	1 Plum	84	.095	L.F.	3.64	4.27		7.91	11
3180	3/4″ diameter		78	.103		5.20	4.60		9.80	13.25
3200	1″ diameter		70	.114		8.40	5.15		13.55	17.65
3220	1-1/4″ diameter		60	.133		11.10	6		17.10	22
3240	1-1/2″ diameter		54	.148		14.20	6.65		20.85	26.50
3260	2″ diameter		44	.182		21	8.15		29.15	37
3280	2-1/2″ diameter	Q-1	64	.250		31.50	10.10		41.60	51
3300	3″ diameter		58	.276		39	11.15		50.15	61.50
3320	3-1/2″ diameter		45	.356		56	14.35		70.35	85
3340	4″ diameter		40	.400		74	16.15		90.15	108
4000	Type DWV tubing, couplings & clevis hanger assemblies 10′ OC									
4100	1-1/4″ diameter	1 Plum	60	.133	L.F.	12.45	6		18.45	23.50
4120	1-1/2″ diameter		54	.148		12	6.65		18.65	24
4140	2″ diameter		44	.182		18.55	8.15		26.70	34
4160	3″ diameter	Q-1	58	.276		24.50	11.15		35.65	45
4180	4″ diameter	″	40	.400		55.50	16.15		71.65	87.50

22 11 13.25 Pipe/Tube Fittings, Copper

	22 11 13.25 Pipe/Tube Fittings, Copper	Crew	Daily Output	Labor-Hours	Unit	Material	Labor	Equipment	Total	Total Incl O&P
0010	**PIPE/TUBE FITTINGS, COPPER**, Wrought unless otherwise noted									
0040	Solder joints, copper x copper									
0070	90° elbow, 1/4″	1 Plum	22	.364	Ea.	3.68	16.30		19.98	31
0100	1/2″		20	.400		1.20	17.95		19.15	31
0120	3/4″		19	.421		2.40	18.90		21.30	33.50
0130	1″		16	.500		6.55	22.50		29.05	44
0140	1-1/4″		15	.533		11.05	24		35.05	51.50
0150	1-1/2″		13	.615		15.35	27.50		42.85	62.50
0160	2″		11	.727		28	32.50		60.50	84
0170	2-1/2″	Q-1	13	1.231		61.50	49.50		111	149
0180	3″		11	1.455		82.50	58.50		141	187
0190	3-1/2″		10	1.600		315	64.50		379.50	455
0200	4″		9	1.778		215	72		287	355
0450	Tee, 1/4″	1 Plum	14	.571		8	25.50		33.50	51
0480	1/2″		13	.615		2.27	27.50		29.77	48
0500	3/4″		12	.667		6.05	30		36.05	55.50
0510	1″		10	.800		16.20	36		52.20	77
0520	1-1/4″		9	.889		23.50	40		63.50	91
0530	1-1/2″		8	1		31.50	45		76.50	109
0540	2″		7	1.143		51	51.50		102.50	140
0550	2-1/2″	Q-1	8	2		115	81		196	259
0560	3″		7	2.286		160	92.50		252.50	325
0580	4″		5	3.200		350	129		479	595
0612	Tee, reducing on the outlet, 1/4″	1 Plum	15	.533		15.60	24		39.60	56.50
0613	3/8″		15	.533		14.90	24		38.90	56
0614	1/2″		14	.571		13.95	25.50		39.45	57.50
0615	5/8″		13	.615		28	27.50		55.50	76.50
0616	3/4″		12	.667		9	30		39	59
0617	1″		11	.727		25	32.50		57.50	81

22 11 13 – Facility Water Distribution Piping

22 11 13.25 Pipe/Tube Fittings, Copper		Crew	Daily Output	Labor-Hours	Unit	Material	2019 Bare Costs Labor	Equipment	Total	Total Incl O&P
0618	1-1/4"	1 Plum	10	.800	Ea.	30	36		66	92
0619	1-1/2"		9	.889		30.50	40		70.50	99
0620	2"		8	1		53	45		98	132
0621	2-1/2"	Q-1	9	1.778		129	72		201	260
0622	3"		8	2		149	81		230	297
0623	4"		6	2.667		287	108		395	490
0624	5"		5	3.200		1,775	129		1,904	2,150
0625	6"	Q-2	7	3.429		2,075	133		2,208	2,525
0626	8"	"	6	4		9,325	156		9,481	10,600
0630	Tee, reducing on the run, 1/4"	1 Plum	15	.533		22	24		46	64
0631	3/8"		15	.533		31	24		55	73.50
0632	1/2"		14	.571		19.20	25.50		44.70	63
0633	5/8"		13	.615		28	27.50		55.50	76.50
0634	3/4"		12	.667		21	30		51	72
0635	1"		11	.727		24.50	32.50		57	80.50
0636	1-1/4"		10	.800		38.50	36		74.50	102
0637	1-1/2"		9	.889		68	40		108	141
0638	2"		8	1		78.50	45		123.50	160
0639	2-1/2"	Q-1	9	1.778		180	72		252	315
0640	3"		8	2		262	81		343	420
0641	4"		6	2.667		555	108		663	785
0642	5"		5	3.200		1,675	129		1,804	2,050
0643	6"	Q-2	7	3.429		2,550	133		2,683	3,025
0644	8"	"	6	4		9,575	156		9,731	10,800
0650	Coupling, 1/4"	1 Plum	24	.333		1.01	14.95		15.96	25.50
0680	1/2"		22	.364		.92	16.30		17.22	28
0700	3/4"		21	.381		2.49	17.10		19.59	30.50
0710	1"		18	.444		4.94	19.95		24.89	38
0715	1-1/4"		17	.471		7	21		28	42
0716	1-1/2"		15	.533		8.70	24		32.70	49
0718	2"		13	.615		14.50	27.50		42	61.50
0721	2-1/2"	Q-1	15	1.067		38.50	43		81.50	113
0726	4"	"	7	2.286		89.50	92.50		182	250
2000	DWV, solder joints, copper x copper									
2030	90° elbow, 1-1/4"	1 Plum	13	.615	Ea.	17.50	27.50		45	65
2050	1-1/2"		12	.667		23	30		53	74.50
2070	2"		10	.800		38	36		74	101
2090	3"	Q-1	10	1.600		85.50	64.50		150	200
2100	4"	"	9	1.778		460	72		532	630
2250	Tee, sanitary, 1-1/4"	1 Plum	9	.889		27	40		67	95
2290	2"	"	7	1.143		52.50	51.50		104	142
2310	3"	Q-1	7	2.286		200	92.50		292.50	370

22 11 13.44 Pipe, Steel		Crew	Daily Output	Labor-Hours	Unit	Material	Labor	Equipment	Total	Total Incl O&P
0010	**PIPE, STEEL**									
0050	Schedule 40, threaded, with couplings, and clevis hanger									
0060	assemblies sized for covering, 10' OC									
0540	Black, 1/4" diameter	1 Plum	66	.121	L.F.	5.80	5.45		11.25	15.25
0560	1/2" diameter		63	.127		3.91	5.70		9.61	13.65
0570	3/4" diameter		61	.131		4.25	5.90		10.15	14.35
0580	1" diameter		53	.151		4.42	6.75		11.17	15.95
0590	1-1/4" diameter	Q-1	89	.180		5.20	7.25		12.45	17.60
0600	1-1/2" diameter		80	.200		5.70	8.10		13.80	19.55

22 11 13.44 Pipe, Steel

22 11 13.44 Pipe, Steel		Crew	Daily Output	Labor-Hours	Unit	Material	Labor	Equipment	Total	Total Incl O&P
						2019 Bare Costs				
0610	2" diameter	Q-1	64	.250	L.F.	11.55	10.10		21.65	29.50
2000	Welded, sch. 40, on yoke & roll hanger assy's, sized for covering, 10' OC									
2040	Black, 1" diameter	Q-15	93	.172	L.F.	4.23	6.95	.60	11.78	16.70
2080	2-1/2" diameter		47	.340		15	13.75	1.19	29.94	40.50
2090	3" diameter		43	.372		16.55	15.05	1.31	32.91	44
2100	3-1/2" diameter		39	.410		19.50	16.55	1.44	37.49	50
2110	4" diameter		37	.432		17	17.45	1.52	35.97	49
2120	5" diameter		32	.500		34.50	20	1.76	56.26	73
2130	6" diameter	Q-16	36	.667		42.50	28	1.56	72.06	94.50
2140	8" diameter		29	.828		69	34.50	1.94	105.44	135
2150	10" diameter		24	1		88	42	2.34	132.34	168
2160	12" diameter		19	1.263		105	53	2.96	160.96	205
2170	14" diameter (two rod roll type hanger for 14" diam. and up)		15	1.600		99.50	67	3.74	170.24	223
2180	16" diameter (two rod roll type hanger)		13	1.846		155	77.50	4.32	236.82	300
2190	18" diameter (two rod roll type hanger)		11	2.182		143	91.50	5.10	239.60	315
2220	24" diameter (two rod roll type hanger)		8	3		194	126	7	327	425

22 11 13.45 Pipe Fittings, Steel, Threaded

Line	22 11 13.45 Pipe Fittings, Steel, Threaded	Crew	Daily Output	Labor-Hours	Unit	Material	Labor	Equipment	Total	Total Incl O&P
0010	**PIPE FITTINGS, STEEL, THREADED**									
0020	Cast iron									
0040	Standard weight, black									
0060	90° elbow, straight									
0100	3/4"	1 Plum	14	.571	Ea.	7.40	25.50		32.90	50
0110	1"	"	13	.615		8.75	27.50		36.25	55
0130	1-1/2"	Q-1	20	.800		17.15	32.50		49.65	72
0140	2"	"	18	.889		27	36		63	88.50
0500	Tee, straight									
0530	1/2"	1 Plum	9	.889	Ea.	11.05	40		51.05	77.50
0540	3/4"		9	.889		12.85	40		52.85	79.50
0550	1"		8	1		11.45	45		56.45	86
0570	1-1/2"	Q-1	13	1.231		27	49.50		76.50	112
0580	2"	"	11	1.455		38	58.50		96.50	138
6000	For galvanized elbows, tees, and couplings, add					20%				

22 11 13.48 Pipe, Fittings and Valves, Steel, Grooved-Joint

Line	22 11 13.48 Pipe, Fittings and Valves, Steel, Grooved-Joint	Crew	Daily Output	Labor-Hours	Unit	Material	Labor	Equipment	Total	Total Incl O&P
0010	**PIPE, FITTINGS AND VALVES, STEEL, GROOVED-JOINT**									
0012	Fittings are ductile iron. Steel fittings noted.									
0020	Pipe includes coupling & clevis type hanger assemblies, 10' OC									
1000	Schedule 40, black									
1040	3/4" diameter	1 Plum	71	.113	L.F.	6.25	5.05		11.30	15.20
1050	1" diameter		63	.127		6.05	5.70		11.75	16
1060	1-1/4" diameter		58	.138		7.30	6.20		13.50	18.20
1070	1-1/2" diameter		51	.157		7.95	7.05		15	20.50
1080	2" diameter		40	.200		9.20	8.95		18.15	25
1090	2-1/2" diameter	Q-1	57	.281		12.10	11.35		23.45	32
1100	3" diameter		50	.320		14.20	12.90		27.10	36.50
1110	4" diameter		45	.356		24	14.35		38.35	50
3990	Fittings: coupling material required at joints not incl. in fitting price.									
3994	Add 1 selected coupling, material only, per joint for installed price.									
4000	Elbow, 90° or 45°, painted									
4030	3/4" diameter	1 Plum	50	.160	Ea.	77	7.20		84.20	96.50
4040	1" diameter		50	.160		41	7.20		48.20	57
4050	1-1/4" diameter		40	.200		41	8.95		49.95	59.50
4060	1-1/2" diameter		33	.242		41	10.85		51.85	63

22 11 13.48 Pipe, Fittings and Valves, Steel, Grooved-Joint		Crew	Daily Output	Labor-Hours	Unit	2019 Bare Costs Material	Labor	Equipment	Total	Total Incl O&P
4070	2" diameter	1 Plum	25	.320	Ea.	41	14.35		55.35	68.50
4080	2-1/2" diameter	Q-1	40	.400		41	16.15		57.15	71.50
4090	3" diameter		33	.485		72	19.60		91.60	112
4100	4" diameter		25	.640		78.50	26		104.50	129
4250	For galvanized elbows, add					26%				
4690	Tee, painted									
4700	3/4" diameter	1 Plum	38	.211	Ea.	82	9.45		91.45	106
4740	1" diameter		33	.242		63.50	10.85		74.35	88
4750	1-1/4" diameter		27	.296		63.50	13.30		76.80	92
4760	1-1/2" diameter		22	.364		63.50	16.30		79.80	97
4770	2" diameter		17	.471		63.50	21		84.50	105
4780	2-1/2" diameter	Q-1	27	.593		63.50	24		87.50	110
4790	3" diameter		22	.727		86.50	29.50		116	144
4800	4" diameter		17	.941		132	38		170	208
4900	For galvanized tees, add					24%				
7790	Valves: coupling material required at joints not incl. in valve price.									
7794	Add 1 selected coupling, material only, per joint for installed price.									

22 11 13.74 Pipe, Plastic

Line	Description	Crew	Daily Output	Labor-Hours	Unit	Material	Labor	Equipment	Total	Total Incl O&P
0010	**PIPE, PLASTIC**									
1800	PVC, couplings 10' OC, clevis hanger assemblies, 3 per 10'									
1820	Schedule 40									
1860	1/2" diameter	1 Plum	54	.148	L.F.	4.76	6.65		11.41	16.15
1870	3/4" diameter		51	.157		5.30	7.05		12.35	17.35
1880	1" diameter		46	.174		9	7.80		16.80	23
1890	1-1/4" diameter		42	.190		9.80	8.55		18.35	25
1900	1-1/2" diameter		36	.222		9.85	9.95		19.80	27
1910	2" diameter	Q-1	59	.271		11.35	10.95		22.30	30.50
1920	2-1/2" diameter		56	.286		10.90	11.55		22.45	31
1930	3" diameter		53	.302		13.10	12.20		25.30	34.50
1940	4" diameter		48	.333		29.50	13.45		42.95	54
4100	DWV type, schedule 40, couplings 10' OC, clevis hanger assy's, 3 per 10'									
4210	ABS, schedule 40, foam core type									
4212	Plain end black									
4214	1-1/2" diameter	1 Plum	39	.205	L.F.	8.25	9.20		17.45	24
4216	2" diameter	Q-1	62	.258		8.80	10.40		19.20	27
4218	3" diameter		56	.286		8.55	11.55		20.10	28.50
4220	4" diameter		51	.314		23.50	12.65		36.15	47
4222	6" diameter		42	.381		19.80	15.40		35.20	47
4240	To delete coupling & hangers, subtract									
4244	1-1/2" diam. to 6" diam.					43%	48%			
5300	CPVC, socket joint, couplings 10' OC, clevis hanger assemblies, 3 per 10'									
5302	Schedule 40									
5304	1/2" diameter	1 Plum	54	.148	L.F.	5.60	6.65		12.25	17.05
5305	3/4" diameter		51	.157		6.60	7.05		13.65	18.80
5306	1" diameter		46	.174		10.85	7.80		18.65	25
5307	1-1/4" diameter		42	.190		12.25	8.55		20.80	27.50
5308	1-1/2" diameter		36	.222		12.15	9.95		22.10	30
5309	2" diameter	Q-1	59	.271		15.40	10.95		26.35	35
5360	CPVC, threaded, couplings 10' OC, clevis hanger assemblies, 3 per 10'									
5380	Schedule 40									
5460	1/2" diameter	1 Plum	54	.148	L.F.	6.45	6.65		13.10	18
5470	3/4" diameter		51	.157		8.15	7.05		15.20	20.50

22 11 13.74 Pipe, Plastic		Crew	Daily Output	Labor-Hours	Unit	Material	Labor	Equipment	Total	Total Incl O&P
						2019 Bare Costs				
5480	1" diameter	1 Plum	46	.174	L.F.	12.40	7.80		20.20	26.50
5490	1-1/4" diameter		42	.190		13.45	8.55		22	29
5500	1-1/2" diameter		36	.222		13.20	9.95		23.15	31
5510	2" diameter	Q-1	59	.271		16.60	10.95		27.55	36
7280	PEX, flexible, no couplings or hangers									
7282	Note: For labor costs add 25% to the couplings and fittings labor total.									
7300	Non-barrier type, hot/cold tubing rolls									
7310	1/4" diameter x 100'				L.F.	.52			.52	.57
7350	3/8" diameter x 100'					.55			.55	.61
7360	1/2" diameter x 100'					.64			.64	.70
7370	1/2" diameter x 500'					.64			.64	.70
7380	1/2" diameter x 1000'					.63			.63	.69
7400	3/4" diameter x 100'					1.01			1.01	1.11
7410	3/4" diameter x 500'					1.14			1.14	1.25
7420	3/4" diameter x 1000'					1.14			1.14	1.25
7460	1" diameter x 100'					1.96			1.96	2.16
7470	1" diameter x 300'					1.96			1.96	2.16
7480	1" diameter x 500'					1.96			1.96	2.16
7500	1-1/4" diameter x 100'					3.33			3.33	3.66
7510	1-1/4" diameter x 300'					3.33			3.33	3.66
7540	1-1/2" diameter x 100'					4.52			4.52	4.97
7550	1-1/2" diameter x 300'					4.50			4.50	4.95
7596	Most sizes available in red or blue									
7700	Non-barrier type, hot/cold tubing straight lengths									
7710	1/2" diameter x 20'				L.F.	.64			.64	.70
7750	3/4" diameter x 20'					1.14			1.14	1.25
7760	1" diameter x 20'					1.96			1.96	2.16
7770	1-1/4" diameter x 20'					3.44			3.44	3.78
7780	1-1/2" diameter x 20'					4.54			4.54	4.99
7790	2" diameter					8.85			8.85	9.75
7796	Most sizes available in red or blue									

22 11 13.76 Pipe Fittings, Plastic		Crew	Daily Output	Labor-Hours	Unit	Material	Labor	Equipment	Total	Total Incl O&P
0010	**PIPE FITTINGS, PLASTIC**									
2700	PVC (white), schedule 40, socket joints									
2760	90° elbow, 1/2"	1 Plum	33.30	.240	Ea.	.51	10.75		11.26	18.25
2770	3/4"		28.60	.280		.59	12.55		13.14	21
2780	1"		25	.320		1.04	14.35		15.39	24.50
2790	1-1/4"		22.20	.360		1.82	16.15		17.97	28.50
2800	1-1/2"		20	.400		1.97	17.95		19.92	31.50
2810	2"	Q-1	36.40	.440		3.08	17.75		20.83	32.50
2820	2-1/2"		26.70	.599		9.90	24		33.90	50.50
2830	3"		22.90	.699		11.20	28		39.20	59
2840	4"		18.20	.879		20	35.50		55.50	80
3180	Tee, 1/2"	1 Plum	22.20	.360		.64	16.15		16.79	27
3190	3/4"		19	.421		.76	18.90		19.66	32
3200	1"		16.70	.479		1.38	21.50		22.88	37
3210	1-1/4"		14.80	.541		2.15	24		26.15	42.50
3220	1-1/2"		13.30	.602		2.61	27		29.61	47.50
3230	2"	Q-1	24.20	.661		3.80	26.50		30.30	48
3240	2-1/2"		17.80	.899		12.50	36.50		49	73.50
3250	3"		15.20	1.053		16.45	42.50		58.95	87.50
3260	4"		12.10	1.322		30	53.50		83.50	121

22 11 13.76 Pipe Fittings, Plastic		Crew	Daily Output	Labor-Hours	Unit	Material	2019 Bare Costs Labor	Equipment	Total	Total Incl O&P
3380	Coupling, 1/2"	1 Plum	33.30	.240	Ea.	.34	10.75		11.09	18.05
3390	3/4"		28.60	.280		.47	12.55		13.02	21
3400	1"		25	.320		.82	14.35		15.17	24.50
3410	1-1/4"		22.20	.360		1.13	16.15		17.28	27.50
3420	1-1/2"		20	.400		1.20	17.95		19.15	31
3430	2"	Q-1	36.40	.440		1.83	17.75		19.58	31
3440	2-1/2"		26.70	.599		4.06	24		28.06	44
3450	3"		22.90	.699		6.35	28		34.35	53.50
3460	4"		18.20	.879		9.20	35.50		44.70	68
4500	DWV, ABS, non pressure, socket joints									
4540	1/4 bend, 1-1/4"	1 Plum	20.20	.396	Ea.	4.73	17.75		22.48	34
4560	1-1/2"	"	18.20	.440		3.65	19.70		23.35	36.50
4570	2"	Q-1	33.10	.483		5.60	19.50		25.10	38
4580	3"		20.80	.769		14.15	31		45.15	66.50
4590	4"		16.50	.970		29	39		68	96
4600	6"		10.10	1.584		124	64		188	242
4650	1/8 bend, same as 1/4 bend									
4800	Tee, sanitary									
4820	1-1/4"	1 Plum	13.50	.593	Ea.	6.30	26.50		32.80	50.50
4830	1-1/2"	"	12.10	.661		5.40	29.50		34.90	54.50
4840	2"	Q-1	20	.800		8.30	32.50		40.80	62
4850	3"		13.90	1.151		22.50	46.50		69	102
4860	4"		11	1.455		40.50	58.50		99	141
4862	Tee, sanitary, reducing, 2" x 1-1/2"		22	.727		7.20	29.50		36.70	56
4864	3" x 2"		15.30	1.046		13.60	42		55.60	84.50
4868	4" x 3"		12.10	1.322		39.50	53.50		93	131
4870	Combination Y and 1/8 bend									
4872	1-1/2"	1 Plum	12.10	.661	Ea.	12.90	29.50		42.40	62.50
4874	2"	Q-1	20	.800		14.35	32.50		46.85	69
4876	3"		13.90	1.151		34	46.50		80.50	114
4878	4"		11	1.455		65	58.50		123.50	168
4880	3" x 1-1/2"		15.50	1.032		34.50	41.50		76	106
4882	4" x 3"		12.10	1.322		51	53.50		104.50	144
4900	Wye, 1-1/4"	1 Plum	13.50	.593		7.25	26.50		33.75	51.50
4902	1-1/2"	"	12.10	.661		8.20	29.50		37.70	57.50
4904	2"	Q-1	20	.800		10.70	32.50		43.20	65
4906	3"		13.90	1.151		25.50	46.50		72	105
4908	4"		11	1.455		51.50	58.50		110	153
4910	6"		6.70	2.388		155	96.50		251.50	330
4918	3" x 1-1/2"		15.50	1.032		20.50	41.50		62	91
4920	4" x 3"		12.10	1.322		40.50	53.50		94	132
4922	6" x 4"		6.90	2.319		123	93.50		216.50	289
4930	Double wye, 1-1/2"	1 Plum	9.10	.879		26.50	39.50		66	93.50
4932	2"	Q-1	16.60	.964		31.50	39		70.50	98.50
4934	3"		10.40	1.538		74	62		136	184
4936	4"		8.25	1.939		145	78.50		223.50	288
4940	2" x 1-1/2"		16.80	.952		29	38.50		67.50	95
4942	3" x 2"		10.60	1.509		51.50	61		112.50	157
4944	4" x 3"		8.45	1.893		115	76.50		191.50	251
4946	6" x 4"		7.25	2.207		165	89		254	325
4950	Reducer bushing, 2" x 1-1/2"		36.40	.440		2.86	17.75		20.61	32
4952	3" x 1-1/2"		27.30	.586		12.60	23.50		36.10	53
4954	4" x 2"		18.20	.879		23.50	35.50		59	84

22 11 13 – Facility Water Distribution Piping

22 11 13.76 Pipe Fittings, Plastic		Crew	Daily Output	Labor-Hours	Unit	Material	2019 Bare Costs Labor	Equipment	Total	Total Incl O&P
4956	6" x 4"	Q-1	11.10	1.441	Ea.	66	58		124	168
4960	Couplings, 1-1/2"	1 Plum	18.20	.440		1.75	19.70		21.45	34.50
4962	2"	Q-1	33.10	.483		2.34	19.50		21.84	34.50
4963	3"		20.80	.769		6.60	31		37.60	58.50
4964	4"		16.50	.970		11.85	39		50.85	77
4966	6"		10.10	1.584		49.50	64		113.50	160
4970	2" x 1-1/2"		33.30	.480		5.05	19.40		24.45	37.50
4972	3" x 1-1/2"		21	.762		14.95	31		45.95	67
4974	4" x 3"		16.70	.958		22.50	38.50		61	88
4978	Closet flange, 4"	1 Plum	32	.250		12.30	11.20		23.50	32
4980	4" x 3"	"	34	.235		14.70	10.55		25.25	33.50
5500	CPVC, Schedule 80, threaded joints									
5540	90° elbow, 1/4"	1 Plum	32	.250	Ea.	12.95	11.20		24.15	32.50
5560	1/2"		30.30	.264		7.75	11.85		19.60	28
5570	3/4"		26	.308		11.25	13.80		25.05	35
5580	1"		22.70	.352		15.80	15.80		31.60	43.50
5590	1-1/4"		20.20	.396		30.50	17.75		48.25	62.50
5600	1-1/2"		18.20	.440		32.50	19.70		52.20	68.50
5610	2"	Q-1	33.10	.483		44	19.50		63.50	80
5620	2-1/2"		24.20	.661		136	26.50		162.50	194
5630	3"		20.80	.769		146	31		177	211
5730	Coupling, 1/4"	1 Plum	32	.250		16.50	11.20		27.70	36.50
5732	1/2"		30.30	.264		13.95	11.85		25.80	35
5734	3/4"		26	.308		22.50	13.80		36.30	47
5736	1"		22.70	.352		25	15.80		40.80	53.50
5738	1-1/4"		20.20	.396		26.50	17.75		44.25	58
5740	1-1/2"		18.20	.440		28.50	19.70		48.20	63.50
5742	2"	Q-1	33.10	.483		33.50	19.50		53	69
5744	2-1/2"		24.20	.661		60	26.50		86.50	110
5746	3"		20.80	.769		70	31		101	128
5900	CPVC, Schedule 80, socket joints									
5904	90° elbow, 1/4"	1 Plum	32	.250	Ea.	12.35	11.20		23.55	32
5906	1/2"		30.30	.264		4.83	11.85		16.68	25
5908	3/4"		26	.308		6.15	13.80		19.95	29.50
5910	1"		22.70	.352		9.80	15.80		25.60	37
5912	1-1/4"		20.20	.396		21	17.75		38.75	52.50
5914	1-1/2"		18.20	.440		23.50	19.70		43.20	58.50
5916	2"	Q-1	33.10	.483		28.50	19.50		48	63.50
5918	2-1/2"		24.20	.661		65.50	26.50		92	117
5920	3"		20.80	.769		74.50	31		105.50	133
5930	45° elbow, 1/4"	1 Plum	32	.250		18.35	11.20		29.55	38.50
5932	1/2"		30.30	.264		5.90	11.85		17.75	26
5934	3/4"		26	.308		8.55	13.80		22.35	32
5936	1"		22.70	.352		13.60	15.80		29.40	41
5938	1-1/4"		20.20	.396		26.50	17.75		44.25	58.50
5940	1-1/2"		18.20	.440		27.50	19.70		47.20	62.50
5942	2"	Q-1	33.10	.483		30.50	19.50		50	65.50
5944	2-1/2"		24.20	.661		63	26.50		89.50	113
5946	3"		20.80	.769		80.50	31		111.50	140
5990	Coupling, 1/4"	1 Plum	32	.250		13.15	11.20		24.35	33
5992	1/2"		30.30	.264		5.10	11.85		16.95	25
5994	3/4"		26	.308		7.15	13.80		20.95	30.50
5996	1"		22.70	.352		9.60	15.80		25.40	36.50

22 11 13 – Facility Water Distribution Piping

22 11 13.76 Pipe Fittings, Plastic		Crew	Daily Output	Labor-Hours	Unit	Material	Labor	Equipment	Total	Total Incl O&P
5998	1-1/4"	1 Plum	20.20	.396	Ea.	14.40	17.75		32.15	45
6000	1-1/2"	↓	18.20	.440		18.10	19.70		37.80	52.50
6002	2"	Q-1	33.10	.483		21	19.50		40.50	55.50
6004	2-1/2"		24.20	.661		47	26.50		73.50	95.50
6006	3"	↓	20.80	.769	↓	51	31		82	107

22 11 19 – Domestic Water Piping Specialties

22 11 19.38 Water Supply Meters		Crew	Daily Output	Labor-Hours	Unit	Material	Labor	Equipment	Total	Total Incl O&P
0010	**WATER SUPPLY METERS**									
2000	Domestic/commercial, bronze									
2020	Threaded									
2080	3/4" diameter, to 30 GPM	1 Plum	14	.571	Ea.	94	25.50		119.50	145
2100	1" diameter, to 50 GPM	"	12	.667	"	143	30		173	206
2300	Threaded/flanged									
2340	1-1/2" diameter, to 100 GPM	1 Plum	8	1	Ea.	350	45		395	460
2360	2" diameter, to 160 GPM	"	6	1.333	"	475	60		535	620
2600	Flanged, compound									
2640	3" diameter, 320 GPM	Q-1	3	5.333	Ea.	2,425	215		2,640	3,000
2660	4" diameter, to 500 GPM	"	1.50	10.667	"	3,875	430		4,305	4,950

22 11 19.42 Backflow Preventers		Crew	Daily Output	Labor-Hours	Unit	Material	Labor	Equipment	Total	Total Incl O&P
0010	**BACKFLOW PREVENTERS**, Includes valves									
0020	and four test cocks, corrosion resistant, automatic operation									
4000	Reduced pressure principle									
4100	Threaded, bronze, valves are ball									
4120	3/4" pipe size	1 Plum	16	.500	Ea.	465	22.50		487.50	550
4140	1" pipe size		14	.571		495	25.50		520.50	585
4150	1-1/4" pipe size		12	.667		950	30		980	1,100
4160	1-1/2" pipe size		10	.800		1,000	36		1,036	1,150
4180	2" pipe size	↓	7	1.143	↓	1,175	51.50		1,226.50	1,375
5000	Flanged, bronze, valves are OS&Y									
5060	2-1/2" pipe size	Q-1	5	3.200	Ea.	5,650	129		5,779	6,400
5080	3" pipe size		4.50	3.556		6,425	144		6,569	7,275
5100	4" pipe size	↓	3	5.333		8,025	215		8,240	9,175
5120	6" pipe size	Q-2	3	8	↓	11,700	310		12,010	13,300

22 11 19.50 Vacuum Breakers		Crew	Daily Output	Labor-Hours	Unit	Material	Labor	Equipment	Total	Total Incl O&P
0010	**VACUUM BREAKERS**									
0013	See also backflow preventers Section 22 11 19.42									
1000	Anti-siphon continuous pressure type									
1010	Max. 150 psi - 210°F									
1020	Bronze body									
1030	1/2" size	1 Stpi	24	.333	Ea.	183	15.35		198.35	226
1040	3/4" size		20	.400		183	18.40		201.40	231
1050	1" size		19	.421		187	19.40		206.40	238
1060	1-1/4" size		15	.533		370	24.50		394.50	445
1070	1-1/2" size		13	.615		455	28.50		483.50	545
1080	2" size	↓	11	.727	↓	465	33.50		498.50	570
1200	Max. 125 psi with atmospheric vent									
1210	Brass, in-line construction									
1220	1/4" size	1 Stpi	24	.333	Ea.	138	15.35		153.35	177
1230	3/8" size	"	24	.333		138	15.35		153.35	177
1260	For polished chrome finish, add				↓	13%				
2000	Anti-siphon, non-continuous pressure type									
2010	Hot or cold water 125 psi - 210°F									

22 11 Facility Water Distribution

22 11 19 – Domestic Water Piping Specialties

22 11 19.50 Vacuum Breakers		Crew	Daily Output	Labor-Hours	Unit	Material	Labor	Equipment	Total	Total Incl O&P
						2019 Bare Costs				
2020	Bronze body									
2030	1/4" size	1 Stpi	24	.333	Ea.	77.50	15.35		92.85	110
2040	3/8" size		24	.333		77.50	15.35		92.85	110
2050	1/2" size		24	.333		87	15.35		102.35	121
2060	3/4" size		20	.400		104	18.40		122.40	144
2070	1" size		19	.421		161	19.40		180.40	209
2080	1-1/4" size		15	.533		282	24.50		306.50	350
2090	1-1/2" size		13	.615		330	28.50		358.50	410
2100	2" size		11	.727		515	33.50		548.50	620
2110	2-1/2" size		8	1		1,475	46		1,521	1,700
2120	3" size		6	1.333		1,975	61.50		2,036.50	2,250
2150	For polished chrome finish, add					50%				
3000	Air gap fitting									
3020	1/2" NPT size	1 Plum	19	.421	Ea.	60	18.90		78.90	97
3030	1" NPT size	"	15	.533		69	24		93	115
3040	2" NPT size	Q-1	21	.762		134	31		165	199
3050	3" NPT size		14	1.143		266	46		312	370
3060	4" NPT size		10	1.600		266	64.50		330.50	400

22 11 19.54 Water Hammer Arresters/Shock Absorbers		Crew	Daily Output	Labor-Hours	Unit	Material	Labor	Equipment	Total	Total Incl O&P
0010	**WATER HAMMER ARRESTERS/SHOCK ABSORBERS**									
0490	Copper									
0500	3/4" male IPS for 1 to 11 fixtures	1 Plum	12	.667	Ea.	29	30		59	81
0600	1" male IPS for 12 to 32 fixtures		8	1		49	45		94	128
0700	1-1/4" male IPS for 33 to 60 fixtures		8	1		49.50	45		94.50	128

22 13 Facility Sanitary Sewerage

22 13 16 – Sanitary Waste and Vent Piping

22 13 16.20 Pipe, Cast Iron		Crew	Daily Output	Labor-Hours	Unit	Material	Labor	Equipment	Total	Total Incl O&P
0010	**PIPE, CAST IRON**, Soil, on clevis hanger assemblies, 5′ OC									
0020	Single hub, service wt., lead & oakum joints 10′ OC									
2120	2" diameter	Q-1	63	.254	L.F.	17.05	10.25		27.30	35.50
2140	3" diameter		60	.267		20	10.75		30.75	39.50
2160	4" diameter		55	.291		34	11.75		45.75	57
2180	5" diameter	Q-2	76	.316		44	12.30		56.30	68
2200	6" diameter	"	73	.329		42	12.80		54.80	67
2320	For service weight, double hub, add					10%				
2340	For extra heavy, single hub, add					48%	4%			
2360	For extra heavy, double hub, add					71%	4%			
2400	Lead for caulking (1#/diam. in.)	Q-1	160	.100	Lb.	1.06	4.04		5.10	7.80
2420	Oakum for caulking (1/8#/diam. in.)	"	40	.400	"	4.32	16.15		20.47	31.50
4000	No hub, couplings 10′ OC									
4100	1-1/2" diameter	Q-1	71	.225	L.F.	16.20	9.10		25.30	33
4120	2" diameter		67	.239		17.30	9.65		26.95	35
4140	3" diameter		64	.250		19.70	10.10		29.80	38
4160	4" diameter		58	.276		33.50	11.15		44.65	55.50
4180	5" diameter	Q-2	83	.289		44.50	11.25		55.75	67.50
4200	6" diameter	"	79	.304		42	11.80		53.80	65.50

22 13 16 – Sanitary Waste and Vent Piping

22 13 16.30 Pipe Fittings, Cast Iron		Crew	Daily Output	Labor-Hours	Unit	Material	Labor	Equipment	Total	Total Incl O&P
						2019 Bare Costs				
0010	**PIPE FITTINGS, CAST IRON**, Soil									
0040	Hub and spigot, service weight, lead & oakum joints									
0080	1/4 bend, 2"	Q-1	16	1	Ea.	22	40.50		62.50	91
0120	3"		14	1.143		29.50	46		75.50	108
0140	4"		13	1.231		46.50	49.50		96	133
0160	5"	Q-2	18	1.333		64.50	52		116.50	156
0180	6"	"	17	1.412		80.50	55		135.50	179
0500	Sanitary tee, 2"	Q-1	10	1.600		31	64.50		95.50	140
0540	3"		9	1.778		50	72		122	173
0620	4"		8	2		63.50	81		144.50	203
0700	5"	Q-2	12	2		122	78		200	262
0800	6"	"	11	2.182		138	85		223	291
5990	No hub									
6000	Cplg. & labor required at joints not incl. in fitting									
6010	price. Add 1 coupling per joint for installed price									
6020	1/4 bend, 1-1/2"				Ea.	10.55			10.55	11.65
6060	2"					11.55			11.55	12.70
6080	3"					16.10			16.10	17.70
6120	4"					24			24	26
6140	5"					62			62	68
6160	6"					57.50			57.50	63.50
6184	1/4 bend, long sweep, 1-1/2"					27			27	29.50
6186	2"					25			25	27.50
6188	3"					30.50			30.50	33.50
6189	4"					48.50			48.50	53.50
6190	5"					94			94	103
6191	6"					107			107	118
6380	Sanitary tee, tapped, 1-1/2"					22.50			22.50	25
6382	2" x 1-1/2"					19.90			19.90	22
6384	2"					19.95			19.95	22
6386	3" x 2"					32			32	35
6388	3"					51			51	56.50
6390	4" x 1-1/2"					28.50			28.50	31
6392	4" x 2"					32.50			32.50	35.50
6393	4"					32.50			32.50	35.50
6394	6" x 1-1/2"					70.50			70.50	77.50
6396	6" x 2"					72			72	79
6459	Sanitary tee, 1-1/2"					14.85			14.85	16.30
6460	2"					15.90			15.90	17.50
6470	3"					19.60			19.60	21.50
6472	4"					37			37	41
6474	5"					90			90	99
6476	6"					88.50			88.50	97.50
8000	Coupling, standard (by CISPI Mfrs.)									
8020	1-1/2"	Q-1	48	.333	Ea.	16.45	13.45		29.90	40
8040	2"		44	.364		16.25	14.70		30.95	42
8080	3"		38	.421		19.70	17		36.70	49.50
8120	4"		33	.485		23	19.60		42.60	57
8160	5"	Q-2	44	.545		56.50	21		77.50	97.50
8180	6"	"	40	.600		59	23.50		82.50	104

22 13 16 – Sanitary Waste and Vent Piping

22 13 16.60 Traps		Crew	Daily Output	Labor-Hours	Unit	Material	Labor	Equipment	Total	Total Incl O&P
						2019 Bare Costs				
0010	**TRAPS**									
0030	Cast iron, service weight									
0050	Running P trap, without vent									
1100	2"	Q-1	16	1	Ea.	162	40.50		202.50	245
1150	4"	"	13	1.231		162	49.50		211.50	260
1160	6"	Q-2	17	1.412		775	55		830	945
3000	P trap, B&S, 2" pipe size	Q-1	16	1		41.50	40.50		82	112
3040	3" pipe size	"	14	1.143		61.50	46		107.50	144
4700	Copper, drainage, drum trap									
4840	3" x 6" swivel, 1-1/2" pipe size	1 Plum	16	.500	Ea.	287	22.50		309.50	350
5100	P trap, standard pattern									
5200	1-1/4" pipe size	1 Plum	18	.444	Ea.	92.50	19.95		112.45	135
5240	1-1/2" pipe size		17	.471		103	21		124	148
5260	2" pipe size		15	.533		158	24		182	214
5280	3" pipe size		11	.727		495	32.50		527.50	600
6710	ABS DWV P trap, solvent weld joint									
6720	1-1/2" pipe size	1 Plum	18	.444	Ea.	11.40	19.95		31.35	45
6722	2" pipe size		17	.471		15	21		36	51
6724	3" pipe size		15	.533		59.50	24		83.50	105
6726	4" pipe size		14	.571		119	25.50		144.50	173
6732	PVC DWV P trap, solvent weld joint									
6733	1-1/2" pipe size	1 Plum	18	.444	Ea.	9.70	19.95		29.65	43
6734	2" pipe size		17	.471		11.85	21		32.85	47.50
6735	3" pipe size		15	.533		40	24		64	83.50
6736	4" pipe size		14	.571		91	25.50		116.50	142
6760	PP DWV, dilution trap, 1-1/2" pipe size		16	.500		265	22.50		287.50	330
6770	P trap, 1-1/2" pipe size		17	.471		66	21		87	107
6780	2" pipe size		16	.500		97.50	22.50		120	144
6790	3" pipe size		14	.571		179	25.50		204.50	239
6800	4" pipe size		13	.615		300	27.50		327.50	375
6860	PVC DWV hub x hub, basin trap, 1-1/4" pipe size		18	.444		52	19.95		71.95	89.50
6870	Sink P trap, 1-1/2" pipe size		18	.444		14.90	19.95		34.85	49
6880	Tubular S trap, 1-1/2" pipe size		17	.471		25.50	21		46.50	63
6890	PVC sch. 40 DWV, drum trap									
6900	1-1/2" pipe size	1 Plum	16	.500	Ea.	37	22.50		59.50	77.50
6910	P trap, 1-1/2" pipe size		18	.444		9.20	19.95		29.15	42.50
6920	2" pipe size		17	.471		12.35	21		33.35	48
6930	3" pipe size		15	.533		42	24		66	85.50
6940	4" pipe size		14	.571		95	25.50		120.50	147
6950	P trap w/clean out, 1-1/2" pipe size		18	.444		15.60	19.95		35.55	49.50
6960	2" pipe size		17	.471		25.50	21		46.50	62.50

22 13 16.80 Vent Flashing and Caps		Crew	Daily Output	Labor-Hours	Unit	Material	Labor	Equipment	Total	Total Incl O&P
0010	**VENT FLASHING AND CAPS**									
0120	Vent caps									
0140	Cast iron									
0180	2-1/2" to 3-5/8" pipe	1 Plum	21	.381	Ea.	51.50	17.10		68.60	84.50
0190	4" to 4-1/8" pipe	"	19	.421	"	75	18.90		93.90	114
0900	Vent flashing									
1000	Aluminum with lead ring									
1040	2" pipe	1 Plum	18	.444	Ea.	5.20	19.95		25.15	38
1060	4" pipe	"	16	.500	"	6.95	22.50		29.45	44.50
1350	Copper with neoprene ring									

22 13 Facility Sanitary Sewerage

22 13 16 – Sanitary Waste and Vent Piping

22 13 16.80 Vent Flashing and Caps		Crew	Daily Output	Labor-Hours	Unit	2019 Bare Costs Material	Labor	Equipment	Total	Total Incl O&P
1430	1-1/2" pipe	1 Plum	20	.400	Ea.	72.50	17.95		90.45	109
1440	2" pipe		18	.444		72.50	19.95		92.45	112
1460	4" pipe		16	.500		87.50	22.50		110	134

22 13 19 – Sanitary Waste Piping Specialties

22 13 19.13 Sanitary Drains		Crew	Daily Output	Labor-Hours	Unit	2019 Bare Costs Material	Labor	Equipment	Total	Total Incl O&P
0010	**SANITARY DRAINS**									
2000	Floor, medium duty, CI, deep flange, 7" diam. top									
2040	2" and 3" pipe size	Q-1	12	1.333	Ea.	244	54		298	355
2080	For galvanized body, add					116			116	127
2120	With polished bronze top					380			380	415
2500	Heavy duty, cleanout & trap w/bucket, CI, 15" top									
2540	2", 3", and 4" pipe size	Q-1	6	2.667	Ea.	7,700	108		7,808	8,650
2560	For galvanized body, add					2,025			2,025	2,225
2580	With polished bronze top					8,700			8,700	9,575

22 13 19.15 Sink Waste Treatment		Crew	Daily Output	Labor-Hours	Unit	2019 Bare Costs Material	Labor	Equipment	Total	Total Incl O&P
0010	**SINK WASTE TREATMENT**, System for commercial kitchens									
0100	includes clock timer & fittings									
0200	System less chemical, wall mounted cabinet	1 Plum	16	.500	Ea.	540	22.50		562.50	630
2000	Chemical, 1 gallon, add					36.50			36.50	40
2100	6 gallons, add					164			164	180
2200	15 gallons, add					445			445	490
2300	30 gallons, add					835			835	915
2400	55 gallons, add					1,425			1,425	1,550

22 13 23 – Sanitary Waste Interceptors

22 13 23.10 Interceptors		Crew	Daily Output	Labor-Hours	Unit	2019 Bare Costs Material	Labor	Equipment	Total	Total Incl O&P
0010	**INTERCEPTORS**									
0150	Grease, fabricated steel, 4 GPM, 8 lb. fat capacity	1 Plum	4	2	Ea.	1,375	89.50		1,464.50	1,675
0200	7 GPM, 14 lb. fat capacity		4	2		1,925	89.50		2,014.50	2,250
1040	15 GPM, 30 lb. fat capacity		4	2		3,350	89.50		3,439.50	3,825
1060	20 GPM, 40 lb. fat capacity		3	2.667		4,100	120		4,220	4,700
3000	Hair, cast iron, 1-1/4" and 1-1/2" pipe connection		8	1		520	45		565	645
3100	For chrome-plated cast iron, add					320			320	350
4000	Oil, fabricated steel, 10 GPM, 2" pipe size	1 Plum	4	2		3,075	89.50		3,164.50	3,550
4100	15 GPM, 2" or 3" pipe size		4	2		4,225	89.50		4,314.50	4,800
4120	20 GPM, 2" or 3" pipe size		3	2.667		5,150	120		5,270	5,875
6000	Solids, precious metals recovery, CI, 1-1/4" to 2" pipe		4	2		780	89.50		869.50	1,000
6100	Dental lab., large, CI, 1-1/2" to 2" pipe		3	2.667		2,725	120		2,845	3,200

22 13 29 – Sanitary Sewerage Pumps

22 13 29.14 Sewage Ejector Pumps		Crew	Daily Output	Labor-Hours	Unit	2019 Bare Costs Material	Labor	Equipment	Total	Total Incl O&P
0010	**SEWAGE EJECTOR PUMPS**, With operating and level controls									
0100	Simplex system incl. tank, cover, pump 15' head									
0500	37 gal. PE tank, 12 GPM, 1/2 HP, 2" discharge	Q-1	3.20	5	Ea.	500	202		702	880
0510	3" discharge		3.10	5.161		545	208		753	940
0600	45 gal. coated stl. tank, 12 GPM, 1/2 HP, 2" discharge		3	5.333		895	215		1,110	1,350
0610	3" discharge		2.90	5.517		930	223		1,153	1,400
0700	70 gal. PE tank, 12 GPM, 1/2 HP, 2" discharge		2.60	6.154		965	248		1,213	1,475
0710	3" discharge		2.40	6.667		1,025	269		1,294	1,575
0730	87 GPM, .7 HP, 2" discharge		2.50	6.400		1,250	258		1,508	1,800
0740	3" discharge		2.30	6.957		1,325	281		1,606	1,900
0760	134 GPM, 1 HP, 2" discharge		2.20	7.273		1,350	294		1,644	1,975
0770	3" discharge		2	8		1,450	325		1,775	2,125

22 13 Facility Sanitary Sewerage

22 13 29 – Sanitary Sewerage Pumps

22 13 29.14 Sewage Ejector Pumps		Crew	Daily Output	Labor-Hours	Unit	2019 Bare Costs Material	Labor	Equipment	Total	Total Incl O&P
1040	Duplex system incl. tank, covers, pumps									
1060	110 gal. fiberglass tank, 24 GPM, 1/2 HP, 2″ discharge	Q-1	1.60	10	Ea.	1,950	405		2,355	2,800
1080	3″ discharge		1.40	11.429		2,025	460		2,485	2,975
1100	174 GPM, .7 HP, 2″ discharge		1.50	10.667		2,500	430		2,930	3,450
1120	3″ discharge		1.30	12.308		2,600	495		3,095	3,700
1140	268 GPM, 1 HP, 2″ discharge		1.20	13.333		2,725	540		3,265	3,875
1160	3″ discharge		1	16		2,825	645		3,470	4,150
1260	135 gal. coated stl. tank, 24 GPM, 1/2 HP, 2″ discharge	Q-2	1.70	14.118		2,000	550		2,550	3,100
2000	3″ discharge		1.60	15		2,125	585		2,710	3,275
2640	174 GPM, .7 HP, 2″ discharge		1.60	15		2,625	585		3,210	3,825
2660	3″ discharge		1.50	16		2,750	620		3,370	4,050
2700	268 GPM, 1 HP, 2″ discharge		1.30	18.462		2,850	720		3,570	4,300
3040	3″ discharge		1.10	21.818		3,025	850		3,875	4,725

22 14 Facility Storm Drainage

22 14 26 – Facility Storm Drains

22 14 26.13 Roof Drains

		Crew	Daily Output	Labor-Hours	Unit	Material	Labor	Equipment	Total	Total Incl O&P
0010	**ROOF DRAINS**									
0140	Cornice, CI, 45° or 90° outlet									
0200	3″ and 4″ pipe size	Q-1	12	1.333	Ea.	380	54		434	505
0260	For galvanized body, add					91.50			91.50	101
0280	For polished bronze dome, add					112			112	124
3860	Roof, flat metal deck, CI body, 12″ CI dome									
3890	3″ pipe size	Q-1	14	1.143	Ea.	355	46		401	465

22 14 29 – Sump Pumps

22 14 29.13 Wet-Pit-Mounted, Vertical Sump Pumps

		Crew	Daily Output	Labor-Hours	Unit	Material	Labor	Equipment	Total	Total Incl O&P
0010	**WET-PIT-MOUNTED, VERTICAL SUMP PUMPS**									
0400	Molded PVC base, 21 GPM at 15′ head, 1/3 HP	1 Plum	5	1.600	Ea.	141	72		213	273
0800	Iron base, 21 GPM at 15′ head, 1/3 HP		5	1.600		154	72		226	287
1200	Solid brass, 21 GPM at 15′ head, 1/3 HP		5	1.600		239	72		311	380
2000	Sump pump, single stage									
2010	25 GPM, 1 HP, 1-1/2″ discharge	Q-1	1.80	8.889	Ea.	3,800	360		4,160	4,800
2020	75 GPM, 1-1/2 HP, 2″ discharge		1.50	10.667		4,025	430		4,455	5,125
2030	100 GPM, 2 HP, 2-1/2″ discharge		1.30	12.308		4,100	495		4,595	5,325
2040	150 GPM, 3 HP, 3″ discharge		1.10	14.545		4,100	585		4,685	5,475
2050	200 GPM, 3 HP, 3″ discharge		1	16		4,350	645		4,995	5,825
2060	300 GPM, 10 HP, 4″ discharge	Q-2	1.20	20		4,700	780		5,480	6,425
2070	500 GPM, 15 HP, 5″ discharge		1.10	21.818		5,350	850		6,200	7,275
2080	800 GPM, 20 HP, 6″ discharge		1	24		6,300	935		7,235	8,450
2090	1,000 GPM, 30 HP, 6″ discharge		.85	28.235		6,925	1,100		8,025	9,425
2100	1,600 GPM, 50 HP, 8″ discharge		.72	33.333		10,800	1,300		12,100	14,000
2110	2,000 GPM, 60 HP, 8″ discharge	Q-3	.85	37.647		11,100	1,525		12,625	14,700
2202	For general purpose float switch, copper coated float, add	Q-1	5	3.200		103	129		232	325

22 31 Domestic Water Softeners

22 31 13 – Residential Domestic Water Softeners

22 31 13.10 Residential Water Softeners		Crew	Daily Output	Labor-Hours	Unit	Material	2019 Bare Costs Labor	Equipment	Total	Total Incl O&P
0010	**RESIDENTIAL WATER SOFTENERS**									
7350	Water softener, automatic, to 30 grains per gallon	2 Plum	5	3.200	Ea.	375	144		519	645
7400	To 100 grains per gallon	"	4	4	"	855	179		1,034	1,225

22 32 Domestic Water Filtration Equipment

22 32 19 – Domestic-Water Off-Floor Cartridge Filters

22 32 19.10 Water Filters

		Crew	Daily Output	Labor-Hours	Unit	Material	Labor	Equipment	Total	Total Incl O&P
0010	**WATER FILTERS**, Purification and treatment									
1000	Cartridge style, dirt and rust type	1 Plum	12	.667	Ea.	78.50	30		108.50	136
1200	Replacement cartridge		32	.250		15.95	11.20		27.15	36
1600	Taste and odor type		12	.667		116	30		146	177
1700	Replacement cartridge		32	.250		38.50	11.20		49.70	61
8000	Commercial, fully automatic or push button automatic									
8200	Iron removal, 660 GPH, 1" pipe size	Q-1	1.50	10.667	Ea.	1,725	430		2,155	2,600
8240	1,500 GPH, 1-1/4" pipe size	"	1	16	"	2,900	645		3,545	4,250

22 33 Electric Domestic Water Heaters

22 33 30 – Residential, Electric Domestic Water Heaters

22 33 30.13 Residential, Small-Capacity Elec. Water Heaters

		Crew	Daily Output	Labor-Hours	Unit	Material	Labor	Equipment	Total	Total Incl O&P
0010	**RESIDENTIAL, SMALL-CAPACITY ELECTRIC DOMESTIC WATER HEATERS**									
1000	Residential, electric, glass lined tank, 5 yr., 10 gal., single element	1 Plum	2.30	3.478	Ea.	445	156		601	740
1060	30 gallon, double element		2.20	3.636		990	163		1,153	1,375
1100	52 gallon, double element		2	4		1,200	179		1,379	1,625

22 33 33 – Light-Commercial Electric Domestic Water Heaters

22 33 33.10 Commercial Electric Water Heaters

		Crew	Daily Output	Labor-Hours	Unit	Material	Labor	Equipment	Total	Total Incl O&P
0010	**COMMERCIAL ELECTRIC WATER HEATERS**									
4000	Commercial, 100° rise. NOTE: for each size tank, a range of									
4010	heaters between the ones shown is available									
4020	Electric									
4100	5 gal., 3 kW, 12 GPH, 208 volt	1 Plum	2	4	Ea.	4,625	179		4,804	5,375
4120	10 gal., 6 kW, 25 GPH, 208 volt		2	4		5,100	179		5,279	5,900
4130	30 gal., 24 kW, 98 GPH, 208 volt		1.92	4.167		8,275	187		8,462	9,400
4136	40 gal., 36 kW, 148 GPH, 208 volt		1.88	4.255		10,800	191		10,991	12,100
4140	50 gal., 9 kW, 37 GPH, 208 volt		1.80	4.444		7,525	199		7,724	8,600
4160	50 gal., 36 kW, 148 GPH, 208 volt		1.80	4.444		11,600	199		11,799	13,000
4200	80 gal., 36 kW, 148 GPH, 208 volt		1.50	5.333		12,200	239		12,439	13,800
4220	100 gal., 36 kW, 148 GPH, 208 volt		1.20	6.667		13,300	299		13,599	15,100
4240	120 gal., 36 kW, 148 GPH, 208 volt		1.20	6.667		13,900	299		14,199	15,800
4280	150 gal., 120 kW, 490 GPH, 480 volt		1	8		45,800	360		46,160	51,000
4300	200 gal., 15 kW, 61 GPH, 480 volt	Q-1	1.70	9.412		35,300	380		35,680	39,400

22 34 Fuel-Fired Domestic Water Heaters

22 34 30 – Residential Gas Domestic Water Heaters

22 34 30.13 Residential, Atmos, Gas Domestic Wtr Heaters		Crew	Daily Output	Labor-Hours	Unit	2019 Bare Costs Material	Labor	Equipment	Total	Total Incl O&P
0010	RESIDENTIAL, ATMOSPHERIC, GAS DOMESTIC WATER HEATERS									
2000	Gas fired, foam lined tank, 10 yr., vent not incl.									
2040	30 gallon	1 Plum	2	4	Ea.	1,775	179		1,954	2,250
2100	75 gallon	"	1.50	5.333	"	2,775	239		3,014	3,450
3000	Tank leak safety, water & gas shut off see 22 05 23.20 8800									

22 34 36 – Commercial Gas Domestic Water Heaters

22 34 36.13 Commercial, Atmos., Gas Domestic Water Htrs.		Crew	Daily Output	Labor-Hours	Unit	Material	Labor	Equipment	Total	Total Incl O&P
0010	COMMERCIAL, ATMOSPHERIC, GAS DOMESTIC WATER HEATERS									
6000	Gas fired, flush jacket, std. controls, vent not incl.									
6040	75 MBH input, 73 GPH	1 Plum	1.40	5.714	Ea.	3,850	256		4,106	4,675
6060	98 MBH input, 95 GPH		1.40	5.714		7,750	256		8,006	8,950
6080	120 MBH input, 110 GPH		1.20	6.667		7,975	299		8,274	9,250
6120	140 MBH input, 130 GPH		1	8		10,200	360		10,560	11,800
6140	155 MBH input, 150 GPH		.80	10		10,300	450		10,750	12,100
6180	200 MBH input, 192 GPH		.60	13.333		10,800	600		11,400	12,900
6220	260 MBH input, 250 GPH	Q-1	.80	20		12,000	810		12,810	14,500
6240	360 MBH input, 360 GPH		.80	20		14,200	810		15,010	16,900
6260	500 MBH input, 480 GPH		.70	22.857		19,900	925		20,825	23,400
6280	725 MBH input, 690 GPH		.60	26.667		23,300	1,075		24,375	27,400
6900	For low water cutoff, add	1 Plum	8	1		375	45		420	485
6960	For bronze body hot water circulator, add	"	4	2		2,200	89.50		2,289.50	2,575

22 34 46 – Oil-Fired Domestic Water Heaters

22 34 46.10 Residential Oil-Fired Water Heaters		Crew	Daily Output	Labor-Hours	Unit	Material	Labor	Equipment	Total	Total Incl O&P
0010	RESIDENTIAL OIL-FIRED WATER HEATERS									
3000	Oil fired, glass lined tank, 5 yr., vent not included, 30 gallon	1 Plum	2	4	Ea.	1,275	179		1,454	1,700
3040	50 gallon	"	1.80	4.444	"	1,500	199		1,699	1,975

22 34 46.20 Commercial Oil-Fired Water Heaters		Crew	Daily Output	Labor-Hours	Unit	Material	Labor	Equipment	Total	Total Incl O&P
0010	COMMERCIAL OIL-FIRED WATER HEATERS									
8000	Oil fired, glass lined, UL listed, std. controls, vent not incl.									
8060	140 gal., 140 MBH input, 134 GPH	Q-1	2.13	7.512	Ea.	26,900	305		27,205	30,100
8080	140 gal., 199 MBH input, 191 GPH		2	8		27,800	325		28,125	31,100
8120	140 gal., 270 MBH input, 259 GPH		1.20	13.333		35,400	540		35,940	39,800
8140	140 gal., 400 MBH input, 384 GPH		1	16		36,300	645		36,945	41,000
8180	140 gal., 720 MBH input, 691 GPH		.92	17.391		38,600	700		39,300	43,700
8200	221 gal., 300 MBH input, 288 GPH		.88	18.182		51,000	735		51,735	57,000
8220	221 gal., 600 MBH input, 576 GPH		.86	18.605		57,000	750		57,750	63,500
8240	221 gal., 800 MBH input, 768 GPH		.82	19.512		57,500	790		58,290	64,500

22 41 Residential Plumbing Fixtures

22 41 13 – Residential Water Closets, Urinals, and Bidets

22 41 13.13 Water Closets		Crew	Daily Output	Labor-Hours	Unit	Material	Labor	Equipment	Total	Total Incl O&P
0010	WATER CLOSETS									
0150	Tank type, vitreous china, incl. seat, supply pipe w/stop, 1.6 gpf or noted									
0200	Wall hung									
0960	For rough-in, supply, waste, vent and carrier	Q-1	2.73	5.861	Ea.	1,250	237		1,487	1,775
0999	Floor mounted									
1020	One piece, low profile	Q-1	5.30	3.019	Ea.	770	122		892	1,050
1100	Two piece, close coupled		5.30	3.019		203	122		325	425
1102	Economy		5.30	3.019		113	122		235	325
1110	Two piece, close coupled, dual flush		5.30	3.019		325	122		447	555

22 41 Residential Plumbing Fixtures

22 41 13 – Residential Water Closets, Urinals, and Bidets

22 41 13.13 Water Closets		Crew	Daily Output	Labor-Hours	Unit	Material	Labor	Equipment	Total	Total Incl O&P
						2019 Bare Costs				
1140	Two piece, close coupled, 1.28 gpf, ADA [G]	Q-1	5.30	3.019	Ea.	325	122		447	560
1960	For color, add					30%				
1980	For rough-in, supply, waste and vent	Q-1	3.05	5.246	Ea.	365	212		577	755

22 41 16 – Residential Lavatories and Sinks

22 41 16.13 Lavatories		Crew	Daily Output	Labor-Hours	Unit	Material	Labor	Equipment	Total	Total Incl O&P
0010	**LAVATORIES**, With trim, white unless noted otherwise									
0500	Vanity top, porcelain enamel on cast iron									
0600	20" x 18"	Q-1	6.40	2.500	Ea.	295	101		396	490
0640	33" x 19" oval	"	6.40	2.500	"	510	101		611	730
0860	For color, add					25%				
1000	Cultured marble, 19" x 17", single bowl	Q-1	6.40	2.500	Ea.	121	101		222	299
1120	25" x 22", single bowl	"	6.40	2.500	"	161	101		262	345
1580	For color, same price									
1900	Stainless steel, self-rimming, 25" x 22", single bowl, ledge	Q-1	6.40	2.500	Ea.	310	101		411	505
1960	17" x 22", single bowl		6.40	2.500		300	101		401	495
2600	Steel, enameled, 20" x 17", single bowl		5.80	2.759		120	111		231	315
2900	Vitreous china, 20" x 16", single bowl		5.40	2.963		205	120		325	420
3200	22" x 13", single bowl		5.40	2.963		211	120		331	430
3580	Rough-in, supply, waste and vent for all above lavatories		2.30	6.957		249	281		530	735
4000	Wall hung									
4040	Porcelain enamel on cast iron, 16" x 14", single bowl	Q-1	8	2	Ea.	440	81		521	620
4180	20" x 18", single bowl	"	8	2	"	240	81		321	395
4580	For color, add					30%				
6000	Vitreous china, 18" x 15", single bowl with backsplash	Q-1	7	2.286	Ea.	153	92.50		245.50	320
6060	19" x 17", single bowl		7	2.286		116	92.50		208.50	279
6960	Rough-in, supply, waste and vent for above lavatories		1.66	9.639		485	390		875	1,175
7000	Pedestal type									
7600	Vitreous china, 27" x 21", white	Q-1	6.60	2.424	Ea.	665	98		763	895
7610	27" x 21", colored		6.60	2.424		850	98		948	1,100
7620	27" x 21", premium color		6.60	2.424		965	98		1,063	1,225
7660	26" x 20", white		6.60	2.424		645	98		743	870
7670	26" x 20", colored		6.60	2.424		825	98		923	1,075
7680	26" x 20", premium color		6.60	2.424		950	98		1,048	1,200
7700	24" x 20", white		6.60	2.424		465	98		563	670
7710	24" x 20", colored		6.60	2.424		590	98		688	810
7720	24" x 20", premium color		6.60	2.424		655	98		753	880
7760	21" x 18", white		6.60	2.424		245	98		343	430
7770	21" x 18", colored		6.60	2.424		271	98		369	460
7990	Rough-in, supply, waste and vent for pedestal lavatories		1.66	9.639		485	390		875	1,175

22 41 16.16 Sinks		Crew	Daily Output	Labor-Hours	Unit	Material	Labor	Equipment	Total	Total Incl O&P
0010	**SINKS**, With faucets and drain									
2000	Kitchen, counter top style, PE on CI, 24" x 21" single bowl	Q-1	5.60	2.857	Ea.	310	115		425	530
2100	31" x 22" single bowl		5.60	2.857		795	115		910	1,075
2200	32" x 21" double bowl		4.80	3.333		380	135		515	640
3000	Stainless steel, self rimming, 19" x 18" single bowl		5.60	2.857		615	115		730	865
3100	25" x 22" single bowl		5.60	2.857		680	115		795	940
3300	43" x 22" double bowl		4.80	3.333		1,150	135		1,285	1,475
4000	Steel, enameled, with ledge, 24" x 21" single bowl		5.60	2.857		525	115		640	770
4100	32" x 21" double bowl		4.80	3.333		545	135		680	820
4960	For color sinks except stainless steel, add					10%				
4980	For rough-in, supply, waste and vent, counter top sinks	Q-1	2.14	7.477		288	300		588	810
5000	Kitchen, raised deck, PE on CI									

22 41 Residential Plumbing Fixtures

22 41 16 – Residential Lavatories and Sinks

22 41 16.16 Sinks		Crew	Daily Output	Labor-Hours	Unit	2019 Bare Costs Material	Labor	Equipment	Total	Total Incl O&P
5100	32" x 21", dual level, double bowl	Q-1	2.60	6.154	Ea.	445	248		693	900
5790	For rough-in, supply, waste & vent, sinks	"	1.85	8.649	"	288	350		638	890

22 41 19 – Residential Bathtubs

22 41 19.10 Baths

		Crew	Daily Output	Labor-Hours	Unit	Material	Labor	Equipment	Total	Total Incl O&P
0010	**BATHS**									
0100	Tubs, recessed porcelain enamel on cast iron, with trim									
0180	48" x 42"	Q-1	4	4	Ea.	2,800	162		2,962	3,375
0300	Mat bottom									
0380	5' long	Q-1	4.40	3.636	Ea.	1,200	147		1,347	1,575
0480	Above floor drain, 5' long	Q-1	4	4	Ea.	845	162		1,007	1,200
0560	Corner 48" x 44"	Q-1	4.40	3.636	Ea.	2,825	147		2,972	3,350
2000	Enameled formed steel, 4'-6" long	Q-1	5.80	2.759	Ea.	505	111		616	740
4600	Module tub & showerwall surround, molded fiberglass									
4610	5' long x 34" wide x 76" high	Q-1	4	4	Ea.	770	162		932	1,100
4750	ADA compliant with 1-1/2" OD grab bar, antiskid bottom									
4760	60" x 32-3/4" x 72" high	Q-1	4	4	Ea.	585	162		747	910
4770	60" x 30" x 71" high with molded seat	Q-1	3.50	4.571	Ea.	760	185		945	1,150
9600	Rough-in, supply, waste and vent, for all above tubs, add	Q-1	2.07	7.729	Ea.	435	310		745	990

22 41 23 – Residential Showers

22 41 23.20 Showers

		Crew	Daily Output	Labor-Hours	Unit	Material	Labor	Equipment	Total	Total Incl O&P
0010	**SHOWERS**									
1500	Stall, with drain only. Add for valve and door/curtain									
1520	32" square	Q-1	5	3.200	Ea.	1,175	129		1,304	1,500
1530	36" square	Q-1	4.80	3.333	Ea.	2,950	135		3,085	3,475
1540	Terrazzo receptor, 32" square	Q-1	5	3.200	Ea.	1,350	129		1,479	1,675
1560	36" square	Q-1	4.80	3.333	Ea.	1,475	135		1,610	1,850
1580	36" corner angle	Q-1	4.80	3.333	Ea.	1,750	135		1,885	2,150
3000	Fiberglass, one piece, with 3 walls, 32" x 32" square	Q-1	5.50	2.909	Ea.	345	117		462	575
3100	36" x 36" square	Q-1	5.50	2.909	Ea.	400	117		517	635
3200	ADA compliant, 1-1/2" OD grab bars, nonskid floor									
3210	48" x 34-1/2" x 72" corner seat	Q-1	5	3.200	Ea.	660	129		789	935
3250	64" x 65-3/4" x 81-1/2" fold. seat, ADA	Q-1	3.80	4.211	Ea.	1,425	170		1,595	1,850
4000	Polypropylene, stall only, w/molded-stone floor, 30" x 30"	Q-1	2	8	Ea.	640	325		965	1,225
4200	Rough-in, supply, waste and vent for above showers	Q-1	2.05	7.805	Ea.	385	315		700	940

22 41 23.40 Shower System Components

		Crew	Daily Output	Labor-Hours	Unit	Material	Labor	Equipment	Total	Total Incl O&P
0010	**SHOWER SYSTEM COMPONENTS**									
4500	Receptor only									
4510	For tile, 36" x 36"	1 Plum	4	2	Ea.	365	89.50		454.50	550
4520	Fiberglass receptor only, 32" x 32"	1 Plum	8	1	Ea.	107	45		152	191
4530	34" x 34"	1 Plum	7.80	1.026	Ea.	123	46		169	212
4540	36" x 36"	1 Plum	7.60	1.053	Ea.	123	47		170	213
4600	Rectangular									
4620	32" x 48"	1 Plum	7.40	1.081	Ea.	151	48.50		199.50	246
4630	34" x 54"	1 Plum	7.20	1.111	Ea.	166	50		216	265
4640	34" x 60"	1 Plum	7	1.143	Ea.	180	51.50		231.50	282
5000	Built-in, head, arm, 2.5 GPM valve	1 Plum	4	2	Ea.	91.50	89.50		181	247
5200	Head, arm, by-pass, integral stops, handles	1 Plum	3.60	2.222	Ea.	290	99.50		389.50	485
5800	Mixing valve, built-in	1 Plum	6	1.333	Ea.	141	60		201	253

22 41 Residential Plumbing Fixtures

22 41 36 – Residential Laundry Trays

22 41 36.10 Laundry Sinks		Crew	Daily Output	Labor-Hours	Unit	Material	2019 Bare Costs Labor	Equipment	Total	Total Incl O&P
0010	**LAUNDRY SINKS**, With trim									
0020	Porcelain enamel on cast iron, black iron frame									
0050	24" x 21", single compartment	Q-1	6	2.667	Ea.	595	108		703	830
3000	Plastic, on wall hanger or legs									
3100	20" x 24", single compartment	Q-1	6.50	2.462	Ea.	165	99.50		264.50	345
3200	36" x 23", double compartment		5.50	2.909		214	117		331	430
9600	Rough-in, supply, waste and vent, for all laundry sinks		2.14	7.477		288	300		588	810

22 41 39 – Residential Faucets, Supplies and Trim

22 41 39.10 Faucets and Fittings

Line	Description	Crew	Daily Output	Labor-Hours	Unit	Material	Labor	Equipment	Total	Total Incl O&P
0010	**FAUCETS AND FITTINGS**									
0150	Bath, faucets, diverter spout combination, sweat	1 Plum	8	1	Ea.	86.50	45		131.50	169
0200	For integral stops, IPS unions, add					111			111	122
0420	Bath, press-bal mix valve w/diverter, spout, shower head, arm/flange	1 Plum	8	1		176	45		221	267
0500	Drain, central lift, 1-1/2" IPS male		20	.400		49.50	17.95		67.45	84
0600	Trip lever, 1-1/2" IPS male		20	.400		60	17.95		77.95	95.50
1000	Kitchen sink faucets, top mount, cast spout		10	.800		84	36		120	151
1100	For spray, add		24	.333		17.15	14.95		32.10	43.50
1300	Single control lever handle									
1310	With pull out spray									
1320	Polished chrome	1 Plum	10	.800	Ea.	197	36		233	276
2000	Laundry faucets, shelf type, IPS or copper unions		12	.667		61.50	30		91.50	117
2100	Lavatory faucet, centerset, without drain		10	.800		67	36		103	133
2120	With pop-up drain		6.66	1.201		50	54		104	144
2210	Porcelain cross handles and pop-up drain									
2220	Polished chrome	1 Plum	6.66	1.201	Ea.	209	54		263	320
2230	Polished brass	"	6.66	1.201	"	315	54		369	435
2260	Single lever handle and pop-up drain									
2280	Satin nickel	1 Plum	6.66	1.201	Ea.	275	54		329	390
2290	Polished chrome		6.66	1.201		197	54		251	305
2800	Self-closing, center set		10	.800		150	36		186	224
4000	Shower by-pass valve with union		18	.444		57.50	19.95		77.45	96
4200	Shower thermostatic mixing valve, concealed, with shower head trim kit		8	1		360	45		405	470
4220	Shower pressure balancing mixing valve									
4230	With shower head, arm, flange and diverter tub spout									
4240	Chrome	1 Plum	6.14	1.303	Ea.	420	58.50		478.50	555
4250	Satin nickel		6.14	1.303		555	58.50		613.50	705
4260	Polished graphite		6.14	1.303		555	58.50		613.50	705
5000	Sillcock, compact, brass, IPS or copper to hose		24	.333		10.60	14.95		25.55	36

22 42 Commercial Plumbing Fixtures

22 42 13 – Commercial Water Closets, Urinals, and Bidets

22 42 13.13 Water Closets

Line	Description	Crew	Daily Output	Labor-Hours	Unit	Material	Labor	Equipment	Total	Total Incl O&P
0010	**WATER CLOSETS**									
3000	Bowl only, with flush valve, seat, 1.6 gpf unless noted									
3100	Wall hung	Q-1	5.80	2.759	Ea.	1,025	111		1,136	1,300
3200	For rough-in, supply, waste and vent, single WC		2.56	6.250		1,300	252		1,552	1,850
3300	Floor mounted		5.80	2.759		325	111		436	545
3370	For rough-in, supply, waste and vent, single WC		2.84	5.634		405	227		632	820
3390	Floor mounted children's size, 10-3/4" high									
3392	With automatic flush sensor, 1.6 gpf	Q-1	6.20	2.581	Ea.	665	104		769	905

22 42 Commercial Plumbing Fixtures

22 42 13 – Commercial Water Closets, Urinals, and Bidets

22 42 13.13 Water Closets		Crew	Daily Output	Labor-Hours	Unit	Material	2019 Bare Costs Labor	Equipment	Total	Total Incl O&P
3396	With automatic flush sensor, 1.28 gpf	Q-1	6.20	2.581	Ea.	620	104		724	850
3400	For rough-in, supply, waste and vent, single WC		2.84	5.634		405	227		632	820

22 42 13.16 Urinals		Crew	Daily Output	Labor-Hours	Unit	Material	Labor	Equipment	Total	Total Incl O&P
0010	**URINALS**									
3000	Wall hung, vitreous china, with self-closing valve									
3100	Siphon jet type	Q-1	3	5.333	Ea.	292	215		507	675
3120	Blowout type		3	5.333		475	215		690	875
3140	Water saving .5 gpf G		3	5.333		560	215		775	975
3300	Rough-in, supply, waste & vent		2.83	5.654		685	228		913	1,125
5000	Stall type, vitreous china, includes valve		2.50	6.400		795	258		1,053	1,300
6980	Rough-in, supply, waste and vent		1.99	8.040		440	325		765	1,025

22 42 16 – Commercial Lavatories and Sinks

22 42 16.13 Lavatories		Crew	Daily Output	Labor-Hours	Unit	Material	Labor	Equipment	Total	Total Incl O&P
0010	**LAVATORIES**, With trim, white unless noted otherwise									
0020	Commercial lavatories same as residential. See Section 22 41 16									

22 42 16.40 Service Sinks		Crew	Daily Output	Labor-Hours	Unit	Material	Labor	Equipment	Total	Total Incl O&P
0010	**SERVICE SINKS**									
6650	Service, floor, corner, PE on CI, 28" x 28"	Q-1	4.40	3.636	Ea.	1,050	147		1,197	1,400
6750	Vinyl coated rim guard, add					63.50			63.50	69.50
6755	Mop sink, molded stone, 22" x 18"	1 Plum	3.33	2.402		530	108		638	760
6760	Mop sink, molded stone, 24" x 36"		3.33	2.402		266	108		374	470
6770	Mop sink, molded stone, 24" x 36", w/rim 3 sides		3.33	2.402		293	108		401	500
6790	For rough-in, supply, waste & vent, floor service sinks	Q-1	1.64	9.756		995	395		1,390	1,750

22 42 33 – Wash Fountains

22 42 33.20 Commercial Wash Fountains		Crew	Daily Output	Labor-Hours	Unit	Material	Labor	Equipment	Total	Total Incl O&P
0010	**COMMERCIAL WASH FOUNTAINS**									
1900	Group, foot control									
2000	Precast terrazzo, circular, 36" diam., 5 or 6 persons	Q-2	3	8	Ea.	7,800	310		8,110	9,075
2100	54" diam. for 8 or 10 persons		2.50	9.600		9,500	375		9,875	11,100
2400	Semi-circular, 36" diam. for 3 persons		3	8		6,350	310		6,660	7,500
2500	54" diam. for 4 or 5 persons		2.50	9.600		9,200	375		9,575	10,700
5610	Group, infrared control, barrier free									
5614	Precast terrazzo									
5620	Semi-circular 36" diam. for 3 persons	Q-2	3	8	Ea.	7,650	310		7,960	8,900
5630	46" diam. for 4 persons		2.80	8.571		8,250	335		8,585	9,625
5640	Circular, 54" diam. for 8 persons, button control		2.50	9.600		10,000	375		10,375	11,600

22 42 39 – Commercial Faucets, Supplies, and Trim

22 42 39.10 Faucets and Fittings		Crew	Daily Output	Labor-Hours	Unit	Material	Labor	Equipment	Total	Total Incl O&P
0010	**FAUCETS AND FITTINGS**									
0910	Dual flush flushometer	1 Plum	12	.667	Ea.	224	30		254	296
0912	Flushometer retrofit kit	"	18	.444	"	17.65	19.95		37.60	52
0920	Urinal									
0930	Exposed, stall	1 Plum	8	1	Ea.	183	45		228	275
0940	Wall (washout)	"	8	1		144	45		189	233
0945	Flush valve, urinal					144			144	159
0950	Pedestal, top spud	1 Plum	8	1		147	45		192	236
0960	Concealed, stall		8	1		170	45		215	260
0970	Wall (washout)		8	1		183	45		228	275
0971	Automatic flush sensor and operator for									
0972	urinals or water closets, standard G	1 Plum	8	1	Ea.	475	45		520	600

22 42 Commercial Plumbing Fixtures

22 42 39 – Commercial Faucets, Supplies, and Trim

22 42 39.10 Faucets and Fittings		Crew	Daily Output	Labor-Hours	Unit	2019 Bare Costs Material	Labor	Equipment	Total	Total Incl O&P
0980	High efficiency water saving									
0984	Water closets, 1.28 gpf G	1 Plum	8	1	Ea.	420	45		465	535
0988	Urinals, .5 gpf G	"	8	1	"	420	45		465	535
2790	Faucets for lavatories									
2800	Self-closing, center set	1 Plum	10	.800	Ea.	150	36		186	224
2810	Automatic sensor and operator, with faucet head		6.15	1.301		485	58.50		543.50	630
3000	Service sink faucet, cast spout, pail hook, hose end		14	.571		76	25.50		101.50	126

22 42 39.30 Carriers and Supports		Crew	Daily Output	Labor-Hours	Unit	Material	Labor	Equipment	Total	Total Incl O&P
0010	**CARRIERS AND SUPPORTS**, For plumbing fixtures									
0500	Drinking fountain, wall mounted									
0600	Plate type with studs, top back plate	1 Plum	7	1.143	Ea.	60	51.50		111.50	150
0700	Top front and back plate	"	7	1.143	"	152	51.50		203.50	252
3000	Lavatory, concealed arm									
3050	Floor mounted, single									
3100	High back fixture	1 Plum	6	1.333	Ea.	630	60		690	795
3200	Flat slab fixture		6	1.333		545	60		605	700
3220	ADA compliant		6	1.333		585	60		645	740
3250	Floor mounted, back to back									
3300	High back fixtures	1 Plum	5	1.600	Ea.	900	72		972	1,100
3400	Flat slab fixtures		5	1.600		1,100	72		1,172	1,350
3430	ADA compliant		5	1.600		800	72		872	1,000
4600	Sink, floor mounted									
4650	Exposed arm system									
4700	Single heavy fixture	1 Plum	5	1.600	Ea.	590	72		662	765
5400	Wall mounted, exposed arms, single heavy fixture		5	1.600		530	72		602	705
6000	Urinal, floor mounted, 2" or 3" coupling, blowout type		6	1.333		670	60		730	840
6100	With fixture or hanger bolts, blowout or washout		6	1.333		480	60		540	630
6300	Wall mounted, plate type system		6	1.333		405	60		465	545
6980	Water closet, siphon jet									
7000	Horizontal, adjustable, caulk									
7040	Single, 4" pipe size	1 Plum	5.33	1.501	Ea.	1,075	67.50		1,142.50	1,275
7050	4" pipe size, ADA compliant		5.33	1.501		805	67.50		872.50	1,000
7100	Double, 4" pipe size		5	1.600		1,725	72		1,797	2,025
7110	4" pipe size, ADA compliant		5	1.600		1,575	72		1,647	1,850
8201	Water closet, residential, vert. centerline, floor mount									
8300	4" copper sweat, 4" vent	1 Plum	6	1.333	Ea.	790	60		850	970
9000	Water cooler (electric), floor mounted									
9100	Plate type with bearing plate, single	1 Plum	6	1.333	Ea.	455	60		515	600

22 43 Healthcare Plumbing Fixtures

22 43 19 – Healthcare Bathtubs

22 43 19.10 Bathtubs		Crew	Daily Output	Labor-Hours	Unit	Material	Labor	Equipment	Total	Total Incl O&P
0010	**BATHTUBS**									
5300	Whirlpool, porcelain enamel on cast iron, 72" x 36"	Q-1	1	16	Ea.	4,550	645		5,195	6,075

22 47 Drinking Fountains and Water Coolers

22 47 16 – Pressure Water Coolers

22 47 16.10 Electric Water Coolers		Crew	Daily Output	Labor-Hours	Unit	2019 Bare Costs Material	Labor	Equipment	Total	Total Incl O&P
0010	**ELECTRIC WATER COOLERS**									
0100	Wall mounted, non-recessed									
0180	8.2 GPH	Q-1	4	4	Ea.	940	162		1,102	1,300
0600	8 GPH hot and cold water		4	4		1,075	162		1,237	1,450
1000	Dual height, 8.2 GPH		3.80	4.211		1,975	170		2,145	2,450
2600	ADA compliant, 8 GPH		4	4		1,000	162		1,162	1,375
4600	Floor mounted, flush-to-wall									
4680	8.2 GPH	1 Plum	3	2.667	Ea.	860	120		980	1,150
9800	For supply, waste & vent, all coolers	"	2.21	3.620	"	214	162		376	500

22 51 Swimming Pool Plumbing Systems

22 51 19 – Swimming Pool Water Treatment Equipment

22 51 19.50 Swimming Pool Filtration Equipment		Crew	Daily Output	Labor-Hours	Unit	Material	Labor	Equipment	Total	Total Incl O&P
0010	**SWIMMING POOL FILTRATION EQUIPMENT**									
0900	Filter system, sand or diatomite type, incl. pump, 6,000 gal./hr.	2 Plum	1.80	8.889	Total	2,200	400		2,600	3,075
1020	Add for chlorination system, 800 S.F. pool	"	3	5.333	Ea.	220	239		459	635

Division 23

Heating, Ventilating & Air Conditioning

Estimating Tips

The labor adjustment factors listed in Subdivision 22 01 02.20 also apply to Division 23.

23 10 00 Facility Fuel Systems

- The prices in this subdivision for above- and below-ground storage tanks do not include foundations or hold-down slabs, unless noted. The estimator should refer to Divisions 3 and 31 for foundation system pricing. In addition to the foundations, required tank accessories, such as tank gauges, leak detection devices, and additional manholes and piping, must be added to the tank prices.

23 50 00 Central Heating Equipment

- When estimating the cost of an HVAC system, check to see who is responsible for providing and installing the temperature control system. It is possible to overlook controls, assuming that they would be included in the electrical estimate.
- When looking up a boiler, be careful on specified capacity. Some manufacturers rate their products on output while others use input.
- Include HVAC insulation for pipe, boiler, and duct (wrap and liner).
- Be careful when looking up mechanical items to get the correct pressure rating and connection type (thread, weld, flange).

23 70 00 Central HVAC Equipment

- Combination heating and cooling units are sized by the air conditioning requirements. (See Reference No. R236000-20 for the preliminary sizing guide.)
- A ton of air conditioning is nominally 400 CFM.
- Rectangular duct is taken off by the linear foot for each size, but its cost is usually estimated by the pound. Remember that SMACNA standards now base duct on internal pressure.
- Prefabricated duct is estimated and purchased like pipe: straight sections and fittings.
- Note that cranes or other lifting equipment are not included on any lines in Division 23. For example, if a crane is required to lift a heavy piece of pipe into place high above a gym floor, or to put a rooftop unit on the roof of a four-story building, etc., it must be added. Due to the potential for extreme variation—from nothing additional required to a major crane or helicopter—we feel that including a nominal amount for "lifting contingency" would be useless and detract from the accuracy of the estimate. When using equipment rental cost data from RSMeans, do not forget to include the cost of the operator(s).

Reference Numbers

Reference numbers are shown at the beginning of some major classifications. These numbers refer to related items in the Reference Section. The reference information may be an estimating procedure, an alternate pricing method, or technical information.

Note: Not all subdivisions listed here necessarily appear. ■

Note: "Powered in part by CINX™, based on licensed proprietary information of Harrison Publishing House, Inc."

Note: Trade Service, in part, has been used as a reference source for some of the material prices used in Division 23.

23 05 Common Work Results for HVAC

23 05 02 – HVAC General

23 05 02.10 Air Conditioning, General		Crew	Daily Output	Labor-Hours	Unit	2019 Bare Costs Material	Labor	Equipment	Total	Total Incl O&P
0010	**AIR CONDITIONING, GENERAL** Prices are for standard efficiencies (SEER 13)									
0020	for upgrade to SEER 14 add					10%				

23 05 05 – Selective Demolition for HVAC

23 05 05.10 HVAC Demolition

		Crew	Daily Output	Labor-Hours	Unit	Material	Labor	Equipment	Total	Total Incl O&P
0010	**HVAC DEMOLITION**									
0100	Air conditioner, split unit, 3 ton	Q-5	2	8	Ea.		330		330	545
0150	Package unit, 3 ton	Q-6	3	8	"		320		320	525
0260	Baseboard, hydronic fin tube, 1/2"	Q-5	117	.137	L.F.		5.65		5.65	9.30
0298	Boilers									
0300	Electric, up thru 148 kW	Q-19	2	12	Ea.		515		515	840
0310	150 thru 518 kW	"	1	24			1,025		1,025	1,675
0320	550 thru 2,000 kW	Q-21	.40	80			3,500		3,500	5,725
0330	2,070 kW and up	"	.30	107			4,650		4,650	7,625
0340	Gas and/or oil, up thru 150 MBH	Q-7	2.20	14.545			605		605	990
0350	160 thru 2,000 MBH		.80	40			1,650		1,650	2,725
0360	2,100 thru 4,500 MBH		.50	64			2,650		2,650	4,350
0370	4,600 thru 7,000 MBH		.30	107			4,425		4,425	7,250
0390	12,200 thru 25,000 MBH		.12	267			11,100		11,100	18,100
1000	Ductwork, 4" high, 8" wide	1 Clab	200	.040	L.F.		1.21		1.21	2.01
1100	6" high, 8" wide		165	.048			1.47		1.47	2.43
1200	10" high, 12" wide		125	.064			1.94		1.94	3.21
1300	12"-14" high, 16"-18" wide		85	.094			2.86		2.86	4.72
1500	30" high, 36" wide		56	.143			4.34		4.34	7.15
2800	Heat pump, package unit, 3 ton	Q-5	2.40	6.667	Ea.		276		276	455
2840	Split unit, 3 ton		2	8			330		330	545
2950	Tank, steel, oil, 275 gal., above ground		10	1.600			66.50		66.50	109
2951	Tank, steel, oil, 550 gal., above ground		5	3.200			133		133	218
2952	Tank, steel, oil, 1000 gal., above ground		2.50	6.400			265		265	435
2960	Remove and reset		3	5.333			221		221	365
5090	Remove refrigerant from system	1 Stpi	40	.200	Lb.		9.20		9.20	15.10

23 05 23 – General-Duty Valves for HVAC Piping

23 05 23.20 Valves, Bronze/Brass

		Crew	Daily Output	Labor-Hours	Unit	Material	Labor	Equipment	Total	Total Incl O&P
0010	**VALVES, BRONZE/BRASS**									
0020	Brass									
1300	Ball combination valves, shut-off and union									
1310	Solder, with strainer, drain and PT ports									
1320	1/2"	1 Stpi	17	.471	Ea.	57	21.50		78.50	98
1330	3/4"		16	.500		62	23		85	107
1340	1"		14	.571		91	26.50		117.50	143
1350	1-1/4"		12	.667		108	30.50		138.50	170
1360	1-1/2"		10	.800		159	37		196	236
1370	2"		8	1		199	46		245	295
1410	Threaded, with strainer, drain and PT ports									
1420	1/2"	1 Stpi	20	.400	Ea.	57	18.40		75.40	92.50
1430	3/4"		18	.444		62	20.50		82.50	102
1440	1"		17	.471		91	21.50		112.50	136
1450	1-1/4"		12	.667		108	30.50		138.50	170
1460	1-1/2"		11	.727		159	33.50		192.50	230
1470	2"		9	.889		199	41		240	286

23 05 Common Work Results for HVAC

23 05 23 – General-Duty Valves for HVAC Piping

23 05 23.30 Valves, Iron Body		Crew	Daily Output	Labor-Hours	Unit	Material	Labor	Equipment	Total	Total Incl O&P
						2019 Bare Costs				
0010	VALVES, IRON BODY									
1650	Gate, 125 lb., N.R.S.									
2150	Flanged									
2240	2-1/2"	Q-1	5	3.200	Ea.	615	129		744	890
2260	3"		4.50	3.556		690	144		834	995
2280	4"		3	5.333		990	215		1,205	1,450
3550	OS&Y, 125 lb., flanged									
3660	3"	Q-1	4.50	3.556	Ea.	450	144		594	730
3680	4"	"	3	5.333		675	215		890	1,100
3700	6"	Q-2	3	8		1,075	310		1,385	1,675
5450	Swing check, 125 lb., threaded									
5500	2"	1 Plum	11	.727	Ea.	540	32.50		572.50	650
5950	Flanged									
6000	2"	1 Plum	5	1.600	Ea.	475	72		547	640
6040	2-1/2"	Q-1	5	3.200		360	129		489	605
6050	3"		4.50	3.556		415	144		559	695
6060	4"		3	5.333		730	215		945	1,150
6070	6"	Q-2	3	8		1,200	310		1,510	1,825

23 07 HVAC Insulation

23 07 13 – Duct Insulation

23 07 13.10 Duct Thermal Insulation

		Crew	Daily Output	Labor-Hours	Unit	Material	Labor	Equipment	Total	Total Incl O&P
0010	DUCT THERMAL INSULATION									
3000	Ductwork									
3020	Blanket type, fiberglass, flexible									
3030	Fire rated for grease and hazardous exhaust ducts									
3060	1-1/2" thick	Q-14	84	.190	S.F.	4.62	7		11.62	16.85
3090	Fire rated for plenums									
3100	1/2" x 24" x 25'	Q-14	1.94	8.247	Roll	172	300		472	700
3110	1/2" x 24" x 25'		98	.163	S.F.	3.43	6		9.43	13.80
3120	1/2" x 48" x 25'		1.04	15.385	Roll	340	565		905	1,325
3126	1/2" x 48" x 25'		104	.154	S.F.	3.39	5.65		9.04	13.25
3140	FSK vapor barrier wrap, .75 lb. density									
3160	1" thick [G]	Q-14	350	.046	S.F.	.23	1.67		1.90	3.07
3170	1-1/2" thick [G]	"	320	.050	"	.27	1.83		2.10	3.38
3210	Vinyl jacket, same as FSK									

23 09 Instrumentation and Control for HVAC

23 09 33 – Electric and Electronic Control System for HVAC

23 09 33.10 Electronic Control Systems

		Crew	Daily Output	Labor-Hours	Unit	Material	Labor	Equipment	Total	Total Incl O&P
0010	ELECTRONIC CONTROL SYSTEMS									
0020	For electronic costs, add to Section 23 09 43.10				Ea.				15%	15%

23 09 43 – Pneumatic Control System for HVAC

23 09 43.10 Pneumatic Control Systems

		Crew	Daily Output	Labor-Hours	Unit	Material	Labor	Equipment	Total	Total Incl O&P
0010	PNEUMATIC CONTROL SYSTEMS									
0011	Including a nominal 50' of tubing. Add control panelboard if req'd.									
0100	Heating and ventilating, split system									
0200	Mixed air control, economizer cycle, panel readout, tubing									

23 09 Instrumentation and Control for HVAC

23 09 43 – Pneumatic Control System for HVAC

23 09 43.10 Pneumatic Control Systems		Crew	Daily Output	Labor-Hours	Unit	Material	2019 Bare Costs Labor	Equipment	Total	Total Incl O&P
0220	Up to 10 tons G	Q-19	.68	35.294	Ea.	4,500	1,500		6,000	7,450
0240	For 10 to 20 tons G		.63	37.915		4,775	1,625		6,400	7,900
0260	For over 20 tons G		.58	41.096		5,175	1,750		6,925	8,550
0300	Heating coil, hot water, 3 way valve,									
0320	Freezestat, limit control on discharge, readout	Q-5	.69	23.088	Ea.	3,350	955		4,305	5,250
0500	Cooling coil, chilled water, room									
0520	Thermostat, 3 way valve	Q-5	2	8	Ea.	1,500	330		1,830	2,200
0600	Cooling tower, fan cycle, damper control,									
0620	Control system including water readout in/out at panel	Q-19	.67	35.821	Ea.	5,925	1,525		7,450	9,025
1000	Unit ventilator, day/night operation,									
1100	freezestat, ASHRAE, cycle 2	Q-19	.91	26.374	Ea.	3,275	1,125		4,400	5,475
3000	Boiler room combustion air, damper to 5 S.F., controls		1.37	17.582		2,975	755		3,730	4,475
3500	Fan coil, heating and cooling valves, 4 pipe control system		3	8		1,325	345		1,670	2,000
3600	Heat exchanger system controls		.86	27.907		2,875	1,200		4,075	5,125
3900	Multizone control (one per zone), includes thermostat, damper									
3910	motor and reset of discharge temperature	Q-5	.51	31.373	Ea.	2,975	1,300		4,275	5,375
4000	Pneumatic thermostat, including controlling room radiator valve	″	2.43	6.593		890	273		1,163	1,425
4060	Pump control system	Q-19	3	8		1,375	345		1,720	2,050
4500	Air supply for pneumatic control system									
4600	Tank mounted duplex compressor, starter, alternator,									
4620	piping, dryer, PRV station and filter									
4640	3/4 HP	Q-19	.64	37.383	Ea.	11,500	1,600		13,100	15,300
4650	1 HP		.61	39.539		12,600	1,700		14,300	16,700
4680	3 HP		.55	43.956		18,300	1,875		20,175	23,300
4690	5 HP		.42	57.143		31,600	2,450		34,050	38,800
4800	Main air supply, includes 3/8″ copper main and labor	Q-5	1.82	8.791	C.L.F.	365	365		730	1,000
4810	If poly tubing used, deduct								30%	30%
8600	VAV boxes, incl. thermostat, damper motor, reheat coil & tubing	Q-5	1.46	10.989	Ea.	1,375	455		1,830	2,275
8610	If no reheat coil, deduct				″				204	204

23 09 53 – Pneumatic and Electric Control System for HVAC

23 09 53.10 Control Components		Crew	Daily Output	Labor-Hours	Unit	Material	Labor	Equipment	Total	Total Incl O&P
0010	**CONTROL COMPONENTS**									
5000	Thermostats									
5030	Manual	1 Stpi	8	1	Ea.	42	46		88	122
5040	1 set back, electric, timed G	″	8	1	″	75.50	46		121.50	159

23 11 Facility Fuel Piping

23 11 23 – Facility Natural-Gas Piping

23 11 23.20 Gas Piping, Flexible (Csst)		Crew	Daily Output	Labor-Hours	Unit	Material	Labor	Equipment	Total	Total Incl O&P
0010	**GAS PIPING, FLEXIBLE (CSST)**									
0100	Tubing with lightning protection									
0110	3/8″	1 Stpi	65	.123	L.F.	3.06	5.65		8.71	12.65
0120	1/2″		62	.129		3.44	5.95		9.39	13.55
0130	3/4″		60	.133		4.48	6.15		10.63	15
0140	1″		55	.145		6.90	6.70		13.60	18.55
0150	1-1/4″		50	.160		8	7.35		15.35	21
0160	1-1/2″		45	.178		14	8.20		22.20	29
0170	2″		40	.200		20	9.20		29.20	37
0200	Tubing for underground/underslab burial									
0210	3/8″	1 Stpi	65	.123	L.F.	4.26	5.65		9.91	14

23 11 23 – Facility Natural-Gas Piping

23 11 23.20 Gas Piping, Flexible (Csst)		Crew	Daily Output	Labor-Hours	Unit	Material	Labor	Equipment	Total	Total Incl O&P
						2019 Bare Costs				
0220	1/2"	1 Stpi	62	.129	L.F.	4.86	5.95		10.81	15.10
0230	3/4"		60	.133		6.15	6.15		12.30	16.85
0240	1"		55	.145		8.30	6.70		15	20
0250	1-1/4"		50	.160		11.15	7.35		18.50	24.50
0260	1-1/2"		45	.178		21	8.20		29.20	36.50
0270	2"	↓	40	.200	↓	25	9.20		34.20	42.50
3000	Fittings									
3010	Straight									
3100	Tube to NPT									
3110	3/8"	1 Stpi	29	.276	Ea.	12.80	12.70		25.50	35
3120	1/2"		27	.296		13.80	13.65		27.45	37.50
3130	3/4"		25	.320		18.80	14.75		33.55	44.50
3140	1"		23	.348		29	16		45	58.50
3150	1-1/4"		20	.400		66	18.40		84.40	103
3160	1-1/2"		17	.471		133	21.50		154.50	182
3170	2"	↓	15	.533	↓	226	24.50		250.50	290
3200	Coupling									
3210	3/8"	1 Stpi	29	.276	Ea.	22.50	12.70		35.20	45.50
3220	1/2"		27	.296		25.50	13.65		39.15	50.50
3230	3/4"		25	.320		35	14.75		49.75	62.50
3240	1"		23	.348		54	16		70	86
3250	1-1/4"		20	.400		120	18.40		138.40	162
3260	1-1/2"		17	.471		252	21.50		273.50	315
3270	2"	↓	15	.533	↓	430	24.50		454.50	515
3300	Flange fitting									
3310	3/8"	1 Stpi	25	.320	Ea.	19.60	14.75		34.35	45.50
3320	1/2"		22	.364		18.75	16.75		35.50	48
3330	3/4"		19	.421		24.50	19.40		43.90	58.50
3340	1"		16	.500		33.50	23		56.50	75
3350	1-1/4"	↓	12	.667	↓	77.50	30.50		108	136
3400	90° flange valve									
3410	3/8"	1 Stpi	25	.320	Ea.	37.50	14.75		52.25	65
3420	1/2"		22	.364		38	16.75		54.75	69.50
3430	3/4"	↓	19	.421	↓	47	19.40		66.40	83.50
4000	Tee									
4120	1/2"	1 Stpi	20.50	.390	Ea.	40	17.95		57.95	73.50
4130	3/4"		19	.421		53	19.40		72.40	90.50
4140	1"	↓	17.50	.457	↓	96	21		117	140
5000	Reducing									
5110	Tube to NPT									
5120	3/4" to 1/2" NPT	1 Stpi	26	.308	Ea.	20	14.15		34.15	45.50
5130	1" to 3/4" NPT	"	24	.333	"	31.50	15.35		46.85	59.50
5200	Reducing tee									
5210	1/2" x 3/8" x 3/8"	1 Stpi	21	.381	Ea.	53	17.55		70.55	87
5220	3/4" x 1/2" x 1/2"		20	.400		49.50	18.40		67.90	84.50
5230	1" x 3/4" x 1/2"		18	.444		82	20.50		102.50	124
5240	1-1/4" x 1-1/4" x 1"		15.60	.513		219	23.50		242.50	280
5250	1-1/2" x 1-1/2" x 1-1/4"		13.30	.602		335	27.50		362.50	415
5260	2" x 2" x 1-1/2"	↓	11.80	.678	↓	505	31		536	605
5300	Manifold with four ports and mounting bracket									
5302	Labor to mount manifold does not include making pipe									
5304	connections which are included in fitting labor.									
5310	3/4" x 1/2" x 1/2"(4)	1 Stpi	76	.105	Ea.	40.50	4.85		45.35	52.50

23 11 Facility Fuel Piping

23 11 23 – Facility Natural-Gas Piping

23 11 23.20 Gas Piping, Flexible (Csst)		Crew	Daily Output	Labor-Hours	Unit	Material	2019 Bare Costs Labor	Equipment	Total	Total Incl O&P
5330	1-1/4" x 1" x 3/4"(4)	1 Stpi	72	.111	Ea.	58	5.10		63.10	72.50
5350	2" x 1-1/2" x 1"(4)	↓	68	.118	↓	73	5.40		78.40	89
5600	Protective striker plate									
5610	Quarter plate, 3" x 2"	1 Stpi	88	.091	Ea.	.92	4.19		5.11	7.85
5620	Half plate, 3" x 7"		82	.098		2.10	4.49		6.59	9.65
5630	Full plate, 3" x 12"	↓	78	.103	↓	3.76	4.72		8.48	11.90

23 13 Facility Fuel-Storage Tanks

23 13 13 – Facility Underground Fuel-Oil, Storage Tanks

23 13 13.09 Single-Wall Steel Fuel-Oil Tanks		Crew	Daily Output	Labor-Hours	Unit	Material	Labor	Equipment	Total	Total Incl O&P
0010	**SINGLE-WALL STEEL FUEL-OIL TANKS**									
5000	Tanks, steel ugnd., sti-p3, not incl. hold-down bars									
5500	Excavation, pad, pumps and piping not included									
5510	Single wall, 500 gallon capacity, 7 ga. shell	Q-5	2.70	5.926	Ea.	1,850	246		2,096	2,425
5520	1,000 gallon capacity, 7 ga. shell	"	2.50	6.400		2,725	265		2,990	3,425
5530	2,000 gallon capacity, 1/4" thick shell	Q-7	4.60	6.957		3,500	288		3,788	4,325
5535	2,500 gallon capacity, 7 ga. shell	Q-5	3	5.333		5,325	221		5,546	6,250
5540	5,000 gallon capacity, 1/4" thick shell	Q-7	3.20	10		11,900	415		12,315	13,800
5560	10,000 gallon capacity, 1/4" thick shell		2	16		10,300	665		10,965	12,400
5610	25,000 gallon capacity, 3/8" thick shell		1.30	24.615		27,500	1,025		28,525	32,000
5630	40,000 gallon capacity, 3/8" thick shell		.90	35.556		43,200	1,475		44,675	49,900
5640	50,000 gallon capacity, 3/8" thick shell	↓	.80	40	↓	48,000	1,650		49,650	55,500

23 13 13.13 Dbl-Wall Steel, Undrgrnd Fuel-Oil, Stor. Tanks		Crew	Daily Output	Labor-Hours	Unit	Material	Labor	Equipment	Total	Total Incl O&P
0010	**DOUBLE-WALL STEEL, UNDERGROUND FUEL-OIL, STORAGE TANKS**									
6200	Steel, underground, 360°, double wall, UL listed,									
6210	with sti-P3 corrosion protection,									
6220	(dielectric coating, cathodic protection, electrical									
6230	isolation) 30 year warranty,									
6240	not incl. manholes or hold-downs.									
6260	1,000 gallon capacity	Q-5	2.25	7.111	Ea.	3,650	295		3,945	4,500
6280	3,000 gallon capacity	Q-7	3.90	8.205		6,725	340		7,065	7,950
6300	5,000 gallon capacity		2.91	10.997		8,575	455		9,030	10,200
6310	6,000 gallon capacity		2.42	13.223		12,100	550		12,650	14,200
6330	10,000 gallon capacity		1.82	17.582	↓	13,400	730		14,130	15,900
6400	For hold-downs 500-2,000 gal., add		16	2	Set	219	83		302	375
6410	For hold-downs 3,000-6,000 gal., add		12	2.667		380	111		491	600
6420	For hold-downs 8,000-12,000 gal., add	↓	11	2.909	↓	450	121		571	695

23 13 13.23 Glass-Fiber-Reinfcd-Plastic, Fuel-Oil, Storage		Crew	Daily Output	Labor-Hours	Unit	Material	Labor	Equipment	Total	Total Incl O&P
0010	**GLASS-FIBER-REINFCD-PLASTIC, UNDERGRND. FUEL-OIL, STORAGE**									
0210	Fiberglass, underground, single wall, UL listed, not including									
0220	manway or hold-down strap									
0230	1,000 gallon capacity	Q-5	2.46	6.504	Ea.	5,075	270		5,345	6,025
0250	4,000 gallon capacity	Q-7	3.55	9.014		9,950	375		10,325	11,500
0270	8,000 gallon capacity		2.29	13.974		13,500	580		14,080	15,800
0280	10,000 gallon capacity	↓	2	16		15,400	665		16,065	18,000
0500	For manway, fittings and hold-downs, add				↓	20%	15%			
2210	Fiberglass, underground, single wall, UL listed, including									
2220	hold-down straps, no manways									
2230	1,000 gallon capacity	Q-5	1.88	8.511	Ea.	5,575	355		5,930	6,700
2250	4,000 gallon capacity	Q-7	2.90	11.034	↓	10,400	455		10,855	12,300

23 13 Facility Fuel-Storage Tanks

23 13 13 – Facility Underground Fuel-Oil, Storage Tanks

23 13 13.23 Glass-Fiber-Reinfcd-Plastic, Fuel-Oil, Storage		Crew	Daily Output	Labor-Hours	Unit	Material	2019 Bare Costs Labor	Equipment	Total	Total Incl O&P
2270	8,000 gallon capacity	Q-7	1.78	17.978	Ea.	14,500	745		15,245	17,200
2280	10,000 gallon capacity		1.60	20		16,400	830		17,230	19,400

23 21 Hydronic Piping and Pumps

23 21 20 – Hydronic HVAC Piping Specialties

23 21 20.46 Expansion Tanks

		Crew	Daily Output	Labor-Hours	Unit	Material	Labor	Equipment	Total	Total Incl O&P
0010	**EXPANSION TANKS**									
1507	Underground fuel-oil storage tanks, see Section 23 13 13									
1512	Tank leak detection systems, see Section 28 33 33.50									
2000	Steel, liquid expansion, ASME, painted, 15 gallon capacity	Q-5	17	.941	Ea.	790	39		829	935
2040	30 gallon capacity		12	1.333		900	55.50		955.50	1,075
2120	100 gallon capacity		6	2.667		1,825	111		1,936	2,200
3000	Steel ASME expansion, rubber diaphragm, 19 gal. cap. accept.		12	1.333		2,775	55.50		2,830.50	3,175
3020	31 gallon capacity		8	2		3,275	83		3,358	3,725
3080	119 gallon capacity		4	4		4,850	166		5,016	5,600

23 21 20.58 Hydronic Heating Control Valves

		Crew	Daily Output	Labor-Hours	Unit	Material	Labor	Equipment	Total	Total Incl O&P
0010	**HYDRONIC HEATING CONTROL VALVES**									
0050	Hot water, nonelectric, thermostatic									
0100	Radiator supply, 1/2" diameter	1 Stpi	24	.333	Ea.	70	15.35		85.35	102
0120	3/4" diameter		20	.400		72.50	18.40		90.90	110
0140	1" diameter		19	.421		89.50	19.40		108.90	131
0160	1-1/4" diameter		15	.533		126	24.50		150.50	179
0500	For low pressure steam, add					25%				
1000	Manual, radiator supply									
1010	1/2" pipe size, angle union	1 Stpi	24	.333	Ea.	65	15.35		80.35	96.50
1020	3/4" pipe size, angle union		20	.400		81.50	18.40		99.90	120
1030	1" pipe size, angle union		19	.421		97	19.40		116.40	139
1140	Balance and stop valve 1/2" size		22	.364		61	16.75		77.75	94.50
1150	3/4" size		20	.400		66	18.40		84.40	103
1160	1" size		19	.421		75	19.40		94.40	115
1170	1-1/4" size		15	.533		97	24.50		121.50	147
8000	System balancing and shut-off									
8020	Butterfly, quarter turn, calibrated, threaded or solder									
8040	Bronze, -30°F to +350°F, pressure to 175 psi									
8060	1/2" size	1 Stpi	22	.364	Ea.	17.30	16.75		34.05	46.50
8070	3/4" size		20	.400		28	18.40		46.40	60.50
8080	1" size		19	.421		34	19.40		53.40	69.50
8090	1-1/4" size		15	.533		55	24.50		79.50	101
8110	2" size		11	.727		90	33.50		123.50	154

23 21 20.78 Strainers, Y Type, Iron Body

		Crew	Daily Output	Labor-Hours	Unit	Material	Labor	Equipment	Total	Total Incl O&P
0010	**STRAINERS, Y TYPE, IRON BODY**									
0100	1/2" pipe size	1 Stpi	20	.400	Ea.	17.40	18.40		35.80	49
0120	3/4" pipe size		18	.444		20	20.50		40.50	55.50
0140	1" pipe size		16	.500		24.50	23		47.50	65
0150	1-1/4" pipe size		15	.533		37.50	24.50		62	82
0180	2" pipe size		8	1		76.50	46		122.50	160
1000	Flanged, 125 lb., 1-1/2" pipe size		11	.727		173	33.50		206.50	245
1020	2" pipe size		8	1		156	46		202	248
1030	2-1/2" pipe size	Q-5	5	3.200		167	133		300	400
1040	3" pipe size		4.50	3.556		182	147		329	440

23 21 Hydronic Piping and Pumps

23 21 20 – Hydronic HVAC Piping Specialties

23 21 20.78 Strainers, Y Type, Iron Body		Crew	Daily Output	Labor-Hours	Unit	2019 Bare Costs Material	Labor	Equipment	Total	Total Incl O&P
1060	4″ pipe size	Q-5	3	5.333	Ea.	310	221		531	710

23 21 23 – Hydronic Pumps

23 21 23.13 In-Line Centrifugal Hydronic Pumps

Line	Description	Crew	Daily Output	Labor-Hours	Unit	Material	Labor	Equipment	Total	Total Incl O&P
0010	**IN-LINE CENTRIFUGAL HYDRONIC PUMPS**									
0600	Bronze, sweat connections, 1/40 HP, in line									
0640	3/4″ size	Q-1	16	1	Ea.	288	40.50		328.50	380
1000	Flange connection, 3/4″ to 1-1/2″ size									
1040	1/12 HP	Q-1	6	2.667	Ea.	755	108		863	1,000
1060	1/8 HP		6	2.667		1,325	108		1,433	1,625
1100	1/3 HP		6	2.667		1,450	108		1,558	1,775
1140	2″ size, 1/6 HP		5	3.200		1,650	129		1,779	2,000
2000	Cast iron, flange connection									
2040	3/4″ to 1-1/2″ size, in line, 1/12 HP	Q-1	6	2.667	Ea.	460	108		568	680
2060	1/8 HP		6	2.667		815	108		923	1,075
2101	Pumps, circulating, 3/4″ to 1-1/2″ size, 1/3 HP		6	2.667		990	108		1,098	1,275
2140	2″ size, 1/6 HP		5	3.200		1,000	129		1,129	1,300
2180	2-1/2″ size, 1/4 HP		5	3.200		1,125	129		1,254	1,425
2220	3″ size, 1/4 HP		4	4		1,150	162		1,312	1,550
2260	1/3 HP		4	4		1,525	162		1,687	1,950
2300	1/2 HP		4	4		1,575	162		1,737	2,000
2340	3/4 HP		4	4		1,825	162		1,987	2,275
2380	1 HP		4	4		2,625	162		2,787	3,175
4000	Close coupled, end suction, bronze impeller									
4040	1-1/2″ size, 1-1/2 HP, to 40 GPM	Q-1	3	5.333	Ea.	2,600	215		2,815	3,225
4090	2″ size, 2 HP, to 50 GPM		3	5.333		3,025	215		3,240	3,675
4100	2″ size, 3 HP, to 90 GPM		2.30	6.957		3,050	281		3,331	3,800
4190	2-1/2″ size, 3 HP, to 150 GPM		2	8		3,500	325		3,825	4,375
4300	3″ size, 5 HP, to 225 GPM		1.80	8.889		4,000	360		4,360	5,025
4410	3″ size, 10 HP, to 350 GPM		1.60	10		5,825	405		6,230	7,075
4420	4″ size, 7-1/2 HP, to 350 GPM		1.60	10		5,975	405		6,380	7,225
5000	Base mounted, bronze impeller, coupling guard									
5190	2-1/2″ size, 3 HP, to 150 GPM	Q-1	1.80	8.889	Ea.	9,125	360		9,485	10,600
5300	3″ size, 5 HP, to 225 GPM		1.60	10		10,500	405		10,905	12,300
5410	4″ size, 5 HP, to 350 GPM		1.50	10.667		11,300	430		11,730	13,200

23 21 29 – Automatic Condensate Pump Units

23 21 29.10 Condensate Removal Pump System

Line	Description	Crew	Daily Output	Labor-Hours	Unit	Material	Labor	Equipment	Total	Total Incl O&P
0010	**CONDENSATE REMOVAL PUMP SYSTEM**									
0020	Pump with 1 gal. ABS tank									
0100	115 V									
0120	1/50 HP, 200 GPH G	1 Stpi	12	.667	Ea.	183	30.50		213.50	253
0140	1/18 HP, 270 GPH G		10	.800		187	37		224	266
0160	1/5 HP, 450 GPH G		8	1		420	46		466	540
0200	230 V									
0240	1/18 HP, 270 GPH	1 Stpi	10	.800	Ea.	202	37		239	283
0260	1/5 HP, 450 GPH G	"	8	1	"	485	46		531	605

23 23 16.16 Refrigerant Line Sets		Crew	Daily Output	Labor-Hours	Unit	2019 Bare Costs Material	Labor	Equipment	Total	Total Incl O&P
0010	REFRIGERANT LINE SETS, Standard									
0100	Copper tube									
0110	1/2" insulation, both tubes									
0120	Combination 1/4" and 1/2" tubes									
0130	10' set	Q-5	42	.381	Ea.	48	15.80		63.80	78.50
0135	15' set		42	.381		69	15.80		84.80	102
0140	20' set		40	.400		74.50	16.60		91.10	109
0150	30' set		37	.432		105	17.90		122.90	146
0160	40' set		35	.457		132	18.95		150.95	176
0170	50' set		32	.500		163	20.50		183.50	214
0180	100' set		22	.727		355	30		385	440
0300	Combination 1/4" and 3/4" tubes									
0310	10' set	Q-5	40	.400	Ea.	53.50	16.60		70.10	86
0320	20' set		38	.421		94.50	17.45		111.95	133
0330	30' set		35	.457		142	18.95		160.95	187
0340	40' set		33	.485		185	20		205	237
0350	50' set		30	.533		233	22		255	293
0380	100' set		20	.800		530	33		563	640
0500	Combination 3/8" & 3/4" tubes									
0510	10' set	Q-5	28	.571	Ea.	61.50	23.50		85	107
0520	20' set		36	.444		95.50	18.40		113.90	135
0530	30' set		34	.471		130	19.50		149.50	175
0540	40' set		31	.516		171	21.50		192.50	223
0550	50' set		28	.571		201	23.50		224.50	260
0580	100' set		18	.889		595	37		632	715
0700	Combination 3/8" & 1-1/8" tubes									
0710	10' set	Q-5	36	.444	Ea.	108	18.40		126.40	148
0720	20' set		33	.485		180	20		200	231
0730	30' set		31	.516		242	21.50		263.50	300
0740	40' set		28	.571		365	23.50		388.50	440
0750	50' set		26	.615		350	25.50		375.50	425
0900	Combination 1/2" & 3/4" tubes									
0910	10' set	Q-5	37	.432	Ea.	65	17.90		82.90	101
0920	20' set		35	.457		115	18.95		133.95	158
0930	30' set		33	.485		173	20		193	224
0940	40' set		30	.533		228	22		250	288
0950	50' set		27	.593		287	24.50		311.50	355
0980	100' set		17	.941		645	39		684	775
2100	Combination 1/2" & 1-1/8" tubes									
2110	10' set	Q-5	35	.457	Ea.	111	18.95		129.95	153
2120	20' set		31	.516		190	21.50		211.50	244
2130	30' set		29	.552		285	23		308	355
2140	40' set		25	.640		385	26.50		411.50	465
2150	50' set		14	1.143		465	47.50		512.50	595
2300	For 1" thick insulation add					30%	15%			
3000	Refrigerant line sets, min-split, flared									
3100	Combination 1/4" & 3/8" tubes									
3120	15' set	Q-5	41	.390	Ea.	68.50	16.20		84.70	102
3140	25' set		38.50	.416		97	17.25		114.25	136
3160	35' set		37	.432		122	17.90		139.90	164
3180	50' set		30	.533		166	22		188	220
3200	Combination 1/4" & 1/2" tubes									
3220	15' set	Q-5	41	.390	Ea.	73	16.20		89.20	107

23 23 Refrigerant Piping

23 23 16 – Refrigerant Piping Specialties

23 23 16.16 Refrigerant Line Sets		Crew	Daily Output	Labor-Hours	Unit	Material	2019 Bare Costs Labor	Equipment	Total	Total Incl O&P
3240	25′ set	Q-5	38.50	.416	Ea.	99.50	17.25		116.75	138
3260	35′ set		37	.432		128	17.90		145.90	171
3280	50′ set		30	.533		173	22		195	227

23 31 HVAC Ducts and Casings

23 31 13 – Metal Ducts

23 31 13.13 Rectangular Metal Ducts

		Crew	Daily Output	Labor-Hours	Unit	Material	Labor	Equipment	Total	Total Incl O&P
0010	**RECTANGULAR METAL DUCTS**									
0020	Fabricated rectangular, includes fittings, joints, supports,									
0021	allowance for flexible connections and field sketches.									
0030	Does not include "as-built dwgs." or insulation.									
0031	NOTE: Fabrication and installation are combined									
0040	as LABOR cost. Approx. 25% fittings assumed.									
0042	Fabrication/Inst. is to commercial quality standards									
0043	(SMACNA or equiv.) for structure, sealing, leak testing, etc.									
0050	Add to labor for elevated installation									
0051	of fabricated ductwork									
0052	10′ to 15′ high						6%			
0053	15′ to 20′ high						12%			
0054	20′ to 25′ high						15%			
0100	Aluminum, alloy 3003-H14, under 100 lb.	Q-10	75	.320	Lb.	3.20	12.95		16.15	25
0140	1,000 to 2,000 lb.		120	.200		1.78	8.10		9.88	15.35
0150	2,000 to 5,000 lb.		130	.185		1.78	7.45		9.23	14.30
0500	Galvanized steel, under 200 lb.		235	.102		.57	4.13		4.70	7.50
0520	200 to 500 lb.		245	.098		.58	3.96		4.54	7.20
0540	500 to 1,000 lb.		255	.094		.56	3.80		4.36	6.90
0560	1,000 to 2,000 lb.		265	.091		.57	3.66		4.23	6.70
0570	2,000 to 5,000 lb.		275	.087		.56	3.53		4.09	6.45

23 31 16 – Nonmetal Ducts

23 31 16.13 Fibrous-Glass Ducts

		Crew	Daily Output	Labor-Hours	Unit	Material	Labor	Equipment	Total	Total Incl O&P
0010	**FIBROUS-GLASS DUCTS**									
3490	Rigid fiberglass duct board, foil reinf. kraft facing									
3500	Rectangular, 1″ thick, alum. faced, (FRK), std. weight	Q-10	350	.069	SF Surf	.90	2.77		3.67	5.60

23 31 16.22 Duct Board

		Crew	Daily Output	Labor-Hours	Unit	Material	Labor	Equipment	Total	Total Incl O&P
0010	**DUCT BOARD**									
3800	Rigid, resin bonded fibrous glass, FSK									
3810	Temperature, bacteria and fungi resistant									
3820	1″ thick	Q-14	150	.107	S.F.	1.63	3.91		5.54	8.40
3830	1-1/2″ thick		130	.123		2.15	4.51		6.66	9.95
3840	2″ thick		120	.133		3.71	4.88		8.59	12.30

23 33 13 – Dampers

	23 33 13.13 Volume-Control Dampers	Crew	Daily Output	Labor-Hours	Unit	Material	Labor (2019 Bare Costs)	Equipment	Total	Total Incl O&P
0010	**VOLUME-CONTROL DAMPERS**									
5990	Multi-blade dampers, opposed blade, 8" x 6"	1 Shee	24	.333	Ea.	32.50	14.45		46.95	59.50
5994	8" x 8"		22	.364		33	15.75		48.75	62.50
5996	10" x 10"		21	.381		36	16.50		52.50	67
6000	12" x 12"		21	.381		45	16.50		61.50	77
6020	12" x 18"		18	.444		60.50	19.25		79.75	98.50
6030	14" x 10"		20	.400		44	17.30		61.30	77
6031	14" x 14"		17	.471		47.50	20.50		68	86
6033	16" x 12"		17	.471		47.50	20.50		68	86
6035	16" x 16"		16	.500		59	21.50		80.50	101
6037	18" x 16"		15	.533		74	23		97	120
6038	18" x 18"		15	.533		78	23		101	125
6070	20" x 16"		14	.571		78	24.50		102.50	127
6072	20" x 20"		13	.615		84.50	26.50		111	137
6074	22" x 18"		14	.571		84.50	24.50		109	134
6076	24" x 16"		11	.727		92.50	31.50		124	154
6078	24" x 20"		8	1		113	43.50		156.50	196
6080	24" x 24"		8	1		122	43.50		165.50	207
6110	26" x 26"		6	1.333		143	57.50		200.50	253
6133	30" x 30"	Q-9	6.60	2.424		212	94.50		306.50	390
6135	32" x 32"	"	6.40	2.500		229	97.50		326.50	415
7500	Variable volume modulating motorized damper, incl. elect. mtr.									
7504	8" x 6"	1 Shee	15	.533	Ea.	180	23		203	237
7506	10" x 6"		14	.571		180	24.50		204.50	239
7510	10" x 10"		13	.615		188	26.50		214.50	251
7520	12" x 12"		12	.667		194	29		223	261
7522	12" x 16"		11	.727		195	31.50		226.50	266
7524	16" x 10"		12	.667		191	29		220	258
7526	16" x 14"		10	.800		200	34.50		234.50	279
7528	16" x 18"		9	.889		206	38.50		244.50	291
7542	18" x 18"		8	1		210	43.50		253.50	305
7544	20" x 14"		8	1		218	43.50		261.50	310
7546	20" x 18"		7	1.143		224	49.50		273.50	330
7560	24" x 12"		8	1		219	43.50		262.50	315
7562	24" x 18"		7	1.143		236	49.50		285.50	340
7564	24" x 24"		6	1.333		246	57.50		303.50	365
7568	28" x 10"		7	1.143		219	49.50		268.50	325
7590	30" x 14"		5	1.600		227	69.50		296.50	365
7610	30" x 24"		3.80	2.105		395	91		486	585
7700	For thermostat, add		8	1		43	43.50		86.50	120
8000	Multi-blade dampers, parallel blade									
8100	8" x 8"	1 Shee	24	.333	Ea.	116	14.45		130.45	152
8160	18" x 12"	"	18	.444	"	160	19.25		179.25	208

23 33 13.16 Fire Dampers

		Crew	Daily Output	Labor-Hours	Unit	Material	Labor	Equipment	Total	Total Incl O&P
0010	**FIRE DAMPERS**									
3000	Fire damper, curtain type, 1-1/2 hr. rated, vertical, 6" x 6"	1 Shee	24	.333	Ea.	34.50	14.45		48.95	62
3020	8" x 6"		22	.364		34.50	15.75		50.25	64
3240	16" x 14"		18	.444		62.50	19.25		81.75	101

23 33 19 – Duct Silencers

23 33 19.10 Duct Silencers

		Crew	Daily Output	Labor-Hours	Unit	Material	Labor	Equipment	Total	Total Incl O&P
0010	**DUCT SILENCERS**									
9000	Silencers, noise control for air flow, duct				MCFM	81.50			81.50	89.50

23 33 Air Duct Accessories

23 33 46 – Flexible Ducts

23 33 46.10 Flexible Air Ducts		Crew	Daily Output	Labor-Hours	Unit	Material	2019 Bare Costs Labor	Equipment	Total	Total Incl O&P
0010	**FLEXIBLE AIR DUCTS**									
1280	Add to labor for elevated installation									
1282	of prefabricated (purchased) ductwork									
1283	10′ to 15′ high						10%			
1284	15′ to 20′ high						20%			
1285	20′ to 25′ high						25%			
1286	25′ to 30′ high						35%			
1287	30′ to 35′ high						40%			
1288	35′ to 40′ high						50%			
1289	Over 40′ high						55%			
1300	Flexible, coated fiberglass fabric on corr. resist. metal helix									
1400	pressure to 12″ (WG) UL-181									
1500	Noninsulated, 3″ diameter	Q-9	400	.040	L.F.	1.21	1.56		2.77	3.92
1520	4″ diameter		360	.044		1.24	1.73		2.97	4.23
1540	5″ diameter		320	.050		1.38	1.95		3.33	4.75
1560	6″ diameter		280	.057		1.61	2.23		3.84	5.45
1580	7″ diameter		240	.067		1.75	2.60		4.35	6.25
1600	8″ diameter		200	.080		2.01	3.12		5.13	7.35
1620	9″ diameter		180	.089		2.22	3.46		5.68	8.20
1640	10″ diameter		160	.100		2.60	3.90		6.50	9.30
1900	Insulated, 1″ thick, PE jacket, 3″ diameter G		380	.042		2.66	1.64		4.30	5.65
1920	5″ diameter G		300	.053		3.01	2.08		5.09	6.75
1940	6″ diameter G		260	.062		3.14	2.40		5.54	7.45
1960	7″ diameter G		220	.073		3.59	2.84		6.43	8.65
1980	8″ diameter G		180	.089		3.96	3.46		7.42	10.10
2000	9″ diameter G		160	.100		4.18	3.90		8.08	11.05
2020	10″ diameter G	↓	140	.114	↓	4.59	4.46		9.05	12.45

23 33 53 – Duct Liners

23 33 53.10 Duct Liner Board		Crew	Daily Output	Labor-Hours	Unit	Material	Labor	Equipment	Total	Total Incl O&P
0010	**DUCT LINER BOARD**									
3490	Board type, fiberglass liner, 3 lb. density									
3940	Board type, non-fibrous foam									
3950	Temperature, bacteria and fungi resistant									
3960	1″ thick G	Q-14	150	.107	S.F.	2.47	3.91		6.38	9.30
3970	1-1/2″ thick G		130	.123		4.23	4.51		8.74	12.25
3980	2″ thick G	↓	120	.133	↓	4.12	4.88		9	12.75

23 34 HVAC Fans

23 34 14 – Blower HVAC Fans

23 34 14.10 Blower Type HVAC Fans		Crew	Daily Output	Labor-Hours	Unit	Material	Labor	Equipment	Total	Total Incl O&P
0010	**BLOWER TYPE HVAC FANS**									
2500	Ceiling fan, right angle, extra quiet, 0.10″ S.P.									
2520	95 CFM	Q-20	20	1	Ea.	297	40.50		337.50	390
2540	210 CFM	″	19	1.053	″	350	42.50		392.50	455

23 34 16 – Centrifugal HVAC Fans

23 34 16.10 Centrifugal Type HVAC Fans		Crew	Daily Output	Labor-Hours	Unit	2019 Bare Costs Material	Labor	Equipment	Total	Total Incl O&P
0010	**CENTRIFUGAL TYPE HVAC FANS**									
5000	Utility set, centrifugal, V belt drive, motor									
5020	1/4" S.P., 1,200 CFM, 1/4 HP	Q-20	6	3.333	Ea.	1,900	134		2,034	2,325
5040	1,520 CFM, 1/3 HP		5	4		2,475	161		2,636	2,975
5060	1,850 CFM, 1/2 HP		4	5		2,450	202		2,652	3,025
5080	2,180 CFM, 3/4 HP		3	6.667		2,850	269		3,119	3,575
7000	Roof exhauster, centrifugal, aluminum housing, 12" galvanized									
7020	curb, bird screen, back draft damper, 1/4" S.P.									
7100	Direct drive, 320 CFM, 11" sq. damper	Q-20	7	2.857	Ea.	805	115		920	1,075
7120	600 CFM, 11" sq. damper		6	3.333		950	134		1,084	1,275
7140	815 CFM, 13" sq. damper		5	4		1,025	161		1,186	1,400
7160	1,450 CFM, 13" sq. damper		4.20	4.762		1,600	192		1,792	2,075
7180	2,050 CFM, 16" sq. damper		4	5		1,825	202		2,027	2,325
7200	V-belt drive, 1,650 CFM, 12" sq. damper		6	3.333		1,475	134		1,609	1,850
7220	2,750 CFM, 21" sq. damper		5	4		1,750	161		1,911	2,200
7230	3,500 CFM, 21" sq. damper		4.50	4.444		1,975	179		2,154	2,475
7240	4,910 CFM, 23" sq. damper		4	5		2,400	202		2,602	2,975

23 34 23 – HVAC Power Ventilators

23 34 23.10 HVAC Power Circulators and Ventilators		Crew	Daily Output	Labor-Hours	Unit	Material	Labor	Equipment	Total	Total Incl O&P
0010	**HVAC POWER CIRCULATORS AND VENTILATORS**									
6000	Propeller exhaust, wall shutter									
6020	Direct drive, one speed, 0.075" S.P.									
6100	653 CFM, 1/30 HP	Q-20	10	2	Ea.	209	80.50		289.50	365
6140	1,323 CFM, 1/15 HP		8	2.500		350	101		451	545
6160	2,444 CFM, 1/4 HP		7	2.857		435	115		550	665
6650	Residential, bath exhaust, grille, back draft damper									
6660	50 CFM	Q-20	24	.833	Ea.	53	33.50		86.50	114
6670	110 CFM	"	22	.909	"	106	36.50		142.50	177

23 34 33 – Air Curtains

23 34 33.10 Air Barrier Curtains		Crew	Daily Output	Labor-Hours	Unit	Material	Labor	Equipment	Total	Total Incl O&P
0010	**AIR BARRIER CURTAINS**, Incl. motor starters, transformers,									
0050	and door switches									
2450	Conveyor openings or service windows									
3000	Service window, 5' high x 25" wide	2 Shee	5	3.200	Ea.	355	139		494	625
3100	Environmental separation									
3110	Door heights up to 8', low profile, super quiet									
3120	Unheated, variable speed									
3130	36" wide	2 Shee	4	4	Ea.	735	173		908	1,100
3134	42" wide		3.80	4.211		755	182		937	1,125
3138	48" wide		3.60	4.444		780	192		972	1,175
3142	60" wide		3.40	4.706		830	204		1,034	1,250
3146	72" wide	Q-3	4.60	6.957		1,000	281		1,281	1,550
3150	96" wide		4.40	7.273		1,700	294		1,994	2,350
3154	120" wide		4.20	7.619		1,475	310		1,785	2,125
3158	144" wide		4	8		1,925	325		2,250	2,650
3200	Door heights up to 10'									
3210	Unheated									
3230	36" wide	2 Shee	3.80	4.211	Ea.	715	182		897	1,075
3234	42" wide		3.60	4.444		710	192		902	1,100
3238	48" wide		3.40	4.706		725	204		929	1,150
3242	60" wide		3.20	5		1,100	217		1,317	1,550

23 34 HVAC Fans

23 34 33 – Air Curtains

23 34 33.10 Air Barrier Curtains		Crew	Daily Output	Labor-Hours	Unit	Material	2019 Bare Costs Labor	Equipment	Total	Total Incl O&P
3246	72″ wide	Q-3	4.40	7.273	Ea.	1,250	294		1,544	1,875
3250	96″ wide		4.20	7.619		1,475	310		1,785	2,125
3254	120″ wide		4	8		2,000	325		2,325	2,725
3258	144″ wide		3.80	8.421		2,175	340		2,515	2,925
3300	Door heights up to 12′									
3310	Unheated									
3334	42″ wide	2 Shee	3.40	4.706	Ea.	1,100	204		1,304	1,550
3338	48″ wide		3.20	5		1,100	217		1,317	1,550
3342	60″ wide		3	5.333		1,125	231		1,356	1,600
3346	72″ wide	Q-3	4.20	7.619		1,875	310		2,185	2,550
3350	96″ wide		4	8		2,175	325		2,500	2,925
3354	120″ wide		3.80	8.421		2,350	340		2,690	3,150
3358	144″ wide		3.60	8.889		3,125	360		3,485	4,050
3400	Door heights up to 16′									
3410	Unheated									
3438	48″ wide	2 Shee	3	5.333	Ea.	1,475	231		1,706	2,000
3442	60″ wide	″	2.80	5.714		1,575	247		1,822	2,125
3446	72″ wide	Q-3	3.80	8.421		2,525	340		2,865	3,325
3450	96″ wide		3.60	8.889		2,800	360		3,160	3,675
3454	120″ wide		3.40	9.412		3,400	380		3,780	4,350
3458	144″ wide		3.20	10		3,500	405		3,905	4,550
3470	Heated, electric									
3474	48″ wide	2 Shee	2.90	5.517	Ea.	2,375	239		2,614	3,000
3478	60″ wide	″	2.70	5.926		2,425	257		2,682	3,100
3482	72″ wide	Q-3	3.70	8.649		4,125	350		4,475	5,125
3486	96″ wide		3.50	9.143		4,350	370		4,720	5,400
3490	120″ wide		3.30	9.697		5,825	390		6,215	7,075
3494	144″ wide		3.10	10.323		5,825	415		6,240	7,075

23 37 Air Outlets and Inlets

23 37 13 – Diffusers, Registers, and Grilles

23 37 13.10 Diffusers		Crew	Daily Output	Labor-Hours	Unit	Material	Labor	Equipment	Total	Total Incl O&P
0010	**DIFFUSERS**, Aluminum, opposed blade damper unless noted									
0100	Ceiling, linear, also for sidewall									
0120	2″ wide	1 Shee	32	.250	L.F.	18.85	10.85		29.70	38.50
0160	4″ wide		26	.308	″	25.50	13.30		38.80	50
0500	Perforated, 24″ x 24″ lay-in panel size, 6″ x 6″		16	.500	Ea.	165	21.50		186.50	217
0520	8″ x 8″		15	.533		173	23		196	229
0530	9″ x 9″		14	.571		176	24.50		200.50	234
0590	16″ x 16″		11	.727		205	31.50		236.50	278
1000	Rectangular, 1 to 4 way blow, 6″ x 6″		16	.500		43	21.50		64.50	83
1010	8″ x 8″		15	.533		61	23		84	106
1014	9″ x 9″		15	.533		53.50	23		76.50	97
1016	10″ x 10″		15	.533		83	23		106	130
1020	12″ x 6″		15	.533		75	23		98	121
1040	12″ x 9″		14	.571		79.50	24.50		104	129
1060	12″ x 12″		12	.667		74.50	29		103.50	130
1070	14″ x 6″		13	.615		81.50	26.50		108	134
1074	14″ x 14″		12	.667		133	29		162	195
1150	18″ x 18″		9	.889		117	38.50		155.50	193
1170	24″ x 12″		10	.800		171	34.50		205.50	247

23 37 Air Outlets and Inlets

23 37 13 – Diffusers, Registers, and Grilles

23 37 13.10 Diffusers		Crew	Daily Output	Labor-Hours	Unit	2019 Bare Costs Material	Labor	Equipment	Total	Total Incl O&P
1180	24″ x 24″	1 Shee	7	1.143	Ea.	284	49.50		333.50	390
1500	Round, butterfly damper, steel, diffuser size, 6″ diameter		18	.444		12.55	19.25		31.80	46
1520	8″ diameter		16	.500		13.25	21.50		34.75	50.50
2000	T-bar mounting, 24″ x 24″ lay-in frame, 6″ x 6″		16	.500		72.50	21.50		94	116
2020	8″ x 8″		14	.571		72	24.50		96.50	121
2040	12″ x 12″		12	.667		88.50	29		117.50	146
2060	16″ x 16″		11	.727		115	31.50		146.50	178
2080	18″ x 18″		10	.800		121	34.50		155.50	191
6000	For steel diffusers instead of aluminum, deduct					10%				

23 37 13.30 Grilles

Code	Description	Crew	Daily Output	Labor-Hours	Unit	Material	Labor	Equipment	Total	Total Incl O&P
0010	**GRILLES**									
0020	Aluminum, unless noted otherwise									
1000	Air return, steel, 6″ x 6″	1 Shee	26	.308	Ea.	21	13.30		34.30	45
1020	10″ x 6″		24	.333		21	14.45		35.45	47
1080	16″ x 8″		22	.364		28.50	15.75		44.25	57.50
1100	12″ x 12″		22	.364		28	15.75		43.75	57
1120	24″ x 12″		18	.444		37.50	19.25		56.75	73.50
1180	16″ x 16″		22	.364		36.50	15.75		52.25	66

23 37 13.60 Registers

Code	Description	Crew	Daily Output	Labor-Hours	Unit	Material	Labor	Equipment	Total	Total Incl O&P
0010	**REGISTERS**									
0980	Air supply									
3000	Baseboard, hand adj. damper, enameled steel									
3012	8″ x 6″	1 Shee	26	.308	Ea.	7.45	13.30		20.75	30
3020	10″ x 6″		24	.333		8.10	14.45		22.55	33
3060	12″ x 6″		23	.348		9.75	15.05		24.80	36
4000	Floor, toe operated damper, enameled steel									
4020	4″ x 8″	1 Shee	32	.250	Ea.	13.60	10.85		24.45	33
4040	4″ x 12″	″	26	.308	″	16	13.30		29.30	39.50
4300	Spiral pipe supply register									
4310	Steel, with air scoop									
4320	4″ x 12″, for 8″ thru 13″ diameter duct	1 Shee	25	.320	Ea.	81	13.85		94.85	112
4330	4″ x 18″, for 8″ thru 13″ diameter duct		18	.444		93.50	19.25		112.75	135
4340	6″ x 12″, for 14″ thru 21″ diameter duct		19	.421		85.50	18.25		103.75	124
4350	6″ x 16″, for 14″ thru 21″ diameter duct		18	.444		97	19.25		116.25	139
4360	6″ x 20″, for 14″ thru 21″ diameter duct		17	.471		107	20.50		127.50	152
4370	6″ x 24″, for 14″ thru 21″ diameter duct		16	.500		129	21.50		150.50	178
4380	8″ x 16″, for 22″ thru 31″ diameter duct		19	.421		101	18.25		119.25	142
4390	8″ x 18″, for 22″ thru 31″ diameter duct		18	.444		109	19.25		128.25	152
4400	8″ x 24″, for 22″ thru 31″ diameter duct		15	.533		136	23		159	189

23 37 23 – HVAC Gravity Ventilators

23 37 23.10 HVAC Gravity Air Ventilators

Code	Description	Crew	Daily Output	Labor-Hours	Unit	Material	Labor	Equipment	Total	Total Incl O&P
0010	**HVAC GRAVITY AIR VENTILATORS**, Includes base									
4200	Stationary mushroom, aluminum, 16″ orifice diameter	Q-9	10	1.600	Ea.	675	62.50		737.50	850
4240	38″ orifice diameter		5	3.200		2,100	125		2,225	2,500
4250	42″ orifice diameter		4.70	3.404		2,775	133		2,908	3,275
4260	50″ orifice diameter		4.44	3.604		3,300	140		3,440	3,850

23 38 Ventilation Hoods

23 38 13 – Commercial-Kitchen Hoods

23 38 13.10 Hood and Ventilation Equipment		Crew	Daily Output	Labor-Hours	Unit	2019 Bare Costs Material	Labor	Equipment	Total	Total Incl O&P
0010	**HOOD AND VENTILATION EQUIPMENT**									
2970	Exhaust hood, sst, gutter on all sides, 4' x 4' x 2'	1 Carp	1.80	4.444	Ea.	4,225	172		4,397	4,925
2980	4' x 4' x 7'	"	1.60	5	"	6,725	194		6,919	7,725

23 41 Particulate Air Filtration

23 41 13 – Panel Air Filters

23 41 13.10 Panel Type Air Filters		Crew	Daily Output	Labor-Hours	Unit	Material	Labor	Equipment	Total	Total Incl O&P
0010	**PANEL TYPE AIR FILTERS**									
2950	Mechanical media filtration units									
3000	High efficiency type, with frame, non-supported G				MCFM	35			35	38.50
3100	Supported type G				"	54			54	59.50
5500	Throwaway glass or paper media type, 12" x 36" x 1"				Ea.	2.27			2.27	2.50

23 41 16 – Renewable-Media Air Filters

23 41 16.10 Disposable Media Air Filters		Crew	Daily Output	Labor-Hours	Unit	Material	Labor	Equipment	Total	Total Incl O&P
0010	**DISPOSABLE MEDIA AIR FILTERS**									
5000	Renewable disposable roll				C.S.F.	5.25			5.25	5.75

23 41 19 – Washable Air Filters

23 41 19.10 Permanent Air Filters		Crew	Daily Output	Labor-Hours	Unit	Material	Labor	Equipment	Total	Total Incl O&P
0010	**PERMANENT AIR FILTERS**									
4500	Permanent washable G				MCFM	25			25	27.50

23 41 23 – Extended Surface Filters

23 41 23.10 Expanded Surface Filters		Crew	Daily Output	Labor-Hours	Unit	Material	Labor	Equipment	Total	Total Incl O&P
0010	**EXPANDED SURFACE FILTERS**									
4000	Medium efficiency, extended surface G				MCFM	6.15			6.15	6.75

23 42 Gas-Phase Air Filtration

23 42 13 – Activated-Carbon Air Filtration

23 42 13.10 Charcoal Type Air Filtration		Crew	Daily Output	Labor-Hours	Unit	Material	Labor	Equipment	Total	Total Incl O&P
0010	**CHARCOAL TYPE AIR FILTRATION**									
0050	Activated charcoal type, full flow				MCFM	650			650	715
0060	Full flow, impregnated media 12" deep					225			225	248
0070	HEPA filter & frame for field erection					440			440	485
0080	HEPA filter-diffuser, ceiling install.				↓	350			350	385

23 43 Electronic Air Cleaners

23 43 13 – Washable Electronic Air Cleaners

23 43 13.10 Electronic Air Cleaners		Crew	Daily Output	Labor-Hours	Unit	Material	Labor	Equipment	Total	Total Incl O&P
0010	**ELECTRONIC AIR CLEANERS**									
2000	Electronic air cleaner, duct mounted									
2150	1,000 CFM	1 Shee	4	2	Ea.	430	86.50		516.50	620
2200	1,200 CFM		3.80	2.105		505	91		596	705
2250	1,400 CFM	↓	3.60	2.222	↓	530	96		626	740

23 51 Breechings, Chimneys, and Stacks

23 51 23 – Gas Vents

23 51 23.10 Gas Chimney Vents		Crew	Daily Output	Labor-Hours	Unit	2019 Bare Costs Material	Labor	Equipment	Total	Total Incl O&P
0010	**GAS CHIMNEY VENTS**, Prefab metal, UL listed									
0020	Gas, double wall, galvanized steel									
0080	3" diameter	Q-9	72	.222	V.L.F.	6.80	8.65		15.45	22
0100	4" diameter		68	.235		8.35	9.15		17.50	24.50
0120	5" diameter		64	.250		9.30	9.75		19.05	26.50
0140	6" diameter		60	.267		11.20	10.40		21.60	29.50
0160	7" diameter		56	.286		22.50	11.15		33.65	43.50
0180	8" diameter		52	.308		23	12		35	45.50
0200	10" diameter		48	.333		44.50	13		57.50	70.50

23 51 26 – All-Fuel Vent Chimneys

23 51 26.30 All-Fuel Vent Chimneys, Double Wall, St. Stl.		Crew	Daily Output	Labor-Hours	Unit	Material	Labor	Equipment	Total	Total Incl O&P
0010	**ALL-FUEL VENT CHIMNEYS, DOUBLE WALL, STAINLESS STEEL**									
7780	All fuel, pressure tight, double wall, 4" insulation, UL listed, 1,400°F.									
7790	304 stainless steel liner, aluminized steel outer jacket									
7800	6" diameter	Q-9	60	.267	V.L.F.	66.50	10.40		76.90	90.50
7806	10" diameter	"	48	.333	"	85	13		98	115
7880	For 316 stainless steel liner, add				L.F.	30%				

23 52 Heating Boilers

23 52 13 – Electric Boilers

23 52 13.10 Electric Boilers, ASME		Crew	Daily Output	Labor-Hours	Unit	Material	Labor	Equipment	Total	Total Incl O&P
0010	**ELECTRIC BOILERS, ASME**, Standard controls and trim									
1000	Steam, 6 KW, 20.5 MBH	Q-19	1.20	20	Ea.	4,300	855		5,155	6,125
1160	60 KW, 205 MBH		1	24		7,075	1,025		8,100	9,450
2000	Hot water, 7.5 KW, 25.6 MBH		1.30	18.462		5,200	790		5,990	7,025
2040	30 KW, 102 MBH		1.20	20		5,600	855		6,455	7,550
2070	60 KW, 205 MBH		1.20	20		5,800	855		6,655	7,800

23 52 19 – Pulse Combustion Boilers

23 52 19.20 Pulse Type Combustion Boilers		Crew	Daily Output	Labor-Hours	Unit	Material	Labor	Equipment	Total	Total Incl O&P
0010	**PULSE TYPE COMBUSTION BOILERS**, High efficiency									
7990	Special feature gas fired boilers									
8000	Pulse combustion, standard controls/trim									
8010	Hot water, DOE MBH output (AFUE %)									
8030	71 MBH (95.2%)	Q-5	1.60	10	Ea.	3,825	415		4,240	4,875
8050	94 MBH (95.3%) G		1.40	11.429		3,825	475		4,300	5,000
8080	139 MBH (95.6%) G		1.20	13.333		4,725	555		5,280	6,100
8090	207 MBH (95.4%)		1.16	13.793		5,425	570		5,995	6,900
8120	270 MBH (96.4%)		1.12	14.286		7,425	590		8,015	9,150
8130	365 MBH (91.7%)		1.07	14.953		8,375	620		8,995	10,300

23 52 23 – Cast-Iron Boilers

23 52 23.20 Gas-Fired Boilers		Crew	Daily Output	Labor-Hours	Unit	Material	Labor	Equipment	Total	Total Incl O&P
0010	**GAS-FIRED BOILERS**, Natural or propane, standard controls, packaged									
1000	Cast iron, with insulated jacket									
2000	Steam, gross output, 81 MBH	Q-7	1.40	22.857	Ea.	2,525	945		3,470	4,325
2080	203 MBH		.90	35.556		3,725	1,475		5,200	6,525
2180	400 MBH		.56	56.838		5,800	2,350		8,150	10,300
2280	1,275 MBH		.34	94.118		15,200	3,900		19,100	23,100
2380	3,060 MBH		.19	172		28,600	7,125		35,725	43,100
2400	3,570 MBH		.18	182		34,800	7,525		42,325	50,500

23 52 23 – Cast-Iron Boilers

23 52 23.20 Gas-Fired Boilers		Crew	Daily Output	Labor-Hours	Unit	2019 Bare Costs Material	Labor	Equipment	Total	Total Incl O&P
3000	Hot water, gross output, 80 MBH	Q-7	1.46	21.918	Ea.	1,950	910		2,860	3,625
3020	100 MBH		1.35	23.704		2,275	985		3,260	4,100
3023	Install medium efficiency propane boiler, 29.3 kW G	Q-5	.68	23.704		2,275	985		3,260	4,100
3025	Install high efficiency gas boiler, 29.3 kW G		.68	23.704		2,275	985		3,260	4,100
3027	Install high efficiency propane boiler, 29.3 kW G		.68	23.704		2,275	985		3,260	4,100
3080	203 MBH	Q-7	1	32		3,725	1,325		5,050	6,275
3180	400 MBH		.64	50		5,150	2,075		7,225	9,075
3280	1,275 MBH		.36	89.888		15,000	3,725		18,725	22,600
3340	2,312 MBH		.22	148		25,800	6,150		31,950	38,500
3400	3,996 MBH		.16	195		34,400	8,100		42,500	51,000
7000	For tankless water heater, add					10%				

23 52 23.40 Oil-Fired Boilers		Crew	Daily Output	Labor-Hours	Unit	Material	Labor	Equipment	Total	Total Incl O&P
0010	**OIL-FIRED BOILERS**, Standard controls, flame retention burner, packaged									
1000	Cast iron, with insulated flush jacket									
2000	Steam, gross output, 109 MBH	Q-7	1.20	26.667	Ea.	2,225	1,100		3,325	4,275
2060	207 MBH		.90	35.556		3,025	1,475		4,500	5,750
2180	1,084 MBH		.38	85.106		10,700	3,525		14,225	17,500
3000	Hot water, same price as steam									

23 52 26 – Steel Boilers

23 52 26.40 Oil-Fired Boilers		Crew	Daily Output	Labor-Hours	Unit	Material	Labor	Equipment	Total	Total Incl O&P
0010	**OIL-FIRED BOILERS**, Standard controls, flame retention burner									
5000	Steel, with insulated flush jacket									
7000	Hot water, gross output, 103 MBH	Q-6	1.60	15	Ea.	1,875	600		2,475	3,025
7020	122 MBH		1.45	16.506		2,025	660		2,685	3,300
7060	168 MBH		1.30	18.405		2,275	735		3,010	3,700
7080	225 MBH		1.22	19.704		3,525	785		4,310	5,175

23 52 26.70 Packaged Water Tube Boilers		Crew	Daily Output	Labor-Hours	Unit	Material	Labor	Equipment	Total	Total Incl O&P
0010	**PACKAGED WATER TUBE BOILERS**									
2000	Packaged water tube, #2 oil, steam or hot water, gross output									
2040	1,200 MBH	Q-7	.50	64	Ea.	25,000	2,650		27,650	31,900
2060	1,600 MBH		.40	80		40,000	3,325		43,325	49,500
2080	2,400 MBH		.30	107		54,000	4,425		58,425	66,500
2100	3,200 MBH		.25	128		61,500	5,300		66,800	76,000
2120	4,800 MBH		.20	160		80,500	6,625		87,125	99,500
2200	Gas fired									
2204	200 MBH	Q-6	1.50	16	Ea.	8,600	640		9,240	10,500
2208	275 MBH		1.40	17.143		9,000	685		9,685	11,000
2212	360 MBH		1.10	21.818		9,225	870		10,095	11,600
2216	520 MBH		.65	36.923		8,475	1,475		9,950	11,800
2220	600 MBH		.60	40		11,000	1,600		12,600	14,700
2224	720 MBH		.55	43.636		10,000	1,750		11,750	13,900
2228	960 MBH		.48	50		18,400	2,000		20,400	23,600
2232	1,220 MBH	Q-7	.50	64		18,900	2,650		21,550	25,200
2236	1,440 MBH		.45	71.111		20,100	2,950		23,050	26,900
2240	1,680 MBH		.40	80		29,700	3,325		33,025	38,200
2244	1,920 MBH		.35	91.429		28,700	3,800		32,500	37,700
2248	2,160 MBH		.33	96.970		30,400	4,025		34,425	40,100
2252	2,400 MBH		.30	107		34,200	4,425		38,625	44,900

23 52 Heating Boilers

23 52 28 – Swimming Pool Boilers

23 52 28.10 Swimming Pool Heaters		Crew	Daily Output	Labor-Hours	Unit	Material	2019 Bare Costs Labor	Equipment	Total	Total Incl O&P
0010	**SWIMMING POOL HEATERS**, Not including wiring, external									
0020	piping, base or pad									
0160	Gas fired, input, 155 MBH	Q-6	1.50	16	Ea.	1,975	640		2,615	3,225
0220	250 MBH		.70	34.286		2,325	1,375		3,700	4,800
0240	300 MBH		.60	40		2,425	1,600		4,025	5,300
0280	500 MBH		.40	60		8,625	2,400		11,025	13,400
0370	1,260 MBH		.21	114		17,500	4,550		22,050	26,800
0400	1,800 MBH		.14	171		20,800	6,850		27,650	34,100
2000	Electric, 12 KW, 4,800 gallon pool	Q-19	3	8		2,125	345		2,470	2,900
2020	15 KW, 7,200 gallon pool		2.80	8.571		2,150	365		2,515	2,950
2040	24 KW, 9,600 gallon pool		2.40	10		2,500	430		2,930	3,450
2100	57 KW, 24,000 gallon pool		1.20	20		3,700	855		4,555	5,450

23 52 39 – Fire-Tube Boilers

23 52 39.13 Scotch Marine Boilers		Crew	Daily Output	Labor-Hours	Unit	Material	Labor	Equipment	Total	Total Incl O&P
0010	**SCOTCH MARINE BOILERS**									
1000	Packaged fire tube, #2 oil, gross output									
1006	15 psi steam									
1020	3,348 MBH, 100 HP	Q-7	.21	152	Ea.	94,500	6,325		100,825	114,500
1120	To fire #6, add		.83	38.554		18,200	1,600		19,800	22,600
1140	To fire #6, and gas, add		.42	76.190		26,800	3,150		29,950	34,700

23 52 88 – Burners

23 52 88.10 Replacement Type Burners		Crew	Daily Output	Labor-Hours	Unit	Material	Labor	Equipment	Total	Total Incl O&P
0010	**REPLACEMENT TYPE BURNERS**									
0990	Residential, conversion, gas fired, LP or natural									
1000	Gun type, atmospheric input 50 to 225 MBH	Q-1	2.50	6.400	Ea.	1,000	258		1,258	1,525
1003	Replace burner		1.92	8.320		1,000	335		1,335	1,650
1020	100 to 400 MBH		2	8		1,675	325		2,000	2,375
1025	Burner, gas, 100 to 400 MBH G					1,675			1,675	1,850
1040	300 to 1,000 MBH	Q-1	1.70	9.412		5,050	380		5,430	6,175
1043	Replace burner G	″	1.30	12.298		5,050	495		5,545	6,375
3000	Flame retention oil fired assembly, input									
3020	.50 to 2.25 GPH	Q-1	2.40	6.667	Ea.	360	269		629	835
3040	2.0 to 5.0 GPH		2	8		340	325		665	900
3060	3.0 to 7.0 GPH		1.80	8.889		580	360		940	1,225
3080	6.0 to 12.0 GPH		1.60	10		930	405		1,335	1,700
4100	Replace burner fireye G	1 Stpi	26.67	.300		249	13.80		262.80	297
4500	Replace burner blower bearing G		4	2		65.50	92		157.50	223
4510	Replace conveyor bearing G		3.60	2.222		73.50	102		175.50	249
4520	Replace fan bearings G		3.40	2.353		189	108		297	385
4525	Fan bearing G					189			189	208
4600	Gas safety, shut off valve, 3/4″ threaded	1 Stpi	20	.400		187	18.40		205.40	236
4610	1″ threaded		19	.421		181	19.40		200.40	231
4620	1-1/4″ threaded		15	.533		204	24.50		228.50	265
4630	1-1/2″ threaded		13	.615		220	28.50		248.50	289
4640	2″ threaded		11	.727		247	33.50		280.50	325
4650	2-1/2″ threaded	Q-1	15	1.067		282	43		325	380
4660	3″ threaded		13	1.231		385	49.50		434.50	505
4670	4″ flanged		3	5.333		2,775	215		2,990	3,400
4680	6″ flanged	Q-2	3	8		6,175	310		6,485	7,300
4700	Replace burner ignition transformer G	1 Stpi	16	.500		99	23		122	147
4705	Burner ignition transformer G					99			99	109

23 52 Heating Boilers

23 52 88 – Burners

23 52 88.10 Replacement Type Burners			Crew	Daily Output	Labor-Hours	Unit	Material	Labor	Equipment	Total	Total Incl O&P
							2019 Bare Costs				
4710	Replace burner ignition electrode	G	1 Stpi	20	.400	Ea.	14.10	18.40		32.50	45.50
4715	Burner ignition electrode	G					14.10			14.10	15.50
4720	Replace burner oil pump	G	1 Stpi	11.43	.700		128	32		160	194
4725	Burner oil pump	G					128			128	141
4730	Replace burner nozzle	G	1 Stpi	26.66	.300		6.35	13.80		20.15	29.50
4735	Burner nozzle	G					6.35			6.35	6.95
4750	Replace burner ignition transformer	G	1 Stpi	14	.571		99	26.50		125.50	152
4755	Burner ignition transformer	G					99			99	109
4760	Replace burner ignition electrode	G	1 Stpi	20	.400		14.10	18.40		32.50	45.50
4765	Burner ignition electrode	G					14.10			14.10	15.50
4770	Replace burner oil pump	G	1 Stpi	10	.800		150	37		187	226
4775	Burner oil pump	G					150			150	165
4780	Replace burner nozzle	G	1 Stpi	24	.333		9.10	15.35		24.45	35
4785	Burner nozzle	G					9.10			9.10	10.05
4800	Replace burner ignition transformer	G	1 Stpi	12	.667		99	30.50		129.50	160
4805	Burner ignition transformer	G					99			99	109
4810	Replace burner ignition electrode	G	1 Stpi	18	.444		14.15	20.50		34.65	49
4815	Burner ignition electrode	G					14.15			14.15	15.55
4820	Replace burner oil pump	G	1 Stpi	8	1		128	46		174	216
4825	Burner oil pump	G					128			128	140
4830	Replace burner nozzle	G	1 Stpi	20	.400		9.10	18.40		27.50	40
4835	Burner nozzle	G					9.10			9.10	10.05
4850	Replace burner oil pump	G	Q-5	6.15	2.602		195	108		303	390
4855	Burner oil pump	G					195			195	215
4860	Replace burner nozzle	G	1 Stpi	14	.571		9.10	26.50		35.60	53
4865	Burner nozzle	G					9.10			9.10	10.05
4880	Replace spreader	G	4 Stpi	.25	130		7,600	6,000		13,600	18,200
4885	Pneumatic coal spreader	G					7,600			7,600	8,375
4900	Replace pump seals/bearings	G	1 Stpi	5.33	1.500		31.50	69		100.50	148
4905	Pump seals/bearings	G					31.50			31.50	35

23 54 Furnaces

23 54 13 – Electric-Resistance Furnaces

23 54 13.10 Electric Furnaces		Crew	Daily Output	Labor-Hours	Unit	Material	Labor	Equipment	Total	Total Incl O&P
0010	**ELECTRIC FURNACES**, Hot air, blowers, std. controls									
0011	not including gas, oil or flue piping									

23 54 16 – Fuel-Fired Furnaces

23 54 16.13 Gas-Fired Furnaces		Crew	Daily Output	Labor-Hours	Unit	Material	Labor	Equipment	Total	Total Incl O&P
0010	**GAS-FIRED FURNACES**									
3000	Gas, AGA certified, upflow, direct drive models									
3040	60 MBH input	Q-9	3.80	4.211	Ea.	695	164		859	1,025
3100	100 MBH input		3.20	5		785	195		980	1,175
3130	150 MBH input		2.80	5.714		1,825	223		2,048	2,375
3140	200 MBH input		2.60	6.154		3,400	240		3,640	4,150
4000	For starter plenum, add		16	1		98	39		137	173

23 54 16.16 Oil-Fired Furnaces		Crew	Daily Output	Labor-Hours	Unit	Material	Labor	Equipment	Total	Total Incl O&P
0010	**OIL-FIRED FURNACES**									
6000	Oil, UL listed, atomizing gun type burner									
6020	56 MBH output	Q-9	3.60	4.444	Ea.	3,125	173		3,298	3,700
6040	95 MBH output		3.40	4.706		2,900	183		3,083	3,475

23 54 Furnaces

23 54 16 – Fuel-Fired Furnaces

23 54 16.16 Oil-Fired Furnaces		Crew	Daily Output	Labor-Hours	Unit	Material	Labor	Equipment	Total	Total Incl O&P
						2019 Bare Costs				
6080	151 MBH output	Q-9	3	5.333	Ea.	3,225	208		3,433	3,900
6100	200 MBH input	↓	2.60	6.154	↓	3,775	240		4,015	4,550

23 55 Fuel-Fired Heaters

23 55 13 – Fuel-Fired Duct Heaters

23 55 13.16 Gas-Fired Duct Heaters

Line	Description	Crew	Daily Output	Labor-Hours	Unit	Material	Labor	Equipment	Total	Total Incl O&P
0010	**GAS-FIRED DUCT HEATERS**, Includes burner, controls, stainless steel									
0020	heat exchanger. Gas fired, electric ignition									
0030	Indoor installation									
0080	100 MBH output	Q-5	5	3.200	Ea.	3,050	133		3,183	3,575
0100	120 MBH output		4	4		3,350	166		3,516	3,950
0130	200 MBH output		2.70	5.926		4,250	246		4,496	5,075
0140	240 MBH output		2.30	6.957		4,425	288		4,713	5,325
0160	280 MBH output		2	8		4,850	330		5,180	5,900
0180	320 MBH output	↓	1.60	10		5,200	415		5,615	6,400
0300	For powered venter and adapter, add				↓	535			535	585

23 55 23 – Gas-Fired Radiant Heaters

23 55 23.10 Infrared Type Heating Units

Line	Description	Crew	Daily Output	Labor-Hours	Unit	Material	Labor	Equipment	Total	Total Incl O&P
0010	**INFRARED TYPE HEATING UNITS**									
0020	Gas fired, unvented, electric ignition, 100% shutoff.									
0030	Piping and wiring not included									
0120	45 MBH	Q-5	5	3.200	Ea.	995	133		1,128	1,325
0160	60 MBH		4	4		1,025	166		1,191	1,400
0180	75 MBH		3	5.333		1,025	221		1,246	1,500
0220	105 MBH	↓	2	8	↓	1,050	330		1,380	1,700
1000	Gas fired, vented, electric ignition, tubular									
1020	Piping and wiring not included, 20′ to 80′ lengths									
1030	Single stage, input, 60 MBH	Q-6	4.50	5.333	Ea.	1,500	213		1,713	2,000
1040	80 MBH		3.90	6.154		1,500	246		1,746	2,050
1050	100 MBH		3.40	7.059		1,500	282		1,782	2,100
1060	125 MBH		2.90	8.276		1,500	330		1,830	2,200
1070	150 MBH		2.70	8.889		1,500	355		1,855	2,225
1080	170 MBH		2.50	9.600		1,500	385		1,885	2,275
1090	200 MBH	↓	2.20	10.909	↓	1,700	435		2,135	2,600
1100	Note: Final pricing may vary due to									
1110	tube length and configuration package selected									
1130	Two stage, input, 60 MBH high, 45 MBH low	Q-6	4.50	5.333	Ea.	1,825	213		2,038	2,375
1140	80 MBH high, 60 MBH low		3.90	6.154		1,825	246		2,071	2,425
1150	100 MBH high, 65 MBH low		3.40	7.059		1,825	282		2,107	2,475
1160	125 MBH high, 95 MBH low		2.90	8.276		1,850	330		2,180	2,575
1170	150 MBH high, 100 MBH low		2.70	8.889		1,850	355		2,205	2,600
1180	170 MBH high, 125 MBH low		2.50	9.600		2,050	385		2,435	2,875
1190	200 MBH high, 150 MBH low	↓	2.20	10.909	↓	2,050	435		2,485	2,975
1220	Note: Final pricing may vary due to									
1230	tube length and configuration package selected									

23 55 Fuel-Fired Heaters

23 55 33 – Fuel-Fired Unit Heaters

23 55 33.16 Gas-Fired Unit Heaters		Crew	Daily Output	Labor-Hours	Unit	2019 Bare Costs Material	Labor	Equipment	Total	Total Incl O&P
0010	**GAS-FIRED UNIT HEATERS**, Cabinet, grilles, fan, ctrls., burner, no piping									
0022	thermostat, no piping. For flue see Section 23 51 23.10									
1000	Gas fired, floor mounted									
1100	60 MBH output	Q-5	10	1.600	Ea.	720	66.50		786.50	905
1140	100 MBH output		8	2		795	83		878	1,000
1180	180 MBH output		6	2.667		1,150	111		1,261	1,425
1500	Rooftop mounted, power vent, stainless steel exchanger									
1520	75 MBH input	Q-6	4	6	Ea.	5,900	240		6,140	6,900
1560	125 MBH input		3.30	7.273		6,425	290		6,715	7,550
1600	175 MBH input		2.60	9.231		7,100	370		7,470	8,400
1620	225 MBH input		2.30	10.435		8,100	415		8,515	9,575
1640	300 MBH input		1.90	12.632		9,350	505		9,855	11,100
1660	350 MBH input		1.40	17.143		10,000	685		10,685	12,100
1720	700 MBH input		.80	30		13,000	1,200		14,200	16,300
1760	1,200 MBH input		.30	80		13,800	3,200		17,000	20,400
2000	Suspension mounted, propeller fan, 20 MBH output	Q-5	8.50	1.882		1,725	78		1,803	2,000
2100	130 MBH output		5	3.200		2,350	133		2,483	2,825
2240	320 MBH output		2	8		4,375	330		4,705	5,375
2500	For powered venter and adapter, add					400			400	440
5000	Wall furnace, 17.5 MBH output	Q-5	6	2.667		630	111		741	875
5020	24 MBH output		5	3.200		690	133		823	975
5040	35 MBH output		4	4		990	166		1,156	1,350

23 56 Solar Energy Heating Equipment

23 56 16 – Packaged Solar Heating Equipment

23 56 16.40 Solar Heating Systems		Crew	Daily Output	Labor-Hours	Unit	Material	Labor	Equipment	Total	Total Incl O&P
0010	**SOLAR HEATING SYSTEMS**									
0020	System/package prices, not including connecting									
0030	pipe, insulation, or special heating/plumbing fixtures									
0500	Hot water, standard package, low temperature									
0540	1 collector, circulator, fittings, 65 gal. tank G	Q-1	.50	32	Ea.	4,050	1,300		5,350	6,575
0580	2 collectors, circulator, fittings, 120 gal. tank G	"	.40	40	"	5,350	1,625		6,975	8,525
0701	Solar energy, medium temperature package									
0720	1 collector, circulator, fittings, 80 gal. tank G	Q-1	.50	32	Ea.	5,300	1,300		6,600	7,950
0740	2 collectors, circulator, fittings, 120 gal. tank G		.40	40		6,975	1,625		8,600	10,300
0780	3 collectors, circulator, fittings, 120 gal. tank G		.30	53.333		7,900	2,150		10,050	12,200
0980	For each additional 120 gal. tank, add G					1,875			1,875	2,075

23 62 Packaged Compressor and Condenser Units

23 62 13 – Packaged Air-Cooled Refrigerant Compressor and Condenser Units

23 62 13.10 Packaged Air-Cooled Refrig. Condensing Units		Crew	Daily Output	Labor-Hours	Unit	Material	Labor	Equipment	Total	Total Incl O&P
0010	**PACKAGED AIR-COOLED REFRIGERANT CONDENSING UNITS**									
0020	Condensing unit									
0030	Air cooled, compressor, standard controls									
0050	1.5 ton	Q-5	2.50	6.400	Ea.	1,075	265		1,340	1,600
0100	2 ton		2.10	7.619		1,050	315		1,365	1,675
0200	2.5 ton		1.70	9.412		1,250	390		1,640	2,025
0300	3 ton		1.30	12.308		1,325	510		1,835	2,275
0400	4 ton		.90	17.778		1,775	735		2,510	3,150

23 62 Packaged Compressor and Condenser Units

23 62 13 – Packaged Air-Cooled Refrigerant Compressor and Condenser Units

23 62 13.10 Packaged Air-Cooled Refrig. Condensing Units		Crew	Daily Output	Labor-Hours	Unit	2019 Bare Costs Material	Labor	Equipment	Total	Total Incl O&P
0500	5 ton	Q-5	.60	26.667	Ea.	2,125	1,100		3,225	4,175
0560	8.5 ton	↓	.53	30.189	↓	3,500	1,250		4,750	5,900
0600	10 ton	↓	.50	32	↓	3,650	1,325		4,975	6,175

23 63 Refrigerant Condensers

23 63 13 – Air-Cooled Refrigerant Condensers

23 63 13.10 Air-Cooled Refrig. Condensers

		Crew	Daily Output	Labor-Hours	Unit	Material	Labor	Equipment	Total	Total Incl O&P
0010	**AIR-COOLED REFRIG. CONDENSERS**									
0080	Air cooled, belt drive, propeller fan									
0220	45 ton	Q-6	.70	34.286	Ea.	10,200	1,375		11,575	13,500
0240	50 ton	↓	.69	34.985	↓	11,000	1,400		12,400	14,300
0260	54 ton	↓	.64	37.795	↓	11,900	1,500		13,400	15,600

23 64 Packaged Water Chillers

23 64 13 – Absorption Water Chillers

23 64 13.13 Direct-Fired Absorption Water Chillers

		Crew	Daily Output	Labor-Hours	Unit	Material	Labor	Equipment	Total	Total Incl O&P
0010	**DIRECT-FIRED ABSORPTION WATER CHILLERS**									
3000	Gas fired, air cooled									
3270	10 ton	Q-5	.40	40	Ea.	22,700	1,650		24,350	27,700

23 64 13.16 Indirect-Fired Absorption Water Chillers

		Crew	Daily Output	Labor-Hours	Unit	Material	Labor	Equipment	Total	Total Incl O&P
0010	**INDIRECT-FIRED ABSORPTION WATER CHILLERS**									
0020	Steam or hot water, water cooled									
0050	100 ton	Q-7	.13	240	Ea.	132,500	9,950		142,450	162,500
0100	148 ton	"	.12	258	"	240,500	10,700		251,200	282,000

23 64 16 – Centrifugal Water Chillers

23 64 16.10 Centrifugal Type Water Chillers

		Crew	Daily Output	Labor-Hours	Unit	Material	Labor	Equipment	Total	Total Incl O&P
0010	**CENTRIFUGAL TYPE WATER CHILLERS**, With standard controls									
0274	Centrifugal, packaged unit, water cooled, not incl. tower									
0275	Chiller, centrifugal									
0278	200 ton	Q-7	.19	172	Ea.	132,500	7,125		139,625	157,000
0279	300 ton	↓	.14	234	↓	151,500	9,675		161,175	183,000
0280	400 ton	↓	.11	283	↓	154,500	11,700		166,200	189,500
0282	450 ton	↓	.11	291	↓	168,500	12,100		180,600	205,500
0290	500 ton	↓	.11	296	↓	189,500	12,300		201,800	228,000
0292	550 ton	↓	.11	305	↓	217,500	12,600		230,100	260,000
0300	600 ton	↓	.10	311	↓	232,500	12,900		245,400	277,000

23 64 19 – Reciprocating Water Chillers

23 64 19.10 Reciprocating Type Water Chillers

		Crew	Daily Output	Labor-Hours	Unit	Material	Labor	Equipment	Total	Total Incl O&P
0010	**RECIPROCATING TYPE WATER CHILLERS**, With standard controls									
0494	Water chillers, integral air cooled condenser									
0980	Water cooled, multiple compressor, semi-hermetic, tower not incl.									
1090	45 ton cooling	Q-7	.29	111	Ea.	25,100	4,600		29,700	35,200
1130	75 ton cooling	↓	.23	139	↓	46,600	5,775		52,375	61,000
1150	95 ton cooling	↓	.19	165	↓	52,500	6,825		59,325	69,000
1160	100 ton cooling	↓	.18	180	↓	50,500	7,450		57,950	67,500
4000	Packaged chiller, remote air cooled condensers not incl.									
4020	15 ton cooling	Q-7	.30	108	Ea.	20,500	4,475		24,975	29,900
4040	25 ton cooling	↓	.25	126	↓	23,200	5,225		28,425	34,100

23 64 Packaged Water Chillers

23 64 19 – Reciprocating Water Chillers

23 64 19.10 Reciprocating Type Water Chillers		Crew	Daily Output	Labor-Hours	Unit	2019 Bare Costs Material	Labor	Equipment	Total	Total Incl O&P
4050	35 ton cooling	Q-7	.24	134	Ea.	28,700	5,550		34,250	40,700
4066	45 ton cooling		.21	150		30,600	6,200		36,800	43,800
4070	50 ton cooling		.21	154		30,900	6,375		37,275	44,500
4080	60 ton cooling		.20	164		33,200	6,800		40,000	47,700

23 64 23 – Scroll Water Chillers

23 64 23.10 Scroll Water Chillers		Crew	Daily Output	Labor-Hours	Unit	Material	Labor	Equipment	Total	Total Incl O&P
0010	**SCROLL WATER CHILLERS**, With standard controls									
0480	Packaged w/integral air cooled condenser									
0482	10 ton cooling	Q-7	.34	94.118	Ea.	15,100	3,900		19,000	23,000
0490	15 ton cooling		.37	86.486		18,000	3,575		21,575	25,700
0500	20 ton cooling		.34	94.118		22,200	3,900		26,100	30,900
0510	25 ton cooling		.34	94.118		25,900	3,900		29,800	34,900

23 64 26 – Rotary-Screw Water Chillers

23 64 26.10 Rotary-Screw Type Water Chillers		Crew	Daily Output	Labor-Hours	Unit	Material	Labor	Equipment	Total	Total Incl O&P
0010	**ROTARY-SCREW TYPE WATER CHILLERS**, With standard controls									
0110	Screw, liquid chiller, air cooled, insulated evaporator									
0120	130 ton	Q-7	.14	229	Ea.	95,500	9,475		104,975	120,500
0124	160 ton		.13	246		122,500	10,200		132,700	151,500
0128	180 ton		.13	250		138,000	10,400		148,400	168,500
0132	210 ton		.12	258		146,000	10,700		156,700	178,000
0136	270 ton		.12	267		173,500	11,100		184,600	209,000
0140	320 ton		.12	276		217,500	11,400		228,900	258,000
0200	Packaged unit, water cooled, not incl. tower									
0240	200 ton	Q-7	.13	252	Ea.	83,500	10,400		93,900	109,000

23 65 Cooling Towers

23 65 13 – Forced-Draft Cooling Towers

23 65 13.10 Forced-Draft Type Cooling Towers		Crew	Daily Output	Labor-Hours	Unit	Material	Labor	Equipment	Total	Total Incl O&P
0010	**FORCED-DRAFT TYPE COOLING TOWERS**, Packaged units									
0070	Galvanized steel									
0080	Induced draft, crossflow									
0100	Vertical, belt drive, 61 tons	Q-6	90	.267	TonAC	198	10.65		208.65	235
0150	100 ton		100	.240		191	9.60		200.60	226
0200	115 ton		109	.220		166	8.80		174.80	197
1000	For higher capacities, use multiples									
5000	Fiberglass tower on galvanized steel support structure									
5010	Draw thru									
5100	100 ton	Q-6	1.40	17.143	Ea.	14,400	685		15,085	16,900
5120	120 ton		1.20	20		16,700	800		17,500	19,700
5140	140 ton		1	24		18,100	960		19,060	21,500
5160	160 ton		.80	30		20,000	1,200		21,200	24,000
5180	180 ton		.65	36.923		22,700	1,475		24,175	27,400
5251	600 ton		.16	150		76,500	6,000		82,500	94,000
5252	1,000 ton		.10	250		127,000	9,975		136,975	156,500
5300	For stainless steel support structure, add					30%				
5360	For higher capacities, use multiples of each size									
6000	Stainless steel									
6010	Induced draft, crossflow, horizontal, belt drive									
6100	57 ton	Q-6	1.50	16	Ea.	30,600	640		31,240	34,700
6120	91 ton		.99	24.242		37,400	970		38,370	42,800

23 65 Cooling Towers

23 65 13 – Forced-Draft Cooling Towers

23 65 13.10 Forced-Draft Type Cooling Towers		Crew	Daily Output	Labor-Hours	Unit	2019 Bare Costs Material	Labor	Equipment	Total	Total Incl O&P
6140	111 ton	Q-6	.43	55.814	Ea.	48,300	2,225		50,525	56,500
6160	126 ton		.22	109		50,500	4,350		54,850	62,500
6170	Induced draft, crossflow, vertical, gear drive									
6180	1,016 ton	Q-6	.15	160	Ea.	180,000	6,375		186,375	208,500

23 74 Packaged Outdoor HVAC Equipment

23 74 33 – Dedicated Outdoor-Air Units

23 74 33.10 Rooftop Air Conditioners

Code	Description	Crew	Daily Output	Labor-Hours	Unit	Material	Labor	Equipment	Total	Total Incl O&P
0010	**ROOFTOP AIR CONDITIONERS**, Standard controls, curb, economizer									
1000	Single zone, electric cool, gas heat									
1140	5 ton cooling, 112 MBH heating	Q-5	.56	28.521	Ea.	4,425	1,175		5,600	6,800
1150	7.5 ton cooling, 170 MBH heating		.50	32.258		5,825	1,325		7,150	8,625
1156	8.5 ton cooling, 170 MBH heating		.46	34.783		7,100	1,450		8,550	10,200
1160	10 ton cooling, 200 MBH heating	Q-6	.67	35.982		9,200	1,425		10,625	12,500
1180	15 ton cooling, 270 MBH heating	"	.57	42.032		13,100	1,675		14,775	17,200
1220	30 ton cooling, 540 MBH heating	Q-7	.47	68.376		35,700	2,825		38,525	44,000
2000	Multizone, electric cool, gas heat, economizer									
2102	15 ton cooling, 360 MBH heating SEER 14 G	Q-7	.61	52.545	Ea.	99,000	2,175		101,175	112,500
2122	20 ton cooling, 360 MBH heating SEER 14 G		.53	60.038		104,500	2,500		107,000	119,000
2142	25 ton cooling, 450 MBH heating SEER 14 G		.45	71.910		128,500	2,975		131,475	146,000
2162	30 ton cooling, 450 MBH heating SEER 14 G		.41	79.012		155,500	3,275		158,775	176,500
2182	35 ton cooling, 540 MBH heating SEER 14 G		.37	85.562		162,000	3,550		165,550	184,500
2202	40 ton cooling, 540 MBH heating SEER 14 G		.28	114		185,000	4,725		189,725	212,000
2222	70 ton cooling, 1,500 MBH heating SEER 14 G		.16	199		247,500	8,250		255,750	285,500
2242	80 ton cooling, 1,500 MBH heating SEER 14 G		.14	229		283,000	9,475		292,475	326,500
2262	90 ton cooling, 1,500 MBH heating SEER 14 G		.13	256		289,500	10,600		300,100	336,000
2282	105 ton cooling, 1,500 MBH heating SEER 14 G		.11	291		322,000	12,100		334,100	374,500

23 81 Decentralized Unitary HVAC Equipment

23 81 13 – Packaged Terminal Air-Conditioners

23 81 13.10 Packaged Cabinet Type Air-Conditioners

Code	Description	Crew	Daily Output	Labor-Hours	Unit	Material	Labor	Equipment	Total	Total Incl O&P
0010	**PACKAGED CABINET TYPE AIR-CONDITIONERS**, Cabinet, wall sleeve,									
0100	louver, electric heat, thermostat, manual changeover, 208 V									
0200	6,000 BTUH cooling, 8,800 BTU heat	Q-5	6	2.667	Ea.	705	111		816	955
0220	9,000 BTUH cooling, 13,900 BTU heat		5	3.200		970	133		1,103	1,300
0240	12,000 BTUH cooling, 13,900 BTU heat		4	4		1,475	166		1,641	1,900
0260	15,000 BTUH cooling, 13,900 BTU heat		3	5.333		1,450	221		1,671	1,975
0320	30,000 BTUH cooling, 10 KW heat		1.40	11.429		2,875	475		3,350	3,925
0340	36,000 BTUH cooling, 10 KW heat		1.25	12.800		2,875	530		3,405	4,025
0360	42,000 BTUH cooling, 10 KW heat		1	16		3,125	665		3,790	4,550
0380	48,000 BTUH cooling, 10 KW heat		.90	17.778		3,325	735		4,060	4,850

23 81 19 – Self-Contained Air-Conditioners

23 81 19.10 Window Unit Air Conditioners

Code	Description	Crew	Daily Output	Labor-Hours	Unit	Material	Labor	Equipment	Total	Total Incl O&P
0010	**WINDOW UNIT AIR CONDITIONERS**									
4000	Portable/window, 15 amp, 125 V grounded receptacle required									
4060	5,000 BTUH	1 Carp	8	1	Ea.	305	39		344	400
4340	6,000 BTUH		8	1		252	39		291	340
4480	8,000 BTUH		6	1.333		355	51.50		406.50	475
4500	10,000 BTUH		6	1.333		600	51.50		651.50	745

23 81 Decentralized Unitary HVAC Equipment

23 81 19 – Self-Contained Air-Conditioners

23 81 19.10 Window Unit Air Conditioners		Crew	Daily Output	Labor-Hours	Unit	Material	2019 Bare Costs Labor	Equipment	Total	Total Incl O&P
4520	12,000 BTUH	L-2	8	2	Ea.	1,950	68		2,018	2,275

23 81 19.20 Self-Contained Single Package

		Crew	Daily Output	Labor-Hours	Unit	Material	Labor	Equipment	Total	Total Incl O&P
0010	**SELF-CONTAINED SINGLE PACKAGE**									
0100	Air cooled, for free blow or duct, not incl. remote condenser									
0110	Constant volume									
0220	5 ton cooling	Q-6	1.20	20	Ea.	4,400	800		5,200	6,150
0221	Self-contained, air cooled, 5 ton cooling, labor only		1.20	20			800		800	1,300
0230	7.5 ton cooling	↓	.90	26.667		5,650	1,075		6,725	7,950
0240	10 ton cooling	Q-7	1	32		7,925	1,325		9,250	10,900
0250	15 ton cooling		.95	33.684		10,300	1,400		11,700	13,600
0260	20 ton cooling		.90	35.556		13,100	1,475		14,575	16,800
0270	25 ton cooling	↓	.85	37.647	↓	19,900	1,550		21,450	24,500
1000	Water cooled for free blow or duct, not including tower									
1010	Constant volume									
1160	20 ton cooling	Q-7	.80	40	Ea.	28,100	1,650		29,750	33,600
1170	25 ton cooling		.75	42.667		33,000	1,775		34,775	39,200
1180	30 ton cooling		.70	45.714		37,200	1,900		39,100	44,000
1200	40 ton cooling	↓	.40	80		42,700	3,325		46,025	52,500
1300	For hot water or steam heat coils, add				↓	12%	10%			

23 81 23 – Computer-Room Air-Conditioners

23 81 23.10 Computer Room Units

		Crew	Daily Output	Labor-Hours	Unit	Material	Labor	Equipment	Total	Total Incl O&P
0010	**COMPUTER ROOM UNITS**									
1000	Air cooled, includes remote condenser but not									
1020	interconnecting tubing or refrigerant									
1080	3 ton	Q-5	.50	32	Ea.	26,600	1,325		27,925	31,500
1120	5 ton		.45	35.556		28,400	1,475		29,875	33,700
1160	6 ton		.30	53.333		54,500	2,200		56,700	63,000
1200	8 ton		.27	59.259		53,000	2,450		55,450	62,500
1240	10 ton		.25	64		55,500	2,650		58,150	65,500
1280	15 ton		.22	72.727		61,000	3,025		64,025	72,000
1290	18 ton	↓	.20	80		70,500	3,325		73,825	83,000
1300	20 ton	Q-6	.26	92.308		73,000	3,675		76,675	86,500
1320	22 ton		.24	100		74,000	4,000		78,000	88,000
1360	30 ton	↓	.21	114	↓	91,000	4,550		95,550	108,000
2200	Chilled water, for connection to									
2220	existing chiller system of adequate capacity									
2260	5 ton	Q-5	.74	21.622	Ea.	19,200	895		20,095	22,700

23 81 29 – Variable Refrigerant Flow HVAC Systems

23 81 29.10 Heat Pump, Gas Driven

		Crew	Daily Output	Labor-Hours	Unit	Material	Labor	Equipment	Total	Total Incl O&P
0010	**HEAT PUMP, GAS DRIVEN**, Variable refrigerant volume (VRV) type									
0020	Not including interconnecting tubing or multi-zone controls									
1000	For indoor fan VRV type AHU see 23 82 19.40									
1010	Multi-zone split									
1020	Outdoor unit									
1100	8 tons cooling, for up to 17 zones	Q-5	1.30	12.308	Ea.	29,700	510		30,210	33,400
1110	Isolation rails		2.60	6.154	Pair	1,050	255		1,305	1,575
1160	15 tons cooling, for up to 33 zones		1	16	Ea.	35,900	665		36,565	40,600
1170	Isolation rails	↓	2.60	6.154	Pair	1,225	255		1,480	1,775
2000	Packaged unit									
2020	Outdoor unit									
2200	11 tons cooling	Q-5	1.30	12.308	Ea.	38,200	510		38,710	42,800

23 81 Decentralized Unitary HVAC Equipment

23 81 29 – Variable Refrigerant Flow HVAC Systems

23 81 29.10 Heat Pump, Gas Driven		Crew	Daily Output	Labor-Hours	Unit	Material	2019 Bare Costs Labor	Equipment	Total	Total Incl O&P
2210	Roof curb adapter	Q-5	2.60	6.154	Ea.	660	255		915	1,150
2220	Thermostat	1 Stpi	1.30	6.154	↓	395	283		678	900

23 81 43 – Air-Source Unitary Heat Pumps

23 81 43.10 Air-Source Heat Pumps		Crew	Daily Output	Labor-Hours	Unit	Material	Labor	Equipment	Total	Total Incl O&P
0010	**AIR-SOURCE HEAT PUMPS**, Not including interconnecting tubing									
1000	Air to air, split system, not including curbs, pads, fan coil and ductwork									
1012	Outside condensing unit only, for fan coil see Section 23 82 19.10									
1060	5 ton cooling, 27 MBH heat @ 0°F	Q-5	.50	32	Ea.	2,700	1,325		4,025	5,150
1080	7.5 ton cooling, 33 MBH heat @ 0°F	″	.45	35.556		4,000	1,475		5,475	6,825
1100	10 ton cooling, 50 MBH heat @ 0°F	Q-6	.64	37.500		6,350	1,500		7,850	9,425
1120	15 ton cooling, 64 MBH heat @ 0°F		.50	48		9,650	1,925		11,575	13,800
1130	20 ton cooling, 85 MBH heat @ 0°F		.35	68.571		18,100	2,725		20,825	24,400
1140	25 ton cooling, 119 MBH heat @ 0°F	↓	.25	96	↓	21,500	3,825		25,325	30,000
6000	Air to water, single package, excluding storage tank and ductwork									
6010	Includes circulating water pump, air duct connections, digital temperature									
6020	controller with remote tank temp. probe and sensor for storage tank.									
6040	Water heating - air cooling capacity									

23 81 46 – Water-Source Unitary Heat Pumps

23 81 46.10 Water Source Heat Pumps		Crew	Daily Output	Labor-Hours	Unit	Material	Labor	Equipment	Total	Total Incl O&P
0010	**WATER SOURCE HEAT PUMPS**, Not incl. connecting tubing or water source									
2000	Water source to air, single package									
2220	5 ton cooling, 29 MBH heat @ 75°F	Q-5	.90	17.778	Ea.	3,375	735		4,110	4,900
2240	7.5 ton cooling, 35 MBH heat @ 75°F		.60	26.667		6,850	1,100		7,950	9,375
2260	10 ton cooling, 50 MBH heat @ 75°F	↓	.53	30.189		7,675	1,250		8,925	10,500
2280	15 ton cooling, 64 MBH heat @ 75°F	Q-6	.47	51.064		14,800	2,050		16,850	19,700
2300	20 ton cooling, 100 MBH heat @ 75°F		.41	58.537		16,100	2,325		18,425	21,500
2310	25 ton cooling, 100 MBH heat @ 75°F		.32	75		22,000	3,000		25,000	29,100
2320	30 ton cooling, 128 MBH heat @ 75°F		.24	102		24,100	4,075		28,175	33,200
2340	40 ton cooling, 200 MBH heat @ 75°F		.21	117		34,000	4,675		38,675	45,100
2360	50 ton cooling, 200 MBH heat @ 75°F	↓	.15	160		38,400	6,375		44,775	52,500
3960	For supplementary heat coil, add				↓	10%				

23 82 Convection Heating and Cooling Units

23 82 19 – Fan Coil Units

23 82 19.10 Fan Coil Air Conditioning		Crew	Daily Output	Labor-Hours	Unit	Material	Labor	Equipment	Total	Total Incl O&P
0010	**FAN COIL AIR CONDITIONING**									
0030	Fan coil AC, cabinet mounted, filters and controls									
0100	Chilled water, 1/2 ton cooling	Q-5	8	2	Ea.	585	83		668	780
0120	1 ton cooling		6	2.667		855	111		966	1,125
0180	3 ton cooling	↓	4	4	↓	1,975	166		2,141	2,450
0262	For hot water coil, add					40%	10%			
0320	1 ton cooling	Q-5	6	2.667	Ea.	1,675	111		1,786	2,025
0940	Direct expansion, for use w/air cooled condensing unit, 1-1/2 ton cooling		5	3.200		650	133		783	930
1000	5 ton cooling	↓	3	5.333		1,125	221		1,346	1,625
1500	For hot water coil, add				↓	40%	10%			
1512	For condensing unit add see Section 23 62									

23 82 Convection Heating and Cooling Units

23 82 19 – Fan Coil Units

23 82 19.20 Heating and Ventilating Units		Crew	Daily Output	Labor-Hours	Unit	2019 Bare Costs Material	Labor	Equipment	Total	Total Incl O&P
0010	**HEATING AND VENTILATING UNITS**, Classroom units									
0020	Includes filter, heating/cooling coils, standard controls									
0080	750 CFM, 2 tons cooling	Q-6	2	12	Ea.	4,575	480		5,055	5,825
0100	1,000 CFM, 2-1/2 tons cooling		1.60	15		5,375	600		5,975	6,900
0120	1,250 CFM, 3 tons cooling		1.40	17.143		5,575	685		6,260	7,250
0140	1,500 CFM, 4 tons cooling	↓	.80	30		5,950	1,200		7,150	8,525
0500	For electric heat, add					35%				
1000	For no cooling, deduct				↓	25%	10%			
23 82 19.40 Fan Coil Air Conditioning										
0010	**FAN COIL AIR CONDITIONING**, Variable refrigerant volume (VRV) type									
0020	Not including interconnecting tubing or multi-zone controls									
0030	For VRV condensing unit see section 23 81 29.10									
0050	Indoor type, ducted									
0100	Vertical concealed									
0130	1 ton cooling G	Q-5	2.97	5.387	Ea.	2,550	223		2,773	3,175
0140	1.5 ton cooling G		2.77	5.776		2,600	239		2,839	3,250
0150	2 ton cooling G		2.70	5.926		2,800	246		3,046	3,475
0160	2.5 ton cooling G		2.60	6.154		2,950	255		3,205	3,675
0170	3 ton cooling G		2.50	6.400		3,000	265		3,265	3,725
0180	3.5 ton cooling G		2.30	6.957		3,150	288		3,438	3,950
0190	4 ton cooling G		2.19	7.306		3,175	305		3,480	4,000
0200	4.5 ton cooling G	↓	2.08	7.692	↓	3,500	320		3,820	4,375
0230	Outside air connection possible									
1100	Ceiling concealed									
1130	0.6 ton cooling G	Q-5	2.60	6.154	Ea.	1,675	255		1,930	2,250
1140	0.75 ton cooling G		2.60	6.154		1,725	255		1,980	2,325
1150	1 ton cooling G	↓	2.60	6.154		1,825	255		2,080	2,425
1152	Screening door G	1 Stpi	5.20	1.538		102	71		173	228
1160	1.5 ton cooling G	Q-5	2.60	6.154		1,875	255		2,130	2,500
1162	Screening door G	1 Stpi	5.20	1.538		118	71		189	246
1170	2 ton cooling G	Q-5	2.60	6.154		2,225	255		2,480	2,875
1172	Screening door G	1 Stpi	5.20	1.538		102	71		173	228
1180	2.5 ton cooling G	Q-5	2.60	6.154		2,575	255		2,830	3,250
1182	Screening door G	1 Stpi	5.20	1.538		177	71		248	310
1190	3 ton cooling G	Q-5	2.60	6.154		2,725	255		2,980	3,425
1192	Screening door G	1 Stpi	5.20	1.538		177	71		248	310
1200	4 ton cooling G	Q-5	2.30	6.957		2,875	288		3,163	3,625
1202	Screening door G	1 Stpi	5.20	1.538		177	71		248	310
1220	6 ton cooling G	Q-5	2.08	7.692		4,750	320		5,070	5,750
1230	8 ton cooling G	"	1.89	8.466	↓	5,350	350		5,700	6,450
1260	Outside air connection possible									
4050	Indoor type, duct-free									
4100	Ceiling mounted cassette									
4130	0.75 ton cooling G	Q-5	3.46	4.624	Ea.	1,975	192		2,167	2,500
4140	1 ton cooling G		2.97	5.387		2,100	223		2,323	2,700
4150	1.5 ton cooling G		2.77	5.776		2,150	239		2,389	2,775
4160	2 ton cooling G		2.60	6.154		2,300	255		2,555	2,950
4170	2.5 ton cooling G		2.50	6.400		2,350	265		2,615	3,025
4180	3 ton cooling G		2.30	6.957		2,525	288		2,813	3,250
4190	4 ton cooling G	↓	2.08	7.692		2,825	320		3,145	3,650
4198	For ceiling cassette decoration panel, add G	1 Stpi	5.20	1.538	↓	305	71		376	450
4230	Outside air connection possible									

23 82 Convection Heating and Cooling Units

23 82 19 – Fan Coil Units

23 82 19.40 Fan Coil Air Conditioning			Crew	Daily Output	Labor-Hours	Unit	Material	Labor	Equipment	Total	Total Incl O&P
							2019 Bare Costs				
4500	Wall mounted										
4530	0.6 ton cooling	G	Q-5	5.20	3.077	Ea.	1,025	128		1,153	1,325
4540	0.75 ton cooling	G		5.20	3.077		1,075	128		1,203	1,375
4550	1 ton cooling	G		4.16	3.846		1,225	159		1,384	1,600
4560	1.5 ton cooling	G		3.77	4.244		1,350	176		1,526	1,775
4570	2 ton cooling	G		3.46	4.624		1,450	192		1,642	1,925
4590	Condensate pump for the above air handlers	G	1 Stpi	6.46	1.238		219	57		276	335
4700	Floor standing unit										
4730	1 ton cooling	G	Q-5	2.60	6.154	Ea.	1,725	255		1,980	2,325
4740	1.5 ton cooling	G		2.60	6.154		1,950	255		2,205	2,575
4750	2 ton cooling	G		2.60	6.154		2,100	255		2,355	2,725
4800	Floor standing unit, concealed										
4830	1 ton cooling	G	Q-5	2.60	6.154	Ea.	1,650	255		1,905	2,250
4840	1.5 ton cooling	G		2.60	6.154		1,875	255		2,130	2,475
4850	2 ton cooling	G		2.60	6.154		2,000	255		2,255	2,625
4880	Outside air connection possible										
8000	Accessories										
8100	Branch divergence pipe fitting										
8110	Capacity under 76 MBH		1 Stpi	2.50	3.200	Ea.	155	147		302	415
8120	76 MBH to 112 MBH			2.50	3.200		183	147		330	445
8130	112 MBH to 234 MBH			2.50	3.200		520	147		667	810
8200	Header pipe fitting										
8210	Max 4 branches										
8220	Capacity under 76 MBH		1 Stpi	1.70	4.706	Ea.	228	217		445	605
8260	Max 8 branches										
8270	76 MBH to 112 MBH		1 Stpi	1.70	4.706	Ea.	465	217		682	865
8280	112 MBH to 234 MBH		"	1.30	6.154	"	740	283		1,023	1,275

23 82 29 – Radiators

23 82 29.10 Hydronic Heating		Crew	Daily Output	Labor-Hours	Unit	Material	Labor	Equipment	Total	Total Incl O&P
0010	**HYDRONIC HEATING**, Terminal units, not incl. main supply pipe									
1000	Radiation									
1100	Panel, baseboard, C.I., including supports, no covers	Q-5	46	.348	L.F.	46	14.40		60.40	74

23 82 36 – Finned-Tube Radiation Heaters

23 82 36.10 Finned Tube Radiation		Crew	Daily Output	Labor-Hours	Unit	Material	Labor	Equipment	Total	Total Incl O&P
0010	**FINNED TUBE RADIATION**, Terminal units, not incl. main supply pipe									
1150	Fin tube, wall hung, 14" slope top cover, with damper									
1200	1-1/4" copper tube, 4-1/4" alum. fin	Q-5	38	.421	L.F.	49.50	17.45		66.95	82.50
1255	2" steel tube, 4-1/4" steel fin		32	.500		45.50	20.50		66	84
1310	Baseboard, pkgd, 1/2" copper tube, alum. fin, 7" high		60	.267		11.75	11.05		22.80	31
1320	3/4" copper tube, alum. fin, 7" high		58	.276		8.10	11.45		19.55	27.50
1340	1" copper tube, alum. fin, 8-7/8" high		56	.286		21.50	11.85		33.35	43
1360	1-1/4" copper tube, alum. fin, 8-7/8" high		54	.296		32	12.30		44.30	55
1500	Note: fin tube may also require corners, caps, etc.									

23 82 39 – Unit Heaters

23 82 39.16 Propeller Unit Heaters		Crew	Daily Output	Labor-Hours	Unit	Material	Labor	Equipment	Total	Total Incl O&P
0010	**PROPELLER UNIT HEATERS**									
3950	Unit heaters, propeller, 115 V 2 psi steam, 60°F entering air									
4000	Horizontal, 12 MBH	Q-5	12	1.333	Ea.	400	55.50		455.50	525
4060	43.9 MBH		8	2		600	83		683	800
4240	286.9 MBH		2	8		1,775	330		2,105	2,500
4260	364 MBH		1.80	8.889		2,250	370		2,620	3,050

23 82 Convection Heating and Cooling Units

23 82 39 – Unit Heaters

23 82 39.16 Propeller Unit Heaters		Crew	Daily Output	Labor-Hours	Unit	Material	Labor	Equipment	Total	Total Incl O&P
4270	404 MBH	Q-5	1.60	10	Ea.	2,350	415		2,765	3,250
4300	Vertical diffuser same price									
4310	Vertical flow, 40 MBH	Q-5	11	1.455	Ea.	605	60.50		665.50	765
4326	131 MBH	"	4	4		895	166		1,061	1,250
4354	420 MBH (460 V)	Q-6	1.80	13.333		2,300	530		2,830	3,400
4358	500 MBH (460 V)		1.71	14.035		3,050	560		3,610	4,300
4362	570 MBH (460 V)		1.40	17.143		4,175	685		4,860	5,725
4366	620 MBH (460 V)		1.30	18.462		4,350	735		5,085	6,000
4370	960 MBH (460 V)	↓	1.10	21.818	↓	8,125	870		8,995	10,400

23 83 Radiant Heating Units

23 83 33 – Electric Radiant Heaters

23 83 33.10 Electric Heating		Crew	Daily Output	Labor-Hours	Unit	Material	Labor	Equipment	Total	Total Incl O&P
0010	**ELECTRIC HEATING**, not incl. conduit or feed wiring									
1100	Rule of thumb: Baseboard units, including control	1 Elec	4.40	1.818	kW	112	83		195	258
1300	Baseboard heaters, 2′ long, 350 watt		8	1	Ea.	27.50	45.50		73	105
1600	4′ long, 1,000 watt		6.70	1.194		39	54.50		93.50	132
2000	6′ long, 1,500 watt		5	1.600		51.50	73		124.50	176
2800	10′ long, 1,875 watt		3.30	2.424		154	111		265	350
5000	Radiant heating ceiling panels, 2′ x 4′, 500 watt		16	.500		395	23		418	470
5050	750 watt		16	.500		500	23		523	585
5300	Infrared quartz heaters, 120 volts, 1,000 watt		6.70	1.194		345	54.50		399.50	470
5350	1,500 watt		5	1.600		345	73		418	500
5400	240 volts, 1,500 watt		5	1.600		395	73		468	555
5450	2,000 watt	↓	4	2	↓	345	91		436	530

23 84 Humidity Control Equipment

23 84 13 – Humidifiers

23 84 13.10 Humidifier Units		Crew	Daily Output	Labor-Hours	Unit	Material	Labor	Equipment	Total	Total Incl O&P
0010	**HUMIDIFIER UNITS**									
0520	Steam, room or duct, filter, regulators, auto. controls, 220 V									
0580	33 lb./hr.	Q-5	4	4	Ea.	3,100	166		3,266	3,700
0620	100 lb./hr.	"	3	5.333	"	5,100	221		5,321	6,000

Division 26 Electrical

Estimating Tips

26 05 00 Common Work Results for Electrical

- Conduit should be taken off in three main categories—power distribution, branch power, and branch lighting—so the estimator can concentrate on systems and components, therefore making it easier to ensure all items have been accounted for.
- For cost modifications for elevated conduit installation, add the percentages to labor according to the height of installation and only to the quantities exceeding the different height levels, not to the total conduit quantities. Refer to subdivision 26 01 02.20 for labor adjustment factors.
- Remember that aluminum wiring of equal ampacity is larger in diameter than copper and may require larger conduit.
- If more than three wires at a time are being pulled, deduct percentages from the labor hours of that grouping of wires.
- When taking off grounding systems, identify separately the type and size of wire, and list each unique type of ground connection.
- The estimator should take the weights of materials into consideration when completing a takeoff. Topics to consider include: How will the materials be supported? What methods of support are available? How high will the support structure have to reach? Will the final support structure be able to withstand the total burden? Is the support material included or separate from the fixture, equipment, and material specified?
- Do not overlook the costs for equipment used in the installation. If scaffolding or highlifts are available in the field, contractors may use them in lieu of the proposed ladders and rolling staging.

26 20 00 Low-Voltage Electrical Transmission

- Supports and concrete pads may be shown on drawings for the larger equipment, or the support system may be only a piece of plywood for the back of a panelboard. In either case, they must be included in the costs.

26 40 00 Electrical and Cathodic Protection

- When taking off cathodic protection systems, identify the type and size of cable, and list each unique type of anode connection.

26 50 00 Lighting

- Fixtures should be taken off room by room using the fixture schedule, specifications, and the ceiling plan. For large concentrations of lighting fixtures in the same area, deduct the percentages from labor hours.

Reference Numbers

Reference numbers are shown at the beginning of some major classifications. These numbers refer to related items in the Reference Section. The reference information may be an estimating procedure, an alternate pricing method, or technical information.

Note: Not all subdivisions listed here necessarily appear. ■

Note: "Powered in part by CINX™, based on liscensed proprietary information of Harrison Publishing House, Inc."

Note: The following companies, in part, have been used as a reference source for some of the material prices used in Division 26:

Electriflex

Trade Service

26 01 02 – Labor Adjustment

	26 01 02.20 Labor Adjustment Factors	Crew	Daily Output	Labor-Hours	Unit	Material	2019 Bare Costs Labor	Equipment	Total	Total Incl O&P
0010	**LABOR ADJUSTMENT FACTORS** (For Div. 26, 27 and 28) R260519-90									
0100	Subtract from labor for Economy of Scale for Wire									
0110	4-5 wires						25%			
0120	6-10 wires						30%			
0130	11-15 wires						35%			
0140	over 15 wires						40%			
0150	Labor adjustment factors (For Div. 26, 27, 28 and 48)									
0200	Labor factors: The below are reasonable suggestions, but									
0210	each project must be evaluated for its own peculiarities, and									
0220	the adjustments be increased or decreased depending on the									
0230	severity of the special conditions.									
1000	Add to labor for elevated installation (above floor level)									
1010	10′ to 14.5′ high						10%			
1020	15′ to 19.5′ high						20%			
1030	20′ to 24.5′ high						25%			
1040	25′to 29.5′ high						35%			
1050	30′ to 34.5′ high						40%			
1060	35′ to 39.5′ high						50%			
1070	40′ and higher						55%			
2000	Add to labor for crawl space									
2010	3′ high						40%			
2020	4′ high						30%			
3000	Add to labor for multi-story building									
3100	For new construction (No elevator available)									
3110	Add for floors 3 thru 10						5%			
3120	Add for floors 11 thru 15						10%			
3130	Add for floors 16 thru 20						15%			
3140	Add for floors 21 thru 30						20%			
3150	Add for floors 31 and up						30%			
3200	For existing structure (Elevator available)									
3210	Add for work on floor 3 and above						2%			
4000	Add to labor for working in existing occupied buildings									
4010	Hospital						35%			
4020	Office building						25%			
4030	School						20%			
4040	Factory or warehouse						15%			
4050	Multi-dwelling						15%			
5000	Add to labor, miscellaneous									
5010	Cramped shaft						35%			
5020	Congested area						15%			
5030	Excessive heat or cold						30%			
6000	Labor factors: the above are reasonable suggestions, but									
6100	each project should be evaluated for its own peculiarities									
6200	Other factors to be considered are:									
6210	Movement of material and equipment through finished areas						10%			
6220	Equipment room min security direct access w/authorization						15%			
6230	Attic space						25%			
6240	No service road						25%			
6250	Poor unloading/storage area, no hydraulic lifts or jacks						20%			
6260	Congested site area/heavy traffic						20%			
7000	Correctional facilities (no compounding division 1 adjustment factors)									
7010	Minimum security w/facilities escort						30%			
7020	Medium security w/facilities and correctional officer escort						40%			

26 01 Operation and Maintenance of Electrical Systems

26 01 02 – Labor Adjustment

26 01 02.20 Labor Adjustment Factors		Crew	Daily Output	Labor-Hours	Unit	Material	2019 Bare Costs Labor	Equipment	Total	Total Incl O&P
7030	Max security w/facilities & correctional officer escort (no inmate contact)						50%			

26 05 Common Work Results for Electrical

26 05 05 – Selective Demolition for Electrical

26 05 05.10 Electrical Demolition

		Crew	Daily Output	Labor-Hours	Unit	Material	Labor	Equipment	Total	Total Incl O&P
0010	**ELECTRICAL DEMOLITION**									
0020	Electrical demolition, conduit to 10′ high, incl. fittings & hangers									
0100	Rigid galvanized steel, 1/2″ to 1″ diameter	1 Elec	242	.033	L.F.		1.51		1.51	2.46
0120	1-1/4″ to 2″	″	200	.040	″		1.82		1.82	2.98
0270	Armored cable (BX) avg. 50′ runs									
0280	#14, 2 wire	1 Elec	690	.012	L.F.		.53		.53	.86
0290	#14, 3 wire		571	.014			.64		.64	1.04
0300	#12, 2 wire		605	.013			.60		.60	.98
0310	#12, 3 wire		514	.016			.71		.71	1.16
0320	#10, 2 wire		514	.016			.71		.71	1.16
0330	#10, 3 wire		425	.019			.86		.86	1.40
0340	#8, 3 wire		342	.023			1.07		1.07	1.74
0350	Non metallic sheathed cable (Romex)									
0360	#14, 2 wire	1 Elec	720	.011	L.F.		.51		.51	.83
0370	#14, 3 wire		657	.012			.56		.56	.91
0380	#12, 2 wire		629	.013			.58		.58	.95
0390	#10, 3 wire		450	.018			.81		.81	1.32
0400	Wiremold raceway, including fittings & hangers									
0420	No. 3000	1 Elec	250	.032	L.F.		1.46		1.46	2.38
0440	No. 4000		217	.037			1.68		1.68	2.74
0460	No. 6000		166	.048			2.20		2.20	3.59
0462	Plugmold with receptacle		114	.070			3.20		3.20	5.20
0465	Telephone/power pole		12	.667	Ea.		30.50		30.50	49.50
0470	Non-metallic, straight section		480	.017	L.F.		.76		.76	1.24
0500	Channels, steel, including fittings & hangers									
0520	3/4″ x 1-1/2″	1 Elec	308	.026	L.F.		1.18		1.18	1.93
0540	1-1/2″ x 1-1/2″		269	.030			1.36		1.36	2.21
0560	1-1/2″ x 1-7/8″		229	.035			1.59		1.59	2.60
1210	Panel boards, incl. removal of all breakers,									
1220	conduit terminations & wire connections									
1720	Junction boxes, 4″ sq. & oct.	1 Elec	80	.100	Ea.		4.56		4.56	7.45
1740	Handy box		107	.075			3.41		3.41	5.55
1760	Switch box		107	.075			3.41		3.41	5.55
1780	Receptacle & switch plates		257	.031			1.42		1.42	2.32
1800	Wire, THW-THWN-THHN, removed from									
1810	in place conduit, to 10′ high									
1830	#14	1 Elec	65	.123	C.L.F.		5.60		5.60	9.15
1840	#12		55	.145			6.65		6.65	10.80
1850	#10		45.50	.176			8		8	13.10
2000	Interior fluorescent fixtures, incl. supports									
2010	& whips, to 10′ high									
2100	Recessed drop-in 2′ x 2′, 2 lamp	2 Elec	35	.457	Ea.		21		21	34
2110	2′ x 2′, 4 lamp		30	.533			24.50		24.50	39.50
2140	2′ x 4′, 4 lamp		30	.533			24.50		24.50	39.50
2180	Surface mount, acrylic lens & hinged frame									
2220	2′ x 2′, 2 lamp	2 Elec	44	.364	Ea.		16.60		16.60	27

26 05 Common Work Results for Electrical

26 05 05 – Selective Demolition for Electrical

26 05 05.10 Electrical Demolition		Crew	Daily Output	Labor-Hours	Unit	2019 Bare Costs Material	Labor	Equipment	Total	Total Incl O&P
2260	2′ x 4′, 4 lamp	2 Elec	33	.485	Ea.		22		22	36
2300	Strip fixtures, surface mount									
2320	4′ long, 1 lamp	2 Elec	53	.302	Ea.		13.75		13.75	22.50
2380	8′ long, 2 lamp	"	40	.400	"		18.25		18.25	30
2600	Exterior fixtures, incandescent, wall mount									
2620	100 Watt	2 Elec	50	.320	Ea.		14.60		14.60	24

26 05 19 – Low-Voltage Electrical Power Conductors and Cables

26 05 19.13 Undercarpet Electrical Power Cables

Line	Description	Crew	Daily Output	Labor-Hours	Unit	Material	Labor	Equipment	Total	Total Incl O&P
0010	UNDERCARPET ELECTRICAL POWER CABLES									
0020	Power System									
0100	Cable flat, 3 conductor, #12, w/attached bottom shield	1 Elec	982	.008	L.F.	5.35	.37		5.72	6.45
0200	Shield, top, steel		1768	.005	"	4.83	.21		5.04	5.65
0250	Splice, 3 conductor		48	.167	Ea.	14.60	7.60		22.20	28.50
0300	Top shield		96	.083		1.35	3.80		5.15	7.70
0350	Tap		40	.200		18.75	9.10		27.85	35.50
0400	Insulating patch, splice, tap & end		48	.167		47	7.60		54.60	64
0450	Fold		230	.035			1.59		1.59	2.59
0500	Top shield, tap & fold		96	.083		1.35	3.80		5.15	7.70
0700	Transition, block assembly		77	.104		68.50	4.74		73.24	83
0750	Receptacle frame & base		32	.250		37.50	11.40		48.90	59.50
0800	Cover receptacle		120	.067		3.17	3.04		6.21	8.45
0850	Cover blank		160	.050		3.75	2.28		6.03	7.85
0860	Receptacle, direct connected, single		25	.320		80.50	14.60		95.10	113
0870	Dual		16	.500		132	23		155	182
0880	Combination high & low, tension		21	.381		97	17.35		114.35	136
0900	Box, floor with cover		20	.400		81.50	18.25		99.75	120
0920	Floor service w/barrier		4	2		230	91		321	400
1000	Wall, surface, with cover		20	.400		53.50	18.25		71.75	89
1100	Wall, flush, with cover		20	.400		37.50	18.25		55.75	71
2500	Telephone System									
2510	Transition fitting wall box, surface	1 Elec	24	.333	Ea.	61.50	15.20		76.70	92.50
2520	Flush		24	.333		61.50	15.20		76.70	92.50
2530	Flush, for PC board		24	.333		61.50	15.20		76.70	92.50
2540	Floor service box		4	2		239	91		330	410
2550	Cover, surface					20.50			20.50	22.50
2560	Flush					20.50			20.50	22.50
2570	Flush for PC board					20.50			20.50	22.50
2700	Floor fitting w/duplex jack & cover	1 Elec	21	.381		46	17.35		63.35	79
2720	Low profile		53	.151		17.20	6.90		24.10	30
2740	Miniature w/duplex jack		53	.151		32.50	6.90		39.40	47.50
2760	25 pair kit		21	.381		49.50	17.35		66.85	83
2780	Low profile		53	.151		17.55	6.90		24.45	30.50
2800	Call director kit for 5 cable		19	.421		91.50	19.20		110.70	133
2820	4 pair kit		19	.421		94	19.20		113.20	135
2840	3 pair kit		19	.421		116	19.20		135.20	159
2860	Comb. 25 pair & 3 conductor power		21	.381		91.50	17.35		108.85	130
2880	5 conductor power		21	.381		90	17.35		107.35	128
2900	PC board, 8 per 3 pair		161	.050		68.50	2.27		70.77	78.50
2920	6 per 4 pair		161	.050		68.50	2.27		70.77	78.50
2940	3 pair adapter		161	.050		62	2.27		64.27	72
2950	Plug		77	.104		3.04	4.74		7.78	11.10
2960	Couplers		321	.025		8.55	1.14		9.69	11.25

26 05 19.13 Undercarpet Electrical Power Cables		Crew	Daily Output	Labor-Hours	Unit	Material	Labor	Equipment	Total	Total Incl O&P
						2019 Bare Costs				
3000	Bottom shield for 25 pair cable	1 Elec	4420	.002	L.F.	.86	.08		.94	1.08
3020	4 pair		4420	.002		.40	.08		.48	.57
3040	Top shield for 25 pair cable		4420	.002	↓	.86	.08		.94	1.08
3100	Cable assembly, double-end, 50′, 25 pair		11.80	.678	Ea.	241	31		272	315
3110	3 pair		23.60	.339		74.50	15.45		89.95	107
3120	4 pair		23.60	.339	↓	91.50	15.45		106.95	126
3140	Bulk 3 pair		1473	.005	L.F.	1.27	.25		1.52	1.80
3160	4 pair	↓	1473	.005	"	1.57	.25		1.82	2.13
3500	Data System									
3520	Cable 25 conductor w/connection 40′, 75 ohm	1 Elec	14.50	.552	Ea.	82.50	25		107.50	132
3530	Single lead		22	.364		226	16.60		242.60	275
3540	Dual lead	↓	22	.364	↓	241	16.60		257.60	292
3560	Shields same for 25 conductor as 25 pair telephone									
3570	Single & dual, none required									
3590	BNC coax connectors, plug	1 Elec	40	.200	Ea.	14.45	9.10		23.55	31
3600	TNC coax connectors, plug	"	40	.200	"	15.55	9.10		24.65	32
3700	Cable-bulk									
3710	Single lead	1 Elec	1473	.005	L.F.	3.50	.25		3.75	4.25
3720	Dual lead	"	1473	.005	"	5.05	.25		5.30	6
3730	Hand tool crimp				Ea.	525			525	580
3740	Hand tool notch				"	19.55			19.55	21.50
3750	Boxes & floor fitting same as telephone									

26 05 19.20 Armored Cable		Crew	Daily Output	Labor-Hours	Unit	Material	Labor	Equipment	Total	Total Incl O&P
0010	**ARMORED CABLE**									
0050	600 volt, copper (BX), #14, 2 conductor, solid	1 Elec	2.40	3.333	C.L.F.	40.50	152		192.50	293
0100	3 conductor, solid		2.20	3.636		63.50	166		229.50	340
0120	4 conductor, solid		2	4		89	182		271	395
0150	#12, 2 conductor, solid		2.30	3.478		41	159		200	305
0200	3 conductor, solid		2	4		70.50	182		252.50	375
0220	4 conductor, solid		1.80	4.444		95.50	203		298.50	435
0250	#10, 2 conductor, solid		2	4		79	182		261	385
0300	3 conductor, solid		1.60	5		108	228		336	490
0320	4 conductor, solid		1.40	5.714		177	261		438	620
0340	#8, 2 conductor, stranded		1.50	5.333		240	243		483	660
0350	3 conductor, stranded		1.30	6.154		240	281		521	725
0370	4 conductor, stranded	↓	1.10	7.273		340	330		670	915
9010	600 volt, copper (MC) steel clad, #14, 2 wire	R-1A	8.16	1.961		42	80.50		122.50	177
9020	3 wire		7.21	2.219		76	91		167	233
9030	4 wire		6.69	2.392		196	98		294	375
9040	#12, 2 wire		7.21	2.219		56	91		147	211
9050	3 wire		6.69	2.392		69	98		167	236
9060	4 wire		6.04	2.649		114	109		223	305
9070	#10, 2 wire		6.04	2.649		101	109		210	288
9080	3 wire		5.65	2.832		131	116		247	335
9090	4 wire		5.23	3.059		212	126		338	440
9100	#8, 2 wire, stranded		3.94	4.061		155	167		322	445
9110	3 wire, stranded		3.40	4.706		189	193		382	525
9120	4 wire, stranded	↓	2.80	5.714	↓	261	235		496	670

26 05 19.55 Non-Metallic Sheathed Cable		Crew	Daily Output	Labor-Hours	Unit	2019 Bare Costs Material	Labor	Equipment	Total	Total Incl O&P
0010	**NON-METALLIC SHEATHED CABLE** 600 volt									
0100	Copper with ground wire (Romex)									
0150	#14, 2 conductor	1 Elec	2.70	2.963	C.L.F.	18.50	135		153.50	241
0200	3 conductor		2.40	3.333		25	152		177	276
0220	4 conductor		2.20	3.636		39	166		205	315
0250	#12, 2 conductor		2.50	3.200		24.50	146		170.50	265
0300	3 conductor		2.20	3.636		39.50	166		205.50	315
0320	4 conductor		2	4		59	182		241	365
0350	#10, 2 conductor		2.20	3.636		41	166		207	315
0400	3 conductor		1.80	4.444		62	203		265	400
0420	4 conductor		1.60	5		88.50	228		316.50	465
0430	#8, 2 conductor		1.60	5		76.50	228		304.50	455
0450	3 conductor		1.50	5.333		116	243		359	525
0500	#6, 3 conductor		1.40	5.714		195	261		456	640
0550	SE type SER aluminum cable, 3 RHW and									
0600	1 bare neutral, 3 #8 & 1 #8	1 Elec	1.60	5	C.L.F.	55.50	228		283.50	430
0650	3 #6 & 1 #6	"	1.40	5.714		66	261		327	500
0700	3 #4 & 1 #6	2 Elec	2.40	6.667		70	305		375	570
0750	3 #2 & 1 #4		2.20	7.273		122	330		452	675
0800	3 #1/0 & 1 #2		2	8		179	365		544	790
0850	3 #2/0 & 1 #1		1.80	8.889		208	405		613	890
0900	3 #4/0 & 1 #2/0		1.60	10		290	455		745	1,075
1450	UF underground feeder cable, copper with ground, #14, 2 conductor	1 Elec	4	2		40.50	91		131.50	194
1500	#12, 2 conductor		3.50	2.286		35	104		139	209
1550	#10, 2 conductor		3	2.667		54	122		176	258
1600	#14, 3 conductor		3.50	2.286		37.50	104		141.50	212
1650	#12, 3 conductor		3	2.667		53.50	122		175.50	257
1700	#10, 3 conductor		2.50	3.200		85	146		231	330
2400	SEU service entrance cable, copper 2 conductors, #8 + #8 neutral		1.50	5.333		103	243		346	510
2600	#6 + #8 neutral		1.30	6.154		132	281		413	605
2800	#6 + #6 neutral		1.30	6.154		154	281		435	630
3000	#4 + #6 neutral	2 Elec	2.20	7.273		202	330		532	760
3200	#4 + #4 neutral		2.20	7.273		223	330		553	785
3400	#3 + #5 neutral		2.10	7.619		335	345		680	935
3600	#3 + #3 neutral		2.10	7.619		269	345		614	860
3800	#2 + #4 neutral		2	8		305	365		670	930
4000	#1 + #1 neutral		1.90	8.421		345	385		730	1,000
4200	1/0 + 1/0 neutral		1.80	8.889		940	405		1,345	1,675
4400	2/0 + 2/0 neutral		1.70	9.412		685	430		1,115	1,450
4600	3/0 + 3/0 neutral		1.60	10		850	455		1,305	1,675
4620	4/0 + 4/0 neutral		1.45	11.034		2,075	505		2,580	3,100
4800	Aluminum 2 conductors, #8 + #8 neutral	1 Elec	1.60	5		49	228		277	425
5000	#6 + #6 neutral	"	1.40	5.714		56.50	261		317.50	490
5100	#4 + #6 neutral	2 Elec	2.50	6.400		65	292		357	545
5200	#4 + #4 neutral		2.40	6.667		64	305		369	565
5300	#2 + #4 neutral		2.30	6.957		79	315		394	605
5400	#2 + #2 neutral		2.20	7.273		92	330		422	640
5450	1/0 + #2 neutral		2.10	7.619		122	345		467	700
5500	1/0 + 1/0 neutral		2	8		127	365		492	735
5550	2/0 + #1 neutral		1.90	8.421		135	385		520	775
5600	2/0 + 2/0 neutral		1.80	8.889		143	405		548	815
5800	3/0 + 1/0 neutral		1.70	9.412		217	430		647	940

26 05 19.55 Non-Metallic Sheathed Cable		Crew	Daily Output	Labor-Hours	Unit	2019 Bare Costs Material	Labor	Equipment	Total	Total Incl O&P
6000	3/0 + 3/0 neutral	2 Elec	1.70	9.412	C.L.F.	183	430		613	900
6200	4/0 + 2/0 neutral		1.60	10		187	455		642	950
6400	4/0 + 4/0 neutral		1.60	10		204	455		659	970
6500	Service entrance cap for copper SEU									
6600	100 amp	1 Elec	12	.667	Ea.	7.95	30.50		38.45	58.50
6700	150 amp		10	.800		10.70	36.50		47.20	71.50
6800	200 amp		8	1		17.25	45.50		62.75	93.50

26 05 19.70 Portable Cord		Crew	Daily Output	Labor-Hours	Unit	Material	Labor	Equipment	Total	Total Incl O&P
0010	**PORTABLE CORD** 600 volt									
0100	Type SO, #18, 2 conductor	1 Elec	980	.008	L.F.	.29	.37		.66	.93
0110	3 conductor		980	.008		.33	.37		.70	.97
0120	#16, 2 conductor		840	.010		.38	.43		.81	1.13
0130	3 conductor		840	.010		4	.43		4.43	5.10
0140	4 conductor		840	.010		.57	.43		1	1.34
0240	#14, 2 conductor		840	.010		.53	.43		.96	1.29
0250	3 conductor		840	.010		.72	.43		1.15	1.50
0260	4 conductor		840	.010		1.07	.43		1.50	1.89
0280	#12, 2 conductor		840	.010		.78	.43		1.21	1.57
0290	3 conductor		840	.010		.91	.43		1.34	1.71
0300	4 conductor		840	.010		1.21	.43		1.64	2.04
0320	#10, 2 conductor		765	.010		.85	.48		1.33	1.72
0330	3 conductor		765	.010		1.27	.48		1.75	2.18
0340	4 conductor		765	.010		1.30	.48		1.78	2.21
0360	#8, 2 conductor		555	.014		2.03	.66		2.69	3.30
0370	3 conductor		540	.015		2.43	.68		3.11	3.77
0380	4 conductor		525	.015		2.14	.69		2.83	3.48
0400	#6, 2 conductor		525	.015		2.96	.69		3.65	4.39
0410	3 conductor		490	.016		3.41	.74		4.15	4.97
0420	4 conductor		415	.019		3.26	.88		4.14	5
0440	#4, 2 conductor	2 Elec	830	.019		8.15	.88		9.03	10.45
0450	3 conductor		700	.023		5.70	1.04		6.74	8
0460	4 conductor		660	.024		6.50	1.11		7.61	8.95
0480	#2, 2 conductor		450	.036		5.20	1.62		6.82	8.35
0490	3 conductor		350	.046		7.15	2.08		9.23	11.25
0500	4 conductor		280	.057		9.05	2.61		11.66	14.20
2000	See 26 27 26.20 for Wiring Devices Elements									

26 05 19.75 Modular Flexible Wiring System		Crew	Daily Output	Labor-Hours	Unit	Material	Labor	Equipment	Total	Total Incl O&P
0010	**MODULAR FLEXIBLE WIRING SYSTEM**									
0020	Commercial system grid ceiling									
0100	Conversion Module	1 Elec	32	.250	Ea.	8	11.40		19.40	27.50
0120	Fixture cable, for fixture to fixture, 3 conductor, 15′ long		40	.200		34	9.10		43.10	52.50
0150	Extender cable, 3 conductor, 15′ long		48	.167		31.50	7.60		39.10	47
0200	Switch drop, 1 level, 9′ long		32	.250		22	11.40		33.40	42.50
0220	2 level, 9′ long		32	.250		30	11.40		41.40	51.50
0250	Power tee, 9′ long		32	.250		27.50	11.40		38.90	48.50
1020	Industrial system open ceiling									
1100	Converter, interface between hardwiring and modular wiring	1 Elec	32	.250	Ea.	21	11.40		32.40	42
1120	Fixture cable, for fixture to fixture, 3 conductor, 21′ long		32	.250		112	11.40		123.40	142
1125	3 conductor, 25′ long		24	.333		147	15.20		162.20	187
1130	3 conductor, 31′ long		19	.421		139	19.20		158.20	185
1150	Fixture cord drop, 10′ long		40	.200		37.50	9.10		46.60	56.50

26 05 Common Work Results for Electrical

26 05 19 – Low-Voltage Electrical Power Conductors and Cables

26 05 19.90 Wire		Crew	Daily Output	Labor-Hours	Unit	Material	Labor	Equipment	Total	Total Incl O&P
						2019 Bare Costs				
0010	**WIRE**, normal installation conditions in wireway, conduit, cable tray									
0020	600 volt, copper type THW, solid, #14	1 Elec	13	.615	C.L.F.	5.80	28		33.80	52.50
0030	#12		11	.727		9.70	33		42.70	64.50
0040	#10		10	.800		15.75	36.50		52.25	77
0050	Stranded, #14		13	.615		7.60	28		35.60	54.50
0100	#12		11	.727		13.65	33		46.65	69
0120	#10		10	.800		21.50	36.50		58	83
0140	#8		8	1		33.50	45.50		79	112
0400	250 kcmil	3 Elec	6	4		345	182		527	680
0420	300 kcmil		5.70	4.211		440	192		632	795
0450	350 kcmil		5.40	4.444		455	203		658	830
0480	400 kcmil		5.10	4.706		530	215		745	935
0490	500 kcmil		4.80	5		800	228		1,028	1,250
0540	600 volt, aluminum type THHN, stranded, #6	1 Elec	8	1		25	45.50		70.50	102
0560	#4	2 Elec	13	1.231		30.50	56		86.50	125
0580	#2		10.60	1.509		42.50	69		111.50	159
0600	#1		9	1.778		62	81		143	200
0620	1/0		8	2		74.50	91		165.50	231
0640	2/0		7.20	2.222		88.50	101		189.50	263
0680	3/0		6.60	2.424		111	111		222	300
0700	4/0		6.20	2.581		123	118		241	325
0720	250 kcmil	3 Elec	8.70	2.759		149	126		275	370
0740	300 kcmil		8.10	2.963		206	135		341	445
0760	350 kcmil		7.50	3.200		208	146		354	465
0780	400 kcmil		6.90	3.478		237	159		396	520
0800	500 kcmil		6	4		267	182		449	590
0850	600 kcmil		5.70	4.211		340	192		532	685
0880	700 kcmil		5.10	4.706		415	215		630	805
0900	750 kcmil		4.80	5		410	228		638	820
0910	1,000 kcmil		3.78	6.349		585	290		875	1,100
0920	600 volt, copper type THWN-THHN, solid, #14	1 Elec	13	.615		6.60	28		34.60	53.50
0940	#12		11	.727		9.80	33		42.80	65
0960	#10		10	.800		14.65	36.50		51.15	75.50
1000	Stranded, #14		13	.615		7.55	28		35.55	54.50
1010	2 wire, #14, install wire only		655	.012	L.F.	.15	.56		.71	1.08
1020	3 wire, #14, install wire only		440	.018	"	.23	.83		1.06	1.60
1200	#12		11	.727	C.L.F.	10.50	33		43.50	65.50
1250	#10		10	.800		16.10	36.50		52.60	77
1300	#8		8	1		28.50	45.50		74	106
1350	#6		6.50	1.231		48.50	56		104.50	145
1400	#4	2 Elec	10.60	1.509		69	69		138	188

26 05 23 – Control-Voltage Electrical Power Cables

26 05 23.20 Special Wires and Fittings		Crew	Daily Output	Labor-Hours	Unit	Material	Labor	Equipment	Total	Total Incl O&P
0010	**SPECIAL WIRES & FITTINGS**									
0100	Fixture TFFN 600 volt 90°C stranded, #18	1 Elec	13	.615	C.L.F.	9.25	28		37.25	56
0150	#16		13	.615		12.80	28		40.80	60
0500	Thermostat, jacket non-plenum, twisted, #18-2 conductor		8	1		9.85	45.50		55.35	85.50
0550	#18-3 conductor		7	1.143		12.10	52		64.10	98.50
0600	#18-4 conductor		6.50	1.231		19.95	56		75.95	114
0650	#18-5 conductor		6	1.333		21.50	61		82.50	123
0700	#18-6 conductor		5.50	1.455		25	66.50		91.50	136
0750	#18-7 conductor		5	1.600		34.50	73		107.50	157

26 05 Common Work Results for Electrical

26 05 23 – Control-Voltage Electrical Power Cables

	26 05 23.20 Special Wires and Fittings	Crew	Daily Output	Labor-Hours	Unit	2019 Bare Costs Material	Labor	Equipment	Total	Total Incl O&P
0800	#18-8 conductor	1 Elec	4.80	1.667	C.L.F.	32	76		108	160

26 05 26 – Grounding and Bonding for Electrical Systems

	26 05 26.80 Grounding	Crew	Daily Output	Labor-Hours	Unit	2019 Bare Costs Material	Labor	Equipment	Total	Total Incl O&P
0010	**GROUNDING**									
0030	Rod, copper clad, 8′ long, 1/2″ diameter	1 Elec	5.50	1.455	Ea.	19.75	66.50		86.25	130
0040	5/8″ diameter		5.50	1.455		22	66.50		88.50	133
0050	3/4″ diameter		5.30	1.509		35.50	69		104.50	151
0080	10′ long, 1/2″ diameter		4.80	1.667		22.50	76		98.50	149
0090	5/8″ diameter		4.60	1.739		23.50	79.50		103	155
0100	3/4″ diameter		4.40	1.818		49.50	83		132.50	189
0130	15′ long, 3/4″ diameter		4	2		49.50	91		140.50	204
0260	Wire ground bare armored, #8-1 conductor		2	4	C.L.F.	69	182		251	375
0270	#6-1 conductor		1.80	4.444		84	203		287	425
0390	Bare copper wire, stranded, #8		11	.727		43	33		76	102
0400	#6		10	.800		41	36.50		77.50	105
0600	#2	2 Elec	10	1.600		89	73		162	217
0800	3/0		6.60	2.424		315	111		426	525
1000	4/0		5.70	2.807		440	128		568	695
1200	250 kcmil	3 Elec	7.20	3.333		445	152		597	740
1800	Water pipe ground clamps, heavy duty									
2000	Bronze, 1/2″ to 1″ diameter	1 Elec	8	1	Ea.	32	45.50		77.50	110
2100	1-1/4″ to 2″ diameter		8	1		29	45.50		74.50	107
2200	2-1/2″ to 3″ diameter		6	1.333		69.50	61		130.50	176
2800	Brazed connections, #6 wire		12	.667		16.15	30.50		46.65	67.50
3000	#2 wire		10	.800		21.50	36.50		58	83.50
3100	3/0 wire		8	1		32	45.50		77.50	110
3200	4/0 wire		7	1.143		36.50	52		88.50	126
3400	250 kcmil wire		5	1.600		42.50	73		115.50	166

26 05 33 – Raceway and Boxes for Electrical Systems

	26 05 33.13 Conduit	Crew	Daily Output	Labor-Hours	Unit	2019 Bare Costs Material	Labor	Equipment	Total	Total Incl O&P
0010	**CONDUIT** To 10′ high, includes 2 terminations, 2 elbows, R260533-22									
0020	11 beam clamps, and 11 couplings per 100 L.F.									
1750	Rigid galvanized steel, 1/2″ diameter	1 Elec	90	.089	L.F.	2.30	4.05		6.35	9.15
1770	3/4″ diameter		80	.100		4.74	4.56		9.30	12.65
1800	1″ diameter		65	.123		7.05	5.60		12.65	16.90
1830	1-1/4″ diameter		60	.133		4.82	6.10		10.92	15.20
1850	1-1/2″ diameter		55	.145		7.75	6.65		14.40	19.35
1870	2″ diameter		45	.178		9.95	8.10		18.05	24
2500	Steel, intermediate conduit (IMC), 1/2″ diameter		100	.080		1.69	3.65		5.34	7.80
2530	3/4″ diameter		90	.089		2.06	4.05		6.11	8.85
2550	1″ diameter		70	.114		2.58	5.20		7.78	11.35
2570	1-1/4″ diameter		65	.123		3.69	5.60		9.29	13.20
2600	1-1/2″ diameter		60	.133		4.67	6.10		10.77	15.05
2630	2″ diameter		50	.160		6.25	7.30		13.55	18.80
2650	2-1/2″ diameter		40	.200		9.25	9.10		18.35	25
2670	3″ diameter	2 Elec	60	.267		12.15	12.15		24.30	33
2700	3-1/2″ diameter		54	.296		17.30	13.50		30.80	41
2730	4″ diameter		50	.320		11.80	14.60		26.40	37
5000	Electric metallic tubing (EMT), 1/2″ diameter	1 Elec	170	.047		.81	2.15		2.96	4.39
5020	3/4″ diameter		130	.062		1.18	2.81		3.99	5.90
5040	1″ diameter		115	.070		1.90	3.17		5.07	7.30
5060	1-1/4″ diameter		100	.080		3.03	3.65		6.68	9.30

26 05 33 – Raceway and Boxes for Electrical Systems

26 05 33.13 Conduit		Crew	Daily Output	Labor-Hours	Unit	2019 Bare Costs Material	Labor	Equipment	Total	Total Incl O&P
5080	1-1/2" diameter	1 Elec	90	.089	L.F.	3.48	4.05		7.53	10.40
5100	2" diameter		80	.100		3.89	4.56		8.45	11.75
5120	2-1/2" diameter		60	.133		4.54	6.10		10.64	14.90
5140	3" diameter	2 Elec	100	.160		5.55	7.30		12.85	18
5160	3-1/2" diameter		90	.178		7.15	8.10		15.25	21
5180	4" diameter		80	.200		12.55	9.10		21.65	28.50
9100	PVC, schedule 40, 1/2" diameter	1 Elec	190	.042		.93	1.92		2.85	4.15
9110	3/4" diameter		145	.055		1.06	2.52		3.58	5.25
9120	1" diameter		125	.064		1.58	2.92		4.50	6.50
9130	1-1/4" diameter		110	.073		1.77	3.32		5.09	7.35
9140	1-1/2" diameter		100	.080		2.08	3.65		5.73	8.25
9150	2" diameter		90	.089		2.85	4.05		6.90	9.75
9160	2-1/2" diameter		65	.123		3.30	5.60		8.90	12.80
9170	3" diameter	2 Elec	110	.145		4.78	6.65		11.43	16.05
9880	Heat bender, to 6" diameter				Ea.	1,500			1,500	1,675
9900	Add to labor for higher elevated installation									
9905	10' to 14.5' high, add						10%			
9910	15' to 20' high, add						10%			
9920	20' to 25' high, add						20%			
9930	25' to 30' high, add						25%			
9940	30' to 35' high, add						30%			
9950	35' to 40' high, add						35%			
9960	Over 40' high, add						40%			

26 05 33.16 Boxes for Electrical Systems		Crew	Daily Output	Labor-Hours	Unit	Material	Labor	Equipment	Total	Total Incl O&P
0010	**BOXES FOR ELECTRICAL SYSTEMS**									
0020	Pressed steel, octagon, 4"	1 Elec	20	.400	Ea.	3.46	18.25		21.71	34
0060	Covers, blank		64	.125		.87	5.70		6.57	10.25
0100	Extension rings		40	.200		4.46	9.10		13.56	19.80
0150	Square, 4"		20	.400		5.20	18.25		23.45	35.50
0200	Extension rings		40	.200		4.59	9.10		13.69	19.95
0250	Covers, blank		64	.125		.81	5.70		6.51	10.20
0300	Plaster rings		64	.125		1.53	5.70		7.23	11
0550	Handy box		27	.296		2.41	13.50		15.91	24.50
0560	Covers, device		64	.125		1.01	5.70		6.71	10.40
0650	Switchbox		27	.296		5.60	13.50		19.10	28
0700	Masonry, 1 gang, 2-1/2" deep		27	.296		8.10	13.50		21.60	31
0710	3-1/2" deep		27	.296		8	13.50		21.50	31
0750	2 gang, 2-1/2" deep		20	.400		17.20	18.25		35.45	49
0760	3-1/2" deep		20	.400		11.40	18.25		29.65	42.50
1100	Concrete, floor, 1 gang		5.30	1.509		123	69		192	248
1400	Cast, 1 gang, FS (2" deep), 1/2" hub		12	.667		23.50	30.50		54	75.50
1410	3/4" hub		12	.667		27	30.50		57.50	79.50
1420	FD (2-11/16" deep), 1/2" hub		12	.667		18.95	30.50		49.45	70.50
1430	3/4" hub		12	.667		24	30.50		54.50	75.50
1450	2 gang, FS, 1/2" hub		10	.800		47.50	36.50		84	112
1460	3/4" hub		10	.800		40.50	36.50		77	104
1470	FD, 1/2" hub		10	.800		57.50	36.50		94	123
1480	3/4" hub		10	.800		60	36.50		96.50	126
1500	3 gang, FS, 3/4" hub		9	.889		79.50	40.50		120	154
1510	Switch cover, 1 gang, FS		64	.125		6.75	5.70		12.45	16.75
1520	2 gang		53	.151		10.35	6.90		17.25	22.50
1530	Duplex receptacle cover, 1 gang, FS		64	.125		6.50	5.70		12.20	16.45

26 05 33 – Raceway and Boxes for Electrical Systems

26 05 33.16 Boxes for Electrical Systems		Crew	Daily Output	Labor-Hours	Unit	2019 Bare Costs Material	Labor	Equipment	Total	Total Incl O&P
1540	2 gang, FS	1 Elec	53	.151	Ea.	10.85	6.90		17.75	23
1542	Weatherproof blank cover, 1 gang		64	.125		1.22	5.70		6.92	10.65
1544	2 gang		53	.151		2.50	6.90		9.40	14
1550	Weatherproof switch cover, 1 gang		64	.125		5.50	5.70		11.20	15.35
1554	2 gang		53	.151		9.75	6.90		16.65	22
1600	Weatherproof receptacle cover, 1 gang		64	.125		7.80	5.70		13.50	17.85
1604	2 gang		53	.151		7.75	6.90		14.65	19.80
1620	Weatherproof receptacle cover, tamper resistant, 1 gang		58	.138		13.60	6.30		19.90	25
1624	2 gang		48	.167		26.50	7.60		34.10	41.50
1750	FSC, 1 gang, 1/2" hub		11	.727		24	33		57	80.50
1760	3/4" hub		11	.727		33.50	33		66.50	90.50
1770	2 gang, 1/2" hub		9	.889		44	40.50		84.50	115
1780	3/4" hub		9	.889		60	40.50		100.50	132
1790	FDC, 1 gang, 1/2" hub		11	.727		27	33		60	84
1800	3/4" hub		11	.727		41.50	33		74.50	99.50
1810	2 gang, 1/2" hub		9	.889		56.50	40.50		97	128
1820	3/4" hub		9	.889		64.50	40.50		105	137
1850	Weatherproof in-use cover, 1 gang		64	.125		24	5.70		29.70	36
1870	2 gang		53	.151		28.50	6.90		35.40	43

26 05 33.17 Outlet Boxes, Plastic		Crew	Daily Output	Labor-Hours	Unit	Material	Labor	Equipment	Total	Total Incl O&P
0010	**OUTLET BOXES, PLASTIC**									
0050	4" diameter, round with 2 mounting nails	1 Elec	25	.320	Ea.	3.09	14.60		17.69	27.50
0100	Bar hanger mounted		25	.320		5.35	14.60		19.95	30
0200	4", square with 2 mounting nails		25	.320		5.30	14.60		19.90	30
0300	Plaster ring		64	.125		2.02	5.70		7.72	11.50
0400	Switch box with 2 mounting nails, 1 gang		30	.267		3.10	12.15		15.25	23.50
0500	2 gang		25	.320		4.21	14.60		18.81	28.50
0600	3 gang		20	.400		5	18.25		23.25	35.50
1400	PVC, FSS, 1 gang, 1/2" hub		14	.571		12.90	26		38.90	56.50
1410	3/4" hub		14	.571		10.75	26		36.75	54.50
1420	FD, 1 gang for variable terminations		14	.571		7.95	26		33.95	51
1450	FS, 2 gang for variable terminations		12	.667		11.20	30.50		41.70	62
1480	Weatherproof blank cover, FS, 1 gang		64	.125		3.68	5.70		9.38	13.35
1500	2 gang		53	.151		4.84	6.90		11.74	16.55
1510	Weatherproof switch cover, FS, 1 gang		64	.125		8.05	5.70		13.75	18.15
1520	2 gang		53	.151		15.25	6.90		22.15	28
1530	Weatherproof duplex receptacle cover, FS, 1 gang		64	.125		11.95	5.70		17.65	22.50
1540	2 gang		53	.151		13.40	6.90		20.30	26
1750	FSC, 1 gang, 1/2" hub		13	.615		8.80	28		36.80	55.50
1760	3/4" hub		13	.615		9.60	28		37.60	56.50
1770	FSC, 2 gang, 1/2" hub		11	.727		16.55	33		49.55	72
1780	3/4" hub		11	.727		14.55	33		47.55	70
1790	FDC, 1 gang, 1/2" hub		13	.615		10.35	28		38.35	57.50
1800	3/4" hub		13	.615		10.40	28		38.40	57.50
1810	Weatherproof, T box w/3 holes		14	.571		10.30	26		36.30	54
1820	4" diameter round w/5 holes		14	.571		9.85	26		35.85	53.50
1850	In-use cover, 1 gang		64	.125		9.50	5.70		15.20	19.75
1870	2 gang		53	.151		12.40	6.90		19.30	25

26 05 33.18 Pull Boxes		Crew	Daily Output	Labor-Hours	Unit	Material	Labor	Equipment	Total	Total Incl O&P
0010	**PULL BOXES**									
0100	Steel, pull box, NEMA 1, type SC, 6" W x 6" H x 4" D	1 Elec	8	1	Ea.	10	45.50		55.50	85.50
0200	8" W x 8" H x 4" D		8	1		12.45	45.50		57.95	88

26 05 33.18 Pull Boxes		Crew	Daily Output	Labor-Hours	Unit	Material	Labor	Equipment	Total	Total Incl O&P
						2019 Bare Costs				
0300	10″ W x 12″ H x 6″ D	1 Elec	5.30	1.509	Ea.	28.50	69		97.50	144
0800	12″ W x 16″ H x 6″ D		4.70	1.702		50	77.50		127.50	182
1000	20″ W x 20″ H x 6″ D	↓	3.60	2.222	↓	119	101		220	295
26 05 33.23 Wireway										
0010	**WIREWAY** to 10′ high									
0100	NEMA 1, screw cover w/fittings and supports, 2-1/2″ x 2-1/2″	1 Elec	45	.178	L.F.	12.30	8.10		20.40	27
0200	4″ x 4″	″	40	.200		13.25	9.10		22.35	29.50
0400	6″ x 6″	2 Elec	60	.267		21	12.15		33.15	43
0600	8″ x 8″	″	40	.400		27.50	18.25		45.75	60.50
4475	NEMA 3R, screw cover w/fittings and supports, 4″ x 4″	1 Elec	36	.222		16.40	10.15		26.55	34.50
4480	6″ x 6″	2 Elec	55	.291		18.20	13.25		31.45	41.50
4485	8″ x 8″		36	.444		32	20.50		52.50	68.50
4490	12″ x 12″	↓	18	.889	↓	57.50	40.50		98	130
26 05 33.25 Conduit Fittings for Rigid Galvanized Steel										
0010	**CONDUIT FITTINGS FOR RIGID GALVANIZED STEEL**									
2280	LB, LR or LL fittings & covers, 1/2″ diameter	1 Elec	16	.500	Ea.	6.95	23		29.95	44.50
2290	3/4″ diameter		13	.615		8.20	28		36.20	55
2300	1″ diameter		11	.727		12.80	33		45.80	68
2330	1-1/4″ diameter		8	1		29.50	45.50		75	107
2350	1-1/2″ diameter		6	1.333		23	61		84	124
2370	2″ diameter		5	1.600		58.50	73		131.50	184
2380	2-1/2″ diameter		4	2		121	91		212	282
2390	3″ diameter		3.50	2.286		166	104		270	355
2400	3-1/2″ diameter		3	2.667		174	122		296	390
2410	4″ diameter		2.50	3.200		190	146		336	445
5280	Service entrance cap, 1/2″ diameter		16	.500		4.90	23		27.90	42.50
5300	3/4″ diameter		13	.615		5.85	28		33.85	52.50
5320	1″ diameter		10	.800		4.79	36.50		41.29	65
5340	1-1/4″ diameter		8	1		5.30	45.50		50.80	80.50
5360	1-1/2″ diameter		6.50	1.231		10.45	56		66.45	103
5380	2″ diameter		5.50	1.455		22	66.50		88.50	132
5400	2-1/2″ diameter		4	2		77	91		168	234
5420	3″ diameter		3.40	2.353		78.50	107		185.50	261
5440	3-1/2″ diameter		3	2.667		125	122		247	335
5460	4″ diameter	↓	2.70	2.963	↓	167	135		302	405
26 05 33.30 Electrical Nonmetallic Tubing (ENT)										
0010	**ELECTRICAL NONMETALLIC TUBING (ENT)**									
0050	Flexible, 1/2″ diameter	1 Elec	270	.030	L.F.	.53	1.35		1.88	2.78
0100	3/4″ diameter		230	.035		.76	1.59		2.35	3.43
0200	1″ diameter		145	.055		1.38	2.52		3.90	5.60
0210	1-1/4″ diameter		125	.064		1.26	2.92		4.18	6.15
0220	1-1/2″ diameter		100	.080		2.05	3.65		5.70	8.20
0230	2″ diameter		75	.107	↓	2.52	4.86		7.38	10.70
0300	Connectors, to outlet box, 1/2″ diameter		230	.035	Ea.	1.07	1.59		2.66	3.77
0310	3/4″ diameter		210	.038		1.88	1.74		3.62	4.90
0320	1″ diameter		200	.040		2.37	1.82		4.19	5.60
0400	Couplings, to conduit, 1/2″ diameter		145	.055		.84	2.52		3.36	5
0410	3/4″ diameter		130	.062		1.10	2.81		3.91	5.80
0420	1″ diameter	↓	125	.064	↓	2.14	2.92		5.06	7.10

26 05 36 – Cable Trays for Electrical Systems

26 05 36.10 Cable Tray Ladder Type		Crew	Daily Output	Labor-Hours	Unit	Material	2019 Bare Costs Labor	Equipment	Total	Total Incl O&P
0010	**CABLE TRAY LADDER TYPE** w/ftngs. & supports, 4" dp., to 15' elev.									
0160	Galvanized steel tray									
0170	4" rung spacing, 6" wide	2 Elec	98	.163	L.F.	13.35	7.45		20.80	27
0800	6" rung spacing, 6" wide		100	.160		12.15	7.30		19.45	25.50
0850	9" wide		94	.170		13.35	7.75		21.10	27.50
0860	12" wide		88	.182		14.05	8.30		22.35	29
0880	24" wide	↓	80	.200	↓	17.95	9.10		27.05	34.50

26 05 39 – Underfloor Raceways for Electrical Systems

26 05 39.30 Conduit In Concrete Slab		Crew	Daily Output	Labor-Hours	Unit	Material	Labor	Equipment	Total	Total Incl O&P
0010	**CONDUIT IN CONCRETE SLAB** Including terminations,									
0020	fittings and supports									
3230	PVC, schedule 40, 1/2" diameter	1 Elec	270	.030	L.F.	.54	1.35		1.89	2.79
3250	3/4" diameter		230	.035		.58	1.59		2.17	3.23
3270	1" diameter		200	.040		.77	1.82		2.59	3.83
3300	1-1/4" diameter		170	.047		1.04	2.15		3.19	4.65
3330	1-1/2" diameter		140	.057		1.27	2.61		3.88	5.65
3350	2" diameter		120	.067		1.58	3.04		4.62	6.70
4350	Rigid galvanized steel, 1/2" diameter		200	.040		2.02	1.82		3.84	5.20
4400	3/4" diameter		170	.047		4.46	2.15		6.61	8.40
4450	1" diameter		130	.062		6.90	2.81		9.71	12.20
4500	1-1/4" diameter		110	.073		4.23	3.32		7.55	10.05
4600	1-1/2" diameter		100	.080		7.25	3.65		10.90	13.95
4800	2" diameter	↓	90	.089	↓	8.95	4.05		13	16.45

26 05 39.40 Conduit In Trench		Crew	Daily Output	Labor-Hours	Unit	Material	Labor	Equipment	Total	Total Incl O&P
0010	**CONDUIT IN TRENCH** Includes terminations and fittings									
0020	Does not include excavation or backfill, see Section 31 23 16									
0200	Rigid galvanized steel, 2" diameter	1 Elec	150	.053	L.F.	8.60	2.43		11.03	13.40
0400	2-1/2" diameter	"	100	.080		11	3.65		14.65	18.05
0600	3" diameter	2 Elec	160	.100		12.85	4.56		17.41	21.50
0800	3-1/2" diameter	"	140	.114	↓	16.75	5.20		21.95	27

26 05 80 – Wiring Connections

26 05 80.10 Motor Connections		Crew	Daily Output	Labor-Hours	Unit	Material	Labor	Equipment	Total	Total Incl O&P
0010	**MOTOR CONNECTIONS**									
0020	Flexible conduit and fittings, 115 volt, 1 phase, up to 1 HP motor	1 Elec	8	1	Ea.	5.70	45.50		51.20	81
0050	2 HP motor		6.50	1.231		10.15	56		66.15	103
0100	3 HP motor		5.50	1.455		9.05	66.50		75.55	118
0110	230 volt, 3 phase, 3 HP motor		6.78	1.180		6.80	54		60.80	95.50
0112	5 HP motor		5.47	1.463		5.85	66.50		72.35	115
0114	7-1/2 HP motor		4.61	1.735		9.35	79		88.35	139
0120	10 HP motor		4.20	1.905		18.25	87		105.25	162
0150	15 HP motor		3.30	2.424		18.25	111		129.25	200
0200	25 HP motor		2.70	2.963		28	135		163	251
1500	460 volt, 5 HP motor, 3 phase		8	1		6.15	45.50		51.65	81.50
1520	10 HP motor		8	1		6.15	45.50		51.65	81.50
1530	25 HP motor	↓	6	1.333	↓	10.95	61		71.95	111

26 05 90.10 Residential Wiring		Crew	Daily Output	Labor-Hours	Unit	2019 Bare Costs Material	Labor	Equipment	Total	Total Incl O&P
0010	**RESIDENTIAL WIRING**									
0020	20' avg. runs and #14/2 wiring incl. unless otherwise noted									
1000	Service & panel, includes 24' SE-AL cable, service eye, meter,									
1010	Socket, panel board, main bkr., ground rod, 15 or 20 amp									
1020	1-pole circuit breakers, and misc. hardware									
1100	100 amp, with 10 branch breakers	1 Elec	1.19	6.723	Ea.	325	305		630	860
1110	With PVC conduit and wire		.92	8.696		360	395		755	1,050
1120	With RGS conduit and wire		.73	10.959		575	500		1,075	1,450
1200	200 amp, with 18 branch breakers	2 Elec	1.80	8.889		1,000	405		1,405	1,750
1220	With PVC conduit and wire		1.46	10.959		1,075	500		1,575	2,000
1230	With RGS conduit and wire		1.24	12.903		1,450	590		2,040	2,550
1800	Lightning surge suppressor	1 Elec	32	.250		75	11.40		86.40	101
2000	Switch devices									
2100	Single pole, 15 amp, ivory, with a 1-gang box, cover plate,									
2110	Type NM (Romex) cable	1 Elec	17.10	.468	Ea.	15.50	21.50		37	52
2120	Type MC cable		14.30	.559		26	25.50		51.50	70
2130	EMT & wire		5.71	1.401		37.50	64		101.50	145
2150	3-way, #14/3, type NM cable		14.55	.550		9.80	25		34.80	52
2170	Type MC cable		12.31	.650		23.50	29.50		53	74.50
2180	EMT & wire		5	1.600		31.50	73		104.50	154
2200	4-way, #14/3, type NM cable		14.55	.550		18.50	25		43.50	61.50
2220	Type MC cable		12.31	.650		32.50	29.50		62	84
2230	EMT & wire		5	1.600		40.50	73		113.50	164
2250	S.P., 20 amp, #12/2, type NM cable		13.33	.600		11.30	27.50		38.80	57
2270	Type MC cable		11.43	.700		21	32		53	75
2280	EMT & wire		4.85	1.649		35.50	75		110.50	162
2290	S.P. rotary dimmer, 600 W, no wiring		17	.471		31.50	21.50		53	69.50
2300	S.P. rotary dimmer, 600 W, type NM cable		14.55	.550		35.50	25		60.50	80
2320	Type MC cable		12.31	.650		46	29.50		75.50	99
2330	EMT & wire		5	1.600		58.50	73		131.50	183
2350	3-way rotary dimmer, type NM cable		13.33	.600		23.50	27.50		51	70
2370	Type MC cable		11.43	.700		34	32		66	89.50
2380	EMT & wire		4.85	1.649		46.50	75		121.50	174
2400	Interval timer wall switch, 20 amp, 1-30 min., #12/2									
2410	Type NM cable	1 Elec	14.55	.550	Ea.	58.50	25		83.50	105
2420	Type MC cable		12.31	.650		64.50	29.50		94	119
2430	EMT & wire		5	1.600		82.50	73		155.50	210
2500	Decorator style									
2510	S.P., 15 amp, type NM cable	1 Elec	17.10	.468	Ea.	20.50	21.50		42	58
2520	Type MC cable		14.30	.559		31.50	25.50		57	76
2530	EMT & wire		5.71	1.401		42.50	64		106.50	151
2550	3-way, #14/3, type NM cable		14.55	.550		15	25		40	57.50
2570	Type MC cable		12.31	.650		29	29.50		58.50	80.50
2580	EMT & wire		5	1.600		37	73		110	160
2600	4-way, #14/3, type NM cable		14.55	.550		23.50	25		48.50	67
2620	Type MC cable		12.31	.650		37.50	29.50		67	90
2630	EMT & wire		5	1.600		45.50	73		118.50	169
2650	S.P., 20 amp, #12/2, type NM cable		13.33	.600		16.55	27.50		44.05	62.50
2670	Type MC cable		11.43	.700		26	32		58	80.50
2680	EMT & wire		4.85	1.649		41	75		116	168
2700	S.P., slide dimmer, type NM cable		17.10	.468		37	21.50		58.50	76
2720	Type MC cable		14.30	.559		47.50	25.50		73	94

26 05 90.10 Residential Wiring		Crew	Daily Output	Labor-Hours	Unit	Material	2019 Bare Costs Labor	Equipment	Total	Total Incl O&P
2730	EMT & wire	1 Elec	5.71	1.401	Ea.	60	64		124	170
2750	S.P., touch dimmer, type NM cable		17.10	.468		48.50	21.50		70	88
2770	Type MC cable		14.30	.559		59	25.50		84.50	107
2780	EMT & wire		5.71	1.401		71.50	64		135.50	183
2800	3-way touch dimmer, type NM cable		13.33	.600		49.50	27.50		77	99
2820	Type MC cable		11.43	.700		60.50	32		92.50	119
2830	EMT & wire		4.85	1.649		73	75		148	203
3000	Combination devices									
3100	S.P. switch/15 amp recpt., ivory, 1-gang box, plate									
3110	Type NM cable	1 Elec	11.43	.700	Ea.	23	32		55	77
3120	Type MC cable		10	.800		33.50	36.50		70	96.50
3130	EMT & wire		4.40	1.818		46	83		129	186
3150	S.P. switch/pilot light, type NM cable		11.43	.700		24	32		56	78.50
3170	Type MC cable		10	.800		35	36.50		71.50	98
3180	EMT & wire		4.43	1.806		47.50	82.50		130	186
3190	2-S.P. switches, 2-#14/2, no wiring		14	.571		13.25	26		39.25	57
3200	2-S.P. switches, 2-#14/2, type NM cables		10	.800		25	36.50		61.50	87
3220	Type MC cable		8.89	.900		40	41		81	111
3230	EMT & wire		4.10	1.951		48	89		137	198
3250	3-way switch/15 amp recpt., #14/3, type NM cable		10	.800		30.50	36.50		67	93
3270	Type MC cable		8.89	.900		44.50	41		85.50	116
3280	EMT & wire		4.10	1.951		52.50	89		141.50	203
3300	2-3 way switches, 2-#14/3, type NM cables		8.89	.900		39	41		80	110
3320	Type MC cable		8	1		60.50	45.50		106	141
3330	EMT & wire		4	2		59.50	91		150.50	214
3350	S.P. switch/20 amp recpt., #12/2, type NM cable		10	.800		39	36.50		75.50	102
3370	Type MC cable		8.89	.900		44.50	41		85.50	116
3380	EMT & wire		4.10	1.951		63	89		152	215
4000	Receptacle devices									
4010	Duplex outlet, 15 amp recpt., ivory, 1-gang box, plate									
4015	Type NM cable	1 Elec	14.55	.550	Ea.	8.70	25		33.70	50.50
4020	Type MC cable		12.31	.650		19.30	29.50		48.80	69.50
4030	EMT & wire		5.33	1.501		30.50	68.50		99	146
4050	With #12/2, type NM cable		12.31	.650		9.85	29.50		39.35	59.50
4070	Type MC cable		10.67	.750		19.40	34		53.40	77.50
4080	EMT & wire		4.71	1.699		34	77.50		111.50	164
4100	20 amp recpt., #12/2, type NM cable		12.31	.650		17.85	29.50		47.35	68
4120	Type MC cable		10.67	.750		27.50	34		61.50	86
4130	EMT & wire		4.71	1.699		42	77.50		119.50	173
4500	Weather-proof cover for above receptacles, add		32	.250		2.06	11.40		13.46	21
4550	Air conditioner outlet, 20 amp-240 volt recpt.									
4560	30′ of #12/2, 2 pole circuit breaker									
4570	Type NM cable	1 Elec	10	.800	Ea.	60	36.50		96.50	126
4580	Type MC cable		9	.889		71	40.50		111.50	145
4590	EMT & wire		4	2		85	91		176	242
4600	Decorator style, type NM cable		10	.800		65	36.50		101.50	131
4620	Type MC cable		9	.889		76	40.50		116.50	150
4630	EMT & wire		4	2		89.50	91		180.50	248
4650	Dryer outlet, 30 amp-240 volt recpt., 20′ of #10/3									
4660	2 pole circuit breaker									
4670	Type NM cable	1 Elec	6.41	1.248	Ea.	54	57		111	153
4680	Type MC cable		5.71	1.401		63.50	64		127.50	174
4690	EMT & wire		3.48	2.299		76	105		181	255

26 05 90.10 Residential Wiring		Crew	Daily Output	Labor-Hours	Unit	Material	Labor	Equipment	Total	Total Incl O&P
						2019 Bare Costs				
4700	Range outlet, 50 amp-240 volt recpt., 30′ of #8/3									
4710	Type NM cable	1 Elec	4.21	1.900	Ea.	82.50	86.50		169	232
4720	Type MC cable		4	2		127	91		218	289
4730	EMT & wire		2.96	2.703		107	123		230	320
4750	Central vacuum outlet, type NM cable		6.40	1.250		53.50	57		110.50	152
4770	Type MC cable		5.71	1.401		67.50	64		131.50	179
4780	EMT & wire	↓	3.48	2.299	↓	88	105		193	268
4800	30 amp-110 volt locking recpt., #10/2 circ. bkr.									
4810	Type NM cable	1 Elec	6.20	1.290	Ea.	63	59		122	165
4820	Type MC cable		5.40	1.481		80	67.50		147.50	198
4830	EMT & wire	↓	3.20	2.500	↓	99	114		213	295
4900	Low voltage outlets									
4910	Telephone recpt., 20′ of 4/C phone wire	1 Elec	26	.308	Ea.	8.55	14.05		22.60	32.50
4920	TV recpt., 20′ of RG59U coax wire, F type connector	"	16	.500	"	17.10	23		40.10	56
4950	Door bell chime, transformer, 2 buttons, 60′ of bellwire									
4970	Economy model	1 Elec	11.50	.696	Ea.	55	31.50		86.50	113
4980	Custom model		11.50	.696		105	31.50		136.50	168
4990	Luxury model, 3 buttons	↓	9.50	.842	↓	188	38.50		226.50	270
6000	Lighting outlets									
6050	Wire only (for fixture), type NM cable	1 Elec	32	.250	Ea.	5.80	11.40		17.20	25
6070	Type MC cable		24	.333		11.65	15.20		26.85	38
6080	EMT & wire		10	.800		22	36.50		58.50	84
6100	Box (4″), and wire (for fixture), type NM cable		25	.320		14.85	14.60		29.45	40.50
6120	Type MC cable		20	.400		20.50	18.25		38.75	53
6130	EMT & wire	↓	11	.727	↓	31.50	33		64.50	88.50
6200	Fixtures (use with line 6050 or 6100 above)									
6210	Canopy style, economy grade	1 Elec	40	.200	Ea.	22.50	9.10		31.60	40
6220	Custom grade		40	.200		52.50	9.10		61.60	72.50
6250	Dining room chandelier, economy grade		19	.421		83	19.20		102.20	123
6260	Custom grade		19	.421		335	19.20		354.20	395
6310	Kitchen fixture (fluorescent), economy grade		30	.267		71.50	12.15		83.65	98.50
6320	Custom grade		25	.320		149	14.60		163.60	188
6350	Outdoor, wall mounted, economy grade		30	.267		29.50	12.15		41.65	52.50
6360	Custom grade		30	.267		120	12.15		132.15	152
6370	Luxury grade		25	.320		243	14.60		257.60	291
6410	Outdoor PAR floodlights, 1 lamp, 150 watt		20	.400		26.50	18.25		44.75	59.50
6420	2 lamp, 150 watt each		20	.400		45	18.25		63.25	79.50
6425	Motion sensing, 2 lamp, 150 watt each		20	.400		110	18.25		128.25	151
6430	For infrared security sensor, add		32	.250		102	11.40		113.40	131
6450	Outdoor, quartz-halogen, 300 watt flood		20	.400		38.50	18.25		56.75	72
6600	Recessed downlight, round, pre-wired, 50 or 75 watt trim		30	.267		66	12.15		78.15	92.50
6610	With shower light trim		30	.267		90	12.15		102.15	119
6620	With wall washer trim		28	.286		92	13.05		105.05	123
6630	With eye-ball trim		28	.286		77	13.05		90.05	107
6700	Porcelain lamp holder		40	.200		3	9.10		12.10	18.20
6710	With pull switch		40	.200		11.30	9.10		20.40	27.50
6750	Fluorescent strip, 2-20 watt tube, wrap around diffuser, 24″		24	.333		48	15.20		63.20	77.50
6760	1-34 watt tube, 48″		24	.333		137	15.20		152.20	176
6770	2-34 watt tubes, 48″		20	.400		158	18.25		176.25	203
6800	Bathroom heat lamp, 1-250 watt		28	.286		31.50	13.05		44.55	56.50
6810	2-250 watt lamps	↓	28	.286	↓	76.50	13.05		89.55	106
6820	For timer switch, see Section 26 05 90.10 line 2400									
6900	Outdoor post lamp, incl. post, fixture, 35′ of #14/2									

	26 05 90.10 Residential Wiring	Crew	Daily Output	Labor-Hours	Unit	Material	Labor	Equipment	Total	Total Incl O&P
						2019 Bare Costs				
6910	Type NM cable	1 Elec	3.50	2.286	Ea.	299	104		403	500
6920	Photo-eye, add		27	.296		28.50	13.50		42	53
6950	Clock dial time switch, 24 hr., w/enclosure, type NM cable		11.43	.700		78.50	32		110.50	139
6970	Type MC cable		11	.727		89	33		122	152
6980	EMT & wire		4.85	1.649		100	75		175	234
7000	Alarm systems									
7050	Smoke detectors, box, #14/3, type NM cable	1 Elec	14.55	.550	Ea.	34.50	25		59.50	79
7070	Type MC cable		12.31	.650		44.50	29.50		74	97.50
7080	EMT & wire		5	1.600		52.50	73		125.50	177
7090	For relay output to security system, add					11.15			11.15	12.25
8000	Residential equipment									
8050	Disposal hook-up, incl. switch, outlet box, 3′ of flex									
8060	20 amp-1 pole circ. bkr., and 25′ of #12/2									
8070	Type NM cable	1 Elec	10	.800	Ea.	28	36.50		64.50	90.50
8080	Type MC cable		8	1		38.50	45.50		84	117
8090	EMT & wire		5	1.600		56	73		129	181
8100	Trash compactor or dishwasher hook-up, incl. outlet box,									
8110	3′ of flex, 15 amp-1 pole circ. bkr., and 25′ of #14/2									
8130	Type MC cable	1 Elec	8	1	Ea.	27.50	45.50		73	105
8140	EMT & wire	"	5	1.600	"	42.50	73		115.50	166
8150	Hot water sink dispenser hook-up, use line 8100									
8200	Vent/exhaust fan hook-up, type NM cable	1 Elec	32	.250	Ea.	5.80	11.40		17.20	25
8220	Type MC cable		24	.333		11.65	15.20		26.85	38
8230	EMT & wire		10	.800		22	36.50		58.50	84
8250	Bathroom vent fan, 50 CFM (use with above hook-up)									
8260	Economy model	1 Elec	15	.533	Ea.	19.45	24.50		43.95	61
8270	Low noise model		15	.533		45.50	24.50		70	89.50
8280	Custom model		12	.667		120	30.50		150.50	182
8300	Bathroom or kitchen vent fan, 110 CFM									
8310	Economy model	1 Elec	15	.533	Ea.	67	24.50		91.50	113
8320	Low noise model	"	15	.533	"	93.50	24.50		118	143
8350	Paddle fan, variable speed (w/o lights)									
8360	Economy model (AC motor)	1 Elec	10	.800	Ea.	131	36.50		167.50	204
8362	With light kit		10	.800		171	36.50		207.50	248
8370	Custom model (AC motor)		10	.800		340	36.50		376.50	435
8372	With light kit		10	.800		380	36.50		416.50	480
8380	Luxury model (DC motor)		8	1		330	45.50		375.50	435
8382	With light kit		8	1		370	45.50		415.50	480
8390	Remote speed switch for above, add		12	.667		34	30.50		64.50	86.50
8500	Whole house exhaust fan, ceiling mount, 36″, variable speed									
8510	Remote switch, incl. shutters, 20 amp-1 pole circ. bkr.									
8520	30′ of #12/2, type NM cable	1 Elec	4	2	Ea.	1,325	91		1,416	1,625
8530	Type MC cable		3.50	2.286		1,350	104		1,454	1,650
8540	EMT & wire		3	2.667		1,375	122		1,497	1,700
8600	Whirlpool tub hook-up, incl. timer switch, outlet box									
8610	3′ of flex, 20 amp-1 pole GFI circ. bkr.									
8620	30′ of #12/2, type NM cable	1 Elec	5	1.600	Ea.	128	73		201	259
8630	Type MC cable		4.20	1.905		135	87		222	290
8640	EMT & wire		3.40	2.353		150	107		257	340
8650	Hot water heater hook-up, incl. 1-2 pole circ. bkr., box;									
8660	3′ of flex, 20′ of #10/2, type NM cable	1 Elec	5	1.600	Ea.	28.50	73		101.50	150
8670	Type MC cable		4.20	1.905		41.50	87		128.50	188
8680	EMT & wire		3.40	2.353		49.50	107		156.50	230

26 05 Common Work Results for Electrical

26 05 90 – Residential Applications

26 05 90.10 Residential Wiring		Crew	Daily Output	Labor-Hours	Unit	2019 Bare Costs Material	Labor	Equipment	Total	Total Incl O&P
9000	Heating/air conditioning									
9050	Furnace/boiler hook-up, incl. firestat, local on-off switch									
9060	Emergency switch, and 40′ of type NM cable	1 Elec	4	2	Ea.	53	91		144	207
9070	Type MC cable		3.50	2.286		68	104		172	245
9080	EMT & wire		1.50	5.333		92	243		335	495
9100	Air conditioner hook-up, incl. local 60 amp disc. switch									
9110	3′ sealtite, 40 amp, 2 pole circuit breaker									
9130	40′ of #8/2, type NM cable	1 Elec	3.50	2.286	Ea.	170	104		274	355
9140	Type MC cable		3	2.667		235	122		357	455
9150	EMT & wire		1.30	6.154		218	281		499	700
9200	Heat pump hook-up, 1-40 & 1-100 amp 2 pole circ. bkr.									
9210	Local disconnect switch, 3′ sealtite									
9220	40′ of #8/2 & 30′ of #3/2									
9230	Type NM cable	1 Elec	1.30	6.154	Ea.	505	281		786	1,025
9240	Type MC cable		1.08	7.407		555	340		895	1,150
9250	EMT & wire		.94	8.511		520	390		910	1,200
9500	Thermostat hook-up, using low voltage wire									
9520	Heating only, 25′ of #18-3	1 Elec	24	.333	Ea.	6.20	15.20		21.40	32
9530	Heating/cooling, 25′ of #18-4	"	20	.400	"	8.15	18.25		26.40	39

26 09 Instrumentation and Control for Electrical Systems

26 09 13 – Electrical Power Monitoring

26 09 13.20 Voltage Monitor Systems		Crew	Daily Output	Labor-Hours	Unit	Material	Labor	Equipment	Total	Total Incl O&P
0010	**VOLTAGE MONITOR SYSTEMS** (test equipment)									
0100	AC voltage monitor system, 120/240 V, one-channel				Ea.	2,725			2,725	2,975
0110	Modem adapter					340			340	375
0120	Add-on detector only					1,425			1,425	1,575
0150	AC voltage remote monitor sys., 3 channel, 120, 230, or 480 V					4,925			4,925	5,425
0160	With internal modem					5,200			5,200	5,725
0170	Combination temperature and humidity probe					765			765	840
0180	Add-on detector only					3,575			3,575	3,925
0190	With internal modem					3,900			3,900	4,275

26 09 13.30 Smart Metering			Crew	Daily Output	Labor-Hours	Unit	Material	Labor	Equipment	Total	Total Incl O&P
0010	**SMART METERING**, In panel										
0100	Single phase, 120/208 volt, 100 amp	G	1 Elec	8.78	.911	Ea.	385	41.50		426.50	495
0120	200 amp	G		8.78	.911		390	41.50		431.50	495
0200	277 volt, 100 amp	G		8.78	.911		415	41.50		456.50	525
0220	200 amp	G		8.78	.911		400	41.50		441.50	510
1100	Three phase, 120/208 volt, 100 amp	G		4.69	1.706		645	78		723	835
1120	200 amp	G		4.69	1.706		850	78		928	1,050
1130	400 amp	G		4.69	1.706		850	78		928	1,050
1140	800 amp	G		4.69	1.706		765	78		843	965
1150	1,600 amp	G		4.69	1.706		775	78		853	980
1200	277/480 volt, 100 amp	G		4.69	1.706		915	78		993	1,125
1220	200 amp	G		4.69	1.706		795	78		873	1,000
1230	400 amp	G		4.69	1.706		820	78		898	1,025
1240	800 amp	G		4.69	1.706		825	78		903	1,025
1250	1,600 amp	G		4.69	1.706		830	78		908	1,025
2000	Data recorder, 8 meters	G		10.97	.729		1,450	33.50		1,483.50	1,650
2100	16 meters	G		8.53	.938		3,600	43		3,643	4,050
3000	Software package, per meter, basic	G					257			257	282

26 09 Instrumentation and Control for Electrical Systems

26 09 13 – Electrical Power Monitoring

26 09 13.30 Smart Metering		Crew	Daily Output	Labor-Hours	Unit	Material	2019 Bare Costs Labor	Equipment	Total	Total Incl O&P
3100	Premium	G			Ea.	680			680	745

26 09 23 – Lighting Control Devices

26 09 23.10 Energy Saving Lighting Devices

			Crew	Daily Output	Labor-Hours	Unit	Material	Labor	Equipment	Total	Total Incl O&P
0010	**ENERGY SAVING LIGHTING DEVICES**										
0100	Occupancy sensors, passive infrared ceiling mounted	G	1 Elec	7	1.143	Ea.	75	52		127	168
0110	Ultrasonic ceiling mounted	G		7	1.143		99.50	52		151.50	194
0120	Dual technology ceiling mounted	G		6.50	1.231		138	56		194	243
0150	Automatic wall switches	G		24	.333		74.50	15.20		89.70	107
0160	Daylighting sensor, manual control, ceiling mounted	G		7	1.143		158	52		210	258
0170	Remote and dimming control with remote controller	G		6.50	1.231		196	56		252	305
0200	Passive infrared ceiling mounted			6.50	1.231		32.50	56		88.50	128
0400	Remote power pack	G		10	.800		32.50	36.50		69	95.50
0450	Photoelectric control, S.P.S.T. 120 V	G		8	1		21	45.50		66.50	98
0500	S.P.S.T. 208 V/277 V	G		8	1		26	45.50		71.50	103
0550	D.P.S.T. 120 V	G		6	1.333		208	61		269	330
0600	D.P.S.T. 208 V/277 V	G		6	1.333		227	61		288	350
0650	S.P.D.T. 208 V/277 V	G		6	1.333		229	61		290	350
0660	Daylight level sensor, wall mounted, on/off or dimming	G	↓	8	1	↓	197	45.50		242.50	292

26 12 Medium-Voltage Transformers

26 12 19 – Pad-Mounted, Liquid-Filled, Medium-Voltage Transformers

26 12 19.20 Transformer, Liquid-Filled

		Crew	Daily Output	Labor-Hours	Unit	Material	Labor	Equipment	Total	Total Incl O&P
0010	**TRANSFORMER, LIQUID-FILLED** Pad mounted									
0020	5 kV or 15 kV primary, 277/480 volt secondary, 3 phase									
0050	225 kVA	R-3	.55	36.364	Ea.	14,800	1,650	236	16,686	19,100
0100	300 kVA		.45	44.444		17,500	2,000	288	19,788	22,900
0200	500 kVA		.40	50		22,200	2,250	325	24,775	28,400
0250	750 kVA	↓	.38	52.632	↓	28,600	2,375	340	31,315	35,800

26 22 Low-Voltage Transformers

26 22 13 – Low-Voltage Distribution Transformers

26 22 13.10 Transformer, Dry-Type

		Crew	Daily Output	Labor-Hours	Unit	Material	Labor	Equipment	Total	Total Incl O&P
0010	**TRANSFORMER, DRY-TYPE**									
0050	Single phase, 240/480 volt primary, 120/240 volt secondary									
0100	1 kVA	1 Elec	2	4	Ea.	395	182		577	735
0300	2 kVA		1.60	5		600	228		828	1,025
0500	3 kVA		1.40	5.714		750	261		1,011	1,250
0700	5 kVA	↓	1.20	6.667		1,225	305		1,530	1,850
0900	7.5 kVA	2 Elec	2.20	7.273		1,400	330		1,730	2,075
1100	10 kVA		1.60	10		1,725	455		2,180	2,650
1300	15 kVA		1.20	13.333		2,125	610		2,735	3,325
1500	25 kVA		1	16		2,600	730		3,330	4,050
1700	37.5 kVA		.80	20		2,825	910		3,735	4,600
1900	50 kVA		.70	22.857		3,425	1,050		4,475	5,450
2100	75 kVA	↓	.65	24.615		4,425	1,125		5,550	6,700
2110	100 kVA	R-3	.90	22.222	↓	5,775	1,000	144	6,919	8,125
2300	3 phase, 480 volt primary, 120/208 volt secondary									
2310	Ventilated, 3 kVA	1 Elec	1	8	Ea.	1,125	365		1,490	1,850
2700	6 kVA	↓	.80	10	↓	1,275	455		1,730	2,150

26 22 Low-Voltage Transformers

26 22 13 – Low-Voltage Distribution Transformers

26 22 13.10 Transformer, Dry-Type		Crew	Daily Output	Labor-Hours	Unit	Material	Labor	Equipment	Total	Total Incl O&P
						2019 Bare Costs				
2900	9 kVA	1 Elec	.70	11.429	Ea.	1,275	520		1,795	2,250
3100	15 kVA	2 Elec	1.10	14.545		1,875	665		2,540	3,150
3300	30 kVA		.90	17.778		1,800	810		2,610	3,300
3500	45 kVA		.80	20		1,925	910		2,835	3,600
3700	75 kVA		.70	22.857		2,575	1,050		3,625	4,525
3900	112.5 kVA	R-3	.90	22.222		4,225	1,000	144	5,369	6,425
4100	150 kVA		.85	23.529		5,150	1,050	153	6,353	7,575
4300	225 kVA		.65	30.769		9,000	1,375	200	10,575	12,400
4500	300 kVA		.55	36.364		9,350	1,650	236	11,236	13,200
4700	500 kVA		.45	44.444		18,800	2,000	288	21,088	24,300
4800	750 kVA		.35	57.143		22,800	2,575	370	25,745	29,700
5380	3 phase, 5 kV primary, 277/480 volt secondary									
5400	High voltage, 112.5 kVA	R-3	.85	23.529	Ea.	17,900	1,050	153	19,103	21,600
5440	500 kVA		.35	57.143		40,600	2,575	370	43,545	49,300
5480	2,000 kVA		.25	80		97,500	3,600	520	101,620	113,500
5590	15 kV primary, 277/480 volt secondary									
5600	High voltage, 112.5 kVA	R-3	.85	23.529	Ea.	29,200	1,050	153	30,403	34,000
5640	500 kVA		.35	57.143		53,500	2,575	370	56,445	63,500
5680	2,000 kVA		.25	80		103,000	3,600	520	107,120	119,500

26 22 16 – Low-Voltage Buck-Boost Transformers

26 22 16.10 Buck-Boost Transformer

		Crew	Daily Output	Labor-Hours	Unit	Material	Labor	Equipment	Total	Total Incl O&P
0010	**BUCK-BOOST TRANSFORMER**									
0100	Single phase, 120/240 V primary, 12/24 V secondary									
0200	0.10 kVA	1 Elec	8	1	Ea.	105	45.50		150.50	190
0400	0.25 kVA		5.70	1.404		147	64		211	266
0600	0.50 kVA		4	2		200	91		291	370
0800	0.75 kVA		3.10	2.581		262	118		380	480
1200	1.5 kVA		1.80	4.444		370	203		573	735
1600	3.0 kVA		1.40	5.714		595	261		856	1,075
2000	3 phase, 240 V primary, 208/120 V secondary, 15 kVA	2 Elec	2.40	6.667		2,275	305		2,580	3,000
2200	30 kVA		1.60	10		2,550	455		3,005	3,550
2400	45 kVA		1.40	11.429		3,050	520		3,570	4,225
2600	75 kVA		1.20	13.333		3,700	610		4,310	5,075
3000	150 kVA	R-3	1.10	18.182		6,125	820	118	7,063	8,225
3200	225 kVA	"	1	20		7,975	900	130	9,005	10,400

26 24 Switchboards and Panelboards

26 24 16 – Panelboards

26 24 16.10 Load Centers

		Crew	Daily Output	Labor-Hours	Unit	Material	Labor	Equipment	Total	Total Incl O&P
0010	**LOAD CENTERS** (residential type)									
0100	3 wire, 120/240 V, 1 phase, including 1 pole plug-in breakers									
0200	100 amp main lugs, indoor, 8 circuits	1 Elec	1.40	5.714	Ea.	98	261		359	535
0300	12 circuits		1.20	6.667		123	305		428	630
0400	Rainproof, 8 circuits		1.40	5.714		126	261		387	565
0500	12 circuits		1.20	6.667		151	305		456	660
0600	200 amp main lugs, indoor, 16 circuits	R-1A	1.80	8.889		203	365		568	820
0700	20 circuits		1.50	10.667		202	440		642	935
0800	24 circuits		1.30	12.308		253	505		758	1,100
1200	Rainproof, 16 circuits		1.80	8.889		245	365		610	865
1300	20 circuits		1.50	10.667		285	440		725	1,025

26 24 16.10 Load Centers		Crew	Daily Output	Labor-Hours	Unit	Material	2019 Bare Costs Labor	Equipment	Total	Total Incl O&P
1400	24 circuits	R-1A	1.30	12.308	Ea.	335	505		840	1,200
1500	30 circuits		1.20	13.333		395	545		940	1,325
1600	40 circuits		.80	20		510	820		1,330	1,900
1800	400 amp main lugs, indoor, 42 circuits		.72	22.222		1,125	910		2,035	2,750
1900	Rainproof, 42 circuits		.72	22.222		1,625	910		2,535	3,300
2200	Plug in breakers, 20 amp, 1 pole, 4 wire, 120/208 volts									
2210	125 amp main lugs, indoor, 12 circuits	1 Elec	1.20	6.667	Ea.	180	305		485	695
2300	18 circuits		.80	10		256	455		711	1,025
2400	Rainproof, 12 circuits		1.20	6.667		225	305		530	745
2500	18 circuits		.80	10		310	455		765	1,100
2600	200 amp main lugs, indoor, 24 circuits	R-1A	1.30	12.308		330	505		835	1,200
2700	30 circuits		1.20	13.333		380	545		925	1,325
2800	36 circuits		1	16		365	655		1,020	1,475
2900	42 circuits		.80	20		495	820		1,315	1,900
3000	Rainproof, 24 circuits		1.30	12.308		268	505		773	1,125
3100	30 circuits		1.20	13.333		300	545		845	1,225
3200	36 circuits		1	16		455	655		1,110	1,575
3300	42 circuits		.80	20		615	820		1,435	2,025

26 24 16.20 Panelboard and Load Center Circuit Breakers		Crew	Daily Output	Labor-Hours	Unit	Material	Labor	Equipment	Total	Total Incl O&P
0010	**PANELBOARD AND LOAD CENTER CIRCUIT BREAKERS**									
0050	Bolt-on, 10,000 amp I.C., 120 volt, 1 pole									
0100	15-50 amp	1 Elec	10	.800	Ea.	18.90	36.50		55.40	80.50
0200	60 amp		8	1		19.05	45.50		64.55	95.50
0300	70 amp		8	1		28	45.50		73.50	105
0350	240 volt, 2 pole									
0400	15-50 amp	1 Elec	8	1	Ea.	50.50	45.50		96	130
0500	60 amp		7.50	1.067		33.50	48.50		82	117
0600	80-100 amp		5	1.600		114	73		187	244
0700	3 pole, 15-60 amp		6.20	1.290		137	59		196	247
0800	70 amp		5	1.600		178	73		251	315
0900	80-100 amp		3.60	2.222		189	101		290	375
1000	22,000 amp I.C., 240 volt, 2 pole, 70-225 amp		2.70	2.963		560	135		695	835
1100	3 pole, 70-225 amp		2.30	3.478		270	159		429	555
2000	Plug-in panel or load center, 120/240 volt, to 60 amp, 1 pole		12	.667		6.15	30.50		36.65	56.50
2004	Circuit breaker, 120/240 volt, 20 A, 1 pole with NM cable		6.50	1.231		11	56		67	104
2006	30 A, 1 pole with NM cable		6.50	1.231		11	56		67	104
2010	2 pole		9	.889		25.50	40.50		66	94
2014	50 A, 2 pole with NM cable		5.50	1.455		30.50	66.50		97	142
2020	3 pole		7.50	1.067		84	48.50		132.50	172
2030	100 amp, 2 pole		6	1.333		103	61		164	212
2040	3 pole		4.50	1.778		115	81		196	259
2050	150-200 amp, 2 pole		3	2.667		257	122		379	480

26 24 16.30 Panelboards Commercial Applications		Crew	Daily Output	Labor-Hours	Unit	Material	Labor	Equipment	Total	Total Incl O&P
0010	**PANELBOARDS COMMERCIAL APPLICATIONS**									
0050	NQOD, w/20 amp 1 pole bolt-on circuit breakers									
0600	4 wire, 120/208 volts, 100 amp main lugs, 12 circuits	1 Elec	1	8	Ea.	990	365		1,355	1,700
0650	16 circuits		.75	10.667		1,150	485		1,635	2,050
0700	20 circuits		.65	12.308		1,300	560		1,860	2,375
0750	24 circuits		.60	13.333		1,250	610		1,860	2,375
0800	30 circuits		.53	15.094		1,625	690		2,315	2,925
0850	225 amp main lugs, 32 circuits	2 Elec	.90	17.778		1,825	810		2,635	3,325
0900	34 circuits		.84	19.048		1,875	870		2,745	3,475

26 24 Switchboards and Panelboards

26 24 16 – Panelboards

26 24 16.30 Panelboards Commercial Applications		Crew	Daily Output	Labor-Hours	Unit	2019 Bare Costs Material	Labor	Equipment	Total	Total Incl O&P
0950	36 circuits	2 Elec	.80	20	Ea.	1,900	910		2,810	3,600
1000	42 circuits		.68	23.529		2,150	1,075		3,225	4,125
1010	400 amp main lugs, 42 circs		.68	23.529		2,150	1,075		3,225	4,125
1200	NEHB, w/20 amp, 1 pole bolt-on circuit breakers									
1250	4 wire, 277/480 volts, 100 amp main lugs, 12 circuits	1 Elec	.88	9.091	Ea.	1,375	415		1,790	2,175
1300	20 circuits	"	.60	13.333		2,025	610		2,635	3,225
1350	225 amp main lugs, 24 circuits	2 Elec	.90	17.778		2,300	810		3,110	3,850
1400	30 circuits		.80	20		2,750	910		3,660	4,525
1448	32 circuits		4.90	3.265		3,200	149		3,349	3,775
1450	36 circuits		.72	22.222		3,200	1,025		4,225	5,175
1500	42 circuits		.60	26.667		3,650	1,225		4,875	6,000
1510	225 amp main lugs, NEMA 7, 12 circuits		.90	17.778		6,050	810		6,860	7,975
1590	24 circuits		.30	53.333		8,075	2,425		10,500	12,900

26 24 19 – Motor-Control Centers

26 24 19.30 Motor Control Center		Crew	Daily Output	Labor-Hours	Unit	Material	Labor	Equipment	Total	Total Incl O&P
0010	**MOTOR CONTROL CENTER** Consists of starters & structures									
0050	Starters, class 1, type B, comb. MCP, FVNR, with									
0100	control transformer, 10 HP, size 1, 12" high	1 Elec	2.70	2.963	Ea.	1,800	135		1,935	2,200
0200	25 HP, size 2, 18" high	2 Elec	4	4		2,050	182		2,232	2,575
0300	50 HP, size 3, 24" high		2	8		3,175	365		3,540	4,100
0350	75 HP, size 4, 24" high		1.60	10		4,225	455		4,680	5,400
0400	100 HP, size 4, 30" high		1.40	11.429		5,575	520		6,095	6,975
0500	200 HP, size 5, 48" high		1	16		8,325	730		9,055	10,400

26 24 19.40 Motor Starters and Controls		Crew	Daily Output	Labor-Hours	Unit	Material	Labor	Equipment	Total	Total Incl O&P
0010	**MOTOR STARTERS AND CONTROLS**									
0050	Magnetic, FVNR, with enclosure and heaters, 480 volt									
0080	2 HP, size 00	1 Elec	3.50	2.286	Ea.	199	104		303	390
0100	5 HP, size 0		2.30	3.478		290	159		449	580
0200	10 HP, size 1		1.60	5		266	228		494	660
0300	25 HP, size 2	2 Elec	2.20	7.273		500	330		830	1,100
0400	50 HP, size 3	"	1.80	8.889		815	405		1,220	1,550
1400	Combination, with fused switch, 5 HP, size 0	1 Elec	1.80	4.444		595	203		798	985
1600	10 HP, size 1	"	1.30	6.154		640	281		921	1,150
1800	25 HP, size 2	2 Elec	2	8		1,025	365		1,390	1,750

26 27 Low-Voltage Distribution Equipment

26 27 13 – Electricity Metering

26 27 13.10 Meter Centers and Sockets		Crew	Daily Output	Labor-Hours	Unit	Material	Labor	Equipment	Total	Total Incl O&P
0010	**METER CENTERS AND SOCKETS**									
0100	Sockets, single position, 4 terminal, 100 amp	1 Elec	3.20	2.500	Ea.	45	114		159	236
0200	150 amp		2.30	3.478		63.50	159		222.50	330
0300	200 amp		1.90	4.211		98	192		290	425
0400	Transformer rated, 20 amp		3.20	2.500		163	114		277	365
0500	Double position, 4 terminal, 100 amp		2.80	2.857		215	130		345	450
0600	150 amp		2.10	3.810		287	174		461	600
0700	200 amp		1.70	4.706		555	215		770	960
1100	Meter centers and sockets, three phase, single pos, 7 terminal, 100 amp		2.80	2.857		118	130		248	340
1200	200 amp		2.10	3.810		228	174		402	535
1400	400 amp		1.70	4.706		840	215		1,055	1,275
2590	Basic meter device									

26 27 13 – Electricity Metering

26 27 13.10 Meter Centers and Sockets		Crew	Daily Output	Labor-Hours	Unit	Material	2019 Bare Costs Labor	Equipment	Total	Total Incl O&P
2600	1P 3W 120/240 V 4 jaw 125A sockets, 3 meter	2 Elec	1	16	Ea.	360	730		1,090	1,600
2610	4 meter		.90	17.778		445	810		1,255	1,825
2620	5 meter		.80	20		530	910		1,440	2,075
2630	6 meter		.60	26.667		555	1,225		1,780	2,600
2640	7 meter		.56	28.571		1,425	1,300		2,725	3,675
2650	8 meter		.52	30.769		1,550	1,400		2,950	4,000
2660	10 meter		.48	33.333		1,925	1,525		3,450	4,600
2680	Rainproof 1P 3W 120/240 V 4 jaw 125A sockets									
2690	3 meter	2 Elec	1	16	Ea.	645	730		1,375	1,900
2700	4 meter		.90	17.778		770	810		1,580	2,175
2710	6 meter		.60	26.667		1,100	1,225		2,325	3,200
2720	7 meter		.56	28.571		1,425	1,300		2,725	3,675
2730	8 meter		.52	30.769		1,550	1,400		2,950	4,000
2750	1P 3W 120/240 V 4 jaw sockets									
2760	with 125A circuit breaker, 3 meter	2 Elec	1	16	Ea.	1,200	730		1,930	2,525
2770	4 meter		.90	17.778		1,525	810		2,335	3,000
2780	5 meter		.80	20		1,900	910		2,810	3,600
2790	6 meter		.60	26.667		2,225	1,225		3,450	4,425
2800	7 meter		.56	28.571		2,725	1,300		4,025	5,125
2810	8 meter		.52	30.769		3,050	1,400		4,450	5,650
2820	10 meter		.48	33.333		3,800	1,525		5,325	6,675
3250	1P 3W 120/240 V 4 jaw sockets									
3260	with 200A circuit breaker, 3 meter	2 Elec	1	16	Ea.	1,800	730		2,530	3,200
3270	4 meter		.90	17.778		2,450	810		3,260	4,025
3290	6 meter		.60	26.667		3,625	1,225		4,850	5,950
3300	7 meter		.56	28.571		4,250	1,300		5,550	6,800
3310	8 meter		.56	28.571		4,900	1,300		6,200	7,500

26 27 16 – Electrical Cabinets and Enclosures

26 27 16.10 Cabinets

		Crew	Daily Output	Labor-Hours	Unit	Material	Labor	Equipment	Total	Total Incl O&P
0010	**CABINETS**									
7000	Cabinets, current transformer									
7050	Single door, 24" H x 24" W x 10" D	1 Elec	1.60	5	Ea.	168	228		396	555
7100	30" H x 24" W x 10" D		1.30	6.154		177	281		458	655
7150	36" H x 24" W x 10" D		1.10	7.273		425	330		755	1,000
7200	30" H x 30" W x 10" D		1	8		245	365		610	865
7250	36" H x 30" W x 10" D		.90	8.889		320	405		725	1,000
7300	36" H x 36" W x 10" D		.80	10		325	455		780	1,100
7500	Double door, 48" H x 36" W x 10" D		.60	13.333		600	610		1,210	1,650
7550	24" H x 24" W x 12" D		1	8		192	365		557	805

26 27 23 – Indoor Service Poles

26 27 23.40 Surface Raceway

		Crew	Daily Output	Labor-Hours	Unit	Material	Labor	Equipment	Total	Total Incl O&P
0010	**SURFACE RACEWAY**									
0090	Metal, straight section									
0100	No. 500	1 Elec	100	.080	L.F.	1.27	3.65		4.92	7.35
0110	No. 700		100	.080		1.43	3.65		5.08	7.50
0200	No. 1000		90	.089		1.46	4.05		5.51	8.20
0400	No. 1500, small pancake		90	.089		2.20	4.05		6.25	9
0600	No. 2000, base & cover, blank		90	.089		2.24	4.05		6.29	9.05
0800	No. 3000, base & cover, blank		75	.107		4.38	4.86		9.24	12.75
1000	No. 4000, base & cover, blank		65	.123		7.45	5.60		13.05	17.35
1200	No. 6000, base & cover, blank		50	.160		12.70	7.30		20	26
2400	Fittings, elbows, No. 500		40	.200	Ea.	2.29	9.10		11.39	17.40

26 27 23 – Indoor Service Poles

26 27 23.40 Surface Raceway		Crew	Daily Output	Labor-Hours	Unit	2019 Bare Costs Material	Labor	Equipment	Total	Total Incl O&P
2800	Elbow cover, No. 2000	1 Elec	40	.200	Ea.	3.67	9.10		12.77	18.95
2880	Tee, No. 500		42	.190		4.23	8.70		12.93	18.80
2900	No. 2000		27	.296		12.20	13.50		25.70	35.50
3000	Switch box, No. 500		16	.500		14.75	23		37.75	53
3400	Telephone outlet, No. 1500		16	.500		17.90	23		40.90	56.50
3600	Junction box, No. 1500		16	.500		11.30	23		34.30	49.50
3800	Plugmold wired sections, No. 2000									
4000	1 circuit, 6 outlets, 3′ long	1 Elec	8	1	Ea.	38.50	45.50		84	117
4100	2 circuits, 8 outlets, 6′ long	"	5.30	1.509	"	61.50	69		130.50	180
9300	Non-metallic, straight section									
9310	7/16″ x 7/8″, base & cover, blank	1 Elec	160	.050	L.F.	1.73	2.28		4.01	5.60
9320	Base & cover w/adhesive		160	.050		1.62	2.28		3.90	5.50
9340	7/16″ x 1-5/16″, base & cover, blank		145	.055		1.90	2.52		4.42	6.20
9350	Base & cover w/adhesive		145	.055		2.23	2.52		4.75	6.55
9370	11/16″ x 2-1/4″, base & cover, blank		130	.062		2.68	2.81		5.49	7.55
9380	Base & cover w/adhesive		130	.062		3.10	2.81		5.91	8
9385	1-11/16″ x 5-1/4″, two compartment base & cover w/screws		80	.100		10.30	4.56		14.86	18.75
9400	Fittings, elbows, 7/16″ x 7/8″		50	.160	Ea.	1.94	7.30		9.24	14.05
9410	7/16″ x 1-5/16″		45	.178		2.01	8.10		10.11	15.45
9420	11/16″ x 2-1/4″		40	.200		2.17	9.10		11.27	17.30
9425	1-11/16″ x 5-1/4″		28	.286		14.05	13.05		27.10	37
9430	Tees, 7/16″ x 7/8″		35	.229		2.47	10.40		12.87	19.70
9440	7/16″ x 1-5/16″		32	.250		2.53	11.40		13.93	21.50
9450	11/16″ x 2-1/4″		30	.267		2.67	12.15		14.82	23
9455	1-11/16″ x 5-1/4″		24	.333		22.50	15.20		37.70	49.50
9460	Cover clip, 7/16″ x 7/8″		80	.100		.50	4.56		5.06	8
9470	7/16″ x 1-5/16″		72	.111		.45	5.05		5.50	8.75
9480	11/16″ x 2-1/4″		64	.125		.95	5.70		6.65	10.35
9484	1-11/16″ x 5-1/4″		42	.190		3.28	8.70		11.98	17.75
9486	Wire clip, 1-11/16″ x 5-1/4″		68	.118		.48	5.35		5.83	9.30
9490	Blank end, 7/16″ x 7/8″		50	.160		.86	7.30		8.16	12.85
9500	7/16″ x 1-5/16″		45	.178		.93	8.10		9.03	14.25
9510	11/16″ x 2-1/4″		40	.200		1.37	9.10		10.47	16.40
9515	1-11/16″ x 5-1/4″		38	.211		5.75	9.60		15.35	22
9520	Round fixture box, 5.5″ diam. x 1″		25	.320		11.25	14.60		25.85	36.50
9530	Device box, 1 gang		30	.267		5.20	12.15		17.35	25.50
9540	2 gang		25	.320		7.60	14.60		22.20	32.50

26 27 26 – Wiring Devices

26 27 26.10 Low Voltage Switching		Crew	Daily Output	Labor-Hours	Unit	Material	Labor	Equipment	Total	Total Incl O&P
0010	**LOW VOLTAGE SWITCHING**									
3600	Relays, 120 V or 277 V standard	1 Elec	12	.667	Ea.	46	30.50		76.50	100
3800	Flush switch, standard		40	.200		11.60	9.10		20.70	27.50
4000	Interchangeable		40	.200		15.20	9.10		24.30	31.50
4100	Surface switch, standard		40	.200		8.25	9.10		17.35	24
4200	Transformer 115 V to 25 V		12	.667		129	30.50		159.50	192
4400	Master control, 12 circuit, manual		4	2		129	91		220	291
4500	25 circuit, motorized		4	2		143	91		234	305
4600	Rectifier, silicon		12	.667		46	30.50		76.50	100
4800	Switchplates, 1 gang, 1, 2 or 3 switch, plastic		80	.100		5.05	4.56		9.61	13
5000	Stainless steel		80	.100		11.40	4.56		15.96	20
5400	2 gang, 3 switch, stainless steel		53	.151		23	6.90		29.90	37

26 27 Low-Voltage Distribution Equipment

26 27 26 – Wiring Devices

26 27 26.20 Wiring Devices Elements		Crew	Daily Output	Labor-Hours	Unit	Material	Labor	Equipment	Total	Total Incl O&P
						2019 Bare Costs				
0010	**WIRING DEVICES ELEMENTS**									
0200	Toggle switch, quiet type, single pole, 15 amp	1 Elec	40	.200	Ea.	.47	9.10		9.57	15.40
1650	Dimmer switch, 120 volt, incandescent, 600 watt, 1 pole G		16	.500		23	23		46	62
2460	Receptacle, duplex, 120 volt, grounded, 15 amp		40	.200		1.51	9.10		10.61	16.55
2470	20 amp		27	.296		9.55	13.50		23.05	32.50
2490	Dryer, 30 amp		15	.533		4.39	24.50		28.89	44.50
2500	Range, 50 amp		11	.727		10.80	33		43.80	66
2600	Wall plates, stainless steel, 1 gang		80	.100		2.58	4.56		7.14	10.30
2800	2 gang		53	.151		4.37	6.90		11.27	16.05
3200	Lampholder, keyless		26	.308		17.45	14.05		31.50	42
3400	Pullchain with receptacle	▼	22	.364	▼	21	16.60		37.60	50

26 28 Low-Voltage Circuit Protective Devices

26 28 16 – Enclosed Switches and Circuit Breakers

26 28 16.10 Circuit Breakers

		Crew	Daily Output	Labor-Hours	Unit	Material	Labor	Equipment	Total	Total Incl O&P
0010	**CIRCUIT BREAKERS** (in enclosure)									
0100	Enclosed (NEMA 1), 600 volt, 3 pole, 30 amp	1 Elec	3.20	2.500	Ea.	515	114		629	750
0200	60 amp		2.80	2.857		620	130		750	900
0400	100 amp		2.30	3.478		710	159		869	1,050
0500	200 amp		1.50	5.333		1,800	243		2,043	2,400
0600	225 amp	▼	1.50	5.333		1,650	243		1,893	2,200
0700	400 amp	2 Elec	1.60	10	▼	2,800	455		3,255	3,850

26 28 16.20 Safety Switches

		Crew	Daily Output	Labor-Hours	Unit	Material	Labor	Equipment	Total	Total Incl O&P
0010	**SAFETY SWITCHES**									
0100	General duty 240 volt, 3 pole NEMA 1, fusible, 30 amp	1 Elec	3.20	2.500	Ea.	85.50	114		199.50	280
0200	60 amp		2.30	3.478		145	159		304	420
0300	100 amp		1.90	4.211		246	192		438	585
0400	200 amp	▼	1.30	6.154		535	281		816	1,050
0500	400 amp	2 Elec	1.80	8.889	▼	1,400	405		1,805	2,175

26 29 Low-Voltage Controllers

26 29 13 – Enclosed Controllers

26 29 13.10 Contactors, AC

		Crew	Daily Output	Labor-Hours	Unit	Material	Labor	Equipment	Total	Total Incl O&P
0010	**CONTACTORS, AC** Enclosed (NEMA 1)									
0050	Lighting, 600 volt 3 pole, electrically held									
0100	20 amp	1 Elec	4	2	Ea.	380	91		471	565
0200	30 amp		3.60	2.222		254	101		355	445
0300	60 amp		3	2.667		545	122		667	800
0400	100 amp		2.50	3.200		835	146		981	1,150
0500	200 amp	▼	1.40	5.714		2,125	261		2,386	2,750
0600	300 amp	2 Elec	1.60	10		4,325	455		4,780	5,500
0800	600 volt 3 pole, mechanically held, 30 amp	1 Elec	3.60	2.222		605	101		706	830
0900	60 amp		3	2.667		1,225	122		1,347	1,550
1000	75 amp		2.80	2.857		1,500	130		1,630	1,875
1100	100 amp		2.50	3.200		1,550	146		1,696	1,975
1200	150 amp		2	4		3,525	182		3,707	4,175
1300	200 amp		1.40	5.714		5,300	261		5,561	6,250
1500	Magnetic with auxiliary contact, size 00, 9 amp	▼	4	2	▼	183	91		274	350

26 29 Low-Voltage Controllers

26 29 13 – Enclosed Controllers

26 29 13.10 Contactors, AC		Crew	Daily Output	Labor-Hours	Unit	Material	Labor	Equipment	Total	Total Incl O&P
						2019 Bare Costs				
1600	Size 0, 18 amp	1 Elec	4	2	Ea.	218	91		309	390
1700	Size 1, 27 amp		3.60	2.222		248	101		349	435
1800	Size 2, 45 amp		3	2.667		460	122		582	705
1900	Size 3, 90 amp		2.50	3.200		740	146		886	1,050
2000	Size 4, 135 amp		2.30	3.478		1,700	159		1,859	2,125
2100	Size 5, 270 amp	2 Elec	1.80	8.889		3,575	405		3,980	4,575
2310	Size 8, 1,215 amp	"	.80	20		21,900	910		22,810	25,500
2500	Magnetic, 240 volt, 1-2 pole, 0.75 HP motor	1 Elec	4	2		147	91		238	310
2520	2 HP motor		3.60	2.222		198	101		299	380
2540	5 HP motor		2.50	3.200		400	146		546	680
2560	10 HP motor		1.40	5.714		660	261		921	1,150
2600	240 volt or less, 3 pole,0 .75 HP motor		4	2		147	91		238	310
2620	5 HP motor		3.60	2.222		183	101		284	365
2640	10 HP motor		3.60	2.222		212	101		313	400
2660	15 HP motor		2.50	3.200		425	146		571	705
2700	25 HP motor		2.50	3.200		425	146		571	705
2720	30 HP motor	2 Elec	2.80	5.714		705	261		966	1,200
2740	40 HP motor		2.80	5.714		705	261		966	1,200
2760	50 HP motor		1.60	10		705	455		1,160	1,525
2800	75 HP motor		1.60	10		1,650	455		2,105	2,575
2820	100 HP motor		1	16		1,650	730		2,380	3,025
2860	150 HP motor		1	16		3,525	730		4,255	5,100
2880	200 HP motor		1	16		3,525	730		4,255	5,100
3000	600 volt, 3 pole, 5 HP motor	1 Elec	4	2		183	91		274	350
3020	10 HP motor		3.60	2.222		212	101		313	400
3040	25 HP motor		3	2.667		425	122		547	665
3100	50 HP motor		2.50	3.200		705	146		851	1,025
3160	100 HP motor	2 Elec	2.80	5.714		1,650	261		1,911	2,250
3220	200 HP motor	"	1.60	10		3,525	455		3,980	4,650

26 29 13.20 Control Stations

		Crew	Daily Output	Labor-Hours	Unit	Material	Labor	Equipment	Total	Total Incl O&P
0010	**CONTROL STATIONS**									
0050	NEMA 1, heavy duty, stop/start	1 Elec	8	1	Ea.	146	45.50		191.50	235
0100	Stop/start, pilot light		6.20	1.290		199	59		258	315
0200	Hand/off/automatic		6.20	1.290		108	59		167	215

26 32 Packaged Generator Assemblies

26 32 13 – Engine Generators

26 32 13.13 Diesel-Engine-Driven Generator Sets

		Crew	Daily Output	Labor-Hours	Unit	Material	Labor	Equipment	Total	Total Incl O&P
0010	**DIESEL-ENGINE-DRIVEN GENERATOR SETS**									
2000	Diesel engine, including battery, charger,									
2010	muffler, & day tank, 30 kW	R-3	.55	36.364	Ea.	11,200	1,650	236	13,086	15,200
2100	50 kW		.42	47.619		20,300	2,150	310	22,760	26,100
2200	75 kW		.35	57.143		23,600	2,575	370	26,545	30,600
2300	100 kW		.31	64.516		27,500	2,900	420	30,820	35,500

26 32 13.16 Gas-Engine-Driven Generator Sets

		Crew	Daily Output	Labor-Hours	Unit	Material	Labor	Equipment	Total	Total Incl O&P
0010	**GAS-ENGINE-DRIVEN GENERATOR SETS**									
0020	Gas or gasoline operated, includes battery,									
0050	charger, & muffler									
0200	7.5 kW	R-3	.83	24.096	Ea.	8,700	1,075	156	9,931	11,500
0300	11.5 kW		.71	28.169		12,300	1,275	183	13,758	15,900

26 32 Packaged Generator Assemblies

26 32 13 – Engine Generators

26 32 13.16 Gas-Engine-Driven Generator Sets	Crew	Daily Output	Labor-Hours	Unit	Material	2019 Bare Costs Labor	Equipment	Total	Total Incl O&P
0400 20 kW	R-3	.63	31.746	Ea.	14,500	1,425	206	16,131	18,600
0500 35 kW	↓	.55	36.364	↓	17,300	1,650	236	19,186	21,900

26 33 Battery Equipment

26 33 43 – Battery Chargers

26 33 43.55 Electric Vehicle Charging	Crew	Daily Output	Labor-Hours	Unit	Material	Labor	Equipment	Total	Total Incl O&P
0010 ELECTRIC VEHICLE CHARGING									
0020 Level 2, wall mounted									
2200 Heavy duty G	R-1A	15.36	1.042	Ea.	2,000	43		2,043	2,275
2210 with RFID G		12.29	1.302		3,375	53.50		3,428.50	3,800
2300 Free standing, single connector G		10.24	1.563		2,200	64		2,264	2,525
2310 with RFID G		8.78	1.822		4,475	75		4,550	5,050
2320 Double connector G		7.68	2.083		4,075	85.50		4,160.50	4,625
2330 with RFID G	↓	6.83	2.343	↓	7,475	96		7,571	8,375

26 33 53 – Static Uninterruptible Power Supply

26 33 53.10 Uninterruptible Power Supply/Conditioner Trans.	Crew	Daily Output	Labor-Hours	Unit	Material	Labor	Equipment	Total	Total Incl O&P
0010 UNINTERRUPTIBLE POWER SUPPLY/CONDITIONER TRANSFORMERS									
0100 Volt. regulating, isolating transf., w/invert. & 10 min. battery pack									
0110 Single-phase, 120 V, 0.35 kVA	1 Elec	2.29	3.493	Ea.	1,200	159		1,359	1,575
0120 0.5 kVA		2	4		1,250	182		1,432	1,675
0130 For additional 55 min. battery, add to 0.35 kVA		2.29	3.493		705	159		864	1,025
0140 Add to 0.5 kVA		1.14	7.018		740	320		1,060	1,325
0150 Single-phase, 120 V, 0.75 kVA		.80	10		1,575	455		2,030	2,500
0160 1.0 kVA	↓	.80	10		2,250	455		2,705	3,250
0170 1.5 kVA	2 Elec	1.14	14.035		3,900	640		4,540	5,325
0180 2 kVA	"	.89	17.978		4,225	820		5,045	6,000
0190 3 kVA	R-3	.63	31.746		5,200	1,425	206	6,831	8,275
0200 5 kVA		.42	47.619		7,425	2,150	310	9,885	12,000
0210 7.5 kVA		.33	60.606		9,475	2,725	395	12,595	15,300
0220 10 kVA		.28	71.429		12,200	3,225	465	15,890	19,200
0230 15 kVA		.22	90.909		15,600	4,100	590	20,290	24,600
0240 3 phase, 120/208 V input 120/208 V output, 20 kVA, incl 17 min. battery		.21	95.238		28,700	4,300	620	33,620	39,300
0242 30 kVA, incl 11 min. battery		.20	100		32,100	4,500	650	37,250	43,400
0250 40 kVA, incl 15 min. battery		.20	100		42,900	4,500	650	48,050	55,500
0260 480 V input 277/480 V output, 60 kVA, incl 6 min. battery		.19	105		48,200	4,750	685	53,635	61,500
0262 80 kVA, incl 4 min. battery		.19	105		57,500	4,750	685	62,935	71,500
0400 For additional 34 min./15 min. battery, add to 40 kVA	↓	.71	28.011	↓	15,000	1,250	182	16,432	18,800
0600 For complex & special design systems to meet specific									
0610 requirements, obtain quote from vendor									

26 35 26 – Harmonic Filters

26 35 26.10 Computer Isolation Transformer		Crew	Daily Output	Labor-Hours	Unit	Material	Labor	Equipment	Total	Total Incl O&P
0010	COMPUTER ISOLATION TRANSFORMER									
0100	Computer grade									
0110	Single-phase, 120/240 V, 0.5 kVA	1 Elec	4	2	Ea.	440	91		531	635
0120	1.0 kVA		2.67	2.996		630	137		767	915
0130	2.5 kVA		2	4		955	182		1,137	1,350
0140	5 kVA		1.14	7.018		1,100	320		1,420	1,725

26 35 26.20 Computer Regulator Transformer		Crew	Daily Output	Labor-Hours	Unit	Material	Labor	Equipment	Total	Total Incl O&P
0010	COMPUTER REGULATOR TRANSFORMER									
0100	Ferro-resonant, constant voltage, variable transformer									
0110	Single-phase, 240 V, 0.5 kVA	1 Elec	2.67	2.996	Ea.	530	137		667	805
0120	1.0 kVA		2	4		725	182		907	1,100
0130	2.0 kVA		1	8		1,250	365		1,615	1,975
0210	Plug-in unit 120 V, 0.14 kVA		8	1		305	45.50		350.50	410
0220	0.25 kVA		8	1		355	45.50		400.50	470
0230	0.5 kVA		8	1		530	45.50		575.50	655
0240	1.0 kVA		5.33	1.501		725	68.50		793.50	910
0250	2.0 kVA		4	2		1,250	91		1,341	1,525

26 35 26.30 Power Conditioner Transformer		Crew	Daily Output	Labor-Hours	Unit	Material	Labor	Equipment	Total	Total Incl O&P
0010	POWER CONDITIONER TRANSFORMER									
0100	Electronic solid state, buck-boost, transformer, w/tap switch									
0110	Single-phase, 115 V, 3.0 kVA, + or - 3% accuracy	2 Elec	1.60	10	Ea.	2,850	455		3,305	3,900
0120	208, 220, 230, or 240 V, 5.0 kVA, + or - 1.5% accuracy	3 Elec	1.60	15		3,700	685		4,385	5,200
0130	5.0 kVA, + or - 6% accuracy	2 Elec	1.14	14.035		3,300	640		3,940	4,675
0140	7.5 kVA, + or - 1.5% accuracy	3 Elec	1.50	16		4,675	730		5,405	6,350
0150	7.5 kVA, + or - 6% accuracy		1.60	15		3,900	685		4,585	5,425
0160	10.0 kVA, + or - 1.5% accuracy		1.33	18.045		6,250	825		7,075	8,225
0170	10.0 kVA, + or - 6% accuracy		1.41	17.021		5,325	775		6,100	7,125

26 35 26.40 Transient Voltage Suppressor Transformer		Crew	Daily Output	Labor-Hours	Unit	Material	Labor	Equipment	Total	Total Incl O&P
0010	TRANSIENT VOLTAGE SUPPRESSOR TRANSFORMER									
0110	Single-phase, 120 V, 1.8 kVA	1 Elec	4	2	Ea.	1,500	91		1,591	1,800
0120	3.6 kVA		4	2		2,350	91		2,441	2,725
0130	7.2 kVA		3.20	2.500		2,800	114		2,914	3,275
0150	240 V, 3.6 kVA		4	2		3,225	91		3,316	3,700
0160	7.2 kVA		4	2		3,800	91		3,891	4,325
0170	14.4 kVA		3.20	2.500		4,925	114		5,039	5,575
0210	Plug-in unit, 120 V, 1.8 kVA		8	1		995	45.50		1,040.50	1,175

26 35 53 – Voltage Regulators

26 35 53.10 Automatic Voltage Regulators		Crew	Daily Output	Labor-Hours	Unit	Material	Labor	Equipment	Total	Total Incl O&P
0010	AUTOMATIC VOLTAGE REGULATORS									
0100	Computer grade, solid state, variable transf. volt. regulator									
0110	Single-phase, 120 V, 8.6 kVA	2 Elec	1.33	12.030	Ea.	6,425	550		6,975	7,975
0120	17.3 kVA		1.14	14.035		7,600	640		8,240	9,400
0130	208/240 V, 7.5/8.6 kVA		1.33	12.030		6,425	550		6,975	7,975
0140	13.5/15.6 kVA		1.33	12.030		7,600	550		8,150	9,250
0150	27.0/31.2 kVA		1.14	14.035		9,575	640		10,215	11,600
0210	Two-phase, single control, 208/240 V, 15.0/17.3 kVA		1.14	14.035		7,600	640		8,240	9,400
0220	Individual phase control, 15.0/17.3 kVA		1.14	14.035		7,600	640		8,240	9,400
0230	30.0/34.6 kVA	3 Elec	1.33	18.045		9,575	825		10,400	11,900
0310	Three-phase single control, 208/240 V, 26/30 kVA	2 Elec	1	16		7,600	730		8,330	9,550
0320	380/480 V, 24/30 kVA	"	1	16		7,600	730		8,330	9,550
0330	43/54 kVA	3 Elec	1.33	18.045		13,800	825		14,625	16,600

26 35 Power Filters and Conditioners

26 35 53 – Voltage Regulators

26 35 53.10 Automatic Voltage Regulators		Crew	Daily Output	Labor-Hours	Unit	2019 Bare Costs Material	Labor	Equipment	Total	Total Incl O&P
0340	Individual phase control, 208 V, 26 kVA	3 Elec	1.33	18.045	Ea.	7,600	825		8,425	9,700
0350	52 kVA	R-3	.91	21.978		9,575	990	143	10,708	12,300
0360	340/480 V, 24/30 kVA	2 Elec	1	16		7,600	730		8,330	9,550
0370	43/54 kVA	"	1	16		9,575	730		10,305	11,700
0380	48/60 kVA	3 Elec	1.33	18.045		13,900	825		14,725	16,700
0390	86/108 kVA	R-3	.91	21.978		15,400	990	143	16,533	18,800
0500	Standard grade, solid state, variable transformer volt. regulator									
0510	Single-phase, 115 V, 2.3 kVA	1 Elec	2.29	3.493	Ea.	2,250	159		2,409	2,725
0520	4.2 kVA		2	4		3,600	182		3,782	4,250
0530	6.6 kVA		1.14	7.018		4,400	320		4,720	5,375
0540	13.0 kVA		1.14	7.018		7,600	320		7,920	8,875
0550	16.6 kVA	2 Elec	1.23	13.008		8,975	595		9,570	10,800
0610	230 V, 8.3 kVA		1.33	12.030		7,600	550		8,150	9,250
0620	21.4 kVA		1.23	13.008		8,975	595		9,570	10,800
0630	29.9 kVA		1.23	13.008		8,975	595		9,570	10,800
0710	460 V, 9.2 kVA		1.33	12.030		7,600	550		8,150	9,250
0720	20.7 kVA		1.23	13.008		8,975	595		9,570	10,800
0810	Three-phase, 230 V, 13.1 kVA	3 Elec	1.41	17.021		7,600	775		8,375	9,625
0820	19.1 kVA		1.41	17.021		8,975	775		9,750	11,200
0830	25.1 kVA		1.23	19.512		8,975	890		9,865	11,300
0840	57.8 kVA	R-3	.95	21.053		16,300	950	137	17,387	19,600
0850	74.9 kVA	"	.91	21.978		16,300	990	143	17,433	19,700
0910	460 V, 14.3 kVA	3 Elec	1.41	17.021		7,600	775		8,375	9,625
0920	19.1 kVA		1.41	17.021		8,975	775		9,750	11,200
0930	27.9 kVA		1.23	19.512		8,975	890		9,865	11,300
0940	59.8 kVA	R-3	1	20		16,300	900	130	17,330	19,500
0950	79.7 kVA		.95	21.053		18,200	950	137	19,287	21,700
0960	118 kVA		.95	21.053		19,200	950	137	20,287	22,800
1000	Laboratory grade, precision, electronic voltage regulator									
1110	Single-phase, 115 V, 0.5 kVA	1 Elec	2.29	3.493	Ea.	1,575	159		1,734	1,975
1120	1.0 kVA		2	4		1,675	182		1,857	2,150
1130	3.0 kVA		.80	10		2,350	455		2,805	3,325
1140	6.0 kVA	2 Elec	1.46	10.959		4,350	500		4,850	5,600
1150	10.0 kVA	3 Elec	1	24		5,650	1,100		6,750	8,000
1160	15.0 kVA	"	1.50	16		6,450	730		7,180	8,300
1210	230 V, 3.0 kVA	1 Elec	.80	10		2,650	455		3,105	3,675
1220	6.0 kVA	2 Elec	1.46	10.959		4,450	500		4,950	5,700
1230	10.0 kVA	3 Elec	1.71	14.035		5,900	640		6,540	7,550
1240	15.0 kVA	"	1.60	15		6,675	685		7,360	8,475

26 36 Transfer Switches

26 36 23 – Automatic Transfer Switches

26 36 23.10 Automatic Transfer Switch Devices		Crew	Daily Output	Labor-Hours	Unit	Material	Labor	Equipment	Total	Total Incl O&P
0010	**AUTOMATIC TRANSFER SWITCH DEVICES**									
0100	Switches, enclosed 480 volt, 3 pole, 30 amp	1 Elec	2.30	3.478	Ea.	3,375	159		3,534	3,975
0200	60 amp	"	1.90	4.211	"	3,375	192		3,567	4,050

26 41 13.13 Lightning Protection for Buildings		Crew	Daily Output	Labor-Hours	Unit	2019 Bare Costs Material	Labor	Equipment	Total	Total Incl O&P
0010	**LIGHTNING PROTECTION FOR BUILDINGS**									
0200	Air terminals & base, copper									
0400	3/8" diameter x 10" (to 75' high)	1 Elec	8	1	Ea.	24	45.50		69.50	101
0500	1/2" diameter x 12" (over 75' high)		8	1		28	45.50		73.50	105
0520	1/2" diameter x 24"		7.30	1.096		37.50	50		87.50	123
0540	1/2" diameter x 60"		6.70	1.194		68.50	54.50		123	165
1000	Aluminum, 1/2" diameter x 12" (to 75' high)		8	1		19.50	45.50		65	96
1020	1/2" diameter x 24"		7.30	1.096		21.50	50		71.50	105
1040	1/2" diameter x 60"		6.70	1.194		28	54.50		82.50	120
1100	5/8" diameter x 12" (over 75' high)		8	1		20.50	45.50		66	97
2000	Cable, copper, 220 lb. per thousand ft. (to 75' high)		320	.025	L.F.	3.05	1.14		4.19	5.20
2100	375 lb. per thousand ft. (over 75' high)		230	.035		5.55	1.59		7.14	8.70
2500	Aluminum, 101 lb. per thousand ft. (to 75' high)		280	.029		.95	1.30		2.25	3.18
2600	199 lb. per thousand ft. (over 75' high)		240	.033		1.40	1.52		2.92	4.02
3000	Arrester, 175 volt AC to ground		8	1	Ea.	146	45.50		191.50	236
3100	650 volt AC to ground		6.70	1.194	"	148	54.50		202.50	251
4000	Air terminals, copper									
4010	3/8" x 10"	1 Elec	50	.160	Ea.	7.70	7.30		15	20.50
4020	1/2" x 12"		38	.211		11.40	9.60		21	28
4030	5/8" x 12"		33	.242		16.10	11.05		27.15	36
4040	1/2" x 24"		30	.267		21	12.15		33.15	43
4050	1/2" x 60"		19	.421		52.50	19.20		71.70	89
4060	Air terminals, aluminum									
4070	3/8" x 10"	1 Elec	50	.160	Ea.	3.12	7.30		10.42	15.35
4080	1/2" x 12"		38	.211		3.41	9.60		13.01	19.40
4090	5/8" x 12"		33	.242		4.37	11.05		15.42	23
4100	1/2" x 24"		30	.267		5.30	12.15		17.45	25.50
4110	1/2" x 60"		19	.421		12.05	19.20		31.25	45
4200	Air terminal bases, copper									
4210	Adhesive bases, 1/2"	1 Elec	15	.533	Ea.	18.05	24.50		42.55	59.50
4215	Adhesive base, lightning air terminal					18.05			18.05	19.85
4220	Bolted bases	1 Elec	9	.889		16.45	40.50		56.95	84
4230	Hinged bases		7	1.143		27	52		79	115
4240	Side mounted bases		9	.889		16.30	40.50		56.80	84
4250	Offset point support		9	.889		27.50	40.50		68	96
4260	Concealed base assembly		5	1.600		83	73		156	211
4270	Tee connector base		13	.615		47	28		75	97.50
4280	Tripod base, 36" for 60" air terminal		7	1.143		39	52		91	128
4290	Intermediate base, 1/2"		15	.533		30	24.50		54.50	72.50
4300	Air terminal bases, aluminum									
4310	Adhesive bases, 1/2"	1 Elec	15	.533	Ea.	9.15	24.50		33.65	49.50
4320	Bolted bases		9	.889		16.10	40.50		56.60	83.50
4330	Hinged bases		7	1.143		19.05	52		71.05	106
4340	Side mounted bases		9	.889		11.85	40.50		52.35	79
4350	Offset point support		9	.889		13.30	40.50		53.80	80.50
4360	Concealed base assembly		5	1.600		42	73		115	166
4370	Tee connector base		13	.615		14.65	28		42.65	62
4380	Tripod base, 36" for 60" air terminal		7	1.143		29.50	52		81.50	118
4390	Intermediate base, 1/2"		15	.533		13.70	24.50		38.20	54.50
4400	Connector cable, copper									
4410	Through wall connector	1 Elec	7	1.143	Ea.	84	52		136	178
4420	Through roof connector		5	1.600		103	73		176	233
4430	Beam connector		7	1.143		27	52		79	115

26 41 Facility Lightning Protection

26 41 13 – Lightning Protection for Structures

26 41 13.13 Lightning Protection for Buildings		Crew	Daily Output	Labor-Hours	Unit	2019 Bare Costs Material	Labor	Equipment	Total	Total Incl O&P
4440	Double bolt connector	1 Elec	38	.211	Ea.	11.70	9.60		21.30	28.50
4450	Bar, copper	↓	40	.200	↓	12.95	9.10		22.05	29
4500	Connector cable, aluminum									
4510	Through wall connector	1 Elec	7	1.143	Ea.	33.50	52		85.50	122
4520	Through roof connector		5	1.600		36.50	73		109.50	159
4530	Beam connector		7	1.143		18.30	52		70.30	105
4540	Double bolt connector	↓	38	.211	↓	6.20	9.60		15.80	22.50
4600	Bonding plates, copper									
4610	I-beam, 8" square	1 Elec	7	1.143	Ea.	21.50	52		73.50	109
4620	Purlin, 8" square		7	1.143		26.50	52		78.50	114
4630	Large heavy duty, 16" square		5	1.600		32	73		105	154
4640	Aluminum, I-beam, 8" square		7	1.143		16.60	52		68.60	103
4650	Purlin, 8" square		7	1.143		16.55	52		68.55	103
4660	Large heavy duty, 16" square		5	1.600		18	73		91	139
4700	Cable support, copper, loop fastener		38	.211		1.11	9.60		10.71	16.85
4710	Adhesive cable holder		38	.211		1.45	9.60		11.05	17.25
4720	Aluminum, loop fastener		38	.211		.72	9.60		10.32	16.45
4730	Adhesive cable holder		38	.211		.78	9.60		10.38	16.50
4740	Bonding strap, copper, 3/4" x 9-1/2"		15	.533		11.40	24.50		35.90	52
4750	Aluminum, 3/4" x 9-1/2"		15	.533		4.85	24.50		29.35	45
4760	Swivel adapter, copper, 1/2"		73	.110		17.15	5		22.15	27
4770	Aluminum, 1/2"	↓	73	.110	↓	9.15	5		14.15	18.20

26 51 Interior Lighting

26 51 13 – Interior Lighting Fixtures, Lamps, and Ballasts

26 51 13.50 Interior Lighting Fixtures			Crew	Daily Output	Labor-Hours	Unit	Material	Labor	Equipment	Total	Total Incl O&P
0010	**INTERIOR LIGHTING FIXTURES** Including lamps, mounting										
0030	hardware and connections										
0100	Fluorescent, C.W. lamps, troffer, recess mounted in grid, RS										
0130	Grid ceiling mount										
0200	Acrylic lens, 1' W x 4' L, two 40 watt		1 Elec	5.70	1.404	Ea.	51.50	64		115.50	161
0300	2' W x 2' L, two U40 watt			5.70	1.404		55	64		119	165
0504	Three 40 watt w/armored cable	G		3.40	2.353		75	107		182	258
0600	2' W x 4' L, four 40 watt			4.70	1.702		62.50	77.50		140	196
0604	Four 40 watt framed w/armored cable	G		3.20	2.500		150	114		264	350
0910	Acrylic lens, 1' W x 4' L, two 32 watt T8	G		5.70	1.404		66.50	64		130.50	177
0930	2' W x 2' L, two U32 watt T8	G		5.70	1.404		96	64		160	210
0940	2' W x 4' L, two 32 watt T8	G		5.30	1.509		76	69		145	196
0950	2' W x 4' L, three 32 watt T8	G		5	1.600		70	73		143	196
0960	2' W x 4' L, four 32 watt T8	G	↓	4.70	1.702	↓	74.50	77.50		152	209
1000	Surface mounted, RS										
1350	1' W x 4' L, sim. wood grain, two 40 watt	G	1 Elec	6.20	1.290	Ea.	148	59		207	258
2000	2' W x 8' L, eight 40 watt		2 Elec	6.20	2.581	"	180	118		298	390
2100	Strip fixture										
2130	Surface mounted										
2150	2' long, two 40 watt, RS	G	1 Elec	8	1	Ea.	33.50	45.50		79	112
2200	4' long, one 40 watt, RS			8.50	.941		31	43		74	104
2300	4' long, two 40 watt, RS			8	1		43.50	45.50		89	123
2304	4' long, 2 lamp & armored cable	G	↓	4.50	1.778		59	81		140	197
2600	8' long, one 75 watt, SL		2 Elec	13.40	1.194		55	54.50		109.50	150
2700	8' long, two 75 watt, SL		"	12.40	1.290	↓	66.50	59		125.50	169

26 51 13 – Interior Lighting Fixtures, Lamps, and Ballasts

26 51 13.50 Interior Lighting Fixtures

	26 51 13.50 Interior Lighting Fixtures		Crew	Daily Output	Labor-Hours	Unit	Material	Labor	Equipment	Total	Total Incl O&P
							2019 Bare Costs				
2704	8′ long, 2 lamp & armored cable	G	1 Elec	3.65	2.192	Ea.	82	100		182	253
2930	4′ long, one 32 watt T8, staggered cove mtd.	G	"	8	1	↓	90	45.50		135.50	174
3000	Strip, pendent mounted, industrial, white porcelain enamel										
3402	4′ long, three 32 watt	G	1 Elec	7	1.143	Ea.	97	52		149	192
3404	4′ long, four 32 watt	G		6.50	1.231		87.50	56		143.50	188
3405	Parabolic louver, 4′ long, continuous row	G		5.70	1.404		99	64		163	213
3414	4′ long, 2 lamp w/cover & armored cable	G		3.70	2.162		102	98.50		200.50	273
3444	8′ long, 2 lamp w/cover & armored cable	G		3.10	2.581		186	118		304	395
3465	Hospital type, 4′ long, two 40 watt, RS	G		4.80	1.667		206	76		282	350
3535	Downlight, recess mounted	G		8	1		148	45.50		193.50	237
3540	Wall wash reflector, recess mounted	G		8	1		110	45.50		155.50	196
3550	Direct/indirect, 4′ long, stl., pendent mtd.	G		5	1.600		162	73		235	298
3560	4′ long, alum., pendent mtd.	G		5	1.600		335	73		408	490
3565	Prefabricated cove, 4′ long, stl. continuous row	G		5	1.600		203	73		276	345
3570	4′ long, alum. continuous row	G	↓	5	1.600	↓	345	73		418	500
4220	Metal halide, integral ballast, ceiling, recess mounted										
4230	prismatic glass lens, floating door										
4240	2′ W x 2′ L, 250 watt		1 Elec	3.20	2.500	Ea.	263	114		377	475
4250	2′ W x 2′ L, 400 watt		2 Elec	5.80	2.759		365	126		491	610
4260	Surface mounted, 2′ W x 2′ L, 250 watt		1 Elec	2.70	2.963		345	135		480	595
4270	400 watt		2 Elec	4.80	3.333	↓	405	152		557	695
4280	High bay, aluminum reflector,										
4290	Single unit, 400 watt		2 Elec	4.60	3.478	Ea.	410	159		569	715
4300	Single unit, 1,000 watt			4	4		595	182		777	950
4310	Twin unit, 400 watt		↓	3.20	5		825	228		1,053	1,275
4320	Low bay, aluminum reflector, 250W DX lamp		1 Elec	3.20	2.500	↓	365	114		479	585
4450	Incandescent, high hat can, round alzak reflector, prewired										
4470	100 watt		1 Elec	8	1	Ea.	68	45.50		113.50	149
4480	150 watt			8	1		104	45.50		149.50	190
4500	300 watt		↓	6.70	1.194	↓	241	54.50		295.50	355
5200	Ceiling, surface mounted, opal glass drum										
5300	8″, one 60 watt lamp		1 Elec	10	.800	Ea.	66.50	36.50		103	133
5400	10″, two 60 watt lamps			8	1		74	45.50		119.50	156
5500	12″, four 60 watt lamps			6.70	1.194		104	54.50		158.50	204
6900	Mirror light, fluorescent, RS, acrylic enclosure, two 40 watt			8	1		113	45.50		158.50	199
6910	One 40 watt			8	1		97	45.50		142.50	182
6920	One 20 watt		↓	12	.667	↓	83.50	30.50		114	141

26 51 13.55 Interior LED Fixtures

	26 51 13.55 Interior LED Fixtures		Crew	Daily Output	Labor-Hours	Unit	Material	Labor	Equipment	Total	Total Incl O&P
0010	**INTERIOR LED FIXTURES** Incl. lamps and mounting hardware										
0100	Downlight, recess mounted, 7.5″ diameter, 25 watt	G	1 Elec	8	1	Ea.	335	45.50		380.50	445
0120	10″ diameter, 36 watt	G		8	1		360	45.50		405.50	470
0160	cylinder, 10 watts	G		8	1		104	45.50		149.50	189
0180	20 watts	G		8	1		129	45.50		174.50	217
1000	Troffer, recess mounted, 2′ x 4′, 3,200 lumens	G		5.30	1.509		138	69		207	263
1010	4,800 lumens	G		5	1.600		178	73		251	315
1020	6,400 lumens	G		4.70	1.702		190	77.50		267.50	335
1100	Troffer retrofit lamp, 38 watt	G		21	.381		69.50	17.35		86.85	105
1110	60 watt	G		20	.400		142	18.25		160.25	186
1120	100 watt	G		18	.444		206	20.50		226.50	260
1200	Troffer, volumetric recess mounted, 2′ x 2′	G		5.70	1.404		320	64		384	455
2000	Strip, surface mounted, one light bar 4′ long, 3,500 K	G		8.50	.941		310	43		353	410
2010	5,000 K	G	↓	8	1	↓	276	45.50		321.50	380

26 51 Interior Lighting

26 51 13 – Interior Lighting Fixtures, Lamps, and Ballasts

26 51 13.55 Interior LED Fixtures			Crew	Daily Output	Labor-Hours	Unit	Material	Labor	Equipment	Total	Total Incl O&P
							2019 Bare Costs				
2020	Two light bar 4' long, 5,000 K	G	1 Elec	7	1.143	Ea.	430	52		482	560
3000	Linear, suspended mounted, one light bar 4' long, 37 watt	G	↓	6.70	1.194		168	54.50		222.50	273
3010	One light bar 8' long, 74 watt	G	2 Elec	12.20	1.311		310	60		370	440
3020	Two light bar 4' long, 74 watt	G	1 Elec	5.70	1.404		335	64		399	470
3030	Two light bar 8' long, 148 watt	G	2 Elec	8.80	1.818		380	83		463	550
4000	High bay, surface mounted, round, 150 watts	G		5.41	2.959		425	135		560	690
4010	2 bars, 164 watts	G		5.41	2.959		385	135		520	645
4020	3 bars, 246 watts	G		5.01	3.197		520	146		666	810
4030	4 bars, 328 watts	G	↓	4.60	3.478		750	159		909	1,075
4040	5 bars, 410 watts	G	3 Elec	4.20	5.716		840	261		1,101	1,350
4050	6 bars, 492 watts	G		3.80	6.324		940	288		1,228	1,500
4060	7 bars, 574 watts	G		3.39	7.075		1,025	325		1,350	1,650
4070	8 bars, 656 watts	G	↓	2.99	8.029		1,050	365		1,415	1,750
5000	Track, lighthead, 6 watt	G	1 Elec	32	.250		55.50	11.40		66.90	79.50
5010	9 watt	G	"	32	.250		63	11.40		74.40	88
6000	Garage, surface mounted, 103 watts	G	2 Elec	6.50	2.462		990	112		1,102	1,275
6100	pendent mounted, 80 watts	G		6.50	2.462		720	112		832	975
6200	95 watts	G		6.50	2.462		855	112		967	1,125
6300	125 watts	G	↓	6.50	2.462	↓	905	112		1,017	1,175

26 52 Safety Lighting

26 52 13 – Emergency and Exit Lighting

26 52 13.10 Emergency Lighting and Battery Units

		Crew	Daily Output	Labor-Hours	Unit	Material	Labor	Equipment	Total	Total Incl O&P
0010	**EMERGENCY LIGHTING AND BATTERY UNITS**									
0300	Emergency light units, battery operated									
0350	Twin sealed beam light, 25 W, 6 V each									
0500	Lead battery operated	1 Elec	4	2	Ea.	144	91		235	305
0700	Nickel cadmium battery operated		4	2		345	91		436	525
0780	Additional remote mount, sealed beam, 25 W 6 V		26.70	.300		29.50	13.65		43.15	55
0790	Twin sealed beam light, 25 W 6 V each	↓	26.70	.300	↓	62.50	13.65		76.15	91

26 52 13.16 Exit Signs

		Crew	Daily Output	Labor-Hours	Unit	Material	Labor	Equipment	Total	Total Incl O&P
0010	**EXIT SIGNS**									
0080	Exit light ceiling or wall mount, incandescent, single face	1 Elec	8	1	Ea.	72.50	45.50		118	155
0100	Double face		6.70	1.194		51	54.50		105.50	145
0120	Explosion proof		3.80	2.105		560	96		656	775
0150	Fluorescent, single face		8	1		84	45.50		129.50	167
0160	Double face	↓	6.70	1.194	↓	76.50	54.50		131	173

26 55 Special Purpose Lighting

26 55 33 – Hazard Warning Lighting

26 55 33.10 Warning Beacons

		Crew	Daily Output	Labor-Hours	Unit	Material	Labor	Equipment	Total	Total Incl O&P
0010	**WARNING BEACONS**									
0015	Surface mount with colored or clear lens									
0100	Rotating beacon									
0110	120V, 40 watt halogen	1 Elec	3.50	2.286	Ea.	107	104		211	287
0120	24V, 20 watt halogen	"	3.50	2.286	"	283	104		387	480
0200	Steady beacon									
0210	120V, 40 watt halogen	1 Elec	3.50	2.286	Ea.	109	104		213	290
0220	24V, 20 watt	↓	3.50	2.286	↓	112	104		216	293

26 55 Special Purpose Lighting

26 55 33 – Hazard Warning Lighting

26 55 33.10 Warning Beacons			Crew	Daily Output	Labor-Hours	Unit	Material	2019 Bare Costs Labor	Equipment	Total	Total Incl O&P
0230	12V DC, incandescent		1 Elec	3.50	2.286	Ea.	104	104		208	284
0300	Flashing beacon										
0310	120V, 40 watt halogen		1 Elec	3.50	2.286	Ea.	106	104		210	287
0320	24V, 20 watt halogen			3.50	2.286		107	104		211	287
0410	12V DC with two 6V lantern batteries			7	1.143		108	52		160	204

26 55 59 – Display Lighting

26 55 59.10 Track Lighting			Crew	Daily Output	Labor-Hours	Unit	Material	Labor	Equipment	Total	Total Incl O&P
0010	**TRACK LIGHTING**										
0080	Track, 1 circuit, 4′ section		1 Elec	6.70	1.194	Ea.	43	54.50		97.50	137
0100	8′ section		2 Elec	10.60	1.509		71	69		140	190
0200	12′ section		"	8.80	1.818		118	83		201	264
0300	3 circuits, 4′ section		1 Elec	6.70	1.194		129	54.50		183.50	231
0400	8′ section		2 Elec	10.60	1.509		133	69		202	258
0500	12′ section		"	8.80	1.818		160	83		243	310
1000	Feed kit, surface mounting		1 Elec	16	.500		14.95	23		37.95	53.50
1100	End cover			24	.333		8.25	15.20		23.45	34
1200	Feed kit, stem mounting, 1 circuit			16	.500		51	23		74	93
1300	3 circuit			16	.500		51	23		74	93
2000	Electrical joiner, for continuous runs, 1 circuit			32	.250		39.50	11.40		50.90	62
2100	3 circuit			32	.250		83	11.40		94.40	110
2200	Fixtures, spotlight, 75 W PAR halogen			16	.500		51.50	23		74.50	94
2210	50 W MR16 halogen			16	.500		173	23		196	227
3000	Wall washer, 250 W tungsten halogen			16	.500		133	23		156	183
3100	Low voltage, 25/50 W, 1 circuit			16	.500		134	23		157	184
3120	3 circuit			16	.500		197	23		220	254

26 56 Exterior Lighting

26 56 13 – Lighting Poles and Standards

26 56 13.10 Lighting Poles			Crew	Daily Output	Labor-Hours	Unit	Material	Labor	Equipment	Total	Total Incl O&P
0010	**LIGHTING POLES**										
6420	Wood pole, 4-1/2" x 5-1/8", 8′ high		1 Elec	6	1.333	Ea.	385	61		446	525
6440	12′ high			5.70	1.404		560	64		624	720
6460	20′ high			4	2		795	91		886	1,025

26 56 19 – LED Exterior Lighting

26 56 19.60 Parking LED Lighting			Crew	Daily Output	Labor-Hours	Unit	Material	Labor	Equipment	Total	Total Incl O&P
0010	**PARKING LED LIGHTING**										
0100	Round pole mounting, 88 lamp watts	G	1 Elec	2	4	Ea.	1,225	182		1,407	1,650

26 56 23 – Area Lighting

26 56 23.10 Exterior Fixtures			Crew	Daily Output	Labor-Hours	Unit	Material	Labor	Equipment	Total	Total Incl O&P
0010	**EXTERIOR FIXTURES** With lamps										
0400	Quartz, 500 watt		1 Elec	5.30	1.509	Ea.	61.50	69		130.50	180
1100	Wall pack, low pressure sodium, 35 watt			4	2		202	91		293	370
1150	55 watt			4	2		240	91		331	415

26 56 23.55 Exterior LED Fixtures			Crew	Daily Output	Labor-Hours	Unit	Material	Labor	Equipment	Total	Total Incl O&P
0010	**EXTERIOR LED FIXTURES**										
0100	Wall mounted, indoor/outdoor, 12 watt	G	1 Elec	10	.800	Ea.	212	36.50		248.50	293
0110	32 watt	G		10	.800		475	36.50		511.50	585
0120	66 watt	G		10	.800		490	36.50		526.50	600
0200	outdoor, 110 watt	G		10	.800		1,200	36.50		1,236.50	1,375

26 56 Exterior Lighting

26 56 23 – Area Lighting

26 56 23.55 Exterior LED Fixtures			Crew	Daily Output	Labor-Hours	Unit	Material	2019 Bare Costs Labor	Equipment	Total	Total Incl O&P
0210	220 watt	G	1 Elec	10	.800	Ea.	1,800	36.50		1,836.50	2,025
0300	modular, type IV, 120 V, 50 lamp watts	G		9	.889		1,175	40.50		1,215.50	1,375
0310	101 lamp watts	G		9	.889		1,325	40.50		1,365.50	1,525
0320	126 lamp watts	G		9	.889		1,650	40.50		1,690.50	1,900
0330	202 lamp watts	G		9	.889		1,925	40.50		1,965.50	2,200
0340	240 V, 50 lamp watts	G		8	1		1,250	45.50		1,295.50	1,450
0350	101 lamp watts	G		8	1		1,375	45.50		1,420.50	1,575
0360	126 lamp watts	G		8	1		1,450	45.50		1,495.50	1,675
0370	202 lamp watts	G		8	1		1,925	45.50		1,970.50	2,200
0400	wall pack, glass, 13 lamp watts	G		4	2		455	91		546	650
0410	poly w/photocell, 26 lamp watts	G		4	2		360	91		451	550
0420	50 lamp watts	G		4	2		780	91		871	1,000
0430	replacement, 40 watts	G		4	2		390	91		481	580
0440	60 watts	G	↓	4	2	↓	400	91		491	590

26 56 26 – Landscape Lighting

26 56 26.20 Landscape Fixtures		Crew	Daily Output	Labor-Hours	Unit	Material	Labor	Equipment	Total	Total Incl O&P
0010	**LANDSCAPE FIXTURES**									
7380	Landscape recessed uplight, incl. housing, ballast, transformer									
7390	& reflector, not incl. conduit, wire, trench									
7420	Incandescent, 250 watt	1 Elec	5	1.600	Ea.	635	73		708	820
7440	Quartz, 250 watt	"	5	1.600	"	605	73		678	785

26 56 26.50 Landscape LED Fixtures		Crew	Daily Output	Labor-Hours	Unit	Material	Labor	Equipment	Total	Total Incl O&P
0010	**LANDSCAPE LED FIXTURES**									
0100	12 volt alum bullet hooded-BLK	1 Elec	5	1.600	Ea.	97	73		170	226
0200	12 volt alum bullet hooded-BRZ		5	1.600		97	73		170	226
0300	12 volt alum bullet hooded-GRN		5	1.600		97	73		170	226
1000	12 volt alum large bullet hooded-BLK		5	1.600		71.50	73		144.50	198
1100	12 volt alum large bullet hooded-BRZ		5	1.600		71.50	73		144.50	198
1200	12 volt alum large bullet hooded-GRN		5	1.600		71.50	73		144.50	198
2000	12 volt large bullet landscape light fixture		5	1.600		71.50	73		144.50	198
2100	12 volt alum light large bullet		5	1.600		71.50	73		144.50	198
2200	12 volt alum bullet light	↓	5	1.600	↓	71.50	73		144.50	198

26 56 33 – Walkway Lighting

26 56 33.10 Walkway Luminaire		Crew	Daily Output	Labor-Hours	Unit	Material	Labor	Equipment	Total	Total Incl O&P
0010	**WALKWAY LUMINAIRE**									
6500	Bollard light, lamp & ballast, 42" high with polycarbonate lens									
7200	Incandescent, 150 watt	1 Elec	3	2.667	Ea.	715	122		837	985

26 56 36 – Flood Lighting

26 56 36.55 LED Floodlights			Crew	Daily Output	Labor-Hours	Unit	Material	Labor	Equipment	Total	Total Incl O&P
0010	**LED FLOODLIGHTS** with ballast and lamp,										
0020	Pole mounted, pole not included										
0100	11 watt	G	1 Elec	4	2	Ea.	435	91		526	630
0110	46 watt	G		4	2		1,475	91		1,566	1,775
0120	90 watt	G		4	2		2,100	91		2,191	2,475
0130	288 watt	G	↓	4	2	↓	1,900	91		1,991	2,250

26 61 Lighting Systems and Accessories

26 61 23 – Lamps Applications

26 61 23.10 Lamps		Crew	Daily Output	Labor-Hours	Unit	2019 Bare Costs Material	Labor	Equipment	Total	Total Incl O&P
0010	LAMPS									
0080	Fluorescent, rapid start, cool white, 2′ long, 20 watt	1 Elec	1	8	C	375	365		740	1,000
0100	4′ long, 40 watt		.90	8.889		248	405		653	935
1000	Metal halide, mogul base, 175 watt		.30	26.667		1,050	1,225		2,275	3,125
1100	250 watt		.30	26.667		1,650	1,225		2,875	3,800
1350	High pressure sodium, 70 watt		.30	26.667		1,575	1,225		2,800	3,725
1370	150 watt	↓	.30	26.667	↓	1,625	1,225		2,850	3,750

Division 27 Communications

Estimating Tips

27 20 00 Data Communications
27 30 00 Voice Communications
27 40 00 Audio-Video Communications

- When estimating material costs for special systems, it is always prudent to obtain manufacturers' quotations for equipment prices and special installation requirements that may affect the total cost.
- For cost modifications for elevated tray installation, add the percentages to labor according to the height of the installation and only to the quantities exceeding the different height levels, not to the total tray quantities. Refer to subdivision 26 01 02.20 for labor adjustment factors.
- Do not overlook the costs for equipment used in the installation. If scissor lifts and boom lifts are available in the field, contractors may use them in lieu of the proposed ladders and rolling staging.

Reference Numbers

Reference numbers are shown at the beginning of some major classifications. These numbers refer to related items in the Reference Section. The reference information may be an estimating procedure, an alternate pricing method, or technical information.

Note: Not all subdivisions listed here necessarily appear. ■

Note: Trade Service, in part, has been used as a reference source for some of the material prices used in Division 27.

27 13 Communications Backbone Cabling

27 13 23 – Communications Optical Fiber Backbone Cabling

27 13 23.13 Communications Optical Fiber		Crew	Daily Output	Labor-Hours	Unit	Material	2019 Bare Costs Labor	Equipment	Total	Total Incl O&P
0010	**COMMUNICATIONS OPTICAL FIBER**									
0040	Specialized tools & techniques cause installation costs to vary.									
0070	Fiber optic, cable, bulk simplex, single mode	1 Elec	8	1	C.L.F.	23	45.50		68.50	99.50
0080	Multi mode		8	1		29.50	45.50		75	107
0090	4 strand, single mode		7.34	1.090		37.50	49.50		87	122
0095	Multi mode		7.34	1.090		50	49.50		99.50	136
0100	12 strand, single mode		6.67	1.199		75.50	54.50		130	173
0150	Jumper				Ea.	36			36	39.50
0200	Pigtail					34			34	37.50
0300	Connector	1 Elec	24	.333		25.50	15.20		40.70	53
0350	Finger splice	"	32	.250		37.50	11.40		48.90	60
1000	Cable, 62.5 microns, direct burial, 4 fiber	R-15	1200	.040	L.F.	.90	1.78	.24	2.92	4.16
1020	Indoor, 2 fiber	R-19	1000	.020		.44	.91		1.35	1.97
1040	Outdoor, aerial/duct	"	1670	.012		.65	.55		1.20	1.61
1060	50 microns, direct burial, 8 fiber	R-22	4000	.009		1.27	.39		1.66	2.04
1080	12 fiber		4000	.009		2.05	.39		2.44	2.90
1100	Indoor, 12 fiber		759	.049		1.99	2.05		4.04	5.55
1120	Connectors, 62.5 micron cable, transmission	R-19	40	.500	Ea.	14.15	23		37.15	53
1140	Cable splice		40	.500		14.90	23		37.90	54
1160	125 micron cable, transmission		16	1.250		14.90	57		71.90	109
1440	Transceiver, 1.9 mile range		5	4		385	183		568	720
1460	1.2 mile range, digital		5	4		395	183		578	735
1480	Cable enclosure, interior NEMA 13		7	2.857		144	131		275	370
1500	Splice w/enclosure encapsulant		16	1.250		221	57		278	335
1510	1/4", black, non-metallic, flexible, tube, liquid tight		8	2.500	C.L.F.	150	114		264	350
1530	1/2"		7.34	2.725		125	125		250	340
1550	1"		7.34	2.725		330	125		455	565
1560	1-1/4"		6.67	2.999		520	137		657	795
1570	1-1/2"		6.67	2.999		570	137		707	855
1580	2"		6.35	3.150		860	144		1,004	1,175

27 15 Communications Horizontal Cabling

27 15 10 – Special Communications Cabling

27 15 10.23 Sound and Video Cables and Fittings

Code	Description	Crew	Daily Output	Labor-Hours	Unit	Material	Labor	Equipment	Total	Total Incl O&P
0010	**SOUND AND VIDEO CABLES & FITTINGS**									
0900	TV antenna lead-in, 300 ohm, #20-2 conductor	1 Elec	7	1.143	C.L.F.	22	52		74	109
0950	Coaxial, feeder outlet		7	1.143		20.50	52		72.50	108
1000	Coaxial, main riser		6	1.333		13.50	61		74.50	114
1100	Sound, shielded with drain, #22-2 conductor		8	1		8.35	45.50		53.85	83.50
1150	#22-3 conductor		7.50	1.067		12.40	48.50		60.90	93
1200	#22-4 conductor		6.50	1.231		19.90	56		75.90	114
1250	Nonshielded, #22-2 conductor		10	.800		10.25	36.50		46.75	71
1300	#22-3 conductor		9	.889		20.50	40.50		61	88.50
1350	#22-4 conductor		8	1		22.50	45.50		68	99.50
1400	Microphone cable		8	1		30	45.50		75.50	108

27 15 13 – Communications Copper Horizontal Cabling

27 15 13.13 Communication Cables and Fittings

Code	Description	Crew	Daily Output	Labor-Hours	Unit	Material	Labor	Equipment	Total	Total Incl O&P
0010	**COMMUNICATION CABLES AND FITTINGS**									
2200	Telephone twisted, PVC insulation, #22-2 conductor	1 Elec	10	.800	C.L.F.	10.30	36.50		46.80	71
2250	#22-3 conductor		9	.889		12.70	40.50		53.20	80

27 15 Communications Horizontal Cabling

27 15 13 – Communications Copper Horizontal Cabling

27 15 13.13 Communication Cables and Fittings		Crew	Daily Output	Labor-Hours	Unit	Material	Labor	Equipment	Total	Total Incl O&P
						2019 Bare Costs				
2300	#22-4 conductor	1 Elec	8	1	C.L.F.	15	45.50		60.50	91
2350	#18-2 conductor	↓	9	.889	↓	21	40.50		61.50	89

27 41 Audio-Video Systems

27 41 33 – Master Antenna Television Systems

27 41 33.10 TV Systems

		Crew	Daily Output	Labor-Hours	Unit	Material	Labor	Equipment	Total	Total Incl O&P
0010	**TV SYSTEMS,** not including rough-in wires, cables & conduits									
0100	Master TV antenna system									
0200	VHF reception & distribution, 12 outlets	1 Elec	6	1.333	Outlet	118	61		179	229
0800	VHF & UHF reception & distribution, 12 outlets		6	1.333	"	251	61		312	375
5000	Antenna, small		6	1.333	Ea.	55.50	61		116.50	160
5100	Large	↓	4	2	"	233	91		324	405

27 51 Distributed Audio-Video Communications Systems

27 51 19 – Sound Masking Systems

27 51 19.10 Sound System

		Crew	Daily Output	Labor-Hours	Unit	Material	Labor	Equipment	Total	Total Incl O&P
0010	**SOUND SYSTEM**, not including rough-in wires, cables & conduits									
2000	Intercom, 30 station capacity, master station	2 Elec	2	8	Ea.	2,325	365		2,690	3,150
2020	10 station capacity	"	4	4		1,375	182		1,557	1,825
2200	Remote station	1 Elec	8	1		199	45.50		244.50	294
2400	Intercom outlets		8	1		117	45.50		162.50	204
2600	Handset		4	2		390	91		481	575
2800	Emergency call system, 12 zones, annunciator		1.30	6.154		1,175	281		1,456	1,725
3000	Bell		5.30	1.509		121	69		190	245
3200	Light or relay		8	1		60.50	45.50		106	141
3400	Transformer		4	2		266	91		357	440
3600	House telephone, talking station		1.60	5		570	228		798	995
3800	Press to talk, release to listen	↓	5.30	1.509		133	69		202	258
4000	System-on button					79.50			79.50	87.50
4200	Door release	1 Elec	4	2		142	91		233	305
4400	Combination speaker and microphone		8	1		242	45.50		287.50	340
4600	Termination box		3.20	2.500		76	114		190	270
4800	Amplifier or power supply		5.30	1.509	↓	875	69		944	1,075
5000	Vestibule door unit		16	.500	Name	161	23		184	214
5200	Strip cabinet		27	.296	Ea.	305	13.50		318.50	355
5400	Directory	↓	16	.500		143	23		166	194
6000	Master door, button buzzer type, 100 unit	2 Elec	.54	29.630		1,450	1,350		2,800	3,800
6020	200 unit	"	.30	53.333	↓	2,750	2,425		5,175	7,000

Division Notes

	CREW	DAILY OUTPUT	LABOR-HOURS	UNIT	BARE COSTS MAT.	LABOR	EQUIP.	TOTAL	TOTAL INCL O&P

Division 28

Electronic Safety & Security

Estimating Tips

- When estimating material costs for electronic safety and security systems, it is always prudent to obtain manufacturers' quotations for equipment prices and special installation requirements that may affect the total cost.
- Fire alarm systems consist of control panels, annunciator panels, batteries with rack, charger, and fire alarm actuating and indicating devices. Some fire alarm systems include speakers, telephone lines, door closer controls, and other components. Be careful not to overlook the costs related to installation for these items. Also be aware of costs for integrated automation instrumentation and terminal devices, control equipment, control wiring, and programming. Insurance underwriters may have specific requirements for the type of materials to be installed or design requirements based on the hazard to be protected. Local jurisdictions may have requirements not covered by code. It is advisable to be aware of any special conditions.
- Security equipment includes items such as CCTV, access control, and other detection and identification systems to perform alert and alarm functions. Be sure to consider the costs related to installation for this security equipment, such as for integrated automation instrumentation and terminal devices, control equipment, control wiring, and programming.

Reference Numbers

Reference numbers are shown at the beginning of some major classifications. These numbers refer to related items in the Reference Section. The reference information may be an estimating procedure, an alternate pricing method, or technical information.

Note: Trade Service, in part, has been used as a reference source for some of the material prices used in Division 28.

28 15 Access Control Hardware Devices

28 15 11 – Integrated Credential Readers and Field Entry Management

28 15 11.11 Standard Card Readers		Crew	Daily Output	Labor-Hours	Unit	Material	2019 Bare Costs Labor	Equipment	Total	Total Incl O&P
0010	**STANDARD CARD READERS**									
0210	Entrance card systems									
0300	Entrance card, barium ferrite				Ea.	4.95			4.95	5.45
0320	Credential					8.15			8.15	8.95
0340	Proximity					3.65			3.65	4.02
0360	Weigand					8.30			8.30	9.15
0700	Entrance card reader, barium ferrite	R-19	4	5		350	229		579	760
0720	Credential		4	5		212	229		441	610
0740	Proximity		4	5		300	229		529	705
0760	Weigand		4	5		495	229		724	920
0850	Scanner, eye retina		4	5		2,650	229		2,879	3,300
0900	Gate opener, cantilever	R-18	3	8.667		2,100	345		2,445	2,875
0950	Switch, tamper	R-19	6	3.333		37	152		189	290
1100	Accessories, electric door strike/bolt		5	4		78.50	183		261.50	385
1120	Electromagnetic lock		5	4		147	183		330	460
1140	Keypad for card reader		4	5		126	229		355	515

28 31 Intrusion Detection

28 31 16 – Intrusion Detection Systems Infrastructure

28 31 16.50 Intrusion Detection		Crew	Daily Output	Labor-Hours	Unit	Material	Labor	Equipment	Total	Total Incl O&P
0010	**INTRUSION DETECTION**, not including wires & conduits									
0100	Burglar alarm, battery operated, mechanical trigger	1 Elec	4	2	Ea.	281	91		372	460
0200	Electrical trigger		4	2		335	91		426	520
0400	For outside key control, add		8	1		87	45.50		132.50	170
0600	For remote signaling circuitry, add		8	1		144	45.50		189.50	234
0800	Card reader, flush type, standard		2.70	2.963		815	135		950	1,125
1000	Multi-code		2.70	2.963		1,225	135		1,360	1,550
1200	Door switches, hinge switch		5.30	1.509		64.50	69		133.50	183
1400	Magnetic switch		5.30	1.509		105	69		174	228
2800	Ultrasonic motion detector, 12 V		2.30	3.478		200	159		359	480
3000	Infrared photoelectric detector		4	2		135	91		226	297
3200	Passive infrared detector		4	2		238	91		329	410
3420	Switchmats, 30" x 5'		5.30	1.509		105	69		174	228
3440	30" x 25'		4	2		206	91		297	375
3460	Police connect panel		4	2		285	91		376	465
3480	Telephone dialer		5.30	1.509		390	69		459	540
3500	Alarm bell		4	2		106	91		197	265
3520	Siren		4	2		148	91		239	310

28 42 Gas Detection and Alarm

28 42 15 – Gas Detection Sensors

28 42 15.50 Tank Leak Detection Systems		Crew	Daily Output	Labor-Hours	Unit	Material	Labor	Equipment	Total	Total Incl O&P
0010	**TANK LEAK DETECTION SYSTEMS** Liquid and vapor									
0100	For hydrocarbons and hazardous liquids/vapors									
0120	Controller, data acquisition, incl. printer, modem, RS232 port									
0140	24 channel, for use with all probes				Ea.	2,775			2,775	3,050
0160	9 channel, for external monitoring				"	1,125			1,125	1,250
0200	Probes									
0210	Well monitoring									

28 42 Gas Detection and Alarm

28 42 15 – Gas Detection Sensors

28 42 15.50 Tank Leak Detection Systems		Crew	Daily Output	Labor-Hours	Unit	2019 Bare Costs Material	Labor	Equipment	Total	Total Incl O&P
0220	Liquid phase detection				Ea.	630			630	695
0230	Hydrocarbon vapor, fixed position					785			785	865
0240	Hydrocarbon vapor, float mounted					560			560	615
0250	Both liquid and vapor hydrocarbon					550			550	605
0300	Secondary containment, liquid phase									
0310	Pipe trench/manway sump				Ea.	725			725	800
0320	Double wall pipe and manual sump					610			610	670
0330	Double wall fiberglass annular space					420			420	460
0340	Double wall steel tank annular space					295			295	325
0500	Accessories									
0510	Modem, non-dedicated phone line				Ea.	269			269	296
0600	Monitoring, internal									
0610	Automatic tank gauge, incl. overfill				Ea.	1,200			1,200	1,325
0620	Product line				"	1,250			1,250	1,375
0700	Monitoring, special									
0710	Cathodic protection				Ea.	675			675	745
0720	Annular space chemical monitor				"	920			920	1,025

28 46 Fire Detection and Alarm

28 46 11 – Fire Sensors and Detectors

28 46 11.21 Carbon-Monoxide Detection Sensors		Crew	Daily Output	Labor-Hours	Unit	Material	Labor	Equipment	Total	Total Incl O&P
0010	**CARBON-MONOXIDE DETECTION SENSORS**									
8400	Smoke and carbon monoxide alarm battery operated photoelectric low profile	1 Elec	24	.333	Ea.	59	15.20		74.20	90
8410	low profile photoelectric battery powered		24	.333		36.50	15.20		51.70	65
8420	photoelectric low profile sealed lithium		24	.333		66.50	15.20		81.70	98.50
8430	Photoelectric low profile sealed lithium smoke and CO with voice combo		24	.333		41.50	15.20		56.70	70.50
8500	Carbon monoxide sensor, wall mount 1Mod 1 relay output smoke & heat		24	.333		505	15.20		520.20	580
8700	Carbon monoxide detector, battery operated, wall mounted		16	.500		46	23		69	87.50
8710	Hardwired, wall and ceiling mounted		8	1		90	45.50		135.50	174
8720	Duct mounted		8	1		350	45.50		395.50	460

28 46 11.27 Other Sensors		Crew	Daily Output	Labor-Hours	Unit	Material	Labor	Equipment	Total	Total Incl O&P
0010	**OTHER SENSORS**									
5200	Smoke detector, ceiling type	1 Elec	6.20	1.290	Ea.	123	59		182	231
5240	Smoke detector, addressable type		6	1.333		227	61		288	350
5420	Duct addressable type		3.20	2.500		525	114		639	760
8300	Smoke alarm with integrated strobe light 120 V, 16DB 60 fpm flash rate		16	.500		112	23		135	160
8310	Photoelectric smoke detector with strobe 120 V, 90 DB ceiling mount		12	.667		205	30.50		235.50	275
8320	120 V, 90 DB wall mount		12	.667		189	30.50		219.50	257

28 46 21 – Fire Alarm

28 46 21.50 Alarm Panels and Devices		Crew	Daily Output	Labor-Hours	Unit	Material	Labor	Equipment	Total	Total Incl O&P
0010	**ALARM PANELS AND DEVICES**, not including wires & conduits									
5600	Strobe and horn	1 Elec	5.30	1.509	Ea.	140	69		209	266
5610	Strobe and horn (ADA type)		5.30	1.509		168	69		237	297
5620	Visual alarm (ADA type)		6.70	1.194		117	54.50		171.50	218
5800	Fire alarm horn		6.70	1.194		55.50	54.50		110	150

28 46 Fire Detection and Alarm

28 46 21 – Fire Alarm

28 46 21.50 Alarm Panels and Devices		Crew	Daily Output	Labor-Hours	Unit	2019 Bare Costs Material	Labor	Equipment	Total	Total Incl O&P
6600	Drill switch	1 Elec	8	1	Ea.	380	45.50		425.50	495
6800	Master box		2.70	2.963		6,575	135		6,710	7,450
7000	Break glass station		8	1		58	45.50		103.50	138
7800	Remote annunciator, 8 zone lamp		1.80	4.444		213	203		416	565
8000	12 zone lamp	2 Elec	2.60	6.154		415	281		696	920
8200	16 zone lamp	"	2.20	7.273		375	330		705	950

Estimating Tips

31 05 00 Common Work Results for Earthwork

- Estimating the actual cost of performing earthwork requires careful consideration of the variables involved. This includes items such as type of soil, whether water will be encountered, dewatering, whether banks need bracing, disposal of excavated earth, and length of haul to fill or spoil sites, etc. If the project has large quantities of cut or fill, consider raising or lowering the site to reduce costs, while paying close attention to the effect on site drainage and utilities.
- If the project has large quantities of fill, creating a borrow pit on the site can significantly lower the costs.
- It is very important to consider what time of year the project is scheduled for completion. Bad weather can create large cost overruns from dewatering, site repair, and lost productivity from cold weather.

Reference Numbers

Reference numbers are shown at the beginning of some major classifications. These numbers refer to related items in the Reference Section. The reference information may be an estimating procedure, an alternate pricing method, or technical information.

Note: Not all subdivisions listed here necessarily appear. ■

31 05 Common Work Results for Earthwork

31 05 13 – Soils for Earthwork

31 05 13.10 Borrow		Crew	Daily Output	Labor-Hours	Unit	Material	2019 Bare Costs Labor	Equipment	Total	Total Incl O&P
0010	**BORROW**									
0020	Spread, 200 HP dozer, no compaction, 2 mile RT haul									
0200	Common borrow	B-15	600	.047	C.Y.	12.60	1.67	4.04	18.31	21

31 05 16 – Aggregates for Earthwork

31 05 16.10 Borrow		Crew	Daily Output	Labor-Hours	Unit	Material	Labor	Equipment	Total	Total Incl O&P
0010	**BORROW**									
0020	Spread, with 200 HP dozer, no compaction, 2 mile RT haul									
0100	Bank run gravel	B-15	600	.047	L.C.Y.	18.65	1.67	4.04	24.36	27.50
0300	Crushed stone (1.40 tons per C.Y.), 1-1/2"		600	.047		27.50	1.67	4.04	33.21	37
0320	3/4"		600	.047		27.50	1.67	4.04	33.21	37
0340	1/2"		600	.047		24.50	1.67	4.04	30.21	34
0360	3/8"		600	.047		16.95	1.67	4.04	22.66	26
0400	Sand, washed, concrete		600	.047		36	1.67	4.04	41.71	47
0500	Dead or bank sand	↓	600	.047	↓	18.10	1.67	4.04	23.81	27

31 11 Clearing and Grubbing

31 11 10 – Clearing and Grubbing Land

31 11 10.10 Clear and Grub Site		Crew	Daily Output	Labor-Hours	Unit	Material	Labor	Equipment	Total	Total Incl O&P
0010	**CLEAR AND GRUB SITE**									
0020	Cut & chip light trees to 6" diam.	B-7	1	48	Acre		1,550	1,675	3,225	4,425
0150	Grub stumps and remove	B-30	2	12			440	1,025	1,465	1,850
0200	Cut & chip medium trees to 12" diam.	B-7	.70	68.571			2,225	2,400	4,625	6,325
0250	Grub stumps and remove	B-30	1	24			885	2,050	2,935	3,700
0300	Cut & chip heavy trees to 24" diam.	B-7	.30	160			5,175	5,625	10,800	14,700
0350	Grub stumps and remove	B-30	.50	48			1,775	4,075	5,850	7,400
0400	If burning is allowed, deduct cut & chip				↓				40%	40%

31 13 Selective Tree and Shrub Removal and Trimming

31 13 13 – Selective Tree and Shrub Removal

31 13 13.10 Selective Clearing		Crew	Daily Output	Labor-Hours	Unit	Material	Labor	Equipment	Total	Total Incl O&P
0010	**SELECTIVE CLEARING**									
0020	Clearing brush with brush saw	A-1C	.25	32	Acre		970	110	1,080	1,725
0100	By hand	1 Clab	.12	66.667			2,025		2,025	3,350
0300	With dozer, ball and chain, light clearing	B-11A	2	8			284	645	929	1,175
0400	Medium clearing	″	1.50	10.667	↓		380	860	1,240	1,575

31 22 Grading

31 22 16 – Fine Grading

31 22 16.10 Finish Grading		Crew	Daily Output	Labor-Hours	Unit	Material	Labor	Equipment	Total	Total Incl O&P
0010	**FINISH GRADING**									
0012	Finish grading area to be paved with grader, small area	B-11L	400	.040	S.Y.		1.42	1.64	3.06	4.15
1050	For small irregular areas	B-32C	2000	.024			.86	1.09	1.95	2.62
1100	Fine grade for slab on grade, machine	B-11L	1040	.015	↓		.55	.63	1.18	1.60
1160	Hand grading	1 Clab	500	.016	S.F.		.49		.49	.80
3300	Finishing grading slopes, gentle	B-11L	8900	.002	S.Y.		.06	.07	.13	.19
3500	Finish grading lagoon bottoms	″	4	4	M.S.F.		142	164	306	415

31 23 16.13 Excavating, Trench

31 23 16.13 Excavating, Trench		Crew	Daily Output	Labor-Hours	Unit	Material	2019 Bare Costs Labor	Equipment	Total	Total Incl O&P
0010	**EXCAVATING, TRENCH**									
0011	Or continuous footing									
0050	1' to 4' deep, 3/8 C.Y. excavator	B-11C	150	.107	B.C.Y.		3.79	2.13	5.92	8.60
0060	1/2 C.Y. excavator	B-11M	200	.080			2.84	1.94	4.78	6.80
0090	4' to 6' deep, 1/2 C.Y. excavator	"	200	.080			2.84	1.94	4.78	6.80
0100	5/8 C.Y. excavator	B-12Q	250	.064			2.31	2.35	4.66	6.40
0300	1/2 C.Y. excavator, truck mounted	B-12J	200	.080			2.89	4.31	7.20	9.50
1352	4' to 6' deep, 1/2 C.Y. excavator w/trench box	B-13H	188	.085			3.08	5	8.08	10.55
1354	5/8 C.Y. excavator	"	235	.068			2.46	4	6.46	8.45
1400	By hand with pick and shovel 2' to 6' deep, light soil	1 Clab	8	1			30.50		30.50	50
1500	Heavy soil	"	4	2			60.50		60.50	100
5050	1' to 4' deep, 3/8 C.Y. tractor loader/backhoe	B-11C	162	.099			3.51	1.98	5.49	7.95
5060	1/2 C.Y. excavator	B-11M	216	.074			2.63	1.80	4.43	6.30
5080	4' to 6' deep, 1/2 C.Y. excavator	"	216	.074			2.63	1.80	4.43	6.30
5090	5/8 C.Y. excavator	B-12Q	276	.058			2.10	2.13	4.23	5.80
5130	1/2 C.Y. excavator, truck mounted	B-12J	216	.074			2.68	3.99	6.67	8.80
5352	4' to 6' deep, 1/2 C.Y. excavator w/trench box	B-13H	205	.078			2.82	4.59	7.41	9.70
5354	5/8 C.Y. excavator	"	257	.062			2.25	3.66	5.91	7.75
6050	1' to 4' deep, 3/8 C.Y. excavator	B-11C	165	.097			3.45	1.94	5.39	7.80
6060	1/2 C.Y. excavator	B-11M	220	.073			2.59	1.76	4.35	6.20
6080	4' to 6' deep, 1/2 C.Y. excavator	"	220	.073			2.59	1.76	4.35	6.20
6090	5/8 C.Y. excavator	B-12Q	275	.058			2.10	2.14	4.24	5.80
6130	1/2 C.Y. excavator, truck mounted	B-12J	220	.073			2.63	3.92	6.55	8.65
6352	4' to 6' deep, 1/2 C.Y. excavator w/trench box	B-13H	209	.077			2.77	4.50	7.27	9.50
6354	5/8 C.Y. excavator	"	261	.061			2.22	3.60	5.82	7.60
7020	Dense hard clay with no sheeting or dewatering included									
7050	1' to 4' deep, 3/8 C.Y. excavator	B-11C	132	.121	B.C.Y.		4.31	2.43	6.74	9.75
7060	1/2 C.Y. excavator	B-11M	176	.091			3.23	2.20	5.43	7.70
7080	4' to 6' deep, 1/2 C.Y. excavator	"	176	.091			3.23	2.20	5.43	7.70
7090	5/8 C.Y. excavator	B-12Q	220	.073			2.63	2.67	5.30	7.25
7130	1/2 C.Y. excavator, truck mounted	B-12J	176	.091			3.29	4.90	8.19	10.80

31 23 16.14 Excavating, Utility Trench

31 23 16.14 Excavating, Utility Trench		Crew	Daily Output	Labor-Hours	Unit	Material	Labor	Equipment	Total	Total Incl O&P
0010	**EXCAVATING, UTILITY TRENCH**									
0011	Common earth									
0050	Trenching with chain trencher, 12 HP, operator walking									
0100	4" wide trench, 12" deep	B-53	800	.010	L.F.		.30	.08	.38	.59
1000	Backfill by hand including compaction, add									
1050	4" wide trench, 12" deep	A-1G	800	.010	L.F.		.30	.07	.37	.57

31 23 16.16 Structural Excavation for Minor Structures

31 23 16.16 Structural Excavation for Minor Structures		Crew	Daily Output	Labor-Hours	Unit	Material	Labor	Equipment	Total	Total Incl O&P
0010	**STRUCTURAL EXCAVATION FOR MINOR STRUCTURES**									
0015	Hand, pits to 6' deep, sandy soil	1 Clab	8	1	B.C.Y.		30.50		30.50	50
0100	Heavy soil or clay		4	2			60.50		60.50	100
1100	Hand loading trucks from stock pile, sandy soil		12	.667			20		20	33.50
1300	Heavy soil or clay		8	1			30.50		30.50	50
1500	For wet or muck hand excavation, add to above								50%	50%

31 23 16.42 Excavating, Bulk Bank Measure

31 23 16.42 Excavating, Bulk Bank Measure		Crew	Daily Output	Labor-Hours	Unit	Material	Labor	Equipment	Total	Total Incl O&P
0010	**EXCAVATING, BULK BANK MEASURE** R312316-40									
0011	Common earth piled									
0020	For loading onto trucks, add								15%	15%
0200	Excavator, hydraulic, crawler mtd., 1 C.Y. cap. = 100 C.Y./hr.	B-12A	800	.020	B.C.Y.		.72	.94	1.66	2.22
0310	Wheel mounted, 1/2 C.Y. cap. = 40 C.Y./hr.	B-12E	320	.050			1.81	1.39	3.20	4.51
1200	Front end loader, track mtd., 1-1/2 C.Y. cap. = 70 C.Y./hr.	B-10N	560	.014			.58	1.06	1.64	2.12

31 23 Excavation and Fill

31 23 16 – Excavation

31 23 16.42 Excavating, Bulk Bank Measure		Crew	Daily Output	Labor-Hours	Unit	Material	2019 Bare Costs Labor	Equipment	Total	Total Incl O&P
1500	Wheel mounted, 3/4 C.Y. cap. = 45 C.Y./hr.	B-10R	360	.022	B.C.Y.		.91	.83	1.74	2.40
5000	Excavating, bulk bank measure, sandy clay & loam piled									
5020	For loading onto trucks, add								15%	15%
5100	Excavator, hydraulic, crawler mtd., 1 C.Y. cap. = 120 C.Y./hr.	B-12A	960	.017	B.C.Y.		.60	.78	1.38	1.85
5610	Wheel mounted, 1/2 C.Y. cap. = 44 C.Y./hr.	B-12E	352	.045	"		1.64	1.26	2.90	4.09
8000	For hauling excavated material, see Section 31 23 23.20									

31 23 19 – Dewatering

31 23 19.20 Dewatering Systems		Crew	Daily Output	Labor-Hours	Unit	Material	Labor	Equipment	Total	Total Incl O&P
0010	**DEWATERING SYSTEMS**									
0020	Excavate drainage trench, 2′ wide, 2′ deep	B-11C	90	.178	C.Y.		6.30	3.56	9.86	14.30
0100	2′ wide, 3′ deep, with backhoe loader	"	135	.119			4.21	2.37	6.58	9.55
0200	Excavate sump pits by hand, light soil	1 Clab	7.10	1.127			34		34	56.50
0300	Heavy soil	"	3.50	2.286			69.50		69.50	115
0500	Pumping 8 hrs., attended 2 hrs./day, incl. 20 L.F.									
0550	of suction hose & 100 L.F. discharge hose									
0600	2″ diaphragm pump used for 8 hrs.	B-10H	4	2	Day		81.50	19.55	101.05	156
0650	4″ diaphragm pump used for 8 hrs.	B-10I	4	2	"		81.50	37.50	119	175
1600	Sump hole construction, incl. excavation and gravel, pit	B-6	1250	.019	C.F.	1.10	.64	.26	2	2.54
1700	With 12″ gravel collar, 12″ pipe, corrugated, 16 ga.	"	70	.343	L.F.	21.50	11.35	4.57	37.42	47.50

31 23 23 – Fill

31 23 23.13 Backfill		Crew	Daily Output	Labor-Hours	Unit	Material	Labor	Equipment	Total	Total Incl O&P
0010	**BACKFILL**									
0015	By hand, no compaction, light soil	1 Clab	14	.571	L.C.Y.		17.35		17.35	28.50
0100	Heavy soil		11	.727	"		22		22	36.50
0300	Compaction in 6″ layers, hand tamp, add to above		20.60	.388	E.C.Y.		11.80		11.80	19.50
0400	Roller compaction operator walking, add	B-10A	100	.080			3.26	1.82	5.08	7.35
0500	Air tamp, add	B-9D	190	.211			6.45	1.42	7.87	12.25
0600	Vibrating plate, add	A-1D	60	.133			4.05	.52	4.57	7.25
0800	Compaction in 12″ layers, hand tamp, add to above	1 Clab	34	.235			7.15		7.15	11.80
1300	Dozer backfilling, bulk, up to 300′ haul, no compaction	B-10B	1200	.007	L.C.Y.		.27	1.08	1.35	1.63
1400	Air tamped, add	B-11B	80	.200	E.C.Y.		6.90	3.59	10.49	15.35

31 23 23.16 Fill By Borrow and Utility Bedding		Crew	Daily Output	Labor-Hours	Unit	Material	Labor	Equipment	Total	Total Incl O&P
0010	**FILL BY BORROW AND UTILITY BEDDING**									
0049	Utility bedding, for pipe & conduit, not incl. compaction									
0050	Crushed or screened bank run gravel	B-6	150	.160	L.C.Y.	21	5.30	2.13	28.43	34.50
0100	Crushed stone 3/4″ to 1/2″		150	.160		27.50	5.30	2.13	34.93	41
0200	Sand, dead or bank		150	.160		18.10	5.30	2.13	25.53	31
0500	Compacting bedding in trench	A-1D	90	.089	E.C.Y.		2.70	.35	3.05	4.84
0600	If material source exceeds 2 miles, add for extra mileage.									
0610	See Section 31 23 23.20 for hauling mileage add.									

31 23 23.17 General Fill		Crew	Daily Output	Labor-Hours	Unit	Material	Labor	Equipment	Total	Total Incl O&P
0010	**GENERAL FILL**									
0011	Spread dumped material, no compaction									
0020	By dozer	B-10B	1000	.008	L.C.Y.		.33	1.29	1.62	1.95
0100	By hand	1 Clab	12	.667	"		20		20	33.50
0500	Gravel fill, compacted, under floor slabs, 4″ deep	B-37	10000	.005	S.F.	.40	.15	.02	.57	.71
0600	6″ deep		8600	.006		.61	.18	.02	.81	.99
0700	9″ deep		7200	.007		1.01	.21	.02	1.24	1.48
0800	12″ deep		6000	.008		1.41	.26	.03	1.70	2.01
1000	Alternate pricing method, 4″ deep		120	.400	E.C.Y.	30.50	12.85	1.27	44.62	56
1100	6″ deep		160	.300		30.50	9.60	.95	41.05	50.50

31 23 23.17 General Fill		Crew	Daily Output	Labor-Hours	Unit	2019 Bare Costs Material	Labor	Equipment	Total	Total Incl O&P
1200	9″ deep	B-37	200	.240	E.C.Y.	30.50	7.70	.76	38.96	47
1300	12″ deep		220	.218		30.50	7	.69	38.19	46

31 23 23.20 Hauling		Crew	Daily Output	Labor-Hours	Unit	2019 Bare Costs Material	Labor	Equipment	Total	Total Incl O&P
0010	**HAULING**									
0011	Excavated or borrow, loose cubic yards									
0012	no loading equipment, including hauling, waiting, loading/dumping									
0013	time per cycle (wait, load, travel, unload or dump & return)									
0014	8 C.Y. truck, 15 MPH avg., cycle 0.5 miles, 10 min. wait/ld./uld.	B-34A	320	.025	L.C.Y.		.87	1.06	1.93	2.60
0016	cycle 1 mile		272	.029			1.02	1.25	2.27	3.05
0018	cycle 2 miles		208	.038			1.34	1.63	2.97	4
0020	cycle 4 miles		144	.056			1.93	2.36	4.29	5.75
0022	cycle 6 miles		112	.071			2.49	3.03	5.52	7.45
0024	cycle 8 miles		88	.091			3.16	3.86	7.02	9.45
0026	20 MPH avg., cycle 0.5 mile		336	.024			.83	1.01	1.84	2.47
0028	cycle 1 mile		296	.027			.94	1.15	2.09	2.81
0030	cycle 2 miles		240	.033			1.16	1.41	2.57	3.47
0032	cycle 4 miles		176	.045			1.58	1.93	3.51	4.72
0034	cycle 6 miles		136	.059			2.05	2.50	4.55	6.10
0036	cycle 8 miles		112	.071			2.49	3.03	5.52	7.45
0044	25 MPH avg., cycle 4 miles		192	.042			1.45	1.77	3.22	4.34
0046	cycle 6 miles		160	.050			1.74	2.12	3.86	5.20
0048	cycle 8 miles		128	.063			2.18	2.65	4.83	6.50
0050	30 MPH avg., cycle 4 miles		216	.037			1.29	1.57	2.86	3.85
0052	cycle 6 miles		176	.045			1.58	1.93	3.51	4.72
0054	cycle 8 miles		144	.056			1.93	2.36	4.29	5.75
0114	15 MPH avg., cycle 0.5 mile, 15 min. wait/ld./uld.		224	.036			1.24	1.52	2.76	3.71
0116	cycle 1 mile		200	.040			1.39	1.70	3.09	4.16
0118	cycle 2 miles		168	.048			1.66	2.02	3.68	4.95
0120	cycle 4 miles		120	.067			2.32	2.83	5.15	6.95
0122	cycle 6 miles		96	.083			2.90	3.54	6.44	8.65
0124	cycle 8 miles		80	.100			3.48	4.25	7.73	10.40
0126	20 MPH avg., cycle 0.5 mile		232	.034			1.20	1.46	2.66	3.58
0128	cycle 1 mile		208	.038			1.34	1.63	2.97	4
0130	cycle 2 miles		184	.043			1.51	1.85	3.36	4.52
0132	cycle 4 miles		144	.056			1.93	2.36	4.29	5.75
0134	cycle 6 miles		112	.071			2.49	3.03	5.52	7.45
0136	cycle 8 miles		96	.083			2.90	3.54	6.44	8.65
0144	25 MPH avg., cycle 4 miles		152	.053			1.83	2.23	4.06	5.45
0146	cycle 6 miles		128	.063			2.18	2.65	4.83	6.50
0148	cycle 8 miles		112	.071			2.49	3.03	5.52	7.45
0150	30 MPH avg., cycle 4 miles		168	.048			1.66	2.02	3.68	4.95
0152	cycle 6 miles		144	.056			1.93	2.36	4.29	5.75
0154	cycle 8 miles		120	.067			2.32	2.83	5.15	6.95
0214	15 MPH avg., cycle 0.5 mile, 20 min. wait/ld./uld.		176	.045			1.58	1.93	3.51	4.72
0216	cycle 1 mile		160	.050			1.74	2.12	3.86	5.20
0218	cycle 2 miles		136	.059			2.05	2.50	4.55	6.10
0220	cycle 4 miles		104	.077			2.68	3.27	5.95	8
0222	cycle 6 miles		88	.091			3.16	3.86	7.02	9.45
0224	cycle 8 miles		72	.111			3.87	4.72	8.59	11.55
0226	20 MPH avg., cycle 0.5 mile		176	.045			1.58	1.93	3.51	4.72
0228	cycle 1 mile		168	.048			1.66	2.02	3.68	4.95
0230	cycle 2 miles		144	.056			1.93	2.36	4.29	5.75

31 23 23.20 Hauling		Crew	Daily Output	Labor-Hours	Unit	Material	2019 Bare Costs Labor	Equipment	Total	Total Incl O&P
0232	cycle 4 miles	B-34A	120	.067	L.C.Y.		2.32	2.83	5.15	6.95
0234	cycle 6 miles		96	.083			2.90	3.54	6.44	8.65
0236	cycle 8 miles		88	.091			3.16	3.86	7.02	9.45
0244	25 MPH avg., cycle 4 miles		128	.063			2.18	2.65	4.83	6.50
0246	cycle 6 miles		112	.071			2.49	3.03	5.52	7.45
0248	cycle 8 miles		96	.083			2.90	3.54	6.44	8.65
0250	30 MPH avg., cycle 4 miles		136	.059			2.05	2.50	4.55	6.10
0252	cycle 6 miles		120	.067			2.32	2.83	5.15	6.95
0254	cycle 8 miles		104	.077			2.68	3.27	5.95	8
0314	15 MPH avg., cycle 0.5 mile, 25 min. wait/ld./uld.		144	.056			1.93	2.36	4.29	5.75
0316	cycle 1 mile		128	.063			2.18	2.65	4.83	6.50
0318	cycle 2 miles		112	.071			2.49	3.03	5.52	7.45
0320	cycle 4 miles		96	.083			2.90	3.54	6.44	8.65
0322	cycle 6 miles		80	.100			3.48	4.25	7.73	10.40
0324	cycle 8 miles		64	.125			4.35	5.30	9.65	13
0326	20 MPH avg., cycle 0.5 mile		144	.056			1.93	2.36	4.29	5.75
0328	cycle 1 mile		136	.059			2.05	2.50	4.55	6.10
0330	cycle 2 miles		120	.067			2.32	2.83	5.15	6.95
0332	cycle 4 miles		104	.077			2.68	3.27	5.95	8
0334	cycle 6 miles		88	.091			3.16	3.86	7.02	9.45
0336	cycle 8 miles		80	.100			3.48	4.25	7.73	10.40
0344	25 MPH avg., cycle 4 miles		112	.071			2.49	3.03	5.52	7.45
0346	cycle 6 miles		96	.083			2.90	3.54	6.44	8.65
0348	cycle 8 miles		88	.091			3.16	3.86	7.02	9.45
0350	30 MPH avg., cycle 4 miles		112	.071			2.49	3.03	5.52	7.45
0352	cycle 6 miles		104	.077			2.68	3.27	5.95	8
0354	cycle 8 miles		96	.083			2.90	3.54	6.44	8.65
0414	15 MPH avg., cycle 0.5 mile, 30 min. wait/ld./uld.		120	.067			2.32	2.83	5.15	6.95
0416	cycle 1 mile		112	.071			2.49	3.03	5.52	7.45
0418	cycle 2 miles		96	.083			2.90	3.54	6.44	8.65
0420	cycle 4 miles		80	.100			3.48	4.25	7.73	10.40
0422	cycle 6 miles		72	.111			3.87	4.72	8.59	11.55
0424	cycle 8 miles		64	.125			4.35	5.30	9.65	13
0426	20 MPH avg., cycle 0.5 mile		120	.067			2.32	2.83	5.15	6.95
0428	cycle 1 mile		112	.071			2.49	3.03	5.52	7.45
0430	cycle 2 miles		104	.077			2.68	3.27	5.95	8
0432	cycle 4 miles		88	.091			3.16	3.86	7.02	9.45
0434	cycle 6 miles		80	.100			3.48	4.25	7.73	10.40
0436	cycle 8 miles		72	.111			3.87	4.72	8.59	11.55
0444	25 MPH avg., cycle 4 miles		96	.083			2.90	3.54	6.44	8.65
0446	cycle 6 miles		88	.091			3.16	3.86	7.02	9.45
0448	cycle 8 miles		80	.100			3.48	4.25	7.73	10.40
0450	30 MPH avg., cycle 4 miles		96	.083			2.90	3.54	6.44	8.65
0452	cycle 6 miles		88	.091			3.16	3.86	7.02	9.45
0454	cycle 8 miles		80	.100			3.48	4.25	7.73	10.40
0514	15 MPH avg., cycle 0.5 mile, 35 min. wait/ld./uld.		104	.077			2.68	3.27	5.95	8
0516	cycle 1 mile		96	.083			2.90	3.54	6.44	8.65
0518	cycle 2 miles		88	.091			3.16	3.86	7.02	9.45
0520	cycle 4 miles		72	.111			3.87	4.72	8.59	11.55
0522	cycle 6 miles		64	.125			4.35	5.30	9.65	13
0524	cycle 8 miles		56	.143			4.97	6.05	11.02	14.85
0526	20 MPH avg., cycle 0.5 mile		104	.077			2.68	3.27	5.95	8
0528	cycle 1 mile		96	.083			2.90	3.54	6.44	8.65

31 23 23.20 Hauling		Crew	Daily Output	Labor-Hours	Unit	Material	2019 Bare Costs Labor	Equipment	Total	Total Incl O&P
0530	cycle 2 miles	B-34A	96	.083	L.C.Y.		2.90	3.54	6.44	8.65
0532	cycle 4 miles		80	.100			3.48	4.25	7.73	10.40
0534	cycle 6 miles		72	.111			3.87	4.72	8.59	11.55
0536	cycle 8 miles		64	.125			4.35	5.30	9.65	13
0544	25 MPH avg., cycle 4 miles		88	.091			3.16	3.86	7.02	9.45
0546	cycle 6 miles		80	.100			3.48	4.25	7.73	10.40
0548	cycle 8 miles		72	.111			3.87	4.72	8.59	11.55
0550	30 MPH avg., cycle 4 miles		88	.091			3.16	3.86	7.02	9.45
0552	cycle 6 miles		80	.100			3.48	4.25	7.73	10.40
0554	cycle 8 miles		72	.111			3.87	4.72	8.59	11.55
1014	12 C.Y. truck, cycle 0.5 mile, 15 MPH avg., 15 min. wait/ld./uld.	B-34B	336	.024			.83	1.69	2.52	3.22
1016	cycle 1 mile		300	.027			.93	1.89	2.82	3.61
1018	cycle 2 miles		252	.032			1.10	2.25	3.35	4.30
1020	cycle 4 miles		180	.044			1.55	3.15	4.70	6
1022	cycle 6 miles		144	.056			1.93	3.94	5.87	7.50
1024	cycle 8 miles		120	.067			2.32	4.73	7.05	9
1025	cycle 10 miles		96	.083			2.90	5.90	8.80	11.25
1026	20 MPH avg., cycle 0.5 mile		348	.023			.80	1.63	2.43	3.11
1028	cycle 1 mile		312	.026			.89	1.82	2.71	3.47
1030	cycle 2 miles		276	.029			1.01	2.05	3.06	3.92
1032	cycle 4 miles		216	.037			1.29	2.63	3.92	5
1034	cycle 6 miles		168	.048			1.66	3.38	5.04	6.45
1036	cycle 8 miles		144	.056			1.93	3.94	5.87	7.50
1038	cycle 10 miles		120	.067			2.32	4.73	7.05	9
1040	25 MPH avg., cycle 4 miles		228	.035			1.22	2.49	3.71	4.75
1042	cycle 6 miles		192	.042			1.45	2.95	4.40	5.65
1044	cycle 8 miles		168	.048			1.66	3.38	5.04	6.45
1046	cycle 10 miles		144	.056			1.93	3.94	5.87	7.50
1050	30 MPH avg., cycle 4 miles		252	.032			1.10	2.25	3.35	4.30
1052	cycle 6 miles		216	.037			1.29	2.63	3.92	5
1054	cycle 8 miles		180	.044			1.55	3.15	4.70	6
1056	cycle 10 miles		156	.051			1.78	3.63	5.41	6.95
1060	35 MPH avg., cycle 4 miles		264	.030			1.05	2.15	3.20	4.09
1062	cycle 6 miles		228	.035			1.22	2.49	3.71	4.75
1064	cycle 8 miles		204	.039			1.36	2.78	4.14	5.30
1066	cycle 10 miles		180	.044			1.55	3.15	4.70	6
1068	cycle 20 miles		120	.067			2.32	4.73	7.05	9
1069	cycle 30 miles		84	.095			3.31	6.75	10.06	12.90
1070	cycle 40 miles		72	.111			3.87	7.90	11.77	15
1072	40 MPH avg., cycle 6 miles		240	.033			1.16	2.36	3.52	4.51
1074	cycle 8 miles		216	.037			1.29	2.63	3.92	5
1076	cycle 10 miles		192	.042			1.45	2.95	4.40	5.65
1078	cycle 20 miles		120	.067			2.32	4.73	7.05	9
1080	cycle 30 miles		96	.083			2.90	5.90	8.80	11.25
1082	cycle 40 miles		72	.111			3.87	7.90	11.77	15
1084	cycle 50 miles		60	.133			4.64	9.45	14.09	18.05
1094	45 MPH avg., cycle 8 miles		216	.037			1.29	2.63	3.92	5
1096	cycle 10 miles		204	.039			1.36	2.78	4.14	5.30
1098	cycle 20 miles		132	.061			2.11	4.30	6.41	8.20
1100	cycle 30 miles		108	.074			2.58	5.25	7.83	10.05
1102	cycle 40 miles		84	.095			3.31	6.75	10.06	12.90
1104	cycle 50 miles		72	.111			3.87	7.90	11.77	15
1106	50 MPH avg., cycle 10 miles		216	.037			1.29	2.63	3.92	5

31 23 23.20 Hauling		Crew	Daily Output	Labor-Hours	Unit	Material	2019 Bare Costs Labor	Equipment	Total	Total Incl O&P
1108	cycle 20 miles	B-34B	144	.056	L.C.Y.		1.93	3.94	5.87	7.50
1110	cycle 30 miles		108	.074			2.58	5.25	7.83	10.05
1112	cycle 40 miles		84	.095			3.31	6.75	10.06	12.90
1114	cycle 50 miles		72	.111			3.87	7.90	11.77	15
1214	15 MPH avg., cycle 0.5 mile, 20 min. wait/ld./uld.		264	.030			1.05	2.15	3.20	4.09
1216	cycle 1 mile		240	.033			1.16	2.36	3.52	4.51
1218	cycle 2 miles		204	.039			1.36	2.78	4.14	5.30
1220	cycle 4 miles		156	.051			1.78	3.63	5.41	6.95
1222	cycle 6 miles		132	.061			2.11	4.30	6.41	8.20
1224	cycle 8 miles		108	.074			2.58	5.25	7.83	10.05
1225	cycle 10 miles		96	.083			2.90	5.90	8.80	11.25
1226	20 MPH avg., cycle 0.5 mile		264	.030			1.05	2.15	3.20	4.09
1228	cycle 1 mile		252	.032			1.10	2.25	3.35	4.30
1230	cycle 2 miles		216	.037			1.29	2.63	3.92	5
1232	cycle 4 miles		180	.044			1.55	3.15	4.70	6
1234	cycle 6 miles		144	.056			1.93	3.94	5.87	7.50
1236	cycle 8 miles		132	.061			2.11	4.30	6.41	8.20
1238	cycle 10 miles		108	.074			2.58	5.25	7.83	10.05
1240	25 MPH avg., cycle 4 miles		192	.042			1.45	2.95	4.40	5.65
1242	cycle 6 miles		168	.048			1.66	3.38	5.04	6.45
1244	cycle 8 miles		144	.056			1.93	3.94	5.87	7.50
1246	cycle 10 miles		132	.061			2.11	4.30	6.41	8.20
1250	30 MPH avg., cycle 4 miles		204	.039			1.36	2.78	4.14	5.30
1252	cycle 6 miles		180	.044			1.55	3.15	4.70	6
1254	cycle 8 miles		156	.051			1.78	3.63	5.41	6.95
1256	cycle 10 miles		144	.056			1.93	3.94	5.87	7.50
1260	35 MPH avg., cycle 4 miles		216	.037			1.29	2.63	3.92	5
1262	cycle 6 miles		192	.042			1.45	2.95	4.40	5.65
1264	cycle 8 miles		168	.048			1.66	3.38	5.04	6.45
1266	cycle 10 miles		156	.051			1.78	3.63	5.41	6.95
1268	cycle 20 miles		108	.074			2.58	5.25	7.83	10.05
1269	cycle 30 miles		72	.111			3.87	7.90	11.77	15
1270	cycle 40 miles		60	.133			4.64	9.45	14.09	18.05
1272	40 MPH avg., cycle 6 miles		192	.042			1.45	2.95	4.40	5.65
1274	cycle 8 miles		180	.044			1.55	3.15	4.70	6
1276	cycle 10 miles		156	.051			1.78	3.63	5.41	6.95
1278	cycle 20 miles		108	.074			2.58	5.25	7.83	10.05
1280	cycle 30 miles		84	.095			3.31	6.75	10.06	12.90
1282	cycle 40 miles		72	.111			3.87	7.90	11.77	15
1284	cycle 50 miles		60	.133			4.64	9.45	14.09	18.05
1294	45 MPH avg., cycle 8 miles		180	.044			1.55	3.15	4.70	6
1296	cycle 10 miles		168	.048			1.66	3.38	5.04	6.45
1298	cycle 20 miles		120	.067			2.32	4.73	7.05	9
1300	cycle 30 miles		96	.083			2.90	5.90	8.80	11.25
1302	cycle 40 miles		72	.111			3.87	7.90	11.77	15
1304	cycle 50 miles		60	.133			4.64	9.45	14.09	18.05
1306	50 MPH avg., cycle 10 miles		180	.044			1.55	3.15	4.70	6
1308	cycle 20 miles		132	.061			2.11	4.30	6.41	8.20
1310	cycle 30 miles		96	.083			2.90	5.90	8.80	11.25
1312	cycle 40 miles		84	.095			3.31	6.75	10.06	12.90
1314	cycle 50 miles		72	.111			3.87	7.90	11.77	15
1414	15 MPH avg., cycle 0.5 mile, 25 min. wait/ld./uld.		204	.039			1.36	2.78	4.14	5.30
1416	cycle 1 mile		192	.042			1.45	2.95	4.40	5.65

31 23 23.20 Hauling		Crew	Daily Output	Labor-Hours	Unit	Material	2019 Bare Costs Labor	Equipment	Total	Total Incl O&P
1418	cycle 2 miles	B-34B	168	.048	L.C.Y.		1.66	3.38	5.04	6.45
1420	cycle 4 miles		132	.061			2.11	4.30	6.41	8.20
1422	cycle 6 miles		120	.067			2.32	4.73	7.05	9
1424	cycle 8 miles		96	.083			2.90	5.90	8.80	11.25
1425	cycle 10 miles		84	.095			3.31	6.75	10.06	12.90
1426	20 MPH avg., cycle 0.5 mile		216	.037			1.29	2.63	3.92	5
1428	cycle 1 mile		204	.039			1.36	2.78	4.14	5.30
1430	cycle 2 miles		180	.044			1.55	3.15	4.70	6
1432	cycle 4 miles		156	.051			1.78	3.63	5.41	6.95
1434	cycle 6 miles		132	.061			2.11	4.30	6.41	8.20
1436	cycle 8 miles		120	.067			2.32	4.73	7.05	9
1438	cycle 10 miles		96	.083			2.90	5.90	8.80	11.25
1440	25 MPH avg., cycle 4 miles		168	.048			1.66	3.38	5.04	6.45
1442	cycle 6 miles		144	.056			1.93	3.94	5.87	7.50
1444	cycle 8 miles		132	.061			2.11	4.30	6.41	8.20
1446	cycle 10 miles		108	.074			2.58	5.25	7.83	10.05
1450	30 MPH avg., cycle 4 miles		168	.048			1.66	3.38	5.04	6.45
1452	cycle 6 miles		156	.051			1.78	3.63	5.41	6.95
1454	cycle 8 miles		132	.061			2.11	4.30	6.41	8.20
1456	cycle 10 miles		120	.067			2.32	4.73	7.05	9
1460	35 MPH avg., cycle 4 miles		180	.044			1.55	3.15	4.70	6
1462	cycle 6 miles		156	.051			1.78	3.63	5.41	6.95
1464	cycle 8 miles		144	.056			1.93	3.94	5.87	7.50
1466	cycle 10 miles		132	.061			2.11	4.30	6.41	8.20
1468	cycle 20 miles		96	.083			2.90	5.90	8.80	11.25
1469	cycle 30 miles		72	.111			3.87	7.90	11.77	15
1470	cycle 40 miles		60	.133			4.64	9.45	14.09	18.05
1472	40 MPH avg., cycle 6 miles		168	.048			1.66	3.38	5.04	6.45
1474	cycle 8 miles		156	.051			1.78	3.63	5.41	6.95
1476	cycle 10 miles		144	.056			1.93	3.94	5.87	7.50
1478	cycle 20 miles		96	.083			2.90	5.90	8.80	11.25
1480	cycle 30 miles		84	.095			3.31	6.75	10.06	12.90
1482	cycle 40 miles		60	.133			4.64	9.45	14.09	18.05
1484	cycle 50 miles		60	.133			4.64	9.45	14.09	18.05
1494	45 MPH avg., cycle 8 miles		156	.051			1.78	3.63	5.41	6.95
1496	cycle 10 miles		144	.056			1.93	3.94	5.87	7.50
1498	cycle 20 miles		108	.074			2.58	5.25	7.83	10.05
1500	cycle 30 miles		84	.095			3.31	6.75	10.06	12.90
1502	cycle 40 miles		72	.111			3.87	7.90	11.77	15
1504	cycle 50 miles		60	.133			4.64	9.45	14.09	18.05
1506	50 MPH avg., cycle 10 miles		156	.051			1.78	3.63	5.41	6.95
1508	cycle 20 miles		120	.067			2.32	4.73	7.05	9
1510	cycle 30 miles		96	.083			2.90	5.90	8.80	11.25
1512	cycle 40 miles		72	.111			3.87	7.90	11.77	15
1514	cycle 50 miles		60	.133			4.64	9.45	14.09	18.05
1614	15 MPH avg., cycle 0.5 mile, 30 min. wait/ld./uld.		180	.044			1.55	3.15	4.70	6
1616	cycle 1 mile		168	.048			1.66	3.38	5.04	6.45
1618	cycle 2 miles		144	.056			1.93	3.94	5.87	7.50
1620	cycle 4 miles		120	.067			2.32	4.73	7.05	9
1622	cycle 6 miles		108	.074			2.58	5.25	7.83	10.05
1624	cycle 8 miles		84	.095			3.31	6.75	10.06	12.90
1625	cycle 10 miles		84	.095			3.31	6.75	10.06	12.90
1626	20 MPH avg., cycle 0.5 mile		180	.044			1.55	3.15	4.70	6

31 23 23.20 Hauling

	31 23 23.20 Hauling	Crew	Daily Output	Labor-Hours	Unit	Material	2019 Bare Costs Labor	Equipment	Total	Total Incl O&P
1628	cycle 1 mile	B-34B	168	.048	L.C.Y.		1.66	3.38	5.04	6.45
1630	cycle 2 miles		156	.051			1.78	3.63	5.41	6.95
1632	cycle 4 miles		132	.061			2.11	4.30	6.41	8.20
1634	cycle 6 miles		120	.067			2.32	4.73	7.05	9
1636	cycle 8 miles		108	.074			2.58	5.25	7.83	10.05
1638	cycle 10 miles		96	.083			2.90	5.90	8.80	11.25
1640	25 MPH avg., cycle 4 miles		144	.056			1.93	3.94	5.87	7.50
1642	cycle 6 miles		132	.061			2.11	4.30	6.41	8.20
1644	cycle 8 miles		108	.074			2.58	5.25	7.83	10.05
1646	cycle 10 miles		108	.074			2.58	5.25	7.83	10.05
1650	30 MPH avg., cycle 4 miles		144	.056			1.93	3.94	5.87	7.50
1652	cycle 6 miles		132	.061			2.11	4.30	6.41	8.20
1654	cycle 8 miles		120	.067			2.32	4.73	7.05	9
1656	cycle 10 miles		108	.074			2.58	5.25	7.83	10.05
1660	35 MPH avg., cycle 4 miles		156	.051			1.78	3.63	5.41	6.95
1662	cycle 6 miles		144	.056			1.93	3.94	5.87	7.50
1664	cycle 8 miles		132	.061			2.11	4.30	6.41	8.20
1666	cycle 10 miles		120	.067			2.32	4.73	7.05	9
1668	cycle 20 miles		84	.095			3.31	6.75	10.06	12.90
1669	cycle 30 miles		72	.111			3.87	7.90	11.77	15
1670	cycle 40 miles		60	.133			4.64	9.45	14.09	18.05
1672	40 MPH avg., cycle 6 miles		144	.056			1.93	3.94	5.87	7.50
1674	cycle 8 miles		132	.061			2.11	4.30	6.41	8.20
1676	cycle 10 miles		120	.067			2.32	4.73	7.05	9
1678	cycle 20 miles		96	.083			2.90	5.90	8.80	11.25
1680	cycle 30 miles		72	.111			3.87	7.90	11.77	15
1682	cycle 40 miles		60	.133			4.64	9.45	14.09	18.05
1684	cycle 50 miles		48	.167			5.80	11.80	17.60	22.50
1694	45 MPH avg., cycle 8 miles		144	.056			1.93	3.94	5.87	7.50
1696	cycle 10 miles		132	.061			2.11	4.30	6.41	8.20
1698	cycle 20 miles		96	.083			2.90	5.90	8.80	11.25
1700	cycle 30 miles		84	.095			3.31	6.75	10.06	12.90
1702	cycle 40 miles		60	.133			4.64	9.45	14.09	18.05
1704	cycle 50 miles		60	.133			4.64	9.45	14.09	18.05
1706	50 MPH avg., cycle 10 miles		132	.061			2.11	4.30	6.41	8.20
1708	cycle 20 miles		108	.074			2.58	5.25	7.83	10.05
1710	cycle 30 miles		84	.095			3.31	6.75	10.06	12.90
1712	cycle 40 miles		72	.111			3.87	7.90	11.77	15
1714	cycle 50 miles		60	.133			4.64	9.45	14.09	18.05
2000	Hauling, 8 C.Y. truck, small project cost per hour	B-34A	8	1	Hr.		35	42.50	77.50	104
2100	12 C.Y. truck	B-34B	8	1			35	71	106	136
2150	16.5 C.Y. truck	B-34C	8	1			35	77.50	112.50	143
2175	18 C.Y. 8 wheel truck	B-34I	8	1			35	88	123	155
2200	20 C.Y. truck	B-34D	8	1			35	79.50	114.50	145
2300	Grading at dump, or embankment if required, by dozer	B-10B	1000	.008	L.C.Y.		.33	1.29	1.62	1.95
2310	Spotter at fill or cut, if required	1 Clab	8	1	Hr.		30.50		30.50	50
9014	18 C.Y. truck, 8 wheels,15 min. wait/ld./uld.,15 MPH, cycle 0.5 mi.	B-34I	504	.016	L.C.Y.		.55	1.40	1.95	2.45
9016	cycle 1 mile		450	.018			.62	1.57	2.19	2.74
9018	cycle 2 miles		378	.021			.74	1.86	2.60	3.26
9020	cycle 4 miles		270	.030			1.03	2.61	3.64	4.57
9022	cycle 6 miles		216	.037			1.29	3.26	4.55	5.70
9024	cycle 8 miles		180	.044			1.55	3.92	5.47	6.85
9025	cycle 10 miles		144	.056			1.93	4.90	6.83	8.60

31 23 23.20 Hauling		Crew	Daily Output	Labor-Hours	Unit	Material	2019 Bare Costs Labor	Equipment	Total	Total Incl O&P
9026	20 MPH avg., cycle 0.5 mile	B-34I	522	.015	L.C.Y.		.53	1.35	1.88	2.37
9028	cycle 1 mile		468	.017			.59	1.51	2.10	2.64
9030	cycle 2 miles		414	.019			.67	1.70	2.37	2.98
9032	cycle 4 miles		324	.025			.86	2.18	3.04	3.80
9034	cycle 6 miles		252	.032			1.10	2.80	3.90	4.90
9036	cycle 8 miles		216	.037			1.29	3.26	4.55	5.70
9038	cycle 10 miles		180	.044			1.55	3.92	5.47	6.85
9040	25 MPH avg., cycle 4 miles		342	.023			.81	2.06	2.87	3.61
9042	cycle 6 miles		288	.028			.97	2.45	3.42	4.28
9044	cycle 8 miles		252	.032			1.10	2.80	3.90	4.90
9046	cycle 10 miles		216	.037			1.29	3.26	4.55	5.70
9050	30 MPH avg., cycle 4 miles		378	.021			.74	1.86	2.60	3.26
9052	cycle 6 miles		324	.025			.86	2.18	3.04	3.80
9054	cycle 8 miles		270	.030			1.03	2.61	3.64	4.57
9056	cycle 10 miles		234	.034			1.19	3.01	4.20	5.25
9060	35 MPH avg., cycle 4 miles		396	.020			.70	1.78	2.48	3.12
9062	cycle 6 miles		342	.023			.81	2.06	2.87	3.61
9064	cycle 8 miles		288	.028			.97	2.45	3.42	4.28
9066	cycle 10 miles		270	.030			1.03	2.61	3.64	4.57
9068	cycle 20 miles		162	.049			1.72	4.35	6.07	7.60
9070	cycle 30 miles		126	.063			2.21	5.60	7.81	9.80
9072	cycle 40 miles		90	.089			3.09	7.85	10.94	13.70
9074	40 MPH avg., cycle 6 miles		360	.022			.77	1.96	2.73	3.42
9076	cycle 8 miles		324	.025			.86	2.18	3.04	3.80
9078	cycle 10 miles		288	.028			.97	2.45	3.42	4.28
9080	cycle 20 miles		180	.044			1.55	3.92	5.47	6.85
9082	cycle 30 miles		144	.056			1.93	4.90	6.83	8.60
9084	cycle 40 miles		108	.074			2.58	6.55	9.13	11.45
9086	cycle 50 miles		90	.089			3.09	7.85	10.94	13.70
9094	45 MPH avg., cycle 8 miles		324	.025			.86	2.18	3.04	3.80
9096	cycle 10 miles		306	.026			.91	2.30	3.21	4.03
9098	cycle 20 miles		198	.040			1.41	3.56	4.97	6.25
9100	cycle 30 miles		144	.056			1.93	4.90	6.83	8.60
9102	cycle 40 miles		126	.063			2.21	5.60	7.81	9.80
9104	cycle 50 miles		108	.074			2.58	6.55	9.13	11.45
9106	50 MPH avg., cycle 10 miles		324	.025			.86	2.18	3.04	3.80
9108	cycle 20 miles		216	.037			1.29	3.26	4.55	5.70
9110	cycle 30 miles		162	.049			1.72	4.35	6.07	7.60
9112	cycle 40 miles		126	.063			2.21	5.60	7.81	9.80
9114	cycle 50 miles		108	.074			2.58	6.55	9.13	11.45
9214	20 min. wait/ld./uld.,15 MPH, cycle 0.5 mi.		396	.020			.70	1.78	2.48	3.12
9216	cycle 1 mile		360	.022			.77	1.96	2.73	3.42
9218	cycle 2 miles		306	.026			.91	2.30	3.21	4.03
9220	cycle 4 miles		234	.034			1.19	3.01	4.20	5.25
9222	cycle 6 miles		198	.040			1.41	3.56	4.97	6.25
9224	cycle 8 miles		162	.049			1.72	4.35	6.07	7.60
9225	cycle 10 miles		144	.056			1.93	4.90	6.83	8.60
9226	20 MPH avg., cycle 0.5 mile		396	.020			.70	1.78	2.48	3.12
9228	cycle 1 mile		378	.021			.74	1.86	2.60	3.26
9230	cycle 2 miles		324	.025			.86	2.18	3.04	3.80
9232	cycle 4 miles		270	.030			1.03	2.61	3.64	4.57
9234	cycle 6 miles		216	.037			1.29	3.26	4.55	5.70
9236	cycle 8 miles		198	.040			1.41	3.56	4.97	6.25

31 23 23.20 Hauling		Crew	Daily Output	Labor-Hours	Unit	Material	2019 Bare Costs Labor	Equipment	Total	Total Incl O&P
9238	cycle 10 miles	B-34I	162	.049	L.C.Y.		1.72	4.35	6.07	7.60
9240	25 MPH avg., cycle 4 miles		288	.028			.97	2.45	3.42	4.28
9242	cycle 6 miles		252	.032			1.10	2.80	3.90	4.90
9244	cycle 8 miles		216	.037			1.29	3.26	4.55	5.70
9246	cycle 10 miles		198	.040			1.41	3.56	4.97	6.25
9250	30 MPH avg., cycle 4 miles		306	.026			.91	2.30	3.21	4.03
9252	cycle 6 miles		270	.030			1.03	2.61	3.64	4.57
9254	cycle 8 miles		234	.034			1.19	3.01	4.20	5.25
9256	cycle 10 miles		216	.037			1.29	3.26	4.55	5.70
9260	35 MPH avg., cycle 4 miles		324	.025			.86	2.18	3.04	3.80
9262	cycle 6 miles		288	.028			.97	2.45	3.42	4.28
9264	cycle 8 miles		252	.032			1.10	2.80	3.90	4.90
9266	cycle 10 miles		234	.034			1.19	3.01	4.20	5.25
9268	cycle 20 miles		162	.049			1.72	4.35	6.07	7.60
9270	cycle 30 miles		108	.074			2.58	6.55	9.13	11.45
9272	cycle 40 miles		90	.089			3.09	7.85	10.94	13.70
9274	40 MPH avg., cycle 6 miles		288	.028			.97	2.45	3.42	4.28
9276	cycle 8 miles		270	.030			1.03	2.61	3.64	4.57
9278	cycle 10 miles		234	.034			1.19	3.01	4.20	5.25
9280	cycle 20 miles		162	.049			1.72	4.35	6.07	7.60
9282	cycle 30 miles		126	.063			2.21	5.60	7.81	9.80
9284	cycle 40 miles		108	.074			2.58	6.55	9.13	11.45
9286	cycle 50 miles		90	.089			3.09	7.85	10.94	13.70
9294	45 MPH avg., cycle 8 miles		270	.030			1.03	2.61	3.64	4.57
9296	cycle 10 miles		252	.032			1.10	2.80	3.90	4.90
9298	cycle 20 miles		180	.044			1.55	3.92	5.47	6.85
9300	cycle 30 miles		144	.056			1.93	4.90	6.83	8.60
9302	cycle 40 miles		108	.074			2.58	6.55	9.13	11.45
9304	cycle 50 miles		90	.089			3.09	7.85	10.94	13.70
9306	50 MPH avg., cycle 10 miles		270	.030			1.03	2.61	3.64	4.57
9308	cycle 20 miles		198	.040			1.41	3.56	4.97	6.25
9310	cycle 30 miles		144	.056			1.93	4.90	6.83	8.60
9312	cycle 40 miles		126	.063			2.21	5.60	7.81	9.80
9314	cycle 50 miles		108	.074			2.58	6.55	9.13	11.45
9414	25 min. wait/ld./uld.,15 MPH, cycle 0.5 mi.		306	.026			.91	2.30	3.21	4.03
9416	cycle 1 mile		288	.028			.97	2.45	3.42	4.28
9418	cycle 2 miles		252	.032			1.10	2.80	3.90	4.90
9420	cycle 4 miles		198	.040			1.41	3.56	4.97	6.25
9422	cycle 6 miles		180	.044			1.55	3.92	5.47	6.85
9424	cycle 8 miles		144	.056			1.93	4.90	6.83	8.60
9425	cycle 10 miles		126	.063			2.21	5.60	7.81	9.80
9426	20 MPH avg., cycle 0.5 mile		324	.025			.86	2.18	3.04	3.80
9428	cycle 1 mile		306	.026			.91	2.30	3.21	4.03
9430	cycle 2 miles		270	.030			1.03	2.61	3.64	4.57
9432	cycle 4 miles		234	.034			1.19	3.01	4.20	5.25
9434	cycle 6 miles		198	.040			1.41	3.56	4.97	6.25
9436	cycle 8 miles		180	.044			1.55	3.92	5.47	6.85
9438	cycle 10 miles		144	.056			1.93	4.90	6.83	8.60
9440	25 MPH avg., cycle 4 miles		252	.032			1.10	2.80	3.90	4.90
9442	cycle 6 miles		216	.037			1.29	3.26	4.55	5.70
9444	cycle 8 miles		198	.040			1.41	3.56	4.97	6.25
9446	cycle 10 miles		180	.044			1.55	3.92	5.47	6.85
9450	30 MPH avg., cycle 4 miles		252	.032			1.10	2.80	3.90	4.90

31 23 23.20 Hauling		Crew	Daily Output	Labor-Hours	Unit	Material	2019 Bare Costs Labor	Equipment	Total	Total Incl O&P
9452	cycle 6 miles	B-34I	234	.034	L.C.Y.		1.19	3.01	4.20	5.25
9454	cycle 8 miles		198	.040			1.41	3.56	4.97	6.25
9456	cycle 10 miles		180	.044			1.55	3.92	5.47	6.85
9460	35 MPH avg., cycle 4 miles		270	.030			1.03	2.61	3.64	4.57
9462	cycle 6 miles		234	.034			1.19	3.01	4.20	5.25
9464	cycle 8 miles		216	.037			1.29	3.26	4.55	5.70
9466	cycle 10 miles		198	.040			1.41	3.56	4.97	6.25
9468	cycle 20 miles		144	.056			1.93	4.90	6.83	8.60
9470	cycle 30 miles		108	.074			2.58	6.55	9.13	11.45
9472	cycle 40 miles		90	.089			3.09	7.85	10.94	13.70
9474	40 MPH avg., cycle 6 miles		252	.032			1.10	2.80	3.90	4.90
9476	cycle 8 miles		234	.034			1.19	3.01	4.20	5.25
9478	cycle 10 miles		216	.037			1.29	3.26	4.55	5.70
9480	cycle 20 miles		144	.056			1.93	4.90	6.83	8.60
9482	cycle 30 miles		126	.063			2.21	5.60	7.81	9.80
9484	cycle 40 miles		90	.089			3.09	7.85	10.94	13.70
9486	cycle 50 miles		90	.089			3.09	7.85	10.94	13.70
9494	45 MPH avg., cycle 8 miles		234	.034			1.19	3.01	4.20	5.25
9496	cycle 10 miles		216	.037			1.29	3.26	4.55	5.70
9498	cycle 20 miles		162	.049			1.72	4.35	6.07	7.60
9500	cycle 30 miles		126	.063			2.21	5.60	7.81	9.80
9502	cycle 40 miles		108	.074			2.58	6.55	9.13	11.45
9504	cycle 50 miles		90	.089			3.09	7.85	10.94	13.70
9506	50 MPH avg., cycle 10 miles		234	.034			1.19	3.01	4.20	5.25
9508	cycle 20 miles		180	.044			1.55	3.92	5.47	6.85
9510	cycle 30 miles		144	.056			1.93	4.90	6.83	8.60
9512	cycle 40 miles		108	.074			2.58	6.55	9.13	11.45
9514	cycle 50 miles		90	.089			3.09	7.85	10.94	13.70
9614	30 min. wait/ld./uld.,15 MPH, cycle 0.5 mi.		270	.030			1.03	2.61	3.64	4.57
9616	cycle 1 mile		252	.032			1.10	2.80	3.90	4.90
9618	cycle 2 miles		216	.037			1.29	3.26	4.55	5.70
9620	cycle 4 miles		180	.044			1.55	3.92	5.47	6.85
9622	cycle 6 miles		162	.049			1.72	4.35	6.07	7.60
9624	cycle 8 miles		126	.063			2.21	5.60	7.81	9.80
9625	cycle 10 miles		126	.063			2.21	5.60	7.81	9.80
9626	20 MPH avg., cycle 0.5 mile		270	.030			1.03	2.61	3.64	4.57
9628	cycle 1 mile		252	.032			1.10	2.80	3.90	4.90
9630	cycle 2 miles		234	.034			1.19	3.01	4.20	5.25
9632	cycle 4 miles		198	.040			1.41	3.56	4.97	6.25
9634	cycle 6 miles		180	.044			1.55	3.92	5.47	6.85
9636	cycle 8 miles		162	.049			1.72	4.35	6.07	7.60
9638	cycle 10 miles		144	.056			1.93	4.90	6.83	8.60
9640	25 MPH avg., cycle 4 miles		216	.037			1.29	3.26	4.55	5.70
9642	cycle 6 miles		198	.040			1.41	3.56	4.97	6.25
9644	cycle 8 miles		180	.044			1.55	3.92	5.47	6.85
9646	cycle 10 miles		162	.049			1.72	4.35	6.07	7.60
9650	30 MPH avg., cycle 4 miles		216	.037			1.29	3.26	4.55	5.70
9652	cycle 6 miles		198	.040			1.41	3.56	4.97	6.25
9654	cycle 8 miles		180	.044			1.55	3.92	5.47	6.85
9656	cycle 10 miles		162	.049			1.72	4.35	6.07	7.60
9660	35 MPH avg., cycle 4 miles		234	.034			1.19	3.01	4.20	5.25
9662	cycle 6 miles		216	.037			1.29	3.26	4.55	5.70
9664	cycle 8 miles		198	.040			1.41	3.56	4.97	6.25

31 23 Excavation and Fill

31 23 23 – Fill

31 23 23.20 Hauling		Crew	Daily Output	Labor-Hours	Unit	Material	2019 Bare Costs Labor	Equipment	Total	Total Incl O&P
9666	cycle 10 miles	B-34I	180	.044	L.C.Y.		1.55	3.92	5.47	6.85
9668	cycle 20 miles		126	.063			2.21	5.60	7.81	9.80
9670	cycle 30 miles		108	.074			2.58	6.55	9.13	11.45
9672	cycle 40 miles		90	.089			3.09	7.85	10.94	13.70
9674	40 MPH avg., cycle 6 miles		216	.037			1.29	3.26	4.55	5.70
9676	cycle 8 miles		198	.040			1.41	3.56	4.97	6.25
9678	cycle 10 miles		180	.044			1.55	3.92	5.47	6.85
9680	cycle 20 miles		144	.056			1.93	4.90	6.83	8.60
9682	cycle 30 miles		108	.074			2.58	6.55	9.13	11.45
9684	cycle 40 miles		90	.089			3.09	7.85	10.94	13.70
9686	cycle 50 miles		72	.111			3.87	9.80	13.67	17.10
9694	45 MPH avg., cycle 8 miles		216	.037			1.29	3.26	4.55	5.70
9696	cycle 10 miles		198	.040			1.41	3.56	4.97	6.25
9698	cycle 20 miles		144	.056			1.93	4.90	6.83	8.60
9700	cycle 30 miles		126	.063			2.21	5.60	7.81	9.80
9702	cycle 40 miles		108	.074			2.58	6.55	9.13	11.45
9704	cycle 50 miles		90	.089			3.09	7.85	10.94	13.70
9706	50 MPH avg., cycle 10 miles		198	.040			1.41	3.56	4.97	6.25
9708	cycle 20 miles		162	.049			1.72	4.35	6.07	7.60
9710	cycle 30 miles		126	.063			2.21	5.60	7.81	9.80
9712	cycle 40 miles		108	.074			2.58	6.55	9.13	11.45
9714	cycle 50 miles	↓	90	.089	↓		3.09	7.85	10.94	13.70

31 23 23.24 Compaction, Structural

		Crew	Daily Output	Labor-Hours	Unit	Material	Labor	Equipment	Total	Total Incl O&P
0010	**COMPACTION, STRUCTURAL**									
0020	Steel wheel tandem roller, 5 tons R312323-30	B-10E	8	1	Hr.		41	19.05	60.05	88
0050	Air tamp, 6" to 8" lifts, common fill	B-9	250	.160	E.C.Y.		4.92	.94	5.86	9.20
0060	Select fill	"	300	.133			4.10	.78	4.88	7.60
0600	Vibratory plate, 8" lifts, common fill	A-1D	200	.040			1.21	.16	1.37	2.18
0700	Select fill	"	216	.037	↓		1.12	.14	1.26	2.02

31 25 Erosion and Sedimentation Controls

31 25 14 – Stabilization Measures for Erosion and Sedimentation Control

31 25 14.16 Rolled Erosion Control Mats and Blankets

		Crew	Daily Output	Labor-Hours	Unit	Material	Labor	Equipment	Total	Total Incl O&P
0010	**ROLLED EROSION CONTROL MATS AND BLANKETS**									
0020	Jute mesh, 100 S.Y. per roll, 4' wide, stapled G	B-80A	2400	.010	S.Y.	.99	.30	.10	1.39	1.70
0100	Plastic netting, stapled, 2" x 1" mesh, 20 mil G	B-1	2500	.010		.24	.30		.54	.75
0120	Revegetation mat, webbed G	2 Clab	1000	.016		3.10	.49		3.59	4.21
0200	Polypropylene mesh, stapled, 6.5 oz./S.Y. G	B-1	2500	.010		1.73	.30		2.03	2.39
0300	Tobacco netting, or jute mesh #2, stapled G	"	2500	.010	↓	.27	.30		.57	.79
0600	Straw in polymeric netting, biodegradable log	A-2	1000	.024	L.F.	6.30	.76	.20	7.26	8.40
1000	Silt fence, install and maintain, remove G	B-62	1300	.018	"	.49	.61	.13	1.23	1.70
1100	Allow 10% per month for maintenance; 6-month max life									

31 31 Soil Treatment

31 31 16 – Termite Control

31 31 16.13 Chemical Termite Control		Crew	Daily Output	Labor-Hours	Unit	2019 Bare Costs Material	Labor	Equipment	Total	Total Incl O&P
0010	**CHEMICAL TERMITE CONTROL**									
0020	Slab and walls, residential	1 Skwk	1200	.007	SF Flr.	.33	.26		.59	.80
0400	Insecticides for termite control, minimum		14.20	.563	Gal.	71	22		93	115
0500	Maximum	↓	11	.727	"	121	28.50		149.50	181

Division Notes

	CREW	DAILY OUTPUT	LABOR-HOURS	UNIT	BARE COSTS MAT.	BARE COSTS LABOR	BARE COSTS EQUIP.	BARE COSTS TOTAL	TOTAL INCL O&P

Estimating Tips

32 01 00 Operations and Maintenance of Exterior Improvements

- Recycling of asphalt pavement is becoming very popular and is an alternative to removal and replacement. It can be a good value engineering proposal if removed pavement can be recycled, either at the project site or at another site that is reasonably close to the project site. Sections on repair of flexible and rigid pavement are included.

32 10 00 Bases, Ballasts, and Paving

- When estimating paving, keep in mind the project schedule. Also note that prices for asphalt and concrete are generally higher in the cold seasons. Lines for pavement markings, including tactile warning systems and fence lines, are included.

32 90 00 Planting

- The timing of planting and guarantee specifications often dictate the costs for establishing tree and shrub growth and a stand of grass or ground cover. Establish the work performance schedule to coincide with the local planting season. Maintenance and growth guarantees can add 20–100% to the total landscaping cost and can be contractually cumbersome. The cost to replace trees and shrubs can be as high as 5% of the total cost, depending on the planting zone, soil conditions, and time of year.

Reference Numbers

Reference numbers are shown at the beginning of some major classifications. These numbers refer to related items in the Reference Section. The reference information may be an estimating procedure, an alternate pricing method, or technical information.

Note: Not all subdivisions listed here necessarily appear. ■

32 01 Operation and Maintenance of Exterior Improvements

32 01 13 – Flexible Paving Surface Treatment

32 01 13.61 Slurry Seal (Latex Modified)		Crew	Daily Output	Labor-Hours	Unit	2019 Bare Costs Material	Labor	Equipment	Total	Total Incl O&P
0010	**SLURRY SEAL (LATEX MODIFIED)**									
3780	Rubberized asphalt (latex) seal	B-45	5000	.003	S.Y.	1.26	.10	.16	1.52	1.74

32 01 13.66 Fog Seal

		Crew	Daily Output	Labor-Hours	Unit	Material	Labor	Equipment	Total	Total Incl O&P
0010	**FOG SEAL**									
0012	Sealcoating, 2 coat coal tar pitch emulsion over 10,000 S.Y.	B-45	5000	.003	S.Y.	.85	.10	.16	1.11	1.29
0030	1,000 to 10,000 S.Y.	"	3000	.005		.85	.17	.27	1.29	1.53
0100	Under 1,000 S.Y.	B-1	1050	.023		.85	.71		1.56	2.11
0300	Petroleum resistant, over 10,000 S.Y.	B-45	5000	.003		1.42	.10	.16	1.68	1.91
0320	1,000 to 10,000 S.Y.	"	3000	.005		1.42	.17	.27	1.86	2.15
0400	Under 1,000 S.Y.	B-1	1050	.023	↓	1.42	.71		2.13	2.73

32 06 Schedules for Exterior Improvements

32 06 10 – Schedules for Bases, Ballasts, and Paving

32 06 10.10 Sidewalks, Driveways and Patios

		Crew	Daily Output	Labor-Hours	Unit	Material	Labor	Equipment	Total	Total Incl O&P
0010	**SIDEWALKS, DRIVEWAYS AND PATIOS** No base									
0020	Asphaltic concrete, 2" thick	B-37	720	.067	S.Y.	6.80	2.14	.21	9.15	11.20
0100	2-1/2" thick	"	660	.073	"	8.60	2.33	.23	11.16	13.55
0300	Concrete, 3,000 psi, CIP, 6 x 6 - W1.4 x W1.4 mesh,									
0310	broomed finish, no base, 4" thick	B-24	600	.040	S.F.	2.22	1.42		3.64	4.76
0350	5" thick		545	.044		2.75	1.56		4.31	5.60
0400	6" thick	↓	510	.047		3.20	1.67		4.87	6.25
0450	For bank run gravel base, 4" thick, add	B-18	2500	.010		.43	.30	.02	.75	.99
0520	8" thick, add	"	1600	.015		.88	.47	.03	1.38	1.76
1000	Crushed stone, 1" thick, white marble	2 Clab	1700	.009		.49	.29		.78	1.01
1050	Bluestone	"	1700	.009		.20	.29		.49	.69
1700	Redwood, prefabricated, 4' x 4' sections	2 Carp	316	.051		4.84	1.96		6.80	8.55
1750	Redwood planks, 1" thick, on sleepers	"	240	.067	↓	4.84	2.58		7.42	9.55

32 06 10.20 Steps

		Crew	Daily Output	Labor-Hours	Unit	Material	Labor	Equipment	Total	Total Incl O&P
0010	**STEPS**									
0011	Incl. excav., borrow & concrete base as required									
0100	Brick steps	B-24	35	.686	LF Riser	17.20	24.50		41.70	59
0200	Railroad ties	2 Clab	25	.640		3.59	19.40		22.99	36
0300	Bluestone treads, 12" x 2" or 12" x 1-1/2"	B-24	30	.800	↓	43	28.50		71.50	94
4025	Steel edge strips, incl. stakes, 1/4" x 5"	B-1	390	.062	L.F.	4.88	1.91		6.79	8.50
4050	Edging, landscape timber or railroad ties, 6" x 8"	2 Carp	170	.094	"	2.32	3.65		5.97	8.60

32 11 Base Courses

32 11 23 – Aggregate Base Courses

32 11 23.23 Base Course Drainage Layers

		Crew	Daily Output	Labor-Hours	Unit	Material	Labor	Equipment	Total	Total Incl O&P
0010	**BASE COURSE DRAINAGE LAYERS**									
0011	For roadways and large areas									
0050	Crushed 3/4" stone base, compacted, 3" deep	B-36C	5200	.008	S.Y.	2.67	.29	.75	3.71	4.23
0100	6" deep		5000	.008		5.35	.30	.78	6.43	7.20
0200	9" deep		4600	.009		8	.33	.84	9.17	10.25
0300	12" deep	↓	4200	.010		10.70	.36	.92	11.98	13.35
0301	Crushed 1-1/2" stone base, compacted to 4" deep	B-36B	6000	.011		7.20	.39	.76	8.35	9.40
0302	6" deep		5400	.012		10.80	.43	.84	12.07	13.55
0303	8" deep		4500	.014		14.40	.52	1.01	15.93	17.80
0304	12" deep	↓	3800	.017	↓	21.50	.61	1.19	23.30	26.50

32 11 Base Courses

32 11 23 – Aggregate Base Courses

32 11 23.23 Base Course Drainage Layers		Crew	Daily Output	Labor-Hours	Unit	Material	2019 Bare Costs Labor	Equipment	Total	Total Incl O&P
0350	Bank run gravel, spread and compacted									
0370	6" deep	B-32	6000	.005	S.Y.	3.63	.20	.36	4.19	4.72
0390	9" deep	↓	4900	.007	↓	5.45	.25	.44	6.14	6.90
0400	12" deep	↓	4200	.008	↓	7.25	.29	.52	8.06	9.05
6900	For small and irregular areas, add						50%	50%		
7000	Prepare and roll sub-base, small areas to 2,500 S.Y.	B-32A	1500	.016	S.Y.		.60	.87	1.47	1.94

32 11 26 – Asphaltic Base Courses

32 11 26.19 Bituminous-Stabilized Base Courses		Crew	Daily Output	Labor-Hours	Unit	Material	Labor	Equipment	Total	Total Incl O&P
0010	**BITUMINOUS-STABILIZED BASE COURSES**									
0020	And large paved areas									
0800	Prime and seal, cut back asphalt	B-45	6000	.003	Gal.	5.40	.09	.14	5.63	6.20
1620	Bituminous stabilized	B-36E	1400	.034	E.C.Y.	77	1.32	2.82	81.14	90
1630	Subbase course, bituminous stabilized	"	1400	.034	"	78.50	1.32	2.82	82.64	92
8900	For small and irregular areas, add						50%	50%		

32 12 Flexible Paving

32 12 16 – Asphalt Paving

32 12 16.14 Paving Asphaltic Concrete		Crew	Daily Output	Labor-Hours	Unit	Material	Labor	Equipment	Total	Total Incl O&P
0010	**PAVING ASPHALTIC CONCRETE**									
0020	6" stone base, 2" binder course, 1" topping	B-25C	9000	.005	S.F.	2.20	.18	.26	2.64	3
0025	2" binder course, 2" topping		9000	.005		2.65	.18	.26	3.09	3.50
0030	3" binder course, 2" topping		9000	.005		3.05	.18	.26	3.49	3.93
0035	4" binder course, 2" topping		9000	.005		3.43	.18	.26	3.87	4.35
0040	1-1/2" binder course, 1" topping		9000	.005		2.01	.18	.26	2.45	2.79
0042	3" binder course, 1" topping		9000	.005		2.59	.18	.26	3.03	3.43
0045	3" binder course, 3" topping		9000	.005		3.50	.18	.26	3.94	4.43
0050	4" binder course, 3" topping		9000	.005		3.89	.18	.26	4.33	4.86
0055	4" binder course, 4" topping		9000	.005		4.34	.18	.26	4.78	5.35
0300	Binder course, 1-1/2" thick		35000	.001		.58	.05	.07	.70	.79
0400	2" thick		25000	.002		.75	.07	.09	.91	1.04
0500	3" thick		15000	.003		1.17	.11	.15	1.43	1.63
0600	4" thick		10800	.004		1.53	.15	.21	1.89	2.16
0800	Sand finish course, 3/4" thick		41000	.001		.31	.04	.06	.41	.47
0900	1" thick	↓	34000	.001		.39	.05	.07	.51	.58
1000	Fill pot holes, hot mix, 2" thick	B-16	4200	.008		.81	.24	.14	1.19	1.44
1100	4" thick		3500	.009		1.19	.29	.16	1.64	1.96
1120	6" thick	↓	3100	.010		1.59	.33	.18	2.10	2.49
1140	Cold patch, 2" thick	B-51	3000	.016		.90	.50	.07	1.47	1.89
1160	4" thick		2700	.018		1.72	.56	.07	2.35	2.89
1180	6" thick	↓	1900	.025	↓	2.68	.79	.10	3.57	4.37

32 13 Rigid Paving

32 13 13 – Concrete Paving

32 13 13.25 Concrete Pavement, Highways		Crew	Daily Output	Labor-Hours	Unit	2019 Bare Costs Material	Labor	Equipment	Total	Total Incl O&P
0010	**CONCRETE PAVEMENT, HIGHWAYS**									
0015	Including joints, finishing and curing									
0020	Fixed form, 12′ pass, unreinforced, 6″ thick	B-26	3000	.029	S.Y.	26	1	1.03	28.03	31.50
0100	8″ thick		2750	.032		35.50	1.09	1.12	37.71	42
0110	8″ thick, small area		1375	.064		35.50	2.17	2.24	39.91	45
0410	Conc. pavement, w/jt., fnsh.& curing, fix form, 24′ pass, unreinforced, 6″T		6000	.015		24.50	.50	.51	25.51	28.50
0430	8″ thick		5500	.016		33.50	.54	.56	34.60	38.50
0520	Welded wire fabric, sheets for rigid paving 2.33 lb./S.Y.	2 Rodm	389	.041		1.43	1.67		3.10	4.34
0530	Reinforcing steel for rigid paving 12 lb./S.Y.		666	.024		6.50	.98		7.48	8.75
0540	Reinforcing steel for rigid paving 18 lb./S.Y.		444	.036		9.75	1.46		11.21	13.10
0620	Slip form, 12′ pass, unreinforced, 6″ thick	B-26A	5600	.016		25	.53	.57	26.10	29
0624	8″ thick		5300	.017		34.50	.56	.60	35.66	39
0640	Slip form, 24′ pass, unreinforced, 6″ thick		11200	.008		24.50	.27	.29	25.06	28
0644	8″ thick		10600	.008		33	.28	.30	33.58	37.50
0700	Finishing, broom finish small areas	2 Cefi	120	.133			4.95		4.95	8

32 14 Unit Paving

32 14 13 – Precast Concrete Unit Paving

32 14 13.18 Precast Concrete Plantable Pavers		Crew	Daily Output	Labor-Hours	Unit	Material	Labor	Equipment	Total	Total Incl O&P
0010	**PRECAST CONCRETE PLANTABLE PAVERS** (50% grass)									
0300	3/4″ crushed stone base for plantable pavers, 6″ depth	B-62	1000	.024	S.Y.	4.15	.80	.17	5.12	6.05
0400	8″ depth		900	.027		5.55	.88	.19	6.62	7.75
0500	10″ depth		800	.030		6.90	.99	.22	8.11	9.50
0600	12″ depth		700	.034		8.30	1.14	.25	9.69	11.25
0700	Hydro seeding plantable pavers	B-81A	20	.800	M.S.F.	11.45	26	18.50	55.95	75.50
0800	Apply fertilizer and seed to plantable pavers	1 Clab	8	1	"	44	30.50		74.50	98.50

32 14 16 – Brick Unit Paving

32 14 16.10 Brick Paving		Crew	Daily Output	Labor-Hours	Unit	Material	Labor	Equipment	Total	Total Incl O&P
0010	**BRICK PAVING**									
0012	4″ x 8″ x 1-1/2″, without joints (4.5 bricks/S.F.)	D-1	110	.145	S.F.	2.55	4.91		7.46	11
0100	Grouted, 3/8″ joint (3.9 bricks/S.F.)		90	.178		2.07	6		8.07	12.30
0200	4″ x 8″ x 2-1/4″, without joints (4.5 bricks/S.F.)		110	.145		2.39	4.91		7.30	10.85
0300	Grouted, 3/8″ joint (3.9 bricks/S.F.)		90	.178		2.07	6		8.07	12.30
0455	Pervious brick paving, 4″ x 8″ x 3-1/4″, without joints (4.5 bricks/S.F.)		110	.145		3.64	4.91		8.55	12.20
0500	Bedding, asphalt, 3/4″ thick	B-25	5130	.017		.64	.57	.52	1.73	2.21
0540	Course washed sand bed, 1″ thick	B-18	5000	.005		.36	.15	.01	.52	.66
0580	Mortar, 1″ thick	D-1	300	.053		.66	1.80		2.46	3.72
0620	2″ thick		200	.080		1.32	2.70		4.02	5.95
1500	Brick on 1″ thick sand bed laid flat, 4.5/S.F.		100	.160		2.91	5.40		8.31	12.20
2000	Brick pavers, laid on edge, 7.2/S.F.		70	.229		4.44	7.70		12.14	17.75

32 14 23 – Asphalt Unit Paving

32 14 23.10 Asphalt Blocks		Crew	Daily Output	Labor-Hours	Unit	Material	Labor	Equipment	Total	Total Incl O&P
0010	**ASPHALT BLOCKS**									
0020	Rectangular, 6″ x 12″ x 1-1/4″, w/bed & neopr. adhesive	D-1	135	.119	S.F.	9.40	4		13.40	17
0100	3″ thick		130	.123		13.15	4.15		17.30	21.50
0300	Hexagonal tile, 8″ wide, 1-1/4″ thick		135	.119		9.40	4		13.40	17
0400	2″ thick		130	.123		13.15	4.15		17.30	21.50
0500	Square, 8″ x 8″, 1-1/4″ thick		135	.119		9.40	4		13.40	17
0600	2″ thick		130	.123		13.15	4.15		17.30	21.50

32 14 Unit Paving

32 14 40 – Stone Paving

32 14 40.10 Stone Pavers		Crew	Daily Output	Labor-Hours	Unit	2019 Bare Costs Material	Labor	Equipment	Total	Total Incl O&P
0010	**STONE PAVERS**									
1100	Flagging, bluestone, irregular, 1" thick,	D-1	81	.198	S.F.	10.50	6.65		17.15	22.50
1150	Snapped random rectangular, 1" thick		92	.174		15.95	5.85		21.80	27.50
1200	1-1/2" thick		85	.188		19.15	6.35		25.50	31.50
1250	2" thick		83	.193		22.50	6.50		29	35.50
1300	Slate, natural cleft, irregular, 3/4" thick		92	.174		9.65	5.85		15.50	20.50
1350	Random rectangular, gauged, 1/2" thick		105	.152		21	5.15		26.15	31.50
1400	Random rectangular, butt joint, gauged, 1/4" thick		150	.107		22.50	3.60		26.10	30.50
1450	For sand rubbed finish, add					9.60			9.60	10.55
1550	Granite blocks, 3-1/2" x 3-1/2" x 3-1/2"	D-1	92	.174		20.50	5.85		26.35	32.50

32 16 Curbs, Gutters, Sidewalks, and Driveways

32 16 13 – Curbs and Gutters

32 16 13.13 Cast-in-Place Concrete Curbs and Gutters

Code	Description	Crew	Daily Output	Labor-Hours	Unit	Material	Labor	Equipment	Total	Total Incl O&P
0010	**CAST-IN-PLACE CONCRETE CURBS AND GUTTERS**									
0290	Forms only, no concrete									
0300	Concrete, wood forms, 6" x 18", straight	C-2	500	.096	L.F.	3	3.31		6.31	8.80
0400	6" x 18", radius	"	200	.240		3.13	8.25		11.38	17.15
0404	Concrete, wood forms, 6" x 18", straight & concrete	C-2A	500	.096		6.45	3.59		10.04	13
0406	6" x 18", radius	"	200	.240		6.60	9		15.60	22

32 16 13.23 Precast Concrete Curbs and Gutters

Code	Description	Crew	Daily Output	Labor-Hours	Unit	Material	Labor	Equipment	Total	Total Incl O&P
0010	**PRECAST CONCRETE CURBS AND GUTTERS**									
0550	Precast, 6" x 18", straight	B-29	700	.069	L.F.	9.35	2.24	1.23	12.82	15.30
0600	6" x 18", radius	"	325	.148	"	10.40	4.82	2.65	17.87	22.50

32 16 13.33 Asphalt Curbs

Code	Description	Crew	Daily Output	Labor-Hours	Unit	Material	Labor	Equipment	Total	Total Incl O&P
0010	**ASPHALT CURBS**									
0012	Curbs, asphaltic, machine formed, 8" wide, 6" high, 40 L.F./ton	B-27	1000	.032	L.F.	1.64	.99	.33	2.96	3.79
0150	Asphaltic berm, 12" W, 3" to 6" H, 35 L.F./ton, before pavement	"	700	.046		.04	1.41	.46	1.91	2.89
0200	12" W, 1-1/2" to 4" H, 60 L.F./ton, laid with pavement	B-2	1050	.038		.02	1.17		1.19	1.97

32 16 13.43 Stone Curbs

Code	Description	Crew	Daily Output	Labor-Hours	Unit	Material	Labor	Equipment	Total	Total Incl O&P
0010	**STONE CURBS**									
1000	Granite, split face, straight, 5" x 16"	D-13	275	.175	L.F.	14.95	6.25	1.35	22.55	28.50
1100	6" x 18"	"	250	.192		19.65	6.90	1.49	28.04	34.50
1300	Radius curbing, 6" x 18", over 10' radius	B-29	260	.185		24	6	3.32	33.32	40
1400	Corners, 2' radius	"	80	.600	Ea.	81	19.55	10.80	111.35	133
1600	Edging, 4-1/2" x 12", straight	D-13	300	.160	L.F.	7.50	5.75	1.24	14.49	19.15
1800	Curb inlets (guttermouth) straight	B-29	41	1.171	Ea.	180	38	21	239	284
2000	Indian granite (Belgian block)									
2100	Jumbo, 10-1/2" x 7-1/2" x 4", grey	D-1	150	.107	L.F.	9.80	3.60		13.40	16.80
2150	Pink		150	.107		8.50	3.60		12.10	15.35
2200	Regular, 9" x 4-1/2" x 4-1/2", grey		160	.100		4.38	3.37		7.75	10.40
2250	Pink		160	.100		5.85	3.37		9.22	12
2300	Cubes, 4" x 4" x 4", grey		175	.091		3.62	3.08		6.70	9.15
2350	Pink		175	.091		3.79	3.08		6.87	9.30
2400	6" x 6" x 6", pink		155	.103		12.90	3.48		16.38	20
2500	Alternate pricing method for Indian granite									
2550	Jumbo, 10-1/2" x 7-1/2" x 4" (30 lb.), grey				Ton	560			560	615
2600	Pink					495			495	545
2650	Regular, 9" x 4-1/2" x 4-1/2" (20 lb.), grey					310			310	340
2700	Pink					410			410	450

32 16 Curbs, Gutters, Sidewalks, and Driveways

32 16 13 – Curbs and Gutters

32 16 13.43 Stone Curbs		Crew	Daily Output	Labor-Hours	Unit	2019 Bare Costs Material	Labor	Equipment	Total	Total Incl O&P
2750	Cubes, 4" x 4" x 4" (5 lb.), grey				Ton	440			440	480
2800	Pink					485			485	535
2850	6" x 6" x 6" (25 lb.), pink					505			505	555
2900	For pallets, add				↓	22			22	24

32 17 Paving Specialties

32 17 23 – Pavement Markings

32 17 23.13 Painted Pavement Markings

		Crew	Daily Output	Labor-Hours	Unit	Material	Labor	Equipment	Total	Total Incl O&P
0010	**PAINTED PAVEMENT MARKINGS**									
0203	Parking striping	B-78	11000	.004	L.F.	.30	.11	.05	.46	.56
0620	Arrows or gore lines	"	2300	.017	S.F.	.21	.53	.22	.96	1.35

32 17 23.14 Pavement Parking Markings

		Crew	Daily Output	Labor-Hours	Unit	Material	Labor	Equipment	Total	Total Incl O&P
0010	**PAVEMENT PARKING MARKINGS**									
0790	Layout of pavement marking	A-2	25000	.001	L.F.		.03	.01	.04	.06
0800	Lines on pvmt., parking stall, paint, white, 4" wide	B-78B	400	.045	Stall	4.85	1.41	.87	7.13	8.60
0825	Parking stall, small quantities	2 Pord	80	.200		9.70	6.55		16.25	21.50
0830	Lines on pvmt., parking stall, thermoplastic, white, 4" wide	B-79	300	.133	↓	18.50	4.20	4.10	26.80	32
1000	Street letters and numbers	B-78B	1600	.011	S.F.	.74	.35	.22	1.31	1.63
1100	Pavement marking letter, 6"	2 Pord	400	.040	Ea.	11.75	1.31		13.06	15.05
1110	12" letter		272	.059		15	1.93		16.93	19.65
1120	24" letter		160	.100		32.50	3.29		35.79	41.50
1130	36" letter		84	.190		43	6.25		49.25	57.50
1140	42" letter		84	.190		52	6.25		58.25	68
1150	72" letter		40	.400		43	13.15		56.15	68.50
1200	Handicap symbol	↓	40	.400		33.50	13.15		46.65	58.50
1210	Handicap parking sign 12" x 18" and post	A-2	12	2	↓	135	63	16.25	214.25	270

32 17 26 – Tactile Warning Surfacing

32 17 26.10 Tactile Warning Surfacing

		Crew	Daily Output	Labor-Hours	Unit	Material	Labor	Equipment	Total	Total Incl O&P
0010	**TACTILE WARNING SURFACING**									
0100	Tactile warning tiles S.F.	2 Clab	400	.040	S.F.	17.95	1.21		19.16	22

32 31 Fences and Gates

32 31 13 – Chain Link Fences and Gates

32 31 13.15 Chain Link Fence

		Crew	Daily Output	Labor-Hours	Unit	Material	Labor	Equipment	Total	Total Incl O&P
0010	**CHAIN LINK FENCE**									
0011	11 ga. wire									
0350	Aluminized steel, 9 ga. wire, 3' high	B-1	185	.130	L.F.	9.05	4.02		13.07	16.60
0400	6' high		115	.209	"	12.65	6.45		19.10	24.50
0450	Add for gate 3' wide, 1-3/8" frame 3' high		12	2	Ea.	155	62		217	273
0490	6' high		10	2.400		187	74.50		261.50	330
0500	Add for gate 4' wide, 1-3/8" frame 3' high		10	2.400		162	74.50		236.50	300
0540	6' high	↓	8	3	↓	201	93		294	375
0860	Tennis courts, 11 ga. wire, 2-1/2" post 10' OC, 1-5/8" top rail									
0900	2-1/2" corner post, 10' high	B-1	95	.253	L.F.	8.85	7.85		16.70	22.50
0920	12' high		80	.300	"	8.70	9.30		18	25
1000	Add for gate 3' wide, 1-5/8" frame 10' high		10	2.400	Ea.	231	74.50		305.50	375
1040	Aluminized, 11 ga. wire 10' high		95	.253	L.F.	11.70	7.85		19.55	26
1100	12' high	↓	80	.300	"	12.70	9.30		22	29.50

32 31 Fences and Gates

32 31 13 – Chain Link Fences and Gates

32 31 13.15 Chain Link Fence		Crew	Daily Output	Labor-Hours	Unit	Material	Labor	Equipment	Total	Total Incl O&P
						2019 Bare Costs				
1140	Add for gate 3′ wide, 1-5/8″ frame, 10′ high	B-1	10	2.400	Ea.	177	74.50		251.50	320
1250	Vinyl covered 11 ga. wire, 10′ high		95	.253	L.F.	10.15	7.85		18	24
1300	12′ high		80	.300	"	12.15	9.30		21.45	29
1400	Add for gate 3′ wide, 1-3/8″ frame, 10′ high		10	2.400	Ea.	335	74.50		409.50	495
32 31 13.80 Residential Chain Link Gate										
0010	**RESIDENTIAL CHAIN LINK GATE**									
0110	Residential 4′ gate, single incl. hardware and concrete	B-80C	10	2.400	Ea.	125	76	20.50	221.50	286
0120	5′		10	2.400		135	76	20.50	231.50	297
0130	6′		10	2.400		146	76	20.50	242.50	310
0510	Residential 4′ gate, double incl. hardware and concrete		10	2.400		223	76	20.50	319.50	395
0520	5′		10	2.400		239	76	20.50	335.50	410
0530	6′		10	2.400		278	76	20.50	374.50	455
32 31 13.82 Internal Chain Link Gate										
0010	**INTERNAL CHAIN LINK GATE**									
0110	Internal 6′ gate, single incl. post flange, hardware and concrete	B-80C	10	2.400	Ea.	273	76	20.50	369.50	450
0120	8′		10	2.400		310	76	20.50	406.50	495
0130	10′		10	2.400		430	76	20.50	526.50	620
0510	Internal 6′ gate, double incl. post flange, hardware and concrete		10	2.400		500	76	20.50	596.50	700
0520	8′		10	2.400		575	76	20.50	671.50	780
0530	10′		10	2.400		725	76	20.50	821.50	945
32 31 13.84 Industrial Chain Link Gate										
0010	**INDUSTRIAL CHAIN LINK GATE**									
0110	Industrial 8′ gate, single incl. hardware and concrete	B-80C	10	2.400	Ea.	465	76	20.50	561.50	665
0120	10′		10	2.400		530	76	20.50	626.50	735
0510	Industrial 8′ gate, double incl. hardware and concrete		10	2.400		730	76	20.50	826.50	950
0520	10′		10	2.400		830	76	20.50	926.50	1,050
32 31 13.88 Chain Link Transom										
0010	**CHAIN LINK TRANSOM**									
0110	Add for, single transom, 3′ wide, incl. components & hardware	B-80C	10	2.400	Ea.	116	76	20.50	212.50	276
0120	Add for, double transom, 6′ wide, incl. components & hardware	"	10	2.400	"	124	76	20.50	220.50	285

32 31 23 – Plastic Fences and Gates

32 31 23.10 Fence, Vinyl		Crew	Daily Output	Labor-Hours	Unit	Material	Labor	Equipment	Total	Total Incl O&P
0010	**FENCE, VINYL**									
0020	Picket, 4″ x 4″ posts @ 6′-0″ OC, 3′ high	B-1	140	.171	L.F.	24.50	5.30		29.80	36
0030	4′ high		130	.185		26.50	5.75		32.25	39
0040	5′ high		120	.200		29	6.20		35.20	42.50
0100	Board (semi-privacy), 5″ x 5″ posts @ 7′-6″ OC, 5′ high		130	.185		27	5.75		32.75	39
0120	6′ high		125	.192		29.50	5.95		35.45	42.50
0200	Basket weave, 5″ x 5″ posts @ 7′-6″ OC, 5′ high		160	.150		26.50	4.65		31.15	36.50
0220	6′ high		150	.160		29.50	4.96		34.46	40.50
0300	Privacy, 5″ x 5″ posts @ 7′-6″ OC, 5′ high		130	.185		26	5.75		31.75	38
0320	6′ high		150	.160		29.50	4.96		34.46	40.50
0350	Gate, 5′ high		9	2.667	Ea.	320	82.50		402.50	490
0360	6′ high		9	2.667		365	82.50		447.50	540
0400	For posts set in concrete, add		25	.960		9.60	30		39.60	59.50
0500	Post and rail fence, 2 rail		150	.160	L.F.	6.10	4.96		11.06	14.90
0510	3 rail		150	.160		7.90	4.96		12.86	16.90
0515	4 rail		150	.160		10.70	4.96		15.66	20

32 31 26 – Wire Fences and Gates

32 31 26.10 Fences, Misc. Metal

	32 31 26.10 Fences, Misc. Metal	Crew	Daily Output	Labor-Hours	Unit	Material	Labor	Equipment	Total	Total Incl O&P
						2019 Bare Costs				
0010	**FENCES, MISC. METAL**									
0012	Chicken wire, posts @ 4′, 1″ mesh, 4′ high	B-80C	410	.059	L.F.	3.48	1.85	.49	5.82	7.40
0100	2″ mesh, 6′ high		350	.069		3.99	2.17	.58	6.74	8.60
0200	Galv. steel, 12 ga., 2″ x 4″ mesh, posts 5′ OC, 3′ high		300	.080		2.76	2.53	.68	5.97	7.95
0300	5′ high		300	.080		3.33	2.53	.68	6.54	8.55
0400	14 ga., 1″ x 2″ mesh, 3′ high		300	.080		3.37	2.53	.68	6.58	8.60
0500	5′ high		300	.080		4.51	2.53	.68	7.72	9.85
1000	Kennel fencing, 1-1/2″ mesh, 6′ long, 3′-6″ wide, 6′-2″ high	2 Clab	4	4	Ea.	500	121		621	750
1050	12′ long		4	4		705	121		826	975
1200	Top covers, 1-1/2″ mesh, 6′ long		15	1.067		136	32.50		168.50	204
1250	12′ long		12	1.333		188	40.50		228.50	274

32 31 29 – Wood Fences and Gates

32 31 29.10 Fence, Wood

	32 31 29.10 Fence, Wood	Crew	Daily Output	Labor-Hours	Unit	Material	Labor	Equipment	Total	Total Incl O&P
0010	**FENCE, WOOD**									
0011	Basket weave, 3/8″ x 4″ boards, 2″ x 4″									
0020	stringers on spreaders, 4″ x 4″ posts									
0050	No. 1 cedar, 6′ high	B-80C	160	.150	L.F.	26	4.74	1.27	32.01	37.50
0860	Open rail fence, split rails, 2 rail, 3′ high, no. 1 cedar		160	.150		9.65	4.74	1.27	15.66	19.85
0870	No. 2 cedar		160	.150		8.40	4.74	1.27	14.41	18.45
0880	3 rail, 4′ high, no. 1 cedar		150	.160		12.20	5.05	1.35	18.60	23.50
0890	No. 2 cedar		150	.160		8.10	5.05	1.35	14.50	18.75
1240	Stockade fence, no. 1 cedar, 3-1/4″ rails, 6′ high		160	.150		13.50	4.74	1.27	19.51	24
1260	8′ high		155	.155		18.70	4.89	1.31	24.90	30
1320	Gate, 3′-6″ wide		8	3	Ea.	89.50	95	25.50	210	283

32 31 29.20 Fence, Wood Rail

	32 31 29.20 Fence, Wood Rail	Crew	Daily Output	Labor-Hours	Unit	Material	Labor	Equipment	Total	Total Incl O&P
0010	**FENCE, WOOD RAIL**									
0012	Picket, No. 2 cedar, Gothic, 2 rail, 3′ high	B-1	160	.150	L.F.	8.40	4.65		13.05	16.95
0050	Gate, 3′-6″ wide	B-80C	9	2.667	Ea.	79	84.50	22.50	186	251
0400	3 rail, 4′ high		150	.160	L.F.	9.45	5.05	1.35	15.85	20
0500	Gate, 3′-6″ wide		9	2.667	Ea.	96	84.50	22.50	203	270
1200	Stockade, No. 2 cedar, treated wood rails, 6′ high		160	.150	L.F.	9.40	4.74	1.27	15.41	19.55
1250	Gate, 3′ wide		9	2.667	Ea.	98.50	84.50	22.50	205.50	273
1300	No. 1 cedar, 3-1/4″ cedar rails, 6′ high		160	.150	L.F.	20	4.74	1.27	26.01	31
1500	Gate, 3′ wide		9	2.667	Ea.	230	84.50	22.50	337	415
2700	Prefabricated redwood or cedar, 4′ high		160	.150	L.F.	16.30	4.74	1.27	22.31	27
2800	6′ high		150	.160		24	5.05	1.35	30.40	36
3300	Board, shadow box, 1″ x 6″, treated pine, 6′ high		160	.150		12.85	4.74	1.27	18.86	23.50
3400	No. 1 cedar, 6′ high		150	.160		25.50	5.05	1.35	31.90	38
3900	Basket weave, No. 1 cedar, 6′ high		160	.150		35	4.74	1.27	41.01	47.50
4200	Gate, 3′-6″ wide	B-1	9	2.667	Ea.	184	82.50		266.50	340

32 32 13 – Cast-in-Place Concrete Retaining Walls

32 32 13.10 Retaining Walls, Cast Concrete		Crew	Daily Output	Labor-Hours	Unit	Material	2019 Bare Costs Labor	Equipment	Total	Total Incl O&P
0010	**RETAINING WALLS, CAST CONCRETE**									
1800	Concrete gravity wall with vertical face including excavation & backfill									
1850	No reinforcing									
1900	6' high, level embankment	C-17C	36	2.306	L.F.	85	92	15.50	192.50	264
2000	33° slope embankment	"	32	2.594	"	105	104	17.45	226.45	305
2800	Reinforced concrete cantilever, incl. excavation, backfill & reinf.									
2900	6' high, 33° slope embankment	C-17C	35	2.371	L.F.	82.50	94.50	15.95	192.95	265

32 32 26 – Metal Crib Retaining Walls

32 32 26.10 Metal Bin Retaining Walls

Code	Description	Crew	Daily Output	Labor-Hours	Unit	Material	Labor	Equipment	Total	Total Incl O&P
0010	**METAL BIN RETAINING WALLS**									
0011	Aluminized steel bin, excavation									
0020	and backfill not included, 10' wide									
0100	4' high, 5.5' deep	B-13	650	.074	S.F.	29	2.41	.90	32.31	37
0200	8' high, 5.5' deep		615	.078		33.50	2.55	.95	37	41.50
0300	10' high, 7.7' deep		580	.083		37	2.70	1	40.70	46.50
0400	12' high, 7.7' deep		530	.091		40	2.95	1.10	44.05	50
0500	16' high, 7.7' deep	↓	515	.093	↓	42.50	3.04	1.13	46.67	52.50

32 32 29 – Timber Retaining Walls

32 32 29.10 Landscape Timber Retaining Walls

Code	Description	Crew	Daily Output	Labor-Hours	Unit	Material	Labor	Equipment	Total	Total Incl O&P
0010	**LANDSCAPE TIMBER RETAINING WALLS**									
0100	Treated timbers, 6" x 6"	1 Clab	265	.030	L.F.	2.54	.92		3.46	4.30
0110	6" x 8'	"	200	.040	"	5.75	1.21		6.96	8.35
0120	Drilling holes in timbers for fastening, 1/2"	1 Carp	450	.018	Inch		.69		.69	1.14
0130	5/8"	"	450	.018	"		.69		.69	1.14
0140	Reinforcing rods for fastening, 1/2"	1 Clab	312	.026	L.F.	.38	.78		1.16	1.71
0150	5/8"	"	312	.026	"	.59	.78		1.37	1.94
0160	Reinforcing fabric	2 Clab	2500	.006	S.Y.	2.17	.19		2.36	2.71
0170	Gravel backfill		28	.571	C.Y.	16.95	17.35		34.30	47
0180	Perforated pipe, 4" diameter with silt sock	↓	1200	.013	L.F.	1.01	.40		1.41	1.78
0190	Galvanized 60d common nails	1 Clab	625	.013	Ea.	.20	.39		.59	.86
0200	20d common nails	"	3800	.002	"	.04	.06		.10	.16

32 32 53 – Stone Retaining Walls

32 32 53.10 Retaining Walls, Stone

Code	Description	Crew	Daily Output	Labor-Hours	Unit	Material	Labor	Equipment	Total	Total Incl O&P
0010	**RETAINING WALLS, STONE**									
0015	Including excavation, concrete footing and									
0020	stone 3' below grade. Price is exposed face area.									
0200	Decorative random stone, to 6' high, 1'-6" thick, dry set	D-1	35	.457	S.F.	73.50	15.40		88.90	107
0300	Mortar set		40	.400		76	13.50		89.50	106
0500	Cut stone, to 6' high, 1'-6" thick, dry set		35	.457		76	15.40		91.40	110
0600	Mortar set		40	.400		77	13.50		90.50	108
0800	Random stone, 6' to 10' high, 2' thick, dry set		45	.356		79	12		91	107
0900	Mortar set		50	.320		82	10.80		92.80	108
1100	Cut stone, 6' to 10' high, 2' thick, dry set		45	.356		79.50	12		91.50	108
1200	Mortar set	↓	50	.320	↓	82.50	10.80		93.30	109

32 33 Site Furnishings

32 33 33 – Site Manufactured Planters

32 33 33.10 Planters		Crew	Daily Output	Labor-Hours	Unit	Material	2019 Bare Costs Labor	Equipment	Total	Total Incl O&P
0010	**PLANTERS**									
0012	Concrete, sandblasted, precast, 48" diameter, 24" high	2 Clab	15	1.067	Ea.	650	32.50		682.50	775
0300	Fiberglass, circular, 36" diameter, 24" high		15	1.067		720	32.50		752.50	845
1200	Wood, square, 48" side, 24" high		15	1.067		1,600	32.50		1,632.50	1,825
1300	Circular, 48" diameter, 30" high		10	1.600		1,125	48.50		1,173.50	1,300
1600	Planter/bench, 72"	↓	5	3.200	↓	3,800	97		3,897	4,325

32 33 43 – Site Seating and Tables

32 33 43.13 Site Seating		Crew	Daily Output	Labor-Hours	Unit	Material	Labor	Equipment	Total	Total Incl O&P
0010	**SITE SEATING**									
0012	Seating, benches, park, precast conc., w/backs, wood rails, 4' long	2 Clab	5	3.200	Ea.	625	97		722	845
0100	8' long		4	4		1,050	121		1,171	1,350
0500	Steel barstock pedestals w/backs, 2" x 3" wood rails, 4' long		10	1.600		1,350	48.50		1,398.50	1,550
0510	8' long		7	2.286		1,700	69.50		1,769.50	2,000
0800	Cast iron pedestals, back & arms, wood slats, 4' long		8	2		435	60.50		495.50	580
0820	8' long		5	3.200		1,050	97		1,147	1,300
1700	Steel frame, fir seat, 10' long	↓	10	1.600	↓	400	48.50		448.50	520

32 84 Planting Irrigation

32 84 23 – Underground Sprinklers

32 84 23.10 Sprinkler Irrigation System		Crew	Daily Output	Labor-Hours	Unit	Material	Labor	Equipment	Total	Total Incl O&P
0010	**SPRINKLER IRRIGATION SYSTEM**									
0011	For lawns									
0800	Residential system, custom, 1" supply	B-20	2000	.012	S.F.	.26	.37		.63	.91
0900	1-1/2" supply	"	1800	.013	"	.43	.41		.84	1.15

32 91 Planting Preparation

32 91 13 – Soil Preparation

32 91 13.16 Mulching		Crew	Daily Output	Labor-Hours	Unit	Material	Labor	Equipment	Total	Total Incl O&P
0010	**MULCHING**									
0100	Aged barks, 3" deep, hand spread	1 Clab	100	.080	S.Y.	3.55	2.43		5.98	7.90
0150	Skid steer loader	B-63	13.50	2.963	M.S.F.	395	90	12.90	497.90	600
0160	Skid steer loader	"	1500	.027	S.Y.	3.55	.81	.12	4.48	5.40
0200	Hay, 1" deep, hand spread	1 Clab	475	.017	"	.46	.51		.97	1.35
0250	Power mulcher, small	B-64	180	.089	M.S.F.	51	2.86	1.86	55.72	63
0350	Large	B-65	530	.030	"	51	.97	.89	52.86	58.50
0400	Humus peat, 1" deep, hand spread	1 Clab	700	.011	S.Y.	3.02	.35		3.37	3.89
0450	Push spreader	"	2500	.003	"	3.02	.10		3.12	3.48
0550	Tractor spreader	B-66	700	.011	M.S.F.	335	.44	.31	335.75	370
0600	Oat straw, 1" deep, hand spread	1 Clab	475	.017	S.Y.	.57	.51		1.08	1.47
0650	Power mulcher, small	B-64	180	.089	M.S.F.	63.50	2.86	1.86	68.22	76.50
0700	Large	B-65	530	.030	"	63.50	.97	.89	65.36	72
0750	Add for asphaltic emulsion	B-45	1770	.009	Gal.	5.85	.29	.46	6.60	7.40
0800	Peat moss, 1" deep, hand spread	1 Clab	900	.009	S.Y.	4.88	.27		5.15	5.80
0850	Push spreader	"	2500	.003	"	4.88	.10		4.98	5.50
0950	Tractor spreader	B-66	700	.011	M.S.F.	540	.44	.31	540.75	595
1000	Polyethylene film, 6 mil	2 Clab	2000	.008	S.Y.	.54	.24		.78	.99
1100	Redwood nuggets, 3" deep, hand spread	1 Clab	150	.053	"	3.16	1.62		4.78	6.15
1150	Skid steer loader	B-63	13.50	2.963	M.S.F.	350	90	12.90	452.90	550
1200	Stone mulch, hand spread, ceramic chips, economy	1 Clab	125	.064	S.Y.	7.05	1.94		8.99	10.95

32 91 Planting Preparation

32 91 13 – Soil Preparation

32 91 13.16 Mulching		Crew	Daily Output	Labor-Hours	Unit	Material	Labor (2019 Bare Costs)	Equipment	Total	Total Incl O&P
1250	Deluxe	1 Clab	95	.084	S.Y.	10.70	2.56		13.26	16
1300	Granite chips	B-1	10	2.400	C.Y.	74	74.50		148.50	205
1400	Marble chips		10	2.400		217	74.50		291.50	360
1600	Pea gravel		28	.857		115	26.50		141.50	170
1700	Quartz	↓	10	2.400	↓	192	74.50		266.50	335
1800	Tar paper, 15 lb. felt	1 Clab	800	.010	S.Y.	.49	.30		.79	1.03
1900	Wood chips, 2" deep, hand spread	"	220	.036	"	1.63	1.10		2.73	3.61
1950	Skid steer loader	B-63	20.30	1.970	M.S.F.	181	60	8.60	249.60	305

32 91 13.26 Planting Beds

Line	Description	Crew	Daily Output	Labor-Hours	Unit	Material	Labor	Equipment	Total	Total Incl O&P
0010	**PLANTING BEDS**									
0100	Backfill planting pit, by hand, on site topsoil	2 Clab	18	.889	C.Y.		27		27	44.50
0200	Prepared planting mix, by hand	"	24	.667			20		20	33.50
0300	Skid steer loader, on site topsoil	B-62	340	.071			2.34	.51	2.85	4.42
0400	Prepared planting mix	"	410	.059			1.94	.43	2.37	3.67
1000	Excavate planting pit, by hand, sandy soil	2 Clab	16	1			30.50		30.50	50
1100	Heavy soil or clay	"	8	2			60.50		60.50	100
1200	1/2 C.Y. backhoe, sandy soil	B-11C	150	.107			3.79	2.13	5.92	8.60
1300	Heavy soil or clay	"	115	.139			4.95	2.78	7.73	11.20
2000	Mix planting soil, incl. loam, manure, peat, by hand	2 Clab	60	.267		46.50	8.10		54.60	64.50
2100	Skid steer loader	B-62	150	.160	↓	46.50	5.30	1.16	52.96	61
3000	Pile sod, skid steer loader	"	2800	.009	S.Y.		.28	.06	.34	.54
4000	Remove sod, F.E. loader	B-10S	2000	.004			.16	.18	.34	.46
4100	Sod cutter	B-12K	3200	.005	↓		.18	.37	.55	.71

32 91 19 – Landscape Grading

32 91 19.13 Topsoil Placement and Grading

Line	Description	Crew	Daily Output	Labor-Hours	Unit	Material	Labor	Equipment	Total	Total Incl O&P
0010	**TOPSOIL PLACEMENT AND GRADING**									
0700	Furnish and place, truck dumped, screened, 4" deep	B-10S	1300	.006	S.Y.	3.49	.25	.27	4.01	4.55
0800	6" deep	"	820	.010	"	4.47	.40	.43	5.30	6.05
0900	Fine grading and seeding, incl. lime, fertilizer & seed,									
1000	With equipment	B-14	1000	.048	S.Y.	.47	1.54	.32	2.33	3.40

32 92 Turf and Grasses

32 92 19 – Seeding

32 92 19.13 Mechanical Seeding

Line	Description	Crew	Daily Output	Labor-Hours	Unit	Material	Labor	Equipment	Total	Total Incl O&P
0010	**MECHANICAL SEEDING** R329219-50									
0020	Mechanical seeding, 215 lb./acre	B-66	1.50	5.333	Acre	575	207	146	928	1,125
0100	44 lb./M.S.Y.	"	2500	.003	S.Y.	.20	.12	.09	.41	.52
0101	44 lb./M.S.Y.	1 Clab	13950	.001	S.F.	.02	.02		.04	.05
0300	Fine grading and seeding incl. lime, fertilizer & seed,									
0310	with equipment	B-14	1000	.048	S.Y.	.46	1.54	.32	2.32	3.40
0400	Fertilizer hand push spreader, 35 lb./M.S.F.	1 Clab	200	.040	M.S.F.	9.90	1.21		11.11	12.90
0600	Limestone hand push spreader, 50 lb./M.S.F.		180	.044		4.87	1.35		6.22	7.60
0800	Grass seed hand push spreader, 4.5 lb./M.S.F.	↓	180	.044	↓	22	1.35		23.35	26.50
1000	Hydro or air seeding for large areas, incl. seed and fertilizer	B-81	8900	.002	S.Y.	.52	.06	.07	.65	.74
1100	With wood fiber mulch added	"	8900	.002	"	2.03	.06	.07	2.16	2.40
1300	Seed only, over 100 lb., field seed, minimum				Lb.	1.92			1.92	2.11
1400	Maximum					1.83			1.83	2.01
1500	Lawn seed, minimum					1.45			1.45	1.60
1600	Maximum				↓	2.55			2.55	2.81
1800	Aerial operations, seeding only, field seed	B-58	50	.480	Acre	625	15.90	64.50	705.40	780

32 92 Turf and Grasses

32 92 19 – Seeding

32 92 19.13 Mechanical Seeding		Crew	Daily Output	Labor-Hours	Unit	Material	2019 Bare Costs Labor	Equipment	Total	Total Incl O&P
1900	Lawn seed	B-58	50	.480	Acre	470	15.90	64.50	550.40	615
2100	Seed and liquid fertilizer, field seed		50	.480		700	15.90	64.50	780.40	865
2200	Lawn seed		50	.480		545	15.90	64.50	625.40	695

32 92 23 – Sodding

32 92 23.10 Sodding Systems

		Crew	Daily Output	Labor-Hours	Unit	Material	Labor	Equipment	Total	Total Incl O&P
0010	**SODDING SYSTEMS**									
0020	Sodding, 1" deep, bluegrass sod, on level ground, over 8 M.S.F.	B-63	22	1.818	M.S.F.	198	55	7.95	260.95	320
0200	4 M.S.F.		17	2.353		249	71.50	10.25	330.75	405
0300	1,000 S.F.		13.50	2.963		325	90	12.90	427.90	520
0500	Sloped ground, over 8 M.S.F.		6	6.667		198	202	29	429	585
0600	4 M.S.F.		5	8		249	243	35	527	715
0700	1,000 S.F.		4	10		325	305	43.50	673.50	905
1000	Bent grass sod, on level ground, over 6 M.S.F.		20	2		262	60.50	8.70	331.20	400
1100	3 M.S.F.		18	2.222		276	67.50	9.70	353.20	425
1200	Sodding 1,000 S.F. or less		14	2.857		305	86.50	12.45	403.95	490
1500	Sloped ground, over 6 M.S.F.		15	2.667		262	81	11.65	354.65	435
1600	3 M.S.F.		13.50	2.963		276	90	12.90	378.90	470
1700	1,000 S.F.		12	3.333		305	101	14.55	420.55	520

32 93 Plants

32 93 13 – Ground Covers

32 93 13.10 Ground Cover Plants

		Crew	Daily Output	Labor-Hours	Unit	Material	Labor	Equipment	Total	Total Incl O&P
0010	**GROUND COVER PLANTS**									
0012	Plants, pachysandra, in prepared beds	B-1	15	1.600	C	84	49.50		133.50	175
0200	Vinca minor, 1 yr., bare root, in prepared beds		12	2	"	72	62		134	182
0600	Stone chips, in 50 lb. bags, Georgia marble		520	.046	Bag	4.79	1.43		6.22	7.60
0700	Onyx gemstone		260	.092		16.25	2.86		19.11	22.50
0800	Quartz		260	.092		16.85	2.86		19.71	23
0900	Pea gravel, truckload lots		28	.857	Ton	15.75	26.50		42.25	61.50

32 93 33 – Shrubs

32 93 33.10 Shrubs and Trees

		Crew	Daily Output	Labor-Hours	Unit	Material	Labor	Equipment	Total	Total Incl O&P
0010	**SHRUBS AND TREES**									
0011	Evergreen, in prepared beds, B&B									
0100	Arborvitae pyramidal, 4'-5'	B-17	30	1.067	Ea.	106	36	22	164	199
0150	Globe, 12"-15"	B-1	96	.250		26.50	7.75		34.25	42.50
0300	Cedar, blue, 8'-10'	B-17	18	1.778		243	59.50	36.50	339	405
0500	Hemlock, Canadian, 2-1/2'-3'	B-1	36	.667		33	20.50		53.50	70.50
0550	Holly, Savannah, 8'-10' H		9.68	2.479		310	77		387	465
0600	Juniper, andorra, 18"-24"		80	.300		52.50	9.30		61.80	73.50
0620	Wiltoni, 15"-18"		80	.300		27	9.30		36.30	45.50
0640	Skyrocket, 4-1/2'-5'	B-17	55	.582		113	19.55	12	144.55	169
0660	Blue pfitzer, 2'-2-1/2'	B-1	44	.545		40	16.90		56.90	72
0680	Ketleerie, 2-1/2'-3'		50	.480		57.50	14.90		72.40	88
0700	Pine, black, 2-1/2'-3'		50	.480		64	14.90		78.90	95
0720	Mugo, 18"-24"		60	.400		56.50	12.40		68.90	83
0740	White, 4'-5'	B-17	75	.427		54.50	14.30	8.80	77.60	93
0800	Spruce, blue, 18"-24"	B-1	60	.400		70.50	12.40		82.90	98
0840	Norway, 4'-5'	B-17	75	.427		89	14.30	8.80	112.10	131
0900	Yew, denisforma, 12"-15"	B-1	60	.400		37.50	12.40		49.90	62
1000	Capitata, 18"-24"		30	.800		35	25		60	79.50

32 93 33 – Shrubs

32 93 33.10 Shrubs and Trees		Crew	Daily Output	Labor-Hours	Unit	2019 Bare Costs Material	Labor	Equipment	Total	Total Incl O&P
1100	Hicksi, 2'-2-1/2'	B-1	30	.800	Ea.	112	25		137	164

32 93 33.20 Shrubs

		Crew	Daily Output	Labor-Hours	Unit	Material	Labor	Equipment	Total	Total Incl O&P
0010	**SHRUBS**									
0011	Broadleaf Evergreen, planted in prepared beds									
0100	Andromeda, 15"-18", cont	B-1	96	.250	Ea.	34.50	7.75		42.25	51
0200	Azalea, 15"-18", cont		96	.250		31	7.75		38.75	47
0300	Barberry, 9"-12", cont		130	.185		19.25	5.75		25	30.50
0400	Boxwood, 15"-18", B&B		96	.250		46	7.75		53.75	64
0500	Euonymus, emerald gaiety, 12"-15", cont		115	.209		25.50	6.45		31.95	38.50
0600	Holly, 15"-18", B&B		96	.250		41	7.75		48.75	58
0900	Mount laurel, 18"-24", B&B		80	.300		75.50	9.30		84.80	98.50
1000	Paxistema, 9"-12" H		130	.185		22.50	5.75		28.25	34
1100	Rhododendron, 18"-24", cont		48	.500		39.50	15.50		55	69
1200	Rosemary, 1 gal. cont		600	.040		19.05	1.24		20.29	23
2000	Deciduous, planted in prepared beds, amelanchier, 2'-3', B&B		57	.421		126	13.05		139.05	160
2100	Azalea, 15"-18", B&B		96	.250		31	7.75		38.75	47
2300	Bayberry, 2'-3', B&B		57	.421		31.50	13.05		44.55	56
2600	Cotoneaster, 15"-18", B&B		80	.300		28	9.30		37.30	46
2800	Dogwood, 3'-4', B&B	B-17	40	.800		34	27	16.50	77.50	99.50
2900	Euonymus, alatus compacta, 15"-18", cont	B-1	80	.300		28	9.30		37.30	46
3200	Forsythia, 2'-3', cont	"	60	.400		18.75	12.40		31.15	41
3300	Hibiscus, 3'-4', B&B	B-17	75	.427		43.50	14.30	8.80	66.60	80.50
3400	Honeysuckle, 3'-4', B&B	B-1	60	.400		28.50	12.40		40.90	52
3500	Hydrangea, 2'-3', B&B	"	57	.421		32	13.05		45.05	57
3600	Lilac, 3'-4', B&B	B-17	40	.800		19.30	27	16.50	62.80	83
3900	Privet, bare root, 18"-24"	B-1	80	.300		15.95	9.30		25.25	33
4100	Quince, 2'-3', B&B	"	57	.421		30	13.05		43.05	54.50
4200	Russian olive, 3'-4', B&B	B-17	75	.427		25	14.30	8.80	48.10	60.50
4400	Spirea, 3'-4', B&B	B-1	70	.343		21	10.65		31.65	40.50
4500	Viburnum, 3'-4', B&B	B-17	40	.800		27	27	16.50	70.50	91.50

32 93 43 – Trees

32 93 43.20 Trees

			Crew	Daily Output	Labor-Hours	Unit	Material	Labor	Equipment	Total	Total Incl O&P
0010	**TREES**										
0011	Deciduous, in prep. beds, balled & burlapped (B&B)										
0100	Ash, 2" caliper	G	B-17	8	4	Ea.	208	134	82.50	424.50	540
0200	Beech, 5'-6'	G		50	.640		221	21.50	13.20	255.70	293
0300	Birch, 6'-8', 3 stems	G		20	1.600		169	53.50	33	255.50	310
0500	Crabapple, 6'-8'	G		20	1.600		144	53.50	33	230.50	283
0600	Dogwood, 4'-5'	G		40	.800		147	27	16.50	190.50	224
0700	Eastern redbud, 4'-5'	G		40	.800		150	27	16.50	193.50	227
0800	Elm, 8'-10'	G		20	1.600		345	53.50	33	431.50	500
0900	Ginkgo, 6'-7'	G		24	1.333		155	45	27.50	227.50	275
1000	Hawthorn, 8'-10', 1" caliper	G		20	1.600		162	53.50	33	248.50	305
1100	Honeylocust, 10'-12', 1-1/2" caliper	G		10	3.200		214	107	66	387	485
1300	Larch, 8'	G		32	1		131	33.50	20.50	185	223
1400	Linden, 8'-10', 1" caliper	G		20	1.600		148	53.50	33	234.50	288
1500	Magnolia, 4'-5'	G		20	1.600		118	53.50	33	204.50	255
1600	Maple, red, 8'-10', 1-1/2" caliper	G		10	3.200		204	107	66	377	475
1700	Mountain ash, 8'-10', 1" caliper	G		16	2		190	67	41	298	365
1800	Oak, 2-1/2"-3" caliper	G		6	5.333		330	179	110	619	775
2100	Planetree, 9'-11', 1-1/4" caliper	G		10	3.200		272	107	66	445	550
2200	Plum, 6'-8', 1" caliper	G		20	1.600		82	53.50	33	168.50	215

32 93 Plants

32 93 43 – Trees

32 93 43.20 Trees			Crew	Daily Output	Labor-Hours	Unit	2019 Bare Costs Material	Labor	Equipment	Total	Total Incl O&P
2300	Poplar, 9′-11′, 1-1/4″ caliper	G	B-17	10	3.200	Ea.	102	107	66	275	360
2500	Sumac, 2′-3′	G		75	.427		45.50	14.30	8.80	68.60	83
2700	Tulip, 5′-6′	G		40	.800		48	27	16.50	91.50	115
2800	Willow, 6′-8′, 1″ caliper	G		20	1.600		99.50	53.50	33	186	235

32 94 Planting Accessories

32 94 13 – Landscape Edging

32 94 13.20 Edging

Code	Description	Crew	Daily Output	Labor-Hours	Unit	Material	Labor	Equipment	Total	Total Incl O&P
0010	**EDGING**									
0050	Aluminum alloy, including stakes, 1/8″ x 4″, mill finish	B-1	390	.062	L.F.	2.17	1.91		4.08	5.55
0051	Black paint		390	.062		2.52	1.91		4.43	5.90
0052	Black anodized		390	.062		2.91	1.91		4.82	6.35
0100	Brick, set horizontally, 1-1/2 bricks per L.F.	D-1	370	.043		1.42	1.46		2.88	3.99
0150	Set vertically, 3 bricks per L.F.	″	135	.119		3.60	4		7.60	10.60
0200	Corrugated aluminum, roll, 4″ wide	1 Carp	650	.012		2.21	.48		2.69	3.22
0250	6″ wide	″	550	.015		2.76	.56		3.32	3.97
0600	Railroad ties, 6″ x 8″	2 Carp	170	.094		2.32	3.65		5.97	8.60
0650	7″ x 9″		136	.118		2.58	4.56		7.14	10.40
0750	Redwood 2″ x 4″		330	.048		2.30	1.88		4.18	5.65
0800	Steel edge strips, incl. stakes, 1/4″ x 5″	B-1	390	.062		4.88	1.91		6.79	8.50
0850	3/16″ x 4″	″	390	.062		3.86	1.91		5.77	7.40

32 94 50 – Tree Guying

32 94 50.10 Tree Guying Systems

Code	Description	Crew	Daily Output	Labor-Hours	Unit	Material	Labor	Equipment	Total	Total Incl O&P
0010	**TREE GUYING SYSTEMS**									
0015	Tree guying including stakes, guy wire and wrap									
0100	Less than 3″ caliper, 2 stakes	2 Clab	35	.457	Ea.	13.30	13.85		27.15	37.50
0200	3″ to 4″ caliper, 3 stakes	″	21	.762	″	19.65	23		42.65	59.50
1000	Including arrowhead anchor, cable, turnbuckles and wrap									
1100	Less than 3″ caliper, 3″ anchors	2 Clab	20	.800	Ea.	21	24.50		45.50	63
1200	3″ to 6″ caliper, 4″ anchors		15	1.067		31.50	32.50		64	88
1300	6″ caliper, 6″ anchors		12	1.333		23.50	40.50		64	93
1400	8″ caliper, 8″ anchors		9	1.778		115	54		169	216

32 96 Transplanting

32 96 23 – Plant and Bulb Transplanting

32 96 23.23 Planting

Code	Description	Crew	Daily Output	Labor-Hours	Unit	Material	Labor	Equipment	Total	Total Incl O&P
0010	**PLANTING**									
0012	Moving shrubs on site, 12″ ball	B-62	28	.857	Ea.		28.50	6.25	34.75	54
0100	24″ ball	″	22	1.091	″		36	7.95	43.95	68.50

32 96 23.43 Moving Trees

Code	Description	Crew	Daily Output	Labor-Hours	Unit	Material	Labor	Equipment	Total	Total Incl O&P
0010	**MOVING TREES**, On site									
0300	Moving trees on site, 36″ ball	B-6	3.75	6.400	Ea.		212	85.50	297.50	445
0400	60″ ball	″	1	24	″		795	320	1,115	1,650

Estimating Tips

33 10 00 Water Utilities
33 30 00 Sanitary Sewerage Utilities
33 40 00 Storm Drainage Utilities

- Never assume that the water, sewer, and drainage lines will go in at the early stages of the project. Consider the site access needs before dividing the site in half with open trenches, loose pipe, and machinery obstructions. Always inspect the site to establish that the site drawings are complete. Check off all existing utilities on your drawings as you locate them. Be especially careful with underground utilities because appurtenances are sometimes buried during regrading or repaving operations. If you find any discrepancies, mark up the site plan for further research. Differing site conditions can be very costly if discovered later in the project.
- See also Section 33 01 00 for restoration of pipe where removal/ replacement may be undesirable. Use of new types of piping materials can reduce the overall project cost. Owners/design engineers should consider the installing contractor as a valuable source of current information on utility products and local conditions that could lead to significant cost savings.

Reference Numbers

Reference numbers are shown at the beginning of some major classifications. These numbers refer to related items in the Reference Section. The reference information may be an estimating procedure, an alternate pricing method, or technical information.

Note: Not all subdivisions listed here necessarily appear. ■

Note: "Powered in part by CINX™, based on liscensed proprietary information of Harrison Publishing House, Inc."

Note: Trade Service, in part, has been used as a reference source for some of the material prices used in Division 33

33 05 Common Work Results for Utilities

33 05 07 – Trenchless Installation of Utility Piping

33 05 07.36 Microtunneling		Crew	Daily Output	Labor-Hours	Unit	2019 Bare Costs Material	Labor	Equipment	Total	Total Incl O&P
0010	**MICROTUNNELING**									
0011	Not including excavation, backfill, shoring,									
0020	or dewatering, average 50'/day, slurry method									
0100	24" to 48" outside diameter, minimum				L.F.				965	965
0110	Adverse conditions, add				"				500	500
1000	Rent microtunneling machine, average monthly lease				Month				97,500	107,000
1010	Operating technician				Day				630	705
1100	Mobilization and demobilization, minimum				Job				41,200	45,900
1110	Maximum				"				445,500	490,500

33 05 61 – Concrete Manholes

33 05 61.10 Storm Drainage Manholes, Frames and Covers

Line	Description	Crew	Daily Output	Labor-Hours	Unit	Material	Labor	Equipment	Total	Total Incl O&P
0010	**STORM DRAINAGE MANHOLES, FRAMES & COVERS**									
0020	Excludes footing, excavation, backfill (See line items for frame & cover)									
0050	Brick, 4' inside diameter, 4' deep	D-1	1	16	Ea.	590	540		1,130	1,550
1110	Precast, 4' ID, 4' deep	B-22	4.10	7.317	"	865	243	47.50	1,155.50	1,400

33 05 63 – Concrete Vaults and Chambers

33 05 63.13 Precast Concrete Utility Structures

Line	Description	Crew	Daily Output	Labor-Hours	Unit	Material	Labor	Equipment	Total	Total Incl O&P
0010	**PRECAST CONCRETE UTILITY STRUCTURES**, 6" thick									
0050	5' x 10' x 6' high, ID	B-13	2	24	Ea.	1,800	785	291	2,876	3,600
0350	Hand hole, precast concrete, 1-1/2" thick									
0400	1'-0" x 2'-0" x 1'-9", ID, light duty	B-1	4	6	Ea.	500	186		686	860
0450	4'-6" x 3'-2" x 2'-0", OD, heavy duty	B-6	3	8	"	1,550	265	107	1,922	2,250

33 11 Groundwater Sources

33 11 13 – Potable Water Supply Wells

33 11 13.10 Wells and Accessories

Line	Description	Crew	Daily Output	Labor-Hours	Unit	Material	Labor	Equipment	Total	Total Incl O&P
0010	**WELLS & ACCESSORIES**									
0011	Domestic									
0100	Drilled, 4" to 6" diameter	B-23	120	.333	L.F.		10.25	22.50	32.75	41.50
1500	Pumps, installed in wells to 100' deep, 4" submersible									
1520	3/4 HP	Q-1	2.66	6.015	Ea.	860	243		1,103	1,350
1600	1 HP	"	2.29	6.987		925	282		1,207	1,500
1800	2 HP	Q-22	1.33	12.030	↓	1,825	485	355	2,665	3,200

33 14 Water Utility Transmission and Distribution

33 14 13 – Public Water Utility Distribution Piping

33 14 13.15 Water Supply, Ductile Iron Pipe

Line	Description	Crew	Daily Output	Labor-Hours	Unit	Material	Labor	Equipment	Total	Total Incl O&P
0010	**WATER SUPPLY, DUCTILE IRON PIPE**									
0020	Not including excavation or backfill									
2000	Pipe, class 50 water piping, 18' lengths									
2020	Mechanical joint, 4" diameter	B-21A	200	.200	L.F.	36	7.40	1.86	45.26	54
3000	Push-on joint, 4" diameter		400	.100		20.50	3.71	.93	25.14	29.50
3020	6" diameter	↓	333.33	.120	↓	23	4.45	1.12	28.57	33.50
8000	Piping, fittings, mechanical joint, AWWA C110									
8006	90° bend, 4" diameter	B-20A	16	2	Ea.	175	71.50		246.50	310
8020	6" diameter		12.80	2.500		256	89.50		345.50	430
8200	Wye or tee, 4" diameter		10.67	2.999		375	108		483	590
8220	6" diameter	↓	8.53	3.751	↓	570	135		705	845

33 14 13 – Public Water Utility Distribution Piping

33 14 13.15 Water Supply, Ductile Iron Pipe		Crew	Daily Output	Labor-Hours	Unit	Material	2019 Bare Costs Labor	Equipment	Total	Total Incl O&P
8398	45° bend, 4″ diameter	B-20A	16	2	Ea.	214	71.50		285.50	355
8400	6″ diameter		12.80	2.500		271	89.50		360.50	445
8405	8″ diameter		10.67	2.999		390	108		498	605
8450	Decreaser, 6″ x 4″ diameter		14.22	2.250		234	80.50		314.50	390
8460	8″ x 6″ diameter	↓	11.64	2.749	↓	360	98.50		458.50	555
8550	Piping, butterfly valves, cast iron									
8560	4″ diameter	B-20	6	4	Ea.	440	124		564	690
8700	Joint restraint, ductile iron mechanical joints									
8710	4″ diameter	B-20A	32	1	Ea.	32	36		68	94.50
8720	6″ diameter		25.60	1.250		38.50	45		83.50	117
8730	8″ diameter		21.33	1.500		57	54		111	151
8740	10″ diameter		18.28	1.751		88.50	63		151.50	201
8750	12″ diameter		16.84	1.900		117	68		185	241
8760	14″ diameter		16	2		145	71.50		216.50	278
8770	16″ diameter		11.64	2.749		180	98.50		278.50	360
8780	18″ diameter		11.03	2.901		254	104		358	450
8785	20″ diameter		9.14	3.501		315	126		441	550
8790	24″ diameter	↓	7.53	4.250		430	152		582	720
9600	Steel sleeve with tap, 4″ diameter	B-20	3	8		425	248		673	880
9620	6″ diameter	″	2	12	↓	495	370		865	1,150

33 14 13.25 Water Supply, Polyvinyl Chloride Pipe		Crew	Daily Output	Labor-Hours	Unit	Material	Labor	Equipment	Total	Total Incl O&P
0010	**WATER SUPPLY, POLYVINYL CHLORIDE PIPE**									
0020	Not including excavation or backfill, unless specified									
2100	PVC pipe, Class 150, 1-1/2″ diameter	Q-1A	750	.013	L.F.	.52	.60		1.12	1.56
2120	2″ diameter		686	.015		.88	.66		1.54	2.05
2140	2-1/2″ diameter	↓	500	.020		1.19	.91		2.10	2.80
2160	3″ diameter	B-20	430	.056	↓	1.62	1.73		3.35	4.64
8700	PVC pipe, joint restraint									
8710	4″ diameter	B-20A	32	1	Ea.	46	36		82	110
8720	6″ diameter		25.60	1.250		56.50	45		101.50	137
8730	8″ diameter		21.33	1.500		83.50	54		137.50	180
8740	10″ diameter		18.28	1.751		149	63		212	267
8750	12″ diameter		16.84	1.900		158	68		226	285
8760	14″ diameter		16	2		223	71.50		294.50	365
8770	16″ diameter		11.64	2.749		300	98.50		398.50	495
8780	18″ diameter		11.03	2.901		370	104		474	575
8785	20″ diameter		9.14	3.501		465	126		591	715
8790	24″ diameter	↓	7.53	4.250	↓	540	152		692	840

33 14 17 – Site Water Utility Service Laterals

33 14 17.15 Tapping, Crosses and Sleeves		Crew	Daily Output	Labor-Hours	Unit	Material	Labor	Equipment	Total	Total Incl O&P
0010	**TAPPING, CROSSES AND SLEEVES**									
4000	Drill and tap pressurized main (labor only)									
4100	6″ main, 1″ to 2″ service	Q-1	3	5.333	Ea.		215		215	355
4150	8″ main, 1″ to 2″ service	″	2.75	5.818	″		235		235	385
4500	Tap and insert gate valve									
4600	8″ main, 4″ branch	B-21	3.20	8.750	Ea.		285	40.50	325.50	515
4650	6″ branch		2.70	10.370			340	48	388	615
4800	12″ main, 6″ branch	↓	2.35	11.915	↓		390	55	445	700

33 31 Sanitary Sewerage Piping

33 31 11 – Public Sanitary Sewerage Gravity Piping

33 31 11.15 Sewage Collection, Concrete Pipe		Crew	Daily Output	Labor-Hours	Unit	Material	2019 Bare Costs Labor	Equipment	Total	Total Incl O&P
0010	**SEWAGE COLLECTION, CONCRETE PIPE**									
0020	See Section 33 42 11.60 for sewage/drainage collection, concrete pipe									

33 31 11.25 Sewage Collection, Polyvinyl Chloride Pipe		Crew	Daily Output	Labor-Hours	Unit	Material	Labor	Equipment	Total	Total Incl O&P
0010	**SEWAGE COLLECTION, POLYVINYL CHLORIDE PIPE**									
0020	Not including excavation or backfill									
2000	20′ lengths, SDR 35, B&S, 4″ diameter	B-20	375	.064	L.F.	1.67	1.99		3.66	5.10
2040	6″ diameter		350	.069		3.67	2.13		5.80	7.55
2080	13′ lengths, SDR 35, B&S, 8″ diameter		335	.072		6.60	2.22		8.82	10.90
2120	10″ diameter	B-21	330	.085		11.40	2.76	.39	14.55	17.55

33 34 Onsite Wastewater Disposal

33 34 13 – Septic Tanks

33 34 13.13 Concrete Septic Tanks		Crew	Daily Output	Labor-Hours	Unit	Material	Labor	Equipment	Total	Total Incl O&P
0010	**CONCRETE SEPTIC TANKS**									
0011	Not including excavation or piping									
0015	Septic tanks, precast, 1,000 gallon	B-21	8	3.500	Ea.	1,050	114	16.20	1,180.20	1,350
0060	1,500 gallon		7	4		1,575	130	18.50	1,723.50	1,950
0100	2,000 gallon		5	5.600		2,275	182	26	2,483	2,825
1150	Leaching field chambers, 13′ x 3′-7″ x 1′-4″, standard	B-13	16	3		505	98	36.50	639.50	760
1420	Leaching pit, precast concrete, 6′ diameter, 3′ deep	B-21	4.70	5.957		840	194	27.50	1,061.50	1,275

33 34 13.33 Polyethylene Septic Tanks		Crew	Daily Output	Labor-Hours	Unit	Material	Labor	Equipment	Total	Total Incl O&P
0010	**POLYETHYLENE SEPTIC TANKS**									
0015	High density polyethylene, 1,000 gallon	B-21	8	3.500	Ea.	1,300	114	16.20	1,430.20	1,625
0020	1,250 gallon		8	3.500		1,300	114	16.20	1,430.20	1,625
0025	1,500 gallon		7	4		1,375	130	18.50	1,523.50	1,750

33 34 16 – Septic Tank Effluent Filters

33 34 16.13 Septic Tank Gravity Effluent Filters		Crew	Daily Output	Labor-Hours	Unit	Material	Labor	Equipment	Total	Total Incl O&P
0010	**SEPTIC TANK GRAVITY EFFLUENT FILTERS**									
3000	Effluent filter, 4″ diameter	1 Skwk	8	1	Ea.	42	39.50		81.50	112
3020	6″ diameter		7	1.143		54.50	45		99.50	135
3030	8″ diameter		7	1.143		250	45		295	350
3040	8″ diameter, very fine		7	1.143		495	45		540	620
3050	10″ diameter, very fine		6	1.333		239	52.50		291.50	350
3060	10″ diameter		6	1.333		274	52.50		326.50	390
3080	12″ diameter		6	1.333		650	52.50		702.50	805
3090	15″ diameter		5	1.600		1,075	63		1,138	1,275

33 34 51 – Drainage Field Systems

33 34 51.10 Drainage Field Excavation and Fill		Crew	Daily Output	Labor-Hours	Unit	Material	Labor	Equipment	Total	Total Incl O&P
0010	**DRAINAGE FIELD EXCAVATION AND FILL**									
2200	Septic tank & drainage field excavation with 3/4 C.Y. backhoe	B-12F	145	.110	C.Y.		3.99	4.70	8.69	11.70
2400	4′ trench for disposal field, 3/4 C.Y. backhoe	"	335	.048	L.F.		1.73	2.03	3.76	5.10
2600	Gravel fill, run of bank	B-6	150	.160	C.Y.	16.95	5.30	2.13	24.38	30
2800	Crushed stone, 3/4″	"	150	.160	"	38	5.30	2.13	45.43	53

33 34 51.13 Utility Septic Tank Tile Drainage Field		Crew	Daily Output	Labor-Hours	Unit	Material	Labor	Equipment	Total	Total Incl O&P
0010	**UTILITY SEPTIC TANK TILE DRAINAGE FIELD**									
0015	Distribution box, concrete, 5 outlets	2 Clab	20	.800	Ea.	93	24.50		117.50	142
0020	7 outlets		16	1		94.50	30.50		125	154
0025	9 outlets		8	2		555	60.50		615.50	710
0115	Distribution boxes, HDPE, 5 outlets		20	.800		77.50	24.50		102	126

33 34 Onsite Wastewater Disposal

33 34 51 – Drainage Field Systems

33 34 51.13 Utility Septic Tank Tile Drainage Field		Crew	Daily Output	Labor-Hours	Unit	2019 Bare Costs Material	Labor	Equipment	Total	Total Incl O&P
0117	6 outlets	2 Clab	15	1.067	Ea.	80.50	32.50		113	142
0118	7 outlets		15	1.067		74.50	32.50		107	136
0120	8 outlets		10	1.600		81.50	48.50		130	170
0240	Distribution boxes, outlet flow leveler	1 Clab	50	.160		2.47	4.86		7.33	10.70

33 41 Subdrainage

33 41 16 – Subdrainage Piping

33 41 16.25 Piping, Subdrainage, Corrugated Metal

		Crew	Daily Output	Labor-Hours	Unit	Material	Labor	Equipment	Total	Total Incl O&P
0010	**PIPING, SUBDRAINAGE, CORRUGATED METAL**									
0021	Not including excavation and backfill									
2010	Aluminum, perforated									
2020	6″ diameter, 18 ga.	B-20	380	.063	L.F.	6.70	1.96		8.66	10.65
2200	8″ diameter, 16 ga.	″	370	.065		8.45	2.01		10.46	12.55
2220	10″ diameter, 16 ga.	B-21	360	.078		10.55	2.53	.36	13.44	16.20
3000	Uncoated galvanized, perforated									
3020	6″ diameter, 18 ga.	B-20	380	.063	L.F.	6.45	1.96		8.41	10.30
3200	8″ diameter, 16 ga.	″	370	.065		7.95	2.01		9.96	12
3220	10″ diameter, 16 ga.	B-21	360	.078		8.40	2.53	.36	11.29	13.85
4000	Steel, perforated, asphalt coated									
4020	6″ diameter, 18 ga.	B-20	380	.063	L.F.	6.60	1.96		8.56	10.55
4030	8″ diameter, 18 ga.	″	370	.065		8.35	2.01		10.36	12.50
4040	10″ diameter, 16 ga.	B-21	360	.078		10.50	2.53	.36	13.39	16.15
4050	12″ diameter, 16 ga.		285	.098		11.80	3.20	.45	15.45	18.75
4060	18″ diameter, 16 ga.		205	.137		18.60	4.45	.63	23.68	28.50

33 42 Stormwater Conveyance

33 42 11 – Stormwater Gravity Piping

33 42 11.40 Piping, Storm Drainage, Corrugated Metal

		Crew	Daily Output	Labor-Hours	Unit	Material	Labor	Equipment	Total	Total Incl O&P
0010	**PIPING, STORM DRAINAGE, CORRUGATED METAL**									
0020	Not including excavation or backfill									
2040	8″ diameter, 16 ga.	B-14	330	.145	L.F.	8.25	4.67	.97	13.89	17.85

33 42 11.60 Sewage/Drainage Collection, Concrete Pipe

		Crew	Daily Output	Labor-Hours	Unit	Material	Labor	Equipment	Total	Total Incl O&P
0010	**SEWAGE/DRAINAGE COLLECTION, CONCRETE PIPE**									
0020	Not including excavation or backfill									
1020	8″ diameter	B-14	224	.214	L.F.	8.45	6.85	1.43	16.73	22
1030	10″ diameter	″	216	.222	″	9.35	7.15	1.48	17.98	23.50
3780	Concrete slotted pipe, class 4 mortar joint									
3800	12″ diameter	B-21	168	.167	L.F.	31	5.45	.77	37.22	44.50
3840	18″ diameter	″	152	.184	″	35.50	6	.85	42.35	50.50
3900	Concrete slotted pipe, Class 4 O-ring joint									
3940	12″ diameter	B-21	168	.167	L.F.	28	5.45	.77	34.22	41
3960	18″ diameter	″	152	.184	″	32	6	.85	38.85	46

33 42 33 – Stormwater Curbside Drains and Inlets

33 42 33.13 Catch Basins

		Crew	Daily Output	Labor-Hours	Unit	Material	Labor	Equipment	Total	Total Incl O&P
0010	**CATCH BASINS**									
0011	Not including footing & excavation									
1600	Frames & grates, C.I., 24″ square, 500 lb.	B-6	7.80	3.077	Ea.	355	102	41	498	605

33 52 Hydrocarbon Transmission and Distribution

33 52 16 – Gas Hydrocarbon Piping

33 52 16.20 Piping, Gas Service and Distribution, P.E.		Crew	Daily Output	Labor-Hours	Unit	Material	Labor	Equipment	Total	Total Incl O&P
						2019 Bare Costs				
0010	PIPING, GAS SERVICE AND DISTRIBUTION, POLYETHYLENE									
0020	Not including excavation or backfill									
1000	60 psi coils, compression coupling @ 100′, 1/2″ diameter, SDR 11	B-20A	608	.053	L.F.	.48	1.89		2.37	3.64
1010	1″ diameter, SDR 11		544	.059		1.13	2.11		3.24	4.71
1040	1-1/4″ diameter, SDR 11		544	.059		1.64	2.11		3.75	5.25
1100	2″ diameter, SDR 11		488	.066		2.54	2.35		4.89	6.65
1160	3″ diameter, SDR 11		408	.078		5.60	2.81		8.41	10.80
1500	60 psi 40′ joints with coupling, 3″ diameter, SDR 11	B-21A	408	.098		6.90	3.64	.91	11.45	14.55
1540	4″ diameter, SDR 11		352	.114		11.70	4.21	1.06	16.97	21
1600	6″ diameter, SDR 11		328	.122		31.50	4.52	1.13	37.15	43
1640	8″ diameter, SDR 11		272	.147		47	5.45	1.37	53.82	62

33 71 Electrical Utility Transmission and Distribution

33 71 16 – Electrical Utility Poles

33 71 16.24 Aluminum Lighting Poles

		Crew	Daily Output	Labor-Hours	Unit	Material	Labor	Equipment	Total	Total Incl O&P
0010	ALUMINUM LIGHTING POLES									
4000	Aluminum, round, tapered seamless shaft, 20′	R-15A	6.85	7.007	Ea.	690	277	41.50	1,008.50	1,250
4020	25′		6	8		1,250	315	47	1,612	1,950
4040	30′		5.35	8.972		1,125	355	53	1,533	1,900

33 71 19 – Electrical Underground Ducts and Manholes

33 71 19.17 Electric and Telephone Underground

		Crew	Daily Output	Labor-Hours	Unit	Material	Labor	Equipment	Total	Total Incl O&P
0010	ELECTRIC AND TELEPHONE UNDERGROUND									
0011	Not including excavation									
0200	backfill and cast in place concrete									
4200	Underground duct, banks ready for concrete fill, min. of 7.5″									
4400	between conduits, center to center									
4601	PVC, type EB, 2 @ 2″ diameter	2 Elec	240	.067	L.F.	2.24	3.04		5.28	7.45
4800	4 @ 2″ diameter		120	.133		4.48	6.10		10.58	14.85
5600	4 @ 4″ diameter		80	.200		7.45	9.10		16.55	23
6200	Rigid galvanized steel, 2 @ 2″ diameter		180	.089		17.25	4.05		21.30	25.50
6400	4 @ 2″ diameter		90	.178		34.50	8.10		42.60	51.50
7400	4 @ 4″ diameter		34	.471		76	21.50		97.50	119

Estimating Tips

- When estimating costs for the installation of electrical power generation equipment, factors to review include access to the job site, access and setting up at the installation site, required connections, uncrating pads, anchors, leveling, final assembly of the components, and temporary protection from physical damage, such as environmental exposure.
- Be aware of the costs of equipment supports, concrete pads, and vibration isolators. Cross-reference them against other trades' specifications. Also, review site and structural drawings for items that must be included in the estimates.
- It is important to include items that are not documented in the plans and specifications but must be priced. These items include, but are not limited to, testing, dust protection, roof penetration, core drilling concrete floors and walls, patching, cleanup, and final adjustments. Add a contingency or allowance for utility company fees for power hookups, if needed.
- The project size and scope of electrical power generation equipment will have a significant impact on cost. The intent of RSMeans cost data is to provide a benchmark cost so that owners, engineers, and electrical contractors will have a comfortable number with which to start a project. Additionally, there are many websites available to use for research and to obtain a vendor's quote to finalize costs.

Reference Numbers

Reference numbers are shown at the beginning of some major classifications. These numbers refer to related items in the Reference Section. The reference information may be an estimating procedure, an alternate pricing method, or technical information.

Note: Trade Service, in part, has been used as a reference source for some of the material prices used in Division 48.

48 15 13 – Wind Turbines

48 15 13.50 Wind Turbines and Components			Crew	Daily Output	Labor-Hours	Unit	2019 Bare Costs Material	Labor	Equipment	Total	Total Incl O&P
0010	WIND TURBINES & COMPONENTS										
0500	Complete system, grid connected										
1000	20 kW, 31′ diam., incl. labor & material	G				System				49,900	49,900
2000	2.4 kW, 12′ diam., incl. labor & material	G				"				18,000	18,000
2900	Component system										
3200	1,000 W, 9′ diam.	G	1 Elec	2.05	3.902	Ea.	1,700	178		1,878	2,175
3400	Mounting hardware										
3500	30′ guyed tower kit	G	2 Clab	5.12	3.125	Ea.	405	95		500	600
3505	3′ galvanized helical earth screw	G	1 Clab	8	1		54.50	30.50		85	110
3510	Attic mount kit	G	1 Rofc	2.56	3.125		216	104		320	425
3520	Roof mount kit	G	1 Clab	3.41	2.346	↓	264	71		335	410
8900	Equipment										
9100	DC to AC inverter for, 48 V, 4,000 W	G	1 Elec	2	4	Ea.	2,300	182		2,482	2,825

Reference Section

All the reference information is in one section, making it easy to find what you need to know . . . and easy to use the data set on a daily basis. This section is visually identified by a vertical black bar on the page edges.

In this Reference Section, we've included Equipment Rental Costs, a listing of rental and operating costs; Crew Listings, a full listing of all crews and equipment, and their costs; Historical Cost Indexes for cost comparisons over time; Location Factors for adjusting costs to the region you are in; Reference Tables, where you will find explanations, estimating information and procedures, or technical data; Change Orders, information on pricing changes to contract documents; and an explanation of all the Abbreviations in the data set.

Table of Contents

Estimating Tips

- This section contains the average costs to rent and operate hundreds of pieces of construction equipment. This is useful information when one is estimating the time and material requirements of any particular operation in order to establish a unit or total cost. Bare equipment costs shown on a unit cost line include, not only rental, but also operating costs for equipment under normal use.

Rental Costs

- Equipment rental rates are obtained from the following industry sources throughout North America: contractors, suppliers, dealers, manufacturers, and distributors.
- Rental rates vary throughout the country, with larger cities generally having lower rates. Lease plans for new equipment are available for periods in excess of six months, with a percentage of payments applying toward purchase.
- Monthly rental rates vary from 2% to 5% of the purchase price of the equipment depending on the anticipated life of the equipment and its wearing parts.
- Weekly rental rates are about 1/3 of the monthly rates, and daily rental rates are about 1/3 of the weekly rate.
- Rental rates can also be treated as reimbursement costs for contractor-owned equipment. Owned equipment costs include depreciation, loan payments, interest, taxes, insurance, storage, and major repairs.

Operating Costs

- The operating costs include parts and labor for routine servicing, such as the repair and replacement of pumps, filters, and worn lines. Normal operating expendables, such as fuel, lubricants, tires, and electricity (where applicable), are also included.
- Extraordinary operating expendables with highly variable wear patterns, such as diamond bits and blades, are excluded. These costs can be found as material costs in the Unit Price section.
- The hourly operating costs listed do not include the operator's wages.

Equipment Cost/Day

- Any power equipment required by a crew is shown in the Crew Listings with a daily cost.
- This daily cost of equipment needed by a crew includes both the rental cost and the operating cost and is based on dividing the weekly rental rate by 5 (the number of working days in the week), then adding the hourly operating cost multiplied by 8 (the number of hours in a day). This "Equipment Cost/Day" is shown in the far right column of the Equipment Rental section.
- If equipment is needed for only one or two days, it is best to develop your own cost by including components for daily rent and hourly operating costs. This is important when the listed Crew for a task does not contain the equipment needed, such as a crane for lifting mechanical heating/cooling equipment up onto a roof.
- If the quantity of work is less than the crew's Daily Output shown for a Unit Price line item that includes a bare unit equipment cost, the recommendation is to estimate one day's rental cost and operating cost for equipment shown in the Crew Listing for that line item.
- Please note, in some cases the equipment description in the crew is followed by a time period in parenthesis. For example: (daily) or (monthly). In these cases the equipment cost/day is calculated by adding the rental cost per time period to the hourly operating cost multiplied by 8.

Mobilization, Demobilization Costs

- The cost to move construction equipment from an equipment yard or rental company to the job site and back again is not included in equipment rental costs listed in the Reference Section. It is also not included in the bare equipment cost of any Unit Price line item or in any equipment costs shown in the Crew Listings.
- Mobilization (to the site) and demobilization (from the site) costs can be found in the Unit Price section.
- If a piece of equipment is already at the job site, it is not appropriate to utilize mobilization or demobilization costs again in an estimate. ■

01 54 33 | Equipment Rental

		01 54 33 Equipment Rental	UNIT	HOURLY OPER. COST	RENT PER DAY	RENT PER WEEK	RENT PER MONTH	EQUIPMENT COST/DAY	
10	0010	**CONCRETE EQUIPMENT RENTAL** without operators R015433-10							10
	0200	Bucket, concrete lightweight, 1/2 C.Y.	Ea.	.88	24.50	74	222	21.80	
	0300	1 C.Y.		.98	28.50	85.50	257	24.95	
	0400	1-1/2 C.Y.		1.24	38.50	115	345	32.90	
	0500	2 C.Y.		1.34	47	141	425	38.90	
	0580	8 C.Y.		6.49	265	795	2,375	210.90	
	0600	Cart, concrete, self-propelled, operator walking, 10 C.F.		2.86	58.50	176	530	58.10	
	0700	Operator riding, 18 C.F.		4.82	98.50	296	890	97.75	
	0800	Conveyer for concrete, portable, gas, 16″ wide, 26′ long		10.64	130	390	1,175	163.10	
	0900	46′ long		11.02	155	465	1,400	181.15	
	1000	56′ long		11.18	162	485	1,450	186.45	
	1100	Core drill, electric, 2-1/2 H.P., 1″ to 8″ bit diameter		1.57	56.50	170	510	46.55	
	1150	11 H.P., 8″ to 18″ cores		5.40	115	345	1,025	112.20	
	1200	Finisher, concrete floor, gas, riding trowel, 96″ wide		9.66	148	445	1,325	166.30	
	1300	Gas, walk-behind, 3 blade, 36″ trowel		2.04	24.50	73.50	221	31	
	1400	4 blade, 48″ trowel		3.07	28	84.50	254	41.50	
	1500	Float, hand-operated (Bull float), 48″ wide		.08	13.85	41.50	125	8.95	
	1570	Curb builder, 14 H.P., gas, single screw		14.04	285	855	2,575	283.30	
	1590	Double screw		15.04	340	1,025	3,075	325.35	
	1600	Floor grinder, concrete and terrazzo, electric, 22″ path		3.04	190	570	1,700	138.30	
	1700	Edger, concrete, electric, 7″ path		1.18	56.50	170	510	43.45	
	1750	Vacuum pick-up system for floor grinders, wet/dry		1.62	98.50	296	890	72.15	
	1800	Mixer, powered, mortar and concrete, gas, 6 C.F., 18 H.P.		7.42	127	380	1,150	135.35	
	1900	10 C.F., 25 H.P.		9.00	150	450	1,350	162	
	2000	16 C.F.		9.36	173	520	1,550	178.85	
	2100	Concrete, stationary, tilt drum, 2 C.Y.		7.23	243	730	2,200	203.85	
	2120	Pump, concrete, truck mounted, 4″ line, 80′ boom		29.88	1,000	3,025	9,075	844.05	
	2140	5″ line, 110′ boom		37.46	1,375	4,150	12,500	1,130	
	2160	Mud jack, 50 C.F. per hr.		6.45	123	370	1,100	125.60	
	2180	225 C.F. per hr.		8.55	145	435	1,300	155.40	
	2190	Shotcrete pump rig, 12 C.Y./hr.		13.97	218	655	1,975	242.75	
	2200	35 C.Y./hr.		15.80	242	725	2,175	271.40	
	2600	Saw, concrete, manual, gas, 18 H.P.		5.54	47.50	143	430	72.90	
	2650	Self-propelled, gas, 30 H.P.		7.90	78.50	235	705	110.15	
	2675	V-groove crack chaser, manual, gas, 6 H.P.		1.64	19	57	171	24.50	
	2700	Vibrators, concrete, electric, 60 cycle, 2 H.P.		.47	9.15	27.50	82.50	9.25	
	2800	3 H.P.		.56	11.65	35	105	11.50	
	2900	Gas engine, 5 H.P.		1.54	16.35	49	147	22.15	
	3000	8 H.P.		2.08	16.50	49.50	149	26.55	
	3050	Vibrating screed, gas engine, 8 H.P.		2.81	95	285	855	79.45	
	3120	Concrete transit mixer, 6 x 4, 250 H.P., 8 C.Y., rear discharge		50.72	615	1,850	5,550	775.80	
	3200	Front discharge		58.88	740	2,225	6,675	916.05	
	3300	6 x 6, 285 H.P., 12 C.Y., rear discharge		58.15	710	2,125	6,375	890.20	
	3400	Front discharge		60.59	750	2,250	6,750	934.70	
20	0010	**EARTHWORK EQUIPMENT RENTAL** without operators R015433-10							20
	0040	Aggregate spreader, push type, 8′ to 12′ wide	Ea.	2.60	27.50	82.50	248	37.30	
	0045	Tailgate type, 8′ wide		2.55	33	99	297	40.15	
	0055	Earth auger, truck mounted, for fence & sign posts, utility poles		13.85	460	1,375	4,125	385.85	
	0060	For borings and monitoring wells		42.65	690	2,075	6,225	756.20	
	0070	Portable, trailer mounted		2.30	33	99	297	38.20	
	0075	Truck mounted, for caissons, water wells		85.39	2,900	8,725	26,200	2,428	
	0080	Horizontal boring machine, 12″ to 36″ diameter, 45 H.P.		22.77	188	565	1,700	295.15	
	0090	12″ to 48″ diameter, 65 H.P.		31.25	330	985	2,950	447	
	0095	Auger, for fence posts, gas engine, hand held		.45	6.35	19	57	7.40	
	0100	Excavator, diesel hydraulic, crawler mounted, 1/2 C.Y. cap.		21.72	450	1,350	4,050	443.75	
	0120	5/8 C.Y. capacity		29.04	590	1,775	5,325	587.30	
	0140	3/4 C.Y. capacity		32.66	700	2,100	6,300	681.25	
	0150	1 C.Y. capacity		41.21	700	2,100	6,300	749.70	

	01 54 33 | Equipment Rental		UNIT	HOURLY OPER. COST	RENT PER DAY	RENT PER WEEK	RENT PER MONTH	EQUIPMENT COST/DAY	
20	0200	1-1/2 C.Y. capacity	Ea.	48.59	865	2,600	7,800	908.70	20
	0300	2 C.Y. capacity		56.58	1,050	3,125	9,375	1,078	
	0320	2-1/2 C.Y. capacity		82.64	1,300	3,900	11,700	1,441	
	0325	3-1/2 C.Y. capacity		120.12	2,150	6,475	19,400	2,256	
	0330	4-1/2 C.Y. capacity		151.63	2,750	8,275	24,800	2,868	
	0335	6 C.Y. capacity		192.39	3,400	10,200	30,600	3,579	
	0340	7 C.Y. capacity		175.20	3,225	9,650	29,000	3,332	
	0342	Excavator attachments, bucket thumbs		3.40	250	750	2,250	177.20	
	0345	Grapples		3.14	215	645	1,925	154.10	
	0346	Hydraulic hammer for boom mounting, 4,000 ft. lb.		13.48	375	1,125	3,375	332.85	
	0347	5,000 ft. lb.		15.95	465	1,400	4,200	407.60	
	0348	8,000 ft. lb.		23.54	675	2,025	6,075	593.35	
	0349	12,000 ft. lb.		25.72	800	2,400	7,200	685.75	
	0350	Gradall type, truck mounted, 3 ton @ 15′ radius, 5/8 C.Y.		43.44	860	2,575	7,725	862.55	
	0370	1 C.Y. capacity		59.40	1,175	3,550	10,700	1,185	
	0400	Backhoe-loader, 40 to 45 H.P., 5/8 C.Y. capacity		11.90	207	620	1,850	219.20	
	0450	45 H.P. to 60 H.P., 3/4 C.Y. capacity		18.02	293	880	2,650	320.20	
	0460	80 H.P., 1-1/4 C.Y. capacity		20.36	375	1,125	3,375	387.85	
	0470	112 H.P., 1-1/2 C.Y. capacity		32.99	590	1,775	5,325	618.90	
	0482	Backhoe-loader attachment, compactor, 20,000 lb.		6.44	150	450	1,350	141.50	
	0485	Hydraulic hammer, 750 ft. lb.		3.68	103	310	930	91.45	
	0486	Hydraulic hammer, 1,200 ft. lb.		6.55	198	595	1,775	171.40	
	0500	Brush chipper, gas engine, 6″ cutter head, 35 H.P.		9.17	110	330	990	139.35	
	0550	Diesel engine, 12″ cutter head, 130 H.P.		23.67	335	1,000	3,000	389.40	
	0600	15″ cutter head, 165 H.P.		26.59	400	1,200	3,600	452.70	
	0750	Bucket, clamshell, general purpose, 3/8 C.Y.		1.40	41	123	370	35.80	
	0800	1/2 C.Y.		1.52	50.50	152	455	42.55	
	0850	3/4 C.Y.		1.64	57.50	173	520	47.75	
	0900	1 C.Y.		1.70	62.50	188	565	51.20	
	0950	1-1/2 C.Y.		2.79	87	261	785	74.50	
	1000	2 C.Y.		2.92	95	285	855	80.40	
	1010	Bucket, dragline, medium duty, 1/2 C.Y.		.82	24.50	74	222	21.35	
	1020	3/4 C.Y.		.78	26	78	234	21.85	
	1030	1 C.Y.		.80	27	81.50	245	22.70	
	1040	1-1/2 C.Y.		1.26	41.50	125	375	35.10	
	1050	2 C.Y.		1.29	45	135	405	37.35	
	1070	3 C.Y.		2.08	66	198	595	56.25	
	1200	Compactor, manually guided 2-drum vibratory smooth roller, 7.5 H.P.		7.22	207	620	1,850	181.75	
	1250	Rammer/tamper, gas, 8″		2.21	46.50	140	420	45.65	
	1260	15″		2.63	53.50	160	480	53.05	
	1300	Vibratory plate, gas, 18″ plate, 3,000 lb. blow		2.13	23.50	70.50	212	31.10	
	1350	21″ plate, 5,000 lb. blow		2.62	32.50	98	294	40.55	
	1370	Curb builder/extruder, 14 H.P., gas, single screw		14.03	283	850	2,550	282.25	
	1390	Double screw		15.04	340	1,025	3,075	325.30	
	1500	Disc harrow attachment, for tractor		.47	79.50	239	715	51.60	
	1810	Feller buncher, shearing & accumulating trees, 100 H.P.		39.20	815	2,450	7,350	803.55	
	1860	Grader, self-propelled, 25,000 lb.		33.35	760	2,275	6,825	721.80	
	1910	30,000 lb.		32.85	660	1,975	5,925	657.85	
	1920	40,000 lb.		51.89	1,275	3,825	11,500	1,180	
	1930	55,000 lb.		66.93	1,650	4,925	14,800	1,520	
	1950	Hammer, pavement breaker, self-propelled, diesel, 1,000 to 1,250 lb.		28.40	475	1,425	4,275	512.20	
	2000	1,300 to 1,500 lb.		42.80	975	2,925	8,775	927.40	
	2050	Pile driving hammer, steam or air, 4,150 ft. lb. @ 225 bpm		12.15	565	1,700	5,100	437.20	
	2100	8,750 ft. lb. @ 145 bpm		14.35	810	2,425	7,275	599.75	
	2150	15,000 ft. lb. @ 60 bpm		14.68	840	2,525	7,575	622.40	
	2200	24,450 ft. lb. @ 111 bpm		15.69	935	2,800	8,400	685.50	
	2250	Leads, 60′ high for pile driving hammers up to 20,000 ft. lb.		3.67	85.50	256	770	80.55	
	2300	90′ high for hammers over 20,000 ft. lb.	↓	5.45	152	455	1,375	134.60	

01 54 33 | Equipment Rental

		01 54 33 Equipment Rental	UNIT	HOURLY OPER. COST	RENT PER DAY	RENT PER WEEK	RENT PER MONTH	EQUIPMENT COST/DAY	
20	2350	Diesel type hammer, 22,400 ft. lb.	Ea.	17.82	475	1,425	4,275	427.55	20
	2400	41,300 ft. lb.		25.68	600	1,800	5,400	565.45	
	2450	141,000 ft. lb.		41.33	950	2,850	8,550	900.65	
	2500	Vib. elec. hammer/extractor, 200 kW diesel generator, 34 H.P.		41.37	690	2,075	6,225	745.95	
	2550	80 H.P.		73.03	1,000	3,000	9,000	1,184	
	2600	150 H.P.		135.14	1,925	5,775	17,300	2,236	
	2800	Log chipper, up to 22" diameter, 600 H.P.		46.26	665	2,000	6,000	770.10	
	2850	Logger, for skidding & stacking logs, 150 H.P.		43.53	835	2,500	7,500	848.25	
	2860	Mulcher, diesel powered, trailer mounted		18.04	220	660	1,975	276.35	
	2900	Rake, spring tooth, with tractor		14.72	360	1,075	3,225	332.75	
	3000	Roller, vibratory, tandem, smooth drum, 20 H.P.		7.81	150	450	1,350	152.45	
	3050	35 H.P.		10.13	252	755	2,275	232.05	
	3100	Towed type vibratory compactor, smooth drum, 50 H.P.		25.27	365	1,100	3,300	422.15	
	3150	Sheepsfoot, 50 H.P.		25.64	375	1,125	3,375	430.15	
	3170	Landfill compactor, 220 H.P.		70.01	1,575	4,750	14,300	1,510	
	3200	Pneumatic tire roller, 80 H.P.		12.92	390	1,175	3,525	338.35	
	3250	120 H.P.		19.39	640	1,925	5,775	540.10	
	3300	Sheepsfoot vibratory roller, 240 H.P.		62.21	1,375	4,125	12,400	1,323	
	3320	340 H.P.		83.82	2,075	6,250	18,800	1,921	
	3350	Smooth drum vibratory roller, 75 H.P.		23.34	635	1,900	5,700	566.70	
	3400	125 H.P.		27.61	715	2,150	6,450	650.90	
	3410	Rotary mower, brush, 60", with tractor		18.79	350	1,050	3,150	360.30	
	3420	Rototiller, walk-behind, gas, 5 H.P.		2.14	89	267	800	70.50	
	3422	8 H.P.		2.81	103	310	930	84.50	
	3440	Scrapers, towed type, 7 C.Y. capacity		6.44	123	370	1,100	125.50	
	3450	10 C.Y. capacity		7.20	165	495	1,475	156.65	
	3500	15 C.Y. capacity		7.40	190	570	1,700	173.20	
	3525	Self-propelled, single engine, 14 C.Y. capacity		133.29	2,575	7,725	23,200	2,611	
	3550	Dual engine, 21 C.Y. capacity		141.37	2,350	7,050	21,200	2,541	
	3600	31 C.Y. capacity		187.85	3,500	10,500	31,500	3,603	
	3640	44 C.Y. capacity		232.68	4,500	13,500	40,500	4,561	
	3650	Elevating type, single engine, 11 C.Y. capacity		61.87	1,125	3,350	10,100	1,165	
	3700	22 C.Y. capacity		114.62	2,325	6,950	20,900	2,307	
	3710	Screening plant, 110 H.P. w/5' x 10' screen		21.13	385	1,150	3,450	399.05	
	3720	5' x 16' screen		26.68	490	1,475	4,425	508.40	
	3850	Shovel, crawler-mounted, front-loading, 7 C.Y. capacity		218.65	3,800	11,400	34,200	4,029	
	3855	12 C.Y. capacity		336.90	5,275	15,800	47,400	5,855	
	3860	Shovel/backhoe bucket, 1/2 C.Y.		2.69	70.50	212	635	63.95	
	3870	3/4 C.Y.		2.67	79.50	238	715	68.90	
	3880	1 C.Y.		2.76	88	264	790	74.85	
	3890	1-1/2 C.Y.		2.95	103	310	930	85.65	
	3910	3 C.Y.		3.44	140	420	1,250	111.50	
	3950	Stump chipper, 18" deep, 30 H.P.		6.90	212	635	1,900	182.20	
	4110	Dozer, crawler, torque converter, diesel 80 H.P.		25.25	450	1,350	4,050	472	
	4150	105 H.P.		34.34	560	1,675	5,025	609.70	
	4200	140 H.P.		41.27	840	2,525	7,575	835.20	
	4260	200 H.P.		63.16	1,300	3,925	11,800	1,290	
	4310	300 H.P.		80.73	1,975	5,925	17,800	1,831	
	4360	410 H.P.		106.74	2,350	7,075	21,200	2,269	
	4370	500 H.P.		133.38	2,900	8,725	26,200	2,812	
	4380	700 H.P.		230.17	5,300	15,900	47,700	5,021	
	4400	Loader, crawler, torque conv., diesel, 1-1/2 C.Y., 80 H.P.		29.54	590	1,775	5,325	591.30	
	4450	1-1/2 to 1-3/4 C.Y., 95 H.P.		30.27	675	2,025	6,075	647.15	
	4510	1-3/4 to 2-1/4 C.Y., 130 H.P.		47.76	1,025	3,100	9,300	1,002	
	4530	2-1/2 to 3-1/4 C.Y., 190 H.P.		57.79	1,225	3,700	11,100	1,202	
	4560	3-1/2 to 5 C.Y., 275 H.P.		71.41	1,475	4,450	13,400	1,461	
	4610	Front end loader, 4WD, articulated frame, diesel, 1 to 1-1/4 C.Y., 70 H.P.		16.62	273	820	2,450	297	
	4620	1-1/2 to 1-3/4 C.Y., 95 H.P.	↓	20.00	320	965	2,900	352.95	

01 54 33 | Equipment Rental

		01 54 33 | Equipment Rental	UNIT	HOURLY OPER. COST	RENT PER DAY	RENT PER WEEK	RENT PER MONTH	EQUIPMENT COST/DAY	
20	4650	1-3/4 to 2 C.Y., 130 H.P.	Ea.	21.07	385	1,150	3,450	398.55	20
	4710	2-1/2 to 3-1/2 C.Y., 145 H.P.		29.53	490	1,475	4,425	531.20	
	4730	3 to 4-1/2 C.Y., 185 H.P.		32.09	515	1,550	4,650	566.70	
	4760	5-1/4 to 5-3/4 C.Y., 270 H.P.		53.19	915	2,750	8,250	975.55	
	4810	7 to 9 C.Y., 475 H.P.		91.17	1,750	5,275	15,800	1,784	
	4870	9 to 11 C.Y., 620 H.P.		131.92	2,625	7,850	23,600	2,625	
	4880	Skid-steer loader, wheeled, 10 C.F., 30 H.P. gas		9.57	163	490	1,475	174.55	
	4890	1 C.Y., 78 H.P., diesel		18.44	410	1,225	3,675	392.50	
	4892	Skid-steer attachment, auger		.75	137	410	1,225	87.95	
	4893	Backhoe		.74	118	355	1,075	76.90	
	4894	Broom		.71	123	370	1,100	79.65	
	4895	Forks		.15	26.50	80	240	17.25	
	4896	Grapple		.72	107	320	960	69.80	
	4897	Concrete hammer		1.05	177	530	1,600	114.40	
	4898	Tree spade		.60	100	300	900	64.80	
	4899	Trencher		.65	108	325	975	70.20	
	4900	Trencher, chain, boom type, gas, operator walking, 12 H.P.		4.18	50	150	450	63.40	
	4910	Operator riding, 40 H.P.		16.69	360	1,075	3,225	348.50	
	5000	Wheel type, diesel, 4′ deep, 12″ wide		68.71	935	2,800	8,400	1,110	
	5100	6′ deep, 20″ wide		87.58	2,150	6,475	19,400	1,996	
	5150	Chain type, diesel, 5′ deep, 8″ wide		16.30	350	1,050	3,150	340.40	
	5200	Diesel, 8′ deep, 16″ wide		89.66	1,875	5,600	16,800	1,837	
	5202	Rock trencher, wheel type, 6″ wide x 18″ deep		47.12	700	2,100	6,300	797	
	5206	Chain type, 18″ wide x 7′ deep		104.70	3,075	9,200	27,600	2,678	
	5210	Tree spade, self-propelled		13.69	390	1,175	3,525	344.55	
	5250	Truck, dump, 2-axle, 12 ton, 8 C.Y. payload, 220 H.P.		23.95	247	740	2,225	339.60	
	5300	Three axle dump, 16 ton, 12 C.Y. payload, 400 H.P.		44.63	350	1,050	3,150	567.05	
	5310	Four axle dump, 25 ton, 18 C.Y. payload, 450 H.P.		50.00	510	1,525	4,575	705	
	5350	Dump trailer only, rear dump, 16-1/2 C.Y.		5.74	147	440	1,325	133.95	
	5400	20 C.Y.		6.20	165	495	1,475	148.60	
	5450	Flatbed, single axle, 1-1/2 ton rating		19.05	71.50	214	640	195.20	
	5500	3 ton rating		23.12	102	305	915	245.95	
	5550	Off highway rear dump, 25 ton capacity		62.86	1,425	4,250	12,800	1,353	
	5600	35 ton capacity		67.10	1,550	4,675	14,000	1,472	
	5610	50 ton capacity		84.12	1,775	5,325	16,000	1,738	
	5620	65 ton capacity		89.83	1,925	5,800	17,400	1,879	
	5630	100 ton capacity		121.61	2,850	8,550	25,700	2,683	
	6000	Vibratory plow, 25 H.P., walking	↓	6.79	60.50	181	545	90.55	
40	0010	**GENERAL EQUIPMENT RENTAL** without operators R015433 -10							40
	0020	Aerial lift, scissor type, to 20′ high, 1200 lb. capacity, electric	Ea.	3.49	50.50	151	455	58.15	
	0030	To 30′ high, 1,200 lb. capacity		3.78	67.50	202	605	70.70	
	0040	Over 30′ high, 1,500 lb. capacity		5.15	122	365	1,100	114.20	
	0070	Articulating boom, to 45′ high, 500 lb. capacity, diesel R015433 -15		9.95	275	825	2,475	244.60	
	0075	To 60′ high, 500 lb. capacity		13.70	510	1,525	4,575	414.60	
	0080	To 80′ high, 500 lb. capacity		16.10	560	1,675	5,025	463.80	
	0085	To 125′ high, 500 lb. capacity		18.40	775	2,325	6,975	612.20	
	0100	Telescoping boom to 40′ high, 500 lb. capacity, diesel		11.27	320	965	2,900	283.15	
	0105	To 45′ high, 500 lb. capacity		12.54	315	945	2,825	289.35	
	0110	To 60′ high, 500 lb. capacity		16.41	535	1,600	4,800	451.25	
	0115	To 80′ high, 500 lb. capacity		21.33	835	2,500	7,500	670.65	
	0120	To 100′ high, 500 lb. capacity		28.80	860	2,575	7,725	745.40	
	0125	To 120′ high, 500 lb. capacity		29.25	840	2,525	7,575	739	
	0195	Air compressor, portable, 6.5 CFM, electric		.91	12.65	38	114	14.85	
	0196	Gasoline		.65	18	54	162	16.05	
	0200	Towed type, gas engine, 60 CFM		9.46	51.50	155	465	106.70	
	0300	160 CFM		10.51	53	159	475	115.85	
	0400	Diesel engine, rotary screw, 250 CFM		12.12	118	355	1,075	167.95	
	0500	365 CFM	↓	16.05	142	425	1,275	213.35	

		01 54 33 | Equipment Rental	UNIT	HOURLY OPER. COST	RENT PER DAY	RENT PER WEEK	RENT PER MONTH	EQUIPMENT COST/DAY	
40	0550	450 CFM	Ea.	20.01	177	530	1,600	266.05	40
	0600	600 CFM		34.20	240	720	2,150	417.60	
	0700	750 CFM		34.72	252	755	2,275	428.80	
	0930	Air tools, breaker, pavement, 60 lb.		.57	10.35	31	93	10.75	
	0940	80 lb.		.57	10.65	32	96	10.90	
	0950	Drills, hand (jackhammer), 65 lb.		.67	17.35	52	156	15.80	
	0960	Track or wagon, swing boom, 4″ drifter		54.83	940	2,825	8,475	1,004	
	0970	5″ drifter		63.49	1,100	3,325	9,975	1,173	
	0975	Track mounted quarry drill, 6″ diameter drill		102.22	1,625	4,875	14,600	1,793	
	0980	Dust control per drill		1.04	25	75.50	227	23.45	
	0990	Hammer, chipping, 12 lb.		.60	27	81	243	21	
	1000	Hose, air with couplings, 50′ long, 3/4″ diameter		.07	10	30	90	6.55	
	1100	1″ diameter		.08	13	39	117	8.40	
	1200	1-1/2″ diameter		.22	35	105	315	22.75	
	1300	2″ diameter		.24	40	120	360	25.90	
	1400	2-1/2″ diameter		.36	56.50	170	510	36.85	
	1410	3″ diameter		.42	37.50	112	335	25.75	
	1450	Drill, steel, 7/8″ x 2′		.08	12.50	37.50	113	8.20	
	1460	7/8″ x 6′		.12	19	57	171	12.30	
	1520	Moil points		.03	4.73	14.20	42.50	3.05	
	1525	Pneumatic nailer w/accessories		.48	31	92.50	278	22.35	
	1530	Sheeting driver for 60 lb. breaker		.04	7.50	22.50	67.50	4.85	
	1540	For 90 lb. breaker		.13	10.15	30.50	91.50	7.15	
	1550	Spade, 25 lb.		.50	7.15	21.50	64.50	8.30	
	1560	Tamper, single, 35 lb.		.59	39.50	119	355	28.55	
	1570	Triple, 140 lb.		.89	59.50	179	535	42.90	
	1580	Wrenches, impact, air powered, up to 3/4″ bolt		.43	12.85	38.50	116	11.10	
	1590	Up to 1-1/4″ bolt		.58	23	69.50	209	18.50	
	1600	Barricades, barrels, reflectorized, 1 to 99 barrels		.03	5.50	16.55	49.50	3.55	
	1610	100 to 200 barrels		.02	4.27	12.80	38.50	2.75	
	1620	Barrels with flashers, 1 to 99 barrels		.03	6.20	18.55	55.50	4	
	1630	100 to 200 barrels		.03	4.95	14.85	44.50	3.20	
	1640	Barrels with steady burn type C lights		.05	8.15	24.50	73.50	5.30	
	1650	Illuminated board, trailer mounted, with generator		3.29	135	405	1,225	107.30	
	1670	Portable barricade, stock, with flashers, 1 to 6 units		.03	6.15	18.50	55.50	4	
	1680	25 to 50 units		.03	5.75	17.25	52	3.70	
	1685	Butt fusion machine, wheeled, 1.5 HP electric, 2″ - 8″ diameter pipe		2.64	203	610	1,825	143.10	
	1690	Tracked, 20 HP diesel, 4″ - 12″ diameter pipe		11.27	525	1,575	4,725	405.15	
	1695	83 HP diesel, 8″ - 24″ diameter pipe		51.47	2,525	7,575	22,700	1,927	
	1700	Carts, brick, gas engine, 1,000 lb. capacity		2.95	69.50	208	625	65.20	
	1800	1,500 lb., 7-1/2′ lift		2.92	65	195	585	62.40	
	1822	Dehumidifier, medium, 6 lb./hr., 150 CFM		1.19	74	222	665	53.95	
	1824	Large, 18 lb./hr., 600 CFM		2.20	140	420	1,250	101.60	
	1830	Distributor, asphalt, trailer mounted, 2,000 gal., 38 H.P. diesel		11.03	340	1,025	3,075	293.20	
	1840	3,000 gal., 38 H.P. diesel		12.90	365	1,100	3,300	323.25	
	1850	Drill, rotary hammer, electric		1.12	27	80.50	242	25.05	
	1860	Carbide bit, 1-1/2″ diameter, add to electric rotary hammer		.03	5.10	15.30	46	3.30	
	1865	Rotary, crawler, 250 H.P.		136.18	2,225	6,650	20,000	2,419	
	1870	Emulsion sprayer, 65 gal., 5 H.P. gas engine		2.78	103	310	930	84.20	
	1880	200 gal., 5 H.P. engine		7.25	173	520	1,550	161.95	
	1900	Floor auto-scrubbing machine, walk-behind, 28″ path		5.64	365	1,100	3,300	265.15	
	1930	Floodlight, mercury vapor, or quartz, on tripod, 1,000 watt		.46	22	66.50	200	16.95	
	1940	2,000 watt		.59	27.50	82	246	21.15	
	1950	Floodlights, trailer mounted with generator, one - 300 watt light		3.56	76	228	685	74.05	
	1960	two 1000 watt lights		4.50	84.50	254	760	86.80	
	2000	four 300 watt lights		4.26	96.50	290	870	92.05	
	2005	Foam spray rig, incl. box trailer, compressor, generator, proportioner		25.54	515	1,550	4,650	514.30	
	2015	Forklift, pneumatic tire, rough terr., straight mast, 5,000 lb., 12′ lift, gas		18.65	212	635	1,900	276.20	

01 54 33 | Equipment Rental

			UNIT	HOURLY OPER. COST	RENT PER DAY	RENT PER WEEK	RENT PER MONTH	EQUIPMENT COST/DAY
40	2025	8,000 lb, 12′ lift	Ea.	22.75	350	1,050	3,150	392
	2030	5,000 lb., 12′ lift, diesel		15.45	237	710	2,125	265.60
	2035	8,000 lb, 12′ lift, diesel		16.75	268	805	2,425	295
	2045	All terrain, telescoping boom, diesel, 5,000 lb., 10′ reach, 19′ lift		17.25	410	1,225	3,675	383
	2055	6,600 lb., 29′ reach, 42′ lift		21.10	400	1,200	3,600	408.80
	2065	10,000 lb., 31′ reach, 45′ lift		23.10	465	1,400	4,200	464.75
	2070	Cushion tire, smooth floor, gas, 5,000 lb. capacity		8.25	76.50	230	690	112
	2075	8,000 lb. capacity		11.37	89	267	800	144.30
	2085	Diesel, 5,000 lb. capacity		7.75	83.50	250	750	112
	2090	12,000 lb. capacity		12.05	130	390	1,175	174.40
	2095	20,000 lb. capacity		17.25	150	450	1,350	228.05
	2100	Generator, electric, gas engine, 1.5 kW to 3 kW		2.58	10.15	30.50	91.50	26.70
	2200	5 kW		3.22	12.65	38	114	33.35
	2300	10 kW		5.93	33.50	101	305	67.65
	2400	25 kW		7.41	84.50	254	760	110.05
	2500	Diesel engine, 20 kW		9.21	74	222	665	118.05
	2600	50 kW		15.95	96.50	289	865	185.35
	2700	100 kW		28.60	133	400	1,200	308.80
	2800	250 kW		54.35	255	765	2,300	587.85
	2850	Hammer, hydraulic, for mounting on boom, to 500 ft. lb.		2.90	90.50	271	815	77.40
	2860	1,000 ft. lb.		4.61	135	405	1,225	117.85
	2900	Heaters, space, oil or electric, 50 MBH		1.46	8	24	72	16.50
	3000	100 MBH		2.72	11.85	35.50	107	28.85
	3100	300 MBH		7.92	40	120	360	87.40
	3150	500 MBH		13.16	45	135	405	132.25
	3200	Hose, water, suction with coupling, 20′ long, 2″ diameter		.02	3.13	9.40	28	2.05
	3210	3″ diameter		.03	4.10	12.30	37	2.70
	3220	4″ diameter		.03	4.80	14.40	43	3.10
	3230	6″ diameter		.11	17	51	153	11.10
	3240	8″ diameter		.27	45	135	405	29.15
	3250	Discharge hose with coupling, 50′ long, 2″ diameter		.01	1.33	4	12	.90
	3260	3″ diameter		.01	2.22	6.65	19.95	1.40
	3270	4″ diameter		.02	3.47	10.40	31	2.25
	3280	6″ diameter		.06	8.65	26	78	5.70
	3290	8″ diameter		.24	40	120	360	25.90
	3295	Insulation blower		.83	6	18.05	54	10.25
	3300	Ladders, extension type, 16′ to 36′ long		.18	31.50	95	285	20.45
	3400	40′ to 60′ long		.64	112	335	1,000	72.10
	3405	Lance for cutting concrete		2.21	64	192	575	56.05
	3407	Lawn mower, rotary, 22″, 5 H.P.		1.06	25	75	225	23.45
	3408	48″ self-propelled		2.90	127	380	1,150	99.20
	3410	Level, electronic, automatic, with tripod and leveling rod		1.05	61.50	185	555	45.40
	3430	Laser type, for pipe and sewer line and grade		2.18	153	460	1,375	109.40
	3440	Rotating beam for interior control		.90	62	186	560	44.40
	3460	Builder's optical transit, with tripod and rod		.10	17	51	153	11
	3500	Light towers, towable, with diesel generator, 2,000 watt		4.27	98	294	880	92.95
	3600	4,000 watt		4.51	102	305	915	97.10
	3700	Mixer, powered, plaster and mortar, 6 C.F., 7 H.P.		2.06	21	62.50	188	28.95
	3800	10 C.F., 9 H.P.		2.24	33.50	101	305	38.15
	3850	Nailer, pneumatic		.48	32.50	97	291	23.25
	3900	Paint sprayers complete, 8 CFM		.85	59.50	179	535	42.60
	4000	17 CFM		1.60	107	320	960	76.80
	4020	Pavers, bituminous, rubber tires, 8′ wide, 50 H.P., diesel		32.02	550	1,650	4,950	586.20
	4030	10′ wide, 150 H.P.		95.91	1,875	5,650	17,000	1,897
	4050	Crawler, 8′ wide, 100 H.P., diesel		87.85	2,000	5,975	17,900	1,898
	4060	10′ wide, 150 H.P.		104.28	2,275	6,825	20,500	2,199
	4070	Concrete paver, 12′ to 24′ wide, 250 H.P.		87.88	1,625	4,900	14,700	1,683
	4080	Placer-spreader-trimmer, 24′ wide, 300 H.P.	↓	117.87	2,475	7,425	22,300	2,428

01 54 33 | Equipment Rental

	01 54 33 Equipment Rental	UNIT	HOURLY OPER. COST	RENT PER DAY	RENT PER WEEK	RENT PER MONTH	EQUIPMENT COST/DAY
40 4100	Pump, centrifugal gas pump, 1-1/2" diam., 65 GPM	Ea.	3.93	52.50	157	470	62.90
4200	2" diameter, 130 GPM		4.99	64.50	194	580	78.75
4300	3" diameter, 250 GPM		5.13	63.50	191	575	79.25
4400	6" diameter, 1,500 GPM		22.31	197	590	1,775	296.45
4500	Submersible electric pump, 1-1/4" diameter, 55 GPM		.40	17.65	53	159	13.80
4600	1-1/2" diameter, 83 GPM		.45	20.50	61	183	15.75
4700	2" diameter, 120 GPM		1.65	25	75.50	227	28.30
4800	3" diameter, 300 GPM		3.04	45	135	405	51.35
4900	4" diameter, 560 GPM		14.80	163	490	1,475	216.35
5000	6" diameter, 1,590 GPM		22.14	200	600	1,800	297.15
5100	Diaphragm pump, gas, single, 1-1/2" diameter		1.13	57.50	172	515	43.45
5200	2" diameter		3.99	70.50	212	635	74.30
5300	3" diameter		4.06	74.50	224	670	77.30
5400	Double, 4" diameter		6.05	155	465	1,400	141.40
5450	Pressure washer, 5 GPM, 3,000 psi		3.88	53.50	160	480	63.05
5460	7 GPM, 3,000 psi		4.95	63	189	565	77.40
5500	Trash pump, self-priming, gas, 2" diameter		3.83	23.50	71	213	44.80
5600	Diesel, 4" diameter		6.70	94.50	284	850	110.40
5650	Diesel, 6" diameter		16.90	165	495	1,475	234.20
5655	Grout pump		18.75	272	815	2,450	313
5700	Salamanders, L.P. gas fired, 100,000 BTU		2.89	14.15	42.50	128	31.65
5705	50,000 BTU		1.67	11.65	35	105	20.35
5720	Sandblaster, portable, open top, 3 C.F. capacity		.60	26.50	79.50	239	20.70
5730	6 C.F. capacity		1.01	39.50	118	355	31.65
5740	Accessories for above		.14	23	69	207	14.90
5750	Sander, floor		.77	19.35	58	174	17.75
5760	Edger		.52	15.85	47.50	143	13.65
5800	Saw, chain, gas engine, 18" long		1.76	22.50	67.50	203	27.55
5900	Hydraulic powered, 36" long		.78	67.50	202	605	46.65
5950	60" long		.78	69.50	208	625	47.85
6000	Masonry, table mounted, 14" diameter, 5 H.P.		1.32	81	243	730	59.20
6050	Portable cut-off, 8 H.P.		1.82	32.50	97.50	293	34
6100	Circular, hand held, electric, 7-1/4" diameter		.23	4.93	14.80	44.50	4.80
6200	12" diameter		.24	7.85	23.50	70.50	6.60
6250	Wall saw, w/hydraulic power, 10 H.P.		3.30	33.50	101	305	46.60
6275	Shot blaster, walk-behind, 20" wide		4.75	272	815	2,450	201
6280	Sidewalk broom, walk-behind		2.25	80	240	720	65.95
6300	Steam cleaner, 100 gallons per hour		3.35	80	240	720	74.80
6310	200 gallons per hour		4.35	96.50	290	870	92.75
6340	Tar kettle/pot, 400 gallons		16.53	76.50	230	690	178.20
6350	Torch, cutting, acetylene-oxygen, 150' hose, excludes gases		.45	14.85	44.50	134	12.50
6360	Hourly operating cost includes tips and gas		20.98				167.85
6410	Toilet, portable chemical		.13	22.50	67	201	14.45
6420	Recycle flush type		.16	27.50	83	249	17.90
6430	Toilet, fresh water flush, garden hose,		.19	33	99	297	21.35
6440	Hoisted, non-flush, for high rise		.16	27	81	243	17.45
6465	Tractor, farm with attachment		17.42	300	900	2,700	319.35
6480	Trailers, platform, flush deck, 2 axle, 3 ton capacity		1.69	21.50	65	195	26.55
6500	25 ton capacity		6.25	138	415	1,250	133
6600	40 ton capacity		8.06	197	590	1,775	182.50
6700	3 axle, 50 ton capacity		8.75	218	655	1,975	200.95
6800	75 ton capacity		11.11	290	870	2,600	262.90
6810	Trailer mounted cable reel for high voltage line work		5.90	282	845	2,525	216.20
6820	Trailer mounted cable tensioning rig		11.71	560	1,675	5,025	428.65
6830	Cable pulling rig		73.99	3,125	9,375	28,100	2,467
6850	Portable cable/wire puller, 8,000 lb. max. pulling capacity		3.71	205	615	1,850	152.70
6900	Water tank trailer, engine driven discharge, 5,000 gallons		7.18	153	460	1,375	149.45
6925	10,000 gallons	▼	9.78	208	625	1,875	203.25

01 54 33 | Equipment Rental

		01 54 33 Equipment Rental	UNIT	HOURLY OPER. COST	RENT PER DAY	RENT PER WEEK	RENT PER MONTH	EQUIPMENT COST/DAY	
40	6950	Water truck, off highway, 6,000 gallons	Ea.	71.97	810	2,425	7,275	1,061	40
	7010	Tram car for high voltage line work, powered, 2 conductor		6.90	152	455	1,375	146.20	
	7020	Transit (builder's level) with tripod		.10	17	51	153	11	
	7030	Trench box, 3,000 lb., 6′ x 8′		.56	93.50	281	845	60.70	
	7040	7,200 lb., 6′ x 20′		.72	120	360	1,075	77.75	
	7050	8,000 lb., 8′ x 16′		1.08	180	540	1,625	116.65	
	7060	9,500 lb., 8′ x 20′		1.21	225	675	2,025	144.65	
	7065	11,000 lb., 8′ x 24′		1.27	212	635	1,900	137.15	
	7070	12,000 lb., 10′ x 20′		1.49	255	765	2,300	164.95	
	7100	Truck, pickup, 3/4 ton, 2 wheel drive		9.26	59.50	179	535	109.90	
	7200	4 wheel drive		9.51	76	228	685	121.70	
	7250	Crew carrier, 9 passenger		12.70	88	264	790	154.35	
	7290	Flat bed truck, 20,000 lb. GVW		15.31	128	385	1,150	199.45	
	7300	Tractor, 4 x 2, 220 H.P.		22.32	208	625	1,875	303.55	
	7410	330 H.P.		32.43	285	855	2,575	430.40	
	7500	6 x 4, 380 H.P.		36.20	330	990	2,975	487.60	
	7600	450 H.P.		44.36	400	1,200	3,600	594.90	
	7610	Tractor, with A frame, boom and winch, 225 H.P.		24.81	283	850	2,550	368.50	
	7620	Vacuum truck, hazardous material, 2,500 gallons		12.83	300	900	2,700	282.65	
	7625	5,000 gallons		13.06	425	1,275	3,825	359.50	
	7650	Vacuum, HEPA, 16 gallon, wet/dry		.85	17.15	51.50	155	17.10	
	7655	55 gallon, wet/dry		.78	26.50	80	240	22.25	
	7660	Water tank, portable		.73	155	465	1,400	98.85	
	7690	Sewer/catch basin vacuum, 14 C.Y., 1,500 gallons		17.37	640	1,925	5,775	523.95	
	7700	Welder, electric, 200 amp		3.83	16	48	144	40.20	
	7800	300 amp		5.56	19.35	58	174	56.10	
	7900	Gas engine, 200 amp		8.97	23	69	207	85.60	
	8000	300 amp		10.16	24	72	216	95.65	
	8100	Wheelbarrow, any size		.06	10.35	31	93	6.70	
	8200	Wrecking ball, 4,000 lb.	↓	2.51	72	216	650	63.25	
50	0010	**HIGHWAY EQUIPMENT RENTAL** without operators R015433-10							50
	0050	Asphalt batch plant, portable drum mixer, 100 ton/hr.	Ea.	88.67	1,500	4,475	13,400	1,604	
	0060	200 ton/hr.		102.29	1,600	4,775	14,300	1,773	
	0070	300 ton/hr.		120.22	1,875	5,600	16,800	2,082	
	0100	Backhoe attachment, long stick, up to 185 H.P., 10-1/2′ long		.37	24.50	74	222	17.75	
	0140	Up to 250 H.P., 12′ long		.41	27.50	83	249	19.90	
	0180	Over 250 H.P., 15′ long		.57	37.50	113	340	27.15	
	0200	Special dipper arm, up to 100 H.P., 32′ long		1.16	77	231	695	55.50	
	0240	Over 100 H.P., 33′ long		1.45	96.50	290	870	69.60	
	0280	Catch basin/sewer cleaning truck, 3 ton, 9 C.Y., 1,000 gal.		35.50	410	1,225	3,675	528.95	
	0300	Concrete batch plant, portable, electric, 200 C.Y./hr.		24.26	540	1,625	4,875	519.05	
	0520	Grader/dozer attachment, ripper/scarifier, rear mounted, up to 135 H.P.		3.16	61.50	184	550	62.10	
	0540	Up to 180 H.P.		4.15	92.50	278	835	88.80	
	0580	Up to 250 H.P.		5.87	148	445	1,325	135.95	
	0700	Pvmt. removal bucket, for hyd. excavator, up to 90 H.P.		2.17	56.50	169	505	51.10	
	0740	Up to 200 H.P.		2.31	72	216	650	61.70	
	0780	Over 200 H.P.		2.53	88.50	265	795	73.20	
	0900	Aggregate spreader, self-propelled, 187 H.P.		50.75	715	2,150	6,450	836	
	1000	Chemical spreader, 3 C.Y.		3.18	46.50	139	415	53.20	
	1900	Hammermill, traveling, 250 H.P.		67.55	2,250	6,750	20,300	1,890	
	2000	Horizontal borer, 3″ diameter, 13 H.P. gas driven		5.43	57.50	172	515	77.85	
	2150	Horizontal directional drill, 20,000 lb. thrust, 78 H.P. diesel		27.66	685	2,050	6,150	631.30	
	2160	30,000 lb. thrust, 115 H.P.		34.00	1,050	3,125	9,375	897	
	2170	50,000 lb. thrust, 170 H.P.		48.74	1,325	3,975	11,900	1,185	
	2190	Mud trailer for HDD, 1,500 gallons, 175 H.P., gas		25.58	157	470	1,400	298.65	
	2200	Hydromulcher, diesel, 3,000 gallon, for truck mounting		17.48	255	765	2,300	292.90	
	2300	Gas, 600 gallon		7.52	107	320	960	124.15	
	2400	Joint & crack cleaner, walk behind, 25 H.P.	↓	3.17	52.50	158	475	56.95	

01 54 33 | Equipment Rental

		01 54 33 | Equipment Rental	UNIT	HOURLY OPER. COST	RENT PER DAY	RENT PER WEEK	RENT PER MONTH	EQUIPMENT COST/DAY	
50	2500	Filler, trailer mounted, 400 gallons, 20 H.P.	Ea.	8.37	220	660	1,975	198.95	50
	3000	Paint striper, self-propelled, 40 gallon, 22 H.P.		6.78	163	490	1,475	152.25	
	3100	120 gallon, 120 H.P.		19.28	410	1,225	3,675	399.25	
	3200	Post drivers, 6" I-Beam frame, for truck mounting		12.45	390	1,175	3,525	334.60	
	3400	Road sweeper, self-propelled, 8' wide, 90 H.P.		36.01	690	2,075	6,225	703.10	
	3450	Road sweeper, vacuum assisted, 4 C.Y., 220 gallons		58.45	650	1,950	5,850	857.60	
	4000	Road mixer, self-propelled, 130 H.P.		46.37	800	2,400	7,200	851	
	4100	310 H.P.		75.23	2,100	6,275	18,800	1,857	
	4220	Cold mix paver, incl. pug mill and bitumen tank, 165 H.P.		95.25	2,250	6,725	20,200	2,107	
	4240	Pavement brush, towed		3.44	96.50	290	870	85.50	
	4250	Paver, asphalt, wheel or crawler, 130 H.P., diesel		94.52	2,200	6,600	19,800	2,076	
	4300	Paver, road widener, gas, 1' to 6', 67 H.P.		46.80	940	2,825	8,475	939.40	
	4400	Diesel, 2' to 14', 88 H.P.		56.55	1,125	3,350	10,100	1,122	
	4600	Slipform pavers, curb and gutter, 2 track, 75 H.P.		58.01	1,225	3,650	11,000	1,194	
	4700	4 track, 165 H.P.		35.79	815	2,450	7,350	776.35	
	4800	Median barrier, 215 H.P.		58.61	1,300	3,900	11,700	1,249	
	4901	Trailer, low bed, 75 ton capacity		10.74	273	820	2,450	249.90	
	5000	Road planer, walk behind, 10" cutting width, 10 H.P.		2.46	33.50	100	300	39.70	
	5100	Self-propelled, 12" cutting width, 64 H.P.		8.28	117	350	1,050	136.25	
	5120	Traffic line remover, metal ball blaster, truck mounted, 115 H.P.		46.70	785	2,350	7,050	843.60	
	5140	Grinder, truck mounted, 115 H.P.		51.04	810	2,425	7,275	893.35	
	5160	Walk-behind, 11 H.P.		3.57	54.50	164	490	61.35	
	5200	Pavement profiler, 4' to 6' wide, 450 H.P.		217.23	3,425	10,300	30,900	3,798	
	5300	8' to 10' wide, 750 H.P.		332.58	4,525	13,600	40,800	5,381	
	5400	Roadway plate, steel, 1" x 8' x 20'		.09	14.65	44	132	9.50	
	5600	Stabilizer, self-propelled, 150 H.P.		41.26	640	1,925	5,775	715.05	
	5700	310 H.P.		76.41	1,700	5,075	15,200	1,626	
	5800	Striper, truck mounted, 120 gallon paint, 460 H.P.		48.89	490	1,475	4,425	686.10	
	5900	Thermal paint heating kettle, 115 gallons		7.73	26.50	80	240	77.85	
	6000	Tar kettle, 330 gallon, trailer mounted		12.31	60	180	540	134.50	
	7000	Tunnel locomotive, diesel, 8 to 12 ton		29.85	600	1,800	5,400	598.75	
	7005	Electric, 10 ton		29.33	685	2,050	6,150	644.70	
	7010	Muck cars, 1/2 C.Y. capacity		2.30	26	77.50	233	33.95	
	7020	1 C.Y. capacity		2.52	33.50	101	305	40.35	
	7030	2 C.Y. capacity		2.66	37.50	113	340	43.90	
	7040	Side dump, 2 C.Y. capacity		2.88	46.50	140	420	51.05	
	7050	3 C.Y. capacity		3.86	51.50	154	460	61.70	
	7060	5 C.Y. capacity		5.64	66.50	199	595	84.90	
	7100	Ventilating blower for tunnel, 7-1/2 H.P.		2.14	51	153	460	47.75	
	7110	10 H.P.		2.43	53.50	160	480	51.40	
	7120	20 H.P.		3.55	69.50	208	625	70	
	7140	40 H.P.		6.16	91.50	275	825	104.25	
	7160	60 H.P.		8.71	98.50	295	885	128.70	
	7175	75 H.P.		10.40	153	460	1,375	175.20	
	7180	200 H.P.		20.85	300	905	2,725	347.75	
	7800	Windrow loader, elevating	↓	54.10	1,350	4,025	12,100	1,238	
60	0010	**LIFTING AND HOISTING EQUIPMENT RENTAL** without operators (R015433-10)							60
	0150	Crane, flatbed mounted, 3 ton capacity	Ea.	14.45	195	585	1,750	232.65	
	0200	Crane, climbing, 106' jib, 6,000 lb. capacity, 410 fpm (R312316-45)		39.84	1,775	5,350	16,100	1,389	
	0300	101' jib, 10,250 lb. capacity, 270 fpm		46.57	2,275	6,800	20,400	1,733	
	0500	Tower, static, 130' high, 106' jib, 6,200 lb. capacity at 400 fpm		45.29	2,075	6,200	18,600	1,602	
	0520	Mini crawler spider crane, up to 24" wide, 1,990 lb. lifting capacity		12.54	535	1,600	4,800	420.30	
	0525	Up to 30" wide, 6,450 lb. lifting capacity		14.57	635	1,900	5,700	496.55	
	0530	Up to 52" wide, 6,680 lb. lifting capacity		23.17	775	2,325	6,975	650.40	
	0535	Up to 55" wide, 8,920 lb. lifting capacity		25.87	860	2,575	7,725	722	
	0540	Up to 66" wide, 13,350 lb. lifting capacity		35.03	1,325	4,000	12,000	1,080	
	0600	Crawler mounted, lattice boom, 1/2 C.Y., 15 tons at 12' radius		37.07	950	2,850	8,550	866.60	
	0700	3/4 C.Y., 20 tons at 12' radius	↓	50.57	1,200	3,625	10,900	1,130	

01 54 33 | Equipment Rental

60		UNIT	HOURLY OPER. COST	RENT PER DAY	RENT PER WEEK	RENT PER MONTH	EQUIPMENT COST/DAY	60
0800	1 C.Y., 25 tons at 12′ radius	Ea.	67.62	1,400	4,225	12,700	1,386	
0900	1-1/2 C.Y., 40 tons at 12′ radius		66.51	1,425	4,300	12,900	1,392	
1000	2 C.Y., 50 tons at 12′ radius		89.03	2,200	6,600	19,800	2,032	
1100	3 C.Y., 75 tons at 12′ radius		75.49	1,875	5,650	17,000	1,734	
1200	100 ton capacity, 60′ boom		86.17	1,975	5,950	17,900	1,879	
1300	165 ton capacity, 60′ boom		106.42	2,275	6,850	20,600	2,221	
1400	200 ton capacity, 70′ boom		138.63	3,175	9,550	28,700	3,019	
1500	350 ton capacity, 80′ boom		182.75	4,075	12,200	36,600	3,902	
1600	Truck mounted, lattice boom, 6 x 4, 20 tons at 10′ radius		39.88	915	2,750	8,250	869	
1700	25 tons at 10′ radius		42.86	1,400	4,175	12,500	1,178	
1800	8 x 4, 30 tons at 10′ radius		45.67	1,475	4,425	13,300	1,250	
1900	40 tons at 12′ radius		48.70	1,550	4,625	13,900	1,315	
2000	60 tons at 15′ radius		53.85	1,650	4,950	14,900	1,421	
2050	82 tons at 15′ radius		59.60	1,775	5,350	16,100	1,547	
2100	90 tons at 15′ radius		66.59	1,950	5,825	17,500	1,698	
2200	115 tons at 15′ radius		75.12	2,175	6,525	19,600	1,906	
2300	150 tons at 18′ radius		81.33	2,275	6,850	20,600	2,021	
2350	165 tons at 18′ radius		87.30	2,425	7,275	21,800	2,153	
2400	Truck mounted, hydraulic, 12 ton capacity		29.59	390	1,175	3,525	471.75	
2500	25 ton capacity		36.46	485	1,450	4,350	581.70	
2550	33 ton capacity		50.82	900	2,700	8,100	946.60	
2560	40 ton capacity		49.62	900	2,700	8,100	937	
2600	55 ton capacity		53.94	915	2,750	8,250	981.50	
2700	80 ton capacity		75.94	1,475	4,400	13,200	1,487	
2720	100 ton capacity		75.18	1,550	4,675	14,000	1,536	
2740	120 ton capacity		103.12	1,825	5,500	16,500	1,925	
2760	150 ton capacity		110.25	2,050	6,125	18,400	2,107	
2800	Self-propelled, 4 x 4, with telescoping boom, 5 ton		15.18	230	690	2,075	259.45	
2900	12-1/2 ton capacity		21.48	335	1,000	3,000	371.85	
3000	15 ton capacity		34.52	535	1,600	4,800	596.20	
3050	20 ton capacity		24.09	660	1,975	5,925	587.75	
3100	25 ton capacity		36.80	615	1,850	5,550	664.40	
3150	40 ton capacity		45.03	650	1,950	5,850	750.25	
3200	Derricks, guy, 20 ton capacity, 60′ boom, 75′ mast		22.81	435	1,300	3,900	442.50	
3300	100′ boom, 115′ mast		36.15	750	2,250	6,750	739.20	
3400	Stiffleg, 20 ton capacity, 70′ boom, 37′ mast		25.48	565	1,700	5,100	543.85	
3500	100′ boom, 47′ mast		39.43	910	2,725	8,175	860.50	
3550	Helicopter, small, lift to 1,250 lb. maximum, w/pilot		99.44	3,525	10,600	31,800	2,916	
3600	Hoists, chain type, overhead, manual, 3/4 ton		.15	.33	1	3	1.35	
3900	10 ton		.79	6	18	54	9.90	
4000	Hoist and tower, 5,000 lb. cap., portable electric, 40′ high		5.14	252	755	2,275	192.10	
4100	For each added 10′ section, add		.12	19.65	59	177	12.75	
4200	Hoist and single tubular tower, 5,000 lb. electric, 100′ high		6.98	350	1,050	3,150	265.80	
4300	For each added 6′-6″ section, add		.21	34.50	103	310	22.25	
4400	Hoist and double tubular tower, 5,000 lb., 100′ high		7.59	385	1,150	3,450	290.75	
4500	For each added 6′-6″ section, add		.23	37.50	113	340	24.40	
4550	Hoist and tower, mast type, 6,000 lb., 100′ high		8.26	400	1,200	3,600	306.10	
4570	For each added 10′ section, add		.13	23	69	207	14.85	
4600	Hoist and tower, personnel, electric, 2,000 lb., 100′ @ 125 fpm		17.55	1,075	3,200	9,600	780.40	
4700	3,000 lb., 100′ @ 200 fpm		20.08	1,200	3,625	10,900	885.65	
4800	3,000 lb., 150′ @ 300 fpm		22.28	1,350	4,075	12,200	993.30	
4900	4,000 lb., 100′ @ 300 fpm		23.05	1,375	4,150	12,500	1,014	
5000	6,000 lb., 100′ @ 275 fpm		24.77	1,450	4,350	13,100	1,068	
5100	For added heights up to 500′, add	L.F.	.01	1.67	5	15	1.10	
5200	Jacks, hydraulic, 20 ton	Ea.	.05	2	6	18	1.60	
5500	100 ton		.40	12	36	108	10.40	
6100	Jacks, hydraulic, climbing w/50′ jackrods, control console, 30 ton cap.		2.17	145	435	1,300	104.40	
6150	For each added 10′ jackrod section, add		.05	3.33	10	30	2.40	

01 54 33 | Equipment Rental

		01 54 33 Equipment Rental	UNIT	HOURLY OPER. COST	RENT PER DAY	RENT PER WEEK	RENT PER MONTH	EQUIPMENT COST/DAY	
60	6300	50 ton capacity	Ea.	3.49	232	695	2,075	166.95	60
	6350	For each added 10′ jackrod section, add		.06	4	12	36	2.90	
	6500	125 ton capacity		9.13	610	1,825	5,475	438.05	
	6550	For each added 10′ jackrod section, add		.62	41	123	370	29.55	
	6600	Cable jack, 10 ton capacity with 200′ cable		1.82	122	365	1,100	87.60	
	6650	For each added 50′ of cable, add	↓	.22	14.65	44	132	10.55	
70	0010	**WELLPOINT EQUIPMENT RENTAL** without operators R015433-10							70
	0020	Based on 2 months rental							
	0100	Combination jetting & wellpoint pump, 60 H.P. diesel	Ea.	15.72	360	1,075	3,225	340.75	
	0200	High pressure gas jet pump, 200 H.P., 300 psi	"	33.93	305	920	2,750	455.45	
	0300	Discharge pipe, 8″ diameter	L.F.	.01	.58	1.74	5.20	.40	
	0350	12″ diameter		.01	.83	2.50	7.50	.60	
	0400	Header pipe, flows up to 150 GPM, 4″ diameter		.01	.53	1.59	4.77	.40	
	0500	400 GPM, 6″ diameter		.01	.62	1.86	5.60	.45	
	0600	800 GPM, 8″ diameter		.01	.83	2.50	7.50	.60	
	0700	1,500 GPM, 10″ diameter		.01	.88	2.64	7.90	.65	
	0800	2,500 GPM, 12″ diameter		.03	1.70	5.10	15.30	1.20	
	0900	4,500 GPM, 16″ diameter		.03	2.18	6.55	19.65	1.55	
	0950	For quick coupling aluminum and plastic pipe, add	↓	.03	2.27	6.80	20.50	1.65	
	1100	Wellpoint, 25′ long, with fittings & riser pipe, 1-1/2″ or 2″ diameter	Ea.	.07	4.52	13.55	40.50	3.25	
	1200	Wellpoint pump, diesel powered, 4″ suction, 20 H.P.		7.02	207	620	1,850	180.20	
	1300	6″ suction, 30 H.P.		9.42	257	770	2,300	229.35	
	1400	8″ suction, 40 H.P.		12.76	350	1,050	3,150	312.10	
	1500	10″ suction, 75 H.P.		18.83	410	1,225	3,675	395.65	
	1600	12″ suction, 100 H.P.		27.32	660	1,975	5,925	613.55	
	1700	12″ suction, 175 H.P.	↓	39.09	725	2,175	6,525	747.75	
80	0010	**MARINE EQUIPMENT RENTAL** without operators R015433-10							80
	0200	Barge, 400 ton, 30′ wide x 90′ long	Ea.	17.69	1,150	3,475	10,400	836.50	
	0240	800 ton, 45′ wide x 90′ long		22.21	1,425	4,275	12,800	1,033	
	2000	Tugboat, diesel, 100 H.P.		29.66	230	690	2,075	375.25	
	2040	250 H.P.		57.58	415	1,250	3,750	710.70	
	2080	380 H.P.		125.36	1,250	3,750	11,300	1,753	
	3000	Small work boat, gas, 16-foot, 50 H.P.		11.38	46.50	139	415	118.85	
	4000	Large, diesel, 48-foot, 200 H.P.	↓	74.90	1,325	3,975	11,900	1,394	

Crews - Open Shop

Crew A-1	Hr.	Daily	Hr.	Daily	Bare Costs	Incl. O&P
1 Building Laborer	$30.35	$242.80	$50.15	$401.20	$30.35	$50.15
1 Concrete Saw, Gas Manual		72.90		80.19	9.11	10.02
8 L.H., Daily Totals		$315.70		$481.39	$39.46	$60.17

Crew A-1A	Hr.	Daily	Hr.	Daily	Bare Costs	Incl. O&P
1 Skilled Worker	$39.45	$315.60	$65.55	$524.40	$39.45	$65.55
1 Shot Blaster, 20"		201.00		221.10	25.13	27.64
8 L.H., Daily Totals		$516.60		$745.50	$64.58	$93.19

Crew A-1B	Hr.	Daily	Hr.	Daily	Bare Costs	Incl. O&P
1 Building Laborer	$30.35	$242.80	$50.15	$401.20	$30.35	$50.15
1 Concrete Saw		110.15		121.17	13.77	15.15
8 L.H., Daily Totals		$352.95		$522.37	$44.12	$65.30

Crew A-1C	Hr.	Daily	Hr.	Daily	Bare Costs	Incl. O&P
1 Building Laborer	$30.35	$242.80	$50.15	$401.20	$30.35	$50.15
1 Chain Saw, Gas, 18"		27.55		30.31	3.44	3.79
8 L.H., Daily Totals		$270.35		$431.51	$33.79	$53.94

Crew A-1D	Hr.	Daily	Hr.	Daily	Bare Costs	Incl. O&P
1 Building Laborer	$30.35	$242.80	$50.15	$401.20	$30.35	$50.15
1 Vibrating Plate, Gas, 18"		31.10		34.21	3.89	4.28
8 L.H., Daily Totals		$273.90		$435.41	$34.24	$54.43

Crew A-1E	Hr.	Daily	Hr.	Daily	Bare Costs	Incl. O&P
1 Building Laborer	$30.35	$242.80	$50.15	$401.20	$30.35	$50.15
1 Vibrating Plate, Gas, 21"		40.55		44.60	5.07	5.58
8 L.H., Daily Totals		$283.35		$445.81	$35.42	$55.73

Crew A-1F	Hr.	Daily	Hr.	Daily	Bare Costs	Incl. O&P
1 Building Laborer	$30.35	$242.80	$50.15	$401.20	$30.35	$50.15
1 Rammer/Tamper, Gas, 8"		45.65		50.22	5.71	6.28
8 L.H., Daily Totals		$288.45		$451.42	$36.06	$56.43

Crew A-1G	Hr.	Daily	Hr.	Daily	Bare Costs	Incl. O&P
1 Building Laborer	$30.35	$242.80	$50.15	$401.20	$30.35	$50.15
1 Rammer/Tamper, Gas, 15"		53.05		58.35	6.63	7.29
8 L.H., Daily Totals		$295.85		$459.56	$36.98	$57.44

Crew A-1H	Hr.	Daily	Hr.	Daily	Bare Costs	Incl. O&P
1 Building Laborer	$30.35	$242.80	$50.15	$401.20	$30.35	$50.15
1 Exterior Steam Cleaner		74.80		82.28	9.35	10.29
8 L.H., Daily Totals		$317.60		$483.48	$39.70	$60.44

Crew A-1J	Hr.	Daily	Hr.	Daily	Bare Costs	Incl. O&P
1 Building Laborer	$30.35	$242.80	$50.15	$401.20	$30.35	$50.15
1 Cultivator, Walk-Behind, 5 H.P.		70.50		77.55	8.81	9.69
8 L.H., Daily Totals		$313.30		$478.75	$39.16	$59.84

Crew A-1K	Hr.	Daily	Hr.	Daily	Bare Costs	Incl. O&P
1 Building Laborer	$30.35	$242.80	$50.15	$401.20	$30.35	$50.15
1 Cultivator, Walk-Behind, 8 H.P.		84.50		92.95	10.56	11.62
8 L.H., Daily Totals		$327.30		$494.15	$40.91	$61.77

Crew A-1M	Hr.	Daily	Hr.	Daily	Bare Costs	Incl. O&P
1 Building Laborer	$30.35	$242.80	$50.15	$401.20	$30.35	$50.15
1 Snow Blower, Walk-Behind		65.95		72.55	8.24	9.07
8 L.H., Daily Totals		$308.75		$473.75	$38.59	$59.22

Crew A-2	Hr.	Daily	Hr.	Daily	Bare Costs	Incl. O&P
2 Laborers	$30.35	$485.60	$50.15	$802.40	$31.60	$52.13
1 Truck Driver (light)	34.10	272.80	56.10	448.80		
1 Flatbed Truck, Gas, 1.5 Ton		195.20		214.72	8.13	8.95
24 L.H., Daily Totals		$953.60		$1465.92	$39.73	$61.08

Crew A-2A	Hr.	Daily	Hr.	Daily	Bare Costs	Incl. O&P
2 Laborers	$30.35	$485.60	$50.15	$802.40	$31.60	$52.13
1 Truck Driver (light)	34.10	272.80	56.10	448.80		
1 Flatbed Truck, Gas, 1.5 Ton		195.20		214.72		
1 Concrete Saw		110.15		121.17	12.72	14.00
24 L.H., Daily Totals		$1063.75		$1587.09	$44.32	$66.13

Crew A-2B	Hr.	Daily	Hr.	Daily	Bare Costs	Incl. O&P
1 Truck Driver (light)	$34.10	$272.80	$56.10	$448.80	$34.10	$56.10
1 Flatbed Truck, Gas, 1.5 Ton		195.20		214.72	24.40	26.84
8 L.H., Daily Totals		$468.00		$663.52	$58.50	$82.94

Crew A-3A	Hr.	Daily	Hr.	Daily	Bare Costs	Incl. O&P
1 Equip. Oper. (light)	$38.75	$310.00	$63.60	$508.80	$38.75	$63.60
1 Pickup Truck, 4x4, 3/4 Ton		121.70		133.87	15.21	16.73
8 L.H., Daily Totals		$431.70		$642.67	$53.96	$80.33

Crew A-3B	Hr.	Daily	Hr.	Daily	Bare Costs	Incl. O&P
1 Equip. Oper. (medium)	$40.75	$326.00	$66.85	$534.80	$37.77	$62.05
1 Truck Driver (heavy)	34.80	278.40	57.25	458.00		
1 Dump Truck, 12 C.Y., 400 H.P.		567.05		623.76		
1 F.E. Loader, W.M., 2.5 C.Y.		531.20		584.32	68.64	75.50
16 L.H., Daily Totals		$1702.65		$2200.88	$106.42	$137.55

Crew A-3C	Hr.	Daily	Hr.	Daily	Bare Costs	Incl. O&P
1 Equip. Oper. (light)	$38.75	$310.00	$63.60	$508.80	$38.75	$63.60
1 Loader, Skid Steer, 78 H.P.		392.50		431.75	49.06	53.97
8 L.H., Daily Totals		$702.50		$940.55	$87.81	$117.57

Crew A-3D	Hr.	Daily	Hr.	Daily	Bare Costs	Incl. O&P
1 Truck Driver (light)	$34.10	$272.80	$56.10	$448.80	$34.10	$56.10
1 Pickup Truck, 4x4, 3/4 Ton		121.70		133.87		
1 Flatbed Trailer, 25 Ton		133.00		146.30	31.84	35.02
8 L.H., Daily Totals		$527.50		$728.97	$65.94	$91.12

Crew A-3E	Hr.	Daily	Hr.	Daily	Bare Costs	Incl. O&P
1 Equip. Oper. (crane)	$41.95	$335.60	$68.85	$550.80	$38.38	$63.05
1 Truck Driver (heavy)	34.80	278.40	57.25	458.00		
1 Pickup Truck, 4x4, 3/4 Ton		121.70		133.87	7.61	8.37
16 L.H., Daily Totals		$735.70		$1142.67	$45.98	$71.42

Crew A-3F	Hr.	Daily	Hr.	Daily	Bare Costs	Incl. O&P
1 Equip. Oper. (crane)	$41.95	$335.60	$68.85	$550.80	$38.38	$63.05
1 Truck Driver (heavy)	34.80	278.40	57.25	458.00		
1 Pickup Truck, 4x4, 3/4 Ton		121.70		133.87		
1 Truck Tractor, 6x4, 380 H.P.		487.60		536.36		
1 Lowbed Trailer, 75 Ton		249.90		274.89	53.70	59.07
16 L.H., Daily Totals		$1473.20		$1953.92	$92.08	$122.12

Crews - Open Shop

Crew No.	Bare Costs		Incl. Subs O&P		Cost Per Labor-Hour	
Crew A-3G	**Hr.**	**Daily**	**Hr.**	**Daily**	**Bare Costs**	**Incl. O&P**
1 Equip. Oper. (crane)	$41.95	$335.60	$68.85	$550.80	$38.38	$63.05
1 Truck Driver (heavy)	34.80	278.40	57.25	458.00		
1 Pickup Truck, 4x4, 3/4 Ton		121.70		133.87		
1 Truck Tractor, 6x4, 450 H.P.		594.90		654.39		
1 Lowbed Trailer, 75 Ton		249.90		274.89	60.41	66.45
16 L.H., Daily Totals		$1580.50		$2071.95	$98.78	$129.50
Crew A-3H	**Hr.**	**Daily**	**Hr.**	**Daily**	**Bare Costs**	**Incl. O&P**
1 Equip. Oper. (crane)	$41.95	$335.60	$68.85	$550.80	$41.95	$68.85
1 Hyd. Crane, 12 Ton (Daily)		708.90		779.79	88.61	97.47
8 L.H., Daily Totals		$1044.50		$1330.59	$130.56	$166.32
Crew A-3I	**Hr.**	**Daily**	**Hr.**	**Daily**	**Bare Costs**	**Incl. O&P**
1 Equip. Oper. (crane)	$41.95	$335.60	$68.85	$550.80	$41.95	$68.85
1 Hyd. Crane, 25 Ton (Daily)		785.85		864.43	98.23	108.05
8 L.H., Daily Totals		$1121.45		$1415.23	$140.18	$176.90
Crew A-3J	**Hr.**	**Daily**	**Hr.**	**Daily**	**Bare Costs**	**Incl. O&P**
1 Equip. Oper. (crane)	$41.95	$335.60	$68.85	$550.80	$41.95	$68.85
1 Hyd. Crane, 40 Ton (Daily)		1280.00		1408.00	160.00	176.00
8 L.H., Daily Totals		$1615.60		$1958.80	$201.95	$244.85
Crew A-3K	**Hr.**	**Daily**	**Hr.**	**Daily**	**Bare Costs**	**Incl. O&P**
1 Equip. Oper. (crane)	$41.95	$335.60	$68.85	$550.80	$38.92	$63.88
1 Equip. Oper. (oiler)	35.90	287.20	58.90	471.20		
1 Hyd. Crane, 55 Ton (Daily)		1299.00		1428.90		
1 P/U Truck, 3/4 Ton (Daily)		140.45		154.50	89.97	98.96
16 L.H., Daily Totals		$2062.25		$2605.40	$128.89	$162.84
Crew A-3L	**Hr.**	**Daily**	**Hr.**	**Daily**	**Bare Costs**	**Incl. O&P**
1 Equip. Oper. (crane)	$41.95	$335.60	$68.85	$550.80	$38.92	$63.88
1 Equip. Oper. (oiler)	35.90	287.20	58.90	471.20		
1 Hyd. Crane, 80 Ton (Daily)		2056.00		2261.60		
1 P/U Truck, 3/4 Ton (Daily)		140.45		154.50	137.28	151.01
16 L.H., Daily Totals		$2819.25		$3438.09	$176.20	$214.88
Crew A-3M	**Hr.**	**Daily**	**Hr.**	**Daily**	**Bare Costs**	**Incl. O&P**
1 Equip. Oper. (crane)	$41.95	$335.60	$68.85	$550.80	$38.92	$63.88
1 Equip. Oper. (oiler)	35.90	287.20	58.90	471.20		
1 Hyd. Crane, 100 Ton (Daily)		2179.00		2396.90		
1 P/U Truck, 3/4 Ton (Daily)		140.45		154.50	144.97	159.46
16 L.H., Daily Totals		$2942.25		$3573.40	$183.89	$223.34
Crew A-3N	**Hr.**	**Daily**	**Hr.**	**Daily**	**Bare Costs**	**Incl. O&P**
1 Equip. Oper. (crane)	$41.95	$335.60	$68.85	$550.80	$41.95	$68.85
1 Tower Crane (monthly)		1181.00		1299.10	147.63	162.39
8 L.H., Daily Totals		$1516.60		$1849.90	$189.57	$231.24
Crew A-3P	**Hr.**	**Daily**	**Hr.**	**Daily**	**Bare Costs**	**Incl. O&P**
1 Equip. Oper. (light)	$38.75	$310.00	$63.60	$508.80	$38.75	$63.60
1 A.T. Forklift, 31' reach, 45' lift		464.75		511.23	58.09	63.90
8 L.H., Daily Totals		$774.75		$1020.03	$96.84	$127.50
Crew A-3Q	**Hr.**	**Daily**	**Hr.**	**Daily**	**Bare Costs**	**Incl. O&P**
1 Equip. Oper. (light)	$38.75	$310.00	$63.60	$508.80	$38.75	$63.60
1 Pickup Truck, 4x4, 3/4 Ton		121.70		133.87		
1 Flatbed Trailer, 3 Ton		26.55		29.20	18.53	20.38
8 L.H., Daily Totals		$458.25		$671.88	$57.28	$83.98

Crew No.	Bare Costs		Incl. Subs O&P		Cost Per Labor-Hour	
Crew A-3R	**Hr.**	**Daily**	**Hr.**	**Daily**	**Bare Costs**	**Incl. O&P**
1 Equip. Oper. (light)	$38.75	$310.00	$63.60	$508.80	$38.75	$63.60
1 Forklift, Smooth Floor, 8,000 Lb.		144.30		158.73	18.04	19.84
8 L.H., Daily Totals		$454.30		$667.53	$56.79	$83.44
Crew A-4	**Hr.**	**Daily**	**Hr.**	**Daily**	**Bare Costs**	**Incl. O&P**
2 Carpenters	$38.75	$620.00	$64.05	$1024.80	$36.78	$60.63
1 Painter, Ordinary	32.85	262.80	53.80	430.40		
24 L.H., Daily Totals		$882.80		$1455.20	$36.78	$60.63
Crew A-5	**Hr.**	**Daily**	**Hr.**	**Daily**	**Bare Costs**	**Incl. O&P**
2 Laborers	$30.35	$485.60	$50.15	$802.40	$30.77	$50.81
.25 Truck Driver (light)	34.10	68.20	56.10	112.20		
.25 Flatbed Truck, Gas, 1.5 Ton		48.80		53.68	2.71	2.98
18 L.H., Daily Totals		$602.60		$968.28	$33.48	$53.79
Crew A-6	**Hr.**	**Daily**	**Hr.**	**Daily**	**Bare Costs**	**Incl. O&P**
1 Instrument Man	$39.45	$315.60	$65.55	$524.40	$38.85	$64.25
1 Rodman/Chainman	38.25	306.00	62.95	503.60		
1 Level, Electronic		45.40		49.94	2.84	3.12
16 L.H., Daily Totals		$667.00		$1077.94	$41.69	$67.37
Crew A-7	**Hr.**	**Daily**	**Hr.**	**Daily**	**Bare Costs**	**Incl. O&P**
1 Chief of Party	$46.85	$374.80	$77.15	$617.20	$41.52	$68.55
1 Instrument Man	39.45	315.60	65.55	524.40		
1 Rodman/Chainman	38.25	306.00	62.95	503.60		
1 Level, Electronic		45.40		49.94	1.89	2.08
24 L.H., Daily Totals		$1041.80		$1695.14	$43.41	$70.63
Crew A-8	**Hr.**	**Daily**	**Hr.**	**Daily**	**Bare Costs**	**Incl. O&P**
1 Chief of Party	$46.85	$374.80	$77.15	$617.20	$40.70	$67.15
1 Instrument Man	39.45	315.60	65.55	524.40		
2 Rodmen/Chainmen	38.25	612.00	62.95	1007.20		
1 Level, Electronic		45.40		49.94	1.42	1.56
32 L.H., Daily Totals		$1347.80		$2198.74	$42.12	$68.71
Crew A-9	**Hr.**	**Daily**	**Hr.**	**Daily**	**Bare Costs**	**Incl. O&P**
1 Asbestos Foreman	$41.20	$329.60	$69.35	$554.80	$40.76	$68.65
7 Asbestos Workers	40.70	2279.20	68.55	3838.80		
64 L.H., Daily Totals		$2608.80		$4393.60	$40.76	$68.65
Crew A-10A	**Hr.**	**Daily**	**Hr.**	**Daily**	**Bare Costs**	**Incl. O&P**
1 Asbestos Foreman	$41.20	$329.60	$69.35	$554.80	$40.87	$68.82
2 Asbestos Workers	40.70	651.20	68.55	1096.80		
24 L.H., Daily Totals		$980.80		$1651.60	$40.87	$68.82
Crew A-10B	**Hr.**	**Daily**	**Hr.**	**Daily**	**Bare Costs**	**Incl. O&P**
1 Asbestos Foreman	$41.20	$329.60	$69.35	$554.80	$40.83	$68.75
3 Asbestos Workers	40.70	976.80	68.55	1645.20		
32 L.H., Daily Totals		$1306.40		$2200.00	$40.83	$68.75
Crew A-10C	**Hr.**	**Daily**	**Hr.**	**Daily**	**Bare Costs**	**Incl. O&P**
3 Asbestos Workers	$40.70	$976.80	$68.55	$1645.20	$40.70	$68.55
1 Flatbed Truck, Gas, 1.5 Ton		195.20		214.72	8.13	8.95
24 L.H., Daily Totals		$1172.00		$1859.92	$48.83	$77.50

Crew No.	Bare Costs		Incl. Subs O&P		Cost Per Labor-Hour	
Crew A-10D	**Hr.**	**Daily**	**Hr.**	**Daily**	**Bare Costs**	**Incl. O&P**
2 Asbestos Workers	$40.70	$651.20	$68.55	$1096.80	$39.81	$66.21
1 Equip. Oper. (crane)	41.95	335.60	68.85	550.80		
1 Equip. Oper. (oiler)	35.90	287.20	58.90	471.20		
1 Hydraulic Crane, 33 Ton		946.60		1041.26	29.58	32.54
32 L.H., Daily Totals		$2220.60		$3160.06	$69.39	$98.75
Crew A-11	**Hr.**	**Daily**	**Hr.**	**Daily**	**Bare Costs**	**Incl. O&P**
1 Asbestos Foreman	$41.20	$329.60	$69.35	$554.80	$40.76	$68.65
7 Asbestos Workers	40.70	2279.20	68.55	3838.80		
2 Chip. Hammers, 12 Lb., Elec.		42.00		46.20	0.66	0.72
64 L.H., Daily Totals		$2650.80		$4439.80	$41.42	$69.37
Crew A-12	**Hr.**	**Daily**	**Hr.**	**Daily**	**Bare Costs**	**Incl. O&P**
1 Asbestos Foreman	$41.20	$329.60	$69.35	$554.80	$40.76	$68.65
7 Asbestos Workers	40.70	2279.20	68.55	3838.80		
1 Trk-Mtd Vac, 14 CY, 1500 Gal.		523.95		576.35		
1 Flatbed Truck, 20,000 GVW		199.45		219.40	11.30	12.43
64 L.H., Daily Totals		$3332.20		$5189.34	$52.07	$81.08
Crew A-13	**Hr.**	**Daily**	**Hr.**	**Daily**	**Bare Costs**	**Incl. O&P**
1 Equip. Oper. (light)	$38.75	$310.00	$63.60	$508.80	$38.75	$63.60
1 Trk-Mtd Vac, 14 CY, 1500 Gal.		523.95		576.35		
1 Flatbed Truck, 20,000 GVW		199.45		219.40	90.42	99.47
8 L.H., Daily Totals		$1033.40		$1304.54	$129.18	$163.07
Crew B-1	**Hr.**	**Daily**	**Hr.**	**Daily**	**Bare Costs**	**Incl. O&P**
1 Labor Foreman (outside)	$32.35	$258.80	$53.45	$427.60	$31.02	$51.25
2 Laborers	30.35	485.60	50.15	802.40		
24 L.H., Daily Totals		$744.40		$1230.00	$31.02	$51.25
Crew B-1A	**Hr.**	**Daily**	**Hr.**	**Daily**	**Bare Costs**	**Incl. O&P**
1 Labor Foreman (outside)	$32.35	$258.80	$53.45	$427.60	$31.02	$51.25
2 Laborers	30.35	485.60	50.15	802.40		
2 Cutting Torches		25.00		27.50		
2 Sets of Gases		335.70		369.27	15.03	16.53
24 L.H., Daily Totals		$1105.10		$1626.77	$46.05	$67.78
Crew B-1B	**Hr.**	**Daily**	**Hr.**	**Daily**	**Bare Costs**	**Incl. O&P**
1 Labor Foreman (outside)	$32.35	$258.80	$53.45	$427.60	$33.75	$55.65
2 Laborers	30.35	485.60	50.15	802.40		
1 Equip. Oper. (crane)	41.95	335.60	68.85	550.80		
2 Cutting Torches		25.00		27.50		
2 Sets of Gases		335.70		369.27		
1 Hyd. Crane, 12 Ton		471.75		518.92	26.01	28.62
32 L.H., Daily Totals		$1912.45		$2696.49	$59.76	$84.27
Crew B-1C	**Hr.**	**Daily**	**Hr.**	**Daily**	**Bare Costs**	**Incl. O&P**
1 Labor Foreman (outside)	$32.35	$258.80	$53.45	$427.60	$31.02	$51.25
2 Laborers	30.35	485.60	50.15	802.40		
1 Telescoping Boom Lift, to 60'		451.25		496.38	18.80	20.68
24 L.H., Daily Totals		$1195.65		$1726.38	$49.82	$71.93
Crew B-1D	**Hr.**	**Daily**	**Hr.**	**Daily**	**Bare Costs**	**Incl. O&P**
2 Laborers	$30.35	$485.60	$50.15	$802.40	$30.35	$50.15
1 Small Work Boat, Gas, 50 H.P.		118.85		130.74		
1 Pressure Washer, 7 GPM		77.40		85.14	12.27	13.49
16 L.H., Daily Totals		$681.85		$1018.28	$42.62	$63.64

Crew No.	Bare Costs		Incl. Subs O&P		Cost Per Labor-Hour	
Crew B-1E	**Hr.**	**Daily**	**Hr.**	**Daily**	**Bare Costs**	**Incl. O&P**
1 Labor Foreman (outside)	$32.35	$258.80	$53.45	$427.60	$30.85	$50.98
3 Laborers	30.35	728.40	50.15	1203.60		
1 Work Boat, Diesel, 200 H.P.		1394.00		1533.40		
2 Pressure Washers, 7 GPM		154.80		170.28	48.40	53.24
32 L.H., Daily Totals		$2536.00		$3334.88	$79.25	$104.22
Crew B-1F	**Hr.**	**Daily**	**Hr.**	**Daily**	**Bare Costs**	**Incl. O&P**
2 Skilled Workers	$39.45	$631.20	$65.55	$1048.80	$36.42	$60.42
1 Laborer	30.35	242.80	50.15	401.20		
1 Small Work Boat, Gas, 50 H.P.		118.85		130.74		
1 Pressure Washer, 7 GPM		77.40		85.14	8.18	8.99
24 L.H., Daily Totals		$1070.25		$1665.88	$44.59	$69.41
Crew B-1G	**Hr.**	**Daily**	**Hr.**	**Daily**	**Bare Costs**	**Incl. O&P**
2 Laborers	$30.35	$485.60	$50.15	$802.40	$30.35	$50.15
1 Small Work Boat, Gas, 50 H.P.		118.85		130.74	7.43	8.17
16 L.H., Daily Totals		$604.45		$933.13	$37.78	$58.32
Crew B-1H	**Hr.**	**Daily**	**Hr.**	**Daily**	**Bare Costs**	**Incl. O&P**
2 Skilled Workers	$39.45	$631.20	$65.55	$1048.80	$36.42	$60.42
1 Laborer	30.35	242.80	50.15	401.20		
1 Small Work Boat, Gas, 50 H.P.		118.85		130.74	4.95	5.45
24 L.H., Daily Totals		$992.85		$1580.73	$41.37	$65.86
Crew B-1J	**Hr.**	**Daily**	**Hr.**	**Daily**	**Bare Costs**	**Incl. O&P**
1 Labor Foreman (inside)	$30.85	$246.80	$51.00	$408.00	$30.60	$50.58
1 Laborer	30.35	242.80	50.15	401.20		
16 L.H., Daily Totals		$489.60		$809.20	$30.60	$50.58
Crew B-1K	**Hr.**	**Daily**	**Hr.**	**Daily**	**Bare Costs**	**Incl. O&P**
1 Carpenter Foreman (inside)	$39.25	$314.00	$64.85	$518.80	$39.00	$64.45
1 Carpenter	38.75	310.00	64.05	512.40		
16 L.H., Daily Totals		$624.00		$1031.20	$39.00	$64.45
Crew B-2	**Hr.**	**Daily**	**Hr.**	**Daily**	**Bare Costs**	**Incl. O&P**
1 Labor Foreman (outside)	$32.35	$258.80	$53.45	$427.60	$30.75	$50.81
4 Laborers	30.35	971.20	50.15	1604.80		
40 L.H., Daily Totals		$1230.00		$2032.40	$30.75	$50.81
Crew B-2A	**Hr.**	**Daily**	**Hr.**	**Daily**	**Bare Costs**	**Incl. O&P**
1 Labor Foreman (outside)	$32.35	$258.80	$53.45	$427.60	$31.02	$51.25
2 Laborers	30.35	485.60	50.15	802.40		
1 Telescoping Boom Lift, to 60'		451.25		496.38	18.80	20.68
24 L.H., Daily Totals		$1195.65		$1726.38	$49.82	$71.93
Crew B-3	**Hr.**	**Daily**	**Hr.**	**Daily**	**Bare Costs**	**Incl. O&P**
1 Labor Foreman (outside)	$32.35	$258.80	$53.45	$427.60	$33.90	$55.85
2 Laborers	30.35	485.60	50.15	802.40		
1 Equip. Oper. (medium)	40.75	326.00	66.85	534.80		
2 Truck Drivers (heavy)	34.80	556.80	57.25	916.00		
1 Crawler Loader, 3 C.Y.		1202.00		1322.20		
2 Dump Trucks, 12 C.Y., 400 H.P.		1134.10		1247.51	48.67	53.54
48 L.H., Daily Totals		$3963.30		$5250.51	$82.57	$109.39
Crew B-3A	**Hr.**	**Daily**	**Hr.**	**Daily**	**Bare Costs**	**Incl. O&P**
4 Laborers	$30.35	$971.20	$50.15	$1604.80	$32.43	$53.49
1 Equip. Oper. (medium)	40.75	326.00	66.85	534.80		
1 Hyd. Excavator, 1.5 C.Y.		908.70		999.57	22.72	24.99
40 L.H., Daily Totals		$2205.90		$3139.17	$55.15	$78.48

Crew B-3B	Bare Costs Hr.	Bare Costs Daily	Incl. Subs O&P Hr.	Incl. Subs O&P Daily	Cost Per Labor-Hour Bare Costs	Cost Per Labor-Hour Incl. O&P
2 Laborers	$30.35	$485.60	$50.15	$802.40	$34.06	$56.10
1 Equip. Oper. (medium)	40.75	326.00	66.85	534.80		
1 Truck Driver (heavy)	34.80	278.40	57.25	458.00		
1 Backhoe Loader, 80 H.P.		387.85		426.63		
1 Dump Truck, 12 C.Y., 400 H.P.		567.05		623.76	29.84	32.82
32 L.H., Daily Totals		$2044.90		$2845.59	$63.90	$88.92

Crew B-3C	Hr.	Daily	Hr.	Daily	Bare Costs	Incl. O&P
3 Laborers	$30.35	$728.40	$50.15	$1203.60	$32.95	$54.33
1 Equip. Oper. (medium)	40.75	326.00	66.85	534.80		
1 Crawler Loader, 4 C.Y.		1461.00		1607.10	45.66	50.22
32 L.H., Daily Totals		$2515.40		$3345.50	$78.61	$104.55

Crew B-4	Hr.	Daily	Hr.	Daily	Bare Costs	Incl. O&P
1 Labor Foreman (outside)	$32.35	$258.80	$53.45	$427.60	$31.43	$51.88
4 Laborers	30.35	971.20	50.15	1604.80		
1 Truck Driver (heavy)	34.80	278.40	57.25	458.00		
1 Truck Tractor, 220 H.P.		303.55		333.90		
1 Flatbed Trailer, 40 Ton		182.50		200.75	10.13	11.14
48 L.H., Daily Totals		$1994.45		$3025.05	$41.55	$63.02

Crew B-5	Hr.	Daily	Hr.	Daily	Bare Costs	Incl. O&P
1 Labor Foreman (outside)	$32.35	$258.80	$53.45	$427.60	$32.83	$54.15
3 Laborers	30.35	728.40	50.15	1203.60		
1 Equip. Oper. (medium)	40.75	326.00	66.85	534.80		
1 Air Compressor, 250 cfm		167.95		184.75		
2 Breakers, Pavement, 60 lb.		21.50		23.65		
2 -50' Air Hoses, 1.5"		45.50		50.05		
1 Crawler Loader, 3 C.Y.		1202.00		1322.20	35.92	39.52
40 L.H., Daily Totals		$2750.15		$3746.65	$68.75	$93.67

Crew B-5A	Hr.	Daily	Hr.	Daily	Bare Costs	Incl. O&P
1 Labor Foreman (outside)	$32.35	$258.80	$53.45	$427.60	$33.69	$55.51
6 Laborers	30.35	1456.80	50.15	2407.20		
2 Equip. Oper. (medium)	40.75	652.00	66.85	1069.60		
1 Equip. Oper. (light)	38.75	310.00	63.60	508.80		
2 Truck Drivers (heavy)	34.80	556.80	57.25	916.00		
1 Air Compressor, 365 cfm		213.35		234.69		
2 Breakers, Pavement, 60 lb.		21.50		23.65		
8 -50' Air Hoses, 1"		67.20		73.92		
2 Dump Trucks, 8 C.Y., 220 H.P.		679.20		747.12	10.22	11.24
96 L.H., Daily Totals		$4215.65		$6408.57	$43.91	$66.76

Crew B-5B	Hr.	Daily	Hr.	Daily	Bare Costs	Incl. O&P
1 Powderman	$39.45	$315.60	$65.55	$524.40	$37.56	$61.83
2 Equip. Oper. (medium)	40.75	652.00	66.85	1069.60		
3 Truck Drivers (heavy)	34.80	835.20	57.25	1374.00		
1 F.E. Loader, W.M., 2.5 C.Y.		531.20		584.32		
3 Dump Trucks, 12 C.Y., 400 H.P.		1701.15		1871.27		
1 Air Compressor, 365 cfm		213.35		234.69	50.95	56.05
48 L.H., Daily Totals		$4248.50		$5658.27	$88.51	$117.88

Crew B-5C	Hr.	Daily	Hr.	Daily	Bare Costs	Incl. O&P
3 Laborers	$30.35	$728.40	$50.15	$1203.60	$34.91	$57.44
1 Equip. Oper. (medium)	40.75	326.00	66.85	534.80		
2 Truck Drivers (heavy)	34.80	556.80	57.25	916.00		
1 Equip. Oper. (crane)	41.95	335.60	68.85	550.80		
1 Equip. Oper. (oiler)	35.90	287.20	58.90	471.20		
2 Dump Trucks, 12 C.Y., 400 H.P.		1134.10		1247.51		
1 Crawler Loader, 4 C.Y.		1461.00		1607.10		
1 S.P. Crane, 4x4, 25 Ton		664.40		730.84	50.93	56.02
64 L.H., Daily Totals		$5493.50		$7261.85	$85.84	$113.47

Crew B-5D	Hr.	Daily	Hr.	Daily	Bare Costs	Incl. O&P
1 Labor Foreman (outside)	$32.35	$258.80	$53.45	$427.60	$33.16	$54.67
3 Laborers	30.35	728.40	50.15	1203.60		
1 Equip. Oper. (medium)	40.75	326.00	66.85	534.80		
1 Truck Driver (heavy)	34.80	278.40	57.25	458.00		
1 Air Compressor, 250 cfm		167.95		184.75		
2 Breakers, Pavement, 60 lb.		21.50		23.65		
2 -50' Air Hoses, 1.5"		45.50		50.05		
1 Crawler Loader, 3 C.Y.		1202.00		1322.20		
1 Dump Truck, 12 C.Y., 400 H.P.		567.05		623.76	41.75	45.92
48 L.H., Daily Totals		$3595.60		$4828.40	$74.91	$100.59

Crew B-6	Hr.	Daily	Hr.	Daily	Bare Costs	Incl. O&P
2 Laborers	$30.35	$485.60	$50.15	$802.40	$33.15	$54.63
1 Equip. Oper. (light)	38.75	310.00	63.60	508.80		
1 Backhoe Loader, 48 H.P.		320.20		352.22	13.34	14.68
24 L.H., Daily Totals		$1115.80		$1663.42	$46.49	$69.31

Crew B-6A	Hr.	Daily	Hr.	Daily	Bare Costs	Incl. O&P
.5 Labor Foreman (outside)	$32.35	$129.40	$53.45	$213.80	$34.91	$57.49
1 Laborer	30.35	242.80	50.15	401.20		
1 Equip. Oper. (medium)	40.75	326.00	66.85	534.80		
1 Vacuum Truck, 5000 Gal.		359.50		395.45	17.98	19.77
20 L.H., Daily Totals		$1057.70		$1545.25	$52.88	$77.26

Crew B-6B	Hr.	Daily	Hr.	Daily	Bare Costs	Incl. O&P
2 Labor Foremen (outside)	$32.35	$517.60	$53.45	$855.20	$31.02	$51.25
4 Laborers	30.35	971.20	50.15	1604.80		
1 S.P. Crane, 4x4, 5 Ton		259.45		285.39		
1 Flatbed Truck, Gas, 1.5 Ton		195.20		214.72		
1 Butt Fusion Mach., 4"-12" diam.		405.15		445.67	17.91	19.70
48 L.H., Daily Totals		$2348.60		$3405.78	$48.93	$70.95

Crew B-6C	Hr.	Daily	Hr.	Daily	Bare Costs	Incl. O&P
2 Labor Foremen (outside)	$32.35	$517.60	$53.45	$855.20	$31.02	$51.25
4 Laborers	30.35	971.20	50.15	1604.80		
1 S.P. Crane, 4x4, 12 Ton		371.85		409.04		
1 Flatbed Truck, Gas, 3 Ton		245.95		270.55		
1 Butt Fusion Mach., 8"-24" diam.		1927.00		2119.70	53.02	58.32
48 L.H., Daily Totals		$4033.60		$5259.28	$84.03	$109.57

Crew B-7	Hr.	Daily	Hr.	Daily	Bare Costs	Incl. O&P
1 Labor Foreman (outside)	$32.35	$258.80	$53.45	$427.60	$32.42	$53.48
4 Laborers	30.35	971.20	50.15	1604.80		
1 Equip. Oper. (medium)	40.75	326.00	66.85	534.80		
1 Brush Chipper, 12", 130 H.P.		389.40		428.34		
1 Crawler Loader, 3 C.Y.		1202.00		1322.20		
2 Chain Saws, Gas, 36" Long		93.30		102.63	35.10	38.61
48 L.H., Daily Totals		$3240.70		$4420.37	$67.51	$92.09

Crew No.	Bare Costs		Incl. Subs O&P		Cost Per Labor-Hour	
Crew B-7A	**Hr.**	**Daily**	**Hr.**	**Daily**	**Bare Costs**	**Incl. O&P**
2 Laborers	$30.35	$485.60	$50.15	$802.40	$33.15	$54.63
1 Equip. Oper. (light)	38.75	310.00	63.60	508.80		
1 Rake w/Tractor		332.75		366.02		
2 Chain Saws, Gas, 18"		55.10		60.61	16.16	17.78
24 L.H., Daily Totals		$1183.45		$1737.84	$49.31	$72.41
Crew B-7B	**Hr.**	**Daily**	**Hr.**	**Daily**	**Bare Costs**	**Incl. O&P**
1 Labor Foreman (outside)	$32.35	$258.80	$53.45	$427.60	$32.76	$54.02
4 Laborers	30.35	971.20	50.15	1604.80		
1 Equip. Oper. (medium)	40.75	326.00	66.85	534.80		
1 Truck Driver (heavy)	34.80	278.40	57.25	458.00		
1 Brush Chipper, 12", 130 H.P.		389.40		428.34		
1 Crawler Loader, 3 C.Y.		1202.00		1322.20		
2 Chain Saws, Gas, 36" Long		93.30		102.63		
1 Dump Truck, 8 C.Y., 220 H.P.		339.60		373.56	36.15	39.76
56 L.H., Daily Totals		$3858.70		$5251.93	$68.91	$93.78
Crew B-7C	**Hr.**	**Daily**	**Hr.**	**Daily**	**Bare Costs**	**Incl. O&P**
1 Labor Foreman (outside)	$32.35	$258.80	$53.45	$427.60	$32.76	$54.02
4 Laborers	30.35	971.20	50.15	1604.80		
1 Equip. Oper. (medium)	40.75	326.00	66.85	534.80		
1 Truck Driver (heavy)	34.80	278.40	57.25	458.00		
1 Brush Chipper, 12", 130 H.P.		389.40		428.34		
1 Crawler Loader, 3 C.Y.		1202.00		1322.20		
2 Chain Saws, Gas, 36" Long		93.30		102.63		
1 Dump Truck, 12 C.Y., 400 H.P.		567.05		623.76	40.21	44.23
56 L.H., Daily Totals		$4086.15		$5502.13	$72.97	$98.25
Crew B-8	**Hr.**	**Daily**	**Hr.**	**Daily**	**Bare Costs**	**Incl. O&P**
1 Labor Foreman (outside)	$32.35	$258.80	$53.45	$427.60	$34.88	$57.42
2 Laborers	30.35	485.60	50.15	802.40		
2 Equip. Oper. (medium)	40.75	652.00	66.85	1069.60		
2 Truck Drivers (heavy)	34.80	556.80	57.25	916.00		
1 Hyd. Crane, 25 Ton		581.70		639.87		
1 Crawler Loader, 3 C.Y.		1202.00		1322.20		
2 Dump Trucks, 12 C.Y., 400 H.P.		1134.10		1247.51	52.10	57.31
56 L.H., Daily Totals		$4871.00		$6425.18	$86.98	$114.74
Crew B-9	**Hr.**	**Daily**	**Hr.**	**Daily**	**Bare Costs**	**Incl. O&P**
1 Labor Foreman (outside)	$32.35	$258.80	$53.45	$427.60	$30.75	$50.81
4 Laborers	30.35	971.20	50.15	1604.80		
1 Air Compressor, 250 cfm		167.95		184.75		
2 Breakers, Pavement, 60 lb.		21.50		23.65		
2 -50' Air Hoses, 1.5"		45.50		50.05	5.87	6.46
40 L.H., Daily Totals		$1464.95		$2290.84	$36.62	$57.27
Crew B-9A	**Hr.**	**Daily**	**Hr.**	**Daily**	**Bare Costs**	**Incl. O&P**
2 Laborers	$30.35	$485.60	$50.15	$802.40	$31.83	$52.52
1 Truck Driver (heavy)	34.80	278.40	57.25	458.00		
1 Water Tank Trailer, 5000 Gal.		149.45		164.40		
1 Truck Tractor, 220 H.P.		303.55		333.90		
2 -50' Discharge Hoses, 3"		2.80		3.08	18.99	20.89
24 L.H., Daily Totals		$1219.80		$1761.78	$50.83	$73.41

Crew No.	Bare Costs		Incl. Subs O&P		Cost Per Labor-Hour	
Crew B-9B	**Hr.**	**Daily**	**Hr.**	**Daily**	**Bare Costs**	**Incl. O&P**
2 Laborers	$30.35	$485.60	$50.15	$802.40	$31.83	$52.52
1 Truck Driver (heavy)	34.80	278.40	57.25	458.00		
2 -50' Discharge Hoses, 3"		2.80		3.08		
1 Water Tank Trailer, 5000 Gal.		149.45		164.40		
1 Truck Tractor, 220 H.P.		303.55		333.90		
1 Pressure Washer		63.05		69.36	21.62	23.78
24 L.H., Daily Totals		$1282.85		$1831.14	$53.45	$76.30
Crew B-9D	**Hr.**	**Daily**	**Hr.**	**Daily**	**Bare Costs**	**Incl. O&P**
1 Labor Foreman (outside)	$32.35	$258.80	$53.45	$427.60	$30.75	$50.81
4 Common Laborers	30.35	971.20	50.15	1604.80		
1 Air Compressor, 250 cfm		167.95		184.75		
2 -50' Air Hoses, 1.5"		45.50		50.05		
2 Air Powered Tampers		57.10		62.81	6.76	7.44
40 L.H., Daily Totals		$1500.55		$2330.01	$37.51	$58.25
Crew B-9E	**Hr.**	**Daily**	**Hr.**	**Daily**	**Bare Costs**	**Incl. O&P**
1 Cement Finisher	$37.15	$297.20	$60.05	$480.40	$33.75	$55.10
1 Laborer	30.35	242.80	50.15	401.20		
1 Chip. Hammers, 12 Lb., Elec.		21.00		23.10	1.31	1.44
16 L.H., Daily Totals		$561.00		$904.70	$35.06	$56.54
Crew B-10	**Hr.**	**Daily**	**Hr.**	**Daily**	**Bare Costs**	**Incl. O&P**
1 Equip. Oper. (medium)	40.75	326.00	66.85	534.80	40.75	66.85
8 L.H., Daily Totals		$326.00		$534.80	$40.75	$66.85
Crew B-10A	**Hr.**	**Daily**	**Hr.**	**Daily**	**Bare Costs**	**Incl. O&P**
1 Equip. Oper. (medium)	$40.75	$326.00	$66.85	$534.80	$40.75	$66.85
1 Roller, 2-Drum, W.B., 7.5 H.P.		181.75		199.93	22.72	24.99
8 L.H., Daily Totals		$507.75		$734.73	$63.47	$91.84
Crew B-10B	**Hr.**	**Daily**	**Hr.**	**Daily**	**Bare Costs**	**Incl. O&P**
1 Equip. Oper. (medium)	$40.75	$326.00	$66.85	$534.80	$40.75	$66.85
1 Dozer, 200 H.P.		1290.00		1419.00	161.25	177.38
8 L.H., Daily Totals		$1616.00		$1953.80	$202.00	$244.22
Crew B-10C	**Hr.**	**Daily**	**Hr.**	**Daily**	**Bare Costs**	**Incl. O&P**
1 Equip. Oper. (medium)	$40.75	$326.00	$66.85	$534.80	$40.75	$66.85
1 Dozer, 200 H.P.		1290.00		1419.00		
1 Vibratory Roller, Towed, 23 Ton		422.15		464.37	214.02	235.42
8 L.H., Daily Totals		$2038.15		$2418.17	$254.77	$302.27
Crew B-10D	**Hr.**	**Daily**	**Hr.**	**Daily**	**Bare Costs**	**Incl. O&P**
1 Equip. Oper. (medium)	$40.75	$326.00	$66.85	$534.80	$40.75	$66.85
1 Dozer, 200 H.P.		1290.00		1419.00		
1 Sheepsft. Roller, Towed		430.15		473.17	215.02	236.52
8 L.H., Daily Totals		$2046.15		$2426.97	$255.77	$303.37
Crew B-10E	**Hr.**	**Daily**	**Hr.**	**Daily**	**Bare Costs**	**Incl. O&P**
1 Equip. Oper. (medium)	$40.75	$326.00	$66.85	$534.80	$40.75	$66.85
1 Tandem Roller, 5 Ton		152.45		167.69	19.06	20.96
8 L.H., Daily Totals		$478.45		$702.50	$59.81	$87.81
Crew B-10F	**Hr.**	**Daily**	**Hr.**	**Daily**	**Bare Costs**	**Incl. O&P**
1 Equip. Oper. (medium)	$40.75	$326.00	$66.85	$534.80	$40.75	$66.85
1 Tandem Roller, 10 Ton		232.05		255.26	29.01	31.91
8 L.H., Daily Totals		$558.05		$790.05	$69.76	$98.76

Crew No.	Bare Costs		Incl. Subs O&P		Cost Per Labor-Hour	
Crew B-10G	**Hr.**	**Daily**	**Hr.**	**Daily**	**Bare Costs**	**Incl. O&P**
1 Equip. Oper. (medium)	$40.75	$326.00	$66.85	$534.80	$40.75	$66.85
1 Sheepsfoot Roller, 240 H.P.		1323.00		1455.30	165.38	181.91
8 L.H., Daily Totals		$1649.00		$1990.10	$206.13	$248.76
Crew B-10H	**Hr.**	**Daily**	**Hr.**	**Daily**	**Bare Costs**	**Incl. O&P**
1 Equip. Oper. (medium)	$40.75	$326.00	$66.85	$534.80	$40.75	$66.85
1 Diaphragm Water Pump, 2"		74.30		81.73		
1 -20' Suction Hose, 2"		2.05		2.25		
2 -50' Discharge Hoses, 2"		1.80		1.98	9.77	10.75
8 L.H., Daily Totals		$404.15		$620.76	$50.52	$77.60
Crew B-10I	**Hr.**	**Daily**	**Hr.**	**Daily**	**Bare Costs**	**Incl. O&P**
1 Equip. Oper. (medium)	$40.75	$326.00	$66.85	$534.80	$40.75	$66.85
1 Diaphragm Water Pump, 4"		141.40		155.54		
1 -20' Suction Hose, 4"		3.10		3.41		
2 -50' Discharge Hoses, 4"		4.50		4.95	18.63	20.49
8 L.H., Daily Totals		$475.00		$698.70	$59.38	$87.34
Crew B-10J	**Hr.**	**Daily**	**Hr.**	**Daily**	**Bare Costs**	**Incl. O&P**
1 Equip. Oper. (medium)	$40.75	$326.00	$66.85	$534.80	$40.75	$66.85
1 Centrifugal Water Pump, 3"		79.25		87.17		
1 -20' Suction Hose, 3"		2.70		2.97		
2 -50' Discharge Hoses, 3"		2.80		3.08	10.59	11.65
8 L.H., Daily Totals		$410.75		$628.02	$51.34	$78.50
Crew B-10K	**Hr.**	**Daily**	**Hr.**	**Daily**	**Bare Costs**	**Incl. O&P**
1 Equip. Oper. (medium)	$40.75	$326.00	$66.85	$534.80	$40.75	$66.85
1 Centr. Water Pump, 6"		296.45		326.10		
1 -20' Suction Hose, 6"		11.10		12.21		
2 -50' Discharge Hoses, 6"		11.40		12.54	39.87	43.86
8 L.H., Daily Totals		$644.95		$885.64	$80.62	$110.71
Crew B-10L	**Hr.**	**Daily**	**Hr.**	**Daily**	**Bare Costs**	**Incl. O&P**
1 Equip. Oper. (medium)	$40.75	$326.00	$66.85	$534.80	$40.75	$66.85
1 Dozer, 80 H.P.		472.00		519.20	59.00	64.90
8 L.H., Daily Totals		$798.00		$1054.00	$99.75	$131.75
Crew B-10M	**Hr.**	**Daily**	**Hr.**	**Daily**	**Bare Costs**	**Incl. O&P**
1 Equip. Oper. (medium)	$40.75	$326.00	$66.85	$534.80	$40.75	$66.85
1 Dozer, 300 H.P.		1831.00		2014.10	228.88	251.76
8 L.H., Daily Totals		$2157.00		$2548.90	$269.63	$318.61
Crew B-10N	**Hr.**	**Daily**	**Hr.**	**Daily**	**Bare Costs**	**Incl. O&P**
1 Equip. Oper. (medium)	$40.75	$326.00	$66.85	$534.80	$40.75	$66.85
1 F.E. Loader, T.M., 1.5 C.Y.		591.30		650.43	73.91	81.30
8 L.H., Daily Totals		$917.30		$1185.23	$114.66	$148.15
Crew B-10O	**Hr.**	**Daily**	**Hr.**	**Daily**	**Bare Costs**	**Incl. O&P**
1 Equip. Oper. (medium)	$40.75	$326.00	$66.85	$534.80	$40.75	$66.85
1 F.E. Loader, T.M., 2.25 C.Y.		1002.00		1102.20	125.25	137.78
8 L.H., Daily Totals		$1328.00		$1637.00	$166.00	$204.63
Crew B-10P	**Hr.**	**Daily**	**Hr.**	**Daily**	**Bare Costs**	**Incl. O&P**
1 Equip. Oper. (medium)	$40.75	$326.00	$66.85	$534.80	$40.75	$66.85
1 Crawler Loader, 3 C.Y.		1202.00		1322.20	150.25	165.28
8 L.H., Daily Totals		$1528.00		$1857.00	$191.00	$232.13

Crew No.	Bare Costs		Incl. Subs O&P		Cost Per Labor-Hour	
Crew B-10Q	**Hr.**	**Daily**	**Hr.**	**Daily**	**Bare Costs**	**Incl. O&P**
1 Equip. Oper. (medium)	$40.75	$326.00	$66.85	$534.80	$40.75	$66.85
1 Crawler Loader, 4 C.Y.		1461.00		1607.10	182.63	200.89
8 L.H., Daily Totals		$1787.00		$2141.90	$223.38	$267.74
Crew B-10R	**Hr.**	**Daily**	**Hr.**	**Daily**	**Bare Costs**	**Incl. O&P**
1 Equip. Oper. (medium)	$40.75	$326.00	$66.85	$534.80	$40.75	$66.85
1 F.E. Loader, W.M., 1 C.Y.		297.00		326.70	37.13	40.84
8 L.H., Daily Totals		$623.00		$861.50	$77.88	$107.69
Crew B-10S	**Hr.**	**Daily**	**Hr.**	**Daily**	**Bare Costs**	**Incl. O&P**
1 Equip. Oper. (medium)	$40.75	$326.00	$66.85	$534.80	$40.75	$66.85
1 F.E. Loader, W.M., 1.5 C.Y.		352.95		388.25	44.12	48.53
8 L.H., Daily Totals		$678.95		$923.04	$84.87	$115.38
Crew B-10T	**Hr.**	**Daily**	**Hr.**	**Daily**	**Bare Costs**	**Incl. O&P**
1 Equip. Oper. (medium)	$40.75	$326.00	$66.85	$534.80	$40.75	$66.85
1 F.E. Loader, W.M., 2.5 C.Y.		531.20		584.32	66.40	73.04
8 L.H., Daily Totals		$857.20		$1119.12	$107.15	$139.89
Crew B-10U	**Hr.**	**Daily**	**Hr.**	**Daily**	**Bare Costs**	**Incl. O&P**
1 Equip. Oper. (medium)	$40.75	$326.00	$66.85	$534.80	$40.75	$66.85
1 F.E. Loader, W.M., 5.5 C.Y.		975.55		1073.11	121.94	134.14
8 L.H., Daily Totals		$1301.55		$1607.91	$162.69	$200.99
Crew B-10V	**Hr.**	**Daily**	**Hr.**	**Daily**	**Bare Costs**	**Incl. O&P**
1 Equip. Oper. (medium)	$40.75	$326.00	$66.85	$534.80	$40.75	$66.85
1 Dozer, 700 H.P.		5021.00		5523.10	627.63	690.39
8 L.H., Daily Totals		$5347.00		$6057.90	$668.38	$757.24
Crew B-10W	**Hr.**	**Daily**	**Hr.**	**Daily**	**Bare Costs**	**Incl. O&P**
1 Equip. Oper. (medium)	$40.75	$326.00	$66.85	$534.80	$40.75	$66.85
1 Dozer, 105 H.P.		609.70		670.67	76.21	83.83
8 L.H., Daily Totals		$935.70		$1205.47	$116.96	$150.68
Crew B-10X	**Hr.**	**Daily**	**Hr.**	**Daily**	**Bare Costs**	**Incl. O&P**
1 Equip. Oper. (medium)	$40.75	$326.00	$66.85	$534.80	$40.75	$66.85
1 Dozer, 410 H.P.		2269.00		2495.90	283.63	311.99
8 L.H., Daily Totals		$2595.00		$3030.70	$324.38	$378.84
Crew B-10Y	**Hr.**	**Daily**	**Hr.**	**Daily**	**Bare Costs**	**Incl. O&P**
1 Equip. Oper. (medium)	$40.75	$326.00	$66.85	$534.80	$40.75	$66.85
1 Vibr. Roller, Towed, 12 Ton		566.70		623.37	70.84	77.92
8 L.H., Daily Totals		$892.70		$1158.17	$111.59	$144.77
Crew B-11A	**Hr.**	**Daily**	**Hr.**	**Daily**	**Bare Costs**	**Incl. O&P**
1 Equipment Oper. (med.)	$40.75	$326.00	$66.85	$534.80	$35.55	$58.50
1 Laborer	30.35	242.80	50.15	401.20		
1 Dozer, 200 H.P.		1290.00		1419.00	80.63	88.69
16 L.H., Daily Totals		$1858.80		$2355.00	$116.18	$147.19
Crew B-11B	**Hr.**	**Daily**	**Hr.**	**Daily**	**Bare Costs**	**Incl. O&P**
1 Equipment Oper. (light)	$38.75	$310.00	$63.60	$508.80	$34.55	$56.88
1 Laborer	30.35	242.80	50.15	401.20		
1 Air Powered Tamper		28.55		31.41		
1 Air Compressor, 365 cfm		213.35		234.69		
2 -50' Air Hoses, 1.5"		45.50		50.05	17.96	19.76
16 L.H., Daily Totals		$840.20		$1226.14	$52.51	$76.63

Crew No.	Bare Costs		Incl. Subs O&P		Cost Per Labor-Hour	
Crew B-11C	**Hr.**	**Daily**	**Hr.**	**Daily**	**Bare Costs**	**Incl. O&P**
1 Equipment Oper. (med.)	$40.75	$326.00	$66.85	$534.80	$35.55	$58.50
1 Laborer	30.35	242.80	50.15	401.20		
1 Backhoe Loader, 48 H.P.		320.20		352.22	20.01	22.01
16 L.H., Daily Totals		$889.00		$1288.22	$55.56	$80.51

Crew B-11K	Hr.	Daily	Hr.	Daily	Bare Costs	Incl. O&P
1 Equipment Oper. (med.)	$40.75	$326.00	$66.85	$534.80	$35.55	$58.50
1 Laborer	30.35	242.80	50.15	401.20		
1 Trencher, Chain Type, 8' D		1837.00		2020.70	114.81	126.29
16 L.H., Daily Totals		$2405.80		$2956.70	$150.36	$184.79

Crew B-11L	Hr.	Daily	Hr.	Daily	Bare Costs	Incl. O&P
1 Equipment Oper. (med.)	$40.75	$326.00	$66.85	$534.80	$35.55	$58.50
1 Laborer	30.35	242.80	50.15	401.20		
1 Grader, 30,000 Lbs.		657.85		723.63	41.12	45.23
16 L.H., Daily Totals		$1226.65		$1659.64	$76.67	$103.73

Crew B-11M	Hr.	Daily	Hr.	Daily	Bare Costs	Incl. O&P
1 Equipment Oper. (med.)	$40.75	$326.00	$66.85	$534.80	$35.55	$58.50
1 Laborer	30.35	242.80	50.15	401.20		
1 Backhoe Loader, 80 H.P.		387.85		426.63	24.24	26.66
16 L.H., Daily Totals		$956.65		$1362.64	$59.79	$85.16

Crew B-11W	Hr.	Daily	Hr.	Daily	Bare Costs	Incl. O&P
1 Equipment Operator (med.)	$40.75	$326.00	$66.85	$534.80	$34.92	$57.46
1 Common Laborer	30.35	242.80	50.15	401.20		
10 Truck Drivers (heavy)	34.80	2784.00	57.25	4580.00		
1 Dozer, 200 H.P.		1290.00		1419.00		
1 Vibratory Roller, Towed, 23 Ton		422.15		464.37		
10 Dump Trucks, 8 C.Y., 220 H.P.		3396.00		3735.60	53.21	58.53
96 L.H., Daily Totals		$8460.95		$11134.97	$88.13	$115.99

Crew B-11Y	Hr.	Daily	Hr.	Daily	Bare Costs	Incl. O&P
1 Labor Foreman (outside)	$32.35	$258.80	$53.45	$427.60	$34.04	$56.08
5 Common Laborers	30.35	1214.00	50.15	2006.00		
3 Equipment Operators (med.)	40.75	978.00	66.85	1604.40		
1 Dozer, 80 H.P.		472.00		519.20		
2 Rollers, 2-Drum, W.B., 7.5 H.P.		363.50		399.85		
4 Vibrating Plates, Gas, 21"		162.20		178.42	13.86	15.24
72 L.H., Daily Totals		$3448.50		$5135.47	$47.90	$71.33

Crew B-12A	Hr.	Daily	Hr.	Daily	Bare Costs	Incl. O&P
1 Equip. Oper. (crane)	$41.95	$335.60	$68.85	$550.80	$36.15	$59.50
1 Laborer	30.35	242.80	50.15	401.20		
1 Hyd. Excavator, 1 C.Y.		749.70		824.67	46.86	51.54
16 L.H., Daily Totals		$1328.10		$1776.67	$83.01	$111.04

Crew B-12B	Hr.	Daily	Hr.	Daily	Bare Costs	Incl. O&P
1 Equip. Oper. (crane)	$41.95	$335.60	$68.85	$550.80	$36.15	$59.50
1 Laborer	30.35	242.80	50.15	401.20		
1 Hyd. Excavator, 1.5 C.Y.		908.70		999.57	56.79	62.47
16 L.H., Daily Totals		$1487.10		$1951.57	$92.94	$121.97

Crew B-12C	Hr.	Daily	Hr.	Daily	Bare Costs	Incl. O&P
1 Equip. Oper. (crane)	$41.95	$335.60	$68.85	$550.80	$36.15	$59.50
1 Laborer	30.35	242.80	50.15	401.20		
1 Hyd. Excavator, 2 C.Y.		1078.00		1185.80	67.38	74.11
16 L.H., Daily Totals		$1656.40		$2137.80	$103.53	$133.61

Crew B-12D	Hr.	Daily	Hr.	Daily	Bare Costs	Incl. O&P
1 Equip. Oper. (crane)	$41.95	$335.60	$68.85	$550.80	$36.15	$59.50
1 Laborer	30.35	242.80	50.15	401.20		
1 Hyd. Excavator, 3.5 C.Y.		2256.00		2481.60	141.00	155.10
16 L.H., Daily Totals		$2834.40		$3433.60	$177.15	$214.60

Crew B-12E	Hr.	Daily	Hr.	Daily	Bare Costs	Incl. O&P
1 Equip. Oper. (crane)	$41.95	$335.60	$68.85	$550.80	$36.15	$59.50
1 Laborer	30.35	242.80	50.15	401.20		
1 Hyd. Excavator, .5 C.Y.		443.75		488.13	27.73	30.51
16 L.H., Daily Totals		$1022.15		$1440.13	$63.88	$90.01

Crew B-12F	Hr.	Daily	Hr.	Daily	Bare Costs	Incl. O&P
1 Equip. Oper. (crane)	$41.95	$335.60	$68.85	$550.80	$36.15	$59.50
1 Laborer	30.35	242.80	50.15	401.20		
1 Hyd. Excavator, .75 C.Y.		681.25		749.38	42.58	46.84
16 L.H., Daily Totals		$1259.65		$1701.38	$78.73	$106.34

Crew B-12G	Hr.	Daily	Hr.	Daily	Bare Costs	Incl. O&P
1 Equip. Oper. (crane)	$41.95	$335.60	$68.85	$550.80	$36.15	$59.50
1 Laborer	30.35	242.80	50.15	401.20		
1 Crawler Crane, 15 Ton		866.60		953.26		
1 Clamshell Bucket, .5 C.Y.		42.55		46.81	56.82	62.50
16 L.H., Daily Totals		$1487.55		$1952.07	$92.97	$122.00

Crew B-12H	Hr.	Daily	Hr.	Daily	Bare Costs	Incl. O&P
1 Equip. Oper. (crane)	$41.95	$335.60	$68.85	$550.80	$36.15	$59.50
1 Laborer	30.35	242.80	50.15	401.20		
1 Crawler Crane, 25 Ton		1386.00		1524.60		
1 Clamshell Bucket, 1 C.Y.		51.20		56.32	89.83	98.81
16 L.H., Daily Totals		$2015.60		$2532.92	$125.97	$158.31

Crew B-12I	Hr.	Daily	Hr.	Daily	Bare Costs	Incl. O&P
1 Equip. Oper. (crane)	$41.95	$335.60	$68.85	$550.80	$36.15	$59.50
1 Laborer	30.35	242.80	50.15	401.20		
1 Crawler Crane, 20 Ton		1130.00		1243.00		
1 Dragline Bucket, .75 C.Y.		21.85		24.04	71.99	79.19
16 L.H., Daily Totals		$1730.25		$2219.03	$108.14	$138.69

Crew B-12J	Hr.	Daily	Hr.	Daily	Bare Costs	Incl. O&P
1 Equip. Oper. (crane)	$41.95	$335.60	$68.85	$550.80	$36.15	$59.50
1 Laborer	30.35	242.80	50.15	401.20		
1 Gradall, 5/8 C.Y.		862.55		948.80	53.91	59.30
16 L.H., Daily Totals		$1440.95		$1900.81	$90.06	$118.80

Crew B-12K	Hr.	Daily	Hr.	Daily	Bare Costs	Incl. O&P
1 Equip. Oper. (crane)	$41.95	$335.60	$68.85	$550.80	$36.15	$59.50
1 Laborer	30.35	242.80	50.15	401.20		
1 Gradall, 3 Ton, 1 C.Y.		1185.00		1303.50	74.06	81.47
16 L.H., Daily Totals		$1763.40		$2255.50	$110.21	$140.97

Crew B-12L	Hr.	Daily	Hr.	Daily	Bare Costs	Incl. O&P
1 Equip. Oper. (crane)	$41.95	$335.60	$68.85	$550.80	$36.15	$59.50
1 Laborer	30.35	242.80	50.15	401.20		
1 Crawler Crane, 15 Ton		866.60		953.26		
1 F.E. Attachment, .5 C.Y.		63.95		70.34	58.16	63.98
16 L.H., Daily Totals		$1508.95		$1975.61	$94.31	$123.48

Crews - Open Shop

Crew No.	Bare Costs		Incl. Subs O&P		Cost Per Labor-Hour	
Crew B-12M	**Hr.**	**Daily**	**Hr.**	**Daily**	**Bare Costs**	**Incl. O&P**
1 Equip. Oper. (crane)	$41.95	$335.60	$68.85	$550.80	$36.15	$59.50
1 Laborer	30.35	242.80	50.15	401.20		
1 Crawler Crane, 20 Ton		1130.00		1243.00		
1 F.E. Attachment, .75 C.Y.		68.90		75.79	74.93	82.42
16 L.H., Daily Totals		$1777.30		$2270.79	$111.08	$141.92
Crew B-12N	**Hr.**	**Daily**	**Hr.**	**Daily**	**Bare Costs**	**Incl. O&P**
1 Equip. Oper. (crane)	$41.95	$335.60	$68.85	$550.80	$36.15	$59.50
1 Laborer	30.35	242.80	50.15	401.20		
1 Crawler Crane, 25 Ton		1386.00		1524.60		
1 F.E. Attachment, 1 C.Y.		74.85		82.33	91.30	100.43
16 L.H., Daily Totals		$2039.25		$2558.93	$127.45	$159.93
Crew B-12O	**Hr.**	**Daily**	**Hr.**	**Daily**	**Bare Costs**	**Incl. O&P**
1 Equip. Oper. (crane)	$41.95	$335.60	$68.85	$550.80	$36.15	$59.50
1 Laborer	30.35	242.80	50.15	401.20		
1 Crawler Crane, 40 Ton		1392.00		1531.20		
1 F.E. Attachment, 1.5 C.Y.		85.65		94.22	92.35	101.59
16 L.H., Daily Totals		$2056.05		$2577.42	$128.50	$161.09
Crew B-12P	**Hr.**	**Daily**	**Hr.**	**Daily**	**Bare Costs**	**Incl. O&P**
1 Equip. Oper. (crane)	$41.95	$335.60	$68.85	$550.80	$36.15	$59.50
1 Laborer	30.35	242.80	50.15	401.20		
1 Crawler Crane, 40 Ton		1392.00		1531.20		
1 Dragline Bucket, 1.5 C.Y.		35.10		38.61	89.19	98.11
16 L.H., Daily Totals		$2005.50		$2521.81	$125.34	$157.61
Crew B-12Q	**Hr.**	**Daily**	**Hr.**	**Daily**	**Bare Costs**	**Incl. O&P**
1 Equip. Oper. (crane)	$41.95	$335.60	$68.85	$550.80	$36.15	$59.50
1 Laborer	30.35	242.80	50.15	401.20		
1 Hyd. Excavator, 5/8 C.Y.		587.30		646.03	36.71	40.38
16 L.H., Daily Totals		$1165.70		$1598.03	$72.86	$99.88
Crew B-12S	**Hr.**	**Daily**	**Hr.**	**Daily**	**Bare Costs**	**Incl. O&P**
1 Equip. Oper. (crane)	$41.95	$335.60	$68.85	$550.80	$36.15	$59.50
1 Laborer	30.35	242.80	50.15	401.20		
1 Hyd. Excavator, 2.5 C.Y.		1441.00		1585.10	90.06	99.07
16 L.H., Daily Totals		$2019.40		$2537.10	$126.21	$158.57
Crew B-12T	**Hr.**	**Daily**	**Hr.**	**Daily**	**Bare Costs**	**Incl. O&P**
1 Equip. Oper. (crane)	$41.95	$335.60	$68.85	$550.80	$36.15	$59.50
1 Laborer	30.35	242.80	50.15	401.20		
1 Crawler Crane, 75 Ton		1734.00		1907.40		
1 F.E. Attachment, 3 C.Y.		111.50		122.65	115.34	126.88
16 L.H., Daily Totals		$2423.90		$2982.05	$151.49	$186.38
Crew B-12V	**Hr.**	**Daily**	**Hr.**	**Daily**	**Bare Costs**	**Incl. O&P**
1 Equip. Oper. (crane)	$41.95	$335.60	$68.85	$550.80	$36.15	$59.50
1 Laborer	30.35	242.80	50.15	401.20		
1 Crawler Crane, 75 Ton		1734.00		1907.40		
1 Dragline Bucket, 3 C.Y.		56.25		61.88	111.89	123.08
16 L.H., Daily Totals		$2368.65		$2921.28	$148.04	$182.58
Crew B-12Y	**Hr.**	**Daily**	**Hr.**	**Daily**	**Bare Costs**	**Incl. O&P**
1 Equip. Oper. (crane)	$41.95	$335.60	$68.85	$550.80	$34.22	$56.38
2 Laborers	30.35	485.60	50.15	802.40		
1 Hyd. Excavator, 3.5 C.Y.		2256.00		2481.60	94.00	103.40
24 L.H., Daily Totals		$3077.20		$3834.80	$128.22	$159.78

Crew No.	Bare Costs		Incl. Subs O&P		Cost Per Labor-Hour	
Crew B-12Z	**Hr.**	**Daily**	**Hr.**	**Daily**	**Bare Costs**	**Incl. O&P**
1 Equip. Oper. (crane)	$41.95	$335.60	$68.85	$550.80	$34.22	$56.38
2 Laborers	30.35	485.60	50.15	802.40		
1 Hyd. Excavator, 2.5 C.Y.		1441.00		1585.10	60.04	66.05
24 L.H., Daily Totals		$2262.20		$2938.30	$94.26	$122.43
Crew B-13	**Hr.**	**Daily**	**Hr.**	**Daily**	**Bare Costs**	**Incl. O&P**
1 Labor Foreman (outside)	$32.35	$258.80	$53.45	$427.60	$32.62	$53.82
4 Laborers	30.35	971.20	50.15	1604.80		
1 Equip. Oper. (crane)	41.95	335.60	68.85	550.80		
1 Hyd. Crane, 25 Ton		581.70		639.87	12.12	13.33
48 L.H., Daily Totals		$2147.30		$3223.07	$44.74	$67.15
Crew B-13A	**Hr.**	**Daily**	**Hr.**	**Daily**	**Bare Costs**	**Incl. O&P**
1 Labor Foreman (outside)	$32.35	$258.80	$53.45	$427.60	$34.88	$57.42
2 Laborers	30.35	485.60	50.15	802.40		
2 Equipment Operators (med.)	40.75	652.00	66.85	1069.60		
2 Truck Drivers (heavy)	34.80	556.80	57.25	916.00		
1 Crawler Crane, 75 Ton		1734.00		1907.40		
1 Crawler Loader, 4 C.Y.		1461.00		1607.10		
2 Dump Trucks, 8 C.Y., 220 H.P.		679.20		747.12	69.18	76.10
56 L.H., Daily Totals		$5827.40		$7477.22	$104.06	$133.52
Crew B-13B	**Hr.**	**Daily**	**Hr.**	**Daily**	**Bare Costs**	**Incl. O&P**
1 Labor Foreman (outside)	$32.35	$258.80	$53.45	$427.60	$33.09	$54.54
4 Laborers	30.35	971.20	50.15	1604.80		
1 Equip. Oper. (crane)	41.95	335.60	68.85	550.80		
1 Equip. Oper. (oiler)	35.90	287.20	58.90	471.20		
1 Hyd. Crane, 55 Ton		981.50		1079.65	17.53	19.28
56 L.H., Daily Totals		$2834.30		$4134.05	$50.61	$73.82
Crew B-13C	**Hr.**	**Daily**	**Hr.**	**Daily**	**Bare Costs**	**Incl. O&P**
1 Labor Foreman (outside)	$32.35	$258.80	$53.45	$427.60	$33.09	$54.54
4 Laborers	30.35	971.20	50.15	1604.80		
1 Equip. Oper. (crane)	41.95	335.60	68.85	550.80		
1 Equip. Oper. (oiler)	35.90	287.20	58.90	471.20		
1 Crawler Crane, 100 Ton		1879.00		2066.90	33.55	36.91
56 L.H., Daily Totals		$3731.80		$5121.30	$66.64	$91.45
Crew B-13D	**Hr.**	**Daily**	**Hr.**	**Daily**	**Bare Costs**	**Incl. O&P**
1 Laborer	$30.35	$242.80	$50.15	$401.20	$36.15	$59.50
1 Equip. Oper. (crane)	41.95	335.60	68.85	550.80		
1 Hyd. Excavator, 1 C.Y.		749.70		824.67		
1 Trench Box		77.75		85.53	51.72	56.89
16 L.H., Daily Totals		$1405.85		$1862.19	$87.87	$116.39
Crew B-13E	**Hr.**	**Daily**	**Hr.**	**Daily**	**Bare Costs**	**Incl. O&P**
1 Laborer	$30.35	$242.80	$50.15	$401.20	$36.15	$59.50
1 Equip. Oper. (crane)	41.95	335.60	68.85	550.80		
1 Hyd. Excavator, 1.5 C.Y.		908.70		999.57		
1 Trench Box		77.75		85.53	61.65	67.82
16 L.H., Daily Totals		$1564.85		$2037.10	$97.80	$127.32
Crew B-13F	**Hr.**	**Daily**	**Hr.**	**Daily**	**Bare Costs**	**Incl. O&P**
1 Laborer	$30.35	$242.80	$50.15	$401.20	$36.15	$59.50
1 Equip. Oper. (crane)	41.95	335.60	68.85	550.80		
1 Hyd. Excavator, 3.5 C.Y.		2256.00		2481.60		
1 Trench Box		77.75		85.53	145.86	160.45
16 L.H., Daily Totals		$2912.15		$3519.13	$182.01	$219.95

Crew No.	Bare Costs		Incl. Subs O&P		Cost Per Labor-Hour	
Crew B-13G	**Hr.**	**Daily**	**Hr.**	**Daily**	**Bare Costs**	**Incl. O&P**
1 Laborer	$30.35	$242.80	$50.15	$401.20	$36.15	$59.50
1 Equip. Oper. (crane)	41.95	335.60	68.85	550.80		
1 Hyd. Excavator, .75 C.Y.		681.25		749.38		
1 Trench Box		77.75		85.53	47.44	52.18
16 L.H., Daily Totals		$1337.40		$1786.90	$83.59	$111.68
Crew B-13H	**Hr.**	**Daily**	**Hr.**	**Daily**	**Bare Costs**	**Incl. O&P**
1 Laborer	$30.35	$242.80	$50.15	$401.20	$36.15	$59.50
1 Equip. Oper. (crane)	41.95	335.60	68.85	550.80		
1 Gradall, 5/8 C.Y.		862.55		948.80		
1 Trench Box		77.75		85.53	58.77	64.65
16 L.H., Daily Totals		$1518.70		$1986.33	$94.92	$124.15
Crew B-13I	**Hr.**	**Daily**	**Hr.**	**Daily**	**Bare Costs**	**Incl. O&P**
1 Laborer	$30.35	$242.80	$50.15	$401.20	$36.15	$59.50
1 Equip. Oper. (crane)	41.95	335.60	68.85	550.80		
1 Gradall, 3 Ton, 1 C.Y.		1185.00		1303.50		
1 Trench Box		77.75		85.53	78.92	86.81
16 L.H., Daily Totals		$1841.15		$2341.03	$115.07	$146.31
Crew B-13J	**Hr.**	**Daily**	**Hr.**	**Daily**	**Bare Costs**	**Incl. O&P**
1 Laborer	$30.35	$242.80	$50.15	$401.20	$36.15	$59.50
1 Equip. Oper. (crane)	41.95	335.60	68.85	550.80		
1 Hyd. Excavator, 2.5 C.Y.		1441.00		1585.10		
1 Trench Box		77.75		85.53	94.92	104.41
16 L.H., Daily Totals		$2097.15		$2622.63	$131.07	$163.91
Crew B-13K	**Hr.**	**Daily**	**Hr.**	**Daily**	**Bare Costs**	**Incl. O&P**
2 Equip. Opers. (crane)	$41.95	$671.20	$68.85	$1101.60	$41.95	$68.85
1 Hyd. Excavator, .75 C.Y.		681.25		749.38		
1 Hyd. Hammer, 4000 ft-lb		332.85		366.13		
1 Hyd. Excavator, .75 C.Y.		681.25		749.38	105.96	116.56
16 L.H., Daily Totals		$2366.55		$2966.49	$147.91	$185.41
Crew B-13L	**Hr.**	**Daily**	**Hr.**	**Daily**	**Bare Costs**	**Incl. O&P**
2 Equip. Opers. (crane)	$41.95	$671.20	$68.85	$1101.60	$41.95	$68.85
1 Hyd. Excavator, 1.5 C.Y.		908.70		999.57		
1 Hyd. Hammer, 5000 ft-lb		407.60		448.36		
1 Hyd. Excavator, .75 C.Y.		681.25		749.38	124.85	137.33
16 L.H., Daily Totals		$2668.75		$3298.91	$166.80	$206.18
Crew B-13M	**Hr.**	**Daily**	**Hr.**	**Daily**	**Bare Costs**	**Incl. O&P**
2 Equip. Opers. (crane)	$41.95	$671.20	$68.85	$1101.60	$41.95	$68.85
1 Hyd. Excavator, 2.5 C.Y.		1441.00		1585.10		
1 Hyd. Hammer, 8000 ft-lb		593.35		652.68		
1 Hyd. Excavator, 1.5 C.Y.		908.70		999.57	183.94	202.33
16 L.H., Daily Totals		$3614.25		$4338.95	$225.89	$271.18
Crew B-13N	**Hr.**	**Daily**	**Hr.**	**Daily**	**Bare Costs**	**Incl. O&P**
2 Equip. Opers. (crane)	$41.95	$671.20	$68.85	$1101.60	$41.95	$68.85
1 Hyd. Excavator, 3.5 C.Y.		2256.00		2481.60		
1 Hyd. Hammer, 12,000 ft-lb		685.75		754.33		
1 Hyd. Excavator, 1.5 C.Y.		908.70		999.57	240.65	264.72
16 L.H., Daily Totals		$4521.65		$5337.10	$282.60	$333.57

Crew No.	Bare Costs		Incl. Subs O&P		Cost Per Labor-Hour	
Crew B-14	**Hr.**	**Daily**	**Hr.**	**Daily**	**Bare Costs**	**Incl. O&P**
1 Labor Foreman (outside)	$32.35	$258.80	$53.45	$427.60	$32.08	$52.94
4 Laborers	30.35	971.20	50.15	1604.80		
1 Equip. Oper. (light)	38.75	310.00	63.60	508.80		
1 Backhoe Loader, 48 H.P.		320.20		352.22	6.67	7.34
48 L.H., Daily Totals		$1860.20		$2893.42	$38.75	$60.28
Crew B-14A	**Hr.**	**Daily**	**Hr.**	**Daily**	**Bare Costs**	**Incl. O&P**
1 Equip. Oper. (crane)	$41.95	$335.60	$68.85	$550.80	$38.08	$62.62
.5 Laborer	30.35	121.40	50.15	200.60		
1 Hyd. Excavator, 4.5 C.Y.		2868.00		3154.80	239.00	262.90
12 L.H., Daily Totals		$3325.00		$3906.20	$277.08	$325.52
Crew B-14B	**Hr.**	**Daily**	**Hr.**	**Daily**	**Bare Costs**	**Incl. O&P**
1 Equip. Oper. (crane)	$41.95	$335.60	$68.85	$550.80	$38.08	$62.62
.5 Laborer	30.35	121.40	50.15	200.60		
1 Hyd. Excavator, 6 C.Y.		3579.00		3936.90	298.25	328.07
12 L.H., Daily Totals		$4036.00		$4688.30	$336.33	$390.69
Crew B-14C	**Hr.**	**Daily**	**Hr.**	**Daily**	**Bare Costs**	**Incl. O&P**
1 Equip. Oper. (crane)	$41.95	$335.60	$68.85	$550.80	$38.08	$62.62
.5 Laborer	30.35	121.40	50.15	200.60		
1 Hyd. Excavator, 7 C.Y.		3332.00		3665.20	277.67	305.43
12 L.H., Daily Totals		$3789.00		$4416.60	$315.75	$368.05
Crew B-14F	**Hr.**	**Daily**	**Hr.**	**Daily**	**Bare Costs**	**Incl. O&P**
1 Equip. Oper. (crane)	$41.95	$335.60	$68.85	$550.80	$38.08	$62.62
.5 Laborer	30.35	121.40	50.15	200.60		
1 Hyd. Shovel, 7 C.Y.		4029.00		4431.90	335.75	369.32
12 L.H., Daily Totals		$4486.00		$5183.30	$373.83	$431.94
Crew B-14G	**Hr.**	**Daily**	**Hr.**	**Daily**	**Bare Costs**	**Incl. O&P**
1 Equip. Oper. (crane)	$41.95	$335.60	$68.85	$550.80	$38.08	$62.62
.5 Laborer	30.35	121.40	50.15	200.60		
1 Hyd. Shovel, 12 C.Y.		5855.00		6440.50	487.92	536.71
12 L.H., Daily Totals		$6312.00		$7191.90	$526.00	$599.33
Crew B-14J	**Hr.**	**Daily**	**Hr.**	**Daily**	**Bare Costs**	**Incl. O&P**
1 Equip. Oper. (medium)	$40.75	$326.00	$66.85	$534.80	$37.28	$61.28
.5 Laborer	30.35	121.40	50.15	200.60		
1 F.E. Loader, 8 C.Y.		1784.00		1962.40	148.67	163.53
12 L.H., Daily Totals		$2231.40		$2697.80	$185.95	$224.82
Crew B-14K	**Hr.**	**Daily**	**Hr.**	**Daily**	**Bare Costs**	**Incl. O&P**
1 Equip. Oper. (medium)	$40.75	$326.00	$66.85	$534.80	$37.28	$61.28
.5 Laborer	30.35	121.40	50.15	200.60		
1 F.E. Loader, 10 C.Y.		2625.00		2887.50	218.75	240.63
12 L.H., Daily Totals		$3072.40		$3622.90	$256.03	$301.91
Crew B-15	**Hr.**	**Daily**	**Hr.**	**Daily**	**Bare Costs**	**Incl. O&P**
1 Equipment Oper. (med.)	$40.75	$326.00	$66.85	$534.80	$35.86	$58.98
.5 Laborer	30.35	121.40	50.15	200.60		
2 Truck Drivers (heavy)	34.80	556.80	57.25	916.00		
2 Dump Trucks, 12 C.Y., 400 H.P.		1134.10		1247.51		
1 Dozer, 200 H.P.		1290.00		1419.00	86.58	95.23
28 L.H., Daily Totals		$3428.30		$4317.91	$122.44	$154.21

Crew No.	Bare Costs		Incl. Subs O&P		Cost Per Labor-Hour	
Crew B-16	**Hr.**	**Daily**	**Hr.**	**Daily**	**Bare Costs**	**Incl. O&P**
1 Labor Foreman (outside)	$32.35	$258.80	$53.45	$427.60	$31.96	$52.75
2 Laborers	30.35	485.60	50.15	802.40		
1 Truck Driver (heavy)	34.80	278.40	57.25	458.00		
1 Dump Truck, 12 C.Y., 400 H.P.		567.05		623.76	17.72	19.49
32 L.H., Daily Totals		$1589.85		$2311.76	$49.68	$72.24

Crew B-17	Hr.	Daily	Hr.	Daily	Bare Costs	Incl. O&P
2 Laborers	$30.35	$485.60	$50.15	$802.40	$33.56	$55.29
1 Equip. Oper. (light)	38.75	310.00	63.60	508.80		
1 Truck Driver (heavy)	34.80	278.40	57.25	458.00		
1 Backhoe Loader, 48 H.P.		320.20		352.22		
1 Dump Truck, 8 C.Y., 220 H.P.		339.60		373.56	20.62	22.68
32 L.H., Daily Totals		$1733.80		$2494.98	$54.18	$77.97

Crew B-17A	Hr.	Daily	Hr.	Daily	Bare Costs	Incl. O&P
2 Labor Foremen (outside)	$32.35	$517.60	$53.45	$855.20	$32.77	$54.22
6 Laborers	30.35	1456.80	50.15	2407.20		
1 Skilled Worker Foreman (out)	41.45	331.60	68.85	550.80		
1 Skilled Worker	39.45	315.60	65.55	524.40		
80 L.H., Daily Totals		$2621.60		$4337.60	$32.77	$54.22

Crew B-17B	Hr.	Daily	Hr.	Daily	Bare Costs	Incl. O&P
2 Laborers	$30.35	$485.60	$50.15	$802.40	$33.56	$55.29
1 Equip. Oper. (light)	38.75	310.00	63.60	508.80		
1 Truck Driver (heavy)	34.80	278.40	57.25	458.00		
1 Backhoe Loader, 48 H.P.		320.20		352.22		
1 Dump Truck, 12 C.Y., 400 H.P.		567.05		623.76	27.73	30.50
32 L.H., Daily Totals		$1961.25		$2745.18	$61.29	$85.79

Crew B-18	Hr.	Daily	Hr.	Daily	Bare Costs	Incl. O&P
1 Labor Foreman (outside)	$32.35	$258.80	$53.45	$427.60	$31.02	$51.25
2 Laborers	30.35	485.60	50.15	802.40		
1 Vibrating Plate, Gas, 21"		40.55		44.60	1.69	1.86
24 L.H., Daily Totals		$784.95		$1274.61	$32.71	$53.11

Crew B-19	Hr.	Daily	Hr.	Daily	Bare Costs	Incl. O&P
1 Pile Driver Foreman (outside)	$41.10	$328.80	$70.00	$560.00	$38.54	$65.06
4 Pile Drivers	39.10	1251.20	66.60	2131.20		
1 Equip. Oper. (crane)	41.95	335.60	68.85	550.80		
1 Building Laborer	30.35	242.80	50.15	401.20		
1 Crawler Crane, 40 Ton		1392.00		1531.20		
1 Lead, 90' High		134.60		148.06		
1 Hammer, Diesel, 22k ft-lb		427.55		470.31	34.90	38.39
56 L.H., Daily Totals		$4112.55		$5792.77	$73.44	$103.44

Crew B-19A	Hr.	Daily	Hr.	Daily	Bare Costs	Incl. O&P
1 Pile Driver Foreman (outside)	$41.10	$328.80	$70.00	$560.00	$38.54	$65.06
4 Pile Drivers	39.10	1251.20	66.60	2131.20		
1 Equip. Oper. (crane)	41.95	335.60	68.85	550.80		
1 Common Laborer	30.35	242.80	50.15	401.20		
1 Crawler Crane, 75 Ton		1734.00		1907.40		
1 Lead, 90' High		134.60		148.06		
1 Hammer, Diesel, 41k ft-lb		565.45		622.00	43.47	47.81
56 L.H., Daily Totals		$4592.45		$6320.65	$82.01	$112.87

Crew B-19B	Hr.	Daily	Hr.	Daily	Bare Costs	Incl. O&P
1 Pile Driver Foreman (outside)	$41.10	$328.80	$70.00	$560.00	$38.54	$65.06
4 Pile Drivers	39.10	1251.20	66.60	2131.20		
1 Equip. Oper. (crane)	41.95	335.60	68.85	550.80		
1 Common Laborer	30.35	242.80	50.15	401.20		
1 Crawler Crane, 40 Ton		1392.00		1531.20		
1 Lead, 90' High		134.60		148.06		
1 Hammer, Diesel, 22k ft-lb		427.55		470.31		
1 Barge, 400 Ton		836.50		920.15	49.83	54.82
56 L.H., Daily Totals		$4949.05		$6712.92	$88.38	$119.87

Crew B-19C	Hr.	Daily	Hr.	Daily	Bare Costs	Incl. O&P
1 Pile Driver Foreman (outside)	$41.10	$328.80	$70.00	$560.00	$38.54	$65.06
4 Pile Drivers	39.10	1251.20	66.60	2131.20		
1 Equip. Oper. (crane)	41.95	335.60	68.85	550.80		
1 Common Laborer	30.35	242.80	50.15	401.20		
1 Crawler Crane, 75 Ton		1734.00		1907.40		
1 Lead, 90' High		134.60		148.06		
1 Hammer, Diesel, 41k ft-lb		565.45		622.00		
1 Barge, 400 Ton		836.50		920.15	58.40	64.24
56 L.H., Daily Totals		$5428.95		$7240.81	$96.95	$129.30

Crew B-20	Hr.	Daily	Hr.	Daily	Bare Costs	Incl. O&P
1 Labor Foreman (outside)	$32.35	$258.80	$53.45	$427.60	$31.02	$51.25
2 Laborers	30.35	485.60	50.15	802.40		
24 L.H., Daily Totals		$744.40		$1230.00	$31.02	$51.25

Crew B-20A	Hr.	Daily	Hr.	Daily	Bare Costs	Incl. O&P
1 Labor Foreman (outside)	$32.35	$258.80	$53.45	$427.60	$35.86	$59.02
1 Laborer	30.35	242.80	50.15	401.20		
1 Plumber	44.85	358.80	73.60	588.80		
1 Plumber Apprentice	35.90	287.20	58.90	471.20		
32 L.H., Daily Totals		$1147.60		$1888.80	$35.86	$59.02

Crew B-21	Hr.	Daily	Hr.	Daily	Bare Costs	Incl. O&P
1 Labor Foreman (outside)	$32.35	$258.80	$53.45	$427.60	$32.58	$53.76
2 Laborers	30.35	485.60	50.15	802.40		
.5 Equip. Oper. (crane)	41.95	167.80	68.85	275.40		
.5 S.P. Crane, 4x4, 5 Ton		129.72		142.70	4.63	5.10
28 L.H., Daily Totals		$1041.93		$1648.10	$37.21	$58.86

Crew B-21A	Hr.	Daily	Hr.	Daily	Bare Costs	Incl. O&P
1 Labor Foreman (outside)	$32.35	$258.80	$53.45	$427.60	$37.08	$60.99
1 Laborer	30.35	242.80	50.15	401.20		
1 Plumber	44.85	358.80	73.60	588.80		
1 Plumber Apprentice	35.90	287.20	58.90	471.20		
1 Equip. Oper. (crane)	41.95	335.60	68.85	550.80		
1 S.P. Crane, 4x4, 12 Ton		371.85		409.04	9.30	10.23
40 L.H., Daily Totals		$1855.05		$2848.64	$46.38	$71.22

Crew B-21B	Hr.	Daily	Hr.	Daily	Bare Costs	Incl. O&P
1 Labor Foreman (outside)	$32.35	$258.80	$53.45	$427.60	$33.07	$54.55
3 Laborers	30.35	728.40	50.15	1203.60		
1 Equip. Oper. (crane)	41.95	335.60	68.85	550.80		
1 Hyd. Crane, 12 Ton		471.75		518.92	11.79	12.97
40 L.H., Daily Totals		$1794.55		$2700.93	$44.86	$67.52

Crew No.	Bare Costs		Incl. Subs O&P		Cost Per Labor-Hour	
Crew B-21C	**Hr.**	**Daily**	**Hr.**	**Daily**	**Bare Costs**	**Incl. O&P**
1 Labor Foreman (outside)	$32.35	$258.80	$53.45	$427.60	$33.09	$54.54
4 Laborers	30.35	971.20	50.15	1604.80		
1 Equip. Oper. (crane)	41.95	335.60	68.85	550.80		
1 Equip. Oper. (oiler)	35.90	287.20	58.90	471.20		
2 Cutting Torches		25.00		27.50		
2 Sets of Gases		335.70		369.27		
1 Lattice Boom Crane, 90 Ton		1698.00		1867.80	36.76	40.44
56 L.H., Daily Totals		$3911.50		$5318.97	$69.85	$94.98
Crew B-22	**Hr.**	**Daily**	**Hr.**	**Daily**	**Bare Costs**	**Incl. O&P**
1 Labor Foreman (outside)	$32.35	$258.80	$53.45	$427.60	$33.20	$54.77
2 Laborers	30.35	485.60	50.15	802.40		
.75 Equip. Oper. (crane)	41.95	251.70	68.85	413.10		
.75 S.P. Crane, 4x4, 5 Ton		194.59		214.05	6.49	7.13
30 L.H., Daily Totals		$1190.69		$1857.15	$39.69	$61.90
Crew B-22A	**Hr.**	**Daily**	**Hr.**	**Daily**	**Bare Costs**	**Incl. O&P**
1 Labor Foreman (outside)	$32.35	$258.80	$53.45	$427.60	$34.89	$57.63
1 Skilled Worker	39.45	315.60	65.55	524.40		
2 Laborers	30.35	485.60	50.15	802.40		
1 Equipment Operator, Crane	41.95	335.60	68.85	550.80		
1 S.P. Crane, 4x4, 5 Ton		259.45		285.39		
1 Butt Fusion Mach., 4"-12" diam.		405.15		445.67	16.61	18.28
40 L.H., Daily Totals		$2060.20		$3036.26	$51.51	$75.91
Crew B-22B	**Hr.**	**Daily**	**Hr.**	**Daily**	**Bare Costs**	**Incl. O&P**
1 Labor Foreman (outside)	$32.35	$258.80	$53.45	$427.60	$34.89	$57.63
1 Skilled Worker	39.45	315.60	65.55	524.40		
2 Laborers	30.35	485.60	50.15	802.40		
1 Equip. Oper. (crane)	41.95	335.60	68.85	550.80		
1 S.P. Crane, 4x4, 5 Ton		259.45		285.39		
1 Butt Fusion Mach., 8"-24" diam.		1927.00		2119.70	54.66	60.13
40 L.H., Daily Totals		$3582.05		$4710.30	$89.55	$117.76
Crew B-22C	**Hr.**	**Daily**	**Hr.**	**Daily**	**Bare Costs**	**Incl. O&P**
1 Skilled Worker	$39.45	$315.60	$65.55	$524.40	$34.90	$57.85
1 Laborer	30.35	242.80	50.15	401.20		
1 Butt Fusion Mach., 2"-8" diam.		143.10		157.41	8.94	9.84
16 L.H., Daily Totals		$701.50		$1083.01	$43.84	$67.69
Crew B-23	**Hr.**	**Daily**	**Hr.**	**Daily**	**Bare Costs**	**Incl. O&P**
1 Labor Foreman (outside)	$32.35	$258.80	$53.45	$427.60	$30.75	$50.81
4 Laborers	30.35	971.20	50.15	1604.80		
1 Drill Rig, Truck-Mounted		2428.00		2670.80		
1 Flatbed Truck, Gas, 3 Ton		245.95		270.55	66.85	73.53
40 L.H., Daily Totals		$3903.95		$4973.74	$97.60	$124.34
Crew B-23A	**Hr.**	**Daily**	**Hr.**	**Daily**	**Bare Costs**	**Incl. O&P**
1 Labor Foreman (outside)	$32.35	$258.80	$53.45	$427.60	$34.48	$56.82
1 Laborer	30.35	242.80	50.15	401.20		
1 Equip. Oper. (medium)	40.75	326.00	66.85	534.80		
1 Drill Rig, Truck-Mounted		2428.00		2670.80		
1 Pickup Truck, 3/4 Ton		109.90		120.89	105.75	116.32
24 L.H., Daily Totals		$3365.50		$4155.29	$140.23	$173.14

Crew No.	Bare Costs		Incl. Subs O&P		Cost Per Labor-Hour	
Crew B-23B	**Hr.**	**Daily**	**Hr.**	**Daily**	**Bare Costs**	**Incl. O&P**
1 Labor Foreman (outside)	$32.35	$258.80	$53.45	$427.60	$34.48	$56.82
1 Laborer	30.35	242.80	50.15	401.20		
1 Equip. Oper. (medium)	40.75	326.00	66.85	534.80		
1 Drill Rig, Truck-Mounted		2428.00		2670.80		
1 Pickup Truck, 3/4 Ton		109.90		120.89		
1 Centr. Water Pump, 6"		296.45		326.10	118.10	129.91
24 L.H., Daily Totals		$3661.95		$4481.39	$152.58	$186.72
Crew B-24	**Hr.**	**Daily**	**Hr.**	**Daily**	**Bare Costs**	**Incl. O&P**
1 Cement Finisher	$37.15	$297.20	$60.05	$480.40	$35.42	$58.08
1 Laborer	30.35	242.80	50.15	401.20		
1 Carpenter	38.75	310.00	64.05	512.40		
24 L.H., Daily Totals		$850.00		$1394.00	$35.42	$58.08
Crew B-25	**Hr.**	**Daily**	**Hr.**	**Daily**	**Bare Costs**	**Incl. O&P**
1 Labor Foreman (outside)	$32.35	$258.80	$53.45	$427.60	$33.37	$55.00
7 Laborers	30.35	1699.60	50.15	2808.40		
3 Equip. Oper. (medium)	40.75	978.00	66.85	1604.40		
1 Asphalt Paver, 130 H.P.		2076.00		2283.60		
1 Tandem Roller, 10 Ton		232.05		255.26		
1 Roller, Pneum. Whl., 12 Ton		338.35		372.19	30.07	33.08
88 L.H., Daily Totals		$5582.80		$7751.44	$63.44	$88.08
Crew B-25B	**Hr.**	**Daily**	**Hr.**	**Daily**	**Bare Costs**	**Incl. O&P**
1 Labor Foreman (outside)	$32.35	$258.80	$53.45	$427.60	$33.98	$55.99
7 Laborers	30.35	1699.60	50.15	2808.40		
4 Equip. Oper. (medium)	40.75	1304.00	66.85	2139.20		
1 Asphalt Paver, 130 H.P.		2076.00		2283.60		
2 Tandem Rollers, 10 Ton		464.10		510.51		
1 Roller, Pneum. Whl., 12 Ton		338.35		372.19	29.98	32.98
96 L.H., Daily Totals		$6140.85		$8541.50	$63.97	$88.97
Crew B-25C	**Hr.**	**Daily**	**Hr.**	**Daily**	**Bare Costs**	**Incl. O&P**
1 Labor Foreman (outside)	$32.35	$258.80	$53.45	$427.60	$34.15	$56.27
3 Laborers	30.35	728.40	50.15	1203.60		
2 Equip. Oper. (medium)	40.75	652.00	66.85	1069.60		
1 Asphalt Paver, 130 H.P.		2076.00		2283.60		
1 Tandem Roller, 10 Ton		232.05		255.26	48.08	52.89
48 L.H., Daily Totals		$3947.25		$5239.65	$82.23	$109.16
Crew B-25D	**Hr.**	**Daily**	**Hr.**	**Daily**	**Bare Costs**	**Incl. O&P**
1 Labor Foreman (outside)	$32.35	$258.80	$53.45	$427.60	$34.30	$56.50
3 Laborers	30.35	728.40	50.15	1203.60		
2.125 Equip. Oper. (medium)	40.75	692.75	66.85	1136.45		
.125 Truck Driver (heavy)	34.80	34.80	57.25	57.25		
.125 Truck Tractor, 6x4, 380 H.P.		60.95		67.05		
.125 Dist. Tanker, 3000 Gallon		40.41		44.45		
1 Asphalt Paver, 130 H.P.		2076.00		2283.60		
1 Tandem Roller, 10 Ton		232.05		255.26	48.19	53.01
50 L.H., Daily Totals		$4124.16		$5475.25	$82.48	$109.50
Crew B-25E	**Hr.**	**Daily**	**Hr.**	**Daily**	**Bare Costs**	**Incl. O&P**
1 Labor Foreman (outside)	$32.35	$258.80	$53.45	$427.60	$34.43	$56.71
3 Laborers	30.35	728.40	50.15	1203.60		
2.250 Equip. Oper. (medium)	40.75	733.50	66.85	1203.30		
.25 Truck Driver (heavy)	34.80	69.60	57.25	114.50		
.25 Truck Tractor, 6x4, 380 H.P.		121.90		134.09		
.25 Dist. Tanker, 3000 Gallon		80.81		88.89		
1 Asphalt Paver, 130 H.P.		2076.00		2283.60		
1 Tandem Roller, 10 Ton		232.05		255.26	48.28	53.11
52 L.H., Daily Totals		$4301.06		$5710.84	$82.71	$109.82

Crew No.	Bare Costs		Incl. Subs O&P		Cost Per Labor-Hour	
Crew B-26	**Hr.**	**Daily**	**Hr.**	**Daily**	**Bare Costs**	**Incl. O&P**
1 Labor Foreman (outside)	$32.35	$258.80	$53.45	$427.60	$33.97	$55.93
6 Laborers	30.35	1456.80	50.15	2407.20		
2 Equip. Oper. (medium)	40.75	652.00	66.85	1069.60		
1 Rodman (reinf.)	40.60	324.80	67.15	537.20		
1 Cement Finisher	37.15	297.20	60.05	480.40		
1 Grader, 30,000 Lbs.		657.85		723.63		
1 Paving Mach. & Equip.		2428.00		2670.80	35.07	38.57
88 L.H., Daily Totals		$6075.45		$8316.43	$69.04	$94.50
Crew B-26A	**Hr.**	**Daily**	**Hr.**	**Daily**	**Bare Costs**	**Incl. O&P**
1 Labor Foreman (outside)	$32.35	$258.80	$53.45	$427.60	$33.97	$55.93
6 Laborers	30.35	1456.80	50.15	2407.20		
2 Equip. Oper. (medium)	40.75	652.00	66.85	1069.60		
1 Rodman (reinf.)	40.60	324.80	67.15	537.20		
1 Cement Finisher	37.15	297.20	60.05	480.40		
1 Grader, 30,000 Lbs.		657.85		723.63		
1 Paving Mach. & Equip.		2428.00		2670.80		
1 Concrete Saw		110.15		121.17	36.32	39.95
88 L.H., Daily Totals		$6185.60		$8437.60	$70.29	$95.88
Crew B-26B	**Hr.**	**Daily**	**Hr.**	**Daily**	**Bare Costs**	**Incl. O&P**
1 Labor Foreman (outside)	$32.35	$258.80	$53.45	$427.60	$34.54	$56.84
6 Laborers	30.35	1456.80	50.15	2407.20		
3 Equip. Oper. (medium)	40.75	978.00	66.85	1604.40		
1 Rodman (reinf.)	40.60	324.80	67.15	537.20		
1 Cement Finisher	37.15	297.20	60.05	480.40		
1 Grader, 30,000 Lbs.		657.85		723.63		
1 Paving Mach. & Equip.		2428.00		2670.80		
1 Concrete Pump, 110' Boom		1130.00		1243.00	43.92	48.31
96 L.H., Daily Totals		$7531.45		$10094.24	$78.45	$105.15
Crew B-26C	**Hr.**	**Daily**	**Hr.**	**Daily**	**Bare Costs**	**Incl. O&P**
1 Labor Foreman (outside)	$32.35	$258.80	$53.45	$427.60	$33.30	$54.84
6 Laborers	30.35	1456.80	50.15	2407.20		
1 Equip. Oper. (medium)	40.75	326.00	66.85	534.80		
1 Rodman (reinf.)	40.60	324.80	67.15	537.20		
1 Cement Finisher	37.15	297.20	60.05	480.40		
1 Paving Mach. & Equip.		2428.00		2670.80		
1 Concrete Saw		110.15		121.17	31.73	34.90
80 L.H., Daily Totals		$5201.75		$7179.17	$65.02	$89.74
Crew B-27	**Hr.**	**Daily**	**Hr.**	**Daily**	**Bare Costs**	**Incl. O&P**
1 Labor Foreman (outside)	$32.35	$258.80	$53.45	$427.60	$30.85	$50.98
3 Laborers	30.35	728.40	50.15	1203.60		
1 Berm Machine		325.30		357.83	10.17	11.18
32 L.H., Daily Totals		$1312.50		$1989.03	$41.02	$62.16
Crew B-28	**Hr.**	**Daily**	**Hr.**	**Daily**	**Bare Costs**	**Incl. O&P**
2 Carpenters	$38.75	$620.00	$64.05	$1024.80	$35.95	$59.42
1 Laborer	30.35	242.80	50.15	401.20		
24 L.H., Daily Totals		$862.80		$1426.00	$35.95	$59.42
Crew B-29	**Hr.**	**Daily**	**Hr.**	**Daily**	**Bare Costs**	**Incl. O&P**
1 Labor Foreman (outside)	$32.35	$258.80	$53.45	$427.60	$32.62	$53.82
4 Laborers	30.35	971.20	50.15	1604.80		
1 Equip. Oper. (crane)	41.95	335.60	68.85	550.80		
1 Gradall, 5/8 C.Y.		862.55		948.80	17.97	19.77
48 L.H., Daily Totals		$2428.15		$3532.01	$50.59	$73.58

Crew No.	Bare Costs		Incl. Subs O&P		Cost Per Labor-Hour	
Crew B-30	**Hr.**	**Daily**	**Hr.**	**Daily**	**Bare Costs**	**Incl. O&P**
1 Equip. Oper. (medium)	$40.75	$326.00	$66.85	$534.80	$36.78	$60.45
2 Truck Drivers (heavy)	34.80	556.80	57.25	916.00		
1 Hyd. Excavator, 1.5 C.Y.		908.70		999.57		
2 Dump Trucks, 12 C.Y., 400 H.P.		1134.10		1247.51	85.12	93.63
24 L.H., Daily Totals		$2925.60		$3697.88	$121.90	$154.08
Crew B-31	**Hr.**	**Daily**	**Hr.**	**Daily**	**Bare Costs**	**Incl. O&P**
1 Labor Foreman (outside)	$32.35	$258.80	$53.45	$427.60	$30.75	$50.81
4 Laborers	30.35	971.20	50.15	1604.80		
1 Air Compressor, 250 cfm		167.95		184.75		
1 Sheeting Driver		7.15		7.87		
2 -50' Air Hoses, 1.5"		45.50		50.05	5.51	6.07
40 L.H., Daily Totals		$1450.60		$2275.06	$36.27	$56.88
Crew B-32	**Hr.**	**Daily**	**Hr.**	**Daily**	**Bare Costs**	**Incl. O&P**
1 Laborer	$30.35	$242.80	$50.15	$401.20	$38.15	$62.67
3 Equip. Oper. (medium)	40.75	978.00	66.85	1604.40		
1 Grader, 30,000 Lbs.		657.85		723.63		
1 Tandem Roller, 10 Ton		232.05		255.26		
1 Dozer, 200 H.P.		1290.00		1419.00	68.12	74.93
32 L.H., Daily Totals		$3400.70		$4403.49	$106.27	$137.61
Crew B-32A	**Hr.**	**Daily**	**Hr.**	**Daily**	**Bare Costs**	**Incl. O&P**
1 Laborer	$30.35	$242.80	$50.15	$401.20	$37.28	$61.28
2 Equip. Oper. (medium)	40.75	652.00	66.85	1069.60		
1 Grader, 30,000 Lbs.		657.85		723.63		
1 Roller, Vibratory, 25 Ton		650.90		715.99	54.53	59.98
24 L.H., Daily Totals		$2203.55		$2910.43	$91.81	$121.27
Crew B-32B	**Hr.**	**Daily**	**Hr.**	**Daily**	**Bare Costs**	**Incl. O&P**
1 Laborer	$30.35	$242.80	$50.15	$401.20	$37.28	$61.28
2 Equip. Oper. (medium)	40.75	652.00	66.85	1069.60		
1 Dozer, 200 H.P.		1290.00		1419.00		
1 Roller, Vibratory, 25 Ton		650.90		715.99	80.87	88.96
24 L.H., Daily Totals		$2835.70		$3605.79	$118.15	$150.24
Crew B-32C	**Hr.**	**Daily**	**Hr.**	**Daily**	**Bare Costs**	**Incl. O&P**
1 Labor Foreman (outside)	$32.35	$258.80	$53.45	$427.60	$35.88	$59.05
2 Laborers	30.35	485.60	50.15	802.40		
3 Equip. Oper. (medium)	40.75	978.00	66.85	1604.40		
1 Grader, 30,000 Lbs.		657.85		723.63		
1 Tandem Roller, 10 Ton		232.05		255.26		
1 Dozer, 200 H.P.		1290.00		1419.00	45.41	49.96
48 L.H., Daily Totals		$3902.30		$5232.29	$81.30	$109.01
Crew B-33A	**Hr.**	**Daily**	**Hr.**	**Daily**	**Bare Costs**	**Incl. O&P**
1 Equip. Oper. (medium)	$40.75	$326.00	$66.85	$534.80	$40.75	$66.85
.25 Equip. Oper. (medium)	40.75	81.50	66.85	133.70		
1 Scraper, Towed, 7 C.Y.		125.50		138.05		
1.250 Dozers, 300 H.P.		2288.75		2517.63	241.43	265.57
10 L.H., Daily Totals		$2821.75		$3324.18	$282.18	$332.42
Crew B-33B	**Hr.**	**Daily**	**Hr.**	**Daily**	**Bare Costs**	**Incl. O&P**
1 Equip. Oper. (medium)	$40.75	$326.00	$66.85	$534.80	$40.75	$66.85
.25 Equip. Oper. (medium)	40.75	81.50	66.85	133.70		
1 Scraper, Towed, 10 C.Y.		156.65		172.32		
1.250 Dozers, 300 H.P.		2288.75		2517.63	244.54	268.99
10 L.H., Daily Totals		$2852.90		$3358.44	$285.29	$335.84

Crews - Open Shop

Crew No.	Bare Costs		Incl. Subs O&P		Cost Per Labor-Hour	
Crew B-33C	**Hr.**	**Daily**	**Hr.**	**Daily**	**Bare Costs**	**Incl. O&P**
1 Equip. Oper. (medium)	$40.75	$326.00	$66.85	$534.80	$40.75	$66.85
.25 Equip. Oper. (medium)	40.75	81.50	66.85	133.70		
1 Scraper, Towed, 15 C.Y.		173.20		190.52		
1.250 Dozers, 300 H.P.		2288.75		2517.63	246.19	270.81
10 L.H., Daily Totals		$2869.45		$3376.65	$286.94	$337.66

Crew B-33D	Hr.	Daily	Hr.	Daily	Bare Costs	Incl. O&P
1 Equip. Oper. (medium)	$40.75	$326.00	$66.85	$534.80	$40.75	$66.85
.25 Equip. Oper. (medium)	40.75	81.50	66.85	133.70		
1 S.P. Scraper, 14 C.Y.		2611.00		2872.10		
.25 Dozer, 300 H.P.		457.75		503.52	306.88	337.56
10 L.H., Daily Totals		$3476.25		$4044.13	$347.63	$404.41

Crew B-33E	Hr.	Daily	Hr.	Daily	Bare Costs	Incl. O&P
1 Equip. Oper. (medium)	$40.75	$326.00	$66.85	$534.80	$40.75	$66.85
.25 Equip. Oper. (medium)	40.75	81.50	66.85	133.70		
1 S.P. Scraper, 21 C.Y.		2541.00		2795.10		
.25 Dozer, 300 H.P.		457.75		503.52	299.88	329.86
10 L.H., Daily Totals		$3406.25		$3967.13	$340.63	$396.71

Crew B-33F	Hr.	Daily	Hr.	Daily	Bare Costs	Incl. O&P
1 Equip. Oper. (medium)	$40.75	$326.00	$66.85	$534.80	$40.75	$66.85
.25 Equip. Oper. (medium)	40.75	81.50	66.85	133.70		
1 Elev. Scraper, 11 C.Y.		1165.00		1281.50		
.25 Dozer, 300 H.P.		457.75		503.52	162.28	178.50
10 L.H., Daily Totals		$2030.25		$2453.53	$203.03	$245.35

Crew B-33G	Hr.	Daily	Hr.	Daily	Bare Costs	Incl. O&P
1 Equip. Oper. (medium)	$40.75	$326.00	$66.85	$534.80	$40.75	$66.85
.25 Equip. Oper. (medium)	40.75	81.50	66.85	133.70		
1 Elev. Scraper, 22 C.Y.		2307.00		2537.70		
.25 Dozer, 300 H.P.		457.75		503.52	276.48	304.12
10 L.H., Daily Totals		$3172.25		$3709.72	$317.23	$370.97

Crew B-33K	Hr.	Daily	Hr.	Daily	Bare Costs	Incl. O&P
1 Equipment Operator (med.)	$40.75	$326.00	$66.85	$534.80	$37.78	$62.08
.25 Equipment Operator (med.)	40.75	81.50	66.85	133.70		
.5 Laborer	30.35	121.40	50.15	200.60		
1 S.P. Scraper, 31 C.Y.		3603.00		3963.30		
.25 Dozer, 410 H.P.		567.25		623.98	297.88	327.66
14 L.H., Daily Totals		$4699.15		$5456.38	$335.65	$389.74

Crew B-34A	Hr.	Daily	Hr.	Daily	Bare Costs	Incl. O&P
1 Truck Driver (heavy)	$34.80	$278.40	$57.25	$458.00	$34.80	$57.25
1 Dump Truck, 8 C.Y., 220 H.P.		339.60		373.56	42.45	46.70
8 L.H., Daily Totals		$618.00		$831.56	$77.25	$103.94

Crew B-34B	Hr.	Daily	Hr.	Daily	Bare Costs	Incl. O&P
1 Truck Driver (heavy)	$34.80	$278.40	$57.25	$458.00	$34.80	$57.25
1 Dump Truck, 12 C.Y., 400 H.P.		567.05		623.76	70.88	77.97
8 L.H., Daily Totals		$845.45		$1081.76	$105.68	$135.22

Crew B-34C	Hr.	Daily	Hr.	Daily	Bare Costs	Incl. O&P
1 Truck Driver (heavy)	$34.80	$278.40	$57.25	$458.00	$34.80	$57.25
1 Truck Tractor, 6x4, 380 H.P.		487.60		536.36		
1 Dump Trailer, 16.5 C.Y.		133.95		147.35	77.69	85.46
8 L.H., Daily Totals		$899.95		$1141.70	$112.49	$142.71

Crew B-34D	Hr.	Daily	Hr.	Daily	Bare Costs	Incl. O&P
1 Truck Driver (heavy)	$34.80	$278.40	$57.25	$458.00	$34.80	$57.25
1 Truck Tractor, 6x4, 380 H.P.		487.60		536.36		
1 Dump Trailer, 20 C.Y.		148.60		163.46	79.53	87.48
8 L.H., Daily Totals		$914.60		$1157.82	$114.33	$144.73

Crew B-34E	Hr.	Daily	Hr.	Daily	Bare Costs	Incl. O&P
1 Truck Driver (heavy)	$34.80	$278.40	$57.25	$458.00	$34.80	$57.25
1 Dump Truck, Off Hwy., 25 Ton		1353.00		1488.30	169.13	186.04
8 L.H., Daily Totals		$1631.40		$1946.30	$203.93	$243.29

Crew B-34F	Hr.	Daily	Hr.	Daily	Bare Costs	Incl. O&P
1 Truck Driver (heavy)	$34.80	$278.40	$57.25	$458.00	$34.80	$57.25
1 Dump Truck, Off Hwy., 35 Ton		1472.00		1619.20	184.00	202.40
8 L.H., Daily Totals		$1750.40		$2077.20	$218.80	$259.65

Crew B-34G	Hr.	Daily	Hr.	Daily	Bare Costs	Incl. O&P
1 Truck Driver (heavy)	$34.80	$278.40	$57.25	$458.00	$34.80	$57.25
1 Dump Truck, Off Hwy., 50 Ton		1738.00		1911.80	217.25	238.97
8 L.H., Daily Totals		$2016.40		$2369.80	$252.05	$296.23

Crew B-34H	Hr.	Daily	Hr.	Daily	Bare Costs	Incl. O&P
1 Truck Driver (heavy)	$34.80	$278.40	$57.25	$458.00	$34.80	$57.25
1 Dump Truck, Off Hwy., 65 Ton		1879.00		2066.90	234.88	258.36
8 L.H., Daily Totals		$2157.40		$2524.90	$269.68	$315.61

Crew B-34I	Hr.	Daily	Hr.	Daily	Bare Costs	Incl. O&P
1 Truck Driver (heavy)	$34.80	$278.40	$57.25	$458.00	$34.80	$57.25
1 Dump Truck, 18 C.Y., 450 H.P.		705.00		775.50	88.13	96.94
8 L.H., Daily Totals		$983.40		$1233.50	$122.93	$154.19

Crew B-34J	Hr.	Daily	Hr.	Daily	Bare Costs	Incl. O&P
1 Truck Driver (heavy)	$34.80	$278.40	$57.25	$458.00	$34.80	$57.25
1 Dump Truck, Off Hwy., 100 Ton		2683.00		2951.30	335.38	368.91
8 L.H., Daily Totals		$2961.40		$3409.30	$370.18	$426.16

Crew B-34K	Hr.	Daily	Hr.	Daily	Bare Costs	Incl. O&P
1 Truck Driver (heavy)	$34.80	$278.40	$57.25	$458.00	$34.80	$57.25
1 Truck Tractor, 6x4, 450 H.P.		594.90		654.39		
1 Lowbed Trailer, 75 Ton		249.90		274.89	105.60	116.16
8 L.H., Daily Totals		$1123.20		$1387.28	$140.40	$173.41

Crew B-34L	Hr.	Daily	Hr.	Daily	Bare Costs	Incl. O&P
1 Equip. Oper. (light)	$38.75	$310.00	$63.60	$508.80	$38.75	$63.60
1 Flatbed Truck, Gas, 1.5 Ton		195.20		214.72	24.40	26.84
8 L.H., Daily Totals		$505.20		$723.52	$63.15	$90.44

Crew B-34M	Hr.	Daily	Hr.	Daily	Bare Costs	Incl. O&P
1 Equip. Oper. (light)	$38.75	$310.00	$63.60	$508.80	$38.75	$63.60
1 Flatbed Truck, Gas, 3 Ton		245.95		270.55	30.74	33.82
8 L.H., Daily Totals		$555.95		$779.35	$69.49	$97.42

Crew B-34N	Hr.	Daily	Hr.	Daily	Bare Costs	Incl. O&P
1 Truck Driver (heavy)	$34.80	$278.40	$57.25	$458.00	$37.77	$62.05
1 Equip. Oper. (medium)	40.75	326.00	66.85	534.80		
1 Truck Tractor, 6x4, 380 H.P.		487.60		536.36		
1 Flatbed Trailer, 40 Ton		182.50		200.75	41.88	46.07
16 L.H., Daily Totals		$1274.50		$1729.91	$79.66	$108.12

Crew No.	Bare Costs		Incl. Subs O&P		Cost Per Labor-Hour	
Crew B-34P	**Hr.**	**Daily**	**Hr.**	**Daily**	**Bare Costs**	**Incl. O&P**
1 Pipe Fitter	$46.05	$368.40	$75.55	$604.40	$40.30	$66.17
1 Truck Driver (light)	34.10	272.80	56.10	448.80		
1 Equip. Oper. (medium)	40.75	326.00	66.85	534.80		
1 Flatbed Truck, Gas, 3 Ton		245.95		270.55		
1 Backhoe Loader, 48 H.P.		320.20		352.22	23.59	25.95
24 L.H., Daily Totals		$1533.35		$2210.76	$63.89	$92.12

Crew B-34Q	Hr.	Daily	Hr.	Daily	Bare Costs	Incl. O&P
1 Pipe Fitter	$46.05	$368.40	$75.55	$604.40	$40.70	$66.83
1 Truck Driver (light)	34.10	272.80	56.10	448.80		
1 Equip. Oper. (crane)	41.95	335.60	68.85	550.80		
1 Flatbed Trailer, 25 Ton		133.00		146.30		
1 Dump Truck, 8 C.Y., 220 H.P.		339.60		373.56		
1 Hyd. Crane, 25 Ton		581.70		639.87	43.93	48.32
24 L.H., Daily Totals		$2031.10		$2763.73	$84.63	$115.16

Crew B-34R	Hr.	Daily	Hr.	Daily	Bare Costs	Incl. O&P
1 Pipe Fitter	$46.05	$368.40	$75.55	$604.40	$40.70	$66.83
1 Truck Driver (light)	34.10	272.80	56.10	448.80		
1 Equip. Oper. (crane)	41.95	335.60	68.85	550.80		
1 Flatbed Trailer, 25 Ton		133.00		146.30		
1 Dump Truck, 8 C.Y., 220 H.P.		339.60		373.56		
1 Hyd. Crane, 25 Ton		581.70		639.87		
1 Hyd. Excavator, 1 C.Y.		749.70		824.67	75.17	82.68
24 L.H., Daily Totals		$2780.80		$3588.40	$115.87	$149.52

Crew B-34S	Hr.	Daily	Hr.	Daily	Bare Costs	Incl. O&P
2 Pipe Fitters	$46.05	$736.80	$75.55	$1208.80	$42.21	$69.30
1 Truck Driver (heavy)	34.80	278.40	57.25	458.00		
1 Equip. Oper. (crane)	41.95	335.60	68.85	550.80		
1 Flatbed Trailer, 40 Ton		182.50		200.75		
1 Truck Tractor, 6x4, 380 H.P.		487.60		536.36		
1 Hyd. Crane, 80 Ton		1487.00		1635.70		
1 Hyd. Excavator, 2 C.Y.		1078.00		1185.80	101.10	111.21
32 L.H., Daily Totals		$4585.90		$5776.21	$143.31	$180.51

Crew B-34T	Hr.	Daily	Hr.	Daily	Bare Costs	Incl. O&P
2 Pipe Fitters	$46.05	$736.80	$75.55	$1208.80	$42.21	$69.30
1 Truck Driver (heavy)	34.80	278.40	57.25	458.00		
1 Equip. Oper. (crane)	41.95	335.60	68.85	550.80		
1 Flatbed Trailer, 40 Ton		182.50		200.75		
1 Truck Tractor, 6x4, 380 H.P.		487.60		536.36		
1 Hyd. Crane, 80 Ton		1487.00		1635.70	67.41	74.15
32 L.H., Daily Totals		$3507.90		$4590.41	$109.62	$143.45

Crew B-34U	Hr.	Daily	Hr.	Daily	Bare Costs	Incl. O&P
1 Truck Driver (heavy)	$34.80	$278.40	$57.25	$458.00	$36.77	$60.42
1 Equip. Oper. (light)	38.75	310.00	63.60	508.80		
1 Truck Tractor, 220 H.P.		303.55		333.90		
1 Flatbed Trailer, 25 Ton		133.00		146.30	27.28	30.01
16 L.H., Daily Totals		$1024.95		$1447.01	$64.06	$90.44

Crew B-34V	Hr.	Daily	Hr.	Daily	Bare Costs	Incl. O&P
1 Truck Driver (heavy)	$34.80	$278.40	$57.25	$458.00	$38.50	$63.23
1 Equip. Oper. (crane)	41.95	335.60	68.85	550.80		
1 Equip. Oper. (light)	38.75	310.00	63.60	508.80		
1 Truck Tractor, 6x4, 450 H.P.		594.90		654.39		
1 Equipment Trailer, 50 Ton		200.95		221.04		
1 Pickup Truck, 4x4, 3/4 Ton		121.70		133.87	38.23	42.05
24 L.H., Daily Totals		$1841.55		$2526.91	$76.73	$105.29

Crew No.	Bare Costs		Incl. Subs O&P		Cost Per Labor-Hour	
Crew B-34W	**Hr.**	**Daily**	**Hr.**	**Daily**	**Bare Costs**	**Incl. O&P**
5 Truck Drivers (heavy)	$34.80	$1392.00	$57.25	$2290.00	$36.69	$60.34
2 Equip. Opers. (crane)	41.95	671.20	68.85	1101.60		
1 Equip. Oper. (mechanic)	42.00	336.00	68.95	551.60		
1 Laborer	30.35	242.80	50.15	401.20		
4 Truck Tractors, 6x4, 380 H.P.		1950.40		2145.44		
2 Equipment Trailers, 50 Ton		401.90		442.09		
2 Flatbed Trailers, 40 Ton		365.00		401.50		
1 Pickup Truck, 4x4, 3/4 Ton		121.70		133.87		
1 S.P. Crane, 4x4, 20 Ton		587.75		646.52	47.59	52.35
72 L.H., Daily Totals		$6068.75		$8113.82	$84.29	$112.69

Crew B-35	Hr.	Daily	Hr.	Daily	Bare Costs	Incl. O&P
1 Labor Foreman (outside)	$32.35	$258.80	$53.45	$427.60	$37.79	$62.32
1 Skilled Worker	39.45	315.60	65.55	524.40		
1 Welder (plumber)	44.85	358.80	73.60	588.80		
1 Laborer	30.35	242.80	50.15	401.20		
1 Equip. Oper. (crane)	41.95	335.60	68.85	550.80		
1 Welder, Electric, 300 amp		56.10		61.71		
1 Hyd. Excavator, .75 C.Y.		681.25		749.38	18.43	20.28
40 L.H., Daily Totals		$2248.95		$3303.89	$56.22	$82.60

Crew B-35A	Hr.	Daily	Hr.	Daily	Bare Costs	Incl. O&P
1 Labor Foreman (outside)	$32.35	$258.80	$53.45	$427.60	$36.46	$60.09
2 Laborers	30.35	485.60	50.15	802.40		
1 Skilled Worker	39.45	315.60	65.55	524.40		
1 Welder (plumber)	44.85	358.80	73.60	588.80		
1 Equip. Oper. (crane)	41.95	335.60	68.85	550.80		
1 Equip. Oper. (oiler)	35.90	287.20	58.90	471.20		
1 Welder, Gas Engine, 300 amp		95.65		105.22		
1 Crawler Crane, 75 Ton		1734.00		1907.40	32.67	35.94
56 L.H., Daily Totals		$3871.25		$5377.81	$69.13	$96.03

Crew B-36	Hr.	Daily	Hr.	Daily	Bare Costs	Incl. O&P
1 Labor Foreman (outside)	$32.35	$258.80	$53.45	$427.60	$34.91	$57.49
2 Laborers	30.35	485.60	50.15	802.40		
2 Equip. Oper. (medium)	40.75	652.00	66.85	1069.60		
1 Dozer, 200 H.P.		1290.00		1419.00		
1 Aggregate Spreader		37.30		41.03		
1 Tandem Roller, 10 Ton		232.05		255.26	38.98	42.88
40 L.H., Daily Totals		$2955.75		$4014.89	$73.89	$100.37

Crew B-36A	Hr.	Daily	Hr.	Daily	Bare Costs	Incl. O&P
1 Labor Foreman (outside)	$32.35	$258.80	$53.45	$427.60	$36.58	$60.16
2 Laborers	30.35	485.60	50.15	802.40		
4 Equip. Oper. (medium)	40.75	1304.00	66.85	2139.20		
1 Dozer, 200 H.P.		1290.00		1419.00		
1 Aggregate Spreader		37.30		41.03		
1 Tandem Roller, 10 Ton		232.05		255.26		
1 Roller, Pneum. Whl., 12 Ton		338.35		372.19	33.89	37.28
56 L.H., Daily Totals		$3946.10		$5456.67	$70.47	$97.44

Crew No.	Bare Costs		Incl. Subs O&P		Cost Per Labor-Hour	
Crew B-36B	**Hr.**	**Daily**	**Hr.**	**Daily**	**Bare Costs**	**Incl. O&P**
1 Labor Foreman (outside)	$32.35	$258.80	$53.45	$427.60	$36.36	$59.80
2 Laborers	30.35	485.60	50.15	802.40		
4 Equip. Oper. (medium)	40.75	1304.00	66.85	2139.20		
1 Truck Driver (heavy)	34.80	278.40	57.25	458.00		
1 Grader, 30,000 Lbs.		657.85		723.63		
1 F.E. Loader, Crl, 1.5 C.Y.		647.15		711.87		
1 Dozer, 300 H.P.		1831.00		2014.10		
1 Roller, Vibratory, 25 Ton		650.90		715.99		
1 Truck Tractor, 6x4, 450 H.P.		594.90		654.39		
1 Water Tank Trailer, 5000 Gal.		149.45		164.40	70.80	77.88
64 L.H., Daily Totals		$6858.05		$8811.58	$107.16	$137.68
Crew B-36C	**Hr.**	**Daily**	**Hr.**	**Daily**	**Bare Costs**	**Incl. O&P**
1 Labor Foreman (outside)	$32.35	$258.80	$53.45	$427.60	$37.88	$62.25
3 Equip. Oper. (medium)	40.75	978.00	66.85	1604.40		
1 Truck Driver (heavy)	34.80	278.40	57.25	458.00		
1 Grader, 30,000 Lbs.		657.85		723.63		
1 Dozer, 300 H.P.		1831.00		2014.10		
1 Roller, Vibratory, 25 Ton		650.90		715.99		
1 Truck Tractor, 6x4, 450 H.P.		594.90		654.39		
1 Water Tank Trailer, 5000 Gal.		149.45		164.40	97.10	106.81
40 L.H., Daily Totals		$5399.30		$6762.51	$134.98	$169.06
Crew B-36E	**Hr.**	**Daily**	**Hr.**	**Daily**	**Bare Costs**	**Incl. O&P**
1 Labor Foreman (outside)	$32.35	$258.80	$53.45	$427.60	$38.36	$63.02
4 Equip. Oper. (medium)	40.75	1304.00	66.85	2139.20		
1 Truck Driver (heavy)	34.80	278.40	57.25	458.00		
1 Grader, 30,000 Lbs.		657.85		723.63		
1 Dozer, 300 H.P.		1831.00		2014.10		
1 Roller, Vibratory, 25 Ton		650.90		715.99		
1 Truck Tractor, 6x4, 380 H.P.		487.60		536.36		
1 Dist. Tanker, 3000 Gallon		323.25		355.57	82.30	90.53
48 L.H., Daily Totals		$5791.80		$7370.46	$120.66	$153.55
Crew B-37	**Hr.**	**Daily**	**Hr.**	**Daily**	**Bare Costs**	**Incl. O&P**
1 Labor Foreman (outside)	$32.35	$258.80	$53.45	$427.60	$32.08	$52.94
4 Laborers	30.35	971.20	50.15	1604.80		
1 Equip. Oper. (light)	38.75	310.00	63.60	508.80		
1 Tandem Roller, 5 Ton		152.45		167.69	3.18	3.49
48 L.H., Daily Totals		$1692.45		$2708.90	$35.26	$56.44
Crew B-37A	**Hr.**	**Daily**	**Hr.**	**Daily**	**Bare Costs**	**Incl. O&P**
2 Laborers	$30.35	$485.60	$50.15	$802.40	$31.60	$52.13
1 Truck Driver (light)	34.10	272.80	56.10	448.80		
1 Flatbed Truck, Gas, 1.5 Ton		195.20		214.72		
1 Tar Kettle, T.M.		134.50		147.95	13.74	15.11
24 L.H., Daily Totals		$1088.10		$1613.87	$45.34	$67.24
Crew B-37B	**Hr.**	**Daily**	**Hr.**	**Daily**	**Bare Costs**	**Incl. O&P**
3 Laborers	$30.35	$728.40	$50.15	$1203.60	$31.29	$51.64
1 Truck Driver (light)	34.10	272.80	56.10	448.80		
1 Flatbed Truck, Gas, 1.5 Ton		195.20		214.72		
1 Tar Kettle, T.M.		134.50		147.95	10.30	11.33
32 L.H., Daily Totals		$1330.90		$2015.07	$41.59	$62.97
Crew B-37C	**Hr.**	**Daily**	**Hr.**	**Daily**	**Bare Costs**	**Incl. O&P**
2 Laborers	$30.35	$485.60	$50.15	$802.40	$32.23	$53.13
2 Truck Drivers (light)	34.10	545.60	56.10	897.60		
2 Flatbed Trucks, Gas, 1.5 Ton		390.40		429.44		
1 Tar Kettle, T.M.		134.50		147.95	16.40	18.04
32 L.H., Daily Totals		$1556.10		$2277.39	$48.63	$71.17

Crew No.	Bare Costs		Incl. Subs O&P		Cost Per Labor-Hour	
Crew B-37D	**Hr.**	**Daily**	**Hr.**	**Daily**	**Bare Costs**	**Incl. O&P**
1 Laborer	$30.35	$242.80	$50.15	$401.20	$32.23	$53.13
1 Truck Driver (light)	34.10	272.80	56.10	448.80		
1 Pickup Truck, 3/4 Ton		109.90		120.89	6.87	7.56
16 L.H., Daily Totals		$625.50		$970.89	$39.09	$60.68
Crew B-37E	**Hr.**	**Daily**	**Hr.**	**Daily**	**Bare Costs**	**Incl. O&P**
3 Laborers	$30.35	$728.40	$50.15	$1203.60	$34.11	$56.16
1 Equip. Oper. (light)	38.75	310.00	63.60	508.80		
1 Equip. Oper. (medium)	40.75	326.00	66.85	534.80		
2 Truck Drivers (light)	34.10	545.60	56.10	897.60		
4 Barrels w/ Flasher		16.00		17.60		
1 Concrete Saw		110.15		121.17		
1 Rotary Hammer Drill		25.05		27.56		
1 Hammer Drill Bit		3.30		3.63		
1 Loader, Skid Steer, 30 H.P.		174.55		192.01		
1 Conc. Hammer Attach.		114.40		125.84		
1 Vibrating Plate, Gas, 18"		31.10		34.21		
2 Flatbed Trucks, Gas, 1.5 Ton		390.40		429.44	15.45	16.99
56 L.H., Daily Totals		$2774.95		$4096.24	$49.55	$73.15
Crew B-37F	**Hr.**	**Daily**	**Hr.**	**Daily**	**Bare Costs**	**Incl. O&P**
3 Laborers	$30.35	$728.40	$50.15	$1203.60	$31.29	$51.64
1 Truck Driver (light)	34.10	272.80	56.10	448.80		
4 Barrels w/ Flasher		16.00		17.60		
1 Concrete Mixer, 10 C.F.		162.00		178.20		
1 Air Compressor, 60 cfm		106.70		117.37		
1 -50' Air Hose, 3/4"		6.55		7.21		
1 Spade (Chipper)		8.30		9.13		
1 Flatbed Truck, Gas, 1.5 Ton		195.20		214.72	15.46	17.01
32 L.H., Daily Totals		$1495.95		$2196.63	$46.75	$68.64
Crew B-37G	**Hr.**	**Daily**	**Hr.**	**Daily**	**Bare Costs**	**Incl. O&P**
1 Labor Foreman (outside)	$32.35	$258.80	$53.45	$427.60	$32.08	$52.94
4 Laborers	30.35	971.20	50.15	1604.80		
1 Equip. Oper. (light)	38.75	310.00	63.60	508.80		
1 Berm Machine		325.30		357.83		
1 Tandem Roller, 5 Ton		152.45		167.69	9.95	10.95
48 L.H., Daily Totals		$2017.75		$3066.72	$42.04	$63.89
Crew B-37H	**Hr.**	**Daily**	**Hr.**	**Daily**	**Bare Costs**	**Incl. O&P**
1 Labor Foreman (outside)	$32.35	$258.80	$53.45	$427.60	$32.08	$52.94
4 Laborers	30.35	971.20	50.15	1604.80		
1 Equip. Oper. (light)	38.75	310.00	63.60	508.80		
1 Tandem Roller, 5 Ton		152.45		167.69		
1 Flatbed Truck, Gas, 1.5 Ton		195.20		214.72		
1 Tar Kettle, T.M.		134.50		147.95	10.04	11.05
48 L.H., Daily Totals		$2022.15		$3071.57	$42.13	$63.99

Crew No.	Bare Costs		Incl. Subs O&P		Cost Per Labor-Hour	
Crew B-37I	**Hr.**	**Daily**	**Hr.**	**Daily**	**Bare Costs**	**Incl. O&P**
3 Laborers	$30.35	$728.40	$50.15	$1203.60	$34.11	$56.16
1 Equip. Oper. (light)	38.75	310.00	63.60	508.80		
1 Equip. Oper. (medium)	40.75	326.00	66.85	534.80		
2 Truck Drivers (light)	34.10	545.60	56.10	897.60		
4 Barrels w/ Flasher		16.00		17.60		
1 Concrete Saw		110.15		121.17		
1 Rotary Hammer Drill		25.05		27.56		
1 Hammer Drill Bit		3.30		3.63		
1 Air Compressor, 60 cfm		106.70		117.37		
1 -50' Air Hose, 3/4"		6.55		7.21		
1 Spade (Chipper)		8.30		9.13		
1 Loader, Skid Steer, 30 H.P.		174.55		192.01		
1 Conc. Hammer Attach.		114.40		125.84		
1 Concrete Mixer, 10 C.F.		162.00		178.20		
1 Vibrating Plate, Gas, 18"		31.10		34.21		
2 Flatbed Trucks, Gas, 1.5 Ton		390.40		429.44	20.51	22.56
56 L.H., Daily Totals		$3058.50		$4408.15	$54.62	$78.72
Crew B-37J	**Hr.**	**Daily**	**Hr.**	**Daily**	**Bare Costs**	**Incl. O&P**
1 Labor Foreman (outside)	$32.35	$258.80	$53.45	$427.60	$32.08	$52.94
4 Laborers	30.35	971.20	50.15	1604.80		
1 Equip. Oper. (light)	38.75	310.00	63.60	508.80		
1 Air Compressor, 60 cfm		106.70		117.37		
1 -50' Air Hose, 3/4"		6.55		7.21		
2 Concrete Mixers, 10 C.F.		324.00		356.40		
2 Flatbed Trucks, Gas, 1.5 Ton		390.40		429.44		
1 Shot Blaster, 20"		201.00		221.10	21.43	23.57
48 L.H., Daily Totals		$2568.65		$3672.72	$53.51	$76.51
Crew B-37K	**Hr.**	**Daily**	**Hr.**	**Daily**	**Bare Costs**	**Incl. O&P**
1 Labor Foreman (outside)	$32.35	$258.80	$53.45	$427.60	$32.08	$52.94
4 Laborers	30.35	971.20	50.15	1604.80		
1 Equip. Oper. (light)	38.75	310.00	63.60	508.80		
1 Air Compressor, 60 cfm		106.70		117.37		
1 -50' Air Hose, 3/4"		6.55		7.21		
2 Flatbed Trucks, Gas, 1.5 Ton		390.40		429.44		
1 Shot Blaster, 20"		201.00		221.10	14.68	16.15
48 L.H., Daily Totals		$2244.65		$3316.32	$46.76	$69.09
Crew B-38	**Hr.**	**Daily**	**Hr.**	**Daily**	**Bare Costs**	**Incl. O&P**
2 Laborers	$30.35	$485.60	$50.15	$802.40	$33.15	$54.63
1 Equip. Oper. (light)	38.75	310.00	63.60	508.80		
1 Backhoe Loader, 48 H.P.		320.20		352.22		
1 Hyd. Hammer (1200 lb.)		171.40		188.54		
1 F.E. Loader, W.M., 4 C.Y.		566.70		623.37		
1 Pvmt. Rem. Bucket		61.70		67.87	46.67	51.33
24 L.H., Daily Totals		$1915.60		$2543.20	$79.82	$105.97
Crew B-39	**Hr.**	**Daily**	**Hr.**	**Daily**	**Bare Costs**	**Incl. O&P**
1 Labor Foreman (outside)	$32.35	$258.80	$53.45	$427.60	$30.68	$50.70
5 Laborers	30.35	1214.00	50.15	2006.00		
1 Air Compressor, 250 cfm		167.95		184.75		
2 Breakers, Pavement, 60 lb.		21.50		23.65		
2 -50' Air Hoses, 1.5"		45.50		50.05	4.89	5.38
48 L.H., Daily Totals		$1707.75		$2692.05	$35.58	$56.08

Crew No.	Bare Costs		Incl. Subs O&P		Cost Per Labor-Hour	
Crew B-40	**Hr.**	**Daily**	**Hr.**	**Daily**	**Bare Costs**	**Incl. O&P**
1 Pile Driver Foreman (outside)	$41.10	$328.80	$70.00	$560.00	$38.54	$65.06
4 Pile Drivers	39.10	1251.20	66.60	2131.20		
1 Building Laborer	30.35	242.80	50.15	401.20		
1 Equip. Oper. (crane)	41.95	335.60	68.85	550.80		
1 Crawler Crane, 40 Ton		1392.00		1531.20		
1 Vibratory Hammer & Gen.		2236.00		2459.60	64.79	71.26
56 L.H., Daily Totals		$5786.40		$7634.00	$103.33	$136.32
Crew B-40B	**Hr.**	**Daily**	**Hr.**	**Daily**	**Bare Costs**	**Incl. O&P**
1 Labor Foreman (outside)	$32.35	$258.80	$53.45	$427.60	$33.54	$55.27
3 Laborers	30.35	728.40	50.15	1203.60		
1 Equip. Oper. (crane)	41.95	335.60	68.85	550.80		
1 Equip. Oper. (oiler)	35.90	287.20	58.90	471.20		
1 Lattice Boom Crane, 40 Ton		1315.00		1446.50	27.40	30.14
48 L.H., Daily Totals		$2925.00		$4099.70	$60.94	$85.41
Crew B-41	**Hr.**	**Daily**	**Hr.**	**Daily**	**Bare Costs**	**Incl. O&P**
1 Labor Foreman (outside)	$32.35	$258.80	$53.45	$427.60	$31.49	$52.00
4 Laborers	30.35	971.20	50.15	1604.80		
.25 Equip. Oper. (crane)	41.95	83.90	68.85	137.70		
.25 Equip. Oper. (oiler)	35.90	71.80	58.90	117.80		
.25 Crawler Crane, 40 Ton		348.00		382.80	7.91	8.70
44 L.H., Daily Totals		$1733.70		$2670.70	$39.40	$60.70
Crew B-42	**Hr.**	**Daily**	**Hr.**	**Daily**	**Bare Costs**	**Incl. O&P**
1 Labor Foreman (outside)	$32.35	$258.80	$53.45	$427.60	$34.36	$56.64
4 Laborers	30.35	971.20	50.15	1604.80		
1 Equip. Oper. (crane)	41.95	335.60	68.85	550.80		
1 Welder	44.85	358.80	73.60	588.80		
1 Hyd. Crane, 25 Ton		581.70		639.87		
1 Welder, Gas Engine, 300 amp		95.65		105.22		
1 Horz. Boring Csg. Mch.		447.00		491.70	20.08	22.09
56 L.H., Daily Totals		$3048.75		$4408.78	$54.44	$78.73
Crew B-43	**Hr.**	**Daily**	**Hr.**	**Daily**	**Bare Costs**	**Incl. O&P**
1 Labor Foreman (outside)	$32.35	$258.80	$53.45	$427.60	$30.75	$50.81
4 Laborers	30.35	971.20	50.15	1604.80		
1 Drill Rig, Truck-Mounted		2428.00		2670.80	60.70	66.77
40 L.H., Daily Totals		$3658.00		$4703.20	$91.45	$117.58
Crew B-44	**Hr.**	**Daily**	**Hr.**	**Daily**	**Bare Costs**	**Incl. O&P**
1 Pile Driver Foreman (outside)	$41.10	$328.80	$70.00	$560.00	$37.52	$63.19
4 Pile Drivers	39.10	1251.20	66.60	2131.20		
1 Equip. Oper. (crane)	41.95	335.60	68.85	550.80		
2 Laborers	30.35	485.60	50.15	802.40		
1 Crawler Crane, 40 Ton		1392.00		1531.20		
1 Lead, 60' High		80.55		88.61		
1 Hammer, Diesel, 15K ft.-lbs.		622.40		684.64	32.73	36.01
64 L.H., Daily Totals		$4496.15		$6348.85	$70.25	$99.20
Crew B-45	**Hr.**	**Daily**	**Hr.**	**Daily**	**Bare Costs**	**Incl. O&P**
1 Building Laborer	$30.35	$242.80	$50.15	$401.20	$32.58	$53.70
1 Truck Driver (heavy)	34.80	278.40	57.25	458.00		
1 Dist. Tanker, 3000 Gallon		323.25		355.57		
1 Truck Tractor, 6x4, 380 H.P.		487.60		536.36	50.68	55.75
16 L.H., Daily Totals		$1332.05		$1751.14	$83.25	$109.45

Crew No.	Bare Costs		Incl. Subs O&P		Cost Per Labor-Hour	
Crew B-46	**Hr.**	**Daily**	**Hr.**	**Daily**	**Bare Costs**	**Incl. O&P**
1 Pile Driver Foreman (outside)	$41.10	$328.80	$70.00	$560.00	$35.06	$58.94
2 Pile Drivers	39.10	625.60	66.60	1065.60		
3 Laborers	30.35	728.40	50.15	1203.60		
1 Chain Saw, Gas, 36" Long		46.65		51.31	0.97	1.07
48 L.H., Daily Totals		$1729.45		$2880.51	$36.03	$60.01
Crew B-47	**Hr.**	**Daily**	**Hr.**	**Daily**	**Bare Costs**	**Incl. O&P**
1 Blast Foreman (outside)	$32.35	$258.80	$53.45	$427.60	$31.35	$51.80
1 Driller	30.35	242.80	50.15	401.20		
1 Air Track Drill, 4"		1004.00		1104.40		
1 Air Compressor, 600 cfm		417.60		459.36		
2 -50' Air Hoses, 3"		51.50		56.65	92.07	101.28
16 L.H., Daily Totals		$1974.70		$2449.21	$123.42	$153.08
Crew B-47A	**Hr.**	**Daily**	**Hr.**	**Daily**	**Bare Costs**	**Incl. O&P**
1 Drilling Foreman (outside)	$32.35	$258.80	$53.45	$427.60	$36.73	$60.40
1 Equip. Oper. (heavy)	41.95	335.60	68.85	550.80		
1 Equip. Oper. (oiler)	35.90	287.20	58.90	471.20		
1 Air Track Drill, 5"		1173.00		1290.30	48.88	53.76
24 L.H., Daily Totals		$2054.60		$2739.90	$85.61	$114.16
Crew B-47C	**Hr.**	**Daily**	**Hr.**	**Daily**	**Bare Costs**	**Incl. O&P**
1 Laborer	$30.35	$242.80	$50.15	$401.20	$34.55	$56.88
1 Equip. Oper. (light)	38.75	310.00	63.60	508.80		
1 Air Compressor, 750 cfm		428.80		471.68		
2 -50' Air Hoses, 3"		51.50		56.65		
1 Air Track Drill, 4"		1004.00		1104.40	92.77	102.05
16 L.H., Daily Totals		$2037.10		$2542.73	$127.32	$158.92
Crew B-47E	**Hr.**	**Daily**	**Hr.**	**Daily**	**Bare Costs**	**Incl. O&P**
1 Labor Foreman (outside)	$32.35	$258.80	$53.45	$427.60	$30.85	$50.98
3 Laborers	30.35	728.40	50.15	1203.60		
1 Flatbed Truck, Gas, 3 Ton		245.95		270.55	7.69	8.45
32 L.H., Daily Totals		$1233.15		$1901.74	$38.54	$59.43
Crew B-47G	**Hr.**	**Daily**	**Hr.**	**Daily**	**Bare Costs**	**Incl. O&P**
1 Labor Foreman (outside)	$32.35	$258.80	$53.45	$427.60	$31.02	$51.25
2 Laborers	30.35	485.60	50.15	802.40		
1 Air Track Drill, 4"		1004.00		1104.40		
1 Air Compressor, 600 cfm		417.60		459.36		
2 -50' Air Hoses, 3"		51.50		56.65		
1 Gunite Pump Rig		313.00		344.30	74.42	81.86
24 L.H., Daily Totals		$2530.50		$3194.71	$105.44	$133.11
Crew B-47H	**Hr.**	**Daily**	**Hr.**	**Daily**	**Bare Costs**	**Incl. O&P**
1 Skilled Worker Foreman (out)	$41.45	$331.60	$68.85	$550.80	$39.95	$66.38
3 Skilled Workers	39.45	946.80	65.55	1573.20		
1 Flatbed Truck, Gas, 3 Ton		245.95		270.55	7.69	8.45
32 L.H., Daily Totals		$1524.35		$2394.55	$47.64	$74.83
Crew B-48	**Hr.**	**Daily**	**Hr.**	**Daily**	**Bare Costs**	**Incl. O&P**
1 Labor Foreman (outside)	$32.35	$258.80	$53.45	$427.60	$32.62	$53.82
4 Laborers	30.35	971.20	50.15	1604.80		
1 Equip. Oper. (crane)	41.95	335.60	68.85	550.80		
1 Centr. Water Pump, 6"		296.45		326.10		
1 -20' Suction Hose, 6"		11.10		12.21		
1 -50' Discharge Hose, 6"		5.70		6.27		
1 Drill Rig, Truck-Mounted		2428.00		2670.80	57.11	62.82
48 L.H., Daily Totals		$4306.85		$5598.57	$89.73	$116.64

Crew No.	Bare Costs		Incl. Subs O&P		Cost Per Labor-Hour	
Crew B-49	**Hr.**	**Daily**	**Hr.**	**Daily**	**Bare Costs**	**Incl. O&P**
1 Labor Foreman (outside)	$32.35	$258.80	$53.45	$427.60	$33.81	$56.25
5 Laborers	30.35	1214.00	50.15	2006.00		
1 Equip. Oper. (crane)	41.95	335.60	68.85	550.80		
2 Pile Drivers	39.10	625.60	66.60	1065.60		
1 Hyd. Crane, 25 Ton		581.70		639.87		
1 Centr. Water Pump, 6"		296.45		326.10		
1 -20' Suction Hose, 6"		11.10		12.21		
1 -50' Discharge Hose, 6"		5.70		6.27		
1 Drill Rig, Truck-Mounted		2428.00		2670.80	46.15	50.77
72 L.H., Daily Totals		$5756.95		$7705.24	$79.96	$107.02
Crew B-50	**Hr.**	**Daily**	**Hr.**	**Daily**	**Bare Costs**	**Incl. O&P**
1 Pile Driver Foreman (outside)	$41.10	$328.80	$70.00	$560.00	$36.11	$60.71
6 Pile Drivers	39.10	1876.80	66.60	3196.80		
1 Equip. Oper. (crane)	41.95	335.60	68.85	550.80		
5 Laborers	30.35	1214.00	50.15	2006.00		
1 Crawler Crane, 40 Ton		1392.00		1531.20		
1 Lead, 60' High		80.55		88.61		
1 Hammer, Diesel, 15K ft.-lbs.		622.40		684.64		
1 Air Compressor, 600 cfm		417.60		459.36		
2 -50' Air Hoses, 3"		51.50		56.65		
1 Chain Saw, Gas, 36" Long		46.65		51.31	25.10	27.61
104 L.H., Daily Totals		$6365.90		$9185.37	$61.21	$88.32
Crew B-51	**Hr.**	**Daily**	**Hr.**	**Daily**	**Bare Costs**	**Incl. O&P**
1 Labor Foreman (outside)	$32.35	$258.80	$53.45	$427.60	$31.31	$51.69
4 Laborers	30.35	971.20	50.15	1604.80		
1 Truck Driver (light)	34.10	272.80	56.10	448.80		
1 Flatbed Truck, Gas, 1.5 Ton		195.20		214.72	4.07	4.47
48 L.H., Daily Totals		$1698.00		$2695.92	$35.38	$56.16
Crew B-52	**Hr.**	**Daily**	**Hr.**	**Daily**	**Bare Costs**	**Incl. O&P**
1 Labor Foreman (outside)	$32.35	$258.80	$53.45	$427.60	$33.31	$55.01
1 Carpenter	38.75	310.00	64.05	512.40		
4 Laborers	30.35	971.20	50.15	1604.80		
.5 Rodman (reinf.)	40.60	162.40	67.15	268.60		
.5 Equip. Oper. (medium)	40.75	163.00	66.85	267.40		
.5 Crawler Loader, 3 C.Y.		601.00		661.10	10.73	11.81
56 L.H., Daily Totals		$2466.40		$3741.90	$44.04	$66.82
Crew B-53	**Hr.**	**Daily**	**Hr.**	**Daily**	**Bare Costs**	**Incl. O&P**
1 Building Laborer	$30.35	$242.80	$50.15	$401.20	$30.35	$50.15
1 Trencher, Chain, 12 H.P.		63.40		69.74	7.92	8.72
8 L.H., Daily Totals		$306.20		$470.94	$38.27	$58.87
Crew B-54	**Hr.**	**Daily**	**Hr.**	**Daily**	**Bare Costs**	**Incl. O&P**
1 Equip. Oper. (light)	$38.75	$310.00	$63.60	$508.80	$38.75	$63.60
1 Trencher, Chain, 40 H.P.		348.50		383.35	43.56	47.92
8 L.H., Daily Totals		$658.50		$892.15	$82.31	$111.52
Crew B-54A	**Hr.**	**Daily**	**Hr.**	**Daily**	**Bare Costs**	**Incl. O&P**
.17 Labor Foreman (outside)	$32.35	$44.00	$53.45	$72.69	$39.53	$64.90
1 Equipment Operator (med.)	40.75	326.00	66.85	534.80		
1 Wheel Trencher, 67 H.P.		1110.00		1221.00	118.59	130.45
9.36 L.H., Daily Totals		$1480.00		$1828.49	$158.12	$195.35
Crew B-54B	**Hr.**	**Daily**	**Hr.**	**Daily**	**Bare Costs**	**Incl. O&P**
.25 Labor Foreman (outside)	$32.35	$64.70	$53.45	$106.90	$39.07	$64.17
1 Equipment Operator (med.)	40.75	326.00	66.85	534.80		
1 Wheel Trencher, 150 H.P.		1996.00		2195.60	199.60	219.56
10 L.H., Daily Totals		$2386.70		$2837.30	$238.67	$283.73

Crew B-54D	Bare Costs Hr.	Bare Costs Daily	Incl. Subs O&P Hr.	Incl. Subs O&P Daily	Cost Per Labor-Hour Bare Costs	Cost Per Labor-Hour Incl. O&P
1 Laborer	$30.35	$242.80	$50.15	$401.20	$35.55	$58.50
1 Equipment Operator (med.)	40.75	326.00	66.85	534.80		
1 Rock Trencher, 6" Width		797.00		876.70	49.81	54.79
16 L.H., Daily Totals		$1365.80		$1812.70	$85.36	$113.29
Crew B-54E	**Hr.**	**Daily**	**Hr.**	**Daily**	**Bare Costs**	**Incl. O&P**
1 Laborer	$30.35	$242.80	$50.15	$401.20	$35.55	$58.50
1 Equipment Operator (med.)	40.75	326.00	66.85	534.80		
1 Rock Trencher, 18" Width		2678.00		2945.80	167.38	184.11
16 L.H., Daily Totals		$3246.80		$3881.80	$202.93	$242.61
Crew B-55	**Hr.**	**Daily**	**Hr.**	**Daily**	**Bare Costs**	**Incl. O&P**
1 Laborer	$30.35	$242.80	$50.15	$401.20	$32.23	$53.13
1 Truck Driver (light)	34.10	272.80	56.10	448.80		
1 Truck-Mounted Earth Auger		756.20		831.82		
1 Flatbed Truck, Gas, 3 Ton		245.95		270.55	62.63	68.90
16 L.H., Daily Totals		$1517.75		$1952.37	$94.86	$122.02
Crew B-56	**Hr.**	**Daily**	**Hr.**	**Daily**	**Bare Costs**	**Incl. O&P**
2 Laborers	$30.35	$485.60	$50.15	$802.40	$30.35	$50.15
1 Air Track Drill, 4"		1004.00		1104.40		
1 Air Compressor, 600 cfm		417.60		459.36		
1 -50' Air Hose, 3"		25.75		28.32	90.46	99.51
16 L.H., Daily Totals		$1932.95		$2394.49	$120.81	$149.66
Crew B-57	**Hr.**	**Daily**	**Hr.**	**Daily**	**Bare Costs**	**Incl. O&P**
1 Labor Foreman (outside)	$32.35	$258.80	$53.45	$427.60	$33.07	$54.55
3 Laborers	30.35	728.40	50.15	1203.60		
1 Equip. Oper. (crane)	41.95	335.60	68.85	550.80		
1 Crawler Crane, 25 Ton		1386.00		1524.60		
1 Clamshell Bucket, 1 C.Y.		51.20		56.32		
1 Centr. Water Pump, 6"		296.45		326.10		
1 -20' Suction Hose, 6"		11.10		12.21		
20 -50' Discharge Hoses, 6"		114.00		125.40	46.47	51.12
40 L.H., Daily Totals		$3181.55		$4226.63	$79.54	$105.67
Crew B-58	**Hr.**	**Daily**	**Hr.**	**Daily**	**Bare Costs**	**Incl. O&P**
2 Laborers	$30.35	$485.60	$50.15	$802.40	$33.15	$54.63
1 Equip. Oper. (light)	38.75	310.00	63.60	508.80		
1 Backhoe Loader, 48 H.P.		320.20		352.22		
1 Small Helicopter, w/ Pilot		2916.00		3207.60	134.84	148.33
24 L.H., Daily Totals		$4031.80		$4871.02	$167.99	$202.96
Crew B-59	**Hr.**	**Daily**	**Hr.**	**Daily**	**Bare Costs**	**Incl. O&P**
1 Truck Driver (heavy)	$34.80	$278.40	$57.25	$458.00	$34.80	$57.25
1 Truck Tractor, 220 H.P.		303.55		333.90		
1 Water Tank Trailer, 5000 Gal.		149.45		164.40	56.63	62.29
8 L.H., Daily Totals		$731.40		$956.30	$91.42	$119.54
Crew B-60	**Hr.**	**Daily**	**Hr.**	**Daily**	**Bare Costs**	**Incl. O&P**
1 Labor Foreman (outside)	$32.35	$258.80	$53.45	$427.60	$34.02	$56.06
3 Laborers	30.35	728.40	50.15	1203.60		
1 Equip. Oper. (crane)	41.95	335.60	68.85	550.80		
1 Equip. Oper. (light)	38.75	310.00	63.60	508.80		
1 Crawler Crane, 40 Ton		1392.00		1531.20		
1 Lead, 60' High		80.55		88.61		
1 Hammer, Diesel, 15K ft.-lbs.		622.40		684.64		
1 Backhoe Loader, 48 H.P.		320.20		352.22	50.32	55.35
48 L.H., Daily Totals		$4047.95		$5347.47	$84.33	$111.41

Crew B-61	Bare Costs Hr.	Bare Costs Daily	Incl. Subs O&P Hr.	Incl. Subs O&P Daily	Cost Per Labor-Hour Bare Costs	Cost Per Labor-Hour Incl. O&P
1 Labor Foreman (outside)	$32.35	$258.80	$53.45	$427.60	$30.75	$50.81
4 Laborers	30.35	971.20	50.15	1604.80		
1 Cement Mixer, 2 C.Y.		203.85		224.24		
1 Air Compressor, 160 cfm		115.85		127.44	7.99	8.79
40 L.H., Daily Totals		$1549.70		$2384.07	$38.74	$59.60
Crew B-62	**Hr.**	**Daily**	**Hr.**	**Daily**	**Bare Costs**	**Incl. O&P**
2 Laborers	$30.35	$485.60	$50.15	$802.40	$33.15	$54.63
1 Equip. Oper. (light)	38.75	310.00	63.60	508.80		
1 Loader, Skid Steer, 30 H.P.		174.55		192.01	7.27	8.00
24 L.H., Daily Totals		$970.15		$1503.20	$40.42	$62.63
Crew B-62A	**Hr.**	**Daily**	**Hr.**	**Daily**	**Bare Costs**	**Incl. O&P**
2 Laborers	$30.35	$485.60	$50.15	$802.40	$33.15	$54.63
1 Equip. Oper. (light)	38.75	310.00	63.60	508.80		
1 Loader, Skid Steer, 30 H.P.		174.55		192.01		
1 Trencher Attachment		70.20		77.22	10.20	11.22
24 L.H., Daily Totals		$1040.35		$1580.43	$43.35	$65.85
Crew B-63	**Hr.**	**Daily**	**Hr.**	**Daily**	**Bare Costs**	**Incl. O&P**
5 Laborers	$30.35	$1214.00	$50.15	$2006.00	$30.35	$50.15
1 Loader, Skid Steer, 30 H.P.		174.55		192.01	4.36	4.80
40 L.H., Daily Totals		$1388.55		$2198.01	$34.71	$54.95
Crew B-63B	**Hr.**	**Daily**	**Hr.**	**Daily**	**Bare Costs**	**Incl. O&P**
1 Labor Foreman (inside)	$30.85	$246.80	$51.00	$408.00	$32.58	$53.73
2 Laborers	30.35	485.60	50.15	802.40		
1 Equip. Oper. (light)	38.75	310.00	63.60	508.80		
1 Loader, Skid Steer, 78 H.P.		392.50		431.75	12.27	13.49
32 L.H., Daily Totals		$1434.90		$2150.95	$44.84	$67.22
Crew B-64	**Hr.**	**Daily**	**Hr.**	**Daily**	**Bare Costs**	**Incl. O&P**
1 Laborer	$30.35	$242.80	$50.15	$401.20	$32.23	$53.13
1 Truck Driver (light)	34.10	272.80	56.10	448.80		
1 Power Mulcher (small)		139.35		153.29		
1 Flatbed Truck, Gas, 1.5 Ton		195.20		214.72	20.91	23.00
16 L.H., Daily Totals		$850.15		$1218.01	$53.13	$76.13
Crew B-65	**Hr.**	**Daily**	**Hr.**	**Daily**	**Bare Costs**	**Incl. O&P**
1 Laborer	$30.35	$242.80	$50.15	$401.20	$32.23	$53.13
1 Truck Driver (light)	34.10	272.80	56.10	448.80		
1 Power Mulcher (Large)		276.35		303.99		
1 Flatbed Truck, Gas, 1.5 Ton		195.20		214.72	29.47	32.42
16 L.H., Daily Totals		$987.15		$1368.70	$61.70	$85.54
Crew B-66	**Hr.**	**Daily**	**Hr.**	**Daily**	**Bare Costs**	**Incl. O&P**
1 Equip. Oper. (light)	$38.75	$310.00	$63.60	$508.80	$38.75	$63.60
1 Loader-Backhoe, 40 H.P.		219.20		241.12	27.40	30.14
8 L.H., Daily Totals		$529.20		$749.92	$66.15	$93.74
Crew B-67	**Hr.**	**Daily**	**Hr.**	**Daily**	**Bare Costs**	**Incl. O&P**
1 Millwright	$41.15	$329.20	$65.55	$524.40	$39.95	$64.58
1 Equip. Oper. (light)	38.75	310.00	63.60	508.80		
1 R.T. Forklift, 5,000 Lb., diesel		265.60		292.16	16.60	18.26
16 L.H., Daily Totals		$904.80		$1325.36	$56.55	$82.83
Crew B-67B	**Hr.**	**Daily**	**Hr.**	**Daily**	**Bare Costs**	**Incl. O&P**
1 Millwright Foreman (inside)	$41.65	$333.20	$66.35	$530.80	$41.40	$65.95
1 Millwright	41.15	329.20	65.55	524.40		
16 L.H., Daily Totals		$662.40		$1055.20	$41.40	$65.95

Crew B-68	Bare Costs Hr.	Bare Costs Daily	Incl. Subs O&P Hr.	Incl. Subs O&P Daily	Cost Per Labor-Hour Bare Costs	Cost Per Labor-Hour Incl. O&P
2 Millwrights	$41.15	$658.40	$65.55	$1048.80	$40.35	$64.90
1 Equip. Oper. (light)	38.75	310.00	63.60	508.80		
1 R.T. Forklift, 5,000 Lb., diesel		265.60		292.16	11.07	12.17
24 L.H., Daily Totals		$1234.00		$1849.76	$51.42	$77.07

Crew B-68A	Hr.	Daily	Hr.	Daily	Bare Costs	Incl. O&P
1 Millwright Foreman (inside)	$41.65	$333.20	$66.35	$530.80	$41.32	$65.82
2 Millwrights	41.15	658.40	65.55	1048.80		
1 Forklift, Smooth Floor, 8,000 Lb.		144.30		158.73	6.01	6.61
24 L.H., Daily Totals		$1135.90		$1738.33	$47.33	$72.43

Crew B-68B	Hr.	Daily	Hr.	Daily	Bare Costs	Incl. O&P
1 Millwright Foreman (inside)	$41.65	$333.20	$66.35	$530.80	$43.55	$70.49
2 Millwrights	41.15	658.40	65.55	1048.80		
2 Electricians	45.60	729.60	74.40	1190.40		
2 Plumbers	44.85	717.60	73.60	1177.60		
1 R.T. Forklift, 5,000 Lb., gas		276.20		303.82	4.93	5.43
56 L.H., Daily Totals		$2715.00		$4251.42	$48.48	$75.92

Crew B-68C	Hr.	Daily	Hr.	Daily	Bare Costs	Incl. O&P
1 Millwright Foreman (inside)	$41.65	$333.20	$66.35	$530.80	$43.31	$69.97
1 Millwright	41.15	329.20	65.55	524.40		
1 Electrician	45.60	364.80	74.40	595.20		
1 Plumber	44.85	358.80	73.60	588.80		
1 R.T. Forklift, 5,000 Lb., gas		276.20		303.82	8.63	9.49
32 L.H., Daily Totals		$1662.20		$2543.02	$51.94	$79.47

Crew B-68D	Hr.	Daily	Hr.	Daily	Bare Costs	Incl. O&P
1 Labor Foreman (inside)	$30.85	$246.80	$51.00	$408.00	$33.32	$54.92
1 Laborer	30.35	242.80	50.15	401.20		
1 Equip. Oper. (light)	38.75	310.00	63.60	508.80		
1 R.T. Forklift, 5,000 Lb., gas		276.20		303.82	11.51	12.66
24 L.H., Daily Totals		$1075.80		$1621.82	$44.83	$67.58

Crew B-68E	Hr.	Daily	Hr.	Daily	Bare Costs	Incl. O&P
1 Struc. Steel Foreman (inside)	$41.95	$335.60	$72.80	$582.40	$41.55	$72.08
3 Struc. Steel Workers	41.45	994.80	71.90	1725.60		
1 Welder	41.45	331.60	71.90	575.20		
1 Forklift, Smooth Floor, 8,000 Lb.		144.30		158.73	3.61	3.97
40 L.H., Daily Totals		$1806.30		$3041.93	$45.16	$76.05

Crew B-68F	Hr.	Daily	Hr.	Daily	Bare Costs	Incl. O&P
1 Skilled Worker Foreman (out)	$41.45	$331.60	$68.85	$550.80	$40.12	$66.65
2 Skilled Workers	39.45	631.20	65.55	1048.80		
1 R.T. Forklift, 5,000 Lb., gas		276.20		303.82	11.51	12.66
24 L.H., Daily Totals		$1239.00		$1903.42	$51.63	$79.31

Crew B-68G	Hr.	Daily	Hr.	Daily	Bare Costs	Incl. O&P
2 Structural Steel Workers	$41.45	$663.20	$71.90	$1150.40	$41.45	$71.90
1 R.T. Forklift, 5,000 Lb., gas		276.20		303.82	17.26	18.99
16 L.H., Daily Totals		$939.40		$1454.22	$58.71	$90.89

Crew B-69	Hr.	Daily	Hr.	Daily	Bare Costs	Incl. O&P
1 Labor Foreman (outside)	$32.35	$258.80	$53.45	$427.60	$33.54	$55.27
3 Laborers	30.35	728.40	50.15	1203.60		
1 Equip. Oper. (crane)	41.95	335.60	68.85	550.80		
1 Equip. Oper. (oiler)	35.90	287.20	58.90	471.20		
1 Hyd. Crane, 80 Ton		1487.00		1635.70	30.98	34.08
48 L.H., Daily Totals		$3097.00		$4288.90	$64.52	$89.35

Crew B-69A	Hr.	Daily	Hr.	Daily	Bare Costs	Incl. O&P
1 Labor Foreman (outside)	$32.35	$258.80	$53.45	$427.60	$33.55	$55.13
3 Laborers	30.35	728.40	50.15	1203.60		
1 Equip. Oper. (medium)	40.75	326.00	66.85	534.80		
1 Concrete Finisher	37.15	297.20	60.05	480.40		
1 Curb/Gutter Paver, 2-Track		1194.00		1313.40	24.88	27.36
48 L.H., Daily Totals		$2804.40		$3959.80	$58.42	$82.50

Crew B-69B	Hr.	Daily	Hr.	Daily	Bare Costs	Incl. O&P
1 Labor Foreman (outside)	$32.35	$258.80	$53.45	$427.60	$33.55	$55.13
3 Laborers	30.35	728.40	50.15	1203.60		
1 Equip. Oper. (medium)	40.75	326.00	66.85	534.80		
1 Cement Finisher	37.15	297.20	60.05	480.40		
1 Curb/Gutter Paver, 4-Track		776.35		853.99	16.17	17.79
48 L.H., Daily Totals		$2386.75		$3500.39	$49.72	$72.92

Crew B-70	Hr.	Daily	Hr.	Daily	Bare Costs	Incl. O&P
1 Labor Foreman (outside)	$32.35	$258.80	$53.45	$427.60	$35.09	$57.78
3 Laborers	30.35	728.40	50.15	1203.60		
3 Equip. Oper. (medium)	40.75	978.00	66.85	1604.40		
1 Grader, 30,000 Lbs.		657.85		723.63		
1 Ripper, Beam & 1 Shank		88.80		97.68		
1 Road Sweeper, S.P., 8' wide		703.10		773.41		
1 F.E. Loader, W.M., 1.5 C.Y.		352.95		388.25	32.19	35.41
56 L.H., Daily Totals		$3767.90		$5218.57	$67.28	$93.19

Crew B-71	Hr.	Daily	Hr.	Daily	Bare Costs	Incl. O&P
1 Labor Foreman (outside)	$32.35	$258.80	$53.45	$427.60	$35.09	$57.78
3 Laborers	30.35	728.40	50.15	1203.60		
3 Equip. Oper. (medium)	40.75	978.00	66.85	1604.40		
1 Pvmt. Profiler, 750 H.P.		5381.00		5919.10		
1 Road Sweeper, S.P., 8' wide		703.10		773.41		
1 F.E. Loader, W.M., 1.5 C.Y.		352.95		388.25	114.95	126.44
56 L.H., Daily Totals		$8402.25		$10316.36	$150.04	$184.22

Crew B-72	Hr.	Daily	Hr.	Daily	Bare Costs	Incl. O&P
1 Labor Foreman (outside)	$32.35	$258.80	$53.45	$427.60	$35.80	$58.91
3 Laborers	30.35	728.40	50.15	1203.60		
4 Equip. Oper. (medium)	40.75	1304.00	66.85	2139.20		
1 Pvmt. Profiler, 750 H.P.		5381.00		5919.10		
1 Hammermill, 250 H.P.		1890.00		2079.00		
1 Windrow Loader		1238.00		1361.80		
1 Mix Paver, 165 H.P.		2107.00		2317.70		
1 Roller, Pneum. Whl., 12 Ton		338.35		372.19	171.16	188.28
64 L.H., Daily Totals		$13245.55		$15820.18	$206.96	$247.19

Crew B-73	Hr.	Daily	Hr.	Daily	Bare Costs	Incl. O&P
1 Labor Foreman (outside)	$32.35	$258.80	$53.45	$427.60	$37.10	$61.00
2 Laborers	30.35	485.60	50.15	802.40		
5 Equip. Oper. (medium)	40.75	1630.00	66.85	2674.00		
1 Road Mixer, 310 H.P.		1857.00		2042.70		
1 Tandem Roller, 10 Ton		232.05		255.26		
1 Hammermill, 250 H.P.		1890.00		2079.00		
1 Grader, 30,000 Lbs.		657.85		723.63		
.5 F.E. Loader, W.M., 1.5 C.Y.		176.47		194.12		
.5 Truck Tractor, 220 H.P.		151.78		166.95		
.5 Water Tank Trailer, 5000 Gal.		74.72		82.20	78.75	86.62
64 L.H., Daily Totals		$7414.27		$9447.86	$115.85	$147.62

Crew No.	Bare Costs		Incl. Subs O&P		Cost Per Labor-Hour	
Crew B-74	**Hr.**	**Daily**	**Hr.**	**Daily**	**Bare Costs**	**Incl. O&P**
1 Labor Foreman (outside)	$32.35	$258.80	$53.45	$427.60	$36.91	$60.69
1 Laborer	30.35	242.80	50.15	401.20		
4 Equip. Oper. (medium)	40.75	1304.00	66.85	2139.20		
2 Truck Drivers (heavy)	34.80	556.80	57.25	916.00		
1 Grader, 30,000 Lbs.		657.85		723.63		
1 Ripper, Beam & 1 Shank		88.80		97.68		
2 Stabilizers, 310 H.P.		3252.00		3577.20		
1 Flatbed Truck, Gas, 3 Ton		245.95		270.55		
1 Chem. Spreader, Towed		53.20		58.52		
1 Roller, Vibratory, 25 Ton		650.90		715.99		
1 Water Tank Trailer, 5000 Gal.		149.45		164.40		
1 Truck Tractor, 220 H.P.		303.55		333.90	84.40	92.84
64 L.H., Daily Totals		$7764.10		$9825.87	$121.31	$153.53
Crew B-75	**Hr.**	**Daily**	**Hr.**	**Daily**	**Bare Costs**	**Incl. O&P**
1 Labor Foreman (outside)	$32.35	$258.80	$53.45	$427.60	$37.21	$61.18
1 Laborer	30.35	242.80	50.15	401.20		
4 Equip. Oper. (medium)	40.75	1304.00	66.85	2139.20		
1 Truck Driver (heavy)	34.80	278.40	57.25	458.00		
1 Grader, 30,000 Lbs.		657.85		723.63		
1 Ripper, Beam & 1 Shank		88.80		97.68		
2 Stabilizers, 310 H.P.		3252.00		3577.20		
1 Dist. Tanker, 3000 Gallon		323.25		355.57		
1 Truck Tractor, 6x4, 380 H.P.		487.60		536.36		
1 Roller, Vibratory, 25 Ton		650.90		715.99	97.51	107.26
56 L.H., Daily Totals		$7544.40		$9432.44	$134.72	$168.44
Crew B-76	**Hr.**	**Daily**	**Hr.**	**Daily**	**Bare Costs**	**Incl. O&P**
1 Dock Builder Foreman (outside)	$41.10	$328.80	$70.00	$560.00	$39.60	$66.62
5 Dock Builders	39.10	1564.00	66.60	2664.00		
2 Equip. Oper. (crane)	41.95	671.20	68.85	1101.60		
1 Equip. Oper. (oiler)	35.90	287.20	58.90	471.20		
1 Crawler Crane, 50 Ton		2032.00		2235.20		
1 Barge, 400 Ton		836.50		920.15		
1 Hammer, Diesel, 15K ft.-lbs.		622.40		684.64		
1 Lead, 60' High		80.55		88.61		
1 Air Compressor, 600 cfm		417.60		459.36		
2 -50' Air Hoses, 3"		51.50		56.65	56.12	61.73
72 L.H., Daily Totals		$6891.75		$9241.41	$95.72	$128.35
Crew B-76A	**Hr.**	**Daily**	**Hr.**	**Daily**	**Bare Costs**	**Incl. O&P**
1 Labor Foreman (outside)	$32.35	$258.80	$53.45	$427.60	$32.74	$53.99
5 Laborers	30.35	1214.00	50.15	2006.00		
1 Equip. Oper. (crane)	41.95	335.60	68.85	550.80		
1 Equip. Oper. (oiler)	35.90	287.20	58.90	471.20		
1 Crawler Crane, 50 Ton		2032.00		2235.20		
1 Barge, 400 Ton		836.50		920.15	44.82	49.30
64 L.H., Daily Totals		$4964.10		$6610.95	$77.56	$103.30
Crew B-77	**Hr.**	**Daily**	**Hr.**	**Daily**	**Bare Costs**	**Incl. O&P**
1 Labor Foreman (outside)	$32.35	$258.80	$53.45	$427.60	$31.50	$52.00
3 Laborers	30.35	728.40	50.15	1203.60		
1 Truck Driver (light)	34.10	272.80	56.10	448.80		
1 Crack Cleaner, 25 H.P.		56.95		62.65		
1 Crack Filler, Trailer Mtd.		198.95		218.85		
1 Flatbed Truck, Gas, 3 Ton		245.95		270.55	12.55	13.80
40 L.H., Daily Totals		$1761.85		$2632.03	$44.05	$65.80

Crew No.	Bare Costs		Incl. Subs O&P		Cost Per Labor-Hour	
Crew B-78	**Hr.**	**Daily**	**Hr.**	**Daily**	**Bare Costs**	**Incl. O&P**
1 Labor Foreman (outside)	$32.35	$258.80	$53.45	$427.60	$30.75	$50.81
4 Laborers	30.35	971.20	50.15	1604.80		
1 Paint Striper, S.P., 40 Gallon		152.25		167.47		
1 Flatbed Truck, Gas, 3 Ton		245.95		270.55		
1 Pickup Truck, 3/4 Ton		109.90		120.89	12.70	13.97
40 L.H., Daily Totals		$1738.10		$2591.31	$43.45	$64.78
Crew B-78B	**Hr.**	**Daily**	**Hr.**	**Daily**	**Bare Costs**	**Incl. O&P**
2 Laborers	$30.35	$485.60	$50.15	$802.40	$31.28	$51.64
.25 Equip. Oper. (light)	38.75	77.50	63.60	127.20		
1 Pickup Truck, 3/4 Ton		109.90		120.89		
1 Line Rem.,11 H.P.,Walk Behind		61.35		67.48		
.25 Road Sweeper, S.P., 8' wide		175.78		193.35	19.28	21.21
18 L.H., Daily Totals		$910.13		$1311.33	$50.56	$72.85
Crew B-78C	**Hr.**	**Daily**	**Hr.**	**Daily**	**Bare Costs**	**Incl. O&P**
1 Labor Foreman (outside)	$32.35	$258.80	$53.45	$427.60	$31.31	$51.69
4 Laborers	30.35	971.20	50.15	1604.80		
1 Truck Driver (light)	34.10	272.80	56.10	448.80		
1 Paint Striper, T.M., 120 Gal.		686.10		754.71		
1 Flatbed Truck, Gas, 3 Ton		245.95		270.55		
1 Pickup Truck, 3/4 Ton		109.90		120.89	21.71	23.88
48 L.H., Daily Totals		$2544.75		$3627.34	$53.02	$75.57
Crew B-78D	**Hr.**	**Daily**	**Hr.**	**Daily**	**Bare Costs**	**Incl. O&P**
2 Labor Foremen (outside)	$32.35	$517.60	$53.45	$855.20	$31.13	$51.41
7 Laborers	30.35	1699.60	50.15	2808.40		
1 Truck Driver (light)	34.10	272.80	56.10	448.80		
1 Paint Striper, T.M., 120 Gal.		686.10		754.71		
1 Flatbed Truck, Gas, 3 Ton		245.95		270.55		
3 Pickup Trucks, 3/4 Ton		329.70		362.67		
1 Air Compressor, 60 cfm		106.70		117.37		
1 -50' Air Hose, 3/4"		6.55		7.21		
1 Breaker, Pavement, 60 lb.		10.75		11.82	17.32	19.05
80 L.H., Daily Totals		$3875.75		$5636.73	$48.45	$70.46
Crew B-78E	**Hr.**	**Daily**	**Hr.**	**Daily**	**Bare Costs**	**Incl. O&P**
2 Labor Foremen (outside)	$32.35	$517.60	$53.45	$855.20	$31.00	$51.20
9 Laborers	30.35	2185.20	50.15	3610.80		
1 Truck Driver (light)	34.10	272.80	56.10	448.80		
1 Paint Striper, T.M., 120 Gal.		686.10		754.71		
1 Flatbed Truck, Gas, 3 Ton		245.95		270.55		
4 Pickup Trucks, 3/4 Ton		439.60		483.56		
2 Air Compressors, 60 cfm		213.40		234.74		
2 -50' Air Hoses, 3/4"		13.10		14.41		
2 Breakers, Pavement, 60 lb.		21.50		23.65	16.87	18.56
96 L.H., Daily Totals		$4595.25		$6696.42	$47.87	$69.75
Crew B-78F	**Hr.**	**Daily**	**Hr.**	**Daily**	**Bare Costs**	**Incl. O&P**
2 Labor Foremen (outside)	$32.35	$517.60	$53.45	$855.20	$30.90	$51.05
11 Laborers	30.35	2670.80	50.15	4413.20		
1 Truck Driver (light)	34.10	272.80	56.10	448.80		
1 Paint Striper, T.M., 120 Gal.		686.10		754.71		
1 Flatbed Truck, Gas, 3 Ton		245.95		270.55		
7 Pickup Trucks, 3/4 Ton		769.30		846.23		
3 Air Compressors, 60 cfm		320.10		352.11		
3 -50' Air Hoses, 3/4"		19.65		21.61		
3 Breakers, Pavement, 60 lb.		32.25		35.48	18.51	20.36
112 L.H., Daily Totals		$5534.55		$7997.89	$49.42	$71.41

Crew B-79	Hr.	Daily	Hr.	Daily	Bare Costs	Incl. O&P
1 Labor Foreman (outside)	$32.35	$258.80	$53.45	$427.60	$31.50	$52.00
3 Laborers	30.35	728.40	50.15	1203.60		
1 Truck Driver (light)	34.10	272.80	56.10	448.80		
1 Paint Striper, T.M., 120 Gal.		686.10		754.71		
1 Heating Kettle, 115 Gallon		77.85		85.64		
1 Flatbed Truck, Gas, 3 Ton		245.95		270.55		
2 Pickup Trucks, 3/4 Ton		219.80		241.78	30.74	33.82
40 L.H., Daily Totals		$2489.70		$3432.67	$62.24	$85.82

Crew B-79B	Hr.	Daily	Hr.	Daily	Bare Costs	Incl. O&P
1 Laborer	$30.35	$242.80	$50.15	$401.20	$30.35	$50.15
1 Set of Gases		167.85		184.63	20.98	23.08
8 L.H., Daily Totals		$410.65		$585.84	$51.33	$73.23

Crew B-79C	Hr.	Daily	Hr.	Daily	Bare Costs	Incl. O&P
1 Labor Foreman (outside)	$32.35	$258.80	$53.45	$427.60	$31.17	$51.47
5 Laborers	30.35	1214.00	50.15	2006.00		
1 Truck Driver (light)	34.10	272.80	56.10	448.80		
1 Paint Striper, T.M., 120 Gal.		686.10		754.71		
1 Heating Kettle, 115 Gallon		77.85		85.64		
1 Flatbed Truck, Gas, 3 Ton		245.95		270.55		
3 Pickup Trucks, 3/4 Ton		329.70		362.67		
1 Air Compressor, 60 cfm		106.70		117.37		
1 -50' Air Hose, 3/4"		6.55		7.21		
1 Breaker, Pavement, 60 lb.		10.75		11.82	26.14	28.75
56 L.H., Daily Totals		$3209.20		$4492.36	$57.31	$80.22

Crew B-79D	Hr.	Daily	Hr.	Daily	Bare Costs	Incl. O&P
2 Labor Foremen (outside)	$32.35	$517.60	$53.45	$855.20	$31.32	$51.72
5 Laborers	30.35	1214.00	50.15	2006.00		
1 Truck Driver (light)	34.10	272.80	56.10	448.80		
1 Paint Striper, T.M., 120 Gal.		686.10		754.71		
1 Heating Kettle, 115 Gallon		77.85		85.64		
1 Flatbed Truck, Gas, 3 Ton		245.95		270.55		
4 Pickup Trucks, 3/4 Ton		439.60		483.56		
1 Air Compressor, 60 cfm		106.70		117.37		
1 -50' Air Hose, 3/4"		6.55		7.21		
1 Breaker, Pavement, 60 lb.		10.75		11.82	24.59	27.04
64 L.H., Daily Totals		$3577.90		$5040.85	$55.90	$78.76

Crew B-79E	Hr.	Daily	Hr.	Daily	Bare Costs	Incl. O&P
2 Labor Foremen (outside)	$32.35	$517.60	$53.45	$855.20	$31.13	$51.41
7 Laborers	30.35	1699.60	50.15	2808.40		
1 Truck Driver (light)	34.10	272.80	56.10	448.80		
1 Paint Striper, T.M., 120 Gal.		686.10		754.71		
1 Heating Kettle, 115 Gallon		77.85		85.64		
1 Flatbed Truck, Gas, 3 Ton		245.95		270.55		
5 Pickup Trucks, 3/4 Ton		549.50		604.45		
2 Air Compressors, 60 cfm		213.40		234.74		
2 -50' Air Hoses, 3/4"		13.10		14.41		
2 Breakers, Pavement, 60 lb.		21.50		23.65	22.59	24.85
80 L.H., Daily Totals		$4297.40		$6100.54	$53.72	$76.26

Crew B-80	Hr.	Daily	Hr.	Daily	Bare Costs	Incl. O&P
1 Labor Foreman (outside)	$32.35	$258.80	$53.45	$427.60	$31.02	$51.25
2 Laborers	30.35	485.60	50.15	802.40		
1 Flatbed Truck, Gas, 3 Ton		245.95		270.55		
1 Earth Auger, Truck-Mtd.		385.85		424.44	26.32	28.96
24 L.H., Daily Totals		$1376.20		$1924.98	$57.34	$80.21

Crew B-80A	Hr.	Daily	Hr.	Daily	Bare Costs	Incl. O&P
3 Laborers	$30.35	$728.40	$50.15	$1203.60	$30.35	$50.15
1 Flatbed Truck, Gas, 3 Ton		245.95		270.55	10.25	11.27
24 L.H., Daily Totals		$974.35		$1474.15	$40.60	$61.42

Crew B-80B	Hr.	Daily	Hr.	Daily	Bare Costs	Incl. O&P
3 Laborers	$30.35	$728.40	$50.15	$1203.60	$32.45	$53.51
1 Equip. Oper. (light)	38.75	310.00	63.60	508.80		
1 Crane, Flatbed Mounted, 3 Ton		232.65		255.91	7.27	8.00
32 L.H., Daily Totals		$1271.05		$1968.32	$39.72	$61.51

Crew B-80C	Hr.	Daily	Hr.	Daily	Bare Costs	Incl. O&P
2 Laborers	$30.35	$485.60	$50.15	$802.40	$31.60	$52.13
1 Truck Driver (light)	34.10	272.80	56.10	448.80		
1 Flatbed Truck, Gas, 1.5 Ton		195.20		214.72		
1 Manual Fence Post Auger, Gas		7.40		8.14	8.44	9.29
24 L.H., Daily Totals		$961.00		$1474.06	$40.04	$61.42

Crew B-81	Hr.	Daily	Hr.	Daily	Bare Costs	Incl. O&P
1 Laborer	$30.35	$242.80	$50.15	$401.20	$32.58	$53.70
1 Truck Driver (heavy)	34.80	278.40	57.25	458.00		
1 Hydromulcher, T.M., 3000 Gal.		292.90		322.19		
1 Truck Tractor, 220 H.P.		303.55		333.90	37.28	41.01
16 L.H., Daily Totals		$1117.65		$1515.30	$69.85	$94.71

Crew B-81A	Hr.	Daily	Hr.	Daily	Bare Costs	Incl. O&P
1 Laborer	$30.35	$242.80	$50.15	$401.20	$32.23	$53.13
1 Truck Driver (light)	34.10	272.80	56.10	448.80		
1 Hydromulcher, T.M., 600 Gal.		124.15		136.57		
1 Flatbed Truck, Gas, 3 Ton		245.95		270.55	23.13	25.44
16 L.H., Daily Totals		$885.70		$1257.11	$55.36	$78.57

Crew B-82	Hr.	Daily	Hr.	Daily	Bare Costs	Incl. O&P
1 Laborer	$30.35	$242.80	$50.15	$401.20	$34.55	$56.88
1 Equip. Oper. (light)	38.75	310.00	63.60	508.80		
1 Horiz. Borer, 6 H.P.		77.85		85.64	4.87	5.35
16 L.H., Daily Totals		$630.65		$995.63	$39.42	$62.23

Crew B-82A	Hr.	Daily	Hr.	Daily	Bare Costs	Incl. O&P
2 Laborers	$30.35	$485.60	$50.15	$802.40	$34.55	$56.88
2 Equip. Opers. (light)	38.75	620.00	63.60	1017.60		
2 Dump Trucks, 8 C.Y., 220 H.P.		679.20		747.12		
1 Flatbed Trailer, 25 Ton		133.00		146.30		
1 Horiz. Dir. Drill, 20k lb. Thrust		631.30		694.43		
1 Mud Trailer for HDD, 1500 Gal.		298.65		328.51		
1 Pickup Truck, 4x4, 3/4 Ton		121.70		133.87		
1 Flatbed Trailer, 3 Ton		26.55		29.20		
1 Loader, Skid Steer, 78 H.P.		392.50		431.75	71.34	78.47
32 L.H., Daily Totals		$3388.50		$4331.19	$105.89	$135.35

Crew B-82B	Hr.	Daily	Hr.	Daily	Bare Costs	Incl. O&P
2 Laborers	$30.35	$485.60	$50.15	$802.40	$34.55	$56.88
2 Equip. Opers. (light)	38.75	620.00	63.60	1017.60		
2 Dump Trucks, 8 C.Y., 220 H.P.		679.20		747.12		
1 Flatbed Trailer, 25 Ton		133.00		146.30		
1 Horiz. Dir. Drill, 30k lb. Thrust		897.00		986.70		
1 Mud Trailer for HDD, 1500 Gal.		298.65		328.51		
1 Pickup Truck, 4x4, 3/4 Ton		121.70		133.87		
1 Flatbed Trailer, 3 Ton		26.55		29.20		
1 Loader, Skid Steer, 78 H.P.		392.50		431.75	79.64	87.61
32 L.H., Daily Totals		$3654.20		$4623.46	$114.19	$144.48

Crew B-82C	Hr.	Daily	Hr.	Daily	Bare Costs	Incl. O&P
2 Laborers	$30.35	$485.60	$50.15	$802.40	$34.55	$56.88
2 Equip. Opers. (light)	38.75	620.00	63.60	1017.60		
2 Dump Trucks, 8 C.Y., 220 H.P.		679.20		747.12		
1 Flatbed Trailer, 25 Ton		133.00		146.30		
1 Horiz. Dir. Drill, 50k lb. Thrust		1185.00		1303.50		
1 Mud Trailer for HDD, 1500 Gal.		298.65		328.51		
1 Pickup Truck, 4x4, 3/4 Ton		121.70		133.87		
1 Flatbed Trailer, 3 Ton		26.55		29.20		
1 Loader, Skid Steer, 78 H.P.		392.50		431.75	88.64	97.51
32 L.H., Daily Totals		$3942.20		$4940.26	$123.19	$154.38

Crew B-82D	Hr.	Daily	Hr.	Daily	Bare Costs	Incl. O&P
1 Equip. Oper. (light)	$38.75	$310.00	$63.60	$508.80	$38.75	$63.60
1 Mud Trailer for HDD, 1500 Gal.		298.65		328.51	37.33	41.06
8 L.H., Daily Totals		$608.65		$837.32	$76.08	$104.66

Crew B-83	Hr.	Daily	Hr.	Daily	Bare Costs	Incl. O&P
1 Tugboat Captain	$40.75	$326.00	$66.85	$534.80	$35.55	$58.50
1 Tugboat Hand	30.35	242.80	50.15	401.20		
1 Tugboat, 250 H.P.		710.70		781.77	44.42	48.86
16 L.H., Daily Totals		$1279.50		$1717.77	$79.97	$107.36

Crew B-84	Hr.	Daily	Hr.	Daily	Bare Costs	Incl. O&P
1 Equip. Oper. (medium)	$40.75	$326.00	$66.85	$534.80	$40.75	$66.85
1 Rotary Mower/Tractor		360.30		396.33	45.04	49.54
8 L.H., Daily Totals		$686.30		$931.13	$85.79	$116.39

Crew B-85	Hr.	Daily	Hr.	Daily	Bare Costs	Incl. O&P
3 Laborers	$30.35	$728.40	$50.15	$1203.60	$33.32	$54.91
1 Equip. Oper. (medium)	40.75	326.00	66.85	534.80		
1 Truck Driver (heavy)	34.80	278.40	57.25	458.00		
1 Telescoping Boom Lift, to 80'		670.65		737.72		
1 Brush Chipper, 12", 130 H.P.		389.40		428.34		
1 Pruning Saw, Rotary		6.60		7.26	26.67	29.33
40 L.H., Daily Totals		$2399.45		$3369.72	$59.99	$84.24

Crew B-86	Hr.	Daily	Hr.	Daily	Bare Costs	Incl. O&P
1 Equip. Oper. (medium)	$40.75	$326.00	$66.85	$534.80	$40.75	$66.85
1 Stump Chipper, S.P.		182.20		200.42	22.77	25.05
8 L.H., Daily Totals		$508.20		$735.22	$63.52	$91.90

Crew B-86A	Hr.	Daily	Hr.	Daily	Bare Costs	Incl. O&P
1 Equip. Oper. (medium)	$40.75	$326.00	$66.85	$534.80	$40.75	$66.85
1 Grader, 30,000 Lbs.		657.85		723.63	82.23	90.45
8 L.H., Daily Totals		$983.85		$1258.43	$122.98	$157.30

Crew B-86B	Hr.	Daily	Hr.	Daily	Bare Costs	Incl. O&P
1 Equip. Oper. (medium)	$40.75	$326.00	$66.85	$534.80	$40.75	$66.85
1 Dozer, 200 H.P.		1290.00		1419.00	161.25	177.38
8 L.H., Daily Totals		$1616.00		$1953.80	$202.00	$244.22

Crew B-87	Hr.	Daily	Hr.	Daily	Bare Costs	Incl. O&P
1 Laborer	$30.35	$242.80	$50.15	$401.20	$38.67	$63.51
4 Equip. Oper. (medium)	40.75	1304.00	66.85	2139.20		
2 Feller Bunchers, 100 H.P.		1607.10		1767.81		
1 Log Chipper, 22" Tree		770.10		847.11		
1 Dozer, 105 H.P.		609.70		670.67		
1 Chain Saw, Gas, 36" Long		46.65		51.31	75.84	83.42
40 L.H., Daily Totals		$4580.35		$5877.31	$114.51	$146.93

Crew B-88	Hr.	Daily	Hr.	Daily	Bare Costs	Incl. O&P
1 Laborer	$30.35	$242.80	$50.15	$401.20	$39.26	$64.46
6 Equip. Oper. (medium)	40.75	1956.00	66.85	3208.80		
2 Feller Bunchers, 100 H.P.		1607.10		1767.81		
1 Log Chipper, 22" Tree		770.10		847.11		
2 Log Skidders, 50 H.P.		1696.50		1866.15		
1 Dozer, 105 H.P.		609.70		670.67		
1 Chain Saw, Gas, 36" Long		46.65		51.31	84.47	92.91
56 L.H., Daily Totals		$6928.85		$8813.06	$123.73	$157.38

Crew B-89	Hr.	Daily	Hr.	Daily	Bare Costs	Incl. O&P
1 Skilled Worker	$39.45	$315.60	$65.55	$524.40	$34.90	$57.85
1 Building Laborer	30.35	242.80	50.15	401.20		
1 Flatbed Truck, Gas, 3 Ton		245.95		270.55		
1 Concrete Saw		110.15		121.17		
1 Water Tank, 65 Gal.		98.85		108.74	28.43	31.28
16 L.H., Daily Totals		$1013.35		$1426.05	$63.33	$89.13

Crew B-89A	Hr.	Daily	Hr.	Daily	Bare Costs	Incl. O&P
1 Skilled Worker	$39.45	$315.60	$65.55	$524.40	$34.90	$57.85
1 Laborer	30.35	242.80	50.15	401.20		
1 Core Drill (Large)		112.20		123.42	7.01	7.71
16 L.H., Daily Totals		$670.60		$1049.02	$41.91	$65.56

Crew B-89B	Hr.	Daily	Hr.	Daily	Bare Costs	Incl. O&P
1 Equip. Oper. (light)	$38.75	$310.00	$63.60	$508.80	$36.42	$59.85
1 Truck Driver (light)	34.10	272.80	56.10	448.80		
1 Wall Saw, Hydraulic, 10 H.P.		46.60		51.26		
1 Generator, Diesel, 100 kW		308.80		339.68		
1 Water Tank, 65 Gal.		98.85		108.74		
1 Flatbed Truck, Gas, 3 Ton		245.95		270.55	43.76	48.14
16 L.H., Daily Totals		$1283.00		$1727.82	$80.19	$107.99

Crew B-90	Hr.	Daily	Hr.	Daily	Bare Costs	Incl. O&P
1 Labor Foreman (outside)	$32.35	$258.80	$53.45	$427.60	$33.81	$55.70
3 Laborers	30.35	728.40	50.15	1203.60		
2 Equip. Oper. (light)	38.75	620.00	63.60	1017.60		
2 Truck Drivers (heavy)	34.80	556.80	57.25	916.00		
1 Road Mixer, 310 H.P.		1857.00		2042.70		
1 Dist. Truck, 2000 Gal.		293.20		322.52	33.60	36.96
64 L.H., Daily Totals		$4314.20		$5930.02	$67.41	$92.66

Crew B-90A	Hr.	Daily	Hr.	Daily	Bare Costs	Incl. O&P
1 Labor Foreman (outside)	$32.35	$258.80	$53.45	$427.60	$36.58	$60.16
2 Laborers	30.35	485.60	50.15	802.40		
4 Equip. Oper. (medium)	40.75	1304.00	66.85	2139.20		
2 Graders, 30,000 Lbs.		1315.70		1447.27		
1 Tandem Roller, 10 Ton		232.05		255.26		
1 Roller, Pneum. Whl., 12 Ton		338.35		372.19	33.68	37.05
56 L.H., Daily Totals		$3934.50		$5443.91	$70.26	$97.21

Crew B-90B	Hr.	Daily	Hr.	Daily	Bare Costs	Incl. O&P
1 Labor Foreman (outside)	$32.35	$258.80	$53.45	$427.60	$35.88	$59.05
2 Laborers	30.35	485.60	50.15	802.40		
3 Equip. Oper. (medium)	40.75	978.00	66.85	1604.40		
1 Roller, Pneum. Whl., 12 Ton		338.35		372.19		
1 Road Mixer, 310 H.P.		1857.00		2042.70	45.74	50.31
48 L.H., Daily Totals		$3917.75		$5249.28	$81.62	$109.36

Crew No.	Bare Costs		Incl. Subs O&P		Cost Per Labor-Hour	
Crew B-90C	**Hr.**	**Daily**	**Hr.**	**Daily**	**Bare Costs**	**Incl. O&P**
1 Labor Foreman (outside)	$32.35	$258.80	$53.45	$427.60	$34.58	$56.94
4 Laborers	30.35	971.20	50.15	1604.80		
3 Equip. Oper. (medium)	40.75	978.00	66.85	1604.40		
3 Truck Drivers (heavy)	34.80	835.20	57.25	1374.00		
3 Road Mixers, 310 H.P.		5571.00		6128.10	63.31	69.64
88 L.H., Daily Totals		$8614.20		$11138.90	$97.89	$126.58
Crew B-90D	**Hr.**	**Daily**	**Hr.**	**Daily**	**Bare Costs**	**Incl. O&P**
1 Labor Foreman (outside)	$32.35	$258.80	$53.45	$427.60	$33.93	$55.90
6 Laborers	30.35	1456.80	50.15	2407.20		
3 Equip. Oper. (medium)	40.75	978.00	66.85	1604.40		
3 Truck Drivers (heavy)	34.80	835.20	57.25	1374.00		
3 Road Mixers, 310 H.P.		5571.00		6128.10	53.57	58.92
104 L.H., Daily Totals		$9099.80		$11941.30	$87.50	$114.82
Crew B-90E	**Hr.**	**Daily**	**Hr.**	**Daily**	**Bare Costs**	**Incl. O&P**
1 Labor Foreman (outside)	$32.35	$258.80	$53.45	$427.60	$34.53	$56.87
4 Laborers	30.35	971.20	50.15	1604.80		
3 Equip. Oper. (medium)	40.75	978.00	66.85	1604.40		
1 Truck Driver (heavy)	34.80	278.40	57.25	458.00		
1 Road Mixer, 310 H.P.		1857.00		2042.70	25.79	28.37
72 L.H., Daily Totals		$4343.40		$6137.50	$60.33	$85.24
Crew B-91	**Hr.**	**Daily**	**Hr.**	**Daily**	**Bare Costs**	**Incl. O&P**
1 Labor Foreman (outside)	$32.35	$258.80	$53.45	$427.60	$36.36	$59.80
2 Laborers	30.35	485.60	50.15	802.40		
4 Equip. Oper. (medium)	40.75	1304.00	66.85	2139.20		
1 Truck Driver (heavy)	34.80	278.40	57.25	458.00		
1 Dist. Tanker, 3000 Gallon		323.25		355.57		
1 Truck Tractor, 6x4, 380 H.P.		487.60		536.36		
1 Aggreg. Spreader, S.P.		836.00		919.60		
1 Roller, Pneum. Whl., 12 Ton		338.35		372.19		
1 Tandem Roller, 10 Ton		232.05		255.26	34.64	38.11
64 L.H., Daily Totals		$4544.05		$6266.18	$71.00	$97.91
Crew B-91B	**Hr.**	**Daily**	**Hr.**	**Daily**	**Bare Costs**	**Incl. O&P**
1 Laborer	$30.35	$242.80	$50.15	$401.20	$35.55	$58.50
1 Equipment Oper. (med.)	40.75	326.00	66.85	534.80		
1 Road Sweeper, Vac. Assist.		857.60		943.36	53.60	58.96
16 L.H., Daily Totals		$1426.40		$1879.36	$89.15	$117.46
Crew B-91C	**Hr.**	**Daily**	**Hr.**	**Daily**	**Bare Costs**	**Incl. O&P**
1 Laborer	$30.35	$242.80	$50.15	$401.20	$32.23	$53.13
1 Truck Driver (light)	34.10	272.80	56.10	448.80		
1 Catch Basin Cleaning Truck		528.95		581.85	33.06	36.37
16 L.H., Daily Totals		$1044.55		$1431.85	$65.28	$89.49
Crew B-91D	**Hr.**	**Daily**	**Hr.**	**Daily**	**Bare Costs**	**Incl. O&P**
1 Labor Foreman (outside)	$32.35	$258.80	$53.45	$427.60	$35.19	$57.92
5 Laborers	30.35	1214.00	50.15	2006.00		
5 Equip. Oper. (medium)	40.75	1630.00	66.85	2674.00		
2 Truck Drivers (heavy)	34.80	556.80	57.25	916.00		
1 Aggreg. Spreader, S.P.		836.00		919.60		
2 Truck Tractors, 6x4, 380 H.P.		975.20		1072.72		
2 Dist. Tankers, 3000 Gallon		646.50		711.15		
2 Pavement Brushes, Towed		171.00		188.10		
2 Rollers Pneum. Whl., 12 Ton		676.70		744.37	31.78	34.96
104 L.H., Daily Totals		$6965.00		$9659.54	$66.97	$92.88

Crew No.	Bare Costs		Incl. Subs O&P		Cost Per Labor-Hour	
Crew B-92	**Hr.**	**Daily**	**Hr.**	**Daily**	**Bare Costs**	**Incl. O&P**
1 Labor Foreman (outside)	$32.35	$258.80	$53.45	$427.60	$30.85	$50.98
3 Laborers	30.35	728.40	50.15	1203.60		
1 Crack Cleaner, 25 H.P.		56.95		62.65		
1 Air Compressor, 60 cfm		106.70		117.37		
1 Tar Kettle, T.M.		134.50		147.95		
1 Flatbed Truck, Gas, 3 Ton		245.95		270.55	17.00	18.70
32 L.H., Daily Totals		$1531.30		$2229.71	$47.85	$69.68
Crew B-93	**Hr.**	**Daily**	**Hr.**	**Daily**	**Bare Costs**	**Incl. O&P**
1 Equip. Oper. (medium)	$40.75	$326.00	$66.85	$534.80	$40.75	$66.85
1 Feller Buncher, 100 H.P.		803.55		883.90	100.44	110.49
8 L.H., Daily Totals		$1129.55		$1418.70	$141.19	$177.34
Crew B-94A	**Hr.**	**Daily**	**Hr.**	**Daily**	**Bare Costs**	**Incl. O&P**
1 Laborer	$30.35	$242.80	$50.15	$401.20	$30.35	$50.15
1 Diaphragm Water Pump, 2"		74.30		81.73		
1 -20' Suction Hose, 2"		2.05		2.25		
2 -50' Discharge Hoses, 2"		1.80		1.98	9.77	10.75
8 L.H., Daily Totals		$320.95		$487.17	$40.12	$60.90
Crew B-94B	**Hr.**	**Daily**	**Hr.**	**Daily**	**Bare Costs**	**Incl. O&P**
1 Laborer	$30.35	$242.80	$50.15	$401.20	$30.35	$50.15
1 Diaphragm Water Pump, 4"		141.40		155.54		
1 -20' Suction Hose, 4"		3.10		3.41		
2 -50' Discharge Hoses, 4"		4.50		4.95	18.63	20.49
8 L.H., Daily Totals		$391.80		$565.10	$48.98	$70.64
Crew B-94C	**Hr.**	**Daily**	**Hr.**	**Daily**	**Bare Costs**	**Incl. O&P**
1 Laborer	$30.35	$242.80	$50.15	$401.20	$30.35	$50.15
1 Centrifugal Water Pump, 3"		79.25		87.17		
1 -20' Suction Hose, 3"		2.70		2.97		
2 -50' Discharge Hoses, 3"		2.80		3.08	10.59	11.65
8 L.H., Daily Totals		$327.55		$494.43	$40.94	$61.80
Crew B-94D	**Hr.**	**Daily**	**Hr.**	**Daily**	**Bare Costs**	**Incl. O&P**
1 Laborer	$30.35	$242.80	$50.15	$401.20	$30.35	$50.15
1 Centr. Water Pump, 6"		296.45		326.10		
1 -20' Suction Hose, 6"		11.10		12.21		
2 -50' Discharge Hoses, 6"		11.40		12.54	39.87	43.86
8 L.H., Daily Totals		$561.75		$752.04	$70.22	$94.01
Crew C-1	**Hr.**	**Daily**	**Hr.**	**Daily**	**Bare Costs**	**Incl. O&P**
2 Carpenters	$38.75	$620.00	$64.05	$1024.80	$34.21	$56.71
1 Carpenter Helper	29.00	232.00	48.60	388.80		
1 Laborer	30.35	242.80	50.15	401.20		
32 L.H., Daily Totals		$1094.80		$1814.80	$34.21	$56.71
Crew C-2	**Hr.**	**Daily**	**Hr.**	**Daily**	**Bare Costs**	**Incl. O&P**
1 Carpenter Foreman (outside)	$40.75	$326.00	$67.35	$538.80	$34.43	$57.13
2 Carpenters	38.75	620.00	64.05	1024.80		
2 Carpenter Helpers	29.00	464.00	48.60	777.60		
1 Laborer	30.35	242.80	50.15	401.20		
48 L.H., Daily Totals		$1652.80		$2742.40	$34.43	$57.13
Crew C-2A	**Hr.**	**Daily**	**Hr.**	**Daily**	**Bare Costs**	**Incl. O&P**
1 Carpenter Foreman (outside)	$40.75	$326.00	$67.35	$538.80	$37.42	$61.62
3 Carpenters	38.75	930.00	64.05	1537.20		
1 Cement Finisher	37.15	297.20	60.05	480.40		
1 Laborer	30.35	242.80	50.15	401.20		
48 L.H., Daily Totals		$1796.00		$2957.60	$37.42	$61.62

Crew C-3	Bare Costs Hr.	Bare Costs Daily	Incl. Subs O&P Hr.	Incl. Subs O&P Daily	Bare Costs	Incl. O&P
1 Rodman Foreman (outside)	$42.60	$340.80	$70.45	$563.60	$36.77	$60.74
3 Rodmen (reinf.)	40.60	974.40	67.15	1611.60		
1 Equip. Oper. (light)	38.75	310.00	63.60	508.80		
3 Laborers	30.35	728.40	50.15	1203.60		
3 Stressing Equipment		31.20		34.32		
.5 Grouting Equipment		77.70		85.47	1.70	1.87
64 L.H., Daily Totals		$2462.50		$4007.39	$38.48	$62.62

Crew C-4	Hr.	Daily	Hr.	Daily	Bare Costs	Incl. O&P
1 Rodman Foreman (outside)	$42.60	$340.80	$70.45	$563.60	$38.54	$63.73
2 Rodmen (reinf.)	40.60	649.60	67.15	1074.40		
1 Building Laborer	30.35	242.80	50.15	401.20		
3 Stressing Equipment		31.20		34.32	0.97	1.07
32 L.H., Daily Totals		$1264.40		$2073.52	$39.51	$64.80

Crew C-4A	Hr.	Daily	Hr.	Daily	Bare Costs	Incl. O&P
2 Rodmen (reinf.)	$40.60	$649.60	$67.15	$1074.40	$40.60	$67.15
4 Stressing Equipment		41.60		45.76	2.60	2.86
16 L.H., Daily Totals		$691.20		$1120.16	$43.20	$70.01

Crew C-5	Hr.	Daily	Hr.	Daily	Bare Costs	Incl. O&P
1 Rodman Foreman (outside)	$42.60	$340.80	$70.45	$563.60	$37.74	$62.32
2 Rodmen (reinf.)	40.60	649.60	67.15	1074.40		
1 Equip. Oper. (crane)	41.95	335.60	68.85	550.80		
2 Building Laborers	30.35	485.60	50.15	802.40		
1 Hyd. Crane, 25 Ton		581.70		639.87	12.12	13.33
48 L.H., Daily Totals		$2393.30		$3631.07	$49.86	$75.65

Crew C-6	Hr.	Daily	Hr.	Daily	Bare Costs	Incl. O&P
1 Labor Foreman (outside)	$32.35	$258.80	$53.45	$427.60	$31.82	$52.35
4 Laborers	30.35	971.20	50.15	1604.80		
1 Cement Finisher	37.15	297.20	60.05	480.40		
2 Gas Engine Vibrators		53.10		58.41	1.11	1.22
48 L.H., Daily Totals		$1580.30		$2571.21	$32.92	$53.57

Crew C-6A	Hr.	Daily	Hr.	Daily	Bare Costs	Incl. O&P
2 Cement Finishers	$37.15	$594.40	$60.05	$960.80	$37.15	$60.05
1 Concrete Vibrator, Elec, 2 HP		9.25		10.18	0.58	0.64
16 L.H., Daily Totals		$603.65		$970.98	$37.73	$60.69

Crew B-89C	Hr.	Daily	Hr.	Daily	Bare Costs	Incl. O&P
1 Cement Finisher	$37.15	$297.20	$60.05	$480.40	$37.15	$60.05
1 Masonry cut-off saw, gas		34.00		37.40	4.25	4.67
8 L.H., Daily Totals		$331.20		$517.80	$41.40	$64.72

Crew C-7	Hr.	Daily	Hr.	Daily	Bare Costs	Incl. O&P
1 Labor Foreman (outside)	$32.35	$258.80	$53.45	$427.60	$33.10	$54.44
5 Laborers	30.35	1214.00	50.15	2006.00		
1 Cement Finisher	37.15	297.20	60.05	480.40		
1 Equip. Oper. (medium)	40.75	326.00	66.85	534.80		
1 Equip. Oper. (oiler)	35.90	287.20	58.90	471.20		
2 Gas Engine Vibrators		53.10		58.41		
1 Concrete Bucket, 1 C.Y.		24.95		27.45		
1 Hyd. Crane, 55 Ton		981.50		1079.65	14.72	16.19
72 L.H., Daily Totals		$3442.75		$5085.51	$47.82	$70.63

Crew C-8	Hr.	Daily	Hr.	Daily	Bare Costs	Incl. O&P
1 Labor Foreman (outside)	$32.35	$258.80	$53.45	$427.60	$34.06	$55.84
3 Laborers	30.35	728.40	50.15	1203.60		
2 Cement Finishers	37.15	594.40	60.05	960.80		
1 Equip. Oper. (medium)	40.75	326.00	66.85	534.80		
1 Concrete Pump (Small)		844.05		928.46	15.07	16.58
56 L.H., Daily Totals		$2751.65		$4055.26	$49.14	$72.42

Crew C-8A	Hr.	Daily	Hr.	Daily	Bare Costs	Incl. O&P
1 Labor Foreman (outside)	$32.35	$258.80	$53.45	$427.60	$32.95	$54.00
3 Laborers	30.35	728.40	50.15	1203.60		
2 Cement Finishers	37.15	594.40	60.05	960.80		
48 L.H., Daily Totals		$1581.60		$2592.00	$32.95	$54.00

Crew C-8B	Hr.	Daily	Hr.	Daily	Bare Costs	Incl. O&P
1 Labor Foreman (outside)	$32.35	$258.80	$53.45	$427.60	$32.83	$54.15
3 Laborers	30.35	728.40	50.15	1203.60		
1 Equip. Oper. (medium)	40.75	326.00	66.85	534.80		
1 Vibrating Power Screed		79.45		87.39		
1 Roller, Vibratory, 25 Ton		650.90		715.99		
1 Dozer, 200 H.P.		1290.00		1419.00	50.51	55.56
40 L.H., Daily Totals		$3333.55		$4388.39	$83.34	$109.71

Crew C-8C	Hr.	Daily	Hr.	Daily	Bare Costs	Incl. O&P
1 Labor Foreman (outside)	$32.35	$258.80	$53.45	$427.60	$33.55	$55.13
3 Laborers	30.35	728.40	50.15	1203.60		
1 Cement Finisher	37.15	297.20	60.05	480.40		
1 Equip. Oper. (medium)	40.75	326.00	66.85	534.80		
1 Shotcrete Rig, 12 C.Y./hr		242.75		267.02		
1 Air Compressor, 160 cfm		115.85		127.44		
4 -50' Air Hoses, 1"		33.60		36.96		
4 -50' Air Hoses, 2"		103.60		113.96	10.33	11.36
48 L.H., Daily Totals		$2106.20		$3191.78	$43.88	$66.50

Crew C-8D	Hr.	Daily	Hr.	Daily	Bare Costs	Incl. O&P
1 Labor Foreman (outside)	$32.35	$258.80	$53.45	$427.60	$34.65	$56.81
1 Laborer	30.35	242.80	50.15	401.20		
1 Cement Finisher	37.15	297.20	60.05	480.40		
1 Equipment Oper. (light)	38.75	310.00	63.60	508.80		
1 Air Compressor, 250 cfm		167.95		184.75		
2 -50' Air Hoses, 1"		16.80		18.48	5.77	6.35
32 L.H., Daily Totals		$1293.55		$2021.22	$40.42	$63.16

Crew C-8E	Hr.	Daily	Hr.	Daily	Bare Costs	Incl. O&P
1 Labor Foreman (outside)	$32.35	$258.80	$53.45	$427.60	$33.22	$54.59
3 Laborers	30.35	728.40	50.15	1203.60		
1 Cement Finisher	37.15	297.20	60.05	480.40		
1 Equipment Oper. (light)	38.75	310.00	63.60	508.80		
1 Shotcrete Rig, 35 C.Y./hr		271.40		298.54		
1 Air Compressor, 250 cfm		167.95		184.75		
4 -50' Air Hoses, 1"		33.60		36.96		
4 -50' Air Hoses, 2"		103.60		113.96	12.01	13.21
48 L.H., Daily Totals		$2170.95		$3254.61	$45.23	$67.80

Crew No.	Bare Costs		Incl. Subs O&P		Cost Per Labor-Hour	
Crew C-9	**Hr.**	**Daily**	**Hr.**	**Daily**	**Bare Costs**	**Incl. O&P**
1 Cement Finisher	$37.15	$297.20	$60.05	$480.40	$34.15	$55.99
2 Laborers	30.35	485.60	50.15	802.40		
1 Equipment Oper. (light)	38.75	310.00	63.60	508.80		
1 Grout Pump, 50 C.F./hr.		125.60		138.16		
1 Air Compressor, 160 cfm		115.85		127.44		
2 -50' Air Hoses, 1"		16.80		18.48		
2 -50' Air Hoses, 2"		51.80		56.98	9.69	10.66
32 L.H., Daily Totals		$1402.85		$2132.66	$43.84	$66.65
Crew C-10	**Hr.**	**Daily**	**Hr.**	**Daily**	**Bare Costs**	**Incl. O&P**
1 Laborer	$30.35	$242.80	$50.15	$401.20	$34.88	$56.75
2 Cement Finishers	37.15	594.40	60.05	960.80		
24 L.H., Daily Totals		$837.20		$1362.00	$34.88	$56.75
Crew C-10B	**Hr.**	**Daily**	**Hr.**	**Daily**	**Bare Costs**	**Incl. O&P**
3 Laborers	$30.35	$728.40	$50.15	$1203.60	$33.07	$54.11
2 Cement Finishers	37.15	594.40	60.05	960.80		
1 Concrete Mixer, 10 C.F.		162.00		178.20		
2 Trowels, 48" Walk-Behind		83.00		91.30	6.13	6.74
40 L.H., Daily Totals		$1567.80		$2433.90	$39.20	$60.85
Crew C-10C	**Hr.**	**Daily**	**Hr.**	**Daily**	**Bare Costs**	**Incl. O&P**
1 Laborer	$30.35	$242.80	$50.15	$401.20	$34.88	$56.75
2 Cement Finishers	37.15	594.40	60.05	960.80		
1 Trowel, 48" Walk-Behind		41.50		45.65	1.73	1.90
24 L.H., Daily Totals		$878.70		$1407.65	$36.61	$58.65
Crew C-10D	**Hr.**	**Daily**	**Hr.**	**Daily**	**Bare Costs**	**Incl. O&P**
1 Laborer	$30.35	$242.80	$50.15	$401.20	$34.88	$56.75
2 Cement Finishers	37.15	594.40	60.05	960.80		
1 Vibrating Power Screed		79.45		87.39		
1 Trowel, 48" Walk-Behind		41.50		45.65	5.04	5.54
24 L.H., Daily Totals		$958.15		$1495.05	$39.92	$62.29
Crew C-10E	**Hr.**	**Daily**	**Hr.**	**Daily**	**Bare Costs**	**Incl. O&P**
1 Laborer	$30.35	$242.80	$50.15	$401.20	$34.88	$56.75
2 Cement Finishers	37.15	594.40	60.05	960.80		
1 Vibrating Power Screed		79.45		87.39		
1 Cement Trowel, 96" Ride-On		166.30		182.93	10.24	11.26
24 L.H., Daily Totals		$1082.95		$1632.33	$45.12	$68.01
Crew C-10F	**Hr.**	**Daily**	**Hr.**	**Daily**	**Bare Costs**	**Incl. O&P**
1 Laborer	$30.35	$242.80	$50.15	$401.20	$34.88	$56.75
2 Cement Finishers	37.15	594.40	60.05	960.80		
1 Telescoping Boom Lift, to 60'		451.25		496.38	18.80	20.68
24 L.H., Daily Totals		$1288.45		$1858.38	$53.69	$77.43
Crew C-11	**Hr.**	**Daily**	**Hr.**	**Daily**	**Bare Costs**	**Incl. O&P**
1 Skilled Worker Foreman	$41.45	$331.60	$68.85	$550.80	$40.09	$66.49
5 Skilled Workers	39.45	1578.00	65.55	2622.00		
1 Equip. Oper. (crane)	41.95	335.60	68.85	550.80		
1 Lattice Boom Crane, 150 Ton		2021.00		2223.10	36.09	39.70
56 L.H., Daily Totals		$4266.20		$5946.70	$76.18	$106.19

Crew No.	Bare Costs		Incl. Subs O&P		Cost Per Labor-Hour	
Crew C-12	**Hr.**	**Daily**	**Hr.**	**Daily**	**Bare Costs**	**Incl. O&P**
1 Carpenter Foreman (outside)	$40.75	$326.00	$67.35	$538.80	$38.22	$63.08
3 Carpenters	38.75	930.00	64.05	1537.20		
1 Laborer	30.35	242.80	50.15	401.20		
1 Equip. Oper. (crane)	41.95	335.60	68.85	550.80		
1 Hyd. Crane, 12 Ton		471.75		518.92	9.83	10.81
48 L.H., Daily Totals		$2306.15		$3546.93	$48.04	$73.89
Crew C-13	**Hr.**	**Daily**	**Hr.**	**Daily**	**Bare Costs**	**Incl. O&P**
2 Struc. Steel Workers	$41.45	$663.20	$71.90	$1150.40	$40.55	$69.28
1 Carpenter	38.75	310.00	64.05	512.40		
1 Welder, Gas Engine, 300 amp		95.65		105.22	3.99	4.38
24 L.H., Daily Totals		$1068.85		$1768.02	$44.54	$73.67
Crew C-14	**Hr.**	**Daily**	**Hr.**	**Daily**	**Bare Costs**	**Incl. O&P**
1 Carpenter Foreman (outside)	$40.75	$326.00	$67.35	$538.80	$34.81	$57.51
3 Carpenters	38.75	930.00	64.05	1537.20		
2 Carpenter Helpers	29.00	464.00	48.60	777.60		
4 Laborers	30.35	971.20	50.15	1604.80		
2 Rodmen (reinf.)	40.60	649.60	67.15	1074.40		
2 Rodman Helpers	29.00	464.00	48.60	777.60		
2 Cement Finishers	37.15	594.40	60.05	960.80		
1 Equip. Oper. (crane)	41.95	335.60	68.85	550.80		
1 Hyd. Crane, 80 Ton		1487.00		1635.70	10.93	12.03
136 L.H., Daily Totals		$6221.80		$9457.70	$45.75	$69.54
Crew C-14A	**Hr.**	**Daily**	**Hr.**	**Daily**	**Bare Costs**	**Incl. O&P**
1 Carpenter Foreman (outside)	$40.75	$326.00	$67.35	$538.80	$38.47	$63.52
16 Carpenters	38.75	4960.00	64.05	8198.40		
4 Rodmen (reinf.)	40.60	1299.20	67.15	2148.80		
2 Laborers	30.35	485.60	50.15	802.40		
1 Cement Finisher	37.15	297.20	60.05	480.40		
1 Equip. Oper. (medium)	40.75	326.00	66.85	534.80		
1 Gas Engine Vibrator		26.55		29.20		
1 Concrete Pump (Small)		844.05		928.46	4.35	4.79
200 L.H., Daily Totals		$8564.60		$13661.26	$42.82	$68.31
Crew C-14B	**Hr.**	**Daily**	**Hr.**	**Daily**	**Bare Costs**	**Incl. O&P**
1 Carpenter Foreman (outside)	$40.75	$326.00	$67.35	$538.80	$38.42	$63.38
16 Carpenters	38.75	4960.00	64.05	8198.40		
4 Rodmen (reinf.)	40.60	1299.20	67.15	2148.80		
2 Laborers	30.35	485.60	50.15	802.40		
2 Cement Finishers	37.15	594.40	60.05	960.80		
1 Equip. Oper. (medium)	40.75	326.00	66.85	534.80		
1 Gas Engine Vibrator		26.55		29.20		
1 Concrete Pump (Small)		844.05		928.46	4.19	4.60
208 L.H., Daily Totals		$8861.80		$14141.66	$42.60	$67.99
Crew C-14C	**Hr.**	**Daily**	**Hr.**	**Daily**	**Bare Costs**	**Incl. O&P**
1 Carpenter Foreman (outside)	$40.75	$326.00	$67.35	$538.80	$36.64	$60.47
6 Carpenters	38.75	1860.00	64.05	3074.40		
2 Rodmen (reinf.)	40.60	649.60	67.15	1074.40		
4 Laborers	30.35	971.20	50.15	1604.80		
1 Cement Finisher	37.15	297.20	60.05	480.40		
1 Gas Engine Vibrator		26.55		29.20	0.24	0.26
112 L.H., Daily Totals		$4130.55		$6802.01	$36.88	$60.73

Crew No.	Bare Costs		Incl. Subs O&P		Cost Per Labor-Hour	
Crew C-14D	**Hr.**	**Daily**	**Hr.**	**Daily**	**Bare Costs**	**Incl. O&P**
1 Carpenter Foreman (outside)	$40.75	$326.00	$67.35	$538.80	$38.32	$63.27
18 Carpenters	38.75	5580.00	64.05	9223.20		
2 Rodmen (reinf.)	40.60	649.60	67.15	1074.40		
2 Laborers	30.35	485.60	50.15	802.40		
1 Cement Finisher	37.15	297.20	60.05	480.40		
1 Equip. Oper. (medium)	40.75	326.00	66.85	534.80		
1 Gas Engine Vibrator		26.55		29.20		
1 Concrete Pump (Small)		844.05		928.46	4.35	4.79
200 L.H., Daily Totals		$8535.00		$13611.66	$42.67	$68.06

Crew C-14E	**Hr.**	**Daily**	**Hr.**	**Daily**	**Bare Costs**	**Incl. O&P**
1 Carpenter Foreman (outside)	$40.75	$326.00	$67.35	$538.80	$37.17	$61.32
2 Carpenters	38.75	620.00	64.05	1024.80		
4 Rodmen (reinf.)	40.60	1299.20	67.15	2148.80		
3 Laborers	30.35	728.40	50.15	1203.60		
1 Cement Finisher	37.15	297.20	60.05	480.40		
1 Gas Engine Vibrator		26.55		29.20	0.30	0.33
88 L.H., Daily Totals		$3297.35		$5425.60	$37.47	$61.65

Crew C-14F	**Hr.**	**Daily**	**Hr.**	**Daily**	**Bare Costs**	**Incl. O&P**
1 Labor Foreman (outside)	$32.35	$258.80	$53.45	$427.60	$35.11	$57.12
2 Laborers	30.35	485.60	50.15	802.40		
6 Cement Finishers	37.15	1783.20	60.05	2882.40		
1 Gas Engine Vibrator		26.55		29.20	0.37	0.41
72 L.H., Daily Totals		$2554.15		$4141.60	$35.47	$57.52

Crew C-14G	**Hr.**	**Daily**	**Hr.**	**Daily**	**Bare Costs**	**Incl. O&P**
1 Labor Foreman (outside)	$32.35	$258.80	$53.45	$427.60	$34.52	$56.28
2 Laborers	30.35	485.60	50.15	802.40		
4 Cement Finishers	37.15	1188.80	60.05	1921.60		
1 Gas Engine Vibrator		26.55		29.20	0.47	0.52
56 L.H., Daily Totals		$1959.75		$3180.80	$35.00	$56.80

Crew C-14H	**Hr.**	**Daily**	**Hr.**	**Daily**	**Bare Costs**	**Incl. O&P**
1 Carpenter Foreman (outside)	$40.75	$326.00	$67.35	$538.80	$37.73	$62.13
2 Carpenters	38.75	620.00	64.05	1024.80		
1 Rodman (reinf.)	40.60	324.80	67.15	537.20		
1 Laborer	30.35	242.80	50.15	401.20		
1 Cement Finisher	37.15	297.20	60.05	480.40		
1 Gas Engine Vibrator		26.55		29.20	0.55	0.61
48 L.H., Daily Totals		$1837.35		$3011.61	$38.28	$62.74

Crew C-14L	**Hr.**	**Daily**	**Hr.**	**Daily**	**Bare Costs**	**Incl. O&P**
1 Carpenter Foreman (outside)	$40.75	$326.00	$67.35	$538.80	$35.98	$59.36
6 Carpenters	38.75	1860.00	64.05	3074.40		
4 Laborers	30.35	971.20	50.15	1604.80		
1 Cement Finisher	37.15	297.20	60.05	480.40		
1 Gas Engine Vibrator		26.55		29.20	0.28	0.30
96 L.H., Daily Totals		$3480.95		$5727.60	$36.26	$59.66

Crew C-14M	**Hr.**	**Daily**	**Hr.**	**Daily**	**Bare Costs**	**Incl. O&P**
1 Carpenter Foreman (outside)	$40.75	$326.00	$67.35	$538.80	$37.18	$61.23
2 Carpenters	38.75	620.00	64.05	1024.80		
1 Rodman (reinf.)	40.60	324.80	67.15	537.20		
2 Laborers	30.35	485.60	50.15	802.40		
1 Cement Finisher	37.15	297.20	60.05	480.40		
1 Equip. Oper. (medium)	40.75	326.00	66.85	534.80		
1 Gas Engine Vibrator		26.55		29.20		
1 Concrete Pump (Small)		844.05		928.46	13.60	14.96
64 L.H., Daily Totals		$3250.20		$4876.06	$50.78	$76.19

Crew No.	Bare Costs		Incl. Subs O&P		Cost Per Labor-Hour	
Crew C-15	**Hr.**	**Daily**	**Hr.**	**Daily**	**Bare Costs**	**Incl. O&P**
1 Carpenter Foreman (outside)	$40.75	$326.00	$67.35	$538.80	$36.02	$59.24
2 Carpenters	38.75	620.00	64.05	1024.80		
3 Laborers	30.35	728.40	50.15	1203.60		
2 Cement Finishers	37.15	594.40	60.05	960.80		
1 Rodman (reinf.)	40.60	324.80	67.15	537.20		
72 L.H., Daily Totals		$2593.60		$4265.20	$36.02	$59.24

Crew C-16	**Hr.**	**Daily**	**Hr.**	**Daily**	**Bare Costs**	**Incl. O&P**
1 Labor Foreman (outside)	$32.35	$258.80	$53.45	$427.60	$34.06	$55.84
3 Laborers	30.35	728.40	50.15	1203.60		
2 Cement Finishers	37.15	594.40	60.05	960.80		
1 Equip. Oper. (medium)	40.75	326.00	66.85	534.80		
1 Gunite Pump Rig		313.00		344.30		
2 -50' Air Hoses, 3/4"		13.10		14.41		
2 -50' Air Hoses, 2"		51.80		56.98	6.75	7.42
56 L.H., Daily Totals		$2285.50		$3542.49	$40.81	$63.26

Crew C-16A	**Hr.**	**Daily**	**Hr.**	**Daily**	**Bare Costs**	**Incl. O&P**
1 Laborer	$30.35	$242.80	$50.15	$401.20	$36.35	$59.27
2 Cement Finishers	37.15	594.40	60.05	960.80		
1 Equip. Oper. (medium)	40.75	326.00	66.85	534.80		
1 Gunite Pump Rig		313.00		344.30		
2 -50' Air Hoses, 3/4"		13.10		14.41		
2 -50' Air Hoses, 2"		51.80		56.98		
1 Telescoping Boom Lift, to 60'		451.25		496.38	25.91	28.50
32 L.H., Daily Totals		$1992.35		$2808.86	$62.26	$87.78

Crew C-17	**Hr.**	**Daily**	**Hr.**	**Daily**	**Bare Costs**	**Incl. O&P**
2 Skilled Worker Foremen (out)	$41.45	$663.20	$68.85	$1101.60	$39.85	$66.21
8 Skilled Workers	39.45	2524.80	65.55	4195.20		
80 L.H., Daily Totals		$3188.00		$5296.80	$39.85	$66.21

Crew C-17A	**Hr.**	**Daily**	**Hr.**	**Daily**	**Bare Costs**	**Incl. O&P**
2 Skilled Worker Foremen (out)	$41.45	$663.20	$68.85	$1101.60	$39.88	$66.24
8 Skilled Workers	39.45	2524.80	65.55	4195.20		
.125 Equip. Oper. (crane)	41.95	41.95	68.85	68.85		
.125 Hyd. Crane, 80 Ton		185.88		204.46	2.29	2.52
81 L.H., Daily Totals		$3415.82		$5570.11	$42.17	$68.77

Crew C-17B	**Hr.**	**Daily**	**Hr.**	**Daily**	**Bare Costs**	**Incl. O&P**
2 Skilled Worker Foremen (out)	$41.45	$663.20	$68.85	$1101.60	$39.90	$66.27
8 Skilled Workers	39.45	2524.80	65.55	4195.20		
.25 Equip. Oper. (crane)	41.95	83.90	68.85	137.70		
.25 Hyd. Crane, 80 Ton		371.75		408.93		
.25 Trowel, 48" Walk-Behind		10.38		11.41	4.66	5.13
82 L.H., Daily Totals		$3654.03		$5854.84	$44.56	$71.40

Crew C-17C	**Hr.**	**Daily**	**Hr.**	**Daily**	**Bare Costs**	**Incl. O&P**
2 Skilled Worker Foremen (out)	$41.45	$663.20	$68.85	$1101.60	$39.93	$66.31
8 Skilled Workers	39.45	2524.80	65.55	4195.20		
.375 Equip. Oper. (crane)	41.95	125.85	68.85	206.55		
.375 Hyd. Crane, 80 Ton		557.63		613.39	6.72	7.39
83 L.H., Daily Totals		$3871.47		$6116.74	$46.64	$73.70

Crew C-17D	**Hr.**	**Daily**	**Hr.**	**Daily**	**Bare Costs**	**Incl. O&P**
2 Skilled Worker Foremen (out)	$41.45	$663.20	$68.85	$1101.60	$39.95	$66.34
8 Skilled Workers	39.45	2524.80	65.55	4195.20		
.5 Equip. Oper. (crane)	41.95	167.80	68.85	275.40		
.5 Hyd. Crane, 80 Ton		743.50		817.85	8.85	9.74
84 L.H., Daily Totals		$4099.30		$6390.05	$48.80	$76.07

Crew No.	Bare Costs		Incl. Subs O&P		Cost Per Labor-Hour	
Crew C-17E	**Hr.**	**Daily**	**Hr.**	**Daily**	**Bare Costs**	**Incl. O&P**
2 Skilled Worker Foremen (out)	$41.45	$663.20	$68.85	$1101.60	$39.85	$66.21
8 Skilled Workers	39.45	2524.80	65.55	4195.20		
1 Hyd. Jack with Rods		104.40		114.84	1.30	1.44
80 L.H., Daily Totals		$3292.40		$5411.64	$41.16	$67.65
Crew C-18	**Hr.**	**Daily**	**Hr.**	**Daily**	**Bare Costs**	**Incl. O&P**
.125 Labor Foreman (outside)	$32.35	$32.35	$53.45	$53.45	$30.57	$50.52
1 Laborer	30.35	242.80	50.15	401.20		
1 Concrete Cart, 10 C.F.		58.10		63.91	6.46	7.10
9 L.H., Daily Totals		$333.25		$518.56	$37.03	$57.62
Crew C-19	**Hr.**	**Daily**	**Hr.**	**Daily**	**Bare Costs**	**Incl. O&P**
.125 Labor Foreman (outside)	$32.35	$32.35	$53.45	$53.45	$30.57	$50.52
1 Laborer	30.35	242.80	50.15	401.20		
1 Concrete Cart, 18 C.F.		97.75		107.53	10.86	11.95
9 L.H., Daily Totals		$372.90		$562.17	$41.43	$62.46
Crew C-20	**Hr.**	**Daily**	**Hr.**	**Daily**	**Bare Costs**	**Incl. O&P**
1 Labor Foreman (outside)	$32.35	$258.80	$53.45	$427.60	$32.75	$53.89
5 Laborers	30.35	1214.00	50.15	2006.00		
1 Cement Finisher	37.15	297.20	60.05	480.40		
1 Equip. Oper. (medium)	40.75	326.00	66.85	534.80		
2 Gas Engine Vibrators		53.10		58.41		
1 Concrete Pump (Small)		844.05		928.46	14.02	15.42
64 L.H., Daily Totals		$2993.15		$4435.67	$46.77	$69.31
Crew C-21	**Hr.**	**Daily**	**Hr.**	**Daily**	**Bare Costs**	**Incl. O&P**
1 Labor Foreman (outside)	$32.35	$258.80	$53.45	$427.60	$32.75	$53.89
5 Laborers	30.35	1214.00	50.15	2006.00		
1 Cement Finisher	37.15	297.20	60.05	480.40		
1 Equip. Oper. (medium)	40.75	326.00	66.85	534.80		
2 Gas Engine Vibrators		53.10		58.41		
1 Concrete Conveyer		186.45		205.10	3.74	4.12
64 L.H., Daily Totals		$2335.55		$3712.30	$36.49	$58.00
Crew C-22	**Hr.**	**Daily**	**Hr.**	**Daily**	**Bare Costs**	**Incl. O&P**
1 Rodman Foreman (outside)	$42.60	$340.80	$70.45	$563.60	$40.90	$67.62
4 Rodmen (reinf.)	40.60	1299.20	67.15	2148.80		
.125 Equip. Oper. (crane)	41.95	41.95	68.85	68.85		
.125 Equip. Oper. (oiler)	35.90	35.90	58.90	58.90		
.125 Hyd. Crane, 25 Ton		72.71		79.98	1.73	1.90
42 L.H., Daily Totals		$1790.56		$2920.13	$42.63	$69.53
Crew C-23	**Hr.**	**Daily**	**Hr.**	**Daily**	**Bare Costs**	**Incl. O&P**
2 Skilled Worker Foremen (out)	$41.45	$663.20	$68.85	$1101.60	$39.74	$65.88
6 Skilled Workers	39.45	1893.60	65.55	3146.40		
1 Equip. Oper. (crane)	41.95	335.60	68.85	550.80		
1 Equip. Oper. (oiler)	35.90	287.20	58.90	471.20		
1 Lattice Boom Crane, 90 Ton		1698.00		1867.80	21.23	23.35
80 L.H., Daily Totals		$4877.60		$7137.80	$60.97	$89.22
Crew C-24	**Hr.**	**Daily**	**Hr.**	**Daily**	**Bare Costs**	**Incl. O&P**
2 Skilled Worker Foremen (out)	$41.45	$663.20	$68.85	$1101.60	$39.74	$65.88
6 Skilled Workers	39.45	1893.60	65.55	3146.40		
1 Equip. Oper. (crane)	41.95	335.60	68.85	550.80		
1 Equip. Oper. (oiler)	35.90	287.20	58.90	471.20		
1 Lattice Boom Crane, 150 Ton		2021.00		2223.10	25.26	27.79
80 L.H., Daily Totals		$5200.60		$7493.10	$65.01	$93.66

Crew No.	Bare Costs		Incl. Subs O&P		Cost Per Labor-Hour	
Crew C-25	**Hr.**	**Daily**	**Hr.**	**Daily**	**Bare Costs**	**Incl. O&P**
2 Rodmen (reinf.)	$40.60	$649.60	$67.15	$1074.40	$32.83	$56.20
2 Rodmen Helpers	25.05	400.80	45.25	724.00		
32 L.H., Daily Totals		$1050.40		$1798.40	$32.83	$56.20
Crew C-27	**Hr.**	**Daily**	**Hr.**	**Daily**	**Bare Costs**	**Incl. O&P**
2 Cement Finishers	$37.15	$594.40	$60.05	$960.80	$37.15	$60.05
1 Concrete Saw		110.15		121.17	6.88	7.57
16 L.H., Daily Totals		$704.55		$1081.96	$44.03	$67.62
Crew C-28	**Hr.**	**Daily**	**Hr.**	**Daily**	**Bare Costs**	**Incl. O&P**
1 Cement Finisher	$37.15	$297.20	$60.05	$480.40	$37.15	$60.05
1 Portable Air Compressor, Gas		16.05		17.66	2.01	2.21
8 L.H., Daily Totals		$313.25		$498.06	$39.16	$62.26
Crew C-29	**Hr.**	**Daily**	**Hr.**	**Daily**	**Bare Costs**	**Incl. O&P**
1 Laborer	$30.35	$242.80	$50.15	$401.20	$30.35	$50.15
1 Pressure Washer		63.05		69.36	7.88	8.67
8 L.H., Daily Totals		$305.85		$470.56	$38.23	$58.82
Crew C-30	**Hr.**	**Daily**	**Hr.**	**Daily**	**Bare Costs**	**Incl. O&P**
1 Laborer	$30.35	$242.80	$50.15	$401.20	$30.35	$50.15
1 Concrete Mixer, 10 C.F.		162.00		178.20	20.25	22.27
8 L.H., Daily Totals		$404.80		$579.40	$50.60	$72.42
Crew C-31	**Hr.**	**Daily**	**Hr.**	**Daily**	**Bare Costs**	**Incl. O&P**
1 Cement Finisher	$37.15	$297.20	$60.05	$480.40	$37.15	$60.05
1 Grout Pump		313.00		344.30	39.13	43.04
8 L.H., Daily Totals		$610.20		$824.70	$76.28	$103.09
Crew C-32	**Hr.**	**Daily**	**Hr.**	**Daily**	**Bare Costs**	**Incl. O&P**
1 Cement Finisher	$37.15	$297.20	$60.05	$480.40	$33.75	$55.10
1 Laborer	30.35	242.80	50.15	401.20		
1 Crack Chaser Saw, Gas, 6 H.P.		24.50		26.95		
1 Vacuum Pick-Up System		72.15		79.36	6.04	6.64
16 L.H., Daily Totals		$636.65		$987.91	$39.79	$61.74
Crew D-1	**Hr.**	**Daily**	**Hr.**	**Daily**	**Bare Costs**	**Incl. O&P**
1 Bricklayer	$36.75	$294.00	$61.25	$490.00	$33.73	$56.23
1 Bricklayer Helper	30.70	245.60	51.20	409.60		
16 L.H., Daily Totals		$539.60		$899.60	$33.73	$56.23
Crew D-2	**Hr.**	**Daily**	**Hr.**	**Daily**	**Bare Costs**	**Incl. O&P**
3 Bricklayers	$36.75	$882.00	$61.25	$1470.00	$34.33	$57.23
2 Bricklayer Helpers	30.70	491.20	51.20	819.20		
40 L.H., Daily Totals		$1373.20		$2289.20	$34.33	$57.23
Crew D-3	**Hr.**	**Daily**	**Hr.**	**Daily**	**Bare Costs**	**Incl. O&P**
3 Bricklayers	$36.75	$882.00	$61.25	$1470.00	$34.54	$57.55
2 Bricklayer Helpers	30.70	491.20	51.20	819.20		
.25 Carpenter	38.75	77.50	64.05	128.10		
42 L.H., Daily Totals		$1450.70		$2417.30	$34.54	$57.55
Crew D-4	**Hr.**	**Daily**	**Hr.**	**Daily**	**Bare Costs**	**Incl. O&P**
1 Bricklayer	$36.75	$294.00	$61.25	$490.00	$31.84	$53.00
3 Bricklayer Helpers	30.70	736.80	51.20	1228.80		
1 Building Laborer	30.35	242.80	50.15	401.20		
1 Grout Pump, 50 C.F./hr.		125.60		138.16	3.14	3.45
40 L.H., Daily Totals		$1399.20		$2258.16	$34.98	$56.45

Crews - Open Shop

Crew No.	Bare Costs		Incl. Subs O&P		Cost Per Labor-Hour	
Crew D-5	**Hr.**	**Daily**	**Hr.**	**Daily**	**Bare Costs**	**Incl. O&P**
1 Block Mason Helper	30.70	245.60	51.20	409.60	30.70	51.20
8 L.H., Daily Totals		$245.60		$409.60	$30.70	$51.20
Crew D-6	**Hr.**	**Daily**	**Hr.**	**Daily**	**Bare Costs**	**Incl. O&P**
3 Bricklayers	$36.75	$882.00	$61.25	$1470.00	$33.73	$56.23
3 Bricklayer Helpers	30.70	736.80	51.20	1228.80		
48 L.H., Daily Totals		$1618.80		$2698.80	$33.73	$56.23
Crew D-7	**Hr.**	**Daily**	**Hr.**	**Daily**	**Bare Costs**	**Incl. O&P**
1 Tile Layer	$35.90	$287.20	$57.95	$463.60	$31.73	$51.20
1 Tile Layer Helper	27.55	220.40	44.45	355.60		
16 L.H., Daily Totals		$507.60		$819.20	$31.73	$51.20
Crew D-8	**Hr.**	**Daily**	**Hr.**	**Daily**	**Bare Costs**	**Incl. O&P**
3 Bricklayers	$36.75	$882.00	$61.25	$1470.00	$34.33	$57.23
2 Bricklayer Helpers	30.70	491.20	51.20	819.20		
40 L.H., Daily Totals		$1373.20		$2289.20	$34.33	$57.23
Crew D-9	**Hr.**	**Daily**	**Hr.**	**Daily**	**Bare Costs**	**Incl. O&P**
3 Bricklayers	$36.75	$882.00	$61.25	$1470.00	$33.73	$56.23
3 Bricklayer Helpers	30.70	736.80	51.20	1228.80		
48 L.H., Daily Totals		$1618.80		$2698.80	$33.73	$56.23
Crew D-10	**Hr.**	**Daily**	**Hr.**	**Daily**	**Bare Costs**	**Incl. O&P**
1 Bricklayer Foreman (outside)	$38.75	$310.00	$64.60	$516.80	$37.04	$61.48
1 Bricklayer	36.75	294.00	61.25	490.00		
1 Bricklayer Helper	30.70	245.60	51.20	409.60		
1 Equip. Oper. (crane)	41.95	335.60	68.85	550.80		
1 S.P. Crane, 4x4, 12 Ton		371.85		409.04	11.62	12.78
32 L.H., Daily Totals		$1557.05		$2376.24	$48.66	$74.26
Crew D-11	**Hr.**	**Daily**	**Hr.**	**Daily**	**Bare Costs**	**Incl. O&P**
2 Bricklayers	$36.75	$588.00	$61.25	$980.00	$34.73	$57.90
1 Bricklayer Helper	30.70	245.60	51.20	409.60		
24 L.H., Daily Totals		$833.60		$1389.60	$34.73	$57.90
Crew D-12	**Hr.**	**Daily**	**Hr.**	**Daily**	**Bare Costs**	**Incl. O&P**
2 Bricklayers	$36.75	$588.00	$61.25	$980.00	$33.73	$56.23
2 Bricklayer Helpers	30.70	491.20	51.20	819.20		
32 L.H., Daily Totals		$1079.20		$1799.20	$33.73	$56.23
Crew D-13	**Hr.**	**Daily**	**Hr.**	**Daily**	**Bare Costs**	**Incl. O&P**
1 Bricklayer Foreman (outside)	$38.75	$310.00	$64.60	$516.80	$35.93	$59.73
2 Bricklayers	36.75	588.00	61.25	980.00		
2 Bricklayer Helpers	30.70	491.20	51.20	819.20		
1 Equip. Oper. (crane)	41.95	335.60	68.85	550.80		
1 S.P. Crane, 4x4, 12 Ton		371.85		409.04	7.75	8.52
48 L.H., Daily Totals		$2096.65		$3275.84	$43.68	$68.25
Crew D-14	**Hr.**	**Daily**	**Hr.**	**Daily**	**Bare Costs**	**Incl. O&P**
3 Bricklayers	$36.75	$882.00	$61.25	$1470.00	$35.24	$58.74
1 Bricklayer Helper	30.70	245.60	51.20	409.60		
32 L.H., Daily Totals		$1127.60		$1879.60	$35.24	$58.74
Crew E-1	**Hr.**	**Daily**	**Hr.**	**Daily**	**Bare Costs**	**Incl. O&P**
2 Struc. Steel Workers	$41.45	$663.20	$71.90	$1150.40	$41.45	$71.90
1 Welder, Gas Engine, 300 amp		95.65		105.22	5.98	6.58
16 L.H., Daily Totals		$758.85		$1255.62	$47.43	$78.48

Crew No.	Bare Costs		Incl. Subs O&P		Cost Per Labor-Hour	
Crew E-2	**Hr.**	**Daily**	**Hr.**	**Daily**	**Bare Costs**	**Incl. O&P**
1 Struc. Steel Foreman (outside)	$43.45	$347.60	$75.40	$603.20	$41.87	$71.97
4 Struc. Steel Workers	41.45	1326.40	71.90	2300.80		
1 Equip. Oper. (crane)	41.95	335.60	68.85	550.80		
1 Lattice Boom Crane, 90 Ton		1698.00		1867.80	35.38	38.91
48 L.H., Daily Totals		$3707.60		$5322.60	$77.24	$110.89
Crew E-3	**Hr.**	**Daily**	**Hr.**	**Daily**	**Bare Costs**	**Incl. O&P**
1 Struc. Steel Foreman (outside)	$43.45	$347.60	$75.40	$603.20	$42.12	$73.07
2 Struc. Steel Workers	41.45	663.20	71.90	1150.40		
1 Welder, Gas Engine, 300 amp		95.65		105.22	3.99	4.38
24 L.H., Daily Totals		$1106.45		$1858.82	$46.10	$77.45
Crew E-3A	**Hr.**	**Daily**	**Hr.**	**Daily**	**Bare Costs**	**Incl. O&P**
1 Struc. Steel Foreman (outside)	$43.45	$347.60	$75.40	$603.20	$42.12	$73.07
2 Struc. Steel Workers	41.45	663.20	71.90	1150.40		
1 Welder, Gas Engine, 300 amp		95.65		105.22		
1 Telescoping Boom Lift, to 40'		283.15		311.46	15.78	17.36
24 L.H., Daily Totals		$1389.60		$2170.28	$57.90	$90.43
Crew E-4	**Hr.**	**Daily**	**Hr.**	**Daily**	**Bare Costs**	**Incl. O&P**
1 Struc. Steel Foreman (outside)	$43.45	$347.60	$75.40	$603.20	$41.95	$72.78
3 Struc. Steel Workers	41.45	994.80	71.90	1725.60		
1 Welder, Gas Engine, 300 amp		95.65		105.22	2.99	3.29
32 L.H., Daily Totals		$1438.05		$2434.01	$44.94	$76.06
Crew E-5	**Hr.**	**Daily**	**Hr.**	**Daily**	**Bare Costs**	**Incl. O&P**
1 Struc. Steel Foreman (outside)	$43.45	$347.60	$75.40	$603.20	$41.73	$71.95
7 Struc. Steel Workers	41.45	2321.20	71.90	4026.40		
1 Equip. Oper. (crane)	41.95	335.60	68.85	550.80		
1 Lattice Boom Crane, 90 Ton		1698.00		1867.80		
1 Welder, Gas Engine, 300 amp		95.65		105.22	24.91	27.40
72 L.H., Daily Totals		$4798.05		$7153.42	$66.64	$99.35
Crew E-6	**Hr.**	**Daily**	**Hr.**	**Daily**	**Bare Costs**	**Incl. O&P**
1 Struc. Steel Foreman (outside)	$43.45	$347.60	$75.40	$603.20	$41.44	$71.38
12 Struc. Steel Workers	41.45	3979.20	71.90	6902.40		
1 Equip. Oper. (crane)	41.95	335.60	68.85	550.80		
1 Equip. Oper. (light)	38.75	310.00	63.60	508.80		
1 Lattice Boom Crane, 90 Ton		1698.00		1867.80		
1 Welder, Gas Engine, 300 amp		95.65		105.22		
1 Air Compressor, 160 cfm		115.85		127.44		
2 Impact Wrenches		37.00		40.70	16.22	17.84
120 L.H., Daily Totals		$6918.90		$10706.35	$57.66	$89.22
Crew E-7	**Hr.**	**Daily**	**Hr.**	**Daily**	**Bare Costs**	**Incl. O&P**
1 Struc. Steel Foreman (outside)	$43.45	$347.60	$75.40	$603.20	$41.73	$71.95
7 Struc. Steel Workers	41.45	2321.20	71.90	4026.40		
1 Equip. Oper. (crane)	41.95	335.60	68.85	550.80		
1 Lattice Boom Crane, 90 Ton		1698.00		1867.80		
2 Welder, Gas Engine, 300 amp		191.30		210.43	26.24	28.86
72 L.H., Daily Totals		$4893.70		$7258.63	$67.97	$100.81
Crew E-8	**Hr.**	**Daily**	**Hr.**	**Daily**	**Bare Costs**	**Incl. O&P**
1 Struc. Steel Foreman (outside)	$43.45	$347.60	$75.40	$603.20	$41.68	$71.94
9 Struc. Steel Workers	41.45	2984.40	71.90	5176.80		
1 Equip. Oper. (crane)	41.95	335.60	68.85	550.80		
1 Lattice Boom Crane, 90 Ton		1698.00		1867.80		
4 Welder, Gas Engine, 300 amp		382.60		420.86	23.64	26.01
88 L.H., Daily Totals		$5748.20		$8619.46	$65.32	$97.95

Crew No.	Bare Costs		Incl. Subs O&P		Cost Per Labor-Hour	
Crew E-9	**Hr.**	**Daily**	**Hr.**	**Daily**	**Bare Costs**	**Incl. O&P**
2 Struc. Steel Foremen (outside)	$43.45	$695.20	$75.40	$1206.40	$41.34	$71.03
5 Struc. Steel Workers	41.45	1658.00	71.90	2876.00		
1 Welder Foreman (outside)	43.45	347.60	75.40	603.20		
5 Welders	41.45	1658.00	71.90	2876.00		
1 Equip. Oper. (crane)	41.95	335.60	68.85	550.80		
1 Equip. Oper. (oiler)	35.90	287.20	58.90	471.20		
1 Equip. Oper. (light)	38.75	310.00	63.60	508.80		
1 Lattice Boom Crane, 90 Ton		1698.00		1867.80		
5 Welder, Gas Engine, 300 amp		478.25		526.08	17.00	18.70
128 L.H., Daily Totals		$7467.85		$11486.28	$58.34	$89.74

Crew E-10	Hr.	Daily	Hr.	Daily	Bare Costs	Incl. O&P
1 Struc. Steel Foreman (outside)	$43.45	$347.60	$75.40	$603.20	$42.12	$73.07
2 Struc. Steel Workers	41.45	663.20	71.90	1150.40		
1 Welder, Gas Engine, 300 amp		95.65		105.22		
1 Flatbed Truck, Gas, 3 Ton		245.95		270.55	14.23	15.66
24 L.H., Daily Totals		$1352.40		$2129.36	$56.35	$88.72

Crew E-11	Hr.	Daily	Hr.	Daily	Bare Costs	Incl. O&P
2 Painters, Struc. Steel	$33.50	$536.00	$59.50	$952.00	$34.02	$58.19
1 Building Laborer	30.35	242.80	50.15	401.20		
1 Equip. Oper. (light)	38.75	310.00	63.60	508.80		
1 Air Compressor, 250 cfm		167.95		184.75		
1 Sandblaster, Portable, 3 C.F.		20.70		22.77		
1 Set Sand Blasting Accessories		14.90		16.39	6.36	7.00
32 L.H., Daily Totals		$1292.35		$2085.91	$40.39	$65.18

Crew E-11A	Hr.	Daily	Hr.	Daily	Bare Costs	Incl. O&P
2 Painters, Struc. Steel	$33.50	$536.00	$59.50	$952.00	$34.02	$58.19
1 Building Laborer	30.35	242.80	50.15	401.20		
1 Equip. Oper. (light)	38.75	310.00	63.60	508.80		
1 Air Compressor, 250 cfm		167.95		184.75		
1 Sandblaster, Portable, 3 C.F.		20.70		22.77		
1 Set Sand Blasting Accessories		14.90		16.39		
1 Telescoping Boom Lift, to 60'		451.25		496.38	20.46	22.51
32 L.H., Daily Totals		$1743.60		$2582.28	$54.49	$80.70

Crew E-11B	Hr.	Daily	Hr.	Daily	Bare Costs	Incl. O&P
2 Painters, Struc. Steel	$33.50	$536.00	$59.50	$952.00	$32.45	$56.38
1 Building Laborer	30.35	242.80	50.15	401.20		
2 Paint Sprayer, 8 C.F.M.		85.20		93.72		
1 Telescoping Boom Lift, to 60'		451.25		496.38	22.35	24.59
24 L.H., Daily Totals		$1315.25		$1943.30	$54.80	$80.97

Crew E-12	Hr.	Daily	Hr.	Daily	Bare Costs	Incl. O&P
1 Welder Foreman (outside)	$43.45	$347.60	$75.40	$603.20	$41.10	$69.50
1 Equip. Oper. (light)	38.75	310.00	63.60	508.80		
1 Welder, Gas Engine, 300 amp		95.65		105.22	5.98	6.58
16 L.H., Daily Totals		$753.25		$1217.21	$47.08	$76.08

Crew E-13	Hr.	Daily	Hr.	Daily	Bare Costs	Incl. O&P
1 Welder Foreman (outside)	$43.45	$347.60	$75.40	$603.20	$41.88	$71.47
.5 Equip. Oper. (light)	38.75	155.00	63.60	254.40		
1 Welder, Gas Engine, 300 amp		95.65		105.22	7.97	8.77
12 L.H., Daily Totals		$598.25		$962.82	$49.85	$80.23

Crew E-14	Hr.	Daily	Hr.	Daily	Bare Costs	Incl. O&P
1 Struc. Steel Worker	$41.45	$331.60	$71.90	$575.20	$41.45	$71.90
1 Welder, Gas Engine, 300 amp		95.65		105.22	11.96	13.15
8 L.H., Daily Totals		$427.25		$680.41	$53.41	$85.05

Crew E-16	Hr.	Daily	Hr.	Daily	Bare Costs	Incl. O&P
1 Welder Foreman (outside)	$43.45	$347.60	$75.40	$603.20	$42.45	$73.65
1 Welder	41.45	331.60	71.90	575.20		
1 Welder, Gas Engine, 300 amp		95.65		105.22	5.98	6.58
16 L.H., Daily Totals		$774.85		$1283.62	$48.43	$80.23

Crew E-17	Hr.	Daily	Hr.	Daily	Bare Costs	Incl. O&P
1 Struc. Steel Foreman (outside)	$43.45	$347.60	$75.40	$603.20	$42.45	$73.65
1 Structural Steel Worker	41.45	331.60	71.90	575.20		
16 L.H., Daily Totals		$679.20		$1178.40	$42.45	$73.65

Crew E-18	Hr.	Daily	Hr.	Daily	Bare Costs	Incl. O&P
1 Struc. Steel Foreman (outside)	$43.45	$347.60	$75.40	$603.20	$41.71	$71.59
3 Structural Steel Workers	41.45	994.80	71.90	1725.60		
1 Equipment Operator (med.)	40.75	326.00	66.85	534.80		
1 Lattice Boom Crane, 20 Ton		869.00		955.90	21.73	23.90
40 L.H., Daily Totals		$2537.40		$3819.50	$63.44	$95.49

Crew E-19	Hr.	Daily	Hr.	Daily	Bare Costs	Incl. O&P
1 Struc. Steel Foreman (outside)	$43.45	$347.60	$75.40	$603.20	$41.22	$70.30
1 Structural Steel Worker	41.45	331.60	71.90	575.20		
1 Equip. Oper. (light)	38.75	310.00	63.60	508.80		
1 Lattice Boom Crane, 20 Ton		869.00		955.90	36.21	39.83
24 L.H., Daily Totals		$1858.20		$2643.10	$77.42	$110.13

Crew E-20	Hr.	Daily	Hr.	Daily	Bare Costs	Incl. O&P
1 Struc. Steel Foreman (outside)	$43.45	$347.60	$75.40	$603.20	$41.07	$70.33
5 Structural Steel Workers	41.45	1658.00	71.90	2876.00		
1 Equip. Oper. (crane)	41.95	335.60	68.85	550.80		
1 Equip. Oper. (oiler)	35.90	287.20	58.90	471.20		
1 Lattice Boom Crane, 40 Ton		1315.00		1446.50	20.55	22.60
64 L.H., Daily Totals		$3943.40		$5947.70	$61.62	$92.93

Crew E-22	Hr.	Daily	Hr.	Daily	Bare Costs	Incl. O&P
1 Skilled Worker Foreman (out)	$41.45	$331.60	$68.85	$550.80	$40.12	$66.65
2 Skilled Workers	39.45	631.20	65.55	1048.80		
24 L.H., Daily Totals		$962.80		$1599.60	$40.12	$66.65

Crew E-24	Hr.	Daily	Hr.	Daily	Bare Costs	Incl. O&P
3 Structural Steel Workers	$41.45	$994.80	$71.90	$1725.60	$41.27	$70.64
1 Equipment Operator (med.)	40.75	326.00	66.85	534.80		
1 Hyd. Crane, 25 Ton		581.70		639.87	18.18	20.00
32 L.H., Daily Totals		$1902.50		$2900.27	$59.45	$90.63

Crew E-25	Hr.	Daily	Hr.	Daily	Bare Costs	Incl. O&P
1 Welder	$41.45	$331.60	$71.90	$575.20	$41.45	$71.90
1 Cutting Torch		12.50		13.75	1.56	1.72
8 L.H., Daily Totals		$344.10		$588.95	$43.01	$73.62

Crew E-26	Hr.	Daily	Hr.	Daily	Bare Costs	Incl. O&P
1 Struc. Steel Foreman (outside)	$43.45	$347.60	$75.40	$603.20	$42.56	$73.20
1 Struc. Steel Worker	41.45	331.60	71.90	575.20		
1 Welder	41.45	331.60	71.90	575.20		
.25 Electrician	45.60	91.20	74.40	148.80		
.25 Plumber	44.85	89.70	73.60	147.20		
1 Welder, Gas Engine, 300 amp		95.65		105.22	3.42	3.76
28 L.H., Daily Totals		$1287.35		$2154.82	$45.98	$76.96

Crew No.	Bare Costs		Incl. Subs O&P		Cost Per Labor-Hour	

Crew E-27	Hr.	Daily	Hr.	Daily	Bare Costs	Incl. O&P
1 Struc. Steel Foreman (outside)	$43.45	$347.60	$75.40	$603.20	$41.76	$71.96
6 Struc. Steel Workers	41.45	1989.60	71.90	3451.20		
1 Equip. Oper. (crane)	41.95	335.60	68.85	550.80		
1 Hyd. Crane, 12 Ton		471.75		518.92		
1 Hyd. Crane, 80 Ton		1487.00		1635.70	30.61	33.67
64 L.H., Daily Totals		$4631.55		$6759.82	$72.37	$105.62

Crew F-3	Hr.	Daily	Hr.	Daily	Bare Costs	Incl. O&P
2 Carpenters	$38.75	$620.00	$64.05	$1024.80	$35.49	$58.83
2 Carpenter Helpers	29.00	464.00	48.60	777.60		
1 Equip. Oper. (crane)	41.95	335.60	68.85	550.80		
1 Hyd. Crane, 12 Ton		471.75		518.92	11.79	12.97
40 L.H., Daily Totals		$1891.35		$2872.13	$47.28	$71.80

Crew F-4	Hr.	Daily	Hr.	Daily	Bare Costs	Incl. O&P
2 Carpenters	$38.75	$620.00	$64.05	$1024.80	$35.49	$58.83
2 Carpenter Helpers	29.00	464.00	48.60	777.60		
1 Equip. Oper. (crane)	41.95	335.60	68.85	550.80		
1 Hyd. Crane, 55 Ton		981.50		1079.65	24.54	26.99
40 L.H., Daily Totals		$2401.10		$3432.85	$60.03	$85.82

Crew F-5	Hr.	Daily	Hr.	Daily	Bare Costs	Incl. O&P
2 Carpenters	$38.75	$620.00	$64.05	$1024.80	$33.88	$56.33
2 Carpenter Helpers	29.00	464.00	48.60	777.60		
32 L.H., Daily Totals		$1084.00		$1802.40	$33.88	$56.33

Crew F-6	Hr.	Daily	Hr.	Daily	Bare Costs	Incl. O&P
2 Carpenters	$38.75	$620.00	$64.05	$1024.80	$36.03	$59.45
2 Building Laborers	30.35	485.60	50.15	802.40		
1 Equip. Oper. (crane)	41.95	335.60	68.85	550.80		
1 Hyd. Crane, 12 Ton		471.75		518.92	11.79	12.97
40 L.H., Daily Totals		$1912.95		$2896.93	$47.82	$72.42

Crew F-7	Hr.	Daily	Hr.	Daily	Bare Costs	Incl. O&P
2 Carpenters	$38.75	$620.00	$64.05	$1024.80	$34.55	$57.10
2 Building Laborers	30.35	485.60	50.15	802.40		
32 L.H., Daily Totals		$1105.60		$1827.20	$34.55	$57.10

Crew G-1	Hr.	Daily	Hr.	Daily	Bare Costs	Incl. O&P
1 Roofer Foreman (outside)	$35.35	$282.80	$63.85	$510.80	$31.26	$56.48
4 Roofers Composition	33.35	1067.20	60.25	1928.00		
2 Roofer Helpers	25.05	400.80	45.25	724.00		
1 Application Equipment		181.15		199.26		
1 Tar Kettle/Pot		178.20		196.02		
1 Crew Truck		154.35		169.79	9.17	10.09
56 L.H., Daily Totals		$2264.50		$3727.87	$40.44	$66.57

Crew G-2	Hr.	Daily	Hr.	Daily	Bare Costs	Incl. O&P
1 Plasterer	$36.30	$290.40	$59.35	$474.80	$32.63	$53.53
1 Plasterer Helper	31.25	250.00	51.10	408.80		
1 Building Laborer	30.35	242.80	50.15	401.20		
1 Grout Pump, 50 C.F./hr.		125.60		138.16	5.23	5.76
24 L.H., Daily Totals		$908.80		$1422.96	$37.87	$59.29

Crew G-2A	Hr.	Daily	Hr.	Daily	Bare Costs	Incl. O&P
1 Roofer Composition	$33.35	$266.80	$60.25	$482.00	$29.58	$51.88
1 Roofer Helper	25.05	200.40	45.25	362.00		
1 Building Laborer	30.35	242.80	50.15	401.20		
1 Foam Spray Rig, Trailer-Mtd.		514.30		565.73		
1 Pickup Truck, 3/4 Ton		109.90		120.89	26.01	28.61
24 L.H., Daily Totals		$1334.20		$1931.82	$55.59	$80.49

Crew G-3	Hr.	Daily	Hr.	Daily	Bare Costs	Incl. O&P
2 Sheet Metal Workers	$43.30	$692.80	$71.80	$1148.80	$36.83	$60.98
2 Building Laborers	30.35	485.60	50.15	802.40		
32 L.H., Daily Totals		$1178.40		$1951.20	$36.83	$60.98

Crew G-4	Hr.	Daily	Hr.	Daily	Bare Costs	Incl. O&P
1 Labor Foreman (outside)	$32.35	$258.80	$53.45	$427.60	$31.02	$51.25
2 Building Laborers	30.35	485.60	50.15	802.40		
1 Flatbed Truck, Gas, 1.5 Ton		195.20		214.72		
1 Air Compressor, 160 cfm		115.85		127.44	12.96	14.26
24 L.H., Daily Totals		$1055.45		$1572.16	$43.98	$65.51

Crew G-5	Hr.	Daily	Hr.	Daily	Bare Costs	Incl. O&P
1 Roofer Foreman (outside)	$35.35	$282.80	$63.85	$510.80	$30.43	$54.97
2 Roofers Composition	33.35	533.60	60.25	964.00		
2 Roofer Helpers	25.05	400.80	45.25	724.00		
1 Application Equipment		181.15		199.26	4.53	4.98
40 L.H., Daily Totals		$1398.35		$2398.07	$34.96	$59.95

Crew G-6A	Hr.	Daily	Hr.	Daily	Bare Costs	Incl. O&P
2 Roofers Composition	$33.35	$533.60	$60.25	$964.00	$33.35	$60.25
1 Small Compressor, Electric		14.85		16.34		
2 Pneumatic Nailers		44.70		49.17	3.72	4.09
16 L.H., Daily Totals		$593.15		$1029.51	$37.07	$64.34

Crew G-7	Hr.	Daily	Hr.	Daily	Bare Costs	Incl. O&P
1 Carpenter	$38.75	$310.00	$64.05	$512.40	$38.75	$64.05
1 Small Compressor, Electric		14.85		16.34		
1 Pneumatic Nailer		22.35		24.59	4.65	5.12
8 L.H., Daily Totals		$347.20		$553.32	$43.40	$69.17

Crew H-1	Hr.	Daily	Hr.	Daily	Bare Costs	Incl. O&P
2 Glaziers	$38.25	$612.00	$62.95	$1007.20	$39.85	$67.42
2 Struc. Steel Workers	41.45	663.20	71.90	1150.40		
32 L.H., Daily Totals		$1275.20		$2157.60	$39.85	$67.42

Crew H-2	Hr.	Daily	Hr.	Daily	Bare Costs	Incl. O&P
2 Glaziers	$38.25	$612.00	$62.95	$1007.20	$35.62	$58.68
1 Building Laborer	30.35	242.80	50.15	401.20		
24 L.H., Daily Totals		$854.80		$1408.40	$35.62	$58.68

Crew H-3	Hr.	Daily	Hr.	Daily	Bare Costs	Incl. O&P
1 Glazier	$38.25	$306.00	$62.95	$503.60	$33.63	$55.77
1 Helper	29.00	232.00	48.60	388.80		
16 L.H., Daily Totals		$538.00		$892.40	$33.63	$55.77

Crew H-4	Hr.	Daily	Hr.	Daily	Bare Costs	Incl. O&P
1 Carpenter	$38.75	$310.00	$64.05	$512.40	$36.22	$59.94
1 Carpenter Helper	29.00	232.00	48.60	388.80		
.5 Electrician	45.60	182.40	74.40	297.60		
20 L.H., Daily Totals		$724.40		$1198.80	$36.22	$59.94

Crew No.	Bare Costs		Incl. Subs O&P		Cost Per Labor-Hour	
Crew J-1	**Hr.**	**Daily**	**Hr.**	**Daily**	**Bare Costs**	**Incl. O&P**
3 Plasterers	$36.30	$871.20	$59.35	$1424.40	$34.28	$56.05
2 Plasterer Helpers	31.25	500.00	51.10	817.60		
1 Mixing Machine, 6 C.F.		135.35		148.88	3.38	3.72
40 L.H., Daily Totals		$1506.55		$2390.89	$37.66	$59.77
Crew J-2	**Hr.**	**Daily**	**Hr.**	**Daily**	**Bare Costs**	**Incl. O&P**
3 Plasterers	$36.30	$871.20	$59.35	$1424.40	$34.92	$56.94
2 Plasterer Helpers	31.25	500.00	51.10	817.60		
1 Lather	38.15	305.20	61.40	491.20		
1 Mixing Machine, 6 C.F.		135.35		148.88	2.82	3.10
48 L.H., Daily Totals		$1811.75		$2882.09	$37.74	$60.04
Crew J-3	**Hr.**	**Daily**	**Hr.**	**Daily**	**Bare Costs**	**Incl. O&P**
1 Terrazzo Worker	$35.40	$283.20	$57.15	$457.20	$33.02	$53.30
1 Terrazzo Helper	30.65	245.20	49.45	395.60		
1 Floor Grinder, 22" Path		138.30		152.13		
1 Terrazzo Mixer		178.85		196.74	19.82	21.80
16 L.H., Daily Totals		$845.55		$1201.67	$52.85	$75.10
Crew J-4	**Hr.**	**Daily**	**Hr.**	**Daily**	**Bare Costs**	**Incl. O&P**
2 Cement Finishers	$37.15	$594.40	$60.05	$960.80	$34.88	$56.75
1 Laborer	30.35	242.80	50.15	401.20		
1 Floor Grinder, 22" Path		138.30		152.13		
1 Floor Edger, 7" Path		43.45		47.80		
1 Vacuum Pick-Up System		72.15		79.36	10.58	11.64
24 L.H., Daily Totals		$1091.10		$1641.29	$45.46	$68.39
Crew J-4A	**Hr.**	**Daily**	**Hr.**	**Daily**	**Bare Costs**	**Incl. O&P**
2 Cement Finishers	$37.15	$594.40	$60.05	$960.80	$33.75	$55.10
2 Laborers	30.35	485.60	50.15	802.40		
1 Floor Grinder, 22" Path		138.30		152.13		
1 Floor Edger, 7" Path		43.45		47.80		
1 Vacuum Pick-Up System		72.15		79.36		
1 Floor Auto Scrubber		265.15		291.67	16.22	17.84
32 L.H., Daily Totals		$1599.05		$2334.16	$49.97	$72.94
Crew J-4B	**Hr.**	**Daily**	**Hr.**	**Daily**	**Bare Costs**	**Incl. O&P**
1 Laborer	$30.35	$242.80	$50.15	$401.20	$30.35	$50.15
1 Floor Auto Scrubber		265.15		291.67	33.14	36.46
8 L.H., Daily Totals		$507.95		$692.87	$63.49	$86.61
Crew J-6	**Hr.**	**Daily**	**Hr.**	**Daily**	**Bare Costs**	**Incl. O&P**
2 Painters	$32.85	$525.60	$53.80	$860.80	$33.70	$55.34
1 Building Laborer	30.35	242.80	50.15	401.20		
1 Equip. Oper. (light)	38.75	310.00	63.60	508.80		
1 Air Compressor, 250 cfm		167.95		184.75		
1 Sandblaster, Portable, 3 C.F.		20.70		22.77		
1 Set Sand Blasting Accessories		14.90		16.39	6.36	7.00
32 L.H., Daily Totals		$1281.95		$1994.70	$40.06	$62.33
Crew J-7	**Hr.**	**Daily**	**Hr.**	**Daily**	**Bare Costs**	**Incl. O&P**
2 Painters	$32.85	$525.60	$53.80	$860.80	$32.85	$53.80
1 Floor Belt Sander		17.75		19.52		
1 Floor Sanding Edger		13.65		15.02	1.96	2.16
16 L.H., Daily Totals		$557.00		$895.34	$34.81	$55.96

Crew No.	Bare Costs		Incl. Subs O&P		Cost Per Labor-Hour	
Crew K-1	**Hr.**	**Daily**	**Hr.**	**Daily**	**Bare Costs**	**Incl. O&P**
1 Carpenter	$38.75	$310.00	$64.05	$512.40	$36.42	$60.08
1 Truck Driver (light)	34.10	272.80	56.10	448.80		
1 Flatbed Truck, Gas, 3 Ton		245.95		270.55	15.37	16.91
16 L.H., Daily Totals		$828.75		$1231.74	$51.80	$76.98
Crew K-2	**Hr.**	**Daily**	**Hr.**	**Daily**	**Bare Costs**	**Incl. O&P**
1 Struc. Steel Foreman (outside)	$43.45	$347.60	$75.40	$603.20	$39.67	$67.80
1 Struc. Steel Worker	41.45	331.60	71.90	575.20		
1 Truck Driver (light)	34.10	272.80	56.10	448.80		
1 Flatbed Truck, Gas, 3 Ton		245.95		270.55	10.25	11.27
24 L.H., Daily Totals		$1197.95		$1897.74	$49.91	$79.07
Crew L-1	**Hr.**	**Daily**	**Hr.**	**Daily**	**Bare Costs**	**Incl. O&P**
.25 Electrician	$45.60	$91.20	$74.40	$148.80	$45.00	$73.76
1 Plumber	44.85	358.80	73.60	588.80		
10 L.H., Daily Totals		$450.00		$737.60	$45.00	$73.76
Crew L-2	**Hr.**	**Daily**	**Hr.**	**Daily**	**Bare Costs**	**Incl. O&P**
1 Carpenter	$38.75	$310.00	$64.05	$512.40	$33.88	$56.33
1 Carpenter Helper	29.00	232.00	48.60	388.80		
16 L.H., Daily Totals		$542.00		$901.20	$33.88	$56.33
Crew L-3	**Hr.**	**Daily**	**Hr.**	**Daily**	**Bare Costs**	**Incl. O&P**
1 Carpenter	$38.75	$310.00	$64.05	$512.40	$40.12	$66.12
.25 Electrician	45.60	91.20	74.40	148.80		
10 L.H., Daily Totals		$401.20		$661.20	$40.12	$66.12
Crew L-3A	**Hr.**	**Daily**	**Hr.**	**Daily**	**Bare Costs**	**Incl. O&P**
1 Carpenter Foreman (outside)	$40.75	$326.00	$67.35	$538.80	$41.60	$68.83
.5 Sheet Metal Worker	43.30	173.20	71.80	287.20		
12 L.H., Daily Totals		$499.20		$826.00	$41.60	$68.83
Crew L-4	**Hr.**	**Daily**	**Hr.**	**Daily**	**Bare Costs**	**Incl. O&P**
1 Skilled Worker	$39.45	$315.60	$65.55	$524.40	$34.23	$57.08
1 Helper	29.00	232.00	48.60	388.80		
16 L.H., Daily Totals		$547.60		$913.20	$34.23	$57.08
Crew L-5	**Hr.**	**Daily**	**Hr.**	**Daily**	**Bare Costs**	**Incl. O&P**
1 Struc. Steel Foreman (outside)	$43.45	$347.60	$75.40	$603.20	$41.81	$71.96
5 Struc. Steel Workers	41.45	1658.00	71.90	2876.00		
1 Equip. Oper. (crane)	41.95	335.60	68.85	550.80		
1 Hyd. Crane, 25 Ton		581.70		639.87	10.39	11.43
56 L.H., Daily Totals		$2922.90		$4669.87	$52.19	$83.39
Crew L-5A	**Hr.**	**Daily**	**Hr.**	**Daily**	**Bare Costs**	**Incl. O&P**
1 Struc. Steel Foreman (outside)	$43.45	$347.60	$75.40	$603.20	$42.08	$72.01
2 Structural Steel Workers	41.45	663.20	71.90	1150.40		
1 Equip. Oper. (crane)	41.95	335.60	68.85	550.80		
1 S.P. Crane, 4x4, 25 Ton		664.40		730.84	20.76	22.84
32 L.H., Daily Totals		$2010.80		$3035.24	$62.84	$94.85
Crew L-5B	**Hr.**	**Daily**	**Hr.**	**Daily**	**Bare Costs**	**Incl. O&P**
1 Struc. Steel Foreman (outside)	$43.45	$347.60	$75.40	$603.20	$42.44	$70.90
2 Structural Steel Workers	41.45	663.20	71.90	1150.40		
2 Electricians	45.60	729.60	74.40	1190.40		
2 Steamfitters/Pipefitters	46.05	736.80	75.55	1208.80		
1 Equip. Oper. (crane)	41.95	335.60	68.85	550.80		
1 Common Laborer	30.35	242.80	50.15	401.20		
1 Hyd. Crane, 80 Ton		1487.00		1635.70	20.65	22.72
72 L.H., Daily Totals		$4542.60		$6740.50	$63.09	$93.62

Crew No.	Bare Costs		Incl. Subs O&P		Cost Per Labor-Hour	
Crew L-6	**Hr.**	**Daily**	**Hr.**	**Daily**	**Bare Costs**	**Incl. O&P**
1 Plumber	$44.85	$358.80	$73.60	$588.80	$45.10	$73.87
.5 Electrician	45.60	182.40	74.40	297.60		
12 L.H., Daily Totals		$541.20		$886.40	$45.10	$73.87

Crew L-7	Hr.	Daily	Hr.	Daily	Bare Costs	Incl. O&P
1 Carpenter	$38.75	$310.00	$64.05	$512.40	$33.28	$55.34
2 Carpenter Helpers	29.00	464.00	48.60	777.60		
.25 Electrician	45.60	91.20	74.40	148.80		
26 L.H., Daily Totals		$865.20		$1438.80	$33.28	$55.34

Crew L-8	Hr.	Daily	Hr.	Daily	Bare Costs	Incl. O&P
1 Carpenter	$38.75	$310.00	$64.05	$512.40	$36.07	$59.78
1 Carpenter Helper	29.00	232.00	48.60	388.80		
.5 Plumber	44.85	179.40	73.60	294.40		
20 L.H., Daily Totals		$721.40		$1195.60	$36.07	$59.78

Crew L-9	Hr.	Daily	Hr.	Daily	Bare Costs	Incl. O&P
1 Skilled Worker Foreman	$41.45	$331.60	$68.85	$550.80	$35.93	$59.73
1 Skilled Worker	39.45	315.60	65.55	524.40		
2 Helpers	29.00	464.00	48.60	777.60		
.5 Electrician	45.60	182.40	74.40	297.60		
36 L.H., Daily Totals		$1293.60		$2150.40	$35.93	$59.73

Crew L-10	Hr.	Daily	Hr.	Daily	Bare Costs	Incl. O&P
1 Struc. Steel Foreman (outside)	$43.45	$347.60	$75.40	$603.20	$42.28	$72.05
1 Structural Steel Worker	41.45	331.60	71.90	575.20		
1 Equip. Oper. (crane)	41.95	335.60	68.85	550.80		
1 Hyd. Crane, 12 Ton		471.75		518.92	19.66	21.62
24 L.H., Daily Totals		$1486.55		$2248.13	$61.94	$93.67

Crew L-11	Hr.	Daily	Hr.	Daily	Bare Costs	Incl. O&P
2 Wreckers	$30.35	$485.60	$51.20	$819.20	$35.35	$58.71
1 Equip. Oper. (crane)	41.95	335.60	68.85	550.80		
1 Equip. Oper. (light)	38.75	310.00	63.60	508.80		
1 Hyd. Excavator, 2.5 C.Y.		1441.00		1585.10		
1 Loader, Skid Steer, 78 H.P.		392.50		431.75	57.30	63.03
32 L.H., Daily Totals		$2964.70		$3895.65	$92.65	$121.74

Crew M-1	Hr.	Daily	Hr.	Daily	Bare Costs	Incl. O&P
3 Elevator Constructors	$61.30	$1471.20	$99.70	$2392.80	$58.24	$94.72
1 Elevator Apprentice	49.05	392.40	79.80	638.40		
5 Hand Tools		49.50		54.45	1.55	1.70
32 L.H., Daily Totals		$1913.10		$3085.65	$59.78	$96.43

Crew M-3	Hr.	Daily	Hr.	Daily	Bare Costs	Incl. O&P
1 Electrician Foreman (outside)	$47.60	$380.80	$77.70	$621.60	$46.70	$76.25
1 Common Laborer	30.35	242.80	50.15	401.20		
.25 Equipment Operator (med.)	40.75	81.50	66.85	133.70		
1 Elevator Constructor	61.30	490.40	99.70	797.60		
1 Elevator Apprentice	49.05	392.40	79.80	638.40		
.25 S.P. Crane, 4x4, 20 Ton		146.94		161.63	4.32	4.75
34 L.H., Daily Totals		$1734.84		$2754.13	$51.02	$81.00

Crew M-4	Hr.	Daily	Hr.	Daily	Bare Costs	Incl. O&P
1 Electrician Foreman (outside)	$47.60	$380.80	$77.70	$621.60	$46.17	$75.40
1 Common Laborer	30.35	242.80	50.15	401.20		
.25 Equipment Operator, Crane	41.95	83.90	68.85	137.70		
.25 Equip. Oper. (oiler)	35.90	71.80	58.90	117.80		
1 Elevator Constructor	61.30	490.40	99.70	797.60		
1 Elevator Apprentice	49.05	392.40	79.80	638.40		
.25 S.P. Crane, 4x4, 40 Ton		187.56		206.32	5.21	5.73
36 L.H., Daily Totals		$1849.66		$2920.62	$51.38	$81.13

Crew Q-1	Hr.	Daily	Hr.	Daily	Bare Costs	Incl. O&P
1 Plumber	$44.85	$358.80	$73.60	$588.80	$40.38	$66.25
1 Plumber Apprentice	35.90	287.20	58.90	471.20		
16 L.H., Daily Totals		$646.00		$1060.00	$40.38	$66.25

Crew Q-1A	Hr.	Daily	Hr.	Daily	Bare Costs	Incl. O&P
.25 Plumber Foreman (outside)	$46.85	$93.70	$76.85	$153.70	$45.25	$74.25
1 Plumber	44.85	358.80	73.60	588.80		
10 L.H., Daily Totals		$452.50		$742.50	$45.25	$74.25

Crew Q-1C	Hr.	Daily	Hr.	Daily	Bare Costs	Incl. O&P
1 Plumber	$44.85	$358.80	$73.60	$588.80	$40.50	$66.45
1 Plumber Apprentice	35.90	287.20	58.90	471.20		
1 Equip. Oper. (medium)	40.75	326.00	66.85	534.80		
1 Trencher, Chain Type, 8' D		1837.00		2020.70	76.54	84.20
24 L.H., Daily Totals		$2809.00		$3615.50	$117.04	$150.65

Crew Q-2	Hr.	Daily	Hr.	Daily	Bare Costs	Incl. O&P
1 Plumber	$44.85	$358.80	$73.60	$588.80	$38.88	$63.80
2 Plumber Apprentices	35.90	574.40	58.90	942.40		
24 L.H., Daily Totals		$933.20		$1531.20	$38.88	$63.80

Crew Q-3	Hr.	Daily	Hr.	Daily	Bare Costs	Incl. O&P
2 Plumbers	$44.85	$717.60	$73.60	$1177.60	$40.38	$66.25
2 Plumber Apprentices	35.90	574.40	58.90	942.40		
32 L.H., Daily Totals		$1292.00		$2120.00	$40.38	$66.25

Crew Q-4	Hr.	Daily	Hr.	Daily	Bare Costs	Incl. O&P
2 Plumbers	$44.85	$717.60	$73.60	$1177.60	$42.61	$69.92
1 Welder (plumber)	44.85	358.80	73.60	588.80		
1 Plumber Apprentice	35.90	287.20	58.90	471.20		
1 Welder, Electric, 300 amp		56.10		61.71	1.75	1.93
32 L.H., Daily Totals		$1419.70		$2299.31	$44.37	$71.85

Crew Q-5	Hr.	Daily	Hr.	Daily	Bare Costs	Incl. O&P
1 Steamfitter	$46.05	$368.40	$75.55	$604.40	$41.45	$68.00
1 Steamfitter Apprentice	36.85	294.80	60.45	483.60		
16 L.H., Daily Totals		$663.20		$1088.00	$41.45	$68.00

Crew Q-6	Hr.	Daily	Hr.	Daily	Bare Costs	Incl. O&P
1 Steamfitter	$46.05	$368.40	$75.55	$604.40	$39.92	$65.48
2 Steamfitter Apprentices	36.85	589.60	60.45	967.20		
24 L.H., Daily Totals		$958.00		$1571.60	$39.92	$65.48

Crew Q-7	Hr.	Daily	Hr.	Daily	Bare Costs	Incl. O&P
2 Steamfitters	$46.05	$736.80	$75.55	$1208.80	$41.45	$68.00
2 Steamfitter Apprentices	36.85	589.60	60.45	967.20		
32 L.H., Daily Totals		$1326.40		$2176.00	$41.45	$68.00

Crew No.	Bare Costs		Incl. Subs O&P		Cost Per Labor-Hour	
Crew Q-8	**Hr.**	**Daily**	**Hr.**	**Daily**	**Bare Costs**	**Incl. O&P**
2 Steamfitters	$46.05	$736.80	$75.55	$1208.80	$43.75	$71.78
1 Welder (steamfitter)	46.05	368.40	75.55	604.40		
1 Steamfitter Apprentice	36.85	294.80	60.45	483.60		
1 Welder, Electric, 300 amp		56.10		61.71	1.75	1.93
32 L.H., Daily Totals		$1456.10		$2358.51	$45.50	$73.70
Crew Q-9	**Hr.**	**Daily**	**Hr.**	**Daily**	**Bare Costs**	**Incl. O&P**
1 Sheet Metal Worker	$43.30	$346.40	$71.80	$574.40	$38.98	$64.63
1 Sheet Metal Apprentice	34.65	277.20	57.45	459.60		
16 L.H., Daily Totals		$623.60		$1034.00	$38.98	$64.63
Crew Q-10	**Hr.**	**Daily**	**Hr.**	**Daily**	**Bare Costs**	**Incl. O&P**
2 Sheet Metal Workers	$43.30	$692.80	$71.80	$1148.80	$40.42	$67.02
1 Sheet Metal Apprentice	34.65	277.20	57.45	459.60		
24 L.H., Daily Totals		$970.00		$1608.40	$40.42	$67.02
Crew Q-11	**Hr.**	**Daily**	**Hr.**	**Daily**	**Bare Costs**	**Incl. O&P**
2 Sheet Metal Workers	$43.30	$692.80	$71.80	$1148.80	$38.98	$64.63
2 Sheet Metal Apprentices	34.65	554.40	57.45	919.20		
32 L.H., Daily Totals		$1247.20		$2068.00	$38.98	$64.63
Crew Q-12	**Hr.**	**Daily**	**Hr.**	**Daily**	**Bare Costs**	**Incl. O&P**
1 Sprinkler Installer	$44.25	$354.00	$72.75	$582.00	$39.83	$65.47
1 Sprinkler Apprentice	35.40	283.20	58.20	465.60		
16 L.H., Daily Totals		$637.20		$1047.60	$39.83	$65.47
Crew Q-13	**Hr.**	**Daily**	**Hr.**	**Daily**	**Bare Costs**	**Incl. O&P**
2 Sprinkler Installers	$44.25	$708.00	$72.75	$1164.00	$39.83	$65.47
2 Sprinkler Apprentices	35.40	566.40	58.20	931.20		
32 L.H., Daily Totals		$1274.40		$2095.20	$39.83	$65.47
Crew Q-14	**Hr.**	**Daily**	**Hr.**	**Daily**	**Bare Costs**	**Incl. O&P**
1 Asbestos Worker	$40.70	$325.60	$68.55	$548.40	$36.63	$61.67
1 Asbestos Apprentice	32.55	260.40	54.80	438.40		
16 L.H., Daily Totals		$586.00		$986.80	$36.63	$61.67
Crew Q-15	**Hr.**	**Daily**	**Hr.**	**Daily**	**Bare Costs**	**Incl. O&P**
1 Plumber	$44.85	$358.80	$73.60	$588.80	$40.38	$66.25
1 Plumber Apprentice	35.90	287.20	58.90	471.20		
1 Welder, Electric, 300 amp		56.10		61.71	3.51	3.86
16 L.H., Daily Totals		$702.10		$1121.71	$43.88	$70.11
Crew Q-16	**Hr.**	**Daily**	**Hr.**	**Daily**	**Bare Costs**	**Incl. O&P**
2 Plumbers	$44.85	$717.60	$73.60	$1177.60	$41.87	$68.70
1 Plumber Apprentice	35.90	287.20	58.90	471.20		
1 Welder, Electric, 300 amp		56.10		61.71	2.34	2.57
24 L.H., Daily Totals		$1060.90		$1710.51	$44.20	$71.27
Crew Q-17	**Hr.**	**Daily**	**Hr.**	**Daily**	**Bare Costs**	**Incl. O&P**
1 Steamfitter	$46.05	$368.40	$75.55	$604.40	$41.45	$68.00
1 Steamfitter Apprentice	36.85	294.80	60.45	483.60		
1 Welder, Electric, 300 amp		56.10		61.71	3.51	3.86
16 L.H., Daily Totals		$719.30		$1149.71	$44.96	$71.86

Crew No.	Bare Costs		Incl. Subs O&P		Cost Per Labor-Hour	
Crew Q-17A	**Hr.**	**Daily**	**Hr.**	**Daily**	**Bare Costs**	**Incl. O&P**
1 Steamfitter	$46.05	$368.40	$75.55	$604.40	$41.62	$68.28
1 Steamfitter Apprentice	36.85	294.80	60.45	483.60		
1 Equip. Oper. (crane)	41.95	335.60	68.85	550.80		
1 Hyd. Crane, 12 Ton		471.75		518.92		
1 Welder, Electric, 300 amp		56.10		61.71	21.99	24.19
24 L.H., Daily Totals		$1526.65		$2219.43	$63.61	$92.48
Crew Q-18	**Hr.**	**Daily**	**Hr.**	**Daily**	**Bare Costs**	**Incl. O&P**
2 Steamfitters	$46.05	$736.80	$75.55	$1208.80	$42.98	$70.52
1 Steamfitter Apprentice	36.85	294.80	60.45	483.60		
1 Welder, Electric, 300 amp		56.10		61.71	2.34	2.57
24 L.H., Daily Totals		$1087.70		$1754.11	$45.32	$73.09
Crew Q-19	**Hr.**	**Daily**	**Hr.**	**Daily**	**Bare Costs**	**Incl. O&P**
1 Steamfitter	$46.05	$368.40	$75.55	$604.40	$42.83	$70.13
1 Steamfitter Apprentice	36.85	294.80	60.45	483.60		
1 Electrician	45.60	364.80	74.40	595.20		
24 L.H., Daily Totals		$1028.00		$1683.20	$42.83	$70.13
Crew Q-20	**Hr.**	**Daily**	**Hr.**	**Daily**	**Bare Costs**	**Incl. O&P**
1 Sheet Metal Worker	$43.30	$346.40	$71.80	$574.40	$40.30	$66.58
1 Sheet Metal Apprentice	34.65	277.20	57.45	459.60		
.5 Electrician	45.60	182.40	74.40	297.60		
20 L.H., Daily Totals		$806.00		$1331.60	$40.30	$66.58
Crew Q-21	**Hr.**	**Daily**	**Hr.**	**Daily**	**Bare Costs**	**Incl. O&P**
2 Steamfitters	$46.05	$736.80	$75.55	$1208.80	$43.64	$71.49
1 Steamfitter Apprentice	36.85	294.80	60.45	483.60		
1 Electrician	45.60	364.80	74.40	595.20		
32 L.H., Daily Totals		$1396.40		$2287.60	$43.64	$71.49
Crew Q-22	**Hr.**	**Daily**	**Hr.**	**Daily**	**Bare Costs**	**Incl. O&P**
1 Plumber	$44.85	$358.80	$73.60	$588.80	$40.38	$66.25
1 Plumber Apprentice	35.90	287.20	58.90	471.20		
1 Hyd. Crane, 12 Ton		471.75		518.92	29.48	32.43
16 L.H., Daily Totals		$1117.75		$1578.93	$69.86	$98.68
Crew Q-22A	**Hr.**	**Daily**	**Hr.**	**Daily**	**Bare Costs**	**Incl. O&P**
1 Plumber	$44.85	$358.80	$73.60	$588.80	$38.26	$62.88
1 Plumber Apprentice	35.90	287.20	58.90	471.20		
1 Laborer	30.35	242.80	50.15	401.20		
1 Equip. Oper. (crane)	41.95	335.60	68.85	550.80		
1 Hyd. Crane, 12 Ton		471.75		518.92	14.74	16.22
32 L.H., Daily Totals		$1696.15		$2530.93	$53.00	$79.09
Crew Q-23	**Hr.**	**Daily**	**Hr.**	**Daily**	**Bare Costs**	**Incl. O&P**
1 Plumber Foreman (outside)	$46.85	$374.80	$76.85	$614.80	$44.15	$72.43
1 Plumber	44.85	358.80	73.60	588.80		
1 Equip. Oper. (medium)	40.75	326.00	66.85	534.80		
1 Lattice Boom Crane, 20 Ton		869.00		955.90	36.21	39.83
24 L.H., Daily Totals		$1928.60		$2694.30	$80.36	$112.26
Crew R-1	**Hr.**	**Daily**	**Hr.**	**Daily**	**Bare Costs**	**Incl. O&P**
1 Electrician Foreman	$46.10	$368.80	$75.25	$602.00	$42.65	$69.59
3 Electricians	45.60	1094.40	74.40	1785.60		
2 Electrician Apprentices	36.50	584.00	59.55	952.80		
48 L.H., Daily Totals		$2047.20		$3340.40	$42.65	$69.59

Crew No.	Bare Costs		Incl. Subs O&P		Cost Per Labor-Hour	
Crew R-1A	**Hr.**	**Daily**	**Hr.**	**Daily**	**Bare Costs**	**Incl. O&P**
1 Electrician	$45.60	$364.80	$74.40	$595.20	$41.05	$66.97
1 Electrician Apprentice	36.50	292.00	59.55	476.40		
16 L.H., Daily Totals		$656.80		$1071.60	$41.05	$66.97

Crew R-1B	Hr.	Daily	Hr.	Daily	Bare Costs	Incl. O&P
1 Electrician	$45.60	$364.80	$74.40	$595.20	$39.53	$64.50
2 Electrician Apprentices	36.50	584.00	59.55	952.80		
24 L.H., Daily Totals		$948.80		$1548.00	$39.53	$64.50

Crew R-1C	Hr.	Daily	Hr.	Daily	Bare Costs	Incl. O&P
2 Electricians	$45.60	$729.60	$74.40	$1190.40	$41.05	$66.97
2 Electrician Apprentices	36.50	584.00	59.55	952.80		
1 Portable Cable Puller, 8000 lb.		152.70		167.97	4.77	5.25
32 L.H., Daily Totals		$1466.30		$2311.17	$45.82	$72.22

Crew R-2	Hr.	Daily	Hr.	Daily	Bare Costs	Incl. O&P
1 Electrician Foreman	$46.10	$368.80	$75.25	$602.00	$42.55	$69.49
3 Electricians	45.60	1094.40	74.40	1785.60		
2 Electrician Apprentices	36.50	584.00	59.55	952.80		
1 Equip. Oper. (crane)	41.95	335.60	68.85	550.80		
1 S.P. Crane, 4x4, 5 Ton		259.45		285.39	4.63	5.10
56 L.H., Daily Totals		$2642.25		$4176.60	$47.18	$74.58

Crew R-3	Hr.	Daily	Hr.	Daily	Bare Costs	Incl. O&P
1 Electrician Foreman	$46.10	$368.80	$75.25	$602.00	$45.07	$73.63
1 Electrician	45.60	364.80	74.40	595.20		
.5 Equip. Oper. (crane)	41.95	167.80	68.85	275.40		
.5 S.P. Crane, 4x4, 5 Ton		129.72		142.70	6.49	7.13
20 L.H., Daily Totals		$1031.13		$1615.30	$51.56	$80.76

Crew R-4	Hr.	Daily	Hr.	Daily	Bare Costs	Incl. O&P
1 Struc. Steel Foreman (outside)	$43.45	$347.60	$75.40	$603.20	$42.68	$73.10
3 Struc. Steel Workers	41.45	994.80	71.90	1725.60		
1 Electrician	45.60	364.80	74.40	595.20		
1 Welder, Gas Engine, 300 amp		95.65		105.22	2.39	2.63
40 L.H., Daily Totals		$1802.85		$3029.22	$45.07	$75.73

Crew R-5	Hr.	Daily	Hr.	Daily	Bare Costs	Incl. O&P
1 Electrician Foreman	$46.10	$368.80	$75.25	$602.00	$39.61	$65.10
4 Electrician Linemen	45.60	1459.20	74.40	2380.80		
2 Electrician Operators	45.60	729.60	74.40	1190.40		
4 Electrician Groundmen	29.00	928.00	48.60	1555.20		
1 Crew Truck		154.35		169.79		
1 Flatbed Truck, 20,000 GVW		199.45		219.40		
1 Pickup Truck, 3/4 Ton		109.90		120.89		
.2 Hyd. Crane, 55 Ton		196.30		215.93		
.2 Hyd. Crane, 12 Ton		94.35		103.79		
.2 Earth Auger, Truck-Mtd.		77.17		84.89		
1 Tractor w/Winch		368.50		405.35	13.64	15.00
88 L.H., Daily Totals		$4685.62		$7048.42	$53.25	$80.10

Crew R-6	Hr.	Daily	Hr.	Daily	Bare Costs	Incl. O&P
1 Electrician Foreman	$46.10	$368.80	$75.25	$602.00	$39.61	$65.10
4 Electrician Linemen	45.60	1459.20	74.40	2380.80		
2 Electrician Operators	45.60	729.60	74.40	1190.40		
4 Electrician Groundmen	29.00	928.00	48.60	1555.20		
1 Crew Truck		154.35		169.79		
1 Flatbed Truck, 20,000 GVW		199.45		219.40		
1 Pickup Truck, 3/4 Ton		109.90		120.89		
.2 Hyd. Crane, 55 Ton		196.30		215.93		
.2 Hyd. Crane, 12 Ton		94.35		103.79		
.2 Earth Auger, Truck-Mtd.		77.17		84.89		
1 Tractor w/Winch		368.50		405.35		
3 Cable Trailers		648.60		713.46		
.5 Tensioning Rig		214.32		235.76		
.5 Cable Pulling Rig		1233.50		1356.85	37.46	41.21
88 L.H., Daily Totals		$6782.05		$9354.49	$77.07	$106.30

Crew R-7	Hr.	Daily	Hr.	Daily	Bare Costs	Incl. O&P
1 Electrician Foreman	$46.10	$368.80	$75.25	$602.00	$31.85	$53.04
5 Electrician Groundmen	29.00	1160.00	48.60	1944.00		
1 Crew Truck		154.35		169.79	3.22	3.54
48 L.H., Daily Totals		$1683.15		$2715.78	$35.07	$56.58

Crew R-8	Hr.	Daily	Hr.	Daily	Bare Costs	Incl. O&P
1 Electrician Foreman	$46.10	$368.80	$75.25	$602.00	$40.15	$65.94
3 Electrician Linemen	45.60	1094.40	74.40	1785.60		
2 Electrician Groundmen	29.00	464.00	48.60	777.60		
1 Pickup Truck, 3/4 Ton		109.90		120.89		
1 Crew Truck		154.35		169.79	5.51	6.06
48 L.H., Daily Totals		$2191.45		$3455.88	$45.66	$72.00

Crew R-9	Hr.	Daily	Hr.	Daily	Bare Costs	Incl. O&P
1 Electrician Foreman	$46.10	$368.80	$75.25	$602.00	$37.36	$61.61
1 Electrician Lineman	45.60	364.80	74.40	595.20		
2 Electrician Operators	45.60	729.60	74.40	1190.40		
4 Electrician Groundmen	29.00	928.00	48.60	1555.20		
1 Pickup Truck, 3/4 Ton		109.90		120.89		
1 Crew Truck		154.35		169.79	4.13	4.54
64 L.H., Daily Totals		$2655.45		$4233.48	$41.49	$66.15

Crew R-10	Hr.	Daily	Hr.	Daily	Bare Costs	Incl. O&P
1 Electrician Foreman	$46.10	$368.80	$75.25	$602.00	$42.92	$70.24
4 Electrician Linemen	45.60	1459.20	74.40	2380.80		
1 Electrician Groundman	29.00	232.00	48.60	388.80		
1 Crew Truck		154.35		169.79		
3 Tram Cars		438.60		482.46	12.35	13.59
48 L.H., Daily Totals		$2652.95		$4023.84	$55.27	$83.83

Crew R-11	Hr.	Daily	Hr.	Daily	Bare Costs	Incl. O&P
1 Electrician Foreman	$46.10	$368.80	$75.25	$602.00	$42.97	$70.26
4 Electricians	45.60	1459.20	74.40	2380.80		
1 Equip. Oper. (crane)	41.95	335.60	68.85	550.80		
1 Common Laborer	30.35	242.80	50.15	401.20		
1 Crew Truck		154.35		169.79		
1 Hyd. Crane, 12 Ton		471.75		518.92	11.18	12.30
56 L.H., Daily Totals		$3032.50		$4623.51	$54.15	$82.56

Crew R-12	Bare Costs Hr.	Bare Costs Daily	Incl. Subs O&P Hr.	Incl. Subs O&P Daily	Cost Per Labor-Hour Bare Costs	Cost Per Labor-Hour Incl. O&P
1 Carpenter Foreman (inside)	$39.25	$314.00	$64.85	$518.80	$36.17	$60.04
4 Carpenters	38.75	1240.00	64.05	2049.60		
4 Common Laborers	30.35	971.20	50.15	1604.80		
1 Equip. Oper. (medium)	40.75	326.00	66.85	534.80		
1 Steel Worker	41.45	331.60	71.90	575.20		
1 Dozer, 200 H.P.		1290.00		1419.00		
1 Pickup Truck, 3/4 Ton		109.90		120.89	15.91	17.50
88 L.H., Daily Totals		$4582.70		$6823.09	$52.08	$77.54

Crew R-13	Hr.	Daily	Hr.	Daily	Bare Costs	Incl. O&P
1 Electrician Foreman	$46.10	$368.80	$75.25	$602.00	$43.67	$71.35
3 Electricians	45.60	1094.40	74.40	1785.60		
.25 Equip. Oper. (crane)	41.95	83.90	68.85	137.70		
1 Equipment Oiler	35.90	287.20	58.90	471.20		
.25 Hydraulic Crane, 33 Ton		236.65		260.32	5.63	6.20
42 L.H., Daily Totals		$2070.95		$3256.82	$49.31	$77.54

Crew R-15	Hr.	Daily	Hr.	Daily	Bare Costs	Incl. O&P
1 Electrician Foreman	$46.10	$368.80	$75.25	$602.00	$44.54	$72.74
4 Electricians	45.60	1459.20	74.40	2380.80		
1 Equipment Oper. (light)	38.75	310.00	63.60	508.80		
1 Telescoping Boom Lift, to 40'		283.15		311.46	5.90	6.49
48 L.H., Daily Totals		$2421.15		$3803.07	$50.44	$79.23

Crew R-15A	Hr.	Daily	Hr.	Daily	Bare Costs	Incl. O&P
1 Electrician Foreman	$46.10	$368.80	$75.25	$602.00	$39.46	$64.66
2 Electricians	45.60	729.60	74.40	1190.40		
2 Common Laborers	30.35	485.60	50.15	802.40		
1 Equip. Oper. (light)	38.75	310.00	63.60	508.80		
1 Telescoping Boom Lift, to 40'		283.15		311.46	5.90	6.49
48 L.H., Daily Totals		$2177.15		$3415.07	$45.36	$71.15

Crew R-18	Hr.	Daily	Hr.	Daily	Bare Costs	Incl. O&P
.25 Electrician Foreman	$46.10	$92.20	$75.25	$150.50	$40.04	$65.33
1 Electrician	45.60	364.80	74.40	595.20		
2 Electrician Apprentices	36.50	584.00	59.55	952.80		
26 L.H., Daily Totals		$1041.00		$1698.50	$40.04	$65.33

Crew R-19	Hr.	Daily	Hr.	Daily	Bare Costs	Incl. O&P
.5 Electrician Foreman	$46.10	$184.40	$75.25	$301.00	$45.70	$74.57
2 Electricians	45.60	729.60	74.40	1190.40		
20 L.H., Daily Totals		$914.00		$1491.40	$45.70	$74.57

Crew R-21	Hr.	Daily	Hr.	Daily	Bare Costs	Incl. O&P
1 Electrician Foreman	$46.10	$368.80	$75.25	$602.00	$45.60	$74.42
3 Electricians	45.60	1094.40	74.40	1785.60		
.1 Equip. Oper. (medium)	40.75	32.60	66.85	53.48		
.1 S.P. Crane, 4x4, 25 Ton		66.44		73.08	2.03	2.23
32.8 L.H., Daily Totals		$1562.24		$2514.16	$47.63	$76.65

Crew R-22	Hr.	Daily	Hr.	Daily	Bare Costs	Incl. O&P
.66 Electrician Foreman	$46.10	$243.41	$75.25	$397.32	$41.77	$68.15
2 Electricians	45.60	729.60	74.40	1190.40		
2 Electrician Apprentices	36.50	584.00	59.55	952.80		
37.28 L.H., Daily Totals		$1557.01		$2540.52	$41.77	$68.15

Crew R-30	Hr.	Daily	Hr.	Daily	Bare Costs	Incl. O&P
.25 Electrician Foreman (outside)	$47.60	$95.20	$77.70	$155.40	$36.37	$59.73
1 Electrician	45.60	364.80	74.40	595.20		
2 Laborers (Semi-Skilled)	30.35	485.60	50.15	802.40		
26 L.H., Daily Totals		$945.60		$1553.00	$36.37	$59.73

Historical Cost Indexes

The table below lists both the RSMeans® historical cost index based on Jan. 1, 1993 = 100 as well as the computed value of an index based on Jan. 1, 2019 costs. Since the Jan. 1, 2019 figure is estimated, space is left to write in the actual index figures as they become available through the quarterly *RSMeans Construction Cost Indexes*.

To compute the actual index based on Jan. 1, 2019 = 100, divide the historical cost index for a particular year by the actual Jan. 1, 2019 construction cost index. Space has been left to advance the index figures as the year progresses.

Year	Historical Cost Index Jan. 1, 1993 = 100		Current Index Based on Jan. 1, 2019 = 100		Year	Historical Cost Index Jan. 1, 1993 = 100	Current Index Based on Jan. 1, 2019 = 100		Year	Historical Cost Index Jan. 1, 1993 = 100	Current Index Based on Jan. 1, 2019 = 100	
	Est.	Actual	Est.	Actual		Actual	Est.	Actual		Actual	Est.	Actual
Oct 2019*					July 2004	143.7	63.2		July 1986	84.2	37.1	
July 2019*					2003	132.0	58.1		1985	82.6	36.3	
April 2019*					2002	128.7	56.6		1984	82.0	36.1	
Jan 2019*	227.3		100.0	100.0	2001	125.1	55.0		1983	80.2	35.3	
July 2018		222.9	98.1		2000	120.9	53.2		1982	76.1	33.5	
2017		213.6	94.0		1999	117.6	51.7		1981	70.0	30.8	
2016		207.3	91.2		1998	115.1	50.6		1980	62.9	27.7	
2015		206.2	90.7		1997	112.8	49.6		1979	57.8	25.4	
2014		204.9	90.1		1996	110.2	48.5		1978	53.5	23.5	
2013		201.2	88.5		1995	107.6	47.3		1977	49.5	21.8	
2012		194.6	85.6		1994	104.4	45.9		1976	46.9	20.6	
2011		191.2	84.1		1993	101.7	44.7		1975	44.8	19.7	
2010		183.5	80.7		1992	99.4	43.7		1974	41.4	18.2	
2009		180.1	79.2		1991	96.8	42.6		1973	37.7	16.6	
2008		180.4	79.4		1990	94.3	41.5		1972	34.8	15.3	
2007		169.4	74.5		1989	92.1	40.5		1971	32.1	14.1	
2006		162.0	71.3		1988	89.9	39.5		1970	28.7	12.6	
2005		151.6	66.7		1987	87.7	38.6		1969	26.9	11.8	

Adjustments to Costs

The "Historical Cost Index" can be used to convert national average building costs at a particular time to the approximate building costs for some other time.

Time Adjustment Using the Historical Cost Indexes:

$$\frac{\text{Index for Year A}}{\text{Index for Year B}} \times \text{Cost in Year B} = \text{Cost in Year A}$$

Example:

Estimate and compare construction costs for different years in the same city.

To estimate the national average construction cost of a building in 1970, knowing that it cost $900,000 in 2019:

INDEX in 1970 = 28.7

INDEX in 2019 = 227.3

$$\frac{\text{INDEX 1970}}{\text{INDEX 2019}} \times \text{Cost 2019} = \text{Cost 1970}$$

$$\frac{28.7}{227.3} \times \$900{,}000 = .126 \times \$900{,}000 = \$113{,}400$$

The construction cost of the building in 1970 was $113,400.

Note: The city cost indexes for Canada can be used to convert U.S. national averages to local costs in Canadian dollars.

Example:

To estimate and compare the cost of a building in Toronto, ON in 2019 with the known cost of $600,000 (US$) in New York, NY in 2019:

INDEX Toronto = 110.1

INDEX New York = 132.1

$$\frac{\text{INDEX Toronto}}{\text{INDEX New York}} \times \text{Cost New York} = \text{Cost Toronto}$$

$$\frac{110.1}{132.1} \times \$600{,}000 = .834 \times \$600{,}000 = \$500{,}076$$

The construction cost of the building in Toronto is $500,076 (CN$).

*Historical Cost Index updates and other resources are provided on the following website:
http://info.thegordiangroup.com/RSMeans.html

Historical Cost Indexes

Year	National	Alabama					Alaska	Arizona		Arkansas		California				
	30 City Average	Birmingham	Huntsville	Mobile	Montgomery	Tuscaloosa	Anchorage	Phoenix	Tuscon	Fort Smith	Little Rock	Anaheim	Bakersfield	Fresno	Los Angeles	Oxnard
Jan 2019	227.3	190E	189.2E	186.5E	188.7E	189.8E	258.5E	196.1E	191.3E	180.4E	183.6E	247.3E	244.6E	248.1E	252.3E	244.9E
2018	217.7	186.4	185.6	183.0	185.1	186.2	253.6	192.4	187.7	177.0	180.1	242.6	240.0	243.4	247.5	240.3
2017	209.4	178.6	178.3	177.9	178.5	179.6	246.2	183.8	181.3	170.6	172.8	233.6	229.2	232.0	237.8	231.9
2016	207.7	185.6	184.2	184.1	183.5	186.3	246.3	181.7	178.1	169.8	174.6	223.6	221.3	224.3	226.2	222.5
2015	204.0	184.8	182.1	185.4	182.7	184.5	244.7	181.0	177.9	167.6	171.1	217.3	214.6	218.8	218.5	215.7
2014	203.0	180.3	175.8	170.6	163.1	165.9	241.0	180.0	177.0	165.1	168.4	214.2	215.2	217.0	217.3	214.8
2013	196.9	173.3	168.7	165.7	158.7	161.6	235.3	174.1	169.2	161.2	163.2	207.0	205.6	210.3	210.7	207.8
2012	194.0	169.1	162.7	163.2	153.8	154.6	232.5	172.2	166.4	158.7	161.0	204.1	202.9	207.3	207.2	204.7
2011	185.7	163.0	156.3	156.9	147.6	148.3	225.1	163.8	159.5	152.0	154.1	196.1	194.6	199.8	199.2	197.0
2010	181.6	159.6	152.9	153.1	144.4	144.9	218.3	160.7	157.3	150.3	152.5	192.8	190.8	195.0	194.9	192.8
2009	182.5	162.7	157.2	155.6	148.8	149.4	222.9	161.3	156.8	149.2	156.3	194.5	192.3	195.0	196.6	194.5
2008	171.0	150.3	146.9	143.7	138.4	138.8	210.8	152.2	148.4	138.6	145.4	182.7	179.8	183.2	184.7	182.6
2007	165.0	146.9	143.3	140.3	134.8	135.4	206.8	147.7	143.1	134.8	141.7	175.1	173.7	176.7	177.5	176.2
2006	156.2	135.7	133.6	126.7	124.0	122.2	196.4	137.9	134.8	123.2	127.2	166.8	164.9	169.8	167.3	167.4
2005	146.7	127.9	125.4	119.3	116.6	114.6	185.6	128.5	124.1	115.4	119.4	156.5	153.0	157.9	157.1	156.4
2004	132.8	115.9	112.9	107.0	104.9	102.9	167.0	116.5	113.6	103.8	108.6	142.1	139.5	143.8	142.0	140.9
2003	129.7	113.1	110.7	104.8	102.6	100.8	163.5	113.9	110.6	101.8	105.8	139.4	137.2	142.1	139.6	138.7
2002	126.7	110.0	103.4	103.6	101.7	99.7	159.5	113.3	110.2	100.4	102.1	136.5	133.4	136.0	136.4	136.3
2001	122.2	106.0	100.5	100.6	98.3	95.9	152.6	109.0	106.4	97.3	98.7	132.5	129.3	132.8	132.4	132.4
2000	118.9	104.1	98.9	99.1	94.2	94.6	148.3	106.9	104.9	94.4	95.7	129.4	125.2	129.4	129.9	129.7
1999	116.6	101.2	97.4	97.7	92.5	92.8	145.9	105.5	103.3	93.1	94.4	127.9	123.6	126.9	128.7	128.2
1998	113.6	96.2	94.0	94.8	90.3	89.8	143.8	102.1	101.0	90.6	91.4	125.2	120.3	123.9	125.8	124.7
1995	105.6	87.8	88.0	88.8	84.5	83.5	138.0	96.1	95.5	85.0	85.6	120.1	115.7	117.5	120.9	120.0
1990	93.2	79.4	77.6	82.7	78.4	75.2	125.8	86.4	87.0	77.1	77.9	107.7	102.7	103.3	107.5	107.4
1985	81.8	71.1	70.5	72.7	70.9	67.9	116.0	78.1	77.7	69.6	71.0	95.4	92.2	92.6	94.6	96.6
1980	60.7	55.2	54.1	56.8	56.8	54.0	91.4	63.7	62.4	53.1	55.8	68.7	69.4	68.7	67.4	69.9
1975	43.7	40.0	40.9	41.8	40.1	37.8	57.3	44.5	45.2	39.0	38.7	47.6	46.6	47.7	48.3	47.0
1970	27.8	24.1	25.1	25.8	25.4	24.2	43.0	27.2	28.5	24.3	22.3	31.0	30.8	31.1	29.0	31.0
1965	21.5	19.6	19.3	19.6	19.5	18.7	34.9	21.8	22.0	18.7	18.5	23.9	23.7	24.0	22.7	23.9
1960	19.5	17.9	17.5	17.8	17.7	16.9	31.7	19.9	20.0	17.0	16.8	21.7	21.5	21.8	20.6	21.7
1955	16.3	14.8	14.7	14.9	14.9	14.2	26.6	16.7	16.7	14.3	14.5	18.2	18.1	18.3	17.3	18.2
▼ 1950	13.5	12.2	12.1	12.3	12.3	11.7	21.9	13.8	13.8	11.8	11.6	15.1	15.0	15.1	14.3	15.1

Year	National	California							Colorado			Connecticut				
	30 City Average	Riverside	Sacramento	San Diego	San Francisco	Santa Barbara	Stockton	Vallejo	Colorado Springs	Denver	Pueblo	Bridgeport	Bristol	Hartford	New Britain	New Haven
Jan 2019	227.3	246.7E	253.7E	245.8E	286.7E	243.9E	252E	259.7E	197.1E	201.4E	196.7E	237E	236.5E	237.4E	236E	238.3E
2018	217.7	242.1	248.9	241.2	281.3	239.3	247.3	254.8	193.4	197.6	193.0	232.5	232.0	232.9	231.6	233.8
2017	209.4	233.2	237.9	229.3	270.8	231.1	236.2	246.8	187.3	187.8	183.1	224.4	222.9	223.9	222.5	225.0
2016	207.7	223.3	227.3	218.3	256.6	221.9	227.9	236.0	189.4	191.0	186.2	226.4	225.6	226.1	225.0	226.9
2015	204.0	217.1	221.0	213.1	250.2	215.7	222.8	230.2	185.1	186.6	183.5	224.1	223.3	224.2	222.8	225.0
2014	203.0	214.1	221.5	211.4	248.0	214.7	218.4	227.9	187.8	189.1	184.7	223.8	222.8	223.9	222.4	224.1
2013	196.9	207.0	215.0	203.2	241.0	207.9	211.8	222.2	180.7	183.4	180.0	217.8	217.2	218.4	216.8	218.5
2012	194.0	204.1	211.9	198.9	237.6	204.8	208.8	218.9	179.4	182.2	177.9	213.9	213.3	214.4	212.9	214.7
2011	185.7	195.8	203.1	192.3	227.9	196.2	201.3	210.3	171.7	174.1	170.7	205.2	204.2	205.1	203.9	205.8
2010	181.6	192.7	196.7	188.2	223.0	192.9	195.9	204.6	170.5	172.6	168.1	198.8	199.1	199.8	198.8	200.7
2009	182.5	193.1	198.1	191.6	224.9	193.6	194.5	202.7	168.4	172.0	166.7	199.0	197.7	198.7	197.4	199.3
2008	171.0	181.3	186.0	179.9	210.6	181.5	183.2	191.6	158.8	161.9	157.9	185.8	184.8	185.7	184.5	186.1
2007	165.0	174.7	179.1	173.6	201.1	175.2	178.2	185.9	152.6	155.9	152.3	179.6	178.3	179.6	178.0	179.5
2006	156.2	166.0	172.2	164.0	191.2	166.4	171.0	177.9	146.1	149.2	145.0	169.5	168.0	169.5	167.7	169.8
2005	146.7	155.4	161.1	153.8	179.7	155.7	160.0	167.2	138.3	141.1	136.4	160.8	159.4	159.6	159.1	160.8
2004	132.8	141.0	147.2	139.0	163.6	141.4	144.2	148.6	125.6	127.2	123.5	143.6	142.9	142.4	142.7	144.3
2003	129.7	138.7	144.9	136.6	162.1	139.2	141.8	146.4	123.1	123.7	120.7	141.3	140.2	140.3	140.0	141.6
2002	126.7	135.3	138.4	134.2	157.9	135.9	136.9	143.8	118.7	121.3	116.7	133.8	133.6	133.1	133.3	134.8
2001	122.2	131.4	135.5	129.7	151.8	131.5	134.1	140.2	113.3	117.2	113.1	128.6	128.5	128.5	128.3	128.6
2000	118.9	128.0	131.5	127.1	146.9	128.7	130.7	137.1	109.8	111.8	108.8	122.7	122.6	122.9	122.4	122.7
1999	116.6	126.5	129.4	124.9	145.1	127.2	127.9	135.3	107.0	109.1	107.1	121.1	121.3	121.2	121.1	121.3
1998	113.6	123.7	125.9	121.3	141.9	123.7	124.7	132.5	103.3	106.5	103.7	119.1	119.4	120.0	119.7	120.0
1995	105.6	119.2	119.5	115.4	133.8	119.0	119.0	122.5	96.1	98.9	96.8	116.0	116.5	116.9	116.3	116.5
1990	93.2	107.0	104.9	105.6	121.8	106.4	105.4	111.4	86.9	88.8	88.8	96.3	95.9	96.6	95.9	96.5
1985	81.8	94.8	92.2	94.2	106.2	93.2	93.7	97.0	79.6	81.0	80.4	86.1	86.2	87.2	86.1	86.3
1980	60.7	68.7	71.3	68.1	75.2	71.1	71.2	71.9	60.7	60.9	59.5	61.5	60.7	61.9	60.7	61.4
1975	43.7	47.3	49.1	47.7	49.8	46.1	47.6	46.5	43.1	42.7	42.5	45.2	45.5	46.0	45.2	46.0
1970	27.8	31.0	32.1	30.7	31.6	31.3	31.8	31.8	27.6	26.1	27.3	29.2	28.2	29.6	28.2	29.3
1965	21.5	23.9	24.8	24.0	23.7	24.1	24.5	24.5	21.3	20.9	21.0	22.4	21.7	22.6	21.7	23.2
1960	19.5	21.7	22.5	21.7	21.5	21.9	22.3	22.2	19.3	19.0	19.1	20.0	19.8	20.0	19.8	20.0
1955	16.3	18.2	18.9	18.2	18.0	18.4	18.7	18.6	16.2	15.9	16.0	16.8	16.6	16.8	16.6	16.8
▼ 1950	13.5	15.0	15.6	15.0	14.9	15.2	15.4	15.4	13.4	13.2	13.2	13.9	13.7	13.8	13.7	13.9

Historical Cost Indexes

Year	National	Connecticut			Delaware	D.C.	Florida						Georgia			
	30 City Average	Norwalk	Stamford	Waterbury	Wilmington	Washington	Fort Lauderdale	Jacksonville	Miami	Orlando	Tallahassee	Tampa	Albany	Atlanta	Columbus	Macon
Jan 2019	227.3	236.7E	243.3E	237.1E	229.9E	211.2E	182.8E	182.6E	183.6E	186E	184.5E	186.2E	190.2E	196E	190.5E	190.3E
2018	217.7	232.2	238.7	232.6	225.6	207.2	179.4	179.2	180.1	182.5	181.0	182.7	186.6	192.3	186.9	186.7
2017	209.4	224.2	230.8	224.2	217.7	199.1	173.7	172.7	173.4	176.5	174.1	177.0	179.4	185.1	1801.1	179.4
2016	207.7	232.7	233.3	225.7	215.7	201.3	177.7	176.3	179.1	180.4	173.4	180.6	176.3	185.0	173.6	172.4
2015	204.0	230.2	230.3	223.8	211.7	197.3	176.1	174.2	178.1	178.3	172.5	178.9	168.6	178.3	171.3	170.0
2014	203.0	230.3	230.5	223.2	210.2	197.0	176.5	171.0	178.4	176.5	164.4	183.2	166.6	177.5	170.2	168.2
2013	196.9	223.1	224.2	217.3	203.6	191.7	173.2	168.4	175.8	174.8	160.8	180.2	163.2	173.3	166.7	164.9
2012	194.0	219.8	220.0	213.4	201.0	189.6	170.9	165.8	173.7	172.8	158.5	177.8	158.3	170.7	160.6	159.3
2011	185.7	211.0	211.2	204.6	193.3	182.3	164.4	159.4	167.0	166.1	151.7	171.4	151.8	164.0	154.3	152.7
2010	181.6	198.3	203.8	199.4	187.7	179.0	160.3	155.4	163.3	162.5	148.1	167.5	148.3	160.1	150.9	149.4
2009	182.5	198.8	204.3	198.4	188.9	181.4	160.7	152.0	165.2	163.9	145.1	165.1	152.0	164.5	155.6	153.5
2008	171.0	185.2	189.2	185.3	177.5	170.4	149.6	142.5	152.9	153.0	135.3	155.6	140.4	153.2	143.9	142.5
2007	165.0	178.6	183.3	179.1	173.4	163.1	145.6	139.0	148.9	148.1	131.7	152.0	135.8	148.0	140.1	138.1
2006	156.2	169.7	173.7	169.1	158.3	153.0	136.0	126.5	136.7	134.3	118.4	136.6	123.6	139.9	126.3	124.3
2005	146.7	160.9	164.1	160.5	149.6	142.7	126.0	119.2	127.1	126.0	110.7	127.7	115.7	131.7	118.9	116.1
2004	132.8	143.6	146.5	143.2	136.1	126.8	114.8	107.8	115.6	113.2	100.6	116.6	102.8	119.6	102.0	105.0
2003	129.7	140.6	145.1	140.8	133.0	124.6	109.0	105.6	110.5	107.8	98.0	103.5	100.5	115.8	99.1	102.7
2002	126.7	133.5	136.2	133.9	129.4	120.3	107.1	104.6	107.4	106.7	97.3	102.8	99.7	113.0	98.3	101.9
2001	122.2	128.2	131.5	128.8	124.8	115.9	104.3	100.8	104.6	103.8	94.8	100.3	96.4	109.1	95.5	99.0
2000	118.9	121.5	126.4	123.2	117.4	113.8	102.4	99.0	101.8	101.2	93.6	98.9	94.9	106.1	94.1	97.0
1999	116.6	120.0	122.0	121.2	116.6	111.5	101.4	98.0	100.8	100.1	92.5	97.9	93.5	102.9	92.8	95.8
1998	113.6	118.8	120.8	120.1	112.3	109.6	99.4	96.2	99.0	98.4	90.9	96.3	91.4	100.8	90.4	93.3
1995	105.6	115.5	117.9	117.3	106.1	102.3	94.0	90.9	93.7	93.5	86.5	92.2	85.0	92.0	82.6	86.8
1990	93.2	96.3	98.9	95.1	92.5	90.4	83.9	81.1	84.0	82.9	78.4	85.0	75.0	80.4	74.0	76.9
1985	81.8	85.3	86.8	85.6	81.1	78.8	76.7	73.4	78.3	73.9	70.6	77.3	66.9	70.3	66.1	68.1
1980	60.7	60.7	60.9	62.3	58.5	59.6	55.3	55.8	56.5	56.7	53.5	57.2	51.7	54.0	51.1	51.3
1975	43.7	44.7	45.0	46.3	42.9	43.7	42.1	40.3	43.2	41.5	38.1	41.3	37.5	38.4	36.2	36.5
1970	27.8	28.1	28.2	28.9	27.0	26.3	25.7	22.8	27.0	26.2	24.5	24.2	23.8	25.2	22.8	23.4
1965	21.5	21.6	21.7	22.2	20.9	21.8	19.8	17.4	19.3	20.2	18.9	18.6	18.3	19.8	17.6	18.0
1960	19.5	19.7	19.7	20.2	18.9	19.4	18.0	15.8	17.6	18.3	17.2	16.9	16.7	17.1	16.0	16.4
1955	16.3	16.5	16.5	17.0	15.9	16.3	15.1	13.2	14.7	15.4	14.4	14.1	14.0	14.4	13.4	13.7
▼ 1950	13.5	13.6	13.7	14.0	13.1	13.4	12.5	11.0	12.2	12.7	11.9	11.7	11.5	11.9	11.0	11.3

Year	National	Georgia	Hawaii	Idaho		Illinois						Indiana				
	30 City Average	Savannah	Honolulu	Boise	Pocatello	Chicago	Decatur	Joliet	Peoria	Rockford	Springfield	Anderson	Evansville	Fort Wayne	Gary	Indianapolis
Jan 2019	227.3	191.7E	266.9E	203.9E	204.5E	266.1E	226.3E	262.3E	232.2E	247.7E	227.9E	199.8E	202.4E	196.8E	227.3E	203.6E
2018	217.7	188.1	261.9	200.1	200.6	261.1	222.0	257.4	227.8	243.0	223.6	196.0	198.6	193.1	223.0	199.8
2017	209.4	180.5	249.7	192.5	193.4	251.7	215.3	247.9	220.4	233.1	216.0	189.8	192.3	185.2	216.8	191.8
2016	207.7	175.1	251.7	191.1	190.2	244.9	212.9	244.3	217.5	229.6	215.1	190.5	192.3	184.4	216.8	192.5
2015	204.0	170.5	250.5	185.4	185.3	238.9	209.0	238.7	213.7	225.3	210.5	186.4	190.1	183.0	212.4	189.6
2014	203.0	167.2	239.7	184.1	184.4	238.1	206.0	236.8	212.5	224.4	208.2	184.5	188.1	180.8	210.1	188.4
2013	196.9	163.5	231.5	179.6	179.9	231.1	200.8	228.7	207.2	217.2	203.2	177.8	182.5	175.9	203.6	182.7
2012	194.0	158.9	227.9	170.4	171.5	226.2	195.7	222.7	199.7	211.8	197.2	175.2	177.8	171.8	197.9	180.6
2011	185.7	151.7	219.0	162.8	163.6	217.6	187.9	216.8	193.4	205.8	189.4	168.4	169.7	165.0	191.0	173.0
2010	181.6	147.9	214.6	161.7	162.4	210.6	181.2	208.3	186.2	197.6	182.2	162.8	165.9	160.0	183.7	168.1
2009	182.5	152.8	218.6	162.6	163.4	209.0	182.2	205.9	186.2	196.5	184.0	164.9	166.9	163.0	185.2	170.3
2008	171.0	141.1	205.5	153.2	152.8	195.7	168.7	185.5	172.2	180.2	168.5	152.4	155.8	151.0	169.0	159.2
2007	165.0	136.9	200.4	148.0	148.1	186.5	160.5	177.7	166.0	175.0	159.2	149.0	152.6	147.7	165.4	154.1
2006	156.2	124.0	191.9	141.6	141.2	177.8	153.8	168.7	155.2	163.3	152.8	139.8	144.4	139.1	154.5	144.6
2005	146.7	116.9	181.2	133.8	132.5	164.4	145.0	158.3	146.5	151.2	145.4	131.4	136.2	131.0	146.4	137.4
2004	132.8	105.6	161.2	122.0	120.5	149.3	129.3	145.1	133.9	138.1	130.3	120.9	123.5	120.1	132.2	125.0
2003	129.7	103.0	159.4	120.2	118.7	146.2	127.4	143.5	132.4	137.3	128.4	119.7	121.4	118.2	131.4	122.5
2002	126.7	102.0	157.2	118.3	117.1	141.2	123.8	138.5	129.6	132.9	125.3	117.5	119.0	116.4	129.1	120.5
2001	122.2	99.0	150.0	114.3	113.4	135.8	120.1	133.7	124.3	127.8	119.8	113.4	115.6	112.1	123.4	116.4
2000	118.9	97.5	144.8	112.9	112.1	131.2	115.1	124.6	119.0	122.2	116.2	109.8	111.5	108.4	117.8	113.2
1999	116.6	96.0	143.0	110.2	109.6	129.6	113.0	122.6	116.4	120.7	113.8	107.1	109.3	106.8	112.8	110.6
1998	113.6	93.7	140.4	107.4	107.0	125.2	110.1	119.8	113.8	115.5	111.1	105.0	107.2	104.5	111.1	108.0
1995	105.6	87.4	130.3	99.5	98.2	114.2	98.5	110.5	102.3	103.6	98.1	96.4	97.2	95.0	100.7	100.1
1990	93.2	77.9	104.7	88.2	88.1	98.4	90.9	98.4	93.7	94.0	90.1	84.6	89.3	83.4	88.4	87.1
1985	81.8	68.9	94.7	78.0	78.0	82.4	81.9	83.4	83.7	83.0	81.5	75.2	79.8	75.0	77.8	77.1
1980	60.7	52.2	68.9	60.3	59.5	62.8	62.3	63.4	64.5	61.6	61.1	56.5	59.0	56.7	59.8	57.9
1975	43.7	36.9	44.6	40.8	40.5	45.7	43.1	44.5	44.7	42.7	42.4	39.5	41.7	39.9	41.9	40.6
1970	27.8	21.0	30.4	26.7	26.6	29.1	28.0	28.6	29.0	27.5	27.6	25.4	26.4	25.5	27.1	26.2
1965	21.5	16.4	21.8	20.6	20.5	22.7	21.5	22.1	22.4	21.2	21.3	19.5	20.4	19.7	20.8	20.7
1960	19.5	14.9	19.8	18.7	18.6	20.2	19.6	20.0	20.3	19.2	19.3	17.7	18.7	17.9	18.9	18.4
1955	16.3	12.5	16.6	15.7	15.6	16.9	16.4	16.8	17.0	16.1	16.2	14.9	15.7	15.0	15.9	15.5
▼ 1950	13.5	10.3	13.7	13.0	12.9	14.0	13.6	13.9	14.0	13.3	13.4	12.3	12.9	12.4	13.1	12.8

Year	National	Indiana			Iowa					Kansas		Kentucky		Louisiana		
	30 City Average	Muncie	South Bend	Terre Haute	Cedar Rapids	Daven-port	Des Moines	Sioux City	Water-loo	Topeka	Wichita	Lexing-ton	Louis-ville	Baton Rouge	Lake Charles	New Orleans
Jan 2019	227.3	198.9E	203.6E	202.4E	206.9E	216.6E	205.7E	199.6E	198.8E	198.7E	192.6E	195.3E	194.4E	188.2E	187.5E	189.8E
2018	217.7	195.2	199.8	198.6	203.0	212.5	201.8	195.8	195.1	195.0	189.0	191.6	190.7	184.7	184.0	186.2
2017	209.4	189.5	192.4	193.1	194.8	205.7	195.0	188.2	188.5	188.1	183.0	186.2	185.0	179.2	178.2	179.3
2016	207.7	191.0	190.7	192.8	191.4	199.9	190.0	184.0	181.0	181.0	191.0	186.0	188.1	180.1	176.7	180.0
2015	204.0	186.0	188.6	190.6	188.8	196.9	189.0	179.1	177.8	175.7	175.1	184.4	185.9	176.6	175.4	177.6
2014	203.0	185.0	185.5	189.2	188.3	195.9	188.5	178.9	177.4	173.6	173.7	183.7	186.1	171.3	170.9	177.5
2013	196.9	179.3	179.3	182.6	183.4	190.8	181.8	173.2	171.3	168.8	167.3	179.1	182.4	166.3	166.2	173.1
2012	194.0	175.8	174.3	178.9	180.3	184.6	179.2	168.8	169.2	162.3	161.4	174.8	177.5	164.4	163.8	171.7
2011	185.7	169.2	168.1	171.3	173.8	178.1	172.7	162.3	162.8	156.3	155.3	167.8	170.8	156.9	156.5	162.1
2010	181.6	163.1	163.5	167.8	170.4	175.6	170.2	159.0	160.2	152.8	151.3	158.7	166.9	153.7	153.4	160.3
2009	182.5	164.5	167.1	168.0	165.9	172.0	162.1	156.3	147.6	154.5	152.4	160.7	167.5	156.9	154.4	161.9
2008	171.0	153.0	153.1	155.9	157.3	163.4	152.4	147.8	139.2	144.9	143.1	152.0	156.8	144.0	141.5	149.3
2007	165.0	149.7	149.5	152.7	152.4	157.6	148.9	144.0	135.0	141.0	138.7	147.6	150.9	139.4	138.0	145.2
2006	156.2	141.2	141.3	144.0	145.8	151.1	142.9	137.3	128.4	134.7	132.9	129.0	143.0	129.4	129.8	135.8
2005	146.7	131.0	133.8	135.2	135.7	142.1	133.5	129.4	120.0	125.9	125.6	121.5	135.1	120.8	117.7	126.2
2004	132.8	120.0	120.6	123.0	122.7	128.6	121.6	116.2	108.3	112.6	113.4	110.1	120.3	105.6	109.0	115.3
2003	129.7	118.7	119.2	121.7	120.9	126.4	120.2	114.6	105.9	109.4	111.3	107.8	118.7	103.3	106.1	112.4
2002	126.7	116.7	117.1	119.9	116.0	122.1	116.7	111.4	103.7	107.8	109.6	106.2	116.4	102.5	105.0	110.6
2001	122.2	112.8	111.6	115.5	112.5	117.5	113.2	107.7	100.6	104.3	105.2	103.3	112.7	99.4	101.3	104.6
2000	118.9	109.1	106.9	110.8	108.8	112.6	108.9	99.0	98.1	101.0	101.1	101.4	109.3	97.8	99.7	102.1
1999	116.6	105.7	103.7	107.9	104.1	109.2	107.6	96.7	96.4	98.7	99.3	99.7	106.6	96.1	97.6	99.5
1998	113.6	103.6	102.3	106.0	102.5	106.8	103.8	95.1	94.8	97.4	97.4	97.4	101.5	94.7	96.1	97.7
1995	105.6	95.6	94.6	97.2	95.4	96.6	94.4	88.2	88.9	91.1	91.0	91.6	94.6	89.6	92.7	91.6
1990	93.2	83.9	85.1	89.1	87.1	86.5	86.9	80.4	80.3	83.2	82.6	82.8	82.6	82.0	84.1	84.5
1985	81.8	75.2	76.0	79.2	75.9	77.6	77.1	72.1	72.3	75.0	74.7	75.5	74.5	74.9	78.5	78.2
1980	60.7	56.1	58.3	59.7	62.7	59.6	61.8	59.3	57.7	58.9	58.0	59.3	59.8	59.1	60.0	57.2
1975	43.7	39.2	40.3	41.9	43.1	41.0	43.1	41.0	40.0	42.0	43.1	42.7	42.5	40.4	40.6	41.5
1970	27.8	25.3	26.2	27.0	28.2	26.9	27.6	26.6	25.9	27.0	25.5	26.9	25.9	25.0	26.7	27.2
1965	21.5	19.5	20.2	20.9	21.7	20.8	21.7	20.5	20.0	20.8	19.6	20.7	20.3	19.4	20.6	20.4
1960	19.5	17.8	18.3	18.9	19.7	18.8	19.5	18.6	18.2	18.9	17.8	18.8	18.4	17.6	18.7	18.5
1955	16.3	14.9	15.4	15.9	16.6	15.8	16.3	15.6	15.2	15.8	15.0	15.8	15.4	14.8	15.7	15.6
1950	13.5	12.3	12.7	13.1	13.7	13.1	13.5	12.9	12.6	13.1	12.4	13.0	12.8	12.2	13.0	12.8

Year	National	Louisiana	Maine		Maryland	Massachusetts									Michigan	
	30 City Average	Shreve-port	Lewis-ton	Portland	Balti-more	Boston	Brockton	Fall River	Law-rence	Lowell	New Bedford	Pitts-field	Spring-field	Wor-cester	Ann Arbor	Dear-born
Jan 2019	227.3	186.5E	201.1E	203.7E	208.7E	253.2E	237.2E	235E	243.7E	242.3E	234.5E	222.9E	225.6E	237E	219.7E	220.7E
2018	217.7	183.0	197.3	199.9	204.8	248.4	232.7	230.6	239.1	237.7	230.1	218.7	221.4	232.5	215.6	216.5
2017	209.4	177.9	192.0	194.7	197.7	240.0	226.4	225.0	233.9	231.3	224.4	211.7	213.5	226.1	208.7	211.1
2016	207.7	173.2	197.2	200.0	193.8	245.0	230.5	229.0	236.9	235.5	228.2	216.1	218.3	230.1	211.3	212.5
2015	204.0	170.9	193.7	196.9	189.2	240.7	228.3	228.2	234.5	233.2	227.4	212.8	214.6	226.8	209.8	211.5
2014	203.0	167.7	192.4	195.4	187.6	239.1	229.1	229.3	234.8	233.9	228.5	213.3	214.8	227.3	206.1	208.4
2013	196.9	159.0	187.9	189.5	183.2	232.9	221.6	221.8	225.3	223.4	221.1	205.9	206.5	219.3	201.2	202.4
2012	194.0	153.2	180.7	182.4	180.8	229.3	217.2	217.8	221.6	219.3	217.0	200.2	201.9	215.2	196.1	199.2
2011	185.7	146.7	167.4	169.1	173.6	219.3	208.2	206.0	211.6	210.1	205.2	191.2	194.5	206.6	186.7	189.4
2010	181.6	145.1	162.1	163.7	168.3	214.3	202.7	200.7	204.9	204.7	200.0	185.9	189.4	201.1	183.3	186.8
2009	182.5	147.6	159.4	161.8	169.1	211.8	199.2	197.6	201.9	201.9	196.9	185.1	187.1	198.2	179.2	186.9
2008	171.0	135.9	149.1	150.9	157.7	198.6	186.4	185.1	188.7	189.1	184.5	171.3	173.7	184.5	169.6	176.4
2007	165.0	132.4	146.4	148.2	152.5	191.8	180.4	179.7	182.0	181.7	179.1	166.8	169.1	180.4	166.8	172.5
2006	156.2	125.2	140.2	139.8	144.9	180.4	171.6	170.4	172.3	173.5	170.4	157.3	159.7	171.6	161.1	165.5
2005	146.7	117.4	131.7	131.4	135.8	169.6	161.3	159.6	162.2	162.3	159.6	147.6	150.7	158.8	147.5	155.8
2004	132.8	105.7	119.8	119.4	121.4	154.1	144.4	143.6	146.4	146.7	143.6	131.9	136.5	143.2	135.9	142.1
2003	129.7	103.9	117.7	117.3	118.1	150.2	139.6	140.6	143.3	143.6	140.5	128.9	133.8	139.0	134.1	139.0
2002	126.7	102.1	117.5	117.1	115.6	145.6	136.0	135.0	136.9	137.3	134.9	124.7	128.3	134.9	131.4	134.3
2001	122.2	98.3	114.6	114.3	111.7	140.9	132.1	131.4	133.1	132.9	131.3	120.3	124.5	130.1	126.8	129.8
2000	118.9	96.2	105.7	105.4	107.7	138.9	129.4	128.4	129.7	130.3	128.3	116.7	121.0	127.3	124.3	125.8
1999	116.6	94.8	105.0	104.7	106.4	136.2	126.7	126.3	127.3	127.4	126.2	115.0	118.8	123.8	117.5	122.7
1998	113.6	92.0	102.8	102.5	104.1	132.8	125.0	124.7	125.0	125.3	124.6	114.2	117.1	122.6	116.0	119.7
1995	105.6	86.7	96.8	96.5	96.1	128.6	119.6	117.7	120.5	119.6	117.2	110.7	112.7	114.8	106.1	110.5
1990	93.2	80.3	88.5	88.6	85.6	110.9	103.7	102.9	105.2	102.8	102.6	98.7	101.4	103.2	93.2	96.5
1985	81.8	73.6	76.7	77.0	72.7	92.8	88.8	88.7	89.4	88.2	88.6	85.0	85.6	86.7	80.1	82.7
1980	60.7	58.7	57.3	58.5	53.6	64.0	63.7	64.1	63.4	62.7	63.1	61.8	62.0	62.3	62.9	64.0
1975	43.7	40.5	41.9	42.1	39.8	46.6	45.8	45.7	46.2	45.7	46.1	45.5	45.8	46.0	44.1	44.9
1970	27.8	26.4	26.5	25.8	25.1	29.2	29.1	29.2	29.1	28.9	29.0	28.6	28.5	28.7	28.5	28.9
1965	21.5	20.3	20.4	19.4	20.2	23.0	22.5	22.5	22.5	22.2	22.4	22.0	22.4	22.1	22.0	22.3
1960	19.5	18.5	18.6	17.6	17.5	20.5	20.4	20.4	20.4	20.2	20.3	20.0	20.1	20.1	20.0	20.2
1955	16.3	15.5	15.6	14.7	14.7	17.2	17.1	17.1	17.1	16.9	17.1	16.8	16.9	16.8	16.7	16.9
1950	13.5	12.8	12.9	12.2	12.1	14.2	14.1	14.1	14.1	14.0	14.1	13.8	14.0	13.9	13.8	14.0

Year	National 30 City Average	Michigan Detroit	Flint	Grand Rapids	Kala-mazoo	Lansing	Sagi-naw	Minnesota Duluth	Minne-apolis	Roches-ter	Mississippi Biloxi	Jackson	Missouri Kansas City	St. Joseph	St. Louis	Spring-field
Jan 2019	227.3	223.8E	208.7E	201.3E	198.5E	208.3E	204.9E	223.8E	237.1E	220.2E	183.6E	184.4E	224.8E	212.8E	227.1E	203E
2018	217.7	219.6	204.8	197.5	194.8	204.4	201	219.6	232.6	216.1	180.1	180.9	220.6	208.8	222.8	199.2
2017	209.4	212.1	199.2	190.9	189.6	198.1	195.3	212.1	221.2	207.6	175.3	175.6	212.9	202.8	213.1	192.9
2016	207.7	212.8	200.2	191.0	189.7	200.9	195.9	217.6	227.2	212.4	173.3	177.7	213.2	205.5	212.1	194.6
2015	204.0	211.8	198.9	191.2	190.4	199.6	194.1	212.9	222.0	209.1	167.3	174.2	210.0	202.3	210.1	191.8
2014	203.0	208.7	196.0	187.9	188.4	195.7	191.7	211.1	220.7	207.5	164.6	169.8	210.8	199.3	208.6	189.5
2013	196.9	203.1	190.6	182.4	182.4	190.2	186.5	205.1	216.3	201.4	160.9	164.3	204.7	193.1	202.4	181.6
2012	194.0	200.2	186.3	176.4	178.9	185.9	182.8	203.1	214.7	199.7	157.6	160.5	200.8	189.0	198.0	177.7
2011	185.7	190.3	178.5	169.9	171.9	178.0	174.6	195.7	208.1	193.8	151.4	154.3	191.1	178.8	190.6	168.3
2010	181.6	187.5	176.2	160.6	167.3	175.8	171.9	193.1	203.8	188.9	148.0	151.0	186.1	174.2	185.9	164.3
2009	182.5	187.8	176.0	156.7	166.3	173.7	168.6	191.8	203.1	188.0	152.8	156.6	185.6	172.3	187.2	163.7
2008	171.0	177.0	164.4	140.2	155.7	161.6	158.7	177.4	190.6	175.0	141.8	147.1	175.5	164.2	176.2	152.9
2007	165.0	172.9	161.7	137.2	152.6	158.7	156.1	174.2	184.5	171.4	137.7	131.2	169.0	157.8	170.6	145.3
2006	156.2	165.8	153.8	131.1	146.0	152.5	150.6	166.2	173.9	161.9	123.3	117.1	162.0	152.4	159.3	139.4
2005	146.7	156.2	145.1	123.1	134.9	142.0	140.6	157.3	164.6	152.8	116.0	109.9	151.3	143.1	149.4	131.7
2004	132.8	141.9	129.6	112.8	122.5	129.7	127.6	139.1	150.1	137.0	105.3	99.3	135.1	126.9	135.5	116.3
2003	129.7	138.7	127.9	110.9	120.8	127.7	126.0	136.7	146.5	134.5	101.4	96.7	131.9	124.8	133.3	113.9
2002	126.7	134.2	126.9	107.4	119.9	125.3	124.4	131.9	139.5	130.0	101.0	96.3	128.4	122.7	129.6	112.5
2001	122.2	129.4	122.4	104.3	115.2	120.1	119.7	131.4	136.1	124.9	98.7	93.9	121.8	114.7	125.5	106.7
2000	118.9	125.3	119.8	102.7	111.6	117.6	115.9	124.1	131.1	120.1	97.2	92.9	118.2	111.9	122.4	104.3
1999	116.6	122.6	113.7	100.9	106.1	111.5	110.1	120.3	126.5	117.1	95.7	91.8	114.9	106.1	119.8	101.1
1998	113.6	119.5	112.3	99.7	104.9	110.1	108.7	117.7	124.6	115.5	92.1	89.5	108.2	104.2	115.9	98.9
1995	105.6	110.1	104.1	91.1	96.4	98.3	102.3	100.3	111.9	102.5	84.2	83.5	99.7	96.1	106.3	89.5
1990	93.2	96.0	91.7	84.0	87.4	90.2	88.9	94.3	100.5	95.6	76.4	75.6	89.5	86.8	94.0	82.6
1985	81.8	81.6	80.7	75.8	79.4	80.0	80.5	85.2	87.9	86.5	69.4	68.5	78.2	77.3	80.8	73.2
1980	60.7	62.5	63.1	59.8	61.7	60.3	63.3	63.4	64.1	64.8	54.3	54.3	59.1	60.4	59.7	56.7
1975	43.7	45.8	44.9	41.8	43.7	44.1	44.8	45.2	46.0	45.3	37.9	37.7	42.7	45.4	44.8	41.1
1970	27.8	29.7	28.5	26.7	28.1	27.8	28.7	28.9	29.5	28.9	24.4	21.3	25.3	28.3	28.4	26.0
1965	21.5	22.1	21.9	20.6	21.7	21.5	22.2	22.3	23.4	22.3	18.8	16.4	20.6	21.8	21.8	20.0
1960	19.5	20.1	20.0	18.7	19.7	19.5	20.1	20.3	20.5	20.2	17.1	14.9	19.0	19.8	19.5	18.2
1955	16.3	16.8	16.7	15.7	16.5	16.3	16.9	17.0	17.2	17.0	14.3	12.5	16.0	16.6	16.3	15.2
1950	13.5	13.9	13.8	13.0	13.6	13.5	13.9	14.0	14.2	14.0	11.8	10.3	13.2	13.7	13.5	12.6

Year	National 30 City Average	Montana Billings	Great Falls	Nebraska Lincoln	Omaha	Nevada Las Vegas	Reno	New Hampshire Man-chester	Nashua	New Jersey Camden	Jersey City	Newark	Pater-son	Trenton	NM Albu-querque	NY Albany
Jan 2019	227.3	199.8E	200.3E	199.5E	200.1E	232.4E	211E	211E	210.2E	248E	251.7E	256.3E	254.7E	250.5E	195.8E	226.3E
2018	217.7	196.0	196.5	195.7	196.3	228.0	207.0	207.0	206.2	243.3	247.0	251.5	249.9	245.8	192.1	222.0
2017	209.4	190.4	191.0	189.6	190.1	220.1	199.8	200.7	200.0	236.3	238.7	242.7	240.8	239.2	185.9	213.5
2016	207.7	191.8	191.1	186.8	187.6	221.1	198.8	200.9	200.5	228.8	229.6	232.9	231.3	229.3	181.6	209.6
2015	204.0	188.1	188.4	183.7	184.7	215.5	194.9	198.7	197.8	224.5	226.1	230.1	228.5	227.6	178.3	208.1
2014	203.0	185.4	186.3	181.1	183.7	210.9	195.7	198.8	197.3	224.6	226.6	230.6	227.8	226.6	178.0	205.1
2013	196.9	180.5	181.1	175.5	180.6	206.8	190.4	193.8	192.5	217.7	219.3	223.6	221.2	218.7	173.6	195.3
2012	194.0	178.6	179.5	169.3	177.1	202.6	186.8	189.2	188.3	213.6	215.2	219.2	217.5	213.7	171.0	191.2
2011	185.7	168.8	170.8	162.8	169.6	195.7	179.8	176.8	176.1	205.5	207.0	210.3	209.2	206.4	163.3	180.9
2010	181.6	166.5	168.6	159.3	165.8	193.7	175.6	172.5	172.0	201.2	203.3	206.3	205.3	200.8	162.5	177.8
2009	182.5	165.3	166.0	161.7	165.0	191.5	176.5	174.0	173.5	196.8	200.5	203.6	202.0	198.6	163.0	178.1
2008	171.0	153.1	154.6	152.0	154.8	176.2	167.0	162.5	161.9	184.0	187.3	189.6	188.7	185.6	152.6	166.0
2007	165.0	147.7	147.8	147.7	150.3	166.7	161.4	159.3	158.7	179.3	183.3	185.2	184.6	181.7	146.7	159.4
2006	156.2	140.5	140.8	132.4	140.7	160.0	154.5	146.7	146.4	167.7	171.0	173.6	172.2	169.9	140.1	151.6
2005	146.7	131.9	132.4	124.1	132.8	149.5	145.4	136.8	136.5	159.0	162.4	163.9	163.0	161.5	130.4	142.2
2004	132.8	118.2	117.9	112.1	119.4	137.1	130.5	124.2	123.9	143.4	145.4	147.2	146.6	146.0	118.3	128.6
2003	129.7	115.8	115.8	110.2	117.3	133.8	128.3	122.4	122.2	142.0	144.3	145.7	145.0	143.0	116.4	126.8
2002	126.7	114.6	114.5	108.2	115.0	131.9	126.4	119.9	119.6	135.9	138.5	140.2	140.1	137.6	114.6	122.6
2001	122.2	117.5	117.7	101.3	111.6	127.8	122.5	116.2	116.2	133.4	136.7	136.9	136.8	136.0	111.4	119.2
2000	118.9	113.7	113.9	98.8	107.0	125.8	118.2	111.9	111.9	128.4	130.5	132.6	132.4	130.6	109.0	116.5
1999	116.6	112.1	112.7	96.6	104.8	121.9	114.4	109.6	109.6	125.3	128.8	131.6	129.5	129.6	106.7	114.6
1998	113.6	109.7	109.1	94.9	101.2	118.1	111.4	110.1	110.0	124.3	127.7	128.9	128.5	127.9	103.8	113.0
1995	105.6	104.7	104.9	85.6	93.4	108.5	105.0	100.9	100.8	107.0	112.2	111.9	112.1	111.2	96.3	103.6
1990	93.2	92.9	93.8	78.8	85.0	96.3	94.5	86.3	86.3	93.0	93.5	93.8	97.2	94.5	84.9	93.2
1985	81.8	83.9	84.3	71.5	77.5	87.6	85.0	78.1	78.1	81.3	83.3	83.5	84.5	82.8	76.7	79.5
1980	60.7	63.9	64.6	58.5	63.5	64.6	63.0	56.6	56.0	58.6	60.6	60.1	60.0	58.9	59.0	59.5
1975	43.7	43.1	43.8	40.9	43.2	42.8	41.9	41.3	40.8	42.3	43.4	44.2	43.9	43.8	40.3	43.9
1970	27.8	28.5	28.9	26.4	26.8	29.4	28.0	26.2	25.6	27.2	27.8	29.0	27.8	27.4	26.4	28.3
1965	21.5	22.0	22.3	20.3	20.6	22.4	21.6	20.6	19.7	20.9	21.4	23.8	21.4	21.3	20.6	22.3
1960	19.5	20.0	20.3	18.5	18.7	20.2	19.6	18.0	17.9	19.0	19.4	19.4	19.4	19.2	18.5	19.3
1955	16.3	16.7	17.0	15.5	15.7	16.9	16.4	15.1	15.0	16.0	16.3	16.3	16.3	16.1	15.6	16.2
1950	13.5	13.9	14.0	12.8	13.0	14.0	13.6	12.5	12.4	13.2	13.5	13.5	13.5	13.3	12.9	13.4

Historical Cost Indexes

Year	National 30 City Average	New York								North Carolina					N. Dakota	Ohio
		Binghamton	Buffalo	New York	Rochester	Schenectady	Syracuse	Utica	Yonkers	Charlotte	Durham	Greensboro	Raleigh	Winston-Salem	Fargo	Akron
Jan 2019	227.3	219.5E	234E	296.1E	223.6E	226.3E	219.6E	215.8E	268.4E	187.3E	191.1E	189.1E	186.8E	188.6E	203.4E	210.3E
2018	217.7	215.4	229.6	290.5	219.4	222.0	215.5	211.7	263.3	183.8	187.5	185.5	183.3	185.0	199.6	206.3
2017	209.4	208.4	220.8	282.9	213.0	213.8	208.9	205.9	256.8	177.0	180.3	178.2	175.8	178.0	192.8	199.4
2016	207.7	202.6	213.7	270.5	208.4	209.8	203.6	200.8	247.9	180.6	182.0	179.8	174.6	179.6	185.5	202.4
2015	204.0	202.4	209.6	268.0	203.5	207.4	200.9	198.1	243.4	173.1	172.3	170.5	167.6	170.6	181.0	199.1
2014	203.0	202.4	207.8	267.7	202.2	204.2	198.8	196.4	245.6	165.5	165.2	163.9	161.6	164.3	177.1	199.2
2013	196.9	195.5	200.7	259.4	194.3	195.5	193.6	188.5	230.0	159.6	158.4	158.7	157.5	159.1	169.6	191.8
2012	194.0	192.9	198.1	256.3	191.2	191.4	189.6	185.2	227.1	156.4	155.6	156.3	155.4	157.2	167.3	187.7
2011	185.7	176.1	188.1	245.6	181.9	181.5	180.8	173.5	218.9	142.6	143.8	142.2	142.8	140.9	160.3	180.8
2010	181.6	173.8	184.0	241.4	179.6	179.0	176.7	170.5	217.1	140.5	141.6	140.0	140.6	138.8	156.2	174.5
2009	182.5	172.9	184.4	239.9	179.6	178.8	176.9	171.3	217.5	145.1	145.8	144.0	145.1	142.7	154.3	175.7
2008	171.0	160.7	173.9	226.8	168.0	167.2	165.9	161.6	202.1	135.8	136.6	134.6	135.3	133.8	145.0	165.4
2007	165.0	155.7	168.6	215.2	163.5	159.8	159.8	155.0	196.5	132.8	133.7	131.7	132.1	131.0	139.8	159.0
2006	156.2	147.5	159.2	204.5	155.2	151.1	150.9	146.4	186.0	125.1	124.6	123.8	124.2	123.0	133.2	152.8
2005	146.7	137.5	149.4	194.0	147.3	142.2	141.9	136.9	176.1	110.5	112.1	112.1	112.1	111.1	125.3	144.4
2004	132.8	123.5	136.1	177.7	132.0	128.4	127.3	124.4	161.8	98.9	100.0	100.1	100.3	99.4	113.0	131.9
2003	129.7	121.8	132.8	173.4	130.3	126.7	124.7	121.7	160.0	96.2	97.5	97.5	97.8	96.8	110.6	129.6
2002	126.7	119.0	128.5	170.1	127.1	123.0	121.8	118.4	154.8	94.8	96.1	96.1	96.3	95.5	106.3	127.2
2001	122.2	116.0	125.2	164.4	123.1	120.1	118.6	115.3	151.4	91.5	92.9	92.9	93.3	92.3	103.1	123.5
2000	118.9	112.4	122.3	159.2	120.0	117.5	115.1	112.4	144.8	90.3	91.6	91.6	91.9	91.0	97.9	117.8
1999	116.6	108.6	120.2	155.9	116.8	114.7	113.7	108.5	140.6	89.3	90.2	90.3	90.5	90.1	96.8	116.0
1998	113.6	109.0	119.1	154.4	117.2	114.2	113.6	108.6	141.4	88.2	89.1	89.2	89.4	89.0	95.2	113.1
1995	105.6	99.3	110.1	140.7	106.6	104.6	104.0	97.7	129.4	81.8	82.6	82.6	82.7	82.6	88.4	103.4
1990	93.2	87.0	94.1	118.1	94.6	93.9	91.0	85.6	111.4	74.8	75.4	75.5	75.5	75.3	81.3	94.6
1985	81.8	77.5	83.2	94.9	81.6	80.1	81.0	77.0	92.8	66.9	67.6	67.7	67.6	67.5	73.4	86.8
1980	60.7	58.0	60.6	66.0	60.5	60.3	61.6	58.5	65.9	51.1	52.2	52.5	51.7	50.8	57.4	62.3
1975	43.7	42.5	45.1	49.5	44.8	43.7	44.8	42.7	47.1	36.1	37.0	37.0	37.3	36.1	39.3	44.7
1970	27.8	27.0	28.9	32.8	29.2	27.8	28.5	26.9	30.0	20.9	23.7	23.6	23.4	23.0	25.8	28.3
1965	21.5	20.8	22.2	25.5	22.8	21.4	21.9	20.7	23.1	16.0	18.2	18.2	18.0	17.7	19.9	21.8
1960	19.5	18.9	19.9	21.5	19.7	19.5	19.9	18.8	21.0	14.4	16.6	16.6	16.4	16.1	18.1	19.9
1955	16.3	15.8	16.7	18.1	16.5	16.3	16.7	15.8	17.6	12.1	13.9	13.9	13.7	13.5	15.2	16.6
1950	13.5	13.1	13.8	14.9	13.6	13.5	13.8	13.0	14.5	10.0	11.5	11.5	11.4	11.2	12.5	13.7

Year	National 30 City Average	Ohio									Oklahoma			Oregon		PA
		Canton	Cincinnati	Cleveland	Columbus	Dayton	Lorain	Springfield	Toledo	Youngstown	Lawton	Oklahoma City	Tulsa	Eugene	Portland	Allentown
Jan 2019	227.3	203.3E	199.4E	213.8E	203.2E	194.7E	207.1E	196.8E	212.5E	204.6E	187.4E	186.2E	184.8E	219.1E	221.7E	226.1E
2018	217.7	199.5	195.6	209.8	199.4	191.0	203.2	193.1	208.5	200.7	183.9	182.7	181.3	215.0	217.5	221.8
2017	209.4	192.2	186.1	202.6	191.6	186.1	196.9	187.8	201.4	194.2	178.0	177.0	174.8	207.3	209.5	213.7
2016	207.7	195.0	190.1	205.9	195.2	189.3	199.9	190.6	206.1	197.9	177.7	180.1	172.0	209.1	209.2	214.8
2015	204.0	191.1	188.1	203.4	192.4	187.9	196.7	188.1	201.0	193.3	175.1	177.1	170.0	203.1	204.3	209.8
2014	203.0	190.8	188.1	203.1	192.4	185.8	196.7	187.2	199.7	193.8	169.3	170.9	166.7	200.4	201.4	209.0
2013	196.9	183.9	181.9	195.5	187.1	180.7	187.6	181.5	193.5	186.5	162.7	165.5	161.7	194.7	195.6	203.2
2012	194.0	180.2	178.6	191.4	183.5	177.6	184.1	178.0	190.4	183.3	158.1	158.0	151.3	190.1	191.8	199.3
2011	185.7	172.1	171.3	184.0	175.7	168.4	176.8	168.4	183.1	176.5	151.7	151.6	144.8	184.9	186.6	192.2
2010	181.6	167.5	166.5	180.2	170.3	163.4	173.0	164.6	176.5	171.7	149.5	149.4	143.1	181.5	182.9	187.9
2009	182.5	168.4	168.1	181.6	171.2	165.3	174.0	166.0	178.5	172.8	152.5	153.3	146.8	181.3	183.6	188.3
2008	171.0	158.4	159.0	170.6	160.2	157.5	163.9	157.5	168.2	161.5	141.3	140.6	136.2	172.3	174.7	175.7
2007	165.0	153.2	152.4	164.9	155.1	149.4	158.1	150.3	161.6	156.7	136.2	135.6	132.5	168.5	169.5	169.4
2006	156.2	147.0	144.9	156.9	146.9	142.9	151.8	143.4	154.5	150.3	129.5	129.5	125.7	160.5	161.7	160.1
2005	146.7	137.7	136.8	147.9	138.0	134.5	143.3	134.8	145.5	141.1	121.6	121.6	117.8	150.9	152.2	150.2
2004	132.8	125.6	124.3	135.6	126.2	121.1	131.5	122.0	132.8	129.3	109.3	109.4	107.3	136.8	137.8	133.9
2003	129.7	123.7	121.3	132.7	123.7	119.4	126.1	120.0	130.4	126.2	107.5	107.8	105.0	134.2	135.9	130.3
2002	126.7	120.9	119.2	130.0	120.9	117.5	124.3	118.0	128.4	123.6	106.3	106.0	104.1	132.2	133.9	128.1
2001	122.2	117.7	115.9	125.9	117.1	114.0	120.4	114.7	123.8	119.9	102.0	102.1	99.7	129.8	131.2	123.5
2000	118.9	112.7	110.1	121.3	112.5	109.0	115.0	109.2	115.7	114.1	98.7	98.9	97.5	126.3	127.4	119.9
1999	116.6	110.6	107.9	118.7	109.5	107.1	112.0	106.4	113.6	111.9	97.5	97.4	96.2	120.9	124.3	117.8
1998	113.6	108.5	105.0	114.8	106.8	104.5	109.4	104.1	111.2	109.0	94.3	94.5	94.5	120.0	122.2	115.7
1995	105.6	98.8	97.1	106.4	99.1	94.6	97.1	92.0	100.6	100.1	85.3	88.0	89.0	112.2	114.3	108.3
1990	93.2	92.1	86.7	95.1	88.1	85.9	89.4	83.4	92.9	91.8	77.1	79.9	80.0	98.0	99.4	95.0
1985	81.8	83.7	78.2	86.2	77.7	76.3	80.7	74.3	84.7	82.7	71.2	73.9	73.8	88.7	90.1	81.9
1980	60.7	60.6	59.8	61.0	58.6	56.6	60.8	56.2	64.0	61.5	52.2	55.5	57.2	68.9	68.4	58.7
1975	43.7	44.0	43.7	44.6	42.2	40.5	42.7	40.0	44.9	44.5	38.6	38.6	39.2	44.6	44.9	42.3
1970	27.8	27.8	28.2	29.6	26.7	27.0	27.5	25.4	29.1	29.6	24.1	23.4	24.1	30.4	30.0	27.3
1965	21.5	21.5	20.7	21.1	20.2	20.0	21.1	19.6	21.3	21.0	18.5	17.8	20.6	23.4	22.6	21.0
1960	19.5	19.5	19.3	19.6	18.7	18.1	19.2	17.8	19.4	19.1	16.9	16.1	18.1	21.2	21.1	19.1
1955	16.3	16.3	16.2	16.5	15.7	15.2	16.1	14.9	16.2	16.0	14.1	13.5	15.1	17.8	17.7	16.0
1950	13.5	13.5	13.3	13.6	13.0	12.5	13.3	12.3	13.4	13.2	11.7	11.2	12.5	14.7	14.6	13.2

Historical Cost Indexes

Year	National	Pennsylvania						RI	South Carolina		South Dakota		Tennessee			
	30 City Average	Erie	Harrisburg	Philadelphia	Pittsburgh	Reading	Scranton	Providence	Charleston	Columbia	Rapid City	Sioux Falls	Chattanooga	Knoxville	Memphis	Nashville
Jan 2019	227.3	209.9E	213.7E	253.5E	224.9E	221E	217.7E	232.7E	189.1E	187.4E	196.5E	200.8E	191.3E	182.8E	191.3E	190.9E
2018	217.7	205.9	209.7	248.7	220.7	216.8	213.6	228.3	185.5	183.9	192.8	197.0	187.7	179.4	187.7	187.3
2017	209.4	199.5	202.4	239.8	213.6	210.7	207.4	218.6	178.4	176.5	185.3	184.8	181.9	173.3	180.9	180.5
2016	207.7	199.7	202.4	238.2	212.2	207.6	207.0	221.6	176.8	176.6	174.7	175.2	177.4	173.9	181.9	181.8
2015	204.0	194.5	200.3	234.7	209.5	204.2	201.7	219.9	171.3	171.9	167.7	168.0	174.3	171.0	177.5	179.9
2014	203.0	191.4	200.3	232.1	207.1	201.4	201.4	220.3	171.0	162.5	168.6	166.8	174.1	170.9	176.5	178.2
2013	196.9	187.2	192.5	223.6	201.4	196.1	196.0	213.4	166.3	157.9	159.6	161.2	167.4	159.3	169.3	172.8
2012	194.0	182.0	187.9	221.8	196.8	190.8	191.5	209.3	155.3	148.1	156.7	158.8	163.8	155.8	166.3	167.5
2011	185.7	176.0	180.9	212.8	187.7	184.3	185.3	196.8	149.4	142.3	149.5	151.3	157.3	149.3	158.4	160.7
2010	181.6	171.0	175.2	209.3	182.3	179.8	180.8	192.8	147.4	140.2	147.3	149.0	152.4	144.7	154.2	156.7
2009	182.5	171.9	176.5	209.5	181.7	180.7	181.3	192.9	150.8	144.3	148.2	151.1	155.7	148.0	157.2	159.8
2008	171.0	161.0	165.7	196.2	169.2	169.3	167.9	178.2	142.1	134.8	138.4	141.3	136.7	133.4	146.0	147.5
2007	165.0	156.0	159.4	188.4	163.2	164.9	163.1	174.4	139.3	131.7	129.5	133.0	133.7	130.6	143.1	144.4
2006	156.2	148.7	150.7	177.8	155.5	155.0	153.9	163.6	122.4	118.7	122.8	126.6	127.0	123.9	136.5	135.5
2005	146.7	140.6	141.5	166.4	146.6	145.5	142.9	155.6	110.2	109.6	115.2	118.5	116.0	115.1	128.6	127.9
2004	132.8	127.0	126.2	148.4	133.1	128.7	129.0	138.8	98.9	98.4	103.9	107.3	105.0	104.5	115.6	115.8
2003	129.7	124.5	123.8	145.5	130.5	126.3	126.0	135.9	96.1	95.7	101.3	104.0	102.8	102.3	111.1	110.9
2002	126.7	122.1	121.1	142.0	127.8	123.4	124.2	131.1	94.8	93.9	100.2	103.2	102.2	101.2	107.6	109.2
2001	122.2	118.9	118.4	136.9	124.1	120.4	121.1	127.8	92.1	91.3	96.8	99.7	98.2	96.5	102.6	104.3
2000	118.9	114.6	114.0	132.1	120.9	115.9	117.7	122.8	89.9	89.1	94.2	98.0	96.9	95.0	100.8	100.8
1999	116.6	113.8	112.8	129.8	119.6	114.8	116.3	121.6	89.0	88.1	92.4	95.8	95.9	93.4	99.7	98.6
1998	113.6	109.8	110.5	126.6	117.1	112.5	113.7	120.3	88.0	87.2	90.7	93.6	94.9	92.4	97.2	96.8
1995	105.6	99.8	100.6	117.1	106.3	103.6	103.8	111.1	82.7	82.2	84.0	84.7	89.2	86.0	91.2	87.6
1990	93.2	88.5	89.5	98.5	91.2	90.8	91.3	94.1	73.7	74.5	77.2	78.4	79.9	75.1	81.3	77.1
1985	81.8	79.6	77.2	82.2	81.5	77.7	80.0	83.0	65.9	66.9	69.7	71.2	72.5	67.7	74.3	66.7
1980	60.7	59.4	56.9	58.7	61.3	57.7	58.4	59.2	50.6	51.1	54.8	57.1	55.0	51.6	55.9	53.1
1975	43.7	43.4	42.2	44.5	44.5	43.6	41.9	42.7	35.0	36.4	37.7	39.6	39.1	37.3	40.7	37.0
1970	27.8	27.5	25.8	27.5	28.7	27.1	27.0	27.3	22.8	22.9	24.8	25.8	24.9	22.0	23.0	22.8
1965	21.5	21.1	20.1	21.7	22.4	20.9	20.8	21.9	17.6	17.6	19.1	19.9	19.2	17.1	18.3	17.4
1960	19.5	19.1	18.6	19.4	19.7	19.0	18.9	19.1	16.0	16.0	17.3	18.1	17.4	15.5	16.6	15.8
1955	16.3	16.1	15.6	16.3	16.5	15.9	15.9	16.0	13.4	13.4	14.5	15.1	14.6	13.0	13.9	13.3
1950	13.5	13.3	12.9	13.5	13.6	13.2	13.1	13.2	11.1	11.1	12.0	12.5	12.1	10.7	11.5	10.9

Year	National	Texas														Utah
	30 City Average	Abilene	Amarillo	Austin	Beaumont	Corpus Christi	Dallas	El Paso	Fort Worth	Houston	Lubbock	Odessa	San Antonio	Waco	Wichita Falls	Ogden
Jan 2019	227.3	182.1E	181.4E	182.3E	186.2E	184.5E	188.6E	181.9E	183.9E	188.7E	184.7E	183.6E	184.5E	180.1E	179.7E	194.3E
2018	217.7	178.7	178.0	178.9	182.7	181.0	185.0	178.5	180.4	185.1	181.2	180.1	181.0	176.7	176.3	190.6
2017	209.4	174.0	172.8	172.6	178.0	178.9	180.0	173.5	174.9	179.2	176.6	175.6	175.5	170.7	170.4	183.3
2016	207.7	173.7	171.0	173.3	178.3	175.5	177.9	172.8	174.1	180.5	175.2	172.6	173.1	170.3	170.1	179.4
2015	204.0	170.8	170.7	172.3	174.0	172.8	174.4	168.8	172.5	176.4	173.3	171.5	171.8	167.3	169.0	177.1
2014	203.0	160.9	167.2	167.5	165.2	167.3	172.4	156.2	169.5	176.0	166.9	159.2	169.5	161.7	160.6	174.2
2013	196.9	155.2	161.7	158.7	160.2	157.7	167.4	151.2	161.7	169.4	159.3	151.5	164.3	156.1	155.9	167.7
2012	194.0	152.5	158.6	154.3	158.2	151.8	165.2	149.1	159.4	167.8	156.6	149.1	160.7	153.8	152.9	165.6
2011	185.7	145.9	151.9	147.6	152.4	145.0	157.9	142.5	152.8	160.8	150.0	142.5	152.6	147.3	146.6	158.7
2010	181.6	144.2	150.1	144.7	150.5	142.1	155.1	140.8	149.6	157.3	148.1	140.7	147.9	145.8	145.0	157.5
2009	182.5	143.5	147.3	146.4	149.8	141.7	155.5	141.1	150.1	160.9	144.6	139.1	150.8	146.7	146.1	155.1
2008	171.0	132.4	137.0	137.5	140.3	133.2	144.1	130.7	138.3	149.1	134.7	128.9	141.0	136.1	136.1	144.8
2007	165.0	128.5	132.2	131.6	137.1	128.8	138.6	126.2	134.8	146.0	130.2	125.0	136.4	132.1	132.1	140.0
2006	156.2	121.6	125.7	125.5	130.1	122.0	131.1	120.3	127.6	138.2	123.2	118.3	129.8	125.2	125.3	133.4
2005	146.7	113.8	117.3	117.9	121.3	114.4	123.7	112.5	119.4	129.0	115.5	110.5	121.3	116.1	117.3	126.0
2004	132.8	102.9	106.3	105.7	108.5	102.9	112.0	101.5	108.4	115.9	104.7	99.9	108.4	104.8	105.1	115.2
2003	129.7	100.5	104.4	103.9	106.1	100.4	109.3	99.5	105.5	113.4	102.3	97.6	104.8	102.7	103.0	112.7
2002	126.7	99.9	101.9	102.4	104.5	98.9	107.9	98.7	104.6	111.5	100.5	95.9	103.7	100.5	101.4	110.8
2001	122.2	93.4	98.4	99.8	102.3	96.6	103.8	95.5	100.9	107.8	97.6	93.4	100.5	97.8	97.3	107.7
2000	118.9	93.4	98.1	99.1	101.6	96.9	102.7	92.4	99.9	106.0	97.6	93.4	99.4	97.3	96.9	104.6
1999	116.6	91.8	94.5	96.0	99.7	94.0	101.0	90.7	97.6	104.6	96.0	92.1	98.0	94.8	95.5	103.3
1998	113.6	89.8	92.7	94.2	97.9	91.8	97.9	88.4	94.5	101.3	93.3	90.1	94.8	92.7	92.9	98.5
1995	105.6	85.2	87.4	89.3	93.7	87.4	91.4	85.2	89.5	95.9	88.4	85.6	88.9	86.4	86.8	92.2
1990	93.2	78.0	80.1	81.3	86.5	79.3	84.5	76.7	82.1	85.4	81.5	78.6	80.7	79.6	80.3	83.4
1985	81.8	71.1	72.5	74.5	79.3	72.3	77.6	69.4	75.1	79.6	74.0	71.2	73.9	71.7	73.3	75.2
1980	60.7	53.4	55.2	54.5	57.6	54.5	57.9	53.1	57.0	59.4	55.6	57.2	55.0	54.9	55.4	62.2
1975	43.7	37.6	39.0	39.0	39.6	38.1	40.7	38.0	40.4	41.2	38.9	37.9	39.0	38.6	38.0	40.0
1970	27.8	24.5	24.9	24.9	25.7	24.5	25.5	23.7	25.9	25.4	25.1	24.6	23.3	24.8	24.5	26.8
1965	21.5	18.9	19.2	19.2	19.9	18.9	19.9	19.0	19.9	20.0	19.4	19.0	18.5	19.2	18.9	20.6
1960	19.5	17.1	17.4	17.4	18.1	17.1	18.2	17.0	18.1	18.2	17.6	17.3	16.8	17.4	17.2	18.8
1955	16.3	14.4	14.6	14.6	15.1	14.4	15.3	14.3	15.2	15.2	14.8	14.5	14.1	14.6	14.4	15.7
1950	13.5	11.9	12.1	12.1	12.5	11.9	12.6	11.8	12.5	12.6	12.2	12.0	11.6	12.1	11.9	13.0

Historical Cost Indexes

Year	National	Utah	Vermont		Virginia					Washington			West Virginia		Wisconsin	
	30 City Average	Salt Lake City	Burlington	Rutland	Alexandria	Newport News	Norfolk	Richmond	Roanoke	Seattle	Spokane	Tacoma	Charleston	Huntington	Green Bay	Kenosha
Jan 2019	227.3	199.5E	203.9E	200.4E	206.2E	189.6E	190.3E	197.5E	194.5E	235.3E	209.2E	228.6E	211.8E	215.7E	218.3E	226.2E
2018	217.7	195.7	200.1	196.6	202.3	186.0	186.7	193.8	190.8	230.9	205.3	224.3	207.8	211.6	214.2	221.9
2017	209.4	188.8	192.8	189.5	195.6	180.7	180.5	182.8	185.1	219.3	198.8	215.8	201.5	204.5	206.9	213.6
2016	207.7	184.7	193.6	190.2	195.6	178.3	178.2	178.9	178.0	213.4	195.1	209.7	204.2	206.5	207.4	212.8
2015	204.0	182.1	191.6	188.5	192.5	175.8	177.4	177.2	175.3	209.9	193.1	207.2	199.9	201.3	200.8	205.4
2014	203.0	177.1	191.3	188.6	190.5	174.4	175.7	176.4	173.5	209.8	190.8	205.2	197.0	199.5	198.9	205.2
2013	196.9	171.1	178.7	176.6	185.0	170.0	171.5	170.7	168.3	203.4	184.3	199.1	187.2	193.2	193.9	201.2
2012	194.0	168.8	172.4	170.5	182.4	168.3	169.7	169.4	166.4	201.9	181.6	194.3	183.4	186.4	190.0	195.5
2011	185.7	161.8	158.8	157.1	174.5	159.8	161.1	159.0	154.2	194.1	175.5	187.6	177.5	180.3	181.4	187.6
2010	181.6	160.9	156.3	154.6	170.9	156.7	158.0	156.6	151.8	191.8	171.5	186.0	171.6	176.1	179.0	185.1
2009	182.5	160.4	158.3	156.7	172.5	160.0	161.9	160.7	155.6	188.5	173.0	186.3	174.6	177.4	175.5	182.5
2008	171.0	149.7	147.2	145.8	161.8	150.8	150.8	150.9	145.9	176.9	162.4	174.9	162.8	165.7	165.6	172.2
2007	165.0	144.6	144.4	143.2	155.0	146.2	146.1	147.3	142.9	171.4	156.7	168.7	158.7	160.1	159.4	164.8
2006	156.2	137.7	131.7	130.7	146.2	133.7	134.6	134.8	129.7	162.9	150.1	160.7	149.4	151.4	152.7	157.1
2005	146.7	129.4	124.6	123.8	136.5	123.5	124.4	125.4	112.2	153.9	141.9	151.5	140.8	141.0	144.4	148.3
2004	132.8	117.8	113.0	112.3	121.5	108.9	110.2	110.9	99.7	138.0	127.6	134.5	124.8	125.6	128.9	132.4
2003	129.7	116.0	110.7	110.1	119.5	104.8	106.2	108.6	97.0	134.9	125.9	133.0	123.3	123.6	127.4	130.9
2002	126.7	113.7	109.0	108.4	115.1	102.9	104.1	106.6	95.2	132.7	123.9	131.4	121.2	120.9	123.0	127.3
2001	122.2	109.1	105.7	105.2	110.8	99.8	100.3	102.9	92.1	127.9	120.3	125.7	114.6	117.5	119.1	123.6
2000	118.9	106.5	98.9	98.3	108.1	96.5	97.6	100.2	90.7	124.6	118.3	122.9	111.5	114.4	114.6	119.1
1999	116.6	104.5	98.2	97.7	106.1	95.6	96.5	98.8	89.8	123.3	116.7	121.6	110.6	113.4	112.1	115.8
1998	113.6	99.5	97.8	97.3	104.1	93.7	93.9	97.0	88.3	119.4	114.3	118.3	106.7	109.0	109.5	112.9
1995	105.6	93.1	91.1	90.8	96.3	86.0	86.4	87.8	82.8	113.7	107.4	112.8	95.8	97.2	97.6	97.9
1990	93.2	84.3	83.0	82.9	86.1	76.3	76.7	77.6	76.1	100.1	98.5	100.5	86.1	86.8	86.7	87.8
1985	81.8	75.9	74.8	74.9	75.1	68.7	68.8	69.5	67.2	88.3	89.0	91.2	77.7	77.7	76.7	77.4
1980	60.7	57.0	55.3	58.3	57.3	52.5	52.4	54.3	51.3	67.9	66.3	66.7	57.7	58.3	58.6	58.3
1975	43.7	40.1	41.8	43.9	41.7	37.2	36.9	37.1	37.1	44.9	44.4	44.5	41.0	40.0	40.9	40.5
1970	27.8	26.1	25.4	26.8	26.2	23.9	21.5	22.0	23.7	28.8	29.3	29.6	26.1	25.8	26.4	26.5
1965	21.5	20.0	19.8	20.6	20.2	18.4	17.1	17.2	18.3	22.4	22.5	22.8	20.1	19.9	20.3	20.4
1960	19.5	18.4	18.0	18.8	18.4	16.7	15.4	15.6	16.6	20.4	20.8	20.8	18.3	18.1	18.4	18.6
1955	16.3	15.4	15.1	15.7	15.4	14.0	12.9	13.1	13.9	17.1	17.4	17.4	15.4	15.2	15.5	15.6
1950	13.5	12.7	12.4	13.0	12.7	11.6	10.7	10.8	11.5	14.1	14.4	14.4	12.7	12.5	12.8	12.9

Year	National	Wisconsin			Wyoming	Canada									
	30 City Average	Madison	Milwaukee	Racine	Cheyenne	Calgary	Edmonton	Hamilton	London	Montreal	Ottawa	Quebec	Toronto	Vancouver	Winnipeg
Jan 2019	227.3	221.1E	228.3E	226E	197.8E	244.2E	245.2E	230.9E	239E	237.4E	241.2E	236.1E	246.1E	234E	221.8E
2018	217.7	216.9	224.0	221.7	194.1	239.6	240.6	226.6	234.5	232.9	236.7	231.7	241.5	229.6	217.6
2017	209.4	207.4	215.2	213.8	185.0	231.4	232.3	227.8	225.0	221.8	227.7	222.9	231.4	221.5	207.7
2016	207.7	207.4	214.4	212.4	178.6	229.1	229.4	221.7	221.3	218.5	223.4	218.9	228.6	219.3	205.3
2015	204.0	202.0	209.4	205.1	175.4	226.6	226.3	221.2	217.0	220.4	220.4	220.4	225.9	217.1	204.1
2014	203.0	202.4	210.2	205.3	173.6	228.2	229.1	222.1	218.9	219.8	214.8	218.3	227.4	218.8	202.9
2013	196.9	197.8	204.6	200.5	167.6	222.7	223.3	216.0	213.7	214.5	214.1	212.6	220.9	216.2	200.2
2012	194.0	191.2	202.2	194.9	165.0	219.1	219.6	211.1	208.2	209.6	209.7	207.5	216.6	214.0	200.7
2011	185.7	183.4	191.9	187.0	155.3	212.6	212.4	204.2	200.1	202.1	202.0	200.6	209.6	207.1	192.6
2010	181.6	181.3	187.1	184.4	154.4	200.6	200.7	197.7	193.7	193.9	195.6	192.7	200.7	192.7	182.9
2009	182.5	180.0	187.3	182.8	155.2	205.4	206.6	203.0	198.9	198.8	200.1	197.4	205.6	199.6	188.5
2008	171.0	168.4	176.3	173.0	146.5	190.2	191.0	194.3	188.6	186.7	188.0	186.8	194.6	184.9	174.4
2007	165.0	160.7	168.9	164.7	141.3	183.1	184.4	186.8	182.4	181.7	182.4	182.0	187.7	180.6	170.1
2006	156.2	152.8	158.8	157.2	128.0	163.6	164.8	169.0	164.9	158.6	163.7	159.0	170.6	169.3	155.9
2005	146.7	145.2	148.4	147.9	118.9	154.4	155.7	160.0	156.2	149.2	155.1	150.1	162.6	159.4	146.9
2004	132.8	129.7	134.2	132.7	104.7	138.8	139.4	142.6	140.4	134.5	141.0	135.7	146.0	141.7	129.7
2003	129.7	128.4	131.1	131.5	102.6	133.6	134.2	139.1	137.0	130.2	137.4	131.4	142.3	137.0	124.4
2002	126.7	124.5	128.2	126.5	101.7	122.5	122.2	136.3	134.1	127.2	134.9	128.4	139.8	134.6	121.4
2001	122.2	120.6	123.9	123.3	99.0	117.5	117.4	131.5	129.1	124.4	130.2	125.5	134.7	130.2	117.2
2000	118.9	116.6	120.5	118.7	98.1	115.9	115.8	130.0	127.5	122.8	128.8	124.1	133.1	128.4	115.6
1999	116.6	115.9	117.4	115.5	96.9	115.3	115.2	128.1	125.6	120.8	126.8	121.9	131.2	127.1	115.2
1998	113.6	110.8	113.4	112.7	95.4	112.5	112.4	126.3	123.9	119.0	124.7	119.6	128.5	123.8	113.7
1995	105.6	96.5	103.9	97.8	87.6	107.4	107.4	119.9	117.5	110.8	118.2	111.5	121.6	116.2	107.6
1990	93.2	84.3	88.9	87.3	79.1	98.0	97.1	103.8	101.4	99.0	102.7	96.8	104.6	103.2	95.1
1985	81.8	74.3	77.4	77.0	72.3	90.2	89.1	85.8	84.9	82.4	83.9	79.9	86.5	89.8	83.4
1980	60.7	56.8	58.8	58.1	56.9	64.9	63.3	63.7	61.9	59.2	60.9	59.3	60.9	65.0	61.7
1975	43.7	40.7	43.3	40.7	40.6	42.2	41.6	42.9	41.6	39.7	41.5	39.0	42.2	42.4	39.2
1970	27.8	26.5	29.4	26.5	26.0	28.9	28.6	28.5	27.8	25.6	27.6	26.0	25.6	26.0	23.1
1965	21.5	20.6	21.8	20.4	20.0	22.3	22.0	22.0	21.4	18.7	21.2	20.1	19.4	20.5	17.5
1960	19.5	18.1	19.0	18.6	18.2	20.2	20.0	20.0	19.5	17.0	19.3	18.2	17.6	18.6	15.8
1955	16.3	15.2	15.9	15.6	15.2	17.0	16.8	16.7	16.3	14.3	16.2	15.3	14.8	15.5	13.3
1950	13.5	12.5	13.2	12.9	12.6	14.0	13.9	13.8	13.5	11.8	13.4	12.6	12.2	12.8	10.9

Location Factors - Commercial

Costs shown in RSMeans cost data publications are based on national averages for materials and installation. To adjust these costs to a specific location, simply multiply the base cost by the factor and divide by 100 for that city. The data is arranged alphabetically by state and postal zip code numbers. For a city not listed, use the factor for a nearby city with similar economic characteristics.

STATE/ZIP	CITY	MAT.	INST.	TOTAL
ALABAMA				
350-352	Birmingham	96.5	71.7	85.9
354	Tuscaloosa	96.6	71.5	85.8
355	Jasper	97.1	71.0	85.9
356	Decatur	96.6	70.1	85.2
357-358	Huntsville	96.5	71.4	85.7
359	Gadsden	96.6	70.6	85.5
360-361	Montgomery	95.6	71.4	85.2
362	Anniston	95.2	67.4	83.3
363	Dothan	95.6	72.9	85.9
364	Evergreen	95.2	71.4	85.0
365-366	Mobile	96.1	68.8	84.4
367	Selma	95.3	72.4	85.5
368	Phenix City	96.1	72.0	85.8
369	Butler	95.5	71.6	85.3
ALASKA				
995-996	Anchorage	117.5	113.6	115.8
997	Fairbanks	118.2	114.7	116.7
998	Juneau	116.5	113.6	115.3
999	Ketchikan	128.4	113.6	122.0
ARIZONA				
850,853	Phoenix	100.0	73.0	88.4
851,852	Mesa/Tempe	98.6	72.0	87.2
855	Globe	99.5	71.8	87.6
856-857	Tucson	97.6	71.4	86.4
859	Show Low	99.7	71.9	87.8
860	Flagstaff	101.0	71.7	88.4
863	Prescott	98.7	71.7	87.1
864	Kingman	97.2	71.8	86.3
865	Chambers	97.2	74.4	87.4
ARKANSAS				
716	Pine Bluff	96.7	64.6	82.9
717	Camden	94.8	59.6	79.7
718	Texarkana	95.6	60.4	80.5
719	Hot Springs	94.2	61.0	79.9
720-722	Little Rock	95.4	65.9	82.8
723	West Memphis	94.3	65.5	81.9
724	Jonesboro	94.7	62.7	81.0
725	Batesville	92.6	60.0	78.7
726	Harrison	94.0	59.1	79.0
727	Fayetteville	91.5	60.7	78.3
728	Russellville	92.7	59.6	78.5
729	Fort Smith	95.1	63.6	81.6
CALIFORNIA				
900-902	Los Angeles	98.7	130.4	112.3
903-905	Inglewood	94.2	129.6	109.4
906-908	Long Beach	95.9	129.6	110.3
910-912	Pasadena	94.8	129.4	109.7
913-916	Van Nuys	97.9	129.4	111.4
917-918	Alhambra	96.8	129.4	110.8
919-921	San Diego	99.6	120.8	108.7
922	Palm Springs	96.9	127.1	109.8
923-924	San Bernardino	94.5	126.8	108.4
925	Riverside	98.8	127.1	110.9
926-927	Santa Ana	96.5	127.0	109.6
928	Anaheim	98.9	126.9	110.9
930	Oxnard	97.2	127.1	110.0
931	Santa Barbara	96.5	126.7	109.4
932-933	Bakersfield	98.0	125.3	109.7
934	San Luis Obispo	97.5	126.6	110.0
935	Mojave	94.7	125.2	107.7
936-938	Fresno	97.7	129.3	111.2
939	Salinas	98.3	135.7	114.3
940-941	San Francisco	106.3	158.1	128.5
942,956-958	Sacramento	100.1	132.7	114.1
943	Palo Alto	98.5	151.4	121.2
944	San Mateo	100.9	150.2	122.1
945	Vallejo	99.5	141.5	117.5
946	Oakland	102.7	151.2	123.5
947	Berkeley	102.3	151.2	123.3
948	Richmond	101.8	145.2	120.4
949	San Rafael	103.9	149.3	123.4
950	Santa Cruz	104.0	136.0	117.7
CALIFORNIA (CONT'D)				
951	San Jose	102.1	151.7	123.4
952	Stockton	100.3	130.8	113.4
953	Modesto	100.2	129.4	112.7
954	Santa Rosa	100.8	148.5	121.2
955	Eureka	102.1	133.2	115.4
959	Marysville	101.2	131.3	114.1
960	Redding	109.1	131.5	118.7
961	Susanville	108.8	131.2	118.4
COLORADO				
800-802	Denver	101.8	74.4	90.0
803	Boulder	97.7	75.9	88.3
804	Golden	99.8	73.7	88.6
805	Fort Collins	101.1	73.7	89.4
806	Greeley	98.3	74.7	88.2
807	Fort Morgan	98.3	72.9	87.4
808-809	Colorado Springs	100.0	73.0	88.4
810	Pueblo	100.5	71.4	88.0
811	Alamosa	102.3	68.8	87.9
812	Salida	102.0	66.6	86.8
813	Durango	102.7	65.0	86.6
814	Montrose	101.4	69.3	87.6
815	Grand Junction	104.5	72.2	90.7
816	Glenwood Springs	102.5	65.4	86.6
CONNECTICUT				
060	New Britain	96.4	118.1	105.7
061	Hartford	97.2	118.5	106.4
062	Willimantic	97.0	118.2	106.1
063	New London	93.6	118.4	104.2
064	Meriden	95.5	118.4	105.4
065	New Haven	98.1	118.5	106.9
066	Bridgeport	97.6	118.4	106.5
067	Waterbury	97.1	118.6	106.3
068	Norwalk	97.0	118.6	106.3
069	Stamford	97.2	125.2	109.2
D.C.				
200-205	Washington	101.0	87.7	95.3
DELAWARE				
197	Newark	98.3	112.3	104.3
198	Wilmington	98.2	112.3	104.2
199	Dover	98.2	112.3	104.2
FLORIDA				
320,322	Jacksonville	96.1	66.5	83.4
321	Daytona Beach	96.3	70.3	85.2
323	Tallahassee	97.3	65.7	83.7
324	Panama City	97.6	64.2	83.3
325	Pensacola	100.3	66.4	85.8
326,344	Gainesville	97.8	64.9	83.7
327-328,347	Orlando	97.9	67.1	84.7
329	Melbourne	99.2	71.2	87.2
330-332,340	Miami	95.8	67.4	83.7
333	Fort Lauderdale	95.3	69.5	84.2
334,349	West Palm Beach	94.3	66.0	82.2
335-336,346	Tampa	96.8	68.1	84.5
337	St. Petersburg	99.0	65.2	84.5
338	Lakeland	96.2	67.1	83.8
339,341	Fort Myers	95.6	68.3	83.9
342	Sarasota	98.7	67.0	85.1
GEORGIA				
300-303,399	Atlanta	99.0	75.6	89.0
304	Statesboro	98.9	65.8	84.7
305	Gainesville	97.4	66.5	84.2
306	Athens	96.8	67.9	84.4
307	Dalton	98.7	71.0	86.8
308-309	Augusta	97.1	72.7	86.6
310-312	Macon	93.9	74.0	85.4
313-314	Savannah	95.6	73.1	86.0
315	Waycross	95.4	68.4	83.8
316	Valdosta	95.3	66.2	82.8
317,398	Albany	95.2	72.6	85.5
318-319	Columbus	95.1	72.7	85.5

Location Factors - Commercial

STATE/ZIP	CITY	MAT.	INST.	TOTAL
HAWAII				
967	Hilo	116.3	118.3	117.1
968	Honolulu	120.9	118.3	119.8
STATES & POSS.				
969	Guam	138.8	52.6	101.9
IDAHO				
832	Pocatello	100.7	78.6	91.2
833	Twin Falls	101.9	77.2	91.3
834	Idaho Falls	99.2	78.2	90.2
835	Lewiston	106.9	85.7	97.8
836-837	Boise	99.7	79.4	91.0
838	Coeur d'Alene	106.8	84.4	97.2
ILLINOIS				
600-603	North Suburban	99.8	144.1	118.8
604	Joliet	99.7	142.2	117.9
605	South Suburban	99.8	144.0	118.7
606-608	Chicago	101.7	146.1	120.7
609	Kankakee	96.4	136.1	113.4
610-611	Rockford	96.8	128.7	110.5
612	Rock Island	94.9	102.1	98.0
613	La Salle	96.1	128.9	110.2
614	Galesburg	95.9	109.5	101.8
615-616	Peoria	97.8	111.5	103.7
617	Bloomington	95.2	112.2	102.5
618-619	Champaign	98.7	109.8	103.5
620-622	East St. Louis	94.4	111.3	101.6
623	Quincy	96.3	105.1	100.1
624	Effingham	95.7	109.2	101.5
625	Decatur	97.2	107.0	101.4
626-627	Springfield	97.9	107.7	102.1
628	Centralia	93.3	113.0	101.8
629	Carbondale	93.1	109.8	100.2
INDIANA				
460	Anderson	95.9	81.8	89.8
461-462	Indianapolis	98.7	83.0	92.0
463-464	Gary	97.2	108.3	102.0
465-466	South Bend	97.4	83.6	91.5
467-468	Fort Wayne	96.4	78.2	88.6
469	Kokomo	94.1	81.0	88.4
470	Lawrenceburg	92.6	78.5	86.5
471	New Albany	93.9	77.2	86.7
472	Columbus	96.1	80.4	89.3
473	Muncie	96.1	80.3	89.3
474	Bloomington	97.7	80.3	90.3
475	Washington	94.6	85.4	90.7
476-477	Evansville	95.4	84.8	90.9
478	Terre Haute	96.2	83.8	90.9
479	Lafayette	95.8	80.1	89.1
IOWA				
500-503,509	Des Moines	96.0	88.4	92.8
504	Mason City	94.7	76.2	86.8
505	Fort Dodge	94.9	74.1	86.0
506-507	Waterloo	96.2	79.3	89.0
508	Creston	95.2	83.1	90.0
510-511	Sioux City	97.1	78.7	89.2
512	Sibley	96.1	63.3	82.1
513	Spencer	97.7	63.6	83.1
514	Carroll	94.9	81.3	89.1
515	Council Bluffs	98.2	80.8	90.7
516	Shenandoah	95.4	83.8	90.4
520	Dubuque	96.5	81.9	90.3
521	Decorah	95.9	75.9	87.3
522-524	Cedar Rapids	97.4	85.8	92.4
525	Ottumwa	95.7	76.3	87.4
526	Burlington	95.1	86.3	91.3
527-528	Davenport	96.5	96.9	96.7
KANSAS				
660-662	Kansas City	96.0	98.6	97.1
664-666	Topeka	97.0	77.7	88.7
667	Fort Scott	94.8	78.2	87.7
668	Emporia	94.9	77.2	87.3
669	Belleville	96.6	72.1	86.1
670-672	Wichita	96.7	70.8	85.6
673	Independence	97.0	77.9	88.8
674	Salina	96.9	72.8	86.6
675	Hutchinson	92.4	73.1	84.1
676	Hays	96.3	73.5	86.5
677	Colby	97.1	75.7	87.9

STATE/ZIP	CITY	MAT.	INST.	TOTAL
KANSAS (CONT'D)				
678	Dodge City	98.3	75.4	88.5
679	Liberal	96.2	73.7	86.5
KENTUCKY				
400-402	Louisville	93.2	77.7	86.6
403-405	Lexington	93.4	79.6	87.5
406	Frankfort	95.5	78.7	88.3
407-409	Corbin	91.1	79.6	86.2
410	Covington	94.4	79.8	88.1
411-412	Ashland	93.2	92.0	92.7
413-414	Campton	94.4	79.9	88.2
415-416	Pikeville	95.8	87.6	92.3
417-418	Hazard	93.8	80.6	88.1
420	Paducah	92.5	83.8	88.8
421-422	Bowling Green	94.5	78.9	87.8
423	Owensboro	94.6	83.1	89.7
424	Henderson	92.2	82.4	88.0
425-426	Somerset	91.7	77.8	85.7
427	Elizabethtown	91.3	76.7	85.1
LOUISIANA				
700-701	New Orleans	99.4	69.5	86.6
703	Thibodaux	96.2	66.4	83.4
704	Hammond	93.8	64.5	81.2
705	Lafayette	95.6	68.9	84.1
706	Lake Charles	95.8	69.8	84.6
707-708	Baton Rouge	96.6	69.7	85.1
710-711	Shreveport	97.8	66.4	84.3
712	Monroe	96.6	65.1	83.1
713-714	Alexandria	96.7	66.6	83.8
MAINE				
039	Kittery	91.3	84.6	88.5
040-041	Portland	96.9	84.1	91.4
042	Lewiston	94.7	84.1	90.1
043	Augusta	96.7	83.1	90.9
044	Bangor	94.3	83.5	89.7
045	Bath	93.2	82.6	88.6
046	Machias	92.6	79.7	87.1
047	Houlton	92.8	79.7	87.2
048	Rockland	91.9	83.1	88.1
049	Waterville	93.1	83.1	88.8
MARYLAND				
206	Waldorf	97.7	87.4	93.3
207-208	College Park	97.7	88.6	93.8
209	Silver Spring	96.9	87.4	92.8
210-212	Baltimore	101.2	85.0	94.2
214	Annapolis	100.5	83.1	93.0
215	Cumberland	97.1	85.5	92.1
216	Easton	98.8	76.0	89.0
217	Hagerstown	97.5	87.0	93.0
218	Salisbury	99.3	68.5	86.1
219	Elkton	96.0	87.6	92.4
MASSACHUSETTS				
010-011	Springfield	96.4	110.4	102.4
012	Pittsfield	96.0	105.1	99.9
013	Greenfield	94.3	110.9	101.4
014	Fitchburg	93.0	118.7	104.0
015-016	Worcester	96.4	118.5	105.9
017	Framingham	92.4	127.0	107.2
018	Lowell	95.8	125.4	108.5
019	Lawrence	96.7	125.8	109.2
020-022, 024	Boston	99.4	133.3	113.9
023	Brockton	96.1	119.2	106.0
025	Buzzards Bay	91.1	117.7	102.5
026	Hyannis	93.4	117.5	103.7
027	New Bedford	95.3	117.6	104.9
MICHIGAN				
480,483	Royal Oak	95.2	100.6	97.5
481	Ann Arbor	97.2	101.7	99.1
482	Detroit	100.2	103.0	101.4
484-485	Flint	96.9	90.8	94.3
486	Saginaw	96.5	87.3	92.6
487	Bay City	96.6	87.5	92.7
488-489	Lansing	97.9	88.7	93.9
490	Battle Creek	95.3	82.5	89.8
491	Kalamazoo	95.6	80.7	89.2
492	Jackson	93.8	93.0	93.5
493,495	Grand Rapids	97.1	81.6	90.5
494	Muskegon	94.1	80.3	88.2

STATE/ZIP	CITY	MAT.	INST.	TOTAL
MICHIGAN (CONT'D)				
496	Traverse City	93.2	78.3	86.8
497	Gaylord	94.3	81.3	88.8
498-499	Iron Mountain	96.2	83.0	90.6
MINNESOTA				
550-551	Saint Paul	97.2	115.8	105.2
553-555	Minneapolis	99.7	116.5	106.9
556-558	Duluth	97.1	102.7	99.5
559	Rochester	97.1	100.3	98.4
560	Mankato	95.1	98.0	96.4
561	Windom	93.7	92.3	93.1
562	Willmar	93.3	99.9	96.1
563	St. Cloud	94.4	114.0	102.8
564	Brainerd	95.0	98.6	96.5
565	Detroit Lakes	96.7	92.2	94.8
566	Bemidji	96.0	95.4	95.8
567	Thief River Falls	95.6	89.4	92.9
MISSISSIPPI				
386	Clarksdale	95.2	54.4	77.7
387	Greenville	98.6	66.0	84.6
388	Tupelo	96.5	58.2	80.1
389	Greenwood	96.5	54.2	78.3
390-392	Jackson	98.0	65.4	84.0
393	Meridian	95.6	65.1	82.5
394	Laurel	97.2	56.9	79.9
395	Biloxi	97.3	64.6	83.3
396	McComb	95.4	54.5	77.9
397	Columbus	97.0	58.4	80.5
MISSOURI				
630-631	St. Louis	99.6	105.8	102.3
633	Bowling Green	97.5	96.0	96.9
634	Hannibal	96.5	94.0	95.4
635	Kirksville	100.2	88.8	95.3
636	Flat River	98.5	96.2	97.5
637	Cape Girardeau	98.1	91.8	95.4
638	Sikeston	96.9	89.4	93.7
639	Poplar Bluff	96.3	89.2	93.3
640-641	Kansas City	98.1	103.6	100.5
644-645	St. Joseph	96.8	92.8	95.1
646	Chillicothe	94.5	95.4	94.9
647	Harrisonville	94.0	102.6	97.7
648	Joplin	95.9	79.7	88.9
650-651	Jefferson City	96.3	90.8	94.0
652	Columbia	96.2	91.3	94.1
653	Sedalia	96.6	93.7	95.4
654-655	Rolla	94.3	96.6	95.3
656-658	Springfield	97.7	81.0	90.5
MONTANA				
590-591	Billings	100.8	75.0	89.7
592	Wolf Point	100.5	77.8	90.8
593	Miles City	98.4	77.9	89.6
594	Great Falls	102.1	74.0	90.1
595	Havre	99.5	75.7	89.3
596	Helena	100.1	76.1	89.8
597	Butte	100.7	76.2	90.2
598	Missoula	97.9	75.7	88.4
599	Kalispell	97.4	76.2	88.3
NEBRASKA				
680-681	Omaha	96.9	80.8	90.0
683-685	Lincoln	97.1	81.1	90.2
686	Columbus	95.6	82.7	90.1
687	Norfolk	96.9	79.2	89.3
688	Grand Island	97.0	78.5	89.1
689	Hastings	96.7	80.0	89.5
690	McCook	96.6	74.8	87.3
691	North Platte	96.6	77.3	88.3
692	Valentine	98.8	70.8	86.8
693	Alliance	98.9	74.7	88.5
NEVADA				
889-891	Las Vegas	102.7	104.1	103.3
893	Ely	101.6	95.6	99.0
894-895	Reno	101.5	84.2	94.1
897	Carson City	100.4	84.2	93.5
898	Elko	100.3	87.4	94.8
NEW HAMPSHIRE				
030	Nashua	95.9	92.8	94.6
031	Manchester	96.2	93.0	94.8

STATE/ZIP	CITY	MAT.	INST.	TOTAL
NEW HAMPSHIRE (CONT'D)				
032-033	Concord	96.1	92.4	94.5
034	Keene	93.0	89.2	91.3
035	Littleton	93.0	80.1	87.4
036	Charleston	92.5	88.6	90.8
037	Claremont	91.7	88.6	90.3
038	Portsmouth	93.3	92.2	92.8
NEW JERSEY				
070-071	Newark	99.6	139.7	116.8
072	Elizabeth	97.3	139.7	115.4
073	Jersey City	96.3	139.4	114.8
074-075	Paterson	97.7	139.5	115.6
076	Hackensack	95.9	139.6	114.6
077	Long Branch	95.6	133.6	111.9
078	Dover	96.1	139.6	114.7
079	Summit	96.2	139.7	114.8
080,083	Vineland	95.8	132.3	111.4
081	Camden	97.5	130.7	111.8
082,084	Atlantic City	96.4	132.6	111.9
085-086	Trenton	99.2	131.4	113.0
087	Point Pleasant	97.7	133.3	113.0
088-089	New Brunswick	98.3	137.7	115.2
NEW MEXICO				
870-872	Albuquerque	97.1	75.0	87.6
873	Gallup	97.3	75.0	87.7
874	Farmington	97.7	75.0	88.0
875	Santa Fe	97.6	75.0	87.9
877	Las Vegas	95.8	75.0	86.9
878	Socorro	95.5	75.0	86.7
879	Truth/Consequences	95.2	71.7	85.2
880	Las Cruces	95.6	71.7	85.4
881	Clovis	98.0	74.9	88.1
882	Roswell	99.5	75.0	89.0
883	Carrizozo	100.2	75.0	89.4
884	Tucumcari	98.7	74.9	88.5
NEW YORK				
100-102	New York	99.0	176.2	132.1
103	Staten Island	94.9	177.8	130.5
104	Bronx	93.3	176.8	129.1
105	Mount Vernon	93.5	151.3	118.3
106	White Plains	93.3	153.5	119.1
107	Yonkers	97.3	151.2	120.5
108	New Rochelle	93.9	146.0	116.2
109	Suffern	93.6	133.5	110.7
110	Queens	99.8	178.9	133.8
111	Long Island City	101.5	178.9	134.7
112	Brooklyn	101.8	178.9	134.9
113	Flushing	102.0	178.9	135.0
114	Jamaica	100.2	178.9	134.0
115,117,118	Hicksville	99.8	160.5	125.8
116	Far Rockaway	102.2	178.9	135.1
119	Riverhead	100.5	156.8	124.6
120-122	Albany	95.0	111.5	102.1
123	Schenectady	95.5	110.6	101.9
124	Kingston	98.9	136.0	114.8
125-126	Poughkeepsie	98.1	140.2	116.1
127	Monticello	97.4	136.9	114.3
128	Glens Falls	90.6	108.7	98.4
129	Plattsburgh	95.6	99.0	97.0
130-132	Syracuse	97.8	100.8	99.1
133-135	Utica	95.9	98.9	97.2
136	Watertown	97.7	99.2	98.3
137-139	Binghamton	97.3	102.8	99.7
140-142	Buffalo	103.0	110.0	106.0
143	Niagara Falls	98.8	110.2	103.7
144-146	Rochester	100.3	101.3	100.8
147	Jamestown	97.8	97.8	97.8
148-149	Elmira	97.6	103.9	100.3
NORTH CAROLINA				
270,272-274	Greensboro	97.7	67.7	84.8
271	Winston-Salem	97.4	67.8	84.7
275-276	Raleigh	96.4	67.1	83.8
277	Durham	99.3	67.7	85.7
278	Rocky Mount	95.3	67.5	83.4
279	Elizabeth City	96.1	69.2	84.6
280	Gastonia	97.0	67.7	84.4
281-282	Charlotte	97.0	67.3	84.3
283	Fayetteville	100.0	67.3	86.0
284	Wilmington	95.9	66.3	83.2
285	Kinston	94.3	67.1	82.7

Location Factors - Commercial

STATE/ZIP	CITY	MAT.	INST.	TOTAL
NORTH CAROLINA (CONT'D)				
286	Hickory	94.7	68.3	83.4
287-288	Asheville	96.3	67.0	83.8
289	Murphy	95.5	66.2	82.9
NORTH DAKOTA				
580-581	Fargo	98.2	80.3	90.5
582	Grand Forks	98.4	79.0	90.1
583	Devils Lake	98.3	80.2	90.5
584	Jamestown	98.3	79.6	90.3
585	Bismarck	98.2	80.0	90.4
586	Dickinson	99.0	79.1	90.5
587	Minot	98.3	79.6	90.3
588	Williston	97.4	79.0	89.6
OHIO				
430-432	Columbus	97.4	83.2	91.3
433	Marion	93.4	88.5	91.3
434-436	Toledo	96.7	92.8	95.0
437-438	Zanesville	94.0	87.1	91.0
439	Steubenville	95.3	92.1	94.0
440	Lorain	98.4	86.8	93.4
441	Cleveland	98.7	92.2	95.9
442-443	Akron	99.5	88.3	94.7
444-445	Youngstown	98.7	84.0	92.4
446-447	Canton	98.9	81.4	91.4
448-449	Mansfield	96.5	86.4	92.1
450	Hamilton	95.9	81.5	89.7
451-452	Cincinnati	96.9	80.7	89.9
453-454	Dayton	95.9	81.3	89.6
455	Springfield	95.9	81.8	89.8
456	Chillicothe	95.4	90.8	93.4
457	Athens	98.2	86.4	93.2
458	Lima	98.3	83.8	92.1
OKLAHOMA				
730-731	Oklahoma City	95.2	68.2	83.6
734	Ardmore	93.8	66.2	82.0
735	Lawton	95.9	66.6	83.3
736	Clinton	95.0	66.1	82.6
737	Enid	95.5	66.5	83.1
738	Woodward	93.9	63.4	80.8
739	Guymon	94.9	64.8	82.0
740-741	Tulsa	96.0	67.1	83.6
743	Miami	92.8	66.5	81.5
744	Muskogee	95.1	64.4	81.9
745	McAlester	92.4	61.9	79.3
746	Ponca City	93.2	65.2	81.2
747	Durant	93.2	66.3	81.7
748	Shawnee	94.6	66.1	82.4
749	Poteau	92.4	66.1	81.1
OREGON				
970-972	Portland	100.2	100.4	100.3
973	Salem	101.7	99.9	100.9
974	Eugene	99.8	98.5	99.3
975	Medford	101.3	97.8	99.8
976	Klamath Falls	101.5	97.8	99.9
977	Bend	100.7	99.8	100.3
978	Pendleton	96.8	101.7	98.9
979	Vale	94.5	87.9	91.7
PENNSYLVANIA				
150-152	Pittsburgh	99.9	102.2	100.9
153	Washington	96.9	102.7	99.4
154	Uniontown	97.2	102.2	99.4
155	Bedford	98.2	94.3	96.5
156	Greensburg	98.2	98.7	98.4
157	Indiana	97.1	100.1	98.4
158	Dubois	98.7	98.3	98.5
159	Johnstown	98.2	94.2	96.5
160	Butler	91.3	103.1	96.4
161	New Castle	91.4	100.3	95.2
162	Kittanning	91.8	100.4	95.5
163	Oil City	91.3	100.3	95.1
164-165	Erie	93.1	96.2	94.5
166	Altoona	93.2	95.1	94.0
167	Bradford	94.9	99.5	96.8
168	State College	94.5	97.0	95.6
169	Wellsboro	95.5	95.2	95.4
170-171	Harrisburg	97.8	94.9	96.5
172	Chambersburg	94.9	91.6	93.4
173-174	York	94.9	95.3	95.1
175-176	Lancaster	93.5	96.8	94.9

STATE/ZIP	CITY	MAT.	INST.	TOTAL
PENNSYLVANIA (CONT'D)				
177	Williamsport	92.1	94.6	93.2
178	Sunbury	94.3	95.0	94.6
179	Pottsville	93.4	98.4	95.6
180	Lehigh Valley	95.0	113.6	103.0
181	Allentown	96.7	108.4	101.7
182	Hazleton	94.4	98.3	96.1
183	Stroudsburg	94.3	107.3	99.9
184-185	Scranton	97.4	98.9	98.1
186-187	Wilkes-Barre	94.1	97.9	95.7
188	Montrose	93.8	99.2	96.1
189	Doylestown	94.0	129.2	109.1
190-191	Philadelphia	100.1	135.3	115.2
193	Westchester	95.8	129.9	110.4
194	Norristown	94.8	130.0	109.9
195-196	Reading	96.5	104.0	99.7
PUERTO RICO				
009	San Juan	117.8	27.2	79.0
RHODE ISLAND				
028	Newport	94.6	113.7	102.8
029	Providence	96.2	113.7	103.7
SOUTH CAROLINA				
290-292	Columbia	96.5	67.9	84.2
293	Spartanburg	96.4	67.9	84.2
294	Charleston	98.0	67.6	84.9
295	Florence	96.2	67.8	84.0
296	Greenville	96.2	67.9	84.1
297	Rock Hill	96.0	67.3	83.7
298	Aiken	96.9	67.9	84.4
299	Beaufort	97.5	56.0	79.7
SOUTH DAKOTA				
570-571	Sioux Falls	96.3	79.2	89.0
572	Watertown	97.1	70.8	85.8
573	Mitchell	96.0	62.4	81.6
574	Aberdeen	98.3	71.5	86.8
575	Pierre	98.7	69.4	86.2
576	Mobridge	96.6	64.4	82.8
577	Rapid City	98.0	73.3	87.4
TENNESSEE				
370-372	Nashville	98.4	71.1	86.7
373-374	Chattanooga	98.9	69.6	86.3
375,380-381	Memphis	97.3	71.1	86.1
376	Johnson City	99.1	59.9	82.3
377-379	Knoxville	95.6	65.2	82.6
382	McKenzie	96.2	56.0	79.0
383	Jackson	97.5	59.7	81.3
384	Columbia	94.7	67.9	83.2
385	Cookeville	96.1	57.3	79.5
TEXAS				
750	McKinney	98.0	65.1	83.9
751	Waxahachie	98.1	65.1	83.9
752-753	Dallas	99.1	67.7	85.6
754	Greenville	98.2	64.5	83.8
755	Texarkana	97.5	63.6	83.0
756	Longview	98.4	62.6	83.0
757	Tyler	98.6	63.2	83.5
758	Palestine	95.1	62.0	80.9
759	Lufkin	95.5	64.2	82.1
760-761	Fort Worth	97.4	64.1	83.1
762	Denton	97.5	64.4	83.3
763	Wichita Falls	94.9	62.7	81.1
764	Eastland	94.0	61.8	80.2
765	Temple	92.5	60.0	78.5
766-767	Waco	94.3	63.9	81.3
768	Brownwood	97.4	60.2	81.4
769	San Angelo	97.1	60.8	81.6
770-772	Houston	100.7	67.9	86.6
773	Huntsville	99.4	64.3	84.4
774	Wharton	100.8	65.8	85.8
775	Galveston	98.5	68.1	85.5
776-777	Beaumont	98.8	67.1	85.2
778	Bryan	95.5	66.4	83.1
779	Victoria	100.5	64.4	85.0
780	Laredo	97.1	62.7	82.4
781-782	San Antonio	99.3	63.5	83.9
783-784	Corpus Christi	99.7	63.0	84.0
785	McAllen	100.6	57.9	82.3
786-787	Austin	97.4	62.7	82.5

STATE/ZIP	CITY	MAT.	INST.	TOTAL
TEXAS (CONT'D)				
788	Del Rio	100.6	62.0	84.0
789	Giddings	97.0	61.9	82.0
790-791	Amarillo	95.7	61.1	80.9
792	Childress	95.8	61.8	81.2
793-794	Lubbock	97.5	64.4	83.3
795-796	Abilene	96.0	62.8	81.8
797	Midland	98.1	63.7	83.3
798-799,885	El Paso	94.6	64.7	81.8
UTAH				
840-841	Salt Lake City	102.0	73.1	89.6
842,844	Ogden	97.5	73.1	87.0
843	Logan	99.4	73.1	88.1
845	Price	100.0	71.7	87.9
846-847	Provo	99.8	72.9	88.3
VERMONT				
050	White River Jct.	95.8	81.0	89.4
051	Bellows Falls	94.2	93.6	94.0
052	Bennington	94.6	90.0	92.6
053	Brattleboro	95.0	93.6	94.4
054	Burlington	99.1	80.6	91.2
056	Montpelier	97.7	85.3	92.4
057	Rutland	96.3	80.5	89.5
058	St. Johnsbury	95.8	80.5	89.3
059	Guildhall	94.5	80.2	88.3
VIRGINIA				
220-221	Fairfax	99.2	84.2	92.8
222	Arlington	100.1	84.3	93.3
223	Alexandria	99.3	84.2	92.8
224-225	Fredericksburg	97.9	80.6	90.5
226	Winchester	98.6	81.3	91.2
227	Culpeper	98.4	84.9	92.6
228	Harrisonburg	98.7	66.0	84.7
229	Charlottesville	99.1	68.0	85.8
230-232	Richmond	98.3	76.4	88.9
233-235	Norfolk	98.8	70.0	86.5
236	Newport News	97.9	69.7	85.8
237	Portsmouth	97.4	63.1	82.7
238	Petersburg	97.9	72.4	87.0
239	Farmville	97.1	62.8	82.4
240-241	Roanoke	99.7	71.6	87.7
242	Bristol	97.9	59.4	81.4
243	Pulaski	97.5	68.3	85.0
244	Staunton	98.3	65.6	84.3
245	Lynchburg	98.4	73.7	87.8
246	Grundy	97.8	59.2	81.2
WASHINGTON				
980-981,987	Seattle	106.0	107.4	106.6
982	Everett	103.4	102.6	103.1
983-984	Tacoma	103.8	102.1	103.1
985	Olympia	100.9	102.3	101.5
986	Vancouver	104.8	96.2	101.1
988	Wenatchee	104.1	91.0	98.5
989	Yakima	104.0	96.6	100.8
990-992	Spokane	99.9	84.5	93.3
993	Richland	99.6	92.2	96.4
994	Clarkston	98.9	83.9	92.5
WEST VIRGINIA				
247-248	Bluefield	96.7	90.2	93.9
249	Lewisburg	98.4	90.5	95.0
250-253	Charleston	96.1	92.5	94.6
254	Martinsburg	96.4	86.1	92.0
255-257	Huntington	97.5	94.3	96.1
258-259	Beckley	94.9	92.1	93.7
260	Wheeling	99.3	91.5	96.0
261	Parkersburg	98.2	90.7	95.0
262	Buckhannon	98.1	90.7	94.9
263-264	Clarksburg	98.7	91.4	95.5
265	Morgantown	98.5	91.5	95.5
266	Gassaway	98.0	90.5	94.8
267	Romney	97.9	89.2	94.2
268	Petersburg	97.7	87.7	93.4
WISCONSIN				
530,532	Milwaukee	99.6	105.9	102.3
531	Kenosha	99.4	104.6	101.6
534	Racine	98.8	105.0	101.5
535	Beloit	98.7	101.4	99.9
537	Madison	99.1	100.3	99.6

STATE/ZIP	CITY	MAT.	INST.	TOTAL
WISCONSIN (CONT'D)				
538	Lancaster	96.9	96.7	96.8
539	Portage	95.3	100.5	97.5
540	New Richmond	94.7	97.0	95.7
541-543	Green Bay	98.7	95.8	97.5
544	Wausau	94.1	95.8	94.8
545	Rhinelander	97.3	94.6	96.1
546	La Crosse	95.2	95.9	95.5
547	Eau Claire	96.8	96.5	96.7
548	Superior	94.5	98.7	96.3
549	Oshkosh	94.8	94.2	94.5
WYOMING				
820	Cheyenne	99.6	72.1	87.8
821	Yellowstone Nat'l Park	97.6	73.2	87.1
822	Wheatland	98.8	68.1	85.6
823	Rawlins	100.3	72.7	88.5
824	Worland	98.3	71.8	87.0
825	Riverton	99.4	69.7	86.6
826	Casper	99.3	69.5	86.5
827	Newcastle	98.1	72.3	87.1
828	Sheridan	100.9	70.6	87.9
829-831	Rock Springs	101.9	71.6	88.9
CANADIAN FACTORS (reflect Canadian currency)				
ALBERTA				
	Calgary	119.0	97.6	109.8
	Edmonton	119.7	97.5	110.2
	Fort McMurray	122.1	90.2	108.4
	Lethbridge	117.2	89.7	105.4
	Lloydminster	111.6	86.5	100.8
	Medicine Hat	111.7	85.8	100.6
	Red Deer	112.2	85.8	100.8
BRITISH COLUMBIA				
	Kamloops	113.0	85.7	101.3
	Prince George	113.9	85.0	101.5
	Vancouver	115.4	91.6	105.2
	Victoria	113.7	84.2	101.0
MANITOBA				
	Brandon	120.9	72.4	100.1
	Portage la Prairie	111.6	70.9	94.1
	Winnipeg	121.1	71.2	99.7
NEW BRUNSWICK				
	Bathurst	110.4	63.9	90.5
	Dalhousie	109.8	64.1	90.2
	Fredericton	117.9	71.2	97.9
	Moncton	110.6	68.1	92.4
	Newcastle	110.5	64.5	90.8
	Saint John	113.0	75.6	96.9
NEWFOUNDLAND				
	Corner Brook	125.0	70.2	101.5
	St. John's	120.8	88.8	107.0
NORTHWEST TERRITORIES				
	Yellowknife	130.4	85.0	110.9
NOVA SCOTIA				
	Bridgewater	111.2	72.2	94.5
	Dartmouth	121.8	72.1	100.5
	Halifax	117.1	85.8	103.7
	New Glasgow	119.9	72.1	99.4
	Sydney	118.1	72.1	98.4
	Truro	110.8	72.2	94.2
	Yarmouth	119.8	72.1	99.4
ONTARIO				
	Barrie	116.0	89.3	104.6
	Brantford	113.1	93.1	104.5
	Cornwall	113.1	89.6	103.0
	Hamilton	116.1	98.1	108.4
	Kingston	114.1	89.7	103.6
	Kitchener	109.4	93.4	102.5
	London	116.8	95.6	107.7
	North Bay	122.8	87.4	107.7
	Oshawa	111.6	92.9	103.6
	Ottawa	116.9	96.4	108.1
	Owen Sound	116.1	87.6	103.9
	Peterborough	113.1	89.4	102.9
	Sarnia	113.2	93.7	104.8

Location Factors - Commercial

STATE/ZIP	CITY	MAT.	INST.	TOTAL
ONTARIO (CONT'D)				
	Sault Ste. Marie	108.6	90.0	100.6
	St. Catharines	107.3	94.3	101.7
	Sudbury	106.9	93.1	101.0
	Thunder Bay	108.5	93.7	102.1
	Timmins	113.2	87.5	102.2
	Toronto	115.8	102.4	110.1
	Windsor	107.8	93.5	101.7
PRINCE EDWARD ISLAND				
	Charlottetown	120.1	61.8	95.1
	Summerside	122.3	59.4	95.3
QUEBEC				
	Cap-de-la-Madeleine	110.7	82.1	98.4
	Charlesbourg	110.7	82.1	98.4
	Chicoutimi	110.3	87.3	100.4
	Gatineau	110.4	81.9	98.2
	Granby	110.6	81.8	98.3
	Hull	110.5	81.9	98.2
	Joliette	110.9	82.1	98.5
	Laval	110.7	81.9	98.3
	Montreal	118.1	89.9	106.0
	Quebec City	116.4	90.3	105.2
	Rimouski	110.3	87.3	100.4
	Rouyn-Noranda	110.4	81.9	98.1
	Saint-Hyacinthe	109.8	81.9	97.8
	Sherbrooke	110.6	81.9	98.3
	Sorel	110.9	82.1	98.5
	Saint-Jerome	110.4	81.9	98.2
	Trois-Rivieres	120.7	82.0	104.1
SASKATCHEWAN				
	Moose Jaw	109.0	65.6	90.4
	Prince Albert	108.2	63.9	89.2
	Regina	122.1	94.2	110.1
	Saskatoon	109.6	90.8	101.5
YUKON				
	Whitehorse	131.9	70.5	105.6

R011105-05 Tips for Accurate Estimating

1. Use pre-printed or columnar forms for orderly sequence of dimensions and locations and for recording telephone quotations.
2. Use only the front side of each paper or form except for certain pre-printed summary forms.
3. Be consistent in listing dimensions: For example, length x width x height. This helps in rechecking to ensure that, the total length of partitions is appropriate for the building area.
4. Use printed (rather than measured) dimensions where given.
5. Add up multiple printed dimensions for a single entry where possible.
6. Measure all other dimensions carefully.
7. Use each set of dimensions to calculate multiple related quantities.
8. Convert foot and inch measurements to decimal feet when listing. Memorize decimal equivalents to .01 parts of a foot (1/8″ equals approximately .01′).
9. Do not "round off" quantities until the final summary.
10. Mark drawings with different colors as items are taken off.
11. Keep similar items together, different items separate.
12. Identify location and drawing numbers to aid in future checking for completeness.
13. Measure or list everything on the drawings or mentioned in the specifications.
14. It may be necessary to list items not called for to make the job complete.
15. Be alert for: Notes on plans such as N.T.S. (not to scale); changes in scale throughout the drawings; reduced size drawings; discrepancies between the specifications and the drawings.
16. Develop a consistent pattern of performing an estimate. For example:
    a. Start the quantity takeoff at the lower floor and move to the next higher floor.
    b. Proceed from the main section of the building to the wings.
    c. Proceed from south to north or vice versa, clockwise or counterclockwise.
    d. Take off floor plan quantities first, elevations next, then detail drawings.
17. List all gross dimensions that can be either used again for different quantities, or used as a rough check of other quantities for verification (exterior perimeter, gross floor area, individual floor areas, etc.).
18. Utilize design symmetry or repetition (repetitive floors, repetitive wings, symmetrical design around a center line, similar room layouts, etc.). Note: Extreme caution is needed here so as not to omit or duplicate an area.
19. Do not convert units until the final total is obtained. For instance, when estimating concrete work, keep all units to the nearest cubic foot, then summarize and convert to cubic yards.
20. When figuring alternatives, it is best to total all items involved in the basic system, then total all items involved in the alternates. Therefore you work with positive numbers in all cases. When adds and deducts are used, it is often confusing whether to add or subtract a portion of an item; especially on a complicated or involved alternate.

R011105-50 Metric Conversion Factors

Description: This table is primarily for converting customary U.S. units in the left hand column to SI metric units in the right hand column. In addition, conversion factors for some commonly encountered Canadian and non-SI metric units are included.

	If You Know		Multiply By		To Find
Length	Inches	x	25.4[a]	=	Millimeters
	Feet	x	0.3048[a]	=	Meters
	Yards	x	0.9144[a]	=	Meters
	Miles (statute)	x	1.609	=	Kilometers
Area	Square inches	x	645.2	=	Square millimeters
	Square feet	x	0.0929	=	Square meters
	Square yards	x	0.8361	=	Square meters
Volume (Capacity)	Cubic inches	x	16,387	=	Cubic millimeters
	Cubic feet	x	0.02832	=	Cubic meters
	Cubic yards	x	0.7646	=	Cubic meters
	Gallons (U.S. liquids)[b]	x	0.003785	=	Cubic meters[c]
	Gallons (Canadian liquid)[b]	x	0.004546	=	Cubic meters[c]
	Ounces (U.S. liquid)[b]	x	29.57	=	Milliliters[c, d]
	Quarts (U.S. liquid)[b]	x	0.9464	=	Liters[c, d]
	Gallons (U.S. liquid)[b]	x	3.785	=	Liters[c, d]
Force	Kilograms force[d]	x	9.807	=	Newtons
	Pounds force	x	4.448	=	Newtons
	Pounds force	x	0.4536	=	Kilograms force[d]
	Kips	x	4448	=	Newtons
	Kips	x	453.6	=	Kilograms force[d]
Pressure, Stress, Strength (Force per unit area)	Kilograms force per square centimeter[d]	x	0.09807	=	Megapascals
	Pounds force per square inch (psi)	x	0.006895	=	Megapascals
	Kips per square inch	x	6.895	=	Megapascals
	Pounds force per square inch (psi)	x	0.07031	=	Kilograms force per square centimeter[d]
	Pounds force per square foot	x	47.88	=	Pascals
	Pounds force per square foot	x	4.882	=	Kilograms force per square meter[d]
Flow	Cubic feet per minute	x	0.4719	=	Liters per second
	Gallons per minute	x	0.0631	=	Liters per second
	Gallons per hour	x	1.05	=	Milliliters per second
Bending Moment Or Torque	Inch-pounds force	x	0.01152	=	Meter-kilograms force[d]
	Inch-pounds force	x	0.1130	=	Newton-meters
	Foot-pounds force	x	0.1383	=	Meter-kilograms force[d]
	Foot-pounds force	x	1.356	=	Newton-meters
	Meter-kilograms force[d]	x	9.807	=	Newton-meters
Mass	Ounces (avoirdupois)	x	28.35	=	Grams
	Pounds (avoirdupois)	x	0.4536	=	Kilograms
	Tons (metric)	x	1000	=	Kilograms
	Tons, short (2000 pounds)	x	907.2	=	Kilograms
	Tons, short (2000 pounds)	x	0.9072	=	Megagrams[e]
Mass per Unit Volume	Pounds mass per cubic foot	x	16.02	=	Kilograms per cubic meter
	Pounds mass per cubic yard	x	0.5933	=	Kilograms per cubic meter
	Pounds mass per gallon (U.S. liquid)[b]	x	119.8	=	Kilograms per cubic meter
	Pounds mass per gallon (Canadian liquid)[b]	x	99.78	=	Kilograms per cubic meter
Temperature	Degrees Fahrenheit	(F-32)/1.8		=	Degrees Celsius
	Degrees Fahrenheit	(F+459.67)/1.8		=	Degrees Kelvin
	Degrees Celsius	C+273.15		=	Degrees Kelvin

[a]The factor given is exact
[b]One U.S. gallon = 0.8327 Canadian gallon
[c]1 liter = 1000 milliliters = 1000 cubic centimeters
1 cubic decimeter = 0.001 cubic meter
[d]Metric but not SI unit
[e]Called "tonne" in England and "metric ton" in other metric countries

R011105-60 Weights and Measures

Measures of Length
1 Mile = 1760 Yards = 5280 Feet
1 Yard = 3 Feet = 36 inches
1 Foot = 12 Inches
1 Mil = 0.001 Inch
1 Fathom = 2 Yards = 6 Feet
1 Rod = 5.5 Yards = 16.5 Feet
1 Hand = 4 Inches
1 Span = 9 Inches
1 Micro-inch = One Millionth Inch or 0.000001 Inch
1 Micron = One Millionth Meter + 0.00003937 Inch

Surveyor's Measure
1 Mile = 8 Furlongs = 80 Chains
1 Furlong = 10 Chains = 220 Yards
1 Chain = 4 Rods = 22 Yards = 66 Feet = 100 Links
1 Link = 7.92 Inches

Square Measure
1 Square Mile = 640 Acres = 6400 Square Chains
1 Acre = 10 Square Chains = 4840 Square Yards = 43,560 Sq. Ft.
1 Square Chain = 16 Square Rods = 484 Square Yards = 4356 Sq. Ft.
1 Square Rod = 30.25 Square Yards = 272.25 Square Feet = 625 Square Lines
1 Square Yard = 9 Square Feet
1 Square Foot = 144 Square Inches
An Acre equals a Square 208.7 Feet per Side

Cubic Measure
1 Cubic Yard = 27 Cubic Feet
1 Cubic Foot = 1728 Cubic Inches
1 Cord of Wood = 4 x 4 x 8 Feet = 128 Cubic Feet
1 Perch of Masonry = 16½ x 1½ x 1 Foot = 24.75 Cubic Feet

Avoirdupois or Commercial Weight
1 Gross or Long Ton = 2240 Pounds
1 Net or Short Ton = 2000 Pounds
1 Pound = 16 Ounces = 7000 Grains
1 Ounce = 16 Drachms = 437.5 Grains
1 Stone = 14 Pounds

Power
1 British Thermal Unit per Hour = 0.2931 Watts
1 Ton (Refrigeration) = 3.517 Kilowatts
1 Horsepower (Boiler) = 9.81 Kilowatts
1 Horsepower (550 ft-lb/s) = 0.746 Kilowatts

Shipping Measure
For Measuring Internal Capacity of a Vessel:
1 Register Ton = 100 Cubic Feet

For Measurement of Cargo:
Approximately 40 Cubic Feet of Merchandise is considered a Shipping Ton, unless that bulk would weigh more than 2000 Pounds, in which case Freight Charge may be based upon weight.

40 Cubic Feet = 32.143 U.S. Bushels = 31.16 Imp. Bushels

Liquid Measure
1 Imperial Gallon = 1.2009 U.S. Gallon = 277.42 Cu. In.
1 Cubic Foot = 7.48 U.S. Gallons

R011110-10 Architectural Fees

Tabulated below are typical percentage fees by project size, for good professional architectural service. Fees may vary from those listed depending upon degree of design difficulty and economic conditions in any particular area.

Rates can be interpolated horizontally and vertically. Various portions of the same project requiring different rates should be adjusted proportionately. For alterations, add 50% to the fee for the first $500,000 of project cost and add 25% to the fee for project cost over $500,000.

Architectural fees tabulated below include Structural, Mechanical and Electrical Engineering Fees. They do not include the fees for special consultants such as kitchen planning, security, acoustical, interior design, etc.

Civil Engineering fees are included in the Architectural fee for project sites requiring minimal design such as city sites. However, separate Civil Engineering fees must be added when utility connections require design, drainage calculations are needed, stepped foundations are required, or provisions are required to protect adjacent wetlands.

	Total Project Size in Thousands of Dollars						
Building Types	100	250	500	1,000	5,000	10,000	50,000
Factories, garages, warehouses, repetitive housing	9.0%	8.0%	7.0%	6.2%	5.3%	4.9%	4.5%
Apartments, banks, schools, libraries, offices, municipal buildings	12.2	12.3	9.2	8.0	7.0	6.6	6.2
Churches, hospitals, homes, laboratories, museums, research	15.0	13.6	12.7	11.9	9.5	8.8	8.0
Memorials, monumental work, decorative furnishings	—	16.0	14.5	13.1	10.0	9.0	8.3

R011110-30 Engineering Fees

Typical **Structural Engineering Fees** based on type of construction and total project size. These fees are included in Architectural Fees.

Type of Construction	Total Project Size (in thousands of dollars)			
	$500	$500-$1,000	$1,000-$5,000	Over $5000
Industrial buildings, factories & warehouses	Technical payroll times 2.0 to 2.5	1.60%	1.25%	1.00%
Hotels, apartments, offices, dormitories, hospitals, public buildings, food stores	↓	2.00%	1.70%	1.20%
Museums, banks, churches and cathedrals	↓	2.00%	1.75%	1.25%
Thin shells, prestressed concrete, earthquake resistive	↓	2.00%	1.75%	1.50%
Parking ramps, auditoriums, stadiums, convention halls, hangars & boiler houses	↓	2.50%	2.00%	1.75%
Special buildings, major alterations, underpinning & future expansion	↓	Add to above 0.5%	Add to above 0.5%	Add to above 0.5%

For complex reinforced concrete or unusually complicated structures, add 20% to 50%.

Typical **Mechanical and Electrical Engineering Fees** are based on the size of the subcontract. The fee structure for both is shown below. These fees are included in Architectural Fees.

Type of Construction	Subcontract Size							
	$25,000	$50,000	$100,000	$225,000	$350,000	$500,000	$750,000	$1,000,000
Simple structures	6.4%	5.7%	4.8%	4.5%	4.4%	4.3%	4.2%	4.1%
Intermediate structures	8.0	7.3	6.5	5.6	5.1	5.0	4.9	4.8
Complex structures	10.1	9.0	9.0	8.0	7.5	7.5	7.0	7.0

For renovations, add 15% to 25% to applicable fee.

R012153-10 Repair and Remodeling

Cost figures are based on new construction utilizing the most cost-effective combination of labor, equipment and material with the work scheduled in proper sequence to allow the various trades to accomplish their work in an efficient manner.

The costs for repair and remodeling work must be modified due to the following factors that may be present in any given repair and remodeling project.

1. Equipment usage curtailment due to the physical limitations of the project, with only hand-operated equipment being used.
2. Increased requirement for shoring and bracing to hold up the building while structural changes are being made and to allow for temporary storage of construction materials on above-grade floors.
3. Material handling becomes more costly due to having to move within the confines of an enclosed building. For multi-story construction, low capacity elevators and stairwells may be the only access to the upper floors.
4. Large amount of cutting and patching and attempting to match the existing construction is required. It is often more economical to remove entire walls rather than create many new door and window openings. This sort of trade-off has to be carefully analyzed.
5. Cost of protection of completed work is increased since the usual sequence of construction usually cannot be accomplished.
6. Economies of scale usually associated with new construction may not be present. If small quantities of components must be custom fabricated due to job requirements, unit costs will naturally increase. Also, if only small work areas are available at a given time, job scheduling between trades becomes difficult and subcontractor quotations may reflect the excessive start-up and shut-down phases of the job.
7. Work may have to be done on other than normal shifts and may have to be done around an existing production facility which has to stay in production during the course of the repair and remodeling.
8. Dust and noise protection of adjoining non-construction areas can involve substantial special protection and alter usual construction methods.
9. Job may be delayed due to unexpected conditions discovered during demolition or removal. These delays ultimately increase construction costs.
10. Piping and ductwork runs may not be as simple as for new construction. Wiring may have to be snaked through walls and floors.
11. Matching "existing construction" may be impossible because materials may no longer be manufactured. Substitutions may be expensive.
12. Weather protection of existing structure requires additional temporary structures to protect building at openings.
13. On small projects, because of local conditions, it may be necessary to pay a tradesman for a minimum of four hours for a task that is completed in one hour.

All of the above areas can contribute to increased costs for a repair and remodeling project. Each of the above factors should be considered in the planning, bidding and construction stage in order to minimize the increased costs associated with repair and remodeling jobs.

R012153-60 Security Factors

Contractors entering, working in, and exiting secure facilities often lose productive time during a normal workday. The recommended allowances in this section are intended to provide for the loss of productivity by increasing labor costs. Note that different costs are associated with searches upon entry only and searches upon entry and exit. Time spent in a queue is unpredictable and not part of these allowances. Contractors should plan ahead for this situation.

Security checkpoints are designed to reflect the level of security required to gain access or egress. An extreme example is when contractors, along with any materials, tools, equipment, and vehicles, must be physically searched and have all materials, tools, equipment, and vehicles inventoried and documented prior to both entry and exit.

Physical searches without going through the documentation process represent the next level and take up less time.

Electronic searches—passing through a detector or x-ray machine with no documentation of materials, tools, equipment, and vehicles—take less time than physical searches.

Visual searches of materials, tools, equipment, and vehicles represent the next level of security.

Finally, access by means of an ID card or displayed sticker takes the least amount of time.

Another consideration is if the searches described above are performed each and every day, or if they are performed only on the first day with access granted by ID card or displayed sticker for the remainder of the project. The figures for this situation have been calculated to represent the initial check-in as described and subsequent entry by ID card or displayed sticker for up to 20 days on site. For the situation described above, where the time period is beyond 20 days, the impact on labor cost is negligible.

There are situations where tradespeople must be accompanied by an escort and observed during the work day. The loss of freedom of movement will slow down productivity for the tradesperson. Costs for the observer have not been included. Those costs are normally born by the owner.

General Requirements | R0129 Payment Procedures

R012909-80 Sales Tax by State

State sales tax on materials is tabulated below (5 states have no sales tax). Many states allow local jurisdictions, such as a county or city, to levy additional sales tax.

Some projects may be sales tax exempt, particularly those constructed with public funds.

State	Tax (%)	State	Tax (%)	State	Tax (%)	State	Tax (%)
Alabama	4	Illinois	6.25	Montana	0	Rhode Island	7
Alaska	0	Indiana	7	Nebraska	5.5	South Carolina	6
Arizona	5.6	Iowa	6	Nevada	6.85	South Dakota	4.5
Arkansas	6.5	Kansas	6.5	New Hampshire	0	Tennessee	7
California	7.25	Kentucky	6	New Jersey	6.625	Texas	6.25
Colorado	2.9	Louisiana	4.45	New Mexico	5.125	Utah	5.95
Connecticut	6.35	Maine	5.5	New York	4	Vermont	6
Delaware	0	Maryland	6	North Carolina	4.75	Virginia	5.3
District of Columbia	5.75	Massachusetts	6.25	North Dakota	5	Washington	6.5
Florida	6	Michigan	6	Ohio	5.75	West Virginia	6
Georgia	4	Minnesota	6.875	Oklahoma	4.5	Wisconsin	5
Hawaii	4	Mississippi	7	Oregon	0	Wyoming	4
Idaho	6	Missouri	4.225	Pennsylvania	6	Average	5.10 %

Sales Tax by Province (Canada)

GST - a value-added tax, which the government imposes on most goods and services provided in or imported into Canada. PST - a retail sales tax, which five of the provinces impose on the prices of most goods and some services. QST - a value-added tax, similar to the federal GST, which Quebec imposes. HST - Three provinces have combined their retail sales taxes with the federal GST into one harmonized tax.

Province	PST (%)	QST (%)	GST(%)	HST(%)
Alberta	0	0	5	0
British Columbia	7	0	5	0
Manitoba	8	0	5	0
New Brunswick	0	0	0	15
Newfoundland	0	0	0	15
Northwest Territories	0	0	5	0
Nova Scotia	0	0	0	15
Ontario	0	0	0	13
Prince Edward Island	0	0	0	15
Quebec	0	9.975	5	0
Saskatchewan	6	0	5	0
Yukon	0	0	5	0

R012909-85 Unemployment Taxes and Social Security Taxes

State unemployment tax rates vary not only from state to state, but also with the experience rating of the contractor. The federal unemployment tax rate is 6.0% of the first $7,000 of wages. This is reduced by a credit of up to 5.4% for timely payment to the state. The minimum federal unemployment tax is 0.6% after all credits.

Social security (FICA) for 2019 is estimated at time of publication to be 7.65% of wages up to $128,400.

R012909-86 Unemployment Tax by State

Information is from the U.S. Department of Labor, state unemployment tax rates.

State	Tax (%)	State	Tax (%)	State	Tax (%)	State	Tax (%)
Alabama	6.74	Illinois	7.75	Montana	6.12	Rhode Island	9.79
Alaska	5.4	Indiana	7.474	Nebraska	5.4	South Carolina	5.46
Arizona	8.91	Iowa	8	Nevada	5.4	South Dakota	9.5
Arkansas	6.0	Kansas	7.6	New Hampshire	7.5	Tennessee	10.0
California	6.2	Kentucky	10.0	New Jersey	5.8	Texas	7.5
Colorado	8.9	Louisiana	6.2	New Mexico	5.4	Utah	7.2
Connecticut	6.8	Maine	5.4	New York	8.5	Vermont	8.4
Delaware	8.0	Maryland	7.50	North Carolina	5.76	Virginia	6.27
District of Columbia	7	Massachusetts	11.13	North Dakota	10.72	Washington	5.7
Florida	5.4	Michigan	10.3	Ohio	8.7	West Virginia	7.5
Georgia	5.4	Minnesota	9.0	Oklahoma	5.5	Wisconsin	12.0
Hawaii	5.6	Mississippi	5.4	Oregon	5.4	Wyoming	8.8
Idaho	5.4	Missouri	9.75	Pennsylvania	10.89	Median	7.47 %

R012909-90 Overtime

One way to improve the completion date of a project or eliminate negative float from a schedule is to compress activity duration times. This can be achieved by increasing the crew size or working overtime with the proposed crew.

To determine the costs of working overtime to compress activity duration times, consider the following examples. Below is an overtime efficiency and cost chart based on a five, six, or seven day week with an eight through twelve hour day. Payroll percentage increases for time and one half and double times are shown for the various working days.

Days per Week	Hours per Day	Production Efficiency					Payroll Cost Factors	
		1st Week	2nd Week	3rd Week	4th Week	Average 4 Weeks	@ 1-1/2 Times	@ 2 Times
5	8	100%	100%	100%	100%	100%	1.000	1.000
	9	100	100	95	90	96	1.056	1.111
	10	100	95	90	85	93	1.100	1.200
	11	95	90	75	65	81	1.136	1.273
	12	90	85	70	60	76	1.167	1.333
6	8	100	100	95	90	96	1.083	1.167
	9	100	95	90	85	93	1.130	1.259
	10	95	90	85	80	88	1.167	1.333
	11	95	85	70	65	79	1.197	1.394
	12	90	80	65	60	74	1.222	1.444
7	8	100	95	85	75	89	1.143	1.286
	9	95	90	80	70	84	1.183	1.365
	10	90	85	75	65	79	1.214	1.429
	11	85	80	65	60	73	1.240	1.481
	12	85	75	60	55	69	1.262	1.524

R013113-40 Builder's Risk Insurance

Builder's risk insurance is insurance on a building during construction. Premiums are paid by the owner or the contractor. Blasting, collapse and underground insurance would raise total insurance costs.

R013113-50 General Contractor's Overhead

There are two distinct types of overhead on a construction project: Project overhead and main office overhead. Project overhead includes those costs at a construction site not directly associated with the installation of construction materials. Examples of project overhead costs include the following:

1. Superintendent
2. Construction office and storage trailers
3. Temporary sanitary facilities
4. Temporary utilities
5. Security fencing
6. Photographs
7. Cleanup
8. Performance and payment bonds

The above project overhead items are also referred to as general requirements and therefore are estimated in Division 1. Division 1 is the first division listed in the CSI MasterFormat but it is usually the last division estimated. The sum of the costs in Divisions 1 through 49 is referred to as the sum of the direct costs.

All construction projects also include indirect costs. The primary components of indirect costs are the contractor's main office overhead and profit. The amount of the main office overhead expense varies depending on the following:

1. Owner's compensation
2. Project managers' and estimators' wages
3. Clerical support wages
4. Office rent and utilities
5. Corporate legal and accounting costs
6. Advertising
7. Automobile expenses
8. Association dues
9. Travel and entertainment expenses

These costs are usually calculated as a percentage of annual sales volume. This percentage can range from 35% for a small contractor doing less than $500,000 to 5% for a large contractor with sales in excess of $100 million.

R013113-55 Installing Contractor's Overhead

Installing contractors (subcontractors) also incur costs for general requirements and main office overhead.

Included within the total incl. overhead and profit costs is a percent mark-up for overhead that includes:

1. Compensation and benefits for office staff and project managers
2. Office rent, utilities, business equipment, and maintenance
3. Corporate legal and accounting costs
4. Advertising
5. Vehicle expenses (for office staff and project managers)
6. Association dues
7. Travel, entertainment
8. Insurance
9. Small tools and equipment

R013113-60 Workers' Compensation Insurance Rates by Trade

The table below tabulates the national averages for workers' compensation insurance rates by trade and type of building. The average "Insurance Rate" is multiplied by the "% of Building Cost" for each trade. This produces the "Workers' Compensation" cost by % of total labor cost, to be added for each trade by building type to determine the weighted average workers' compensation rate for the building types analyzed.

Trade	Insurance Rate (% Labor Cost)		% of Building Cost			Workers' Compensation		
	Range	Average	Office Bldgs.	Schools & Apts.	Mfg.	Office Bldgs.	Schools & Apts.	Mfg.
Excavation, Grading, etc.	2.3 % to 19.2%	8.5%	4.8%	4.9%	4.5%	0.41%	0.42%	0.38%
Piles & Foundations	3.9 to 27.3	13.4	7.1	5.2	8.7	0.95	0.70	1.17
Concrete	3.5 to 25.6	11.8	5.0	14.8	3.7	0.59	1.75	0.44
Masonry	3.6 to 50.4	13.8	6.9	7.5	1.9	0.95	1.04	0.26
Structural Steel	4.9 to 43.8	21.2	10.7	3.9	17.6	2.27	0.83	3.73
Miscellaneous & Ornamental Metals	2.8 to 27.0	10.6	2.8	4.0	3.6	0.30	0.42	0.38
Carpentry & Millwork	3.7 to 31.9	13.0	3.7	4.0	0.5	0.48	0.52	0.07
Metal or Composition Siding	5.1 to 113.7	19.0	2.3	0.3	4.3	0.44	0.06	0.82
Roofing	5.7 to 103.7	29.0	2.3	2.6	3.1	0.67	0.75	0.90
Doors & Hardware	3.1 to 31.9	11.0	0.9	1.4	0.4	0.10	0.15	0.04
Sash & Glazing	3.9 to 24.3	12.1	3.5	4.0	1.0	0.42	0.48	0.12
Lath & Plaster	2.7 to 34.4	10.7	3.3	6.9	0.8	0.35	0.74	0.09
Tile, Marble & Floors	2.2 to 21.5	8.7	2.6	3.0	0.5	0.23	0.26	0.04
Acoustical Ceilings	1.9 to 29.7	8.5	2.4	0.2	0.3	0.20	0.02	0.03
Painting	3.7 to 36.7	11.2	1.5	1.6	1.6	0.17	0.18	0.18
Interior Partitions	3.7 to 31.9	13.0	3.9	4.3	4.4	0.51	0.56	0.57
Miscellaneous Items	2.1 to 97.7	11.2	5.2	3.7	9.7	0.58	0.42	1.09
Elevators	1.4 to 11.4	4.7	2.1	1.1	2.2	0.10	0.05	0.10
Sprinklers	2.0 to 16.4	6.7	0.5	—	2.0	0.03	—	0.13
Plumbing	1.5 to 14.6	6.3	4.9	7.2	5.2	0.31	0.45	0.33
Heat., Vent., Air Conditioning	3.0 to 17.0	8.3	13.5	11.0	12.9	1.12	0.91	1.07
Electrical	1.9 to 11.3	5.2	10.1	8.4	11.1	0.53	0.44	0.58
Total	1.4 % to 113.7%	—	100.0%	100.0%	100.0%	11.71%	11.15%	12.52%
		Overall Weighted Average 11.79%						

Workers' Compensation Insurance Rates by States

The table below lists the weighted average Workers' Compensation base rate for each state with a factor comparing this with the national average of 11.8%.

State	Weighted Average	Factor	State	Weighted Average	Factor	State	Weighted Average	Factor
Alabama	13.9%	129	Kentucky	11.3%	105	North Dakota	7.4%	69
Alaska	11.0	102	Louisiana	20.0	185	Ohio	6.2	57
Arizona	9.1	84	Maine	9.8	91	Oklahoma	8.7	81
Arkansas	5.7	53	Maryland	11.2	104	Oregon	9.0	83
California	23.6	219	Massachusetts	9.1	84	Pennsylvania	24.9	231
Colorado	8.1	75	Michigan	8.0	74	Rhode Island	10.6	98
Connecticut	17.0	157	Minnesota	16.2	150	South Carolina	20.3	188
Delaware	10.7	99	Mississippi	11.4	106	South Dakota	10.3	95
District of Columbia	9.1	84	Missouri	14.0	130	Tennessee	7.9	73
Florida	11.3	105	Montana	7.8	72	Texas	6.2	57
Georgia	32.9	305	Nebraska	13.4	124	Utah	7.4	69
Hawaii	9.2	85	Nevada	8.9	82	Vermont	10.8	100
Idaho	9.2	85	New Hampshire	11.3	105	Virginia	7.4	69
Illinois	20.1	186	New Jersey	15.4	143	Washington	8.9	82
Indiana	3.7	34	New Mexico	13.7	127	West Virginia	4.7	44
Iowa	12.1	112	New York	19.1	177	Wisconsin	13.3	123
Kansas	6.6	61	North Carolina	17.1	158	Wyoming	6.3	58
			Weighted Average for U.S. is	11.8% of payroll = 100%				

The weighted average skilled worker rate for 35 trades is 11.8%. For bidding purposes, apply the full value of Workers' Compensation directly to total labor costs, or if labor is 38%, materials 42% and overhead and profit 20% of total cost, carry 38/80 x 11.8% = 6.0% of cost (before overhead and profit) into overhead. Rates vary not only from state to state but also with the experience rating of the contractor.

Rates are the most current available at the time of publication.

R013113-80 Performance Bond

This table shows the cost of a Performance Bond for a construction job scheduled to be completed in 12 months. Add 1% of the premium cost per month for jobs requiring more than 12 months to complete. The rates are "standard" rates offered to contractors that the bonding company considers financially sound and capable of doing the work. Preferred rates are offered by some bonding companies based upon financial strength of the contractor. Actual rates vary from contractor to contractor and from bonding company to bonding company. Contractors should prequalify through a bonding agency before submitting a bid on a contract that requires a bond.

Contract Amount	Building Construction Class B Projects	Highways & Bridges	
		Class A New Construction	Class A-1 Highway Resurfacing
First $ 100,000 bid	$25.00 per M	$15.00 per M	$9.40 per M
Next 400,000 bid	$ 2,500 plus $15.00 per M	$ 1,500 plus $10.00 per M	$ 940 plus $7.20 per M
Next 2,000,000 bid	8,500 plus 10.00 per M	5,500 plus 7.00 per M	3,820 plus 5.00 per M
Next 2,500,000 bid	28,500 plus 7.50 per M	19,500 plus 5.50 per M	15,820 plus 4.50 per M
Next 2,500,000 bid	47,250 plus 7.00 per M	33,250 plus 5.00 per M	28,320 plus 4.50 per M
Over 7,500,000 bid	64,750 plus 6.00 per M	45,750 plus 4.50 per M	39,570 plus 4.00 per M

General Requirements R0154 Construction Aids

R015423-10 Steel Tubular Scaffolding

On new construction, tubular scaffolding is efficient up to 60′ high or five stories. Above this it is usually better to use a hung scaffolding if construction permits. Swing scaffolding operations may interfere with tenants. In this case, the tubular is more practical at all heights.

In repairing or cleaning the front of an existing building the cost of tubular scaffolding per S.F. of building front increases as the height increases above the first tier. The first tier cost is relatively high due to leveling and alignment.

The minimum efficient crew for erecting and dismantling is three workers. They can set up and remove 18 frame sections per day up to 5 stories high. For 6 to 12 stories high, a crew of four is most efficient. Use two or more on top and two on the bottom for handing up or hoisting. They can also set up and remove 18 frame sections per day. At 7' horizontal spacing, this will run about 800 S.F. per day of erecting and dismantling. Time for placing and removing planks must be added to the above. A crew of three can place and remove 72 planks per day up to 5 stories. For over 5 stories, a crew of four can place and remove 80 planks per day.

The table below shows the number of pieces required to erect tubular steel scaffolding for 1000 S.F. of building frontage. This area is made up of a scaffolding system that is 12 frames (11 bays) long by 2 frames high.

For jobs under twenty-five frames, add 50% to rental cost. Rental rates will be lower for jobs over three months duration. Large quantities for long periods can reduce rental rates by 20%.

Description of Component	Number of Pieces for 1000 S.F. of Building Front	Unit
5′ Wide Standard Frame, 6′-4″ High	24	Ea.
Leveling Jack & Plate	24	
Cross Brace	44	
Side Arm Bracket, 21″	12	
Guardrail Post	12	
Guardrail, 7′ section	22	
Stairway Section	2	
Stairway Starter Bar	1	
Stairway Inside Handrail	2	
Stairway Outside Handrail	2	
Walk-Thru Frame Guardrail	2	▼

Scaffolding is often used as falsework over 15′ high during construction of cast-in-place concrete beams and slabs. Two foot wide scaffolding is generally used for heavy beam construction. The span between frames depends upon the load to be carried with a maximum span of 5′.

Heavy duty shoring frames with a capacity of 10,000#/leg can be spaced up to 10′ O.C. depending upon form support design and loading.

Scaffolding used as horizontal shoring requires less than half the material required with conventional shoring.

On new construction, erection is done by carpenters.

Rolling towers supporting horizontal shores can reduce labor and speed the job. For maintenance work, catwalks with spans up to 70′ can be supported by the rolling towers.

R015423-20 Pump Staging

Pump staging is generally not available for rent. The table below shows the number of pieces required to erect pump staging for 2400 S.F. of building frontage. This area is made up of a pump jack system that is 3 poles (2 bays) wide by 2 poles high.

Item	Number of Pieces for 2400 S.F. of Building Front	Unit
Aluminum pole section, 24′ long	6	Ea.
Aluminum splice joint, 6′ long	3	↓
Aluminum foldable brace	3	↓
Aluminum pump jack	3	↓
Aluminum support for workbench/back safety rail	3	↓
Aluminum scaffold plank/workbench, 14″ wide x 24′ long	4	↓
Safety net, 22′ long	2	↓
Aluminum plank end safety rail	2	↓

The cost in place for this 2400 S.F. will depend on how many uses are realized during the life of the equipment.

R015433-10 Contractor Equipment

Rental Rates shown elsewhere in the data set pertain to late model high quality machines in excellent working condition, rented from equipment dealers. Rental rates from contractors may be substantially lower than the rental rates from equipment dealers depending upon economic conditions; for older, less productive machines, reduce rates by a maximum of 15%. Any overtime must be added to the base rates. For shift work, rates are lower. Usual rule of thumb is 150% of one shift rate for two shifts; 200% for three shifts.

For periods of less than one week, operated equipment is usually more economical to rent than renting bare equipment and hiring an operator.

Costs to move equipment to a job site (mobilization) or from a job site (demobilization) are not included in rental rates, nor in any Equipment costs on any Unit Price line items or crew listings. These costs can be found elsewhere. If a piece of equipment is already at a job site, it is not appropriate to utilize mob/demob costs in an estimate again.

Rental rates vary throughout the country with larger cities generally having lower rates. Lease plans for new equipment are available for periods in excess of six months with a percentage of payments applying toward purchase.

Rental rates can also be treated as reimbursement costs for contractor-owned equipment. Owned equipment costs include depreciation, loan payments, interest, taxes, insurance, storage, and major repairs.

Monthly rental rates vary from 2% to 5% of the cost of the equipment depending on the anticipated life of the equipment and its wearing parts. Weekly rates are about 1/3 the monthly rates and daily rental rates are about 1/3 the weekly rates.

The hourly operating costs for each piece of equipment include costs to the user such as fuel, oil, lubrication, normal expendables for the equipment, and a percentage of the mechanic's wages chargeable to maintenance. The hourly operating costs listed do not include the operator's wages.

The daily cost for equipment used in the standard crews is figured by dividing the weekly rate by five, then adding eight times the hourly operating cost to give the total daily equipment cost, not including the operator. This figure is in the right hand column of the Equipment listings under Equipment Cost/Day.

Pile Driving rates shown for the pile hammer and extractor do not include leads, cranes, boilers or compressors. Vibratory pile driving requires an added field specialist during set-up and pile driving operation for the electric model. The hydraulic model requires a field specialist for set-up only. Up to 125 reuses of sheet piling are possible using vibratory drivers. For normal conditions, crane capacity for hammer type and size is as follows.

Crane Capacity	Hammer Type and Size		
	Air or Steam	Diesel	Vibratory
25 ton	to 8,750 ft.-lb.		70 H.P.
40 ton	15,000 ft.-lb.	to 32,000 ft.-lb.	170 H.P.
60 ton	25,000 ft.-lb.		300 H.P.
100 ton		112,000 ft.-lb.	

Cranes should be specified for the job by size, building and site characteristics, availability, performance characteristics, and duration of time required.

Backhoes & Shovels rent for about the same as equivalent size cranes but maintenance and operating expenses are higher. The crane operator's rate must be adjusted for high boom heights. Average adjustments: for 150′ boom add 2% per hour; over 185′, add 4% per hour; over 210′, add 6% per hour; over 250′, add 8% per hour and over 295′, add 12% per hour.

Tower Cranes of the climbing or static type have jibs from 50′ to 200′ and capacities at maximum reach range from 4,000 to 14,000 pounds. Lifting capacities increase up to maximum load as the hook radius decreases.

Typical rental rates, based on purchase price, are about 2% to 3% per month.

Erection and dismantling run between 500 and 2000 labor hours. Climbing operation takes 10 labor hours per 20′ climb. Crane dead time is about 5 hours per 40′ climb. If crane is bolted to side of the building add cost of ties and extra mast sections. Climbing cranes have from 80′ to 180′ of mast while static cranes have 80′ to 800′ of mast.

Truck Cranes can be converted to tower cranes by using tower attachments. Mast heights over 400′ have been used.

A single 100′ high material **Hoist and Tower** can be erected and dismantled in about 400 labor hours; a double 100′ high hoist and tower in about 600 labor hours. Erection times for additional heights are 3 and 4 labor hours per vertical foot respectively up to 150′, and 4 to 5 labor hours per vertical foot over 150′ high. A 40′ high portable Buck hoist takes about 160 labor hours to erect and dismantle. Additional heights take 2 labor hours per vertical foot to 80′ and 3 labor hours per vertical foot for the next 100′. Most material hoists do not meet local code requirements for carrying personnel.

A 150′ high **Personnel Hoist** requires about 500 to 800 labor hours to erect and dismantle. Budget erection time at 5 labor hours per vertical foot for all trades. Local code requirements or labor scarcity requiring overtime can add up to 50% to any of the above erection costs.

Earthmoving Equipment: The selection of earthmoving equipment depends upon the type and quantity of material, moisture content, haul distance, haul road, time available, and equipment available. Short haul cut and fill operations may require dozers only, while another operation may require excavators, a fleet of trucks, and spreading and compaction equipment. Stockpiled material and granular material are easily excavated with front end loaders. Scrapers are most economically used with hauls between 300′ and 1-1/2 miles if adequate haul roads can be maintained. Shovels are often used for blasted rock and any material where a vertical face of 8′ or more can be excavated. Special conditions may dictate the use of draglines, clamshells, or backhoes. Spreading and compaction equipment must be matched to the soil characteristics, the compaction required and the rate the fill is being supplied.

R015433-15 Heavy Lifting

Hydraulic Climbing Jacks

The use of hydraulic heavy lift systems is an alternative to conventional type crane equipment. The lifting, lowering, pushing, or pulling mechanism is a hydraulic climbing jack moving on a square steel jackrod from 1-5/8″ to 4″ square, or a steel cable. The jackrod or cable can be vertical or horizontal, stationary or movable, depending on the individual application. When the jackrod is stationary, the climbing jack will climb the rod and push or pull the load along with itself. When the climbing jack is stationary, the jackrod is movable with the load attached to the end and the climbing jack will lift or lower the jackrod with the attached load. The heavy lift system is normally operated by a single control lever located at the hydraulic pump.

The system is flexible in that one or more climbing jacks can be applied wherever a load support point is required, and the rate of lift synchronized.

Economic benefits have been demonstrated on projects such as: erection of ground assembled roofs and floors, complete bridge spans, girders and trusses, towers, chimney liners and steel vessels, storage tanks, and heavy machinery. Other uses are raising and lowering offshore work platforms, caissons, tunnel sections and pipelines.

R015436-50 Mobilization

Costs to move rented construction equipment to a job site from an equipment dealer's or contractor's yard (mobilization) or off the job site (demobilization) are not included in the rental or operating rates, nor in the equipment cost on a unit price line or in a crew listing. These costs can be found consolidated in the Mobilization section of the data and elsewhere in particular site work sections. If a piece of equipment is already on the job site, it is not appropriate to include mob/demob costs in a new estimate that requires use of that equipment. The following table identifies approximate sizes of rented construction equipment that would be hauled on a towed trailer. Because this listing is not all-encompassing, the user can infer as to what size trailer might be required for a piece of equipment not listed.

3-ton Trailer	20-ton Trailer	40-ton Trailer	50-ton Trailer
20 H.P. Excavator	110 H.P. Excavator	200 H.P. Excavator	270 H.P. Excavator
50 H.P. Skid Steer	165 H.P. Dozer	300 H.P. Dozer	Small Crawler Crane
35 H.P. Roller	150 H.P. Roller	400 H.P. Scraper	500 H.P. Scraper
40 H.P. Trencher	Backhoe	450 H.P. Art. Dump Truck	500 H.P. Art. Dump Truck

R024119-10 Demolition Defined

Whole Building Demolition - Demolition of the whole building with no concern for any particular building element, component, or material type being demolished. This type of demolition is accomplished with large pieces of construction equipment that break up the structure, load it into trucks and haul it to a disposal site, but disposal or dump fees are not included. Demolition of below-grade foundation elements, such as footings, foundation walls, grade beams, slabs on grade, etc., is not included. Certain mechanical equipment containing flammable liquids or ozone-depleting refrigerants, electric lighting elements, communication equipment components, and other building elements may contain hazardous waste, and must be removed, either selectively or carefully, as hazardous waste before the building can be demolished.

Foundation Demolition - Demolition of below-grade foundation footings, foundation walls, grade beams, and slabs on grade. This type of demolition is accomplished by hand or pneumatic hand tools, and does not include saw cutting, or handling, loading, hauling, or disposal of the debris.

Gutting - Removal of building interior finishes and electrical/mechanical systems down to the load-bearing and sub-floor elements of the rough building frame, with no concern for any particular building element, component, or material type being demolished. This type of demolition is accomplished by hand or pneumatic hand tools, and includes loading into trucks, but not hauling, disposal or dump fees, scaffolding, or shoring. Certain mechanical equipment containing flammable liquids or ozone-depleting refrigerants, electric lighting elements, communication equipment components, and other building elements may contain hazardous waste, and must be removed, either selectively or carefully, as hazardous waste, before the building is gutted.

Selective Demolition - Demolition of a selected building element, component, or finish, with some concern for surrounding or adjacent elements, components, or finishes (see the first Subdivision (s) at the beginning of appropriate Divisions). This type of demolition is accomplished by hand or pneumatic hand tools, and does not include handling, loading, storing, hauling, or disposal of the debris, scaffolding, or shoring. "Gutting" methods may be used in order to save time, but damage that is caused to surrounding or adjacent elements, components, or finishes may have to be repaired at a later time.

Careful Removal - Removal of a piece of service equipment, building element or component, or material type, with great concern for both the removed item and surrounding or adjacent elements, components or finishes. The purpose of careful removal may be to protect the removed item for later re-use, preserve a higher salvage value of the removed item, or replace an item while taking care to protect surrounding or adjacent elements, components, connections, or finishes from cosmetic and/or structural damage. An approximation of the time required to perform this type of removal is 1/3 to 1/2 the time it would take to install a new item of like kind. This type of removal is accomplished by hand or pneumatic hand tools, and does not include loading, hauling, or storing the removed item, scaffolding, shoring, or lifting equipment.

Cutout Demolition - Demolition of a small quantity of floor, wall, roof, or other assembly, with concern for the appearance and structural integrity of the surrounding materials. This type of demolition is accomplished by hand or pneumatic hand tools, and does not include saw cutting, handling, loading, hauling, or disposal of debris, scaffolding, or shoring.

Rubbish Handling - Work activities that involve handling, loading or hauling of debris. Generally, the cost of rubbish handling must be added to the cost of all types of demolition, with the exception of whole building demolition.

Minor Site Demolition - Demolition of site elements outside the footprint of a building. This type of demolition is accomplished by hand or pneumatic hand tools, or with larger pieces of construction equipment, and may include loading a removed item onto a truck (check the Crew for equipment used). It does not include saw cutting, hauling or disposal of debris, and, sometimes, handling or loading.

R024119-20 Dumpsters

Dumpster rental costs on construction sites are presented in two ways.

The cost per week rental includes the delivery of the dumpster; its pulling or emptying once per week, and its final removal. The assumption is made that the dumpster contractor could choose to empty a dumpster by simply bringing in an empty unit and removing the full one. These costs also include the disposal of the materials in the dumpster.

The Alternate Pricing can be used when actual planned conditions are not approximated by the weekly numbers. For example, these lines can be used when a dumpster is needed for 4 weeks and will need to be emptied 2 or 3 times per week. Conversely the Alternate Pricing lines can be used when a dumpster will be rented for several weeks or months but needs to be emptied only a few times over this period.

R024119-30 Rubbish Handling Chutes

To correctly estimate the cost of rubbish handling chute systems, the individual components must be priced separately. First choose the size of the system; a 30-inch diameter chute is quite common, but the sizes range from 18 to 36 inches in diameter. The 30-inch chute comes in a standard weight and two thinner weights. The thinner weight chutes are sometimes chosen for cost savings, but they are more easily damaged.

There are several types of major chute pieces that make up the chute system. The first component to consider is the top chute section (top intake hopper) where the material is dropped into the chute at the highest point. After determining the top chute, the intermediate chute pieces called the regular chute sections are priced. Next, the number of chute control door sections (intermediate intake hoppers) must be determined. In the more complex systems, a chute control door section is provided at each floor level. The last major component to consider is bolt down frames; these are usually provided at every other floor level.

There are a number of accessories to consider for safe operation and control. There are covers for the top chute and the chute door sections. The top chute can have a trough that allows for better loading of the chute. For the safest operation, a chute warning light system can be added that will warn the other chute intake locations not to load while another is being used. There are dust control devices that spray a water mist to keep down the dust as the debris is loaded into a dumpster. There are special breakaway cords that are used to prevent damage to the chute if the Dumpster is removed without disconnecting from the chute. There are chute liners that can be installed to protect the chute structure from physical damage from rough abrasive materials. Warning signs can be posted at each floor level that is provided with a chute control door section.

In summary, a complete rubbish handling chute system will include one top section, several intermediate regular sections, several intermediate control door (intake hopper) sections and bolt down frames at every other floor level starting with the top floor. If so desired, the system can also include covers and a light warning system for a safer operation. The bottom of the chute should always be above the Dumpster and should be tied off with a breakaway cord to the Dumpster.

R028213-20 Asbestos Removal Process

Asbestos removal is accomplished by a specialty contractor who understands the federal and state regulations regarding the handling and disposal of the material. The process of asbestos removal is divided into many individual steps. An accurate estimate can be calculated only after all the steps have been priced.

The steps are generally as follows:

1. Obtain an asbestos abatement plan from an industrial hygienist.
2. Monitor the air quality in and around the removal area and along the path of travel between the removal area and transport area. This establishes the background contamination.
3. Construct a two part decontamination chamber at entrance to removal area.
4. Install a HEPA filter to create a negative pressure in the removal area.
5. Install wall, floor and ceiling protection as required by the plan, usually 2 layers of fireproof 6 mil polyethylene.
6. Industrial hygienist visually inspects work area to verify compliance with plan.
7. Provide temporary supports for conduit and piping affected by the removal process.
8. Proceed with asbestos removal and bagging process. Monitor air quality as described in Step #2. Discontinue operations when contaminate levels exceed applicable standards.
9. Document the legal disposal of materials in accordance with EPA standards.
10. Thoroughly clean removal area including all ledges, crevices and surfaces.
11. Post abatement inspection by industrial hygienist to verify plan compliance.
12. Provide a certificate from a licensed industrial hygienist attesting that contaminate levels are within acceptable standards before returning area to regular use.

R028319-60 Lead Paint Remediation Methods

Lead paint remediation can be accomplished by the following methods.

1. Abrasive blast
2. Chemical stripping
3. Power tool cleaning with vacuum collection system
4. Encapsulation
5. Remove and replace
6. Enclosure

Each of these methods has strengths and weaknesses depending on the specific circumstances of the project. The following is an overview of each method.

1. **Abrasive blasting** is usually accomplished with sand or recyclable metallic blast. Before work can begin, the area must be contained to ensure the blast material with lead does not escape to the atmosphere. The use of vacuum blast greatly reduces the containment requirements. Lead abatement equipment that may be associated with this work includes a negative air machine. In addition, it is necessary to have an industrial hygienist monitor the project on a continual basis. When the work is complete, the spent blast sand with lead must be disposed of as a hazardous material. If metallic shot was used, the lead is separated from the shot and disposed of as hazardous material. Worker protection includes disposable clothing and respiratory protection.
2. **Chemical stripping** requires strong chemicals to be applied to the surface to remove the lead paint. Before the work can begin, the area under/adjacent to the work area must be covered to catch the chemical and removed lead. After the chemical is applied to the painted surface it is usually covered with paper. The chemical is left in place for the specified period, then the paper with lead paint is pulled or scraped off. The process may require several chemical applications. The paper with chemicals and lead paint adhered to it, plus the containment and loose scrapings collected by a HEPA (High Efficiency Particulate Air Filter) vac, must be disposed of as hazardous material. The chemical stripping process usually requires a neutralizing agent and several wash downs after the paint is removed. Worker protection includes a neoprene or other compatible protective clothing and respiratory protection with face shield. An industrial hygienist is required intermittently during the process.
3. **Power tool cleaning** is accomplished using shrouded needle blasting guns. The shrouding with different end configurations is held up against the surface to be cleaned. The area is blasted with hardened needles and the shroud captures the lead with a HEPA vac and deposits it in a holding tank. An industrial hygienist monitors the project. Protective clothing and a respirator are required until air samples prove otherwise. When the work is complete the lead must be disposed of as hazardous material.
4. **Encapsulation** is a method that leaves the well bonded lead paint in place after the peeling paint has been removed. Before the work can begin, the area under/adjacent to the work must be covered to catch the scrapings. The scraped surface is then washed with a detergent and rinsed. The prepared surface is covered with approximately 10 mils of paint. A reinforcing fabric can also be embedded in the paint covering. The scraped paint and containment must be disposed of as hazardous material. Workers must wear protective clothing and respirators.
5. **Removing and replacing** are effective ways to remove lead paint from windows, gypsum walls and concrete masonry surfaces. The painted materials are removed and new materials are installed. Workers should wear a respirator and tyvek suit. The demolished materials must be disposed of as hazardous waste if it fails the TCLP (Toxicity Characteristic Leachate Process) test.
6. **Enclosure** is the process that permanently seals lead painted materials in place. This process has many applications such as covering lead painted drywall with new drywall, covering exterior construction with tyvek paper then residing, or covering lead painted structural members with aluminum or plastic. The seams on all enclosing materials must be securely sealed. An industrial hygienist monitors the project, and protective clothing and a respirator are required until air samples prove otherwise.

All the processes require clearance monitoring and wipe testing as required by the hygienist.

R031113-10 Wall Form Materials

Aluminum Forms

Approximate weight is 3 lbs. per S.F.C.A. Standard widths are available from 4″ to 36″ with 36″ most common. Standard lengths of 2′, 4′, 6′ to 8′ are available. Forms are lightweight and fewer ties are needed with the wider widths. The form face is either smooth or textured.

Metal Framed Plywood Forms

Manufacturers claim over 75 reuses of plywood and over 300 reuses of steel frames. Many specials such as corners, fillers, pilasters, etc. are available. Monthly rental is generally about 15% of purchase price for first month and 9% per month thereafter with 90% of rental applied to purchase for the first month and decreasing percentages thereafter. Aluminum framed forms cost 25% to 30% more than steel framed.

After the first month, extra days may be prorated from the monthly charge. Rental rates do not include ties, accessories, cleaning, loss of hardware or freight in and out. Approximate weight is 5 lbs. per S.F. for steel; 3 lbs. per S.F. for aluminum.

Forms can be rented with option to buy.

Plywood Forms, Job Fabricated

There are two types of plywood used for concrete forms.

1. Exterior plyform which is completely waterproof. This is face oiled to facilitate stripping. Ten reuses can be expected with this type with 25 reuses possible.
2. An overlaid type consists of a resin fiber fused to exterior plyform. No oiling is required except to facilitate cleaning. This is available in both high density (HDO) and medium density overlaid (MDO). Using HDO, 50 reuses can be expected with 200 possible.

Plyform is available in 5/8″ and 3/4″ thickness. High density overlaid is available in 3/8″, 1/2″, 5/8″ and 3/4″ thickness.

5/8″ thick is sufficient for most building forms, while 3/4″ is best on heavy construction.

Plywood Forms, Modular, Prefabricated

There are many plywood forming systems without frames. Most of these are manufactured from 1-1/8″ (HDO) plywood and have some hardware attached. These are used principally for foundation walls 8′ or less high. With care and maintenance, 100 reuses can be attained with decreasing quality of surface finish.

Steel Forms

Approximate weight is 6-1/2 lbs. per S.F.C.A. including accessories. Standard widths are available from 2″ to 24″, with 24″ most common. Standard lengths are from 2′ to 8′, with 4′ the most common. Forms are easily ganged into modular units.

Forms are usually leased for 15% of the purchase price per month prorated daily over 30 days.

Rental may be applied to sale price, and usually rental forms are bought. With careful handling and cleaning 200 to 400 reuses are possible.

Straight wall gang forms up to 12′ x 20′ or 8′ x 30′ can be fabricated. These crane handled forms usually lease for approx. 9% per month.

Individual job analysis is available from the manufacturer at no charge.

R031113-40 Forms for Reinforced Concrete

Design Economy

Avoid many sizes in proportioning beams and columns.

From story to story avoid changing column dimensions. Gain strength by adding steel or using a richer mix. If a change in size of column is necessary, vary one dimension only to minimize form alterations. Keep beams and columns the same width.

From floor to floor in a multi-story building vary beam depth, not width, as that will leave the slab panel form unchanged. It is cheaper to vary the strength of a beam from floor to floor by means of a steel area than by 2″ changes in either width or depth.

Cost Factors

Material includes the cost of lumber, cost of rent for metal pans or forms if used, nails, form ties, form oil, bolts and accessories.

Labor includes the cost of carpenters to make up, erect, remove and repair, plus common labor to clean and move. Having carpenters remove forms minimizes repairs.

Improper alignment and condition of forms will increase finishing cost. When forms are heavily oiled, concrete surfaces must be neutralized before finishing. Special curing compounds will cause spillages to spall off in first frost. Gang forming methods will reduce costs on large projects.

Materials Used

Boards are seldom used unless their architectural finish is required. Generally, steel, fiberglass and plywood are used for contact surfaces. Labor on plywood is 10% less than with boards. The plywood is backed up with 2 x 4′s at 12″ to 32″ O.C. Walers are generally 2 - 2 x 4′s. Column forms are held together with steel yokes or bands. Shoring is with adjustable shoring or scaffolding for high ceilings.

Reuse

Floor and column forms can be reused four or possibly five times without excessive repair. Remember to allow for 10% waste on each reuse.

When modular sized wall forms are made, up to twenty uses can be expected with exterior plyform.

When forms are reused, the cost to erect, strip, clean and move will not be affected. 10% replacement of lumber should be included and about one hour of carpenter time for repairs on each reuse per 100 S.F.

The reuse cost for certain accessory items normally rented on a monthly basis will be lower than the cost for the first use.

After the fifth use, new material required plus time needed for repair prevent the form cost from dropping further; it may go up. Much depends on care in stripping, the number of special bays, changes in beam or column sizes and other factors.

Costs for multiple use of formwork may be developed as follows:

2 Uses

$$\frac{\text{(1st Use + Reuse)}}{2} = \text{avg. cost/2 uses}$$

3 Uses

$$\frac{\text{(1st Use + 2 Reuses)}}{3} = \text{avg. cost/3 uses}$$

4 Uses

$$\frac{\text{(1st use + 3 Reuses)}}{4} = \text{avg. cost/4 uses}$$

R031113-60 Formwork Labor-Hours

Item	Hours Required				Total Hours	Multiple Use		
	Unit	Fabricate	Erect & Strip	Clean & Move	1 Use	2 Use	3 Use	4 Use
Beam and Girder, interior beams, 12″ wide	100 S.F.	6.4	8.3	1.3	16.0	13.3	12.4	12.0
Hung from steel beams		5.8	7.7	1.3	14.8	12.4	11.6	11.2
Beam sides only, 36″ high		5.8	7.2	1.3	14.3	11.9	11.1	10.7
Beam bottoms only, 24″ wide		6.6	13.0	1.3	20.9	18.1	17.2	16.7
Box out for openings		9.9	10.0	1.1	21.0	16.6	15.1	14.3
Buttress forms, to 8′ high		6.0	6.5	1.2	13.7	11.2	10.4	10.0
Centering, steel, 3/4″ rib lath			1.0		1.0			
3/8″ rib lath or slab form			0.9		0.9			
Chamfer strip or keyway	100 L.F.		1.5		1.5	1.5	1.5	1.5
Columns, fiber tube 8″ diameter			20.6		20.6			
12″			21.3		21.3			
16″			22.9		22.9			
20″			23.7		23.7			
24″			24.6		24.6			
30″			25.6		25.6			
Columns, round steel, 12″ diameter			22.0		22.0	22.0	22.0	22.0
16″			25.6		25.6	25.6	25.6	25.6
20″			30.5		30.5	30.5	30.5	30.5
24″			37.7		37.7	37.7	37.7	37.7
Columns, plywood 8″ x 8″	100 S.F.	7.0	11.0	1.2	19.2	16.2	15.2	14.7
12″ x 12″		6.0	10.5	1.2	17.7	15.2	14.4	14.0
16″ x 16″		5.9	10.0	1.2	17.1	14.7	13.8	13.4
24″ x 24″		5.8	9.8	1.2	16.8	14.4	13.6	13.2
Columns, steel framed plywood 8″ x 8″			10.0	1.0	11.0	11.0	11.0	11.0
12″ x 12″			9.3	1.0	10.3	10.3	10.3	10.3
16″ x 16″			8.5	1.0	9.5	9.5	9.5	9.5
24″ x 24″			7.8	1.0	8.8	8.8	8.8	8.8
Drop head forms, plywood		9.0	12.5	1.5	23.0	19.0	17.7	17.0
Coping forms		8.5	15.0	1.5	25.0	21.3	20.0	19.4
Culvert, box			14.5	4.3	18.8	18.8	18.8	18.8
Curb forms, 6″ to 12″ high, on grade		5.0	8.5	1.2	14.7	12.7	12.1	11.7
On elevated slabs		6.0	10.8	1.2	18.0	15.5	14.7	14.3
Edge forms to 6″ high, on grade	100 L.F.	2.0	3.5	0.6	6.1	5.6	5.4	5.3
7″ to 12″ high	100 S.F.	2.5	5.0	1.0	8.5	7.8	7.5	7.4
Equipment foundations		10.0	18.0	2.0	30.0	25.5	24.0	23.3
Flat slabs, including drops		3.5	6.0	1.2	10.7	9.5	9.0	8.8
Hung from steel		3.0	5.5	1.2	9.7	8.7	8.4	8.2
Closed deck for domes		3.0	5.8	1.2	10.0	9.0	8.7	8.5
Open deck for pans		2.2	5.3	1.0	8.5	7.9	7.7	7.6
Footings, continuous, 12″ high		3.5	3.5	1.5	8.5	7.3	6.8	6.6
Spread, 12″ high		4.7	4.2	1.6	10.5	8.7	8.0	7.7
Pile caps, square or rectangular		4.5	5.0	1.5	11.0	9.3	8.7	8.4
Grade beams, 24″ deep		2.5	5.3	1.2	9.0	8.3	8.0	7.9
Lintel or Sill forms		8.0	17.0	2.0	27.0	23.5	22.3	21.8
Spandrel beams, 12″ wide		9.0	11.2	1.3	21.5	17.5	16.2	15.5
Stairs			25.0	4.0	29.0	29.0	29.0	29.0
Trench forms in floor		4.5	14.0	1.5	20.0	18.3	17.7	17.4
Walls, Plywood, at grade, to 8′ high		5.0	6.5	1.5	13.0	11.0	9.7	9.5
8′ to 16′		7.5	8.0	1.5	17.0	13.8	12.7	12.1
16′ to 20′		9.0	10.0	1.5	20.5	16.5	15.2	14.5
Foundation walls, to 8′ high		4.5	6.5	1.0	12.0	10.3	9.7	9.4
8′ to 16′ high		5.5	7.5	1.0	14.0	11.8	11.0	10.6
Retaining wall to 12′ high, battered		6.0	8.5	1.5	16.0	13.5	12.7	12.3
Radial walls to 12′ high, smooth		8.0	9.5	2.0	19.5	16.0	14.8	14.3
2′ chords		7.0	8.0	1.5	16.5	13.5	12.5	12.0
Prefabricated modular, to 8′ high		—	4.3	1.0	5.3	5.3	5.3	5.3
Steel, to 8′ high		—	6.8	1.2	8.0	8.0	8.0	8.0
8′ to 16′ high		—	9.1	1.5	10.6	10.3	10.2	10.2
Steel framed plywood to 8′ high		—	6.8	1.2	8.0	7.5	7.3	7.2
8′ to 16′ high		—	9.3	1.2	10.5	9.5	9.2	9.0

R032110-10 Reinforcing Steel Weights and Measures

Bar Designation No.**	Nominal Weight Lb./Ft.	U.S. Customary Units			SI Units			
		Nominal Dimensions*			Nominal Dimensions*			
		Diameter in.	Cross Sectional Area, in.^{2}	Perimeter in.	Nominal Weight kg/m	Diameter mm	Cross Sectional Area, cm^2	Perimeter mm
3	.376	.375	.11	1.178	.560	9.52	.71	29.9
4	.668	.500	.20	1.571	.994	12.70	1.29	39.9
5	1.043	.625	.31	1.963	1.552	15.88	2.00	49.9
6	1.502	.750	.44	2.356	2.235	19.05	2.84	59.8
7	2.044	.875	.60	2.749	3.042	22.22	3.87	69.8
8	2.670	1.000	.79	3.142	3.973	25.40	5.10	79.8
9	3.400	1.128	1.00	3.544	5.059	28.65	6.45	90.0
10	4.303	1.270	1.27	3.990	6.403	32.26	8.19	101.4
11	5.313	1.410	1.56	4.430	7.906	35.81	10.06	112.5
14	7.650	1.693	2.25	5.320	11.384	43.00	14.52	135.1
18	13.600	2.257	4.00	7.090	20.238	57.33	25.81	180.1

* The nominal dimensions of a deformed bar are equivalent to those of a plain round bar having the same weight per foot as the deformed bar.
** Bar numbers are based on the number of eighths of an inch included in the nominal diameter of the bars.

R032110-20 Metric Rebar Specification - ASTM A615-81

Grade 300 (300 MPa* = 43,560 psi; +8.7% vs. Grade 40)				
Grade 400 (400 MPa* = 58,000 psi; –3.4% vs. Grade 60)				
Bar No.	Diameter mm	Area mm^2	Equivalent in.^{2}	Comparison with U.S. Customary Bars
10M	11.3	100	.16	Between #3 & #4
15M	16.0	200	.31	#5 (.31 in.^{2})
20M	19.5	300	.47	#6 (.44 in.^{2})
25M	25.2	500	.78	#8 (.79 in.^{2})
30M	29.9	700	1.09	#9 (1.00 in.^{2})
35M	35.7	1000	1.55	#11 (1.56 in.^{2})
45M	43.7	1500	2.33	#14 (2.25 in.^{2})
55M	56.4	2500	3.88	#18 (4.00 in.^{2})

* MPa = megapascals

R032110-40 Weight of Steel Reinforcing Per Square Foot of Wall (PSF)

Reinforced Weights: The table below suggests the weights per square foot for reinforcing steel in walls. Weights are approximate and will be the same for all grades of steel bars. For bars in two directions, add weights for each size and spacing.

C/C Spacing in Inches	Bar Size #3 Wt. (PSF)	#4 Wt. (PSF)	#5 Wt. (PSF)	#6 Wt. (PSF)	#7 Wt. (PSF)	#8 Wt. (PSF)	#9 Wt. (PSF)	#10 Wt. (PSF)	#11 Wt. (PSF)
2″	2.26	4.01	6.26	9.01	12.27				
3″	1.50	2.67	4.17	6.01	8.18	10.68	13.60	17.21	21.25
4″	1.13	2.01	3.13	4.51	6.13	8.10	10.20	12.91	15.94
5″	.90	1.60	2.50	3.60	4.91	6.41	8.16	10.33	12.75
6″	.752	1.34	2.09	3.00	4.09	5.34	6.80	8.61	10.63
8″	.564	1.00	1.57	2.25	3.07	4.01	5.10	6.46	7.97
10″	.451	.802	1.25	1.80	2.45	3.20	4.08	5.16	6.38
12″	.376	.668	1.04	1.50	2.04	2.67	3.40	4.30	5.31
18″	.251	.445	.695	1.00	1.32	1.78	2.27	2.86	3.54
24″	.188	.334	.522	.751	1.02	1.34	1.70	2.15	2.66
30″	.150	.267	.417	.600	.817	1.07	1.36	1.72	2.13
36″	.125	.223	.348	.501	.681	.890	1.13	1.43	1.77
42″	.107	.191	.298	.429	.584	.753	.97	1.17	1.52
48″	.094	.167	.261	.376	.511	.668	.85	1.08	1.33

R032110-50 Minimum Wall Reinforcement Weight (PSF)

This table lists the approximate minimum wall reinforcement weights per S.F. according to the specification of .12% of gross area for vertical bars and .20% of gross area for horizontal bars.

Location	Wall Thickness	Bar Size	Horizontal Steel Spacing C/C	Sq. In. Req'd per S.F.	Total Wt. per S.F.	Bar Size	Vertical Steel Spacing C/C	Sq. In. Req'd per S.F.	Total Wt. per S.F.	Horizontal & Vertical Steel Total Weight per S.F.
Both Faces	10″	#4	18″	.24	.89#	#3	18″	.14	.50#	1.39#
	12″	#4	16″	.29	1.00	#3	16″	.17	.60	1.60
	14″	#4	14″	.34	1.14	#3	13″	.20	.69	1.84
	16″	#4	12″	.38	1.34	#3	11″	.23	.82	2.16
	18″	#5	17″	.43	1.47	#4	18″	.26	.89	2.36
One Face	6″	#3	9″	.15	.50	#3	18″	.09	.25	.75
	8″	#4	12″	.19	.67	#3	11″	.12	.41	1.08
	10″	#5	15″	.24	.83	#4	16″	.14	.50	1.34

R032110-70 Bend, Place and Tie Reinforcing

Placing and tying by rodmen for footings and slabs run from nine hrs. per ton for heavy bars to fifteen hrs. per ton for light bars. For beams, columns, and walls, production runs from eight hrs. per ton for heavy bars to twenty hrs. per ton for light bars. The overall average for typical reinforced concrete buildings is about fourteen hrs. per ton. These production figures include the time for placing accessories and usual inserts, but not their material cost (allow 15% of the cost of delivered bent rods). Equipment handling is necessary for the larger-sized bars so that installation costs for the very heavy bars will not decrease proportionately.

Installation costs for splicing reinforcing bars include allowance for equipment to hold the bars in place while splicing as well as necessary scaffolding for iron workers.

R032110-80 Shop-Fabricated Reinforcing Steel

The material prices for reinforcing, shown in the unit cost sections of the data set, are for 50 tons or more of shop-fabricated reinforcing steel and include:

1. Mill base price of reinforcing steel
2. Mill grade/size/length extras
3. Mill delivery to the fabrication shop
4. Shop storage and handling
5. Shop drafting/detailing
6. Shop shearing and bending
7. Shop listing
8. Shop delivery to the job site

Both material and installation costs can be considerably higher for small jobs consisting primarily of smaller bars, while material costs may be slightly lower for larger jobs.

R032205-30 Common Stock Styles of Welded Wire Fabric

This table provides some of the basic specifications, sizes, and weights of welded wire fabric used for reinforcing concrete.

New Designation		Old Designation		Steel Area per Foot				Approximate Weight per 100 S.F.	
Spacing — Cross Sectional Area (in.) — (Sq. in. 100)		Spacing — Wire Gauge (in.) — (AS & W)		Longitudinal		Transverse			
				in.	cm	in.	cm	lbs	kg
Rolls	6 x 6 — W1.4 x W1.4	6 x 6 — 10 x 10		.028	.071	.028	.071	21	9.53
	6 x 6 — W2.0 x W2.0	6 x 6 — 8 x 8	1	.040	.102	.040	.102	29	13.15
	6 x 6 — W2.9 x W2.9	6 x 6 — 6 x 6		.058	.147	.058	.147	42	19.05
	6 x 6 — W4.0 x W4.0	6 x 6 — 4 x 4		.080	.203	.080	.203	58	26.91
	4 x 4 — W1.4 x W1.4	4 x 4 — 10 x 10		.042	.107	.042	.107	31	14.06
	4 x 4 — W2.0 x W2.0	4 x 4 — 8 x 8	1	.060	.152	.060	.152	43	19.50
	4 x 4 — W2.9 x W2.9	4 x 4 — 6 x 6		.087	.227	.087	.227	62	28.12
	4 x 4 — W4.0 x W4.0	4 x 4 — 4 x 4		.120	.305	.120	.305	85	38.56
Sheets	6 x 6 — W2.9 x W2.9	6 x 6 — 6 x 6		.058	.147	.058	.147	42	19.05
	6 x 6 — W4.0 x W4.0	6 x 6 — 4 x 4		.080	.203	.080	.203	58	26.31
	6 x 6 — W5.5 x W5.5	6 x 6 — 2 x 2	2	.110	.279	.110	.279	80	36.29
	4 x 4 — W1.4 x W1.4	4 x 4 — 4 x 4		.120	.305	.120	.305	85	38.56

NOTES: 1. Exact W—number size for 8 gauge is W2.1
2. Exact W—number size for 2 gauge is W5.4

The above table was compiled with the following excerpts from the WRI Manual of Standard Practices, 7th Edition, Copyright 2006. Reproduced with permission of the Wire Reinforcement Institute, Inc.:

1. Chapter 3, page 7, Table 1 Common Styles of Metric Wire Reinforcement (WWR) With Equivalent US Customary Units
2. Chapter 6, page 19, Table 5 Customary Units
3. Chapter 6, Page 23, Table 7 Customary Units (in.) Welded Plain Wire Reinforcement
4. Chapter 6, Page 25 Table 8 Wire Size Comparison
5. Chapter 9, Page 30, Table 9 Weight of Longitudinal Wires Weight (Mass) Estimating Tables
6. Chapter 9, Page 31, Table 9M Weight of Longitudinal Wires Weight (Mass) Estimating Tables
7. Chapter 9, Page 32, Table 10 Weight of Transverse Wires Based on 62″ lengths of transverse wire (60″ width plus 1″ overhand each side)
8. Chapter 9, Page 33, Table 10M Weight of Transverse Wires

R033053-10 Spread Footings

General: A spread footing is used to convert a concentrated load (from one superstructure column or substructure grade beams) into an allowable area load on supporting soil.

Because of punching action from the column load, a spread footing is usually thicker than strip footings which support wall loads. One or two story commercial or residential buildings should have no less than 1′ thick spread footings. Heavier loads require no less than 2′ thick. Spread footings may be square, rectangular or octagonal in plan.

Spread footings tend to minimize excavation and foundation materials, as well as labor and equipment. Another advantage is that footings and soil conditions can be readily examined. They are the most widely used type of footing, especially in mild climates and for buildings of four stories or under. This is because they are usually more economical than other types, if suitable soil and site conditions exist.

They are used when suitable supporting soil is located within several feet of the surface or line of subsurface excavation. Suitable soil types include sands and gravels, gravels with a small amount of clay or silt, hardpan, chalk, and rock. Pedestals may be used to bring the column base load down to the top of the footing. Alternately, undesirable soil between the underside of the footing and the top of the bearing level can be removed and replaced with lean concrete mix or compacted granular material.

Depth of footing should be below topsoil, uncompacted fill, muck, etc. It must be lower than frost penetration but should be above the water table. It must not be at the ground surface because of potential surface erosion. If the ground slopes, approximately three horizontal feet of edge protection must remain. Differential footing elevations may overlap soil stresses or cause excavation problems if clear spacing between footings is less than the difference in depth.

Other footing types are usually used for the following reasons:

A. Bearing capacity of soil is low.
B. Very large footings are required, at a cost disadvantage.
C. Soil under footing (shallow or deep) is very compressible, with probability of causing excessive or differential settlement.
D. Good bearing soil is deep.
E. Potential for scour action exists.
F. Varying subsoil conditions within building perimeter.

Cost of spread footings for a building is determined by:

1. The soil bearing capacity.
2. Typical bay size.
3. Total load (live plus dead) per S.F. for roof and elevated floor levels.
4. The size and shape of the building.
5. Footing configuration. Does the building utilize outer spread footings or are there continuous perimeter footings only or a combination of spread footings plus continuous footings?

Soil Bearing Capacity in Kips per S.F.

Bearing Material	Typical Allowable Bearing Capacity
Hard sound rock	120 KSF
Medium hard rock	80
Hardpan overlaying rock	24
Compact gravel and boulder-gravel; very compact sandy gravel	20
Soft rock	16
Loose gravel; sandy gravel; compact sand; very compact sand-inorganic silt	12
Hard dry consolidated clay	10
Loose coarse to medium sand; medium compact fine sand	8
Compact sand-clay	6
Loose fine sand; medium compact sand-inorganic silts	4
Firm or stiff clay	3
Loose saturated sand-clay; medium soft clay	2

R033053-60 Maximum Depth of Frost Penetration in Inches

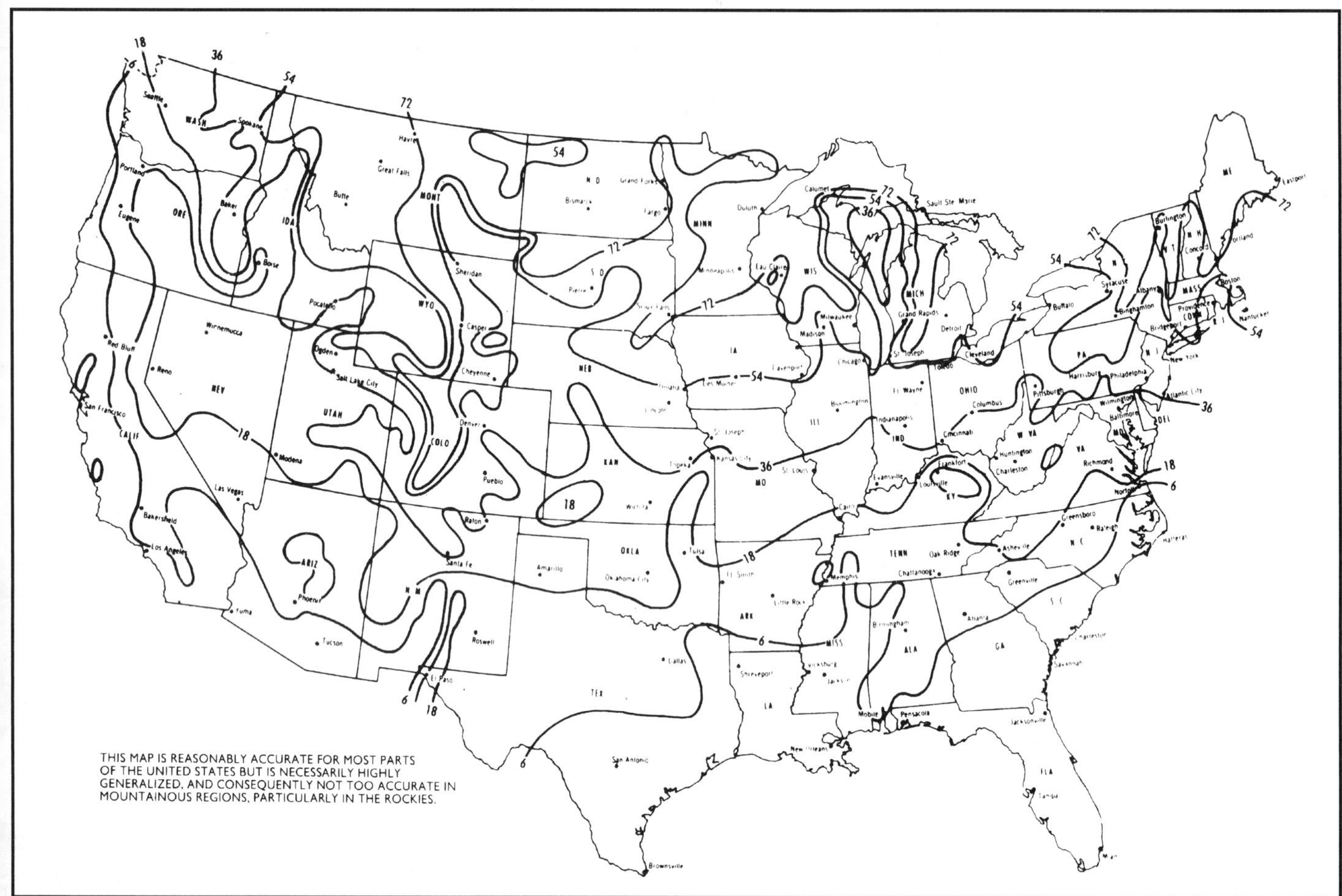

THIS MAP IS REASONABLY ACCURATE FOR MOST PARTS OF THE UNITED STATES BUT IS NECESSARILY HIGHLY GENERALIZED, AND CONSEQUENTLY NOT TOO ACCURATE IN MOUNTAINOUS REGIONS, PARTICULARLY IN THE ROCKIES.

R033105-10 Proportionate Quantities

The tables below show both quantities per S.F. of floor areas as well as form and reinforcing quantities per C.Y. Unusual structural requirements would increase the ratios below. High strength reinforcing would reduce the steel weights. Figures are for 3000 psi concrete and 60,000 psi reinforcing unless specified otherwise.

Type of Construction	Live Load	Span	Per S.F. of Floor Area				Per C.Y. of Concrete		
			Concrete	Forms	Reinf.	Pans	Forms	Reinf.	Pans
Flat Plate	50 psf	15 Ft.	.46 C.F.	1.06 S.F.	1.71lb.		62 S.F.	101 lb.	
		20	.63	1.02	2.40		44	104	
		25	.79	1.02	3.03		35	104	
	100	15	.46	1.04	2.14		61	126	
		20	.71	1.02	2.72		39	104	
		25	.83	1.01	3.47		33	113	
Flat Plate (waffle construction) 20″ domes	50	20	.43	1.00	2.10	.84 S.F.	63	135	53 S.F.
		25	.52	1.00	2.90	.89	52	150	46
		30	.64	1.00	3.70	.87	42	155	37
	100	20	.51	1.00	2.30	.84	53	125	45
		25	.64	1.00	3.20	.83	42	135	35
		30	.76	1.00	4.40	.81	36	160	29
Waffle Construction 30″ domes	50	25	.69	1.06	1.83	.68	42	72	40
		30	.74	1.06	2.39	.69	39	87	39
		35	.86	1.05	2.71	.69	33	85	39
		40	.78	1.00	4.80	.68	35	165	40
Flat Slab (two way with drop panels)	50	20	.62	1.03	2.34		45	102	
		25	.77	1.03	2.99		36	105	
		30	.95	1.03	4.09		29	116	
	100	20	.64	1.03	2.83		43	119	
		25	.79	1.03	3.88		35	133	
		30	.96	1.03	4.66		29	131	
	200	20	.73	1.03	3.03		38	112	
		25	.86	1.03	4.23		32	133	
		30	1.06	1.03	5.30		26	135	
One Way Joists 20″ Pans	50	15	.36	1.04	1.40	.93	78	105	70
		20	.42	1.05	1.80	.94	67	120	60
		25	.47	1.05	2.60	.94	60	150	54
	100	15	.38	1.07	1.90	.93	77	140	66
		20	.44	1.08	2.40	.94	67	150	58
		25	.52	1.07	3.50	.94	55	185	49
One Way Joists 8″ x 16″ filler blocks	50	15	.34	1.06	1.80	.81 Ea.	84	145	64 Ea.
		20	.40	1.08	2.20	.82	73	145	55
		25	.46	1.07	3.20	.83	63	190	49
	100	15	.39	1.07	1.90	.81	74	130	56
		20	.46	1.09	2.80	.82	64	160	48
		25	.53	1.10	3.60	.83	56	190	42
One Way Beam & Slab	50	15	.42	1.30	1.73		84	111	
		20	.51	1.28	2.61		68	138	
		25	.64	1.25	2.78		53	117	
	100	15	.42	1.30	1.90		84	122	
		20	.54	1.35	2.69		68	154	
		25	.69	1.37	3.93		54	145	
	200	15	.44	1.31	2.24		80	137	
		20	.58	1.40	3.30		65	163	
		25	.69	1.42	4.89		53	183	
Two Way Beam & Slab	100	15	.47	1.20	2.26		69	130	
		20	.63	1.29	3.06		55	131	
		25	.83	1.33	3.79		43	123	
	200	15	.49	1.25	2.70		41	149	
		20	.66	1.32	4.04		54	165	
		25	.88	1.32	6.08		41	187	

R033105-10 Proportionate Quantities (cont.)

4000 psi Concrete and 60,000 psi Reinforcing—Form and Reinforcing Quantities per C.Y.						
Item	Size		Forms	Reinforcing	Minimum	Maximum
Columns (square tied)	10″ x 10″		130 S.F.C.A.	#5 to #11	220 lbs.	875 lbs.
	12″ x 12″		108	#6 to #14	200	955
	14″ x 14″		92	#7 to #14	190	900
	16″ x 16″		81	#6 to #14	187	1082
	18″ x 18″		72	#6 to #14	170	906
	20″ x 20″		65	#7 to #18	150	1080
	22″ x 22″		59	#8 to #18	153	902
	24″ x 24″		54	#8 to #18	164	884
	26″ x 26″		50	#9 to #18	169	994
	28″ x 28″		46	#9 to #18	147	864
	30″ x 30″		43	#10 to #18	146	983
	32″ x 32″		40	#10 to #18	175	866
	34″ x 34″		38	#10 to #18	157	772
	36″ x 36″		36	#10 to #18	175	852
	38″ x 38″		34	#10 to #18	158	765
	40″ x 40″		32 ▼	#10 to #18	143	692
Item	Size	Form	Spiral	Reinforcing	Minimum	Maximum
Columns (spirally reinforced)	12″ diameter	34.5 L.F.	190 lbs.	#4 to #11	165 lbs.	1505 lb.
		34.5	190	#14 & #18	—	1100
	14″	25	170	#4 to #11	150	970
		25	170	#14 & #18	800	1000
	16″	19	160	#4 to #11	160	950
		19	160	#14 & #18	605	1080
	18″	15	150	#4 to #11	160	915
		15	150	#14 & #18	480	1075
	20″	12	130	#4 to #11	155	865
		12	130	#14 & #18	385	1020
	22″	10	125	#4 to #11	165	775
		10	125	#14 & #18	320	995
	24″	9	120	#4 to #11	195	800
		9	120	#14 & #18	290	1150
	26″	7.3	100	#4 to #11	200	729
		7.3	100	#14 & #18	235	1035
	28″	6.3	95	#4 to #11	175	700
		6.3	95	#14 & #18	200	1075
	30″	5.5	90	#4 to #11	180	670
		5.5	90	#14 & #18	175	1015
	32″	4.8	85	#4 to #11	185	615
		4.8	85	#14 & #18	155	955
	34″	4.3	80	#4 to #11	180	600
		4.3	80	#14 & #18	170	855
	36″	3.8	75	#4 to #11	165	570
		3.8	75	#14 & #18	155	865
	40″	3.0	70	#4 to #11	165	500
		3.0	70	#14 & #18	145	765

R033105-10 Proportionate Quantities (cont.)

3000 psi Concrete and 60,000 psi Reinforcing—Form and Reinforcing Quantities per C.Y.						
Item	**Type**	**Loading**	**Height**	**C.Y./L.F.**	**Forms/C.Y.**	**Reinf./C.Y.**
Retaining Walls	Cantilever	Level Backfill	4 Ft.	0.2 C.Y.	49 S.F.	35 lbs.
			8	0.5	42	45
			12	0.8	35	70
			16	1.1	32	85
			20	1.6	28	105
		Highway Surcharge	4	0.3	41	35
			8	0.5	36	55
			12	0.8	33	90
			16	1.2	30	120
			20	1.7	27	155
		Railroad Surcharge	4	0.4	28	45
			8	0.8	25	65
			12	1.3	22	90
			16	1.9	20	100
			20	2.6	18	120
	Gravity, with Vertical Face	Level Backfill	4	0.4	37	None
			7	0.6	27	↓
			10	1.2	20	↓
		Sloping Backfill	4	0.3	31	↓
			7	0.8	21	↓
			10	1.6	15	↓

Item	Span	Live Load in Kips per Linear Foot							
		Under 1 Kip		2 to 3 Kips		4 to 5 Kips		6 to 7 Kips	
		Forms	Reinf.	Forms	Reinf.	Forms	Reinf.	Forms	Reinf.
Beams	10 Ft.	—	—	90 S.F.	170 #	85 S.F.	175 #	75 S.F.	185 #
	16	130 S.F.	165 #	85	180	75	180	65	225
	20	110	170	75	185	62	200	51	200
	26	90	170	65	215	62	215	—	—
	30	85	175	60	200	—	—	—	—

Item	Size	Type	Forms per C.Y.	Reinforcing per C.Y.
Spread Footings	Under 1 C.Y.	1,000 psf soil	24 S.F.	44 lbs.
		5,000	24	42
		10,000	24	52
	1 C.Y. to 5 C.Y.	1,000	14	49
		5,000	14	50
		10,000	14	50
	Over 5 C.Y.	1,000	9	54
		5,000	9	52
		10,000	9	56
Pile Caps (30 Ton Concrete Piles)	Under 5 C.Y.	shallow caps	20	65
		medium	20	50
		deep	20	40
	5 C.Y. to 10 C.Y.	shallow	14	55
		medium	15	45
		deep	15	40
	10 C.Y. to 20 C.Y.	shallow	11	60
		medium	11	45
		deep	12	35
	Over 20 C.Y.	shallow	9	60
		medium	9	45
		deep	10	40

R033105-10 Proportionate Quantities (cont.)

3000 psi Concrete and 60,000 psi Reinforcing — Form and Reinforcing Quantities per C.Y.						
Item	Size	Pile Spacing	50 T Pile	100 T Pile	50 T Pile	100 T Pile
Pile Caps (Steel H Piles)	Under 5 C.Y.	24″ O.C.	24 S.F.	24 S.F.	75 lbs.	90 lbs.
		30″	25	25	80	100
		36″	24	24	80	110
	5 C.Y. to 10 C.Y.	24″	15	15	80	110
		30″	15	15	85	110
		36″	15	15	75	90
	Over 10 C.Y.	24″	13	13	85	90
		30″	11	11	85	95
		36″	10	10	85	90

Item		8″ Thick		10″ Thick		12″ Thick		15″ Thick	
	Height	Forms	Reinf.	Forms	Reinf.	Forms	Reinf.	Forms	Reinf.
Basement Walls	7 Ft.	81 S.F.	44 lbs.	65 S.F.	45 lbs.	54 S.F.	44 lbs.	41 S.F.	43 lbs.
	8	↓	44	↓	45	↓	44	↓	43
	9	↓	46	↓	45	↓	44	↓	43
	10	↓	57	↓	45	↓	44	↓	43
	12	↓	83	↓	50	↓	52	↓	43
	14	↓	116	↓	65	↓	64	↓	51
	16	↓		↓	86	↓	90	↓	65
	18	↓		↓		↓	106	↓	70

R033105-20 Materials for One C.Y. of Concrete

This is an approximate method of figuring quantities of cement, sand and coarse aggregate for a field mix with waste allowance included.

With crushed gravel as coarse aggregate, to determine barrels of cement required, divide 10 by total mix; that is, for 1:2:4 mix, 10 divided by 7 = 1-3/7 barrels.

If the coarse aggregate is crushed stone, use 10-1/2 instead of 10 as given for gravel.

To determine tons of sand required, multiply barrels of cement by parts of sand and then by 0.2; that is, for the 1:2:4 mix, as above, 1-3/7 x 2 x .2 = .57 tons.

Tons of crushed gravel are in the same ratio to tons of sand as parts in the mix, or 4/2 x .57 = 1.14 tons.

1 bag cement = 94#
4 bags = 1 barrel

1 C.Y. sand or crushed gravel = 2700#
1 ton sand or crushed gravel = 20 C.F.

1 C.Y. crushed stone = 2575#
1 ton crushed stone = 21 C.F.

Average carload of cement is 692 bags; of sand or gravel is 56 tons.

Do not stack stored cement over 10 bags high.

R033105-30 Metric Equivalents of Cement Content for Concrete Mixes

94 Pound Bags per Cubic Yard	Kilograms per Cubic Meter	94 Pound Bags per Cubic Yard	Kilograms per Cubic Meter
1.0	55.77	7.0	390.4
1.5	83.65	7.5	418.3
2.0	111.5	8.0	446.2
2.5	139.4	8.5	474.0
3.0	167.3	9.0	501.9
3.5	195.2	9.5	529.8
4.0	223.1	10.0	557.7
4.5	251.0	10.5	585.6
5.0	278.8	11.0	613.5
5.5	306.7	11.5	641.3
6.0	334.6	12.0	669.2
6.5	362.5	12.5	697.1

a. If you know the cement content in pounds per cubic yard, multiply by .5933 to obtain kilograms per cubic meter.

b. If you know the cement content in 94 pound bags per cubic yard, multiply by 55.77 to obtain kilograms per cubic meter.

R033105-40 Metric Equivalents of Common Concrete Strengths (to convert other psi values to megapascals, multiply by 0.006895)

U.S. Values psi	SI Values Megapascals	Non-SI Metric Values kgf/cm^2*
2000	14	140
2500	17	175
3000	21	210
3500	24	245
4000	28	280
4500	31	315
5000	34	350
6000	41	420
7000	48	490
8000	55	560
9000	62	630
10,000	69	705

* kilograms force per square centimeter

R033105-50 Quantities of Cement, Sand and Stone for One C.Y. of Concrete per Various Mixes

This table can be used to determine the quantities of the ingredients for smaller quantities of site mixed concrete.

Concrete (C.Y.)	Mix = 1:1:1-3/4			Mix = 1:2:2.25			Mix = 1:2.25:3			Mix = 1:3:4		
	Cement (sacks)	Sand (C.Y.)	Stone (C.Y.)	Cement (sacks)	Sand (C.Y.)	Stone (C.Y.)	Cement (sacks)	Sand (C.Y.)	Stone (C.Y.)	Cement (sacks)	Sand (C.Y.)	Stone (C.Y.)
1	10	.37	.63	7.75	.56	.65	6.25	.52	.70	5	.56	.74
2	20	.74	1.26	15.50	1.12	1.30	12.50	1.04	1.40	10	1.12	1.48
3	30	1.11	1.89	23.25	1.68	1.95	18.75	1.56	2.10	15	1.68	2.22
4	40	1.48	2.52	31.00	2.24	2.60	25.00	2.08	2.80	20	2.24	2.96
5	50	1.85	3.15	38.75	2.80	3.25	31.25	2.60	3.50	25	2.80	3.70
6	60	2.22	3.78	46.50	3.36	3.90	37.50	3.12	4.20	30	3.36	4.44
7	70	2.59	4.41	54.25	3.92	4.55	43.75	3.64	4.90	35	3.92	5.18
8	80	2.96	5.04	62.00	4.48	5.20	50.00	4.16	5.60	40	4.48	5.92
9	90	3.33	5.67	69.75	5.04	5.85	56.25	4.68	6.30	45	5.04	6.66
10	100	3.70	6.30	77.50	5.60	6.50	62.50	5.20	7.00	50	5.60	7.40
11	110	4.07	6.93	85.25	6.16	7.15	68.75	5.72	7.70	55	6.16	8.14
12	120	4.44	7.56	93.00	6.72	7.80	75.00	6.24	8.40	60	6.72	8.88
13	130	4.82	8.20	100.76	7.28	8.46	81.26	6.76	9.10	65	7.28	9.62
14	140	5.18	8.82	108.50	7.84	9.10	87.50	7.28	9.80	70	7.84	10.36
15	150	5.56	9.46	116.26	8.40	9.76	93.76	7.80	10.50	75	8.40	11.10
16	160	5.92	10.08	124.00	8.96	10.40	100.00	8.32	11.20	80	8.96	11.84
17	170	6.30	10.72	131.76	9.52	11.06	106.26	8.84	11.90	85	9.52	12.58
18	180	6.66	11.34	139.50	10.08	11.70	112.50	9.36	12.60	90	10.08	13.32
19	190	7.04	11.98	147.26	10.64	12.36	118.76	9.84	13.30	95	10.64	14.06
20	200	7.40	12.60	155.00	11.20	13.00	125.00	10.40	14.00	100	11.20	14.80
21	210	7.77	13.23	162.75	11.76	13.65	131.25	10.92	14.70	105	11.76	15.54
22	220	8.14	13.86	170.05	12.32	14.30	137.50	11.44	15.40	110	12.32	16.28
23	230	8.51	14.49	178.25	12.88	14.95	143.75	11.96	16.10	115	12.88	17.02
24	240	8.88	15.12	186.00	13.44	15.60	150.00	12.48	16.80	120	13.44	17.76
25	250	9.25	15.75	193.75	14.00	16.25	156.25	13.00	17.50	125	14.00	18.50
26	260	9.64	16.40	201.52	14.56	16.92	162.52	13.52	18.20	130	14.56	19.24
27	270	10.00	17.00	209.26	15.12	17.56	168.76	14.04	18.90	135	15.02	20.00
28	280	10.36	17.64	217.00	15.68	18.20	175.00	14.56	19.60	140	15.68	20.72
29	290	10.74	18.28	224.76	16.24	18.86	181.26	15.08	20.30	145	16.24	21.46

R033105-70 Placing Ready-Mixed Concrete

For ground pours allow for 5% waste when figuring quantities.

Prices in the front of the data set assume normal deliveries. If deliveries are made before 8 A.M. or after 5 P.M. or on Saturday afternoons add 30%. Negotiated discounts for large volumes are not included in prices in front of the data set.

For the lower floors without truck access, concrete may be wheeled in rubber-tired buggies, conveyer handled, crane handled or pumped. Pumping is economical if there is top steel. Conveyers are more efficient for thick slabs.

At higher floors the rubber-tired buggies may be hoisted by a hoisting tower and wheeled to the location. Placement by a conveyer is limited to three floors and is best for high-volume pours. Pumped concrete is best when the building has no crane access. Concrete may be pumped directly as high as thirty-six stories using special pumping techniques. Normal maximum height is about fifteen stories. The b

est pumping aggregate is screened and graded bank gravel rather than crushed stone.

Pumping downward is more difficult than pumping upward. The horizontal distance from pump to pour may increase preparation time prior to pour. Placing by cranes, either mobile, climbing or tower types, continues as the most efficient method for high-rise concrete buildings.

R033105-80 Slab on Grade

General: Ground slabs are classified on the basis of use. Thickness is generally controlled by the heaviest concentrated load supported. If load area is greater than 80 sq. in., soil bearing may be important. The base granular fill must be a uniformly compacted material of limited capillarity, such as gravel or crushed rock. Concrete is placed on this surface of the vapor barrier on top of base.

Ground slabs are either single or two course floors. Single course are widely used. Two course floors have a subsequent wear resistant topping.

Reinforcement is provided to maintain tightly closed cracks.

Control joints limit crack locations and provide for differential horizontal movement only. Isolation joints allow both horizontal and vertical differential movement.

Use of Table: Determine appropriate type of slab (A, B, C, or D) by considering type of use or amount of abrasive wear of traffic type.

Determine thickness by maximum allowable wheel load or uniform load, opposite 1st column, thickness. Increase the controlling thickness if details require, and select either plain or reinforced slab thickness and type.

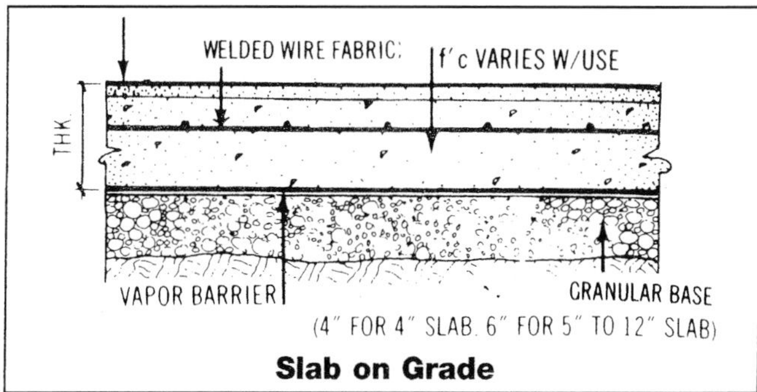

Slab on Grade

Thickness and Loading Assumptions by Type of Use

SLAB THICKNESS (IN.)	TYPE	A	B	C	D	◄ Slab I.D.
		Non	Light	Normal	Heavy	◄ Industrial
		Little	Light	Moderate	Severe	◄ Abrasion
		Foot Only	Pneumatic Wheels	Solid Rubber Wheels	Steel Tires	◄ Type of Traffic
		Load* (K)	Load* (K)	Load* (K)	Load* (K)	Max. Uniform Load to Slab ▼ (PSF)
4"	Reinf. Plain	4K				100
5"	Reinf. Plain	6K	4K			200
6"	Reinf. Plain		8K	6K	6K	500 to 800
7"	Reinf. Plain			9K	8K	1,500
8"	Reinf. Plain				11K	* Max. Wheel Load in Kips (incl. impact)
10"	Reinf. Plain				14K	
12"	Reinf. Plain					
DESIGN ASSUMPTIONS	Concrete, Chuted	f'c = 3.5 KSI	4 KSI	4.5 KSI	Slab @ 3.5 KSI	ASSUMPTIONS BY SLAB TYPE
	Toppings			1" Integral	1" Bonded	
	Finish	Steel Trowel	Steel Trowel	Steel Trowel	Screed & Steel Trowel	
	Compacted Granular Base	4" deep for 4" slab thickness 6" deep for 5" slab thickness & greater				ASSUMPTIONS FOR ALL SLAB TYPES
	Vapor Barrier	6 mil polyethylene				
	Forms & Joints	Allowances included				
	Rein-forcement	WWF as required ≥ 60,000 psi				

R033105-85 Lift Slabs

The cost advantage of the lift slab method is due to placing all concrete, reinforcing steel, inserts and electrical conduit at ground level and in reduction of formwork. Minimum economical project size is about 30,000 S.F. Slabs may be tilted for parking garage ramps.

It is now used in all types of buildings and has gone up to 22 stories high in apartment buildings. The current trend is to use post-tensioned flat plate slabs with spans from 22′ to 35′. Cylindrical void forms are used when deep slabs are required. One pound of prestressing steel is about equal to seven pounds of conventional reinforcing.

To be considered cured for stressing and lifting, a slab must have attained 75% of design strength. Seven days are usually sufficient with four to five days possible if high early strength cement is used. Slabs can be stacked using two coats of a non-bonding agent to insure that slabs do not stick to each other. Lifting is done by companies specializing in this work. Lift rate is 5′ to 15′ per hour with an average of 10′ per hour. Total areas up to 33,000 S.F. have been lifted at one time. 24 to 36 jacking columns are common. Most economical bay sizes are 24′ to 28′ with four to fourteen stories most efficient. Continuous design reduces reinforcing steel cost. Use of post-tensioned slabs allows larger bay sizes.

R033543-10 Polished Concrete Floors

A polished concrete floor has a glossy mirror-like appearance and is created by grinding the concrete floor with finer and finer diamond grits, similar to sanding wood, until the desired level of reflective clarity and sheen are achieved. The technical term for this type of polished concrete is bonded abrasive polished concrete. The basic piece of equipment used in the polishing process is a walk-behind planetary grinder for working large floor areas. This grinder drives diamond-impregnated abrasive discs, which progress from coarse- to fine-grit discs.

The process begins with the use of very coarse diamond segments or discs bonded in a metallic matrix. These segments are coarse enough to allow the removal of pits, blemishes, stains, and light coatings from the floor surface in preparation for final smoothing. The condition of the original concrete surface will dictate the grit coarseness of the initial grinding step which will generally end up being a three- to four-step process using ever finer grits. The purpose of this initial grinding step is to remove surface coatings and blemishes and to cut down into the cream for very fine aggregate exposure, or deeper into the fine aggregate layer just below the cream layer, or even deeper into the coarse aggregate layer. These initial grinding steps will progress up to the 100/120 grit. If wet grinding is done, a waste slurry is produced that must be removed between grit changes and disposed of properly. If dry grinding is done, a high performance vacuum will pick up the dust during grinding and collect it in bags which must be disposed of properly.

The process continues with honing the floor in a series of steps that progresses from 100-grit to 400-grit diamond abrasive discs embedded in a plastic or resin matrix. At some point during, or just prior to, the honing step, one or two coats of stain or dye can be sprayed onto the surface to give color to the concrete, and two coats of densifier/hardener must be applied to the floor surface and allowed to dry. This sprayed-on densifier/hardener will penetrate about 1/8″ into the concrete to make the surface harder, denser and more abrasion-resistant.

The process ends with polishing the floor surface in a series of steps that progresses from resin-impregnated 800-grit (medium polish) to 1500-grit (high polish) to 3000-grit (very high polish), depending on the desired level of reflective clarity and sheen.

The Concrete Polishing Association of America (CPAA) has defined the flooring options available when processing concrete to a desired finish. The first category is aggregate exposure, the grinding of a concrete surface with bonded abrasives, in as many abrasive grits necessary, to achieve one of the following classes:

A. Cream – very little surface cut depth; little aggregate exposure

B. Fine aggregate (salt and pepper) – surface cut depth of 1/16″; fine aggregate exposure with little or no medium aggregate exposure at random locations

C. Medium aggregate – surface cut depth of 1/8″; medium aggregate exposure with little or no large aggregate exposure at random locations

D. Large aggregate – surface cut depth of 1/4″; large aggregate exposure with little or no fine aggregate exposure at random locations

The second CPAA defined category is reflective clarity and sheen, the polishing of a concrete surface with the minimum number of bonded abrasives as indicated to achieve one of the following levels:

1. Ground – flat appearance with none to very slight diffused reflection; none to very low reflective sheen; using a minimum total of 4 grit levels up 100-grit
2. Honed – matte appearance with or without slight diffused reflection; low to medium reflective sheen; using a minimum total of 5 grit levels up to 400-grit
3. Semi-polished – objects being reflected are not quite sharp and crisp but can be easily identified; medium to high reflective sheen; using a minimum total of 6 grit levels up to 800-grit
4. Highly-polished – objects being reflected are sharp and crisp as would be seen in a mirror-like reflection; high to highest reflective sheen; using a minimum total of up to 8 grit levels up to 1500-grit or 3000-grit

The CPAA defines reflective clarity as the degree of sharpness and crispness of the reflection of overhead objects when viewed 5′ above and perpendicular to the floor surface. Reflective sheen is the degree of gloss reflected from a surface when viewed at least 20′ from and at an angle to the floor surface. These terms are relatively subjective. The final outcome depends on the internal makeup and surface condition of the original concrete floor, the experience of the floor polishing crew, and the expectations of the owner. Before the grinding, honing, and polishing work commences on the main floor area, it might be beneficial to do a mock-up panel in the same floor but in an out of the way place to demonstrate the sequence of steps with increasingly fine abrasive grits and to demonstrate the final reflective clarity and reflective sheen. This mock-up panel will be within the area of, and part of, the final work.

R034105-30 Prestressed Precast Concrete Structural Units

Type	Location	Depth	Span in Ft.		Live Load Lb. per S.F.
Double Tee 8' to 10'	Floor	28″ to 34″	60 to 80		50 to 80
	Roof	12″ to 24″	30 to 50		40
	Wall	Width 8′	Up to 55′ high		Wind
Multiple Tee 8'	Roof	8″ to 12″	15 to 40		40
	Floor	8″ to 12″	15 to 30		100
Plank	Roof		Roof	Floor	
		4″	13	12	40 for Roof
		6″	22	18	
or	or	8″	26	25	
		10″	33	29	100 for Floor
	Floor	12″	42	32	
Single Tee 8' to 10'	Roof	28″	40		40
		32″	80		
		36″	100		
		48″	120		
AASHO Girder	Bridges	Type 4	100		Highway
		5	110		
		6	125		
Box Beam 4'	Bridges	15″ 27″ 33″	40 to 100		Highway

The majority of precast projects today utilize double tees rather than single tees because of speed and ease of installation. As a result casting beds at manufacturing plants are normally formed for double tees. Single tee projects will therefore require an initial set up charge to be spread over the individual single tee costs.

For floors, a 2″ to 3″ topping is field cast over the shapes. For roofs, insulating concrete or rigid insulation is placed over the shapes.

Member lengths up to 40′ are standard haul, 40′ to 60′ require special permits and lengths over 60′ must be escorted. Excessive width and/or length can add up to 100% on hauling costs.

Large heavy members may require two cranes for lifting which would increase erection costs by about 45%. An eight man crew can install 12 to 20 double tees, or 45 to 70 quad tees or planks per day.

Grouting of connections must also be included.

Several system buildings utilizing precast members are available. Heights can go up to 22 stories for apartment buildings. The optimum design ratio is 3 S.F. of surface to 1 S.F. of floor area.

R034136-90 Prestressed Concrete, Post-Tensioned

In post-tensioned concrete the steel tendons are tensioned after the concrete has reached about 3/4 of its ultimate strength. The cableways are grouted after tensioning to provide bond between the steel and concrete. If bond is to be prevented, the tendons are coated with a corrosion-preventative grease and wrapped with waterproofed paper or plastic. Bonded tendons are usually used when ultimate strength (beams & girders) are controlling factors.

High strength concrete is used to fully utilize the steel, thereby reducing the size and weight of the member. A plasticizing agent may be added to reduce water content. Maximum size aggregate ranges from 1/2″ to 1-1/2″ depending on the spacing of the tendons.

The types of steel commonly used are bars and strands. Job conditions determine which is best suited. Bars are best for vertical prestresses since they are easy to support. The trend is for steel manufacturers to supply a finished package, cut to length, which reduces field preparation to a minimum.

Bars vary from 3/4″ to 1-3/8″ diameter. The table below gives time in labor-hours per tendon for placing, tensioning and grouting (if required) a 75′ beam. Tendons used in buildings are not usually grouted; tendons for bridges usually are grouted. For strands the table indicates the labor-hours per pound for typical prestressed units 100′ long. Simple span beams usually require one- end stressing regardless of lengths. Continuous beams are usually stressed from two ends. Long slabs are poured from the center outward and stressed in 75′ increments after the initial 150′ center pour.

Labor Hours per Tendon and per Pound of Prestressed Steel						
Length	**100′ Beam**		**75′ Beam**		**100′ Slab**	
Type Steel	**Strand**		**Bars**		**Strand**	
Diameter	**0.5″**		**3/4″**	**1-3/8″**	**0.5″**	**0.6″**
Number	**4**	**12**	**1**	**1**	**1**	**1**
Force in Kips	**100**	**300**	**42**	**143**	**25**	**35**
Preparation & Placing Cables	3.6	7.4	0.9	2.9	0.9	1.1
Stressing Cables	2.0	2.4	0.8	1.6	0.5	0.5
Grouting, if required	2.5	3.0	0.6	1.3		
Total Labor Hours	8.1	12.8	2.3	5.8	1.4	1.6
Prestressing Steel Weights (Lbs.)	215	640	115	380	53	74
Labor-hours per Lb. Bonded	0.038	0.020	0.020	0.015		
Non-bonded					0.026	0.022

Flat slab construction — 4000 psi concrete with span-to-depth ratio between 36 and 44. Two way post-tensioned steel averages 1.0 lb. per S.F. for 24′ to 28′ bays (usually strand) and additional reinforcing steel averages .5 lb. per S.F.

Pan and joist construction — 4000 psi concrete with span-to-depth ratio between 28 to 30. Post-tensioned steel averages .8 lb. per S.F. and reinforcing steel about 1.0 lb. per S.F. Placing and stressing average 40 hours per ton of total material.

Beam construction — 4000 to 5000 psi concrete. Steel weights vary greatly.

Labor cost per pound goes down as the size and length of the tendon increase. The primary economic consideration is the cost per kip for the member.

Post-tensioning becomes feasible for beams and girders over 30′ long; for continuous two-way slabs over 20′ clear; and for transferring upper building loads over longer spans at lower levels. Post-tension suppliers will provide engineering services at no cost to the user. Substantial economies are possible by using post-tensioned lift slabs.

R040130-10 Cleaning Face Brick

On smooth brick a person can clean 70 S.F. an hour; on rough brick 50 S.F. per hour. Use one gallon muriatic acid to 20 gallons of water for 1000 S.F. Do not use acid solution until wall is at least seven days old, but a mild soap solution may be used after two days.

Time has been allowed for clean-up in brick prices.

R040513-10 Cement Mortar (material only)

Type N - 1:1:6 mix by volume. Use everywhere above grade except as noted below. - 1:3 mix using conventional masonry cement which saves handling two separate bagged materials.

Type M - 1:1/4:3 mix by volume, or 1 part cement, 1/4 (10% by wt.) lime, 3 parts sand. Use for heavy loads and where earthquakes or hurricanes may occur. Also for reinforced brick, sewers, manholes and everywhere below grade.

Mix Proportions by Volume and Compressive Strength of Mortar

Where Used	Allowable Proportions by Volume					Compressive Strength @ 28 days
	Mortar Type	Portland Cement	Masonry Cement	Hydrated Lime	Masonry Sand	
Plain Masonry	M	1 1	1 —	— 1/4	6 3	2500 psi
	S	1/2 1	1 —	— 1/4 to 1/2	4 4	1800 psi
	N	— 1	1 —	— 1/2 to 1-1/4	3 6	750 psi
	O	— 1	1 —	— 1-1/4 to 2-1/2	3 9	350 psi
	K	1	—	2-1/2 to 4	12	75 psi
Reinforced Masonry	PM	1	1	—	6	2500 psi
	PL	1	—	1/4 to 1/2	4	2500 psi

Note: The total aggregate should be between 2.25 to 3 times the sum of the cement and lime used.

The labor cost to mix the mortar is included in the productivity and labor cost of unit price lines in unit cost sections for brickwork, blockwork and stonework.

The material cost of mixed mortar is included in the material cost of those same unit price lines and includes the cost of renting and operating a 10 C.F. mixer at the rate of 200 C.F. per day.

There are two types of mortar color used. One type is the inert additive type with about 100 lbs. per M brick as the typical quantity required. These colors are also available in smaller-batch-sized bags (1 lb. to 15 lb.) which can be placed directly into the mixer without measuring. The other type is premixed and replaces the masonry cement. Dark green color has the highest cost.

R040519-50 Masonry Reinforcing

Horizontal joint reinforcing helps prevent wall cracks where wall movement may occur and in many locations is required by code. Horizontal joint reinforcing is generally not considered to be structural reinforcing and an unreinforced wall may still contain joint reinforcing.

Reinforcing strips come in 10′ and 12′ lengths and in truss and ladder shapes, with and without drips. Field labor runs between 2.7 to 5.3 hours per 1000 L.F. for wall thicknesses up to 12″.

The wire meets ASTM A82 for cold drawn steel wire and the typical size is 9 ga. sides and ties with 3/16″ diameter also available. Typical finish is mill galvanized with zinc coating at .10 oz. per S.F. Class I (.40 oz. per S.F.) and Class III (.80 oz. per S.F.) are also available, as is hot dipped galvanizing at 1.50 oz. per S.F.

R042110-10 Economy in Bricklaying

Have adequate supervision. Be sure bricklayers are always supplied with materials so there is no waiting. Place experienced bricklayers at corners and openings.

Use only screened sand for mortar. Otherwise, labor time will be wasted picking out pebbles. Use seamless metal tubs for mortar as they do not leak or catch the trowel. Locate stack and mortar for easy wheeling.

Have brick delivered for stacking. This makes for faster handling, reduces chipping and breakage, and requires less storage space. Many dealers will deliver select common in 2′ x 3′ x 4′ pallets or face brick packaged. This affords quick handling with a crane or forklift and easy tonging in units of ten, which reduces waste.

Use wider bricks for one wythe wall construction. Keep scaffolding away from the wall to allow mortar to fall clear and not stain the wall.

On large jobs develop specialized crews for each type of masonry unit.

Consider designing for prefabricated panel construction on high rise projects.

Avoid excessive corners or openings. Each opening adds about 50% to the labor cost for area of opening.

Bolting stone panels and using window frames as stops reduce labor costs and speed up erection.

R042110-20 Common and Face Brick

Common building brick manufactured according to ASTM C62 and facing brick manufactured according to ASTM C216 are the two standard bricks available for general building use.

Building brick is made in three grades: SW, where high resistance to damage caused by cyclic freezing is required; MW, where moderate resistance to cyclic freezing is needed; and NW, where little resistance to cyclic freezing is needed. Facing brick is made in only the two grades SW and MW. Additionally, facing brick is available in three types: FBS, for general use; FBX, for general use where a higher degree of precision and lower permissible variation in size than FBS are needed; and FBA, for general use to produce characteristic architectural effects resulting from non-uniformity in size and texture of the units.

In figuring the material cost of brickwork, an allowance of 25% mortar waste and 3% brick breakage was included. If bricks are delivered palletized with 280 to 300 per pallet, or packaged, allow only 1-1/2% for breakage. Packaged or palletized delivery is practical when a job is big enough to have a crane or other equipment available to handle a package of brick. This is so on all industrial work but not always true on small commercial buildings.

The use of buff and gray face is increasing, and there is a continuing trend to the Norman, Roman, Jumbo and SCR brick.

Common red clay brick for backup is not used that often. Concrete block is the most usual backup material with occasional use of sand lime or cement brick. Building brick is commonly used in solid walls for strength and as a fire stop.

Brick panels built on the ground and then crane erected to the upper floors have proven to be economical. This allows the work to be done under cover and without scaffolding.

R042110-50 Brick, Block & Mortar Quantities

Running Bond						For Other Bonds Standard Size Add to S.F. Quantities in Table to Left		
Number of Brick per S.F. of Wall - Single Wythe with 3/8″ Joints				C.F. of Mortar per M Bricks, Waste Included				
Type Brick	Nominal Size (incl. mortar) L H W	Modular Coursing	Number of Brick per S.F.	3/8″ Joint	1/2″ Joint	Bond Type	Description	Factor
Standard	8 x 2-2/3 x 4	3C=8″	6.75	8.1	10.3	Common	full header every fifth course	+20%
Economy	8 x 4 x 4	1C=4″	4.50	9.1	11.6		full header every sixth course	+16.7%
Engineer	8 x 3-1/5 x 4	5C=16″	5.63	8.5	10.8	English	full header every second course	+50%
Fire	9 x 2-1/2 x 4-1/2	2C=5″	6.40	550 # Fireclay	—	Flemish	alternate headers every course	+33.3%
Jumbo	12 x 4 x 6 or 8	1C=4″	3.00	22.5	29.2		every sixth course	+5.6%
Norman	12 x 2-2/3 x 4	3C=8″	4.50	11.2	14.3	Header = W x H exposed		+100%
Norwegian	12 x 3-1/5 x 4	5C=16″	3.75	11.7	14.9	Rowlock = H x W exposed		+100%
Roman	12 x 2 x 4	2C=4″	6.00	10.7	13.7	Rowlock stretcher = L x W exposed		+33.3%
SCR	12 x 2-2/3 x 6	3C=8″	4.50	21.8	28.0	Soldier = H x L exposed		—
Utility	12 x 4 x 4	1C=4″	3.00	12.3	15.7	Sailor = W x L exposed		-33.3%

Concrete Blocks Nominal Size	Approximate Weight per S.F.		Blocks per 100 S.F.	Mortar per M block, waste included	
	Standard	Lightweight		Partitions	Back up
2″ x 8″ x 16″	20 PSF	15 PSF	113	27 C.F.	36 C.F.
4″	30	20		41	51
6″	42	30		56	66
8″	55	38		72	82
10″	70	47		87	97
12″	85	55		102	112

Brick & Mortar Quantities
©Brick Industry Association. 2009 Feb. Technical Notes on Brick Construction 10:
Dimensioning and Estimating Brick Masonry. Reston (VA): BIA. Table 1 Modular Brick Sizes and Table 4 Quantity Estimates for Brick Masonry.

R042210-20 Concrete Block

The material cost of special block such as corner, jamb and head block can be figured at the same price as ordinary block of equal size. Labor on specials is about the same as equal-sized regular block.

Bond beams and 16″ high lintel blocks are more expensive than regular units of equal size. Lintel blocks are 8″ long and either 8″ or 16″ high.

Use of a motorized mortar spreader box will speed construction of continuous walls.

Hollow non-load-bearing units are made according to ASTM C129 and hollow load-bearing units according to ASTM C90.

R050516-30 Coating Structural Steel

On field-welded jobs, the shop-applied primer coat is necessarily omitted. All painting must be done in the field and usually consists of red oxide rust inhibitive paint or an aluminum paint. The table below shows paint coverage and daily production for field painting.

See Division 09 97 13.23 for hot-dipped galvanizing and for field-applied cold galvanizing and other paints and protective coatings.

See Division 05 01 10.51 for steel surface preparation treatments such as wire brushing, pressure washing and sand blasting.

Type Construction	Surface Area per Ton	Coat	One Gallon Covers		In 8 Hrs. Person Covers		Average per Ton Spray	
			Brush	Spray	Brush	Spray	Gallons	Labor-hours
Light Structural	300 S.F. to 500 S.F.	1st	500 S.F.	455 S.F.	640 S.F.	2000 S.F.	0.9 gals.	1.6 L.H.
		2nd	450	410	800	2400	1.0	1.3
		3rd	450	410	960	3200	1.0	1.0
Medium	150 S.F. to 300 S.F.	All	400	365	1600	3200	0.6	0.6
Heavy Structural	50 S.F. to 150 S.F.	1st	400	365	1920	4000	0.2	0.2
		2nd	400	365	2000	4000	0.2	0.2
		3rd	400	365	2000	4000	0.2	0.2
Weighted Average	225 S.F.	All	400	365	1350	3000	0.6	0.6

R050521-20 Welded Structural Steel

Usual weight reductions with welded design run 10% to 20% compared with bolted or riveted connections. This amounts to about the same total cost compared with bolted structures since field welding is more expensive than bolts. For normal spans of 18′ to 24′ figure 6 to 7 connections per ton.

Trusses — For welded trusses add 4% to weight of main members for connections. Up to 15% less steel can be expected in a welded truss compared to one that is shop bolted. Cost of erection is the same whether shop bolted or welded.

General — Typical electrodes for structural steel welding are E6010, E6011, E60T and E70T. Typical buildings vary between 2# to 8# of weld rod per ton of steel. Buildings utilizing continuous design require about three times as much welding as conventional welded structures. In estimating field erection by welding, it is best to use the average linear feet of weld per ton to arrive at the welding cost per ton. The type, size and position of the weld will have a direct bearing on the cost per linear foot. A typical field welder will deposit 1.8# to 2# of weld rod per hour manually. Using semiautomatic methods can increase production by as much as 50% to 75%.

R050523-10 High Strength Bolts

Common bolts (A307) are usually used in secondary connections (see Division 05 05 23.10).

High strength bolts (A325 and A490) are usually specified for primary connections such as column splices, beam and girder connections to columns, column bracing, connections for supports of operating equipment or of other live loads which produce impact or reversal of stress, and in structures carrying cranes of over 5-ton capacity.

Allow 20 field bolts per ton of steel for a 6 story office building, apartment house or light industrial building. For 6 to 12 stories allow 25 bolts per ton, and above 12 stories, 30 bolts per ton. On power stations, 20 to 25 bolts per ton are needed.

R051223-10 Structural Steel

The bare material prices for structural steel, shown in the unit cost sections of the data set, are for 100 tons of shop-fabricated structural steel and include:

1. Mill base price of structural steel
2. Mill scrap/grade/size/length extras
3. Mill delivery to a metals service center (warehouse)
4. Service center storage and handling
5. Service center delivery to a fabrication shop
6. Shop storage and handling
7. Shop drafting/detailing
8. Shop fabrication
9. Shop coat of primer paint
10. Shop listing
11. Shop delivery to the job site

In unit cost sections of the data set that contain items for field fabrication of steel components, the bare material cost of steel includes:

1. Mill base price of structural steel
2. Mill scrap/grade/size/length extras
3. Mill delivery to a metals service center (warehouse)
4. Service center storage and handling
5. Service center delivery to the job site

R051223-20 Steel Estimating Quantities

One estimate on erection is that a crane can handle 35 to 60 pieces per day. Say the average is 45. With usual sizes of beams, girders, and columns, this would amount to about 20 tons per day. The type of connection greatly affects the speed of erection. Moment connections for continuous design slow down production and increase erection costs.

Short open web bar joists can be set at the rate of 75 to 80 per day, with 50 per day being the average for setting long span joists.

After main members are calculated, add the following for usual allowances: base plates 2% to 3%; column splices 4% to 5%; and miscellaneous details 4% to 5%, for a total of 10% to 13% in addition to main members.

The ratio of column to beam tonnage varies depending on type of steels used, typical spans, story heights and live loads.

It is more economical to keep the column size constant and to vary the strength of the column by using high strength steels. This also saves floor space. Buildings have recently gone as high as ten stories with 8″ high strength columns. For light columns under W8X31 lb. sections, concrete filled steel columns are economical.

High strength steels may be used in columns and beams to save floor space and to meet head room requirements. High strength steels in some sizes sometimes require long lead times.

Round, square and rectangular columns, both plain and concrete filled, are readily available and save floor area, but are higher in cost per pound than rolled columns. For high unbraced columns, tube columns may be less expensive.

Below are average minimum figures for the weights of the structural steel frame for different types of buildings using A36 steel, rolled shapes and simple joints. For economy in domes, rise to span ratio = .13. Open web joist framing systems will reduce weights by 10% to 40%. Composite design can reduce steel weight by up to 25% but additional concrete floor slab thickness may be required. Continuous design can reduce the weights up to 20%. There are many building codes with different live load requirements and different structural requirements, such as hurricane and earthquake loadings, which can alter the figures.

Structural Steel Weights per S.F. of Floor Area

Type of Building	No. of Stories	Avg. Spans	L.L. #/S.F.	Lbs. Per S.F.	Type of Building	No. of Stories	Avg. Spans	L.L. #/S.F.	Lbs. Per S.F.
Steel Frame Mfg.	1	20′x20′	40	8	Apartments	2-8	20′x20′	40	8
		30′x30′		13		9-25			14
		40′x40′		18	Office	to 10	Various	80	10
Parking garage	4	Various	80	8.5		20			18
Domes	1	200′	30	10		30			26
(Schwedler)*		300′		15		over 50			35

R051223-25 Common Structural Steel Specifications

ASTM A992 (formerly A36, then A572 Grade 50) is the all-purpose carbon grade steel widely used in building and bridge construction.

The other high-strength steels listed below may each have certain advantages over ASTM A992 structural carbon steel, depending on the application. They have proven to be economical choices where, due to lighter members, the reduction of dead load and the associated savings in shipping cost can be significant.

ASTM A588 atmospheric weathering, high-strength, low-alloy steels can be used in the bare (uncoated) condition, where exposure to normal atmosphere causes a tightly adherant oxide to form on the surface, protecting the steel from further oxidation. ASTM A242 corrosion-resistant, high-strength, low-alloy steels have enhanced atmospheric corrosion resistance of at least two times that of carbon structural steels with copper, or four times that of carbon structural steels without copper. The reduction or elimination of maintenance resulting from the use of these steels often offsets their higher initial cost.

Steel Type	ASTM Designation	Minimum Yield Stress in KSI	Shapes Available
Carbon	A36	36	All structural shape groups, and plates & bars up through 8″ thick
	A529	50	Structural shape group 1, and plates & bars up through 2″ thick
		50	All structural shape groups
High-Strength Low-Alloy	A913	60	
Quenched & Self-Tempered		65	
		70	
		42	All structural shape groups, and plates & bars up through 6″ thick
High-Strength Low-Alloy	A572	50	All structural shape groups, and plates & bars up through 4″ thick
Columbium-Vanadium		55	Structural shape groups 1 & 2, and plates & bars up through 2″ thick
		60	Structural shape groups 1 & 2, and plates & bars up through 1-1/4″ thick
		65	Structural shape group 1, and plates & bars up through 1-1/4″ thick
High-Strength Low-Alloy	A992	50	All structural shape groups
Columbium-Vanadium			
Weathering		42	Structural shape groups 4 & 5, and plates & bars over 1-1/2″ up through 4″ thick
High-Strength Low-Alloy	A242	46	Structural shape group 3, and plates & bars over 3/4″ up through 1-1/2″ thick
		50	Structural shape groups 1 & 2, and plates & bars up through 3/4″ thick
Weathering		42	Plates & bars over 5″ up through 8″ thick
High-Strength Low-Alloy	A588	46	Plates & bars over 4″ up through 5″ thick
		50	All structural shape groups, and plates & bars up through 4″ thick
Quenched and Tempered	A852	70	Plates & bars up through 4″ thick
Low-Alloy			
Quenched and Tempered	A514	90	Plates & bars over 2-1/2″ up through 6″ thick
Alloy		100	Plates & bars up through 2-1/2″ thick

R051223-30 High Strength Steels

The mill price of high strength steels may be higher than A992 carbon steel, but their proper use can achieve overall savings through total reduced weights. For columns with L/r over 100, A992 steel is best; under 100, high strength steels are economical. For heavy columns, high strength steels are economical when cover plates are eliminated. There is no economy using high strength steels for clip angles or supports or for beams where deflection governs. Thinner members are more economical than thick.

The per ton erection and fabricating costs of the high strength steels will be higher than for A992 since the same number of pieces, but less weight, will be installed.

R053100-10 Decking Descriptions

General - All Deck Products

A steel deck is made by cold forming structural grade sheet steel into a repeating pattern of parallel ribs. The strength and stiffness of the panels are the result of the ribs and the material properties of the steel. Deck lengths can be varied to suit job conditions, but because of shipping considerations, are usually less than 40 feet. Standard deck width varies with the product used but full sheets are usually 12″, 18″, 24″, 30″, or 36″. The deck is typically furnished in a standard width with the ends cut square. Any cutting for width, such as at openings or for angular fit, is done at the job site.

The deck is typically attached to the building frame with arc puddle welds, self-drilling screws, or powder or pneumatically driven pins. Sheet to sheet fastening is done with screws, button punching (crimping), or welds.

Composite Floor Deck

After installation and adequate fastening, a floor deck serves several purposes. It (a) acts as a working platform, (b) stabilizes the frame, (c) serves as a concrete form for the slab, and (d) reinforces the slab to carry the design loads applied during the life of the building. Composite decks are distinguished by the presence of shear connector devices as part of the deck. These devices are designed to mechanically lock the concrete and deck together so that the concrete and the deck work together to carry subsequent floor loads. These shear connector devices can be rolled-in embossments, lugs, holes, or wires welded to the panels. The deck profile can also be used to interlock concrete and steel.

Composite deck finishes are either galvanized (zinc coated) or phosphatized/painted. Galvanized deck has a zinc coating on both the top and bottom surfaces. The phosphatized/painted deck has a bare (phosphatized) top surface that will come into contact with the concrete. This bare top surface can be expected to develop rust before the concrete is placed. The bottom side of the deck has a primer coat of paint.

A composite floor deck is normally installed so the panel ends do not overlap on the supporting beams. Shear lugs or panel profile shapes often prevent a tight metal to metal fit if the panel ends overlap; the air gap caused by overlapping will prevent proper fusion with the structural steel supports when the panel end laps are shear stud welded.

Adequate end bearing of the deck must be obtained as shown on the drawings. If bearing is actually less in the field than shown on the drawings, further investigation is required.

Roof Deck

A roof deck is not designed to act compositely with other materials. A roof deck acts alone in transferring horizontal and vertical loads into the building frame. Roof deck rib openings are usually narrower than floor deck rib openings. This provides adequate support of the rigid thermal insulation board.

A roof deck is typically installed to endlap approximately 2″ over supports. However, it can be butted (or lapped more than 2″) to solve field fit problems. Since designers frequently use the installed deck system as part of the horizontal bracing system (the deck as a diaphragm), any fastening substitution or change should be approved by the designer. Continuous perimeter support of the deck is necessary to limit edge deflection in the finished roof and may be required for diaphragm shear transfer.

Standard roof deck finishes are galvanized or primer painted. The standard factory applied paint for roof decks is a primer paint and is not intended to weather for extended periods of time. Field painting or touching up of abrasions and deterioration of the primer coat or other protective finishes is the responsibility of the contractor.

Cellular Deck

A cellular deck is made by attaching a bottom steel sheet to a roof deck or composite floor deck panel. A cellular deck can be used in the same manner as a floor deck. Electrical, telephone, and data wires are easily run through the chase created between the deck panel and the bottom sheet.

When used as part of the electrical distribution system, the cellular deck must be installed so that the ribs line up and create a smooth cell transition at abutting ends. The joint that occurs at butting cell ends must be taped or otherwise sealed to prevent wet concrete from seeping into the cell. Cell interiors must be free of welding burrs, or other sharp intrusions, to prevent damage to wires.

When used as a roof deck, the bottom flat plate is usually left exposed to view. Care must be maintained during erection to keep good alignment and prevent damage.

A cellular deck is sometimes used with the flat plate on the top side to provide a flat working surface. Installation of the deck for this purpose requires special methods for attachment to the frame because the flat plate, now on the top, can prevent direct access to the deck material that is bearing on the structural steel. It may be advisable to treat the flat top surface to prevent slipping.

A cellular deck is always furnished galvanized or painted over galvanized.

Form Deck

A form deck can be any floor or roof deck product used as a concrete form. Connections to the frame are by the same methods used to anchor floor and roof decks. Welding washers are recommended when welding a deck that is less than 20 gauge thickness.

A form deck is furnished galvanized, prime painted, or uncoated. A galvanized deck must be used for those roof deck systems where a form deck is used to carry a lightweight insulating concrete fill.

R061110-30 Lumber Product Material Prices

The price of forest products fluctuates widely from location to location and from season to season depending upon economic conditions. The bare material prices in the unit cost sections of the data set show the National Average material prices in effect Jan. 1 of this data year. It must be noted that lumber prices in general may change significantly during the year.

Availability of certain items depends upon geographic location and must be checked prior to firm-price bidding.

R061636-20 Plywood

There are two types of plywood used in construction: interior, which is moisture-resistant but not waterproofed, and exterior, which is waterproofed.

The grade of the exterior surface of the plywood sheets is designated by the first letter: A, for smooth surface with patches allowed; B, for solid surface with patches and plugs allowed; C, which may be surface plugged or may have knot holes up to 1″ wide; and D, which is used only for interior type plywood and may have knot holes up to 2-1/2″ wide. "Structural Grade" is specifically designed for engineered applications such as box beams. All CC & DD grades have roof and floor spans marked on them.

Underlayment-grade plywood runs from 1/4″ to 1-1/4″ thick. Thicknesses 5/8″ and over have optional tongue and groove joints which eliminate the need for blocking the edges. Underlayment 19/32″ and over may be referred to as Sturd-i-Floor.

The price of plywood can fluctuate widely due to geographic and economic conditions.

Typical uses for various plywood grades are as follows:

AA-AD Interior — cupboards, shelving, paneling, furniture

BB Plyform — concrete form plywood

CDX — wall and roof sheathing

Structural — box beams, girders, stressed skin panels

AA-AC Exterior — fences, signs, siding, soffits, etc.

Underlayment — base for resilient floor coverings

Overlaid HDO — high density for concrete forms & highway signs

Overlaid MDO — medium density for painting, siding, soffits & signs

303 Siding — exterior siding, textured, striated, embossed, etc.

R073126-20 Roof Slate

16″, 18″ and 20″ are standard lengths, and slate usually comes in random widths. For standard 3/16″ thickness use 1-1/2″ copper nails. Allow for 3% breakage.

R075113-20 Built-Up Roofing

Asphalt is available in kegs of 100 lbs. each; coal tar pitch in 560 lb. kegs. Prepared roofing felts are available in a wide range of sizes, weights and characteristics. However, the most commonly used are #15 (432 S.F. per roll, 13 lbs. per square) and #30 (216 S.F. per roll, 27 lbs. per square).

Inter-ply bitumen varies from 24 lbs. per sq. (asphalt) to 30 lbs. per sq. (coal tar) per ply, MF4@ 25%. Flood coat bitumen also varies from 60 lbs. per sq. (asphalt) to 75 lbs. per sq. (coal tar), MF4@ 25%. Expendable equipment (mops, brooms, screeds, etc.) runs about 16% of the bitumen cost. For new, inexperienced crews this factor may be much higher.

A rigid insulation board is typically applied in two layers. The first is mechanically attached to nailable decks or spot or solid mopped to non-nailable decks; the second layer is then spot or solid mopped to the first layer. Membrane application follows the insulation, except in protected membrane roofs, where the membrane goes down first and the insulation on top, followed with ballast (stone or concrete pavers). Insulation and related labor costs are NOT included in prices for built-up roofing.

R075213-30 Modified Bitumen Roofing

The cost of modified bitumen roofing is highly dependent on the type of installation that is planned. Installation is based on the type of modifier used in the bitumen. The two most popular modifiers are atactic polypropylene (APP) and styrene butadiene styrene (SBS). The modifiers are added to heated bitumen during the manufacturing process to change its characteristics. A polyethylene, polyester or fiberglass reinforcing sheet is then sandwiched between layers of this bitumen. When completed, the result is a pre-assembled, built-up roof that has increased elasticity and weatherability. Some manufacturers include a surfacing material such as ceramic or mineral granules, metal particles or sand.

The preferred method of adhering SBS-modified bitumen roofing to the substrate is with hot-mopped asphalt (much the same as built-up roofing). This installation method requires a tar kettle/pot to heat the asphalt, as well as the labor, tools and equipment necessary to distribute and spread the hot asphalt.

The alternative method for applying APP and SBS modified bitumen is as follows. A skilled installer uses a torch to melt a small pool of bitumen off the membrane. This pool must form across the entire roll for proper adhesion. The installer must unroll the roofing at a pace slow enough to melt the bitumen, but fast enough to prevent damage to the rest of the membrane.

Modified bitumen roofing provides the advantages of both built-up and single-ply roofing. Labor costs are reduced over those of built-up roofing because only a single ply is necessary. The elasticity of single-ply roofing is attained with the reinforcing sheet and polymer modifiers. Modifieds have some self-healing characteristics and because of their multi-layer construction, they offer the reliability and safety of built-up roofing.

Thermal & Moist. Protec. R0784 Firestopping

R078413-30 Firestopping

Firestopping is the sealing of structural, mechanical, electrical and other penetrations through fire-rated assemblies. The basic components of firestop systems are safing insulation and firestop sealant on both sides of wall penetrations and the top side of floor penetrations.

Pipe penetrations are assumed to be through concrete, grout, or joint compound and can be sleeved or unsleeved. Costs for the penetrations and sleeves are not included. An annular space of 1″ is assumed. Escutcheons are not included.

A metallic pipe is assumed to be copper, aluminum, cast iron or similar metallic material. An insulated metallic pipe is assumed to be covered with a thermal insulating jacket of varying thickness and materials.

A non-metallic pipe is assumed to be PVC, CPVC, FR Polypropylene or similar plastic piping material. Intumescent firestop sealants or wrap strips are included. Collars on both sides of wall penetrations and a sheet metal plate on the underside of floor penetrations are included.

Ductwork is assumed to be sheet metal, stainless steel or similar metallic material. Duct penetrations are assumed to be through concrete, grout or joint compound. Costs for penetrations and sleeves are not included. An annular space of 1/2″ is assumed.

Multi-trade openings include costs for sheet metal forms, firestop mortar, wrap strips, collars and sealants as necessary.

Structural penetrations joints are assumed to be 1/2″ or less. CMU walls are assumed to be within 1-1/2″ of the metal deck. Drywall walls are assumed to be tight to the underside of metal decking.

Metal panel, glass or curtain wall systems include a spandrel area of 5′ filled with mineral wool foil-faced insulation. Fasteners and stiffeners are included.

R081313-20 Steel Door Selection Guide

Standard steel doors are classified into four levels, as recommended by the Steel Door Institute in the chart below. Each of the four levels offers a range of construction models and designs to meet architectural requirements for preference and appearance, including full flush, seamless, and stile & rail. Recommended minimum gauge requirements are also included.

For complete standard steel door construction specifications and available sizes, refer to the Steel Door Institute Technical Data Series, ANSI A250.8-98 (SDI-100), and ANSI A250.4-94 Test Procedure and Acceptance Criteria for Physical Endurance of Steel Door and Hardware Reinforcements.

Level		Model	Construction	For Full Flush or Seamless		
				Min. Gauge	Thickness (in)	Thickness (mm)
I	Standard Duty	1 2	Full Flush Seamless	20	0.032	0.8
II	Heavy Duty	1 2	Full Flush Seamless	18	0.042	1.0
III	Extra Heavy Duty	1 2 3	Full Flush Seamless *Stile & Rail	16	0.053	1.3
IV	Maximum Duty	1 2	Full Flush Seamless	14	0.067	1.6

*Stiles & rails are 16 gauge; flush panels, when specified, are 18 gauge

Openings R0853 Plastic Windows

R085313-20 Replacement Windows

Replacement windows are typically measured per United Inch.

United Inches are calculated by rounding the width and height of the window opening up to the nearest inch, then adding the two figures.

The labor cost for replacement windows includes removal of sash, existing sash balance or weights, parting bead where necessary and installation of new window.

Debris hauling and dump fees are not included.

R087110-10 Hardware Finishes

This table describes hardware finishes used throughout the industry. It also shows the base metal and the respective symbols in the three predominate systems of identification. Many of these are used in pricing descriptions in Division Eight.

US″	BMHA*	CDN^	Base	Description
US P	600	CP	Steel	Primed for Painting
US 1B	601	C1B	Steel	Bright Black Japanned
US 2C	602	C2C	Steel	Zinc Plated
US 2G	603	C2G	Steel	Zinc Plated
US 3	605	C3	Brass	Bright Brass, Clear Coated
US 4	606	C4	Brass	Satin Brass, Clear Coated
US 5	609	C5	Brass	Satin Brass, Blackened, Satin Relieved, Clear Coated
US 7	610	C7	Brass	Satin Brass, Blackened, Bright Relieved, Clear Coated
US 9	611	C9	Bronze	Bright Bronze, Clear Coated
US 10	612	C10	Bronze	Satin Bronze, Clear Coated
US 10A	641	C10A	Steel	Antiqued Bronze, Oiled and Lacquered
US 10B	613	C10B	Bronze	Antiqued Bronze, Oiled
US 11	616	C11	Bronze	Satin Bronze, Blackened, Satin Relieved, Clear Coated
US 14	618	C14	Brass/Bronze	Bright Nickel Plated, Clear Coated
US 15	619	C15	Brass/Bronze	Satin Nickel, Clear Coated
US 15A	620	C15A	Brass/Bronze	Satin Nickel Plated, Blackened, Satin Relieved, Clear Coated
US 17A	621	C17A	Brass/Bronze	Nickel Plated, Blackened, Relieved, Clear Coated
US 19	622	C19	Brass/Bronze	Flat Black Coated
US 20	623	C20	Brass/Bronze	Statuary Bronze, Light
US 20A	624	C20A	Brass/Bronze	Statuary Bronze, Dark
US 26	625	C26	Brass/Bronze	Bright Chromium
US 26D	626	C26D	Brass/Bronze	Satin Chromium
US 20	627	C27	Aluminum	Satin Aluminum Clear
US 28	628	C28	Aluminum	Anodized Dull Aluminum
US 32	629	C32	Stainless Steel	Bright Stainless Steel
US 32D	630	C32D	Stainless Steel	Stainless Steel
US 3	632	C3	Steel	Bright Brass Plated, Clear Coated
US 4	633	C4	Steel	Satin Brass, Clear Coated
US 7	636	C7	Steel	Satin Brass Plated, Blackened, Bright Relieved, Clear Coated
US 9	637	C9	Steel	Bright Bronze Plated, Clear Coated
US 5	638	C5	Steel	Satin Brass Plated, Blackened, Bright Relieved, Clear Coated
US 10	639	C10	Steel	Satin Bronze Plated, Clear Coated
US 10B	640	C10B	Steel	Antique Bronze, Oiled
US 10A	641	C10A	Steel	Antiqued Bronze, Oiled and Lacquered
US 11	643	C11	Steel	Satin Bronze Plated, Blackened, Bright Relieved, Clear Coated
US 14	645	C14	Steel	Bright Nickel Plated, Clear Coated
US 15	646	C15	Steel	Satin Nickel Plated, Clear Coated
US 15A	647	C15A	Steel	Nickel Plated, Blackened, Bright Relieved, Clear Coated
US 17A	648	C17A	Steel	Nickel Plated, Blackened, Relieved, Clear Coated
US 20	649	C20	Steel	Statuary Bronze, Light
US 20A	650	C20A	Steel	Statuary Bronze, Dark
US 26	651	C26	Steel	Bright Chromium Plated
US 26D	652	C26D	Steel	Satin Chromium Plated

* - BMHA Builders Hardware Manufacturing Association
″ - US Equivalent
^ - Canadian Equivalent
Japanning is imitating Asian lacquer work

R088110-10 Glazing Productivity

Some glass sizes are estimated by the "united inch" (height + width). The table below shows the number of lights glazed in an eight-hour period by the crew size indicated, for glass up to 1/4″ thick. Square or nearly square lights are more economical on a S.F. basis. Long slender lights will have a high S.F. installation cost. For insulated glass reduce production by 33%. For 1/2″ float glass reduce production by 50%. Production time for glazing with two glaziers per day averages: 1/4″ float glass 120 S.F.; 1/2″ float glass 55 S.F.; 1/2″ insulated glass 95 S.F.; 3/4″ insulated glass 75 S.F.

Glazing Method	United Inches per Light							
	40″	60″	80″	100″	135″	165″	200″	240″
Number of Men in Crew	1	1	1	1	2	3	3	4
Industrial sash, putty	60	45	24	15	18	—	—	—
With stops, putty bed	50	36	21	12	16	8	4	3
Wood stops, rubber	40	27	15	9	11	6	3	2
Metal stops, rubber	30	24	14	9	9	6	3	2
Structural glass	10	7	4	3	—	—	—	—
Corrugated glass	12	9	7	4	4	4	3	—
Storefronts	16	15	13	11	7	6	4	4
Skylights, putty glass	60	36	21	12	16	—	—	—
Thiokol set	15	15	11	9	9	6	3	2
Vinyl set, snap on	18	18	13	12	12	7	5	4
Maximum area per light	2.8 S.F.	6.3 S.F.	11.1 S.F.	17.4 S.F.	31.6 S.F.	47 S.F.	69 S.F.	100 S.F.

R092000-50 Lath, Plaster and Gypsum Board

Gypsum board lath is available in 3/8″ thick x 16″ wide x 4′ long sheets as a base material for multi-layer plaster applications. It is also available as a base for either multi-layer or veneer plaster applications in 1/2″ and 5/8″ thick–4′ wide x 8′, 10′ or 12′ long sheets. Fasteners are screws or blued ring shank nails for wood framing and screws for metal framing.

Metal lath is available in diamond mesh patterns with flat or self-furring profiles. Paper backing is available for applications where excessive plaster waste needs to be avoided. A slotted mesh ribbed lath should be used in areas where the span between structural supports is greater than normal. Most metal lath comes in 27″ x 96″ sheets. Diamond mesh weighs 1.75, 2.5 or 3.4 pounds per square yard, slotted mesh lath weighs 2.75 or 3.4 pounds per square yard. Metal lath can be nailed, screwed or tied in place.

Many **accessories** are available. Corner beads, flat reinforcing strips, casing beads, control and expansion joints, furring brackets and channels are some examples. Note that accessories are not included in plaster or stucco line items.

Plaster is defined as a material or combination of materials that when mixed with a suitable amount of water, forms a plastic mass or paste. When applied to a surface, the paste adheres to it and subsequently hardens, preserving in a rigid state the form or texture imposed during the period of elasticity.

Gypsum plaster is made from ground calcined gypsum. It is mixed with aggregates and water for use as a base coat plaster.

Vermiculite plaster is a fire-retardant plaster covering used on steel beams, concrete slabs and other heavy construction materials. Vermiculite is a group name for certain clay minerals, hydrous silicates or aluminum, magnesium and iron that have been expanded by heat.

Perlite plaster is a plaster using perlite as an aggregate instead of sand. Perlite is a volcanic glass that has been expanded by heat.

Gauging plaster is a mix of gypsum plaster and lime putty that when applied produces a quick drying finish coat.

Veneer plaster is a one or two component gypsum plaster used as a thin finish coat over special gypsum board.

Keenes cement is a white cementitious material manufactured from gypsum that has been burned at a high temperature and ground to a fine powder. Alum is added to accelerate the set. The resulting plaster is hard and strong and accepts and maintains a high polish, hence it is used as a finishing plaster.

Stucco is a Portland cement based plaster used primarily as an exterior finish.

Plaster is used on both interior and exterior surfaces. Generally it is applied in multiple-coat systems. A three-coat system uses the terms scratch, brown and finish to identify each coat. A two-coat system uses base and finish to describe each coat. Each type of plaster and application system has attributes that are chosen by the designer to best fit the intended use.

Gypsum Plaster Quantities for 100 S.Y.	2 Coat, 5/8″ Thick		3 Coat, 3/4″ Thick		
	Base 1:3 Mix	Finish 2:1 Mix	Scratch 1:2 Mix	Brown 1:3 Mix	Finish 2:1 Mix
Gypsum plaster	1,300 lb.		1,350 lb.	650 lb.	
Sand	1.75 C.Y.		1.85 C.Y.	1.35 C.Y.	
Finish hydrated lime		340 lb.			340 lb.
Gauging plaster		170 lb.			170 lb.

Vermiculite or Perlite Plaster Quantities for 100 S.Y.	2 Coat, 5/8″ Thick		3 Coat, 3/4″ Thick		
	Base	Finish	Scratch	Brown	Finish
Gypsum plaster	1,250 lb.		1,450 lb.	800 lb.	
Vermiculite or perlite	7.8 bags		8.0 bags	3.3 bags	
Finish hydrated lime		340 lb.			340 lb.
Gauging plaster		170 lb.			170 lb.

Stucco-Three-Coat System Quantities for 100 S.Y.	On Wood Frame	On Masonry
Portland cement	29 bags	21 bags
Sand	2.6 C.Y.	2.0 C.Y.
Hydrated lime	180 lb.	120 lb.

R092910-10 Levels of Gypsum Board Finish

In the past, contract documents often used phrases such as "industry standard" and "workmanlike finish" to specify the expected quality of gypsum board wall and ceiling installations. The vagueness of these descriptions led to unacceptable work and disputes.

In order to resolve this problem, four major trade associations concerned with the manufacture, erection, finish, and decoration of gypsum board wall and ceiling systems developed an industry-wide *Recommended Levels of Gypsum Board Finish.*

The finish of gypsum board walls and ceilings for specific final decoration is dependent on a number of factors. A primary consideration is the location of the surface and the degree of decorative treatment desired. Painted and unpainted surfaces in warehouses and other areas where appearance is normally not critical may simply require the taping of wallboard joints and 'spotting' of fastener heads. Blemish-free, smooth, monolithic surfaces often intended for painted and decorated walls and ceilings in habitated structures, ranging from single-family dwellings through monumental buildings, require additional finishing prior to the application of the final decoration.

Other factors to be considered in determining the level of finish of the gypsum board surface are (1) the type of angle of surface illumination (both natural and artificial lighting), and (2) the paint and method of application or the type and finish of wallcovering specified as the final decoration. Critical lighting conditions, gloss paints, and thin wall coverings require a higher level of gypsum board finish than heavily textured surfaces which are subsequently painted or surfaces which are to be decorated with heavy grade wall coverings.

The following descriptions were developed by the Association of the Wall and Ceiling Industries-International (AWCI), Ceiling & Interior Systems Construction Association (CISCA), Gypsum Association (GA), and Painting and Decorating Contractors of America (PDCA) as a guide.

Level 0: Used in temporary construction or wherever the final decoration has not been determined. Unfinished. No taping, finishing or corner beads are required. Also could be used where non-predecorated panels will be used in demountable-type partitions that are to be painted as a final finish.

Level 1: Frequently used in plenum areas above ceilings, in attics, in areas where the assembly would generally be concealed, or in building service corridors and other areas not normally open to public view. Some degree of sound and smoke control is provided; in some geographic areas, this level is referred to as "fire-taping," although this level of finish does not typically meet fire-resistant assembly requirements. Where a fire resistance rating is required for the gypsum board assembly, details of construction should be in accordance with reports of fire tests of assemblies that have met the requirements of the fire rating acceptable.

All joints and interior angles shall have tape embedded in joint compound. Accessories are optional at specifier discretion in corridors and other areas with pedestrian traffic. Tape and fastener heads need not be covered with joint compound. Surface shall be free of excess joint compound. Tool marks and ridges are acceptable.

Level 2: It may be specified for standard gypsum board surfaces in garages, warehouse storage, or other similar areas where surface appearance is not of primary importance.

All joints and interior angles shall have tape embedded in joint compound and shall be immediately wiped with a joint knife or trowel, leaving a thin coating of joint compound over all joints and interior angles. Fastener heads and accessories shall be covered with a coat of joint compound. Surface shall be free of excess joint compound. Tool marks and ridges are acceptable.

Level 3: Typically used in areas receiving heavy texture (spray or hand applied) finishes before final painting, or where commercial-grade (heavy duty) wall coverings are to be applied as the final decoration. This level of finish should not be used where smooth painted surfaces or where lighter weight wall coverings are specified. The prepared surface shall be coated with a drywall primer prior to the application of final finishes.

All joints and interior angles shall have tape embedded in joint compound and shall be immediately wiped with a joint knife or trowel, leaving a thin coating of joint compound over all joints and interior angles. One additional coat of joint compound shall be applied over all joints and interior angles. Fastener heads and accessories shall be covered with two separate coats of joint compound. All joint compounds shall be smooth and free of tool marks and ridges. The prepared surface shall be covered with a drywall primer prior to the application of the final decoration.

Level 4: This level should be used where residential grade (light duty) wall coverings, flat paints, or light textures are to be applied. The prepared surface shall be coated with a drywall primer prior to the application of final finishes. Release agents for wall coverings are specifically formulated to minimize damage if coverings are subsequently removed.

The weight, texture, and sheen level of the wall covering material selected should be taken into consideration when specifying wall coverings over this level of drywall treatment. Joints and fasteners must be sufficiently concealed if the wall covering material is lightweight, contains limited pattern, has a glossy finish, or has any combination of these features. In critical lighting areas, flat paints applied over light textures tend to reduce joint photographing. Gloss, semi-gloss, and enamel paints are not recommended over this level of finish.

All joints and interior angles shall have tape embedded in joint compound and shall be immediately wiped with a joint knife or trowel, leaving a thin coating of joint compound over all joints and interior angles. In addition, two separate coats of joint compound shall be applied over all flat joints and one separate coat of joint compound applied over interior angles. Fastener heads and accessories shall be covered with three separate coats of joint compound. All joint compounds shall be smooth and free of tool marks and ridges. The prepared surface shall be covered with a drywall primer like Sheetrock® First Coat prior to the application of the final decoration.

Level 5: The highest quality finish is the most effective method to provide a uniform surface and minimize the possibility of joint photographing and of fasteners showing through the final decoration. This level of finish is required where gloss, semi-gloss, or enamel is specified; when flat joints are specified over an untextured surface; or where critical lighting conditions occur. The prepared surface shall be coated with a drywall primer prior to the application of the final decoration.

All joints and interior angles shall have tape embedded in joint compound and be immediately wiped with a joint knife or trowel, leaving a thin coating of joint compound over all joints and interior angles. Two separate coats of joint compound shall be applied over all flat joints and one separate coat of joint compound applied over interior angles. Fastener heads and accessories shall be covered with three separate coats of joint compound.

A thin skim coat of joint compound shall be trowel applied to the entire surface. Excess compound is immediately troweled off, leaving a film or skim coating of compound completely covering the paper. As an alternative to a skim coat, a material manufactured especially for this purpose may be applied such as Sheetrock® Tuff-Hide primer surfacer. The surface must be smooth and free of tool marks and ridges. The prepared surface shall be covered with a drywall primer prior to the application of the final decoration.

R096613-10 Terrazzo Floor

The table below lists quantities required for 100 S.F. of 5/8″ terrazzo topping, either bonded or not bonded.

Description	Bonded to Concrete 1-1/8″ Bed, 1:4 Mix	Not Bonded 2-1/8″ Bed and 1/4″ Sand
Portland cement, 94 lb. Bag	6 bags	8 bags
Sand	10 C.F.	20 C.F.
Divider strips, 4′ squares	50 L.F.	50 L.F.
Terrazzo fill, 50 lb. Bag	12 bags	12 bags
15 Lb. tarred felt		1 C.S.F.
Mesh 2 x 2 #14 galvanized		1 C.S.F.
Crew J-3	0.77 days	0.87 days

2′ x 2′ panels require 1.00 L.F. divider strip per S.F.
3′ x 3′ panels require 0.67 L.F. divider strip per S.F.
4′ x 4′ panels require 0.50 L.F. divider strip per S.F.
5′ x 5′ panels require 0.40 L.F. divider strip per S.F.
6′ x 6′ panels require 0.33 L.F. divider strip per S.F.

Finishes — R0972 Wall Coverings

R097223-10 Wall Covering

The table below lists the quantities required for 100 S.F. of wall covering.

Description	Medium-Priced Paper	Expensive Paper
Paper	1.6 dbl. rolls	1.6 dbl. rolls
Wall sizing	0.25 gallon	0.25 gallon
Vinyl wall paste	0.6 gallon	0.6 gallon
Apply sizing	0.3 hour	0.3 hour
Apply paper	1.2 hours	1.5 hours

Most wallpapers now come in double rolls only.
To remove old paper, allow 1.3 hours per 100 S.F.

Finishes — R0991 Painting

R099100-10 Painting Estimating Techniques

Proper estimating methodology is needed to obtain an accurate painting estimate. There is no known reliable shortcut or square foot method. The following steps should be followed:

- List all surfaces to be painted, with an accurate quantity (area) of each. Items having similar surface condition, finish, application method and accessibility may be grouped together.
- List all the tasks required for each surface to be painted, including surface preparation, masking, and protection of adjacent surfaces. Surface preparation may include minor repairs, washing, sanding and puttying.
- Select the proper Means line for each task. Review and consider all adjustments to labor and materials for type of paint and location of work. Apply the height adjustment carefully. For instance, when applying the adjustment for work over 8′ high to a wall that is 12′ high, apply the adjustment only to the area between 8′ and 12′ high, and not to the entire wall.

When applying more than one percent (%) adjustment, apply each to the base cost of the data, rather than applying one percentage adjustment on top of the other.

When estimating the cost of painting walls and ceilings remember to add the brushwork for all cut-ins at inside corners and around windows and doors as a LF measure. One linear foot of cut-in with a brush equals one square foot of painting.

All items for spray painting include the labor for roll-back.

Deduct for openings greater than 100 SF or openings that extend from floor to ceiling and are greater than 5′ wide. Do not deduct small openings.

The cost of brushes, rollers, ladders and spray equipment is considered part of a painting contractor's overhead, and should not be added to the estimate. The cost of rented equipment such as scaffolding and swing staging should be added to the estimate.

R099100-20 Painting

Item	Coat	One Gallon Covers			In 8 Hours a Laborer Covers			Labor-Hours per 100 S.F.		
		Brush	Roller	Spray	Brush	Roller	Spray	Brush	Roller	Spray
Paint wood siding	prime	250 S.F.	225 S.F.	290 S.F.	1150 S.F.	1300 S.F.	2275 S.F.	.695	.615	.351
	others	270	250	290	1300	1625	2600	.615	.492	.307
Paint exterior trim	prime	400	—	—	650	—	—	1.230	—	—
	1st	475	—	—	800	—	—	1.000	—	—
	2nd	520	—	—	975	—	—	.820	—	—
Paint shingle siding	prime	270	255	300	650	975	1950	1.230	.820	.410
	others	360	340	380	800	1150	2275	1.000	.695	.351
Stain shingle siding	1st	180	170	200	750	1125	2250	1.068	.711	.355
	2nd	270	250	290	900	1325	2600	.888	.603	.307
Paint brick masonry	prime	180	135	160	750	800	1800	1.066	1.000	.444
	1st	270	225	290	815	975	2275	.981	.820	.351
	2nd	340	305	360	815	1150	2925	.981	.695	.273
Paint interior plaster or drywall	prime	400	380	495	1150	2000	3250	.695	.400	.246
	others	450	425	495	1300	2300	4000	.615	.347	.200
Paint interior doors and windows	prime	400	—	—	650	—	—	1.230	—	—
	1st	425	—	—	800	—	—	1.000	—	—
	2nd	450	—	—	975	—	—	.820	—	—

Special Construction | R1311 Swimming Pools

R131113-20 Swimming Pools

Pool prices given per square foot of surface area include pool structure, filter and chlorination equipment, pumps, related piping, ladders/steps, maintenance kit, skimmer and vacuum system. Decks and electrical service to equipment are not included.

Residential in-ground pool construction can be divided into two categories: vinyl lined and gunite. Vinyl lined pool walls are constructed of different materials including wood, concrete, plastic or metal. The bottom is often graded with sand over which the vinyl liner is installed. Vermiculite or soil cement bottoms may be substituted for an added cost.

Gunite pool construction is used both in residential and municipal installations. These structures are steel reinforced for strength and finished with a white cement limestone plaster.

Municipal pools will have a higher cost because plumbing codes require more expensive materials, chlorination equipment and higher filtration rates.

Municipal pools greater than 1,800 S.F. require gutter systems to control waves. This gutter may be formed into the concrete wall. Often a vinyl/stainless steel gutter or gutter/wall system is specified, which will raise the pool cost.

Competition pools usually require tile bottoms and sides with contrasting lane striping, which will also raise the pool cost.

Plumbing | R2201 Operation & Maintenance of Plumbing

R220102-20 Labor Adjustment Factors

Labor Adjustment Factors are provided for Divisions 21, 22, and 23 to assist the mechanical estimator account for the various complexities and special conditions of any particular project. While a single percentage has been entered on each line of Division 22 01 02.20, it should be understood that these are just suggested midpoints of ranges of values commonly used by mechanical estimators. They may be increased or decreased depending on the severity of the special conditions.

The group for "existing occupied buildings" has been the subject of requests for explanation. Actually there are two stages to this group: buildings that are existing and "finished" but unoccupied, and those that also are occupied. Buildings that are "finished" may result in higher labor costs due to the workers having to be more careful not to damage finished walls, ceilings, floors, etc. and may necessitate special protective coverings and barriers. Also corridor bends and doorways may not accommodate long pieces of pipe or larger pieces of equipment. Work above an already hung ceiling can be very time consuming. The addition of occupants may force the work to be done on premium time (nights and/or weekends), eliminate the possible use of some preferred tools such as pneumatic drivers, powder charged drivers, etc. The estimator should evaluate the access to the work area and just how the work is going to be accomplished to arrive at an increase in labor costs over "normal" new construction productivity.

R236000-20 Air Conditioning Requirements

BTU's per hour per S.F. of floor area and S.F. per ton of air conditioning.

Type of Building	BTU/Hr per S.F.	S.F. per Ton	Type of Building	BTU/Hr per S.F.	S.F. per Ton	Type of Building	BTU per S.F.	S.F. per Ton
Apartments, Individual	26	450	Dormitory, Rooms	40	300	Libraries	50	240
Corridors	22	550	Corridors	30	400	Low Rise Office, Exterior	38	320
Auditoriums & Theaters	40	300/18*	Dress Shops	43	280	Interior	33	360
Banks	50	240	Drug Stores	80	150	Medical Centers	28	425
Barber Shops	48	250	Factories	40	300	Motels	28	425
Bars & Taverns	133	90	High Rise Office—Ext. Rms.	46	263	Office (small suite)	43	280
Beauty Parlors	66	180	Interior Rooms	37	325	Post Office, Individual Office	42	285
Bowling Alleys	68	175	Hospitals, Core	43	280	Central Area	46	260
Churches	36	330/20*	Perimeter	46	260	Residences	20	600
Cocktail Lounges	68	175	Hotel, Guest Rooms	44	275	Restaurants	60	200
Computer Rooms	141	85	Corridors	30	400	Schools & Colleges	46	260
Dental Offices	52	230	Public Spaces	55	220	Shoe Stores	55	220
Dept. Stores, Basement	34	350	Industrial Plants, Offices	38	320	Shop'g. Ctrs., Supermarkets	34	350
Main Floor	40	300	General Offices	34	350	Retail Stores	48	250
Upper Floor	30	400	Plant Areas	40	300	Specialty	60	200

*Persons per ton

12,000 BTU = 1 ton of air conditioning

R260519-90 Wire

Wire quantities are taken off by either measuring each cable run or by extending the conduit and raceway quantities times the number of conductors in the raceway. Ten percent should be added for waste and tie-ins. Keep in mind that the unit of measure of wire is C.L.F. not L.F. as in raceways so the formula would read:

$$\frac{\text{(L.F. Raceway x No. of Conductors) x 1.10}}{100} = \text{C.L.F.}$$

Price per C.L.F. of wire includes:
1. Setting up wire coils or spools on racks
2. Attaching wire to pull in means
3. Measuring and cutting wire
4. Pulling wire into a raceway
5. Identifying and tagging

Price does not include:
1. Connections to breakers, panelboards, or equipment
2. Splices

Job Conditions: Productivity is based on new construction to a height of 15′ using rolling staging in an unobstructed area. Material staging is assumed to be within 100′ of work being performed.

Economy of Scale: If more than three wires at a time are being pulled, deduct the following percentages from the labor of that grouping:

4-5 wires	25%
6-10 wires	30%
11-15 wires	35%
over 15	40%

If a wire pull is less than 100′ in length and is interrupted several times by boxes, lighting outlets, etc., it may be necessary to add the following lengths to each wire being pulled:

Junction box to junction box	2 L.F.
Lighting panel to junction box	6 L.F.
Distribution panel to sub panel	8 L.F.
Switchboard to distribution panel	12 L.F.
Switchboard to motor control center	20 L.F.
Switchboard to cable tray	40 L.F.

Measure of Drops and Riser: It is important when taking off wire quantities to include the wire for drops to electrical equipment. If heights of electrical equipment are not clearly stated, use the following guide:

	Bottom A.F.F.	Top A.F.F.	Inside Cabinet
Safety switch to 100A	5′	6′	2′
Safety switch 400 to 600A	4′	6′	3′
100A panel 12 to 30 circuit	4′	6′	3′
42 circuit panel	3′	6′	4′
Switch box	3′	3′6″	1′
Switchgear	0′	8′	8′
Motor control centers	0′	8′	8′
Transformers - wall mount	4′	8′	2′
Transformers - floor mount	0′	12′	4′

R260533-22 Conductors in Conduit

The table below lists the maximum number of conductors for various sized conduit using THW, TW or THWN insulations.

Copper Wire Size	Conduit Size ½″			¾″			1″			1-¼″			1-½″			2″			2-½″			3″		3-½″		4″	
	TW	THW	THWN	TW	THW	THWN	TW	THW	THWN	TW	THW	THWN	TW	THW	THWN	TW	THW	THWN	TW	THW	THWN	THW	THWN	THW	THWN	THW	THWN
#14	9	6	13	15	10	24	25	16	39	44	29	69	60	40	94	99	65	154	142	93		143		192			
#12	7	4	10	12	8	18	19	13	29	35	24	51	47	32	70	78	53	114	111	76	164	117		157			
#10	5	4	6	9	6	11	15	11	18	26	19	32	36	26	44	60	43	73	85	61	104	95	160	127		163	
#8	2	1	3	4	3	5	7	5	9	12	10	16	17	13	22	28	22	36	40	32	51	49	79	66	106	85	136
#6		1	1		2	4		4	6		7	11		10	15		16	26		23	37	36	57	48	76	62	98
#4		1	1		1	2		3	4		5	7		7	9		12	16		17	22	27	35	36	47	47	60
#3		1	1		1	1		2	3		4	6		6	8		10	13		15	19	23	29	31	39	40	51
#2		1	1		1	1		2	3		4	5		5	7		9	11		13	16	20	25	27	33	34	43
#1					1	1		1	1		3	3		4	5		6	8		9	12	14	18	19	25	25	32
1/0					1	1		1	1		2	3		3	4		5	7		8	10	12	15	16	21	21	27
2/0					1	1		1	1		1	2		3	3		5	6		7	8	10	13	14	17	18	22
3/0					1	1		1	1		1	1		2	3		4	5		6	7	9	11	12	14	15	18
4/0						1		1	1		1	1		1	2		3	4		5	6	7	9	10	12	13	15
250 kcmil								1	1		1	1		1	1		2	3		4	4	6	7	8	10	10	12
300								1	1		1	1		1	1		2	3		3	4	5	6	7	8	9	11
350									1		1	1		1	1		1	2		3	3	4	5	6	7	8	9
400											1	1		1	1		1	1		2	3	4	5	5	6	7	8
500											1	1		1	1		1	1		1	2	3	4	4	5	6	7
600												1		1	1		1	1		1	1	3	3	4	4	5	5
700														1	1		1	1		1	1	2	3	3	4	4	5
750 ▼														1	1		1	1		1	1	2	2	3	3	4	4

R312316-40 Excavating

The selection of equipment used for structural excavation and bulk excavation or for grading is determined by the following factors.

1. Quantity of material
2. Type of material
3. Depth or height of cut
4. Length of haul
5. Condition of haul road
6. Accessibility of site
7. Moisture content and dewatering requirements
8. Availability of excavating and hauling equipment

Some additional costs must be allowed for hand trimming the sides and bottom of concrete pours and other excavation below the general excavation.

When planning excavation and fill, the following should also be considered.

1. Swell factor
2. Compaction factor
3. Moisture content
4. Density requirements

A typical example for scheduling and estimating the cost of excavation of a 15′ deep basement on a dry site when the material must be hauled off the site is outlined below.

Assumptions:

1. Swell factor, 18%
2. No mobilization or demobilization
3. Allowance included for idle time and moving on job
4. No dewatering, sheeting, or bracing
5. No truck spotter or hand trimming

Number of B.C.Y. per truck = 1.5 C.Y. bucket x 8 passes = 12 loose C.Y.

$$= 12 \times \frac{100}{118} = 10.2 \text{ B.C.Y. per truck}$$

Truck Haul Cycle:

Load truck, 8 passes	=	4 minutes
Haul distance, 1 mile	=	9 minutes
Dump time	=	2 minutes
Return, 1 mile	=	7 minutes
Spot under machine	=	1 minute
		23 minute cycle

Fleet Haul Production per day in B.C.Y.

$$4 \text{ trucks} \times \frac{50 \text{ min. hour}}{23 \text{ min. haul cycle}} \times 8 \text{ hrs.} \times 10.2 \text{ B.C.Y.}$$

$$= 4 \times 2.2 \times 8 \times 10.2 = 718 \text{ B.C.Y./day}$$

Add the mobilization and demobilization costs to the total excavation costs. When equipment is rented for more than three days, there is often no mobilization charge by the equipment dealer. On larger jobs outside of urban areas, scrapers can move earth economically provided a dump site or fill area and adequate haul roads are available. Excavation within sheeting bracing or cofferdam bracing is usually done with a clamshell and production is low, since the clamshell may have to be guided by hand between the bracing. When excavating or filling an area enclosed with a wellpoint system, add 10% to 15% to the cost to allow for restricted access. When estimating earth excavation quantities for structures, allow work space outside the building footprint for construction of the foundation and a slope of 1:1 unless sheeting is used.

R312316-45 Excavating Equipment

The table below lists theoretical hourly production in C.Y./hr. bank measure for some typical excavation equipment. Figures assume 50 minute hours, 83% job efficiency, 100% operator efficiency, 90° swing and properly sized hauling units, which must be modified for adverse digging and loading conditions. Actual production costs in the front of the data set average about 50% of the theoretical values listed here.

Equipment	Soil Type	B.C.Y. Weight	% Swell	1 C.Y.	1-1/2 C.Y.	2 C.Y.	2-1/2 C.Y.	3 C.Y.	3-1/2 C.Y.	4 C.Y.
Hydraulic Excavator "Backhoe" 15′ Deep Cut	Moist loam, sandy clay	3400 lb.	40%	85	125	175	220	275	330	380
	Sand and gravel	3100	18	80	120	160	205	260	310	365
	Common earth	2800	30	70	105	150	190	240	280	330
	Clay, hard, dense	3000	33	65	100	130	170	210	255	300
Power Shovel Optimum Cut (Ft.)	Moist loam, sandy clay	3400	40	170	245	295	335	385	435	475
				(6.0)	(7.0)	(7.8)	(8.4)	(8.8)	(9.1)	(9.4)
	Sand and gravel	3100	18	165	225	275	325	375	420	460
				(6.0)	(7.0)	(7.8)	(8.4)	(8.8)	(9.1)	(9.4)
	Common earth	2800	30	145	200	250	295	335	375	425
				(7.8)	(9.2)	(10.2)	(11.2)	(12.1)	(13.0)	(13.8)
	Clay, hard, dense	3000	33	120	175	220	255	300	335	375
				(9.0)	(10.7)	(12.2)	(13.3)	(14.2)	(15.1)	(16.0)
Drag Line Optimum Cut (Ft.)	Moist loam, sandy clay	3400	40	130	180	220	250	290	325	385
				(6.6)	(7.4)	(8.0)	(8.5)	(9.0)	(9.5)	(10.0)
	Sand and gravel	3100	18	130	175	210	245	280	315	375
				(6.6)	(7.4)	(8.0)	(8.5)	(9.0)	(9.5)	(10.0)
	Common earth	2800	30	110	160	190	220	250	280	310
				(8.0)	(9.0)	(9.9)	(10.5)	(11.0)	(11.5)	(12.0)
	Clay, hard, dense	3000	33	90	130	160	190	225	250	280
				(9.3)	(10.7)	(11.8)	(12.3)	(12.8)	(13.3)	(12.0)
				Wheel Loaders				Track Loaders		
				3 C.Y.	4 C.Y.	6 C.Y.	8 C.Y.	2-1/4 C.Y.	3 C.Y.	4 C.Y.
Loading Tractors	Moist loam, sandy clay	3400	40	260	340	510	690	135	180	250
	Sand and gravel	3100	18	245	320	480	650	130	170	235
	Common earth	2800	30	230	300	460	620	120	155	220
	Clay, hard, dense	3000	33	200	270	415	560	110	145	200
	Rock, well-blasted	4000	50	180	245	380	520	100	130	180

R312323-30 Compacting Backfill

Compaction of fill in embankments, around structures, in trenches, and under slabs is important to control settlement. Factors affecting compaction are:

1. Soil gradation
2. Moisture content
3. Equipment used
4. Depth of fill per lift
5. Density required

Production Rate:

$$\frac{1.75' \text{ plate width x 50 F.P.M. x 50 min./hr. x .67' lift}}{27 \text{ C.F. per C.Y.}} = 108.5 \text{ C.Y./hr.}$$

Production Rate for 4 Passes:

$$\frac{108.5 \text{ C.Y.}}{4 \text{ passes}} = 27.125 \text{ C.Y./hr. x 8 hrs.} = 217 \text{ C.Y./day}$$

Example:

Compact granular fill around a building foundation using a 21″ wide x 24″ vibratory plate in 8″ lifts. Operator moves at 50 F.P.M. working a 50 minute hour to develop 95% Modified Proctor Density with 4 passes.

R329219-50 Seeding

The type of grass is determined by light, shade and moisture content of soil plus intended use. Fertilizer should be disked 4″ before seeding. For steep slopes disk five tons of mulch and lay two tons of hay or straw on surface per acre after seeding. Surface mulch can be staked, lightly disked or tar emulsion sprayed. Material for mulch can be wood chips, peat moss, partially rotted hay or straw, wood fibers and sprayed emulsions. Hemp seed blankets with fertilizer are also available. For spring seeding, watering is necessary. Late fall seeding may have to be reseeded in the spring. Hydraulic seeding, power mulching, and aerial seeding can be used on large areas.

Utilities R3311 Water Utility Distribution Piping

R331113-80 Piping Designations

There are several systems currently in use to describe pipe and fittings. The following paragraphs will help to identify and clarify classifications of piping systems used for water distribution.

Piping may be classified by schedule. Piping schedules include 5S, 10S, 10, 20, 30, Standard, 40, 60, Extra Strong, 80, 100, 120, 140, 160 and Double Extra Strong. These schedules are dependent upon the pipe wall thickness. The wall thickness of a particular schedule may vary with pipe size.

Ductile iron pipe for water distribution is classified by Pressure Classes such as Class 150, 200, 250, 300 and 350. These classes are actually the rated water working pressure of the pipe in pounds per square inch (psi). The pipe in these pressure classes is designed to withstand the rated water working pressure plus a surge allowance of 100 psi.

The American Water Works Association (AWWA) provides standards for various types of **plastic pipe.** C-900 is the specification for polyvinyl chloride (PVC) piping used for water distribution in sizes ranging from 4″ through 12″. C-901 is the specification for polyethylene (PE) pressure pipe, tubing and fittings used for water distribution in sizes ranging from 1/2″ through 3″. C-905 is the specification for PVC piping sizes 14″ and greater.

PVC pressure-rated pipe is identified using the standard dimensional ratio (SDR) method. This method is defined by the American Society for Testing and Materials (ASTM) Standard D 2241. This pipe is available in SDR numbers 64, 41, 32.5, 26, 21, 17, and 13.5. A pipe with an SDR of 64 will have the thinnest wall while a pipe with an SDR of 13.5 will have the thickest wall. When the pressure rating (PR) of a pipe is given in psi, it is based on a line supplying water at 73 degrees F.

The National Sanitation Foundation (NSF) seal of approval is applied to products that can be used with potable water. These products have been tested to ANSI/NSF Standard 14.

Valves and strainers are classified by American National Standards Institute (ANSI) Classes. These Classes are 125, 150, 200, 250, 300, 400, 600, 900, 1500 and 2500. Within each class there is an operating pressure range dependent upon temperature. Design parameters should be compared to the appropriate material dependent, pressure-temperature rating chart for accurate valve selection.

Change Orders

Change Order Considerations

A change order is a written document usually prepared by the design professional and signed by the owner, the architect/engineer, and the contractor. A change order states the agreement of the parties to: an addition, deletion, or revision in the work; an adjustment in the contract sum, if any; or an adjustment in the contract time, if any. Change orders, or "extras", in the construction process occur after execution of the construction contract and impact architects/engineers, contractors, and owners.

Change orders that are properly recognized and managed can ensure orderly, professional, and profitable progress for everyone involved in the project. There are many causes for change orders and change order requests. In all cases, change orders or change order requests should be addressed promptly and in a precise and prescribed manner. The following paragraphs include information regarding change order pricing and procedures.

The Causes of Change Orders

Reasons for issuing change orders include:

- Unforeseen field conditions that require a change in the work
- Correction of design discrepancies, errors, or omissions in the contract documents
- Owner-requested changes, either by design criteria, scope of work, or project objectives
- Completion date changes for reasons unrelated to the construction process
- Changes in building code interpretations, or other public authority requirements that require a change in the work
- Changes in availability of existing or new materials and products

Procedures

Properly written contract documents must include the correct change order procedures for all parties—owners, design professionals, and contractors—to follow in order to avoid costly delays and litigation.

Being "in the right" is not always a sufficient or acceptable defense. The contract provisions requiring notification and documentation must be adhered to within a defined or reasonable time frame.

The appropriate method of handling change orders is by a written proposal and acceptance by all parties involved. Prior to starting work on a project, all parties should identify their authorized agents who may sign and accept change orders, as well as any limits placed on their authority.

Time may be a critical factor when the need for a change arises. For such cases, the contractor might be directed to proceed on a "time and materials" basis, rather than wait for all paperwork to be processed—a delay that could impede progress. In this situation, the contractor must still follow the prescribed change order procedures including, but not limited to, notification and documentation.

Lack of documentation can be very costly, especially if legal judgments are to be made, and if certain field personnel are no longer available. For time and material change orders, the contractor should keep accurate daily records of all labor and material allocated to the change.

Owners or awarding authorities who do considerable and continual building construction (such as the federal government) realize the inevitability of change orders for numerous reasons, both predictable and unpredictable. As a result, the federal government, the American Institute of Architects (AIA), the Engineers Joint Contract Documents Committee (EJCDC), and other contractor, legal, and technical organizations have developed standards and procedures to be followed by all parties to achieve contract continuance and timely completion, while being financially fair to all concerned.

Pricing Change Orders

When pricing change orders, regardless of their cause, the most significant factor is when the change occurs. The need for a change may be perceived in the field or requested by the architect/engineer *before* any of the actual installation has begun, or may evolve or appear *during* construction when the item of work in question is partially installed. In the latter cases, the original sequence of construction is disrupted, along with all contiguous and supporting systems. Change orders cause the greatest impact when they occur *after* the installation has been completed and must be uncovered, or even replaced. Post-completion changes may be caused by necessary design changes, product failure, or changes in the owner's requirements that are not discovered until the building or the systems begin to function.

Specified procedures of notification and record keeping must be adhered to and enforced regardless of the stage of construction: *before, during,* or *after* installation. Some bidding documents anticipate change orders by requiring that unit prices including overhead and profit percentages—for additional as well as deductible changes—be listed. Generally these unit prices do not fully take into account the ripple effect, or impact on other trades, and should be used for general guidance only.

When pricing change orders, it is important to classify the time frame in which the change occurs. There are two basic time frames for change orders: *pre-installation change orders*, which occur before the start of construction, and *post-installation change orders*, which involve reworking after the original installation. Change orders that occur between these stages may be priced according to the extent of work completed using a combination of techniques developed for pricing *pre-* and *post-installation* changes.

Factors To Consider When Pricing Change Orders

As an estimator begins to prepare a change order, the following questions should be reviewed to determine their impact on the final price.

General

- *Is the change order work* pre-installation *or* post-installation?

  Change order work costs vary according to how much of the installation has been completed. Once workers have the project scoped in their minds, even though they have not started, it can be difficult to refocus. Consequently they may spend more than the normal amount of time understanding the change. Also, modifications to work in place, such as trimming or refitting, usually take more time than was initially estimated. The greater the amount of work in place, the more reluctant workers are to change it. Psychologically they may resent the change and as a result the rework takes longer than normal. Post-installation change order estimates must include demolition of existing work as required to accomplish the change. If the work is performed at a later time, additional obstacles, such as building finishes, may be present which must be protected. Regardless of whether the change occurs pre-installation or post-installation, attempt to isolate the identifiable factors and price them separately. For example, add shipping costs that may be required pre-installation or any demolition required post-installation. Then analyze the potential impact on productivity of psychological and/or learning curve factors and adjust the output rates accordingly. One approach is to break down the typical workday into segments and quantify the impact on each segment.

Change Order Installation Efficiency

The labor-hours expressed (for new construction) are based on average installation time, using an efficiency level. For change order situations, adjustments to this efficiency level should reflect the daily labor-hour allocation for that particular occurrence.

- *Will the change substantially delay the original completion date?*

  A significant change in the project may cause the original completion date to be extended. The extended schedule may subject the contractor to new wage rates dictated by relevant labor contracts. Project supervision and other project overhead must also be extended beyond the original completion date. The schedule extension may also put installation into a new weather season. For example, underground piping scheduled for October installation was delayed until January. As a result, frost penetrated the trench area, thereby changing the degree of difficulty of the task. Changes and delays may have a ripple effect throughout the project. This effect must be analyzed and negotiated with the owner.

- *What is the net effect of a deduct change order?*

  In most cases, change orders resulting in a deduction or credit reflect only bare costs. The contractor may retain the overhead and profit based on the original bid.

Materials

- *Will you have to pay more or less for the new material, required by the change order, than you paid for the original purchase?*

  The same material prices or discounts will usually apply to materials purchased for change orders as new construction. In some instances, however, the contractor may forfeit the advantages of competitive pricing for change orders. Consider the following example:

  A contractor purchased over $20,000 worth of fan coil units for an installation and obtained the maximum discount. Some time later it was determined the project required an additional matching unit. The contractor has to purchase this unit from the original supplier to ensure a match. The supplier at this time may not discount the unit because of the small quantity, and he is no longer in a competitive situation. The impact of quantity on purchase can add between 0% and 25% to material prices and/or subcontractor quotes.

- *If materials have been ordered or delivered to the job site, will they be subject to a cancellation charge or restocking fee?*

  Check with the supplier to determine if ordered materials are subject to a cancellation charge. Delivered materials not used as a result of a change order may be subject to a restocking fee if returned to the supplier. Common restocking charges run between 20% and 40%. Also, delivery charges to return the goods to the supplier must be added.

Labor

- *How efficient is the existing crew at the actual installation?*

  Is the same crew that performed the initial work going to do the change order? Possibly the change consists of the installation of a unit identical to one already installed; therefore, the change should take less time. Be sure to consider this potential productivity increase and modify the productivity rates accordingly.

- *If the crew size is increased, what impact will that have on supervision requirements?*

  Under most bargaining agreements or management practices, there is a point at which a working foreman is replaced by a nonworking foreman. This replacement increases project overhead by adding a nonproductive worker. If additional workers are added to accelerate the project or to perform changes while maintaining the schedule, be sure to add additional supervision time if warranted. Calculate the hours involved and the additional cost directly if possible.

- *What are the other impacts of increased crew size?*

  The larger the crew, the greater the potential for productivity to decrease. Some of the factors that cause this productivity loss are: overcrowding (producing restrictive conditions in the working space) and possibly a shortage of any special tools and equipment required. Such factors affect not only the crew working on the elements directly involved in the change order, but other crews whose movements may also be hampered. As the crew increases, check its basic composition for changes by the addition or deletion of apprentices or nonworking foreman, and quantify the potential effects of equipment shortages or other logistical factors.

- *As new crews, unfamiliar with the project, are brought onto the site, how long will it take them to become oriented to the project requirements?*

  The orientation time for a new crew to become 100% effective varies with the site and type of project. Orientation is easiest at a new construction site and most difficult at existing, very restrictive renovation sites. The type of work also affects orientation time. When all elements of the work are exposed, such as concrete or masonry work, orientation is decreased. When the work is concealed or less visible, such as existing electrical systems, orientation takes longer. Usually orientation can be accomplished in one day or less. Costs for added orientation should be itemized and added to the total estimated cost.

- *How much actual production can be gained by working overtime?*

  Short term overtime can be used effectively to accomplish more work in a day. However, as overtime is scheduled to run beyond several weeks, studies have shown marked decreases in output. The following chart shows the effect of long term overtime on worker efficiency. If the anticipated change requires extended overtime to keep the job on schedule, these factors can be used as a guide to predict the impact on time and cost. Add project overhead, particularly supervision, that may also be incurred.

Days per Week	Hours per Day	Production Efficiency					Payroll Cost Factors	
		1st Week	2nd Week	3rd Week	4th Week	Average 4 Weeks	@ 1-1/2 Times	@ 2 Times
5	8	100%	100%	100%	100%	100%	100%	100%
	9	100	100	95	90	96.25	105.6	111.1
	10	100	95	90	85	92.50	110.0	120.0
	11	95	90	75	65	81.25	113.6	127.3
	12	90	85	70	60	76.25	116.7	133.3
6	8	100	100	95	90	96.25	108.3	116.7
	9	100	95	90	85	92.50	113.0	125.9
	10	95	90	85	80	87.50	116.7	133.3
	11	95	85	70	65	78.75	119.7	139.4
	12	90	80	65	60	73.75	122.2	144.4
7	8	100	95	85	75	88.75	114.3	128.6
	9	95	90	80	70	83.75	118.3	136.5
	10	90	85	75	65	78.75	121.4	142.9
	11	85	80	65	60	72.50	124.0	148.1
	12	85	75	60	55	68.75	126.2	152.4

Effects of Overtime

Caution: Under many labor agreements, Sundays and holidays are paid at a higher premium than the normal overtime rate.

The use of long-term overtime is counterproductive on almost any construction job; that is, the longer the period of overtime, the lower the actual production rate. Numerous studies have been conducted, and while they have resulted in slightly different numbers, all reach the same conclusion. The figure above tabulates the effects of overtime work on efficiency.

As illustrated, there can be a difference between the *actual* payroll cost per hour and the *effective* cost per hour for overtime work. This is due to the reduced production efficiency with the increase in weekly hours beyond 40. This difference between actual and effective cost results from overtime work over a prolonged period. Short-term overtime work does not result in as great a reduction in efficiency and, in such cases, effective cost may not vary significantly from the actual payroll cost. As the total hours per week are increased on a regular basis, more time is lost due to fatigue, lowered morale, and an increased accident rate.

As an example, assume a project where workers are working 6 days a week, 10 hours per day. From the figure above (based on productivity studies), the average effective productive hours over a 4-week period are:

$$0.875 \times 60 = 52.5$$

Depending upon the locale and day of week, overtime hours may be paid at time and a half or double time. For time and a half, the overall (average) *actual* payroll cost (including regular and overtime hours) is determined as follows:

$$\frac{40 \text{ reg. hrs.} + (20 \text{ overtime hrs.} \times 1.5)}{60 \text{ hrs.}} = 1.167$$

Based on 60 hours, the payroll cost per hour will be 116.7% of the normal rate at 40 hours per week. However, because the effective production (efficiency) for 60 hours is reduced to the equivalent of 52.5 hours, the effective cost of overtime is calculated as follows:

For time and a half:

$$\frac{40 \text{ reg. hrs.} + (20 \text{ overtime hrs.} \times 1.5)}{52.5 \text{ hrs.}} = 1.33$$

The installed cost will be 133% of the normal rate (for labor).

Thus, when figuring overtime, the actual cost per unit of work will be higher than the apparent overtime payroll dollar increase, due to the reduced productivity of the longer work week. These efficiency calculations are true only for those cost factors determined by hours worked. Costs that are applied weekly or monthly, such as equipment rentals, will not be similarly affected.

Equipment

- *What equipment is required to complete the change order?*

  Change orders may require extending the rental period of equipment already on the job site, or the addition of special equipment brought in to accomplish the change work. In either case, the additional rental charges and operator labor charges must be added.

Summary

The preceding considerations and others you deem appropriate should be analyzed and applied to a change order estimate. The impact of each should be quantified and listed on the estimate to form an audit trail.

Change orders that are properly identified, documented, and managed help to ensure the orderly, professional, and profitable progress of the work. They also minimize potential claims or disputes at the end of the project.

Back by customer demand!

You asked and we listened. For customer convenience and estimating ease, we have made the 2019 Project Costs available for download at **RSMeans.com/2019books**. You will also find sample estimates, an RSMeans data overview video, and a book registration form to receive quarterly data updates throughout 2019.

Estimating Tips

- The cost figures available in the download were derived from hundreds of projects contained in the RSMeans database of completed construction projects. They include the contractor's overhead and profit. The figures have been adjusted to January of the current year.
- These projects were located throughout the U.S. and reflect a tremendous variation in square foot (S.F.) costs. This is due to differences, not only in labor and material costs, but also in individual owners' requirements. For instance, a bank in a large city would have different features than one in a rural area. This is true of all the different types of buildings analyzed. Therefore, caution should be exercised when using these Project Costs. For example, for courthouses, costs in the database are local courthouse costs and will not apply to the larger, more elaborate federal courthouses.
- None of the figures "go with" any others. All individual cost items were computed and tabulated separately. Thus, the sum of the median figures for plumbing, HVAC, and electrical will not normally total up to the total mechanical and electrical costs arrived at by separate analysis and tabulation of the projects.
- Each building was analyzed as to total and component costs and percentages. The figures were arranged in ascending order with the results tabulated as shown. The 1/4 column shows that 25% of the projects had lower costs and 75% had higher. The 3/4 column shows that 75% of the projects had lower costs and 25% had higher. The median column shows that 50% of the projects had lower costs and 50% had higher.
- Project Costs are useful in the conceptual stage when no details are available. As soon as details become available in the project design, the square foot approach should be discontinued and the project should be priced as to its particular components. When more precision is required, or for estimating the replacement cost of specific buildings, the current edition of *Square Foot Costs with RSMeans data* should be used.
- In using the figures in this section, it is recommended that the median column be used for preliminary figures if no additional information is available. The median figures, when multiplied by the total city construction cost index figures (see City Cost Indexes) and then multiplied by the project size modifier at the end of this section, should present a fairly accurate base figure, which would then have to be adjusted in view of the estimator's experience, local economic conditions, code requirements, and the owner's particular requirements. There is no need to factor in the percentage figures, as these should remain constant from city to city.
- The editors of this data would greatly appreciate receiving cost figures on one or more of your recent projects, which would then be included in the averages for next year. All cost figures received will be kept confidential, except that they will be averaged with other similar projects to arrive at square foot cost figures for next year.

See the website above for details and the discount available for submitting one or more of your projects.

50 17 | Project Costs

	50 17 00 | Project Costs		UNIT	UNIT COSTS			% OF TOTAL			
				1/4	MEDIAN	3/4	1/4	MEDIAN	3/4	
01	0000	**Auto Sales with Repair**	S.F.							01
	0100	Architectural		104	116	126	59.8%	59.2%	63.7%	
	0200	Plumbing		8.70	9.10	12.15	5%	5%	5%	
	0300	Mechanical		11.65	15.60	17.20	6.7%	6.7%	8.6%	
	0400	Electrical		17.90	22	27.50	10.3%	10.3%	12.1%	
	0500	Total Project Costs	↓	174	182	187				
02	0000	**Banking Institutions**	S.F.							02
	0100	Architectural		157	192	234	60.4%	60.4%	65.5%	
	0200	Plumbing		6.30	8.80	12.25	2.4%	2.4%	3%	
	0300	Mechanical		12.55	17.35	20.50	4.8%	4.8%	5.9%	
	0400	Electrical		30.50	37	57	11.7%	11.7%	12.6%	
	0500	Total Project Costs	↓	260	293	360				
03	0000	**Court House**	S.F.							03
	0100	Architectural		82.50	162	162	54.6%	54.5%	58.3%	
	0200	Plumbing		3.12	3.12	3.12	2.1%	2.1%	1.1%	
	0300	Mechanical		19.50	19.50	19.50	12.9%	12.9%	7%	
	0400	Electrical		25	25	25	16.6%	16.6%	9%	
	0500	Total Project Costs	↓	151	278	278				
04	0000	**Data Centers**	S.F.							04
	0100	Architectural		187	187	187	68%	67.9%	68%	
	0200	Plumbing		10.20	10.20	10.20	3.7%	3.7%	3.7%	
	0300	Mechanical		26	26	26	9.5%	9.4%	9.5%	
	0400	Electrical		24.50	24.50	24.50	8.9%	9%	8.9%	
	0500	Total Project Costs	↓	275	275	275				
05	0000	**Detention Centers**	S.F.							05
	0100	Architectural		173	183	194	59.2%	59.2%	59%	
	0200	Plumbing		18.25	22	27	6.3%	6.2%	7.1%	
	0300	Mechanical		23	33	39.50	7.9%	7.9%	10.6%	
	0400	Electrical		38	45	58.50	13%	13%	14.5%	
	0500	Total Project Costs	↓	292	310	365				
06	0000	**Fire Stations**	S.F.							06
	0100	Architectural		95	121	171	47%	50%	52.4%	
	0200	Plumbing		9.95	13.50	15.65	4.9%	4.9%	5.8%	
	0300	Mechanical		13.45	18.40	25.50	6.7%	6.6%	8%	
	0400	Electrical		22.50	28.50	32.50	11.1%	11%	12.3%	
	0500	Total Project Costs	↓	202	231	300				
07	0000	**Gymnasium**	S.F.							07
	0100	Architectural		86.50	114	114	64.6%	64.4%	57.3%	
	0200	Plumbing		2.12	6.95	6.95	1.6%	1.6%	3.5%	
	0300	Mechanical		3.25	29	29	2.4%	2.4%	14.6%	
	0400	Electrical		10.65	20.50	20.50	7.9%	7.9%	10.3%	
	0500	Total Project Costs	↓	134	199	199				
08	0000	**Hospitals**	S.F.							08
	0100	Architectural		105	172	187	42.9%	43%	47.1%	
	0200	Plumbing		7.70	14.70	32	3.1%	3.1%	4%	
	0300	Mechanical		51	57.50	74.50	20.8%	20.8%	15.8%	
	0400	Electrical		23	46.50	60.50	9.4%	9.5%	12.7%	
	0500	Total Project Costs	↓	245	365	395				
09	0000	**Industrial Buildings**	S.F.							09
	0100	Architectural		44	70	227	56.4%	56.3%	68.6%	
	0200	Plumbing		1.70	6.40	12.95	2.2%	2.2%	6.3%	
	0300	Mechanical		4.71	8.95	42.50	6%	6%	8.8%	
	0400	Electrical		7.20	8.20	68.50	9.2%	9.2%	8%	
	0500	Total Project Costs	↓	78	102	425				
10	0000	**Medical Clinics & Offices**	S.F.							10
	0100	Architectural		87.50	120	158	53.7%	51%	56.3%	
	0200	Plumbing		8.50	12.85	20.50	5.2%	5.2%	6%	
	0300	Mechanical		14.20	22	45	8.7%	8.8%	10.3%	
	0400	Electrical		19.55	26	36.50	12%	11.9%	12.2%	
	0500	Total Project Costs	↓	163	213	286				

50 17 00 | Project Costs

		50 17 00 | Project Costs	UNIT	UNIT COSTS			% OF TOTAL		
				1/4	MEDIAN	3/4	1/4	MEDIAN	3/4
11	0000	**Mixed Use**	S.F.						
	0100	Architectural		86.50	126	207	47%	47.5%	59.4%
	0200	Plumbing		6	9.15	11.55	3.3%	3.2%	4.3%
	0300	Mechanical		14.80	24	46.50	8%	7.8%	11.3%
	0400	Electrical		15.25	24	40.50	8.3%	8.3%	11.3%
	0500	Total Project Costs	↓	184	212	335			
12	0000	**Multi-Family Housing**	S.F.						
	0100	Architectural		75	125	167	60.5%	67.6%	56.6%
	0200	Plumbing		6.40	12.75	15.10	5.2%	5.6%	5.8%
	0300	Mechanical		6.95	11.70	37.50	5.6%	6.3%	5.3%
	0400	Electrical		10.10	18	22.50	8.1%	8.8%	8.1%
	0500	Total Project Costs	↓	124	221	282			
13	0000	**Nursing Home & Assisted Living**	S.F.						
	0100	Architectural		70	92	116	58.3%	58.4%	59%
	0200	Plumbing		7.55	11.35	12.50	6.3%	5.9%	7.3%
	0300	Mechanical		6.20	9.15	17.95	5.2%	5.2%	5.9%
	0400	Electrical		10.25	16.20	22.50	8.5%	8.6%	10.4%
	0500	Total Project Costs	↓	120	156	188			
14	0000	**Office Buildings**	S.F.						
	0100	Architectural		92.50	126	177	60.1%	60%	64.6%
	0200	Plumbing		4.98	7.85	15.40	3.2%	3.1%	4%
	0300	Mechanical		10.75	17.65	25.50	7%	6.8%	9.1%
	0400	Electrical		12.35	21	34	8%	7.9%	10.8%
	0500	Total Project Costs	↓	154	195	285			
15	0000	**Parking Garage**	S.F.						
	0100	Architectural		31	38	39.50	82.7%	82.1%	82.6%
	0200	Plumbing		1.02	1.07	2	2.7%	2.7%	2.3%
	0300	Mechanical		.79	1.22	4.62	2.1%	2.1%	2.7%
	0400	Electrical		2.72	2.98	6.25	7.3%	7.1%	6.5%
	0500	Total Project Costs	↓	37.50	46	49.50			
16	0000	**Parking Garage/Mixed Use**	S.F.						
	0100	Architectural		100	110	112	61%	61.2%	64.3%
	0200	Plumbing		3.22	4.22	6.45	2%	2%	2.5%
	0300	Mechanical		13.80	15.50	22.50	8.4%	8.4%	9.1%
	0400	Electrical		14.45	21	21.50	8.8%	8.8%	12.3%
	0500	Total Project Costs	↓	164	171	177			
17	0000	**Police Stations**	S.F.						
	0100	Architectural		113	127	160	53.3%	54%	48.5%
	0200	Plumbing		15	18	18.10	7.1%	7%	6.9%
	0300	Mechanical		34	47.50	49	16%	16.1%	18.1%
	0400	Electrical		25.50	28	29.50	12%	12.1%	10.7%
	0500	Total Project Costs	↓	212	262	297			
18	0000	**Police/Fire**	S.F.						
	0100	Architectural		110	110	340	67.9%	68.2%	65.9%
	0200	Plumbing		8.65	9.15	34	5.3%	5.5%	5.5%
	0300	Mechanical		13.55	21.50	77.50	8.4%	8.4%	12.9%
	0400	Electrical		15.40	19.70	88.50	9.5%	9.6%	11.8%
	0500	Total Project Costs	↓	162	167	610			
19	0000	**Public Assembly Buildings**	S.F.						
	0100	Architectural		115	156	218	62.5%	63%	61.7%
	0200	Plumbing		5.95	8.75	12.90	3.2%	3%	3.5%
	0300	Mechanical		13.60	22.50	34.50	7.4%	8%	8.9%
	0400	Electrical		18.60	25.50	40.50	10.1%	10.5%	10.1%
	0500	Total Project Costs	↓	184	253	360			
20	0000	**Recreational**	S.F.						
	0100	Architectural		108	170	231	56.3%	55.7%	59.2%
	0200	Plumbing		8.35	15.35	24.50	4.3%	4.6%	5.3%
	0300	Mechanical		12.90	19.60	31	6.7%	6.9%	6.8%
	0400	Electrical		15.80	28	39	8.2%	7.7%	9.8%
	0500	Total Project Costs	↓	192	287	435			

	50 17 00 | Project Costs		UNIT	UNIT COSTS 1/4	UNIT COSTS MEDIAN	UNIT COSTS 3/4	% OF TOTAL 1/4	% OF TOTAL MEDIAN	% OF TOTAL 3/4	
21	0000	**Restaurants**	S.F.							21
	0100	Architectural		123	198	245	60.6%	77.9%	59.1%	
	0200	Plumbing		9.95	31	39	4.9%	14.6%	9.3%	
	0300	Mechanical		14.55	19.30	47	7.2%	11.2%	5.8%	
	0400	Electrical		14.45	30.50	51	7.1%	17.9%	9.1%	
	0500	Total Project Costs	↓	203	335	420				
22	0000	**Retail**	S.F.							22
	0100	Architectural		54	60.50	109	73%	59.9%	64.4%	
	0200	Plumbing		5.65	7.85	9.95	7.6%	6.2%	8.4%	
	0300	Mechanical		5.15	7.40	9.05	7%	5.6%	7.9%	
	0400	Electrical		7.20	11.55	18.50	9.7%	7.9%	12.3%	
	0500	Total Project Costs	↓	74	94	148				
23	0000	**Schools**	S.F.							23
	0100	Architectural		94	120	160	58%	58.8%	55.6%	
	0200	Plumbing		7.50	10.40	15.15	4.6%	4.7%	4.8%	
	0300	Mechanical		17.85	24.50	36.50	11%	11.1%	11.3%	
	0400	Electrical		17.45	24	30.50	10.8%	11%	11.1%	
	0500	Total Project Costs	↓	162	216	286				
24	0000	**University, College & Private School Classroom & Admin Buildings**	S.F.							24
	0100	Architectural		121	150	188	60.2%	61%	54%	
	0200	Plumbing		6.90	10.70	15.10	3.4%	3.4%	3.8%	
	0300	Mechanical		26	37.50	45	12.9%	12.9%	13.5%	
	0400	Electrical		19.50	27.50	33.50	9.7%	9.8%	9.9%	
	0500	Total Project Costs	↓	201	278	370				
25	0000	**University, College & Private School Dormitories**	S.F.							25
	0100	Architectural		79	139	147	67.5%	67.1%	62.6%	
	0200	Plumbing		10.45	14.80	22	8.9%	8.9%	6.7%	
	0300	Mechanical		4.69	19.95	31.50	4%	4%	9%	
	0400	Electrical		5.55	19.35	29.50	4.7%	4.8%	8.7%	
	0500	Total Project Costs	↓	117	222	263				
26	0000	**University, College & Private School Science, Eng. & Lab Buildings**	S.F.							26
	0100	Architectural		136	144	188	48.7%	49.1%	50.5%	
	0200	Plumbing		9.35	14.20	26	3.4%	3.4%	5%	
	0300	Mechanical		42.50	67	68.50	15.2%	15.3%	23.5%	
	0400	Electrical		27.50	32	37.50	9.9%	9.8%	11.2%	
	0500	Total Project Costs	↓	279	285	320				
27	0000	**University, College & Private School Student Union Buildings**	S.F.							27
	0100	Architectural		108	283	283	50.9%	60%	54.4%	
	0200	Plumbing		16.25	16.25	24	7.7%	4.3%	3.1%	
	0300	Mechanical		31	50	50	14.6%	9.7%	9.6%	
	0400	Electrical		27	47	47	12.7%	13.3%	9%	
	0500	Total Project Costs	↓	212	520	520				
28	0000	**Warehouses**	S.F.							28
	0100	Architectural		46	72	171	66.7%	67.4%	58.5%	
	0200	Plumbing		2.40	5.15	9.90	3.5%	3.5%	4.2%	
	0300	Mechanical		2.84	16.20	25.50	4.1%	4.1%	13.2%	
	0400	Electrical	S.F.	5.15	19.40	32.50	7.5%	7.5%	15.8%	
	0500	Total Project Costs	S.F.	69	123	238				

Square Foot Project Size Modifier

One factor that affects the S.F. cost of a particular building is the size. In general, for buildings built to the same specifications in the same locality, the larger building will have the lower S.F. cost. This is due mainly to the decreasing contribution of the exterior walls plus the economy of scale usually achievable in larger buildings. The Area Conversion Scale shown below will give a factor to convert costs for the typical size building to an adjusted cost for the particular project.

The Square Foot Base Size lists the median costs, most typical project size in our accumulated data, and the range in size of the projects.

The Size Factor for your project is determined by dividing your project area in S.F. by the typical project size for the particular Building Type. With this factor, enter the Area Conversion Scale at the appropriate Size Factor and determine the appropriate cost multiplier for your building size.

Example: Determine the cost per S.F. for a 152,600 S.F. Multi-family housing.

$$\frac{\text{Proposed building area} = 152{,}600\text{ S.F.}}{\text{Typical size from below} = 76{,}300\text{ S.F.}} = 2.00$$

Enter Area Conversion scale at 2.0, intersect curve, read horizontally the appropriate cost multiplier of .94. Size adjusted cost becomes .94 x $194.00 = $182.36 based on national average costs.

Note: For Size Factors less than .50, the Cost Multiplier is 1.1
For Size Factors greater than 3.5, the Cost Multiplier is .90

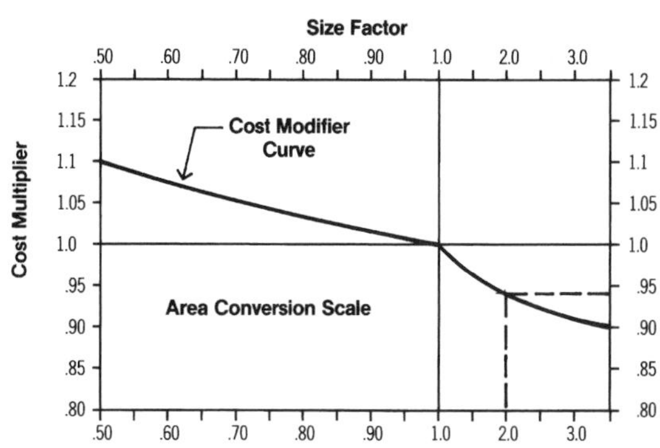

System	Median Cost (Total Project Costs)	Typical Size Gross S.F. (Median of Projects)	Typical Range (Low – High) (Projects)
Auto Sales with Repair	$182.00	24,900	4,700 – 29,300
Banking Institutions	293.00	9,300	3,300 – 38,100
Detention Centers	310.00	37,800	12,300 – 183,300
Fire Stations	231.00	12,300	6,300 – 29,600
Hospitals	365.00	87,100	22,400 – 410,300
Industrial Buildings	$102.00	22,100	5,100 – 200,600
Medical Clinics & Offices	213.00	22,500	2,300 – 327,000
Mixed Use	212.00	27,200	7,200 – 109,800
Multi-Family Housing	221.00	54,700	2,500 – 1,161,500
Nursing Home & Assisted Living	156.00	38,200	1,500 – 242,600
Office Buildings	195.00	20,600	1,100 – 930,000
Parking Garage	46.00	151,800	99,900 – 287,000
Parking Garage/Mixed Use	171.00	254,200	5,300 – 318,000
Police Stations	262.00	28,500	15,400 – 88,600
Public Assembly Buildings	253.00	22,600	2,200 – 235,300
Recreational	287.00	19,900	1,000 – 223,800
Restaurants	335.00	6,100	5,500 – 42,000
Retail	94.00	28,700	5,200 – 84,300
Schools	216.00	73,500	1,300 – 410,800
University, College & Private School Classroom & Admin Buildings	278.00	48,300	9,400 – 196,200
University, College & Private School Dormitories	222.00	28,900	1,500 – 126,900
University, College & Private School Science, Eng. & Lab Buildings	285.00	73,400	25,700 – 117,600
Warehouses	123.00	10,400	600 – 303,800

Abbreviations

A	Area Square Feet; Ampere
AAFES	Army and Air Force Exchange Service
ABS	Acrylonitrile Butadiene Stryrene; Asbestos Bonded Steel
A.C., AC	Alternating Current; Air-Conditioning; Asbestos Cement; Plywood Grade A & C
ACI	American Concrete Institute
ACR	Air Conditioning Refrigeration
ADA	Americans with Disabilities Act
AD	Plywood, Grade A & D
Addit.	Additional
Adh.	Adhesive
Adj.	Adjustable
af	Audio-frequency
AFFF	Aqueous Film Forming Foam
AFUE	Annual Fuel Utilization Efficiency
AGA	American Gas Association
Agg.	Aggregate
A.H., Ah	Ampere Hours
A hr.	Ampere-hour
A.H.U., AHU	Air Handling Unit
A.I.A.	American Institute of Architects
AIC	Ampere Interrupting Capacity
Allow.	Allowance
alt., alt	Alternate
Alum.	Aluminum
a.m.	Ante Meridiem
Amp.	Ampere
Anod.	Anodized
ANSI	American National Standards Institute
APA	American Plywood Association
Approx.	Approximate
Apt.	Apartment
Asb.	Asbestos
A.S.B.C.	American Standard Building Code
Asbe.	Asbestos Worker
ASCE	American Society of Civil Engineers
A.S.H.R.A.E.	American Society of Heating, Refrig. & AC Engineers
ASME	American Society of Mechanical Engineers
ASTM	American Society for Testing and Materials
Attchmt.	Attachment
Avg., Ave.	Average
AWG	American Wire Gauge
AWWA	American Water Works Assoc.
Bbl.	Barrel
B&B, BB	Grade B and Better; Balled & Burlapped
B&S	Bell and Spigot
B.&W.	Black and White
b.c.c.	Body-centered Cubic
B.C.Y.	Bank Cubic Yards
BE	Bevel End
B.F.	Board Feet
Bg. cem.	Bag of Cement
BHP	Boiler Horsepower; Brake Horsepower
B.I.	Black Iron
bidir.	bidirectional
Bit., Bitum.	Bituminous
Bit., Conc.	Bituminous Concrete
Bk.	Backed
Bkrs.	Breakers
Bldg., bldg	Building
Blk.	Block
Bm.	Beam
Boil.	Boilermaker
bpm	Blows per Minute
BR	Bedroom
Brg., brng.	Bearing
Brhe.	Bricklayer Helper
Bric.	Bricklayer
Brk., brk	Brick
brkt	Bracket
Brs.	Brass
Brz.	Bronze
Bsn.	Basin
Btr.	Better
BTU	British Thermal Unit
BTUH	BTU per Hour
Bu.	Bushels
BUR	Built-up Roofing
BX	Interlocked Armored Cable
°C	Degree Centigrade
c	Conductivity, Copper Sweat
C	Hundred; Centigrade
C/C	Center to Center, Cedar on Cedar
C-C	Center to Center
Cab	Cabinet
Cair.	Air Tool Laborer
Cal.	Caliper
Calc	Calculated
Cap.	Capacity
Carp.	Carpenter
C.B.	Circuit Breaker
C.C.A.	Chromate Copper Arsenate
C.C.F.	Hundred Cubic Feet
cd	Candela
cd/sf	Candela per Square Foot
CD	Grade of Plywood Face & Back
CDX	Plywood, Grade C & D, exterior glue
Cefi.	Cement Finisher
Cem.	Cement
CF	Hundred Feet
C.F.	Cubic Feet
CFM	Cubic Feet per Minute
CFRP	Carbon Fiber Reinforced Plastic
c.g.	Center of Gravity
CHW	Chilled Water; Commercial Hot Water
C.I., CI	Cast Iron
C.I.P., CIP	Cast in Place
Circ.	Circuit
C.L.	Carload Lot
CL	Chain Link
Clab.	Common Laborer
Clam	Common Maintenance Laborer
C.L.F.	Hundred Linear Feet
CLF	Current Limiting Fuse
CLP	Cross Linked Polyethylene
cm	Centimeter
CMP	Corr. Metal Pipe
CMU	Concrete Masonry Unit
CN	Change Notice
Col.	Column
CO_2	Carbon Dioxide
Comb.	Combination
comm.	Commercial, Communication
Compr.	Compressor
Conc.	Concrete
Cont., cont	Continuous; Continued, Container
Corkbd.	Cork Board
Corr.	Corrugated
Cos	Cosine
Cot	Cotangent
Cov.	Cover
C/P	Cedar on Paneling
CPA	Control Point Adjustment
Cplg.	Coupling
CPM	Critical Path Method
CPVC	Chlorinated Polyvinyl Chloride
C.Pr.	Hundred Pair
CRC	Cold Rolled Channel
Creos.	Creosote
Crpt.	Carpet & Linoleum Layer
CRT	Cathode-ray Tube
CS	Carbon Steel, Constant Shear Bar Joist
Csc	Cosecant
C.S.F.	Hundred Square Feet
CSI	Construction Specifications Institute
CT	Current Transformer
CTS	Copper Tube Size
Cu	Copper, Cubic
Cu. Ft.	Cubic Foot
cw	Continuous Wave
C.W.	Cool White; Cold Water
Cwt.	100 Pounds
C.W.X.	Cool White Deluxe
C.Y.	Cubic Yard (27 cubic feet)
C.Y./Hr.	Cubic Yard per Hour
Cyl.	Cylinder
d	Penny (nail size)
D	Deep; Depth; Discharge
Dis., Disch.	Discharge
Db	Decibel
Dbl.	Double
DC	Direct Current
DDC	Direct Digital Control
Demob.	Demobilization
d.f.t.	Dry Film Thickness
d.f.u.	Drainage Fixture Units
D.H.	Double Hung
DHW	Domestic Hot Water
DI	Ductile Iron
Diag.	Diagonal
Diam., Dia	Diameter
Distrib.	Distribution
Div.	Division
Dk.	Deck
D.L.	Dead Load; Diesel
DLH	Deep Long Span Bar Joist
dlx	Deluxe
Do.	Ditto
DOP	Dioctyl Phthalate Penetration Test (Air Filters)
Dp., dp	Depth
D.P.S.T.	Double Pole, Single Throw
Dr.	Drive
DR	Dimension Ratio
Drink.	Drinking
D.S.	Double Strength
D.S.A.	Double Strength A Grade
D.S.B.	Double Strength B Grade
Dty.	Duty
DWV	Drain Waste Vent
DX	Deluxe White, Direct Expansion
dyn	Dyne
e	Eccentricity
E	Equipment Only; East; Emissivity
Ea.	Each
EB	Encased Burial
Econ.	Economy
E.C.Y	Embankment Cubic Yards
EDP	Electronic Data Processing
EIFS	Exterior Insulation Finish System
E.D.R.	Equiv. Direct Radiation
Eq.	Equation
EL	Elevation
Elec.	Electrician; Electrical
Elev.	Elevator; Elevating
EMT	Electrical Metallic Conduit; Thin Wall Conduit
Eng.	Engine, Engineered
EPDM	Ethylene Propylene Diene Monomer
EPS	Expanded Polystyrene
Eqhv.	Equip. Oper., Heavy
Eqlt.	Equip. Oper., Light
Eqmd.	Equip. Oper., Medium
Eqmm.	Equip. Oper., Master Mechanic
Eqol.	Equip. Oper., Oilers
Equip.	Equipment
ERW	Electric Resistance Welded

E.S.	Energy Saver
Est.	Estimated
esu	Electrostatic Units
E.W.	Each Way
EWT	Entering Water Temperature
Excav.	Excavation
excl	Excluding
Exp., exp	Expansion, Exposure
Ext., ext	Exterior; Extension
Extru.	Extrusion
f.	Fiber Stress
F	Fahrenheit; Female; Fill
Fab., fab	Fabricated; Fabric
FBGS	Fiberglass
F.C.	Footcandles
f.c.c.	Face-centered Cubic
f'c.	Compressive Stress in Concrete; Extreme Compressive Stress
F.E.	Front End
FEP	Fluorinated Ethylene Propylene (Teflon)
F.G.	Flat Grain
F.H.A.	Federal Housing Administration
Fig.	Figure
Fin.	Finished
FIPS	Female Iron Pipe Size
Fixt.	Fixture
FJP	Finger jointed and primed
Fl. Oz.	Fluid Ounces
Flr.	Floor
Flrs.	Floors
FM	Frequency Modulation; Factory Mutual
Fmg.	Framing
FM/UL	Factory Mutual/Underwriters Labs
Fdn.	Foundation
FNPT	Female National Pipe Thread
Fori.	Foreman, Inside
Foro.	Foreman, Outside
Fount.	Fountain
fpm	Feet per Minute
FPT	Female Pipe Thread
Fr	Frame
F.R.	Fire Rating
FRK	Foil Reinforced Kraft
FSK	Foil/Scrim/Kraft
FRP	Fiberglass Reinforced Plastic
FS	Forged Steel
FSC	Cast Body; Cast Switch Box
Ft., ft	Foot; Feet
Ftng.	Fitting
Ftg.	Footing
Ft lb.	Foot Pound
Furn.	Furniture
FVNR	Full Voltage Non-Reversing
FVR	Full Voltage Reversing
FXM	Female by Male
Fy.	Minimum Yield Stress of Steel
g	Gram
G	Gauss
Ga.	Gauge
Gal., gal.	Gallon
Galv., galv	Galvanized
GC/MS	Gas Chromatograph/Mass Spectrometer
Gen.	General
GFI	Ground Fault Interrupter
GFRC	Glass Fiber Reinforced Concrete
Glaz.	Glazier
GPD	Gallons per Day
gpf	Gallon per Flush
GPH	Gallons per Hour
gpm, GPM	Gallons per Minute
GR	Grade
Gran.	Granular
Grnd.	Ground
GVW	Gross Vehicle Weight
GWB	Gypsum Wall Board
H	High Henry
HC	High Capacity
H.D., HD	Heavy Duty; High Density
H.D.O.	High Density Overlaid
HDPE	High Density Polyethylene Plastic
Hdr.	Header
Hdwe.	Hardware
H.I.D., HID	High Intensity Discharge
Help.	Helper Average
HEPA	High Efficiency Particulate Air Filter
Hg	Mercury
HIC	High Interrupting Capacity
HM	Hollow Metal
HMWPE	High Molecular Weight Polyethylene
HO	High Output
Horiz.	Horizontal
H.P., HP	Horsepower; High Pressure
H.P.F.	High Power Factor
Hr.	Hour
Hrs./Day	Hours per Day
HSC	High Short Circuit
Ht.	Height
Htg.	Heating
Htrs.	Heaters
HVAC	Heating, Ventilation & Air-Conditioning
Hvy.	Heavy
HW	Hot Water
Hyd.; Hydr.	Hydraulic
Hz	Hertz (cycles)
I.	Moment of Inertia
IBC	International Building Code
I.C.	Interrupting Capacity
ID	Inside Diameter
I.D.	Inside Dimension; Identification
I.F.	Inside Frosted
I.M.C.	Intermediate Metal Conduit
In.	Inch
Incan.	Incandescent
Incl.	Included; Including
Int.	Interior
Inst.	Installation
Insul., insul	Insulation/Insulated
I.P.	Iron Pipe
I.P.S., IPS	Iron Pipe Size
IPT	Iron Pipe Threaded
I.W.	Indirect Waste
J	Joule
J.I.C.	Joint Industrial Council
K	Thousand; Thousand Pounds; Heavy Wall Copper Tubing, Kelvin
K.A.H.	Thousand Amp. Hours
kcmil	Thousand Circular Mils
KD	Knock Down
K.D.A.T.	Kiln Dried After Treatment
kg	Kilogram
kG	Kilogauss
kgf	Kilogram Force
kHz	Kilohertz
Kip	1000 Pounds
KJ	Kilojoule
K.L.	Effective Length Factor
K.L.F.	Kips per Linear Foot
Km	Kilometer
KO	Knock Out
K.S.F.	Kips per Square Foot
K.S.I.	Kips per Square Inch
kV	Kilovolt
kVA	Kilovolt Ampere
kVAR	Kilovar (Reactance)
KW	Kilowatt
KWh	Kilowatt-hour
L	Labor Only; Length; Long; Medium Wall Copper Tubing
Lab.	Labor
lat	Latitude
Lath.	Lather
Lav.	Lavatory
lb.; #	Pound
L.B., LB	Load Bearing; L Conduit Body
L. & E.	Labor & Equipment
lb./hr.	Pounds per Hour
lb./L.F.	Pounds per Linear Foot
lbf/sq.in.	Pound-force per Square Inch
L.C.L.	Less than Carload Lot
L.C.Y.	Loose Cubic Yard
Ld.	Load
LE	Lead Equivalent
LED	Light Emitting Diode
L.F.	Linear Foot
L.F. Hdr	Linear Feet of Header
L.F. Nose	Linear Foot of Stair Nosing
L.F. Rsr	Linear Foot of Stair Riser
Lg.	Long; Length; Large
L & H	Light and Heat
LH	Long Span Bar Joist
L.H.	Labor Hours
L.L., LL	Live Load
L.L.D.	Lamp Lumen Depreciation
lm	Lumen
lm/sf	Lumen per Square Foot
lm/W	Lumen per Watt
LOA	Length Over All
log	Logarithm
L-O-L	Lateralolet
long.	Longitude
L.P., LP	Liquefied Petroleum; Low Pressure
L.P.F.	Low Power Factor
LR	Long Radius
L.S.	Lump Sum
Lt.	Light
Lt. Ga.	Light Gauge
L.T.L.	Less than Truckload Lot
Lt. Wt.	Lightweight
L.V.	Low Voltage
M	Thousand; Material; Male; Light Wall Copper Tubing
M^2CA	Meters Squared Contact Area
m/hr.; M.H.	Man-hour
mA	Milliampere
Mach.	Machine
Mag. Str.	Magnetic Starter
Maint.	Maintenance
Marb.	Marble Setter
Mat; Mat'l.	Material
Max.	Maximum
MBF	Thousand Board Feet
MBH	Thousand BTU's per hr.
MC	Metal Clad Cable
MCC	Motor Control Center
M.C.F.	Thousand Cubic Feet
MCFM	Thousand Cubic Feet per Minute
M.C.M.	Thousand Circular Mils
MCP	Motor Circuit Protector
MD	Medium Duty
MDF	Medium-density fibreboard
M.D.O.	Medium Density Overlaid
Med.	Medium
MF	Thousand Feet
M.F.B.M.	Thousand Feet Board Measure
Mfg.	Manufacturing
Mfrs.	Manufacturers
mg	Milligram
MGD	Million Gallons per Day
MGPH	Million Gallons per Hour
MH, M.H.	Manhole; Metal Halide; Man-Hour
MHz	Megahertz
Mi.	Mile
MI	Malleable Iron; Mineral Insulated
MIPS	Male Iron Pipe Size
mj	Mechanical Joint
m	Meter
mm	Millimeter
Mill.	Millwright
Min., min.	Minimum, Minute

Abbreviations

Misc.	Miscellaneous
ml	Milliliter, Mainline
M.L.F.	Thousand Linear Feet
Mo.	Month
Mobil.	Mobilization
Mog.	Mogul Base
MPH	Miles per Hour
MPT	Male Pipe Thread
MRGWB	Moisture Resistant Gypsum Wallboard
MRT	Mile Round Trip
ms	Millisecond
M.S.F.	Thousand Square Feet
Mstz.	Mosaic & Terrazzo Worker
M.S.Y.	Thousand Square Yards
Mtd., mtd., mtd	Mounted
Mthe.	Mosaic & Terrazzo Helper
Mtng.	Mounting
Mult.	Multi; Multiply
MUTCD	Manual on Uniform Traffic Control Devices
M.V.A.	Million Volt Amperes
M.V.A.R.	Million Volt Amperes Reactance
MV	Megavolt
MW	Megawatt
MXM	Male by Male
MYD	Thousand Yards
N	Natural; North
nA	Nanoampere
NA	Not Available; Not Applicable
N.B.C.	National Building Code
NC	Normally Closed
NEMA	National Electrical Manufacturers Assoc.
NEHB	Bolted Circuit Breaker to 600V.
NFPA	National Fire Protection Association
NLB	Non-Load-Bearing
NM	Non-Metallic Cable
nm	Nanometer
No.	Number
NO	Normally Open
N.O.C.	Not Otherwise Classified
Nose.	Nosing
NPT	National Pipe Thread
NQOD	Combination Plug-on/Bolt on Circuit Breaker to 240V.
N.R.C., NRC	Noise Reduction Coefficient/ Nuclear Regulator Commission
N.R.S.	Non Rising Stem
ns	Nanosecond
NTP	Notice to Proceed
nW	Nanowatt
OB	Opposing Blade
OC	On Center
OD	Outside Diameter
O.D.	Outside Dimension
ODS	Overhead Distribution System
O.G.	Ogee
O.H.	Overhead
O&P	Overhead and Profit
Oper.	Operator
Opng.	Opening
Orna.	Ornamental
OSB	Oriented Strand Board
OS&Y	Outside Screw and Yoke
OSHA	Occupational Safety and Health Act
Ovhd.	Overhead
OWG	Oil, Water or Gas
Oz.	Ounce
P.	Pole; Applied Load; Projection
p.	Page
Pape.	Paperhanger
P.A.P.R.	Powered Air Purifying Respirator
PAR	Parabolic Reflector
P.B., PB	Push Button
Pc., Pcs.	Piece, Pieces
P.C.	Portland Cement; Power Connector
P.C.F.	Pounds per Cubic Foot
PCM	Phase Contrast Microscopy
PDCA	Painting and Decorating Contractors of America
P.E., PE	Professional Engineer; Porcelain Enamel; Polyethylene; Plain End
P.E.C.I.	Porcelain Enamel on Cast Iron
Perf.	Perforated
PEX	Cross Linked Polyethylene
Ph.	Phase
P.I.	Pressure Injected
Pile.	Pile Driver
Pkg.	Package
Pl.	Plate
Plah.	Plasterer Helper
Plas.	Plasterer
plf	Pounds Per Linear Foot
Pluh.	Plumber Helper
Plum.	Plumber
Ply.	Plywood
p.m.	Post Meridiem
Pntd.	Painted
Pord.	Painter, Ordinary
pp	Pages
PP, PPL	Polypropylene
P.P.M.	Parts per Million
Pr.	Pair
P.E.S.B.	Pre-engineered Steel Building
Prefab.	Prefabricated
Prefin.	Prefinished
Prop.	Propelled
PSF, psf	Pounds per Square Foot
PSI, psi	Pounds per Square Inch
PSIG	Pounds per Square Inch Gauge
PSP	Plastic Sewer Pipe
Pspr.	Painter, Spray
Psst.	Painter, Structural Steel
P.T.	Potential Transformer
P. & T.	Pressure & Temperature
Ptd.	Painted
Ptns.	Partitions
Pu	Ultimate Load
PVC	Polyvinyl Chloride
Pvmt.	Pavement
PRV	Pressure Relief Valve
Pwr.	Power
Q	Quantity Heat Flow
Qt.	Quart
Quan., Qty.	Quantity
Q.C.	Quick Coupling
r	Radius of Gyration
R	Resistance
R.C.P.	Reinforced Concrete Pipe
Rect.	Rectangle
recpt.	Receptacle
Reg.	Regular
Reinf.	Reinforced
Req'd.	Required
Res.	Resistant
Resi.	Residential
RF	Radio Frequency
RFID	Radio-frequency Identification
Rgh.	Rough
RGS	Rigid Galvanized Steel
RHW	Rubber, Heat & Water Resistant; Residential Hot Water
rms	Root Mean Square
Rnd.	Round
Rodm.	Rodman
Rofc.	Roofer, Composition
Rofp.	Roofer, Precast
Rohe.	Roofer Helpers (Composition)
Rots.	Roofer, Tile & Slate
R.O.W.	Right of Way
RPM	Revolutions per Minute
R.S.	Rapid Start
Rsr	Riser
RT	Round Trip
S.	Suction; Single Entrance; South
SBS	Styrene Butadiere Styrene
SC	Screw Cover
SCFM	Standard Cubic Feet per Minute
Scaf.	Scaffold
Sch., Sched.	Schedule
S.C.R.	Modular Brick
S.D.	Sound Deadening
SDR	Standard Dimension Ratio
S.E.	Surfaced Edge
Sel.	Select
SER, SEU	Service Entrance Cable
S.F.	Square Foot
S.F.C.A.	Square Foot Contact Area
S.F. Flr.	Square Foot of Floor
S.F.G.	Square Foot of Ground
S.F. Hor.	Square Foot Horizontal
SFR	Square Feet of Radiation
S.F. Shlf.	Square Foot of Shelf
S4S	Surface 4 Sides
Shee.	Sheet Metal Worker
Sin.	Sine
Skwk.	Skilled Worker
SL	Saran Lined
S.L.	Slimline
Sldr.	Solder
SLH	Super Long Span Bar Joist
S.N.	Solid Neutral
SO	Stranded with oil resistant inside insulation
S-O-L	Socketolet
sp	Standpipe
S.P.	Static Pressure; Single Pole; Self-Propelled
Spri.	Sprinkler Installer
spwg	Static Pressure Water Gauge
S.P.D.T.	Single Pole, Double Throw
SPF	Spruce Pine Fir; Sprayed Polyurethane Foam
S.P.S.T.	Single Pole, Single Throw
SPT	Standard Pipe Thread
Sq.	Square; 100 Square Feet
Sq. Hd.	Square Head
Sq. In.	Square Inch
S.S.	Single Strength; Stainless Steel
S.S.B.	Single Strength B Grade
sst, ss	Stainless Steel
Sswk.	Structural Steel Worker
Sswl.	Structural Steel Welder
St.; Stl.	Steel
STC	Sound Transmission Coefficient
Std.	Standard
Stg.	Staging
STK	Select Tight Knot
STP	Standard Temperature & Pressure
Stpi.	Steamfitter, Pipefitter
Str.	Strength; Starter; Straight
Strd.	Stranded
Struct.	Structural
Sty.	Story
Subj.	Subject
Subs.	Subcontractors
Surf.	Surface
Sw.	Switch
Swbd.	Switchboard
S.Y.	Square Yard
Syn.	Synthetic
S.Y.P.	Southern Yellow Pine
Sys.	System
t.	Thickness
T	Temperature; Ton
Tan	Tangent
T.C.	Terra Cotta
T & C	Threaded and Coupled
T.D.	Temperature Difference
TDD	Telecommunications Device for the Deaf
T.E.M.	Transmission Electron Microscopy
temp	Temperature, Tempered, Temporary
TFFN	Nylon Jacketed Wire

Abbreviations

TFE	Tetrafluoroethylene (Teflon)
T. & G.	Tongue & Groove; Tar & Gravel
Th.,Thk.	Thick
Thn.	Thin
Thrded	Threaded
Tilf.	Tile Layer, Floor
Tilh.	Tile Layer, Helper
THHN	Nylon Jacketed Wire
THW.	Insulated Strand Wire
THWN	Nylon Jacketed Wire
T.L.,TL	Truckload
T.M.	Track Mounted
Tot.	Total
T-O-L	Threadolet
tmpd	Tempered
TPO	Thermoplastic Polyolefin
T.S.	Trigger Start
Tr.	Trade
Transf.	Transformer
Trhv.	Truck Driver, Heavy
Trlr	Trailer
Trlt.	Truck Driver, Light
TTY	Teletypewriter
TV	Television
T.W.	Thermoplastic Water Resistant Wire
UCI	Uniform Construction Index
UF	Underground Feeder
UGND	Underground Feeder
UHF	Ultra High Frequency
U.I.	United Inch
U.L., UL	Underwriters Laboratory
Uld.	Unloading
Unfin.	Unfinished
UPS	Uninterruptible Power Supply
URD	Underground Residential Distribution
US	United States
USGBC	U.S. Green Building Council
USP	United States Primed
UTMCD	Uniform Traffic Manual For Control Devices
UTP	Unshielded Twisted Pair
V	Volt
VA	Volt Amperes
VAT	Vinyl Asbestos Tile
V.C.T.	Vinyl Composition Tile
VAV	Variable Air Volume
VC	Veneer Core
VDC	Volts Direct Current
Vent.	Ventilation
Vert.	Vertical
V.F.	Vinyl Faced
V.G.	Vertical Grain
VHF	Very High Frequency
VHO	Very High Output
Vib.	Vibrating
VLF	Vertical Linear Foot
VOC	Volatile Organic Compound
Vol.	Volume
VRP	Vinyl Reinforced Polyester
W	Wire; Watt; Wide; West
w/	With
W.C., WC	Water Column; Water Closet
W.F.	Wide Flange
W.G.	Water Gauge
Wldg.	Welding
W. Mile	Wire Mile
W-O-L	Weldolet
W.R.	Water Resistant
Wrck.	Wrecker
WSFU	Water Supply Fixture Unit
W.S.P.	Water, Steam, Petroleum
WT., Wt.	Weight
WWF	Welded Wire Fabric
XFER	Transfer
XFMR	Transformer
XHD	Extra Heavy Duty
XHHW	Cross-Linked Polyethylene Wire
XLPE	Insulation
XLP	Cross-linked Polyethylene
Xport	Transport
Y	Wye
yd	Yard
yr	Year
Δ	Delta
%	Percent
~	Approximately
Ø	Phase; diameter
@	At
#	Pound; Number
<	Less Than
>	Greater Than
Z	Zone

A

C

D

G

H

I

J

K

L

N

O

P

Q

R

S

U

V

W

X

Y

Z

Notes

Notes

Notes

Notes

Notes

Division Notes

		CREW	DAILY OUTPUT	LABOR-HOURS	UNIT	BARE COSTS MAT.	BARE COSTS LABOR	BARE COSTS EQUIP.	BARE COSTS TOTAL	TOTAL INCL O&P

Division Notes

	CREW	DAILY OUTPUT	LABOR-HOURS	UNIT	BARE COSTS MAT.	BARE COSTS LABOR	BARE COSTS EQUIP.	BARE COSTS TOTAL	TOTAL INCL O&P

Division Notes

I		CREW	DAILY OUTPUT	LABOR-HOURS	UNIT	BARE COSTS MAT.	LABOR	EQUIP.	TOTAL	TOTAL INCL O&P

Division Notes

		DAILY	LABOR-		BARE COSTS				TOTAL
	CREW	OUTPUT	HOURS	UNIT	MAT.	LABOR	EQUIP.	TOTAL	INCL O&P

Division Notes

		CREW	DAILY OUTPUT	LABOR-HOURS	UNIT	BARE COSTS MAT.	BARE COSTS LABOR	BARE COSTS EQUIP.	BARE COSTS TOTAL	TOTAL INCL O&P

Other Data & Services

A tradition of excellence in construction cost information and services since 1942

Table of Contents

For more information visit our website at RSMeans.com

Unit prices according to the latest MasterFormat®

Cost Data Selection Guide

The following table provides definitive information on the content of each cost data publication. The number of lines of data provided in each unit price or assemblies division, as well as the number of crews, is listed for each data set. The presence of other elements such as reference tables, square foot models, equipment rental costs, historical cost indexes, and city cost indexes, is also indicated. You can use the table to help select the RSMeans data set that has the quantity and type of information you most need in your work.

Unit Cost Divisions	Building Construction	Mechanical	Electrical	Commercial Renovation	Square Foot	Site Work Landsc.	Green Building	Interior	Concrete Masonry	Open Shop	Heavy Construction	Light Commercial	Facilities Construction	Plumbing	Residential
1	584	406	427	531	0	516	200	326	467	583	521	273	1056	416	178
2	779	278	86	735	0	995	207	397	218	778	737	479	1222	285	274
3	1744	340	230	1265	0	1536	1041	354	2273	1744	1929	537	2027	316	444
4	961	21	0	921	0	726	180	615	1159	929	616	534	1176	0	448
5	1889	158	155	1093	0	852	1787	1106	729	1889	1025	979	1906	204	746
6	2453	18	18	2111	0	110	589	1528	281	2449	123	2141	2125	22	2661
7	1596	215	128	1634	0	580	763	532	523	1593	26	1329	1697	227	1049
8	2140	80	3	2733	0	255	1140	1813	105	2142	0	2328	2966	0	1552
9	2107	86	45	1931	0	309	455	2193	412	2048	15	1756	2356	54	1521
10	1089	17	10	685	0	232	32	899	136	1089	34	589	1180	237	224
11	1097	201	166	541	0	135	56	925	29	1064	0	231	1117	164	110
12	548	0	2	298	0	219	147	1551	14	515	0	273	1574	23	217
13	744	149	158	253	0	366	125	254	78	720	267	109	760	115	104
14	273	36	0	223	0	0	0	257	0	273	0	12	293	16	6
21	130	0	41	37	0	0	0	296	0	130	0	121	668	688	259
22	1165	7559	160	1226	0	1573	1063	849	20	1154	1682	875	7506	9416	719
23	1198	7001	581	940	0	157	901	789	38	1181	110	890	5240	1918	485
25	0	0	14	14	0	0	0	0	0	0	0	0	0	0	0
26	1512	491	10455	1293	0	811	644	1159	55	1438	600	1360	10236	399	636
27	94	0	447	101	0	0	0	71	0	94	39	67	388	0	56
28	143	79	223	124	0	0	28	97	0	127	0	70	209	57	41
31	1511	733	610	807	0	3266	289	7	1218	1456	3282	605	1570	660	614
32	838	49	8	905	0	4475	355	406	315	809	1891	440	1752	142	487
33	1248	1080	534	252	0	3040	38	0	239	525	3090	128	1707	2089	154
34	107	0	47	4	0	190	0	0	31	62	221	0	136	0	0
35	18	0	0	0	0	327	0	0	0	18	442	0	84	0	0
41	62	0	0	33	0	8	0	22	0	61	31	0	68	14	0
44	75	79	0	0	0	0	0	0	0	0	0	0	75	75	0
46	23	16	0	0	0	274	261	0	0	23	264	0	33	33	0
48	8	0	36	2	0	0	21	0	0	8	15	8	21	0	8
Totals	26136	19092	14584	20692	0	20952	10322	16446	8340	24902	16960	16134	51148	17570	12993

Assem Div	Building Construction	Mechanical	Elec-trical	Commercial Renovation	Square Foot	Site Work Landscape	Assemblies	Green Building	Interior	Concrete Masonry	Heavy Construction	Light Commercial	Facilities Construction	Plumbing	Asm Div	Residential
A		15	0	188	164	577	598	0	0	536	571	154	24	0	1	378
B		0	0	848	2554	0	5661	56	329	1976	368	2094	174	0	2	211
C		0	0	647	954	0	1334	0	1641	146	0	844	251	0	3	588
D		1057	941	712	1858	72	2538	330	824	0	0	1345	1104	1088	4	851
E		0	0	86	261	0	301	0	5	0	0	258	5	0	5	391
F		0	0	0	114	0	143	0	0	0	0	114	0	0	6	357
G		527	447	318	312	3378	792	0	0	535	1349	205	293	677	7	307
															8	760
															9	80
															10	0
															11	0
															12	0
Totals		1599	1388	2799	6217	4027	11367	386	2799	3193	2288	5014	1851	1765		3923

Reference Section	Building Construction Costs	Mechanical	Electrical	Commercial Renovation	Square Foot	Site Work Landscape	Assem.	Green Building	Interior	Concrete Masonry	Open Shop	Heavy Construction	Light Commercial	Facilities Construction	Plumbing	Resi.
Reference Tables	yes	yes	yes	yes	no	yes	yes	yes	yes	yes	yes	yes	yes	yes	yes	yes
Models					111			25					50			28
Crews	582	582	582	561		582		582	582	582	560	582	560	561	582	560
Equipment Rental Costs	yes	yes	yes	yes		yes		yes	yes	yes	yes	yes	yes	yes	yes	yes
Historical Cost Indexes	yes	yes	yes	yes	yes	yes	yes	yes	yes	yes	yes	yes	yes	yes	yes	no
City Cost Indexes	yes	yes	yes	yes	yes	yes	yes	yes	yes	yes	yes	yes	yes	yes	yes	yes

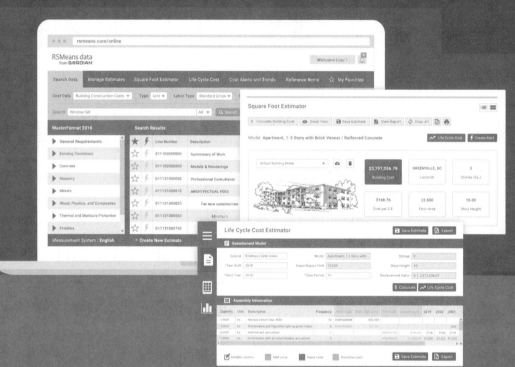

RSMeans Data Online **Core**

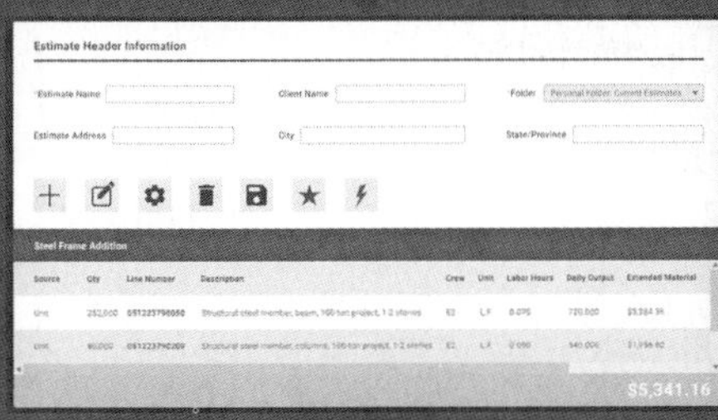

rsmeans.com/core

The Core tier of RSMeans Data Online provides reliable construction cost data along with the tools necessary to quickly access costs at the material or task level. Users can create unit line estimates with RSMeans data from Gordian and share them with ease from the web-based application.

Key Features:

- Search
- Estimate
- Share

Data Available:

16 datasets and packages curated by project type

RSMeans Data Online **Complete**

The Complete tier of RSMeans Data Online is designed to provide the latest in construction costs with comprehensive tools for projects of varying scopes. Unit, assembly or square foot price data is available 24/7 from the web-based application. Harness the power of RSMeans data from Gordian with expanded features and powerful tools like the square foot model estimator and cost trends analysis.

Key Features:

- Alerts
- Square Foot Estimator
- Trends

Data Available:

20 datasets and packages curated by project type

rsmeans.com/complete

RSMeans Data Online **Complete Plus**

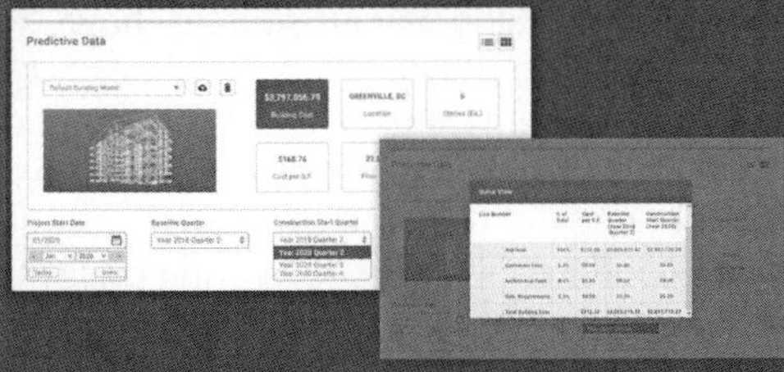

rsmeans.com/completeplus

The Complete Plus tier of RSMeans Data Online was developed to take project planning and estimating to the next level with predictive cost data. Users gain access to an exclusive set of tools and features and the most comprehensive database in the industry. Leverage everything RSMeans Data Online has to offer with the all-inclusive Complete Plus tier.

Key Features:

- Predictive Cost Data
- Life Cycle Costing
- Full History

Data Available:

Full library and renovation models

2019 Seminar Schedule

877-620-6245

Note: call for exact dates, locations, and details as some cities are subject to change.

Location	Dates
Seattle, WA	January and August
Dallas/Ft. Worth, TX	January
Austin, TX	February
Jacksonville, FL	February
Anchorage, AK	March and September
Las Vegas, NV	March
Washington, DC	April and September
Charleston, SC	April
Toronto	May
Denver, CO	May
San Francisco, CA	June
Bethesda, MD	June
Dallas, TX	September
Raleigh, NC	October
Baltimore, MD	November
Orlando, FL	November
San Diego, CA	December
San Antonio, TX	December

Gordian also offers a suite of online RSMeans data self-paced offerings.
Check our website RSMeans.com/products/training.aspx for more information.

Facilities Construction Estimating

In this two-day course, professionals working in facilities management can get help with their daily challenges to establish budgets for all phases of a project.

Some of what you'll learn:

- Determining the full scope of a project
- Identifying the scope of risks and opportunities
- Creative solutions to estimating issues
- Organizing estimates for presentation and discussion
- Special techniques for repair/remodel and maintenance projects
- Negotiating project change orders

Who should attend: facility managers, engineers, contractors, facility tradespeople, planners, and project managers.

Mechanical & Electrical Estimating

This two-day course teaches attendees how to prepare more accurate and complete mechanical/electrical estimates, avoid the pitfalls of omission and double-counting, and understand the composition and rationale within the RSMeans mechanical/electrical database.

Some of what you'll learn:

- The unique way mechanical and electrical systems are interrelated
- M&E estimates—conceptual, planning, budgeting, and bidding stages
- Order of magnitude, square foot, assemblies, and unit price estimating
- Comparative cost analysis of equipment and design alternatives

Who should attend: architects, engineers, facilities managers, mechanical and electrical contractors, and others who need a highly reliable method for developing, understanding, and evaluating mechanical and electrical contracts.

Construction Cost Estimating: Concepts and Practice

This one or two day introductory course to improve estimating skills and effectiveness starts with the details of interpreting bid documents and ends with the summary of the estimate and bid submission.

Some of what you'll learn:

- Using the plans and specifications to create estimates
- The takeoff process—deriving all tasks with correct quantities
- Developing pricing using various sources; how subcontractor pricing fits in
- Summarizing the estimate to arrive at the final number
- Formulas for area and cubic measure, adding waste and adjusting productivity to specific projects
- Evaluating subcontractors' proposals and prices
- Adding insurance and bonds
- Understanding how labor costs are calculated
- Submitting bids and proposals

Who should attend: project managers, architects, engineers, owners' representatives, contractors, and anyone who's responsible for budgeting or estimating construction projects.

Assessing Scope of Work for Facilities Construction Estimating

This two-day practical training program addresses the vital importance of understanding the scope of projects in order to produce accurate cost estimates for facility repair and remodeling.

Some of what you'll learn:

- Discussions of site visits, plans/specs, record drawings of facilities, and site-specific lists
- Review of CSI divisions, including means, methods, materials, and the challenges of scoping each topic
- Exercises in scope identification and scope writing for accurate estimating of projects
- Hands-on exercises that require scope, take-off, and pricing

Who should attend: corporate and government estimators, planners, facility managers, and others who need to produce accurate project estimates.

Maintenance & Repair Estimating for Facilities

This two-day course teaches attendees how to plan, budget, and estimate the cost of ongoing and preventive maintenance and repair for existing buildings and grounds.

Some of what you'll learn:

- The most financially favorable maintenance, repair, and replacement scheduling and estimating
- Auditing and value engineering facilities
- Preventive planning and facilities upgrading
- Determining both in-house and contract-out service costs
- Annual, asset-protecting M&R plan

Who should attend: facility managers, maintenance supervisors, buildings and grounds superintendents, plant managers, planners, estimators, and others involved in facilities planning and budgeting.

Practical Project Management for Construction Professionals

In this two-day course, acquire the essential knowledge and develop the skills to effectively and efficiently execute the day-to-day responsibilities of the construction project manager.

Some of what you'll learn:

- General conditions of the construction contract
- Contract modifications: change orders and construction change directives
- Negotiations with subcontractors and vendors
- Effective writing: notification and communications
- Dispute resolution: claims and liens

Who should attend: architects, engineers, owners' representatives, and project managers.

Life Cycle Cost Estimating for Facility Asset Managers

Life Cycle Cost Estimating will take the attendee through choosing the correct RSMeans database to use and then correctly applying RSMeans data to their specific life cycle application. Conceptual estimating through RSMeans new building models, conceptual estimating of major existing building projects through RSMeans renovation models, pricing specific renovation elements, estimating repair, replacement and preventive maintenance costs today and forward up to 30 years will be covered.

Some of what you'll learn:

- Cost implications of managing assets
- Planning projects and initial & life cycle costs
- How to use RSMeans data online

Who should attend: facilities owners and managers and anyone involved in the financial side of the decision making process in the planning, design, procurement, and operation of facility real assets.

Please bring a laptop with ability to access the internet.

Building Systems and the Construction Process

This one-day course was written to assist novices and those outside the industry in obtaining a solid understanding of the construction process - from both a building systems and construction administration approach.

Some of what you'll learn:

- Various systems used and how components come together to create a building
- Start with foundation and end with the physical systems of the structure such as HVAC and Electrical
- Focus on the process from start of design through project closeout

This training session requires you to bring a laptop computer to class.

Who should attend: building professionals or novices to help make the crossover to the construction industry; suited for anyone responsible for providing high level oversight on construction projects.

Training for our Online Estimating Solution

Construction estimating is vital to the decision-making process at each state of every project. Our online solution works the way you do. It's systematic, flexible and intuitive. In this one-day class you will see how you can estimate any phase of any project faster and better.

Some of what you'll learn:

- Customizing our online estimating solution
- Making the most of RSMeans "Circle Reference" numbers
- How to integrate your cost data
- Generating reports, exporting estimates to MS Excel, sharing, collaborating and more

Also offered as a self-paced or on-site training program!

Training for our CD Estimating Solution

This one-day course helps users become more familiar with the functionality of the CD. Each menu, icon, screen, and function found in the program is explained in depth. Time is devoted to hands-on estimating exercises.

Some of what you'll learn:

- Searching the database using all navigation methods
- Exporting RSMeans data to your preferred spreadsheet format
- Viewing crews, assembly components, and much more
- Automatically regionalizing the database

This training session requires you to bring a laptop computer to class.

When you register for this course you will receive an outline for your laptop requirements.

Also offered as a self-paced or on-site training program!

Site Work Estimating with RSMeans data

This one-day program focuses directly on site work costs. Accurately scoping, quantifying, and pricing site preparation, underground utility work, and improvements to exterior site elements are often the most difficult estimating tasks on any project. Some of what you'll learn:

- Evaluation of site work and understanding site scope including: site clearing, grading, excavation, disposal and trucking of materials, backfill and compaction, underground utilities, paving, sidewalks, and seeding & planting.
- Unit price site work estimates—Correct use of RSMeans site work cost data to develop a cost estimate.
- Using and modifying assemblies—Save valuable time when estimating site work activities using custom assemblies.

Who should attend: Engineers, contractors, estimators, project managers, owner's representatives, and others who are concerned with the proper preparation and/or evaluation of site work estimates.

Please bring a laptop with ability to access the internet.

Facilities Estimating Using the CD

This two-day class combines hands-on skill-building with best estimating practices and real-life problems. You will learn key concepts, tips, pointers, and guidelines to save time and avoid cost oversights and errors.

Some of what you'll learn:

- Estimating process concepts
- Customizing and adapting RSMeans cost data
- Establishing scope of work to account for all known variables
- Budget estimating: when, why, and how
- Site visits: what to look for and what you can't afford to overlook
- How to estimate repair and remodeling variables

This training session requires you to bring a laptop computer to class.

Who should attend: facility managers, architects, engineers, contractors, facility tradespeople, planners, project managers, and anyone involved with JOC, SABRE, or IDIQ.

Registration Information

Register early to save up to $100!!!

Register 45+ days before date of a class and save $50 off each class. This savings cannot be combined with any other promotional or discounting of the regular price of classes!

How to register

By Phone

Register by phone at 877-620-6245

Online

Register online at RSMeans.com/products/seminars.aspx

Note: Purchase Orders or Credits Cards are required to register.

Two-day seminar registration fee - $1,200.

One-Day Construction Cost Estimating or Building Systems and the Construction Process - $765.

Government pricing

All federal government employees save off the regular seminar price. Other promotional discounts cannot be combined with the government discount. Call 781-422-5115 for government pricing.

CANCELLATION POLICY:

If you are unable to attend a seminar, substitutions may be made at any time before the session starts by notifying the seminar registrar at 1-781-422-5115 or your sales representative.

If you cancel twenty-one (21) days or more prior to the seminar, there will be no penalty and your registration fees will be refunded. These cancellations must be received by the seminar registrar or your sales representative and will be confirmed to be eligible for cancellation.

If you cancel fewer than twenty-one (21) days prior to the seminar, you will forfeit the registration fee.

In the unfortunate event of an RSMeans cancellation, RSMeans will work with you to reschedule your attendance in the same seminar at a later date or will fully refund your registration fee. RSMeans cannot be responsible for any non-refundable travel expenses incurred by you or another as a result of your registration, attendance at, or cancellation of an RSMeans seminar.

Any on-demand training modules are not eligible for cancellation, substitution, transfer, return or refund.

AACE approved courses

Many seminars described and offered here have been approved for 14 hours (1.4 recertification credits) of credit by the AACE International Certification Board toward meeting the continuing education requirements for recertification as a Certified Cost Engineer/Certified Cost Consultant.

AIA Continuing Education

We are registered with the AIA Continuing Education System (AIA/CES) and are committed to developing quality learning activities in accordance with the CES criteria. Many seminars meet the AIA/CES criteria for Quality Level 2. AIA members may receive 14 learning units (LUs) for each two-day RSMeans course.

Daily course schedule

The first day of each seminar session begins at 8:30 a.m. and ends at 4:30 p.m. The second day begins at 8:00 a.m. and ends at 4:00 p.m. Participants are urged to bring a hand-held calculator since many actual problems will be worked out in each session.

Continental breakfast

Your registration includes the cost of a continental breakfast and a morning and afternoon refreshment break. These informal segments allow you to discuss topics of mutual interest with other seminar attendees. (You are free to make your own lunch and dinner arrangements.)

Hotel/transportation arrangements

We arrange to hold a block of rooms at most host hotels. To take advantage of special group rates when making your reservation, be sure to mention that you are attending the RSMeans Institute data seminar. You are, of course, free to stay at the lodging place of your choice. (Hotel reservations and transportation arrangements should be made directly by seminar attendees.)

Important

Class sizes are limited, so please register as soon as possible.

Note: Pricing subject to change.